CDX Learning Systems™

We Support
ASE Education Foundation

FUNDAMENTALS OF

Automotive Maintenance and Light Repair

SECOND EDITION

Preparation for ASE G1

Kirk VanGelder
ASE Certified Master Technician & LI
ASE Evaluation Team Leader
Certified Automotive Service Instructor
Vancouver, Washington, USA

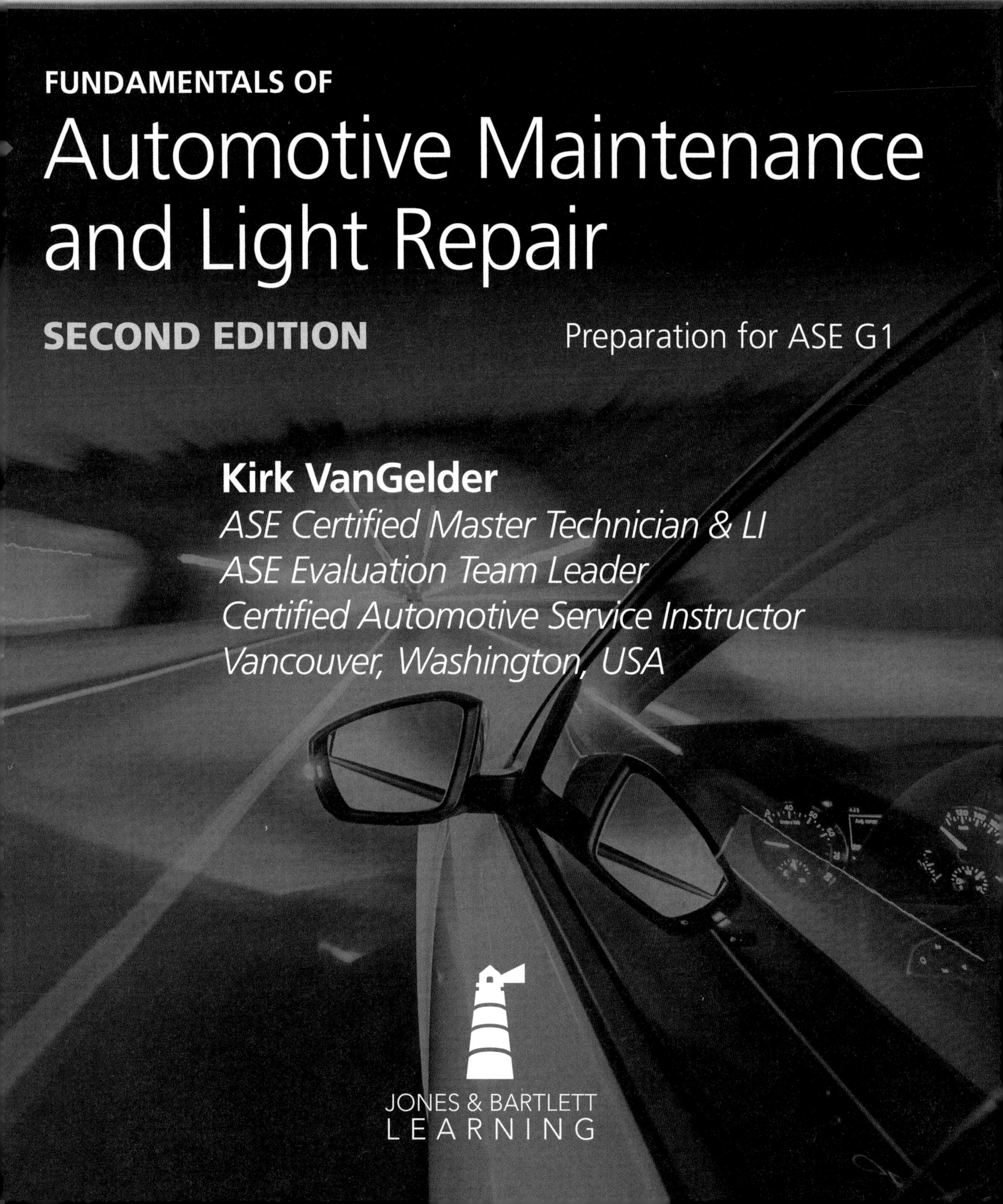

JONES & BARTLETT
LEARNING

World Headquarters
Jones & Bartlett Learning
5 Wall Street
Burlington, MA 01803
978-443-5000
info@jblearning.com
www.jblearning.com

Jones & Bartlett Learning books and products are available through most bookstores and online booksellers. To contact Jones & Bartlett Learning directly, call 800-832-0034, fax 978-443-8000, or visit our website, www.jblearning.com.

Substantial discounts on bulk quantities of Jones & Bartlett Learning publications are available to corporations, professional associations, and other qualified organizations. For details and specific discount information, contact the special sales department at Jones & Bartlett Learning via the above contact information or send an email to specialsales@jblearning.com.

Production Credits
General Manager: Kimberly Brophy
VP, Product Development: Christine Emerton
Product Owner: Kevin Murphy
Senior Managing Editor: Donna Gridley
Senior Content Development Editor: Amanda Brandt
Director, Project Management: Jenny Corriveau
Project Specialist: Brooke Haley
Marketing Manager: Amanda Banner
VP, Manufacturing and Inventory Control: Therese Connell
Composition: Integra Software Services Pvt. Ltd.
Project Management: Integra Software Services Pvt. Ltd.
Cover Design: Scott Moden
Text Design: Scott Moden
Cover Image: Front cover: © gehringj/E+/Getty Images; Back cover: © blue jean images RF/Getty Images
Printing and Binding: LSC Communications
Cover Printing: LSC Communications

ISBN: 978-1-284-14339-3

Library of Congress Cataloging-in-Publication Data unavailable at time of printing.

6048

Printed in the United States of America
23 22 21 20 10 9 8 7 6 5 4 3 2

BRIEF CONTENTS

SECTION 1 Safety and Foundation . . . 3

CHAPTER 1 Careers in Automotive Technology . . . 4
CHAPTER 2 Introduction to Automotive Technology . . . 18
CHAPTER 3 Introduction to Automotive Safety . . . 40
CHAPTER 4 Personal Safety . . . 80
CHAPTER 5 Vehicle Service Information and Diagnostic Process . . . 97
CHAPTER 6 Hand and Measuring Tools . . . 120
CHAPTER 7 Power Tools and Equipment . . . 170
CHAPTER 8 Fasteners and Thread Repair . . . 203
CHAPTER 9 Vehicle Protection and Jack and Lift Safety . . . 228
CHAPTER 10 Vehicle Maintenance Inspection . . . 250
CHAPTER 11 Communication and Employability Skills . . . 281

SECTION 2 Engine Repair . . . 309

CHAPTER 12 Motive Power Theory—SI Engines . . . 310
CHAPTER 13 Engine Mechanical Testing . . . 342
CHAPTER 14 Lubrication System Theory . . . 362
CHAPTER 15 Servicing the Lubrication System . . . 387
CHAPTER 16 Cooling System Theory . . . 402
CHAPTER 17 Servicing the Cooling System . . . 432

SECTION 3 Automatic Transmissions . . . 459

CHAPTER 18 Automatic Transmission Fundamentals . . . 460
CHAPTER 19 Maintaining the Automatic Transmission/Transaxle . . . 480
CHAPTER 20 Hybrid and Continuously Variable Transmissions . . . 493

SECTION 4 Manual Transmissions . . . 509

CHAPTER 21 Manual Transmission/Transaxle Principles . . . 510
CHAPTER 22 The Clutch System . . . 529
CHAPTER 23 Driveshafts, Axles, and Final Drives . . . 547

SECTION 5 Steering and Suspension . . . 575

CHAPTER 24 Wheels and Tires Theory . . . 576
CHAPTER 25 Servicing Wheels and Tires . . . 597
CHAPTER 26 Steering Systems Theory . . . 633
CHAPTER 27 Servicing Steering Systems . . . 659
CHAPTER 28 Suspension Systems Theory . . . 679

CHAPTER 29 Servicing Suspension Systems 709
CHAPTER 30 Wheel Alignment 732

SECTION 6 Brakes 757
CHAPTER 31 Principles of Braking 758
CHAPTER 32 Hydraulics and Power Brakes Theory 776
CHAPTER 33 Servicing Hydraulic Systems and Power Brakes 804
CHAPTER 34 Disc Brake Systems Theory 831
CHAPTER 35 Servicing Disc Brakes 849
CHAPTER 36 Drum Brake Systems Theory 871
CHAPTER 37 Servicing Drum Brakes 889
CHAPTER 38 Wheel Bearings 905
CHAPTER 39 Electronic Brake Control 927

SECTION 7 Electric 949
CHAPTER 40 Principles of Electrical Systems 950
CHAPTER 41 Electrical Components and Repair 973
CHAPTER 42 Meter Usage and Circuit Diagnosis 997
CHAPTER 43 Battery Systems 1035
CHAPTER 44 Starting and Charging Systems 1060
CHAPTER 45 Lighting Systems 1092
CHAPTER 46 Body Electrical System 1117

SECTION 8 Heating and Air Conditioning 1145
CHAPTER 47 Principles of Heating and Air-Conditioning Systems 1146

SECTION 9 Engine Performance 1169
CHAPTER 48 Ignition Systems 1170
CHAPTER 49 Gasoline Fuel Systems 1200
CHAPTER 50 Engine Management System 1237
CHAPTER 51 On-Board Diagnostics 1261
CHAPTER 52 Induction and Exhaust 1275
CHAPTER 53 Emission Control 1299
CHAPTER 54 Alternative Fuel Systems 1320

Appendix A 2017 ASE Education Foundation Automobile Accreditation Task List Correlation Guide 1336

Appendix B ASE Education Foundation Integrated Applied Academic Skills Correlation Guide 1343

Glossary 1345

Index 1374

CONTENTS

SECTION 1 Safety and Foundation

CHAPTER 1 Careers in Automotive Technology 4
Introduction 5
A Brief History of the Automobile 5
Careers in the Automotive Sector 7
Types of Shops 12
Automotive Industry Certification 14
Ready for Review 16
Key Terms 16
Review Questions 16
ASE Technician A/Technician B Style Questions 17

CHAPTER 2 Introduction to Automotive Technology 18
Introduction 19
Body Designs 19
Overview of Vehicle Systems 24
Drivetrain Layouts 28
Torque 32
Ready for Review 37
Key Terms 37
Review Questions 38
ASE Technician A/Technician B Style Questions 38

CHAPTER 3 Introduction to Automotive Safety 40
Introduction 41
Work Environment - Safety Features 47
Preventing and Fighting Fires 56
Hazardous Materials Safety 61
Vehicle System Safety 72
Ready for Review 77
Key Terms 77
Review Questions 78
ASE Technician A/Technician B Style Questions 78

CHAPTER 4 Personal Safety 80
Personal Safety Overview 81
Hand Protection 86
Head Protection 88
Ready for Review 95
Key Terms 95
Review Questions 95
ASE Technician A/Technician B Style Questions 96

CHAPTER 5 Vehicle Service Information and Diagnostic Process 97
Vehicle Information Overview 98
VIN and Production Date Code, and Vehicle Information Labels 107
Repair Order Information 110
Strategy-Based Diagnosis 113
Ready for Review 118
Key Terms 118
Review Questions 118
ASE Technician A/Technician B Style Questions 119

CHAPTER 6 Hand and Measuring Tools 120
Hand Tool Overview 121
Basic Hand Tools 125
Other Hand Tools 134
Hammers and Struck Tools 141
Taps and Dies 145
Precision Measuring Tools 154
Vernier Calipers 159
Ready for Review 165
Key Terms 166
Review Questions 168
ASE Technician A/Technician B Style Questions 168

CHAPTER 7 Power Tools and Equipment 170
Introduction 171
Battery Charging and Jump-Starting 171
Air Tools 177
Electric Power Tools 184
Cleaning Tools 193
Ready for Review 200
Key Terms 200
Review Questions 200
ASE Technician A/Technician B Style Questions 201

CHAPTER 8 Fasteners and Thread Repair 203
Fastener Identification Overview 204
Non-Threaded Fasteners 213
Replace Threaded Fasteners 217
Repair Damaged Fastener Threads 220

Ready for Review. 226
Key Terms. 226
Review Questions 226
ASE Technician A/Technician B Style Questions . . 227

CHAPTER 9 Vehicle Protection and Jack and Lift Safety 228
Introduction. 229
Preventing Vehicle Damage 229
Lifting Equipment. 234
Vehicle Lifts 242
Ready for Review. 248
Key Terms. 248
Review Questions 248
ASE Technician A/Technician B Style Questions . . 249

CHAPTER 10 Vehicle Maintenance Inspection 250
Vehicle Inspection Preliminaries 251
Underhood Fluid Inspection 255
Engine Drive Belts 263
Under-Vehicle Inspection. 267
Exterior Vehicle Inspections. 273
Ready for Review. 278
Key Terms. 278
Review Questions 278
ASE Technician A/Technician B Style Questions . . 279

CHAPTER 11 Communication and Employability Skills .281
Introduction. 282
Active Listening 282
The Art of Speaking 285
Employability Skills. 288
Effective Reading 293
Effective Writing. 299
Ready for Review. 304
Key Terms. 305
Review Questions 305
ASE Technician A/Technician B Style Questions . . 305

SECTION 2 Engine Repair

CHAPTER 12 Motive Power Theory— SI Engines. 310
Heat Engine Overview. 311
Principles of Engine Operation (Physics). 313
Force, Work, and Power. 315
Four-Stroke Spark-Ignition Engines 318
Components of the Spark-Ignition Engine 323
Two-Stroke Spark-Ignition Engines 335
Ready for Review. 339
Key Terms. 339
Review Questions 340
ASE Technician A/Technician B Style Questions . . 340

CHAPTER 13 Engine Mechanical Testing . . .342
Engine Mechanical Testing Overview. 343
Diagnosing Engine Noise and Vibrations 345
Engine Vacuum Tests 347
Cylinder Power Balance Test Overview 349
Cranking and Running Compression Tests 353
Cylinder Leakage Test Overview. 355
Ready for Review. 359
Key Terms. 359
Review Questions 359
ASE Technician A/Technician B Style Questions . . 360

CHAPTER 14 Lubrication System Theory . . 362
Introduction. 363
Oil 363
Types of Oil 364
Lubrication System Components 368
Oil-Certifying Bodies and Their Rating Standards . . 374
Oil Indicators. 378
Types of Lubrication Systems 381
Ready for Review. 383
Key Terms. 383
Review Questions 385
ASE Technician A/Technician B Style Questions . . 385

CHAPTER 15 Servicing the Lubrication System 387
Lubrications System - Maintenance and Repair. . . 388
Oil and Filter Change: Draining the Engine Oil . . . 392
Lubrication System Diagnosis 398
Ready for Review. 399
Key Term 400
Review Questions 400
ASE Technician A/Technician B Style Questions . . 400

CHAPTER 16 Cooling System Theory 402
Cooling System Purpose 403
Vehicle Coolant 405
Centrifugal Force Is Used to Circulate Coolant . . 410
Overview of Cooling System Components 413

Thermostat and Housing . . . 416
Cooling Fan . . . 419
Radiator Hoses . . . 421
Water Jackets . . . 425
Ready for Review . . . 428
Key Terms . . . 429
Review Questions . . . 429
ASE Technician A/Technician B Style Questions . . 430

CHAPTER 17 Servicing the Cooling System . . 432
Introduction . . . 433
Preventive Maintenance . . . 433
Checking and Adjusting Coolant . . . 436
Inspecting and Adjusting an Accessory Drive Belt . . . 439
Removing and Replacing a Thermostat . . . 445
Inspecting and Testing the Cooling Fan . . . 445
Cooling System Diagnosis . . . 451
Testing the Cooling System for Leaks . . . 454
Ready for Review . . . 456
Key Terms . . . 456
Review Questions . . . 456
ASE Technician A/Technician B Style Questions . . 457

SECTION 3 Automatic Transmissions

CHAPTER 18 Automatic Transmission Fundamentals . . . 460
Introduction . . . 461
Function of an Automatic Transmission . . . 461
Torque Converters . . . 464
Torque Converter Operation . . . 466
Lock-Up Converters . . . 468
Gear Train—Principles of Operation . . . 470
Holding/Driving Gears . . . 472
Ready for Review . . . 476
Key Terms . . . 477
Review Questions . . . 477
ASE Technician A/Technician B Style Questions . . 478

CHAPTER 19 Maintaining the Automatic Transmission/Transaxle . . . 480
Introduction . . . 481
General Transmission Maintenance . . . 481
Replacing Fluid and Filters . . . 482
In-Vehicle Transmission Repair . . . 486
Ready for Review . . . 490
Key Terms . . . 490
Review Questions . . . 491
ASE Technician A/Technician B Style Questions . . 491

CHAPTER 20 Hybrid and Continuously Variable Transmissions . . . 493
Introduction . . . 494
Hybrid Drive Systems . . . 494
Hybrid Electric Vehicle Models . . . 497
Continuously Variable Transmission (CVT) . . . 501
Ready for Review . . . 504
Key Terms . . . 505
Review Questions . . . 505
ASE Technician A/Technician B Style Questions . . 506

SECTION 4 Manual Transmissions

CHAPTER 21 Manual Transmission/Transaxle Principles . . . 510
Introduction . . . 511
The History of Manual Transmissions . . . 511
Fundamentals of Manual Transmissions . . . 512
Manual Transmission Drivetrain Layout . . . 514
Manual Transmission Components . . . 515
Clutch System . . . 517
Differential and Final Drive . . . 519
Preventive Maintenance . . . 521
Ready for Review . . . 526
Keyterms . . . 526
Review Questions . . . 527
ASE Technician A/Technician B Style Questions . . 528

CHAPTER 22 The Clutch System . . . 529
Introduction . . . 530
Clutch Principles . . . 530
Clutch Components . . . 532
Pressure Plates . . . 534
Throw-Out Bearing and Clutch Fork . . . 536
Clutch Operating Mechanisms . . . 537
Clutch Maintenance . . . 539
Ready for Review . . . 544
Key Terms . . . 544
Review Questions . . . 544
ASE Technician A/Technician B Style Questions . . 545

CHAPTER 23 Driveshafts, Axles, and Final Drives . . . 547
Drive Train Layout . . . 548
Driveline Subassemblies and Components . . . 550

Final Drives/Differentials 555
Axles and Half-Shafts 560
Inspecting and Replacing Wheel Studs and Lug Nuts 563
Joints and Couplings 563
Ready for Review 571
Key Terms 571
Review Questions 572
ASE Technician A/Technician B Style Questions . . 573

SECTION 5 Steering and Suspension

CHAPTER 24 Wheels and Tires Theory 576
Introduction 577
Tire and Wheel Physics 577
Wheels 579
Tires 584
Tire Markings 586
Tire Safety Features 590
Ready for Review 594
Key Terms 594
Review Questions 595
ASE Technician A/Technician B Style Questions . . 596

CHAPTER 25 Servicing Wheels and Tires . . 597
Tire Maintenance Preliminaries 598
Proper Tire Inflation 600
Checking for Tire Wear Patterns 604
Wheel Balance 608
Dismounting a Tire 613
Dismounting, Inspecting, and Remounting a Tire on a Wheel Equipped with a TPMS Sensor 616
Replacing a Valve Stem 619
Servicing TPMS sensors 620
Tire Diagnosis 624
Measuring Wheel, Tire, Axle Flange, and Hub Runout 626
Ready for Review 630
Key Terms 631
Review Questions 631
ASE Technician A/Technician B Style Questions . . 631

CHAPTER 26 Steering Systems Theory 633
Introduction 634
Steering System Overview 634
Steering Geometry 635
Parallelogram Steering Linkage 637
Steering Columns 640
Steering Boxes 643
Worm Gearbox 645
Power Steering 647
Electric Power Steering 651
Four-Wheel Steering Systems 654
Ready for Review 655
Key Terms 656
Review Questions 657
ASE Technician A/Technician B Style Questions . . 657

CHAPTER 27 Servicing Steering Systems . . 659
Introduction 660
Diagnosing Steering Systems 663
Maintenance and Repair 664
Perform Rack-and-Pinion Service 667
Perform Parallelogram Steering Linkage Service . . 668
Inspecting and Testing Electric Power-Assist Steering 673
Disabling and Enabling the SRS 674
Ready for Review 677
Key Terms 677
Review Questions 677
ASE Technician A/Technician B Style Questions . . 678

CHAPTER 28 Suspension Systems Theory . . 679
Suspension System Principles 680
Suspension System Components 684
Shock Absorbers and Struts 688
Adjustable Shock Absorbers 690
Control Arms and Rods 692
Types of Suspension Systems 696
Front Suspension 698
Rear Suspension 700
Active and Adaptive Suspension Systems 704
Ready for Review 705
Key Terms 706
Review Questions 707
ASE Technician A/Technician B Style Questions . . 708

CHAPTER 29 Servicing Suspension Systems 709
Suspension System Service Preliminaries 710
Diagnosis 710
Measure Ride Height 714
Unloading a Suspension to Measure Ball Joint Play 716
Removing, Inspecting, and Installing Stabilizer Components 716

Removing, Inspecting, and Installing SLA Suspension Coil Springs and Spring Insulators 719
Removing and Inspecting Upper and Lower Control Arms and Components 721
Install SLA Suspension Components 722
Inspecting the Strut Cartridge or Assembly 722
Inspecting Strut Rods and Bushings 727
Ready for Review 730
Key Terms 730
Review Questions 730
ASE Technician A/Technician B Style Questions . . 731

CHAPTER 30 Wheel Alignment 732
Wheel Alignment Fundamentals 733
Toe-Out on Turns 735
Steering Axis Inclination 737
Thrust Angle, Centerlines, and Setback (Tracking) 739
Performing a Wheel Alignment 742
Wheel Alignment Preliminaries 743
Tools 745
Performing Four-Wheel Alignment 748
Checking Secondary Alignment Angles 749
Ready for Review 752
Key Terms 752
Review Questions 753
ASE Technician A/Technician B Style Questions . . 753

SECTION 6 Brakes

CHAPTER 31 Principles of Braking 758
Introduction 759
The History of Brakes 759
Service Brakes and Parking Brakes 761
Kinetic Energy 763
Friction and Friction Brakes 765
Rotational Force 767
Types of Brake Systems 770
Ready for Review 773
Key Terms 773
Review Questions 774
ASE Technician A/Technician B Style Questions . . 774

CHAPTER 32 Hydraulics and Power Brakes Theory 776
Introduction 777
Principles of Hydraulics 777
Hydraulic Components 779
Master Cylinder 780
Quick Take-up Master Cylinders 784
Brake Pedals 785
Brake Lines 787
Hydraulic Braking System Control 790
Metering Valves 793
Brake Warning Light and Stop Lights 796
Power Brakes 797
Hydraulic Brake Booster 800
Ready for Review 801
Key Terms 801
Review Questions 802
ASE Technician A/Technician B Style Questions . . 803

CHAPTER 33 Servicing Hydraulic Systems and Power Brakes 804
Servicing Brake Hydraulic Systems 805
Brake Fluid Handling 807
Bleeding Brake Systems 810
Inspecting the Brake Pedal 813
Hydraulic System Component Diagnosis 816
Inspecting the Master Cylinder 816
Diagnosing Power Brake Systems 818
Inspecting Brake Lines and Hoses 821
Diagnosing the Brake Warning Lamp 825
Ready for Review 828
Key Terms 829
Review Questions 829
ASE Technician A/Technician B Style Questions . . 830

CHAPTER 34 Disc Brake Systems Theory . . 831
Disc Brake Fundamentals 832
Disc Brake Calipers 834
Disc Brake Pads and Friction Materials 838
Disc Brake Rotors 842
Parking Brakes on Disc Brakes 845
Ready for Review 846
Key Terms 846
Review Questions 847
ASE Technician A/Technician B Style Questions . . 847

CHAPTER 35 Servicing Disc Brakes 849
Servicing Disc Brakes 850
Maintain and Repair Disc Brakes 851
Disassembling Calipers 856
Inspecting and Measuring Disc Brake Rotors 858
Refinishing Rotors on Vehicle 863
Inspecting and Replacing Wheel Studs 864

Ready for Review . . . 869
Key Terms . . . 869
Review Questions . . . 869
ASE Technician A/Technician B Style Questions . . 870

CHAPTER 36 Drum Brake Systems Theory . . . 871
Drum Brake Fundamentals . . . 872
Types of Drum Brake Systems . . . 874
Drum Brake Components . . . 875
Wheel Cylinders . . . 877
Brake Shoes and Linings . . . 879
Drum Brake Springs . . . 882
Self-Adjusters . . . 883
Ready for Review . . . 886
Key Terms . . . 886
Review Questions . . . 887
ASE Technician A/Technician B Style Questions . . 888

CHAPTER 37 Servicing Drum Brakes . . . 889
Servicing Drum Brakes . . . 890
Maintain and Repair Drum Brakes . . . 891
Refinishing Brake Drums . . . 894
Removing, Cleaning, and Inspecting Brake Shoes and Hardware . . . 896
Removing, Inspecting, and Installing Wheel Cylinders . . . 898
Pre-Adjusting Brakes and Installing Drums . . . 900
Ready for Review . . . 903
Key Terms . . . 903
Review Questions . . . 903
ASE Technician A/Technician B Style Questions . . 904

CHAPTER 38 Wheel Bearings . . . 905
Wheel Bearing Theory . . . 906
Wheel Bearing Types . . . 907
Grease Seals and Axle Seals . . . 910
Wheel Bearing Arrangements for Rear Drive Axles . . . 913
Diagnosis . . . 914
Maintenance and Repair . . . 915
Removing and Reinstalling Sealed Wheel Bearings . . . 922
Ready for Review . . . 923
Key Terms . . . 924
Review Questions . . . 924
ASE Technician A/Technician B Style Questions . . 925

CHAPTER 39 Electronic Brake Control . . . 927
Evolution of Electronic Brake Control Systems . . . 928
Antilock Braking System Overview . . . 930
ABS Master Cylinder . . . 932
Wheel Speed Sensors . . . 936
ABS Electronic Brake Control Module (EBCM) . . 940
Traction Control System (TCS) Overview . . . 940
Electronic Stability Control (ESC) Overview . . . 943
Ready for Review . . . 945
Key Terms . . . 945
Review Questions . . . 946
ASE Technician A/Technician B Style Questions . . 947

SECTION 7 Electric

CHAPTER 40 Principles of Electrical Systems . . . 950
Importance of Learning Electrical Theory . . . 951
Electrical Fundamentals . . . 952
Movement of Free Electrons . . . 953
Volts, Amps, and Ohms . . . 955
Sources of Electricity . . . 957
Effects of Electricity . . . 959
Ohm's Law . . . 961
Electrical Power and the Power Equation . . . 963
Series Circuits . . . 964
Parallel Circuits . . . 965
Direct Current and Alternating Current . . . 967
Using Basic Electrical Concepts to Solve Problems . . . 968
Ready for Review . . . 969
Key Terms . . . 970
Review Questions . . . 971
ASE Technician A/Technician B Style Questions . . 971

CHAPTER 41 Electrical Components and Repair . . . 973
Electrical Component Preliminaries . . . 974
Circuit Protection Devices . . . 975
Relays . . . 977
Motors . . . 979
Resistors . . . 980
Wires . . . 982
Wiring Harnesses . . . 984
Wiring Diagram Fundamentals . . . 986
Wire Maintenance and Repair . . . 990
Soldering Wires and Terminals . . . 991

Ready for Review . . . 994
Key Terms . . . 994
Review Questions . . . 995
ASE Technician A/Technician B Style Questions . . 996

CHAPTER 42 Meter Usage and Circuit Diagnosis . . . 997
Introduction to Multimeters . . . 998
Digital Multimeter Purpose . . . 999
Min/Max and Hold Setting . . . 1001
Measuring Volts, Ohms, and Amps . . . 1003
Voltage Exercises . . . 1006
Resistance Exercises . . . 1008
Current Exercises . . . 1009
Perform Series Circuit Measurements . . . 1011
Perform Parallel Circuit Measurements . . . 1013
Perform Series-Parallel Circuit Measurements . . . 1015
Variable Resistors . . . 1018
Electrical Circuit Testing . . . 1018
Using a DMM to Measure Voltage and Voltage Drop . . . 1022
Locating Opens, Shorts, Grounds, and High Resistance . . . 1025
Checking Circuits with a Test Light . . . 1028
Inspecting and Testing Switches, Connectors, Relays, Solenoid Solid-State Devices, and Wires . . . 1031
Ready for Review . . . 1031
Key Terms . . . 1032
Review Questions . . . 1032
ASE Technician A/Technician B Style Questions . . . 1033

CHAPTER 43 Battery Systems . . . 1035
Introduction to the Battery . . . 1036
Low-Maintenance and Maintenance-Free Batteries and Cells . . . 1037
Battery Configurations . . . 1039
Battery Ratings . . . 1041
Battery Life . . . 1042
Battery Maintenance . . . 1044
Inspecting, Cleaning, Filling, and Replacing the Battery and Cables . . . 1046
Charging the Battery . . . 1047
Testing Battery State of Charge and Specific Gravity . . . 1050
Testing Battery Capacity Preliminaries . . . 1052
Identifying Modules That Lose Their Initialization During Battery Removal . . . 1053
Measuring Parasitic Draw . . . 1054
Ready for Review . . . 1056
Key Terms . . . 1057
Review Questions . . . 1057
ASE Technician A/Technician B Style Questions . 1058

CHAPTER 44 Starting and Charging Systems . . . 1060
Introduction to Starting Systems . . . 1061
Starter Motor Construction . . . 1063
Starter Motor Engagement . . . 1064
Armature Windings and Commutator . . . 1065
Starter Drives and the Ring Gear . . . 1066
Solenoid Operation . . . 1067
Starter Control Circuit . . . 1068
Starter Draw Testing . . . 1070
Testing Starter High-Current Circuit Voltage Drop . . . 1072
Inspecting and Testing Relays and Solenoids . . . 1074
Removing and Installing a Starter . . . 1074
Idle–Stop/Start–Stop Systems . . . 1076
Charging Systems . . . 1077
Alternator Component Overview . . . 1079
Stator . . . 1080
Rectification . . . 1082
Voltage Regulation . . . 1083
Charging System Output Test . . . 1085
Testing Charging System Circuit Voltage Drop . . 1086
Replacing an Alternator . . . 1088
Ready for Review . . . 1089
Key Terms . . . 1090
Review Questions . . . 1090
ASE Technician A/Technician B Style Questions . 1091

CHAPTER 45 Lighting Systems . . . 1092
Lighting System Introduction . . . 1093
Types of Lamps . . . 1093
Lamp/Lightbulb Configurations . . . 1096
Types of Lighting Systems . . . 1097
Driving Lights . . . 1098
Brake Lights and CHMSL . . . 1099
Turn Signal Lights . . . 1100
Headlights . . . 1102
Lighting Circuit Testing and Service . . . 1107
Courtesy Lights . . . 1107

Checking and Changing a Headlight Bulb 1111
Ready for Review 1114
Key Terms 1114
Review Questions 1114
ASE Technician A/Technician B Style Questions . 1115

CHAPTER 46 Body Electrical System 1117
Vehicle Networks 1118
Vehicle Communications Networks 1119
Using a Scan Tool 1122
Electric Accessory Motors 1124
Power Door Locks 1127
Electric Lock and Keyless Entry Systems 1128
Horn Systems 1130
Wiper/Washer System 1131
Supplemental Restraint Systems 1135
Ready for Review 1140
Key Terms 1141
Review Questions 1141
ASE Technician A/Technician B Style Questions . 1142

SECTION 8 Heating and Air Conditioning

CHAPTER 47 Principles of Heating and Air-Conditioning Systems 1146
HVAC Introduction 1147
HVAC Principles 1148
Refrigerant Principles 1151
Air-Conditioning Components and Operation . . 1153
Types of Refrigerant 1155
Heating and Ventilation System Overview 1158
Defroster 1160
Performance Testing 1161
Ready for Review 1165
Key Terms 1165
Review Questions 1166
ASE Technician A/Technician B Style Questions . 1166

SECTION 9 Engine Performance

CHAPTER 48 Ignition Systems 1170
Ignition System Introduction 1171
Primary and Secondary Circuits 1172
Required Voltage Versus Available Voltage 1174
Spark Timing 1175
Components Common to All Ignition Systems . . 1176
Spark Plug 1178
Types of Ignition Systems 1180
Engine Spark Timing 1184
Electronic Ignition Systems 1185
Components of Distributorless-Type Systems 1189
Ignition System Maintenance 1192
Ready for Review 1195
Key Terms 1196
Review Questions 1197
ASE Technician A/Technician B Style Questions . 1198

CHAPTER 49 Gasoline Fuel Systems 1200
Introduction 1201
Gasoline Fuel System Principles 1202
Gasoline Fuel 1203
Controlling Fuel Burn 1204
Fuel/Air Requirements for Internal Combustion 1205
Fuel 1205
Fuel Delivery System Components 1207
Fuel Pump 1209
Fuel Pump Relay 1210
Fuel Tank Sending Unit 1210
Fuel Lines 1211
Fuel Filter 1211
Fuel Rail 1212
Fuel Pressure Regulation 1213
Fuel Injectors 1214
Types of EFI Systems 1215
Gasoline Direct Injection Systems 1217
GDI Drawbacks 1220
Carbureted Fuel Systems 1221
Carburetor Operation 1222
Carburetor Circuits 1223
Float Circuit 1223
Idle and Off-Idle Circuits 1223
Main Metering Circuit 1223
Power Circuit 1224
Accelerator Pump Circuit 1224
The Choke 1224
Carburetor Barrels 1224
Computer-Controlled Carburetors 1226
Maintenance and Repair 1226
Inspecting and Testing Fuel Pumps 1229
Checking Fuel for Contaminants and Quality 1230

Inspecting and Testing Fuel Injectors (Non-GDI) 1231
Ready for Review 1233
Key Terms 1233
Review Questions 1235
ASE Technician A/Technician B Style Questions . 1235

CHAPTER 50 Engine Management System 1237
Introduction 1238
Digital and Analog Signals 1238
Engine Management Sensors 1240
Potentiometers 1240
Throttle Position Sensor 1240
Accelerator Pedal Position Sensor 1241
Thermistors 1241
Crankshaft Position Sensor 1242
Camshaft Position Sensor 1243
Ignition Pickup-Style Position Sensor 1243
Vehicle Speed Sensor (VSS) 1244
Oxygen Sensor 1244
Air Supply 1246
Manifold Absolute Pressure Sensor 1248
Barometric Pressure Sensor 1248
Fuel Pressure Sensor 1249
Knock Sensor 1249
Switches 1250
Powertrain Control Module 1250
Controlled Devices 1252
Relays 1252
Solenoids 1253
Control Modules 1253
Electric Motor Actuators 1255
Electronically Controlled Throttle 1255
Feedback and Looping 1255
Short- and Long-Term Fuel Trim 1257
Fuel Shutoff Mode and Clear Flood Mode 1257
Ready for Review 1258
Key Terms 1258
Review Questions 1259
ASE Technician A/Technician B Style Questions . 1260

CHAPTER 51 On-Board Diagnostics 1261
Introduction 1262
Reasons for Onboard Diagnostic Systems 1262
Onboard Diagnostic Systems 1263
OBD Terminology 1264
Diagnostic Trouble Codes 1265
Malfunction Indicator Lamp (MIL) Operation . . . 1267
Drive Cycles 1267
System Readiness Monitors 1268
Scan Tools 1269
Retrieving and Recording DTCs, OBD Monitor Status, and Freeze-Frame Data 1270
Ready for Review 1272
Key Terms 1272
Review Questions 1272
ASE Technician A/Technician B Style Questions . 1273

CHAPTER 52 Induction and Exhaust 1275
Introduction 1276
The Intake System 1276
Air Cleaner 1277
Ducting 1279
Intake Air Heating 1280
Intake Manifolds 1281
Variable Intake Systems 1282
Volumetric Efficiency 1283
Turbocharger Systems 1285
The Exhaust System 1287
Exhaust System Components 1288
Catalytic Converter 1289
The Muffler System 1291
Variable-Flow Exhaust 1292
Inspecting the Air Filter 1293
Inspecting the Induction System 1294
Inspecting the Exhaust System 1295
Ready for Review 1296
Key Terms 1297
Review Questions 1297
ASE Technician A/Technician B Style Questions . 1298

CHAPTER 53 Emission Control 1299
Introduction 1300
Composition of Air 1300
Hydrocarbons 1301
Carbon Monoxide 1301
Toxicity 1301
Oxides of Nitrogen 1302
Sulfur Dioxide 1303
Controlling Emissions 1304
Importance of Controlled Combustion 1304
Combustion Chamber Design 1305
Stoichiometric Ratio 1305
Controlling Air–Fuel Ratios 1306

Precombustion/Postcombustion Treatment 1307
Emission Control System Evolution 1307
Catalytic Converters 1308
Crankcase Emission Control 1310
Types of PCV Systems 1310
Exhaust Gas Recirculation System 1311
Purpose and Operation 1312
Evaporative Emission Control 1313
Heated Air Intake Systems 1314
Diagnosing PCV-Related Concerns 1315
Inspecting and Servicing the PCV System 1315
Ready for Review 1317
Key Terms 1317
Review Questions 1318
ASE Technician A/Technician B Style Questions . 1319

CHAPTER 54 Alternative Fuel Systems ... 1320
Introduction 1321
Alternative Fuels 1321
Energy Security 1322
Environmental Concerns 1322
Economic Concerns 1322
Vehicle Emissions and Standards 1323
Emission Standards 1323
Battery Electric Vehicles 1323
Batteries 1325
Types of Hybrids and Electric Vehicles 1325
Power Sources 1326
Plug-in Hybrid Electric Vehicles 1327
Extended-Range Electric Vehicles 1327
Fuel Cell (Electric) Vehicles 1327
Hybrid Electric Vehicles 1327
Hybrid Drive Configurations 1328
Hybrid Vehicle Efficiency Enhancements 1329
Hybrid Vehicle Operation 1329
Hybrid and Electric Vehicle Service Precautions 1329
Identifying and Disabling the High-Voltage System 1331
Hybrid Auxiliary (12V) Battery Service 1333
Ready for Review 1333
Key Terms 1334
Review Questions 1334
ASE Technician A/Technician B Style Questions . 1335

Appendix A 2017 ASE Education Foundation Automobile Accreditation Task List Correlation Guide 1336

Appendix B ASE Education Foundation Integrated Applied Academic Skills Correlation Guide 1343

Glossary 1345

Index 1374

NOTE TO STUDENTS

This book was created to help you on your path to a career in the transportation industry. Employability basics covered early in the text will help you get and keep a job in the field. Essential technical skills are built in cover to cover and are the core building blocks of an advanced technician's skill set. This book also introduces "strategy-based diagnostics," a method used to solve technical problems correctly on the first attempt. The text covers every task the industry standard recommends for technicians, and will help you on your path to a successful career.

As you navigate this textbook, ask yourself, "What does a technician need to know and be able to do at work?"

This book is set up to answer that question. Each chapter starts by listing the technicians' tasks that are covered within the chapter. These are your objectives. Each chapter ends by reviewing those things a technician needs to know. The content of each chapter is written to explain each objective. As you study, continue to ask yourself that question. Gauge your progress by imagining yourself as the technician. Do you have the knowledge, and can you perform the tasks required at the beginning of each chapter? Combining your knowledge with hands-on experience is essential to becoming a Master Technician.

During your training, remember that the best thing you can do as a technician is learn to learn. This will serve you well because vehicles keep advancing, and good technicians never stop learning.

Stay curious. Ask questions. Practice your skills, and always remember that one of the best resources you have for learning is right there in your classroom . . . your instructor.

Best wishes and enjoy!
The CDX Learning Systems Team

ACKNOWLEDGMENTS

Editorial Board

Contributors

Aims Community College
Greeley, Colorado

Addison Trail High School
Addison, Illinois

Larry Baker
Aims Community College
Greeley, Colorado

Ron Beaumont
Brisbane, Queensland
Australia

Roy Belding
King Limousine Service
King of Prussia, Pennsylvania

Michael Broud
Heritage High School
Palm Bay, Florida

Walter Brueggeman
Tidewater Community College
Chesapeake, Virginia

Casey's Independent Auto Repair
Vancouver, Washington

Kent Chambers
Northwest Technical Institute
Lowell, Arkansas

Sean Chesney
Vancouver, Washington

CJC Auto Parts
Lombard, Illinois

Clark College
Vancouver, Washington

Clark County Skills Center
Vancouver, Washington

Roger Duvall
Grayson County Technology Center
Leitchfield, Kentucky

John Frala
Rio Hondo Community College
Whittier, California

Fraser Automotive
Plainfield, Illinois

Brandon Fryman
Denver, Colorado

Tony Gumushian
Prairie State High School
Chicago, Illinois

Haggerty Automotive Group
Glen Ellyn, Illinois

Kaz Harris
Brisbane, Queensland
Australia

Ed Heim
Battleground High School
Yacolt, Washington

Kelly Herbert
WarrenTech
Arvada, Colorado

Nancy Hoffman
Dover, New Hampshire

Jim Hunnicutt
F. H. Peterson Academies of Technology
Jacksonville, Florida

Matthew Lamperd
Brisbane, Queensland
Australia

Les Schwab Tire Centers
Battle Ground, Washington
Orchards, Washington

McCord's Vancouver Toyota
Vancouver, Washington

Tom Millard
WarrenTech
Arvada, Colorado

Jennifer Miller
Port St. Lucie, Florida

Jesse Mitchell
Stark State College
North Canton, Ohio

Joe Moore
Southern Maine Community College
Portland, Maine

Kevin Murphy
Stark State College
North Canton, Ohio

Jeffrey Rehkhopf
Florida State College
Jacksonville, Florida

Ron's Automotive
Vancouver, Washington

David Sitchler
Burlington County Institute of Technology
Westampton, New Jersey

Emma Spencer
Brisbane, Queensland
Australia

Russ Strayline
Abington, Pennsylvania

Warren Tech
Lakewood, Colorado

▶ Reviewers

Michael A. Broud
Heritage High School
Palm Bay, Florida

Timothy Campbell
Wenatchee Valley Technical Center
Wenatchee, Washington

Matt Carpenter
Southern Alberta Institute of Technology
Calgary, Alberta
Canada

Frederick Cole
Tidewater Community College
Chesapeake, Virginia

Brett Colston
Oconee Fall Line Technical College
Dublin, Georgia

Al Cox
Metropolitan Community College
Omaha, Nebraska

Joe Cruz
Tennessee Department of Education/CTE
Nashville, Tennessee

Seth DeArmond
Indianapolis Public Schools Career Technology Magnet
Indianapolis, Indiana

Ken Dunn
Liverpool Community College
Everton, Liverpool
United Kingdom

Robert William Evans
Yankton Senior High School
Yankton, South Dakota

Hervey Forward
OCM BOCES, McEvoy Campus
Cortland, New York

Jerry Friesen
Brooks Composite High School
Brooks, Alberta
Canada

Joshua E. George
The Billings Career Center
Billings, Montana

Joe Glassford
Vested, LLC
Sunriver, Oregon

Curtis J. Goodwin
Northwest Kansas Technical College
Goodland, Kansas

Alan Grant
Westlake High School
Westlake Village, California

Allan Haberman
Blue Streak-Hygrade Motor Products
Winnipeg, Manitoba
Canada

Ben Haggeman
Southern Alberta Institute of Technology
Calgary, Alberta
Canada

Wade Hansma
West Central High School
Rocky Mountain House, Alberta
Canada

Kaz Harris
Brisbane, Queensland
Australia

Chance Henderson
Dauphin Regional Comprehensive Secondary School
Dauphin, Manitoba
Canada

Reginald Hildebrand
Assiniboine Community College
Brandon, Manitoba
Canada

Todd Hills
Northeast Iowa Community College
Calmar, Iowa

Robert Holm
Walla Walla University
College Place, Washington

David Howell
Tidewater Community College Regional
Chesapeake, Virginia

Kevin Human
Tennessee Technology Center at Harriman
Harriman, Tennessee

Tim Isaac
Foothills Composite High School
Okotoks, Alberta
Canada

Robert Jackson
Copper Mountain College
Joshua Tree, California

Shawn Klemm
Iowa Lakes Community College
Algona, Iowa
Algona High School
Algona, Iowa

Mark P. Lammers
Ivy Tech Community College
Evansville, Indiana

Harry Lewis
Niagara College
Niagara Falls, Ontario
Canada

Robbie Lindhorst
Southeastern Illinois College
Harrisburg, Illinois

Katherine Luhman
Montana State University
Billings College of Technology
Billings, Montana

David Macholz
Suffolk County Community College
Eastern Suffolk BOCES, Edward J. Milliken
Technical Center
Oakdale, New York

Andy Murray
Motherwell College
Motherwell, United Kingdom

Michael Myrowich
Red River College
Winnipeg, Manitoba
Canada

Larry Nobles
Tidewater Community College
Portsmouth, Virginia

Raymond H. Oviyach
Consultant, Automotive Vocational &
Technical Education
Kingwood, Texas

Joseph Palazzolo
GKN Driveline
Commerce Township, Michigan

Katherine Pfau
Montana State University
Billings, Montana

Jeffrey Rehkopf
Florida State College
Jacksonville, Florida

Mark Ridgeway
Gates Corporation
Denver, Colorado

Jon Severson
Sioux Falls Career and Technical Education Academy
Sioux Falls, South Dakota

Matthew Shanahan
College of DuPage
Bartlett, Illinois

Mark Spisak
Central Piedmont Community College
Joe Hendrick Center for Automotive Excellence
Charlotte, North Carolina

Jim Stafford
Tennessee Technology Center at Newbern
Newbern, Tennessee

Don Sykora
Morton College
Cicero, Illinois

Shane Taplin
Wellington Institute of Technology
Lower Hutt, Wellington
New Zealand

Donald Thompson
Florida State College
Jacksonville, Florida

Rob Thompson
South-Western Career Academy
Grove City, Ohio

Joe Wash
Clifton High School
Clifton, New Jersey
Lincoln Technical Institute
Union, New Jersey

Bill Weber
Seminole State College of Florida
Sanford, Florida

David W. Wharf
Thompson Rivers University
Kamloops, British Columbia
Canada

Kenneth Wurster
SUNY Canton
Canton, New York

Peter Zifovich
Campbelltown, New South Wales
Australia

Marty Zuzens
Assiniboine Community College
Brandon, Manitoba
Canada

Jeffrey Murray
ASE Master Certified
Jay M. Robinson High School

Lee Berger
ASE Master Technician
B.S. in Career and Technical Education

William Andrews
ASE Master Technician

HOW TO USE THIS TEXT

The second edition of *Fundamentals of Automotive Maintenance and Light Repair* is a comprehensive resource that covers the foundational theory and skills necessary to prepare entry-level technicians to maintain and repair today's light duty vehicles.

Chapters in the *Second Edition* have been broken down into theory and procedure. This format ensures that material can be tailored to each learning environment, leading to a more user-friendly text. Additionally, several features are included in the text to facilitate student learning. Instructors are encouraged to incorporate these features and activities into their lessons.

▶ Learning Objectives

Learning Objectives are skills, knowledge, and behaviors that translate to on-the-job requirements. Instructors should align Learning Objectives to the outcomes defined by accreditors and advisory boards. This will help instructors ensure that they have provided the training necessary for safety and competence on the job. Learning Objectives are listed at the beginning of the chapter and highlighted when supporting material appears in the chapter content. Organizing each chapter by the Learning Objectives makes the content more manageable for students and focuses their attention on the relevant information.

▶ You Are the Automotive Technician

Each chapter includes a You Are the Automotive Technician scenario and open-ended questions intended to provide relevance to the content students are about to learn. Instructors can use this feature to stimulate classroom discussion, capture students' attention, and provide an overview of key topics in the chapter.

▶ Skill Drills

Skill Drills offer a step-by-step portrayal, in words and images, of necessary skills. These are particularly helpful as students prepare to perform the task themselves for the first time, or if they need reference information in the lab. Instructors can use Skill Drills when discussing difficult steps or preparing students for the nuances of a procedure. Breaking down these processes into individual steps helps students internalize the importance of each step. The visual component further assists the student in determining what needs to be done, and how it needs to be done, at each step.

▶ Applied Math

The Applied Math feature provides a practical scenario for specific math skills in the shop. Instructors who are required to address STEM and similar GLOs will find this feature especially helpful as both a skill review and a reference tool. After the scenario is presented, the student is guided through the steps to solve the problem. This feature pulls the student's existing math skills into the automotive context and helps students apply these skills to real-world automotive situations.

▶ Technician Tips

Technician Tips add extra background information, details, and suggestions students will find helpful in both their studies and their work in the shop. These details provide an insight into the topic from technicians with years of experience.

▶ Safety Tips

Safety Tips draw attention to specific safety concerns and how to avoid injury in the shop. Instructors know that reinforcement is key for many of these very important, but less exciting, practices. Safety Tips alert students to potential hazards and remind instructors to address the dangers in class ahead of exposure in the shop.

▶ Wrap-Up

The Wrap-Up at the end of each chapter contains several components to help pull together the information learned in the chapter. In addition to highlighting key topics, the Wrap-Up gives students an opportunity to test their knowledge of the material they have learned. Instructors can use the questions in this section as a homework assignment, an in-class (individual or group) activity, or the basis of a class discussion. Students can develop their critical-thinking and problem-solving skills in the context of automotive maintenance and light repair—skills essential for success in the field.

SECTION I
Safety and Foundation

- ▶ CHAPTER 1 Careers in Automotive Technology
- ▶ CHAPTER 2 Introduction to Automotive Technology
- ▶ CHAPTER 3 Introduction to Automotive Safety
- ▶ CHAPTER 4 Personal Safety
- ▶ CHAPTER 5 Vehicle Service Information & Diagnostic Process
- ▶ CHAPTER 6 Hand and Measuring Tools
- ▶ CHAPTER 7 Power Tools and Equipment
- ▶ CHAPTER 8 Fasteners and Thread Repair
- ▶ CHAPTER 9 Vehicle Protection and Jack and Lift Safety
- ▶ CHAPTER 10 Vehicle Maintenance Inspection
- ▶ CHAPTER 11 Communication and Employability Skills

CHAPTER 1

Careers in Automotive Technology

Learning Objectives

- **LO 1-01** Outline the history of the automobile.
- **LO 1-02** Describe the careers in the automotive service sector.
- **LO 1-03** Describe each type of repair facility.
- **LO 1-04** Describe the importance of automotive industry certification and ongoing training.

ASE Education Foundation Tasks

See Appendix A to view the 2017 ASE Education Foundation Automobile Accreditation Task List Correlation Guide.

▶ Introduction

LO 1-01 Outline the history of the automobile.

Early vehicles were very basic machines. Their engines were started by manually operating a crank handle. They also needed almost continual tinkering and maintenance. As vehicle technology developed, the maintenance requirements have evolved as well (**FIGURE 1-1**). Early vehicles have many of the same basic systems as today's vehicles. This includes the engine, ignition, cooling, lubrication, suspension, and drivetrain. However, the systems on modern vehicles are much more sophisticated and reliable. This means that modern vehicles travel much further between maintenance visits than earlier models. As opposed to visits every 1000 miles, it's now every 7500–10,000 miles. In a few years, it could be up to 25,000 miles between scheduled services.

Maintenance requirements along with major repairs have decreased over time. Engines used to last 100,000 miles on average. Now they routinely last well over 200,000 miles. Better metals, machining processes, and lubricants all extend the life of the vehicle's parts. They still wear out or need to be replaced—just not as often as in years past.

Vehicles also have many more safety and convenience systems and features. This makes them much more complex. At the same time, the amount of service information has grown to cover the new systems. This requires that the technician have more knowledge to perform repairs on these vehicles. So, strong reading skills are needed to keep up with the demands of technology. It also requires a solid understanding of electrical and electronic theory and diagnosis.

FIGURE 1-1 Typical dealership shop.

▶ A Brief History of the Automobile

In the late 1800s, several engineers were working on the concept and design of the automobile. Karl Benz is generally acknowledged to have invented the modern automobile around 1885 (**FIGURE 1-2**). The concept of the automobile continued to develop in those early days. Many inventors produced various models. The early versions of the automobile were like hand-built horse carriages converted into automobiles, but with engines. Being hand-built, these early automobiles tended to lack uniformity. This made them unreliable, and expensive to buy and maintain. Therefore, they were considered a novelty that only the wealthy could afford.

FIGURE 1-2 Karl Benz is generally acknowledged to have invented the modern automobile around 1885.

You Are the Automotive Technician

A customer brings her 2016 V6 Dodge Minivan to the dealership for its 15,000-mile oil change. You pull the vehicle into the bay and set the vehicle safely on the hydraulic lift. Next, you reference the computer to check the service history and any technical service bulletins (TSBs) or recalls for the vehicle. You can see that all previous services were done on time, and there are no TSBs or recalls for the vehicle. You then verify the manufacturer's scheduled maintenance recommendations for the mileage on the vehicle. You find that the tires also need to be rotated and the air filter inspected. After you complete the tire rotation, you use the hydraulic hoist to raise the vehicle and then proceed to perform the oil and filter change. First, you drain and dispose of the old oil, remove and replace the filter, and then you add 4.5 quarts (4.32 liters) of new oil. When finished, you check all of the fluids, belts, hoses, and the air filter. You find them to be in good condition. You reset the oil life monitor on the vehicle, clean your area, and return tools to the proper location. You process the customer's invoice with notes from your inspection, tire rotation, and oil change. The service advisor reviews the work and invoice with the customer. She thanks the customer for her business and provides her with a reminder card to return for the next scheduled maintenance appointment.

1. If you had noticed a worn belt or hose during the oil change, which type of technician would you have asked to look at the vehicle?
2. In the shop, who is responsible for initially filling out a new repair order?
3. What are the job duties of a lot attendant?
4. Of the types of shops listed in this chapter, which would you prefer to work in and why?

▶ TECHNICIAN TIP

Service or maintenance intervals are also influenced by the severity of operating conditions. The more severe the conditions, the more frequent the required maintenance. Most service information gives both a normal-duty and a severe-duty maintenance schedule.

In the early 1900s, the advent of mass production made automobiles available to more people. Henry Ford applied two concepts that helped make the Model T affordable for the masses. The first was the concept of "interchangeability." This meant each part was made to the same specification so it would fit properly with its related parts. Parts could now be stockpiled, and ready for later use.

Henry Ford's second concept was the assembly line, which brought the car to the worker (**FIGURE 1-3**). This approach made assembly much more efficient. It increased the number of vehicles that could be built in a shift and lowered production costs. As the prices dropped, more powerful and reliable automobiles were produced. The popularity, and lower cost, allowed the automobile to become the preferred mode of personal transportation.

Vehicle Manufacturing

Vehicles were manufactured by small independent companies in the late 1800s. They relied on large amounts of labor and limited automation. Now, manufacturers have large-scale production lines which use extensive automation. The globalization of the automotive industry has seen manufacturers sharing models. This allows them to make vehicles that are sold across the world.

Modern assembly lines require large-scale investments. Because of this, manufacturers must be confident that consumers will buy the new vehicle model. Otherwise, the billions of investment dollars would be wasted. This is due to the need for expensive retooling of the production line (**FIGURE 1-4**).

Assembly lines use robots for many of the assembly processes, including welding seams. Assemblers continue to work up and down the assembly line, doing tasks that are still too complicated for robots.

Vehicle manufacturing is a high-volume business. Everything needs to work in the correct timing and sequence. This occurs from the supply of the required parts, down to the speed at which the production line runs. The use of sophisticated technology allows for the mass production of high-quality, affordable vehicles.

Mass production uses "just-in-time" manufacturing. With this system, vehicle manufacturers schedule (days or weeks in advance) the order that vehicles will be produced. Large-scale parts manufacturers then preassemble various parts into unit assemblies. Those units are delivered in the proper order to the vehicle manufacturer shortly before assembly. They go into the assembly lines to meet up with their specified vehicle, at just the right time. This means that the manufacturer doesn't have to store large quantities of parts on site.

Technology in Vehicles

Technology in vehicles continues to adapt and change. Consumers expect increased comfort, entertainment, and reduced impact on the environment. This adds more complex electrical, electronic, and mechanical systems. Future consumer demands and environmental pressures will continue to increase this trend. A few of the recent technological trends are:

- Hybrid Electric Vehicles (HEVs)
- Gas Direct Injection (GDI) Engines
- Electric vehicles (EVs) (**FIGURE 1-5**)
- Lane departure and blind spot warning
- Autonomous vehicles (AVs) (self-driving) (**FIGURE 1-6**)

The next 5–10 years are going to be exciting times as technology moves forward!

FIGURE 1-3 The assembly line was essential in the mass production of vehicles.

FIGURE 1-4 Modern assembly lines require large-scale investments in high-tech equipment.

FIGURE 1-5 Fully electric Tesla Model X vehicle.

FIGURE 1-6 Autonomous vehicles are likely to radically change the way we travel between places, and even own vehicles.

▶ Careers in the Automotive Sector

LO 1-02 Describe the careers in the automotive service sector.

The automotive sector provides for many career choices within the manufacturing, service, and retail industries. The manufacturing sector provides career choices from factory workers and assemblers to design engineers and senior administrators (**FIGURE 1-7**). In the service and retail sectors, jobs range from maintenance/service technicians, light vehicle technicians, and heavy vehicle technicians to service advisors and service managers. As technical complexity has grown in modern vehicles, so has the need for more specialized job roles. For example, there is now a need for hybrid technicians. They are trained to service the systems on hybrid vehicles.

Lot Attendant

One common entry-level position is as a **lot attendant**. Lot attendants work with both the sales and service side of the new or used car dealership. They are primarily responsible for keeping the vehicles in the lot organized, clean, and prepared for sale (**FIGURE 1-8**). This means that they constantly move vehicles around the lot as vehicles are testdriven and sold. They are also responsible for keeping the vehicles clean, gassed up, and charged. This keeps them ready for a testdrive or sale. When new vehicles are delivered to the dealership, lot attendants may be responsible for checking the vehicles in. This involves inspecting them for damage and missing accessories. Each vehicle's vehicle identification number is compared with the invoice.

Lot attendants must have a valid driver's license and a clean driving record. It is also important that they be cautious drivers who don't take risks, which could result in an accident. Knowing that vehicles are packed into tight spaces at most car lots means that there is not much room for error (**FIGURE 1-9**). This means that good driving skills and situational awareness are important. The vehicles typically cost between $20,000 and $80,000 each, so an accident would be costly to the dealership.

Lot attendants must also be able to take directions from many different people. They must also be able to prioritize these requests in order to achieve the goals of the dealership. This means they must be able to think on their feet, juggle a number of tasks, have a good memory, and be able to create and modify plans on the run.

Finally, being a lot attendant gives valuable experience in learning how a dealership runs. Many lot attendants move up to other jobs inside the dealership. So, becoming a lot attendant is a great way to get your foot in the door, even if you have your eye on another job in the dealership.

Lube Technician

Lube technicians carry out all aspects of manufacturer-scheduled maintenance activities. And they do it on a range of vehicle systems (**FIGURE 1-10**). Lube technicians:

- change oil and filters,
- carry out fluid inspection and fluid service,
- resetting of maintenance reminder systems, and
- and tire rotations.

While performing these duties, lube technicians also perform a visual inspection of the entire vehicle. They look for any other service needs such as worn belts, hoses, tires, and suspension system parts. When servicing vehicles, lube technicians are required to safely raise and support vehicles using hydraulic hoists or jacks. They also routinely use hand and air tools.

In addition, lube technicians typically enter time, materials, and the maintenance tasks they performed into the shop's computerized repair order system. They may also be required to assist other types of technicians with their work. Finally, lube technicians are responsible for keeping their workspace, and tools, clean and organized.

Light Line Technician

Light line technicians diagnose and replace the mechanical and electrical components of motor vehicles. Common examples include:

- gaskets,
- belts,
- hoses,

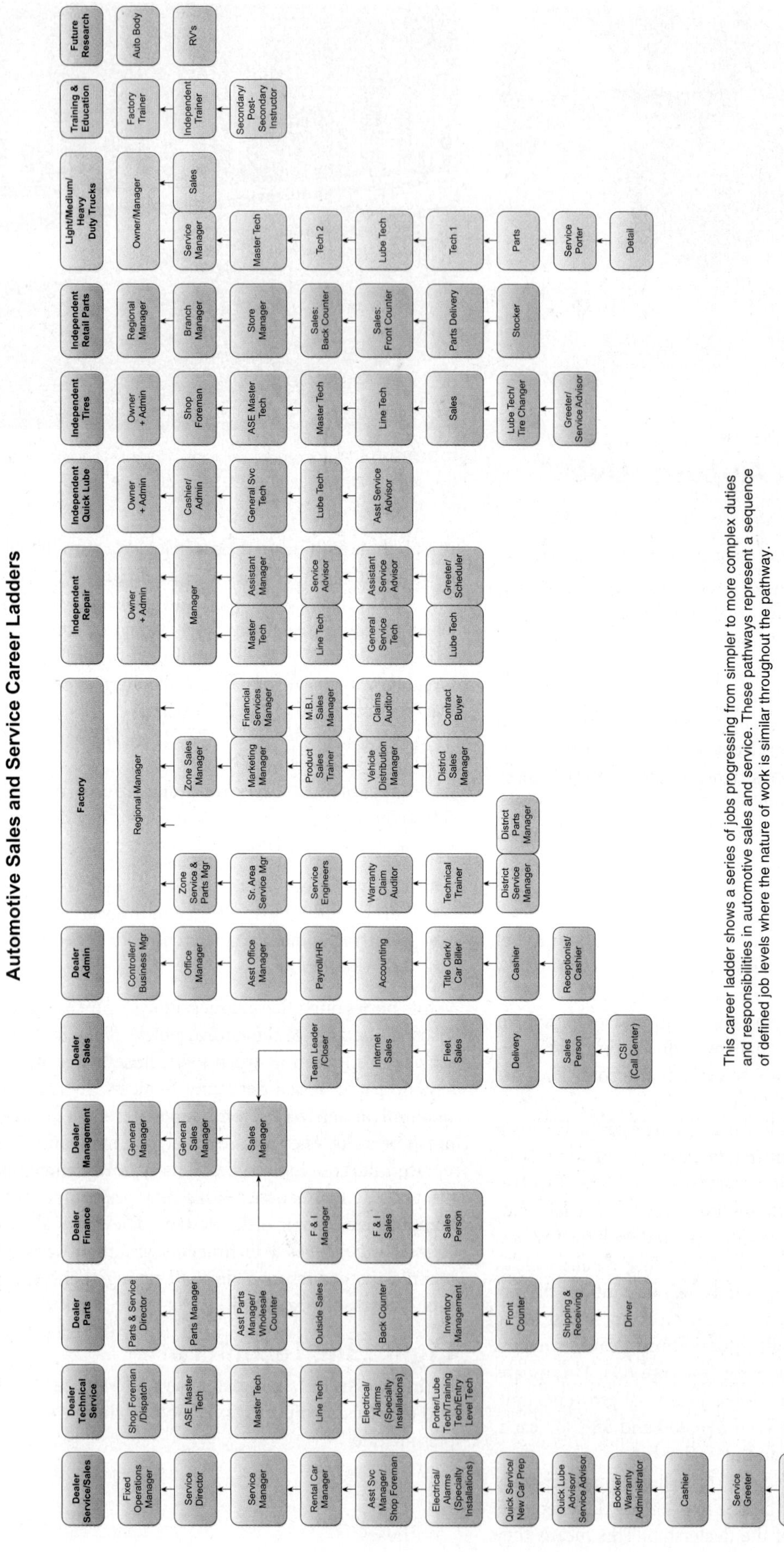

FIGURE 1-7 The opportunities for challenging and rewarding careers are endless in the automotive field.

Courtesy of Shoreline Community College, Seattle, Washington.

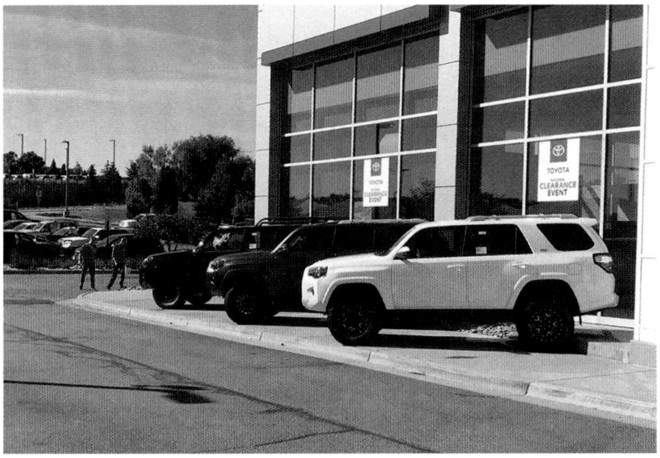

FIGURE 1-8 Lot attendants are responsible for keeping the vehicles organized, clean, and prepared for sale.

FIGURE 1-9 New car lots are notoriously packed tight, so caution when moving vehicles is very important.

- timing belts,
- water pumps,
- radiators,
- alternators, and
- starters (**FIGURE 1-11**).

In doing their job, light line technicians may be required to:

- discuss problems with vehicle owners,
- operate special test equipment, and
- testdrive vehicles to identify faults.

They also need to be able to research service information and interpret wiring diagrams. This information is used to diagnose and make repairs.

In addition, light line technicians reassemble, test, clean, and adjust repaired or replaced parts or assemblies. They use various instruments to make sure the parts are working properly. They also test and repair electrical systems such as lighting, instrumentation, vehicle sensors, and engine management systems. Finally, light line technicians inspect vehicles. They may issue state safety certificates or list the work required before a certificate can be issued.

FIGURE 1-10 Lube technicians are responsible for carrying out all aspects of the vehicle's scheduled maintenance.

FIGURE 1-11 Light line technicians diagnose and replace the mechanical and electrical components of vehicles.

Heavy Line Technician

Heavy line technicians undertake major engine, transmission, and differential overhaul and repair. They may diagnose, overhaul, repair, or replace parts and assemblies (**FIGURE 1-12**). They must be able to research service information and use the information to help determine the cause of the problem. They also reassemble, test, clean, and adjust repaired or replaced parts or assemblies. This involves using various test and measuring instruments to make sure the parts are working properly.

Some heavy line technicians are more generalized and work on a broad range of vehicles. Others specialize in particular areas by working on specific makes and models. Heavy line technicians may also specialize in particular vehicle systems, such as engines, transmissions, or final drives.

Chassis and Brake Technician

Chassis and brake technicians specialize and work primarily on the chassis and brakes of vehicles. This includes steering and suspension system repairs (**FIGURE 1-13**). They inspect, diagnose, and service these systems. In many cases, this also

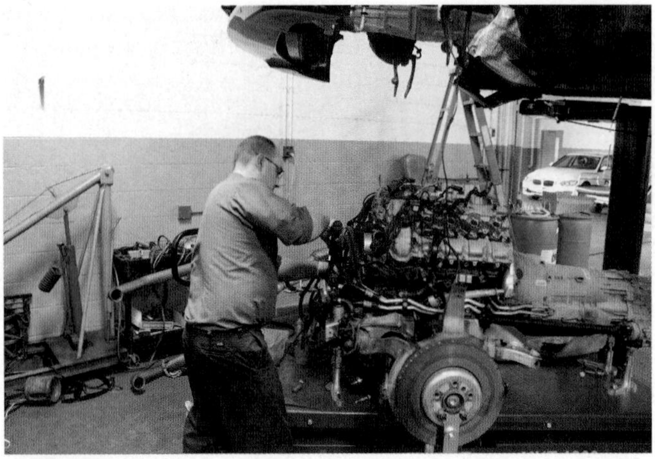

FIGURE 1-12 A heavy line technician removing an engine.

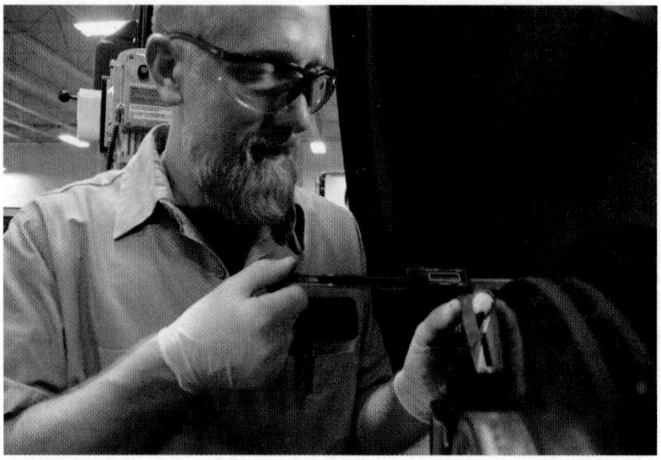

FIGURE 1-14 Brake technicians diagnose faults and repair those faults, replace or overhaul brake systems, and test the components of disc, drum, or power brake systems.

FIGURE 1-13 A chassis and brake technician performing suspension repairs.

FIGURE 1-15 Electrical technicians install, maintain, identify faults with, and repair electrical wiring and computer-based equipment in vehicles.

includes performing wheel alignments once any repairs have been performed.

Chassis technicians generally perform this work on all types of vehicles. They diagnose faults in steering and suspension systems. This begins with understanding the vehicle owner's concern and testdriving the vehicle. They can then note the vehicle's performance and compare that with their knowledge of how the components and systems function. After diagnosis, they replace faulty components. This work could include replacing bushings and servicing wheel bearings. Or it could involve checking and replacing shock absorbers or steering joints and knuckles. They also perform wheel alignments.

Brake technicians diagnose and repair faults, and replace or overhaul brake systems. This involves testing the components of disc, drum, and power brake systems (**FIGURE 1-14**). They begin diagnosis by speaking with the vehicle's owner and testdriving the vehicle. This allows them to note the vehicle's performance and compare that with their knowledge of how the components and system functions. This information is used with the service information to properly diagnose and service the vehicle.

Brake technicians also visually inspect brake units for wear, damage, or possible failure. They then repair or replace the components as required. Brake technicians can measure brake drums and disc rotors to the nearest 0.0001 (0.00254 mm). This is used to determine whether the wear or finished size meets specifications. Often, brake technicians replace leaky brake cylinders, machine rotors, and drums, when necessary. They also ensure that brake systems are filled with the correct brake fluid, are bled or flushed, and are functioning properly.

Electrical/Drivability Technician

The roles of the electrical and drivability technician may be performed by a single person. Or they may be performed by technicians who specialize in only one of the areas. For example, in larger shops, roles could be assigned to separate electrical and drivability technicians. In smaller shops, one technician could perform both roles. Often, roles cross over. For example, an electrical technician needs to understand drivability. And a drivability technician needs to understand electrical systems.

Electrical Technician

Electrical technicians diagnose, replace, maintain, and repair electrical wiring and electronic components (**FIGURE 1-15**). They diagnose charging and starting system faults as well as work with body electric systems. This includes the electrical portions of accessories such as:

- power windows/door locks,
- radios,
- air conditioning systems,
- lighting systems, and
- antitheft systems.

They may also perform drivability diagnosis and repair on computer-controlled engine management system faults. This includes faults in the fuel, ignition, and emission systems.

Often, electrical technicians use meters, oscilloscopes, and test instruments to diagnose electrical faults. They refer to service information and wiring diagrams to understand how circuits operate and are controlled. Once they have a good understanding, they perform tests to locate the cause of the fault. This requires a strong understanding of electrical and electronic theory as well as problem-solving skills. It is said that on electrical repair, the diagnosis is by far the largest part of the job, while the actual repair is usually much smaller. Electrical technicians often use solder equipment and special terminal tools when repairing electrical faults.

Drivability Technician

Drivability technicians diagnose mechanical and electrical faults that affect the performance and emissions of vehicles. They carry out maintenance activities, replace parts, and repair computer-based systems in vehicles. They work with computer-controlled engine management systems to diagnose and repair faults on electronically controlled vehicle systems. Some of these include the fuel injection, ignition, and emission control systems. (**FIGURE 1-16**).

Often, drivability technicians use electronic test equipment and circuit wiring diagrams to locate electrical, fuel, and emission systems faults. Common test equipment include scan tools, pressure transducers, exhaust gas analyzers, lab scopes, and meters. They may be required to reprogram (reflash) powertrain control modules (PCMs) using computerized equipment on a wide variety of vehicles. In doing so, they use updates supplied by the manufacturer to ensure they run properly, and within acceptable emissions limits.

Transmission Specialist

With the increasing complexity of modern transmissions, comes the need for transmission specialists. They are also required to use specialized service equipment to repair them. **Transmission specialists** diagnose, overhaul, and repair transmission units (**FIGURE 1-17**). They work on various types of manual and automatic transmissions. Transmission specialists usually specialize in either light vehicle or heavy truck transmissions.

Transmission specialists may also work on the other components of the drivetrain. Common components include driveshafts and differentials. They testdrive vehicles and listen to customer concerns. They use many of the hand tools that heavy line technicians use. They also use specialized tools to measure tolerances, check electrical circuits, and measure hydraulic pressures.

Shop Foreman

A **shop foreman** is the supervisor in a shop. Shop foremen oversee the work of all types of technicians and staff. They also communicate with customers and external suppliers. They handle the various administrative duties involved with running a shop. Some shop foremen are responsible for hiring and training new workers. They may also provide regular performance reviews (**FIGURE 1-18**). They oversee technicians' work to ensure that customers receive quality repair work. Finally, the shop foreman is responsible for enforcing safety procedures at all times. This is to avoid accidental injuries to technicians or damage to vehicles.

Service Consultant

Service consultants (advisors) work with both customers and technicians (**FIGURE 1-19**). They are the first point of contact for the customer. They provide advice and assistance to customers concerning their vehicles. Service consultants book customer

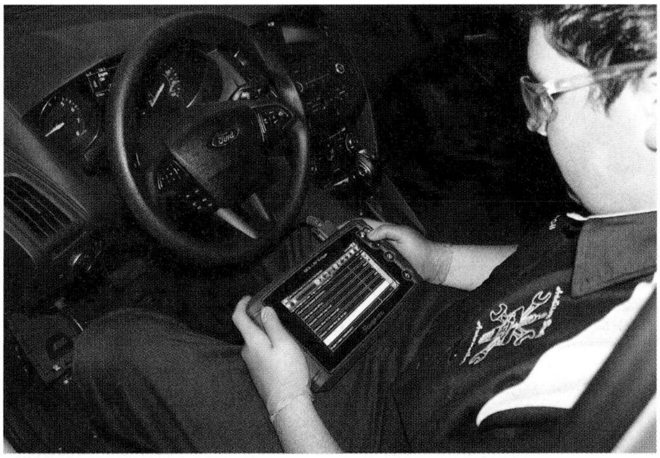

FIGURE 1-16 A drivability technician using a scan tool to diagnose a vehicle.

FIGURE 1-17 A transmission specialist rebuilding an automatic transmission.

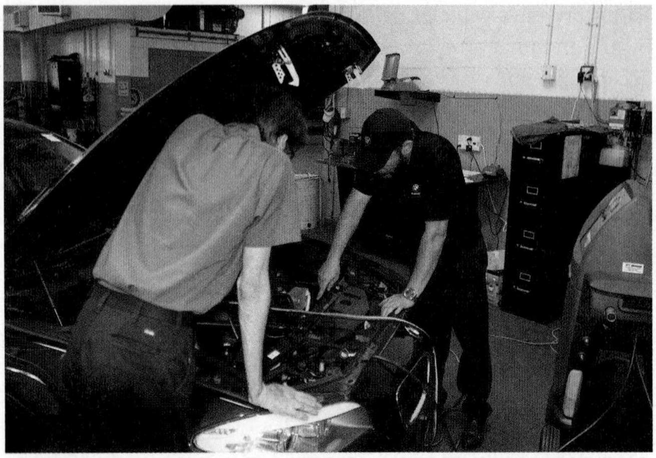

FIGURE 1-18 A shop foreman training a new technician on how to perform a procedure.

FIGURE 1-19 Service consultants work with both customers and technicians.

work into the shop. They fill out repair orders, price repairs, invoice, and keep track of work being performed. They build customer relations in order to provide a high level of customer satisfaction. They are the interface between the technician and the customer. Good communication and organizational skills are essential. A service consultant can advance to become a service manager.

Service Manager

The role of a service manager is very demanding and challenging. Service managers are responsible for the functioning of the entire service department. This career requires well-established communication skills and the ability to motivate people. They are critical in creating a positive work environment. The service manager often hires and supervises employees in the service department. They deal with any customer complaints. They are also accountable for the overall performance of the shop (**FIGURE 1-20**).

This job requires a high level of:

- personal commitment and focus,
- exceptional people skills,

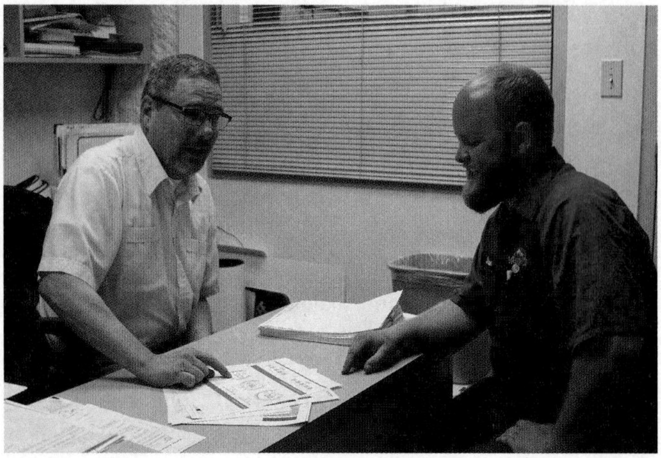

FIGURE 1-20 The service manager is accountable for the overall performance of the shop.

- excellent leadership qualities, and
- a high level of business knowledge.

Service managers may have worked their way up through the various roles within a shop. For example, the service manager may have once been a technician or a service consultant. Others may enter the position with backgrounds and college degrees in business management. In smaller shops, the service manager typically reports directly to the business owner. In larger shops, the service manager reports to the service director, who in turn reports to the business owner.

We have only described some of the most common jobs in the automotive sector. Please refer back to Figure 1-7 to identify any other jobs that you may have an interest in. Then, research their duties, requirements, and pay on the Internet.

▶ Types of Shops

LO 1-03 Describe each type of repair facility.

There are many shops that cater to specific customer needs. They can be broken down into the following types of repair facilities:

- Dealership
- Independent shop
- Specialty shop
- Franchise/retailer
- Fleet shop

Each type of shop caters to a particular segment of the industry. For example, dealerships are affiliated with a specific vehicle manufacturer. The dealership sells new and used vehicles. Technicians perform maintenance, service, and warranty repairs on that manufacturer's vehicles (**FIGURE 1-21**). Customers often only bring their vehicles back to the dealership for these services while the vehicle is in warranty.

Dealership technicians work on the latest vehicles which are at the cutting edge of vehicle technology. This can make diagnosis and repair more difficult. Because of this, manufacturers provide their technicians with additional resources. The first is providing factory training for their technicians. They attend

FIGURE 1-21 Dealership technicians perform maintenance, service, and warranty repairs on vehicles sold by a particular manufacturer.

FIGURE 1-22 Independent shops are not affiliated with vehicle manufacturers.

FIGURE 1-23 Specialty shops focus on one type of service.

FIGURE 1-24 Typical quick lube franchise business.

these training classes to learn how to maintain, diagnose, and repair these vehicles. Also, dealership technicians have instant access to the manufacturer's service information. This information is much more complete than generic service information. Finally, when dealership technicians run into a difficult diagnostic situation, they have direct access to the manufacturer's service representatives. These representatives have additional training and experience. So, they are great resources to dealership technicians.

Independent shops are not affiliated with vehicle manufacturers. So, they have limited access to new vehicle technology training (**FIGURE 1-22**). In many cases, they service a broad range of vehicles. But it is common for independent shops to limit their clientele. They usually specialize in one of the following: European, Asian, or domestic vehicles. Some shops may specialize in a particular system of the vehicle, such as brakes and alignment. Independent technicians often work on vehicles that are out of warranty. While the technicians need to be familiar with diagnosing vehicles from several manufacturers, the technology has been around for awhile. This helps offset the challenge of needing to know a variety of vehicles.

Specialty shops are usually independent shops that focus on one type of service. Common examples are transmission service, electrical system repair, or tires and wheel alignment (**FIGURE 1-23**). They usually become very experienced with the system they work on. In many cases, this allows them to provide excellent service. They typically work on a variety of vehicles from a variety of manufacturers. A shop that specializes in one area, such as air-conditioning, may experience a slow period during certain times of the year. This is appealing to some owners who enjoy vacation time.

Franchises are similar to specialty shops, but they are connected to a larger parent organization. This can help with marketing. It also provides a mechanism for warranty claims that are honored at related franchise shops across the country. Some examples include Goodyear Tire Company, AAMCO Transmissions, and Jiffy Lube (**FIGURE 1-24**).

A fleet shop works at maintaining and repairing a specific fleet of vehicles. They could be a private business that services

FIGURE 1-25 Fleet shop.

its own vehicles in-house (**FIGURE 1-25**). It could be a government agency, such as a city or county, that services its own vehicles and equipment.

It is important that all vehicles get serviced on a regular basis to prevent downtime. This is especially true for commercial vehicles that are depended upon for work. This means that most of the fleet shop work is at the maintenance and light repair level. If larger repairs are needed, some fleet shops send the vehicle to either a dealership or specialty shop. Fleet shops are typically a bit slower paced than other types of shops. This is due to the fact that fleet technicians are generally paid on an hourly basis. At the same time, most dealership and independent technicians are paid on a flat rate basis.

▶ Automotive Industry Certification

LO 1-04 Describe the importance of automotive industry certification and ongoing training.

The automotive service industry in the United States usually does not require technicians to be licensed, although some localities do. This means that a technician generally does not have to pass a licensure test in order to work in the industry. At the same time, many technicians become **Automotive Service Excellence (ASE)** certified (**FIGURE 1-26**). This certification helps technicians to demonstrate their skill. It also helps create a professional workplace. ASE is an independent, nonprofit organization. They are dedicated to improving vehicle repair by testing and certifying automotive professionals. A few localities and many shops require technicians to be ASE certified. But overall, it is a voluntary certification for technicians.

To earn ASE certification, technicians are required to:

- pass one or more ASE certification tests and
- have two years of qualifying work experience as a technician.

ASE certification needs to be renewed every five years. This is done by taking and passing recertification tests. There are currently about 300,000 certified ASE technicians in the United States.

ASE®

WE SUPPORT
PROFESSIONAL CERTIFICATION
THROUGH THE
National Institute for
AUTOMOTIVE
SERVICE
EXCELLENCE

FIGURE 1-26 ASE certifies technicians.

Courtesy of National Institute for Automotive Service Excellence (ASE).

ASE | We Support
Education Foundation

FIGURE 1-27 ASE Education Foundation certifies automotive training programs.

Courtesy of National Automotive Technicians Education Foundation (NATEF).

The **ASE Education Foundation** is an accrediting body. It certifies secondary and post-secondary automotive, diesel, and collision repair programs (**FIGURE 1-27**). The ASE Education Foundation is an independent, nonprofit organization under the umbrella of ASE. In order for programs to be accredited by the ASE Education Foundation, they have to demonstrate their compliance to a rigorous set of standards. The standards are developed by the automotive industry.

Program instructors must also maintain ASE certification in the areas they teach. Instructors must also complete at least 20 hours of technical update training each year. ASE Education Foundation accreditation is valid for five years. Then, the program must go through a reaccreditation process. Mid-way

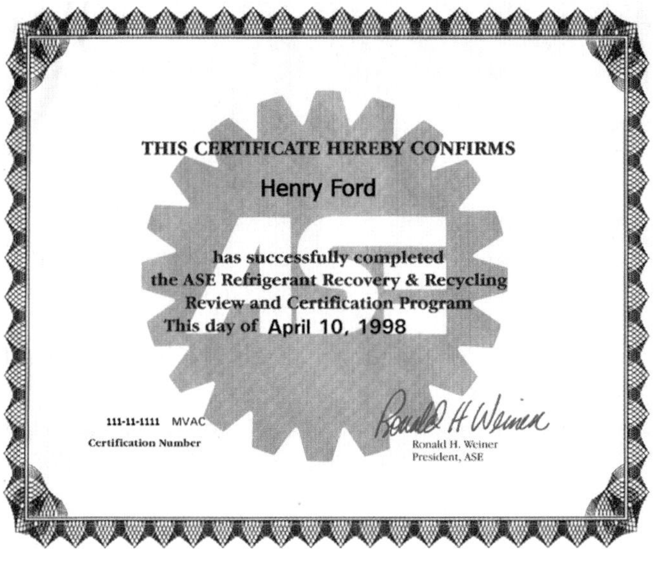

THIS CERTIFICATE HEREBY CONFIRMS
Henry Ford
has successfully completed
the ASE Refrigerant Recovery & Recycling
Review and Certification Program
This day of April 10, 1998
111-11-1111 MVAC
Certification Number
Ronald H. Weiner
President, ASE

FIGURE 1-28 EPA 609 certificate.

through the five-year accreditation cycle, the program advisory committee conducts a compliance review. The results of the review are then approved or denied by the ASE Education Foundation.

The ASE Education Foundation has recently brought another organization under its umbrella. The organization is known as **Automotive Youth Educational Systems (AYES)**. The goals of AYES have now been integrated into the mission of the ASE Education Foundation. Their mission states: "Our mission is to educate, prepare and inspire a new kind of automotive service workforce. One that embraces innovation, today's workforce demands and critical thinking." They do this by establishing Business and Education partnerships between:

- qualified high school automotive programs,
- qualified automotive dealers and aftermarket service employers, and
- qualified high school automotive technology students.

These programs may receive access to new vehicle technology and manufacturer service information. This can help prepare students for working on today's vehicles. It is also a great way for students to successfully enter the automotive workforce.

One part of this model is a 320-hour internship opportunity. Internships usually take place during the summer between the student's junior and senior year. During this internship, students work alongside a trained and qualified mentor. This is usually an ASE-certified Master Technician. By working this way, students get first-hand experience in working in a real shop. This helps prepare them for entry-level career positions. It also allows them to pursue advanced studies in automotive technology. It is a great way for students to practice what they are learning in the training program. Shops and manufacturers also benefit from working with

FIGURE 1-29 Dealership technicians at a regional training class.

highly motivated students who are preparing to go to work for them.

Special Certification

The automotive industry has two special certification requirements. These only apply to certain automotive technician specialists. To diagnose and repair vehicle emission failures in some localities, technicians must be certified by the appropriate agency. In many cases, the technician must have ASE Advanced Engine Performance certification. In other cases, the technician must have passed an approved emission specialist's course.

Second, technicians who handle refrigerants or work on air-conditioning systems need special Environmental Protection Agency (EPA) Section 609 certification. This is obtained by taking a training course and passing the 609 exam. Once certified, technicians are legally able to handle refrigerants and repair AC systems. All students should obtain their 609 certification during their training program. You can then show it to potential employers when you graduate (**FIGURE 1-28**).

Ongoing Training

Even though most technicians have graduated from automotive technology programs, they are never finished learning. Being an automotive technician means you will always be learning new things. As discussed, vehicle technology continues to change at a rapid rate. In fact, it is easy to fall behind. Training classes help technicians stay up to date.

Dealership technicians typically have access to the manufacturer's online training classes. They also get the chance to attend regional training classes (**FIGURE 1-29**). Independent technicians usually have a more difficult time accessing training. But there are both online and face-to-face classes that technicians can attend.

Also, organizations like the Automobile Service Association (ASA) hold yearly regional training conferences. These conference sessions are popular and attract some of the best minds in the industry. Because of this, they are well attended. No matter what, successful automotive technicians are always learning. By doing the same, you are preparing for an exciting and rewarding journey.

▶ Wrap-Up

Ready for Review

- Karl Benz is generally acknowledged to have invented the modern automobile around 1885. The advent of mass production made automobiles available to more people in the early 1900s.
  - The popularity and lower cost allowed the automobile to become the preferred mode of personal transportation.
- The automotive sector provides many career choices within the manufacturing, service, and retail industries. Different positons include, but are not limited to: lot attendant, lube technician, light line technicians, heavy line technicians, chassis and brake technicians, electrical/drivability technician transmission specialist, shop foreman, service consultant, and service manager.
- Shops that cater to specific customer needs can be categorized as dealership, independent shop, specialty shop, franchise/retailer, and fleet shop.
- Automotive Service Excellence (ASE) certification helps technicians demonstrate their skill and helps create a professional workplace. As vehicle technology continues to change at a rapid rate, training classes help technicians stay up to date and prepare them for an exciting and rewarding journey.

Key Terms

ASE Education Foundation The portion of ASE that evaluates and accredits entry-level automotive technology programs. It also brings together education, workplace experience, and mentorship.

Automotive Service Excellence (ASE) An independent, nonprofit organization dedicated to the improvement of vehicle repair through the testing and certification of automotive professionals.

Automotive Youth Educational Systems (AYES) An independent, nonprofit organization that is a partnership among automotive manufacturers, their dealerships, and affiliated secondary automotive programs.

Chassis and brake technicians Person who services the chassis and brake systems.

drivability technician A technician who diagnoses and identifies mechanical and electrical faults that affect vehicle performance and emissions.

electrical technician A technician who diagnoses, replaces, maintains, identifies fault with, and repairs electrical wiring and computer-based equipment in vehicles.

heavy line technician A technician who undertakes major engine, transmission, and differential overhaul and repair.

light line technician A technician who diagnoses and replaces the mechanical and electrical components of motor vehicles.

lot attendant. Person who is responsible for keeping vehicles in salable condition on a car lot.

lube technician A technician who carries out scheduled maintenance activities on a range of mechanical and related vehicle components.

service consultant/advisor A customer service worker who works with both customers and technicians; the first point of contact for customers seeking vehicle repairs.

shop foreman The supervisor in a shop who oversees the work of technicians and staff, and communicates with customers and external suppliers.

transmission specialist A technician who diagnoses, overhauls, and repairs transmissions.

Review Questions

1. When compared with early vehicles, modern vehicles do not have:
   - **a.** As much sophistication and reliability
   - **b.** As much speed
   - **c.** As many maintenance requirements
   - **d.** As many convenience systems
2. Which of the following statements is true?
   - **a.** "Interchangeability" meant that parts did not have to be custom built to match a particular car.
   - **b.** Assembly line involved workers moving to the car as it was being assembled rather than bringing the car to the worker.
   - **c.** The advent of mass production made automobiles available to only the wealthy.
   - **d.** Efficiency in mass production is increased when parts are produced where the vehicle is being assembled.
3. In "Just in Time" manufacturing:
   - **a.** There is no need for scheduling.
   - **b.** The manufacturer doesn't have to store large quantities of parts.
   - **c.** Parts are delivered a few months before assembly and in the order the supplier prefers.
   - **d.** Technology use is reduced in response to environmental pressures.
4. Which of the following would be the most likely to program or reprogram (reflash) powertrain control modules (PCMs) using computerized equipment?
   - **a.** Transmission specialists
   - **b.** Brake technicians
   - **c.** Drivability technicians
   - **d.** Light line technicians
5. All of the statements below are true EXCEPT:
   - **a.** Service consultants work only with customers
   - **b.** A shop foreman is the supervisor of the shop
   - **c.** Service managers are responsible for the functioning of the entire service department
   - **d.** A shop foreman may also be responsible for hiring and training new workers

6. A technician wants to be at the cutting edge of technology. Which type of repair facility would be the best fit for him or her?
   a. Franchise
   b. Fleet shop
   c. Independent shop
   d. Dealership
7. Jiffy Lube outlets are examples of:
   a. Dealerships
   b. Franchises
   c. Fleet shops
   d. Independent shops
8. Which of the following type of shop is a bit slower paced?
   a. Dealerships
   b. Franchises
   c. Fleet shops
   d. Independent shops
9. Special certification is required for technicians who handle which of the following systems?
   a. Refrigerants and AC systems
   b. Computer-controlled systems
   c. Automatic transmission systems
   d. Engine management systems
10. Which of the following organizations certify technicians in all areas of a vehicle
   a. EPA
   b. ASE
   c. ASA
   d. AYES

ASE Technician A/Technician B Style Questions

1. Technician A says that newer vehicles require less maintenance compared with older vehicles. Technician B says that major repairs are needed more frequently on newer engines. Who is correct?
   a. Technician A
   b. Technician B
   c. Both A and B
   d. Neither A nor B
2. Technician A says that Henry Ford is known for the invention of the gasoline engine. Technician B says that Carl Benz is credited with the invention of the automobile. Who is correct?
   a. Technician A
   b. Technician B
   c. Both A and B
   d. Neither A nor B
3. Technician A says that the production of vehicles today requires a mix of robotic and human assembly. Technician B says that most of the parts on a vehicle are preassembled into unit assemblies before they reach the assembly line. Who is correct?
   a. Technician A
   b. Technician B
   c. Both A and B
   d. Neither A nor B
4. Technician A says that only a certain few people will find jobs in the automotive industry. Technician B says that the automotive industry offers numerous career choices. Who is correct?
   a. Technician A
   b. Technician B
   c. Both A and B
   d. Neither A nor B
5. Technician A says that a technician can specialize in different areas based on his or her interest and ability. Technician B says that most specialty shops, such as transmission shops, primarily work on only one manufacturer's vehicles. Who is correct?
   a. Technician A
   b. Technician B
   c. Both A and B
   d. Neither A nor B
6. Technician A says that the shop foreman is the first point of contact for customers. Technician B says that the service consultant typically reports directly to the business owner. Who is correct?
   a. Technician A
   b. Technician B
   c. Both A and B
   d. Neither A nor B
7. Technician A says that dealership technicians generally have access to manufacturers' training courses. Technician B says that an independent shop works on a wide variety of vehicles that requires a broad skill level in technicians. Who is correct?
   a. Technician A
   b. Technician B
   c. Both A and B
   d. Neither A nor B
8. Technician A says that EPA Section 609 certification is required for all technicians. Technician B says that having ASE certifications can make it easier to get a job. Who is correct?
   a. Technician A
   b. Technician B
   c. Both A and B
   d. Neither A nor B
9. Technician A says that franchises are connected to a larger parent organization. Technician B says that independent shops usually work on vehicles that are still in warranty. Who is correct?
   a. Technician A
   b. Technician B
   c. Both A and B
   d. Neither A nor B
10. Technician A says that a technician can progress to different jobs within the industry. Technician B says that careers in the automotive industry include parts manager. Who is correct?
   a. Technician A
   b. Technician B
   c. Both A and B
   d. Neither A nor B

CHAPTER 2

Introduction to Automotive Technology

Learning Objectives

- **LO 2-01** Identify vehicle body types and their characteristics.
- **LO 2-02** List the functions of common vehicle systems and describe general vehicle operation.
- **LO 2-03** Describe drivetrain layouts and their major components.
- **LO 2-04** Describe torque and identify engine configurations.

ASE Education Foundation Tasks

See Appendix A to view the 2017 ASE Education Foundation Automobile Accreditation Task List Correlation Guide.

▶ Introduction

As you prepare to work in the automotive repair industry, you will need to learn a whole new vocabulary. Some of these words are "hygroscopic," "asymmetrical," "reciprocating motion," and "volumetric efficiency" (**FIGURE 2-1**). Look at healthcare professionals, for example. They need to know how each of the body's systems functions, the organs that make up the system, and how to diagnose an issue. Automotive technicians need to have the same level of understanding about vehicles. The good news is that automotive names are not in Latin, but it may seem like they are at times.

Knowing automotive terminology and concepts will help you fit into your new work environment. It will also help you communicate accurately with customers, suppliers, and fellow employees. You will also be able to complete repair orders, parts requisitions, and warranty paperwork as you maintain, diagnose, and repair vehicles. This chapter will help you start the process of learning automotive terminology related to vehicle types, drivetrain layouts, engine configurations, and axle arrangements.

▶ Body Designs

LO 2-01 Identify vehicle body types and their characteristics.

Vehicle bodies come in a variety of designs depending on the intended function of the vehicle. But they are also designed in such a way to accommodate style, aesthetics, and, most importantly, safety. Vehicle body design has changed over time to accommodate the owners' lifestyles and personal tastes.

Glossary Excerpt	
Amp	Ampere
Amp/hour	Amperes per hour. A standard measure for a rate of current flow
Amperage	An amount of current, expressed as amperes
Ampere	Usually called an amp, the unit for measuring electrical current
Ampere-turns	The unit of measurement for electrical magnetic field strength
Analog instrument	An instrument that displays measurements with a needle on a dial
Analog signal	An electrical signal that varies in amplitude within a given parameter
Anchor	A mounting point on a vehicle for a stressed, non-structural component such as a seat or seat belt
Anchor end	The end of a brake shoe that is attached to a fixed point on the backing plate
Anchor pin	The steel pin attached to the backing plate of drum brakes. Return springs are attached to the anchor pin and to the brake shoes to hold the shoes against the anchor pin in a non-applied position. In an applied position it prevents the shoes from rotating with the drums
Anion	A negative ion. Alkali, molten carbonate, and solid oxide fuel cells are "anion-mobile" cells
Anode	The positively charged electrode in an electrolytic cell toward which current flows
Anodize	An electrochemical process that coats and hardens the surface of aluminum
ANSI	American National Standards Institute. A privately funded organization that promotes uniform standards in areas such as measurements
Antenna	A conductive metallic structure used for radiating or receiving electromagnetic signals, such as those for radio transmissions and television signals
Anti-foam agent	An additive that reduces foaming caused by the churning action of the crankshaft in the engine oil
Antifreeze	A liquid that mixes easily with water and is used to cool the engine. It also lowers the coolant's freezing point and increases its boiling point. Coolant is typically mixed at a ratio of 50% anitfreeze and 50% water.

FIGURE 2-1 A sample list of automotive vocabulary terms.

You Are the Automotive Technician

The sales manager asks you to provide training for his/her new vehicle sales associates in your recently expanded dealership. The first training lesson takes place in the dealership showroom where several styles of vehicles are on display. You explain the concept of the unibody design and compare it to the full-frame design. You also show examples of various drivetrain layouts and their pros and cons. The second lesson is in the shop area designated for engine repair. You show the trainees the various engine classifications and configurations.

1. What are the common vehicle body types and their characteristics?
2. How does the unibody design differ from the full-frame vehicle?
3. What are the benefits of the unibody design?
4. What are the various engine configurations, and how do they differ?
5. How is a transaxle different from a transmission?

FIGURE 2-2 Vehicle body design continues to evolve to accommodate the owners' lifestyles and personal tastes. **A.** Car from the 1950s. **B.** Car from the 2010s.

A. © William Attard McCarthy/Shutterstock. **B.** © Teddy Leung/Shutterstock.

FIGURE 2-3 A sedan has an enclosed body, with a maximum of four doors.

© Maksim Toome/Shutterstock.

FIGURE 2-4 Traditionally, the coupe has two standard-size seats in front, with two smaller seats behind.

© G Fiume /Getty Images Sport/Getty Images.

An example is the Scion XB or the M-Benz Smart fortwo (**FIGURE 2-2**). Manufacturers also use vehicle body design in advertising to tempt buyers to purchase their vehicles.

Common types of body design cater to both passenger and light commercial use. Terms to describe various body designs have become part of common automotive language. However, names describing the same body design type can vary from country to country. For example, a sedan in the United States is called a saloon in the United Kingdom. Other types of body designs include:

- station wagons,
- hatchbacks,
- convertibles,
- coupes,
- vans and minivans,
- pickups, and
- sport utility vehicles (SUVs).

Sedan

A **sedan** has an enclosed body, with a maximum of four doors to allow access to the passenger compartment (**FIGURE 2-3**). The sedan design also allows for storage of luggage or other items in a trunk. It is located in the rear of the vehicle and accessible from a trunk lid. A sedan traditionally has a fixed roof. However, there are soft-top versions of sedans, which have only two doors.

Coupe

A **coupe** has only two doors. Reducing the number of doors to the passenger compartment makes the vehicle structurally more rigid. Traditionally, a coupe has two standard-size seats in front and possibly two smaller seats behind (**FIGURE 2-4**). Coupes are available in both a fixed-roof and a convertible style. They also are equipped with a trunk for storage purposes. In most cases, they are usually on the small side.

Hatchback

Hatchbacks are available in three-door and five-door designs. The odd-numbered door is a hatch that lifts up at the rear of the vehicle. This gives access to the luggage area. The rear seats usually fold down to increase the luggage area (**FIGURE 2-5**). Often the rear seat is split. This provides more flexibility because only one side needs to be folded down if the other seat is required for

FIGURE 2-5 Rear seats in hatchbacks usually fold down to increase the luggage area.

FIGURE 2-7 Hardtop convertibles have a hard roof that makes the vehicle look more like a conventional fixed-roof vehicle.

FIGURE 2-6 A convertible is an automobile that can convert from having an enclosed top to having an open top by means of a roof that can be removed, retracted, or folded away.

FIGURE 2-8 A station wagon has increased luggage capacity and a large rear door for access.

a passenger. Hatchbacks are versatile vehicles, combining some of the benefits of both sedans and station wagons.

Convertible

A **convertible** is an automobile with a roof that can be removed, retracted, or folded away (**FIGURE 2-6**). The roof is most often a flexible fabric such as canvas or vinyl. Most convertibles use electric motors that retract and raise the top. In some vehicles, known as hardtop convertibles, the roof is made up of folding or fixed steel or fiberglass panels. When in place, the hard roof makes such vehicles look more like conventional fixed-roof coupe vehicles (**FIGURE 2-7**). In other vehicles, only a smaller section of the roof area is convertible, such as a T-top. The term "roadster" was applied to a vehicle with no permanent roof covering or side windows. Nowadays, the term is most often used to describe any convertible sports car.

> ▶ **TECHNICIAN TIP**
>
> A convertible is commonly known as a cabriolet in Europe.

Station Wagon

A **station wagon** has an extended roof that goes all the way to the rear of the vehicle. It is similar to a van, but not as tall. The extra length in the roof increases the luggage capacity. In some cases, the passenger capacity is increased with extra seats in the very rear of the vehicle. Station wagons have a large rear door for easy access. And the rear seats can usually be folded to increase the storage capacity even further (**FIGURE 2-8**). Station wagons usually have fixed roofs.

Pickup

The **pickup**, or **truck**, carries and tows cargo. Usually, it has heavier-duty chassis and suspension components than a passenger car. This is used to support greater loads. Traditionally, pickups had only a single cab with two doors. This limited the number of passengers they could carry. Today's pickups have options for extended cabs or four-door versions to carry more passengers (**FIGURE 2-9**). In some cases, the four-door pickup has a reduced cargo-carrying space. This is to accommodate the extra seating in the cab.

FIGURE 2-9 More recent versions of pickups have options for extended cabs or four-door versions to carry more passengers.

FIGURE 2-10 Minivans can be configured for maximum cargo space or passengers.

Minivan

Minivans are usually lighter-duty vehicles. They have suspension systems similar to passenger cars. Alternatively, full-size vans use heavy-duty pickup truck–type suspension systems. Minivans can be configured in two different ways. The first way maximizes the number of seats for passengers. The second way maximizes the cargo space (**FIGURE 2-10**). Also, because they are of light duty, the fuel economy of minivans is substantially better than that of full-size vans.

Sport Utility Vehicle

SUVs are popular in the United States. They can easily be used to carry out functions that would otherwise require several different vehicles (**FIGURE 2-11**). They act like both a full-size van and a pickup. They have a heavy-duty chassis so they can carry heavier loads. This load can be in the form of passengers, luggage, or cargo. They can also tow moderately heavy loads. This makes them a great vehicle for family outings, as they can pull a trailer while still carrying a number of passengers and luggage.

FIGURE 2-11 Sport utility vehicles (SUVs) are designed for flexible use while being heavier duty than minivans.

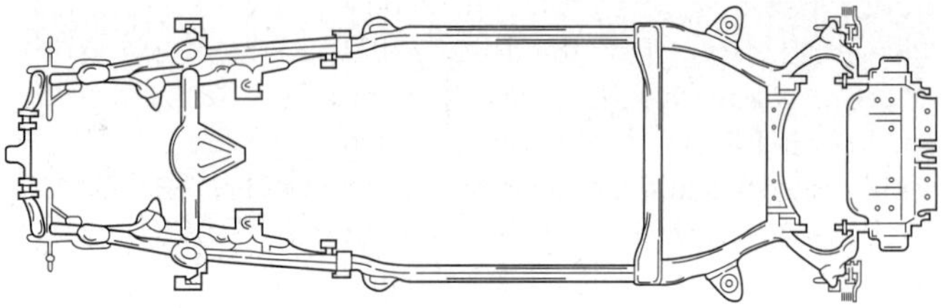

FIGURE 2-12 Steel ladder-frame chassis.

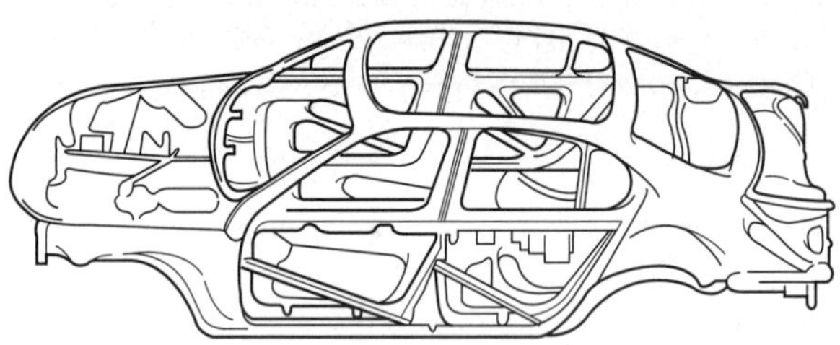

FIGURE 2-13 The unibody design.

Vehicle Chassis

A **chassis** is an underlying supporting structure for vehicles where other components are mounted. It is similar to the skeleton of a human. In a vehicle, a traditional chassis gives the vehicle structural strength. It also has a platform to mount the engine, wheels, transmission, and all the other mechanical components. Also bolted onto this frame is the body. Originally made of wood, vehicle chassis were soon changed to an open steel ladder-frame structure. This is easier to manufacture and is longer lasting (**FIGURE 2-12**).

"Body-on-frame" is the term used when a vehicle body is mounted on a rigid frame or chassis. It was the preferred way of building passenger vehicles. Manufacturers did not need to retool the structural components to release new models of vehicles with different body styles. However, by the 1960s, most manufacturers switched to vehicle designs that integrated the bodywork into a single unit with the chassis. The vehicle body became part of the vehicle structure rather than just an external skin. This is the **unibody design**, or single-shell design (**FIGURE 2-13**). The unibody design is constructed of a large number of steel sheet metal panels. They are precisely formed in presses and spot-welded together into a structural unit.

The unibody design was first used in aircraft and then spread to automobiles. This is because with less of a chassis component, it was quicker to manufacture and lighter in weight. The lighter weight meant less cost in both material and labor. Another benefit of being lighter was that the vehicles became more fuel-efficient.

Vehicle Closure Designs

A vehicle body contains many openings apart from the vehicle doors. These include:

- the engine compartment hood,
- hatch and tailgate openings,
- the fuel door, and
- in some cases a battery access cover.

All of these openings have to be secured and may require a remote switch or lever to be activated. In some cases, the access door is opened mechanically by the driver pulling a lever. This moves a cable and releases a latch. Other doors may use electric- or vacuum-operated solenoids to release the latch. In this case, the driver pushes a switch that sends an electric or vacuum signal to the release mechanism, which releases the door. Some rear hatch doors have a hinged window incorporated. This window offers easy access to the storage space without opening the entire back door.

Engine compartment hoods on modern-day vehicles usually have a remote release lever. It prevents unauthorized access to the engine compartment. This is mainly for security reasons. The release lever is usually located inside the passenger compartment:

- under the dash,
- in the glove compartment, or
- on a doorjamb (**FIGURE 2-14**).

Once the hood is open, it stays open by one of the following three methods:

- large springs on the hinges,
- pressurized gas strut assemblies, or
- a prop rod (**FIGURE 2-15**).

FIGURE 2-14 The engine compartment release may be located inside the passenger compartment under the dash, in the glove compartment, or on a doorjamb.

FIGURE 2-15 Mechanisms for holding hoods open. **A.** Springs. **B.** Gas strut.

FIGURE 2-15 Mechanisms for holding hoods open. **C.** Prop rod.

▶ Overview of Vehicle Systems

LO 2-02 List the functions of common vehicle systems and describe general vehicle operation.

A vehicle's components are arranged in systems that are designed to perform specific functions (**FIGURE 2-16**). Each of the systems is critical to the operation, passenger safety, and environmental impact of the vehicle. Knowing all the systems and their purposes will aid you when we explore them further. Here is a list of the major vehicle systems:

- Powertrain system—One of the largest systems on the vehicle. It has several smaller subsystems to get its job done. It powers the vehicle down the road and provides all on-vehicle power.
- Engine system—The power plant of the vehicle. This system converts gasoline energy or electrical energy into mechanical energy.
- Lubrication system—Lubricates the internal components of the engine for long life and quiet operation.
- Cooling system—Regulates the temperature of the engine so that it operates at the ideal temperature for efficiency and long life.
- Fuel system—Stores and delivers the proper amount of fuel to the engine's combustion chambers.
- Ignition system—Creates and delivers high-voltage sparks at the right time. This ignites the air–fuel mixture in the combustion chamber.
- Emission control system—Controls and reduces the amount of harmful emissions that the powertrain creates.
- Transmission system—Extends the vehicle's operating range by increasing either the engine's torque or the engine speed.
- Electrical system—The nerve center of the vehicle. Electricity is used in all of the vehicle's systems. It can be used to both control, and possibly power, circuits.
- Starting system—Used to crank the engine over so that it will start and run. This is usually performed by an electric starter motor.
- Charging system—Used to charge the vehicle's battery as well as run the entire vehicle's electrical system when the engine is running.
- Lighting system—Supplies lighting both outside and inside of the vehicle, including warning lights.
- Entertainment system—Provides entertainment such as audio, video, and Internet to vehicle occupants.
- Safety system—An important system on the vehicle that is designed to protect the vehicle's occupants. In some vehicles, it also provides a level of accident avoidance.
- SRS system—Supplemental restraint system, designed to restrain and cushion people in an accident.
- TPMS system—Tire pressure monitoring system, monitors tire pressure and warns the driver of tires with low pressure.
- Crash avoidance system—Monitors the area around the vehicle and predicts potential accidents. It then either alerts the driver or takes evasive actions to avoid the accident.
- Climate control system—Maintains a comfortable temperature and humidity inside the vehicle. Also clears the windshield of fog and frost.
- Braking system—Used to slow or stop the vehicle. Must be capable of performing this quickly and on all types of surfaces.
- Steering system—Allows the vehicle to track straight while making it easy for the wheels to be steered by the driver, without losing control.
- Suspension system—Keeps the tires in contact with the road surface while absorbing road harshness. It also reduces vehicle body sway and dive during vehicle maneuvers.

We cover each of these systems and more in the following chapters. This is so that you will be prepared to maintain, diagnose, and repair them. But for now, sit back and see how some of these systems work together to drive the vehicle down the road.

Vehicle Operation Overview

For a vehicle to operate, a lot of almost magical things need to happen. First, it requires a means of converting stored energy

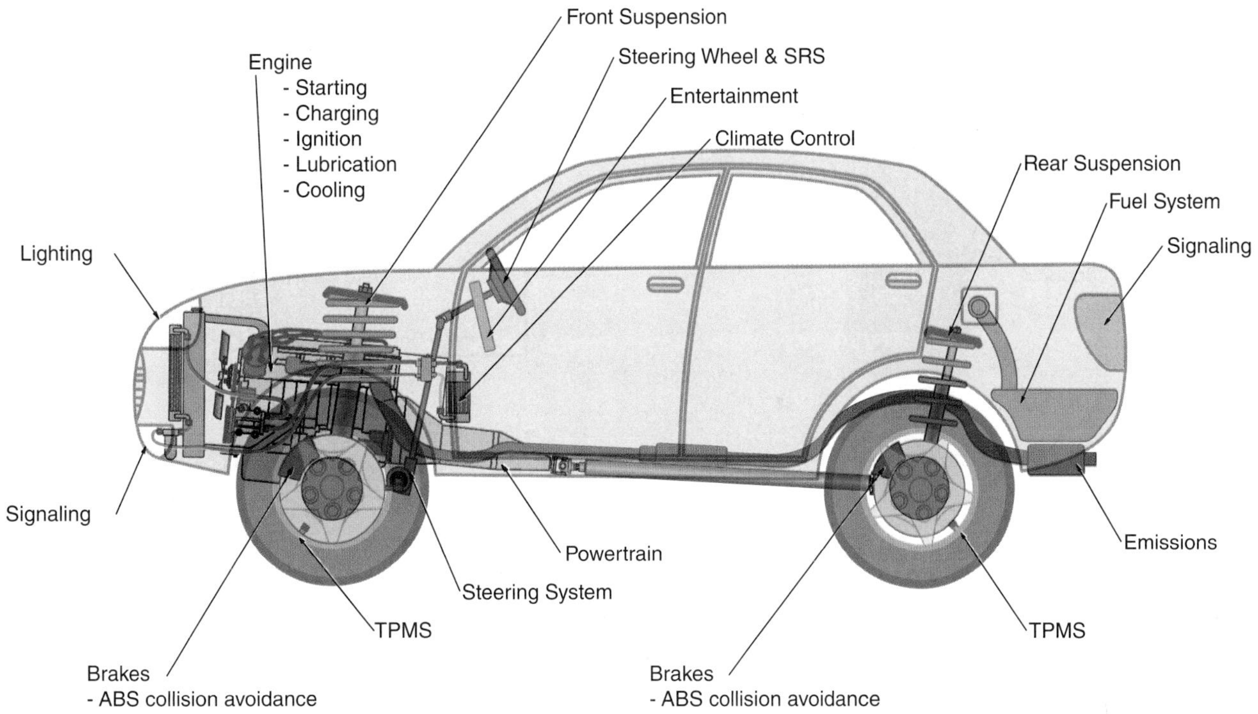

FIGURE 2-16 Vehicles are made of a variety of systems.

into a form of energy that can turn the wheels. In most vehicles, the stored energy is in the chemical form of gasoline or diesel fuel. Nowadays, natural gas, alcohol, biodiesel, hydrogen, and battery acid are used in vehicles as their stored chemical energy source.

Chemical energy can be converted into mechanical energy in two primary ways. The first is through the operation of an internal combustion engine. The second is through the operation of a battery and electric motor (**FIGURE 2-17**). Both of these methods take energy in a chemical form and convert it into mechanical energy by causing a shaft to rotate. In an internal combustion engine, it is the crankshaft that rotates. In an electric motor, it is the armature. The shaft then provides mechanical energy to move the vehicle. It also powers all the other accessories on the vehicle.

The vehicle's internal combustion engine is an engineering marvel. It combines fuel with air to create a combustible mixture. This mixture is compressed by the pistons and ignited by a spark plug in the engine cylinders (**FIGURE 2-18**). The burning, expanding gases create high pressure. This rapidly pushes the pistons down the cylinders and spins the crankshaft.

The driver initiates the starting process by turning the ignition key to run (or using a smart key). Power is sent from the

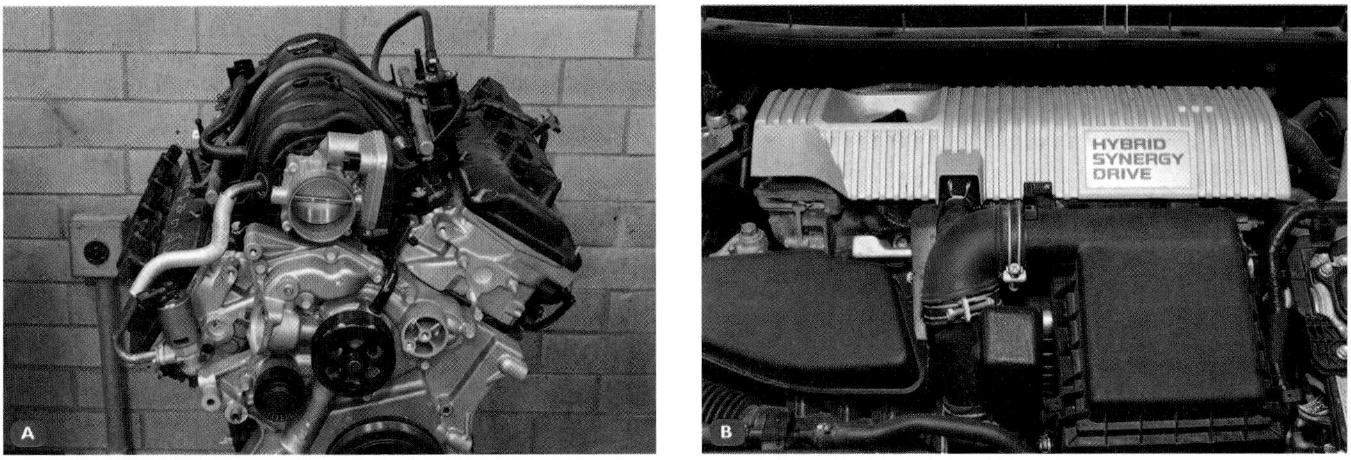

FIGURE 2-17 Stored chemical energy is converted to mechanical energy to propel the vehicle down the road. **A.** Internal combustion engine. **B.** Hybrid engine with internal combustion engine and electric motor.

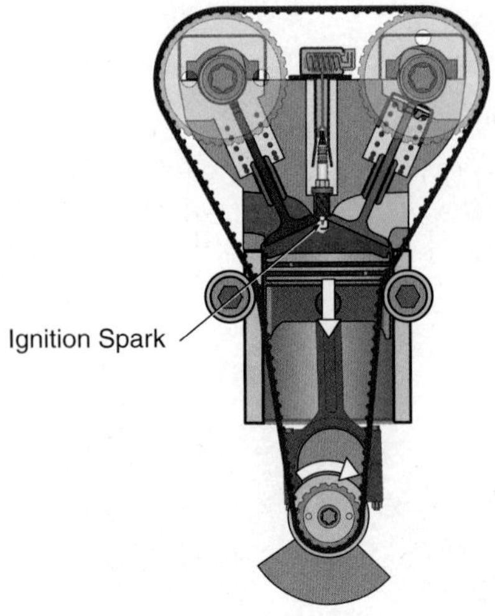

FIGURE 2-18 Internal combustion engine.

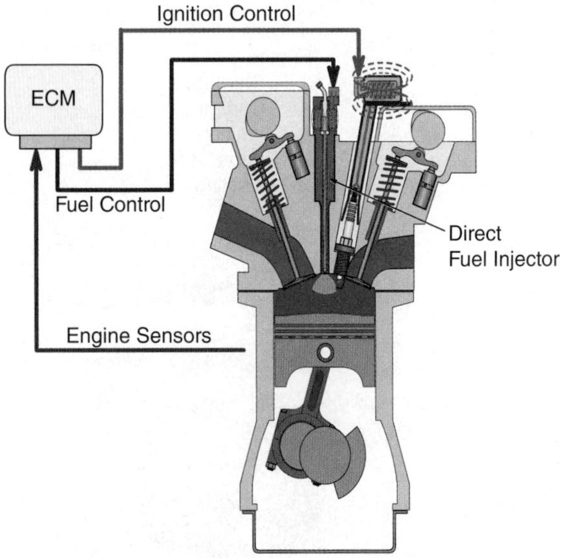

FIGURE 2-20 The engine management system controls the amount of fuel injected.

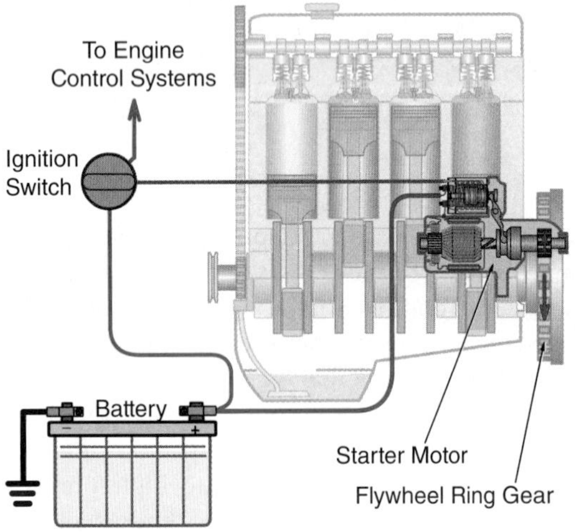

FIGURE 2-19 The starter cranks the engine over.

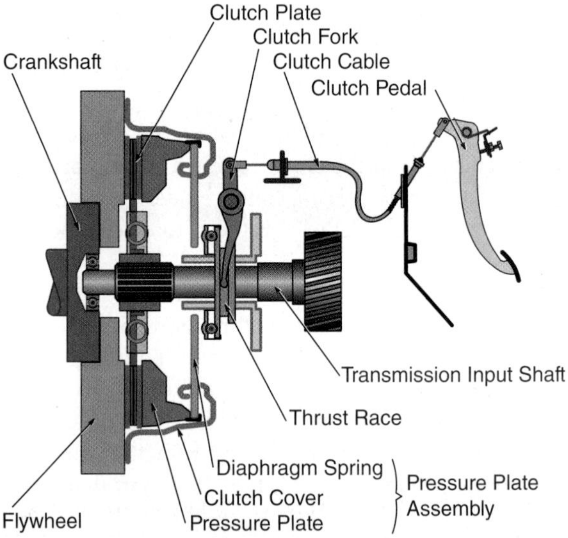

FIGURE 2-21 The crankshaft rotates the flywheel or flexplate, which sends power to the transmission.

battery to a variety of vehicle circuits, including energizing of the fuel pump. This pressurizes the fuel system in preparation for the engine to start. When the key is moved just a bit further to the crank position (or the start button is activated), the starter motor is energized. It cranks over the engine by turning the crankshaft (**FIGURE 2-19**).

The rotation of the crankshaft moves the pistons up and down. This draws air and fuel into the cylinder, which is ignited when the piston reaches the top of the cylinder. Once the cylinders start to fire off, the engine runs by itself and the starter motor is disengaged. The engine continues to run while the ignition key is in the run position. Depressing the accelerator pedal allows more air and fuel to enter the engine. This greatly increases the engine speed and power.

As the engine cranks, the engine management system monitors the many engine sensors. It makes critical decisions such as the correct timing to fire the spark plugs and when and how much fuel to be injected (**FIGURE 2-20**). The fuel mixture is ignited by a high-voltage spark (up to 100,000 volts) that is created by the ignition system. This causes the burning gases to expand very rapidly. In fact, it almost explodes. This expansion forces each piston down its cylinder, which in turn causes the crankshaft to rotate.

As the pistons move up and down, they rotate the crankshaft. This turns the **flywheel** or flexplate, which is bolted to the engine crankshaft (**FIGURE 2-21**). In a manual transmission, the flywheel transmits the engine output through a clutch. In an automatic transmission, the flexplate transmits the engine output through a torque convertor.

The vehicle's transmission takes the torque from the engine and multiplies it. This happens by running the torque through a

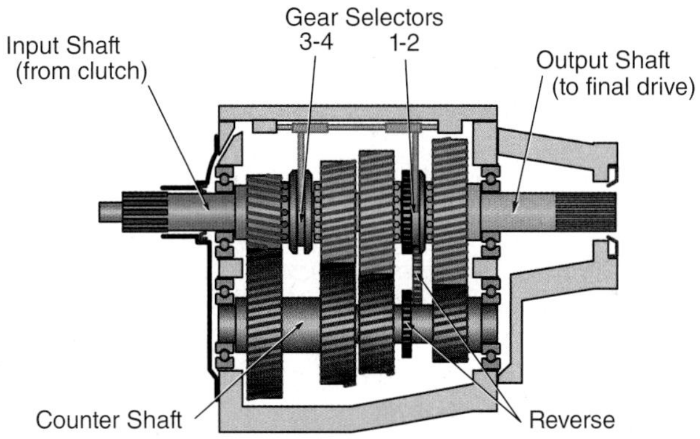

FIGURE 2-22 Transmissions transmit the engine's power through various gears.

FIGURE 2-23 Rear axle assembly sending torque from the driveshaft to the rear wheels.

number of different gear ratios. They modify the torque to allow the vehicle to pull quickly away from a stop or cruise effortlessly down the highway (**FIGURE 2-22**). The transmission connects to a final drive assembly. It divides up the power and sends it through axles to the drive wheels (**FIGURE 2-23**).

Now that the vehicle is moving at a fast speed, we need to be able to stop it. Brake assemblies are attached to the wheels. They provide a means of slowing or stopping the vehicle.

The driver operates the brakes by stepping on a brake pedal. It applies the brakes hydraulically through a master cylinder. The power booster amplifies the force from the brake pedal to the master cylinder. This increases the hydraulic pressure to the wheel brake units. The brake units apply brake friction pads to metal discs or drums connected to the wheels. This creates friction, which transforms the vehicle's kinetic energy into heat energy, slowing the wheel's rotation. The harder the driver presses on the brake pedal, the firmer the brakes are applied and the quicker the vehicle slows down (**FIGURE 2-24**).

The vehicle suspension system suspends the body above the wheels through flexible springs and pivoting links. This allows the wheels to follow uneven roads while isolating the passengers from the bumps and dips. It also maintains the orientation of the wheels so that the vehicle drives in a stable and predictable manner (**FIGURE 2-25**).

FIGURE 2-24 Brakes stopping a vehicle.

FIGURE 2-25 The suspension system absorbs road shock.

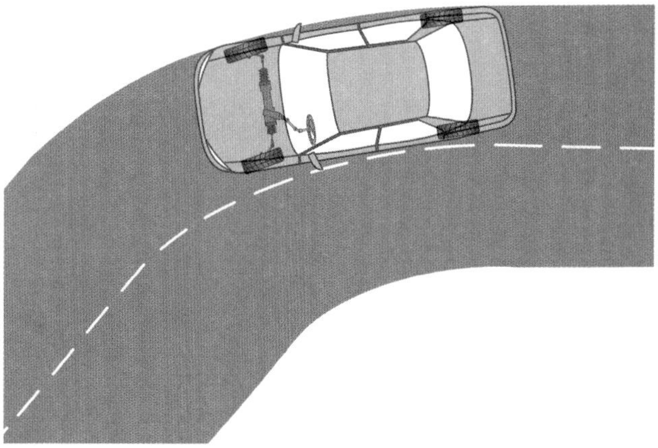

FIGURE 2-26 The steering system allows the driver to control the intended direction of travel.

The steering system connects the steering wheel to the road wheels. This occurs so that the driver can point the wheels in the intended direction of travel (**FIGURE 2-26**). Power steering systems assist the driver in steering the vehicle's wheels. They use either an engine-driven pump or electric motor to power the system.

The electrical system is the nerve system of the vehicle. It is interconnected to virtually all of the other systems. It includes the battery, which stores a supply of electricity. This is for starting the vehicle and operating the electrical accessories (**FIGURE 2-27**). A charging system, driven by the engine, creates electrical energy to feed the battery. It also runs all the other parts of the electrical system. The electrical system is part of virtually every other system on the vehicle. This includes the:

- powertrain control system,
- lighting system,

FIGURE 2-27 The electrical system operates all the systems on the vehicle.

- accessory systems,
- safety systems,
- passenger comfort systems, and
- entertainment systems.

All these systems work together to provide a safe, efficient, and enjoyable form of transportation. But because customers have different needs and desires, not all vehicles are designed to look and perform the same way. We will explore the differences in other topics.

▶ Drivetrain Layouts

LO 2-03 Describe drivetrain layouts and their major components.

The **drivetrain** includes the major assemblies that power the vehicle down the road. This includes the engine, transmission/transaxle, differential, axles, and wheels (**FIGURE 2-28**). Drivetrains are designed in different layouts based on the vehicle use and manufacturer's preferences. For example, the drivetrain layout used in pickups facilitates driving up muddy mountain roads. But the layout is different in a high-performance sports car that runs around a smooth asphalt track. Differences in configuration between each of the drivetrains' major assemblies define the drivetrain layout.

The drivetrain layout includes three main engine-mounting positions: front, mid, and rear (**FIGURE 2-29**). The front engine design is the most common in everyday vehicles. It has the engine mounted between the front wheels. Mid-engine vehicles have the engine mounted in front of the rear wheels. It provides a more equal weight distribution between the front and rear wheels. However, this occupies the passenger or cargo space. Rear-engine vehicles have the engine mounted between or behind the rear wheels.

Mid- and rear-engine designs are usually reserved for performance-type vehicles. There have been some exceptions, however, such as the Volkswagen Beetle. Manufacturers also mount engines in one of two orientations. **Longitudinal** (front to back) and **transverse** (side to side) are options. The orientation

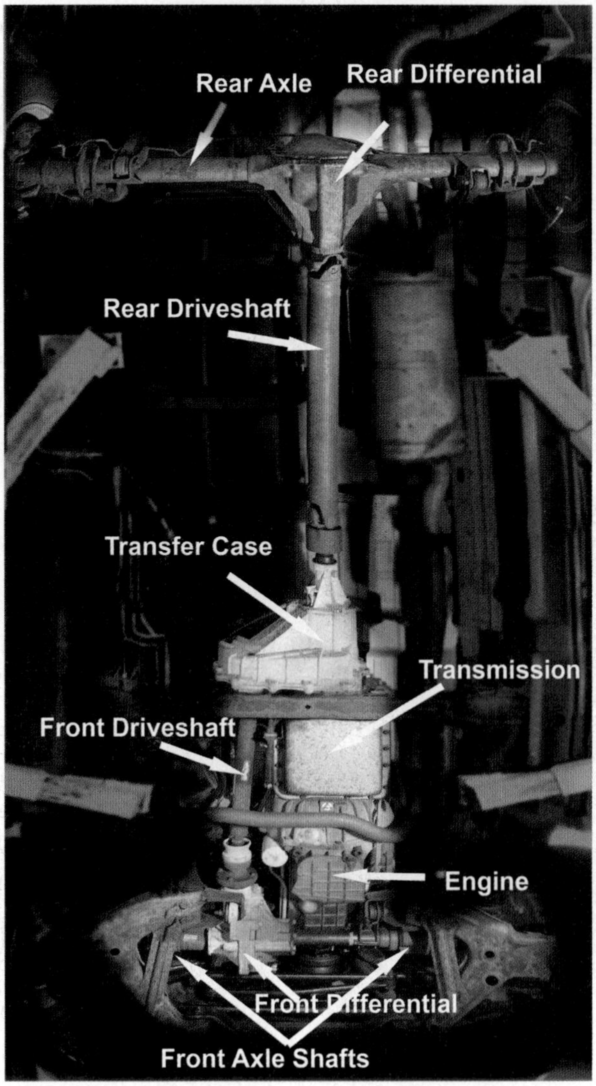

FIGURE 2-28 The drivetrain encompasses the engine, transmission, differential, axles, and wheels.

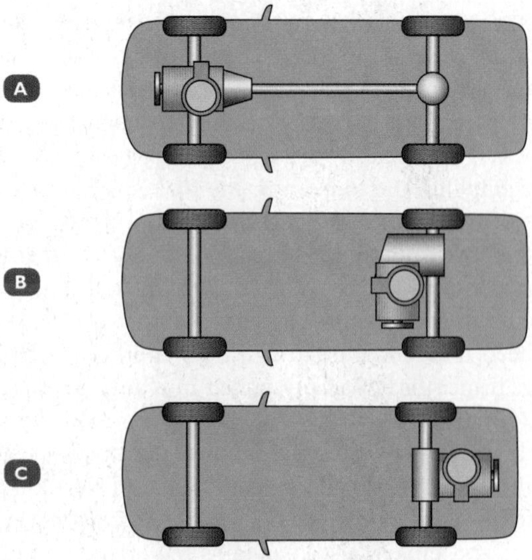

FIGURE 2-29 The three engine-mounting positions. **A.** Front. **B.** Mid. **C.** Rear.

depends on which design best fits the vehicle and the rest of the drivetrain (**FIGURE 2-30**).

The engine does not necessarily drive all four wheels. Drivetrain layouts accommodate four common drive wheel arrangements. **Front-wheel drive (FWD)** is very common in modern vehicles. In this arrangement, only the front wheels are driven by the engine. **Rear-wheel drive (RWD)** is when the engine drives only the rear wheels. Both of these arrangements are called two-wheel drive (2WD) vehicles. Becoming increasingly popular is **all-wheel drive (AWD)**, with all four wheels driven by the engine all the time. The final arrangement is **four-wheel drive (4WD)**, which is slightly different from AWD. In a 4WD vehicle, the driver can select between 2WD and 4WD.

The drivetrain layout can be defined by a combination of the following:

- engine position,
- engine orientation, and
- type of drive.

For example, using the different variations can give the following drivetrain layouts:

- Front-engine, front-wheel drive
- Front-engine, rear-wheel drive
- Front-engine, all-wheel drive/4WD
- Rear-engine, rear-wheel drive
- Rear-engine, all-wheel drive
- Mid-engine, rear-wheel drive
- Mid-engine, all-wheel drive

Transmission and Axle Configurations

In most vehicles, the engine is bolted firmly to either a transmission or a transaxle (**FIGURE 2-31**). A transmission transmits engine power to the driveshaft, final drive and differential gears, and driving axles. Transmissions are usually used in front-engine, RWD vehicles. Alternatively, a transaxle is a self-contained unit. It has the transmission, final drive gears, and differential located in one casing. It is usually used on front-engine, FWD vehicles or rear-engine, RWD vehicles. Some sports cars with front-engine, RWD have them. These vehicles have the transaxle connected to the engine by a driveshaft (**FIGURE 2-32**).

Live and Dead Axles

Vehicles can be described by the number of axles and driven wheels. Each axle typically has one wheel on each end of the axle. **Axles** come in two configurations: live axle and dead axle.

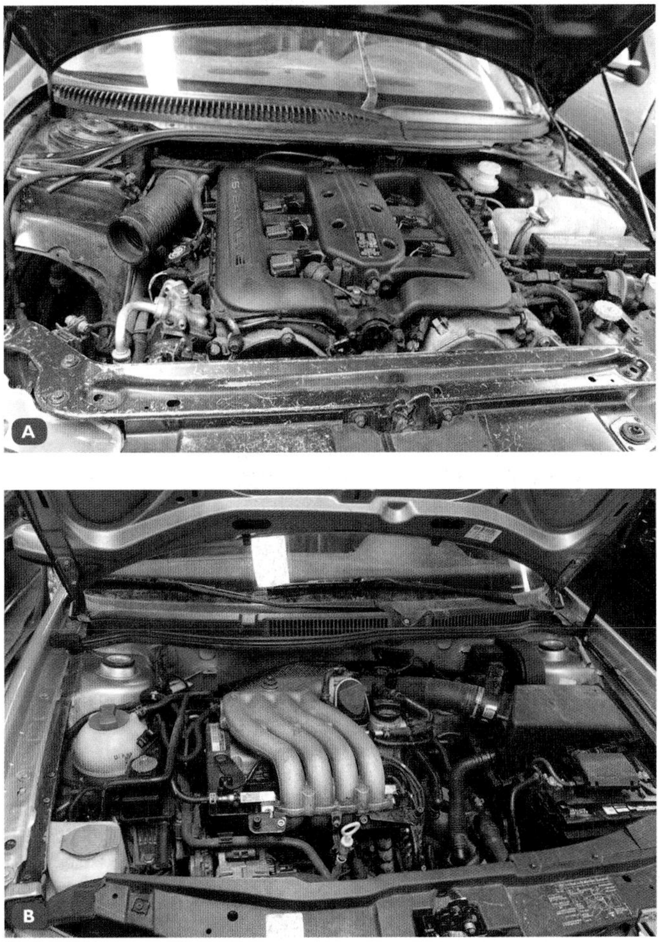

FIGURE 2-30 A. Longitudinal engine orientation. **B.** Transverse engine orientation.

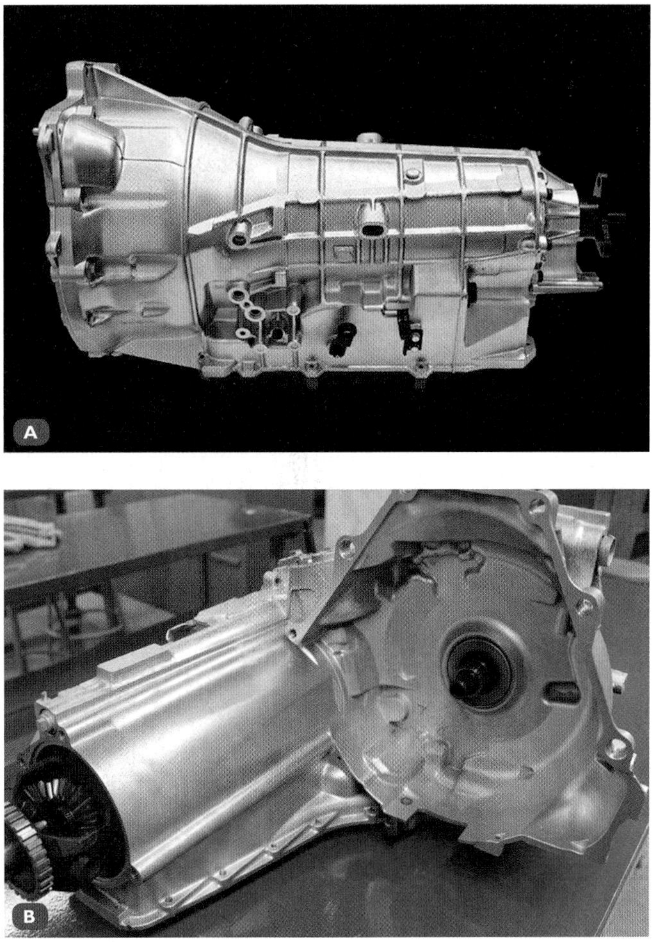

FIGURE 2-31 A. Transmission. **B.** Transaxle.

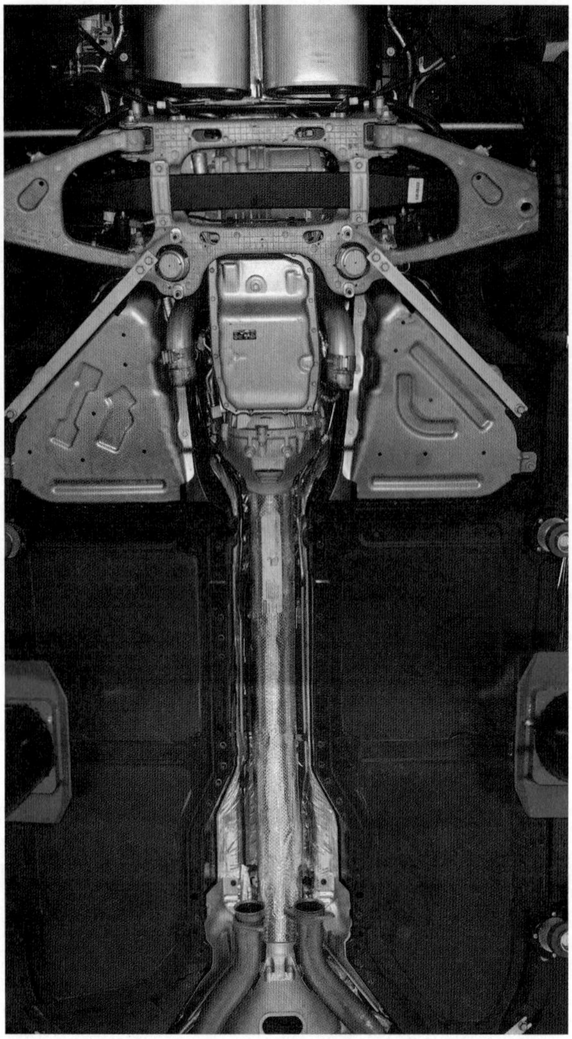

FIGURE 2-32 Front-engine, rear-transaxle arrangement used in some Corvettes.

Live axles use the engine's torque to turn the wheels (drive the vehicle) and at the same time support the weight of the vehicle. The wheels and axles on a live axle are also called drive wheels and drive axles because they propel the vehicle. Dead axles support the weight of the vehicle only while allowing the wheels to rotate freely on the axle. The wheels on dead axles are not considered drive wheels because they support only the vehicle's weight.

Most light vehicles have only two axles: one live axle and one dead axle (**FIGURE 2-33**). On commercial vehicles, the load carried on a single axle is limited by law, so vehicles with extra axles are common. A heavy vehicle may have six wheels (three axles) to support the vehicle, but only two wheels (one axle) may actually drive it. The extra axle at the rear may be used only to support the weight of the vehicle and is sometimes called a lazy axle. This vehicle is called a 6 × 2 (expressed as "6-by-2") vehicle. If the lazy axle is changed to a drive axle, this becomes a 6 × 4 vehicle. Some heavy transport vehicles have an extra steering axle, which allows even more weight to be carried.

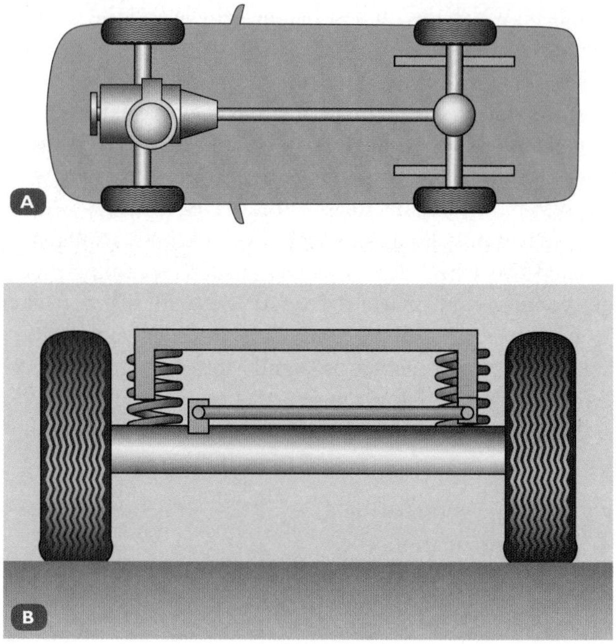

FIGURE 2-33 Most light vehicles have only two axles. **A.** Live axle. **B.** Dead axle.

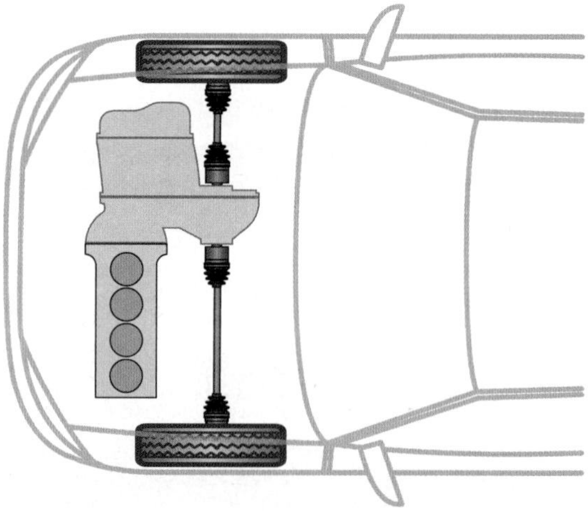

FIGURE 2-34 Front-wheel drive arrangement.

Location of Live Axles

The location of the live axle determines if the vehicle is classified as RWD, FWD, 4WD, or AWD. If the live axle is at the front of the vehicle, it is considered an FWD vehicle (**FIGURE 2-34**). FWD vehicles use the front wheels to pull the vehicle along. In light passenger vehicles, a live front axle gives it lighter body weight and increased interior room. The engine and transaxle are at the front. They can be mounted transversely, with the engine parallel to the front axle. Or they can be mounted longitudinally, where the engine is in line with the centerline of the vehicle. FWD gives good traction on the front wheels, but less traction on the rear wheels. This is especially true when braking or taking evasive maneuvers.

If the live axle is at the rear of the vehicle, it is considered a RWD vehicle. The rear wheels push the vehicle along (**FIGURE 2-35**). With the engine in front, this spreads the weight of the drivetrain assemblies throughout the vehicle. This puts more of the weight on the rear wheels. Some RWD vehicles have the engine at the rear, driving the wheels through a transaxle. Moving the engine to the rear allows a lower hood line, which may improve the aerodynamics. The increase in weight over the rear wheels can improve their traction, but at the expense of less traction on the front wheels.

The previous examples would be called 2WD vehicles, or 4 × 2. This means that there are four wheels, but only two drive wheels. On 4WD and AWD vehicles, both axles can be live, and all four wheels can drive (4 × 4) the vehicle (**FIGURE 2-36**). Driving all four wheels requires additional components and complexity, which add weight and cost. Most 4 × 4 vehicles are more expensive and less fuel-efficient than similar 4 × 2 vehicles.

A 4WD vehicle can drive all four wheels. It has a driveshaft, a final drive and differential gears, and axles for both the front and rear axle assemblies. The transfer case controls the drive to the front and rear axles. Part-time 4WD means the vehicle is usually driven in 2WD and switched to 4WD when needed. This is accomplished by engaging the transfer case. It has a selector lever to select 2WD or 4WD. The transfer case locks the driveshafts together and directs the torque to both axles. Some transfer cases also allow for selection of high and low ranges, which changes the gear ratios to the wheels. This is helpful when driving in rough off-road conditions, where vehicle speed is very slow.

AWD vehicles also provide drive to all four wheels of the vehicle. AWD systems should not be confused with part-time 4WD vehicles. They cannot normally be manually disconnected or deselected. The AWD vehicle can be used on hard pavement, with drive to all wheels. This is because the transfer case employs a center differential unit. It allows the front and rear axles to rotate at slightly different speeds.

There are various methods of splitting the drive between the front and rear wheels. Some transfer cases use an electronically controlled multi-plate clutch. Others use a **viscous coupling**. Both of these devices allow a small difference in speed between the front and rear axles when the vehicle is turning, but not much. They are not designed to handle large differences in speed, such as when one or more of the wheels lose traction. The driver can still temporarily lock the front and rear axles together. This is accomplished by moving a separate lever, as in a conventional 4WD, or by moving the main gear selector. A device called a differential lock is used.

Some AWD sedans use a front engine and transaxle, with a driveshaft connected to drive the rear wheels (**FIGURE 2-37**). These cars are lighter and less rugged than conventional off-road types and usually operate at higher speeds. The drive to all four wheels provides better-balanced handling and traction for cornering in slippery conditions.

Regardless of the type of system fitted, the aim is the same. It is to provide drive to all four wheels constantly while the vehicle is in motion. The power is not necessarily split 50/50 between the front and rear wheels. Torque is usually split 60% to front wheels and 40% to rear wheels. Some vehicles include sophisticated traction and torque control systems. These are used to maintain effective traction of the wheels to the road under all driving conditions.

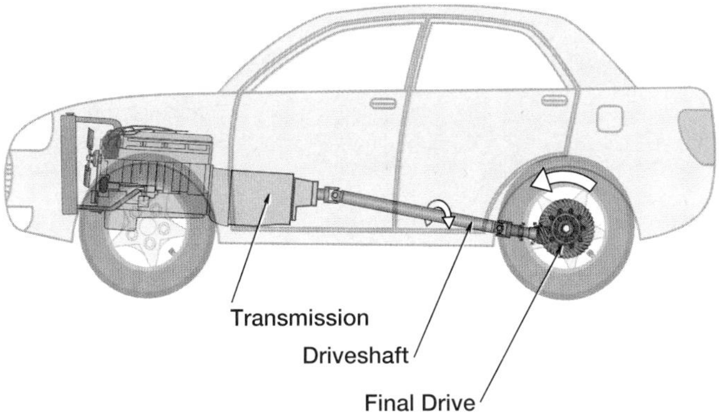

FIGURE 2-35 Rear-wheel drive arrangement.

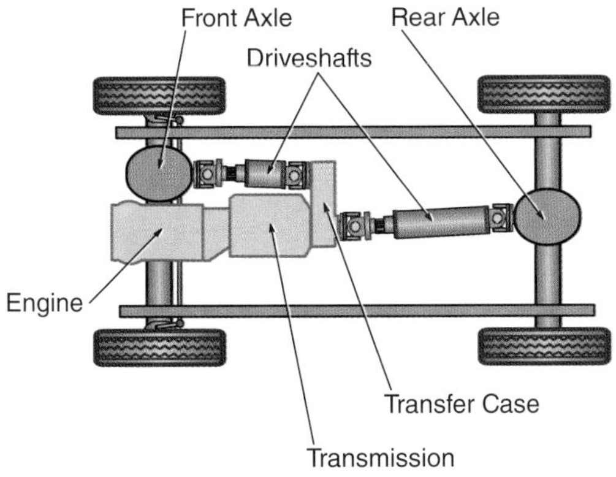

FIGURE 2-36 Four-wheel drive vehicle.

FIGURE 2-37 All-wheel drive layout using a modified transaxle.

Transmissions and Final Drives

A vehicle with a manual transmission uses a clutch to engage and disengage the engine from the transmission (**FIGURE 2-38**). Engine torque is transmitted through the clutch to the transmission or transaxle. The transmission or transaxle contains sets of gears that increase or decrease the torque. This allows the vehicle to have more pulling power in low gears and higher road speed in higher gears. Drive from the transmission is then transmitted to the rest of the drivetrain. The lower the gear selected, the higher the torque transmitted. A vehicle starting from rest needs a lot of torque. But once it is moving, it can maintain the speed with only a relatively small amount of torque. A higher gear can then be selected and engine speed reduced.

A front engine, rear wheel drive vehicle uses a **driveshaft**. It is used to transmit torque from the transmission to the final drive. The **final drive** provides a final gear reduction to multiply the torque before applying it to the drive axles (**FIGURE 2-39**). On front-engine, RWD vehicles, the driveshaft is fitted down the centerline of the vehicle. The final drive at the rear of the vehicle changes the direction of the drive by 90 degrees. It changes it from the center of the vehicle out to the wheels via the axles. Inside the final drive, a **differential gear set** divides the torque to the axles. It allows for the difference in speed of each wheel when cornering.

Axles transmit the torque to the driving wheels. In a RWD vehicle, the axles can be solid or contain joints to allow for movement of the suspension. For a transaxle-equipped vehicle, each driveshaft has movable joints. They allow for suspension and steering movement.

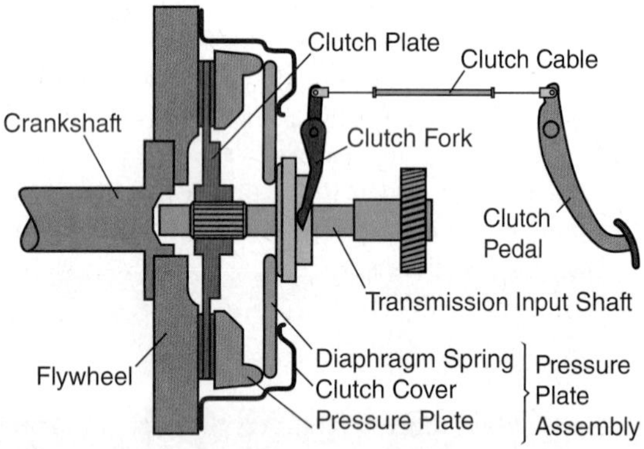

FIGURE 2-38 A vehicle with a manual transmission uses a clutch to engage and disengage the engine from the drivetrain.

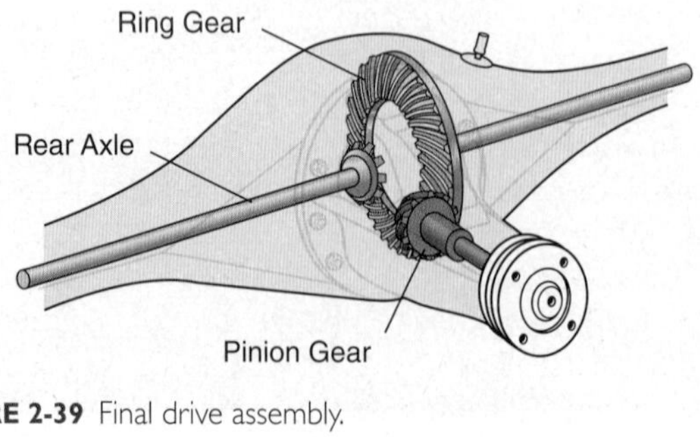

FIGURE 2-39 Final drive assembly.

Note: Differential gears and bearings removed from the center of ring gear for illustration purposes.

An automatic transmission performs similar functions to a manual transmission. The differences are that gear selection is automatically controlled either hydraulically or electronically. The automatic transmission uses a **torque converter** (instead of a clutch). This acts as a hydraulic coupling. It transfers the drive from the engine to the transmission hydraulically (**FIGURE 2-40**). Automatic transmissions will be covered in more depth in the Automatic Transmission section.

▶ Torque

LO 2-04 Describe torque and identify engine configurations.

Torque is the twisting force applied to a shaft. In a vehicle, torque is used to drive the vehicle down the road (**FIGURE 2-41**). Torque is created in the engine when combustion of the air–fuel mixture causes high pressure to push the pistons down the cylinders. This applies a twisting force to the crankshaft. The torque in the crankshaft is then used to twist the gears in the transmission. It then twists the gears in the final drive assembly. Finally, it is used to twist the axles, and therefore to twist the wheels and tires. They, in turn, drive the vehicle down the road.

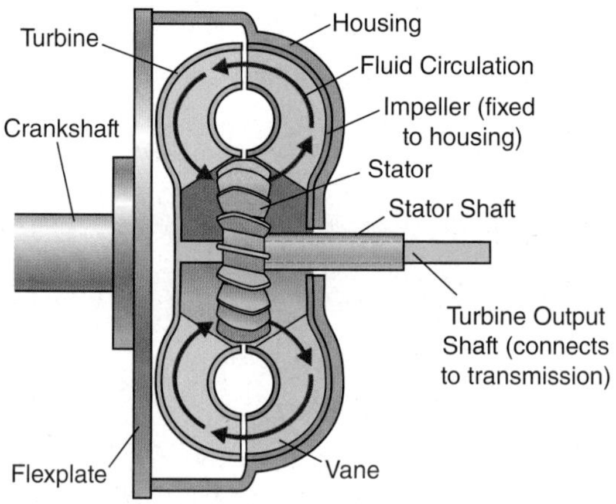

FIGURE 2-40 Torque converter.

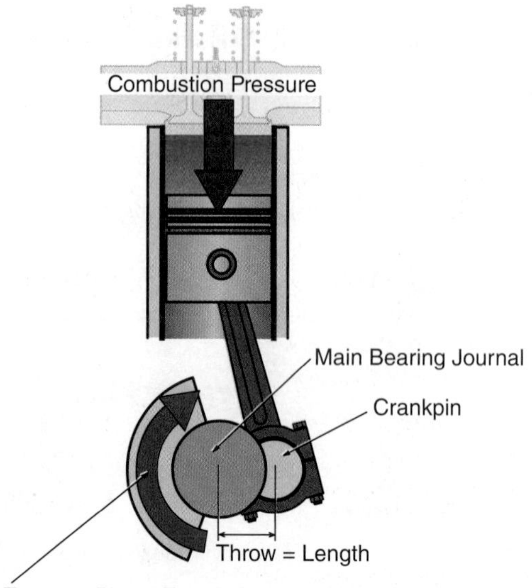

FIGURE 2-41 Torque applied to a shaft.

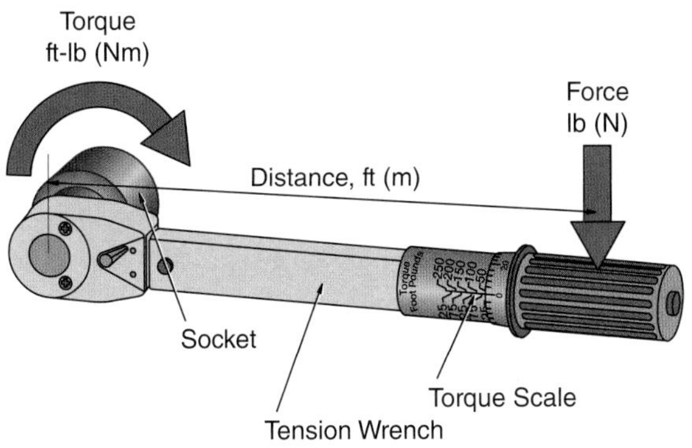

FIGURE 2-42 Torque applied to a bolt.

TABLE 2-1 Torque Conversions

12 in-lb	1 ft-lb
1 in-lb	0.08 ft-lb
1 Nm	0.74 ft-lb
1 Nm	8.8 in-lb

Torque is also used by technicians when tightening bolts and nuts (**FIGURE 2-42**). Most of the bolts and nuts on the vehicle have a manufacturer-specified torque that they must be tightened to. This is accomplished using a special wrench that can measure torque. Thus, torque is an important concept for technicians to understand. Torque in the imperial system is measured by the foot-pound (ft-lb) and inch-pound (in-lb). In the metric system, it is measured by the Newton meter (Nm). **TABLE 2-1** lists standard torque conversions.

Torque Measurement

The measure of torque is based on the equivalent twisting force exerted by an amount of weight (mass) applied to a perpendicular lever of a given length. The following designations are used in measuring torque:

- A foot-pound (ft-lb) is the twisting force applied to a shaft by a lever 1-foot long with a 1-pound mass on the end.
- An inch-pound (in-lb) is the twisting force applied to a shaft by a lever 1-inch (25.4 mm) long with a 1-pound (0.45 kg) mass on the end.
- A Newton meter (Nm) is the twisting force applied to a shaft by a lever 1-meter (3.3 ft) long with a force of 1 Newton applied to the end of the lever. (1 N is equivalent to the force applied by a mass of 102 g, or 0.102 kg.)

Classifying Engines

Engines create torque to drive the vehicle and power the accessories. Engines come in a variety of shapes and sizes, but they all create torque. Engines are classified by:

- type,
- cylinder arrangement,
- number of cylinders/rotors, and
- total engine displacement in cubic inches or liters.

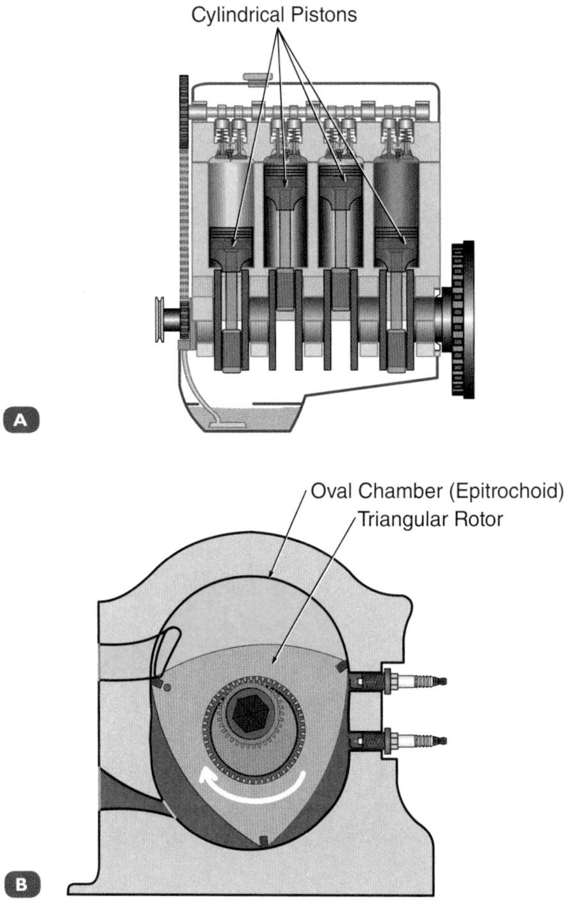

FIGURE 2-43 A. Piston engine. **B.** Rotary engine.

Each of these factors affects the torque that the engine produces. Two common types of engines are piston and rotary (**FIGURE 2-43**). The piston engine uses cylindrical pistons that move up and down in cylinder bores. The rotary engine uses a roughly triangular rotor that turns inside a roughly oval chamber. Most automotive engines are of the piston type rather than the less common rotary type. This is because they have a longer life and produce fewer emissions.

Piston Engines

In a **piston engine**, the way engine cylinders are arranged is called the **engine configuration**. Multi-cylinder internal combustion automotive engines are produced in four common configurations:

- In-line—The pistons are all in one bank on one side of a common crankshaft (**FIGURE 2-44**).
- Horizontally opposed—The pistons are in two banks on both sides of a common crankshaft (**FIGURE 2-45**).
- V—The pistons are in two banks on opposite sides, forming a deep V with a common crankshaft at the base of the V (**FIGURE 2-46**).
- VR and W—In a VR engine, the pistons are in one bank but form a shallow V within the bank. The W engine consists of two VR banks in a deep V configuration with each other (**FIGURE 2-47**).

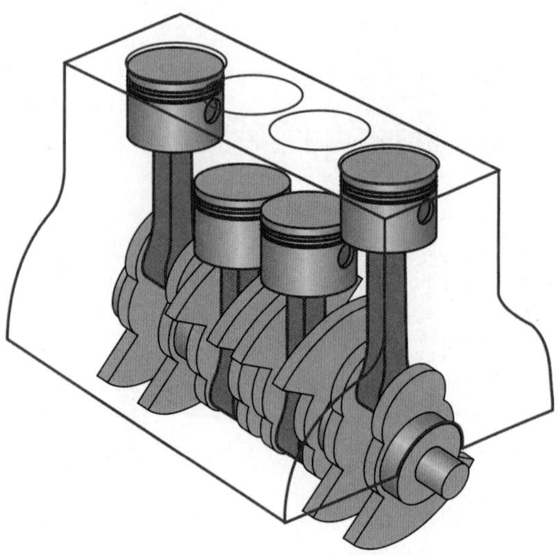

FIGURE 2-44 In-line engine.

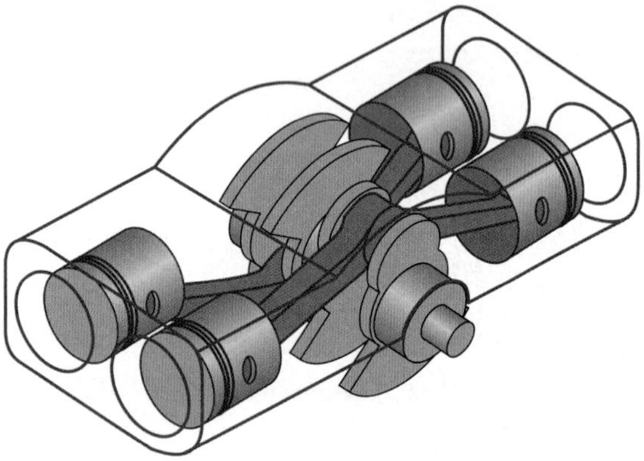

FIGURE 2-45 Horizontally opposed engine.

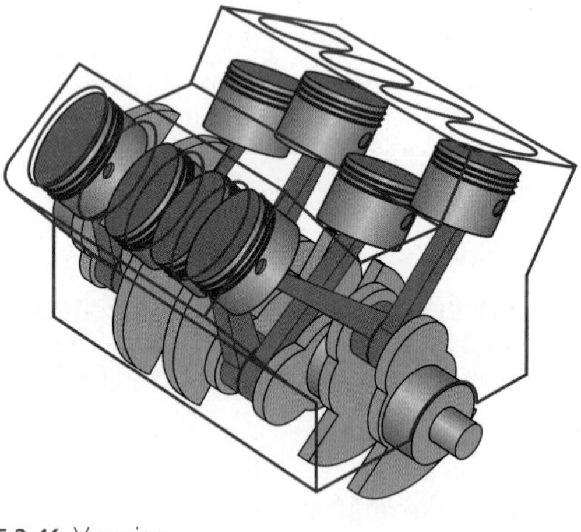

FIGURE 2-46 V engine.

Engineers design engines with tilted cylinder banks to reduce engine height. This can reduce the height of the hood as well, which allows a more streamlined hood line (**FIGURE 2-48**).

Tilting can be carried to an extreme. A flat engine is designed with the engine completely on its side in the case of a horizontally opposed engine. This greatly reduces the engine height.

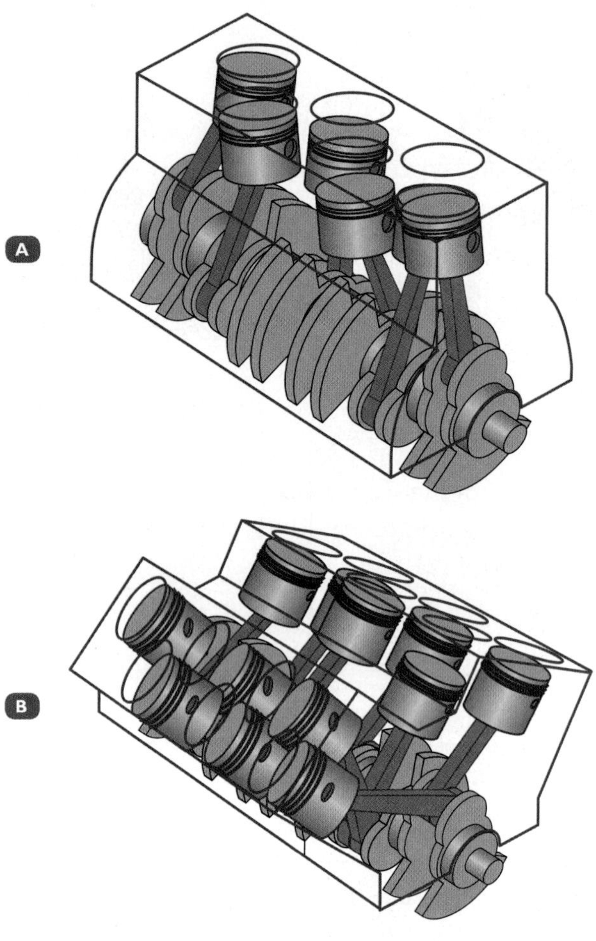

FIGURE 2-47 **A.** VR engine. **B.** W engine.

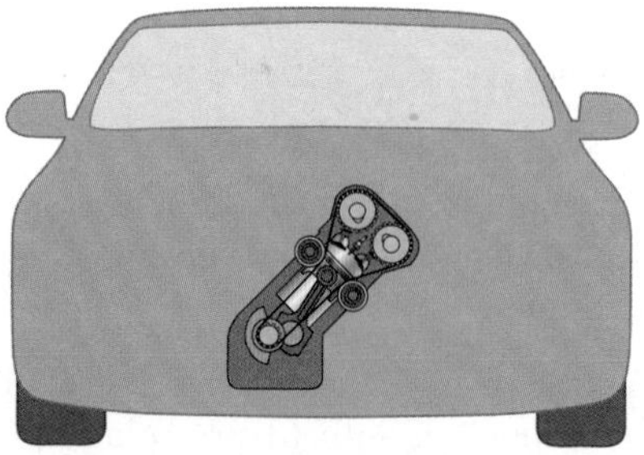

FIGURE 2-48 Tilting cylinder banks can reduce both engine height and hood height, which allows a more streamlined hood line.

As the number of cylinders increases, the length of the engine block and the crankshaft also increases. This can become a problem structurally and space-wise. One way to avoid this problem is by having more than one row of cylinders, as in a horizontally opposed, V, or W configuration. These designs make the engine block and the crankshaft shorter and more rigid. In vehicle applications, the number of cylinders can vary, usually from a minimum of 3, up to 12. Some examples are listed here:

- In-line 4—Compact, fairly inexpensive, easier to work on, and better fuel economy than other types.
- V8—Approximately twice as powerful as the in-line 4, but with less than twice the space. Good power for its size and not overly complicated.

- Flat 6—Low-profile engine that fits well in vehicles with very low hood lines.
- W12—Most powerful compared to its overall dimensions, but more complicated and expensive than the other engines.

Common angles between the banks of cylinders are:

- 180 degrees (horizontally opposed type),
- 90 degrees (V type),
- 60 degrees (V type), and
- 15 degrees (VR type).

Angles vary due to the number of cylinders and the manufacturer's design considerations.

In-line

Cylinders can be arranged side-by-side in a single row. These **in-line engines** can be found in three-, four-, five-, and six-cylinder configurations (**FIGURE 2-49**). There have been in-line eight-cylinder engines, but they are too long to fit into the engine bay of a conventional modern car. In-line engines can be mounted longitudinally (lengthwise) or transversely (sideways) in the engine bay. In-line engines are generally less complicated to design and manufacture. This is because they do not have to share components with a second bank of cylinders. This lack of shared components can mean extra working room in the engine compartment when performing repairs. As a general rule of thumb, in-line engines are easier to work on than the other cylinder arrangements.

Horizontally Opposed

Horizontally opposed engines are sometimes referred to as "flat" engines. They are commonly found in four- and six-cylinder configurations (**FIGURE 2-50**). They are shorter lengthwise than a comparable in-line engine, but wider than a V type. Horizontally opposed engines have two banks of cylinders. They are 180 degrees apart, on opposite sides of the crankshaft. It is a useful design when little vertical space is available. A horizontally opposed engine is only fitted longitudinally.

FIGURE 2-49 Cylinders arranged side-by-side in a single row identify the in-line engine.

V

V engines have two banks of cylinders sitting side-by-side in a V arrangement sharing a common crankshaft (**FIGURE 2-51**). This compact design allows for about twice the power output from a V engine than an in-line engine of the same length. In automotive applications, V engines can typically be found in 6-, 8-, 10-, and 12-cylinder configurations. A V6 has two banks of three cylinders; a V8, two banks of four cylinders; etc. The angle of the V varies depending on the number of cylinders. It can be found by dividing 720 degrees (two rotations of the crankshaft, which equals one complete cycle) by the number of cylinders. The natural angle for a V8 is 90 degrees. The natural angle for a V6 is 120 degrees, for a V10 is 72 degrees, and for a V12 is 60 degrees.

Designing the engine around the natural angle means the engine can have a shorter length. This is because each crankshaft throw can be shared between two cylinders. Some manufacturers vary their angles from those natural angles. They may use 90 degrees (or 60 degrees) for a V6 and 15 degrees for a VR6. This is because of convenience or design requirements. Varying away

FIGURE 2-50 Horizontally opposed engines are commonly found in four- and six-cylinder configurations.

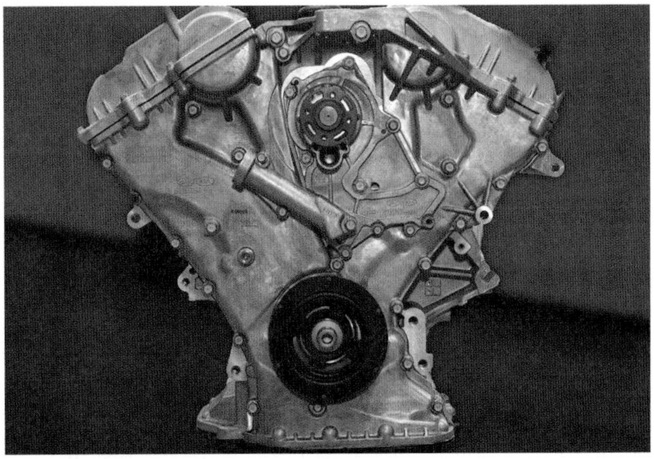

FIGURE 2-51 V engines are shorter than in-line engines of equivalent capacity.

FIGURE 2-52 **A.** Individual crankshaft journals. **B.** Splayed crankshaft journals.

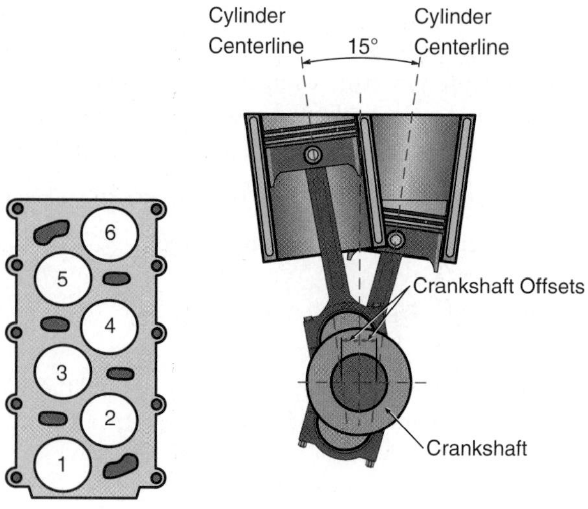

FIGURE 2-53 VR cylinder arrangement.

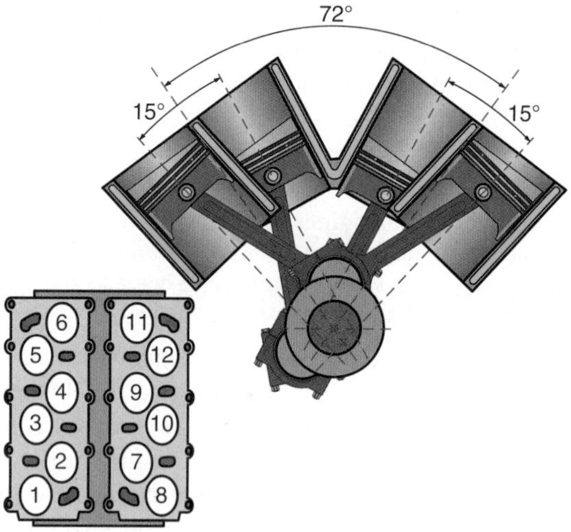

FIGURE 2-54 W engine.

from the natural angle means that the crankshaft must have one crank throw per cylinder. This makes the engine slightly longer. This can be accomplished by having completely separate crankshaft journals or by using splayed journals (FIGURE 2-52).

VR and W

The **VR engine** uses a single bank of cylinders, but the cylinders are staggered at a shallow 15-degree V within the bank (FIGURE 2-53). This design not only allows the engine to be shorter than an in-line engine, but also makes it narrower than a typical V engine. The **W engine** consists of two VR cylinder banks in a deeper V arrangement to each other. This gives a very compact, yet powerful, engine design (FIGURE 2-54).

Rotary Engines

Rotary engines are very powerful for their size. They do not use conventional pistons that slide back and forth inside a straight cylinder. Instead, a rotary engine uses a roughly triangular rotor that turns inside a roughly oval-shaped housing (FIGURE 2-55). The rotary engine has three combustion events for each rotation of the rotor. As the rotor turns, it carries the air–fuel mixture around the chambers. The chambers are created between the tips of the rotor and the chamber wall. The rotor compresses the

FIGURE 2-55 Rotary engine.

air–fuel mixture, and spark plugs ignite the mixture, just like a conventional engine.

The rotary engine does not have intake and exhaust valves like a traditional piston engine. Instead, it has exhaust and intake ports that are covered and uncovered by the rotating rotor in the chamber. This design reduces the number of parts and makes it less complicated than a piston engine. Rotary engines generally have more than one rotor, with two being the most common. The design of the rotary engine provides a very compact power unit.

▶ Wrap-Up

Ready for Review

- Vehicle bodies come in a variety of designs depending on the intended function of the vehicle and continue to evolve to accommodate the owners' lifestyles and personal tastes.
  - Different types of body designs include: sedan, coupe, hatchback, convertible, station wagon, pickup, minivan, and SUV.
- Various vehicle systems are designed to perform specific functions that are critical to the operation, passenger safety, and environmental impact of the vehicle. All of the systems work together to provide a safe, efficient, and enjoyable form of transportation.
- A drivetrain includes the engine, transmission/transaxle, differential, axles, and wheels that are used to power the vehicle down the road.
  - The drivetrain layout can be defined by a combination of: engine position, engine orientation, and type of drive.
- Torque is the twisting force applied to a shaft and its measurement is based on the equivalent twisting force exerted by an amount of weight (mass) applied to a perpendicular lever of a given length.
  - Engines can be configured based on type, cylinder arrangement, number of cylinders/rotors, and total engine displacement in cubic inches or liters.

Key Terms

all-wheel drive (AWD) A drivetrain arrangement in which all of the wheels drive the vehicle.

axle A shaft connected to wheels that transmits the driving torque to the wheels.

chassis The main support frame in a vehicle. It includes the running gear, such as suspension, the engine, and the drivetrain.

convertible A vehicle that converts from having an enclosed top to having an open top by a roof that can be removed, retracted, or folded away.

coupe A two-door vehicle that has seating for two people and may have a small rear seat.

differential gears A gear arrangement that splits the available torque equally between two wheels while allowing them to turn at different speeds when required.

driveshaft The shaft or tube fitted with universal couplings that is connected between the transmission and other drivetrain components, to transmit torque and rotation.

drivetrain A term used to identify the engine, transmission/transaxle, differential, axles, and wheels.

engine configuration The way engine cylinders are arranged—for example, V, flat, or in-line.

final drive A component that provides a final gear reduction and allows for the difference in speed of each wheel when cornering.

flywheel A heavy, round metal disc attached to the end of the crankshaft to smooth out vibrations from the crankshaft assembly and provide one of the friction surfaces for a clutch disc used on a manual transmission.

four-wheel drive (4WD) A drivetrain layout in which the engine drive has either two wheels or four wheels, depending on which mode is selected by the driver.

front-wheel drive (FWD) A drivetrain layout in which the engine drives the front wheels.

hatchback A vehicle that has a shared passenger and cargo area; it typically is available in three- and five-door arrangements.

horizontally opposed engine An engine with two banks of cylinders, 180 degrees apart, on opposite sides of the crankshaft. It is also called a flat engine or a boxer engine.

in-line engine An engine in which the cylinders are arranged side by side in a single row.

longitudinal The orientation of the engine in which the front of the engine is facing the front of the vehicle. It is most commonly found in rear-wheel-drive vehicles.

minivan A lighter-duty van used for carrying six to eight occupants or light cargo.

pickup A vehicle that carries cargo; it has stronger chassis components and suspension than a sedan.

piston engine An internal combustion engine that uses cylindrical pistons, moving back and forth in a cylinder, to extract mechanical energy from chemical energy.

rear-wheel drive (RWD) A drivetrain layout in which the engine drives the rear wheels.

rotary engine An engine that uses a triangular rotor turning in a housing instead of conventional pistons.

sedan A vehicle configuration that has an enclosed body, with a maximum of four doors to allow access to the passenger compartment.

sport utility vehicle (SUV) A passenger vehicle built on a light-truck chassis; it is usually equipped with four-wheel drive and capable of hauling heavier loads than typical passenger vehicles.

station wagon A vehicle configuration with four doors, with a roof line that continues into the rear cargo area and a rear door for access.

torque Twisting force applied to a shaft that may or may not result in motion.

torque converter A type of fluid coupling that is also capable of multiplying torque. It is turned by the crankshaft and transmits torque to the input shaft of an automatic transmission.

transverse A term used to describe the side-to-side engine orientation when mounted in the engine compartment.

truck A large, heavy vehicle for carrying cargo.

unibody design A vehicle design that does not use a rigid frame to support the body. The body panels are designed to provide the strength for the vehicle.

V engine A term used to describe an engine configuration that has two banks of cylinders sitting side by side in a V arrangement and sharing a common crankshaft.

viscous coupling A silicone clutch assembly used in AWD differentials to provide a slight amount of differential action for control of axle rotational speeds.

VR engine A term used to describe an engine configuration that uses a single bank of cylinders staggered at a shallow 15-degree V.

W engine A term used to describe an engine configuration consisting of two VR cylinder banks in a deep V arrangement.

Review Questions

1. A station wagon has which characteristic?
   a. 2 doors
   b. Increased luggage capacity
   c. Short roof lines
   d. Convertible top
2. The sport utility vehicle is a combination of which vehicles?
   a. Van and truck
   b. Convertible and sedan
   c. Van and station wagon
   d. Coupe and van
3. Chemical energy can be converted into mechanical energy through which system?
   a. Emission system
   b. Exhaust system
   c. Transmission system
   d. Engine system
4. What function does the TPMS perform?
   a. Warn of engine failure
   b. Keep occupants safe in accident
   c. Monitor tire pressure
   d. Regulate passenger temperature
5. Which vehicle drivetrain configuration drives all four wheels all the time?
   a. Front wheel drive
   b. All wheel drive
   c. 4-Wheel drive
   d. Rear-wheel drive
6. Which method is NOT used to control power between the front and rear wheels?
   a. Viscous coupling
   b. Transfer case
   c. Differential lock
   d. Master cylinder
7. Which torque value is used to determine the tightening of fasteners?
   a. Newton kilograms
   b. Inch meters
   c. Foot pounds
   d. Centigram newtons
8. An engine can be identified by which feature?
   a. Weight
   b. Gear ratio
   c. Head bolt torque
   d. Number of cylinders
9. A flat (boxer) engine has cylinder angles of:
   a. 45 Degrees
   b. 72 Degrees
   c. 118 Degrees
   d. 180 Degrees
10. How does a rotary engine function?
   a. Triangular rotor rotates inside an oval housing
   b. Oval rotor rotates in a circular cylinder housing
   c. Triangular rotor rotates in a circular housing
   d. Oval rotor rotates in a triangular cylinder housing

ASE Technician A/Technician B Style Questions

1. Technician A says that an inline engine has the cylinders in a single row. Technician B says that inline engines can be found in 3, 4, 5, and 6 cylinder configurations. Who is correct?
   a. Technician A
   b. Technician B
   c. Both A and B
   d. Neither A nor B
2. Technician A says that tilting the cylinder bank can lower hood heights. Technician B says that tilting the cylinder banks increases the engine power output. Who is correct?
   a. Technician A
   b. Technician B
   c. Both A and B
   d. Neither A nor B
3. Technician A says that torque is the distance the piston travels. Technician B says that one customary unit of measure for torque is foot-pounds. Who is correct?
   a. Technician A
   b. Technician B
   c. Both A and B
   d. Neither A nor B
4. Technician A says that the higher the gear selected the more torque is available. Technician B says that a transaxle contains gears that increase or decrease torque. Who is correct?
   a. Technician A
   b. Technician B

c. Technician A and B
d. Neither A nor B

5. Technician A says that the location of the live axle will determine the drive configuration. Technician B says that a live axle just supports the wheel. Who is correct?
   a. Technician A
   b. Technician B
   c. Both A and B
   d. Neither A nor B
6. Technician A says that the electrical system helps with starting the vehicle. Technician B says that the electrical system is the nerve system of the vehicle. Who is correct?
   a. Technician A
   b. Technician B
   c. Both A and B
   d. Neither A nor B
7. Technician A says that when starting the engine, the crankshaft moves the pistons up and down. Technician B says that once the engine is running, the flywheel moves the pistons up and down. Who is correct?
   a. Technician A
   b. Technician B
   c. Both A and B
   d. Neither A nor B
8. Technician A says that the engine cooling system controls the temperature inside the vehicle. Technician B says that the engine cooling system regulates the engine temperature for efficiency. Who is correct?
   a. Technician A
   b. Technician B
   c. Both A and B
   d. Neither A nor B
9. Technician A says that the uni-body is mounted on a rigid frame. Technician B says that the body-on-frame chassis integrates the bodywork into a single unit with the chassis. Who is correct?
   a. Technician A
   b. Technician B
   c. Both A and B
   d. Neither A nor B
10. Technician A says that hatchback vehicles have 3 or 5 doors. Technician B says that a convertible vehicle can remove or retract the top.
   a. Technician A
   b. Technician B
   c. Both A and B
   d. Neither A nor B

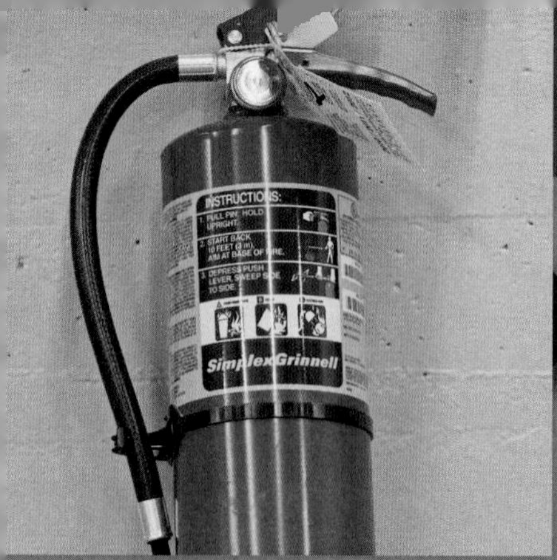

CHAPTER 3

Introduction to Automotive Safety

Learning Objectives

- **LO 3-01** Adhere to workplace safety guidelines.
- **LO 3-02** Locate shop safety features and equipment.
- **LO 3-03** Prevent fires and operate fire safety equipment.
- **LO 3-04** Research material data using safety data sheets (SDS).
- **LO 3-05** Work near hazardous on-vehicle systems.

ASE Education Foundation Tasks

See Appendix A to view the 2017 ASE Education Foundation Automobile Accreditation Task List Correlation Guide.

▶ Introduction

LO 3-01 Adhere to workplace safety guidelines.

There are potential hazards in most workplaces, especially in automotive shops. Occupational safety and health is very important so that everyone can work without being injured. Governments normally have legislation in place regarding safety measures. Significant penalties can be imposed on those who do not follow the safety practices in the workplace. It is important to learn about hazards so that you can identify them. Once identified, you can take actions to protect yourself and your coworkers (**FIGURE 3-1**).

Some hazards are obvious, such as vehicles falling from hoists, or tires exploding during inflation. Other hazards are less obvious, such as the long-term effects of breathing the fumes from solvents. There are many things to learn about safety in the automotive shop. Therefore, it is impossible to cover every situation you will encounter. One of the most important skills to learn is the ability to recognize unsafe practices or equipment. Once recognized, take measures to prevent injuries from happening.

FIGURE 3-1 Identifying hazards.

Occupational safety and health is everyone's responsibility. You have a responsibility to ensure that you work safely and take care not to put others at risk by acting in an unsafe manner. Your employer also has a responsibility to provide a safe working environment.

To ensure the safety of yourself and others, make sure you are aware of the correct safety procedures at your workplace. This means listening very carefully to safety information provided by your employer. Also ask for clarification, help, or instructions if you are unsure how to perform a task safely. Always think about how you are performing shop tasks. And be on the lookout for unsafe equipment and work practices. Be sure to wear the correct **personal protective equipment (PPE)**. PPE refers to items of safety equipment like:

- safety footwear,
- gloves, clothing,
- protective eyewear, and
- hearing protection (**FIGURE 3-2**).

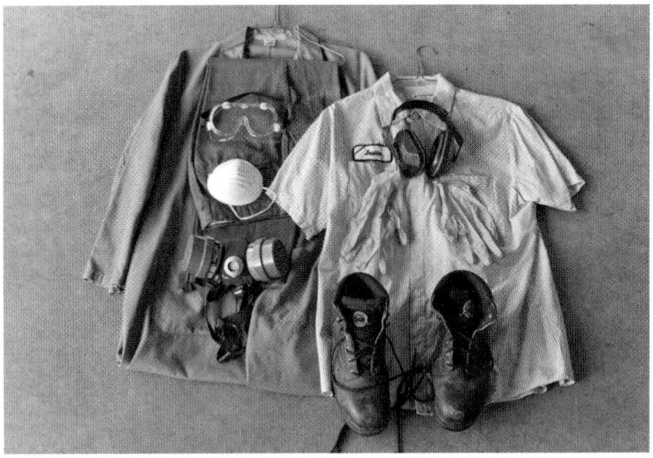

FIGURE 3-2 Personal protective equipment (PPE) are items of safety equipment like safety footwear, gloves, clothing, protective eyewear, and hearing protection.

You Are the Automotive Technician

You are changing the oil on a new type of vehicle for the first time. The oil pan has the drain plug on its side instead of the bottom. You place the drain pan directly under the drain plug like you normally do. Unfortunately, when the plug comes out, the oil shoots sideways right over the side of the drain pan. You reposition it quickly, but not before a large puddle is on the floor.

1. Why is it important to review the SDS before cleaning up a spill?
2. What is the minimum PPE that should be worn to manage this spill?
3. What are some of the health hazards of coming into frequent or prolonged contact with used engine oil?
4. If the oil were to catch fire, which fire extinguisher should be used to put the fire out?

Safety Overview

Motor vehicle servicing is one of the most common vocations worldwide. Hundreds of thousands of shops service millions of vehicles every day. That means at any given time, very many people are servicing vehicles. Because of this, there is great potential for things to go wrong (**FIGURE 3-3**). It is up to you and your workplace to make sure all work activities are conducted safely. Accidents are rarely caused by properly maintained tools; accidents are generally caused by people.

FIGURE 3-3 Accident in the shop.

▶ TECHNICIAN TIP

Whenever you perform a task in the shop, you must use personal protective clothing and equipment that are appropriate for the task. They also must conform to your local safety regulations and policies. Among other items, these may include:

- Work clothing, such as coveralls and steel-capped footwear
- Eye protection, such as safety glasses and face masks
- Ear protection, such as earmuffs and earplugs
- Hand protection, such as gloves and barrier cream
- Respiratory equipment, such as face masks and valved respirators

If you are not certain what is appropriate or required, ask your supervisor.

Don't Underestimate the Dangers

Because vehicle servicing and repair are so commonplace, it is easy to overlook many potential risks related to this field. Think carefully about what you are doing and how you are doing it. Think through the steps, trying to anticipate things that may go wrong and taking steps to prevent them. Also be wary of taking shortcuts. In most cases, the time saved by taking a shortcut is nothing compared to the time spent recovering from an accident (**FIGURE 3-4**).

FIGURE 3-4 Technician recovering from a broken arm due to a slip or fall.

Accidents and Injuries Can Happen at Any Time

There is the possibility of an accident occurring whenever work is undertaken. For example, fires and explosions are a constant hazard wherever there are flammable fuels. Also, electricity can kill quickly as well as cause painful shocks and burns. Heavy equipment and machinery can easily cause injuries such as broken bones or crushed fingers and toes. Hazardous solvents and other chemicals can burn or lead to blindness as well as contribute to many kinds of illness. Oil spills and tools left lying around can cause slips, trips, and falls. Poor lifting and handling techniques can cause chronic strain injuries, particularly to your back (**FIGURE 3-5**).

FIGURE 3-5 Poor lifting and handling techniques can cause chronic strain injuries, particularly to your back.

Accidents and Injuries Are Avoidable

Almost all accidents are avoidable or preventable by taking a few precautions. Think of nearly every accident you have witnessed or heard about. In most cases, someone made a mistake. Accidents caused by mistakes like horseplay, neglecting maintenance on tools, or using tools improperly can lead to injury (**FIGURE 3-6**).

Most accidents can be prevented if people follow policies and develop a "safety first" attitude. By following regulations and safety procedures, you can make your workplace much safer. Learn and follow all of the correct safety procedures for your workplace. Always wear the right PPE, and stay alert and aware of what is happening around you. Think about what you are doing, how you are doing it, and its effect on others. You also need to know what to do in case of an emergency. Document and report all accidents and injuries whenever they happen. Then take the proper steps to make sure they never happen again.

OSHA and EPA

OSHA stands for the **Occupational Safety and Health Administration (OSHA)**. It is a U.S. government agency that was created to provide national leadership in occupational safety and health. It works toward finding the most effective ways to help prevent worker fatalities and workplace injuries and illnesses. OSHA can

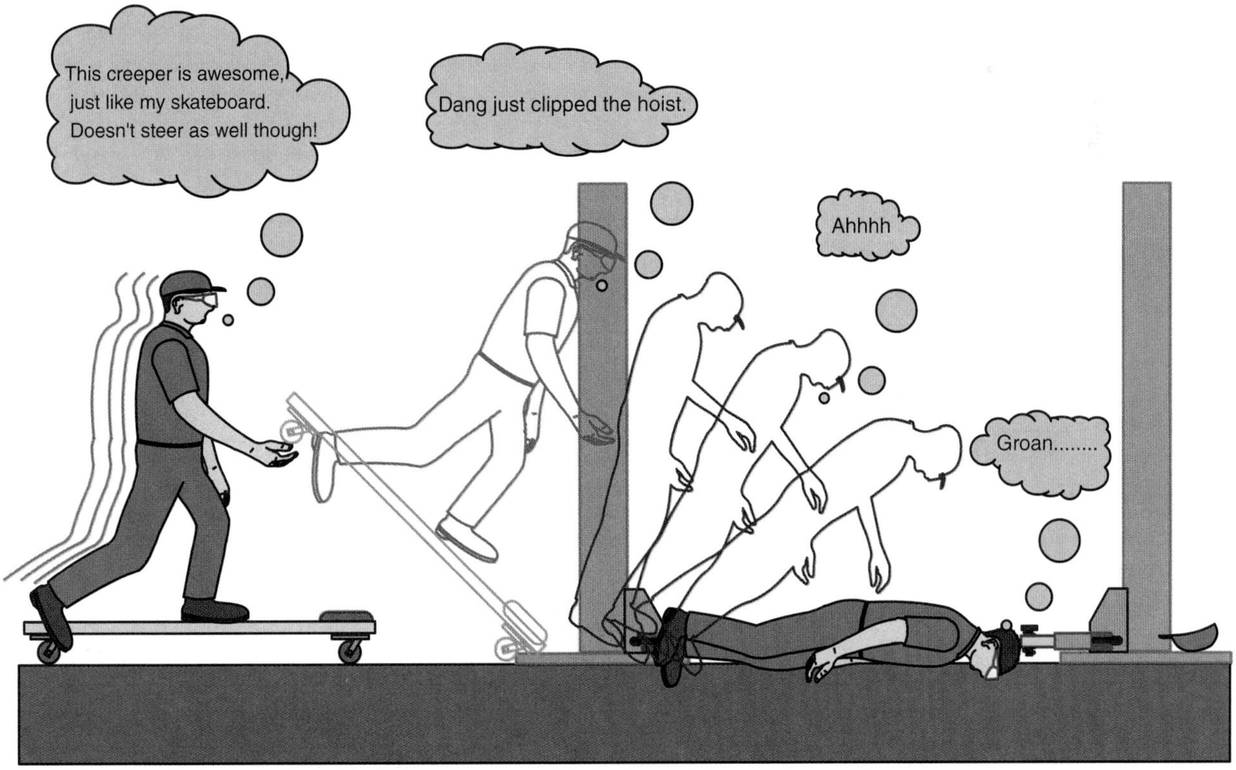

FIGURE 3-6 Accidents are usually caused by mistakes.

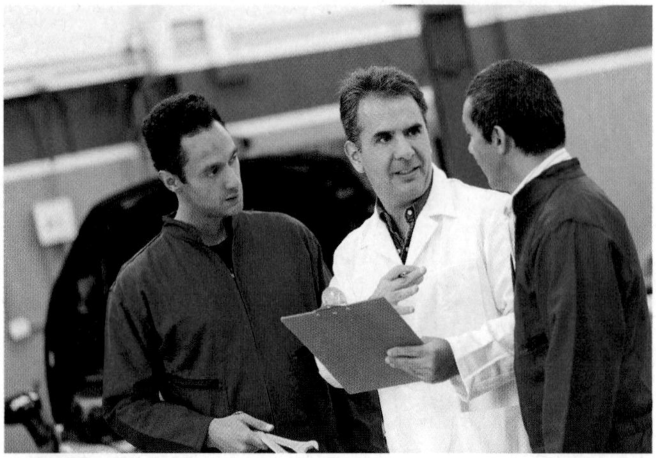

FIGURE 3-7 OSHA inspection.

FIGURE 3-8 Correct chemical storage.

conduct workplace inspections and fine employers and workplaces if they violate its regulations and procedures (**FIGURE 3-7**). For example, a fine may be imposed on the employer or workplace if a worker is electrocuted by a piece of faulty machinery that has not been regularly tested and maintained.

EPA stands for the **Environmental Protection Agency**. This federal government agency deals with issues related to environmental safety. The EPA conducts research and monitoring, sets standards, and conducts workplace inspections. It also holds companies legally accountable in order to protect the environment. Shop activities need to comply with EPA laws and regulations. This ensures that waste products are disposed of in an environmentally responsible way. It also makes sure chemicals and fluids are stored correctly, and work practices do not harm the environment (**FIGURE 3-8**).

Shop Policies and Procedures

Shop policies and procedures are a set of documents that outline how tasks and activities in the shop are to be conducted and managed. They also ensure that the shop operates according to

Applied Science

AS-2: Environmental Issues: The technician develops and maintains an understanding of all federal, state, and local rules and regulations regarding environmental issues related to the work of the automobile technician.

You need to keep up to date with local laws and regulations regarding environmental issues, or you will have large fines to pay. You can log on to local state websites to check the latest laws and regulations. The U.S. Environmental Protection Agency is known as the EPA. Its website also has up-to-date information on environmental regulations (www.epa.gov/lawsregs/). More information on the specific regulations for the automotive industry can be found at www.epa.gov/lawsregs/sectors/automotive.html. This site has a full listing of laws and regulations, compliance measures, and enforcement tactics. For additional laws that are enforced by your local state environmental agencies, go to www.epa.gov/epahome/state.htm.

AS-3: Environmental Issues: The technician uses such things as government impact statements, media information, and general knowledge of pollution and waste management to correctly use and dispose of the products that result from the performance of a repair task.

When you complete a job, you must be able to identify what to do with any waste products. This could be as simple as knowing where and how to recycle cardboard boxes. Or it could be as complex as knowing what to do with brake components that may contain asbestos. Normally, there is a table posted in the garage that details the correct measures for disposing of and storing waste material. A sample table follows:

Component	Material/Parts	Removal and Safety Information	Recommended Storage
Air-conditioning gases (refrigerant)	R12, R134a	Use approved evacuation and collection equipment required. *Note:* Requires AC license.	Use approved storage containers—reused or recycled.
Batteries	Plastic, rubber, lead, sulfuric acid	Avoid contact between sulfuric acid and your skin, clothing, or eyes.	Store off the ground in a covered area for collection by recycler.
Brake fluid	Diethylene and polyethylene glycol-monoalkyl ethers	These are corrosive and highly toxic to the environment. Drain into pan or tray.	Store in a drum in a covered area for collection by a licensed operator.
Brake shoes and pads pre-2004	Asbestos	Fibers are dangerous if inhaled.	Put in a plastic bag in a sealed container for collection by contractor.

Component	Material/Parts	Removal and Safety Information	Recommended Storage
Coolant	Phosphoric acid, hydrazine ethylene glycol, alcohols	Radiator coolant can be toxic to the environment.	Store in sealed drums in a covered area for recycling or collection by a licensed operator.
Coverings for plastic parts, and plastic bags and containers for parts shipping	Plastic-made components	Plastic components that can be recycled have a recycling symbol: this symbol has a number inside telling the recycling company what the product is made of.	If the plastic container has a recycle code, then put it in the recycle bin. If it does not, then put in general waste. Plastic oil containers cannot be recycled.
Fuel	Unleaded, diesel	Avoid fumes. Fire hazard; keep well away from ignition sources. Siphon from tank to avoid spillage.	Store in a drum in a covered area.
Metal	Brake discs, housings made of metal (gearbox/engine case and components), metal cuttings	Some metal products can be heavy, so lift with care.	All metal components can be recycled; keep waste metal in a separate recycle bin for sale or disposal.
Oil	Engine, transmission, and differential oils	Fire hazard; keep well away from ignition sources.	Store in a drum/container in a covered area for collection by a licensed operator.
Oil filters	Steel paper fiber	—	Drain filter, then crush and store in a leak-proof drum for collection.
Parts boxes and paperwork	Paper, cardboard	—	Store in a recycling bin to be taken away for recycling.
Tires	Rubber, steel, fabric	Keep away from ignition sources.	Store in a fenced area for collection by recycler.
Trim, plastic fittings, and seats	Plastic, metal, cloth	—	Store racked or binned for reuse, sale, or recycling.
Tubes and rubber components	Rubber hoses, mounts, etc.	Keep away from ignition sources.	Store in a collection bin, to be collected and recycled.
Undeployed airbags	Plastics, metals, igniters, explosives	Recommended specific training on airbags before attempting removal. Handle with care; accidental deployment can cause serious harm. If a unit is to be scrapped, ensure that it is safely deployed first.	Store face up in a secure area.

OSHA and EPA laws and regulations. A **policy** is a guiding principle that sets the shop direction. A **procedure** is a list of the steps required to get the same result each time a task or activity is performed (**FIGURE 3-9**). An example of a policy is an OSHA document that describes how the shop complies with legislation. It can simply be a sign saying, "Safety glasses must be worn at all times in the shop." An example of a procedure is a document that describes the steps required to safely use the vehicle hoist.

Each shop has its own set of policies and procedures. It also has a system in place to make sure the policies and procedures are regularly reviewed and updated. Regular reviews ensure that new policies and procedures are developed and old ones are modified in case something has changed. For example, if the shop moves to a new building, then a review of policies and procedures is needed. This will ensure that they relate to the new shop, its layout, and equipment. In general, the policies and procedures are written to guide shop practice. They help ensure compliance with laws, statutes, and regulations; and reduce the risk of injury. Always follow your shop policies and procedures. This helps reduce the risk of injury to your coworkers and yourself and to prevent damage to property.

It is everyone's responsibility to know and follow the rules. Locate the general shop rules and procedures for your workplace. Look through the contents or index pages to familiarize yourself with the contents. Discuss the policy and the shop rules and procedures with your supervisor. Ask questions to ensure that you understand how the rules and procedures should be applied. Also understand your role in making sure they are followed.

Air Quality

Managing air quality in shops helps protect you from potential harm. It also protects the environment. There are many shop activities, stored liquids, and other hazards that can reduce the quality of air in shops. Some of these are listed here:

- Dangerous fumes from running engines
- Welding (gas and electric)
- Painting
- Liquid storage areas
- Air conditioning servicing
- Dust particles from brake servicing

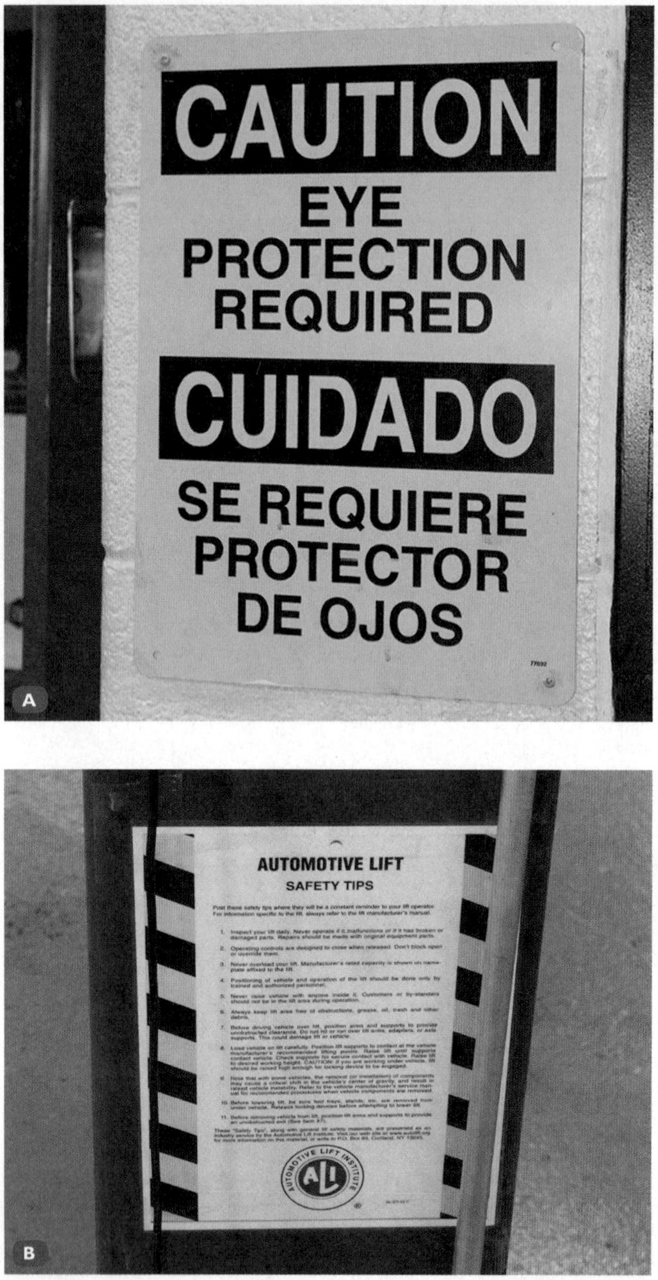

FIGURE 3-9 **A.** Policy. **B.** Procedure.

FIGURE 3-10 Exhaust hoses should be vented to a place from where the fumes will not be drawn back indoors.

Running Engines

Running engines produce dangerous exhaust gases including carbon monoxide (CO) and carbon dioxide (CO_2). Carbon monoxide in small concentrations can kill or cause serious injuries. Carbon dioxide is a greenhouse gas, and vehicles are a major source of CO_2 in the atmosphere. Exhaust gases also contain hydrocarbons (HC) and oxides of nitrogen (NO_x). These gases can form smog and also cause breathing problems for some people.

Carbon monoxide in particular is extremely dangerous. It is odorless and colorless and can build up to toxic levels very quickly in confined spaces. In fact, it does not take very much CO to pose a danger. The maximum exposure limit is regulated by the following agencies:

- OSHA permissible exposure limit (PEL) is 50 parts per million (ppm) of air for an eight-hour period.
- National Institute for Occupational Safety and Health has established a recommended exposure limit of 35 ppm for an eight-hour period.

The reason the PEL is so low is because CO attaches itself to red blood cells much more easily than does oxygen. And it never leaves the blood cell. This prevents the blood cells from carrying as much oxygen. And if enough CO has been inhaled, it effectively asphyxiates the person. Always follow the correct safety precautions when running engines indoors or in a confined space. This includes service pits, as gases can accumulate there.

The best solution when running engines in an enclosed space (shop) is to directly connect the vehicle's exhaust pipe to an exhaust extraction system hose. This ventilates the fumes to the outside air. The extraction system should be vented to a place well away from other people. It should also be far enough away from where the fumes will not be drawn back indoors (**FIGURE 3-10**).

Do not assume that an engine fitted with a catalytic converter can be run safely indoors; it cannot. Catalytic converters are fitted into the exhaust system. They help control exhaust emissions through chemical reactions. To operate efficiently, they require high temperatures. They allow pollutants to flow unconverted when the exhaust gases are relatively cool. This happens when the engine is only idling or being run intermittently. A catalytic convertor can never substitute for adequate ventilation or exhaust extraction equipment. Even if the catalytic converter were working at 100% efficiency, the exhaust would contain large amounts of CO_2 and very low amounts of oxygen. Neither of these conditions can sustain life.

▶ Work Environment - Safety Features

LO 3-02 Locate shop safety features and equipment.

The work environment can be described as anywhere you work. The condition of the work environment plays an important role in making the workplace safer. A safe work environment goes a long way toward preventing accidents, injuries, and illnesses. There are many ways to describe a safe work environment. Generally, it would:

- contain a well-organized shop layout,
- use of shop policies and procedures,
- safe equipment,
- safety equipment,
- safety training,
- employees who work safely,
- good supervision, and
- a workplace culture that supports safe work practices.

Conversely, a shop that is cluttered with junk, poorly lit, and full of safety hazards is unsafe (**FIGURE 3-11**).

Evacuation Routes

Evacuation routes are a safe way of escaping danger and gathering in a prearranged safe place. This is so that everyone can be accounted for in the event of an emergency. It is important to have more than one evacuation route. This helps to avoid any mishap in case any single route is blocked during the emergency. Your shop should have an evacuation procedure that clearly identifies the evacuation routes (**FIGURE 3-12**).

Often the evacuation routes are marked with colored lines painted or taped on the floors. Exits should be highlighted with signs that may be illuminated. Exits should never be chained closed or obstructed (**FIGURE 3-13**).

Always make sure you are familiar with the evacuation routes for the shop. Before conducting any task, identify which route you will take if an emergency occurs.

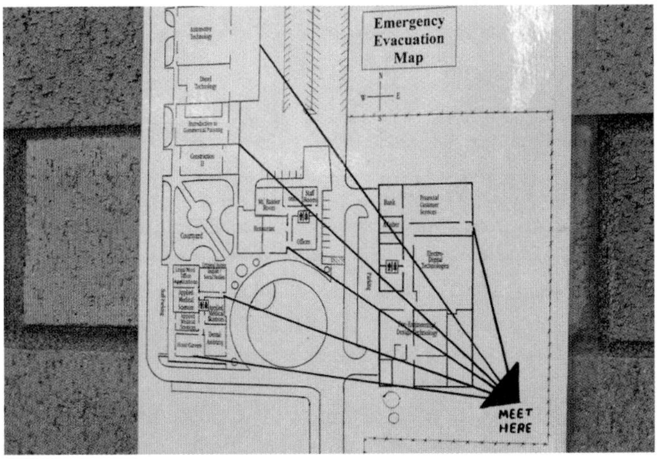

FIGURE 3-12 Your shop may have an evacuation procedure that clearly identifies the evacuation routes.

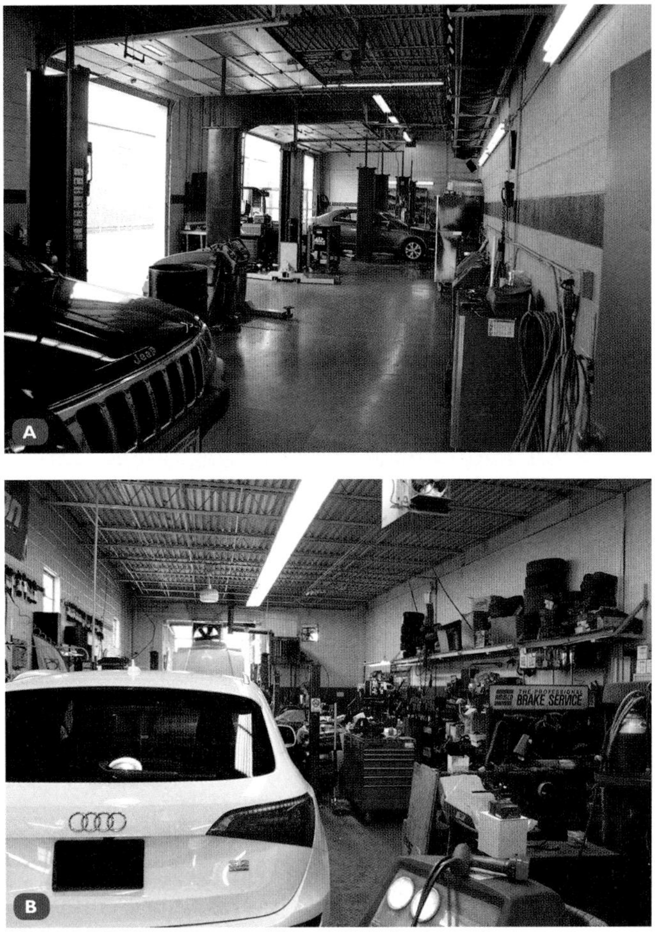

FIGURE 3-11 **A.** Relatively safe shop. **B.** Relatively unsafe shop.

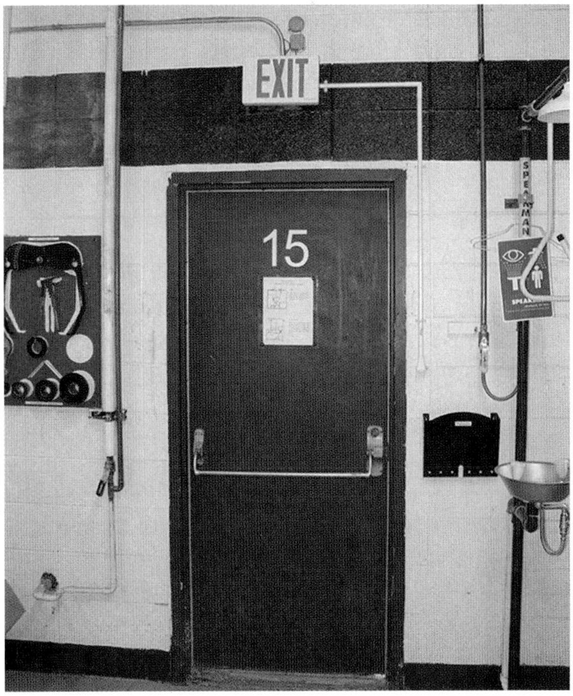

FIGURE 3-13 Exits should be marked, clear of obstructions, and not chained closed.

> ▶ TECHNICIAN TIP
>
> Never place anything in the way of evacuation routes, including equipment, tools, parts, cleaning supplies, or vehicles.

Standard Safety Measures

Signs

Always remember that a shop is a hazardous environment. To make people more aware of specific shop hazards, legislative bodies have developed a series of safety signs. These signs are designed to give adequate warning of an unsafe situation. Each sign has four components:

- **Signal word:** There are three signal words—danger, warning, and caution. *Danger* indicates an immediately hazardous situation, which, if not avoided, will result in death or serious injury. Danger is usually indicated by white text with a red background. *Warning* indicates a potentially hazardous situation, which, if not avoided, could result in death or serious injury. The sign is usually in black text with an orange or yellow background. *Caution* indicates a potentially hazardous situation, which, if not avoided, may result in minor or moderate injury. It may also be used to alert against unsafe practices. This sign is usually in black text with a yellow background (**FIGURE 3-14**).
- **Background color:** The choice of background color also draws attention to potential hazards. It is used to provide contrast so that the letters or images stand out. For example, a red background is used to identify a definite hazard; yellow indicates caution for a potential hazard. A green background is used for emergency-type signs, such as for first aid, fire protection, and emergency equipment. A blue background is used for general information signs.
- **Text:** The sign will sometimes include explanatory text. This is to provide additional safety information. Some signs are designed to convey a personal safety message.
- **Pictorial message:** In symbol signs, a pictorial message appears alone or is combined with explanatory text. This type of sign allows the safety message to be conveyed to people who are illiterate or who do not speak the local language.

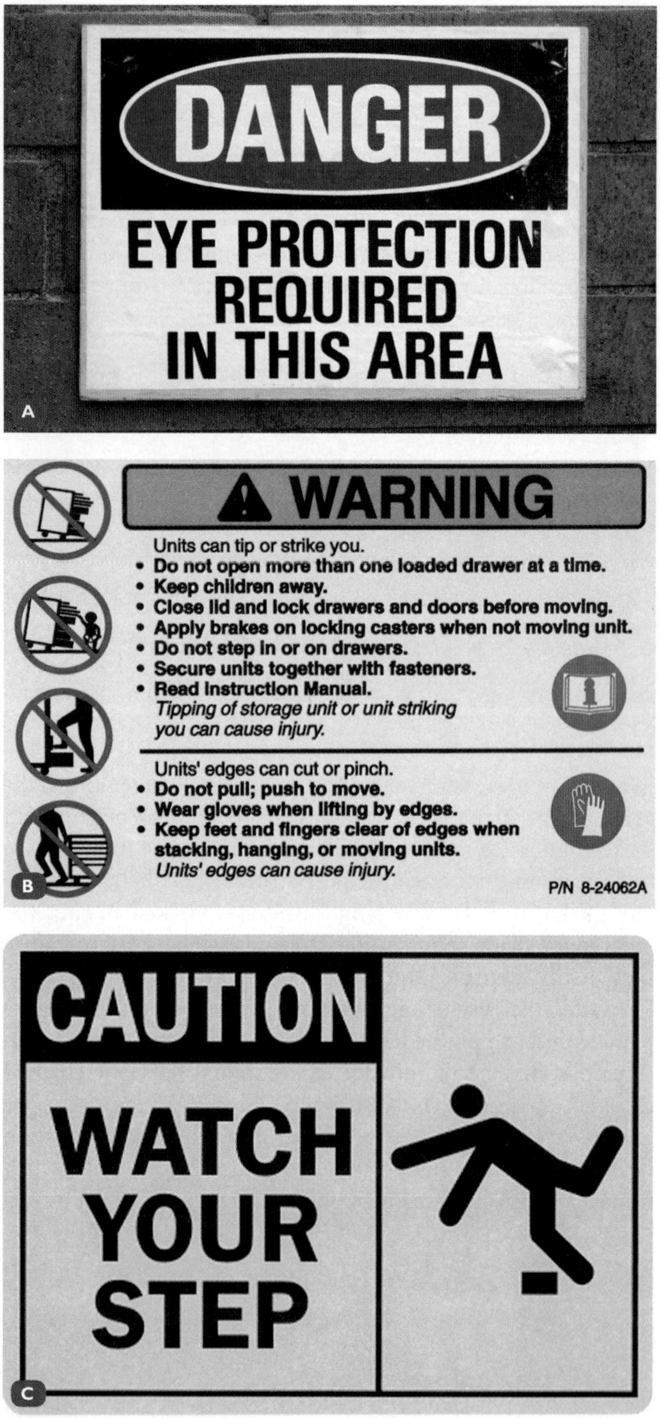

FIGURE 3-14 Signs. **A.** Danger is usually indicated by white text on a red background. **B.** Warning is usually in black text with an orange background. **C.** Caution is usually in black text with a yellow background.

Identifying Hazardous Environments

A **hazardous environment** is a place where hazards exist. A **hazard** is anything that could hurt you or someone else. Hazards are present in most of the workplaces. It is almost impossible to remove all hazards. But it is important to identify hazards and work to reduce their potential for causing harm by putting specific measures in place. For example, operating a bench grinder poses a number of hazards. Although it is not possible to eliminate the hazards of using the bench grinder, by putting specific measures in place, the risk of those hazards can be reduced. A risk analysis of a bench grinder would identify the following hazards and risks:

- a high-velocity particle that could damage your eyesight or that of someone working nearby;
- the grinding wheel breaking apart, damaging eyesight or causing cuts and abrasion;
- electrocution if electrical parts are faulty;

- a risk to your hands from heat or high-velocity particles;
- a risk to your hearing due to excessive noise;
- a risk of fire if flammables are present, and
- a risk of entrapment of clothing or body parts through rotating machinery.

To reduce the risk of these hazards, several measures can be taken. Position the bench grinder in a safe area away from where others work. Make sure electrical items are regularly checked for electrical and mechanical safety. When operating the equipment, wear PPE such as:

- protective eyewear,
- gloves,
- hearing protection,
- hairnets, or caps;
- make sure no flammables are nearby.
- Also, do not wear loose clothing that can be caught in the grinder.

To identify hazardous environments, follow the steps in **SKILL DRILL 3-1**.

SKILL DRILL 3-1 Identifying Hazardous Environments

1. Familiarize yourself with the shop layout. There are special work areas that are defined by painted lines. These lines show the hazardous zone around certain machines and areas. If you are not working on the machines, you should stay outside the marked area.

2. Study the various warning signs around your shop. Understand the meaning of the signal word, the colors, the text, and the symbols or pictures on each sign. Ask your supervisor if you do not fully understand any part of the sign.

3. Identify exits. Find out where each door, window, and gate is and whether it is usually open or locked. Plan your escape route, should you need to exit in a hurry. Also know the designated gathering point and go there in an emergency.

4. Check for air quality. There should be good ventilation and very little chemical fumes or smell. Locate the extractor fans or ventilation outlets, and make sure they are not obstructed in any way. Locate and observe the operation of the exhaust extraction hoses, pump, and outlets that are used with the vehicle's exhaust pipes.

5. Check the location and types of fire extinguishers in your shop. Be sure you know when and how to use each type of fire extinguisher.

6. Identify flammable hazards. Find out where flammable materials are kept, and make sure they are stored properly.

SKILL DRILL 3-1 Identifying Hazardous Environments (Continued)

7. Check the hoses and fittings on the air compressor and air guns for any damage or excessive wear. You must be particularly careful when troubleshooting air guns. Always disconnect an air gun from the air line before inspecting it—if you do not, severe eye damage can result.

8. Identify caustic chemicals and acids associated with activities in your shop. Ask your supervisor for information on any special hazards in your particular shop and any special avoidance procedures, which may apply to you and your working environment.

Safety Equipment

Shop safety equipment is designed to help make the work environment more safe, and includes items such as:

- **Handrails:** Handrails are used to separate walkways and pedestrian traffic from work areas. They provide a physical barrier that directs pedestrian traffic. It also offers protection from vehicle movements (**FIGURE 3-15**).
- **Machinery guards:** Machinery guards and yellow lines prevent people from accidentally walking into the operating equipment. It may also indicate that a safe distance should be kept from the equipment (**FIGURE 3-16**).
- **Painted lines:** Large, fixed machinery such as lathes and milling machines present a hazard to the operator and others working in the area. To prevent accidents, a machinery guard or a painted yellow line on the floor usually borders this equipment (**FIGURE 3-17**).
- **Sound-insulated rooms:** Sound-insulated rooms are usually used when operating equipment makes a lot of noise. A chassis dynamometer is an example. A vehicle operating on a

FIGURE 3-15 Handrail.

FIGURE 3-16 Machinery guard.

FIGURE 3-17 Painted safety lines.

FIGURE 3-18 Sound-insulated room.

FIGURE 3-19 Exhaust extraction and shop ventilation equipment.

dynamometer produces a lot of noise from its tires, exhaust, and engine. To protect other shop users from the noise, the dynamometer is usually placed in a sound-insulated room. This keeps shop noise to a minimum (**FIGURE 3-18**).

- **Adequate ventilation:** Exhaust gases and chemical vapors are serious health hazards in the shop. Whenever a vehicle's engine is running, toxic gases are emitted from its exhaust. To prevent an excess of toxic gas buildup, a well-ventilated work area is needed. In addition, a method of directly venting the vehicle's exhaust to the outside is needed. It may only take a minute or two for a poorly running vehicle to fill the shop with enough CO to affect people's health. Chemical vapors are also a hazard and need to be vented outside (**FIGURE 3-19**).
- **Doors and gates:** Doors and gates are used for the same reason as machinery guards and painted lines. A doorway is a physical barrier that can be locked and sealed. It separates a hazardous environment from the rest of the shop or a general work area from an office or specialist work area (**FIGURE 3-20**).
- **Temporary barriers:** In the day-to-day operation of a shop, there is often a reason to temporarily separate one

FIGURE 3-20 Sign on the door, used to control access.

work bay from others. If a welding machine or an oxyacetylene cutting torch is in use, it may be necessary to place a temporary screen or barrier around the work area. This is to protect other shop users from welding flash or injury (**FIGURE 3-21**).

▶ TECHNICIAN TIP

Stay alert for hazards or anything that might be dangerous. If you see, hear, or smell anything odd, take steps to fix it, or tell your supervisor about the problem.

Electrical Safety

Many people are injured by electricity in shops. Poor electrical safety practices can cause shocks and burns as well as fires and explosions. Make sure you know where the electrical shutoffs or panels for your shop are located. All circuit breakers and fuses should be clearly labeled. This is so that you know which circuits and functions they control (**FIGURE 3-22**).

FIGURE 3-21 Portable welding curtain.

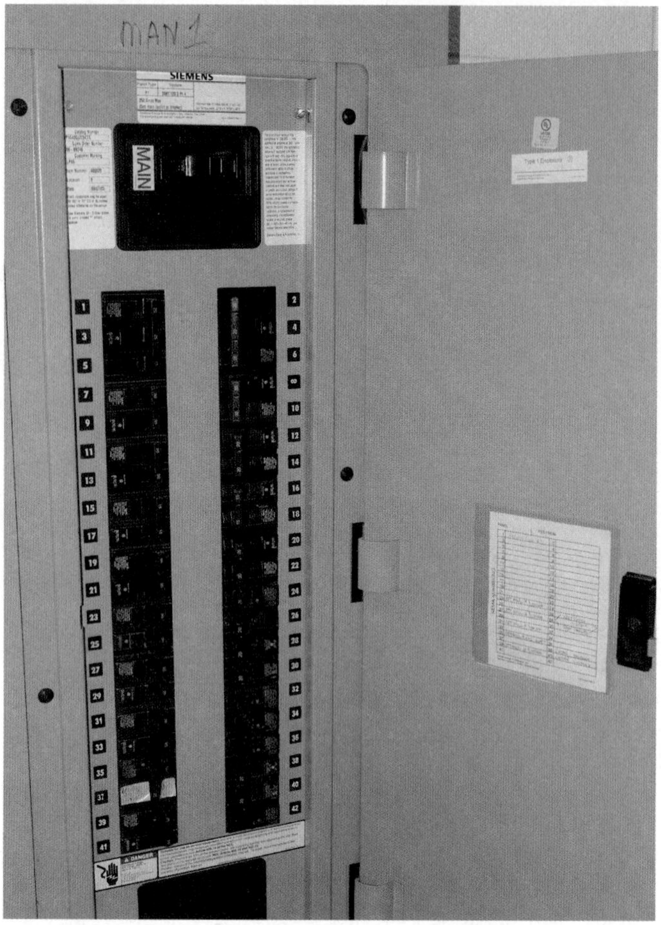

FIGURE 3-22 All electrical switches and fuses should be clearly labeled. This is so that you know which circuits and functions they control.

In the case of an emergency, you may need to know how to shut off the electricity supply to a work area or to your entire shop. Keep the circuit breaker and/or electrical panel covers closed. This is needed to keep them in good condition, prevent unauthorized access, and prevent accidental contact with the electricity supply. It is important that you do not block or obstruct access to this electrical panel. Keep equipment and tools well away so that emergency access is not hindered. In some localities, 3' (0.91 m) of unobstructed space must be maintained around the panel at all times.

There should be a sufficient number of electrical receptacles in your work area for all your needs. Do not connect multiple appliances to a single receptacle with a simple double adapter. If necessary, use a multi-outlet safety strip that has a built-in circuit breaker. Electric receptacles should be at least 3' (0.91 m) above floor level. This reduces the risk of igniting spilled fuel vapors or other flammable liquids.

Portable Electrical Equipment

If you need to use an extension cord, make sure it is made of flexible wiring—not the stiffer type of house wiring. Also make sure that it is fitted with a ground wire. The cord should

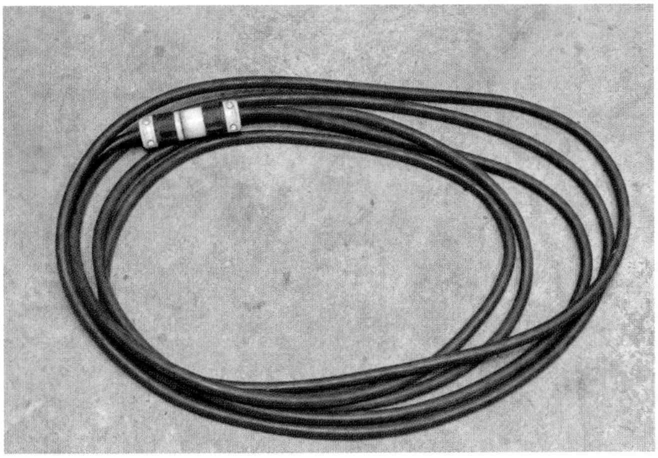

FIGURE 3-23 The extension cord should be neoprene-covered.

be neoprene-covered, as this material resists oil damage (**FIGURE 3-23**).

Always check cords for cuts, abrasions, or other damage. Be careful how you place the extension cord, so it does not cause a tripping hazard. Also avoid rolling equipment or vehicles over it, as doing so can damage the cord. Never use an extension cord in wet conditions or around flammable liquids. Portable electric tools that operate at 240 volts are often sources of serious shock and burn accidents. Be particularly careful when using these items. Always inspect the security of the attached plug before connecting the item to the power supply. Use 110-volt or lower voltage tools if they are available. All electric tools must be equipped with a ground prong or be **double insulated** (**FIGURE 3-24**).

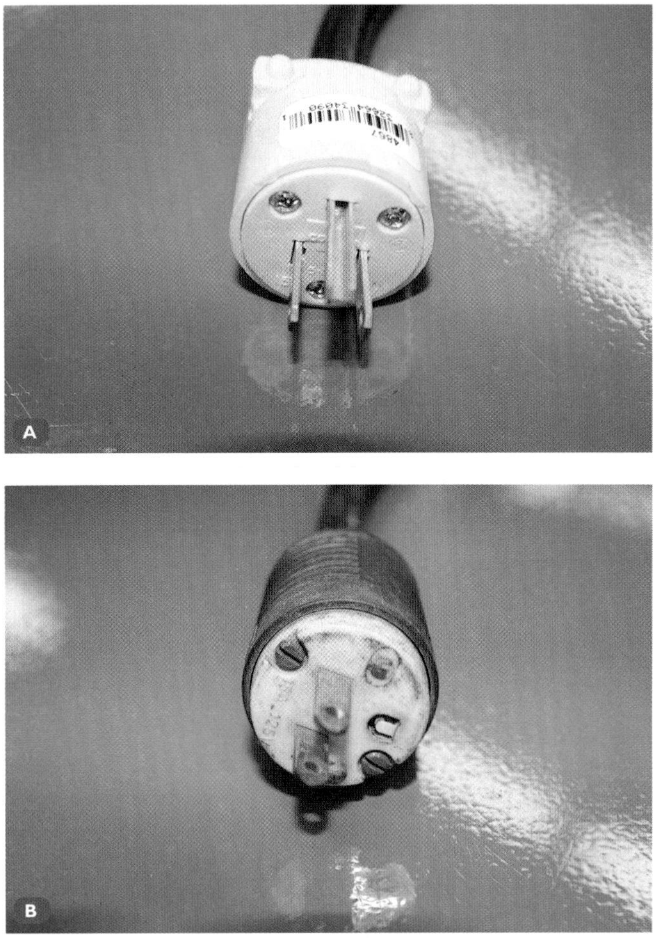

FIGURE 3-24 Ground prong. **A.** Okay. **B.** Missing.

If electrical cords do not have the ground prong or are not double insulated, do *not* use them. Never use any high-voltage tool in a wet environment. In contrast, air-operated tools cannot give you an electric shock. This is because they operate on air pressure instead of electricity. Therefore, they are safer to use in a wet environment.

Portable Shop Lights

Portable shop lights/droplights have been the cause of many accidents over the years. Incandescent bulbs get extremely hot and can cause burns. They are also prone to shatter, which can cut the skin and damage the eyes. They can also cause fires and electric shock. In fact, in some places, incandescent portable shop lights cannot be used in automotive shops. They must be replaced with less hazardous lights (**FIGURE 3-25**).

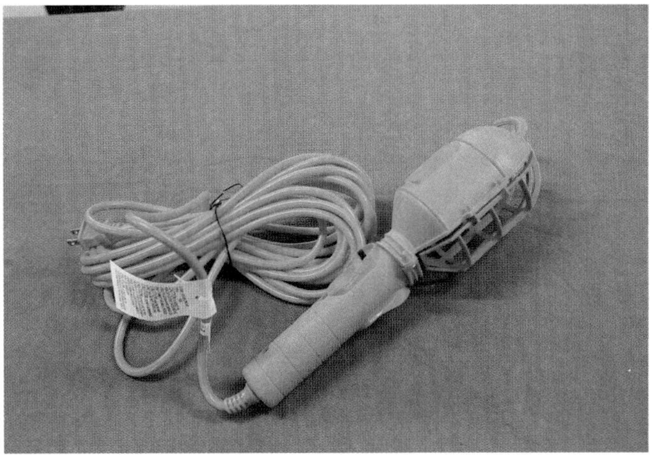

FIGURE 3-25 Incandescent droplight which should *not* be used in a shop.

One such light is the fluorescent droplight. Although this type of light stays much cooler than incandescent lights, it can still shatter, causing cuts or damage to eyes. Because of this, droplights should be fully enclosed in a clear, insulating case. They may also contain mercury, which becomes dispersed when the bulb is shattered. This creates a hazardous-materials situation (**FIGURE 3-26**).

The safest portable shop light that has come to market is the LED (light-emitting diode) shop light. It uses much lower voltage, and the LED is much less prone to shattering, so it is much safer to use. It also uses a small amount of electricity to produce a large amount of light, so many of them are cordless. They may also include a magnetic base, so the light can be attached to any steel surface. It can then be adjusted to shine where needed (**FIGURE 3-27**).

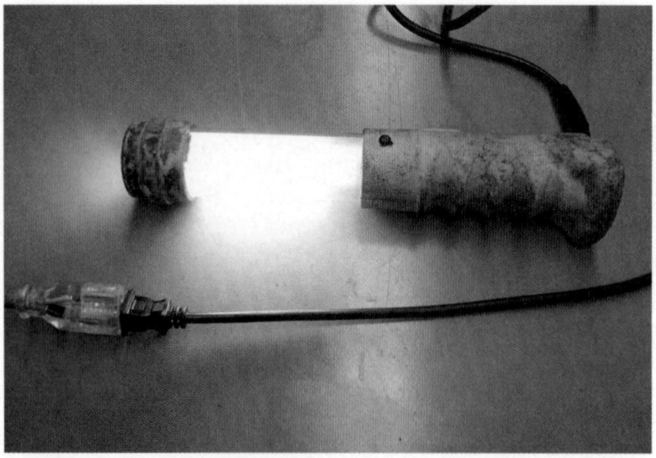

FIGURE 3-26 Fluorescent droplight.

FIGURE 3-27 Cordless LED light.

> ▶ TECHNICIAN TIP
>
> Try out the new LED flashlights or cordless LED shop lights. Good ones provide a lot of light, and the batteries last a long time. It is not unusual to hear about an LED flashlight that got dropped behind a toolbox while it was on. Days later, it was still illuminated.

Shop Layout

The shop should have a layout that is efficient and safe, with clearly defined working areas and walkways. Customers should not be allowed to wander through work areas unescorted. A good shop layout can be achieved by thinking about:

- how the work is to be done,
- how equipment is used, and
- what traffic movements, both pedestrian and vehicular, occur within the shop.

A well-planned shop should have clearly defined areas for various activities. These areas include:

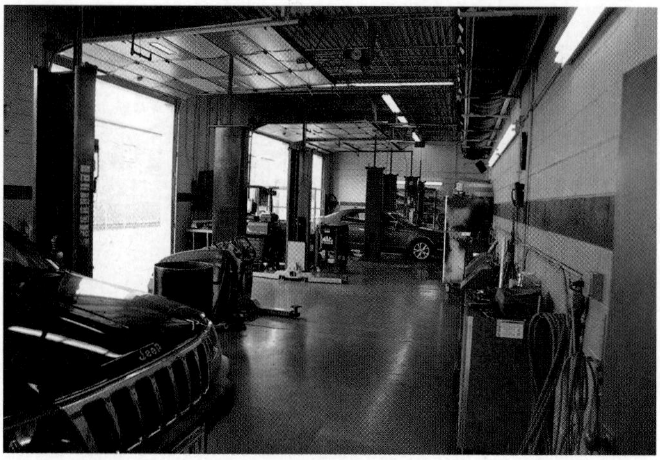

FIGURE 3-28 Shop layout can minimize accidents as well as improve work efficiency.

- parts cleaning,
- parts storage,
- tool storage,
- flammable liquid storage,
- jacking or lifting,
- tire service, and
- painting.

Safety equipment should be located where they are easily accessible. These include eyewash stations, fire extinguishers, and first aid kits. All flammable items should be kept in an approved fireproof storage container or cabinet. Firefighting equipment should be close at hand. And finally, supplies for cleaning the shop need to be strategically located so that it is easy to keep the shop clean and orderly. This includes garbage and recycling containers, brooms, mops, dust pans, and even floor scrubbers (**FIGURE 3-28**).

Eyewash Stations and Emergency Showers

Hopefully you will never need to use an eyewash station or emergency shower. The best treatment is prevention. So make sure you wear all the PPE required for each specific task, to avoid as many eye injuries as possible. Eyewash stations are used to flush the eyes with clean water or sterile liquid in the event that you get foreign liquid or particles in your eyes. There are different types of eyewash stations. The two main types are disposable eyewash packs and fixed eyewash stations. Some emergency or deluge showers also have an eyewash station built in (**FIGURE 3-29**).

When an individual gets chemicals in their eyes, they typically need assistance in reaching the eyewash station. Do this by taking their arm and leading them to it. They may not want to open their eyes, even in the water. So encourage them to use their fingers to pull their eyelids open. If a chemical splashed in their eyes, encourage them to rinse their eyes for 15 minutes. While they are rinsing their eyes, call for medical assistance.

FIGURE 3-29 Some emergency showers have an eyewash station built in.

One thing to note is that eyewash stations are used for flushing chemicals or debris from the surface of the eyes. An eyewash should not be used for punctures or cuts to the eye. In this case, cover the eye with sterile gauze, if available, and seek medical assistance. To flush eyes with a fixed eyewash station, follow the steps in **SKILL DRILL 3-2**. If flushing the eyes with disposable eyewash packs, follow the steps in **SKILL DRILL 3-3**.

Shop Safety Inspections

Shop safety inspections are valuable ways of identifying unsafe equipment, materials, or activities. They can be corrected to reduce the risk of accidents or injuries. The inspection can be formalized by using inspection sheets to check specific items. Or they can be general walk-arounds where you consciously look for and document problems that can be corrected (**FIGURE 3-30**).

Here are some of the common things to look for:

- Items blocking emergency exits or walkways
- Poor safety signage
- Unsafe storage of flammable goods
- Tripping or slipping hazards
- Faulty or unsafe equipment or tools
- Missing equipment guards

SKILL DRILL 3-2 Using a Fixed Eyewash Station

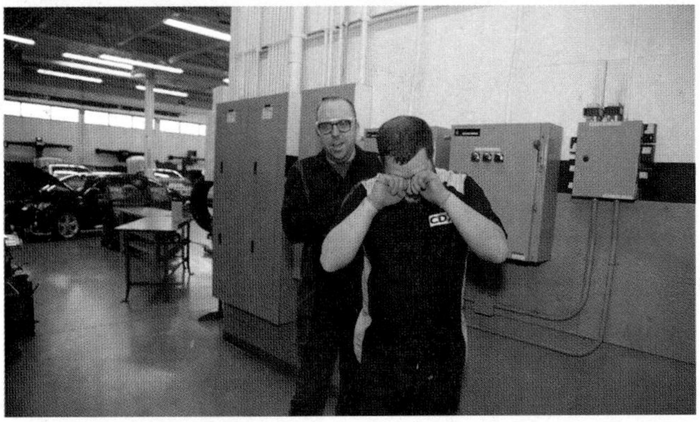

1. Guide the injured person to the eyewash station

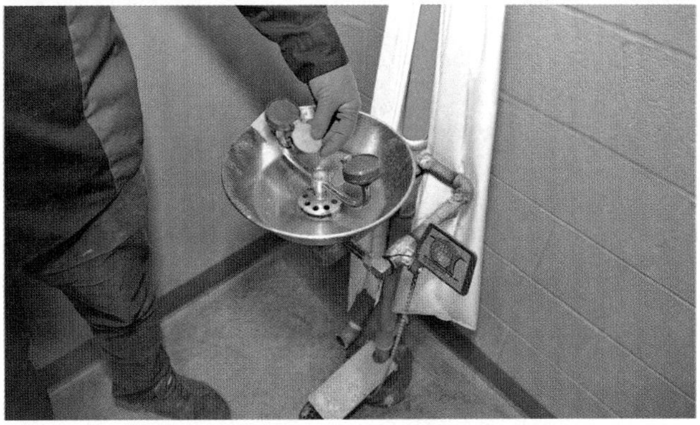

2. Remove any covers/caps from the eyewash nozzles

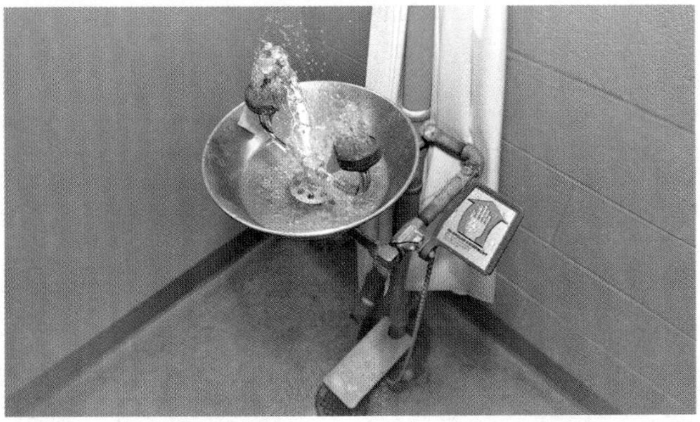

3. Turn the eyewash station on.

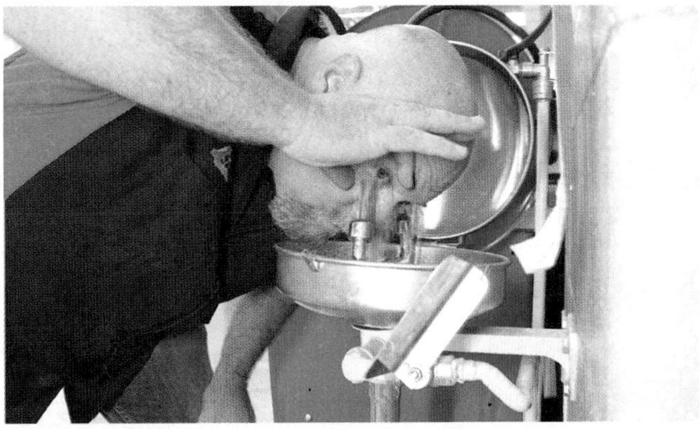

4. Place eye/s in front of eyewash nozzles and force them open. Flush for 15 minutes while medical assistance is called.

SKILL DRILL 3-3 Using Disposable Eyewash Packs

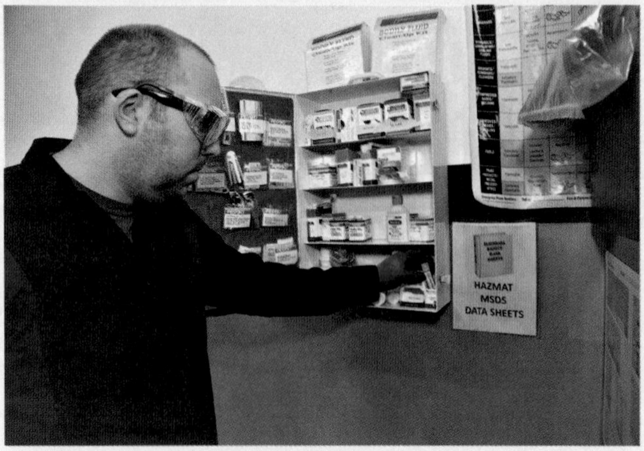

1. Get the eyewash pack to the injured person.

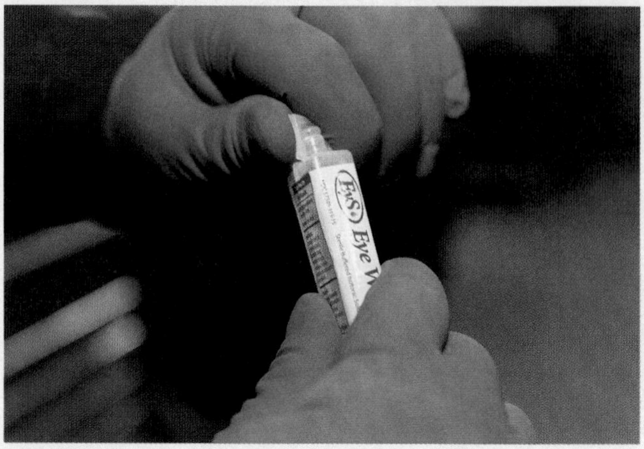

2. Remove any caps from the eyewash pack.

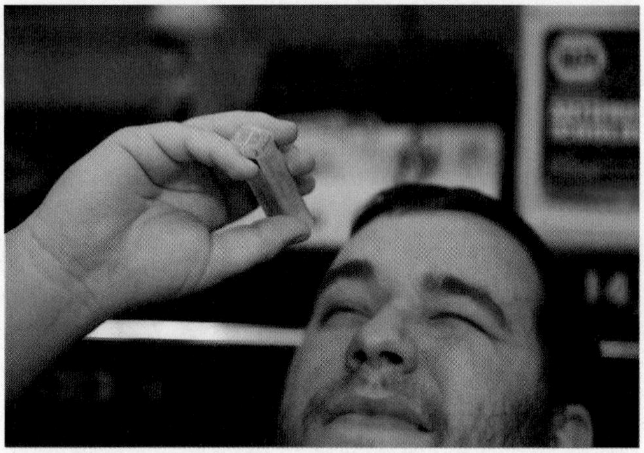

3. Tilt the person's head so the water can flow across the eye and out the side

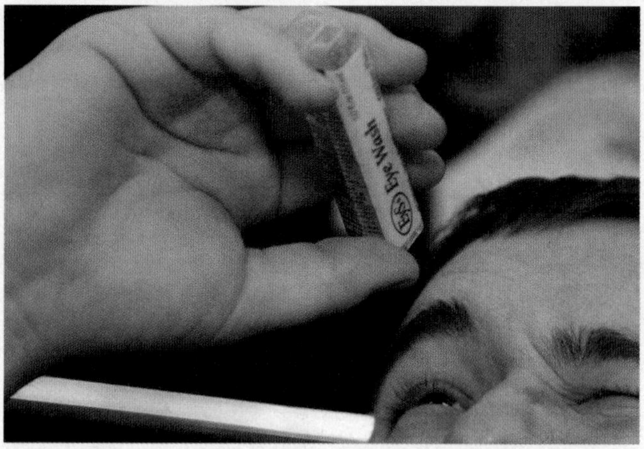

4. Flush the eye/s for 15 minutes, or until the eyewash runs out while medical assistance is called

- Misadjusted bench grinder tool rests
- Missing or expired fire extinguishers
- Clutter, spills, unsafe shop practices
- People not wearing the correct PPE

Formal and informal safety inspections should be held regularly. For example, an inspection sheet might be used weekly or monthly to formally evaluate the shop. Informal inspections might be held daily to catch issues that are of a more immediate nature. Never ignore or put off a safety issue.

▶ Preventing and Fighting Fires

LO 3-03 Prevent fires and operate fire safety equipment.

The danger of a gasoline fire is always present in an automotive shop. Most automobiles carry a fuel tank, often with large quantities of fuel on board. This is more than sufficient to cause a large, very destructive, and potentially explosive fire (**FIGURE 3-31**).

In fact, 1 gallon (3.8 liter) of gasoline has the same amount of energy as 20 sticks of dynamite. So take precautions to make

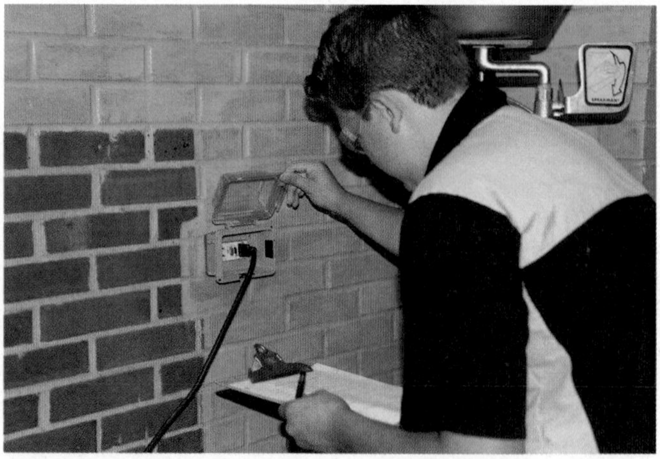

FIGURE 3-30 Shop safety inspection using a checklist.

FIGURE 3-31 Shop fire.

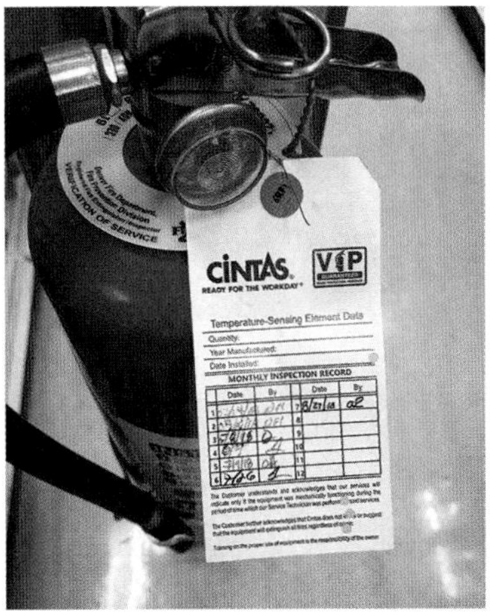

FIGURE 3-32 Fire extinguisher inspection tag with dates and sign offs.

sure that fires don't get started in the first place. And if a fire does break out, you need to be prepared by having the correct type, size, and quantity of extinguishers on hand.

Also, most localities require that each of the shop's fire extinguishers be checked on a periodic schedule, such as monthly. This is to make sure they are ready to be used and are not past their service date. There is usually an inspection tag connected to the fire extinguisher. It is used to record the date of inspection and the signature of the person inspecting it (**FIGURE 3-32**).

One of the simplest ways of preventing fires is to make sure you clean up spills immediately. Also avoid ignition sources, like sparks, near flammable liquids or gases. Also, being aware of the following topics will help you know how to minimize the risk of fires in the shop.

> ▶ **TECHNICIAN TIP**
>
> It is not a matter of "if" but "when" a shop is going to have a fire. When shops do catch on fire, the fire can grow quickly out of control. So preventing a fire is the best action you can take. If a fire starts, react appropriately. In some cases, you just need to get out safely. In other cases, it may be safe to try to fight the fire quickly before it spreads too far. Always be aware of the potential for a fire, and plan ahead by thinking through the task you are about to undertake. Know where firefighting equipment is kept and how it works.

Fuel Vapor

Liquid fuel vaporizes rapidly, especially when spilled, and the vapor is extremely easy to ignite. Fuel vapor is invisible and three to four times heavier than air. This allows it to spread unseen across a wide area, filling any low spots such as service pits. So a source of ignition can be quite some distance from the original spill and still ignite it. Fuel can even vaporize from paper towels or rags used to wipe up liquid spills. These materials should be allowed to dry in the open air, not held in front of a heater element or other ignition source. Any spark

FIGURE 3-33 Spill-proof gas can.

or naked flame, even a lit cigarette, can start an explosive fire. Always keep flammable liquids in spill-proof containers. If you use gas cans, use the ones that automatically close when not in use (**FIGURE 3-33**).

Always be aware of the possibility of fuel vapor buildup. If you smell gasoline or other flammable vapor, stop work and investigate the source. This may mean having coworkers stop all work and assist you. Do not ignore the presence of fuel vapors.

Spillage Risks

Spills frequently occur when technicians remove and replace fuel filters. Disconnecting the hoses without properly releasing the fuel pressure causes liquid fuel to spray a great deal. So always check the manufacturer's procedure for depressurizing the fuel system before opening it up.

Spills can also occur during removal of a fuel pump or fuel tank sending unit. This is because they are commonly located on the inside of the fuel tank. Unless the tank can be opened from the top without removing the tank from the vehicle, it is best to first empty the tank safely, to avoid

FIGURE 3-34 Spill response kit.

spills. Spills also can occur when fuel is being drained into unsuitable containers. Avoid spills by following the manufacturer's specified procedure when removing fuel system components. Also, keep a spill response kit nearby to deal with any spills quickly. Spill kits should contain absorbent material and barrier dams to contain moderate-sized spills (**FIGURE 3-34**).

And remember that fuel vapors are heavier than air and can travel a long way from the spill. So all sources of ignition need to be removed from the immediate vicinity. This can be more than 50' away.

Draining Fuel

If there is a possibility of fuel spillage while working on a vehicle, then you should first remove the fuel safely. Do this only in a well-ventilated, level space, preferably outside in the open air. Make sure all potential sources of ignition have been removed from the area, and disconnect the battery on the vehicle. Do not drain fuel from a vehicle over an inspection pit. Make sure the container you are draining into is an approved fuel storage container (fuel retriever). Also make sure that it is large enough to contain all of the fuel in the system being drained. A fuel retriever minimizes the chance of sudden large spills occurring (**FIGURE 3-35**).

You may need to use a narrow diameter hose to bypass the anti-spillage device in the filler neck (**FIGURE 3-36**).

If removing fuel lines or fuel filters, the fuel system must be depressurized. This helps to prevent fuel from spraying when the hoses are disconnected. Check the service manual for details on how best to depressurize or drain the fuel from the vehicle you are working on.

FIGURE 3-35 Fuel retriever.

FIGURE 3-36 Fuel retriever use.

Extinguishing Fires

Three elements must be present at the same time for a fire to occur: fuel, oxygen, and heat (**FIGURE 3-37**). The secret of firefighting involves the removal of at least one of these elements. If a fire occurs in the shop, it is usually the oxygen or the heat that is removed to extinguish the fire. For example, a fire blanket when applied correctly removes the oxygen. A water extinguisher removes heat from the fire. Fire extinguishers are used to extinguish the majority of small fires in a shop. Never hesitate to call the fire department if you cannot extinguish a fire quickly and safely.

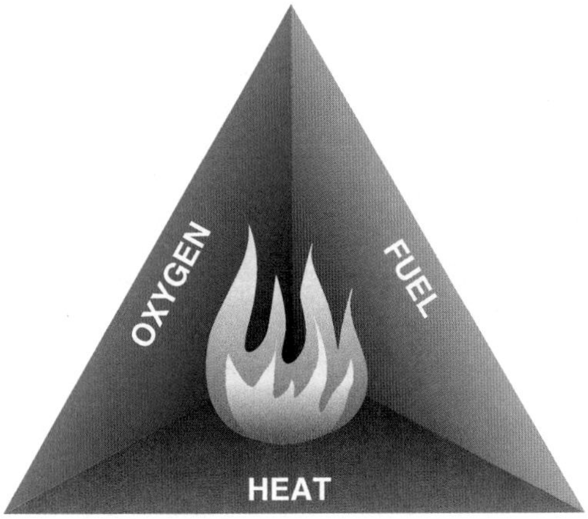

FIGURE 3-37 Fire triangle.

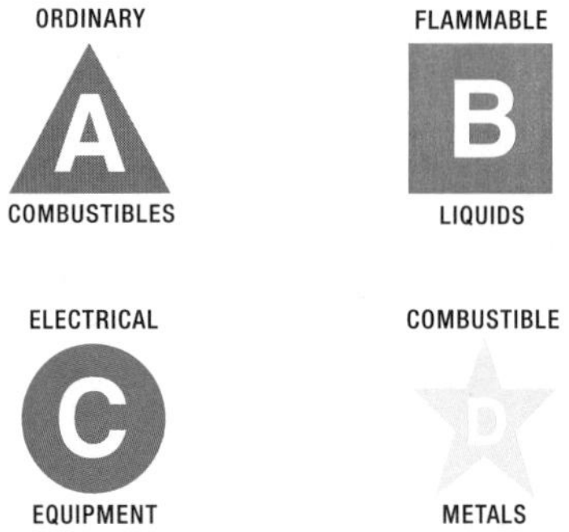

FIGURE 3-38 Traditional labels on fire extinguishers often incorporate a shape as well as a letter.

Fire Classifications

In the United States, fire is classified into five types:

- Class A fires involve ordinary combustibles such as wood, paper, or cloth.
- Class B fires involve flammable liquids or gaseous fuels.
- Class C fires involve electrical equipment.
- Class D fires involve combustible metals such as sodium, titanium, and magnesium.
- Class K fires involve cooking oil or fat.

Fire Extinguisher Types

Fire extinguishers are marked with pictograms depicting the types of fires that the extinguisher is approved to fight (**FIGURE 3-38**):

- Class A: Green triangle
- Class B: Red square
- Class C: Blue circle
- Class D: Yellow pentagram
- Class K: Black hexagon

FIGURE 3-39 If there is any doubt that you can extinguish the fire, pull the fire alarm to get the fire department here as soon as possible.

Fire Extinguisher Operation

Unless the fire is very small, *ALWAYS* sound the alarm before attempting to fight a fire (**FIGURE 3-39**). It is better to get the fire department coming right away and end up not needing them, than to risk the fire getting out of control. Fire generally spreads quickly. If you cannot fight the fire safely, leave the area while you wait for backup. Close any doors and windows, if you have time. This will help slow the fire down.

You need to size up the fire before you make the decision to fight it with a fire extinguisher. Identify what sort of material is burning, the extent of the fire, and the likelihood of it spreading. Also, if the fire is in electrical wires or equipment, make sure you won't be electrocuted while trying to extinguish it. To operate a fire extinguisher, follow the acronym for fire extinguisher use: PASS (pull, aim, squeeze, sweep).

- *Pull* out the pin that locks the handle at the top of the fire extinguisher to prevent accidental use. Carry the fire extinguisher in one hand, and use your other hand to:
- *Aim* the nozzle at the base of the fire. Stand about 8–12 feet (2.4–3.7 m) away from the fire, and
- *Squeeze* the handle to discharge the fire extinguisher. Remember that if you release the handle on the fire extinguisher, it will stop discharging.
- *Sweep* the nozzle from side-to-side at the base of the fire. Continue to watch the fire. Although it may appear to be extinguished, it may suddenly reignite. Portable fire extinguishers only operate for about 10–25 seconds before they are empty. So use them effectively (**FIGURE 3-40**).

If the fire is indoors, you should be standing between the fire and the nearest safe exit (**FIGURE 3-41**). If the fire is outside, you should stand facing the fire, with the wind on your back. This is so that the smoke and heat are being blown away from you (**FIGURE 3-42**). If possible, get an assistant to guide you and inform you of the fire's progress. Again, make sure you have a means of escape, should the fire get out of control.

When you are certain that the fire is out, report it to your supervisor. Also report what actions you took to put out the fire. Allow the circumstances of the fire to be investigated. Once your supervisor or the fire department has given you the all clear, clean up the debris. Also submit the used fire extinguisher for inspection and service.

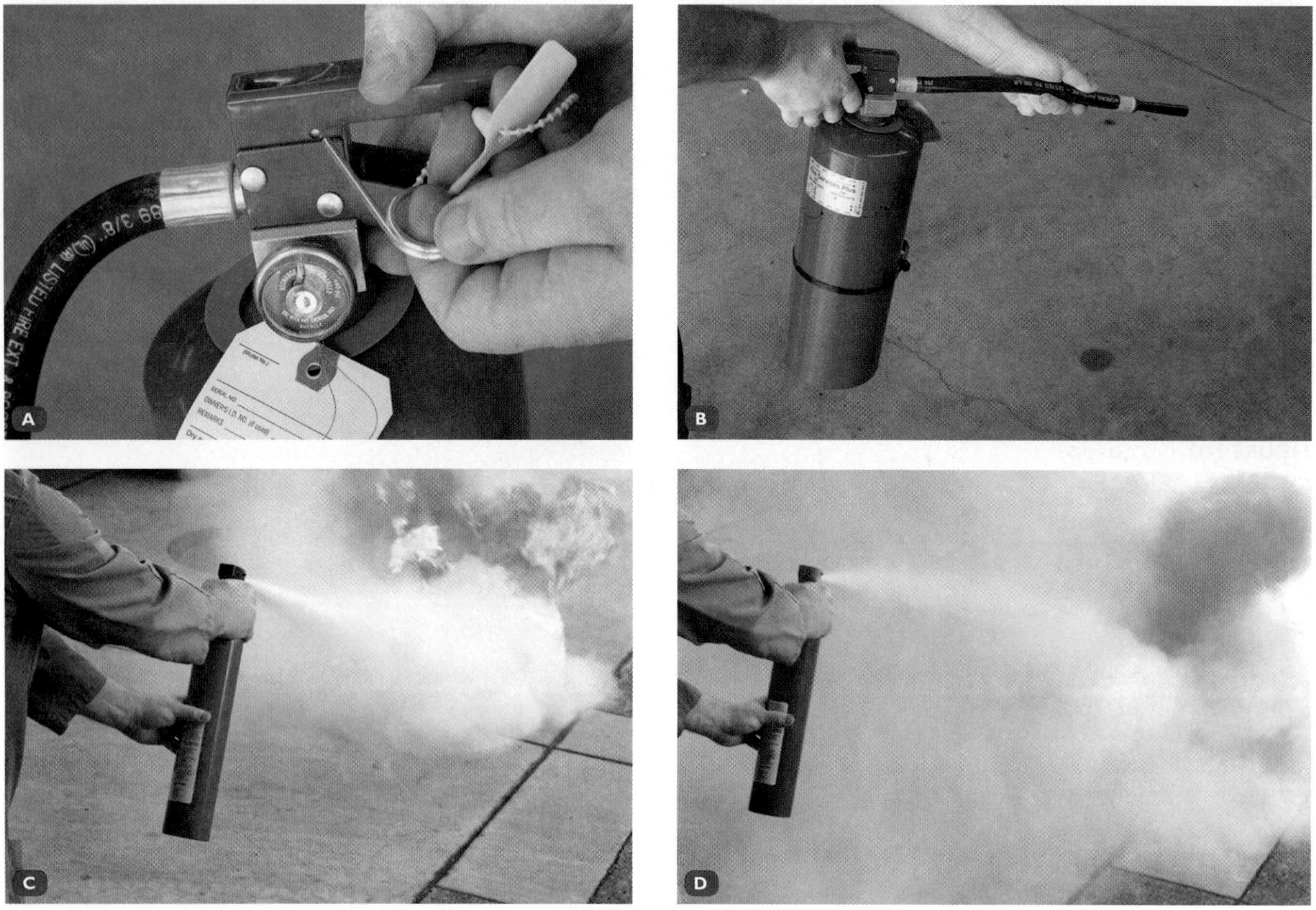

FIGURE 3-40 To operate a fire extinguisher, follow PASS. **A.** Pull. **B.** Aim. **C.** Squeeze. **D.** Sweep.

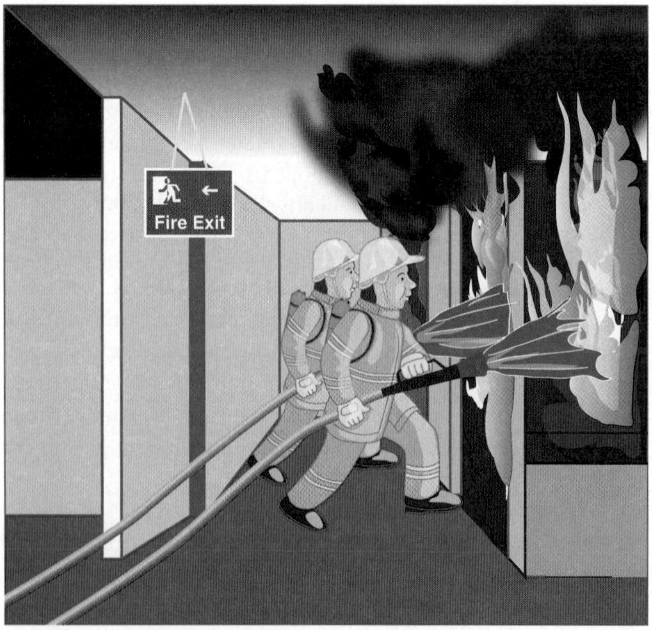

FIGURE 3-41 When fighting indoor fires, stand between the fire and the nearest safe exit.

FIGURE 3-42 When fighting outdoor fires, stand facing the fire, with the wind to your back, making sure you have a clear exit path.

FIGURE 3-43 Fire blanket in storage container.

FIGURE 3-44 Hazardous materials.

Fire Blankets

Fire blankets are designed to smother a small fire (**FIGURE 3-43**). They are ideal for use in situations where a fire extinguisher could cause damage. For example, if there is a small fire under the hood of a vehicle, a fire blanket might be able to smother the fire without running the risk of getting powder from the fire extinguisher down the intake system. They are also very useful in putting out a fire on a person. The fire blanket can be thrown around the person, smothering any fire. The blanket should be tucked snugly around the person and may need to stay on for up to 30 minutes. This is to prevent the fire from restarting when removed.

Obtain a fire blanket and study the instructions on the packaging. If instructions are not provided, research how to use a fire blanket, or ask your supervisor. You may require instruction from an authorized person in using the fire blanket. If you do use a fire blanket, make sure you replace it with a new one.

▶ Hazardous Materials Safety

LO 3-04 Research material data using safety data sheets (SDS).

A **hazardous material** is any material that poses an unreasonable risk of damage or injury to persons, property, or the environment. It is dangerous if it is not properly controlled during:

- handling,
- storage,
- manufacture,
- processing,
- packaging,
- use and disposal, or
- transportation.

These materials can be solids, liquids, or gases. Most technicians use hazardous materials daily. Examples are cleaning solvents, gasket cement, brake fluid, and coolant. In fact, there are likely to be hundreds of hazardous materials in a typical shop (**FIGURE 3-44**). Hazardous materials must be properly handled, labeled, and stored in the shop.

Safety Data Sheets

Hazardous materials are used daily and may make you very sick if they are not used properly. **Safety data sheets (SDS)** contain detailed information about hazardous materials. They help you understand:

- how they should be safely used,
- any health effects relating to them,
- how to treat a person who has been exposed to them, and
- how to deal with them in a fire situation.

SDS can be obtained from the manufacturer of the material. The shop should have an SDS for each hazardous substance or dangerous product. In the United States, it is required that workplaces have an SDS for every chemical that is on site.

Safety data sheets (SDS) were formerly called material safety data sheets (MSDS). In 2012, OSHA changed the requirements for the hazard communication system (HCS). It now conforms to the United Nations Globally Harmonized System of Classification and Labeling of Chemicals (GHS). The MSDS needed to change its name and its format to fit the new standards. While the original MSDS had 8 sections, safety data sheets are required to have 16 sections. Each section provides additional details and makes it easier to find specific data when needed. In addition, GHS requires all employers to train their employees. They must be trained in the new chemical labeling requirements and the new format for the safety data sheets.

Whenever you deal with a potentially hazardous product, you should consult the SDS to learn how to use that product safely. If you are using more than one product, make sure you consult all the SDS for those products. Be aware that certain combinations of products can be more dangerous than any of them separately. SDS are usually kept in a clearly marked binder. They should be regularly updated as chemicals come into the workplace. As of June 1, 2015, the HCS requires new SDS to be in a uniform format. It must include the section numbers, the headings, and associated information under **TABLE 3-1**. See **FIGURE 3-45** for a sample SDS for a common brake-cleaning chemical.

To identify information found on an SDS, follow the steps in **SKILL DRILL 3-4**.

Cleaning Hazardous Dust Safely

Toxic dust is any dust that may contain fine particles that could be harmful to humans or the environment. If you are unsure as to the toxicity of any particular dust, then you should always treat it as toxic. Take the precautions identified in the SDS or shop procedures. Brake and clutch dust are potential toxic dusts that automotive shops must manage. The dust is made up of very fine particles that can easily spread and contaminate an area.

One of the more common sources of toxic dust is inside drum brakes and manual transmission bell housings. It is a good idea to avoid all dust if possible, whether it is classified as toxic or not. If you do have to work with dust, never use compressed air to blow it from components or parts. And always use PPE such as face masks, eye protection, and gloves when working around it.

Various tools have been developed to clean toxic dust from vehicle components. The most common one is the brake wash station (**FIGURE 3-46**). It uses an aqueous solution to wet down and wash the dust into a collection basin. The basin needs periodic maintenance to properly dispose of the accumulated sludge. This tool is probably the simplest way to effectively deal with hazardous dust. This is because it is easy to set up, use, and store. A high-efficiency particulate air (HEPA) dust collection system uses a HEPA filter to trap very small dust particles. It is no longer approved for use as a hazardous dust collection system.

After completing a service or repair task on a vehicle, there is often dirt and dust left behind. The chemicals present in this dirt usually contain toxic chemicals that can build up and cause health problems. To keep the levels of dirt and dust to a minimum, clean it up immediately after the task is complete. The vigorous action of sweeping causes the dirt and dust to rise. When sweeping the floor, use a soft broom that pushes, rather than flicks, the dirt and dust forward. Create smaller piles and dispose of them frequently.

Another successful way of cleaning shop dirt and dust is to use a water hose. The wastewater must be caught in a settling pit and not run into a storm water drain. Many shops also have floor scrubbers that use a water/soap solution to clean the floor (**FIGURE 3-47**). These shops usually vacuum up the dirty water and store it in a tank until it can be disposed of properly.

TABLE 3-1 SDS Section Descriptions

Section 1, Identification, includes product identifier; manufacturer or distributor name, address, and phone number; emergency phone number; recommended use; restrictions on use.
Section 2, Hazard(s) identification, includes all hazards regarding the chemical; required label elements.
Section 3, Composition/information on ingredients, includes information on chemical ingredients; trade secret claims.
Section 4, First-aid measures, includes important symptoms/effects, acute, delayed; required treatment.
Section 5, Firefighting measures, lists suitable extinguishing techniques, equipment; chemical hazards from fire.
Section 6, Accidental release measures, lists emergency procedures; protective equipment; proper methods of containment and cleanup.
Section 7, Handling and storage, lists precautions for safe handling and storage, including incompatibilities.
Section 8, Exposure controls/personal protection, lists OSHA's permissible exposure limits (PELs); threshold limit values (TLVs); appropriate engineering controls; personal protective equipment (PPE).
Section 9, Physical and chemical properties, lists the chemical's characteristics.
Section 10, Stability and reactivity, lists chemical stability and possibility of hazardous reactions.
Section 11, Toxicological information, includes routes of exposure; related symptoms, acute and chronic effects; numerical measures of toxicity.
Section 12, Ecological information*
Section 13, Disposal considerations*
Section 14, Transport information*
Section 15, Regulatory information*
Section 16, Other information, includes the date of preparation or last revision.

**Note*: Since other agencies regulate this information, OSHA will not be enforcing Sections 12 through 15 (29 CFR 1910.1200(g)(2)).

Source: https://www.osha.gov/Publications/HazComm_QuickCard_SafetyData.html.

CRC

SAFETY DATA SHEET

Section 1: Product & Company Identification

Product Name: **Brakleen® Brake Parts Cleaner** (aerosol)

Product Number (s): **05089, 05089-6, 05089T, 75089, 85089, 85089AZ**

Product Use: Brake parts cleaner

Manufactured / Supplier Contact Information:

In United States:
CRC Industries, Inc.
885 Louis Drive
Warminster, PA 18974
www.crcindustries.com
1-215-674-4300(General)
(800) 521-3168 (Technical)
(800) 272-4620 (Customer Service)

In Canada:
CRC Canada Co.
2-1246 Lorimar Drive
Mississauga, Ontario L5S 1R2
www.crc-canada.ca
1-905-670-2291

In Mexico:
CRC Industries Mexico
Av. Benito Juárez 4055 G
Colonia Orquídea
San Luís Potosí, SLP CP 78394
www.crc-mexico.com
52-444-824-1666

24-Hr Emergency – CHEMTREC: (800) 424-9300 or (703) 527-3887

Section 2: Hazards Identification

Emergency Overview

DANGER: Vapor Harmful. Contents Under Pressure.
As defined by OSHA's Hazard Communication Standard, this product is hazardous.
Appearance & Odor: Colorless liquid, irritating odor at high concentrations

Potential Health Effects:

ACUTE EFFECTS:

EYE: May cause slight temporary eye irritation. Vapors may irritate the eyes at concentrations of 100 ppm.

SKIN: Short single exposures may cause skin irritation. Prolonged exposure may cause severe skin irritation, even a burn. A single prolonged exposure is not likely to result in the material being absorbed through skin in harmful amounts.

INHALATION: Dizziness may occur at concentrations of 200 ppm. Progressively higher levels may also cause nasal irritation, nausea, incoordination, and drunkenness. Very high levels or prolonged exposure could lead to unconsciousness and death.

INGESTION: Single dose oral toxicity is considered to be extremely low. Swallowing large amounts may cause injury if aspirated into the lungs. This may be rapidly absorbed through the lungs and result in injury to other body systems.

CHRONIC EFFECTS: Repeated contact with skin may cause drying or flaking of skin. Excessive or long term exposure to vapors may increase sensitivity to epinephrine and increase myocardial irritability.

TARGET ORGANS: Central nervous system. Possibly liver and kidney.

Medical Conditions Aggravated by Exposure: None known.

See Section 11 for toxicology and carcinogenicity information on product ingredients.

Page 1 of 7

FIGURE 3-45 Continued

Product Name: Brakleen® Brake Parts Cleaner (aerosol)
Product Number (s): 05089,05089-6, 05089T, 75089, 85089, 85089AZ

Section 3: Composition/Information and Ingredients

COMPONENT	CAS NUMBER	% by Wt.
Tetrachloroethylene (PERC)	127-18-4	> 95
Carbon Dioxide	124-38-9	< 5

Section 4: First Aid Measures

Eye Contact: Immediately flush with plenty of water for 15 minutes. Call a physician if irritation persists.

Skin Contact: Remove contaminated clothing and wash affected area with soap and water. Call a physician if irritation persists. Wash contaminated clothing prior to re-use.

Inhalation: Remove person to fresh air. Keep person calm. If not breathing, give artificial respiration. If breathing is difficult give oxygen. Call a physician.

Ingestion: Do NOT induce vomiting. Call a physician immediately.

Note to Physicians: Because rapid absorption may occur through lungs if aspirated and cause systemic effects, the decision of whether to induce vomiting or not should be made by a physician. If lavage is performed, suggest endotracheal and/or esophageal control. If burn is present, treat as any thermal burn, after decontamination. Exposure may increase myocardial irritability. Do not administer sympathomimetic drugs unless absolutely necessary. No specific antidote.

Section 5: Fire-Fighting Measures

Flammable Properties: This product is nonflammable in accordance with aerosol flammability definitions. (See 16 CFR 1500.3(c)(6))

Flash Point: None (TCC)
Autoignition Temperature: None
Upper Explosive Limit: None
Lower Explosive Limit: None

Fire and Explosion Data:

Suitable Extinguishing Media: This material does not burn. Use extinguishing agent suitable for surrounding fire.

Products of Combustion: Hydrogen chloride, trace amounts of phosgene and chlorine

Explosion Hazards: Aerosol containers, when exposed to heat from fire, may build pressure and explode.

Protection of Fire-Fighters: Firefighters should wear self-contained, NIOSH-approved breathing apparatus for protection against suffocation and possible toxic decomposition products. Proper eye and skin protection should be provided. Use water spray to keep fire-exposed containers cool and to knock down vapors which may result from product decomposition.

Section 6: Accidental Release Measures

Personal Precautions: Use personal protection recommended in Section 8. Do not breathe vapors.

Environmental Precautions: Take precautions to prevent contamination of ground and surface waters. Do not flush into sewers or storm drains.

Methods for Containment & Clean-up: Dike area to contain spill. Ventilate the area with fresh air. If in confined space or limited air circulation area, clean-up workers should wear appropriate

FIGURE 3-45 Continued

Product Name: Brakleen® Brake Parts Cleaner (aerosol)
Product Number (s): 05089,05089-6, 05089T, 75089, 85089, 85089AZ

respiratory protection. Recover or absorb spilled material using an absorbent designed for chemical spills. Place used absorbents into proper waste containers.

Section 7: Handling and Storage

Handling Procedures: Vapors of this product are heavier than air and will collect in low areas. Make sure ventilation removes vapors from low areas. Do not eat, drink or smoke while using this product. Use caution around energized equipment. The metal container will conduct electricity if it contacts a live source. This may result in injury to the user from electrical shock and/or flash fire. For product use instructions, please see the product label.

Storage Procedures: Store in a cool dry area out of direct sunlight. Aerosol cans must be maintained below 120 F to prevent cans from rupturing.

Aerosol Storage Level: I

Section 8: Exposure Controls/Personal Protection

Exposure Guidelines:

	OSHA		ACGIH		OTHER		
COMPONENT	TWA	STEL	TWA	STEL	TWA	SOURCE	UNIT
Tetrachloroethylene	100	N.E.	25	100	N.E.		ppm
Carbon dioxide	5000	30000 v	5000	30,000	N.E.		ppm
N.E. – Not Established (c) – ceiling (s) – skin (v) – vacated							

Controls and Protection:

Engineering Controls: Area should have ventilation to provide fresh air. Local exhaust ventilation is generally preferred because it can control the emissions of the contaminant at the source, preventing dispersion into the general work area. Use mechanical means if necessary to maintain vapor levels below the exposure guidelines. If working in a confined space, follow applicable OSHA regulations.

Respiratory Protection: None required for normal work where adequate ventilation is provided. If engineering controls are not feasible or if exposure exceeds the applicable exposure limits, use a NIOSH-approved cartridge respirator with organic vapor cartridge. Air monitoring is needed to determine actual employee exposure levels. Use a self-contained breathing apparatus in confined spaces and for emergencies.

Eye/face Protection: For normal conditions, wear safety glasses. Where there is reasonable probability of liquid contact, wear splash-proof goggles.

Skin Protection: Use protective gloves such as PVA, Teflon, or Viton. Also, use full protective clothing if there is prolonged or repeated contact of liquid with skin.

Section 9: Physical and Chemical Properties

Physical State: liquid
Color: colorless
Odor: irritating odor
Odor Threshold: 50 ppm
Specific Gravity: 1.619

FIGURE 3-45 Continued

Product Name: Brakleen® Brake Parts Cleaner (aerosol)
Product Number (s): 05089,05089-6, 05089T, 75089, 85089, 85089AZ

Initial Boiling Point: 250 F
Freezing Point: ND
Vapor Pressure: 13 mmHg @ 68 F
Vapor Density: 5.76 (air = 1)
Evaporation Rate: very fast
Solubility: 0.015 g/ 100 g @ 77 F in water
Coefficient of water/oil distribution (log P_{ow}): 2.88
pH: NA
Volatile Organic Compounds: wt %: 0 g/L: 0 lbs./gal: 0

Section 10: Stability and Reactivity

Stability: Stable

Conditions to Avoid: Avoid direct sunlight or ultraviolet sources. Avoid open flames, welding arcs, and other high temperature sources which induce thermal decomposition.

Incompatible Materials: Avoid contact with metals such as: aluminum powders, magnesium powders, potassium, sodium, and zinc powder. Avoid unintended contact with amines. Avoid contact with strong bases and strong oxidizers.

Hazardous Decomposition Products: Hydrogen chloride, trace amounts of chlorine and phosgene

Possibility of Hazardous Reactions: No

Section 11: Toxicological Information

Long-term toxicological studies have not been conducted for this product. The following information is available for components of this product.

Acute Toxicity:

Component	Oral LD50 (rat)	Dermal LD50 (rabbit)	Inhalation LC50 (rat)
Tetrachloroethylene	2629 mg/kg	> 10 g/kg	5200 mg/kg/4H
Carbon dioxide	No data	No data	470,000 ppm/30M

Chronic Toxicity:

Component	OSHA Carcinogen	IARC Carcinogen	NTP Carcinogen	Irritant	Sensitizer
Tetrachloroethylene	No	Group 2A	Reasonably Anticipated to be a Carcinogen	E (mild) / S (severe)	No
Carbon dioxide	No	No	No	None	No

E – Eye S – Skin R - Respiratory

Reproductive Toxicity: No information available
Teratogenicity: No information available
Mutagenicity: Tetrachloroethylene: in vitro studies were negative
animal studies were negative
Synergistic Effects: No information available

Section 12: Ecological Information

Ecological studies have not been conducted for this product. The following information is available for components of this product.

FIGURE 3-45 Continued

Ecotoxicity:	Tetrachloroethylene --	96 Hr LC50 Rainbow Trout: 5.28 mg/L (static) 96 Hr LC50 Fathead minnow: 13.4 mg/L (flow-through)
Persistence / Degradability:		Biodegradation under aerobic conditions is below detectable limits. Biodegradation may occur under anaerobic conditions. Biodegradation rate may increase in soil and/or water with acclimation.
Bioaccumulation / Accumulation:		Bioconcentration potential is low (BCF less than 100).
Mobility in Environment:		Potential for mobility in soil is medium.

Section 13: Disposal considerations

Waste Classification: The dispensed liquid product is a RCRA hazardous waste for toxicity with the following potential waste codes: U210, F001, F002, D039. Pressurized containers are a D003 reactive waste. (See 40 CFR Part 261.20 – 261.33)
Empty aerosol containers may be recycled. Any liquid product should be managed as a hazardous waste.

All disposal activities must comply with federal, state, provincial and local regulations. Local regulations may be more stringent than state, provincial or national requirements.

Section 14: Transport Information

US DOT (ground):	Consumer Commodity, ORM-D
ICAO/IATA (air):	Consumer Commodity, ID8000, 9
IMO/IMDG (water):	Aerosols, UN1950, 2.2, Limited Quantity
Special Provisions:	None

Section 15: Regulatory Information

U.S. Federal Regulations:

Toxic Substances Control Act (TSCA):
All ingredients are either listed on the TSCA inventory or are exempt.

Comprehensive Environmental Response, Compensation and Liability Act (CERCLA):
Reportable Quantities (RQ's) exist for the following ingredients: Tetrachloroethylene (100 lbs)

Spills or releases resulting in the loss of any ingredient at or above its RQ require immediate notification to the National Response Center (800-424-8802) and to your Local Emergency Planning Committee.

Superfund Amendments Reauthorization Act (SARA) Title III:
Section 302 Extremely Hazardous Substances (EHS): None

Section 311/312 Hazard Categories:		
	Fire Hazard	No
	Reactive Hazard	No
	Release of Pressure	Yes
	Acute Health Hazard	Yes
	Chronic Health Hazard	Yes

Section 313 Toxic Chemicals: This product contains the following substances subject to the reporting requirements of Section 313 of Title III of the Superfund Amendments and Reauthorization Act of 1986 and 40 CFR Part 372:

FIGURE 3-45 Continued

Brakleen® Brake Parts Cleaner (aerosol)

Tetrachloroethylene (97.7%)

Clean Air Act:
Section 112 Hazardous Air Pollutants (HAPs): | Tetrachloroethylene

U.S. State Regulations:

California Safe Drinking Water and Toxic Enforcement Act (Prop 65):
This product may contain the following chemicals known to the state of California to cause cancer, birth defects or other reproductive harm: Tetrachloroethylene

Consumer Products VOC Regulations: This product cannot be sold for use in California and New Jersey. In other states with Consumer Products VOC regulations, this product is compliant as a Brake Cleaner.

State Right to Know:

New Jersey:	127-18-4, 124-38-9
Pennsylvania:	127-18-4, 124-38-9
Massachusetts:	127-18-4, 124-38-9
Rhode Island :	127-18-4, 124-38-9

Canadian Regulations:

Canadian DSL Inventory: All ingredients are either listed on the DSL Inventory or are exempt.

WHMIS Hazard Class: A, D1B, D2A, D2B

European Union Regulations:

RoHS Compliance: This product is compliant with Directive 2002/95/EC of the European Parliament and of the Council of 27 January 2003. This product does not contain any of the restricted substances as listed in Article 4(1) of the RoHS Directive.

Additional Regulatory Information: None

Section 16: Other Information

HMIS® (II)	
Health:	2
Flammability:	0
Reactivity:	0
PPE:	B

Ratings range from 0 (no hazard) to 4 (severe hazard)

NFPA

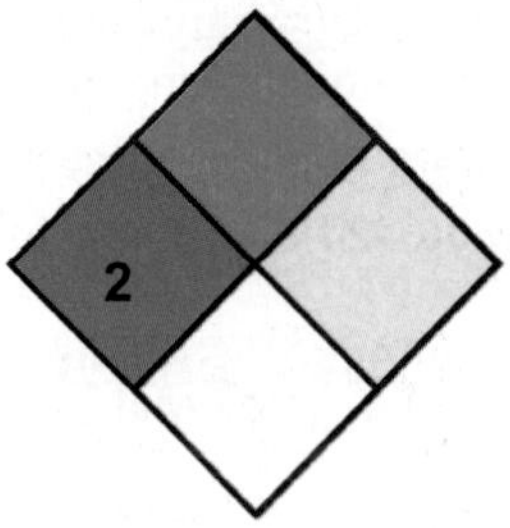

Prepared By: Michelle Rudnick
CRC #: 491G
Revision Date: 01/25/2010

Changes since last revision: MSDS reformatted to meet the requirements of the Canadian Controlled Products Regulations.

FIGURE 3-45 Continued

Product Name: Brakleen® Brake Parts Cleaner (aerosol)
Product Number (s): 05089,05089-6, 05089T, 75089, 85089, 85089AZ

The information contained in this document applies to this specific material as supplied. It may not be valid for this material if it is used in combination with any other materials. This information is accurate to the best of CRC Industries' knowledge or obtained from sources believed by CRC to be accurate. Before using any product, read all warnings and directions on the label. For further clarification of any information contained on this SDS consult your supervisor, a health & safety professional, or CRC Industries.

ACGIH:	American Conference of Governmental Industrial Hygienists
CAS:	Chemical Abstract Service
CFR:	Code of Federal Regulations
DOT:	Department of Transportation
DSL:	Domestic Substance List
g/L:	grams per Liter
HMIS:	Hazardous Materials Identification System
IARC:	International Agency for Research on Cancer
IATA:	International Air Transport Association
ICAO:	International Civil Aviation Organization
IMDG:	International Maritime Dangerous Goods
IMO:	International Maritime Organization
lbs./gal:	pounds per gallon
LC:	Lethal Concentration
LD:	Lethal Dose
NA:	Not Applicable
ND:	Not Determined
NIOSH:	National Institute of Occupational Safety & Health
NFPA:	National Fire Protection Association
NTP:	National Toxicology Program
OSHA:	Occupational Safety and Health Administration
PMCC:	Pensky-Martens Closed Cup
PPE:	Personal Protection Equipment
ppm:	Parts per Million
RoHS:	Restriction of Hazardous Substances
STEL:	Short Term Exposure Limit
TCC:	Tag Closed Cup
TWA:	Time Weighted Average
WHMIS:	Workplace Hazardous Materials Information System

Canadian Regulations

Canadian DSL Inventory: All ingredients are either listed on the DSL Inventory or are exempt.

FIGURE 3-45 Example of an SDS of a common brake-cleaning chemical found in most shops.

SKILL DRILL 3-4 Identifying Information in an SDS

1. Once you have studied the information on the container label, find the SDS for that particular material. Always check the revision date to ensure that you are reading the most recent update.
2. Note the chemical and trade names for the material, its manufacturer, and the emergency telephone number to call.
3. Find out why this material is potentially hazardous. It may be flammable, it may explode, or it may be poisonous if inhaled or touched with your bare skin. Check the **threshold limit values (TLVs)**. The concentration of this material in the air you breathe in your shop must not exceed the TLVs. There could be physical symptoms associated with breathing harmful chemicals. Find out what will happen to you if you suffer overexposure to the material. This can be either through breathing it or by coming into physical contact with it. This helps you to take safety precautions, such as wearing eye, face, or skin protection, or a mask or respirator, while using the material, or like washing your skin afterward.
4. Note the flash point for this material so that you know at what temperature it may catch fire. Also note what kind of fire extinguisher you would use to fight a fire involving this material. The wrong fire extinguisher could make the emergency even worse.
5. Study the reactivity for this material to identify the physical conditions or other materials that you should avoid when using this material. It could be heat, moisture, or some other chemical.
6. Find out what special precautions you should take when working with this material. These include personal protection for your skin, eyes, and lungs, and proper storage and use of the material.
7. Be sure to refresh your knowledge of your SDS from time to time. Be confident that you know how to handle and use the material and what action to take in an emergency, should one occur.

FIGURE 3-46 Aqueous brake wash station uses a water-based system to wet down and wash dust into a collection basin.

FIGURE 3-47 Many shops use a floor washer, often referred to as a Zamboni, to wash the floor and pick up dust and spills.

▶ TECHNICIAN TIP

- Some vehicle components, including brake and clutch linings, contain asbestos. Despite having very good heat properties, they are toxic. Asbestos dust causes lung cancer. Complications from breathing the dust may not show until decades after exposure.
- Airborne dust in the shop can also cause breathing problems such as asthma and throat infections.
- Never cause dust from vehicle components to be blown into the air. It can stay floating for many hours, meaning that other people will breathe the dust unknowingly.
- Wear protective gloves whenever using solvents.
- If you are unfamiliar with a solvent or a cleaner, refer to the SDS for information about its correct use and applicable hazards.
- Always wash your hands thoroughly with soap and water after performing repair tasks on brake and clutch components.
- Always wash work clothes separately from other clothes so that toxic dust does not transfer from one garment to another.
- Always wear protective clothing and the appropriate safety equipment.

▶ TECHNICIAN TIP

Whenever using an atomizer with solvents and cleaners, make sure there is adequate ventilation. Wear appropriate breathing apparatus and eye protection.

To safely clean brake dust, follow the steps in **SKILL DRILL 3-5**.

Used Engine Oil and Fluids

Used engine oil and fluids are liquids that have been drained from the vehicle. This usually occurs during servicing operations. Used oil and fluids often contain dangerous chemicals. They can also contain impurities such as heavy metals. This means that these fluids need to be safely recycled if possible. If not, then they must be disposed of in an environmentally friendly way (**FIGURE 3-48**).

There are laws and regulations that control the way in which waste fluids are to be handled and disposed. The shop will have policies and procedures that describe how you should handle and dispose of these fluids. Be careful not to mix incompatible fluids such as used engine oil and used coolant. Doing

SKILL DRILL 3-5 Safely Cleaning Brake Dust

1. When performing any cleaning tasks on brake or clutch components, always wear a face mask, gloves, and eye protection.

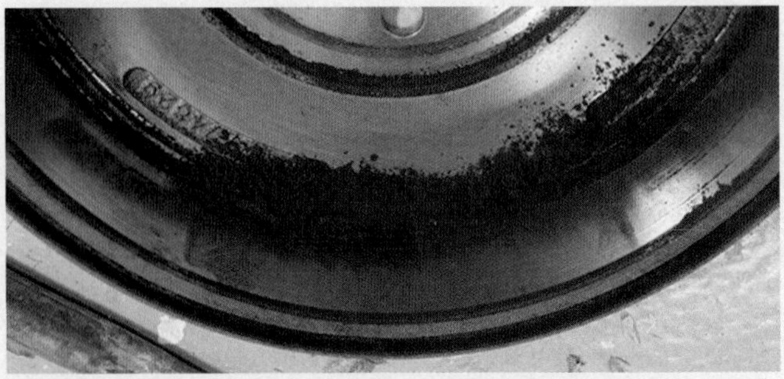

2. Position the brake wash station under the bottom of the backing plate. When cleaning drum brakes, remove the brake drum and check for the presence of dust and brake fluid. When cleaning a clutch, position the wash station underneath the bell housing.

SKILL DRILL 3-5 Safely Cleaning Brake Dust (Continued)

3. Turn on the wash station pump, and paint the solution over the components to wet and clean the components and remove the dust. Any toxic dust will be washed down and caught in the wash station.

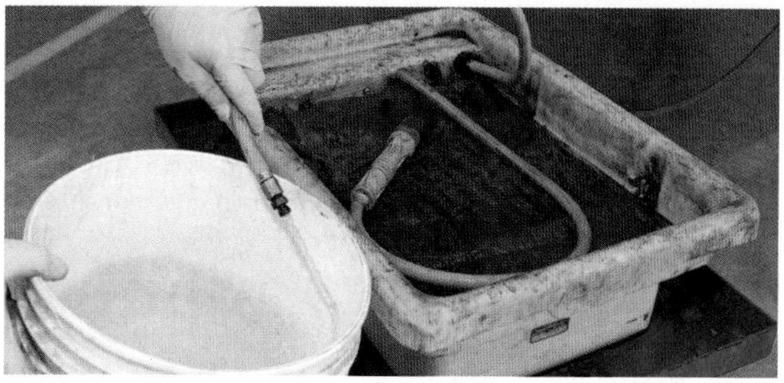

4. Periodically dispose of the residue in an approved manner.

so makes the hazardous materials very much more expensive to dispose of. Generally speaking, petroleum products can be mixed together. Follow your local, state, and federal regulations when disposing of waste fluids.

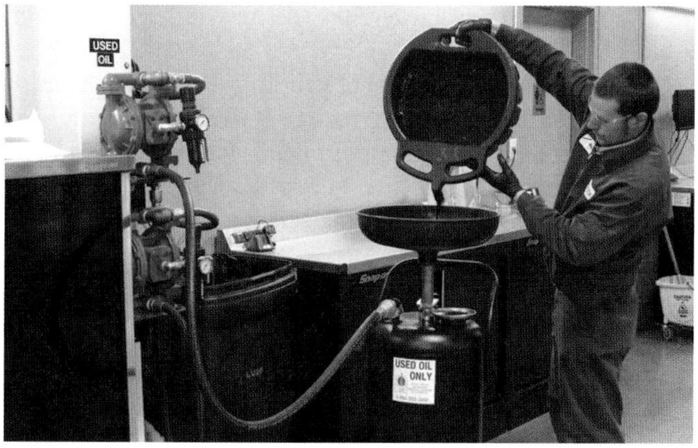

FIGURE 3-48 Used oil and fluids often contain dangerous chemicals and need to be safely recycled or disposed of in an environmentally friendly way.

Used engine oil is a hazardous material. It contains many impurities that can damage your skin. Coming into frequent or prolonged contact with used engine oil can cause dermatitis and other skin disorders. This includes some forms of cancer. Avoid direct contact as much as possible by always using impervious gloves and other protective clothing. These should be cleaned or replaced regularly. Using a barrier-type hand lotion also helps protect your hands. And it makes cleaning them much easier.

Always follow safe work practices, which minimize the possibility of accidental spills. Keeping a high standard of personal hygiene and cleanliness is important. Get into the habit of washing off harmful materials as soon as possible after contact. If you have been in periodic contact with used engine oil, you should regularly inspect your skin for signs of damage or deterioration. If you have any concerns, consult your doctor.

Applied Science

AS-4: Waste Management: The technician identifies the waste products resulting from a repair task.

The most important part of identifying waste is to first determine what products are waste and what can be reused or recycled. The second is to identify what type of material the waste is. Then determine what type of disposal needs to take place with the waste material.

If you replace a set of front disc brake rotors and pads, then you will have to determine where to put the waste material resulting from this job. You will have to refer to the waste disposal table posted in the shop to determine where each item of waste should go. The disc brake rotors should be put in the metal scrap bin for recycling. The disc brake pads may also be put in the metal bin for recycling unless they were manufactured before 2004. If so, they may contain asbestos and need to be put in a plastic airtight bag for removal by a licensed contractor.

AS-5: Waste Management: The technician handles the disposal of materials such as automotive lubricants in accordance with applicable federal, state, and local rules and regulations.

When you carry out a major service, you may have to replace several fluids. Typically this includes the engine oil and oil filter, coolant, and automatic transmission fluid. You will have to determine where to put the waste material resulting from this job. Consult the waste disposal guidelines posted in the shop. For example, refer to the sample waste disposal table shown in a previous Applied Science box. You can see that engine oil and transmission fluid need to be collected and stored in a container. It can then be removed by a licensed contractor. The oil filter needs to be drained and then crushed and kept in a leak-proof container. It too can be collected by a licensed contractor. The coolant needs to be kept in a separate container and collected to be recycled by a licensed operator. Cardboard boxes that come with the oil or air filters can be put in the recycling bin for recycling.

▶ Vehicle System Safety

LO 3-05 Work near hazardous on-vehicle systems.

Technicians need to understand that many dangers and hazardous conditions are located in the vehicle itself. Some examples are:

- high-voltage circuits that can result in electrocution
- automatic controls that operate vehicle systems. They may potentially be activated while the technician is working on the vehicle.
- accidental SRS deployment
- high-pressure liquids that can penetrate your skin or cause eye damage
- very high or very low temperatures that can cause burns or freezing

High-Voltage Safety

There are many things to be aware of when working on:

- hybrid electric vehicles (HEVs),
- plug-in hybrid electric vehicles (PHEVs), and
- all-electric vehicles (EVs).

These vehicles have high-voltage electrical systems that range from 100 to 600 volts. OSHA considers 50 volts or higher as hazardous. So, working on these vehicles is also hazardous. At the same time, there are plenty of opportunities to service hybrid and electric vehicles without venturing into the high-voltage areas. An example is changing oil.

But just because you are not working on the main traction motor doesn't mean that you won't come into contact with other high-voltage circuits. In many hybrid and electric vehicles, high voltage is used to operate a variety of accessories. Examples include power steering and air-conditioning. You need to be aware that high-voltage wires can be located nearly anywhere in the vehicle. These high-voltage wires are recognized by orange (usually) convolute or wiring (**FIGURE 3-49**). Some vehicles use other colors, such as blue, to designate high-voltage wiring. Check the service information to verify the high-voltage wire color used.

There is a critical rule to remember when working on a vehicle that has high voltage circuits. Never work on this type of vehicle without the proper training. Also, if you do have the proper training, you also need a responsible person in your work environment to check on you as you work around high voltage.

Automatic Controls Safety

Besides the high-voltage dangers of a hybrid or electric vehicle, there are other safety precautions you need to be aware of. One of the first is that the vehicle should be completely powered down before working on it. The internal combustion engine (ICE) can start at any time the vehicle's powertrain control module (PCM) deems necessary.

Never assume that the vehicle is turned off because it is silent or still. The vehicle may appear to be turned off. This is because the ICE has been shut down, and the vehicle is silent. But the system can still be in "ready" mode (most vehicle models show this on the vehicle dash display panel). The PCM is then capable of starting the ICE or the drive motor at any moment (**FIGURE 3-50**). If a technician is changing a belt or working around the ICE, the engine could start up. He or she could then be injured or killed.

Also prior to service, the vehicle should be in park, with the parking brake applied. For many service procedures, the keyless fob, if the vehicle is so equipped, should be stored more than 15' (4.6 m) from the vehicle. This will prevent the vehicle from accidentally being powered on.

Due to the unique nature of hybrid and electric vehicles, it is very important to read the manufacturer's service information. Also, complete all required manufacturer's training prior to working on these vehicles. Taking these steps will instruct you in how to properly follow all safety procedures related to hybrid vehicles.

Hybrid and electric vehicles are not the only types of vehicles with automatic controls. Other systems can be powered automatically as well, such as:

- Radiator fan—Can be turned on automatically by the PCM as needed.
- High-pressure brake fluid accumulator—Can be turned on automatically if brake pressure falls.
- Electronically controlled braking systems—Can apply the brake pads automatically after vehicle shut down.

FIGURE 3-49 High-voltage wiring.

FIGURE 3-50 System-ready indicator showing that the vehicle is powered on.

- Automatic ride-height systems—Can raise or lower suspension components, such as when being lifted on a hoist, or with a jack.
- And others—Any device with an electric motor is potentially capable of being turned on by its control module.

There are potential risks of injury with devices that can be operated automatically. It is important that you research the service information for safety precautions. Do this prior to carrying out any maintenance, diagnosis, or repairs to these vehicles.

Accidental SRS Deployment

The supplemental restraint system (SRS) is made up of several devices. They are designed to deploy with great force in the event of an accident. These devices include (**FIGURE 3-51**):

- Driver's side airbag
- Passenger side airbag
- Knee-airbags
- Side curtain airbags
- Ballistic seat belts
- Pedestrian airbags

Most of these devices are activated by a small electric signal. It is possible to accidentally deploy them during diagnosis or repair. This could injure or kill the technician because of the large deployment force created by the ignited propellant. Manufacturers specify that the SRS system or device be disabled when carrying out a servicing procedure on them. If not properly disabled, the airbag could be deployed accidentally. The airbag assembly and other required components need to be replaced whenever they have been deployed. This can be expensive.

SAFETY TIP

Although an airbag is powerful enough to counteract the weight of a moving body during an accident, it does not take much electricity to ignite one. In fact, a fresh 9-volt battery will provide more than enough electrical energy to set it off. In the same way, an ordinary test lamp connected to 12 volts can ignite the airbag if the wrong wire is probed. Always be careful when working around SRS systems.

Generally, to disable an SRS system or device, the correct SRS fuse must be located and removed (**FIGURE 3-52**). Then, disablement needs to be verified by turning the ignition switch to the run position and observing the SRS light. It should remain on (**FIGURE 3-53**). Once the fuse is removed, the negative battery terminal gets disconnected (**FIGURE 3-54**).

The vehicle should sit for about 15 minutes to discharge the capacitors. They store power to deploy the SRS devices in the event power is severed in an accident. Note that a memory minder *should not* be used when working on, or around, the SRS system. Also, do not use an ohmmeter, test light, or any other type of test equipment not specified in the service manual on any airbag, or airbag circuit. This is important in order to avoid accidental deployment of the airbag. Failure to follow this could result in injury or death.

Some manufacturers instruct you to disconnect the SRS device's electrical connector. Disconnecting this connector

FIGURE 3-52 Remove the SRS fuse.

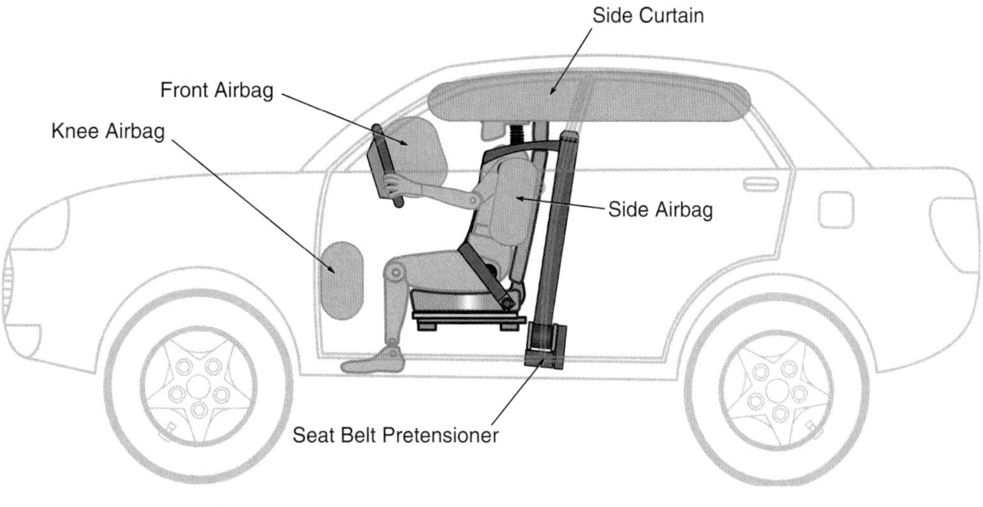

FIGURE 3-51 Typical SRS injury reduction devices.

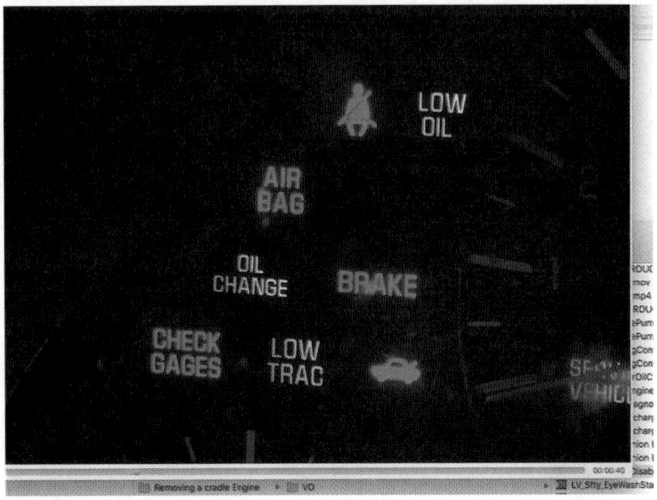

FIGURE 3-53 Turn the key on and verify that the SRS or air bag warning light is illuminated.

FIGURE 3-54 Remove the negative battery terminal and wait for 15 minutes.

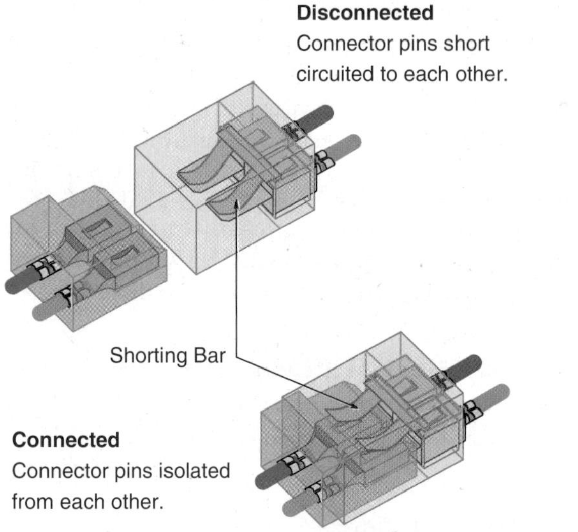

FIGURE 3-55 A shorting bar inside most air bag connectors shorts out the terminals when the connector is disconnected. This helps avoid accidental deployment when the air bag is removed from the vehicle.

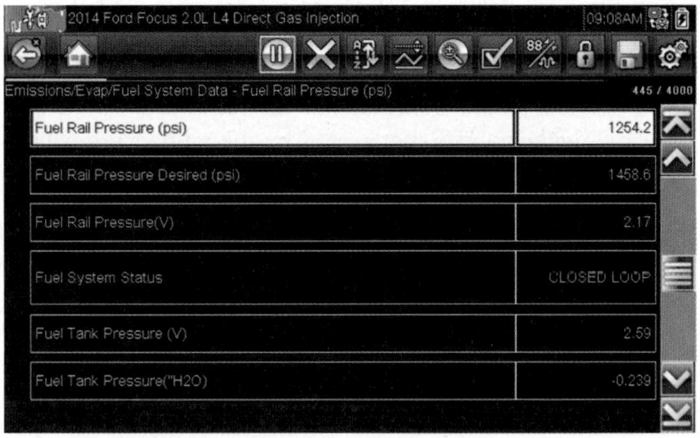

FIGURE 3-56 GDI fuel pressure as shown on a scan tool.

generally activates a shorting bar in the device's side of the connector. This makes it much harder for the device to accidentally deploy (**FIGURE 3-55**). Many manufacturers use yellow color to denote the wiring for the SRS system, but don't assume all do. Accidental deployment of the airbag system could happen if you inadvertently probe the wrong wire. Always be aware of the system/circuit you are working on. And make sure you properly disable any SRS devices as directed by the service information.

With the SRS system or devices disabled, you can carry out any needed diagnosis or repair on the vehicle. Just remember to enable the system and devices once service is completed. Follow the vehicle's specific service information to properly disable and enable the SRS system and devices.

SAFETY TIP

Never lean over or in front of an armed airbag when working on the dash, instrument panel, or airbag system. Doing so can cause serious injury or death if it accidentally deploys. Always follow the manufacturer's procedure when disarming and arming the SRS system and devices.

SAFETY TIP

Just because an airbag has been deployed doesn't mean it is safe. Dual-stage airbags may be designed with two ballistic charges, and only one of them may have been deployed. Airbags must have all charges deployed before disposal. But only do so according to the manufacturer's service information.

Working on Pressurized Systems Safely

Fluids can be under relatively low pressure such as in cooling systems (typically under 25 psi (172 kPa). And yet, they can cause hot coolant to be sprayed on skin or into eyes. Fluids can also be under high pressure such as fuel in a Gas Direct Injected engine (2,900 psi or 200 bar) (**FIGURE 3-56**). The pressure in this case is high enough to penetrate the skin. Hazardous chemicals that penetrate the skin can cause physical damage and infection.

Many systems maintain pressure even when the vehicle is shut down. This makes it dangerous to work on even then. So, first you will need to know which systems are pressurized. Second, you will need to know how to work on them safely. And third, you will need to know how to properly handle the fluids as hazardous materials.

Here are some of the pressurized systems and components that you will need to work around:

- Cooling systems—retain pressure on the coolant when the system is hotter than the ambient temperature.
- Fuel systems—retain pressure on the fuel when the vehicle is off.
- Air conditioning systems—retain pressure in the system when the vehicle is off.
- Electronically controlled braking systems—retain pressure in the high-pressure portion of the system.
- Tires—retain pressure continuously.
- Power steering systems—are pressurized only during operation.
- Automatic transmissions—are pressurized primarily during operation. May maintain pressure when the engine is off on some vehicles.
- Electronically controlled suspension systems—retain pressure when off.

These are many, but not all, of the systems that have hazardous pressure. Be sure to research the service information to see if the vehicle system you are working on is pressurized. If it is, research the procedure for safely releasing the pressure before working on it.

Working Near Extreme Temperatures

Unfortunately, much of the time technicians need to work on vehicles that are at operational temperature. This exposes them to temperatures high enough to cause dangerous burns. At the same time, objects and liquids can be cold enough to freeze skin and eyeballs. Both of these conditions can cause severe injuries.

Even after the vehicle has been shut down for extended periods of time, technicians can still encounter extreme temperatures. For example, some Toyota Priuses are equipped with a coolant heat storage system. Coolant can be kept very hot for up to three days. So working on it, even if the vehicle hasn't been run for a few days, could still end up burning you. The following are systems or components that can burn you:

- Virtually anything on the engine, but especially the exhaust pipes and catalytic converter (**FIGURE 3-57**).
- The cooling system components (**FIGURE 3-58**).
- Brake drums and rotors after the vehicle has been driven recently (**FIGURE 3-59**).
- Automatic transmission (**FIGURE 3-60**).

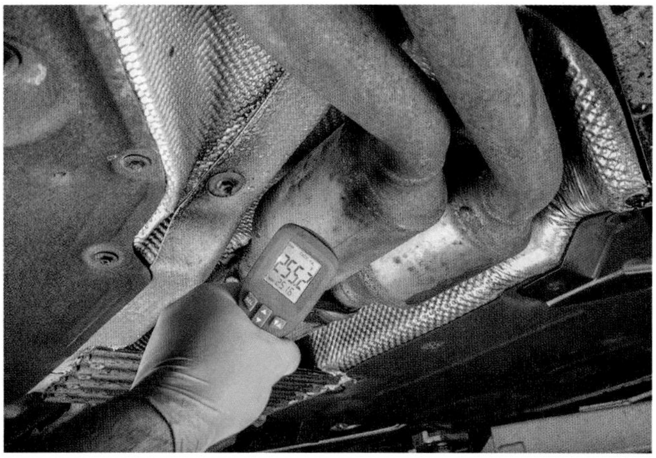

FIGURE 3-57 Exhaust components can be extremely hot and cause severe burns.

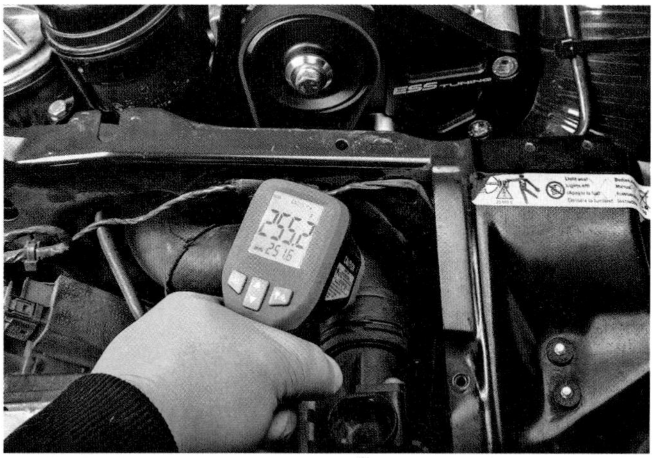

FIGURE 3-58 Cooling system components and coolant can be extremely hot and cause severe burns.

FIGURE 3-59 Brake components can be extremely hot and cause severe burns.

- The hot side of the air conditioning system including the compressor, condenser, and tubing (**FIGURE 3-61**).
- The cold side of the air conditioning system, including the evaporator and tubing, can cause freeze burns. Also, any

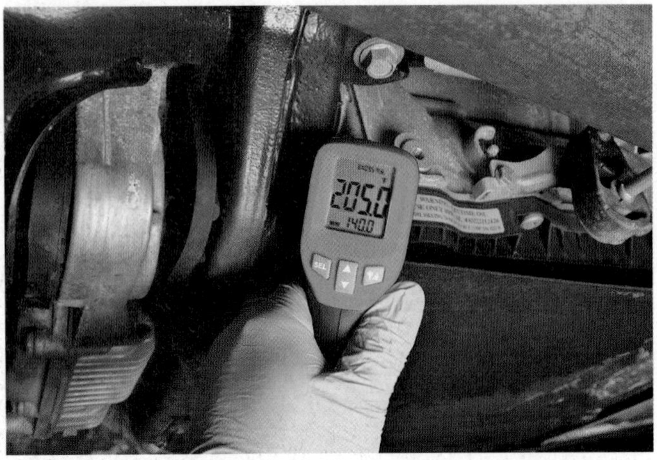

FIGURE 3-60 Automatic transmission components can be extremely hot and cause severe burns.

FIGURE 3-61 Air conditioning components can be extremely hot and cause severe burns.

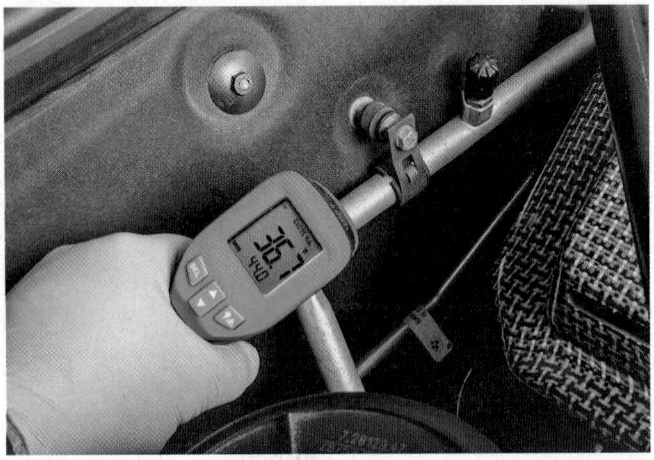

FIGURE 3-62 AC components can be extremely cold and cause severe freeze burns.

FIGURE 3-63 Blower motor resistors can be extremely hot and cause severe burns.

FIGURE 3-64 Shock absorbers can be extremely hot and cause severe burns.

refrigerant leak can cause freeze burns to the skin if allowed to come in contact with the escaping refrigerant (**FIGURE 3-62**).

- Blower motor resistors get extremely hot when the blower motor is operating (**FIGURE 3-63**).
- Shock absorbers and struts can get extremely hot if used in rough conditions (**FIGURE 3-64**).
- Other components that have been operating near their full limit. This may be an alternator that has been trying to charge a battery that has a shorted cell (**FIGURE 3-65**).
- These are just some examples of situations where burns are possible. As you can see, an infrared temp gun is a handy tool. Use it to check the temperature of a surface before grabbing it and burning yourself. If you don't have this tool, then using the back side of your hand. You can usually feel heat coming off of hot surfaces, before touching them (**FIGURE 3-66**).

FIGURE 3-65 Alternators can be extremely hot and cause severe burns.

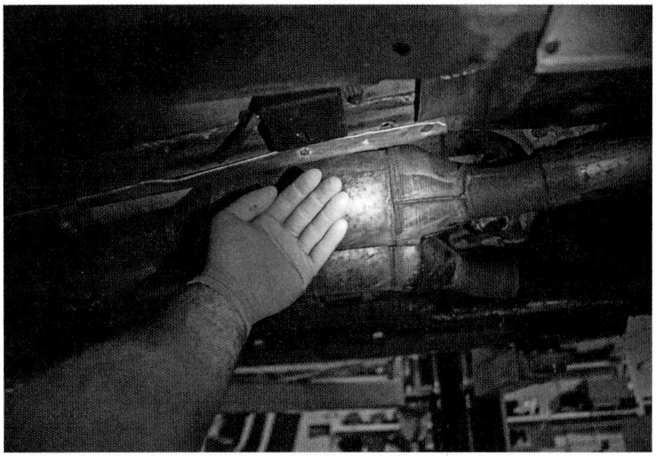

FIGURE 3-66 Using the back side of your hand allows you to feel if a surface is hot or not.

Wrap-Up

Ready for Review

- Use of correct personal protective equipment can ensure the safety of yourself and others in a work environment. A safe work environment goes a long way toward preventing accidents, injuries, and illnesses.
  - Occupational Safety and Health Administration works toward finding the most effective ways to help prevent worker fatalities and workplace injuries and illnesses. Environmental Protection Agency conducts research and monitoring, sets standards, and conducts workplace inspections.
- The condition of a work environment plays an important role in making the workplace safer. Evacuation routes, standard safety measures, and identification of hazardous environments ensure a safe work environment.
  - Shop safety equipment help make the work environment more safe, and includes items such as: handrails, machinery guards, painted lines, sound-insulated rooms, adequate ventilation, doors and gates, and temporary barriers.
- Firefighting involves the removal of at least one of the elements essential for a fire to occur: fuel, oxygen, and heat. Fire extinguishers are used to extinguish the majority of small fires in a shop and categorized into classes: A, B, C, D, and K.
- Safety data sheets (SDS) contain detailed information about hazardous materials and help a technician understand the safe use, health effects, and treatment of hazardous materials.
- High-voltage circuits, automatic controls that operate vehicle systems, accidental SRS deployment, high-pressure liquids, and very high or very low temperatures are hazardous conditions in a vehicle.

Key Terms

Adequate ventilation Ventilation system or procedures that are designed to keep the air in the shop at a safe level.

Background color Background color used on a sign to indicate the type of warning.

Doors and gates Used to separate areas and control access.

double insulated Tools or appliances that are designed in such a way that no single failure can result in a dangerous voltage coming into contact with the outer casing of the device.

Environmental Protection Agency (EPA) A US federal government agency that deals with issues related to environmental safety.

Handrails Used to separate walkways from hazardous areas.

hazard Anything that could hurt you or someone else.

hazardous environment A place where hazards exist.

hazardous material Any material that poses an unreasonable risk of damage or injury to persons, property, or the environment if it is not properly controlled during handling, storage, manufacture, processing, packaging, use and disposal, or transportation.

Machinery guard Used to help prevent accidental contact with dangerous objects.

Occupational Safety and Health Administration (OSHA) The agency that ensures safe and healthy working conditions by setting and enforcing standards and by providing training, outreach, education, and assistance.

Painted line Used to indicate a safe walking distance from a piece of equipment.

personal protective equipment (PPE) Safety equipment designed to protect the technician, such as safety boots, gloves, clothing, protective eyewear, and hearing protection.

Pictorial message Symbol used on a sign to convey a warning.

policy A guiding principle that sets the shop direction.

procedure A list of the steps required to get the same result each time a task or activity is performed.

safety data sheet (SDS) A sheet that provides information about handling, use, and storage of a material that may be hazardous. Previously called material safety data sheets.

Signal word Specific words used on signs in a shop to indicate a warning; danger, warning, and caution.

Sound-insulated rooms A room designed to isolate noise from the rest of the shop.

Temporary barrier Used to provide specific protection for specific areas, such as a welding screen.

Text Information on a warning sign to provide additional information.

toxic dust Any dust that may contain fine particles that could be harmful to humans or the environment.

Review Questions

1. Which piece of Personal Protective Equipment is an essential part of working in a shop?
   - **a.** Jewelry
   - **b.** Reading glasses
   - **c.** Hearing protection
   - **d.** Street shoes
2. Which of the following is a type of PPE?
   - **a.** Respirators
   - **b.** Headphones
   - **c.** Finger splints
   - **d.** Colored glasses
3. Which of the following help reduce accidents and injuries?
   - **a.** Poor supervision
   - **b.** Cluttered shop floor
   - **c.** Policies and procedures
   - **d.** Dim lighting
4. Carbon monoxide is an exhaust gas that can cause injury. What is OSHA'spermissible exposure limit over an 8-hour period?
   - **a.** 25 ppm
   - **b.** 75 ppm
   - **c.** 100 ppm
   - **d.** 50 ppm
5. Electrical tools used in a shop must either have a cord with a ground prong or:
   - **a.** be in a safety cage
   - **b.** be double insulated
   - **c.** be used outside
   - **d.** be made of metal
6. In a shop, fires can occur. Which fire extinguisher would be used for flammable liquids?
   - **a.** Class B
   - **b.** Class K
   - **c.** Class C
   - **d.** Class A
7. What is the acronym used when operating a fire extinguisher?
   - **a.** SPOT
   - **b.** SPAN
   - **c.** PASS
   - **d.** PAST
8. Where should a technician find information on the safe use of chemicals in the shop?
   - **a.** Safety Materials Manual
   - **b.** Safety Material Dictionary
   - **c.** Safety Databases
   - **d.** Safety Data Sheets
9. The Hazard Communication System requires uniformity for SDS. How many sections are required for the new standards?
   - **a.** 10
   - **b.** 8
   - **c.** 16
   - **d.** 18
10. When working in a shop, hazardous dust can be present. What precaution should be taken?
    - **a.** Use an aqueous solution to wash it into a collection basin
    - **b.** Blow it off with an air nozzle to break it up
    - **c.** Sweep it up with a stiff broom and dust pan
    - **d.** Hold breath while using a hammer to dislodge it

ASE Technician A/Technician B Style Questions

1. Technician A says that a brake wash station is used to clean brake dust. Technician B says that the SDS should be referenced if unsure about the correct use of any brake cleaning chemicals. Who is correct?
   - **a.** Technician A
   - **b.** Technician B
   - **c.** Both A and B
   - **d.** Neither A nor B
2. Technician A says that Safety Data Sheets contain detailed health information about hazardous materials. Technician B says that Safety Data Sheets can be obtained from the manufacturer of the material. Who is correct?
   - **a.** Technician A
   - **b.** Technician B
   - **c.** Technician A and B
   - **d.** Neither A nor B
3. Technician A says that fire blankets should not be used to put out a fire on a car. Technician B says that a Type ABC fire extinguisher can be used to put out a fire on a car. Who is correct?
   - **a.** Technician A
   - **b.** Technician B
   - **c.** Both A and B
   - **d.** Neither A nor B

4. Technician A says that when working with fuel the vapors can easily ignite. Technician B says to make sure work that involves fuel is done in a tightly closed shop. Who is correct?
   a. Technician A
   b. Technician B
   c. Both A and B
   d. Neither A nor B
5. Technician A says that shop inspections are only required for older shops. Technician B says that shop inspections are valuable to identify unsafe equipment. Who is correct?
   a. Technician A
   b. Technician B
   c. Both A and B
   d. Neither A nor B
6. Technician A says that eyewash stations are used to flush out debris or chemicals from the eyes. Technician B says that the eyewash isused to flush cuts or punctures to the eye. Who is correct?
   a. Technician A
   b. Technician B
   c. Both A and B
   d. Neither A nor B
7. Technician A says that fluorescent droplights are safe to use in a shop because of lower operating temperatures. Technician B says that the LED in LED shop lights is less prone to shattering than most other bulbs. Who is correct?
   a. Technician A
   b. Technician B
   c. Both A and B
   d. Neither A nor B
8. Technician A says that it is ok to set things in front of the circuit breaker panel as long as they are movable. Technician B says that all circuit breakers and fuses should be labeled. Who is correct?
   a. Technician A
   b. Technician B
   c. Both A and B
   d. Neither A nor B
9. Technician A says that a bench grinder may damage hearing due to excessive noise. Technician B says that a bench grinder poses a risk of entrapment for clothing or body parts. Who is correct?
   a. Technician A
   b. Technician B
   c. Both A and B
   d. Neither A nor B
10. Technician A says that a vehicle fitted with a catalytic converter is safe to run indoors. Technician B says that a catalytic converter can never be a substitute for adequate exhaust extraction equipment. Who is correct?
    a. Technician A
    b. Technician B
    c. Both A and B
    d. Neither A nor B

CHAPTER 4

Personal Safety

Learning Objectives

- **LO 4-01** Dress for the workplace.
- **LO 4-02** Comply with hand protection guidelines.
- **LO 4-03** Comply with protective head gear guidelines.

ASE Education Foundation Tasks

See Appendix A to view the 2017 ASE Education Foundation Automobile Accreditation Task List Correlation Guide.

▶ Personal Safety Overview

LO 4-01 Dress for the workplace.

Personal safety is not something to take lightly. Accidents cause injury and death every day in workplaces across the world (**FIGURE 4-1**). Even if accidents do not result in death, they can be very costly. They result in lost productivity, disability, rehabilitation, and litigation costs. Workplace safety affects people and society heavily. Because of this, government looks to limit accidents and promote safe working environments. The primary federal agency for workplace safety is the Occupational Safety and Health Administration (OSHA). States have their own agencies that administer the federal guidelines. They also create additional regulations that apply to their state.

OSHA standard 29 CFR 1910.132 requires employers to assess the workplace. Employers need to determine if hazards are present, or are likely to be present. In addition, OSHA regulations require employers to protect their employees from workplace hazards. These hazards include machines, work procedures, and hazardous substances that can cause injury. To do that, employers must institute all feasible **engineering controls** and **work practice controls**. This helps eliminate and reduce hazards even before using personal protective equipment (PPE).

Employers have a responsibility not only to assess safety issues but to eliminate or reduce those that can be mitigated, and then provide PPE and training for those hazards that cannot be mitigated. At the same time, if you are injured, you are the one suffering the consequences. Accidents can result in being out of work for a period of time, permanent disability, or death. So it is in your best interest to take responsibility for your own safety while on the job.

Engineering controls means the employer has physically changed the machine or work environment. This is done to prevent employee exposure to the potential hazard. An example

FIGURE 4-1 Accidents are costly.

You Are the Automotive Technician

It's your first day on the job, and you are asked to report to the main office, where your new supervisor gives you your PPE. Before you can begin working on the shop floor, you are given training on the proper use of PPE. Here are some of the questions you must be able to answer.

1. Which type of gloves should be worn when handling solvents and cleaners?
2. Why must safety glasses be worn at all times in the shop?
3. Why should rings, watches, and jewelry never be worn in the shop?
4. When should hearing protection be worn?
5. For what types of tasks should a face shield be worn?
6. Why must hair be tied up or restrained in the shop?
7. Which type of eye protection should be worn when using or assisting a person using an oxyacetylene welder?

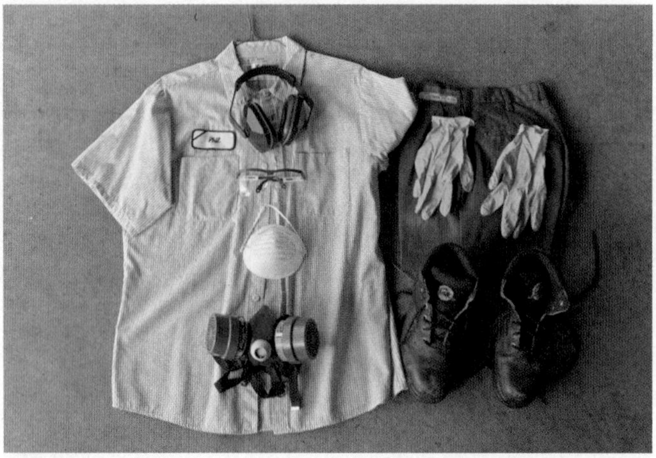

FIGURE 4-2 Personal protective equipment (PPE) includes clothing, shoes, safety glasses, hearing protection, masks, and respirators.

might be adding a ventilation system to an area where solvent tanks are used. Work practice controls means the employer changed the way employees do their jobs in order to reduce exposure to the hazard. An example of this might be requiring employees to use a brake wash station when working on brake systems.

Personal protective equipment (PPE) is various types of specialized equipment. It can block the entry of hazardous materials into the body or protect the body from injury. PPE includes:

- clothing,
- shoes,
- eye protection,
- face protection,
- head protection,
- hearing protection,
- gloves,
- masks, and
- respirators (**FIGURE 4-2**).

Before you undertake any activity, consider all potential hazards. Then, select the correct PPE based on the risk associated with the activity. For example, if you are going to change hydraulic brake fluid, put on some impervious gloves. This will protect your skin from chemicals.

As you go through this chapter, you will learn how to identify the correct PPE for a given activity. It is important that the PPE you use fits correctly and is appropriate for the task you are undertaking. For example, if the task requires you to use a full-face shield, do not try to cut corners and only wear safety glasses. You also need to make sure the PPE you are using is worn correctly. For example, a hairnet that does not capture all of your hair is not protecting you adequately.

Protective Clothing

Protective clothing includes items like shirts, vests, pants, and shoes. These items are your first line of defense against injuries and accidents. Clothing appropriate for the task must be worn when performing any work. Always make sure protective

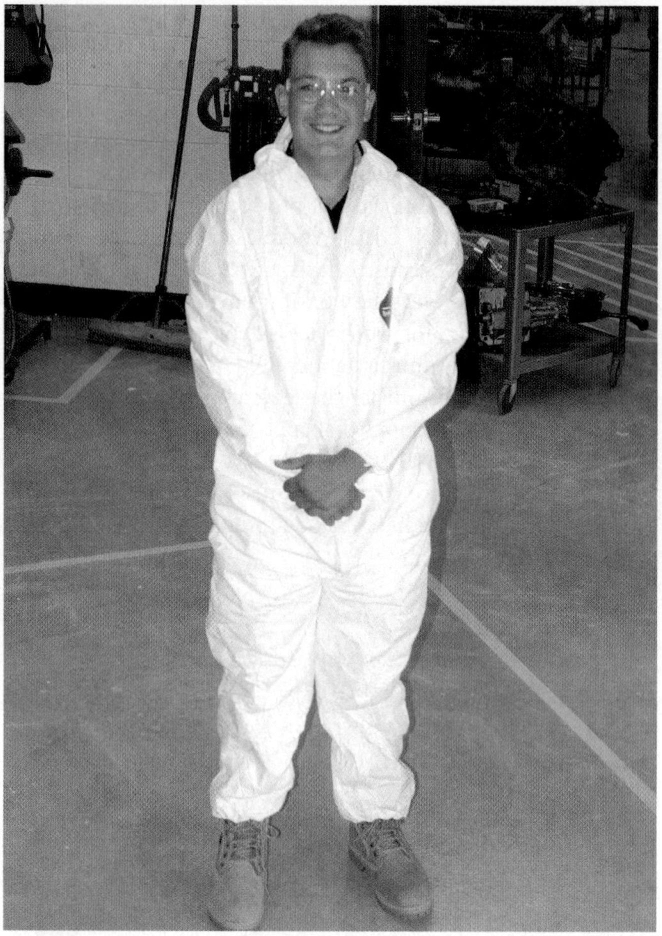

FIGURE 4-3 Disposable fiber suits are used for jobs like painting.

clothing is kept clean and in good condition. You should replace any clothing that is not in good condition, as it is no longer able to fully protect you. Types of protective clothing materials and their uses are as follows:

- **Paper-like fiber:** Disposable suits made of this material provide protection against dust and splashes (**FIGURE 4-3**).
- **Treated wool and cotton:** Adapts well to changing workplace temperatures. Comfortable and fire-resistant. Protects against dust, abrasion, and rough and irritating surfaces (**FIGURE 4-4**).
- **Duck:** Protects employees against cuts and bruises while they handle heavy, sharp, or rough materials (**FIGURE 4-5**).
- **Leather:** Often used against dry heat and flame (**FIGURE 4-6**).
- **Rubber, rubberized fabrics, neoprene, and plastics:** Provides protection against certain acids and other chemicals (**FIGURE 4-7**).

▶ TECHNICIAN TIP

Each shop activity requires specific clothing, depending on its nature. Research and identify what specific type of clothing is required for every activity you undertake. Wear appropriate clothing for the activity you will be involved in, according to the shop's policies and procedures.

FIGURE 4-4 Treated wool and cotton uniform are used for jobs in dry conditions.

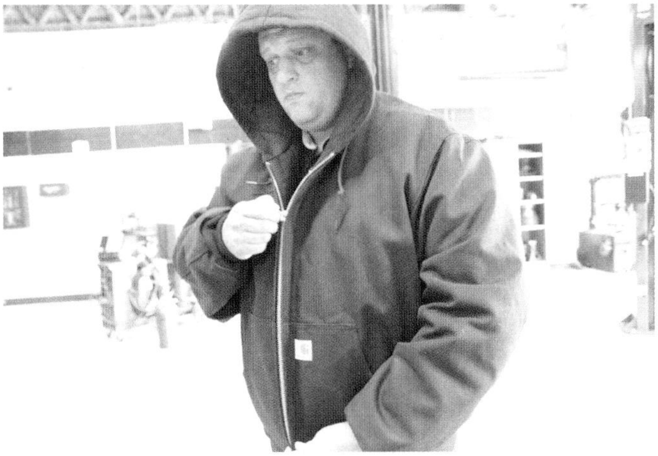

FIGURE 4-5 Duck material work clothes are used for jobs in wet conditions.

Work Clothing

Always wear appropriate work clothing. This can be a one-piece coverall/overall or a separate shirt and pants combination (**FIGURE 4-8**). Either type should be comfortable enough to allow you freedom of movement, without being loose enough to catch on machinery. Also, the sleeves and legs need to be the proper length. They should not be too short (and don't protect you) or too long and bunch up which can catch sparks, get caught on machinery, or trip you up. The material must be flame retardant and strong enough that it cannot be easily torn. A flap must cover metal buttons, snaps, or zippers. If you wear a long-sleeve shirt, the cuffs must be close fitting, without being tight. Pants should not have cuffs so that hot debris cannot become trapped in the fabric.

Wearing work clothing properly is just as important as having the correct clothing. When showing up to work, you should be dressed in your work clothes, ready to work, about 10–15 minutes before your scheduled start time. Most shops will have a changing room, so you can either come to work ready-to-go or you can show up early to get changed. Just make sure you have all of your work clothing with you when you arrive. Most shops have lockers that you can keep your work clothing in, so you probably don't have to take it back and forth to work with you.

Let's start with one-piece coveralls. These are worn over your clothing and helps protect it from spills and damage.

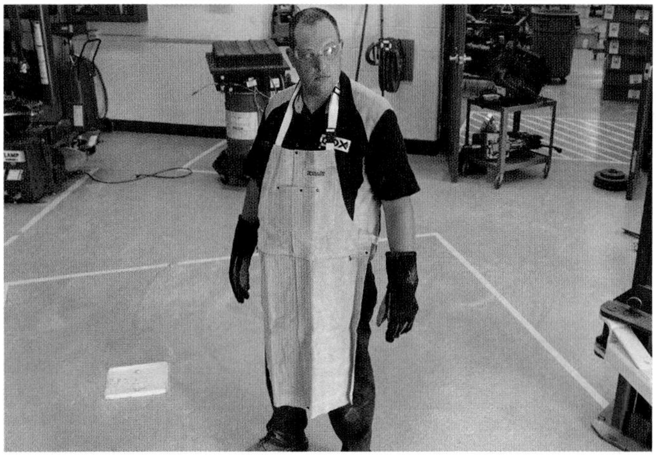

FIGURE 4-6 Leather apron is used for jobs with sparks and welding splatter.

FIGURE 4-7 Neoprene is used in gloves and aprons to protect from certain chemicals.

FIGURE 4-8 **A.** One-piece coverall. **B.** Shirt and pants combination.

FIGURE 4-9 Keep an extra set of work clothes at work in case you need them.

FIGURE 4-10 Work shoes provide support for your feet when working on hard surfaces.

Coveralls need to be worn all the way up. They should also be zipped, or buttoned, all the way up. The one exception is that you can generally leave the top button undone, unless you are welding or grinding and need to keep sparks out. Do not unzip it and tie the arms around your waist. This is unsafe. Also, pant legs should normally be kept out of boots, since that could catch sparks and direct them into your boots.

When wearing separate shirt and pants, the shirt should be buttoned up fully if you are welding or grinding. The top button can be left unbuttoned if not. Pants need to be worn just above the hip bone. No sagging allowed. Typically, a belt should be worn. But the belt needs to be of the safety belt type that can break away if it gets caught on something. It also needs to have a covered buckle that won't scratch vehicles.

Care of Clothing

Always wash your work clothes separately from your other clothes to prevent contaminating your regular clothes. Start a new working day with clean work clothes, and change out of contaminated clothing as soon as possible. It is a good idea to keep a spare set of work clothes in the shop to use in case the ones you are wearing become overly dirty or a toxic or corrosive fluid is spilled on them (**FIGURE 4-9**).

Footwear

The proper footwear provides protection against situations or conditions:

- items falling on your feet
- chemicals
- cuts, abrasions, and punctures
- slips.

They also provide good support for your feet, especially when working on hard surfaces like concrete (**FIGURE 4-10**). The soles of your shoes must be acid- and slip-resistant, and the uppers must be made from a puncture-proof material such as leather. Some shops and technicians prefer safety shoes with a steel toe cap to protect the toes (**FIGURE 4-11**). Always wear shoes that comply with your local shop standards. Also, keep the shoe laced up and tied to avoid tripping hazards.

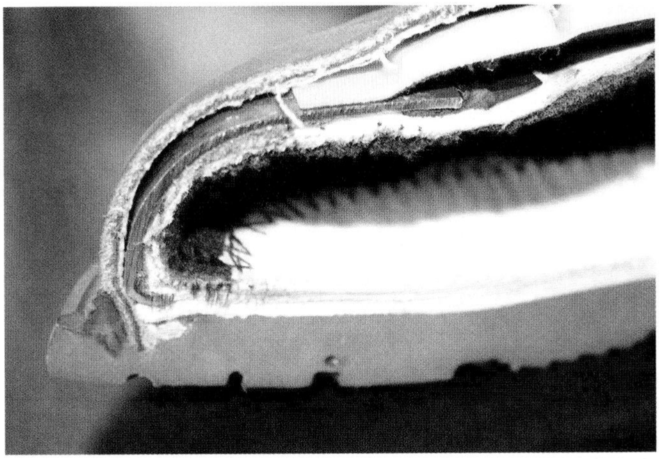

FIGURE 4-11 Steel-toed boots protect your toes from falling objects.

Jewelry

When in a workshop environment, jewelry such as watches, rings, necklaces, and dangling earrings present a number of hazards. They can get caught in rotating machinery, leading to ripping of skin or even strangulation. Also, because they are mainly constructed from metal, they can conduct electricity. This can result in severe burns, or even electrocution. Imagine leaning over a running engine with a dangling necklace. It could get caught in the accessory belt and pull you into the rotating parts if it doesn't break. Not only will it get destroyed but it could seriously injure you. The same goes with dangling earrings. Not only is it safer to remove these items but your valuables will not get damaged or lost. So as part of your preparation for work, make sure you remove, or properly secure, any unsafe jewelry (**FIGURE 4-12**).

- Watches and bracelets—Watches and bracelets with metal bands or leather bands with buckles should never be worn in the shop. If the band is non-conductive and has an easy break-away feature, it MAY be safe to wear in the workshop, if it meets the shop's safety policies.
- Necklaces—In general, necklaces should never be worn in the shop, no matter the type.
- Corded nametags—Nametags that hang from a cord around the neck are not to be worn in the shop unless they are certified to meet the OSHA breakaway requirements.
- Neckties—In general, neckties should never be worn in the shop. However, clip-on ties may be acceptable in some situations.
- Rings—In general, rings should never be worn in shop, no matter the type (**FIGURE 4-13**).
- Earrings—Because virtually all earrings can catch on things, it is best to not wear any earrings in the shop. But some shops allow close fitting stud style earrings. All other earrings should not be worn in the shop.
- Face and head piercings—In general, jewelry associated with head and face piercings should not be worn in the shop. Some shops MAY allow close fitting studs to be worn in piercings, but certainly not all of them will.

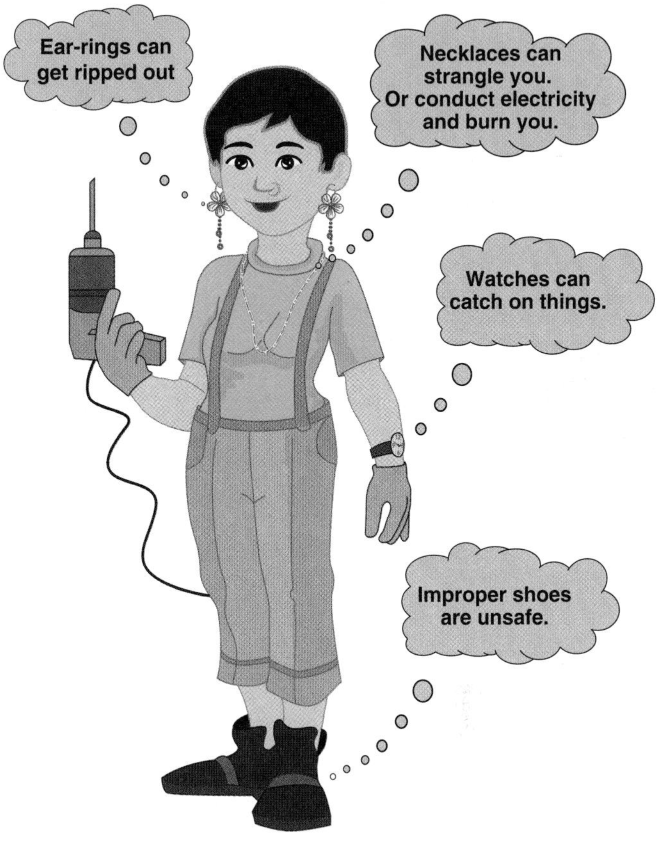

FIGURE 4-12 Jewelry presents a number of safety issues.

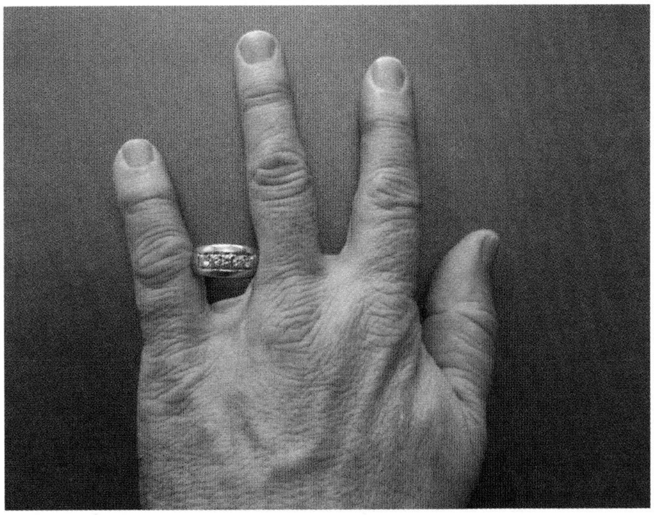

FIGURE 4-13 Finger missing because the wedding ring caught on rotating machinery.

- Earbuds/headphones—Earbuds and headphones pose a strangulation risk, as well as a distraction. They should never be worn in the shop, unless they are specifically designed for, and used, where this type of hearing protection is required.
- Glasses/contacts—This topic is covered in the: "Comply with protective head gear guidelines" objective.

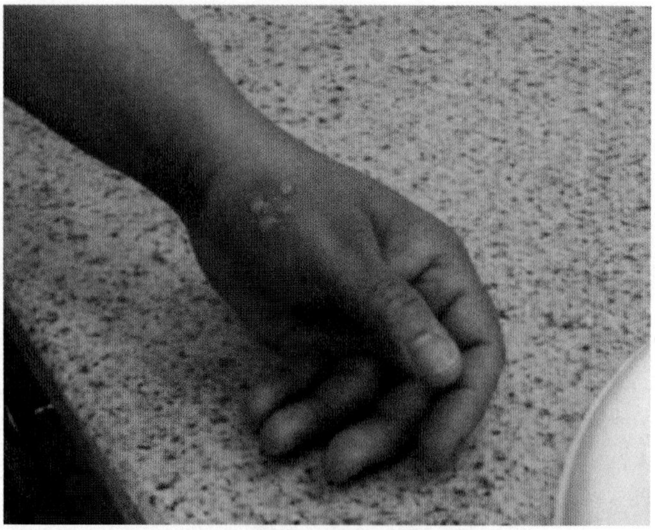

FIGURE 4-14 Example of severe dermatitis due to prolonged chemical exposure.

FIGURE 4-15 Removing gloves before working on a drill press.

▶ Hand Protection

LO 4-02 Comply with hand protection guidelines.

Hands are a very complex and sensitive part of the body, with many nerves, tendons, and blood vessels. They are susceptible to injury and damage. Exposure to chemicals can also lead to severe dermatitis (FIGURE 4-14). Nearly every activity performed on vehicles requires the use of your hands, which provides many opportunities for injury. Whenever required, wear gloves to protect your hands. There are many types of gloves available, and their applications vary greatly as you will see below. It is important to wear the correct type of glove for the various activities you perform. In fact, when working on rotating equipment, it may be necessary for you to remove your gloves. This is so that they don't get caught in the machinery and pull you into it (FIGURE 4-15).

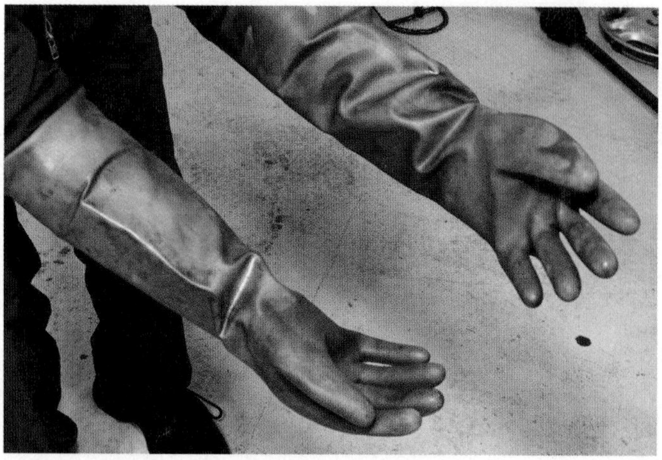

FIGURE 4-16 Chemical gloves should extend to the middle of your forearm to reduce the risk of chemical burns.

FIGURE 4-17 Leather gloves protect your hands from burns when welding or handling hot components.

Chemical Gloves

Heavy-duty and impenetrable chemical gloves should always be worn when using solvents and cleaners. They should also be worn when working on batteries. Chemical gloves should extend to the middle of your forearm to reduce the risk of chemicals splashing onto your skin (FIGURE 4-16). Always inspect chemical gloves for holes or cracks before using them, and replace them when they become worn. Some chemical gloves are also slightly heat-resistant. This type of chemical glove is suitable for use when removing radiator caps and mixing coolant.

Leather Gloves

Leather gloves protect your hands from burns when welding or handling hot components (FIGURE 4-17). You should also use them when removing steel from a storage rack and when handling sharp objects. When using leather gloves for handling hot components, be aware of the potential for **heat buildup**. Heat buildup occurs when the leather glove can no longer absorb or reflect heat. Heat is then transferred to the inside of the leather glove. At this point, the leather gloves' ability to protect you from the heat is reduced. You will need to stop work, remove the leather gloves, and allow them to cool down before continuing

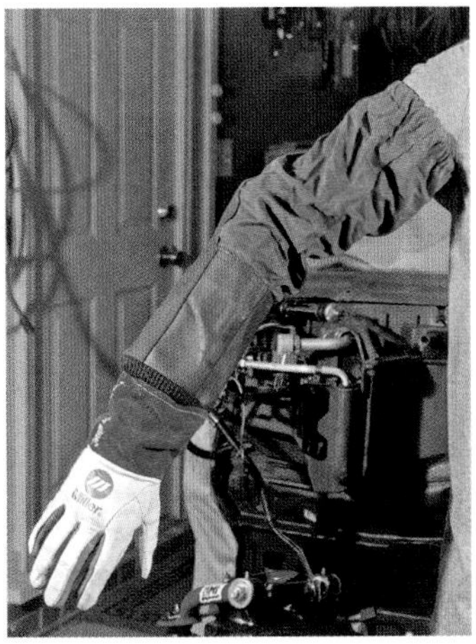

FIGURE 4-18 Heat-resistant sleeve.

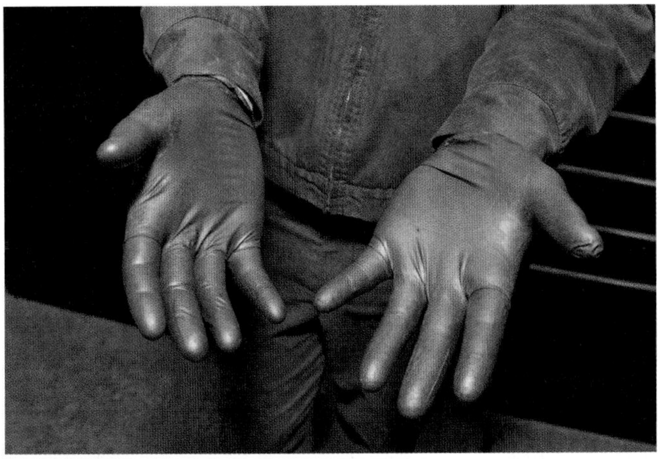

FIGURE 4-19 Light-duty gloves should be used to protect your hands from exposure to greases and oils.

to work. Also avoid picking up very hot metal with leather gloves. This is, because it causes the leather to harden, making it less flexible during use. If very hot metal must be moved, it would be better to use an appropriate pair of pliers.

Some technicians use heat-resistant arm sleeves. They help protect their arms when working in engine bays around hot exhaust systems or even sharp edges. They typically have a thumb hole to keep the sleeve from sliding up the arm (**FIGURE 4-18**).

Light-Duty Gloves

Light-duty gloves should be used to protect your hands from exposure to greases and oils (**FIGURE 4-19**). Light-duty gloves are typically disposable and can be made from different materials, such as nitrile, latex, and even plastic (**FIGURE 4-20**). Some people have allergies to these materials. If you have an allergic reaction when wearing these gloves, try using a glove made from a different material.

General-Purpose Cloth Gloves

Cloth gloves are designed to be worn in cold temperatures, particularly during winter. They help keep your hands warm when using cold tools. They also keep cold tools from freezing

FIGURE 4-20 Common types of gloves. **A.** Latex, **B.** Nitrile, **C.** PVC.

to your skin (**FIGURE 4-21**). Over time, cloth gloves accumulate dirt and grime, so you need to wash them regularly. Regularly inspect cloth gloves for damage and wear, and replace them when required. Cloth gloves are not an effective barrier against chemicals or oils, so never use them for that purpose.

Barrier Cream

Barrier cream looks and feels like a moisturizing cream, but it has a special use. Its formula is designed to provide extra

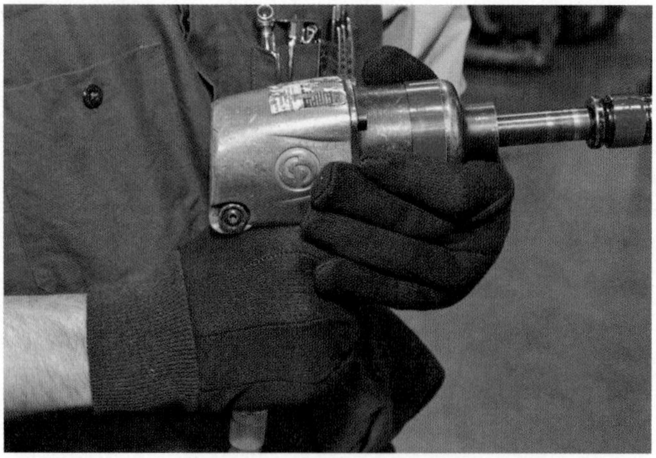

FIGURE 4-21 Cloth gloves work well in cold temperatures, particularly during winter, so that cold tools do not stick to your skin.

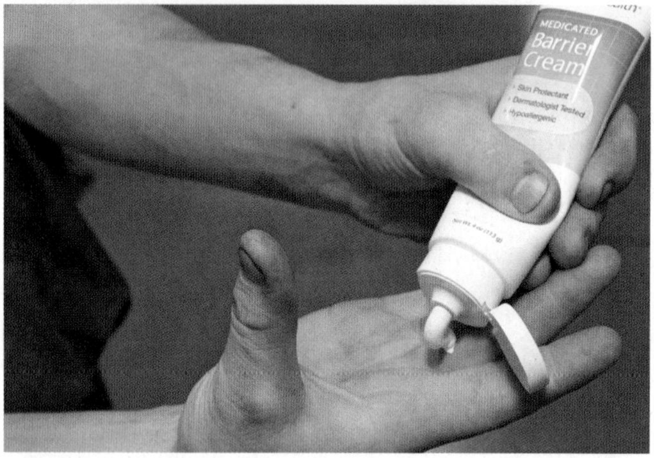

FIGURE 4-22 Barrier cream helps prevent chemicals from being absorbed into your skin and is applied to your hands before you begin work.

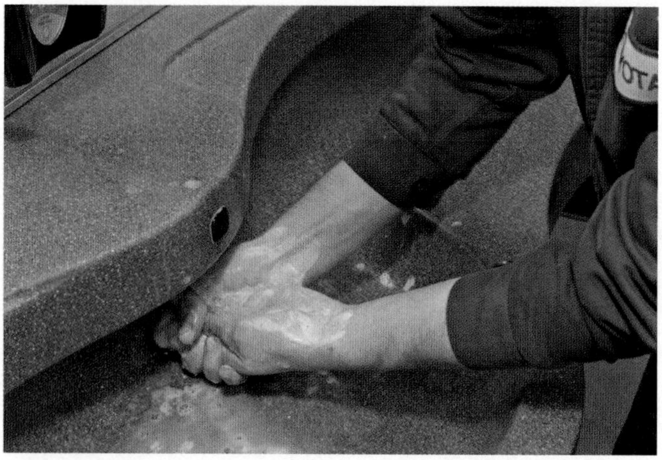

FIGURE 4-23 When cleaning your hands, use only specialized hand cleaners, which protect your skin.

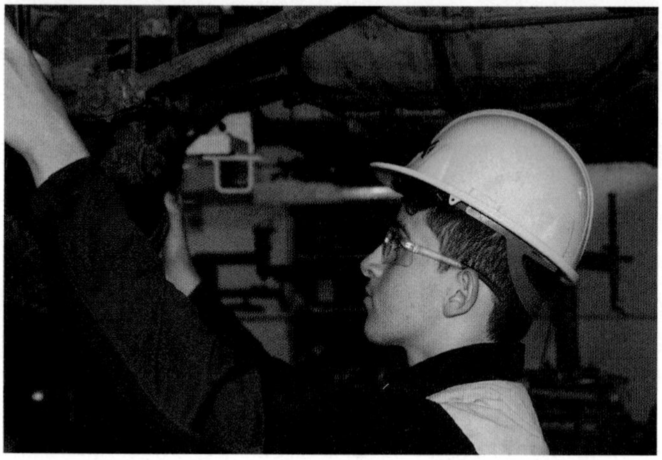

FIGURE 4-24 Using head protection, when required, will help keep you safe.

protection from chemicals and oils. Barrier cream prevents chemicals from being absorbed into your skin. So they should be applied to your hands before you begin work (**FIGURE 4-22**). Even the slightest exposure to certain chemicals can lead to dermatitis, a painful skin irritation. Never use a standard moisturizer as a replacement for proper barrier cream. Barrier cream also makes it easier to clean your hands. This is because it prevents fine particles from adhering to your skin.

Cleaning Your Hands

When cleaning your hands, use only specialized hand cleaners, which protect your skin (**FIGURE 4-23**). Your hands are porous and easily absorb liquids on contact. Never use solvents such as gasoline or kerosene to clean your hands. This is because they can be absorbed into the bloodstream. They also remove the skin's natural protective oils.

Also, cleaning your hands helps prevent transferring grime or residue from one surface to another. This is especially important in two situations. First, it minimizes the possibility of transferring grime to painted surfaces or upholstery on the vehicle. This would require time-consuming cleaning, or upsetting the customer. The second is transferring grime to food you eat. Ingesting chemicals due to dirty hands can create health issues over time. So use the appropriate gloves whenever you can, and clean your hands regularly.

▶ Head Protection

LO 4-03 Comply with protective head gear guidelines.

Your head is a very complex and sensitive part of your body. It has many nerves and blood vessels, and of course, your eyes and ears, all of which require protection. Protection can be accomplished by removing unsafe items, securing items, or wearing PPE (**FIGURE 4-24**). This is complicated by the fact that you still need to use your eyes, ears, and sense of smell to work safely. So proper preparation and PPE is very important to help ensure your safety. For example, as part of preparation, you need to make sure to remove anything that would endanger you. Common items are jewelry, some hats, and headphones. We will explore the preparation process, and the equipment needed for proper head protection, in the following topics.

Headgear

Headgear includes items like hairnets, caps, and hard hats. OSHA requires hair to be covered and secured. This helps to prevent it from getting tangled in rotating or moving parts. Hairnets are typically an effective way of securing hair (**FIGURE 4-25A**). They are simple to use yet effective on long hair. Some technicians wear a ball cap to contain hair that reaches a shirt collar (**FIGURE 4-25B**). Ball caps also help to keep their hair clean when working under vehicles. Some caps are designed specifically with additional padding on the top. They provide a little bit of protection against bumps.

If hair is longer than can be contained in a cap, then technicians should use a hairnet. Although it is acceptable to tie long hair back into a bun or knot (**FIGURE 4-25C**). A pony tail might be acceptable in some situations. Just make sure it is done in such a way that it meets the requirement of being secured (**FIGURE 4-25D**). Long pony tails can still get caught in machinery. So it may still require using a hair net along with the pony tail.

Hard hats protect your head from knocks or bumps. It is easy to hit your head on vehicle parts or the vehicle hoist when working under a vehicle (**FIGURE 4-26**). Head wounds tend to bleed a lot, so hard hats can prevent the need for visiting an emergency room for stitches. Hard hats need to be adjusted properly for them to work correctly. The shell of the hard hat must be suspended above your head an appropriate distance (**FIGURE 4-27**). And the head band must be tensioned properly so that the hard hat does not slip or fall off easily. See the hard hat manufacturer's instruction manual for proper adjustment.

Eye and Face Protection

Eyes are very sensitive organs, and they need to be protected against damage and injury. There are many things in the workshop environment that can damage or injure the eyes. Examples are high-velocity particles coming from a grinder. Or chemicals that get splashed or sprayed into your eyes. High-intensity light coming from a welder is also a potential injury.

In fact, the American National Standards Institute (ANSI) reports that approximately 2,000 workers per day suffer on-the-job eye injuries. Because of this, OSHA requires that: "The employer shall ensure that each affected employee uses appropriate eye or face protection when exposed to eye or face hazards from flying particles, molten metal, liquid chemicals, acids or caustic liquids, chemical gases or vapors, or potentially injurious light radiation." This means that you will always need to select

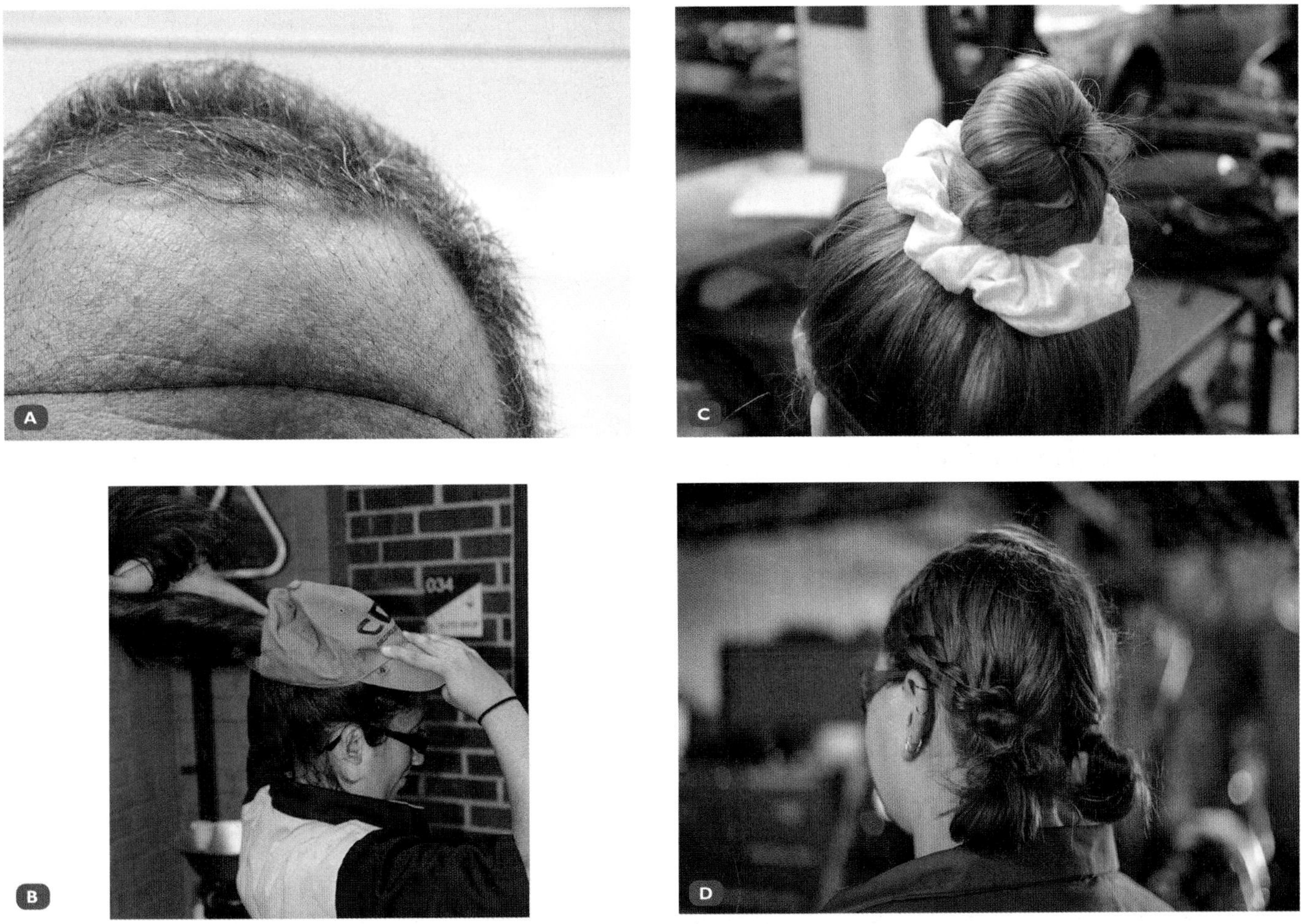

FIGURE 4-25 Containing hair. **A.** Hair net being used. **B.** Ball cap. **C.** Hair put up in a bun. **D.** Hair put up in a pony tail.

FIGURE 4-26 Hard hats prevent head wounds.

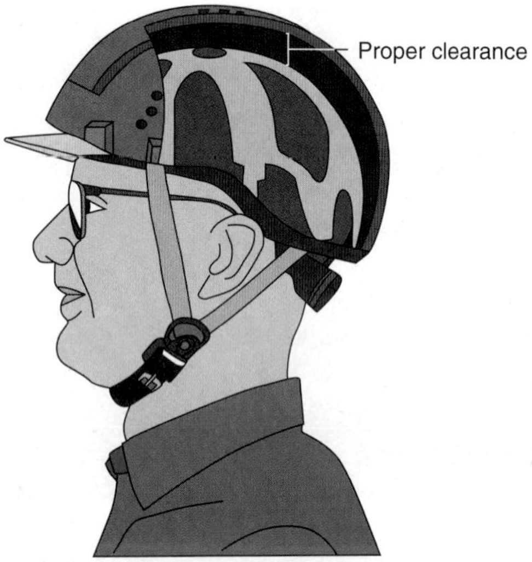

FIGURE 4-27 Hard hat properly adjusted to maintain the specified clearance above the head.

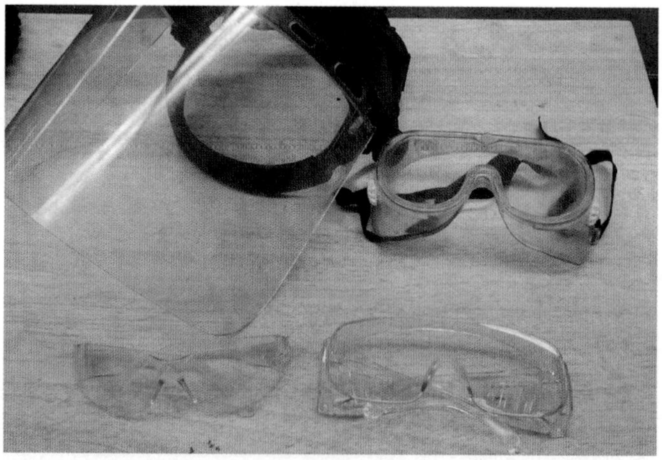

FIGURE 4-28 Various eye and face PPE.

the appropriate eye protection for the work you are undertaking (**FIGURE 4-28**). Sometimes, this may mean that more than one type of protection is required. For example, when grinding, you should wear a pair of safety glasses underneath your face shield for added protection. The following sections explain each type of eye protection equipment.

Safety Glasses

OSHA requires that "The employer shall ensure that each affected employee uses eye protection that provides side protection when there is a hazard from flying objects. Detachable side protectors (e.g. clip-on or slide-on side shields) meeting the pertinent requirements of this section are acceptable." Automotive workshops are subject to flying objects, so safety glasses with built-in, or detachable, side shields are required. The most common type of eye protection for automotive shops is a pair of safety glasses. They must be clearly marked with ANSI "Z87.1" (**FIGURE 4-29**). The ANSI Z87.1 standard was issued initially in 2003, and was revised in 2010 and 2015. The 2015 standard mandated easy-to-understand lens and frame markings that indicate specific ratings beyond the Z87.1 standard (**TABLE 4-1**). Approved safety glasses meeting the appropriate standard must be worn whenever you are in a workshop. (**FIGURE 4-30**).

When in the workshop, the only time safety glasses should be removed is when you are using other eye protection equipment. Prescription and tinted safety glasses are also available. Tinted safety glasses are designed to be worn outside in bright sunlight conditions. So, never wear them indoors or in low-light conditions as they may reduce your ability to see clearly. Safety glasses specifically made for low light can be worn. For people who wear prescription glasses, there are three options that OSHA approves of:

- Prescription glasses, with side shields and protective lenses meeting the requirements of ANSI Z87.1
- Goggles, or specific safety glasses. These need to fit comfortably over corrective eyeglasses without disturbing the alignment of either glasses
- Goggles that incorporate corrective lenses mounted behind protective lenses

TABLE 4-1 Z87.1 – 2015 Ratings Beyond the Basic Rating

Type of Rating	Description of Rating
Impact	Z87.1 = Basic impact rating; Z87.1+ = High-velocity impact rating
Splash and dust	D3 = Splash rating; D4 = Dust
Fine dust	D5 = Fine dust
Welding	W and the shade number
Ultraviolet light	U and the scale number
Infrared light	R and the scale number
Visible light filter	L and the scale number
Prescription glasses	Z87-2 on both the frame and arms
Head size	H = Designed for smaller head size
Other	V = Photochromic; S = Special lens tint

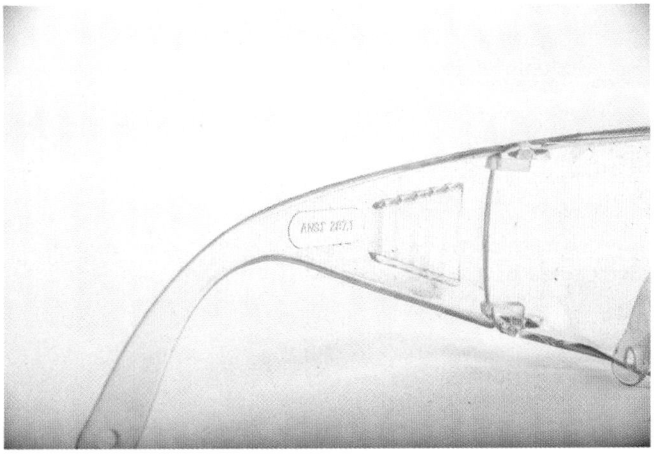

FIGURE 4-29 Safety glasses with the Z87.1-2015 marking.

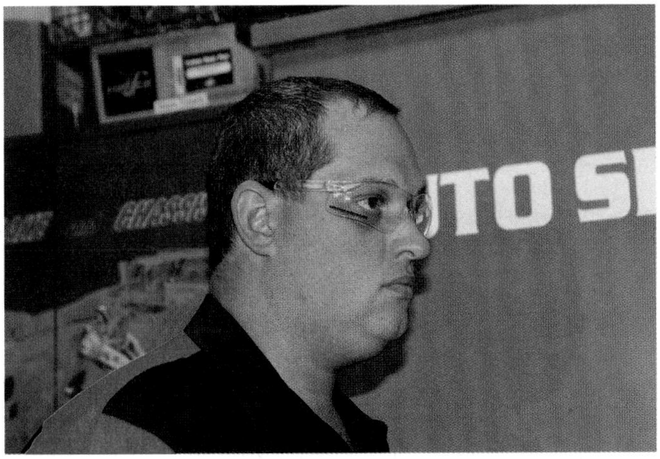

FIGURE 4-30 Safety glasses are designed to protect your eyes from direct impact or debris damage.

SAFETY TIP

You might be tempted to take your safety glasses off while you are doing a nonhazardous task in the shop, like a former student of mine. He was doing paperwork in his stall while his best friend was driving pins out of the tracks of a bulldozer. Unfortunately, the head of the punch his friend was using was mushroomed. And on one hit, a fragment broke off, flew across the stall, and hit the student in the eye, blinding him permanently in that eye. So always wear safety glasses while in a work area, even if you are not working.

Safety Goggles

Safety goggles may or may not provide as much eye protection as safety glasses, depending on their rating. But they do give additional protection against chemical splashes and foreign particles (**FIGURE 4-31**). Wear splash goggles for the best protection against chemicals. Splash goggles are either non-vented, or indirectly vented. In either case, they are designed so that chemical splashes cannot reach the eyes. Goggles are prone to fogging up when in use. Buy them with anti-fog lenses. Or you can apply one of the special anti-fog cleaning liquids or cloths to the inside surface of the lens.

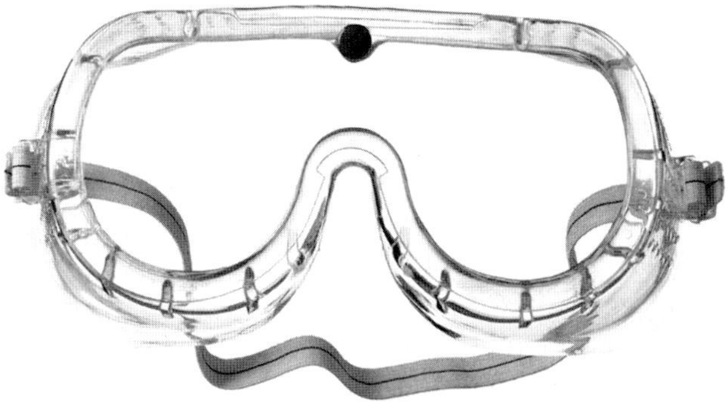

FIGURE 4-31 Safety goggles not only provide a similar eye protection as safety glasses, but with added protection against harmful fluids that may find their way behind the lenses.

Source: © Picsfive/ShutterStock, Inc.

Wear safety goggles whenever you are working with chemicals. Some examples are when servicing batteries, air-conditioning systems and fuel systems. Also wear them when servicing other systems that contain pressurized gas.

▶ TECHNICIAN TIP

Each lab/shop activity requires at least the safe use of safety glasses, clothing, and shoes, depending on its nature. Research and identify whether any additional safety devices are required for every activity you undertake.

Full-Face Shield

Safety glasses and goggles help protect your eyes. A full-face shield helps protect your face from sparks or chemical splashes (**FIGURE 4-32**). The clear mask of the face shield allows you to see all that you are doing while protecting your face. It is also recommended that you use a full-face shield combined with safety goggles when using any kind of grinder or wire wheel equipment.

Welding Helmet

The light from a welding arc is very bright. It also contains high levels of ultraviolet radiation. A **welding helmet** helps protect your eyes when using, or assisting a person using, an electric welder. The lens on a welding helmet has heavily shaded glass to reduce the intensity of the light from the welding arc. This allows you to see the task you are performing more clearly (**FIGURE 4-33**).

Lenses come in a variety of ratings depending on the type of welding you are doing. Always make sure you are using a properly rated lens for the welder you are using (**TABLE 4-2**). The remainder of the helmet is made from a durable material. It blocks any other light, which can burn your skin similar to a sunburn, from reaching your face. It also protects you from welding sparks.

Photosensitive welding helmets that darken automatically when an arc is struck are also available (**FIGURE 4-34**).Their big

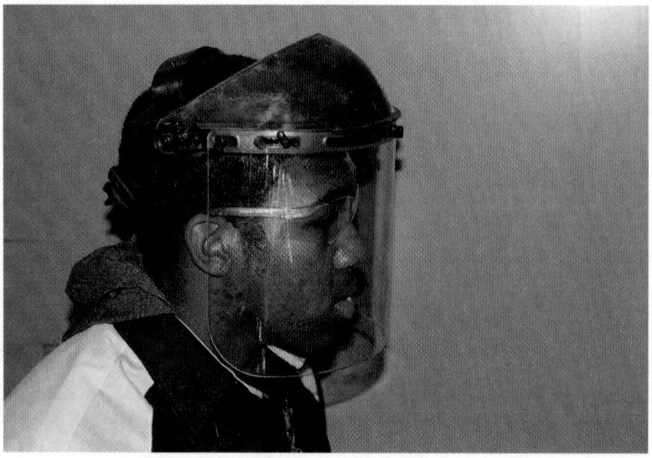

FIGURE 4-32 Full-face shield.

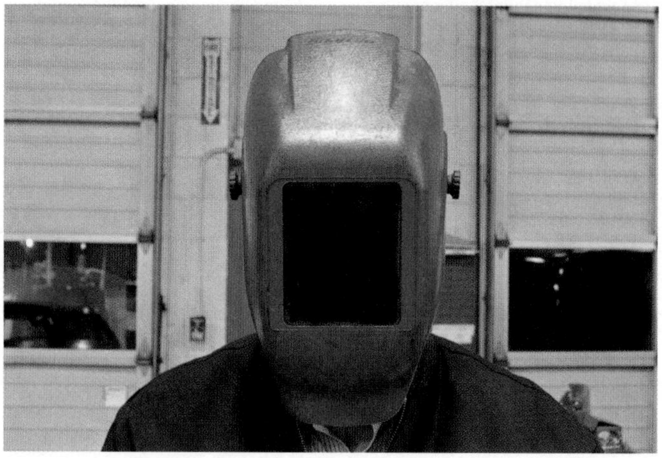

FIGURE 4-33 The lens on a welding helmet has heavily tinted glass to reduce the intensity of the light from the welding tip, allowing you to see what you are doing.

TABLE 4-2 Filter Lens Shade Numbers for Protection Against Radiant Energy

Welding Operation	Shade Number
Shielded metal-arc welding 1/18-, 3/32-, 1/8-, 5/32-inch-diameter electrodes	10
Gas-shielded arc welding (nonferrous) 1/16-, 3/32-, 1/8-, 5/32-inch diameter electrodes	11
Gas-shielded arc welding (ferrous) 1/16-, 3/32-, 1/8-, 5/32-inch diameter electrodes	12
Shielded metal-arc welding 3/16-, 7/32-, 1/4-inch diameter electrodes	12
5/16-, 3/8-inch diameter electrodes	12
Atomic hydrogen welding	10–14
Carbon-arc welding	14
Soldering	2
Torch brazing	3 or 4
Light cutting, up to 1 inch	3 or 4
Medium cutting, 1 inch to 6 inches	4 or 5
Heavy cutting, over 6 inches	5 or 6
Gas welding (light), up to 1/8 inch	4 or 5
Gas welding (medium), 1/8 inch to ½ inch	5 or 6
Gas welding (heavy), over ½ inch	6 or 8

Source: Reprinted from PPE Assessment, OSHA Office of Training and Education, Tab. 2, http://www.osha.gov/dte/library/ppe_assessment/ppe_assessment.html.

advantage is that you do not have to lift and lower the helmet by hand while welding. It was once considered unsafe to wear contact lenses while welding. This has been proven to be a myth. When welding while wearing contact lenses, just make sure that you are wearing the proper welding goggles or helmet for the job.

FIGURE 4-34 Photosensitive welding helmets darken automatically as soon as an arc is struck.

FIGURE 4-35 Gas welding goggles can be worn instead of a welding helmet when using or assisting a person using an oxyacetylene welder.

SAFETY TIP

Be aware that the ultraviolet radiation can burn your skin like a sunburn. So always wear the appropriate welding apparel to protect yourself from this hazard.

Gas Welding Goggles

Gas welding goggles can be worn instead of a welding mask when using or assisting a person using an oxyacetylene welder (**FIGURE 4-35**). The eyepieces are available in heavily shaded versions. But they are not as shaded as those used in an electric welding helmet. There is much less ultraviolet radiation from an oxyacetylene flame, so a welding helmet is not required. However, the flame is bright enough to damage your eyes. So always use goggles of the correct shade rating for the type of welding you are performing. Gas welding goggles come in two common models: single eye piece and separate eye pieces.

Ear Protection

Ear protection is worn to save hearing (hearing conservation). OSHA requires engineering controls to be utilized when the sound level in the shop exceeds those listed in **TABLE 4-3**. If the engineering controls fail to reduce the sound to those levels, then hearing protection shall be provided by the employer. Alternatively, OSHA requires employers to provide hearing protection if the sound levels are 85 decibels (dB) or higher. OSHA also requires that hearing protection be worn when sound levels are 90 dB or higher (Table 4-3). Exposure to impulsive or impact noise should never exceed 140 dB peak sound pressure level, even with hearing protection. This is true even though it is only for a short burst.

Adequate hearing protection is defined as that which lowers the sound to a time-weighted average of 90 dB or lower. For noise levels of common sounds, see (**TABLE 4-4**). Here is one easy way to know if the sound level is high enough to require hearing protection. Must you raise your voice to be heard by a person who is 2 feet (60 cm) away from you? If so, then the sound level is likely to be 85 dB or more.

TABLE 4-3 Permissible Noise Exposures

Duration Per Day, Hours	Sound Level dBA—Slow Response
8	90
6	92
4	95
3	97
2	100
1½	102
1	105
½	110
¼ or less	115

Note: Impulsive or impact noise should not exceed 140 dB peak sound pressure level even with appropriate hearing protection.

TABLE 4-4 Noise Level of Common Sounds

Approximate Decibel Level	Examples
0 dB	The quietest sound that can be heard by a normal human ear
30 dB	Whisper, quiet library
60 dB	Normal conversation, sewing machine, typewriter
90 dB	Lawnmower, shop tools, truck traffic
100 dB	Chainsaw, pneumatic drill, snowmobile
115 dB	Sandblasting, loud rock concert, vehicle horn
140 dB	Gun muzzle blast, jet engine; noise causes pain and even brief exposure injures unprotected ears; maximum noise level allowed with hearing protection

Ear protection comes in three forms (**FIGURE 4-36**):

- Ear muff style—Covers the entire ear.
- Single-use earplugs—These are self-forming inserts that expand to fill the ear canal.
- Pre-formed earplugs—These are custom fit to the wearer's ear canal and are reusable.

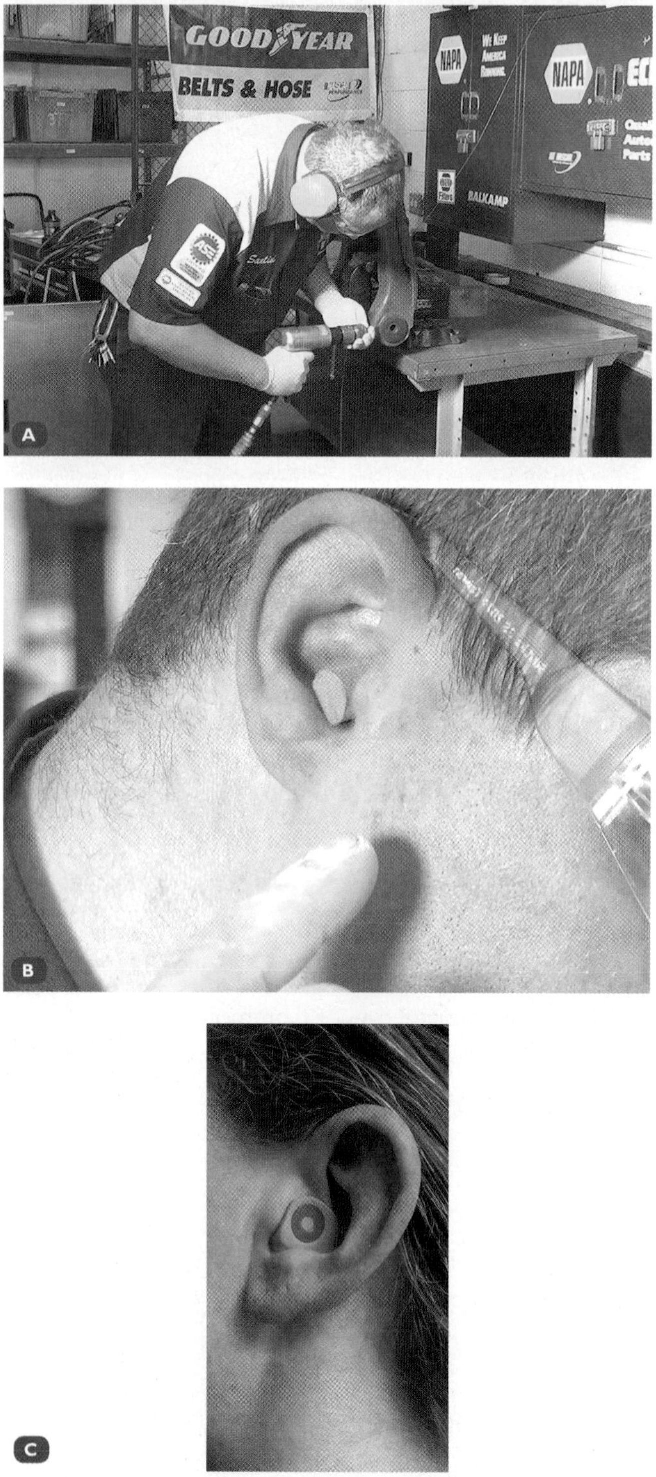

FIGURE 4-36 Ear protection comes in three forms: **A.** Ear muff type covers the entire ear. **B.** Single-use in-ear type. **C.** Pre-formed in-ear type.

All three types reduce the noise level approximately 15–30 dB if fitted correctly. Generally speaking, the in-ear styles have higher noise-reduction ratings than the types that cover the ear. If the noise is not excessively loud, either type of protection will work. If you are in an extremely loud environment (above about 105 dB), you will want to verify that the option you choose is rated to match the noise level. Or in some cases, you may need to use both types at the same time.

When using the ear muff style of hearing protection, first make sure that it is in good working condition, including any padding. Fit it properly to your head by adjusting the bands or straps so that the cups are centered over your ears. They must push equally against your head (**FIGURE 4-37**). If the cups don't fit properly, they won't seal and you will need to try another pair.

When using the in-the-ear style (foam insert) hearing protection, first roll the tip of the insert between your fingers (**FIGURE 4-38**). Then quickly, yet carefully, insert it in one ear, as far as you comfortably can (**FIGURE 4-39**). The foam insert will expand to fill the space in your ear canal (**FIGURE 4-40**). Do the same with the other insert for your other ear.

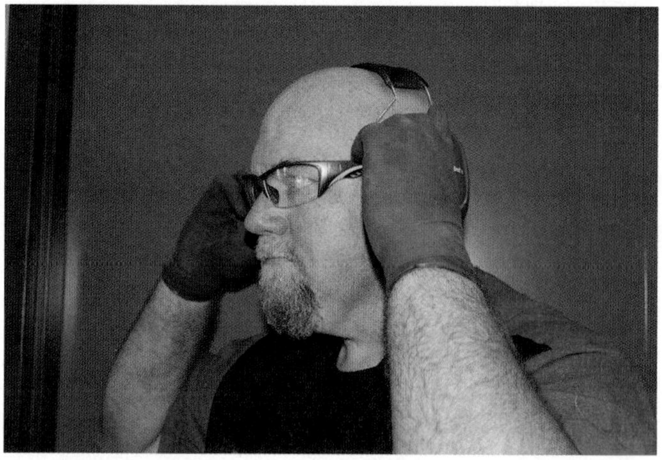

FIGURE 4-37 Adjust the bands on the hearing protector so that the cups are centered on your ears and provide equal pressure around each seal.

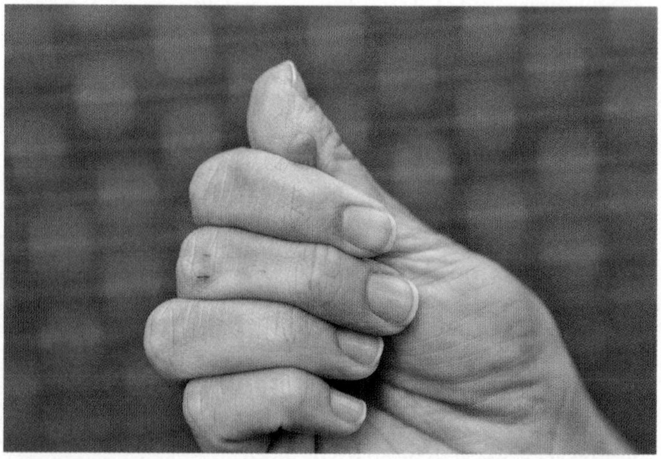

FIGURE 4-38 Roll the tip of the foam insert to shrink its size.

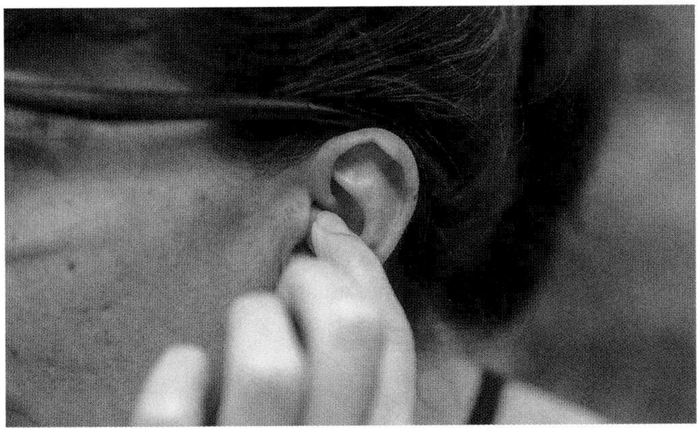

FIGURE 4-39 Insert the foam insert into your ear canal as far as it is comfortable.

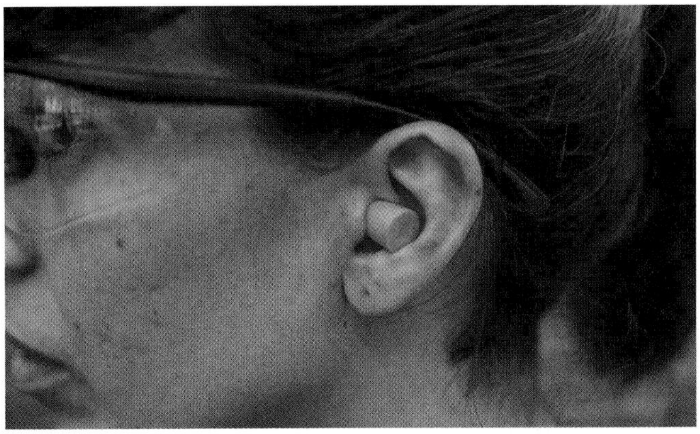

FIGURE 4-40 The foam insert should expand and fill your ear canal completely.

▶ Wrap-Up

Ready for Review

- ▶ Clothing appropriate for the task must be worn when performing any work. Personal protective equipment (PPE) includes clothing, shoes, safety glasses, hearing protection, masks, and respirators.
- ▶ Nearly every activity performed on vehicles requires the use of your hands and need to be covered with gloves to protect them. Different types of gloves that can be worn are: chemical, leather, and light-duty gloves.
- ▶ Head protection can be accomplished by removing unsafe items, securing items, or wearing PPE.
  - Head protection PPE include headgear, eye and face protection, safety glasses/goggles, face shield/helmet, and ear protection.

Key Terms

barrier cream A cream that looks and feels like a moisturizing cream but has a specific formula to provide extra protection from chemicals and oils.

Duck A heavy plain woven cotton fabric used in some shop clothing.

ear protection Protective gear worn when the sound levels exceed 85 decibels, when working around operating machinery for any period of time or when the equipment you are using produces loud noise.

engineering and work practice controls Systems and procedures required by OSHA and put in place by employers to protect their employees from hazards.

gas welding goggles Protective gear designed for gas welding; they provide protection against foreign particles entering the eye and are tinted to reduce the glare of the welding flame.

headgear Protective gear that includes items like hairnets, caps, or hard hats.

heat buildup A dangerous condition that occurs when the glove can no longer absorb or reflect heat, and heat is transferred to the inside of the glove.

Leather Shop PPE made from animal skins. Used to protect from dry heat and flame.

Paper-like fiber Used in disposable suits for protection against dust and splashes.

Rubber, rubberized fabrics, neoprene, and plastics Shop PPE made from these materials. Used to protect from chemicals.

Treated wool and cotton Fire-resistant material used in some shop clothing.

welding helmet Protective gear designed for arc welding; it provides protection against foreign particles entering the eye, and the lens is tinted to reduce the glare of the welding arc.

Review Questions

1. If you are injured in an accident, who primarily suffers the consequences?
   a. Customer
   b. You
   c. Employer
   d. Other employees
2. Employers are required to protect their employees from hazards in the workplace. Which agency regulates employee safety?
   a. SAE
   b. API
   c. NIOAPA
   d. OSHA
3. Using the appropriate clothing for the task to be performed is important. What type of clothing would provide the best protection against acids and other chemicals?
   a. Treated wool
   b. Leather
   c. Rubberized fabric
   d. Duck-treated material
4. Proper footwear can help prevent injuries and provide protection. Which safety feature in footwear reduces a crushing injury?
   a. Slip-resistant sole
   b. Steel toe cap
   c. Leather toe cap
   d. Puncture proof material

5. Which piece of personal protection equipment would provide the best protection from burns when welding?
   a. Puncture proof gloves
   b. Cotton fiber gloves
   c. Reinforced wool gloves
   d. Leather gloves
6. Working in a shop can be hazardous to the technician if not properly prepared. Which of the following is OK to wear in the shop?
   a. Dangling earrings and necklaces
   b. Watches with non-conductive, break-away bands
   c. Rings
   d. Watches with metal bands
7. At what decibel level in a shop is hearing protection required to be used?
   a. 90 dB
   b. 140 dB
   c. 115 dB
   d. 80 dB
8. Cleaning your hands regularly helps protect your skin. How else can dirty hands lead to contamination?
   a. Through breathing
   b. Through shoes
   c. Through food
   d. Through hair
9. What type of PPE would provide the best protection to reduce head injuries?
   a. Hard hat
   b. Bump cap
   c. Hair net
   d. Baseball cap
10. If a technician is performing a task, more than one type of PPE may be required. What is required if a technician is grinding?
   a. Safety glasses and a face shield
   b. Safety glasses and a welding mask
   c. Safety glasses and a bump cap
   d. Safety glasses and a rubberized apron

ASE Technician A/Technician B Style Questions

1. Technician A says to make sure work clothes are in good condition to protect you. Technician B says that clothing appropriate for the task helps protect against injuries. Who is correct?
   a. Technician A
   b. Technician B
   c. Both A and B
   d. Neither A nor B
2. Technician A says that coveralls can help protect your clothing from spills. Technician B says that coveralls can be worn around your waist if more comfortable. Who is correct?
   a. Technician A
   b. Technician B
   c. Both A and B
   d. Neither A nor B
3. Technician A says that any standard footwear will provide adequate protection in the shop. Technician B says that proper foot support is needed when working on hard surfaces. Who is correct?
   a. Technician A
   b. Technician B
   c. Both A and B
   d. Neither A nor B
4. Technician A says that gloves rated for chemicals should be worn when working with chemicals. Technician B says to inspect chemical gloves for damage before use. Who is correct?
   a. Technician A
   b. Technician B
   c. Both A and B
   d. Neither A nor B
5. Technician A says that barrier creams and moisturizer creams perform the same task. Technician B says that barrier creams should be applied once your hands are dirty. Who is correct?
   a. Technician A
   b. Technician B
   c. Both A and B
   d. Neither A nor B
6. Technician A says that long hair should be secured to avoid entanglement. Technician B says that hairnets are NOT an effective means to secure hair. Who is correct?
   a. Technician A
   b. Technician B
   c. Both A and B
   d. Neither A nor B
7. Technician A says that safety glasses only need to be worn in the shop when doing something hazardous. Technician B says that safety glasses used in auto shops must provide side protection. Who is correct?
   a. Technician A
   b. Technician B
   c. Both A and B
   d. Neither A nor B
8. Technician A says that OSHA requires hearing protection when noise levels exceed 75 decibels. Technician B says that sound levels in a shop should never exceed 100 decibels. Who is correct?
   a. Technician A
   b. Technician B
   c. Both A and B
   d. Neither A nor B
9. Technician A says that the pre-formed earplugs are custom-fit to a user's ear canal. Technician B says that the earmuff style can NOT be used in auto shops. Who is correct?
   a. Technician A
   b. Technician B
   c. Both A and B
   d. Neither A nor B
10. Technician A says that a welding helmet lens glass is heavily tinted to reduce the intensity of light from a welder. Technician B says that welding shade glass is rated based on the task being performed. Who is correct?
   a. Technician A
   b. Technician B
   c. Both A and B
   d. Neither A nor B

CHAPTER 5

Vehicle Service Information and Diagnostic Process

Learning Objectives

- **LO 5-01** Utilize information systems.
- **LO 5-02** Identify vehicle information.
- **LO 5-03** Complete a repair order.
- **LO 5-04** Explain Strategy-Based Diagnosis and the 3 C's.

ASE Education Foundation Tasks

See Appendix A to view the 2017 ASE Education Foundation Automobile Accreditation Task List Correlation Guide.

▶ Vehicle Information Overview

LO 5-01 Utilize information systems.

Over the last 100 years, motor vehicles have become increasingly comfortable and reliable. This is due to the application of technology and improved manufacturing processes. For example, the modern motor vehicle has complex computer-controlled electrical systems. Older vehicles had very basic systems with no electronic components. This increase in computer controls allowed more features to be built into vehicles.

With this increase in technology and expanded range of makes and models, there are some challenges. One of the most challenging is finding relevant, complete, and accurate service information. This is critical when performing any kind of maintenance, diagnosis, or repair activities.

Researching the service information starts with gathering the vehicle information. Vehicle information can come from several sources. These include vehicle identification plates, emission control labels, or even from the repair order. The repair order also includes the customer information. Once the vehicle information is known, it can be used to find the correct service information. That could be any of the following: owner's manuals, shop manuals, and technical service bulletins.

The various sources of information are increasingly available through online services. They can also be obtained through printed books or manuals. It is important that you know how to research and apply this information correctly. This is so you can accurately service vehicles (**FIGURE 5-1**).

FIGURE 5-1 A technician researching service information.

Owner's Manual

Manufacturers supply a vehicle **owner's manual**. It comes with virtually every new vehicle purchased and is usually kept in the vehicle's glove box. Secondhand vehicles may or may not have the owner's manual in the glove box. The information contained in the owner's manual varies for each manufacturer. A typical owner's manual includes:

- an overview of the controls and features of the vehicle
- the proper operation, care, and maintenance of the vehicle
- owner service procedures and
- specifications or technical data (**FIGURE 5-2**).

The owner's manual also details such elements as:

- vehicle security PIN (personal identification number) codes;
- warranty and service information;
- fuel, lubricant, and coolant capacities;
- tire-changing specifications;
- jacking and towing information; and
- a list of service facilities.

The layout and amount of detail in an owner's manual varies according to the manufacturer and age of the vehicle. Newer vehicles tend to have more extensive owner's manuals. This is due to the increased amount of accessories the vehicle is equipped with.

Using an Owner's Manual

To locate the specifications for servicing a vehicle, follow the steps in **SKILL DRILL 5-1**.

You Are the Automotive Technician

A well-known vehicle manufacturer has recalled over 400,000 of its vehicles for a brake issue. The issue can be repaired with a software update. It reprograms the vehicles' powertrain control module. But not all of the suspect vehicles need the software update. The technical service bulletin describes the steps needed to verify whether the vehicle has the fault. The manufacturer has notified their customers of a product recall through both mail and email. The dealership that you work for has been receiving many calls from customers to set up recall appointments. Today, you are working on the first vehicle for this recall. You verify that it meets the criteria for the update. You update the software and then drive the vehicle to verity that the issue is resolved.

1. Where would you locate the technical service bulletins (TSBs), service campaigns, and recalls in the dealership?
2. Why is it important to verify whether the TSB has been issued for that particular vehicle?
3. What sources would you use to look up the scheduled maintenance chart for the vehicle?
4. Using the above scenario, write up what you would put on the repair order for each of the "3 C's."

CONTENTS

Brief . . . 2
Dash Board . . . 3
First Drive . . . 4
Vehicle Features . . . 17
Performance/Maintenance . . . 21

Keys/Doors/Windows . . . 23
RFID Keys/Locks . . . 24
Doors . . . 25
Security Systems . . . 25
Mirrors
Exterior . . . 25
Interior . . . 26
Windows . . . 27
Roof . . . 30

Seats/Safety Systems . . . 31
Head Rests . . . 31
Seats
Front . . . 33
Rear . . . 33
Safety Belts . . . 34
Airbag System . . . 35
LATCH (Child Restraints) . . . 35

Storage . . . 36
Compartments . . . 36
Additional Storage . . . 37

Instruments/Controls . . . 38
Controls . . . 38
Warnings, Gauges, and Indicators . . . 39
HUD (Heads-Up-Display) . . . 40
Instrument Messages . . . 41
Instrument Personalization . . . 44

Lighting Systems . . . 47
Exterior . . . 47
Interior . . . 46
Additional Lighting . . . 46

Technology Systems . . . 47
Introduction . . . 47
Navigation . . . 48
Audio . . . 49
Blue Tooth Sync . . . 50

Climate Controls . . . 51
HVAC Systems . . . 51
Vents . . . 52

Driving/Operating . . . 53
Driving Information . . . 53
Starting and Operating . . . 54
Exhaust . . . 55
CV Transmission . . . 56
Regenerative Brakes . . . 57
Boost Performance Systems . . . 58
Cruise Control . . . 60
Fuel . . . 62
Towing . . . 64
Conversions and Upgrades . . . 64

Vehicle Care . . . 65
General Information . . . 65
Vehicle Checks . . . 66
Headlamp Aligning . . . 67
LED Replacements . . . 68
Electrical System . . . 69
Wheels and Tires . . . 70

1

FIGURE 5-2 A typical owner's manual includes: an overview, a list of the controls and features, operation care and maintenance information, owner service procedures, and specifications or technical data.

SKILL DRILL 5-1 Using an Owner's Manual

1. Decide what information you need to know about the job and about the vehicle. For example, your job is to change the engine oil. Make sure you know the make, model, and year of manufacture of the vehicle and the type and size of the engine. In order to change the oil, you need to know the specifications for the engine oil; what type of oil to use, and how much oil is needed.
2. Locate the appropriate manual. This kind of information is most readily found in the vehicle's owner's manual. Open the owner's manual to the table of contents to help you quickly find the information you need.
3. Locate the vehicle specifications, and identify the correct type of motor oil for this vehicle. The table of contents lists a page number for the fuel and lubricant capacities and another page number for the refill capacities. First, turn to the page listing each of the vehicle's lubricant specifications. Find the correct specifications for the engine oil, and make a note. Next, turn to the page listing the refill capacities. You find that this eight-cylinder engine requires 5 quarts (4.7 liters) of oil.
4. Another way of finding information is to refer to the index at the back of the owner's manual. For example, look under E for engine, and find "Engine Oil," or look under L for lubricants or O for oil. Each item should refer you to the same page.
5. Once you have the specific requirements of the vehicle, you are ready to begin servicing the vehicle.

Shop Manual

Shop or service manuals are available for just about every make and model ever made. Service manuals come in two types—factory and after-market (**FIGURE 5-3**). Factory manuals are produced by vehicle manufacturers. They specify the procedures to maintain, repair, and diagnose their vehicles. Usually, a factory service manual is specific to one year and make of vehicle, such as a 2017 Ford Mustang.

After-market service manuals are published by independent companies for the same purpose. Usually, after-market manuals are not as detailed as factory manuals. They may not cover topics such as trim and entertainment systems. They are arranged in one of two ways. They either cover a range of years for a particular vehicle, such as 2012 to 2015 Ford Mustangs. Or they cover a range of vehicles for a single year, such as 2011 General Motors vehicles. Paper shop manuals have become less

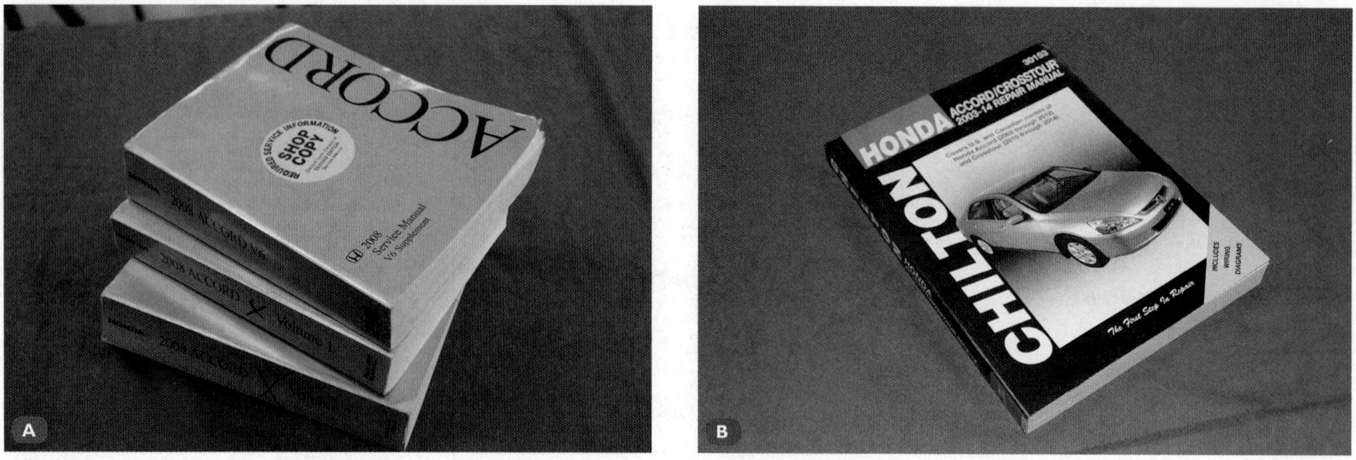

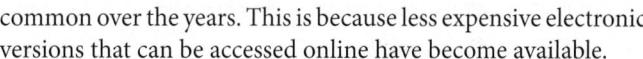

FIGURE 5-3 **A.** Factory manual. **B.** After-market manual.

common over the years. This is because less expensive electronic versions that can be accessed online have become available.

> **▶ TECHNICIAN TIP**
>
> Factory manuals are usually more complete and ideal to have. But it can be very costly to maintain a library of individual manuals for each of the vehicles serviced by a shop. After-market manuals help make the cost of service information more affordable. But they are generally not as complete. The modern vehicle is becoming very complex with all the technology now fitted to it. This means shop manuals have also grown in complexity and size. The result is a significant expense and larger storage requirements. This is another reason that manufacturers have increasingly gone to electronic service manuals.

> **▶ TECHNICIAN TIP**
>
> Here is an example of how much more a technician of today needs to know versus 70 years ago. In 1947, MOTOR, an organization devoted to supplying automotive data, produced a repair manual. It covered the repair procedures and specifications for all of the vehicles produced by more than 20 manufacturers from 1935 to 1946. This book was only about 1,100 pages long. Today it takes MOTOR two to three 1,100-page books to cover the same information for a single vehicle for one manufacturer for one year. This shows why technicians have to rely on service information more now than ever before.

Online versions are becoming very popular. They allow shops to access the information they need without having to pay for and store large numbers of paper shop manuals. Also, it is easier for the publisher to update information as changes or corrections. As a result, the information is generally more accurate than printed materials, which need periodic updates. These online service manual subscriptions are usually provided on a daily, monthly, or yearly basis.

Typical paper and online shop manuals are broken into a number of sections that relate to systems within the vehicle. Engine, transmission, drivetrain, suspension, and electrical are some examples. The sections of the shop manuals are further divided into topics or subject areas. For example, in the engine section, topics could be general description, engine diagnosis, and on-vehicle service. A typical shop manual page has a task description. It is broken into steps and diagrams or pictures to aid the technician (**FIGURE 5-4**). It is important to know that all service manuals arrange the content in their own way. Using a variety of different manuals helps you become familiar with the process of finding the information you are looking for.

Using a Shop Manual

Manufacturers or after-market publishers develop paper shop manuals. They provide you with correct information on performing all service and repair tasks on the vehicles referenced. The information found in shop manuals provides systematic procedures. They also identify special tools, safety precautions, and specifications relevant to the task.

Shop manuals are organized according to vehicle systems. And they have indexes for quick referencing. Knowing that a water pump is part of the cooling system and that the cooling system is part of the engine system (or in some cases, the HVAC system) is important. It helps you locate the specific information you need.

To identify correct service procedures using a shop manual, follow the steps in **SKILL DRILL 5-2**.

Using a Service Information Program

Service information programs are computer applications. They are used to provide technical information for the repair and maintenance of vehicles (**FIGURE 5-5**). This information was typically provided in paper service manuals. In most cases, manufacturers have moved to online delivery of their service information systems. They do this through subscriptions that shops and schools purchase. The information is typically laid out similarly to paper manuals. But because it is electronic, it can be searched much more easily. Also, publishers can link in full-size illustrations, pictures, and charts.

To use a service information program, you may be required to log in to both the computer and the service information program.

MAINTENANCE/SPECIFICATIONS

CHANGING YOUR WIPERS

The wiper arms can be manually moved when the ignition is disabled. This allows for ease of blade replacement and cleaning under the blades.

1. Disable the ignition before removing the blade.
2. Pull the arm away from the glass.
3. Left leading edge retaining block to release the blade. Swing the blade, away from with the arm, to remove it.
4. Swing the new blade toward the arm and snap it into place. Replace the retaining block at the leading edge of the wiper arm. Lower the wiper arm back to the windshield. The wiper arms will automatically return to their normal position the next time the ignition is enabled.

Refresh wiper blades at least twice a year for premium performance.

Poor preforming wipers quality can be improved by cleaning the blades and the windshield. See *Windows and wiper blades* in the *Cleaning* chapter.

To extend the life of wiper blades, scrape off the ice on the windshield BEFORE turning on the wipers. The ice has many sharp edges and will damage and shred the cleaning edge of your wiper blade.

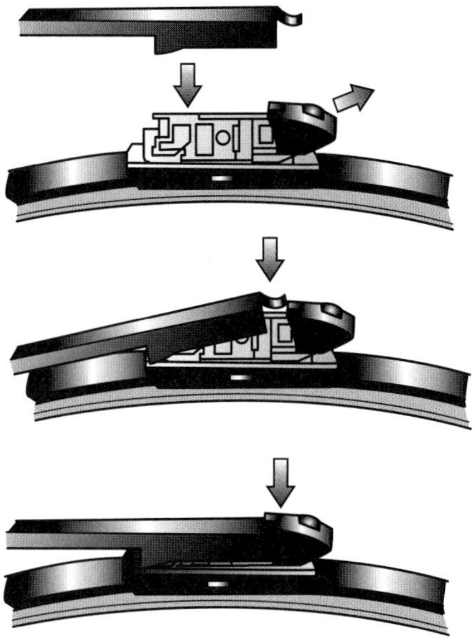

FIGURE 5-4 A typical shop manual page has a task description broken into steps and diagrams or pictures to aid the technician.

SKILL DRILL 5-2 Using a Shop Manual

1. Gather the information you need:
- Year, make, model
- Type and size of engine
- Vehicle identification number (VIN)
- Identify the specific information you need to know

2. Find the appropriate shop manual for the make, model, and year of the vehicle you are working on.

3. Locate the correct section that contains the information you need. The first few pages of the shop manual usually contain the table of contents.

4. Locate the service procedures in the proper section. For example, if you were performing brake repairs, you would turn to the Brakes section. The text and the pictures describe how to properly perform a procedure and tell you the tools to use.

5. Locate the vehicle specifications by consulting the specifications page in the proper section. You may need to know the VIN of the vehicle to find the correct specifications for the vehicle you are working on.

Applied Science

AS-7: Maps/Charts/Tables/Graphs: The technician uses the information in service manual charts, tables, or graphs to determine the manufacturer's specifications for system(s) operation(s). AS-8: Maps/Charts/Tables/Graphs: The technician uses the information in service manual charts, tables, or graphs to determine the appropriate repair/replacement procedure and/or part.

Most service manuals have a chart that you need to consult before performing preventative maintenance. The chart lists what needs to be performed according to the vehicle's mileage. As you can see, there is a big difference between a 43,000 mile (minor) service and a 52,000-mile (major) service. On the 52,000-mile service, far more items are inspected and/or replaced. If the vehicle came into the workshop for a 43,000-mile service, you can easily read from the service chart what needs to be done.

Mileage	43,000 Miles	52,000 Miles
Maintenance Items	I: Drive belts	I: Drive belts
	R:Engine oil	R:Engine oil
	R:Engine oil filter	R: Engine oil filter
	I: Battery	I: Cooling and heater system
	I:Engine air cleaner filter	I: Engine coolant
	I:Brake pedal and parking brake	I: Exhaust pipe and mountings
	I:Brake pads and discs	I:Battery
	I:Brake fluid	R: Engine air cleaner filter
	I:Clutch fluid	I: Brake pedal and parking
	I:Brake pipes and hope	Brake
	I:Power steering fluid	I:Parking brake linings and drums
	I:Steening wheel	I: Brake pads and discs
	I:Drive shaft boots	I: Brake fluid
	I:Suspension ball joint and dust covers	I: Brake pipes and hoses
	I:Tires and psi	I:Power steering fluid
	I:Rotate Wheels	I: Steering wheel and linkage
	I:Seatbelt, webbing condition, budde, and retractor mechanism operation	I: Front and rear suspension
	C: Air conditioner filter	I: Lights, horns, wipers, and washers
	I: Refrigerant amount of air conditioner	I: Seatbelt, webbing condition, buckle, and retractor mechanism operation
	I:Valve clearance	C: Air conditioner filter

Note: T= Tighten, R = Replace, I = Inspect, A = Adjust, L = Lubricate, and C= Clean

FIGURE 5-5 Computer databases provide information on procedures, parts, and service problems.

Make sure you have the username and login available before you start. A printer is also helpful to print copies of the information. You will likely need to use it when performing service and repairs. If you don't print it off, you may need to take notes.

To obtain the correct information, you need the same vehicle identification information as with a service manual. The repair order may provide you with this information. Or you may have to research vehicle identification information from the vehicle. You may also need to perform some initial diagnosis of the fault to further identify the specific information needed.

You can usually obtain information by searching online for the vehicle. Then select from the list of systems. For example, brakes or maintenance, followed by subsystems such as disc brakes or fluid capacities. A keyword search may also be available. For example, use the keyword "service interval" to obtain a list of scheduled service intervals. Using a generic word like "engine" may return a very large list. If this occurs, the search can be narrowed further. Do this by entering more specific criteria, such as "engine oil," "water pump," or "camshaft."

The information is displayed on pages that have a mixture of text and diagrams along with explanations. Some of the diagrams may have detailed views, so you can see how parts fit together. Links may be provided to other relevant information such as a schematic diagram, or diagnostic tests. Most systems contain help menus, or training guides with examples, to assist you in using the system, if required.

To use a service information program, follow the steps in **SKILL DRILL 5-3.**

SKILL DRILL 5-3 Using a Service Information Program

1. If necessary, start the computer and select the service information program.
2. Log in to the application, using the appropriate username and password.
3. Enter the vehicle identification information into the system in the appropriate places: year, make, model, engine, and possibly VIN.
4. Search for the information you require to perform the service or repair.
5. The search engine will provide a list of possible matches for you to select from. If the initial search does not produce what you are looking for, try changing the search criteria. Keep searching until you find the information.
6. Finally, once the general details for the item are displayed, gather the specific information on the specifications or repairs. You may need more than one piece of information.
7. Print out or write down the information needed. Put this on a clipboard, and take it with you to perform the diagnosis, service, or repair.

Applied Science

AS-11 Information Processing: The technician can use computer databases to input and retrieve customer information for billing, warranty work, and other record-keeping purposes.

Dealership service departments have access to databases run by manufacturers. They can:

- access warranty information,
- track vehicle servicing and warranty repair history, and
- log warranty repair jobs for payment by the manufacturer.

When customers present their vehicle for a warranty repair, the customer service staff begin by consulting the database. They can confirm that the vehicle is within its warranty period. They can also confirm that the warranty has not been invalidated for any reason. The repair order is then passed to the technician for diagnosis and repair. Any parts required for the warranty repair must be labeled by the technician. They are also stored for possible recall by the manufacturer.

For example, a young man comes in complaining that his vehicle is "running rough." The customer service staff confirms that the vehicle is nine months old and only has 14,500 miles. They see that it is within the manufacturer's three-year/100,000-mile warranty period. They check the manufacturer's database to confirm that the vehicle's warranty has not been invalidated. The repair order is then handed on to the technician. The technician diagnoses the fault as a defective ignition coil and fills out a warranty parts form.

Once the repair has been completed and the parts labeled, the warranty parts form and any repair order paperwork are passed back to administrative staff for processing. Processing includes:

- billing the manufacturer for the correct, preapproved amount of time,
- logging the repair on the database for payment, and
- ensuring that all documentation is correct for auditing purposes.

Warranty Parts Form

Customer concern	Vehicle running rough	Vehicle Information	
Cause	#6 ignition coil open circuit on primary winding	VIN:	1G112345678910111
Correction	Replaced #6 ignition coil	RO Number:	123456
Parts description	#6 ignition coil	Date of repair:	10/04/2016

Technical Service Bulletins

Technical service bulletins (TSBs) are issued by manufacturers to provide information to technicians. They can also describe:

- unexpected problems,
- updated parts, or
- changes to repair procedures that may occur with a particular vehicle system, part, or component (**FIGURE 5-6**).

The typical TSB contains step-by-step procedures and diagrams on how to identify whether there is a fault. If there is a fault, it describes how to perform an effective repair.

At the time of production, manufacturers prepare service and technical information. They attempt to anticipate the information the technicians need to perform service and repairs. Once the vehicle is in use, situations can arise when particular components or repair procedures may need either additional information or changes. This is where TSBs are most useful. For example, a change may be needed to update the procedure that bleeds air from the cooling system. In this situation, the manufacturer would issue a service bulletin. They would explain the problem. And then list the updated procedure to bleed air from the cooling system.

Using TSBs

To use a TSB, follow these guidelines:

1. Locate where the TSBs are kept in your shop or look them up in the electronic service information system.

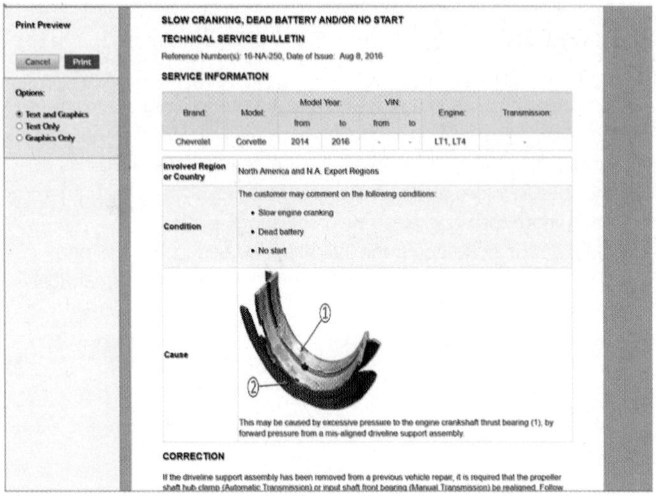

Print Preview

Cancel Print

Options:

- Text and Graphics
- Text Only
- Graphics Only

SLOW CRANKING, DEAD BATTERY AND/OR NO START

TECHNICAL SERVICE BULLETIN

Reference Number(s): 16-NA-250, Date of Issue: Aug 8, 2016

SERVICE INFORMATION

Brand:	Model:	Model Year: from	Model Year: to	VIN: from	VIN: to	Engine:	Transmission:
Chevrolet	Corvette	2014	2016	-	-	LT1, LT4	-

Involved Region or Country	North America and N.A. Export Regions
Condition	The customer may comment on the following conditions: • Slow engine cranking • Dead battery • No start
Cause	(1) (2) This may be caused by excessive pressure to the engine crankshaft thrust bearing (1), by forward pressure from a mis-aligned driveline support assembly.

CORRECTION

If the driveline support assembly has been removed from a previous vehicle repair, it is required that the propeller shaft hub clamp (Automatic Transmission) or input shaft front bearing (Manual Transmission) be realigned. Follow

FIGURE 5-6 Technical service bulletin.

2. Prior to performing repairs, look through the TSBs and get to know the type of information contained in them.
3. Before working on a vehicle, it is a good practice to check whether a TSB has been issued for that vehicle and type of fault or repair. This can help prevent wastage of time.
4. Compare the information contained in the TSB to that found in the shop manual. Note the differences, and if necessary, copy the TSB and take it with you to perform the repair.
5. Perform the repair following the TSB where appropriate while also referring to the shop manual.
6. If required in your shop policy, note the details of the service bulletin in the appropriate area on the repair order.

Service Campaigns and Recalls

Service campaigns and recalls are usually conducted by manufacturers. They do this when a safety issue is discovered with a particular vehicle. Recalls are costly for manufacturers. They can require the repair of an entire model or production run of vehicles. Potentially, this could involve many thousands of vehicles. Depending on the nature of the problem, recalls can be mandatory and enforced by law. Or, manufacturers may choose to voluntarily conduct a recall to ensure the safe operation of the vehicle. They can also chose to do this to minimize damage to their business and product image.

SAFETY TIP

Each country has specific laws regarding product recalls. Find out the laws in your jurisdiction.

An example of a mandatory recall is a vehicle with an unintentional, built-in fault within the airbag system. It results in the airbag not deploying, or deploying when it should not. It could be caused by the airbag wiring harness that is built slightly short. Because it is short, the airbag wire flexes and breaks prematurely. In this case, the manufacturer would identify the

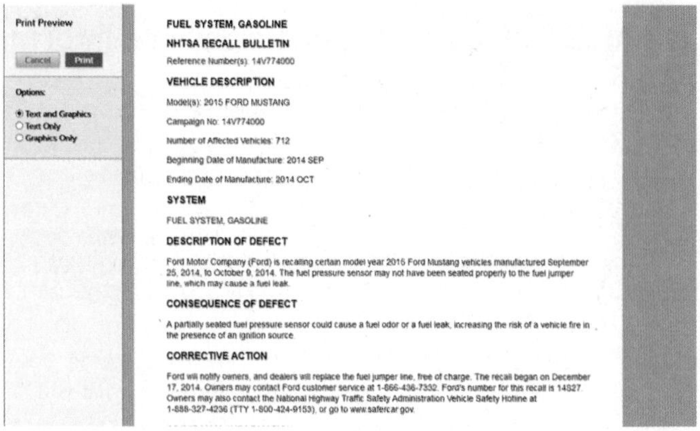

Print Preview

Cancel Print

Options:

- Text and Graphics
- Text Only
- Graphics Only

FUEL SYSTEM, GASOLINE

NHTSA RECALL BULLETIN

Reference Number(s): 14V774000

VEHICLE DESCRIPTION

Model(s): 2015 FORD MUSTANG

Campaign No: 14V774000

Number of Affected Vehicles: 712

Beginning Date of Manufacture: 2014 SEP

Ending Date of Manufacture: 2014 OCT

SYSTEM

FUEL SYSTEM, GASOLINE

DESCRIPTION OF DEFECT

Ford Motor Company (Ford) is recalling certain model year 2015 Ford Mustang vehicles manufactured September 25, 2014, to October 9, 2014. The fuel pressure sensor may not have been seated properly to the fuel jumper line, which may cause a fuel leak.

CONSEQUENCE OF DEFECT

A partially seated fuel pressure sensor could cause a fuel odor or a fuel leak, increasing the risk of a vehicle fire in the presence of an ignition source.

CORRECTIVE ACTION

Ford will notify owners, and dealers will replace the fuel jumper line, free of charge. The recall began on December 17, 2014. Owners may contact Ford customer service at 1-866-436-7332. Ford's number for this recall is 14S27. Owners may also contact the National Highway Traffic Safety Administration Vehicle Safety Hotline at 1-888-327-4236 (TTY 1-800-424-9153), or go to www.safercar.gov.

FIGURE 5-7 Typical manufacturer recall announcement.

problem (broken wire). Then they would find its cause (wire harness not designed correctly). They would also determine the vehicles affected (certain model/s within a certain production range). Next, they would identify any new components (revised wire harness) needed. Then they would determine the repair procedure (how the new harness needs to be replaced, routed, and secured). Finally, they would identify the recertification requirements (process to verify the repair corrects the fault).

A recall would then be issued and advertised in popular media. Letters indicating that the vehicles should be returned for repair would be sent from the manufacturer. They are sent to known owners of the particular vehicle (**FIGURE 5-7**). Usually, all costs associated with the recall are paid by the manufacturer.

Using Service Campaign Information

To utilize service campaigns or recall information, follow these guidelines. Locate where the special service messages, service campaigns/recalls, and vehicle/service warranty applications can be accessed. Look through the TSBs, service recalls, and service warranty applications. Get to know the type of information that is contained in them. Identify how they could be used in your daily tasks.

When working on vehicles, check to see if a TSB has been issued for that vehicle and type of fault. Perform the repairs following the special service messages, service campaigns/recalls, and vehicle/service warranty applications. Fill in the required documentation as required in your shop policies. Note the details of the special service messages, service campaigns/recalls, and vehicle/service warranty applications. Place them in the appropriate area on the repair order.

▶ TECHNICIAN TIP

Customer satisfaction ratings are very important to dealerships and shops. One way to impress customers is to check for recalls on every vehicle that comes in for service. Inform the customer if you find anything. Some manufacturers flag a vehicle for any outstanding recalls when the VIN is entered into the dealership's computerized repair system. This pops up whenever the vehicle is brought in for service.

Report Receipt Date: JAN 24, 2013

NHTSA Campaign Number: 13V023000

Component(s): AIR BAGS

All Products Associated with this Recall expand

Manufacturer: General Motors LLC

SUMMARY: General Motors LLC (GM) is recalling certain model year 2012 Chevrolet Camaro, Cruze, and Sonic, and model year 2012 Buick Verano vehicles. The driver side frontal air bag has a shorting bar which may intermittently contact the air bag terminals.

CONSEQUENCE: If the bar and terminals are contacting each other at the time of a crash necessitating deployment of the driver's frontal air bag, that air bag will not deploy, increasing the driver's risk of injury.

REMEDY: GM will notify owners, and dealers will replace the steering wheel air bag coil, free of charge. The safety recall began on February 13, 2013. Owners may contact General Motors at 1-800-521-7300.

NOTES: This is General Motors recall number 12261 and is an expansion of NHTSA recall 12V-522. Owners may also contact the National Highway Traffic Safety Administration Vehicle Safety Hotline at 1-888-327-4236 (TTY 1-800-424-9153), or go to www.safercar.gov.

Labor Guide

Labor guides list how much time will be involved in performing a standard or warranty-related service or repair. They are regularly updated as new models are released into the market. They provide a basis for making job time and cost estimates for the customer. Flat-rate servicing costs are usually derived from a labor guide. For example, a customer wants to know how much it will cost to replace a leaking intake manifold gasket on a particular vehicle. A technician can look up this procedure in a labor guide. They can find the expected amount of repair time required for that particular repair on the specific vehicle.

With the advent of technology and the Internet, most labor guides are now accessed online. The online versions are paid for by subscription. This is usually obtained with a monthly or annual fee to access the information. Having access to online labor guides means the shop does not have to wait for a new version of the print publication to become available. Online versions of labor guides also can be updated as new models of vehicles are released. They are also updated when the manufacturer releases updated information (**FIGURE 5-8**).

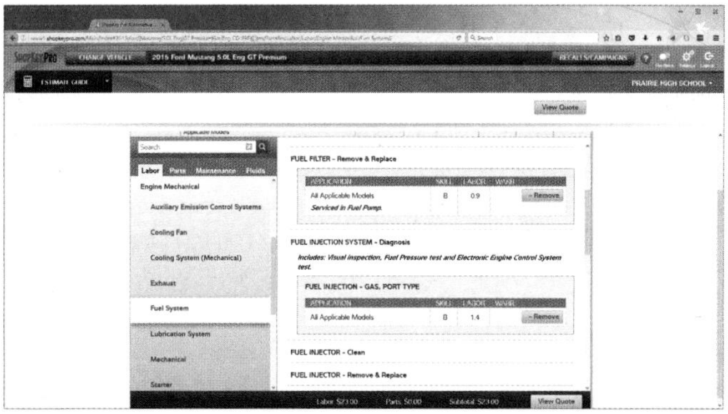

FIGURE 5-8 Online labor guide.

Using a Labor Guide

The labor guide indicates how quickly an average technician can complete tasks. Experienced technicians can usually perform the job more quickly than the labor guide specifies. This is true if the technician has performed a task many times. It also helps if the technician has invested in tools to help them work more efficiently.

But because each task and vehicle has its small differences, the time given does not always match the reality. Sometimes, bolts break, or are severely rusted, and require more time to perform. It is also possible that the vehicle has modifications that make the repair job more intensive. For example, the vehicle has after-market wheels that have wheel locks installed. It will then take longer to remove the wheels when doing a brake job. These extras are sometimes listed in the flat-rate manual as "Additional Time." Either way, the additional time needs to be added to the estimated price of the repair.

The information contained within a labor guide is organized by vehicle year, make, model, and engine size. Once that is found, there are menus that you can use. They narrow down the information to the specific job you are researching.

To use a labor guide, follow the steps in **SKILL DRILL 5-4**.

SKILL DRILL 5-4 Using a Labor Guide

1. Decide what specific labor operations you need to locate. Make sure you know the year, make, model, engine, and any other pertinent details of the vehicle.
2. Log in to the labor estimating system.
3. Enter the vehicle information into the system.
4. Find the labor operation either by working your way through the menu tree or by typing a keyword into the search bar.
5. Once you locate the labor operation, there are usually two columns that list the time. The first one is "warranty time." It is the amount of time the manufacturer would pay the shop for the operation under a customer warranty. The second time listed is the "customer pay" time. That is the amount of time a customer would be billed for and is usually 20–40% longer than warranty time. The length of time is usually listed in tenths of an hour. So 0.6 hours represent 36 minutes. Every tenth of an hour equals six minutes.
6. Check for any "combination" time that would need to be added to the base job when a related job is also being completed. Combination time recognizes that combined tasks in many cases save a lot of time over individual tasks because the customer

SKILL DRILL 5-4 Using a Labor Guide (Continued)

is already being charged for part of the job in the first task. This could be time needed to flush the brakes when the main operation is to replace the brake pads.

7. Check for an "additional" time. This is extra time needed to deal with situations that occur on a relatively common basis, such as vehicle-installed options that are not common to all vehicles, like wheel locks. If you are replacing the brake pads on a vehicle with wheel locks, the customer should be charged for the extra time it takes to find the lock key and to remove and install the wheel locks (finding the lock key can itself add substantial time). Thus, "additional" time needs to be added to the base operation time if the vehicle meets the criteria for the particular "additional" situation.
8. Calculate the total time and multiply it by the shop's hourly labor rate. You now have the correct figure to estimate the charge for the particular service.

Parts Program

Parts programs are the modern-day version of parts manuals. They are essentially an electronic version of a parts manual. Parts programs are typically accessed over the Internet, but some shops use DVDs or even paper manuals. **Parts specialists** are the individuals working at the parts counter. They use these programs to identify parts and find part numbers for ordering purposes.

Parts programs are available for virtually all makes and models of vehicles. They are essentially a catalog of all the commonly available parts that make up a vehicle. The parts are arranged by systems. For example, brake, engine, and transmission systems would include parts diagrams for each of the major assemblies within the system. Diagrams of each part in an assembly are shown along with a part number. That is a unique identifying number for that particular part.

▶ TECHNICIAN TIP

Dealership technicians have one advantage over most independent technicians. They have an onsite parts department that stocks many of the parts needed for repairs. This can save time by not having to wait for parts to be delivered. Many independent shops maintain a relatively small inventory of high-demand parts, such as filters, belts, and light bulbs. This minimizes wait times. For less common parts, they typically use several local parts houses to supply those parts. This can result in delays waiting for parts to arrive.

Using a Parts Program

A parts program is a computer application that is used to identify part numbers for vehicle components. Part numbers have to be identified so that correct replacement components can be ordered to replace faulty parts. The software is typically accessed via the Internet using a browser.

To use a parts program, you need to have a basic understanding of how to start and use a computer. Usernames and passwords may be required to log in to the computer and the parts program. So make sure you have those available before you start. A printer is also helpful to print out copies of the information so that you can use it when ordering parts. Alternatively, you may need to take notes.

To identify the correct part, you need to know:

- where on the vehicle the part is installed,
- what system or subsystem it comes from, and
- vehicle identification information, such as date of manufacture, model, engine size, and VIN numbers.

Make sure you have this information on hand before you use the system. Searches can be conducted by keywords. If the part is for the brake system, in the search criteria box, enter "brake." Using a generic word like "brake" may return a very large list of components. If this occurs, the search can be narrowed further by entering more specific criteria such as "disc brake."

The parts are displayed in diagrams that are labeled. They show individual parts in exploded view, making it easier to identify the parts (**FIGURE 5-9**). The diagrams may number the parts and have a key on the page for reference to part numbers. Or arrows may point to listed part numbers on the page. Most systems contain help menus or training guides with examples to assist you in using the software, if required.

To locate parts information on the computer, follow the steps in **SKILL DRILL 5-5**.

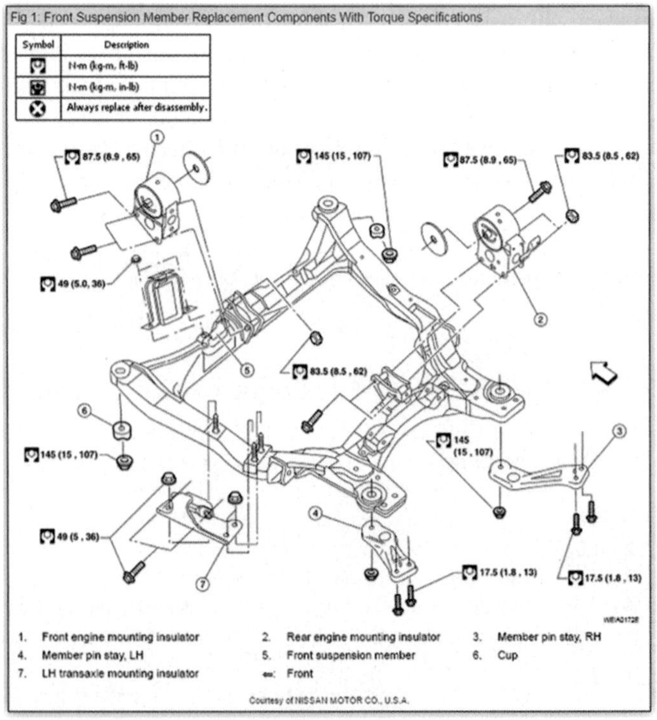

FIGURE 5-9 Typical exploded parts diagram.

SKILL DRILL 5-5 Using a Parts System

1. Log in to the application, using the appropriate username and password.
2. Enter the year, make, model, engine size, and/or VIN number information into the system in the appropriate places.
3. Search for the parts you require to conduct the service or repair.
4. The search engine will provide a list of possible matches for you to select from. If the initial search does not produce what you are looking for, try changing the search criteria. Keep searching until you find the information.
5. Gather information on the identified parts, including part numbers, location, availability, and cost.
6. Print, write down, or directly place an order for the desired parts.

▶ VIN and Production Date Code, and Vehicle Information Labels

LO 5-02 Identify vehicle information.

Every day, large numbers of vehicles are produced across the world. They have many variations of makes and models with different equipment levels. To accurately diagnose faults, order the correct parts, and properly install them require being able to accurately identify the vehicle. Vehicle information labels provide that information. **VIN** is a unique serial number. It is assigned to every vehicle produced. This means that no two vehicles have the same VIN (**FIGURE 5-10**). The VIN is designed to identify motor vehicles of all kinds. This includes cars, trucks, buses, motorcycles, etc. It was originally defined in the International Organization for Standardization (ISO) Standard 3779 in 1977. It was revised in 1983.

Since 1981, the VIN has been made up of 17 characters. It is usually located behind the front left corner of the windshield. It is also inscribed on the engine, the transmission, both front fenders, the hood, the doors, both bumpers, both rear quarter panels, and the trunk or hatchback. The VIN can be used to check the service history of a vehicle. It is also used for ordering the correct replacement parts. Labeling the vehicle and major vehicle parts with the VIN also deters auto theft. This is because it provides an easy way of uniquely identifying and tracing the major vehicle parts back to a specific vehicle. If the vehicle was reported stolen, then the parts would be associated to that vehicle.

FIGURE 5-10 A sample VIN.

Whenever a vehicle is registered or a registered vehicle is sold, a record of the VIN is kept. From this registry, information about the vehicle can be accessed. This includes the **title history**, which can tell you who has owned the vehicle. The registry may reveal a **salvage title**. It tells you that the vehicle has been wrecked and suffered economically unrepairable damage. It also can tell you if a **lemon law buyback** has occurred on the vehicle. Lemon laws exist in some states. They protect consumers from purchasing vehicles that have undergone several unsuccessful attempts to repair the same fault. Or from purchasing vehicles in which repair of the defects has caused the vehicle to be out of service for an extended time. The registry can also indicate if the vehicle has had an odometer rollback (mileage reduction). This provides evidence of odometer tampering.

Production date codes provide a critical piece of information when identifying vehicles. The production date is the date the vehicle was manufactured. It is indicated by month and year. This may be different from the model year designation of the vehicle.

Vehicle manufacturers do not introduce model years according to the regular calendar year (January 1st through December 31st). They release model years by their own calendar (typically in late summer or fall). In fact, most manufacturers introduce new model year vehicles in August or even July. This means that if a vehicle has a production date of October, 2018, it would be a 2019 model year. This is very important to know when ordering parts. If you order parts according to the production date code (October 2018), you could end up receiving the wrong parts. This is because it is really a 2019 vehicle.

Other information labels are fitted to the vehicle to provide ready access to additional information. For example, the following information can be found on labels:

- tire inflation pressures,
- vehicle weight and, load-carrying capacity,
- emission calibrations,
- engine coolant requirements, and
- air-conditioning refrigerant type and quantity.

These information labels are used regularly by technicians to identify vehicles, perform service on them, and order parts.

Locating the VIN and Production Date Code

The VIN is usually located behind the front left corner of the windshield. It is also inscribed on the major components of the vehicle. The production date code is usually located on the driver's side door pillar (B pillar). It will usually be listed as "Date of manufacture" or "Manufacture date." It is typically given as month/year.

To locate the VIN and production date code, follow the steps in **SKILL DRILL 5-6**.

Decoding a VIN

There are two different, but essentially compatible, 17-character VIN standards: the North American VIN system, and the ISO Standard 3779, which is used in most of the rest of the world. (**FIGURE 5-11**) shows how the numbers are structured.

To decode a North American VIN, follow the steps in **SKILL DRILL 5-7**.

So a vehicle with the VIN 1G1YN3DE-A5100001 can be identified as a 2010 Chevrolet Corvette ZR1 Convertible 7.0 L, V-8 LS7, SFI, and it was the first one off the production line in the Bowling Green assembly plant that year.

SKILL DRILL 5-6 Locating the VIN and Production Date Code

1. Locate the make of the vehicle from the body nameplate, which is usually found on the front or rear of the vehicle. Now locate the model from the body trim. The model may be a name, number, letter, or combination of these elements.

2. Next locate the VIN, usually found on a plate in the upper left dashboard and often visible through the windshield. In some instances, the plate may be mounted in a different location. If the plate is not visible through the windshield, check under the hood to see if it is mounted in the engine bay area. Note each letter and number exactly as it appears on the VIN plate.

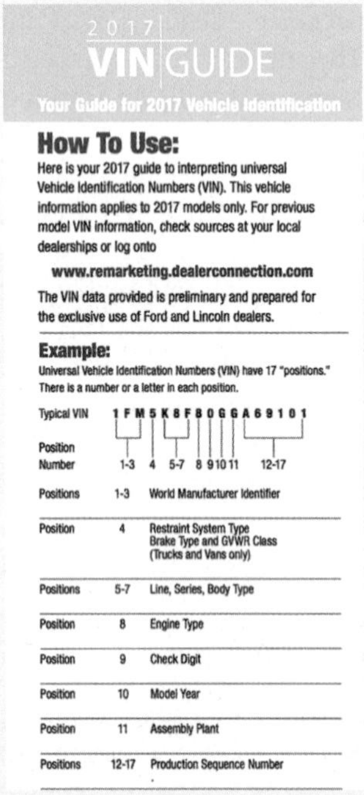

2017 VIN GUIDE

Your Guide for 2017 Vehicle Identification

How To Use:

Here is your 2017 guide to interpreting universal Vehicle Identification Numbers (VIN). This vehicle information applies to 2017 models only. For previous model VIN information, check sources at your local dealerships or log onto

www.remarketing.dealerconnection.com

The VIN data provided is preliminary and prepared for the exclusive use of Ford and Lincoln dealers.

Example:

Universal Vehicle Identification Numbers (VIN) have 17 "positions." There is a number or a letter in each position.

Typical VIN 1 F M 5 K 8 F 8 0 G G A 6 9 1 0 1

Position Number 1-3 4 5-7 8 9 10 11 12-17

Positions	1-3	World Manufacturer Identifier
Position	4	Restraint System Type Brake Type and GVWR Class (Trucks and Vans only)
Positions	5-7	Line, Series, Body Type
Position	8	Engine Type
Position	9	Check Digit
Position	10	Model Year
Position	11	Assembly Plant
Positions	12-17	Production Sequence Number

3. Decode the VIN and note the information. Each manufacturer provides a VIN decoding chart for its vehicles in its shop service information. This chart is normally found in the "general information" section of the manual or electronic service information system. Using the VIN decoding chart, write down the information or print it for later use when locating specifications or parts.

4. Look on the driver's door pillar for the label showing the production date code, and write down the month and year. If the label indicates what model year the vehicle conforms to, list that as well.

1	2	3	4	5	6	7	8	9	10	11	12	13	14	15	16	17
Manufacturer Identifier			Vehicle Attributes					Check Character	Model Year	Plant Code	Sequential Number					

A

1	2	3	4	5	6	7	8	9	10	11	12	13	14	15	16	17
World Manufacturer Identifier (WMI)			Vehicle Descriptor System (VDS)						Vehicle Identifier System (VIS)							

B

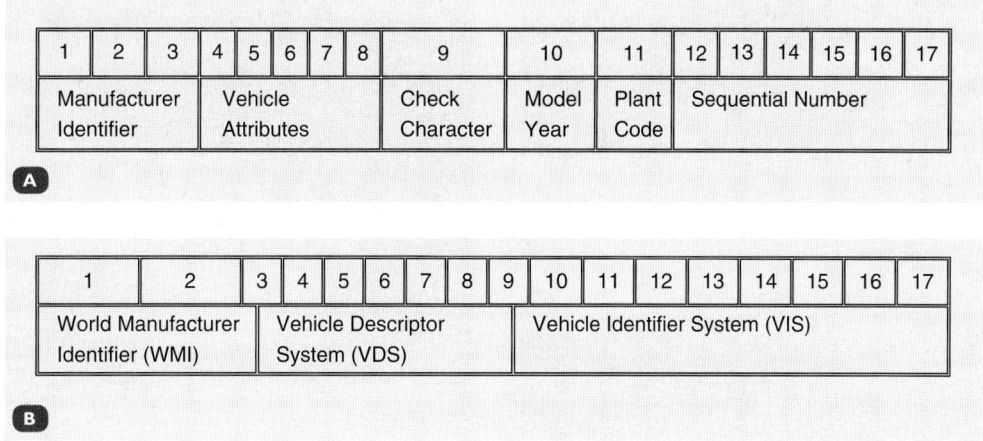

FIGURE 5-11 **A.** North American VIN system. **B.** ISO Standard 3779.

SKILL DRILL 5-7 Decoding a North American VIN

1. The VIN is **1G1YN3DE-A5100001**. Refer to (**FIGURE 5-12**). The first character is the country of origin. This number or letter tells you where the vehicle was manufactured. For instance, a "1" means that the vehicle was made in the United States; a "2" is for Canada; a "J" means Japan, etc.
2. The second character is usually a letter; it tells you the name of the manufacturer. "G" stands for "General Motors."
3. The third character tells you the division that made the vehicle. It could be a Pontiac, an Oldsmobile, or a GMC truck, for example. The number "1" means this is a Chevrolet.

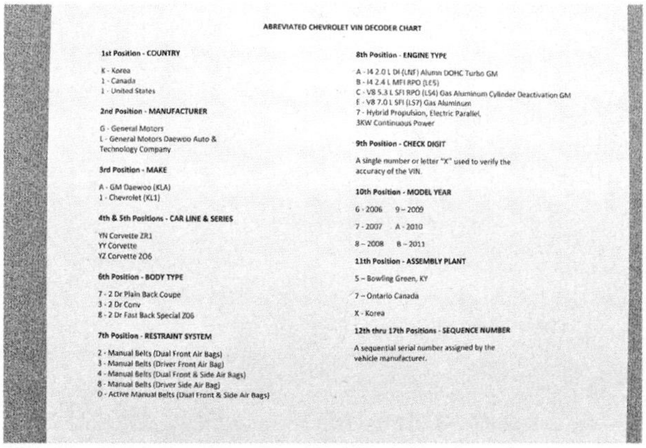

ABREVIATED CHEVROLET VIN DECODER CHART

1st Position - COUNTRY

K - Korea
1 - Canada
1 - United States

2nd Position - MANUFACTURER

G - General Motors
L - General Motors Daewoo Auto & Technology Company

3rd Position - MAKE

A - GM Daewoo (KLA)
1 - Chevrolet (KL1)

4th & 5th Positions - CAR LINE & SERIES

YN Corvette ZR1
YY Corvette
YZ Corvette Z06

6th Position - BODY TYPE

7 - 2 Dr Plain Back Coupe
3 - 2 Dr Conv
8 - 2 Dr Fast Back Special Z06

7th Position - RESTRAINT SYSTEM

2 - Manual Belts (Dual Front Air Bags)
3 - Manual Belts (Driver Front Air Bag)
4 - Manual Belts (Dual Front & Side Air Bags)
8 - Manual Belts (Driver Side Air Bag)
0 - Active Manual Belts (Dual Front & Side Air Bags)

8th Position - ENGINE TYPE

A - I4 2.0 L DI (LNF) Alumn DOHC Turbo GM
B - I4 2.4 L MFI RPO (LE5)
C - V8 5.3 L SFI RPO (LS4) Gas Aluminum Cylinder Deactivation GM
E - V8 7.0 L SFI (LS7) Gas Aluminum
7 - Hybrid Propulsion, Electric Parallel, 3KW Continuous Power

9th Position - CHECK DIGIT

A single number or letter "X" used to verify the accuracy of the VIN.

10th Position - MODEL YEAR

6 - 2006 9 – 2009
7 - 2007 A - 2010
8 – 2008 B – 2011

11th Position - ASSEMBLY PLANT

5 – Bowling Green, KY
7 – Ontario Canada
X - Korea

12th thru 17th Positions - SEQUENCE NUMBER

A sequential serial number assigned by the vehicle manufacturer.

FIGURE 5-12 2010 GM VIN Chart.

4. The fourth and fifth characters give you the model, or series, of the vehicle. You will need a decoding chart for the details. Here the YN means we have a Corvette ZR1 Custom 3ZR Manual.
5. The sixth character describes the body type: two-door, four-door, coupe, sedan, etc. Here we have a two-door convertible.
6. The seventh character tells you the type of seat restraints fitted to the vehicle. In this case, it is active manual seat belts, airbags front (driver and passenger), and front seat side.
7. The eighth character is the engine code, which provides details of the engine type, size, or displacement, and where the engine was made. Here we have a 7.0 L, LS7, gas, eight-cylinder, SFI, aluminum, GM.
8. The ninth character is the check character. It is used internally by the manufacturer.
9. The tenth character tells you the model year of manufacture. You can decode this character according to a model year identification chart, which in this example shows us that the vehicle was assembled for the 2010 model year.
10. The 11th character tells you the assembly plant or factory where the vehicle was put together. In this case, it's Bowling Green, Kentucky.
11. The final six numbers make up the sequential number of the vehicle as it comes off the assembly line, starting at a base number, which is usually 100,000. So the first vehicle to be produced usually, but not always, has the number 100001. In our example, the vehicle was the first to come off the assembly line in that year.

Using Other Vehicle Information Labels

Vehicle Emission Control Information (VECI) Label

The **vehicle emission control information (VECI) label** is used by technicians to identify engine and emission control information for the vehicle (**FIGURE 5-13**). It is usually located in the engine compartment on either the hood, strut tower, or radiator support. It typically includes the following information:

- Engine family and displacement
- Model year the vehicle conforms to
- Spark plug part number and gap
- Evaporative emission system family
- Emission control system schematic
- Certification application

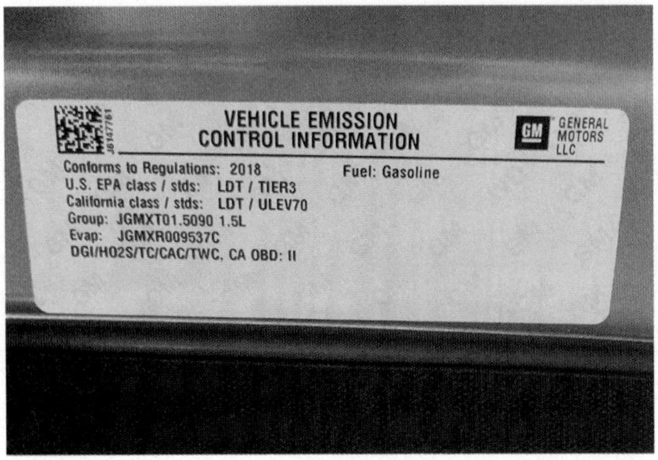

FIGURE 5-13 VECI label.

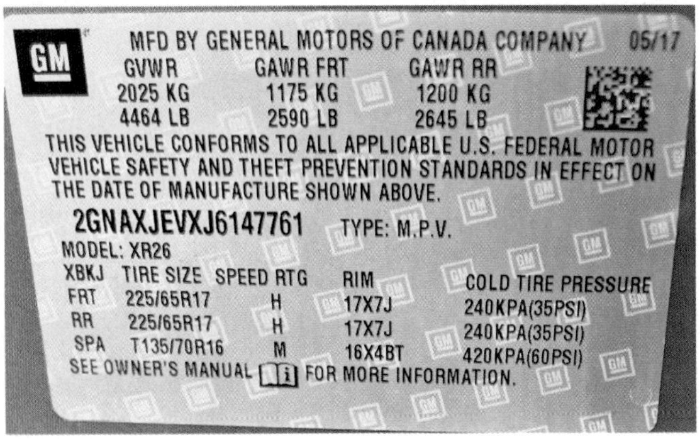

FIGURE 5-14 VSC label.

Vehicle Safety Certification (VSC) Label

The **vehicle safety certification (VSC) label** certifies that the vehicle meets the Federal Motor Vehicle Safety, Bumper, and Theft Prevention Standards in effect at the time of manufacture (**FIGURE 5-14**). It is used by technicians to identify some basic types of information about the vehicle. This includes the month and year of manufacture, gross vehicle weight rating (GVWR), and on some vehicles, tire information. It is usually affixed to the driver's side door pillar or on the side of the door next to the pillar. It typically includes the following information:

- Month and year of manufacture
- GVWR and gross axle weight rating
- VIN
- Recommended tire sizes
- Recommended tire inflation pressures
- Paint and trim codes

Other Labels

Other labels include the refrigerant label, the coolant label, and the belt routing label. The **refrigerant label** lists the type and total capacity of refrigerant that is installed in the A/C system (**FIGURE 5-15**). The **coolant label** lists the type of coolant installed in the cooling system (**FIGURE 5-16**). The **belt routing label** lists a diagram of the serpentine belt routing for the engine accessories (**FIGURE 5-17**).

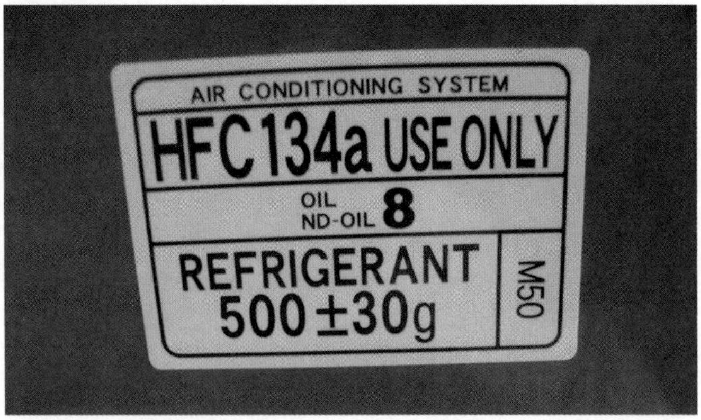

FIGURE 5-15 Refrigerant label.

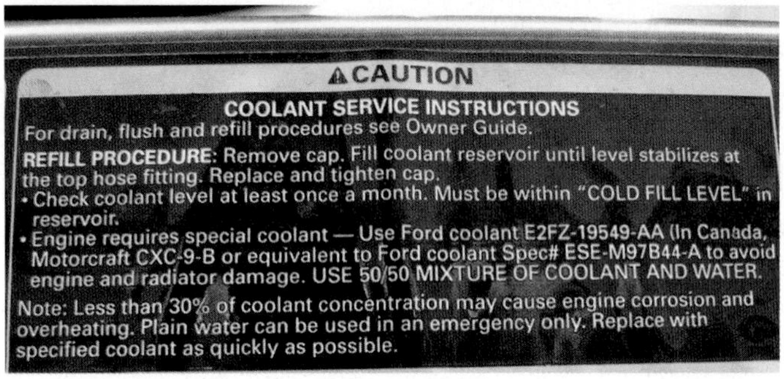

FIGURE 5-16 Coolant label.

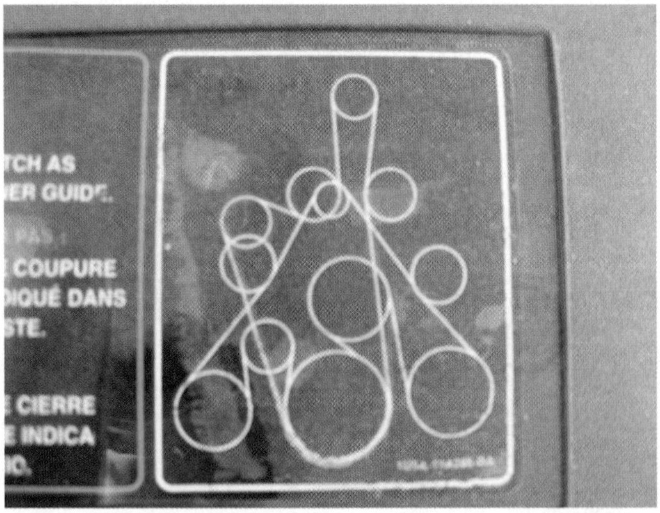

FIGURE 5-17 Belt routing label.

▶ Repair Order Information

LO 5-03 Complete a repair order.

A **repair order**, or work order, is used by shops to document the details regarding the repair of a vehicle (**FIGURE 5-18**). They can be paper, but more often shops use electronic repair orders.

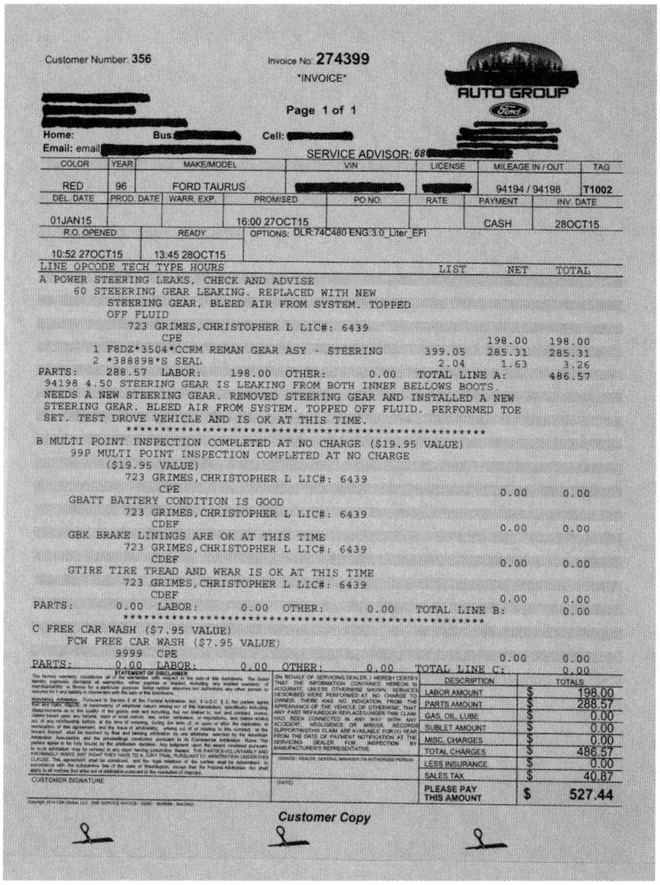
Customer Number: 356 Invoice No: 274399
INVOICE
[redacted] AUTO GROUP
Page 1 of 1
Home: Bus: [redacted] Cell: [redacted]
Email: email [redacted]
SERVICE ADVISOR: 68 [redacted]

COLOR	YEAR	MAKE/MODEL	VIN	LICENSE	MILEAGE IN / OUT	TAG
RED	96	FORD TAURUS	[redacted]	[redacted]	94194 / 94198	T1002

DEL. DATE	PROD. DATE	WARR. EXP.	PROMISED	PO NO.	RATE	PAYMENT	INV. DATE
01JAN15			16:00 27OCT15			CASH	28OCT15

R.O. OPENED	READY	OPTIONS: DLR:74C480 ENG:3.0_Liter_EFI
10:52 27OCT15	13:45 28OCT15	

```
LINE OPCODE TECH TYPE HOURS                              LIST     NET    TOTAL
A POWER STEERING LEAKS, CHECK AND ADVISE
    60 STEEERING GEAR LEAKING. REPLACED WITH NEW
       STEERING GEAR, BLEED AIR FROM SYSTEM. TOPPED
       OFF FLUID
          723 GRIMES,CHRISTOPHER L LIC#: 6439
              CPE                                              198.00   198.00
     1 F8DZ*3504*CCRM REMAN GEAR ASY - STEERING     399.05   285.31   285.31
     2 *388898*S SEAL                                  2.04     1.63     3.26
PARTS:   288.57  LABOR:   198.00  OTHER:     0.00  TOTAL LINE A:      486.57
 94198 4.50 STEERING GEAR IS LEAKING FROM BOTH INNER BELLOWS BOOTS.
 NEEDS A NEW STEERING GEAR. REMOVED STEERING GEAR AND INSTALLED A NEW
 STEERING GEAR. BLEED AIR FROM SYSTEM. TOPPED OFF FLUID. PERFORMED TOE
 SET. TEST DROVE VEHICLE AND IS OK AT THIS TIME.
          ****************************************************
B MULTI POINT INSPECTION COMPLETED AT NO CHARGE ($19.95 VALUE)
    99P MULTI POINT INSPECTION COMPLETED AT NO CHARGE
       ($19.95 VALUE)
          723 GRIMES,CHRISTOPHER L LIC#: 6439
              CPE                                                0.00     0.00
    GBATT BATTERY CONDITION IS GOOD
          723 GRIMES,CHRISTOPHER L LIC#: 6439
              CDEF                                               0.00     0.00
    GBK BRAKE LININGS ARE OK AT THIS TIME
          723 GRIMES,CHRISTOPHER L LIC#: 6439
              CDEF                                               0.00     0.00
    GTIRE TIRE TREAD AND WEAR IS OK AT THIS TIME
          723 GRIMES,CHRISTOPHER L LIC#: 6439
              CDEF                                               0.00     0.00
PARTS:     0.00  LABOR:     0.00  OTHER:       0.00  TOTAL LINE B:        0.00
          ****************************************************
C FREE CAR WASH ($7.95 VALUE)
    FCW FREE CAR WASH ($7.95 VALUE)
          9999  CPE                                              0.00     0.00
PARTS:     0.00  LABOR:     0.00  OTHER:       0.00  TOTAL LINE C:        0.00
```

STATEMENT OF DISCLAIMER

CUSTOMER SIGNATURE

(DATE)

DESCRIPTION	TOTALS
LABOR AMOUNT	$ 198.00
PARTS AMOUNT	$ 288.57
GAS, OIL, LUBE	$ 0.00
SUBLET AMOUNT	$ 0.00
MISC. CHARGES	$ 0.00
TOTAL CHARGES	$ 486.57
LESS INSURANCE	$ 0.00
SALES TAX	$ 40.87
PLEASE PAY THIS AMOUNT	$ 527.44

Customer Copy

FIGURE 5-18 A print out of an electronic repair order.

Initial information for the repair order includes customer and vehicle details. It also lists a brief description of the customer's concern(s). The repair order provides the technician with a summary of the fault or needed service on the vehicle. So it is critical that the customer service staff get an accurate, complete, and yet concise record of the customer's concern(s). This should also include any information and details about the concern.

The technician and customer service staff complete the repair order. They document the exact cause of the concern and the correction required to fix it. This provides a detailed record that can be reviewed as part of the vehicle's service history. It is also helpful when the technician is faced with a new or recurring issue on the vehicle. We will cover the 3 C's (Concern, Cause, and Correction) in a later section.

Detailed information on the repair order includes the shop, customer, and vehicle details such as:

- Shop information
- Service advisor
- Customer name, address, and contact info
- Customer authorization signature
- vehicle make, model, and year
- VIN, license plate, odometer reading
- date
- customer concern information
- the cause of the concern(s)
- the correction for the concern(s)
- the hours of labor and
- the parts used for the repair.

The repair order should always include all of the information pertaining to the customer, vehicle, and cost of repair. Repair orders are legal documents that can be used as evidence in the event of a lawsuit. Make sure the information is complete and accurate whenever filling out a repair order. Store paper repair orders in an organized safe place, such as in a file cabinet. Electronic repair orders should be stored on a secure computer network.

To identify the information needed and the service requested on a repair order, follow the steps in **SKILL DRILL 5-8**.

> **▶ TECHNICIAN TIP**
>
> Repair orders are also used to inform the customer of needed repairs or service. This usually results in the customer authorizing the needed repair by initialing, or signing, the repair order. But if the customer does not agree to the repair, the shop should have the customer initial, or sign, a statement on the repair order. This signifies that they are declining the repair and understand the safety issues involved. This may help prevent the shop from being held liable in the event of an accident. Also, any additional needed work must be authorized by the customer before those repairs are started.

Completing a Repair Order

When completing a repair order, make sure you fill it out completely since it is a legal document. The repair order is broken

SKILL DRILL 5-8 Using Repair Orders

1. Locate a repair order used in your shop.
2. Familiarize yourself with the repair order, and identify the following information on the repair order:
   - **a.** Date
   - **b.** Customer details: name and address, daytime phone number
   - **c.** Vehicle details: year, make, model, color, odometer reading, VIN, and license plate number
   - **d.** Customer signature
   - **e.** Customer concern details
3. Note any additional information that is required on your shop's repair order.
4. Following the shop procedures, determine the workflow for the tasks that are listed.
5. Use the repair order to carry out the requested service or repair.
6. Complete repair order with details of the cause of the customer concern(s) and the correction(s) needed or performed.

up into sections; customer, vehicle, requested service, parts, and accounting. Each section needs to be filled out appropriately. The accounting section contains information about the methods of payment. This can be check, cash, credit card, or on account for later billing. To work out the total cost of the service, you need to know:

- The cost of labor
- The cost of parts
- The amount of tax
- The cost of gas and consumables you used to service the vehicle
- Any environmental or disposal fees

You also need to have the customer's authorization to carry out the service. After the initial authorization, if there are any additional repair costs beyond a certain percent or dollar amount, you need to receive and document the customer's approval. Failure to do so may mean that the customer doesn't have to pay the extra cost.

TECHNICIAN TIP

Always obtain the customer's authorization before servicing his/her vehicle. Also make sure you have the customer's authorization before making any changes to the repair order that would cost the customer additional money.

Service History

Service history is a complete list of all the servicing and repairs that have been performed on a vehicle (**FIGURE 5-19**). The scheduled service history can be recorded in a service booklet or owner's manual. These are typically kept in the glove compartment. The service history can provide valuable information to technicians conducting repairs. It also can provide potential new owners of used vehicles an indication of how well the vehicle was maintained. A vehicle with a regular service history is a good indication that all of the vehicle's systems have been well maintained. The vehicle will often be worth more during resale.

Applied Communications

AC-23: Repair Orders: The technician writes a repair order containing customer vehicle information, customer complaints, parts and materials used (including prices), services performed, labor hours, and suggested repairs/maintenance.

A repair order is a key document used to communicate with both your customers and coworkers. It is a legal contract between the service provider and the customer. It contains details of the services to be provided by you and the authorization from the customer. To make sure everyone understands clearly what is involved, a repair order should contain the following information:

- Your company or service providers—The service provider section contains the company name, address, and contact details; the name of a service advisor who is overseeing the job; and the amount of time the service technician needs to service the vehicle.
- The customer—The customer section contains the customer's name, address, and contact phone numbers.

  The customer's vehicle—The vehicle section includes details about the vehicle to be serviced. Check the vehicle's license plate before starting work. The license plate numbers are usually unique within a country. You should also record information about the vehicle's make, model, and color. This information makes it easier for you to locate the vehicle on the parking lot. You need to know the manufacture date of the vehicle to be able to order the right parts. The odometer reading and the date help keep track of how much distance the vehicle travels and the time period between each visit to the shop. The VIN is designed to be unique worldwide and contains specific information about the vehicle. Many shops do a "walk-around" with the customer to note any previous damage to the vehicle and to look for any obvious faults such as worn tires, rusted-out exhaust pipes, or torn wiper blades.
- The service operations—This section contains the details of the service operations and parts.

  The first part is the service operation details. For example, the vehicle is in for a 150,000-mile (240,000 km) service, which can be done in three hours, resulting in approximately a $300 labor cost. The information about the chargeable labor time to complete a specific task can be found in a labor guide manual. In some workplaces, this information is built into the computer system and is automatically displayed.

  The second part of this section is the details of parts used in the service, including the descriptions, quantities, codes, and prices. The codes for each service and part are normally abbreviations that are used for easy reference in the shop. Some shops may have their own reference code system.

  As you do the vehicle inspection, you may discover other things that need to be replaced or repaired. These additional services can be recorded in another section. It is essential that you check with your customers and obtain their approval before carrying out any additional services.
- The parts requirements—This section lists the parts required to perform the repair, and their cost. It can also include specific supplies used to perform the job, such as cleaners. Or it can be a general supply charge that is used to cover the cost of commonly used supplies. If the parts come with a warranty, it is helpful to list the length of the warranty in the part description. For example, a new fuel pump may come with a 24-month warranty.
- Repair orders also contain accounting information so that they can be used to invoice the customer. Most customers pay upon completion of the job. But some customers have an account set up with the shop to pay monthly for accumulated repair costs. Shops commonly set this up for companies that have a fleet of vehicles that are regularly brought in for repairs and maintenance. When a vehicle on an account system comes in, you need to also record the account name and number on the repair order. If the fleet numbers their vehicles, you will need to record the company's vehicle number as well.

FIGURE 5-19 Print outs of completed repair orders as saved in the online repair order system.

Most manufacturers store all service history performed in their dealerships (based on the VIN) on a corporate server. This makes it accessible from any of their dealerships. They also use this vehicle service history when it comes to evaluating warranty claims. A vehicle that does not have a complete service history may not be eligible for warranty claims.

Independent shops generally keep records of the repairs they perform. If a vehicle is repaired at multiple shops, repair history is much more difficult to track. This may result in a denial of warranty claims.

To review the vehicle service history, follow the steps in **SKILL DRILL 5-9**.

▶ TECHNICIAN TIP

A vehicle's service history is valuable for several reasons:

It can provide helpful information to the technician when performing repairs.

It allows potential new owners of the vehicle to know how well the vehicle and its systems were maintained.

Manufacturers use the history to evaluate warranty claims.

▶ Strategy-Based Diagnosis

LO 5-04 Explain Strategy-Based Diagnosis and the 3 C's.

Customers experience issues with their vehicles on a regular basis. Parts wear out or malfunction and cause the vehicle to not operate properly. Technicians are called on to investigate and identify the cause of the malfunction. Then they determine what the appropriate fix is. That is called diagnosing the problem. A diagnosis is the conclusion a technician comes to after investigating the problem.

Diagnosing problems can be very challenging to perform. Especially in a timely and efficient manner. Technicians find that having a proven plan in place ahead of time greatly simplifies the process of strategically solving problems. The plan must be simple to remember and consistent in its approach. It must also work for the entire range of diagnostic problems that technicians will encounter. In this way, technicians can have one single plan to approach any diagnostic situation they may encounter. This will give them confidence in their ability to resolve it. This problem-solving plan is called **strategy-based diagnosis** and involves the following steps:

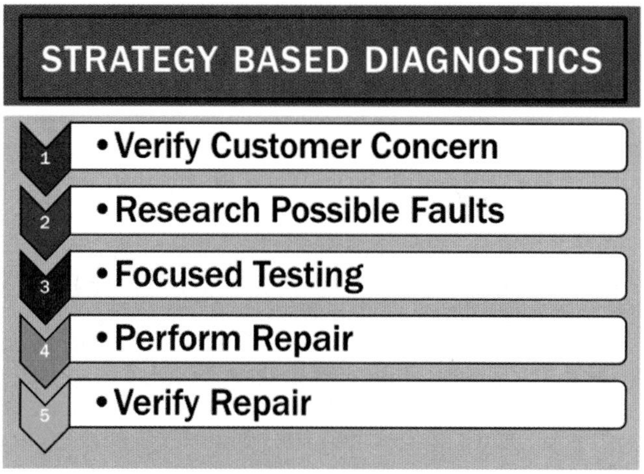

Let's look at a simple example of the diagnostic process in action. A customer dropped off her car the day before with the following concern: "The tire pressure monitoring system (TPMS) warning lamp is on and the car pulls to the left." The technician takes the first step of "verifying the concern," when walking up to the vehicle. He does this by looking at the tires. He notices that the driver's side front tire appears to be quite low on air pressure.

The technician then turns the ignition key to the "Run" position and observes the TPMS warning system. The display shows a warning message indicating low pressure in the left front tire. This verifies the customer concern that the TPMS warning lamp is on. It also gives the technician a clue about the cause of the vehicle pulling left. This is because low tire pressure

SKILL DRILL 5-9 Reviewing Service History

1. Locate the service history for the vehicle. This may be in shop records or in the service history booklet within the vehicle glove compartment. Some shops may keep the vehicle's service history on a computer.
2. Familiarize yourself with the service history of the vehicle.
   - **a.** On what date was the vehicle first serviced?
   - **b.** On what date was the vehicle last serviced?
   - **c.** What was the most major service performed?
   - **d.** Was the vehicle ever serviced for the same problem more than once?
3. Compare the vehicle service history to the manufacturer's scheduled maintenance requirements, and list any discrepancies.
   - **a.** Have all the services been performed?
   - **b.** Have all the items been checked?
   - **c.** Are there any outstanding items?

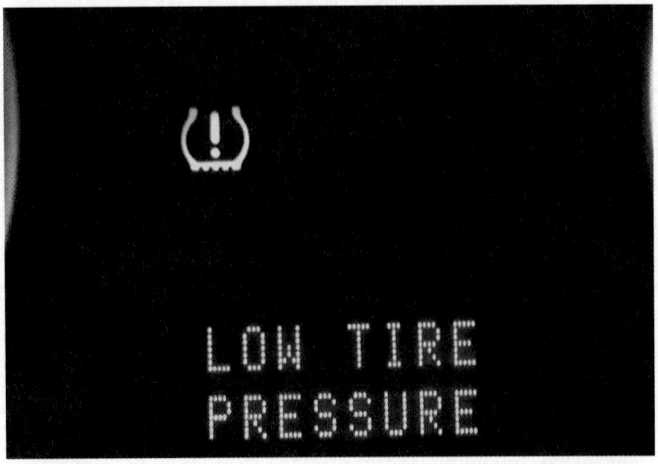

FIGURE 5-20 TPMS warning system showing the left front tire low.

will cause a vehicle to pull toward the side with low pressure. This completes Step 1 (**FIGURE 5-20**).

Step 2 is, "Researching Possible Faults and Gathering Information." The technician knows from training that the low-pressure tire can cause the vehicle to pull toward the side that is low. They also know that the TPMS warning lamp is designed to warn the driver of low-pressure tires and faults in the system. Checking the tire placard reveals that the specified tire pressure is 32 psi (220.63 kPa) for this vehicle.

Researching the service information, the technician finds two relevant pieces of information. First, that the air pressure needs to be adjusted after the vehicle has been parked for at least four hours. And second, that pressures under a certain point will activate the TPMS. The technician then knows that if the tire pressure is below that specified psi (kPa), the TPMS was working properly. They will also know that there is another problem causing the low tire pressure. And that will need to be identified. So now it is time to perform focused testing on the vehicle (**FIGURE 5-21**).

Step 3 is, "Focused Testing." The technician follows the specified procedure for properly testing the tire pressure, and finds that it reads 21 psi (144.79 kPa) (**FIGURE 5-22**). At this point, the technician needs to locate the cause of the low tire

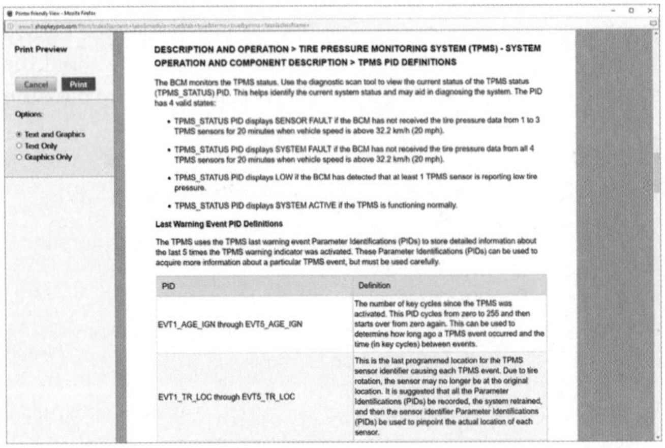

Print Preview

Cancel Print

Options:

- Text and Graphics
- Text Only
- Graphics Only

DESCRIPTION AND OPERATION > TIRE PRESSURE MONITORING SYSTEM (TPMS) - SYSTEM OPERATION AND COMPONENT DESCRIPTION > TPMS PID DEFINITIONS

The BCM monitors the TPMS status. Use the diagnostic scan tool to view the current status of the TPMS status (TPMS_STATUS) PID. This helps identify the current system status and may aid in diagnosing the system. The PID has 4 valid states:

- TPMS_STATUS PID displays SENSOR FAULT if the BCM has not received the tire pressure data from 1 to 3 TPMS sensors for 20 minutes when vehicle speed is above 32.2 km/h (20 mph).
- TPMS_STATUS PID displays SYSTEM FAULT if the BCM has not received the tire pressure data from all 4 TPMS sensors for 20 minutes when vehicle speed is above 32.2 km/h (20 mph).
- TPMS_STATUS PID displays LOW if the BCM has detected that at least 1 TPMS sensor is reporting low tire pressure.
- TPMS_STATUS PID displays SYSTEM ACTIVE if the TPMS is functioning normally.

Last Warning Event PID Definitions

The TPMS uses the TPMS last warning event Parameter Identifications (PIDs) to store detailed information about the last 5 times the TPMS warning indicator was activated. These Parameter Identifications (PIDs) can be used to acquire more information about a particular TPMS event, but must be used carefully.

PID	Definition
EVT1_AGE_IGN through EVT5_AGE_IGN	The number of key cycles since the TPMS was activated. This PID cycles from zero to 255 and then starts over from zero again. This can be used to determine how long ago a TPMS event occurred and the time (in key cycles) between events.
EVT1_TR_LOC through EVT5_TR_LOC	This is the last programmed location for the TPMS sensor identifier causing each TPMS event. Due to tire rotation, the sensor may no longer be at the original location. It is suggested that all the Parameter Identifications (PIDs) be recorded, the system retrained, and then the sensor identifier Parameter Identifications (PIDs) be used to pinpoint the actual location of each sensor.

FIGURE 5-21 Service information listing the enable criteria that activate the TPMS warning system.

FIGURE 5-22 Technician measuring tire pressure, showing 21 psi (144.79 kPa).

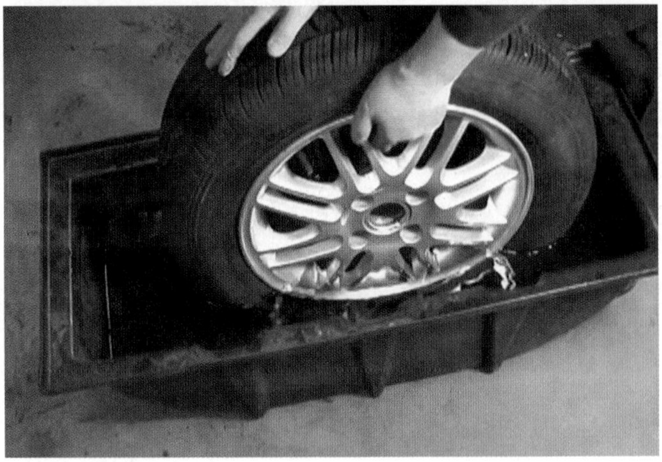

FIGURE 5-23 Tire being tested for a leak and leaking from the center of the tread.

pressure. The technician does so by removing the wheel and tire from the vehicle. The tire is then placed in a tire dunk tank to locate any leaks. Bubbles are coming from a small nail embedded near the center of the tread (**FIGURE 5-23**). The technician informs the customer of the need to plug and patch the tire, and then verify that the TPMS warning light is no longer illuminated. The customer authorizes the repair.

Step 4 is, "Performing the Repair." The technician follows the proper steps to plug and patch the hole in the tire (**FIGURE 5-24**). The tire is remounted on the wheel. It is then inflated to the recommended 32 psi (220.63 kPa). It is rechecked to make sure the original leak is repaired, and there are no other leaks as well. They then balance the wheel/tire assembly, and remount the wheel on the vehicle, being sure to tighten the lug nuts to the proper torque.

Step 5 is, "Verify the Repair." The technician turns the ignition key to the "Run" position and watches the TPMS warning light. It initially comes on for a few seconds as the system goes through its initial check. After that, the TPMS warning lamp turns off, indicating that the system is monitoring the pressure correctly (**FIGURE 5-25**). Now, all the technician needs to do is test-drive the vehicle to verify that it is no longer pulling to the

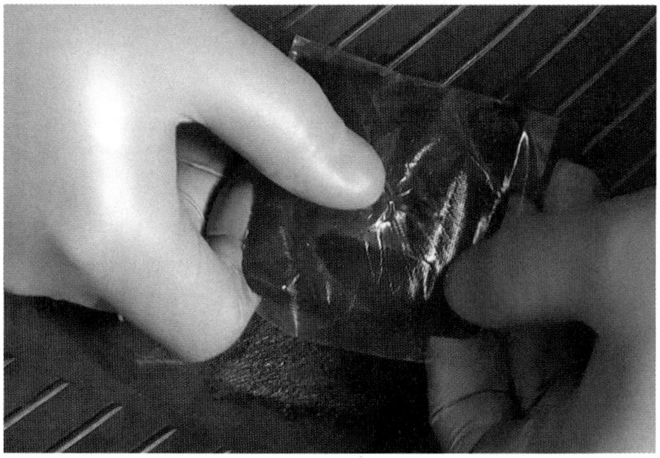

FIGURE 5-24 Tire being plug patched.

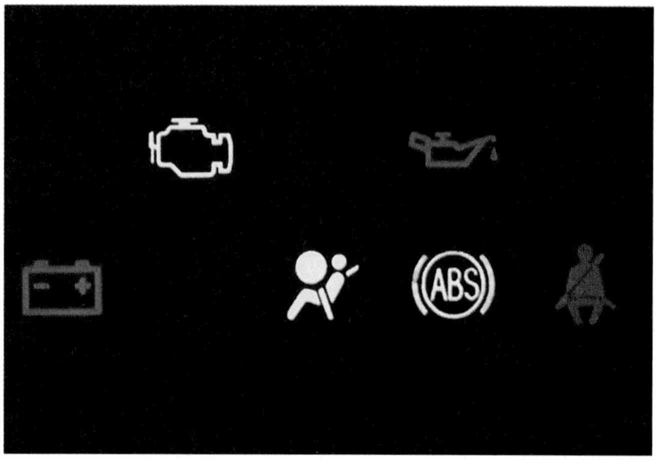

FIGURE 5-25 Technician verifying that the TPMS warning light is off.

left, which he does. He then finishes up the repair order so the customer can pay the bill and get on his/her way.

This strategy-based diagnostic process is used by technicians to help them logically diagnose problems. We used a very simple example of a technician using that process. We will continue to expand on the theory of the diagnostic strategy as you progress through this course. We will also present more complicated diagnostic procedures and processes. But for now, we wanted to introduce the process. That way, you can start to use it from the early portions of this training, and build your experience as you go. Stay tuned for more details on this in future topics.

3 C's

Now that you understand how the diagnostic process works, you need a record of the repair. This is so you can inform the customer on what work was performed and then pay the bill. The shop also uses information off the repair order to know how much they will have to pay you. It also provides a record for other warranty and liability issues. The information recorded on repair orders is based on the "3 C's." They represent "concern," "cause," and "correction" (**FIGURE 5-26**).

The 3 C's provide a summary of the repair, so the write up of the 3 C's is not quite like writing an English paper or novel. At the same time, it needs to contain all of the needed information so that people reading it will have a good understanding of what is included. I like to say that the write up of the 3 C's needs to be clear, concise, and complete. So that means to be clear, you need to use the right words that accurately describe the situation. Concise means that it needs to be as short as possible, so don't use complete sentences. And finally, it needs to be complete, which means that there is enough of a description that everything that needs to be covered is covered. As you can see, there needs to be a balance between concise and complete. Here are some examples of bad and good write-ups of the 3 C's.

"Concern" is the description of the customer's understanding of the problem with the vehicle. When a customer experiences an issue with their vehicle, they attempt to communicate that to the service advisor. The advisor helps the customer put the concern into words that make sense and that

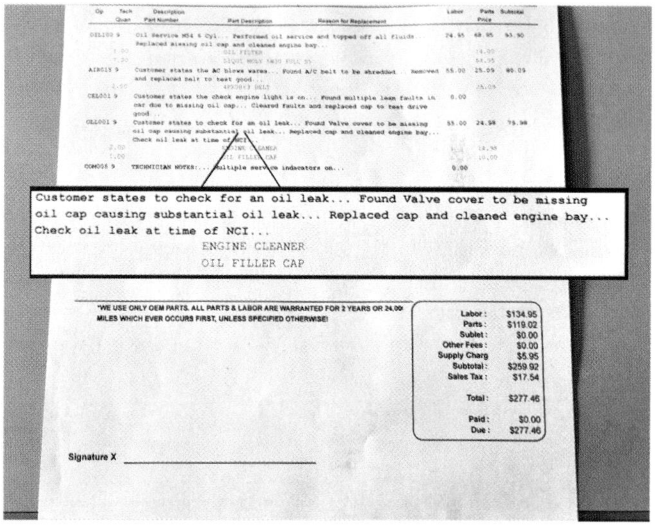

FIGURE 5-26 3 C's as listed on a repair order.

accurately describe the actual problem. Once the service advisor has a clear understanding of the customer's concern, they write it up. We will use the example from the Strategy-Based Diagnosis scenario about the TPMS warning lamp illuminated (**FIGURE 5-27**).

The service advisor also helps to evaluate the customer's concern. They help determine if it is one concern with related symptoms, or if it is separate concerns that are not likely to be related. If not, they will need to be diagnosed or performed separately. For example, a customer may say:

> My blower motor doesn't work on high speed and my steering wheel shakes when I apply the brakes.

In this case, the service advisor would know that each of these concerns are likely caused by separate faults. So they would list each concern separately on the repair order.

For gaining an understanding of the customer concern, the service advisor will need to carefully listen to and communicate with the customer. For example, on a vehicle that won't start, the customer may initially state that the engine won't crank over. With some questioning, the service advisor helps the customer

Example of a write-up that is not clear:

Correction—Got power back to the injector. Replaced broken components.

Notice that it doesn't say what was done to get power to the injector, nor does it say which components were replaced, making it difficult to know what exactly was done.

Example of a write up that is clear:

Correction—Repaired the broken Red/White power wire at the #6 fuel injector with solder and heat shrink, also replaced the #6 fuel injector harness connector.

Example of a write up that is not concise:

Correction—I inspected all of the wires under the hood and found that the power wire for the #6 fuel injector, which is red with a white stripe, broken in two. I found a soldering iron and solder and soldered the wires back together. After I soldered it, I insulated it with heat shrink tubing. I also noticed that the #6 fuel injector connector looked like it was melted from something. So I went to the parts department and they didn't have one in stock, so they ordered it. It came in that afternoon and I installed it.

Notice that there is a LOT of information that is not needed. Again, look at the example of the clear write-up to see what is more appropriate.

Example of a write up that is not complete:

Correction—Repaired the fuel injector wire, replaced the connector.

Notice that it doesn't say which wire, which injector, or which connector.

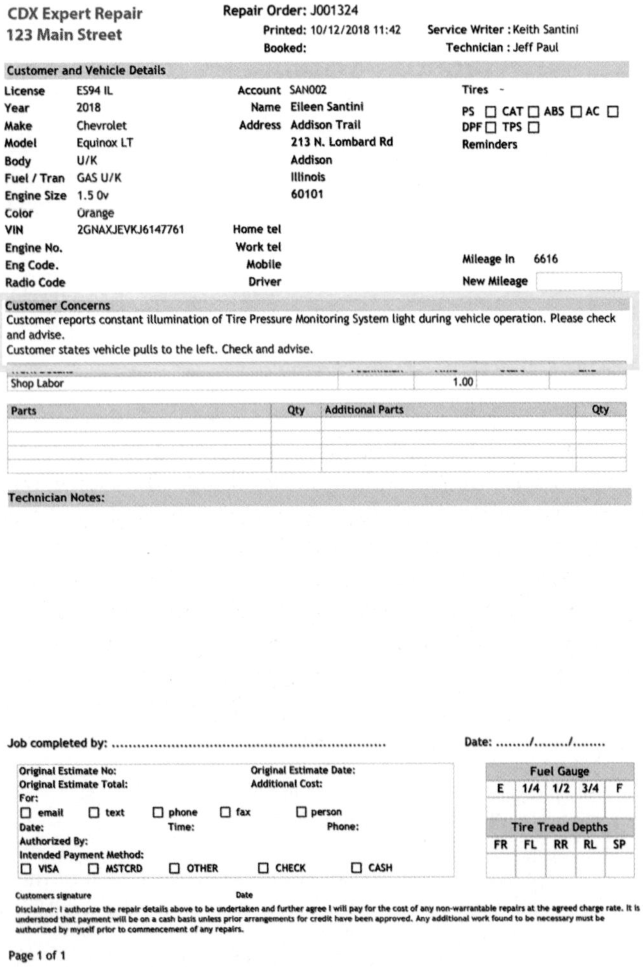

CDX Expert Repair
123 Main Street

Repair Order: J001324
Printed: 10/12/2018 11:42 Service Writer : Keith Santini
Booked: Technician : Jeff Paul

Customer and Vehicle Details

License	ES94 IL	Account	SAN002	Tires	-
Year	2018	Name	Eileen Santini	PS ☐ CAT ☐ ABS ☐ AC ☐	
Make	Chevrolet	Address	Addison Trail	DPF ☐ TPS ☐	
Model	Equinox LT		213 N. Lombard Rd	Reminders	
Body	U/K		Addison		
Fuel / Tran	GAS U/K		Illinois		
Engine Size	1.5 0v		60101		
Color	Orange				
VIN	2GNAXJEVKJ6147761	Home tel			
Engine No.		Work tel			
Eng Code.		Mobile		Mileage In	6616
Radio Code		Driver		New Mileage	

Customer Concerns
Customer reports constant illumination of Tire Pressure Monitoring System light during vehicle operation. Please check and advise.
Customer states vehicle pulls to the left. Check and advise.

Shop Labor		1.00		

Parts	Qty	Additional Parts	Qty

Technician Notes:

Job completed by: .... Date: ..../......../........

Original Estimate No: Original Estimate Date:
Original Estimate Total: Additional Cost:
For:
☐ email ☐ text ☐ phone ☐ fax ☐ person
Date: Time: Phone:
Authorized By:
Intended Payment Method:
☐ VISA ☐ MSTCRD ☐ OTHER ☐ CHECK ☐ CASH

Fuel Gauge				
E	1/4	1/2	3/4	F

Tire Tread Depths				
FR	FL	RR	RL	SP

Customers signature Date

Disclaimer: I authorize the repair details above to be undertaken and further agree I will pay for the cost of any non-warrantable repairs at the agreed charge rate. It is understood that payment will be on a cash basis unless prior arrangements for credit have been approved. Any additional work found to be necessary must be authorized by myself prior to commencement of any repairs.

Page 1 of 1

FIGURE 5-27 Customer concern listed on the repair order.

to better describe the problem. During that process, the advisor discovers that the engine is cranking over, but not starting. So the actual concern is that the "engine cranks, but won't start." For a technician diagnosing the problem, that is a vastly different situation than the "engine will not crank." By having a clear description of the actual concern, the technician will avoid wasting time looking for a nonexistent problem.

In order to be more thorough, many shops use a paper or electronic symptom questionnaire. This guides the customer through a series of questions that will give the technician as much information as possible. In some cases, they will have several questionnaires. Each questionnaire targets a specific type of problem, such as brake-related, drivability-related, and suspension-related.

"Cause" is where the actual fault that is causing the concern is documented. To determine that, you need to fully diagnose the concern, making sure you find the root cause of the concern. For example, in the previous topic, the TPMS warning lamp was on, which was the customer's concern. As we saw, the left front tire had low air pressure. This could definitely cause the TPMS lamp to be lit. If the technician had inflated the tire to the proper pressure, it would have turned the TPMS warning lamp off and solved the customer's problem. But it would have only fixed it temporarily until the air pressure in the tire leaked out again. So in that case, the technician did not stop the diagnosis when he found low tire pressure. He continued the diagnosis and investigated why the tire pressure was low. This was caused by a nail that had punctured the tire causing the leak. So in this situation, the root cause of the customer's concern was a nail in the tire, causing low tire pressure, which turned the TPMS warning lamp on. This then needs to be written up clearly, but concisely (**FIGURE 5-28**).

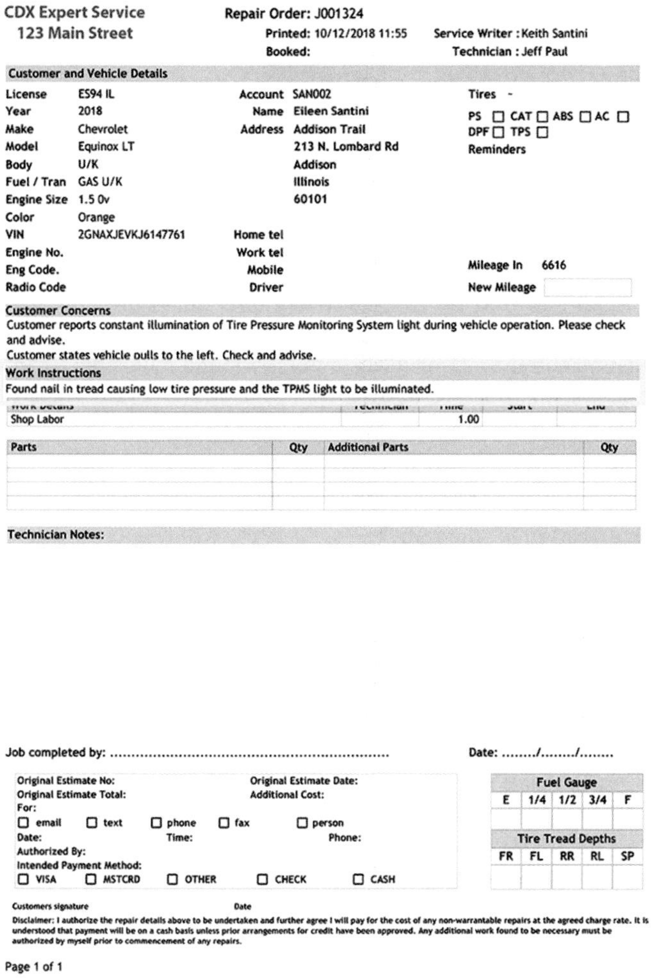
CDX Expert Service
123 Main Street

Repair Order: J001324
Printed: 10/12/2018 11:55 Service Writer : Keith Santini
Booked: Technician : Jeff Paul

Customer and Vehicle Details

License	ES94 IL	Account	SAN002	Tires	-
Year	2018	Name	Eileen Santini	PS ☐ CAT ☐ ABS ☐ AC ☐	
Make	Chevrolet	Address	Addison Trail	DPF ☐ TPS ☐	
Model	Equinox LT		213 N. Lombard Rd	Reminders	
Body	U/K		Addison		
Fuel / Tran	GAS U/K		Illinois		
Engine Size	1.5 0v		60101		
Color	Orange				
VIN	2GNAXJEVKJ6147761	Home tel			
Engine No.		Work tel			
Eng Code.		Mobile		Mileage In	6616
Radio Code		Driver		New Mileage	

Customer Concerns
Customer reports constant illumination of Tire Pressure Monitoring System light during vehicle operation. Please check and advise.
Customer states vehicle pulls to the left. Check and advise.

Work Instructions
Found nail in tread causing low tire pressure and the TPMS light to be illuminated.

Work Details	Technician	Time	Start	End
Shop Labor		1.00		

Parts	Qty	Additional Parts	Qty

Technician Notes:

Job completed by: .... Date: ..../......../........

Original Estimate No: Original Estimate Date:
Original Estimate Total: Additional Cost:
For:
☐ email ☐ text ☐ phone ☐ fax ☐ person
Date: Time: Phone:
Authorized By:
Intended Payment Method:
☐ VISA ☐ MSTCRD ☐ OTHER ☐ CHECK ☐ CASH

Fuel Gauge				
E	1/4	1/2	3/4	F

Tire Tread Depths				
FR	FL	RR	RL	SP

Customers signature Date
Disclaimer: I authorize the repair details above to be undertaken and further agree I will pay for the cost of any non-warrantable repairs at the agreed charge rate. It is understood that payment will be on a cash basis unless prior arrangements for credit have been approved. Any additional work found to be necessary must be authorized by myself prior to commencement of any repairs.

Page 1 of 1

FIGURE 5-28 Cause of the concern written up on the repair order.

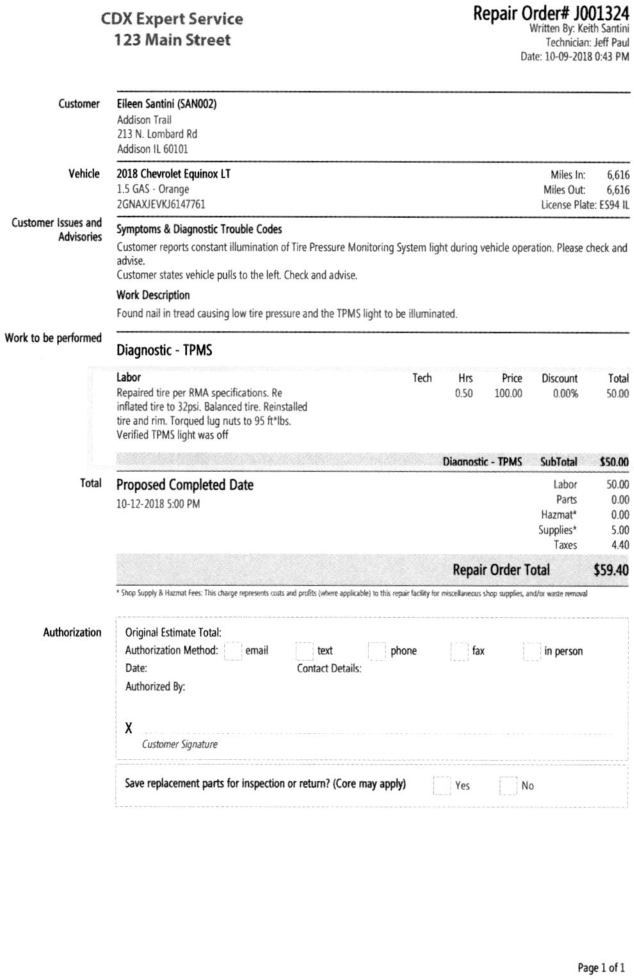
CDX Expert Service
123 Main Street

Repair Order# J001324
Written By: Keith Santini
Technician: Jeff Paul
Date: 10-09-2018 0:43 PM

Customer: Eileen Santini (SAN002)
Addison Trail
213 N. Lombard Rd
Addison IL 60101

Vehicle: 2018 Chevrolet Equinox LT
1.5 GAS - Orange
2GNAXJEVKJ6147761
Miles In: 6,616
Miles Out: 6,616
License Plate: ES94 IL

Customer Issues and Advisories: Symptoms & Diagnostic Trouble Codes
Customer reports constant illumination of Tire Pressure Monitoring System light during vehicle operation. Please check and advise.
Customer states vehicle pulls to the left. Check and advise.
Work Description
Found nail in tread causing low tire pressure and the TPMS light to be illuminated.

Work to be performed: Diagnostic - TPMS

Labor	Tech	Hrs	Price	Discount	Total
Repaired tire per RMA specifications. Re inflated tire to 32psi. Balanced tire. Reinstalled tire and rim. Torqued lug nuts to 95 ft*lbs. Verified TPMS light was off		0.50	100.00	0.00%	50.00

Diagnostic - TPMS SubTotal $50.00

Total: Proposed Completed Date
10-12-2018 5:00 PM

Labor	50.00
Parts	0.00
Hazmat*	0.00
Supplies*	5.00
Taxes	4.40

Repair Order Total $59.40

* Shop Supply & Hazmat Fees: This charge represents costs and profits (where applicable) to this repair facility for miscellaneous shop supplies, and/or waste removal

Authorization: Original Estimate Total:
Authorization Method: ☐ email ☐ text ☐ phone ☐ fax ☐ in person
Date: Contact Details:
Authorized By:
X
Customer Signature

Save replacement parts for inspection or return? (Core may apply) ☐ Yes ☐ No

Page 1 of 1

FIGURE 5-29 Actions performed to correct the fault.

Technicians always need to determine the root cause of a concern if the problem is to be fixed permanently. And the way you determine the root cause is by repeatedly asking yourself what caused the particular finding until you get to the root. And yes, you can go too far. In the leaking tire scenario, you could ask yourself, "Why is there a nail in the tire?," and the answer might be because the customer drove through a construction zone. Most of the time, you don't have to go that far unless the customer will encounter that situation regularly. You might consider the deeper questions if the customer comes in within a few weeks with another nail in a tire. At that point, telling him that this is the second roofing nail in his tire, he may decide that he should drive a different route home than through the construction zone.

"Correction" is where the description of what is needed (or was performed) to fix the customer's concern is documented. In the preceding situation, the correction would be: "Repaired tire with plug patch, inflated tire to 32 psi (220.63 kPa). Balanced wheel assembly, torqued lug nuts to 95 ft/lb (130 N-m). Verified that the TPMS lamp is off and the vehicle doesn't pull" (**FIGURE 5-29**). This step occurs once the problem and faults are fully understood and the fault assessed. It is done so that the correct parts can be replaced to complete a successful repair. Don't forget that to ensure a successful repair, the system should be thoroughly checked to verify that the fault has been corrected. It must be working properly before returning the vehicle to the customer. To apply the 3 C's, follow the steps in **SKILL DRILL 5-10**.

SKILL DRILL 5-10 Using the 3 C's to Document the Repair

1. Identify and document the customer's concern/s on the repair order. Obtain as much information as possible, as this will help the technician understand the problem. Ask questions like, "How long has the problem been occurring?" or "Does it occur at any particular time or temperature?" Clearly, concisely, and completely list the customer's concern/s.
2. Once the diagnosis has been completed and the root cause of the customer concern is identified, clearly, concisely, and completely list the exact cause of the fault, along with any readings performed which verify the cause.
3. List the actions needed (or performed) that will correct (or have corrected) the fault which is causing the customer concern.

▶ Wrap-Up

Ready for Review

- Vehicle information can come from several sources and are increasingly available through online services. Different information systems include owner's manual, shop or service manual, technical service bulletin, labor guide, and parts program.
- Vehicle information number is a unique serial number designed to identify motor vehicles of all kinds. The VIN is usually located behind the front left corner of the windshield and is made up of 17 characters.
  - North American VIN system, ISO Standard 3779
- A repair order provides the technician with a summary of the fault or needed service on the vehicle. Customer, vehicle, requested service, parts, and accounting sections should be filled to complete a repair order.
- Strategy-based diagnosis is a problem-solving plan that includes: verifying the customer's concern, researching possible faults and gathering information, focused testing, performing the repair, and verifying the repair.
  - The "3 C's" "concern," "cause," and "correction" provide a summary of the repair.

Key Terms

belt routing label A label that lists a diagram of the routing of the belt(s) for the engine accessories.

coolant label A label that lists the type of coolant installed in the cooling system.

labor guide A guide that provides information to make estimates for repairs.

lemon law A consumer protection law used in some states to identify a new vehicle that has undergone several unsuccessful attempts to repair the same fault.

lemon law buyback A vehicle that has been bought back by the manufacturer due to inability to repair the same fault within a specified number of attempts.

owner's manual An informational guide supplied by the manufacturer; it contains basic vehicle operating information.

parts program A computer software program for identifying and ordering replacement vehicle parts.

parts specialist The person who serves customers at the parts counter.

refrigerant label A label that lists the type and total capacity of refrigerant that is installed in the A/C system.

repair order A form used by shops to collect information regarding a vehicle coming in for repair; also referred to as a work order.

salvage title A record that a vehicle has been severely damaged or deemed a total loss by an insurance company; also called a branded title.

service campaign and recall A corrective measure conducted by manufacturers when a safety issue is discovered with a particular vehicle.

service history A complete list of all the servicing and repairs that have been performed on a vehicle.

shop or service manual Manufacturer or aftermarket information on the repair and service of vehicles.

strategy-based diagnosis A best practice diagnostic process that utilizes the same process every time.

technical service bulletins (TSBs) A bulletin from the vehicle manufacturer that includes commonly found faults in a particular system.

title history A detailed account of a vehicle's past.

vehicle emission control information (VECI) label A label used by technicians to identify engine and emission control information for the vehicle.

vehicle safety certification (VSC) label A label certifying that the vehicle meets the Federal Motor Vehicle Safety, Bumper, and Theft Prevention Standards in effect at the time of manufacture.

Review Questions

1. Owner's manuals contain important information about a vehicle. Which item can be found in the owner manual?
   a. A list of common faults
   b. Engine rebuilding specifications
   c. Maintenance information
   d. Emission control diagnosis
2. Shop manuals are broken into sections based on the vehicle systems. Which section would contain information on replacing the water pump?
   a. Engine system
   b. Diagnostic procedures
   c. Electrical systems
   d. Fluids and capacities
3. A vehicle has after-market wheels with wheel locks. When looking up the removal time in a labor guide, how would it be listed?
   a. Hazard time
   b. Optional time
   c. Standard time
   d. Additional time
4. The Vehicle Identification Number is essential to determine what options are on a vehicle. What else can the VIN provide?
   a. Engine identification
   b. Mileage information
   c. Owner information
   d. Service location identification
5. The Vehicle Safety Certification certifies that the vehicle meets FMVS standards. What other information does the VSC typically identify?
   a. Fluid capacities
   b. Maintenance recommendations
   c. Tire inflation pressures
   d. Crash test certification

6. A repair order contains a great deal of information about the vehicle and the customer. What else can be found on the work order?
   a. Listings for other local shops
   b. Accounting and payment information
   c. Customer satisfaction survey
   d. Shop safety procedures
7. Which of the following is NOT a step in a strategy-based diagnosis process?
   a. Focused testing
   b. Perform the repair
   c. Verify the repair
   d. Substitute parts
8. Labor guides generally provide information on:
   a. Repair specifications
   b. Standard or warranty-related repair time
   c. Service intervals
   d. Hourly rates for different shops
9. The repair order is a legal document and may be used in court. What information is required on the work order to be legal?
   a. Date of vehicle purchase
   b. Dealership of original purchase
   c. Warranty history from other dealers
   d. Complete vehicle information
10. Many technicians use a consistent format, 3 C's, for documentation of information on work orders. What does the 3 C's represent?
   a. Correction, concern, comments
   b. Concern, cause, correction
   c. Customer, cause, concern
   d. Concern, correction, clear

ASE Technician A/Technician B Style Questions

1. Technician A says that in the past, service information was only listed in books. Technician B says that online service information is usually better because publishers can update it when needed. Who is correct?
   a. Technician A
   b. Technician B
   c. Both A and B
   d. Neither A nor B
2. Technician A says that scheduled service charts list specifications such as spark plug gap. Technician B says that scheduled service charts list the service required at various mileage intervals. Who is correct?
   a. Technician A
   b. Technician B
   c. Both A and B
   d. Neither A nor B
3. Technician A says that the VIN provides information related to seat restraints. Technician B says that the VIN can be used to check the service history for the vehicle. Who is correct?
   a. Technician A
   b. Technician B
   c. Both A and B
   d. Neither A nor B
4. Technician A says that a repair order contains information documenting the work performed on a vehicle. Technician B says that the repair order lists the steps a technician will need to follow to diagnose the vehicle. Who is correct?
   a. Technician A
   b. Technician B
   c. Both A and B
   d. Neither A nor B
5. Technician A says that looking at the vehicle service history can help diagnose a vehicle fault. Technician B says that service history can be found by looking it up in shop manuals. Who is correct?
   a. Technician A
   b. Technician B
   c. Botha A and B
   d. Neither A nor B
6. Technician A says that a strategy-based diagnostic procedure means you only have to do one test to determine the problem. Technician B says that using a consistent process to diagnose a vehicle can simplify the diagnosis. Who is correct?
   a. Technician A
   b. Technician B
   c. Both A and B
   d. Neither A nor B
7. Technician A says to verify the repair is complete before returning the vehicle. Technician B says that the root cause of the problem should be determined. Who is correct?
   a. Technician A
   b. Technician B
   c. Both A and B
   d. Neither A nor B
8. Technician A says that a Technical Service Bulletin may contain updated service or repair procedures. Technician B says that TSBs are only issued once the vehicle is out of warranty. Who is correct?
   a. Technician A
   b. Technician B
   c. Both A and B
   d. Neither A nor B
9. Technician A says to have a customer initial or sign the repair order if repairs are declined. Technician B says that if a customer authorizes repairs, document it on the work order. Who is correct?
   a. Technician A
   b. Technician B
   c. Both A and B
   d. Neither A nor B
10. Technician A says that the production date is the date the vehicle was manufactured. Technician B says that the model year may be different from the production date. Who is correct?
   a. Technician A
   b. Technician B
   c. Both A and B
   d. Neither A nor B

CHAPTER 6

Hand and Measuring Tools

Learning Objectives

- **LO 6-01** Identify basic hand tool safety.
- **LO 6-02** Identify basic wrenches and sockets.
- **LO 6-03** Identify other basic hand tools.
- **LO 6-04** Identify basic hammers and struck tools.
- **LO 6-05** Identify basic taps, dies, and specialty tools.
- **LO 6-06** Measure precisely using measuring tools.
- **LO 6-07** Measure precisely using other measuring tools.

ASE Education Foundation Tasks

See Appendix A to view the 2017 ASE Education Foundation Automobile Accreditation Task List Correlation Guide.

▶ Hand Tool Overview

LO 6-01 Identify basic hand tool safety.

In this section, we explore a variety of hand tools, equipment, and precision-measuring tools that are fundamental to your success as an automotive technician. They provide the means for work to be undertaken on vehicles. This includes lifting, diagnosing, removing, cleaning, inspecting, installing, and adjusting. Nearly all shop tasks involve the use of some sort of tool or a piece of equipment (**FIGURE 6-1**). This makes their purchase, use, and maintenance very important to the overall performance of the shop.

In fact, most tool purchases are considered an investment because they generate income when they are used. For example, a tool you buy for $100 might save you two hours of working time every time you use it. You only need to use it a couple of times in order for it to pay for itself. Then, every time after that, it pays you a bonus. This means that you need to treat tools like your own personal moneymakers.

One way to treat tools right is to always use them in the way they are designed to be used. Don't abuse them. Think about the task at hand and identify the most effective tools to do the task. Inspect the tool before using it and use it correctly. When finished, clean it, inspect it, and store it in the correct location. Doing all of these things ensures that your tools will be available the next time you want to use them. They will also last a long time.

FIGURE 6-1 Technicians rely on tools to perform work.

▶ TECHNICIAN TIP

A sticker on the dash of a work truck I once used read: "Treat me well—your paycheck depends on it. If I can't work, neither can you!" That goes for all tools and equipment. Your ability to do work as a technician will either be enhanced or reduced by the condition of the tools and equipment you use.

General Safety Guidelines

Although it is important to be trained on the safe use of tools and equipment, it is even more critical to have a safe attitude. A safe attitude will help you avoid being involved in an accident. Students who think they will never be involved in an accident will not be as aware of unsafe situations as they should be. And that can lead to accidents. So while we are covering the various tools and equipment you will encounter in the shop, pay close attention to the safety and operation procedures. Tools are a technician's best friend, but if used improperly, they can injure or kill (**FIGURE 6-2**).

Work Safe and Stay Safe

Whenever using tools, always think safety first. There is nothing more important than your personal safety. If tools (both hand and power) are used incorrectly, you can potentially injure yourself and others. Always follow tool/equipment instructions and all shop policies. This includes using the recommended personal protective equipment (PPE). Accidents only take a second or two to happen but can take a lifetime to recover from. You are ultimately responsible for your safety, so remember to work safe and stay safe.

Safe Handling and Use of Tools

Tools must be safely handled and used to prevent injury and damage. Always inspect tools prior to use, and never use damaged tools. Check the manufacturer and the shop procedures, or ask your supervisor if you are uncertain about how to use tools. Inspect and clean the tools once you finish using them. Always return the tools to their correct storage location.

To safely handle and use appropriate tools, follow **SKILL DRILL 6-1**.

You Are the Automotive Technician

After finishing work on the last vehicle of the day, you are required to return your workstation back to order. You clean, inspect, and return tools and equipment to their designated place. You wipe up any spills, according to the shop procedure, and clear the floor of any debris, to avoid slips and falls. During your workspace inspection, you determine that the insulation on the droplight cord is frayed. There are also some tools that need to be cleaned and put away.

1. What needs to happen with the droplight?
2. What should you do to with a micrometer before storing it?
3. How do you check a micrometer for accuracy?
4. Describe a double flare and how it is different from a single flare.

FIGURE 6-2 Tools used improperly can injure or kill.

SKILL DRILL 6-1 Safe Handling and Use of Tools

1. Select the correct tool(s) to undertake tasks.

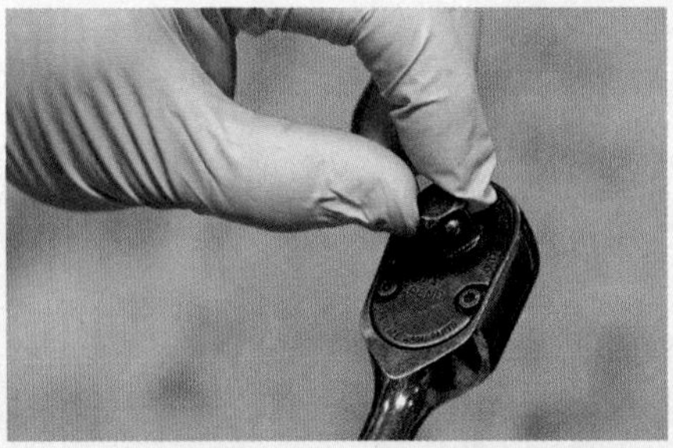

2. Inspect tools prior to use to ensure they are in good working order. If tools are faulty, remove them from service according to shop procedures.

3. Clean tools prior to use if necessary.

4. Use tools to complete the task while ensuring manufacturer and shop procedures are followed. Always use tools safely to prevent injury and damage.

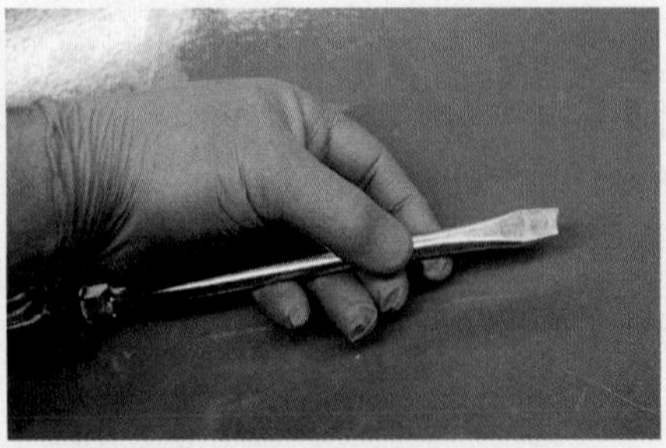

5. Ensure tools are clean and in good working order after use. Report and tag damaged tools, and remove them from service, following shop procedures.

6. Return tools to correct storage locations.

SKILL DRILL 6-2 Safe Procedures for Handling Tools and Equipment

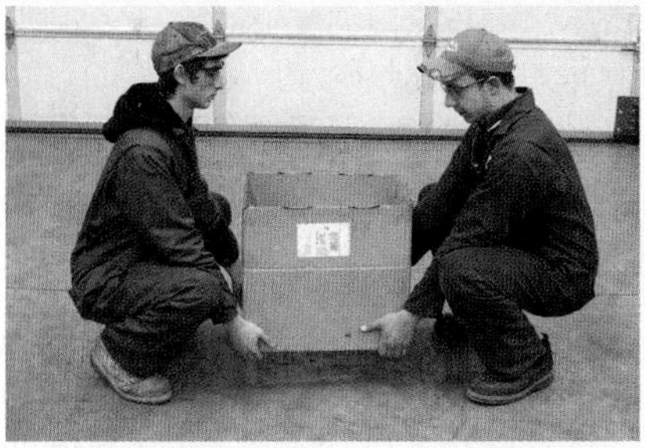

1. Seek assistance if tools and equipment are too heavy or too awkward to be managed by a single person.

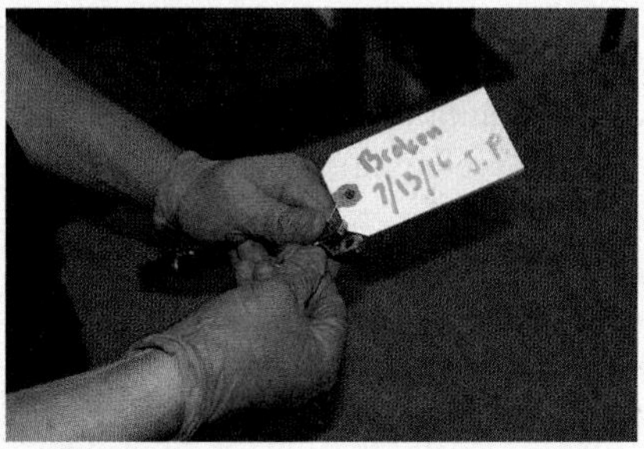

2. Inspect the tools and equipment for possible defects before starting work. Report and/or tag faulty tools and equipment according to shop procedures.

3. Select and wear appropriate PPE for the tools and equipment being used.

4. Use tools and equipment safely.

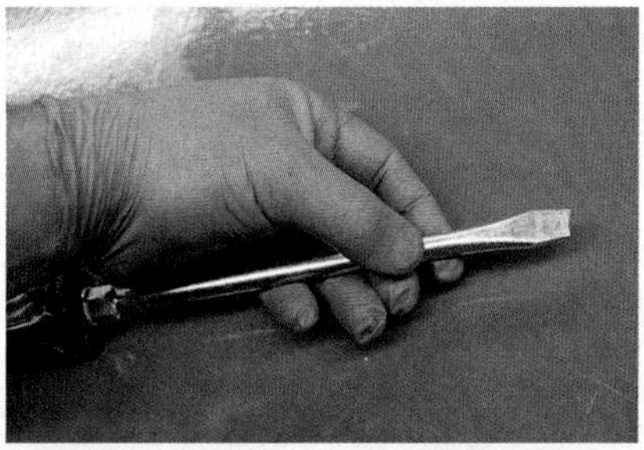

5. Check tools for faults after using them and report and/or tag faulty tools and equipment according to shop procedures.

6. Clean and return the tools and equipment to correct storage locations when tasks are completed.

Safe Procedures for Handling Tools and Equipment

Some tools are heavy or awkward to use, so seek assistance if required. Be sure to use correct handling techniques when using tools. To utilize safe procedures for handling tools and equipment, follow the steps in **SKILL DRILL 6-2**.

Tool Usage

Tools extend our abilities to perform many tasks. For example, jacks, stands, and hoists extend our ability to lift and hold heavy objects. Hand tools extend our ability to perform fundamental tasks like gripping, turning, tightening, measuring, and cutting (**FIGURE 6-3**). Electrical meters enable us to measure things we cannot see, feel, or hear. Power and air tools multiply our strength by performing tasks quickly and efficiently. As you are working, always think about what tool can make the job easier, safer, or more efficient. As you become familiar with more tools, your productivity, quality of work, and effectiveness will improve.

Every tool is designed to be used in a certain way to do the job safely. It is critical to use a tool in the way it was designed to be used and to do so safely. For example, a screwdriver is designed to tighten and loosen screws, not to be used as a chisel. Ratchets are designed to turn sockets, not to be used as a hammer. Think about the task you are undertaking, select the correct tools for the task, and use each tool as it was designed.

FIGURE 6-3 Tools extend our abilities.

> ▶ **TECHNICIAN TIP**
>
> Using tools can make you much more efficient and effective in performing your job. Without tools, it would be very difficult to carry out vehicle repairs and servicing. It is also the reason that many technicians invest well over $20,000 in their personal tools. If purchased wisely, tools can help you perform more work in a shorter period of time. This makes you more money. So think of your tools as an investment that pays for themselves over time.

Lockout/Tag-Out

Lockout/tag-out is an umbrella term that describes a set of safety practices and procedures. They are intended to reduce the risk to technicians who might inadvertently use unsafe tools, equipment, or materials. This means that any tools/equipment that are determined to be potentially unsafe, or that are being serviced, need to be locked out and/or tagged out.

An example of lockout would be physically securing a broken, unsafe, or out-of-service tool so that it cannot be used by a technician. In many cases, the item is also tagged out so it is not inadvertently placed back into service. An example of tag-out would be affixing a clear and obvious label to a piece of equipment. It describes the:

- fault found,
- the name of the person who found the fault,
- the date that the fault was found, and
- warns not to use the equipment (**FIGURE 6-4**).

Never attempt to use a tool or any equipment that has been tagged out or locked out.

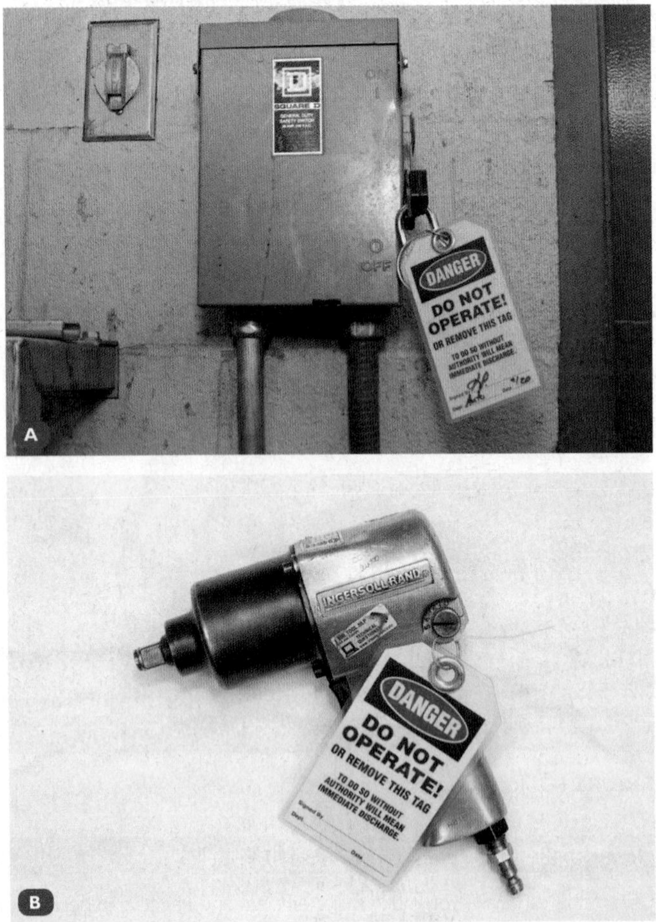

FIGURE 6-4 **A.** An example of lockout would be physically locking out a tool or piece of equipment so that it cannot be accessed and used by someone who may be unaware of the potential danger of doing so. **B.** An example of tag-out would be affixing a clear and obvious label to a piece of equipment that describes the fault found and that warns not to use the equipment.

SAFETY TIP

Standardized lockout/tag-out procedures are a mandatory part of workplace safety regulations in most countries. Familiarize yourself with the specific lockout/tag-out practices that apply in your workplace.

Tool Storage

Typically, technicians have a selection of their own tools. These include various hand tools, air tools, measuring tools, and electrical meters (**FIGURE 6-5**). Often they add tools to their toolbox over time. These tools are kept in a toolbox that can be rolled around as needed. Tool boxes are typically kept in the technician's work stall.

Speaking of toolboxes, technicians need to invest in a quality toolbox to keep their tools secure. They should have enough capacity to hold all current and future tools. They should be built to last because a toolbox is in continuous use throughout the workday. The drawers should have enough capacity to handle the weight of the tools in it. Also, it should open and close easily. Drawer slides that use bearings make them much easier to open and close (**FIGURE 6-6**). The toolbox should also be

FIGURE 6-5 Typical technician's toolbox.

FIGURE 6-6 Toolbox drawers need to open and close easily.

SKILL DRILL 6-3 Manufacturer's Special Tools

To identify tools and their usage in automotive applications, follow these steps:

1. Create a list of tools in your toolbox and identify their application for automotive repair and service.
2. Look through the shop's tool storage areas, and create a list of the tools found in each storage area.
3. Identify their application for automotive repair and service.

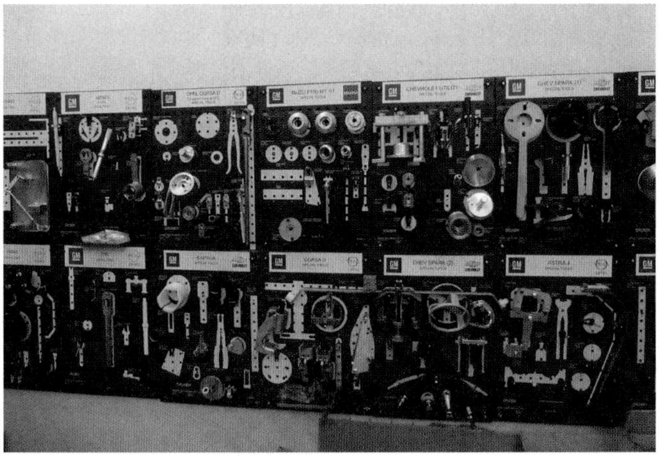

FIGURE 6-7 Manufacturer's special tools.

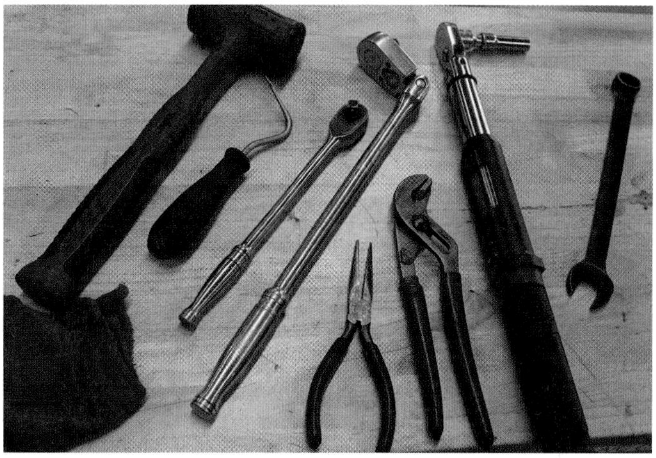

FIGURE 6-8 Hand tools.

equipped with an easy-to-use locking system to secure the tools when they are not being used.

The shop usually has a selection of specialty tools and equipment that are available for technicians to use on a shared basis. These tools are located in specific areas around the shop. Typically they are placed in a centralized location so that they are relatively easy to access (**FIGURE 6-7**). They include:

- specialized manufacturer tools, such as pullers or installation tools;
- high-cost tools such as factory scan tools; and
- tools that are not portable, such as hoists and tire machines.

Specialty tools and equipment are shared by everyone in the shop. So it is critical that they are inspected and stored properly after each use. This way, they will be available for the next technician who needs them. Shared tools can cause a real problem in the shop if people do not treat them well or return them to their proper place (Figure 6-7).

▶ Basic Hand Tools

LO 6-02 Identify basic wrenches and sockets.

Like all tools, hand tools extend our ability to do work. Hand tools come in a variety of shapes, sizes, and functions (**FIGURE 6-8**). A large percentage of your personal tools will be hand tools. Over the years, manufacturers have introduced new parts that require their own different types of hand tools. Some examples of parts that manufacturers are changing are fasteners, wire harness terminals, and quick-connect fittings. This means that technicians need to continually add updated tools to their toolbox.

▶ TECHNICIAN TIP

Invest in quality tools. Because tools extend your abilities, poor-quality tools affect the quality and quantity of your work. Price is not always the best indicator of quality, but it plays a role. As you learn the purpose and function of the tools in this chapter, you should be able to identify high-quality tools versus poor-quality tools. Do this by looking at them, handling them, and putting them to work.

Standard and Metric Designations

Tools, measuring instruments, and fasteners come in two size designations. The first is the United States customary system (USCS), commonly referred to as "standard." The second uses the metric system. Tools and measuring instruments can be identified as standard or metric by their size markings. New **fasteners** have their designation identified on the packaging. Other fasteners may have to be measured to identify their designation. Manufacturer's charts showing thread and fastener sizing assist in determining if bolts are standard or metric.

To identify standard and metric designation, follow these steps:

1. Examine the component, tool, or fastener to see if any marking identifies it as standard or metric (**FIGURE 6-9**).
2. If not, refer to manufacturer specifications to identify components as standard or metric.

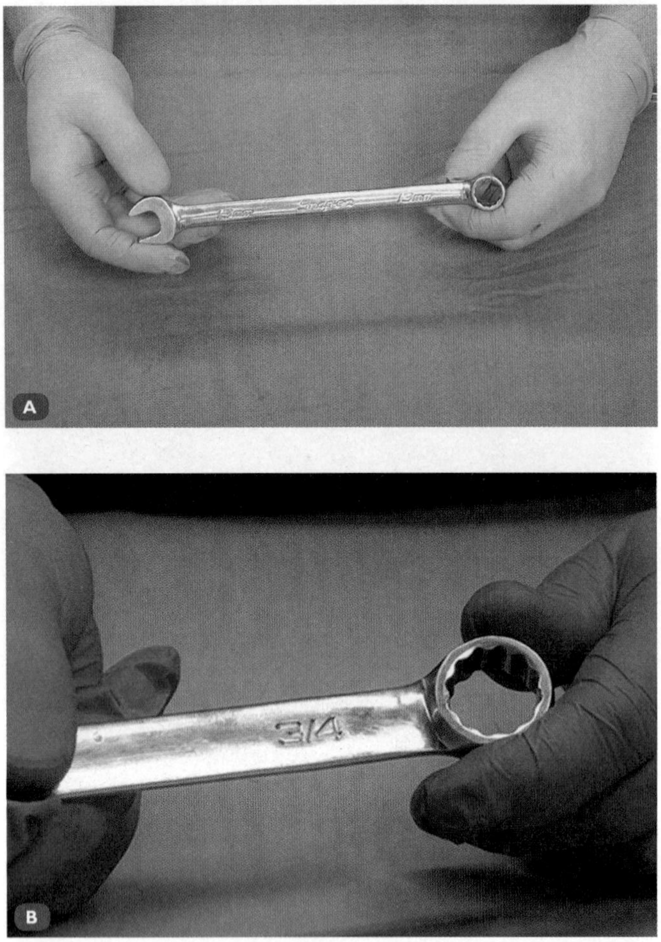

FIGURE 6-9 Tools come in both **A.** Metric and **B.** Standard sizes.

3. If no markings or specifications are available, use measuring devices to gauge the size of the item. Then compare to thread and fastener charts to identify the sizing. Inch-to-metric conversion charts assist in identifying component designation (**FIGURE 6-10**).

Wrenches

Wrenches are used to tighten and loosen nuts and **bolts**, which are two types of fasteners. There are three commonly used wrenches: the **box-end wrench**, the **open-end wrench**, and the **combination wrench** (**FIGURE 6-11**). The box-end wrench fits fully around the head of the bolt or nut. It grips each of the six points at the corners, just like a socket. This is exactly the sort of grip needed if a nut or bolt is very tight. It makes the box-end wrench less likely to round off the points on the head of the bolt than the open-end wrench.

The ends of box-end wrenches are bent or offset. They are easier to grip and have different-sized heads at each end (**FIGURE 6-12**). One disadvantage of the box-end wrench is that it can be awkward to use once the nut or bolt has been loosened a bit. This is because you have to lift it off the head of the fastener and move it to each new position.

The open-end wrench is open on the end, and the two parallel flats only grip two points of the fastener. Open-end wrenches either have different-sized heads on each end of the wrench, or they have the same size, but with different angles (**FIGURE 6-13**). The head is at an angle to the handle and is not bent or offset, so it can be flipped over and used on both sides. This is a good wrench to use in very tight spaces as you can flip it over and get a new angle. This allows the head to catch new points on the fastener.

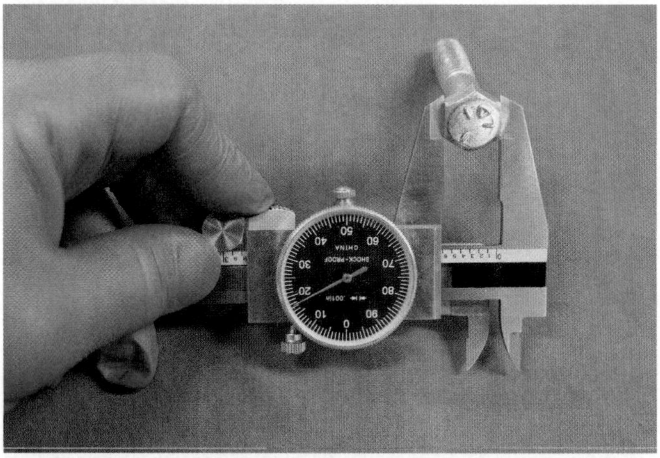

FIGURE 6-10 If no markings or specifications are available, use measuring devices to gauge the size of the item. Then compare thread and fastener to charts to identify the sizing.

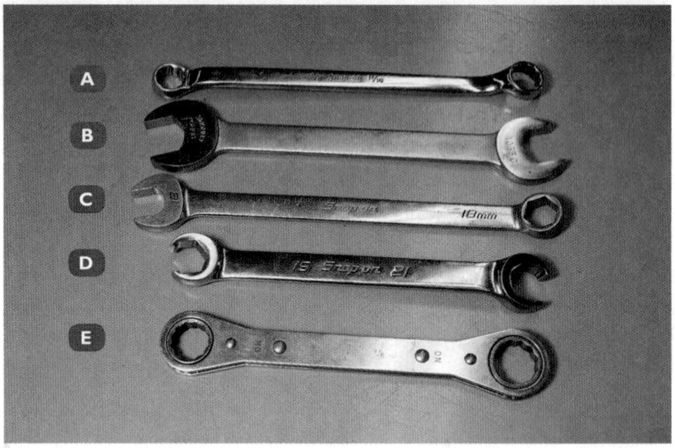

FIGURE 6-11 **A.** Box-end wrench. **B.** Open-end wrench. **C.** Combination wrench. **D.** Flare nut wrench. **E.** Ratcheting box-end wrench.

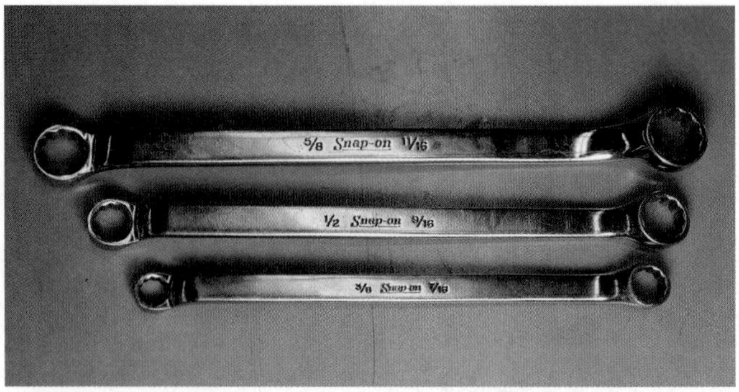

FIGURE 6-12 Box-end wrenches.

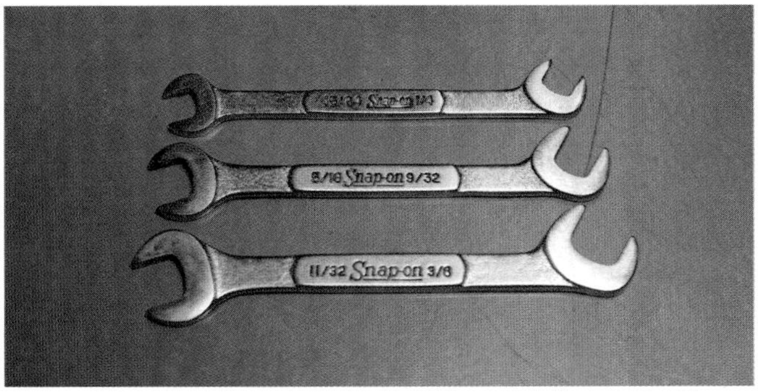

FIGURE 6-13 Types of open-end wrenches.

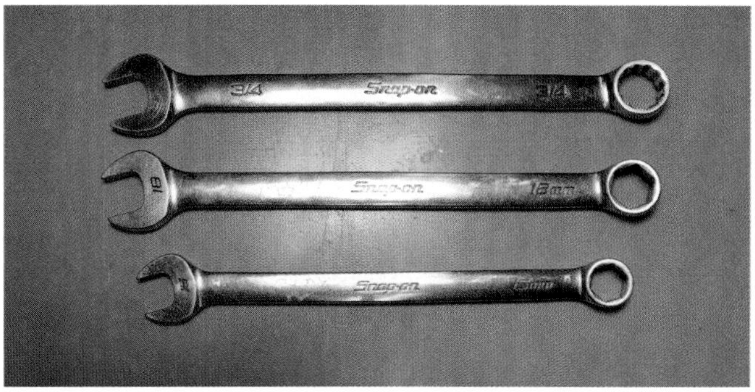

FIGURE 6-14 Combination wrenches.

Although an open-end wrench often gives the best access to a fastener, if the fastener is extremely tight, the open-end wrench should not be used. Since this type of wrench only grips two points, the wrench can suddenly slip if the jaws flex. This slippage can round off the points of the fastener. The best way to approach a tight fastener is to use a box-end wrench to break the bolt or nut free, and then use the open-end wrench to finish the job. The open-end wrench should only be used on fasteners that are no more than moderately tightened.

▶ TECHNICIAN TIP

Wrenches (which are also known as spanners in some countries) will only do a job properly if they are the right size for the given nut or bolt head. The size used to describe a wrench is the distance across the flats of the nut or bolt. There are two systems in common use—standard (in inches) and metric (in millimeters). Each system provides a range of sizes. They are identified by a fraction, which indicates fractions of an inch for the standard system. Or they are identified by a number, which indicates millimeters for the metric system. In most cases, standard and metric tools cannot be used interchangeably on a particular fastener. Even though a metric tool fits a standard sized fastener, it may not grip as tight, which could cause it to round the head of the fastener off.

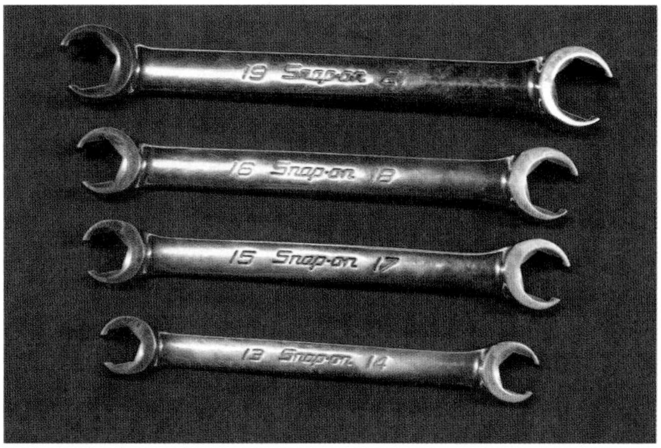

FIGURE 6-15 Flare nut wrenches.

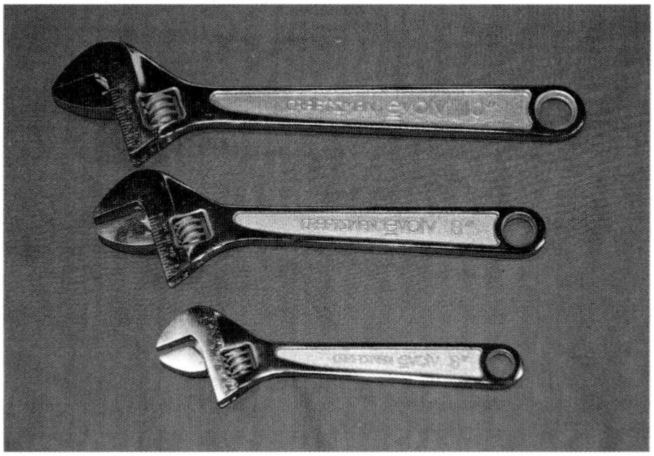

FIGURE 6-16 Open-end adjustable wrench.

The combination wrench has an open-end head on one end and a box-end head on the other end (**FIGURE 6-14**). Both ends are usually of the same size. That way the box-end wrench can be used to break the bolt loose, and the open end can be used for turning the bolt. Because of its versatility, this is probably the most popular wrench for technicians.

A variation on the open-end wrench is the **flare nut wrench**, also called a flare tubing wrench, or line wrench. It gives a better grip than the open-end wrench because it grabs five of the six points of the fastener, not two (**FIGURE 6-15**). However, because it is open on the end, it is not as strong as a box-end wrench. The partially open side lets the wrench be placed over the tubing or pipe so the wrench can be used to turn the tube fittings. Do not use the flare nut wrench on extremely tight fasteners as the jaws may still spread, damaging the nut.

Another open-end wrench is the open-end adjustable wrench, or crescent wrench. This wrench has a movable jaw that can be adjusted by turning an adjusting screw to fit any fastener within its range (**FIGURE 6-16**). It should only be used if other wrenches are not available. Because it is not as strong as a fixed wrench, it can slip off and damage the head of tight bolts or nuts. Still, it is a handy tool to have because it can be adjusted to fit most any fastener size.

A **ratcheting box-end wrench** is a useful tool in some applications because it does not require removal of the tool to reposition it. It has an inner piece that fits over and grabs the

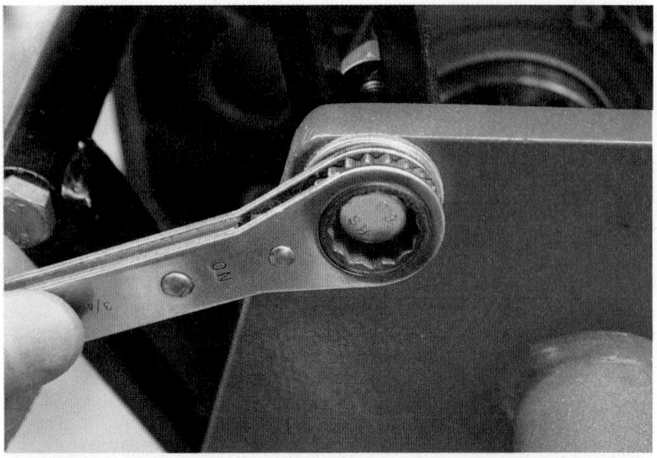

FIGURE 6-17 Ratcheting box-end wrench.

FIGURE 6-18 Ratcheting open-end wrench.

points on the fastener. The inner piece is able to rotate within the outer housing. A ratcheting mechanism lets it rotate in one direction and lock in the other direction. In some cases, the wrench just needs to be flipped over to be used in the opposite direction. In other cases, it has a lever that changes the direction from clockwise to counterclockwise (**FIGURE 6-17**). Just be careful to not overstress this tool by using it to tighten or loosen very tight fasteners, as the outer housing is not very strong.

There is also a ratcheting open-end wrench, but it uses no moving parts. One of the sides is partially removed so that only the bottom one-third remains to catch a point on the bolt (**FIGURE 6-18**). When it is used, the normal side works just like a standard open-end wrench. The shorter side of the open-end wrench catches the point on the fastener so it can be turned. When moving the wrench to get a new bite, the wrench is pulled slightly outward. This disengages the short side while leaving the long side to slide along the faces of the bolt. The wrench is then rotated to the new position and pushed back in so that the short side engages the next point. This wrench, like other open-end wrenches, is not designed to tighten or loosen tight fasteners. But it does work well in blind places where a socket or ratcheting box-end wrench cannot be used.

FIGURE 6-19 Pipe wrench.

FIGURE 6-20 Oil filter wrenches.

Special Wrenches

Specialized wrenches such as the **pipe wrench** grip pipes and can exert a lot of force to turn them (**FIGURE 6-19**). Because the handle pivots slightly, the more pressure put on the handle to turn the wrench, the more the grip tightens. The jaws are hardened and serrated. So increasing the pressure also increases the risk of marking or even gouging metal from the pipe. The jaw is adjustable, so it can be threaded in or out to fit different pipe sizes. Also, they come in different lengths, allowing you to increase the leverage applied to the pipe.

A specialized wrench called an **oil filter wrench** grabs the filter to allow you to remove it. These wrenches are available in various designs and sizes (**FIGURE 6-20**). Some are adjustable to fit many filter sizes. Also, note that an oil filter wrench should be used *only* to remove an oil filter, never to install it. Almost all oil filters should be installed and tightened by hand.

Using Wrenches Correctly

Choosing the correct wrench for a job usually depends on two things. The first is how tight the fastener is. The second is, how much room there is to get the wrench onto the fastener, and then to turn it. When being used, it is always possible that a wrench could slip. Before putting a lot of tension on the wrench, try to anticipate what will happen if it does slip. If possible, it is usually better to pull a wrench toward you than to push it away (**FIGURE 6-21**). If you have to push, use an open palm to push, so your knuckles won't get crushed if the wrench slips. If pulling toward yourself, make sure your face is not close to your hand or in line with your pulling motion. Many technicians have punched themselves in the face when the wrench slipped.

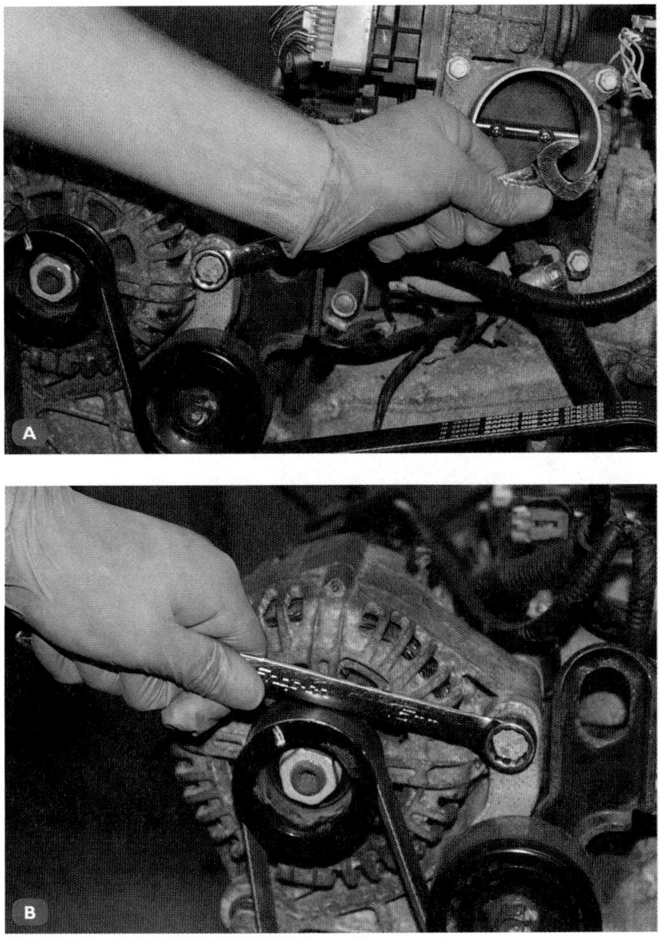

FIGURE 6-21 A. Pulling on a wrench is generally better than pushing on a wrench. **B.** Pushing regularly results in bruised or broken knuckles.

Sockets

Sockets are very popular because of their adaptability and ease of use (**FIGURE 6-22**). Sockets are a good choice where the top of the fastener is reasonably accessible. The **socket** fits onto the fastener snugly and grips it on all six corners. This provides the type of grip needed on any nut or bolt that is extremely tight. Sockets come in a variety of configurations, and technicians usually have a lot of sockets so they can get in a variety of tight places. Individual sockets fit a particular size nut or bolt, so they are usually purchased in sets.

Sockets are classified by the following characteristics:

- Standard or metric
- Size of drive used to turn them: 1/2" (12.7 mm), 3/8" (9.525 mm), and 1/4" (6.35 mm) are most common; 1" (25.4 mm) and 3/4" (19.05 mm) are less common.
- Number of points: 6 and 12 are most common; 4 and 8 are less common.
- Depth of socket: Standard and deep are most common; shallow is less common.
- Thickness of wall: Standard and impact are most common; thin wall is less common.

Sockets are built with a recessed square drive that fits the square drive of the ratchet or other driver (**FIGURE 6-23**). The size of the drive determines how much twisting force can be applied to the socket. The larger the drive, the larger the twisting force. Small fasteners usually only need a small torque, so having too large of a drive may result in a situation that the socket cannot gain access to the bolt.

For fasteners that are really tight, an impact wrench exerts a lot more torque on a socket than turning it by hand. Impact sockets are usually thicker walled than standard wall sockets. They have six points so they can withstand the forces generated by the impact wrench. It also allows them to grip the fastener securely (**FIGURE 6-24**).

Six- and 12-point sockets fit the heads of hexagonal-shaped fasteners. Four- and 8-point sockets fit the heads of square-shaped

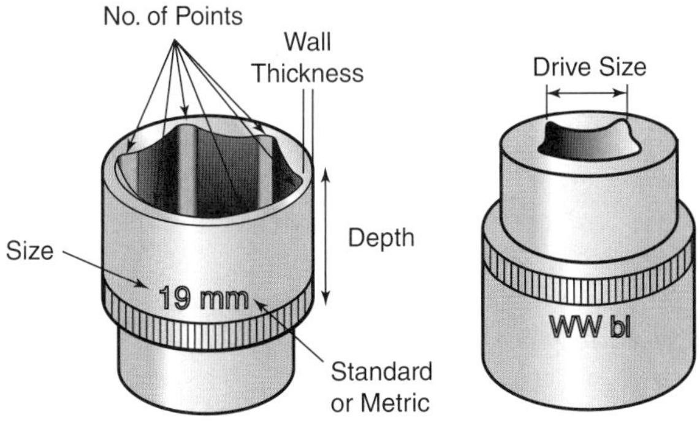

FIGURE 6-22 The anatomy of a socket.

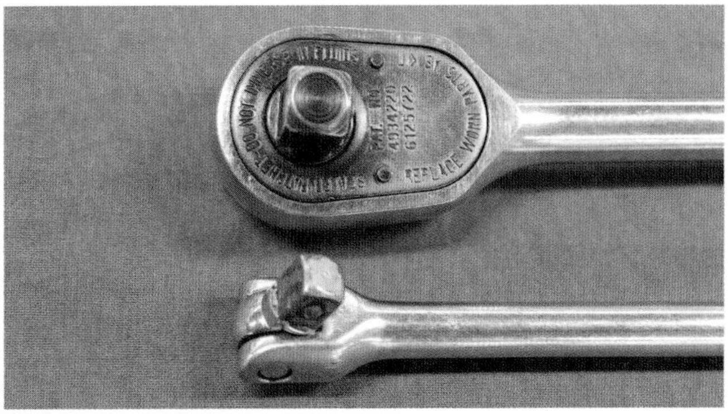

FIGURE 6-23 Sockets are designed to fit a matching drive on a ratchet.

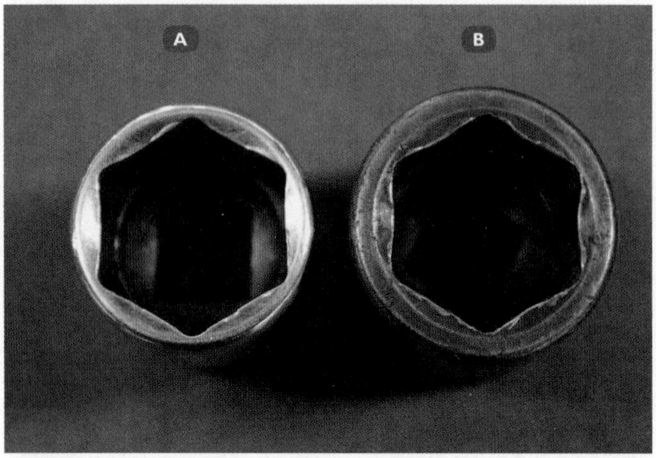

FIGURE 6-24 **A.** Standard wall socket. **B.** Impact socket.

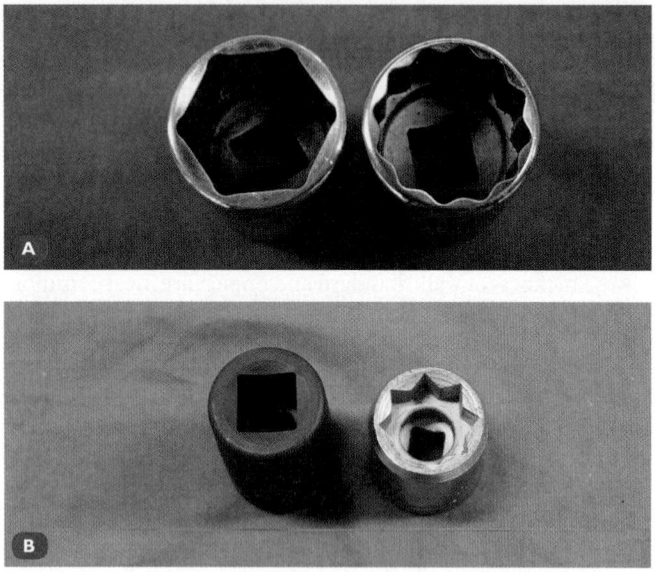

FIGURE 6-25 **A.** Six- and 12-point sockets. **B.** Four- and 8-point sockets.

FIGURE 6-26 **A.** Deep socket. **B.** Standard-length socket.

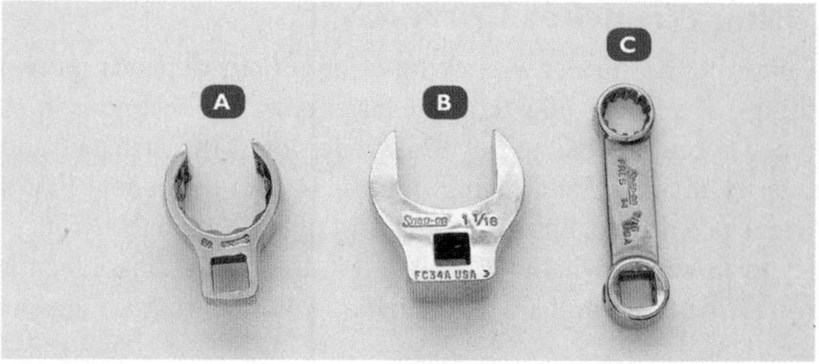

FIGURE 6-27 Crow's foot wrenches. **A.** Open-end. **B.** Box-end. **C.** Flare-nut.

FIGURE 6-28 **A.** Allen socket. **B.** Torx socket (Internal, external, and tamper-proof). **C.** Screwdriver socket.

fasteners (**FIGURE 6-25**). Because 6-point and 4-point sockets fit the exact shape of the fastener, they have the strongest grip on the fastener. They only fit on the fastener in half as many positions as a 12- or 8-point socket, though. This makes them harder to fit onto the fastener in places where the ratchet handle is restricted.

Another factor in accessing a fastener is the depth of the socket. If a nut is threaded quite a way down a stud or bolt, then a standard-length socket will not fit far enough over the stud to reach the nut (**FIGURE 6-26**). In this case, a deep socket will usually reach the nut.

Other tools are made with a square drive and designed to be used in place of sockets. Some of those are crow's foot wrenches, which can be open-end, box-end, or flare-nut wrenches (**FIGURE 6-27**). Other accessory sockets include Allen sockets, Torx sockets, and screwdriver sockets (**FIGURE 6-28**).

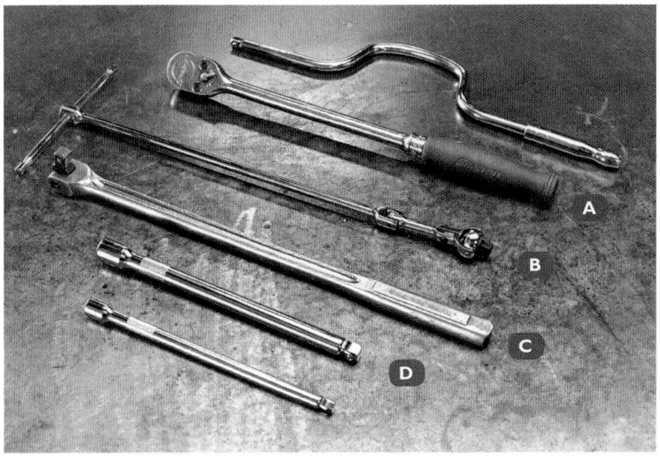

FIGURE 6-29 Tools to turn sockets. **A.** Ratchet. **B.** T-handle. **C.** Breaker bar. **D.** Extension.

FIGURE 6-30 A course-tooth ratchet is stronger than a fine-tooth ratchet, but takes more room to swing the handle for it to catch the next tooth.

Sockets are usually purchased in sets, with each set providing a different capability. You can see why technicians could have several hundred sockets in their toolbox. Having a variety of sockets allows the technician to do jobs easier and quicker. This makes them a good investment.

Socket Drivers

Turning a socket requires a handle and other accessories (**FIGURE 6-29**). The most common socket handle, the **ratchet**, makes the work of tightening or loosening a nut easy where not a lot of force is involved. It can be set to turn in either direction and does not need much room to swing. Although built to be conveniently used, it is not super strong, so too much force could strip the ratchet mechanism. To help overcome this, ratchet mechanisms come in two types. One is a coarse-tooth ratchet which is strong, but takes more room to swing the handle for it to catch the next tooth. And another is a fine-tooth ratchet which can be used when the room to swing the handle is small (**FIGURE 6-30**).

FIGURE 6-31 Extensions can be used together to gain access needed for tight situations.

For heavier tightening or loosening, a breaker bar gives the most leverage. When that is not available, a **sliding T-handle** may be more useful. With this tool, both hands can be used, and the position of the T-piece is adjustable to clear any obstructions when turning it. The connection between the socket and the accessory is made by a square drive. The larger the drive, the heavier and bulkier the socket will be. The 1/4" (6.35 mm) drive is for small work in difficult areas. The 3/8" (9.525 mm) drive accessories handle a lot of general work where torque requirements are not too high. The 1/2" (12.7 mm) drive is required for all-around service. The 3/4" (19.05 mm) and 1" (25.4 mm) drives are required for large work with high torque settings.

Many fasteners are located in positions where access can be difficult. There are many different lengths of extensions available. This allows the socket to be on the fastener while extending the drive point out to where a handle can be attached. Because extensions come in various lengths, they can be connected together. This allows you to get just the right length needed for a particular situation (**FIGURE 6-31**).

If an object is in the way of getting a socket on a fastener, a flexible joint can be used. This applies the turning force to the socket through an angle. The socket can still be turned, even though you are no longer directly in line with the fastener. There are five common types of flexible joints: U-joint style, CV style, wobble extension, cable extension, and flex socket (**FIGURE 6-32**). The flex socket has the universal joint built into it, so its overall length is shorter. This makes it able to get into tighter spaces. In some situations, you may need to use more than one flexible joint to get around objects that are in the way. This is especially true when removing some bell housing bolts on some transmissions.

A **speed brace** or speeder handle is the fastest way to spin a fastener on or off a thread by hand. It cannot apply much torque to the fastener, though. It is mainly used to remove a fastener that has already been loosened. It can also be used to run the fastener onto the thread until it begins to tighten (**FIGURE 6-33**).

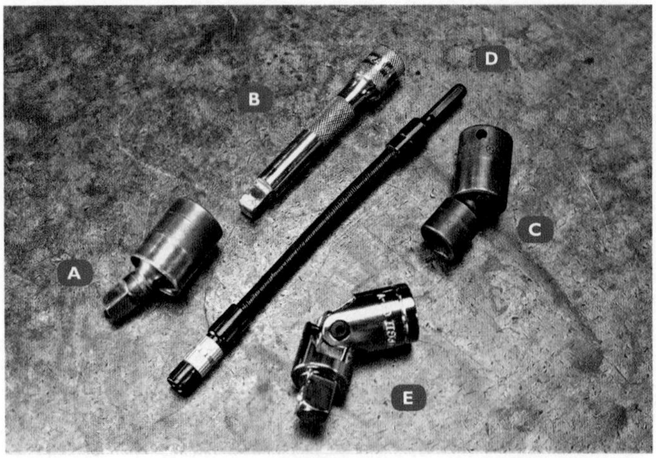

FIGURE 6-32 Flexible extensions. **A.** U-joint style. **B.** CV joint style. **C.** Wobble extension style. **D.** Cable extension style. **E.** Flex socket style.

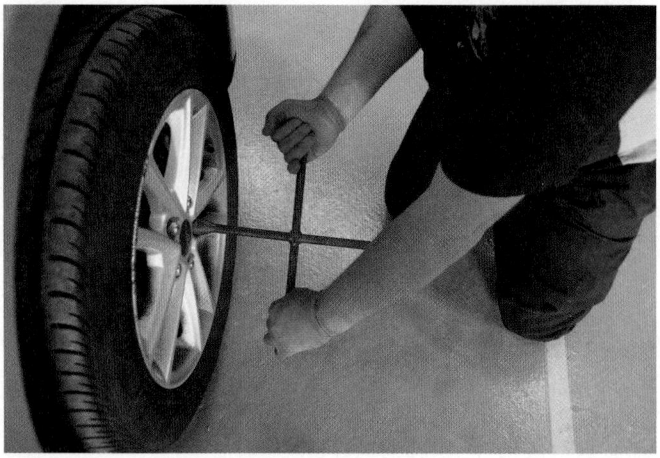

FIGURE 6-34 Lug wrench.

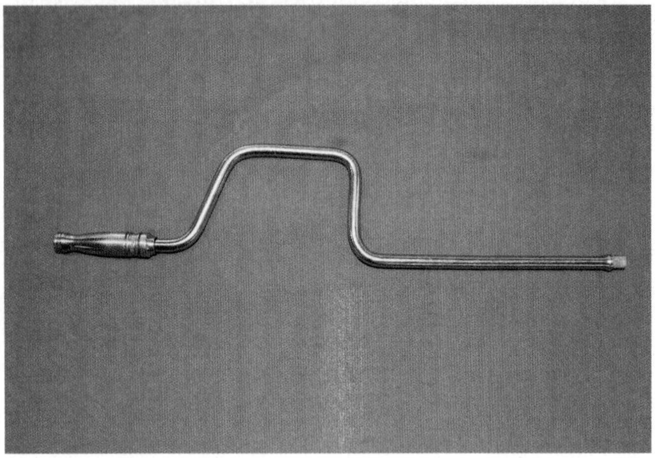

FIGURE 6-33 Speed brace.

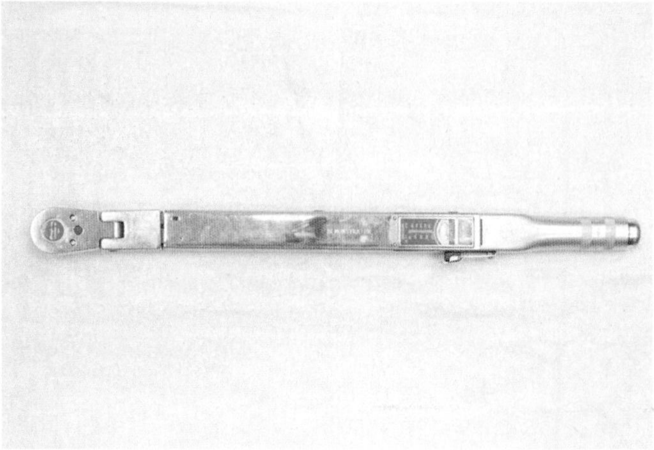

FIGURE 6-35 A torque wrench.

A **lug wrench** has special-sized lug nut sockets permanently attached to it. One common model has four different-sized sockets, one on each arm (**FIGURE 6-34**). Never hit or jump on a lug wrench when loosening the lug nuts. If the lug wrench could not remove them, you should use an impact wrench. The impact wrench provides a hammering effect in conjunction with rotation to help loosen tight fasteners. *Never* use an impact wrench to tighten lug fasteners. Torque all lug fasteners to the proper torque with a properly calibrated torque wrench. And do it in the proper sequence.

Torque Wrenches

A **torque wrench** is also known as a tension wrench (**FIGURE 6-35**). It is used to tighten fasteners to a predetermined torque. The drive on the end fits any socket and accessory of the same drive size found in ordinary socket sets. Manufacturers do not specify torque settings for every nut and bolt. When they do, though, it is important to follow the specifications. For example, manufacturers specify a torque for head bolts. The torque specified ensures that the bolt provides the proper clamping pressure and will not come loose. However, it will not be so tight as to risk breaking the bolt or stripping the threads. The torque value is specified in foot-pounds (ft-lb), inch-pounds (in-lb), or Newton meters (Nm).

Torque wrenches come in various types: beam style, clicker, dial, and electronic (**FIGURE 6-36**). The simplest and least expensive is the beam-style torque wrench. It uses a spring steel beam that flexes under tension. A smaller fixed rod then indicates the amount of torque on a scale mounted to the beam. The amount of deflection of the beam coincides with the amount of torque on the scale. One drawback of this design is that you have to be positioned directly above the scale so you can read it accurately. That can be a problem when working in awkward positions.

The clicker-style torque wrench uses an adjustable clutch inside that slips (clicks) when the preset torque is reached. You can set it for a particular torque on the handle (**FIGURE 6-37**). As the bolt is tightened, and once the preset torque is reached, the torque wrench clicks. The higher the torque, the louder the click; the lower the torque, the quieter

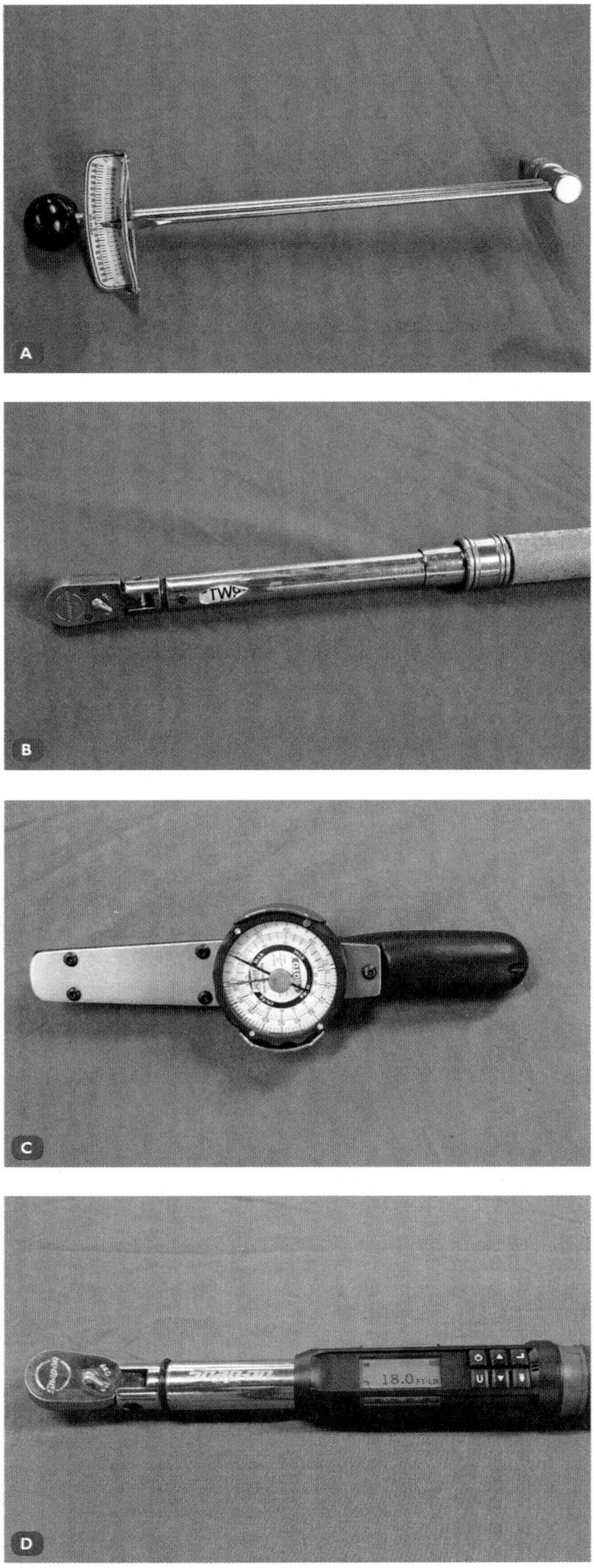

FIGURE 6-36 Torque wrenches. **A.** Beam style. **B.** Clicker style. **C.** Dial. **D.** Electronic.

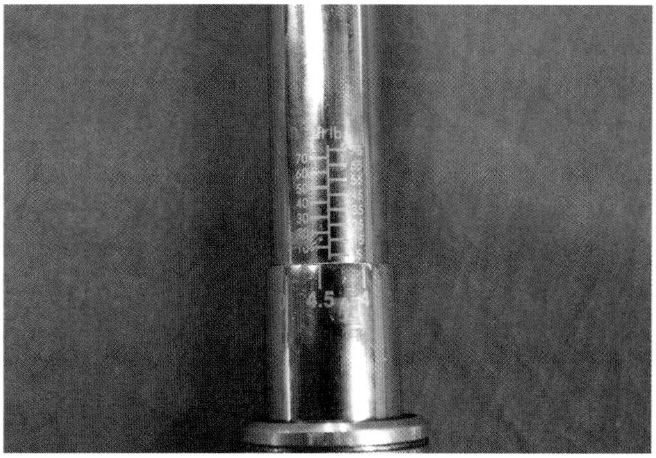

FIGURE 6-37 Torque setting scale on the handle.

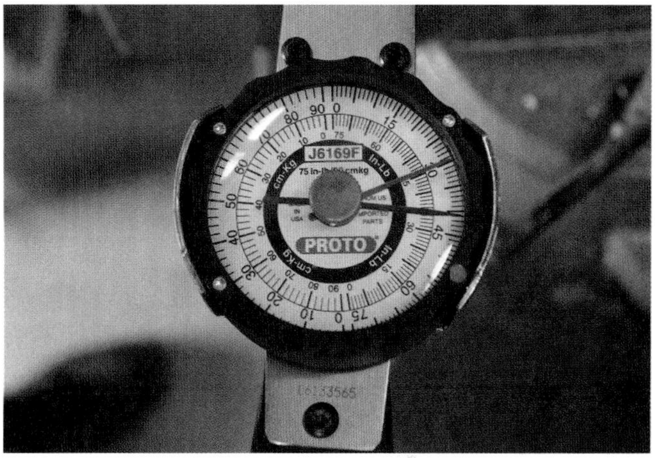

FIGURE 6-38 Dial torque wrench reading torque.

the click. Be careful when using this style of torque wrench, especially at lower torque settings. It is easy to miss the click and overtighten, break, or strip the bolt. Once the torque wrench clicks, stop turning it. Continuing to turn the torque wrench will continue to tighten the fastener if you turn it past the click point.

The dial torque wrench turns a dial that indicates the torque based on the torque being applied. Like the beam-style torque wrench, you have to see the dial to know how much torque is being applied (**FIGURE 6-38**). Many dial torque wrenches have a movable indicator that is moved by the dial and stays at the highest reading. That way you can double-check the torque achieved once the torque wrench is released. Once the proper torque is reached, the indicator can be moved back to zero for the next fastener being torqued.

The digital torque wrench usually uses a spring steel bar with an electronic strain gauge to measure the amount of torque being applied. The torque wrench can be preset to the desired torque. It will then display the torque as the fastener is being tightened (**FIGURE 6-39**). When it reaches the preset torque, it usually gives an audible signal, such as a beep. This makes it useful in situations where a scale or dial cannot be read.

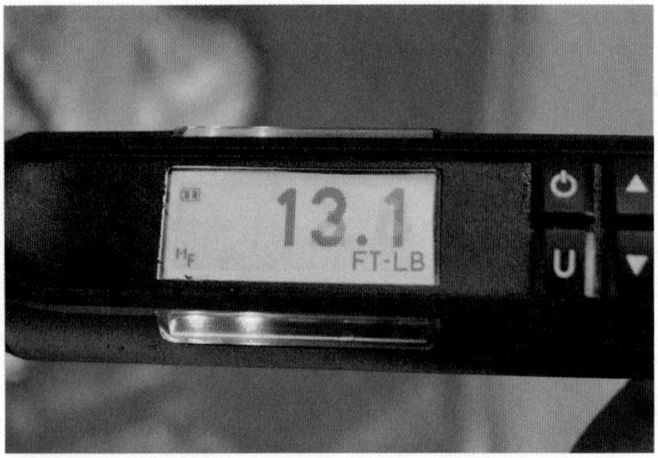

FIGURE 6-39 Digital torque wrench displaying torque.

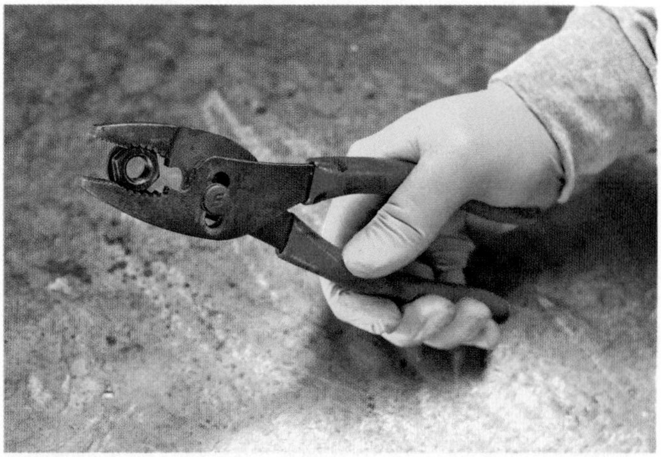

FIGURE 6-40 Pliers are used for grasping and cutting. These are slip-joint pliers.

▶ Other Hand Tools

LO 6-03 Identify other basic hand tools.

Pliers

Pliers are a hand tool designed to hold, cut, or compress materials (**FIGURE 6-40**). They are usually made out of two pieces of strong steel joined at a fulcrum point. They have jaws and cutting surfaces at one end and handles designed to provide leverage at the other. There are many types of pliers, including slip-joint, combination, arc joint, needle-nose, and flat-nose (**FIGURE 6-41**).

Quality combination pliers are one of the most commonly used pliers in a shop. They pivot together so that any force applied to the handles is multiplied in the strong jaws. Most combination pliers are designed so that they have surfaces to both grip and cut. Combination pliers offer two gripping surfaces, one for gripping flat objects and one for gripping rounded objects. They also have one or two pairs of cutters. The cutters in the jaws should be used for softer materials that will not damage the blades. The cutters next to the pivot can shear through thin materials, like steel wire or pins.

Most pliers are limited by their size in what they can grip. Beyond a certain point, the handles are spread too wide, or the jaws cannot open wide enough. **Arc joint pliers** overcome this limitation with a moveable pivot. Often, these are called Channellocks™, named after the company that first made them. These pliers have parallel jaws that allow you to increase or decrease the size of the jaws by selecting a different set of channels (**FIGURE 6-42**). They are useful for a wider grip and a tighter squeeze on parts too big for conventional pliers.

Another type of pliers is **needle-nose pliers**. They have long, pointed jaws and can reach into tight spots or hold small items that other pliers cannot (**FIGURE 6-43**). For example, they can pick up a small bolt that has fallen into a tight spot. **Flat-nose pliers** have an end or nose that is flat and square. In contrast, combination pliers have a rounded end. A flat nose makes it possible to bend wire or even a thin piece of sheet steel accurately along a straight edge.

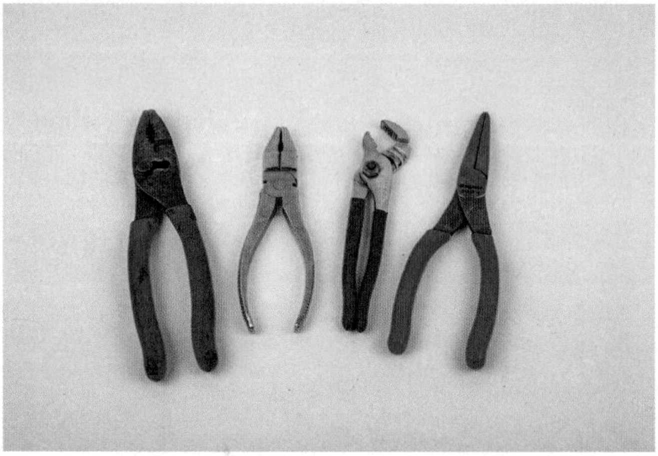

FIGURE 6-41 Variety of pliers.

FIGURE 6-42 Arc joint pliers.

Diagonal cutting pliers are used for cutting wire or cotter pins (**FIGURE 6-44**). Diagonal cutters are the most common cutters in the toolbox. But they should not be used on hard or heavy-gauge materials because the cutting surfaces will be damaged. End-cutting pliers, also called **nippers**, have a cutting edge at right angles to their length (**FIGURE 6-45**). They are designed to cut through soft metal objects sticking out from a surface.

Snap ring pliers have metal pins that fit in the holes of a snap ring (**FIGURE 6-46**). Snap rings can be of the internal or external type. If internal, then internal snap ring pliers compress the snap ring. This allows the snap ring to be removed and installed in its internal groove. If external, then external snap ring pliers are used to expand the snap ring. This allows the snap ring to be removed and installed in its external groove. Not all snap rings have holes in them for pliers. Some snap rings have ends that are cut on an angle and require the use of flat-nose snap ring pliers (**FIGURE 6-47**). Always wear safety glasses when working with snap rings because they can easily slip off the snap ring pliers and fly off at tremendous speeds, possibly causing severe eye injuries.

Locking pliers are also called vice grips. They are general-purpose pliers used to clamp and hold one or more objects (**FIGURE 6-48**). Locking pliers are helpful for freeing up one or more of your hands when you are working. They can clamp

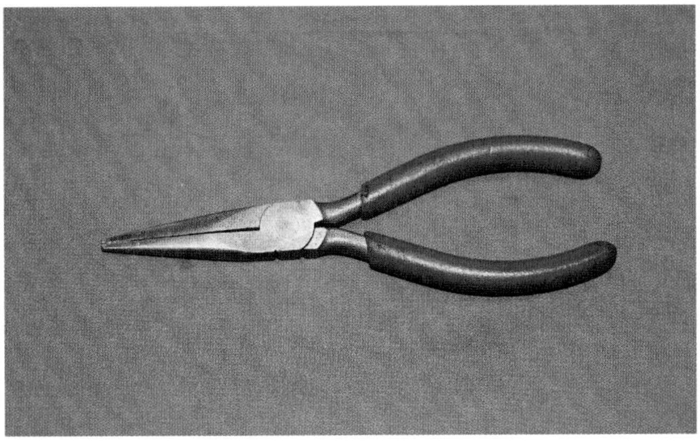

FIGURE 6-43 Needle-nose pliers.

FIGURE 6-44 Diagonal side cutters.

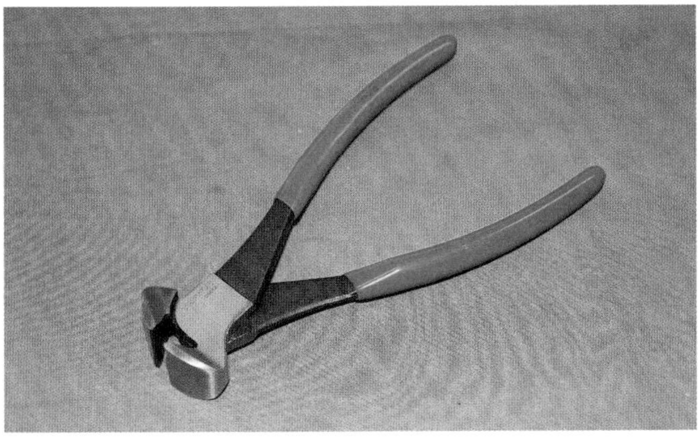

FIGURE 6-45 Nippers, or end-cutting, pliers.

FIGURE 6-46 Snap ring pliers and snap ring.

FIGURE 6-47 Flat-nose snap ring pliers and snap ring.

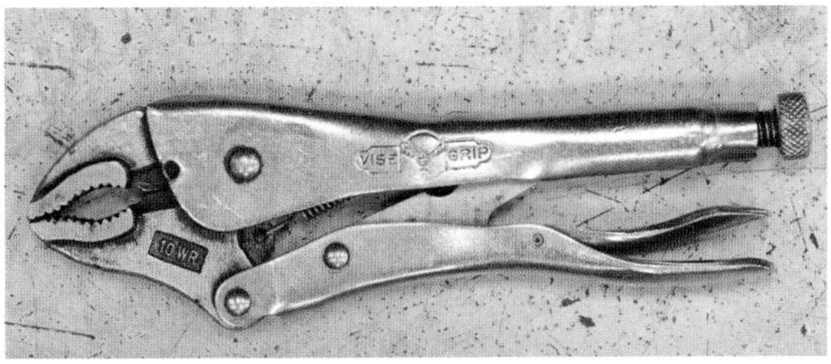

FIGURE 6-48 Locking pliers.

something and lock themselves in place to hold it for you. They are also adjustable, so they can be used for a variety of tasks. To clamp an object with locking pliers, put the object between the jaws, and turn the screw until the handles are almost closed. Then, squeeze them together to lock them shut. You can increase or decrease the gripping force with the adjustment screw. To release the object, squeeze the release lever, and they will open right up.

SAFETY TIP

When applying pressure to pliers, make sure your hands are not greasy, or they might slip. Select the right type and size of pliers for the job. As with most tools, if you have to exert almost all your strength to get something done, then you are using either the wrong tool or the wrong technique. If the pliers slip, you will get hurt. At the very least, you will damage the tool and what you are working on. Pliers get a lot of hard use in the shop, so they do get worn and damaged. If they are worn or damaged, they will be inefficient and can be dangerous. Always check the condition of all shop tools on a regular basis, and replace any that are worn or damaged.

Allen Wrenches

Allen wrenches are sometimes called Allen or hex keys. They are tools designed to tighten and loosen fasteners with Allen heads (**FIGURE 6-49**). The Allen head fastener has an internal hexagonal recess that the Allen wrench fits in snugly. Allen wrenches come in sets, and there is a correct wrench size for every Allen head. They give the best grip on a screw or bolt of all the drivers, and their shape makes them good at getting into tight spots. Care must be taken to make sure the correct size of Allen wrench is used, or else the wrench and/or socket head will be rounded off.

The traditional Allen wrench is a hexagonal bar with a right-angle bend at one end. They are made in various metric and standard sizes. As their popularity has increased, so too has the number of tool variations. Now Allen sockets are available, as are T-handle Allen keys (**FIGURE 6-50**).

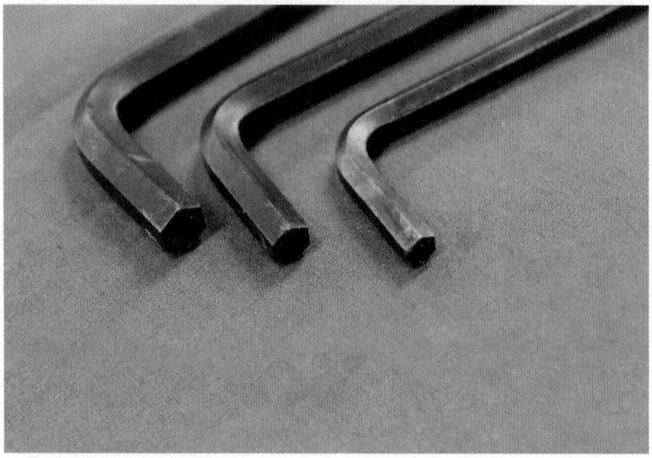

FIGURE 6-49 Typical Allen wrench head.

Screwdrivers

The correct screwdriver to use depends on the type of slot or recess in the head of the screw or bolt, and how accessible it is (**FIGURE 6-51**). It is very important to match the tip of the screwdriver exactly with the slot or recess in the head of a fastener. Otherwise, the tool might slip. This would damage the fastener or the tool and possibly injure you.

Most screwdrivers cannot grip as securely as wrenches. So always check where the screwdriver blade can end up if it slips off the head of the screw. Many technicians learned this the hard way when they stabbed a screwdriver into, or through, their hand. Not only is this painful, but it can become infected or damage the nerves.

The most common screwdriver has a flat tip, or blade, which gives it the name **flat-blade screwdriver**. The blade should be almost as wide and thick as the slot in the fastener. This allows the twisting force of the screwdriver to be applied right out to the edges of the head, where it has most effect. The blade should be a snug fit in the slot of the screw head. The twisting force is then applied evenly along the sides of the slot. This guards against the screwdriver suddenly chewing a piece out of the slot and slipping just when the most force is being exerted.

Flat-blade screwdrivers come in a variety of sizes and lengths, so find the right one for the job. If viewed from the side, the blade should taper slightly to the very end where the flat tip fits into the slot. If the tip of the blade is not clean and square, it should be reshaped or replaced. When you use a flat-blade

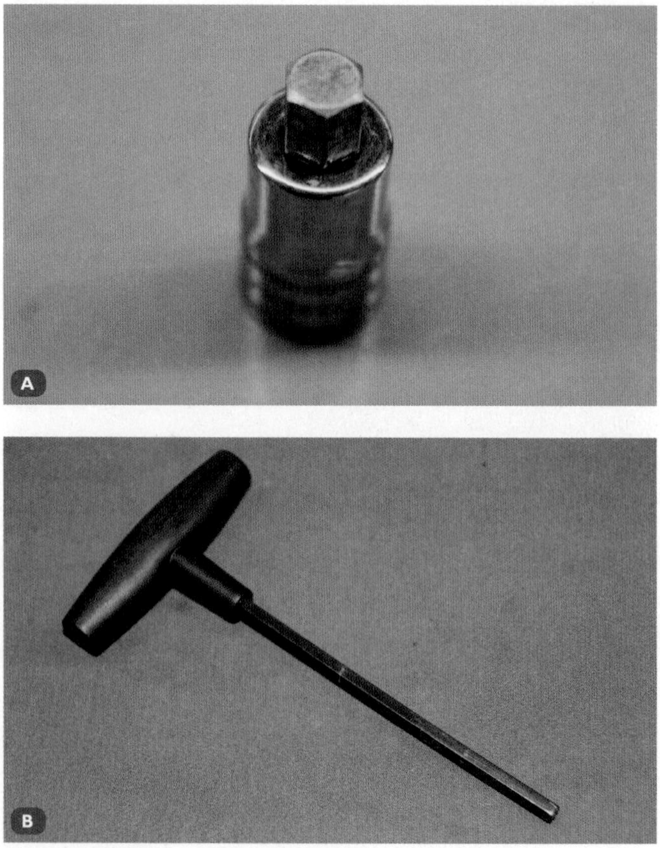

FIGURE 6-50 **A.** Allen socket. **B.** T-handle Allen wrench.

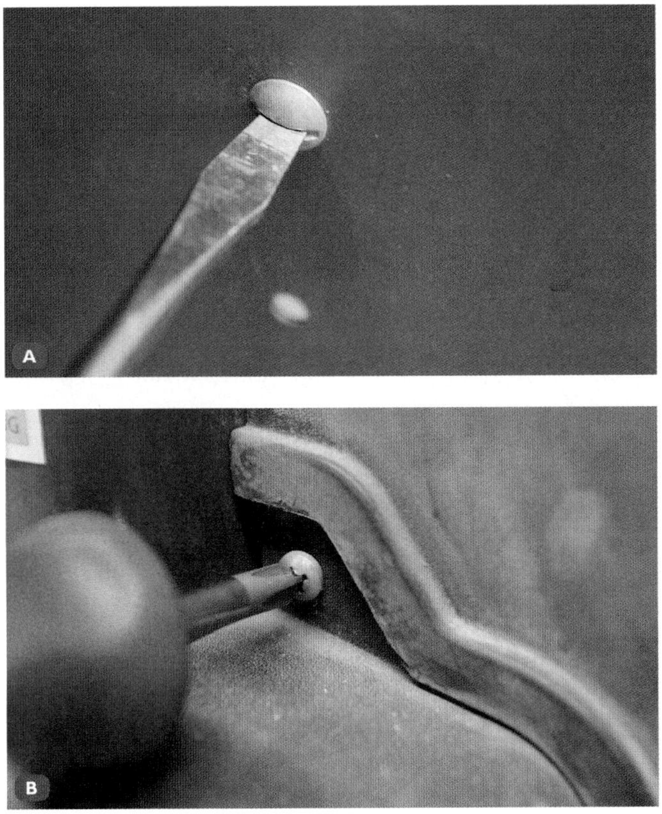

FIGURE 6-51 **A.** Slotted screw and screw driver. **B.** Phillips screw and screw driver.

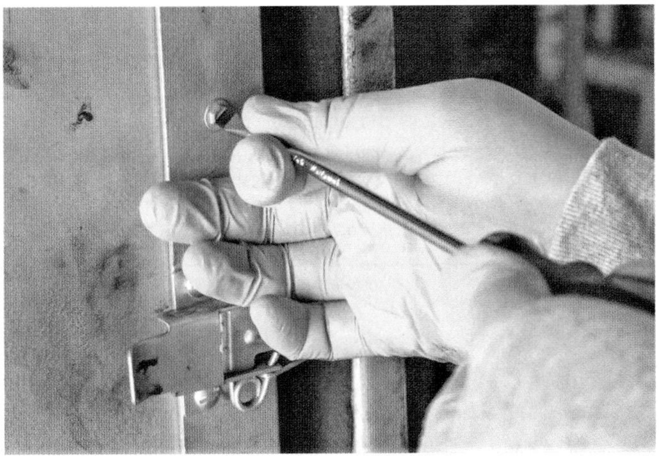

FIGURE 6-52 Support the screw driver so it stays centered in the slot.

screwdriver, support the shaft with your free hand as you turn it (but keep it behind the tip). This helps keep the blade square in the slot and centered (**FIGURE 6-52**). Screwdrivers that slip are a common source of damage and injury in shops.

A screw or bolt with a cross-shaped recess requires a **Phillips head screwdriver** or a Pozidriv screwdriver (**FIGURE 6-53**). The cross-shaped slot on both types holds the tip of the screwdriver centered on the fastener. The Phillips tip fits a tapered recess in the head of a Phillips screw. The Pozidriv tip fits into slots with parallel sides in the head of a Pozidriv screw. Again,

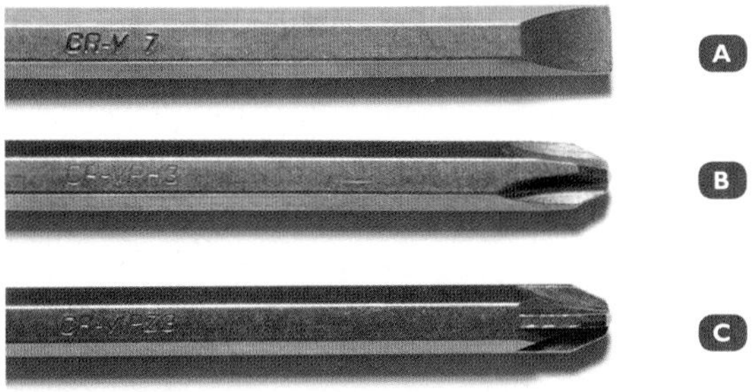

FIGURE 6-53 **A.** Flat-blade screwdrivers. **B.** Phillips screwdriver. **C.** Pozidriv screwdriver.

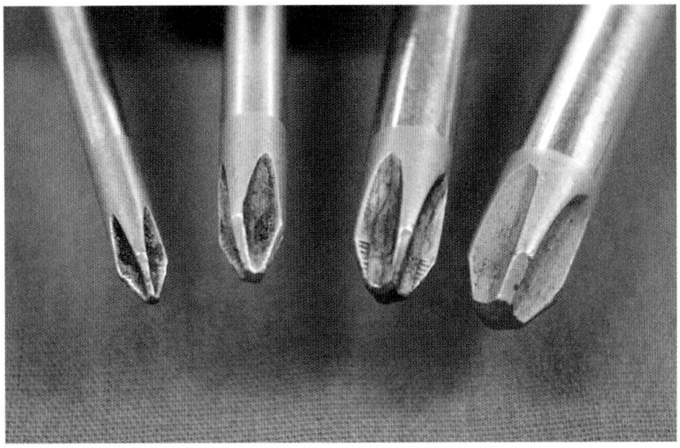

FIGURE 6-54 Four sizes of Phillips screwdrivers.

the screwdriver must be the right size for the screw. The fitting process is simplified with these two types of screwdrivers. Four sizes are enough to fit almost all fasteners with these sorts of screw heads (**FIGURE 6-54**).

The **offset screwdriver** fits into spaces where a straight screwdriver cannot and is useful where there is not much room to turn it (**FIGURE 6-55**). The two tips look identical, but one is set at 90 degrees to the other. This is because sometimes there is only room to make a quarter turn of the driver. Thus, the driver has two blades on opposite ends so that offset ends of the screwdriver can be used alternately.

The **ratcheting screwdriver** is a popular screwdriver handle. It usually comes with a selection of removable flat and Phillips tips. It engages in one direction and freewheels in the other direction, depending on how the slider is set. When set for loosening, a screw can be undone without removing the tip from the head of the screw. When set for tightening, a screw can be installed just as easily.

An **impact driver** is used when a screw or a bolt is rusted/corroded in place or overtightened. An impact driver is a tool that can apply more force than the other members of this family. Screw slots can easily be stripped with the use of a standard screwdriver. The force of the hammer forcing the bit into the screw, while twisting it, makes it more likely that the screw will break loose.

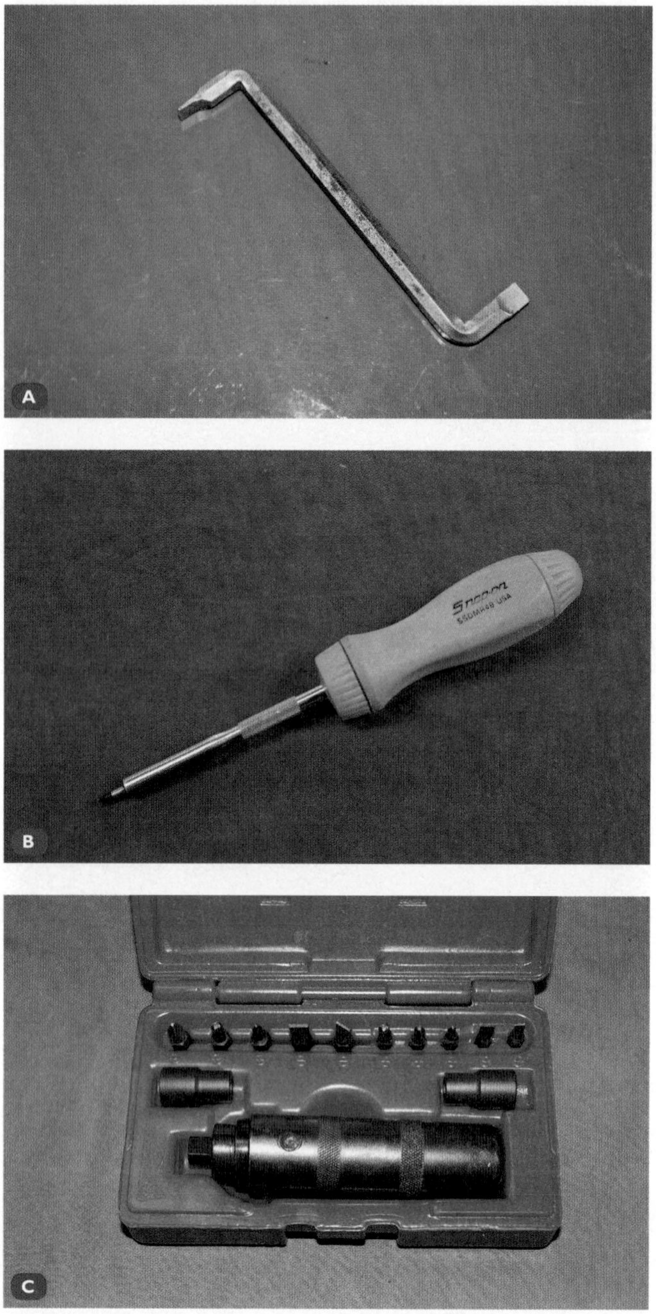

FIGURE 6-55 **A.** Offset screwdriver. **B.** Ratcheting screwdriver. **C.** Impact driver.

The impact driver accepts a variety of special impact tips. Choose the right one for the screw head. Fit the tip in place, and then tension it in the direction it has to turn. A sharp blow with the hammer breaks the screw free. It can then be unscrewed normally.

Magnetic Pickup Tools and Mechanical Fingers

Magnetic pickup tools and mechanical fingers are very useful for grabbing items in tight spaces (**FIGURE 6-56**). A magnetic pickup tool typically has a magnet attached to the end of a telescoping, swivel joint. The magnet is strong enough to pick up

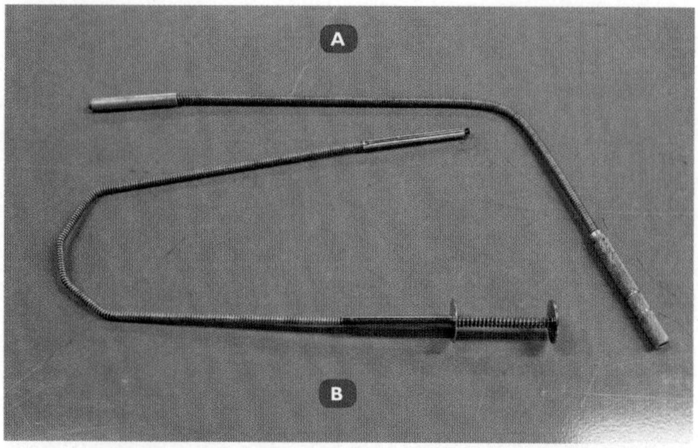

FIGURE 6-56 **A.** Magnetic pickup tools. **B.** Mechanical fingers.

screws, bolts, sockets, and other ferrous (containing iron, making it magnetic) metals. For example, if a screw is dropped into a tight space, a magnetic pickup tool can be used to extract it.

Mechanical fingers are also designed to extract or insert objects in tight spaces. Because they actually grab the object, they can pick up nonmagnetic objects. This makes them handy for picking up rubber or plastic parts. They use a flexible body and come in different lengths, but typically are about 12–18" (305–457 mm) long. They have expanding grappling fingers on one end to grab items. The other end has a push mechanism to expand the fingers and a retracting spring to contract the fingers.

▶ TECHNICIAN TIP

It may be challenging getting the magnet down inside some areas because the magnet wants to keep sticking to other objects. One trick in this situation is to roll up a piece of paper so that a tube is created. Stick that down into the area of the dropped part. Then, slide the magnet down the tube, which helps it get past the magnetic objects. Once the magnet is down, you may want to remove the roll of paper. Just remember two things: First, patience is important when using this tool, and second, do your best to NEVER drop anything!

Cutting Tools

Bolt cutters cut heavy wire, non-hardened rods, and bolts (**FIGURE 6-57**). Their compound joints and long handles give the leverage and cutting pressure that is needed for heavy gauge materials. **Tin snips** are the nearest thing in the toolbox to a pair of scissors. They can cut thin-sheet metal, and lighter versions make it easy to follow the outline of gaskets. Most snips come with straight blades. But if there is an unusual shape to cut, there is a pair with left- or right-handed curved blades. **Aviation snips** are designed to cut soft metals. They are easy to use because the handles are spring-loaded in the open position. They are also double pivoted for extra leverage.

Pry Bars

Pry bars are tools constructed of strong metal and are used as a lever to move, adjust, or pry (**FIGURE 6-58**). Pry bars come in a variety of shapes and sizes. Many have a tapered end that is slightly bent, with a plastic handle on the other end. This design works well for applying force to tension belts or for moving parts into alignment.

Another type of pry bar is the **roll bar**. One end is sharply curved and tapered, which is used for prying. The other end is tapered to a dull point and is used to align larger holes such as transmission bell housings or engine motor mounts. Because pry bars are made of hardened steel, care should be taken when using them on softer materials. This will help avoid any damage.

Gasket Scrapers

A typical **gasket scraper** has a hardened, sharpened blade. It is designed to remove a gasket on cast iron surfaces and non-critical aluminum sealing surfaces such as thermostat housings. When removing gaskets from critical sealing surfaces, such as the aluminum head gasket sealing surface, plastic gasket scrapers are required. They won't damage the sealing surface when used properly (**FIGURE 6-59**).

On one end, gasket scrapers have a comfortable handle like a screwdriver handle. On the other end, a blade is fitted

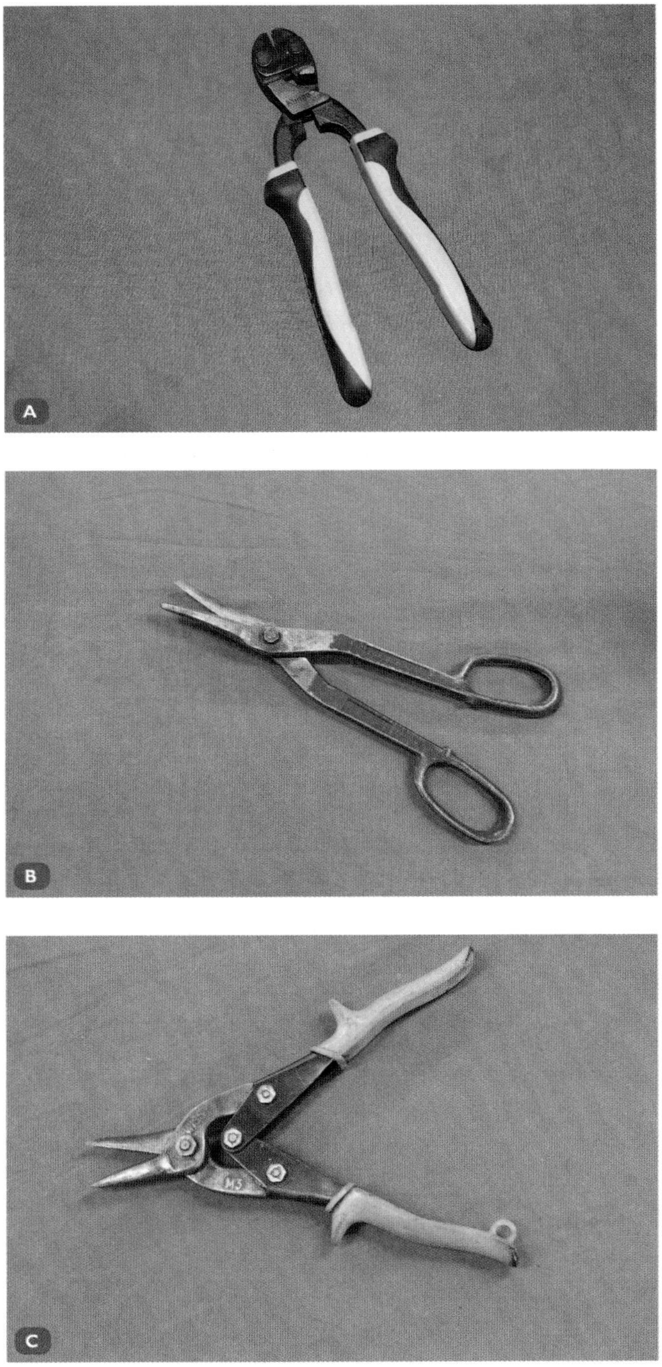

FIGURE 6-57 **A.** Bolt cutters. **B.** Tin snips. **C.** Aviation snips.

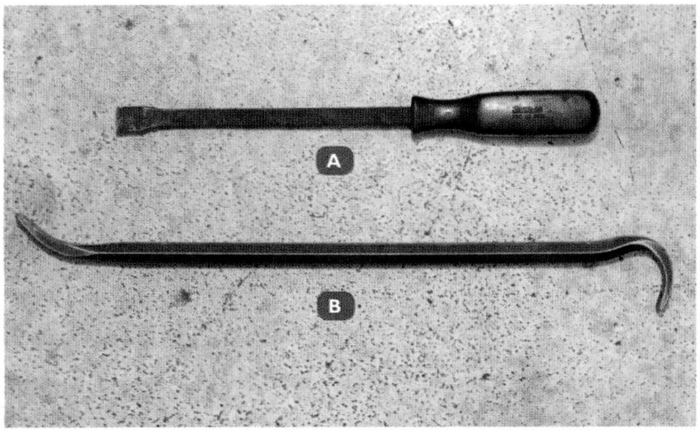

FIGURE 6-58 **A.** Pry bar. **B.** Roll bar.

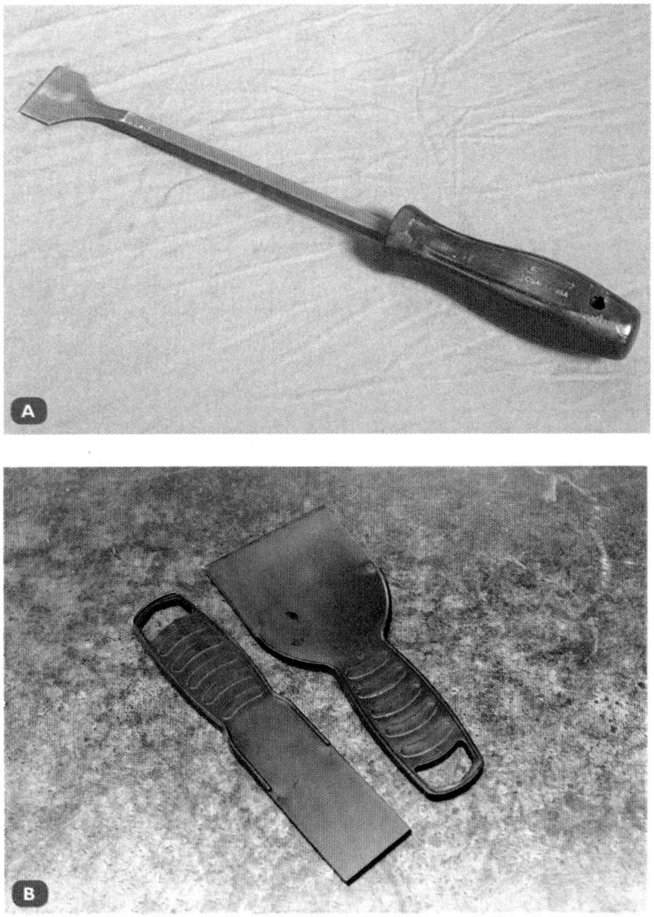

FIGURE 6-59 **A.** Metal gasket scraper. **B.** Plastic gasket scraper.

with a sharp edge to assist in the removal of gaskets. The gasket scraper should be kept sharp and straight. This makes it easier to remove all traces of the old gasket and sealing compounds. The blades come in different sizes, with a typical size being 1" (25 mm) wide. Whenever you use a gasket scraper, be very careful not to nick or damage the surface being cleaned (**FIGURE 6-60**).

> ▶ TECHNICIAN TIP
>
> Many engine components are made of aluminum. Because aluminum is quite soft, it is critical that you use the gasket scraper very carefully so as not to damage the surface. This can be accomplished by keeping the gasket scraper at a fairly flat angle to the surface. Also, the gasket scraper should only be used by hand, not with a hammer. Some manufacturers require the use of plastic gasket scrapers on certain aluminum components such as cylinder heads and blocks.

Files

Files are hand tools designed to remove small amounts of material from the surface of a workpiece (**FIGURE 6-61**). Files come in a variety of shapes, sizes, and coarseness depending on the material being worked and the size of the job. Files have a pointed tang on one end that is fitted to a handle. Files are often sold without handles, but they should not be used until a handle of the right size has been fitted. A correctly sized handle fits snugly without working loose when the file is being used. Always check the handle before using the file. If the handle is loose, give it a sharp rap to tighten it up, or if it is the threaded type, screw it on tighter. If it fails to fit snugly, you must use a different-size handle.

FIGURE 6-60 Use the gasket scraper very carefully to avoid damaging the surface.

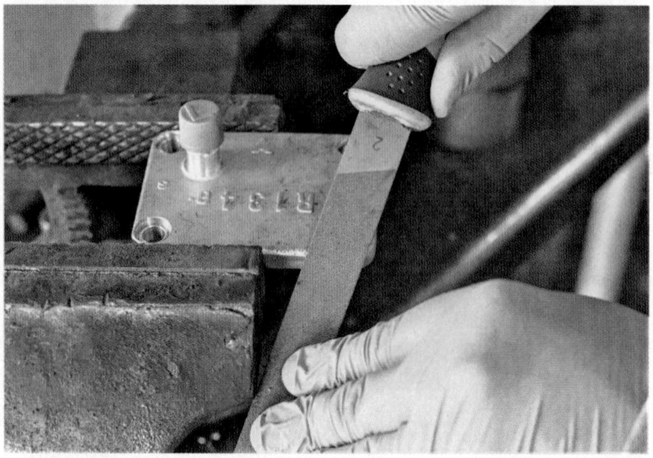

FIGURE 6-61 Files are used to remove materials from the surface of a workpiece.

What makes one file different from another is not just the shape but how much material it is designed to remove with each stroke. The teeth on the file determine how much material will be removed (**FIGURE 6-62**). Because the teeth face one direction only, the file cuts in one direction only. Dragging the file backward over the surface of the metal only dulls the teeth and wears them out quickly.

Teeth on a coarse-grade file are larger, with a greater space between them. A coarse-grade file working on a piece of soft metal removes a lot of material with each stroke. But it leaves a rough finish. A smooth-grade file has smaller teeth cut more closely together. It removes much less material on each stroke, but the finish is much smoother. On many jobs, the coarse file is used first to remove the material quickly. Then, a smoother file gently removes the remaining material and leaves a smoother finish to the work. The full list of grades in flat files, from rough to smooth, follows:

- Rough files have the coarsest teeth, with approximately 20 teeth per inch. They are used when a lot of materials must be removed quickly. They leave a very rough finish and have to be followed by the use of finer files to produce a smooth final finish.
- Coarse bastard files are still a coarse file, with approximately 30 teeth per inch, but they are not as course as the rough file. They are also used to rough out or remove the material quickly.
- Second-cut files have approximately 40 teeth per inch and provide a smoother finish than the rough or coarse bastard file. They are good all-round intermediary files and leave a reasonably smooth finish.
- Smooth files have approximately 60 teeth per inch and are a finishing file used to provide a smooth final finish.
- Dead smooth files have 100 teeth per inch or more and are used where a very fine finish is required.

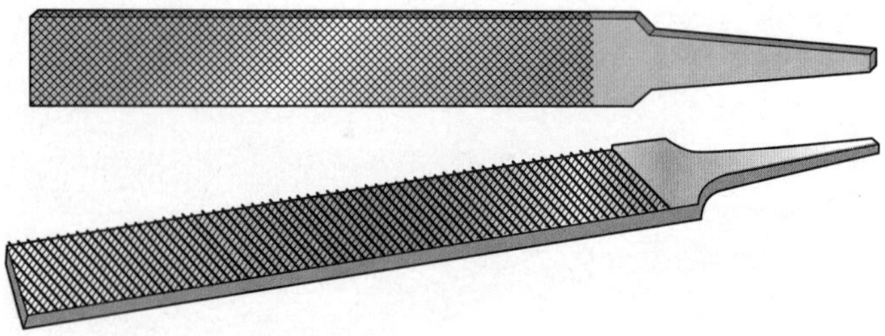

Description:
The teeth on a file.

FIGURE 6-62 The teeth on a file determine how much material will be removed from the object being filed.

Some flat files are available with one smooth edge (no teeth), called safe edge files. They allow filing up to an edge without damaging it. Flat files work well on straightforward jobs, but some jobs require special files. A **warding file** is thinner than other files and comes to a point; it is used for working in narrow slots (**FIGURE 6-63**). A **square file** has teeth on all four sides, so you can use it in a square or rectangular hole. A square file can make the right shape for a squared metal key to fit in a slot.

SAFETY TIP

Hands should always be kept away from the surface of the file and the metal that is being worked on. It is easy for skin to get caught between the edge of the file and the piece you are working. This creates a scissor-like condition that can result in severe cuts. Also, filing can produce small slivers of metal that penetrate your skin and can be difficult to remove.

A **triangular file** has three sides. It is triangular, so it can get into internal corners easily. It is able to cut right into a corner without removing material from the sides. **Curved files** are typically either half-round or round. A half-round file has a shallow convex surface that can file in a concave hollow or in an acute internal corner (**FIGURE 6-64**). The fully round file, sometimes called a rat-tail file, can make holes bigger. It can also file inside a concave surface with a tight radius.

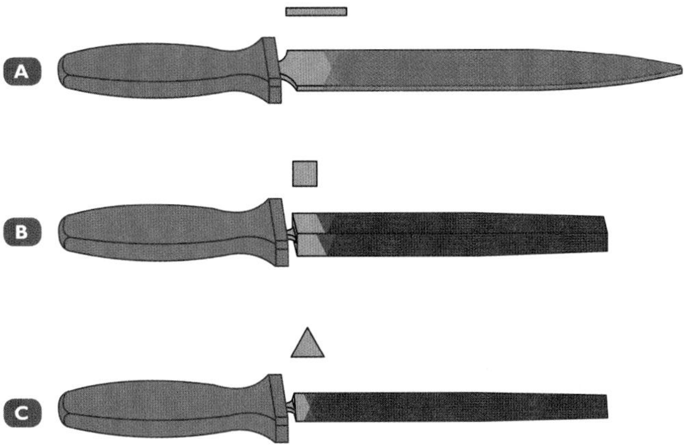

FIGURE 6-63 **A.** Warding file. **B.** Square file. **C.** Triangular file.

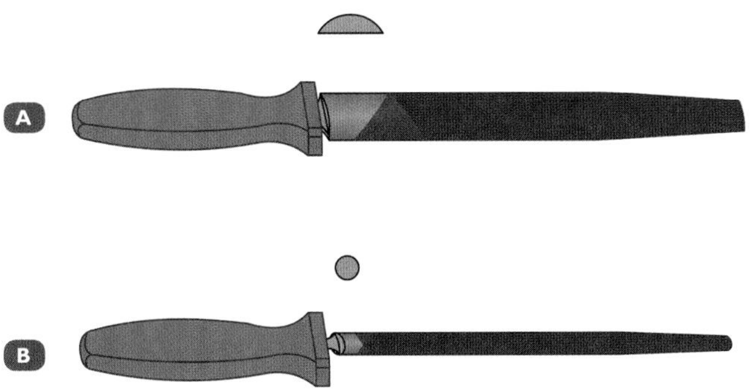

FIGURE 6-64 **A.** Half-round file. **B.** Round, or rat-tail, file.

The **thread file** cleans clogged or distorted threads on bolts and studs. Thread files come in either standard or metric configurations, so make sure you use the correct file. Each file has eight different surfaces that match different thread dimensions. So the correct face must be used (**FIGURE 6-65**).

Files should be cleaned after each use. If they are clogged, they can be cleaned by using a file card, or file brush (**FIGURE 6-66**). This tool has short steel bristles that clean out the small particles that clog the teeth of the file. Rubbing a piece of chalk over the surface of the file prior to filing makes it easier to clean.

▶ Hammers and Struck Tools

LO 6-04 Identify basic hammers and struck tools.

Hammers are a vital part of the shop tool collection, and a variety of hammers are commonly used (**FIGURE 6-67**). The most common hammer in an automotive shop is the **ball-peen (engineer's) hammer**. Like most hammers, its head is hardened steel. A punch or a chisel can be driven with the flat face. Its name comes from the ball peen or rounded face. This end is usually used for flattening or **peening** a rivet. The hammer should always match the size of the job. But it is usually better to use one that is a little too big than too small.

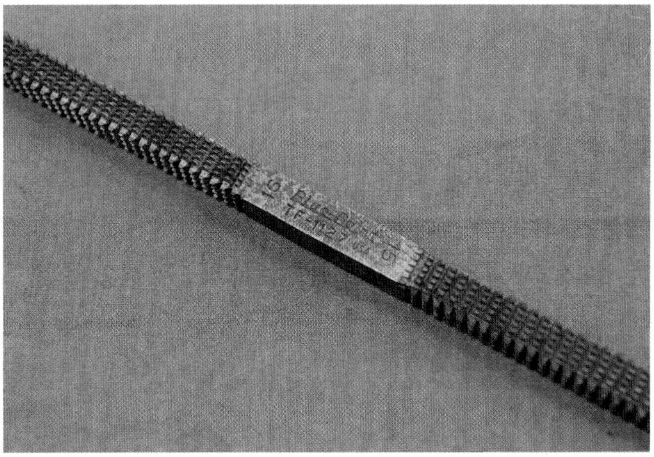

FIGURE 6-65 Thread file.

FIGURE 6-66 File card.

Hitting chisels with a **steel hammer** is fine, but sometimes you only need to tap a component to position it. A steel hammer might mark or damage the part, especially if it is made of a softer metal, such as aluminum. In such cases, a soft-faced hammer should normally be used for the job. Soft-faced hammers range from very soft with rubber or plastic heads to slightly harder with brass or copper heads (**FIGURE 6-68**).

When a large chisel needs a really strong blow, it is time to use a **sledgehammer**. The sledgehammer is like a small mallet, with two square faces made of high-carbon steel. It is the heaviest type of hammer that can be used one-handed. The sledgehammer is used in conjunction with a chisel to cut off a bolt where corrosion has made it impossible to remove the nut.

A **dead blow hammer** is designed not to bounce back when it hits something. A rebounding hammer can be dangerous or destructive. A dead blow hammer can be made with a lead head or, more commonly, a hollow polyurethane head filled with lead shot. The head absorbs the blow when the hammer makes contact, reducing any bounce-back or rebounding. This hammer is ideal for dislodging stuck parts.

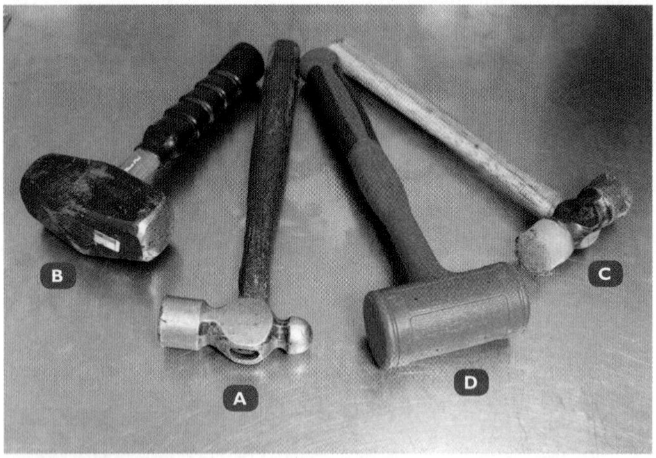

FIGURE 6-67 A. Ball peen hammer. **B.** Sledge hammer. **C.** Soft-faced hammer. **D.** Dead blow hammer.

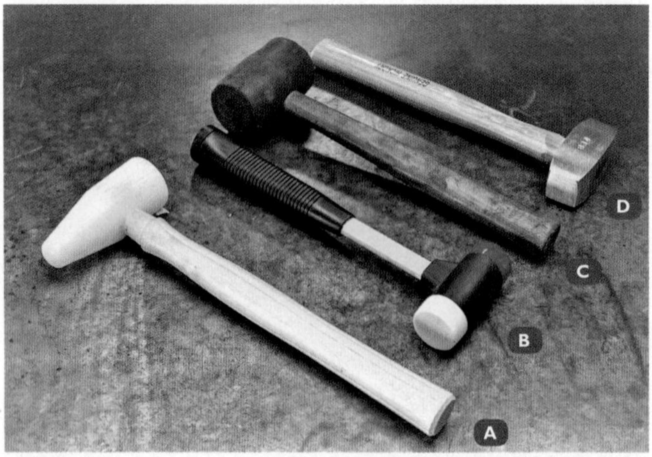

FIGURE 6-68 Soft-faced hammers. **A.** Rubber. **B.** Plastic. **C.** Rubber/Plastic. **D.** Brass.

A **hard rubber mallet** is a special-purpose tool and has a head made of hard rubber. It is often used for moving things into place where it is important not to damage the item being moved. For example, it can be used to install a hubcap or to break a gasket seal on an aluminum housing.

SAFETY TIP

The hammer you use depends on the part you are striking. Hammers with a metal face should almost always be harder than the part you are hammering. Never strike two hardened tools together, as this can cause the hardened parts to shatter.

SAFETY TIP

When using hammers and chisels, always wear safety glasses. Also be aware that safety glasses by themselves need to be form fitting. They must not be allowed to slide down your nose, for them to be effective. It may be prudent to wear either a face shield or safety goggles for extra protection.

Chisels

The most common kind of chisel is a **cold chisel** (**FIGURE 6-69**). It gets its name from the fact it is used to cut cold metals rather than heated metals. It has a flat blade made of high-quality steel and a cutting angle of approximately 70 degrees. The cutting end is tempered and hardened because it has to be harder than the metals it is cutting. The head of the chisel needs to be softer so it will not chip when it is hit with a hammer. Technicians sometimes use a cold chisel to remove bolts whose heads have rounded off.

A variation of the cold chisel is a spring-loaded cold chisel (**FIGURE 6-70**). This chisel works really well in tight spaces where a hammer cannot be swung. The chisel is made up of three parts: the chisel, the weighted hammerhead, and a spring in tension. It holds the other two components together. It is operated by first holding the chisel end against the part you are working on, and then pulling back on the hammerhead and

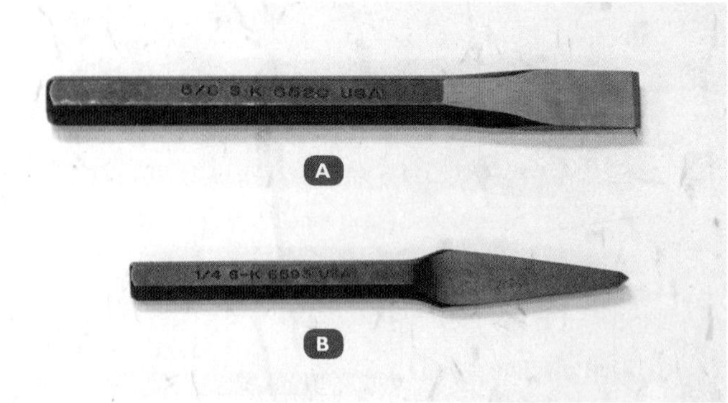

FIGURE 6-69 A. Cold chisel. **B.** Cross-cut chisel.

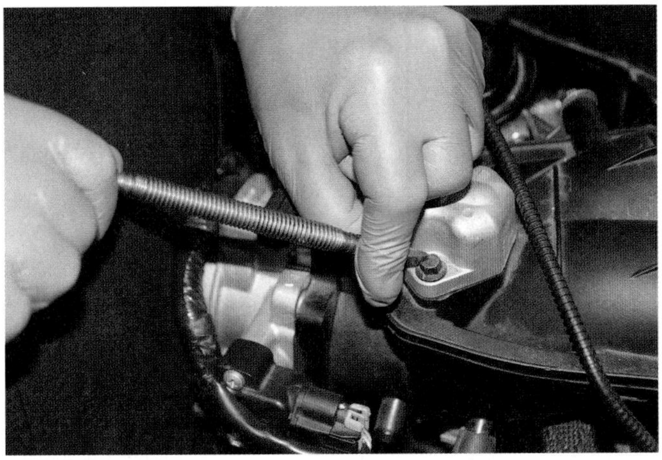

FIGURE 6-70 Spring-loaded chisels.

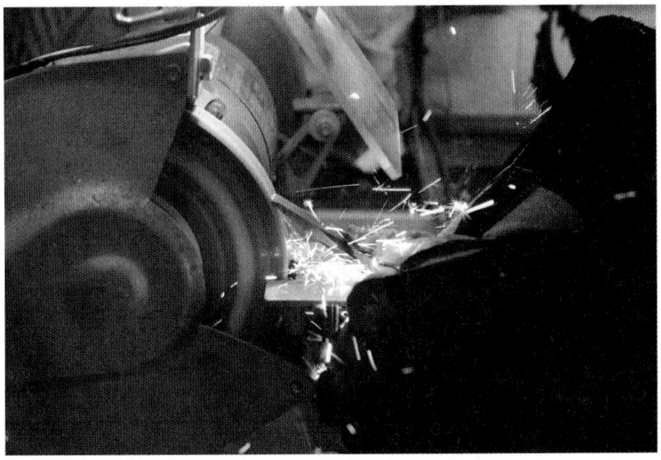

FIGURE 6-71 Dressing a chisel.

releasing it. This allows the spring to rapidly slam the hammerhead into the end of the chisel. The force of the hammerhead hitting the end of the chisel transfers a lot of energy to the chisel.

A **cross-cut chisel** is so named because the sharpened edge is across the blade width. This chisel narrows down along the stock, so it is good for getting into grooves. It is used for cleaning out or even making key ways. The flying chips of metal should always be directed away from the user.

SAFETY TIP

Chisels and punches are designed with a softer striking end than hammers. Over time, this softer metal "mushrooms." Small fragments are prone to breaking off when hammered. These fragments can cause eye injury or other penetrative injuries to people in the area. Always inspect chisels and punches for mushrooming. Dress them on a grinder when necessary (**FIGURE 6-71**).

Punches

Punches are used when the head of the hammer is too large to strike the object being hit without causing damage to adjacent parts. A punch transmits the hammer's striking power from the soft upper end down to the tip that is made of hardened high-carbon steel. A punch transmits an accurate blow from the hammer at exactly one point. This is something that cannot be guaranteed when using a hammer on its own.

Four of the most common punches are the prick punch, center punch, drift punch, and pin punch (**FIGURE 6-72**). When marks need to be drawn on an object like a steel plate, a **prick punch** can be used to mark the points so they will not rub off. They can also be used to scribe intersecting lines between given points. The prick punch's point is very sharp, so a gentle tap leaves a clear indentation. The **center punch** is not as sharp as a prick punch and is usually bigger. It makes a bigger indentation that centers a drill bit at the point where a hole is required to be drilled.

Most center punches are used with a hammer. Some center punches, though, operate automatically when the punch is pressed tightly up against the part you are punching. This type of punch has a spring and weighted hammer inside of the back end of the center punch. It is machined in such a way that pushing the center punch against a surface compresses a spring behind a movable weight (**FIGURE 6-73**). When pushed far enough, the weight is released, and the spring forces it against the center rod in the punch. This causes the punch to indent the work surface.

A **drift punch** is also named a starter punch because you should always use it first to get a pin moving. It has a tapered shank, and the tip is slightly hollow so it does not spread the end of a pin and make it an even tighter fit. Once the starter drift has gotten the pin moving, a suitable pin punch will drive the pin out or in. A drift punch also works well for aligning holes on two mating objects, such as a valve cover and cylinder head. Forcing the drift punch in the hole aligns both components for easier installation of the remaining bolts.

A **pin punch** has a long slender shaft that has straight sides. It is used to drive out pins or rivets (**FIGURE 6-74**). A lot of components are either held together or accurately located by pins. Pins can be pretty tight, and a group of pin punches of various diameters is specially designed to deal with them.

Special punches with hollow ends are called **wad punches** or **hollow punches** (**FIGURE 6-75**). They are the most efficient tool to make a hole in soft materials. They are commonly used on soft sheet material like shim steel, plastic, leather, or gasket material. When being used, there should always be a soft surface under the work area. Ideally, this would be the end grain of a wooden block. If a hollow punch loses its sharpness or has nicks around its edge, it will make a mess instead of a hole.

Number and letter punches are used to mark engine components such as connecting rods. Number and letter punches come in boxed sets (**FIGURE 6-76**). The rules for using a number or letter punch set are the same as for all punches. The punch must be square with the surface being worked on, not on an angle. And the hammer must hit the top of the punch squarely.

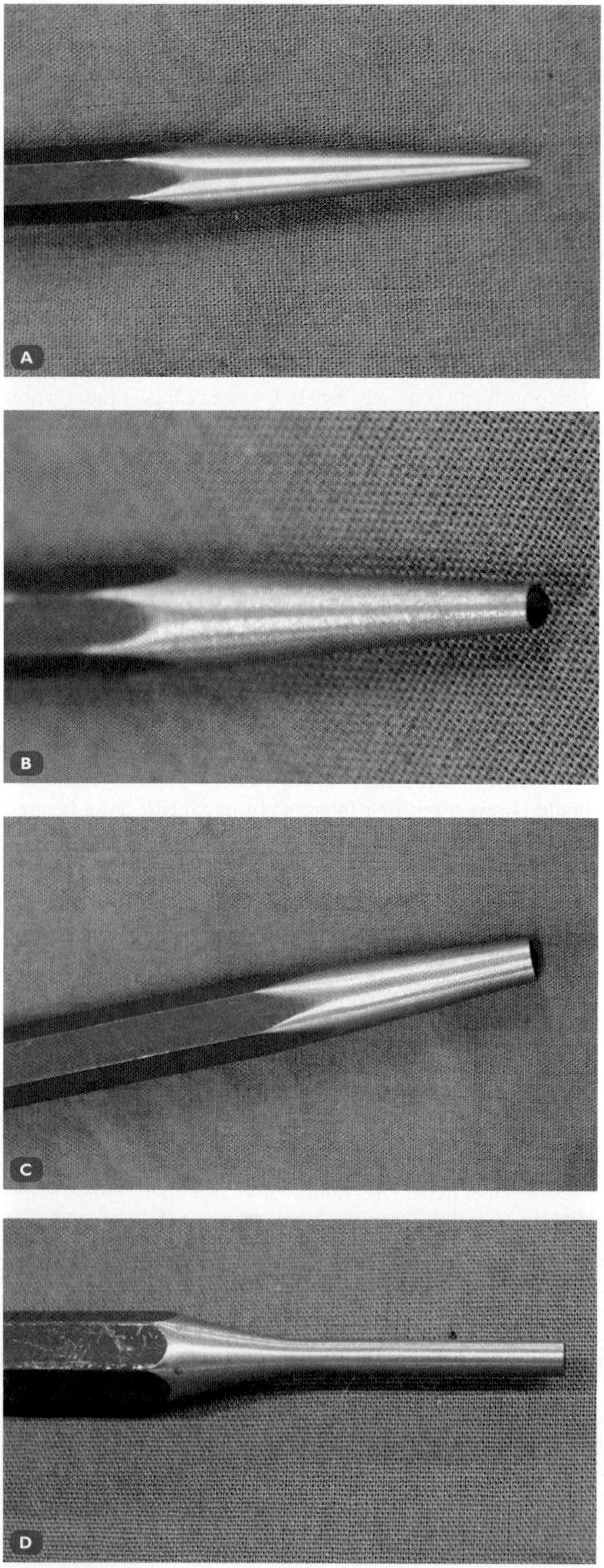

FIGURE 6-72 **A.** Prick punch. **B.** Center punch. **C.** Drift punch. **D.** Pin punch.

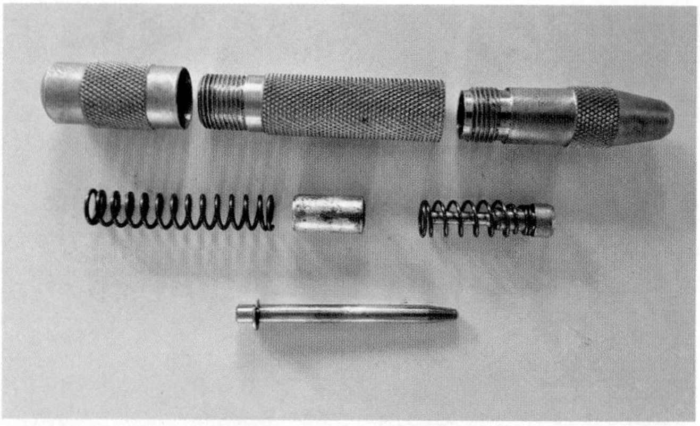

FIGURE 6-73 Internal workings of an automatic center punch.

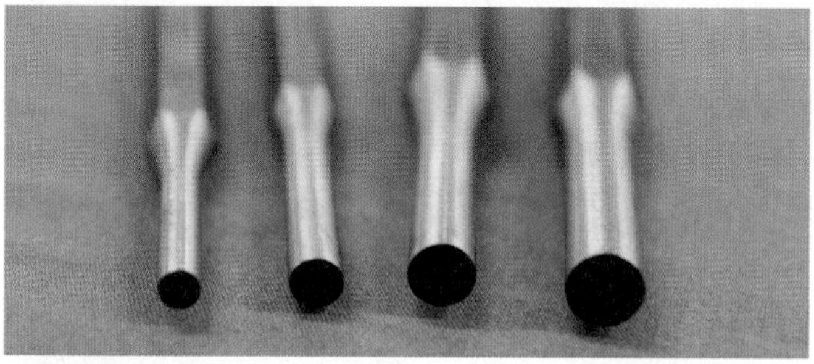

FIGURE 6-74 Various pin punches.

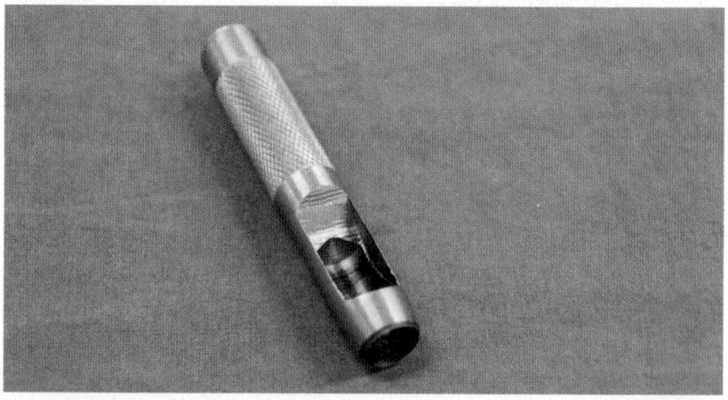

FIGURE 6-75 Wad punch.

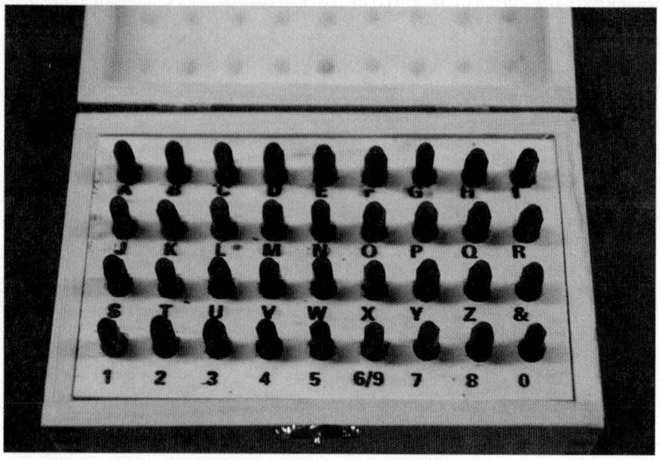

FIGURE 6-76 Number and letter punches.

▶ Taps and Dies

LO 6-05 Identify basic taps, dies, and specialty tools.

Taps and dies are used to form threads in metal so that they can be fastened together (**FIGURE 6-77**). The tap cuts female threads in a component so that a fastener can be screwed into it. The die is used to cut male threads on a bolt so that it can be screwed into the female threads created by the tap. The tap and die are companion tools that create matching threads so that they can both be used to fasten things together.

Taps

Various types of taps are designed based on what you want to do and the material you are working with. The most common taps are the taper tap, intermediate tap (also called a plug tap), bottoming tap (also called a flat-bottomed tap), and the thread chaser (**FIGURE 6-78**). These are explored in more depth below.

Taper Tap: A **taper tap** narrows at the tip, which makes it easier to start straight when cutting threads in a new hole. It also makes it less likely to break the tip off the tap because it removes metal in a less aggressive manner. Taps are very hard, which gives them good wear resistance but also makes them very brittle. So they can break off easily. If a tap breaks in the hole, it can be very hard to remove. This makes taper taps good to use when starting a hole.

Intermediate Tap: The second type of tap is an **intermediate tap**, also known as a plug tap. It is more aggressive than a taper tap, but not as aggressive as a bottoming tap. This is the most common tap used by technicians. Although it is a bit more aggressive than the taper tap, it can be used as a starter tap for a new hole.

Bottoming Tap: The **bottoming tap** is used when you need to cut threads to the very bottom of a blind hole. It has a flat bottom and the threads are the same all the way to the end. This type of tap is virtually impossible to use in a new, unthreaded hole.

Thread Chaser: **Thread chasers** are used to clean up the threads of a hole to make sure they are free from debris and dirt. They do not cut threads, but just clean up the existing threads so that the bolt does not encounter any excessive resistance.

Dies

A **die** is used to cut external threads on a metal shank or bolt. The threads in a die create the male companion to the female threads made by a tap. Usually, dies come with a setscrew that allows the user to slightly adjust the size of the die. This allows the threads to be cut to the right fit that matches the threaded hole. Loose-fitting threads strip more easily than they should and are not as secure. Tight-fitting threads increase the torque required to turn the bolt, thereby reducing the clamping force relative to bolt torque. So adjusting thread fit is typically accomplished by adjusting the die. Dies are hardened, which makes them very wear-resistant, but very brittle. Die nuts do not have the split like the threading die does because they are used to clean up threads like thread chasers (**FIGURE 6-79**).

Proper Use of Taps and Dies

The proper use of taps and dies is a major issue often overlooked in repairing a thread on a bolt or threads in a hole. Improper

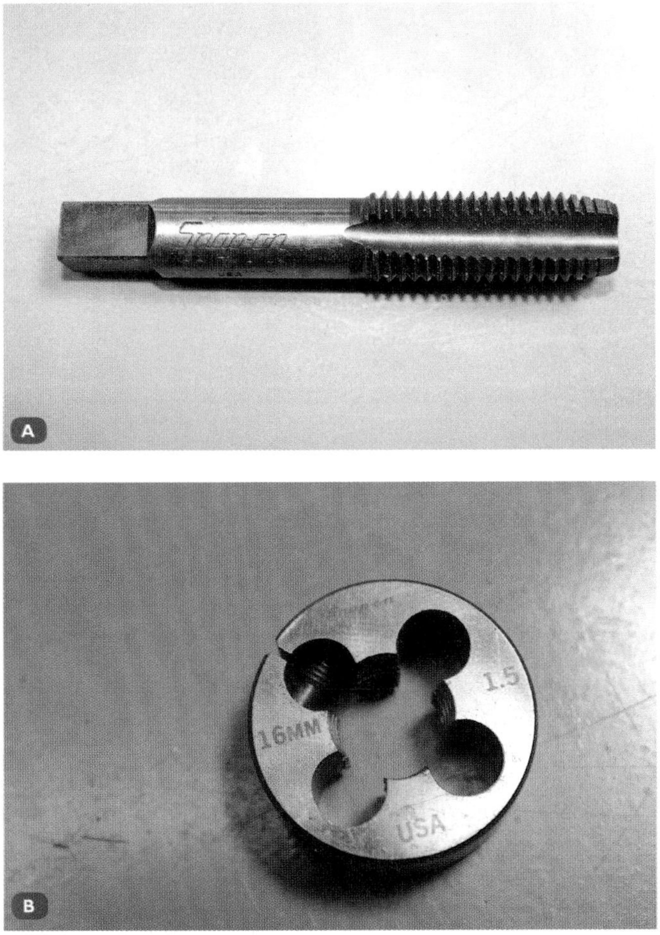

FIGURE 6-77 **A.** Tap. **B.** Die.

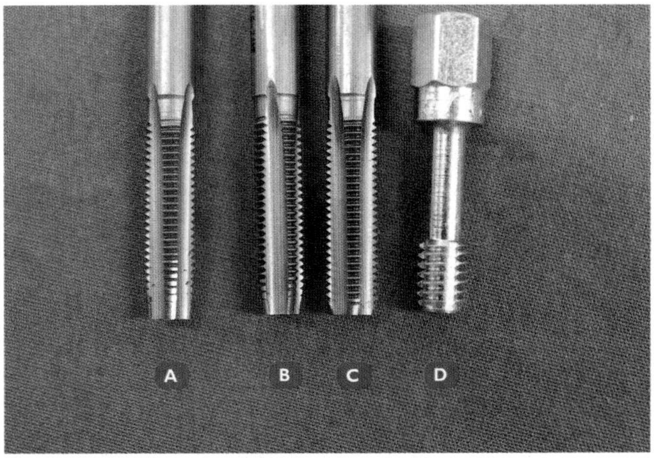

FIGURE 6-78 **A.** Taper tap. **B.** Intermediate tap. **C.** Bottoming tap. **D.** Thread chaser.

use of these two items causes the bolt not to thread into the hole properly. It could even break the tool, which is something to avoid. If they are used improperly, they have a tendency to facture and break.

Because taps and dies both need to be rotated, special tools are used to turn them. A **tap handle** is used to turn the taps. It has a right-angled jaw that matches the squared end of the taps (**FIGURE 6-80**). The jaws are designed to hold the tap securely. And the handles provide the leverage for the operator to comfortably rotate the tap to cut the thread. To cut a thread in an awkward space, a T-shaped tap handle is very convenient (**FIGURE 6-81**). Its handle is not as long, so it fits into tighter spaces; however, it is harder to turn and to guide accurately.

To cut a brand new thread on a blank rod or shaft, a die held in a **die stock** is used (**FIGURE 6-82**). The die fits into the octagonal recess in the die stock. It is usually held in place by a thumb screw. The die may be split so that it can be adjusted more tightly onto the work with each pass of the die. This allows you to get a good fit between the external threads on the shaft and the threads in the hole.

When tapping a hole, the diameter of the hole is determined by a tap drill chart. This chart shows what drill size is needed for a particular tap. Taps are made for any given bolt size (**FIGURE 6-83**).

Just remember that if you are drilling a 1/4" (6 mm) or larger hole, use a smaller pilot drill first. Once the properly

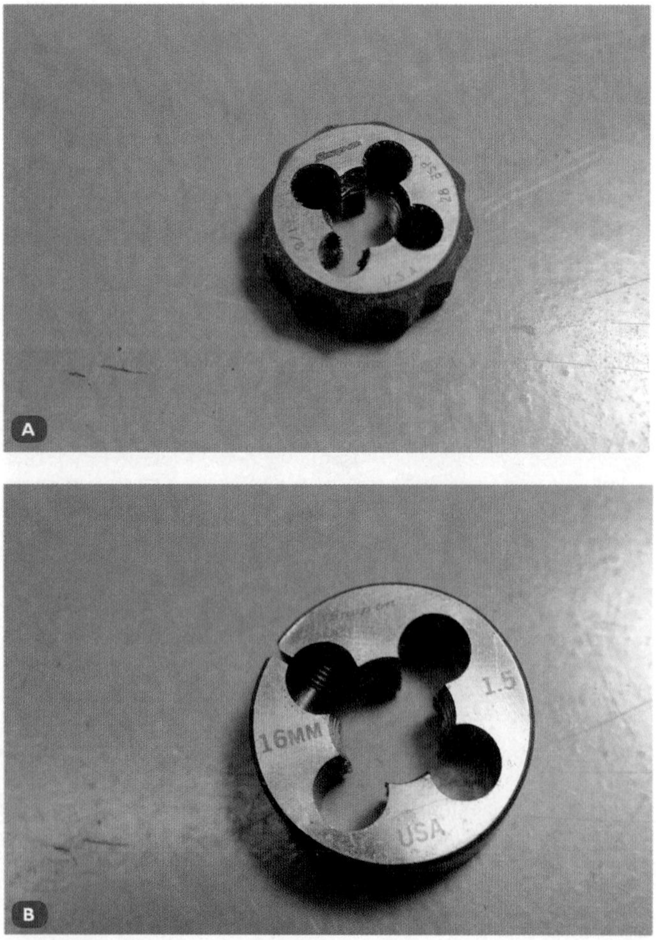

FIGURE 6-79 **A.** Die nut. **B.** Split die.

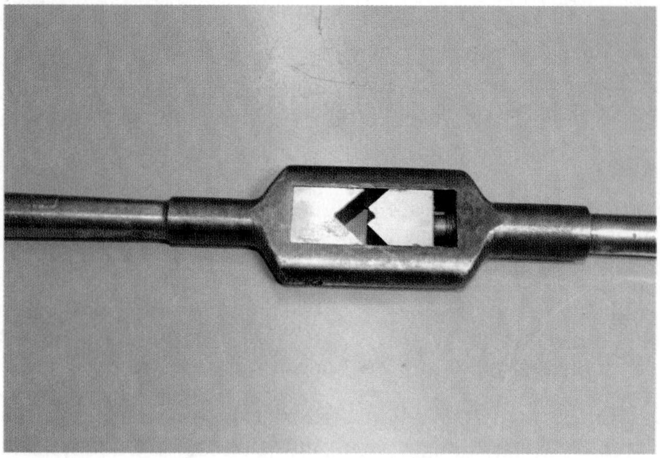

FIGURE 6-80 Tap handle.

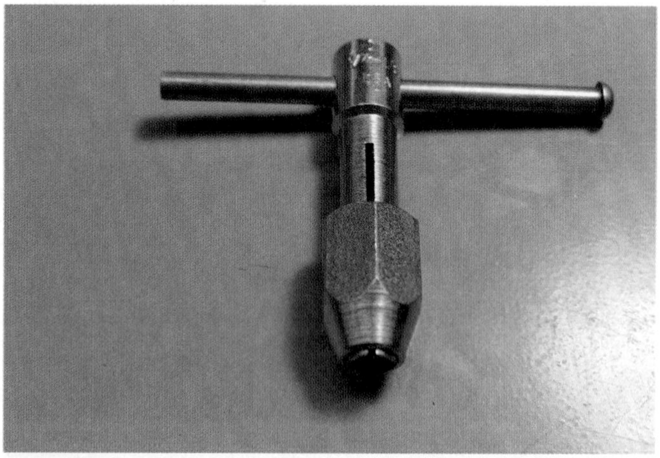

FIGURE 6-81 T-shaped tap handle.

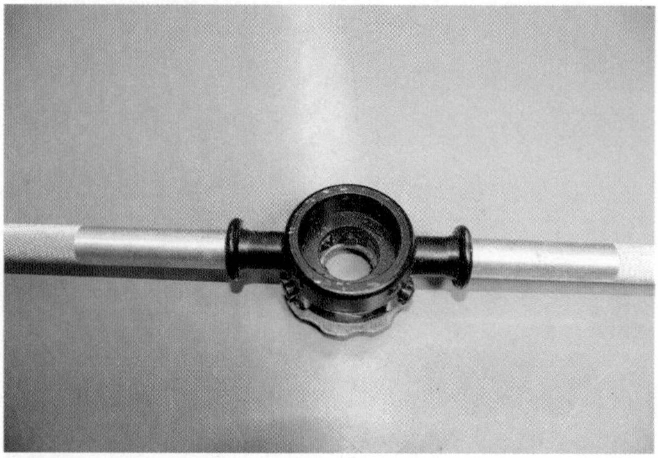

FIGURE 6-82 Die stock.

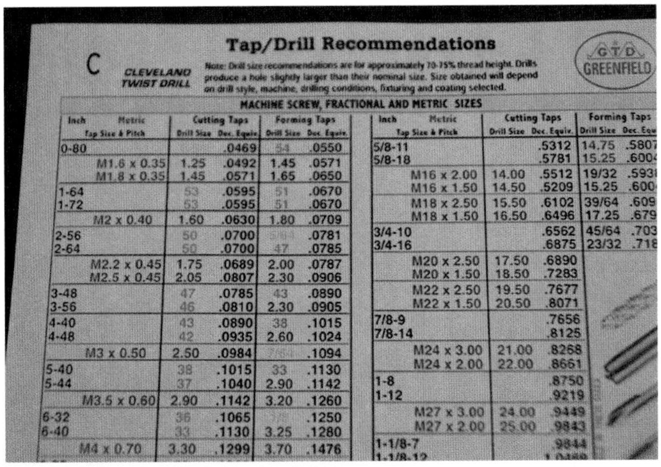

Tap/Drill Recommendations

C CLEVELAND TWIST DRILL

Note: Drill size recommendations are for approximately 70-75% thread height. Drills produce a hole slightly larger than their nominal size. Size obtained will depend on drill style, machine, drilling conditions, fixturing and coating selected.

GTD GREENFIELD

MACHINE SCREW, FRACTIONAL AND METRIC SIZES

Inch Tap Size & Pitch	Metric Tap Size & Pitch	Cutting Taps Drill Size	Cutting Taps Dec. Equiv.	Forming Taps Drill Size	Forming Taps Dec. Equiv.
0-80			.0469	54	.0550
	M1.6 x 0.35	1.25	.0492	1.45	.0571
	M1.8 x 0.35	1.45	.0571	1.65	.0650
1-64		53	.0595	51	.0670
1-72		53	.0595	51	.0670
	M2 x 0.40	1.60	.0630	1.80	.0709
2-56		50	.0700	[illegible]	.0781
2-64		50	.0700	47	.0785
	M2.2 x 0.45	1.75	.0689	2.00	.0787
	M2.5 x 0.45	2.05	.0807	2.30	.0906
3-48		47	.0785	43	.0890
3-56		46	.0810	2.30	.0905
4-40		43	.0890	38	.1015
4-48		42	.0935	2.60	.1024
	M3 x 0.50	2.50	.0984	[illegible]	.1094
5-40		38	.1015	33	.1130
5-44		37	.1040	2.90	.1142
	M3.5 x 0.60	2.90	.1142	3.20	.1260
6-32		36	.1065	[illegible]	.1250
6-40		33	.1130	3.25	.1280
	M4 x 0.70	3.30	.1299	3.70	.1476

Inch Tap Size & Pitch	Metric Tap Size & Pitch	Cutting Taps Drill Size	Cutting Taps Dec. Equiv.	Forming Taps Drill Size	Forming Taps Dec. Equiv.
5/8-11			.5312	14.75	.5807
5/8-18			.5781	15.25	.6004
	M16 x 2.00	14.00	.5512	19/32	.593
	M16 x 1.50	14.50	.5209	15.25	.600
	M18 x 2.50	15.50	.6102	39/64	.609
	M18 x 1.50	16.50	.6496	17.25	.679
3/4-10			.6562	45/64	.703
3/4-16			.6875	23/32	.718
	M20 x 2.50	17.50	.6890		
	M20 x 1.50	18.50	.7283		
	M22 x 2.50	19.50	.7677		
	M22 x 1.50	20.50	.8071		
7/8-9			.7656		
7/8-14			.8125		
	M24 x 3.00	21.00	.8268		
	M24 x 2.00	22.00	.8661		
1-8			.8750		
1-12			.9219		
	M27 x 3.00	24.00	.9449		
	M27 x 2.00	25.00	.9843		
1-1/8-7			.9844		
1-1/8-12			[illegible]		

FIGURE 6-83 Tap drill chart.

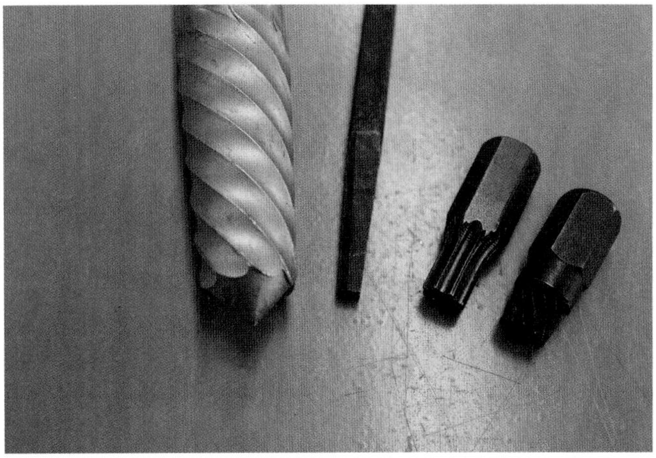

FIGURE 6-84 Screw extractors.

sized hole has been drilled, the taper tap or intermediate tap can be started in the hole. Make sure to use the proper lubricant for the metal you are tapping. Also be sure to start the tap straight. The best way to do that is to start the tap about one turn, stop, and then use a square to check the position of the tap in two places 90 degrees apart. If it is not perfectly straight, you can usually straighten it while turning it another half turn.

Again stop and verify that it is straight in two positions, 90 degrees apart. If the tap is straight, turn the tap about one full turn, and then back it off about a quarter turn. Continue cutting and backing off the tap until either the tap turns easily or you are at the bottom of the hole. Remove the tap. If you are cutting threads in a blind hole, you need to use a bottoming tap to finish the threads off. If not, clean the threaded hole and test-fit a bolt in the hole to check the threads. The bolt should turn smoothly by hand.

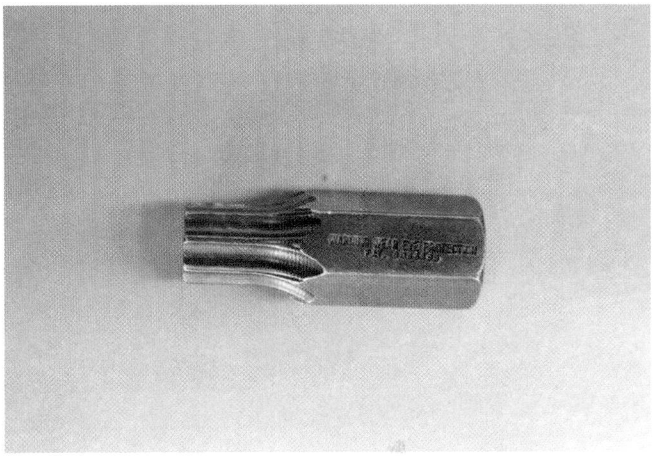

FIGURE 6-85 Straight-sided screw extractor.

Screw Extractors

Screw extractors are devices designed to remove screws, studs, or bolts that have broken off in threaded holes. A common type of extractor uses a coarse left-hand tapered thread formed on its hardened body (**FIGURE 6-84**). Normally, a hole is drilled in the center of the broken screw, and then the extractor is screwed into the hole. The left-hand thread grips the broken part of the bolt and unscrews it.

The extractor is marked with two sizes: one showing the size range of screws it is designed to remove, and the other, the size of the hole that needs to be drilled. It is important to carefully drill the hole in the center of the bolt or stud in case you end up having to drill the bolt out. If you drill the hole off center, you will not be able to drill it out all the way to the inside diameter of the threads. This makes removal of the broken bolt much harder.

Some screw extractors use a hardened, tapered square shank. It is hammered into a hole drilled into the center of the broken off bolt. The square edges of the screw extractor cut into the bolt and are used to grip it so it can be removed. Another type of screw extractor is the straight-sided, vertically splined, round shaft. The sides of the shaft have straight splines running the length of the extractor (**FIGURE 6-85**). It does not taper, so it is strong up and down its length. A straight hole of the correct diameter is drilled into the broken-off bolt. Then, the extractor is driven into the newly drilled hole. The vertical splines, being larger than the hole, grab onto the bolt, allowing it to be unthreaded.

Clamps, Vices, and Pullers

There are many types of vices, clamps, and pullers available. The **bench vice** is a useful tool for holding anything that can fit into its jaws (**FIGURE 6-86**). Some common uses include sawing, filing, or chiseling. The jaws are serrated to give extra grip. They are also very hard, which means that when the vice is tightened, the jaws can mar whatever they are gripping. To prevent this, a pair of soft jaws can be fitted whenever the danger of damage arises. They are usually made of aluminum or some other soft metal or can have a rubber-type surface applied to them.

When materials are too awkward to grip vertically in a plain vice, it may be easier to use an **offset vice**. The offset vice

FIGURE 6-86 Bench vice.

FIGURE 6-87 Offset vice being used to hold a pipe.

has its jaws set to one side to allow long components to be held vertically (**FIGURE 6-87**). For example, a long bar can be held vertically in an offset vice so it can be cut with a die.

A **drill vice** is designed to hold material on a drill worktable (**FIGURE 6-88**). The drill worktable has slots cut into it to allow the vice to be bolted down on the table. This helps to hold the material securely. The vice can be moved on the bed until the precise drilling point is located. It is then tightened down by bolts to hold the drill vice in place during drilling.

The name for the **C-clamp** comes from its shape (**FIGURE 6-89**). It can hold parts together while they are being assembled, drilled, or welded. It can reach around awkwardly shaped pieces that do not fit in a vice. It is also commonly used to retract disc brake caliper pistons. This clamp is portable, so it can be taken to where it is needed.

Pullers

Pullers are a very common universal tool that can be used for removing bearings, bushings, pulleys, and gears (**FIGURE 6-90**). Specialized pullers are also available for specific tasks where a

FIGURE 6-88 Drill vice on a drill worktable.

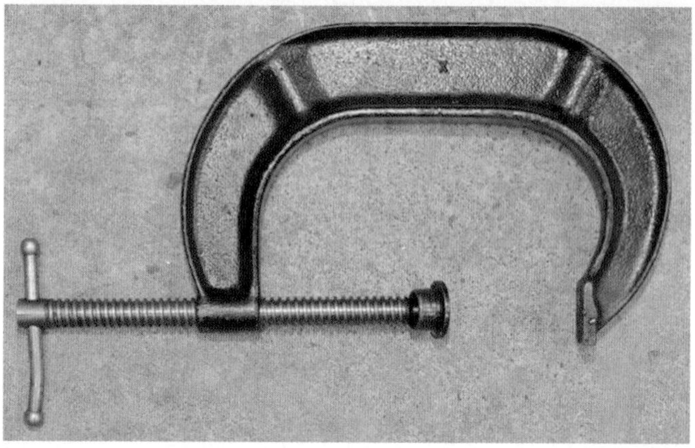

FIGURE 6-89 C-Clamp.

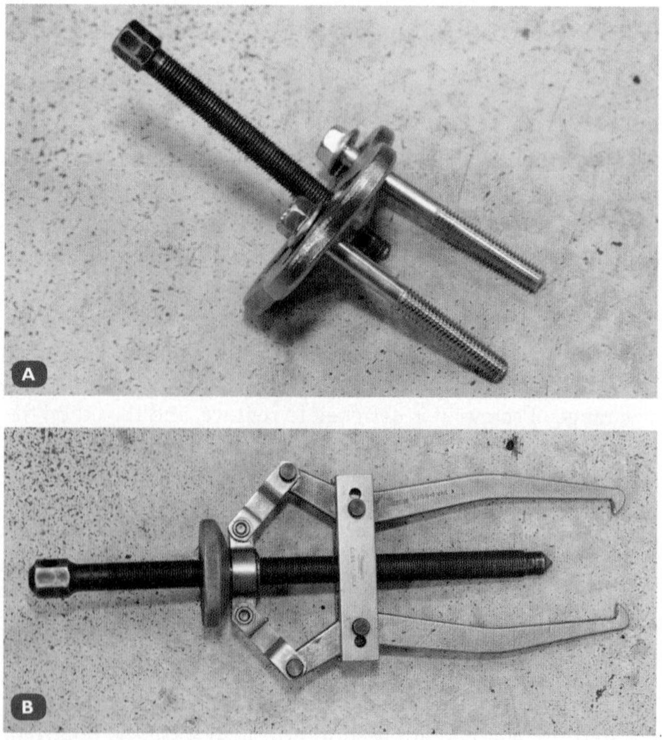

FIGURE 6-90 A. Puller. **B.** Gear puller.

standard puller would not work. The most common pullers have two or three legs that grip or push on the part to be removed. A center bolt, called a forcing screw, is then screwed in, producing a pulling action, which extracts the part. **Gear pullers** come in a range of sizes and shapes, all designed for particular applications. They consist of three main parts: jaws, a cross-arm, and a forcing screw. They are designed to connect to the component either externally or internally.

The **forcing screw** is a long, fine-threaded bolt that is applied to the center of the cross-arm. When the forcing screw is turned, it applies a very large force to the component you are removing. The forcing screw typically has interchangeable feet (**FIGURE 6-91**). A tapered cone-style foot does a good job of centering the puller. But it also creates a very large wedging effect, which can distort the end of the shaft. A flat-style foot is very good for pushing against the end of a shaft, but does not center itself. In either situation, the wrong foot size can push on the internal threads of the shaft, damaging them.

The cross-arm attaches the jaws to the forcing screw. If the **cross-arm** has four arms, three of the arms are spaced 120 degrees apart. The fourth arm is positioned 180 degrees apart from one arm (**FIGURE 6-92**). This allows the cross-arm to be used as either a two- or a three-arm puller.

Using Gear Pullers

Gear and bearing pullers are designed for hundreds of applications. Their main purpose is to remove a component such as a gear, pulley, or bearing from a shaft. They are also used to remove a component from inside a hole. Normally these components are pressed onto the shaft or into the hole. So removing them requires considerable force. To select, install, and use a gear puller to remove a pulley, follow the steps in **SKILL DRILL 6-4**.

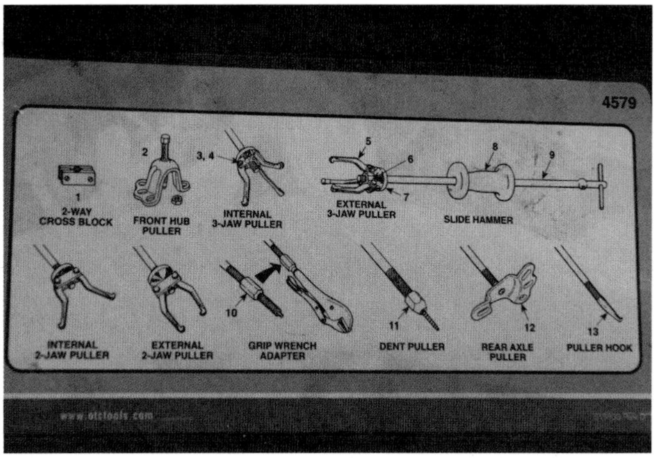

FIGURE 6-91 Interchangeable feet for the gear puller.

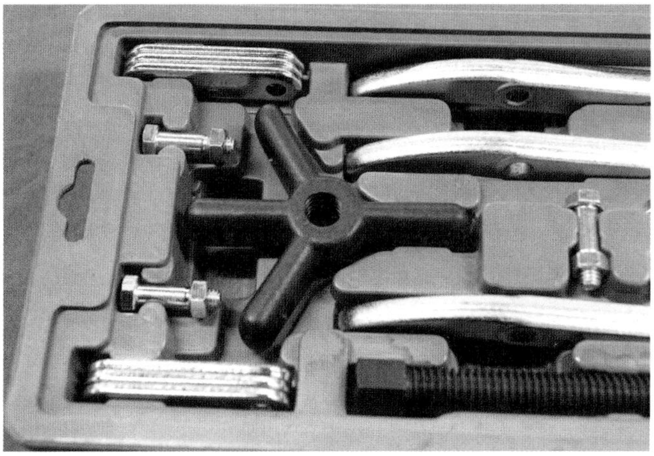

FIGURE 6-92 Four-arm puller.

SKILL DRILL 6-4 Using Gear Pullers

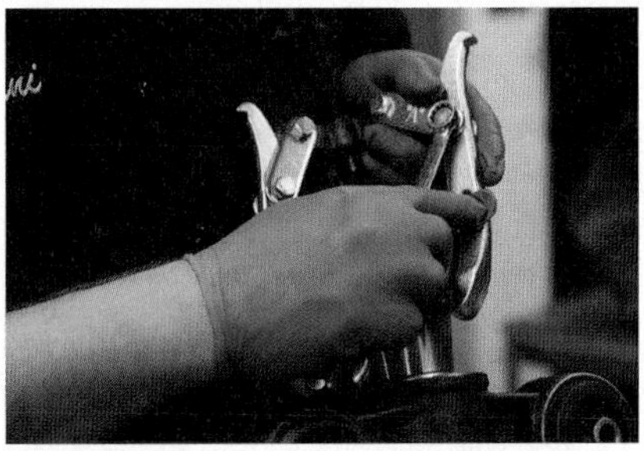

1. Examine the gear puller you have selected for the job. Identify the jaws; there may be two or three of them, and they must fit the part you want to remove. The cross-arm enables you to adjust the diameter of the jaws. The forcing screw should fit snugly onto the part you are removing. Finally, select the right wrench or socket size to fit the nut on the end of the forcing screw.

2. Adjust and fit the puller. Adjust the jaws and cross-arms of the puller so that it fits tightly around the part to be removed. The arms of the jaws should be pulling against the component at close to right angles.

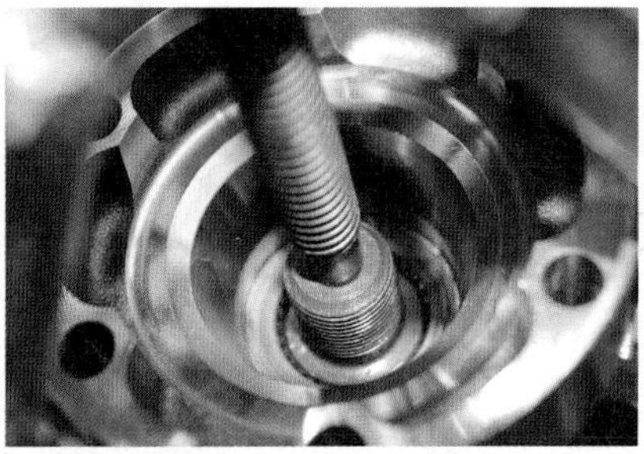

3. Position the forcing screw. Use the appropriate wrench to run the forcing screw down to touch the shaft. Check that the point of the forcing screw is centered on the shaft. If not, adjust the jaws and cross-arms until the point is in the center of the shaft. Also, be careful to use the correct foot on the end of the puller.

SKILL DRILL 6-4 Using Gear Pullers (Continued)

4. Tighten the forcing screw slowly and carefully onto the shaft. Check that the puller is not going to slip off center or off the pulley. Readjust the puller if necessary.

5. If the forcing screw and puller jaws remain in the correct position, tighten the forcing screw, and pull the part off the shaft.

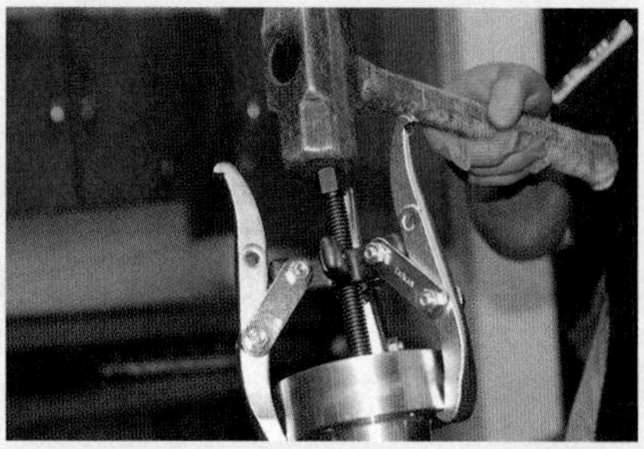

6. You may sometimes have to use a hammer to hit directly on the end of the forcing screw to help break the part loose.

SAFETY TIP

Always wear eye protection when using a gear puller.

Make sure the puller is located correctly on the workpiece. If the jaws cannot be fitted correctly on the part, then select a more appropriate puller. Do not use a puller that does not fit the job.

Flaring Tools

Flaring and riveting tools are only used on an occasional basis, but you still need to be familiar with them. A **tube flaring tool** is used to flare the end of a tube so it can be connected to another tube or component. One example of this is where the brake line screws into a wheel cylinder. The flared end is compressed between two threaded parts so that it seals the joint and withstands high pressures. There are three common types of flares. The **single flare** is for tube systems with low pressures like a fuel line. The **double flare** is for higher pressures such as in a brake system. The **ISO flare** (sometimes called a bubble flare) is the metric version used in brake systems (**FIGURE 6-93**).

Flaring tools have two main parts. One is a set of bars with holes that match the diameter of the tube end that is being shaped. The second is a yoke that drives a cone into the mouth of the tube (**FIGURE 6-94**). To make a single flare, the end of the tube is placed level with the surface of the top of the flaring bars. With the clamp screw firmly tightened, the feed screw flares the end of the tube.

Making a double flare is similar, but an extra step is added before flaring the tube. Also, more of the tube is exposed to allow for folding the flare over into a double flare. A double flaring button is placed into the end of the tube. The feed

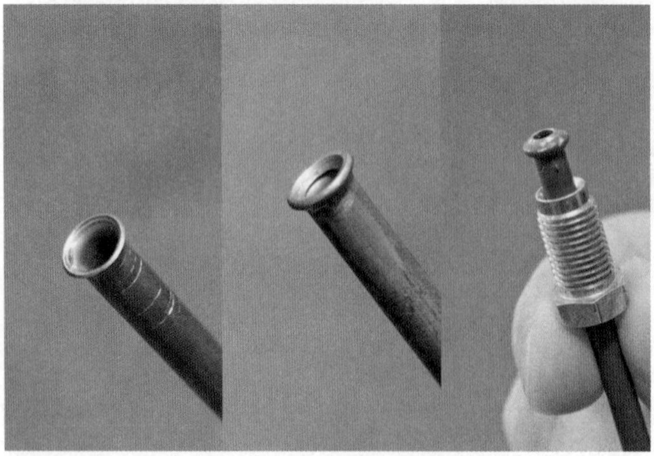

FIGURE 6-93 Single flare, double flare, and ISO flare.

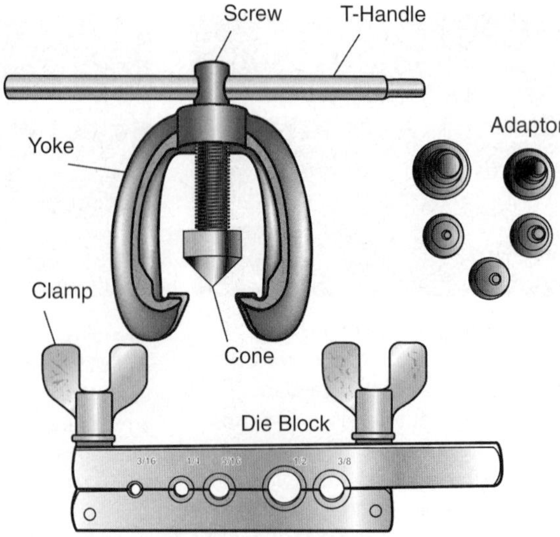

FIGURE 6-94 Components of a flare tool.

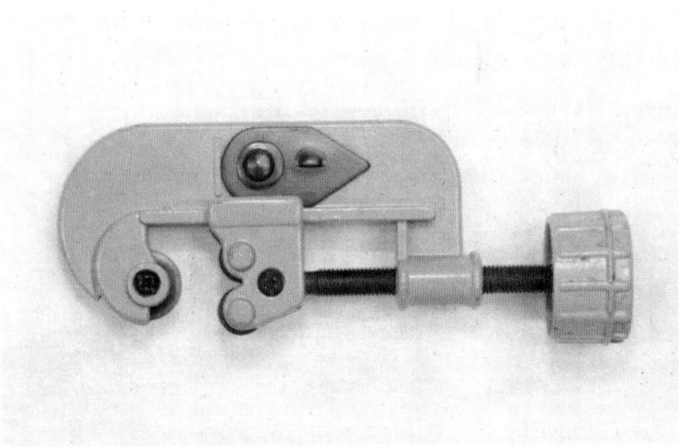

FIGURE 6-95 Tubing cutter.

screw then pushes the flaring button into the tube. When the flaring button is removed after tightening, the tube looks like a bubble. Placing the cone and yoke over the bubble allows you to force the bubble to fold in on itself, forming the double flare.

An ISO flare uses a flaring tool made specifically for that type of flare. The process is similar to that of the double-flare but stops after using the ISO button. An ISO flare does not get doubled back on itself. It should resemble a bubble shape when the work is finished.

A **tubing cutter** is more convenient and neater than a saw when cutting pipes and metal tubing (**FIGURE 6-95**). The sharpened wheel does the cutting. As the tool turns around the pipe, the screw increases the pressure on the cutting wheel. This forces the wheel deeper and deeper into the pipe until it finally cuts through. There is a larger version that is used for cutting exhaust pipes.

SAFETY TIP

A flaring tool is used to produce a pressure seal for sealing brake lines and fuel system tubing. Make sure you test the flared joint for leaks before completing the repair. Otherwise, the brakes could fail or the leaking fuel could catch on fire.

Using Flaring Tools

To make a successful flare, it is important to have the correct length of tube protruding through the tool before clamping. A higher protruding end will result in the tube folding over and thereby leaving the hole too small for the fluid to pass through. In case of too little end, there won't be enough tube to fold over properly, and the joint won't have full surface contact.

If you are making a double flare or ISO flare, make sure you use the correctly sized button for the tubing size. The button is also used to measure the amount of tube required to protrude from the tool prior to forming it. To prevent the tool from slipping on the tube, make sure the tool is clamped sufficiently tight around the tube. Do this before starting to create the flare. To create a flared fitting, follow the steps in **SKILL DRILL 6-5**.

SKILL DRILL 6-5 Using Flaring Tools

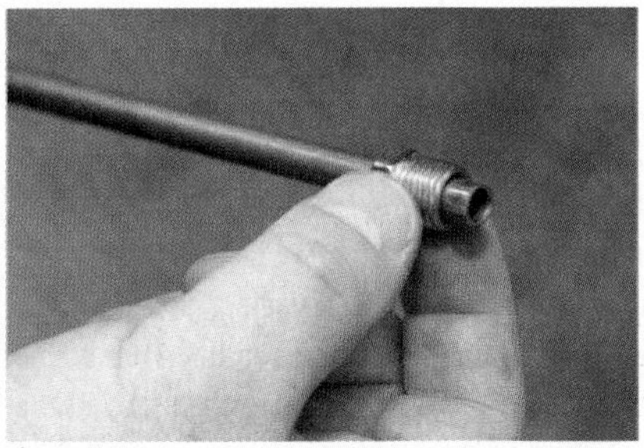

1. Choose the tube you will use to make the flare, and put the flare nut on the tube before creating the flare.

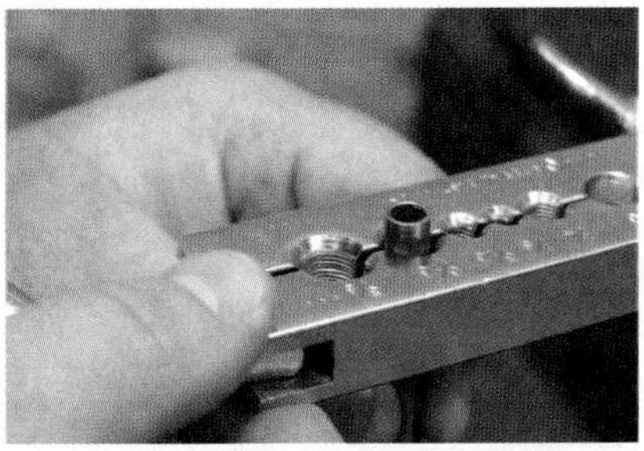

2. Match the size of the tube to the correct hole in the tubing clamp.

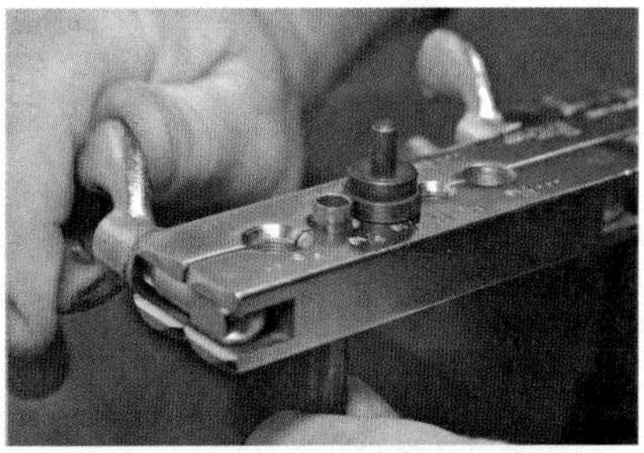

3. Holding the flaring tool, put the tube into the clamp. Position the tube so that the correct length is showing through the tool. If you are conducting a double flare, use the correctly sized button to ensure the proper length of the tube is sticking up above the top of the clamp. Tighten the two halves of the clamp together using the wing nuts. Make sure the tool is tight enough to clamp the tube so it will not slip.

SKILL DRILL 6-5 Using Flaring Tools (Continued)

4. Put the cone and forming tool over the clamp, and turn the handle to make the flare. If you are doing a double flare or ISO flare, place the button in the end of the tube, install the cone and forming tool, and turn the handle to make the bubble. Remove the button from the tube.

5. If this is an ISO flare, inspect it to see if it is properly formed. If it is a double flare, put the cone and forming tool back on the clamp, and tighten the forming tool handle to create the double flare.

6. Remove the forming tool.

7. Remove the tube from the clamp, and check the flare to ensure it is free of burrs and is correctly formed.

Riveting Tools

There are many applications for blind rivets. Various rivet types and tools may be used to do the riveting. Rivets are used in many places in automotive applications. They can be used where there is a need for a fastener that does not have to be easily removed. Some window regulators, body panels, trim pieces, and even some suspension members use rivets to keep them attached without vibrating loose. **Pop rivet guns** are convenient for occasional riveting of light materials (**FIGURE 6-96**).

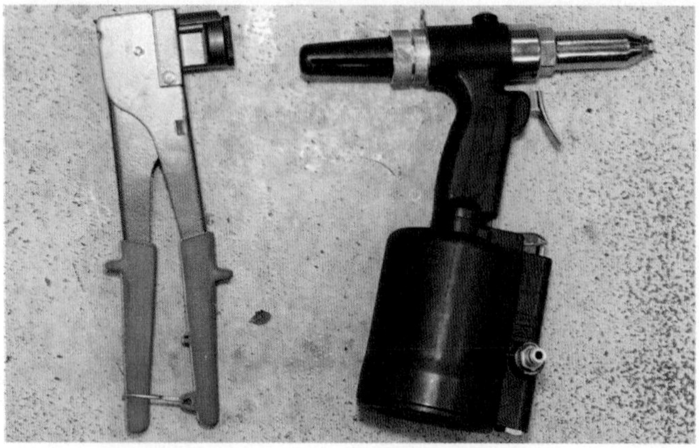

FIGURE 6-96 Pop rivet guns.

A typical pop, or **blind rivet**, has two main pieces. First, a body that forms the **finished rivet**. And second, a mandrel which is discarded when the riveting is completed (**FIGURE 6-97**). It is called a blind rivet because there is no need to access the other side of the hole in which the rivet is installed. In some types, the rivet is plugged shut so that it is waterproof or pressure proof.

The rivet is inserted into the riveting tool. When squeezed, it pulls the end of the **mandrel** back through the body of the rivet. Because the **mandrel head** is bigger than the hole through the body of the rivet, it swells tightly against the hole. Finally, the

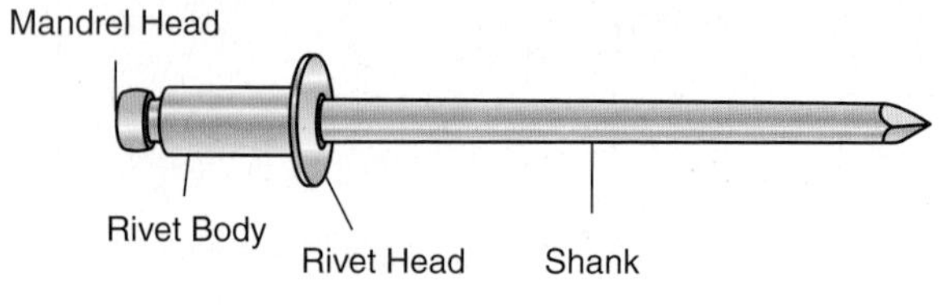

FIGURE 6-97 Anatomy of a rivet.

mandrel head will snap off under the pressure and fall out. This leaves the rivet body gripping the two pieces of material together.

Using Riveting Tools

Rivet tools are used to join two pieces of metal or other material together. For example, sheet metal that needs to be attached to a stiffening frame. To perform a riveting operation, you need a rivet gun, rivets, a drill, the right-sized drill bit, and the materials to be riveted.

Rivets come in various diameters and lengths for different sizes of jobs. They are made of various types of metals to suit the job at hand. When selecting rivets for a job, consider the diameter, length, and rivet material. Larger diameter rivets should be used for jobs that require more strength. The rivet length should be sufficient to protrude past the materials being riveted by about 1.6 times the diameter of the rivet stem. Typically, you should select rivets that are made from the same material as that being riveted. For example, stainless steel rivets should be used for riveting stainless steel. And aluminum rivets should be used to rivet aluminum.

Pilot holes must be drilled through the metal to be riveted. Ensure that the hole is just large enough for the rivet to comfortably pass through it. Do not make it so the rivet fits loosely. If the hole is too large, the rivet will be loose and will not hold the materials securely together. When drilling holes for rivets, stay back from the material's edge. This helps ensure that the rivets do not break through the edge of the materials being riveted. A good rule of thumb is to allow at least twice the diameter of the rivet stem as clearance from any edge.

Most rivet tools are capable of riveting various sizes of rivets. They have a number of nosepiece sizes to work with different sizes of rivets. Make sure you select the proper nosepiece for the rivet you are using.

▶ TECHNICIAN TIP

A rivet is a single-use fastener. Unlike a nut and bolt, which can normally be disassembled and reused, a rivet cannot. The metal shell that makes up a pop rivet is crushed into place so that it holds the parts firmly together. If it ever needs to be removed, it must be drilled out or cut off.

▶ SAFETY TIP

When compressing the rivet handles, be careful not to place your fingers between the handles. When the rivet "pops," they could end up pinching your fingers.

To use a riveting tool to rivet two pieces of material together, follow the steps in **SKILL DRILL 6-6**.

SKILL DRILL 6-6 Using Riveting Tools

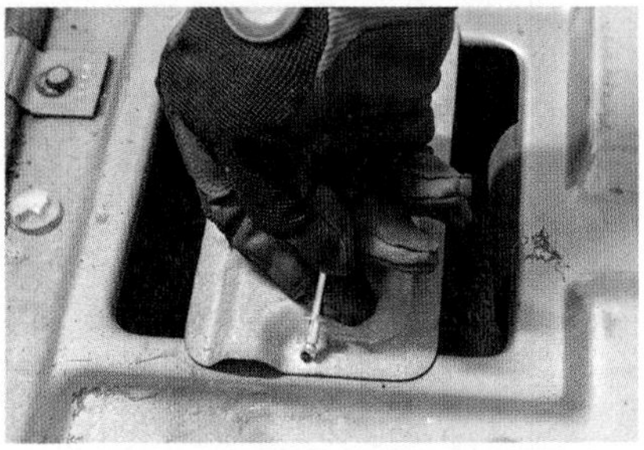

1. Select the correct rivet for the components you are riveting. Make sure the rivet is of the correct length.

2. Verify the pilot holes in the material is of the proper size. Remove any burrs.

3. Make sure the correctly sized nosepiece for the rivet size is fitted to the rivet tool.

4. Insert the rivet into the gun, and push the rivet through the materials to be riveted. Hold firm pressure while pushing the rivet into the work.

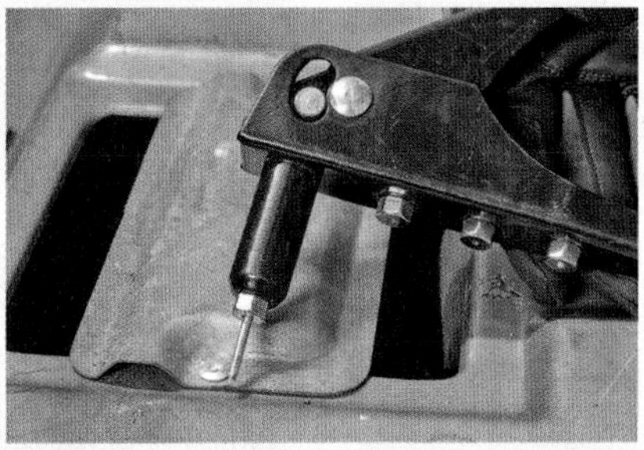

5. Operate the rivet tool to compress the rivet. Continue this process until the rivet stem or shank breaks away from the rivet head.

6. Check the rivet joint to ensure the pieces are firmly held together.

▶ Precision Measuring Tools

LO 6-06 Measure precisely using measuring tools.

Technicians are required to perform a variety of measurements while carrying out their job. This requires knowledge of what measuring tools are available and how to use them. Measuring tools can generally be classified according to what type of measurements they can make. A measuring tape is useful for measuring longer distances. It is accurate to a millimeter or fraction of an inch (**FIGURE 6-98**). A steel rule is capable of accurate measurements on shorter lengths, down to a millimeter or a fraction of an inch. Precision-measuring tools are accurate to much smaller dimensions, such as a micrometer. In some cases, they can accurately measure down to 1/10,000 of an inch (0.0001") or 1/1000 of a millimeter (0.001 mm).

Measuring Tapes

Measuring tapes are a flexible type of ruler and are a common measuring tool. The most common type found in shops is a thin metal strip about 0.5"–1" (13–25 mm) wide that is rolled up inside a housing with a spring return mechanism. Measuring tapes can be of various lengths, with 16" or 25" (5 or 8 m) being very common. The measuring tape is pulled from the housing to measure items, and a spring return winds it back into the housing. The housing usually has a built-in locking mechanism. It holds the extended measuring tape against the spring return mechanism. The hooked end can be placed over the edge of the object you are measuring and pulled against spring tension to take the measurement (**FIGURE 6-99**).

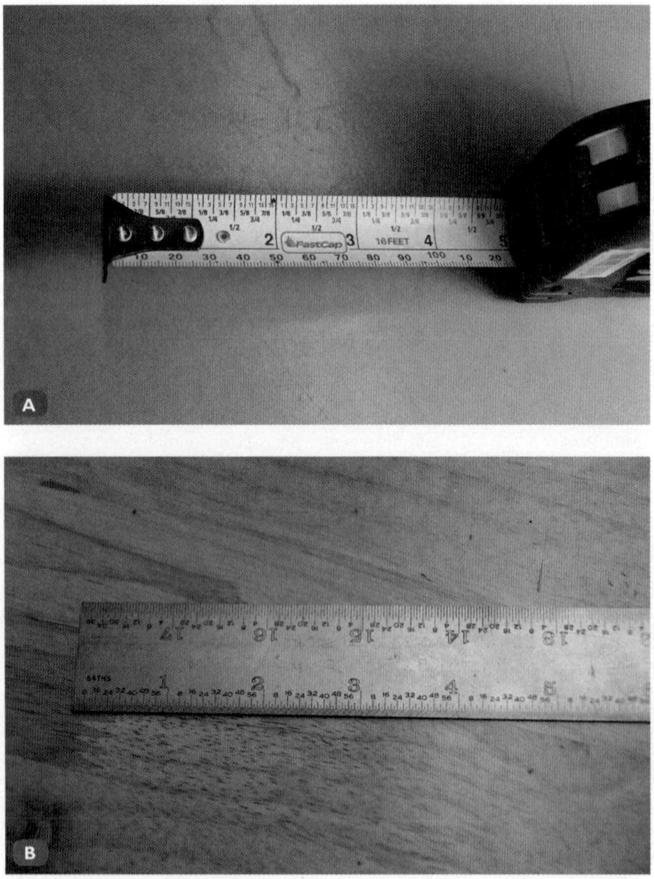

FIGURE 6-98 A. Measuring tape. **B.** Steel rule.

> **▶ TECHNICIAN TIP**
>
> The USCS, also called the standard system, and the metric system are two sets of standards for quantifying weights and measurements. Each system has defined units. For example, the standard system uses inches, feet, and yards. The metric system uses millimeters, centimeters, and meters. Conversions can be undertaken from one system to the other. For example, 1 inch is equal to 25.4 millimeters, and 1 foot is equal to 304.8 millimeters.
>
> Tools that make use of a measuring system come in both standard and metric measurements. These include wrenches, sockets, drill bits, micrometers, rulers, and many others. To work on modern vehicles, an understanding of both systems and their conversion is required. Conversion tables can be used to convert from one system to the other. The more you work with both systems, the easier it will be to understand how they relate to each other.

Steel Rulers

As the name suggests, a **steel rule** is a ruler that is made from steel. Steel rules commonly come in 12", 24", and 36" lengths. They are used like any ruler to measure and mark out items. A steel rule is a very strong ruler, has precise markings, and resists damage. When using a steel rule, you can rest it on its edge so that the markings are closer to the material being measured. This helps to mark the work more precisely (**FIGURE 6-100**). Always protect the steel rule from damage by storing it carefully. A damaged ruler will not give an accurate measurement. Never take measurements from the very end of a damaged steel rule, as damaged ends may affect the accuracy of your measurements.

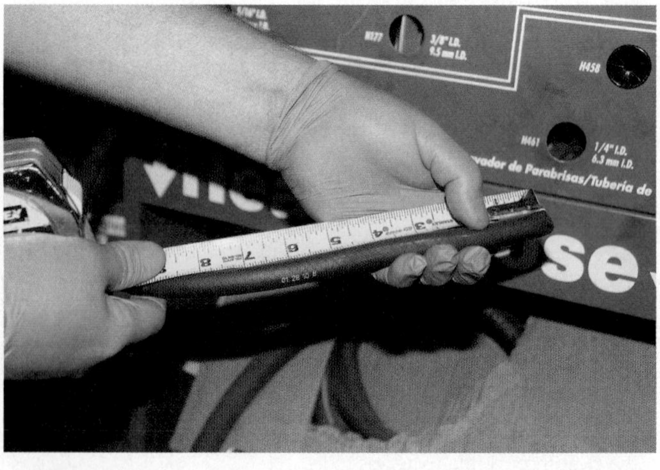

FIGURE 6-99 The hook makes it easy to position the end to take a measurement.

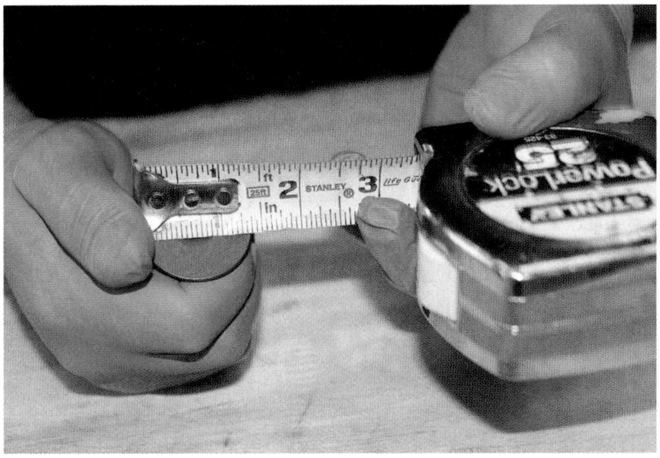

FIGURE 6-100 Tipping a steel rule on its side to get a more accurate reading.

> **▶ TECHNICIAN TIP**
>
> If the end of a rule is damaged, you may be able to measure from the 1" mark and subtract an inch from the measurement.

Outside, Inside, and Depth Micrometers

Micrometers are precise measuring tools. They are designed to measure small distances and are available in both inch and millimeter (mm) calibrations. Typically, a standard micrometer can measure down to a resolution of 1/1000 of an inch (0.001"). A metric micrometer can measure down to 1/100 of a millimeter (0.01 mm). Vernier micrometers equipped with a vernier scale can measure down to 1/10,000 of an inch (0.0001") or 1/1000 of a millimeter (0.001 mm).

The most common types of micrometers are the outside, inside, and depth micrometers (**FIGURE 6-101**). As the name suggests, an **outside micrometer** measures the outside dimensions of an item. For example, it could measure the diameter of a valve stem. The **inside micrometer** measures the inside dimensions. For example, the inside micrometer could measure the cylinder bore of an engine. **Depth micrometers** measure the depth of an item such as how much clearance a piston has below the surface of the block.

The most common micrometer is the outside micrometer and is made up of several parts (**FIGURE 6-102**). The horseshoe-shaped part is the frame. It is built to make sure the micrometer holds its shape. Some frames have plastic finger pads so that body heat is not transferred to the metal frame as easily. This prevents body heat from causing the metal to expand slightly and affecting the reading. On one end of the frame is the anvil, which contacts one side of the part being measured. The other contact point is the spindle. The micrometer measures the distance between the anvil and spindle, so that is where the part being measured fits.

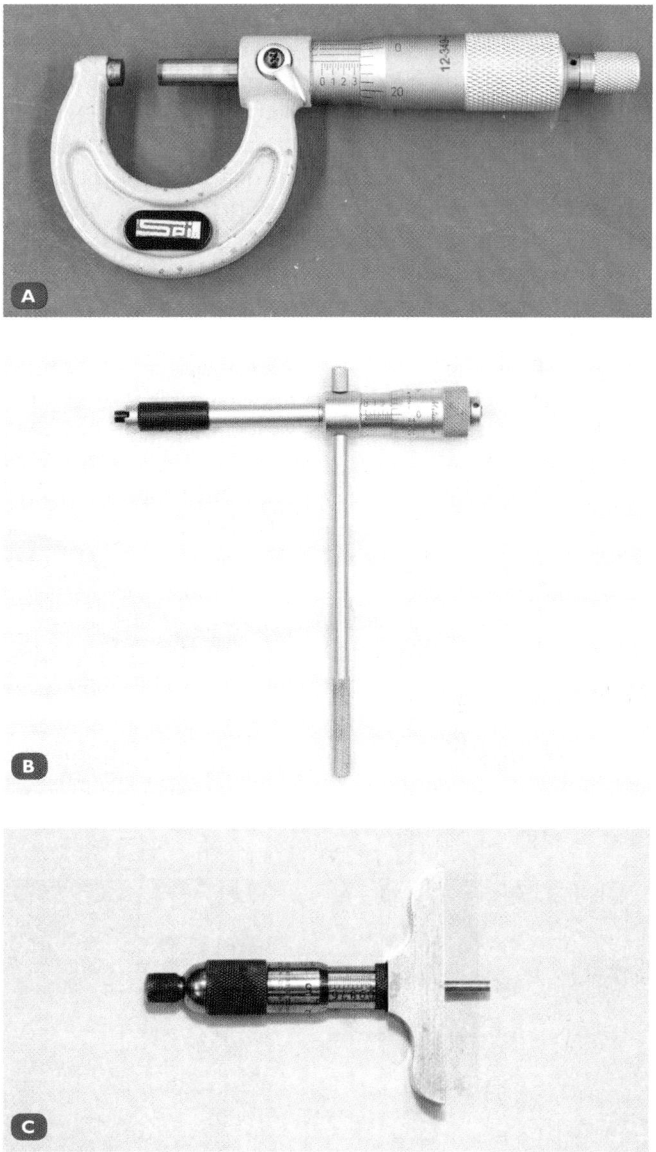

FIGURE 6-101 **A.** Outside micrometer. **B.** Inside micrometer. **C.** Depth micrometer.

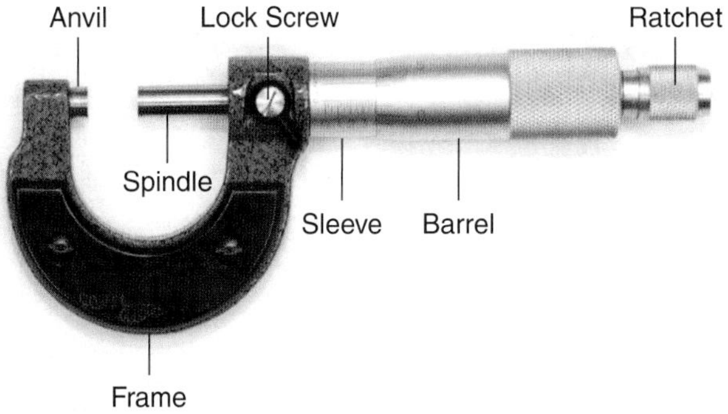

FIGURE 6-102 Parts of an outside micrometer.

The measurement is read on the sleeve/barrel and thimble. The sleeve/barrel is stationary and has the linear markings on it. The thimble fits over the sleeve and has graduated markings on it. The thimble is connected directly to the spindle, and both turn as a unit. Because the spindle and sleeve/barrel have matching threads, the thimble rotates the spindle inside of the sleeve/barrel. The thread moves the spindle inward and outward. The thimble usually incorporates either a ratchet or a clutch mechanism, which is turned lightly by finger. This prevents overtightening of the micrometer thimble when taking a reading.

A lock nut, lock ring, or lock screw is used on most micrometers to lock the thimble in place while you read the micrometer. Standard micrometers use a specific thread of 40 TPI (threads per inch) on the spindle and sleeve. This means that the thimble rotates exactly 40 turns in 1" of travel. Every complete rotation moves the spindle 1/40th of an inch, or 0.025" (1 ÷ 40 = 0.025). In four rotations, the spindle moves 0.100" (0.025 × 4 = 0.100). The linear markings on the sleeve show each of the 0.100" marks between 0 and 1 inch as well as each of the 0.025" marks (**FIGURE 6-103**). Because the thimble has graduated marks from 0 to 24 (each mark representing 0.001"), every complete turn of the thimble uncovers another one of the 0.025" marks on the sleeve. If the thimble stops short of any complete turn, it will indicate the number of 0.001" marks past the zero line on the sleeve (**FIGURE 6-104**).

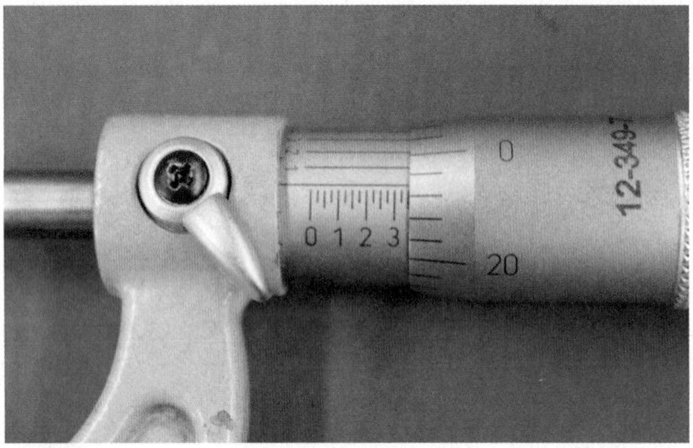

FIGURE 6-103 0.100" and 0.025" markings on the sleeve.

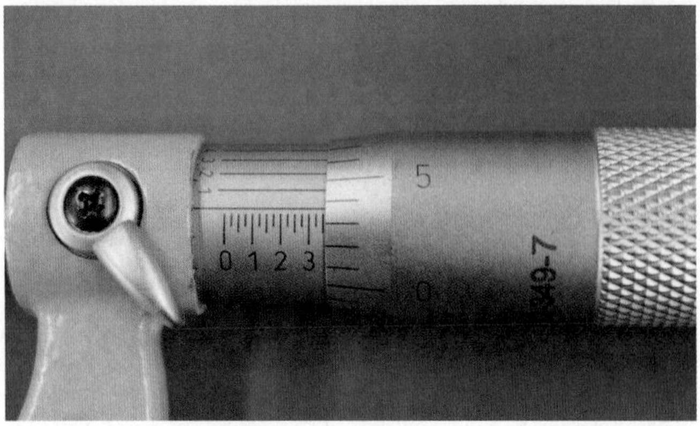

FIGURE 6-104 Markings on the thimble.

So reading a micrometer is as simple as adding up the numbers as shown below.

> ▶ **TECHNICIAN TIP**
>
> Micrometers are precision-measuring instruments and must be handled and stored with care. They should always be stored with a gap between the spindle and anvil so that metal contraction and expansion do not interfere with their calibration.

To read a standard micrometer, perform the following steps (**FIGURE 6-105**):

1. Verify that the micrometer is properly calibrated.
2. Verify what size of micrometer you are using. If it is a 0–1" micrometer, start with 0.000. If it is a 1–2" micrometer, start with 1.000". A 2–3" micrometer would start with 2.000", etc. (To give an example, let's say it is 2.000".)
3. Read how many 0.100" marks the thimble has uncovered (e.g., 0.300").
4. Read how many 0.025" marks the thimble has uncovered past the 0.100" mark in step 3 (e.g., 2 × 0.025 = 0.050").
5. Read the number on the thimble that lines up with the zero line on the sleeve (e.g., 13 × 0.001 = 0.013").
6. Finally, total all of the individual readings (e.g., 2.000 + 0.300 + 0.050 + 0.013 = 2.363").

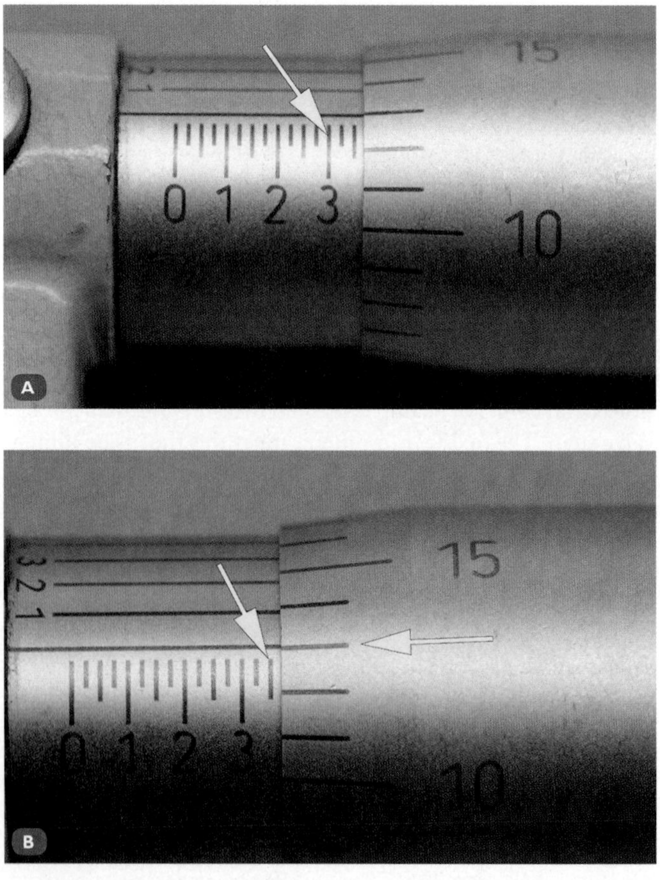

FIGURE 6-105 **A.** Read how many 0.100" and 0.025" marks the thimble has uncovered. **B.** Read the number on the thimble that lines up with the zero line on the sleeve.

FIGURE 6-106 Metric markings on the sleeve.

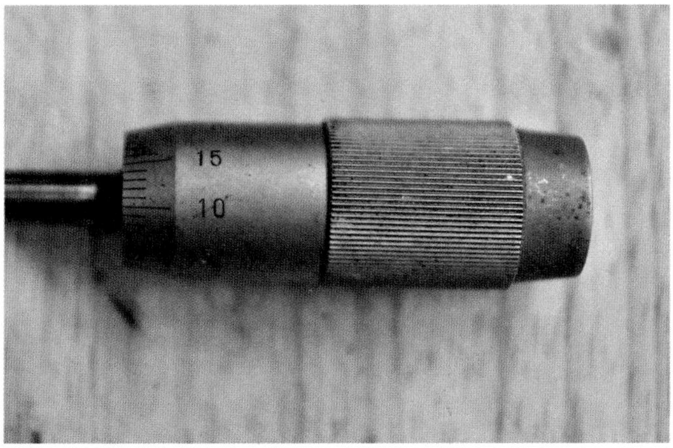

FIGURE 6-107 Markings on the thimble.

A metric micrometer uses the same components as the standard micrometer. However, it uses a different **thread pitch** on the spindle and sleeve. It uses a 0.5-mm thread pitch (2.0 threads per millimeter) and opens up approximately 25 mm. Each rotation of the thimble moves the spindle 0.5 mm. It therefore takes 50 rotations of the thimble to move the full 25 mm distance. The sleeve/barrel is labeled with individual millimeter marks and half-millimeter marks from the beginning to the end (**FIGURE 6-106**). The thimble has graduated marks from 0 to 49 (**FIGURE 6-107**).

▶ TECHNICIAN TIP

All micrometers need to be checked for calibration (also called "zeroing") before each use. A 0–1" or 0–25 mm outside micrometer can be lightly closed all of the way. If the anvil and spindle are clean, the micrometer should read 0.000. This indicates that the micrometer is calibrated correctly. If the micrometer is bigger than 1", or 25 mm, then a "standard" is used to verify the calibration. A standard is a hardened, machined rod of a precise length, such as 2", or 50 mm. When inserted in the same-sized micrometer, the reading should be exactly the same as listed on the standard. If a micrometer is not properly calibrated, it will give you incorrect readings. So it should not be used until it is recalibrated. See the tool's instruction manual for the calibration procedure.

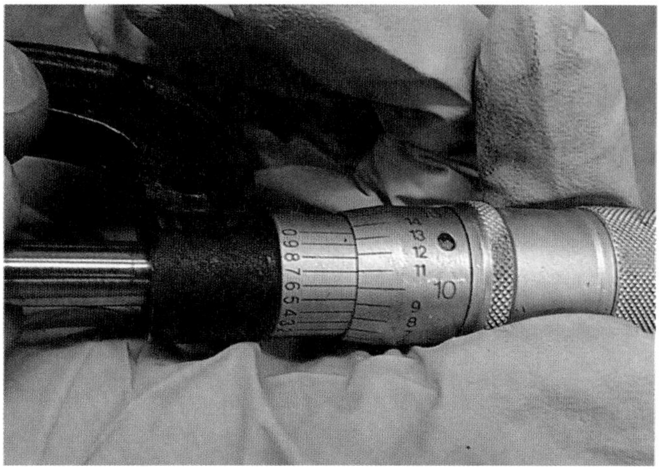

FIGURE 6-108 Vernier scale on a micrometer showing 7 on the sleeve lined up the best with a line on the thimble.

Reading a metric micrometer involves the following steps:

1. Read the number of full millimeters the thimble has passed (to give an example, let's say it is 23.00 mm).
2. Check to see if it passed the 0.5 mm mark (e.g., 0.50 mm).
3. Check to see which mark on the thimble lines up with or is just passed (e.g., 37 × 0.01 mm = 0.37 mm).
4. Total all of the numbers (e.g., 23.00 mm + 0.50 mm + 0.37 mm = 23.87 mm).

If the micrometer is equipped with a vernier gauge, meaning it can read down to 1/10,000 of an inch (0.0001") or 1/1000 of a millimeter (0.001 mm), you need to complete one more step. Identify which of the vernier lines is closest to one of the lines on the thimble (**FIGURE 6-108**). Sometimes, it is hard to determine which is the closest. So decide which of the three are the closest, and then use the center line. At the frame side of the sleeve will be a number that corresponds to the vernier line, numbered 1–0. Add the vernier to the end of your reading. For example, 2.363 + 0.0007 = 2.3637", and 23.77 + 0.007 = 23.777 mm.

For inside measurements, the inside micrometer works on the same principles as the outside micrometer. The depth micrometer works similarly, but the sleeve is backward, so be careful when reading it.

Using Micrometers

To maintain accuracy of measurements, it is important that both the micrometer and the items to be measured are clean and free of any dirt or debris. Also, make sure the micrometer is zeroed before taking any measurements. Never overtighten a micrometer. Never store it with its measuring surfaces touching. This may damage the tool and affect its accuracy. When measuring, make sure the item can pass through the micrometer surfaces snugly and squarely. This is best accomplished by using the ratchet to tighten the micrometer. Always take the measurement a number of times and compare the results with each other. This will help ensure you have measured accurately. To correctly measure using an outside micrometer, follow the steps in **SKILL DRILL 6-7**.

SKILL DRILL 6-7 Using Micrometers

1. Select the correct size of micrometer. Verify that the anvil and spindle are clean and that it is calibrated properly. Clean the surface of the part you are measuring.

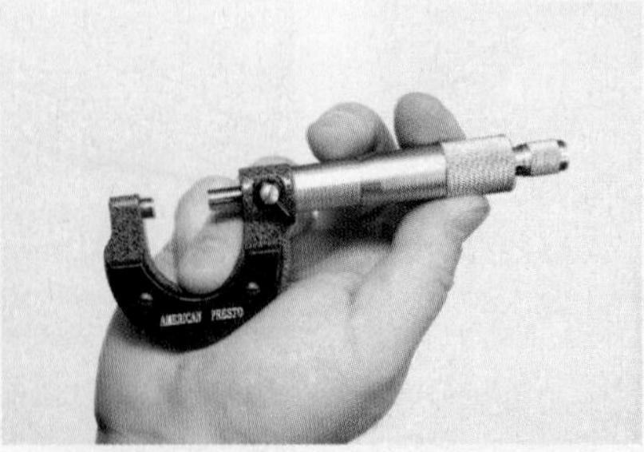

2. In your right hand, hold the frame of the micrometer between your pinky, ring finger, and the palm of your hand, with the thimble between your thumb and forefinger.

3. With your left hand, hold the part you are measuring, and place the micrometer over it.

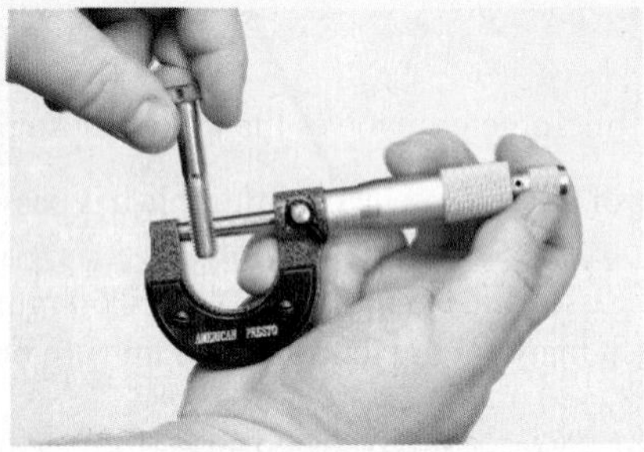

4. Using your thumb and forefinger, lightly tighten the ratchet. It is important that the correct amount of force is applied to the spindle when taking a measurement. The spindle and anvil should just touch the component with a slight amount of drag when the micrometer is removed from the measured piece. Be careful that the part is square in the micrometer so the reading is correct. Try rocking the micrometer in all directions to make sure it is square.

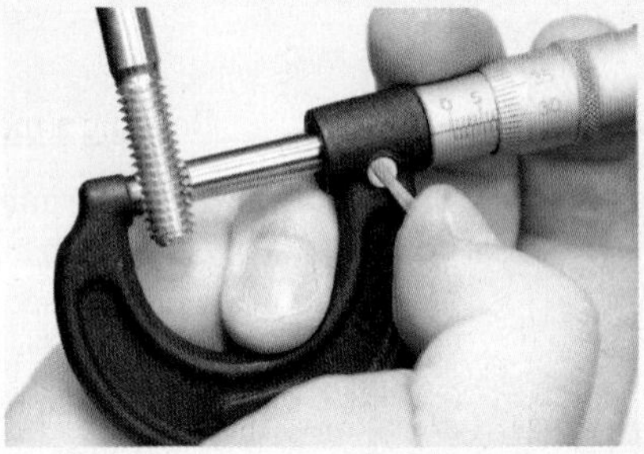

5. Once the micrometer is properly snug, tighten the lock mechanism so the spindle will not turn. Read the micrometer and record your reading.

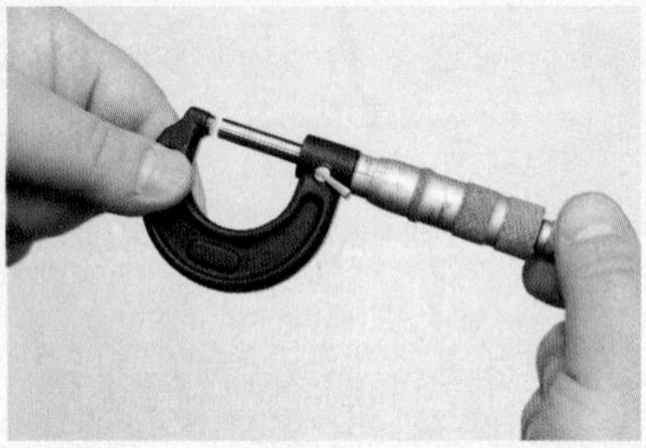

6. When all readings are finished, clean the micrometer, position the spindle so it is backed off from the anvil, and return it to its protective case.

Telescoping Gauges

Sometimes, you may need to measure distances in awkward spots like the bottom of a deep cylinder. **Telescoping gauges** can do this. They have spring-loaded plungers that are released with a screw on the handle. This allows the plungers to expand inside of a cylinder. The screw on the handle then locks the plungers in position. This allows the telescoping gauge to be removed. The distance across the plungers can be measured with an outside micrometer (**FIGURE 6-109**). The measurement on the micrometer indicates the diameter of the cylinder at that point. Telescoping gauges come in a variety of sizes to fit various sizes of holes and bores.

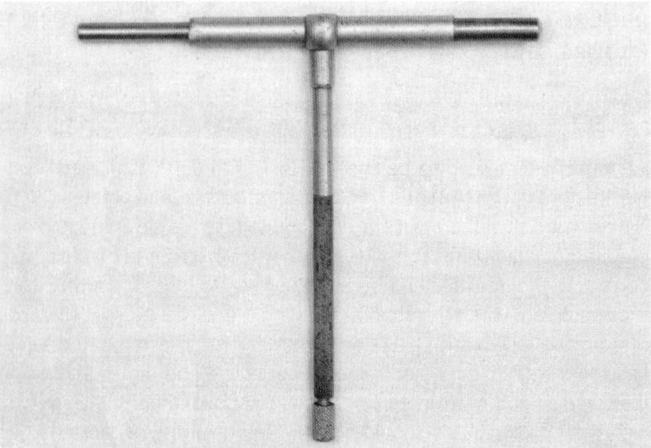

FIGURE 6-109 Telescoping gauge.

Split Ball Gauges

A **split ball gauge** or small hole gauge is good for measuring small holes where telescoping gauges cannot fit (**FIGURE 6-110**).

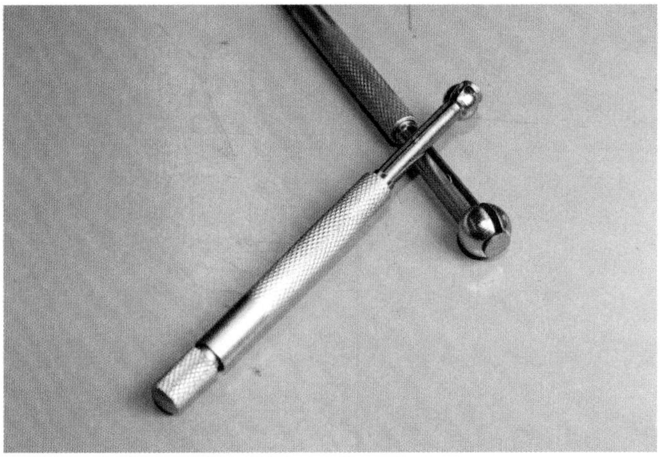

FIGURE 6-110 Split ball gauges.

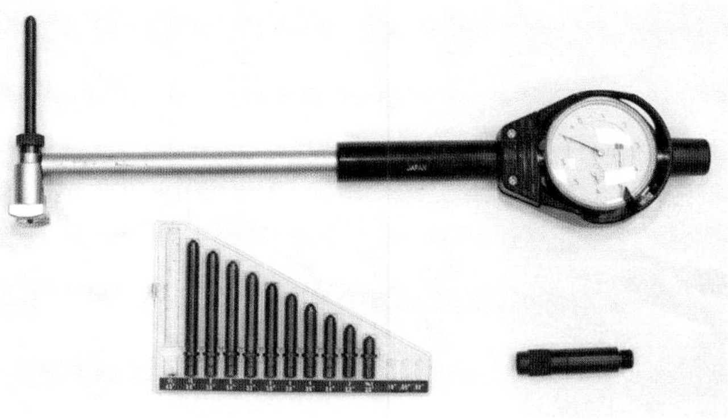

FIGURE 6-111 A dial bore gauge set.

They use a similar principle to the telescoping gauge. But the measuring head uses a split ball mechanism that allows it to fit into very small holes. Split ball gauges are ideal for measuring valve guides on a cylinder head for wear. A split ball gauge can be fitted in the bore and expanded until there is a slight drag. Then, it can be retracted and measured with an outside micrometer.

Dial Bore Gauges

A **dial bore gauge** measures the inside diameter of bores with a high degree of accuracy and speed (**FIGURE 6-111**). The dial bore gauge can measure a bore directly. It uses telescoping pistons on a T-handle with a dial mounted on the handle. The dial bore gauge combines a telescoping gauge and dial indicator in one instrument. A dial bore gauge can measure if the cylinder is worn, tapered, or out-of-round. The resolution of a dial bore gauge is typically accurate to 5/10,000 of an inch (0.0005") or 1/100 of a millimeter (0.01 mm).

Using Dial Bore Gauges

Bore gauges are available in different ranges of size. It is important to select a gauge with the correct range for the bore you are measuring. To use a dial bore gauge, select an appropriate-sized adapter to fit the internal diameter of the bore. Install it to the measuring head. Many dial bore gauges also have a fixture to calibrate the tool to the size you desire (**FIGURE 6-112**). The fixture is set to the size desired, and the dial bore gauge is placed in it. The dial bore gauge is then adjusted to the proper reading.

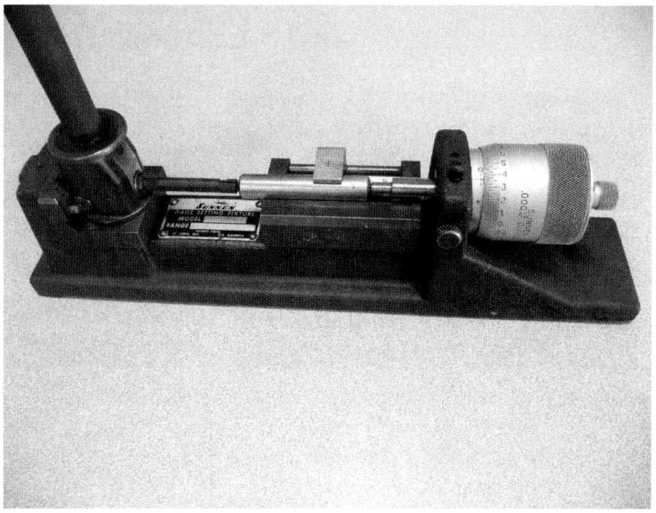

FIGURE 6-112 Dial bore gauge being calibrated to a predetermined size.

Once it is calibrated, the dial bore gauge can be inserted inside the bore to be measured. Hold the gauge in line with the bore, and slightly rock it to ensure it is centered. Make sure the gauge is at a 90-degree angle to the bore and read the dial. Always take the measurement a number of times and compare the results to ensure you have measured accurately. It takes a bit of practice to get accurate readings.

Store a bore gauge carefully in its storage box and ensure the locking mechanism is released while in storage.

To correctly measure using a dial bore gauge, follow the steps in **SKILL DRILL 6-8**.

▶ Vernier Calipers

LO 6-07 Measure precisely using other measuring tools.

Vernier calipers are a precision instrument. They are used for measuring outside, inside, and depth dimensions, all in one tool (**FIGURE 6-113**). They have a graduated bar with markings like a ruler. On the bar, a sliding sleeve with a jaw is mounted for taking inside or outside measurements. Measurements are taken by comparing the scales on the sliding sleeve to the graduated bar.

Some versions of vernier calipers have dial scales (**FIGURE 6-114**). The dial caliper has the main scale on the graduated bar. The fractional measurements are taken from a dial with a rotating needle. These tend to be easier to read than straight vernier calipers. More recently, digital readouts on vernier calipers have become commonplace (**FIGURE 6-115**). The principle of their use is the same as any vernier caliper. However, they have a digital scale that reads the measurement directly.

Using Vernier Calipers

Always store vernier calipers in a storage box to protect them. Ensure the measuring surfaces are kept clean for accurate

SKILL DRILL 6-8 Using Dial Bore Gauges

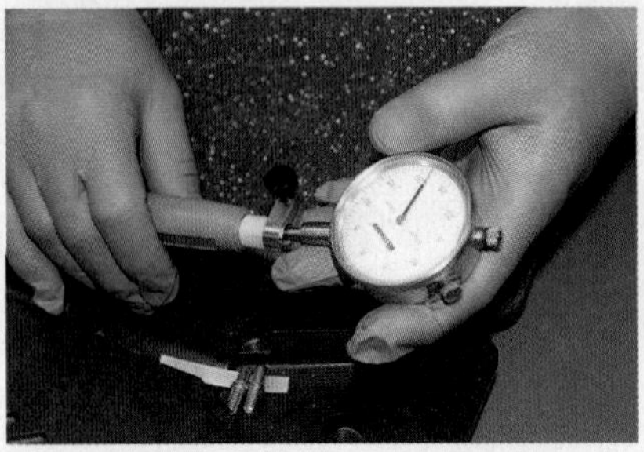

1. Select the correct size of the dial bore gauge you will use, and fit any adapters to it. Check the calibration and adjust it as necessary. Insert the dial bore gauge into the bore. The accurate measurement will be at exactly 90 degrees to the bore. To find the accurate measurement, rock the dial bore gauge handle slightly back and forth until you find the centered position.

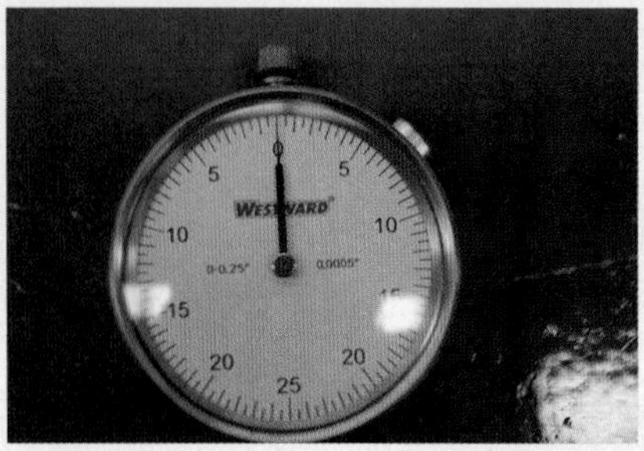

2. Check the calibration and adjust it as necessary.

3. Insert the dial bore gauge into the bore. The accurate measurement will be at exactly 90 degrees to the bore. To find the accurate measurement, rock the dial bore gauge handle slightly back and forth until you find the centered position.

4. Read the dial to determine the bore measurement.

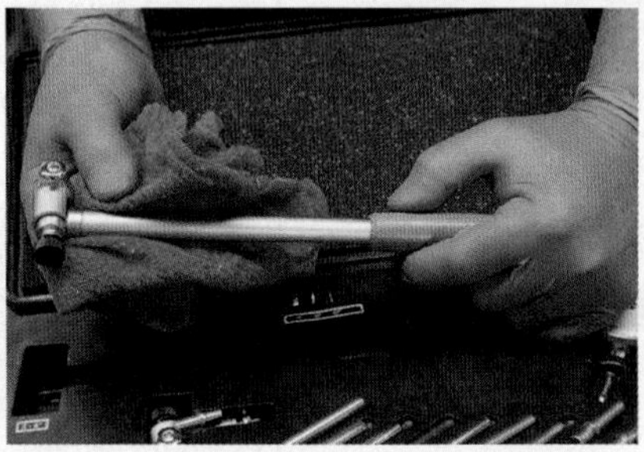

5. Always clean the dial bore gauge, and return it to its protective case when you have finished using it.

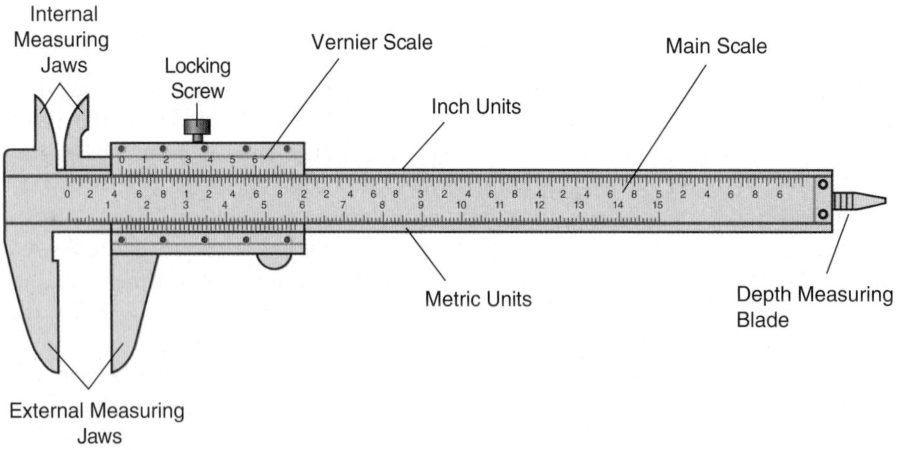

FIGURE 6-113 Vernier calipers can take three types of readings.

measurement. When making a measurement, make sure the caliper is at right angles to the surfaces to be measured. Make sure that all slack is taken up between the part being measured and the jaws of the caliper. This means that you will have to hold some pressure on the jaws, pushing them into the object being measured. You should always repeat the measurement a

FIGURE 6-114 Dial caliper.

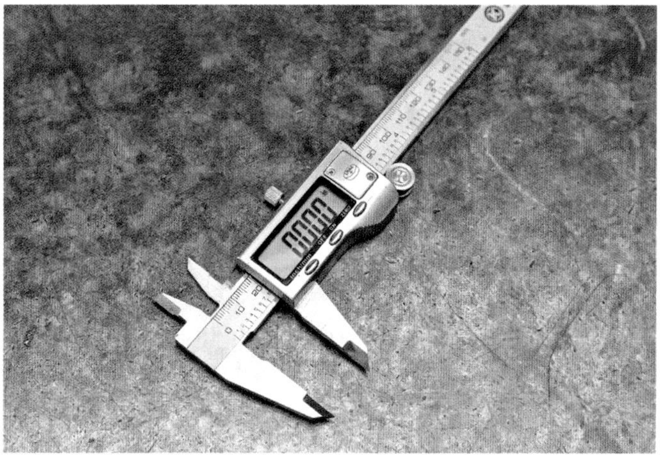

FIGURE 6-115 Digital caliper.

number of times. Then, compare your results to make you have measured accurately. To correctly measure using digital vernier calipers, follow the steps in **SKILL DRILL 6-9**.

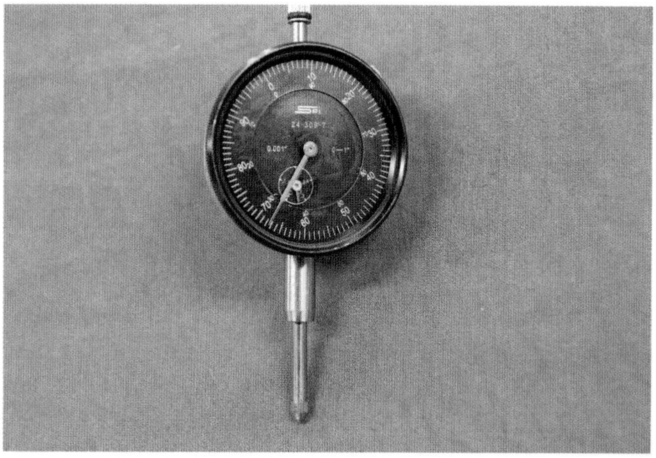

FIGURE 6-116 Dial indicator.

Dial Indicators

Dial indicators can also be known as dial gauges. As the name suggests, they have a dial and needle where measurements are read. They have a measuring plunger with a pointed or rounded contact end that is spring loaded. The plunger is connected via gears in the housing to the dial needle (**FIGURE 6-116**). The dial accurately displays movement of the plunger in and out as it rests against an object. For example, they can be used to measure the trueness of a rotating disc brake rotor.

A dial indicator can also measure how round something is, such as a **crankshaft**, which can be rotated in a set of **V blocks** (**FIGURE 6-117**). If the crankshaft is bent, it will show as movement on the dial indicator as the crankshaft is rotated. The dial indicator senses slight movement at its tip and magnifies it into a measurable swing on the dial.

Dial indicators normally have either one or two indicator needles. The large needle indicates the fine reading of

SKILL DRILL 6-9 Using Vernier Calipers

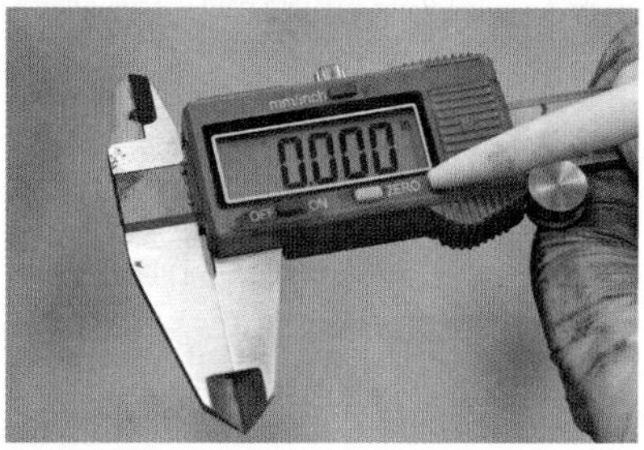

1. Verify that the vernier caliper is calibrated (zeroed) before using it.

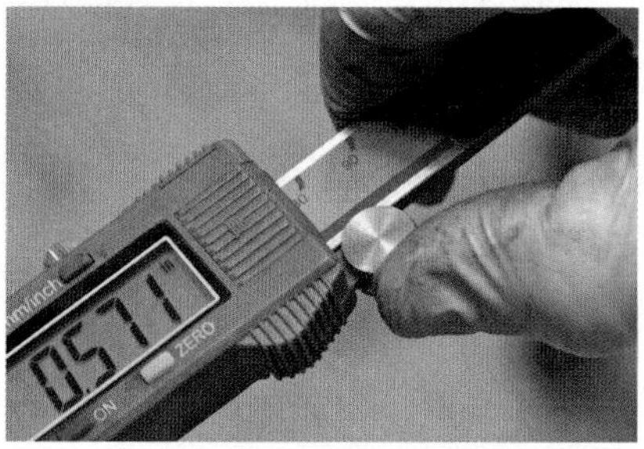

2. Position the caliper correctly for the measurement you are making. Internal and external readings are normally made with the vernier caliper positioned at 90 degrees to the face of the component to be measured. Length and depth measurements are usually made parallel to or in line with the object being measured. Use your thumb to press or withdraw the sliding jaw to measure the outside or inside of the part.

3. Read the scale of the vernier caliper, being careful not to change the position of the moveable jaw. If using a non-digital caliper, always read the dial or face straight on. A view from the side can give a considerable **parallax error**. Parallax error is a visual error caused by viewing measurement markers at an incorrect angle.

FIGURE 6-117 A dial indicator being used to measure crankshaft runout.

thousandths of an inch. If it has a second needle, it will be smaller and indicates the coarse reading of tenths of an inch. The large needle is able to move numerous times around the outer scale. One full turn may represent 0.100" or 1 mm. The small inner scale indicates how many times the outer needle has moved around its scale. In this way, the dial indicator is able to read movement of up to 1" or 2 cm. Dial indicators can typically measure with an accuracy of 0.001" or 0.01 mm.

The type of dial indicator you use is determined by the amount of movement you expect from the component you are measuring. The indicator must be set up so that there is no gap between the dial indicator and the component to be measured. It also must be set perpendicular and centered to the part being measured. Most dial indicator sets contain various attachments and support arms. This allows them to be configured specifically for a variety of measuring tasks.

Using Dial Indicators

Dial indicators are used in many types of service jobs. They are particularly useful in determining runout on rotating shafts and surfaces. Runout is the side-to-side variation of movement when a component is turned. When attaching a dial indicator:

- Keep support arms as short as possible.
- Make sure all attachments are tightened to prevent unnecessary movement between the indicator and the component.
- Make sure the dial indicator plunger is positioned at 90 degrees to the face of the component to be measured.
- Always read the dial face straight on, as a view from the side can give a considerable parallax error.

The outer face of the dial indicator is designed so it can be rotated so that the zero mark can be positioned directly under the pointer. This is how a dial indicator is zeroed. To correctly measure using a dial indicator, follow the steps in **SKILL DRILL 6-10**.

SKILL DRILL 6-10 Using Dial Indicators

1. Select the gauge type, size, attachment, and bracket that fit the part you are measuring. Mount the dial indicator firmly to keep it stationary.

2. Adjust the indicator so that the plunger is at 90 degrees to the part you are measuring, and lock it in place.

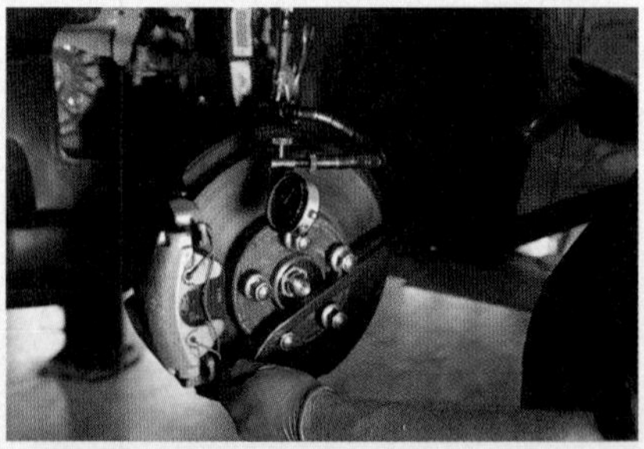

3. Rotate the part one complete turn, and locate the low spot. Zero the indicator.

4. Find the point of maximum height and note the reading. This indicates the runout value.

5. Continue the rotation, making sure the needle does not go below zero. If it does, re-zero the indicator and remeasure the point of maximum variation.

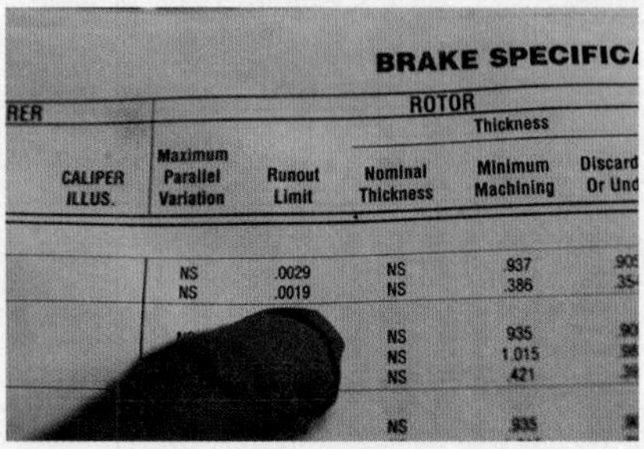

6. Check your readings against the manufacturer's specifications. If the deviation is greater than the specifications allow, consult your supervisor.

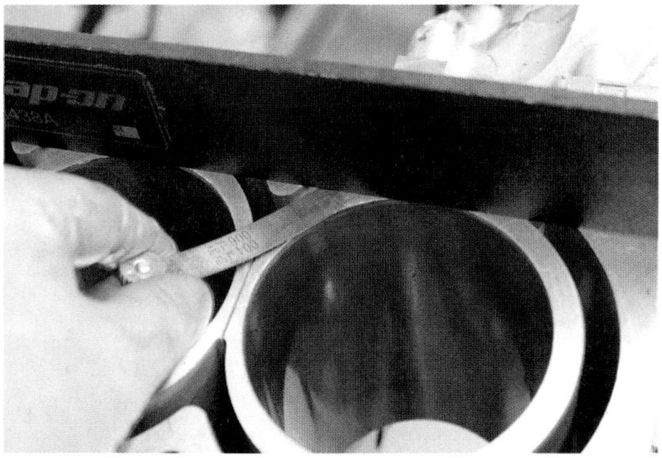

FIGURE 6-118 Straight edge being used to measure the flatness of a surface.

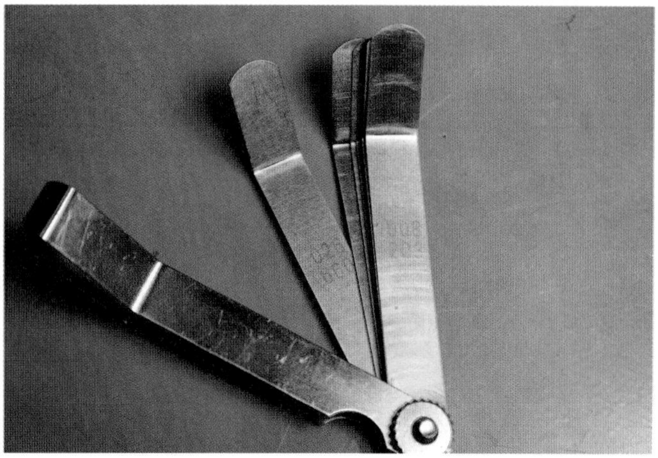

FIGURE 6-120 Bent feeler gauge set.

FIGURE 6-119 Feeler blade set with blades of varying thickness.

FIGURE 6-121 Feeler gauge set with one brass feeler gauge.

Straight Edges

Straight edges are usually made from hardened steel and are machined so that the edge is perfectly straight. A straight edge is used to check the flatness of a surface. It is placed on its edge against the surface to be checked (FIGURE 6-118). The gap between the straight edge and the surface can be measured by using feeler gauges. Sometimes, the gap can be seen easily if light is shone from behind the surface being checked. Straight edges are often used to measure the amount of warpage the surface of a cylinder head has.

Feeler Gauges

Feeler gauges (also called feeler blades) are used to measure the width of gaps. One example is the clearance between valves and rocker arms. Feeler gauges are flat metal strips of varying thicknesses (FIGURE 6-119). The thickness of each feeler gauge is clearly marked on it. They are sized from fractions of an inch or fractions of a millimeter. They usually come in sets and are available in standard and metric measurements. Some feeler gauges come in a bent arrangement to be more easily inserted in cramped spaces (FIGURE 6-120).

Some sets contain feeler gauges made of brass (FIGURE 6-121). These are used to take measurements between components that are magnetic. If steel gauges are used, the drag caused by the magnetism would mimic the drag of a proper clearance. Brass gauges are not subject to magnetism, so they work well in that situation.

Some feeler blades come in a stepped version. For example, the end of the gauge might be 0.010" thick, while the rest of the gauge is 0.012" thick (FIGURE 6-122). This works well for adjusting valve clearance. If the specification is 0.010", then the 0.010 section can be placed in the gap. If the 0.012 section slides into the gap, then the valve needs to be readjusted. If it stops at the lip of the 0.012" section, then the gap is correct (FIGURE 6-123).

Two or more non-stepped feeler gauges can be stacked together to make up a desired thickness. For example, to measure a thickness of 0.029 of an inch, a 0.017 and a 0.012 feeler gauge could be used together to make up the size. If you want to measure an unknown gap, you can interchange feeler gauges until you find the one or more that fits snugly into the gap. Then, total their thickness to determine the measurement of the gap.

Wire feeler gauges are made with hardened wire of specified thickness (FIGURE 6-124).They are used to measure gaps that may not be completely parallel, such as spark plug gaps. Because they are round, they can find the minimum gap much more accurately (FIGURE 6-125).

SAFETY TIP

Never use feeler gauges on operating machinery.

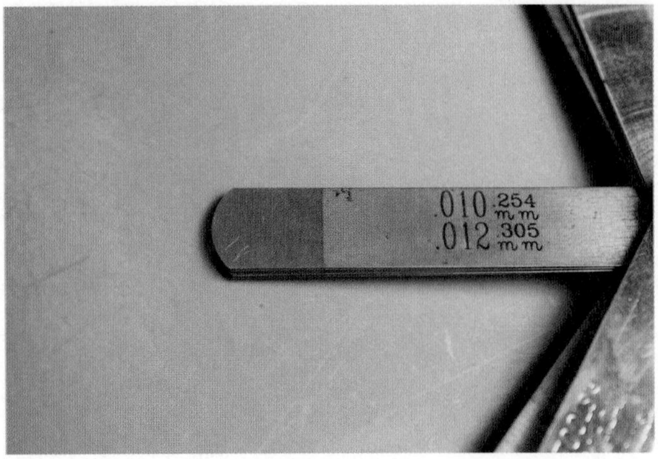

FIGURE 6-122 Stepped feeler gauge set.

FIGURE 6-124 Wire-type feeler gauge,

FIGURE 6-123 Stepped feeler blade being used during a valve adjustment.

FIGURE 6-125 Wire feeler gauges are used for surfaces that are not perfectly parallel.

SAFETY TIP

Feeler gauges are strips of hardened metal that have been ground or rolled to a precise thickness. They can be very thin and will cut through skin if not handled correctly.

Using Feeler Gauges

If the feeler gauge feels too loose when measuring a gap, select the next larger size, and measure the gap again. Repeat this procedure until the feeler gauge has a slight drag between both parts. If the feeler gauge is too tight, select a smaller size until the feeler gauge fits properly. When measuring a spark plug gap, flat feeler gauges should not be used because the spark plug electrodes are not perfectly parallel. So it is preferable to use wire feeler gauges when measuring spark plug gaps (**FIGURE 6-126**). Wire feeler gauges use accurately machined pieces of wire instead of flat metal strips. To select and use feeler gauge sets, follow the steps in **SKILL DRILL 6-11**.

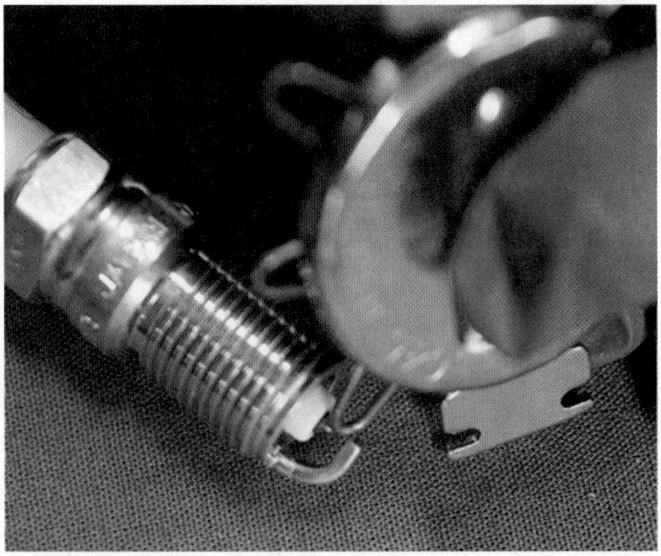

FIGURE 6-126 Wire feeler gauge used to check a spark plug gap.

SKILL DRILL 6-11 Using Feeler Gauges

1. Select the appropriate type and size feeler gauge set for the job you are working on.

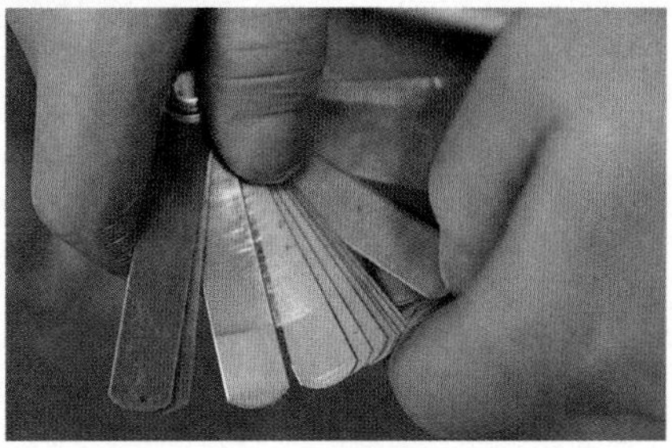

2. Inspect the feeler gauges to make sure they are clean, rust-free, and undamaged, but slightly oiled for ease of movement.

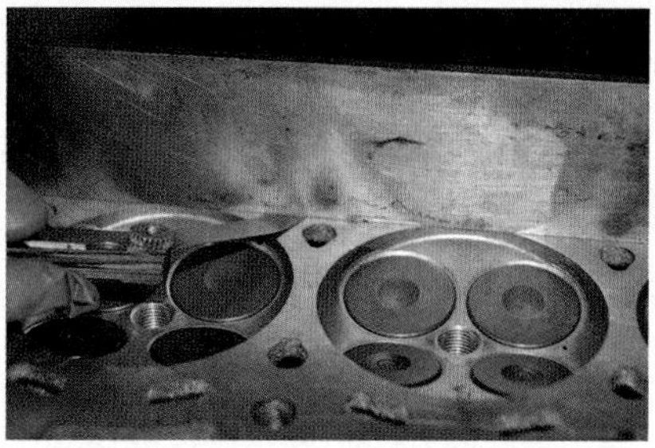

3. Choose one of the smaller wires or blades, and try to insert it in the gap on the part. If it slips in and out easily, choose the next size up. When you find one that touches both sides of the gap and slides with only gentle pressure, then you have found the exact width of that gap.

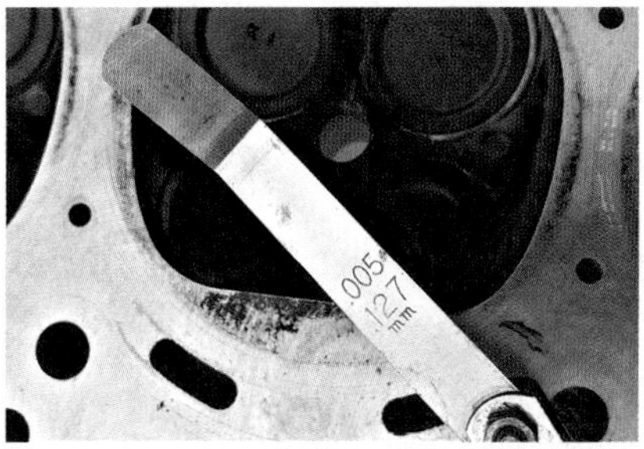

4. Read the markings on the wire or blade, and check these against the manufacturer's specifications for this component. If gap width is outside the tolerances specified, inform your supervisor.

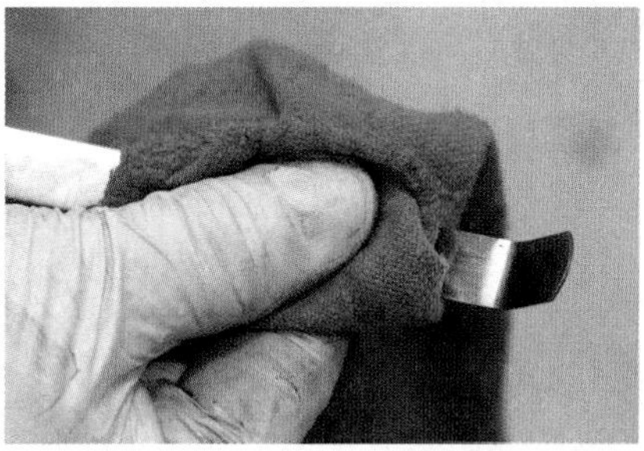

5. Clean the feeler gauge set with an oily cloth before storage to prevent rust.

▶ Wrap-Up

Ready for Review

- Tools must be safely handled and used to prevent injury and damage. Lockout/tag-out describes a set of safety practices and procedures for tool usage. A quality toolbox should be used to keep tools secure.
- Box-end wrench, the open-end wrench, and combination wrench are commonly used. Sockets are classified by: standard or metric, size of drive used to turn them, number of points, depth of socket, and thickness of wall.
- Basic hand tools include: pliers, Allen wrenches, screwdrivers, magnetic pickup tools and mechanical fingers, cutting tools, pry bars, gasket scrapers, and files.
- Different types of hammer include: ball peen hammer, sledge hammer, soft-faced hammer, and dead blow hammer. Chisels and punches can also be used as struck tools.
- Taps and dies are used to form threads in metal. The most common taps are the taper tap, intermediate tap, bottoming tap, and the thread chaser.
    - Screw extractors, gear pullers, flaring tools, ad riveting tools are other specialized tools.
- Micrometers are precise measuring tools and are designed to measure small distances. Most common types of micrometers are the outside, inside, and depth micrometers.
- Vernier calipers are used for measuring outside, inside, and depth dimensions, all in one tool.
    - Measurements are taken by comparing the scales on the sliding sleeve to the graduated bar.

Key Terms

Allen wrenches A tool that fits into a fastener with an internal hexagonal recess.

arc joint pliers Pliers with parallel slip jaws that can increase in size; also called Channellocks.

aviation snips A scissorlike tool for cutting sheet metal.

ball-peen (engineer's) hammer A hammer with one flat face and one rounded face.

bench vice Tool used to hold parts while working on them.

blind rivet A rivet that can be installed from its insertion side.

bolt A type of threaded fastener with a thread on one end and a hexagonal head on the other.

bottoming tap A thread-cutting tap designed to cut threads to the bottom of a blind hole.

box-end wrench A wrench or spanner with a closed or ring end to grip bolts and nuts.

C-clamp A clamp shaped like the letter C; it comes in various sizes and can clamp various items.

center punch Less sharp than a prick punch, the center punch makes a bigger indentation that centers a drill bit at the point where a hole is required to be drilled.

cold chisel The most common type of chisel, used to cut cold metals. The cutting end is tempered and hardened so that it is harder than the metals that need to be cut.

combination wrench A type of wrench that has an open end on one end and a closed-end wrench on the other.

crankshaft A vehicle engine component that transfers the reciprocating movement of pistons into rotary motion.

cross-arm A description for an arm that is set at right angles, or 90 degrees, to another component.

cross-cut chisel A type of chisel for metal work that cleans out or cuts key ways.

curved file A type of file that has a curved surface for filing holes.

dead blow hammer A type of hammer that has a cushioned head to reduce the amount of head bounce.

depth micrometer A micrometer that measures the depth of an item such as how far a piston is below the surface of the block.

diagonal cutting pliers Cutting pliers for small wire or cable.

dial bore gauge A gauge that is used to measure the inside diameter of bores with a high degree of accuracy and speed.

dial indicator A device for precision measurements used to measure small variations, such as end play, movement in a bearing, or run-out.

die stock Tool used to hold a die during use

die Used to cut external threads on a metal shank or bolt.

double flare A seal that is made at the end of metal tubing or pipe.

drift punch A type of punch used to start pushing roll pins to prevent them from spreading.

drill vice Tool used to hold parts while they are being drilled on a drill press

fasteners Devices that securely hold items together, such as screws, cotter pins, rivets, and bolts.

feeler gauge Flat metal strips used to measure the width of gaps, such as the clearance between valves and rocker arms. Also called feeler blades.

finished rivet A rivet after the completion of the riveting process.

flare nut wrench A type of box-end wrench that has a slot in the box section to allow the wrench to slip through a tube or pipe. Also called a flare tubing wrench.

flat-blade screwdriver A screwdriver that has a flat tip or blade.

flat-nose pliers Pliers that are flat and square at the end of the nose.

forcing screw The center screw on a gear, bearing, or pulley puller. Also called a jacking screw.

gasket scraper A broad, sharp, flat blade to assist in removing gaskets and glue.

Gear pullers Tool used to remove press-fit gears from shafts.

hard rubber mallet A special-purpose tool with a head made of hard rubber; often used for moving things into place where it is important not to damage the item being moved.

hollow punch A punch with a center hollow for cutting circles in thin materials such as gaskets.

impact driver A tool that is struck with a hammer to provide an impact turning force to remove tight fasteners.

inside micrometer A micrometer that measures inside dimensions.

intermediate tap One of a series of taps designed to cut an internal thread. Also called a plug tap.

ISO flare Metric system flare commonly called a bubble flare

locking pliers A type of pliers where the jaws can be set and locked into position.

Lockout/tag-out A safety tag system to ensure that faulty equipment or equipment in the middle of repair is not used.

lug wrench A tool designed to remove wheel lugs nuts and commonly shaped like a cross.

mandrel head The head of the pop rivet that connects to the shaft and causes the rivet body to flare.

mandrel The shaft of a pop rivet.

measuring tape A thin measuring blade that rolls up and is contained in a spring-loaded dispenser.

micrometer An accurate measuring device for internal and external dimensions. Commonly abbreviated as *mic*.

needle-nose pliers Pliers with long tapered jaws for gripping small items and getting into tight spaces.

nippers Another name for end-cutting pliers.

offset screwdriver A screwdriver with a 90-degree bend in the shaft for working in tight spaces.

offset vice. Tool used to hold long components such as pipes and axles.

oil-filter wrench A wrench used to grip and loosen an oil filter. Not to be used for tightening an oil filter.

open-end wrench A wrench with open jaws to allow side entry to a nut or bolt.

outside micrometer A precision measuring instrument meant to measure the outside dimensions of components. It is usually accurate to 0.0001 inch (0.0025 mm).

parallax error A visual error caused by viewing measurement markers at an incorrect angle.

peening A term used to describe the action of flattening a rivet through a hammering action.

Phillips head screwdriver A type of screwdriver that fits a head shaped like a cross in screws. Also called a Phillips screwdriver.

pin punch A type of punch in various sizes with a straight or parallel shaft.

pipe wrench A wrench that grips pipes and can exert a lot of force to turn them. Because the handle pivots slightly, the more pressure put on the handle to turn the wrench, the more the grip tightens.

pliers A hand tool with gripping jaws.

pop rivet gun A hand tool for installing pop rivets.

prick punch A pinch with a sharp point for accurately marking a point on metal.

pry bar A high-strength carbon steel rod with offsets for levering and prying. Also called a crowbar.

Puller Tool used to grab and pull press-fit parts apart.

punches A generic term to describe high-strength carbon steel shafts with a blunt point for driving. Center and prick punches are exceptions and have a sharp point for marking or making an indentation.

ratcheting box-end wrench A wrench with an inner piece that is able to rotate within the outer housing, allowing it to be repositioned without being removed.

ratcheting screwdriver A screwdriver with a selectable ratchet mechanism built into the handle that allows the screwdriver tip to ratchet as it is being used.

ratcheting screwdriver A screwdriver with a selectable ratchet mechanism built into the handle that allows the screwdriver tip to ratchet as it is being used.

ratchet A generic term to describe a handle for sockets that allows the user to select direction of rotation. It can turn sockets in restricted areas without the user having to remove the socket from the fastener.

roll bar Another type of pry bar, with one end used for prying and the other end for aligning larger holes, such as engine motor mounts.

screw extractor A tool for removing broken screws or bolts.

single flare A sealing system made on the end of metal tubing.

sledgehammer A heavy hammer with two flat faces

sliding T-handle A handle, fitted at 90 degrees to the main body, that can be slid from side to side.

snap ring pliers A pair of pliers for installing and removing internal or external snap rings.

socket An enclosed metal tube, commonly with 6 or 12 points, used to remove and install bolts and nuts.

speed brace A U-shaped socket wrench that allows high-speed operation; also called a speeder handle.

split ball gauge A gauge that is good for accurately measuring small holes where telescoping gauges cannot fit; also called a small hole gauge.

square file A type of file with a square cross section.

steel hammer A hammer with a head made of hardened steel.

steel rule An accurate measuring ruler made of steel, or stainless steel.

straight edge A measuring device, generally made of steel, to check how flat a surface is.

taper tap A tap with a taper; it is usually the first of three taps used when cutting internal threads.

tap handle A tool designed to securely hold taps for cutting internal threads.

Taps and dies Tools used to create internal and external threads

telescoping gauge A gauge that expands and locks to the internal diameter of bores; a caliper or outside micrometer is used to measure its size.

thread chaser A device similar to a die that cleans up rusty or damaged threads.

thread file A type of file that cleans clogged or distorted threads on bolts and studs.

thread pitch The coarseness or fineness of a thread as measured by either the threads per inch or the distance from the peak of one thread to the next. Metric fasteners are measured in millimeters.

tin snips Cutting device for sheet metal; works in a similar fashion to scissors.

torque wrench A tool used to measure the rotational or twisting force applied to fasteners.

triangular file A type of file with three sides so it can get into internal corners.

tube flaring tool A tool that makes a sealing flare on the end of metal tubing.

tubing cutter A hand tool for cutting pipe or tubing squarely.

V blocks Tools used to set round objects in while measuring them.

wad punch A type of punch that is hollow for cutting circular shapes in soft materials such as gaskets.

warding file A type of thin, flat file with a tapered end.

wrench A generic term to describe tools that tighten and loosen fasteners with hexagonal heads.

Review Questions

1. Technician tool purchases are considered:
   a. an investment
   b. a liability
   c. a waste of money
   d. a last-resort option
2. Which of the following is NOT a type of socket?
   a. Impact-rated socket
   b. Adjustable socket
   c. Shallow socket
   d. Six-point socket
3. Which of these would you use for a wider grip and a tighter squeeze on parts too big for conventional pliers?
   a. Diagonal cutting pliers
   b. Combination pliers
   c. Arc joint pliers
   d. Snap ring pliers
4. Which tool would be best used to remove a threaded fastener that has a cross-shaped recess on the head?
   a. Straight screwdriver
   b. Offset slotted screwdriver
   c. Phillips screwdriver
   d. Allen driver
5. Which tool would be best used to make external threads on a steel rod?
   a. A tap
   b. A thread maker
   c. A thread file
   d. A die
6. All of the following are steps for properly using a micrometer EXCEPT:
   a. Verify that the micrometer is zeroed.
   b. That the measuring surfaces are clean.
   c. Use the ratchet to run the micrometer down.
   d. Turn the thimble an additional 0.005" to take up any slack.
7. When would a technician use a dial indicator?
   a. Checking for runout
   b. Checking the diameter of a part
   c. Measuring spark plug gap
   d. Measuring angles on parts
8. Which style of torque wrench is the simplest and least expensive?
   a. Clicker
   b. Dial
   c. Beam
   d. Electronic
9. The depth of blind holes in housings can best be measured using a(n):
   a. measuring tape.
   b. steel rule.
   c. dial bore gauge.
   d. vernier calipers.
10. When a large chisel needs a really strong blow, use a:
    a. sledgehammer.
    b. hard rubber hammer.
    c. dead blow hammer.
    d. ball-peen hammer.

ASE Technician A/Technician B Style Questions

1. Technician A says that proper maintenance and inspection is required on tools regularly. Technician B says that tools don't wear out, so extra time is not needed for inspection. Who is correct?
   a. Technician A
   b. Technician B
   c. Both A and B
   d. Neither A nor B
2. Two techs are measuring a cylinder bore down to 0.001". Technician A says that a tape measure will work. Technician B says that a telescopic gauge and micrometer will work. Who is correct?
   a. Technician A
   b. Technician B
   c. Both A and B
   d. Neither A nor B
3. Two technicians are discussing hammers. Technician A says a dead blow hammer reduces rebound of the hammer. Technician B says that a dead blow hammer should be used with a chisel to cut the head of a bolt off. Who is correct?
   a. Technician A
   b. Technician B
   c. Both A and B
   d. Neither A nor B
4. Two techs are threading a hole in a steel part. Technician A says to consult a tap drill chart to determine the proper sized drill bit. Technician B says that the tap needs to be lubricated when cutting threads. Who is correct?
   a. Technician A
   b. Technician B
   c. Both A and B
   d. Neither A nor B
5. Two techs are about to measure runout on a rotating surface. Technician A says to use a micrometer and telescoping gauge. Technician B says to use a Vernier caliper. Who is correct?
   a. Technician A
   b. Technician B
   c. Both A and B
   d. Neither A nor B
6. Technician A says that files are made to be used without handles. Technician B says that thread files are used for cutting threads in new holes. Who is correct?
   a. Technician A
   b. Technician B

c. Both A and B
d. Neither A nor B

7. Technician A says that a Vernier Caliper is a precision measuring tool if used properly. Technician B says a Vernier Caliper can measure outside diameter, inside diameter, and depth. Who is correct?
a. Technician A
b. Technician B
c. Both A and B
d. Neither A nor B

8. When using tools in the shop. Technician A says a safe attitude will help you avoid accidents. Technician B says that while tools are a technician's best friend, if used improperly, they can injure or kill you. Who is correct?
a. Technician A
b. Technician B
c. Both A and B
d. Neither A nor B

9. Technician A says that a roll bar is sharply curved on one end and tapered to a dull point on the other end. Technician B says that metal gasket scrapers should be used on aluminum surfaces. Who is correct?
a. Technician A
b. Technician B
c. Both A and B
d. Neither A nor B

10. Technician A says that a beam-style torque wrench tightens the bolt automatically by itself. Technician B says a digital torque wrench usually gives an audible signal when it reaches the preset torque. Who is correct?
a. Technician A
b. Technician B
c. Both A and B
d. Neither A nor B

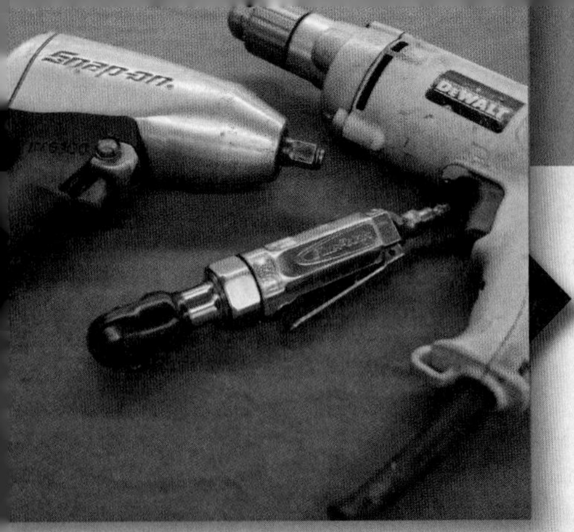

CHAPTER 7

Power Tools and Equipment

Learning Objectives

- **LO 7-01** Identify and safely operate battery charging and jump-starting equipment.
- **LO 7-02** Identify and safely operate air tools.
- **LO 7-03** Use cutting and grinding tools.
- **LO 7-04** Perform a solder repair.
- **LO 7-05** Operate cleaning equipment.

ASE Education Foundation Tasks

See Appendix A to view the 2017 ASE Education Foundation Automobile Accreditation Task List Correlation Guide.

▶ Introduction

Power tools and shop equipment use electric or air power to increase the amount of work that technicians can perform in a given amount of time (**FIGURE 7-1**). This results in more income for both the shop and the technician. It also comes with risk of damage to the vehicle or injury to the technician if the power tool or equipment is not used properly. So it is important that you learn how to use these tools safely.

The following topics introduce the safe working procedures of some of the common power tools and equipment used in the shop. But we can't cover all of the differences between makes and models. Always refer to the manufacturer's operator's manual. And get your instructor's permission before using any power tools or equipment. We will begin with battery charging and jump-starting equipment.

▶ Battery Charging and Jump-Starting

LO 7-01 Identify and safely operate battery charging and jump-starting equipment.

Batteries are common in shops and are used in vehicles and rechargeable tools. Extreme caution should be taken when working around or with batteries. This includes both when they are being charged, and when they are not. Batteries produce dangerous and explosive hydrogen gases. Because of this, sparks and short circuits near the battery should be avoided.

Always wear appropriate personal protective equipment (PPE), such as goggles, gloves, and protective clothing, when working around batteries. Also, some regulatory agencies require that an eyewash station be located near the battery charging station. This is required in case of a battery explosion. Check with your local authorities to determine any distance requirements.

There are many different types of **battery chargers**, and each is designed for a particular purpose and application. Battery chargers can be **fast chargers**, with high current output to charge a battery quickly. Or they can be **slow chargers** which have lower current outputs. They put less stress on the battery, which is ideal if time is not a consideration.

Smart chargers incorporate microprocessors to check and control the charge rate. This is important so the battery receives the correct amount of charge depending on its state of charge (**FIGURE 7-2**). These types of chargers are becoming more popular and ensure that the battery receives the optimal charge. This promotes longer battery life. Even though most motor vehicle batteries are typically 12 volts, they store a lot of energy. The high current supply from a battery can be very dangerous. Remember, batteries have to deliver enough power to crank over a cold engine.

FIGURE 7-1 Power tools increase a technician's productivity.

FIGURE 7-2 Smart battery charger.

You Are the Automotive Technician

The shop just agreed to accept an automotive student intern from the local college for the semester. Because you remember what it was like to be a new person in the shop, you volunteered to supervise the student and help get him/her up to speed. Because he/she will need to be familiar with certain equipment in the shop, you want to make sure you train him/her properly. To prepare for the training, you ask yourself the following questions. You write down the answers so you will make sure to cover all of the important details with the intern.

1. What are the precautions when charging batteries?
2. What are the precautions when working around compressed air?
3. What safety requirements are needed when using a bench grinder?
4. What are the steps for using a spray wash cabinet?

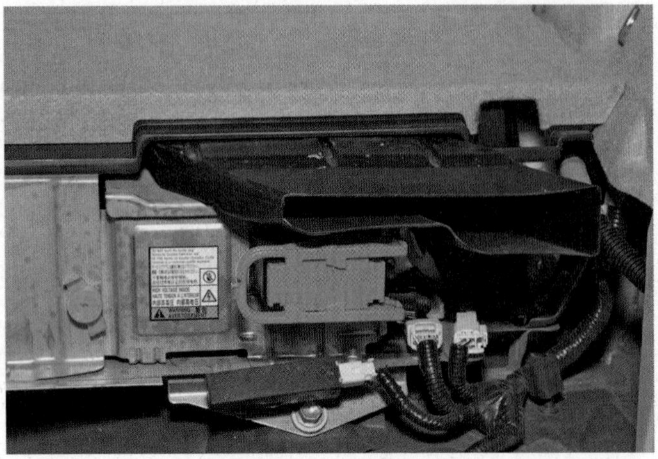

FIGURE 7-3 High-voltage battery packs, like those fitted to hybrid vehicles, are extremely dangerous. They have the potential for high voltage and current. Take special precautions for dealing with high-voltage systems.

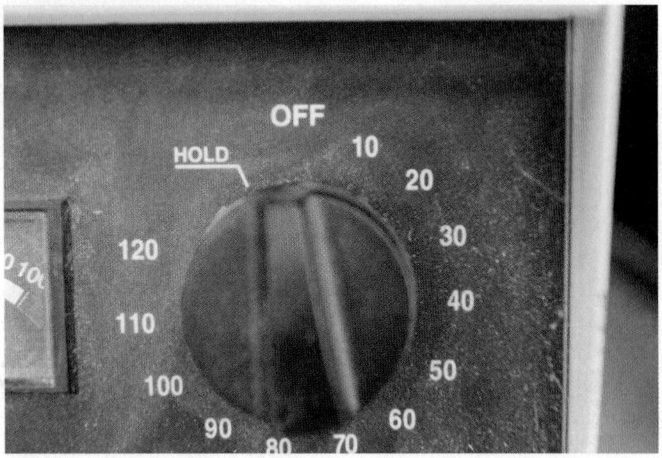

FIGURE 7-4 Switch off the charger before connecting it to or disconnecting it from the battery.

Batteries also produce enough power to melt a metal rod resting across the terminals. High-voltage battery packs, like those fitted to hybrid vehicles, are even more dangerous. They have the potential for high voltage and current. Special precautions for dealing with high-voltage systems must be taken (**FIGURE 7-3**). Always treat batteries with care and respect.

▶ TECHNICIAN TIP

Vehicle batteries are usually lead acid types, often in a 12-volt configuration. One 24-volt battery or two 12-volt batteries connected in series can provide 24 volts.

Technology changes in vehicles have increased the need for more electrical power. In the future, this will drive the need for higher voltage battery systems. Hybrid vehicles are an example of this. Their operating voltages are typically 200–600 volts. Generally, the higher the system voltage, the more efficient the system is. This is because the electrical current can be lower for a given amount of power. And because current determines wire size, the wires can be smaller. However, higher voltages also create a greater shock hazard, so bear this in mind when working around and with batteries.

SAFETY TIP

High voltages used in a hybrid vehicle are extremely dangerous. The voltage and current flow is many times greater than that needed to kill a person. Most hybrid manufacturers require their technicians to undergo special factory training. This is needed before they will allow them to service a hybrid vehicle. Also, they usually allow only very experienced technicians to undergo the training, not novices. In fact, one of the tools that Toyota requires of their shops for working on a hybrid vehicle is a nonconductive shepherd's hook. This can be used to drag a technician away from high voltage if the technician is electrocuted while working on the vehicle.

SAFETY TIP

Always remove your hand, wrist, and neck jewelry before working with batteries and electrical systems. If any of these come into contact with the battery terminals or power wire, it can cause a short circuit. You will receive painful skin burns from the very rapid heating of the metal you are wearing. You can also receive flash burns from an arcing current. A wristwatch or ring is much harder to take off when it is red- or white-hot and burned onto your skin!

SAFETY TIP

Batteries give off hydrogen gas while they are being charged and discharged. Hydrogen is a light and highly explosive gas that is easily ignited by a simple spark. Batteries are filled with **sulfuric acid**. If the hydrogen creates an explosion, the battery case can then rupture. It could spray everything and everyone nearby with this dangerous and corrosive liquid. Be very careful not to create a spark when you are connecting or disconnecting battery cables or hooking up a charger to the battery terminals. Switch off the charger before connecting and disconnecting it from the battery (**FIGURE 7-4**).

Do not try to charge a battery faster than the battery manufacturer recommends. And never use a battery load tester immediately after charging a battery. This is because both charging and rapidly discharging a battery generate heat and hydrogen. If you load-test a battery after charging it without waiting for it to cool down, you will increase the risk of distorting the plates inside the battery. This would increase the risk of explosion if the plates end up touching each other.

How to Charge Batteries

Batteries go dead for a variety of reasons. A common cause is the driver forgetting to turn off the headlights when exiting the vehicle. Or maybe the owner went on vacation for a month, and the battery discharged slowly over that time. Because the battery only stores electricity, anything that stays on when the vehicle is not running drains the battery. The more discharged a battery is, the more it needs to be recharged.

When charging a battery, slow charging a battery is less stressful on the battery than fast charging. So, if possible, slow charge a battery instead of fast charging it. Removing the negative battery terminal while changing a battery reduces the risk of burning up any electronic devices on the vehicle. This prevents any excessive voltage from the battery charger from being applied to the vehicle's electrical system. This is especially true with today's electronically intensive cars. However, disconnecting the vehicle's battery risks losing information. This includes the radio presets and other learned data. Using a **memory saver (memory minder)** prevents this. It provides backup power to retain electronic memory settings in the vehicle's computer systems (**FIGURE 7-5**).

In some vehicles, manufacturers install more than one battery to provide additional power. Examples of this include diesel pickup trucks and SUVs. Knowing how the batteries are connected together helps you determine how you need to connect a battery charger.

Batteries can be connected in series or in parallel. When connected in series, they are connected in line with each other. This means that the positive terminal of one 12-volt battery is connected to the negative terminal of the other battery (**FIGURE 7-6**). They have a nominal output voltage of 24 volts across the most negative and most positive battery terminals.

If you have a 24-volt battery charger, you can charge both batteries at once. You do this by connecting the battery charger to the most negative and positive terminals. If you only have a 12-volt charger, you will have to charge one battery at a time. Or you can reconnect them so they are connected in parallel. Just make sure you charge both batteries fully, which could take more than 12 hours each if slow charging.

When connected in parallel, they are connected side by side. This means that the positive terminal of one battery is connected to the positive of the other battery. And the negative to negative (**FIGURE 7-7**). The output voltage will be equal to the voltage of one battery. It is taken from the positive and negative posts of either battery.

FIGURE 7-5 Memory savor being hooked up to a vehicle.

For example, two 12-volt batteries connected in parallel have an output voltage of about 12 volts. In this situation, a 12-volt charger can be used to charge both batteries at the same time while the batteries are connected together. But it is likely to take about twice as long as it would if only charging one battery.

After charging and reinstalling a battery, it is good practice to clean the battery terminals and posts. This is done with a battery terminal cleaner. To correctly charge a battery using battery charging equipment, follow the steps in **SKILL DRILL 7-1**.

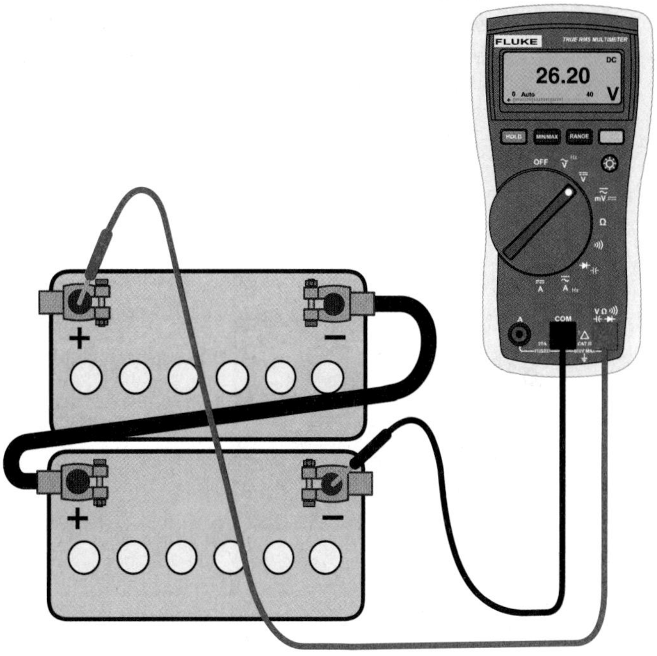

FIGURE 7-6 Batteries connected in series.

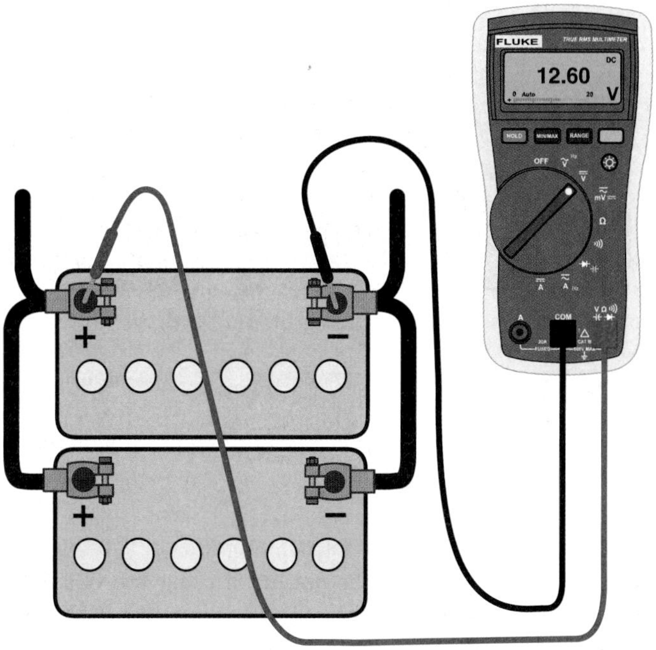

FIGURE 7-7 Batteries connected in parallel.

SKILL DRILL 7-1 Charging Batteries

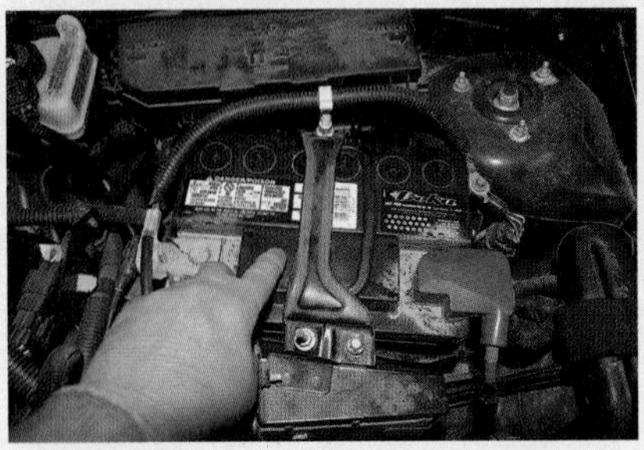

1. Determine the voltage of the system that needs charging. If you are charging a 12-volt battery, use the 12-volt setting on the charger. If you are charging a 24-volt battery, or two 12-volt batteries connected in series, use the 24-volt setting on the charger, if it has one.

2. Identify the positive and negative terminals. Never simply use the color of the cables to determine the positive or negative terminals; use the + and − or the "Pos" and "Neg" marks.

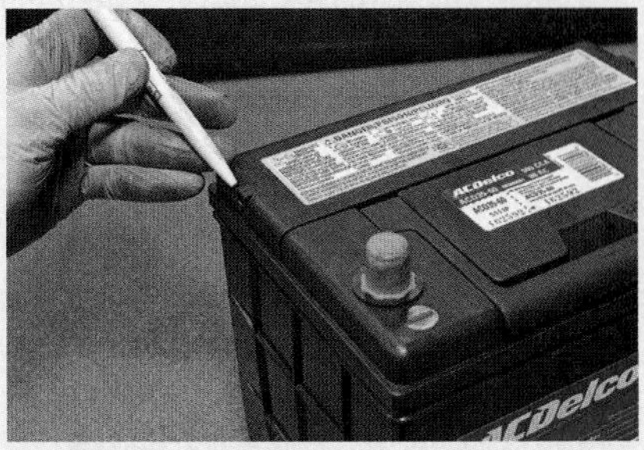

3. Inspect the battery by carrying out a visual inspection of the battery to ensure there are no cracks, holes, or damage to the casing.

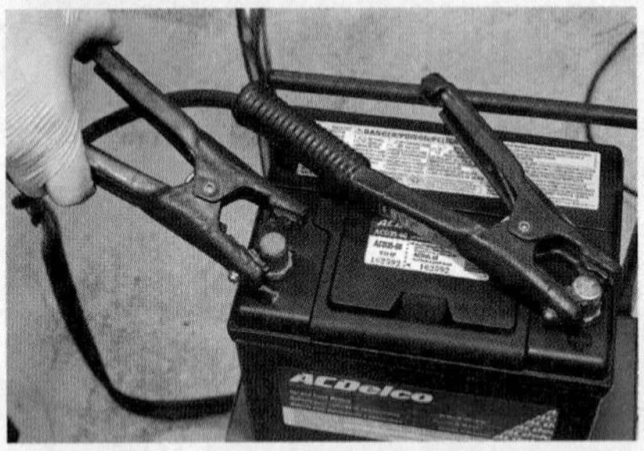

4. Verify that the charger is unplugged from the wall and turned off. Connect the red lead from the charger to the positive battery terminal. Connect the black lead from the charger to the negative battery terminal.

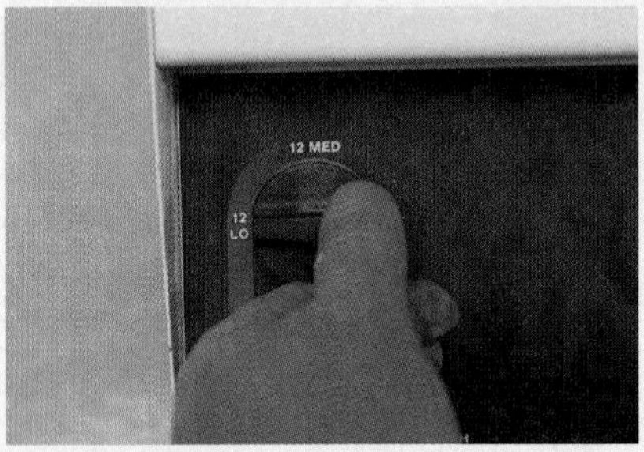

5. Check the settings on the charger, and verify that they are correct for what you are charging. Turn the charger on, and select the automatic setting, if equipped. Select the rate of charge. A fast charge should be carried out only under constant supervision.

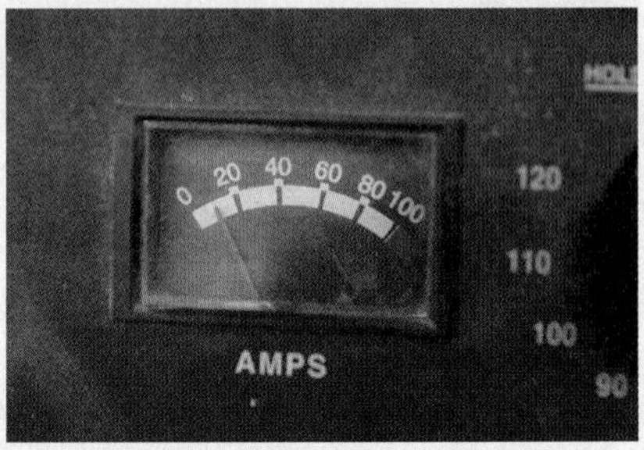

6. Verify that the voltage and amperage the charger is putting out are proper.

7. Once the battery is charged, turn the charger off. Disconnect the black lead from the negative battery terminal and the red lead from the positive battery terminal.

8. Allow the battery to stand for at least five minutes before testing the battery. Using a load tester or hydrometer, test the charged state of the battery.

Jump-Starting Vehicles

Jump-starting a vehicle is the process of using one vehicle to start another. It involves using a vehicle with a charged battery to provide electrical energy to the vehicle with a discharged or dead battery. Jump-starting a vehicle can put stresses on both vehicles. This is because starting a vehicle requires a high amount of electrical energy. When the discharged vehicle is being cranked, the battery voltage tends to fall very low. This is because the battery is already discharged. This causes the alternator on the running vehicle to put out its maximum current, which puts the alternator under heavy load. But as soon as the jumped vehicle stops cranking, the voltage shoots up quickly. It is potentially high enough to damage the electronic components in either vehicle. The same voltage spike can happen when the jumper cables are being disconnected.

Some vehicle manufacturers are now recommending that their vehicles should not be jump-started. Instead, they recommend that the battery be charged or replaced. Some towing companies have policies stating that they will not jump-start certain vehicles. Instead, they will only replace the battery or tow the vehicle to a shop to be recharged. If you do decide to jump-start a vehicle, *always* read the owner's manual for both vehicles. And follow all of their jump-starting guidelines. Also, never attempt to jump-start a frozen battery.

It is usually best to let the running vehicle charge the battery on the other vehicle for 5–10 minutes before trying to start the vehicle. It is also good practice to place a load on the charged battery by turning on an accessory such as the headlights. Do this prior to trying to crank the dead vehicle over. This helps absorb any sudden rise in voltage that may occur once the starter stops cranking. Once the dead vehicle's engine has started, let both vehicles come to an idle for a moment. Do not turn off the vehicle with the discharged battery. It needs an extended amount of time to recharge.

There is another method of reducing the risk of damage to sensitive electronic devices when jump-starting a vehicle. You can use jumper cables that have a built-in or auxiliary **surge protector** (**FIGURE 7-8**). This will help protect against high-voltage spikes. To start a vehicle with a discharged battery, using jumper leads and a second vehicle, follow the steps in **SKILL DRILL 7-2**.

▶ TECHNICIAN TIP

Alternators are not generally designed to charge a battery when it is highly discharged. If a battery is dead, it is always best to recharge the battery using a battery charger. That way, the alternator will not have to work so hard. Some technicians say it takes only 15 minutes to burn up an alternator when charging a dead battery. If you must drive the vehicle after being jump-started, turn off as many accessories as possible. This will lessen the load on the alternator.

SAFETY TIP

When connecting jumper cables, a spark will almost always occur on the last connection you make. That is why it is critical to make the last connection away from the battery. The engine block is generally good for this. A spark also occurs when you disconnect the first jumper cable connection. So the connection at the engine block needs to be the first cable to be disconnected.

SAFETY TIP

Secure the hood with a hood prop before going under it. Otherwise, it could fall and injure someone. It can also short out the jumper cables, causing a spark, which could result in the battery blowing up.

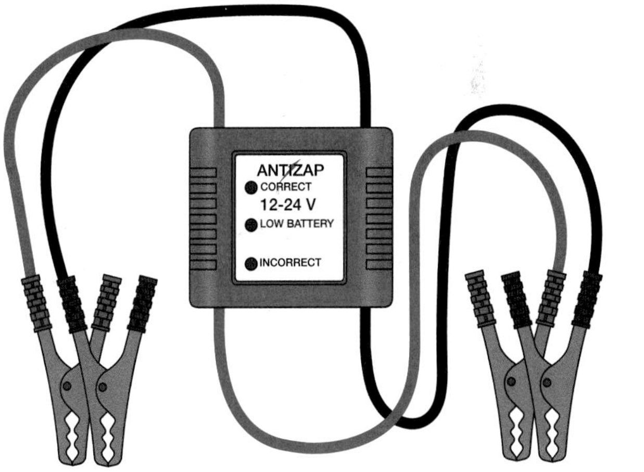

FIGURE 7-8 Jumper cables with built-in surge protection.

SKILL DRILL 7-2 Jump-Starting Vehicles with Jumper Cables

1. Position the charged battery close enough to the discharged battery that it is within comfortable range of your jumper cables. If the charged battery is in another vehicle, make sure the two vehicles are not touching.

2. First, connect the red jumper lead to the positive terminal of the discharged battery in the vehicle you are trying to start.

3. Next, connect the other end of this lead to the positive terminal of the charged battery or the remote terminal.

SKILL DRILL 7-2 Jump-Starting Vehicles with Jumper Cables (Continued)

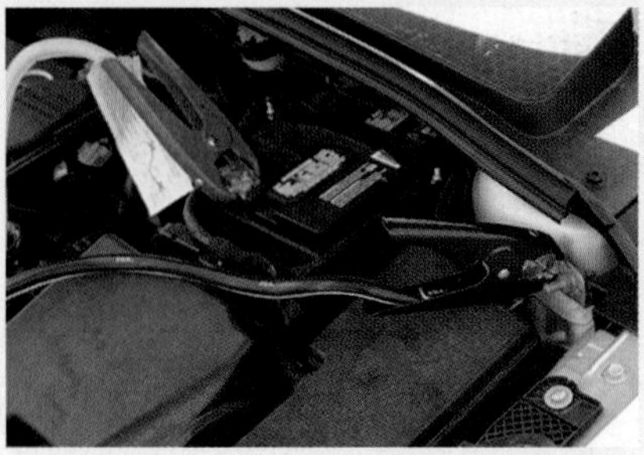

4. Then, connect the black jumper lead to the negative terminal of the charged battery or the battery remote terminal.

5. Connect the other end of the negative lead to a good ground on the engine block of the vehicle with the discharged battery, and as far away as possible from the battery.

6. Do not connect the lead to the negative terminal of the discharged battery itself; doing so may cause a dangerous spark. Also, do not connect the negative lead to the body or chassis as the ground wire from the body back to the negative battery terminal is usually too small to carry the current needed for jump-starting the vehicle.

7. Try to start the vehicle with the discharged battery. If the booster battery does not have enough charge or the jumper cables are too small in diameter to do this: start the engine in the booster vehicle, and allow it to partially charge the discharged battery for several minutes. Turn on the headlights on the booster vehicle to reduce the possibility of a voltage spike damaging electronic equipment, and try starting the discharged vehicle again.

8. Disconnect the leads in the reverse order of connecting them, starting with the negative clamp.

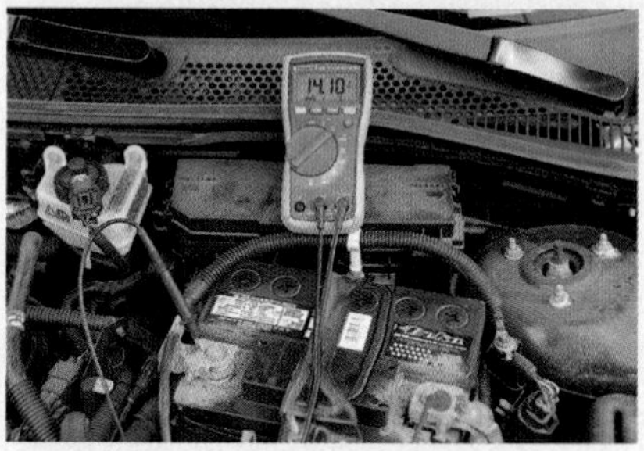

9. If the charging system is working correctly and the battery is in good condition, the battery will be recharged while the engine is running, although it could end up overheating and damaging the alternator.

SAFETY TIP

Keep your face and body as far back as you can while connecting jumper leads.

Do not connect the negative cable to the discharged battery, because the spark may blow up the battery.

Use only specially designed heavy-duty jumper cables to start a vehicle with a dead battery. Do not try to connect the batteries with any other type of cable.

Always make sure you wear the appropriate PPE before starting the job. Remember, batteries contain sulfuric acid, and it is very caustic.

Always follow any manufacturer's personal safety instructions to prevent damage to the vehicle you are servicing.

To jump-start vehicles without the need for another vehicle, many shops use jump boxes instead of jumper cables. The jump box has a built-in, high discharge battery and two short cables with battery clamps. Some jump boxes have on/off switches while others do not. If the jump box has a switch, then the switch should be turned off before connecting the battery clamps. If it does not have a switch, then it is best to first connect the red battery clamp to the battery positive terminal. Then, connect the black battery clamp to a good engine ground, if the cable will reach.

It is important to connect the battery clamps securely to the battery terminals and/or engine ground. By doing this, the maximum current can flow out of the jump box, and into the vehicle's electrical system. Also, the jump box needs to be placed in a

SKILL DRILL 7-3 Jump-Starting Vehicles with a Jump Box

1. Position the jump box in a secure position that it is within comfortable range of the dead battery.

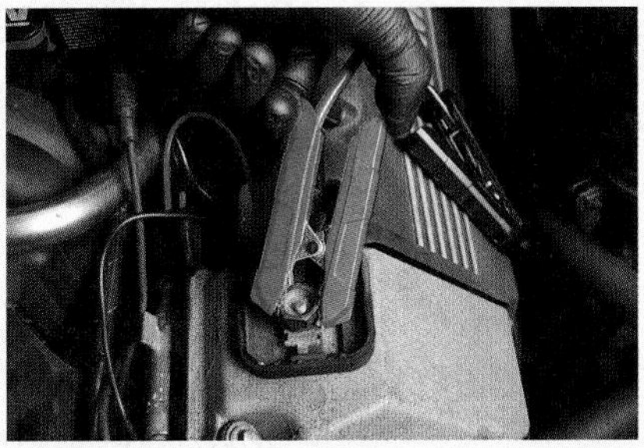

2. Turn the switch on the jumper box off, if equipped.

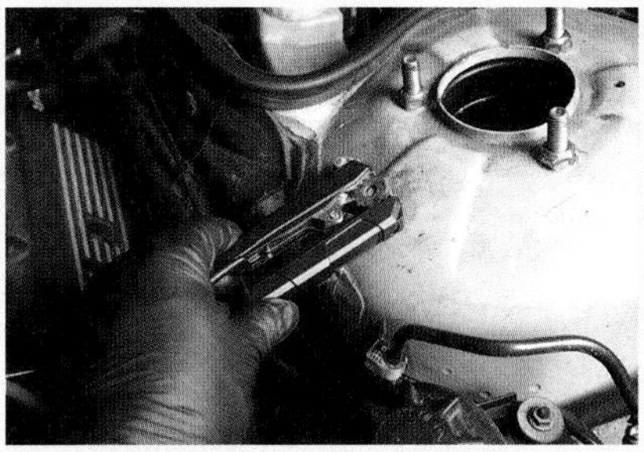

3. Connect the red battery clamp to the positive terminal of the discharged battery.

4. Connect the black battery clamp to the negative terminal of the charged battery or the battery remote terminal. Do not connect the black clamp to the body as the ground wire from the body back to the negative battery terminal is usually too small to carry the current needed for jump-starting the vehicle.

5. Turn the switch on the jump box on and try to start the vehicle. If the engine won't crank over properly, reposition the battery clamps and try starting the vehicle again.

6. Disconnect the battery clamps in the reverse order of connecting them.

secure position and away from belts and pulleys. This will help prevent it from falling over when the vehicle starts. In many cases, this is either right on top of the dead battery or next to it, but away from the engine.

To start a vehicle with a discharged battery, using a jump box, follow the steps in **SKILL DRILL 7-3**.

▶ Air Tools

LO 7-02 Identify and safely operate air tools.

Powered tools allow technicians to perform work faster than doing it by hand. Power tools can be powered by various power sources. The most common power sources are air, electric, and battery. Powering tools with air pressure has pros and cons. One pro is that air pressure doesn't normally create sparks. So it is safer to use air tools in potentially explosive environments than electric tools. They also don't present as much of a shock hazard either. In many cases, air tools are lighter and more powerful than electric tools. Because of these reasons, air tools are common in automotive shops.

Air tools also have disadvantages. These include having to be attached to an air hose, which can impede mobility. Air tools can also be louder than their electric counter parts, so ear protection is normally required. Finally, most air tools require daily lubrication. So for long life they require maintenance.

Air Compressors and Equipment

Compressed air powers a wide range of tools and equipment. Tools that use compressed air include impact wrenches, drills, grinders, pumps, grease guns, air nozzles, and hoists (**FIGURE 7-9**). The compressed air system is made up of a compressor, a pressure regulator, air hose or fixed piping, and the actual tool or item that is powered by the compressed air. The air compressor has a storage tank and is driven by an electric motor or, for more portability, a gasoline engine. In many shops, the air compressor is housed in a separate room to help isolate the noise.

Standard compressors have been, and many still are, based on piston-type compressors. They operate similarly to a piston engine, where each piston compresses the air on the compression stroke. But in the case of a compressor, the piston forces the air out of the cylinder and into a storage tank under pressure.

Many newer air compressors are of the scroll compressor type. It uses a pair of rotating scrolls to compress the air and is much quieter and more efficient. With either type of compressor, a pressure regulator controls the air pressure supplied to the distribution system. The air hose or lines transport the compressed air from the compressor to the tool.

SAFETY TIP

Always respect compressed air: Wear appropriate PPE, and never use compressed air inappropriately, such as using it to blow dust off yourself.

Compressed Air Safety

Serious, sometimes fatal, injuries can be caused by compressed air. This is why some states specify a maximum pressure that at shop air systems cannot exceed. Even with that, air can still be injected into the body through the skin. It can also be injected into a body opening, such as your mouth or ear. Internal human blood vessels and organs will rupture at much lower pressures than the maximum allowed. For these reasons, do not play with air equipment, such as blowing air at another person or yourself.

Always handle air equipment carefully and with respect. Be extra careful when working with air equipment in a confined or awkward space. Examples include being under a vehicle, and when clearing or cleaning the equipment. The air pressure can blow dirt, debris, or liquids back at you with high force. Another issue to be careful of is air hoses that are starting to balloon. That means that the internal webbing in the hose has broken and is in danger of rupturing. If that happens when you are holding it, air can be injected into your skin.

SAFETY TIP

If air, grease, or any other substance is injected into someone's body, it is not always easy to tell how much damage has been done. Always provide initial first aid, and then seek medical advice if any kind of penetration injury occurs.

Air Driers and Automatic Oilers

Air driers are fitted to compressed air systems. They remove the moisture or water from the compressed air. Water in the compressed air is a result of compressing air from the atmosphere. In the atmosphere, water is in the form of humidity. So it gets compressed along with the air.

If water gets into the air lines, it may damage the inside workings of air tools. An air drier can be a stand-alone device fitted to the compressed air system or can be incorporated into a filter/regulator system. The combination filter/regulator system removes water from the air, filters any debris that may come from the compressor tank, and regulates the line pressure from the tank.

A simple air drier can be nothing more than an inline water trap. It catches water droplets that have condensed out of the air as it cooled down (**FIGURE 7-10**). This type of air drier usually needs to be drained manually on a periodic basis. However, some devices are equipped with an automatic drain system.

Another type of air drier is the chiller type. This uses the principles of air conditioning to chill the hot compressed air and force condensation to occur at a higher rate (**FIGURE 7-11**). This system is much more expensive but does a much more effective job of removing all traces of moisture from the compressed air. You are likely to find this system in a body shop, where moisture in the compressed air must be removed. Otherwise, water would be mixed in with the paint as it is sprayed onto the surface of the vehicle.

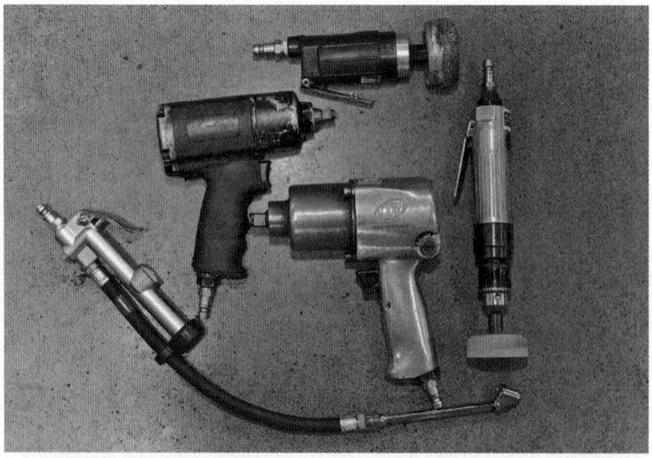

FIGURE 7-9 A wide variety of tools use compressed air, including drills, grinders, pumps, grease guns, jacks, and impact wrenches.

FIGURE 7-10 Water trap air drier.

FIGURE 7-11 Chiller-type air drier.

Compressed air tools and equipment require a regular application of a lubricating oil to reduce wear and tear. This is typically done by adding a few drops of air tool oil to the air fitting prior to use each day. This will lubricate the internal workings of the air tool (**FIGURE 7-12**).

Automatic oilers are designed to regularly oil an air tool or air equipment so it does not have to be done manually. Automatic oilers are usually fitted in the air hose near the air tool or equipment. They regularly supply small amounts of oil into the stream of compressed air. The oil is then transported along with air to the tool or equipment (**FIGURE 7-13**). Automatic oilers need periodic inspection to make sure they deliver the correct amount of oil. The built-in oil reservoir also needs to be refilled with air tool oil on a regular schedule.

Air Tools

Air tools use compressed air at high pressure to operate (**FIGURE 7-14**). Air compressors in automotive shops typically run at greater than 90 psi (621 kPa), so caution needs to be exercised when working with them. Air tools have quick-connect fittings so that various air tools can be used on the same air hose (**FIGURE 7-15**). There are several styles of quick-connect fittings, and a shop will usually use one style throughout the entire shop.

SAFETY TIP

Air tools generally produce more noise than electric tools, so wear ear protection when using air tools.

Always use impact sockets with impact guns.

The most common air tool in an automotive shop is the **air impact wrench**. It is sometimes called an impact gun or **rattle gun**, which is easy to understand why when you hear one. Taking the wheels off a car to replace the tires is a typical application for this air tool (**FIGURE 7-16**). Removing lug nuts often requires a lot of torque to free the lug nuts, and air impact wrenches work well for that.

The air impact wrench can be set to spin in either direction, and a valve roughly controls how much torque it applies. But it should never be used for final tightening of lug nuts. There is a danger in overtightening the lug nuts, as it can cause the lug studs to fail and the wheel to separate from the vehicle while it is moving.

Another rule with the air impact wrench is that you have to use special hardened impact sockets, extensions, and joints. The tools are thicker and stronger than standard duty sockets and extensions. They are designed to withstand the hammering force that the impact wrench subjects them to.

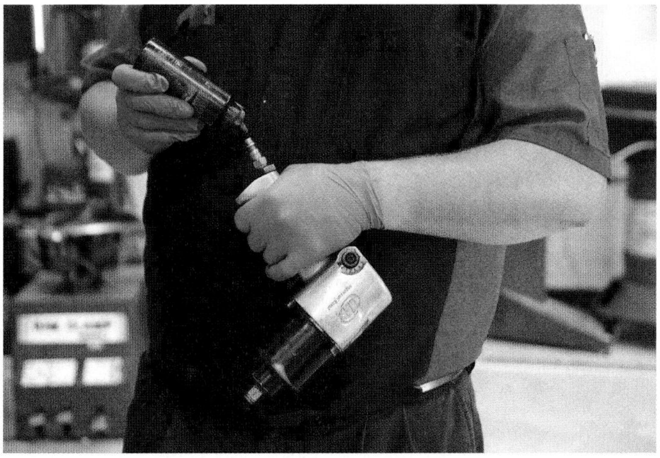

FIGURE 7-12 Manually lubricating air tools.

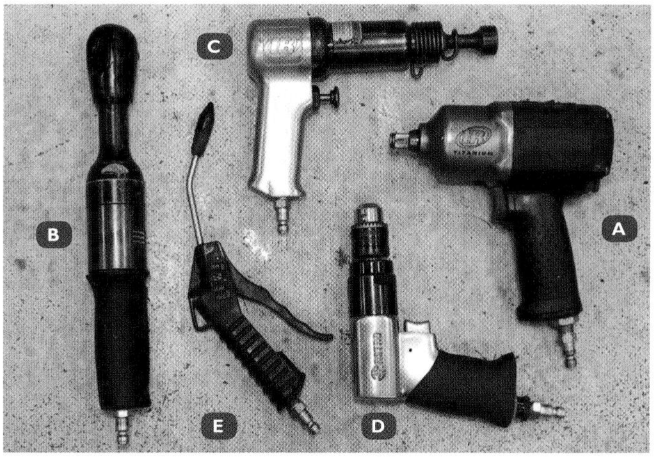

FIGURE 7-14 **A.** Air impact wrench. **B.** Air ratchet. **C.** Air hammer. **D.** Air drill. **E.** Blowgun or air nozzle.

FIGURE 7-13 Compressed air automatic oiler.

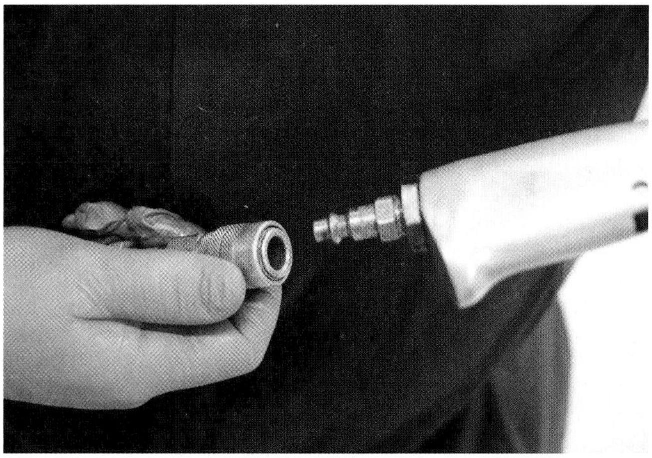

FIGURE 7-15 Quick disconnect fitting.

An **air ratchet** uses the force of compressed air to turn a ratchet drive (**FIGURE 7-17**). It is used on smaller nuts and bolts. Once the nut is loosened, the air ratchet spins it off in a fraction of the time it would take by hand. Since it is reversible, it can be used to install nuts and bolts, especially where there is not much room to swing a ratchet handle.

An **air hammer**, sometimes called an air chisel, is useful for driving and cutting (**FIGURE 7-18**). The extra force that is generated by the compressed air makes it more efficient than a hand chisel and hammer. Just as there are many chisels, there are many bits that fit into the air hammer, depending on the job at hand.

An **air drill** has some important advantages over the more common electric power drill. With the right attachment, it can drill holes, grind, polish, and clean parts. Unlike the electric drill, it does not run the risk of producing sparks. This is very important when working around flammable liquids or batteries. An air drill does not drag a live electric cable behind it that could be cut, possibly causing shock and burns. It also does not get hot with heavy use.

A blowgun, or **air nozzle**, is probably the simplest air tool. It controls the flow of compressed air. It is controlled by a lever or valve that is used to blast debris and dirt out of confined spaces. Blasting debris and dirt can be dangerous, so eye protection must be worn whenever this tool is used. Noise levels are usually high, so ear protection should also be worn.

An air nozzle should always be directed away from the user and anyone else working nearby. It is also dangerous to use an air nozzle to clean yourself off. Note that Occupational Safety and Health Administration (OSHA)-approved nozzles lower the tip pressure by venting some of the air for safety reasons (**FIGURE 7-19**). Only use OSHA-approved nozzles for general blowing purposes; otherwise, the shop could be liable for a substantial fine.

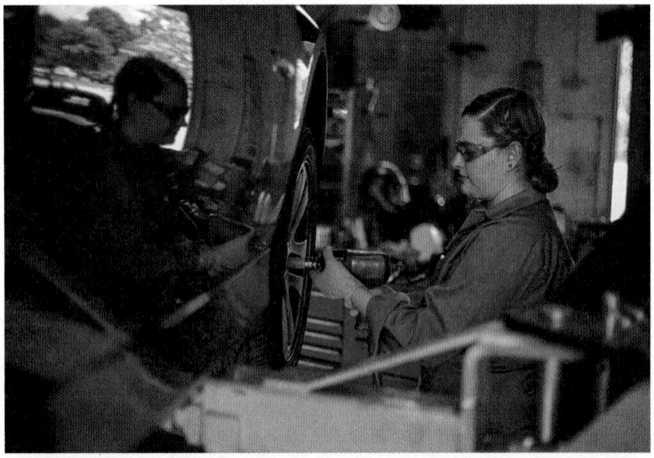

FIGURE 7-16 Impact wrench being used to remove lug nuts.

FIGURE 7-17 Air ratchet being used.

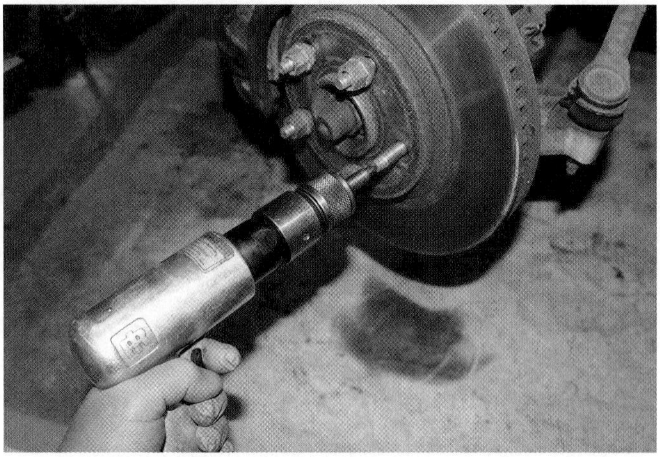

FIGURE 7-18 Air hammer being used.

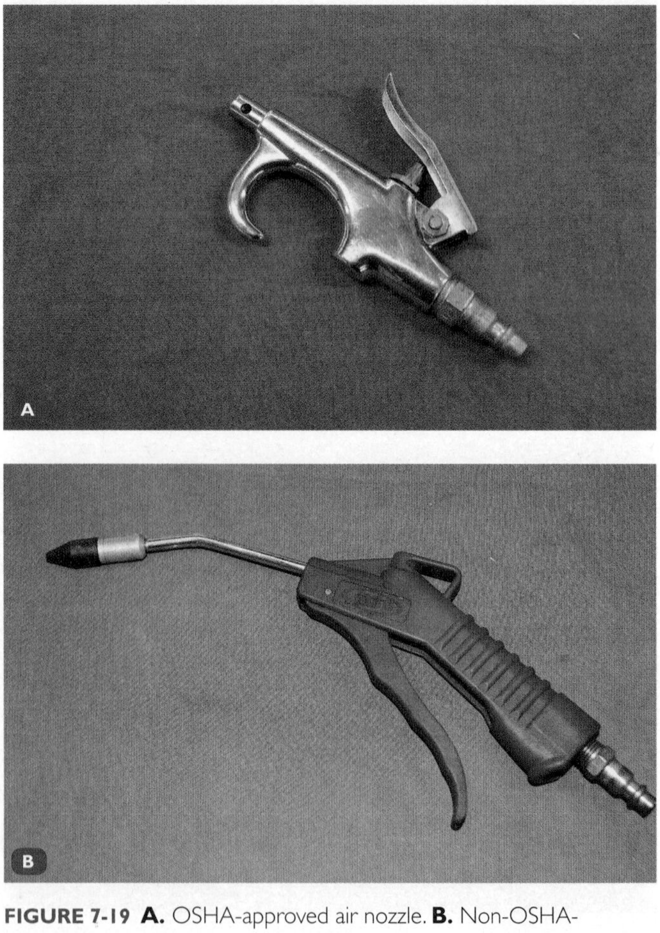

FIGURE 7-19 A. OSHA-approved air nozzle. **B.** Non-OSHA-approved air nozzle.

SKILL DRILL 7-4 Using Air Nozzles

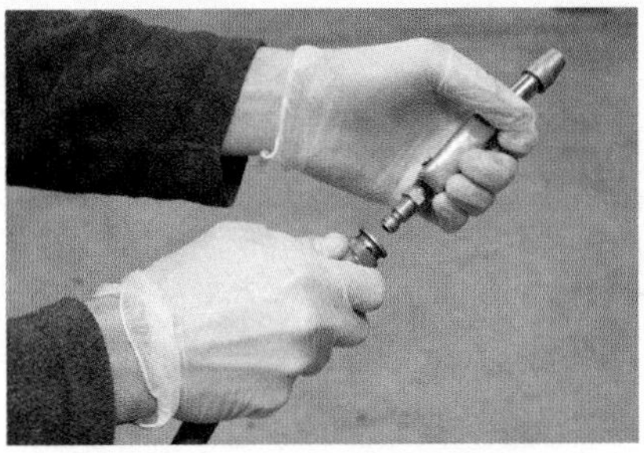

1. Fit an OSHA-approved air nozzle to the end of the air hose. Make sure there are no air leaks.

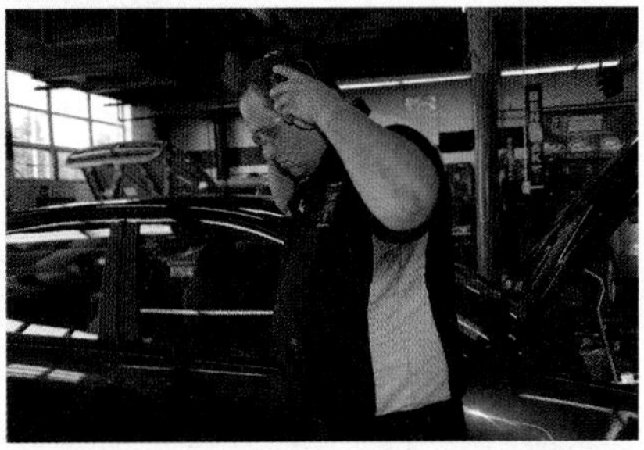

2. The air nozzle is used to blast dirt and debris out of confined spaces. To avoid injury, be sure to wear eye and ear protection whenever you use the air nozzle.

3. Do *not* use the air nozzle to dust yourself off because you risk injury. Be sure to direct the air jet away from yourself and away from anyone else who may be working nearby.

Using Air Nozzles

An air nozzle can be a handy tool for blowing dirt and debris out of holes, for example, around spark plugs prior to removing them. It is helpful to activate the air nozzle away from the area you intend to blow off before doing it. This way you can get a feel for how the valve operates. To operate an air nozzle, pull the trigger gently, and modulate the flow of air through the nozzle. If too much air is allowed through, you may blow dirt particles back at you, so always wear your safety glasses. Also, keep your mouth closed, as it helps prevent eating a bunch of dirt! To correctly operate an air nozzle, follow the steps in **SKILL DRILL 7-4**.

SAFETY TIP

Do not use the air nozzle to clean brake dust from brake components. It will disperse the hazardous dust.

Do not use a high-pressure air nozzle to disperse liquid solvents or fuels. A low-pressure blowing action can help these volatile materials to evaporate more quickly, but a high-pressure air jet could atomize the liquid, allowing it to form a flammable mixture.

Do not point the air nozzle at other people.

Never use the air nozzle to blow air over yourself or other people.

Always wear eye protection when using air tools.

Do not drive over air hoses.

Make sure the hoses are in good condition before using.

Using Air Impact Wrenches

The amount of torque an air impact wrench can produce is determined by the tool and the pressure in the air system feeding it. Because of these variables, there is no way of knowing how much torque an impact wrench is applying to a fastener. It is easy to overtighten or undertighten fasteners. An air impact wrench can generally be used to take up the looseness in a nut or bolt. The final tightening, however, must be performed using a torque wrench set to the manufacturer's specifications. Every impact wrench has a control mechanism that allows it to change direction. It also loosely controls the amount of torque the wrench develops.

Always use impact sockets and extensions when using an air impact wrench. To correctly operate an air impact wrench, follow the steps in **SKILL DRILL 7-5**.

Using Air Drills

An air drill is another tool operated by compressed air. Since they are powered by air and not electricity, they are safer to use

SKILL DRILL 7-5 Using Air Impact Wrenches

1. Select the properly sized impact gun and socket, and inspect them for damage.

2. Lubricate the gun if an automatic oiler is not installed in the system.

3. Adjust the direction of spin—forward or backward—with the selector

SKILL DRILL 7-5 Using Air Impact Wrenches (Continued)

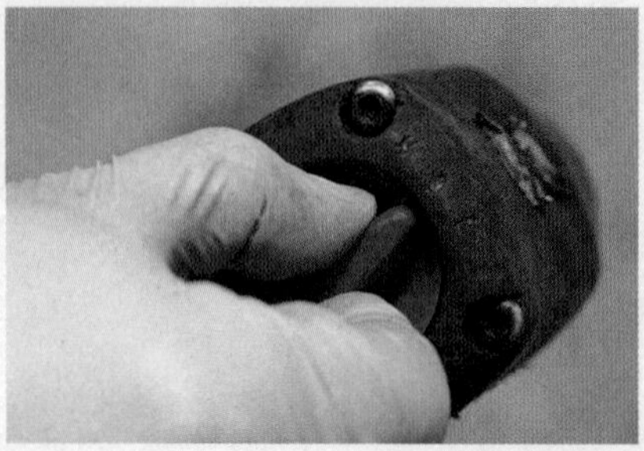

4. Turn the valve to increase or reduce the torque to match the needs of the fastener. If removing lug nuts, adjust it toward the upper middle torque setting. If running a nut back on a stud, adjust it toward the lowest torque setting. You can always readjust it if you end up needing more torque.

5. Place the impact wrench fully over the bolt or nut and give the trigger a quick squeeze to verify that the impact wrench is turning the correct direction.

6. Continue to remove or install the fastener by squeezing the trigger only long enough to get the job done. Release the trigger before getting to the end as it takes time for the impact wrench to slow down. Otherwise, the nut could fly off the end of the stud, or if tightening, you could end up overtightening it.

in an environment where flammable materials are present. The amount of torque an air drill can produce is less than most electric drills, but will be determined by the pressure in the air system feeding it. Air drills are smaller and typically turn at slower speeds than electric drills. At the same time, air drills operate similarly to their electric counterparts and are equipped with the same type of drill chuck. Most air drills are of the 90-degree angle style and fit in places a typical electric drill might not. To correctly operate an air drill, follow the steps in **SKILL DRILL 7-6**.

Using Air Hammers

Air hammers act in a manner similar to a jackhammer. However, their size makes their cycling rate faster. There are a variety of attachments, so use the correct attachment for the task you are performing.

SKILL DRILL 7-6 Using Air Drills

1. Select the properly sized air drill and drill bit, inspect them for damage, and lubricate the gun if an automatic oiler is not installed in the system.

2. Adjust the direction of spin—forward or backward—with the selector.

3. Turn the valve to increase or reduce the torque to match the needs of the job.

4. Place the air drill into position and hold it firmly. Give the trigger a quick squeeze to verify that the air drill is turning the correct direction and does not slip.

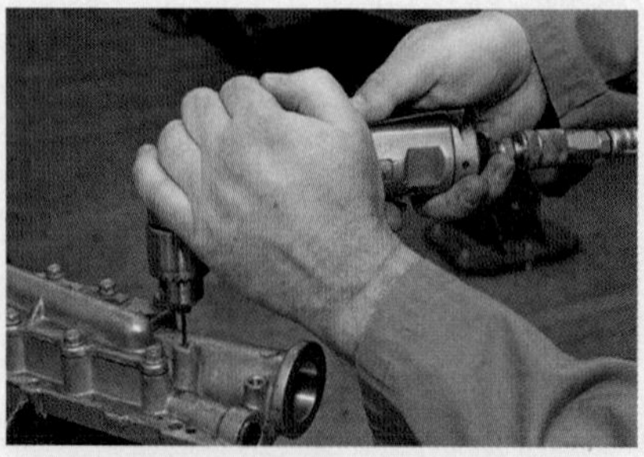

5. Continue to operate the air drill by squeezing the trigger only long enough to get the job done.

This could be a chisel for cutting a bolt or a punch for driving out a broken lug stud. The attachment is held in place by a coil spring that tends to break over time. So inspect the air hammer before use, and never use it if it is broken. Place the tool bit against the workpiece. Hold it firmly before you pull the trigger, as it is likely to jump around. To use an air hammer, follow the steps in **SKILL DRILL 7-7**.

SAFETY TIP

An air hammer's operation produces a noise level that exceeds the maximum safe exposure level for human ears. Always wear ear protection when using an air hammer.

Hydraulic Press

In an automobile, some parts are press fit together. This is accomplished when the engineers design an interference fit between the parts to hold them together. For example, some rear axle bearings are designed with a diameter that is about 0.001" smaller than the outside diameter of the axle it fits over. Once installed, the bearing is clamped tightly to the axle, staying firmly in place. Over time, if the bearing fails, then it will need to be pressed off the axle shaft, using several tons of force (**FIGURE 7-20**). Depending on the size, hydraulic presses can typically develop many tons of force, and hence they are used to perform this kind of task.

Because hydraulic presses develop so much force, they can be very dangerous to operate. The extreme forces generated can cause components to shatter with great force, resulting in shrapnel injuries or death. Parts that do not shatter can slip and be thrown a great distance at very high speed. Also, it is possible to get a body part caught under the ram. If so, it is easy for bones to be broken and crushed. So never operate a hydraulic press without specific training by and the permission of your instructor. Here are some general guidelines when using a hydraulic press:

- Never remove any guards or safety devices.
- Always wear safety glasses and a face shield.
- Never wear loose clothing. But it is good practice to wear thick clothing or a snug-fitting leather apron.
- Keep the area clear of obstacles and tripping hazards.
- Make sure all pins or locks are securely in place.
- Only use approved press plates and adapters.

SKILL DRILL 7-7 Using Air Hammers

1. Select the properly sized air hammer and attachment, inspect them for damage, and lubricate the gun if an automatic oiler is not installed in the system.

2. Fit the appropriate bit into the nose of the air hammer, and ensure the spring is installed correctly.

3. Place the air hammer into position and hold it firmly. Give the trigger a quick squeeze to see how the air hammer will react.

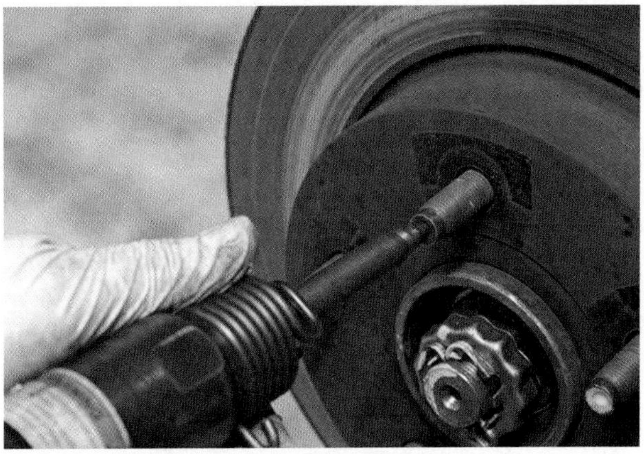

4. Continue to operate the air hammer by squeezing the trigger and watching the progress carefully. Allow the air hammer to do the work, and work slowly around the item.

FIGURE 7-20 Hydraulic press being used to press off an axle bearing.

- Only press in line with the hydraulic ram; don't press at an angle.
- Inspect the press for damage regularly.
- If the part does not move with reasonable pressure, verify that all keepers, bolts, or other fasteners have been removed from the part.

▶ Electric Power Tools

LO 7-03 Use cutting and grinding tools.

Electric power tools are common in automotive shops. They provide additional power for getting tasks done. Electric power tools can be operated on the shop's electrical power. They can also be battery operated, which is lower power and less likely to be a shock hazard. Electric power tools should be inspected before use. Inspect for frayed cords and missing ground terminals. Also, never use electric tools in an explosive environment. This includes gasoline, solvent, or other flammable gases. Lastly, never use electric tools until you have been fully trained on their safe use.

Drills and Drill Bits

There are several types of drills and drill bits (**FIGURE 7-21**). A portable drill can be corded or cordless. A corded drill has a cord that you have to plug into an electrical supply. The operating voltage of a drill depends on the country's electric supply. Corded drills are a good choice when moderate power is needed or if extended drilling is required. Cordless drills use their own internal batteries. When you cannot bring the work to the drill, you can take the drill to the work. But do not expect a cordless drill to be able to drill large holes through hard metal. Although they are very versatile, they are limited to the amount of work they can do by their power rating.

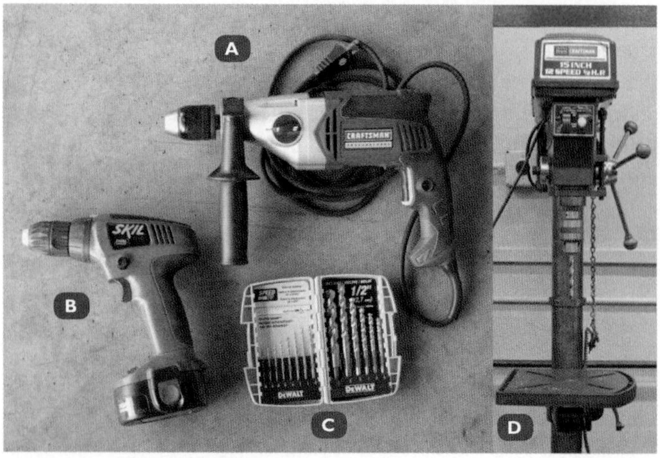

FIGURE 7-21 **A.** Corded drill. **B.** Cordless drill. **C.** Drill bits. **D.** Drill press.

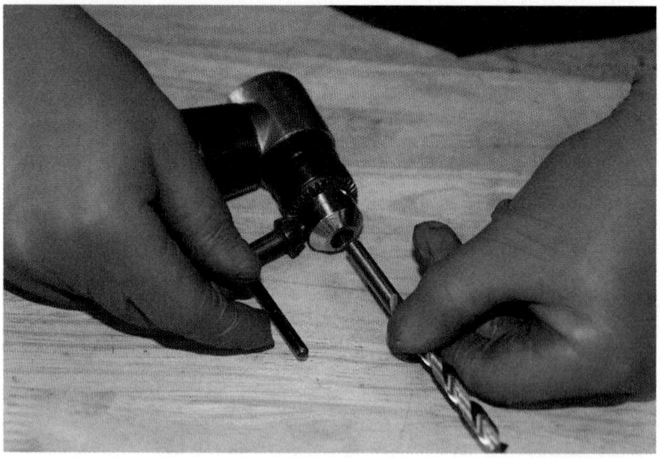

FIGURE 7-22 Twist drill bit and drill chuck.

The biggest drill bit that will fit into the chuck of these drills is usually marked on the body of the drill or chuck, along with the speeds at which it turns. Most drills have a drill chuck that grips different sized drill bits. Portable drills are commonly sized according to their chuck size such as a 3/8" or 1/2" drill. The chuck is tightened using a chuck key which have teeth that mesh with the teeth on the chuck (**FIGURE 7-22**). However, some drills have a keyless chuck which can be tightened by hand (**FIGURE 7-23**).

Some portable drills have only two operating speeds. But most portable drills have a variable speed rating that is determined by how far the trigger is moved. In this case, it can be set to any speed within the drill's range.

▶ TECHNICIAN TIP

Drills are also used to drive other accessories such as rotary files, screwdriver bits, and sockets.

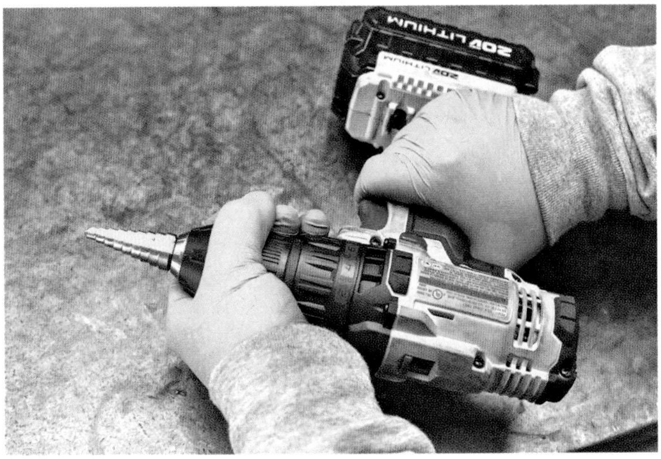

FIGURE 7-23 Drill with keyless chuck.

FIGURE 7-24 Morse taper drill bit.

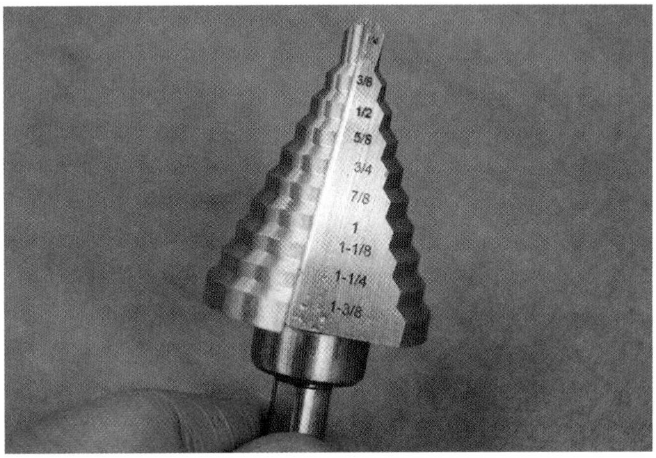

FIGURE 7-25 Multi-fluted tapered hole drill.

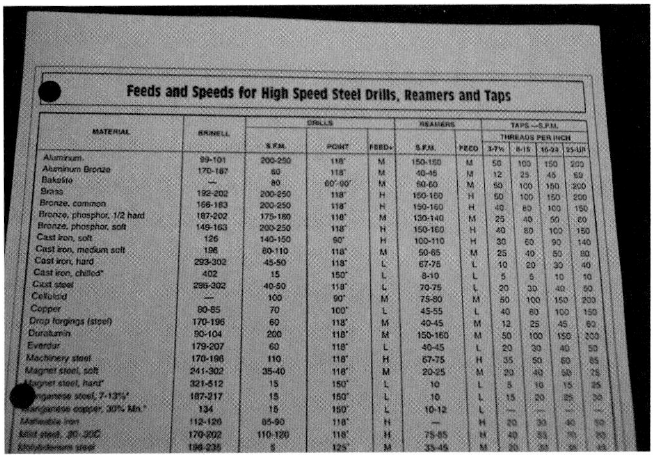

Feeds and Speeds for High Speed Steel Drills, Reamers and Taps

MATERIAL	BRINELL	DRILLS			REAMERS		TAPS —S.F.M. THREADS PER INCH			
		S.F.M.	POINT	FEED*	S.F.M.	FEED	3-7½	8-15	16-24	25-UP
Aluminum.	99-101	200-250	118°	M	150-160	M	50	100	150	200
Aluminum Bronze	170-187	60	118°	M	40-45	M	12	25	45	60
Bakelite	—	80	60°-90°	M	50-60	M	50	100	150	200
Brass	192-202	200-250	118°	H	150-160	H	50	100	150	200
Bronze, common	166-183	200-250	118°	H	150-160	H	40	80	100	150
Bronze, phosphor, 1/2 hard	187-202	175-180	118°	M	130-140	M	25	40	50	80
Bronze, phosphor, soft	149-163	200-250	118°	H	150-160	H	40	80	100	150
Cast iron, soft	126	140-150	90°	H	100-110	H	30	60	90	140
Cast iron, medium soft	196	80-110	118°	M	50-65	M	25	40	50	80
Cast iron, hard	293-302	45-50	118°	L	67-75	L	10	20	30	40
Cast iron, chilled*	402	15	150°	L	8-10	L	5	5	10	10
Cast steel	286-302	40-50	118°	L	70-75	L	20	30	40	50
Celluloid	—	100	90°	M	75-80	M	50	100	150	200
Copper	80-85	70	100°	L	45-55	L	40	60	100	150
Drop forgings (steel)	170-196	60	118°	M	40-45	M	12	25	45	60
Duralumin	90-104	200	118°	M	150-160	M	50	100	150	200
Everdur	179-207	60	118°	L	40-45	L	20	30	40	50
Machinery steel	170-196	110	118°	H	67-75	H	35	50	60	85
Magnet steel, soft	241-302	35-40	118°	M	20-25	M	20	40	50	75
agnet steel, hard*	321-512	15	150°	L	10	L	5	10	15	25
nganese steel, 7-13%*	187-217	15	150°	L	10	L	15	20	25	30
nganese copper, 30% Mn.*	134	15	150°	L	10-12	L	—	—	—	—
Malleable iron	112-126	85-90	118°	H	—	H	20	30	40	50
Mild steel, 20-30C	170-202	110-120	118°	H	75-85	H	40	[illegible]	[illegible]	[illegible]
[illegible] steel	196-235	[illegible]	125°	M	35-45	M	[illegible]	[illegible]	[illegible]	[illegible]

FIGURE 7-26 Drill speed reference chart.

A **drill press** allows for accurate drilling with more control than is offered by a portable drill. A mounted drill can feed the drill bit at a controlled rate. The worktable on the drill press typically has a vise to secure the job at a constant angle to the drill bit. Also, this drill can be set to operate at different drilling speeds.

Morse taper is a system for securing drill bits to drills. The Morse taper size changes according to drill size. The shank of the drill bit is tapered and looks like the tang of a file (**FIGURE 7-24**). It fits snugly into the drill spindle, which has a similar taper on its inner surface. The tang on the drill bit is placed in the spindle, and it drives the drill bit. It is a quick way to change drills without constantly adjusting the chuck.

When a hole already drilled in sheet metal needs enlarging, a multi-fluted tapered hole drill will do the job quickly (**FIGURE 7-25**). This drill bit can drill several sizes of holes depending on how deep the bit is moved into the material.

A drilling speed chart is usually supplied with the drill press and should be kept nearby for handy reference (**FIGURE 7-26**). It compares drill sizes and metals to show the proper drill speed. For example, to drill a 0.375" (10 mm) hole through a piece of aluminum, the drill speed should be 1800 rpm.

Drilling metals is also best performed with the aid of a lubricant. The lubricant helps cool the cutting edges of the drill bit, as well as lubricate it. Each metal requires its own type of lubricant, so check a drilling guide for the metal you are working on.

Bench and Angle Grinders

Power grinders come in different sizes and speeds (**FIGURE 7-27**). The size of a power grinder is normally determined by the diameter of the largest grinding wheel or disc that can be fitted to it. Some grinders are fixed to a bench or pedestal, and the work is brought to the grinder. Others are portable devices that can be taken to the work. Bench or pedestal grinders tend to be powered by electricity. Portable ones can be electric- or air-powered.

A maximum safe operating speed is usually printed on **grinding wheels and discs**. This maximum speed must never be exceeded, or the wheel or disc could disintegrate. Every well-equipped shop has a solidly mounted grinder. It is either mounted on a pedestal bolted to the shop floor, or securely attached to the workbench. Appropriate eye protection must be worn when grinders are being used. Also, the wheel guards and shields must be positioned correctly and firmly in place.

A **bench grinder (pedestal grinder)** normally has a rating listing the size of the grinding wheel it can take (**FIGURE 7-28**). Do not try to install a grinding wheel larger or smaller than it is rated for. Grinding wheels come in grades from coarse to very fine, depending on the size of the abrasive grains that are bonded together to make the wheel. They also range in hardness, depending on the abrasive used and the material that

FIGURE 7-27 **A.** Bench grinder. **B.** Angle grinder.

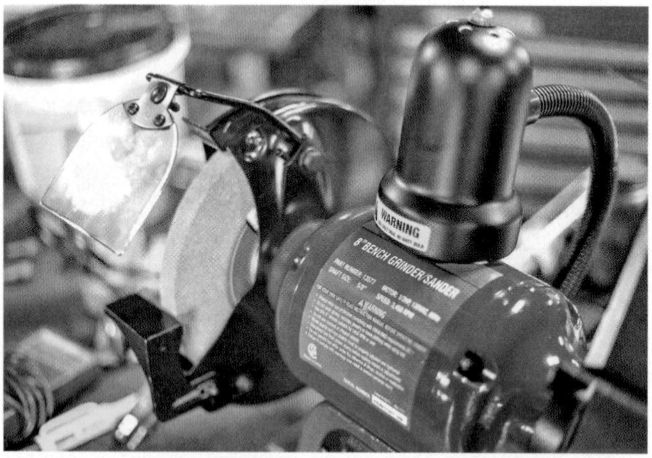

FIGURE 7-28 Bench grinders are typically rated according to the specified diameter of the grinding wheel.

FIGURE 7-29 Hand-held cutoff wheel.

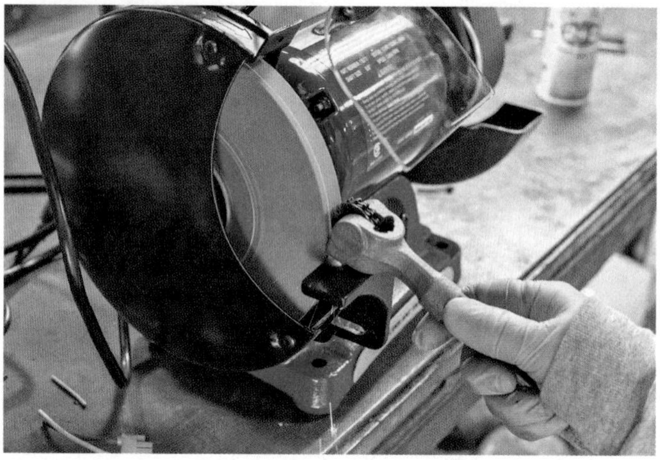

FIGURE 7-30 Dressing an abrasive wheel with a dressing tool.

bonds the particles together. Check to determine the most suitable grinding wheel for a particular grinding application.

An **angle grinder** is usually needed when the bench grinder is not appropriate. The angle grinder uses discs rather than wheels. During grinding, the face of the disc is used instead of the edge. An angle grinder can throw sparks many feet, so direct the sparks in a safe direction. Or set up a guard to catch them. Also use hearing protection whenever grinding, as it is very noisy and can damage your ears. Although not as common in an automotive shop, the **straight grinder** takes conventional grinding wheels, just like the stationery grinders. However, the grinding wheel diameter is limited to about 4.75" (126 mm). In many cases, the grinder has a long shaft that moves the grinding wheel away from the motor. This makes it handy to get into recessed areas.

Hand-held cutoff wheels can be powered by electricity or air (**FIGURE 7-29**). A special thin grinding disc enables them to cut. The edge of the wheel is used for cutting, and they are useful for jobs that cannot be reached with a hacksaw.

Using Bench Grinders

When you are grinding metal, it must not be allowed to overheat, because this will adversely affect its hardness. If the metal becomes too hot and is allowed to cool slowly, it may become soft. If it is cooled quickly (quenched), it may become brittle. As you grind the metal, stop and dip it regularly into the water pot attached to the base of the grinder. This will prevent the metal from getting too hot. Some bench grinders are not supplied with a water pot. If this is the case, you need to have a water can located near the grinder so that you can cool the piece you are grinding.

When using a bench grinder, the face of the abrasive wheel must be kept square. This is done with a dressing tool, which removes some of the abrasive compound. If the abrasive wheel is not square, use a dressing tool to square it up (**FIGURE 7-30**). Then, readjust the tool rest to the proper clearance.

As the abrasive wheel wears down, or the wheel is dressed, the gap between the wheel and the tool rest increases. This creates a dangerous situation. Once the gap enlarges to the thickness of the metal you are grinding, the metal can be pulled into the gap. This can cause the metal to be thrown from the grinder with much force. So always make sure the tool rest gap is set properly. This is usually about 1/8" (3.2 mm) or no more than half the thickness of the metal you are grinding, whichever is smaller.

Also, most bench grinders are equipped with a spark deflector, sometimes called a "shatter guard." This reduces sparks being thrown forward and can help contain pieces of the grinding wheel if it shatters. The shatter guard is located at the top of each grinding wheel guard opening. It needs to be periodically adjusted as the grinder wheel wears down. Refer to the manufacturer's specifications, which typically call for a gap of between 1/16" and 1/4" (1.59 and 6.35 mm).

To set up, adjust, and use a bench grinder, follow the steps in **SKILL DRILL 7-8**.

SKILL DRILL 7-8 Using Bench Grinders

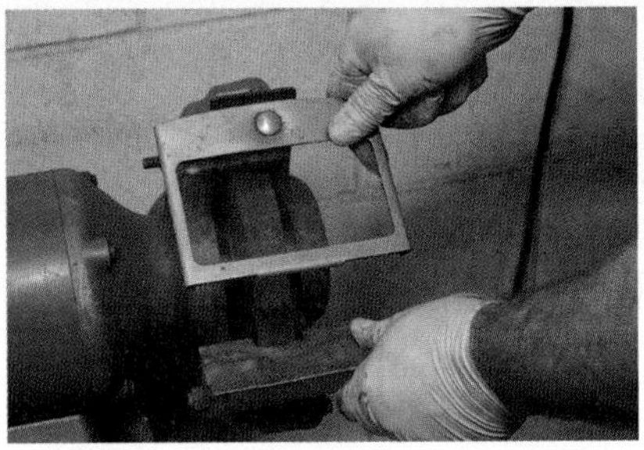

1. Before you start using the bench grinder, inspect the wheels, and check to see if the tool rest, safety shield, and shatter guard are adjusted properly. There should also be water in the pot.

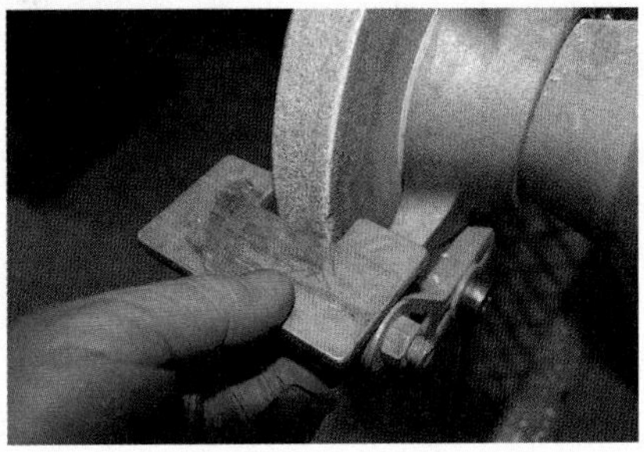

2. If needed, adjust the tool rest so it has a maximum gap of 0.125" (3.2 mm) (or no more than half the thickness of the metal you are grinding), whichever is smaller, between the tool rest and wheel. If you are unsure of how to do this, ask your supervisor.

3. If needed, adjust the shatter guard to a maximum gap of 0.0625" (1.6 mm) between the guard and wheel.

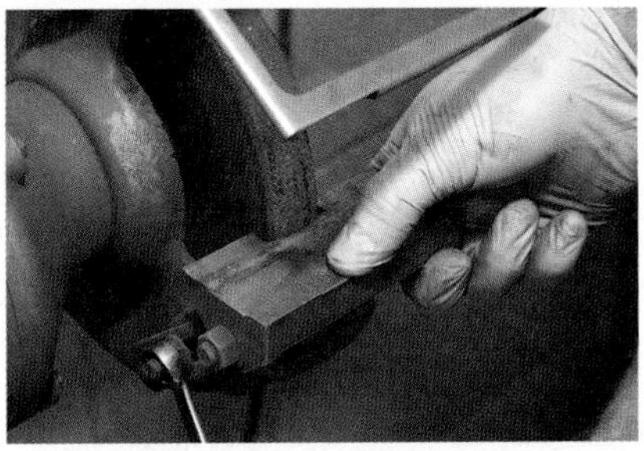

4. The tool rest should be slightly below the center of the wheel. To adjust the tool rest, locate the adjusting bolt, and loosen it with a box-end wrench. Set the tool rest at the right height and distance from the wheel, and then tighten the adjusting bolt.

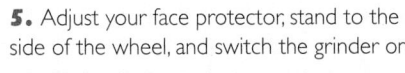

5. Adjust your face protector, stand to the side of the wheel, and switch the grinder on.

6. After the grinder is fully up to speed, move to the front of the wheel. Hold the part you are going to grind firmly on the tool rest, and move it slowly and gently forward until it comes into contact with the wheel. The grinding wheel removes the metal it contacts. Use the full face of the grinding wheel to prevent wearing one area of the wheel.

7. Occasionally, dip the part into the water to keep it cool.

8. When you have finished, turn off the power and unplug the grinder.

SAFETY TIP

Make all adjustments with the grinder stopped and unplugged.

Stand to the side of the grinder when starting the electric motor.

Never use a cracked or gouged grinding wheel.

Do not operate a grinder unless it is securely mounted to the bench or floor.

Do not grind on the side of the wheel because it may cause the wheel to shatter.

Maintain a 3-foot (0.9 m) perimeter around the grinder free of flammables, clutter, and people.

Using Angle Grinders

The angle grinder uses an electric motor to drive an abrasive disc at a high speed (**FIGURE 7-31**). The grinder disc is turned at speeds that range from 5000 to 12000 rpm. The turning disc is used to grind or cut metal. The grinder size relates to the diameter of the cutting disc, which can range from 4" to 9" (102 to 229 mm). The size of grinder you use depends on the type of job you are doing. The smaller the grinder, the higher the speed it turns. Sanding discs and wire wheels can be fitted on the grinder, making it a versatile electric tool. The handle can be fitted to either the left, right, and sometimes the top of the head to make it easy to use for left-handed as well as right-handed people.

FIGURE 7-31 A. The angle grinder uses an electric motor to drive an abrasive disc at a high speed. **B.** The handle is removable and can be installed in either side of the grinder.

The **abrasive disc**, or cutting wheel, is attached to the grinder by a flange and nut. The nut is specially designed to fit in a recess in the center of the pad or wheel. It is tightened by a tool that is provided with the grinder when purchased. Do not lose this wrench because it is the only tool that can tighten the nut properly. When using the grinder with cutting discs, you should always use the edge of the disc rather than the face.

▶ TECHNICIAN TIP

Do not confuse a grinder with a **sander/polisher**. The sander/polisher turns at lower speeds, typically 600–3000 rpm (**FIGURE 7-32**). It is commonly used to sand and polish paint. The pads these tools use cannot be turned at a high speed. If the polish pad were attached to an angle grinder, the higher rotational speed would cause the polishing pad to burn the paint and cause the polish pad to fly apart.

SAFETY TIP

- Always wear impact-resistant protective glasses, ear protection, and a full-face shield when using an angle grinder.
- Wear safety shoes, leather gloves, and an apron to protect your body from flying metal chips.
- Use the correct type of disc.
- Make sure the grinder handles are secure.
- Use the correct flange or spindle nut for the type of disc being used. If you do not, the disc can shatter at high speeds and injure you.
- Angle grinders, like all portable grinding tools, need to be equipped with safety guards to protect you from flying fragments in case the disc breaks apart.
- Always make sure the spindle wheel does not exceed the abrasive disc maximum speed.
- Make sure there are no obvious defects or damage to the disc before you use it.

To correctly use an angle grinder, follow the steps in **SKILL DRILL 7-9**.

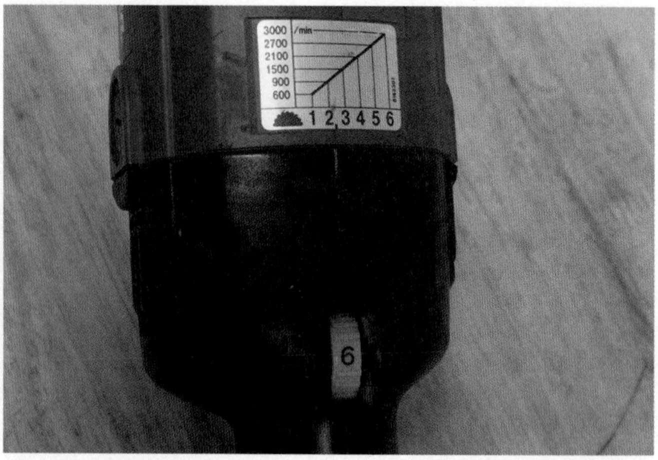

FIGURE 7-32 Sander/polishers are rated at lower rpm than grinders.

SKILL DRILL 7-9 Using Angle Grinders

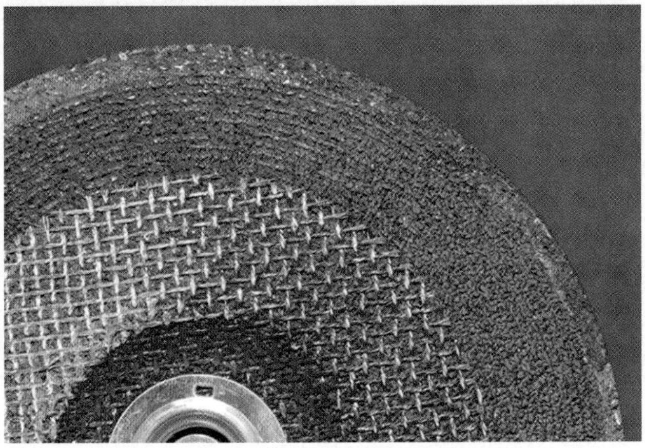

1. Inspect the grinding disc for any cracks or damage.

2. Check the area for any flammables and to determine where the sparks will fly. Take any precautions necessary.

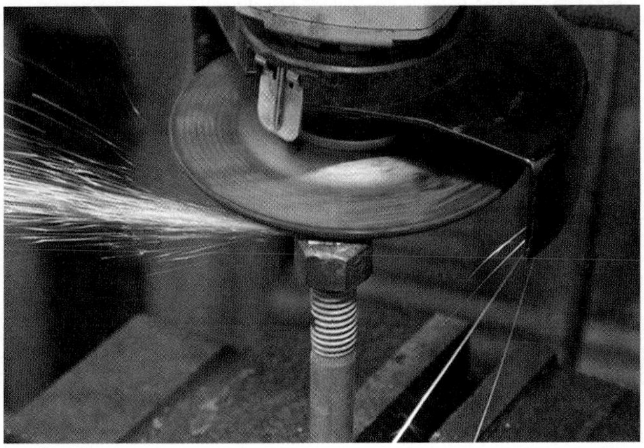

3. Hold the grinder firmly with the face of the disc, not the edge, against the work. Be careful that the motor's torque does not cause the grinder to slip out of your hand. Do not press too hard. Let the grinder do the work.

4. If using the grinder with a cutting disc, use the edge of the disc, not the face.

Soldering Tools

LO 7-04 Perform a solder repair.

Solder is a mixture of metals with low melting points and is used to join metals together. Tin–lead solder has been used for soldering wires and other metals together for decades. Tin–lead solder for automotive applications consists of approximately 60% tin and 40% lead, and melts at approximately 370°F (188°C). With the environmental hazards of lead becoming an issue over the past 40 years, lead-free solder was introduced in the past decade (**FIGURE 7-33**). Lead-free solder is made up of tin and copper, or of tin, copper, and silver, and usually has a melting point about 20–30°F higher than tin-lead solder. Because solder is a relatively soft compound, it is not used to make joints in situations where high stresses are involved.

In automotive applications, solder generally comes in the form of a wire. It can be solid, requiring an external **flux** cleaning agent of **rosin** if soldering electrical connections or an acid if soldering non-electrical connections (**FIGURE 7-34**). The solder can also be hollow, with the rosin or acid in the core. In this case, it would be referred to as rosin-core solder or acid-core solder (**FIGURE 7-35**). Make sure you use rosin flux with electrical connections, and acid with all other connections.

The process of soldering involves heating the metals (wires) hot enough so that the solder melts and fills the spaces between the metals. When the solder cools, the solder holds the parts together. It also transmits electricity, if used in electrical circuits. The temperature of the soldering operation is critical. If it is not hot enough, the solder does not flow very well. It does not

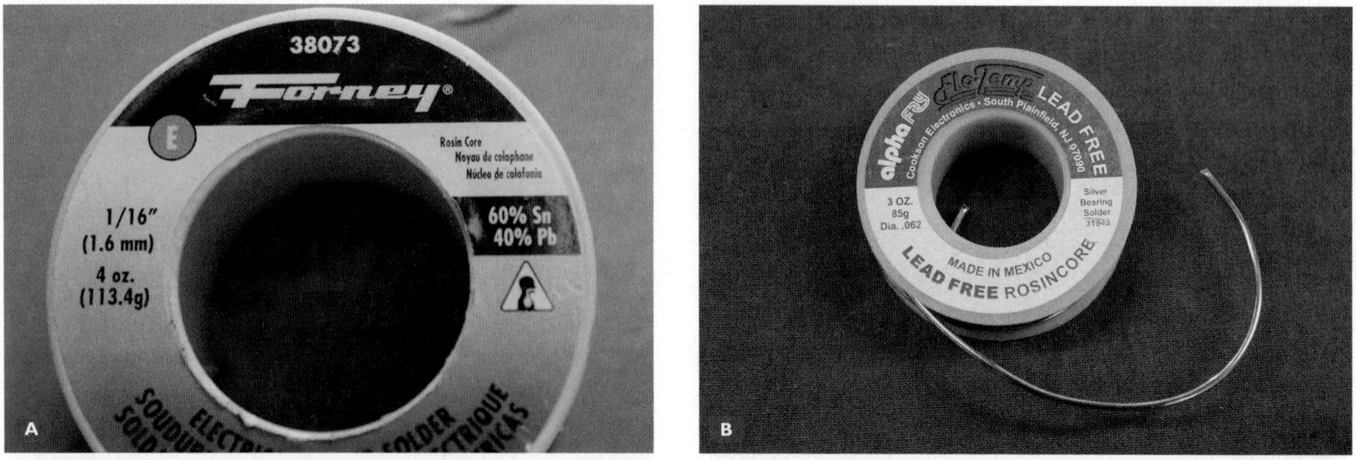

FIGURE 7-33 **A.** Tin–lead solder. **B.** Lead-free solder.

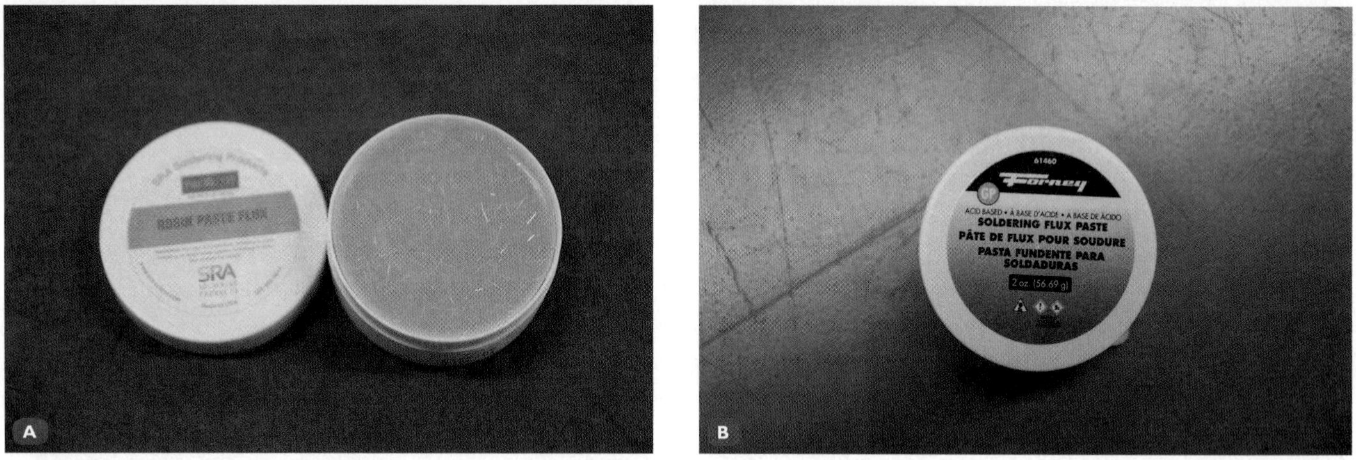

FIGURE 7-34 **A.** Rosin flux. **B.** Acid flux.

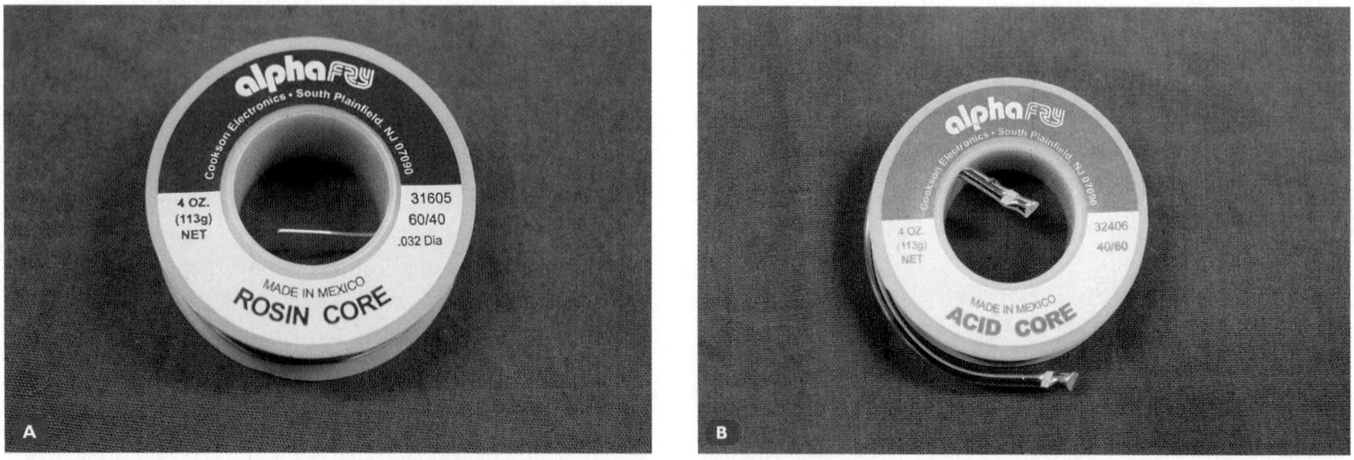

FIGURE 7-35 **A.** Rosin core solder. **B.** Acid core solder.

make good contact with the metal surfaces, and tends to glob up (**FIGURE 7-36**). This leads to a weak joint and poor electrical conductivity.

If it is too hot, the solder tends to run off the joint and overheat the components being soldered (**FIGURE 7-37**). In the case of electronic components, overheating can make them inoperative. Only heat the components enough to melt the solder and cause it to flow (**FIGURE 7-38**). You can also protect electronic components by using a heat dam. A heat dam absorbs some of the heat from the wires being soldered and prevents it from traveling on to the electronic component (**FIGURE 7-39**).

The heat is provided by an electric **soldering iron** or gun, or a butane soldering iron (**FIGURE 7-40**). A typical soldering iron has a handle that is thermally insulated from the tip. Then, heat is transferred by metal-to-metal contact from the tip into the metal to be soldered. Basic soldering irons are heated manually by a gas flame. More sophisticated soldering irons are electrically operated and have thermal tips that are controlled by thermostats to maintain more accurate tip temperatures. Soldering irons can either have a fixed tip size or tips that can be interchanged with different sizes for different kinds of jobs.

Sometimes, when soldering, you have to remove solder from a joint before resoldering a new part into the circuit. In

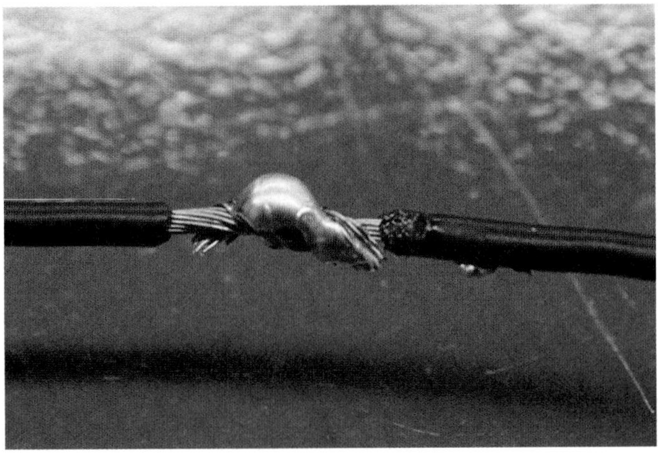

FIGURE 7-36 Solder joint that was too cold.

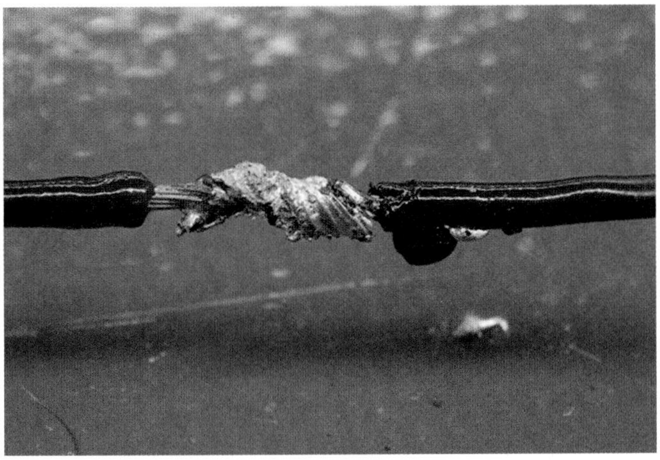

FIGURE 7-37 Solder joint that was too hot.

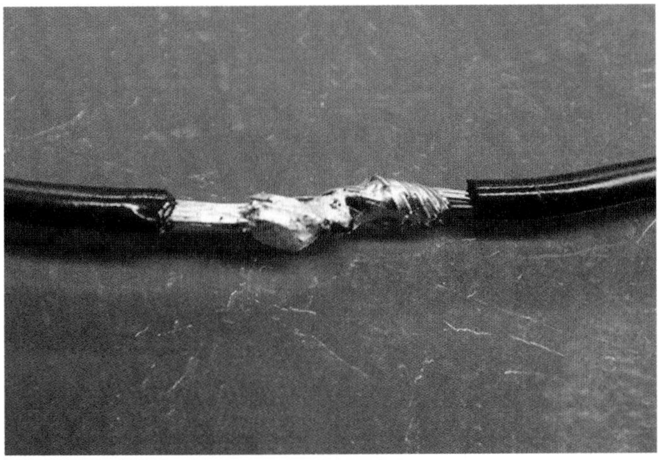

FIGURE 7-38 Good solder joint.

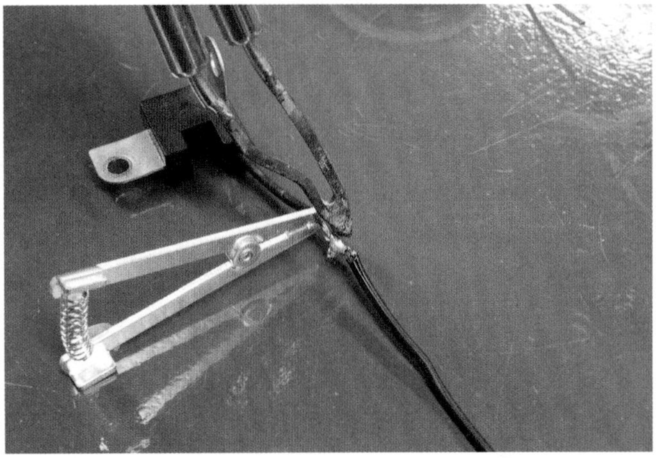

FIGURE 7-39 Soldering while using a heat dam.

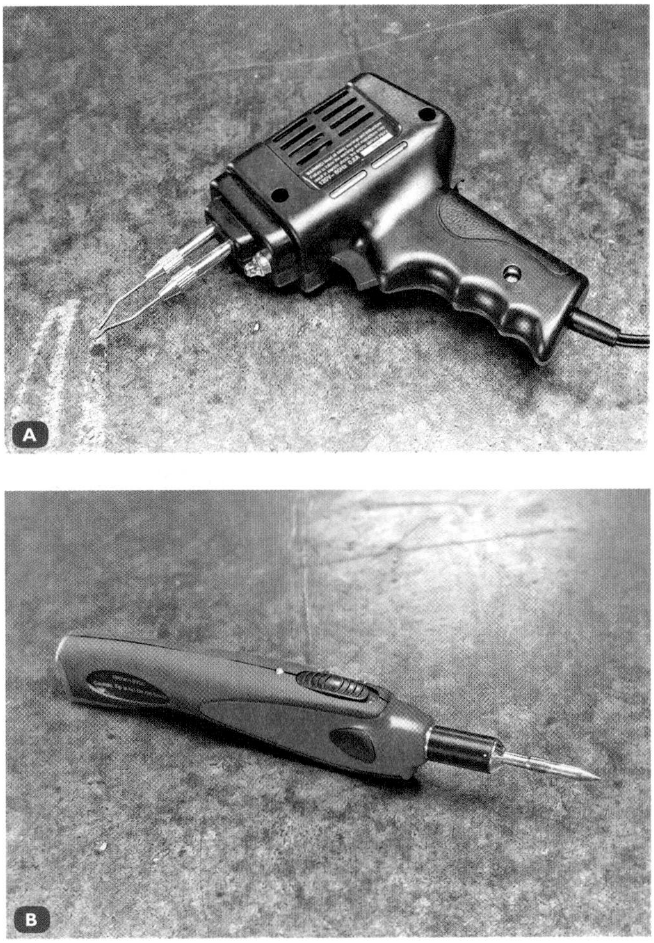

FIGURE 7-40 **A.** Soldering gun. **B.** Soldering iron.

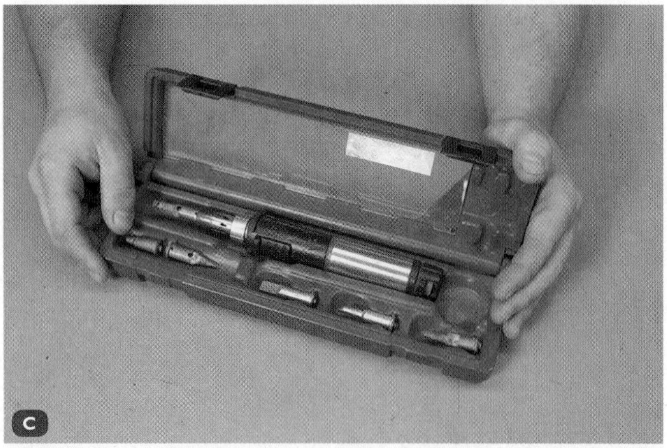

FIGURE 7-40 C. Butane soldering iron.

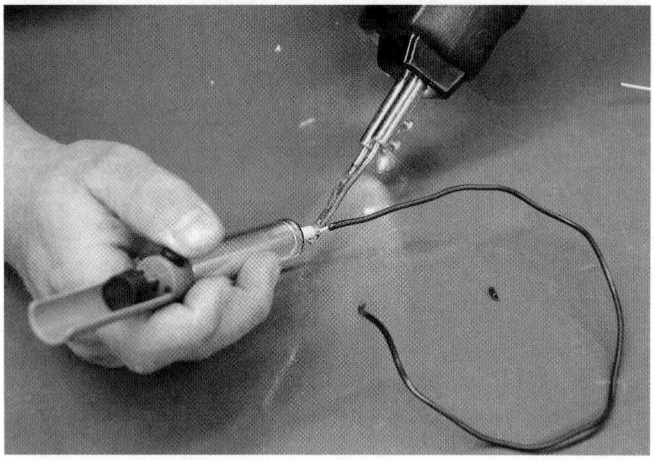

FIGURE 7-41 Desoldering tool being used.

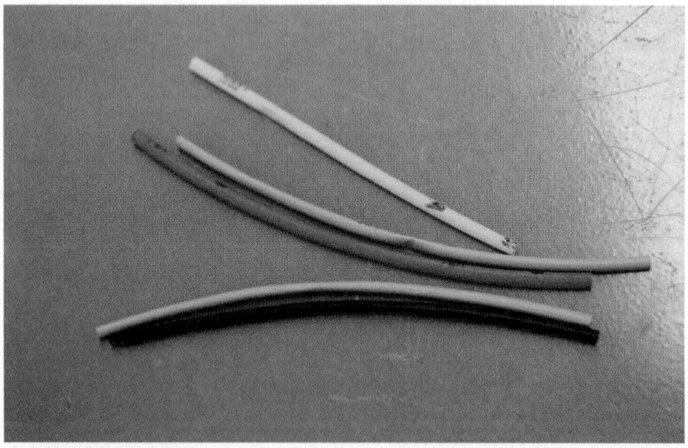

FIGURE 7-42 Heat shrink tubing.

this case, a desoldering tool can be used to suck up the solder once it is melted. This makes it much easier to disassemble the joint (**FIGURE 7-41**).

Once a good solder joint has been created, you need to protect it. One of the best ways to do so is with heat shrink tubing (**FIGURE 7-42**). Heat shrink tubing is a hollow tube of insulating plastic that shrinks when it is heated. When the properly sized

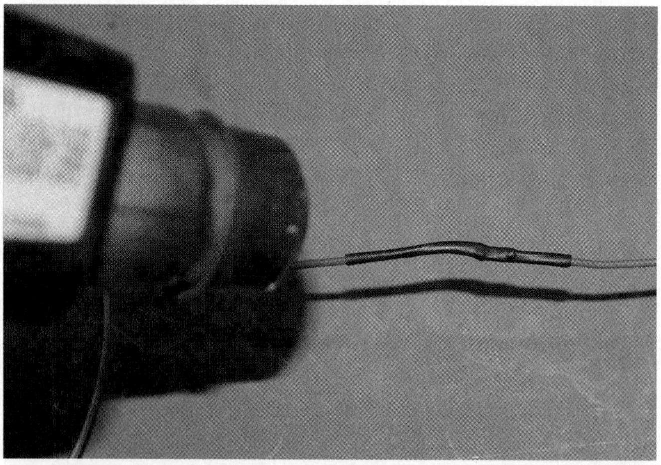

FIGURE 7-43 Heat shrink tubing being heated.

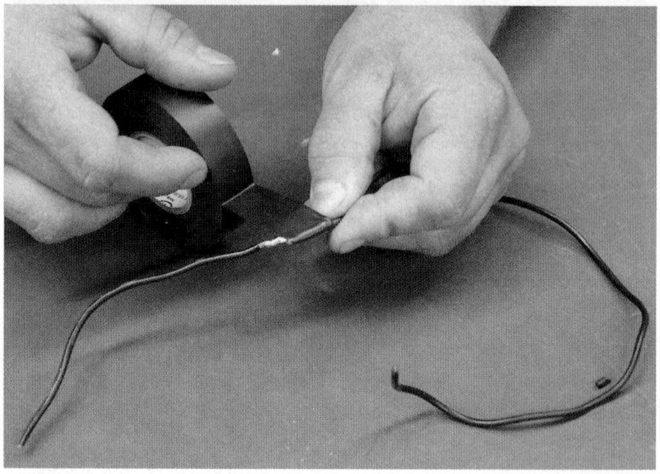

FIGURE 7-44 Wire joint being wrapped with electrical tape.

heat shrink tube is placed over the solder joint and heated with a heat gun, it will shrink tightly around the joint (**FIGURE 7-43**). Some heat shrink tubing contains a small amount of sealer that melts when heated. It flows around the joint, and seals the tubing to the wire insulation.

Another way to protect solder joints is with electrical tape. When installed correctly, the tape insulates and protects the joint. The tape should be wrapped around the wire in a spiral manner while being lightly pulled tight in overlapping wraps. In most cases, the tape is first wrapped down the wire (and joint), overlapping each layer by about 50%, and then back up the wire in the same fashion. Additional layers can be added for extra protection as needed (**FIGURE 7-44**).

Using Soldering Tools

Apply flux to the joint if **cored solder** is not being used. Always remove any excess flux when finished. Ensure that the joint is held steady during and after the solder is applied. A dull solder surface indicates a cold, high-resistance joint that needs to be resoldered. Select the correct tip size that will heat the joint and melt the solder within a few seconds. To use a soldering iron to solder two pieces of wire or metal together, follow the steps in **SKILL DRILL 7-10**.

SKILL DRILL 7-10 Using Soldering Tools

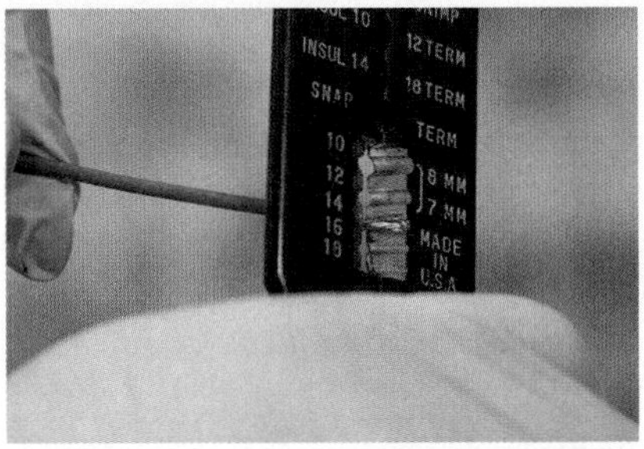

1. Prepare the materials to be soldered. Strip wires or clean metal parts before soldering.

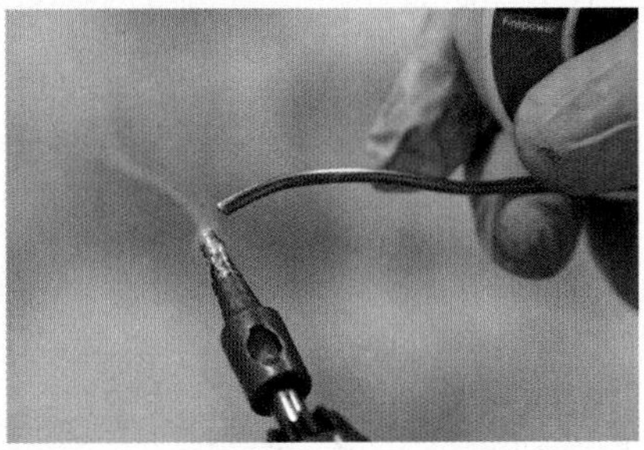

2. Prepare the soldering iron by ensuring the correctly sized tip is fitted and is clean (you may have to use a file to reshape and clean it). Tin the soldering iron tip by melting some solder to it and wiping any excess from the tip with a rag.

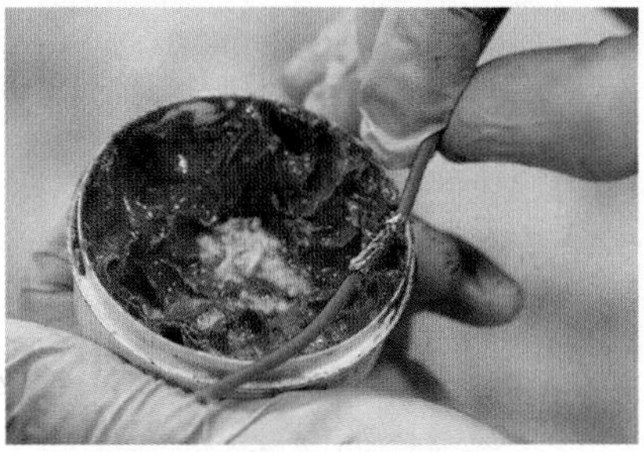

3. Apply flux to the wires or metal to be soldered. This may not be necessary if you are using cored solder.

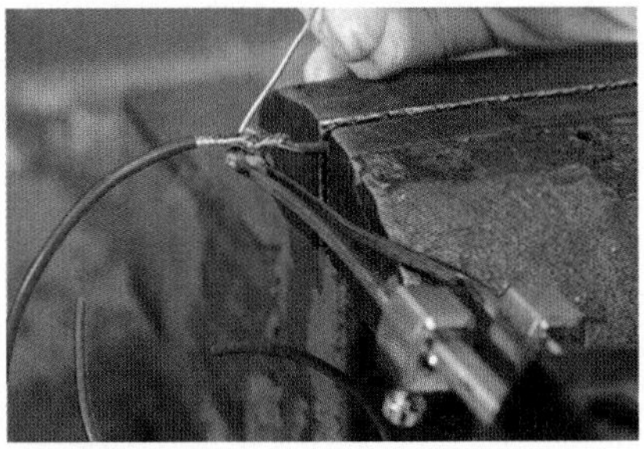

4. Apply the hot solder iron tip to heat the joint, and then apply solder to the joint (not the soldering iron). If the solder does not melt within a few seconds, remove it and allow the joint to heat further before reapplying.

5. Once the solder has been applied, ensure the joint does not move until the solder has cooled sufficiently to set. Once cooled, inspect the joint; it should be shiny and firm.

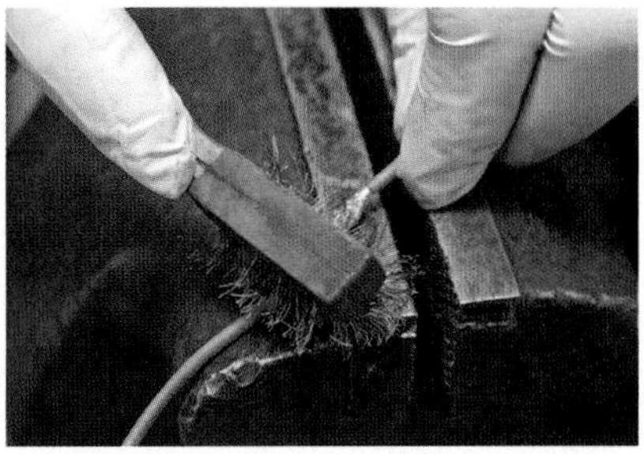

6. Clean any excess flux from the joint.

SAFETY TIP

Always wear eye and hand protection when soldering.

Always wipe excess solder from the iron; never flick the iron to remove excess solder.

Make sure the tool is clean, in good condition, and suitable for the type of material you are going to solder.

Check electrical leads and plugs for damage prior to use.

▶ Cleaning Tools

LO 7-05 Operate cleaning equipment.

Pressure Washers and Cleaners

Pressure washers are valuable tools for cleaning engine compartments and components (**FIGURE 7-45**). They can be powered by an electric motor or a gasoline engine fitted to a high-pressure pump. The pressure washer takes water at normal pressure and boosts it substantially. The **cleaning gun** has a high-pressure nozzle that focuses the high-pressure water (possibly over 2000 psi [13,790 kPa]). It quickly cleans accumulated dirt and grease from components. Some pressure washers have a provision for detergent to be injected into the water stream to help remove the grime. Others have the ability to heat the water, in some cases hot enough to turn it to steam. Hot water and steam help loosen oil and grease buildup.

FIGURE 7-45 Pressure washer being used to clean an engine compartment.

Pressure washers are dangerous because of their high pressure and possibly high temperature. Always wear appropriate PPE when working with pressure cleaners. This includes goggles or face shield, protective gloves, close-fitting clothes with long sleeves and full-length pants, and leather-type boots or shoes.

Using Pressure Washers

Pressure washers used in automotive applications are available in a range of makes and types depending on the application. There are fixed and mobile pressure washers. Familiarize yourself with the equipment before using it. Incorrect handling can result in damage to the washer or to the vehicle or components you are cleaning. There are also health risks to yourself and your coworkers. The biggest advantage of using fluid to clean vehicles, components, and spaces is that it wets the dirt or contaminants, so no dust is created. However, the waste products must be caught and disposed of properly in either a catch basin or a wastewater settling system. The waste materials must not be released into a storm water drain.

It is critical that you only use the pressure washer for its approved purposes. Pressure washers with very high pressure, or steam cleaners, can actually remove paint. So be familiar with the appropriate use and operation of the pressure washer. If in doubt, clean the area by hand using a clean sponge and clean water.

It is very important to note the type of solvent or detergent used in a pressure washer. Some vehicle components are damaged by some solutions. Therefore, in all cases, they should only be cleaned in wash tanks containing the correct cleaning fluid. When using high-pressure washers, it is always important not to spray in areas where water can have a harmful effect. Examples include fuse boxes, relays, control units, and open air cleaners. The damage may not become clear for some time after the cleaning process. This can cause system failures, which are difficult to diagnose.

If you need to use a pressure washer in sensitive areas, take precautionary measures to protect the units from water damage. Cover them with sturdy plastic bags. If the wheel brake units get wet, drive the vehicle for a short distance with the brakes slightly applied. This will remove any residual water from the brake shoes or pads through heat transfer and evaporation. Depending on the outside temperature, it usually only takes a few seconds to return the brakes to their proper operation.

On completion of the job, the proper disposal of contaminated materials is required. Shops may be subject to prosecution for the incorrect disposal of waste materials. So always know the regulations and follow all policies for your jurisdiction.

SAFETY TIP

- Always wear a face shield and gloves when using cleaning and washing equipment.
- Always wear safety shoes when using any washing equipment, to prevent slips on slippery surfaces.
- Always be aware of the location of safety switches located on equipment and of eyewash and first aid stations, should an accident occur.
- Do not place your hand or any other part of your body in the stream of water from the high-pressure wand. Many pressure washers generate enough pressure to instantly cut the skin and muscle to the bone.
- Do not aim the high-pressure wand at another person.
- Always test the temperature of the wand and the hose before you pick it up. The handle of the pressure wand is insulated to protect the user from heat, but the wand extension and the hose are not.
- If the pressure cleaner uses a heating element, turn the heater off. Allow water to flow through the wand until it has cooled, before you turn the unit off.
- If you are unfamiliar with a solvent or a cleaning agent, refer to the safety data sheet (SDS). It has information about its correct use and applicable hazards.

To safely use pressure washers, follow the steps in **SKILL DRILL 7-11**.

SKILL DRILL 7-11 Using Pressure Washers

1. Before using the pressure washer, locate the position of safety switches, and put on a face shield, work apron, and gloves. Note the location of the eyewash and first aid stations.

2. If the pressure washer is not permanently connected to the water supply, follow the manufacturer's instructions to do so. Connect the electrical plug to a power outlet protected by a ground fault circuit interrupter (GFCI). Make sure electrical connections and adjacent areas are protected from water spray by using appropriate protective shields.

3. It may be necessary to apply a degreasing agent with a spray bottle and hand brush. This will penetrate and soften excess dirt before you operate the pressure washer.

SKILL DRILL 7-11 Using Pressure Washers (Continued)

4. Turn on the water supply first, but not the power switch. Make sure water is flowing through the washer unit by testing the flow through the pressure wand before turning the power on. It should flow freely, but not at high pressure.

5. Turn the power on, and you will hear the motor engage. Point the wand toward the ground, and test that the water now flows at high pressure.

6. Pull the trigger, and using either a circular or a sweeping motion, direct the high-pressure water onto the area to be cleaned. Avoid getting the high-pressure spray on the exterior paintwork of the vehicle by placing the wand close to the area to be cleaned.

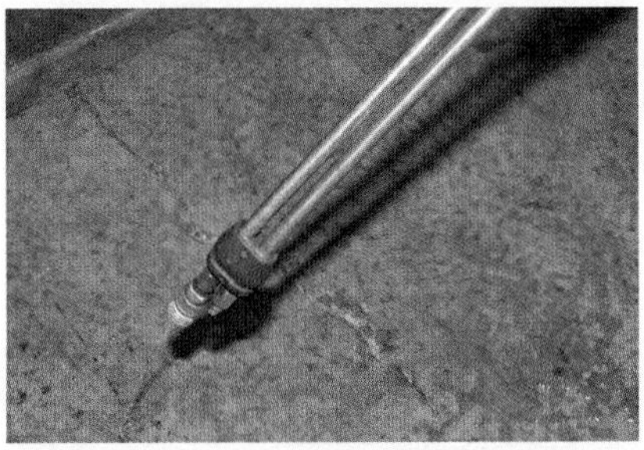

7. When the contaminants have been removed, release the trigger, and remove the wand from the cleaning area. Turn the electrical power off, and then turn the water supply off. If it is a hot water or steam cleaner, let the water run until it cools off so that you do not overheat the water in the heating coils.

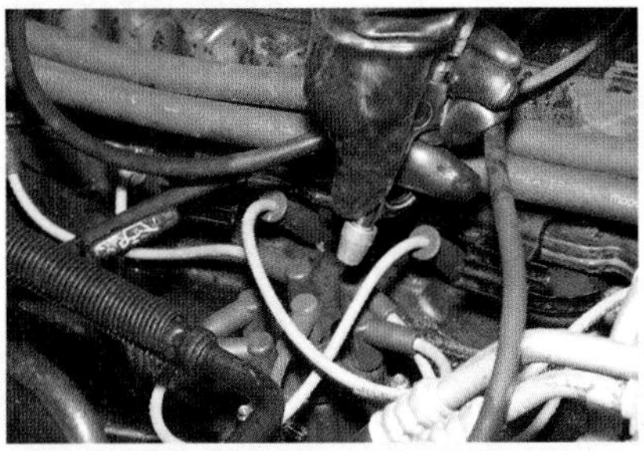

8. Use an air nozzle to disperse any residual water from electrical components.

9. Start the vehicle and let it run for a few moments to dry. This will help remove any residual water.

10. Clean up any residual material, and place it in a bin or an environmental waste container.

Spray Wash Cabinets

Spray wash cabinets spray high-temperature, high-pressure cleaning solutions onto parts inside a sealed cabinet. They are automated and act like a dishwasher for parts (**FIGURE 7-46**). This reduces the labor required to clean parts. Once the door is closed and the unit is turned on, the technician is free to move onto other tasks. Spray wash cabinets are available in a variety of sizes to cater to different-sized parts. They provide a high level of cleaning performance. The cleaning solution is designed to clean without leaving dirty residue on the parts. Most spray cabinets are fitted with a filtering system to reduce the frequency of cleaning solution changes.

FIGURE 7-46 Spray wash cabinet.

Using Spray Wash Cabinets

Spray wash cabinets are available in a range of makes and types depending on the application. It is important that you read the instructions before operating spray wash cabinet. Spray wash cabinets incorporate either a built-in waste recovery system, or capture the contaminated fluid for later disposal. Always follow recommended safety procedures. Some spray wash cabinets use dangerous chemicals at very high pressure to clean away the contaminants. So always know what chemical is being used as a cleaning agent so that you can look up any precautions in the SDS.

It is also very important to note that some cleaning agents can damage some vehicle components. If in doubt about the particular parts you are cleaning, ask your supervisor. On completion of the job, the correct disposal of any contaminated materials is required. Shops may be subject to prosecution for the incorrect disposal of waste materials. To use a spray wash cabinet to clean components, follow the steps in **SKILL DRILL 7-12**.

SAFETY TIP

Do not operate the spray wash cabinet without the door securely closed. The spray wash cabinet uses high-pressure, high-temperature cleaning fluid. Parts will be hot after being in the spray cabinet.

Solvent Tanks

A **solvent tank** is a cleaning tank that contains a suitable solvent to clean parts by removing oil, grease, dirt, and grime. Solvent tanks are available in different sizes. Many solvent tanks have a pump that pushes solvent out a nozzle where it can be directed to the parts being cleaned. A brush, either on the nozzle or separate from it, can be used to loosen the grease and grime. The solvent falls back into the bottom of the solvent tank, where the heavier residue settles to the bottom.

Other solvent tanks are designed so that parts can be immersed in the tank on racks or suspended on pieces of wire. Then, they are slowly lowered and soaked in the tank for a period of time. Some solvent tanks may have an agitation system or use a heated cleaning fluid to speed up the process. They may also have a circulation system and filters to remove debris in the solvent. This extends the life of the solvent between changes.

SAFETY TIP

Whenever using a tank-type cleaner with solvents, make sure there is adequate ventilation. Also, wear appropriate breathing apparatus and eye protection.

SKILL DRILL 7-12 Using Spray Wash Cabinets

1. Always refer to the manufacturer's manual for specific operating instructions. Before using the equipment, locate the position of safety switches, and put on a face shield and gloves.

2. Check the spray wash cabinet to ensure it is operating correctly with enough cleaning fluid.

3. Open the spray wash cabinet. Place the components to be cleaned into the wash tray. Seek assistance if parts are too heavy to be handled by one person. Distribute them so each part will be cleaned effectively.

4. Close the spray wash cabinet and start the cleaning cycle.

5. Make sure the cleaning cycle has finished before you open the spray wash cabinet door. Ensure you are wearing appropriate eye and hand protection.

6. When opening the door, be careful because parts will be so hot to touch. Examine the parts to ensure they have been completely cleaned. In some models, you need to rinse the components after removing them from the washer. Follow the directions for the model you are using.

Using Solvent Tanks

Familiarize yourself with the equipment prior to use. Incorrect use can result in damage to components that you are cleaning. Solvent tanks may use potentially poisonous and/or flammable chemicals to clean parts. It is very important to note the type of solvent being used. This will allow you to look up all the necessary precautions as listed on the SDS. Also, some vehicle components can be damaged by some solvents. So ask your supervisor about what can be safely washed and what cannot.

On completion of the job, properly dispose of all contaminated materials. Shops may be subject to prosecution for the incorrect disposal of waste materials. To use a solvent tank to clean components, follow the steps in **SKILL DRILL 7-13**.

SAFETY TIP

- If equipped, keep lids closed as much as possible, as many solvents are flammable and evaporate.
- The cleaning solution may be hot.
- Make sure there is adequate ventilation.
- Wear appropriate eye and hand protection and, if necessary, appropriate breathing apparatus.

Brake Washers

Brake washers are used to wash brake dust from wheel brake units and their components. It is possible that the brake dust may contain asbestos, which is a cancer-causing agent. Dust in general is a lung irritant. Brake washers are designed to capture the brake dust before it enters the shop environment. It does so by wetting down the dust on the brake parts and then washing it into the cleaning tray. Brake washers store contaminated washer fluid for later disposal in an environmentally friendly manner.

Brake washers are normally designed to operate at low pressure and use a range of cleaning agents. The most popular agent is an aqueous solution made up of water and a water-soluble detergent. A low-pressure air nozzle may be provided to blow excess fluid from the component back into the catch tray. Avoid using solvent when cleaning brake components. It contaminates friction materials and may cause seals to swell. Soapy water is a good cleaning agent for most brake components.

Using Brake Washers

Brake washers are a handy piece of shop equipment to deal with hazardous dust in a quick and relatively easy manner. This is especially true if the washer uses an aqueous (water and

SKILL DRILL 7-13 Using Solvent Tanks

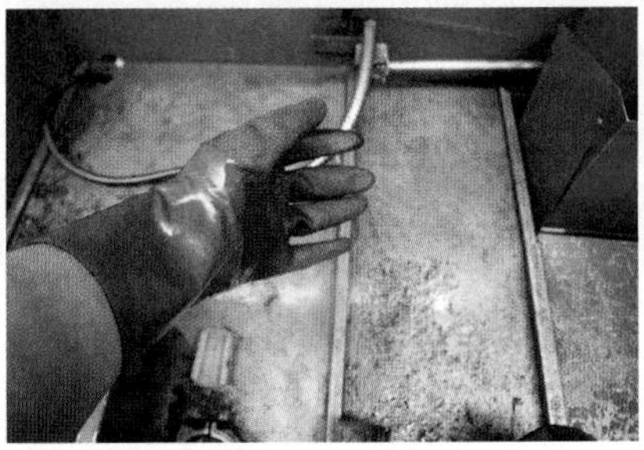

1. Always refer to the manufacturer's manual for specific operating instructions. Before using the equipment, locate the position of safety switches and put on a face shield and gloves. Note the location of the eyewash and first aid stations.

2. Lift the components to be cleaned into the wash tray or lower them into the solvent tank. Seek assistance if parts are too heavy to be handled by one person. Distribute them so each part will be cleaned effectively.

3. Start the circulating or agitating pump if equipped.

4. Soak or run the solvent onto the components in the solvent tank for the appropriate amount of time. You may need to use a brush to clean the parts. Be careful not to cause excessive splashing of cleaning fluid.

5. Examine the parts to ensure they have been completely cleaned.

6. Remove parts carefully from the solvent tank. This allows excess cleaning solvent to drip back into the tank. Ensure contaminated cleaning fluids can be captured to enable disposal in an environmentally friendly manner.

detergent) cleaning solution. The one thing that this solution is not suited for is cleaning greasy residue. If you need to remove grease, you should first use a paper towel or grease rag to wipe it off. If that is not enough, then a chemical like BrakeKleen can be used to dissolve the grease. Once the grease is gone, any remaining brake dust can be cleaned with the brake washer. To use a brake washer to clean components, follow the steps in **SKILL DRILL 7-14**.

Sand or Bead Blasters

Sand or bead blasters use high pressure to blast small abrasive particles to clean the surface of parts. The most common method of propelling the sand or glass beads is with compressed air. Sand or bead blasting can occur in a specially designed cabinet, or on portable models that are used in open-air situations. The cabinets contain the blasting operation in a controlled safe environment. They are best for smaller parts that can fit into the cabinet. Portable systems that do not operate within a cabinet can blast larger parts. But they do require more protection for the operator and surrounding environment.

The sand or bead blaster cabinet is fitted with a hand-operated blasting nozzle, a viewing port, and foot-operated control switch. It has openings with tough rubber gloves sealed

SKILL DRILL 7-14 Using Brake Washers

1. Make sure all the washing solution is contained within the cleaning tray and returns to the reservoir. Washing solution must not enter the environment. Before using the component cleaner, make sure the solution is compatible with the component to be cleaned. An aqueous solution is an environmentally friendly solution for most materials.

2. Put on gloves and safety glasses or a face shield. Move the brake washer under the wheel brake unit to be cleaned. Make sure the waste drain is not blocked and the low-pressure air nozzle, if equipped, is operational.

3. Using a semi-stiff brush, paint the solution over the components to wet the components. Then, scrub the components and flush them with the cleaning solution. Continue to do so until the components are clean.

4. Use the low-pressure air nozzle to dry any excess solution from the components. Make sure the waste materials are caught in the cleaning tray and drained back into the tank below.

into them. This allows a technician's hands to be inside the cabinet while being protected from the abrasive sand or beads. Wet sand or bead blasters are also available. They have the added advantage of reducing the amount of dust created during blasting.

Using Sand or Bead Blasters

Technicians use sand or bead blasters to clean paint, corrosion, or dirt from metal parts. Because the sand or beads are abrasive, they can remove metal from the surface of the components. So be careful of what you use it on. Softer materials can be damaged very easily. Sand or bead blasters used in automotive applications are available in a range of makes and models, depending on application. To use a sand or bead blaster to clean components, follow the steps in **SKILL DRILL 7-15**.

SAFETY TIP

Sand or bead blasters use very fine particles at high pressure to clean away the contaminants. Also, the sand breaks down into silica particles which should not be inhaled. Silica can cause lung damage, called silicosis, over an extended period of time.

SAFETY TIP

- Do not operate the sand or bead blaster without the door securely closed.
- Wear appropriate breathing apparatus or a dust mask.
- Always wear eye protection to prevent injury from flying particles or escaping compressed air.
- On completion of the job, the correct disposal of contaminated materials is a top priority.

SKILL DRILL 7-15 Using Sand or Bead Blasters

1. Always refer to the manufacturer's manual for specific operating instructions. Before using the equipment, locate the position of safety switches. If not using a cabinet, then a face shield, dust mask, and gloves are required. Note the location of the eyewash and first aid stations.

2. Check the sand or bead blaster to ensure it is operating correctly with enough cleaning sand or beads.

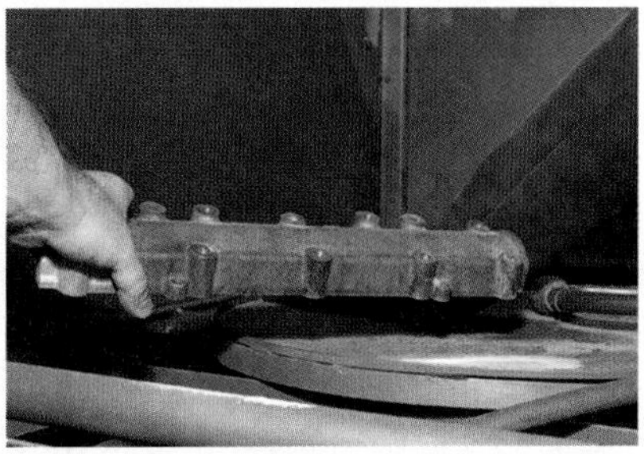

3. Open the sand or bead blaster. Lift the components to be blasted into the cabinet. Seek assistance if parts are too heavy to be handled by one person.

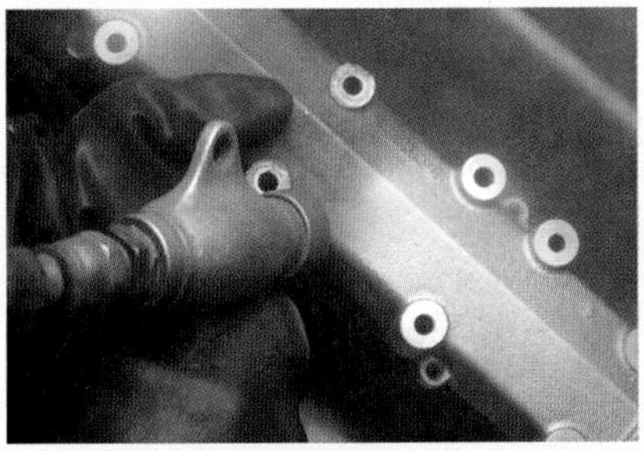

4. Close the sand or bead blaster, and start sand or bead blasting. Direct the blaster nozzle toward the parts to be cleaned. Avoid accidentally blasting the viewing port, as doing so will permanently reduce the transparency of the port.

5. Open the door and examine the parts to ensure they have been completely cleaned. If the parts are not sufficiently cleaned, continue blasting.

6. Remove the parts and clean, preferably with a liquid cleaning solution or water, to ensure all abrasive materials are removed.

▶ Wrap-Up

Ready for Review

- Battery chargers can be fast chargers or slow chargers. Slow charging is less stressful on a battery than fast charging. Always follow all of the jump-starting guidelines to avoid unnecessary stresses on the vehicles.
- Power tools can be powered by air, electric, and battery. Wear appropriate PPE, and never use compressed air inappropriately.
- Drill and drill bits and bench and angle grinders provide additional power for getting tasks done.
- The process of soldering involves heating the metals (wires) hot enough so that the solder melts and fills the spaces between the metals. Heat shrink tube is placed over the solder joint and heated with a heat gun, to encase tightly around the joint.
- Pressure washers and cleaners, spray wash cabinets, solvent tanks, brake washers, sand or bead blasters are valuable tools for cleaning engine compartments and components.

Key Terms

abrasive discs (cutting wheels) Abrasive wheels or flat discs fitted to bench, pedestal, and portable grinders.

air drill A compressed air–powered drill.

Air driers A device fitted to compressed air lines to remove moisture.

air hammer A tool powered by compressed air with various hammer, cutting, punching, or chisel attachments; also called an air chisel.

air impact wrench An impact tool powered by compressed air, designed to undo tight fasteners.

air nozzle A compressed air device that emits a fine stream of compressed air for drying or cleaning parts.

air ratchet A ratchet tool for use with sockets, powered by compressed air.

angle grinder A portable grinder for grinding or cutting metal.

automatic oiler A device fitted to compressed air systems to oil air tools.

Batteries Device used to store electricity in chemical form.

battery charger A device that charges a battery, reversing the discharge process.

bench grinder (pedestal grinder) A grinder that is fixed to a bench or pedestal.

cleaning gun A device with a nozzle controlled by a trigger fitted to the outlet of pressure cleaners.

cored solder Solder that is in the form of a hollow wire. The center is filled with flux, which is used as a cleaning agent while the solder is being applied to the metal surfaces.

drill press A device that incorporates a fixed drill with multiple speeds and an adjustable worktable. It can be free standing or fixed to a bench.

fast chargers A type of battery charger that charges batteries quickly.

flux A material that is used during brazing and soldering operations to prevent oxidation and remove impurities from the metals.

grinding wheels and discs Abrasive wheels or flat discs fitted to bench, pedestal, and portable grinders.

memory saver (memory minder) Battery backup device for vehicle computer systems.

Morse taper A system for quickly changing and securing drill bits to drills.

pressure washer/cleaner A cleaning machine that boosts low-pressure tap water to a high-pressure output.

rattle gun A term used to describe an air impact wrench, based on the noise it makes.

rosin A type of liquid or paste (flux) that, in solid form, is contained within solder and is used to prevent oxidization.

sand or bead blasters A cleaning system that uses high pressure and fine particles of glass bead or sand.

sander/polisher A power tool with a rotating disc or head to which polishing or sanding discs can be attached.

slow charger A battery charger that charges at low current.

smart charger A battery charger with microprocessor-controlled charging rates and times.

solder A metal with a low melting temperature that is used to fuse metal components.

soldering irons A heating tool to heat solder and wires to produce a low-resistance joint.

solvent tank A tank containing solvents to clean vehicle parts.

spray wash cabinet A cleaning cabinet that sprays cleaning solution, under pressure, to clean vehicle parts.

straight grinder A powered grinder with the wheel set at 90 degrees to the shaft.

sulfuric acid A type of acid that, when mixed with pure water, forms the basis of battery acid or electrolyte.

surge protector An electrical protection device for preventing electrical surges.

Review Questions

1. When charging a battery you should do all of the following EXCEPT:
   a. Disconnect the negative battery terminal while charging.
   b. Slow charge it if possible.
   c. Charge a frozen battery to unfreeze it.
   d. Use a memory saver to retain electronic memory settings.

2. When jump-starting a vehicle:
   a. Make the last connection to a good ground on the engine block.
   b. Connect the Red jumper lead to the negative battery terminals.
   c. Make sure the bumpers on both cars are firmly touching.
   d. Connect the Black jumper lead to the positive battery terminals.
3. Which of these is used for driving and cutting?
   a. Air drill
   b. Air ratchet
   c. Air hammer
   d. Air nozzle
4. All of the following are true about electric power tools except:
   a. They can be battery operated.
   b. The should be lubricated daily.
   c. They should be inspected for missing ground terminals.
   d. They should never be used in explosive environments.
5. What should be done before using an impact wrench?
   a. Add a few drops of air tool oil to the air fitting.
   b. Fill the impact wrench with cooling fluid.
   c. Install a standard chrome socket on the drive.
   d. Wrap the socket with duct tape to prevent shattering.
6. When grinding metal on a bench grinder:
   a. Make sure there is no gap between the wheel and tool rest.
   b. Stand off to the side the entire time.
   c. Grind on both sides of the wheel.
   d. Stop and dip it regularly in the water pot.
7. Which solder should be used on electrical solder repairs?
   a. Acid core
   b. Rosin Core
   c. Solid solder
   d. Hollow solder
8. After a solder repair is completed, how should the repair be protected?
   a. No protection is required
   b. Dielectric grease and RTV sealant
   c. Heat shrink tubing with sealant
   d. Duct tape
9. Which of the following is a true statement about spray wash cabinets?
   a. They reduce the labor required to clean parts.
   b. They use low-temperature, low-pressure cleaning solutions.
   c. They leave a dirty residue on the parts.
   d. They are not sealed, so they can make a mess on the floor.
10. Pressure washers use a high-powered spray to remove dirt and grime from parts. All of the following are safety precautions EXCEPT:
    a. Be careful where you point the nozzle.
    b. Test the temperature of the wand before you pick it up.
    c. Use of a cleaning agent requires referencing an SDS.
    d. Use the pressure washer for getting grease off your hands.

ASE Technician A/Technician B Style Questions

1. Technician A says solvent tanks may use flammable solvents or chemicals. Technician B says there must be adequate ventilation when using solvent tanks. Who is correct?
   a. Technician A
   b. Technician B
   c. Both A and B
   d. Neither A nor B
2. Technician A says when performing a brake job, it is acceptable to blow off the brake dust. Technician B says to use a brake washer to catch the brake dust for proper disposal. Who is correct?
   a. Technician A
   b. Technician B
   c. Both A and B
   d. Neither A nor B
3. Technician A says that when soldering electrical wires, rosin flux must be used. Technician B says that a heat dam is used to prevent heat from traveling to the electronic component when soldering. Who is correct?
   a. Technician A
   b. Technician B
   c. Both A and B
   d. Neither A nor B
4. Technician A says to have a firm grip on an angle grinder when you turn it on to compensate for torque. Technician B says to check the area where the sparks are going to fly for flammables. Who is correct?
   a. Technician A
   b. Technician B
   c. Both A and B
   d. Neither A nor B
5. Technician A says that heat shrink tubing is used to protect a solder joint. Technician B says that a desoldering tool uses acid to remove solder. Who is correct?
   a. Technician A
   b. Technician B
   c. Both A and B
   d. Neither A nor B
6. When using a bench grinder, if the protective guard is in the way, Technician A says that the guard may be removed. Technician B says that the guard is to protect the operator if the grinding wheel shatters. Who is correct?
   a. Technician A
   b. Technician B
   c. Both A and B
   d. Neither A nor B
7. Technician A says that air tools and equipment require regular application of lubricating oil to reduce wear and tear. Technician B says that some compressed air systems use an inline water trap that needs to be drained periodically. Who is correct?
   a. Technician A
   b. Technician B
   c. Both A and B
   d. Neither A nor B

8. Technician A says that when using an impact wrench, the maximum torque setting will provide the proper torque. Technician B says that impact wrenches are acceptable for taking up the looseness in a nut or bolt, but a torque wrench is required for final tightening. Who is correct?
    a. Technician A
    b. Technician B
    c. Both A and B
    d. Neither A nor B
9. Technician A says that after charging and reinstalling a vehicle battery, it is good practice to clean the battery terminals and posts. Technician B says that fast charging a vehicle battery extends its life. Who is correct?
    a. Technician A
    b. Technician B
    c. Both A and B
    d. Neither A nor B
10. Technician A says that when jump-starting a vehicle, a spark may ignite the hydrogen gas. Technician B says to use an engine ground away from the battery to minimize the possibility of an explosion. Who is correct?
    a. Technician A
    b. Technician B
    c. Both A and B
    d. Neither A nor B

CHAPTER 8

Fasteners and Thread Repair

Learning Objectives

- LO 8-01 Identify threaded fasteners.
- LO 8-02 Identify non-threaded fasteners.
- LO 8-03 Replace threaded fasteners.
- LO 8-04 Repair damaged fastener threads.

ASE Education Foundation Tasks

See Appendix A to view the 2017 ASE Education Foundation Automobile Accreditation Task List Correlation Guide.

▶ Fastener Identification Overview

LO 8-01 Identify threaded fasteners.

Fasteners come in two common types: threaded fasteners and non-threaded fasteners (**FIGURE 8-1**). Both types of **fasteners** are designed to secure parts. **Threaded fasteners** are primarily designed to clamp objects together. Non-threaded fasteners are designed to hold parts together. For example, preventing a component from falling off a shaft by using C-clips, cotter pins, roll pins, or other retainers. Getting to know the various kinds of fasteners will help you know how to disassemble and reassemble component assemblies. We cover the major types of threaded and non-threaded fasteners below.

The idea of threaded fasteners has been around since approximately 240 B.C. That's when Archimedes invented the screw conveyor. The screw design was adapted to a straight pin to attach different materials together. As time progressed, the materials that the screw was made from became stronger. The sizing also became more precise.

Thread profiles also changed over time. Whitworth from England, in 1841, created a thread profile that had a rounded crest and rounded root (**FIGURE 8-2**). Then, in 1864, in the United States, Sellers created a thread profile that had a flat crest and flat root (**FIGURE 8-3**).

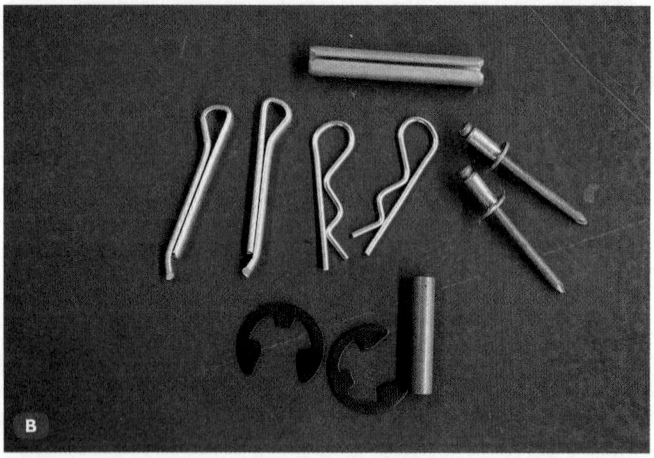

FIGURE 8-1 A. Threaded fasteners. **B.** Non-threaded fasteners.

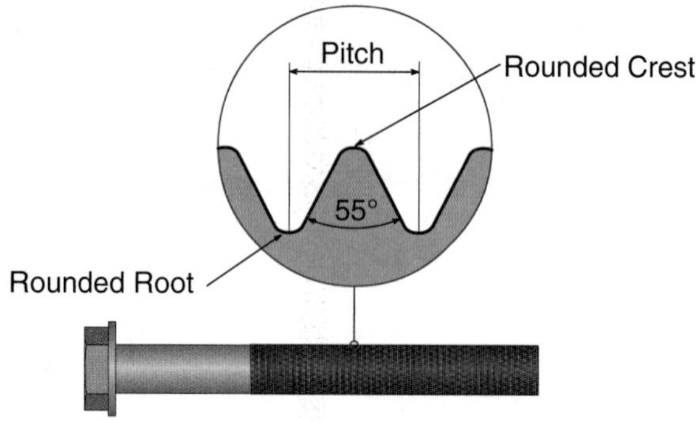

FIGURE 8-2 Whitworth thread.

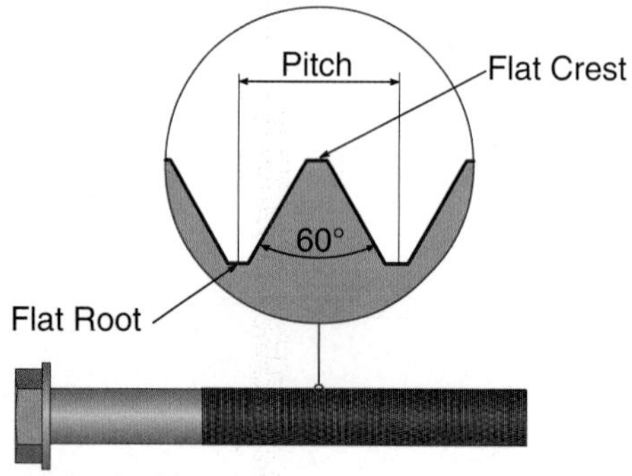

FIGURE 8-3 Sellers thread.

You Are the Automotive Technician

The new intern was helping you disassemble a leaking water pump. All was going well until one of the bolts broke off while removing it. It is evident that the shank of the bolt had been narrowed considerably due to rust and corrosion over time. You know that removing this bolt could turn the job into a real nightmare if not performed properly and carefully. It is also an opportunity to teach the intern how to perform this important task. As you are thinking about this task, several questions come to mind.

1. If part of the bolt extends past the surface it is threaded into, what can be done to try to remove it without using a bolt extractor?
2. What processes can be tried on bolts that are stuck, before they are broken off?
3. Is it okay to use a replacement bolt of a different grade from the original bolt? Why or why not?
4. What is the purpose of a thread-locking compound? Give an example of when it should be used.
5. What is the purpose of antiseize? Give examples of when it should and shouldn't be used?

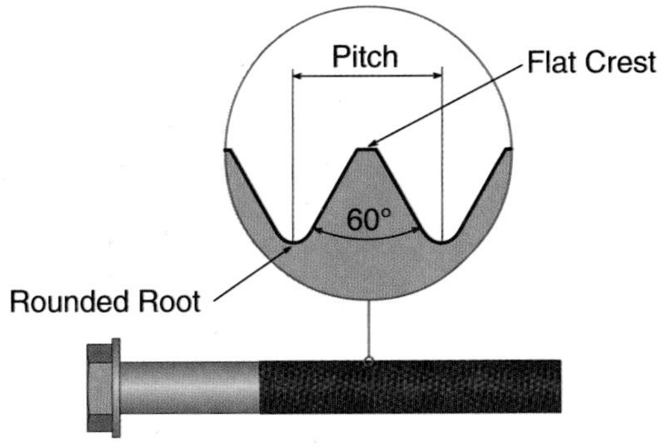

FIGURE 8-4 Current thread.

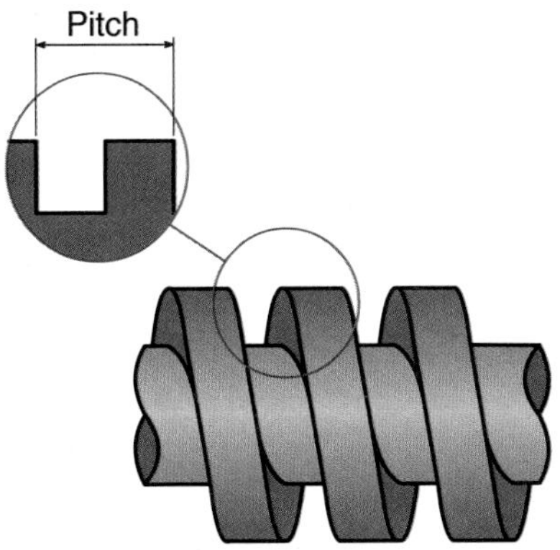

FIGURE 8-5 Square thread as used in a vice.

In order to create the threaded fasteners we use today, the flat crest and rounded root is used along with a 60-degree pitch angle (**FIGURE 8-4**). Metric and standard size bolts use the same shape of threads, excluding specialty-use bolts. Specialty-use bolts are made for a particular purpose, such as the lead screw for a vice or gear puller. The square edges of this specialty thread (**square thread**) tend to exert force more in line with the screw. A V-thread tends to push at a 90-degree angle to the thread, not the screw (**FIGURE 8-5**).

Fastener Standardization

There are many different groups that monitor and set the standards that make the automotive industry conform to one universally recognized specification set. There are three main groups to know about:

- The American Society for Testing and Materials (ASTM) is a nongovernmental controlled group that tests and sets standards in all areas of industry.
- The International Organization for Standardization (ISO) is an independent developer of standards that are created to ensure a reliable, safe, and quality product.
- The Society of Automotive Engineers (SAE) was initially started to standardize automotive production and has since grown to encompass a lot of engineering disciplines creating standards and best practices literature.

FIGURE 8-6 Metric bolts can be identified by the grade number on the top of the bolt head.

Metric Bolts

The metric bolt came about from the ISO standard 898. It defines mechanical and physical properties of metric fasteners. The ISO is an independent developer of voluntary international standards. In other words, any one government or group does not control them. The ISO develops standards that apply to many different industries in many countries. They use an unbiased logical approach.

The metric decimal system has been around since the late 1600s. It is sequential so it is easy to use. Many countries have adopted this standard for their weights and measure systems. But some countries have also retained their own customary standards. For industries that require more precision, the metric system is used almost exclusively.

The automotive industry is one that deals with precise measurement in every aspect of automobile design. Because of this, manufacturers have embraced the metric measuring system. Most automotive companies are using this worldwide. This also leads to efficiency and conformity. While most vehicle manufacturers use metric fasteners almost exclusively, they don't always do so entirely. So you need to be able to distinguish between metric and standard fasteners. Most metric fasteners can be identified by the grade number cast into the head of the bolt (**FIGURE 8-6**).

Standard Bolts

The standard measurement of bolts is a combination of the Imperial and U.S. customary measurement systems. SAE standard J429 covers the Mechanical and Material Requirements. They govern this standard for externally threaded fasteners.

Based on the British units of measure, units have been developed over the past millennia to arrive at today's standards. Instead of using a decimal number–based system like the Metric units, they are fractional based. People unfamiliar with fractions can find using it difficult. Standard bolts can generally be

FIGURE 8-7 Standard bolts can be identified by the hash marks on top of the bolt head.

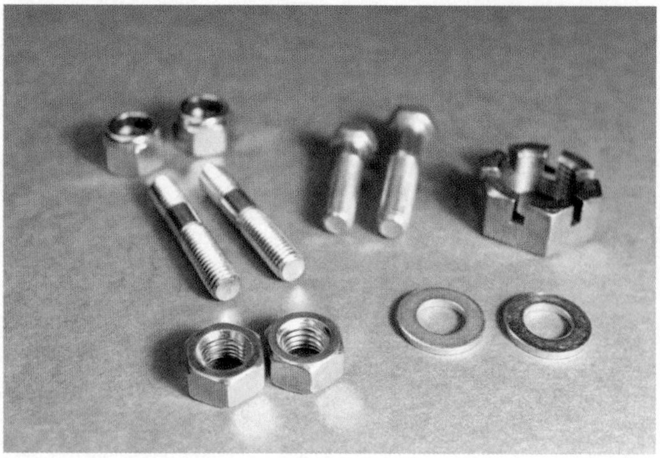

FIGURE 8-8 Bolts, studs, and nuts.

identified by hash marks cast into the head of the bolt, indicating the bolt's grade (**FIGURE 8-7**). We will cover bolt grade in just a bit.

Bolts, Studs, and Nuts

Bolts, studs, and nuts are threaded fasteners designed for jobs requiring a fastener heavier than a screw. They are mostly made of metal alloys, making them stronger (**FIGURE 8-8**). Typical **bolts** are cylindrical pieces of metal with a hexagonal head on one end and a thread cut into the shaft at the other end. The thread acts as an inclined plane. As the bolt is turned, it is drawn into or out of the mating thread. Hexagonal **nuts** thread onto the bolt thread. The hexagonal heads for the bolt and nut are designed to be turned by tools such as wrenches and sockets. Note that other bolt head designs are used as well. These require specially shaped tools as covered in the Hand Tools chapter.

A **stud** does not have a fixed hexagonal head. Rather, it has a thread cut on each end. It is threaded into one part where it stays. The mating part is then slipped over it, and a nut is threaded onto the other end of the stud to secure the part. Studs are commonly used to attach one component to another, such as a throttle body to the intake manifold. Studs can have different threads on each end, to work best with the material they are threaded into. Coarse threads work well in aluminum. In such cases, one end of the stud may use a coarse thread to grip the threads in an aluminum intake manifold and to position the throttle body. On the other end, there may be a fine thread for pulling everything together tightly with a steel nut.

Bolts, nuts, and studs have either standard or metric threads. They are designated by their thread diameter, thread pitch, length, and grade. The diameter is measured across the outside of the threads. It is measured in fractions of an inch for standard-type fasteners. And millimeters for metric-type fasteners. So, a 3/8" (9.5 mm) bolt has a thread diameter of 3/8" (9.5 mm). It is important to note that a 3/8" (9.5 mm) bolt does not have a 3/8" (9.5 mm) bolt head.

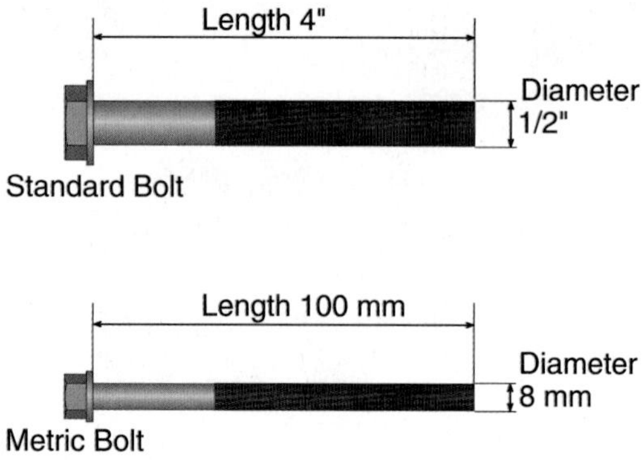

FIGURE 8-9 Bolt diameter and length.

Sizing Bolts

Bolts are sized either by metric or standard measuring systems. A 1/2" bolt does not mean that the bolt has a head that fits into a 1/2" socket. It means that the distance across the outside diameter of the bolt's threads measures 1/2" in diameter. The same goes for the metric equivalent. An 8 mm bolt is the diameter of the outside diameter of the bolt's threads.

The length of a bolt is fairly straightforward. It is measured from the end of the bolt to the bottom of the head and is listed in inches or millimeters (**FIGURE 8-9**). Note that the length of bolt does not include the thickness of the head.

Thread Pitch

The coarseness of any thread is called its **thread pitch** (**FIGURE 8-10**). In the standard system, bolts, studs, and nuts are measured in threads per inch (TPI). To determine the TPI, simply count the number of threads there are in 1" (25.4 mm). Each bolt diameter in the standard system can typically have one of two thread pitches. One being the **Unified National Coarse Thread (UNC)**. The other is the **Unified National Fine Thread (UNF)**. For example, a 3/8–16 is coarse, and a 3/8–24 is fine.

In the metric system, the thread pitch is measured in millimeters by the distance between the peaks of the threads. So course threads have larger distances, and fine threads have smaller distances. Each bolt diameter in the metric system can have up to four thread pitches (**FIGURE 8-11**). Consult a metric thread pitch chart, because there is no clear pattern of thread pitches for metric fasteners. These charts can be found in tap and die sets or on the Internet.

Thread Pitch Gauge

Thread pitch gauges make identifying the thread pitch on any bolt quick and easy. Without a thread pitch gauge, technicians must measure the number of threads per inch, or the distance from peak to peak in millimeters. Each thread pitch gauge matches a particular thread pitch that is stamped or engraved on the gauge. To use a thread pitch gauge, find the gauge that fits perfectly into the threads you are measuring. Just make sure you keep the gauge parallel to the shaft of the bolt; otherwise, you will get a wrong match (**FIGURE 8-12**).

Grading of Bolts

What is bolt grading? Bolt grade or class means that the fastener meets the strength requirements of that grade or classification. There are a variety of grades, so it is important to use the specified grade of bolt for each component. When bolts are tested for grade, the bolt is first put in a testing tool. Then it is stressed to the point of breakage so that the maximum readings can be determined to properly grade the fastener. We need to know a bolt's grade so that we are able to use the correct fastener for the application. This helps avoid bolt failures down the road.

Standard bolts are typically grade 1 to grade 8, with grade 8 being the strongest. The grade can be identified by counting the hash marks on the head of the bolt and adding two to that number. For example, there are three hash marks on the head of a grade 5 bolt, and six hash marks on the head of a grade 8 bolt (**FIGURE 8-13**).

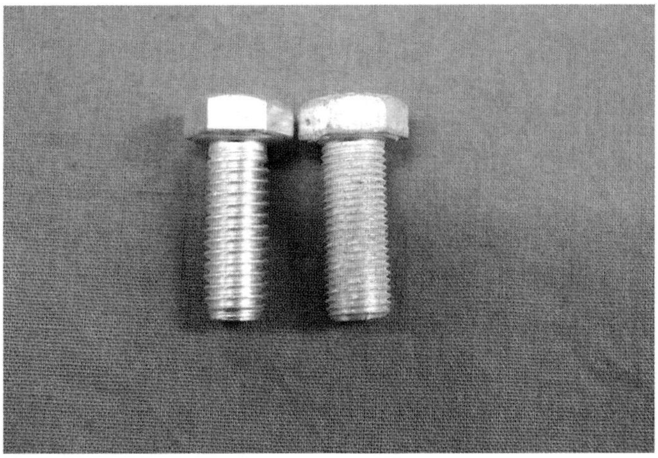

FIGURE 8-10 UNC and UNF standard bolt thread pitch.

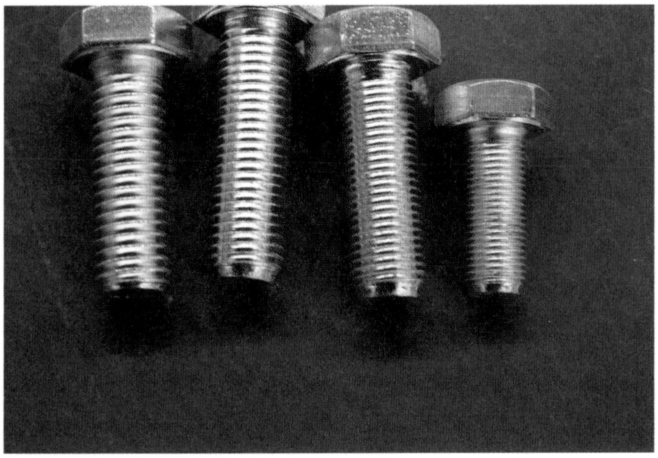

FIGURE 8-11 Metric thread pitches.

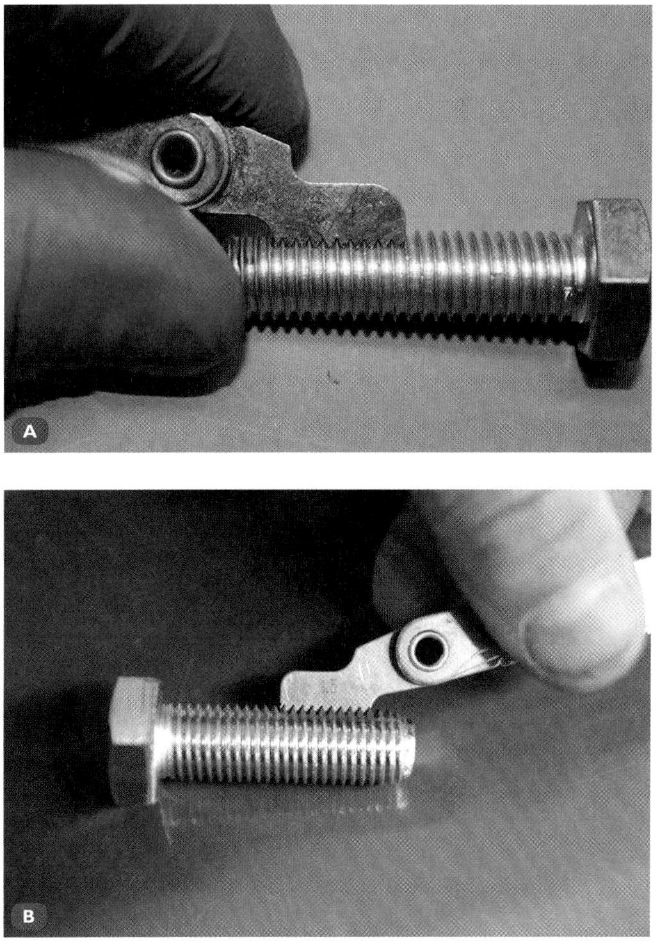

FIGURE 8-12 Thread pitch gauge being used. **A.** Standard. **B.** Metric.

FIGURE 8-13 **A.** Grade 5 standard bolt. **B.** Grade 8 standard bolt.

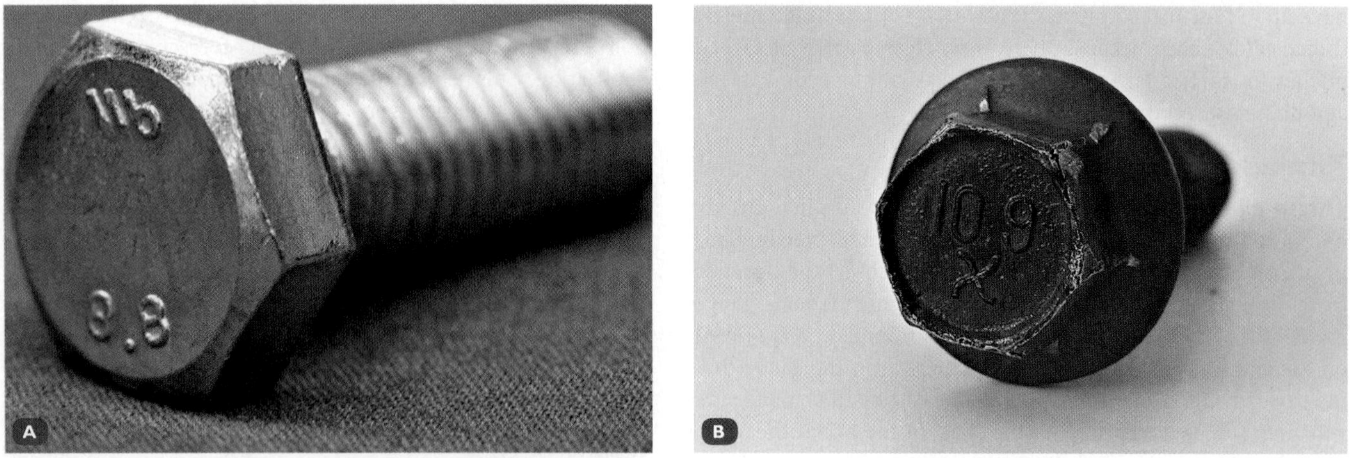

FIGURE 8-14 A. Grade 8.8 metric bolt. **B.** Grade 10.9 metric bolt.

Metric bolt grades are typically grade 4.6 to grade 12.9, with grade 12.9 being the strongest. In metric bolts, the grade number is cast into the top of the bolt head. These numbers on metric bolts have specific meanings. The number before the decimal indicates its tensile strength in megapascals (MPa). This is found by multiplying the number by 100. For example, a metric 12.9-grade bolt would have a tensile strength of 1200 MPa. The number to the right of the decimal when multiplied by 10 indicates the yield point as a percentage of the tensile strength (**FIGURE 8-14**).

Just because the higher grades are stronger doesn't mean you should use those in all applications. There are times when lower graded bolts are more appropriate for particular applications. For example, as with some head bolts that need to have a little bit of give in them. This allows for expansion and contraction of the cylinder head when the head gasket is clamped in place. Otherwise, the head gasket seal could fail. In other situations, using the wrong grade of bolt could cause either the bolt to fail or the threads in the threaded hole to fail. So you should always use the specified grade bolt when replacing bolts with new ones.

Strength of Bolts

Using the right bolt includes using one of the correct strength. If the bolt is not strong enough, it may break during installation. It may also break once it is put in service. If the bolt is too strong, it may damage the threads on the mating part if overtightened. It may also shatter under certain types of loads. There are a variety of ways that bolt strength is measured. We will cover the main ones below.

Tensile Strength

Tensile strength is the maximum tension the fastener can withstand without being pulled apart. Tensile strength is determined by the strength of the material and the size of the stress area. When a typical threaded fastener fails in pure tension, it usually fractures at the threaded portion. That is the weakest area because it is also the smallest area of the bolt (**FIGURE 8-15**). Generally speaking, the higher the grade of bolt, the higher the tensile strength.

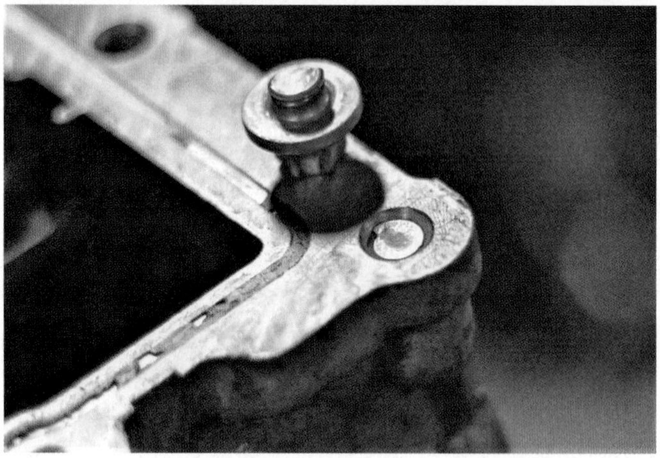

FIGURE 8-15 Bolt failure (tension).

Shear Strength

Shear strength is defined as the maximum load that can be supported prior to fracture, when applied at a right angle to the fastener's axis. A load occurring in one transverse plane is known as single shear (**FIGURE 8-16**). Double shear is a load applied in two planes, where the fastener could be cut into three pieces. For most standard threaded fasteners, shear strength is not specified. This is true even if the fastener may be commonly used in shear applications. Generally speaking, the higher the grade of bolt, the higher the shear strength.

Proof Load

The proof load represents the usable strength range for certain standard fasteners. By definition, the proof load is an applied tensile load that the fastener must support without exceeding the elastic phase. This is the point up to which the bolt returns to its original point when tension is removed.

Fatigue Strength

A fastener subjected to repeated cyclic loads can suddenly and unexpectedly break. This can happen even if the loads are well

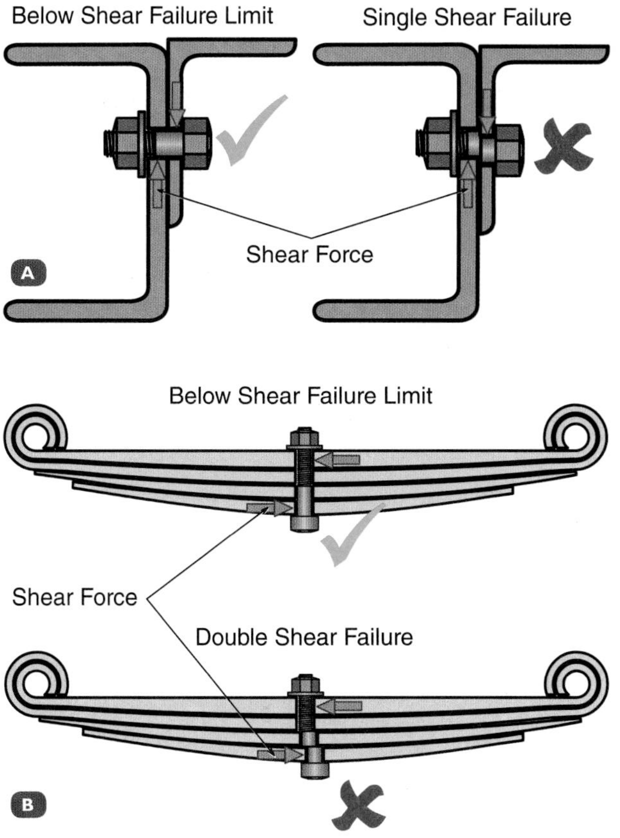

FIGURE 8-16 **A.** Single shear. **B.** Double shear.

below the strength of the material. The repeated cyclic loading weakens the fastener over time. It fails from fatigue. The fatigue strength is the maximum stress a fastener can withstand for a specified number of repeated cycles before failing. Connecting rod bolts are an example where fatigue strength would need to withstand millions of cyclic loads.

Torsional Strength

Torsional strength is a measure of a material's ability to withstand a twisting load. This is usually expressed in terms of torque, in which the fastener fails by being twisted off about its axis. A fastener that fails due to low torsional strength typically fails during installation or removal. Generally speaking, the higher the grade, the higher the torsional strength of the fastener.

Ductility

Ductility is the ability of a material to deform before it fractures. A material that experiences very little or no plastic deformation before fracturing is considered brittle. Think of a piece of flat glass. If you try to bend it, it breaks easily. This is an example of a material with low ductility.

Most automotive fasteners need to have some measure of ductility to avoid catastrophic failure. A reasonable indication of a fastener's ductility is the ratio of its specified minimum yield strength to the minimum tensile strength. The lower this ratio, the more ductile the fastener. This means that the more ductile the bolt, the more it can flex or stretch without breaking. Generally speaking, the higher the grade, the lower the ductility of

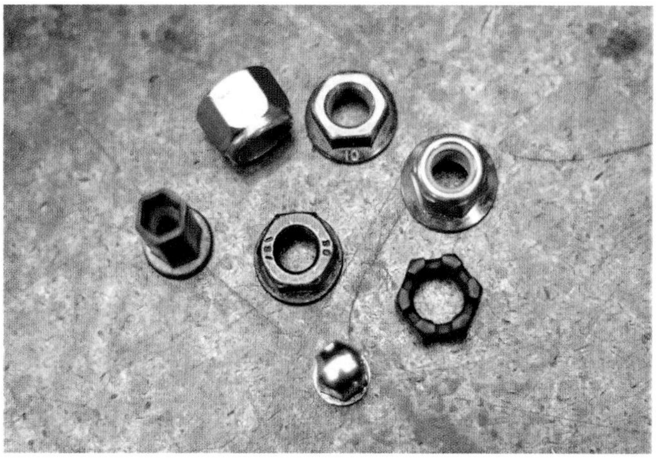

FIGURE 8-17 Variety of nuts.

the fastener. So a balance between tensile strength and ductility is needed that is based on the requirements of the application.

Toughness

Toughness is defined as a material's ability to absorb impact or shock loading. Impact strength toughness is rarely a specified requirement for automotive applications. There is a specification for various aerospace industry fasteners as well as ASTM A320. This specification for alloy steel bolts for low-temperature service is one of the few specifications that requires impact testing on certain grades.

Nuts

Nuts are screwed onto the threads of a bolt or stud to hold something together. They have internal threads that match the threads on the same size bolt. The many different kinds of nuts are mostly application-specific (**FIGURE 8-17**). We cover some of the more popular ones below.

When you replace a nut with a new nut on a vehicle, you need to match the nut to the grade of bolt that it is going to be screwed onto. For example, if the bolt is a grade 8, you need a grade 8 nut. By matching the grade of fasteners, you are keeping the intended integrity of the manufacturer's fastening system. If you mix different grades of bolts and nuts, one could prematurely fail, causing catastrophic failure of the component.

Locking Nuts

Locking nuts are used when there is a chance of the nut vibrating loose. Two common types of locking nuts are used in automotive applications. One is the nylon insert lock nut and the other is the deformed lock nut (**FIGURE 8-18**). The nylon lock nut has a nylon insert with a smaller diameter hole than the threads in the nut. As the nut is threaded onto the bolt, the nylon insert is deformed by the bolt threads. This tends to lock the nut onto the thread of the bolt or stud so that it can't vibrate loose.

In a deformed lock nut, the top thread of the nut is deformed, thus pinching the bolt threads so the nut can't vibrate loose. Both of these types of lock nuts are designed for single

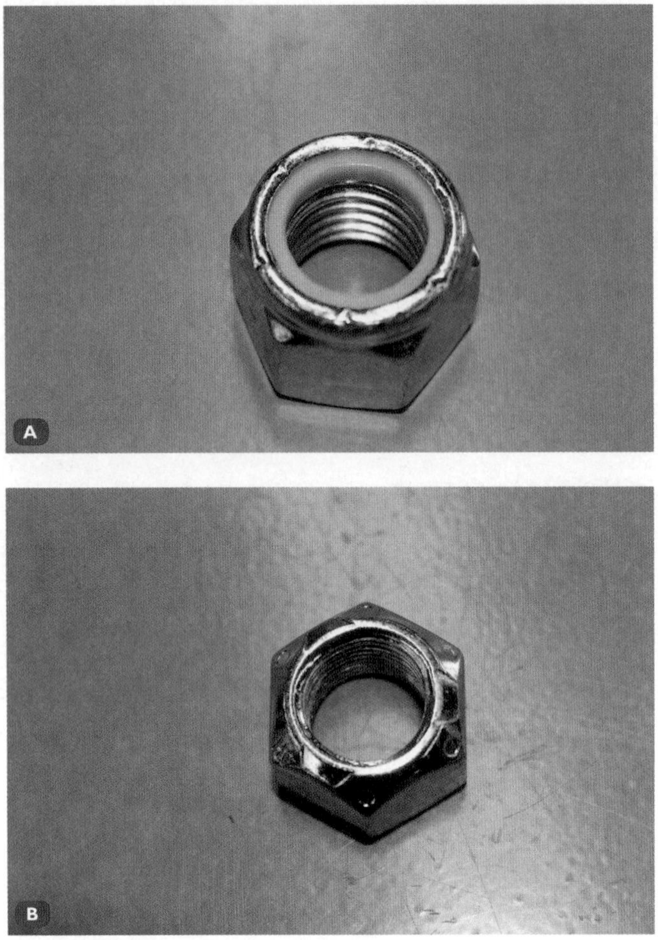

FIGURE 8-18 **A.** Nylon lock nut. **B.** Deformed lock nut.

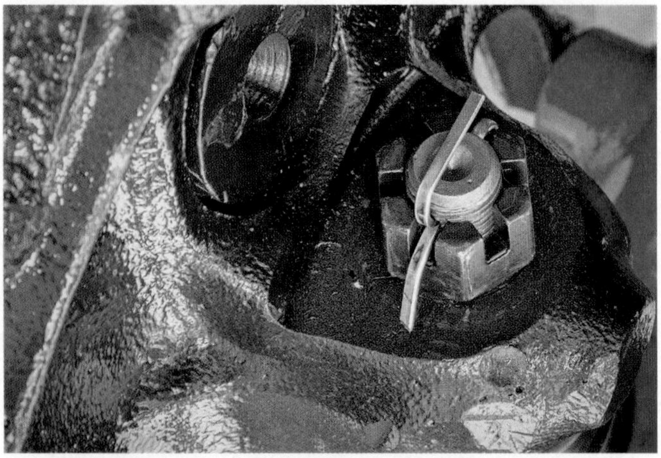

FIGURE 8-19 Castle nut used on a ball joint.

FIGURE 8-20 New cotter pins installed in the castle nuts.

FIGURE 8-21 A variety of specialty nuts.

use. This means they should be replaced with new ones once they have been removed.

Castle Nuts

Castle nuts, sometimes called castellated nuts, are used with a cotter pin to keep the nuts from moving once they are installed. Protrusions and notches are machined into the top of the nut. The notches then can be lined up with a hole in the threaded bolt or stud (FIGURE 8-19).

A cotter pin passes through the notch on one side of the nut, then through the hole in the stud and out the notch in the other side of the nut. The cotter pin physically prevents the castle nut from backing off the stud. These nuts are usually used on suspension components or other critical components where it is important they don't fall off. It is good practice to always use new cotter pins. You also need to visually verify if they are installed in the castle nuts before finishing up a job (FIGURE 8-20).

Specialty Nuts

In the automotive industry, there are lot of specialty nuts with special characteristics. For example, an extended shoulder, a special head size, or a special material—required for a particular application. You must use the correct nut for the correct application (FIGURE 8-21). Sheet metal nuts are a cheap alternative to a regular nut. Sometimes called J nuts, these nuts are a folded piece of sheet metal that fits over a hole in a piece of sheet metal. They are threaded to accept a fastener. They usually do not need to be held as they are clipped to the component. Failure to use the proper fasteners for the application usually results in failure of the component.

Washers

Why are washers used underneath bolt heads and nuts? There are three main purposes of a washer:

- To distribute the force exerted on the component that it is pressing against evenly
- To prevent the surface around the hole from being worn down due to tightening of the head of the bolt or nut against a surface
- To provide a measure of locking force to keep the bolt or nut from loosening due to vibration

Not all applications require washers that provide all three functions, so make sure you know what the application requires. When selecting a washer for a bolt or nut, make sure you use the correct type, size, and grade.

Grades of Washers

Just as with nuts, you want to match the grade of washer to the nut and bolt. This maximizes the clamping force on the component (**FIGURE 8-22**). Using washers that have different hardness characteristics can cause the surface of the softer metal to wear away. This reduces the clamping force over time, which can cause the bolt or the component to fail. Not all washers have grade ratings on them. So be careful to select the proper washer when replacing a damaged or missing washer with a new one.

FIGURE 8-22 Washer with its grade rating marked on it.

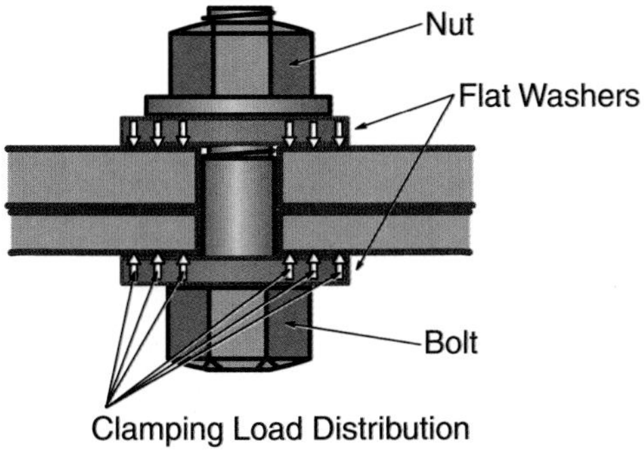

FIGURE 8-23 Flat washer spreading out the clamping force.

Flat Washers

A flat washer is a piece of steel that has been cut in a circular shape. It has a hole in the middle for the bolt or stud to fit through. Washers are sized and paired with a metric or standard bolt. This allows the pressure created by torquing the bolt down to spread out on the component it is being used on (**FIGURE 8-23**). They also prevent marring of the surface of the component around the bolt hole. And they act as a bearing surface so that the bolt achieves the expected clamping force at the specified torque. Flat washers are graded just like bolts. This means you need to match the grade on the washer with the fasteners that you are using them on.

Lock Washers

Lock washers are made from spring steel and have an angled slit cut in them. This allows for the spring action to hold tension against the nut so that the nut will not loosen. Also, the angled slit has sharp edges on the top and bottom. The sharp edges bite into the bolt head and surface of the component if the bolt tries to loosen (**FIGURE 8-24**). Lock washers are used with conventional nuts. They are ideal in applications that experience a lot of vibration. Lock washers are usually considered to be single use. So they should always be replaced rather than reused.

Star Washers

Star washers, also called toothed lock washers, are used to stop the rotation of the nut. They work in a similar way as lock washers in that the teeth are spring steel and are at an angle. This causes them to bite into the bottom of the bolt head and the surface being clamped (**FIGURE 8-25**). If these washers are required for your application, they need to be installed before the nut is installed. They also should be replaced with new ones rather than being reused.

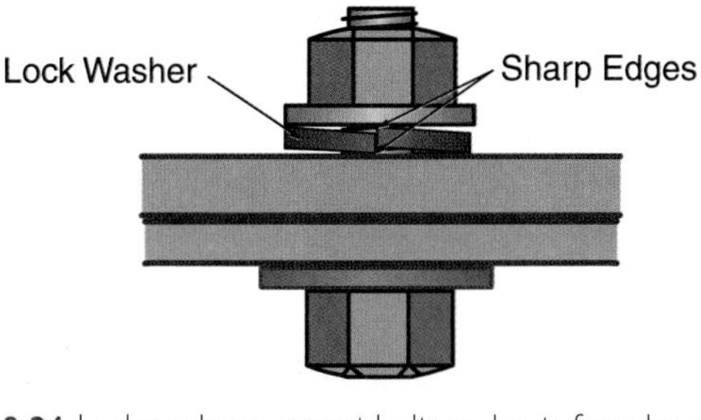

FIGURE 8-24 Lock washers prevent bolts and nuts from loosening.
(*Note:* This nut and bolt is not fully tightened to show the action of the lock washer)

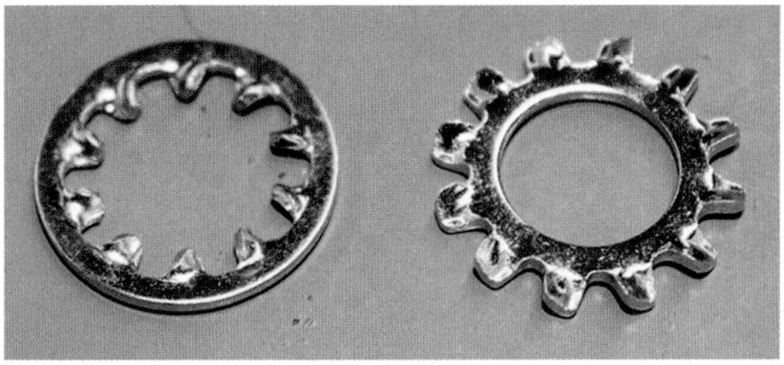

FIGURE 8-25 Star washers prevent nuts and bolts from loosening.

Threadlocker and Anti-seize

Thread-locking compound is a liquid that is put on the threads of a bolt or stud to lock the nut in place. The thread-locking compound acts like a very strong glue. Once the compound between the nut and bolt cures, it bonds the two together so that they do not move. A popular thread-locking compound (threadlocker) that most automotive technicians prefer is Loctite.

The two main strengths of locking compounds that are used in the automotive industry are identified as blue and red (**FIGURE 8-26**). The blue is a medium strength which allows for relatively easy removal of the nut or bolt with a socket and ratchet. The red is much stronger. The red version usually requires a fair amount of heat to soften the compound so the nut or bolt can be removed.

There is also a green type of threadlocker that has penetrating qualities. It is used after the nut is tightened. Thread-locking compound is a safety item that technicians use on critical parts they do not want to come loose. In fact, the manufacturer may specify its use on certain fasteners. So, be aware of situations that require a thread-locking compound, and use the one with specified strength.

Anti-seize compound is the opposite of thread-locking compound (**FIGURE 8-27**). It keeps threaded fasteners from becoming corroded together or seized to each other. One common use of anti-seize compound is on black steel spark plug threads when installed into an aluminum cylinder head. Anti-seize compound is a coating that prevents rust and provides lubrication so that the fastener can be removed in the future. Because of that, it should *not* be used on most fasteners in a vehicle. Otherwise, they could vibrate loose, causing damage or an accident. When you do use anti-seize compound on specified fasteners, use a very light coating, as a little bit goes a very long way (**FIGURE 8-27**). In fact, for a technician, a small can of anti-seize could last for several years of use.

FIGURE 8-26 Thread-locking compound.

FIGURE 8-27 Anti-seize compound.

Screws

Many different types of screws are found in automotive applications (**FIGURE 8-28**). A screw is very similar to a bolt except for a couple of differences. They are not hardened. And they tend to be used in light-duty applications where they need to hold together components that have a low shear or tensile strength.

Machine Screws

Similar to a bolt, a machine screw is used to fasten components together. But they are usually driven by a screwdriver with a Philips or slotted head. Machine screws are usually driven into a nut or threaded hole that has been tapped to the thread pitch of the screw. As vehicles are becoming more complex, manufacturers have started to introduce new types of driving tools for these screws. Torx, square, and Allen head screws are becoming common in this fastener group (**FIGURE 8-29**).

FIGURE 8-28 Variety of screws.

FIGURE 8-29 Machine screw heads.

Self-Tapping Screws

Self-tapping screws are designed to create their own holes in the material they are being driven into. They have a fluted tip so that they drill a hole into the base material, without needing a pilot hole (**FIGURE 8-30**). The threads on a self-tapping screw are very similar to those on a sheet metal screw. Usually, they have a cap screw–type head so they can be driven with a screw gun. But they can have Phillips, square, Allen, or Torx heads as well.

Trim Screws

Trim screws are used in applications where there is a need to hold plastic and metal trim to the vehicle. They are also used to hold door panels, trim pieces, and other small components to the vehicle. They are basic screws that are usually painted either the trim color or black. This is so that they blend in with the material that they are used with (**FIGURE 8-31**).

Sheet Metal Screws

Sheet metal screws are used to attach things to sheet metal. They have a chip on the tip of the screw to cut threads as it is installed (**FIGURE 8-32**). These screws are usually threaded all the way up to the head so that they can be run down all the way, clamping the pieces together.

▶ Non-Threaded Fasteners

LO 8-02 Identify non-threaded fasteners.

Non-threaded fasteners also hold parts in place, but not tightly clamped in place, like a threaded fastener would do. Some examples of where they would be used are:

- to hold a bearing in place. (**FIGURE 8-33**)
- to hold a u-joint in place. (**FIGURE 8-34**)
- to hold a bracket in place. (**FIGURE 8-35**)
- to retain a nut and prevent it from loosening up. (**FIGURE 8-36**)
- to retain a piece of trim. (**FIGURE 8-37**)

Snap Rings

Let's look a bit more in depth at some of the common types of non-threaded fasteners starting with snap rings. Snap rings are shaped like the letter "C," but with a much

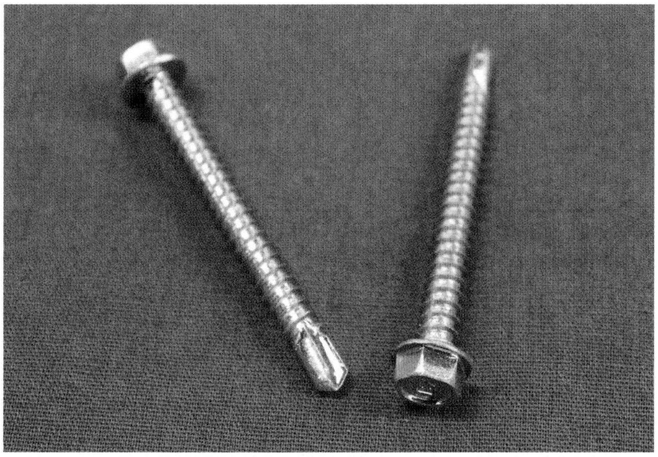

FIGURE 8-30 Self-tapping screws.

FIGURE 8-32 Sheet metal screws.

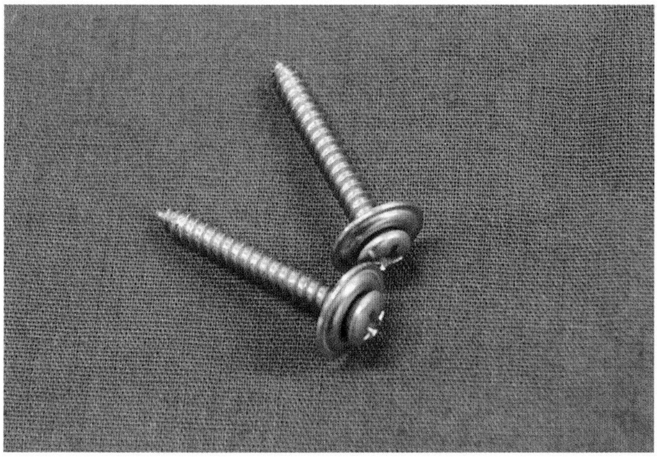

FIGURE 8-31 Trim screws.

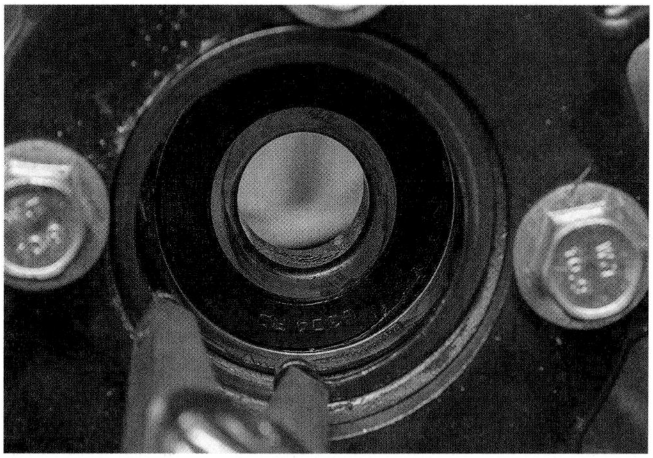

FIGURE 8-33 Snap ring holding a bearing in place.

FIGURE 8-34 U-joint retainer clip holding a U-joint cap in place.

FIGURE 8-35 Rivet holding a bracket in place.

FIGURE 8-36 Cotter pin holding a castellated nut in place.

FIGURE 8-37 Push clip holding a piece of trim in place.

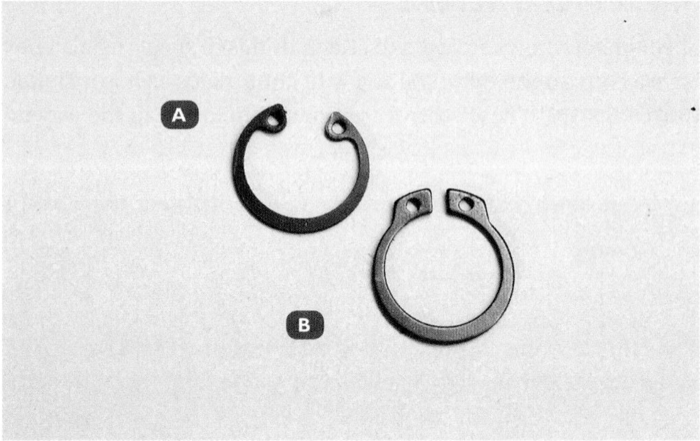

FIGURE 8-38 **A.** Internal snap ring. **B.** External snap ring.

FIGURE 8-39 Internal snap ring holding a valve in place.

smaller opening. They are made of fairly thin spring steel. They can be either internal or external types of snap rings (**FIGURE 8-38**). Internal snap rings fit inside the groove in a hole (**FIGURE 8-39**). External snap rings fit in grooves on the outside of a shaft or housing. Both types of snap rings either hold a part in place, or prevent the part from traveling beyond a certain point. Snap rings typically have a hole in each end of the snap ring. This is so snap ring pliers can be used to either expand or contract the snap ring during installation and removal (**FIGURE 8-40**).

SAFETY TIP

There is one big caution when installing or removing snap rings and other types of spring steel clips. Realize that if they slip off the tool, they will fly at a great speed and distance. This creates a very dangerous eye hazard, so safety glasses are required. In addition, since snap rings are generally small, they can be hard to find after they have flown across the shop. Laying a shop rag across the snap ring and snap ring pliers during removal and installation can help keep it from flying away.

Similar to snap rings are circlips. Circlips are also made of spring steel, and they can be both internal and external types. Circlips typically do not have holes for snap ring pliers. So they use either angled or stepped ends (**FIGURE 8-41**). The angled ends are used on external circlips and can typically be expanded using paddle-tip-style snap ring pliers (**FIGURE 8-42**). The stepped ends are used on internal circlips and can typically be removed with a screw driver or small pry bar (**FIGURE 8-43**).

Another type of clip is an "e" clip. It is shaped similar to a lower case "e," but with the top open (**FIGURE 8-44**). e-Clips fit into an external groove on small diameter shafts and keep parts from sliding off the shaft.

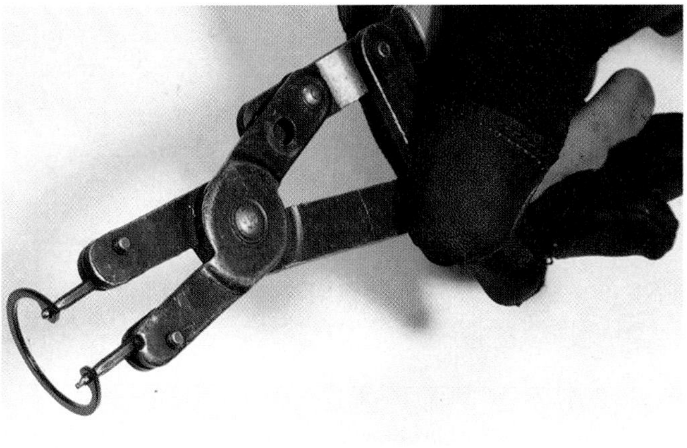

FIGURE 8-40 Snap ring pliers fit into the small holes in each end of the snap ring.

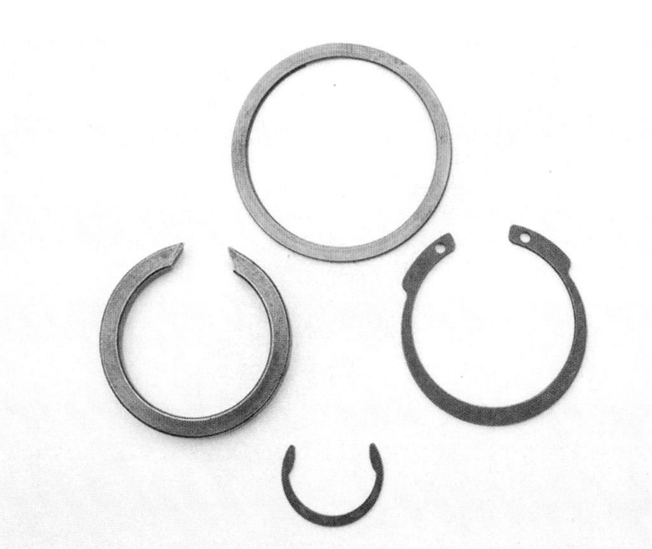

FIGURE 8-41 Various types of circlips.

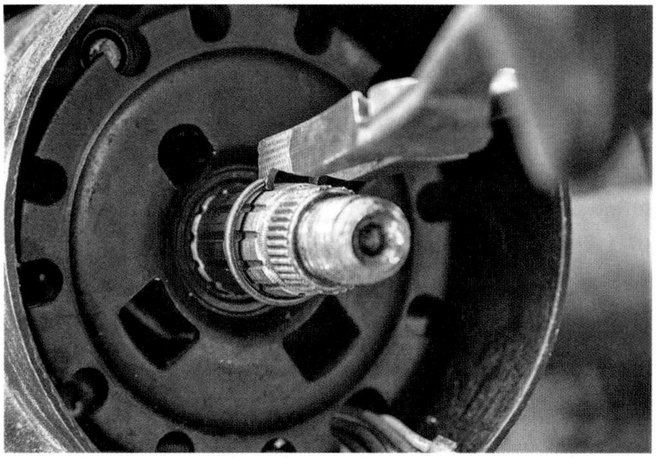

FIGURE 8-42 Paddle-style snap ring pliers being used to remove an external circlip.

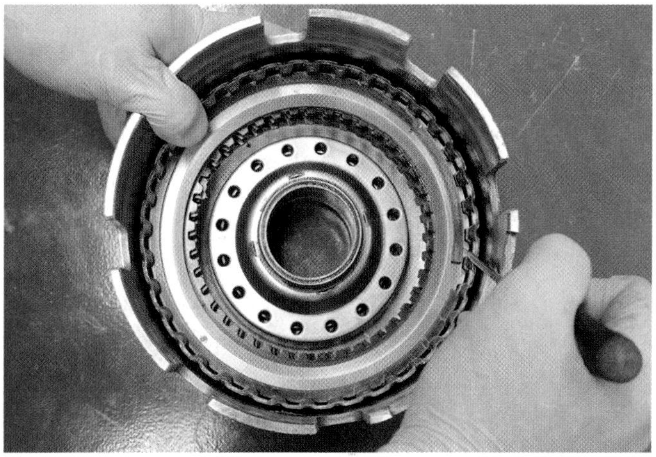

FIGURE 8-43 Screw driver being used to remove an internal circlip.

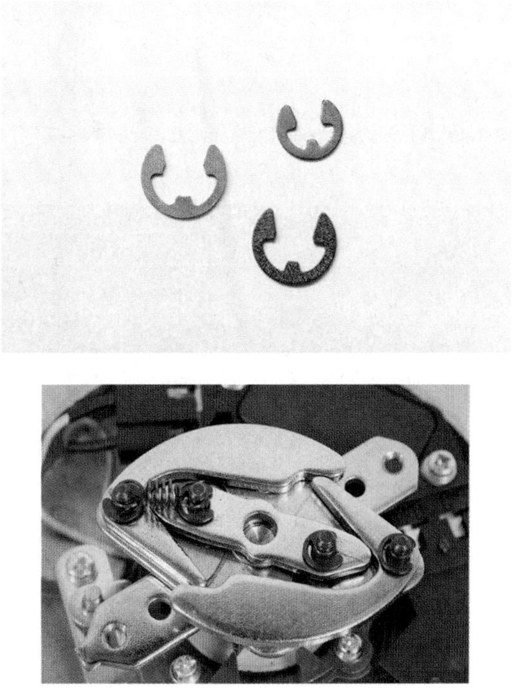

FIGURE 8-44 An e-clip is an external type of clip.

Rivets are another type of non-threaded fastener and were covered in the Tools chapter.

Pins

Various types of pins are used to retain parts: cotter pins, roll pins, and spiral pins (**FIGURE 8-45**). Cotter pins are commonly used to retain nuts on the steering and suspension system (**FIGURE 8-46**). Cotter pins are made of soft pliable metal. This allows them to be put through the recess in the nut and the hole in the stud and then bent in such a way that the nut cannot loosen (**FIGURE 8-47**). The long end is then cut off (**FIGURE 8-48**). When removing the nut, the cotter pin needs to be removed first. Generally, this is done by cutting it with diagonal pliers and pulling it out of the stud.

Roll pins and spiral pins are friction fit pins that firmly hold one part to another, typically a gear to a shaft (**FIGURE 8-49**). Roll pins are rolled into a fairly tight "C" shape (**FIGURE 8-50A**). Spiral pins are rolled into a spiral (**FIGURE 8-50B**). Sometimes, the hole in the gear and shaft is centered, which allows the gear to be installed in two positions. Other times, the hole is drilled slightly off-center so that the parts can only be assembled in one position. This is ideal for parts that require exact positions, such as timing components.

Roll pins and spiral pins are installed with a hammer once the parts are lined up and the holes between the two are completely lined up. To remove these types of pins, a pin punch (straight-sided

FIGURE 8-47 Cotter pins are bent into place.

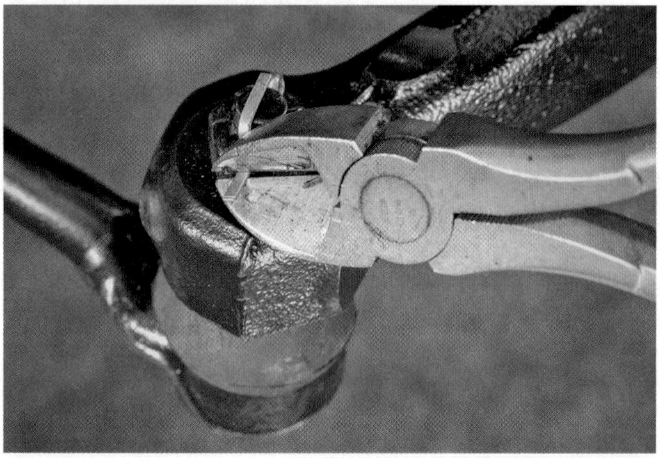

FIGURE 8-48 The remaining end is then cut off flush with the outside of the nut.

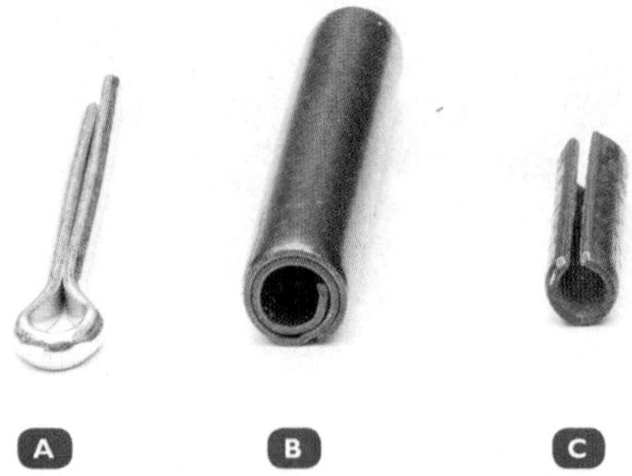

FIGURE 8-45 **A.** Cotter pin. **B.** Spiral pin. **C.** Roll pin.

FIGURE 8-46 Cotter pin used to retain a tie-rod nut.

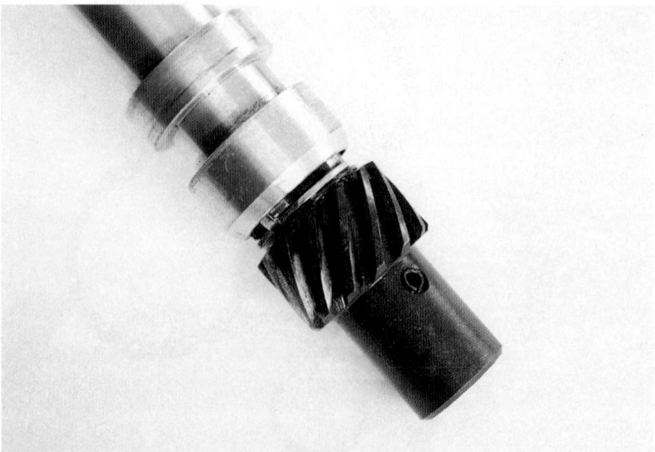

FIGURE 8-49 A roll pin holding a gear onto a shaft.

punch) of the appropriate diameter is used. The pin punch is driven by a hammer and pushes the pin out of the hole.

Clips

The last type of non-threaded fasteners we will look at are various types of clips. This includes push clips, christmas trees, and plastic retainers (**FIGURE 8-51**). These types of fasteners are designed to be installed quickly and without great precision in placement. In other words, they can accommodate slight differences in alignment between parts. This makes them ideal for retaining trim pieces.

The push clip has two parts: an outer housing with fingers, and a center pin. The outer housing with fingers fits through hole in both parts. The outer housing is placed firmly in the hole. A center pin is placed in the center housing which expands the fingers outward, securing one part to the other.

Christmas trees are used extensively to retain door panels and other trim pieces. They are plastic pins with thin, flexible rings around them (**FIGURE 8-52**) that grip the hole when pushed through. As they are installed, some of the rings get pushed through the hole. They grip the sides of the hole and hold the trim piece in place. To remove these clips, a special tool is used to pry the clips back out of the hole. Many times this process breaks or damage the clips, making replacement necessary.

Plastic retainers come in a variety of shapes, sizes, and purposes (**FIGURE 8-53**). Some plastic retainers are designed to be reused, while others are made only for single use. When removing plastic retainers, look at them to determine the best way to remove them. But do know that in some cases, it is best to cut the plastic retainer so you don't damage the piece that it is holding in place. Also, if you are having difficulty removing a plastic retainer, don't hesitate to ask another more experienced technician for assistance. They may have the finesse needed to remove the retainer.

▶ Replace Threaded Fasteners

LO 8-03 Replace threaded fasteners.

Bolt tension is what keeps a bolt from loosening. It is also what holds parts together with the proper clamping force. **Torque** is the twisting force used to create bolt tension. The torque value is

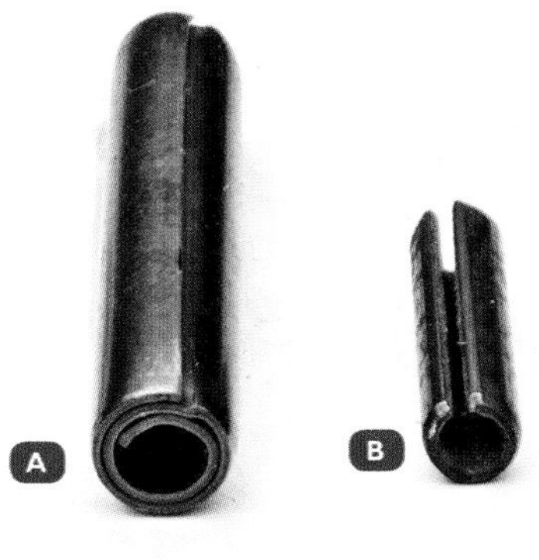

FIGURE 8-50 **A.** A spiral pin is rolled into a spiral shape. **B.** A roll pin is "C"-shaped spring steel.

FIGURE 8-52 Christmas tree type of clip showing the rings that grip the hole.

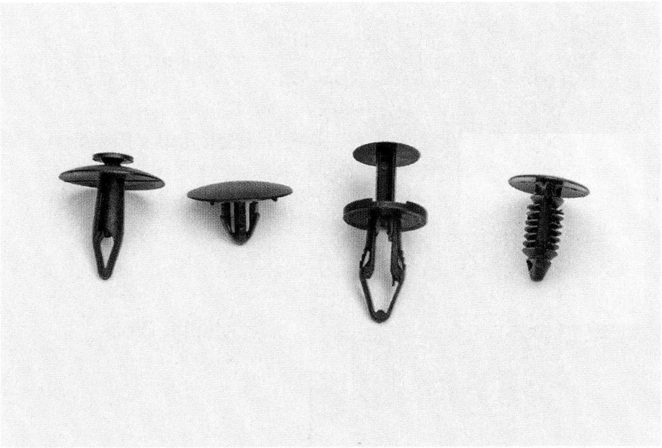

FIGURE 8-51 Various types of clips.

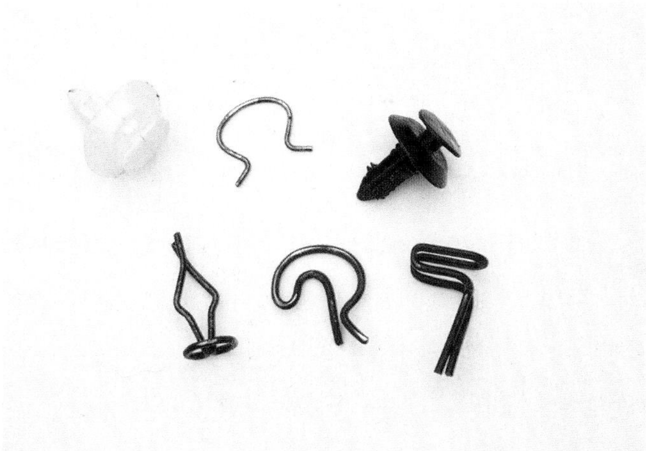

FIGURE 8-53 Retaining clips come in a variety of shapes and sizes.

the amount of twisting force applied to a fastener by the torque wrench. A foot-pound is described as the amount of twisting force applied to a shaft by a perpendicular lever 1 ft (30 cm) long with a weight of 1 lb (0.45 kg) placed on the outer end (**FIGURE 6-54**). A torque value of 100 ft-lb is the same as a 100 lb (45.36 kg) weight placed at the end of a 1' (2.5 cm)-long lever. A torque value of an inch-pound is 1 lb (0.45 kg) placed at the end of a 1' (2.5 cm)-long lever. This means that 12 in-lb equals 1 ft-lb and vice versa.

A Newton meter (Nm) is described as the amount of twisting force applied to a shaft by a perpendicular lever 1 meter long with a force of 1 Newton applied to the outer end. A torque value of 100 Nm is the same as applying a 100 Newton force to the end of a 1-m-long lever. One ft-lb is equal to 1.35 Nm. So either designation of torque is the measurement of twisting force.

Torque Charts

Torque specifications for bolts and nuts are usually found in the service information. Bolt, nut, and stud manufacturers also produce torque charts. They contain the information you need to determine the maximum torque of bolts or nuts. For example, most charts include the following:

- bolt diameter,
- threads per inch,
- grade, and
- maximum torque setting for both dry and lubricated bolts and nuts (**FIGURE 8-55**).

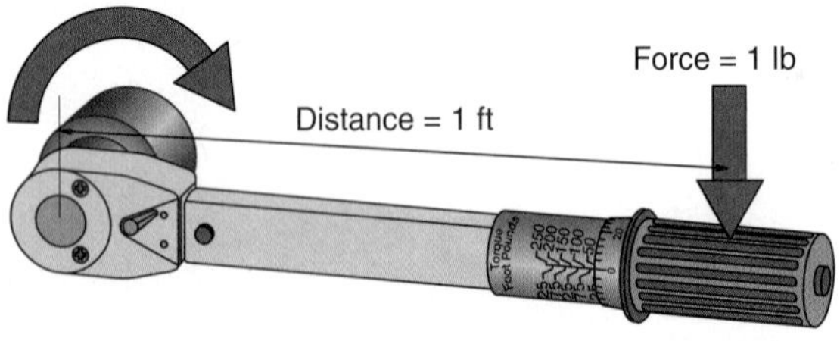

FIGURE 8-54 Torque is the measurement of twisting force. 1 ft-lb of torque.

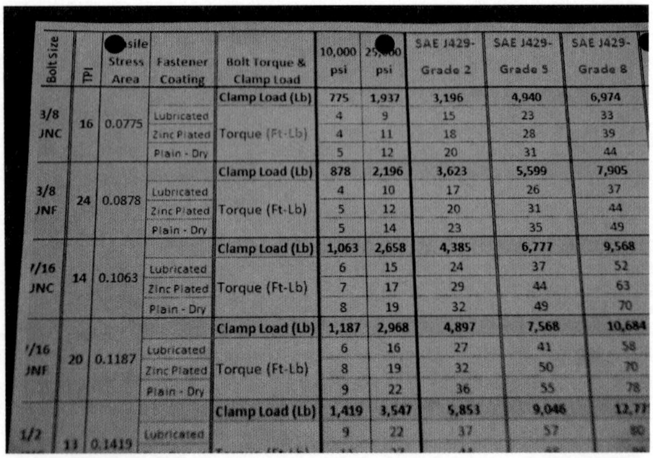

Bolt Size	TPI	[illegible]sile Stress Area	Fastener Coating	Bolt Torque & Clamp Load	10,000 psi	2[illegible]0 psi	SAE J429-Grade 2	SAE J429-Grade 5	SAE J429-Grade 8
3/8 JNC	16	0.0775		Clamp Load (Lb)	775	1,937	3,196	4,940	6,974
			Lubricated	Torque (Ft-Lb)	4	9	15	23	33
			Zinc Plated		4	11	18	28	39
			Plain - Dry		5	12	20	31	44
3/8 JNF	24	0.0878		Clamp Load (Lb)	878	2,196	3,623	5,599	7,905
			Lubricated	Torque (Ft-Lb)	4	10	17	26	37
			Zinc Plated		5	12	20	31	44
			Plain - Dry		5	14	23	35	49
7/16 JNC	14	0.1063		Clamp Load (Lb)	1,063	2,658	4,385	6,777	9,568
			Lubricated	Torque (Ft-Lb)	6	15	24	37	52
			Zinc Plated		7	17	29	44	63
			Plain - Dry		8	19	32	49	70
7/16 JNF	20	0.1187		Clamp Load (Lb)	1,187	2,968	4,897	7,568	10,684
			Lubricated	Torque (Ft-Lb)	6	16	27	41	58
			Zinc Plated		8	19	32	50	70
			Plain - Dry		9	22	36	55	78
1/2	13	0.1419		Clamp Load (Lb)	1,419	3,547	5,853	9,046	12,77
			Lubricated		9	22	37	57	80

FIGURE 8-55 Bolt torque chart.

A lubricated bolt and nut reach their maximum clamping force at a lower torque setting than if they are dry. In practice, most torque specifications call for the nuts and bolts to have dry threads prior to tightening. There are some exceptions. Close examination of the torque specification is critical. Also remember that the bolt manufacturer's torque chart is a maximum recommended torque. It is not necessarily the torque required by the vehicle manufacturer for the specific application that the bolt is used for.

Threaded Fasteners and Torque

Threaded fasteners are designed to secure parts that are under various tension and sheer stresses. The nature of the stresses placed on parts and threaded fasteners depends on their use and location. For example, head bolts withstand tension stresses by clamping the head gasket between the cylinder head and block. This is so that combustion pressures can be contained within the cylinder (**FIGURE 8-56**). The bolts must withstand the very high combustion pressures trying to push the head off the top of the block. In this situation, the pressure tries to stretch the head bolts, which means they are under tension.

Some fasteners must withstand sheer stresses, which are sideways forces trying to shear the bolt into two (**FIGURE 8-57**). An example of fasteners withstanding sheer stresses is wheel lug studs and lug nuts. They clamp the wheel assembly to the suspension system, and the weight of the vehicle tries to sheer the lug studs. If this were to happen, the wheel would fall off the vehicle, likely leading to an accident.

Threaded fasteners are designed to be tightened to a specified torque depending on several factors. Examples include the job at hand, the grade or hardness of the material they are made from, their size, and the thread type. If a fastener is overtightened, it could become weakened or break. If it is undertightened, it could work loose over time. This means that fasteners must be tightened to a specific point, which is accomplished by using a torque wrench.

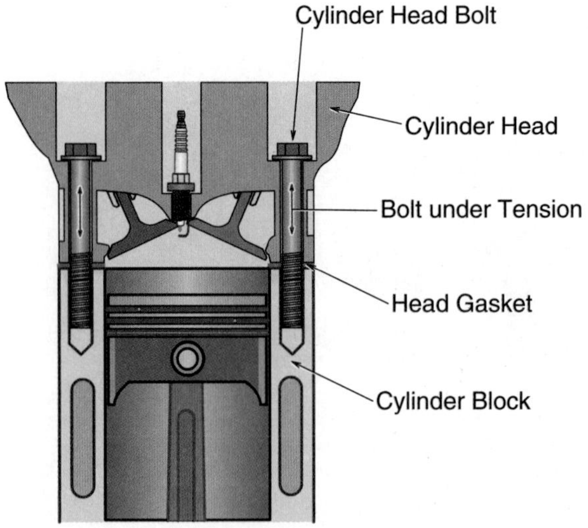

FIGURE 8-56 Bolts clamp parts together.

Torque wrenches fall out of calibration over time or if they are not used properly. They should be checked and calibrated on a periodic basis (**FIGURE 8-58**). This can be performed in the shop if the proper calibration equipment is available. Or the torque wrench can be sent to a qualified service center. Most quality torque wrench manufacturers provide a recalibration service for their customers.

Using Torque Wrenches

The torque wrench is used to apply a specified amount of torque to a fastener. There are various torque wrench designs used by torque wrench manufacturers. Each type indicates the torque applied. Some give signal when a pre-set torque is reached. This can be an audible signal, such as a click or a beep; or a visual signal such as a light turning on or a pin clicking out. Some provide a scale and needle that must be observed while you are torqueing the fastener.

To help ensure that the proper amount of torque gets from the torque wrench to the bolt, support the head of the torque wrench with one hand (**FIGURE 8-59**). When using a torque wrench, it is best not to use extensions. Extensions make it harder to support the head, which can end up absorbing some of the torque. If possible, use a deep socket instead.

Torque Sticks

Torque sticks are socket extensions made of spring steel. They are designed to absorb a certain amount of the impulses of an impact wrench. This is so that fasteners (typically lug nuts) will be tightened to a certain torque. Torque sticks come in a variety of torque ratings. A set of torque sticks is required to meet the torque specifications of different manufacturers (**FIGURE 8-60**).

Torque sticks are somewhat of a controversial tool in automotive shops. They are dependent on several things being right: strength of the impact wrench (and settings), air pressure in the shop, and method used by the technician. This is why some manufacturers and tire retailers do not recommend the use of torque sticks when performing the final torque on lug nuts. Some manufacturers allow a lower rated torque stick be used to run the lug nuts down. Then, they require the use of a calibrated torque wrench to perform the final torque.

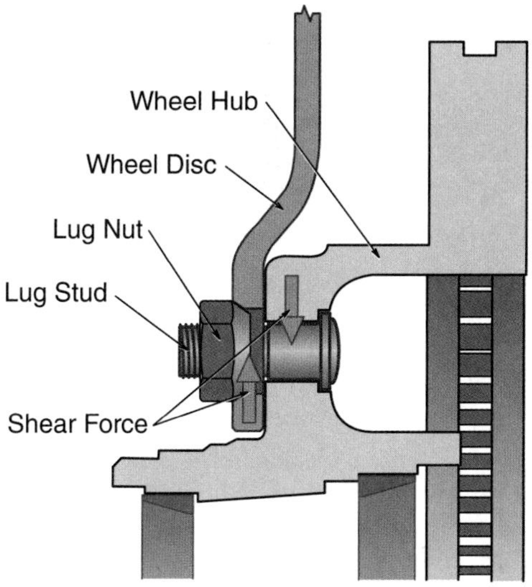

FIGURE 8-57 Bolt under sheer stresses.

FIGURE 8-59 It may be necessary to support the torque wrench when using extensions.

FIGURE 8-58 Checking torque wrench calibration.

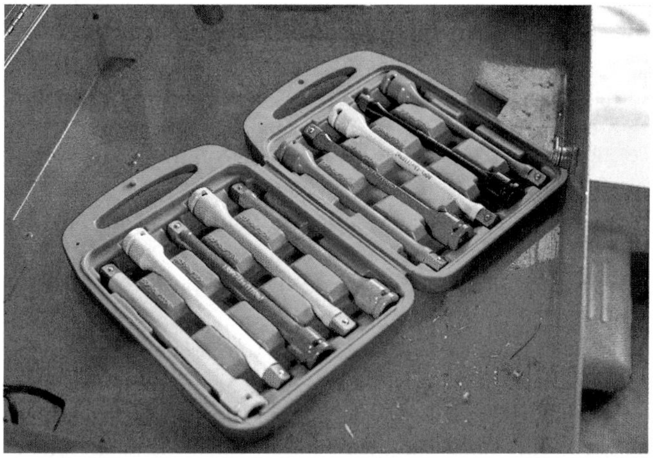

FIGUE 8-60 Torque sticks of various torque ratings.

Torque-to-Yield and Torque Angle

Torque is not always the best method of ensuring that a bolt is tightened so as to give the proper amount of clamping force. If the threads are rusty, rough, or damaged in any way, the amount of twisting force required to tighten the fastener increases. For example, there is more clamping force applied in a smooth fastener torqued the same amount as a rusty fastener. This also brings up the question of whether threads should be lubricated. In most automotive cases, the torque values specified are for dry, non-lubricated threads. But always check the manufacturer's specifications.

When bolts are tightened, they are also stretched. As long as they are not tightened too much, they will return to their original length when loosened. This is called **elasticity**. If they continue to be tightened and stretch beyond their point of elasticity, they will not return to their original length when loosened. This is called the **yield point**. As threaded fasteners are torqued, they go through the following phases:

Rundown Phase: Free-running fastener (may or may not have any torque).

Alignment Phase: Fastener and joint mating surfaces are drawn into alignment.

Elastic Phase: *This is the third and final stage for normal bolts!* The slope of the torque/angle curve is constant. The fastener is elongated but will return to original length upon loosening.

Plastic or Yield Phase: *Over-torqued condition for normal bolts. Torque-to-yield (TTY) bolts are tightened just into the beginning points of this phase.* Permanent deformation and elongation of the fastener and/or joint occur. Necking of the fastener occurs as the bolt is stretched further into the yield phase.

Torque-to-yield means that a fastener is torqued to, or just beyond, its yield point. With the changes in engine metallurgy that manufacturers are using in today's vehicles, bolt technology also had to change. Manufacturers have adopted **torque-to-yield (TTY) bolts**. This helps prevent bolts from loosening over time. It also helps maintain an adequate clamping force when the engine is both cold and hot. TTY bolts are designed to provide a consistent clamping force when torqued to their yield point or just beyond. The challenge is that the torque does not increase very much, or at all, once the yield point is reached. So using a torque wrench by itself will not indicate the point at which the manufacturer wants the bolt tightened.

TTY bolts generally require a new torquing procedure called torque angle. In virtually all cases, TTY bolts cannot be reused. They have been stretched into their yield zone and would very likely fail if retorqued (**FIGURE 8-61**).

Torque angle is considered a more precise method to tighten TTY bolts. It is essentially a multistep process. Bolts are first torqued to a required moderate torque setting, in the required pattern. They are then further tightened one or more additional specified angles (torque angle) using an angle gauge. This provides further tightening, which tightens the bolt to, or just beyond, their yield point.

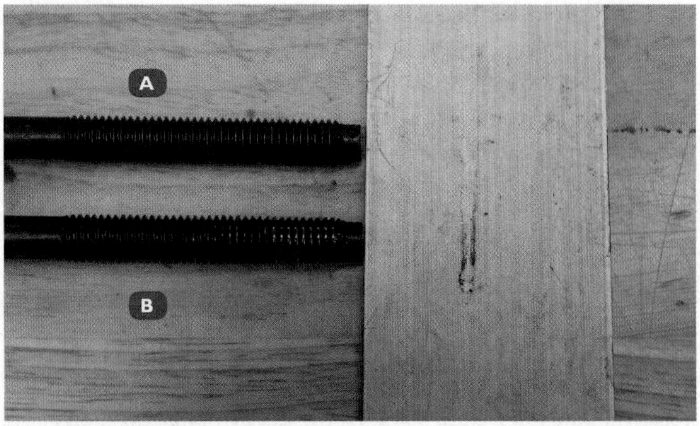

FIGURE 8-61 **A.** New TTY bolt. **B.** Used TTY bolt.

FIGURE 8-62 Left-hand lug nuts are installed on some vehicles.

In some cases, the manufacturer wants the bolts torqued in a particular sequence, then, detorqued in a particular sequence. Then, retorqued once again in a particular sequence. And finally tightened at an additional specified angle. So always check the manufacturer's specifications and procedure before torquing TTY bolts.

To use a torque angle gauge in conjunction with a torque wrench, follow the steps in **SKILL DRILL 8-1**.

▶ Repair Damaged Fastener Threads

LO 8-04 Repair damaged fastener threads.

Broken bolts can turn a routine job into a nightmare job by adding hours dealing with the broken bolt. In most cases, it is worth taking a bit of extra time doing whatever is needed to avoid breaking a bolt. One of the first things to do is to make sure that you are not dealing with a left-handed bolt (**FIGURE 8-62**). Left-hand bolts loosen by turning them clockwise. So trying to loosen them by turning them counterclockwise will actually tighten them, making it likely you will break them off.

Another cause of broken bolts is fasteners that are rusted or corroded in place. Although it is impossible to prevent

SKILL DRILL 8-1 Using a Torque Angle Gauge with a Torque Wrench

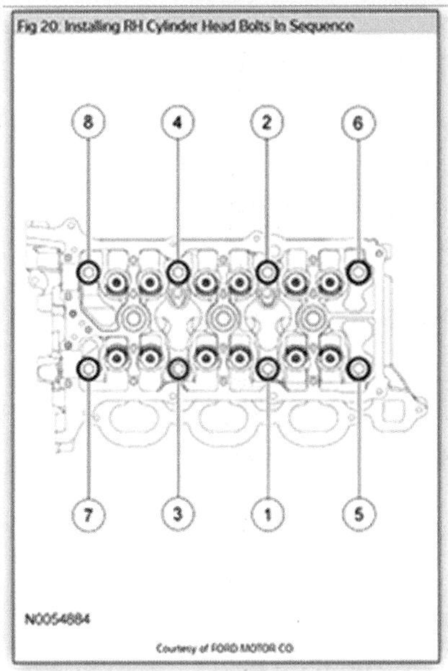

1. Check the specifications. Determine the correct torque value (ft-lb or N·m) and sequence for the bolts or fasteners you are using. Also, check the torque angle specifications for the bolt or fastener, and whether it involves one step or more than one step.

2. Tighten the bolt to the specified torque. If the component requires multiple bolts or fasteners, make sure to tighten them all to the same torque value in the sequence and steps that are specified by the manufacturer.

3. Install the torque angle gauge over the head of the bolt, and then put the torque wrench on top of the gauge and zero it, if necessary.

4. Turn the torque wrench the specified number of degrees indicated on the angle gauge.

5. If the component requires multiple bolts or fasteners, make sure to tighten them all to the same torque angle in the sequence that is specified by the manufacturer.

6. Some torquing procedures could call for four or more steps to complete the torquing process properly.

breakage of all bolts in this case, there are several things you can do to minimize the number of bolts you break. One of the first things to do is to use a good penetrating oil. Penetrating oil actually penetrates the rust between the nut and bolt. It softens the rust, making it easier to break the nut or bolt loose (**FIGURE 8-63**). Once the penetrating oil has been given time to soak into the threads of the fastener, you may attempt to loosen the fastener.

Note that you may want to try both loosening and tightening the bolt to help break it loose. It may be good to try using an impact wrench as well, because the hammering effect might assist in breaking it loose. Just try it a little at a time in each direction, and not with too much force.

If the fastener is still not loosening, you may need to try heating the component. Heat can help break up the rust or corrosion. The heat not only expands the metal components, it also expands the space between the nut and bolt. This also helps break the nut or bolt loose.

The bolt or nut can be heated up in two primary ways. The first is with an oxyacetylene torch (**FIGURE 8-64**). But

FIGURE 8-63 Penetrating oil used to break down the rust on rusty fasteners.

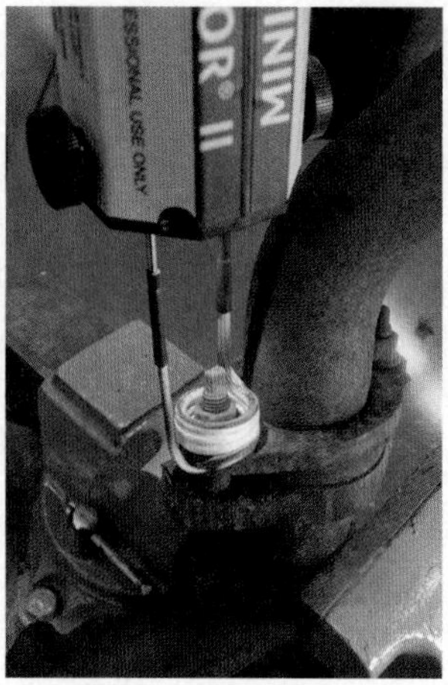
FIGURE 8-65 Heating a stuck nut with an induction heater.

FIGURE 8-64 An oxyacetylene torch being used to heat up a nut being removed.

FIGURE 8-66 Repaired thread.

be careful because the flame from the torch also burns everything in its path. When heating up a bolt on a vehicle, there are almost always other things in the way that could be affected by the heat. So the torch method risks damaging other parts.

The second way to heat up a bolt is with an inductive heater. It uses electrical induction to heat up any ferrous metal that is placed in the inductive coil (**FIGURE 8-65**). The induction heater is preferred by many technicians, as it does not have a flame and it only heats up what is in the induction coil. Most induction heaters come with several sizes of induction coils which makes them handy for most applications.

When heating a fastener to remove it, you usually heat it up to a dull orange color and then let it cool a bit before attempting to remove it. Once a heated fastener is removed, it must be replaced with a new nut or bolt. The heat weakens the fastener.

Thread Repair

Thread repair is used in situations where it is not feasible to replace a damaged thread. This may be because the damaged thread is located in a large expensive component, such as the engine block or cylinder head of a vehicle. Or, it could be because replacement parts are not available. The aim of thread repair is to restore the thread to a condition that restores the fastening integrity (**FIGURE 8-66**). It is most often performed on internal threads, such as in a housing, engine block, or cylinder head. But external threads, such as on a bolt, can sometimes be repaired. However, it is usually easier to replace a bolt if the threads become damaged, rather than repairing it.

Types of Thread Repair

Many different tools and methods can be used to repair a thread. The least invasive method is to reshape the threads. If the threads on a bolt are not too badly damaged, then a thread

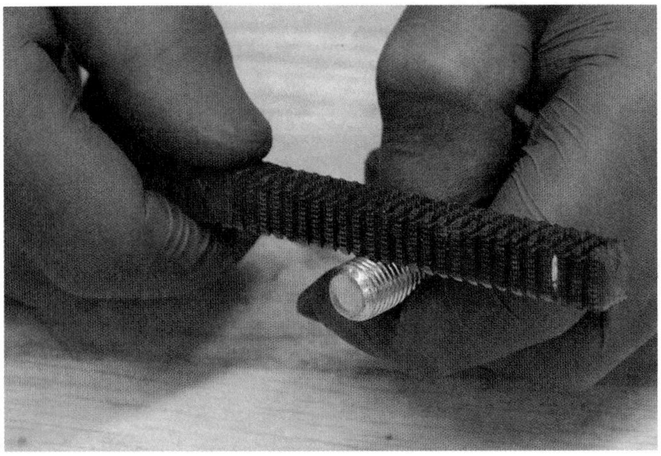

FIGURE 8-67 Threads being cleaned up with a thread file.

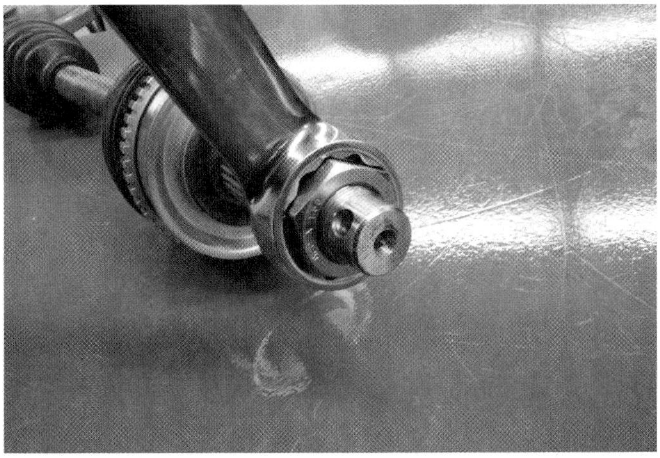

FIGURE 8-68 Thread-restoring tool being used.

file can be used to clean them up (**FIGURE 8-67**). Each thread file has eight different sets of file teeth that match various thread pitches. Select the set that matches the bolt you are working on, and file the bolt in line with the threads. The file removes any distorted metal from the threads. Only file until the bad spot is reshaped.

The thread-restoring tool looks like an ordinary tap and die set. But, instead of cutting the threads, it reshapes the damaged portion of the thread (**FIGURE 8-68**). The type that looks like a tap restores threads in internal holes. The type that looks like a die restores external threads on bolts and studs. Threads that have substantial damage require other methods of repair.

A common method for replacing damaged internal threads is a thread insert. A number of manufacturers make thread inserts, and they all work in a similar fashion. There are two main types of thread inserts: helical thread inserts and sleeve-type thread inserts (**FIGURE 8-69**). A helical thread insert replaces the damaged threads with new threads of the

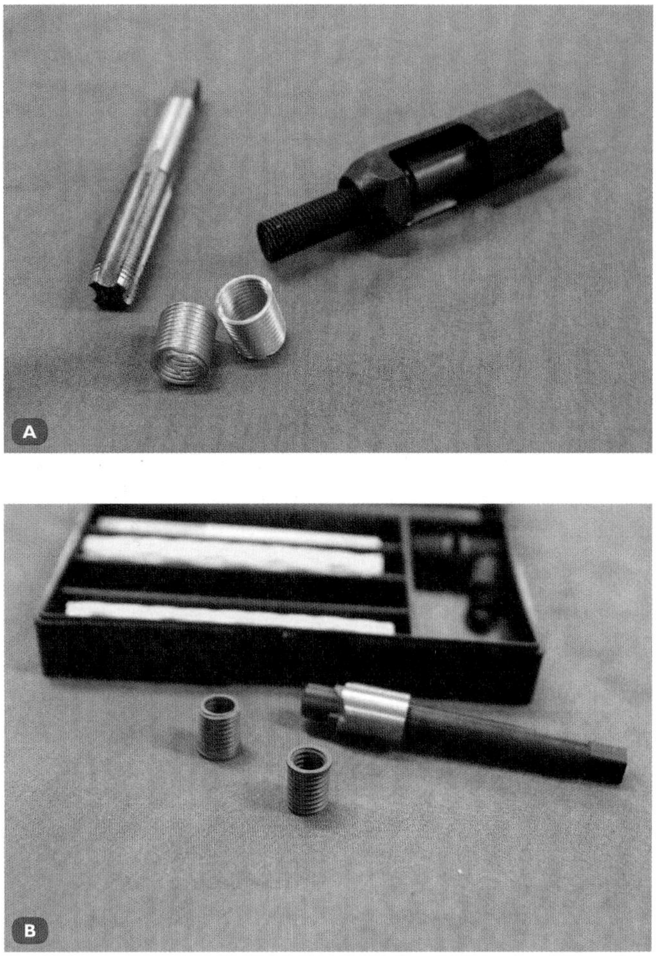

FIGURE 8-69 Thread inserts. **A.** Heli-Coil. **B.** Time-Sert.

same internal thread pitch as the original threads. The damaged threaded hole is drilled out larger and tapped with a special tap that matches the outside size of the helical insert. Then, the threaded coil insert is screwed into place. The insert provides a brand new internal thread that matches the original size. A common manufacturer for this type of insert is Heli-Coil.

The solid sleeve type of insert also replaces the damaged thread with the same size thread as the original. The difference between this type of insert compared to the helical insert is that this one is a slightly thicker insert. It resembles a sleeve or bushing, but with threads on both the inside and outside. It may also have a small flange at the top of the insert that prevents it from being installed too deeply in the hole. Like the previous insert, the damaged hole needs to be drilled out to the proper oversize. Then, a special tap is used to cut new threads that match the external thread on the insert. The insert is then installed in the hole and locked into place with either thread-locking compound or keys that get driven down into the base metal. A common manufacturer for this type of insert is Time-Sert.

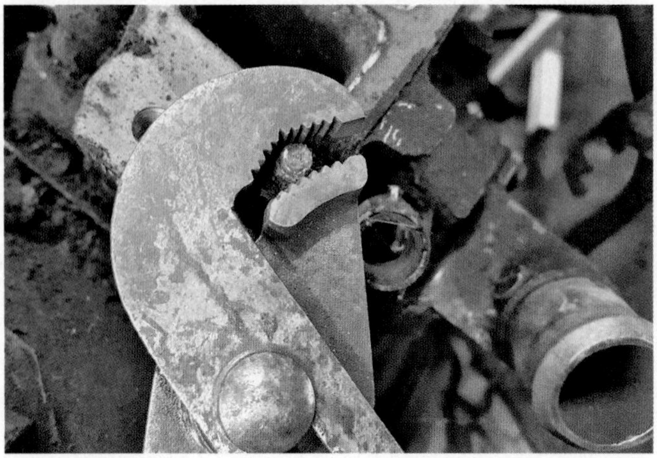

FIGURE 8-70 Using pliers to remove a broken bolt.

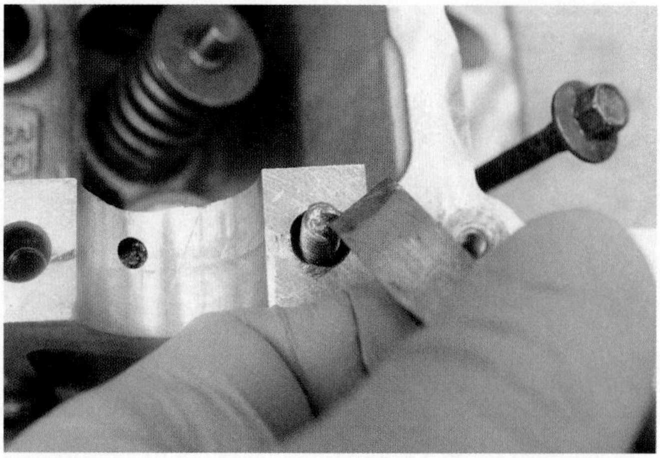

FIGURE 8-71 Removing a broken bolt with a center punch or cold chisel.

Self-tapping inserts are a type of insert that is screwed in, similar to a self-tapping screw. It makes its own threads and at the same time locks the insert into the base material. Bushing inserts are threaded inserts that are pressed into blind holes with an arbor press. These types of inserts can be inserted in any material. You must be aware of the material that you are inserting them in. Self-tapping inserts have internal threads, so bolts and screws can be threaded into them.

The process of repairing a thread should first start with attempting to remove the broken bolt without damaging the threads. If accomplished, then you will likely save time overall. You may also avoid the possibility of making the problem worse, such as breaking off an easy out in the broken bolt. If the bolt can't be removed without damaging the threads, then the use of a thread insert to repair an internal thread is probably needed.

To remove a broken bolt, inspect the site. If enough of the bolt is sticking out of the surface, then a pair of pliers or locking pliers may be enough to turn and remove the bolt. The author has had quite a bit of luck using a pair of special curved jaw Channellock pliers. The curved jaw tends to get a better grip on the broken-off bolt than regular straight jaw Channellock pliers (**FIGURE 8-70**). Using a penetrant or heat may help coax the bolt out.

What if the bolt is broken off flush with the surface? If the bolt is large enough in diameter, then you may be able to use a small center punch to turn the bolt. You do this by tapping on the outside diameter of the bolt, but in the reverse direction (**FIGURE 8-71**). If that doesn't work, or the bolt is too small in diameter, then try a screw or bolt extraction tool.

To use a screw extraction tool, use a center punch to mark the center of the bolt. This assists with centering the drill bit as it starts drilling. Select the correct size **screw extractor**, and drill the designated hole size in the center of the broken bolt to accommodate the extractor. Once the hole is drilled, insert the extractor and turn it counterclockwise (**FIGURE 8-72**). The flutes on the extractor should grab the inside of the bolt and hopefully enable you to back it out. Be careful not to exert too much force on the extractor if the bolt is extremely stuck in place. If the extractor breaks, it is almost impossible to remove it, as it is made of hardened steel that cannot be cut by most drill bits.

FIGURE 8-72 Removing a bolt with a screw extractor.

After removing the broken bolt, run a lubricated tap or thread-restoring tool of the correct size and thread pitch through the hole. This helps to clean up any rust or damage. If the screw extractor can't remove the broken bolt, then it will need to be drilled out. Once it is drilled out, check to see if the internal threads were damaged during the removal process. If so, the thread will have to be repaired with a thread insert, as described earlier. To conduct thread repair, follow the steps in **SKILL DRILL 8-2**.

SKILL DRILL 8-2 Conducting Thread Repair

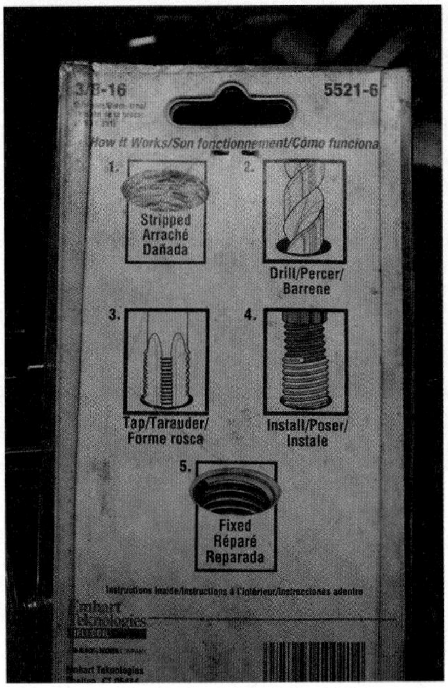

1. Always refer to the manufacturer's manual for specific operating instructions.

2. Inspect the condition of the threads, and determine the repair method.

3. Determine the type and size of the thread to be repaired. Thread pitch gauges and vernier calipers may be used to measure the thread.

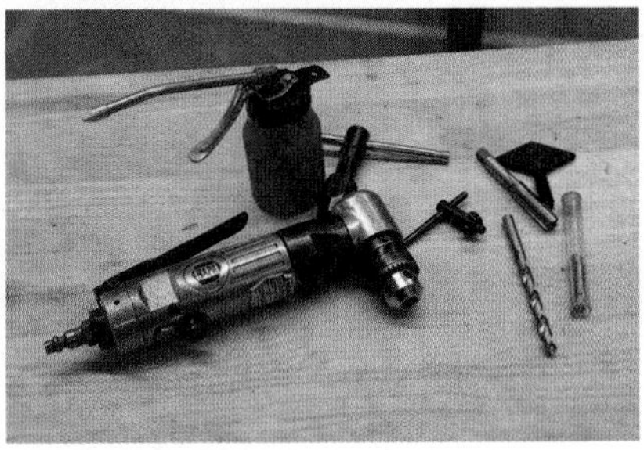

4. Prepare materials for conducting the repair: dies and taps or a drill bit and drill; cutting oil, if required; and inserts.

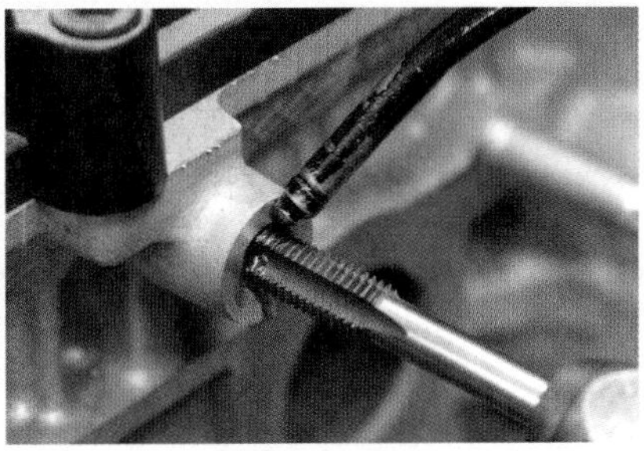

5. Select the correctly sized tap or die if conducting a minor repair. Run the die or tap through or over the thread; be sure to use cutting lubricant.

6. If using inserts, select the correctly sized insert, based on the original bolt size. Drill the damaged hole, ensuring the drill is in perfect alignment with the hole.

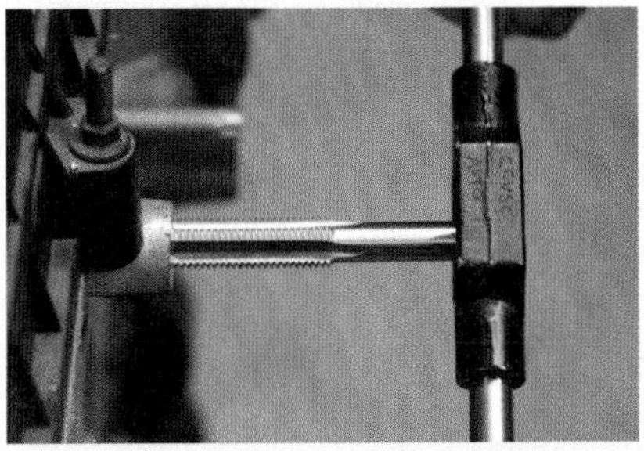

7. Cut the new thread, using the proper size tap. Make sure you use cutting lubricant if required.

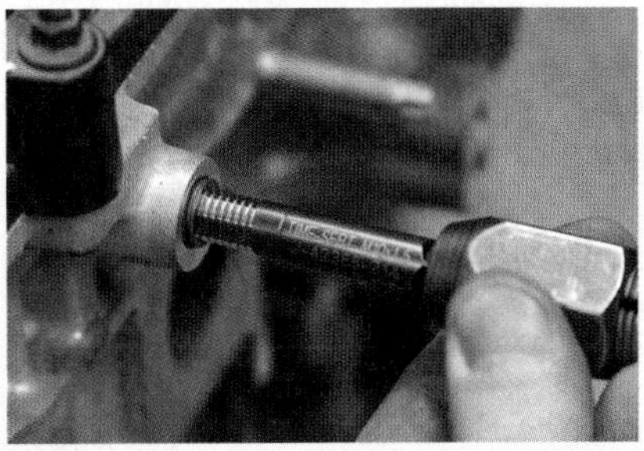

8. Using the insert-installing tool, install the insert by screwing it into the newly cut threads. Make sure the insert is secure or locked into the hole using the method specified by the manufacturer. Some inserts use a locking tab, whereas others use a liquid thread locker that hardens and holds the insert in place.

9. Test the insert to ensure it is secure and that the bolt will screw all the way in.

Wrap-Up

Ready for Review

- Threaded fasteners need to meet the standards set by The American Society for Testing and Materials, The International Organization for Standardization, or The Society of Automotive Engineers.
  - Standards include: sizing bolts, thread pitch, thread pitch gauge, grading of bolts, strength of bolts, tensile strength, shear strength, proof load, fatigue strength, toughness, ductility, and torsional strength.
- Non-threaded fasteners hold parts in place, but are not tightly clamped in place. Fasteners include snap rings, pins, and clips.
- Threaded fasteners are designed to be tightened to a specified torque depending on several factors. They should be checked and calibrated on a periodic basis with a proper calibration equipment.
- Heat can help break a nut or bolt loose. A thread-restoring tool reshapes the damaged portion of the thread. A common method for replacing damaged internal threads is a thread insert.
  - Helical thread inserts and sleeve-type thread inserts

Key Terms

bolt A type of threaded fastener with a thread on one end and a hexagonal head on the other.

elasticity The amount of stretch or give a material has.

fasteners Devices that securely hold items together, such as screws, cotter pins, rivets, and bolts.

nut A fastener with a hexagonal head and internal threads for screwing on bolts.

screw extractor A tool for removing broken screws or bolts.

square thread A thread type with square shoulders used to translate rotational to lateral movement.

stud A type of threaded fastener with a thread cut on each end rather than having a bolt head on one end.

tensile strength The amount of force required before a material deforms or breaks when being pulled apart.

thread pitch The coarseness or fineness of a thread as measured by either the threads per inch or the distance from the peak of one thread to the next. Metric fasteners are measured in millimeters.

thread repair A generic term to describe a number of processes that can be used to repair threads.

threaded fasteners Bolts, studs, and nuts designed to secure parts that are under various tension and sheer stresses.

torque Twisting force applied to a shaft that may or may not result in motion.

torque angle A method of tightening bolts or nuts based on angles of rotation.

torque specifications Supplied by manufacturers, who describe the amount of twisting force required for a fastener or provide a specification showing the twisting force from an engine crankshaft.

torque-to-yield (TTY) A method of tightening bolts close to their yield point or the point at which they will not return to their original length.

torque-to-yield (TTY) bolts Bolts that are tightened using the torque-to-yield method.

UNC (Unified National Coarse) Used to describe thread pitch.

Unified National Coarse Thread (UNC) Used in the standard bolt system. A bolt with fewer threads per inch for a given diameter.

Unified National Fine Used to describe thread pitch.

Unified National Fine Thread (UNF) Used in the standard bolt system. A bolt with more threads per inch for a given diameter.

yield point The point at which a bolt is stretched so much that it will not return to its original length when loosened; it is measured in pounds per square inch or kilopascals of bolt cross-section.

Review Questions

1. What are the three groups that establish the standards to establish fasteners specifications?
   a. ISO, ASE, ASTM
   b. ASTM, ISO, EIO
   c. SAE, ISO, ASTM
   d. SAE, EIO, ASE
2. How can a person distinguish a metric fastener?
   a. Grade number on the head
   b. Number of threads per inch
   c. Hash marks on the bolt head
   d. Color code on one side
3. What are the two main thread pitch designations used for SAE fasteners?
   a. SAE, UAE
   b. UNF, UNC
   c. UNC, ISO
   d. UNF, AST
4. Snap rings:
   a. Fit under nuts to spread out the clamping force
   b. Are used to retain plastic trim pieces in a vehicle
   c. Are threaded into place
   d. Fit inside a groove
5. When replacing a threaded fastener, where would you most likely find torque specifications?
   a. Service manual
   b. Owners manual
   c. Go by rule-of-thumb
   d. Another technician

6. When torqueing fasteners it is critical to:
   a. Torque it well past its yield point
   b. Lubricate the threads
   c. Always use new bolts
   d. Use the correct grade bolt
7. Which threaded fasteners cannot be reused?
   a. Torque-to-stretch
   b. Torque–to-yield
   c. Torque-to-specification
   d. Dry torque fasteners
8. All of the following are ways to avoid breaking a stuck bolt EXCEPT:
   a. Use penetrating oil and let it sit for a while
   b. Try both tightening and loosening the bolt
   c. Use a breaker bar and just keep increasing the torque
   d. Use an inductive heater to heat the bolt up
9. When using a thread, insert:
   a. A bolt of at least one size larger diameter will need to be used
   b. The hole will need to be drilled about twice as deep as the original
   c. A nut must be used to secure the thread insert in the hole
   d. The thread insert replaces the damaged internal threads
10. All of the following can be used to remove a broken bolt EXCEPT:
   a. Use a new bolt to push the old bolt out
   b. Use a screw extractor
   c. Use a hammer and small center punch
   d. Use curved jaw channel lock pliers

ASE Technician A/Technician B Style Questions

1. Technician A says that a thread insert can be used to repair damaged internal threads. Technician B says a thread file may be used to repair lightly damaged external threads. Who is correct?
   a. Technician A
   b. Technician B
   c. Both A and B
   d. Neither A nor B
2. Technician A says to use heat to help remove a broken fastener. Technician B says that broken bolts mean the part must be replaced. Who is correct?
   a. Technician A
   b. Technician B
   c. Both A and B
   d. Neither A nor B
3. Technician A says that a grade 5 bolt has five hash marks on the head. Technician B says that the diameter of a bolt is measured across the outside diameter of the threads. Who is correct?
   a. Technician A
   b. Technician B
   c. Both A and B
   d. Neither A nor B
4. Technician A says when heating a broken fastener, induction heating works best. Technician B says the oxy-acetylene torch may cause damage to other components. Who is correct?
   a. Technician A
   b. Technician B
   c. Both A and B
   d. Neither A nor B
5. Technician A says torque angle means that a fastener is torqued to a specified torque and then turned a certain number of degrees further. Technician B says torque-to-yield bolts can be reused because they are made to stretch. Who is correct?
   a. Technician A
   b. Technician B
   c. Both A and B
   d. Neither A nor B
6. Technician A says fasteners should always be lubricated when tightened. Technician B says lubricating the fastener will change the clamp load of the fastener. Who is correct?
   a. Technician A
   b. Technician B
   c. Both A and B
   d. Neither A nor B
7. Technician A says non-threaded fasteners are designed to firmly clamp parts together. Technician B says a cotter pin is designed to be reusable. Who is correct?
   a. Technician A
   b. Technician B
   c. Both A and B
   d. Neither A nor B
8. Technician A says it is essential to wear safety glasses when working with snap rings. Technician B says that roll pins are typically used to hold one part to another. Who is correct?
   a. Technician A
   b. Technician B
   c. Both A and B
   d. Neither A nor B
9. Technician A says that lock washers are usually considered to be reusable. Technician B says washers are graded the same as nuts and bolts. Who is correct?
   a. Technician A
   b. Technician B
   c. Both A and B
   d. Neither A nor B
10. Technician A says a lower grade bolt can be used in place of a higher grade bolt. Technician B says nuts used should be the same grade as the bolt. Who is correct?
   a. Technician A
   b. Technician B
   c. Both A and B
   d. Neither A nor B

CHAPTER 9

Vehicle Protection and Jack and Lift Safety

Learning Objectives

- **LO 9-01** Prepare vehicle for service, and return to customer.
- **LO 9-02** Operate jacks, engine hoists, and stands.
- **LO 9-03** Operate hoists and describe use of inspection pits.

ASE Education Foundation Tasks

See Appendix A to view the 2017 ASE Education Foundation Automobile Accreditation Task List Correlation Guide.

▶ Introduction

Whenever a shop conducts repairs, its professionalism is on display. Customers expect their vehicle to be treated with respect, repaired, and returned clean and free of any additional damage (**FIGURE 9-1**). A professional shop always takes precautions to do this. It also ensures that all work is conducted safely and efficiently.

Before you undertake a task, pause for a moment to identify good work practices. Make sure you prevent accidental damage to a customer's vehicle. For example, before starting work, protect the vehicle from accidental spills and scratches. Installing protective covers to seats, carpets, and steering wheels are one way to do this.

To prevent vehicle damage or personal injury, it is important to inspect lifting equipment before each use. This includes vehicle lifts, jacks, jack stands, engine hoists, slings, and chains. Also make sure they are well maintained. Some states require annual certification inspections of vehicle lifts to help ensure their safety.

▶ Preventing Vehicle Damage

LO 9-01 Prepare vehicle for service, and return to customer.

Vehicle Walk-Around

You want to do what you can to avoid customer dissatisfaction. It is good practice for the service advisor to perform a vehicle walk-around with the customer. During this time, notes are made of any existing damage or missing components on the vehicle. These should be noted on the check-in sheet or repair order and discussed with the customer (**FIGURE 9-2**). This helps prevent the customer from coming back to the shop and complaining that the vehicle was damaged in the shop.

Also, this is a good time for the service advisor to visually look the vehicle over for any needed maintenance or repairs. For example, worn tires and wiper blades. The service advisor should also check with the customer to make sure any valuables, medications, or firearms are not left in the vehicle. Alternatively, the shop may allow those items be stored properly, such as in the trunk. You, as the technician, should also be on the lookout for any damage, needed repairs, and valuables. If found, inform your supervisor or the service advisor. And always follow

FIGURE 9-1 Prior to returning the vehicle to the customer, make sure the vehicle is clean and free of any additional damages.

FIGURE 9-2 Vehicle walk-around with the customer.

You Are the Automotive Technician

A customer brings her 2014 BMW 325i into the auto dealership for routine service. During the vehicle walk-around, you notice a missing BMW emblem from the hood. The customer says it was there when she washed the car the past weekend. You also notice that the rear tires appear to be excessively worn. Using a tread-depth indicator, you demonstrate to the customer that the tires are worn below the legal limit. You escort her to the sales department to choose new tires and wait for her vehicle. Before moving the vehicle for service, you install seat covers, a steering wheel cover, and floor mats.

1. What type of hoist is best for rotating and changing vehicle tires?
2. Why is it important to check the safe working load (SWL) rating of the lifting equipment before using it?
3. A vehicle is about to be removed from the vehicle inspection pit; what are some safety precautions you must perform before it can be removed?
4. What are some steps you need to take before returning the vehicle to the customer?

company policy. Typically, the service advisor installs at least a floor cover during the walk-around. So make sure they are installed before getting into the vehicle.

▶ TECHNICIAN TIP

The preservice vehicle walk-around with the customer can be very valuable. Occasionally, a customer is unaware of damage to his or her vehicle that was sustained before dropping the vehicle off at the shop. For example, maybe the customer stopped by the store to pick up a couple of things on the way to the shop. While in the store, another car sideswiped the passenger side of the vehicle. Because it was on the passenger side, the driver may not have noticed it. However, when picking up the car from the shop and looking it over, the customer sees that the passenger side has been sideswiped. They immediately blame the shop, saying, "My car wasn't that way when I dropped it off." A good preservice walk-around can prevent this kind of misunderstanding.

Fender, Seat, Carpet, and Steering Wheel Covers

Working on vehicles requires the use of tools, equipment, and chemicals. It also involves the physical process of testing and replacing parts. All of these activities have the potential to damage the vehicle if proper precautions are not taken. For example, a dropped screwdriver or wrench may scratch the paint if fender covers are not used.

The vehicle and its components tend to be more prone to accidental damage in areas that have higher levels of service activity. Prime areas for damage are the engine bay and passenger compartment. The engine bay is the center of activity where many tools are used, batteries serviced, and oil changed. The chance of an accident in the engine bay is high if caution and care are not applied.

Protective equipment for vehicles is used to prevent damage to these areas while servicing the vehicle (**FIGURE 9-3**). Fender covers are a protective layer used to cover the fenders when work is conducted around the engine bay. They are usually made from either a durable fabric blanket or a flexible energy-absorbing foam compound about a 0.250" (6.35 mm) thick. They are designed to fit across the top and down the side of the fender.

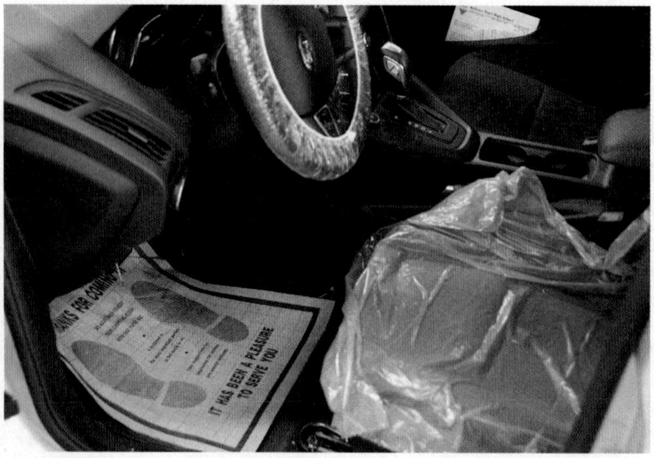

FIGURE 9-3 Always use seat and steering wheel covers, floor mats, and fender covers when working on a vehicle.

▶ TECHNICIAN TIP

Many of today's vehicles use very thin sheet metal in their fenders. This makes them easily dented if you lean against them or put your body weight against them. Fender covers *cannot* protect the vehicle from this kind of damage.

Seats, carpets, and steering wheels are made from materials that are sensitive to marks and damage from grease, oil, and dirt. Use protective covers to prevent accidental damage to these surfaces. They are made from materials that provide a barrier to oil, grease, and dirt. Examples include waxed or plastic laminated carpet protectors, plastic or fabric seat protectors, and plastic steering wheel covers.

▶ TECHNICIAN TIP

Covers for fenders, carpet, seat, and the steering wheel should be the first things on and the last things removed when working on vehicles.

▶ TECHNICIAN TIP

Never place tools in your back pocket. If you put a screwdriver in your back pocket and then sit in the driver's seat, the screwdriver would likely poke a hole in the seat upholstery, even if you used a seat cover.

Corrosives and Greases

Corrosives and greases have the potential to cause damage to customers' vehicles. Corrosives, such as battery acids, can corrode metals, etch painted surfaces, and eat holes in fabrics. Greases, such as wheel bearing grease, can cause staining of fabric and leather materials. Corrosives and greases can also transfer from contaminated surfaces of the vehicle onto the customer's clothing. This results in ruining their clothing as well. Be careful to prevent corrosives and greases from coming into contact with areas of the vehicle that are not designed for them. This means keeping your work area clean, including your hands and uniform.

Because your hands become dirty while working, it is critical that you learn not to put your hands on clean surfaces (**FIGURE 9-4**). Novice technicians have a bad habit of putting their hands on painted surfaces while working. This transfers grease and oil to the painted surface. Removing dirty fingerprints can be time-consuming, especially on oxidized paint surfaces.

▶ TECHNICIAN TIP

Brake fluid is another liquid that can damage a vehicle. Most brake fluids soften paint quickly if allowed to get on the surface. Brake fluid on your hands can easily get onto painted surfaces if you are not careful. If brake fluid gets on a painted surface, immediately remove it with a water-soaked, clean rag, but don't scrub it as the paint may wipe off.

FIGURE 9-4 Putting your hands on painted surfaces is a habit that needs to be stopped.

FIGURE 9-5 Special absorbent materials in granular form can be used to absorb some liquid spills, such as engine oil.

Also, be diligent to prevent spills. If they do occur, be sure to clean them up thoroughly using appropriate methods. These can usually be found in the safety data sheets (SDS) for each material. Damage caused by corrosive agents may not be immediately visible and may lead to other problems. Examples include rust or corrosion, if the corrosive agents are not cleaned up and neutralized immediately.

If you are handling corrosive materials and believe they have spilled, clean the area immediately, and use a neutralizer. For example, battery electrolyte contains sulfuric acid. This material can be cleaned up by neutralizing it with an alkaline such as common baking soda and flushing the area with fresh water. Special absorbent materials in granular form can also be used to absorb liquid spills such as engine oil (**FIGURE 9-5**). Once the engine oil is absorbed, the granules can be swept up for disposal in an environmentally safe way.

Greases and liquids can cause stains at any time, so be sure to work as cleanly as possible, and use protective covers to prevent damage. If a liquid spill does occur, clean up any excess with an absorbent material. Then, use a cleaner suitable for the type of spill and the material being cleaned. For example, use upholstery cleaner for spills on vehicle seats, or carpet cleaner for carpets.

Using Protective Covers

Proper Use of Fender Covers, Seat Covers, Floor Mats, and Steering Wheel Covers

As the purpose of protective covers is to prevent damage of the vehicle, you should always inspect them prior to use. Ensure that fender covers and floor mats are clean on both sides. This means that they do not have any metal or hard objects stuck to them. Ensure that they fit securely and provide adequate protection.

To properly apply fender covers, seat covers, floor mats, and steering wheel covers, follow the steps in **SKILL DRILL 9-1**.

SKILL DRILL 9-1 Applying Fender Covers, Seat Covers, Floor Mats, and Steering Wheel Covers

1. Select appropriate protective covers for the vehicle and type of repair. Inspect the covers for rocks, metal, or fluids that would damage the vehicle.

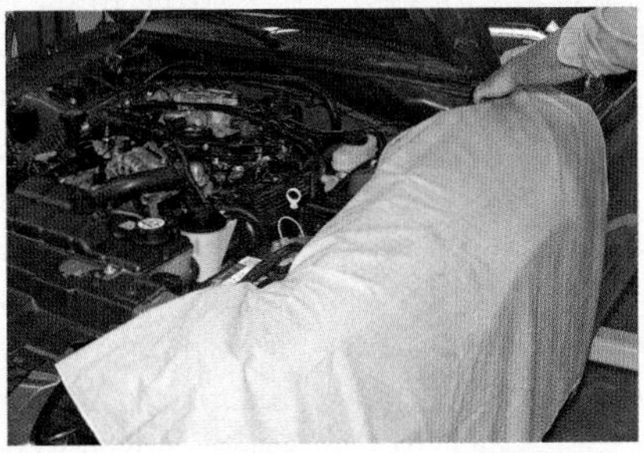

2. Position the fender covers so they provide adequate protection. Ensure that fender covers stay in position, providing protection while the vehicle is in the shop.

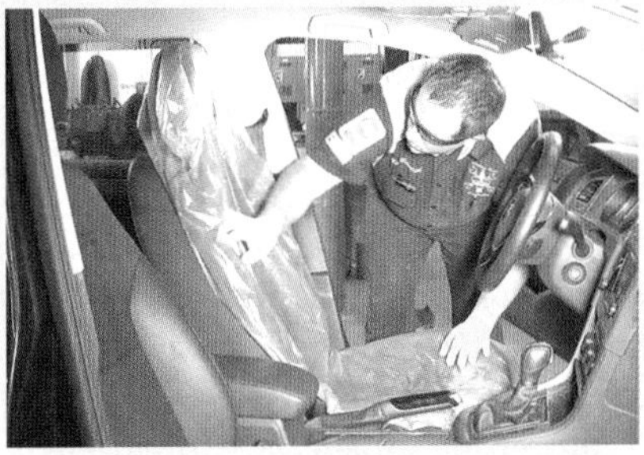

3. Position the seat cover so it covers both the back and bottom of the seat. Make sure it hangs over all of the edges and will stay in place.

SKILL DRILL 9-1 Applying Fender Covers, Seat Covers, Floor Mats, and Steering Wheel Covers (Continued)

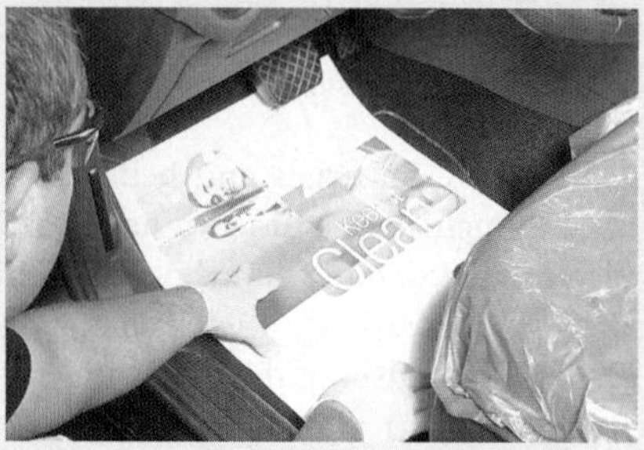

4. Install the floor mat on the floor so that it covers as much of the floor as possible. Also, make sure it does not interfere with the throttle, brake, or clutch pedals.

5. Install the steering wheel cover so that it covers as much of the steering wheel as possible without interfering with any controls.

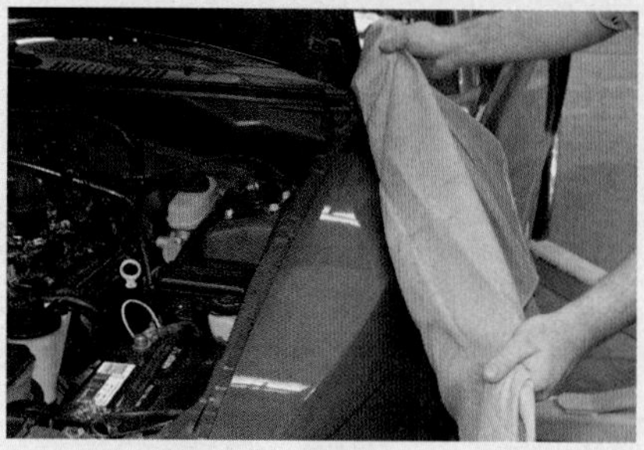

6. Remove all protective covers prior to customer pickup.

▶ TECHNICIAN TIP

Some fender covers are made with a magnetic strip in them, which is designed to help hold the fender cover in place. Unfortunately, the magnet attracts metal particles, which can be held between the cover and the fender. The particles can then scratch the paint. Always check these types of covers very thoroughly for metal particles.

Preparation for Customer Pickup

An important part of the customer's experience with your shop is the customer's impression of their vehicle once they get it back. Remember that they have just paid the bill, which can be quite large. They might not be in a good overall mood. At the same time, they are going to expect that not only is their vehicle fixed properly, but that it is in at least as good of shape as when they left it at the shop.

Assuming that the repairs were made correctly, there are other things that could lead to an unsatisfied customer. Examples include greasy fingerprints or scratches on the paint, greasy stains on the upholstery, or even the radio station being changed. This means that you play a critical role in the level of satisfaction that the customer experiences. It is also why you need to actively prepare the vehicle for delivery back to the customer.

One of the first things to do is check to make sure that all tools and old parts have been removed from the vehicle and stored properly. Make sure all vehicle protection covers are removed prior to releasing the vehicle to the customer. After removing the vehicle protection covers, visually inspect the vehicle. Verify that no damage to the vehicle has occurred during repair. If damage is discovered, alert your supervisor and get instructions on what steps need to be taken to solve this issue.

Check the vehicle for cleanliness by ensuring that all trash has been removed and no oil or grease marks are on the vehicle. Check that all windows are crystal clear. The dashboard, knobs, steering wheel, and center console should be spotless and clean. The floor mats should have no grime. There should be no fingerprints on the door latches, fenders, or the backs of the mirrors (**FIGURE 9-6**).

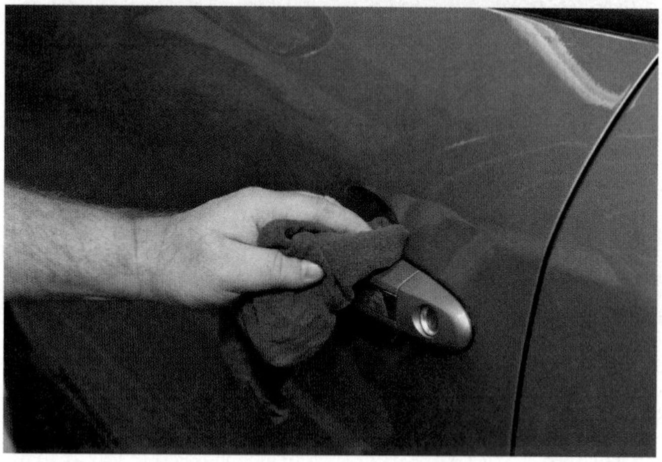

FIGURE 9-6 Preparing a vehicle for a customer by removing any finger prints.

If cleaning is needed, wear appropriate personal protective gear. This includes safety glasses and gloves when working with cleaning materials. Some shops include running the vehicle through a car wash before returning it to the customer. This helps to ensure customer satisfaction.

Verify that the radio, seats, and mirrors are set the same as when the vehicle came in. If any of the customer's belongings were moved during the repair, move them back into their original places. Take a look to make sure that all of the vehicle parts you removed to perform the repair have been reinstalled. Some common items that technicians forget to reinstall are the wheel covers, tire valve caps, engine trim pieces, and under-vehicle skid plates. A good habit to form is prior to removing the vehicle from your stall, look around your tool box for any of these types of items.

SKILL DRILL 9-2 Preparing to Return Vehicle to Customer

1. Check for any forgotten tools or parts.

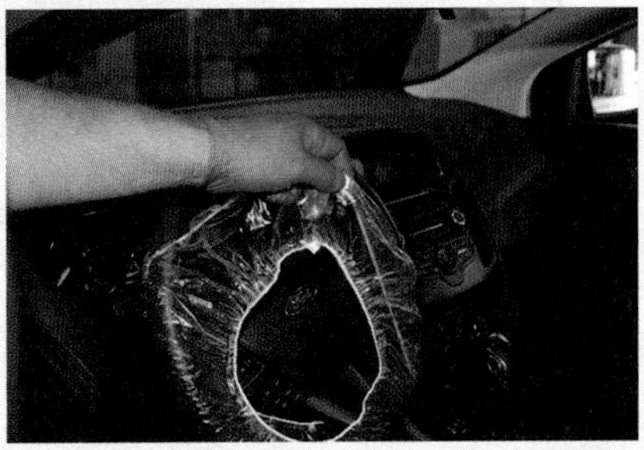

2. Remove fender, seat, and steering wheel covers.

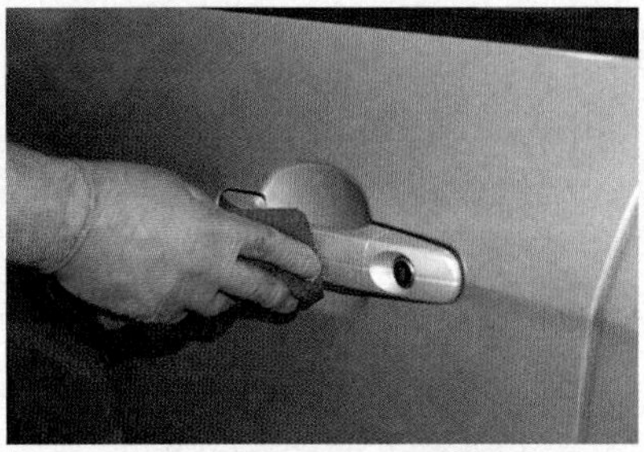

3. Clean any fingerprints and other grime from the vehicle.

4. Reset the radio, seats, and mirrors to their original settings.

5. Make sure that all of the vehicle parts you removed have been reinstalled.

6. Verify that all warning lights are off.

To ensure that the vehicle is prepared to return to the customer per school/company policy, follow the steps in **SKILL DRILL 9-2**.

Preventing Mechanical Damage

The possibility for accidental mechanical damage is always present when a vehicle is in the shop. It can occur easily; for example, an incorrectly lifted vehicle may damage the suspension or body. Another example is running into another vehicle while bringing a vehicle into the tight bays of the shop. The best way to minimize accidental mechanical damage is to:

- think carefully about the task or work you are performing
- create a plan that avoids having an accident
- follow shop and manufacturer procedures
- use tools and equipment correctly and
- seek advice if you are not sure

For example, if you need to raise a vehicle that is equipped with running boards, it is possible that the hoist could contact the running boards and damage them. To avoid damage, you should proceed more carefully than normal. Stop and look for potential problems regularly. If unsure, ask a more experienced technician to assist you. Let's look at hazards when moving vehicles.

Moving and Road Testing Vehicles

Vehicles often need to be road tested or simply moved from one place to another in, or out of, the shop. Accidents are likely to occur at this time, especially if the driver involved is inexperienced or unqualified. Only authorized, fully trained, and licensed drivers should be allowed to move vehicles. Only the most skilled, responsible, and experienced drivers available should be allowed to test-drive performance vehicles.

Have someone outside the vehicle supervise and guide any vehicle movement inside the shop (commonly called a spotter). This is especially important in restricted spaces, when backing up, and when nearing blind corners. Partially roll the driver's side window down so that you can hear the person directing you (**FIGURE 9-7**).

When moving a vehicle in the shop, and in tight parking lots, you should drive no faster than a normal walking speed. When you get close to an object, such as another vehicle, hoist, or person, you should slow down to a crawl. The person directing you while maneuvering the vehicle should use hand signals and their voice to guide you. Make sure you understand what the hand and voice signals mean by discussing them before moving the vehicle.

If you are the person guiding the driver, use large hand signals, and make sure you stand where the driver can clearly see you. But this should NOT be directly in front of or behind the vehicle (**FIGURE 9-8**). Also, use a loud voice so that the driver will be able to hear you from inside the vehicle. Make sure you

FIGURE 9-7 Partially roll down the driver's side window so that you can hear your spotter.

FIGURE 9-8 If you are the person guiding the driver, use large hand signals and loud voice commands. Also, make sure you stand where the driver can clearly see you.

give the signal to stop in plenty of time. Realize that it takes a second or two for the driver to respond. So stop them well short of running into something.

▶ TECHNICIAN TIP

Make sure customers are aware of your rules for moving cars. Keep the keys for all vehicles secure and away from the vehicles when not in use.

SAFETY TIP

Vehicles do not always run properly when they are brought in for service, which can lead to an accident. For example, a car with a hesitation problem will not move until the throttle is pressed down farther than normal. When the car does move, it lurches forward. In the tight spaces of a shop, this could cause the vehicle to run into something or someone. You have two options in this situation. The safest option is to push the vehicle instead of driving it. The second is a bit controversial—two-footing the pedals. Two-footing means that you put your left foot on the brake pedal, with the brakes lightly applied, and your right foot on the throttle. Thus, as you are driving, you can use the brakes to keep the vehicle moving very slow. This method needs practice and goes against most driver training. But when the engine is not running properly, the standard method of using the right foot for both the throttle and brake puts the vehicle at a severe risk of an accident.

FIGURE 9-9 Some examples of lifting equipment are vehicle lifts, floor jacks, jack stands, engine and component hoists, chains, slings, and shackles.

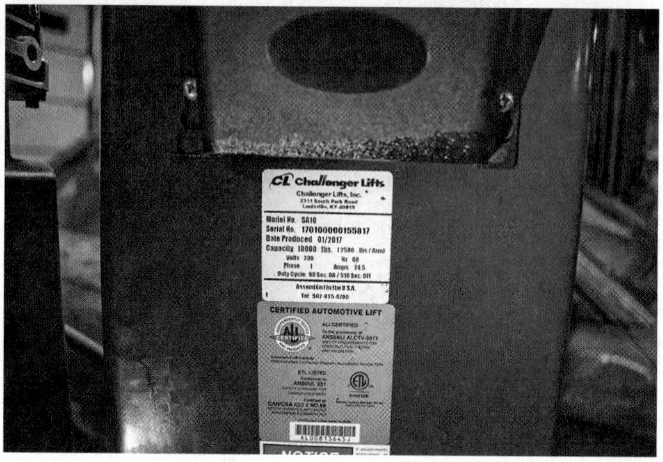

FIGURE 9-10 Typical manufacturer's certificate listing the lift capacity and serial number.

▶ Lifting Equipment

LO 9-02 Operate jacks, engine hoists, and stands.

Many different types of lifting equipment may be used in a shop. Some examples include vehicle lifts, floor jacks, jack stands, engine and component hoists, chains, slings, and shackles (**FIGURE 9-9**). Lifting equipment is designed to lift and securely hold loads.

Each piece of lifting equipment is designed for a specific purpose and has an operating capacity. The operating capacity is listed on the manufacturer's certificate located on the lifting equipment (**FIGURE 9-10**). The capacity is usually expressed

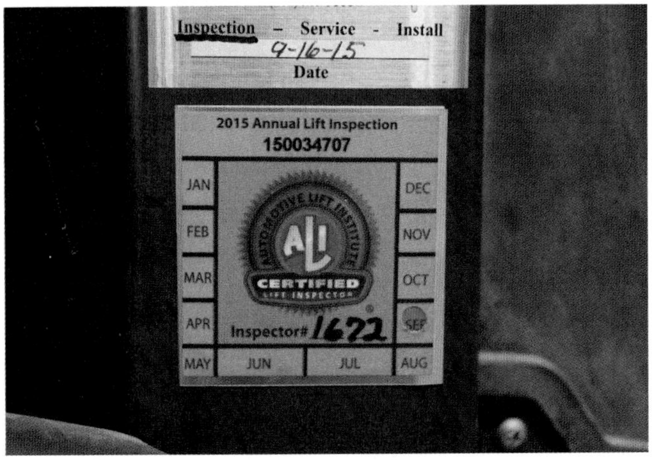

FIGURE 9-11 An annual vehicle lift inspection certificate.

FIGURE 9-12 The operator should inspect the lifting equipment daily for damage.

as the **safe working load (SWL)**. For example, if the SWL is 2,000 lb (907 kg), the equipment can safely lift up to 2,000 lb, or 1 ton. When using lifting equipment, never exceed its capacity, and always maintain some reserve capacity as an extra safety margin. Using lifting equipment incorrectly may lead to equipment failure. This can cause serious injury or death, and damage to the vehicle and surrounding equipment.

SAFETY TIP

When multiple pieces of lifting equipment are used, the SWL is limited to the lowest rated piece of equipment. For example, if a chain with an SWL of 2 tons is used with a 5-ton SWL shackle and a 3-ton SWL engine hoist, then the maximum amount of weight that can be safely lifted is 2 tons.

Inspection of Lifting Equipment

Lifting equipment should be periodically checked and tested to make sure it is safe. The testing should be recorded for each piece of lifting equipment in accordance with local regulations. And the equipment should be tagged with its inspection date. Inspections should identify any damage, such as cracks, dents, cuts, and abrasions. This may prevent the lifting equipment from performing as it is designed. In case of any damage found, it must be repaired before use. Refer to the operator's manual to determine how often maintenance inspections are recommended. The time frame is usually every 12 months in the case of hoists and lifts. It may be longer for items of lifting equipment such as chains and slings. Always check local regulations to determine the requirements for periodic testing of lifting equipment.

In some countries, lifting equipment is subject to periodic statutory inspection and certification. If this is the case, the **inspection certificate** should be attached to or displayed near the lifting equipment (**FIGURE 9-11**). Before using a piece of lifting equipment, make sure the most recent inspection recorded is within the prescribed time limit. If it is not, the test certificate has expired, and you should notify your supervisor.

Lifting equipment should also be visually inspected on a daily basis by the operator. Make sure everything appears to be in good working order. You should be aware of the condition of hoses, cables, latches, stops, pins, and controls (**FIGURE 9-12**). If anything out of the ordinary is found, report it to your supervisor.

Jacks and Jack Stands

Jacks and jack stands are always used together. Jacks are used to lift objects. And jack stands are used to support the raised object. Always support a vehicle with a jack stand when it is raised off the ground, even for a short task. Many people have been killed by going under a vehicle that was supported only by a jack when it slipped and fell on them. We will explore the safe use of jacks and jack stands in the following sections.

Jacks

A **vehicle jack** is a lifting tool used to raise part of a vehicle from the ground. It can also be used to raise heavy components into position. The vehicle's emergency jack can be used to raise and support the vehicle while changing a wheel on the side of the road. Vehicle jacks must not be used to support the weight of the vehicle during any task that requires you to get underneath any part of the vehicle. For those shop tasks, a vehicle jack should be used only to raise the vehicle. It can then be lowered onto suitably rated and carefully positioned jack stands (**FIGURE 9-13**).

There are three main types of mechanisms that provide the lifting action for vehicle jacks: **hydraulic jacks**, **pneumatic jacks**, and **mechanical jacks**. Hydraulic and pneumatic jacks are the most common types of vehicle jacks. They can be mounted on slides or on a wheeled trolley. In hydraulic jacks, pressurized oil acts on a piston to provide the lifting action. In pneumatic jacks, compressed air acts on a piston to lift the vehicle. In mechanical jacks, a screw or gears provide the mechanical leverage required for lifting.

Different jacks are available for different purposes (**FIGURE 9-14**):

- Floor jacks are a common type of hydraulic jack that is mounted on four wheels. Two of which swivel to provide a steering mechanism. The floor jack has a long handle that is used both to operate the jacking mechanism and to move and position the jack. Floor jacks have a low profile, making them suitable to position under vehicles.
- Sliding bridge jacks are usually fitted in pairs to four-post lifts. They act as an accessory to lift the vehicle off the drive-on lift runways. They are operated by a hydraulic mechanism or compressed air. Since they sit between the runways, they are used to lift one end of the vehicle. Thus, it makes it more convenient to work on wheels and brakes.
- Transmission jacks are specialized jacks for lifting and lowering transmissions during removal and installation. Transmission jacks are usually mounted on a trolley with wheels. They have a large flat plate area on which the transmission can be strapped or chained securely. They are usually operated by a hydraulic mechanism but can also be powered by compressed air.

FIGURE 9-13 A vehicle jack should be used only to raise the vehicle so that it can then be lowered onto suitably rated and carefully positioned stable jack stands.

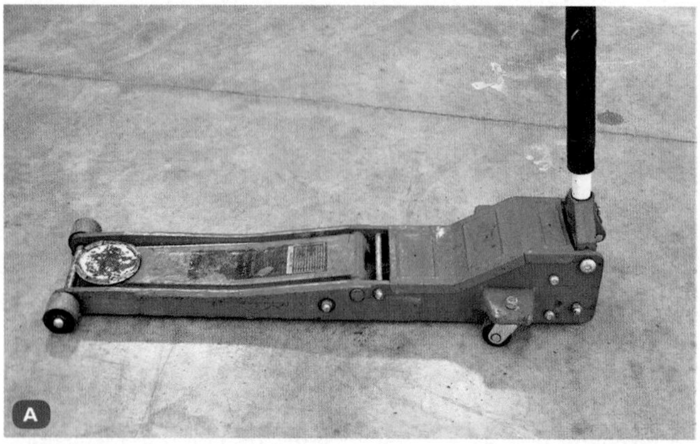

FIGURE 9-14 Jacks. **A.** Floor jack.

FIGURE 9-14 Jacks. **B.** Sliding bridge jack. **C.** Transmission jack. **D.** High-lift (or farm) jacks. **E.** Air jack.

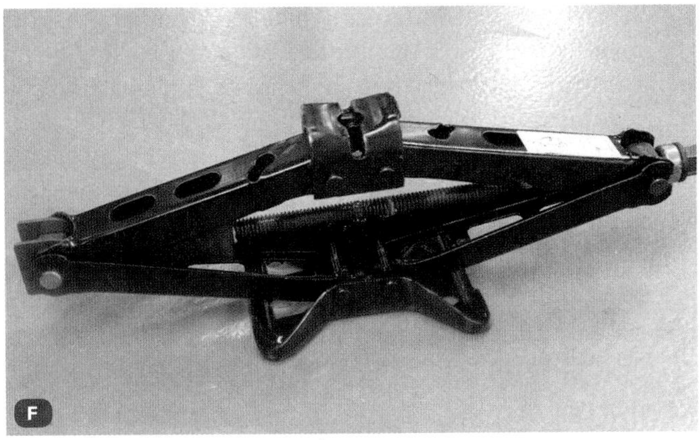

FIGURE 9-14 Jacks. **F.** Scissor jack. **G.** Bottle jack.

- High-lift (or farm) jacks are a versatile type of jack designed to lift, winch, clamp, pull, and push. They have a mechanical mechanism and are designed to provide high lift capability—for example, 36" (0.91 m) or more. Because of their high lift capability, they are often used on farms or on four-wheel drive vehicles.
- Air jacks use compressed air to either operate a large ram or inflate an expandable air bag to lift the vehicle. Often the air jack is fitted to a moveable platform with a long handle. Air jacks are used to lift vehicles as an alternative to floor jacks. Because a compressed air supply is required, air jacks are usually used in the shop and not for mobile operations.
- Scissor jacks are one of the most common types of jacks provided as part of the vehicle tool kit. They are used to jack a vehicle to lift one wheel at a time. They employ a scissor action that is controlled by turning a long horizontal screw that acts on levers to raise or lower the scissor jack.
- Bottle jacks are portable and usually have either a mechanical screw or a hydraulic ram mechanism. It rises vertically from the center of the jack as the handle is operated. They are relatively inexpensive and may be provided with vehicles for the purpose of changing flat tires.

SAFETY TIP

All vehicle jacks must always be of the correct capacity and used in accordance with the manufacturer's instructions. They should be inspected on a regular basis to ensure that they are in safe working order.

FIGURE 9-15 Jack stands normally come in matched pairs and should always be used as a pair.

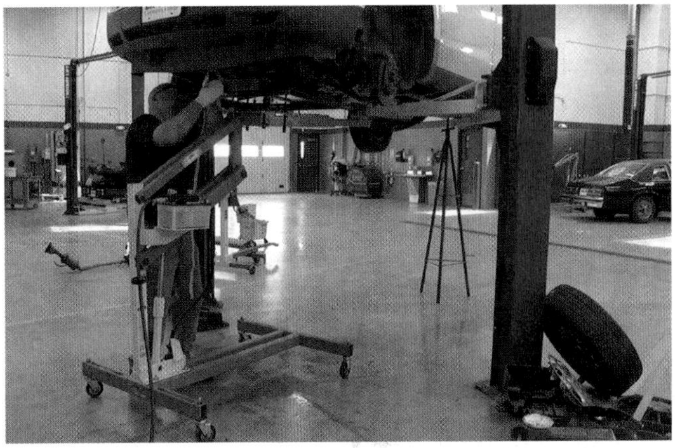

FIGURE 9-16 Tall jack stands being used to support the vehicle on a lift while a heavy component is being removed.

Jack Stands

Jack stands (axle stands) are adjustable supports that are used with vehicle jacks. They are designed to support the weight of the vehicle after the vehicle has been raised. Jack stands are mechanical devices. They mechanically lock in place at the height selected. This makes them very dependable if they are rated strong enough for the load they are holding. To be safe they must be used properly.

Lifting devices are also lowering devices, so it is unsafe to work underneath a vehicle that is supported only by a vehicle jack. Jacks can give way or be accidentally lowered. Jack stands normally come in matched pairs and should always be used as a pair (**FIGURE 9-15**). Jack stands are load rated and should only be used for loads less than the rating indicated on the jack stand.

Some shops have tall jack stands that are used along with a vehicle lift. They are much taller than standard jack stands. Tall jack stands are used to stabilize a vehicle when it is up on a lift. This is especially true if a heavy component, such as a transaxle, is being removed or installed (**FIGURE 9-16**). Never try to lower the vehicle with the tall jack stands still in place. Doing so can cause the vehicle to slip off the lift.

Using Lifting Equipment

Identify Jack Points

If a vehicle is lifted from, or supported on, the wrong places, it is likely to damage components. It can even cause the vehicle to fall. Both situations are costly and dangerous. Also, with the push for lighter weight vehicles, components that were once safe to lift from are no longer safe. As an example, we used to place a jack under the differential on rear wheel drive vehicles, and lift the rear of the vehicle by that. Some manufacturers warn that this can damage the differential housing. This means that before a vehicle can be lifted, you need to research the manufacturer's specified lift points for the vehicle you are lifting. These specifications can be found on a vehicle lift chart, or in the service information (**FIGURE 9-17**).

Once you identify the lift points in the vehicle lift chart, you will need to locate those points on the vehicle (**FIGURE 9-18**).

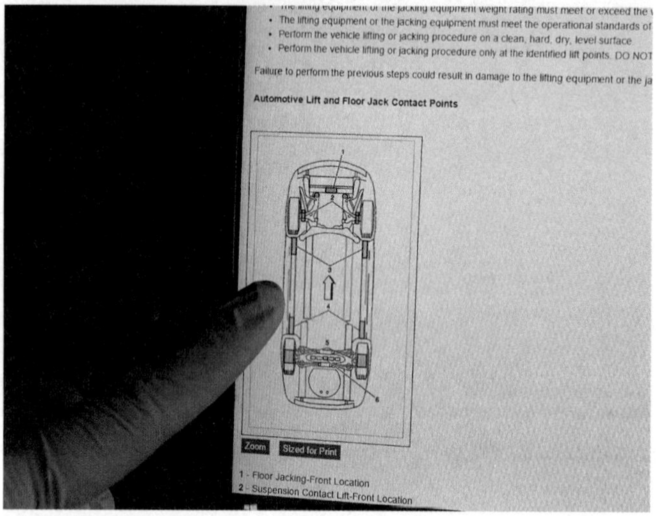

FIGURE 9-17 Use a vehicle lift chart or the service information to identify the lift points for the vehicle you are lifting, and the type of lift you are using.

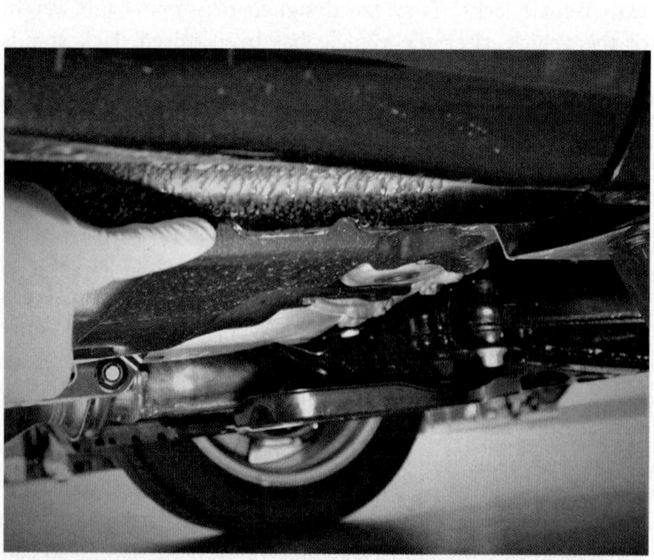

FIGURE 9-18 Common lift points on a vehicle.

Newer vehicles sometimes use symbols or words to identify the points. Others use a reinforced section of metal on the lift point. At times, it can be confusing to locate the actual lift point on the vehicle due to lack of detail on the lift chart. No matter what, the lift point needs to be strong enough to support the vehicle. So always evaluate the position you believe to be the lift point. If it is thin sheet metal, or a cast part that may not be strong enough, stop and ask for help. It is far better to get a second opinion than to have a vehicle fall, or become damaged. Here are a few points that you should never jack or lift the vehicle from:

- Oil pan
- Transmission pan
- Gas tank
- Exhaust pipes
- Axle half shaft
- Fender or door

Using Vehicle Jacks and Stands

The capacity of vehicle jack you use is determined by the weight of the vehicle you want to lift. Most shops have a vehicle jack that has a lifting capacity of about 2.5 tons (2,500 kg). If the end of the vehicle is heavier than that, or if the vehicle is loaded, you need to use a vehicle jack with a larger lifting capacity.

Make sure the jack stands are also rated to support the weight you are lifting. They must be in good condition before you use them to support the vehicle. If they are cracked or bent, they will not support the vehicle safely. Always use matched pairs of jack stands that are in good shape.

Never support a vehicle on anything other than jack stands. Do not use wood or steel blocks to support the vehicle. They may slide or split under the weight of the vehicle. Do not use bricks or concrete blocks to support the vehicle. They may crumble under its weight.

Jack stands must be used on solid, hard surfaces that are level. Using them on soft surfaces like dirt or even asphalt can cause the jack stand to sink in and tip over. Unlevel surfaces can cause the vehicle to roll which could cause it to slide off the jack or stands, or tip them over. When used properly, jack stands provide a stable support for a raised vehicle that is safer than the jack. This is because the vehicle cannot be accidentally lowered while the jack stands are in place.

When placing the jack or jack stands, never place them under fuel lines, brake lines, or wiring harnesses. The jacks and jack stands must be placed on the manufacturer's specified lift points. If the vehicle is heavily rusted, make sure the lift points are solid enough to support the vehicle. If in doubt, ask your supervisor.

Always grip jack stands by the sides to move them. Never grip them by the top or the bottom to move them, as the center post can slip and pinch or injure you. Check that the base of the stand is flat on the ground before lowering the vehicle onto it. Otherwise, the stand might tip over, causing the vehicle to slip off. When positioned correctly, the vehicle can be lowered onto the jack stands, and the vehicle jack can be moved out of the way. To lower a vehicle that is on jack stands, it must be raised again so that the jack stands can be removed. To lift and secure a vehicle with a floor jack and jack stands, follow the steps in **SKILL DRILL 9-3**.

SAFETY TIP

Be very careful to not climb under the vehicle when placing or removing the jack stands. This can be difficult to do on some vehicles. Use your arms to place the jack stands while keeping your body out from under the vehicle.

SKILL DRILL 9-3 Lifting and Securing a Vehicle with a Vehicle Jack and Jack Stands

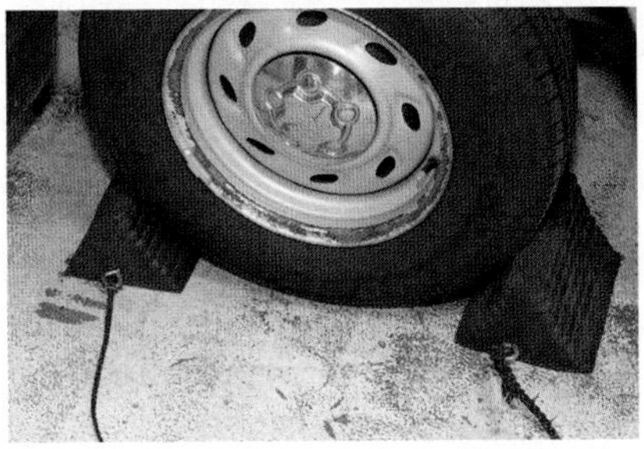

1. Position the vehicle on a flat, solid surface. Put the vehicle into neutral or park and set the parking brake. Place wheel chocks in front of and behind the wheels that are not going to be raised off the ground.

2. Check the manufacturer's labels on the jack and stands. Make sure they are rated higher than the weight you are lifting. If in doubt, ask your instructor.

3. Place one jack stand on each side of the vehicle at the same point, and adjust them so that they are both the same height.

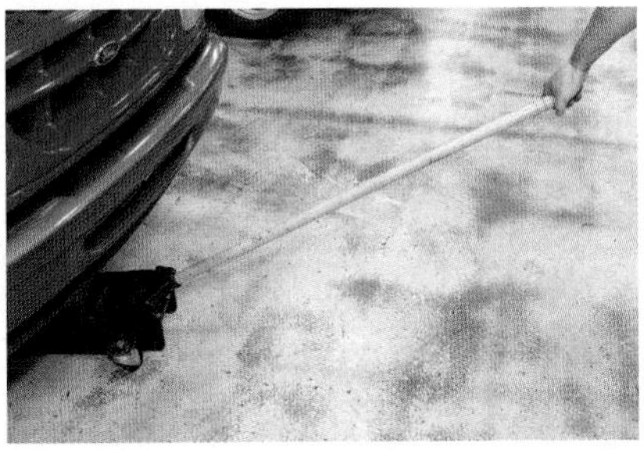

4. Roll the floor jack under the vehicle, and position the lifting pad correctly under the frame or cross member. Turn the jack handle clockwise, and begin pumping the handle up and down until the lifting pad touches and begins to lift the vehicle.

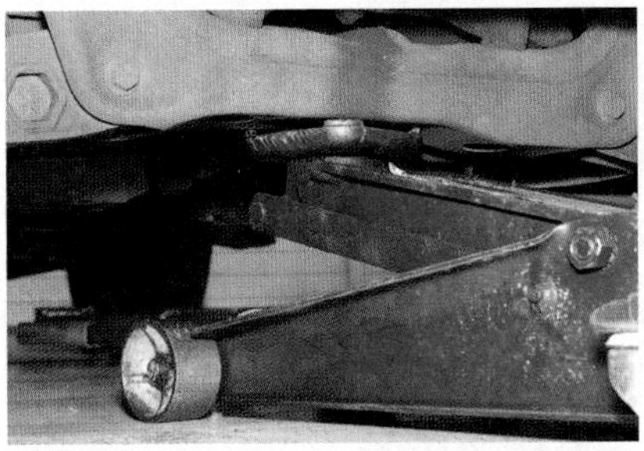

5. Stop and check the placement of the lifting pad under the vehicle to make sure there is no danger of slipping. Double-check the position of the wheel chocks to make sure they have not moved. If the vehicle is stable, continue lifting it until it is at the height at which you can safely work under it.

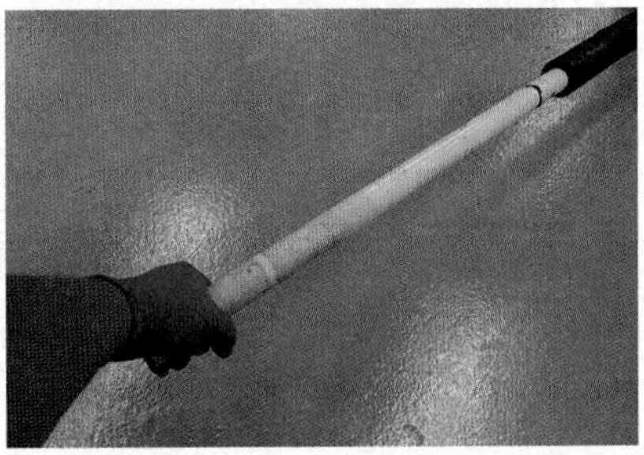

6. Slide the two jack stands underneath the vehicle and position them to support the vehicle's weight. Slowly turn the jack handle counterclockwise to open the release valve, and gently lower the vehicle onto the jack stands.

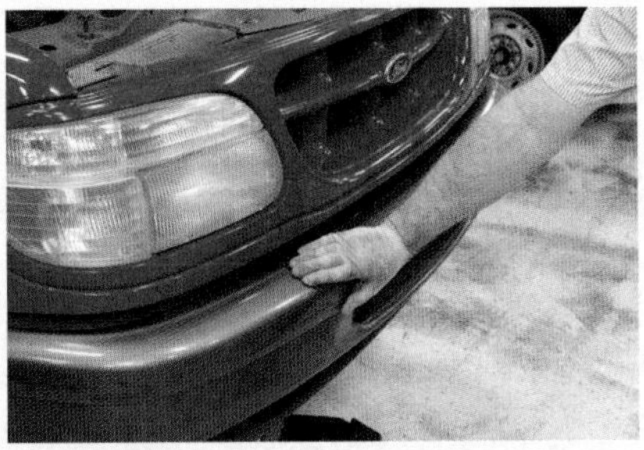

7. When the vehicle has settled onto the jack stands, lower the vehicle jack completely, and remove it from under the vehicle. Gently push the vehicle sideways to make sure it is secure. Repeat this process to lift the other end of the vehicle.

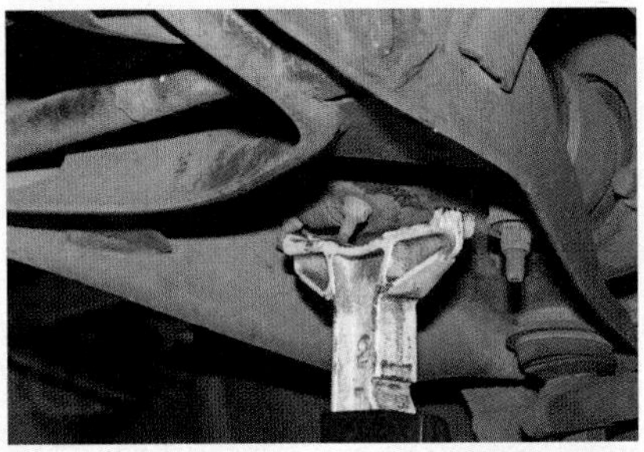

8. When the repairs are complete, use the jack to raise the vehicle off the jack stands. Slide the jack stands from under the vehicle. Make sure no one goes under the vehicle or puts any body parts under the vehicle, as the jack could fail or slip.

9. Slowly turn the jack handle counterclockwise to gently lower the vehicle to the ground. Return the jack, jack stands, and wheel chocks to their storage area before you continue working on the vehicle.

SAFETY TIP

Some independent suspension components are not strong enough to support the weight of the vehicle. Make sure you always use the specified lift points to lift the vehicle.

Engine Hoists and Stands

Engine hoists, or mobile floor cranes, are capable of lifting very heavy objects. Examples include engines, while they are being removed from or installed in a vehicle. The lifting arm of the engine hoist is moved by a hydraulic cylinder and is adjustable for length. However, extending the lifting arm reduces its lifting capacity because it moves the load farther away from the supporting frame. The supporting legs can be extended for stability. But the more the lifting arm and the legs are extended, the lower the lifting capacity of the engine hoist. The safe lifting capacity at various extensions is normally marked on the lifting arm (**FIGURE 9-19**).

The engine or component to be lifted is attached to the lifting arm by a sling or a lifting chain (**FIGURE 9-20**). The sling and lifting chain must be rated for weight in excess of the engine being lifted. It must be firmly attached before the engine hoist is raised. This means that any bolts holding the lifting chain are a minimum of grade 5 and threaded in a minimum of five complete turns. When the engine or other component has been lifted and slowly and carefully moved away from the vehicle, it should be lowered as soon as possible. This is because the farther off the ground an engine is lifted, the less stable the engine hoist becomes as it is rolled around. Also, it is good to mount the engine onto an engine stand or set it onto the floor as soon as practical.

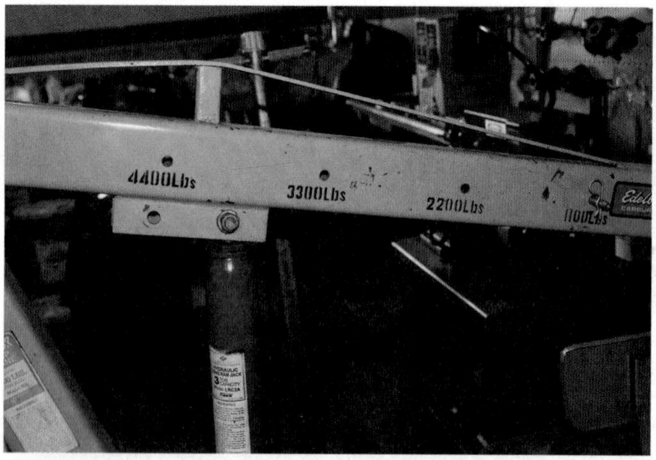

FIGURE 9-19 Lifting capacity at various extensions.

FIGURE 9-20 A lifting chain being used to lift an engine out of the engine compartment.

SAFETY TIP

Never use a hoist to lift any weight greater than the lifting capacity of the hoist, sling, chains, or bolts.

Engine stands are a convenient device to support an engine when it is being worked on while out of the vehicle (**FIGURE 9-21**). Engine stands come in a variety of ratings, some of which are very light duty and should not be used with any but the smallest engines. When mounting an engine on the stand, use a minimum of grade 5 bolts. Ensure that they thread into the hole at least five complete turns. You should also mount the arms of the stand so that the weight of the engine is centered on the mounting head pivot. Otherwise, the engine could be top-heavy and rotate quickly, injuring you.

Using Engine Hoists and Stands

Engine hoists are capable of lifting very heavy objects, which make them suitable for lifting engines. Make sure the hoist is rated for the weight being lifted. In the same way, the engine hoist must be configured to lift that weight. Make sure the lifting chain and the fasteners attaching the lifting chain, or sling, have a tensile strength in excess of the weight being lifted. To keep from overstressing the sling, leave enough length in the sling. This is determined by the angle at the top of the sling, which should be close to 45 degrees and not exceed 90 degrees (**FIGURE 9-22**).

If removing an engine from an engine bay, lower the engine so that it is close to the ground after removal. If the engine is lifted high in the air, the engine hoist will be unstable. When

FIGURE 9-21 Engine stand with engine attached.

FIGURE 9-22 To be safe, the lifting chain angle should be between 45 and 90 degrees.

moving a suspended engine, move the engine hoist slowly. Do not change direction quickly. The engine will swing and may cause the whole apparatus to tumble. To use an engine hoist and choose the correct attachments to lift an engine, follow the steps in **SKILL DRILL 9-4**.

▶ TECHNICIAN TIP

- The load rating of the engine hoist must be greater than the weight of the object to be lifted.
- Never leave an unsupported engine hanging on an engine hoist. Secure the engine on an engine stand, or on the ground, before starting to work on it.
- If using an engine stand, make sure it is designed to support the weight of the engine. And make sure that you have the correct number of bolts to hold the engine to the stand.
- Always extend the legs of the engine hoist in relation to the lifting arm to ensure adequate stability.

SKILL DRILL 9-4 Using Engine Hoists and Stands

1. Prepare to use the engine hoist. Lower the lifting arm, and position the lifting end and chain over the center of the engine.

2. Inspect the chain, steel cable or sling, and bolts to make sure they are in good condition. Look carefully around the engine to determine if it has lifting eyes or other anchor points.

3. If the engine has lifting eyes, attach the sling with D-shackles or chain hooks. If you need to screw in bolts and spacer washers to lift the engine, make sure you use the correct bolt and spacer size for the chain or cable. Screw in the bolts until the sling is held tight against the engine.

4. Attach the hook of the hoist under the center of the sling, and raise the engine hoist just enough to lift the engine an inch or two (0.025–0.05 mm). Double-check the sling and attachment points for safety. The center of gravity of the engine should be directly under the hook of the engine hoist, and there should be no twists or kinks in the chain or sling.

5. Raise the engine hoist until the engine is clear of the ground and any obstacles. Slowly and gently move the engine hoist and engine to the new location.

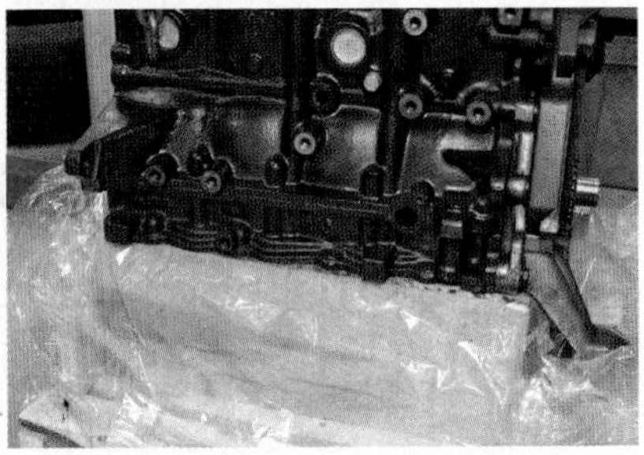

6. Make sure the engine is positioned correctly. You may need to place blocks under the engine to stabilize it. Once you are sure the engine is stable, lower the engine hoist, and remove the sling and any securing fasteners. Finally, return the equipment to its storage area.

▶ Vehicle Lifts

LO 9-03 Operate hoists and describe use of inspection pits.

Vehicle lifts raise whole vehicles off the ground so that a technician can easily work on the underside of the vehicle. The vehicle lift is also useful for raising the vehicle to a height that removes the need for the technician to bend down (**FIGURE 9-23**). For example, when changing tires, the vehicle can be raised to waist height to avoid excessive bending.

Vehicle lifts have a number of different designs. They also come in a range of sizes and configurations to meet the particular needs of the shop. For instance, some vehicle lifts are mobile. Others are designed for use where the ceiling height is limited. Some vehicle lifts can be linked together electronically so they can be used on longer vehicles such as trucks and buses (**FIGURE 9-24**).

Vehicle lifts are typically classified as in-ground and above-ground (**FIGURE 9-25**). In-ground lifts have the working mechanism in and below the concrete floor. Because most of the working parts of the lift are below floor level, they don't have large sections above ground that can damage vehicles or impede foot traffic. One potential problem with in-ground lifts is they can leak hydraulic fluid into the ground. Because of this, they need to have a sealed containment chamber around the cylinders and lines to catch any fluid.

Above-ground lifts have their working parts above ground. These are easier and less costly to install, but they do impede

FIGURE 9-23 Vehicle lifts allow technicians to more easily perform work under a vehicle.

FIGURE 9-24 Vehicle lift linked together electronically for large trucks and busses.

FIGURE 9-25 **A.** In-ground lift. **B.** Above-ground lift.

traffic. And negotiating vehicles around them can lead to accidents. Technicians need to be familiar with the operation of both types of lifts because they are both in common use.

The most common types of vehicle lifts in general use are single-post, two-post, and four-post lifts. Other types of lifts include scissor lifts, parallelogram lifts, and mobile or specialty lifts.

Safety Locks

Every vehicle lift in the shop must have a built-in mechanical locking device. This makes it so the vehicle lift can be secured at the chosen height once the vehicle is raised (**FIGURE 9-26**). This locking device prevents the vehicle from being accidentally lowered. It is like a jack stand, and it holds the vehicle mechanically in place. This will happen even if the hydraulic lifting mechanism fails. You should never physically go under a raised vehicle for any reason unless the safety locking mechanism has been activated.

Ratings and Inspections

All vehicle lifts are rated for a maximum weight and/or type of usage. This maximum weight rating is listed on a label attached to the lift (**FIGURE 9-27**). Never use the vehicle lifts to lift or support a weight greater than their rated limit. This means you will need to know the weight of the vehicle being lifted. Also, never use them for any task other than that recommended by the manufacturer.

In many countries, lifts are required to be periodically inspected and certified. This is typically required every year for vehicle hoists. Before you use a lift, check the identification plate for its maximum weight rating and compare that to the vehicle being lifted. Also, make sure it has a current certification label (**FIGURE 9-28**).

Single-Post Lift

A **single-post lift** raises the vehicle on a platform supported by a single solid shaft located centrally under the vehicle. This type of lift is very compact and leaves the perimeter of the vehicle easily accessible. However, the central post obscures part of the underside of the vehicle. This makes jobs such as transmission removal difficult or impossible.

The single-post lift is an in-ground lift. This means that the central post and hydraulic cylinder are in the ground as part of the concrete floor (**FIGURE 9-29**). This also means that very little of the lift is above ground to damage vehicles or impede foot traffic in the shop. Even so, technicians need to drive over them carefully when pulling vehicles onto the lift.

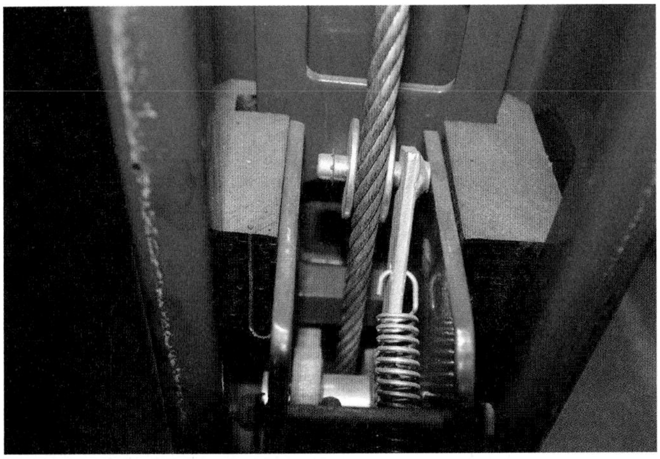

FIGURE 9-26 Typical mechanical lock on a two-post lift.

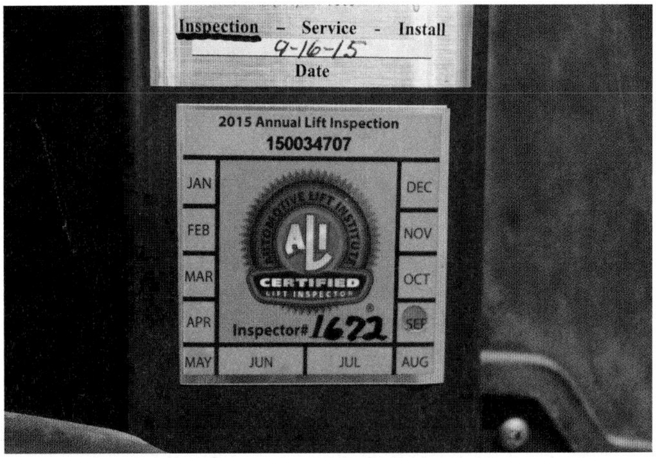

FIGURE 9-28 Typical certification label for a lift.

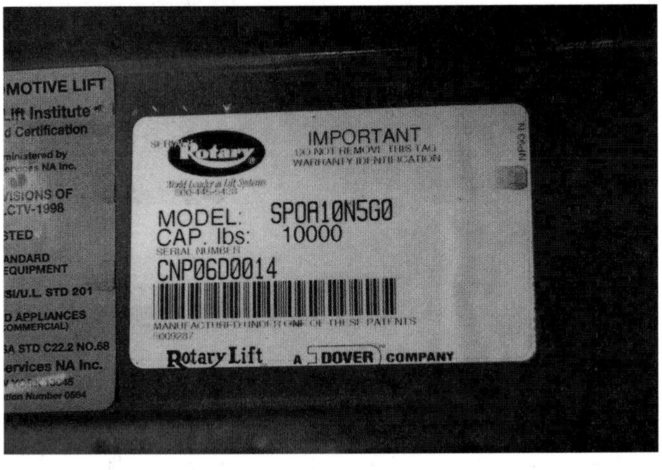

FIGURE 9-27 Label showing the weight capacity of a vehicle hoist.

FIGURE 9-29 In-ground lift.

Two-Post Lift

Two-post lifts can either be in-ground or above-ground. They come in two configurations—symmetrical and asymmetrical (**FIGURE 9-30**). Symmetrical two-post lifts have arms that are of approximately equal length. This roughly centers the vehicle lengthwise between the posts. This positioning creates a challenge because the posts are usually right in the way of the vehicle's front doors. This makes it harder to get into and out of the vehicle.

Asymmetrical lifts have shorter arms in front of the posts than the arms behind the posts. This allows the vehicle to be positioned farther back on the lift, allowing better access to the doors on the vehicle. Both types of two-post lifts leave the underside of the vehicle easily accessible. It also allows a technician to remove the wheels while the vehicle is raised.

Two-post lifts and single-post lifts require careful positioning of the lift arms. They need to be positioned under the appropriate lifting points, two on each side of the vehicle (**FIGURE 9-31**). The service information lists the lifting points so that the vehicle can be raised without causing any structural damage.

> **TECHNICIAN TIP**
>
> Some vehicles are equipped with an automatic leveling system, which must be disabled before lifting the vehicle. Always check the service information to see if this applies to the vehicle you are lifting.

FIGURE 9-30 **A.** Symmetrical lift. **B.** Asymmetrical lift.

Four-Post Lift

A **four-post lift** is very easy to use with most vehicles. It has two long narrow platforms that the vehicle is driven onto. There is one platform on each side of the vehicle (**FIGURE 9-32**). The platforms are then raised, taking the vehicle with them. The underside of the vehicle is accessible to the technician. Because the vehicle rests on its wheels on this type of lift, the wheels cannot be removed unless the lift is fitted with sliding bridge jacks. The sliding bridge jacks are used to raise the wheels off the platform.

Four-post lifts are ideal for jobs such as wheel alignments, oil changes, and exhaust work. They are typically quick to use since technicians can drive up on the lift, and immediately raise the vehicle in the air. At the same time, there are some cautions to be aware of. Most four-post lifts are at an angle to the center lane in the shop. This makes it difficult to get the vehicle lined up straight with the platforms when driving onto and off the lift. It is easy to accidentally drive off the side of the hoist with one or more wheels. This can damage the car and lift. It can also make it extremely difficult to get the vehicle back on the

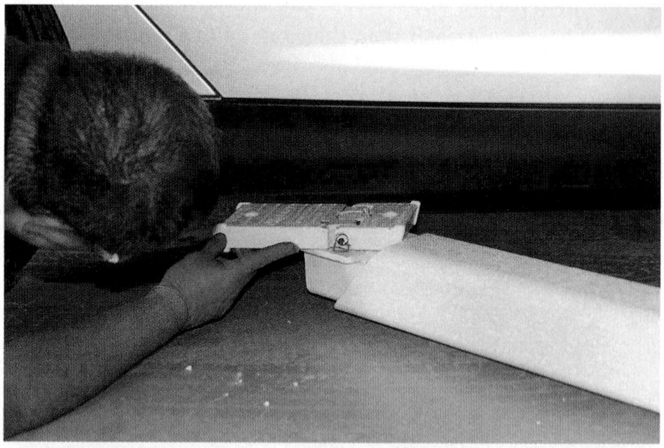

FIGURE 9-31 A two-post lift requires careful positioning of the lift arms so that they are under appropriate lifting points, two on each side of the vehicle.

FIGURE 9-32 A four-post lift.

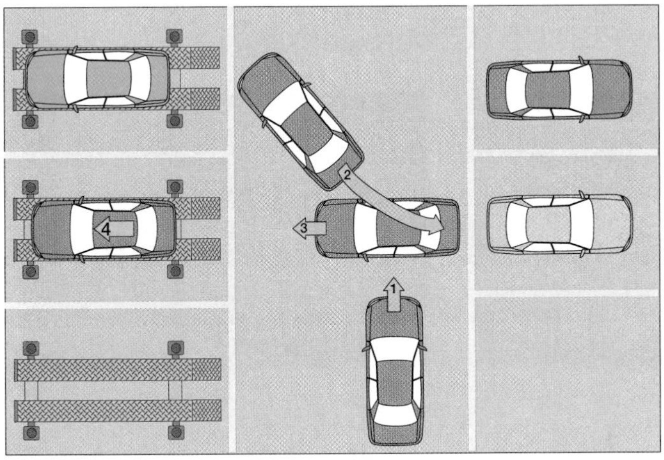

FIGURE 9-33 When pulling a vehicle onto a four-post lift, follow the path to line the vehicle up with the platforms.

platforms. To line the vehicle up for the four-post lift, follow the path listed in **FIGURE 9-33**.

Using Two-Post Lifts

Before lifting any vehicle, make sure the frame is structurally sound. If you see rust or signs of major repair, lifting the vehicle with a vehicle lift may cause damage to the vehicle, or may be dangerous to you. Make sure you know exactly how to operate the vehicle lift. Take particular care that you know exactly where the stop control is so that you can use it quickly in an emergency. Refer to the operator's manual for the correct operating procedure.

Check the amount of clearance under the vehicle. If any of the lifting mechanism is designed so that the vehicle is driven over it, verify that the vehicle has enough clearance. Driving a low-slung vehicle over the lifting mechanism may result in damage to the underside of the vehicle. These vehicles may require shallow ramps that raise the vehicle's wheels so it can go onto the lift.

The lifting points on a vehicle are typically located at the same place as the jacking points. Check the vehicle's service information if you are not sure where the lift points are. The lifting arms must be positioned under the center of the lift points so that the weight of the vehicle is distributed evenly. Check whether the vehicle requires rubber pads to protect the undercoating. If the vehicle is equipped with running boards, verify that the lift arms will not contact the running boards before contacting the lift points. You may be able to set the lift pads to a higher setting so that the running boards clear the lift arms.

Make sure there will be adequate headroom above the vehicle once it is raised. Taller vehicles, especially those fitted with roof racks, may need more headroom than you think. Ask your supervisor if there is the least bit of doubt! The vehicle lift should be raised so you can comfortably work under it. Lock the lift in place before moving underneath or working on the vehicle. To lift a vehicle using a two-post lift, follow the steps in **SKILL DRILL 9-5**.

Using Four-Post Lifts

Four-post lifts are often used to lift a vehicle for wheel alignment services and oil changes. Make sure you know how to operate the four-post lift. Take particular care to know where the stop control is so that you can use it quickly in an emergency. Once the vehicle is at working height, make sure that the lift is either lowered onto the mechanical stops, or the safety device is engaged. When a four-post lift is lowered, pay careful attention to all four corners of the lift. Sometimes, one corner may not get released from the mechanical stop. This causes the vehicle to tip dangerously as it is lowered. This can also damage the hoist. Always refer to the operator's manual for the correct procedure for using the four-post lift. To lift a vehicle using a four-post lift, follow the steps in **SKILL DRILL 9-6**.

SKILL DRILL 9-5 Lifting a Vehicle Using a Two-Post Lift

1. Prepare to use the two-post lift. Check the lift and check the vehicle clearance. Carefully drive the vehicle so that it is centered between the two posts, left and right.

2. Also ensure that it is positioned properly, front to back, for the type of lift and vehicle you are using. Leave the vehicle in neutral, and apply the emergency brake.

3. Position the lifting pads under the vehicle lifting points. Make sure the lifting pads are adjusted to the same height for both sides of the vehicle. Move to the operating controls, and raise the two-post lift just far enough to come into contact with the vehicle.

SKILL DRILL 9-5 Lifting a Vehicle Using a Two-Post Lift (Continued)

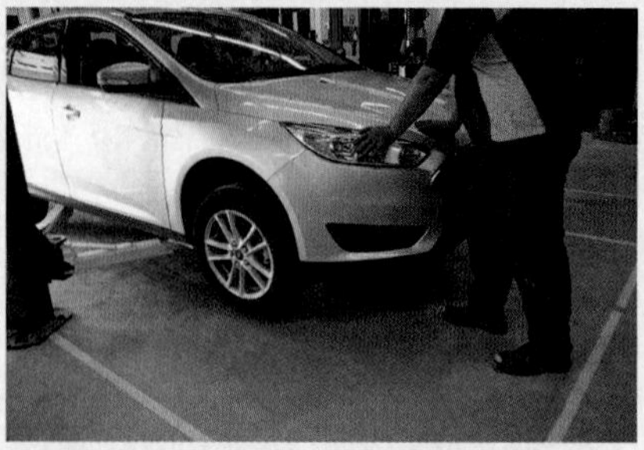

4. Make sure no one is near the vehicle, and then raise the vehicle just until the wheels are a couple of inches off the floor. Check the position of the lifting pads, and shake the vehicle gently to confirm that it is stable.

5. Lift the vehicle to slightly above the working height, and then lower it onto the locks or safety device.

6. Before the two-post lift is lowered, remove all tools and equipment from the area, and wipe up any spilled fluids. Raise the lift to unlock the lift before lowering it. Make sure no one is near the vehicle before lowering it. Once the vehicle is on the ground, remove the lifting arms and drive it away.

SKILL DRILL 9-6 Using Four-Post Lifts to Lift a Vehicle

1. Prepare to safely use the vehicle lift. With the aid of an assistant guiding the driver, or a large mirror in front of the lift, drive the vehicle slowly and carefully onto the four-post lift, and position it centrally. If the lift has front-wheel restraints, drive the vehicle forward until the wheels lock into the brackets.

2. Get out of the vehicle, and check that it is correctly positioned on the platform. If it is, apply the emergency brake, and select first gear for a manual transmission or park for an automatic.

3. Make sure the four-post lift area is clear. Move to the controls, and lift the vehicle until it reaches the appropriate work height. If the four-post lift has a manual safety mechanism, lock it in place to engage whatever safety device is used.

4. Before the four-post lift is lowered, remove all tools and equipment from the area, and wipe up any spilled fluids.

5. Remove the safety device or unlock the lift before lowering it. Make sure no one is near the area.

6. Once the four-post lift is fully lowered, with the help of a guide, you can carefully back the vehicle off the lift.

Vehicle Inspection Pits

Vehicle inspection pits are usually built into the floor area of a vehicle repair bay (**FIGURE 9-34**). The vehicle inspection pit allows the technician to access the underside of the vehicle without the need for a lift or jack. Vehicle inspection pits are wide enough to allow a technician to move along. But narrow enough to fit between the wheels of a vehicle.

Stairs or steps are usually positioned at one end of the inspection pit to allow access for the technician. Lighting is installed in the inspection pit area for greater visibility. The depth of the inspection pit is fixed and provides a general working height to the underside of the vehicle.

Many people have been injured falling into vehicle inspection pits. This happens even if they have been working in or near the inspection pit for a long time. When there is no vehicle over the inspection pit, you should fence it or place the safety net over the top. This will help to stop people from accidentally stumbling into it. You should also restrict general access to the immediate work area.

FIGURE 9-34 Vehicle inspection pit with safety net.

▶ TECHNICIAN TIP

Even low-voltage shop lights can set off an explosion. Whenever possible, use an explosion-proof lamp that will not break, even if it is dropped 6" (1.8 m). Lighting units in an inspection pit should be sealed behind toughened glass or polycarbonate. They should be set into the inspection pit walls so that falling objects will not damage them.

SAFETY TIP

Fuel, paints, and solvents can all give off flammable vapors that are heavier than air. Like water, these vapors tend to flow down and accumulate in places like inspection pits. This increases the risk of explosion and polluting the breathable air. Unless bulkier and more expensive explosion-protected power tools are available in your shop, the only safe power tools to use in an inspection pit are air-powered. They are less likely to cause sparks that could ignite flammable gases.

Using Vehicle Inspection Pits

When using a vehicle inspection pit, have a responsible person guide you as you drive over the pit. Make sure you understand each other's hand signals. Before using the inspection pit, check its lighting. Ensure that there are no obstacles in the inspection pit or its surrounding area. Do not drive or move the vehicle over an inspection pit if someone is in the inspection pit.

To use a vehicle inspection pit, follow the steps in **SKILL DRILL 9-7**.

SKILL DRILL 9-7 Using Vehicle Inspection Pits

1. Ensure that the vehicle inspection pit and surrounding area are clear of obstructions and you have a clear pathway to drive the vehicle over the vehicle inspection pit. Turn the lights on in the inspection pit.

2. Get someone to help guide you as you drive the vehicle over the vehicle inspection pit. Drive the vehicle slowly and carefully, and position it centrally.

3. Check that the vehicle is correctly positioned on the platform. If it is, apply the emergency brake, and select first gear or park. Enter the pit.

SKILL DRILL 9-7 Using Vehicle Inspection Pits (Continued)

4. Perform the needed work on the vehicle, being sure to work in a safe manner, using tools approved for confined spaces.

5. Before driving the vehicle off the vehicle inspection pit, remove all tools and equipment from the vehicle inspection pit area, and wipe up any spilled fluids.

6. Make sure no one is near the vehicle or in the vehicle inspection pit. Have someone guide you off the pit area.

▶ Wrap-Up

Ready for Review

- A professional shop always takes precautions to treat a customer's vehicle with respect, repair, and return it clean and free of any additional damage to do this. Fender, seat, carpet, and steering wheel covers are to be used to avoid unintentional damage to vehicle.
  - Accidental mechanical damage can be avoided by following shop and manufacturer procedures.
- Lifting equipment such as vehicle lifts, floor jacks, jack stands, engine and component hoists, chains, slings, and shackles are designed to lift and securely hold loads. The operating capacity of each equipment is listed on the manufacturer's certificate located on the lifting equipment.
  - Identify the lift points in the vehicle lift chart, to avoid damage to vehicle during lifting.
- A vehicle inspection pit allows the technician to access the underside of a vehicle without the need for a lift or jack. Before using the inspection pit, check its lighting and ensure that there are no obstacles in the inspection pit or its surrounding area.

Key Terms

engine hoist A small crane used to lift engines.

four-post lift Vehicle lift with four lifting posts. One near each corner of the vehicle.

hydraulic jack A type of vehicle jack that uses oil under pressure to lift vehicles.

inspection certificate Label applied to lifting equipment showing when it was last inspected and certified.

jack stands Metal stands with adjustable height to hold a vehicle once it has been jacked up.

mechanical jack A type of vehicle jack that uses mechanical leverage to lift a vehicle.

pneumatic jack A type of vehicle jack that uses compressed gas or air to lift a vehicle.

safe working load (SWL) The maximum safe lifting load for lifting equipment.

Two-post lifts Vehicle lift with two centrally located lifting posts.

vehicle inspection pit A trench permanently fitted into the floor of the shop to allow easy work access to the vehicle's underside.

vehicle jack A tool for lifting a vehicle.

Vehicle lifts Equipment designed to lift the entire vehicle off the ground

Review Questions

1. All of the following are good practices when a customer drops a vehicle for service EXCEPT:
   a. Do a walk around to look for damage
   b. Install seat covers
   c. Use floor mats to protect the carpet
   d. Set the radio to your preferred station
2. What is used to protect the vehicle when working under the hood?
   a. Shop rags
   b. Oil drip cloth
   c. Fender covers
   d. A bright light
3. Before returning the vehicle to the customer, the technician should?
   a. Clean the vehicle
   b. Clean the tools
   c. Wash the bay
   d. Reset the trip meter
4. When using a two-post lift:
   a. Raise the vehicle to working height and shake the vehicle to see if it is stable
   b. Raise the vehicle a few inches off the floor and shake the vehicle gently to see if it is stable
   c. The arms automatically align themselves with the vehicle lift points
   d. The vehicle is driven onto two long narrow platforms

5. When using a jack to lift a vehicle so a tire can be removed, which piece of equipment should also be used?
   a. A second jack
   b. A rated lifting pad
   c. Jack stands
   d. Fender covers
6. When multiple pieces of lifting equipment are used, the SWL is:
   a. That of the piece with the lowest rating
   b. The sum of the SWLs of the individual pieces
   c. That of the piece with the highest rating
   d. The average of the SWLs of the individual pieces
7. When fastening an engine to a stand, what grade bolt should be used?
   a. Grade 2
   b. Grade 3
   c. Grade 5
   d. Does not matter
8. What must be built into every vehicle lift in the shop?
   a. Computer controls
   b. Air-powered cylinder
   c. Color-coded controls
   d. Mechanical locking device
9. Four-post lifts are ideal for all of the following types of work EXCEPT:
   a. Alignments
   b. Oil changes
   c. Exhaust work
   d. Brake work
10. When driving a vehicle over an inspection pit
   a. Have a spotter guide you
   b. Make sure the lights in the pit are off
   c. Drive as quickly as you can
   d. Make sure a person is in the pit

ASE Technician A/Technician B Style Questions

1. Technician A says that harmful vapors can collect in an inspection pit. Technician B says that electric powered tools should be used in inspection pits. Who is correct?
   a. Technician A
   b. Technician B
   c. Both A and B
   d. Neither A nor B
2. Technician A says that vehicle positioning is not very critical on two-post lifts. Technician B says to make sure all the locks on a four-post lift disengage when lowering. Who is correct?
   a. Technician A
   b. Technician B
   c. Both A and B
   d. Neither A nor B
3. Technician A says that vehicle lifts should be inspected and certified every year. Technician B says that lifts should be raised slightly to undo the safety locks when lowering a vehicle. Who is correct?
   a. Technician A
   b. Technician B
   c. Both A and B
   d. Neither A nor B
4. Technician A says lifting chains should be rated higher than the object being lifted. Technician B says that extending the lifting arm on an engine hoist increases its capacity. Who is correct?
   a. Technician A
   b. Technician B
   c. Both A and B
   d. Neither A nor B
5. Technician A says that symmetrical two-post lifts allow the vehicle to be positioned farther back on the lift. Technician B says to use the lift points specified by the manufacturer. Who is correct?
   a. Technician A
   b. Technician B
   c. Both A and B
   d. Neither A nor B
6. Technician A says when using jack stands, make sure they are level on the ground. Technician B says to make sure the jack stands are on a solid, hard surface. Who is correct?
   a. Technician A
   b. Technician B
   c. Both A and B
   d. Neither A nor B
7. Technician A says to return the vehicle to the customer in a better shape than when it was dropped off. Technician B says to use seat covers, steering wheel covers, and floor mats on customer vehicles. Who is correct?
   a. Technician A
   b. Technician B
   c. Both A and B
   d. Neither A nor B
8. Technician A says to do a walk around the vehicle with the customer to note any problems. Technician B says to check for dirty finger prints before returning the vehicle to the customer. Who is correct?
   a. Technician A
   b. Technician B
   c. Both A and B
   d. Neither A nor B
9. Technician A says some brake fluids can cause damage to painted surfaces. Technician B says that it is generally best to clean up any brake fluid spills on painted surfaces after it has sat for a while. Who is correct?
   a. Technician A
   b. Technician B
   c. Both A and B
   d. Neither A nor B
10. Technician A says that fender covers with magnetic strips can end up scratching the paint. Technician B says that your back pocket is a good place to keep a screw driver. Who is correct?
   a. Technician A
   b. Technician B
   c. Both A and B
   d. Neither A nor B

CHAPTER 10
Vehicle Maintenance Inspection

Learning Objectives

- **LO 10-01** Perform in-vehicle inspection.
- **LO 10-02** Perform fluid inspection.
- **LO 10-03** Perform belt, hose, and air filter/cabin air filter inspection.
- **LO 10-04** Perform under-vehicle inspection.
- **LO 10-05** Perform exterior vehicle inspection.

ASE Education Foundation Tasks

See Appendix A to view the 2017 ASE Education Foundation Automobile Accreditation Task List Correlation Guide.

▶ Vehicle Inspection Preliminaries

LO 10-01 Perform in-vehicle inspection.

Vehicles are made up of thousands of parts, all of which work together to provide a safe and reliable means of transportation. But those parts don't last forever. They need periodic inspection to assess their condition. In fact, in the United States alone, unperformed maintenance totals approximately $66 billion. This means that there are a lot of vehicles on the road that need maintenance. This creates a safety issue for motorist due to accidents and breakdowns. It also has a financial cost seeing that regular maintenance can extend the useful life of a vehicle by more than 100,000 miles (160,100 km). Because vehicle inspections are so important, technicians of all levels perform these inspections (**FIGURE 10-1**).

There are several levels of inspections that can be performed. Some of the common ones are:

- Basic Inspection—Such as a 30-point inspection—performed during oil changes or any time a vehicle is in the shop.
- In-depth Inspection—Such as a 128-point inspection—Commonly performed when the vehicle hits certain milestones, such as at 100,000 miles (160,000 km).
- State safety inspection—This is usually required every 1–3 years. Verifies that the vehicle meets minimum safety requirements.
- Certified used car inspection—Many car dealers sell certified used cars. They use a specific inspection form. This certifies that the vehicle meets certain minimum specifications as listed in the form. Areas that didn't meet the specifications are brought up to those specifications.
- Vehicle pre-purchase Inspection—Performed when a customer wants to purchase a noncertified used vehicle to make sure it is in good condition.

Most shops use inspection forms to ensure a thorough and efficient inspection. It also allows the results to be documented. These can be paper forms (**FIGURE 10-2**) that the technician fills out by hand. But this makes storage for long-term reference a challenge. A more efficient solution is to use electronic forms which are filled out on a tablet or computer (**FIGURE 10-3**). They can be stored electronically and recalled at any time. The information is then used to inform the customer of their vehicle's condition, and the cost to repair any issues found (**FIGURE 10-4**).

In researching electronic inspection systems for this chapter, I ran across a very innovative product. It is called Cherry Inspect (www.drivecherry.com). This is a purpose-built electronic inspection system. It allows a shop to customize their inspection forms to fit their needs. Because it is electronic, it includes a bunch of time-saving features. It speeds up the inspection time approximately 40% over using paper forms. This allows for increased productivity because more work can be performed in a shorter amount of time.

FIGURE 10-1 Regular inspection of a vehicle's systems helps ensure safe and reliable operation.

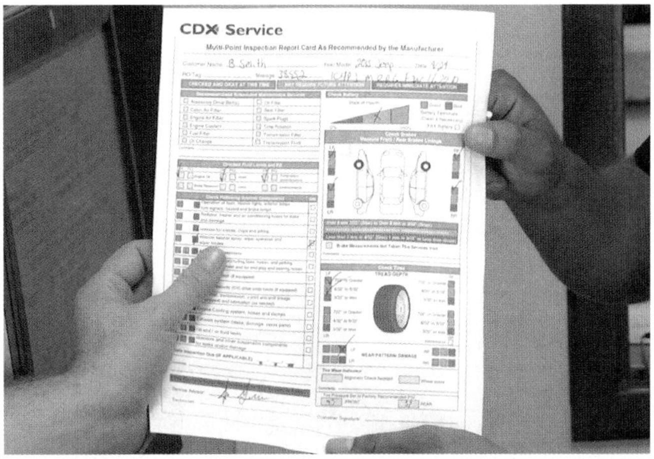

FIGURE 10-2 Sample paper inspection form.

You Are the Automotive Technician

You work for a car rental company that maintains their vehicles in-house. Your main task today is to perform preventive maintenance on a vehicle in the fleet that is scheduled for its next service. To perform a routine maintenance check, you follow a procedural checklist. That includes changing the oil and filter, rotating the tires and inspecting brake, steering, and suspension systems. It also involves checking/inspecting other important parts. During the inspection, you notice that the brake fluid level is low. You also notice that the wiper blades smear badly. Finally, you see that the left rear brake light is not working, which requires a bulb replacement.

1. What two conditions can low brake fluid level indicate?
2. How do you determine what engine oil to use?
3. What must be done after an oil change to remind the customer of the next change?
4. What is used to check the oil level after it has been changed?

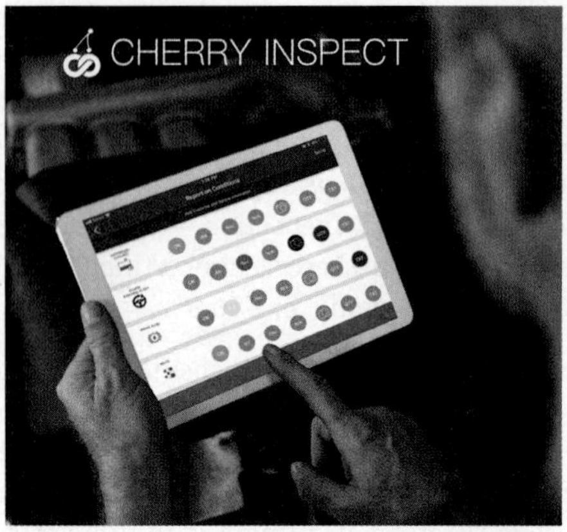

FIGURE 10-3 Sample electronic inspection form on a tablet.

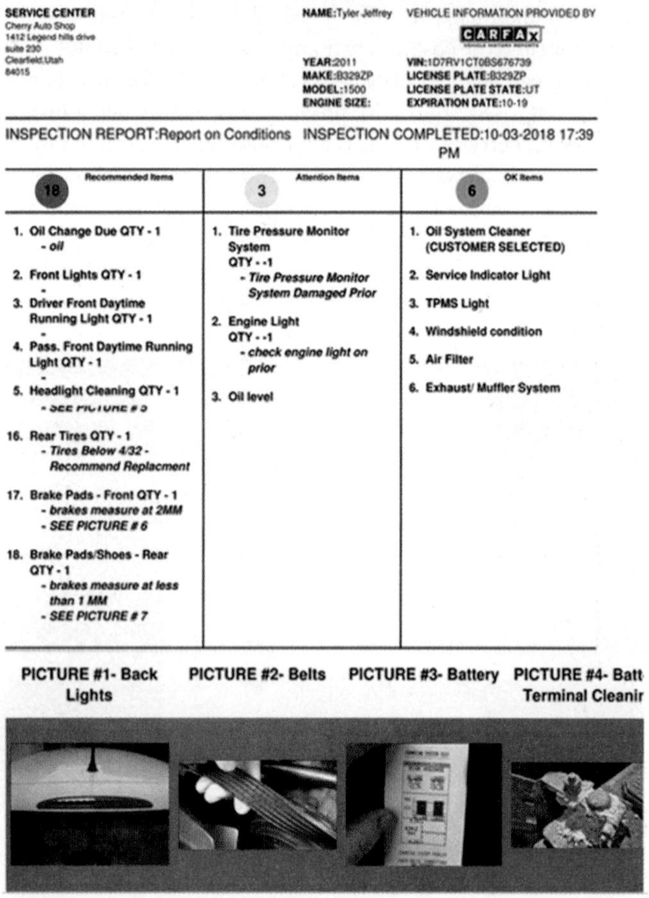

SERVICE CENTER
Cherry Auto Shop
1412 Legend hills drive
suite 230
Clearfield,Utah
84015

NAME:Tyler Jeffrey
VEHICLE INFORMATION PROVIDED BY CARFAX

YEAR:2011
MAKE:B329ZP
MODEL:1500
ENGINE SIZE:

VIN:1D7RV1CT0BS676739
LICENSE PLATE:B329ZP
LICENSE PLATE STATE:UT
EXPIRATION DATE:10-19

INSPECTION REPORT:Report on Conditions INSPECTION COMPLETED:10-03-2018 17:39 PM

18 Recommended Items	3 Attention Items	6 OK Items
1. Oil Change Due QTY - 1 - *oil* 2. Front Lights QTY - 1 - 3. Driver Front Daytime Running Light QTY - 1 - 4. Pass. Front Daytime Running Light QTY - 1 - 5. Headlight Cleaning QTY - 1 - *SEE PICTURE # 5* 16. Rear Tires QTY - 1 - *Tires Below 4/32 - Recommend Replacment* 17. Brake Pads - Front QTY - 1 - *brakes measure at 2MM* - *SEE PICTURE # 6* 18. Brake Pads/Shoes - Rear QTY - 1 - *brakes measure at less than 1 MM* - *SEE PICTURE # 7*	1. Tire Pressure Monitor System QTY - -1 - *Tire Pressure Monitor System Damaged Prior* 2. Engine Light QTY - -1 - *check engine light on prior* 3. Oil level	1. Oil System Cleaner (CUSTOMER SELECTED) 2. Service Indicator Light 3. TPMS Light 4. Windshield condition 5. Air Filter 6. Exhaust/ Muffler System

PICTURE #1- Back Lights PICTURE #2- Belts PICTURE #3- Battery PICTURE #4- Batt Terminal Cleanir

FIGURE 10-4 Sample summary sheet from electronic inspection system (Cherry Inspect).

The system allows the technician to also include written notes and pictures. This better presents the situation to the customer (**FIGURE 10-5**). It provides very valuable information to customers as it gives them solid evidence to base their repair decisions on. Plus, this system provides evidence that the customer was notified of the condition. This is important in the event that a customer refuses the recommended repairs and has an accident. It becomes evidence if there is a liability lawsuit filed.

When you perform an inspection, it is important that you are accurate and complete. Being accurate is important. It means that you are performing the inspection according to the manufacturer's specified procedure. It also means comparing the results to their specifications. This usually means taking measurements and using objective criteria for your evaluation. You should not guess. For example, it is best to measure the tread depth or brake pad thickness instead of guessing the percentage of life left in the component. Once you have the objective results, they can be compared to the manufacturer's specifications. Together, these will give you valuable information. You can then inform the customer on the condition of their vehicle and what is needed to maintain it.

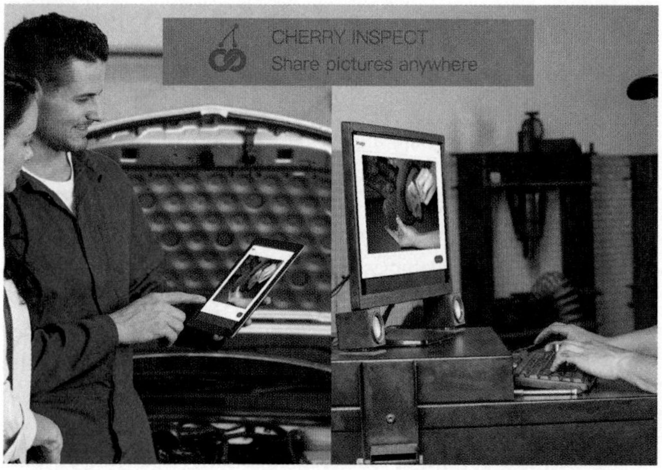

FIGURE 10-5 Electronic Inspection system being used to show a customer a picture of the issue with their vehicle.

For an inspection to be complete, it needs to include everything that should be inspected at the level of inspection being performed. Nothing should be missed. This mean that the inspection needs to be organized in a manner that prevents missed items. This chapter explains what to look for, and how to perform these important inspections.

In-Vehicle Inspections

It would be easy to assume that customers would be aware of any in-vehicle concerns in their vehicle, and therefore report them. But many drivers either ignore them, or don't realize they indicate a problem. So, do not overlook this critical inspection. In fact, certain in-vehicle inspections should be made as you drive the vehicle into the service bay.

Pay attention to the instrument cluster and warning lights. Also, note how the pedals feel and how the vehicle sounds when first started, and as you drive it into the stall. If the Malfunction Indicator Lamp (MIL) is illuminated while driving the vehicle into the stall, you will likely need to retrieve any stored Diagnostic Trouble Codes (DTCs) from the onboard computer. This requires a scan tool to perform. As always, document any unusual findings and give them to your supervisor or service advisor.

▶ TECHNICIAN TIP

The customer may have become used to the way the vehicle drives and feels. Yet, there may be concerns such as a squishy brake pedal, or play in the steering wheel that the customer is unaware of. So, it is up to the technician to detect faults when the customer may think the fault is just a normal condition.

FIGURE 10-6 Always feel the brake pedal before starting the vehicle.

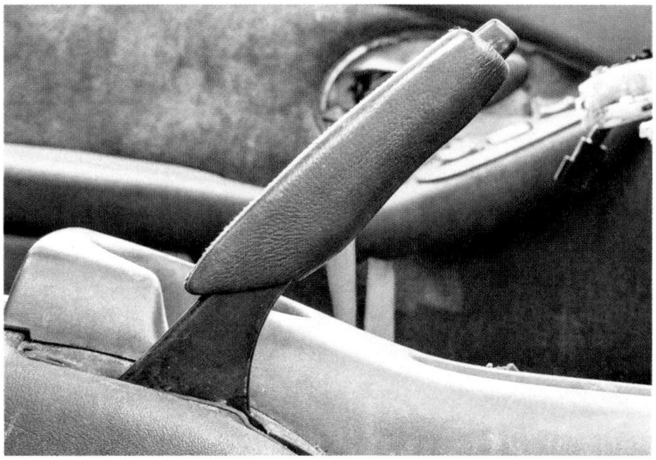

FIGURE 10-7 This parking brake goes too far before it is fully applied requiring further attention.

Checking the Brake Pedal

The brake pedal acts as a lever to increase the force applied to the brake assemblies by the driver. Changes to how far the pedal travels or to its resistance—if it feels harder or softer than normal—can indicate a problem in the braking system. Once you start the engine, always check the brake pedal feel and travel before driving the vehicle into or out of the shop (**FIGURE 10-6**). If the pedal is low, extremely squishy or hard, do NOT drive the vehicle.

Also, listen for unusual brake noises when driving the vehicle into the shop. High-pitched scraping noises or heavy grinding noises could indicate a worn brake lining. The vehicle owner may be used to the feel of a low pedal or the sound of noisy brakes, whereas you will recognize it as an indication of a problem. See the chapter on Hydraulics and Power Brakes for more information on how to service the braking system.

Checking the Parking Brake

All vehicles must be manufactured with a service brake and a parking brake. The service brake is designed to slow or stop the vehicle when driving. It uses a foot-operated pedal to hydraulically apply brake units at all four wheels. The parking brake is designed to hold the vehicle when it is parked. It uses either a hand-operated or a foot-operated lever to mechanically apply the brake units at only two wheels.

The parking brake should be checked as part of a routine safety or vehicle inspection. This level of inspection involves applying the parking brake and observing how far it travels before engaging. If it goes all the way to its stop, or near its stop, it fails this test and will need further attention (**FIGURE 10-7**).

In climates with below-freezing temperatures, the parking brake cable can freeze in the applied position. This makes it so the parking brake will not release. In freezing conditions, it is probably best not to test the parking brake operation. See the chapter on Disc Brake Systems for more information on servicing the parking brake.

> ▶ **TECHNICIAN TIP**
>
> If the vehicle does not have a hand- or foot-operated parking brake, check for a "P" button on the console or dash. The vehicle may be equipped with an electronically controlled parking brake.

Checking the Instrument Panel Warning Lamps

The many **instrument panel warning lamps** can mean faults with various systems on the vehicle. If a fault is detected, the appropriate warning lamp will be illuminated. Warning lamps can also mean proper operation of the system. Each warning system performs a self-check each time the ignition is switched on or the engine cranked.

You should observe the action of the warning lights when starting the vehicle to determine whether any more service may be required. For example, the amber anti-lock brake system (ABS) warning lamp will come on, then stay on for a few seconds, and then go off. This means that the ABS control module has successfully completed a preliminary self-check. The same thing happens with the Onboard Diagnostic System II (OBDII). It turns on the MIL during the bulb check, and then turns it off once the engine starts and no DTCs are stored in memory. To check instrument panel warning lamps, follow the steps in **SKILL DRILL 10-1**.

Retrieving Diagnostic Trouble Codes

The OBDII system tests and analyzes each of the engine's systems while the vehicle is being operated. The computer will set and store DTCs if a fault is detected. When performing an in-vehicle inspection, check to see if the MIL is illuminated. If it is, it is important to retrieve any DTCs stored in the vehicle's computer memory. These codes will indicate which faults have been detected in the system. They will be useful when informing the customer of the status of their vehicle. They also give the technician valuable information when diagnosing the fault.

> ▶ **TECHNICIAN TIP**
>
> Never erase the DTCs until all testing and repairs have been completed. Even then it may be best to allow the computer to automatically clear them. This will be done after it performs a specified number of drive cycles after the problem has been corrected.

SKILL DRILL 10-1 Checking Instrument Panel Warning Lamps

1. Perform an instrument panel self-test. When the key is switched on (before starting the engine), most of the dash warning lamps will light up as a bulb check. Note any that do not light up as expected.

2. Perform an engine running check. Start the engine and observe the warning lamps. All should go off after a few seconds as the related control module runs a self-check and then commands the lamp to go off.

DTCs can be retrieved with a scan tool. The scan tool plugs into the vehicles Data Link Connector (DLC) which allows the scan tool to communicate with the vehicle's computer. The technician selects menu items to get to the "retrieve codes" option. Once selected, the scan tool will display any current and pending DTCs. You will need to document those codes on the inspection system. To retrieve and record DTCs, follow the steps in **SKILL DRILL 10-2**.

SKILL DRILL 10-2 Retrieving and Recording DTCs

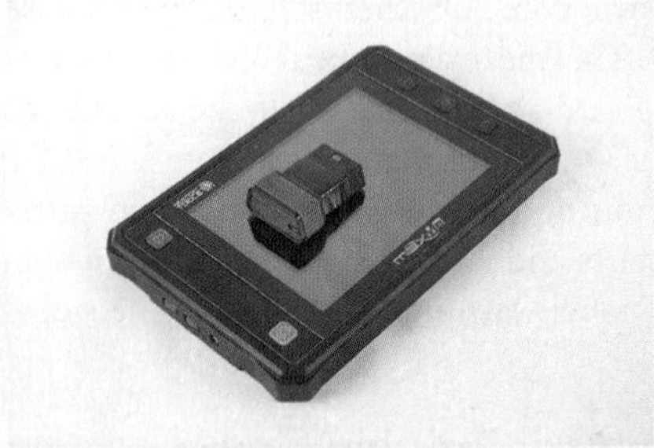

1. Select the scan tool to provide the best coverage for the type and make of vehicle.

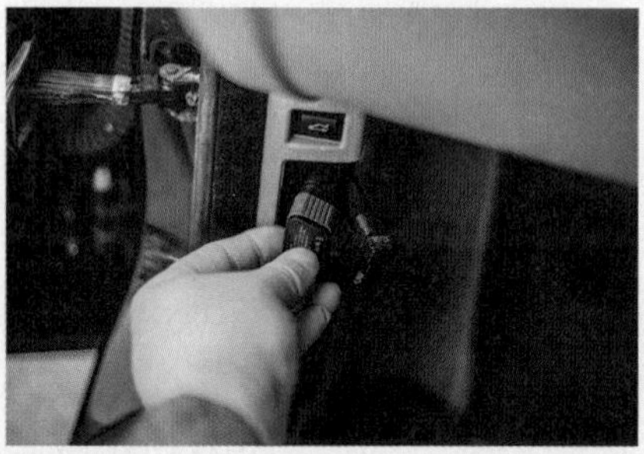

2. Locate the DLC and connect the scan tool.

3. Power on the scan tool.

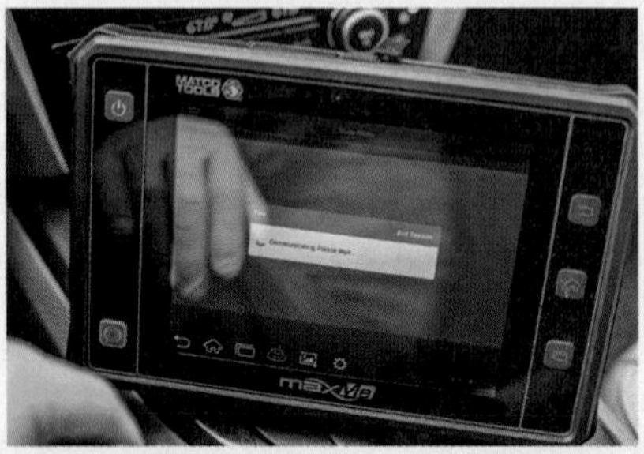

4. Turn the vehicle's ignition on which should establish scan tool communication with the vehicle's computer.

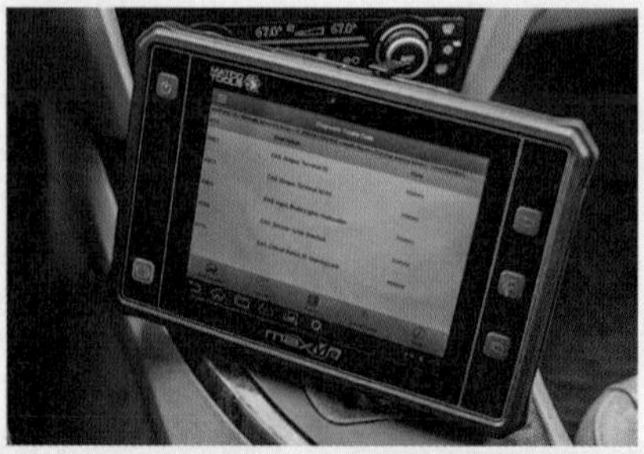

5. Retrieve and record the DTCs.

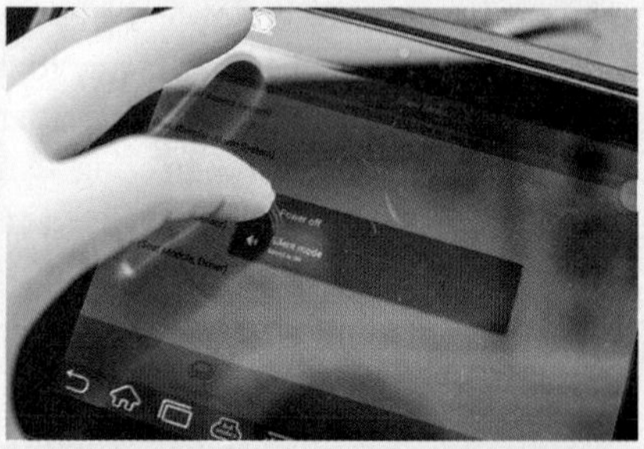

6. Power off the scan tool, turn the ignition off, and disconnect the scan tool.

Checking the Horn

The vehicle horn is usually operated by either a relay or by the vehicle **body control module (BCM)**. There may be a single horn or a pair of horns, depending on the vehicle. With two horns, one sounds at a lower pitch than the other. The horns are located at the front of the vehicle, behind the grill or bumper. The horn can easily be checked before driving the vehicle into the shop. Do NOT operate the horn in an annoying manner.

To perform a horn check, follow the steps in **SKILL DRILL 10-3**.

Inspecting the Interior Lights

The interior lights provide illumination to the inside of the vehicle. This includes the courtesy lights, dome lights, vanity lights, and map lights. Activating the key fob or opening the driver's door should activate the courtesy lights and dome light/s. Some of these lights are on a timer and stay on for a specified amount of time before they shut off. This gives the driver time to insert the key into the ignition.

Once the door is closed, and the lights are off, you should be able to turn the dome light/s on by either a switch on the dome light, or a switch on the dash. Vanity lights are typically on the back side of the sun visor. A switch will turn it on and off. Map lights typically turn on automatically when they are moved into a position they can be used. To inspect the interior lights, follow the steps in **SKILL DRILL 10-4**.

▶ Underhood Fluid Inspection

LO 10-02 Perform fluid inspection.

A check under the hood is important to the life and operation of the vehicle (**FIGURE 10-8**). The inspection should be performed at the manufacturer's recommended intervals and also prior to any long trip. Component damage or failure is often caused by a lack of maintenance or low fluid level in the related system. For example, low oil level in the engine can cause major damage to the engine bearings and crankshaft. Future problems also can be prevented by a thorough inspection of the underhood fluids. The discovery of a low fluid level may help to avoid a breakdown on the highway.

Fluids

Some of the fluids used in a vehicle, such as engine oil, are needed to keep the mechanical systems lubricated and functioning correctly. Other fluids may be safety related, such as windshield washer fluid and brake fluid. No matter what the fluid, they all need to be at the proper level (**FIGURE 10-9**). Always use the manufacturer's recommended type and amount of fluid when checking these items.

For the engine, the engine oil and coolant levels must be checked. The brake, hydraulic clutch, and ABS all depend on the proper level of brake fluid. Power steering fluid and transmission fluid need to be at the recommended level for these systems to operate properly. The windshield washer fluid level should be checked and topped off as needed.

SKILL DRILL 10-3 Performing a Horn Check

1. Check the vehicle horn. Turn on the ignition, and press the horn button. The horn should sound.
2. If the horn is not working, locate it under the hood with the help of the manufacturer's service information. Check the wiring to make sure it is connected securely.

SKILL DRILL 10-4 Inspecting the Interior Lights

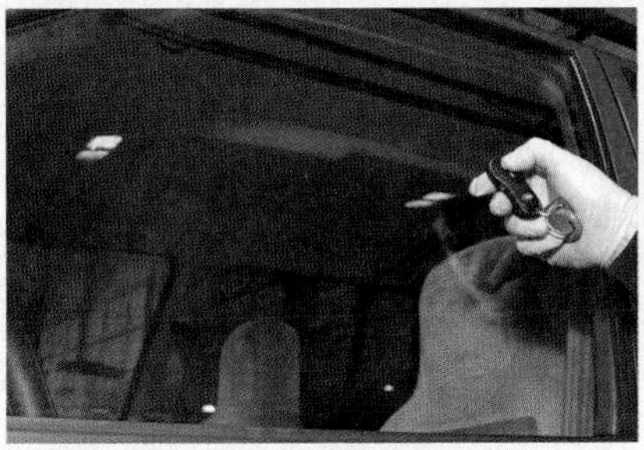

1. Using the remote key fob or door key, unlock the doors. On most vehicles, this causes the interior lights to come on. Check that each light works as intended. On some vehicles, the lights do not come on until the door is actually opened. Check each door on the vehicle.

2. Enter the vehicle and close the door. Wait to see that the lights go off after a time. Repeat this check with each door on the vehicle.

3. Operate the courtesy lights from any other switches. Check the vanity lights and map lights, if they exist.

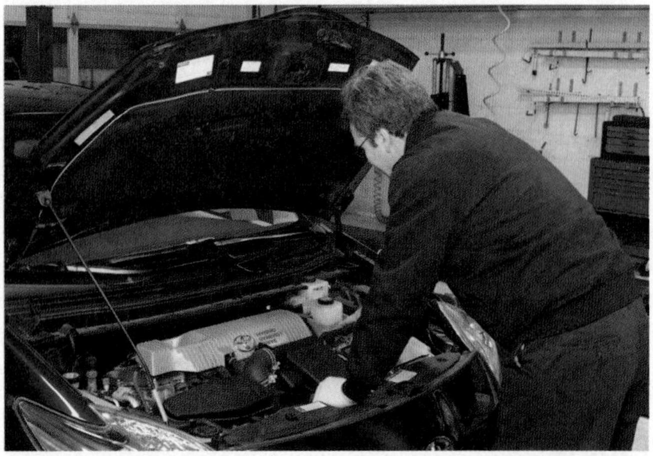

FIGURE 10-8 A check under the hood is vital to the life and operation of the vehicle.

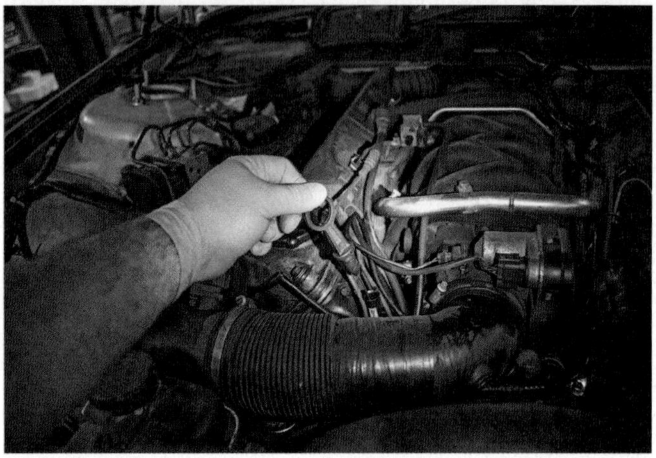

FIGURE 10-9 All fluids need to be checked for the proper level.

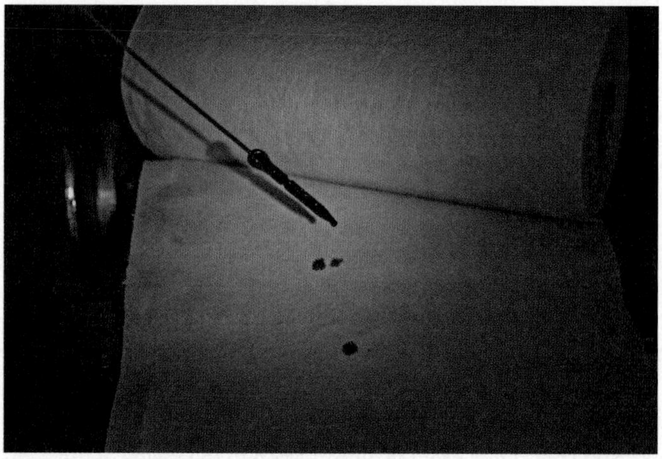

FIGURE 10-10 Checking fluid appearance.

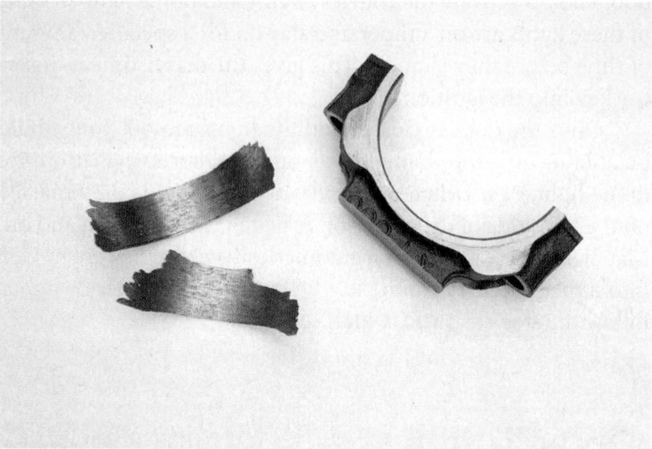

FIGURE 10-11 Parts damaged by lack of lubrication.

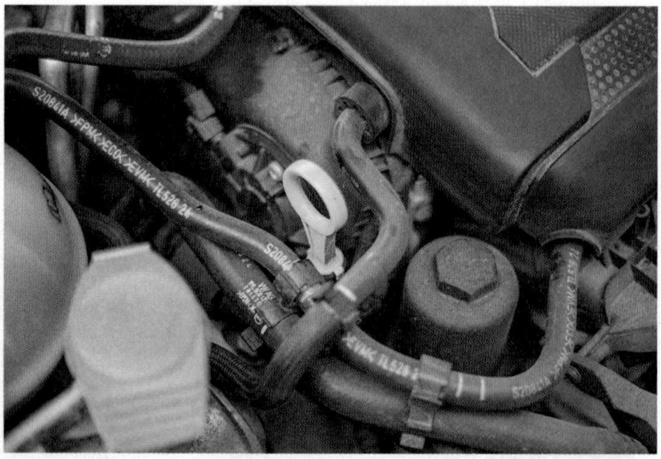

FIGURE 10-12 Oil dipstick is usually marked "OIL" or brightly colored.

Just as the level of the fluid is important, so too is the quality of the fluid (**FIGURE 10-10**). Nearly all of the fluids in a vehicle get old and wear out, requiring replacement; the one exception being some automatic transmission/transaxle fluids. So, whenever you check the level of any fluid, check the quality of the fluid too. You might notice a change in color, a change in consistency, a mix of fluids, or a change in smell, such as burnt transmission/transaxle fluid. Each fluid shows its age differently, so become familiar with how to identify both good and bad fluids.

Engine Oil

The level of the oil in the engine's lubrication system is critical to the engine's operation. The engine oil is picked up by the oil pump, filtered through the oil filter, and then sent under pressure to the crankshaft and camshaft bearings. If the level is too low, the oil pump will starve for oil. If the level is too high, the oil will be struck by the crankshaft, churning it into foam. Either condition causes a lack of lubrication, which can damage the engine bearings and other internal engine parts (**FIGURE 10-11**).

The engine oil level should be checked periodically, usually at every other fuel stop as part of a preventive maintenance plan. Always check the oil level when the vehicle is on a level surface, not on a hill or slope. The oil can be checked with the dipstick, which is usually marked "oil" or brightly colored (**FIGURE 10-12**). The marks on the bottom of the oil dipstick usually have lines that indicate "full" and "add," or "min" and

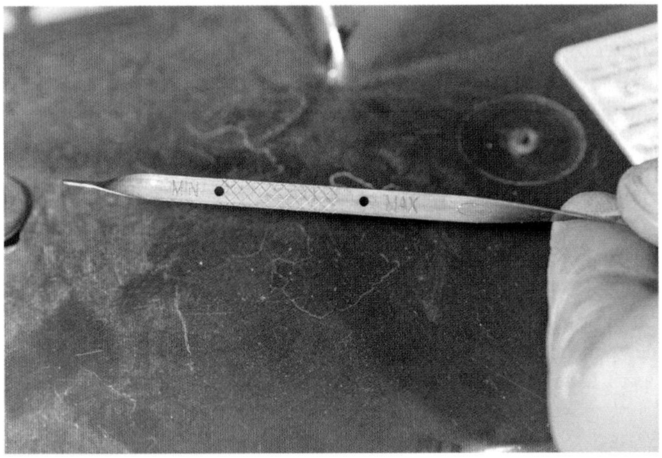

FIGURE 10-13 Oil dipsticks typically are labeled either "FULL" and "ADD" or "MIN" and "MAX."

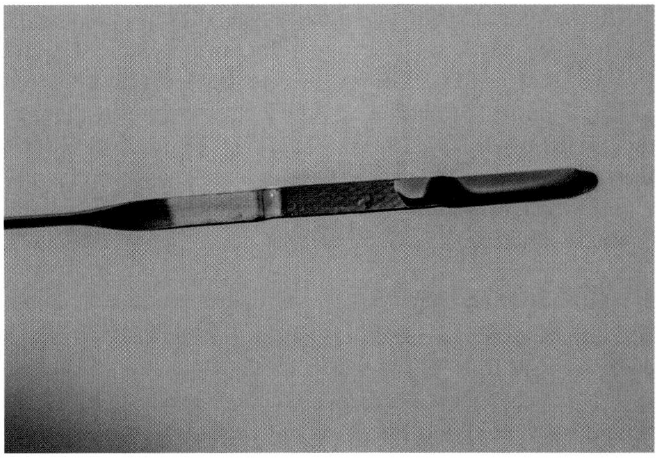

FIGURE 10-15 Flip the dipstick over and compare the reading on both sides. The lowest reading is the most accurate.

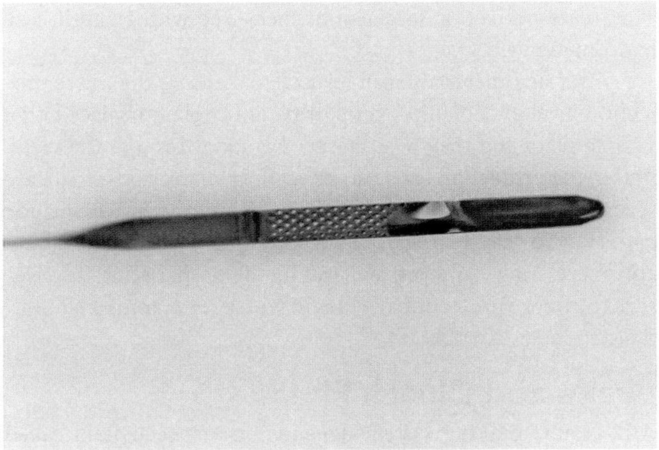

FIGURE 10-14 Hold the dipstick level when reading it.

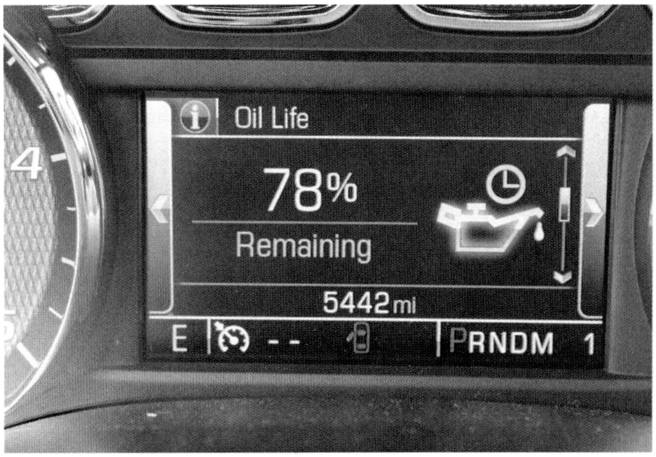

FIGURE 10-16 Oil-life monitor.

"max" (**FIGURE 10-13**). The difference in quantity between the add and full marks on an engine oil dipstick is typically 1 quart (0.9 liters). But can be as much as 2 quarts (1.9 liters) on some vehicles. This can be verified in the vehicle owner's manual or the manufacturer's published service information.

Always check engine oil with the engine off. Wipe off the dipstick, identify the marks on the dipstick, and reinsert it fully in the tube. Then, pull it out and hold it horizontally to read it (**FIGURE 10-14**). It is also a good idea to flip the dipstick over and compare the oil level on the back side with the level shown on the front side. The lowest reading on either side is likely the most accurate (**FIGURE 10-15**).

When checking the oil level, also consider checking whether it is time for an oil change. There are two ways to determining this. If the vehicle is equipped with an oil-life monitoring (OLM) system, it will show the percentage of oil life remaining (**FIGURE 10-16**). If not equipped with an OLM, the service information will show the recommended mileage or time interval between changes (**FIGURE 10-17**).

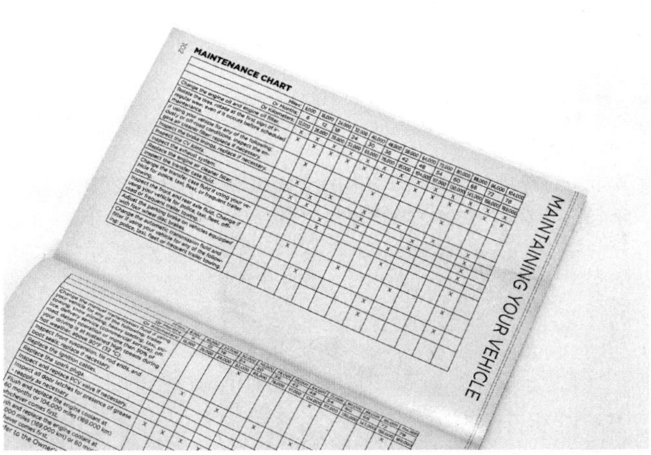

FIGURE 10-17 Service information listing the recommended oil and filter change mileage or time interval.

If the vehicle is due for an oil change, the level should still be checked first. A low reading on the dipstick could indicate that a seal or gasket is leaking, or that the engine is using oil. Either of these situations requires further investigation. Refer to the service information to select the correct oil for the vehicle. If the vehicle requires an oil change, refer to the chapter on Engine Lubrication service. It includes skill drills on how to change the engine oil and filter.

Engine Coolant

The engine cooling system depends on the coolant to transfer excessive heat from the engine to the radiator. Proper coolant level is critical to this. The engine coolant level and condition should be checked whenever the oil level is checked. Coolant levels that are low indicate a coolant leak that must be diagnosed and repaired.

To check the coolant level, some vehicles have a transparent reservoir or surge tank marked with "hot" and "cold." This allows checking the coolant without removing the cap (**FIGURE 10-18**). Some vehicles also may have an overflow tank with a tube leading from the radiator cap filler neck to a transparent overflow tank. The level can usually be seen through the side of the tank and compared to the marks, which indicate the hot or cold fluid levels. Older vehicles may not have an overflow tank. They are checked by removing the radiator cap and checking that the coolant level in the radiator is about 1.5" (38 mm) below the filler neck.

FIGURE 10-18 Checking the coolant level on a clear reservoir.

SAFETY TIP

Do not remove the pressure cap (radiator cap) when the engine is warm or hot. The system is under pressure, and removing the cap could allow the coolant to immediately boil and spray out, causing severe burns. Always allow the system to cool before removing the cap.

Because the coolant also provides freeze protection, the coolant's freeze point also should be checked. A weak antifreeze solution could allow the coolant to freeze during cold temperatures. Water expands when frozen. Freezing of the coolant could crack the engine block, radiator, or other cooling system parts. The freeze protection level can be measured with an antifreeze hydrometer **SKILL DRILL 10-5**.

A refractometer also can be used to measure the freeze protection level of coolant. A drop or two of coolant is placed on a sample plate and the cover closed. Looking through the viewfinder, the protection level can be read. In many cases, there are several scales you will be able to see. Common scales include battery acid, ethylene glycol antifreeze, and propylene glycol antifreeze. Make sure you are reading the proper scale. To measure the freeze protection of coolant with a refractometer, follow the step in **SKILL DRILL 10-6**.

Brake and Clutch Fluid

A hydraulic braking system depends on a special fluid called brake fluid. Brake fluid is stored in a reservoir attached to or near to the brake system master cylinder (**FIGURE 10-19**). If the brake fluid level gets too low, air can be pulled into the hydraulic system, which causes the brake pedal to be soft and too low. This causes the brakes to work poorly or not at all.

SKILL DRILL 10-5 Using a Hydrometer to Measure the Freeze Protection Level

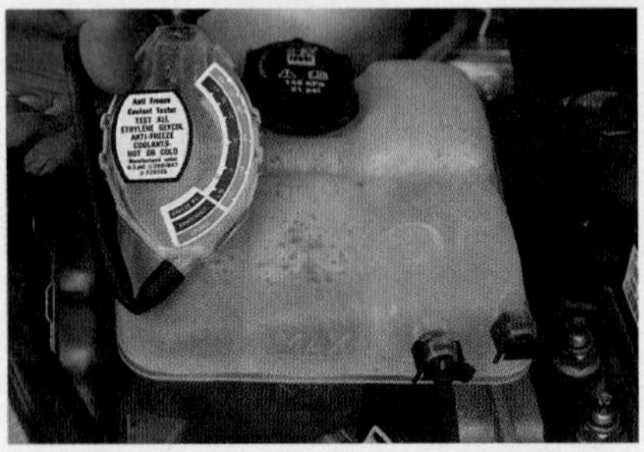

1. Select a hydrometer that is designed to be used with the type of antifreeze being tested.

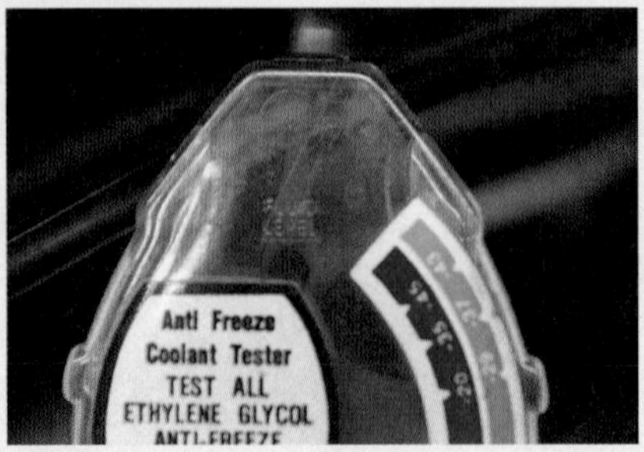

2. Draw enough coolant into the hydrometer to bring it up to the "fill" line.

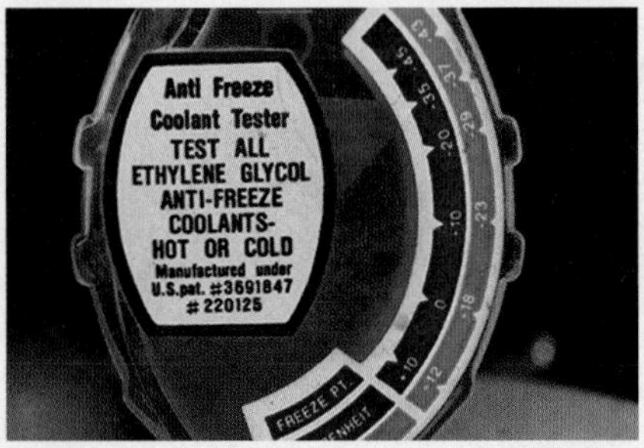

3. Hold the hydrometer vertically and read the freeze protection level.

SKILL DRILL 10-6 Using a Refractometer to Measure the Freeze Protection Level of Coolant

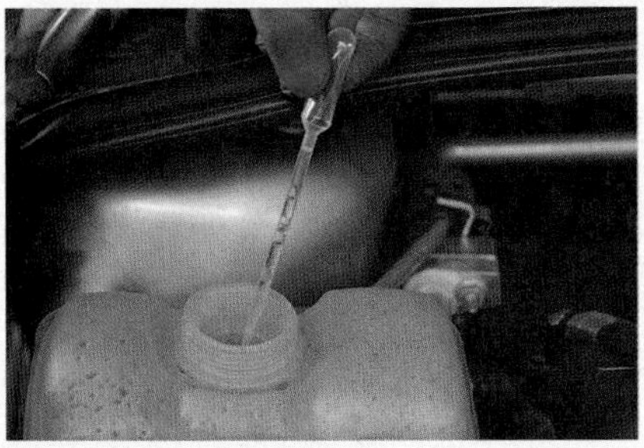

1. Remove the radiator cap and draw out a few drops of coolant.

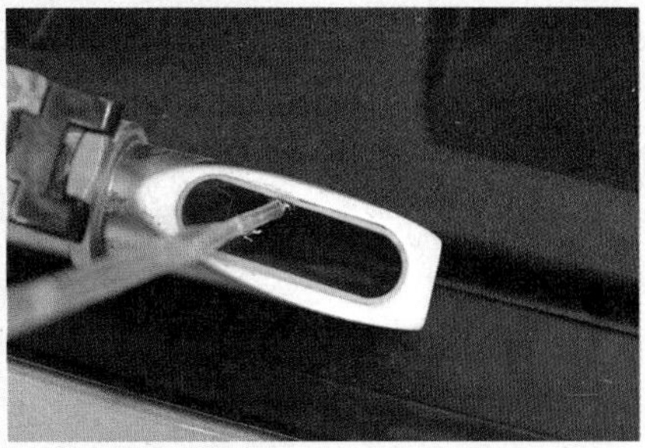

2. Place a drop or two of the coolant on the sample plate.

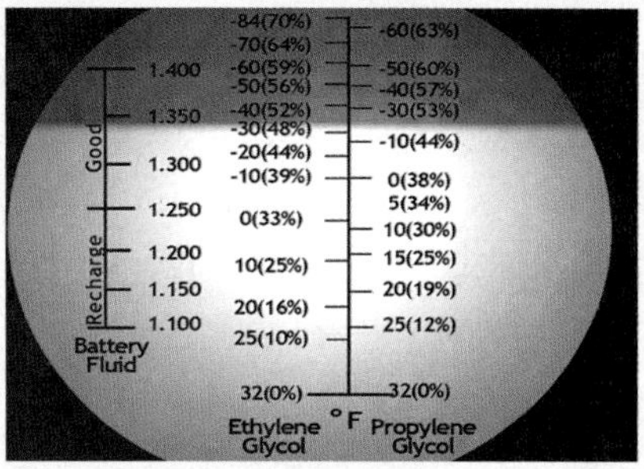

3. Close the diffuser plate, hold the refractometer level, look through the viewfinder, and read the proper scale.

Applied Math

AM-1: Ratios/Percentages: The technician can convert test readings in decimal or fractional form to a ratio or percentage form for comparison with the manufacturer's specifications, and vice versa.

A technician is servicing the cooling system of a vehicle with a 12-quart (11.52 liters) capacity per service information. The location has a cold climate. The shop supervisor informed the technician to fill the cooling system with a 60/40 mix of antifreeze and distilled water. Considering that the total capacity of the system is 12 quarts (11.52 liters), the technician needs to convert the percentages to a fractional form.

This can be done by multiplying the total quantity by each percentage. So, 12 quarts × 60% is 12 × 0.60, which equals 7.2 quarts (6.91 liters) of antifreeze. And 12 quarts × 40% is 12 × 0.40, which equals 4.8 quarts (4.60 liters) of distilled water.

Most climates require a 50/50 mix of antifreeze and water. This can be converted to a ratio or percentage form. Six quarts of distilled water and 6 quarts (5.76 liters) of antifreeze would be a 1:1 ratio. Concerning the percentage form, we have 50% distilled water and 50% antifreeze. This is often referred to as a 50/50 mix.

FIGURE 10-19 The brake fluid is stored in a reservoir attached to or near to the brake system master cylinder.

The brake fluid should be checked whenever the oil level is checked, or at least monthly. Also, check the color of the brake fluid. Most brake fluid is very clear. If it is getting dark, it is most likely becoming oxidized or contaminated and needs to be changed. Most brake fluids are **hygroscopic**, meaning they absorb water from the atmosphere. Because of this, most manufacturers recommend changing brake fluid every two to four years. However, the best process is to check the condition of the brake fluid with either a brake fluid tester or brake fluid test strips. Check the service information or the vehicle owner's manual for the specified fluid type.

▶ TECHNICIAN TIP

Brake fluid test strips use leached copper content in the brake fluid to show moisture contamination. The copper leaches from the steel brake lines. Because of this, hydraulic clutch systems that use plastic master cylinders and clutch lines cannot be tested with test strips. A brake fluid tester should be used in this situation.

Vehicles with a manual transmission may have a hydraulic clutch system that uses brake fluid. Depending on the vehicle, the brake master cylinder reservoir also may supply the clutch master cylinder. Or the clutch may have a separate fluid reservoir (**FIGURE 10-20**). Be sure to check both of these reservoirs for the proper fluid level. Refer to the chapter on Hydraulics and Power Brakes for an explanation of the different types of brake fluid.

FIGURE 10-20 Some hydraulic clutches use a reservoir separate from the brake master cylinder.

> ▶ **TECHNICIAN TIP**
>
> Low master cylinder brake fluid levels usually indicate one of two possible issues with the system: Either there is a brake fluid leak in the system, or the disc brake pads are worn. If the brake fluid level is low, inform your supervisor; it may be necessary to perform a brake inspection.

FIGURE 10-21 The power steering fluid reservoir. **A.** Mounted on the engine-driven hydraulic pump. **B.** Mounted separately.

FIGURE 10-22 Checking power steering fluid.

Power Steering Fluid

Most vehicles are equipped with power-assisted steering systems. The power for the system usually comes from an engine-driven hydraulic pump. Some vehicles use an electric motor to drive a hydraulic pump or directly operate the steering linkage. These are typically found on hybrid-electric vehicles.

The pump delivers fluid under pressure to the power unit at the steering box, or rack-and-pinion. The fluid reservoir can be mounted as part of the engine-driven hydraulic pump, or it can be a separate container (**FIGURE 10-21**). The power steering fluid level must be at the proper level to avoid drawing air into the hydraulic system. It also must prevent fluid overflow when the engine is hot.

In most cases, the power steering fluid level can be checked with a dipstick connected to the filler cap. The engine should be idling and the fluid hot. In many cases, the dipstick lists both a "cold" and a "hot" level, or a "safe" level (**FIGURE 10-22**). When checking the level, you also should check the appearance of the fluid. Dark or black fluid usually means the fluid is old and needs to be changed. The power steering fluid level should be checked as a normal part of the underhood inspection. Refer to the chapter on Servicing Steering Systems for more information on power steering fluid types and how to service them.

Automatic Transmission/Transaxle Fluid

The correct automatic transmission/transaxle fluid level (referred to as automatic transmission fluid from here on) is critical to the effective operation of the transmission. If the level is too low, slipping and shift timing faults can result. Because reverse gear usually requires a larger volume of fluid, a delayed shift into reverse could be an early indication that the fluid is low. If the fluid level is too high, the transmission fluid will churn and aerate. This can lead to low pressures, resulting in slipping clutches. Always make sure the transmission fluid level is correct.

Automatic transmission fluid level is usually checked with a dipstick located under the hood. This is usually near the front of

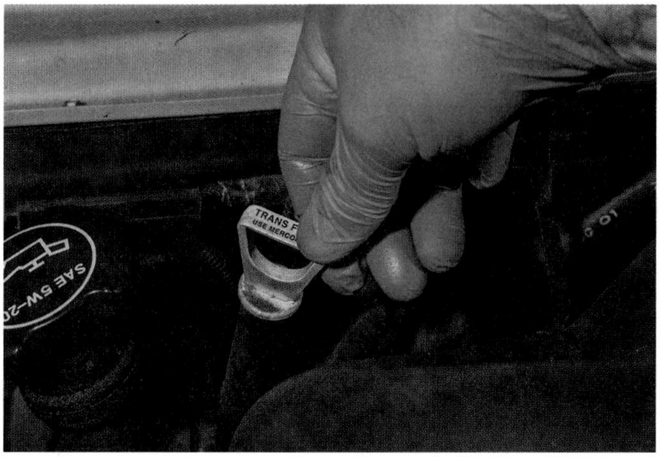

FIGURE 10-23 The transmission dipstick is usually located near the front of the transmission.

FIGURE 10-24 A few automatic transmissions are checked using a fill plug, or level plug, on the side of the transmission.

FIGURE 10-25 Wipe off the dipstick.

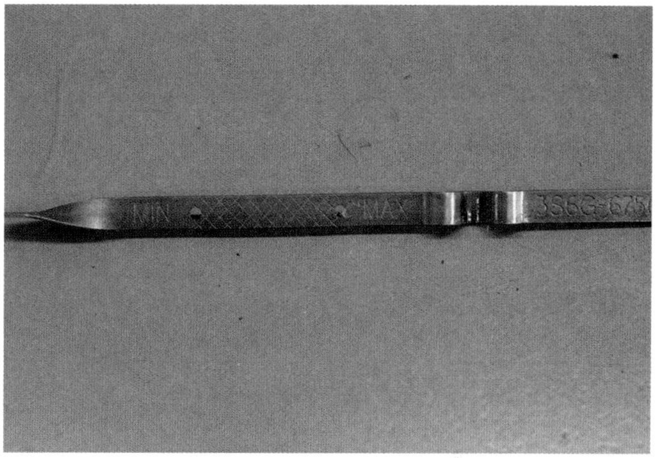

FIGURE 10-26 Observe the markings.

the transmission (**FIGURE 10-23**). Some automatic transmissions do not have a dipstick; they are checked using a fill plug, or level plug, on the side of the transmission (**FIGURE 10-24**).

If the transmission does use a dipstick, the fluid level is usually checked with:

- the engine running,
- the transmission warmed up fully, and
- the gear selector in park or neutral, depending on the vehicle.

When checking transmission fluid:

1. Wipe off the dipstick (**FIGURE 10-25**).
2. Observe the markings (**FIGURE 10-26**).
3. Reinsert the dipstick fully, remove it once again, hold it horizontal, and read it (**FIGURE 10-27**).
4. Turn the dipstick over and read the back side. The lowest reading between the front and back is likely the most accurate reading (**FIGURE 10-28**).

If adding transmission fluid, make sure you select the correct fluid. There are a number of different transmission fluids

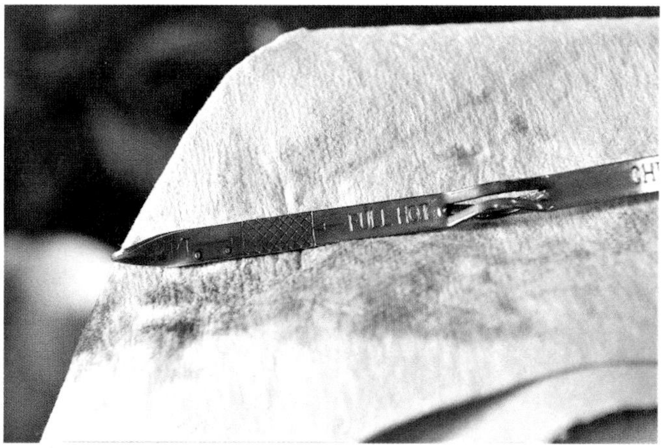

FIGURE 10-27 Reinsert the dipstick fully, remove it once again, hold it horizontal, and read it.

specified for various vehicles. Fluid is added through the dipstick tube, so use a funnel. The lines between "full" and "add" are usually only about 1 pint (0.5 liters), so add only a small amount of fluid at a time, checking the fluid level regularly.

In recent years, vehicle manufacturers have been eliminating the transmission dipstick on some of their vehicles. This prevents the vehicle owner from installing the incorrect fluid or overfilling the transmission. It also provides one less entry point for dirt and contaminants to enter the transmission.

On transmissions without a dipstick, it is critical to check the service information for the proper procedure for checking the fluid level. Many late-model Ford and Toyota vehicles have what looks like a drain plug installed in the transmission pan. With the vehicle running, this plug is removed. Fluid is forced up into the transmission through this hole. The hole has a tube attached to it inside the transmission (**FIGURE 10-29**). Once the fluid level is at full, excess fluid will simply drain back out. The plug can then be reinstalled.

General Motors vehicles often have a threaded plug, located on the transmission case. The plug is removed while the vehicle is running. Again, fluid is added until transmission fluid flows out of the hole, and then the plug is reinstalled (**FIGURE 10-30**).

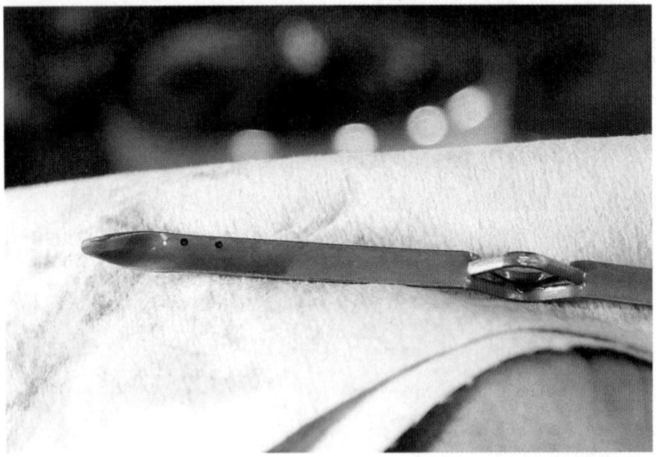

FIGURE 10-28 Turn the dipstick over and read the back side. Use the lowest reading.

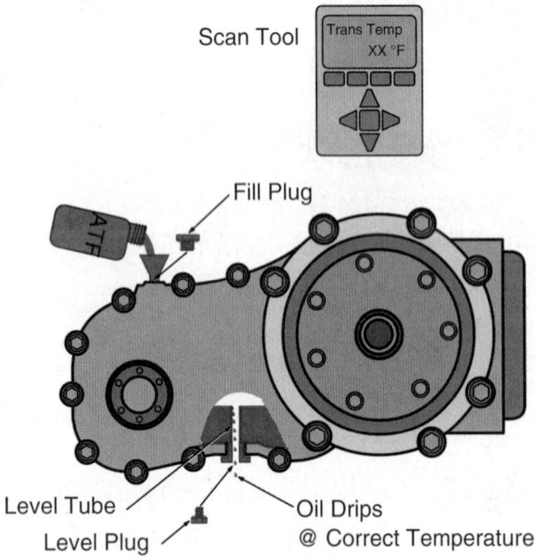

FIGURE 10-29 On transmissions with the fill tube on the bottom of the pan, once the fluid is at the correct level, additional fluid spills out.

> **TECHNICIAN TIP**
>
> Many newer vehicles—typically European models and some Asian imports—do not have a specified method of checking the transmission fluid level. The transmissions are considered sealed and lubricated for the life of the vehicle. The manufacturers have determined that the transmission fluid will not be low as long as there are no leaks. Any leak requires that the transmission be repaired.

Manual Transmission/Transaxle Fluid

In most cases, transmission fluid level on manual transmissions/transaxles, transfer cases, and differentials can only be checked from under the vehicle. So, it will be covered in the under-vehicle inspection section of this chapter.

Diesel Exhaust Fluid

Some late-model diesel-powered vehicles use a fluid called **diesel exhaust fluid** (DEF), one example being AdBlue™.

DEF is injected into the exhaust stream to reduce oxides of nitrogen during certain driving conditions. Because DEF is consumed over time, it has to be replenished periodically, ideally during oil changes. The DEF filler cap is often located under the hood and may be colored blue (**FIGURE 10-31**).

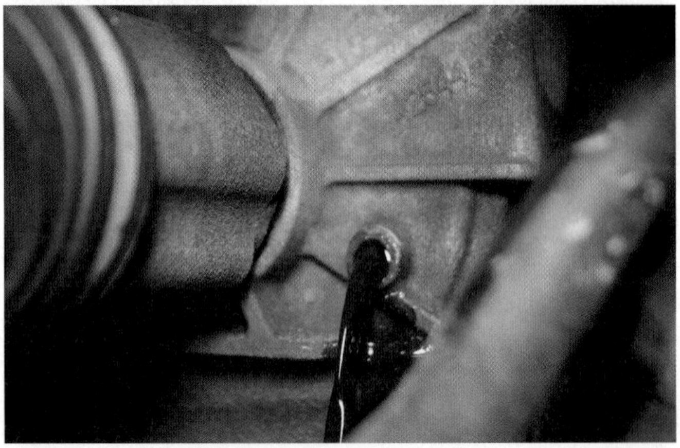

FIGURE 10-30 On transmissions with a fill plug on the side of the transmission, fill until transmission fluid just starts to come out.

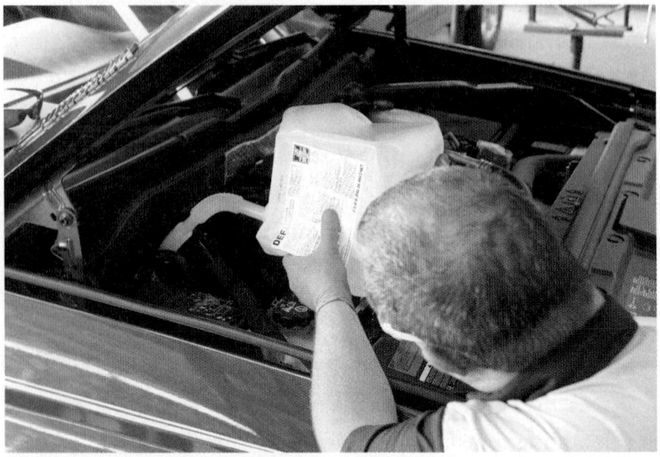

FIGURE 10-31 Adding DEF to the reservoir.

Do not make the expensive mistake of putting washer fluid (or any other fluid) in the DEF tank.

Windshield Washer Fluid

The windshield washer system on any vehicle is an important safety feature. Driving in muddy conditions, light mist, or fog requires that the washer system be ready to work when needed. The washer fluid reservoir is normally located under the hood of the vehicle (**FIGURE 10-32**). Some vehicles also may have a separate washer reservoir for the rear wiper. It is typically located somewhere in the rear hatch or trunk area. Refer to the owner's manual or service information to locate the filler cap.

The windshield washer fluid should be checked whenever the oil is changed, or at each vehicle service. Driving in dusty conditions may require that the windshield washer fluid be checked more often. It is important to use properly formulated and mixed washer fluid, especially in freezing weather. The fluid is normally purchased premixed in the proper ratio that protects against freezing.

To check and add to the windshield washer fluid, follow the steps in **SKILL DRILL 10-7**.

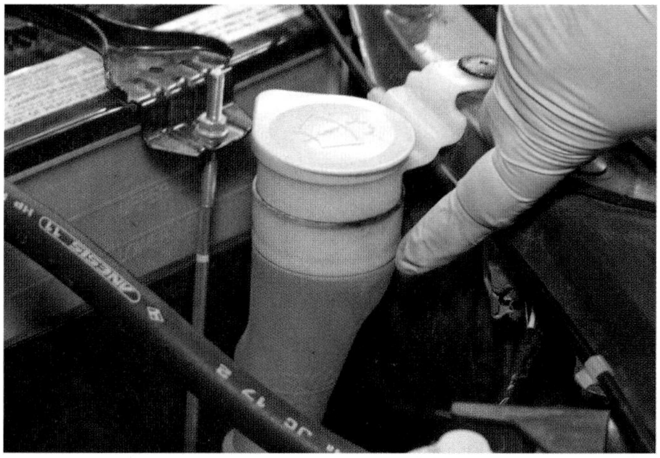

FIGURE 10-32 The washer reservoir is usually located under the hood.

▶ TECHNICIAN TIP

Never use laundry or dishwashing detergent to top off the reservoir, as the chemicals in the detergent can damage the vehicle's paint.

▶ Engine Drive Belts

LO 10-03 Perform belt, hose, and air filter/cabin air filter inspection.

Engine drive belts are used to operate the various accessories on the engine. Examples include the water pump, power steering pump, air conditioner compressor, and alternator. As the vehicle ages, these belts wear, along with their associated idler and tensioner pulleys. Some manufacturers recommend that the belts be replaced at about five years of age, as part of a preventive maintenance program.

A newer belt technology is becoming more common. These are called stretchy belts or, as one manufacturer named theirs, Stretch-Fit belts. They do not use a tensioner for tensioning the belt. Their stretchiness applies an appropriate amount of tension to the belt over its useful life. Some vehicle manufacturers claim this can be up to 150,000 miles (241,400 km). So, always check the manufacturer's scheduled service guide for belt replacement intervals.

There are two main types of accessory drive belts: the V-type and the serpentine type (**FIGURE 10-33**). A V-type belt sits inside a deep V-shaped groove in the pulley. The sides of the V-belt contact and wedge in the sides of the pulley. Serpentine-type belts have a flat profile with a number of grooves running lengthwise along the belt. These grooves are the exact reverse of the grooves in the outer diameter of the pulleys. They increase the contact surface area and prevent the belt from slipping off the drive pulley as it rotates.

Visually inspect the drive belts whenever the hood is opened for service. The engine will quickly overheat if the belt breaks or comes off. This is because the water pump is the most important component driven by the belt.

If the belts are more than five to six years old, check the manufacturer's replacement schedule in the service

SKILL DRILL 10-7 Checking and Refilling Windshield Washer Fluid

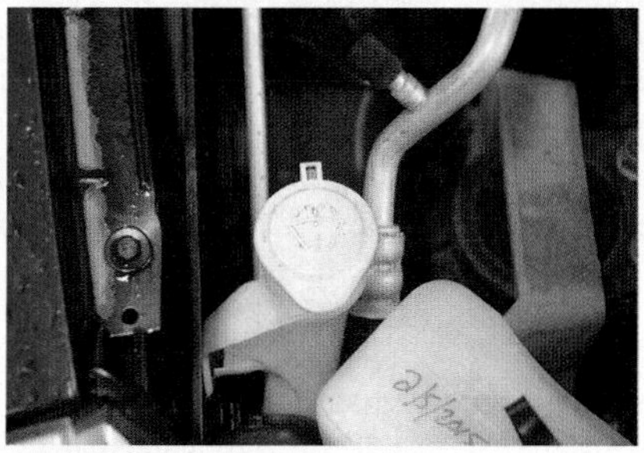

1. Locate the front windshield wiper fluid container.

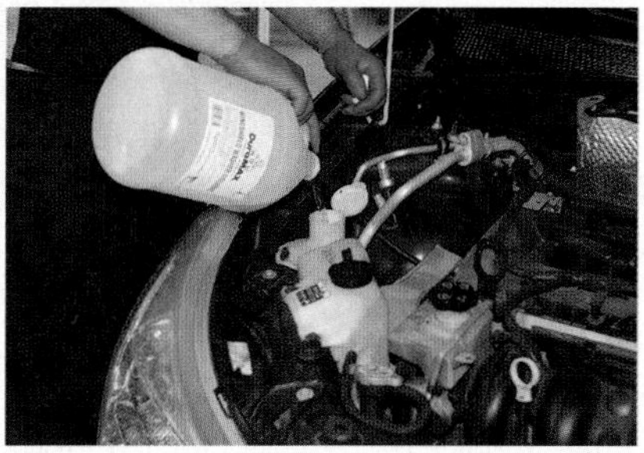

2. Check the windshield washer fluid level. If the level is low, refill the reservoir with the appropriate washer fluid.

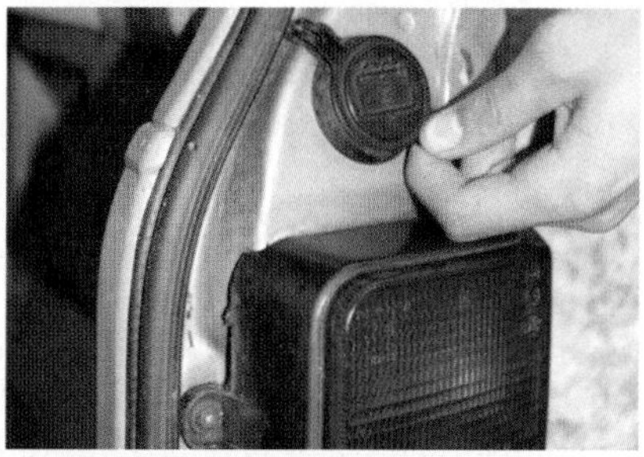

3. If equipped, check and fill the rear window washer reservoir.

FIGURE 10-33 Drive belts. **A.** A V-type belt fits into the V of the drive pulley. **B.** Serpentine-type belts have a flat profile with a number of grooves running lengthwise along the belt.

information. Most vehicles using a serpentine belt also have a spring-operated tensioner and pulley. This tensioner may have a built-in damper that reduces noise and vibration. Some manufacturers recommend that the tensioner be replaced along with the belt.

> ▶ TECHNICIAN TIP
>
> Some belts, called stretchy belts or Stretch-Fit belts, do not use a method for tensioning the belt. When removing them, they usually need to be cut with side cutting pliers.

The belts should be checked for the following:

- Cracks—Cracks that exceed a certain number per inch in a belt indicate that the belt may soon fail and should be replaced (**FIGURE 10-34**).
- Oil soaking—A belt that has been soaked in oil will not grip properly on the pulleys and will slip. If the oil contamination is severe enough for this to happen, replace the belt (**FIGURE 10-35**).
- Glazing—Glazing is shininess on the surface of the belt, which comes in contact with the pulley (**FIGURE 10-36**). If the belt is extremely worn, the glazing could be due to the belt bottoming out, and it should be replaced. If it is not old and worn, glazing could indicate that the belt is not tight enough. Tightening the belt may be all that is necessary, depending on how bad the glazing is.

FIGURE 10-34 A serpentine belt with excessive cracks should be replaced.

FIGURE 10-35 Oil-soaked belt.

FIGURE 10-36 Glazed belt.

- Tears—Torn or split belts are unserviceable and should be replaced (**FIGURE 10-37**).
- Bottoming out—When a V-type belt or serpentine belt becomes extremely worn, the bottom of the V may contact the bottom of the groove in the pulley, preventing the sides of the belt from making good contact with the sides of the pulley grooves. This reduced friction causes slippage. A belt worn enough to bottom out should be replaced (**FIGURE 10-38**). Serpentine belts are checked for wear by use of a small tool that fits into the grooves on the belt. The tool should sit higher than the ridges in the belt if the belt is still serviceable (**FIGURE 10-39**).

Hoses

The vehicle usually has two large radiator hoses and some smaller heater hoses that carry hot coolant through the system (**FIGURE 10-40**). At the radiator, there is a large upper radiator hose and a large lower radiator hose. The smaller heater hoses (usually two) run from the engine block, manifold, or water pump to connections at the heater assembly (near the firewall).

The engine should be cool when inspecting the hoses. A hot engine has pressure in the cooling system that may make a soft hose feel stiff, when it really may need to be replaced. If the engine is hot, look for bulging in the hoses (**FIGURE 10-41**).

FIGURE 10-37 Torn belt.

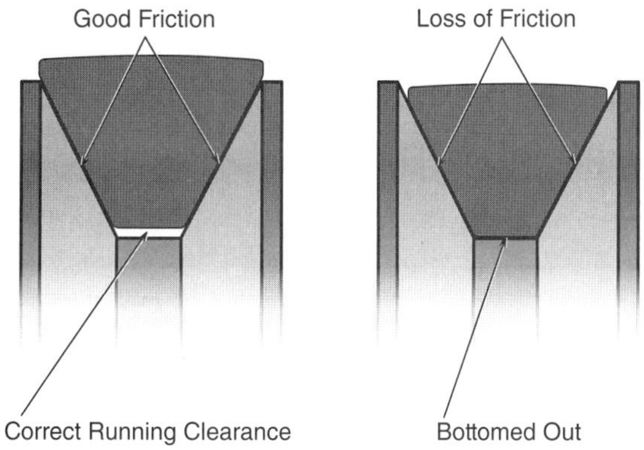

FIGURE 10-38 A V-belt worn enough to bottom out should be replaced.

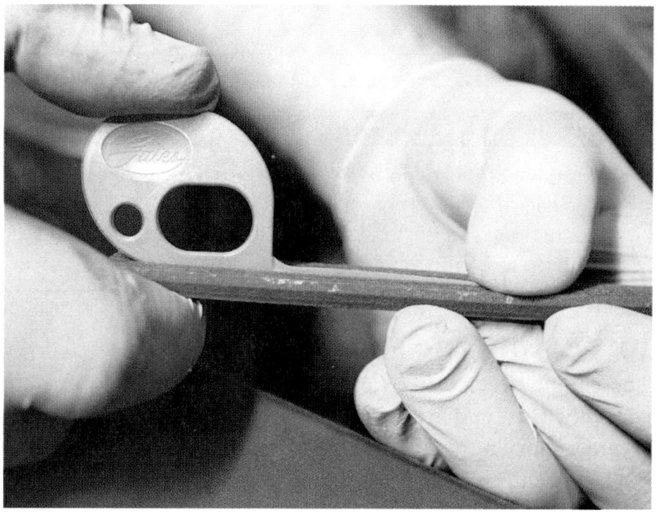

FIGURE 10-39 A serpentine belt wear tool should fit into each groove and still stick up higher than the ridges in the belt.

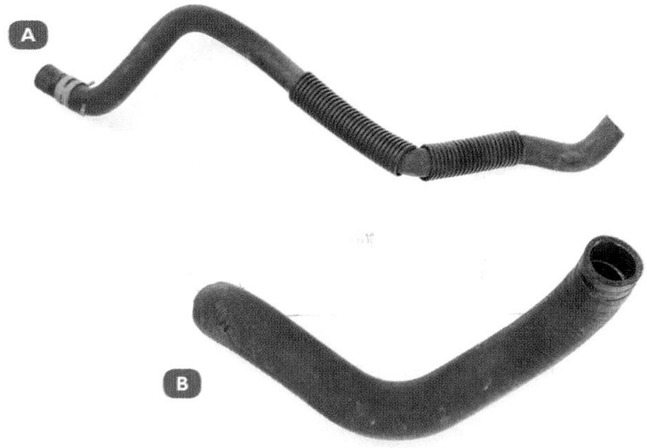

FIGURE 10-40 **A.** Heater hose. **B.** Radiator hose.

FIGURE 10-41 Inspecting coolant hoses.

SAFETY TIP

Never remove the radiator cap on an engine that has been partially or fully warmed up. Doing so can cause severe burns due to the pressurized coolant. If the radiator cap needs to be removed, wait until the engine cools.

▶ TECHNICIAN TIP

In most cases, hoses need to be felt to determine their condition. They should be neither too hard nor too soft. Also, the relative stiffness should be consistent over the length of the hose. If you feel differences in stiffness, the hose should be replaced.

Inspecting the Air Filter

The engine needs a free flow of clean air in order to operate correctly. Dust and grit in the air can be very abrasive and will severely shorten the life of the engine if not filtered out. If the filter element is not fitted correctly and does not seal properly, dirty air can bypass the filter and enter the engine.

The location of the air filter varies depending on the type of fuel system on the vehicle. Check the service information for the exact procedure. Some air filters are mounted to the top of the engine, usually found on older vehicles using a carburetor or throttle body fuel injection. The air cleaner on a multiport fuel-injected vehicle is typically located in a rectangular box within the air induction system. While inspecting the air filter, take a look at the air cleaner housing and ductwork for cracks or holes. These conditions would allow unfiltered air to enter the engine.

▶ TECHNICIAN TIP

The paper filter element actually becomes more efficient at filtering dirt particles the more it is used. This is because the passageways become smaller as dirt is caught in them. Smaller and smaller dirt particles are caught over time. However, if the filter becomes too clogged, it will restrict air, which reduces engine power output.

Once the air filter is removed, it is fairly easy to inspect. First, inspect it for any damage to the sealing surfaces. If they are bent or damaged, they won't seal dirt out and therefore require replacement. If the filter is in good shape, then inspect it for clogging. This is best done by holding the filter up to light and looking through it. If it is bright, then it is not clogged. If little to no light comes through, then it is clogged and needs to be replaced. To inspect and change the air filter, follow the steps in **SKILL DRILL 10-8**.

▶ TECHNICIAN TIP

Just opening the filter housing up and looking at the top of the filter is not a proper way to inspect the filter. In most vehicles, the air flows UP through the filter, so it is the bottom side that is the dirty side. Looking only at the top side of the filter will give you inaccurate information about the filter. Remove it and hold it up to the light.

Cabin Air Filter

Many vehicles now include a cabin air filter in the heating, ventilation, and air conditioning (HVAC) system to filter the air before it enters the cabin. The filter is housed in the air box and can be accessed from one of a variety of positions, depending on the vehicle. The access may be from under the hood near the firewall, under the windshield, or behind the glove box (**FIGURE 10-42**). It is usually fairly easy to remove and replace once you find the access cover.

The cabin air filter should be inspected during every service and replaced according to the manufacturer's specified interval. Typically, this is once a year or every 12,000 to 15,000

SKILL DRILL 10-8 Inspecting and Changing an Air Filter

1. On fuel-injected engines, unlatch or unscrew the filter housing fasteners to remove the air filter. It may be necessary to loosen the clamps and hoses on the induction tubing to remove the filter housing cover.

2. On carbureted or throttle body–injected engines, remove the top of the air filter by unscrewing the wing nut, and remove the air filter.

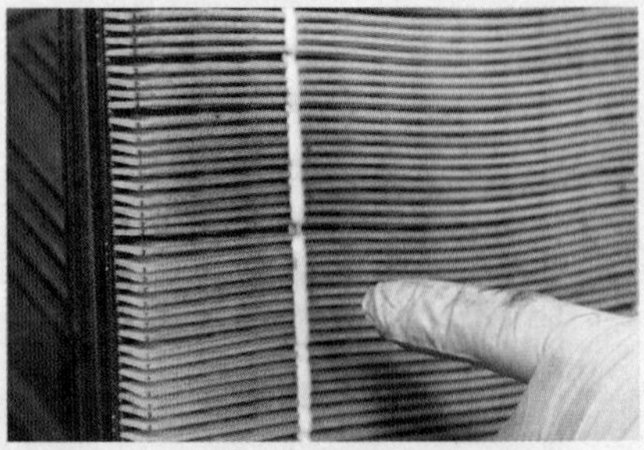

3. Inspect the air cleaner element by holding the filter element up to light and looking through it. If it is bright with no tears or cracks, it can be reused. If it is dark or damaged in any way, it needs to be replaced.

SKILL DRILL 10-8 Inspecting and Changing an Air Filter (Continued)

4. Clean the inside of the air filter housing, and inspect it and any ducts for cracks. If the air filter is being replaced, obtain a new air filter and compare it with the old one to ensure that they are exactly the same.

5. Place the new air filter inside the filter housing, making sure it is aligned properly on both sides.

6. Replace the cover of the air filter housing and tighten the latches, screws, or wing nut until completely closed. Reinstall any induction tubing or clamps.

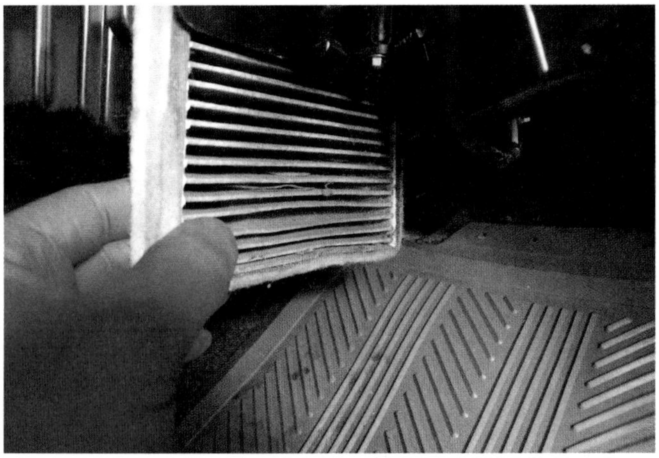

FIGURE 10-42 A typical location of a cabin filter.

miles (19,000 to 24,000 km). When inspecting the cabin air filter, use the same guidelines as for an engine air filter. Hold it up to a light, and look at it to see if the light shines through, or if it is blocked from being dirty. Also, check it for any cracks, tears, or deformities that would cause it to be ineffective (**FIGURE 10-43**).

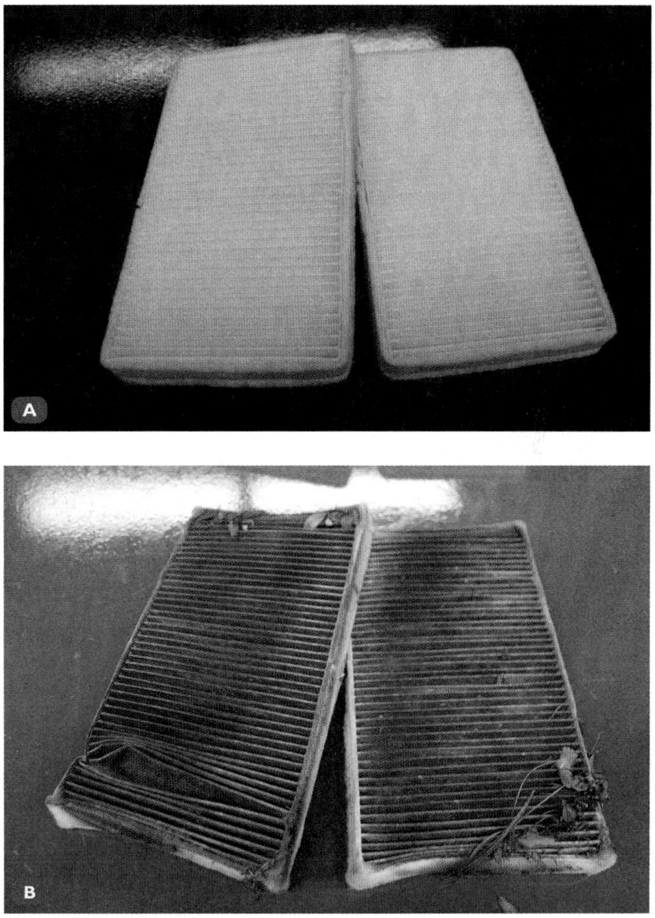

FIGURE 10-43 Cabin filters. **A.** Clean. **B.** Dirty.

▶ Under-Vehicle Inspection

LO 10-04 Perform under-vehicle inspection.

The under-vehicle inspection is a systematic visual inspection of all major vehicle systems that can be accessed from below. Since these parts are not accessible from above, they often get overlooked. Also, because they are exposed to the elements, they are prone to additional wear and damage.

With the vehicle safely lifted on a hoist, an under-vehicle inspection is a good way to get a feel for the overall condition of the vehicle (**FIGURE 10-44**). Tire issues, leaks, worn parts, and structural damage can be found relatively easy. This allows you to inform the customer of any areas of concern before they become a real problem. Like all inspections, the results need to be documented on an inspection form. This can be done by paper or electronically, and given to your supervisor or service advisor to review with the customer.

Tire Inspection

The tires and their condition are one of the most important safety considerations on the vehicle. They are checked for pressure, wear patterns, damage, and tread depth. Tires that are underinflated or overinflated do not grip the road fully, and they tend to wear out sooner. Since tires loose air pressure over time, they need to be inspected and aired up periodically. This should commonly be performed every 3 months, or at every oil change.

Normal tire pressures vary from vehicle to vehicle. Recommended tire pressures for the vehicle are located on the vehicle manufacturer's tire placard. This is typically placed on the driver's side door pillar (**FIGURE 10-45**). The maximum tire pressure, located on the tire sidewall, is the maximum pressure for that tire. It is NOT the recommended pressure for the vehicle (**FIGURE 10-46**). At the same time, never inflate the tire above the specified maximum pressure, as the tire may explode. Tires are inflated using compressed air or nitrogen (**FIGURE 10-47**).

Adequate tread depth and even tread wear are also important for safety. A tire worn to a minimum tread depth may work fine on dry pavement but be dangerous in wet or snowy weather. Keep this in mind as you measure the tread depth. Also, tires don't always wear evenly. If this is the case, there is likely a steering, suspension, or tire inflation issue that needs to be identified. So, always measure at least three places across the tread to identify uneven wear in the tread (**FIGURE 10-48**).

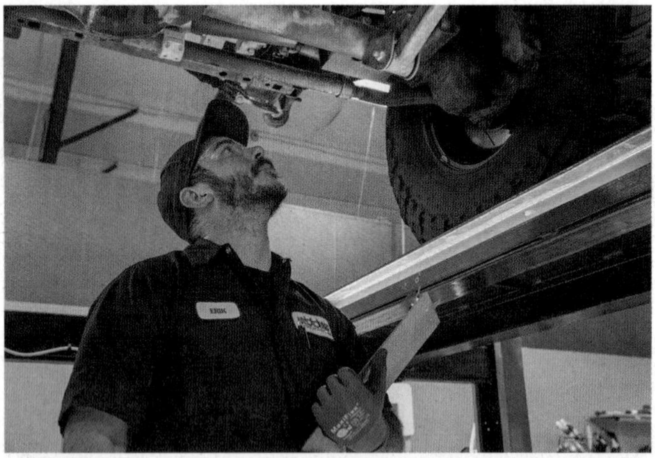

FIGURE 10-44 An under-vehicle inspection is a good way to get a feel for the overall condition of the vehicle.

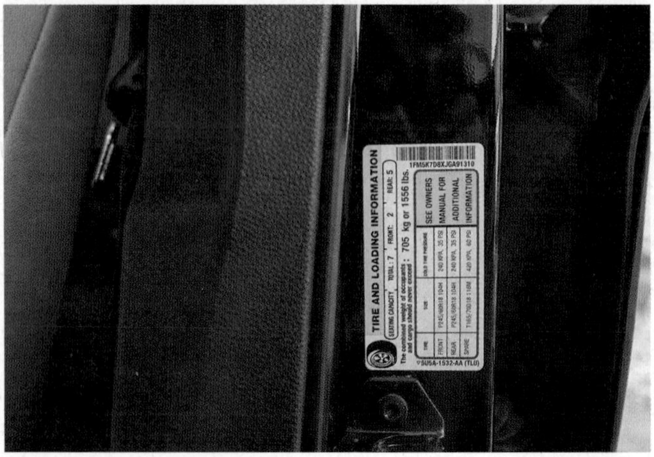

FIGURE 10-45 Typical tire placard.

FIGURE 10-46 Maximum tire inflation pressure. (Do not exceed!)

FIGURE 10-47 Adding air to a tire with a tire inflator.

FIGURE 10-48 Tire being checked for wear patterns, damage, and tread depth.

Finally, inspect the tire for damage, bulges, and age. Damage shows up as deep cuts, gashes, or objects that penetrate the tire. Deep cuts and gashes typically require tire replacement. Tires that have been penetrated with an object can be repaired with a plug-patch. But the hole must be no bigger than 1/4" (6 mm) in diameter. It must also be located within the tire's tread (not on the sidewall).

Bulges happen when belts break or separate in the tire's carcass, and require replacement of the tire. They can be located by running your hand over the tread and sidewalls of the tire and feeling for bulges. Just be aware that often times steel cords will be exposed on tires with bulges (**FIGURE 10-49**). These can be very sharp. So, always visually inspect the tire for exposed cords before running your hand over the tire.

Tire age is another important factor to consider when inspecting tires. Tires don't last forever. They have a useful service life that is typically between 6 and 10 years, depending on the manufacturer and the environment. It is hazardous to use a tire that has outlived its useful service life.

Tire age can be determined by locating the tire date code on the sidewall of the tire (**FIGURE 10-50**). Most tires use a four-digit code where the first two digits are the week of manufacture, starting in January. The last two digits indicate the year. Therefore, a tire with a date code of "3011" means that the tire was manufactured on the 30th week of 2011. Refer to the chapter on Servicing Wheels for more information on wheels and tires.

FIGURE 10-49 Because bulges in tires are due to broken belts, often the steel cords stick out of the tread. Be careful to not run your hand across them.

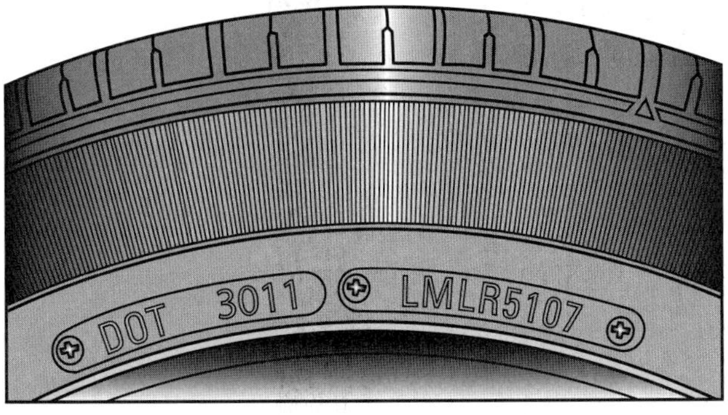

FIGURE 10-50 Four-digit DOT date code showing this tire was manufactured on the 30th week of 2011.

Checking Manual Transmission/Transaxle/Transfer Case Fluid Level

As a preventive maintenance task, fluid level of the manual transmission, transaxle, and/or transfer case should be checked. In most cases, fluid level on these components can only be checked from under the vehicle. The vehicle must be level. The level is usually accessed through a fill plug on the side of the transmission/transaxle. If fluid comes out of the fill hole when the plug is removed, allow any extra fluid to drain out, and then reinsert the fill plug. If no fluid comes out of the fill hole, carefully stick out a finger in the hole and bend your finger down to feel the level of the fluid (**FIGURE 10-51**). If the level is within a 0.25" (0.0006 mm) of the bottom of the fill hole, the level is okay, and the fill plug can be reinstalled. If the fluid level is lower than that, the proper fluid must be added until the level is even with the bottom of the fill hole. Also, not all manual transmissions use gear lube. Some use engine oil of a specified viscosity. Others even use a specified type of automatic transmission fluid. So, make sure you check the service information to determine the correct type of fluid required **SKILL DRILL 10-9**.

SAFETY TIP

Do not rotate the wheels or engine with your finger in the fill hole, as transmission parts could pinch or sever your finger.

Fluid level and condition can tell a technician a lot about what is going on inside. For example, consider metal filings found floating inside the transmission fluid. If this happens, then the technician knows that the transmission will have to be pulled apart or replaced. In the event of a suspected leak, the leak will need to be identified and repaired. Once repaired, the transmission fluid level should be checked and topped off. This will help avoid catastrophic failure. Make it a habit to visually inspect the transmission every time it is in for an oil change or other service. This will help you catch any issues before they become severe.

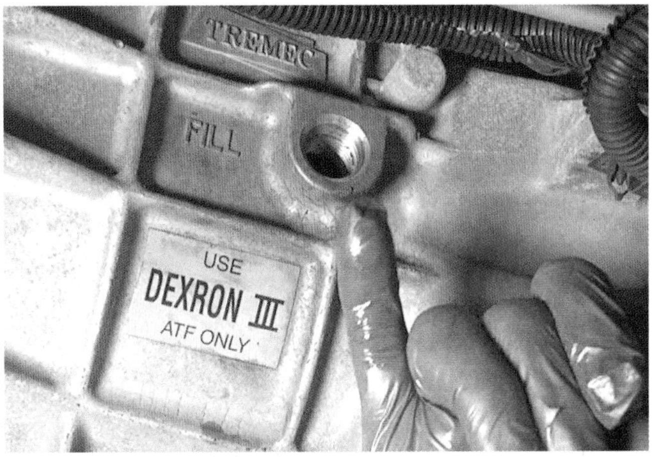

FIGURE 10-51 Checking the transmission fluid level in a manual transmission.

SKILL DRILL 10-9 Checking the Fluid Level of a Manual Transmission/Transaxle

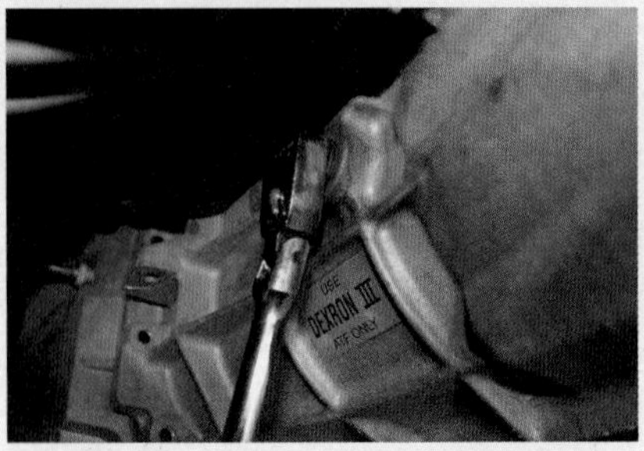

1. Safely raise and support the vehicle on the lift so that it is level. Inspect the transmission for leaks. Remove the filler plug, using the proper tool. Inspect the filler plug and fill hole for thread damage, and replace or repair if necessary.

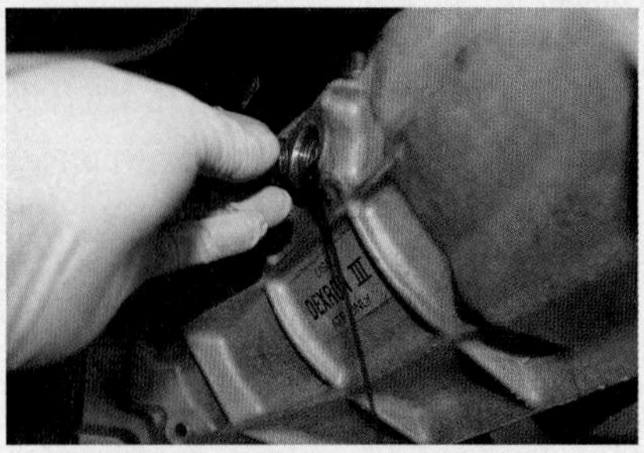

2. If the gearbox fluid begins to run out as the filler plug is removed, let the gearbox fluid seek its own level before reinstalling the filler plug. The gearbox fluid level should be at the bottom of the fill plug hole.

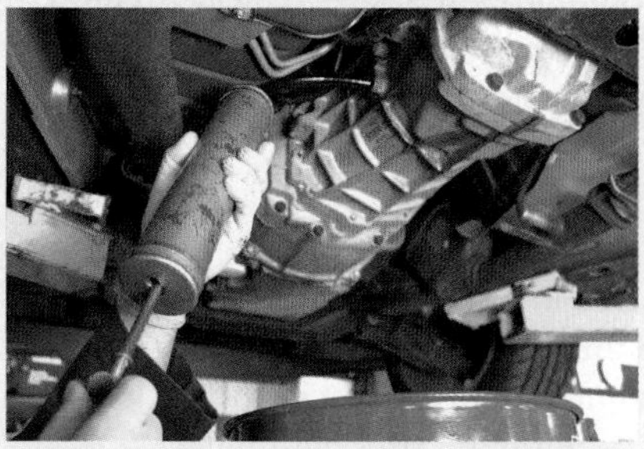

3. If the fluid level is low, refill with the specified fluid, reinstall the filler plug, and wipe the area around the filler plug hole with a clean shop towel. Tighten the filler plug to the specified torque.

Checking Differential Fluid

Checking and adjusting the differential/transfer case fluid level is very similar to checking the transmission fluid level. But there are some differences. Examples include limited-slip and posi-traction assemblies which require specially designed additives or fluid. Transfer cases may also require special lubricants as specified by the manufacturer. Always check the service information for the vehicle you are working on.

Differentials and transfer cases generally have fill plugs that can be used when checking the fluid level in the same way a transmission does. However, they often do not have a drain plug. In these cases, either a specific bolt may need to be removed, or a cover may need to be unbolted and removed. Follow the service information **SKILL DRILL 10-10**.

Checking for Fluid Leaks

The service bay should be clean and dry before driving the vehicle into the shop. This will help you to locate any fluid leaks that may be present on the vehicle. Active leaks leave telltale drips or puddles on the clean floor, making it easier to identify what may be leaking. Vehicles have a variety of fluids that can leak. Some fluids come in a variety of colors, such as red, which can be used for both automatic transmission fluid and antifreeze. Become familiar with each fluid's distinctive colors, feel, or smell:

- Brake fluid—May appear clear or light amber for DOT 3 and DOT 4. Whereas DOT 5 is usually purple, slightly slippery, and has an unpleasant, slightly acid-type smell.
- Automatic transmission fluid and some manual transmission fluid—Normally reddish in color, although some manufacturers may use a clear or amber color; it is very slippery and oily and has an oily smell.
- Power steering fluid—Typically red if it uses automatic transmission fluid, or clear in color if it uses power steering fluid.
- Coolant—Normally green, orange, or yellow in color; some manufacturers (e.g., General Motors) use a coolant that is red or light red. It is slippery, and has a sweet smell, like syrup.

SKILL DRILL 10-10 Checking and Adjusting the Differential Fluid Level

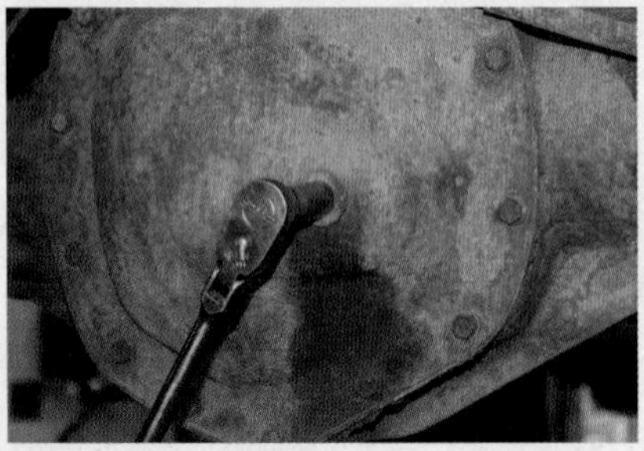

1. Safely raise and support the vehicle. Inspect the differential and transfer case for leaks. Position a clean drain pan under the filler plug. Remove the filler plug, using the proper tool. Inspect the filler plug threads for thread, and replace if necessary.

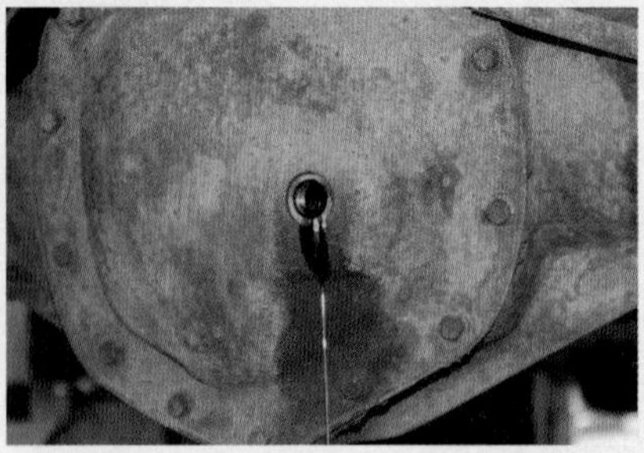

2. If the fluid begins to run out as the filler plug is removed, let the fluid seek its own level before reinstalling the filler plug. The fluid level should be at the bottom of the filler plug hole.

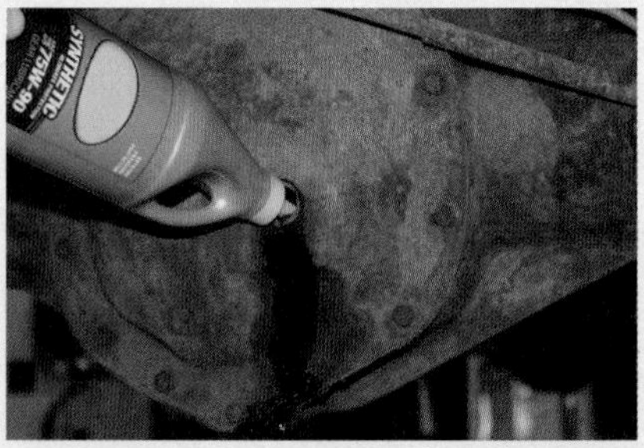

3. If the fluid level is low, refill with the specified fluid, reinstall the filler plug, and wipe the area around the filler plug hole with a clean shop towel. Tighten the filler plug to the specified torque.

- Engine oil—Clear, brown or black in color, very slippery and a bit thick, with an oily smell.
- Gear Oil—Light brown in color, very slippery and thick (like syrup) with an oily smell.
- Gasoline—Clear in color; evaporates easily, and has a distinctive gas odor.
- Diesel—Dirty clear in color, thin, has an oily smell.

Checking for leaks can be done as part of the under-vehicle inspection. The leaks are normally more visible on a warmed-up vehicle, although some coolant leaks only appear when the engine is cold. Discuss with the customer whether there are any unusual smells when the vehicle is running.

To check for leaks, drive the vehicle into a clean, well-lit work stall. Ideally, the stall will also have a lift so that the vehicle can be further inspected if a leak is suspected. Safely raise and support the vehicle. Use a flashlight or shop light to inspect the underside of the vehicle for any drips or wet areas. You may be able to identify the cause of the leak at this point.

Try to identify the type of fluid that is leaking and the area from which it is leaking. Remember that gravity tends to pull any leaking fluids down. So, always look toward the top of the wet area to help determine the source of the leak. If fluid leaks onto a moving part, it can be thrown a good distance, so check for a common source. Finally, a leak under pressure, such as coolant, can be sprayed a good distance from a small hole. So, always look for a stream, using a good light to help identify the location of the leak. Finding the leak's source will tell you which component on the vehicle is likely leaking.

If the location of an engine leak cannot be determined with the engine off, start the engine and wait a few minutes. Then carefully look to see if any leaks appear. If so, carefully inspect the components to identify the location of the leak. Remember to stay away from moving or hot parts. If you still can't see a leak, turn the engine off and inspect the engine again. Some leaks only occur shortly after the engine is shut down. To locate and identify fluid leaks, follow **SKILL DRILL 10-11**.

SKILL DRILL 10-11 Locate and Identify Fluid Leaks

1. Safely raise and support the vehicle on a lift. Use a light to check for fluid leaks at any place that fluids can leak.

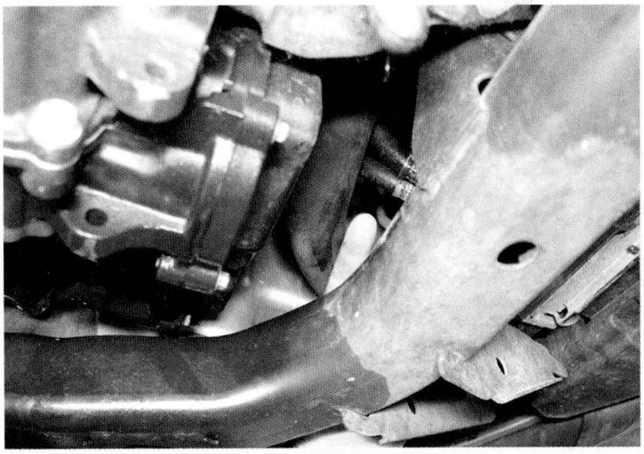

2. Once leaky fluid is found, follow it to its source.

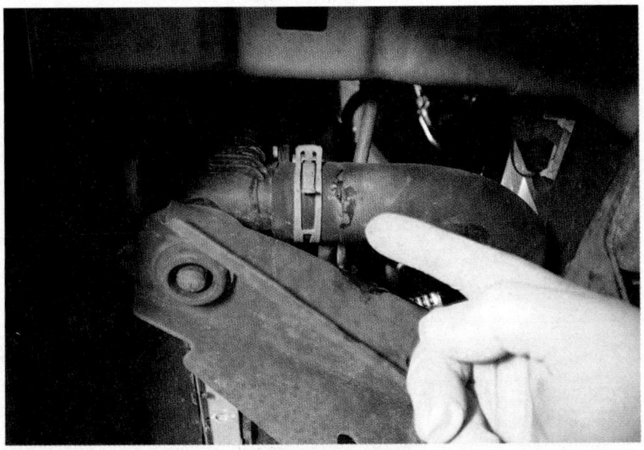

3. If it is hard to identify exactly where the leak is coming from, you may need to make it leak by pressurizing the cooling system, applying the brakes, or operating the leaking system.

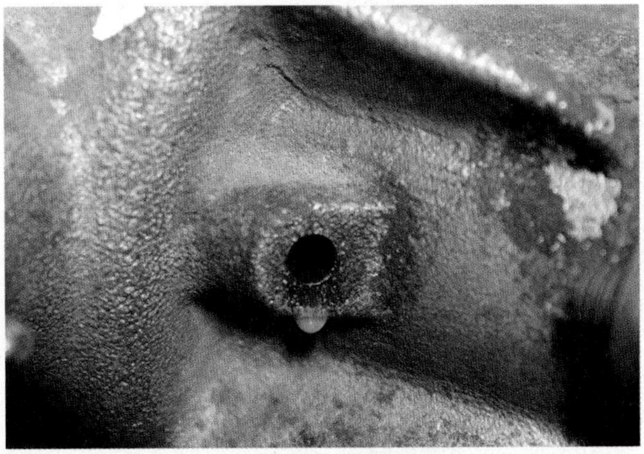

4. If you still can't make it leak, try releasing the pressure, or turning the system off. Sometimes, leaks happen only when the system is shut off.

Steering and Suspension Inspection

The steering and suspension systems manage the movement of the wheels and body as the vehicle is being driven. All of the joints, bushings, springs, and other components are subject to extreme forces, which cause them to wear. Worn parts cannot do their job like they should, and need to be identified and replaced. When inspecting the steering system, look for wear and other types of damage (**FIGURE 10-52**). This can occur in tie-rods and tie-rod ends, idler arm, pitman arm, and the steering gear or rack-and-pinion. The best way to locate parts with wear is to wiggle one of the front wheels side-to-side while feeling for play in each of the steering joints (**FIGURE 10-53**). Do this for each wheel.

When inspecting the suspension system, some components are under very high load. So, you may not be able to move the component by hand to feel any play in it. In this case, you will visually inspect the appearance of the components. In many cases, there will be indications of worn parts such as torn boots, or wear marks that indicate excessive movement. While inspecting the suspension system, also inspect:

- the bushings for tears or displacement (**FIGURE 10-54**)
- the grease fittings that need to be greased, if any
- the shock absorbers for leaks
- the brake hoses for cracks and bulging
- the brake lines for kinks and rust
- the wheel bearings for excessive play.

Drive axles and Driveshafts

Drive axles and driveshafts are equipped with joints that can wear out, and the axles themselves can be bent or damaged. The following list describes what to look for when inspecting these components.

- Constant-velocity joints (**CV joints**) and dust boots. Look for cracked, torn, or leaking boots (**FIGURE 10-55**).
- The driveshaft: Check for any excess movement in driveshaft universal joints. Look for any dents or bends in the shaft.
- The differential, rear axle, and rear suspension area: The rear axle includes the differential and axle shafts. Look for leaks around the differential, and check the rear shock absorbers, leaf springs, brake hoses, and lines.

FIGURE 10-52 Extremely worn steering components (tie rod end) can't hold the wheels in the proper position.

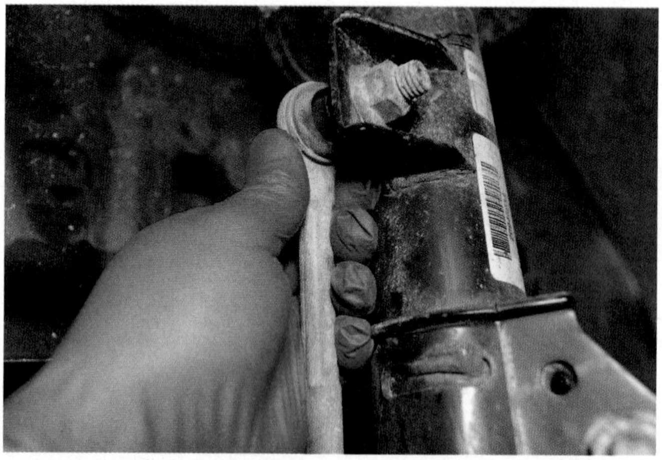

FIGURE 10-54 Extremely worn suspension components can't hold the wheels in the proper position.

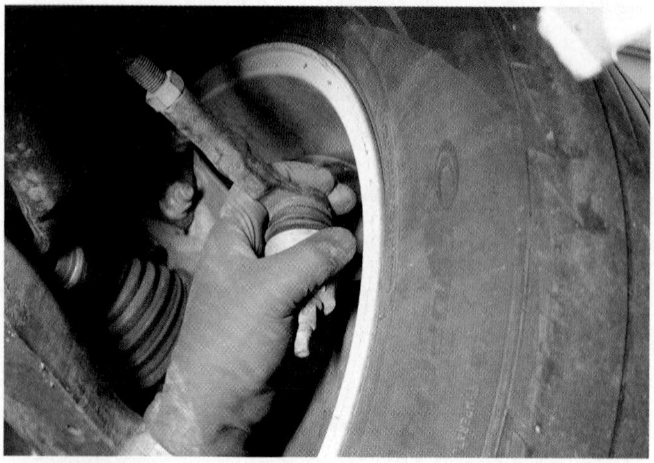

FIGURE 10-53 Wiggle one front wheel side-to-side while feeling for play in the joint.

FIGURE 10-55 A torn CV boot found during the inspection.

Engine, Transmission, and Exhaust

The engine, transmission, and exhaust have a number of items that need to be visually inspected. Inspect the following items as described.

- The engine: Look for torn or cracked motor mounts, coolant hoses, and belts (**FIGURE 10-56**).
- The transmission: Look for torn or cracked mounts, and faults or looseness in the clutch mechanism or shift linkage.
- The exhaust system: Check for signs of exhaust leaks, corrosion, or deterioration, including the exhaust hanger mounts. A good way to check the integrity of the exhaust pipe is to try to squeeze it along its length with a pair of arc joint pliers. If it is squishy, then it needs to be replaced. Check the condition of any heat shields.

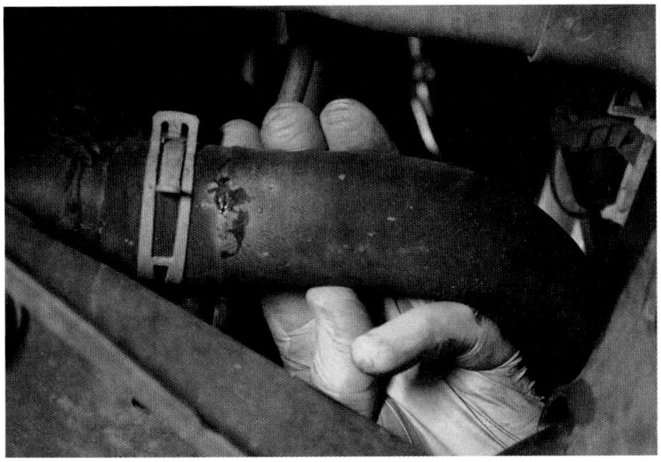

FIGURE 10-56 Lower radiator hoses get neglected and need to be inspected.

Other Items

- The parking brake cables: The cables are encased in a housing that attaches the parking brake lever or pedal to the rear brakes. Check for rusted, frozen, broken, or crushed cables.
- The fuel tank: The fuel tank is metal or plastic, depending on the vehicle. Inspection should include the filler tube and hose, the vent and fuel delivery lines, and the fuel tank straps and protective shields. The fuel tank must be secure and fuel lines inspected for damage or abrasion.

▶ TECHNICIAN TIP

When checking the fuel tank, any odor of gasoline indicates a leak. Keep checking until you find it. It is not normal for modern-day vehicles to have any odor of gasoline.

To perform an under-vehicle inspection, follow the steps in **SKILL DRILL 10-12.**

▶ Exterior Vehicle Inspections

LO 10-05 Perform exterior vehicle inspection.

A periodic inspection of the vehicle's exterior can prevent troubles that may cause safety or operational concerns. It is much better to discover a problem such as a worn tire or broken taillight bulb during an inspection than when the car is broken down on the side of the road, or the driver is pulled over by the police. A small percentage of owners check their own vehicle for problems, but most depend on the service technician to do it for them. Any time the vehicle is in the shop for maintenance or repair, you should perform this inspection.

SKILL DRILL 10-12 Performing an Under-Vehicle Inspection

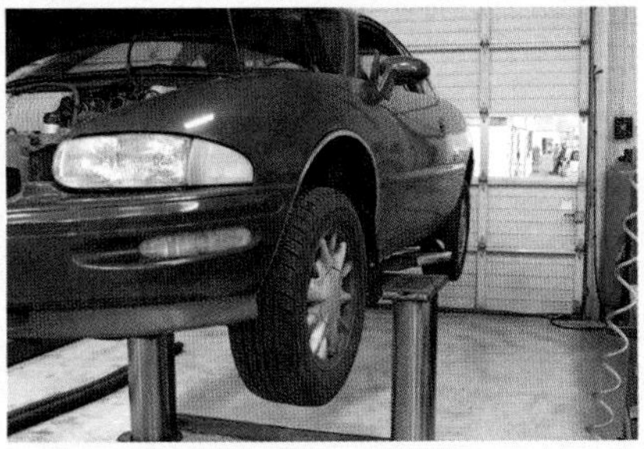

1. Safely raise and secure the vehicle at a comfortable working height. Work systematically. Pay particular attention to any fluid leaks.

2. Check the steering parts. Grasp the front and rear of the tire and pivot it to detect wear in the steering components. Grasp the tire at the top and bottom and pivot it to detect movement in the wheel bearings or ball joint. Look for missing or torn rubber boots around the tie-rod ends and steering rack. Inspect the rack bushings and other rubber suspension bushings.

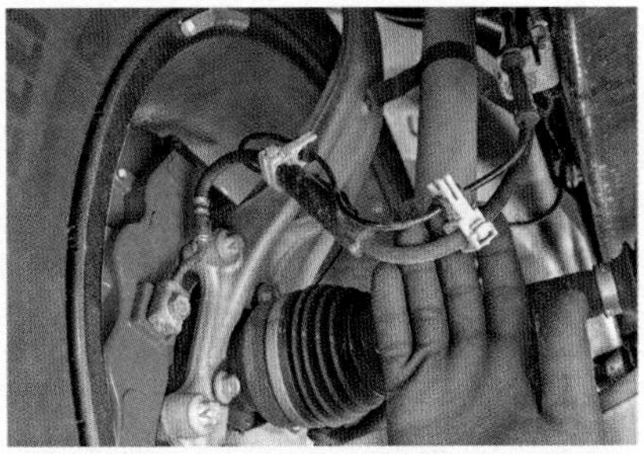

3. Inspect all four shock absorbers for signs of damage or leaks. Check the brake hoses and lines for signs of cracking, abrasions, or bulging.

SKILL DRILL 10-12 Performing an Under-Vehicle Inspection (Continued)

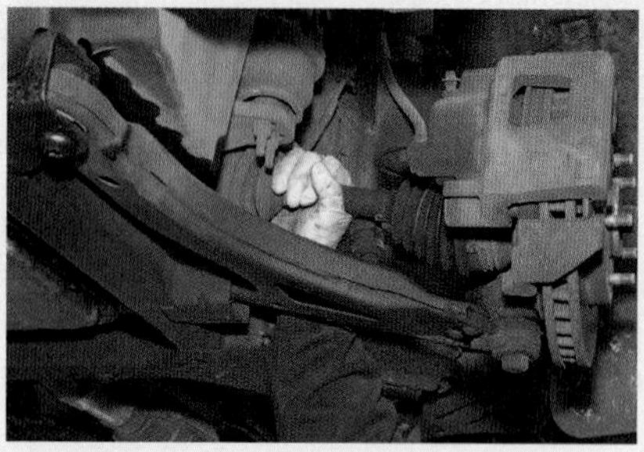

4. Check the front-wheel drive axles. On vehicles with front-wheel drive, examine the inner and outer CV boots for cracks or tears.

5. Check the transmission area. Trace and record the source of fluid leaks. Check the mount. With a manual transmission, check the clutch operating mechanism for damage. For an automatic transmission, check the shift linkage for any damage. If the transmission is electronically controlled, check the wiring for damage.

6. Check the exhaust system. Examine the catalytic converter, muffler, and resonator for signs of corrosion or deterioration. Inspect the heat shields. Check the tailpipe for corrosion, and check for damaged or missing hangers.

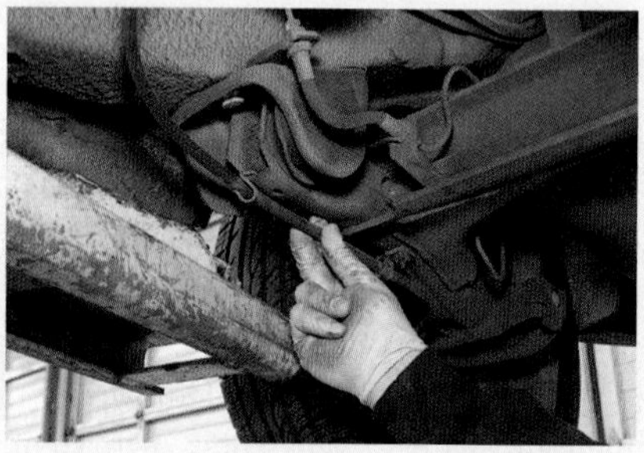

7. Check the parking brake cables. Inspect the parking brake cable to make sure it is not frayed, damaged, or binding. Look for rusted or swollen cable housings. Pull on the cables, and check that the parking brake applies. Lubricate the cable if specified.

8. Check the drive shaft. On rear-wheel drive vehicles, inspect the drive shaft universal joints for signs of excess movement or rust. To check for wear, rotate the shaft and flange in opposite directions (there should be no movement). On four-wheel drive vehicles, inspect the front drive shaft universals.

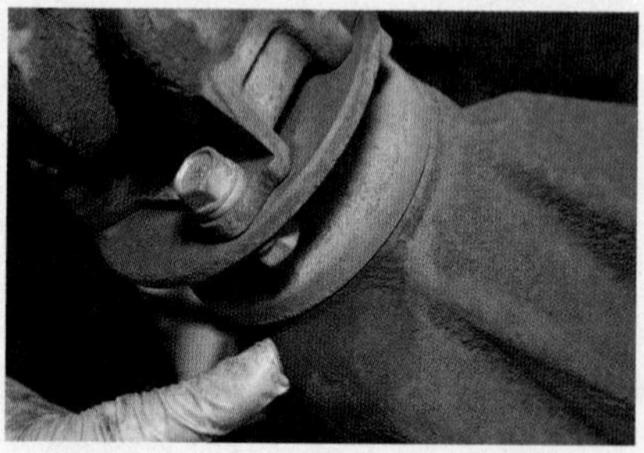

9. Check the differential and rear suspension area. Inspect the pinion shaft oil seal for any signs of leakage. Inspect the suspension mounting bushings for signs of deterioration or damage. If the vehicle is fitted with leaf springs, inspect them for any cracks or misalignment. Inspect the brake hoses for signs of cracking, abrasion, and bulges.

10. Check the fuel tank area. Check all the fuel lines and brake lines for signs of damage, abrasions, leaks, or rust.

> ▶ TECHNICIAN TIP
>
> The customer or owner of the vehicle can perform an inspection while washing the vehicle. Bad wiper blades, loose trim, and other items may be noticed that are otherwise not seen.

Performing a Visual Inspection of the Vehicle's Exterior

The vehicle exterior should be checked periodically for overall roadworthiness. The vehicle owner may perform this inspection themselves. But more often, the service technician does it during oil changes. The inspection should also be made any time the vehicle is in the shop for service work. While doing this inspection, the technician should be sure to work in a systematic manner. Using an inspection sheet helps ensure that faults are not missed.

To perform a visual inspection of the vehicle's exterior, follow the steps in **SKILL DRILL 10-13**.

Inspecting Shock Absorbers

Shock absorbers and struts are located near each wheel. They dampen body movement from bumps in the road. A common reason for testing shock absorbers is unusual tire wear, such as tires having a cupped appearance. Also, the driver may complain of a soft or bouncy ride. In some cases, a shock absorber can bind up, creating a very stiff ride. If a vehicle has adjustable shock absorbers, make sure the shock absorber adjustments are the same for the left- and right-hand sides. Some shock absorbers contain pressurized gas, which can leak out. This can cause uneven ride height and shock absorber performance issues.

Many of today's vehicles are equipped with a strut-type suspension instead of conventional shock absorbers. Testing either type of system involves the same procedure, a bounce test. Basically, while the vehicle is stationary, push up and down on a strong point at each corner of the vehicle (not the fenders as they can be dented) several times. Watch how the vehicle responds after you release it. Typically, if you let go at the bottom, a good shock will allow the corner of the vehicle to rise and then settle back into position. On some vehicles with softer suspensions, it may allow the corner to rise, fall, and rise back into position. More oscillations would than that would indicate worn shock absorbers.

Pay particular attention to the top strut mounting during the bounce test. Place your hand on top of the mounting during the bounce test. Any noise or movement in the mounting could indicate the need to replace the strut mount. Also, have someone turn the steering wheel from lock-to-lock while feeling and listening to the top strut mount. If you do feel movement or hear noise, report it to your supervisor. Visually inspect the shock absorber mounting points for security and corrosion. Note any wet-looking patches on the sides of the shock absorbers. A wet patch is a common indicator that the strut or shock absorber needs replacement because of a fluid leak. Slight dampness on the shock is typically normal, but drips on the shock are not normal and indicate excessive leaking. To inspect shock absorbers, follow the steps in **SKILL DRILL 10-14**.

Inspecting the Exterior Lights

The lighting system allows the driver to see the road when driving at night or in poor-visibility conditions. They also provide signals of your intentions to other drivers. These lights need to be checked periodically as they do burn out on occasion. The exterior lighting system includes the following:

- headlights,
- taillights,
- turn signals,
- side markers,
- brake lights,
- license plate lights, and
- back-up lights.

Some vehicles may have cornering lights, driving lights, or fog lights. Note that the rear lights may have three or more bulbs per side. So be sure to check that all of them are in working condition. To inspect the exterior lights, follow the steps in **SKILL DRILL 10-15**.

Checking and Replacing the Wiper Blades

The windshield wiper blades and arms are an important safety system on every vehicle. Many states with a vehicle inspection program fail a vehicle if the wiper blades are missing, torn, or worn out. The blades, along with the washer system, help the driver to see clearly under all driving conditions.

The wiper blades should be checked as part of the exterior inspection. Usually, any wiper blade that is more than a year old is ready for replacement. This is especially true if the vehicle is parked outside. Both the blade and the wiper arm should be checked.

The wiper blade should be flexible and not torn. The wiper arm should flex at the hinge and be held firmly against the

SKILL DRILL 10-13 Performing a Visual Inspection on the Vehicle's Exterior

1. Prepare the vehicle. Park the vehicle in a well-lit area. Turn the engine off, and unlock the doors and trunk or rear hatch.
2. Walk around the vehicle, observing any obvious items that need attention.
3. Check exterior component and system operation. Check the body condition to make sure all the body components are secure. Look for loose plastic trim.
4. Open and close doors to check that they are operating correctly.
5. Push and pull on the bumpers and fenders to ensure that they are secure.
6. Inspect the external mirrors to ensure that they are secure and not broken.

SKILL DRILL 10-14 Inspecting Shock Absorbers

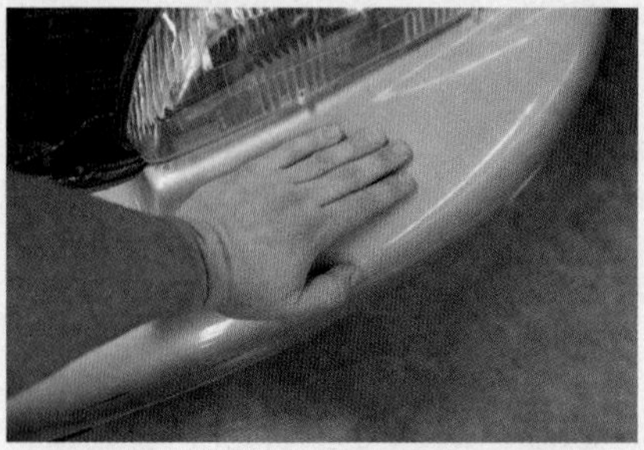

1. Place your weight on a bumper, and begin to bounce the vehicle until it reaches its maximum amount of travel produced by your weight. Stop bouncing at the bottom of the bounce. If the shock absorbers are performing well, the vehicle will rebound once, and then return to its original position.

2. Pay particular attention to the top strut mounting during the bounce test. Place your hand on top of the mounting during the bounce test. Any noise or looseness in the mounting could indicate the need to replace the mount.

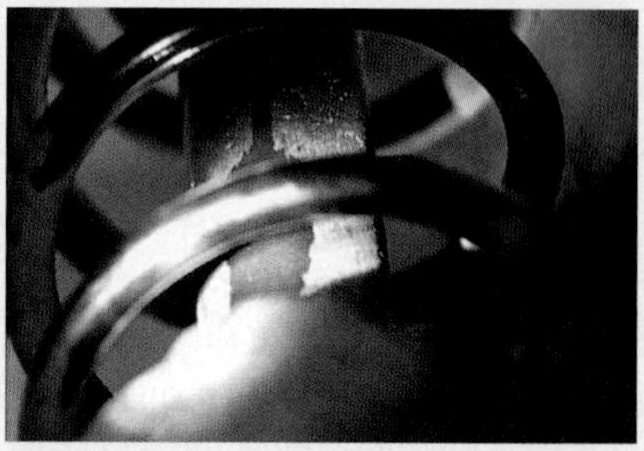

3. Visually inspect the shock absorber mounting points for security and corrosion, and note any wet-looking patches on the sides of the shock absorbers. Slight dampness on the shock is typically normal, but a drip on the shock indicates leaking.

SKILL DRILL 10-15 Inspecting the Exterior Lights

1. Have someone sit in the vehicle, turn the ignition "ON," and switch the light switch to the park light position. Check that the taillights, side markers, and rear license plate lights come on.

2. Put the turn signal switch in the left-turn and then in the right-turn position, and check that the signals flash equally on each side.

3. With the ignition key "ON," and the engine "OFF," place the transmission in reverse and observe the back-up lights.

4. Depress the brake pedal to make sure the brake lights work. Check that the third (center) brake light works.

5. Make sure the high and low headlight beams work properly.

6. Make sure the park lights, side markers, turn indicators, and daytime running lamps (if equipped) are all working properly

windshield by the wiper arm spring. The rear wiper blade and arm are checked in the same way.

Never operate the wipers when they are dry because this may damage the blades or scratch the surface of the windshield. Never bend the arms to make better contact with the windshield. The arms are pre-tensioned by the manufacturer, and damage could result. If the arms seem to have lost their spring tension, obtain a suitable replacement.

To check and replace windshield wiper blades, follow the steps in **SKILL DRILL 10-16**.

> ▶ TECHNICIAN TIP
>
> Many a windshield has been broken while inspecting or replacing windshield wiper blades (**FIGURE 10-57**). The spring holds the wiper arm firmly against the windshield. If you drop the wiper arm while holding it away from the windshield, the spring will snap it against the windshield. There is enough force to potentially break the windshield, especially if the wiper blade is removed from the arm. You can prevent a broken windshield by making sure the arm is never allowed to slip. Another option is to place a folded-up fender cover on the windshield where the wiper arm would hit.

FIGURE 10-57 A windshield broken during wiper blade replacement.

Inspecting the Windshield

The windshield should be inspected during the wiper blade inspection process. Scratched, scored, or pitted glass will not wipe clean, even with new blades. Windshields may become etched or pitted, causing the wipers to function poorly. This is a safety hazard and should be repaired. The glass in some cases may be polished to repair the condition, or it may have to be replaced. Some small chips or cracks can be repaired with special resins and tools. These services can be performed by an automotive glass repair service or at some collision repair shops. Large chips or cracks longer than 3" (76 mm) may require replacement of the windshield. To inspect the windshield, follow the steps in **SKILL DRILL 10-17**.

SKILL DRILL 10-16 Checking and Replacing the Windshield Wiper Blades

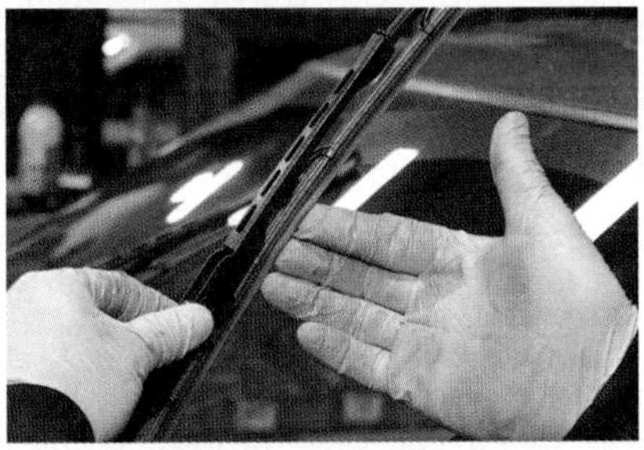

1. Check the windshield wiper blades. Lift the wiper arm away from the windshield and inspect the condition of the blades. Look for damage or loss of resilience in the material.

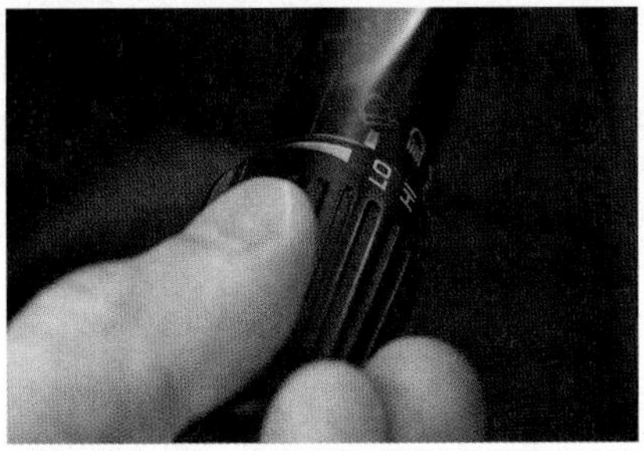

2. Wet the windshield with a hose or with the washers and switch the wipers on. If the windshield is being wiped cleanly, do not replace the wiper blades. If the wiper blades are not wiping the glass evenly or are smearing, replace the blades.

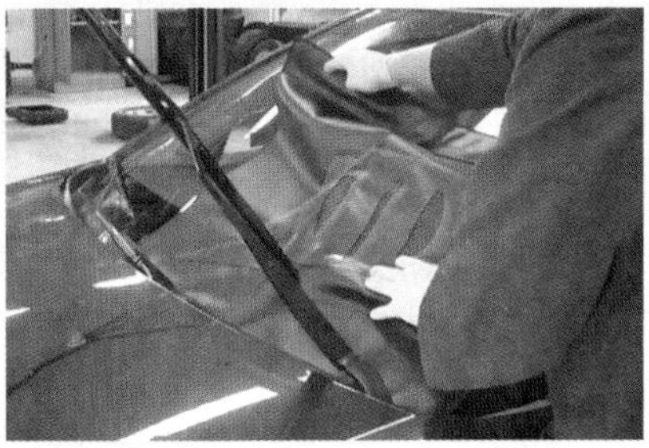

3. Place a folded-up fender cover under the wiper blade you are working on to protect the windshield.

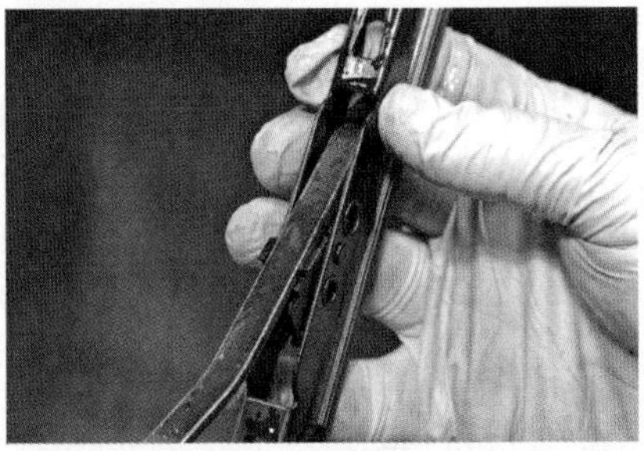

4. Remove the blade assembly.

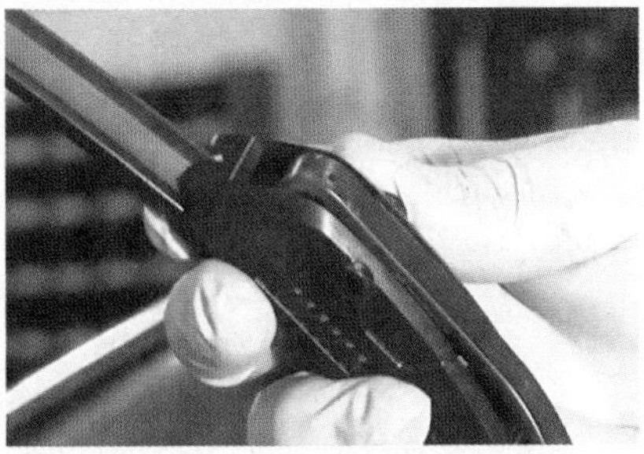

5. Obtain and install the appropriate replacement blades.

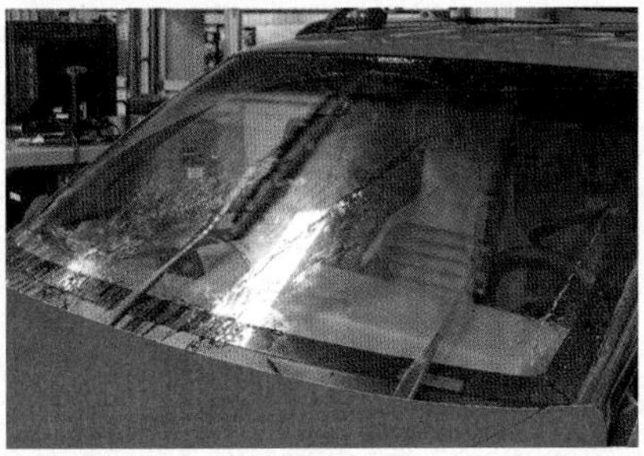

6. Once the wiper blade is installed, test them for proper operation.

SKILL DRILL 10-17 Inspecting the Windshield

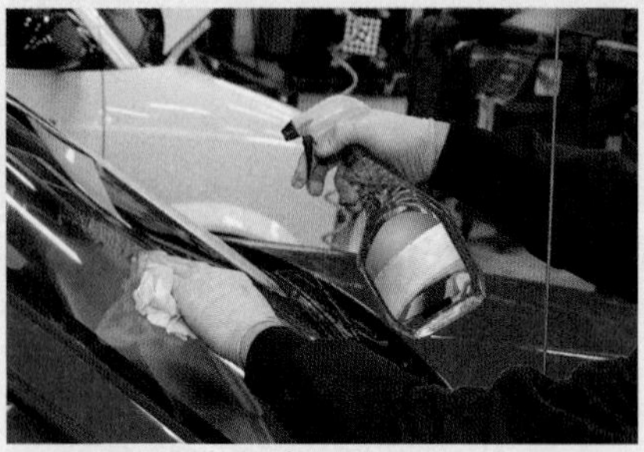

1. Prepare the windshield. First, use glass cleaner to clean the windshield thoroughly.

2. Inspect the glass. Look closely at the surface of the glass. It may help to use a flashlight or trouble light at an angle while inspecting.

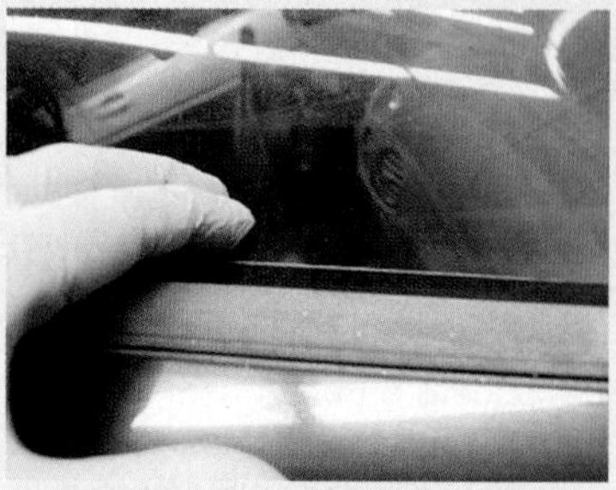

3. Look for any delamination conditions present.

▶ Wrap-Up

Ready for Review

- Different levels of in-vehicle inspections include: basic inspection, in-depth inspection, state safety inspection, certified used car inspection, and vehicle pre-purchase Inspection.
- An underhood fluid inspection should be performed at the manufacturer's recommended intervals and also prior to any long trip. Important fluids to be checked include: engine oil, engine coolant, brake and clutch fluid, power steering fluid, automatic transmission/transaxle fluid, manual transmission/transaxle fluid, and windshield washer fluid.
- V-type and the serpentine type belts should be visually inspected whenever the hood is opened for service. The belts should be checked for cracks, oil soaking, glazing, tears, and bottoming out.
  - When inspecting hoses, the engine should be cool. While inspecting the air filter, the air cleaner housing and ductwork should be checked for cracks or holes. Looking only at the top side of the filter will give you inaccurate information about the filter.
- An under-vehicle inspection should be carried out with the vehicle safely lifted on a hoist to get a feel for the overall condition of the vehicle. Inspection includes: tire inspection, transfer case fluid level, fluid leaks, steering and suspension inspection, drive axles and driveshafts, and engine, transmission, and exhaust.
- The vehicle exterior should be checked periodically. Shock absorbers, exterior lights, wiper blades, and windshield should be inspected to prevent troubles that may cause safety or operational concerns.

Key Terms

body control module (BCM) An onboard computer that controls many vehicle functions, including the vehicle interior and exterior lighting, horn, door locks, power seats, and windows.

CV joints A joint used to transmit torque through wider angles and without the change of velocity that occurs in U-joints.

diesel exhaust fluid (DEF) A mixture of urea and water that is injected into the exhaust system of a late-model diesel-powered vehicle to reduce exhaust oxides of nitrogen emissions.

hygroscopic A property of a substance or liquid that causes it to attract and absorb moisture (water), as a sponge absorbs water. Brake fluid absorbs water out of the air; thus it is hygroscopic.

instrument panel warning lamps Lamps that illuminate to warn a driver of a fault in a system.

Review Questions

1. Most shops use inspection forms when inspecting vehicles because:
   a. They slow the technician down
   b. They ensure a thorough and efficient inspection
   c. Technicians are typically lazy
   d. Technicians can't write very well
2. If the Malfunction Indicator Lamp is illuminated while the vehicle is being operated:
   a. Shut the engine off immediately
   b. Try revving the engine to see if the lamp goes off
   c. Retrieve any DTCs with a scan tool
   d. Step on the brake pedal very firmly for at least 10 seconds
3. When checking engine oil, it is important that the level be:
   a. Between the minimum and maximum lines
   b. Higher than the maximum line
   c. At least 1/4" below the minimum line
   d. Not touching the bottom of the dipstick

4. What is the main reason that brake fluid needs to be changed periodically?
   a. The boiling point increases too high
   b. It becomes too thin (viscosity)
   c. It absorbs water (Hygroscopic)
   d. It ruins the brake lining
5. When checking a serpentine belt for bottoming out:
   a. Use a straight edge across the belt ridges and measure the depth of the grooves
   b. Twist the belt 90 degrees and see if cracks appear in the ridges
   c. Measure the width of the belt to see if it has stretched too far
   d. Use a small tool that should sit higher than the ridges in the belt
6. A good way to test the integrity of exhaust pipes is to:
   a. Heat it up with a propane torch
   b. Squeeze the pipe with arc joint pliers
   c. Use a pry bar to see if the pipe will bend
   d. Fill it with water and see if any leaks out
7. How should you inspect an air filter to see if it needs to be replaced?
   a. Wash it in the sink and see how water flows through it
   b. Use an air nozzle to blow air through it
   c. You can't see light through it very well
   d. Air filters can't be inspected, just replace them
8. All of the following systems should be checked as part of an undervehicle inspection EXCEPT:
   a. The steering system
   b. The suspension system
   c. The exhaust system
   d. The electrical system
9. All of the following statements are true EXCEPT:
   a. The wiper blade should be flexible and not torn
   b. The wiper blades should be checked as part of an exterior inspection
   c. While checking, operate the wipers when they are dry
   d. Never bend the arms to make better contact with the windshield
10. Which of these is normally reddish in color?
   a. Engine oil
   b. Brake fluid
   c. Automatic transmission fluid
   d. Gear lube

ASE Technician A/Technician B Style Questions

1. Technician A says very wet fluid on struts or shocks is typically normal. Technician B says that shock absorbers can be tested with a bounce test. Who is correct?
   a. Technician A
   b. Technician B
   c. Both A and B
   d. Neither A nor B
2. Technician A says that improper handling of a windshield wiper can lead to a broken windshield. Technician B says that when testing wipers, if they wipe cleanly, replace them anyway. Who is correct?
   a. Technician A
   b. Technician B
   c. Both A and B
   d. Neither A nor B
3. Two technicians are discussing steering system inspection. Technician A says to wiggle the front tires side-to-side while feeling for play in the joints. Technician B says that there are usually wear marks on the parts that indicate excessive movement. Who is correct?
   a. Technician A
   b. Technician B
   c. Both A and B
   d. Neither A nor B
4. Technician A says that when the engine is started, the amber anti-lock brake system (ABS) warning lamp should come on, stay on for a few seconds, and then go off, indicating a successfully completed preliminary self-check. Technician B says that if a fault is detected in the system, the warning lamp will blink five times. Who is correct?
   a. Technician A
   b. Technician B
   c. Both A and B
   d. Neither A nor B
5. Technician A says that coolant freeze protection can be measured with a hydrometer. Technician B says that coolant freeze protection can be measured with a refractometer. Who is correct?
   a. Technician A
   b. Technician B
   c. Both A and B
   d. Neither A nor B
6. Technician A says that operating the brake pedal can indicate a hydraulic brake problem. Technician B says that high-pitched scraping noises could indicate worn brake lining. Who is correct?
   a. Technician A
   b. Technician B
   c. Both A and B
   d. Neither A nor B
7. Tire A says the date code of the tire can be found on the tire sidewall. Technician B says the recommended tire pressure can be found on the tire sidewall.
   a. Technician A
   b. Technician B
   c. Technician A and B
   d. Neither A nor B
8. Technician A says the cabin air filter can be inspected similar to an engine air filter. Technician B says the cabin air filter should be changed about every 4 or 5 years. Who is correct?
   a. Technician A
   b. Technician B
   c. Both A and B
   d. Neither A nor B

9. Technician A says that Stretch-Fit belts use a spring-loaded tensioner to keep them tight. Technician B says that a bottomed-out belt can be fixed by tightening it. Who is correct?
   a. Technician A
   b. Technician B
   c. Both A and B
   d. Neither A nor B

10. Technician A says Diesel Exhaust Fluid helps to control oxides of nitrogen. Technician B says that Diesel Exhaust Fluid should be replenished during oil changes. Who is correct?
   a. Technician A
   b. Technician B
   c. Both A and B
   d. Neither A nor B

CHAPTER 11
Communication and Employability Skills

Learning Objectives

- **LO 11-01** Demonstrate active listening skills.
- **LO 11-02** Demonstrate effective speaking skills.
- **LO 11-03** Demonstrate employability skills.
- **LO 11-04** Demonstrate effective reading and researching skills.
- **LO 11-05** Demonstrate effective writing skills.

ASE Education Foundation Tasks

See Appendix A to view the 2017 ASE Education Foundation Automobile Accreditation Task List Correlation Guide.

▶ Introduction

We have been communicating all of our lives. But most of us are not aware of the listening, reading, writing, and speaking skills needed to be a good communicator. It requires very little effort to communicate well, once you know the principles behind each communication skill. Learning and applying good communication skills will save you time. It will also help you avoid or get through tricky situations. These skills build over time. You will find that you learn something new every day as you encounter new situations or meet new people. It is a lifelong learning process to perfect your communication skills (**FIGURE 11-1**).

Communication is a necessary workplace skill for a person to be successful in a service facility. So this entire chapter is dedicated to communication and employability skills. This chapter describes the steps to becoming an effective communicator. It offers tips on how to be a good listener—the first step in good communication. It also explains how to speak to both customers and your coworkers. Along the way, we discuss the writing and documentation requirements in the workplace.

Note: The majority of NATEF's Applied Academic Skills for Communication are covered in the context of this chapter.

FIGURE 11-1 A service advisor listening to a customer describe their vehicle's concern.

The applied science and math boxes found in other chapters provide examples of how to apply concepts to everyday activities in the shop. This chapter provides examples of how to apply the concepts of good communication in everyday life.

▶ Active Listening

LO 11-01 Demonstrate active listening skills.

Active listening is an essential skill. We may hear what someone is saying. But are we truly understanding what the person is trying to communicate? The listening process can be difficult to perfect, but is one of the most important skills to possess. This is especially true when gathering information from a customer, or any other person. The active listener focuses all of his or her attention on the speaker, including verbal and nonverbal messages. When appropriate, the active listener encourages the speaker to further communicate and clarify details that may have otherwise been left out.

Applied Communication

AC-15: Listening/Reading/Speaking/Writing: The technician identifies and uses effective strategies for listening, reading, speaking, and writing when dealing with customers, coworkers, and supervisors.
AC-34: Listening: The technician adapts a listening strategy that will obtain the information required for solving the problem.

The Listening Process

To be a good listener, we need to be aware of barriers that can disrupt the listening process. These barriers can be mental and physical. Mental barriers are thoughts and feelings that interfere with our listening. This can be our own assumptions, emotions, and prejudices. To fully absorb what someone is telling us, we need to learn to set these feelings aside. It takes effort and is not always easy, but it is important to keep an open mind throughout the listening process. For example, when listening to a customer who is describing a concern about his or her vehicle, it is good practice to allow the customer to fully complete the thought. This is true even if you believe you have all the information you need. Keeping an open mind is also an important first step in practicing empathy, which will be discussed later.

You Are the Automotive Technician

During your second week on the job, your supervisor has pulled you off the shop floor and assigned you to the service department office. You will be shadowing the service advisor, Bob, to gain firsthand experience in communicating effectively with customers. The first customer has an appointment with the service department for a safety recall issue. Bob maintains eye contact while gathering all the necessary information from the customer. After the customer has answered all of Bob's questions, Bob reviews with the customer his understanding of the concern, to be sure it is accurate. As a new service technician, you practice writing up a repair order for Bob to critique after the customer has been taken care of. Bob writes up a repair order and reviews the concern and agreed-on course of action with the customer. He then guides the customer to the service department lounge to wait for his vehicle.

1. What are the three types of "right questions" to ask when gathering information?
2. What are the three Cs all work orders should contain?
3. Explain why empathy is an important skill to have when dealing with upset customers.

As a listener and active participant, we can encourage the flow of communication. This is done by welcoming the speaker, showing that we want to understand the message, and acknowledging the speaker's feelings and concerns. To do this, give the speaker your undivided attention by practicing these steps:

- Stop what you are doing.
- Remove as many distractions as possible.
- Focus on him or her.
- Look the speaker in the eye.

As the person is speaking, you also need to provide listening feedback. This indicates to the speaker that you are engaged in what he or she is saying. Feedback can be nonverbal (e.g., facial expressions, body posture) and verbal. Nonverbal language, such as nodding while listening and gesturing while speaking, is used to reinforce or add emphasis.

If, while speaking, you send a conflicting message, such as looking annoyed while stating that you value the customer's opinion, the person may tend to believe the nonverbal message instead of the verbal one. For example, a customer has their car towed in to your shop because it won't start. They tell you that the cause is a bad headlight. You may be tempted to roll your eyes at this claim. But regardless of your personal opinion, you must act as a professional. You must maintain a sincere and attentive attitude toward the customer.

Nonverbal communication is perceived to be more spontaneous and less conscious than verbal messages. So most people tend to believe the nonverbal message more than the actual words expressed. In this case, rolling your eyes would likely override any "correct" words you might say. So always be careful of the nonverbal messages you send.

▶ TECHNICIAN TIP

In many situations, nonverbal communication can be more important than the verbal message itself. So make sure that your nonverbal message matches the verbal message you are communicating.

Empathy

Empathy can help us to avoid selective hearing. To empathize with someone is to attempt to see the situation from that person's point of view. Empathy requires good listening skills. This includes using verbal and nonverbal listening feedback effectively. You also have to take in and consider the message without applying your own biases.

True empathy means breaking down or putting aside existing mental barriers. For example, when customers are faced with an expensive repair that is not in their budget, we can come across as uncaring if we present them with a take-it-or-leave-it attitude. Instead, express your understanding of the customer's situation. Seek to explore valid options to resolve the issue. These can help build trust with the customer. Remember, we can empathize with people even if we do not agree with them. Also, we don't have to take responsibility for their problem. But we can at least attempt to help them find the best possible solution that will work for them. You will know that you have demonstrated empathy when no viable options were found, but the customer still thanks you sincerely for your assistance.

Empathy helps you understand the message better. It also helps you to be less reactive in a negative way as you realize you could feel the same way in that person's situation. It may also help motivate you to find a better solution to the problem, knowing that it could be you in the very same circumstances.

Nonverbal Feedback

Applied Communication

AC-35: Nonverbal and Verbal Cues: The technician uses verbal and nonverbal cues in discussion to help identify, verify, and solve problems.

Nonverbal feedback can be a very useful tool when listening. Your body position, eye contact, and facial expression can all set the direction of a conversation. We use body language to help emphasize our message. And when listening, we can use it to provide listening feedback. When listening to a speaker, try to sit or stand upright, while making eye contact (**FIGURE 11-2**). Try to avoid folding your arms, as this can be perceived as either an aggressive or defensive stance. At the same time, do not act too casual, for example, by standing with your hands in your pockets. It may seem harmless, but a customer or supervisor would likely see such a posture as a sign of disrespect or disinterest.

Imagine if you were talking to a service manager during a job interview, and the manager leaned back and put their feet up on the desk. That one gesture would send you a message about the manager's level of professionalism, and their respect for you. In turn, it would affect the way you viewed the meeting.

Maintaining some eye contact during the conversation is important. Not looking down at something or someone else also lets the speaker know you are paying attention. If you are taking notes, be sure to look up periodically and make eye contact with the person.

Just as the speaker is noting our facial expressions, remember to pay attention to the speaker's facial expressions. This will give you a clearer sense of the message being delivered. If, for example, a customer scrunches up his/her face while talking about a squealing noise, you can probably assume the customer would be happy if you make that noise go away.

When we say that space is a part of nonverbal language, we mean the physical space between ourselves and the people we are communicating with. This is commonly known as "personal space." Everyone has their preferred amount of personal space, which varies depending on circumstances. Some of the variables include familiarity, gender, status, and culture. Because of this, people will have different requirements for their preferred personal space. But typically, 2½ feet (0.76 m) is considered a good rule of thumb in most professional settings. Be aware of space as a nonverbal message, and adjust accordingly (**FIGURE 11-3**).

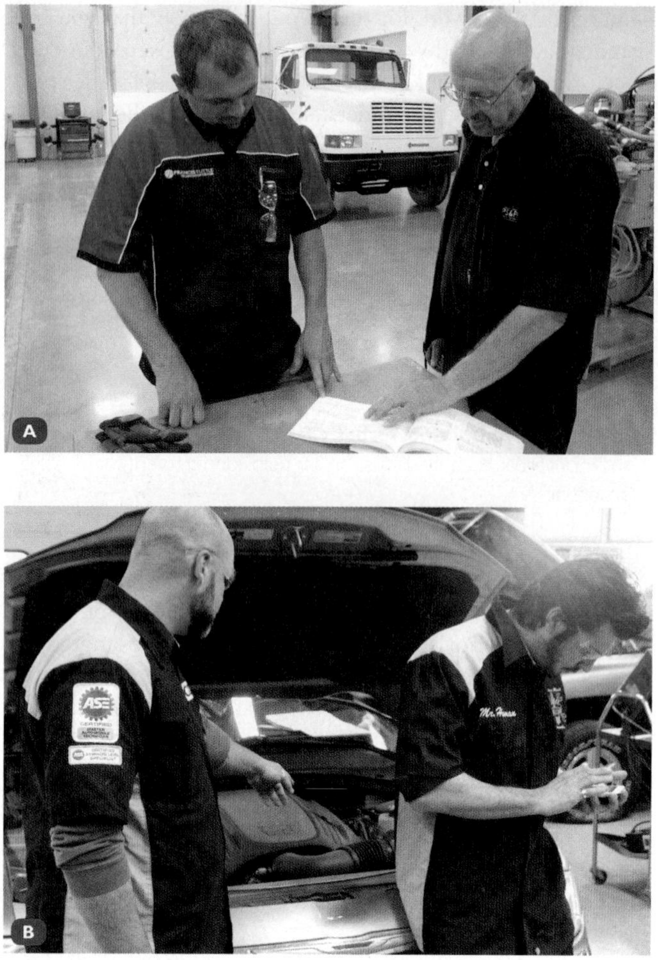

FIGURE 11-2 Nonverbal communication. **A.** Good. **B.** Bad.

FIGURE 11-3 Personal space. **A.** About right. **B.** Too close.

Verbal Feedback

Verbal feedback includes very simple signals that enhance the conversation. They let the person know you understand, but may need additional information. Some examples are the use of validating statements and supporting statements. These are types of clarifying statements or clarifying questions.

> A **validating statement** shows common interest in the topic being discussed. A phrase such as "I see" or "That is strange" indicates that you are paying attention. A validating statement also helps show empathy for the person speaking. It can be a simple "I understand" or "That must be frustrating for you."
>
> A **supporting statement** can urge the speaker to elaborate on a particular topic. Statements like "Give me an example" and "What else do you notice when it happens?" let the speaker know you would like more detail because you are genuinely interested in finding a solution. By asking for more details, you are likely to obtain more thorough information. This gives you a better starting point for solving the problem. For example, imagine you are talking with a customer about a noise his car made while going over a bump. If you were to say, "When else does it happen?" the customer would know that you understood but wanted a little more information.

As an example, a customer brings in a car that will not crank. Technician A speaks with this customer and gets only enough information to fill out the work order. He did not ask any clarifying questions, so he did not gather any further details. Technician B gathers the same information, but also uses supporting statements to gather additional information.

What is the end result for the two technicians? Technician B found out that the customer had recently gotten a new key cut for the vehicle. This led the technician to check whether the key had been programmed to the security system. It had not been programmed, causing the no-start condition, which is exactly what was wrong. Technician A was able to come to the same conclusion, but only after making several checks that led to the unprogrammed key. Both Technician A and Technician B were able to diagnose the fault. But Technician B was able to come to a conclusion much more quickly than Technician A. This is because of the extra information received by asking supporting questions.

▶ The Art of Speaking

LO 11-02 Demonstrate effective speaking skills.

Applied Communication

AC-28: Speaking: The technician employs communication strategies for customers, supervisors, and coworkers that will yield high-quality information for use in problem solving.
AC-29: Information–Oral: When analyzing a problem, the technician evaluates the usefulness of oral information provided by customers and coworkers.
AC-30: Information–Oral: The technician makes logical inferences and recommends solutions to problems, based on discussions with customers, coworkers, and supervisors.
AC-31: Information–Oral: The technician comprehends information gathered during discussions with customers, supervisors, and coworkers regarding problem symptoms and possible solutions.
AC-32: Information–Oral/Written: The technician analyzes information based on discussions, notes, observations, personal experiences, and data searches that will assist in solving the problem.
AC-33: Information Supplying: The technician clarifies information to customers, associates, parts suppliers, and supervisors.

Speaking is often referred to as an art. This is because there are so many facets involved in effectively communicating. Whether you are asking a customer about a vehicle, or your boss for a raise, knowing how to "artfully speak" will benefit you.

Speaking is a three-step process:

1. *Think* about the message.
2. Accurately *present* the message.
3. *Check* whether the message is correctly understood. If it is not, you have to respond by rethinking and re-presenting the message, and rechecking with your listener. This process continues until you are satisfied that the listener correctly understands your message.

Think before you speak is easier said than done, but it is not impossible. In some situations, you have time to think and plan beforehand. Other situations require you to think on your feet. Thinking on your feet simply means you may be called upon to answer or handle a difficult situation quickly. Having to get things right in a fast-paced setting can result in rash decisions, but there are a few tactics you can use to make these situations easier:

1. *Relax.* This is not easy to do in an urgent situation. But attempting to remain calm is extremely beneficial, keeping your mind clear and helping you to embody the confidence needed in such a situation.
2. *Listen.* If the situation requires you to answer a tough question, make sure you fully understand the question before answering. You can always ask the questioner to repeat or, better yet, rephrase the question. This gives you more time to think about your answer. It also provides another chance to read into the intent of the question. Remember that if someone is asking you a question, then it means this person is showing interest. Interest is a good thing!
3. *Pause.* Silence is golden—when used properly. Most people are uncomfortable with silence. But it is perfectly acceptable to take slight pauses to organize your thoughts before speaking. This also gives you the advantage of controlling the pace of the conversation. If a situation that you have been put in charge of is slipping out of control, regain control with slight pauses and thoughtful answers. It may be beneficial to say, "Let me think about that for a second." Or you may take a few seconds to restate your understanding of the problem and the person's question. Doing so can help the other person understand that you are devoting effort to the question, while also giving you a bit of time to think through an answer.

Before speaking, take a moment to consider that your tone of voice reveals a lot about your feelings and adds significant meaning to your message. The tone of voice includes:

- how high or low the pitch of your voice is,
- how fast or slow you speak,
- how soft or loud your voice is.

These voice characteristics give listeners indications of your current emotional state.

Even if you disagree with what is being asserted, try to first use empathizing statements, before offering a solution. This can be as simple as "I understand" or "I can imagine how frustrating that would be." For example, a regular customer brought his vehicle in last week for an illuminated malfunctioning indicator lamp (MIL). The vehicle was repaired and the codes were erased. Now the same customer is back with the MIL on again, and furious that they had to return. Regardless of whether the MIL is on for the same cause or not, you still have an upset customer to deal with.

Most times, people who are angry respond well to a statement like "I can imagine how frustrating it is for you to have to return." When you say this, remember to keep a calm and steady tone of voice. Ultimately, the goal is to de-escalate the customer's anger. Try not to argue with the customer. Rather, support him or her, and offer a solution that will work for both of you. In a situation like the one given above, you could say:

> I can imagine how frustrating that is for you. Let's bring the car in and see if it's the same problem as before, or not. If it's the same problem, our one-year warranty will cover it. Can I get you a cup of coffee in our waiting room while we look into it?

So, when encountering an upset customer, the goal is to first de-escalate the anger. Then, guide the conversation back to rational communication, and come to an agreement about how the problem will be resolved. Always try to empathize and de-escalate the situation. Do not argue with the customer. Focus the discussion on finding a mutually agreeable solution, if possible.

After thinking about what we want to say, and how to say it, we can then use the second step of the communication process. And that is to accurately present the message using verbal and nonverbal language. Remember that when we say "verbal language," we mean the actual words that are being spoken. Nonverbal language includes how we speak those words—our tone of voice, body language, and expression—and takes into consideration the environment. Imagine you are with a friend in a quiet café, drinking coffee, and she is telling you how much she values your opinion. Now imagine that she is clenching her fists, scowling, and screaming those same words to you. It would change the message a little, wouldn't it?

The last step of the speaking process is to make sure the listener correctly understands your message. Look to see that the listener is making eye contact with you. If you perceive a lack of understanding or confusion, be prepared to repeat yourself. Or try to explain it in another way, or by using an example. Make sure you suppress any outward irritation you may feel at needing to do so. In stressful situations, the parties involved are often not as open-minded as in calmer moments. Therefore, they are not as receptive to your words. This can also end with misunderstandings and expectations that go unmet. So, take the extra step and ask clarifying questions of the listener, or rephrase your message.

TECHNICIAN TIP

A calm and happy customer or coworker is normally easy to communicate with. Most people struggle with how to handle an angry one. In almost all cases, keeping a calm and steady tone of voice and a pleasant overall demeanor is the best bet. Never lose your temper. Doing so will only escalate the situation and lead to an unwanted outcome.

Asking Questions

Questioning is an important speaking skill that will help keep you out of a lot of trouble. People speak to deliver a message, but many times that message is not completely received. More information is needed, or details need to be confirmed. Asking questions to gather more information can help you to make good decisions. Or, once you come to an agreement, you can use questions to confirm those details. In either of these situations, you avoid trouble from miscommunication. In the first case, you gather enough information to avoid a bad decision. In the second case, you confirm the expectations regarding the agreement. This helps to avoid disappointment and loss of trust.

To use questions effectively, you need to know how to ask the right questions. You can ask three types of questions:

- Open questions
- Closed questions
- Yes-or-no questions

Each type of question is beneficial when used appropriately. Good communicators know when and how to use the appropriate type. If you have mastered each of these types of questions, you can keep a conversation going while speaking very little. These questions can be used in all kinds of situations, from casual conversations with new acquaintances, to detailed conversations with customers.

In dealing with customers, you should usually start with open questions. This allows you to gather general information about the issue. Then, use closed questions to find out specific details. Use yes-or-no questions to further check or confirm the listener's responses. They are also useful to gain the customer's agreement to authorize a repair or diagnostic procedure.

Open Questions

An open question encourages people to speak freely so you can gather facts, insights, and opinions from them. It's a good way to start a conversation with a new acquaintance or even an established customer. Open questions usually begin with these words:

What
How
Why
Could you tell me

For example, if you were questioning a customer about his/her visit to a repair facility, you could ask, "What type of service did you receive from XYZ Automotive?" This question opens the topic up for discussion and allows the customer to give details about his/her visit.

Closed Questions

If you want to know more information, you can use closed questions to establish facts and details. Closed questions usually begin with the words:

When
Where
Which
Who
How many
How much

This type of question requires a specific answer, and there is usually only one answer. For example, you could ask a customer, "How many times have you visited XYZ Automotive?" or, "When did you start going to them?" These closed questions allow for only one answer, without much room for discussion. These questions help you narrow the topic and guide the discussion in the direction in which you would like it to go. This is especially helpful when talking with a customer.

Yes-or-No Questions

Yes-or-no questions allow individuals to answer with a simple yes or no. This type of question is useful for checking or confirming a person's responses. For example, "Would you recommend them?" gets to the point and helps clarify information. Generally, you should not start out with yes-or-no questions, because they discourage further explanation. However, ending with yes-or-no questions is a good way to get confirmation. For example, after explaining the need for replacing the customer's water pump and answering the customer's questions about the job, it would be very appropriate to ask, "Can we go ahead and replace that leaky water pump for you?"

Telephone Skills

We have learned about the important aspects of the speaking process. A phone conversation presents some different challenges. Because we can't see each other, we can rely only upon verbal messages and some nonverbal cues, such as the tone of voice. Other nonverbal cues that we miss are body language, appearance, and the environment.

On the phone, we are limited in how we present our messages, so we need to think about our words and tone more carefully perhaps than in person. It is a good idea to plan and even write down each of the points you want to say before even picking up the phone.

Phone communication consists of three parts: greeting, exchanging messages, and finishing the call. We should always answer a phone call by first saying hello, identifying ourselves, and our place of business, succinctly and clearly. An example could be, "Hello, this is John with XYZ Automotive. How may I help you?" Try to keep your greeting friendly and short.

The second part of the phone conversation is exchanging messages. This requires concise, clear communication. It includes using clarifying questions and summarizing any main points. At all times, you should be polite and considerate. And remember that the most important nonverbal clue sent out over the phone is your tone of voice. It reveals a lot about your feelings. Here are a few tips to create a good impression:

- Do not sound bored. Try to keep some inflection in your voice; do not speak in a monotone.
- Sound calm and in control, even if you were caught at a busy moment.
- No matter what, never lose your temper or patience.
- If there is a need to keep someone on hold for an extended time, offer to call the person back, and then do it.

Finish a call by confirming the actions both you and the caller will take to ensure that the messages on both sides were accurately received. Then, end with a pleasant and friendly goodbye (**FIGURE 11-4**). Remember to thank all customers for their business, and invite them to come back. When taking a phone message for someone else, make sure you get all the information. This includes the caller's name and organization, contact details, the date and time of the call, and a summary of the caller's message.

FIGURE 11-4 Always end calls with a friendly goodbye.

When making a phone call, use these same skills. And always have necessary information available to give to the person you are calling. For example, if you need to order parts, relevant vehicle information should be shared with your parts supplier. You usually need to have vehicle make, model, year, engine size, and transmission type. Many times you need the vehicle identification number (VIN) as well.

Giving and Receiving Instructions

Applied Communication

AC-14: Directions/Task: The technician follows all written and oral directions that relate to the applicable task or system.

A critical aspect of an efficient and well-run workshop is the ability to give and receive clear, logical instructions. Usually, these should contain information about who, what, when, where, and why, and direction on how a task should be completed. What is the job that needs to be done? Who should do it? When should the job be done? Apart from knowing what information we should include in our instructions, we also need to consider the following:

- The instructions should be clear, complete, and concise.
- The instructions shouldn't be more than the person can remember; otherwise, they need to be written down.
- Don't ramble.
- Check for understanding.

When receiving instructions, we should make sure we understand the instructions completely. For example, the instruction "Use the tire machine to mount a set of four tires" indicates what to do. Ask follow-up questions, such as, "Should the white lettering or white walls be on the outside?" "Should I install new valve stems?" "Do you want me to install them back on the vehicle once they are mounted?" These help ensure we understood completely. Such questions may not be needed every time an instruction is given. But by clarifying the instructions, the quality of work will be closer to the customer's expectations. It will also save time and hassles in the long run.

Communication in a Team

Being part of a team can make working an enjoyable experience or a horrible experience, depending on the team. In large part, the ability of each team member to communicate effectively determines the success of the team. A high-performing team can accomplish much. As the saying goes, the sum is greater than its parts. We can achieve more when we effectively work together. Being part of a team allows us to achieve these goals:

- Learn new things from other team members.
- Share ideas, knowledge, and resources.
- Complement each other's strengths and weaknesses.
- Feel a sense of belonging.

A poorly functioning team spends a lot of time and energy bickering and blaming. And not enough time and energy being productive. Transitioning from a poorly functioning team to a high-performing team requires commitment to the team, as well as self-discipline. When all team members are committed to a set of common goals, they can contribute in positive ways to the success of the team.

Developing such a team requires good leadership skills and good followership skills. Good leadership skills involve setting a clear vision of the goals, empowering team members to contribute their best efforts, and recognizing each team member's strengths and weaknesses. Good leadership also provides training or mentoring to address any weaknesses in the team.

Followership is just as important as leadership. Not much would get done if everyone were a leader. Good followership skills involve being fully engaged in the team and its goals, stepping in and performing the work that needs to be done, participating fully in all decision-making, and giving honest feedback.

We need to be committed to the team and team goals to make teamwork successful. Each team member should have a defined role and responsibilities that go with that role. Each member is then able to rely on the others to do their part. Think of a team as a chain; one broken link can break the whole chain. A good team player is someone who commits to being a part of that team. They contribute to its success by fulfilling his or her role.

Some shops are set up to work on the team system. They typically have three to six members on a team, who all work to accomplish the work assigned to the team. Generally, the team consists of a service advisor, shop team leader, and one or two top-level techs, one or two mid-level techs, and one or two entry-level techs.

The team gets paid for the number of hours of work the team produces. They split that money, based on the number of hours each person worked. So if the team works efficiently, the number of worked hours goes down, and the pay for each team member goes up. If the team does not work efficiently, the amount of work completed goes down, as does each team member's pay. Effective communication plays a big part in efficient productivity in a team.

▶ TECHNICIAN TIP

To be able to fulfill the team commitment, people need to know what their roles and responsibilities are and what is expected of them. Leaders help facilitate that. Good communication is then required for the team to come together and work efficiently.

▶ Employability Skills

LO 11-03 Demonstrate employability skills.

Employability skills are sometimes referred to as "job-keeping skills." They are likely the most important skills you need to acquire as a future technician. The biggest reason that technicians get fired is due to poor employability skills (**FIGURE 11-5**). Job-keeping skills can be simple. Some examples are showing up to work on time, every day, and keeping your driver's license free of tickets. Learning and practicing good employability skills will pay off over and over (**FIGURE 11-6**).

FIGURE 11-5 Technician getting fired for regularly showing up late.

FIGURE 11-6 Good employability skills pay off.

Employment Requirements

Working as a technician typically requires that you be above a certain age, usually 18 years. Being an automotive technician is considered a hazardous occupation. Operating vehicle lifts and vehicles themselves can be very dangerous. However, in some states, 16-year-olds can be employed if they are part of a formal training program.

One of the biggest hurdles to being hired, or even being allowed to continue working, is the requirement of a driver's license. This also means having a clean driving record (**FIGURE 11-7**). Shops cannot employ a technician they cannot insure. That typically means no more than one or two minor moving violations within the last three years. Major violations such as reckless driving or driving under the influence will likely lead to losing your job. Because most technicians are car people who like to go fast, speeding tickets are not uncommon. Unfortunately, they can derail your career as a technician very fast. So work hard to keep your driving record clean.

Another requirement is to be drug and alcohol free. Most employers require you to pass a drug test before being hired (**FIGURE 11-8**). They also may perform random drug tests. If an accident happens, the employer may require a mandatory drug test. So staying away from illicit drugs is a requirement for keeping your job as a technician. This can be a problem for technicians in states that have legalized marijuana. Although it may be legal in those states, it is currently illegal in the eyes of the federal government. This can cause confusion for employers and insurance companies. If you live in a state where marijuana is legal, know your employer's policy so that you don't get caught by surprise. Just because marijuana may be legal in your state doesn't mean you won't lose your job if you use it. Alcohol is another trap that can catch young and old technicians alike, and is detected by drug tests.

Honesty, integrity, reliability, and quality are also requirements of being a valued technician. They help you build a strong positive reputation that people can trust. And trust means people don't have to worry or wonder whether leaving their vehicle in your care was the right thing to do.

- **Honesty**—Merriam-Webster defines "honesty" as a refusal to lie, steal, or deceive in any way. This can be as simple as telling the truth about situations. It can also mean charging people for the job you actually did, not for the job you told them you did but actually didn't. So practice honesty; it pays off in the long run.
- **Integrity**—Merriam-Webster defines "integrity" as an adherence to moral and ethical principles. This takes honesty to a higher level because it indicates an overarching commitment to being ethical in character. I could be honest by saying that the part broke while I was pressing it apart. When, in fact, it broke because I didn't follow the service information, which said that a retaining ring had to be removed before pressing the component apart. A good simple summary of integrity is "doing the right thing." So, if you always do the "right" thing, you won't go wrong.
- **Reliability**—Merriam-Webster defines "reliability" as the quality state of or being fit to be trusted or relied on. It is vital to earn trust with both supervisors and customers. For most people, that means that you have to prove yourself to them by upholding your word. You do this by doing the things you agree to do, in the time you say you will do

DRIVING RECORD HISTORY

```
TYPE VIOL/SUS CONV/REI HISTORY ENTRY                         PTS
---- ---------- ---------- ------------------------------------------- -----
00   00            NONE CURRENT                              0. 0

                                   ** END OF RECORD **
```

A

```
TYPE VIOL/SUS CONV/REI HISTORY ENTRY                         PTS
---- ---------- ---------- ------------------------------------------- -----
VIOL 04/11/1998 05/18/1998 RECKLESS OPERATION                6.0
                City/Location.......: OH
                Event Type..........: VIOLATION
                State Code..........: E01
                Court/Agency........: UNKNOWN
                Order/Viol #........: C0126765301
                Miscellaneous.......: REK OP
                CODE: 3210RO NO
                IMMEDIATE IMPOUND

REIN 09/25/1995 09/25/1995 FAIL TO PAY - FINE AND COSTS
                State Code..........: D53
                County/Location.....: OH
                Event Type..........: REINSTATEMENT
                Eligible Date.......: 09/25/95
                Miscellaneous.......: WITHDRAWAL EXTENT:ALL
                PRIVILEGES WITHDRAWAL
                LOC REFN:PENDING REIN
                FEE WITHDRAWAL STATE
```

B

FIGURE 11-7 A. Good driving report. **B.** Bad driving report.

FIGURE 11-8 Passing a drug test is typically required as part of the hiring process, or in the event of an accident.

it. This includes showing up to work on time and working diligently during work time. So, if you agree to do something, then do it. And do it without having to be reminded. Be someone that others can rely on.

- **Quality**—Merriam-Webster defines "quality" as "superiority in kind." This trait should be a mark of everything you do. The old saying is true: "If it's worth doing, it's worth doing well." Since customers have high expectations that shops will provide them with quality work, shop owners expect that from their employees. Therefore, learning to do quality work will stand out to both employers and customers. This will allow you to earn money to pay for your desired lifestyle.

In many training programs, learning employability skills is often overlooked. Instead, the emphasis is often placed on learning the technical skills. At the same time, employers have a hard time finding people with good employability skills. And while employers say they can teach the technical skills, they cannot teach the employability skills. So, mastering these skills, along with technical skills, will make you very valuable to employers.

FIGURE 11-9 Personal hygiene is important when presenting a professional image.

Appearance and Environment

Appearance is the image you present of yourself to the public. All aspects of your physical appearance, including personal hygiene, clothes, jewelry, hairstyle, posture, and outward demeanor, combine to create the first impression made in any encounter. That first impression often informs the judgment others make about you. This in turn affects the level of respect and trust you achieve. It is much harder to convey your message effectively if your audience is distracted by some aspect of your appearance. Or if they do not take you seriously.

When working in a professional environment, always do your best to look professional. This starts with personal hygiene. No one enjoys being around stinky, dirty people. So, personal hygiene is an important habit for all people to develop. Technicians work in somewhat dirty conditions, and perform a fair amount of physical work. So it is common to become dirty, sweaty, and stinky if you do not clean up regularly. Technicians should typically shower daily and use deodorant so as not to offend fellow employees and customers (**FIGURE 11-9**). Personal hygiene also includes grooming. For men, this includes neatly shaving or trimming facial hair as well as maintaining a professional hairstyle. Women should maintain a professional hairstyle.

FIGURE 11-10 When working in a professional environment, always do your best to look professional.

Remember that while you are at work, you embody the image of your company. Shorts, improper footwear, or untucked or filthy clothes can all send negative signals to a customer. Customers are much more willing to have their vehicle serviced by someone who looks well put together than someone who does not. And that means wearing clothing that meets or exceeds your company policy.

If you are a technician, that typically means wearing the shop's uniform. If that is the case, then wear it properly (**FIGURE 11-10**). That means tucked in, buttoned up, shirts and clean pants that are not drooping. If the company doesn't have an official clothing policy, then wear clothing appropriate for your job. If you are a technician, then ensure clean, close-fitting, but not tight, work pants and shirt are in order. Also consider wearing a belt, if appropriate. Avoid wearing metal belt buckles that could damage a vehicle if you lean up against it.

If you are an employee who deals with customers, such as a service advisor, you may need to step up the level of your clothing. This may mean a shirt and tie, or polo shirt, with dress pants or slacks (**FIGURE 11-11**). If you work the parts counter, you could be required to wear either a work uniform or dress shirt and dress pants.

You should also keep an extra change of clothes at the shop in case you need to change them if they get excessively dirty. Most shops provide lockers for employees to keep their uniforms

FIGURE 11-11 Customer facing employees typically have a higher level of dress.

FIGURE 11-12 Showing up on time means showing up early, getting ready to work, and starting work at the appointed time.

and street clothes in during work hours. Professional appearance means that you also take care of your clothes. Most shops provide a weekly laundry service to make sure you have clean uniforms at all times. But you have to manage getting them turned in on time so they can be picked up and cleaned. Once you get the clean uniforms back, make sure you store them properly. Don't throw them in your trunk, where they can get all wrinkled and dirty. Hang them up neatly so they will be ready to wear.

The surrounding work environment is also worth consideration. It can affect the image you are portraying. A disorganized, cluttered, and dirty area leaves a negative impression with most customers. This can lead them to believe that you don't care about appearances or quality. Look around you and evaluate the housekeeping. Is it clean, organized, and inviting? Or is it neglected, dirty, and gross? A little housekeeping goes a long way toward making customers feel confident in letting you work on their vehicle.

Another issue that affects the work environment is distractions. This could be excessive background noise from a blaring radio, or inappropriate coworker conversations. Interruptions by coworkers, phone calls, or text messages can also hinder productivity. Whenever it is in your power to do so, work to keep these distractions at a minimum.

Be aware of dangers around you as well. Although most insurance policies prohibit customers in the work area, there may be occasional times when they need to see a particular issue. If you have to take customers into the shop, always escort them, and be sure to keep them safe. For example, do they need to wear safety glasses? Also, escort them back out of the shop as soon as possible. You can continue the conversation in the customer write-up area.

Time Management

We know that in the work environment, time is money. This is especially true in most repair shop environments. Every minute is costing somebody something—the customer, the shop owner, and/or the technician. Punctuality is an important nonverbal message in a business environment. When you are punctual, you demonstrate a good work ethic and professionalism. Punctual means showing up on time (typically 5–10 minutes early to get ready to start work at the appointed time) (**FIGURE 11-12**).

While you should always be on time to work, it is possible that an occasional unexpected delay can prevent that. If you cannot make it into work on time, make sure you call your supervisor as soon as possible. Let him or her know what happened and when you will be in. The same applies if you need to call in sick. Do so as soon as you can, so other arrangements can be made. Also, being late or sick should be very rare.

Remember that customers are depending on the shop to complete their work on time. And the shop depends on you to get it done. Any employee delays or absences cause scheduling trouble for the management and customers. It doesn't take very many absences or late arrivals before management decides that it would be better to hire a dependable replacement.

Punctuality also means you complete the job when you say you will. If you tell a customer their vehicle will be finished at 3:00 p.m., you should plan on finishing it before then. That way, if something goes wrong, you will have a bit of a time cushion. Occasionally, there are unforeseen events that keep you from completing a job on time. But those should be rare exceptions. If there is a delay, then you need to communicate that to your customers as quickly as possible. That way they can make other arrangements.

Finally, you should be giving a full measure of work for the time you are being paid. Using time for things that are not work related is stealing from your employer. Some of the more common ways employees steal time from their employer are:

- Routinely texting your friends
- Taking personal calls during work hours
- Surfing the web
- Getting in non-work-related conversations with other employees
- Extended lunch or break times

These are just a few examples of stealing time from your employer. Always use your work time efficiently, just as you would want your own employees to do.

Customer Service

Good customer service is vital in today's competitive business environment. The quality of customer service influences people to choose us over our competitors. Good service makes people feel good about continuing to buy our products or services. This is how the business gets the money to pay your wages. So customer service has a direct impact on the ability of employers to hire employees, provide wages, and offer promotions. In other words, providing good customer service benefits you and others.

In fact, vehicle manufacturers place high importance on the customer satisfaction index (CSI) rating at their dealerships (**FIGURE 11-13**). The CSI rating is gathered from virtually all the customers who have service work done. The CSI rating is reported each month. It is used to evaluate individual technicians and the entire service facility. If the CSI rating is high, bonuses can be paid to everyone who contributed to that success. If the CSI rating slips, bonuses can be withheld. It is also likely that new processes may be implemented to help improve the CSI rating.

To be able to provide good customer service, we must first understand who our customers are. Once we know who they are, we can then identify their needs. There are two types of customers, internal and external. Examples of internal customers would be parts counter people and sales people. External customers are people outside of the business such as vehicle owners and other shops.

FIGURE 11-13 Customer Satisfaction Index report.

Internal customers are as important as external customers. By helping our coworkers with their jobs, it ultimately helps our external customers. It is helpful to remember that it is almost always external customers who bring money into your shop. This occurs when they trade their money for your service or product. At the same time, internal customers help us meet the needs of external customers. So treat them as respected members of the team.

Being focused on customer service means you are fully engaged in providing the highest level of service that you can. This includes clear, friendly communication to help prevent misunderstandings while establishing achievable expectations (**FIGURE 11-14**). It also means simple things like respecting the customer's vehicle. This means not getting grease on the vehicle's steering wheel, upholstery, or paint. Or you can take extra steps such as washing the vehicle, or vacuuming it out. That way it is cleaner than when you started working on it.

Keep in mind, one of the most important parts of customer service is fixing the problem in the vehicle the very first time. Customers hate bringing a vehicle back for the same problem. So, fixing their vehicle goes a long way toward earning your customer's satisfaction.

Identifying Our Customers' Needs

Different customers have different needs. The same customer may have different needs at different times. In most cases, we can divide customers into three categories:

- those who are more concerned about getting things right
- those who want to get things done and
- those who just want to get along with people.

Each of these customers is motivated by different factors. The person who wants the job done right is usually not interested in lesser quality parts and wants you to take the time to make sure the job is not rushed. They will generally appreciate using top-quality parts, so lead with that option. If you assume

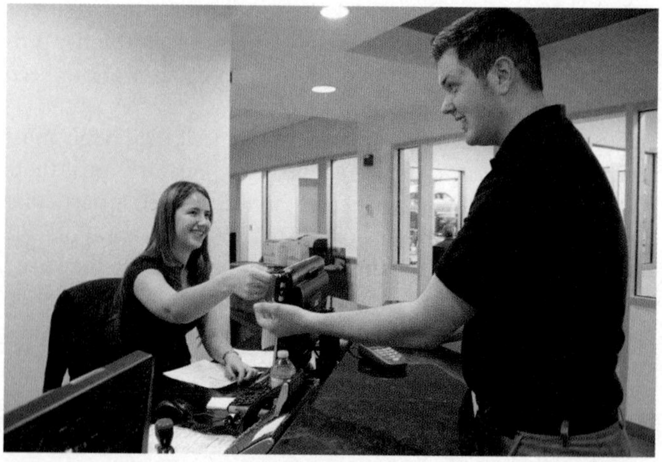

FIGURE 11-14 Focus on customer service and communicate clearly.

that this customer wants lower quality parts used, and install them, they will likely be unhappy. If the diagnosis was rushed and turned out to be incorrect, they will likely be very unhappy.

People who are most interested in getting things done will probably be very keen to have the repairs finished on schedule. A good example is if the vehicle breaks down while on vacation. The customer will likely want the vehicle back quickly, and may not be as interested in the highest quality parts. But if the vehicle is not finished on schedule, this person will likely be unhappy. So make sure the estimated time of delivery is accurate and reasonable.

For the person who just wants to get along with people, it is important that a level of trust is maintained. Trust is broken in several ways. One way is if you don't follow through on your word. If the repair takes longer than you said, or if it costs more, trust is broken. Another way trust is broken is if the customer catches you in a lie. Telling them that you did something that you didn't do, is a lie. If trust is broken, it will be hard to regain.

So, when working with customers, you need to identify their greatest need, so you can serve them properly. At the same time, it is best to address all three factors, as best you can, each time service is performed. That will help keep customer satisfaction high.

If you are the one in charge of gathering information from the customer about the concern, be sure to get specific details from the customer about:

- what is happening
- when it happens
- how often it happens and
- how long it's been happening.

Take notes that will help to diagnose the problem. Then, when relaying this information to someone else, be sure that they understand the symptoms with the same detail. For example, a customer is concerned about a particular noise in their vehicle. If the noise doesn't happen all the time, then ask when the noise happens. Also ask under what conditions the noise happens. You may need to prompt the customer with follow-up questions like: "Does it happen when it is cold or hot? When going around a corner or over bumps? Or when braking or accelerating?" It may be helpful to ask where the noise seems to be coming from, and what it sounds like. Of course, if the noise can be reproduced, you may need to test drive the vehicle. This may mean that you need to ride with the customer so that they can identify the noise for you.

TECHNICIAN TIP

Many shops have Customer Concern sheets that prompt the customer to answer questions about the vehicle concern. These sheets can be system- or symptom-specific (**FIGURE 11-15**). For example, there might be one for concerns with the braking system. Another might be for unusual noises. The customer then answers the questions to the best of his or her ability. This gives the technician valuable information when planning out the steps of diagnosis or repair.

Applied Communication

AC-25: Notes: The technician makes notes regarding symptoms, possible causes, and other data that aid in diagnosing and solving the problem.

AC-36: Information Requests: The technician requests specific symptom information from the customer and discusses solutions with supervisors and associates.

▶ Effective Reading

LO 11-04 Demonstrate effective reading and researching skills.

Every day we are faced with interpreting written information. This includes service information, emails, voicemails, and work orders (**FIGURE 11-16**). And yet, understanding written messages is more difficult than verbal messages. In written communication, we do not receive the verbal and nonverbal clues that come with face-to-face communication.

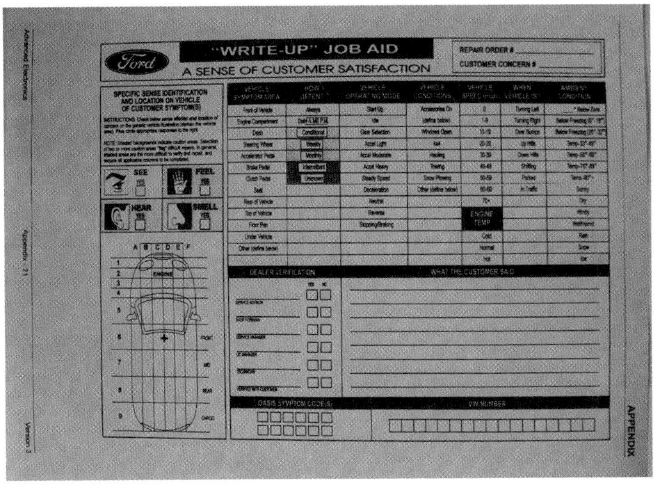

FIGURE 11-15 Customer Concern Sheet.

FIGURE 11-16 Technicians need solid reading skills.

Nor do we have the ability to ask clarifying questions of the text. So, learning effective reading strategies will help you to overcome this. It will allow you to find, understand, and use information much more efficiently. And that will help you diagnose and repair vehicles.

Applied Communication

AC-1: Reading: The technician adapts a reading strategy for all written materials (e.g., customer's notes, service information, and computer/data readouts) to help identify the solution to the problem.
AC-3: Abbreviations/Acronyms: The technician identifies and uses written abbreviations and acronyms in diagnosing and solving problems.
AC-6: Charts/Tables/Graphs: The technician consults charts, tables, and graphs to determine the manufacturer's specifications to identify out-of-tolerance system components.
AC-7: Sequence: The technician consults service information to determine the appropriate sequence of procedures required for solving a specific problem.
AC-13: Skimming/Scanning: The technician reviews service information to identify problems and applies that information to appropriate repair procedures.

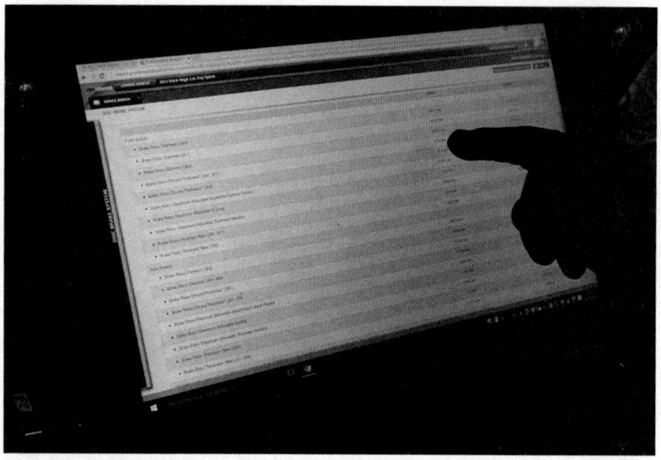

FIGURE 11-17 The selective reading method is used to find specific bits of information.

FIGURE 11-18 The comprehending reading method is used when you need to understand a block of information, such as performing a diagnostic test.

Reading Comprehension

Technicians are required to read a lot of information. This, includes repair orders, service information, technical service bulletins (TSBs), and training materials. Many of these reading materials can be written at high-grade levels. For example, many service manuals and TSBs are written at grade level 14 and higher. Technicians need to be proficient readers to be able to comprehend this information. Diagnosing and repairing vehicles depend on it.

Before you start reading anything in particular, you need to know your goal for reading it. In reading, your goal may be any, or all, of the following:

- Access a specific piece of information quickly, such as a particular specification.
- Understand the information, such as how to perform a particular series of tests on a system you are diagnosing.
- Remember the information, such as learning about a new technology that is being introduced by a manufacturer, how it works, and how to test it during diagnosis.

Once we know our goal in reading the information, we can choose the suitable reading method. The three common reading methods are as follows: selective, comprehending, and absorbing. *Selective* reading is reading to find small bits of information we need to know (**FIGURE 11-17**). In fact, this method is useful when looking for a particular piece of information. The quickest way to use selective reading is to skim through the following resources. Start with the first one and continue until you find what you are looking for:

- table of contents,
- index,
- introduction,
- headings, and
- the body of the text.

When you use the *comprehending* method of reading, you need to interpret and understand medium-sized chunks of information. A good example of reading for comprehension is reading the instructions for performing a diagnostic test (**FIGURE 11-18**). Understanding the information requires careful attention to the structure of the sentences and paragraphs. You need to pay attention to the punctuation and sentence structure so you can determine the meaning of the information. In fact, you may need to read each sentence more than once, so you understand what is being communicated in the text.

When you need to remember larger chunks of information for a longer period of time, you use the *absorbing* reading method. A good example would be reading about how a new type of system operates (**FIGURE 11-19**). This method allows you to absorb the information into your long-term memory. This type of reading takes longer than the previous two methods. The reader needs to read through the information more slowly. This includes stopping to ponder the information to see how it fits together. You will likely have to reread it several times until it fully makes sense. You know it makes sense when you can visualize the concept in the form of a picture.

To use an *absorbing* reading method:

- Read the information slowly and methodically.
- Stop and visualize what is being said in written form.
- Reread the sentence or paragraph as needed.
- Absorb the pictured information into our memory.
- Review the information regularly so you will retain it over time.

When we talk about reading, we should not be concerned only about how fast we can read. We also need to be concerned about how well we understand and retain what we have read. Unfortunately, the faster we try to read, the less we are likely to concentrate on the meaning. So these two objectives (speed and comprehension) work against each other. To combat that, practice reading. The more we practice, the faster and more accurate we can read.

Two of these reading methods are typically used when researching a particular problem with a vehicle. For example, a vehicle that is hard to start when it is cold outside. You might first check to see if there are any related TSBs. You can use the selective reading method to quickly skim through the TSBs, looking for specific hard starting symptoms.

When the relevant TSB is found, use the comprehending reading method to evaluate the information. Read it carefully to understand the information presented. This can include the symptoms, cause, test procedures, and corrective actions necessary for repairing the vehicle. Once you have a good understanding, you are ready to diagnose the fault.

▶ TECHNICIAN TIP

Charts, tables, and graphs help you to find information in a timely manner. An example is torque specifications. This information is usually presented as a chart showing the component and the required torque, usually listed in Newton meters (N·m) as well as foot-pounds (ft-lb).

▶ TECHNICIAN TIP

The Society of Automotive Engineers (SAE) has a well-known list of J1930 terms, which are abbreviations used by the automotive manufacturers. In addition, each manufacturer may have additional abbreviations that differ from the SAE terms. For example, the term "GEN" (generator) is used by the SAE, and the term "ALT" (alternator) is used by several automotive manufacturers.

Researching and Using Information Sources

Technicians spend a lot of time researching information (**FIGURE 11-20**). This could be as simple as looking for a specification needed to perform a specific task. Or it could be researching how a manufacturer designed the system, so you

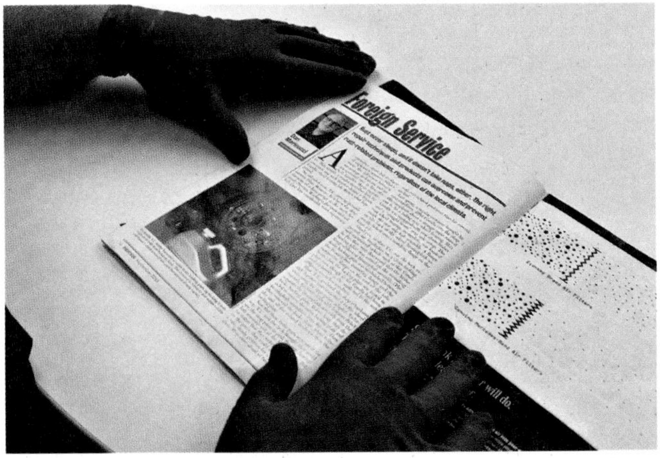

FIGURE 11-19 The absorbing reading method is used when you are trying to learn a larger chunk of information such as the theory of a new type of system and how to diagnose it.

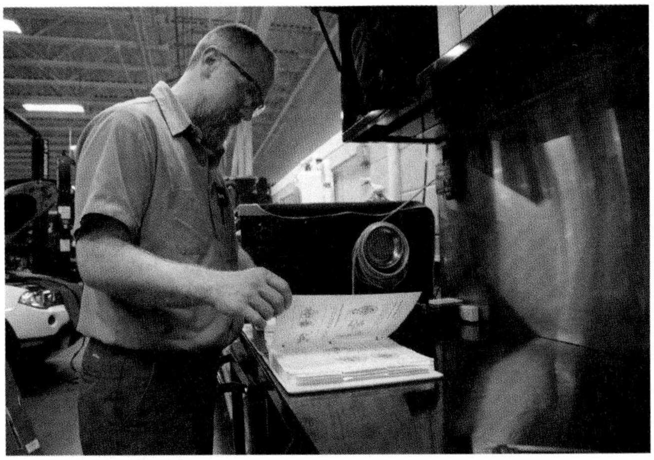

FIGURE 11-20 Technicians spend a lot of time researching information.

can determine whether it is operating correctly or not. In some cases, research is needed to see what previous work has been performed on the vehicle.

Researching is like conducting an investigation. It can be done in three steps. When we first come across a problem, we have to define what the real issue is. That requires us to know how the system or part is meant to work. This is usually found in the *Description and Operation* section of service information (**FIGURE 11-21**). Next, we look for information or clues that can help us solve the problem. This is usually found in the *Testing and Diagnosis* section of service information (**FIGURE 11-22**). Wiring diagrams can also be helpful at this step if dealing with electrical problems. Finally, we put the pieces together to determine the best solution. This is when the diagnostic plan is formulated.

If using a book for reference, start with the table of contents to find the section you are looking for. If you come across a word you do not understand, look up the definition. Most books have a dictionary or glossary of terms toward the end of the text. When using an online resource, familiarize yourself with the site's navigation. Take note of all drop-down menus, shortcut options, and search features.

While using these types of resources, you can use the skimming technique to find information quickly. Try skimming for key words. For example, you need to locate information on oil specifications for a particular vehicle. You would just skim through the resource's text, looking for the words "oil," "lubrication," or "specifications." If included in the reference material, charts, tables, and graphs are usually easy to skim as well. Keep looking until you locate the needed information.

Most automotive resources include diagnostic trouble charts. These provide a great way to narrow down faults without missing any steps in the procedure. Trouble charts also use the cause-and-effect approach to diagnosis (**FIGURE 11-23**). For example, if a fuse is blown, you can use the trouble tree to help you work through the step-by-step diagnosis. This will help you locate what caused the fuse to blow. As you perform the diagnostic tests and gather results, you should take notes so that you can refer back to them. Notes are also a good resource when you complete the repair order after finishing the job.

As you gain experience in the automotive industry, you will find that many vehicles have similar faults. Don't hesitate to use your prior knowledge to assist you in a new diagnostic situation. Just don't assume that all similar problems are caused by the same fault. You will want to test and verify your assumptions before suggesting a course of action to the customer.

Always save the manuals for shop equipment and tools, and organize them in handy places for future reference. Refer to these manuals to periodically maintain the equipment and tools (**FIGURE 11-24**). For example, most air tools require a drop

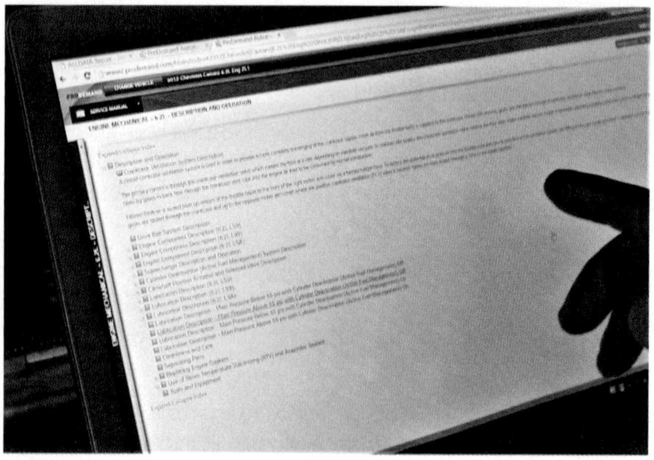

FIGURE 11-21 The Description and Operation section of service information helps us to determine if the system we are diagnosing is operating correctly or not.

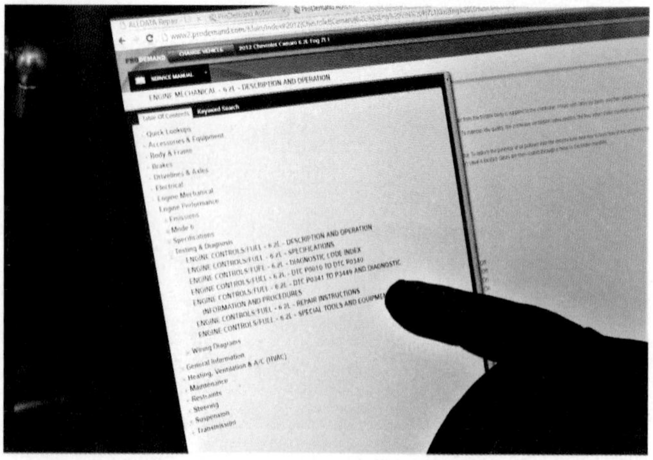

FIGURE 11-22 The Testing and Diagnosis section of service information gives us the steps to diagnose a problem.

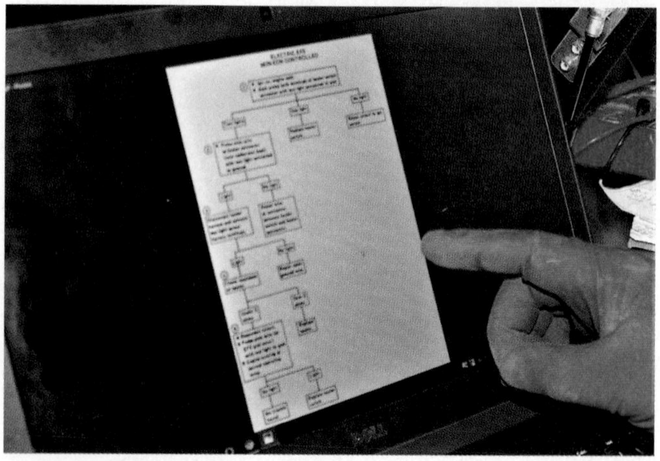

FIGURE 11-23 Service information provides diagnostic trouble charts for guiding you through a problem.

FIGURE 11-24 Manuals for tools and equipment are very handy when needed. Keep them in a safe place.

FIGURE 11-25 Compile primary and secondary information to define the problem.

or two of hydraulic oil every day. They also require the use of a water separator in the air line that may need to be drained periodically.

Applied Communication

AC-8: Dictionary: The technician refers to a dictionary to check spelling and define unfamiliar terms.
AC-9: Text Resources: The technician uses glossaries, indexes, database menus, and tables of contents to gather the information needed for diagnosis and repair.
AC-10: Database: The technician uses databases to obtain service information.
AC-11: Operator's Manual: The technician comprehends and applies information from accompanying manuals in order to use and maintain automotive tools and equipment.
AC-12: Service (Shop) Manual: The technician uses service information in both database and print formats to identify potential malfunctions.
AC-16: Study Habits/Methods: The technician uses proven research methods when consulting the manufacturer's service information (e.g., shop manuals, service bulletins, and computer databases).
AC-17: Prior Knowledge: The technician uses prior knowledge of similar problems to determine the specific cause(s) of problems.
AC-18: Cause/Effect Relationships: The technician comprehends and uses cause-and-effect relationships presented in service information problem-solving trees.
AC-19: Definitions: The technician applies industry definitions to solve problems in automotive components and systems.

Problem Solving

Before you spend much time researching information, it is important that you define the problem. As you are defining the problem, make sure you narrow it down as accurately as possible. For example, if the car won't start, is this because it does not crank over at all? Or does it crank over but not start? Narrowing the problem down in this way helps you filter out irrelevant information. If the problem is very broad, it is better to break it into smaller chunks. Then, conduct a separate research activity on each one.

The next step is to look for the information that may help solve the problem. There is plenty of information readily available. But to determine how useful and reliable the information is, we need to know the sources of that information. We can obtain information from two sources: primary and secondary. The primary sources of information are people who have direct experience with the same or a similar problem. We can obtain information by interviewing those primary sources directly. However, information from primary sources can be subjective. This is because their information is narrowly focused. So unless you have the exact same issue, caused by the same exact fault, they are likely to lead you astray. Therefore, primary sources are not always reliable.

Information from secondary sources, or secondhand information, is compiled from a variety of sources. It is usually more reliable because it is more generic, objective, and broad-based. This secondhand information is available in various formats, including print, audiovisual, and computer-based (**FIGURE 11-25**). We can divide this information into different content categories, including the following:

- Vehicle service information
- Automotive training and educational sources
- Technical assistance

Vehicle Service Information

The first place to look for information about a vehicle is in the shop. Computer service information, shop manuals, aftermarket manuals, and owner manuals are good resources for basic service information. They include information on:

- vehicle systems and how to operate them
- the locations of major components and lifting points

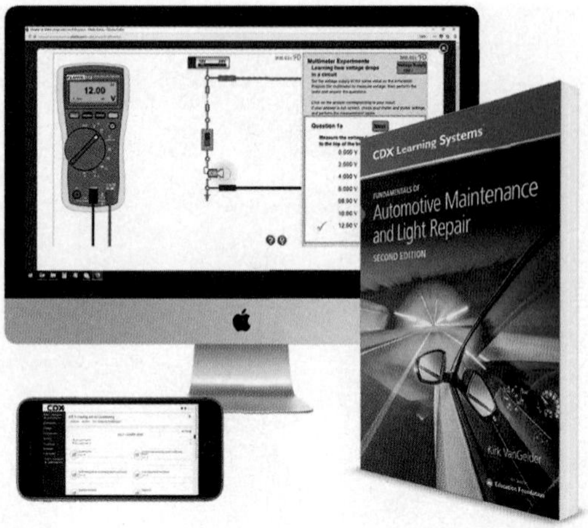

FIGURE 11-26 Paper and online training materials are available from publishers.

FIGURE 11-27 Training class in session.

- vehicle care and maintenance information
- vehicle specifications and
- diagnostic and repair information

The manufacturer's service information will guide the technician in the proper sequence of procedures to diagnose a given fault. The basic idea is to check the simple things first. In most cases, this involves a visual inspection for obvious issues such as a vacuum line or electrical connector that is loose or disconnected. If no problem is found, then the technician should go to the next step in the service procedure. This will likely involve testing the system or components affected. The service information will continue to guide the technician through additional steps as necessary.

Educational Training Sources

Publishers provide extensive materials covering the basic theory of operation of automotive systems and components. These materials come in a variety of forms. Textbooks and workbooks come as paper-based materials (**FIGURE 11-26**). Also, there are computer-based materials. They provide more options for interactivity and visual engagement of the student. Although there is free information available, most of these providers require a subscription to enroll in their courses.

For dealership technicians, manufacturers usually provide training classes (**FIGURE 11-27**). These can be classroom-based or online classes. They cover diagnostic and repair procedures for their particular models. Similarly, aftermarket and parts suppliers provide similar information about their products. This information is also typically available by either attending a physical class, or as online training.

Technical Assistance Services

For troubleshooting, there is help available on the phone and on the Internet. To use either resource, you must subscribe as a member. This usually involves paying a fee. Technical assistance hotlines put you in contact with professionals who assist you in diagnosing a particularly difficult problem over the phone (**FIGURE 11-28**). The technical assistant has access to a variety of technical service information. Some of this information is gathered while helping other technicians with their issues. Thus, a large database of information can be accumulated. This can save a technician a lot of time when dealing with a difficult diagnostic situation.

FIGURE 11-28 Technical assistance is only a phone call away, but it does cost to use them.

For free information, there are chat rooms and forums where questions can be posted to other technicians online; however, there is no guarantee of the accuracy and availability of the resources. The websites of automotive hobbyists, enthusiasts, and car clubs are often good resources as well.

One very good online reference that is available only to professional automotive technicians and students is the International Automotive Technicians Network (www.iatn.net). This organization consists of over 85,000 active members. They have more than 2 million years of combined experience. They share their knowledge with each other on over a dozen forums. They recently have begun granting free student accounts to students of member instructors. These free student accounts allow students to monitor the forums. This is a great way to learn from them while not allowing the students to interact with the technicians.

TECHNICIAN TIP

Have you ever worked on a vehicle and seemed to have hit a dead end in diagnosis? Don't be afraid to discuss the symptoms with other technicians or supervisors around you. Sometimes, others have seen the same issue or have a bit more experience and will be able to help. Or maybe they will look at the problem from a slightly different perspective, which can give you something new to try.

Effective Writing

LO 11-05 Demonstrate effective writing skills.

Not only are technicians required to be able to read on the job, they also must be able to write. One of the biggest writing tasks involves repair orders. Technicians fill out repair orders to document their findings, conclusions, and any repairs undertaken. This requires an accurate, short, but complete summary of the work completed. In fact, the shop and ultimately the technicians get paid based on the quality of the write-up of the repair order. If items get left off of the repair order, then the customer will not be charged the full amount. And you, the technician will not be paid your full wage.

The repair order is a legal document which can be used as evidence in court if there is ever a problem in the future.

Applied Communication

AC-2: Information–Written: The technician can comprehend and apply the available written information needed to diagnose, analyze, and solve a problem.

AC-4: Information–Written: The technician evaluates the usefulness of available written information clearly and thoroughly when analyzing a problem.

AC-5: Information–Written: The technician makes logical inferences and recommendations based on information provided on the repair order.

AC-20: Summaries: The technician uses appropriate grammar and sentence structure when summarizing problems in reports.

AC-21: Sentences: The technician uses conventional sentence structure, spelling, capitalization, and punctuation when composing sentences for warranty reports.

AC-22: Writing: The technician adapts a writing strategy that is most appropriate for the intended audience (e.g., customers, supervisor, or fellow employees) when documenting repairs.

AC-24: Purpose: The technician adapts speaking and/or writing styles that are consistent with the purpose of the communication.

AC-26: Paragraphs: The technician composes complete paragraphs, with appropriate details, presenting accurate information regarding symptoms, diagnosis, and results when preparing warranty claims and work orders.

AC-27: Diction/Structure: The technician adapts diction and structure to the context of all verbal and written communication based on the audience, purpose, and specific situation.

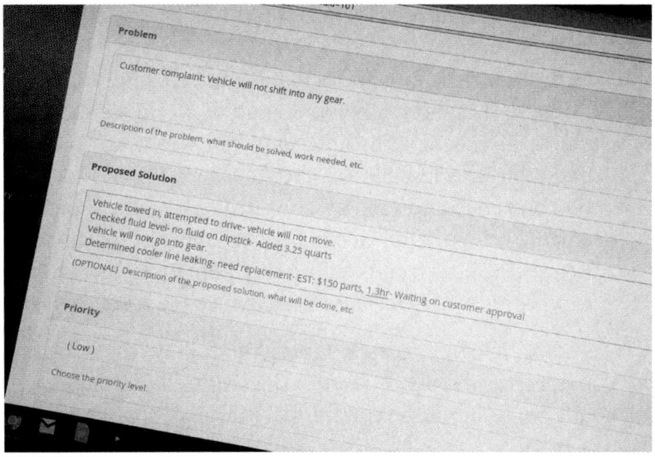

FIGURE 11-29 Technician completing an online repair order.

Thus, if the repair order is poorly written, incomplete, or inaccurate, the shop will be much more likely to lose the court case. This could put it at risk of having to pay thousands of dollars. If it resulted in an injury accident, it could easily be in the hundreds of thousands of dollars. So, writing is one of the most important tasks you as a technician perform on a daily basis (**FIGURE 11-29**).

Writing Business Correspondence

Writing for technicians generally involves completing a write-up of diagnosis and repair conclusions on repair orders. It also includes filling out parts requests. However, technicians may sometimes need to undertake a more formal business correspondence using complete paragraphs. As with any type of writing, first think about what you want to say and draft your message. And then refine and finalize your message.

Just as it is important to think before speaking, you also need to think about what you want to achieve before composing a written message. Using your notes, sketch out a short outline of major points; this is a great place to start when writing. Then, use the outline to assist you in the writing. Use complete sentences. Remember to carefully proofread your writing so that you catch and fix any spelling or grammatical errors. You should use spell-check, but don't rely on it to catch everything. Spell-checkers can't tell you when you have used "two" and "to" incorrectly. Or "there" and "their." Or when you meant to use "they" and only typed "the", just to mention a few examples. Once you have completed the draft, it may be necessary to create a final version of the write-up.

When writing a business letter, after creating an outline of the main points, you must then organize and put them together. Here are a few tips:

- Structure your letter logically so readers can follow your flow of thoughts; check the content by reading it out loud.
- Use plain English when addressing a customer, avoiding complicated words or technical jargon. On the other

hand, when addressing a manufacturer or warranty clerk, be very specific.

- Check your spelling and punctuation. This is easily done using spell-check or by looking up the term(s). Be aware that spell-check can misinterpret your words, so you need to double-check spelling corrections.
- Be courteous and empathize with the readers, especially when writing a letter of complaint or delivering bad news.
- Be precise, direct, and to the point. Repetition or unnecessary words waste time and can confuse the readers.

There is a simple but necessary format to follow when creating a business letter (**FIGURE 11-30**). The letter should have your company name and address at the top. Most business letterheads have these details preprinted. Next include the date, followed by the recipient's name and address.

Usually, you should start a letter with a greeting like "Dear Sir or Madam," or when we know the receiver's name, "Dear Mr. Mortenson." "Re:" or "Reference:" is followed by the subject or purpose of the letter. The next part is the content that you composed in your outline. A business letter usually concludes with a complimentary closing, such as "Sincerely." Finally, a letter needs to be signed and followed by the sender's typewritten name and position.

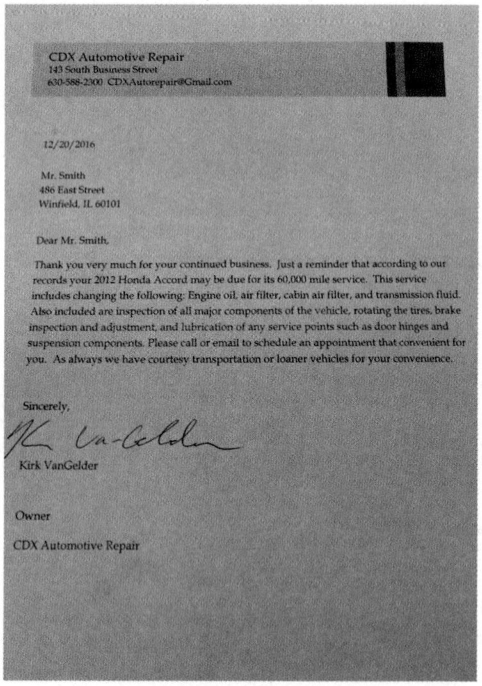

CDX Automotive Repair
143 South Business Street
630-588-2300 CDXAutorepair@Gmail.com

12/20/2016

Mr. Smith
486 East Street
Winfield, IL 60101

Dear Mr. Smith,

Thank you very much for your continued business. Just a reminder that according to our records your 2012 Honda Accord may be due for its 60,000 mile service. This service includes changing the following: Engine oil, air filter, cabin air filter, and transmission fluid. Also included are inspection of all major components of the vehicle, rotating the tires, brake inspection and adjustment, and lubrication of any service points such as door hinges and suspension components. Please call or email to schedule an appointment that convenient for you. As always we have courtesy transportation or loaner vehicles for your convenience.

Sincerely,

Kirk VanGelder

Owner

CDX Automotive Repair

FIGURE 11-30 Typical business letter.

Completing a Repair Order

When you are writing up your findings, conclusions, and repairs performed on a repair order, one thing to remember is that the information should be concise, yet complete. This is a difficult skill to develop but will improve with practice.

A work order should contain the following information:

- The customer's concern: Describe the concern, how you verified the concern, and any related issues found.
- The cause of the concern: Describe what specifically is the root cause of the concern.
- The action taken, or needed to be taken, to correct the concern: Describe specifically how the concern was, or could be, resolved.

These elements constitute what is called the three "Cs": concern, cause, and correction. Remember that each of these sections needs to be written up as concisely yet completely as possible (**FIGURE 11-31**). For example:

Concern: The vehicle overheats within about 10 minutes of vehicle operation from a cold start, no matter whether it is hot or cold outside.

Cause: Coolant is full and protected to –30°F (–34.4°C). Started vehicle, monitored engine temperature with temp gun. Temperature exceeded specified 195°F (90.5°C) thermostat opening temperature, and thermostat was still not open at

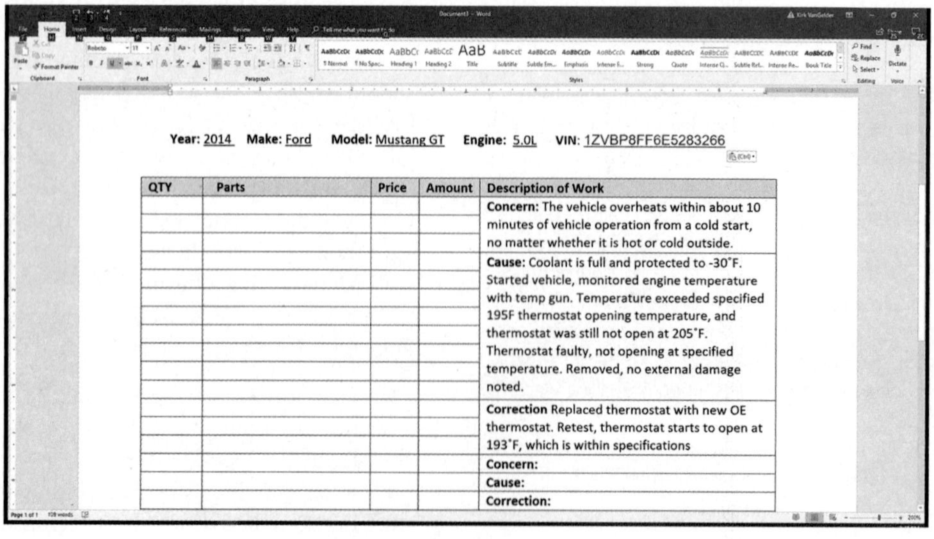

Year: 2014 **Make:** Ford **Model:** Mustang GT **Engine:** 5.0L **VIN:** 1ZVBP8FF6E5283266

QTY	Parts	Price	Amount	Description of Work
				Concern: The vehicle overheats within about 10 minutes of vehicle operation from a cold start, no matter whether it is hot or cold outside.
				Cause: Coolant is full and protected to -30°F. Started vehicle, monitored engine temperature with temp gun. Temperature exceeded specified 195F thermostat opening temperature, and thermostat was still not open at 205°F. Thermostat faulty, not opening at specified temperature. Removed, no external damage noted.
				Correction Replaced thermostat with new OE thermostat. Retest, thermostat starts to open at 193°F, which is within specifications
				Concern:
				Cause:
				Correction:

FIGURE 11-31 The 3 Cs filled out as listed in the text.

205°F (96.1°C). Thermostat faulty, not opening at specified temperature. Removed, no external damage noted.

Correction: Replaced thermostat with new OE thermostat. Retested, thermostat starts to open at 193°F (89.4°C), which is within specifications.

A repair order should also contain the customer's contact information, vehicle information (e.g., make/model/VIN), parts, prices, labor time, taxes, and any recommended service that is found while repairing the vehicle. Remember that a repair order is a legal document, so treat it as if your job depends on it; it does.

While in training, you document tasks performed in the shop on forms referred to as job sheets or tasksheets. You can use these forms to develop your ability to accurately complete the three Cs. Doing so will prepare you for the repair orders you will complete when working out in the industry.

Completing a Shop Safety Inspection Form

Most shops will have a safety inspection form that has to be completed on a regular basis. This is typically weekly or monthly, although some tasks may have to be performed daily (**FIGURE 11-32**). Inspection forms guide you to visually inspect critical items in the shop such as:

- automotive lifts, overhead doors, hydraulic equipment, pneumatic equipment, hoses and cords, fire extinguishers, and emergency exits.

The importance of these inspections should not be taken lightly, as neglect can cause safety issues and premature break downs.

Safety inspection forms are also a way of keeping up on routine maintenance tasks. Common types of maintenance include:

- changing the oil or replacing the belt on an air compressor
- refilling the tire machine automatic oiler and
- draining any water traps in the compressed air system.

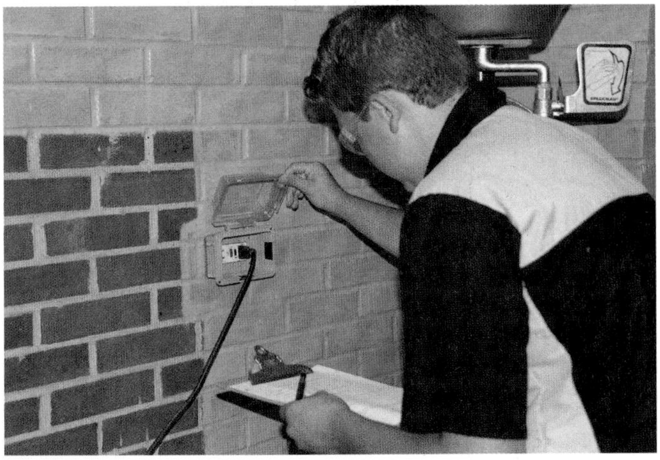

FIGURE 11-32 Shop safety inspection form being used.

If you are the person completing the inspection, make sure you are trained to evaluate the safety and condition of the items you are inspecting. Inspect the items carefully and thoroughly so you don't miss any issues. Be sure to accurately document any issues found, and inform your manager.

Completing a Defective Equipment Report

One way to help ensure safety in the workshop is to inspect equipment regularly. And arrange for repair or replacement whenever it does not meet the safety standards. Keeping equipment well maintained will help avoid equipment downtime to provide a safer work environment.

Whenever you come across any defective equipment, you should do the following:

- Tag the defective items and either place them in a secured area or secure them so no one else can use them by mistake (**FIGURE 11-33**).
- Complete a defective equipment report (**FIGURE 11-34**).
- Notify your supervisor.

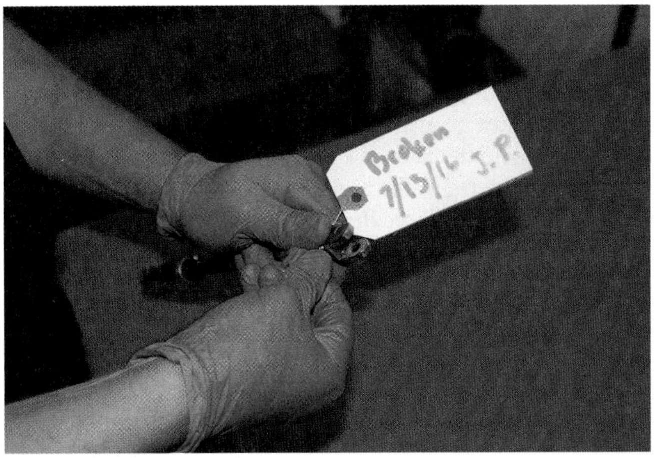

FIGURE 11-33 Tagging out and securing defective tools.

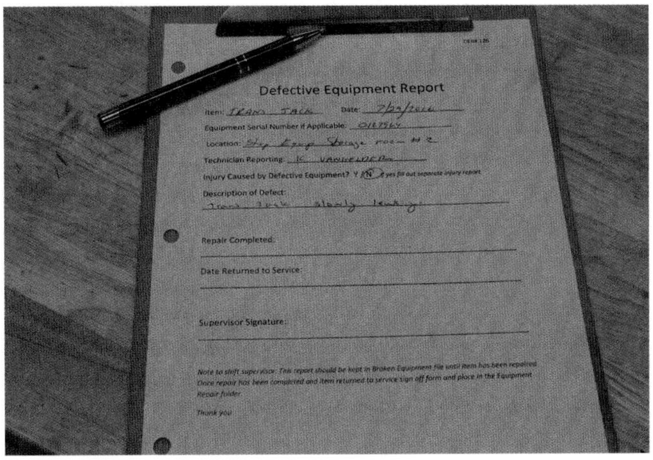

FIGURE 11-34 Completing a defective equipment report.

By tagging out faulty equipment, your immediate actions can help protect coworkers from accidents and injuries. The report will provide details on why the equipment or tools were tagged out. On the report, record:

- the date when the defect was detected
- the location of the defective equipment
- the name of the equipment and the serial number (if possible) and
- a description of the defect and the action you have taken.
- Then, sign your name as the reporter.
- Finally, notify your supervisor.

If an accident occurs due to defective equipment, you will need to complete both an accident report and a defective equipment report.

Lockout/Tagout

There are many dangers in the automotive repair facility that may need to be properly documented and tagged. One example could be a defective automotive lift. If a problem is noticed with a piece of equipment, the lockout/tagout procedure should be followed. Lockout/tagout procedures have been developed to prevent avoidable and unnecessary workshop accidents. These procedures have many functions:

- The tag notifies other users that the tool or component is dangerous to use. Any equipment that is found to be faulty must be identified so that other users are not put at risk. Write the fault, the date, and your name on the tag. Attach the tag to the tool. Smaller equipment should also be tagged and placed in a location where it is not forgotten. Notify your supervisor so that repairs or a replacement can be arranged.
- If a machine is faulty, the lockout procedure is used. Most large workshop equipment is permanently wired to the electrical supply and usually will have an isolation switch that disconnects the electrical power. A lockout tag should be placed on the isolation switch as well as the equipment. Turn the machine off at the power and master switches, attaching the lockout tag in a manner that prevents the switches from being turned on. Once again, notify your supervisor so repairs can be arranged.
- The lockout/tagout procedure is also used to notify other technicians that a vehicle is not drivable.

Your workshop will have a procedure for vehicle lockout/tagout. It may involve the technician filling out a "defective vehicle" label listing the nature of the defect, name of the technician, and date and time of the defect.

If you remove the vehicle keys, do not keep them in your pocket or on your workbench. Attach a label or tag to the keys that identify the vehicle they belong to. Then, store them in a secure key organizer.

If a vehicle is going through a relearn process (i.e., the vehicle's computer communicating with another computer), it may be necessary to leave the ignition on for many hours. In this case, tag the vehicle with instructions to leave the ignition on. Otherwise, a passing technician may turn it off in an effort to be helpful.

If a vital component has been removed for service, it may not be obvious to a casual observer, so it is necessary to tag the vehicle. The best place to tag the vehicle is on the steering wheel or driver's window. This is where others trying to operate the vehicle will see it. Also, remove the ignition key and store it in a safe place. Ask your supervisor to demonstrate the lockout/tagout process used in your workshop and to show you the location of the key organizer.

To properly identify faulty equipment, follow the steps in **SKILL DRILL 11-1**.

SKILL DRILL 11-1 Identifying Faulty Equipment

1. Basic workshop tools that are broken or worn should be replaced. Make sure you tag the tool as faulty or broken, and do not use it until you buy a replacement. Then discard the tool.

2. Power tools that have been identified as faulty, due to failure of parts, should also be tagged and set aside. The tools can only be used again after an authorized agent has made the repair.

SKILL DRILL 11-1 Identifying Faulty Equipment (Continued)

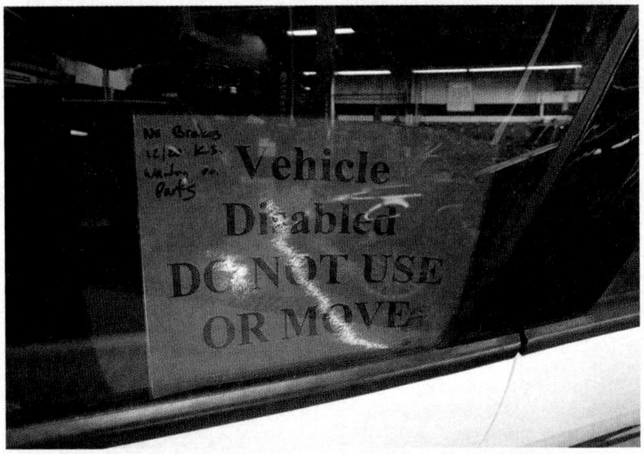

3. Isolation tags are also used on disabled vehicles or vehicles undergoing a repair. In this case, you must locate and complete the "Disabled Vehicle" warning notice. Write the license number of the vehicle and the nature of the defect. Write your name and then the date and time you completed the notice. Attach the notice to the steering wheel or driver's window.

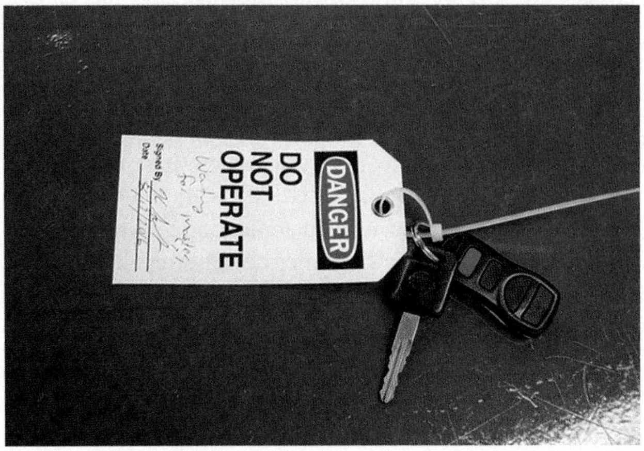

4. Remove the keys and lock the vehicle, if appropriate. Attach a tag to the keys that identifies the vehicle they belong to. Store the keys in the key organizer, and notify your supervisor.

Completing an Accident Report

Safety is the most important issue in the workplace. We must make a conscious decision to work safely and act responsibly to protect others and ourselves. Unfortunately, accidents do happen. When one occurs, an accident report should be completed by those involved, both the victim and witnesses, if possible (**FIGURE 11-35**). The information in the report is used for several purposes. First, to protect current employees and employers against false claims. It also protects future employees by calling attention to a situation that caused an accident. This allows steps to be taken to prevent it in the future.

To ensure that the information on the report is accurate, the accident report should be completed as soon as practically possible. Over time, peoples' memories fade. So, completing the report while the facts of the accident are still fresh in everyone's memory makes it more accurate.

A typical accident report includes:

- the date and time when the accident happened,
- the location where the accident happened,
- the name of the person who was injured,
- the name of any witnesses,
- the details of the accident,
- any first aid treatment provided, and
- any medical assistance rendered.

It is also important to note whether the accident will be subject to, or covered by, any insurance claims. Finally, the form needs to be signed by the person reporting the accident, and turned in to the supervisor.

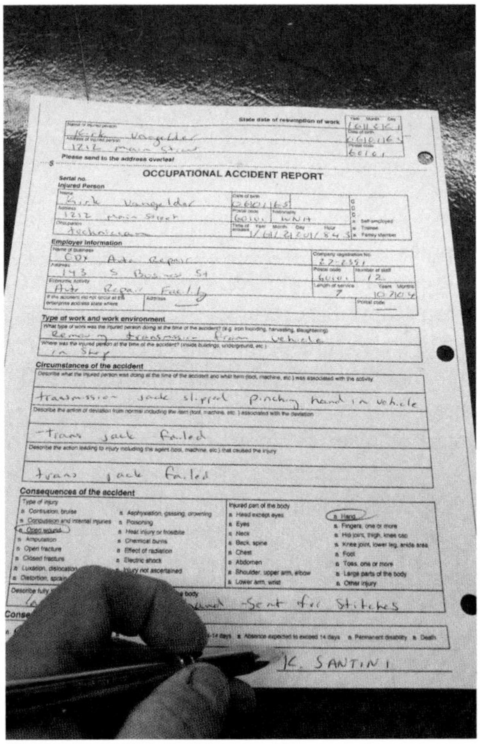

FIGURE 11-35 Technician filling out an accident report.

Completing a Vehicle Inspection Form

Technicians provide customers with a very beneficial service when performing a thorough inspection on their vehicles. When performing an inspection, check that all major components and systems are operational, secured, and safe. This should be in accordance with the vehicle manufacturer's specifications. This means testing the operation of the electrical and mechanical systems, such as lights and brakes, and visual inspections of the components such as tires and glass.

An inspection form is a useful guide when conducting a vehicle inspection. By following the checklist, a technician can test all the components in a systematic way. This will help ensure that is fully operational (**FIGURE 11-36**). It also becomes a record for the shop to bring a customer's attention to needed service. In the event that a customer declines repairs, it shows that the shop made the customer aware of them. Many shops use their own inspection form that lists all items tested, which are known as points. Depending on the number of points covered, the inspection may be called a 30-point inspection, a 72-point inspection, etc.

To complete the inspection form, you should inspect all the components and systems on the checklist, including the following:

- Fluids, belts, and hoses
- Steering and suspension system
- Brakes
- Drive line
- Fuel system
- Exhaust system
- Tires
- Lighting system
- Electrical system
- Visibility
- Seat belts
- General components

Once the inspection is complete, the results should be presented to the customer. They then decide if any repairs are to be made. This is where verbal communication comes back into play. The customer must have a good understanding of what the vehicle needs. They also must know why it is important to have it repaired and the consequences of not repairing it. This conversation is easier if you have a trusting relationship with the customer.

The trusting relationship is determined by their past experience with you and the shop. Assuming the level of trust is strong, your communication skills will come into play. They include good verbal, nonverbal, listening, clarifying, writing, and presentation skills as well.

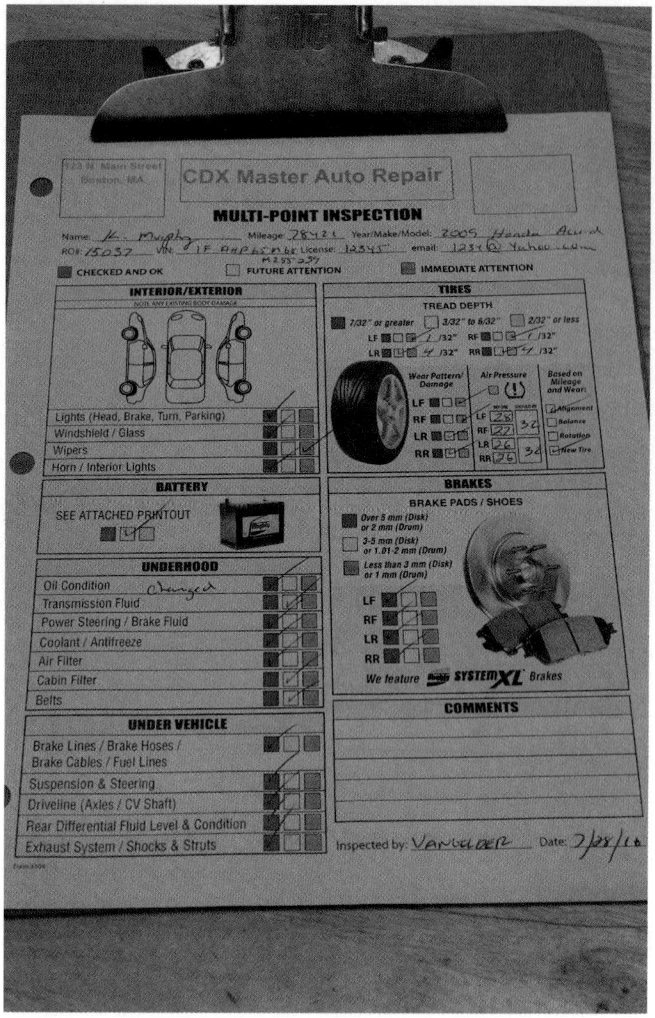

FIGURE 11-36 Technician using a vehicle inspection form.

In conclusion, communication is a critical component of a successful business and efficient workplace. In our current digital world, face-to-face contact is getting less and less common. This makes it increasingly important to familiarize yourself with a wide variety of communication skills. Using and practicing these skills can be a key factor in how far you go in the automotive industry. It will also determine how much money you make. And the good news is that effective communication does not even require expensive tools!

▶ Wrap-Up

Ready for Review

- ▶ Active listening skills is one of the most important skills to gather information from a customer, or any other person. It should include empathy, nonverbal feedback, and verbal feedback.
- ▶ Speaking is a three-step process that consists of thinking about a message, presenting a message, and checking whether the message is correctly understood. Relaxing, listening, and pausing can help ease situations.
  - Open-ended questions, close-ended questions, yes/no questions can be used to obtain the necessary information.
- ▶ Employability skills can be as simple as showing up to work on time, every day, and keeping your driver's license

free of tickets. Honesty, integrity, reliability, and quality are also requirements of being a valued technician.
  - Professional appearance, time management, and effective customer service are also vital in today's competitive business environment.
- Before reading anything in particular, know your goal for reading it. Selective, comprehending, and absorbing reading methods are necessary to understand information properly.
  - Computer service information, shop manuals, aftermarket manuals, and owner manuals are good resources for basic service information.
- Effective writing of a repair order should be accurate, short, but complete summary of the work completed.
  - Proper structure; language; spelling and punctuation; and being courteous, precise, and direct to the point are main points to be considered for effective writing.

Key Terms

Cause A concise, but complete, description of the root cause of the customer's concern.

Concern The first C in the three Cs, documenting the original concern that the customer came into the shop with. This documentation will go on the repair order, invoice, and service history.

Correction The third C in the three Cs, documenting the repair that solved the vehicle fault. This documentation goes on the repair order, invoice, and service history and must include the procedure used as well as a brief description of the correction.

supporting statement A statement that urges the speaker to elaborate on a particular topic.

validating statement A statement that shows common interest in the topic being discussed.

Review Questions

1. The active listener should focus their attention:
   - **a.** On taking notes
   - **b.** Formulating a response
   - **c.** On the customer
   - **d.** On customer posture
2. When listening to a customer problem, how should the service advisor gather more information?
   - **a.** Research the situation on the Internet
   - **b.** Tell the customer what you think the problem is
   - **c.** Have the customer describe it to another service advisor
   - **d.** Ask open-ended questions
3. All of the following are types of questions EXCEPT:
   - **a.** Sequential questions
   - **b.** Open questions
   - **c.** Closed questions
   - **d.** Yes-or-no questions
4. All of the following are steps in the speaking process EXCEPT:
   - **a.** Think about the message
   - **b.** Say the first thing that comes to mind
   - **c.** Accurately present the message
   - **d.** Check whether the message was understood
5. When a shop hires a technician, one of the largest hurdles to overcome is
   - **a.** Lack of required tools
   - **b.** The new technician is too young
   - **c.** Having a clean driving record
   - **d.** Reliable transportation
6. Which method of reading is used when reading the instructions for a diagnostic test?
   - **a.** Selective
   - **b.** Comprehending
   - **c.** Absorbing
   - **d.** Reasoning
7. What is one of the most important tasks a technician performs on a daily basis?
   - **a.** Cleaning up the work area
   - **b.** Avoiding excessive breaks
   - **c.** Writing
   - **d.** Watching
8. When would selective reading be used?
   - **a.** When looking up a specification
   - **b.** When looking up diagnostic information
   - **c.** When reading a chapter in a textbook
   - **d.** When learning how a new system works
9. When a technician encounters faulty equipment, what is the procedure for dealing with it?
   - **a.** Repair it
   - **b.** Use it till it breaks
   - **c.** Leave it until later
   - **d.** Lock out/Tag out
10. All of the following are employment requirements for a technician EXCEPT:
    - **a.** Being drug and alcohol free
    - **b.** Having all of your ASE certifications
    - **c.** Having a clean driving record
    - **d.** Having honesty, integrity, reliability, and quality

ASE Technician A/Technician B Style Questions

1. Technician A says a vehicle inspection form is used to bring attention to the customer for needed repairs. Technician B says that by following an inspection form the technician can test all the components in a systematic way. Who is correct?
   - **a.** Technician A
   - **b.** Technician B
   - **c.** Both A and B
   - **d.** Neither A nor B
2. Technician A says that most people believe nonverbal communication more than the actual words. Technician B says that verbal feedback is not important as long as eye contact is maintained. Who is correct?
   - **a.** Technician A
   - **b.** Technician B
   - **c.** Both A and B
   - **d.** Neither A nor B

3. Technician A says that it is ok to lose your temper while on the phone because the person can't see you. Technician B says that if there is a need to keep a person on hold for an extended time, offer to call them back, and then do it. Who is correct?
   a. Technician A
   b. Technician B
   c. Both A and B
   d. Neither A nor B
4. Technician A says that with supporting statements, the service writer can demonstrate they understand, but want more information. Technician B says that with validating statements, the service writer can show empathy for the person speaking. Who is correct?
   a. Technician A
   b. Technician B
   c. Both A and B
   d. Neither A nor B
5. Technician A says that technical assistance hotlines send professionals to your shop to diagnose difficult problems. Technician B says that technical assistance hotlines can save a technician a lot of time when diagnosing difficult problems. Who is correct?
   a. Technician A
   b. Technician B
   c. Both A and B
   d. Neither A nor B
6. Technician A says that open-ended questions encourage people to speak freely so you can gather insights and opinions. Technician B says that closed-ended questions will provide facts and details. Who is correct?
   a. Technician A
   b. Technician B
   c. Both A and B
   d. Neither A nor B

7. Technician A says that taking personal texts and phone calls during work is OK. Technician B says that showing up to work on time every day is an employability skill. Who is correct?
   a. Technician A
   b. Technician B
   c. Both A and B
   d. Neither A nor B
8. Technician A says that maintaining a professional hairstyle is an important employability trait. Technician B says that taking showers daily is an important employability trait. Who is correct?
   a. Technician A
   b. Technician B
   c. Both A and B
   d. Neither A nor B
9. Technician A says that repair orders are legal documents so they have to be filled out accurately and carefully. Technician B says that ensuring the repair order is well written, clear, and concise promotes a professional reputation. Who is correct?
   a. Technician A
   b. Technician B
   c. Both A and B
   d. Neither A nor B
10. Technician A says that most information technicians have to read is written at a very high level. Technician B says that technicians spend a lot of time researching information. Who is correct?
   a. Technician A
   b. Technician B
   c. Both A and B
   d. Neither A nor B

BRAKE
ABS
DTC
SERVICE ENGINE SOON
SERVICE
P
mls -7000
mph
km/h
1/min

SECTION 2
Engine Repair

▶ CHAPTER 12 Motive Power Theory—SI Engines

▶ CHAPTER 13 Engine Mechanical Testing

▶ CHAPTER 14 Lubrication System Theory

▶ CHAPTER 15 Servicing the Lubrication System

▶ CHAPTER 16 Cooling System Theory

▶ CHAPTER 17 Servicing the Cooling System

CHAPTER 12

Motive Power Theory—SI Engines

Learning Objectives

- **LO 12-01** Demonstrate knowledge of heat engines.
- **LO 12-02** Demonstrate knowledge of the physics of engine operation.
- **LO 12-03** Demonstrate knowledge of force, work, and power.
- **LO 12-04** Demonstrate knowledge of four-stroke engine arrangement, operation, and measurement.
- **LO 12-05** Demonstrate knowledge of spark-ignition engine components.
- **LO 12-06** Demonstrate knowledge of two-stroke and rotary engine operation.

ASE Education Foundation Tasks

See Appendix A to view the 2017 ASE Education Foundation Automobile Accreditation Task List Correlation Guide.

▶ Heat Engine Overview

LO 12-01 Demonstrate knowledge of heat engines.

The internal combustion engine (ICE) is an irreplaceable part of modern society. We rely on it to haul food and water, deliver passengers to their destinations, and even save lives. Over time, the ICE has seen many changes. However, the basics have remained similar (**FIGURE 12-1**). At the same time, they have been highly refined.

FIGURE 12-1 **A.** Antique flat-head engine. **B.** Modern dual overhead cam engine.

In this chapter, we:

- cover the types of spark-ignition engines that are available
- identify the major components that make up the engine and
- describe how these components operate together.

We also compare them briefly with diesel, two-stroke, and rotary engines.

Principles of Thermodynamic (Heat) Engines

Thermodynamics is generally defined as the branch of physical science that deals with heat and its relation to other forms of energy. These forms of energy can be mechanical, electrical, or chemical. In this chapter, we discuss how we use heat energy to produce power and make work happen.

Heat causes things to expand. If they are contained, they create pressure as they expand. This pressure is used to move mechanical parts, which in turn can be used to power devices. In automotive applications, the useful effect of heat is to create motive (moving) power. This power drives a vehicle down the road as well as provides power for all of the onboard systems.

Heat is produced when air and fuel are burned. This process is called combustion. Where the heat is created determines, in part, how engines are classified. Engines used for motive power may be classified as external combustion or internal combustion engines.

External combustion means that the fuel is being burned outside of the engine. Internal combustion means the fuel is burned inside of the engine. Both internal and external combustion engines are called "heat engines" because they run off of heat energy. This is why a basic understanding of thermodynamics is critical to understand how engines operate.

Two examples of the external combustion engine are the steam engine and the Stirling engine. At one time, external combustion engines, such as steam engines, were used to power almost all equipment (**FIGURE 12-2**). Steam engines are the best example of external engines powering equipment such as:

- farm tractors
- railroad trains
- automobiles

You Are the Automotive Technician

You are working in the back shop when a salesman from the new car sales department asks if you can answer some questions for a customer. The customer's previous car was totaled in a parking lot accident, and he is very interested in a couple of cars on the lot. He has some technical questions that need to be answered. And the service manager selected you to answer the questions. You greet the customer, and he asks you the following questions:

1. "I see a nice diesel pickup truck that I could use on the farm. How does a compression-ignition engine operate differently from a spark-ignition engine?"
2. "My son would like me to buy that RX8. How does a rotary engine operate differently from a piston engine?"
3. "As I am looking at specifications on the vehicles, what is the difference between horsepower and torque?"
4. "What is meant by overhead cam?"
5. "What is an interference engine?"

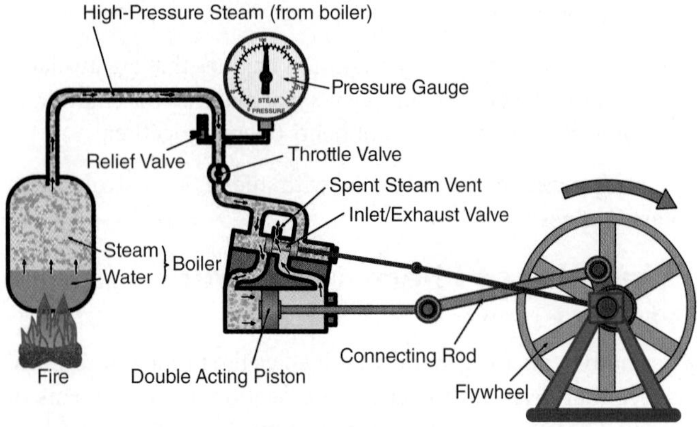

FIGURE 12-2 External combustion engine—Steam engine.

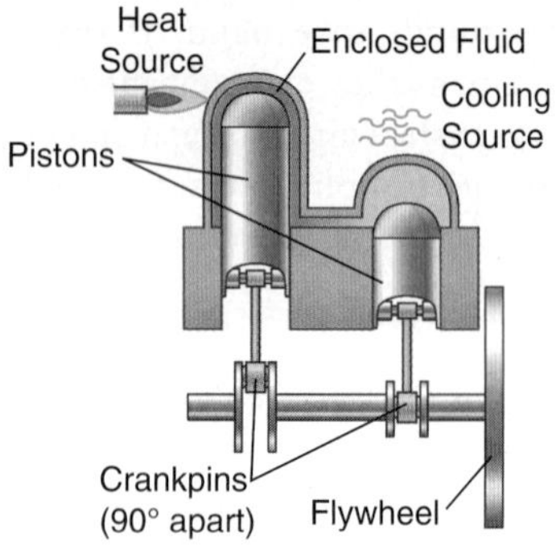

FIGURE 12-3 External combustion engine – Stirling engine.

- boats and ships, and more.
- Even a steam-powered airplane was produced, although it never became popular.

In a steam engine, steam is created in an external boiler (external combustion). As the steam is created, it builds pressure. The high-pressure steam is used to push a piston back and forth in a cylinder. Most steam engines applied steam alternately to each side of the piston. This powered the piston in both directions. One problem with steam engines is that they take a relatively long time to generate steam pressure. So you could not just hop in a steam-powered car and take off. The boilers also presented an explosion hazard if they generated too much pressure. Another explosion risk was if the boiler was weakened due to rust.

The Stirling engine is also an external combustion engine. This engine uses two cylinders with a passage way that connects both cylinders (**FIGURE 12-3**). One cylinder houses the power piston and, the other, the displacer piston. The power cylinder is heated, and the air expands, pushing the piston down. The hot air is then pushed into the displacer cylinder, where it is cooled. Air moves back and forth between the cylinders to expand and cool, which drives the engine.

Stirling engines hold promise as an alternative source of power. But they have not become popular for transportation because their output cannot be easily varied. Solar-powered Stirling engines are gaining popularity as home power sources because they operate on free solar energy. This also means there are no byproducts of combustion to worry about. Stirling engines can run almost silently and therefore can be used near people without disturbing them.

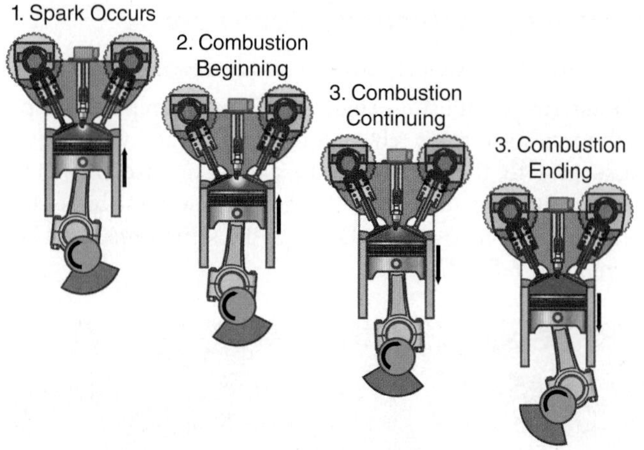

FIGURE 12-4 Combustion in an SI engine.

The **ICE** has almost completely replaced the external combustion engine. The ICE is still the favored mode of power for the transportation industry. For on-road applications, it is used in cars, trucks, buses, etc. For off-road applications, it is used in tractors, trains, and ships. This chapter focuses on spark-ignition (SI) ICEs, as they are the most widely used engines in modern automobiles.

NOTE: Refer to the Alternative Fuel Systems chapter for more information on hybrid vehicles, flex-fuel vehicles, "pure" battery-electric vehicles, fuel cell vehicles, and other alternative-fueled vehicles.

Gasoline (spark-ignition) and diesel (compression-ignition)engines are prime examples of ICEs. Fuel is burned inside the ICE. According to Charles's law, when a gas is heated, it expands. Because fuel contains energy in chemical form, it creates high pressure when it is burned in a sealed combustion chamber. The pressure pushes a moveable piston. The moving piston produces power to do work.

The ICE can be classified in two ways. It can either be a reciprocating piston engine, or a rotary engine. In reciprocating piston engines, pistons move back and forth in their cylinders. Piston engines use a crankshaft to convert the reciprocating movement of the pistons into rotary motion at the crankshaft. The rotary engine uses a rotating motion rather than reciprocating motion. Both the piston and the rotary engine are described in greater detail later in this chapter.

Piston engines are either SI engines or CI engines. In SI engines, liquid fuels are compressed in the combustion chamber. Liquid fuels can be gasoline, ethanol, methanol, or butanol. The compressed fuel is ignited by an electrical spark. This happens when the spark jumps across the air gap of a spark plug inside the combustion chamber. Timing of combustion is controlled by when the spark jumps across the electrodes of the spark plug (**FIGURE 12-4**).

In diesel engines, air in the combustion chamber is compressed so tightly that it becomes extremely hot. In fact, it is hot enough to ignite the fuel as soon as it is injected into the combustion chamber. Timing of the combustion process is controlled by when the fuel is injected. Because of compression ignition, CI engines do not use spark plugs (**FIGURE 12-5**).

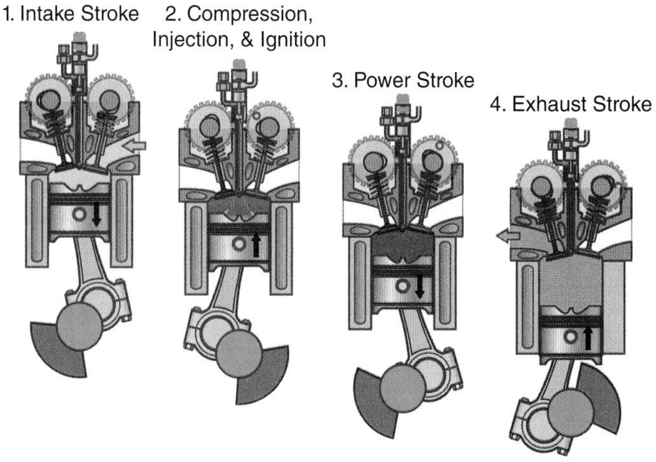

FIGURE 12-5 Combustion in a Compression Ignition (CI) engine.

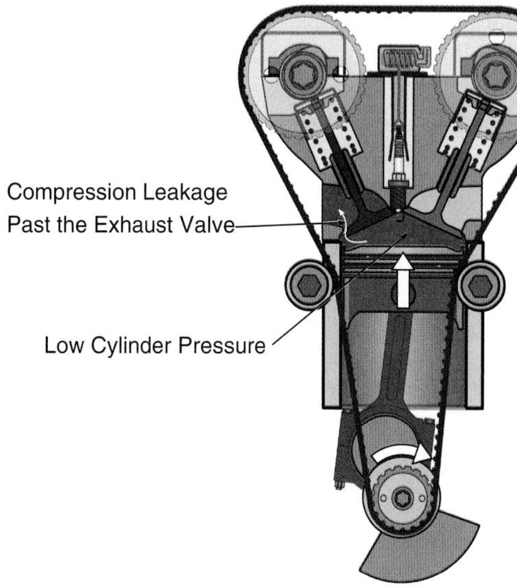

FIGURE 12-6 If the cylinder has a leak, the pressure will not rise, making combustion inefficient.

▶ Principles of Engine Operation (Physics)

LO 12-02 Demonstrate knowledge of the physics of engine operation.

Engines operate according to the unchanging laws of physics and thermodynamics. Understanding the physics involved with an engine will help you diagnose engine problems. For example, knowing that pressure rises when the volume of a sealed container is reduced will help you understand that if the pressure doesn't rise properly, then there is a leak present. The diagnosis will then focus on locating the source of the leak.

Valves seal the ports in the engine's cylinder head(s). If the valves leak, the pressure in the cylinder will not rise as it should during the compression stroke of the piston (**FIGURE 12-6**). Too little pressure squeezing the air–fuel mixture together in the cylinder causes inefficient combustion when it is ignited. This results in a rough-running engine.

A good example of the importance of compression is the burning of black powder in open air. It produces fire and smoke, but no explosion. If the same powder is wrapped tightly, it becomes a firecracker (or even a stick of dynamite)—exploding the wrapping and producing a big bang. In the same way, more power is produced when air and fuel are compressed into a tightly packed space.

Pressure and Temperature

According to Charles's law, the pressure and temperature of a gas are directly related to each other. As pressure rises, so does temperature. As pressure decreases, so does temperature. For example, think of a cylinder with a moveable plunger at one end (**FIGURE 12-7**). This plunger seals tightly in the cylinder so that no air can escape past the plunger. Installed on the other end of the cylinder are a pressure gauge and a thermometer.

As the plunger is pushed in, the air molecules are squeezed together, and air pressure rises. As the air molecules are squeezed together more tightly, the pressure and the temperature rise. This is caused by friction between the air molecules as they bounce off each other with greater force. The air in the cylinder heats up as this happens, and thus, the temperature rises.

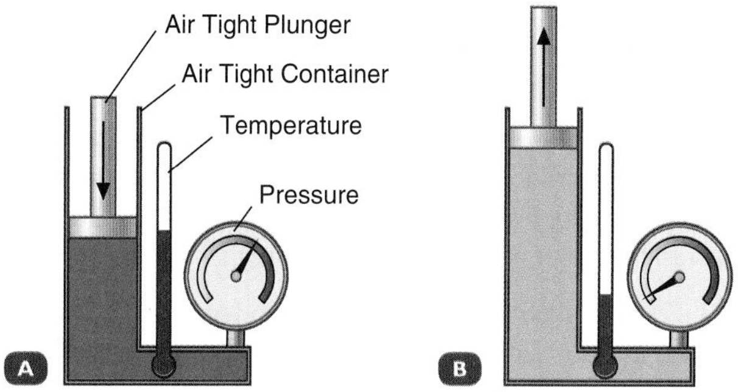

FIGURE 12-7 Pressure changes temperature. **A.** As pressure goes up, temperature goes up. **B.** As pressure goes down, temperature goes down.

▶ TEST YOUR UNDERSTANDING

When the plunger is pushed into the cylinder, the air pressure and temperature are increased. What will happen to the temperature if the plunger is held steady at that point? If you said it would stay the same because the pressure stayed the same, you would be wrong. Heat travels from hot areas to cold areas. The heat would slowly transfer to the surrounding air until the temperatures are equalized. And yes, that would lower the pressure inside the cylinder slightly, as the temperature goes down. Once the cylinder is cool, what happens to the temperature of the air inside it if the plunger is pulled way back so that there is a vacuum inside? If you say it would be colder than the outside air, you would be correct. Reducing pressure reduces temperature, even below the outside temperature. And that is a principle we use in an air-conditioning system. To find out more, turn to the HVAC Principles chapter.

A diesel engine uses this principle to ignite the fuel injected into an engine cylinder. The air is compressed so tightly that it becomes hot enough to ignite the fuel when it is sprayed into the cylinder. This is why diesel engines are called CI engines.

Again, using the example of a plunger in a cylinder, what happens when the plunger is pulled outward? Pulling out the plunger reduces the gas pressure and gives the molecules more room to move. They bump into each other less and temperature decreases. Thus, a drop in pressure produces a lower temperature.

Looking at a related scenario, what happens to pressure when the gas temperature changes? When a gas is heated, its molecules start to move more quickly and require more space. Heating a gas in a sealed container increases the pressure in the container (thermal expansion) (**FIGURE 12-8**). Cooling a gas has the opposite effect. As the molecules slow down, they demand less space. As a result of cooling a gas in a sealed container, the pressure drops.

Temperature and Energy

The temperature of a gas is one measure of how much energy it has. The more energy a gas has, the more work it can do. The heating of gas particles makes them move faster, which produces more pressure. This pressure exerts more force on the container in which the gas is located, which is how an ICE functions.

Pressure is first raised quite a bit through compression. It is then raised much higher through combustion of the air–fuel mixture. Burning the air–fuel mixture increases the temperature inside the container tremendously. And that creates the necessary pressure to produce work. The more energy the air–fuel mixture has, the more force it exerts on the piston. And the more work the piston can do (**FIGURE 12-9**). This principle takes place during the power stroke and pushes the piston down the sealed container.

Latent (stored) heat energy exists in various kinds of fuels. It is released to do work when the fuel is ignited and burned. Types of fuels that contain latent heat energy include:

- liquid fuels such as gasoline, diesel, and ethanol
- gaseous fuels such as natural gas, propane, and hydrogen and
- solid fuels such as gunpowder, wood, and coal.

Latent heat energy is often measured and expressed as British thermal units, abbreviated Btu. One Btu equals the heat required to raise the temperature of 1 pound (lb) (0.45 kg) of water by 1°F (17.22°C). Gasoline has a comparatively high Btu per gallon rating of around 14,000 Btu. Diesel fuel is even more energy dense, however, at around 25,000 Btu per gallon. Coal has a much lower Btu rating. This is one reason why both the home heating and the transportation (rail) industries have moved from coal to petroleum. It has better energy density. It also takes up less space for storage and transport.

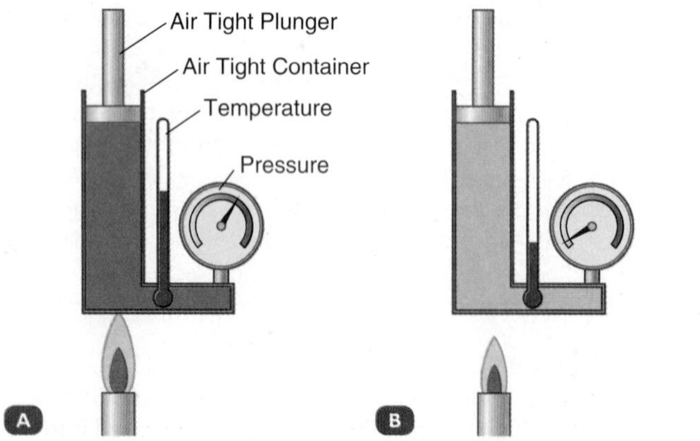

FIGURE 12-8 Temperature changes pressure. **A.** As temperature goes up, pressure goes up. **B.** As temperature goes down, pressure goes down.

Pressure and Volume

Pressure and volume are inversely related. As one rises, the other falls. A cylinder with a pressure gauge and movable piston is a good example. It contains air, and as the piston is pushed in, the inside air is forced into a smaller volume. At the same time, the pressure gauge shows an increase in pressure. It is this increase in pressure that allows the pump to do its work. When the piston is pulled out, the volume occupied by the gas grows larger, and the pressure drops. A larger volume has less gas pressure. When the volume is reduced, the gas pressure rises. Keep in mind that larger pressures are desirable to increase the amount of work done in an engine (**FIGURE 12-10**).

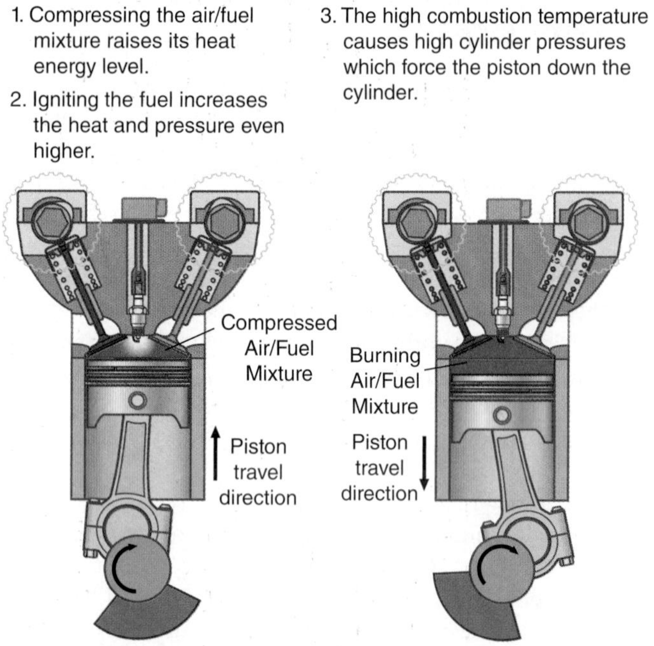

FIGURE 12-9 Burning a compressed gas increases the temperature, producing more pressure and increased force.

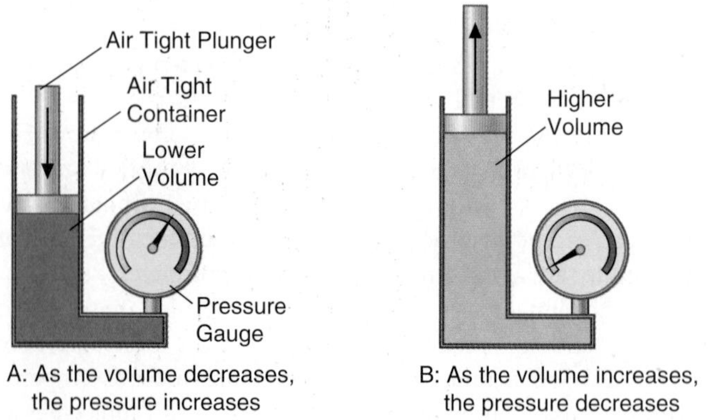

FIGURE 12-10 Volume affects pressure.

▶ Force, Work, and Power

LO 12-03 Demonstrate knowledge of force, work, and power.

Effort to produce a push or pull action is referred to as force (**FIGURE 12-11**). A compressed spring applies force to cause, or resist, movement. A tensioned lifting cable applies force to hold a vehicle in place on a lift. Force is measured in pounds, kilograms, or Newtons.

When force causes movement, work is performed. For example, when the compressed spring or the tensioned lifting cable causes movement, work is performed. Without movement, work cannot be performed even if force is applied.

Work is equal to distance moved multiplied by the force applied. If the lifting cable of a hoist lifts a 250 lb (112.5 kg) engine 4" (10 cm) in the air, the amount of work done is equal to 4" (10 cm) times 250 lb (112.5 kg), or 1,000 foot-pounds of work. Work is measured in foot-pounds (ft-lb), watts (W), or joules (J). Work can only be accomplished when something is moved. When force causes movement, work is performed (**FIGURE 12-12**).

FIGURE 12-11 Force is a push or pull that may or may not produce movement.

▶ TECHNICIAN TIP

Holding a heavy starter motor in place while trying to get the bolts started is not technically work; it is force. Although it seems like you are working hard, you are not performing work in the true physics sense. Even if your arm muscles twitch as you stand under the vehicle and hold the part in place, it is not a work. It takes movement, along with force, to qualify as work. So, lifting the starter from the ground to its position on the engine is work. But holding it there is not work. Understanding the difference between these two terms will give you a good foundation for understanding power.

The rate or speed at which work is performed is called power (**FIGURE 12-13**). The more power that can be produced, the more work can be performed in a given amount of time. Power is measured in ft-lb per second or ft-lb per minute. If an electric motor can lift a 600 lb (270 kg) weight 20" (50 cm) in 10 seconds, the power used would be equal to:

$$20 \text{ (feet)} \times 600 \text{ (lb)} \div 10 \text{ (seconds)} = 12{,}000 \text{ ft-lb per } 10 \text{ seconds} = 1{,}200 \text{ ft-lb per second.}$$

One horsepower equals 550 ft-lb per second, or 33,000 ft-lb per minute. So, 1,200 ft-lb per second equals:

$$1{,}200 \text{ (ft-lb)} \div 550 \text{ (ft-lb per second)} = 2.18 \text{ horsepower}$$

The watt or kilowatt (1,000 watts) is the metric unit of measurement for power, where 746 watts equals 1 horsepower. So, the electric motor in the example would be developing:

$$2.18 \text{ (horsepower)} \times 746 \text{ (watts)} = 1627 \text{ watts, or } 1.627 \text{ kilowatts, of power}$$

Power in Rotating Components

Determining power in rotating components is not as straightforward as when parts are moved in straight lines. Torque is described as a twisting force. Because torque is a force, movement does not have to occur to have torque. Torque is applied before or during

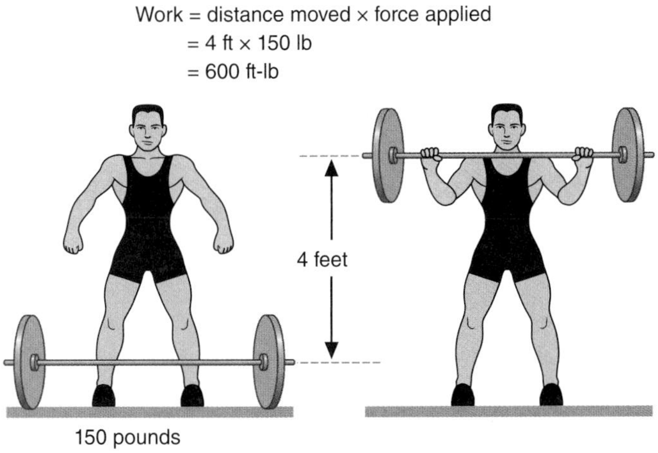

FIGURE 12-12 Work.

*Note: 1 Joule = a force of 1 Newton which moves an object 1 meter. Also called a Newton meter (Nm).

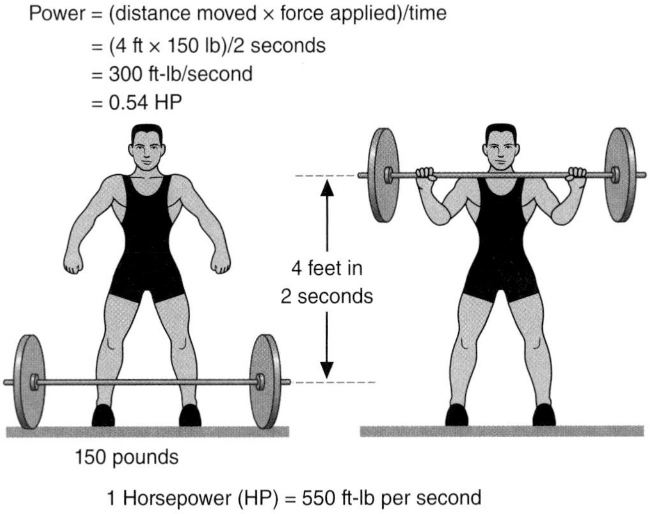

FIGURE 12-13 Formula for power.

movement. When a twist cap on a water bottle is removed, maximum torque is applied just before the cap starts to turn. When the same cap is tightened, maximum torque is applied once the cap starts to get tight. The concept of "twisting force" should always come to mind when the term "torque" is used.

When a piston is pushed down a cylinder during the power stroke, it applies force to a connecting rod linking the piston and crankshaft. This causes the crankshaft to rotate (**FIGURE 12-14**). The rotational force applied to the crankshaft is called torque. The crankshaft then applies that twisting force to other components such as the transmission and front pulley.

The unit of measurement for torque in the imperial system is ft-lb. In the metric system, it is Newton meters (Nm). If a force of 100 lb (45 kg) is applied to the end of a 1" (30.48 cm)-long wrench (lever) attached to a bolt, the resulting torque applied to the wrench will be 100 ft-lb. If no movement occurred, torque was still applied as a force, but no work was performed.

Work only occurs when movement happens. So, applying torque (twisting force) to something while it is turning means that work is being performed. This means that rotating shafts are capable of performing work.

Power means that work is being done at a certain rate, which involves time. So, power is the amount of work performed in a certain amount of time. The measurement of engine power is calculated from the amount of torque at the crankshaft and the speed at which it is turning in rpm. The formula for engine horsepower is:

$$\text{Horsepower} = \text{rpm} \times \text{ft-lb} \div 5252$$

For example, if an engine creates 500 ft-lb of torque at 4000 rpm, then the amount of horsepower is:

$$500 \times 4{,}000 \div 5{,}252, \text{ or } 380 \text{ horsepower at } 4000 \text{ rpm}$$

Because horsepower would change with rpm, it is necessary to express not only the power value but also the engine speed, in rpm, at which it occurs.

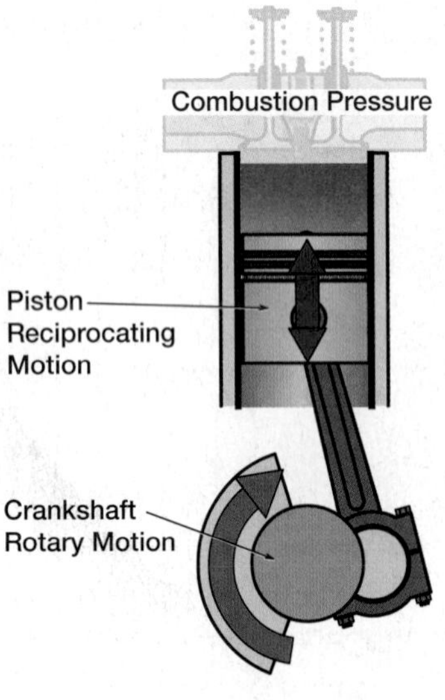

FIGURE 12-14 Crankshaft converts reciprocating motion of the piston into rotary motion.

▶ TECHNICIAN TIP

Power can also be expressed in kilowatts. A kilowatt (1,000 watts) is equivalent to 1,000 Newtons per meter per second.

▶ WANT TO KNOW MORE?

The formula for computing an engine's horsepower might seem confusing. As we said earlier, 1 horsepower equals 33,000 ft-lb per minute. Yet, we divide the product of torque × rpm by 5,252. Why do we use 5,252 instead of 33,000? The reason is a bit complicated, having to do with the definition of ft-lb. When relating to work, ft-lb means force times distance moved. That definition works well in a lifting situation, but not so well in a twisting (torque) situation.

In a torque situation, ft-lb means a twisting force applied to a shaft—that is, applied force times lever distance of the applied force. Thus, it is possible for torque to result in no movement, only an applied force. And we label that force in ft-lb even though for work or power to happen, movement must take place.

To calculate the twisting power of a shaft, we need a way to add distance moved and time taken to the equation so that power can be represented. We can't just say that 33,000 ft-lb = 1 horsepower when talking about torque. That is why rpm is included in the formula. Because rpm refers to a certain number of revolutions per minute, it includes both time and distance. But how much distance is 1 rpm? The answer relates to radians.

A radian describes how many radius distances there are in the circumference of a circle (**FIGURE 12-15**). Because there are 3.14 diameters in the circumference of a circle, there are twice as many, or 6.28, radius distances (radians) in the circumference of any circle. The larger the circle, the longer the radians, but still only 6.28 of them fit within the total circumference. And that is where we get our distance. Torque relates to an equivalent amount of force a certain distance from the rotational center (radius). There are 6.28 radians in the circumference of a circle. Therefore, every revolution equals a distance of 6.28 radians.

To convert torque ft-lb to work ft-lb, there has to be movement. We can either multiply the rpm by 6.28 and use the 33,000 (ft-lb per horsepower) factor, or divide the 33,000 ft-lb by 6.28 (distance around a circle in radians) and come up with a new factor, which is 5,252. We can then use that with the original "torque × rpm" numbers for calculating engine horsepower. To keep the numbers more manageable, most people go with the 5,252 factor. Thus, if we multiply the ft-lb of torque × rpm and divide that number by 5,252, we will have calculated the engine's horsepower at that particular rpm.☺

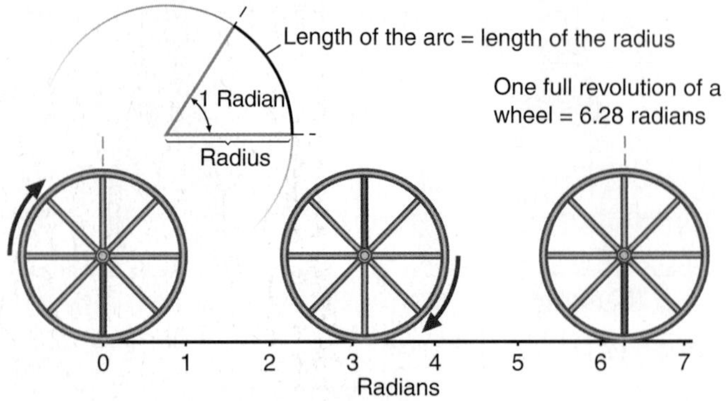

FIGURE 12-15 Radians in the circumference of a circle.

Torque Versus Horsepower

As torque is a twisting or turning force, horsepower is the rate (in time and distance) at which that force (torque) is produced. Torque alone does not mean work has been accomplished. It takes movement and time to accomplish a given amount of work in a given amount of time (horsepower). At the same time, for an engine to produce torque, it has to be running. So, the rotation of the crankshaft (torque × distance) in a running engine means that work and power are both occurring. This is because torque × rpm gives us power (**FIGURE 12-16**).

An engine puts out varying amounts of torque and power as it is operating. Thus, if it is producing torque, it is producing power. Factors that affect an engine's torque output include:

- engine volumetric efficiency (the rate of air intake and exhaust) at various crankshaft speeds and
- internal component friction (parasitic losses).

In a naturally aspirated engine (nonpressurized intake system), torque peaks at the rpm where the engine's cylinders fill the most with air. Torque starts to drop as engine speed increases past peak torque rpm (**FIGURE 12-17**).

FIGURE 12-16 An automotive chassis dynamometer measures power output of the engine at the wheels of the vehicle.

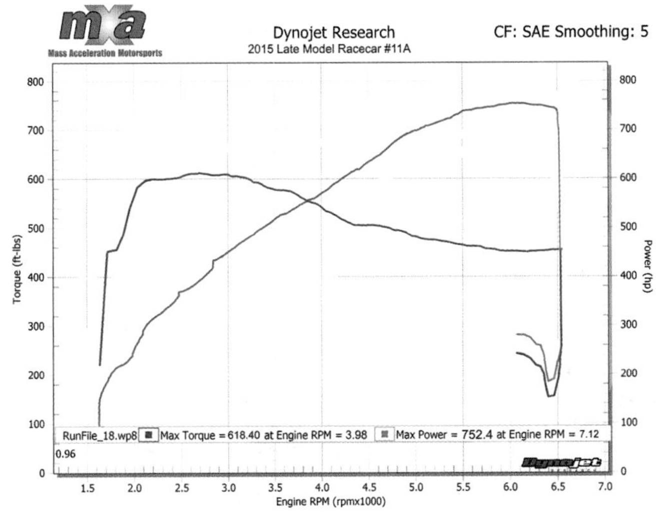

FIGURE 12-17 The relationship between torque and horsepower.

Actually, in a naturally aspirated (naturally breathing, nonpressurized) engine, air almost never completely fills the combustion chamber while the engine is running. This is due to the resistance of air flowing through the intake system. Peak engine torque rpm occurs at the peak volumetric efficiency (somewhere around 85–90% in an unmodified engine). Peak torque usually occurs at some low-to-mid-rpm engine speed. The main factors that affect the rpm at which peak torque occurs are:

- bore and stroke of the engine
- intake and exhaust port size and
- valve timing.

Engine rpm tends to rise faster than torque falls off (above peak volumetric efficiency). This causes an engine's maximum horsepower occur at a higher rpm than the peak torque rpm (**FIGURE 12-18**). At some point in the rpm range, the torque on the crankshaft drops so low that the crankshaft can no longer do additional work. This causes the horsepower to start to decrease. Remember, it is horsepower that does the work, but torque makes it happen.

Engine torque (and therefore horsepower) increases can be achieved through any engine modifications that improve volumetric efficiency. In fact, a turbocharger or supercharger increases an engine's volumetric efficiency well above 100%. For example, a 1.6-liter Volkswagen diesel engine when turbocharged to its rated 11 psi (75 kPa) of boost would be theoretically equivalent, in terms of horsepower, to a naturally aspirated 2.8-liter engine.

Applied Science

AS-50: Work: The technician can explain the relationship between torque and horsepower.

Two technicians are discussing the relationship between torque and horsepower during their break time. Al believes that torque is the turning force at the engine's crankshaft. Bob says that horsepower is the rate at which work is produced. Both technicians are correct as

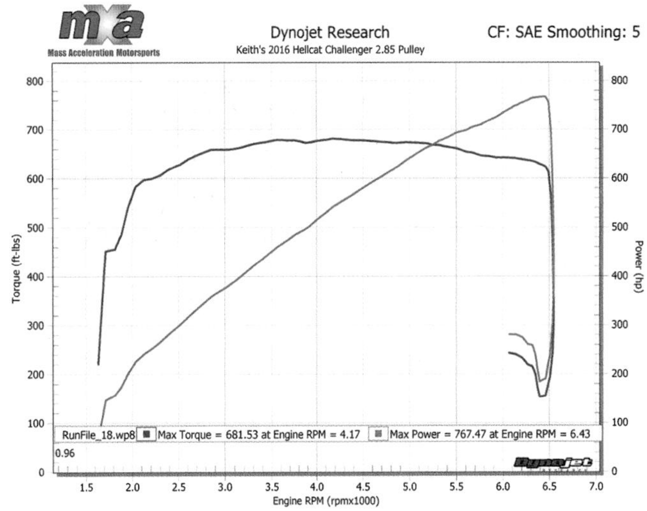

FIGURE 12-18 Horsepower typically peaks at a higher rpm than torque peaks at.

they discuss the unique relationship of torque and horsepower. One cannot exist without the other.

When we compare a race car to a bulldozer, we can see two different applications of torque and horsepower. A bulldozer has lots of torque, but the engine may be operating at 2000 rpm. A race car may be operating at 8000 rpm with a lower amount of torque. Both the race car and bulldozer create the same amount of horsepower (**FIGURE 12-19**). But that power occurs at different rpms. Low-speed torque is needed in some situations. And high rpm horsepower is needed in some others. (Additional information on the relationship between torque and horsepower is described in this chapter.)

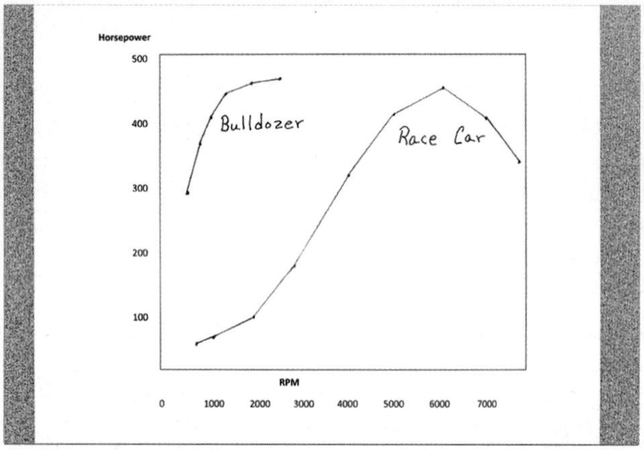

FIGURE 12-19 Same horsepower, different applications.

Generator Load Factor Sample

Continuous Base Load:

Power available for continuous full load operation.Overload of 10% is permitted for 1 hour in every 12 hours of operation

Prime Power:

Power available at variable load with an average load factor not exceeding 80% of the Prime Power rating. Overload of 10% is permitted for 1 hour for every 12 hours of operation.

Standby Power:

Power available at variable load in the event of main power network failure for 500 hours max in one year. No overload permitted!

FIGURE 12-20 Example of a load factor for an engine.

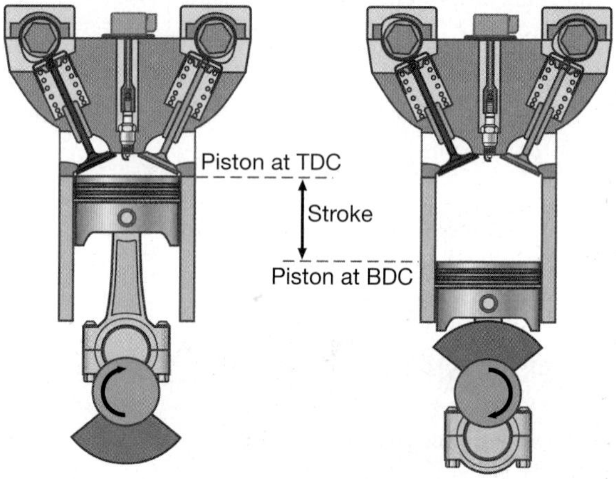

FIGURE 12-21 Piston movement from TDC to BDC or from BDC to TDC is one stroke.

Engine Load Factor

Load factor or power range is typically used in regard to diesel engines and describes how long an engine can produce its maximum power output. An engine that is quite powerful may not be designed to produce its maximum power over long periods. A gasoline engine in a car may give good acceleration in short bursts. But it can overheat and/or fail if operated at maximum speed and power for too long. In contrast, a heavy-duty diesel engine is designed to operate near or at its maximum speed and load for long periods without damage.

One way to describe the load factor of an engine is to give its power as an average over a certain period. This is stated as a percentage and is called load factor (**FIGURE 12-20**). An engine required to operate at maximum power over 10 hours is said to have a load factor of 100%. Using that as a standard, most car engines usually only operate at about 20% to 30% during a typical daily drive. The engine is never run at 100% load for an extended time. In fact, it may never reach 100% even for a short time.

▶ Four-Stroke Spark-Ignition Engines

LO 12-04 Demonstrate knowledge of four-stroke engine arrangement, operation, and measurement.

The SI engine used in modern-day vehicles operates on the four-stroke cycle principle. It takes four strokes of the piston to complete one cycle. This complete cycle is called the Otto cycle and is named after German engineer Nikolaus Otto. When the piston in a cylinder is at the position farthest away from the crankshaft, it is said to be at top dead center (TDC). When the piston in a cylinder is at a position closest to the crankshaft, it is said to be at bottom dead center (BDC). When the piston moves from TDC to BDC or from BDC to TDC, one stroke has occurred (**FIGURE 12-21**). Two or more strokes of the piston are called reciprocating motion. This can be an up-and-down or back-and-forth motion within the cylinder. Piston engines are therefore referred to as reciprocating engines.

Piston engines can be simple, single-piston engines such as those on lawn mowers. Or they can be much more complicated, multi-piston engines. They are used in automobiles, trucks, and heavy equipment. Multi-cylinder engines come in various cylinder arrangements (**FIGURE 12-22**):

- Inline engines with pistons one behind another
- Flat engines with horizontally opposed cylinders, called boxer engines
- V-type engines with the pistons forming a V configuration

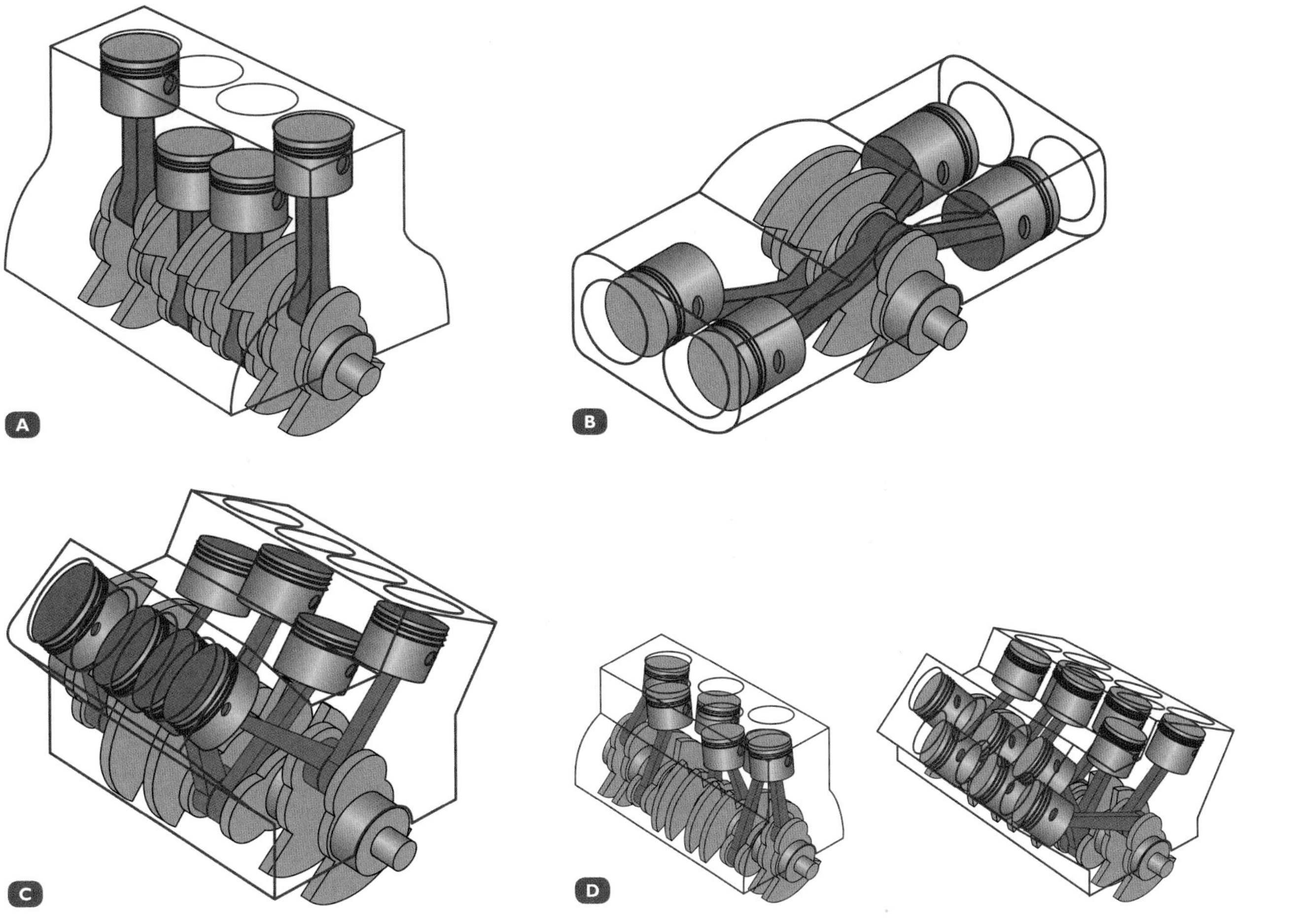

FIGURE 12-22 **A.** Inline. **B.** Flat. **C.** V-type. **D.** VR and W types.

- VR or W-type engines, a modified version of the V-type engine, with cylinders in a narrower angle to one another and housed in the same cylinder bank

Basic Four-Stroke Operation

In a single four-stroke cycle, only one stroke out of four delivers new energy to turn the crankshaft. The four strokes must include the five key events common to all ICEs (**FIGURE 12-23**):

- Intake
- Compression
- Ignition
- Power
- Exhaust

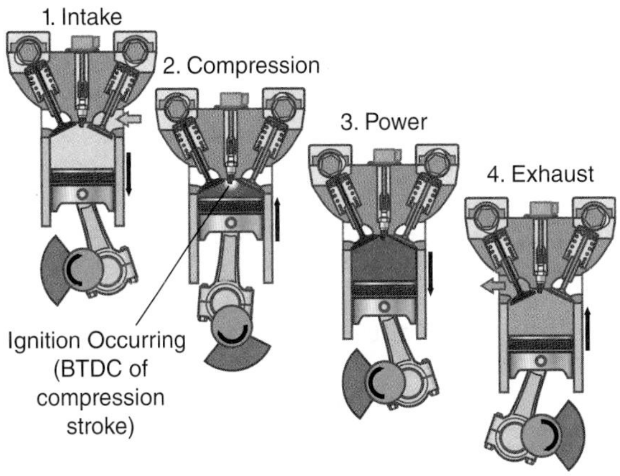

FIGURE 12-23 The basic four-stroke cycle.

These events occur in the same order every time and make up one complete cycle that repeats over and over.

The intake stroke starts with the piston at TDC. The exhaust valve is closed, with the intake valve(s) starting to open. As the piston moves down from TDC to BDC, the volume above the top of the piston increases. This lowers the pressure inside the cylinder compared to the pressure outside the cylinder. Higher outside air pressure forces air (usually with fuel) into the cylinder. As the piston reaches BDC, the intake valve(s) closes, and the intake stroke ends.

The compression stroke starts near BDC when the intake valve(s) closes. The piston moves from BDC to TDC. As the piston moves up, the air–fuel mixture is compressed into a smaller and smaller volume. Compression causes the temperature of the air–fuel charge to rise. This makes ignition easier and the combustion (burning of fuel) more complete and efficient.

As the piston reaches TDC of the compression stroke, ignition occurs. The air–fuel mixture is ignited and burns rapidly at up to about 4,500°F (2,482°C). The heat of combustion causes the burning gases to expand greatly (thermal expansion). This creates very high pressure in the combustion chamber. The pressure is applied to the entire cylinder, including the top of the piston.

The power stroke occurs as this extreme force pushes the piston from TDC to BDC. Both valves remain closed during this event. The exhaust valve(s) starts to open near BDC.

At the end of the power stroke, the exhaust valve(s) opens and the exhaust stroke occurs. The piston moves from BDC to TDC, which pushes the burned gases out of the cylinder through the open exhaust valve(s). When the piston nears TDC, the exhaust valve(s) starts to close and the intake valve(s) starts to open. This completes the four-stroke cycle which starts over from the beginning. Note that the crankshaft has completed two full rotations during the four-stroke cycle. Thus, four complete strokes make one complete cycle. And the crankshaft has made two complete rotations in that cycle.

Engine Measurement—Size

ICEs are designated by the amount of space (volume) their pistons displace as they move from TDC to BDC. This is called engine displacement. So, a 5.4-liter V8 engine has eight cylinders that displace a total volume of 5.4 liters (330 cubic inches). Displacement can be listed in cubic centimeters, liters, or cubic inches. To find an engine's displacement, you need to know the bore, stroke, and number of cylinders for a particular engine.

The cylinder bore is the diameter of the engine cylinder. The bore is measured across the cylinder, parallel with the block deck. Cylinder bore sizes on automotive applications typically run around 3" to 4". The distance the piston travels from TDC to BDC, or from BDC to TDC, is called the piston stroke.

Piston stroke is determined by the offset portion of the crankshaft called the throw. The crankshaft is described in greater detail later in this chapter. Piston stroke also varies from less than 3" to more than 4". Engine specifications typically list the bore size first and the stroke length second (bore vs. stroke).

▶ TECHNICIAN TIP

Generally speaking, the longer the stroke, the greater the engine torque produced. But it usually has a lower maximum rpm. In the same way, a shorter stroke generally enables the engine to run at higher rpm. This allows the engine to create higher horsepower.

The volume that a piston displaces from BDC to TDC is called piston displacement. Increasing the diameter of the bore or increasing the length of the stroke produces a larger piston displacement. The formula for calculating piston displacement (**FIGURE 12-24**) is:

Cylinder bore squared × 0.785 × the piston stroke

This formula works for calculating both the standard displacement in cubic inches and the metric displacement in cubic centimeters (ccs) or liters.

For example, a 5.4-liter (329-cubic inch) V8 truck engine has a 3.55" bore, a 4.16" stroke, and eight cylinders. Using the formula for displacement:

3.55 × 3.55 (bore squared) = 12.6025 × 0.785 (constant) = 9.893
= 9.893 × 4.16 (stroke) = 41.155-cubic-inch piston displacement

Once you know the piston displacement, multiply the piston displacement with the number of cylinders in the engine (**FIGURE 12-25**). Continuing from the previous example:

41.155-cubic-inch piston displacement ×8 (number of cylinders) = 329.24-cubic-inch engine displacement

The displacement of an engine (also called engine size) can be altered by changing cylinder bore (diameter), piston stroke (length), or the number of cylinders.

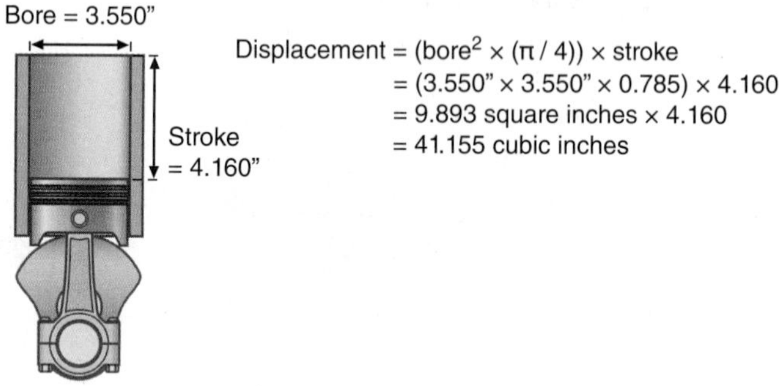

Bore = 3.550"
Stroke
= 4.160"
Displacement = (bore² × (π / 4)) × stroke
= (3.550" × 3.550" × 0.785) × 4.160
= 9.893 square inches × 4.160
= 41.155 cubic inches

FIGURE 12-24 Piston displacement.

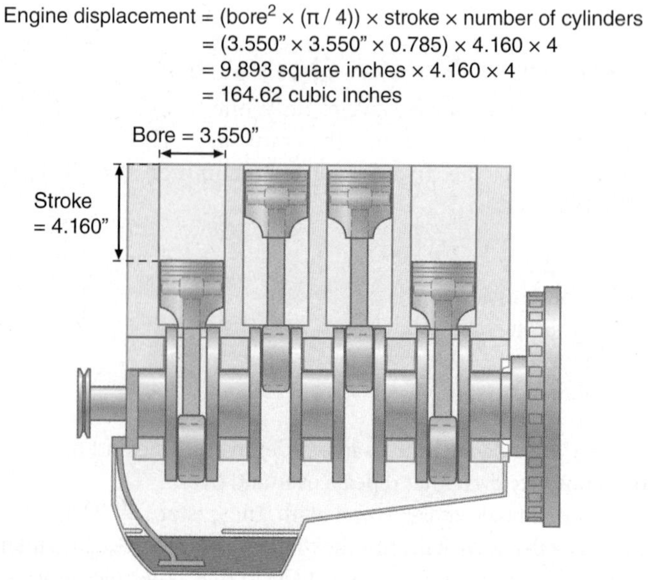

FIGURE 12-25 Engine displacement.

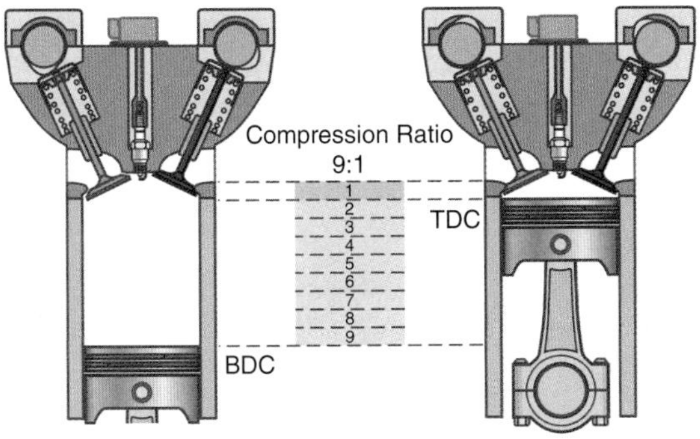

FIGURE 12-26 The compression ratio of an engine is found by taking the volume of the cylinder at BDC and comparing it to the volume at TDC. In this example, a 9:1 compression ratio is found.

> **TECHNICIAN TIP**
>
> An engine with the same size bore and stroke is referred to as a square engine. An engine with a larger bore than stroke is called an oversquare engine (short-stroke engine). An engine with a bore smaller than the stroke is called an undersquare engine (long-stroke engine). Oversquare engines tend to make their power at higher rpm. Undersquare engines tend to make their power at lower rpm.

Compression ratio compares the volume of the cylinder when the piston is at BDC with the volume at TDC (**FIGURE 12-26**). Maximum cylinder volume is at BDC, and minimum cylinder volume is at TDC. The ratio is given as two numbers. A compression ratio listed as 8:1 (8 to 1) means that the maximum cylinder volume is eight times larger than the minimum cylinder volume. Or in other words, the air–fuel mixture is compressed into 1/8th of its original size. Compression ratio is affected by:

- changing the size and shape of the top of the piston
- changing the size of the combustion chamber in the cylinder head or piston or
- altering valve timing.

The higher the compression ratio, the higher the compression pressures within the combustion chamber. This increases the thermal expansion during combustion. And that makes the engine more fuel-efficient. But a compression ratio that is too high can cause the air–fuel mixture to be ignited before the spark occurs. The high temperature created by the high compression ignites the air–fuel mixture. This preignition can cause damage to the engine bearings and pistons. Therefore, manufacturers design their engines with an ideal compression ratio to avoid this.

Atkinson and Miller Cycle Engines

The Atkinson cycle engine and the Miller cycle engine are variations of the traditional four-stroke engine. These engines operate more efficiently. But they produce lower power outputs for the same displacement.

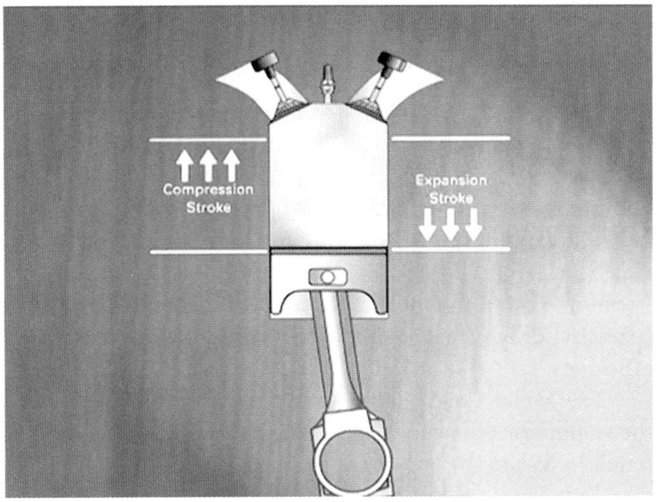

FIGURE 12-27 Conventional engine—compression and expansion strokes are of the same length.

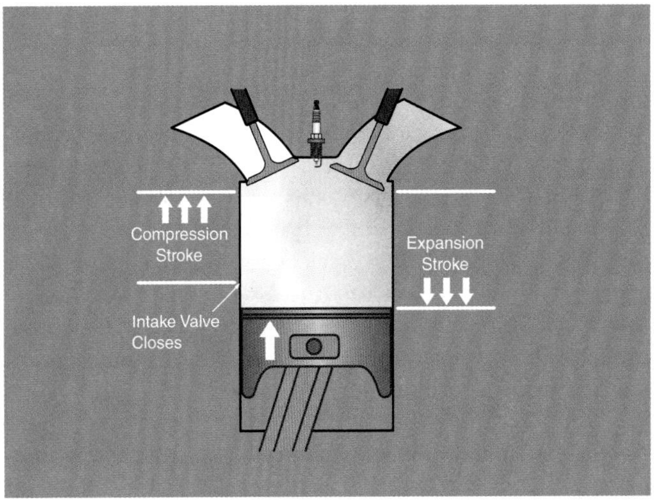

FIGURE 12-28 Miller cycle and Atkinson cycle engines—compression and expansion strokes are of different effective lengths due to delayed intake valve closing.

In a conventional four-stroke cycle, the compression and the power (expansion) strokes are of the same length (**FIGURE 12-27**). Increasing the stroke and raising the expansion ratio also raise the compression ratio. There is a limit to how high the compression ratio can be. Raising it too much results in engine-damaging detonation.

The Miller and the Atkinson cycles have overcome the problem of having the same compression and expansion ratios. They use valve timing variations to make the effective compression stroke shorter than the expansion stroke (**FIGURE 12-28**). The effective compression stroke is shortened by delaying the closing of the intake valve at the beginning of the compression stroke. This shortens the compression stroke which maintains the proper compression ratio. Thus, the compression pressure is still typically the same as that of a conventional four-stroke engine. But the expansion stroke is lengthened.

The effective expansion stroke is also lengthened by delaying the opening of the exhaust valve until closer to BDC. This allows the pressure created by the expansion of the burning gases to act on the piston longer. It also applies pressure to the crankshaft for a longer time, which increases the efficiency.

Some of the intake gases are pushed back from the cylinder into the intake manifold. So, Miller and Atkinson engines use a larger throttle opening for a given amount of power. Having a larger throttle opening results in lower manifold vacuum. Lower manifold vacuum reduces pumping losses (parasitic drag) during the intake stroke. This increases fuel efficiency.

The Miller cycle engine adds a supercharger to increase the volumetric efficiency. It also boosts the power output when required. When the engine is operating at low load and speed, the supercharger is not needed. A clutch disengages the drive, and hence there is no unnecessary drag on the engine. When extra power is required, the clutch is engaged. This activates the supercharger, which boosts the amount of air drawn into the engine, supercharging the cylinders.

The Atkinson cycle engine is efficient within a specific operating range (the so-called engine "sweet spot" of peak torque rpm). This is typically between 2000 and 4500 rpm. But its overall power output and torque are lower than a conventional ICE of the same displacement. This type of engine is less useful as the primary power source. But it is ideal in applications such as a series-parallel hybrid vehicle. In this way, it can work in tandem with a battery-driven electric motor. This engine, combined with an electric motor, provides more torque than it can produce by itself. It is also well suited to be used to charge the high-voltage battery, as it can do that within the most efficient engine rpm range.

Also, the lower maximum operating rpm allows engine components to be of lighter construction and weight as compared to a conventional ICE. Lighter and smaller components reduce friction and increase engine efficiency. In addition, the crankshaft is mounted slightly off-center from the cylinder bores (**FIGURE 12-29**). This position reduces the thrust load on the piston and cylinder wall during the power stroke. It results in reduced power loss due to friction.

Scavenging

Scavenging is the process of using a column of moving air to create a low-pressure area behind it. This assists in removing any remaining burned gases from the combustion chamber. It also replaces these gases with a new charge. As the exhaust stroke ends and the intake stroke begins, both valves are open for a short time. The time that both valves are open is called **valve overlap**.

As the exhaust gases leave the combustion chamber, it creates a column of moving gases through the exhaust pipe. Once moving, the flow tends to continue. This creates a low pressure behind it that helps to draw any remaining gases from the cylinder (**FIGURE 12-30**). It also draws new air and fuel into the cylinder. At the same time, the air and fuel being pushed (by atmospheric pressure) into the combustion chamber also helps to push the remaining exhaust gases out.

The flow effect during this valve overlap is called scavenging. Valve overlap has a desirable effect during high power/high rpm demand as more air and fuel are able to be pulled into the engine. However, during engine idling, valve overlap produces a rougher idle. The vacuum in the intake manifold tends to pull exhaust gases back into the intake manifold, diluting the incoming air. The rpm at which the most efficient scavenging occurs contributes to peak volumetric efficiency. This produces peak engine torque. Better exhaust scavenging and induction system (intake) breathing work together to improve volumetric efficiency. This is achieved by:

- smoothing intake and exhaust passages
- using tuned intake and exhaust runners (to maximize ram effect and scavenging)
- using a low back-pressure exhaust.

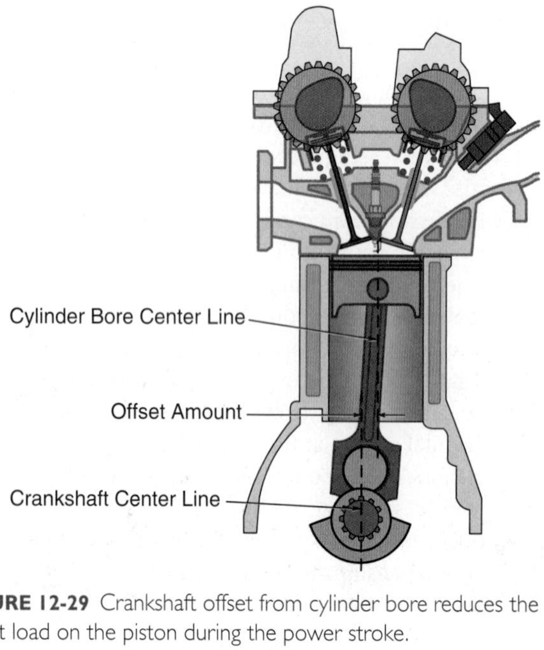

FIGURE 12-29 Crankshaft offset from cylinder bore reduces the thrust load on the piston during the power stroke.

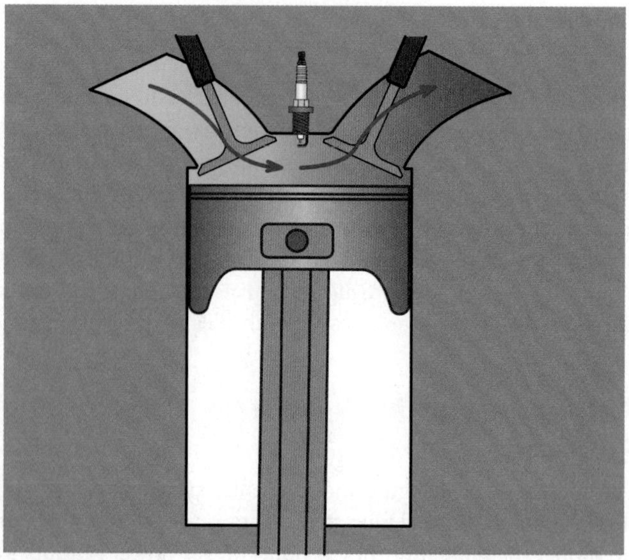

FIGURE 12-30 Scavenging helps remove all exhaust gases from the combustion chamber. It also helps draw a larger amount of air and fuel into the cylinder.

▶ Components of the Spark-Ignition Engine

LO 12-05 Demonstrate knowledge of spark-ignition engine components.

The SI engine is the most widely used engine to power passenger vehicles in the United States. In the rest of the world, CI engines are more common. But both types of engines share similar components.SI engines have evolved over their 125-year life, but the fundamental principles are still the same:

- an air–fuel mixture is compressed in the cylinder to increase its energy,
- it is ignited by a high-voltage spark,
- and the mixture burns rapidly, causing the thermal expansion needed to push the piston down, and
- the exhaust gases are pushed out of the cylinder.

Manufacturers have made incredible gains in the manufacturing of engines and engine components. Many of these gains are due to new technologies. Engine blocks and cylinder heads are commonly manufactured from lightweight aluminum. Valve covers and intake manifolds are being made of durable plastic materials. Pistons are made of newer aluminum alloys. And in some cases, connecting rods are manufactured from powdered metals. This makes them lighter and stronger.

The engine can be divided into a couple of main assemblies: the bottom end and the top end (**FIGURE 12-31**). The engine's so-called bottom end is where the crankshaft, bearings, connecting rods, piston assemblies, and oilpump reside. The crankshaft, connecting rods, flywheel, and harmonic balancer make up the "rotating assembly."

The so-called top end is where the cylinder heads and combustion chambers reside. The cylinder head(s) contains the overhead valves and valve train. If of the cam-in-head configuration (overhead cam), it also contains the camshaft. Each of these assemblies and components are explored further below.

Short Block and Long Block

If a rebuilt engine is needed, an engine subassembly may be purchased. A short block replacement includes all of the parts in the engine block from below the head gasket to above the oil pan (**FIGURE 12-32**). A cam-in-block engine also includes the camshaft and timing gears. An overhead-cam short block does not include the camshaft or timing gears.

A long block replacement engine includes the short block, plus the cylinder head(s), new or reconditioned valve train, camshaft and timing chain, and/or gears (or timing belt) (**FIGURE 12-33**). A long block engine still requires swapping parts from the original engine to the long block. These typically include:

- the intake and exhaust manifolds
- fuel injection system
- starter and alternator
- power steering pump and air-conditioning compressor.

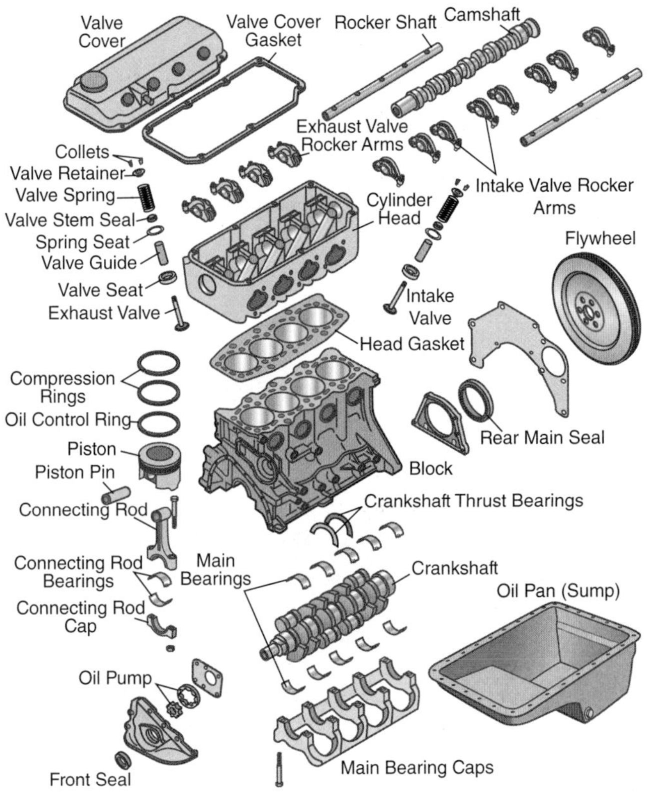

FIGURE 12-31 The engine contains many parts that work together to power the vehicle.

FIGURE 12-32 Short block assembly.

FIGURE12-33 Long block assembly.

Cylinder Block, Crankshaft, and Flywheel

The cylinder block is the single largest part of the engine. The block can be made of cast iron or aluminum, which is much lighter. The block casting includes the cylinder bore openings, also known as cylinders. The cylinders are machined into the block to allow for the fitting of pistons (**FIGURE 12-34**). The block deck is the top of the block and is machined flat. The cylinder head bolts to the block deck.

Passages for the flow of coolant and lubrication are machined or cast into the block. Holes are machined into the bottom of the block and are called main bearing bores. They have removable main caps and are used to hold the crankshaft in place. Each cap is held in place with two or more bolts.

On most modern engines, the main bearing caps are now a part of the engine girdle, also called a bed plate. The use of a girdle provides an even stronger design as all main bearing caps are connected and reinforce each other.

Reinforcements for strength and attachment points for related parts are also machined into the block. The lowest portion of the block is called the crankcase because it houses the crankshaft. The oil pan completes the crankcase.

The crankshaft can be made of cast iron or forged steel. Or it can be machined out of a solid piece of steel. The crankshaft has main bearing journals and connecting rod journals machined and polished into it. Main bearing journals ride in main bearings installed in the block. And the connecting rod journals provide a connection point for the connecting rods (**FIGURE 12-35**).

The crankshaft is supported by Babbitt-lined main bearing inserts. The inserts fit in the main bearing saddles of the block. End play of the crankshaft is limited by a thrust bearing at one of the main bearings (**FIGURE 12-36**).

Offset from the crankshaft centerline are the rod journals, also called throws (**FIGURE 12-37**). They are essentially the levers of the crankshaft. The longer the throw of the crankshaft, the longer the stroke of the piston. A longer stroke produces more torque from the engine. The rod journals are also machined and polished.

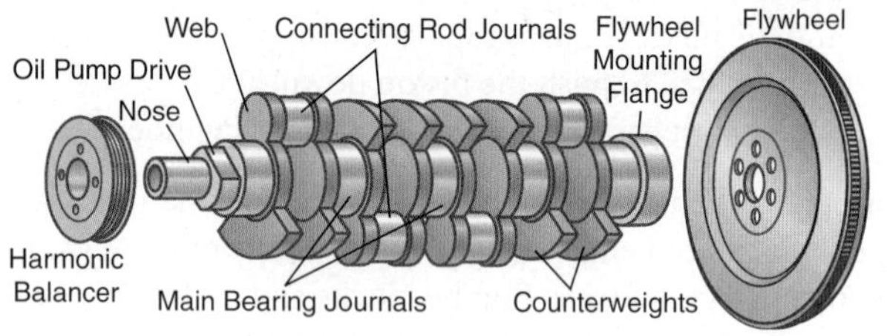

FIGURE 12-35 The basic parts of the crankshaft.

FIGURE 12-36 One of the main bearings has a thrust bearing that limits forward and backward movement of the crankshaft.

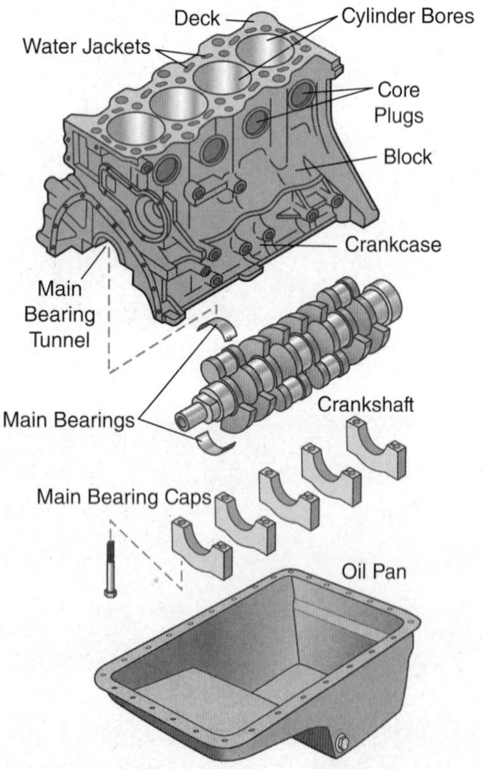

FIGURE 12-34 The block is the single largest part of the engine, with other components attached to it.

FIGURE 12-37 Crankshaft rod journals (throws).

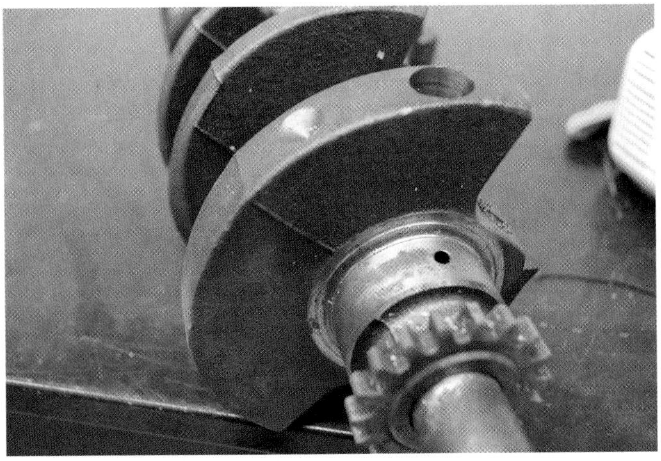

FIGURE 12-38 Crankshaft counterweights.

FIGURE 12-39 **A.** Flexplate. **B.** Flywheel.

As the crankshaft turns, the rod journals' "big ends" circle around the centerline of the crankshaft. To prevent vibration, counterweights are formed on the crankshaft (**FIGURE 12-38**). The counterweights balance the weight of the piston assembly, connecting rods, and rod journals.

At the front of the crankshaft is a "snout." It provides a mount for a vibration damper (harmonic balancer) and drive gears, sprockets, or pulleys. The back of the crankshaft has a flange where the flywheel is connected by bolts or studs.

The flywheel is a weighted assembly that stores kinetic energy from each power stroke. It helps keep the crankshaft turning through non-power strokes. Vehicles with manual transmissions have a clutch assembly attached to the flywheel. Vehicles with automatic transmissions use a flexplate and torque converter assembly. The effect on the crankshaft is the same with either assembly. They store energy and keep the crankshaft rotating smoothly (**FIGURE 12-39**).

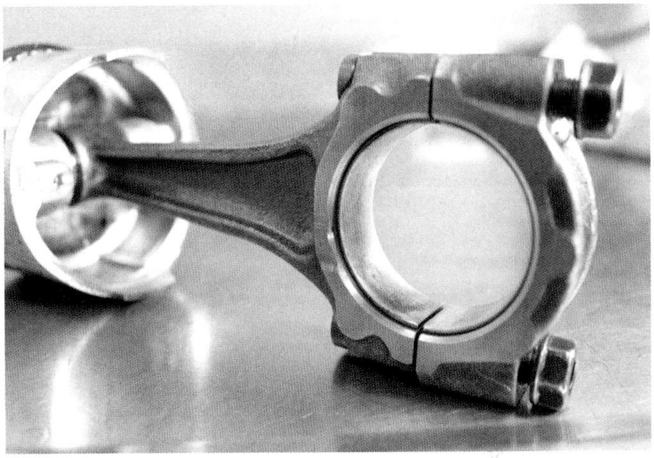

FIGURE 12-40 Connecting rod connected to a piston.

Connecting Rod and Piston

The connecting rod connects the piston to the crankshaft. It transfers piston movement and combustion pressure to the crankshaft rod journals. It has a "small end" and a "big end." The piston end (small end) of the connecting rod follows the reciprocal movement of the piston. The piston pin or "wrist pin" attaches the piston to the connecting rod (**FIGURE 12-40**). It fits into a one-piece bushing in the small end of the rod. The piston pivots on the "wrist pin" attachment.

The other end (large end) of the connecting rod attaches to the rod journal on the crankshaft. It has a removable rod cap that bolts to the end of the connecting rod body. The connecting rod and cap are machined to allow the placement of the connecting rod bearing.

Many connecting rods are made of cast iron or steel. However, some race cars and exotic sports cars use aluminum or titanium connecting rods. This allows the engine to rev more quickly and higher, producing more power.

The powdered metal rod is the most recent technology. This type of rods is prepared by mixing a precise amount of various powdered metals. The powder includes iron, copper, and carbon, and other agents; the mixture is placed in a die and forged (pressed) under high pressure. It is then put through a sintering process, where it is heated up. It thereby almost melts to form an alloy rod. This alloy is stronger than the original individual metals.

Powdered metal rods have several benefits. They are lighter than the typical cast and forged rod. They also use approximately 40% less metal, compared to a standard forged rod. The powdered metal rod meets or exceeds the tensile and yield strength of forged rods. And the cost is considerably less.

The powdered metal rod is then fracture-split at the big end. This gives more adhesion at the parting line (**FIGURE 12-41**) than a machined cap and rod. It is also a time saver. It eliminates the machining involved on the parting line between the rod and the rod cap.

With the fracture-split design, the rod cap is a precision fit on the rod. The rod bolt is not relied upon to align the cap to the rod. Because the fractured surfaces would be ruined by machining, fracture rods cannot be rebuilt if they are out of specification. The powdered metal rod is used on most newer engines.

The piston transfers combustion pressures to the crankshaft through the connecting rod. It is typically made of lightweight aluminum and possibly synthetic material. Pistons must be light because they change direction many times a second (**FIGURE 12-42**).

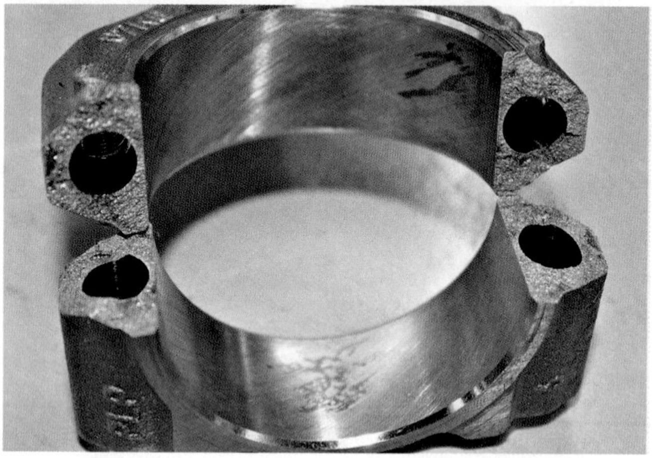

FIGURE 12-41 Mating surfaces of a powdered metal fracture-split rod.

Consider that an engine idling at 750 rpm, the piston changes direction 25 times each second. We can imagine that piston movement would be hard to see even at idle. And yet at a redline (maximum allowed) speed of 6000 rpm, each piston changes direction 200 times per second. Can you imagine the stress placed on the pistons, connecting rods, and bearings? What's more, a 12000-rpm redline sports car or motorcycle engine would experience 400 piston reversals per second!

The top of the piston is called the piston head. It is exposed to extremes of heat and pressure during combustion. Pistons can come in many styles of crowns, such as flat top, dished, domed, or recessed (**FIGURE 12-43**). Piston crown shape allows

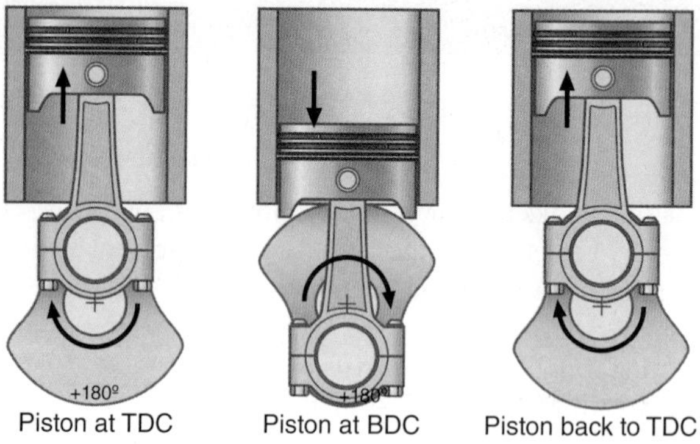

FIGURE 12-42 The piston moves two strokes during one revolution.

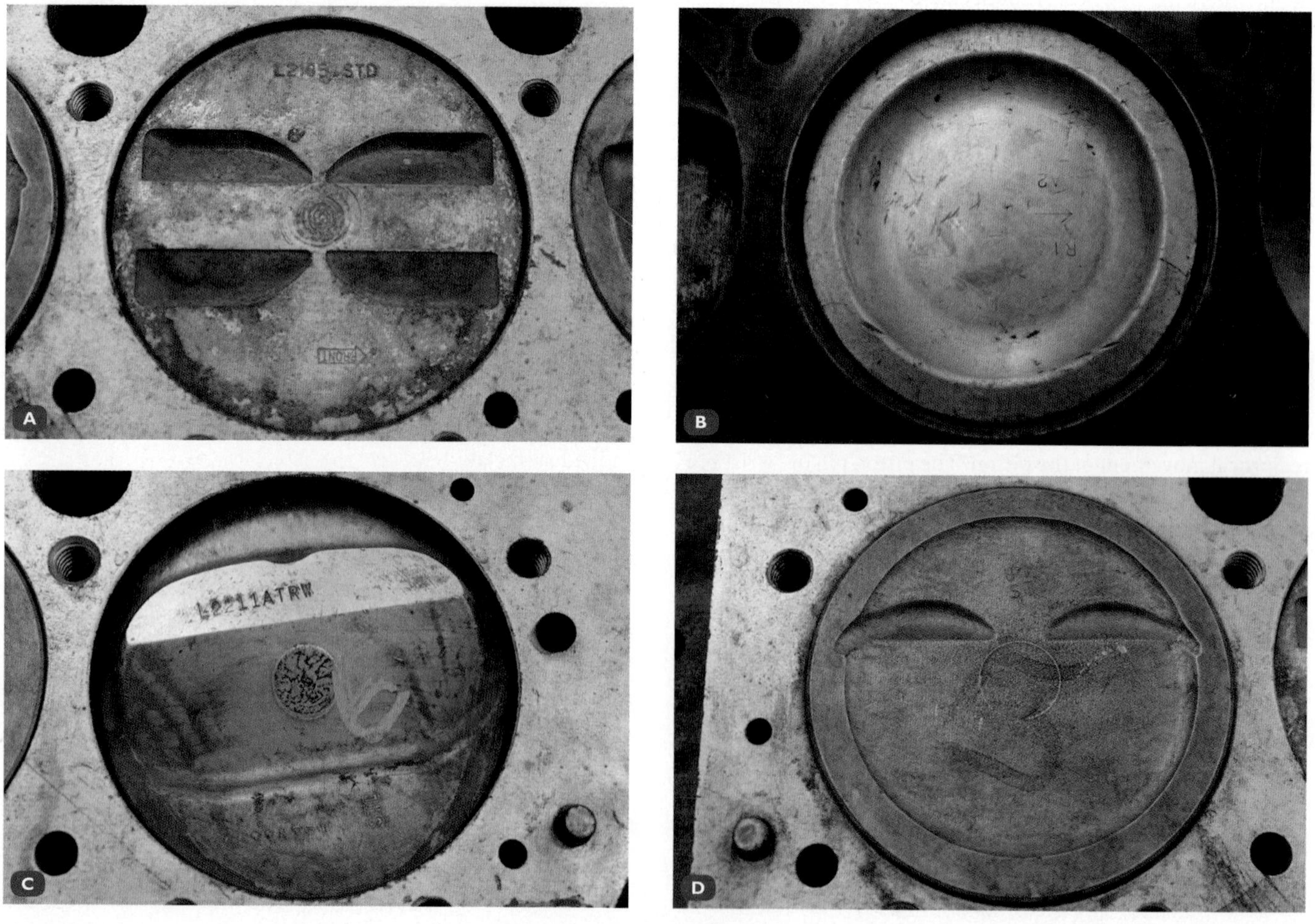

FIGURE 12-43 Piston crown shapes. **A.** Flat top. **B.** Dished. **C.** Domed. **D.** Recessed.

FIGURE 12-44 Piston coatings. **A.** Moly coating. **B.** Ceramic coating.

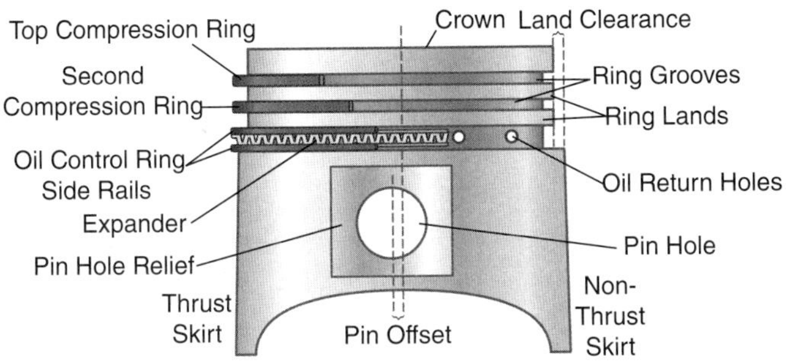

FIGURE 12-45 Piston and piston rings.

Applied Science

AS-99: Rotational: The technician can explain how rotational motion can be converted to linear motion and why balance is important in rotating systems.

When working in an automotive shop, it is important that the technician understand basic engine operation. The text has explained much regarding each component in the engine block and cylinder head. The technician should know that reciprocating motion is a linear (back-and-forth) motion. One example being the pistons in a cylinder. The crankshaft converts this reciprocating motion into rotational motion.

It is also important to know that the camshaft lobe converts rotary motion into linear motion. As described in the text, the lobe is used to lift the lifter and open the valve. Opening the valve requires linear motion on all automotive engines. The text describes a cam as an egg-shaped piece (lobe) mounted on the camshaft. The egg shape of the cam lobe is designed to lift the valve open, hold it open, and let it close. This is the operation that converts rotating motion into linear motion.

Balance is extremely important in a rotating system for a number of reasons. It gives a smooth-running engine, with long engine life and dependability. To help maintain balance, a vibration damper or harmonic balancer is mounted on the front of the crankshaft. This important component is carefully engineered to meet the needs of a specific engine regarding proper balance.

the engine designer to alter the compression ratio as desired. Manufacturers are now using special coatings on the pistons. They provide better lubrication qualities and reduced heat absorption (**FIGURE 12-44**).

Below the piston head are grooves machined into the piston. They hold circular piston sealing rings. The piston rings provide a seal between the outside of the piston and the inside of the cylinder wall. They have to seal this space as the piston moves up and down the cylinder (**FIGURE 12-45**).

Usually, a total of three piston rings are used. The upper two rings are compression rings. They prevent combustion pressure, called blowby gas, from leaking past the pistons into the crankcase. The lower piston ring is an oil control ring. It keeps lubricating oil on the cylinder wall and out of the combustion chamber.

The Cylinder Head

The head forms the top of the cylinder and is sealed in place with the use of a head gasket. The cylinder head has a combustion chamber either cast or machined into it. Combustion chambers in the cylinder head come in several different designs. Common designs are the wedge, hemispherical, and pent-roof combustion chambers (**FIGURE 12-46**).

The cylinder head is constructed of cast iron or aluminum. Most engines are now constructed using an aluminum cylinder head, which reduces the weight of the engine. The cylinder head contains the valves and valve train (valve actuating components) of the engine. The head also includes intake and exhaust ports to which intake and exhaust manifolds are attached.

Valves

The ICE uses so-called poppet valves. These are somewhat mushroom-shaped parts that slide up and down in the valve guides (**FIGURE 12-47**). The valves, under pressure from the valve springs, rest on seats of hardened materials such as Stellite®.

Valves are used to open and close ports in a cylinder head. The intake valve controls the flow of air and/or fuel into the combustion chamber. The exhaust valve controls the flow of exhaust gases out of the combustion chamber. The exposed intake port area usually is larger than the exhaust port area. This is because engine vacuum created by the piston on the intake stroke is not as effective at moving air into the engine as the

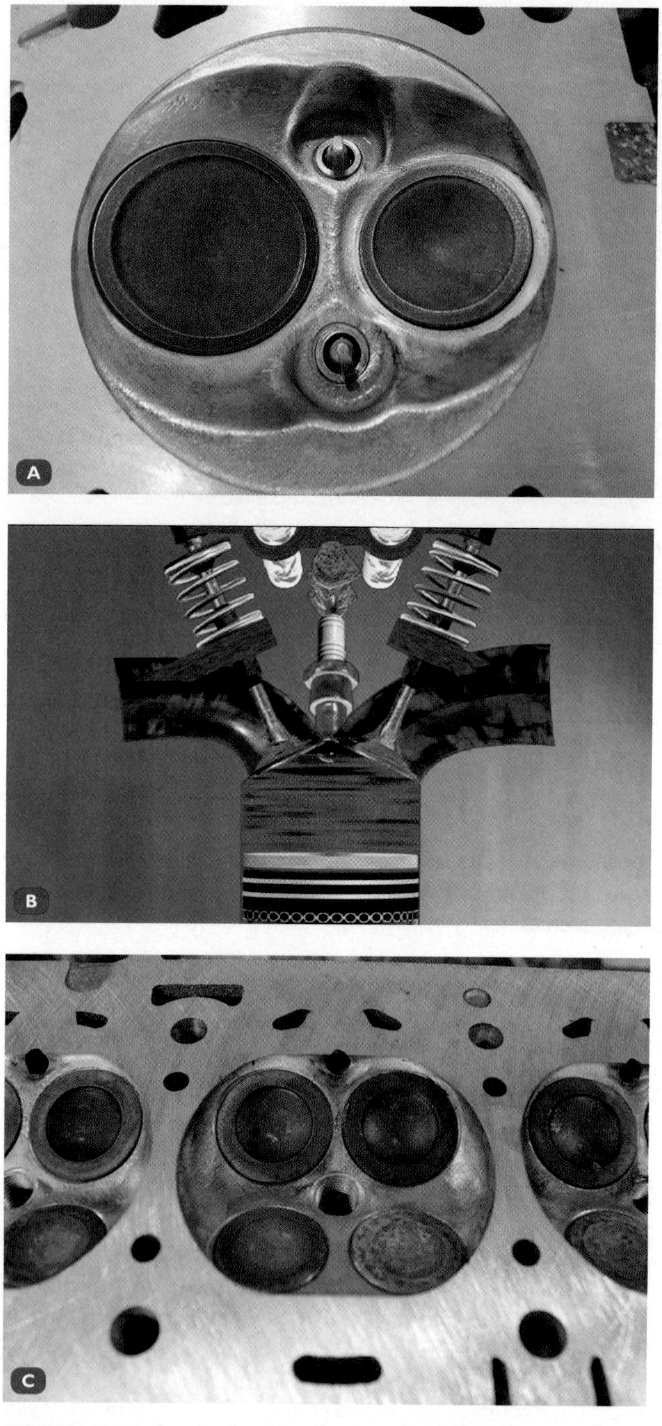

FIGURE 12-46 Combustion chambers can be designed in several configurations. **A.** Wedge. **B.** Hemispherical. **C.** Pent-roof

pressure created by the piston on the exhaust stroke is at pushing exhaust gases out of the engine.

The exposed intake port area can be increased by making the intake valve larger than the exhaust valve. Or, the manufacturer can use multiple intake valves in each cylinder. In the case of a three-valve engine, there would be two intake valves and one exhaust valve. In this case, the intake valves would be smaller than the exhaust valve.

FIGURE 12-47 Poppet valves in a cylinder head.

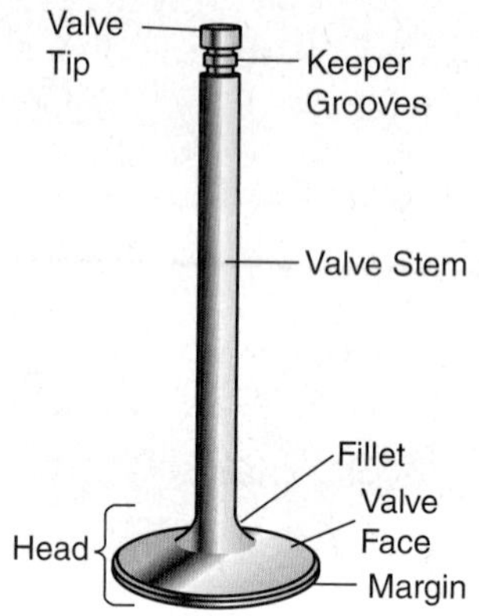

FIGURE 12-48 The parts of a valve.

The valve head is disc shaped, and the top of the valve head faces the combustion chamber. A machined surface on the back of the valve head is the valve face (**FIGURE 12-48**).The valve face seals on a hardened valve seat in the cylinder head. Located between the valve head and the valve face is a flat surface on the outer edge of the valve head, called the valve margin. The margin helps to prevent the valve head from melting under the heat and pressure of combustion.

The shaft attached to the valve head is the valve stem. The stem operates in a valve guide, which is in the cylinder head. The valve stem and guide work together to maintain valve alignment as the valve is opened and closed. Grooves are machined into the top end of the stem in order to receive locking pieces. The locking pieces are referred to as valve keepers. They lock the valve spring retainer and spring on the valve stem. Keepers prevent the retainer from coming loose while under tension from the valve spring. Engines may have two to five valves per cylinder.

Engine Cam and Camshaft

Valves need a system to make them open and close. Control of the valves is accomplished through the use of cams on a common shaft. A cam is an egg-shaped piece (lobe) mounted on

FIGURE 12-49 Cam lobes on a camshaft.

FIGURE 12-51 Overhead cam (OHC) engine.

FIGURE 12-50 Cam-in-block arrangement.

the camshaft (**FIGURE 12-49**). The egg shape of the cam lobe is designed to lift the valve open, hold it open, and let it close.

The camshaft is timed to the rotation of the crankshaft. This means that the valves open and close when the pistons are at specific positions. Timing the valve opening and closing to the piston position is critical to ensure proper operation of the engine. The camshaft is driven by the crankshaft through gears, a toothed belt, or a chain.

Up until the 1950s, many engines had their valves installed in the engine block. Such engines are called flat-head engines. Some manufacturers still place the camshaft in the center of the block (**FIGURE 12-50**), but the valves are installed in the cylinder head(s). These so-called cam-in-block engines use tappets, or "lifters," which ride on the camshaft lobes. The lifters follow the cam lobe and actuate the pushrods, rocker arms, and valves.

In most automotive engines today, however, the camshaft is mounted on top of the cylinder head. These engines are called overhead cam (OHC) engines (**FIGURE 12-51**). Intake and exhaust valves may all be actuated by a single camshaft, called a single overhead cam (SOHC) engine. Or there may be two camshafts per head, called a dual overhead cam (DOHC) engine. One camshaft is used to actuate all of the intake valves and another to actuate all of the exhaust valves. When separate intake and exhaust camshafts are used, the camshafts typically sit right on top of the valves. This means there is no need for rocker arms. Most manufacturers use a lifter called a bucket lifter, placed right on top of the valve and valve spring. This actuates the valve directly from the camshaft.

Camshaft lobes are designed to open the valve, hold it open, and allow it to close. The timing of these events is critical to proper engine operation. The shape and profile of the cam lobes determine valve timing. The challenge is that an engine needs different valve timing at different rpms.

Automotive engines do not operate at a single rpm. So a camshaft must be designed to provide the best balance for all speeds. High-performance engines built for racing use camshafts designed for high rpm power. But those would not work well for use on the street, where engines rarely stay above 3000 rpm. Newer engine designs have overcome some of these limitations by using variable valve timing.

Camshaft Specifications

As seen previously, the base circle of the cam lobe is the rounded bottom part of the lobe. This is where the lifters rest when the valves are closed (**FIGURE 12-52**). The opening ramp opens the valve, and the closing ramp closes it. The clearance ramps take up any clearance in the valve train. One clearance ramp extends from the base circle to the opening ramp. The other extends from the end of the closing ramp back to the base circle. The shape of the camshaft lobes affects the power characteristics of the engine.

One specification of the cam lobe is the lift of the cam lobe. Lift is the amount the lifter or cam follower moves with the cam. The more the valve is lifted off its seat, the more air that can enter the engine. However, more lift creates more pressure on the valve train components. The valve gets pushed further, and the spring gets compressed further.

Too much lift can create coil bind in the valve springs. Coil bind occurs when the coils of the spring touch each other. This can cause the spring to break. It also causes the camshaft and

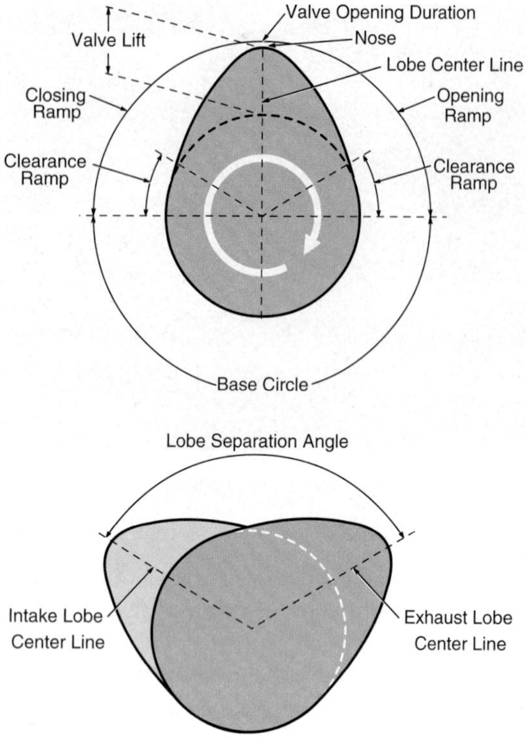

FIGURE 12-52 The camshaft lobe controls the opening and closing of the valves and affects performance greatly.

Specifications for a 2000- to 5000-rpm street camshaft:

	Intake		Exhaust
Lift	0.440		0.440
Duration at 0.050"	222°		222°
Cam lobe centerline	109° ATDC		116° BTDC
Lobe separation angle (camshaft degrees)		113° between lobe centerlines	
Degree valve opens at 0.050"	2° BTDC		48° BBDC
Degree valve closes at 0.050"	40° ABDC		–6° ATDC

Specifications for the same camshaft to be used up to 6500 rpm:

	Intake		Exhaust
Lift	0.500		0.515
Duration at 0.050"	256°		270°
Cam lobe centerline	108° ATDC		116° BTDC
Lobe separation angle (camshaft degrees)		112° between lobe centerlines	
Degree valve opens at 0.050"	20° BTDC		71° BBDC
Degree valve closes at 0.050"	56° ABDC		19° ATDC

FIGURE 12-53 An aftermarket performance camshaft includes a specifications card that is used during engine assembly to ensure correct timing. The top chart shows the specifications of a street camshaft, and the bottom chart a higher performance camshaft. Notice the differences in valve lift and duration.

lifter to wear excessively. Another problem with too much valve lift is too little valve-to-piston clearance. Too much valve lift can cause the valve to hit the piston and ruin the engine.

Duration is another specification engineers use when designing a cam lobe. Duration is the amount of time the valve stays open, given in degrees of crankshaft rotation (not camshaft rotation). The longer the valve is kept open, the more air will be able to move into and out of the engine.

Cam lobe centerline is where the center of the cam lobe is located in relation to the piston, in degrees. There is a cam lobe centerline specification for both the intake and exhaust lobes. Cam centerline and duration can be used to determine when the valves will open in degrees of crankshaft rotation.

Cam lobe separation refers to how far the centerline of the intake lobe is offset from the centerline of the exhaust lobe. Generally speaking, the wider the separation, the better the camshaft is for building power at lower rpm. Narrowing the separation builds power at higher rpm. The smaller the number of degrees of lobe separation, the more valve overlap there is. Valve overlap simply means that the intake and exhaust valves are open at the same time.

The duration and lobe separation are specifications measured in crankshaft degrees, not in camshaft degrees. Lift, duration, lobe separation, and overlap are features determined by the grind (profile) of the camshaft. The bigger the lift, and the longer the duration, the more power is created at high rpm.

High-performance engines have more degrees of valve overlap than street engines because of longer cam duration. Increased valve overlap and lift increase top-end rpm power due to column inertia (ram effect). But they reduce low-speed power and idle quality, again due to a loss of column inertia. Valve overlap specifications will not be given if you have a DOHC engine. This is because the exhaust and intake lobes run on separate shafts.

The so-called street cams, three-quarter race, and full race cams have progressively more lift and duration than stock cams. They are tailored to meet the operating conditions of the engine. For street engines, too much lift or duration may not be beneficial. The resulting lower intake manifold vacuum causes the engine to idle roughly.

When comparing cam specifications, it is important to know that there are two listings for duration specifications: advertised duration and duration at 0.050" lift (**FIGURE 12-53**). Advertised duration measures the duration from where the lifter first starts to move. Duration at 0.050" lift measures the duration starting when the valve opens 0.050" and ending when it returns to 0.050". This specification is a better indicator of how the camshaft's profile will affect the operation of the engine. By 0.050" lift, the cam is well into its lift profile. So, it is a more accurate method of comparing duration specifications between camshafts.

Mechanical and Hydraulic Valve Train

The valve train is the combination of parts that work together to open and close the engine's valves. The valve train operates off the camshaft, and the part that rides against the cam lobe is the valve lifter. The lifter transfers motion from the cam lobe to a pushrod, or it may directly act on the valve.

The valve lifter works as a mechanical spacer, providing a hardened bearing surface that slides on the cam lobe. If the lifter is a mechanically solid piece, it is said to be a mechanical lifter. And therefore, the engine is said to have a mechanical valve train. If the lifter has a hydraulic plunger in its center, it is a hydraulic lifter. And the engine is said to have a hydraulic valve train. The hydraulic plunger is designed to take up slack in the valve train. The slack is due to the expansion and contraction of components during engine warm-up and cool-down.

With a hydraulic valve train, valve adjustment is made by the hydraulic lifter, which takes up any clearance automatically. If a mechanical valve train is used, the valves will need to be adjusted at periodic intervals. This will ensure proper clearance is maintained as parts in the valve train wear.

Roller Rockers and Lifters

Roller-equipped rocker arms are used on many new engines because they reduce friction and increase engine efficiency. They may also be used to replace the cast or stamped steel rocker arms that have been used for many years as part of a standard valve train.

A typical rocker arm has a fulcrum, a half-round bearing, or a shaft that the rocker moves on as a bearing surface. One end of the rocker sits on the top of the valve stem. It pushes the valve open and then lets it close. The other end of the rocker arm connects with the pushrod.

The rocker's sliding motion across the tip of the valve stem and center pivot results in friction. Friction creates a loss of power and wear. To overcome this, a roller may be added to the end of the rocker where it contacts the valve stem. Needle bearings may also be added to the pivot point where the fulcrum is (**FIGURE 12-54**). This system greatly reduces friction and results in a more reliable valve train.

Rocker arms are designed to increase the amount of lift designed into the cam lobe. A typical rocker arm has the center pivot closer to the pushrod end. This design causes the valve end of the rocker arm to move more than the pushrod end. Many rocker arms have a 1.5:1 lift ratio, which means that the valve moves 1.5 times farther than the pushrod. High-performance rockers have even higher ratios, such as a 1.6:1 or 1.7:1. These rockers increase the lift of the valve even more than a standard rocker arm.

The other friction loss point in the valve train is at the lifter as it rides on the cam lobe. To reduce this friction loss, rollers with needle bearings are added to the base of the lifter on some engines (**FIGURE 12-55**). The lifter now rolls on the cam lobe profile rather than sliding on it. This modification further reduces friction in the valve train. This results in more power delivered to the engine flywheel, and less power wasted as heat.

The use of rollers on the lifters also allows the use of more aggressive cam lift profiles. Such profiles would rapidly wear an ordinary flat lifter and cam lobe. This difference means flat lifters should never be run on camshafts designed for roller lifters. Also, the roller lifter must be held in position so that it does not turn in the lifter bore and scrape on and wear the cam lobe. To prevent this, a retainer keeps the lifter from turning sideways (**FIGURE 12-56**).

Valve Clearance

Valve clearance is the amount of slack between the tip of the valve stem and its mating part. This could be a rocker arm, cam

FIGURE 12-54 A. Standard rocker arm. **B.** Roller rocker arm.

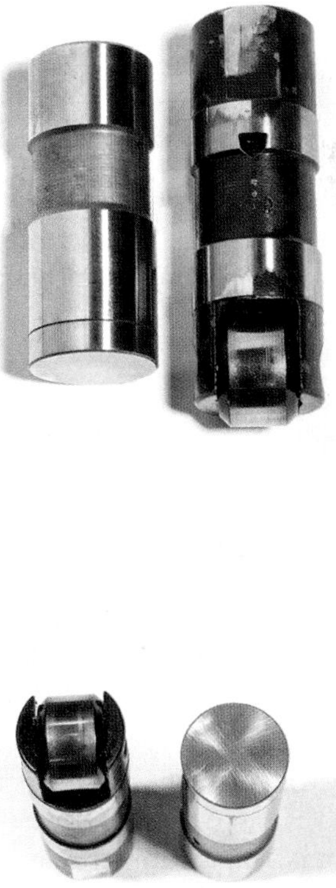

FIGURE 12-55 Roller lifter compared to a standard lifter.

FIGURE 12-56 Roller lifters need a provision to keep them from turning in their bore.

follower, or bucket-style lifter. If valve clearance is too large, the valves will tick and make enough noise to irritate the operator. It also increases wear on the valve train.

If valve clearance is too small, the valve can be held open longer than it should be. As the cylinder head and valve train parts heat, they expand. As they expand, the clearance decreases. So adequate clearance is needed to allow for this expansion. Insufficient valve clearance could result in burned valves.

Some valves are adjustable through the use of adjusting screws, nuts, or metal shims (**FIGURE 12-57**). Other valves use nonadjustable hydraulic components. On these, the parts are simply bolted in place and the hydraulic system automatically compensates.

On systems that use adjusting screws or nuts, they can be adjusted to provide the proper clearance. On bucket lifter systems, shims are installed in the lifter to provide the proper clearance. These shims are used to preset the proper valve clearance during cylinder head assembly or during a major tune-up.

Valve Train Drives

The valve train is operated by the camshaft, which is driven by a crankshaft-driven chain or belt (**FIGURE 12-58**). In older engine designs, the camshaft was driven by a gear-to-gear arrangement like that found in small lawn mower engines. The trouble with a gear-to-gear design is that it tends to be a bit noisy compared to a chain or belt. It also requires that the camshaft be relatively close to the crankshaft.

In any four-cycle engine design, the camshaft must rotate at half of crankshaft speed. This is accomplished by using a camshaft gear or pulley with twice the number of teeth as the crankshaft gear. The ratio of the crank to cam gear is 2:1. Thus, it takes two turns of the crankshaft to turn the camshaft one turn.

The timing chain style of camshaft drive used in a cam-in-block engine is very different from that used in the OHC engine. The cam-in-block engine uses a chain behind a timing chain cover located on the front of the engine (**FIGURE 12-59**). The chain is splash-lubricated from oiling system. The timing chain

FIGURE 12-57 Valve adjustments. **A.** Rocker arm screw and locknut. **B.** Rocker arm center bolt..

is fairly short because the camshaft is close to the crankshaft. Most of these engines do not use a chain tensioner.

The OHC engine requires a longer timing chain or belt. It also requires the use of one or more tensioners (**FIGURE 12-60**). The timing chain in an OHC engine typically has hard plastic-type guides for the chain to slide on. They also assist the tensioner(s) with correct tracking and tension of the timing chain. The OHC timing chain runs in oil to ensure that the chain is lubricated. Without oil, the chain would wear out rapidly.

A timing belt system uses a toothed or cogged belt to turn toothed or cogged pulleys on the camshaft. It runs dry and requires no oil. The belt is a scheduled maintenance replacement item. It needs to be replaced at the mileage or time recommended by the manufacturer (e.g., 60,000 miles [100,000 km] or five years, whichever occurs first).

The timing chain drive is louder than a belt, but the belt will not last as long as the timing chain. If exposed to fluids or dirt, it will wear out even more quickly. The timing chain must have a constant supply of engine oil to lubricate the chain. This keeps it from wearing out quickly. Timing gears rarely jump time. But if the timing chain or belt breaks, serious engine damage could result. This will depend on whether it is a freewheeling engine or an interference engine.

The freewheeling engine has enough clearance between the pistons and the valves that they can never make contact. In the event the timing belt breaks, any valve that is hanging all the way open will not be hit by the piston. This prevents major engine damage (**FIGURE 12-61**). On a freewheeling engine, a broken timing belt will be an inconvenience to the customer. The engine dies and will not restart. But the good news is that no mechanical engine damage usually occurs.

By contrast, the interference engine has no clearance between the piston and valves if the valve were to be fully open when the piston is at TDC. This means there is minimal clearance between the valves and pistons during normal operation. When the timing belt breaks, the pistons hit the valves. This leads to bent or broken-off valves. It can also punch holes in the pistons and bend connecting rods. It can even ruin the main and connecting rod bearings.

A broken timing belt on an interference engine can mean a huge repair expense. And in some cases, the entire engine must be replaced. The manufacturer's service information will normally list if the engine is an interference engine.

FIGURE 12-58 Cam drives. **A.** Belt-driven OHC. **B.** Chain-driven OHC.

FIGURE 12-59 Cam in-block timing chain and gears.

FIGURE 12-60 Timing chain, gears, and tensioner system.

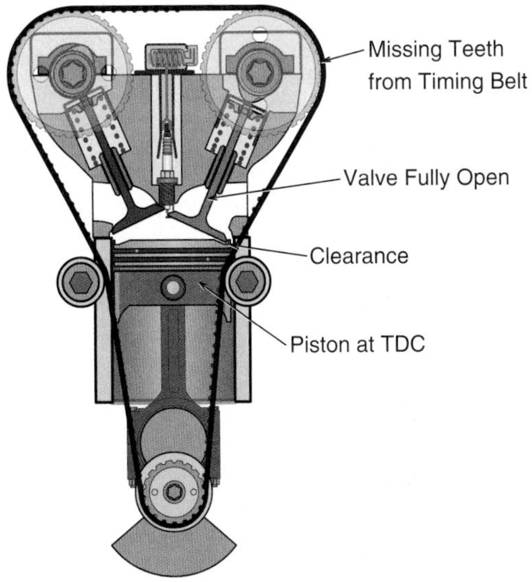

FIGURE 12-61 Freewheeling engine.

Intake Manifold

The intake manifold is part of the engine's air intake (induction) system. It sits between the throttle body and the cylinder head(s). On a V-type engine, it usually is located between the cylinder heads. For an inline engine, it bolts to the side of the head. Intake manifolds deliver air (air with fuel on carbureted or throttle body–injected engines) to the cylinder head.

Intake manifold designs have seen many changes since early engines. Once rectangular and boxy in shape, they are now sleek. The materials from which they are made have also changed. Intake manifolds were first manufactured in cast iron, and later in lightweight aluminum. Now, most manufacturers are using even lighter thermoplastic materials (**FIGURE 12-62**). This reduces the weight of the engine and leads to better fuel economy.

Intake manifolds contribute greatly to an engine's performance. Some of the variables are its inside diameter, its length, and how smooth the inside is. Each of these affects how an intake manifold will contribute to engine performance. Some sportier vehicles use variable-length "tuned" intake runners. They switch between a shorter or longer path for the air to flow, depending on engine rpm. This topic is discussed further in the Induction and Exhaust chapter.

FIGURE 12-62 Intake manifolds. **A.** Cast iron. **B.** Aluminum. **C.** Thermoplastic.

Exhaust Manifold

The exhaust manifold is the output side of the engine's breathing apparatus. With a crossflow head, the exhaust manifold bolts to the cylinder head opposite the intake manifold. This allows the air and exhaust to flow in a straighter line. As with intake manifolds, exhaust manifold design can have a big impact on the performance of the engine due to scavenging. The so-called tuned headers, for example, use equal-length tubes (runners). They scavenge (extract) the exhaust from neighboring cylinders (**FIGURE 12-63**). This topic is discussed further in the Induction and Exhaust chapter.

FIGURE 12-63 Exhaust manifolds. **A.** Cast iron. **B.** Tuned headers.

▶ Two-Stroke Spark-Ignition Engines

LO 12-06 Demonstrate knowledge of two-stroke and rotary engine operation.

Two-stroke engines are notable for their ability to produce a large power-to-weight ratio. The power capabilities of this engine come from the fact that it has a power stroke for every two strokes. That is twice as often as a four-stroke engine that has one power stroke for every four strokes. However, there are drawbacks to this engine design—namely that it is a high-emissions engine. Almost all car manufacturers have moved away from two-stroke production. But it is used extensively in small equipment. Let's look at its basic operation.

Basic Two-Stroke Cycle Engine

Principles

The two-stroke SI engine is different from the four-stroke SI engine. In a **two-stroke engine**, the inlet and exhaust ports are opened and closed by the movement of the piston. There are no poppet valves like those used in the four-stroke cycle engine. The two-stroke engine is still an ICE and shares the five events common to all SI engines. What is different is the method of air induction and exhaust used:

- Intake occurs in two parts. First, the air–fuel mixture is drawn into the crankcase as the piston moves up. (Different)
- It is then transferred from the crankcase to above the piston when the piston moves down. (Different)
- During compression, the mixture in the cylinder is forced into a small volume as the piston moves up. (Same)
- During ignition, the spark from the spark plug ignites the mixture and it burns. (Same)
- During the power stroke, energy released by combustion generates the force that pushes the piston down and turns the crankshaft. (Same)
- During exhaust, leftover gases are expelled from the cylinder when the piston is near BDC. (Same, but through a port, not a valve)

As in all ICEs, expanding gases drive the piston down and turn the crankshaft and flywheel. The flywheel continues rotating, which pushes the piston back up to TDC in the cylinder (**FIGURE 12-64**).

With the two-stroke cycle engine, the crankshaft makes one revolution (two strokes) for every complete cycle. In one revolution of the crankshaft, two piston strokes occur: one down and one up. Each downward stroke of the piston is a power stroke and turns the crankshaft.

The two-stroke cycle engine differs from the four-stroke cycle engine because both the top and bottom of the piston move air and fuel. The upward piston movement creates suction below the piston in the crankcase. This suction pulls the air–fuel mixture into the crankcase (**FIGURE 12-65**). The air and fuel sit in the crankcase until the piston begins to move down. The downward movement of the piston creates a small pressure in the crankcase. This is referred to as crankcase compression (**FIGURE 12-66**). As the piston moves downward, it uncovers a transfer port. The air and fuel rush into the cylinder from the crankcase.

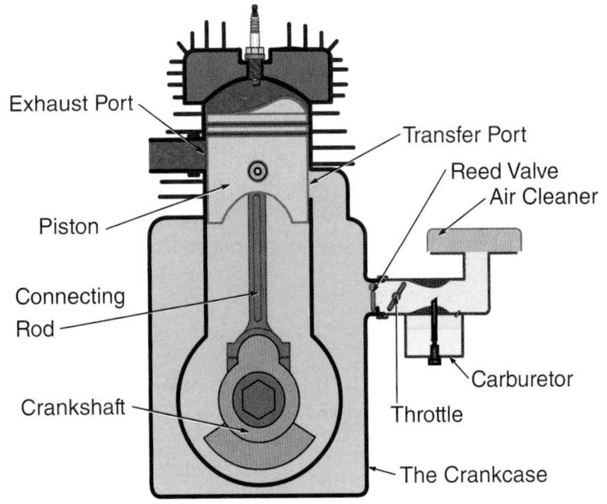

FIGURE 12-64 Two-stroke engine.

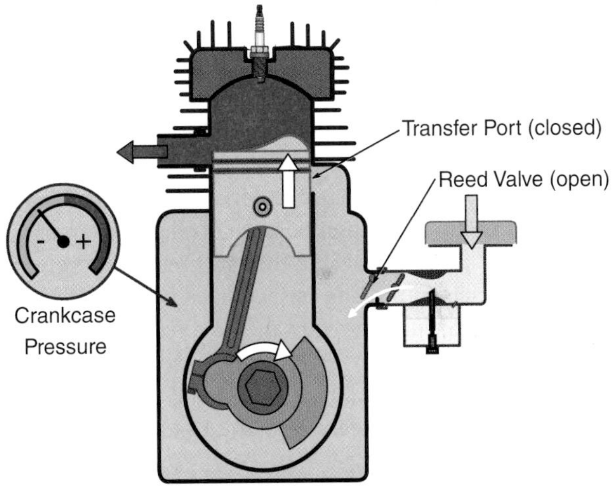

FIGURE 12-65 Two stroke—crankcase intake.

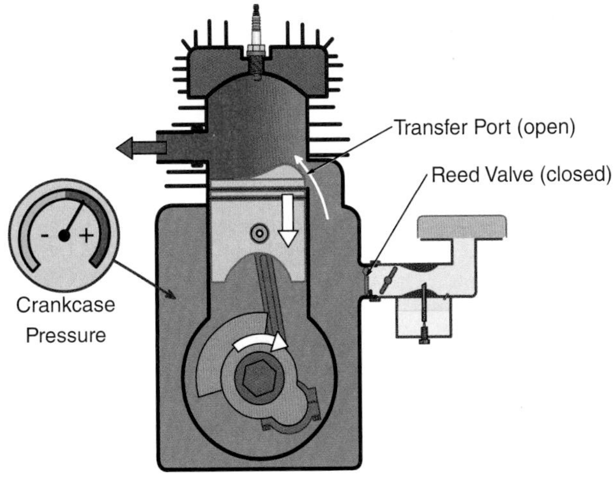

FIGURE 12-66 Two stroke—crankcase compression.

The piston begins moving up toward TDC. This draws air and fuel into the crankcase. The air and fuel above the piston get squeezed in the combustion chamber. The spark plug ignites the mixture. As the piston moves down, it also uncovers the exhaust port, and exhaust gases exit the cylinder. As the piston moves farther down, it uncovers the transfer port again. The crankcase compression pushes new air and fuel into the cylinder.

To summarize, with upward movement of the piston:

- compression of the air–fuel mixture happens in the cylinder above the piston and
- intake of new air and fuel happens in the crankcase below the piston.

When the piston moves down:

- power is being applied to the crankshaft
- exhaust occurs toward the bottom and
- crankcase compression builds to push air and fuel through the transfer port into the cylinder.

The placement of the ports makes all these processes possible. It also eliminates the use of valves to control air into and out of the cylinder.

Two-Stroke Intake System

The intake system of a two-stroke engine is called a piston port intake system. This is because the piston acts as a valve to cover and uncover the ports, allowing flow into and out of the cylinder. The piston uses pressure changes to move air and fuel into and out of the crankcase. Because of this, there must be a valve between the crankcase and the carburetor. The valve prevents air and fuel from being pushed back out the carburetor. Manufacturers use either a reed valve or a rotary valve for this purpose.

A **reed valve** is a small flexible metal plate that covers the inlet port in the crankcase (**FIGURE 12-67**). The reed valve is made of spring metal and can be attached to the crankcase or to the inlet port. The reed valve acts like a one-way check valve and opens and closes automatically, according to changes in pressure in the crankcase. For performance applications, such as in off-road motorcycle racing, a different reed valve may be used. It changes the performance characteristics of the engine.

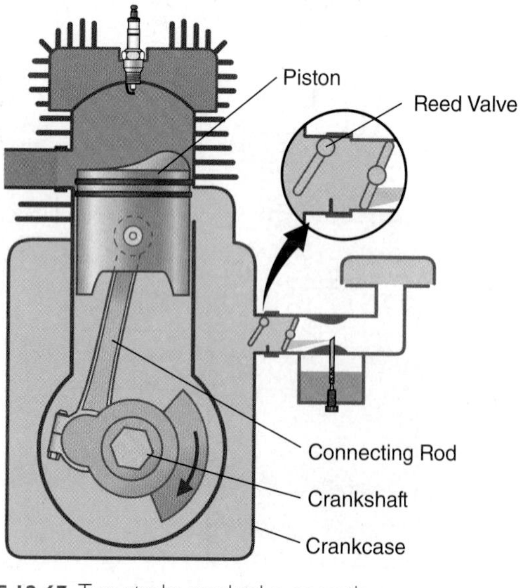

FIGURE 12-67 Two-stroke reed valve operation.

Some two-stroke intake systems use a rotary valve instead of a reed valve. The rotary valve covers and uncovers the port from the carburetor. It is mounted on the crankshaft. As the crankshaft turns, the rotary valve rotates with it, opening and closing the inlet port. Like the reed valve, this ensures that pressurized air and fuel does not get pushed backward into the carburetor.

Four- and Two-Stroke Engine Differences

In a two-stroke gasoline engine, lubrication is usually performed by mixing oil with the gasoline. The oil lubricates the moving parts within the engine. But it also burns in the combustion chamber. This oil causes smoke to be emitted from the exhaust. This is one reason why two-stroke engines are no longer used in automobiles. They also tend to create excessive hydrocarbon exhaust emissions.

Two-stroke engines do have some attractive benefits.

- First is their lightweight, compact, and powerful package. These features make them ideal for handheld equipment such as weed cutters and chain saws.
- They can be run in any position because there is no worry of oil leaking out of the sump or fuel out of the carburetor.
- They are also simple, using fewer parts, so they are easier and less expensive to build.
- Pressurized oil and oil passages are not needed for the crankshaft and connecting rod bearings in two-stroke engines. Most two-stroke engines use roller bearings on the crankshaft, which are splash-lubricated by the oil mixed in with the gas as it churns in the crankcase.
- Many two-stroke engines are air-cooled, thus eliminating the water-based cooling system of the four-stroke engine.
- There is no oil pan with an oil pump or oil in it, nor a radiator with coolant in it. The two-stroke engine does not use any of the valve train components used by four-stroke engines, so even more weight is shed from the engine package.

These are some of the benefits of the two-stroke engine.

Again, the downfall to the two-stroke engine design is the pollution released from the engine due to burning oil in the fuel. Another issue is the amount of unburned fuel exhausted from the engine. This is due to scavenging of the exhaust. It is estimated that 25% of the incoming fuel and oil leave the exhaust unburned. The oil and fuel represent hydrocarbons released to the atmosphere, and wasted energy.

The four-stroke engine releases fewer hydrocarbons because it controls inlet and outlet flow better. The release of less pollution makes the four-stroke engine the choice for most manufacturers. It allows them to meet the pollution requirements of the Environmental Protection Agency. However, companies do continue to produce two-stroke engines. They have found ways to make them pollute less. But much research is still needed to improve this engine design.

Rotary Combustion Spark-Ignition Engine

As we have seen from the two-stroke engine, the fewer parts that are used, the better the power production and the smaller the engine can be. The rotary engine fits into the same category of using fewer parts to produce power. The rotary combustion (RC) engine has found its way into automobiles, planes, helicopters, boats, motorcycles, lawn mowers, and other applications. Displacement has varied from tiny air-cooled models to much larger liquid-cooled, multirotor units.

The rotary engine is also called the "Wankel" engine. It got its name because Felix Wankel improved it for automotive use in the 1940s. The RC engine was commercially released in 1964 in the NSU Wankel Spider model. In 1967, the NSU Ro80 with a two-rotor engine was launched. Under license from NSU, Mazda successfully used the rotary engine in several vehicles. It started producing RC engines from the late 1960s all the way through the RX series (**FIGURE 12-68**). The engines were a redefining period for engine development. But were never really a success for any of the other companies.

Basic Principles of the Rotary Engine

The rotary engine is not as common as the four-stroke or two-stroke cycle engines. But its basic principle is well accepted. The rotary engine layout is vastly different from that of a reciprocating engine. The piston engine is called a reciprocating engine because the pistons move back and forth over the same path. This reciprocating motion is converted to rotary motion at the crankshaft.

In contrast, a rotary engine does not use a piston that reciprocates. Rather, it has a rotor that—you guessed it—rotates. This means that the rotary engine does not need to convert inefficient reciprocating motion to rotary motion. In the reciprocating engine, the piston must stop at the top and bottom of each stroke, many times a second. The stopping and starting of the piston assembly puts tremendous pressure on the connecting rod and rod bolts. Due to inertia, the piston tries to move out of the top and bottom of the cylinder bore.

The rotary engine does not have to stop/start its "piston" as it rotates. The rotor is roughly triangular in shape and turns inside of a housing. The housing works on a geometric principle called an epitrochoid curve. An epitrochoid curve is the circular movement around the perimeter of another circle. The rotor moves in a unique pattern to ensure that the rotor tips follow the oblong shape of the housing (**FIGURE 12-69**).

Because the rotor spins, rather than moving up and down, engine operation is relatively smooth and vibration free. Each rotor and housing is akin to that of a three-cylinder two-stroke engine. This is because of the rotor's three-sided shape. The rotor has three working chambers. Thus, for each rotation of the rotor, we get three power pulses. Low-end torque is improved to the point that (though not recommended) a rotary vehicle can be accelerated from a standstill in fourth gear with minimal lugging. Rotary engines can be made with one, two, or even three or more rotor housings stacked side by side.

Let's look at the basic principles of a rotary engine. Although it appears different, the rotary engine is still an ICE. Recall the five events common to all ICEs: intake, compression, ignition, power, and exhaust.

The rotary engine's intake cycle occurs when one tip of the rotor passes the intake port. It draws the air–fuel mixture into the working chamber through the inlet port (**FIGURE 12-70**). The turning rotor then carries the mixture around the spark plugs. Along the way, the volume of the working chamber decreases and compresses the mixture. The mixture is ignited, and combustion occurs. Expanding gases produce a power pulse, driving the rotor farther around. When the exhaust port is uncovered, the rotor sweeps the burned gases out of the housing. Each face of the rotor is a separate working chamber. So, three combustion events occur for each single revolution of the rotor.

FIGURE 12-68 Wankel rotary engine.

FIGURE 12-69 Cutaway of a rotary engine.

Basic Components of the Rotary Engine

The rotor is mounted in an oval-shaped housing. The housing is made of aluminum alloy, but the curved interior surface has hard chromium plating. This surface has to put up with the wear and tear of the rotor seals sliding against it (**FIGURE 12-71**). Cooling passages machined into the housing allow coolant to flow around the outside surface of the housing. This cools the engine.

There are usually two spark plugs fixed to the housing that enable combustion to occur. These are referred to as the leading and trailing spark plugs. The combustion chamber shape is a long trough. This can contribute to incomplete combustion due to the large surface area of the chamber. It also means a longer distance for the flame to travel in the combustion chamber. Designers found that, by installing a leading and trailing plug, a more efficient combustion process could be produced. The rotor housing has an intake port to let air and fuel into the combustion chamber. It also has an exhaust port to expel burned gases.

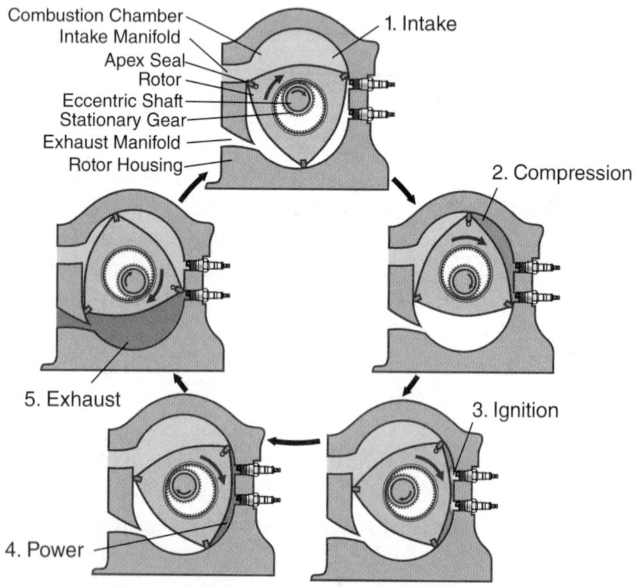

FIGURE 12-70 Operation of a rotary engine.

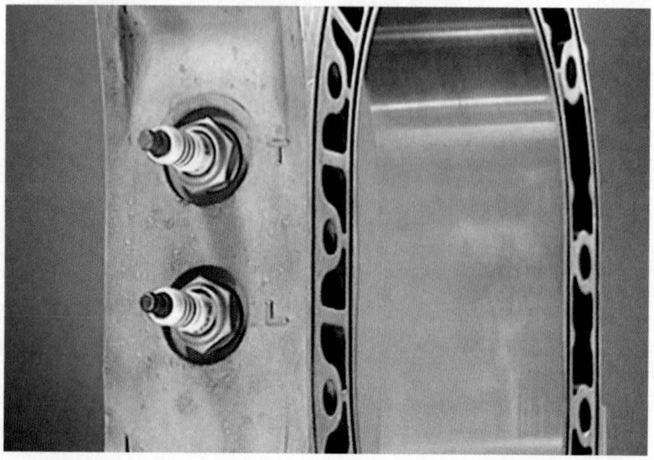

FIGURE 12-71 Rotary engine housing.

The rotor has three apexes, or points. Each apex has a seal which seals the rotor and rotor housing (**FIGURE 12-72**). These seals work like piston rings in a reciprocating piston engine. Side and corner seals create a seal between the rotor and the side housing. This prevents combustion gases from leaking around the rotor at the apex seals. The rotor also has oil seals on the side of the rotor. They keep oil that is inside the rotor from finding its way into the combustion chamber.

The combustion chamber is formed by hollows in the flanks of the rotor. These hollows are sometimes called bathtubs. Front and rear housings, or side housings, are bolted to each side of the rotor housing. If it is a two-rotor engine, there is an intermediate housing between the two rotors.

An internal gear in each rotor meshes with a corresponding stationary gear in the front and rear housings (**FIGURE 12-73**).

When combustion occurs, the meshing of the teeth forces the rotor to walk around the stationary gear. The teeth being in mesh combines with an eccentric shaft to make the rotor follow the curved surface of the housing. This gives the rotor somewhat of a planetary motion. The rotor is attached to the eccentric shaft at points called rotor journals. This eccentric shaft is like a crankshaft in a piston engine, but with the journals off-center (**FIGURE 12-74**). Because the rotor is off-center, the force applied to the shaft is off-center too. The whole shaft is supported by main journals, so the final output is smooth rotary motion.

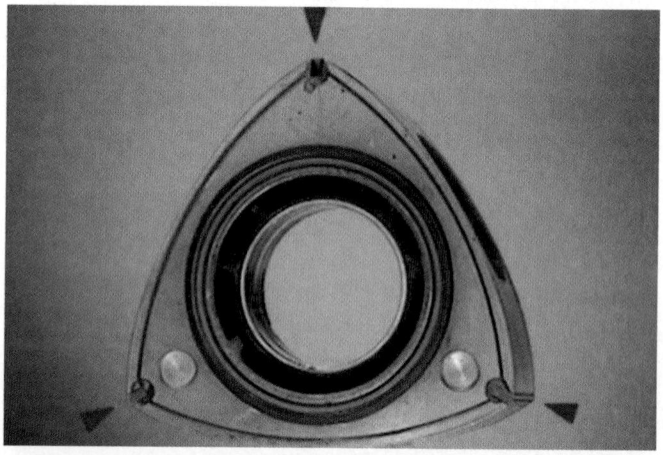

FIGURE 12-72 Rotary engine rotor and seals.

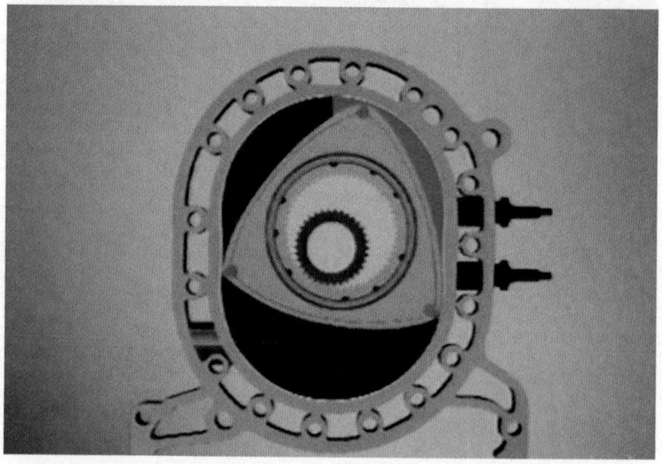

FIGURE 12-73 Stationary gear and internal rotor gear.

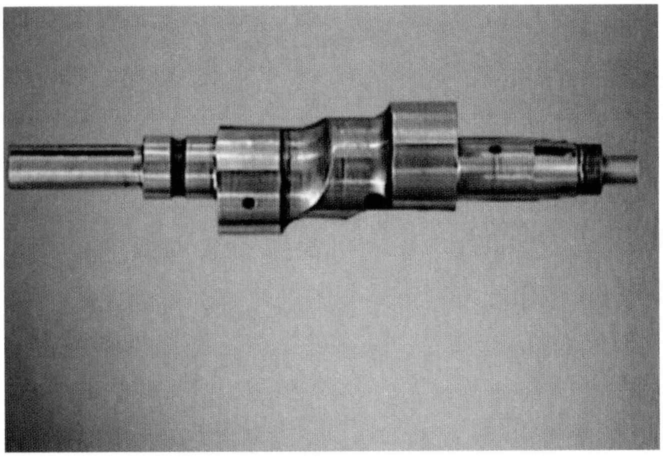

FIGURE 12-74 Rotary engine crankshaft.

Engine Power Pulses

A single-rotor rotary engine produces three power pulses per rotor rotation. This means one power pulse per eccentric shaft rotation. In a four-stroke engine, there is one power pulse for every two crankshaft revolutions. And in a two-stroke engine, there is one power pulse for each crankshaft revolution.

Because of the ratio of the gears in the housings and rotor, the eccentric shaft makes one revolution for each power phase. That is the same as three eccentric shaft revolutions for each rotation of the rotor. So, the eccentric shaft turns at three times the speed of the rotor. A standard rotary engine typically has two rotors. Each rotor is offset from the other, in separate chambers. So, a two-rotor engine ends up with two power pulses per revolution of the eccentric shaft.

▶ Wrap-Up

Ready for Review

- Engines used for motive power may be classified as external combustion or internal combustion engines. Internal combustion engines can be either a reciprocating piston engine or a rotary engine.
  - Piston engines use a crankshaft to convert the reciprocating movement of the pistons into rotary motion at the crankshaft. Rotary engines use a rotating motion rather than reciprocating motion.
- Engines operate according to the unchanging laws of physics and thermodynamics and are dependent on the relationship between pressure and temperature, temperature and energy, and pressure and volume.
- Effort to produce a push or pull action is referred to as force, work is equal to distance moved multiplied by the force applied, and the rate or speed at which work is performed is called power.
  - As torque is a twisting or turning force, horsepower is the rate (in time and distance) at which that force (torque) is produced.
- Multi-cylinder engines come in various cylinder arrangements: inline engines, boxer engines, V-type engines, and VR or W-type engines. A four stroke engine makes use of intake, compression, power, and exhaust strokes to complete one cycle.
  - ICEs are designated by the amount of space (volume) their pistons displace as they move from TDC to BDC. The displacement of an engine can be altered by changing cylinder bore, piston stroke, or the number of cylinders.
- SI engine can be divided into a couple of main assemblies. The bottom end houses the crankshaft, bearings, connecting rods, piston assemblies, and oil pump. The top end houses the cylinder heads and combustion chambers. The cylinder head(s) contains the overhead valves and valve train.
- In two-stroke cycle engine, the crankshaft makes one revolution (two strokes) for every complete cycle. The upward stroke of piston initiates compression and intake, while the downward stroke initiates combustion and exhaust stroke.
  - A rotary engine makes use of a rotor for intake, compression, ignition, power, and exhaust strokes. Three combustion events occur for each single revolution of the rotor.

Key Terms

compression ratio (CR) The volume of the cylinder with the piston at bottom dead center as compared to the volume of the cylinder at top dead center, given in a ratio such as 9:1 CR.

ICE Internal combustion engine. A device that burns fuel inside itself to generate power.

reed valve A small flexible metal plate that covers the inlet port of a two-stroke engine and opens and closes to let air and fuel into the crankcase.

scavenging The process of removing burned gases from the cylinder through the use of moving airflow pulling or extracting the gases out.

two-stroke engine An engine that uses only two strokes to complete its running cycle.

valve overlap The portion of time that both valves are open at the same time.

Review Questions

1. When air and fuel are burned in an engine, the process is called combustion. What type of combustion process is used in modern automobiles?
   a. Internal combustion
   b. Reserve combustion
   c. External combustion
   d. Spontaneous combustion
2. Which type of engine uses a power piston and a displacer piston?
   a. Steam engine
   b. Deutz engine
   c. Stirling engine
   d. Wankel engine
3. Power is:
   a. A twisting or bending force
   b. A push or pull action
   c. Equal to the distance moved multiplied by the force applied
   d. The rate at which work is performed
4. The more energy a fuel has, the more work it can do. What is the unit of measure used to determine the amount of energy a fuel has?
   a. Cetane rating
   b. British thermal unit
   c. Octane rating
   d. Thermal expansion coefficient
5. What is the correct term used to define a twisting force?
   a. Rotational force
   b. Force vector application
   c. Torque
   d. Tension
6. When the piston reaches its lowest point in the cylinder, what position is the piston said to be?
   a. Top dead center
   b. Bottom dead center
   c. Lowest stroke point
   d. Stroke position
7. Which engine stroke pushes the piston from TDC to BDC?
   a. Intake
   b. Compression
   c. Power
   d. Exhaust
8. In a sealed container, when temperature increases:
   a. Pressure increases
   b. Pressure decreases
   c. Volume decreases
   d. Energy stays the same
9. Which of these components convert reciprocating movement of the pistons into rotary motion?
   a. Camshaft
   b. Crankshaft
   c. Flywheel
   d. Piston rings
10. At what speed does the crankshaft rotate as compared to the camshaft?
    a. Same speed
    b. Twice as fast
    c. Half speed
    d. One-third

ASE Technician A/Technician B Style Questions

1. Technician A says that understanding the laws of thermal dynamics and physics can help in engine diagnostics. Technician B says that pressure and temperature rise are directly proportional. Who is correct?
   a. Technician A
   b. Technician B
   c. Both A and B
   d. Neither A nor B
2. Technician A says that gasoline has a rating of 25,000 Btu. Technician B says that diesel fuel has a 14,000 Btu rating. Who is correct?
   a. Technician A
   b. Technician B
   c. Both A and B
   d. Neither A nor B
3. Technician A says that work is performed any time an object is moved. Technician B says that any time force is applied to an object, work is performed. Who is correct?
   a. Technician A
   b. Technician B
   c. Both A and B
   d. Neither A nor B
4. Technician A says that the engine with the same size bore and stroke is referred to as an interference engine. Technician B says that stroke is the distance the piston travels from TDC to BDC. Who is correct?
   a. Technician A
   b. Technician B
   c. Both A and B
   d. Neither A nor B
5. Technician A says that engine displacement can be defined by cubic inches, cubic centimeters, or liters. Technician B says to find displacement you need to know the bore, stroke, and number of cylinders. Who is correct?
   a. Technician A
   b. Technician B
   c. Both A and B
   d. Neither A nor B
6. Technician A says that the compression stroke is also called the expansion stroke. Technician B says that the compression ratio compares the cylinder volume at BDC to TDC. Who is correct?
   a. Technician A
   b. Technician B
   c. Both A and B
   d. Neither A nor B

7. Technician A says that the connecting rod transfers movement between the crankshaft and the piston. Technician B says that the connecting rod can be made of cast iron or powdered metal. Who is correct?
   a. Technician A
   b. Technician B
   c. Both A and B
   d. Neither A nor B
8. Technician A says that a freewheeling engine has enough clearance between the pistons and the valves so that they can never make contact. Technician B says that an interference engine has pistons that will hit each other if the timing belt breaks. Who is correct?
   a. Technician A
   b. Technician B
   c. Both A and B
   d. Neither A nor B
9. Technician A says that spark ignition engines use the heat of compression to ignite the air–fuel mixture. Technician B says that compression ignition engines use spark plugs to ignite the air–fuel mixture. Who is correct?
   a. Technician A
   b. Technician B
   c. Both A and B
   d. Neither A nor B
10. Technician A says that the camshaft can be belt-driven. Technician B says that the camshaft can be chain-driven. Who is correct?
    a. Technician A
    b. Technician B
    c. Both A and B
    d. Neither A nor B

CHAPTER 13

Engine Mechanical Testing

Learning Objectives

- **LO 13-01** Describe the overview of engine mechanical testing.
- **LO 13-02** Isolate engine noises and vibrations.
- **LO 13-03** Evaluate the results of engine vacuum tests.
- **LO 13-04** Evaluate the results of cylinder power balance tests.
- **LO 13-05** Evaluate the results of cylinder cranking compression tests.
- **LO 13-06** Evaluate the results of cylinder leakage tests.

ASE Education Foundation Tasks

See Appendix A to view the 2017 ASE Education Foundation Automobile Accreditation Task List Correlation Guide.

▶ Engine Mechanical Testing Overview

LO 13-01 Describe the overview of engine mechanical testing.

For an engine to operate efficiently and effectively, the mechanical condition of the engine must be in good working order. The pistons, piston rings, cylinder walls, head gasket, and valves must seal properly. If they do not, then the engine will not operate correctly, and all the tuning in the world will not be able to fix it. This makes assessing the condition of the engine a critical step in the diagnostic process (**FIGURE 13-1**). And it needs to be performed before tune-up parts are replaced.

There are few things more dreaded than having to tell a customer that the $650 tune-up you just performed did not fix the engine's misfiring problem. And in fact, the fault was caused by a burned valve. So, the vehicle really needs a $2,500 valve job or a $5,000 engine replacement. First, the customer will not be happy that the vehicle is not fixed. Second, the repair will now cost substantially more money than the customer expected. Third, the customer now has very good reason to doubt your competence. They will wonder if you are correct this time, when you were wrong earlier. This is a no-win situation. But it can be avoided by always diagnosing the problem, instead of throwing parts at it.

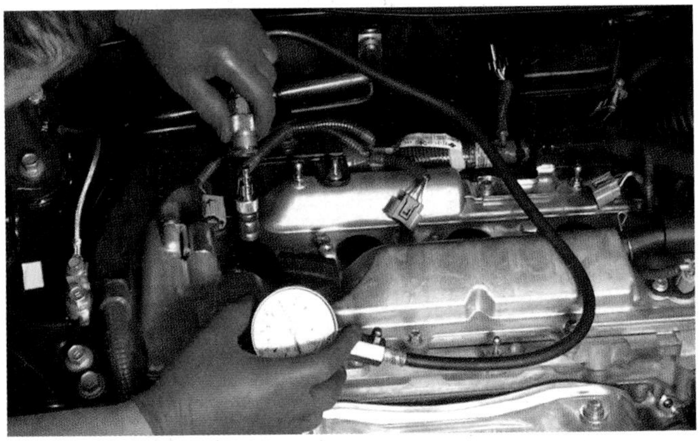

FIGURE 13-1 Always determine the root cause of an engine fault. Never assume it may be something simple. Engine mechanical tests help you determine the root cause.

Engine mechanical testing is also performed to diagnose what major engine work is needed. Some examples are:

- replacing a head gasket,
- rebuilding of the cylinder heads, or
- performing a full engine replacement.

Understanding exactly what is wrong will allow you to better talk with the customer about the needed repairs. It will also help you build credibility with your customers by having a full understanding the situation. And then clearly communicating it to them in a professional manner. This chapter will help you learn the skills and procedures used in performing engine mechanical tests.

Engine Mechanical Testing

Engine mechanical testing uses a series of tests to assess the mechanical condition of the engine. The tests start off broad, and then narrow down as each test is performed. This process will help you to first identify the location or type of fault. From there, you will determine the cause of the fault. Having a good understanding of engine theory will help you evaluate the results of each test. It will also help to guide you down the diagnostic path after each step.

Mechanical testing starts with following the strategy-based diagnostic process (**FIGURE 13-2**) covered earlier in this book. In this chapter, we mainly focus on only the first three steps. The last two steps are covered in the *Fundamentals of Automotive Technology* book.

So, always start by verifying the customer's concern. This involves getting as much information from the customer as possible. With a clear understanding of the concern, you will

You Are the Automotive Technician

A customer comes into the dealership complaining that his vehicle is not running as smoothly as before. They also say that the check engine light is flashing. You know that this can show a potentially catalyst-damaging fault. You explain to the customer that they did the right thing by bringing the vehicle in for diagnosis. First, you locate the customer's vehicle history and write up a repair order including the customer's concerns. Second, you use a scan tool to retrieve the diagnostic trouble codes (DTCs) and find a P0304-cylinder 4 misfire detected code. Third, you hold the throttle to the floor, with the ignition switch in the off position, and then crank the engine over. The engine exhibits an uneven cranking sound, indicating a compression-related fault. Next, you research the technical service bulletins (TSBs) and find one that relates to this code and a low compression condition. The TSB indicates possible soft camshaft lobes, which can wear down and not open the valve(s) fully. To verify whether this is the case, the TSB directs you to perform cranking and running compression tests on any misfiring cylinders. The TSB then lists the acceptable minimum pressures for each test.

1. Why did the engine have an uneven cranking sound?
2. What does a power balance test indicate?
3. What does the running compression test indicate?
4. How would you determine which cylinder is misfiring if the engine computer doesn't have that capability?

need to observe the fault. A visual inspection of the gauges will give you helpful, general indications of the engine's condition. Issues such as low oil pressure, overheating, and illuminated malfunction indicator lamp (MIL) give you clues to the cause of the fault.

A good time to observe the gauges and warning indicators is when you are either test-driving the vehicle, or as you pull it into the shop. Use that time to investigate the information provided through those indicators. One thing to remember is that the indicators can be wrong. So, you may need to verify that the indicator is working properly and providing you with accurate information. If you assume it is correct when it is not, you will be led down the wrong path. You may be able to verify the warning indicators by other types of observation. For example, if the oil pressure is truly low, then there will likely be tapping or knocking noises coming from the engine that support this conclusion.

Starting the engine and listening to its operation can indicate a host of problems, from loose belts to worn main bearings. It can also reveal whether the engine is misfiring, and how steady the misfire is. A visual inspection of the engine for leaks will affirm the ability of the seals and gaskets to contain each of the engine's fluids. It can also determine other issues with the engine.

After verifying the concern and giving the vehicle a good initial inspection, it is time to research the vehicle's service history. Check to see if the new concern is related to any previous work. If so, that may give you some clues as to where to start looking for the fault. If the vehicle service history doesn't appear to be related to the concern, you will need to gather additional information. This usually starts with researching the service information. Look up any related TSBs and the manufacturer's diagnostic flowcharts. In doing so, use that information along with your knowledge of the affected system and symptoms to create a focused testing plan. This begins with determining a logical starting point to begin testing.

Once you have a starting point, continue to reference the service information. Use it and the diagnostic procedure to determine the follow-up tests needed to find the problem. That is what diagnosis is all about. You need to gather information, sort it all out, and build a focused plan of attack. Then, you need to test and interpret the results. If needed, decide what the next step is. But first, let's look at the tools and equipment that are used to assess the mechanical condition of the engine.

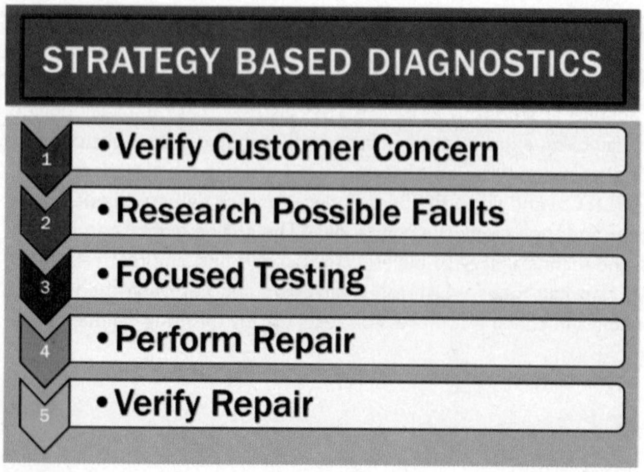

FIGURE 13-2 Strategy-based diagnosis.

Testing Tools

Testing tools have come a long way over time. Some seasoned mechanics have been known to use bubble gum and its wrapper to diagnose a burned valve (see the Technician Tip below). Today, a variety of tools and equipment are available. They allow technicians to fully assess the mechanical condition of modern engines. Each of the tools introduced here has a specific function. Their use will be further described in the Skill Drills later in this chapter:

- Compression tester: Used when a technician suspects a cylinder may have low compression. It measures the amount of compression pressure a cylinder can generate (**FIGURE 13-3**).
- Vacuum gauge: Measures the amount of vacuum an engine can generate during various operating conditions (**FIGURE 13-4**).

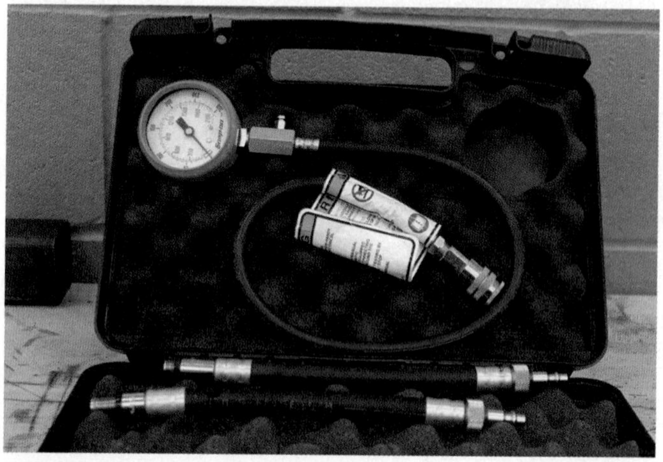

FIGURE 13-3 A compression tester.

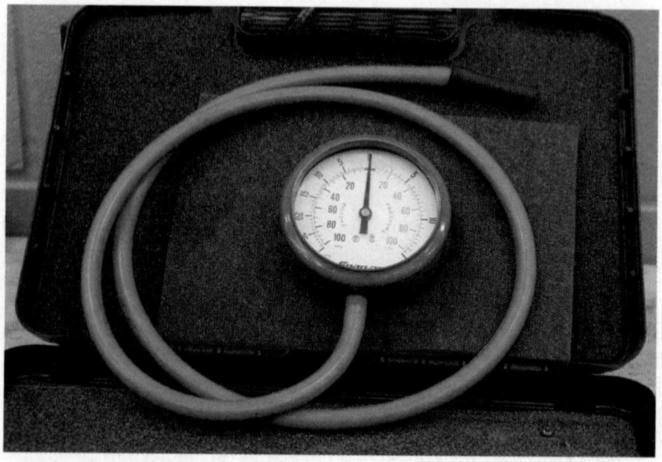

FIGURE 13-4 A vacuum gauge.

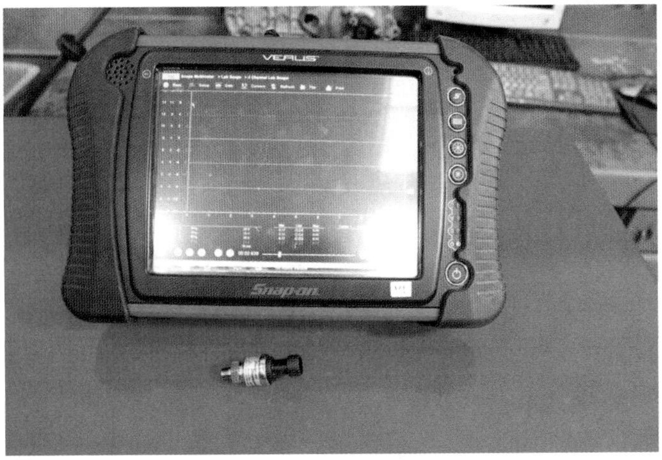

FIGURE 13-5 A pressure transducer and lab scope.

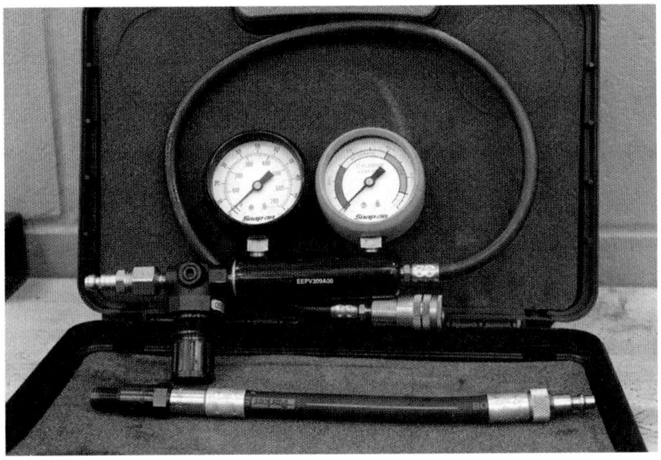

FIGURE 13-6 A cylinder leakage tester.

FIGURE 13-7 A scan tool.

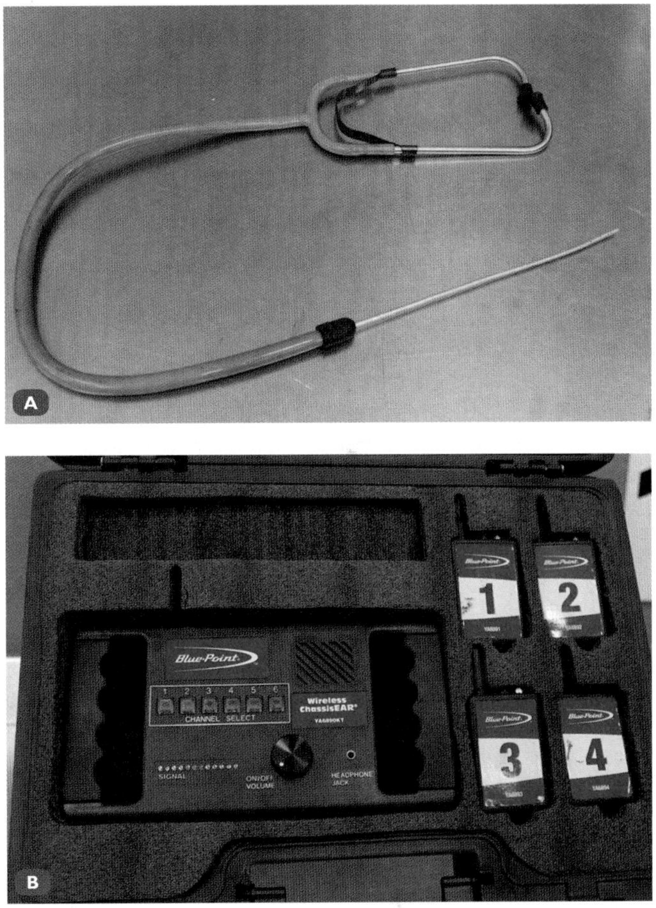

FIGURE 13-8 **A.** A standard stethoscope. **B.** An electronic stethoscope.

- Pressure transducer and lab scope: The **pressure transducer** measures engine vacuum or pressure and displays it graphically on a lab scope. It is very accurate. It creates a detailed trace on the lab scope. This trace can be compared to a known good trace and used to determine mechanical issues in the engine (**FIGURE 13-5**).
- Cylinder leakage tester: Directs air into the cylinder and measures the percentage of air that is leaking from the cylinder. The technician can determine where the pressurized air is leaking from the cylinder by looking, listening, or feeling (**FIGURE 13-6**).
- Scan tool: Communicates to the vehicle's computers through the **data link connector** (DLC). It displays the readings from the various sensors, retrieves trouble codes, freeze-frame data, and system-monitor data. On some vehicles, it performs output tests, such as a cylinder power balance test, or commands other output devices to operate (**FIGURE 13-7**).
- Stethoscope—standard and electronic: Used by technicians to listen to unusual noises in the vehicle. Stethoscopes come in both standard and electronic versions (**FIGURE 13-8**).

▶ Diagnosing Engine Noise and Vibrations

LO 13-02 Isolate engine noises and vibrations.

Running engines are fairly quiet considering all of the mechanical activity that happens within them. But if something starts to go wrong, noises can be one of the first indicators. Noise issues can indicate something as simple to fix as replacing a worn accessory belt. Or they can be as complicated as a spun connecting rod bearing. This would generally require rebuilding the entire engine.

When diagnosing engine noises, you need a good foundation to start with. That begins with a solid understanding of engine theory which we covered previously. Many sounds can be pinpointed through applying that knowledge to what you are hearing. This builds experience that will help you diagnose noises in the future. So, investigate unusual noises and build your experience.

Let's look at some of the engine noises that are caused by engine faults. A loud knocking noise could be from a worn or spun main bearing or connecting rod bearing. A main bearing noise is generally deeper sounding than that of a rod bearing. Also, a main bearing makes an evenly spaced single knock, whereas a rod bearing generally makes a double knock.

A light ticking noise could be a valve lifter problem, which can be heard near the camshaft or valve cover area. A normal source of a light ticking noise is the fuel injectors. They make a ticking sound as they open and close. This is normal.

You might hear a light knocking noise that comes under slight rocking of the throttle. That could be caused by a collapsed piston skirt. A whirring noise can be caused by worn bearings in the alternator, water pump, AC compressor, power steering pump, or belt tensioner.

One way to help locate any type of engine noise is to use a mechanic's stethoscope. Mechanical stethoscopes are the most common. But many electronic stethoscopes have settings that enhance selected sound frequencies while filtering out others. This makes them very handy. Place the stethoscope only against stationary engine components. Place them in a variety of positions around the engine, and listen to the noises. Generally, the louder the noise, the closer you are to its source.

Depending on the noise, a stethoscope may not be appropriate. In the case of a squeaky belt, spray water on one belt at a time. If the noise goes away temporarily, you have identified the one that is squeaking. For squeaks and creaks that come from linkage and joints, spraying them with a lubricant, one at a time, and then operating them will typically identify the offending joint.

Vibrations can be difficult to pinpoint. Vibrations can come from the engine or the drivetrain. The best clue is to determine the frequency of the vibration. You can then compare that to the various components on the engine and drivetrain to identify the fault. If the frequency matches the engine rpm, then you would be looking for a fault in the crankshaft, or parts connected to it. If the frequency matches tire speed, then look for a fault in the tires or axles.

If the vibration occurs only when the vehicle is moving, suspect a component within the drivetrain or driveline. This could be the U-joints, CV joints, driveshaft, axles, or even tire balance.

A good way to isolate the drivetrain is to drive the vehicle up to the speed at which the vibration occurs. Then, place the transmission in neutral and allow the engine to idle. If the vibration is still present, then the issue is probably associated with the wheels, tires, or axles. If the vibration goes away, then raise the engine rpm while still in neutral. If the vibration reoccurs, then the issue is most likely with the engine. If there is no vibration, reengage the appropriate drive gear and accelerate moderately. If the vibration reappears, then the issue is likely with the driveshaft, U-joints, or CV joints.

Cranking Sound Diagnosis Overview

Engine noises can give a technician valuable insight into the condition of the engine. As you have learned, compression is one of the five critical events for each cylinder to operate properly. If the compression in one or more cylinders is too low, then the affected cylinders will not create as much power as cylinders with the proper amount of compression. This causes the engine to run rough. In many cases, this misleads the customer to request a "tune-up." Yet, that will not fix the low-compression issue, and the engine will still run poorly after the tune-up. This results in an unhappy customer as well as technician.

To help avoid that situation, a cranking sound diagnosis should be performed. It can identify whether the compression is similar across each of the cylinders. If it is not similar across all cylinders, a tune-up will not fix the problem. If it is similar across all cylinders, then compression issues are not likely causing the engine to run rough. In this case, further diagnosis of the fuel and ignition system is needed.

When you perform a cranking sound diagnosis, you must disable the engine so that it will not start. Then crank the engine over, using the key, and listen to the cranking sounds. It might take a little time to orient yourself toward the sound. Any noises you hear could be from:

- a misaligned starter
- the knock of a spun crankshaft bearing
- an uneven cranking sound that a low-compression cylinder gives
- a fast cranking sound from a low-compression condition that is due to bent valves caused by a broken timing belt or
- a whole slew of other mechanical possibilities.

Just know that most engine problems have an associated noise. This means that noises provide clues to identify and locate the fault. Your job is to gain experience in distinguishing between normal noises and abnormal noises. Let's see how a cranking sound diagnosis is performed.

Diagnosing Cranking Sound

The sound of the starter motor cranking the engine over can give us an indication of the engine's compression. The starter will make an even cranking sound if the compression is similar across the cylinders. As each piston comes up on compression, it loads down the starter motor and slows the cranking speed. If the compression is not the same at each cylinder, the starter will make an uneven cranking sound. Train your ear to pick up the different compression-related sounds using a vehicle with known good compression. First, disable the engine so that it will not start. You can disable the engine in four possible ways:

- Use the "clear flood" mode if the vehicle is equipped with it. To do this: With the ignition key off, hold the throttle to the floor, then crank the engine. If it starts, let up on the throttle and turn the key off. Then, use one of the two other methods.
- Disable the fuel injectors by disconnecting their electrical connectors.
- Disable the fuel pump by pulling the fuel pump fuse or relay, starting the engine, and waiting for it to die.
- Disable the ignition system (see service manual for proper procedure) so that it will crank, but not start. Realize that the injectors will still spray fuel, which can wash down the cylinder walls, so this is not the best way to disable the cylinders.

No matter which method you use, hold the throttle to the floor and crank the engine. Listen to the sounds it makes. Next perform the same task on a vehicle with known bad compression, and listen to the difference in cranking sounds. To perform a cranking sound diagnosis, follow the steps in **SKILL DRILL 13-1**.

▶ Engine Vacuum Tests

LO 13-03 Evaluate the results of engine vacuum tests.

Intake manifold vacuum readings can be used to indicate an engine's general condition. The typical vacuum reading for a properly running engine at idle is a steady 17" to 21" of mercury (57.6 to 71.1 kPa) (**FIGURE 13-9**). Manifold vacuum is affected by a few major factors (**TABLE 13-1**). The first is the overall tune of the engine. This includes ignition timing and valve timing. But it could also be caused by a restricted exhaust system. Problems with these factors generally create low, but steady vacuum readings.

The second major factor that affects vacuum readings is related to the mechanical condition of the engine. This can cause the needle to either oscillate rhythmically, or be too low. A sharp back-and-forth motion in the needle can be from conditions such as a burnt or sticking valve.

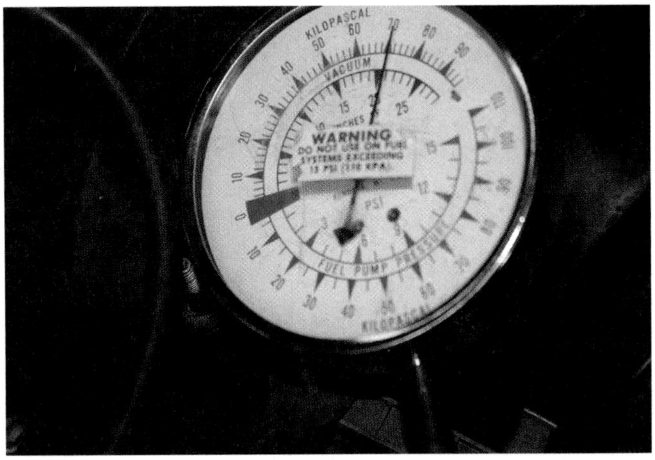

FIGURE 13-9 A vacuum reading on an engine with good idle vacuum.

TABLE 13-1 Chart of Vacuum Readings—Vacuum Gauge

Reading	Indication
17" to 21", steady needle at idle	Good reading
Needle oscillates back and forth about 4" to 8".	Burned or constantly leaking valve
A low and steady reading	Possible late valve timing or ignition timing
Snap acceleration needle drops to zero and then only reads 20" to 23" of vacuum; should be much higher, around 27".	Possible worn rings
Good reading at idle. As engine speeds up to a steady 2500 rpm, the needle slowly goes down and may continue to drop	Possible restricted exhaust system

SKILL DRILL 13-1 Performing a Cranking Sound Diagnosis

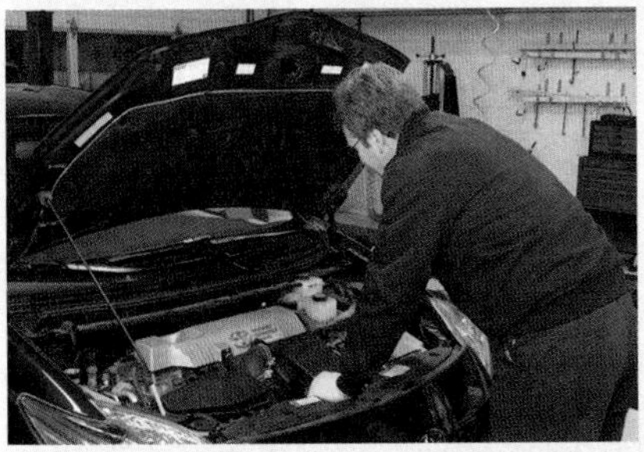

1. Raise the hood and make sure it is secure.

2. Disable the engine by one of the methods described above.

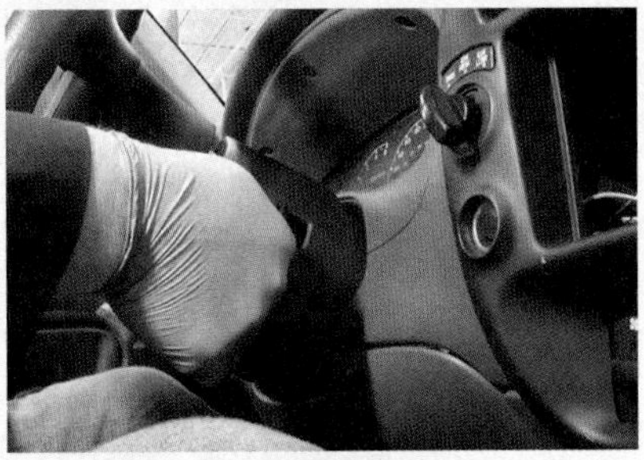

3. Hold the throttle to the floor, and crank the engine over. Listen to the sound the engine makes while cranking. Determine any necessary actions.

If the reading is too low, it can be caused by worn piston rings, cylinder walls, or camshaft lobes. It could also be caused by an intake manifold leak. A low reading once the engine reaches a higher speed, such as 2500 rpm, could be caused by several faults. Common faults would be:

- a restricted exhaust system,
- ignition timing that is not advancing,
- faulty variable valve timing system, or
- faulty exhaust gas recirculation (EGR) valve.

As you can see, the vacuum test can provide you with helpful information, when diagnosing an engine problem. At the same time, many of these conditions will need to be verified by further testing. Those tests are covered later in this section.

Intake Manifold Vacuum Testing

A vacuum gauge is used to determine the engine's general condition. The vacuum gauge reading shows the difference in pressure between the outside atmospheric pressure and intake manifold vacuum. Because vacuum is any pressure below atmospheric pressure, this test is also called a manifold pressure test. The vacuum gauge is installed in a vacuum port on the intake manifold. Always select the largest and most centrally located vacuum port (**FIGURE 13-10**). This may require using a vacuum tee. The vacuum tee allows the existing component to operate while measuring the vacuum.

Vacuum is typically measured in inches of mercury, or kPa. Some vacuum gauges use a small orifice to dampen the needle movement. To see if the gauge has this dampening orifice, remove the vacuum hose from the gauge. Then, look in the opening of the gauge to see if the opening is completely clear (**FIGURE 13-11**). It may have a small orifice inside. If it has a dampening orifice, the vacuum gauge will not be able to display rapid needle movements. So, this type of vacuum gauge is not useful in diagnosing valve issues. But it can be used for diagnosing other issues.

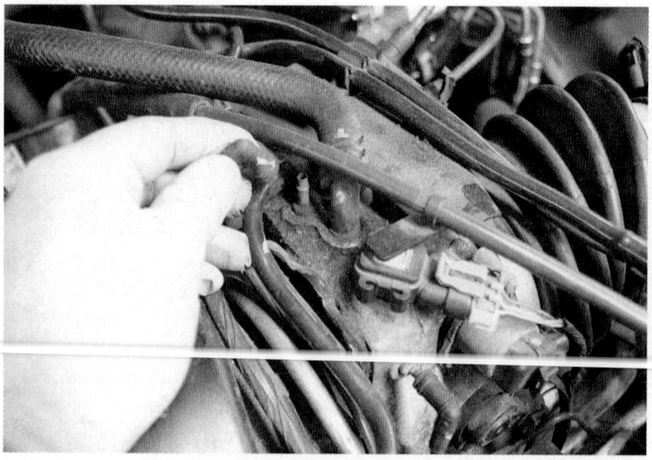

FIGURE 13-10 A vacuum gauge needs to be hooked up to a centrally located, large vacuum port.

Testing Engine Vacuum Using a Vacuum Gauge

During an engine vacuum test, all vacuum gauges are calibrated to sea level pressure. When testing above sea level, you will need to compensate for the altitude. A general rule of thumb is that for every 1,000 feet (305m) above sea level, atmospheric pressure drops 1" (3.4 kPa). For instance, if the gauge reads 18" (61 kPa) of vacuum at sea level, it would drop to 17" (57.6 kPa) at 1,000 feet (305 m) above sea level.

The vacuum gauge connected to the intake manifold shows the vacuum (pressure) inside. The average inches of vacuum for a good running engine at idle is a steady 17" to 21" (57.6–71.1 kPa) of vacuum. The average vacuum reading at idle for a performance camshaft is around 12"–15" (40.6–50.8 kPa).

When testing intake manifold vacuum, the engine should be at operating temperature. The engine should be idling to start with. At this point, the gauge can be read, and the behavior of the needle observed. Compare your readings with the chart above.

To attain the highest vacuum possible, the throttle should be snapped open and closed quickly. Note that vehicles with throttle control may not be able to do the snap throttle test. This test checks the integrity of the piston rings and cylinder walls. The vacuum reading should be approximately 27" (91.5 kPa).

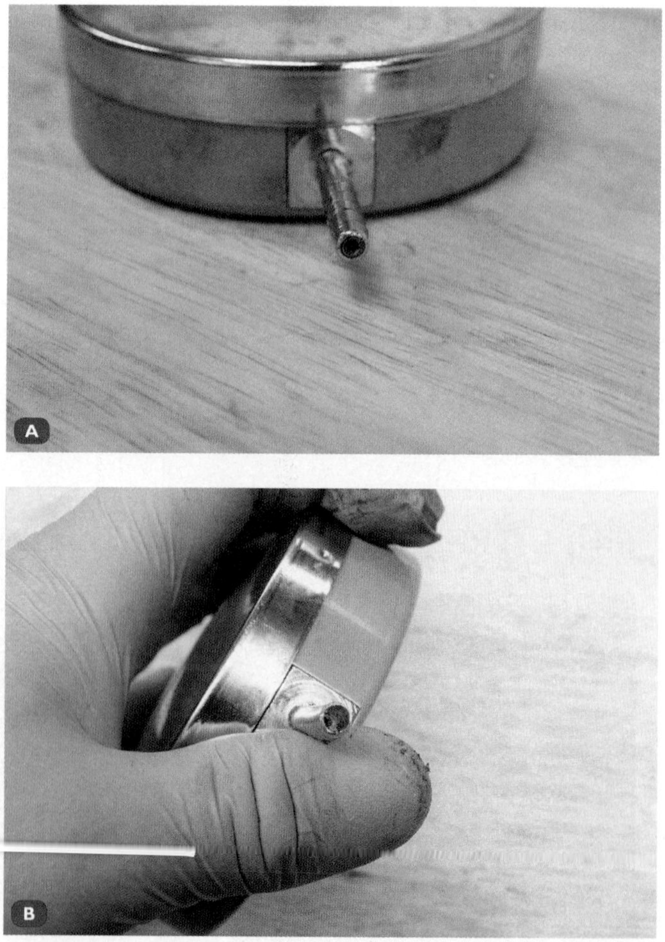

FIGURE 13-11 A. Vacuum gauge with a dampening orifice. **B.** Vacuum gauge without a dampening orifice.

SKILL DRILL 13-2 Testing Engine Vacuum

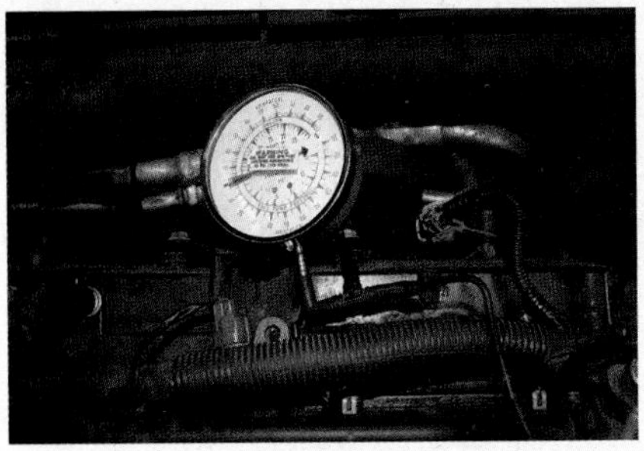

1. Connect the vacuum gauge to the intake manifold. You might have to use a vacuum tee. Make sure the engine is at operating temperature. Start the engine, take a reading at idle, and record the reading.

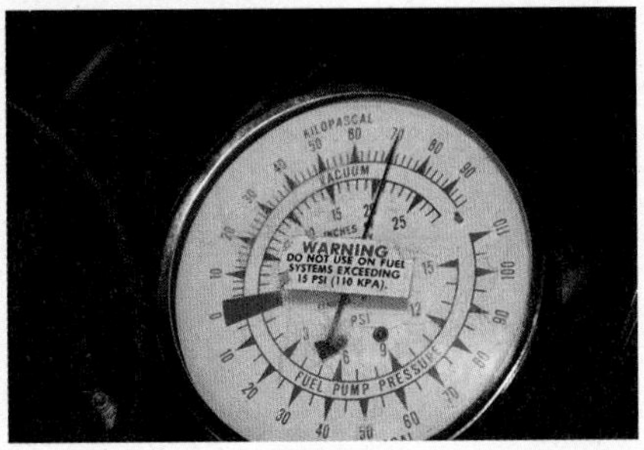

2. Snap accelerate the engine by quickly opening and closing the throttle; record the highest vacuum reading attained.

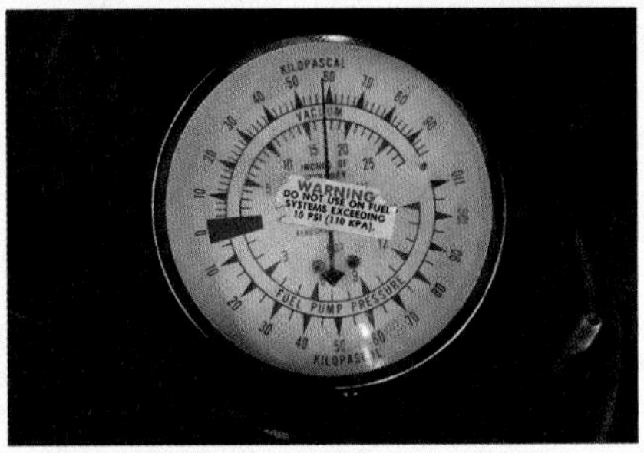

3. Hold the throttle steady at 2500 rpm and record your reading.

A reading below about 23" (77.9 kPa) indicates potentially worn piston rings.

Finally, the throttle should then be opened to achieve a steady 2,500 rpm and the vacuum reading recorded. To test engine vacuum using a vacuum gauge, follow the steps in **SKILL DRILL 13-2**.

Testing Engine Vacuum Using a Pressure Transducer

With the advent of lab scopes, many technicians are using pressure transducers to measure the engine's manifold pressure. This allows the vacuum to be displayed graphically on a lab scope. The pressure transducer is a very accurate measuring tool. So, it can indicate small changes in manifold pressure. This allows it to detect issues with individual valves and cylinders.

The pressure transducer is connected to a lab scope and the intake manifold port. That way the manifold pressure is displayed as a trace. The trace is created by each piston's intake stroke (**FIGURE 13-12**). If the ignition pattern for cylinder 1 is monitored on a separate trace, the vacuum pulses can be tied to specific cylinders. Technicians can view these traces and see how well each cylinder is contributing to the vacuum.

Using a pressure transducer and lab scope is similar to using a vacuum gauge. But it is much more accurate and allows you to look at the vacuum graphically. By using a pressure transducer on a variety of vehicles with various issues, you will become familiar with common faults. To use a pressure transducer and lab scope for testing engine vacuum, follow the steps in **SKILL DRILL 13-3**.

> ▶ TECHNICIAN TIP
>
> A pressure transducer is a very handy tool. It can be adapted to fit in the spark plug hole to create a pressure trace of running and cranking compression. It can also be used in the exhaust system to measure the exhaust pulses. Pressure transducers are used more and more commonly when diagnosing engine problems.

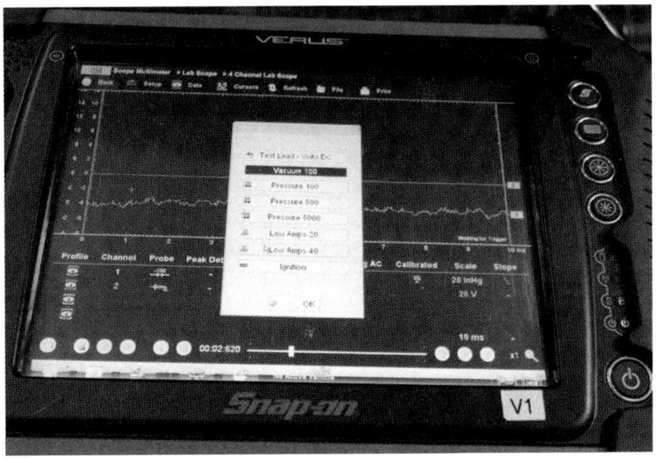

FIGURE 13-12 Vacuum trace.

▶ Cylinder Power Balance Test Overview

LO 13-04 Evaluate the results of cylinder power balance tests.

A cylinder power balance test (also called a power balance test, for short) is used for two purposes. First, after failing the previous tests, it identifies which cylinder(s) is not operating properly. Second, it is used as a general indication of each cylinder's overall health. Every cylinder in the engine should contribute equally to the engine's power output. When a mechanical, electrical, or fuel problem occurs within an engine, the affected cylinders will not produce as much, if any, power when compared to the other cylinders.

The cylinder power balance test allows us to measure each cylinder's output. It is found by disabling one cylinder at a time and measuring the rpm drop (**FIGURE 13-13**). Disabling a cylinder that is not operating correctly will not produce much, if any, rpm drop. Disabling a cylinder that is working properly, however, will produce a much larger rpm

SKILL DRILL 13-3 Testing Engine Vacuum Using a Pressure Transducer

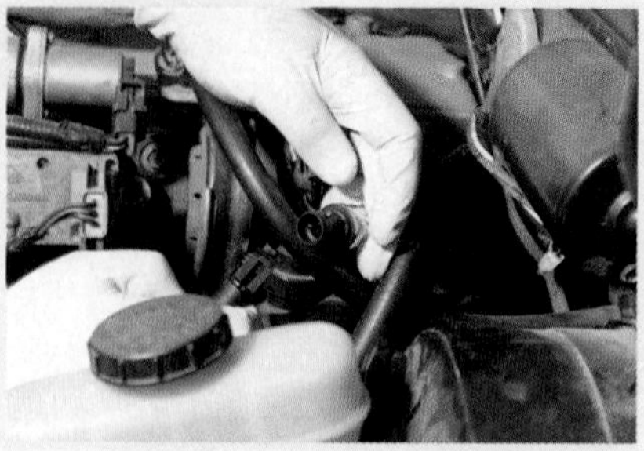

1. Connect the pressure transducer to the intake manifold and the lab scope.

2. Connect the second channel of the lab scope to the ignition system so that it can identify cylinder 1.

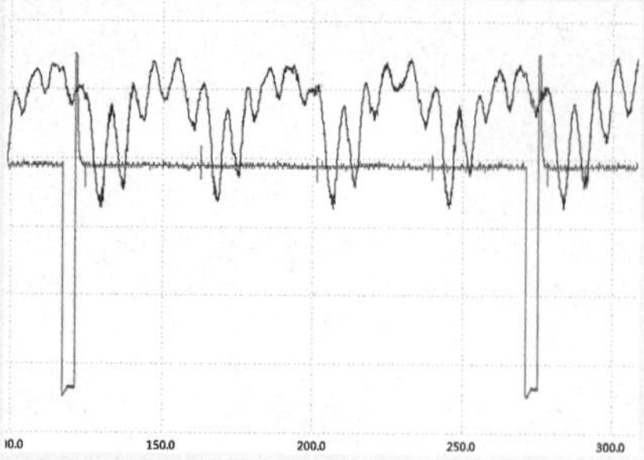

3. Start the engine and let it idle. Adjust the lab scope so that it captures the vacuum pulses for all cylinders. Observe the vacuum trace.

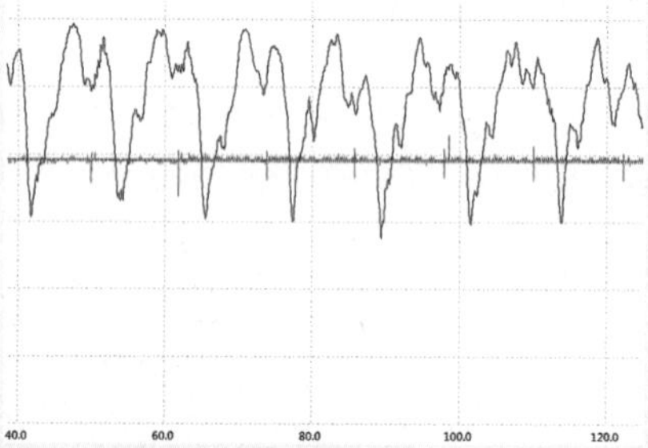

4. Snap accelerate the engine by opening and closing the throttle, and compare the trace to known good readings.

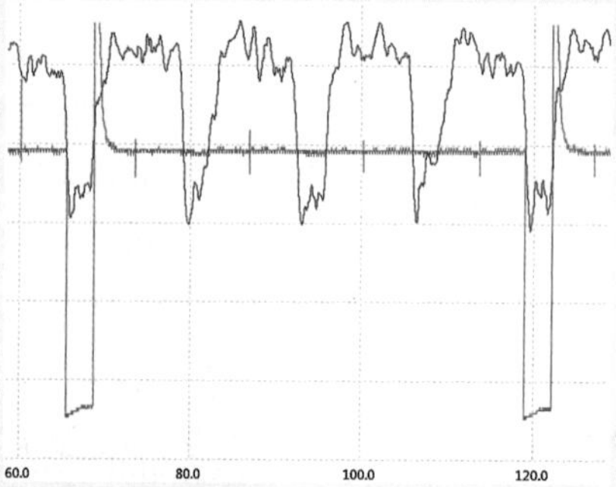

5. Hold the throttle steady at 1200–1500 rpm, and compare the trace to known good readings.

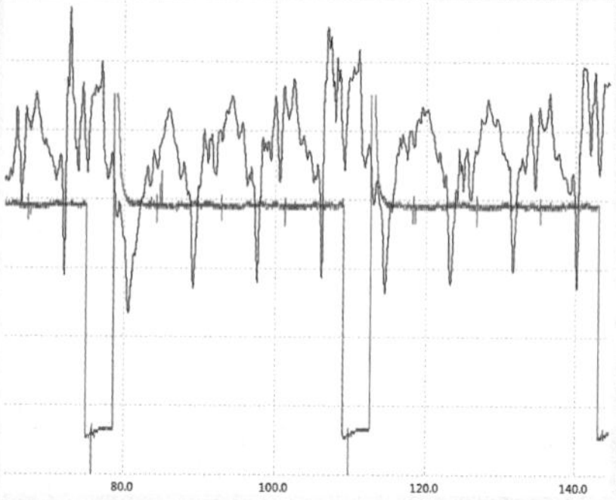

6. Then, hold the throttle steady at 2500 rpm, and compare the trace to known good readings.

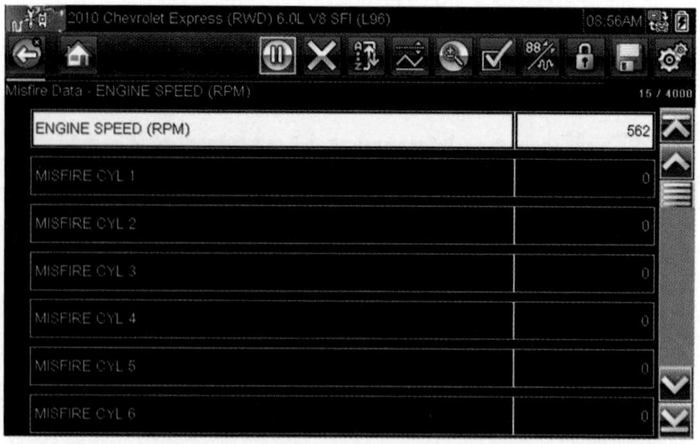

FIGURE 13-13 Cylinder power balance results from a good engine.

drop. The greater the difference in rpm drop, the greater the difference between each cylinder's ability to produce its share of the engine's power.

Many newer vehicles control the idle speed electronically through the powertrain control module (PCM). In these vehicles, if something loads the engine down, such as the air-conditioning compressor turning on, the PCM will compensate. It will allow more air and fuel into the engine so that the idle speed stays relatively the same. In this case, disabling one of the cylinders will not produce a drop in rpm because the PCM compensates for it.

In vehicles which automatically maintain idle rpm, you will need to use the PCM's built-in power balance feature, if equipped. Alternatively, in some vehicles, you can disconnect the electrical connector on the idle speed control system. This prevents the PCM from adjusting the idle speed. Unfortunately, in some vehicles, if you disconnect the idle speed control system, the engine will die. On these vehicles, you can usually perform the power balance test slightly above idle. This is because the PCM only controls engine speed at idle. Raising the rpm can be accomplished in two ways. First, by wedging an appropriate tool between the throttle body and the throttle linkage. And second by having an assistant hold the throttle steady while you perform the test.

Once the faulty cylinders are identified, you will need to take the next step to determine what is causing them to be low. There are two main causes. The first is a mechanical issue related to compression. And the second is an ignition- or fuel-related issue. But first, let's learn how to perform the power balance test.

Performing a Cylinder Power Balance Test

There are several ways that a power balance test can be performed. Knowing each of the options will allow you to choose the easiest one for the vehicle you are diagnosing. For many onboard diagnostic system generation II (OBDII) vehicles, you can identify which cylinders are misfiring by using a scan tool. The scan tool accesses any stored DTCs in the PCM. The DTCs will indicate which cylinders are misfiring, and how severe the misfire is (**FIGURE 13-14**).

Another option is to use the scan tool to access mode 6 data in the PCM. This will give you information on how many misfires happen on each cylinder (**FIGURE 13-15**). In some cases, you can also use the scan tool to command the PCM to perform an automatic power balance test. Once the test is complete, it even reports the results right on the scan tool. On most newer vehicles, using the scan tool is by far the best method of performing a power balance test.

If the system is not set up to perform the test automatically, you will have to do it manually. You will need to determine whether to disable the ignition or the fuel to each individual cylinder. If the engine has port fuel injectors that are accessible, disconnecting the electrical connector from each injector, one at a time, is the preferred method (**FIGURE 13-16**). This is because it stops fuel from being injected into the cylinder being tested. Otherwise, if the ignition system is disabled on a cylinder, fuel may still be injected. But it will not be burned in the cylinder. However, it will burn in the catalytic converter. This can cause it to overheat and possibly be damaged. Therefore, shutting down the fuel injector is preferable, if it is an option on the engine you are working on.

If the engine has individual ignition coils on each spark plug, you can disconnect the primary electrical connector from each ignition coil (**FIGURE 13-17**). This shuts off the spark to

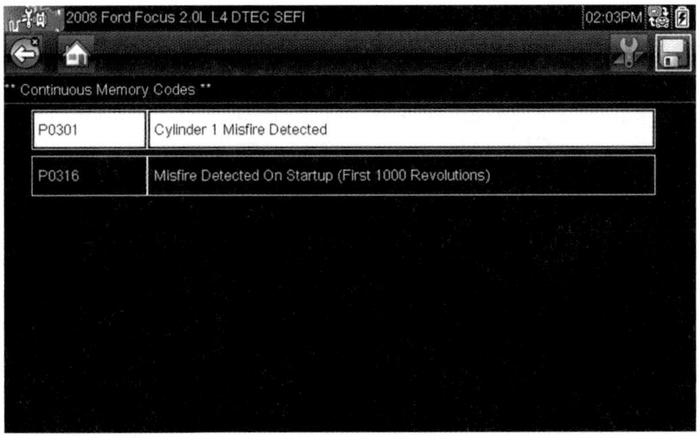

FIGURE 13-14 A scan tool can retrieve DTCs that indicate which cylinder is misfiring and the severity.

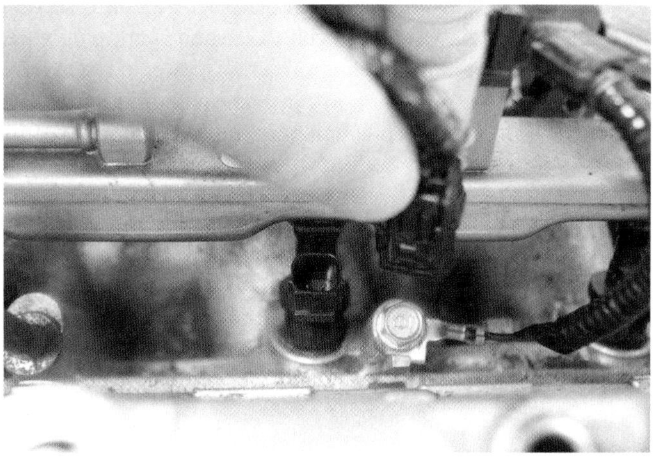

FIGURE 13-16 Disconnecting each fuel injector harness is a good way to disable cylinders one at a time.

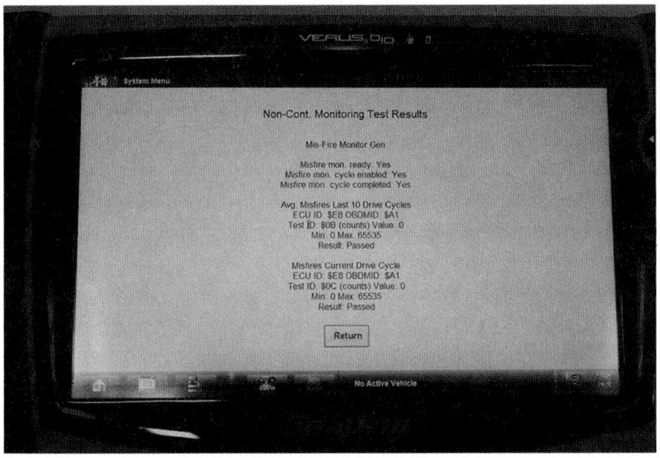

FIGURE 13-15 Using a scan tool to access mode6 showing cylinder misfires.

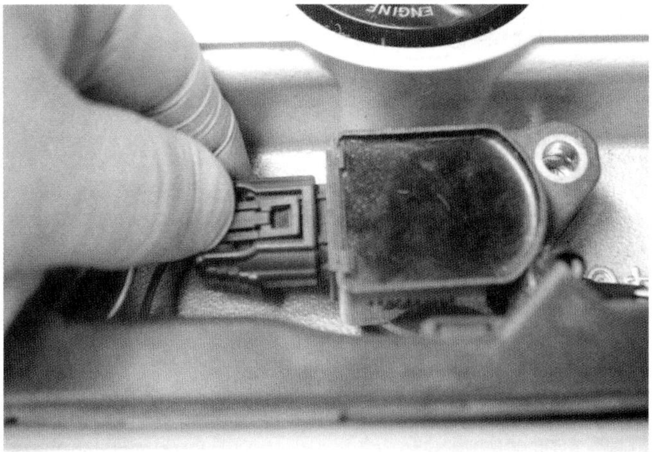

FIGURE 13-17 Disconnecting each ignition coil harness is a good way to disable cylinders one at a time.

FIGURE 13-18 Using a test light and vacuum hose to short out individual cylinders on waste spark ignition systems.

FIGURE 13-19 Disconnecting one spark plug wire at a time on a distributor can be used to disable cylinders.

that spark plug. If the vehicle has coils that share cylinders (waste spark system), then you can place a 1" (25 mm) section of vacuum hose between each coil tower and spark plug wire. Then, connect the alligator clip from a non-powered test light to a good engine ground. To short out each spark plug, touch the tip of the test light to each length of vacuum hose one at a time (**FIGURE 13-18**).

If the vehicle is equipped with a distributor, disconnect one spark plug wire at a time from the distributor cap (it is good to use a test lead to ground the spark at the distributor cap terminal to prevent the spark from damaging the ignition module inside the distributor) (**FIGURE 13-19**). This shuts down the spark for the cylinder being tested. To perform a cylinder power balance test, follow the steps in **SKILL DRILL 13-4**.

SKILL DRILL 13-4 Performing a Cylinder Power Balance Test

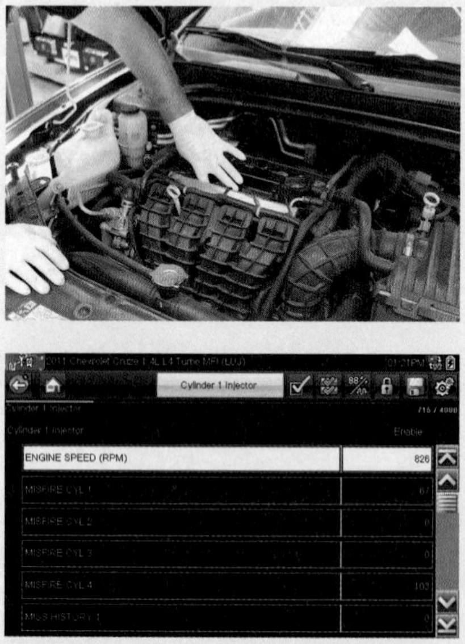

1. Visually inspect the engine to determine the best method to disable the cylinders. If necessary, disable the idle control system. Start the engine and allow it to idle. Record the idle rpm.

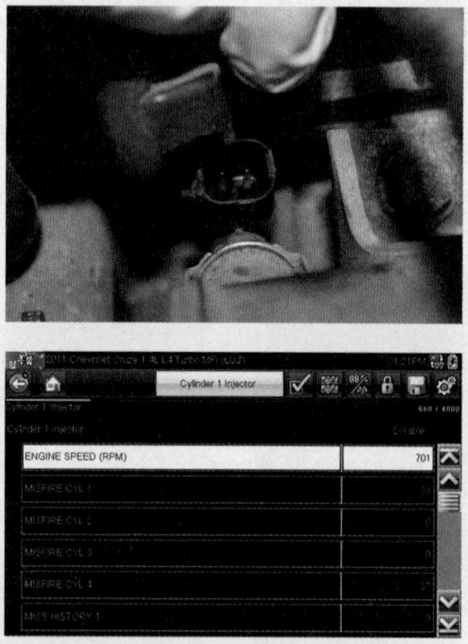

2. Using the method chosen to disable cylinders, disable the first cylinder, and record the rpm. (Do not leave the cylinder disabled for more than a few seconds.)

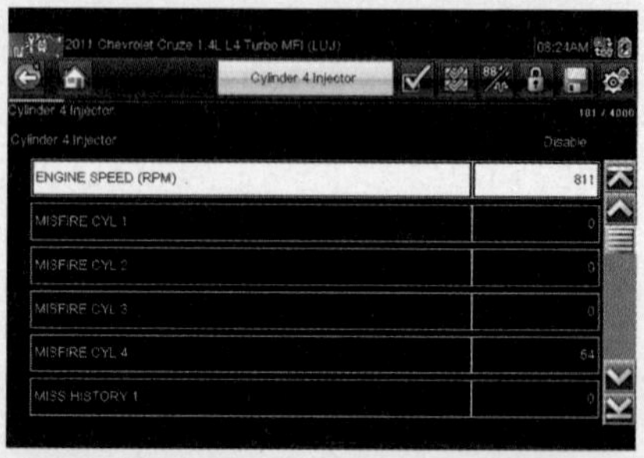

3. Reactivate the cylinder, and allow the engine to run for 10 seconds to stabilize. Repeat the steps on each of the cylinders, and record your readings. Determine any necessary action.

▶ Cranking and Running Compression Tests

LO 13-05 Evaluate the results of cylinder cranking compression tests.

The cranking compression test is performed to determine if a misfiring is caused by a compression problem. A compression test measures the air pressure as it is compressed in the cylinder (**FIGURE 13-20**). If the combustion chamber is sealed properly, the compression pressures in all cylinders should be within specifications. Typically, they should measure within 10% to 15% of each other, but check the specifications.

A cranking compression test indicates how well each combustion chamber is sealed. A running compression test indicates the engine's ability to breathe. This is also referred to as volumetric efficiency. You should always check factory specifications before performing these tests so that you know what results to expect.

In a cranking or a running compression test, a compression tester is installed in a spark plug hole. This is usually done by hand-threading a high-pressure hose into one spark plug hole. The hose is then connected to the compression gauge. The engine needs to be cranked over or started, depending on the test that is being performed. The compression gauge reads the amount of pressure produced in the cylinder. Once the reading is recorded, the pressure is vented from the compression gauge. And the gauge is ready for the next cylinder.

Performing a Cranking Compression Test

During a cranking compression test, the engine is cranked over, but not started. As the piston moves up, it compresses the air in the cylinder. To get an accurate reading, the engine should crank until at least five compression pulses are observed on the compression gauge. If the final reading is low, there could be a problem with the valves, rings, pistons, or head gasket. All of these are considered major engine problems.

FIGURE 13-20 Compression tester showing good compression.

For the compression test to be completely accurate, the following conditions should be met:

- The engine is at operating temperature.
- All of the spark plugs are removed.
- The battery is fully charged.
- The throttle is held wide open.
- At least five compression pulses are made on each cylinder (the same number of pulses for each cylinder).

Once you have accurate readings, they will need to be compared to specifications. Manufacturers list the specifications differently; some examples follow:

- As the minimum pressure allowed in psi or kPa, such as 125 psi (861.8 kPa) minimum
- As the maximum difference between cylinders in psi or kPa, such as maximum variation—35 psi (241.3 kPa)
- As the maximum difference between cylinders in percent, such as maximum variation—15%

The above situation is designed to test the compression in each of the engine's cylinders. But what if only one or two cylinders are suspected of having low compression (due to failing the power balance test)? Then, only the suspect spark plugs should be removed. A compression test is performed only on those cylinders. If the compression is reasonably close to the specifications, then the compression is not causing the issue. It is likely caused by a fuel or ignition fault.

When a low-compression cylinder is found, you should perform a wet compression test. This test is performed by placing a couple of squirts of clean engine oil in the low cylinder. And then retesting the compression. If the compression pressure rises significantly, the problem is typically worn piston rings in that cylinder. This is due to the short-term sealing ability of the added oil.

Worn piston rings are a substantial problem. In fact, it typically requires rebuilding or replacement of the engine. If the oil did not make the compression rise, then the problem is likely to be a leaky valve, head gasket, or a hole in the piston. A cylinder leakage test can determine which fault is causing the problem, and will be covered later.

A typical diagnostic test would be checking the compression of an engine after suspecting low compression. In this scenario, a customer is concerned about a four-cylinder engine that is running rough at idle. Fuel and ignition problems have been ruled out. A compression gauge will be used to measure the compression on each cylinder. The technician records the readings from each cylinder on the following chart. The technician then consults the manufacturer's specifications. From that, the technician will be able to determine whether the compression meets specifications. To perform a cranking compression test, follow the steps in **SKILL DRILL 13-5**.

Applied Math

AM-16: Charts/Tables/Graphs: The technician can construct a chart, table, or graph that depicts and compares a range of performance characteristics of various system operational conditions.

Cylinder	Cranking Compression in psi	Running Compression in psi
Cylinder1		
Cylinder2		
Cylinder3		
Cylinder4		

▶ TECHNICIAN TIP

Another way to measure compression is with a relative compression test. This test uses an inductive ammeter connected to a lab scope to measure the current flow for the starter as the engine is being cranked. As each cylinder comes up on the compression stroke, the engine is harder to turn. Because of this, the starter works harder, and the current flow is higher. If every cylinder has the same relative compression, then the current spikes will be similar. If a cylinder has low compression, then that current spike will be lower than the rest. This is a quick way of checking to see if there is a compression-related problem.

▶ TECHNICIAN TIP

When performing a compression test, the final reading is not the only thing to observe. If the piston rings are sealing well, then the first compression pulse on the gauge should be at least half as much pressure as the final reading.

Performing a Running Compression Test

In a running compression test, the engine will be running during the compression test. While the cranking compression test checks the sealing capability of the cylinder. The running compression test checks the engine's ability to move air into and out of the cylinder. This is referred to as the engine's ability

SKILL DRILL 13-5 Performing a Cranking Compression Test

1. Remove any spark plug wires or ignition coils connected directly to the spark plugs. Disable the ignition primary system, or ground the coil wire(s).

2. Disable the injectors, or remove the fuel pump fuse or relay.

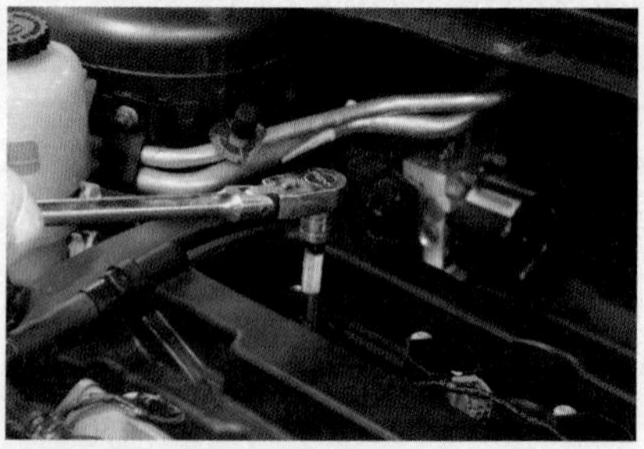

3. Remove all spark plugs.

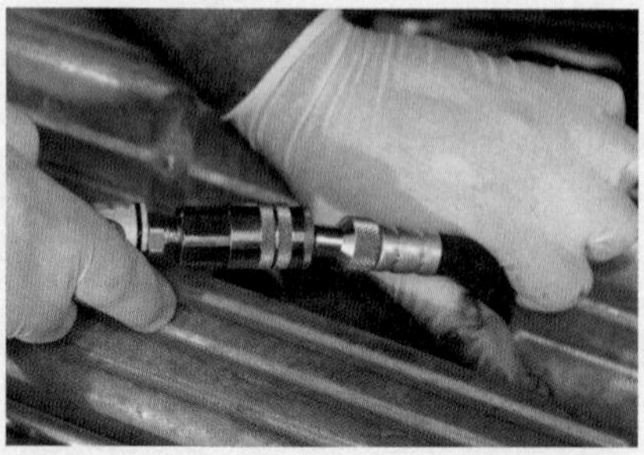

4. Connect the compression tester to the spark plug hole to be tested.

5. Hold the throttle down, and crank the engine over at least five pulses. Record the first and last needle readings. Repeat this procedure on the other cylinders.

6. Perform a wet test on any cylinders with low compression. Determine any necessary action.

to breathe. For example, if a camshaft lobe is badly worn, less air will enter or exit the cylinder, depending on which valve is affected. The running compression test helps a technician to evaluate this process.

The test is performed in two parts: idle and snap throttle. At idle, the running compression pressure will be approximately half of the cranking compression pressure. This is because the throttle is relatively closed, restricting airflow into the engine. The second part of the test is a snap throttle test. With the engine idling, the throttle is snapped open and then closed very quickly (about 1 second). This allows a big rush of air into the cylinders. The idea is to not make the rpm change very much during the test.

If the intake and exhaust system are operating correctly, then the compression tester needle will jump to about 80% of the cranking compression pressure. If the intake side of the system is restricted, then the reading will be lower than the 80% threshold. If the exhaust side of the system is restricted, then the pressure will be substantially higher than the 80% threshold.

The running compression test is performed by leaving all of the spark plugs in the engine except for the one in the cylinder that you are testing. Most technicians leave the Schrader valve in the compression tester while performing the running compression test. This holds pressure in the gauge, but it is hard on the Schrader valves. So, always have a couple of spares handy. Also, know that compression tester Schrader valves use lighter weight springs than tire Schrader valves. So do not interchange them.

This test can detect a number of faults. Some examples include flat cam lobes, broken valve springs or rocker arms, carboned-up valves, or restricted intake and exhaust passageways in general. If the running compression test indicates a fault, perform a visual inspection of the suspect parts.

To perform a running compression test, follow the steps in **SKILL DRILL 13-6**.

Applied Science

AS-6: Operational: The technician can relate scientific terms to automotive system diagnosis, service, and repair.

The use of pressure gauges is common in most automotive repair facilities. When performing engine mechanical tests, the technician uses compression gauges and cylinder leakage testers. These read in pounds per square inch (psi) or kilopascals (kPa). Psi refers to the primary units of measure for pressure in the United States. The metric unit is the kilopascal, or kPa. 10 psi is equal to approximately 68.95 kPa. Most manufacturers will supply both units as technical information. If necessary, the technician can consult conversion charts to obtain the unit that is needed.

▶ Cylinder Leakage Test Overview

LO 13-06 Evaluate the results of cylinder leakage tests.

The cylinder leakage test is performed on a cylinder with low compression. It is used to determine the severity of the compression leak and where the leak is located. Compressed air is applied to the cylinder through a tester. The tester is calibrated to show the amount of cylinder leakage as a percent of air entering the cylinder. An ideal reading is close to 0% (**FIGURE 13-21**). But because piston rings have a small end gap to allow for expansion as the engine heats up, a cylinder will not be sealed 100%. There is almost always at least a small amount of leakage past the piston ring gaps. Typically, manufacturers consider up to 20% cylinder leakage past the piston rings acceptable. But the smaller the leakage, the better the rings and cylinder wall.

SKILL DRILL 13-6 Performing a Running Compression Test

1. Remove the spark plug on the cylinder that you are testing, and ground the spark plug wire.

2. Install the proper hose and compression tester into the spark plug hole. Start the engine, allow it to idle, press and release the bleed valve, and record the reading.

3. Have your partner quickly snap the throttle open for about 1 second and then quickly close it. (Make sure the key can be turned off quickly if the throttle sticks.) Record the reading. Repeat the process on the other cylinders. Determine any necessary action.

Although it is okay to have a small amount of leakage past the piston rings, it is not okay to have any leakage past one of the valves or the head gasket. Leaks at these places mean the engine likely has a major mechanical engine issue. And it will likely need a substantial repair to fix it.

The point of this test is to do two things. The first is to measure how much air is leaking out of the cylinder. The second is to determine the location of the leak. The gauge tells you the percentage of air leaking from the cylinder, so that is straightforward.

Determining where the air is leaking from is a bit more challenging. Air leaking past the piston rings leaks to the crankcase. This can be heard by removing the oil fill cap and listening in the valve cover. Because there is always some air leaking past the piston rings, you will always be able to hear some air leaking there (**FIGURE 13-22**). If that is the only leak that you end up diagnosing, then the gauge will indicate if the leakage past the piston rings is excessive.

If an exhaust valve is burnt or warped, then you will hear leakage out of the exhaust pipe. If the intake valve is burnt or warped, then you will hear leakage out the intake system. There should be no leakage past either of the valves (**FIGURE 13-23**). So, any leak into the exhaust or intake is a bad leak.

If the head gasket is leaking, then you will hear air coming out from one of two places. If the head gasket is leaking to an adjacent cylinder, you will hear air from that spark plug hole. The head gasket could also be blown to a cooling system passage. If so, you will see the coolant level rise, or bubbles in the coolant, when looking in the surge tank or radiator.

It is also possible that the cylinder is leaking into the cooling system through a crack. The crack could be in the head or the block. But both of them show bubbles in the coolant. The problem is that cracks like that cannot be fixed in the car. So the engine would have to be rebuilt or replaced. This makes it important that if you see bubbles in the coolant during this test that you don't assume it is a blown head gasket. The customer will be unhappy if the problem is not solved after replacing the head gasket. They also won't be happy to hear that there are more problems than just a blown head gasket. Especially, if you didn't warn them of the possibility.

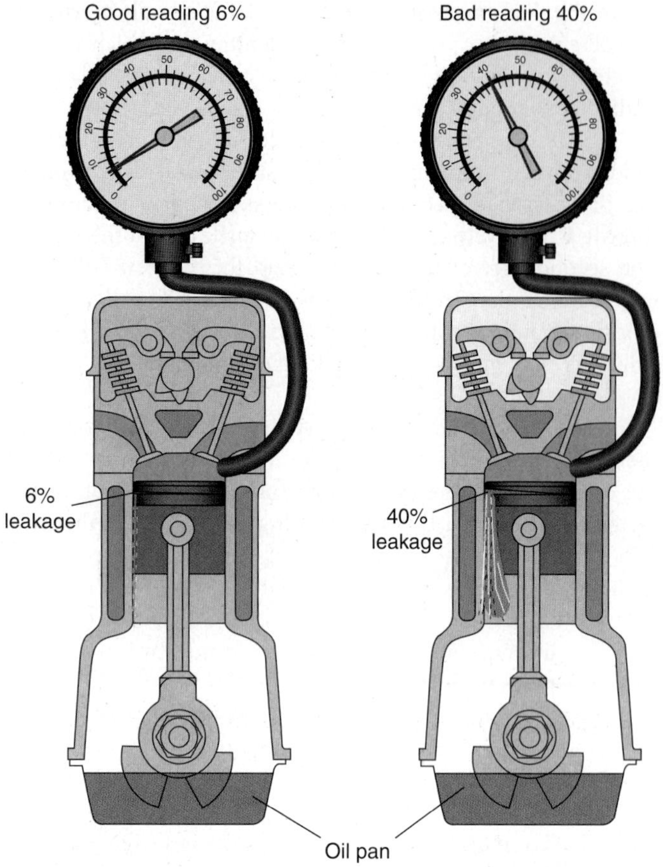

FIGURE 13-22 Cylinder leakage past the piston rings can be heard out the oil fill hole.

Before you can use the cylinder leakage tester, it needs to be calibrated. You generally do so by connecting only the air supply to the tester. Do NOT connect the hose going to the cylinder. Once the air supply is connected, adjust the knob so the gauge reads zero. That calibrates the gauge to the air system.

One mistake that students regularly make is to use the wrong adapter hose. It is easy to mistake a compression tester hose for a cylinder leakage tester hose. Unfortunately, they are different. The compression tester hose has a Shrader valve (**FIGURE 13-24**). The cylinder leakage hose does not have one. If you use a compression tester hose, the Shrader valve will not let air past it into the cylinder. This will give a cylinder leakage reading of zero. Since we know that piston rings leak, any reading of zero is suspect.

The way to verify that you have the correct hose is to unscrew it from the cylinder and hook it up to the cylinder leakage tester. It should read 100% leakage. If not, look to see if it has a Shrader valve in the spark plug end of the hose. If so, it is a compression tester hose and cannot be used.

FIGURE 13-21 A typical cylinder leakage reading on a good engine.

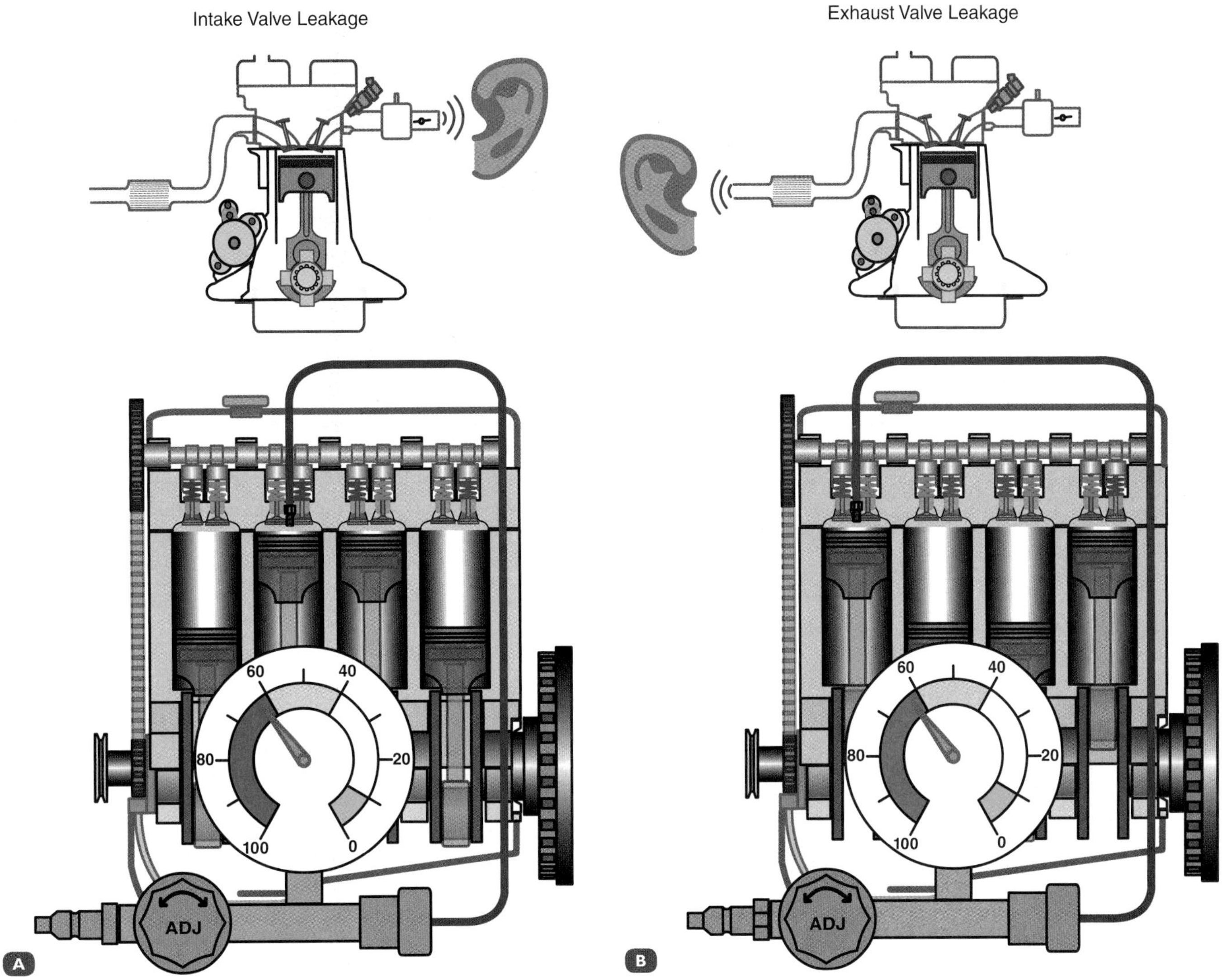

FIGURE 13-23 Leakage past the: **A.** Intake valve can be heard out of the throttle body. **B.** Exhaust valve can be heard out the exhaust pipe.

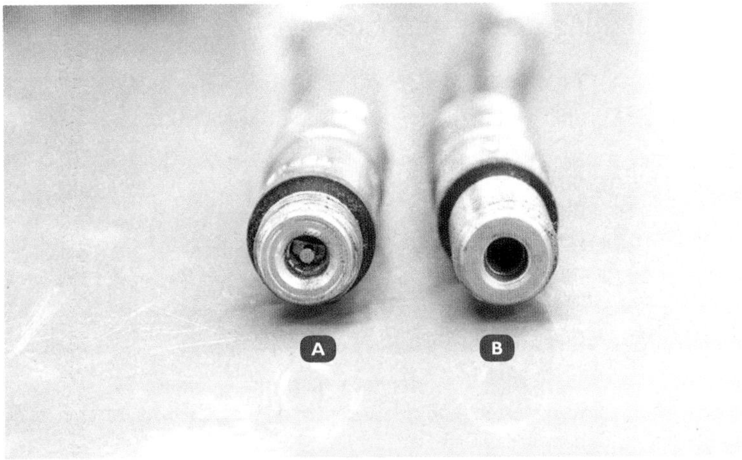

FIGURE 13-24 **A.** Compression tester hose has a Shrader valve. **B.** Cylinder leakage hose does not have a Shrader valve.

▶ TECHNICIAN TIP

There are a couple of ways that a technician can get in trouble when interpreting the results of a cylinder leakage test. First, if a valve is being held open, then the problem could be misdiagnosed. Valves can be held open by a piece of carbon, or not enough valve clearance.

To verify if the valve is being held open, remove the valve cover, and verify that the valve has the proper valve clearance. Also, try tapping the valve open with a soft hammer while watching the cylinder leakage gauge. If tapping on the valve stops the leakage, there was likely some carbon holding the valve slightly open. If the valve is being held open by the valve train, try adjusting the valve. Then, retest the valve for leakage. You should know that a valve that had too little valve lash is likely to be burnt in a very short amount of time. The longer it was operated with too little clearance, the more likely it is burnt.

Performing a Cylinder Leakage Test

There are a couple of critical steps needed to make sure the cylinder leakage test is accurate. First, the engine should be near operating temperature. This will ensure that oil has been circulated to the piston rings to help them seal.

Next, it is helpful to loosen each of the spark plugs about one turn for the cylinders you will be testing. Then, run the engine at 1500 rpm for 10 to 15 seconds with the plugs loose. This process helps blow out any chunks of carbon that break off when the spark plugs are removed. If you do not do this, it is possible for one of these chunks of carbon to get stuck between a valve and valve seat. This holds the valve open slightly and produces a false diagnosis of a burnt valve.

The cylinder leakage test is usually only performed on a cylinder with low compression. So, only remove the spark plug for the cylinder you are testing, and any spark plug next to that one. If the suspect cylinder is in the middle of the bank, then you will need to remove the spark plug on either side. If the suspect cylinder is at the end of the bank, then you will need to remove only the nearest spark plug. It also helps to remove the air cleaner assembly, oil filler cap, and radiator cap for listening purposes during the test.

Cylinder leakage is measured when the piston is on top dead center on the compression stroke. This means that you will have to turn the crankshaft to place the piston in this position before pressurizing the cylinder. This can be very challenging. The piston, connecting rod, and crankshaft throw must be in near-perfect alignment. Otherwise, the pressure on the piston from the cylinder leakage tester will push the piston down. This turns the crankshaft. If this happens, then the intake or exhaust valve will open, depending on which way the piston ends up turning the crankshaft.

The hardest part is getting the piston exactly on top dead center. There are two primary ways to do so. One is to screw the cylinder leakage tester hose into the spark plug hole. Then, slowly turn the engine over by hand while lightly floating your thumb over the end of the hose. You should be able to feel pressure and vacuum as the piston moves slowly up and down. When you feel the transition from pressure to vacuum, turn the engine in the opposite direction slightly. Stop right as the pressure stops and before the vacuum begins. It takes experience to get the feel for this.

The second way is to use a plastic straw that fits down the spark plug hole. The straw gets pushed up by the piston without damaging the cylinder or piston. Rotate the engine by hand until the piston is as high as it will go, as indicated by the plastic straw. While the piston is on top dead center, you will not know if it is on the top of the compression stroke or the exhaust stroke. You will need to pressurize the cylinder to see if it is leaking out of the intake and exhaust valves. If it does, turn the engine one complete revolution and try it again.

To perform a cylinder leakage test, follow the steps in **SKILL DRILL 13-7**.

SKILL DRILL 13-7 Performing a Cylinder Leakage Test

1. Remove the spark plug of the low-compression cylinder and any adjacent spark plugs. Install the cylinder leakage tester adapter hose into the spark plug hole.

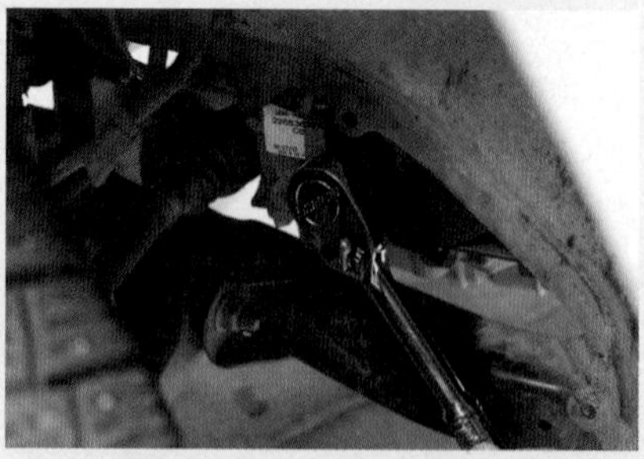

2. Position the piston for the cylinder being tested at top dead center on the compression stroke.

3. Connect the compressed air hose to the tester, and adjust the tester so it reads zero.

SKILL DRILL 13-7 Performing a Cylinder Leakage Test (Continued)

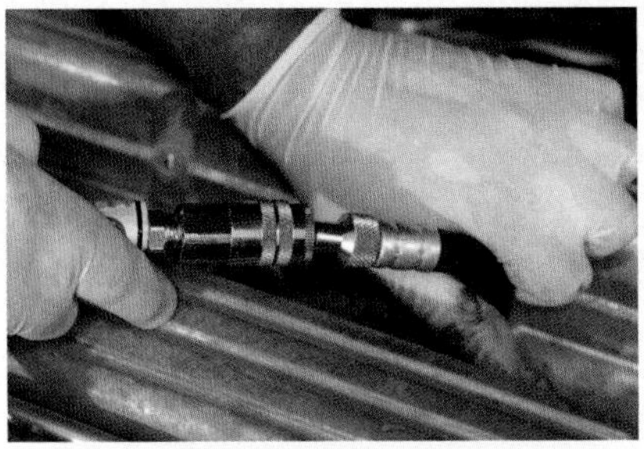

4. Connect the cylinder leakage adapter hose to the tester. Make sure the engine does not turn over. If the engine turns over, you will need to reset the piston back to the compression stroke. Record the reading.

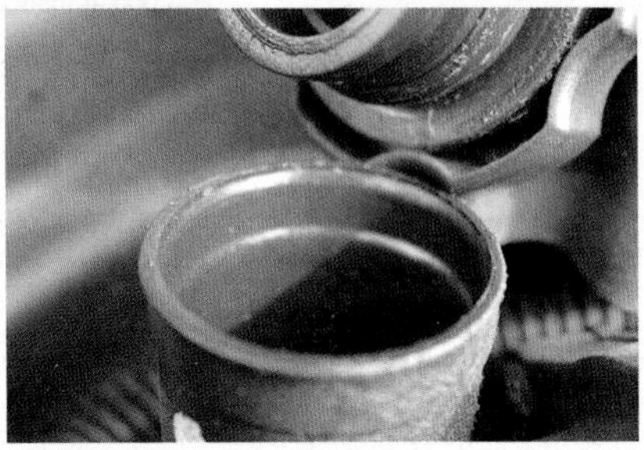

5. Listen for leakage from the oil fill hole, the throttle body, and the exhaust pipe.

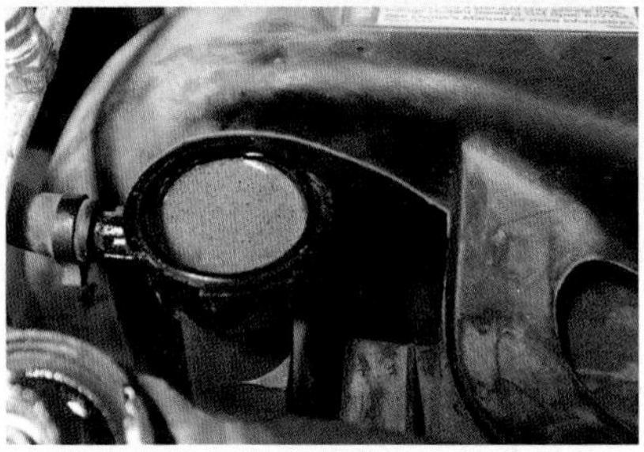

6. Look for bubbles in the surge tank or radiator. Determine any necessary action.

▶ Wrap-Up

Ready for Review

- Engine mechanical testing uses a series of tests to assess the mechanical condition of the engine. Mechanical testing starts with following the strategy-based diagnostic process: verifying customer's concern, researching possible faults and gathering information, focused testing, performing the repair, and verifying the repair.
- Engine noises can indicate the source of underlying problems. A mechanical or electronic stethoscope can help locate any type of engine noise.
  - Vibrations can be difficult to pinpoint. The best clue is to determine the frequency of the vibration.
- Intake manifold vacuum readings can be used to indicate an engine's general condition. A typical vacuum reading for a properly running engine at idle is a steady 17" to 21" of mercury.
  - A pressure transducer is a very accurate measuring tool and allows the vacuum to be displayed graphically on a lab scope.
- A cylinder power balance test allows the measurement of each cylinder's output by disabling one cylinder at a time and measuring the rpm drop. A scan tool can be used to perform a cylinder power balance test.
- A compression test measures the air pressure as it is compressed in the cylinder to determine if a misfiring is caused by a compression problem. A running compression test indicates the engine's ability to breathe.
  - For a compression test to be completely accurate: the engine should be at operating temperature, all of the spark plugs should be removed, the battery should be fully charged, the throttle should be held wide open, at least five compression pulses should be made on each cylinder.
- A cylinder leakage test determines the severity and location of the compression leak. Compressed air is applied to the cylinder through a tester to determine leakage.

Key Terms

data link connector (DLC) The under-dash connector through which the scan tool communicates to the vehicle's computers. The scan tool displays the readings from the various sensors and can retrieve trouble codes, freeze-frame data, and system monitor data.

pressure transducer An electrical device that creates an electrical signal based on a pressure input and displays it graphically on a lab scope.

Review Questions

1. What is the basic process a technician should use when diagnosing concerns?
   a. Input vs output process
   b. Strategy-based diagnostic process
   c. Research-based diagnostic process
   d. Concern-based process
2. When diagnosing an engine mechanical concern, there are a number of tools to aid a technician. What tool will help rule out a cylinder compression problem?
   a. Scan tool
   b. Vacuum gauge
   c. Stethoscope
   d. Pressure transducer

3. Engine vacuum can be a useful indicator of the engine's general condition. What would be the typical engine vacuum at engine idle?
   a. 17" to 21"
   b. 15" to 17"
   c. 12" to 15"
   d. 23" to 25"
4. When performing a cranking sound diagnosis, you can put some vehicles into "clear flood mode" by:
   a. Disconnecting one fuel injector at a time while the engine is idling.
   b. With your foot on the brake, turn the key on, off, then immediately crank the engine.
   c. With the key off, hold the throttle to the floor, then crank the engine.
   d. Disconnecting one ignition coil at a time while the engine is idling.
5. Which test would be used to help determine the cause of low compression on one cylinder?
   a. Wet compression test
   b. Power balance test
   c. Vacuum test
   d. Cranking sound diagnosis test
6. A cylinder leakage test can help a technician pinpoint the source of a leak. If air is heard leaking out the throttle body, which part would be leaking?
   a. Head gasket
   b. Piston rings
   c. Intake valve
   d. Exhaust valve
7. A cylinder leakage test is being performed on a cylinder with low compression. What position does the piston need to be in when performing the test?
   a. Bottom dead center
   b. Top dead center
   c. Intake stroke
   d. Any position
8. When performing a power balance test, each cylinder needs to be disabled individually to observe RPM drop. On most newer vehicles, which is the best method to use to do this?
   a. Use a scan tool
   b. Disable the injectors
   c. Disable the spark
   d. Use a compression tester
9. Performing a vacuum test on an engine can reveal a great deal about its condition. Which tool would provide the most accurate vacuum results?
   a. Cylinder leakage tester
   b. Analog vacuum gauge
   c. Pressure transducer
   d. Refractometer
10. When disabling a cylinder during a power balance test, what indicates a correctly operating cylinder?
    a. No RPM drop
    b. The engine speeds up
    c. A substantial RPM drop
    d. The engine runs smoothly

ASE Technician A/Technician B Style Questions

1. Technician A says that a cylinder leakage tester can help diagnose an engine mechanical fault. Technician B says that a compression tester can help determine an engine mechanical fault. Who is correct?
   a. Technician A
   b. Technician B
   c. Both A and B
   d. Neither A nor B
2. Technician A says that if the engine is running rough, listen to the starter when cranking the engine over. Technician B says that the starter will make an even cranking sound if a compression fault is present. Who is correct?
   a. Technician A
   b. Technician B
   c. Both A and B
   d. Neither A nor B
3. Technician A says that low vacuum at idle could be caused by a restricted air filter. Technician B says that a restricted exhaust could cause a low vacuum reading at 2500 rpm. Who is correct?
   a. Technician A
   b. Technician B
   c. Both A and B
   d. Neither A nor B
4. Technician A says that a running compression test determines the ability of the engine to move air in and out. Technician B says that a relative compression test uses the current draw of the starter while cranking to determine relative engine compression. Who is correct?
   a. Technician A
   b. Technician B
   c. Both A and B
   d. Neither A nor B
5. When performing a cylinder leakage test, a technician notices bubbles in the coolant. Technician A says that this could be an indication of a cracked head. Technician B says that this could be an indication of a bad head gasket. Who is correct?
   a. Technician A
   b. Technician B
   c. Both A and B
   d. Neither A nor B

6. A technician could perform a wet compression test to narrow down the cause of low compression. Technician A says that if the compression stays the same, there could be worn piston rings. Technician B says that if the compression rises, the issue is with the valves. Who is correct?
   a. Technician A
   b. Technician B
   c. Both A and B
   d. Neither A nor B
7. When performing a power balance test, the cylinder being tested must be disabled. Technician A says to disconnect the fuel injector for the cylinder being tested. Technician B Says to disable the spark for the cylinder being tested. Who is correct?
   a. Technician A
   b. Technician B
   c. Both A and B
   d. Neither A nor B
8. Two techs are discussing running compression tests. Technician A says that the test involves quickly snapping the throttle open while watching the compression gauge. Technician B says that the test involves squirting oil in the cylinder just before performing the test. Who is correct?
   a. Technician A
   b. Technician B
   c. Both A and B
   d. Neither A nor B
9. Two techs are discussing a compression test. Technician A says that the throttle plate should be held wide open during the test. Technician B says that the vehicle's battery should be fully charged. Who is correct?
   a. Technician A
   b. Technician B
   c. Both A and B
   d. Neither A nor B
10. Two techs are discussing a cylinder leakage test. Technician A says that some leakage past the valves is normal. Technician B says that some leakage past the piston rings is normal. Who is correct?
   a. Technician A
   b. Technician B
   c. Both A and B
   d. Neither A nor B

CHAPTER 14

Lubrication System Theory

Learning Objectives

- **LO 14-01** Describe the functions of lubricating oil.
- **LO 14-02** Describe the common types of oil and their additives.
- **LO 14-03** Identify and describe lubrication system components.
- **LO 14-04** Describe oil-certifying bodies and their rating standards.
- **LO 14-05** Describe oil indicators and warning systems.
- **LO 14-06** Identify and describe the types of lubrication systems.

ASE Education Foundation Tasks

See Appendix A to view the 2017 ASE Education Foundation Automobile Accreditation Task List Correlation Guide.

▶ Introduction

Machinery like our automobiles relies on lubrication to keep the moving parts from wearing out quickly. Lubricating oil is used for that purpose. It is processed from crude oil in a refinery. The process of refining crude oil also creates gasoline, diesel, and many other beneficial products. Lubricating oil is much more than simply crude oil dumped into our engine's crankcase. It is heavily processed to remove impurities. Its lubricating qualities are also enhanced by blending in many different additives.

Each moving part in the engine needs lubricating oil (**FIGURE 14-1**). The system that moves the oil through the engine is called the lubrication system. This chapter covers the theory and components of lubrication systems. It provides a solid foundation for the next chapter. That is where we will discuss servicing the lubrication system.

▶ Oil

LO 14-01 Describe the functions of lubricating oil.

Oil originates from the ground as **crude oil** (**FIGURE 14-2**). Crude oil varies in color from a dirty yellow to dark brown, to black. It can be thin like gasoline or a thick oil, or tarlike substance. Crude oil is pumped from the ground and processed into many products such as fuel for use in diesel and gasoline vehicles. Crude oil is also broken down into other products. These are used in plastics, kerosene, aviation fuel, asphalt, cosmetics, pharmaceuticals, and many other products.

Many of the products refined from crude oil are used in the transportation industry. For example, **lubricating oil** is distilled from the crude oil and used as a base stock. Additives are added to the base stock to make the lubricating oil useful in engines. Other additives, such as thickening agents, are added to the base stocks to make lubricating grease used in bearings. The additives that are added to the base stock perform a variety of tasks. A few of those are:

- preventing acid formations,
- cutting down the oxidation process, and
- maintaining the correct viscosity over a broad temperature range.

We cover those qualities in more depth later.

▶ TECHNICIAN TIP

Lubricating oil has been used since the invention of machinery. When metal moves on another piece of metal, the parts wear quickly in absence of lubrication. Lubricating oil also helps to quiet the moving parts and remove heat from metal surfaces.

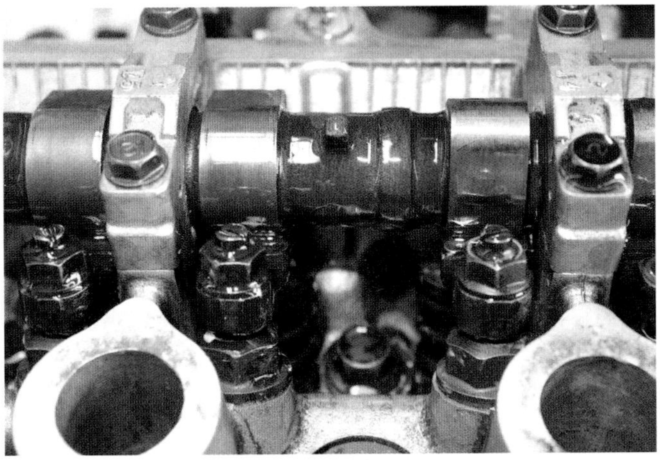

FIGURE 14-1 Oil lubricates components inside the engine.

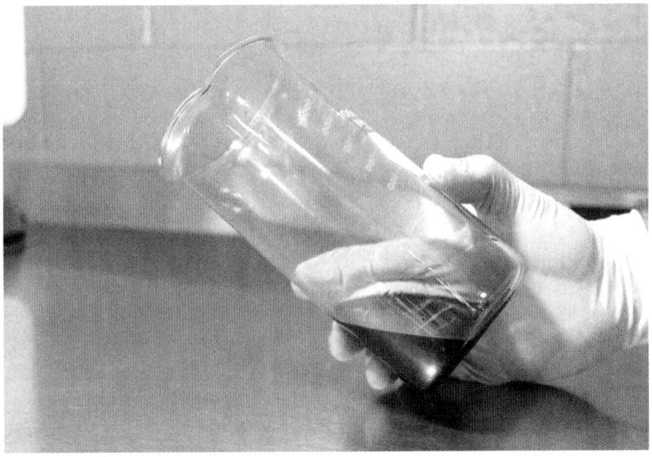

FIGURE 14-2 Crude oil straight from the ground.

You Are the Automotive Technician

A customer comes in needing a their patched on their vehicle. When the vehicle is brought into the shop, an inspection is performed, and the oil doesn't read on the dipstick. The customer tells you that a friend added two quarts to it when he borrowed it last week. It regularly needs that much or more oil. The customer says that they haven't been concerned about the oil level because the oil light is not on. They wants to know why the oil light didn't come on. They also want to know what oil does that is so important. How would you answer her following questions?

1. What functions does the oil perform inside an engine?
2. Why didn't the oil pressure light indicate that the oil was below the "add" line?
3. What is viscosity?
4. What does "W" stand for in 5W30 oil?

Functions of Lubricating Oil

Lubricating oil performs five main functions:

- Lubricates
- Cushions
- Cools
- Cleans
- Seals

Lubrication involves:

- reducing friction
- protecting against corrosion and
- preventing metal-to-metal contact between the moving surfaces.

Friction occurs between all surfaces that come into contact with each other. When moving surfaces come together, friction tends to slow them down. Friction can be useful, such as in a brake system. In the moving parts of engines, friction is a bad thing and will lead to serious damage. Friction can make metal parts so hot they melt and fuse together. When this happens, an engine is said to have seized and will need to be rebuilt or replaced.

Lubrication is the first function of oil. It reduces unwanted friction as well as wear on moving parts. Clearances, such as those between the crankshaft journal and crankshaft bearing, fill with lubricating oil. This allows the engine parts move or float on layers of oil instead of directly on each other (**FIGURE 14-3**). By reducing friction, less power is needed to move these components. And more of the engine's power can be used to turn the crankshaft instead of being wasted as heat. The result is increased power to move the vehicle, along with better fuel economy.

The second function of oil is cushioning. How long an engine lasts depends mostly on how well it is lubricated. This is especially true at the points of **extreme loading**, or high-wear areas, such as between the cam lobe and cam follower. At the same time, the connecting rod and crankshaft bearings take large amounts of stress. The pistons transfer thousands of pounds of force to the crankshaft each time a cylinder fires. The lubricating oil between the surfaces helps to cushion these shock loads. It is similar to the way a shock absorber absorbs a bump in the road.

Lubricating oil also helps cool an engine, which is the third function of oil. The lubricating oil collects heat from the engine's components and then returns to the **oil pan**, where it cools. The heat from the lubricating oil is picked up by the air moving over the oil pan. Many heavy-duty and high-performance vehicles have cooling fins on their oil pan. Others have a separate oil cooler to extract more heat from the oil. This helps the oil do an even better job of cooling critical engine components (**FIGURE 14-4**). When inspecting an engine for leaks, don't forget to check the oil cooler, if equipped.

Lubricating oil also works as a cleaning agent, which is the fourth function of oil. There are additives in the lubricating oil that allow it to collect particles of metal and carbon and carry them back to the oil pan. Larger pieces fall to the bottom of the oil pan. Smaller pieces are suspended in the oil and are removed when the oil moves through the oil filter. When oil is changed, most of the particles are removed with the oil filter and old oil.

The last function of oil is that it seals. It plays a key role in sealing the piston rings to the cylinder walls. Without a small film of oil between the rings and cylinder walls, blowby gases would be much higher. This results in diluted oil, lower compression, lower power, and lower fuel economy.

Knowing the five functions of oil will allow you to better diagnose and maintain customer's vehicles. So, remember them as we explore how the system works, as well as each component.

▶ Types of Oil

LO 14-02 Describe the common types of oil and their additives.

There are two main types of oil; conventional and synthetic. These two types can be blended together and are called synthetic blends. Each of these types has its pros and cons. So understanding the differences will be important for you to know as a technician.

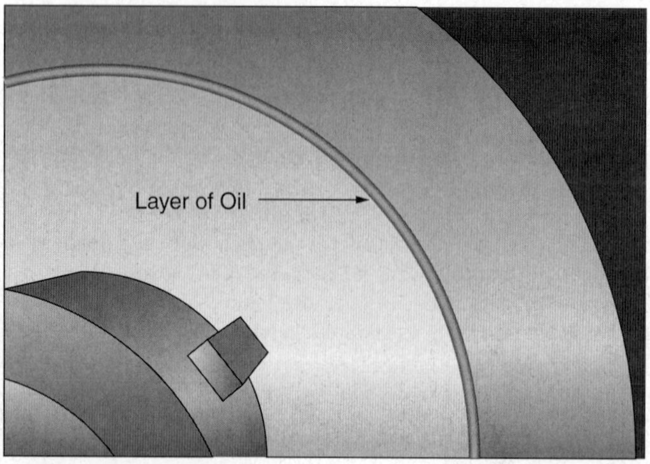

FIGURE 14-3 Clearances fill with lubricating oil so that engine parts move or float on layers of oil instead of directly on each other.

FIGURE 14-4 An oil cooler assists oil to do its job of cooling internal engine components.

Conventional Oil

Conventional oil is processed from crude oil pumped from the ground (**FIGURE 14-5**). The crude oil contains many impurities that are removed during the refining process. One of the impurities found in all crude oil is wax. This wax is removed during refining and is used for candle wax. It also serves as an additive in some food and candy. Wax is not a good thing in oil because it creates a thickening effect when it gets cold. If cold enough, it becomes too thick to flow through the engine.

Crude oil is broken down into mineral oil. It is then combined with additives to enhance the lubricating qualities. Without the additives, conventional oil would not work well. It would foam easily, break down quickly, and corrode the engine parts after being in the engine for a short time.

Synthetic Oil

There are two main categories of synthetic lubricating oils: type 3, which is not a true synthetic. And type 4 (polyalphaolefin [PAO]), which is a true **synthetic oil**. Both types of synthetics are more costly to manufacture. The base stocks of type 3 oils are more highly refined than conventional oil. And since the type 4 synthetic oils are developed in a lab, they are also more costly.

Synthetic lubricants have a number of advantages over conventional oils. They offer better protection against engine wear. They can operate at the higher temperatures needed by performance engines. They have better low-temperature viscosity. This allows the oil to be circulated through the engine more quickly during low-temperature engine start-ups.

FIGURE 14-5 Conventional oil is processed from crude oil pumped from the ground.

Synthetics have fewer wax impurities that coagulate at low temperatures. They are chemically more stable so they don't oxidize as easily. And they are generally thinner, and hence allow for closer tolerances in engine components without loss of lubrication. Modern high-performance engines run much tighter tolerances. This requires a thinner oil that is able to hold up under higher temperatures.

Some synthetics also last considerably longer, extending oil change intervals to 20,000 miles (30,000 km) or more. This means fewer oil changes, which benefits the environment by reducing the used oil stream. It also reduces the need for finding new sources of oil.

True synthetic oils are based on artificially made hydrocarbons, commonly **PAO** oil. This is an artificially made oil base stock—meaning it is not refined from crude oil. Synthetic oils were developed in Germany during World War II due to the lack of crude oil. Synthetic oil was used primarily in jet engines because of their high heat demands. Normal conventional oil created heavy carbon deposits on bearings due to the extreme heat. This led to bearing, and ultimately engine, failures. Amsoil was the first synthetic to be approved by the American Petroleum Institute (API) in 1972. Many companies now offer synthetic oils.

Very few synthetic oils on the market are full PAO synthetic oils (**FIGURE 14-6**). Many of the oils allowed to be labeled as synthetic are in fact blends of highly processed **mineral oil** and PAO. Or in some cases, just highly refined base stock that possesses lubrication qualities similar to PAOs (**FIGURE 14-7**).

Synthetic Blends

Synthetic blends are a mix of conventional high-quality oil and full synthetic oil (**FIGURE 14-8**). They provide some of the benefits of the full synthetic oil, and some of the cost-effectiveness of conventional oil. These oils need to be changed sooner than a full synthetic but less frequently than conventional oil. The purer the base stock is after the refinement process, the longer the oil will last in the engine.

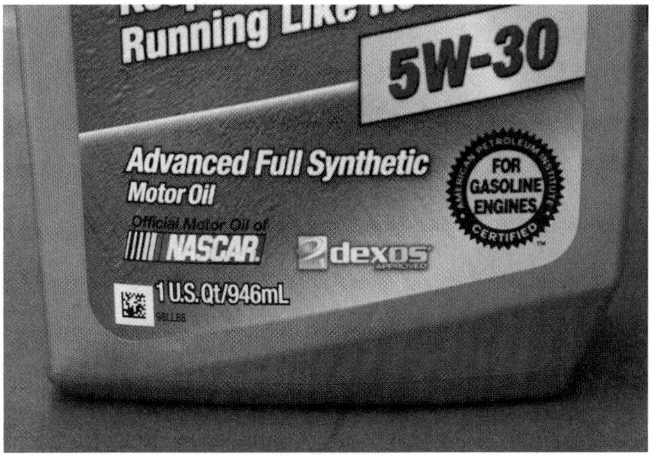

FIGURE 14-6 Full PAO synthetic oil.

FIGURE 14-7 Type 3 synthetic.

FIGURE 14-8 A typical synthetic blend motor oil.

Some manufacturers are now recommending synthetic blend oil over conventional oil. This is due to the better protection and performance of these types of oils. Because it is half synthetic, half conventional, the full benefit of the thinner, higher performance pure synthetic is diluted. But in turn, the conventional half is improved by adding oil that has no impurities. If the vehicle will be used hard, such as for towing, synthetic blends will perform better than conventional oil. This is because of the ability of the synthetic oil to stand up to the higher heat and load placed on the engine.

> **▶ TECHNICIAN TIP**
>
> Be sure to use at least the minimum recommended oil by the manufacturer. Manufacturers of engines spend a lot of money and time designing engines that are efficient and long-lasting. But that can only happen if their recommendations are followed.

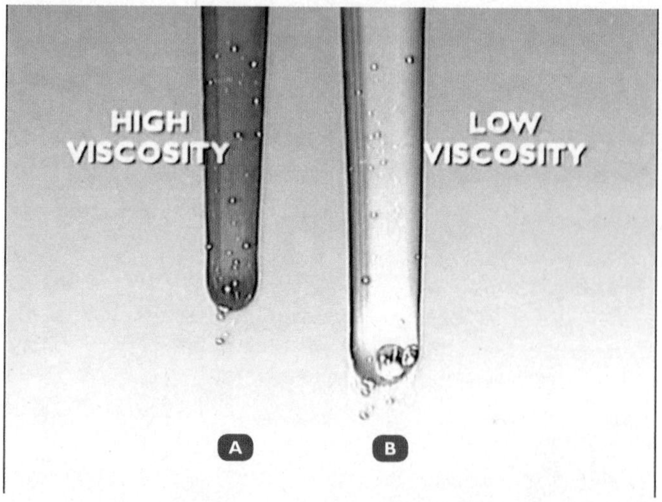

FIGURE 14-9 **A.** Oil with high viscosity is relatively thick. **B.** Oil with low viscosity is relatively thin.

Viscosity

For an oil to be able to cushion the engine parts, it needs to have the proper viscosity. **Viscosity** is a measure of how easily a liquid flows (**FIGURE 14-9**). Low-viscosity liquid is thin and flows easily. High-viscosity liquid is thick and flows slowly. Lubricating oil must be thin enough to circulate easily between moving parts, but not so thin that it will be squeezed out easily. If it is too viscous, it moves too slowly to get to the moving

Applied Science

AS-1 Inhibitors: The technician can explain the need for additives in automobile lubricants.

Automotive lubricants are composed of base stock plus an additive package. Additives are also used in lubricants for manual and automatic transmissions as well as differentials.

Oil additives are a very necessary part of modern lubricants. The improvements in additives are one of the factors that allow vehicles to last longer than ever before. Oil additives consist of chemical compounds that have many beneficial functions.

Detergents are additives that help keep the oil clean. Corrosion (or rust) inhibiting additives work to prevent oxidation of engine parts. According to Wikipedia, a corrosion inhibitor is a chemical compound that, when added to a liquid or gas, decreases the corrosion rates of a material. This applies typically to a metal or an alloy.

FIGURE 14-10 5W-30 multiviscosity engine oil.

parts, especially in a cold engine. As engine machining and metal technology have become more advanced, the clearances between lubricated parts have decreased. As a result, engine manufacturers have specified thinner oils for their engines. This is so that oil can flow into the smaller clearances. The thinner oil also flows more easily, which reduces drag and increases fuel economy.

Oil Additives

Special chemicals called additives are added to the base oil by the oil companies. Different combinations of these additives allow the oil to perform the different functions in an engine. A description of common additives follows:

Extreme-pressure additives coat parts with a protective layer so that the oil resists being forced out under heavy load. This improves the cushioning effect.

Oxidation inhibitors stop very hot oil from combining with oxygen in the air. This process produces a sticky tarlike material that coats parts and clogs the oil **galleries** and drain-back passages.

Corrosion inhibitors help stop acids from forming that cause corrosion, especially on bearing surfaces. Corrosion due to acid etches into bearing surfaces and causes premature wear of the bearings.

Antifoaming agents reduce the effect of oil churning in the crankcase and minimize foaming. Foaming allows air bubbles to form in the engine oil. This reduces the lubricating quality of oil and contributes to breakdown of the oil due to oxidation. Because air is compressible, oil with foam reduces the ability of the oil to keep the moving parts separated. This leads to more wear and friction.

Detergents reduce carbon deposits on parts such as piston rings and valves.

Dispersants collect particles that can block the system, separate them from each other, and keep them moving. They will be removed when the oil and filter are changed.

Pour point depressants keep oil from forming wax particles under cold-temperature operation. When wax crystals form, they result in the **gelling** of the oil. This makes it harder for oil to keep flowing during cold start-up conditions. Base stock derived from crude oil will not retain its viscosity (it thickens) if the temperature gets cold enough. So, viscosity improvers are used to modify the viscosity.

A **viscosity index improver** is an additive that helps to reduce the change in viscosity as the temperature of the oil changes. Viscosity index improvers primarily keep the engine oil from becoming too thin during hot operation.

Initially, oil was produced with a specific viscosity rating. For example, a common oil was 30 weight oil. That worked well when the engine was fully warm. But it was too thick to flow well when the engine was cold. Cold engines need oil with a lower viscosity (say 5 weight oil) so that the oil can flow well. Unfortunately, that oil is too thin once it, and the engine, warms up.

Multiviscosity oils, such as 5W-30 oils, overcome this issue (**FIGURE 14-10**). Viscosity improvers are added to a relatively thin oil (5 weight oil). With viscosity improvers, the oil doesn't thin out as much as it heats up. Once it is hot, it is similar to hot 30 weight oil. In other words, it is thin (like it should be) when it is cold. And doesn't thin out much when it heats up.

▶ TECHNICIAN TIP

Before multiviscosity oils, it was a normal practice for engines to need one grade of lubricating oil for summer and another for winter.

As stated above, one of the functions of lubrication is corrosion protection. Acids build up in the engine due to the accumulation of combustion by-products and moisture. These combustion by-products come from blowby gases which contain chemicals and moisture. The chemicals react and form acids. When the engine is turned off, it begins to cool. The moisture then condenses into droplets that fall into the oil and form acids. The acids attack the internal components, causing unnecessary damage. The oil contains anticorrosion additives that coat the engine surfaces. This helps protect them from the effects of the acids and water.

▶ Lubrication System Components

LO 14-03 Identify and describe lubrication system components.

The **lubrication system** is a series of engine components that work together to keep the moving parts inside an engine lubricated (**FIGURE 14-11**). Proper lubrication ensures that the engine:

- runs cooler
- produces maximum power and
- gets maximum fuel efficiency.

Lubrication also ensures that the engine will last for a long time. The lubrication system has many components that work together to deliver the oil to the correct locations in the engine. A typical lubrication system consists of an:

- **oil sump**, also called an oil pan
- an **oil pump strainer** (also called a pickup tube)
- an **oil pump**
- an **oil pressure regulator**
- **oil galleries**
- an oil filter and
- a low-pressure warning system.

The oil is stored in the oil sump, or oil pan. Oil is drawn from the oil pan through the oil pump strainer by the oil pump. The oil travels from the oil pump to the oil filter, which removes small particles from the oil. Oil moves from the filter to the oil galleries, which are small passages in the cylinder block and head(s). They direct oil to the moving parts. Oil that has been pumped to the crankshaft main bearings travels through oilways to the connecting rods. Oil may also be splashed from the connecting rods onto the cylinder walls. The circulation of the oil assists with the cooling of the internal parts.

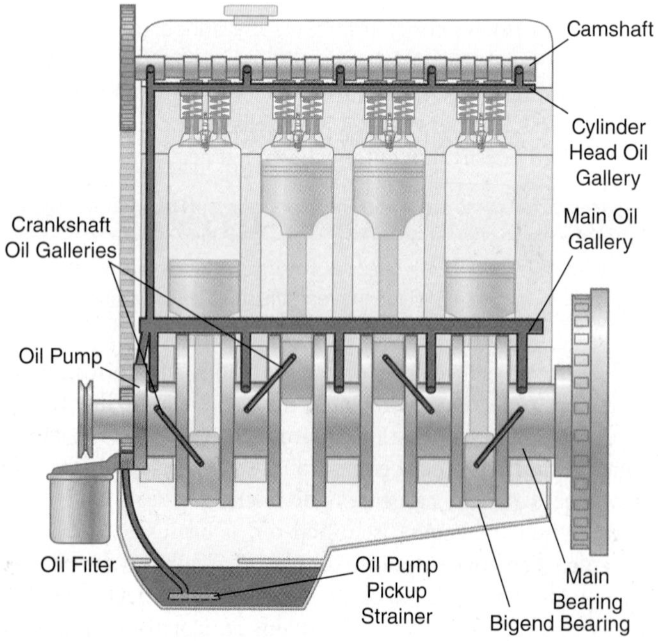

FIGURE 14-11 The lubrication system.

Wet Sump and Dry Sump Systems

There are two types of oil storage systems—the wet sump system and the dry sump system (**FIGURE 14-12**). In a wet sump lubrication system, the oil pan is a reservoir or storage container for the engine lubricating oil. It collects oil returning from the lubrication system. The oil pan is bolted to the bottom of the block and forms the lower portion of the crankcase. The oil pan is sealed to the engine with silicone sealer or an oil pan gasket. The oil pan is equipped with a drain plug. It allows the oil to be drained from the engine during oil changes (**FIGURE 14-13**).

The oil pan is formed from either stamped sheet metal or cast aluminum. It is shaped to ensure that oil will return to its deepest section. The oil pickup tube and strainer are located in this deep section. This ensures they stay submerged in oil and prevent air from being drawn into the oil pump. The oil pan's large external surface area helps heat transfer from the oil to the outside air. In some designs, the oil pan is an aluminum alloy casting with fins and ribs to assist in this heat transfer. A wet

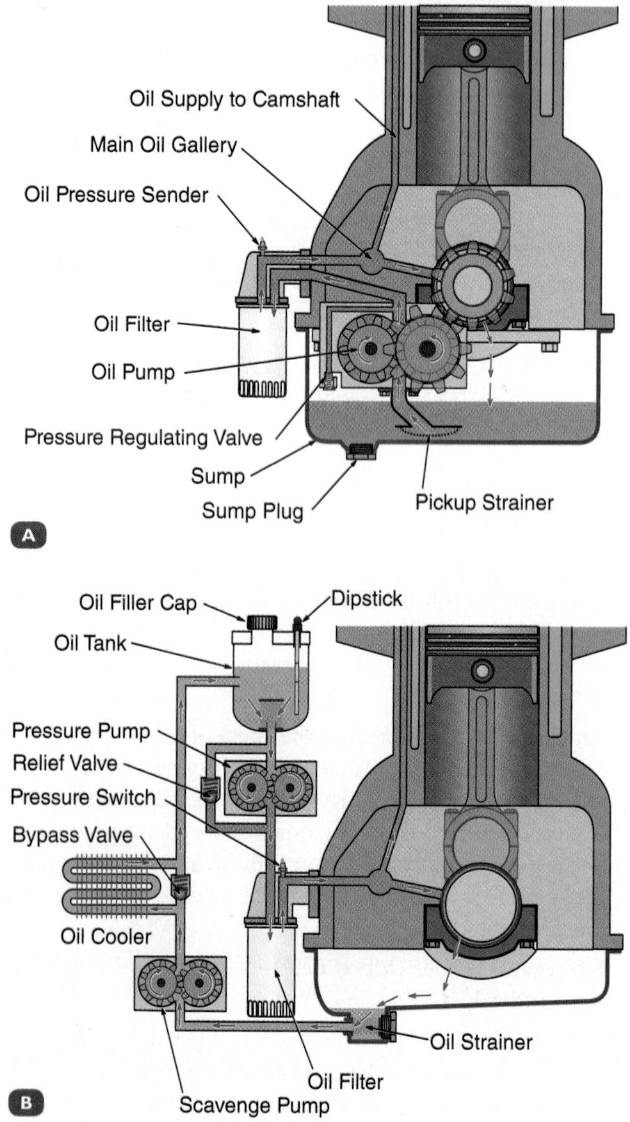

FIGURE 14-12 A. Wet sump. **B.** Dry sump.

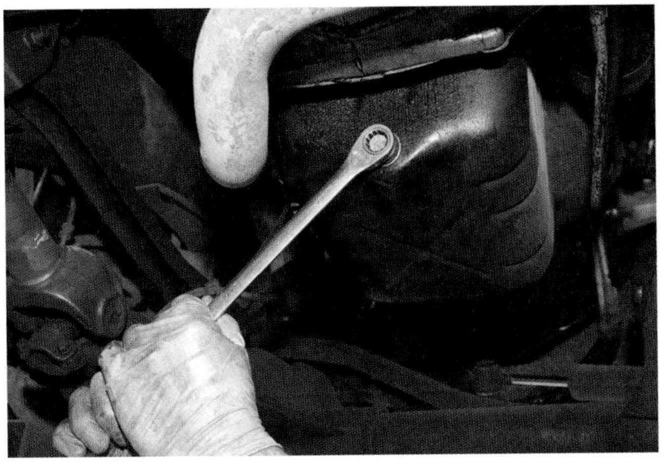

FIGURE 14-13 A drain plug allows oil to be drained during oil changes.

FIGURE 14-14 A windage tray.

sump system is used on most production vehicles because of its low cost and simplicity.

Typically, a dry sump system is used in high-performance applications where ground clearance or extra capacity is needed. A dry sump system uses a shallow oil collection pan, which is used only to collect oil, not store it. This shallower pan allows the engine to sit lower in the engine compartment. This lowers the vehicle's center of gravity for improved handling. The dry sump system also does a better job of providing oil to the oil pump during aggressive driving conditions.

In a dry sump system, the oil is not stored under the engine in an oil pan. The **scavenge pump** pulls oil from the collection pan and moves it to the oil reservoir located outside the engine. A pressure pump pulls engine oil from the oil tank, or reservoir, and delivers it to the engine at the proper pressure. Because no oil is stored in the engine in this system, a dipstick that is normally used to check oil levels in the engine is not needed. At the same time, the oil tank requires either a dipstick or a sight glass so that the oil level can be checked there.

Some high-performance vehicles have a **windage tray**, which is a shallow tray located close to the crankshaft, inside the oil pan. It is installed to prevent churning of the oil by the rotation of the crankshaft (**FIGURE 14-14**). The windage tray can be made from stamped sheet metal with slots cut in it. Or it can be made of a one-way mesh screen that allows oil to flow down through the mesh, but not back again. This design helps keep oil away from the rotating crank as much as possible. Drag on the spinning crankshaft is also reduced. It also prevents aeration of the oil.

Baffles are also used in some oil pans on high-performance vehicles (**FIGURE 14-15**). Baffles are flat pieces of sheet metal placed around the oil pump pickup in the oil pan. They prevent oil from surging away from the pickup during cornering, braking, and accelerating. Baffles can act as doors for oil flow if they are allowed to pivot. When the baffle moves one way, oil flows toward the pickup. When the baffle is moved the other way, the baffle closes and holds oil near the pickup. Baffles are commonly used on vehicles that experience strong g-forces. Examples are rally cars, stock cars, and drag-racing cars.

FIGURE 14-15 Oil pan with baffles to keep oil near the pickup screen.

Applied Math

AM-1: Volume: The technician can use various measurement techniques to determine the volume as applicable.

Two technicians are examining an engine oil pan as it is being cleaned. It is a standard automotive pan from a V6 engine and has a rectangular shape. Randy says that the volume of the pan could be calculated by multiplying the area of the base times the height. Tom agrees with this and adds that in the U.S. system of measurement, the result would be in cubic inches, which can be converted into quarts.

At break time, they decide to measure the pan and calculate its volume in quarts. The pan measures 6" wide, 9" long, and 5.35" deep. The area of the base is 6 × 9, which is 54 square inches. Next, they multiplied by the height, 5.35", to obtain 288.9 square inches. This is approximately 5 quarts (considering one quart is equal to 57.750 cubic inches).

To calculate the example in metric units, the process would be very similar except that we would be using metric units. Centimeters would be used for the linear units to determine the base × height. The cubic centimeters would then be converted to liters.

FIGURE 14-16 Pickup tube.

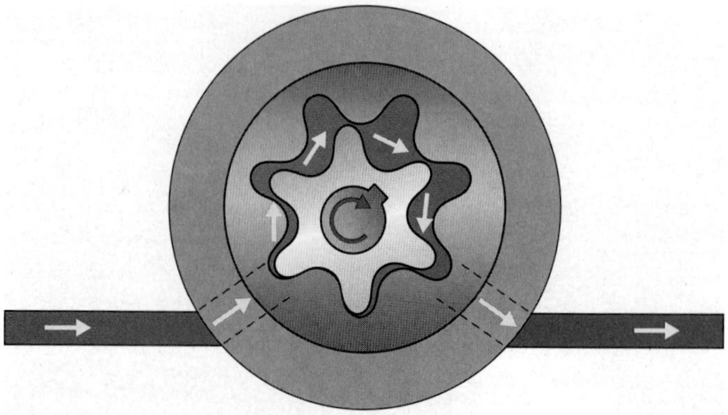

FIGURE 14-17 Rotor-type oil pump.

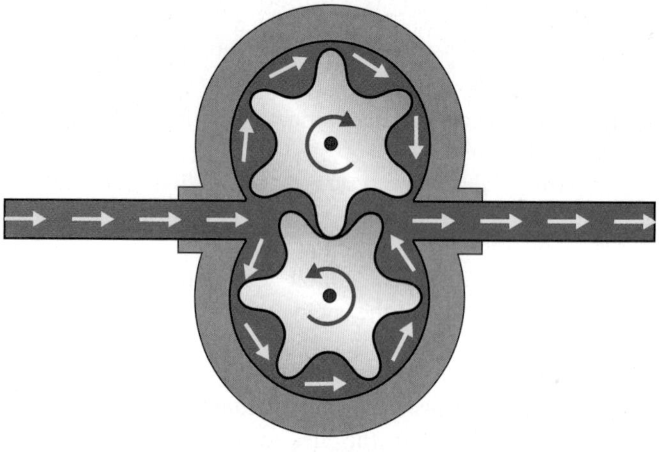

FIGURE 14-18 Geared oil pump.

FIGURE 14-19 Crescent pump.

Pickup Tube

Between the oil pan and oil pump is a **pickup tube** with a flat cup and a wire mesh strainer immersed in the oil (**FIGURE 14-16**). The pickup tube pulls oil from the oil pan by suction of the oil pump and atmospheric pressure. A strainer on the pickup tube stops large particles of debris from entering the oil pump and damaging it. The pickup tube leads to the inlet of the oil pump on the low-pressure side. The pickup tube is either bolted or press-fit to the oil pump. The joint must be sealed so that air is not drawn into the oil pump. The pickup tube commonly bolts to the block to hold it in place and resist vibration. If the pickup tube were to fall out, the engine would not receive oil. The pump would not reach down into the oil pan from which oil is drawn.

Oil Pump

Oil pumps force oil under pressure through the lubrication system. Pressurized oil is delivered to the parts that require lubrication. Oil pumps may be driven from the camshaft or the crankshaft. There are three main types of oil pumps: rotor style, gear style, and crescent style.

In a **rotor-type oil pump**, an inner rotor drives an outer rotor. As they turn, the volume between them increases (**FIGURE 14-17**). The larger volume created between the rotors lowers the pressure at the pump inlet, creating a vacuum. Outside atmospheric pressure, which is higher, forces oil into the pump. The oil fills the spaces between the **rotor lobes**. As the lobes of the inner rotor move into the spaces in the outer rotor, oil is squeezed out through the outlet port. In other words, oil is drawn into the spaces between the lobes on the inlet side and travels around with the lobes. The oil cannot get back to the inlet side because the lobes come together. Oil is therefore forced out of the pump outlet.

In a **geared oil pump**, the driving gear meshes with a second gear (**FIGURE 14-18**). As both gears turn, their teeth separate, creating a low-pressure area. Higher atmospheric pressure outside forces the oil up through the inlet. This fills the spaces between the gear teeth with oil. As the gears rotate, they carry oil around the chamber. As the teeth mesh again, oil is forced from the outlet into the oil gallery.

The **crescent pump** uses a similar principle (**FIGURE 14-19**). It is usually mounted on the front of the cylinder block and straddles the front of the crankshaft. The inner gear is then driven by the crankshaft directly. An external toothed gear meshes with the inner gear. Some gear teeth are meshed, but others are separated by the crescent-shaped part of the pump housing. The increasing volume between gear teeth causes pressure to fall, creating a vacuum. Atmospheric pressure pushes oil into the pump. Oil is then carried around between the gears and crescent. The gear teeth come together preventing oil from flowing past them. Oil is then forced out of the outlet port.

As you see, oil pumps move oil from one side of the pump to the other, and in doing so, force oil into the oil gallery. Most oil pumps are of the positive displacement design. This means that they move a given amount of oil from the inlet to the outlet each revolution. The faster the pump turns, the more the oil is pumped. But too high a speed can cause the pressure to be too high. This can erode bearings or cause oil leaks. So, oil pressure must be controlled within specifications. Oil pressure is determined by two factors:

1. The size of the leaks in the system. In the case of an engine, this means the amount of clearance between the bearings and the journals, and the diameter of any holes in the system (**FIGURE 14-20**).
2. The amount of oil flowing in the system, which is directly affected by the speed and size of the pump (**FIGURE 14-21**).

As you can imagine, the bearings provide a fairly consistent set of leaks. When the engine is new, the leaks are fairly small. When the engine has acquired many miles, the leaks are larger. This is why, over the life of the engine, engine oil pressure falls. In fact, low oil pressure can mean one of three things (other than a bad oil pressure gauge):

1. the oil leaks inside the engine have gotten excessive (e.g., worn bearings);
2. the oil pump is worn out and not creating as much flow as it needs to; or
3. the oil is thinner than it should be (e.g., saturated with gasoline from a leaky fuel injector), which causes it to pass through the leaks faster than it should.

So, anytime you run across low oil pressure, you should consider it to be a serious condition that needs to be diagnosed. Testing procedures are covered in the next chapter. But if the leaks inside the engine only allow a certain amount of oil to leak past them, what happens when the engine speeds up and the oil pump turns faster? If you said that the oil pressure increases, you would be correct. The next question is: What prevents the oil pressure from going too high? Keeping the engine speed low

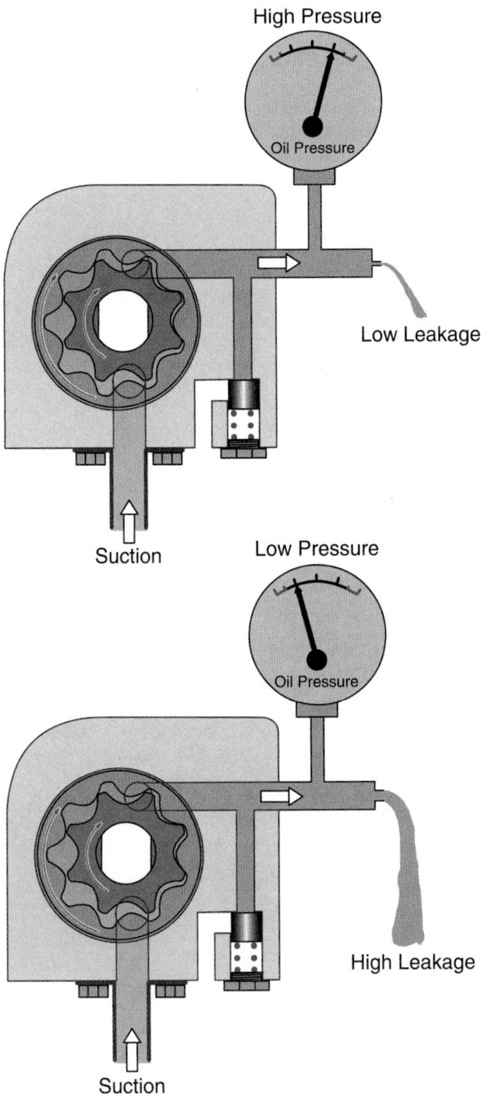

FIGURE 14-20 Oil pressure is affected by the size of the leaks in a system.

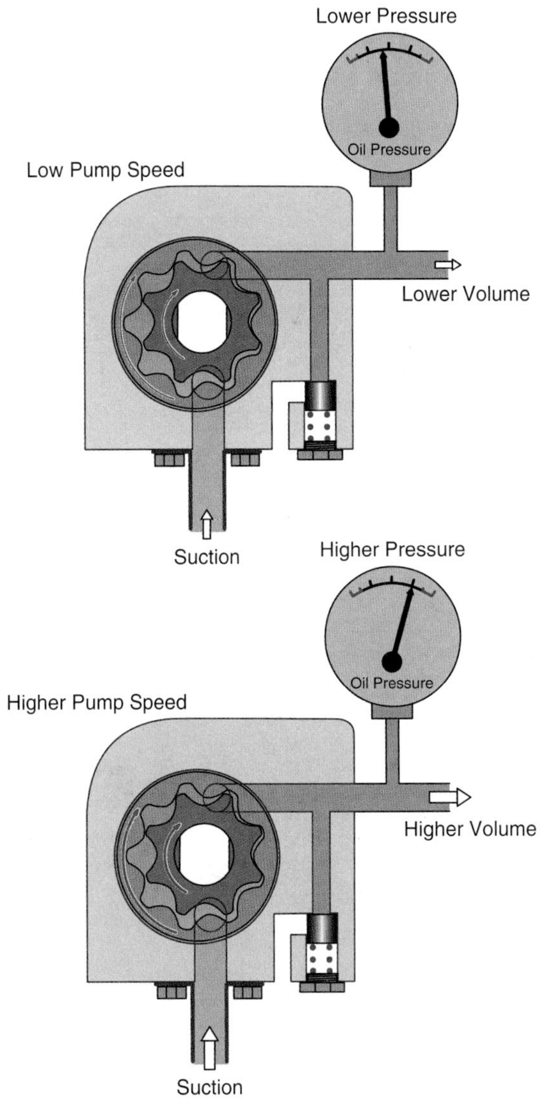

FIGURE 14-21 Oil pressure is affected by changes in the volume of oil flowing.

would be one answer, but this solution is not very practical. Stay tuned: We cover that issue next.

Oil Pressure Relief Valve

A normal oil pump is capable of delivering more oil than an engine needs. Extra volume provides a safety measure to ensure the engine is never starved for oil. As the oil pump rotates with increased engine speed, the volume of oil delivered also increases. The fixed clearances between the moving parts of the engine limit the amount of oil escaping. This causes the pressure to build up in the lubrication system.

An **oil pressure relief valve** stops excess pressure from developing. It is like a controlled leak. It releases just enough oil back to the oil pan to regulate the pressure in the whole system. The oil pressure relief valve contains a spring that is calibrated to a specific pressure (**FIGURE 14-22**). Oil pressure climbs and pushes on the relief valve. At the specified pressure, the oil pressure relief valve opens just enough to bleed sufficient oil back to the pan to maintain the regulated pressure. If the engine speed and oil flow increase, the pressure relief valve will open a bit farther. This allows more oil to escape back to the sump. If the engine speed and oil flow decrease, the pressure relief valve will close an appropriate amount.

> **▶ TECHNICIAN TIP**
>
> So, what is the difference between a high-pressure oil pump and a high-volume oil pump? A high-pressure pump has a stiffer pressure relief spring. It allows higher oil pressure to build in the system. A high-volume pump has greater volume between the rotor or gear teeth. This is usually accomplished by making both the rotor/gear and the oil pump housing deeper. More oil can then be drawn into the pump during each revolution. And, therefore, more oil is forced out of the pump each revolution. Generally speaking, a high-volume pump is more beneficial. Because it can pump more oil, the pressure will not fall as quickly as the engine experiences wear and tear.

Oil Filters

There are two basic oil-filtering systems: full-flow and bypass (**FIGURE 14-23**). The most common **full-flow filter systems** are designed to filter all of the oil before delivering it to the gallery.

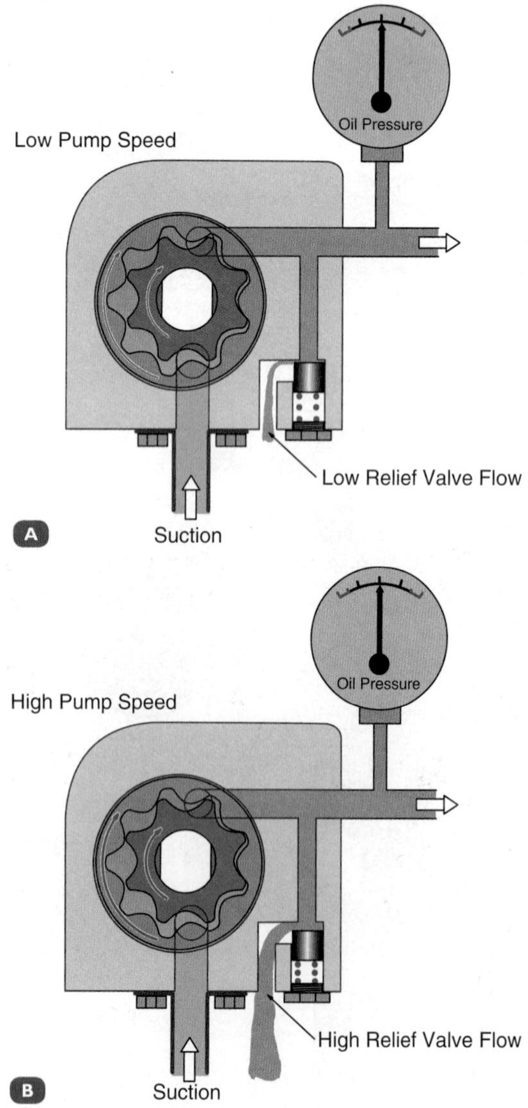

FIGURE 14-22 Oil pressure relief valve at: **A.** Low oil flow. **B.** High oil flow.

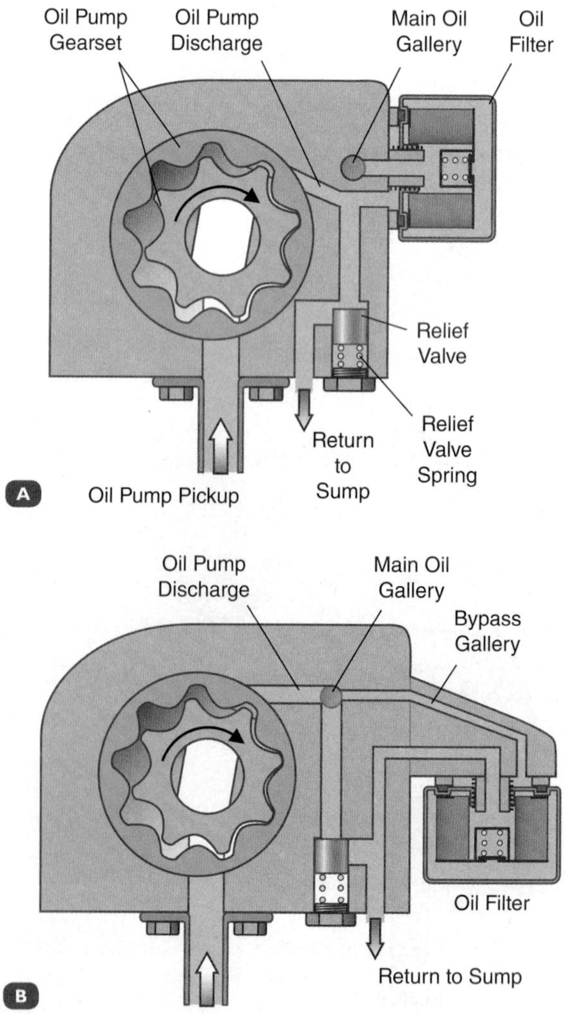

FIGURE 14-23 **A.** Full-flow filtering system. **B.** Bypass filtering system.

FIGURE 14-24 Pleated oil filter paper.

The filter is located right after the oil pump. This ensures that all of the oil is filtered before it is sent on. The bypass filtering system is more common on diesel engines. It is used in conjunction with a full-flow filtering system. The **bypass filter system** is discussed later in this section. Note that a bypass filtering system is different than the bypass valve in a regular oil filter.

Oil filters use a pleated filter paper for the filtering medium (**FIGURE 14-24**). Oil flows through the paper, and as it does so, it filters out particles in the oil. Most full-flow oil filters catch particles down to 30 microns. A micron is 0.000039" (0.001 mm)—a very small particle. A human hair's thickness can be as small as 50 microns, for example. As the oil filter catches these fine particles, the paper filter element begins to clog, making it harder for the oil to flow through.

During cold engine starts, or if the filter becomes clogged, oil flows through a bypass valve. The valve opens to let unfiltered oil flow past the filter to the lubricated components. The manufacturers believe it is better to have unfiltered oil flow to components than no oil at all. To prevent excessive engine wear, it is critical to change the oil filter at the manufacturer's recommended interval.

There are two common types of oil filters: spin-on and cartridge (**FIGURE 14-25**). The spin-on type is the most common. It uses a one-piece filter assembly with a crimped housing and threaded base. The pleated paper filter element is formed into the inside of the crimped housing. This kind of filter spins off with the use of an oil filter wrench and tightens by hand force only.

A square-cut rubber O-ring fits into a groove in the base of the filter. The groove holds the O-ring in place so that it can seal the base of the filter to the engine block. A new O-ring comes with the filter, so it gets replaced with the filter. Be aware, though, that the old O-ring may stick to the filter adapter on the engine block. If you do not notice this, the old O-ring will be in the way of the new one. The old O-ring will not stay in place because it is not in the filter groove. As a result, it will get pushed out of place when the engine is started. This will cause a large oil leak. Most, if not all, of the engine oil will be pumped out onto the ground. Always check for the old O-ring when removing the oil filter.

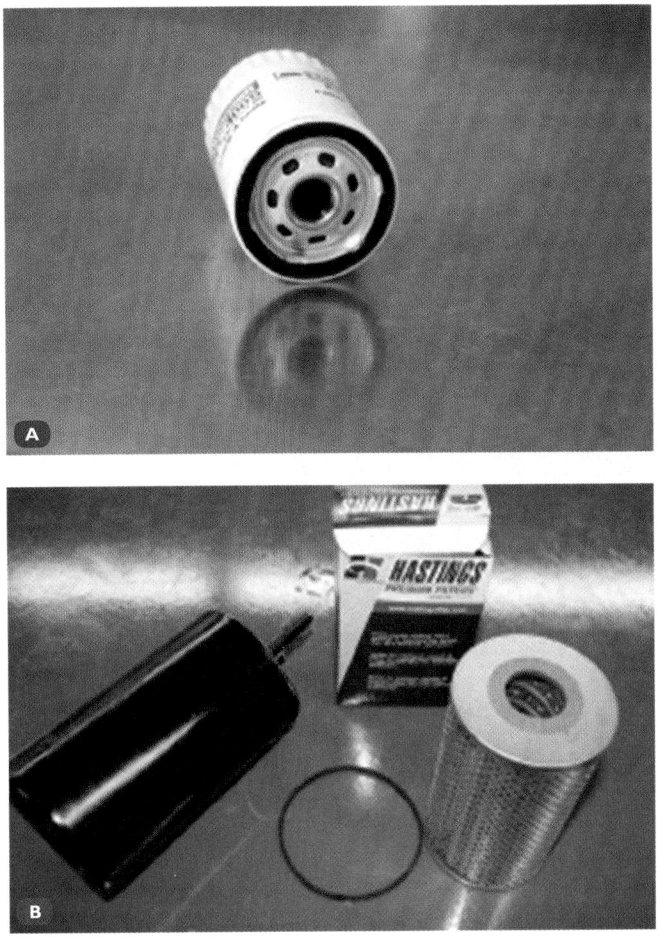

FIGURE 14-25 **A.** Spin-on filter and O-ring. **B.** Cartridge paper filter, housing, and bolt.

The cartridge style of oil filter uses a separate reusable metal or plastic housing and a replaceable filter cartridge. It is typically held together in one of two ways: a threaded center bolt or screw-on housing. If it uses a center bolt, it will have a sealing washer between the bolt and the housing to prevent oil from leaking out. There is also a seal that fits in one of two ways. Either between the cylindrical housing and the filter adapter on the block. Or between the cylindrical housing and the end cap. Cartridge filters must be disassembled during oil changes. Once disassembled, the housing is cleaned, and the paper filter element and any O-rings or seals are replaced with new ones.

Most oil filters on diesel engines are larger than those on similar gasoline engines. And some diesel engines have two oil filters. Diesel engines produce more carbon particles than gasoline engines. Because of this, engines can have a full-flow element to trap larger impurities and a bypass element to collect sludge and carbon soot.

In a bypass system, the bypass element filters only some of the oil from the oil pump. An oil line is tapped into the oil gallery. It collects finer particles than a full-flow filter. After the oil is filtered, it is returned to the oil pan. If the bypass filter were to clog and stop oil flow, the flow of oil lubricating the engine components would not be affected.

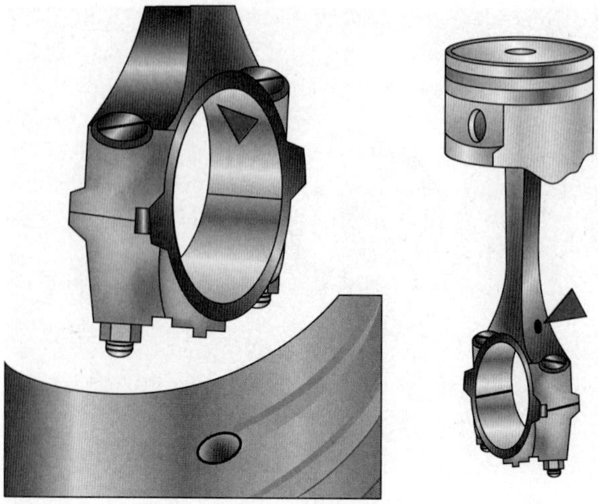

FIGURE 14-26 Some connecting rods have oil spurt holes, which are positioned to receive oil from similar holes in the crankshaft.

> **TECHNICIAN TIP**
>
> Magnets are also used as a type of filter. They attract ferrous metal particles and hold them in place until they can be cleaned off. Some manufacturers use magnetic drain plugs, which then must be inspected and cleaned off as part of an oil change. Others place a magnet to the inside or outside of the oil pan. Although this style cannot be readily cleaned, it does hold the magnetic particles in place so they cannot travel freely.

Spurt Holes and Oil Galleries

Pistons, rings, and pins are lubricated by oil thrown onto the cylinder walls from the connecting rod bearings. Some connecting rods have **oil spurt holes** that are positioned to receive oil from similar holes in the crankshaft (**FIGURE 14-26**). Oil can then spurt out at the point in the engine cycle when the largest area of cylinder wall is exposed. This oil lubricates the cylinder walls and piston wrist pin. It may also help cool the underside of the piston.

Engines with timing chains typically use one or more spurt holes that spray oil directly on the chain. Some heavy-duty engines have oil nozzles that spray oil up onto the bottom side of the piston. This is used to cool the piston and prevent it from overheating and melting down. Oil nozzles have fairly small holes that can become plugged by old, dirty oil fairly easily. So, regular oil changes are critical.

Oil is fed to the cylinder head through oil galleries and on to the camshaft bearings and valve train. When oil reaches the top of the cylinder head, it lubricates the valve train. At this point, it has completed its pressurized journey. The oil drains back to the oil pan through oil drain-back holes located in the cylinder head and engine block.

> **TECHNICIAN TIP**
>
> In many cases, the ends of the oil galleries are plugged with a threaded pipe plug or a small soft plug. During an engine rebuild, these plugs are normally removed and the galleries cleaned with stiff wire brushes. It is very important to make sure the plugs have been reinstalled with a sealer and tightened properly so that they do not leak. Many a technician has started up a newly rebuilt engine and had oil pour out of the bell housing. This being due to one or more of the oil gallery plugs that were not reinstalled.

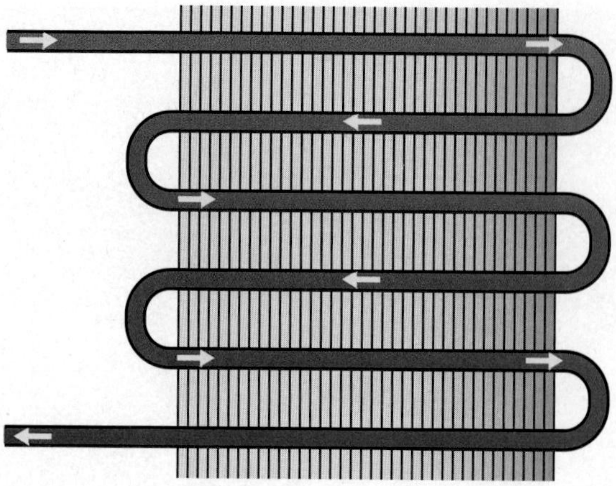

FIGURE 14-27 An oil cooler system.

Oil Coolers

Engines that operate under severe conditions may use an **oil cooler** to cool the oil in the engine. There are two types of oil coolers: oil-to-water coolers and oil-to-air coolers. In some engines, an oil-to-water oil cooler and the oil filter are on the same mounting on the cylinder block (**FIGURE 14-27**).

An oil-to-water oil cooler acts as a heat exchanger. It works by transferring heat from the oil to the coolant in the cooling system. One fluid circulates through tubes in the cooler, and the other fluid surrounds the tubes. As the coolant circulates, heat is removed from the oil.

In the oil-to-air design, the oil cooler is mounted in the airstream at the front of the vehicle. This type of oil cooler uses the flow of air passing across its fins to cool the oil circulating through it, similar to a radiator. Most vehicles equipped with a tow package will have one of these two types of oil coolers. High-performance vehicles that will be operated at high speeds and/or temperatures can also have a factory-installed oil cooler. They can also be added as an aftermarket accessory.

▶ Oil-Certifying Bodies and Their Rating Standards

LO 14-04 Describe oil-certifying bodies and their rating standards.

There are several certifying bodies for engine oil, each with its own standards. The three most common are:

- the American Petroleum Institute (API),
- the American Society of Automotive Engineers (SAE), and
- the International Lubricant Standardization and Approval Committee (ILSAC).

However, there are three others that technicians must be aware of:

- the Japanese Automotive Standards Organization (JASO);
- the Association des Constructeurs Européensd' Automobiles (ACEA), also called the European Automobile Manufacturers Association; and
- the original equipment manufacturers' (OEM) own standards.

Let's look at them one at a time.

American Petroleum Institute (API)

The API sets minimum performance standards for lubricants, including engine oils. The API has a two-part classification: service class and service standard. The API service class has two general classifications:

- S for spark ignition engines and
- C for compression ignition engines, also referred to as "commercial."

This is important to know because S-rated oil cannot be used in compression ignition engines unless it also carries the appropriate C rating. Be careful to use the correct service class of oil in a particular engine.

The API symbol is the donut symbol located on the back of the oil bottle (**FIGURE 14-28**). In the top half of the symbol is the service class—S or C—and the service standard that the oil meets. The center part carries the SAE viscosity rating for the oil. The API symbol may also carry a "Resource-Conserving" or "Energy-Conserving" designation if it is a fuel-saving oil. Be sure to use oil that has a correct API rating and also an energy-conserving designation in all North American vehicles.

The API also created a "starburst" symbol. It can be placed on oil containers that meet the current engine protection standard and fuel economy requirements of the ILSAC. ILSAC is a joint effort of U.S. and Japanese automobile manufacturers. See later topics.

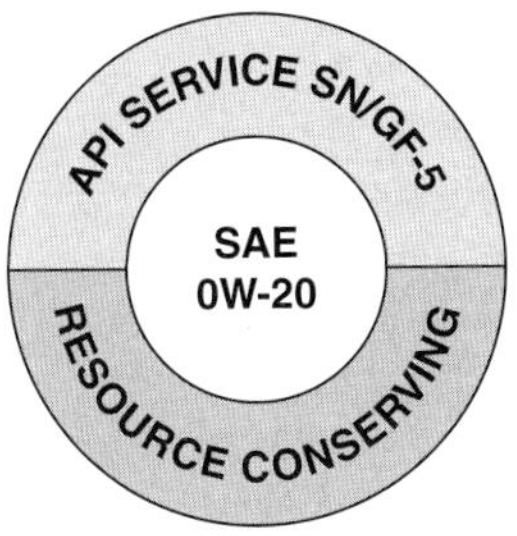

FIGURE 14-28 The API donut shows the API service class and service standard, the viscosity, the ILSAC performance rating, and the energy-conserving designation.

The API service standards have changed over time. The first service standard was SA (spark engine oil – level A). It was used in engines up until 1930. This is pure mineral oil without any additives. As engine manufacturers improved engine technology, or as government regulations changed, engine oil with new qualities was required. This meant that the API would introduce a new service standard. The API SN level was added in October 2010 for 2011 gasoline vehicles.

API CJ-4 was added in 2010 to meet four-stroke diesel engine requirements. New CK-4 (backward compatible) and FA-4 (not backward compatible) ratings came into effect in December 2016. FA-4 is intended only for on-highway diesel engines. The API FA-4 donut features a shaded section to set it apart from CK-4 oils (**FIGURE 14-29**).

> **TECHNICIAN TIP**
>
> In most cases, higher rated engine oils are backward compatible. This means you can use SM oil in a vehicle that requires SL. But there is one exception that some technicians have found. SN-rated oil has very low levels of phosphorus and zinc, which aids in flat tappet camshaft lubrication. So, if you are working on an older engine that uses flat tappets, you probably do not want to use SN-rated oil, but SM instead.

The API classifies oils into five groups:

1. Group 1 oils are produced by simple distillation of crude oil. The components of the oil are separated by their boiling point and by the use of solvents. This is used to extract sulfur, nitrogen, and oxygen compounds. This method was the only commercial refinement process until the early 1970s. The bulk of commercial oil products on the market are still produced by this process, such as conventional engine oils.
2. Group 2 and group 3 oils are refined with hydrogen. They are refined at much higher temperatures and pressures, in a process known as **hydrocracking**. This process results in a base mineral oil with many of the higher performance characteristics of synthetic oils.

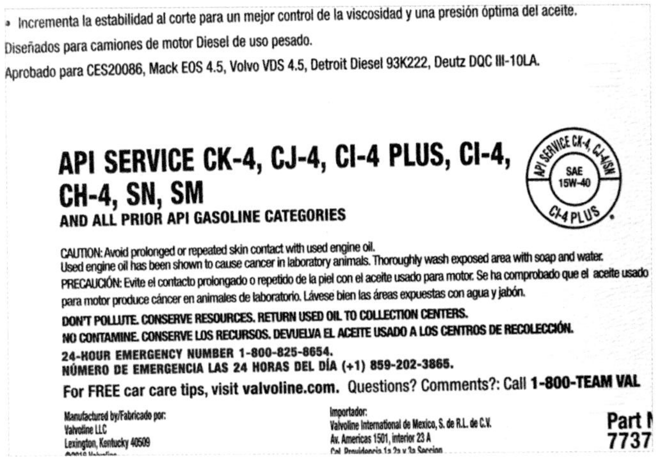

FIGURE 14-29 API CK-4 donut.

3. The more heavily hydrocracked group 3 oils have a very high viscosity index (above 120). They have many, but not all, of the higher performance characteristics of a full PAO synthetic oil. Although not fully synthetic, these oils can be sold as synthetic oil in North America. And because of their lower cost, many of the oils billed as "synthetic" are of group 3 base stock.
4. Group 4 oils are all of the full synthetic PAO group (most common true synthetic). This synthetic oil has more robust qualities and benefits over group 3 oils.
5. Group 5 includes all other types of synthetic oil. Typically, group 5 oils are more expensive than group 4 oils. Although they are very thermally stable, they are more commonly used in aviation and industrial applications.

▶ TECHNICIAN TIP

Over time, lubricating oil breaks down by reacting with dissolved atmospheric oxygen. It is mostly the impurities present in the oil that react in this way. In most refineries, impurities are removed by using solvent. **Hydrogenating** is a newer process that is more effective at removing impurities. It uses hydrogen during the refining process. By hydrogenating the oil, along with oxidation-inhibiting additives, the deterioration rate can be slowed by more than a hundredfold. Hydrogenating also reduces the presence of aromatic hydrocarbons. This improves the oxidation inhibitor action which minimizes sludge and varnish deposits.

American Society of Automotive Engineers (SAE)

Engine oil producers must also meet the SAE viscosity rating for each particular oil. Engine oil with an SAE number of 30 has a higher viscosity, or is thicker, than an SAE 5 oil. Oils with low viscosity ratings, such as SAE 0W, 5W, and 10W (the "W" stands for winter viscosity), are tested at a low temperature—around 0°F (−17.8°C). These ratings indicate how the oil will flow during engine start-ups in cold climate conditions.

Oils with high viscosity ratings, such as SAE 20, 30, 40, and 50, are tested at a high temperature—around 210°F (98.9°C). These ratings indicate how the oil will flow when the engine is being used under loaded conditions in hotter situations. Oil with higher viscosity will flow slower than low-viscosity oil.

Modern oils are blends of oils that combine these properties. The oils are blended with viscosity index improvers to form multigrade, or multiviscosity, oils. They provide better lubrication over a wider range of climatic conditions than monograde oils. These oils are classified by a two-part designation, such as SAE 0W-20 (**FIGURE 14-30**). In this example, when the oil was tested at 0°F (−17.8°C), it met the specifications for a viscosity of 0W weight oil. When the same oil was tested at 210°F (98.9°C), it met the viscosity specifications for 20 weight oil.

Multiviscosity oils flow easily during cold engine start-up. But they do not thin out very much as the engine comes up to operating temperature. These properties allow the oil to get to the components more quickly during start-up. And at the same time, it maintains the ability to cushion components when it is hot. Multiviscosity oils extend the operating temperature range of the engine. But always refer to the vehicle's service information to determine the correct oil viscosity to use. Check it for the climate the engine will be operated in.

FIGURE 14-30 Example of multiviscosity oil.

In 2015, regulations were adopted requiring engines to meet the significantly higher fuel economy standard of 54.5 mpg by 2025. One way to meet this is by using thinner oils that operate under more extreme conditions. Because of this, SAE, in January of 2015, introduced a new viscosity grade called SAE 16 (**FIGURE 14-31**). The following year, two more grades, SAE 12 and SAE 8,have been added. These are high-temperature viscosity grades, *not* the cold-temperature winter viscosity grades. For example, that would allow for engine oils such as 5W16 and 0W12. Note that these engine oils are extremely thin and should not be used in engines that don't specify them. They are unlikely to provide adequate wear protection for engines not designed for their use.

FIGURE 14-31 SAE introduced grade SAE 16 oil recently.

Applied Science

AS-2: Viscosity: The technician can demonstrate an understanding of fluid viscosity as a measurement and explain how it impacts engine performance.

Viscosity is the measurement of a liquid's resistance to flow. This concept is often best understood by example. Imagine you have a small funnel that you fill with honey. You will find that the funnel drains quite slowly. If you filled the funnel with water, it would drain almost instantly. The difference is because honey has a higher viscosity than water.

International Lubricant Standardization and Approval Committee (ILSAC)

ILSAC works in conjunction with the API. Together they create new specifications for gasoline engine oils. However, ILSAC requires that the oil provide increased fuel economy over a base lubricant. These oils should reduce vehicle owners' fuel costs a small amount. This is true when compared to an oil that does not meet the ILSAC standard.

Like the API standard, ILSAC issues sequentially higher rating levels each time the standards are updated. ILSAC GF-5 replaced GF-4 and became the standard in October 2010. Engine oils that meet the highest ILSAC standard (currently GF-5) can display the API starburst symbol. API created this symbol to verify that the oil meets the highest ILSAC standard (**FIGURE 14-32**).

The next ILSAC standard is scheduled to be called GF-6. It is split into two categories, GF-6A and GF-6B. GF-6A will address traditional viscosities and will be backwards compatible. GF-6Bwill represent viscosity grades 0W-16 and lower. They are known as ultra-low viscosity (ULV) lubricants. They will not be backwards compatible. Oils meeting the standard will be released by September 2018. This higher standard will allow engine manufacturers to keep enhancing their engines. This helps in meeting the 2025 fuel economy standard of 54.4 mpg.

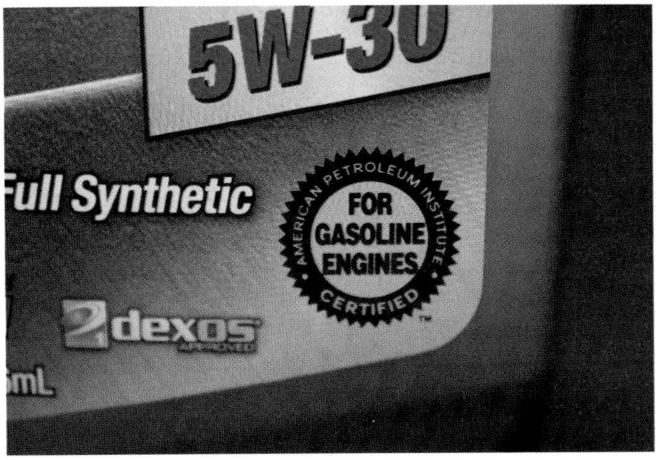

FIGURE 14-32 ILSAC starburst symbol.

Association des Constructeurs Européensd' Automobiles (ACEA)

The ACEA classifications are formulated for engine oils used in European vehicles. They are much more stringent than the API and ILSAC standards. Some of the characteristics the ACEA-rated oil must score high on are:

- soot thickening,
- water, sludge, and piston deposits,
- oxidative thickening,
- fuel economy, and
- after-treatment compatibility.

Although some of these may be tested by the API and ILSAC, the standards are set high to achieve ACEA certification ratings. This means that the engine oil provides additional protection or characteristics that API- or ILSAC-rated oils may not match. If you are servicing a European vehicle, it is advised that you do not go by any API recommendations. Instead, make sure the oil meets the recommended ACEA rating specified by the manufacturer. Or in some cases, the manufacturer's own specification rating (**FIGURE 14-33**).

FIGURE 14-33 Oil bottle showing ACEA rating.

FIGURE 14-34 Oil bottle showing JASO diesel engine oil rating.

FIGURE 14-35 Manufacturer-specific oil rating.

Japanese Automotive Standards Organization (JASO)

The JASO standards set the classification for motorcycle engines. This is for both two-stroke and four-stroke engines, as well as Japanese automotive diesel engines. For four-cycle motorcycle engines, the JASO T 903:2011 came into effect in October 2011. It designates different ratings for wet clutch (MA) and dry clutch (MB). For two-stroke motorcycles, JASO M 35:2003 came into effect in October 2003. And for automotive diesel engines, JASO M355:2015 came into effect in October 2015 (**FIGURE 14-34**).

OEM-Specific Standards

Engine manufacturers continue to design new features or longer drain intervals into their engines. This is sometimes faster than some of the oil-rating organizations could (or would) change their standards. So, some engine manufacturers came up with their own standards. These standards are specific to individual manufacturers. And in some cases, they are specific to individual engines of a particular manufacturer. A few examples follow.

Oil meeting Volkswagen's VW 506.00 standard is suitable for use on diesel engines (not with single injector pump). They have an extended service interval of up to 31,000 miles or two years.

Oil meeting General Motor's dexos1™ was specified for use starting with all 2011 GM gasoline-powered vehicles. It was backward compatible in all older GM vehicles. General Motor's dexos2™ is specified for all GM vehicles equipped with Duramax diesel engines (**FIGURE 14-35**). Its viscosity is SAE 5W30 and meets the ACEA A3/B3 standard. It has a service interval of up to 18,600 miles.

Oil meeting BMW's Longlife-04 standard is approved for fully synthetic long-life oil. It is usually required for BMWs equipped with a diesel particulate filter.

As you can see, it is important to understand the oil requirements for the vehicle you are working on. This will help ensure that you use the specified oil. Using the wrong oil can result in severe damage to the engine. Furthermore, using the wrong oil can void the customer's warranty. This can leave the customer, or your shop, responsible for repairs. Long gone are the days of grabbing five bottles of any 10W-30 oil off the shelf and putting it into any car that rolls through the door. So, always research the specified oil requirements for the vehicle you are servicing before selecting an oil to use.

▶ TECHNICIAN TIP

Be sure to check the owner's manual or service manual of the vehicle to ensure that you are using the correct oil rating and viscosity for the engine. This is especially true for ACEA and manufacturer-specific oils. Do not use just any oil that is sitting on the parts shelf.

▶ Oil Indicators

LO 14-05 Describe oil indicators and warning systems.

A lubrication system failure can be catastrophic to the engine. Because of this, an oil pressure warning system is installed. It lets the driver know the system has failed. If oil pressure falls too low, a pressure sensor threaded into the oil gallery can activate a warning device. There are three types of warning devices:

- low oil pressure warning light,
- oil pressure gauge,
- low oil pressure warning message (**FIGURE 14-36**).

The pressure sensor is also commonly called a sending unit. This is because it sends a signal to the light, gauge, or message center in the dash. Sending units come in two types: pressure switch style and pressure sensor style. If the sending unit is designed as part of a warning lamp system, it is a pressure switch style. It is made up of a spring-loaded diaphragm and a set of switch contacts (**FIGURE 14-37**). Oil pressure is on the engine side of the diaphragm, and a spring is on the other side.

With the engine off and the ignition switch in the run position, the oil pressure is zero. The spring holds the diaphragm toward the engine, and the switch contacts are closed (i.e., making contact with each other). Current now flows from the warning

FIGURE 14-36 If oil pressure falls too low, a pressure sensor in a gallery can: **A.** activate a warning light, **B.** register on a gauge, or **C.** turn on a warning message.

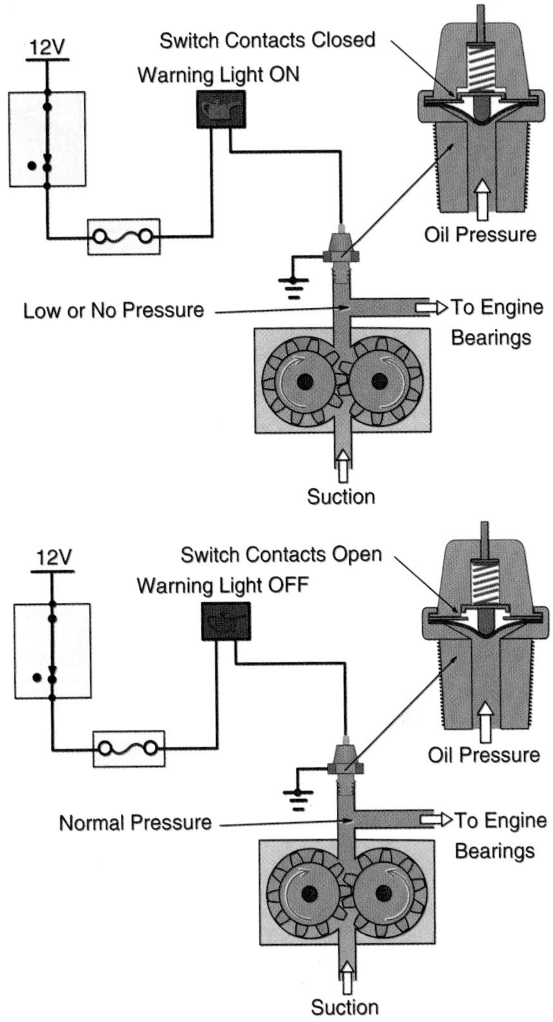

FIGURE 14-37 Oil pressure switch.

lamp in the dash through the closed switch contacts. This completes the circuit to ground, which turns on the warning light.

When the engine is started, oil pressure increases above spring pressure. The diaphragm is pushed away from the engine, opening the switch contacts. Current flow stops, and turns off the warning light. This is how the system should work when everything is working normally. If the oil pressure drops below spring pressure while the engine is running, the warning light will come on. In some vehicles, the sending unit sends the electrical signal to the body control module (BCM). The BCM is programmed to turn the light on below a certain pressure.

On systems with an oil pressure gauge, a pressure sensor is used. It uses a variable resistor within the sending unit to indicate the amount of pressure. The variable resistor is moved by the oil pressure moving the diaphragm against the spring pressure (**FIGURE 14-38**). As the pressure increases, the diaphragm is moved further. This changes the resistance of the variable resistor. This in turn changes the amount of current flowing through the oil pressure gauge, causing it to read higher. If the engine oil pressure drops, then the spring pushes the diaphragm toward the engine. The resistance of the variable resistor changes, again changing the current flowing through the oil pressure gauge. This decreases the pressure reading on the gauge.

In many instances, a driver may not notice that the oil pressure gauge reading has dropped. If this happens and they keep driving the vehicle, the engine may be damaged. Some oil pressure gauges include a warning light to warn the driver of low oil pressure. The warning light is designed to catch the driver's attention. That way he or she can stop the vehicle and investigate the cause of the low oil pressure.

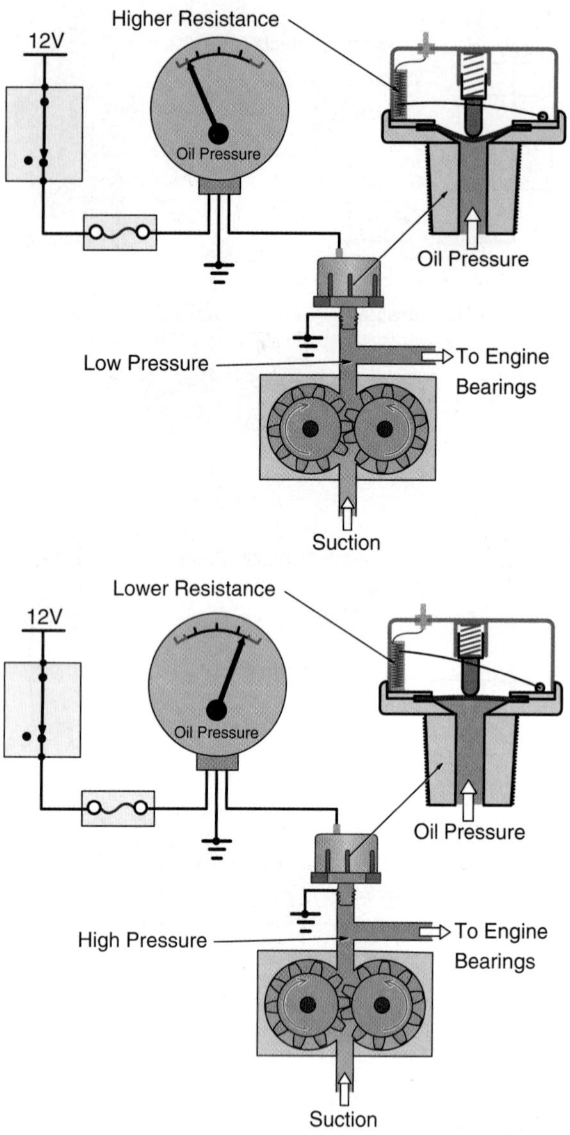

FIGURE 14-38 Oil pressure sending unit.

If the sending unit is part of a driver information system, then the sending unit could be of the switch type or the pressure sensor type. It may also include a sensor for monitoring the oil level. Some systems include an oil temperature monitor. You should investigate various manufacturers' driver information systems in the service information. This will help you become familiar with the different systems and strategies each manufacturer uses.

▶ TECHNICIAN TIP

Some people erroneously refer to the low oil pressure warning system as a low oil level warning system. They think this because, if the oil level gets really low, then the oil pump will draw air into the lubrication system. The oil pressure will then fall, turning on the low oil pressure warning light. Unfortunately, if the oil is allowed to get that low, it is doing damage to the engine. A true low oil level warning system is designed to alert the driver when the oil level approaches the "add" mark on the dipstick. Which is well before engine damage is being done.

▶ TECHNICIAN TIP

Many vehicle manufacturers have moved to using a switch-type sending unit with a gauge. The switch is spring loaded so that it will come ON below a predetermined pressure. But it also allows current to flow through a resistor that is in series with the switch. When the oil pressure is above the spring pressure, the resistor causes the oil pressure gauge to stay mid-scale. When oil pressure falls below this setting, the contacts open, and the gauge reads low. This can confuse drivers as the oil pressure remains very steady for many years. Then, all of a sudden, it drops toward zero. As the engine clearances become larger, or if oil becomes thin, oil pressure drops. Once it falls low enough that spring pressure overcomes oil pressure, the gauge will read low or zero.

Oil Monitoring Systems

Oil monitoring systems are used to inform the driver when the oil needs to be changed. There are several types of **oil monitoring systems**. Some oil systems are simply timers that keep track of mileage. Once a predetermined mileage is reached, it activates a warning light. This notifies the driver that it is time to change the engine oil (**FIGURE 14-39**).

Other oil monitoring systems are much more sophisticated. They analyze the conductivity of the oil through a sensor in the oil pan. This monitors changes in the oil that indicate the oil needs to be changed. Once the system determines that the life of the oil used up, the change oil light or warning message is displayed.

Another monitoring system is called an oil-life monitor. It calculates the expected life of the oil and displays it to the driver (**FIGURE 14-40**). The computer receives inputs from several sensors that take into account:

- the number of start-ups,
- mileage,
- driving habits/conditions,
- temperature,
- length of run time, and
- other data to calculate the remaining oil life.

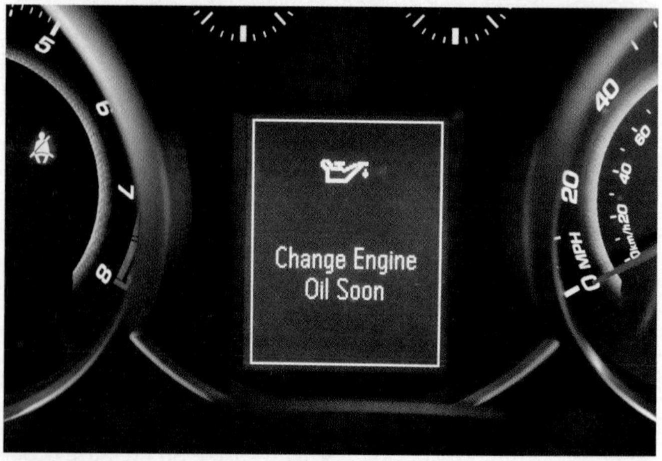

FIGURE 14-39 Oil change needed warning light.

FIGURE 14-40 Oil-life monitor showing the percentage of life remaining.

The calculated remaining oil life is displayed as a percentage. When the oil is freshly changed, and the oil-life monitor is reset—it will say the oil life is 100%. As the oil wears out, the monitor will read closer to 0% oil life. Thus, informing the driver of the condition of the oil. It also allows the driver to know when to schedule an oil change.

This system monitors the conditions the oil is operating under. Because of this, the life of the oil can change drastically depending on the conditions. For example, if the vehicle were only driven in moderate temperatures for long distances, the oil would last thousands of miles longer than a vehicle driven in stop-and-go traffic. That makes this system much more accurate at predicting when the oil truly needs to be changed. When changing oil, you need to reset the oil monitoring system, if equipped. Each vehicle has a specific reset procedure to turn the light or message off after an oil change. We will cover this in the oil change section.

▶ TECHNICIAN TIP

As a technician, you have to be able to find the procedure to turn the light off and reset the oil monitoring system. Refer to the service manual or owner's manual for resetting procedures. The reset procedures can be as simple as pushing a button. Or harder, having to go through a set of steps. In some cases, you have to use a dedicated tool.

▶ Types of Lubrication Systems

LO 14-06 Identify and describe the types of lubrication systems.

Pressure-Fed System

Modern vehicle engines use a **pressure, or force-feed, lubrication system**. In this system, the oil is forced throughout the engine under pressure (**FIGURE 14-41**). In gasoline engines, oil will not flow up into the engine by itself. So, the oil pump collects it through a pickup tube and forces it through an oil filter. It then flows into passageways, called galleries, in the engine block. The galleries allow oil to be fed to the crankshaft bearings first. Oil then flows through holes drilled in the crankshaft to the connecting rods.

The oil also moves from the galleries onto the camshaft bearings and the valve mechanism. After circulating through the engine, the oil drains back to the oil sump to cool. This design is called a wet sump lubrication system.

Diesel engines are lubricated in much the same way as gasoline engines. But there are a few differences. Diesel engines typically operate at the top end of their power range. So, their internal operating temperatures are usually higher than those in gasoline engines. Thus, the parts in diesel engines are usually more stressed.

Because diesel fuel is ignited by the heat of compression, the compression pressures (and compression ratio) are much higher than in gasoline engines. Diesel fuel has more British thermal units (Btus) of heat energy than gasoline, so it produces more heat when it is ignited. This places more stress on the engine's moving parts. And the parts have to be much heavier. With heavier parts, oil must be able to handle higher shear forces. As a result, diesel oils need a different range of properties than oil for gasoline engines. This means that they are classified differently.

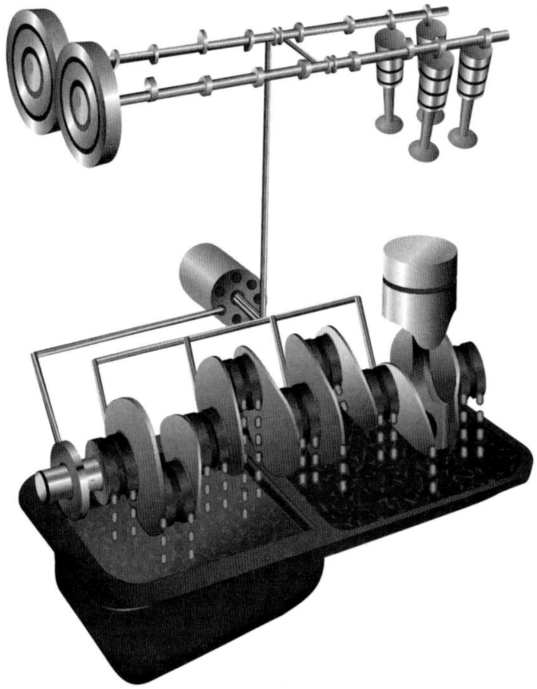

FIGURE 14-41 A pressure, or force-feed, lubrication system.

While spark ignition engine oil is "S" rated, diesel engine oil is "C" rated. The "C" rating stands for compression ignition (**FIGURE 14-42**). To help overcome the high oil temperatures, it is also common for many diesel engines to use an oil-to-water cooler. Diesel engines use engine coolant to cool the oil in the engine. The cooler and oil filter are usually on the same mounting on the block.

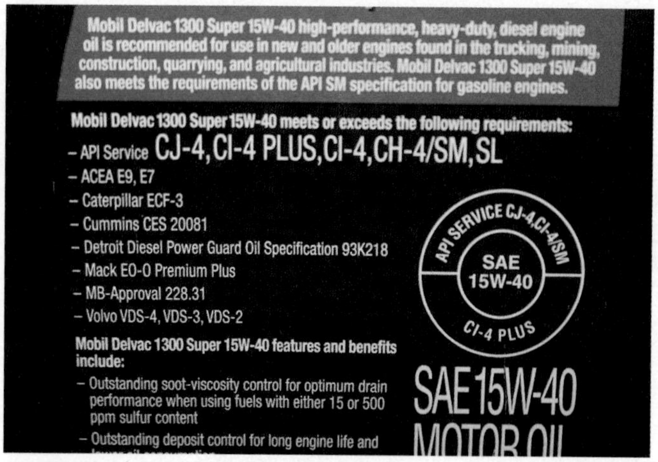

FIGURE 14-42 Diesel-rated oil as designated by the "C."

Applied Science

AS-3: Friction: The technician can explain the need for lubrication to minimize friction.

Oil is a good lubricant in an engine because it has a low coefficient of friction. It creates a protective layer between two metal components, which both have a high coefficient of friction. The high coefficient of friction produces heat. Heat causes the metal to expand, potentially creating engine wear and damage. Oil keeps the two metal components from rubbing against each other, thus preventing damage.

Splash Lubrication

Not all lubricated engine components are lubricated by the pressure-feed system. Some are lubricated by the **splash lubrication** method (**FIGURE 14-43**). In this method, the oil is thrown around and gets into spaces that need lubrication. Automotive and diesel engines use splash lubrication for lubricating the:

- cylinder walls,
- pistons,
- wrist pin,
- valve guides, and
- sometimes the timing chain.

The oil that is splashed around usually comes from moving parts that are pressure-fed. As the oil leaks out of those parts as designed, it is thrown around. In this way, it provides splash lubrication to the needed components.

Most small four-stroke gasoline engines use only splash lubrication to lubricate all of the parts on the engine. This includes the crankshaft bearings, camshaft, lifters, and valves. On horizontal-crankshaft engines, a **dipper** on the bottom of the connecting rod scoops up oil from the crankcase. It throws it all around the crankcase, lubricating the parts. The dipper is also able to splash oil up to the valve mechanism.

Alternatively, an **oil slinger** can be driven by the crankshaft or camshaft. A slinger is a device that runs half-submerged in

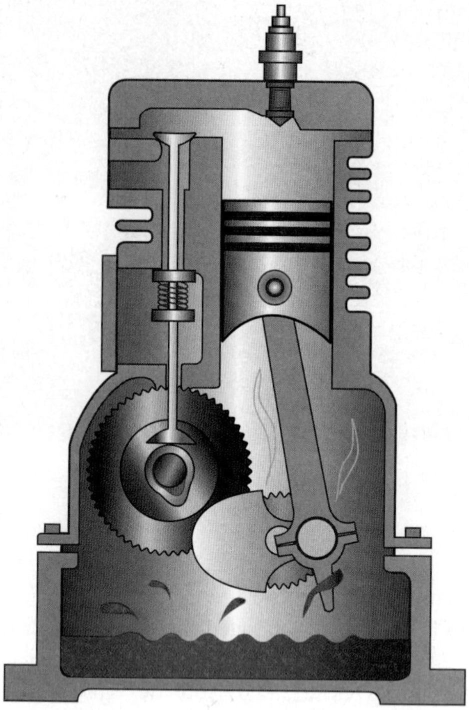

FIGURE 14-43 A splash lubrication method.

FIGURE 14-44 Two-stroke engine requiring a gasoline–oil pre-mix at 50:1.

the engine oil. The oil is slung from the slinger upward by centrifugal force to lubricate moving parts. A similar system is used in most small vertical-crankshaft engines. Oil is also splashed up to the valve mechanism from the centrifugal force of the slinger.

Two-Stroke Engine Premix Fuel Systems

Most two-stroke gasoline engines use a specified gasoline–oil mixture for lubrication. Two-stroke oil is different than engine oil. The two cannot be substituted for each other. They must be used according to the manufacturer's specifications. For many small two-stroke engines, the oil and fuel are premixed to the engine manufacturer's specification. For example, an engine may require a 50:1 mixture, which is 50 parts gasoline to 1 part oil (**FIGURE 14-44**).

The mixture of air, oil, and fuel passes through a sealed crankcase on its way to the combustion chamber. The crankcase thus becomes part of the fuel intake system and cannot be used as an oil sump. As the air, fuel, and oil enter the crankcase, the fuel evaporates. This leave behind enough oil to keep parts coated and lubricated.

The crankshaft bearings in two-stroke engines typically are ball or roller types. Because the components of this kind of bearing only roll over each other, they require a lesser amount of lubrication. So, pressure lubrication is not needed.

Two-Stroke Engine Oil Injection Systems

Some two-stroke gasoline engines use an oil injection system. This system does not require that the oil and gasoline be mixed manually. A small engine-driven oil pump takes oil from a tank and pumps a measured amount directly into the engine. It then mixes with the fuel and lubricates the internal engine parts (**FIGURE 14-45**). The oil pump is designed to deliver the correct amount of oil for the engine speed and the throttle setting. However, because only a small amount of oil is needed, the orifice that the oil is sprayed out of is very tiny. It is easily plugged by any debris that is allowed to fall into the oil tank when it is filled.

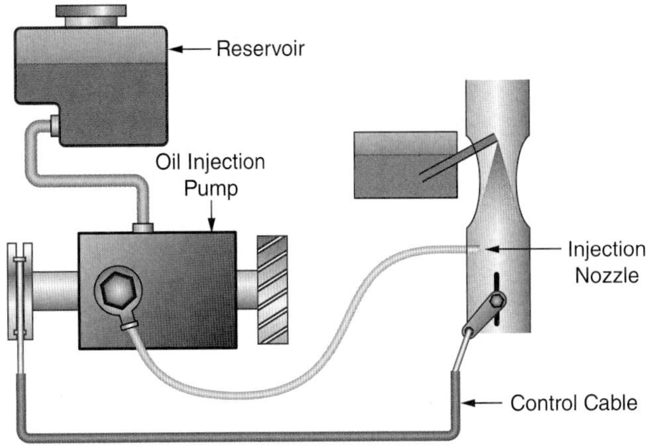

FIGURE 14-45 An oil injection system in a two-stroke engine.

▶ TECHNICIAN TIP

When fueling a two-stroke engine, it is easy to forget that the gasoline must be mixed with oil. If you forget and put only gasoline in it, you will ruin the engine very quickly. To make matters worse, a new engine cannot tolerate this fueling error as well as an older engine. There are therefore quite a few worn-out two-cycle engines sitting around, retired long before their time.

▶ Wrap-Up

Ready for Review

- Lubricating oil lubricates, cushions, cools, cleans, and seals various components and reduces friction, protects against corrosion, and preventing metal-to-metal contact between moving surfaces.
- Conventional oil is processed from crude oil pumped from the ground and type 3 and type 4 are two main categories of synthetic lubricating oils.
  - The following additives allow the oil to perform the different functions in an engine: extreme-pressure additives, oxidation inhibitors, corrosion inhibitors, antifoaming agents, detergents, dispersants, pour point depressants, and viscosity index improver.
- A typical lubrication system consists of oil sump, an oil pump strainer, an oil pump, an oil pressure regulator, oil galleries, an oil filter, and a low pressure warning system.
- The most common certifying bodies for engine oil are: The American Petroleum Institute (API), the American Society of Automotive Engineers (SAE), and the International Lubricant Standardization and Approval Committee (ILSAC).
- The warning devices of an oil pressure warning system include low oil pressure warning light, oil pressure gauge, and low oil pressure warning message.
- Pressure-fed system and splash lubrication are two main lubrications systems used in an engine. Two-stroke engine use either a specified gasoline–oil mixture for lubrication or an oil injection system.

Key Terms

antifoaming agents Oil additives that keep oil from foaming.

Baffles A part used to prevent sloshing of liquids. Used in oil pans on high performance vehicles.

bypass filter An oil filter system that only filters some of the oil.

conventional oil Type of oil that is processed from crude oil to the desired viscosity, after which additives are added to increase wear resistance.

corrosion inhibitors Oil additives that keep acid from forming in the oil.

crescent pump An oil pump that uses a crescent-shaped part to separate the oil pump gears from each other, allowing oil to be moved from one side of the pump to the other.

crude oil Oil that originates from the ground; it is then refined into a usable substance.

detergents Reduce carbon deposits on parts such as piston rings and valves.

dipper A part used to splash oil inside the crankcase for lubrication purposes.

dispersants Oil additives that keep contaminants held in suspension in the oil, to be removed by the filter or when the oil is changed.

extreme loading Large pressure placed on two bearing surfaces. Extreme loading will try to press oil from between bearing surfaces.

extreme-pressure additives Coat parts with a protective layer so that the oil resists being forced out under heavy load.

full-flow filter Type of filters that are designed to filter all of the oil before delivering it to the engine.

galleries Passageways drilled or cast into the engine block or head(s), which carry pressurized lubricating oil to various moving parts in the engine, such as the camshaft bearings.

geared oil pump An oil pump that has two gears running side by side together to move oil from one side of the pump gears to the other.

gelling The thickening of oil to a point that it will not flow through the engine; it becomes close to a solid in extreme cold temperatures.

hydrocracking Refining crude oil with hydrogen, resulting in a base oil that has the higher performance characteristics of synthetic oils.

hydrogenating A process used during refining of crude oil. Hydrogen is added to crude oil to create a chemical reaction to take out impurities such as sulfur.

lubricating oil Processed crude oil with additives to help it perform well in the engine.

lubrication system A system of parts that work together to deliver lubricating oil to the various moving parts of the engine.

mineral oil Base stock processed from crude oil in a refinery, used as the base material of all conventional oil.

oil cooler A device that transfers heat away from oil by passing it near either engine coolant or outside air. Cooling the oil helps to keep it from overheating and breaking down.

oil galleries Main passageways for transporting lubrication in an engine.

oil monitoring system A system that monitors or calculates the condition of the oil and alerts the driver when it is time to be changed. These systems have to be reset for the customer after an oil change is performed.

oil pan The metal pan located at the bottom of the engine; it usually covers the crankshaft and rods, commonly where the oil sump is located on a wet sump oil system.

oil pressure regulator A spring operated valve which limits the maximum pressure in an engine lubrications system

oil pressure relief valve A calibrated, spring-loaded valve that allows for pressure bleed-off if the oil pump creates too much pressure.

oil pump A positive-displacement pump that produces oil flow within the engine and lubricates the internal moving components.

oil pump strainer A screen located on the oil pump pickup that keeps debris from being picked up by the oil pump.

oil slinger A device used to fling oil up onto moving engine parts.

oil spurt holes Holes drilled into the connecting rod that spray oil up onto the cylinder walls and the piston wrist pins.

oil sump The lower part of the oil pan that collects and holds lubricating oil for the engine. The oil pickup screen sits in this low point.

oxidation inhibitor An oil additive that helps keep hot oil from combining with oxygen to produce sludge or tar, which clogs oil galleries and drain-back passages.

pickup tube A tube connected to the oil pump that acts like a straw for the oil pump to pull oil from the sump of the oil pan.

polyalphaolefin (PAO) oil An artificially made base stock (synthetic) that is not refined from crude oil. Oil used in R-12 systems and those converted from R-12. Oil molecules are more consistent in size, and no impurities are found in this oil, as it is made in a lab.

pour point depressants Oil additives that keep wax crystals from forming and causing the oil to gel during cold operation.

pressure, or force-feed, lubrication system A lubrication system that has a pump to pressurize the lubricating oil and push it through the engine to moving parts.

rotor lobes Lobes or rounded edges on rotors that squeeze oil and create pressure.

rotor-type oil pump A pump with an inner rotor driving an outer one; as they turn, the volume between them increases. The larger volume created between the rotors lowers the pressure at the pump inlet, drawing fluid in and filling the spaces. As the lobes of the inner rotor move into the spaces in the outer rotor, oil is squeezed out through the outlet.

scavenge pump A pump used with a dry sump oiling system to pull oil from the dry sump pan and move it to an oil tank outside the engine.

splash lubrication A lubrication system that relies on oil being splashed onto moving parts by rotating engine parts striking the oil. These systems are typically used in small engines.

synthetic oil Synthetic oil that, in its pure form, uses artificially made base stocks and is not derived from crude oil. This oil lasts longer and performs better than normal oil. The base stock additives are similar to those in conventional oils.

synthetic blend A blend of conventional engine oil and pure synthetic oil.

viscosity The measurement of how easily a liquid flows; the most common organization that rates lubricating fluids is SAE.

viscosity index improver An oil additive that resists a change in viscosity over a range of temperatures.

windage tray Component usually made out of sheet metal or plastic that bolts onto the bottom of the main bearing saddles; it prevents the churning of the oil by the rotation of the crankshaft.

Review Questions

1. Engine oil performs all of the following functions EXCEPT:
   a. Cleans
   b. Powers
   c. Lubricates
   d. Cools
2. Synthetic oils have many advantages over conventional oils. Which of the following is a benefit of using a synthetic oil?
   a. Lower cost
   b. Oxidize easily
   c. Higher wax content
   d. Extended oil changes
3. Engine oils have to operate over a wide temperature range. What additive is used to allow oil to function over the required wide temperature range?
   a. Extreme pressure additive
   b. Antifoaming agents
   c. Viscosity index improver
   d. Pour point depressants
4. Choose the correct statement?
   a. A dry sump system stores oil in a tank outside of the engine
   b. A wet sump system uses a scavenge pump and pressure pump
   c. The wet sump pump system is the best fit during aggressive driving conditions
   d. The dry sump system is the best fit in low-performance applications
5. Which of these is most likely to cause low engine oil pressure over time?
   a. Oil is thicker than it should be
   b. Oil pump is creating excessive flow
   c. Oil leaks inside the engine have become excessive
   d. Oil pump is of the wrong type
6. Which oil certifying organization provides stringent classifications formulated for engine oils used in European vehicles?
   a. API
   b. SAE
   c. JASO
   d. ACEA
7. If oil pressure is lost in an engine, catastrophic damage may occur. Which component warns the driver of an oil pressure problem?
   a. Low oil pressure warning light
   b. Flashing malfunction indicator lamp
   c. High oil pressure warning light
   d. Low oil level warning light
8. Oil life monitoring systems can be simple or more complex. All of the following factors are used to determine engine oil life EXCEPT:
   a. Number of start-ups
   b. Driving habits/conditions
   c. Brand of oil used
   d. Length of run time
9. Internal engine components can be lubricated by oil pressure or splash lubrication. Which component would be pressure lubricated?
   a. Valve guides
   b. Crankshaft
   c. Cylinder walls
   d. Pistons
10. Which component is only splash lubricated?
    a. Cylinder walls
    b. Crankshaft
    c. Camshaft
    d. Connecting rods

ASE Technician A/Technician B Style Questions

1. Technician A says the engine oil can act as a cooling agent by removing heat. Technician B says the engine oil acts as a cleaning agent. Who is correct?
   a. Technician A
   b. Technician B
   c. Both A and B
   d. Neither A nor B
2. Since oils have to function over a wide temperature range, multi-weight (5W-30) oils were developed. Technician A says the "W" stands for "weight." Technician B says the "W" stands for "winter." Who is correct?
   a. Technician A
   b. Technician B
   c. Both A and B
   d. Neither A nor B
3. Two technicians are discussing oil pans. Technician A says high-performance vehicles have a windage tray to reduce oil churning. Technician B says high-performance vehicles use baffles to prevent oil from surging away from the oil pick-up. Who is correct?
   a. Technician A
   b. Technician B
   c. Both A and B
   d. Neither A nor B
4. Oil pressure and flow are essential to the continued operation of an engine. Technician A says oil pressure is affected by the clearances inside the engine. Technician B says the volume of oil flowing in the engine affects oil pressure. Who is correct?
   a. Technician A
   b. Technician B
   c. Both A and B
   d. Neither A nor B
5. Two techs are discussing oil filters. Technician A says that spin-on filters use an O-ring to seal the base of the filter to the engine block. Technician B says that cartridge-type oil filters use paper filters that can be cleaned and reused. Who is correct?
   a. Technician A
   b. Technician B
   c. Both A and B
   d. Neither A nor B

6. As the demand on engines increases so does oil temperatures. Technician A says an oil-to-water cooler transfers heat to the coolant. Technician B says the oil-to-air oil cooler transfers heat to the atmosphere. Who is correct?
   a. Technician A
   b. Technician B
   c. Both A and B
   d. Neither A nor B
7. An oil pressure switch may provide a signal to turn on an oil pressure light. Technician A says the oil pressure switch will close the contacts to complete a circuit when there is no oil pressure. Technician B says the BCM may turn on the oil pressure light if the pressure drops. Who is correct?
   a. Technician A
   b. Technician B
   c. Both A and B
   d. Neither A nor B
8. Technician A says that API's "S" rating means it should only be used in supercharged engines. Technician B says that "C"-rated oil is certified for use in cars, and not trucks. Who is correct?
   a. Technician A
   b. Technician B
   c. Both A and B
   d. Neither A nor B
9. Technician A says that the higher the viscosity number, the thicker the oil. Technician B says that most modern vehicles use single weight oil. Who is correct?
   a. Technician A
   b. Technician B
   c. Both A and B
   d. Neither A nor B
10. Technician A says that one function of engine oil is to seal. Technician B says that one function of engine oil is to cushion. Who is correct?
    a. Technician A
    b. Technician B
    c. Both A and B
    d. Neither A nor B

CHAPTER 15

Servicing the Lubrication System

Learning Objectives

- **LO 15-01** Checking oil level and condition.
- **LO 15-02** Changing engine oil.
- **LO 15-03** Diagnose engine lubrication system issues.

ASE Education Foundation Tasks

See Appendix A to view the 2017 ASE Education Foundation Automobile Accreditation Task List Correlation Guide.

▶ Lubrications System - Maintenance and Repair

LO 15-01 Checking oil level and condition.

The engine lubrication system protects the internal parts of an engine (**FIGURE 15-1**). Because of this, periodic inspection, maintenance, diagnosis, and repair are required. Inspection involves checking the level, condition, and service life of the oil. This should be performed by the operator every couple of fuel tank fill-ups or monthly at a minimum. It also involves technicians inspecting the engine for oil leaks during oil changes.

Maintenance involves changing the oil, oil filter, and resetting the oil-life monitor, if applicable. This is performed according to the manufacturer's maintenance schedule. Diagnosis involves:

- testing the oil pressure,
- determining the cause of low or high oil pressure,
- and identifying the cause of oil leaks, and
- diagnosing oil pressure and level warning system faults.

We cover each of these tasks in this chapter.

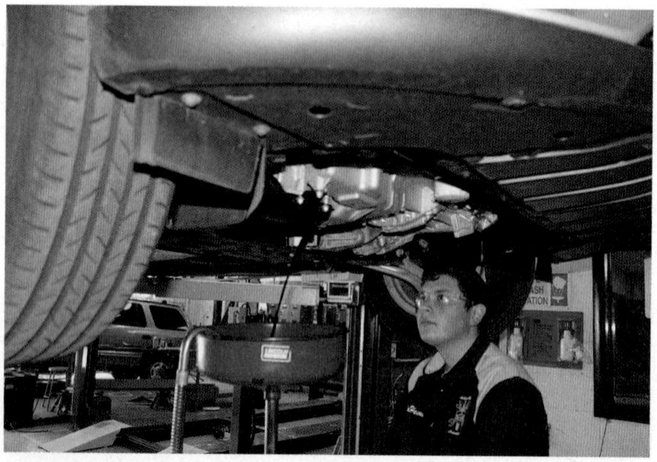

FIGURE 15-1 Entry-level technician performing an oil and filter change.

When working on the lubrication system, always wear appropriate personal protective equipment to help protect you. Also, remember that waste oil is a hazardous material. It needs to be stored and disposed of properly. Review your state and local requirements for disposing of waste oil and oil filters. Engine oil can be extremely hot, leading to severe burns. And if it spills on the floor, it only takes a dime-sized spot to cause slips and falls. These are some of the most common injuries in a shop. So, always clean up spills immediately.

Tools

The tools for lubrication system maintenance and repair include (**FIGURE 15-2**):

- A variety of oil filter wrenches to remove the oil filter
- A set of wrenches and a socket set to remove the oil drain plug and any engine covers
- A mirror and a quality light for leak testing
- A special socket to remove many of the oil pressure switches or sensors
- A digital multimeter (DMM) to test wiring and sensors of the oil warning light or gauge
- A pressure gauge for testing oil pressure
- A sensor substitution box

Checking the Engine Oil

There is a strong danger of damaging the engine if the engine oil level drops too low. This makes the job of checking engine oil level regularly, critical. It should be performed at least at every other fuel fill-up, depending on the age and condition of the vehicle. If the engine oil level is low, you will need to determine why. The oil level can be low for two main reasons. Either there is an oil leak, or the engine is consuming oil by burning it. Burning engine oil could be due to:

- worn piston rings,
- worn valve guides, or
- a malfunction with the positive crankcase ventilation (PCV) system.

Oil leaks can occur at the various gaskets, seals, and oil-pressurized components. One common leak point is the oil

You Are the Automotive Technician

A customer brings a five-year-old vehicle into your shop for an oil and filter change. The vehicle is right at the recommended 7,500-mile interval. The customer has noticed some oil spots on the garage floor. And the oil level was a bit below the "add" mark this morning when checked. The customer is concerned that it has an oil leak. As you start the vehicle to pull it onto the hoist, the oil light comes on when the ignition key is turned on. It goes off once the engine starts, proving that the oil pressure warning light circuit is working. With the engine running, you find that a drop of oil forms every so often on the end of the oil pressure switch. This indicates that it is leaking and needs to be replaced. None of the engine seals and gaskets show any signs of leakage.

1. What must you look for when removing a spin-on oil filter?
2. When performing oil and filter changes, what must you do with most oil drain plug gaskets?
3. What is the process for tightening spin-on oil filters?
4. What things must be done after changing the oil and verifying there are no oil leaks?

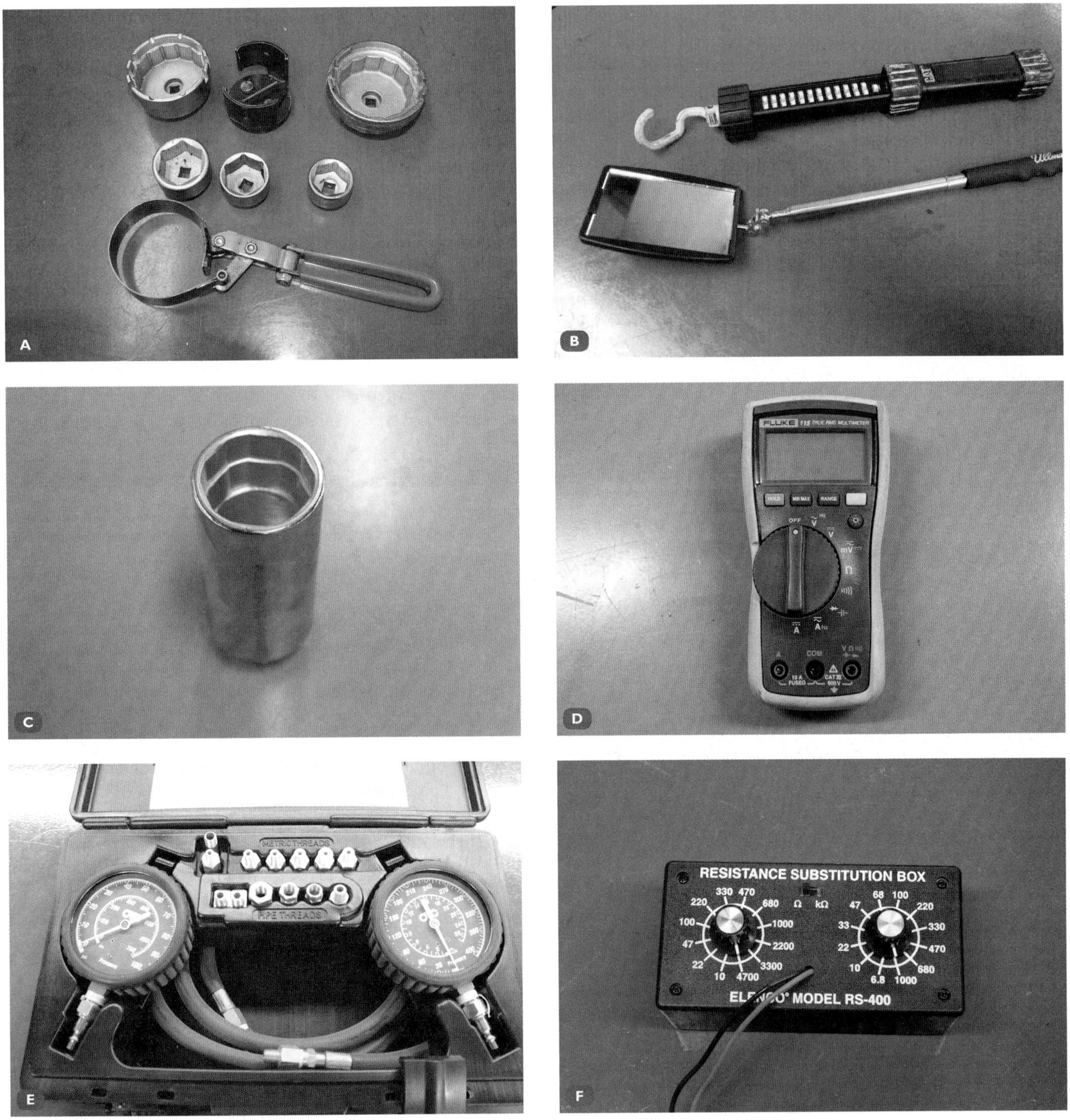

FIGURE 15-2 Tools for lubrication system maintenance. **A.** Oil filter wrenches. **B.** Mirror and quality light. **C.** Oil pressure sensor socket. **D.** Digital multimeter (DMM). **E.** Pressure gauge. **F.** Sensor substitution box..

pressure sending unit. Checking oil level should also be part of any pre-trip check. It should also be a part of a pre-delivery inspection on a new car at the dealership. This happens before handing over the vehicle to the customer.

Oil level is typically checked with a dipstick located under the hood. The dipstick sits firmly in a tube so that the lower end of it is positioned in the oil in the oil pan. It's marked so that when the engine is filled to the proper level, oil will coat the dipstick up to the "full" line (**FIGURE 15-3**). A "safe" or "add" line is usually also included. It is typically at the point where the engine needs 1 quart (0.96 liter) of oil added to make it full. The oil level should always be maintained between the two lines.

When checking oil, always make sure the vehicle is on a level surface and the engine is off before taking a reading. If you do not, you will get inaccurate readings. Also, remove the dipstick and wipe it off before reinserting it to take the reading.

When reading it, hold the dipstick horizontal so the oil will not run down the stick. Typically, the amount of oil needed to raise the oil level from the bottom of the "safe" mark to the "full" mark is about one quart (0.96 liter). This amount may vary. So, always check the service information to determine the correct quantity.

Some newer vehicles are not equipped with an engine oil dipstick. Instead, they may use an oil level sensor that reports the oil level to an indicator on the dash (**FIGURE 15-4**). In this way, the driver can monitor the oil level from the driver's seat, without even opening the hood.

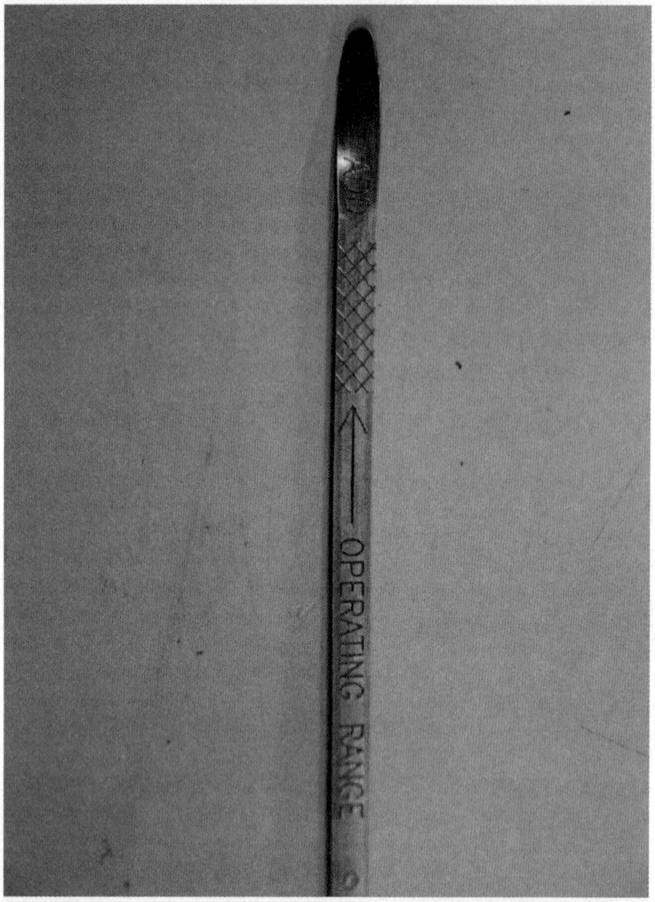

FIGURE 15-3 A dipstick showing "full" and "add" marks.

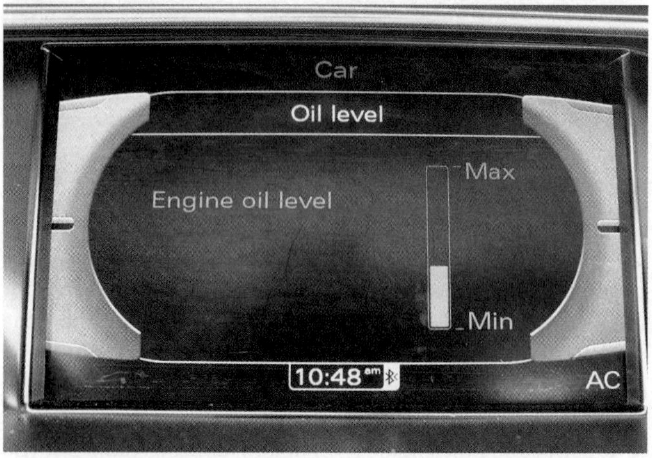

FIGURE 15-4 Instead of a dipstick, some vehicles use an oil level sensor and display the level on the instrument panel.

When checking the oil level, it is usually difficult to assess the condition of engine oil simply by its color. Fresh oil is translucent and oil that needs to be replaced often looks black and dirty. But oil loses its clean, fresh look fairly quickly. So, it still may have a lot of life left in it.

The best guide for knowing when to change the oil is the vehicle's oil-life monitor, if the vehicle is so equipped. It will tell you the percentage of oil life remaining before a change is needed. If the vehicle is not equipped with an oil-life monitor, you will have to determine it another way. This can be done by checking the oil change sticker on the windshield. You may also be able to look up the last oil change in the repair order system. Doing this can help ensure that the oil is not changed too often, avoiding an unnecessary expense for the customer. When topping off the oil, always use the type of oil specified in the service information.

> ▶ TECHNICIAN TIP
>
> Our job as technicians is to provide high-quality work only when it is truly needed by the customer. Part of being a professional is letting customers know when they do and when they do not need a service performed. This helps to build trust with the customer. It also often results in the customer returning to your shop for future repairs.

When checking the level of the oil, also visually check the condition of the oil. Use the dipstick to wipe some oil onto a clean white paper towel. Hold a light over the towel. Move the towel back and forth to see if light reflects from metal. If metal is in the oil, it indicates substantial wear happening inside the engine. This will require further engine diagnosis and major repairs.

If the oil on the dipstick looks milky gray, it is possible that there is water (or coolant) being mixed into the oil (**FIGURE 15-5**). This could indicate a serious problem somewhere inside the engine. Common causes of this are leaking

FIGURE 15-5 Oil that looks milky gray could indicate that coolant or water is being mixed into the oil. This could be caused by a bad head gasket, cracked head, or leaky oil cooler.

head gaskets, a cracked head, or faulty oil cooler. If you find milky coolant, report this to your supervisor immediately.

Engine operating conditions can also influence the oil's condition. For instance, very short operating cycles can cause condensation to build up inside the engine. An extreme case of this will cause very rapid oil deterioration. Preventing this requires more frequent oil changes.

The oil of a vehicle that is running too rich, or that has a leaking fuel injector, will smell like fuel and will be very thin. These problems can ruin an engine quickly, as the diluted oil will not adequately lubricate the engine parts. Finally, if you had to add oil to the engine, do not forget to reinstall the filler cap after topping off the oil.

To check the engine oil, follow the steps in **SKILL DRILL 15-1**.

SKILL DRILL 15-1 Checking the Engine Oil

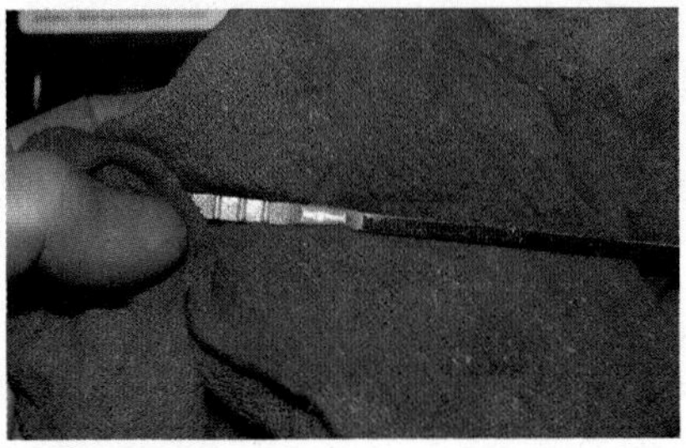

1. Locate the dipstick. With the engine off, remove the dipstick, catching any drops of oil on a rag, and wipe it clean. Observe the markings on the lower end of the stick, which indicate the "full" and "add" marks or specify the "safe" zone.

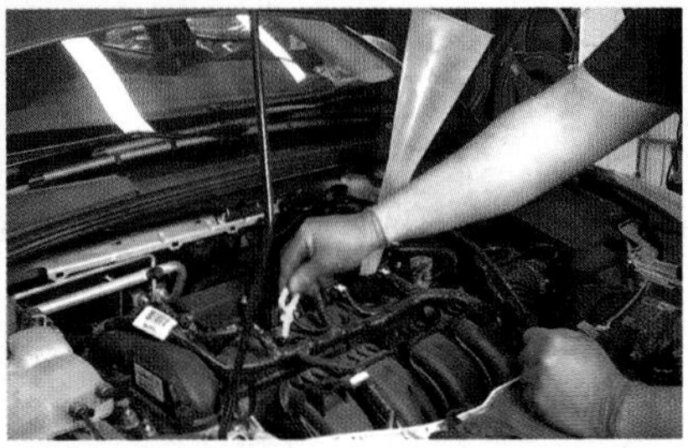

2. Replace the dipstick and push it back down into the sump as far as it will go.

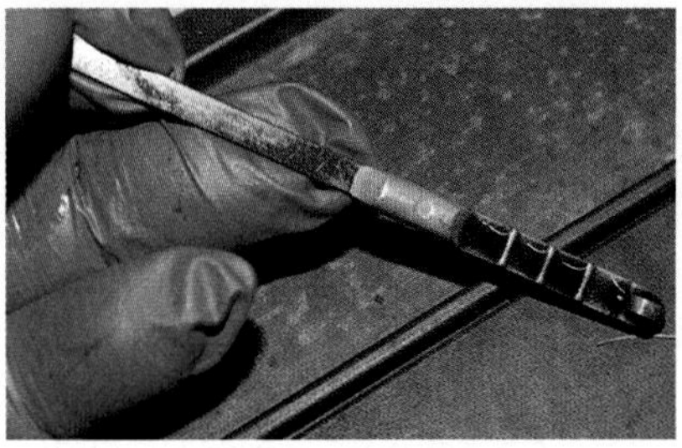

3. Remove it again, and hold it horizontal while checking the level indicated on the bottom of the stick. If the level is near or below the "add" mark, then you will need to determine if the engine just needs topped up with fresh oil or replaced with new oil and oil filter.

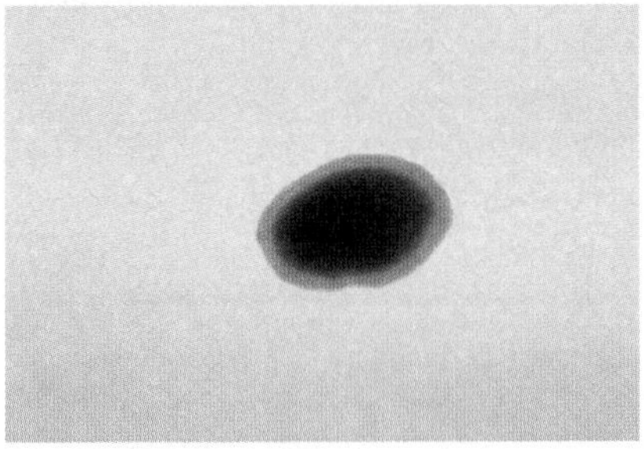

4. Check the oil for any conditions such as unusual color or texture. Report these to your supervisor.

5. Check the oil-monitoring system, oil sticker, or service record to determine if the oil needs to be changed. (Some oil-monitoring systems show the percentage of life left in the oil.)

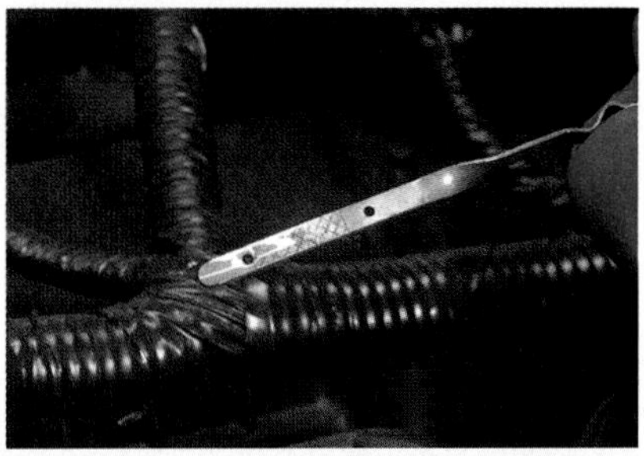

6. If additional oil is needed, estimate the amount by checking the service information guide to the dipstick markings. Remove the filler cap at the top of the engine.

7. Using a funnel to avoid spillage, turn the oil bottle so the spout is on the high side of the bottle, and gently pour the oil into the engine.

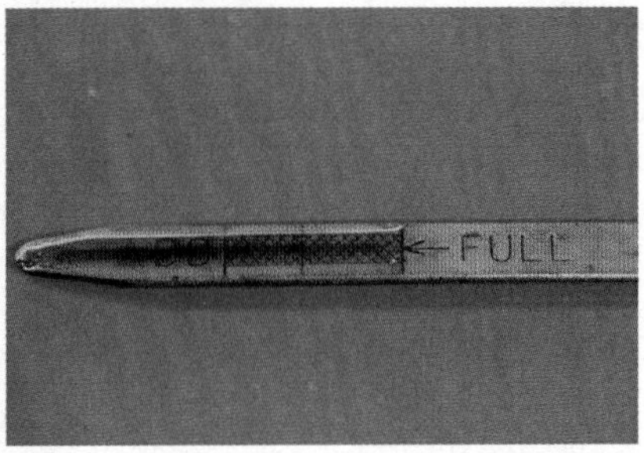

8. Recheck the oil level.

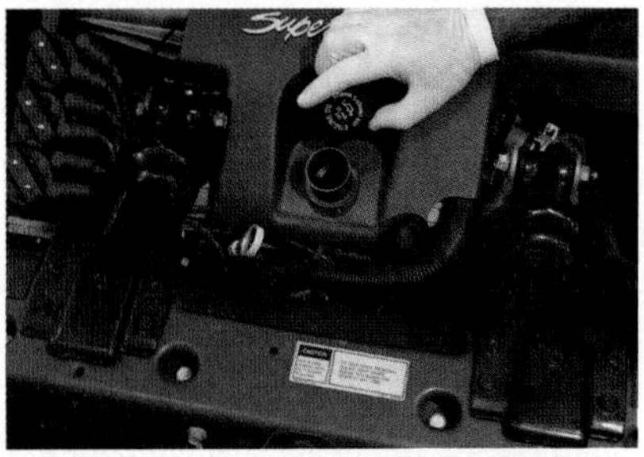

9. Replace the oil filler cap, and check the dipstick again to make sure the oil level is now correct.

▶ TECHNICIAN TIP

Make sure the hood is secure with a hood prop rod, if necessary. Always make sure you wear the appropriate personal protection equipment before starting the job. It is very easy to think that nothing can happen on a basic job like checking the oil level.

SAFETY TIP

If the engine has been running, be careful not to burn your hand or arm on the exhaust manifold or any other hot part of the engine when reaching for the dipstick. Remember, the dipstick and the oil on it will also be hot. Dripping oil from the dipstick will smoke or burn if it falls on any hot engine surfaces.

Oil Analysis

Oil will suspend particles as the engine wears. Analysis of the engine oil is a useful way to see what parts are wearing in the engine. The military as well as many companies use oil analysis to ensure that the engine oil is changed at the appropriate interval. A small tube is slipped down the oil dipstick tube all the way to the oil pan. A vacuum device is used to pull a sample of oil from the pan to a collection container. This sample is labeled with the vehicle information and sent to a lab. The lab thoroughly analyzes the oil and reports the findings back to the shop. The report lists the physical properties such as:

- viscosity,
- condensed water,
- fuel dilution,
- antifreeze,
- acids,
- metal content, and
- oil additives.

Each of these attributes can be used to determine the condition (remaining life) of the oil as well as the engine. When oil analysis is performed for a particular engine on a regular basis, issues can usually be addressed before they become catastrophic. Oil analysis is typically used in heavy vehicle applications because they may use 3 or more gallons (11.4 or more liters) of engine oil. Some race teams also analyze engine oil to ensure that the engine is not being damaged.

▶ Oil and Filter Change: Draining the Engine Oil

LO 15-02 Changing engine oil.

Draining the engine oil is performed any time the oil and filter are to be changed. It also needs to be drained whenever the oil pan needs to be removed for service work. Over time, the oil becomes dirty and the additives wear out. Also, as the engine wears, very small pieces of metal get suspended in the oil. Removing the old contaminated oil helps to make the engine last longer. Always follow the manufacturer's oil change interval. Also, remember that normal use and severe use have different oil change intervals.

When draining the oil, several precautions are necessary. First, engine oil is normally changed after the engine is fully warmed up. This helps to stir up any contaminants, making them easier to flush out with the draining of the oil. However, that means that the oil can be 200–300°F (93–149°C), so use disposable gloves to reduce the risks of being burned (**FIGURE 15-6**).

Second, make sure you locate the correct drain plug—or in some cases, the correct two drain plugs. Some vehicles use an oil pan with a hump in the middle. This requires removing two drain plugs, one in each part of the pan. Some vehicles have a drain plug on the transmission/transaxle that can be mistaken for the engine oil drain plug. If in doubt, look it up or ask your supervisor to point out the correct drain plug.

Third, many drain plugs are either angled off the bottom radius of the oil pan or almost sideways at the bottom side of the pan. This means when the drain plug is removed, hot oil will shoot sideways. Always make sure you take into consideration what path the oil will take (**FIGURE 15-7**). Oil can

FIGURE 15-6 Use disposable gloves to protect from contaminants and heat.

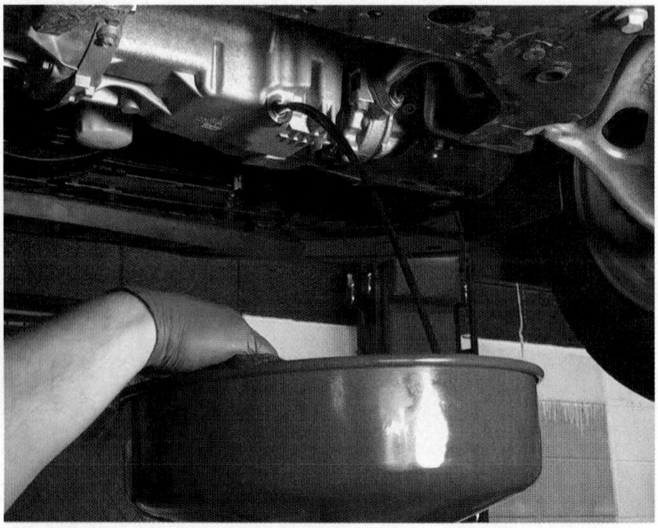

FIGURE 15-7 When removing drain plugs that are located in the side of the pan, be aware that the oil can shoot out sideways a considerable distance.

shoot out pretty far. The lower the drain pan is compared to the drain plug, the harder it is to judge the distance the oil will spray.

Fourth, the drain plug gasket needs to be inspected and possibly replaced. The gasket can be a single-use gasket made of plastic, aluminum, or fiber (**FIGURE 15-8**). This type of gasket should be replaced during every oil change. In use, it is crushed to conform to any irregularities of the pan and drain plug. Because it is crushed, it will not conform as easily the next time it is tightened. This means that someone may think it needs to be tightened excessively to prevent leakage. Overtightening can strip the threads on the oil pan, especially if it is made of aluminum. It can also strip the threads on the drain plug itself. Replace a non-silicone drain plug gasket every time it is removed.

Some drain plugs have an integrated long-life silicone seal (**FIGURE 15-9**). The silicone seal retains its ability to seal for quite a few oil changes as long as it has not been abused. So, this type of seal rarely needs to be replaced.

FIGURE 15-8 Single-use drain plug gaskets can be made of plastic, aluminum, or paper.

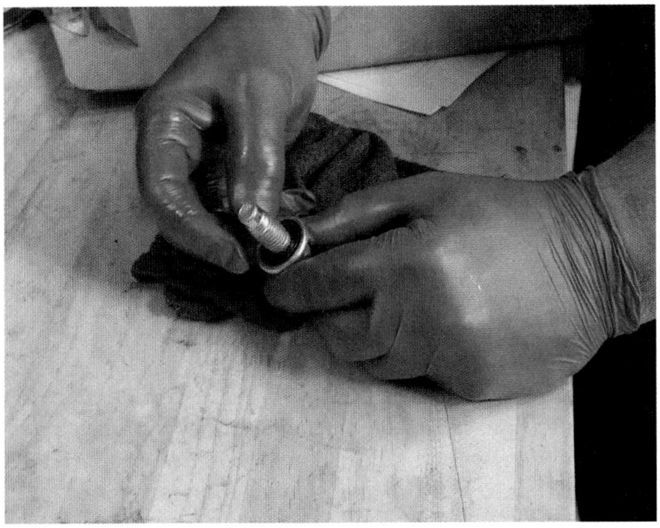

FIGURE 15-9 Oil drain plug with long-life silicone seal.

Fifth, drain plugs need to be tightened to the proper torque. If a drain plug were to loosen and fall out, all of the oil would leak out pretty quickly. The engine would be ruined if the vehicle is not stopped immediately. At the same time, if drain plugs are tightened too tight, the threads in the oil pan can be stripped (**FIGURE 15-10**). This may require replacement of the oil pan, especially if it is an aluminum oil pan. In some cases, the threads can be replaced with a thread insert or oversize drain plug. But neither of those is ideal as they are costly for the shop or customer.

To drain the engine oil, follow the steps in **SKILL DRILL 15-2**.

▶ TECHNICIAN TIP

Refer the customer to the service information to determine when the oil should be changed. That could be based on a mileage interval, time interval, or oil-life monitor. You may also need to verify whether the vehicle is operated under normal or severe conditions. Severe conditions may include driving in low temperatures, stop-and-go driving, dusty conditions, short trips, towing, etc. Intervals for severe driving conditions are sooner than normal driving intervals.

Oil and Filter Change: Replacing Oil Filters

Oil filters are designed to filter out particles that find their way into the oil. The filter catches particles that result from a variety of sources. This can be carbon from combustion or small metal flakes that result from normal engine wear. The engine oil suspends some of the particles while the heavier particles fall to the bottom of the oil pan. The oil filter is designed to catch the particles that are flowing with the oil. It is critical to change the oil filter at the manufacturer's recommended mileage. This helps ensure that it does not become clogged. Clogged filters restrict the oil flow.

When changing a spin-on oil filter, be careful of a few things. First, an oil filter wrench is only used to remove an oil filter. It is never used to install an oil filter. Installation must be performed by hand.

FIGURE 15-10 Stripped threads in an aluminum oil pan.

SKILL DRILL 15-2 Draining the Engine Oil

1. Before you begin, obtain the oil drain container (and make sure it has enough room for the oil to be drained), verify that you have the correct oil and oil filter, and ensure that the engine oil is up to operating temperature.

2. Identify the location of the oil drain plug. Some vehicles have two drain plugs, draining separate sump areas. Position the drain pan so it will catch the oil.

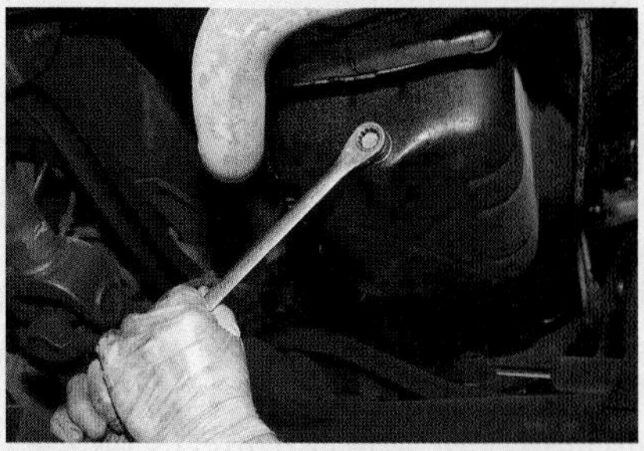

3. Use a box wrench or socket to remove the drain bolt. Be careful not to remove the transmission drain plug by mistake.

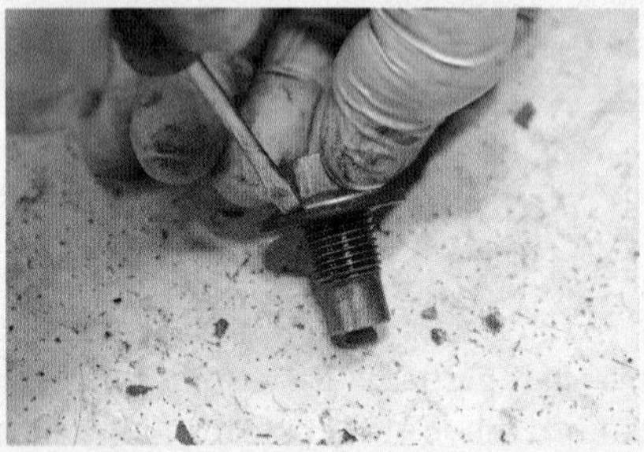

4. Inspect the drain plug and gasket; replace as necessary. Allow the oil to drain while you are dealing with the drain plug, gasket, and oil filter (see SKILL DRILL 15-3).

5. Screw in the drain plug all the way by hand, and then tighten it to the torque specified by the manufacturer. Wipe any drips from the underside of the engine.

6. Safely dispose of the drained oil according to all local regulations.

Second, the O-ring that seals the spin-on oil filter to the block tends to stick on the engine block when the filter is being removed. This can lead to double-gasketing. This occurs when the new O-ring in the new filter is installed over the old O-ring. Because the groove in the filter is only deep enough to hold the new O-ring, the old O-ring is not held in place (**FIGURE 15-11**). Once the engine is started, the oil pressure pushes the old O-ring out of place. Oil is then pumped very quickly out of the engine and onto the floor. This makes a huge mess and wastes good oil. But more importantly, if you don't realize it has happened, the engine could be damaged in only a minute or two of running.

Always check that the old O-ring was removed with the old filter. Do this by comparing the old filter to the new filter. If the old O-ring is not on the old filter, reach up and peel it off the filter mounting adapter.

Third, when installing the spin-on oil filter, smear a bit of oil on the surface of the O-ring. Doing so lubricates the O-ring so that it will spin with the oil filter as it is being tightened. Failure to lube the O-ring can cause it to bind and roll out of the oil filter groove when the filter is being tightened. This again causes a substantial oil leak.

Finally, when installing the spin-on oil filter, it can be hard to see how it is going on the threaded filter adapter. If you get it cross-threaded, it will leak just like a double-gasketed O-ring. Plus you are likely to damage the threads on the adapter. This will make it harder to install a new oil filter in the future.

To prevent cross-threading the filter, always start by turning it with your fingers. Once you suspect it is started, stop and try to lift the filter off the adapter. If it comes off, it has not started. Try to start it by finger again; then try to lift it off. If it does not lift off, then it has started onto the threads. Now count the turns that the filter spins on. If it is not cross-threaded, it should go on at least five full turns before the gasket contacts the filter mounting surface.

If the filter does not spin on at least five full turns by hand, unscrew it and try it again. Once you get it started properly and run down by finger, you can then tighten it by hand the appropriate amount. Typically, the filter manufacturer specifies about three-quarters to one turn after the gasket contacts the adapter.

Increasingly, manufacturers are returning to using cartridge filters. This is because it is easier to properly dispose of

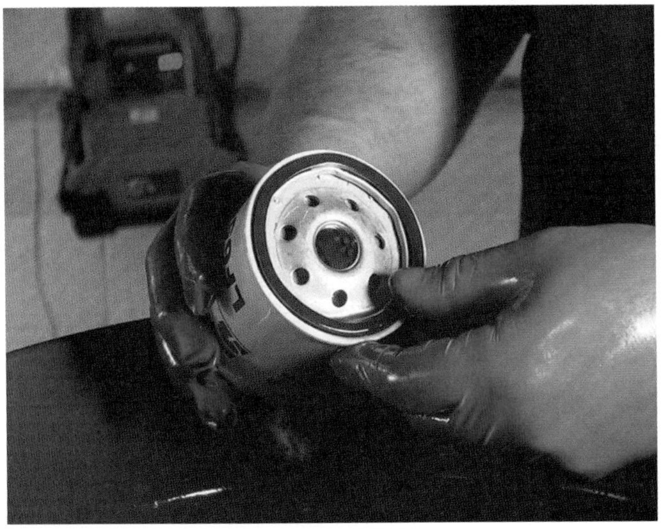

FIGURE 15-11 Oil filter oil ring and groove it fits in.

FIGURE 15-12 Many cartridge filters use an end cap that screws onto the housing. Always use the correct tool when removing and installing it.

the oil. It also reduces the amount of waste generated from used spin-on filters. When changing a cartridge filter, be aware of the following situations. First, the only parts that get replaced are the paper filter cartridge and the O-rings or seals. All of the other parts are reused, so do not damage them or throw them away during disassembly.

Second, some cartridge filters are near the bottom of the engine, whereas others are on top. Use the service information to help you locate the filter. If the filter is on the top of the engine, there is a good chance that you will need to prefill the cartridge before installing the filter end cap. Again, check the service information for the vehicle you are working on.

Third, be careful with tightening a cartridge filter. It is easy to crack or damage the housing—especially if it is plastic. Always follow the manufacturer's torque procedure (**FIGURE 15-12**).

Finally, because the filter housing is being reused, it is important that it is clean before being reinstalled. You may need to wash it out in a clean solvent tank. Just be sure that you also remove any solvent residue before reinstalling it.

Replacing Spin-On and Cartridge Filters

To replace a spin-on filter, follow the steps in **SKILL DRILL 15-3**. To replace a cartridge filter, follow the steps in **SKILL DRILL 15-4**.

Oil and Filter Change: Refilling Engine Oil

Refilling an engine's oil supply is necessary when performing an oil and filter change. It may also be necessary to refill the engine oil after a lubrication system part has been replaced, if the oil was drained. Always determine the oil you are adding meets the manufacturer's specifications. Also, use the recommended amount of oil and grade of oil listed in the service information.

SKILL DRILL 15-3 Replacing a Spin-On Filter

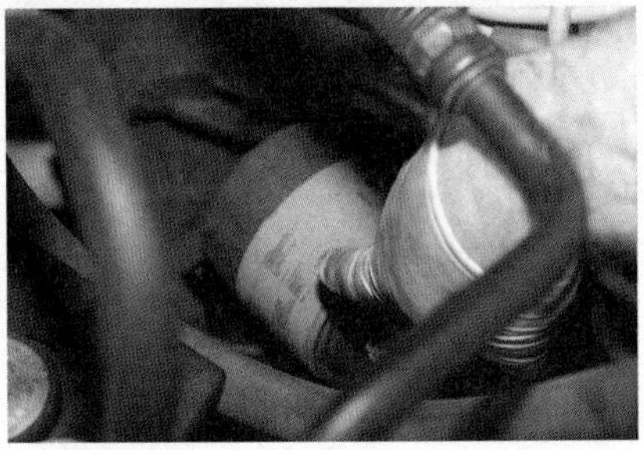

1. Check for new filter availability. Locate the filter being changed. It will usually be located on the side of the engine block or at an angle underneath the engine. Select the proper oil filter wrench.

2. Position a drain pan to catch any oil that will leak from the filter.

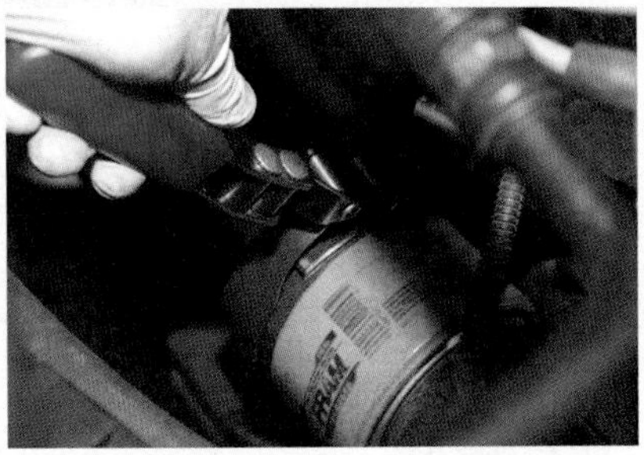

3. Remove the filter.

SKILL DRILL 15-3 Replacing a Spin-On Filter (Continued)

4. Clean the seating area on the oil filter adapter so that its surface can seal properly. Make sure the O-ring from the removed filter is not still stuck to the mounting surface like in this picture.

5. Confirm you have the correct replacement filter. Smear a little oil on the surface of the new O-ring.

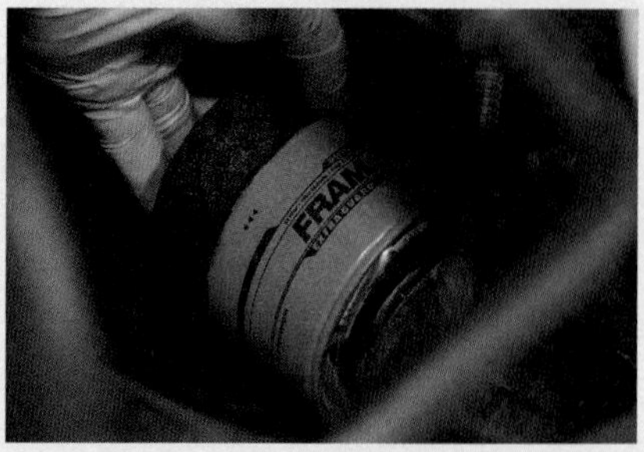

6. Screw in the filter until it just starts, and ensure that it cannot be pulled off. Then, turn the filter by hand until the filter lightly contacts the base. Then, tighten the specified amount. Be careful not to cross-thread the oil filter.

SKILL DRILL 15-4 Replacing a Cartridge Filter (Replaceable Element)

1. Before removing a cartridge-style oil filter, make sure a suitable replacement filter is available.

2. The filter may be located on the side of the engine block, at an angle underneath the engine, or on the top of the engine. If the filter is located in a plastic end cap housing, be sure to use the special tool that the service information calls for.

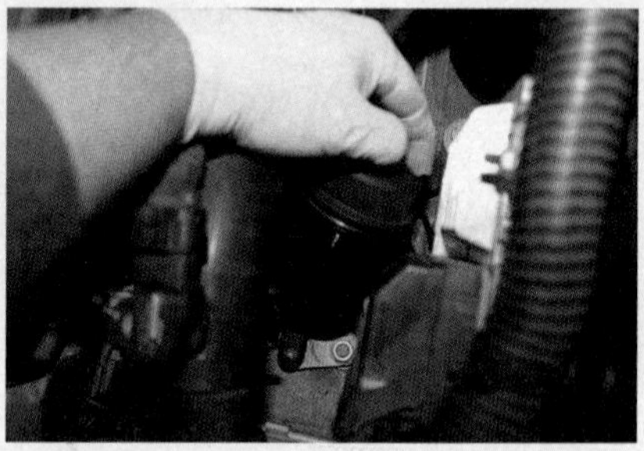

3. Position a drain pan to catch any oil that leaks from the filter. Unscrew the filter cartridge retaining bolt or oil filter cap, and remove the cartridge filter and housing or end cap.

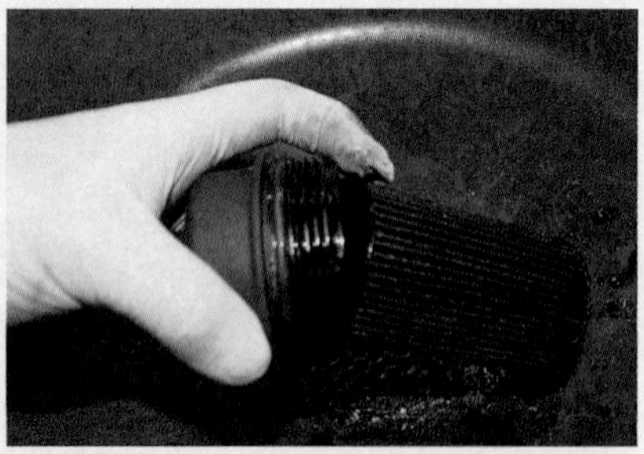

4. Remove the filter and clean the housing or end cap as necessary, and replace any O-rings or gaskets on the assembly.

5. Smear a little oil on the surface of any new O-rings. Install the new filter cartridge back into its housing.

6. If the cartridge is on the top of the engine, the service information may direct you to pour a specified amount of oil into the filter cavity before the end cap is installed. Screw the cartridge bolt or end cap back in place, and tighten to the specified torque. Be careful not to cross-thread the oil filter bolt or end cap.

After adding the required amount of oil, be sure to verify there are no leaks. Do this by starting the engine to fill the oil filter and build oil pressure. Then, shut the engine off to check for leaks and verify the level on the dipstick. Fill the oil to the max line and no farther.

To refill the engine oil, follow the steps in **SKILL DRILL 15-5**.

SKILL DRILL 15-5 Refilling Engine Oil

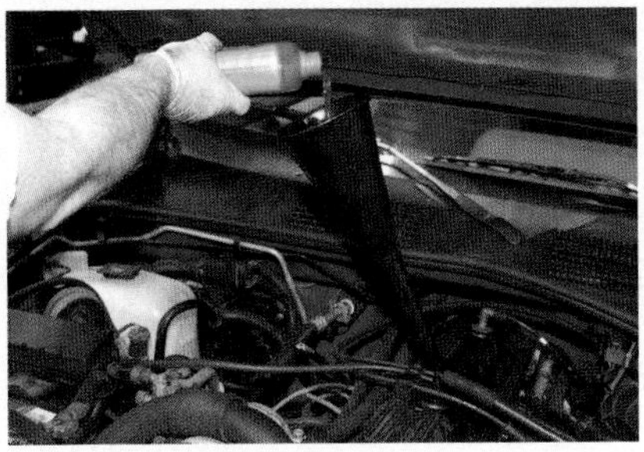

1. Research the correct grade and the quantity of oil you will need. Position the spout on the high side of the bottle. Pour the oil into the funnel slowly enough to avoid the risk of blowback or overflow. Fill the engine only to the level indicated on the engine dipstick. Replace the filler cap.

2. Start the engine and check the oil pressure indicator on the dash. If the oil pressure is inadequate, stop. Do not continue to run the engine.

3. If the oil pressure is good, turn the engine off and check underneath the vehicle to make sure no oil is leaking from the oil filter or drain plug.

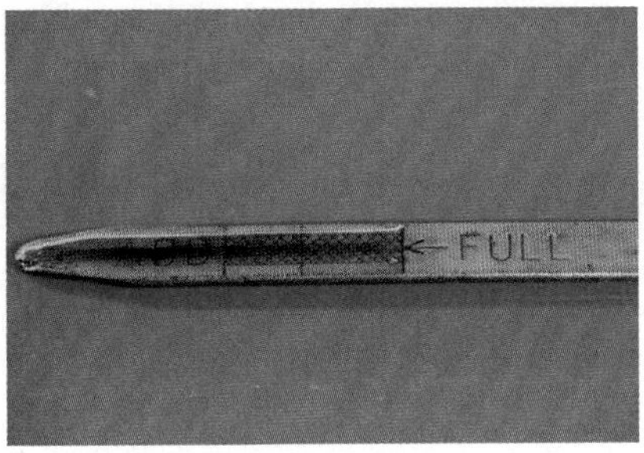

4. With the vehicle level, check the oil level again with the dipstick. It may be necessary to top off the engine by adding a small quantity of oil to compensate for the amount absorbed by the new filter. Do not overfill.

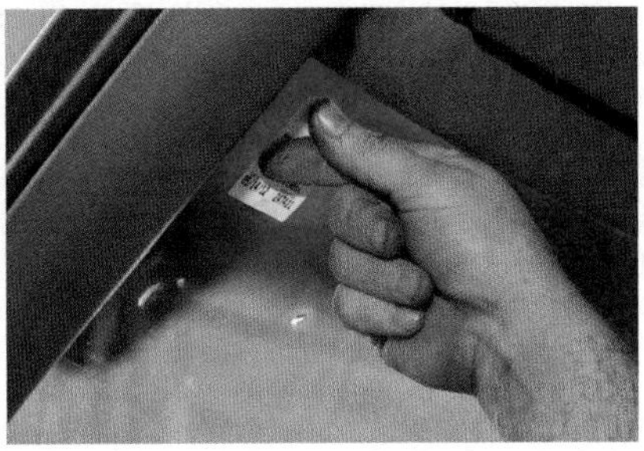

5. Refer to the service information, and install a static oil change reminder sticker.

6. Reset the maintenance reminder system to remind the owner when the next oil change is due.

Applied Math

AM-1: Volume: The technician can determine if the existing volume is within the manufacturer's recommended tolerance.

A technician was instructed to change the engine oil and filter on a vehicle. In addition to this, he was also to change the automatic transmission fluid and filter. The service information stated that 4.5 quarts (4.32 liters) of oil would be needed for an engine oil change with filter replacement. The service information specified 8.5 quarts (8.16 liters) for the automatic transmission fluid and filter change.

The technician begins working by draining the engine oil and filter. As the new oil is put into the engine, the technician counts the number of quart containers. He discovers that after the oil change was complete, it takes a total of 4.5 quarts (4.32 liters) to bring the oil level to the exact full mark on the dipstick. By this method, the technician determined that the existing volume was within the manufacturer's tolerance.

At this point, the technician drains the transmission fluid and changes the filter. As before, the technician counts the number of quart containers necessary for filling the system properly. He pours in 1 quart (0.96 liters) at a time in order to take an accurate count. The technician observes that it takes 8.5 quarts (8.16 liters) to fill the system to the full line, with the engine running in park. By this method, the technician verified that the existing volume is within the manufacturer's tolerance.

▶ Lubrication System Diagnosis

LO 15-03 Diagnose engine lubrication system issues.

In this section we cover simple lubrication system diagnosis and common issues. Issues such as oil leaks and improper oil level. Since this content is focused on maintenance and light repair, we will not cover diagnosis of low oil pressure issues. Also, we will not cover internal lubrication system issues. The skills taught below are focused toward entry level technicians. Let's get started.

Common Issues

Some of the common issues found in lubrication systems tend to stem from the same common variable—the customer. Changing the engine oil at the recommended interval is the most important maintenance task the customer can have performed. Failure to change oil results in sludge formation. Sludge clogs the oil passageways and keeps oil from lubricating critical parts. When bearings fail to get enough lubricant, they begin to wear and then will seize. Seized bearings result in the need for a new or rebuilt engine. An engine replacement is vastly more expensive than the relatively inexpensive oil changes the customer skipped. As technicians, we need to educate customers on the importance of oil changes.

Another common issue customers allow to happen is letting the oil level fall too low. Low oil level wears the oil out sooner, which reduces engine life. If it gets low enough, then air will be drawn into the lubrication system. Air does not provide the protection that oil does, so the engine experiences extreme wear. Often an engine will be damaged beyond use if it loses oil pressure for only a minute or two. Because of this, some engines use an automatic cut-out that turns the engine off if oil pressure falls too low.

The oil level can be low for a couple of reasons. It may have leaked away, or it may have burnt off. An engine can leak its lubricating oil from seals or gaskets that are intended to seal the oil in. The most common oil leaks are found at leaking:

- Oil drain plug gasket
- Oil filter seal
- Valve cover gaskets
- Oil pan gaskets
- Crankshaft seals
- Camshaft seals
- Timing cover gasket
- Oil cooler
- Oil pressure sensor

Since oil leaks leave evidence, customers will ask you to diagnose where that puddle of oil on the garage floor is coming from. Diagnosing oil leaks is primarily done by visually inspecting the engine looking for signs of the leakage. Oil will run down the engine from valve covers or the camshaft seals and may appear to be coming from a lower point. Remember to inspect all of the components listed above.

Oil leaking from gaskets will generally just leave trails that run down toward the ground. But if the oil hits a horizontal flange, it can run sideways before running down again. Oil leaking from seals on rotating parts will typically sling oil outward. For example, a crankshaft seal will sling oil around the front of the engine in a circular pattern. Gaskets and seals may only leak when the engine is running. Or they may only leak when the engine is off. You may need to perform the leak inspection both ways to locate the leak.

Another common area to look for leaks is on the oil pressure sending unit. The sending unit uses a diaphragm, which can develop a hole over time. When this happens, oil will typically leak out of the body either at the crimp or at the terminal end. This typically occurs while the engine is running. If you watch it while the engine is running, you will usually see it drip. Just be aware that if the sending unit is at a downward angle, then oil from another location can possibly run down to it and then drip off of it. So, always look above the sending unit. If there is little to no oil above it, the sending unit is likely faulty.

On engines that are coated with oil, it can be difficult finding the source of the leak. In this case, you may need to do some preparation work. One option is to steam-clean the engine to remove the oil and grime. This will make it easier to see where the oil is actually coming from. The second option is to use a fluorescent dye and ultraviolet light. The dye is added to the engine oil and the engine is run for an appropriate amount of time. Then, the engine is inspected for leaks using the ultraviolet light. The fluorescent dye shows up under the light making it easier to determine the leak, even if it is small.

▶ TECHNICIAN TIP

Remember that oil will sometimes blow upward when the vehicle is driving down the highway. It can also blowback under the vehicle, so always try to imagine how oil is being thrown around. The use of fluorescent dye may be necessary if a leak is covering more than one area of the engine. It may also be necessary to pressure-wash the undercarriage and retest for a large leak.

▶ TECHNICIAN TIP

In the case of finding a leak, always remember that you need to find the root cause of the problem. Gaskets and seals are only designed to withstand a certain amount of pressure before they leak. If the engine has excessive blowby or the PCV system is not working properly, pressure can build up in the crankcase. This can cause oil to leak past gaskets and seals. A quick way to check for crankcase pressure is to start the engine. Then, remove the crankcase fresh air intake tube or hose from the engine and cover the engine side of it. If a vacuum develops, the crankcase is being vented. If pressure develops, further diagnosis of the PCV system and piston rings is necessary.

As mentioned above, the oil level can also be low due to oil being burned in the engine. Here are the most common causes of engines that burn oil:

- Worn/sticking piston rings
- worn valve guides/seals

- intake manifold gasket leak
- turbo seals, if equipped or
- by an improperly operating PCV system.

Each of these conditions allows oil into the combustion chamber, where it is burned. If the amount of oil that is burning is small, then the falling oil level may be the only indicator. As the amount of burning oil increases, it will begin producing smoke out of the exhaust pipe. This is especially noticeable at higher engine rpm or when decelerating.

Burning oil also leaves deposits on combustion chamber and exhaust system components. One of the easier places to see it is on the combustion chamber side of the spark plug insulators. It also coats the oxygen sensors and catalytic converter. In each of these cases, the deposits hinder the operation of the component.

Diagnosing the cause of burning oil starts with verifying that the engine is not leaking oil externally. Once that is determined, the rate at which the engine burns oil must be measured. This is done by performing an **oil consumption test**. Since all engines burn oil, most manufacturers specify the maximum rate of consumption. This typically varies from about a quart every 3,000 miles (4,828 km) to a quart every 1,000 miles (1,610 km).

The oil consumption test consists of the following steps:

- Confirm there are no external oil leaks.
- Top off the oil to the full line.
- Seal the oil fill cap, dipstick, and drain plug.
- Customer drives the vehicle for the specified distance.
- Vehicle is returned to the shop and the oil level measured.
- The oil consumption is calculated and compared to specifications.

If the vehicle fails the oil consumption test, some manufacturers will repair the vehicle under warranty. But the vehicle must have less than a certain number of miles on it to qualify.

Too little oil in the engine is a problem, but so is too much oil. With excess oil, the crankshaft can whip it into foam and cause leaks by flooding the seals. Another problem with excessive oil is the drag produced when the crankshaft hits the oil in the oil pan. The oil slows the crankshaft down. This condition will cause a loss of power and result in more fuel consumption. Additionally, the foaming caused by the crankshaft hitting the oil will result in starving the engine of precious lubricating oil. The pump will send these air bubbles to the moving parts which cannot provide the lubrication they need.

Tapping noises are another lubrication issue found in engines. Low oil pressure can result in valve lifters not pumping up. This creates a tapping noise in the valve train. Many times, this noise is due to either low oil level, or thinned-out oil. In this case, the oil level and condition should be checked first. If the oil level and condition are good, then use an oil pressure gauge to measure the oil pressure.

- If the pressure is good, then suspect faulty lifters or flat camshaft.
- If the pressure is bad, then suspect a faulty oil pump, excessive bearing clearance, or internal oil leak.

Knocking noises occur when the oil cannot cushion the parts it separates. This can be due to several issues. The first issue can be worn parts with improper running clearances. Second, the oil pressure may be too low. And third, the oil may be too thin. In this case, the oil level and condition should be checked first. If the oil level and condition are good, then use an oil pressure gauge to measure the oil pressure.

- If the oil pressure is normal, then use the sound of the noise to locate the component with excessive clearance. A single deep knock is likely a faulty main bearing. A sharper double rap is likely a connecting rod bearing. Knocking noises that only happen when the engine is cold are likely to be from a collapsed piston skirt.
- If the oil pressure is lower than specifications, then suspect a faulty oil pump, excessive bearing clearance, or internal oil leak.

▶ Wrap-Up

Ready for Review

- ▶ Oil level is typically checked with a dipstick located under the hood. A vehicle's oil-life monitor is the best guide for knowing when to change the oil. Oil analysis of the engine oil lists physical properties such as viscosity, condensed water, fuel dilution, antifreeze, acids, metal content, and oil additives.
- ▶ Always follow the manufacturer's engine oil change interval. Normal use and severe use have different oil change intervals.
  - When draining the engine oil, use disposable gloves to reduce the risks of being burned from hot engine oil, locate the correct drain plug, consider the path the oil will take from the drain plug, inspect the drain plug gasket, and tighten the drain plugs to proper torque.
- ▶ Most common oil leaks are found at leaking: oil drain plug gasket, oil filter seal, valve cover gaskets, oil pan gasket, crankshaft seals, camshaft seals, timing cover gasket, oil cooler, and oil pressure sensor.
  - Worn/sticking piston rings, worn valve guides/seals, intake manifold gasket leak, turbo seals, or an improperly operating PCV system can cause an engine to burn oil.

Key Term

oil consumption test A test designed to measure the amount of oil used in a certain amount of miles.

Review Questions

1. How much oil does it typically take to raise the oil level from the bottom of the "Safe" mark to the "Full" mark on a passenger car dipstick?
   a. 1 pint
   b. 1 gallon
   c. 1 quart
   d. 1 pound
2. If the oil level is low, which of the following is likely to be the source of the leak?
   a. Leaking exhaust manifold gasket
   b. Leaking oil pressure sending unit
   c. Misadjusted valves
   d. Loose spark plugs
3. If the oil on the dipstick looks milky gray, what could cause this?
   a. Engine is running too rich
   b. Oil is worn out
   c. Water in the oil
   d. Engine is misfiring
4. Which of the following is the best way to determine if the engine oil needs to be changed?
   a. Time and mileage
   b. Oil type
   c. Appearance and condition
   d. Oil life monitor
5. Which statement is correct with respect to changing engine oil?
   a. The engine should be fully warmed up
   b. The engine should be cold
   c. The drain plug should be tightened until it stops leaking
   d. Normal use and severe use have the same change interval
6. When performing an oil change, the filter must be replaced. What must you do when installing a spin-on filter?
   a. Make sure the O-ring is installed dry
   b. Put thread sealer on the filter threads
   c. Make sure the old O-ring is removed
   d. Tighten it with an oil filter wrench
7. All of the following could cause oil to be burned in the engine EXCEPT:
   a. Worn/stuck piston rings
   b. Worn valve guides/seals
   c. Improperly operating PCV system
   d. Leaky valve cover gaskets
8. When changing a cartridge oil filter (Replaceable element):
   a. Tighten the end cap to the specified torque
   b. Only change it when the engine is cold
   c. Always replace the filter housing with a new one
   d. Reuse the seals and O-rings
9. If oil pressure is lower than specifications, all of the following could be the cause EXCEPT:
   a. Faulty oil pump
   b. Leaky valve cover gaskets
   c. Excessive bearing clearance
   d. Internal oil leak
10. After refilling the engine with oil, all of the following should be done EXCEPT:
    a. Start the engine and check for leaks
    b. Shut the engine off and recheck the engine oil level
    c. Tighten the oil filter an additional half turn
    d. Fill the oil to the max (Full) line

ASE Technician A/Technician B Style Questions

1. When checking the engine oil level, the vehicle should be on a level surface. Technician A says to wipe the dipstick off and then recheck the level. Technician B says that when reading the dipstick, hold it horizontal. Who is correct?
   a. Technician A
   b. Technician B
   c. Both A and B
   d. Neither A nor B
2. Technician A says to use time and mileage to determine engine oil change interval. Technician B says to check if the oil is still slippery and if not change it. Who is correct?
   a. Technician A
   b. Technician B
   c. Both A and B
   d. Neither A nor B
3. When changing engine oil. Technician A says to let the vehicle sit for one hour to drain. Technician B says that some engines have more than one oil drain plug. Who is correct?
   a. Technician A
   b. Technician B
   c. Both A and B
   d. Neither A nor B
4. Performing an oil change requires the replacement of the oil and the filter. Technician A says that the oil filter may be a spin-on design. Technician B says that the oil filter may be a cartridge design. Who is correct?
   a. Technician A
   b. Technician B
   c. Both A and B
   d. Neither A nor B
5. Engine oil leaks can be difficult to diagnose. Tracing the oil trail back will lead to the source. Technician A says that oil is slung outwards from a rotating component. Technician B says that oil leaks from gaskets generally spray pretty far. Who is correct?
   a. Technician A
   b. Technician B
   c. Both A and B
   d. Neither A nor B

6. If a customer complains that an engine is using oil, further diagnosis is required. Upon inspection, no signs of an external oil leak are discovered. Technician A says to add dye to the oil and use a black light for closer analysis. Technician B says to perform an oil consumption test. Who is correct?
   a. Technician A
   b. Technician B
   c. Both A and B
   d. Neither A nor B
7. Engine oil level is critical to the correct operation of an engine. Technician A says that too much oil may cause increased fuel consumption. Technician B says that too little oil may cause valve train noises. Who is correct?
   a. Technician A
   b. Technician B
   c. Both A and B
   d. Neither A nor B
8. Technician A says that after changing the engine oil and filter, you should apply a static sticker to remind the owner when the next oil change is due. Technician B says that you should reset the maintenance reminder system to remind the owner when the next oil change is due. Who is correct?
   a. Technician A
   b. Technician B
   c. Both A and B
   d. Neither A nor B
9. Large fleets may use oil analysis to schedule preventive maintenance. Technician A says that the oil analysis will show the metal content of the oil. Technician B says that the oil analysis will determine remaining oil life. Who is correct?
   a. Technician A
   b. Technician B
   c. Both A and B
   d. Neither A nor B
10. During an oil change, the filter should be replaced. Technician A says the new screw-on filter should be installed and tightened by hand. Technician B says the new filter gasket should be lubricated before installation. Who is correct?
   a. Technician A
   b. Technician B
   c. Both A and B
   d. Neither A nor B

CHAPTER 16
Cooling System Theory

Learning Objectives

- **LO 16-01** Describe the methods of heat transfer, cooling system configurations, and their operation.
- **LO 16-02** Describe engine coolant and its required properties.
- **LO 16-03** Describe coolant flow in engines.
- **LO 16-04** Describe the radiator and its associated components.
- **LO 16-05** Describe the operation of the thermostat and water pump.
- **LO 16-06** Describe the operation of cooling fans.
- **LO 16-07** Describe hoses, belts, and tensioners.
- **LO 16-08** Describe miscellaneous cooling system components.

ASE Education Foundation Tasks

See Appendix A to view the 2017 ASE Education Foundation Automobile Accreditation Task List Correlation Guide.

▶ Cooling System Purpose

LO 16-01 Describe the methods of heat transfer, cooling system configurations, and their operation.

Cooling systems play a critical role in the life span of the engine (**FIGURE 16-1**). Typically, a great deal of focus is placed on the maintenance of the engine's lubrication system. But little focus is placed on the engine's cooling system. And yet, the Department of Transportation has stated that cooling system failure is the leading cause of mechanical breakdowns on the highway. So, it deserves our attention. The cooling system also can have a huge effect on:

- the lubrication system
- engine emissions and
- fuel economy, thus making its role a significant one.

In fact, a myth regarding the cooling system is that it keeps the engine cool. That is not true. The purpose of the cooling system is to:

- allow the engine to reach its ideal operating temperature as quickly as possible and
- maintain it during all operating conditions.

A cooling system can fail in two ways. The most common failure most people are familiar with is allowing the engine to overheat. This is usually evident by steam pouring out of the engine compartment. Overheating can lead quickly to major engine damage.

FIGURE 16-1 Cooling systems are the leading cause of mechanical breakdowns on the highway.

The second way the cooling system fails is by preventing the engine from reaching its ideal operating temperature, either at all or too slowly. This may be evident by the heater not getting very warm or the defrost not working very well. Although this does not cause damage immediately, it does cause the engine to wear out much more quickly. Low engine temperatures also cause poor fuel economy and high vehicle emissions.

This chapter explains how the principles of heat transfer are used to keep the engine within its ideal temperature range. We also explore how each of the cooling system components contributes to that goal.

Heat Transfer

Heat is thermal energy. It cannot be destroyed; it can only be transferred. It always moves from areas of higher temperature to areas of lower temperature (**FIGURE 16-2**). This principle is applied to the transfer of heat energy from engine parts to **ambient** air. Coolant is used as a medium to carry the heat. To control this heat transfer, it is necessary to understand how heat behaves. Heat travels in only three ways (**FIGURE 16-3**):

1. From one solid to another, by a process called **conduction**.
2. Through liquids and gases, by a process called **convection**, whereby heat follows paths called convection currents.
3. Through space, by **radiation**.

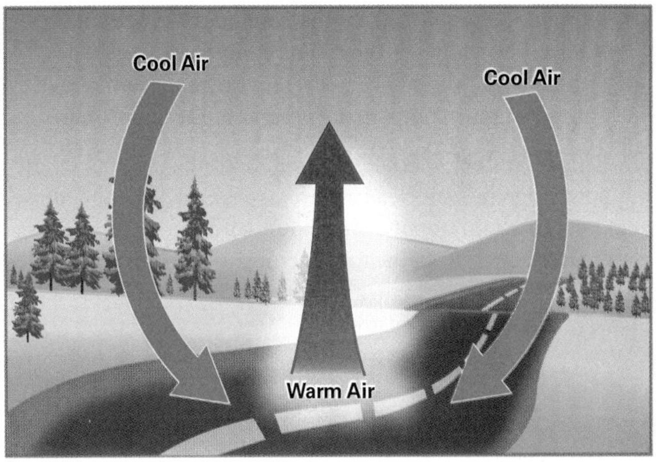

FIGURE 16-2 Heat always moves from hot areas to cooler areas.

You Are the Automotive Technician

You are discussing cooling system theory with a fellow technician. She said that her teacher taught the cooling system theory in a way that made sense to her, and she feels confident that she knows how they operate. You told her that you feel confident as well, and she should try to stump you with some questions. Answer the following questions she asks you:

1. What are the three ways that heat transfers from one place to another?
2. How does a thermostat regulate engine temperature?
3. How does pressure affect the boiling point of coolant?
4. What are the different types of fans, and how do they operate?

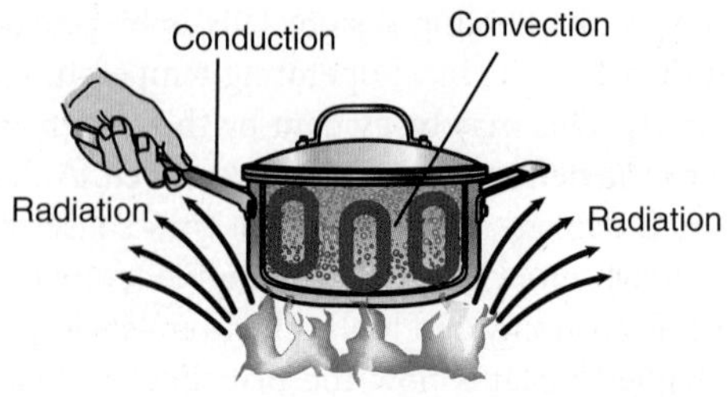

FIGURE 16-3 Conduction is the movement of heat energy through solids; convection is the movement of heat through liquids or gases; and radiation is the movement of heat through space.

Heat Transfer in Internal Combustion Engines

The internal combustion engine relies on the heat of combustion to produce torque. Torque is used to move the vehicle and power the accessories. Unfortunately, much of the heat produced during combustion is wasted. No matter how efficient an engine is, the heat energy never completely transforms into kinetic energy (**FIGURE 16-4**). Excess heat must be removed to avoid overheating of the engine. Excess heat energy is typically wasted in three ways:

- About 33% is wasted by being dumped straight out of the exhaust to the atmosphere (some of this wasted energy can be recovered by a turbocharger).
- About 33% is wasted by the cooling system, which prevents overheating of the engine components.
- About 5% is wasted by internal friction and from radiating off hot engine components straight to the atmosphere.

This leaves only about 25% to 30% of the original energy for powering the vehicle. However, turbocharged and diesel engines can attain about 35% to 40% efficiency. Turbochargers recover some of the heat energy lost out the exhaust. Diesel engines operate at higher compression ratios. These factors increase engine efficiency greatly.

Applied Science

AS-27: Heat: The technician can demonstrate an understanding of the effect of heat on automotive systems.

The extreme heat created by burning fuel in the internal combustion engine causes a rapid rise of pressure. This drives the pistons. Some of the heat from the combustion process must be dissipated so that it does not overheat or melt the engine parts. This is the job of the engine cooling system. It helps the engine to quickly come up to its specified operating temperature. It also helps maintain that temperature in spite of ambient weather conditions or engine load.

If the engine is not allowed to warm up properly, other systems will not operate properly. For example, the engine temperature greatly affects the air–fuel mixture (fuel trim). Fuel trim affects fuel economy and emissions. Operation of the evaporative emission system and other systems are also affected by engine temperature. The cooling system must be maintained so it operates correctly.

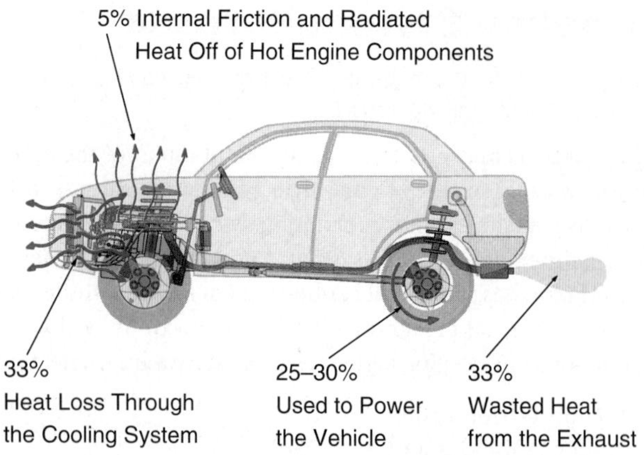

FIGURE 16-4 Approximate heat loss in a gasoline engine.

Principles of Engine Cooling

From the second law of thermodynamics, we know that heat always moves from hot to cold. The automotive cooling system provides a means of transferring heat from the hot parts of the engine to ambient air. This is accomplished through a group of parts working together to circulate coolant throughout a sealed system. They carry heat away from those engine parts. Heat can then be released in one of two places. (1) To outside air away from the vehicle. (2) Or to air entering the passenger compartment, which can be used for comfort and safety purposes.

Manufacturers use various cooling system configurations. Regardless of the design used by the manufacturer, the job is the same; maintain the ideal operating temperature. The ideal operating temperature for most engines is somewhere around 200°F (93°C), give or take 20°F (11°C), depending on the vintage of the vehicle.

There are two types of cooling systems: liquid cooled and air cooled. Most automotive engines are liquid cooled. A liquid-cooled system uses coolant to transfer heat throughout the system. Coolant is made up of special antifreeze and anticorrosion chemicals mixed with water. In general terms, a water pump causes coolant to circulate through **water jackets** in the engine block and cylinder head. The coolant picks up heat along the way. It then gives up much of that heat when it passes through a radiator (**FIGURE 16-5**).

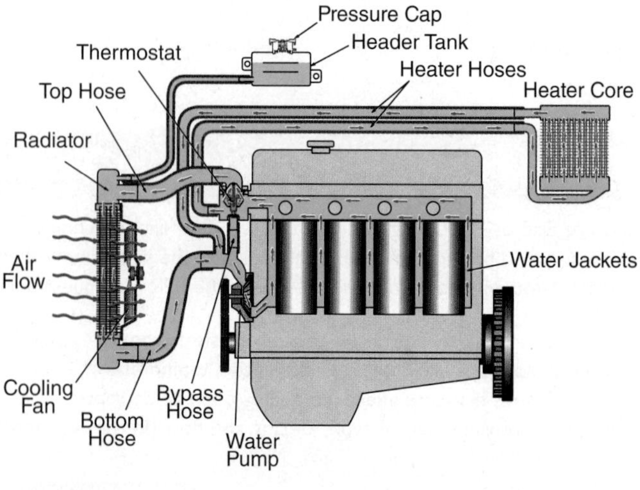

FIGURE 16-5 Simple cooling system.

As air moves over and through the radiator, the heat energy is transferred from the coolant to the ambient air. The lower temperature coolant is then returned to the engine to absorb more heat energy. And the cycle continues.

In modern vehicles, radiators are low and wide to allow the hood to sit lower to the ground. This improves aerodynamics. Because of this design, modern vehicles use a water pump to circulate coolant through the system.

The circulation of the coolant is controlled by the thermostat. It opens and closes to control coolant flow from the engine to the radiator (**FIGURE 16-6**). When the engine is cold, coolant circulates through the engine (and heater core) only. Once the engine warms up, the thermostat opens. It allows the coolant from the engine to flow through the radiator. In this manner, a thermostat is used to regulate the coolant temperature.

The thermostat initially blocks coolant flow to the radiator. This keeps the coolant circulating in the engine, where it heats up quickly. Once operating temperature is reached, the thermostat starts to open and allows the coolant to flow to the radiator. The thermostat does not normally open completely. It operates somewhere between its fully closed or fully open position. This helps maintain an optimum engine operating temperature regardless of engine load.

Some newer vehicles use a coolant heat storage system. This system uses a vacuum-insulated container, similar to a thermos bottle. The storage container holds an amount of hot coolant. It maintains the temperature for up to three days after the engine is shut down (**FIGURE 16-7**). As the engine is started the next time, a small electric water pump circulates this hot coolant through the engine. The hot coolant preheats the engine. This greatly reduces hydrocarbon exhaust emissions during start-up. And it substantially shortens the warm-up time.

Air cooling is common on smaller internal combustion engines. Most of these engines use cooling fins around the cylinders and heads. Their design makes the exposed surface area as large as possible. Many air-cooled engines also use a fan to direct air through and over the cooling fins. This increases the cooling capacity.

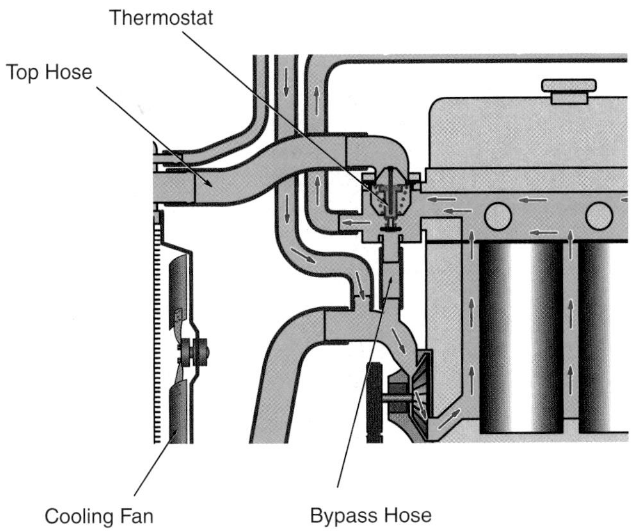

FIGURE 16-6 The thermostat controls the flow of coolant.

Air-cooled engines have fallen out of favor in passenger vehicles because they are harder to maintain at a stable temperature. This causes increased emissions output. When air-cooled engines are used in automobiles, they are usually not exposed to the air. Rather, the engine is housed in an enclosed engine bay. Still, for a vehicle moving at highway speed, airflow over the engine may be high enough to prevent overheating. At lower speeds or during idling, heat may build up and overheat the engine.

One way to remove excess heat from the engine is to use a fan, along with shrouds and ducts. This directs air over, or even between, the cylinders. An engine that uses these components is called a "forced draft" air-cooled engine. Some air-cooled engines are "open draft" air-cooled. This type requires the engine to be moving through the air to have sufficient airflow (**FIGURE 16-8**).

▶ Vehicle Coolant

LO 16-02 Describe engine coolant and its required properties.

Coolant is a mixture of water and antifreeze solution, which is used to carry heat from the engine. If an engine did not have the heat removed from it, it would fail very quickly. Coolant absorbs the heat from the engine by convection. Because coolant contacts the hot metal directly, it is a very effective heat sink. The coolant is important for three reasons:

1. It prevents an engine from overheating while in use.
2. It keeps the engine from freezing while not in use in cold climates.
3. It prevents corrosion of the parts in the cooling system.

Water alone is a better coolant than antifreeze. It can absorb a larger amount of heat than most other liquids. But water has some drawbacks. It freezes if its temperature drops below 32°F (0°C) (the temperature at which water becomes a solid). As water freezes, it expands approximately 10% as it becomes a solid (**FIGURE 16-9**). If it expands in the coolant passages inside the engine, these passages—typically made of cast aluminum or cast iron—cannot flex to allow expansion and will break. This causes blocks and heads to crack. The cracks allow coolant to leak internally inside the engine. In most cases, the engine is unrepairable.

Another thing to realize about only using water as a coolant is that water is corrosive. It causes metal to rust and corrode. Think about a piece of unpainted metal that is lying outside in the rain. Rust and corrosion build up quickly on that metal. This is simply because of the reaction (oxidation) of the metal, water, and oxygen.

Antifreeze prevents corrosion and rusting through anti-corrosion additives mixed into the solution. Another important note on water is that water contains minerals. Minerals potentially leave excessive deposits even when added to antifreeze. Because of this, most manufacturers recommend using distilled water for cooling systems.

Antifreeze is made from one of two base chemicals—ethylene glycol or propylene glycol. A small amount of additives are used to protect against corrosion and foaming.

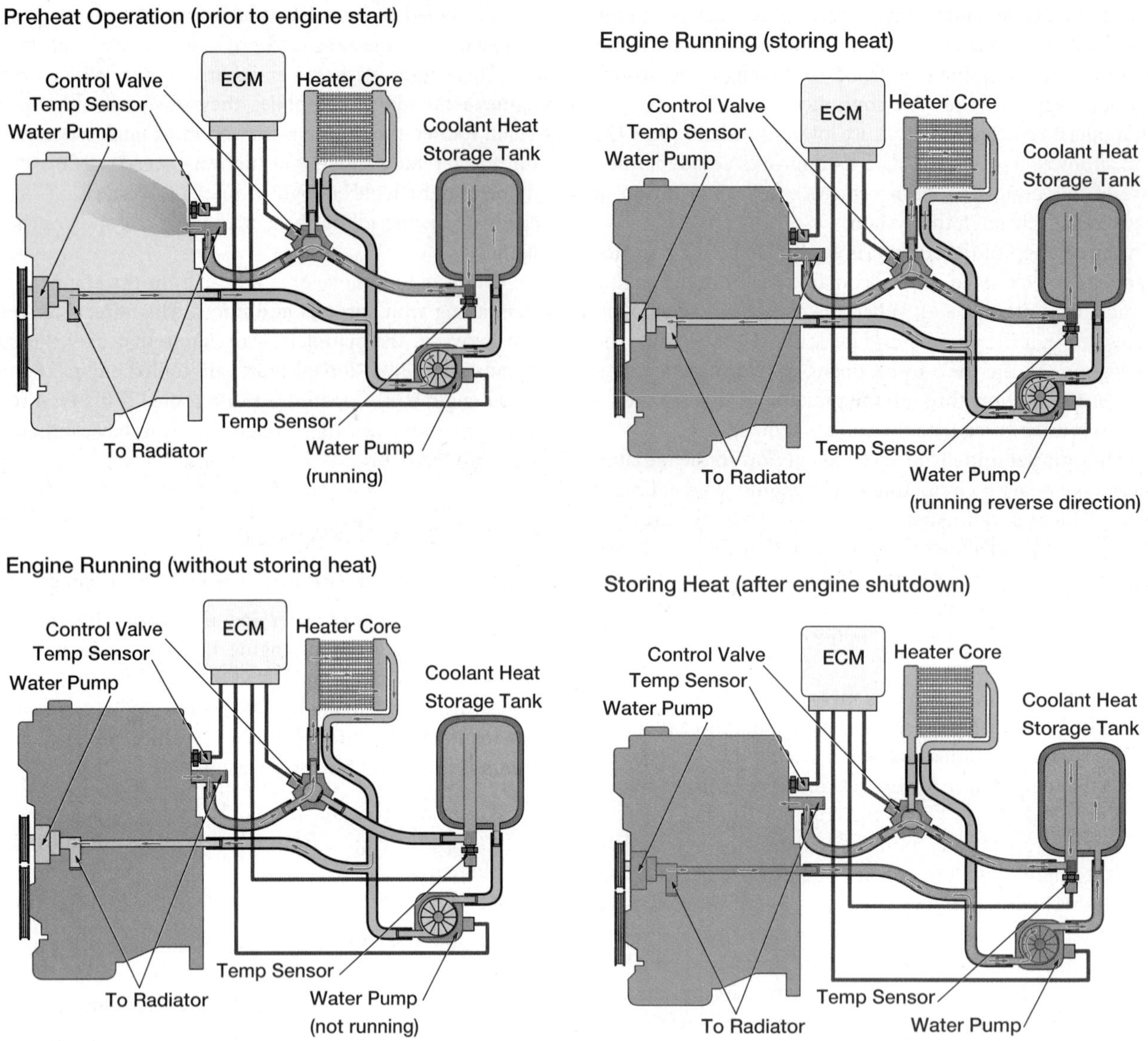

FIGURE 16-7 Toyota coolant heat storage system.

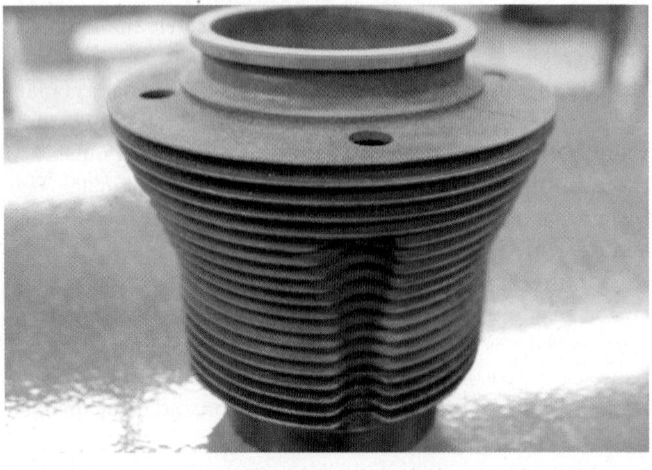

FIGURE 16-8 A cylinder from an air-cooled Volkswagen engine. Note the cooling fins around the cylinder.

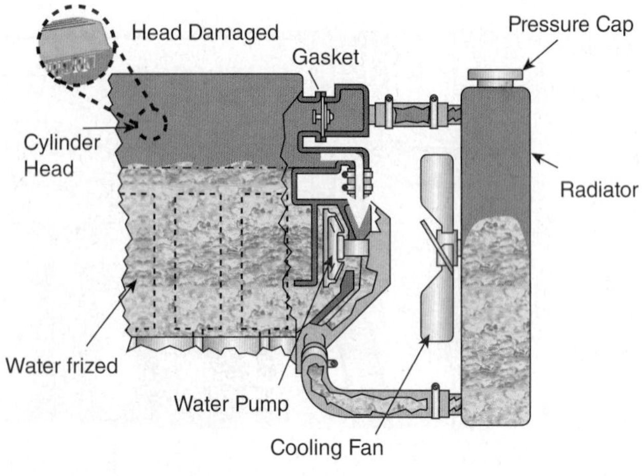

FIGURE 16-9 Water expands as it freezes. This can crack blocks and heads.

(**FIGURE 16-10**). Ethylene and propylene glycol may achieve a maximum very low freezing point of −70°F (−57°C) when mixed with the appropriate amount of water. **Ethylene glycol** is very toxic to humans and animals. **Propylene glycol** is not toxic and is used in nontoxic antifreezes. Both of these antifreezes actually freeze around 0°F (−18°C) if not mixed with water. So, water actually lowers the freeze point of antifreeze. It also absorbs more heat than antifreeze. So it is a very necessary part of coolant.

FIGURE 16-10 A. Ethylene glycol. **B.** Propylene glycol.

The freezing point of coolant will vary depending upon how much water is added to the antifreeze (**FIGURE 16-11**). Because antifreeze does not absorb heat as effectively as water, it should not be mixed at a ratio higher than 65% antifreeze and 35% water. Using a higher proportion of antifreeze will actually reduce the cooling capacity of the mixture. This will raise the operating temperature of the engine.

The most common coolant mixture is a 50/50 balance of water and antifreeze. This is an ideal coolant for most hot and cold climates. It also provides adequate corrosion protection. Also, at a 50% mixture, the boiling point increases from 212°F (100°C) to around 228°F (109°C). As you can see, these are extremely beneficial characteristics of antifreeze. This is even more true as manufacturers continue to build engines that are more powerful, create more heat, and operate at higher temperatures.

Antifreeze can be purchased as full strength (100%) or as a 50/50 premix with water. Full-strength antifreeze that you buy from the dealer or parts store consists of three parts:

- glycol (around 96%),
- corrosion inhibitors and additives (around 2–3%), and
- water (around 2%).

Glycol, as discussed previously, keeps the freezing point low and the boiling point high. Corrosion inhibitors and additives:

- prevent corrosion and erosion,
- resist foaming,
- ensure coolant is compatible with cooling system component materials and hard water,
- resist sedimentation,
- lubricate the water pump seals, and
- balance the acid-to-alkaline content of the antifreeze.
- Water is added to blend the inhibitors with the glycol.

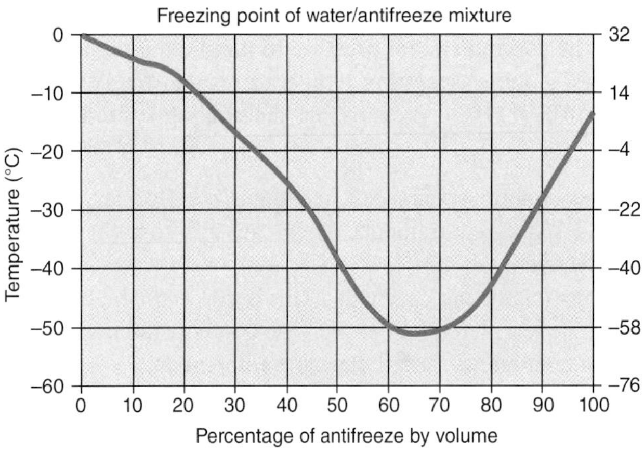

FIGURE 16-11 Freezing point of antifreeze and water solution.

Antifreeze is an amazing chemical that performs a monumental task in the operation of our vehicles. It works so well that it is often overlooked for maintenance by the customer. However, the additives wear out and become less effective over time. So, coolant does need to be changed at recommended intervals. Doing so reduces the possibility of engine damage and failure.

Boiling Point and Pressure

The **boiling point** of a liquid is the temperature at which it begins to change from a liquid to a gas. Water at sea level atmospheric pressure of 14.7 psi (101.4 kPa or 1 atmosphere [atm]) boils at 212°F (100°C). Atmospheric pressure decreases as elevation is increased. Because atmospheric pressure is lower at higher elevations (such as in the mountains), the boiling temperature of a liquid is lower. Think of it as *lower pressure = lower boiling point.* Conversely, raising the pressure has the opposite effect. It raises the boiling point of a liquid. Stated another way:

- Liquid in a vacuum = lower boiling point.
- Liquid under pressure = higher boiling point.

Thus, the boiling point of a liquid varies depending upon the surrounding environmental pressure.

Raising the pressure enables the coolant to remain a liquid at temperatures well above the normal boiling point of 212°F (100°C). The pressurized coolant boils at a higher temperature than it would outside the system. It's the same if the system has a leak. This is because for every psi the pressure is raised on coolant, the boiling point is raised about 3°F (1.67°C).

SAFETY TIP

You should *never* remove a radiator pressure cap on a hot engine. If the coolant is above 212°F (110°C), all of the coolant will turn to steam as soon as the pressure is released. This pushes superheated steam out of the radiator, potentially scalding you.

Over the years, manufacturers have intentionally raised the operating temperature of their engines. Higher temperatures improve efficiency and reduce emissions. This required raising the cooling system pressure to handle the higher temperatures. Engine operating temperatures can be as high as about 250°F (121°C). Pressurizing the cooling system means that:

- The cooling system can be downsized (for less weight and space requirement). They can still cool the engine effectively.
- The radiator can be smaller. This is due to the higher temperature differential between the cooling system's operating temperature and the outside ambient air.

In today's vehicles, most automotive cooling systems are pressurized at 13–21 psi (89.6–144.8 kPa). However, some factory radiator caps are rated as high as 24 psi (165.5 kPa). And high-performance systems can go as high as 34 psi (234.4 kPa) (**TABLE 16-1**).

TABLE 16-1 Boiling Point of Water at Various Pressures

(Boiling point increases 3°F per each psi of increased pressure.)
13 psi: 212 + 39 = 251°F (121.6°C)
15 psi: 212 + 45 = 257°F (125°C)
19 psi: 212 + 57 = 269°F (131.6°C)
21 psi: 212 + 63 = 275°F (13°5C)
24 psi: 212 + 72 = 284°F (140°C)

▶ TECHNICIAN TIP

The term "psig" is another way of saying the pressure that is read on a normal pressure gauge. The "g" stands for gauge. So, gauge pressure is the amount of pressure measured above atmospheric pressure (1 atm). A compound pressure gauge typically reads both above and below atmospheric pressure. It measures psig above zero, and inches of vacuum (or mm of mercury) below zero.

In contrast, the term "psia" means *absolute pressure*, which does not start at atmospheric pressure. In other words, a psia gauge would read 14.7 psig at sea level. If you want to know the absolute pressure (in psia), just add gauge pressure (psig) to atmospheric pressure (usually considered 14.7 psi at sea level).

Applied Science

AS-30: Fusion/Vaporization: The technician can demonstrate an understanding of how heat causes a change in the state of matter.

Vaporization and fusion are two ways in which the application of heat can cause matter to change states. Vaporization refers to the transitional phase from liquid to vapor. Fusion refers to the transitional phase from solid to liquid. Both concepts have applications in the shop.

Pressurized cooling systems in motor vehicles are designed to prevent vaporization; in this case, from boiling. Water boils at 212°F (100°C) at atmospheric pressure at sea level. When coolant in the cooling system is pressurized, its boiling point is raised. This allows the system to continue to cool the engine at temperatures above 212°F (100°C). Vaporization must be avoided because, unlike liquids, vapors do not absorb as much heat from engine components. Many cooling systems have pressure limits of 14 to 15 pounds per square inch (psi) (97 to 103 kilopascals [kPa]). This pressure effectively raises the boiling point of the coolant by approximately 45°F (25°C).

In contrast, water turns from liquid to solid (freezes) at 32°F (0°C). When water freezes, it expands approximately 10%. Pressure from ice is strong enough to break cast iron heads and blocks. This can cause leaks in the cooling system and very expensive repairs. Antifreeze is used to lower the freezing point of water to prevent freezing of the water. The antifreeze and water mixture is called coolant.

Electrolysis

Electrolysis is the process of pulling chemicals (materials) apart by using electricity. It is also the process of creating electricity through the use of chemicals and dissimilar metals. Electrolysis is used in manufacturing to create some metals, gases, and chemicals. You may have performed electrolysis in a science class. A common experiment is to pass electrical current through water to break the hydrogen out of the water.

Electrolysis is what takes place in an automotive battery. Two dissimilar metals are submerged in an acid solution, resulting in the reaction that is called electricity. Another common experiment in science class is the potato battery. It is made from two different types of metal stuck into the potato. It is used to power a digital clock or a lightbulb.

Electrolysis can occur in places where it is not desired. This has undesirable effects in automotive cooling systems. Electrolysis is possible when the coolant breaks down and becomes more acidic. Many types of metals are used in the engine, such as cast iron, aluminum, copper, and brass.

Introducing an acid solution into a mix of metals produces electricity. When this electricity is produced, the movement of the electrons begins to erode away the metals in the system (**FIGURE 16-12**). Eventually, pinholes are created in the thinner, softer cooling system components. Most commonly, this occurs in the heater core, radiator core, or aluminum cylinder head. To combat electrolysis, customers should have the old coolant flushed out and replaced with new coolant.

Electricity can also appear in the cooling system due to faulty grounds. This can occur on accessories or even the starter motor circuit. Electricity is known to follow the path of least resistance and will be partially carried through the coolant. It can then erode metals, if that path is easier than its intended path. We discuss how to test for electrolysis and also revisit the topic of bad grounds later in this chapter.

Coolant Types

As mentioned earlier, water-cooled engines must be protected from freezing, boiling, and corrosion. Water absorbs a larger amount of heat than most other liquids. But it freezes at a relatively high temperature. It is also corrosive. Mixing antifreeze with water provides an adequate coolant solution. It lowers the water's freeze point, raises the boiling point, and provides anticorrosive properties.

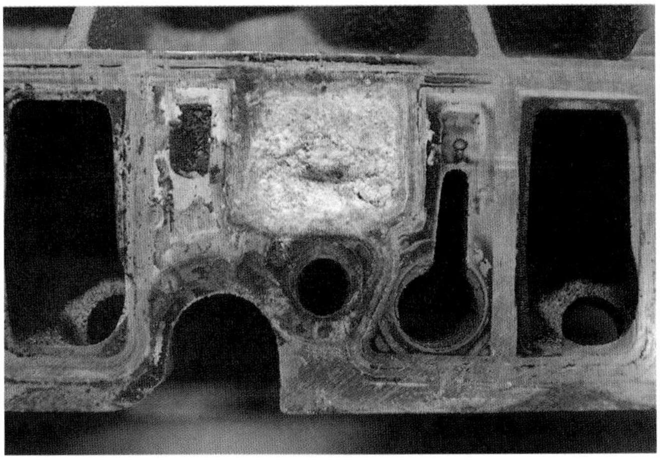

FIGURE 16-12 Erosion caused by electrolysis.

There are several types of coolants available for use in liquid-cooled engines. The recommended coolant depends on the original equipment manufacturer's (OEM) recommendation. This is influenced by:

- the metallurgy of the engine parts and
- how long the coolant is designed to last.

It is important to note that brands and types of coolant (antifreeze) differ from one manufacturer to another. Some believe coolant can be identified according to its color, which may be anything from green or purple to yellow/gold, orange, blue, or pink. OEM cooling system designs and coolant recommendations have changed in recent years. So, now, the color of coolant is no longer a reliable way to identify a particular type of antifreeze.

Mixing types of antifreeze can cause a reaction that turns the chemicals in antifreeze to sludge. This sludge can plug up the passages in the system, including the radiator and heater core. Always read the container label and follow OEM coolant recommendations.

Most coolant types start with a base of ethylene glycol. Specific corrosion inhibitors, lubricants, and other chemicals are added. The corrosion inhibitors determine the type of coolant it is. Maintaining the proper coolant acid/alkaline pH balance is also critical to cooling system life. This means that pH testing can help determine when to perform coolant replacement.

Ethylene glycol is a toxic chemical that works very well as an antifreeze. Ethylene glycol mixes well with water and has a low viscosity. This allows it to circulate easily through the cooling system. Propylene glycol performs essentially the same as ethylene glycol and is not toxic. In fact, propylene glycol antifreeze is sold as a nontoxic coolant.

The types of ethylene glycol or propylene glycol antifreeze available are based on the type of corrosion inhibitors. The most common types of corrosion inhibitors are as follows:

- Inorganic acid technology (IAT)
- Organic acid technology (OAT)
- Hybrid organic acid technology (HOAT)
- Poly organic acid technology (POAT)

The first category of coolant, IAT, became available in the 1930s and was green in color (**FIGURE 16-13**). This coolant is still in use today. It contains phosphate and silicate as corrosion inhibitors. Phosphate protects iron and steel parts, and silicate keeps aluminum from corroding. IAT coolant needs to be changed every two years or 24,000 miles (39,000 km), as the additives break down over that time.

The second category, OAT, is a longerlasting coolant. Also called extended-life coolant, it is designed to be changed at five years or 150,000 miles (241,000 km). This is a giant increase in the change interval from IAT coolant. OAT coolant was introduced in North America around 1994. One example of OAT is Dex-Cool, the orange coolant used by GM (**FIGURE 16-14**). The anticorrosion additives in OAT coolants do not break down as quickly, which explains the longer service time. The primary

FIGURE 16-13 IAT coolant is typically green in color.

FIGURE 16-14 OAT coolant such as Dex-Cool.

FIGURE 16-15 HOAT coolant.

FIGURE 16-16 POAT coolant.

additives in OAT coolants are organic acids, such as sebacate. These coolants do not use the additives used in IAT coolants.

The third category, HOAT, is a coolant that contains a mixture of inorganic and organic additives. This type of coolant can use silicate and organic acid. This gives the best of both worlds of coolants. Some manufacturers found that without silicate, problems arose if the system was not serviced. Oxidation of the coolant occurred, leading to a breakdown of the corrosion inhibitors. This ultimately caused failure of cooling system parts and gaskets. Other tests indicate that silicates cause premature water pump failures. Less silicate is present in HOAT coolant than in IAT coolants. But it is still considered to be an extended life coolant which has a five year or 150,000 mile (241,000 km) change interval. An example of this type of coolant is the yellow coolant used by Ford (**FIGURE 16-15**).

The fourth category, POAT, is a relatively new coolant that contains a proprietary blend of corrosion inhibitors (**FIGURE 16-16**). It is a very long-life coolant, providing up to seven years or 250,000 miles (402,000 km) of protection. It is claimed that it is compatible with most other types of coolant, but check the manufacturer's current service information.

Within the OAT or HOAT classifications, manufacturers may specify different corrosion additives and colored dyes. It is important that you service the vehicle with the coolant that is called for by the manufacturer. In most cases, you should never mix two or more types of coolants. Nor should you refill with a type other than that originally used in a vehicle. Doing so will likely compromise the cooling system's service life and cause premature failure of the system.

▶ Centrifugal Force Is Used to Circulate Coolant

LO 16-03 Describe coolant flow in engines.

Centrifugal force is a force pulling outward on a rotating body. For example, take a tennis ball, tie it to a string, and swing it around you. Centrifugal force pulls the tennis ball outward, making the string taut. Another example of centrifugal force occurs when a vehicle turns a corner. Centrifugal force resists the turning of the vehicle. It tries to keep the vehicle moving in a straight line. In fact, it can create a sliding condition if centrifugal force is great enough.

Centrifugal force can be useful in some cases, such as in the water pump. Coolant enters the center of the water pump. As the internal rotor spins, centrifugal force moves the liquid outward toward the outlet (**FIGURE 16-17**). Centrifugal force pushes the coolant into the passageways that surround the cylinders. The coolant then travels through the radiator to be cooled.

Coolant Flow—Normal and Reverse Flow

In a water-cooled cooling system with normal flow, the flow of coolant starts at the water pump. Cold coolant is moved through the engine by the water pump. The coolant starts to warm up as the engine is allowed to run. The coolant flows around the cylinders where combustion is taking place and picks up excess heat. It then moves upward through the cylinder head. It passes over the top of the combustion chamber (the hottest part of the engine). All the while, it picks up heat as it passes those hotter surfaces.

From the head, it moves to the thermostat. The thermostat works like a trapdoor. If it is closed, the coolant will continue to circulate within the engine. It flows through a bypass hose or passage to move back down to the water pump. This is back to the starting point of its journey again.

The thermostat is meanwhile sensing the coolant's temperature. At a specified temperature, it begins to open (**FIGURE 16-18**). As the thermostat opens, the coolant begins to flow through the radiator hose and to the radiator. Coolant enters the radiator's inlet tank and then the radiator core. It flows through small tubes in the core that have heat-dissipating fins on the outside. The tubes and fins transfer the coolant's heat to the outside air. As it cools, coolant moves to the cool side of the radiator. From there, it continues its journey back to the engine through the other radiator hose. The coolant's return flow to the engine is aided by the suction created by the water pump impeller.

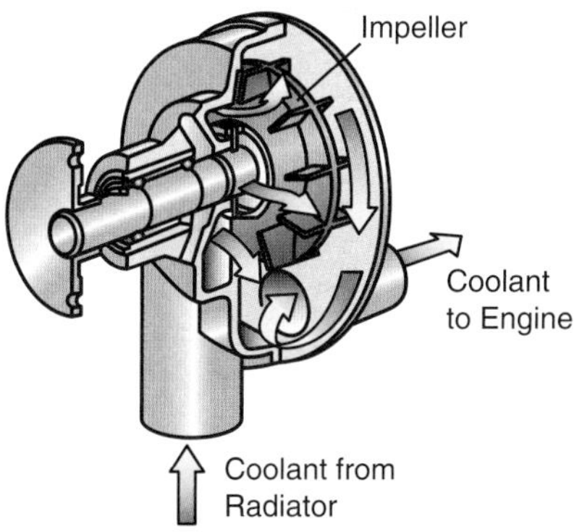

FIGURE 16-17 When coolant enters the center of this pump and the internal rotor spins, centrifugal force moves the liquid outward.

One problem with the normal-flow system was discovered on race cars. Race car engines had issues with cylinder heads overheating. They were failing prematurely. Since cylinder heads generally run hotter than engine blocks, under race conditions, the heads would get too hot and fail. High cylinder head temperatures also increase detonation which is harmful to engines.

Another issue with high cylinder head temperature is head gasket failure. As a cylinder head heats up, it expands more than the block. That causes it to slide further across the head gasket each time. When the head does this repeatedly, the gasket is more likely to fail.

In the normal-flow cooling system, coolant goes through the engine block first. It then moves to the hottest part—the cylinder head. Engineers found that if they reversed the flow of coolant, they could keep the cylinder head and block closer to the same temperature. This is done by directing coolant through the head first, and then to the block (**FIGURE 16-19**). This evens out the temperature and makes the head, valves, and head gaskets last longer.

In the typical reverse-flow design, coolant is moved by the water pump to the cylinder head. From there it flows down through the block to the thermostat. The thermostat is located between the inlet and outlet side of the engine to help regulate cold coolant more closely. It also helps to reduce temperature shock to engine components. When the coolant is cold, it circulates back to the water pump.

Once the engine heats up, the thermostat directs the coolant back to the radiator through the radiator hose. The coolant is cooled in the radiator and then sent back to the water pump through the other radiator hose. In general, all of the components are generally the same between the two types of systems.

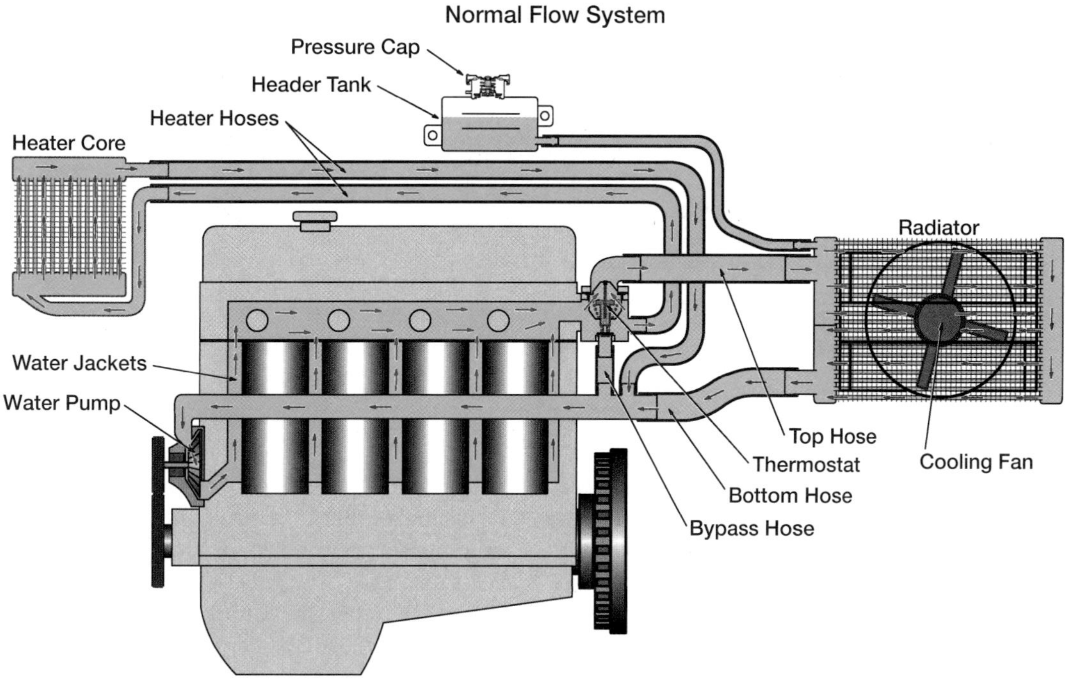

FIGURE 16-18 Coolant flow in a normal flow system.

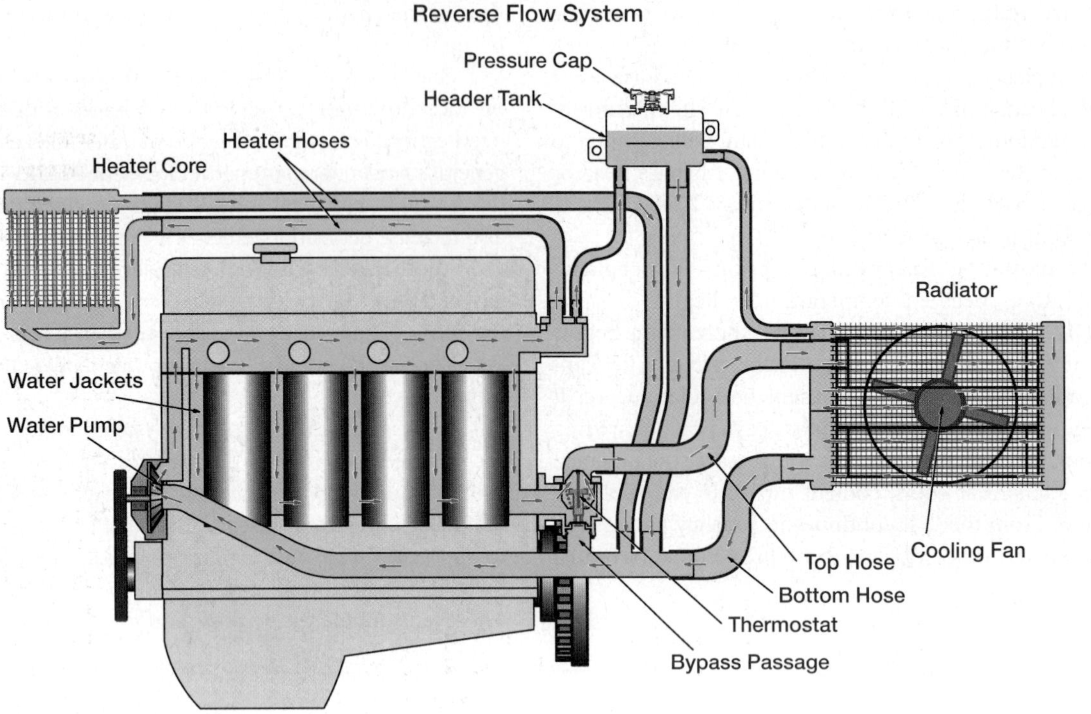

FIGURE 16-19 Coolant flow in a reverse flow system.

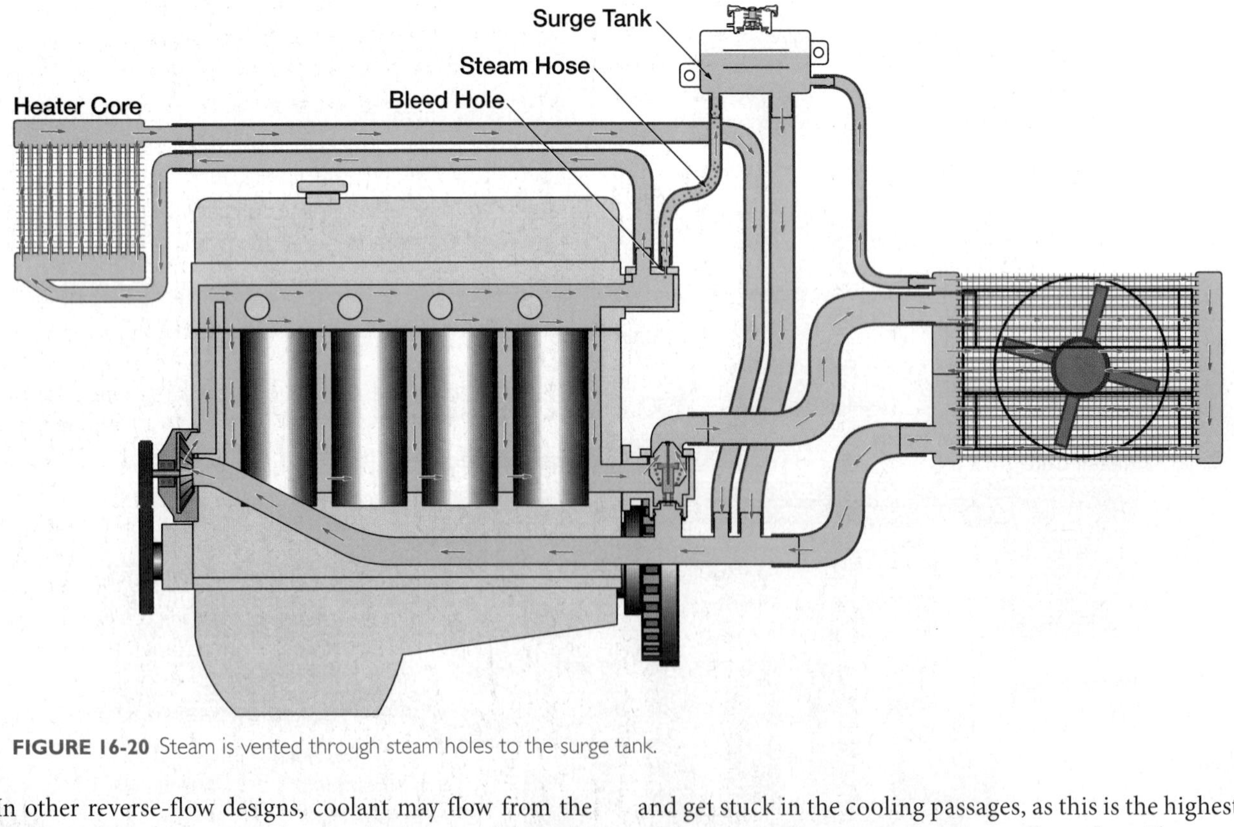

FIGURE 16-20 Steam is vented through steam holes to the surge tank.

In other reverse-flow designs, coolant may flow from the thermostat in the lower radiator hose to the bottom of the radiator. Then up to the top of the radiator and back to the cylinder heads. Then down through the block, to the thermostat. And finally back to the lower radiator.

One problem with the reverse-flow design was identified. As coolant moved through the head, steam tended to form and get stuck in the cooling passages, as this is the highest part of the engine. Because the engine will overheat if gas pockets build up, the solution was to drill holes in the head for steam to escape. The steam holes direct steam through a tube from the head back to a **surge tank** (**FIGURE 16-20**). The steam turns back into liquid and is recycled through the system. The surge tank is discussed in greater detail later in the chapter.

Rotary Engine Cooling System

The cooling system of the rotary engine is not much different from the piston engine's cooling system. The cooling system uses a standard radiator, thermostat, and radiator hoses (**FIGURE 16-21**). Coolant flows from the radiator to the water pump. The water pump sends coolant into the **rotor housing**. The rotor housing has passageways (water jackets) cast into it, which go around the rotor housing. They allow heat to be pulled from around the rotor housing (**FIGURE 16-22**). After coolant moves through the water jackets, it is sent to the thermostat, where it returns to the radiator.

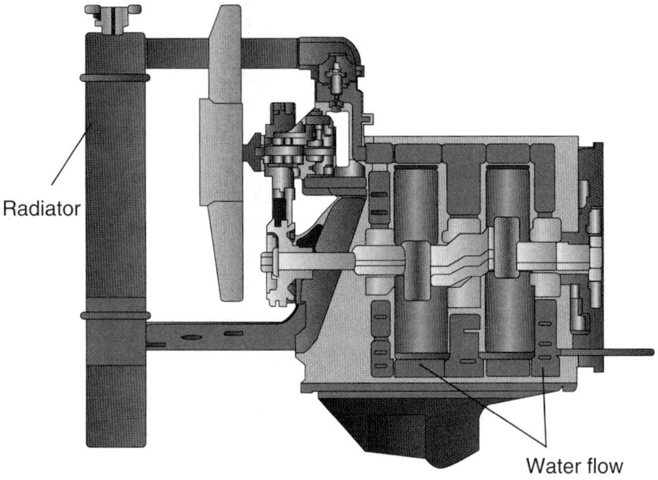

FIGURE 16-21 Rotary engine cooling system.

FIGURE 16-22 Rotary engine coolant passages.

▶ Overview of Cooling System Components

LO 16-04 Describe the radiator and its associated components.

The primary components of a vehicle cooling system and their purpose are:

- *Coolant*: Coolant is the liquid used to prevent freezing, overheating, and corrosion of the engine.
- *Radiator*: The radiator is usually made of copper, brass, or aluminum tubes. The top or side tanks are made of copper, brass, aluminum, or plastic for coolant to collect in. Air is drawn through the radiator to transfer heat energy to ambient air. The fins on the tubes of the radiator give more surface area for **heat dissipation**—the spreading of heat over a large area to ease heat transfer.
- *Thermostat*: The thermostat regulates coolant flow to the radiator. It opens at a predetermined temperature to allow coolant flow to the radiator for cooling. It also enables the engine to reach operating temperature more quickly for reduced emissions and wear.
- *Recovery system*: The recovery system uses an **overflow tank** to catch any coolant that is released from the pressure cap when the coolant heats up. It works like a catch can.
- *Surge tank*: This pressurized tank is piped into the cooling system. Coolant constantly moves through it. It is used when the radiator is not the highest part of the cooling system. It is also used to capture steam and condense it back into liquid. Remember, air collects at the highest point in the cooling system.
- *Water pump*: This pump is used to move coolant throughout the cooling system in order to transfer heat energy. The water pump is typically driven off the engine timing belt or accessory belt. On some engines, it is driven by the camshaft timing chain.
- *Cooling fan*: This fan forces air through the radiator for faster heat transfer. Cooling fans can be driven by a belt or by an electric or hydraulic motor. The fan can be controlled by viscous fluid or thermostatic sensors, switches, and relays.
- *Radiator hoses*: These hoses are used to connect the radiator to the water pump and engine. They are usually made of formed, nylon-reinforced rubber. Some lower radiator hoses use coiled wire inside them to prevent hose collapse during high engine rpm.
- *Heater hoses*: These hoses connect the water pump and engine to the heater core. They carry heated coolant to and from the heater core.
- *Drive belts*: These belts provide power to drive the water pump and other accessories on the front of the engine. Four types are used: V-belts, serpentine (also called multi-groove) belts, stretch-fit belts, and toothed belts.
- *Temperature indicators*: Temperature indicators provide information to the operator about engine temperature. The temperature gauge indicates engine temperature continuously. A temperature warning indicator comes on only when the engine is overheating. It is used to warn the

operator that engine damage will occur if the vehicle is driven much farther.

- *Water* jackets: Water jackets are passages surrounding the cylinders and head on the engine where coolant can flow to pick up excess heat. They are sealed by replaceable core plugs.
- *Heater core*: The heater core is a small radiator used to provide heat to the passenger compartment from the hot coolant passing through it. The amount of heat can be controlled by a heater control valve.
- *Auxiliary coolers*: Auxiliary coolers are used to cool automatic transmission fluid, power steering fluid, exhaust gas recirculation gases, and compressed intake air. Each of these coolers transmits heat. This is done either to the cooling system or directly to the atmosphere. There is a wide variety of auxiliary coolers. Refer to the manufacturer's service information to learn about each type of cooler.

Radiator

The **radiator** is located in the front part of the vehicle, where maximum airflow can pass through it. It is typically located behind the grill and under the front of the hood. Its actual location under the hood depends on several factors:

- the engine configuration,
- the available space, and
- the shape or hood line itself.

The radiator consists of top and bottom tanks, or side tanks and a core. The radiator core allows the coolant to pass through it while extracting heat. The radiator core conducts the heat generated by the engine to the atmosphere.

The materials used in the radiator must be good heat conductors, such as brass, copper, or aluminum. Brass or copper is often used for tanks when combined with a brass or copper core. Modern vehicles often use plastic tanks combined with an aluminum core. This design saves weight and cost while still providing good heat transfer.

The core consists of a number of cooling tubes that carry coolant between the two tanks. The tubes can be in a horizontal (cross-flow) design or a vertical (down-flow) design (**FIGURE 16-23**). In a **down-flow radiator**, the cooling tubes run top to bottom, with the tanks on the top and bottom. This design is fairly tall.

In a **cross-flow radiator**, the cooling tubes are arranged horizontally, with one tank on each side. Because of this arrangement, the same amount of cooling area can be achieved without the need for a very tall radiator. So they tend to be short and wide. This design feature allows the hood profile to be lower, allowing for better aerodynamics. Other benefits include improved fuel economy and increased forward vision by the driver. The function of both types of radiator configurations is the same, which is to cool the coolant before it reenters the engine.

In both the cross-flow and the down-flow designs, the core is built of the same components. The cooling tubes are designed with heat-dissipating fins. The fins increase the surface area so heat can be dissipated to the air (**FIGURE 16-24**). Coolant transfers heat to the tube walls. The tube walls transfer heat to the fins. And the fins transfer heat to the air. Air rushing by the fins carries the heat away. Liquid coolant emerges cooler at the outlet of the radiator. In this way, coolant was able to give up much of its heat to the atmosphere.

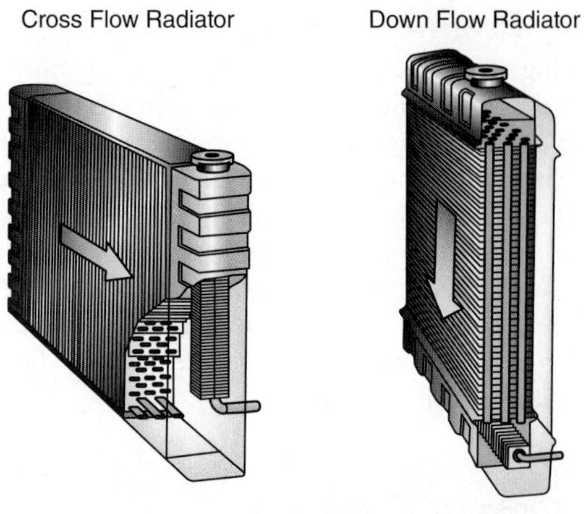

FIGURE 16-23 Cross-flow radiators have cooling tubes mounted horizontally, whereas down-flow radiators have tubes mounted vertically, requiring a taller hood profile.

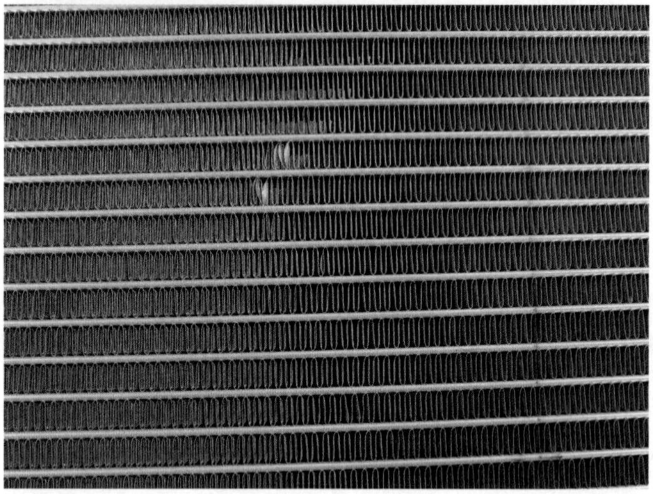

FIGURE 16-24 Radiator fins help dissipate heat.

Most radiators come with a method of draining coolant from them. This drain plug can take several forms. Older vehicles used a petcock (**FIGURE 16-25**). This usually has a T-shaped metal knob that needs to be turned "IN" to open it up. Many a new technician has tried to turn it "OUT" only to break it off, creating a problem that is not easy to fix.

Some drain plugs are plastic quarter-turn valves. If turned too far, these drain plugs will break off. These drain plugs cannot always be found as a replacement part, and replacement of the entire radiator may be necessary.

The last type of common drain valves are simple plastic threaded plugs. They typically are only tightened by finger and use a rubber sealing washer. Remember that radiator petcocks and drain plugs can be sources of coolant leaks, so don't overlook them.

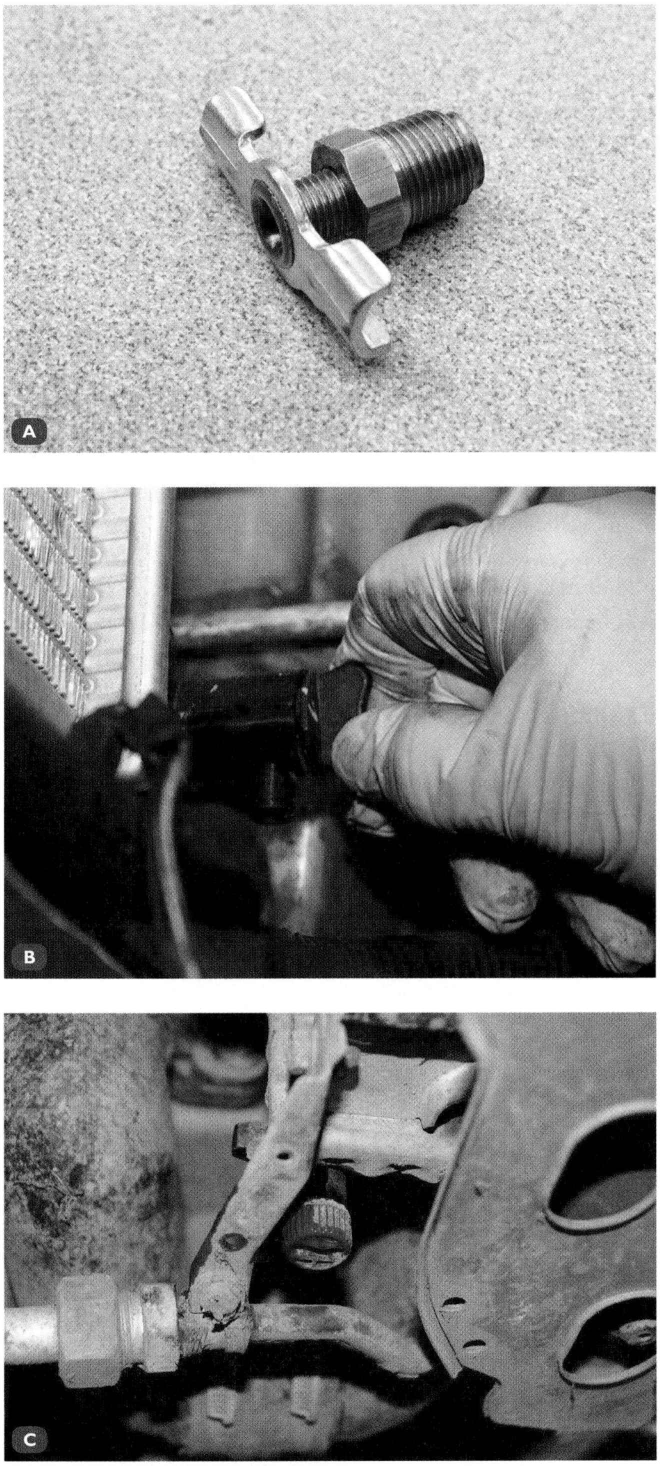

FIGURE 16-25 Radiator drain plugs. **A.** Petcock. **B.** Plastic quarter-turn valve. **C.** Plastic drain plug.

Radiator Shrouding

In the interest of fuel economy (less drag coefficient), vehicle hood lines have become more streamlined. And engine compartments are smaller and more crowded. These changes in turn make it more difficult for air to flow through the radiator. Airflow is therefore dependent on shrouding above, behind, and below the radiator. Shrouds direct ram air to the radiator,

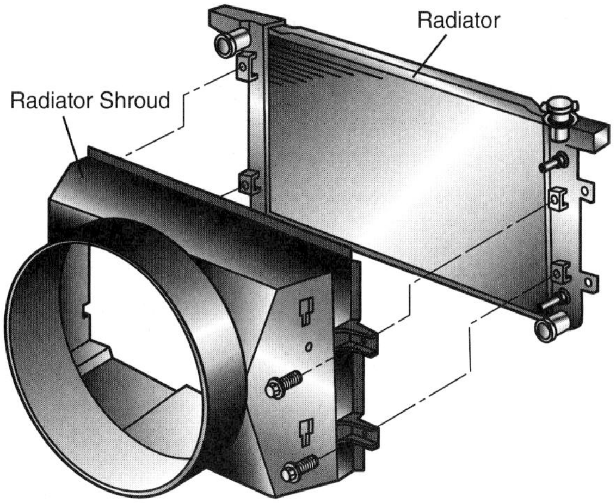

FIGURE 16-26 Radiator shrouding.

and from it (**FIGURE 16-26**). Because of this, shrouds are very important parts of the cooling system.

Shrouds are also used around the radiator fan to help:

- draw air through the entire radiator core, not just in front of the fan blades.
- prevent air from simply circulating around the tips of the fan blades instead of being drawn through the radiator.

If the shrouding is removed and not replaced, the engine will not be cooled efficiently. It will likely overheat, especially on hot days.

Applied Science

AS-31: Insulation: The technician can explain the role of insulation in maintaining temperatures.

Certain parts of the vehicle's heating and cooling system must be insulated in order to perform properly. Air-conditioning refrigerant lines, heating ducts, and other areas are insulated to prevent the loss of, or absorption of, heat. Even the hood itself may be insulated. This not only reduces the amount of heat radiated from the engine. But it also quiets the noise of the engine.

Applied Science

AS-32: Radiation: The technician can demonstrate an understanding of heat transfer that involves infrared rays.

At one end of the light spectrum, just beyond visibility, lies the infrared portion of light. Infrared is a form of heat energy, which can be detected by special instruments. Thermography involves the use of heat-sensing instruments to detect heat sources. The amount of radiation emitted by an object increases with temperature. Therefore, thermography allows one to see variations in temperature.

In the automotive trade, an infrared "temp gun" is pointed at objects to determine their temperature. This tool is useful for detecting cylinders that are not contributing full power to the engine. It can also find leaks in air-conditioning systems.

On the dashboard of many vehicles lies an infrared-sensing "sun-load sensor." It helps automatic heating, ventilation, and air conditioning (HVAC) systems regulate cabin temperature. It also enables automatic headlight dimming.

Radiator Pressure Cap

If coolant boils, it can be as damaging to an engine as having it freeze. One way to prevent coolant from boiling is to use a radiator pressure cap. Increased pressure on coolant raises its boiling point. The radiator cap allows pressure to build up to a specified amount. Once the pressure builds to that point, excess pressure is vented.

Here is how the radiator cap works. As coolant temperature rises, the coolant expands, and pressure in the cooling system rises. The increased pressure raises the boiling point of the coolant. Engine temperature keeps rising, and the coolant expands further. Pressure builds against a spring-loaded valve in the radiator cap. At a preset pressure, the valve opens and releases a small amount of coolant.

When the valve in the radiator cap opens, the coolant leaves the radiator through the overflow tube. It flows to the overflow container (**FIGURE 16-27**). In past designs, the overflow tube would drop the coolant onto the ground, which was an environmental hazard. By releasing coolant into a coolant overflow container, the coolant can be pulled back into the system once it cools.

Coolant in the radiator cools when the engine is off. When coolant cools, it contracts, and with this contraction, it creates a vacuum (low pressure) in the radiator. A vacuum valve is located in the radiator cap. When vacuum causes it to open, it allows coolant to be pulled back into the radiator from the overflow container. Because of this design, no coolant is lost as with the older systems that dropped coolant onto the ground. The vacuum valve in the cap also stops low pressure from developing in the radiator as the coolant cools. A vacuum would cause the radiator hoses to collapse as a result of atmospheric pressure on the outside of the hose.

Recovery System

A coolant recovery system maintains coolant in the system at all times. The recovery system consists of:

- an overflow bottle,
- a sealed radiator pressure cap, and
- a small hose connecting the bottle to the radiator neck (**FIGURE 16-28**).

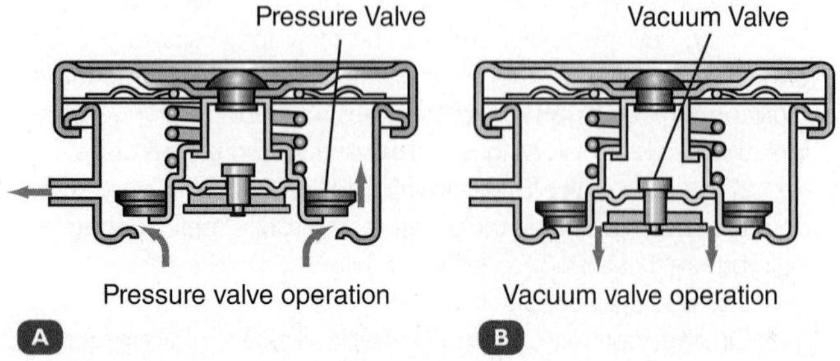

FIGURE 16-27 A. Pressure valve being forced open. **B.** Vacuum valve being pulled open.

As engine temperatures rise, the coolant expands. Pressure builds against a valve in the radiator cap until, at a preset pressure, the valve opens. Hot coolant flows out of the radiator, through the connecting hose, into an overflow bottle. As the engine cools, the coolant contracts, and pressure in the cooling system drops below atmospheric pressure. Atmospheric pressure in the overflow bottle opens the vacuum valve in the radiator cap. Atmospheric pressure pushes the coolant back into the radiator.

Like water, air contains oxygen, which reacts with metals to form corrosion. Oxygen can also cause oxidation of the coolant which turns it into sludge. With use of a recovery system, no coolant is lost, and excess air is kept out of the system. This helps maintain the life of both the engine and coolant.

Surge Tank

Some vehicles are equipped with a surge tank. The surge tank has coolant constantly running through it. It is usually located higher than the top of the radiator. Thus, any gas in the system will make its way to this tank, ensuring that only liquid coolant circulates through the system.

The surge tank has at least one line in and one line out. It is usually made of hardened plastic. This allows for a visual inspection of the fluid level through the plastic. This tank is usually where the cooling system is filled or topped off with coolant. In many vehicles with a surge tank, the pressure cap is mounted on the surge tank instead of the radiator (**FIGURE 16-29**).

▶ Thermostat and Housing

LO 16-05 Describe the operation of the thermostat and water pump.

The **thermostat** is located under the thermostat housing. The thermostat regulates the flow of coolant. It allows coolant to flow from the engine to the radiator when the engine is running at its

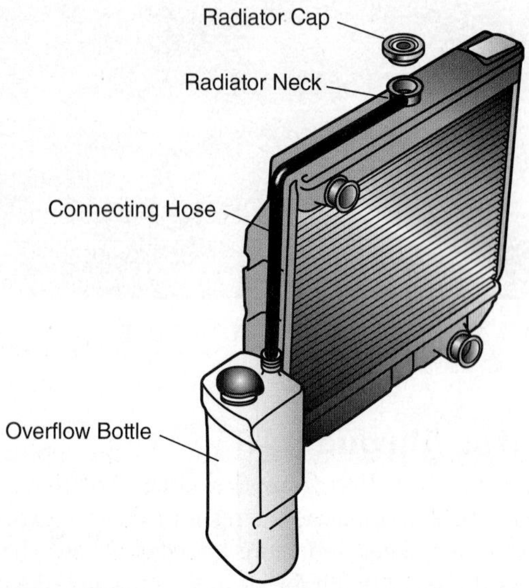

FIGURE 16-28 A coolant recovery system.

FIGURE 16-29 A surge tank removes air and gases from the coolant. This is where you normally fill the cooling system.

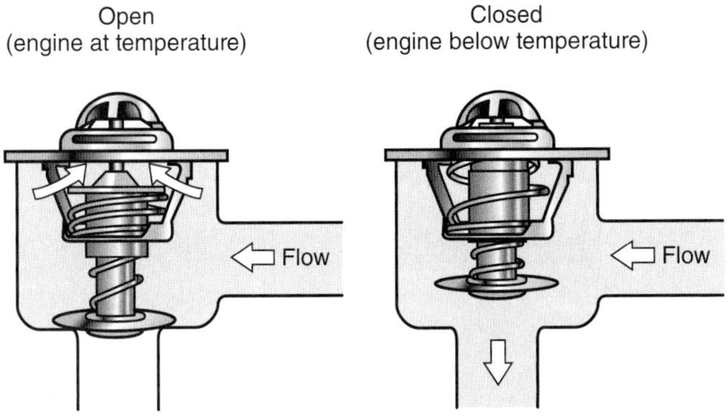

FIGURE 16-30 The thermostat has a moving valve that is controlled by a wax pellet. When the wax is cool, the valve stays closed; as temperature increases, the wax melts and forces the valve open.

operating temperature. The thermostat prevents coolant from flowing to the radiator when the engine is cold. This allows the engine to warm up more rapidly, thus reducing engine wear and emissions.

The thermostat is a spring-loaded valve that is controlled by a wax pellet located inside the valve (FIGURE 16-30). As the temperature of the coolant rises, the wax pellet melts and expands. This forces the spring-loaded valve to open at a preset temperature. As the valve opens, coolant is allowed to flow through it.

The thermostat works like a door to control the movement of coolant. When the engine is cold, the door is closed. When hot, the door opens. You can sometimes see a slight swing of the temperature gauge as the thermostat cycles open or closed.

Some engines are designed such that the coolant bypass passage is directly under the thermostat. In those situations, the thermostat may have a flat disc attached to the bottom of it. The disc moves with movement of the thermostat. When the thermostat fully opens, the flat disc blocks off the bypass passage so that all coolant must flow through the radiator (FIGURE 16-31).

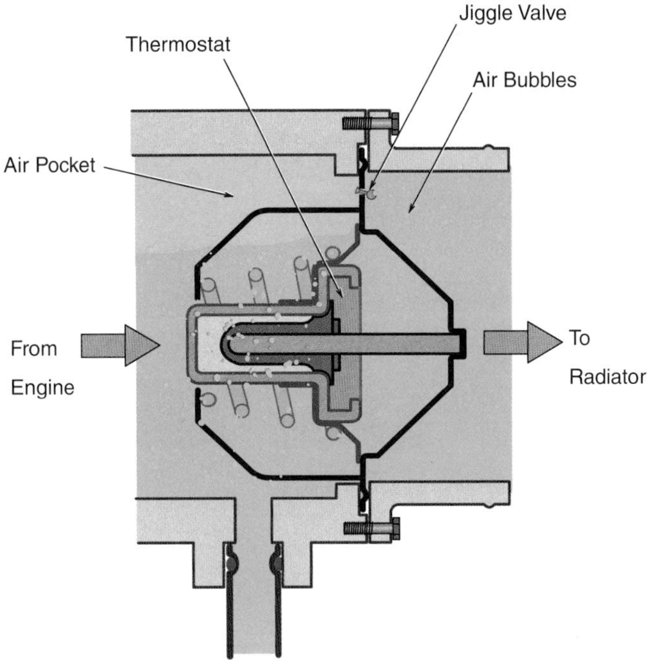

FIGURE 16-31 Thermostat with a jiggle valve for bleeding air from the system when refilled..

Yet, when the thermostat partially closes, the bypass passage will be partially open. This helps give more effective cooling when the thermostat is fully open. This is because all of the coolant flows through the radiator.

Most thermostats have a small hole on one side of the thermostat valve. This hole allows any air in the system to move past the closed thermostat when the valve is closed. This is especially helpful when the cooling system has been drained and is being filled. The hole usually contains a little pin, called a jiggle valve or jiggle pin. It helps break the surface tension of the coolant and allows any air to flow slowly through the hole. Air trapped in the cooling system is thus able to slowly find its way past the thermostat, where it can be bled. When installing the thermostat, the jiggle valve should be in the uppermost position. That way, as much air as possible can be bled from the system.

The thermostat and housing are normally located on the outlet side of the coolant flow from the engine. However, on some engines, they are located on the inlet side of the engine. The reason for this is to better control the amount of cold water that rushes into the engine. Large swings in temperature can create a temperature shock to the engine.

Some vehicles include a manual bleed valve to bleed the system. It is typically located on the thermostat housing or on a high part of the cooling system (FIGURE 16-32). After the vehicle cooling system has been serviced and refilled, the air needs to be bled from the system. The technician uses the bleed valve to vent any trapped air to the atmosphere. It is good practice to bleed the system again once the engine has warmed up.

FIGURE 16-32 Coolant bleed valve.

FIGURE 16-33 Typical water pump and impeller.

▶ TECHNICIAN TIP

Preventing overheating is one function of the cooling system. It also helps the engine reach its best operating temperature as soon as possible. Every engine has a temperature at which it operates best. If operated below or above this temperature, ignition and combustion problems may occur. Because of this, the MIL (malfunction indicator lamp) may illuminate if the thermostat has failed. The powertrain control module (PCM) determines this because the engine failed to reach its normal operating temperature within a specified time. Thus, a faulty thermostat can cause the MIL to illuminate.

Water Pump

The water pump is usually located at the front of the engine block. A hose typically connects it to the output of the radiator. So, relatively cool coolant enters the water pump at the center of the pump. As the impeller rotates, it catches the coolant and flings it outward with centrifugal force (**FIGURE 16-33**). This type of water pump is called a centrifugal pump. It creates movement of the coolant due to centrifugal force. But it does not cause much buildup of pressure. Coolant is driven through the outlet into the water jackets of the engine block and cylinder heads.

Coolant can be directed to critical hot spots, such as around the exhaust ports in the cylinder head. This helps to stop localized overheating. The cylinder head gasket has holes of various sizes to help control how much coolant flows to these and other locations of the head (**FIGURE 16-34**). If the head gasket is not properly installed, the holes being in the wrong place will restrict coolant flow.

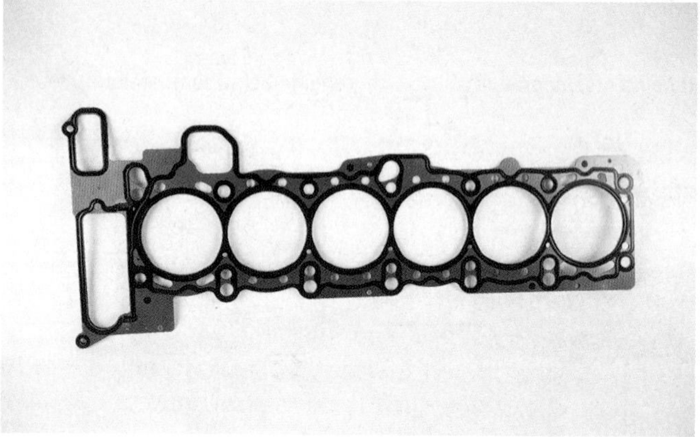

FIGURE 16-34 Head gaskets have holes that control the flow of coolant to the cylinder head.

The water pump is usually belt-driven from a pulley on the front of the crankshaft. The engine drives the water pump using either an accessory belt or the timing belt (**FIGURE 16-35**). Some newer vehicles use an electrically driven water pump.

FIGURE 16-35 Water pumps are typically driven by timing belts.

Internally, the water pump has fanlike blades on an impeller or rotor. The impeller is turned by a shaft connected to the pulley. The shaft rides on a heavy-duty double-row ball bearing. This is needed for long life and to withstand belt tension. The shaft is sealed where it enters the pump chamber. Most water pumps have a small weep hole that sits between the seal and the bearing. It vents any coolant that leaks past the seal to the outside of the engine (**FIGURE 16-36**). Checking this hole for signs of coolant leakage is part of the process of inspecting the engine for leaks.

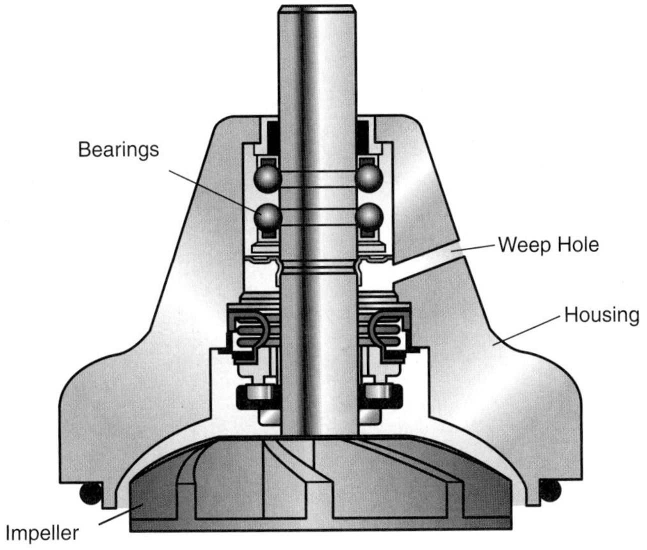

FIGURE 16-36 Cutaway view of water pump and weep hole.

▶ Cooling Fan

LO 16-06 Describe the operation of cooling fans.

Cooling fans are used to provide airflow through the radiator core for engine cooling. This is most needed during slow driving or stop-and-go driving. There are two main categories of cooling fans: engine driven and electric. Engine-driven cooling fans are typically located on the water pump shaft. In a few cases, they may be attached directly to the engine crankshaft. Either way, engine horsepower is needed to directly drive the fan. This requires extra fuel even when the fan is not needed. Such fans are also noisy and dangerous to work around. Electric fans avoid most of these issues.

It takes a fair amount of energy to turn a fan, yet the fan is not always needed. It is not needed during cold engine operation. It is also not normally needed when the vehicle is traveling above about 35 mph (miles per hour; 56 kph [km per hour]). Above those speeds, the airflow is strong enough to cool the radiator without using the fan.

For years, vehicle manufacturers have sought ways to reduce the amount of energy the fan uses. One design is called a flex fan. It uses flexible steel or plastic blades. As engine speed increases, the blades straighten out and lessen their **pitch** (**FIGURE 16-37**). This design increases fuel economy and reduces noise.

Another type of engine-driven fan uses a **viscous coupler.** It connects the water pump pulley to the cooling fan. The viscous coupler is called a fan clutch. It can engage and disengage the fan from the pulley (**FIGURE 16-38**). The fan clutch typically uses two discs that have closely fitted interwoven rings and grooves. When viscous silicone oil is allowed to fill the small spaces between the rings and the grooves, it transmits torque from one disc to the other. This transmits torque across the two halves of the hub.

A bimetallic spring on the front is attached to a valve. As air temperature from the radiator rises, the bimetallic spring opens the valve. The valve allows silicone oil to flow into the coupler. This transmits more of the pulley's speed to the fan to move more air.

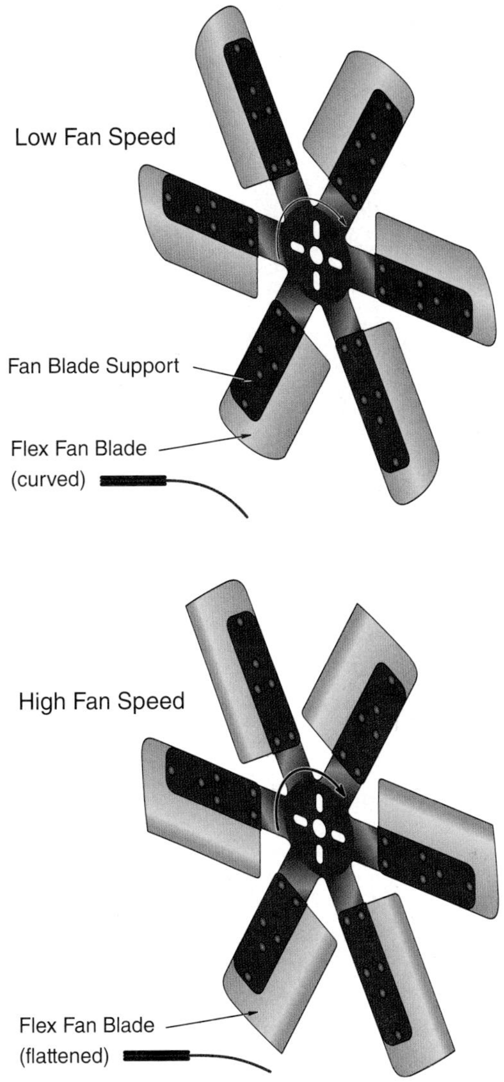

FIGURE 16-37 Flex fan. Blades flatten out at higher rpm.

As air temperature from air flowing through the radiator decreases, the bimetallic spring closes the valve. The silicone oil moves out of the coupler back to the reservoir located in the clutch body. This causes the fan to slow and move less air. The pulley for this type of fan is driven at all times by the accessory belt. The benefit of using a viscous clutch is that it increases fuel economy by being able to cycle the fan on and off.

A variation of the clutch fan is the solenoid-controlled fan clutch. It operates in the same manner as the thermostatic fan clutch. The clutch uses oil to control the speed of the fan (allows slip to slow fan down) and volume of air moving across the radiator. The only difference in the operation is that the bimetallic spring is replaced with an electric solenoid on the front of the clutch. This solenoid is controlled by the PCM. As temperature increases and decreases, the PCM controls the solenoid that controls the oil flow to the viscous coupler. On this type of fan, a feedback sensor provides actual fan speed, or rpm information, back to the PCM.

Over time, fan designs have evolved to increase efficiency. Also, more vehicles moved to front-wheel drive, which turns the front of the engine away from the radiator. This makes

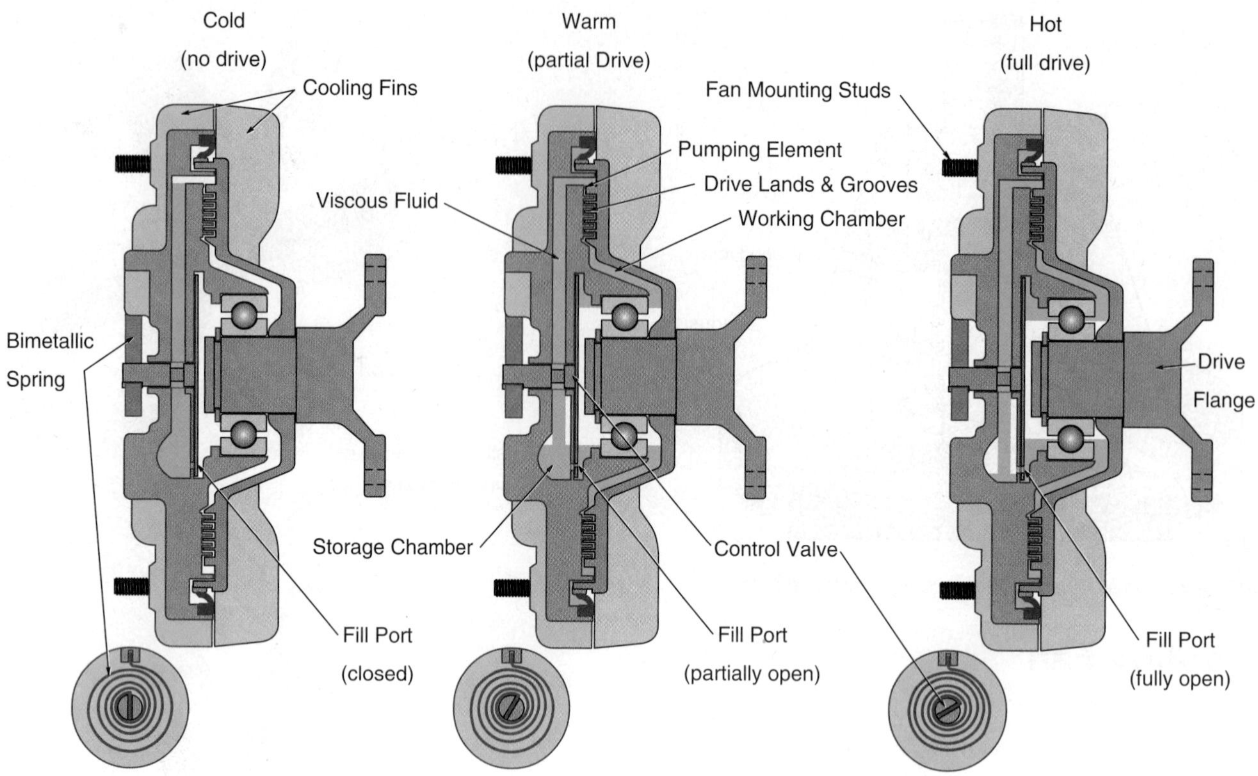

FIGURE 16-38 Viscous fan clutch.

engine-driven fans not very useful. So the electric fan was introduced. Now it is by far the most common and simple type of cooling fan. They are mounted directly to the radiator.

An electric fan can be turned on and off easily whenever it is needed (**FIGURE 16-39**). It can also operate at full speed even though the engine is idling. This versatility makes it very efficient. It only runs when the engine is above the ideal operating temperature. In addition, because it can run at full speed independent of the engine, it can move plenty of air to cool the engine effectively.

Electric fans use an electric motor to turn an attached fan. In most cases, the fan blades are made of plastic, making them safer than metal blades. The blades can be designed to either push or pull air through the radiator. This increases the mounting options. The electric fan is controlled by one of the following two methods:

- A control module, such as the PCM, is used to energize a fan relay to turn the fan on and off.
- A thermo-control switch to either directly turn the fan on and off, or through a fan relay.

The PCM knows the temperature of the engine by means of the engine coolant temperature (ECT) sensor. The PCM can then supply either power or ground to the fan relay, energizing the relay and fan (**FIGURE 16-40**).

The **thermo-control switch** is a temperature-sensitive switch. It is mounted into a coolant passage on the engine, or into the radiator. When engine temperature gets hot enough (say, 215°F [102°C]), the thermo-control switch closes. This either sends power directly to the fan or causes a relay to be activated, which turns on the fan. Once the engine coolant temperature cools back down, the switch opens and the fan stops.

FIGURE 16-39 Electric cooling fan.

Thermo-control switches often operate on the bimetallic strip principle. These consist of two different metals or alloys laminated back to back. As different metals and alloys heat and cool, they expand and contract at different rates. Since two different metals are joined, when heated, the greater expansion of one forces the whole strip to flex into a curved shape. As the strip changes shape, it can be designed to complete an electric circuit by closing a switch. The switch then turns the fan on (**FIGURE 16-41**).

With an electric fan, the electricity to run it comes from the alternator. The alternator gets its energy from the engine, which

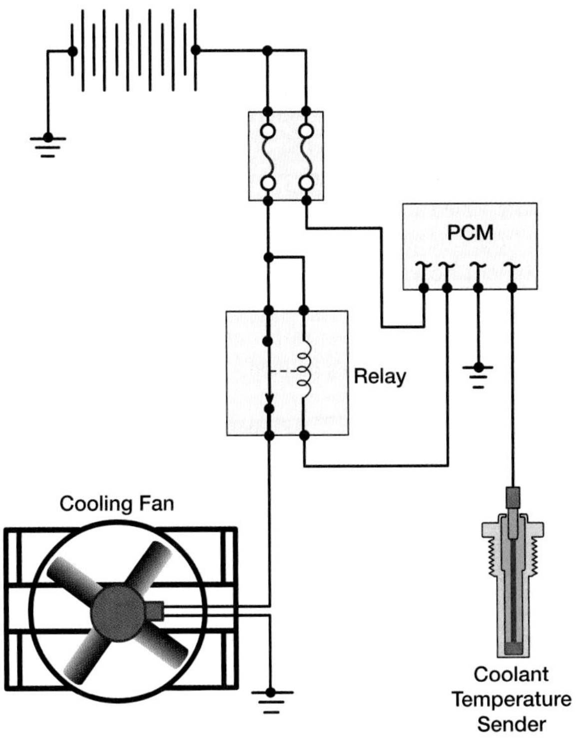

FIGURE 16-40 PCM-controlled fans typically use a relay and coolant temperature sensor to control fan operation.

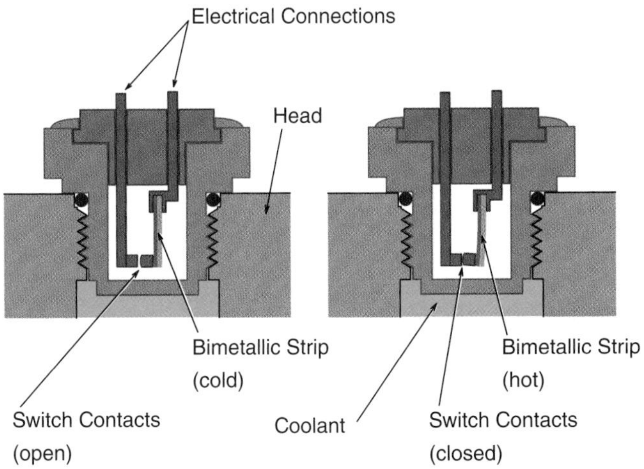

FIGURE 16-41 Bimetallic strip.

comes from the gasoline fuel. Because an electric fan only needs to operate part of the time, fuel is saved whenever it is off. For further fuel savings, some manufacturers use multiple fans and control them separately (**FIGURE 16-42**). Others use multispeed fans to provide just enough fan operation to keep the coolant at the proper temperature.

Another type of cooling fan is the hydraulically operated fan. In many cases, it uses power steering fluid from the power steering pump to power the fan (**FIGURE 16-43**). Because the power steering pump can create substantial power, hydraulically driven fans are powerful. They can be used to draw a large

FIGURE 16-42 Some vehicles use more than one electric fan.

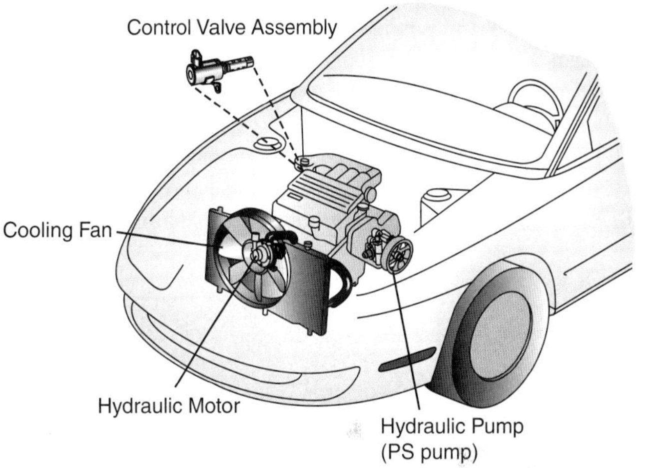

FIGURE 16-43 Hydraulically operated cooling fan.

amount of air through the radiator. This makes them ideal for use on vehicles with heavy trailer towing capacities.

Hydraulically controlled fan systems consist of the power steering pump, a fluid control device, the hydraulic fan motor, and high-pressure connecting hoses. The fan is typically controlled by a pulse-width-modulated solenoid valve. The solenoid controls how much hydraulic fluid is directed to the fan motor. The more fluid, the faster the fan turns. The PCM varies the signal to the solenoid valve.

One benefit of the hydraulically controlled fan is that it can be operated at near full speed and force, even at idle. This is similar to an electric fan. Yet it can tap into more engine power than the electric fan, so it is more heavy duty.

▶ Radiator Hoses

LO 16-07 Describe hoses, belts, and tensioners.

On most vehicles, there are two radiator hoses: the upper hose and the lower hose. They are also called the inlet hose and the outlet hose. **Radiator hoses** are rubber hoses that are subject to pressure and high temperature. They are reinforced with a layer of fabric, typically nylon. This gives them strength and prevents them from ballooning, and yet they are still flexible. The reason radiator hoses need to be flexible is to allow movement between the engine and body. Because the engine is mounted on flexible mounts, the hoses must be flexible.

Radiator hoses are often molded into a special shape to suit the particular vehicle (**FIGURE 16-44**). Some radiator hoses, especially lower hoses, have a spiral wire inside. This keeps the hose from collapsing during heavy acceleration when the water pump is drawing a lot of water from the radiator.

The top radiator hose is typically attached to the thermostat housing, and the inlet side of the radiator. Heated coolant is sent from the thermostat to the radiator. The bottom or lower radiator hose is connected between the outlet of the radiator and the inlet of the water pump. The radiator hoses are held in position by clamps. These can be spring clamps, wire wound clamps, or worm drive clamps (**FIGURE 16-45**).

Radiator hoses deteriorate over time and use. They can also be damaged by oil or fuel leaking on them. Thus, they need to be inspected and changed periodically. Many vehicle manufacturers recommend radiator and heater hose replacement approximately every four to five years or 48,000 to 60,000 miles (77,000 to 97,000 km).

When replacing hoses, be sure to reinstall the clamps correctly; otherwise, the seal between the hose and the component will leak. Every component has a raised ridge built into it. The hose clamp has to clamp on the inside of this ridge, not on top of it. If the hose clamp is on top of the ridge, the hose is likely to pop off once the cooling system becomes pressurized (**FIGURE 16-46**).

There are two hoses for the heater core: one inlet hose and one outlet hose. The heater hose sends hot coolant to and from the heater core. This provides heat to the passenger compartment. The heater hoses are sealed and retained by the use of a hose clamp. As with radiator hoses, be sure to install hose clamps correctly.

Some coolant hoses are made of silicone (**FIGURE 16-47**). They are designed to withstand heat better and last longer than the standard rubber hose. Be sure to inspect hoses whenever you are servicing the customer's vehicle.

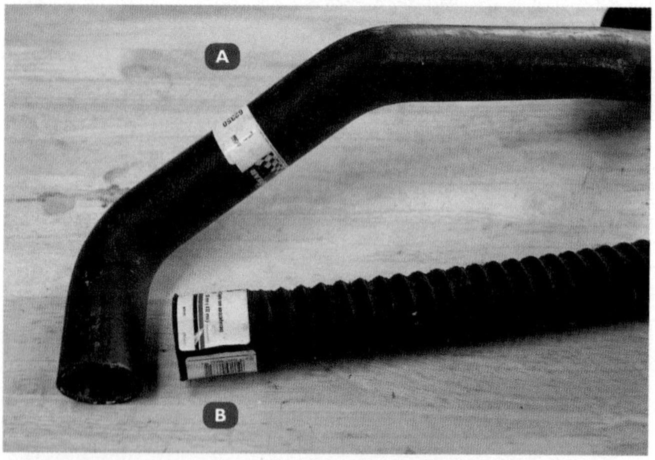

FIGURE 16-44 Radiator hoses. **A.** Molded. **B.** Flexible.

FIGURE 16-45 Hose clamps.

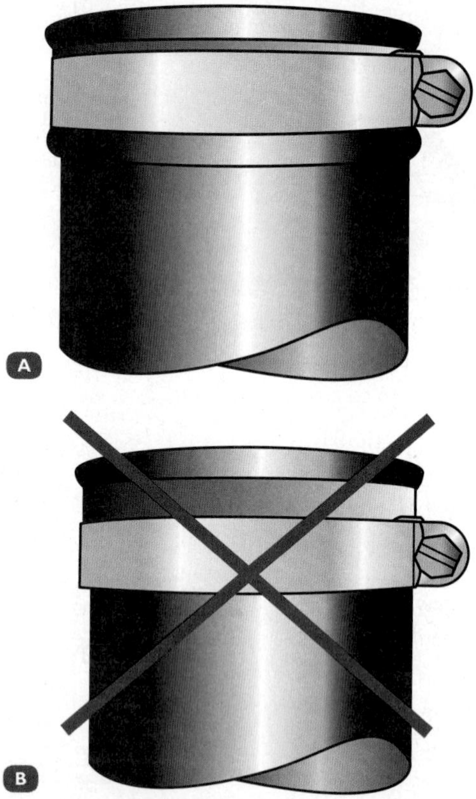

FIGURE 16-46 **A.** Proper installation of hose clamp. **B.** Improper installation of hose clamp

FIGURE 16-47 Silicone coolant hose.

Other Coolant Hoses

There are additional coolant hoses that carry hot coolant through the cooling system. Some examples are the bypass hose, steam vent hoses, and hoses to the throttle body (**FIGURE 16-48**). They are often ignored when checking the radiator hoses and other cooling system components. It is critical that these hoses be checked. Any break in the system will dump the coolant and cause the engine to quickly overheat.

One of the additional hoses is the bypass hose. The bypass hose is typically located on the water pump. It is also connected to the intake manifold on many V-configured engines, such as a V6 or V8 (**FIGURE 16-49**). This hose allows the water pump to circulate the coolant in the engine when the thermostat is closed. This hose is made of the same materials as the radiator hoses and heater hoses.

Another coolant hose that may be used is a throttle body coolant line. This hose runs from the intake up to the throttle body to heat the throttle body (**FIGURE 16-50**). It helps prevent icing of the throttle body during cold outside temperatures when there is high moisture content in the air. The cold wet air being pulled through the throttle plate may create ice. The ice can restrict airflow and cause the throttle to stick. Running hot coolant through the throttle body eliminates this problem.

There are many possible applications of additional cooling system hoses. Some examples are for remote oil coolers, turbochargers, or even some alternators. In a compressed natural gas vehicle, coolant hoses are routed to the CNG pressure regulator to keep it warm. A similar setup may be used for the engine's idle air control motor and more. The point is to know where all the coolant hoses are on a vehicle so none will be missed during an inspection.

Drive Belts

Drive belts are used to drive a variety of accessories on the front of the engine. Belts normally need to be tensioned properly. If they are too loose, they squeal and slip. If they are too tight, they put excessive wear on the bearings in the accessory or tensioner. We will discuss belt tension below. There are four types of drive belts:

- *V-type*: A V-type belt has a wedge-shaped interior and sits inside a corresponding groove in the pulley (**FIGURE 16-51**). The sides of the V-belt wedge in the sides of the pulley.
- *Serpentine*: A serpentine-type belt is also called a multi-groove V-belt. It has a flat profile with a number of small

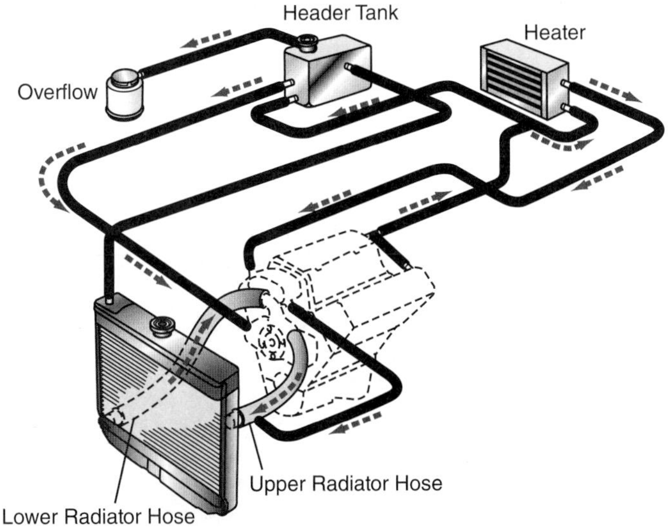

FIGURE 16-48 The cooling system may have multiple flexible coolant hoses, as seen in this cooling system schematic.

FIGURE 16-49 Typical thermostat bypass hose.

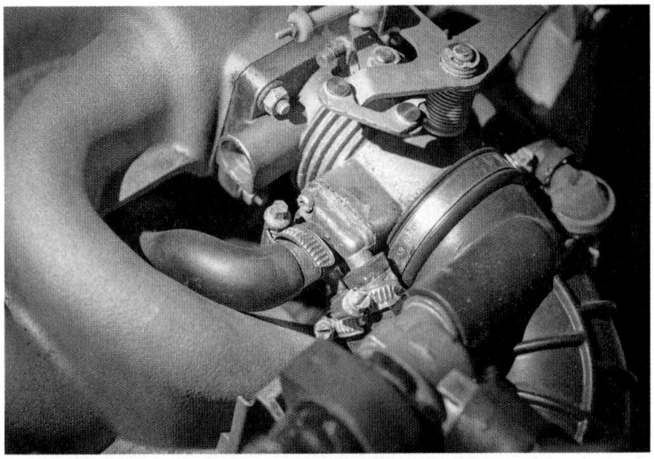

FIGURE 16-50 Coolant lines are used to warm up the throttle body on some vehicles.

V-shaped grooves running lengthwise along the inside of the belt (**FIGURE 16-52**). These grooves are the exact reverse of the grooves in the outer edge of the pulleys. The increased number of grooves increases the contact surface area. They also prevent the belt from slipping off the pulley as it rotates. The serpentine belt is used to drive multiple accessories and therefore saves underhood space forward of the engine. It winds its way around the crankshaft pulley and some, or all, of the engine-driven accessories. Most serpentine belts use a spring-loaded automatic tensioner to maintain proper tightness of the belt.

- *Stretch-fit belt*: A stretch-fit belt looks like an ordinary serpentine belt. But is found on vehicles without a tensioner. It is made of a special material that allows it to stretch just enough to be installed over the pulleys (**FIGURE 16-53**). It then shrinks back to its original size, which is shorter than the distance around the pulleys. This stretchiness keeps the belt properly tensioned over its life. Stretch belts require special tools to install and usually must be cut off when being removed.
- *Toothed belt*: The toothed belt has teeth on the inside that are perpendicular to the belt. The teeth on the belt fit inside the teeth of a gear (**FIGURE 16-54**). Timing belts are always toothed belts to keep the camshaft running in sync with the crankshaft. To save on labor cost, these belts are generally replaced whenever a water pump replacement is required. Likewise, the water pump is generally replaced whenever the timing belt is changed.

Belt Tension

The technician must be careful to properly tension belts when installing or adjusting them. If excessive tension is placed on the belt, the bearings on the accessories or tensioner can be overloaded. This causes them to get too hot, and fail due to excessive working load. If the belt tension is too low, the belt will squeal and slip. A slipping belt will quickly glaze the surface of the belt and pulley. This makes it harder for the belt to grip the pulley, even when it is tensioned properly.

Tensioners

Tensioners are used to keep the drive belt at the proper tension. Proper tension ensures the least amount of slippage without causing damage to component bearings. Tensioners can be either manual or automatic. Manual tensioners come in a wide

FIGURE 16-51 V-type belt and pulley.

FIGURE 16-53 Stretch-fit belt and pulley.

FIGURE 16-52 Serpentine belt and pulley.

FIGURE 16-54 Toothed belt and pulley.

variety of configurations. One type uses a pulley that is adjusted by turning a tensioning bolt (**FIGURE 16-55**). When the bolt is tightened, the pulley moves against the belt with increased tension. If the bolt is rotated the other direction, tension decreases. The tensioner is locked in place by tightening the nut either on the front of the pulley or on the adjustment slot.

A spring-loaded automatic tensioner is typically used with serpentine accessory belts (**FIGURE 16-56**). This type of tensioner adjusts itself, so there is no chance of getting it too tight. However, be sure to note the routing of the serpentine belt when servicing it. There are many pulleys to route the belt around. And it can become confusing if you do not have a routing picture to refer to.

Automatic tensioners can wear out and lose spring tension. This usually causes belt slippage or unusual noises. Tensioner pivot points also wear out. This causes misalignment of the tensioner pulley on the belt. The belt will wear out more quickly, slip, or make noises. Also, a seized accessory or tensioner bearing can cause a no-crank or slow-crank engine condition. In fact, it is easy to mistake that for a seized engine. Simply loosening or removing the belt will allow you to determine if it is an accessory that is causing the fault.

If the water pump is driven by the timing belt, one of two types of tensioners may be used. One type is the spring-loaded tensioner. In this type, a spring sets the tension and a bolt locks the tensioner into position (**FIGURE 16-57**). The other type is the oil-actuated tensioner (**FIGURE 16-58**). The oil-actuated tensioner uses oil pressure from the engine to provide additional tension on the belt. Regardless of the style of tensioner, belt tension can be checked with a belt tension tool. The tension can be compared to specifications, which are typically published in the service information.

▶ Water Jackets

LO 16-08 Describe miscellaneous cooling system components.

Coolant passages such as water jackets are cast into the block and heads during the manufacturing process. They are designed to allow coolant to circulate around the tops and sides of the cylinders. Water jackets are critical for the transfer of excess heat energy (**FIGURE 16-59**). Buildup and scale can restrict heat transfer and coolant flow. Proper maintenance of draining, flushing, and refilling the system can help avoid this damage. Follow the recommended maintenance intervals. (Also see earlier reference to head gasket design.)

FIGURE 16-55 Manual tensioner.

FIGURE 16-57 Timing belt tensioner—spring style.

FIGURE 16-56 A spring-loaded automatic tensioner is typically used with serpentine accessory belts.

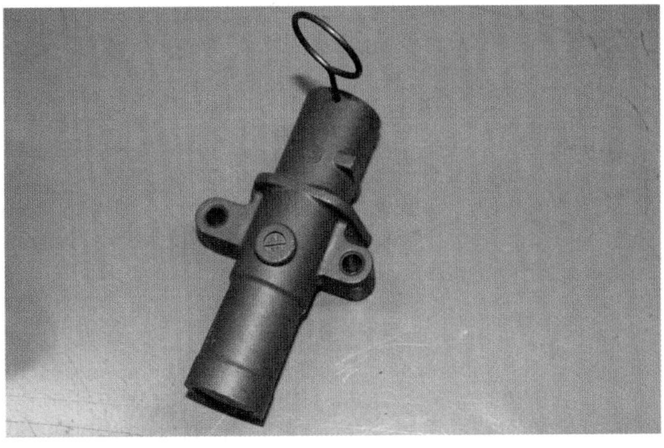

FIGURE 16-58 Timing belt tensioner—oil-actuated style.

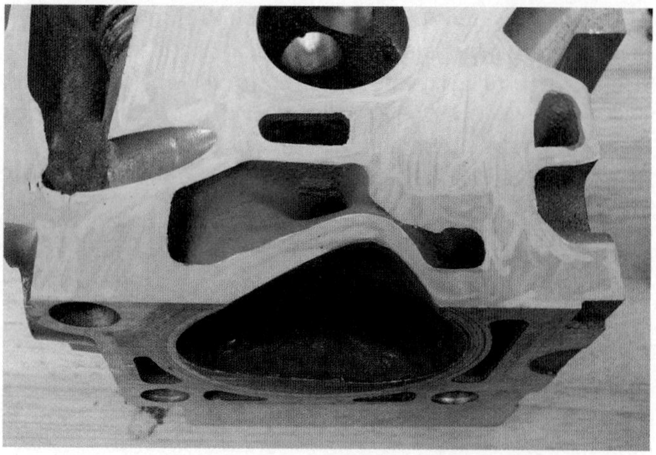

FIGURE 16-59 Cutaway view of the water jacket in the cylinder head.

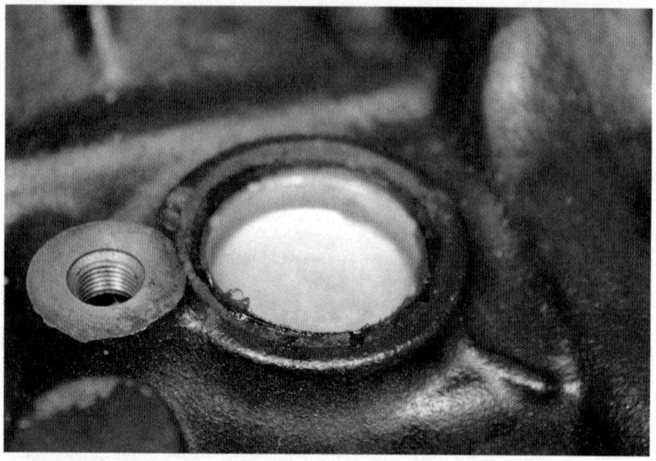

FIGURE 16-60 Core plugs are designed to seal the openings left from the casting process of the block and heads.

Core Plugs

Core plugs are also known as soft plugs or expansion plugs. They are plugs made of aluminum, brass, or steel. They are designed to seal the openings to the water jackets that were left from the casting process when the sand was removed (**FIGURE 16-60**).

Under some conditions, the core plugs *might* pop out if the engine coolant is allowed to freeze. That is, if the proper mixture of antifreeze was not used. Because water expands when it freezes, the block or heads can crack internally or externally near the coolant passages. Sometimes, the core plug will be pushed out and the coolant will leak out before the block cracks. However, that is not what they are designed for. So, do not rely on soft plugs to protect the engine from freezing. Also, core plugs can rust out and start leaking coolant. So, do not forget to inspect them when trying to locate a coolant leak.

Heater Core

The heater core is simply a small radiator that is mounted inside the heater box in the passenger compartment. As air is blown past the fins of the core, heat energy is radiated to the air. This heat is used to heat the passenger compartment for comfort (**FIGURE 16-61**).

FIGURE 16-61 Heater core.

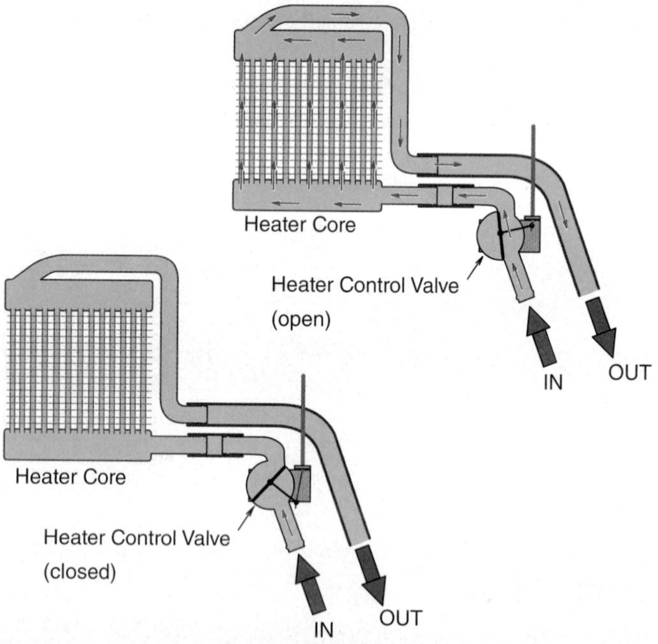

FIGURE 16-62 Heater control valve.

The heater core connects to the engine's cooling system. It is supplied with hot coolant by circulation of the water pump.

Typically, hot coolant enters the bottom of the heater core and exits the top. Thus, the hot water flows from the bottom to the top of the heater core. This allows heat to flow up through the heater core. In this way, more heat can be pulled from the coolant. Heater cores are typically constructed of aluminum, brass, or copper. Because they are made of metal, they can corrode and leak coolant.

Heater Control Valve

If used, the heater control valve is mounted in one of the heater hoses that supply coolant to the heater core. This valve controls the flow of coolant to the heater core. The more coolant allowed to flow, the warmer the air flowing out of the heater (**FIGURE 16-62**). The climate control panel, which is adjusted by the operator, controls this valve. This feature is discussed further in the Electronic Climate Control chapter.

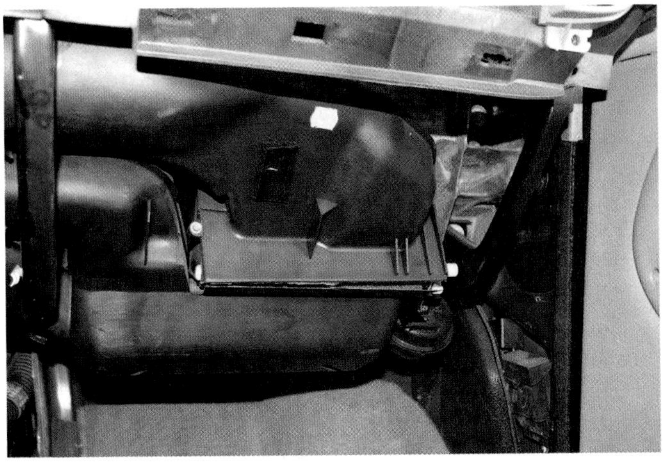
FIGURE 16-63 Typical heater box and air doors.

Air Doors and Actuators

The heater box consists of many air doors. These air doors are plastic or metal flaps that seal off parts of the air box to control airflow (FIGURE 16-63). The air doors are moved by one of three methods: cable, vacuum actuator, or electric actuator, called a stepper motor (FIGURE 16-64). The **actuator** is a device that is electrically or vacuum controlled. It is used to physically move doors within the heater box to control airflow. This system is discussed in detail in the Electronic Climate Control chapter.

The layout and function of the doors depend on the design of the system. Most systems flow air through the evaporator at all times. But air is typically diverted around the heater core when heat is not required. Also, for best defrost operation, the air is directed first over the evaporator to remove any moisture from the air. It is then directed over the heater core to warm it up. This process results in dry, warm air to more quickly defog or defrost the windshield.

Temperature Indicators

Temperature indicators can come in two types: a temperature gauge or a temperature warning light. Both are located in the instrument cluster (FIGURE 16-65). The two types are sometimes used in conjunction. Overheating can heavily damage an engine, so a warning indicator is necessary.

A temperature warning light is a good indicator of an overheating condition. But it cannot indicate a condition where the engine stays below operating temperature. This is a problem since low operating temperatures cause excess engine wear, increased emission output, and decreased fuel economy.

A temperature gauge indicates engine temperature to the driver. Typically, whether the temperature is normal, below normal, or above normal. But drivers can forget to monitor it. And that means that the engine could overheat without the driver noticing. Thus, a warning light in addition to a temperature gauge affords the best assurance.

Temperature gauges and warning lights both operate from a signal sent from a coolant temperature sensor. The sensor is located on the engine in a coolant passage. Therefore, the sensor will provide an accurate reading, as long as the coolant level is not low. If the coolant is low, then it is likely to read a much lower temperature than it actually is.

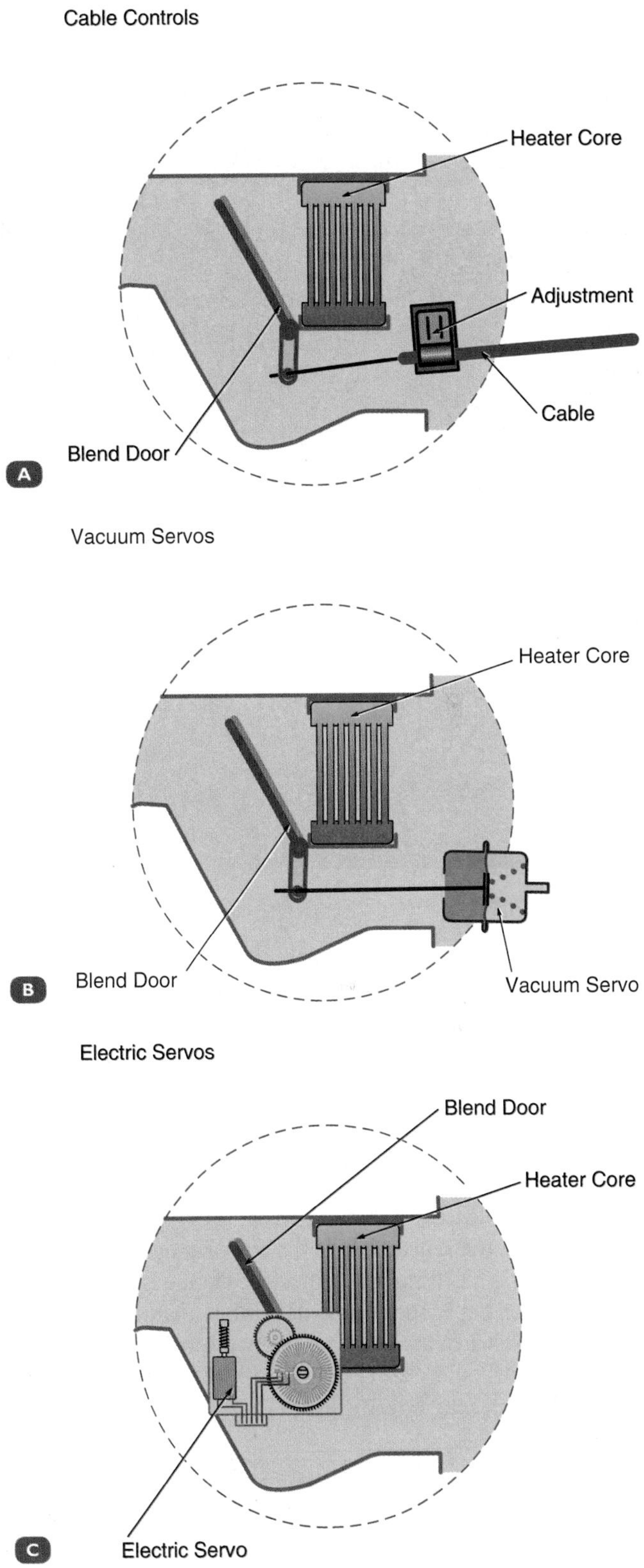

FIGURE 16-64 Air door operating mechanisms. **A.** Cable. **B.** Vacuum actuator. **C.** Electric actuator.

FIGURE 16-65 **A.** Temperature gauge. **B.** Temperature warning lamp.

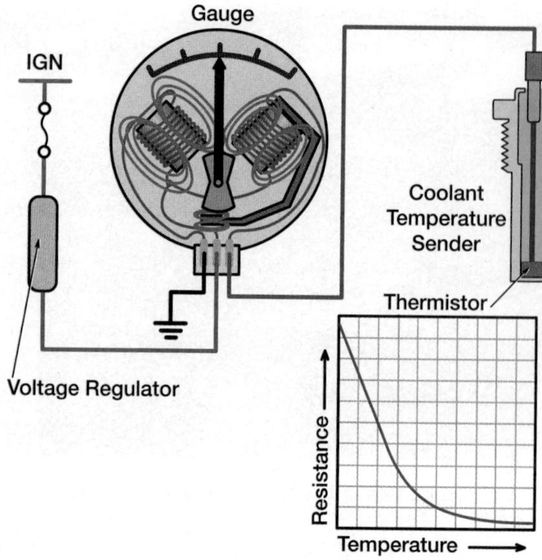

FIGURE 16-66 Temperature gauge wiring diagram.

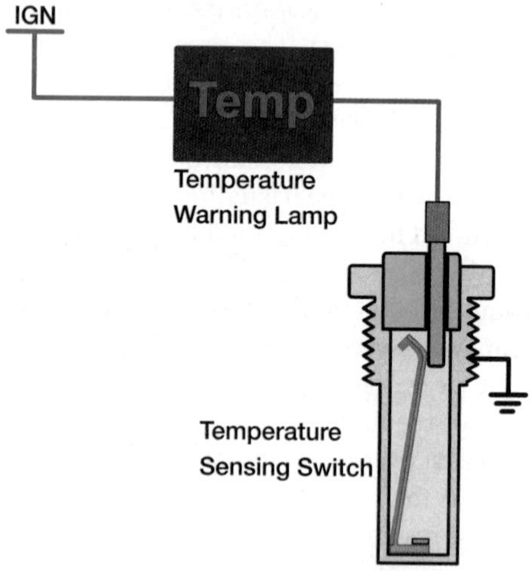

FIGURE 16-67 Temperature warning light diagram.

The temperature gauge uses a sensor which is designed to continuously indicate the temperature of the engine. It uses a thermistor that changes resistance with temperature. As temperature changes, so does the resistance in the sensor. This changes the amount of current flowing through the circuit and gauge. And that changes the reading on the gauge (**FIGURE 16-66**). We will discuss the various types of sensors and how to test them in the next chapter.

On a warning light system, when engine coolant gets hotter than it should, the sensor causes the warning light or message to turn on. This alerts the driver that the engine is overheating. This type of sensor is a thermoswitch. It has a bimetallic spring that closes the switch at a specified temperature (**FIGURE 16-67**). Current is then allowed to flow across the switch contacts, which illuminates the warning light.

There may also be a low coolant indicator that shows when engine coolant level is low. This system works by having a low-level sensor in the surge tank or overflow bottle. It turns on a warning light in the instrument cluster if the coolant level falls below an acceptable level. If the low coolant indicator illuminates, there is a good chance that the cooling system has a leak that must be located.

▶ Wrap-Up

Ready for Review

- Heat transfer occurs through: conduction, convection, and radiation.
    - Cooling systems maintain the ideal operating temperature and can be liquid cooled or air cooled.
- Coolant is a mixture of water and antifreeze solution used to carry heat from the engine. A coolant should prevent corrosion and erosion, resist foaming, be compatible with cooling system component materials, resist sedimentation, and lubricate the water pump seals.
- Coolant circulation makes use of centrifugal force and the flow can be normal flow or reverse flow. Primary components of a vehicle cooling system include: coolant, radiator, thermostat, recovery system, surge tank, water

pump, cooling fan, radiator hoses, heater hoses, drive belts, temperature indicators, water jackets, heater core, and auxiliary coolers.

- A radiator transfers heat energy from the engine to ambient air. It can be either down-flow or cross-flow.
- A thermostat is a is a spring-loaded valve that regulates the flow of coolant to maintain optimum engine operating temperature.
  - A water pump is usually belt-driven from a pulley on the front of the crankshaft and is used to move coolant throughout the cooling system in order to transfer heat energy.
- Cooling fans can be either engine driven or electric to provide airflow through the radiator core for engine cooling.
  - A flex fan uses flexible steel or plastic blades; a viscous coupler connects the water pump pulley to the cooling
- Radiator hoses connect the radiator to the water pump and engine and are usually made of formed, nylon-reinforced rubber; drive belts provide power to drive the water pump and other accessories; and tensioners are used to keep the drive belt at the proper tension.
- Water jackets allow coolant to circulate around the tops and sides of the cylinders; core plugs seal the openings to the water jackets; a heater core is a small radiator mounted inside the heater box in the passenger compartment; heater control valve controls the flow of coolant to the heater core; and temperature indicators provide information to the operator about engine temperature.

Key Terms

actuators Electrical devices that are used to control fan speed, heater core coolant flow, air-conditioning clutch activation, compressor displacement, blend doors, and fresh-air/recirculation doors.

ambient Relates to the immediate surroundings. For example, the temperature of the atmospheric air surrounding a vehicle.

boiling point The temperature at which a substance begins to change from a liquid to a gas.

centrifugal force The apparent force by which a rotating mass tries to move outward, away from its axis of rotation.

conduction The process of transferring heat through matter by the movement of heat energy through solids from one particle to another.

convection The process of transferring heat by the circulatory movement that occurs in a gas or fluid as areas of differing temperatures exchange places due to variations in density and the action of gravity.

cross-flow radiator Type of radiator that has the coolant flow from left to right and is more conformable to the low hood designs of today's vehicles.

down-flow radiator A radiator in which the coolant flows from the top to the bottom.

electrolysis The process of pulling metals apart by using electricity or by creating electricity through the use of chemicals and dissimilar metals.

ethylene glycol A chemical used as antifreeze that provides the lower freezing point of coolant and raises the boiling point. It is a toxic antifreeze.

heat dissipation The spreading of heat over a large area to increase heat transfer.

overflow tank A tank used to catch any coolant that is released from the radiator cap (works like a catch can).

pitch The angle of something.

propylene glycol A chemical used as antifreeze. It is labeled as a nontoxic antifreeze.

radiation The movement of energy through space, such as the movement of energy from the sun to the earth.

radiator A device that transfers heat from a fluid within to a location outside.

radiator hoses Rubber hoses that connect the radiator to the engine. Because they are subject to pressure, they are reinforced with a layer of fabric, typically nylon. Some radiator hoses use coiled wire inside them to prevent hose collapse as the coolant cools.

rotor housing Houses the rotors in a Wankel/rotary engine. This is the base for the engine, similar to the engine block.

surge tank A pressurized tank that is piped into the cooling system. Coolant constantly moves through it. It is used when the radiator is not the highest part of the cooling system. (Remember, air collects at the highest point in the cooling system.)

thermo-control switch A temperature-sensitive switch that is mounted into a coolant passage on the engine or into the radiator to control electric fan operation.

thermostat Regulates coolant flow to the radiator. It opens at a predetermined temperature to allow coolant flow to the radiator for cooling. It also enables the engine to reach operating temperature more quickly for reduced emissions and wear.

viscous coupler Called a fan clutch, a hub that connects the water pump drive to the cooling fan, using a temperature-sensitive viscous fluid to cause the fan to turn faster as the temperature of the air pulled through the radiator increases.

water jacket Passages surrounding the cylinders and head on the engine where coolant can flow to pick up excess heat. They are sealed by replaceable core plugs.

Review Questions

1. When heat energy transfers through a liquid or gas, what is it called?
   a. Conduction
   b. Convection
   c. Radiation
   d. Induction
2. When heat energy transfers through a solid, what is it called?
   a. Convection
   b. Radiation
   c. Ambient equalization
   d. Conduction

3. What would operating an engine with 100% water in the cooling system cause?
   a. Corrosion of the metal parts in the cooling system
   b. Premature hose deterioration
   c. Engine overheating
   d. Loss of heat in the passenger compartment
4. What should be done to prevent excessive deposits and buildup within the cooling system?
   a. Replace the coolant filter often
   b. Flush or replace coolant every year
   c. Use cold tap water when mixing antifreeze and water
   d. Use distilled water when mixing antifreeze and water
5. In a reverse flow cooling system, which major component is engine coolant directed to in the engine after leaving the radiator?
   a. The thermostat
   b. The engine block
   c. The cylinder head
   d. The throttle body
6. Which component is primarily responsible for the pressure in a cooling system?
   a. The thermostat
   b. The water pump
   c. The radiator cap
   d. The surge tank
7. What is the primary purpose of a thermostat jiggle valve?
   a. To act as a bypass for coolant
   b. To relieve water pump pressure under load
   c. To ensure coolant flows the correct direction
   d. To aid in bleeding air from the system
8. How are mechanical cooling system clutch fans often engaged?
   a. A ratcheting mechanism
   b. A viscous fluid
   c. A pneumatic actuator
   d. A pressure switch
9. An engine may be equipped with all the following belts, except?
   a. A serpentine belt
   b. A shrink belt
   c. A toothed timing belt
   d. A V-type belt
10. Which of the following is most responsible for providing or blocking heat to the passenger compartment?
    a. The heater control valve
    b. The water pump
    c. The thermostat
    d. The radiator cap

ASE Technician A/Technician B Style Questions

1. Automotive cooling system problems are being discussed. Technician A states that engine overheating is a common failure. Technician B states that an engine remaining below proper engine operating temperature is a common failure. Who is correct?
   a. Technician A only
   b. Technician B only
   c. Both Technicians A and B
   d. Neither Technician A nor B
2. A vehicle's cooling system is being discussed. Technician A states that engine coolant should not be mixed at a ratio higher than 65% coolant and 35% water. Technician B states that a 50% coolant and 50% water is adequate for most climates. Who is correct?
   a. Technician A only
   b. Technician B only
   c. Both Technicians A and B
   d. Neither Technician A nor B
3. A vehicle is having its water pump replaced. When the water pump is removed, severe corrosion is found on the water pump impeller, and inside the engine. Technician A states that the radiator cap may be overpressurizing the system. Technician B states that the vehicle may have been run with 100% water in the cooling system for some time. Who is correct?
   a. Technician A only
   b. Technician B only
   c. Both Technicians A and B
   d. Neither Technician A nor B
4. A vehicle has had several radiator replacements, and each time there has been pinhole leaks in the aluminum. Technician A states that the engine coolant may need to be exchanged more often. Technician B states that the engine grounds should be inspected closely since this could be the cause. Who is correct?
   a. Technician A only
   b. Technician B only
   c. Both Technicians A and B
   d. Neither Technician A nor B
5. An overheating vehicle is being discussed. Technician A states that coolant in a normal flow cooling system flows from the radiator to the heater core to warm up the passenger compartment quickly. Technician B states that in a reverse flow cooling system, coolant flows to the cylinder heads first, which keeps engine components at a more uniform temperature. Who is correct?
   a. Technician A only
   b. Technician B only
   c. Both Technicians A and B
   d. Neither Technician A nor B
6. A technician is inspecting a vehicle and finds the upper radiator hose is collapsed, and the cooling system pulled into a vacuum when cold. Technician A states that a leak in the surge tank could be the cause. Technician B states that the radiator cap is most likely defective. Who is correct?
   a. Technician A only
   b. Technician B only
   c. Both Technicians A and B
   d. Neither Technician A nor B
7. An overheating vehicle is being discussed. Technician A states that the thermostat is open until the engine fully warms up. Technician B states that the vehicle may be

overheating because the thermostat is stuck closed. Who is correct?

a. Technician A only
b. Technician B only
c. Both Technicians A and B
d. Neither Technician A nor B

8. An overheating vehicle is being discussed. Technician A states that the vehicle may be overheating because the heater control valve is stuck closed. Technician B states that the vehicle may be overheating because the PCM in not energizing the cooling fan relay. Who is correct?
a. Technician A only
b. Technician B only
c. Both Technicians A and B
d. Neither Technician A nor B

9. Coolant hoses are being discussed. Technician A states that many manufacturers recommend coolant hose replacement on a maintenance schedule. Technician B states that coolant hoses can fail prematurely due to contamination by oil. Who is correct?
a. Technician A only
b. Technician B only
c. Both Technicians A and B
d. Neither Technician A nor B

10. A vehicle comes in with no coolant and a core plug popped out of the engine block. Technician A states that the water pump is likely to be overpressurizing the system. Technician B states that the coolant may not have a low enough freezing point. Who is correct?
a. Technician A only
b. Technician B only
c. Both Technicians A and B
d. Neither Technician A nor B

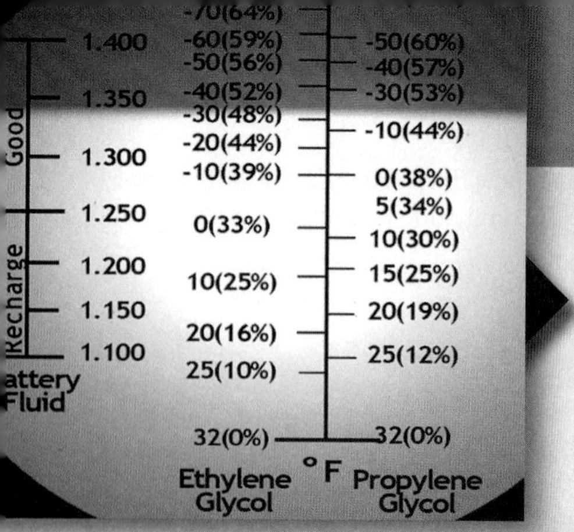

CHAPTER 17

Servicing the Cooling System

Learning Objectives

- **LO 17-01** Test engine coolant.
- **LO 17-02** Service engine coolant.
- **LO 17-03** Service drive belts.
- **LO 17-04** Service coolant hoses.
- **LO 17-05** Service thermostat and bypass.
- **LO 17-06** Service fan, clutch, shroud, and water pump.
- **LO 17-07** Diagnose cooling system performance.
- **LO 17-08** Diagnose cooling system leaks and verify engine operating temperature.

ASE Education Foundation Tasks

See Appendix A to view the 2017 ASE Education Foundation Automobile Accreditation Task List Correlation Guide.

▶ Introduction

Now that you understand the function and operation of the cooling system we can move on to cooling system service. This chapter covers the maintenance required to keep the cooling system function properly. And if the cooling system does fail, you will be prepared to diagnose the root cause of the fault. This chapter will also prepare you to repair cooling systems on a variety of vehicles.

▶ Preventive Maintenance

LO 17-01 Test engine coolant.

Preventive maintenance of the cooling system is critical. It allows for long life and reliability of the engine. Failure to perform required maintenance will result in cooling system failure. Extreme failures can lead to breakdowns and major engine damage. Knowing this will allow you to better recommend needed maintenance to customers.

Manufacturers publish the required cooling system maintenance in the service information (**FIGURE 17-1**). The maintenance schedule lists what services are due at a particular mileage or date. With standard IAT coolant, the maintenance schedule will say "every two years or 30,000 miles, whichever comes first." Other coolants will have a different requirement. Belts and hoses also have inspection and maintenance requirements. These service intervals are critical to follow. So, always check the maintenance schedule for the vehicle you are working on.

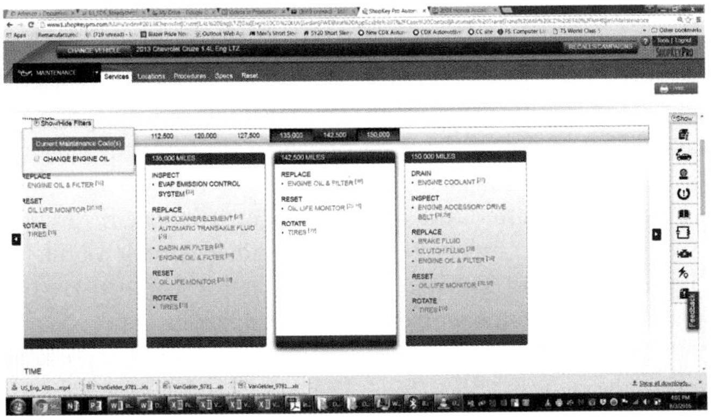

FIGURE 17-1 Maintenance schedule for the cooling system on a sample vehicle.

EPA Guidelines

Ethylene glycol antifreeze, the most common antifreeze, is highly toxic to humans and animals. Because of this, the Environmental Protection Agency (EPA) has strict regulations for the handling and disposal of vehicle coolant. Coolant should never be dumped into a storm drain or down a shop floor drain. Coolant should only be poured into an approved container. It can then either be recycled in-house or removed by a licensed recycler (**FIGURE 17-2**). Shops can be fined for violating this requirement.

Care must be taken when a spill occurs while servicing a vehicle. It should be cleaned up promptly according to EPA regulations. Taking a little extra time to place a catch pan under the component being removed can prevent a spill and save time in the long run. Also, coolant should never be mixed with oil or other liquids. And separate catch pans should always be used and marked accordingly.

Checking Coolant Level

The engine cooling system depends on the coolant to transfer excessive heat from the engine to the radiator. Proper coolant level is critical to this. The engine coolant level and condition should be checked whenever the oil level is checked. Coolant levels that are low indicate a coolant leak that must be diagnosed and repaired.

To check coolant level, some vehicles have a transparent reservoir or surge tank marked with "hot" and "cold." This allows checking the coolant without removing the cap (**FIGURE 17-3**). The coolant level should not be below the lower mark when the vehicle is cold. It should be near the upper mark when the coolant is hot. Older vehicles may not have an overflow tank. They are checked by removing the radiator cap and checking that the coolant level in the radiator is about 1.5" (38 mm) below the filler neck.

SAFETY TIP

Do not remove the pressure cap (radiator cap) when the engine is warm or hot. The system is under pressure, and removing the cap could allow the coolant to immediately boil and spray out, causing severe burns. Always allow the system to cool before removing the cap.

You Are the Automotive Technician

A customer brings a 2012 Honda Accord into your dealership. The concern is that the coolant boils over on hot days. The engine gauge gets higher than normal, but never to the red zone, when steam comes out from under the hood. You ask some clarifying questions, and get the following type of answers: "It happens most during stop-and-go traffic on the way home from work, which is at the hot time of the day." They are not sure if the electric cooling fan is coming on or not since they are focused on traffic. A quick lube shop takes care of the fluids in the vehicle. So, it is not clear for the customer if the vehicle is using coolant or not. The customer agrees to let you diagnose it. You take it back to the shop and verify that the coolant level is near the full mark. You then take a minute to remember the cooling system theory your auto instructor taught you and plan your steps for diagnosis.

1. What are the possible causes for coolant that boils over below the red zone?
2. How can engine operating temperature be verified?
3. What two factors determine the boiling point of engine coolant?
4. What are some of the most likely causes of coolant that boils over when in the red zone?

FIGURE 17-2 Used coolant being stored for recycling.

FIGURE 17-3 Checking the coolant level on a clear reservoir.

Measuring Freeze Protection

The concentration of antifreeze in coolant must be strong enough to prevent it from freezing. It also must be strong enough to prevent corrosion of the cooling system parts. In most climates, the ratio should be 50/50 (50% antifreeze and 50% water). This gives a freeze point of approximately −34°F (−37°C). To make sure the freeze protection level is correct, it needs to be checked in a couple of situations. First, the customer may request this service as part of a winterization package. This makes sure the coolant is strong enough for freezing weather. Second, it should be verified any time coolant is replaced in the cooling system.

A hydrometer or refractometer is used to measure the freeze protection of an engine's coolant. Both tools measure the specific gravity of a liquid. When coolant is drawn into the hydrometer, a float will rise to a certain level depending upon the density of the coolant. Antifreeze has a higher specific gravity than water. So, the higher the float rises in the liquid, the greater the percentage of antifreeze in the mix.

One drawback to hydrometers is that they are typically antifreeze-specific. That means you need one for ethylene glycol and one for propylene glycol. This is because the specific gravities of both fluids are different. Another drawback is that as the temperature of the coolant goes up, the specific gravity goes down. To compensate, some hydrometers have a built-in thermometer and a chart. This allows you to compensate the reading based on the temperature of the coolant.

To use a hydrometer to test the freeze point of the coolant, follow the steps in **SKILL DRILL 17-1**.

SKILL DRILL 17-1 Using a Hydrometer to Test the Freeze Point of the Coolant

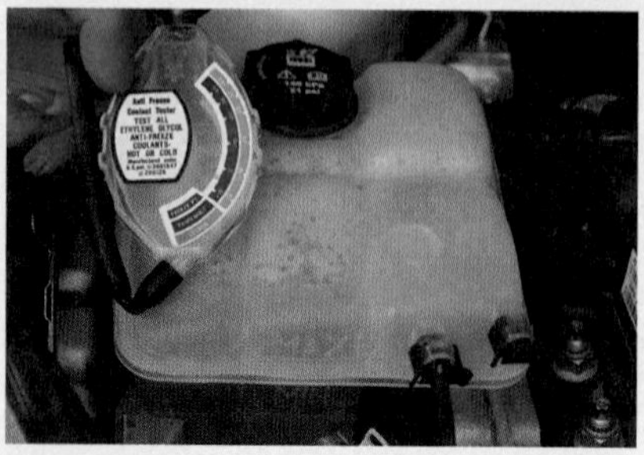

1. Be sure the cooling system is cool. Remove the pressure cap. Determine the type of antifreeze, and verify that the hydrometer is designed to be used with it. Place the hydrometer tube in the coolant, and squeeze the bulb on top.

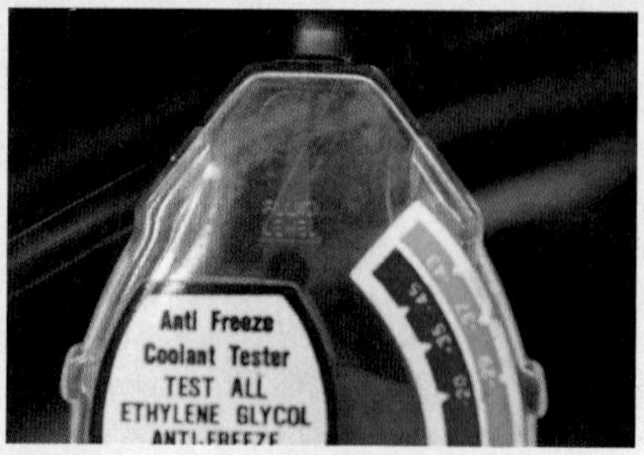

2. Release the bulb to pull a coolant sample into the hydrometer. Verify it is above the minimum level in the tester.

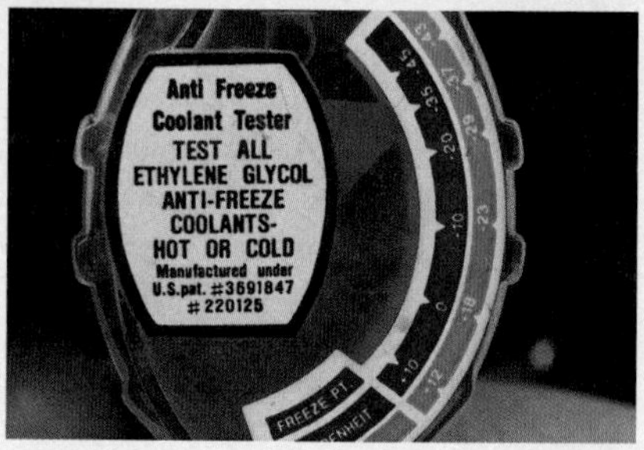

3. Hold the tool vertical. Read the scale to verify the freeze protection of the coolant. Return coolant sample to the radiator or surge tank.

A refractometer can also tell the proportions of antifreeze and water in the coolant mix (or the level of freeze protection). It works by allowing light to shine through a small patch of the fluid. The light bends in accordance with the particular liquid's specific gravity. The bending of the light displays on a scale inside the tool. The line between the light and dark indicates the specific gravity of the fluid. One nice thing about a refractometer is that it has a scale for both types of antifreeze. So, it accurately reads the freeze point for both types.

To use a refractometer to test the freeze point of the coolant, follow the steps in **SKILL DRILL 17-2**.

Testing the Coolant pH

Just because the coolant's freeze protection is adequate, it doesn't mean that the additives are still good. The pH testing of coolant is intended to indicate the life left in the coolant. As corrosion inhibitors break down over time, the solution of water and antifreeze becomes more acidic. As the acid level builds, so does corrosion and electrolysis.

At the same time, pH testing is subject to a fair amount of speculation. There is not any hard and fast specification for an acceptable pH level. Vehicle manufacturers, coolant suppliers, and coolant test strip companies seem to have a wide range of opinion of what pH is acceptable and what is not. The best we can say is that most entities suggest that the pH should be somewhere between 9.5 and 10.5 when installed new in the vehicle. And they recommend a flush below approximately 8.5. At the same time, if the pH level exceeds 11.0, aluminum components in the cooling system start to corrode. So there is a happy range of pH level you should shoot for.

Measuring the pH can be accomplished in two ways. The first way is with test strips that turn color based on the level of acidity in the coolant. The second way is with electronic testers that measure the pH of the coolant directly. pH testing of coolant can help determine if the corrosion inhibitors are still working in the antifreeze. But the pH level of the water you use and the antifreeze chosen will affect the pH level. So, you need to take that into consideration.

To test the coolant pH, follow the steps in **SKILL DRILL 17-3**.

SKILL DRILL 17-2 Using a Refractometer to Test the Freeze Point of the Coolant

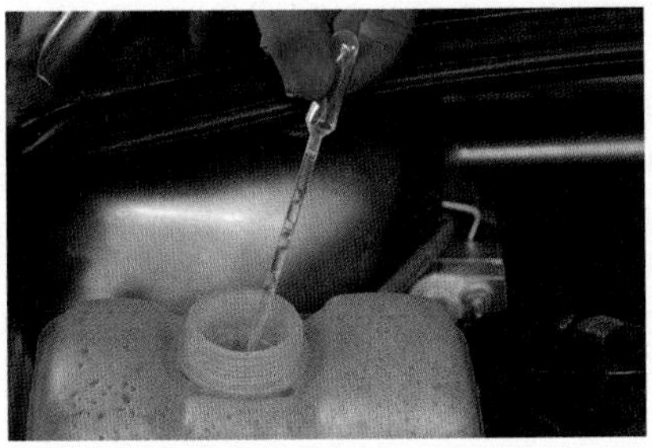

1. Be sure the cooling system is cool. Remove the pressure cap.

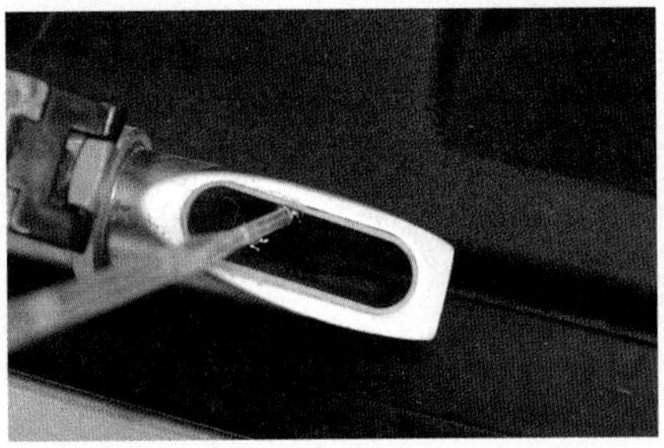

2. Determine the type of antifreeze, and verify that the refractometer is designed to be used with it. Place a few drops of coolant on the sample plate on the top of the tool.

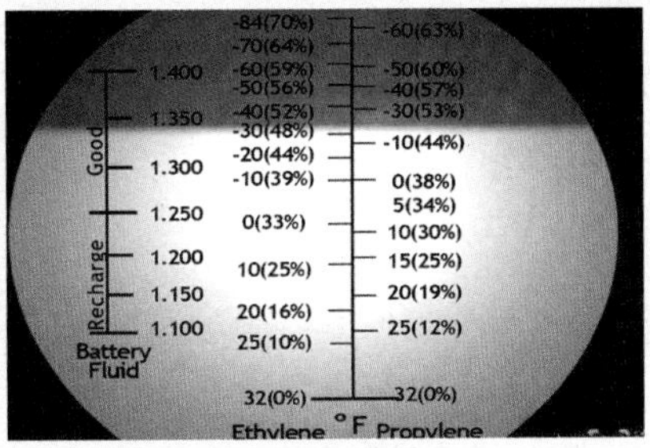

3. Hold the refractometer roughly level under a light, look through the viewfinder, and read the scale to verify the freeze protection of the coolant.

SKILL DRILL 17-3 Testing the Coolant pH

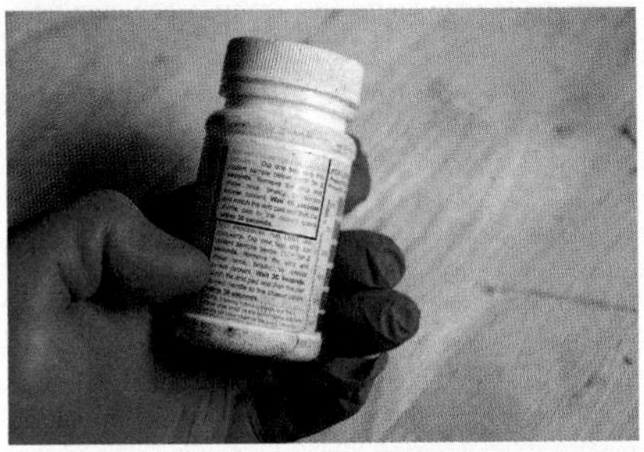

1. Ensure that the coolant is relatively cool before removing the radiator cap. If using a pH test strip kit, read the instructions. Know how long to dip the strip and how long to wait to compare it to the scale. Some strips measure the freeze point, so know the time for that as well.

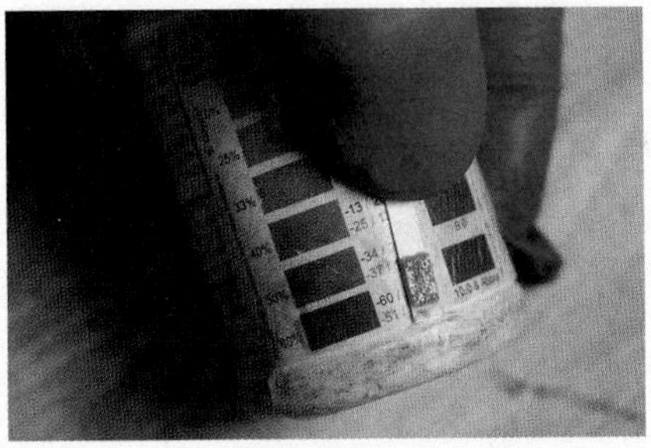

2. Dip test strip into the coolant, and wait the amount of time directed by the instructions. Compare the color of the test strip to the color scale on the kit.

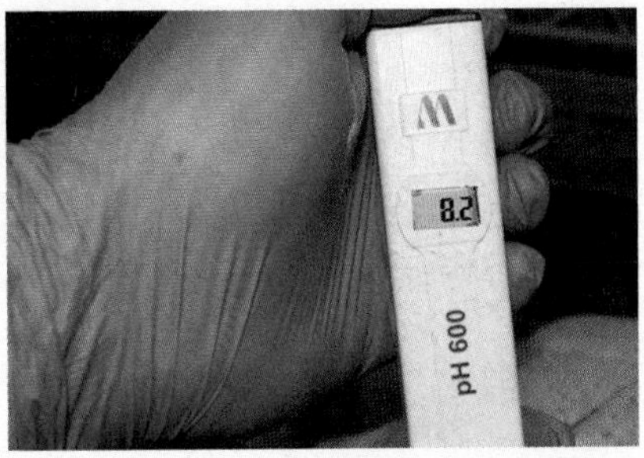

3. If using an electronic pH tester, turn on the tester, and immerse it in the coolant. Read the meter. Some cooling system experts suggest a coolant flush if the pH level is below 8.5.

Testing for Electrolysis

The search for greater fuel efficiency continues. Lighter weight conductive and nonconductive materials used in engines introduce new problems. As a result, there is the need for training new technicians. **Electrolysis** is the reaction of different metals to an acid solution which produces electricity. Electrolysis in the cooling system occurs when the coolant breaks down and becomes more acidic.

When electricity is produced, the movement of the electrons begins to erode the metals in the system (**FIGURE 17-4**). Eventually, pinholes are created in the thinner, softer cooling system components. Most commonly, this occurs in the heater core, radiator core, or aluminum cylinder head. To combat electrolysis, the old coolant should be flushed out and replaced with new coolant.

Electrolysis can also be due to faulty grounds in the electrical system. If a circuit has a faulty ground, the current will try to find its way back to the battery in other ways. And that can include sending some of the current flow through the coolant.

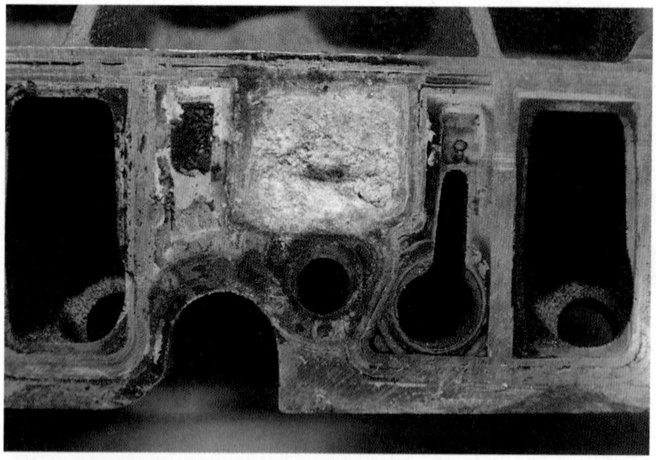

FIGURE 17-4 Erosion caused by electrolysis.

Current flowing through the coolant can erode metal surfaces in the cooling system. In this case, the coolant is not at fault, and flushing will have little to no effect on the condition. Repairing the failed ground is the solution in this case.

To verify whether electricity is finding its way into the cooling system, voltage can be measured in the coolant. This is performed when various electrical loads are operated. If voltages are over 0.3 volts when the load or loads are activated, there is a problem with the ground circuit. You will then need to perform a voltage drop test on the ground circuit being operated.

To perform electrolysis testing for a suspected coolant problem, follow the steps in **SKILL DRILL 17-4**.

▶ Checking and Adjusting Coolant

LO 17-02 Service engine coolant.

Checking coolant implies three separate tasks. First, is the level at the full mark? Second, is the level of freeze protection appropriate for the climate the vehicle is operated in? And third, is the pH level correct?

Checking coolant level should be part of every oil change. This way, any leaks can be identified before they become more serious. You may also need to check or adjust coolant level if the customer's low coolant indicator comes on. There are usually two correct level marks on the reservoir, one for hot and one for cold. The coolant level should not be below the lower mark when the vehicle is cold. It should be near the upper mark when the coolant is hot. If the coolant is indeed low, testing for a coolant leak will be necessary.

The freeze protection level can be tested with a coolant hydrometer or refractometer. Testing of the coolant's pH can be performed with a test strip or electronic tester. If either of these tests shows that the coolant does not meet specifications, it will have to be flushed and replaced.

To check and adjust coolant, follow the steps in **SKILL DRILL 17-5**.

SKILL DRILL 17-4 Performing Electrolysis Testing

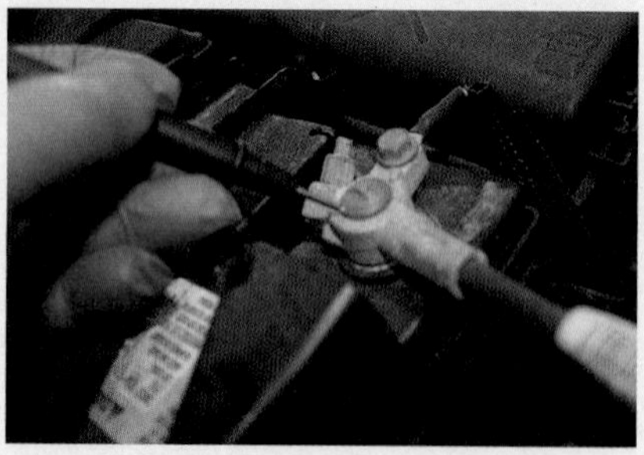

1. Connect the black lead of the voltmeter to a good engine ground.

2. Hold the red lead of the voltmeter in the coolant in the radiator.

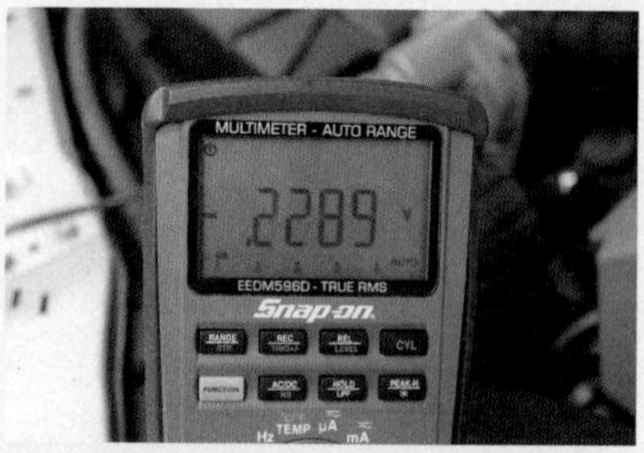

3. Observe the voltage reading. If it is greater than 0.3 volts (300 millivolts), flush the cooling system, refill, and retest. If less than 0.3 volts, operate electrical circuits and read the meter again. If greater than 0.3 volts, check the ground connections.

SKILL DRILL 17-5 Checking and Adjusting Coolant

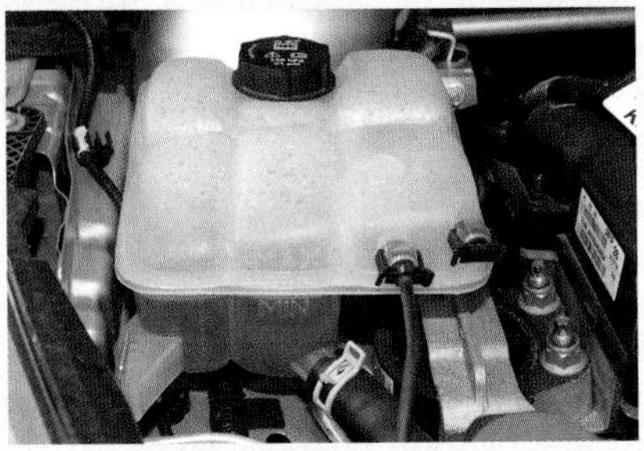

1. Check the level of coolant in this reservoir; if the engine is hot, the level should be visible near the upper mark. If the engine is cold, it should be at or above the lower mark.

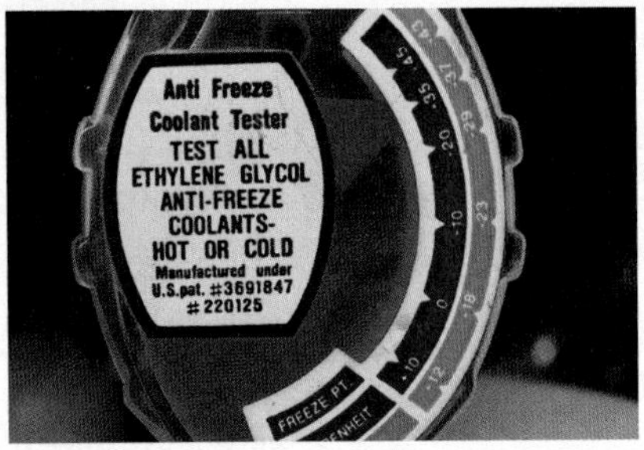

2. Before adding new coolant, check the specific gravity of the coolant in the system with a coolant hydrometer or refractometer.

3. Depending on the freeze protection of the coolant, add either 50/50 premixed coolant, straight antifreeze, or straight water as appropriate to bring it to the proper level. If the level was low, find the cause of the loss.

Adjusting Coolant Freeze Protection Level

Getting the correct freeze protection level sounds fairly simple, but it can be challenging. Let's do the math on a 50/50 mix in a system that holds 16 quarts (15.36 liters). It would require 8 quarts (7.68 liters) of antifreeze mixed with 8 quarts (7.68 liters) of distilled water. A 10-quart (9.6 liters) system would require 5 quarts (4.8 liters) of each, and so forth. Simple, right? Not quite. Let's look further.

The challenge is that the cooling system never drains out completely. Even if drained all the way, there is probably about 25%–30% of the coolant still in the system. If that is pure water, or highly diluted coolant, the final mix will be hard to get right. To overcome this, the technician would first add the entire 50% of antifreeze to the system. Then, top the system off with distilled water. That should bring the system to a 50/50 mix.

As an example, let's look at a 16-quart (15.36 liters) system. If it is drained completely, it is likely that about 4–5 quarts (3.8–4.8 liters) still remain in it. So, we would need to pour all 8 quarts (7.68 liters) of antifreeze into the radiator first. Then, the system will likely take about 3–4 quarts (2.8–3.8 liters) of distilled water to top it off. This will result in the 8 quarts (7.68 liters) of antifreeze, 4–5 quarts (3.8–4.8 liters) of undrained water, and 3–4 quarts (2.8–3.8 liters) of new water. Note that if not all of the water is needed to top it off, it will still end up at a 50/50 mix.

A more difficult challenge is bringing a weak antifreeze solution to a 50/50 mix. Say you want to bring the coolant from 25% antifreeze–75% water mix to a 50/50 mix. In this case, you will need to drain some coolant to make room for the new coolant. But how much do you drain out? This is where some math comes in. Let's say the system is a 16-quart (15.36 liters) system.

- 25% of 16 quarts (15.36 liters) is 4 quarts (3.8 liters). (So, there is about 4 quarts (3.8 liters) of antifreeze in the system.)
- 4 more quarts (3.8 liters) of antifreeze are needed to make it a 50/50 mix.
- But if we drain 4 quarts (3.8 liters) of coolant, we will be draining 1 quart (0.96 liters) of antifreeze with it.
- So, we need to drain an additional quart of coolant to make up for that.
- So, we need to drain about 5 quarts (4.8 liters) of coolant and add 5 quarts (4.8 liters) of antifreeze.
- That will bring us pretty close to a 50/50 mix.
- Note that if the existing percentage of antifreeze is different, or if you need to end up with a different percentage, you will need to use those numbers to determine the above amounts.

If the coolant was flushed with a machine using a 50/50 mix, then a 50/50 premix can be used to top it off. After topping off the cooling system, be sure to run the engine to circulate the coolant. Then, check the mixture's freeze protection level with a hydrometer or refractometer.

Applied Science

AS-1: Proportion Mixtures: The technician can correctly mix fluids using proportions.

As discussed earlier, engine coolant is composed of water and antifreeze. The normally recommended mixture of these two liquids is 50/50—50% water and 50% antifreeze—which provides freeze protection to about −34°F (−37°C). Automotive cooling system capacity typically is given in quarts or liters. Smaller vehicles require perhaps 5 quarts (4.8 liters). Larger SUVs with air conditioning require up to 20 or more quarts (19.2 or more liters). Coolant can now be purchased premixed for convenience, but you will be paying for 50% water, plus the cost of shipping it, so it is usually less expensive to mix your own coolant.

Draining and Refilling Coolant

Draining and refilling coolant is necessary for several reasons:

- if the customer requests a preventive maintenance service to flush the cooling system.
- if the coolant pH is not within specifications
- if the coolant was contaminated by an incorrect fluid being added to the radiator, such as by mistakenly adding power steering fluid to the coolant reservoir.
- When a part of the cooling system must be removed, the coolant will need to be drained and refilled.

When draining the coolant, it will drain faster if you remove the radiator cap before opening the petcock. To drain the cooling system further than the petcock can do itself, some engines are equipped with block drains. They are located near the bottom of the block. This allows more coolant to be drained from the block.

When refilling the system, it is critical that any trapped air be bled from behind the thermostat. This can be accomplished by loosening a bleed screw if so equipped. If not, then loosen the highest hose in the cooling system. Fill the radiator or surge tank until coolant comes out the hose. Finally, always dispose of used coolant in an environmentally approved manner.

Your shop may have a coolant flushing machine, which will change the steps of the following skill drill. If you are using a flushing machine, follow the directions for the machine you are using.

To drain and refill coolant, follow the steps in **SKILL DRILL 17-6**.

Flushing the Coolant

Coolant flushing is necessary as a preventive maintenance service. It is generally performed because the coolant is worn out or contaminated. Flushing the coolant removes virtually all of the old coolant so that new coolant can be added. In some instances, a flushing chemical is used before the flush is performed.

Flushing of the coolant is performed in one of two ways: manually or with a flushing machine. Most shops have a coolant flushing machine. If you are using this machine, follow the directions on the machine as they are all different. In this example, we perform the manual flush.

To flush the coolant, follow the steps in **SKILL DRILL 17-7**.

SKILL DRILL 17-6 Draining and Refilling Coolant

1. Remove the radiator cap. Locate the radiator drain plug, if equipped, and place a catch pan marked for coolant underneath the drain plug. Drain the radiator into the catch pan.

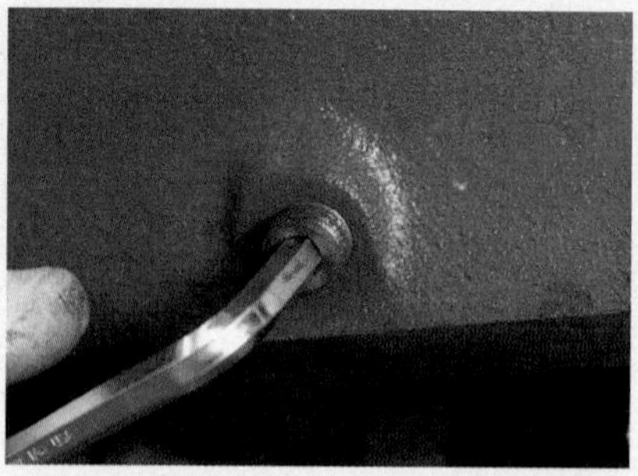

2. Remove the block drain plugs, and allow the coolant to drain into the pan.

3. Refill the cooling system with the proper coolant mix after closing the drain plugs. Start the engine and verify the proper level. Dispose of coolant in an approved way.

SKILL DRILL 17-7 Flushing the Coolant

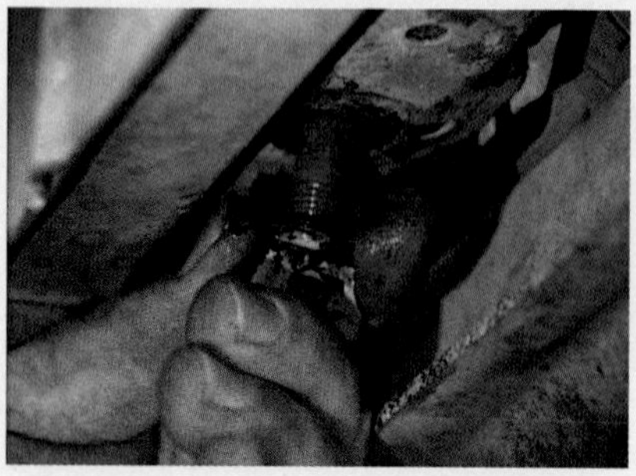

1. Remove the radiator cap. Locate the radiator drain plug, if installed. Drain the coolant into the catch pan.

2. Remove any engine drain plugs in the block and allow the block to drain into the catch pan.

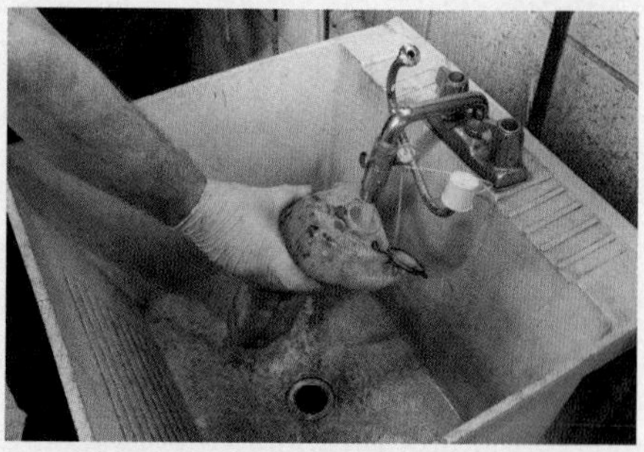

3. Remove the surge tank or overflow tank, and clean thoroughly with hot water.

SKILL DRILL 17-7 Flushing the Coolant (Continued)

4. Reinstall the tank and all drain plugs, and refill the radiator with clean water, and if desired, a coolant flushing additive. Start the engine and allow it to warm fully.

5. Drain water from the radiator and engine block. Replace all drain plugs.

6. Refill the radiator with the proper mix of antifreeze and water. Bleed any air from the system. This can be done using a vacuum bleeding system or by running the engine with the cap off and topping it off.

▶ Inspecting and Adjusting an Accessory Drive Belt

LO 17-03 Service drive belts.

Inspecting the engine drive belt should be part of any maintenance inspection. Drive belts stretch over time with use and may require adjustment. Adjusting the drive belt may be necessary if the vehicle is not equipped with an automatic tensioner. When adjusting belts, if they are too loose, they will squeal or chirp and slip. If the belt is too tight, it will put extra force against the bearings on the accessories being turned, which can cause them to wear out prematurely. Conditions to look for on a drive belt include:

- *Cracks:* Cracks in a belt used to indicate that immediate replacement was needed. With today's belts, many manufacturers tolerate a certain number of cracks per inch. Check the manufacturer's specifications before recommending a belt be replaced (**FIGURE 17-5**).
- *Oil contamination:* A belt that has been soaked in oil will not grip properly on the pulleys and will slip. Diagnose and repair the cause of the oil contamination, and replace the belt.
- *Glazing:* Glazing is shininess on the surface of the belt, which comes in contact with the pulley. If the belt is worn, the glazing could be caused by the belt "bottoming out" in the pulley, and it should be replaced. If it is not old and worn, glazing could simply indicate that the belt is not tight enough. Tightening the belt may be all that is necessary, depending on how bad the glazing is (**FIGURE 17-6**).
- *Tears:* Torn or split belts are unserviceable and should be replaced immediately (**FIGURE 17-7**).

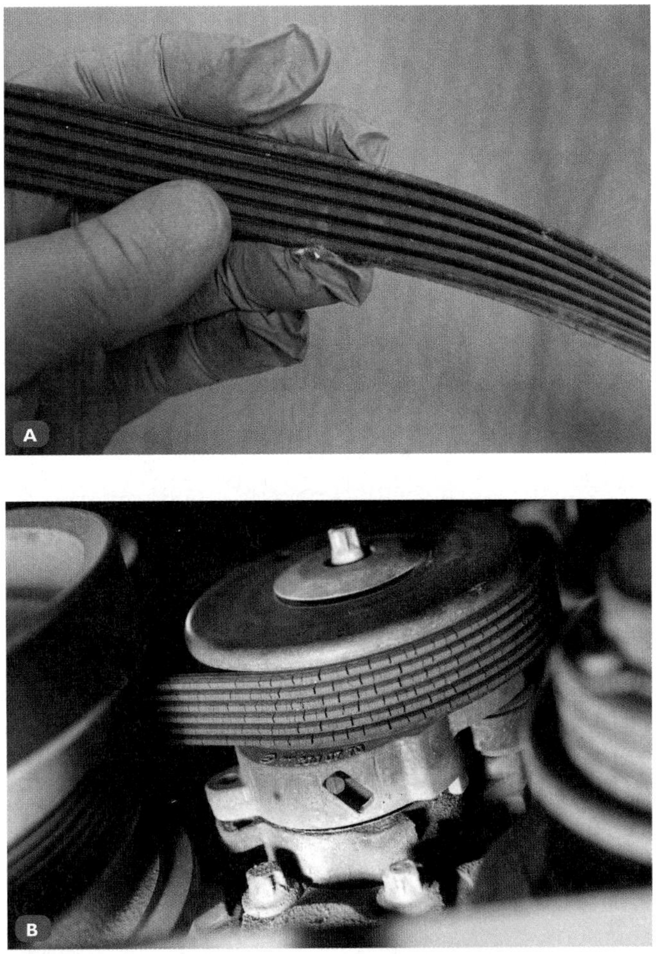

FIGURE 17-5 A. Acceptable belt. **B.** Unacceptable belt.

- *Bottoming out:* When V-type or serpentine-type belts become extremely worn, the bottom of the V-shape may contact the bottom of the groove in the pulley. This prevents the sides of the belt from making good contact with the sides of the pulley groove. This reduced friction causes slippage. A belt worn enough to bottom out should be replaced. V-belts can be inspected for the depth of the belt in the pulley visually (**FIGURE 17-8**). A serpentine belt can be checked with a plastic wear gauge. If the plastic tool is even with the top of the belt ribs, or lower, the belt needs to be replaced (**FIGURE 17-9**).
- *Pulley wear:* Always inspect the pulley when inspecting the belt. A worn pulley will slip and squeal.
- *Too wide*: When a V-belt that is too wide is used, the belt sits above the edges of the pulley. As the sides of the belt wear, a step develops in the side of the belt. After a while, this step runs on the top of the pulley, reducing the grip on the sides of the pulley. This can commonly cause the belt to squeal even though it is tight. Replacing the belt with one of the proper width will solve the issue (**FIGURE 17-10**).

SAFETY TIP

Never try to inspect belts with the engine running. Fan belts have very little stretch in them. If your fingers get pulled between the belt and pulley, it will break the bones. It may even pinch your fingers off.

FIGURE 17-8 Bottomed-out V-belt.

FIGURE 17-6 Glazed belt.

FIGURE 17-9 Bottomed-out serpentine belt.

FIGURE 17-7 Torn belt.

FIGURE 17-10 V-belt that is too wide for the pulleys.

To inspect and adjust an accessory drive belt, follow the steps in **SKILL DRILL 17-8**.

> **▶ TECHNICIAN TIP**
>
> Some vehicles require the technician to manually adjust the tension on the belt. Other vehicles have an automatic spring tensioning system. Tension can be checked with a belt tension gauge. There are several different types, so follow the operating instructions for the tool. If you do not have a tension gauge, you can estimate the tension by pushing the belt inward with your thumb. If it is correctly tensioned, you should be able to deflect the belt only about half an inch for each foot of belt span between pulleys.

Replacing an Accessory Drive Belt

Replacement of an engine drive belt may be necessary when the belt is cracked, glazed, separating, and getting ready to fail. Another reason to replace belts is if it has met the replacement time in the scheduled maintenance chart. Verifying that the pulley system is not damaged or misaligned will be the first step if noise is a problem. Checking pulley alignment is done with a straightedge across the face of the pulleys. It can also be performed with a special laser that fits in the grooves of the pulley. If it is a serpentine belt system, the pulley edges should be within 1/16" (1.6 mm) of alignment with each other.

Before removing the belt, you need to verify two things. First, verify that the replacement belt is the correct belt. Compare both the width, length, and number of grooves of the belt to specifications. And second, verify the routing of the belt.

Some serpentine belts wrap around about 10 different pulleys. So it is easy to get the routing wrong. This can cause accessories to turn backward. Most vehicles have a belt routing label under the hood. Find that and compare it to the routing of the existing belt. If the belt routing diagram is not available or is unreadable, create your own diagram on a piece of paper. That can save you from a big problem of trying to figure out the correct routing by trial and error.

When replacing serpentine belts, it is common that the belt will not seat in the proper grooves in one or more pulleys. If this is not corrected, the belt will wear out quickly, or even be thrown off the pulley. So, always check each pulley to make sure the belt is centered in the pulley. In many cases, you will have to use your fingers to feel how the belt is positioned in the pulley.

To replace a standard accessory drive belt, follow the steps in **SKILL DRILL 17-9**.

Replacing a Stretch Fit Belt

Although Stretch Fit belts can last up to 10 years or 150,000 miles, you should still be familiar with the procedure to remove and replace them. First, because they are designed to be operated

SKILL DRILL 17-8 Inspecting and Adjusting an Accessory Drive Belt

1. If a V-belt, twist the belt so that you can see the underside of the V shape.

2. If a serpentine belt, check the ribs with a serpentine belt gauge.

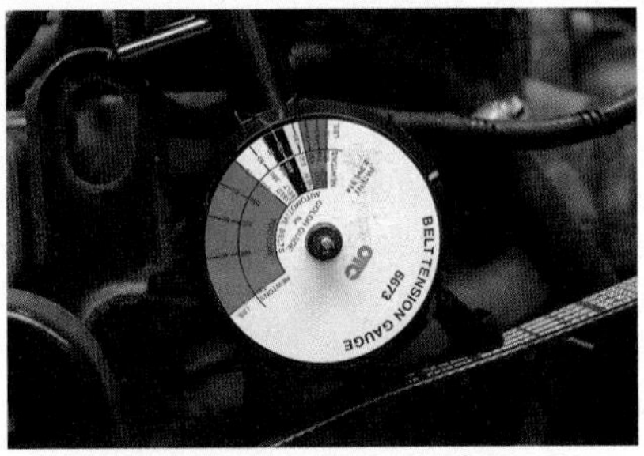

3. Check the belt tension by attaching the belt tension gauge to the belt, and measure the tension. On manually adjusted belts, loosen the locking fastener.

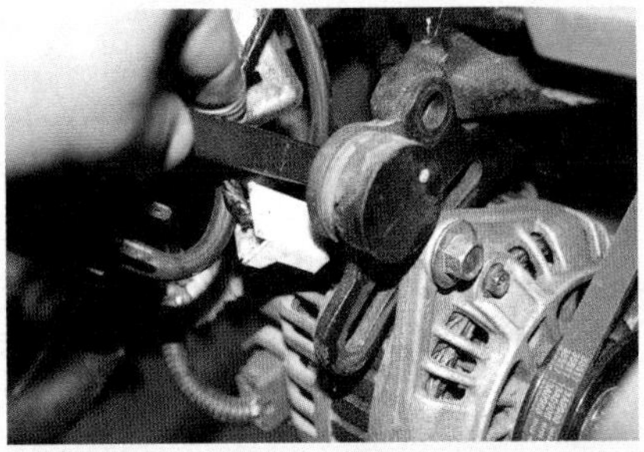

4. For the slotted style, use a pry bar to carefully pry the adjustable component until the belt tension is adjusted. Tighten the locking fastener.

5. On the style that uses a tensioning screw, tighten the tensioning screw until the belt is properly tensioned. Tighten the locking fastener.

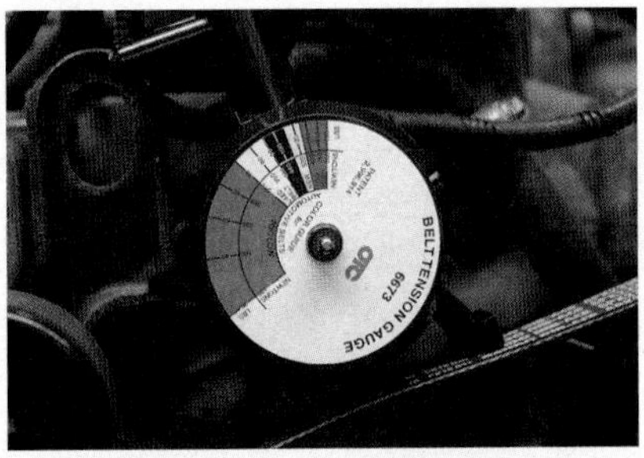

6. Check belt tension with the belt tension gauge. Remove the gauge. Start the engine and verify proper operation.

SKILL DRILL 17-9 Replacing an Accessory Drive Belt

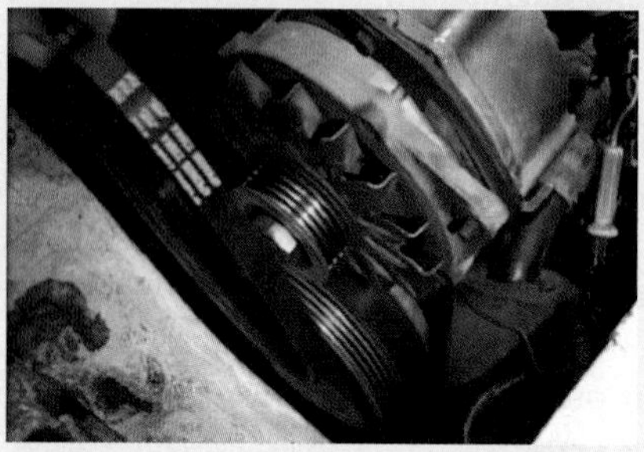

1. On a manually adjusted belt, locate and loosen the adjustment locking fastener. Loosen the belt and remove it.

2. On an automatically adjusted belt, locate the spring-loaded idler pulley, install an adjuster tool, and pry the tensioner back. Remove the belt.

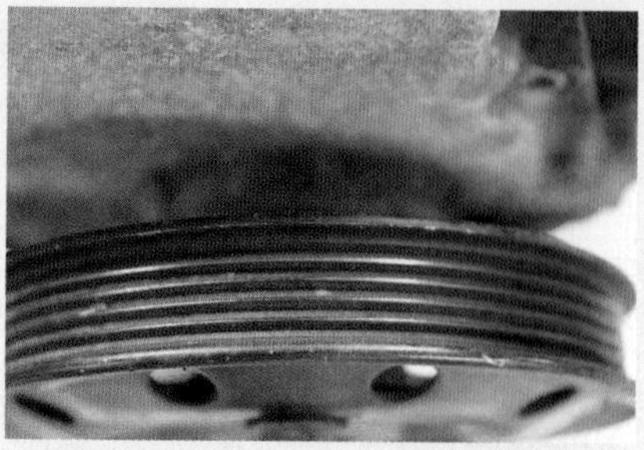

3. Inspect the belts and pulleys for wear and damage.

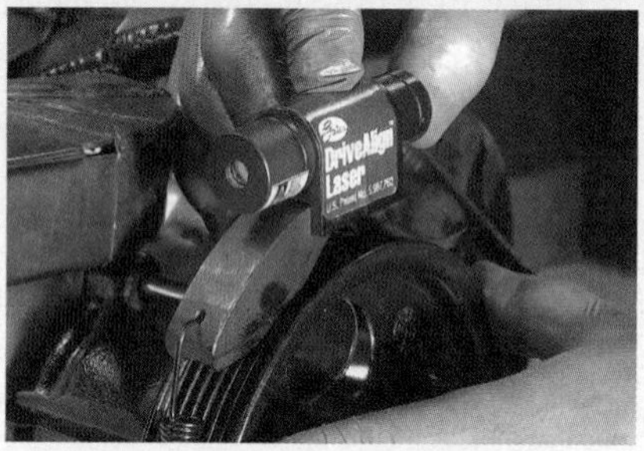

4. Check all drive pulleys for alignment using a straightedge or laser.

5. Select the correct replacement belt. Install the V-belt or serpentine belt. Make sure the belt is properly routed around the pulleys and fully seated in each pulley.

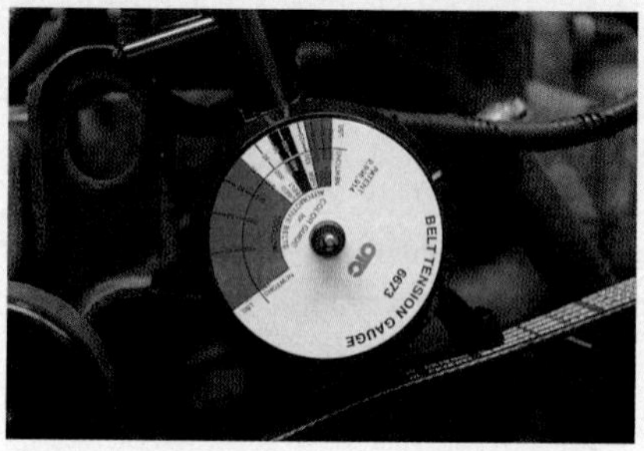

6. Correctly tension the new belt. Start the engine and verify proper operation. Stop the engine and recheck the tension.

without a tensioner, there is no way to loosen the belt to remove it. So, Stretch Fit belts are typically cut with diagonal side cutters to remove them. This means that you need to replace any Stretch Fit belt with a new one whenever the belt is removed for any reason.

To install a new Stretch Fit belt, you need a special installation tool. This tool grabs onto the crankshaft pulley and levers the belt over the side of the pulley as the crankshaft is turned. This presents several safety concerns. First, you need to turn the engine over by hand because the tool is on the crankshaft pulley. It would be thrown if the engine happened to start while cranking it over. So, don't use the starter to crank the engine to install the belt.

Installing a Stretch Fit belt by hand means you *must* have the ignition switch OFF when turning the engine. You don't want the engine to start, or even kick when you are turning it by hand. Use a socket and ratchet on the front crankshaft bolt to slowly turn the crankshaft while installing the belt.

To install a Stretch Fit drive belt, follow the steps in **SKILL DRILL 17-10**.

SKILL DRILL 17-10 Installing a Stretch Fit Drive Belt

1. Gather the correct replacement belt and installation tool.

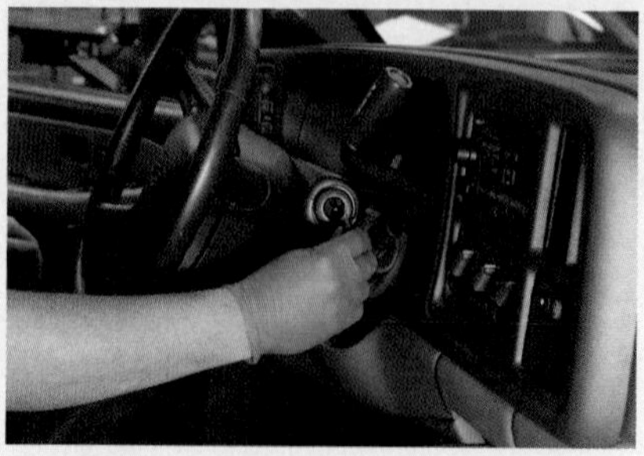

2. Make sure the ignition key is OFF.

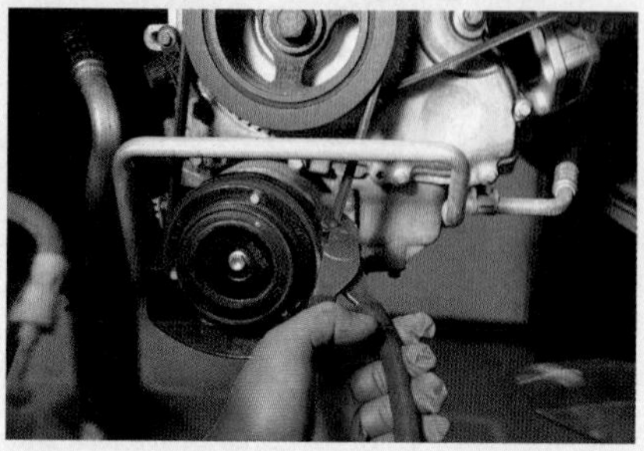

3. Cut the old belt with a pair of diagonal side cutters.

SKILL DRILL 17-10 Installing a Stretch Fit Drive Belt (Continued)

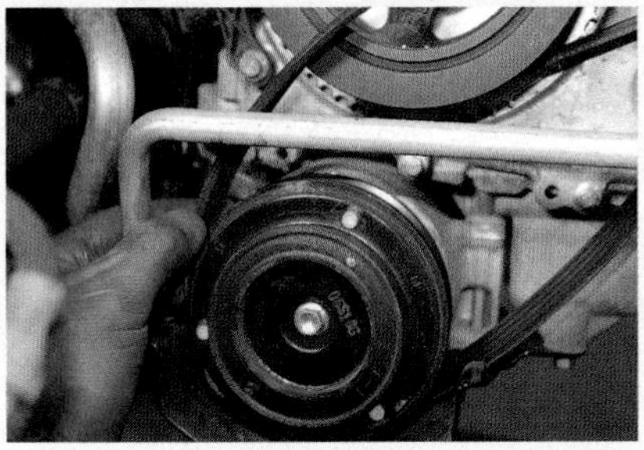

4. Position the belt fully on all of the pulleys except the crankshaft pulley.

5. Position the installation tool and belt so that the belt is snug against the pulley and tool. Using a socket, turn the crankshaft slowly to stretch the belt over the side of the pulley.

6. Verify that the belt is properly centered in all of the pulleys.

Inspecting and Replacing a Coolant Hose

LO 17-04 Service coolant hoses.

Preventing vehicle breakdowns is a very important job of a technician. Cooling system failures can easily lead to breakdowns, and hose failures are a likely cause. If the hose is deteriorating, it will eventually burst. Coolant will be pumped out of the engine, resulting in overheating.

Coolant hoses should be checked anytime the vehicle is in the shop for other work. If you find one deteriorated coolant hose, chances are that the other hose(s) may be deteriorating in the same way. For this reason, most technicians generally replace both radiator hoses at once as a sensible precaution.

Do not forget that there may be many hoses in the cooling system. You need to inspect all of them to ensure proper function of the cooling system. Some hoses are buried and hard to inspect, so a flashlight and mirror may be helpful. Coolant hose problems include:

- *Swollen hose:* This hose has lost its reinforcement and is swelling under pressure. It may soon rupture; typically, you will see a bubble protruding from the side of this hose. Replace this hose immediately (**FIGURE 17-11**).
- *Hardened hose:* This hose has become brittle and will break and leak. Verify hardening by squeezing the hose and comparing it to a known good hose (**FIGURE 17-12**).
- *Cracked hose:* This hose has cracked and will soon start to leak. Verify cracking by a visual inspection (**FIGURE 17-13**).
- *Soft hose:* This hose has become very weak and is in danger of ballooning or bursting. Verify softening by squeezing the hose and comparing it to a known good hose (**FIGURE 17-14**).

▶ TECHNICIAN TIP

"You can't diagnose bad if you don't know good." This saying is very appropriate when checking radiator hoses. It is a good idea to check the feel of radiator and heater hoses on a variety of new vehicles to become familiar with what new hoses feel like. Then, it will be easy to identify the bad ones.

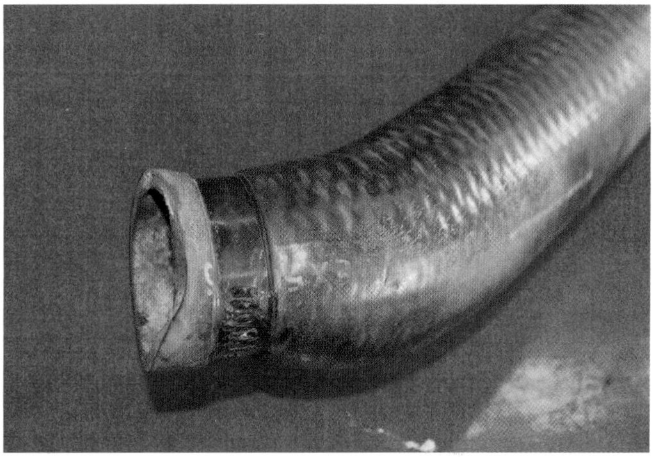

FIGURE 17-11 Swollen hose due to broken cords.

FIGURE 17-12 Hardened/brittle hose due to old age and heat.

Hose clamps come in several forms and require different tools to properly remove or replace. These clamps secure the hose to the component it is connected to. Be sure to follow proper installation instructions to prevent future leaks from this seal. Types of clamps include:

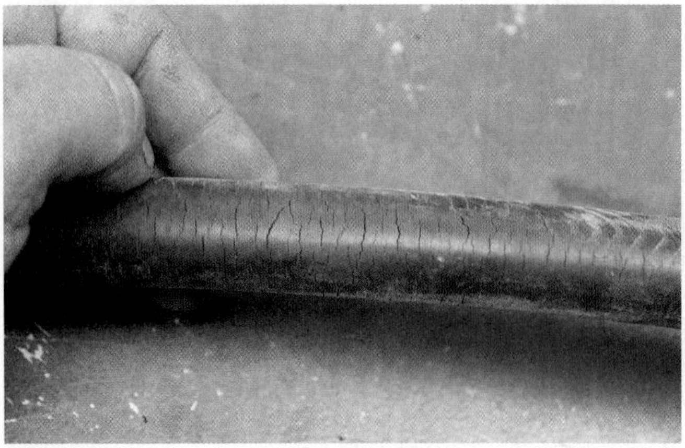

FIGURE 17-13 Cracked hose due to old age.

FIGURE 17-14 Soft hose due to old age or contamination from oil or gas.

- *Wire or spring clamp:* It is not adjustable. It is fitted and removed with special hose clamp pliers, which have grooved jaws (**FIGURE 17-15**).
- *Banded or screw clamp:* Adjust with a screwdriver.
- *Gear or worm clamp:* Adjust with a screwdriver or nut driver.

Clamps are relatively inexpensive. So, it is good practice to install new clamps at the same time as new hoses. Even if

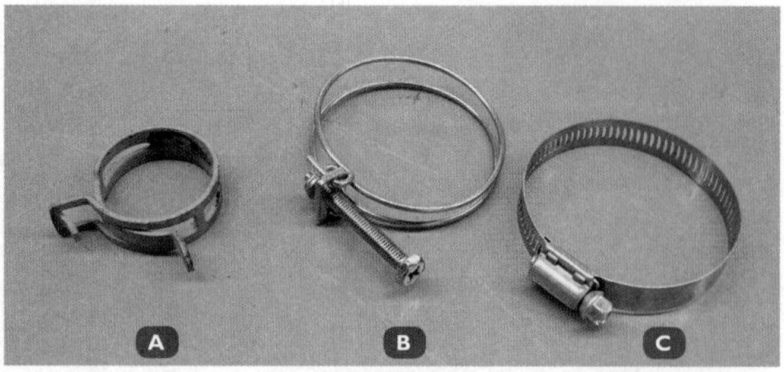

FIGURE 17-15 **A.** Spring clamp. **B.** Banded or screw clamp. **C.** Gear or worm clamp.

not corroded, the old clamps may have become distorted when being removed. The time and money associated with replacing clamps is very small compared to the hassle the customer would experience if a used hose clamp fails. So use new clamps when replacing hoses.

When removing hoses from the radiator and heater core, be very careful that you don't damage either one. The tubes are usually either thin metal or plastic. If metal, then the tubes are soldered to the radiator or heater core. If you try to twist the hose off, you can twist the tube out of the solder joint.

You also shouldn't try to pry the hose off with a screw driver as that can dent the tube, which can cause leaks. So, the best way to remove hoses is to slit them. Then, carefully peel them off if they are stuck and don't come off easily.

When repositioning hoses, make sure they are fully installed on the fitting. There is usually a small hump on the surface of the fitting. The hose needs to be installed past that hump far enough that the clamp can be placed on the hose-end side of the hump. In other words, the clamp should be installed between the end of the hose and the hump. Do NOT install the clamp on top of the hump. This will NOT hold the hose securely. The hose will likely blow off the fitting once the cooling system pressurizes.

To check and replace a coolant hose, follow the steps in **SKILL DRILL 17-11**.

SKILL DRILL 17-11 Checking and Replacing a Coolant Hose

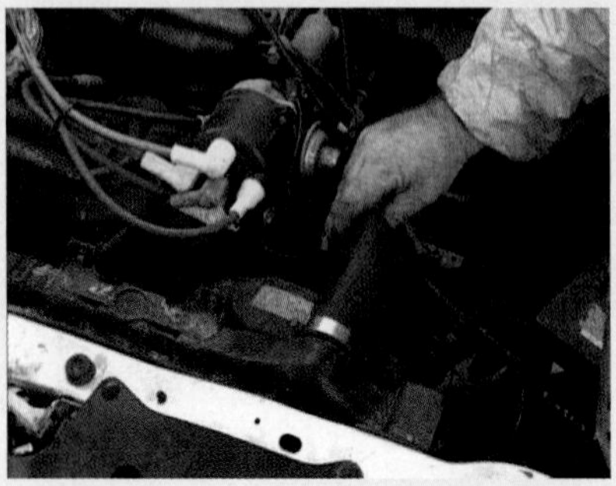

1. Inspect the hoses by squeezing them, and visually inspect the clamps. Remove the hose clamp. Carefully pull or cut the hose off the fittings.

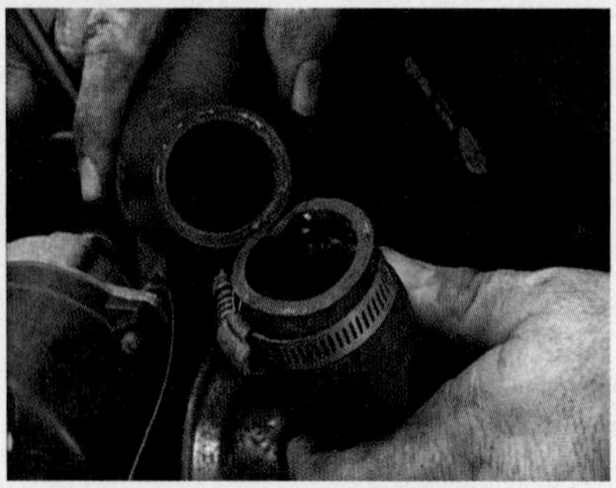

2. Verify that the replacement hose is correct. Cut to length if necessary. Reinstall the new hose all the way into position. Install new clamps.

3. Refill coolant, verify correct level, and pressure check for leaks.

SAFETY TIP

Never try to assess the serviceability of a coolant hose while the engine is hot. Turn the engine off, and let it cool so that you can handle the hoses comfortably and safely. It is easy to hurt yourself, even when the most exhaustive protection measures are taken.

▶ Removing and Replacing a Thermostat

LO 17-05 Service thermostat and bypass.

It will be necessary to remove and replace a thermostat if the thermostat is found to be faulty. Faulty thermostats create either an overheating concern or an underheating concern. It is good practice to suggest a thermostat replacement whenever cooling system service is needed.

Allow sufficient time for the system to cool adequately before opening the pressurized system. Drain at least 50% of the coolant in the system to avoid spills. Follow the specified procedure to remove the thermostat. Clean any old gasket material and corrosion that has built up on both sealing surfaces where the thermostat seats.

For installation, always use a new thermostat gasket, O-ring seal, or rubber seal (**FIGURE 17-16**). Properly position the thermostat air bleed valve (if equipped). Make sure the thermostat is facing the correct direction (sensing bulb toward the engine). It must also be fully seated in the groove and stay there, before installing the housing. In most cases, if the thermostat falls out of its recessed groove, one of the thermostat housing ears will break off when the bolts are tightened. This will require replacement of the housing. Also, the thermostat itself is likely to be damaged. Tighten the housing bolts to the specified torque. Follow the manufacturer's procedure to properly bleed all air from the cooling system.

To remove and replace a thermostat, follow the steps in **SKILL DRILL 17-12**.

Inspecting the Thermostat Bypass

The thermostat bypass sends coolant back to the water pump inlet when the thermostat is closed. That way coolant can circulate through the engine, picking up heat. The bypass can be a simple passageway directly under the thermostat. You need to inspect the bypass port to make sure it is not restricted by corrosion or foreign materials. If so, clean it out or replace the bypass port as needed.

In some cases, the thermostat will have a disc on one end. The disc opens and closes the bypass port as the thermostat opens and closes. Make sure the new thermostat has this disc if the system is designed to use it. Installing a thermostat without the disc can cause overheating since hot coolant will be bypassing the radiator. Also, make sure that the disc is at the correct height so that it will be able to block off the bypass port when the thermostat opens.

Other bypass systems use a rubber hose to connect the thermostat housing to the water pump inlet (**FIGURE 17-17**). This hose usually gets neglected, so you should always make sure to inspect it along with the other hoses. It is usually either a short piece of heater hose or formed hose. So replacement hoses are usually available.

FIGURE 17-16 Type of thermostat seals. **A.** Gasket. **B.** O-ring. **C.** Rubber seal.

▶ Inspecting and Testing the Cooling Fan

LO 17-06 Service fan, clutch, shroud, and water pump.

The cooling fan is a critical part in the cooling system. When coolant moves through the engine, it picks up heat. Even though the coolant is moving heat to the radiator, air still has to move

SKILL DRILL 17-12 Removing and Replacing a Thermostat

1. Drain the coolant from the cooling system, using an approved catch container. Remove the thermostat.

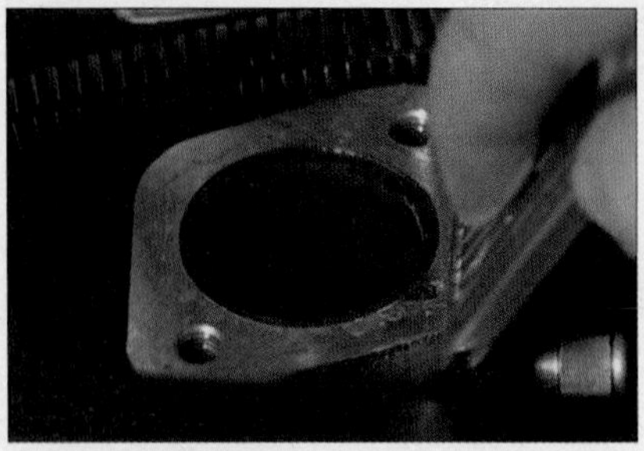

2. Inspect the thermostat housing, and clean any gasket material from the mating surfaces. Clean and inspect the thermostat groove.

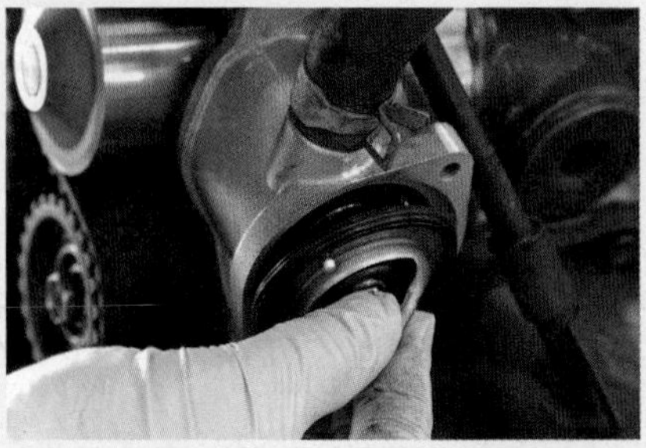

3. Verify the correct new thermostat. Position it facing the correct direction and with the bleed valve in the correct position.

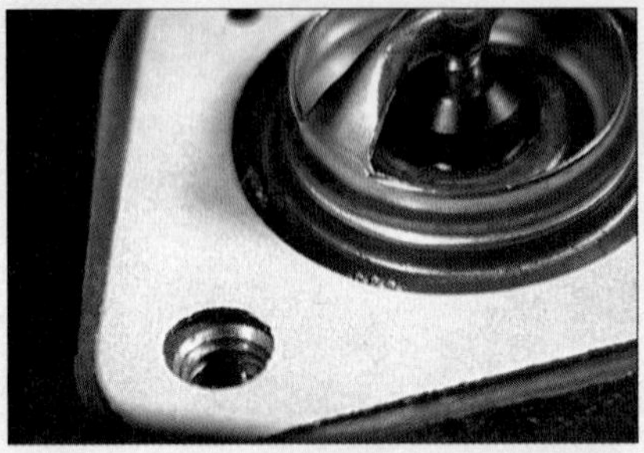

4. Install the gasket, O-ring, or rubber seal and thermostat housing.

5. Torque to housing specifications.

6. Refill the cooling system to the proper level.

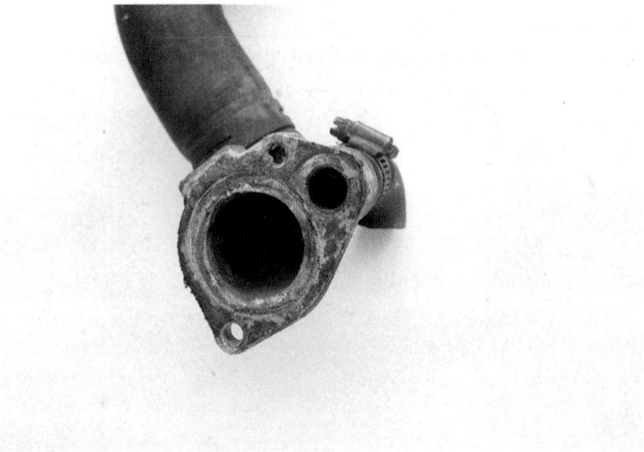

FIGURE 17-17 Rubber hose style of thermostat bypass.

FIGURE 17-18 Temperature warning lamp illuminated.

through the radiator for heat to be removed from the coolant. The cooling fan is used to pull air through the radiator whenever airflow is needed.

If the cooling fan stops working and the vehicle sits in traffic, engine heat will continue to build. In short order, the coolant's boiling point is reached. This will cause the pressure relief valve in the radiator cap to vent. This releases steam from the cooling system and produces a hissing sound. The customer may note that a temperature warning light has come on (**FIGURE 17-18**). If the vehicle has a temperature gauge, it has likely moved toward its full hot position. If the engine is operated in this condition, it will be damaged in a short amount of time. It is even possible to damage the automatic transmission if the engine gets too hot. Since the transmission fluid is typically cooled through the cooler located in the radiator, it can also overheat.

When testing the fan, start by determining what type of fan the vehicle is equipped with. It can be a mechanical fan, clutch fan, electric fan, or hydraulic fan. Mechanical fans are the easiest to inspect, typically requiring only a visual inspection. Check the condition and tension of the drive belt, the condition of the pulley, and the condition of the fan blades themselves. Especially look for cracks in any metal fan blades, as they can cause the blade to be ripped off and thrown with deadly force. It is also possible that someone previously installed the fan blades backward. This lowers the effectiveness of the fan a lot (**FIGURE 17-19**).

A clutch fan is inspected in the same way as the mechanical fan, but the clutch also must be tested. First, with the engine off, the bearings in the fan clutch can be tested by trying to move the blades frontward and backward. Then, rotate them to see if there is any play in the bearings in any direction (**FIGURE 17-20**). To test the operation of the fan clutch, some manufacturers specify placing cardboard in front of the radiator to block most of the airflow. The engine is then started with the vehicle in the shop. Monitor the engine temperature and watch and listen to see if the fan clutch engages at the proper engine temperature. If it does not, the fan clutch most likely has to be replaced. But

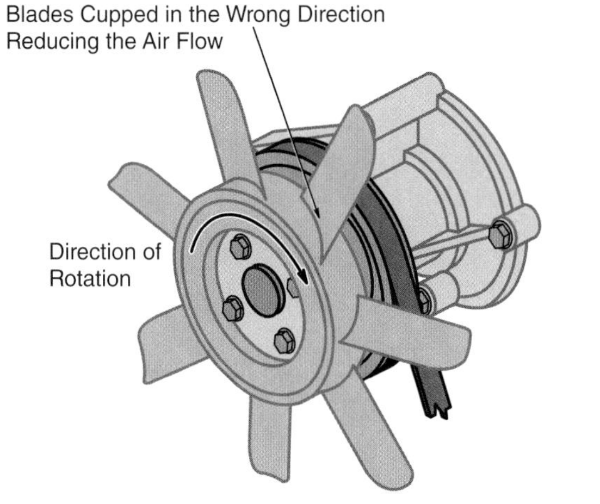

FIGURE 17-19 The mechanical fan has been installed backward on this vehicle, making it much less efficient at moving air.

FIGURE 17-20 Testing a fan clutch for excessive play.

some fan clutches can be refilled with silicone oil and put back in service.

Electric fans are quite reliable. If they do cause trouble, it is usually because they don't come on when they are supposed to. Because electric fans are electrically operated, diagnosing them is quite different from diagnosing mechanical fans. Be aware that some electric fans can come on automatically, even when the key is off. So, be careful when working around them. In fact, you should unplug the fan's electrical connector before physically inspecting the fan assembly. Just be sure to reconnect it when you are finished.

Once the fan is disconnected, check that the fan is free to turn and does not catch or bind on any of the shrouding. If it is free to rotate, then the electrical circuit and the fan motor itself will have to be tested. Remember, the electric fan only operates when the engine is at the upper end of the engine's operating temperature, so it shouldn't run all the time.

To test the electric fan, you will need to research the wiring diagram. This will let you know how the circuit is controlled and wired (**FIGURE 17-21**). Most electric fans are activated by a relay. The relay is controlled by either a temperature sending unit or the powertrain control module (PCM). Use the wiring diagram and your electrical diagnosis skills to determine what is causing the fault.

To inspect and test the electric cooling fan, follow the steps in **SKILL DRILL 17-13**.

Inspecting and Testing Fans, Fan Clutch, Fan Shroud, and Air Dams

The need to inspect and test fans and fan shrouds will arise when the customer complains of overheating. Air dams are located under the vehicle. They are designed to direct airflow into the radiator as the vehicle is moving down the road. If the air dam is damaged or missing, the engine temperature may be higher than originally designed. Air dams are easily damaged by customers who hit curbs when parking. So, always be on the lookout for damaged air dams.

If the vehicle only overheats while stopped, it could be an issue with the fan shroud, since it directs airflow through the radiator. The shroud is what surrounds the cooling fan and allows a maximum amount of air to be moved by the cooling fan (**FIGURE 17-22**). If the shroud is missing, air will be pulled from around the fan blades rather than pulling air through the radiator.

Sometimes, a customer will complain about a drumming noise when accelerating or going uphills. This could be caused by the engine fan hitting the fan shroud because of broken motor mounts. If this is the case, motor mount replacement is required. A visual inspection of these components will verify whether they are in proper position and operating as designed.

A mechanical or electric fan can fail, which will also result in air not being pulled through the radiator. A quick test to verify that air is moving through the radiator is to hold a piece of paper in front of the radiator. If suction pulls the paper in, it's working. If it does not, then air is not being pulled through the

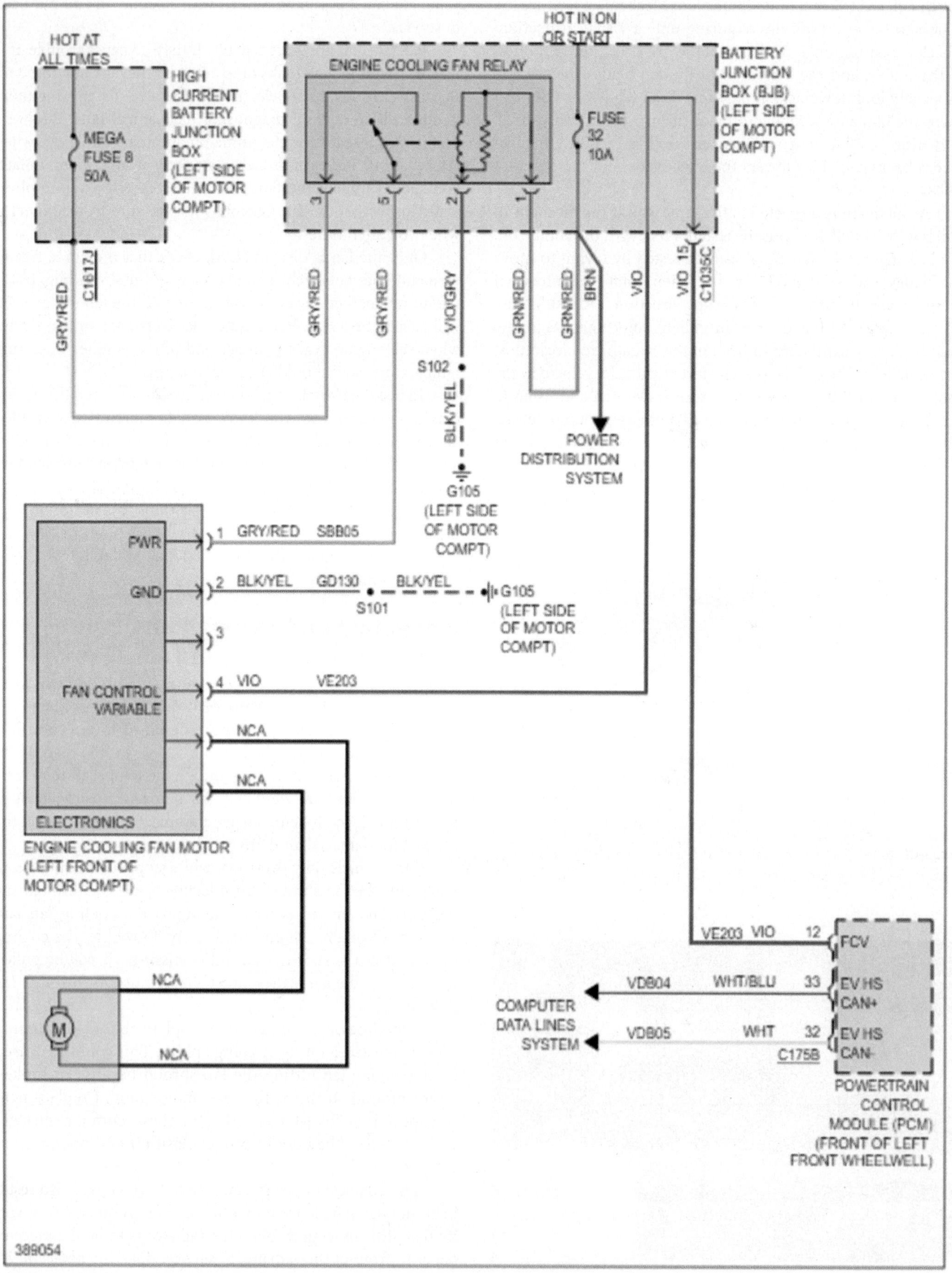

FIGURE 17-21 Typical cooling fan wiring diagram.

SKILL DRILL 17-13 Inspecting and Testing the Cooling Fan

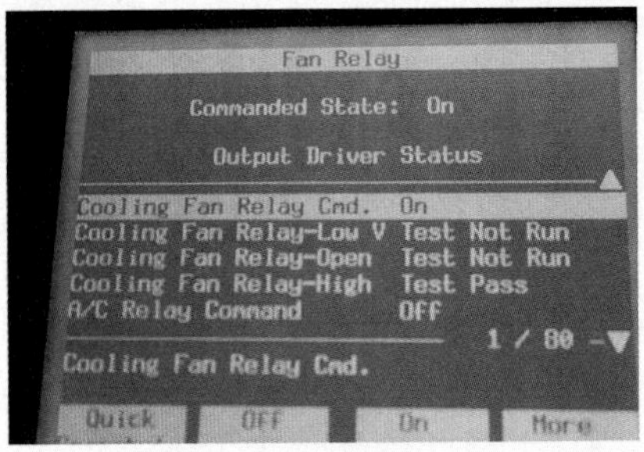

1. Install the scan tool. Find the bidirectional controls, and command the fan on. Verify fan operation. If it comes on, move to step 2. If not, move to step 3.

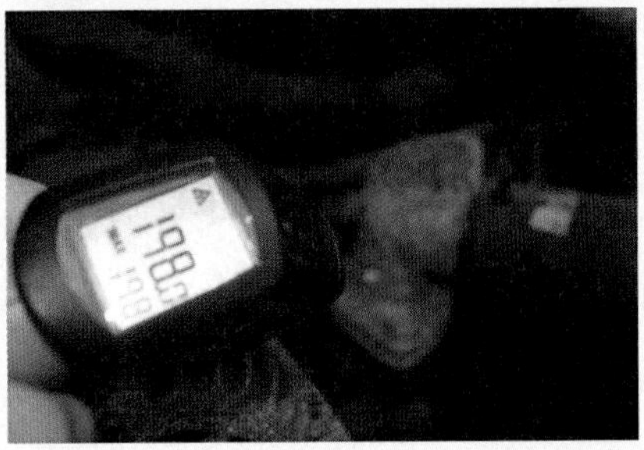

2. Use a temperature gun to measure the operating temperature, and compare its reading to the coolant temperature displayed on the scan tool. If substantially different, diagnose the temperature gauge issue and determine any necessary actions.

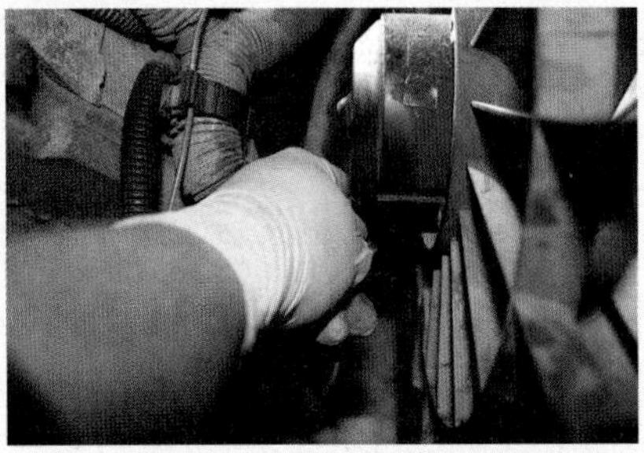

3. If the fan does not operate, wiggle test the fan wires.

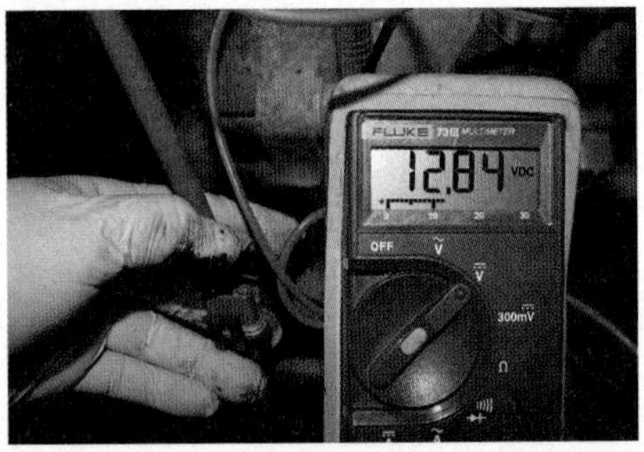

4. If it still does not work, measure the voltage to the input terminal of the fan. If substantially less than battery voltage is present, use a wiring diagram of the fan circuit to diagnose the cause.

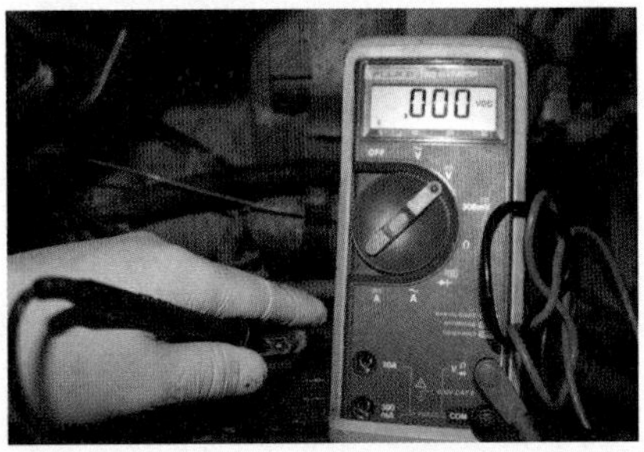

5. If voltage is present to the input terminal, measure the voltage drop on the ground side of the fan.

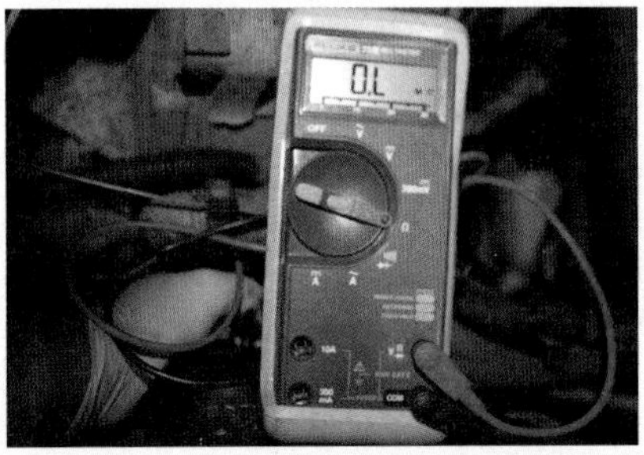

6. If the voltage drop is acceptable on the ground side, measure the continuity of the fan motor with an ohmmeter. If out of specs, replace the fan motor.

FIGURE 17-22 Fan shrouds allow maximum air to be pulled through the radiator.

radiator. Follow the service information when diagnosing the fan equipped on your vehicle.

To inspect the shroud/air dam and test the electric cooling fan, follow the steps in **SKILL DRILL 17-14**.

Replacing a Water Pump

Replacing a water pump is a fairly common task. It used to be performed mostly as a result of noise or leak issues on accessory-belt-driven water pumps (**FIGURE 17-23**). When the water pump seal failed, it would be evidenced by water leaking out of the weep hole (**FIGURE 17-24**). But now, many water pumps are driven by timing belts. So, it is common to replace the water pump along with the timing belt at the belt's recommended replacement schedule (**FIGURE 17-25**). When changing this type of pump, the timing belt has to be removed. So, following the manufacturer's procedure is necessary to prevent damage to the valves and pistons. Once the drive belt is removed from either style of water pump, the removal of the pump involves:

- unbolting of the pump,
- carefully prying it off the mating surface,
- cleaning the mounting surface and bolt holes.

Once all of the cleanup is finished, it is ready to reinstall the water pump.

- Put it back in place with the appropriate gasket, gasket sealer, or O-ring.

SKILL DRILL 17-14 Inspecting and Testing Fans, Fan Clutch, Fan Shroud, and Air Dams

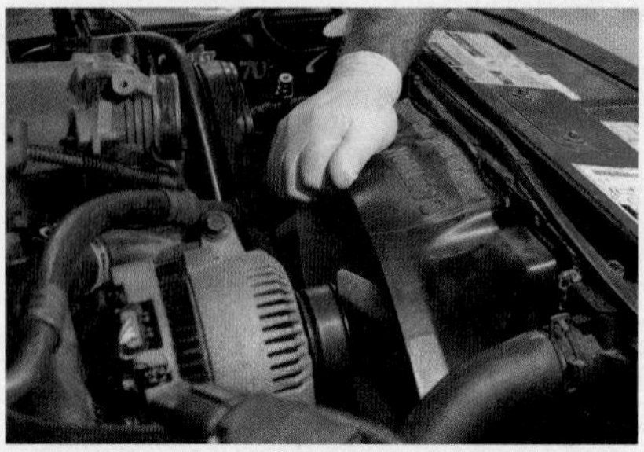

1. Visually inspect all shrouds, ducts, and air dams around the radiator and fan for damaged or missing pieces. If equipped with a fan clutch, rotate the fan and check it for looseness, play, or binding. If bad, replace the fan clutch.

2. If the fan clutch feels free, place cardboard or a fender cover in front of most of the radiator. Start the engine, monitor the engine temperature with a temperature gun, and verify that the fan cycles on at the proper temperature. If not, replace the fan clutch.

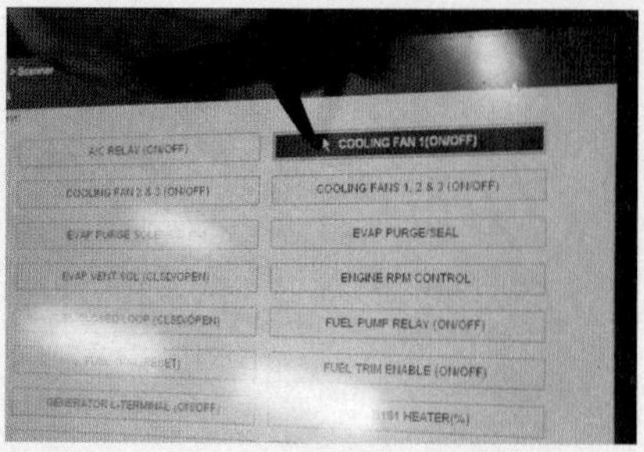

3. If equipped with an electric fan, connect a scan tool to the data link connector under the dash. Locate the bidirectional command menu, and command the cooling fan on. This activates the relay and should send power to the fan.

4. Verify that the fan is running properly. Verify that air is flowing across the radiator. If the fan is not operating, locate the cooling fan relay, remove the relay, and check for the presence of power and ground at the appropriate terminals in the relay block. With the key on and the horn button pushed, two terminals should have battery positive voltage, and two terminals should have battery negative voltage. If not, diagnose the faulty circuit.

5. If all terminals read properly, test the relay winding for continuity. If out of specifications, replace the relay.

6. If okay, install a relay socket tester, reinstall the relay into the relay box, and then activate the electric fan. Measure the voltage drop on each leg of the circuit and the relay contacts themselves. Trace the path that indicated a problem, and locate the voltage drop or defective component. Determine the necessary actions based on your findings.

FIGURE 17-23 Replacing a water pump.

FIGURE 17-24 Water leaking from the weep hole indicates a worn water pump seal.

FIGURE17-25 Water pumps are typically driven by timing belts.

FIGURE 17-26 Cracked head due to freezing.

Applied Science

AS-2: Centrifugal/Centripetal: The technician can explain the relationship of centrifugal/centripetal force to the functioning or a failure of a rotating system.

An automotive instructor is presenting a class on the vehicle's cooling system. He has an assortment of new and used water pumps to serve as training devices. The instructor begins by stating that the water pump operates on the principle of centrifugal force. As our text indicates, **centrifugal force** is a force pulling outward on a rotating body. *Centripetal* force is just the opposite. Wikipedia states that *centrifugal* is from a Latin word meaning "to flee the center." *Centripetal* is from a word meaning "to seek the center." These are forces which are opposite in direction.

The impeller is the rotating portion of the water pump. It is equipped with fins to direct the coolant outward, based on the principle of centrifugal force. The correct flow rate is based on the design of the water pump's impeller and the restrictions in the cooling system, namely the thermostat.

During class, new and used water pumps are inspected to find out the condition of the impeller. Erosion damage is visible on several of the used pumps in the area of the impeller fins. Erosion is a physical wearing away of the metal caused by contaminants in the coolant. These contaminants include dirt, grit, or sediment circulating in the coolant. The instructor stresses proper cooling system maintenance. This will prevent premature failure of the water pump and other system components.

- Tighten the bolts to the specified torque.
- Reinstall the drive belt and tension it properly.
- Refill the cooling system and bleed any air out.

▶ Cooling System Diagnosis

LO 17-07 Diagnose cooling system performance.

The cooling system is often overlooked. It does its job very effectively and is rarely thought of until problems arise. The lack of scheduled maintenance of the cooling system is the leading reason that cooling systems fail. Properly performed cooling system maintenance will ensure that the cooling system continues to operate as designed.

Common failures of the cooling system can require repairs ranging from simple to complex. An example of a simple repair is a radiator hose clamp that was not installed properly and is creating a leak. An example of a more complex repair is a vehicle that needs a new head and block because they cracked due to freezing from low antifreeze protection (**FIGURE 17-26**).

Corrosion can build up to the point that leads to an internal coolant leak. Corrosion can also coat and/or block coolant passages. It can also erode the fins off the water pump impellers, which reduces coolant flow. Either of these situations can cause the engine to run hot and overheat.

A common customer complaint is that the heater does not produce heat when turned on. This issue can be the result of several problems that will have to be diagnosed. The most common causes are:

- If the coolant is too low due to a leak. There won't be enough coolant in the engine to pick up the heat. All leaks will need to be identified and repaired.
- If the coolant level is correct, it could be a stuck open thermostat that will not let the engine come up to operating temperature. The thermostat will need to be replaced.
- A plugged-up heater core or faulty heater control valve. Both of these situations prevent coolant from flowing through the heater core. So, it cannot heat up the air flowing through it.
- A broken, stuck, or misadjusted air door in the heater box. This will prevent air from flowing through the heater core. So the air won't warm up.

Another common cooling system complaint is that the engine overheats. Engine overheating conditions can have several causes, such as:

- Low coolant level due to a leak,
- If the thermostat sticks closed, the engine will overheat because coolant cannot get to the radiator to be cooled.
- A faulty radiator cap can result in boiling coolant at normal operating temperatures.
- Clogged radiator tubes or fins won't transfer heat from the coolant as well, which results in overheating.

- An inoperative cooling fan will not pull air through the radiator, so the coolant will overheat.
- A water pump impeller that is eroded or slipping on the shaft will not circulate coolant fast enough to take heat away from the engine.
- A blown head gasket will put air into the cooling system which will cause it to overheat.

These are some of the more common issues that cause overheating. Understanding how the cooling system works is necessary to successfully diagnose and repair these faults. You will also need to use your senses to help determine what is wrong. Your sense of touch can allow you to feel for hot and cold parts in the system. For example, the engine is overheating, but the upper radiator hose is cold to the touch (**FIGURE 17-27**). This could indicate a thermostat that is stuck closed or a water pump that is not circulating coolant.

Your sense of smell could pick up the scent of antifreeze leaking into the passenger compartment from a leaky heater core (**FIGURE 17-28**). Your sense of hearing can tell you that the accessory belt is loose and slipping (**FIGURE 17-29**). And your sense of sight could see the telltale stream of coolant leaking out of the water pump weep hole (**FIGURE 17-30**). It can also identify a rusted-through soft plug. However, it is advisable not to use your sense of taste when diagnosing cooling system issues.

If the problem cannot be determined by your senses alone, you may have to resort to tools or equipment to help you locate the issue. A cooling system pressure tester allows you to pressurize the cooling system and radiator cap. This will help make leaks visible.

Starting the engine and using an exhaust analyzer to sniff the vapors coming out of the neck of the radiator can indicate a small combustion chamber leak. Combustion gases should never leak into the cooling system. Using a cylinder leakage tester and pressurizing each cylinder can help locate which cylinder has the combustion leak. However, you will still have to remove the head to determine if it is either a blown head gasket or a cracked head. If the leak is a small external leak and the source cannot be located easily, you may need to use a fluorescent dye and ultraviolet light to make it stand out better.

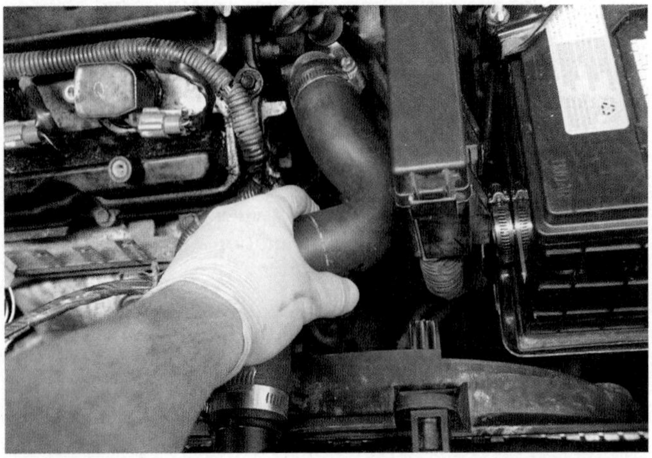

FIGURE 17-27 Your sense of touch allows you to feel temperature differences.

An infrared temperature gun is a useful tool that can be used to measure the operating temperature of the engine. This will verify whether the engine really is overheating or not. It can also be used to check the radiator for cold spots. Cold spots indicate blockage within the core of the radiator.

FIGURE 17-28 Your sense of smell can detect a heater core leak.

FIGURE 17-29 Your sense of hearing can detect a loose belt.

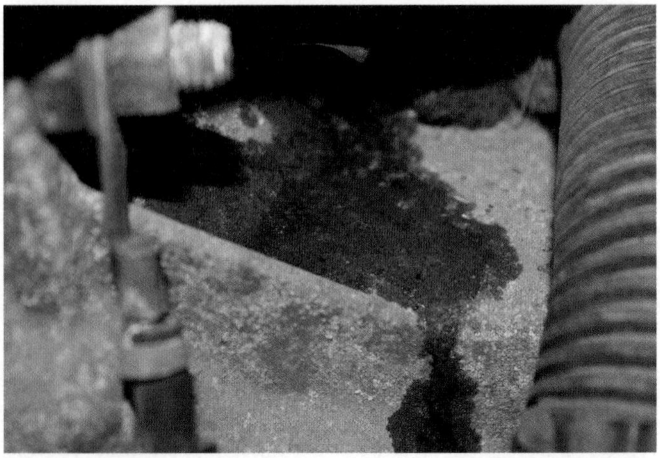

FIGURE 17-30 Your eyes can detect coolant leaking from various places.

A scan tool provides a quick way to do several things. It can monitor the operating temperature, verify any cooling system diagnostic trouble codes (DTCs), or command the electric fan to come on. As you can see, diagnosis is dependent on a few things. First, understanding how cooling systems operate. Second, observing what is happening. And third, having the right tools and equipment available, and knowing how to use them. Having a good grasp of those things will take you a long way toward diagnosing customers' cooling system concerns.

Tools

Special tools and equipment are used for diagnosing and servicing the engine cooling system, including the following (**FIGURE 17-31**):

- *Coolant system pressure tester:* Used to apply pressure to the cooling system to diagnose leakage complaints. Under pressure, coolant may leak internally to the combustion chamber, intake or exhaust system, or the engine lubrication system. It can also leak externally to the outside of the engine.
- *Hydrometer:* Used to test coolant mixture and freeze protection by testing the specific gravity of the coolant. You must use a hydrometer specifically designed for the antifreeze you are testing.

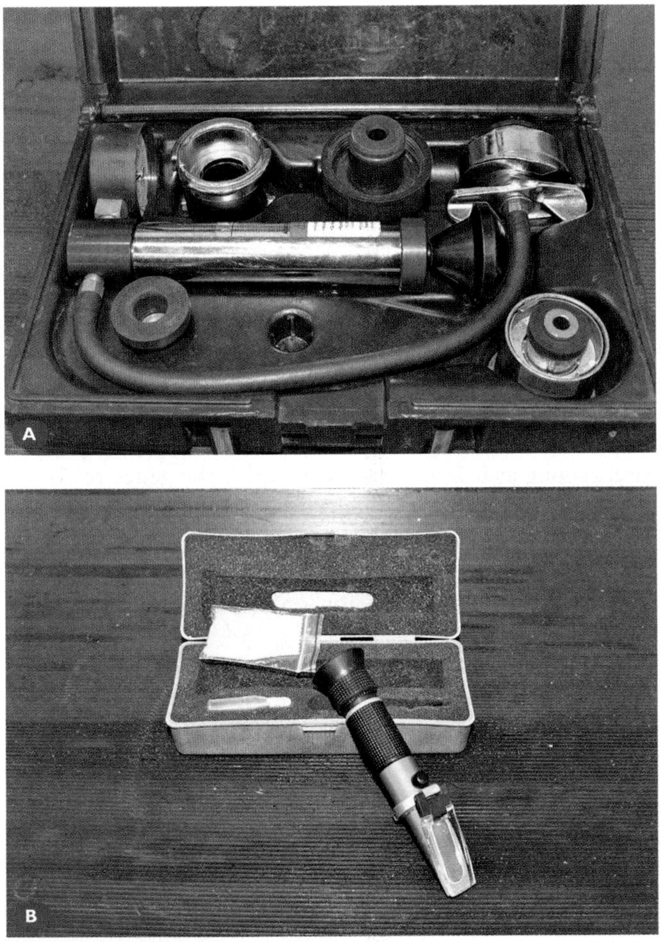

FIGURE 17-31 Common tools needed to service cooling systems. **A.** Pressure tester. **B.** Refractometer.

- *Refractometer:* Used to test coolant mixture and freeze protection by testing the fluid's ability to bend light. This tester can be used with any type of antifreeze.
- *Coolant pH test strips:* Used to test the acid-to-alkalinity balance of the coolant.
- *Coolant dye kit:* Used to aid leak detection by adding dye to coolant and using an ultraviolet light source (black light) to trace the source of the leak. The dye glows fluorescent when an ultraviolet light is shined on it.
- *Infrared temperature sensor:* A noncontact thermometer. It is used to check actual temperatures and variations of temperature throughout the cooling system. Helps pinpoint faulty parts and system blockages.
- *Thermometer:* Used to check the temperature of air exiting the heating ducts.
- *Voltmeter:* Used to check for electrical problems such as electric cooling fan and temperature gauge issues.
- *Belt tension gauge:* Used to check belt tension.
- *Serpentine belt wear gauge:* Used to check whether the serpentine belt grooves are worn past their specifications.
- *Hose clamp pliers:* Used to safely remove spring-type radiator clamps.
- *Borescope:* Used for examining internal passages for evidence of a coolant leak.
- *Scan tool:* Used to activate the cooling fan and air door actuators through bidirectional controls for testing. Monitors coolant temperature sensor operation. And reads DTCs related to cooling system operation.
- *Cooling system flush machine:* Used to flush coolant backward through the system with cleaners that remove corrosion buildup and old coolant. Most of these machines have their own pump, so the vehicle does not have to run to perform the flush.
- *Exhaust gas analyzer:* Used to detect exhaust gases that are finding their way into the cooling system. This can be due to a leaking head gasket or damaged head or block. Be careful not to allow liquid coolant to be picked up by the analyzer probe.

Visual Inspection

Many times a visual inspection of the cooling system will give you a good indication of any issues. Start by checking the level of the coolant in both the overflow bottle and radiator. At the same time, check the condition of the coolant to see if it is cloudy or contaminated. Also check the belt condition and for the proper tension, and check hoses for any leaks or wear.

If the sweet aroma of coolant is detected, check for leaking coolant at the heater box drain. This could indicate a possible cracked or rotted leaking heater core. Check the engine exhaust for excessive white smoke. A head gasket failure or a cracked head or valve seat may cause coolant to leak into the combustion chamber. This appears as white smoke in the exhaust, especially when accelerating.

Also, start the engine with the radiator cap off, and look for bubbles from combustion in the radiator. Bubbles would indicate a leak in the combustion chamber, likely at the head gasket. Disassembly and inspection would need to occur to verify that it is the head gasket and not a cracked head or block.

SAFETY TIP

When working around the cooling system, care must be taken. This is particularly true if the engine is at operating temperature, as the coolant is hot enough to scald. Always allow the system to cool before removing the radiator or pressure cap. If you must remove the radiator cap from a hot system, wear protective gloves and eyewear. Place a cloth fender cover on top of the radiator cap before releasing it slowly, to the first (safety) point. This will help keep the pressure inside from erupting.

▶ Testing the Cooling System for Leaks

LO 17-08 Diagnose cooling system leaks and verify engine operating temperature.

Pressure testing the cooling system is usually an effective way to locate leaks. Pressurizing the system causes coolant to leak out much more quickly. This makes leaks easier to locate. But before you pressurize the system, make sure it is topped off with coolant or water. Otherwise, the leak could be above the coolant level, and only air may get leaked, which is much harder to observe. Topping off also allows you to pressurize the system quicker because it removes air, which is compressible. Also, if the system is full of liquid, the pressure reading on the gauge will fall faster if there is a leak in the system.

When pressure is applied, only apply the amount shown on the radiator cap. Do not exceed that pressure or else components could burst. With pressure applied, leaks generally show up easily, as identified by coolant leaking from the source. The use of a droplight and a mirror may be necessary to see behind the engine or in tight areas. Don't forget to check all of the hoses, soft plugs, water pump weep hole, and the heater core in the passenger compartment. Also, some extended vehicles have a second heater core near the rear of the passenger compartment.

If the pressure gauge is losing pressure, but no leaks are found, ensure that the tester is installed correctly. If it still loses pressure, ensure that the tool itself isn't leaking. Once the tool is verified and you cannot find an external leak, you can test to see if the cooling system is leaking internally into the engine. Check engine oil for evidence of coolant. It will have a milky appearance on the dipstick.

If coolant is not leaking into the oil, it could be leaking into the cylinder. Remove the spark plugs and look for evidence of coolant being burned in the cylinders. This can show up as a color-stained spark plug insulator. A combustion chamber experiencing a coolant leak will appear to be steam cleaned. A borescope can be used to inspect the cylinders through the spark plug holes.

Normally the engine should be off when carrying out any visual inspection of the system. It should also be off when you connect test equipment such as the pressure tester. However, it is possible that the leak only occurs when the engine is running. This is most likely to happen around the water pump seal, making it necessary to run the engine while testing. If you do have to run the engine after the tester has been installed and pressurized, there is a big caution. Make sure to watch the pressure gauge, and release excess pressure as the engine heats up. Failure to do so can lead to excessive pressure being created in the cooling system. This can rupture hoses, the radiator, or the heater core. So, never let the pressure exceed the pressure listed on the radiator cap. When the engine is running, make sure you keep well away from any rotating or hot parts.

Also, remember to pressure test the radiator cap, as a leak at the cap prevents the cooling system from building pressure. This lowers the boiling point, which can cause the coolant to boil at normal operating temperatures. Many a technician has been fooled into thinking that the vehicle still had a serious overheating problem after changing cooling system components. Actually, all it needed was a new radiator cap.

Most pressure testers are hand-operated and come with a number of adapters to fit a variety of cooling systems. Adapters are used to connect the tester to the radiator, surge tank, or radiator cap.

To pressure test the cooling system, follow the steps in **SKILL DRILL 17-15**.

▶ TECHNICIAN TIP

If you need to replace a pressure cap, use only a cap with the correct recommended pressure. If a cap with a lower pressure rating is installed, it could cause the coolant to boil over. Alternatively, a higher rated cap will increase the pressure in the system and could result in a hose, radiator, or heater core bursting.

Verifying Engine Operating Temperature

Technicians need to verify engine operating temperature for several issues. First, if the customer complains about an overheating issue. Second, if there is an underheating issue such as inadequate heat from the heater. Third, if the temperature gauge is not providing a normal reading. And finally, if the vehicle is experiencing poor fuel economy.

Engine operating temperature can be verified in a couple of ways. First is the vehicle's dash-mounted temperature gauge. It provides a good indication of the engine's temperature, as long as it is working correctly. If the gauge reads too cold or too hot, you will need a second opinion. The same is needed if it is reading normally but you suspect the engine temperature is not as indicated. You can get a second opinion by using either an infrared noncontact temperature gun or a scan tool.

The temperature gun, as described earlier, measures the amount of heat energy (temperature) of an object. Just realize that some objects do not conduct heat as well as others. If you can point the temperature gun at a metal component, it will produce a more accurate reading. Point it toward the engine's thermostat housing (or next to it), with the engine fully warmed up. A close approximation of the engine's operating temperature can be measured and compared to the specifications.

Another way to verify the engine temperature is to use a scan tool to access the data for the engine coolant temperature sensor. The temperature reading it gives can be compared to

SKILL DRILL 17-15 Pressure Testing the Cooling System

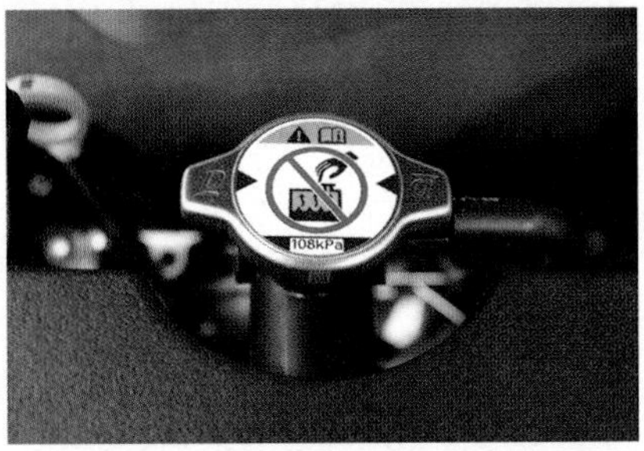

1. Verify specified cooling system pressure. This can normally be taken from the radiator cap. But be aware that the wrong cap may be installed.

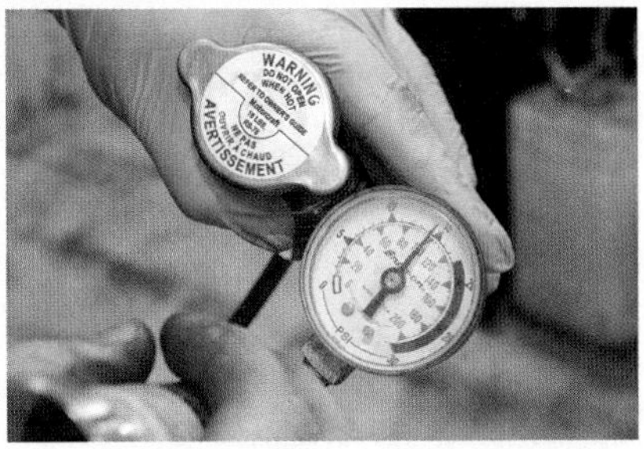

2. Install the radiator cap on the pressure tester, and pressurize the cap to the correct pressure. It should hold pressure at approximately the rated pressure and vent at slightly above the rated pressure.

3. Top off the radiator with coolant or water

4. Install the tester on radiator. Pressurize the system to the specified cap pressure.

5. Watch the pressure reading for a drop while performing a visual check for any leaks.

6. Check hoses, soft plugs, and any heater cores; determine necessary action.

both the dash gauge and the temp gun readings. Comparing all three readings is good practice. If they agree, this does three things. It will verify the engine operating temperature. It will verify the temperature reading on the temperature gauge on the dash. And it will verify the coolant temperature signal that the vehicle's PCM receives. If they are not in agreement, you will need to determine which one is not accurate.

Note that a vehicle with an electric fan usually has two listed temperatures. One temperature at which the fan should turn on and another temperature at which it should turn off. The normal operating temperature is anywhere between those temperatures in this situation.

To verify the engine operating temperature, follow the steps in **SKILL DRILL 17-16**.

SKILL DRILL 17-16 Verifying Engine Operating Temperature

1. Verify that the coolant level is correct before starting the engine. If low, check for the presence of a leak before measuring the operating temperature.

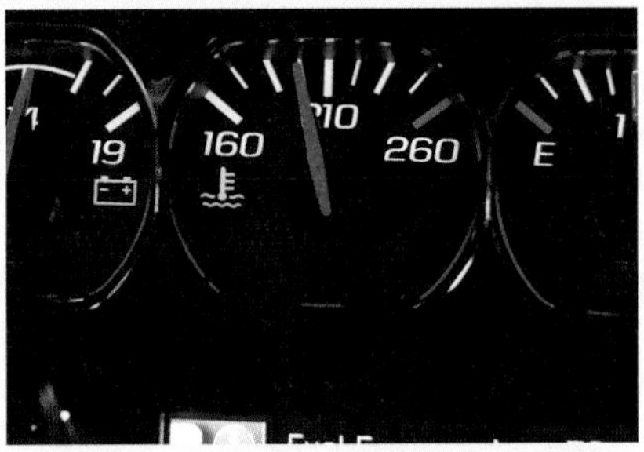

2. Start the engine and allow it to reach full operational temperature, monitoring the temperature along the way in case it starts to overheat.

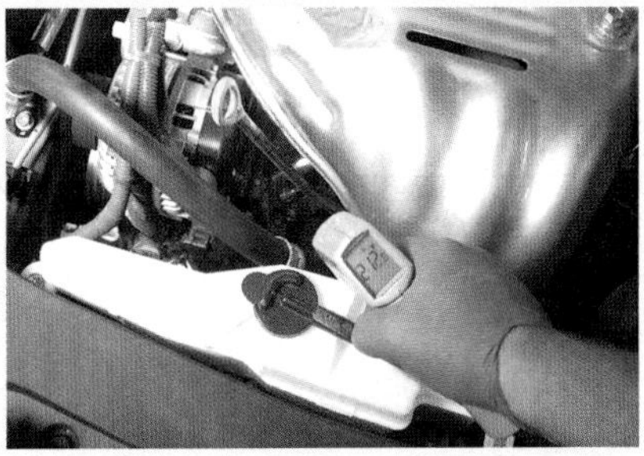

3. Using an infrared temperature gun, test the temperature of the engine near the location of the thermostat or coolant temperature sensor. Compare to specifications.

SKILL DRILL 17-16 Verifying Engine Operating Temperature (Continued)

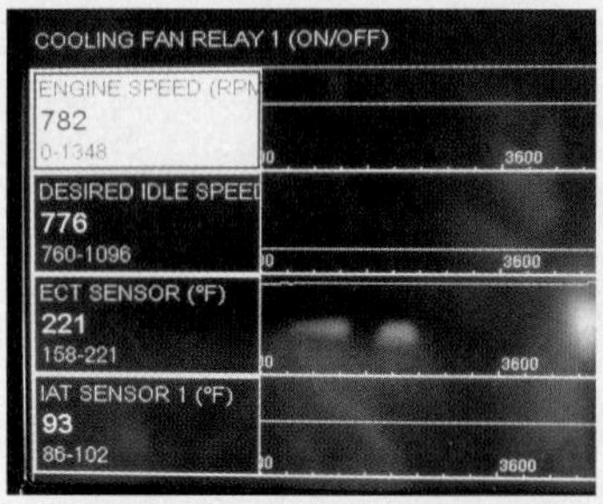

4. Using a scan tool, retrieve the engine coolant temperature sensor temperature reading.

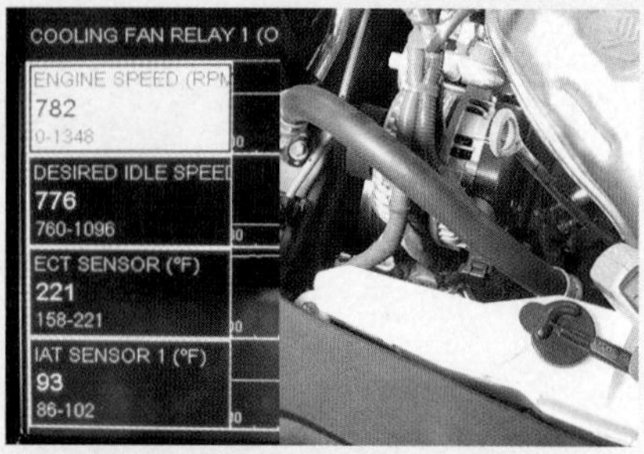

5. Be aware that there will be some temperature difference because the temperature sensor is sitting in the coolant, and the temperature gun is measuring the surface temperature near the sensor.

6. Compare the results from the temperature gun, the coolant temperature sensor, and the vehicle's temperature gauge; this gives you a good idea of the engine's operating temperature.

▶ Wrap-Up

Ready for Review

- Engine coolant can be tested with test strips and electronic testers for pH, a refractometer can be used to test the freeze point, and a voltmeter can be used to perform electrolysis testing.
- During coolant service: the level should be at the full mark, the level of freeze protection should be appropriate, and the pH level should be correct.
- Drive belt service should include checks for: cracks, oil contamination, glazing, tears, bottoming out, pulley wear, and belt width.
- Coolant problems include: swollen hose, hardened hose, cracked hose, and soft hose.
- A thermostat should be replaced whenever cooling system service is needed. The thermostat bypass should be inspected for restrictions due to corrosion or foreign materials.
- To test a fan, determine the type of fan the vehicle is equipped with. A damaged air dam leads to higher engine temperature. It is common to replace the water pump along with the timing belt at the belt's recommended replacement schedule.
- Special tools and equipment used for diagnosing and servicing the engine cooling system include: coolant system pressure tester, hydrometer, refractometer, coolant pH test strips, coolant dye kit, infrared temperature sensor, borescope, scan tool, cooling system flush machine, and exhaust gas analyzer.
- Pressure testing the cooling system is an effective way to locate leaks. Engine operating temperature can be verified by the vehicle's dash-mounted temperature gauge or with a scan tool to access the data for the engine coolant temperature sensor.

Key Terms

centrifugal force The apparent force by which a rotating mass tries to move outward, away from its axis of rotation.

electrolysis The process of pulling metals apart by using electricity or by creating electricity through the use of chemicals and dissimilar metals.

ethylene glycol A chemical used as antifreeze that provides the lower freezing point of coolant and raises the boiling point. It is a toxic antifreeze.

Review Questions

1. A vehicle is tested for electrolysis. Further electrical testing is required if the voltage reading is greater than:
   a. 300 millivolts
   b. 150 millivolts
   c. 3 millivolts
   d. 1.5 millivolts
2. Coolant condition can be tested with all the following tools EXCEPT:
   a. Hydrometer
   b. Barometer
   c. Refractometer
   d. Electronic tester
3. When replacing an accessory drive belt, all the following are important considerations EXCEPT:
   a. Pulley alignment
   b. Crankshaft to camshaft timing
   c. Belt dimensions
   d. Belt routing
4. What is the best option for removing a stuck coolant hose?
   a. Twisting the hose off with pliers
   b. Prying the hose off with a screwdriver
   c. Leave the portion stuck to fitting and install a larger diameter hose over it
   d. Slit the hose and peel it away from the fitting
5. When replacing a thermostat, all these statements are true EXCEPT:
   a. Clean the mating surfaces
   b. Inspect the bypass port for corrosion

c. Replace old gaskets with an O-ring
d. Tighten the bolts to the specified torque

6. What causes coolant to leak from the water pump weep hole?
a. A faulty thermostat
b. A faulty radiator cap
c. A faulty bypass hose
d. A faulty pump shaft seal

7. What part is often changed with a water pump?
a. A crankshaft pulley
b. A timing belt
c. An alternator
d. A starter motor

8. What is the correct tool to examine internal passages for evidence of a coolant leak?
a. An exhaust gas analyzer
b. A refractometer
c. A hydrometer
d. A borescope

9. Coolant leaking into the combustion chamber produces what color smoke from the exhaust?
a. White
b. Blue
c. Black
d. Red

10. Milky looking engine oil is most likely caused by:
a. A faulty radiator
b. A leaking heater hose
c. A blown head gasket
d. Corroded water pump impeller fins

ASE Technician A/Technician B Style Questions

1. Technician A says that the same hydrometer can be used for both ethylene glycol and propylene glycol coolants. Technician B says that some hydrometers include a thermometer to compensate the reading based on coolant temperature. Who is correct?
a. Technician A only
b. Technician B only
c. Both A and B
d. Neither A nor B

2. Two technicians are discussing a coolant drain and refill. Technician A says that draining the coolant removes all coolant from the system. Technician B says that any trapped air in the system will bleed itself out automatically. Who is correct?
a. Technician A only
b. Technician B only
c. Both A and B
d. Neither A nor B

3. Technician A says that some small cracks in the accessory drive belt may be within tolerance. Technician B says that serpentine belts can be checked with a wear gauge. Who is correct?
a. Technician A only
b. Technician B only
c. Both A and B
d. Neither A nor B

4. Technician A says that coolant hoses should be squeezed to check for hardening or softening. Technician B says that coolant hoses should be visually inspected for swelling and cracks. Who is correct?
a. Technician A only
b. Technician B only
c. Both A and B
d. Neither A nor B

5. Technician A says that installing a thermostat incorrectly can cause the housing to break. Technician B says that it is OK to install a thermostat in either direction. Who is correct?
a. Technician A only
b. Technician B only
c. Both A and B
d. Neither A nor B

6. Technician A says that a faulty radiator fan relay can cause an engine to overheat. Technician B says that bidirectional control is NOT used to test an electric cooling fan. Who is correct?
a. Technician A only
b. Technician B only
c. Both A and B
d. Neither A nor B

7. Technician A says that broken engine mounts can cause the fan to hit the shroud. Technician B says that missing or broken air dams can cause an engine to overheat. Who is correct?
a. Technician A only
b. Technician B only
c. Both A and B
d. Neither A nor B

8. Technician A says that you may be able to smell antifreeze from a leaking heater core. Technician B says that you may be able to hear a loose belt when the engine is running. Who is correct?
a. Technician A only
b. Technician B only
c. Both A and B
d. Neither A nor B

9. Technician A says to use a scan tool to check engine operating temperature. Technician B says to use an infrared temperature gun to check the engine operating temperature. Who is correct?
a. Technician A only
b. Technician B only
c. Both A and B
d. Neither A nor B

10. When performing a coolant pressure test, Technician A says to pressurize the system to the pressure shown on the radiator cap. Technician B says that the coolant should be drained before performing the test. Who is correct?
a. Technician A only
b. Technician B only
c. Both A and B
d. Neither A nor B

SECTION 3
Automatic Transmissions

▶ CHAPTER 18 Automatic Transmission Fundamentals

▶ CHAPTER 19 Maintaining the Automatic Transmission/Transaxle

▶ CHAPTER 20 Hybrid and Continuously Variable Transmissions

CHAPTER 18

Automatic Transmission Fundamentals

Learning Objectives

- **LO 18-01** Describe the functions and types of automatic transmissions.
- **LO 18-02** Describe torque converter construction.
- **LO 18-03** Describe torque converter operation.
- **LO 18-04** Describe lock-up converters and heat exchangers.
- **LO 18-05** Describe gear train operation.
- **LO 18-06** Describe methods of holding–driving gears.
- **LO 18-07** Describe automatic transmission components.

ASE Education Foundation Tasks

See Appendix A to view the 2017 ASE Education Foundation Automobile Accreditation Task List Correlation Guide.

▶ Introduction

Automatic transmissions were once considered an expensive option on automobiles. Today, the vast majority of vehicles sold in the United States include automatic transmissions. They are now considered standard equipment. In fact, it now mostly comes down to how many speeds a customer prefers. It is not whether they want an automatic or manual transmission. Some passenger vehicle manufacturers produce automatic transmissions with up to nine forward speeds. It is likely that even more speeds will be added in the future (**FIGURE 18-1**).

It is necessary to have a basic knowledge of how automatic transmissions operate. Even if most technicians never will need to rebuild one. This knowledge is needed for diagnostic purposes due to the interrelated nature of the transmission with the rest of the vehicle.

Technicians may encounter malfunctions that at first appear to be transmission problems. It is only after transmission replacement that they find that the problem is still present. Further diagnosis shows it was caused by an entirely different system. For example, a plugged catalytic converter can cause a transmission to shift erratically. And a bad vehicle speed sensor (VSS) can prevent a transmission from shifting at all. Getting this wrong is very frustrating and costly to all who are involved.

In some cases, an automatic transmission can be easily repaired without a complete teardown and rebuild. Having a good understanding of how a transmission operates will allow you to determine that (**FIGURE 18-2**). And that will save the customer money. It will also help build trust with the customer.

Unfortunately, though, technicians will often immediately condemn a transmission and replace it with a rebuilt unit. This may seem like an easy option to the technician. It's just replacing the transmission, and the problem is likely to go away. But that can be much more expensive than a smaller targeted repair for the customer. Also, a customer who is able to drive away with a $300 repair is likely to become a much more loyal customer than one who is sold a complete transmission for $3,000. It's even worse when they discover later that he or she was taken advantage of. For these reasons, the more knowledge and experience you have with transmissions, the better you will be able to diagnose and repair vehicles for customers. This leads to more job security and satisfaction. In this ASE area, we will cover the theory of operation and maintenance of automatic transmissions.

▶ Function of an Automatic Transmission

LO 18-01 Describe the functions and types of automatic transmissions.

An automatic transmission has two major functions that separate it from a manual transmission. First, the transmission can select and shift gears without input from the driver. This function is accomplished within the gear train of the transmission. It is aided by the hydraulic and electronic control systems.

FIGURE 18-1 Nine-speed ZF transmission.

FIGURE 18-2 Sometimes transmission concerns can be fixed with simple repairs.

You Are the Automotive Technician

It is your second week as a new technician. You overhear two older technicians discussing automatic transmissions. Both of the technicians have been in the industry for a long time, but they have had very little training. Most of what they have learned has been learned on the job. They seem to agree on much of their understanding about how automatic transmissions operate. But they have several points of disagreement. Because you have been through an automotive technology training program, they ask you to explain some things to them.

1. How does a torque converter transmit more torque to the transmission than the engine creates?
2. How can the planetary gears in an automatic transmission always be in mesh, and yet the transmission can change from one gear ratio to other gear ratios?
3. How do multidisc clutches, bands, one-way clutches, and dog clutches operate?

Second, the transmission can automatically couple and uncouple from the engine when needed. This is much like a clutch on a manual transmission vehicle, but without any input from the driver. The torque converter performs this function. Without the ability to automatically connect and disconnect the transmission from the engine, a vehicle would stall when it comes to a stop. A manual transmission vehicle will do this if the driver forgets to depress the clutch pedal. These two functions separate automatic transmissions from manual transmissions. We will explore these concepts further in this chapter.

Types of Automatic Transmissions

There are several types or classifications of automatic transmissions:

- conventional transmissions,
- transaxles,
- dual clutch,
- continuously variable,
- hybrid transmissions, and
- a Honda/Saturn type of transmission.

Each of these types of automatic transmissions has elements common to the conventional automatic transmission. We will explain the major differences below.

A conventional automatic transmission uses one or more planetary gear sets. These gear sets create several gear ratios needed to drive the vehicle. Conventional automatic transmissions are typically connected to the engine through a torque converter. The torque converter is covered in detail later in the chapter. The gears inside a conventional transmission are constantly meshed to each other. Holding devices, such as clutches or bands, will either stop or drive the rotation of gears in a planetary gear set. This creates the needed gear ratios.

On most rear-wheel drive (RWD) vehicles, the differential and final drive gear are located in the rear axle assembly. The differential allows for the vehicle to make turns without scrubbing or dragging the tires. The final drive also increases the torque to the wheels. In the 1970s, most manufacturers developed front-wheel drive (FWD) vehicles for several reasons:

- improved traction over RWD vehicles,
- less overall vehicle weight, and
- increased passenger compartment space.

Driving the front wheels required a different arrangement. That arrangement is the **transaxle**. It is a combination of a transmission, differential, and final drive gear all in one compact unit (**FIGURE 18-3**).

In most cases, the transaxle bolts directly to the engine. However, some RWD vehicles, such as late-model Chevy Corvettes, use a transaxle assembly located in the rear of the vehicle. And yet the engine is located in the front. The Corvette uses this transaxle arrangement to better distribute the weight between the front and rear wheels to improve handling. A fairly long shaft connects the engine to the transaxle.

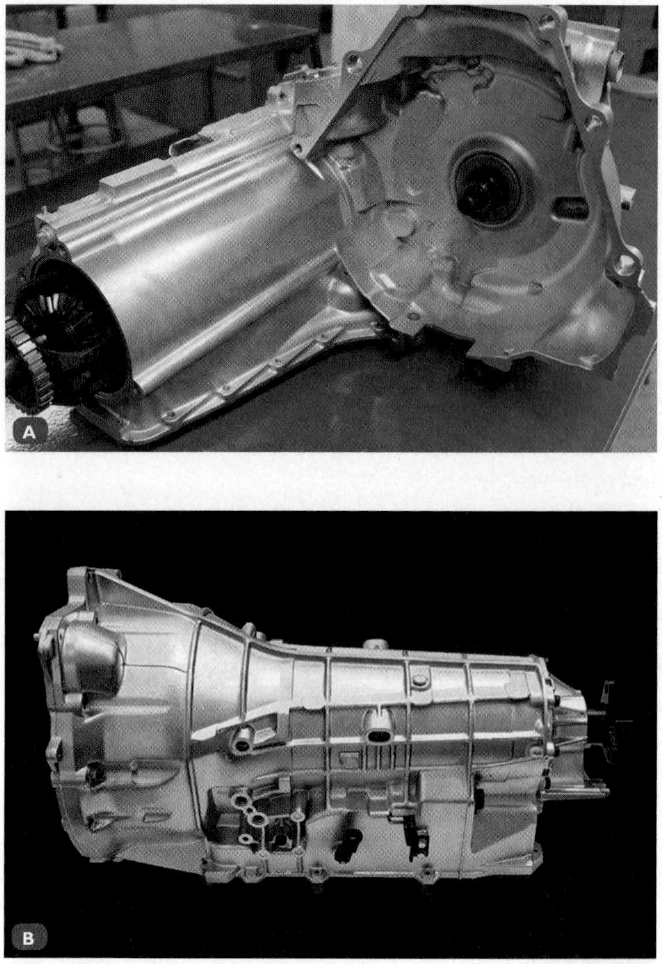

FIGURE 18-3 **A.** Transaxle. **B.** Transmission.

Dual-clutch transmissions are a newer type of automatic transmission. They use two clutches, either wet or dry, in place of a **standard torque converter**. A wet clutch is one that is immersed in hydraulic oil or transmission fluid. And a dry clutch is like a conventional manual transmission clutch.

Each clutch is connected to a shaft driving several gears inside the transmission. Clutch 1 is connected to gears 1, 3, and 5. Clutch 2 is connected to gears 2, 4, and 6 (**FIGURE 18-4**). Each of these gears is in constant mesh with a matching gear. The transmission operates by locking one gear at a time to the output shaft. The gears have different ratios so they change the speed of the output shaft. Their operation is similar in some ways to that of a manual transmission. But the changing of gears is controlled by the vehicle computer, not the driver.

Continuously variable transmissions (CVTs) do not use typical gears as in other transmissions. Rather, CVTs commonly use two pulleys that change diameter in response to vehicle load and speed. There is a large, heavy metal belt connecting the two pulleys (**FIGURE 18-5**). When the vehicle is starting from a stop, the input pulley, which is coupled to the engine, has a small diameter. The output pulley, which is coupled to the drive wheels, has a large diameter.

FIGURE 18-4 A Ford Focus dual-clutch transmission. Gears 1, 3, and 5 are on one shaft and 2, 4, and 6 are on the other shaft. Note that this is more like a manual transmission than an automatic transmission.

FIGURE 18-5 The pulleys and metal belt of a CVT transmission.

The size difference between the pulleys creates a large amount of torque multiplication. Typically, the smaller input pulley turns approximately three times for every one turn of the output pulley. When the vehicle reaches cruising speed, the input pulley has compressed. This causes the diameter of the pulley to increase. At the same time, the output pulley is reduced in diameter. In this position, the output pulley spins up to three times for every revolution of the input pulley.

Using this system, the transmission does not have fixed gear ratios like a conventional transmission. Instead, it is able to vary the gear ratio infinitely within the limits of the changing diameters of the pulleys. This variability allows the engine to operate in its most efficient rpm range to save fuel.

Hybrid vehicles use drivetrains that can be categorized as series or parallel hybrid drivetrains. In a series drivetrain, the power from the engine is supplemented with power from an electric motor. In a **series hybrid drivetrain**, the electric motor is not typically able to propel the vehicle on its own. This electric motor is often placed between the engine and the transmission.

FIGURE 18-6 A power-splitting transmission (PST) uses a planetary gear set to obtain an infinitely adjustable transmission.

A **parallel hybrid drivetrain** also has two or more devices to power the transmission. The engine can mechanically send power through the transmission to the wheels. Or one or more electric motors can send power through the transmission to the wheels. Either the electric motors or the gasoline engine can propel the vehicle along, individually or in combination.

A **series-parallel hybrid drivetrain** is designed so that it can function as both a series hybrid and a parallel hybrid. It uses what is called a **power-splitting transmission (PST)** (**FIGURE 18-6**). This transmission is sometimes referred to as a type of electronic CVT (e-CVT). But it uses a planetary gear set to obtain an infinitely adjustable transmission. It allows power to be applied to the planetary gear set (epicyclic gearing) through three sources:

- the internal combustion engine (ICE) and
- two motor/generators.

This design is an ingenious application of the planetary gear set. In its normal use, one member is held and another is driven. But in the case of a hybrid vehicle, where you have more than one power source, two components can be driven. Moreover, they can be driven at different speeds. And in the case of the electric motors, they can be driven in reverse. This ability to drive more than one member at various speeds allows the planetary gear to act like a CVT.

In this configuration, the planetary gear set acts as a power divider. It uses all three power sources, or two generators, to balance the overall needs of the system. This includes:

- vehicle speed,
- traction motor battery charge, and
- ICE efficiency.

This system is operated by sophisticated computer controls. Hybrids and CVT transmissions are covered in greater detail in the chapter on that topic.

Many Honda and Saturn vehicles use an automatic transmission. But the transmission more closely resembles a manual transmission. These transmissions are sometimes called

dual-shaft transmissions. They have a main shaft and a countershaft like a manual transmission (**FIGURE 18-7**). They do not use planetary gear sets. Instead of using synchronizer assemblies like in a manual transmission, they use multidisc clutch packs. The clutch packslock the individual gears onto the main shaft.

▶ Torque Converters

LO 18-02 Describe torque converter construction.

Torque converters transmit engine torque to the transmission. They allow the engine to idle at a stop even though the transmission is still in gear. As engine speed increases, they transmit greater amounts of torque, driving the vehicle. We explain their operation in depth a bit later.

Torque converters are generally robust components that don't give a lot of trouble. But if they do, it is important to have a basic working knowledge of them in order to diagnose problems with them. And when you do find one that is faulty, you will typically only replace it. Since they are usually welded together from the factory, they can't be repaired without special equipment. So, most of the time torque converters only get replaced when the transmission is being replaced or rebuilt. However, occasionally, the torque converter itself fails.

It is important to understand how the torque converter operates. It will help you successfully diagnose problems in them. When lock-up torque converters came out in the early 1980s, one manufacturer had a very large percentage of these torque converters coming in for warranty replacement. The problem, however, was not corrected by replacing the torque converter. After careful diagnosis, the problem turned out to be a faulty spark plug wire. The problem would only appear once the lock-up torque convertor engaged. This led technicians to blame the torque converter. The manufacturer released several technical service bulletins (TSBs) to address this concern. If the technicians had been better able to diagnose the system, they might have accurately diagnosed the problem sooner. And it would have cost much less money.

On some heavy-duty vehicles, the torque converters can be disassembled and rebuilt in a regular shop. This is because the two halves of the housing are bolted together instead of welded (**FIGURE 18-8**). But on light-duty vehicles, most shops simply order a remanufactured torque converter.

Torque Converter Principles

The purpose of a vehicle transmission is to transmit engine torque to the driving wheels. In a manual transmission, engine torque is controlled by the driver. This is done by opening or closing the throttle. The driver also manually selects the appropriate gear with the shifter mechanism.

In an automatic transmission, the driver still controls the engine torque with the throttle. But in this case, gear selection is automatically controlled by the transmission. The transmission also works in concert with the torque converter to modify the engine's torque. We will explain that later (**FIGURE 18-9**).

The torque converter is mounted between the engine and the transmission. It is in the same place as a manual transmission clutch (**FIGURE 18-10**). The torque converter effectively connects the engine flywheel to the input shaft of the transmission. The torque converter does the same job as a manual clutch. It transmits engine torque to the input shaft of the transmission.

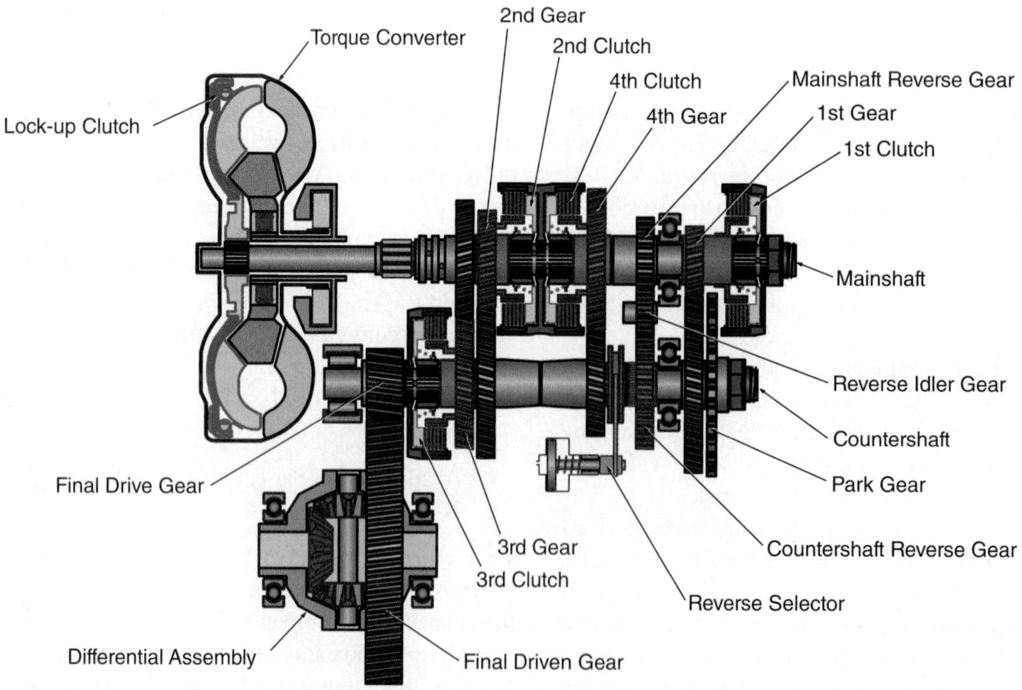

FIGURE 18-7 A Honda automatic transmission with a main shaft and countershaft.

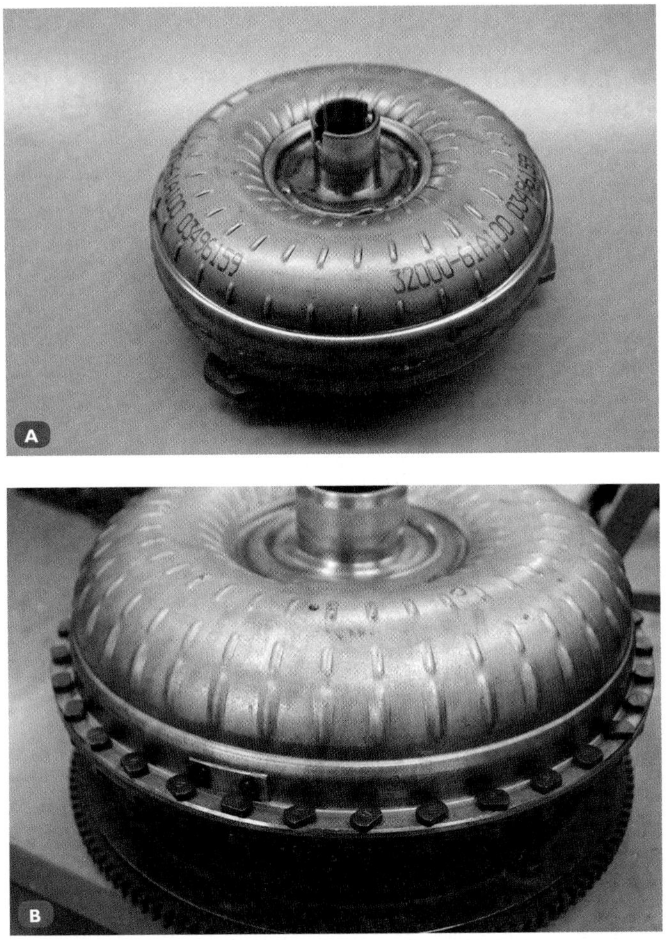

FIGURE 18-8 **A.** A light-vehicle torque converter that is welded together. **B.** A heavy-duty truck torque converter that is bolted together.

FIGURE 18-9 A typical torque converter that has been removed from a vehicle.

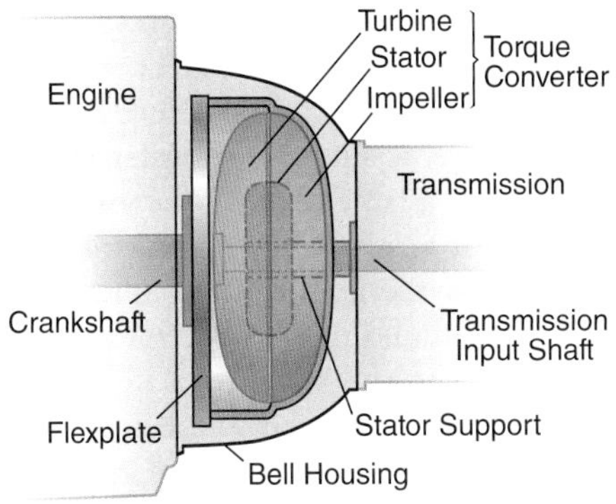

FIGURE 18-10 A torque converter mounted between the engine and the transmission.

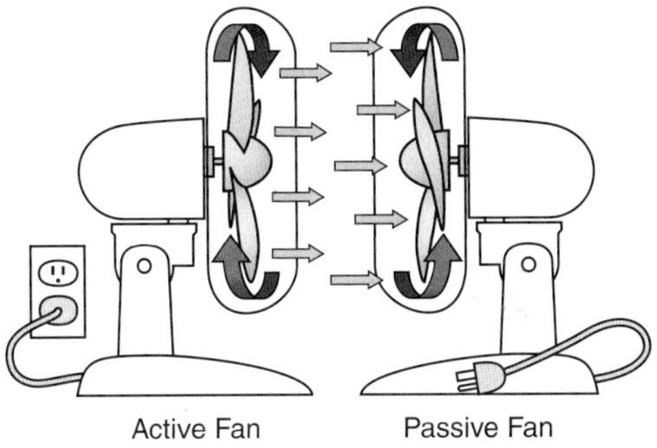

FIGURE 18-11 One fan driving a second fan functions much like a fluid coupler, with air as the medium rather than transmission fluid.

But it can do one thing a manual transmission clutch cannot do. It can multiply torque from the engine to the input shaft of the transmission under certain driving conditions. This is one reason that vehicles equipped with automatic transmissions make good tow vehicles. The torque converter can multiply the engine torque when starting from a standstill. This is just what is needed when pulling a boat trailer out of the water on a steep boat ramp. How can it do that, you ask? Well, keep reading. We will explain how each of the torque converter components work together to multiply torque under certain conditions.

Torque Converter Components

Early automatic transmissions used a **fluid coupler**, which was the precursor to a torque converter. A fluid coupler is basically two fans facing each other (**FIGURE 18-11**). One fan is the driving fan and is driven by the engine. The second fan is attached to the input shaft of the transmission. When the fluid is thrown off the driving fan, it hits the driven fan, causing the second fan to begin to spin. When the second fan spins, the input shaft of the transmission also begins to spin.

The device is called a fluid coupler because none of the converter components are physically connected to the others. Rather, the force of the fluid flow causes the transfer of power. This fluid coupler acts as an automatic clutch. At engine idle speeds, it allows the engine to operate while the vehicle is stopped in a drive range. And it can do so without the driver having to shift the transmission to neutral, or step on a clutch pedal. Unfortunately, fluid couplers were very inefficient, so they haven't been used since the 1950s.

The principle of the torque converter operation is similar to the fluid coupler. But it has an extra component. In its simplest form, a single-stage torque converter has three elements: the impeller, the turbine, and the stator (**FIGURE 18-12**). All three have angled or curved vanes and are contained in a single housing. They are separated from each other by thrust bearings, but are close together for efficient torque transfer.

The impeller has a large number of vanes attached to the converter housing. It forms the driving member (like the fan that is plugged in for the fluid coupler) (**FIGURE 18-13**). The vanes rotate with the housing. Because the housing is bolted to the engine's flexplate, the impeller rotates at engine speed. Each impeller vane has a slight curvature and is set radially in the case. The impeller causes fluid to flow inside the converter when the engine is running.

The turbine is similar in construction to the impeller, but with more vanes and a greater curvature. The direction of curvature of the turbine vanes is opposite to that of the impeller vanes. The turbine is free to rotate in the housing. The center hub has splines that mate with splines on the input shaft of the transmission (**FIGURE 18-14**). The turbine is turned by fluid leaving the impeller. This then rotates the transmission input shaft.

Both the impeller and the turbine are fitted with a guide ring. It helps to secure the vanes in position. It also reduces turbulence as fluid is flowing from the impeller to the turbine and back to the impeller again. This improves torque converter efficiency. So far, we have what could be called a fluid coupler. The next component (stator) is what turns this assembly into a torque converter.

The stator has a small set of curved blades attached to a central hub. It is positioned between the impeller and the turbine. The center hub is mounted on a one-way clutch. It is splined to the stator support shaft (**FIGURE 18-15**). The stator support shaft is attached firmly to the transmission case and does not rotate. The one-way clutch allows the stator to rotate only in the same direction as the impeller. Trying to rotate the stator in the opposite direction locks the stator on the support shaft and holds it stationary.

▶ Torque Converter Operation

LO 18-03 Describe torque converter operation.

When the engine starts, the transmission pump rapidly fills the converter with transmission fluid. The impeller is driven by the engine and turns at crankshaft speed. The turbine is splined

FIGURE 18-12 A torque converter that has been cut open to reveal the impeller, stator, and turbine.

FIGURE 18-14 A turbine mounted inside the converter case. Note that the turbine is free to spin inside the housing on bearings.

FIGURE 18-13 An impeller brazed to the housing of the torque converter.

FIGURE 18-15 A stator removed from a torque converter.

to the transmission input shaft. When a gear is selected by the driver, the input shaft becomes locked to the output shaft of the transmission. This is due to the various gears, bands, and clutches that are applied in the transmission. The spinning impeller uses centrifugal force to throw fluid outward. The fluid follows the case and strikes the turbine vanes transmitting power to the turbine. Fluid then goes back around the guide ring in a forward direction. This is because of the shape of the housing and the curvature of the vanes.

With the engine idling, little torque is transferred from the impeller to the turbine. This is because the fluid flow is too slow. When the engine accelerates, higher impeller speed discharges the fluid against the turbine vanes with greater force. This causes the turbine to begin to spin and transfers torque to the input shaft of the transmission.

The fluid, still at high velocity, now flows between the turbine vanes. The fluid leaves the turbine in a direction opposite to impeller rotation due to the curvature of the turbine vanes. In a fluid coupler, this fluid flow would then strike the impeller in the opposite direction of impeller rotation. That would tend to slow it down, reducing efficiency and torque.

In a torque converter, the stator redirects the fluid so that it reenters the impeller in the same direction as impeller rotation (**FIGURE 18-16**). The energy that was left in the fluid after turning the turbine is recycled and helps to spin the impeller. This allows the engine to turn faster. This is the process that creates torque multiplication. During torque multiplication, the stator remains stationary while the turbine and impeller spin.

Torque Multiplication

Torque multiplication exists only when there is a difference in speed between the impeller and the turbine. The amount of torque multiplication depends on this difference. When the turbine is stalled, it has a maximum torque multiplication of about 2.5:1 for most passenger vehicles. **Stall** is an operating condition where the turbine is stationary and the engine throttle is wide open. This makes the rotational speed of the impeller as high as possible.

Stall can be approximated when a vehicle moves from rest, up an incline, towing a heavy trailer. Instantaneously, as the

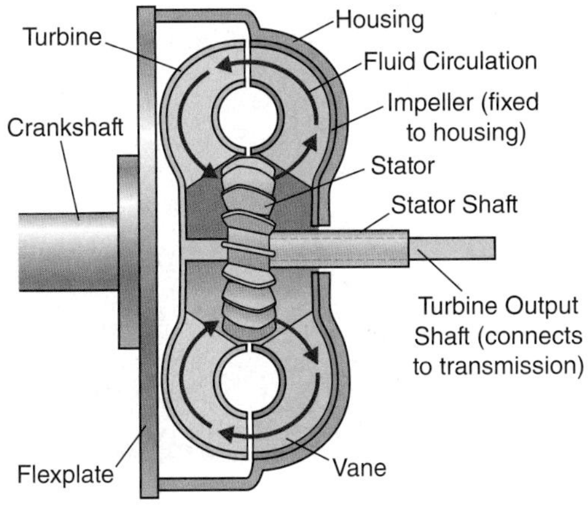

FIGURE 18-16 Fluid flow through a torque converter.

vehicle begins to move, maximum torque multiplication occurs. This is because there is a large difference between the speed of the impeller and the turbine. At a stall speed of 2100 rpm, if the torque at that speed is 200 ft-lb (271.16 N·m), the torque input to the transmission could be as high as 500 ft-lb (677.91 N·m). This is about 2.5 times the input torque. This torque enters the transmission, where it is multiplied further by the planetary gears. It is multiplied again at the final drive assembly before being applied to the vehicle's wheels. This torque multiplication is what gives a vehicle its pulling power moving away from a stop.

SAFETY TIP

Stall speed can be tested by performing a stall test (brake torque) in the shop. To perform this test, place the transmission in low gear, apply the brake pedal fully, and then carefully push the throttle open all the way. The maximum rpm the engine achieves is the stall speed. This can be a dangerous test and should only be performed under your supervisor's direct guidance.

This multiplication tapers off as the turbine speed begins to match the speed of the impeller. When turbine speed reaches around 90% of impeller speed, torque multiplication falls to zero. Torque multiplication is then about 1:1. This is known as the coupling point. At the coupling point, fluid flow from the turbine vanes is relatively low. But it is at high speed, following the direction of the torque converter rotation. The rapidly turning turbine discharges its fluid against the back of the stator blades. This force unlocks the stator's one-way clutch, allowing the stator to rotate with the fluid. The stator, turbine, and impeller all rotate as one unit at this time. Unlocking the stator prevents the stator from being in the way of the fluid flow.

Fluid Flow

The rotating impeller carries the fluid with it inside the converter housing. The fluid will be rotating around with each of the three converter components. This is known as **rotary flow**. At the same time, centrifugal force moves the fluid outward, away from the converter axis. During torque multiplication, the shape of the converter case makes the fluid flow in a circular motion, through the impeller, turbine, and stator. This is known as **vortex flow**. Combining these two fluid flows produces a progressive circular, or spiraling, motion. This is known as the spiral flow (**FIGURE 18-17**).

In a stalled converter, fluid flows at high velocity. It flows from the revolving impeller through the stationary turbine and stator. This condition results in a fast-moving vortex flow and high torque multiplication. When the turbine starts to rotate and increases in speed, the centrifugal force on the fluid in the turbine opposes the high-velocity flow from the impeller. This reduces vortex flow and torque multiplication. At coupling point, the vortex flow of fluid is slight, and there is no torque multiplication. The converter now acts as a fluid coupler.

During hill climbing, the turbine slows. At the same time, the driver increases engine power, causing the impeller to speed

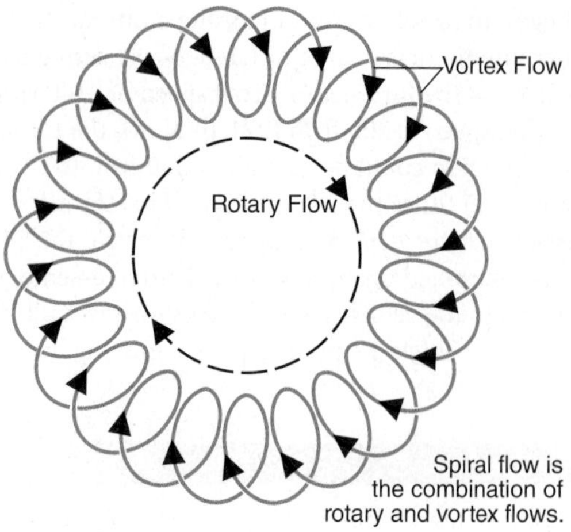

FIGURE 18-17 Rotary fluid flow, vortex flow, and spiral flow.

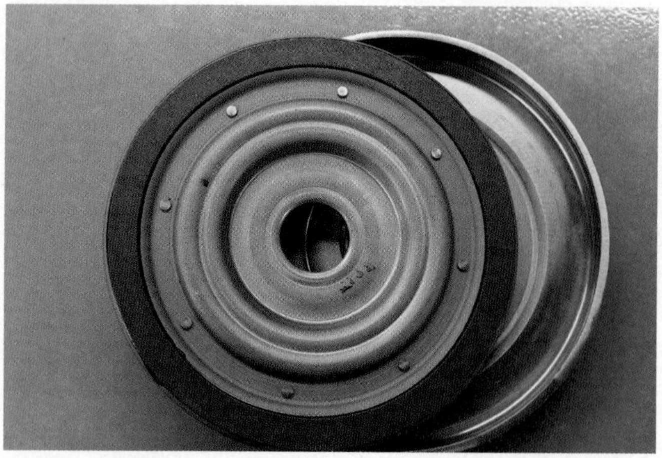

FIGURE 18-18 A lock-up torque converter piston with friction material bonded to the back of the piston.

up. The combination of these two factors causes an increase in vortex flow. This again causes torque multiplication. It continues until either the turbine speeds up or the impeller slows down. The converter automatically adjusts its output, within design limits, to meet driving requirements.

▶ Lock-Up Converters

LO 18-04 Describe lock-up converters and heat exchangers.

Under a drive condition, the impeller and turbine generally never achieve a 1:1 speed ratio. The turbine almost always spins slightly slower than the impeller. This means that the converter is typically slipping approximately 5% to 10%. This is seen when comparing engine speed to transmission input speed. During the late 1970s and early 1980s, manufacturers added a lock-up function to the torque converter in their vehicles. This dealt with this slippage condition and increased the vehicle's fuel economy.

In a lock-up converter, the impeller and turbine are locked together when conditions are suitable. This provides a 1:1 drive from the engine to the transmission input shaft. Lock-up normally occurs at higher road speeds when the vehicle is under light load.

Inside the front of the torque converter, there is a large piston. It has friction material bonded to its surface near its outside diameter (**FIGURE 18-18**). When the piston is hydraulically applied, the piston and friction material are pressed against the front housing of the torque converter. This locks the turbine to the converter housing (**FIGURE 18-19**). A 1:1 connection is provided from the engine to transmission. So, there is no slippage between the torque converter components. The lock-up torque converter also helps to reduce transmission fluid temperatures. In some cases, this requires an automatic transmission fluid (ATF) heater system.

Torsional damper springs are built into the piston assembly (**FIGURE 18-20**). When the clutch is engaged, these springs dampen engine torsional vibrations. Without these damper springs, every power pulse of the pistons would be felt by the driver. Many drivers would complain about excessive engine vibration.

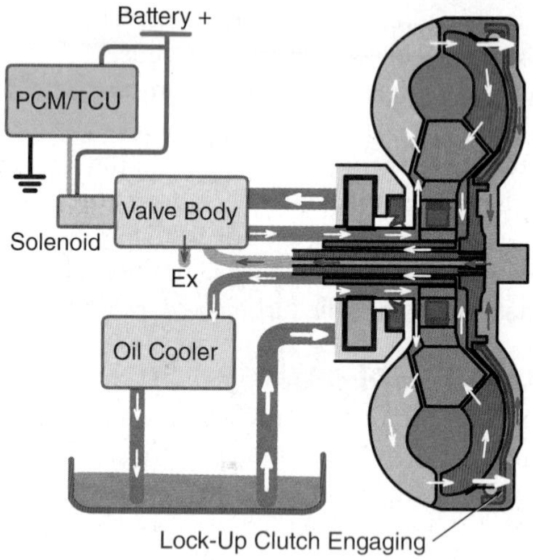

FIGURE 18-19 The torque converter clutch is hydraulically applied and forced against the front housing of the torque converter, locking the two together.

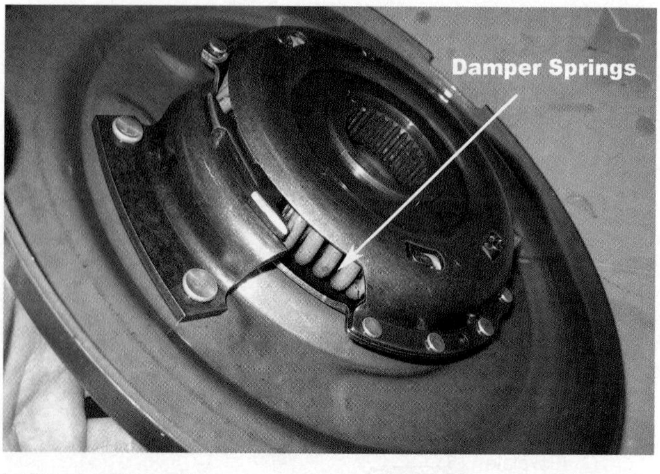

FIGURE 18-20 Lock-up pistons use torsional springs to absorb power pulses from the engine.

Heat Exchanger

Torque converter slip and loss of power through the transmission produce excessive heat. This heat must be dissipated. At stall, a lot of engine output is converted into heat. It can easily bring the transmission fluid operating temperature close to its boiling point. Excessive temperature can produce cavitation bubbles in the fluid, which reduce converter efficiency. It can also shorten the life of the transmission fluid greatly. Too hot and it oxidizes; too cold and it does not lubricate as well.

Most automatic transmission vehicles use a heat exchanger to cool the transmission fluid. It is sometimes called a transmission cooler. Most of the time, it is located in the outlet tank of the radiator. In this way, the transmission fluid can be cooled. But the engine coolant at the outlet side of the radiator will prevent it from overcooling the transmission fluid (**FIGURE 18-21**).

Fluid flows from the pump to the valve body. Part of the fluid flows to the converter and the rest to operate the bands and clutches. Because the converter creates a lot of heat, the fluid then flows through the heat exchanger before returning to the transmission. It is then used to lubricate the planetary gears and bushings in the transmission. Heavy-duty vehicles or those with towing packages have an extra external transmission cooler. It is used to help dissipate the added heat from towing (**FIGURE 18-22**). Typically, it is installed in front of the radiator.

When using an external transmission fluid cooler, it is generally agreed that it is best to run the transmission fluid through the external cooler first. Then, route it back through the cooler in the radiator. Doing so prevents overcooling of the transmission fluid.

Another method of improving cooling is to install a finned aluminum transmission pan. This finned pan allows air that is passing under the vehicle to draw heat away from the fluid. Often a pan of this type is a deeper pan, allowing the transmission to hold an extra quart or two of fluid (**FIGURE 18-23**).

Some manufacturers incorporate a transmission fluid warmer into their vehicles. They use engine coolant to warm up the ATF quickly and keep it from cooling too much. These warmers are often found on vehicles with CVTs. They help reduce the viscosity of the ATF, which reduces the friction losses of the thicker ATF.

FIGURE 18-21 Cutaway showing the transmission cooler inside the radiator tank.

The warmer assembly uses a thermostatic valve. It directs the flow of ATF through the heated warmer assembly when the ATF is cold. The valve then bypasses the heated warmer assembly if the ATF is hot (**FIGURE 18-24**).

FIGURE 18-22 An external transmission cooler (the smaller one behind the tubes) is typically used on vehicles equipped with towing packages.

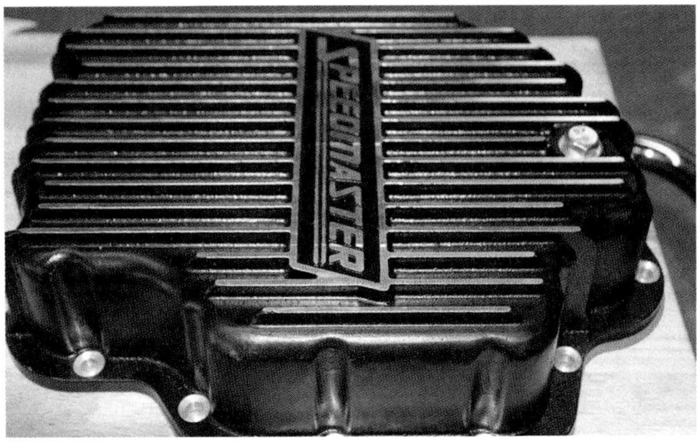

FIGURE 18-23 A deep sump finned transmission pan.

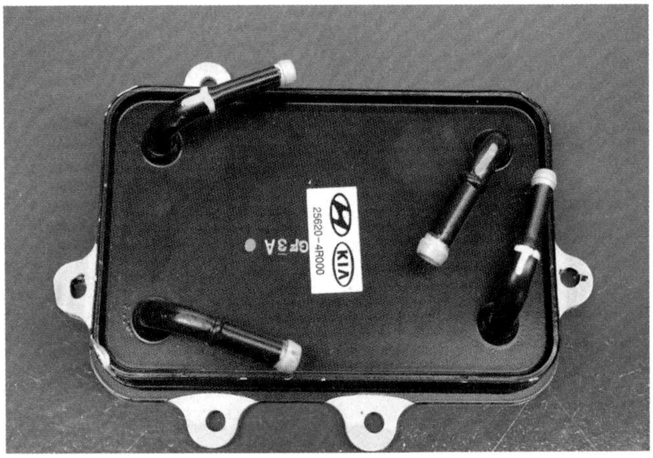

FIGURE 18-24 A transmission fluid warmer from a Hyundai Sonata.

▶ Gear Train—Principles of Operation

LO 18-05 Describe gear train operation.

The gear train of a modern automatic transmission generally consists of the following:

- planetary gears,
- shafts,
- transfer chains or gears,
- bearings and bushings, and
- the final drive unit on a transaxle.

These components are often referred to as the "hard parts" of a transmission. They would generally not be replaced during a traditional rebuild unless they show signs of wear. The only exceptions are some of the bearings and bushings.

Most automatic transmissions use planetary gear sets to obtain several gear ratios. A planetary gear set is a device with several intermeshed gears assembled in a compact design. The planetary gear set has a **sun gear** in the center. Smaller **planet gears** revolve around the sun gear (**FIGURE 18-25**). These planet gears are held together in a **planet carrier**. The planet gears revolve inside a larger **ring gear**. The ring gear wraps around the outside of the whole planetary gear set.

Various gear ratios can be created in a planetary gear by holding and driving the different members. Hydraulic pressure is used to apply individual clutches and bands. The bands and clutches either hold or drive one or more of the planetary gear components. We explain that in more depth later in the chapter. Combining two or more planetary gear sets (compound planetary gear sets) creates additional gear ratios.

Gear Ratio/Torque Multiplication

Gear ratio refers to the difference in diameter between the driving (input) gear and the driven (output) gear. When you loosen a bolt, what tool do you use? Do you use a socket alone? No, that would not give you much leverage. Instead, you place a long ratchet or breaker bar onto the end of the socket to increase the amount of leverage. This increase in leverage is the same as torque multiplication in a gear ratio.

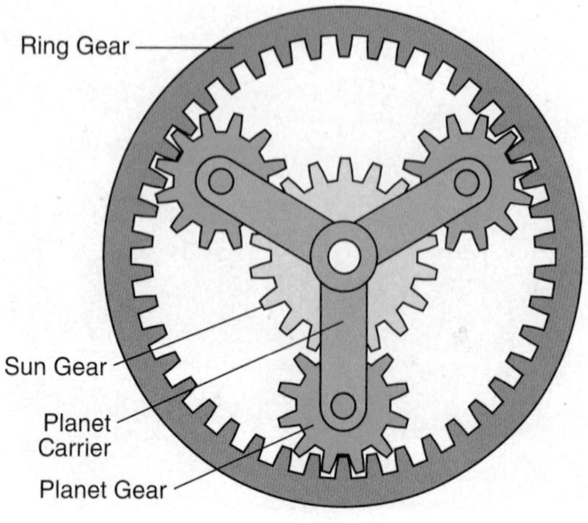

FIGURE 18-25 Simple planetary gear set.

FIGURE 18-26 shows a gear with 8 teeth driving a gear with 24 teeth. The smaller drive gear will be able to turn three times before the large driven gear turns once. In this scenario, we have created a gear ratio of three to one (3:1). This gear ratio will create three times more torque. But the driven gear will turn at three times less speed and distance. This is known as gear reduction. If we were to put 100 ft-lb (136 N·m) of torque into this gear system, we would get 300 ft-lb (407 N·m) of torque out, but at one-third the speed and distance.

If we have a gear with 15 teeth driving a gear with only 10 teeth, we will have a gear ratio of 0.67:1. This is an **overdrive** ratio (**FIGURE 18-27**), and it results in less torque being transmitted to the wheels but will increase wheel speed. Any ratio of less than 1:1 is considered an overdrive ratio. Overdrive ratios are typically used to increase fuel economy while traveling at highway speeds. It does this by allowing the engine to operate at a lower speed.

A typical automatic transmission has four or more forward gear ratios and one reverse. Transmission gear ratios vary by transmission and manufacturer, but are often approximately as listed here:

- First gear—3:1, resulting in a large increase in torque, but a major decrease in speed

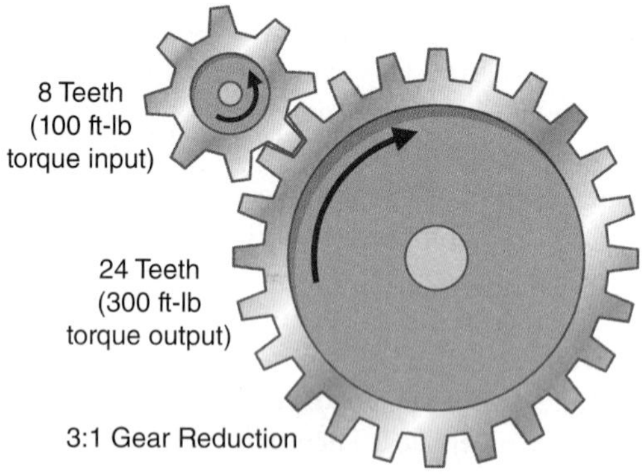

FIGURE 18-26 Having a smaller input gear with 8 teeth driving a larger output gear with 24 teeth results in a 3:1 gear reduction.

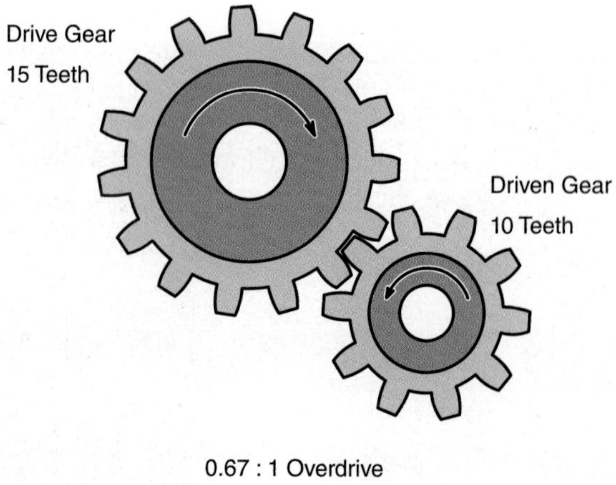

FIGURE 18-27 Having a larger input gear with 15 teeth and a smaller output gear with 10 teeth results in a 0.67:1 overdrive.

- Second gear—1.75:1, still resulting in a torque increase with a smaller speed decrease
- Third gear—typically 1:1, or direct, ratio, meaning that power is sent directly through the transmission with no torque increase or speed increase
- Fourth gear—0.80:1, resulting in an overdrive ratio in order to increase fuel economy
- Reverse gear—2:1, resulting in a torque increase with a speed reduction

Early transmissions were simple two-speed automatics with a low gear and a direct gear. As engine size in vehicles decreased, the two-gear ratios were no longer adequate. Three-speed automatic transmissions became the norm through the early 1980s. After that, four-speed automatics were installed in vehicles to improve fuel economy. Now they have even greater need to increase fuel economy. Manufacturers have added more gear ratios to their transmissions. Many vehicles are leaving the factory today with six- and seven-speed transmissions. And a few manufacturers use nine-speed automatics. Even with all of these gears, the basic principle of gear reduction and overdrive remains the same.

Gear Set Styles

Individual gears in automatic transmissions are typically one of three types: helical cut, spur (also known as straight-cut gears), or hypoid gears (**FIGURE 18-28**). **Spur gears** were used in early transmissions due to ease of manufacturing and lower cost. Spur gears also tend to be much louder and do not offer as much strength as the other types of gears. For this reason, manufacturers usually use helical gears for passenger vehicles. At the same time, spur gears don't create thrust forces. That makes them useful in high-torque applications, such as heavy off-road diesel equipment.

Helical-cut gears (or simply "helical gears") are cut on a spiral around the axis of the shaft. Helical gears offer the benefit of always having more than one gear tooth in contact with the driving gear. This increases the strength. Helical gears are quieter than spur gears but have more thrust motion due to the spiral cut of the gear teeth. Because of this increased thrust, helical gears must have a method to control the amount of thrust movement. This is provided by **thrust washers** or thrust bearings.

Hypoid gears are often used in the final drive assemblies. Hypoid gears are often used to change the direction of power flow by 90 degrees. They are a type of helical gear in which the axes of the two gears are not aligned (**FIGURE 18-29**). These gears therefore have the thrust action of helical gears along with a scraping action between the teeth. Because of this scraping action and very high pressure on the gear teeth, hypoid gears require special lubricants to prevent damage.

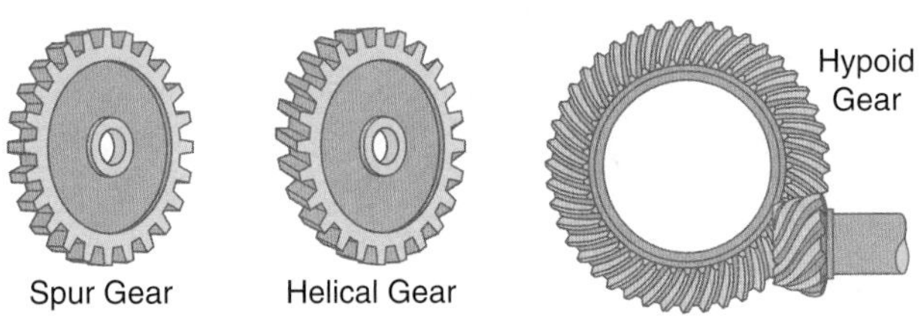

FIGURE 18-28 Spur, helical, and hypoid gears.

Planetary Gear Sets

A simple planetary gear set is the backbone of most modern automatic transmissions. One of their biggest benefits is that they are in constant mesh. And because they can provide multiple gear ratios, a variety of speeds can be obtained without having to take them out of mesh. Planetary gear sets use either helical gears or straight-cut gears. A simple planetary gear set contains (**FIGURE 18-30**):

- a sun gear in the center, with multiple revolving planet gears around it.
- The planet gears are held in place by the planet carrier.
- The outside edge of the planet gears is meshed with the ring gear.

By holding one of the components (planet carrier, sun, or ring) and driving one of the other components, we can create up to six separate gear ratios. Four gear ratios in the forward direction (two reduction and two overdrive). And two in the reverse direction (one reduction and one overdrive). If any two components are locked together, we end up with another speed—1:1, called direct drive.

Just because a planetary gear can obtain four forward gears and two reverse gears does not mean the manufacturer can use each of those ratios. Manufacturers have to design the planetary

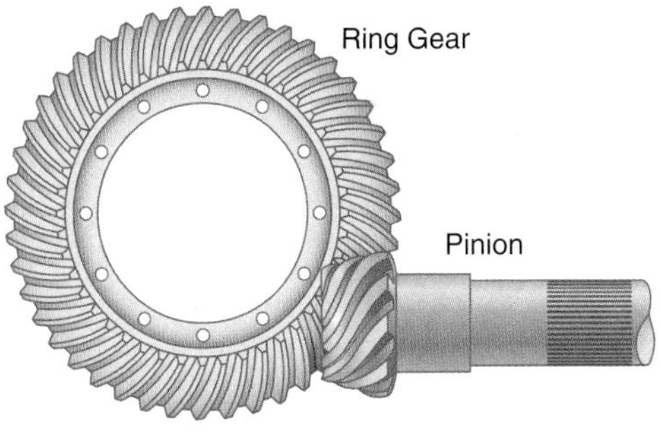

FIGURE 18-29 A hypoid gear arrangement.

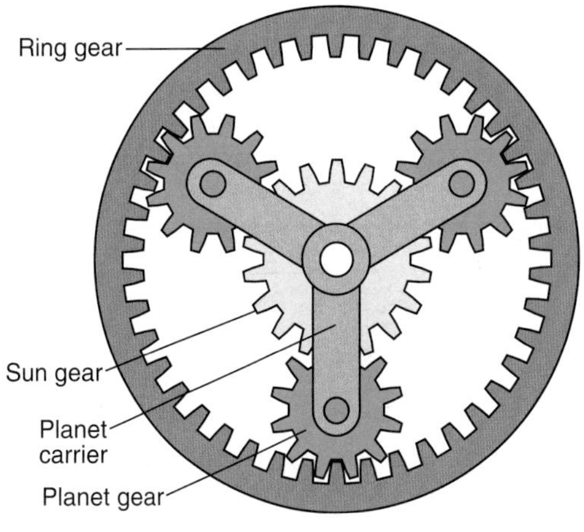

FIGURE 18-30 A simple planetary gear set.

gear sets to get the required number of gear ratios as well as useful gear ratio splits. The benefit of designing transmissions with planetary gears is that you can add planetary gear sets to create additional ratios (**FIGURE 18-31**).

▶ Holding/Driving Gears

LO 18-06 Describe methods of holding–driving gears.

As we stated earlier, to create a gear ratio, we need to hold one member of a planetary gear set and drive another. There are four basic holding devices used in transmissions today:

- bands,
- multidisc clutches,
- one-way clutches, and
- dog clutches.

These holding devices will either lock two members of the planetary gear set together, or lock one member to the transmission housing. This prevents it from spinning.

Applied Math

AM-1: Direct/Inverse Variation: The technician can solve problems requiring the use of fractions, decimals, ratios, and percentages.

When working out gear combinations, it may be necessary to calculate gear ratios based on counting the number of gear teeth. This is simple in the case of differentials and manual transmissions. But the planetary gear sets used in automatic transmissions require more involved calculations.

In the case of a four-speed automatic transmission with an overdrive top gear obtained by driving the planet carrier, the sun gear (30 teeth) held, and the ring gear (72 teeth) driven, the formula to calculate the gear ratio would be:

$1/(1 + \text{sun/ring})$

The effective ratio is therefore calculated as:

$1/(1 + 30/72)$, or 0.71:1.

Different combinations of driving, driven, and held gears require different formulae to calculate ratios.

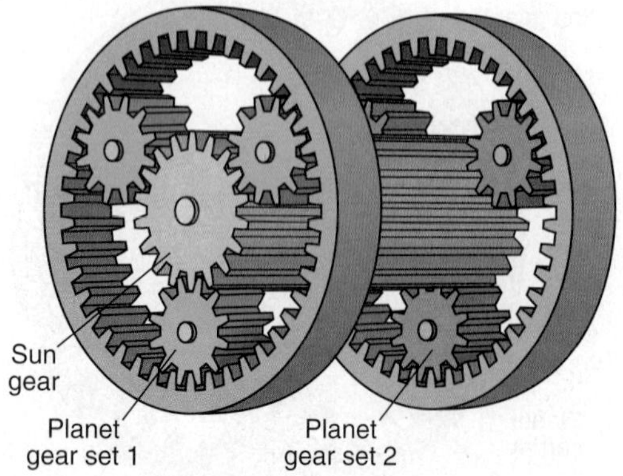

FIGURE 18-31 Compound planetary gear made by combing simple planetary gears.

The **band** is a friction-lined steel belt that wraps around the outside of a drum (**FIGURE 18-32**). When a servo applies pressure to one end of the band, it closes the diameter of the band. The band then squeezes the drum and prevents it from turning. The drum is then locked to one member of a planetary gear set.

Multidisc clutches are unique in that they not only can be used to hold a member of the planetary gear set, but they are also used to drive a member of the planetary gear set (**FIGURE 18-33**). In a multidisc clutch, individual friction discs, often called frictions, are stacked between smooth steel plates, often called steels. Both the frictions and the steels have a hollow center. This allows them to wrap around the outside of a hub or drum which is connected to a member of the planetary gear set.

The steels typically have splines, or teeth, on the outside edge. They mesh with splines, or teeth inside a drum. The frictions have splines or teeth on the inside edge. They are meshed to a hub or drum connected to a member of the planetary gear set. When the steels and frictions are forced together by a hydraulic piston, the separately revolving steels and frictions are locked together. This locks the drum or hub to the member of the planetary gear set.

One-way clutches can be either one-way rollers or sprags. Both the one-way roller and the sprag operate on the same

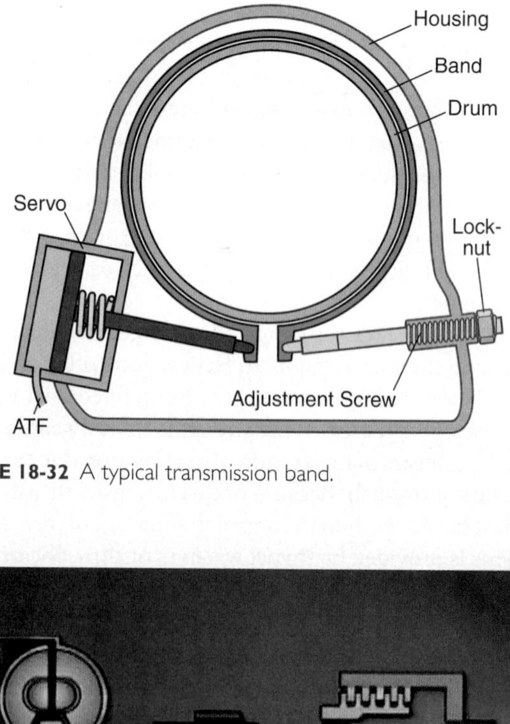

FIGURE 18-32 A typical transmission band.

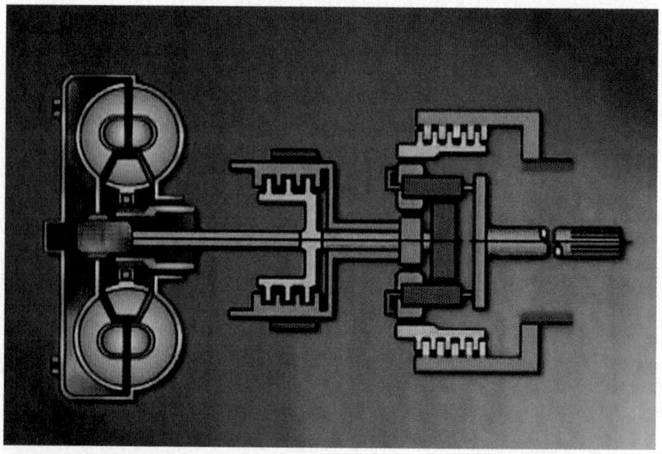

FIGURE 18-33 Two multi-disc clutches shown here.

principle. But sprags have a greater torque-holding capacity. On both the one-way roller and the sprag, an inside hub is allowed to spin freely in one direction. But turning the hub in the opposite direction causes the rollers or sprags to become wedged in place (**FIGURE 18-34**). This wedging action prevents rotation of the hub.

In the one-way roller clutch, each roller has a waved compression spring that pushes the roller toward the narrow end of the wedge. Rotating the hub toward the wide end of the wedge allows the hub to rotate freely. Rotating the hub in the opposite direction locks the outer race onto the rollers wedged between the two races. Thus, a one-way clutch allows rotation in one direction but prevents rotation in the opposite direction.

> **TECHNICIAN TIP**
>
> It is beneficial on some transmissions to increase the number of individual rollers or sprags to increase the holding power. Common performance transmissions have multiple versions of one-way rollers and sprags.

Dog clutches are two-piece mechanisms that can be engaged or disengaged from each other. Their interlocking teeth can hold or drive components (**FIGURE 18-35**). They do not rely on friction to lock up. They provide a direct mechanical connection between the two components. When decoupling them, one is moved away from the other.

Automatic Transmission Fluids

LO 18-07 Describe automatic transmission components.

ATF is a specialized fluid that has been designed for a specific job. The ATF must be able to transfer heat from the internal components of the transmission to the transmission cooler. This helps prevent damage to the internal seals and bushings of the transmission. The fluid must also lubricate the internal gears, bearings, and bushings of the transmission. It also must have enough of a **coefficient of friction** to allow the clutches to hold and not slip. Coefficient of friction is the force required to move two sliding surfaces over each other.

Some ATF fluids are typically dyed red for easy identification. But there are other colors used as well. At the same time, all ATF contains additives such as those listed here:

- *Rust and corrosion inhibitors:* These additives prevent the internal parts of the transmission from developing rust and corrosion. Rust and corrosion can affect the shift quality and longevity of the transmission. As rust particles break off, they become an abrasive in the fluid, causing increased wear.
- *Friction modifiers:* Manufacturers add friction modifiers to the fluid to ensure that the fluid has the proper coefficient of friction. This helps to produce the desired shift quality. A fluid with a lower coefficient of friction produces softer, longer shifts. This is due to an increase in clutch slippage. A fluid with a higher coefficient of friction causes shorter, harsher shifts. This is due to reduced clutch slippage. It also increases driveline shock.
- *Seal conditioners:* These additives are designed to help protect the seals inside a transmission. It keeps the seals soft and makes them to swell slightly to help prevent leaks and clutch slippage.
- *Detergents:* ATF has a large amount of detergent to prevent dirt and other foreign particles from becoming trapped inside the transmission. The detergent causes the dirt and other particles to be attracted to the fluid so they flow with the fluid. When the fluid passes through the filter, the large particles become trapped in the filter. The small particles are removed during the next transmission fluid change.
- *Antifoam:* These additives help prevent foaming of the transmission fluid. When moving parts spin through a fluid, they tend to produce air bubbles. These air bubbles can quickly multiply and become foam. Foam is compressible and can cause a transmission to slip because insufficient pressure is applied to the clutches.
- *Viscosity modifiers:* These additives are similar to the engine oil additives that allow us to have multiviscosity engine oils such as 5W-30. The additives allow the fluid to remain thin when the temperature is cold. And they prevent the fluid from becoming too thin as the transmission fluid warms up.

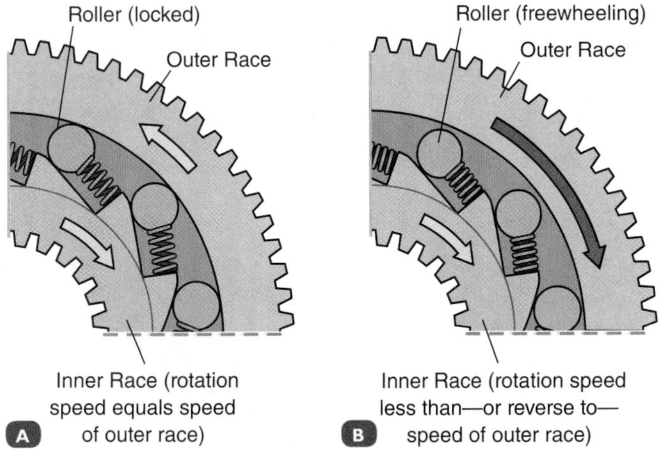

FIGURE 18-34 One-way clutches. **A.** A one-way roller in the locked position. **B.** A one-way roller in its freewheeling position.

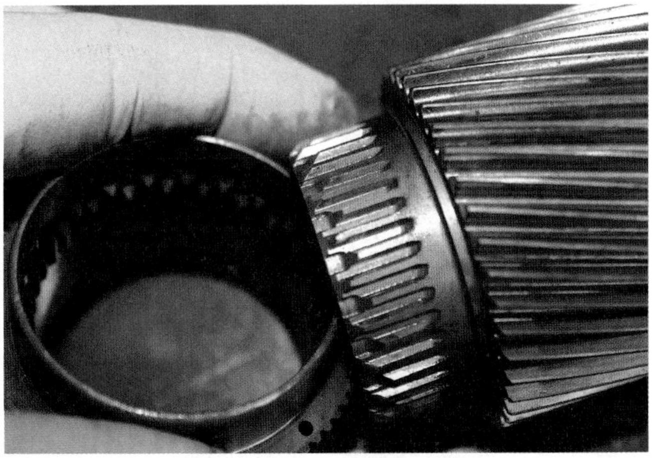

FIGURE 18-35 Dog clutches have teeth on each side that interlock with each other, providing a direct mechanical connection between the two.

ATFs can be mineral oil based or a synthetic lubricant (**FIGURE 18-36**). Many late-model vehicles recommend a specific synthetic ATF. Initially, most manufacturers used and recommended General Motors and Ford Motor Company fluids. In the past few decades, most auto manufacturers have developed their own fluids for use in their transmissions. This practice has required repair facilities and quick lube shops to stock many different types of fluids.

Some shops carry a few major types of fluids and use those in every vehicle. But that is not recommended. It may cause undesirable transmission operation and may void the transmission warranty. Several vehicle manufacturers have published TSBs related to incorrect fluid use and the negative effects on the transmission. Research the service information to determine the correct fluid for a vehicle.

On most vehicles, ATF needs to be changed periodically to remove dirt and contaminants from the transmission. Technicians used to tell if the transmission fluid needed to be changed by looking at and smelling it. If the fluid was dark or smelled burnt, it needed to be changed. You cannot determine the condition of modern ATF by looks and smell. Some newer synthetic fluids have a burnt smell when they are brand new and are often darker than mineral-based ATF.

Most manufacturers specify the recommended change interval in miles or months. But some manufacturers have specified no fluid changes for the life of the transmission. They may not even provide a method of checking the fluid level.

Flexplate and Ring Gear

On an automatic transmission, there is no flywheel. Rather, there is a thin lightweight steel flexplate. The flexplate bolts to the rear crankshaft flange in the same manner as a flywheel would. The torque converter is bolted to the flexplate with three or more bolts. Because there is no thrust force as in a manual transmission, the flexplate can be significantly lighter than a traditional flywheel (**FIGURE 18-37**). Also, the large mass of the torque converter dampens the engine pulses similar to that of a flywheel.

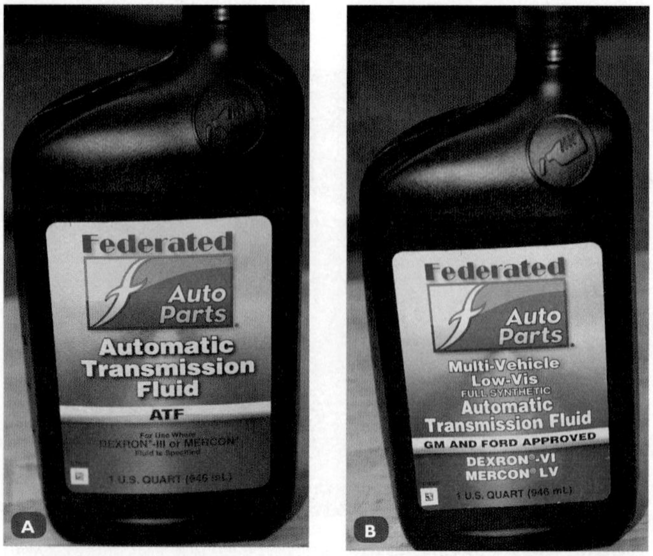

FIGURE 18-36 A. Mineral-based ATF. **B.** Synthetic-based ATF.

Wrapped around the outside edge of the flexplate is a ring gear. It is used when starting the vehicle. When the driver turns the ignition switch to the crank position, the starter pinion gear moves out to contact the ring gear. Once engaged, the starter motor cranks the engine over. This ring gear is not typically a serviceable, separate part of the flexplate. If the ring gear wears out, the entire flexplate will typically have to be replaced. Ring gears on flywheels can often be replaced. In some applications, manufacturers weld the ring gear to the torque converter instead of the flexplate (**FIGURE 18-38**).

▶ TECHNICIAN TIP

When a faulty engine or transmission is suspected because of knocking noises, double-check that the flexplate bolts are tight. Loose bolts on the flexplate can sound like a knocking engine bearing. Also, a cracked flexplate can cause a knocking noise that can sound like a defective rod bearing. In some cases, the entire center of the flexplate can break free of the rest of the flexplate. This prevents the engine from cranking over when the starter is engaged. It is easy for a technician to see that the front pulley does not turn when the engine is cranked and think that the engine has a broken crankshaft. In reality, the outer portion of the flexplate is spinning on the broken center piece.

FIGURE 18-37 A. An automatic transmission flexplate and ring gear. Note the ring gear around the outside edge of the flexplate that is used when starting the vehicle. **B.** A manual transmission flywheel and ring gear.

FIGURE 18-38 Torque converter with a welded-on ring gear.

Case, Extension Housing, and Pan

Modern automatic transmission cases are made of lightweight aluminum and alloys. Early transmission cases were sometimes made of cast iron. When a transmission is new, the manufacturer typically leaves the transmission case uncoated. When the transmission is rebuilt, the rebuilders often paint the transmission a particular color. This serves two purposes:

- It helps identify a transmission that was previously rebuilt.
- It helps seal the transmission after it has been cleaned. Aluminum can become porous after excessive cleaning of the case with detergents and solvents. The paint helps to fill in any case porosity.

On the rear or output side of the transmission, there is an aluminum housing called the **extension housing**. The extension housing is usually bolted to the transmission case. A **gasket** seals the housing to the case. The extension housing may contain

- the VSS drive gear or
- speedometer drive gear and/or
- a governor assembly to measure vehicle speed (**FIGURE 18-39**)
- an extension housing bushing and seal.

Inside the end of the extension housing is the extension housing bushing. It supports the driveshaft. The extension housing on most four-wheel drive pickups and SUVs has a surface that bolts the transmission up to the transfer case.

On the bottom of most transmissions is a sheet metal or aluminum pan. Inside this pan, the transmission filter and possibly the transmission valve body are located. Most FWD transaxles have a second pan called the side pan, which houses the valve body (**FIGURE 18-40**). The bottom transmission pan is often removed during routine transmission fluid service. The fluid and filter are changed, and the magnet cleaned, if so equipped. The magnet catches small particles of steel and iron that are floating in the fluid. The transmission pan holds about half the transmission fluid when the vehicle is off. Much of the rest of the fluid is contained in the torque converter.

Gaskets and Seals

Gaskets are used throughout the automatic transmission. Most gaskets are made from paper, fiber, or cork material. These types of gaskets must be replaced every time a component is removed. Other gaskets are a reusable type with neoprene inserted into a plastic housing. Reusable gaskets are fairly common for transmission pan gaskets where the part is frequently removed for routine service. Some reusable gaskets have built-in torque limiters. These torque limiters are metal sleeves built into the gasket around the bolt holes. This helps prevent overtightening of the gasket (**FIGURE 18-41**).

There are many types of internal seals for an automatic transmission (**FIGURE 18-42**). These internal seals are often the root cause of a transmission slipping and eventually failing. All seals need to be properly lubricated when assembling the transmission. This helps with installation and prevents premature failure. Seals are lubricated with ATF, petroleum jelly, or automatic transmission assembly lubricant. Do not use grease to lubricate the seals, as grease can clog up the internal passages

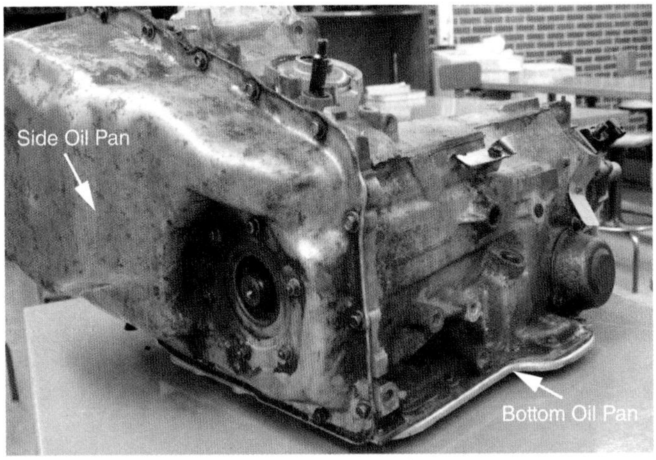

FIGURE 18-40 A common FWD transaxle showing the bottom oil pan and side pan.

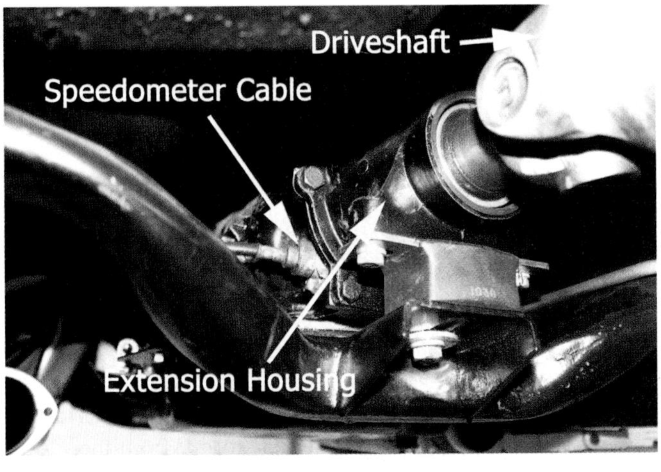

FIGURE 18-39 A typical extension housing with the speedometer cable connected.

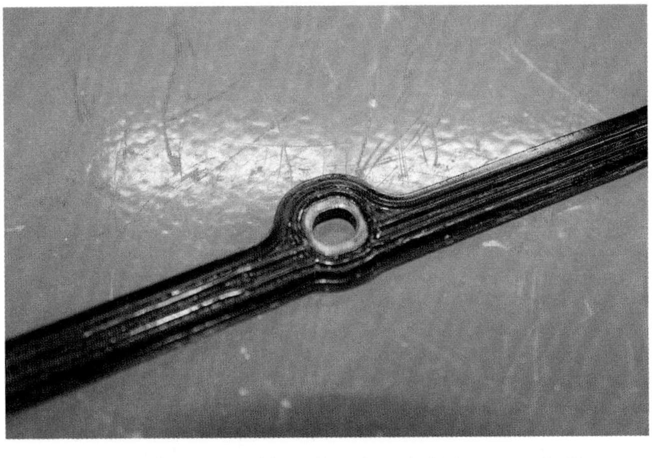

FIGURE 18-41 Some reusable gaskets have built-in torque limiters that prevent overtightening of the gasket.

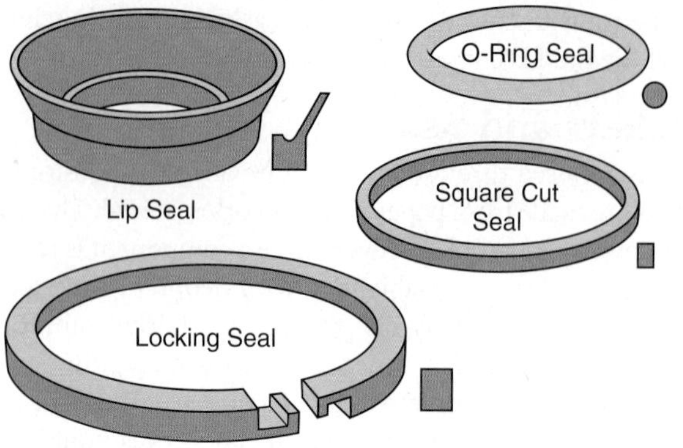

FIGURE 18-42 Typical seals used in an automatic transmission.

of the transmission. A description of seals used in a transmission system follows:

- *Square-cut seals:* Square-cut seals can be made from neoprene rubber or Teflon. Teflon seals require special handling and are discussed separately. Square-cut seals are similar to the seal found inside a brake caliper. They are sometimes located on the piston of a multidisc clutch.
- *Lip seals:* Lip seals are made from neoprene. A lip seal has greater sealing ability than a square-cut seal. When pressure is applied to the back of the seal, it forces the seal out against the inside of the clutch drum. These seals are similar to the cup seals located inside a drum brake wheel cylinder. Lip seals can be easily damaged by careless installation and infrequent fluid and filter changes. They are typically used inside multidisc clutch assemblies on the piston.
- *Locking seal:* The locking seal is made from cast iron and is similar to a piston compression ring except that the ends overlap and lock together. Locking seals are typically used between the front pump and the rotating clutch drums.
- *O-rings:* O-rings are made from neoprene rubber and are often found sealing external parts of the transmission. Examples are the speedometer cable assembly or governor cover. O-rings can also be used on accumulators and servo covers.
- *Teflon seals:* Teflon seals come in several styles—continuous, butt cut, scarf cut, and step joint. The continuous seal provides the best sealing action and should be used whenever possible. But it does require special tools for proper installation. Teflon seals have replaced the locking seals on most modern transmissions. They are found sealing the front pump to input shafts and clutch drums.

TECHNICIAN TIP

Often, Teflon seals are placed in boiling water prior to assembly to help expand the seal. The component on which the seal is being installed may be placed in the freezer to contract the part.

Parking Pawl Assembly

Most automatic transmissions for passenger vehicles include an internal parking mechanism. When the driver shifts the transmission into park, a lever, called a parking pawl, is activated. It is forced into notches cut into a hardened steel drum on the output shaft of the transmission (**FIGURE 18-43**). This parking pawl prevents rotation of the output shaft when it is engaged.

The parking pawl is often operated through a spring. This is in case the driver accidentally places the vehicle into park while it is still moving. In this scenario, the vehicle will make a loud ratcheting noise as the parking pawl bounces against the rotating notches of the output shaft drum. The spring is designed to help minimize damage to the parking pawl and the output shaft.

The spring also allows the driver to place the vehicle shift selector into park even when the parking paw does not quite align with the notches in the drum. If this occurs, the gearshift is placed in park. But the vehicle may need to roll an inch or two for the parking pawl to engage the notch.

FIGURE 18-43 A parking pawl assembly on a late-model CVT transmission.

Wrap-Up

Ready for Review

- An automatic transmission can select and shift gears without driver input and can automatically couple and uncouple from the engine when needed.
  - Transmission types: conventional transmissions, transaxles, dual clutch, continuously variable, hybrid transmissions, and a Honda/Saturn type of transmission
- A single-stage torque converter has three elements: the impeller, the turbine, and the stator. The torque converter is mounted between the engine and the transmission.
- A transmission pump rapidly fills the torque converter with transmission fluid to multiply torque from the engine to the input shaft of the transmission under certain driving conditions.

- In a lock-up converter, the impeller and turbine are locked together to provide a 1:1 drive from the engine to the transmission input shaft. A heat exchanger is used to cool the transmission fluid and the engine coolant at the outlet side of the radiator will prevent it from overcooling the transmission fluid.
- A gear train consists of planetary gears, shafts, transfer chains or gears, bearings and bushings, and the final drive unit on a transaxle.
- A holding device will either lock two members of the planetary gear set together, or lock one member to the transmission housing. Four basic holding devices used in transmissions: bands, multidisc clutches, one-way clutches, and dog clutches.
- Automatic transmission components include, but are not limited to: flexplate and ring gear; case, extension housing, and pan; gaskets and seals; and parking pawl assembly.

Key Terms

band A metal band with friction material bonded to one side. The band is contracted around a drum to stop the drum from spinning.

coefficient of friction (CoF) The amount of force required to move an object while in contact with another, divided by its weight.

CVT (continuously variable transmission) A type of transmission that lacks the fixed gears found in conventional transmissions; it adjusts gear ratios infinitely within the design of the transmission.

dual-clutch transmission A transmission with two input shafts controlled by two separate clutches.

dual-shaft transmission An automatic transmission that more closely resembles a manual transmission, as it does not use planetary gearsets.

extension housing A component of the automatic transmission housing that covers the output shaft of the transmission. The extension housing also supports the end of the driveshaft and may hold components such as the vehicle speed sensor, speedometer drive assembly, and governor assembly.

fluid coupler A type of hydraulic coupling used on vintage vehicles to connect and transfer power from the engine to the transmission.

gasket A rubber, cork, or paper spacer that goes between two parts to seal the gap between the parts.

gear ratio The relationship between two gears in mesh as a comparison to input versus output.

Helical-cut gears Gears that are cut on a helix, or spiral

hypoid gearing A type of spiral bevel gearset that mounts the pinion gear below the centerline of the crown gear.

multidisc clutch A type of holding device used by an automatic transmission to stop the movement of one component of a planetary gearset. It uses several thin friction discs and thin steel plates that are squeezed together when hydraulic pressure is applied to a piston in the clutch.

one-way clutch A type of holding device that allows free rotation in one direction but will lock up in the opposite direction; also called an over-running clutch.

parallel hybrid drivetrain A type of hybrid transmission in which power can flow from a gasoline engine, an electric motor, or any combination of the two.

planet carrier The device that holds the planet gears in place, keeping them equally spaced.

planet gears Gears that mesh with the sun gear and ring gear in planetary gear sets.

power-splitting transmission (PST) A type of hybrid transmission that splits the power flow going to the wheels from one or more electric motors and an internal combustion engine.

ring gear The outer gear in a planetary gear. Also the gear that meshes with the pinion gear in a final drive.

rotary flow A type of fluid flow in a torque converter in which fluid flows around the centerline of the torque converter in a circle.

series hybrid drivetrain A type of hybrid transmission in which power flows from the engine through an electric motor. The electric motor supplements the power from the engine to the wheels.

series-parallel hybrid drivetrain A type of hybrid drivetrain that can function as both a series hybrid and parallel hybrid, meaning that the gasoline engine can turn a generator that can be used to power an electric motor. The gasoline engine can also drive the vehicle directly through the transmission, and the electric motor can work in parallel with the gasoline engine to drive the vehicle.

spur gear A gear with teeth cut parallel to its axis of rotation.

standard torque converter A hydraulic coupling device consisting of an impeller, turbine, stator, and housing; located between the engine and the transmission.

sun gear The center gear of a planetary gearset around which the other gears rotate.

thrust washers Washers that provide a bearing surface between parts.

transaxle A type of transmission, typically used in FWD vehicles, in which the transmission also includes the differential and final drive gear assembly.

vortex flow State in which the fluid in the torque converter is traveling from the impeller, through the turbine, through the stator, and back to the impeller.

Review Questions

1. What does a conventional automatic transmission use to create its gear ratios?
   a. A main shaft
   b. A counter shaft
   c. A pair of variable pullies
   d. One or more planetary gearsets

2. Which type of automatic transmission does not change gears by shifting?
   a. A continuously variable transmission
   b. A dual clutch transmission
   c. A conventional RWD transmission
   d. A conventional FWD transmission
3. Which torque converter component is attached to the transmission input shaft?
   a. The stator
   b. The impeller
   c. The turbine
   d. The transmission pump
4. Which torque converter component is responsible for redirecting fluid under heavy loads, resulting in multiplied torque?
   a. The impeller
   b. The turbine
   c. The stator
   d. The lock-up clutch
5. When a torque converter impeller is spinning, but the turbine is stopped, what operating condition is the torque converter in?
   a. Stall
   b. Torque stratification
   c. Coupling
   d. Rotary flow
6. When is a lock-up torque converter most likely to be locked 1:1?
   a. Under heavy load
   b. At highway speed
   c. At low speed
   d. At stall speed
7. What type of gearset is most likely to be used for a ring and pinion in a rear differential assembly?
   a. A spiral helix
   b. A helical
   c. A spur
   d. A hypoid
8. Which holding device would be most likely to utilize a servo?
   a. A band
   b. A one-way clutch
   c. A multidisc clutch pack
   d. A dog clutch
9. Which of the following is a type of one-way clutch?
   a. Diode
   b. Sprag
   c. Check flow
   d. Locker
10. Many automatic tramissions utilize a pawl, what is its primary function?
   a. To activate a clutch pack
   b. To lock a band to a drum
   c. To shift into reverse gear
   d. To act as a parking mechanism

ASE Technician A/Technician B Style Questions

1. A vehicle with an automatic transmission is not shifting correctly. Technician A states that this could be a result of an electrical problem such as a malfunctioning vehicle speed sensor. Technician B states that this could be a result of an engine performance problem such as a restricted exhaust. Who is correct?
   a. Technician A only
   b. Technician B only
   c. Both Technicians A and B
   d. Neither Technician A nor B
2. Torque converters are being discussed. Technician A states that most torque converters can be disassembled by removing bolts. Technician B states that most torque converters are welded together and require a specialty shop to repair. Who is correct?
   a. Technician A only
   b. Technician B only
   c. Both Technicians A and B
   d. Neither Technician A or B
3. A torque converter is being discussed. Technician A states that the impeller is driven by the engine and turns at crankshaft speed. Technician B states that the turbine drives the transmission input shaft. Who is correct?
   a. Technician A only
   b. Technician B only
   c. Both Technicians A and B
   d. Neither Technician A nor B
4. A torque converter is being discussed. Technician A states that the turbine is driven by the engine and turns at engine RPM. Technician B states the impeller drives the transmission input shaft. Who is correct?
   a. Technician A only
   b. Technician B only
   c. Both Technicians A and B
   d. Neither Technician A nor B
5. Lock-up torque converters are being discussed. Technician A states that the torsional damper springs are used to help release the clutch when exiting lock-up condition. Technician B states that the torsional damper springs are used to absorb the engine pulses when in a lock-up condition. Who is correct?
   a. Technician A only
   b. Technician B only
   c. Both Technicians A and B
   d. Neither Technician A nor B
6. A Planetary gearset is being discussed. Technician A states that the single planetary gearset can provide an underdrive and an overdrive gear ratio. Technician B states that the single planetary gearset can be used to produce reverse. Who is correct?
   a. Technician A only
   b. Technician B only
   c. Both Technicians A and B
   d. Neither Technician A nor B

7. An automatic transmission band is being discussed. Technician A states that a sprag is used to apply the band. Technician B states that a servo is used to apply the band. Who is correct?
   a. Technician A only
   b. Technician B only
   c. Both Technicians A and B
   d. Neither Technician A nor B
8. An automatic transmission is equipped with a multidisc clutch. Technician A states that a piston is used to apply the clutch. Technician B states that the clutch discs grip steel plates when activated. Who is correct?
   a. Technician A only
   b. Technician B only
   c. Both Technicians A and B
   d. Neither Technician A nor B
9. A vehicle is brought into the shop for routine maintenance. Technician A states that all automatic transmissions have a dipstick for checking fluid level. Technician B states that some vehicles may not require ATF service, which is expected to last the life of the vehicle. Who is correct?
   a. Technician A only
   b. Technician B only
   c. Both Technicians A and B
   d. Neither Technician A nor B
10. A transmission requires removal of the oil pan. Technician A states that if it has a paper gasket, the gasket must be replaced. Technician B states that the gasket may be re-used, if it is a re-usable type and in good condition. Who is correct?
    a. Technician A only
    b. Technician B only
    c. Both Technicians A and B
    d. Neither Technician A nor B

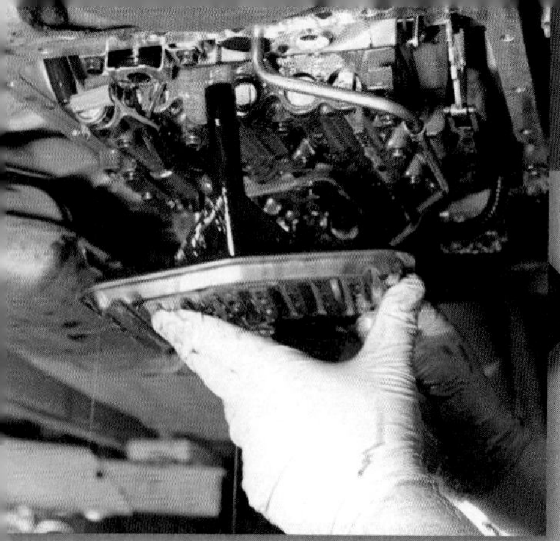

CHAPTER 19

Maintaining the Automatic Transmission/Transaxle

Learning Objectives

- **LO 19-01** Check automatic transmission fluid level and inspect for leaks.
- **LO 19-02** Perform automatic transmission fluid service.
- **LO 19-03** Perform the in-vehicle transmission service tasks.

ASE Education Foundation Tasks

See Appendix A to view the 2017 ASE Education Foundation Automobile Accreditation Task List Correlation Guide.

▶ Introduction

Most automotive technicians will not perform complete rebuilding of automatic transmissions and transaxles in their shops. But it is very important for technicians to have a good understanding of how the transmission operates and the correct maintenance procedures. A technician who improperly services an automatic transmission can cause it to fail prematurely. This can lead to internal damage and a large expense for either the customer or the shop.

A technician who does not have the ability to diagnose a transmission failure may be doomed to transmission replacement. Or worse, this technician may replace the transmission only to find that the original problem still exists.

Say that a customer brings a vehicle in for a transmission problem and is told that the vehicle needs a new transmission for $3500. That customer may then take the vehicle to another shop for a second opinion. What if that other shop actually knows what it is doing? They find that the transmission is only leaking from a gasket and is low on fluid. Or they may find that an easily replaced shift solenoid is faulty. The total bill may be less than $400, and the customer has now found a new, reliable shop. Not only that, but the customer may tell other potential customers not to take their vehicles to that particular shop. All because of that one critical experience.

*Note: When we say "transmission," we use the term generically to include transaxles also. When we use the term "transaxle," we generally only mean transaxle. Readers should use their understanding of transmission and transaxle theory to understand the intended meaning. For example, if we are discussing ring and pinion backlash, that can only apply to a transaxle.

▶ General Transmission Maintenance

LO 19-01 Check automatic transmission fluid level and inspect for leaks.

Most automatic transmissions require some periodic preventive maintenance. This is designed to reduce transmission failure. It also saves the customer money in the long run. It should include a check of the transmission for proper fluid level, and a visual inspection for leaks.

Periodic maintenance often includes transmission fluid and filter replacement. This allows for the inspection of debris in the bottom of the transmission pan. On older vehicles, the bands needed periodic adjustment. Always check the service information for the list of required maintenance.

Checking Transmission Fluid

On transmissions with a dipstick, transmission fluid should be checked at least at every oil change. Checking the fluid level is one of the first steps in diagnosing an automatic transmission problem. Without the proper fluid level, the transmission will not shift properly. It also may suffer internal damage when the fluid level is too low or too high. If the fluid level is low, it is important to properly identify the source of fluid loss. We will cover that in more depth shortly.

For most, but not all, automatic transmissions, the fluid level is checked with the vehicle idling in park or neutral. The transmission should be at operating temperature. The vehicle must be on level ground as well. When reading a transmission dipstick, it usually only takes a pint of fluid to raise the level from the bottom of the crosshatched area or add mark to the full mark (**FIGURE 19-1**). An engine oil dipstick typically requires 1 quart (0.96 liters) between marks.

Vehicle manufacturers have been eliminating the transmission dipstick on some of their vehicles. This prevents the vehicle owner from installing the incorrect fluid. Customers might also overfill the transmission. It also provides one less entry point for dirt and contaminants to enter the transmission.

On transmissions without a dipstick, it is critical to check the service information for the proper procedure for checking the fluid level. Many late-model Ford and Toyota vehicles have what looks like a drain plug installed in the transmission pan. With the vehicle running, this plug is removed. Fluid is forced up into the transmission through this hole. The hole has a tube attached to it inside the transmission (**FIGURE 19-2**). Once the fluid level is at full, excess fluid will simply drain back out. The plug can then be reinstalled.

You Are the Automotive Technician

Today a customer visits your shop with a transmission concern in a 2011 Chevy Impala. He tells you that it takes about five seconds for it to go into gear the first thing in the morning. The engine rpm sometimes flairs on an upshift. You start by checking the fluid level and find it well below the safe mark. The fluid leaves a moderately dark center when placed on a white paper towel, but no metal filings. You top it off with about 1.5 quarts (1.44 liters) of the proper transmission fluid to get it ready for a testdrive. During the testdrive, you notice that the malfunction indicator light (MIL) is illuminated. Also, the torque converter clutch doesn't engage. The customer now says the flaring on upshifts is gone. Everything else appears normal.

1. What might a visual inspection reveal on this vehicle?
2. What test(s) should be performed based on the results of the testdrive?
3. Assuming the tests reveal that the transmission does not need to be rebuilt, what service(s) should be recommended to the customer?

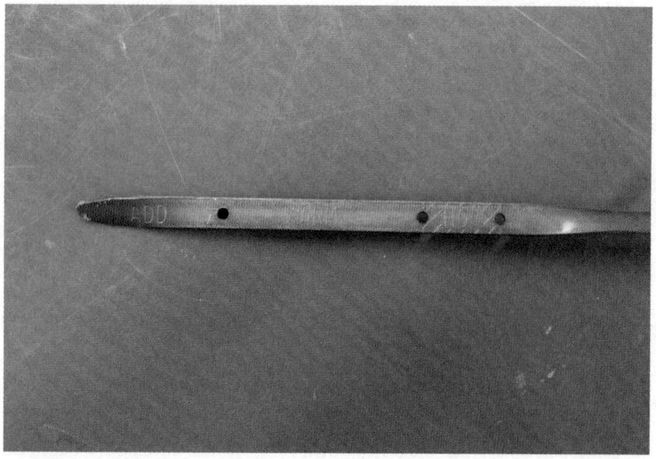

FIGURE 19-1 A typical transmission dipstick. This transmission typically requires 1 pint (0.47 liters) to bring the fluid level from the bottom of the crosshatched area to the full mark.

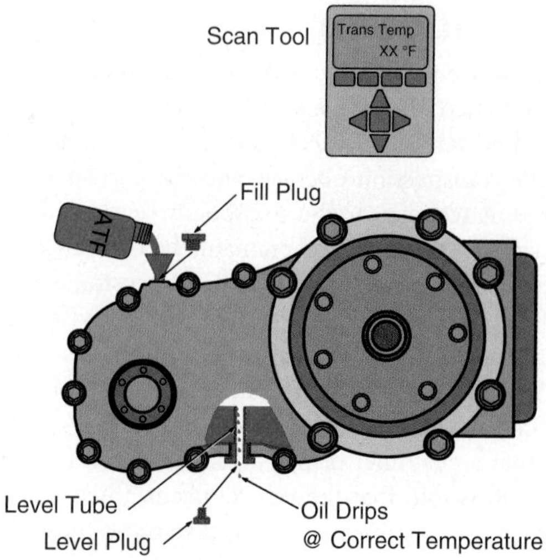

FIGURE 19-2 On transmissions with the fill tube on the bottom of the pan, once the fluid is at the correct level, additional fluid spills out.

General Motors vehicles often have a threaded plug, located on the transmission case. The plug is removed while the vehicle is running. Again, fluid is added until transmission fluid flows out of the hole, and then the plug is reinstalled. (**FIGURE 19-3**).

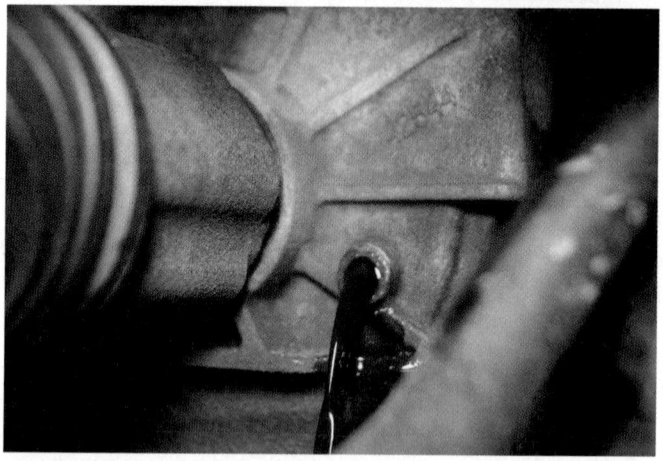

FIGURE 19-3 On transmissions with a fill plug on the side of the transmission, fill until transmission fluid just starts to come out.

Locating Leaks

When fluids leak, they travel downward due to gravity. Air flow from driving tends to blow fluid toward the back of the vehicle. This fluid travel can make finding a leak more challenging, especially if the vehicle has been leaking for a while.

If the transmission has a large amount of oil and dirt on it, locating the leak can be difficult. Cleaning the transmission case with an engine degreaser or pressure washer can help. Another option if the leak is hard to locate is to place a leak detection dye in the transmission fluid. After running the transmission for a few minutes, any leaking dye will be visible with an ultraviolet light.

Transmissions can leak from a variety of places. Some places to check are the:

- transmission pan,
- area around the entrance of the filler tube to the transmission,
- extension housing gasket,
- output shaft seals,
- speedometer/vehicle speed sensor seal,
- selector shaft seal,
- area around the electrical connectors that go into the transmission case,
- front pump seals,
- fluid cooler lines, and fittings.

If the vehicle has a vacuum modulator, remove the vacuum hose from the modulator and see if there is any transmission fluid in the hose. If there is, the modulator is bad. Also be sure to remove the radiator cap (with vehicle cold). Check for any transmission fluid in the radiator indicating a leaky cooler.

To diagnose fluid loss and condition concerns, follow the steps in **SKILL DRILL 19-1**.

▶ Replacing Fluid and Filters

LO 19-02 Perform automatic transmission fluid service.

The most common transmission work that the average technician performs is a visual inspection. This, along with fluid and filter replacement. At the minimum, transmission fluid should be changed according to the manufacturer's maintenance schedule. If the vehicle is used for towing or operates in dusty environments, the fluid should be changed more often. Some manufacturers recommend fluid and filter changes every 25,000 miles (40,000 km). Others advise against changing the fluid for

SKILL DRILL 19-1 Checking Fluid Level and Inspecting Fluid Loss

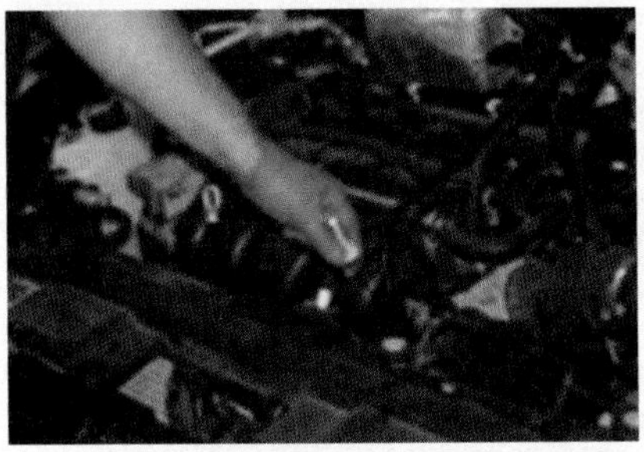

1. Look up the procedure for checking the transmission fluid level in the appropriate service information, and check the level.

2. Inspect the transmission for signs of leakage.

3. If the transmission has a large amount of transmission fluid or engine oil covering it, pour a leak detection dye in the transmission fluid, and use an ultraviolet light to look for fresh transmission fluid leaking.

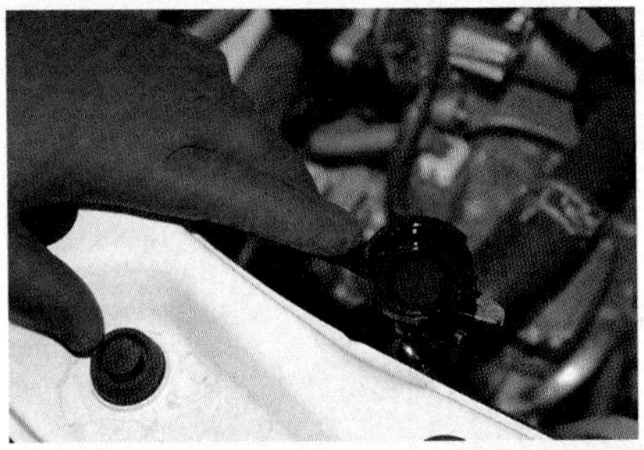

4. Remove the radiator cap and check for the presence of transmission fluid in the coolant.

the life of the vehicle because the transmission is sealed. You will find many recommendations in between these extremes.

As discussed in the Automatic Transmission Fundamentals chapter, it used to be possible to check the quality of transmission fluid by looking at its color and smelling it. If the fluid was a darkish red or had a burnt smell, it was time for fluid replacement. Some modern transmission fluids are a darker red when they are brand new and even have a slightly burnt smell to them. For this reason, it is important to check the fluid in a new way. Take a few drops of transmission fluid and place them on a clean paper towel. The fluid will disperse on the paper towel, but any contaminants will remain. If the fluid spot on the paper towel has a darker center, it is a sign that there are contaminants in the transmission fluid. This indicates that the fluid should be changed (**FIGURE 19-4**). If metal particles are found, further diagnosis is needed.

When changing the fluid and filter, there are a few things to be careful of:

- Don't lose any parts. On some transmissions, checkballs and springs can be held in place by the filter. Removing the filter frees these parts to fall on the floor or in the drain pan.
- Tighten any filter and pan bolts to the specified torque to avoid stripping the aluminum threads or breaking the bolts.
- Research the proper transmission fluid type and capacity (which is approximated for a filter change).

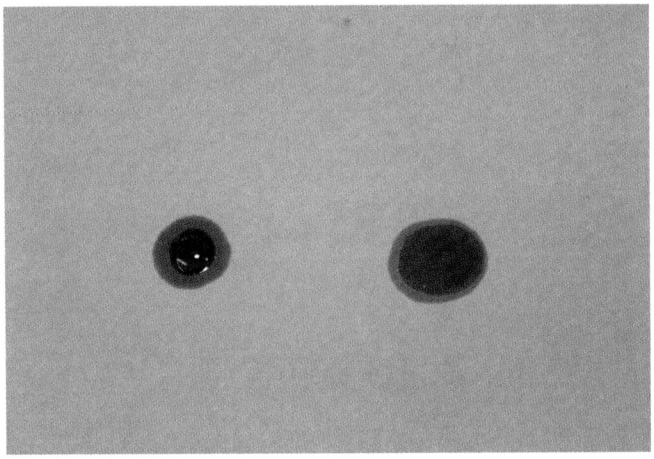

FIGURE 19-4 Place a drop of transmission fluid on a clean paper towel. If the fluid is dirty, the center of the drop will appear darker.

▶ TECHNICIAN TIP

Some manufacturers install a replaceable in-line filter in the cooler line. Be aware of that when you research the replacement procedure for the transmission filter. Also, some transmissions have external filters that look similar to an engine oil filter.

To replace the fluid and internal filter, follow the steps in **SKILL DRILL 19-2**.

SKILL DRILL 19-2 Draining and Replacing Fluid and Filter

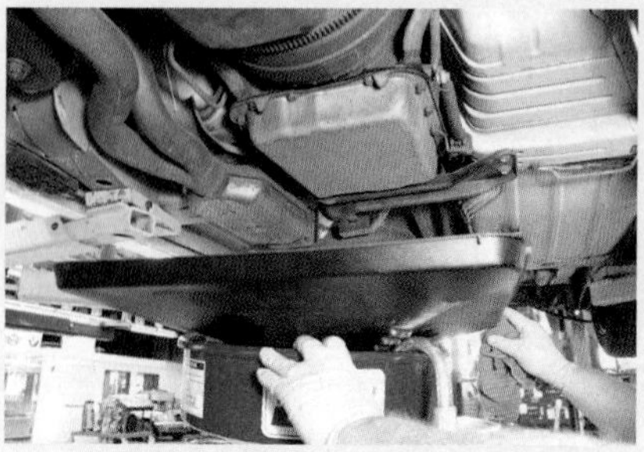

1. Safely raise and support the vehicle on the hoist. Place a drain pan with a large transmission drain funnel under the transmission pan. If the transmission has a drain plug, remove it, and place it to the side to prevent losing it. *Be careful: The transmission fluid will be hot.* If the transmission does not have a drain plug, loosen and remove all of the bolts except for two or three that are next to each other in a corner.

2. Hold the pan against the transmission with one hand while loosening the remaining bolts a few turns. Slowly allow the transmission pan to tilt downward so that transmission fluid flows into the drain pan. Once most of the fluid has been drained, remove the remaining bolts.

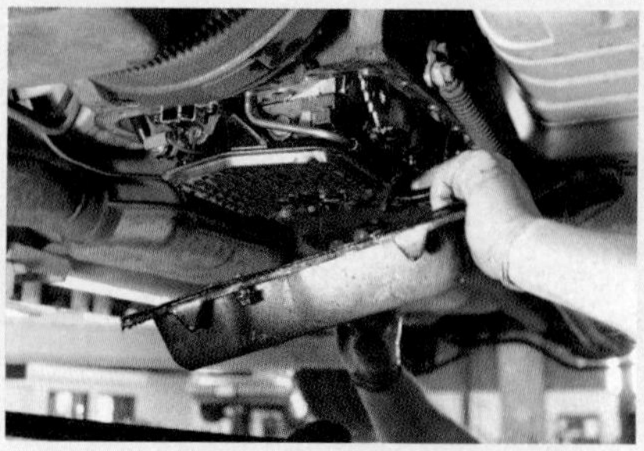

3. Lower the pan and inspect it for nonferrous metal and old clutch material. Metal is sparkly, and clutch material is a dark residue in the bottom of the pan. The presence of some clutch material is normal. Also, inspect the magnet in the pan (if equipped). A small amount of metal is normal.

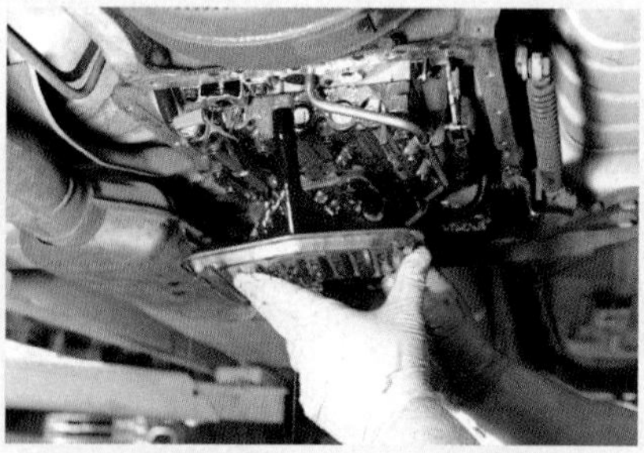

4. If equipped, remove any bolts or clips holding the transmission filter in place, and lower the filter.

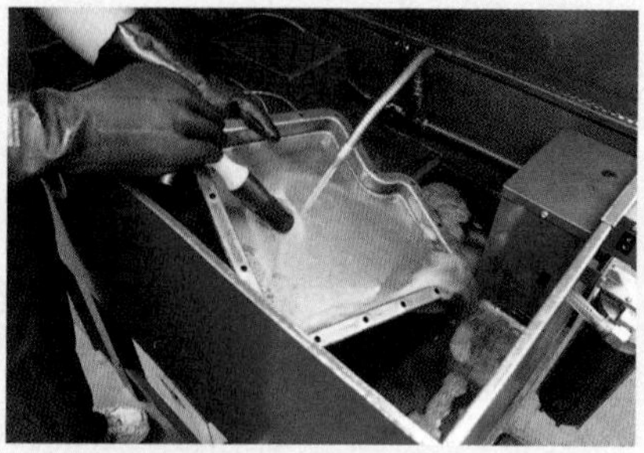

5. Clean the pan and the magnet thoroughly in a parts washer. Dry the pan with a lint-free shop towel or compressed air. Check that the sealing surface of the pan is flat. See your supervisor if it is not.

6. Compare the new filter and transmission gasket to the old one to make sure they are correct. If the filter kit came with a new filter gasket or grommet, install the new one to prevent air leaks and possible transmission damage.

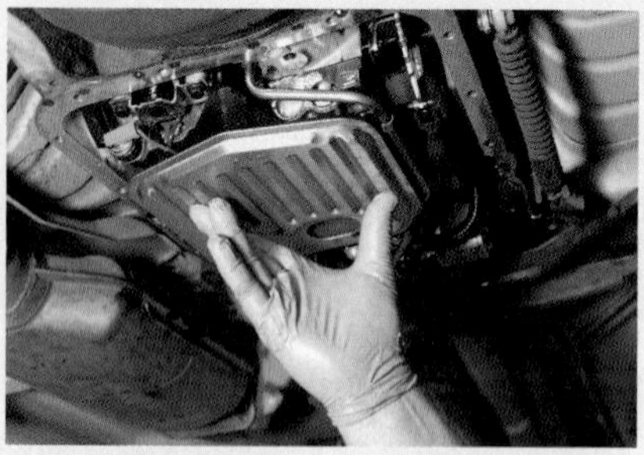

7. Install the new filter, and torque any retaining bolts to the manufacturer's specifications.

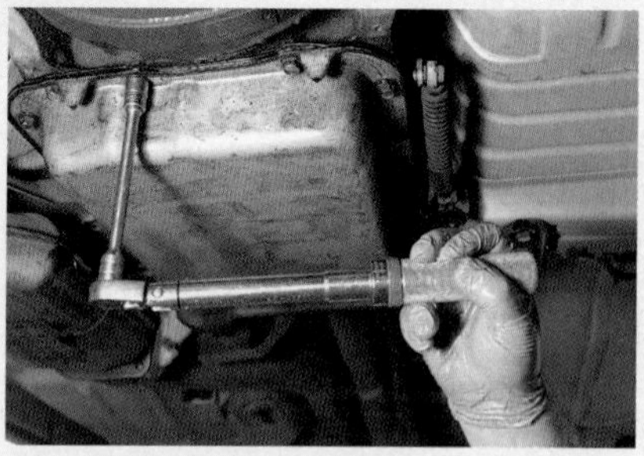

8. Put the new gasket on the transmission pan, and place the pan onto the transmission. Start all of the bolts before tightening any of them. Make sure the gasket remains in the correct place. Torque all of the bolts to specifications.

9. Lower the vehicle and install about 75% of the correct new fluid. Start the vehicle and check the fluid level. Add fluid as necessary to bring it to the bottom of the safe or add mark. While your foot is firmly depressing the brake pedal, place the gear selector in each of the gear ranges. Check the fluid level and top off as necessary. Raise the vehicle, and check for any leaks.

Inspecting and Cleaning the Transmission Cooler

If the transmission fluid was contaminated, you may need to clean and inspect the transmission fluid cooler and lines. If not cleaned, contaminated fluid will mix with the new fluid as soon as the vehicle is started. The transmission cooler and lines are cleaned by thoroughly flushing them out.

Some transmissions have a separate filter located in the cooler lines. These should be replaced, not cleaned. Also, be aware that some manufacturers install a check valve in one of the cooler lines. This is to prevent drainage of the fluid in the converter when the engine is off. This valve can catch debris, which can cause a restriction or even stick the check valve closed. If this happens, no transmission fluid will get to the planetary gears for lubrication, ruining them fairly quickly. There is debate about whether these check valves can be cleaned adequately, since the check valves prevent the lines from being reverse flushed. So, follow the manufacturer's recommendations.

Transmission coolers can be cleaned in one of two ways. The first method is to use a dedicated cleaning system with a pump that will pulse clean solvent through the transmission cooler and lines. Dirt and debris are removed and returned to the cleaning system with the solvent. The machine's filter will then remove the contaminants.

The second method involves using an aerosol flushing kit available at many parts stores. The aerosol can is a one-time use system. So, a new one is needed for every transmission. When performing either method, hook up to the transmission cooler so that the solvent will flow backward through the cooler. This helps dislodge any stuck debris. Cooler lines with check valves cannot be reverse flushed.

A transmission fluid cooler needs to be inspected. If it is leaking, it could cause the transmission to fail from low fluid level. If it is an internal leak, it could also cause antifreeze to be drawn into the cooler from the radiator. This would contaminate the transmission fluid. In the same way, transmission fluid will leak into the radiator. This causes the antifreeze to become contaminated with transmission fluid. It will also cause the transmission to suffer from low fluid level.

To inspect, leak test, and flush or replace transmission cooler lines and fittings, follow the steps in **SKILL DRILL 19-3**.

SKILL DRILL 19-3 Inspecting and Flushing Cooler Lines

1. Look up the recommended transmission cooler service method. Remove the fluid cooler lines from the transmission if the transmission is still in the vehicle.

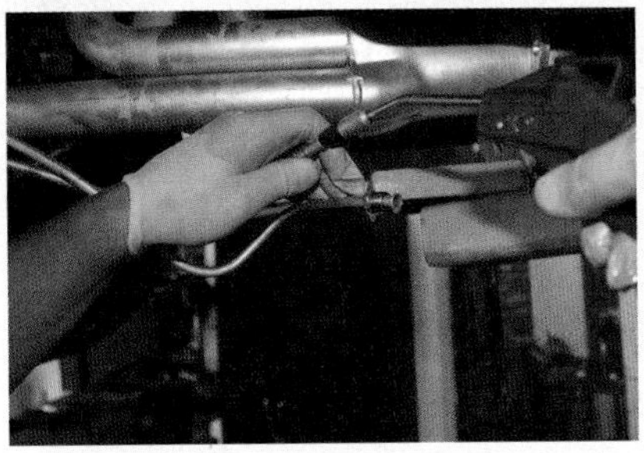

2. Using compressed air, blow into one cooler line while catching the residue in a container as it comes out the other line. Switch directions and repeat.

3. Install the cooler flush machine or aerosol can lines onto the transmission cooler lines so that the flow is in the reverse direction.

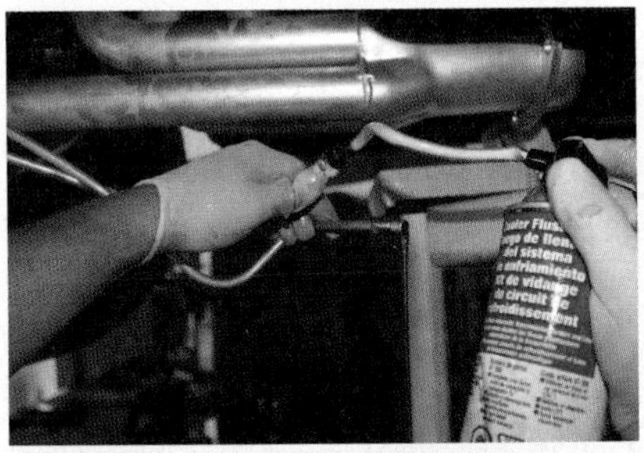

4. Start the flush machine or aerosol can, and allow it to run the recommended time. If necessary, switch directions on the lines so they can be flushed in the other direction.

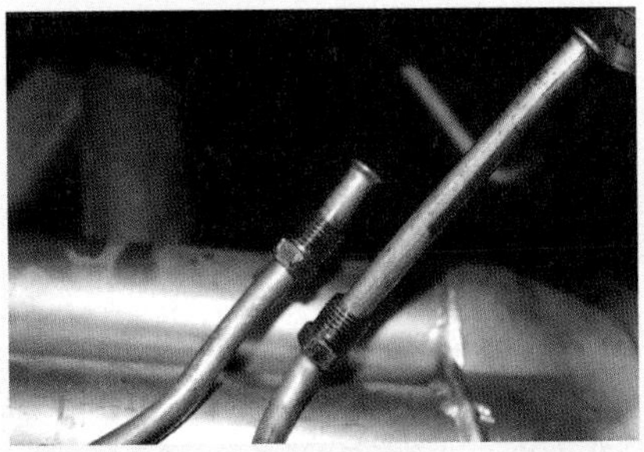

5. Remove the flush machine, and blow out the lines again so that no residue remains inside the lines.

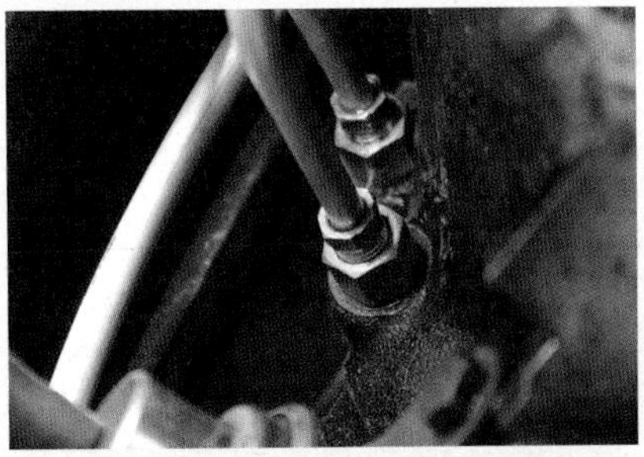

6. Reinstall the lines onto the transmission. After properly filling the transmission with fluid, start the vehicle and inspect the lines and fittings for any signs of leakage. Check inside the radiator for signs of the transmission cooler leaking into the radiator.

▶ In-Vehicle Transmission Repair

LO 19-03 Perform the in-vehicle transmission service tasks.

Many automatic transmission repairs can be completed without removing the transmission from the vehicle. This saves money since the transmission doesn't have to be removed. Examples of typical in-vehicle transmission repairs include the following:

- replacement of external gaskets and seals,
- replacement of vehicle speed sensor,
- replacement of shift solenoids,
- replacement of extension housing bushing, and powertrain mounts,
- repair of the valve body, adjustment of linkage, and
- adjustment of some servos and bands.

Inspecting and Replacing Gaskets and Seals

Transmissions have many external seals and gaskets that can be replaced without removing the transmission. Examples include the extension housing seal, the vehicle speed sensor seal, the pan gasket, and the extension housing gasket. A transmission fluid leak can result in failure of the transmission due to a low fluid level. As a technician, it is important to fix any external leaks on the transmission to prevent this failure. In the case of a leaky gasket (pan or extension housing), the leak may be caused by loose bolts. If so, retighten them to the specified torque. But *never* tighten them beyond specifications. If retightening them doesn't stop the leak, then replace the gasket.

Some seals, such as the extension housing seal, may require the replacement of the bushing behind the seal. With a worn bushing, the driveshaft may move up and down excessively. Flexing of the seal may stretch it out and allow transmission fluid to leak. Replacement of an extension housing bushing requires the removal of the driveshaft and the extension housing. The bushing can then be driven out using a bushing driver set.

Most seals on the outside of a transmission require specialized tools to be removed and replaced. Some of these specialty tools are universal types. Others are specific for a particular model of transmission.

To inspect for leakage and replace external seals, gaskets, and bushings, follow the steps in **SKILL DRILL 19-4**.

SKILL DRILL 19-4 Inspecting and replacing the extension housing bushing

1. Safely raise and support the vehicle on a hoist. Carefully inspect the front pump, output shaft, selector shaft seals, pan gasket, side pan gasket (if equipped), and extension housing gasket for leaks.

2. Inspect the extension housing bushing by moving the driveshaft up and down. If there is excessive movement in the driveshaft, the extension housing bushing must be replaced. Place a drain pan under the extension housing and remove the driveshaft from the vehicle.

3. Remove the extension housing from the vehicle. On some older vehicles, the seal and bushing can be replaced while it is still in the vehicle, using a specially designed puller. Remove the extension housing seal using the correct tool.

4. Use the correct-sized bushing driver to remove the extension housing bushing.

5. Use the correct-sized bushing driver to carefully drive the bushing into place. Take note of any lubrication holes that need to be lined up before installation.

6. Use the correct seal installation tool to install the new seal in the housing. Lubricate the edge of the seal with clean transmission fluid. Reinstall the extension housing and the driveshaft.

Scanning the Transmission Control Module/Powertrain Control Module

Some transmission faults will set a diagnostic trouble code (DTC). These can be a big help in determining the cause of the fault. A high-quality or factory scan tool can be an invaluable asset when diagnosing DTCs. A generic scan tool can read generic codes, which can be helpful. But a factory scan tool can also view factory codes and data that the generic tool may not be able to access.

Further, many factory scan tools give the technician **bidirectional control** of the transmission. This means the technician is able to command different solenoids and actuators on and off to check their operation. For example, the torque converter clutch (TCC) could be commanded on while the engine is idling in drive with the brakes on. This should kill the engine if the TCC solenoid is operating and the clutch is in good shape. Activating this solenoid and observing the reaction give you valuable information. It helps determine the next steps in the diagnostic process.

Any trouble codes should be researched in the appropriate service information. This will help you to find the correct diagnostic procedure. Following the diagnostic procedure step by step is critical in a successful repair of the vehicle. The diagnostic procedure often requires the use of a digital volt-ohmmeter (DVOM). It will aid you in the diagnosis of wiring, switches, solenoids, sensors, and actuators.

To retrieve transmission DTCs, follow the steps in **SKILL DRILL 19-5**.

Inspecting and Repairing Powertrain Mounts

The **powertrain mounts** hold the engine and transmission in the proper position in the vehicle. If they are allowed to move out of position, all of the related components go out of alignment. In this case, the shift linkage or throttle linkage, could become bound or out of place. With the shift linkage out of place, the gear selector indicator might indicate one gear, but the transmission is in another gear. This could cause an accident, especially if the driver expects the vehicle to be in neutral or reverse when it is actually in drive. Worn powertrain mounts can also cause drive axles to be out of alignment, causing a vibration. Worn powertrain mounts should always be replaced with the correct part for the vehicle.

▶ TECHNICIAN TIP

A technician was once called by a customer who was on a trip. He was towing his large boat, and was experiencing a strange phenomenon. It seemed that the vehicle had developed a mind of its own. As soon as the vehicle would start up a hill, the gas pedal would drop to the floor, and the vehicle would go to full throttle. If the driver turned the engine off at the top of the hill and restarted the engine, everything worked fine until the next hill. Once again, the gas pedal would drop to the floor. The technician walked the customer through the process of testing the powertrain mounts. One of the powertrain mounts was found to be torn. It allowed the engine to move and pulled the throttle linkage open. This made the pedal dropped to the floor. Replacing the motor mount solved the problem.

Most powertrain mounts are of the rubber style. They are molded from precisely engineered rubber, with specific shapes and voids (**FIGURE 19-5**). Rubber mounts are subject to cracks and tears that occur over time due to their constant flexing. Their life is also reduced if saturated with oil, power steering fluid, or gasoline. These chemicals soften the rubber.

Powertrain mounts can also be of the hydraulic style (**FIGURE 19-6**). Hydraulic mounts use silicone hydraulic fluid that is squeezed back and forth between two chambers, similar to the design of a shock absorber. They use a metered orifice between the chambers to control the flow of fluid, which dampens the engine pulsations.

Some hydraulic mounts use a computer-controlled electronic control valve to control the amount of fluid that can flow inside the mount. This is similar to a shock absorber.

SKILL DRILL 19-5 Scanning the TCM/PCM

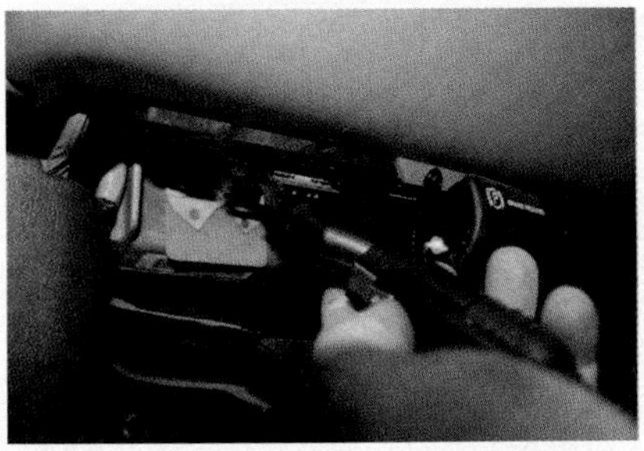

1. Install a scan tool onto the vehicle's data link connector (DLC). For the location of the DLC, consult the service information.

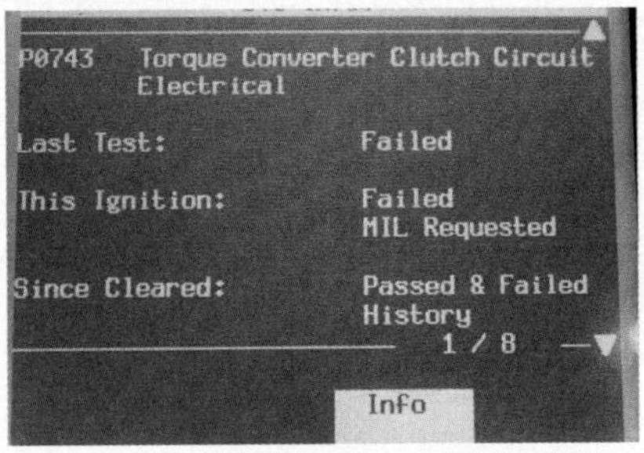

2. Retrieve any DTCs from the vehicle. Record the codes.

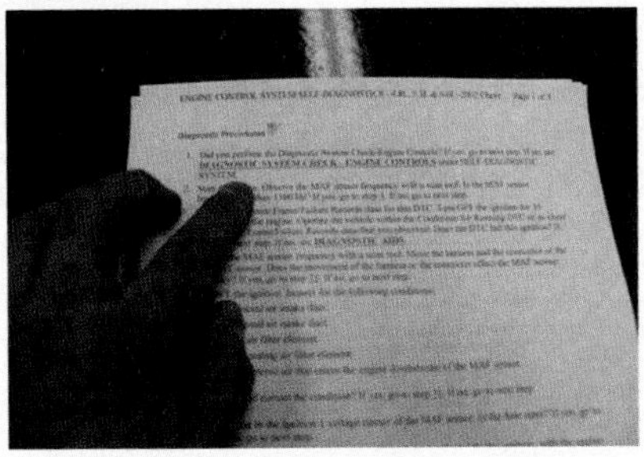

3. Using the service information, research the diagnostic procedure for any codes found. Follow the diagnostic procedure step by step until you have completed the diagnostic procedure and found the cause of the DTC.

FIGURE 19-5 Rubber style powertrain mounts.

FIGURE 19-6 Hydraulic-style powertrain mount.

The control valve opens an alternate passageway or varies the size of the orifice. This changes the rate of flow between the chambers. Thereby allowing the computer to control the amount of dampening that is needed for various operating conditions.

Hydraulic mounts should be inspected for leaks and excessive movement. The solenoid valve can also be tested for continuity. It can also be commanded to operate with a scan tool. Hydraulic-style powertrain mounts are typically much more expensive than rubber mounts. And may be available only through the manufacturer.

Often, in the event of a broken powertrain mount, customers complain of a loud thump. The noise typically occurs when they apply the accelerator in drive or reverse, or when they apply the brake pedal. This loud thump is caused by the engine and/or transmission being pulled up from the broken mount due to the engine's torque. And then another thump when it is set back down when the brake is applied.

Powertrain mounts wear out over time, so they need to be inspected periodically. Watch for leaks that can drip on mounts and soften the rubber. If found, it is wise to encourage customers to have leaks repaired before they damage the mounts.

To inspect, replace, and align powertrain mounts, follow the steps in **SKILL DRILL 19-6**.

Adjusting the Shift Linkage and Transmission Range Sensor

At times, it may be necessary to adjust the shift linkage along with the transmission range sensor. If the linkage is not adjusted properly, the transmission position indicator (PRNDL indicator) will show the wrong gear. Also, the backup, or reverse, lights may not work.

If the neutral safety switch is out of adjustment the engine won't start in park or neutral. But it may start in one of the other gears, which is very dangerous. Make sure the brake pedal is firmly applied when checking that the vehicle starts only in the park and neutral positions.

To inspect, adjust, and replace the manual valve shift linkage, transmission range sensor/switch, and park/neutral position switch, follow the steps in **SKILL DRILL 19-7**.

SKILL DRILL 19-6 Inspecting, Replacing, and Aligning Powertrain Mounts

1. Look up the correct procedure for checking and replacing powertrain mounts in the service information. Safely raise and support the vehicle on a hoist to inspect the powertrain mounts under the vehicle.

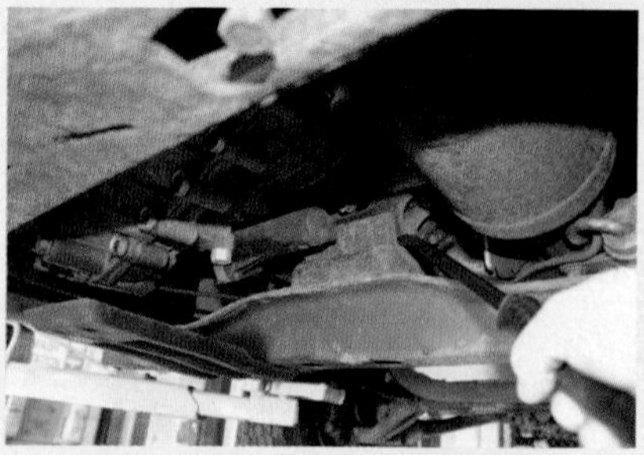

2. Use a pry bar to carefully push up on the engine and transmission while watching the powertrain mounts. If the rubber section of the powertrain mount separates from the metal bracket, or if the rubber is torn, replace the mount.

3. Lower the vehicle, apply the brake, and start the engine. Place the vehicle into gear, and apply the throttle slowly. Watch for excessive engine movement on the mounts. Repeat in reverse to check the opposite mounts.

SKILL DRILL 19-6 Inspecting, Replacing, and Aligning Powertrain Mounts (Continued)

4. Use an engine support fixture, engine hoist, or transmission jack to raise the engine or transmission just far enough that the weight is off the powertrain mount. Be *very* careful not to cause the vehicle to shift on the hoist!

5. Remove the bolts securing the mount to the transmission or engine, and then the bolts securing the mount to the frame of the vehicle.

6. Remove the old mount, and compare it with the new mount.

7. Place the new mount in the correct position, and lower the jack slightly. Be careful to keep your fingers away from any pinch points. Reinstall the bolts and torque them to specifications.

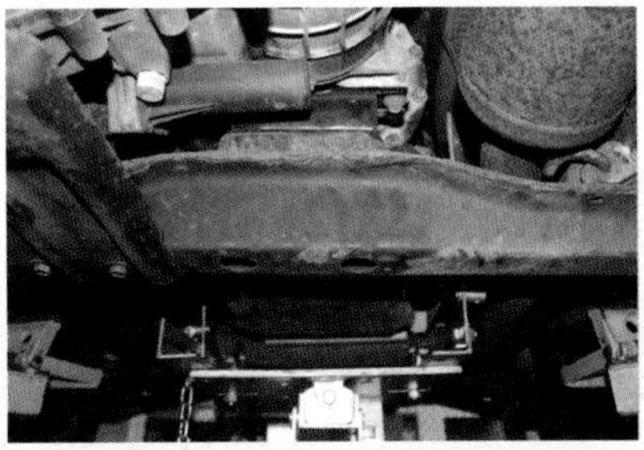

8. Lower the engine or transmission back down. Reinstall all components that were removed to access the powertrain mount.

SKILL DRILL 19-7 Inspecting, Adjusting, and Replacing Shift Linkage

1. Look up the proper service procedure in the appropriate service information. Place the gear selector in the park position.

2. Safely raise and support the vehicle on a hoist (if necessary to access the transmission range switch and manual valve linkage). Disconnect the shift linkage from the transmission.

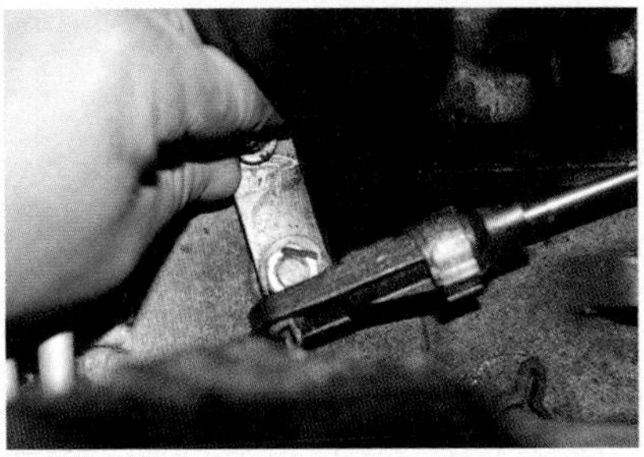

3. Check that the manual lever is in the park position. The lever should snap into position.

SKILL DRILL 19-7 Inspecting, Adjusting, and Replacing Shift Linkage (Continued)

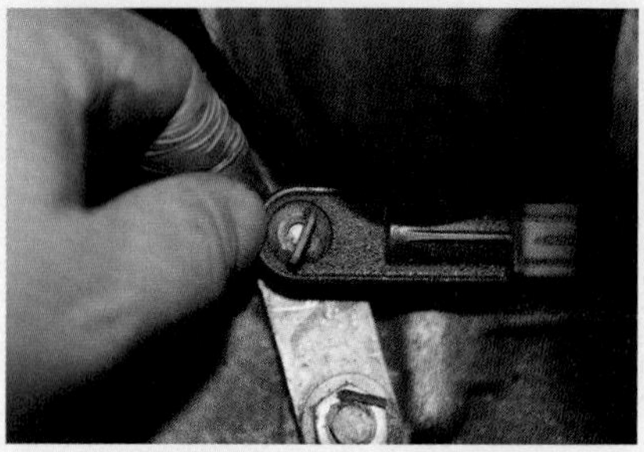

4. The shift linkage should fit right onto the manual valve with no pulling on the linkage or the manual valve.

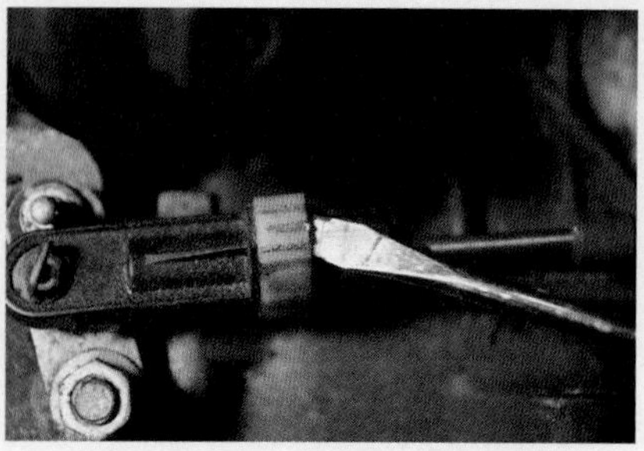

5. If the linkage does not line up, loosen the adjustment on the shift linkage and adjust the linkage so that it will install properly on the manual valve. Tighten the adjustment on the shift linkage.

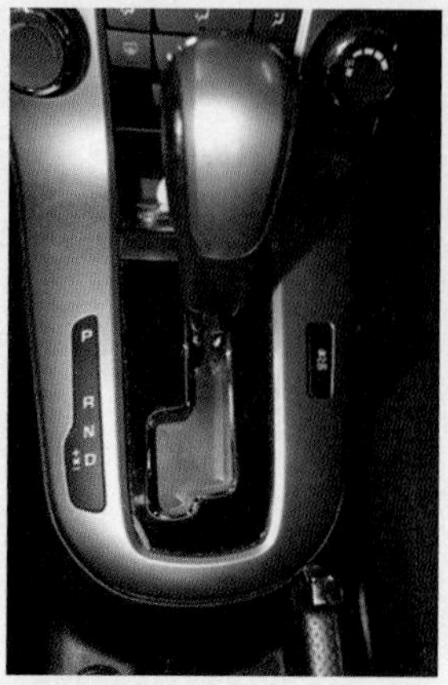

6. Double-check that the PRNDL indicator still indicates the vehicle is in park.

7. Use an ohmmeter to check that the neutral safety switch/transmission range switch has continuity on the correct terminals.

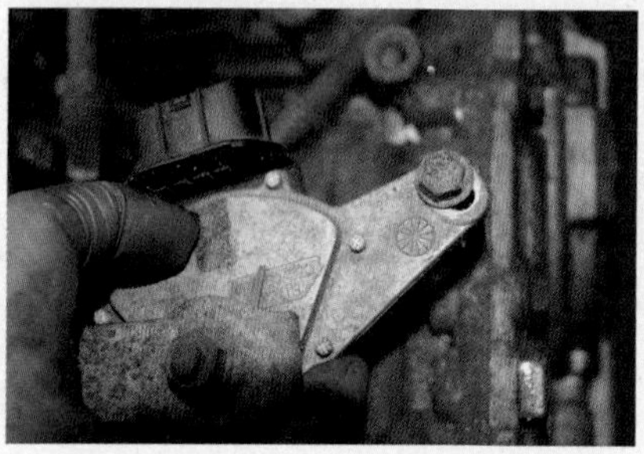

8. If not, loosen the switch and adjust its position, if possible. If continuity is never obtained or is obtained in every gear, replace the switch.

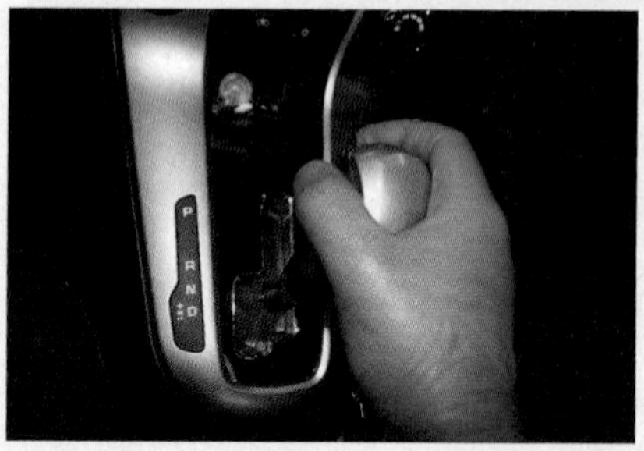

9. Run the shifter through all of the gear ranges, checking for proper operation.

▶ Wrap-Up

Ready for Review

- A typical transmission dipstick requires 1 pint (0.47 liters) to bring the fluid level from the bottom of the crosshatched area to the full mark. A leak detection dye in the transmission fluid can help detect leaks.
  - Transmissions can leak from: transmission pan, area around the entrance of the filler tube to the transmission, extension housing gasket, output shaft seals, speedometer/vehicle speed sensor seal, selector shaft seal, area around the electrical connectors that go into the transmission case, front pump seals, fluid cooler lines, and fittings.
- A few drops of transmission can be placed on a clean paper towel to check for contaminants. When changing the fluid and filter: don't lose any parts, tighten any filter and pan bolts to the specified torque, and research the proper transmission fluid type and capacity.
- Typical in-vehicle transmission repairs include: replacement of external gaskets and seals, replacement of vehicle speed sensor, replacement of shift solenoids, replacement of extension housing bushing, and powertrain mounts, repair of the valve body, adjustment of linkage, and adjustment of some servos and bands.

Key Terms

bidirectional control The ability to command different solenoids and actuators on and off to check their operation.

powertrain mount A rubber or metal bracket used to secure the engine and transmission into the vehicle. Some vehicles use hydraulic or electrohydraulic powertrain mounts.

Review Questions

1. How often should a transmission with a dipstick have its fluid level checked?
   a. Every day
   b. Every fuel tank fill-up
   c. Every oil change
   d. Never
2. A transmission without a dipstick:
   a. Cannot be refilled
   b. May require a special tool or procedure to fill
   c. Should not have the fluid level checked
   d. Are overfilled to allow some seepage over time
3. All of these are common external transmission fluid leak points EXCEPT:
   a. Transmission pan gasket
   b. Front pump seal
   c. Torque converter clutch seal
   d. Output shaft seal
4. ____________ can be used to find transmission fluid leaks.
   a. Fluid dye
   b. A smoke machine
   c. A digital multimeter
   d. A stethoscope
5. When inspecting the transmission pan magnet, all the following conditions are normal EXCEPT:
   a. Fine particles that look and feel smooth
   b. A small amount of metal filings
   c. A small amount of clutch material
   d. A few roller bearings
6. How often should a sealed transmission have its fluid and filter changed?
   a. Every 25,000 miles
   b. Every 50,000 miles
   c. Every 90,000 miles
   d. Never
7. Which of these repairs commonly requires transmission removal?
   a. Replacement of the torque converter
   b. Replacement of shift solenoids
   c. Valve body removal and repair
   d. Clutch band adjustment
8. What transmission concern can be caused by failed powertrain mounts?
   a. Leaking extension housing gasket
   b. Burned fluid
   c. The transmission is in a different gear than indicated
   d. A stuck cooler line check valve
9. Electronic powertrain mounts change tension by:
   a. Thickening the rubber material between the upper and lower plates
   b. Adjusting a torsion bar back and forth
   c. Moving the location of the fulcrum
   d. Changing the size of the orifice to allow more or less fluid flow
10. Inspect a powertrain mount by:
   a. Prying up and down with a pry bar
   b. Spraying it with white lithium grease
   c. Measuring its runout with a dial indicator
   d. Check its resistance with an ohm meter

ASE Technician A/Technician B Style Questions

1. Technician A says that the engine should be idling when checking transmission fluid. Technician B says to add a quart of fluid when the level is at the low line on the dipstick. Who is correct?
   a. Technician A only
   b. Technician B only
   c. Both A and B
   d. Neither A nor B
2. Technician A says that milky pink transmission fluid could indicate a leaking transmission fluid cooler in the radiator. Technician B says that a faulty vacuum modulator can cause transmission fluid to be burned in the engine. Who is correct?
   a. Technician A only
   b. Technician B only
   c. Both A and B
   d. Neither A nor B
3. Technician A says that transmission fluid should be serviced less often in dusty conditions. Technician B says that fluid condition is checked on a dirty shop rag. Who is correct?
   a. Technician A only
   b. Technician B only
   c. Both A and B
   d. Neither A nor B
4. While filling a transmission after a service, technician A says you need to start the engine before adding any transmission fluid. Technician B says that you need to top off the fluid after operating the transmission through all the gear ranges. Who is correct?
   a. Technician A only
   b. Technician B only
   c. Both A and B
   d. Neither A nor B
5. Technician A says that small parts may fall out when the transmission filter is removed. Technician B says that transmission pan bolt hole threads are commonly damaged by over-torquing. Who is correct?
   a. Technician A only
   b. Technician B only
   c. Both A and B
   d. Neither A nor B
6. Technician A says to clean the transmission fluid cooler if the fluid is contaminated. Technician B says that a restricted

in-line transmission fluid cooler filter can damage the planetary gears. Who is correct?
a. Technician A only
b. Technician B only
c. Both A and B
d. Neither A nor B

7. Technician A says that you should torque the transmission pan bolts until the gasket stops leaking. Technician B says that some extension housings have a bushing that can be replaced. Who is correct?
a. Technician A only
b. Technician B only
c. Both A and B
d. Neither A nor B

8. Technician A says that there is not much difference between a factory scan tool and a generic code reader. Technician B says bidirectional control means reprograming a powertrain control module (PCM). Who is correct?
a. Technician A only
b. Technician B only
c. Both A and B
d. Neither A nor B

9. Technician A says that trouble codes tell you what part to replace. Technician B says that trouble codes can help you find the correct diagnostic procedure. Who is correct?
a. Technician A only
b. Technician B only
c. Both A and B
d. Neither A nor B

10. Technician A says that the life of rubber powertrain mounts can be reduced by oil contamination. Technician B says that some powertrain mounts can leak a silicone-based fluid. Who is correct?
a. Technician A only
b. Technician B only
c. Both A and B
d. Neither A nor B

CHAPTER 20

Hybrid and Continuously Variable Transmissions

Learning Objectives

- **LO 20-01** Describe hybrid vehicle functions.
- **LO 20-02** Describe various hybrid vehicle models.
- **LO 20-03** Describe CVT operation.

ASE Education Foundation Tasks

See Appendix A to view the 2017 ASE Education Foundation Automobile Accreditation Task List Correlation Guide.

▶ Introduction

There is a growing need for vehicles with increased fuel economy and lower carbon dioxide emissions. Manufacturers have been working to meet these goals by creating more efficient powertrains. This is where the hybrid vehicles come in.

Hybrid vehicles have been at the forefront of this development for more than 20 years. And continuously variable transmissions (CVT) have been around even longer (**FIGURE 20-1**). These vehicles are showing up in repair facilities in greater numbers every day. It is important that technicians become familiar with their design and operation. This is because of their unique safety and operational characteristics.

▶ Hybrid Drive Systems

LO 20-01 Describe hybrid vehicle functions.

A **hybrid drive system** is defined as a system that uses two or more power sources. A common example is an internal combustion engine (ICE) and an electric motor. These are used to move the vehicle down the road (**FIGURE 20-2**). There are many different styles and systems of hybrid vehicles being sold today. We cover the operation of several of the more common hybrid vehicles on the market at the time of this writing. These are primarily gasoline–electric hybrids. In the future, we could expect various other hybrid arrangements, such as diesel–electric hybrids and fuel cell–electric hybrids.

Before exploring how a hybrid operates, let's discuss the functions many hybrid drive systems have. Hybrid vehicles have several different functions that separate them from conventional vehicles. The most common ones are:

- idle stop,
- torque smoothing,
- regenerative braking,
- torque assist, and
- electric only propulsion.

Not all hybrids have every one of these functions, but often they have most of them. Hybrids may be considered mild hybrids or full hybrids based on the number of hybrid functions they use. Knowing what functions a hybrid vehicle is equipped with and how the vehicle operates is necessary for two reasons. First, so you will know whether it is operating correctly. And second, so you will know what precautions you need to take in order to work on it safely.

Idle Stop

Idle stop is one of the most common functions of a hybrid vehicle. When used with an integrated starter/generator, the

FIGURE 20-1 Inside view of a continuously variable transmission.

FIGURE 20-2 Inside view of both electric motors from a power-splitting transmission out of a Toyota Prius.

You Are the Automotive Technician

Recently, the Express Delivery company you work for has converted its fleet to hybrid vehicles. Hybrid vehicles have lower carbon dioxide emissions, increased fuel economy, and a smaller global footprint. These vehicles often use very high-voltage batteries that present a severe shock hazard. This can easily lead to death if handled improperly. Your supervisor has asked you to complete a scheduled maintenance service on one of the hybrid vehicles.

1. Why is it important to fully understand the hybrid system and its operation before attempting any repairs on a hybrid vehicle?
2. Describe "regeneration."
3. Describe the operation of a variable-diameter pulley CVT.
4. When checking a rear backup bulb, why would it be unsafe to turn the ignition key on, place in "R," and then go check to see if the bulb is working?

vehicle qualifies for mild hybrid status. When the driver stops the vehicle, such as at a stoplight, the powertrain control module (PCM) shuts off the engine. Shutting off the engine reduces fuel consumption and carbon dioxide emissions. By using high-voltage electric motors, the PCM is able to crank the vehicle over very quickly and smoothly. An almost instant start is created as soon as the driver presses down on the accelerator, or lets off the brake. This cranking speed can be as high as 1000 rpm.

Stopping the engine at stoplights or when idling creates several problems that need to be addressed. Stopping the engine stops the operation of the heating and cooling systems. In a conventional vehicle, the water pump and air-conditioning compressor are driven by an engine belt. This is not possible on a vehicle with the idle stop function (**FIGURE 20-3**).

Many idle stop vehicles have a small, electric hydraulic pump in the transmission. This hydraulic pump prevents a delay in the engagement of the transmission when the engine is restarted.

Refer to the Principles of Heating and Air-Conditioning Systems chapter for more information on hybrid heating and cooling systems.

Torque Smoothing

Torque smoothing refers to the ability of the hybrid vehicle to help smooth out the power pulses of the ICE. This creates a flatter torque output curve. The system uses an electric motor to provide the torque smoothing function. This feature is especially helpful on small three- and four-cylinder engines. These engines tend to produce stronger engine pulsations at lower engine speeds.

When the ICE is on a compression stroke, the electric motor can be used to increase the amount of torque (**FIGURE 20-4**). As the ICE is on a power stroke, the electric motor creates less torque. This smoothes out the variations in crankshaft speed during each revolution. As the ICE's rpm increases, the torque smoothing pulses are needed less and less.

Regenerative Braking

Accelerating a vehicle from a stoplight requires a large amount of energy. Yet, when a conventional vehicle is being stopped, most of the kinetic energy of the vehicle's movement is converted into

FIGURE 20-3 A picture of the driver information center on a Toyota Prius, showing the vehicle in an idle stop situation. The engine is not running in this mode.

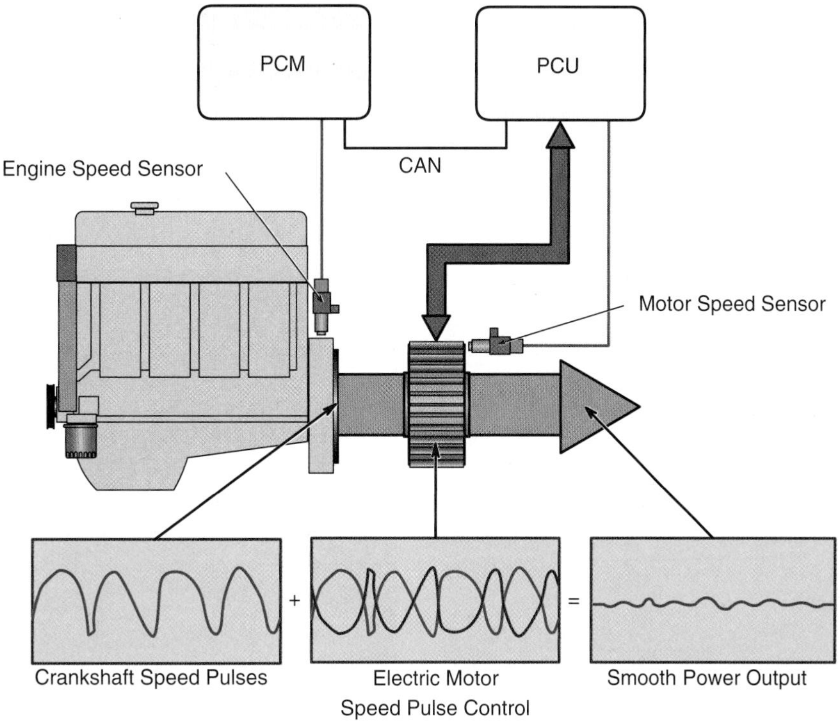

FIGURE 20-4 Torque smoothing uses the electric motor to smooth out the compression and power pulses.

heat. The friction of the brake pads against the brake rotors creates the heat. Therefore, all of this braking energy is wasted on a conventional vehicle. Most hybrid vehicles use **regenerative braking**. It helps to ensure that some of the energy that was used to accelerate the vehicle is recaptured during braking.

On a hybrid vehicle, when the driver initiates a stop, the electric motor becomes a generator. The kinetic energy of the vehicle's movement is used to turn the generator and create electricity. The electricity is used to charge the high-voltage battery. The harder the driver steps on the brake pedal, the more electricity is generated. It is important to remember that the more electricity that a generator is required to produce, the harder it is to turn the generator. This action causes the vehicle to slow down and eventually stop. If needed, the conventional braking system can provide additional braking (**FIGURE 20-5**).

Regenerative brakes can only develop a certain amount of stopping power. If the driver needs more braking power than the regenerative brakes can supply, mechanical brakes are available as backup. However, they cannot recapture any of the kinetic energy. Judicious use of the brakes on a hybrid vehicle results in the best fuel economy. This is because the mechanical brakes are not even applied.

Continuous driving does not allow the regenerative braking system to recharge the battery. For example, being on the interstate. It is during stop-and-go driving that hybrids benefit most from this system. In fact, hybrid vehicles often achieve significantly higher gas mileage (mpg) during city driving than on the highway. This means that regenerative braking systems are ideal for vehicles that operate in stop-and-go conditions. Some examples are delivery vehicles, taxis, and city transit buses.

Torque Assist

An electric motor is capable of creating its maximum torque as soon as it begins spinning. This torque is ideal to help propel the vehicle from a stop. The electric motor creates **torque assist** for the ICE, which creates its maximum torque much higher in the rpm range. By using an electric motor with an ICE, the overall displacement of the ICE can be reduced. For example, a vehicle that needed a 2.0-liter engine to operate adequately might only need a 1.4-liter engine when it is combined with the electric motor. This reduction in size of the ICE results in a fuel savings in both city driving and on the highway (**FIGURE 20-6**).

Electric-Only Propulsion

On many full hybrids, the vehicle can operate at low speeds using the electric motor only. This is especially helpful in slow-moving or stop-and-go traffic. It greatly lowers the emissions output of the vehicle. This cuts down the amount of air pollution that builds up around cities. When a driver presses on the accelerator pedal, the PCM commands the high-voltage battery pack to supply electricity to the electric motor. It does so until the vehicle reaches speeds up to approximately 40 mph (64 kph) (depending on the model). At that point, the PCM starts the ICE to continue accelerating the vehicle (**FIGURE 20-7**).

FIGURE 20-6 A picture of the driver information center from a Toyota Prius showing the ICE and the electric motor being used to propel the vehicle (torque assist).

FIGURE 20-5 The driver information screen from a Toyota Prius, showing power flow from the electric motor to the battery (regeneration).

FIGURE 20-7 A driver information center on a Toyota Prius, showing the electric motor alone propelling the vehicle.

SAFETY TIP

Extreme caution should be exercised when working around hybrid vehicles. These vehicles often use very high voltages that can cause serious harm or death. Only begin working on a hybrid vehicle after you have had the proper safety training. This will help you thoroughly understand the system and its operation. It will also orient you to the proper safety equipment. Always check with the appropriate service information before working on a hybrid vehicle. In addition to the high voltages, many systems have very strong permanent magnets. These magnets attract any ferrous metal that gets close to them. Your fingers or hand can become pinched between the part and the magnet. The magnets often cannot be removed without a special tool.

▶ Hybrid Electric Vehicle Models

LO 20-02 Describe various hybrid vehicle models.

Currently, hybrid vehicles are available from almost every vehicle manufacturer. And new models are being added every year as the demand for more fuel-efficient vehicles increases (**FIGURE 20-8**). In this section, we cover some of the more common hybrid vehicle types and models available.

Belt Alternator Starter

Belt alternator starter (BAS) vehicles have been produced primarily by General Motors (**FIGURE 20-9**). The BAS unit is a belt-driven alternator/starter motor. Conventional starter motors and alternators operate on nominal 12 volts. The BAS system uses a 42-volt battery.

When the engine is running, the belt spins the BAS. Since it is an alternator, it produces electricity, like a conventional alternator. The electricity charges the high-voltage battery. When the engine is stopped, the BAS can act as a starter motor. But because it operates on high voltage, it can crank the engine over very quickly through the belt drive. This design allows the vehicle to use the fuel savings of the idle stop function.

As the vehicle is slowing down, the BAS is used for regenerative braking. It can produce a large amount of current to charge the battery while also slowing the vehicle down. This allows the vehicle to capture some of the fuel savings of regenerative braking. Note that this system is not designed to power the vehicle down the road independent of the ICE.

The transaxle used in vehicles with the BAS systems is similar to that used in non-hybrid vehicles. The only major difference, other than PCM programming, is the addition of an electrically operated hydraulic pump. This pump allows a smooth acceleration as the engine restarts after an idle stop. Without this electric pump, there would be a momentary lag as the transaxle reengages after the engine was restarted.

Honda Integrated Motor Assist

Early Honda hybrids used a system called the **integrated motor assist (IMA)**. The IMA system uses a thin electric motor in place of a conventional flywheel or flexplate. This electric motor is used to supplement the gasoline engine's power when accelerating. The Honda IMA system is classified as a parallel hybrid. This is because the electric motor is operating at the same time as the gasoline engine.

Currently, Honda IMA hybrids cannot drive using the electric motor only because the IMA is directly coupled to the gasoline engine. Therefore, when the motor is operating, the gasoline engine must be spinning. The Honda IMA system uses between 144 and 158 volts, depending on the specific model.

The IMA system has regeneration capability. So, the electric motor is used to charge the high-voltage battery pack when the vehicle is decelerating. This converts some of the vehicle's kinetic energy into electrical energy, rather than wasting the energy as heat. The electric motor is also used to rapidly restart the engine after the driver releases the brake pedal.

The Honda IMA system also has a conventional 12-volt starter. It is used to initially start the vehicle when it is cold out, or when the high-voltage battery is not charged. The transaxle used in combination with the IMA system can be a:

- conventional automatic transmission,
- a manual transmission, or
- more typically, a CVT.

FIGURE 20-8 Toyota Prius is the most common hybrid vehicle.

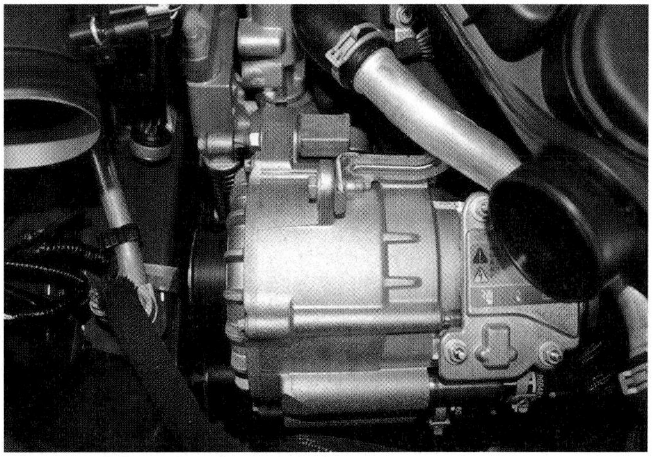

FIGURE 20-9 A BAS system installed on a Chevrolet Malibu.

All three types of transmissions have been modified so they can fit the IMA assembly. The IMA assembly fits between the engine and the transmission. The automatic transmission uses an auxiliary hydraulic pump to keep the clutches engaged when the engine is off. Other transaxle modifications include a stronger lock-up torque converter. This allows the drive wheels to power the IMA assembly during regeneration.Finally, gear ratios are designed to take advantage of the operation of the IMA assembly. This results in higher fuel savings and better performance (**FIGURE 20-10**).

FIGURE 20-10 An electric motor from a Honda IMA system.

Honda Intelligent Multi-Mode Drive (i-MMD) Two-Motor Hybrid Powertrain

The Honda two-motor hybrid system uses two electric motors (propulsion motor and generator motor) and an ICE. It is considered a full hybrid that is classified as a series-parallel hybrid. Either the gasoline ICE or the electric propulsion motor can propel the vehicle. Additionally, both can be used together, either in series or parallel.

One especially unique feature of this powertrain is that there is no real transmission. It is a direct drive powertrain with no shifts. Nor does it have a continuously variable geartrain. It utilizes a single-speed gearbox, where both power sources come together to drive the vehicle (**FIGURE 20-11**). To do that, the propulsion motor is permanently coupled to the direct drive system. But the ICE is only coupled to the drive system by a lock-up clutch. The generator motor is permanently coupled to the ICE. It acts as a generator used to both charge the high-voltage battery and power the propulsion motor. It also acts as a starter motor for the ICE during idle/stop operation.

Operation at low vehicle speeds is provided entirely by the propulsion motor. It is powered electrically by either the high-voltage battery, or power from the generator motor. It can be powered by both, if needed. At slow vehicle speeds, the ICE will operate as needed to charge the high-voltage battery and provide power to the propulsion motor. Above approximately 40 mph (64 kph), the ICE is more efficient. The generator motor

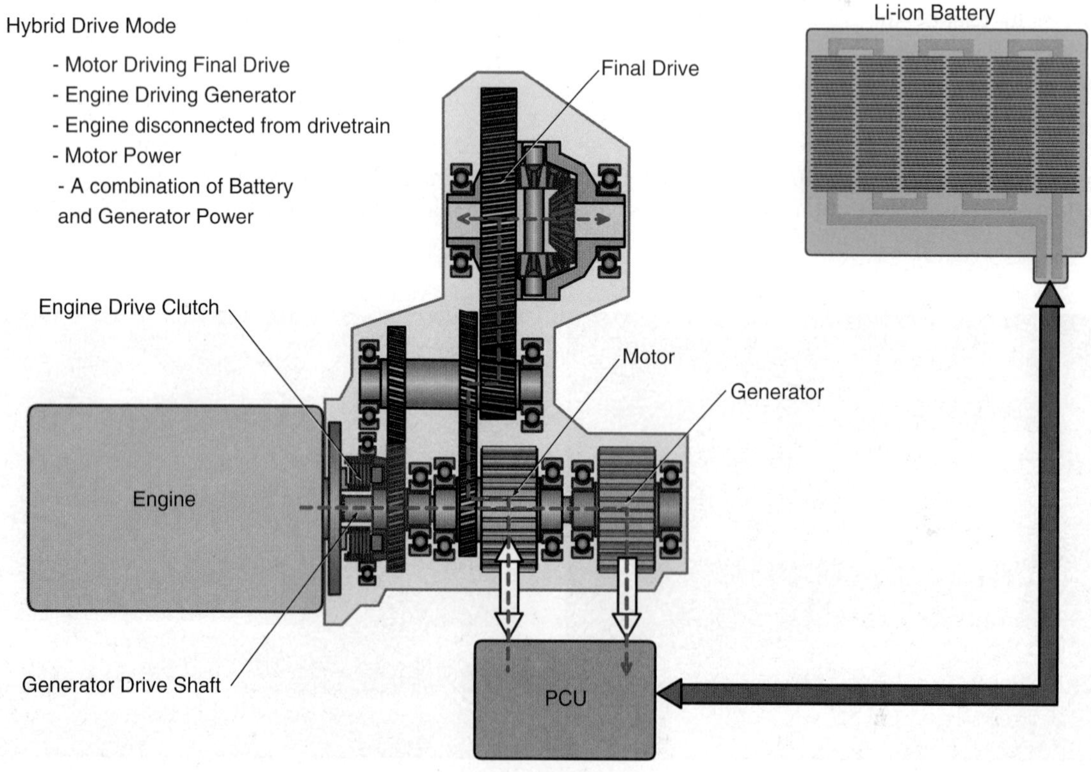

FIGURE 20-11 Honda two-motor hybrid powertrain schematic.

spins the ICE up to speed, and it is coupled to the drive system through a lock-up clutch. Above this speed, the ICE can propel the vehicle by itself. It will also power the generator motor to charge the battery or power the propulsion motor. The control system determines what is needed at the time.

During deceleration, the propulsion motor is turned by the wheels and regenerates power back to the battery. This recaptures braking energy that would otherwise be wasted. Reverse is obtained simply by reversing the direction of the propulsion motor. It then drives the vehicle in reverse. This reduces the need for additional parts and complexity.

Toyota and Lexus Hybrids

Toyota and Lexus hybrids use a power-splitting device that is considered a type of CVT. These hybrids use voltages from 200 to 650 volts, depending on the application. The Toyota and Lexus hybrids are full hybrids that can be classified as a series-parallel hybrid. Either the gasoline ICE or both electric motor/generators can propel the vehicle, or they all can be used together.

SAFETY TIP

It is very important that you follow the manufacturer's procedure when testing the vehicle. This is because the vehicle can be driven by the electric motor only. Also, the ICE can be started by the PCM at any time the system is powered up. This situation could be illustrated by a simple backup bulb inspection. You would normally turn the ignition switch to the run position, place the gear selector in "R," and then look at the backup bulbs. Unfortunately, with a hybrid system, the vehicle could start on its own. If it did so, it could travel across the shop, putting others, and the vehicle, in danger. Follow the manufacturer's procedures exactly when working on a hybrid vehicle.

The transaxle contains:

- two electric motor/generators,
- a final drive gear set,
- a differential,
- and a planetary gear set.

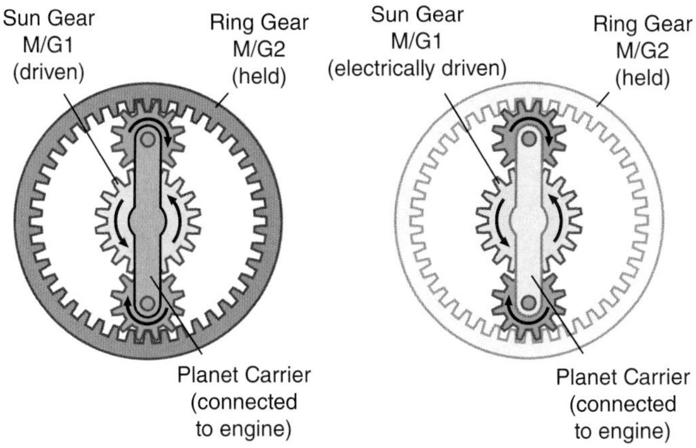

FIGURE 20-12 The planetary operation as the engine is started.

Each of the electric motor/generators and the ICE are connected to a separate part of the planetary gear set, as shown in (**FIGURE 20-12**).

The ring gear of the planetary gear set is connected to the final drive. Some use a chain, while others use gears (**FIGURE 20-13**). The ring gear is also directly connected to motor/generator 2 (M/G2). M/G2 is used to propel the vehicle. The sun gear of the planetary gear set is attached to the motor/generator 1 (M/G1). M/G1 is used as a generator to charge the high-voltage battery pack and to crank the engine over. The planet carrier of the planetary gear set is attached to the ICE.

When starting the gasoline engine, M/G1 spins the planetary carrier, which is attached to the ICE. With the vehicle stopped, M/G2 is basically locked and prevented from turning. These two conditions allow the engine to be cranked over.

When the engine starts, M/G1 begins spinning. M/G1 switches to generator mode and begins charging the high-voltage battery pack. M/G2 then begins to spin, and the power from the ICE is added to M/G2 to spin the ring gear and propel the vehicle (**FIGURE 20-14**).

Toyota and Lexus hybrids are able to propel the vehicle in electric-only mode at low speeds. During electric-only mode, the ICE is shut down. This effectively holds the planet carrier

FIGURE 20-13 The ring gear is attached to MG2 and the drive chain. The sun gear is attached to MG1.

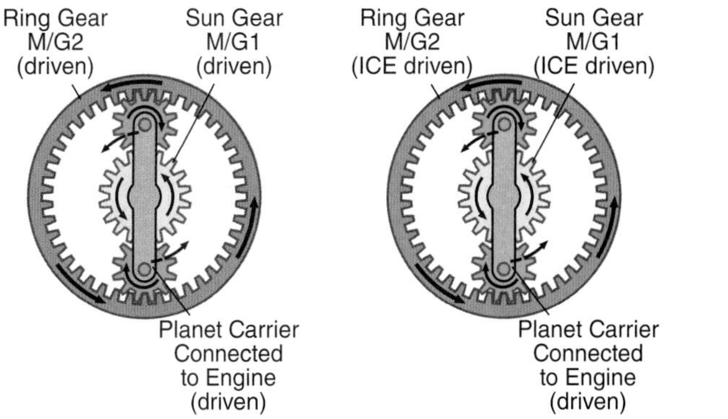

FIGURE 20-14 The planetary operation as the vehicle is moving and M/G2 and the ICE are moving the vehicle.

from spinning. High voltage is supplied to M/G2, which spins the ring gear and drives the wheels. M/G1 is allowed to spin freely (**FIGURE 20-15**). In reverse, the current flow to M/G2 is reversed and causes the motor to spin in the opposite direction. In reverse, the ICE is not used to propel the vehicle.

During deceleration at low speeds, the engine is shut off to hold the carrier. M/G1 is allowed to freewheel while M/G2 is driven by the wheels. M/G2 is switched to generator mode, and the power generated is used to recharge the high-voltage battery pack (**FIGURE 20-16**).

The vehicle's PCM controls the amount of current produced from M/G2 during deceleration to prevent wheel lock-up and skidding. During electric-only operation, the PCM monitors the battery voltage and restarts the ICE as needed. There is no torque converter or clutch mechanism. There is, however, a damper assembly that is used to cushion the power pulses from the engine. It also reduces the drive line shock.

Ford Motor Company Hybrids

The Ford hybrids are very similar in operation to the Toyota system. The Ford transaxle also uses two electric motor/generators that, in conjunction with an ICE, split power through a planetary gear set. There is only one major difference between the Ford and the Toyota systems. The Toyota system has the two electric motor/generators directly connected to the ring gear and the sun gear. On the Ford system, they are attached through a set of transfer gears (**FIGURE 20-17**).

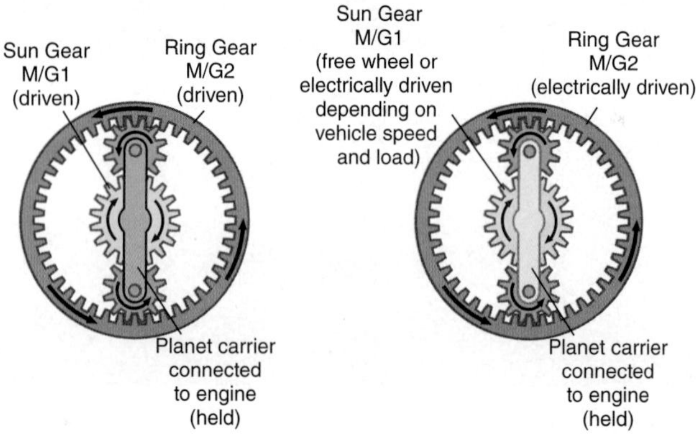

FIGURE 20-15 The planetary operation as M/G1 is propelling the vehicle without the aid of the ICE.

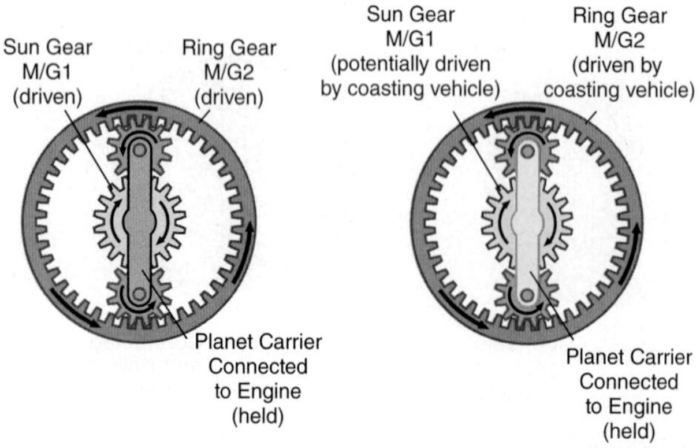

FIGURE 20-16 The planetary gear operation during deceleration.

Two-Mode Hybrid

The **two-mode hybrid** was a joint venture by the General Motors, Chrysler, and BMW companies. Their goal was to create a hybrid system that could be used on trucks and larger luxury vehicles. The system is mostly housed in what looks like a conventional truck transmission housing (**FIGURE 20-18**). It uses a system voltage of 300 volts to power the two electric motor/generators housed inside the transmission case. It has either three or four planetary gear sets, depending on the application. The electric motor/generators and ICE operate in a continuously variable fashion. While it is somewhat similar to the Toyota power-splitting transmission, the Toyota only uses one planetary gear set.

The transmission gets its name from the two operational modes (**FIGURE 20-19**). The first mode is used at low speeds and light engine load. During mode 1, the vehicle can be propelled by the ICE, the electric motor/generator, or a combination of the ICE and electric motor/generator. This functionality qualifies it as a full hybrid system. When operating in mode 1, one of the electric motor/generators is used to keep the high-voltage battery charged. The second motor/generator is used to assist in

FIGURE 20-17 A cutaway of a Ford hybrid transmission. Notice M/G1 and M/G2.

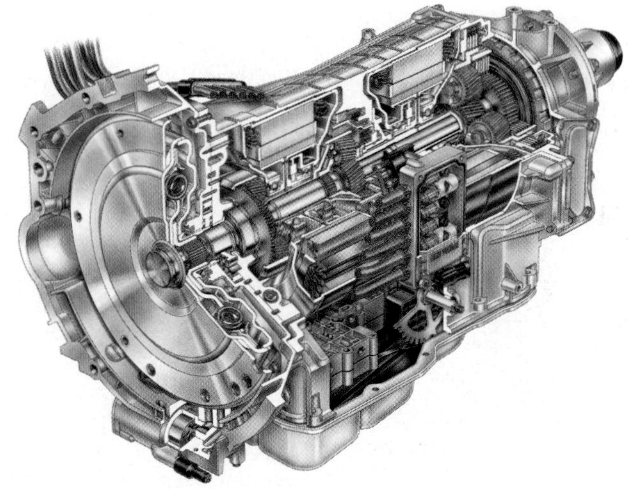

FIGURE 20-18 A typical two-mode hybrid transmission.

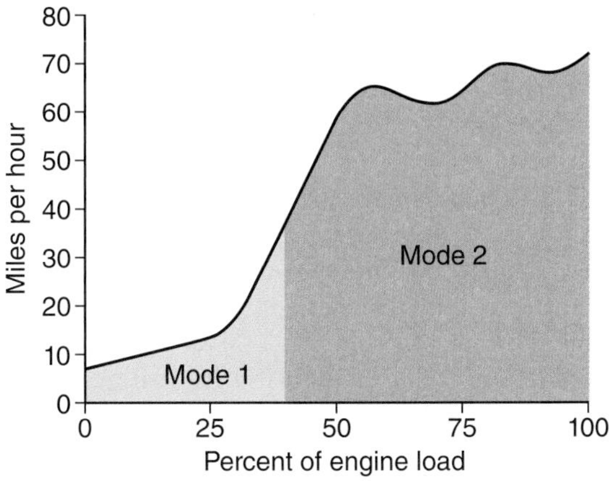

FIGURE 20-19 A chart showing the operation of the two modes of a two-mode hybrid.

propelling the vehicle. The vehicle can be driven by the electric motor only. This happens only if the PCM determines that the high-voltage battery has enough charge to operate the vehicle. If at any time the PCM determines that the battery voltage is too low, or if the driver demands more power, the ICE will be restarted automatically. One of the motor/generators cranks the engine over.

In mode 1, the electric motor/generators, in combination with the planetary gear sets, allow the transmission to operate as a type of CVT. They keep the engine operating in its most efficient rpm range. During deceleration, both electric motor/generators are used to recharge the battery.

The second mode is used for higher vehicle speed and/or load. During the second mode, the gasoline engine runs to propel the vehicle. When maximum power is needed, both electric motor/generators are used to help propel the vehicle. When less power is needed, or the PCM determines that the battery voltage is low, one of the motor/generators is used to generate power. The PCM phases in and out each of the planetary gear sets to create four distinct gear ratios. This also keeps the electric motor/generators operating in their efficient rpm range.

During the second mode, the PCM employs different fuel-saving techniques. Examples include shutting down individual cylinders, and using variable cam timing. When some cylinders are shut down to save fuel, the electric motor/generators are used to smooth out the power demands (torque smoothing). This helps compensate for the reduced number of power pulses from the ICE.

▶ Continuously Variable Transmission (CVT)

LO 20-03 Describe CVT operation.

A CVT has no fixed gear ratios. The transmission can infinitely change the gear ratio within its operational design. Changes to the gear ratio occur in a smooth stepless progression to suit speed and load conditions. This design allows the engine to operate in its most efficient rpm operating range for fuel economy or performance. For fuel economy, the system will keep the engine rpm relatively low. For sportier performance, the system will allow the engine rpm to be higher. This means that the engine can operate within fairly narrow rpm ranges. This happens even though the vehicle speed is being varied substantially.

Types of CVTs

There are three basic types of CVTs commonly used in production vehicles (**FIGURE 20-20**). The first type is

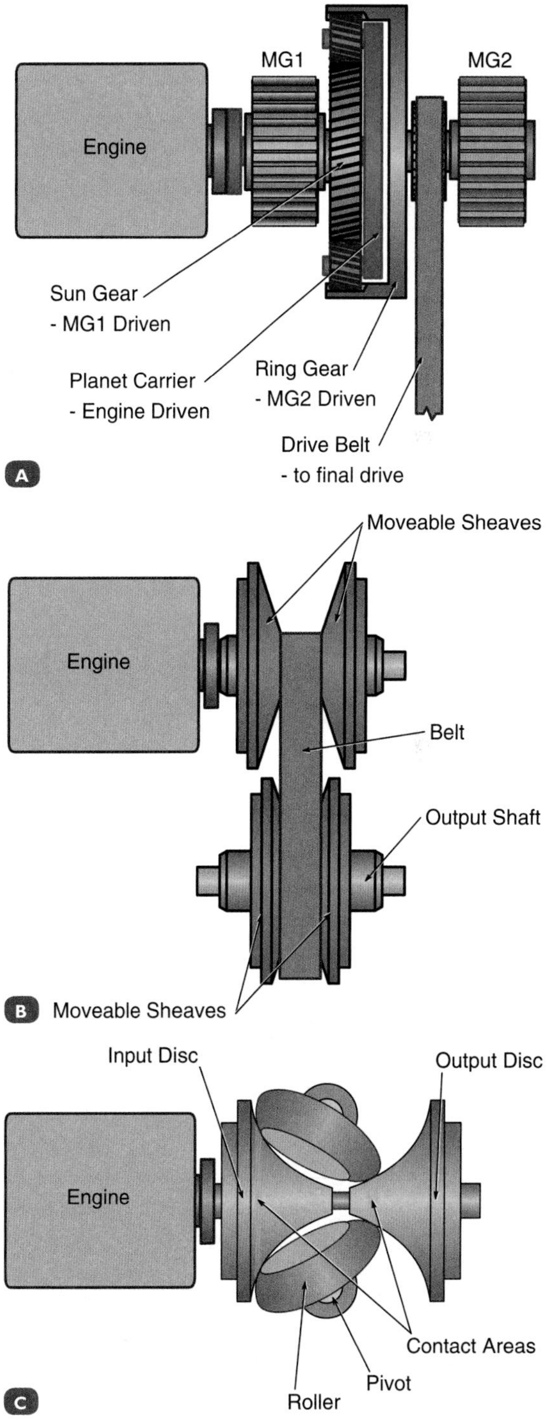

FIGURE 20-20 **A.** Electronic continuously variable transmission (ECVT). **B.** Variable-diameter pulley or Reeves drive CVT. **C.** Toroidal or roller-based CVT.

often referred to as an **electronic continuously variable transmission (ECVT)**. It is commonly found in full hybrid vehicles. The second type is the most common CVT, called a **variable-diameter pulley (VDP)** or Reeves drive CVT. It is used on both hybrid and standard vehicles. The last type of CVT found in automobiles is a toroidal or roller-based CVT. It is much less common than the other two.

We have already covered the ECVT. It is the transmission/transaxle found in Toyota/Ford hybrid vehicles. The General Motors two-mode transmission also uses planetary gears in a partial continuously variable manner. This is accomplished by using planetary gear sets in combination with electric motor/generators and an ICE. By doing this, the manufacturer is able to create an infinite number of gear ratios, within the design limits.

Variable-Diameter Pulley CVT

The **VDP** CVT is one of the most common types of CVTs being used in vehicles today. The VDP system operates using two variable diameter pulleys with either a steel or a rubber belt between them. Some of the early CVTs in Europe used a stiff rubber belt, as do many snowmobiles and all-terrain vehicles. The rubber belt has a limited service life in an automobile, so most manufacturers use a steel drive belt.

Each of the pulleys has one or two movable drive faces, called sheaves. These sheaves can be moved inward or outward, relative to each other. This changes the effective diameter of each pulley (**FIGURE 20-21**).

When the vehicle starts from a stop, the input pulley has a small diameter while the output pulley has a large diameter. This size difference creates a high torque multiplication to start the vehicle from a stop (low speed). The drive ratio from a stop is approximately 3:1 through the pulleys. As the vehicle gains speed, the input pulley diameter increases while the output pulley diameter decreases (high speed).

The input pulley diameter is changed by applying hydraulic oil to one of the pulley sheaves. This pushes the two sheaves closer together. The belt is forced to ride higher on the faces of the pulley. The effective diameter of the pulley increases, as shown in (**FIGURE 20-22**).

When the two pulleys have the same diameter, the gear ratio of the pulleys is 1:1. As vehicle speed continues to increase, the input pulley diameter continues to increase. This causes the output pulley to decrease in diameter. An overdrive ratio of 1:2, or even greater is created on some transmissions. Remember that this ratio does not include any final drive gearing, which reduces the actual gear ratio further.

Belt tension is maintained by large springs in the output shaft pulley. The springs cause the sheaves of the output pulley to be close together. This creates a large-diameter pulley. As the input pulley diameter increases, the springs in the output pulley are compressed. This decreases the diameter of the output pulley. The diameter of the pulleys is controlled to keep the engine operating at its intended rpm range. The PCM monitors the speed of the two pulleys to maintain the correct gear ratio. Some CVTs also incorporate a sensor to measure the position of one of the input pulley sheaves. This allows for greater control and monitoring of the CVT.

Some manufacturers have programmed regular shift points into the PCM. This moves the pulley sheaves to predetermined positions. It creates the feel of separate shifts. Some manufacturers allow the driver to manually shift through these predetermined shift points while driving.

The steel belt is made up of hundreds of transversely mounted steel plates. They are held in place with several steel bands running longitudinally around their edges. The transverse steel plates grab the sides of the pulley and in turn are pushed by the pulleys. The longitudinal steel bands are used to hold the belt together and keep the steel plates lined up. These bands hold the plates closely together in a very strong and flexible ring. The plates are guided by the bands, but not attached

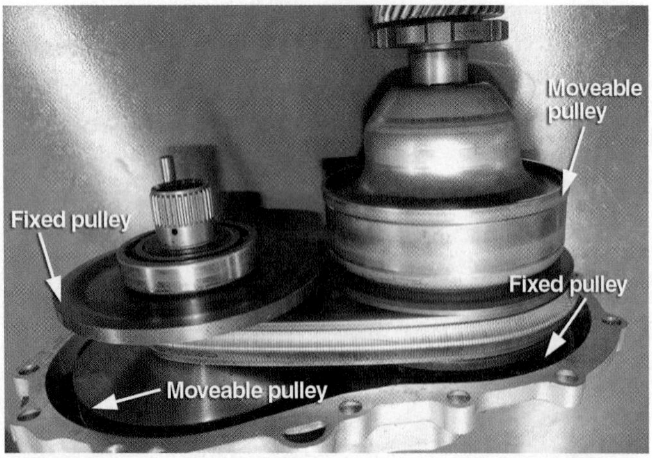

FIGURE 20-21 A VDP CVT. Note the small-diameter input pulley on the left and the large-diameter output pulley on the right. The two sheaves that are moveable are labeled. The other two sheaves remain stationary.

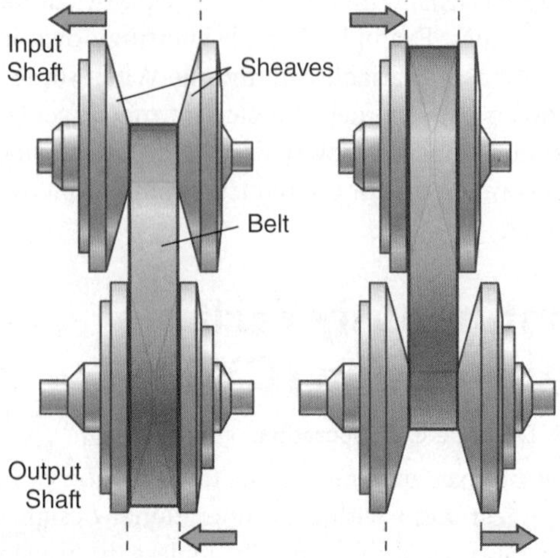

FIGURE 20-22 An illustration of the changing sizes of the input and output pulleys.

to them. Drive is transmitted by compressing the plate elements rather than relying on tension in the band. Each block leaving the primary pulley pushes the blocks ahead of it to the secondary pulley. The bands keep the blocks in contact with the pulley faces. The blocks are compressed on the drive side and float loosely along the bands on the return side.

The large number of plates that are in contact with the pulleys keeps the surface pressures low on the belt. This allows high torque to be transmitted. Both the plates and pulleys are made of very hard materials to resist wearing out quickly. The belt and pulleys are lubricated by transmission fluid sprayed directly onto them at high pressure. A typical steel belt is shown in (**FIGURE 20-23**).

FIGURE 20-23 A steel CVT belt that has been removed from a Nissan transmission.

CVTs require special transmission fluid. And some CVTs have a special transmission oil heater built into them. This heater is used to warm up the transmission fluid so that it can operate properly during cold weather. Filter and fluid change intervals must be adhered to in order to prevent damage to the transmission.

Currently, most manufacturers do not recommend repairing internal components of a CVT. Instead, they recommend the transmission be replaced with a new or rebuilt unit. CVTs can be rebuilt, but many parts are difficult and costly to obtain. Specialty tools are required to compress the output pulley and reinstall the belt.

Low and Reverse

Most CVT manufacturers use multidisc clutch packs and a planetary gear set to allow the transmission to operate in reverse. These clutch packs and planetary gears are typically located on the input pulley shaft. They change the direction of rotation of the pulleys and belt (**FIGURE 20-24**).

Ford Motor Company also uses these planetary gear sets to create a low ratio when it is requested by the driver (**FIGURE 20-25**). Ford advertises their CVTs as two-speed CVTs. When the driver selects low, the planetary gear set is used to create a greater torque multiplication. This increases the pulling power. This is helpful when driving in rough terrain or when starting from a stop on a steep hill.

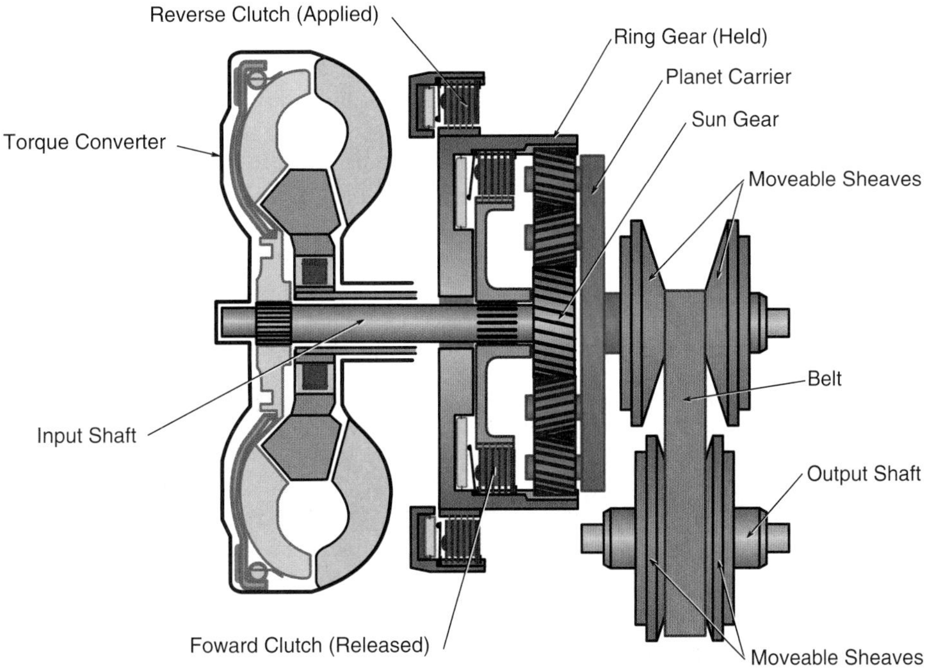

FIGURE 20-24 A planetary gear controlled by clutches allows some CVTs to have reverse and low gears.

Applied Science

AS-1: Simple Machines: The technician can demonstrate an understanding of how cams, pulleys, and levers are used to multiply forces or change the direction of force in a mechanical system.

Simple machines, such as cams, pulleys, and levers, have become much improved and refined over the years. Looking back into history, these three mechanical units have transformed the way we perform many tasks.

Cams have the ability to change rotary motion to linear motion; for example, a camshaft in an engine. Pulleys can be used to change the direction of a force. As an example, raising a flag on a flagpole requires the use of a pulley. It can also provide a mechanical advantage when multiple pulleys are used. Levers are one of the basic tools that may have been used in prehistoric times. The Greek mathematician, Archimedes, described the use of levers in 260 B.C. Examples of levers include pry bars, pliers, claw hammers, and tongs.

As described in our text, the VDP CVT is one of the most common types of CVT being used in vehicles today. The VDP system operates using two variable diameter pulleys with either a steel or rubber belt between them. The rubber belt has a limited service life in an automobile, so most manufacturers use a steel drive belt. Each of the pulleys has one or two movable drive faces, called sheaves. These sheaves can be moved inward or outward, relative to each other. This changes the effective diameter of the pulleys.

A CVT is a transmission that can change through an infinite number of gear ratios. This infinite number is limited only between the minimum and maximum capabilities of the particular unit. This contrasts with other transmissions that offer a fixed number of gear ratios. Fuel economy is enhanced by the engine being allowed to run at its most efficient rpm for a range of vehicle speeds.

FIGURE 20-25 Control switch for selecting low gear ratio.

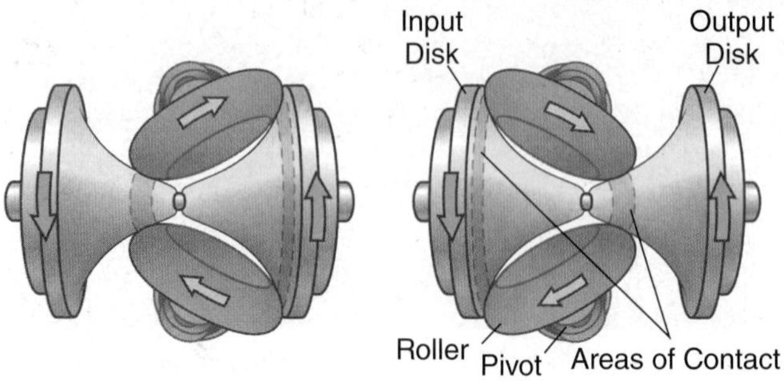

FIGURE 20-26 An illustration of a toroidal CVT. The transmission on the left is in a high torque multiplication ratio, and the transmission on the right is in an overdrive ratio.

Toroidal CVT

The **toroidal CVT** design is currently limited because of high manufacturing costs. The toroidal, or roller, design uses two curved discs—an input disc and an output disc. Power is transferred from the input disc to the output disc through a set of variable-angle rollers (**FIGURE 20-26**).

When the rollers are tilted down toward the center axis of the input disc, the opposite side of the roller moves up toward the outer edge of the output disc. This positioning creates a high gear ratio of approximately 4:1 to begin accelerating the vehicle. As the vehicle accelerates, the roller angle changes. When the rollers are perpendicular to the input and output discs, the gear ratio is 1:1. As the vehicle continues to accelerate, the rollers angle in the opposite direction. When the roller is near the center axis of the output disc, its opposite side is near the outer edge of the input disc. This creates a gear ratio of approximately 1:4. Remember that these ratios do not include any final drive gearing in their calculations.

The toroidal CVT is currently more costly to manufacture than a belt-style CVT. But it can handle larger amounts of torque. A special transmission fluid is needed for these transmissions. It prevents wear to the rollers and discs. At the same time, it provides the correct coefficient of friction for the rollers to operate.

▶ Wrap-Up

Ready for Review

- Most common functions of hybrid vehicles include: idle stop, torque smoothing, regenerative braking, torque assist, and electric only propulsion.
- Belt alternator starter (BAS) vehicles, Honda integrated motor assist, Honda intelligent multi-mode drive (i-MMD) two-motor hybrid powertrain, Toyota and Lexus hybrids, Ford Motor Company hybrids, and two-mode hybrid are some of the common hybrid vehicle types and models available today. New models are being added every year as the demand for more fuel-efficient vehicles increases.

- A continuously variable transmission (CVT) has no fixed gear ratios and allows the engine to operate in its most efficient rpm operating range for fuel economy or performance.
    - Three basic types of CVTs: electronic continuously variable transmission, variable-diameter pulley or Reeves drive CVT, toroidal or roller-based CVT

Key Terms

hybrid drive system A drive system that uses two or more propulsion systems, such as electric motors and an ICE.

torque smoothing A process that uses an electric motor to smooth out engine power pulses when an ICE is operating at low rpm or when the vehicle is using fuel management techniques such as cylinder deactivation.

belt alternator starter (BAS) A type of hybrid drive system that uses a belt-driven alternator/starter that operates on 42 volts.

electronic continuously variable transmission A type of hybrid transmission that often uses two electric motors in combination with an ICE. The two electric motors and the ICE transfer power through a planetary gearset, allowing an infinite number of gear ratios.

IMA (integrated motor assist) A Honda hybrid drive system that uses a moderate-sized electric motor installed between the engine and the transmission.

regenerative braking A type of braking in which the kinetic energy of the vehicle's motion is captured rather than being lost to heat as it is in a conventional braking system. This is accomplished by using the drive motors as generators, which recharge the traction batteries.

toroidal CVT A type of CVT that uses moveable rollers in contact with input and output drive discs. The rollers transfer power from one drive disc to the other. Their position determines the effective gear ratio.

torque assist Use of an electric motor to supplement the engine's torque whenever additional torque is needed, allowing for a smaller ICE to be used.

two-mode hybrid A type of hybrid drive system in which there are two distinct modes of operation. In one mode, the electric motor can propel the vehicle and be used for regenerative braking; in the second mode, the electric motors can be used to assist the engine while the engine uses fuel management techniques such as cylinder deactivation.

variable-diameter pulley A pulley that can change its diameter by moving closer or further apart.

variable-diameter pulley (VDP) A type of CVT that uses two pulleys with moveable sheaves, allowing the effective diameter of the pulleys to change, resulting in variable gear ratios.

Review Questions

1. How is a hybrid drive system defined?
    a. A vehicle with high voltage
    b. A system with more than one battery
    c. A vehicle with an internal combustion engine
    d. A system with more than one power source
2. A technician performing diagnosis on a hybrid vehicle mentions the ICE, what is most likely being referred to?
    a. The high-voltage air-conditioning system
    b. The internal combustion engine
    c. The hybrid inverter cooling equipment
    d. The electric motors inside the transaxle
3. All of the following are common benefits of a hybrid powertrain EXCEPT?
    a. The ability to turn off the engine at stop lights
    b. Brake pads and brake shoes often last longer
    c. There is no longer a need to change engine oil
    d. The power delivery is more smooth
4. Some hybrid vehicles were equipped with a Belt Alternator Starter (BAS). What was the typical voltage on this system?
    a. 24 volts
    b. 42 volts
    c. 120 volts
    d. 240 volts
5. On a Honda Intelligent Multi-Mode Drive (i-MMD) system, what type of transmission is used?
    a. A constantly variable transmission (CVT)
    b. A hybrid transaxle
    c. A conventional hydraulic transmission
    d. A single speed gearbox
6. To drive the wheels, Toyota and Lexus hybrids use the gasoline engine, along with MG1 and ______.
    a. MG2
    b. SG1
    c. DD2
    d. MGG
7. Modern Toyota and Lexus hybrid systems can operate on up to ____?
    a. 240 volts
    b. 300 volts
    c. 450 volts
    d. 650 volts
8. The GM, Chrysler, and BMW two-mode hybrid system:
    a. uses a variably pulley CVT
    b. operates on over 600 volts
    c. can drive the vehicle at slow speeds
    d. can travel up to 200 miles on battery power
9. A customer recently purchased a new vehicle with a Continuously Variable Transmission (CVT). They complain that the transmission seems to be slipping and does not upshift. What is the most likely cause?
    a. This is a normal feeling for CVT
    b. The CVT should be replaced
    c. The CVT belt needs adjustment
    d. The CVT fluid and cooler should be flushed
10. In some vehicles equipped with a continuously variable transmission (CVT), how are manufacturers giving the feeling of shifts?
    a. By adding steps to the CVT pulleys that grab and release
    b. By replacing the rubber belt with an updated steel belt
    c. By falsely fluctuating the vehicle tachometer under acceleration
    d. By programming the PCM and having predetermined sheave positions

ASE Technician A/Technician B Style Questions

1. Hybrid vehicles are being discussed. Technician A states that all hybrid vehicles can drive on electricity alone, without the engine running. Technician B states that some hybrids use a 42-volt motor to assist the engine, or to start it when stopped. Who is correct?
   a. Technician A only
   b. Technician B only
   c. Both Technicians A and B
   d. Neither Technician A nor B
2. Hybrid regenerative braking is being discussed. Technician A states that hybrids that utilize regenerative braking will often have better MPG in stop-and-go traffic, than at highway speeds. Technician B states that vehicles with regenerative braking often do not apply the conventional brakes at all when the brake pedal is lightly depressed. Who is correct?
   a. Technician A only
   b. Technician B only
   c. Both Technicians A and B
   d. Neither Technician A nor B
3. A hybrid vehicle utilizing a small displacement engine is being discussed. Technician A states that on this vehicle, an electric motor is most useful at highway speeds. Technician B states that an electric motor produces peak torque at low rpms. Who is correct?
   a. Technician A only
   b. Technician B only
   c. Both Technicians A and B
   d. Neither Technician A nor B
4. The Honda IMA system is being discussed. Technician A states that the system is considered a parallel hybrid. Technician B states that this system can move the vehicle on electricity only, without the engine running. Who is correct?
   a. Technician A only
   b. Technician B only
   c. Both Technicians A and B
   d. Neither Technician A nor B
5. The GM two-mode hybrid system is being discussed. Technician A states that the electric motors on this system are located inside the bellhousing of the transmission. Technician B states that this system uses electric assist primarily at low speeds, although they can provide assist at high speeds. Who is correct?
   a. Technician A only
   b. Technician B only
   c. Both Technicians A and B
   d. Neither Technician A nor B
6. The Toyota/Lexus hybrid transaxle is being discussed. Technician A states that MG1 is primarily used to start the engine and charge the high-voltage battery. Technician B states that MG2 is used primarily to drive the wheels through its connection to the differential. Who is correct?
   a. Technician A only
   b. Technician B only
   c. Both Technicians A and B
   d. Neither Technician A nor B

7. The Toyota/Lexus hybrid vehicle is being discussed. Technician A states that the torque converter allows this vehicle to come to a complete stop without stalling. Technician B states that this system uses a damper assembly to reduce drive line shock. Who is correct?
   a. Technician A only
   b. Technician B only
   c. Both Technicians A and B
   d. Neither Technician A nor B
8. A CVT is being discussed. Technician A states that a CVT can deliver maximum fuel efficiency since it can keep the engine rpm at its peak efficiency while vehicle speed can vary substantially. Technician B states that a CVT can maximize power by keeping engine rpm relatively low under high engine load conditions. Who is correct?
   a. Technician A only
   b. Technician B only
   c. Both Technicians A and B
   d. Neither Technician A nor B
9. A CVT is being discussed. Technician A states that most CVTs use conventional automatic transmission fluid. Technician B states that a CVT may have a fluid heater to help the CVT operate normally in cold climates. Who is correct?
   a. Technician A only
   b. Technician B only
   c. Both Technicians A and B
   d. Neither Technician A nor B
10. A vehicle comes into the shop needing a CVT repair; it has suffered an internal failure. Technician A states that the CVT will likely need to be replaced with a new or remanufactured unit. Technician B states that there may be expensive special tools to service the unit, but parts may not be readily available. Who is correct?
    a. Technician A only
    b. Technician B only
    c. Both Technicians A and B
    d. Neither Technician A nor B

SECTION 4
Manual Transmissions

▶ CHAPTER 21 Manual Transmission/Transaxle Principles

▶ CHAPTER 22 The Clutch System

▶ CHAPTER 23 Driveshafts, Axles, and Final Drives

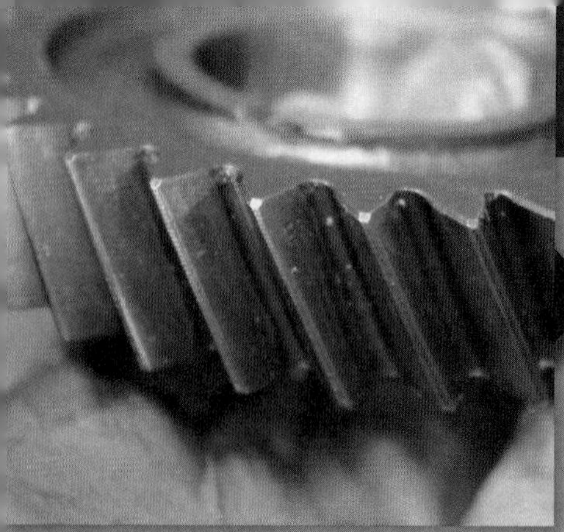

CHAPTER 21

Manual Transmission/ Transaxle Principles

Learning Objectives

- **LO 21-01** Describe the history of manual transmissions.
- **LO 21-02** Describe manual transmission fundamentals.
- **LO 21-03** Describe manual transmission drivetrain layout and operation.
- **LO 21-04** Describe manual transmission gears, shafts, and bearings.
- **LO 21-05** Describe the purpose of clutches, transmission/ transaxles, and transfer cases.
- **LO 21-06** Describe the purpose of differentials, final drives, and axles.
- **LO 21-07** Perform drivetrain fluid maintenance.

ASE Education Foundation Tasks

See Appendix A to view the 2017 ASE Education Foundation Automobile Accreditation Task List Correlation Guide.

▶ Introduction

In this chapter, we explore the history and principles of the modern manual transmission **drivetrain** system. Some of those principles involve certain aspects of physics. Examples include mechanical advantage and gear ratios. For other principles, you need an understanding of the theory of operation of the major drivetrain assemblies. It is important to have a solid understanding of these principles. They will give you a framework on which to hang the higher level theory covered in later chapters.

A good starting place is to understand what is meant by "drivetrain." The drivetrain consists of the component assemblies that transmit power from the engine all the way to the drive wheels. The manual transmission is at the heart of the drivetrain. It receives power from the engine by way of the clutch assembly. The clutch is operated by the driver to disconnect and connect the transmission from the engine. The transmission provides a range of shiftable gears for the driver to select when operating the vehicle.

In a front-engine, rear-wheel drive (RWD) vehicle, the transmission sends power to the final drive assembly. The final drive assembly changes the direction of the twisting force 90 degrees so it can be sent out the axles. The axles power the wheels and tires.

In a front-engine, front-wheel drive (FWD) vehicle, the transaxle contains both the transmission and the final drive assembly in a common unit. It sends the power out the axles to the front wheels and tires (**FIGURE21-1**). In all-wheel drive vehicles (AWD), the transmission or transaxle sends power to all four wheels and tires.

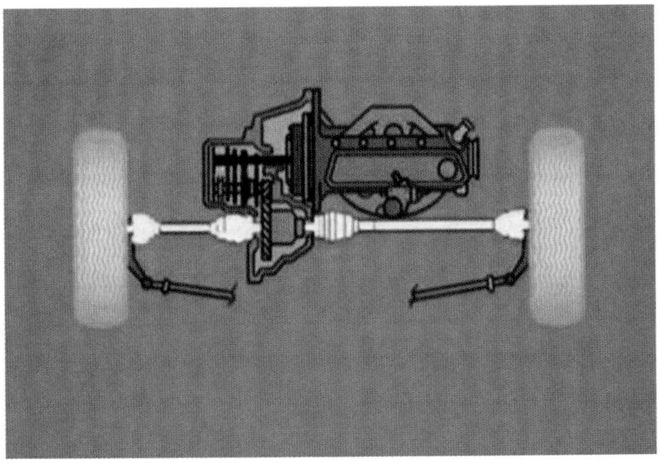

FIGURE 21-1 Front-wheel drive layout.

▶ The History of Manual Transmissions

LO 21-01 Describe the history of manual transmissions.

In 1877, a patent for an FWD carriage with a one-cylinder engine was obtained by George Selden. The "claim to fame" of that vehicle was the transmission (**FIGURE 21-2**). The power from the engine drove a set of bevel gears, which in turn drove a shaft and a pulley. Leather belts were used on a pulley, and a geared wheel on the axle to make it move. One small wheel on the engine got the car going by meshing with the ring gear on one of the drive wheels. The big wheel on the axle then made the car move along at a staggering—at the time—20 miles per hour.

This early engine had one belt-driven high gear for speed. It also had another belt-driven low gear for increased torque. If the car needed to climb a hill, the driver had to stop, get out, and change the belt to the lower gear. This setup was similar to a 10-speed bicycle, where the shifting mechanism physically moves a chain from one gear to another.

FIGURE 21-2 The engine and transmission, which were on the front axle, drove this early vehicle.

You Are the Automotive Technician

A Boy Scout troop is working on a merit badge regarding vehicles and vehicle maintenance. They come into your shop to ask you questions that their scoutmaster cannot answer. You greet them, show them around the shop, and introduce them to your fellow employees, who tell them about their specialties. Once that is done, they gather together to ask you some specific questions about terms they heard but didn't understand. How would you answer their questions?

1. What is the difference between a transmission and transaxle?
2. What is the purpose of a final drive?
3. What is a transfer case, and what does it do?
4. What are spur gears and helical gears?

FIGURE 21-3 Panhard and Levassor's attempt at a front drive axle.

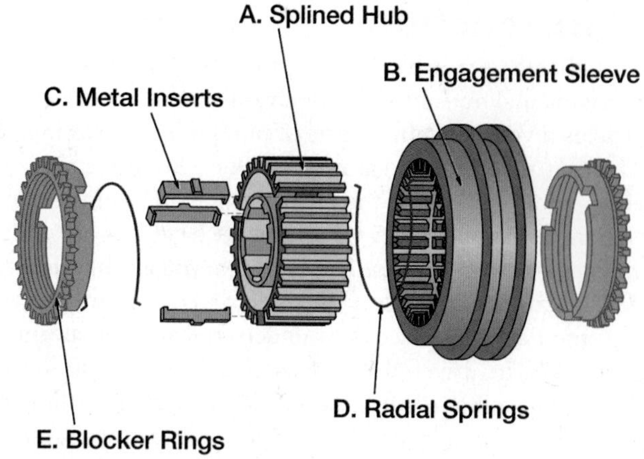

FIGURE 21-4 Cone-style synchronizer.

In 1894, a couple of Frenchmen, Louis René Panhard and Émile Levassor, designed an FWD multi-gear manual transmission (**FIGURE 21-3**). When they tried to put on a demonstration of their new transmission, the engine in their demo vehicle encountered problems. They were unable to make it move under its own power.

One year later, Panhard and Levassor were able to demonstrate their new multi-gear transmission. Their transmission and clutch arrangement is the prototype for most manual transmissions today. It used a clutch-driven, three-speed, sliding-gear transmission. When the driver wanted to shift gears, he or she would push the clutch pedal to disengage the engine. The driver would then move the shifter lever. This slid the previous gears out of mesh and the newly selected gears into mesh. The process allowed the vehicle to move at higher and lower speeds. This sliding-gear arrangement became the basis for nearly all manual transmissions since then. And manufacturers have continued to develop enhancements to the design.

Before 1898, vehicles were either belt- or chain-driven. Because the chain or belt was exposed to the elements, this design required frequent maintenance. It could also be dangerous if someone got too close to it. In 1898, Louis Renault connected an engine to a transmission and created a live rear axle. The metal axle shaft was supported by bushings. Renault then adapted a differential-type rear axle that was based on an idea that American C. E. Duryea had back in 1893. The differential assembly had a number of gears arranged in such a way as to allow each wheel to turn at its own speed when going around a corner. This solved the problems of loss of traction and rapid tire wear due to scuffing of tires when making turns with a solid axle.

By 1904, most of the carmakers had adopted the sliding-gear manual transmission design in one form or another. Many improvements have been made since then. One of the biggest improvements was the invention of the **gear synchronizer** (syncromesh). It applies a friction device between the gear and the shaft to match the gear speed to the shaft speed. As a result, the gear and shaft spin at the same speed. This allows gear selection to be made without "grinding" of the gears.

In 1928, Cadillac introduced the first **synchromesh transmission**. Porsche improved the design in 1952 by using moly-coated steel rings. They were more efficient and reliable and were licensed by many of the manufacturers of the time. In the early 1960s, Borg Warner introduced the modern cone-style synchronizer. It is still widely used today (**FIGURE 21-4**).

> ▶ **TECHNICIAN TIP**
>
> In the time span between the sliding-gear transmission and the synchromesh transmission, there were other attempts to make it easier for the driver to change gears. In 1907, the Ford Model T was equipped with a transmission that used constantly meshed planetary gear sets. These sets consist of a central gear—called a sun gear. They are meshed with multiple planetary pinion gears, which are meshed inside a ring gear. Although this gear arrangement is not common in manual transmissions, it is widely used in automatic transmissions today.

> ▶ **TECHNICIAN TIP**
>
> All transmission designs since Panhard and Levassor have had one goal in mind: to make shifting easier for the driver. The easiest transmission to shift is the automatic transmission. It is strictly an American innovation.

▶ Fundamentals of Manual Transmissions

LO 21-02 Describe manual transmission fundamentals.

Mechanical Advantage

Mechanical advantage is defined as the increase of the input force by trading distance moved for greater output force. For example, imagine using a 10"(25 cm)-lever to move a large rock. If you place a short section of log (the pivot point) 2"(5 cm) from

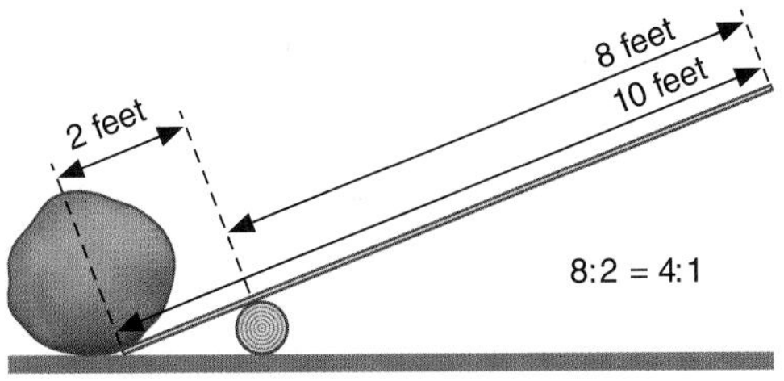

FIGURE 21-5 Using a 10" (25 cm)-lever to move a rock, with the pivot point 2" (5 cm) from the end nearest the rock, gives a mechanical advantage of 8 to 2 (4:1).

the end of the lever nearest the rock, you would have a mechanical advantage of 8 to 2 ratio (**FIGURE 21-5**). Ratios are normally expressed as the equivalent of the first number compared to 1. Thus, 8 to 2 would be expressed as 4 to 1, and is written as 4:1. Assuming no frictional loss, a lever with this mechanical advantage would exert four times as much force against the rock as the amount of force being applied to the other end of the lever. So we could say the lever gives a mechanical advantage of 4:1.

Given this 4:1 mechanical advantage, if a person pushes the lever down with a force of 100 lb (45 kg), the force the lever generates against the rock is 400 lb (180 kg). By using a lever, the person achieves more output force than he or she puts in. However, the person also is moving the long side of the lever four times as far as the rock side of the lever. So, although the output force is four times greater than the input force, the input distance moved is four times greater than the output distance. Also, the input speed is four times faster than the output speed. Thus, the total amount of work on each end of the lever is the same (work = force × distance). It is simply rearranged to give an increase of the output force. The lesson here is that mechanical advantage can be used to generate a larger output force. But it requires an increase in the input distance and speed.

This concept applies directly to manual transmissions. Each set of gears, called a **gear set**, provides a certain amount of mechanical advantage. It also has a specific speed differential associated with it. For example, in low gear, the smallest gear in the transmission turns the largest gear in the transmission. One complete turn of the small gear turns the larger gear only a small portion of a full turn. This provides mechanical advantage and is called gear reduction. At the same time, because the small gear is turning much faster than the output gear, the output speed is slow compared with the input speed.

The driver can select the transmission gear set that best matches the operating conditions under which he or she is driving. When driving on rough and hilly off-road terrain, a lower gear works well. When driving on a smooth, straight, and level freeway, a higher gear is best.

Gear Ratios

How do gear sets create mechanical advantage? They do so by having different ratios. Gears with the same number of teeth have a ratio of 1:1, meaning that this set of gears has no mechanical advantage. At the same time, there is no difference of speed between these gears. They each turn at the same speed because they have the same number of teeth.

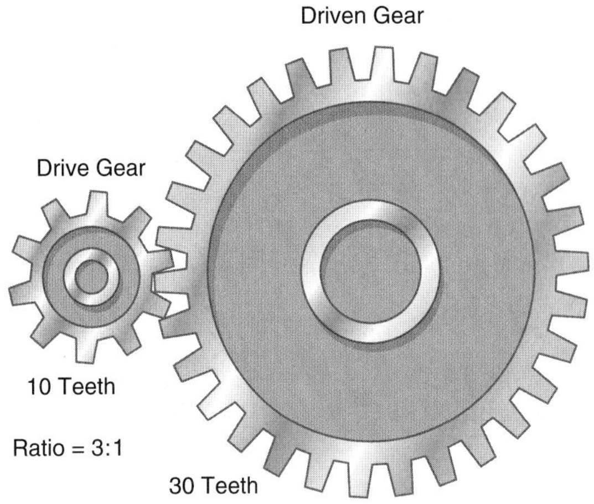

FIGURE 21-6 The mechanical advantage of two gears with a different ratio, which provides torque to get moving.

A gear set in which the input gear has half as many teeth as the output gear has a ratio of 2:1. Gear ratios can be found by comparing the number of teeth each gear has (**FIGURE 21-6**). If the drive gear (the input gear) has fewer teeth than the driven gear (the output gear), then the gear set will have mechanical advantage. In this case, the smaller drive gear has to rotate faster than the larger driven gear, giving an increase in output force.

All gear ratios are calculated by the following formula:

$$\frac{\text{Driven}}{\text{Drive}},$$

Written out, gear ratio is the number of driven gear teeth divided by the number of drive gear teeth. For example, if the drive gear has 10 teeth and the driven gear has 30 teeth, then the mechanical advantage is 3:1. In this case, the driven gear has three times the output force (torque), but only one-third of the **rotational speed**. In other words, if the drive gear is turning at 100 revolutions per minute (rpm), the driven gear is turning at 33.3 rpm, but with three times the torque.

If you were to make the drive gear larger by increasing it to 15 teeth and then you connected it to a driven gear with 30 teeth, the gear ratio would drop to 2:1. But the speed of the driven gear would increase to one-half as fast as the drive gear. In other words, if the drive gear is turning at 100 rpm, the driven gear is now turning at 50 rpm. But it would only have two times the torque.

What if you were to make the drive gear larger by increasing it to 30 teeth, and connected it to a driven gear with 24 teeth? The output gear would be turning faster than the input gear. This is called **overdrive**. The gear ratio for this arrangement would be 0.8:1.

As you can see, as the gear ratio decreases, the output speed increases. This is what happens inside a transmission each time the driver selects a higher gear. The output torque and speed can be varied as necessary, based on the speed and load of the vehicle. When the vehicle speed is low, the driver selects a lower gear (which has a high gear ratio). That way, the engine can be used to accelerate the vehicle. As the engine speed increases, the driver selects the next gear so that the vehicle can be accelerated further. This continues to happen until the vehicle reaches the desired speed, or runs out of higher gears.

Speed and torque output is the result of the selected gear ratio. Different gear ratios allow the driver to tailor the engine speed and power to the driving conditions. The ultimate task of the manual transmission is to allow the following. First, it allows the vehicle to easily move away from a stop. It does so by multiplying the torque through use of a low gear, which has a high gear ratio. And second, once moving, it allows the driver to sequentially select higher gears (lower gear ratios) for higher vehicle speeds. This helps to keep the engine speed appropriate.

> ▶ TECHNICIAN TIP
>
> Following manufacturer-recommended shifting speed guidelines will ensure good fuel economy. In fact, many vehicles have a shift indicator on the dash. It signals the driver when an upshift or downshift would provide better fuel economy.

Power Flow

Power flow is defined as the path in which power is transmitted through a series of components. Power flows through the entire drivetrain. The engine in the vehicle is typically the main source of all mechanical power for the vehicle. The power that the engine creates flows from the crankshaft to the flywheel and to the clutch assembly. When the clutch pedal is released (in the up position), the clutch transmits power from the flywheel to the input shaft of the transmission/transaxle. Power leaves the transmission/transaxle output shaft and flows to the final drive assembly. It changes direction and flows through the axles and out to the wheels. This is a very simple and generic explanation of the path in which power flows from the engine to the tires.

When it comes to the manual transmission, power flows into the transmission through the input shaft. A gear on the input shaft transfers power to a meshed gear on the countershaft. The countershaft has a number of gears that mesh with gears on the output shaft (**FIGURE 21-7**). The gear the driver selects determines the gear ratio from the countershaft to the output shaft. Because of the various gear sets and driver selection, the transmission has multiple gear ratios through which the power can flow.

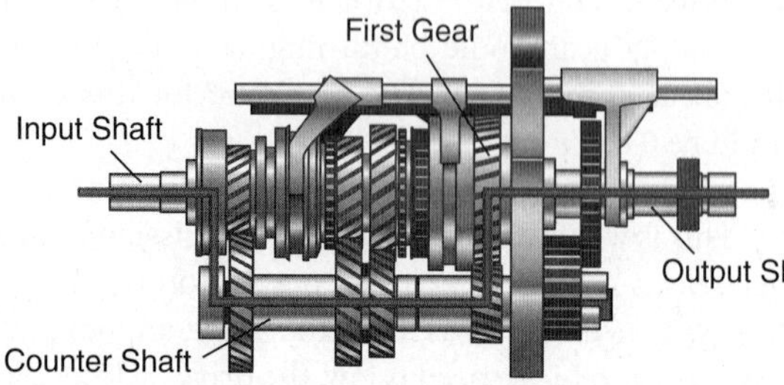

FIGURE 21-7 Power flow in a transmission.

▶ Manual Transmission Drivetrain Layout

LO 21-03 Describe manual transmission drivetrain layout and operation.

The manual transmission allows the driver to control the power flow in the drivetrain. The **clutch system** is part of the manual transmission. It is the medium by which the driver can connect and disconnect the engine from the transmission.

On the driver's side of the passenger compartment, there are usually three pedals on the floor (**FIGURE 21-8**). The pedal on the far right is the **accelerator pedal**. It is used for controlling the power output of the engine. The center pedal is the **brake pedal**. It is used to stop or slow the vehicle's motion. And the far left is the **clutch pedal**. It operates the clutch system. Pushing down on the clutch pedal disconnects the clutch components. This allows the driver to select one of the forward or reverse gears within the transmission. Letting the clutch pedal back up reengages the clutch components. Power can again be transmitted from the engine to the transmission.

The term **transmission** refers to layouts where the transmission sends power to the final drive assembly. Transmissions are usually used in RWD vehicle configurations. The term transaxle refers to layouts where the transmission and final drive are integrated into a common assembly. They are usually used on FWD vehicles.

The **final drive assembly** gives the final gear reduction to the drivetrain. It powers the drive wheels through axles (**FIGURE 21-9**). The final drive assembly increases the torque from the transmission by means of mechanical advantage. The amount of mechanical advantage is determined by the gear ratio of the final drive. It affects the amount of torque applied by the wheels to the ground. It also impacts the speed of the engine, and the vehicle's fuel economy.

If the final drive has a high gear ratio (higher numerical number), then the vehicle will have more pulling power. But the engine will operate at a higher rpm at any vehicle speed, thereby decreasing the fuel economy. Manufacturers match the final drive gear ratio to the vehicle based on the engine, vehicle weight, and expected use of the vehicle.

FIGURE 21-8 Vehicles equipped with a manual transmission have three pedals that the driver controls.

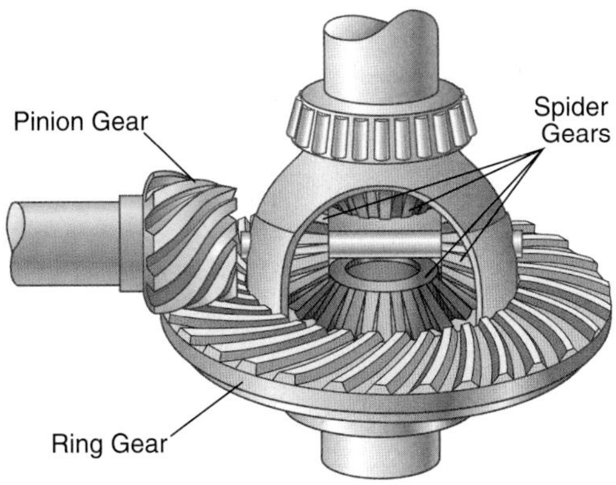

FIGURE 21-9 Typical final drive assembly.

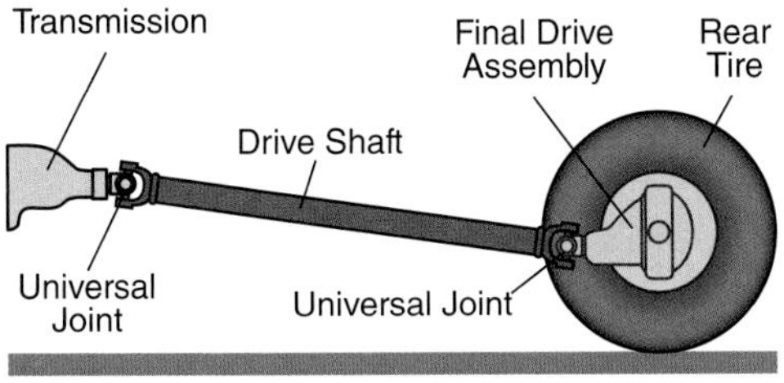

FIGURE 21-10 A typical driveshaft.

When a vehicle travels around a corner, the outside wheel travels a greater distance than the inside wheel. It also turns at a faster rate. If both wheels were connected to a solid axle, the tires would scuff on the road when turning a corner. In modern automobiles, the final drive assembly sits between the ends of two separate axles. The final drive assembly incorporates a set of **differential gears**. One gear is connected to each axle. This allows each axle to rotate at its own speed while going around a corner. At the same time, the final drive assembly powers both axles through the differential gear assembly.

On RWD vehicles, the **driveshaft** transmits power from the transmission to the final drive assembly (**FIGURE 21-10**). The driveshaft uses **universal joints**. They allow the driveshaft to change angles due to the movement of the suspension relative to the body. Because FWD vehicles incorporate the final drive in the transaxle, no external driveshaft is needed.

Both FWD and RWD vehicles use **drive axles** to power the wheels. The axles can be solid, as is the case with many RWD vehicles. Or they can use flexible **half-shafts**, which use **constant-velocity (CV) joints**, as is the case with FWD vehicles (**FIGURE 21-11**). FWD axles must be able to change length as the vehicle goes over bumps and dips. They must also allow the wheels to be steered. This is provided by different types of CV joints.

Operation

The driver steps on the clutch pedal, which releases the clutch and disconnects the engine from the transmission. Depressing the clutch pedal also closes the contacts of the clutch safety switch. The closed clutch safety switch allows the starter motor to crank the engine over when the driver turns the key to the "crank" position. The engine then starts and idles. With the clutch pedal fully depressed, the driver moves the gearshift lever to select first gear or reverse gear. Once in gear, the driver can slowly release the clutch pedal. At the same time, the driver lightly steps on the gas pedal. The clutch progressively connects the engine to the transmission. This allows the vehicle to start moving.

With the transmission in first gear, the transmission is in the lowest gear. In this gear, the vehicle cannot travel very fast, but it provides the best pulling power. Torque is then sent to the final drive, where its speed undergoes a further gear reduction. This increases the torque further before it is sent to the axles. The axles turn the wheels, which are connected to the ground and cause the vehicle to move.

When the driver wants to change gears, he or she pushes the clutch pedal with their foot. The gearshift lever is then moved by hand to the next gear position. Once the gear is selected, the driver again slowly releases the clutch pedal. This again reconnects the engine to the transmission so the vehicle can accelerate.

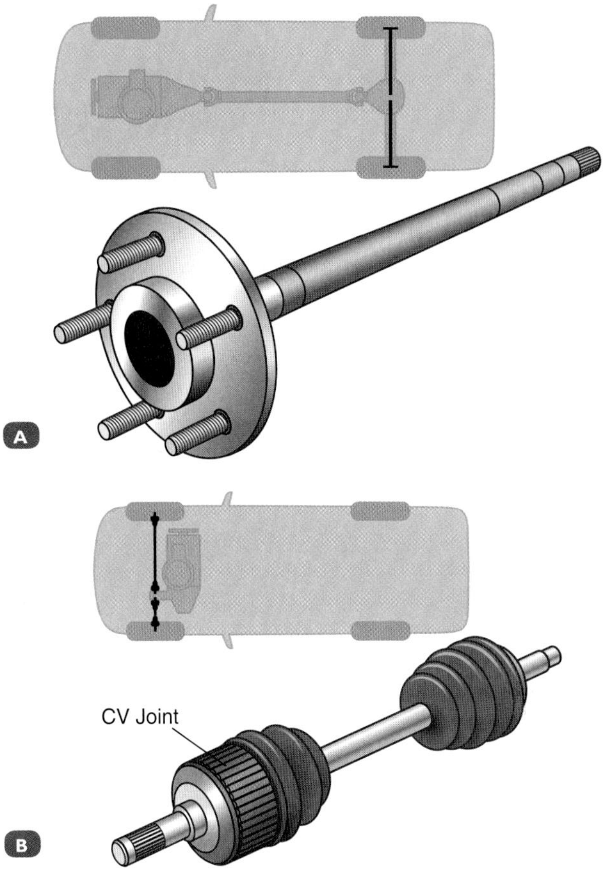

FIGURE 21-11 Drive axles. **A.** Solid drive axle. **B.** Half-shaft drive axle.

▶ Manual Transmission Components

LO 21-04 Describe manual transmission gears, shafts, and bearings.

Today's manual drivetrains are precisely machined, mechanical wonders. Each of the subassemblies plays a vital role in transmitting power from the engine to the wheels (**FIGURE 21-12**). Understanding the main components of the unit assemblies will

FIGURE 21-12 Typical manual drivetrain from an AWD vehicle.

FIGURE 21-13 Manual transmissions are mechanical wonders with many precisely machined gears and shafts.

give you a good overview of their operation. Also, having an understanding of the internal workings of the subassemblies will help you diagnose drivetrain problems. We now cover some of the main components of the drivetrain.

Shafts, Gears, and Bearings

Inside a manual transmission is a series of shafts, gears, and bearings. They make it possible for the driver to select the preferred gear for the road conditions (**FIGURE 21-13**). The availability

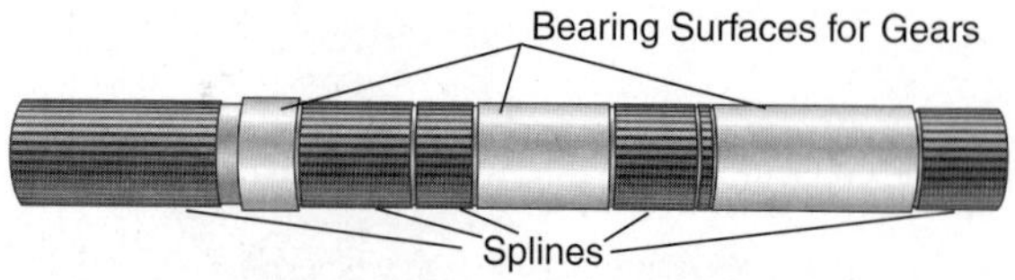

FIGURE 21-14 Shafts are used to support gears and are machined precisely to accommodate bearings and individual gears.

FIGURE 21-15 Grooves are cut in shafts to hold gears and bearings in place.

of several different gear ratios allows the vehicle to operate at a variety of speeds. They make it so the vehicle can accelerate from a dead stop almost effortlessly. It also allows travel at a variety of speeds without over-revving the engine.

Shafts are used to support gears. They are machined precisely to accommodate bearings and individual gears. The surfaces are smooth, allowing the bearings and gears to rotate on a thin film of oil (**FIGURE 21-14**). This reduces metal-to-metal friction and prevents catastrophic failure from overheating. Some surfaces on the shaft are **splined** (parallel grooves in a shaft that mate with a component with matching grooves). The grooves allow other parts, such as synchronizers, to be held stationary on the shaft.

Other machining may create raised areas on the shafts such as shoulders. Shoulders provide a firm support for gears and bearings to butt up against. Alternately, grooves may be cut into the shafts to accommodate **snap rings**. Snap rings are used for holding the gears in the proper position on the shaft once the gears are installed (**FIGURE 21-15**).

The ends of the shafts have machined surfaces to hold the bearings. Bearings are used to center and hold the shafts in alignment. This allows the gears rotate smoothly when in mesh with their mating gears. Friction bearings (bushings) provide sliding metal-to-metal contact between components. They are usually made of brass or bronze. Nonfriction bearings provide rolling metal-to-metal contact between components. They are usually of the ball or roller type.

Gears are round parts with teeth cut on the outside perimeter. They mate with one or more other gears to achieve a specific gear ratio. The mated gears are called a gear set. Each set is cut on the same angle or pitch to allow for proper contact of the gears with each other.

Manual transmission gear teeth are usually of two configurations: spur or helical. Spur gears have straight teeth that are parallel to the shaft the gear turns on (**FIGURE 21-16A**). They can be noisy during operation. However, they don't create axial forces like helical gears do. Helical gears have teeth that are cut on an angle to the shaft (**FIGURE 21-16B**). Because of this, they place high stress on the bearings in the case. So, spur gears are used in heavy-duty vehicles such as off-road equipment. Helical gears operate more quietly, and are more commonly used in passenger vehicles.

Bearings are necessary to maintain the position and alignment of the shafts and gears while under two types of loads: radial and axial. **Radial loads** are perpendicular to the shaft. They occur because the gears have a tendency to push each other apart as torque is transmitted between them (**FIGURE 21-17**).

Axial loads are in line with the shaft. Axial loads occur in transmissions because of the cut of the helical gear teeth. Since they are cut at an angle to each other, they cause a fore-and-aft load (**FIGURE 21-18**).

Bearings also allow shafts and gears to have greater rotational speeds while also minimizing metal-to-metal friction. This also contributes to increased fuel economy and longer transmission life. Bearings can be of the friction (bushing) type, or nonfriction (roller bearing) type. Both are used in transmissions to resist overheating and keep shafts aligned.

FIGURE 21-16 A. Spur gears have straight teeth. They can be noisy during operation. **B.** Helical gears are cut on an angle for better coverage. They operate quietly.

> **TECHNICIAN TIP**
>
> Manual transmission failures occur. Proper preventive maintenance is critical to the life of all parts inside the transmission. Proper oil levels, types, and viscosities contribute to longevity of these units.

▶ Clutch System

LO 21-05 Describe the purpose of clutches, transmission/transaxles, and transfer cases.

In a manual transmission vehicle, the clutch locks the engine and transmission together. The engine is connected to the clutch, and the clutch is connected to the transmission. The clutch is designed so it can connect and disconnect. Imagine a wall-mounted dimmer switch for a ceiling light in a house. When the dimmer switch is in the off position, no electrical power is transmitted to the light. This results in no illumination of the room.

When the clutch pedal is in the depressed or down position, no power is transmitted to the transmission. And there is no movement of the vehicle. When the dimmer switch is in the full on position, electrical power is transmitted to the lightbulb,

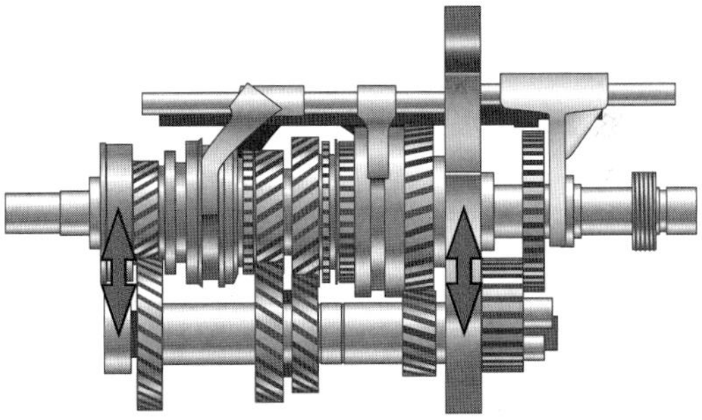

FIGURE 21-17 Radial-loaded bearing.

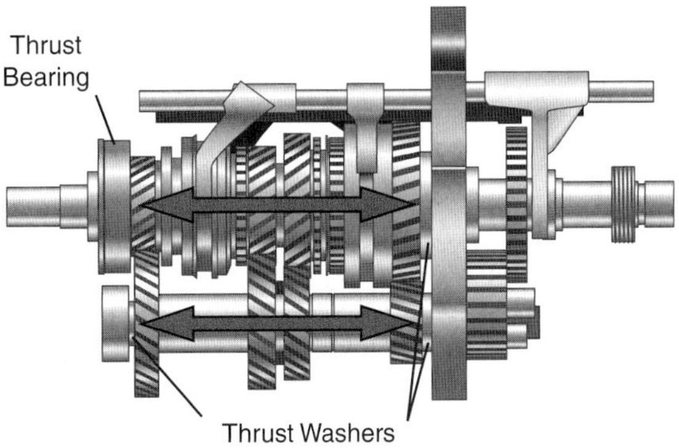

FIGURE 21-18 Axial-loaded thrust bearings/thrust washers.

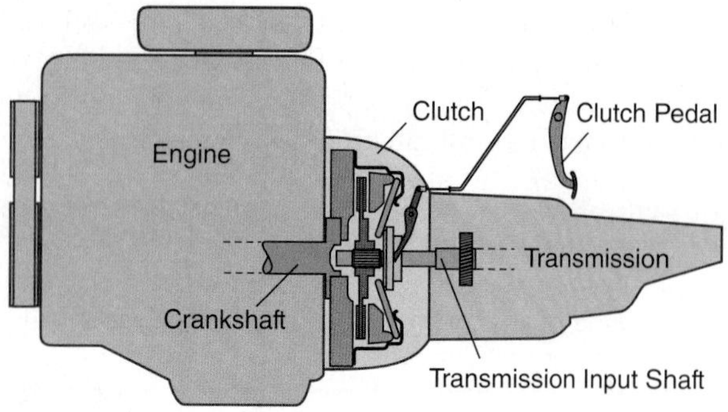

FIGURE 21-19 Clutch components are designed to connect and disconnect the power to the transmission.

resulting in illumination. Similarly, when the clutch pedal is in the released or up position, power is transmitted to the transmission. And the vehicle moves.

Much like the dimmer switch, the clutch pedal provides movement to operate the clutch (**FIGURE 21-19**). The movement causes engagement and disengagement of the clutch. During engagement, slippage between the clutch components occurs. This slippage can cause the clutch components to wear (just as wear can be present in the dimmer switch over time).

When a dimmer switch is turned on, there is a drag of amperage on the power source as the light draws power. When the clutch is released, the engine is running but the vehicle is not moving. The clutch must allow some slippage until the vehicle starts moving. Otherwise, the engine would die. This slippage creates heat and is detrimental to both the dimmer switch and the clutch system.

If the clutch is only partially released, there will be a large amount of slippage in the clutch. This is just like power being reduced by the dimmer switch when it is left in a partially on position. The difference is that the dimmer switch is designed to handle the resulting heat, but the clutch is not.

> ▶ TECHNICIAN TIP
>
> The clutch system may have to be inspected and maintained at regular intervals depending on the driver's habits. Customers who tend to slip the clutch, rest their foot on the clutch pedal, or drive aggressively put more wear on the clutch components. It is always a good practice to inspect the clutch pedal action to catch a problem before it becomes serious.

Transmission/Transaxle

Since the invention of the automobile, the need for a device to manage engine power and torque has been satisfied by the transmission. As vehicle design has evolved over the decades, the transaxle was developed (**FIGURE 21-20**). The only difference between a transmission and a transaxle is that a transaxle has the final drive assembly built in. This makes the transmission and final drive a single unit. In a vehicle with a transmission, the final drive is a separate unit.

FIGURE 21-20 Transmissions/transaxles. **A.** Typical RWD transmission. **B.** Typical FWD transaxle.

> ▶ TECHNICIAN TIP
>
> Most manual transmissions/transaxles are manually shifted, but some are electronically controlled. In these systems, a computer plays a part in how and when the shift will occur. These electronic systems require the use of a scan tool to diagnose faults in the transmission. Scan tools can also be used to program the parameters for shifting.

> ▶ TECHNICIAN TIP
>
> The manufacturer determines the gear ratios in the transmission/transaxle based on many things. These include the power-to-weight ratio, the vehicle size, the application, and the intended use. This concept is discussed in depth in the Manual Transmissions/Transaxles Basic Diagnosis and Maintenance chapter.

The manufacturers rate transmissions/transaxles. They are rated for how much twisting force (torque) they can handle. The rating is usually measured in ft-lb. It helps ensure that the transmission/transaxle is strong enough for the engine it is used with. This includes passenger vehicles, pickup trucks, or sport utility vehicles (SUVs). Transmissions must operate in a manner

FIGURE 21-21 A typical transfer case used for four-wheel drive (4WD).

that meets customer expectations. Sometimes customers have a choice between a five-speed and a six-speed transmission, or between a six-speed and a seven-speed transmission, or even more.

Transfer Case

The **transfer case** is typically mounted to the rear of the transmission. Its purpose is to transfer power to both the front and rear axles. This provides a four-wheel drive (4WD) so the vehicle can move with more stability in inclement weather and compromised road conditions. Steering ability is also increased, making the vehicle track better in slippery conditions.

The transfer case takes the power from the transmission and directs it to one or both axles, depending on the mode selected (**FIGURE 21-21**). This system requires that the front wheels be powered. Thus, a front differential and two front axles were added. A second driveshaft is necessary to provide power to the front axle assembly.

The transfer case and the manual transmission are similar in operation. They both have a series of shafts, gears, and bearings and include shift mechanisms. The transfer case can be made of cast iron to provide strength, or magnesium or aluminum to lessen the weight. It is also filled with lubricant to reduce friction. This increases the life span, and maintains quiet operation of the gears by providing a thin film of oil between them.

Transfer cases can be operated manually or electronically (**FIGURE 21-22**). Mechanically shifted transfer cases have a shift lever. This is usually next to the transmission gearshift. The transfer case shift lever is usually shorter. It is not used as frequently as the transmission gearshift. Electronically shifted transfer cases may use an electronically controlled vacuum actuator. Or, they may use an electric motor to shift the transfer case. Sensors may be used to operate a light that informs the driver which mode is selected.

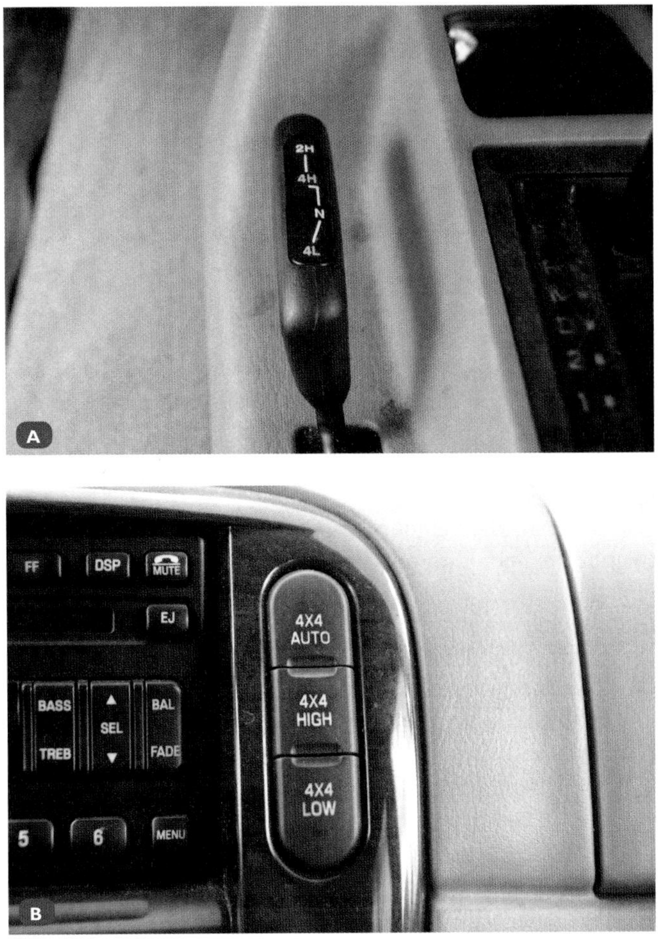

FIGURE 21-22 Transfer case control can be: **A.** manually controlled or **B.** electronically controlled.

> ▶ **TECHNICIAN TIP**
>
> There are two types of 4WD systems—full-time 4WD and part-time 4WD. These will be explained in greater depth in the Drivetrain Components chapter. Most manufacturers of SUVs incorporate some form of 4WD or AWD as a selling/safety feature to the public. Early 4WD vehicles were not very fuel-efficient due to heavy and bulky designs. Over time, manufacturers have made fuel efficiency and safety priorities.

▶ Differential and Final Drive

LO 21-06 Describe the purpose of differentials, final drives, and axles.

The term "differential" is used in two different ways. The first is the most technically accurate. It refers to the components inside the differential housing that allow the axles to turn at different speeds when the vehicle is cornering (**FIGURE 21-23**). This is called the differential assembly. Because the outside tire must travel farther when cornering than the inside tire, it must rotate at a faster speed. The differential assembly allows this to happen. Otherwise, the tires would bind, skip, hop, and slide when going around a corner. This would cause them to wear out quickly and could easily cause a loss of vehicle control.

Technicians also use the term "differential" as another name for the final drive assembly (**FIGURE 21-24**). The final drive provides the final gear reduction necessary for drivetrain operation. The gear reduction happens because the pinion gear (drive gear) is much smaller than the ring gear (driven gear). This difference in size results in an increase in mechanical advantage.

In addition, the final drive assembly takes the power from the transmission. It turns the power 90 degrees and sends it out the axles to the tires. The differential assembly allows the tires to travel around corners without binding. It does so as the final drive is powering them.

Because the differential is housed within the final drive assembly, many technicians call the final drive assembly, the differential. Any time you hear the term "differential," you need to ask whether they mean the final drive assembly, or the differential assembly.

Differential assemblies also come in two varieties. First, there is the **open differential assembly**. With the open differential assembly, power is supplied to both wheels equally only when each tire maintains traction. If one wheel is stuck in the snow or ice, the other wheel cannot supply power. Because of the nature of the open differential assembly, power is only supplied to the slipping wheel.

The second type is the **limited slip differential assembly**. It allows both of the rear wheels to supply power to the ground, even if one wheel loses traction. This allows the wheel with traction to drive the vehicle forward or backward.

There are two types of drive axle assemblies used in modern vehicles: the solid axle and the independent axle. The **solid axle** uses a solid axle housing, which means movement of one wheel affects the other wheel (**FIGURE 21-25**). As the wheel on one side of the vehicle hits a bump, the wheel on the other side pivots. This affects the ride of the vehicle and the grip of the tire to the road.

An **independent axle** allows each wheel to move independently of the other wheel. This results in a more comfortable ride and better handling (**FIGURE 21-26**). In a rear wheel drive, independent axle, the final drive assembly is firmly bolted to the vehicle frame or unibody. The final drive uses flexible half shafts to send power to the independently operated wheels.

▶ TECHNICIAN TIP

If the vehicle is equipped with an open differential, traction control can help overcome the loss of traction on one wheel. The traction control system applies the brakes to the wheel that is slipping. This results in more power being sent to the wheel that is not slipping.

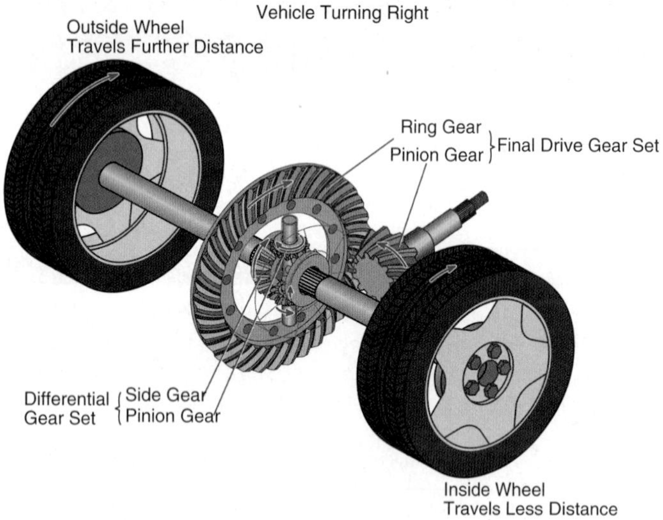

FIGURE 21-23 The differential assembly allows the wheels on an axle to rotate at different speeds when cornering.

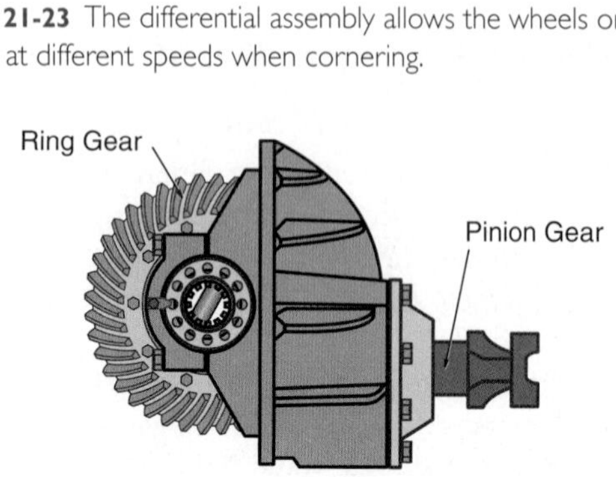

FIGURE 21-24 A final drive assembly sometimes inaccurately referred to as a differential.

FIGURE 21-25 Solid rear axle assembly.

FIGURE 21-26 Independent rear axle assembly.

▶ TECHNICIAN TIP

Final drives are filled with special lubricants and additives. They help create a long life and maintain quiet operation of the gears. Visual and physical inspection of the lubricant is part of required maintenance. It helps prevent premature failure of the gears. Maintenance schedules are helpful for technicians to know the required maintenance intervals.

Drive Axle

The drive axle is the component that supplies the power from the final drive to the wheels. It is part of the **drive axle assembly**. There are two types of axles—live and dead. A **live axle** powers the wheels attached to it (**FIGURE 21-27**). **Dead axles** allow the wheels to freely rotate on the axle assembly, but do not drive the wheels (**FIGURE 21-28**).

There are two types of live axles. The first type is the **independent suspension drive axle** (**FIGURE 21-29**). It uses one half-shaft axle for each of the two wheels. These axles must be able to flex to allow suspension movement. Because of this, they incorporate what are called CV joints. CV joints connect the transaxle final drive to the wheels. CV joints are known for their flexibility in allowing greater operating angles than universal joints. This allows them to be used for steering the front wheels while also supplying them with power. Rubber boots contain the lubricant inside each CV joint to maintain its integrity.

The second type of live axle in use today is the **solid axle** (**FIGURE 21-30**). These axles are fitted within a strong nonflexible housing. The outer ends of the axles have flanges with studs and nuts that are used to mount wheels and tires. The other end of the axle has splines that slide into the differential side gears inside the differential housing. This is where the solid axle gets its power. There are three types of solid live axles used today in RWD vehicles: the semi-floating axle, the three-quarter floating axle, and the full floating axle. All three types are discussed in detail in the Drivetrain Components chapter.

FIGURE 21-27 Live axles drive wheels.

FIGURE 21-28 Dead axles hold wheels in position only; they don't drive the wheels.

▶ TECHNICIAN TIP

A good preventive maintenance program will keep on top of all leaks and fluid losses. Always consult manufacturer specifications for correct fluid types.

▶ Preventive Maintenance

LO 21-07 Perform drivetrain fluid maintenance.

Preventive maintenance helps extend the service life of the drivetrain components. Some newer vehicles have fewer maintenance requirements. Regardless, regular inspection is still critical. Vehicles that need maintenance should adhere to the manufacturer's schedule. This ensures a long life. It usually consists of:

- checking and replacing lubricants on a regular basis,
- lubricating linkage pivot points,
- performing adjustments such as the clutch adjustment, and
- visually inspecting all drivetrain components for leaks, damage, or wear.

Failure to perform required maintenance can result in severe damage to the internal components. This can lead to expensive repairs.

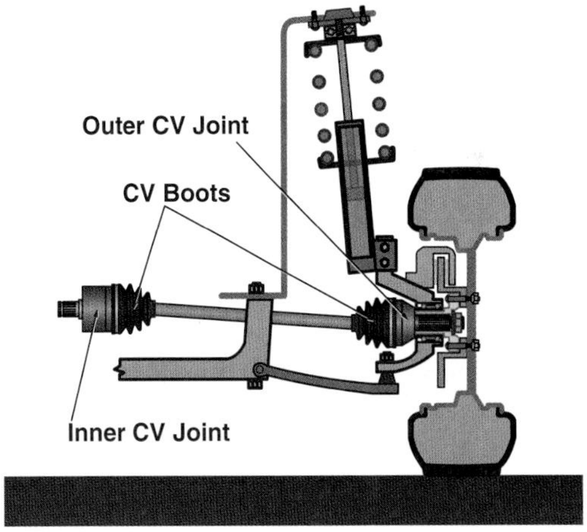

FIGURE 21-29 Independent suspension drive axle.

FIGURE 21-30 Solid axle fits inside a solid axle housing.

FIGURE 21-31 Checking the fluid level.

FIGURE 21-32 Typical gear lube used in manual transmissions.

For example, never checking the transmission fluid level, even though it has a small leak, can lead to a substantially low fluid level. The low fluid level can ruin the transmission long before its time.

Maintenance also improves the reliability of the vehicle and helps prevent vehicle breakdowns. Although most vehicle owners understand this, not all of them choose to have preventive maintenance performed on a regular basis. Many owners do not know the extent of the maintenance needed, or how it is performed. This is where you come in. By learning what maintenance is needed, and how to perform it, you can help get the maximum reliability and life span out of the vehicle (**FIGURE 21-31**).

Lubrication

Virtually every drivetrain component requires lubrication according to manufacturer specifications. This continues even after the warranty runs out. The clutch, transmission, final drive, driveshafts, and drive axles need to be inspected or lubricated on a regular basis.

Proper lubricants will ensure that all drivetrain components last a long time. There has been a variety of lubricants for drivetrains over the years. The most common manual transmission lubricant is gear lube (**FIGURE 21-32**). Some manufacturers specify a specific weight of engine oil be used. Some manufacturers even specify automatic transmission fluid be used in their manual transmissions. It is important to maintain all lubricants on a regular basis. This is because of the amount of stress put on the drivetrain.

The transmission/transaxle and differential lubricant/oil levels keep all rotating parts cool. The lubricant absorbs heat and transfers it to the case. The case then dissipates the heat to the atmosphere. The lubricant also supplies a thin film of oil between close tolerances of gears, shafts, and bearings. This reduces metal-to-metal contact. It also prolongs the life of these parts. Improper lubricant/oil levels could cause overheating and unwanted failures to occur.

Components such as CV joints require different types of semisolid lubricants, or grease. Special lead-based grease is used in many applications. This is because there is a lot of torque that CV joints must endure (**FIGURE 21-33**). Since lead is a hazardous material, follow all precautionary measures. This includes using gloves and washing your hands thoroughly after working with it.

FIGURE 21-33 Some CV joints use lead-based grease.

Grease possesses a property called **thixotropy**. This term refers to its ability to be a solid but, while under stress, to flow in order to lubricate properly. Many types of semisolid lubricants are in use today. Always remember to consult the manufacturer's service information for the correct application.

Checking Transmission Fluid

Checking the fluid level of the manual transmission is a preventive maintenance task. It should be checked according to the manufacturer's maintenance schedule. Routinely, this occurs when other tasks are performed, such as an engine oil and filter change. Also, not all manual transmissions use gear lube. Some use engine oil of a specified viscosity. Others even use a specified type of automatic transmission fluid. So, make sure you check the service information to determine the correct type and quantity of fluid.

When checking fluid level, the engine should be off. The vehicle should also be level. Always place a drain pan under the fill plug before removing the fill plug. This will allow you to catch any fluid that spills out. Locate and remove the fill plug. Typically, it is a threaded plug. But some transmissions use a rubber plug that pushes in. Fluid may drain out of the fill plug if the fluid level is a bit too high. The level should be right at the bottom of the fill hole, or within ¼" (6 mm) (**FIGURE 21-34**). If

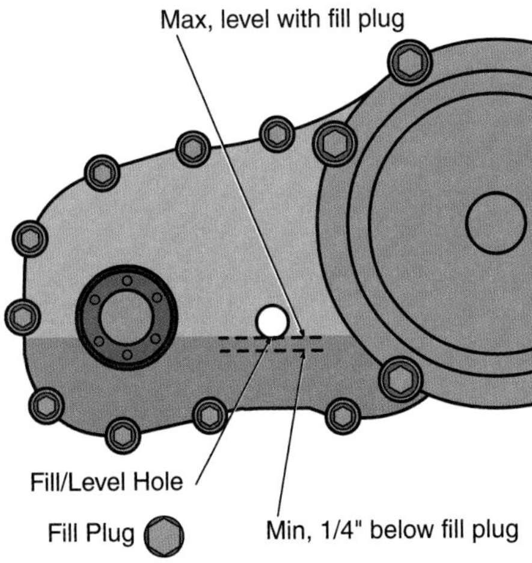

FIGURE 21-34 Proper fluid level for a manual transmission

it is lower than that, inspect the transmission for a leak. In the event of a leak, once the repair is complete, the fluid should be topped off with the specified fluid.

Fluid level and condition should be visually inspected anytime a transmission fault is suspected. Condition and level of the transmission fluid can reveal a lot about what is going on inside. For example, what if metal is found in the transmission fluid? Then, the technician knows that the transmission will have to be rebuilt or replaced. Make it a habit to visually inspect the transmission every time the vehicle is in for an oil change or other service. That way you can catch issues before they become severe.

> ▶ TECHNICIAN TIP
>
> It is always a good idea to keep up with preventive maintenance on a regular basis. Remind the customer that preventive maintenance is the road to long life for the vehicle. It can also help save thousands of dollars in repairs.

To check and adjust the fluid level, follow the steps in **SKILL DRILL 21-1**.

Identifying the Cause of Fluid Loss

Performing preventive maintenance on a regular basis helps ensure that fluid leaks are addressed before they become a major problem. Diagnosing leaks and concerns requires that you raise the vehicle on a hoist or jack stands. Noises and shifting concerns can be signs of fluid leaks or problems. So, always inspect the transmission if these symptoms are present. If a leak is found, identify the exact cause of the leak. In some cases, repairing fluid leaks can be done with the transmission in the vehicle. Other times, it requires removal of the transmission.

Fluid leaks may be a result of the following conditions:

- Too much fluid in the transmission
- Leaking seals, including the following:
  - Half-shaft axle seals
  - Input shaft retainer O-rings or lip seals
  - Speed sensor seals or O-ring
  - Backup light switch O-ring
  - Shifting lever shaft seals
  - Transmission rear seal
  - Side covers or access plates
- Improper fluid type
- Transmission case porosity or cracks
- Missing, loose, or stripped case bolts
- Damaged gaskets for case halves
- Loose drain or fill plugs

To identify the cause of fluid loss in a transmission, follow the steps in **SKILL DRILL 21-2**.

Checking and Adjusting Final Drive/Transfer Case Fluid

Checking and adjusting the final drive and transfer case fluid level is like checking the transmission fluid level. But there are some differences. First, some differential/final drive assemblies require specially designed additives or fluids. Examples include

SKILL DRILL 21-1 Checking the Fluid Level of a Manual Transmission

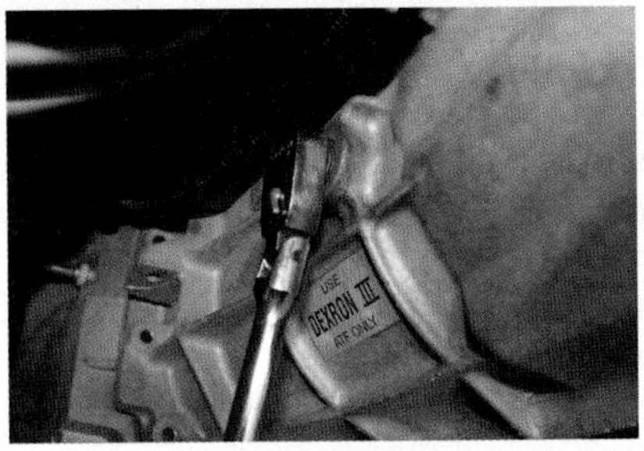

1. Safely raise and support the vehicle on the lift so that it is level. Inspect the transmission for leaks. Remove the filler plug, using the proper tool. Inspect the filler plug and fill hole for thread damage, and replace or repair if necessary.

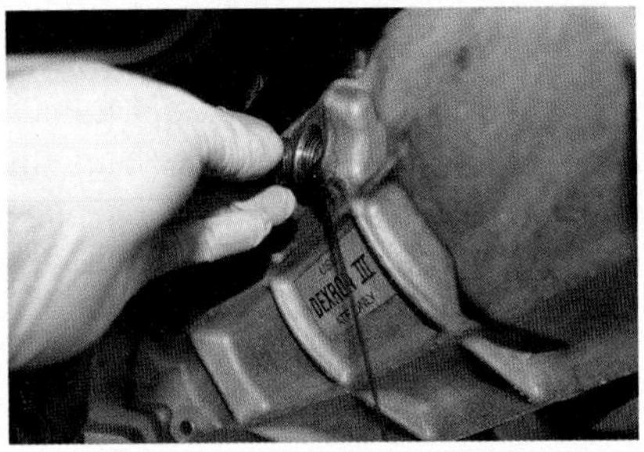

2. If the transmission fluid begins to run out as the filler plug is removed, let the transmission fluid seek its own level before reinstalling the filler plug. The transmission fluid level should be at the bottom of the filler plug hole.

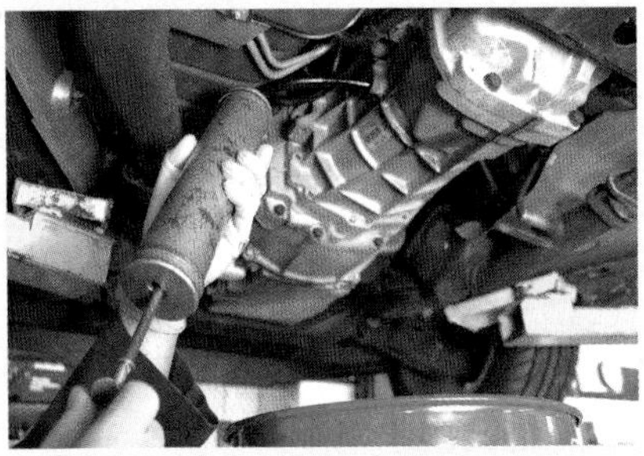

3. If the fluid level is low, refill with the specified fluid, reinstall the filler plug, and wipe the area around the filler plug hole with a clean shop towel. Tighten the filler plug to the specified torque.

SKILL DRILL 21-2 Identifying the Cause of Fluid Loss in a Transmission

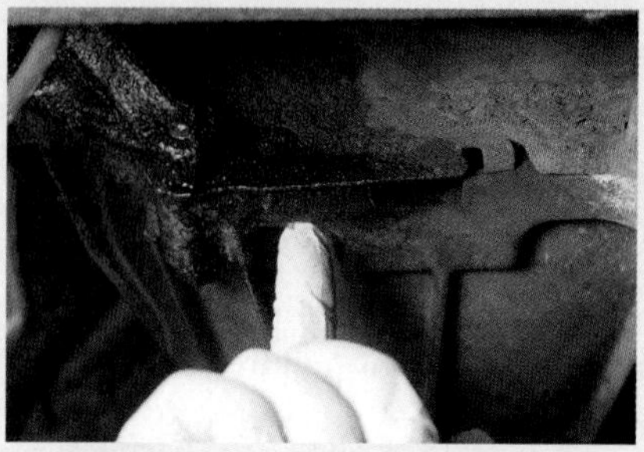

1. Safely raise and support the vehicle. Look for leaks in the transmission bell housing to the engine block (front seal).

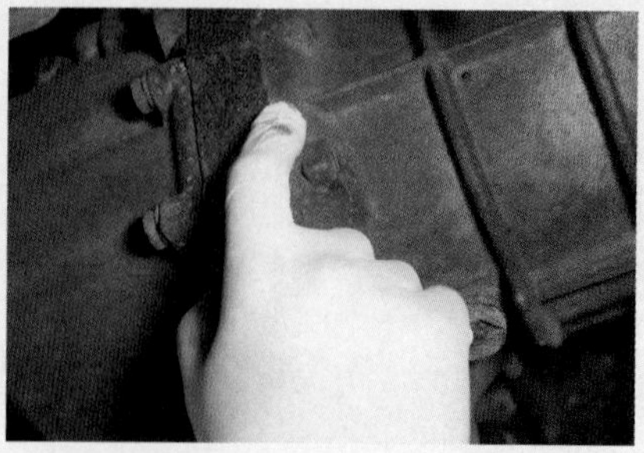

2. Look for leaks in the transmission breather outlet. Look for leaks in all case gasket areas.

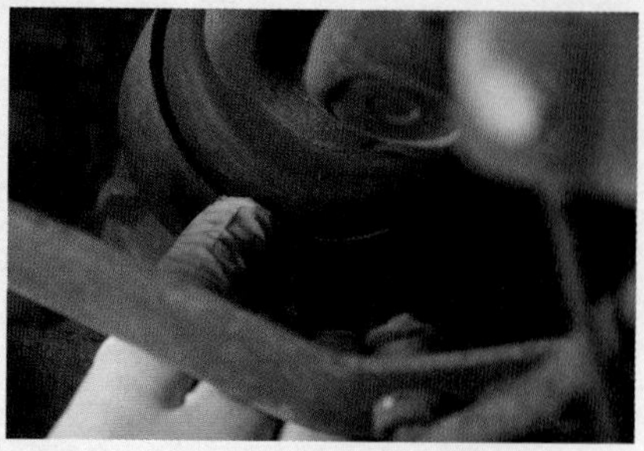

3. Look for leaks in the rear tail shaft seal.

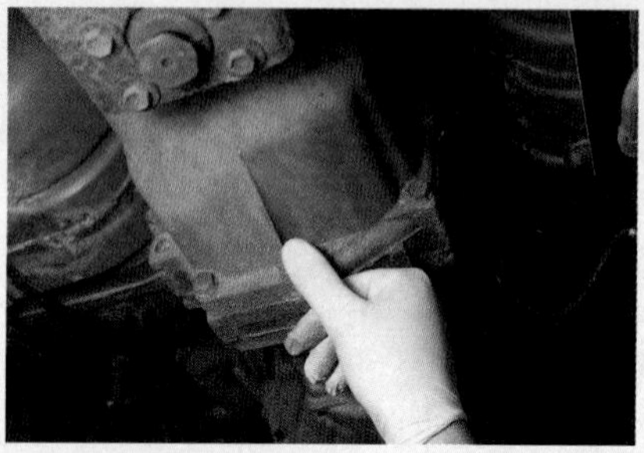

4. Inspect for transmission case defects, such as cracks and porosity.

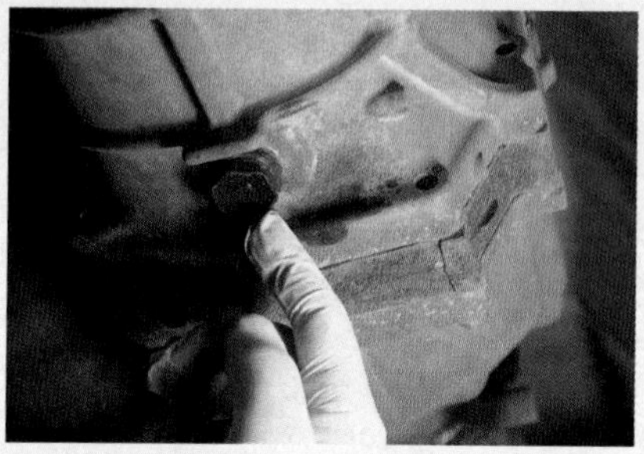

5. Look for leaks at the drain and fill plugs.

6. Check the gear fluid level. If excess level is found, let the excess drain into a container. If the fluid is more than ¼" (6 mm) below the bottom of the threaded hole, add new fluid of the correct viscosity and type.

limited-slip and positraction assemblies. Transfer cases may also require special lubricants as specified by the manufacturer. Always check the service information for the vehicle you are working on.

Final drives and transfer cases generally have fill plugs that can be used for checking the fluid level in the same way as a transmission. However, some applications do not have a drain plug. In these cases, either a specific bolt may need to be removed. Or a cover may need to be unbolted and removed for the fluid to be drained. Follow the procedure listed in the service information.

To check and adjust the differential/transfer case fluid level, follow the steps in **SKILL DRILL 21-3**.

Changing Manual Transmission and Final Drive Fluid

Most manufacturers specify a change interval for the transmission fluid. Transmission fluid left unchanged will break down from heat. This causes it to lose its capacity to lubricate all rotating parts. This will cause premature bearing and gear failure. Regular change intervals help ensure that none of these conditions occur.

Manual transmissions use fluids specified by the manufacturer. Do not assume that all transmissions use the same fluid. They don't. Some use gear lube of differing viscosities. Some use engine oil of a specified viscosity. And some even specify automatic transmission fluid in the transmission. Therefore, always verify that you are using the specified fluid in a particular vehicle. There is one more thing to keep in mind. Some transaxles use a separate reservoir for the final drive, which is separate from the transaxle reservoir. And in some cases, they use different types of fluid. Always check the manufacturer's specifications before changing out the fluids.

To prepare for a transmission fluid change, it is a good idea to have the transmission fluid warm. When it is warm, it will flow better to ensure proper draining. If possible, drive the vehicle until the engine is at operating temperature. Then change the transmission fluid. Do not touch the transmission fluid. It will be hot and could cause burns.

To change the transmission fluid, follow the steps in **SKILL DRILL 21-4**.

SKILL DRILL 21-3 Checking and Adjusting the Differential/Transfer Case Fluid Level

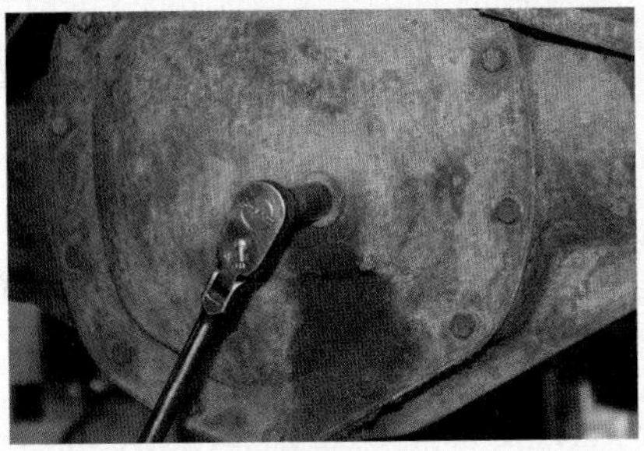

1. Safely raise and support the vehicle. Inspect the differential and transfer case for leaks. Position a clean drain pan under the filler plug. Remove the filler plug, using the proper tool. Inspect the filler plug threads for thread, and replace if necessary.

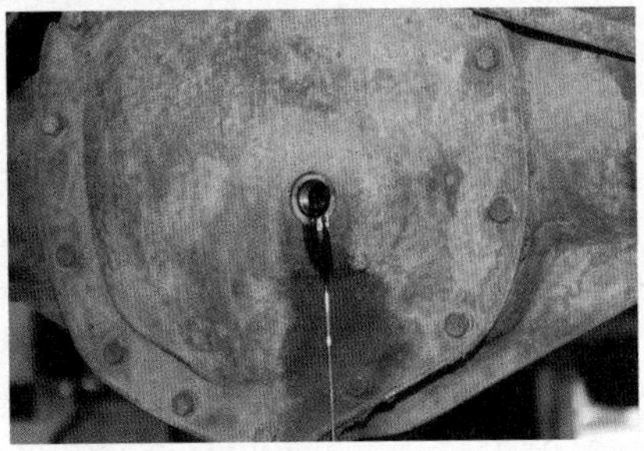

2. If the fluid begins to run out as the filler plug is removed, let the fluid seek its own level before reinstalling the filler plug. The fluid level should be at the bottom of the filler plug hole.

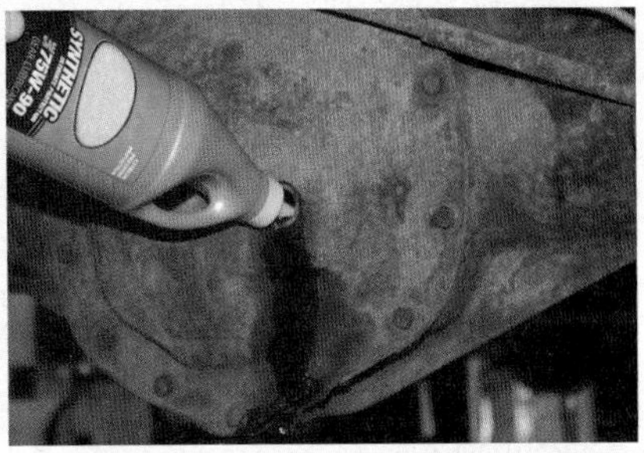

3. If the fluid level is low, refill with the specified fluid, reinstall the filler plug, and wipe the area around the filler plug hole with a clean shop towel. Tighten the filler plug to the specified torque.

SKILL DRILL 21-4 Changing the Transmission Fluid

1. Safely raise and support the vehicle, using an approved lift. Obtain a clean drain pan to put the used fluid in.

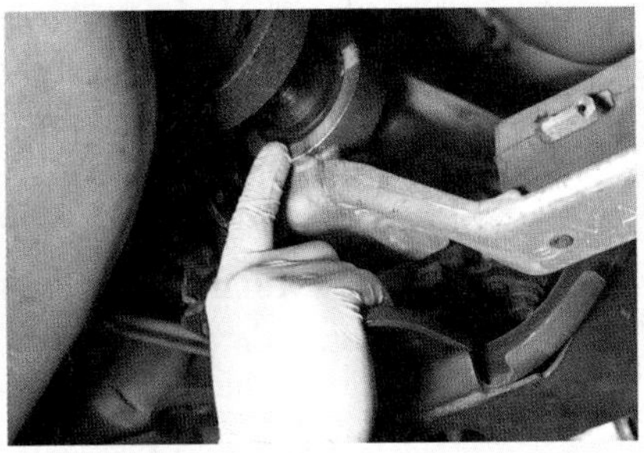

2. Inspect the transmission for leaks.

3. Remove the drain plug from the bottom of the transmission, being careful of the hot transmission fluid. Let transmission fluid drain until it has stopped running. If necessary, drain the final drive assembly.

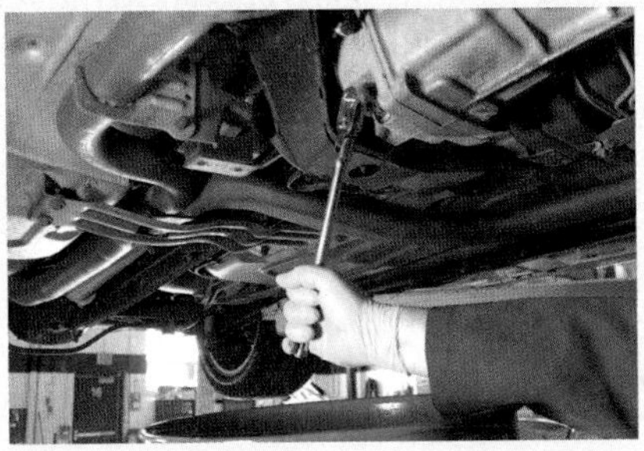

4. Replace the drain plug(s) and tighten to specification, and remove the fill plug(s).

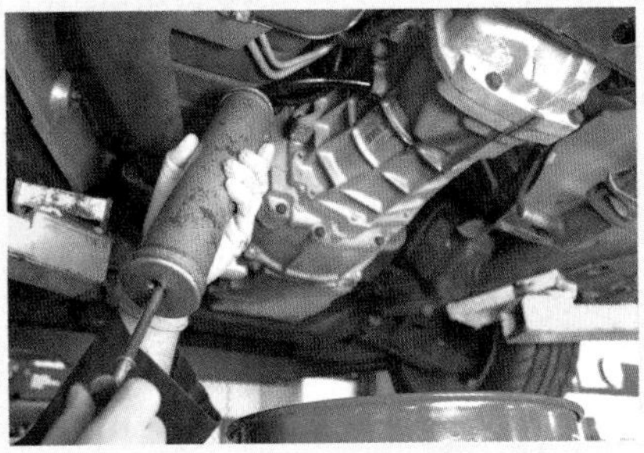

5. Refill the transmission and final drive to the proper level, using manufacturer-approved transmission fluid. Replace the fill plug(s), and tighten to the proper torque. Wipe away any spillage. Road test the vehicle. If necessary, put the vehicle back on the lift, and check for any leaks.

▶ Wrap-Up

Ready for Review

- The one-cylinder engine with an FWD carriage patented by George Selden in 1877 made use of leather belts on a pulley and a geared wheel on the axle to make the engine move. In 1894, Louis René Panhard and Émile Levassor designed an FWD multi-gear manual transmission that used a clutch-driven, three-speed, sliding-gear transmission.
  - This sliding-gear arrangement became the basis for nearly all manual transmissions since then.
- In a manual transmission, different gear ratios allow the driver to tailor the engine speed and power to the driving conditions.
  - Power flows into the transmission through the input shaft, then to the countershaft, and finally to the output shaft through various gears.
- Transmission refers to layouts where the transmission sends power to the final drive assembly. Transaxle refers to layouts where the transmission and final drive are integrated into a common assembly.
  - Power from transmission passes through universal joint, drive shaft, universal joint, final drive assembly, and finally to the rear tires.
- In a manual transmission, shafts accommodate bearings and individual gears, gears mate with one or more other gears to achieve a specific gear ratio, and bearings maintain the position and alignment of the shafts and gears under load.
- A clutch is used to transmit power to the transmission as and when needed. Transmissions/transaxles are rated for how much twisting force (torque) they can handle. A transfer case transfers power to both the front and rear axles.
- A differential refers to the components inside the differential housing that allow the axles to turn at different speeds when the vehicle is cornering, a final drive provides the final gear reduction necessary for drivetrain operation, a drive axle supplies the power from the final drive to the wheels.
  - Differentials assemblies can be either open differential assembly or limited slip differential assembly.
- Drivetrain maintenance consists of checking and replacing lubricants on a regular basis, lubricating linkage pivot points, performing adjustments such as the clutch adjustment, and visually inspecting all drivetrain components for leaks, damage, or wear.

Keyterm

accelerator pedal The foot-operated pedal used by the driver to increase and decrease the amount of power the engine develops.

axial load The load applied in line with a shaft. It can be controlled with thrust bearings.

brake pedal The foot-operated lever which provides leverage to the brake system.

clutch pedal The foot-operated pedal used by the driver to engage and disengage the clutch.

clutch system A mechanically operated assembly that connects and disconnects the engine from the transmission.

constant-velocity (CV) joints A flexible joint used to transmit torque in axles and driveshafts. This style maintains a constant velocity no matter the angle.

Dead axles An axle that supports only supports the wheel. It doesn't drive it.

differential gears A gear arrangement that splits the available torque equally between two wheels while allowing them to turn at different speeds when required.

drive axles A very strong part used to turn the wheels.

drive axle assembly The components that make up the drive axle, including the axles, final drive assembly, bearings, and axle housing.

driveshaft The shaft or tube fitted with universal couplings that is connected between the transmission and other drivetrain components, to transmit torque and rotation.

drivetrain A term used to identify the engine, transmission/transaxle, differential, axles, and wheels.

final drive assembly An assembly used to power the drive wheels and allow the wheels to rotate at different speeds as the vehicle turns.

gear A relatively round, rotating part with internal or external teeth that are designed to mesh with another gear for the purpose of transmitting torque.

gear set Gears that are in mesh with each other.

gear synchronizer An assembly in the transmission that is used to bring two unequally spinning shafts or gears to the same speed when upshifting or downshifting.

half-shaft An axle that has CV joints on each end and that fits between the transaxle and wheel. Typically, one is used on each side of a vehicle.

independent suspension drive axle A type of suspension that allows each wheel on a drive axle to move independently of the other.

limited slip differential assembly A differential assembly that uses a clutch assembly or gear assembly to allow a limited amount of slip between the two axles. It is used to increase drive wheel traction in slippery conditions.

live axle The axle that drives the machine by turning the power from the driveshaft 90 degrees to deliver it to the wheels and providing the final gear reduction in the drivetrain; also known as a drive axle.

mechanical advantage Occurs when we give up either speed or torque to increase either torque or speed through a machine.

open differential assembly A differential assembly that allows both axles to turn at their own speed when turning a corner, but is dependent on the traction of the tires to deliver torque to the ground. If one wheel has no traction, all of the engine's torque will be used at that wheel, causing it to simply spin.

power flow The path that power takes from the beginning of an assembly to the end.

radial load The load that is perpendicular to a shaft, usually controlled by bearings or bushings.

rotational speed The speed at which an object rotates, measured in revolutions per minute (rpm).

shaft The long, narrow component that carries one or more gears or has gears machined into it.

snap ring The spring steel, C-shaped ring that is fitted in a groove and holds gears, bearings, and shafts in place.

solid axle A single piece of steel that provides a simple means of mounting the hub and wheel units; also called beam axle or straight axle.

solid axle A single piece of steel that provides a simple means of mounting the hub and wheel units; also called beam axle or straight axle.

splined Typically, a shaft and gear that have parallel grooves machined in them so they mate with each other and lock together rotationally.

synchromesh transmission A modern transmission that uses gear synchronizers to match the speeds of gears and shafts during upshifts and downshifts.

thixotropy The ability of a semisolid grease to flow when agitated or stressed.

transfer case A gearbox arrangement that allows the torque from the transmission to be split between the front and rear driving axles of a vehicle.

transmission An assembly that houses a variety of gearsets that allow the vehicle to be driven at a wider range of speeds and terrain conditions than would be possible without a transmission.

universal joints A flexible cross-shaped joint used to transmit torque.

Review Questions

1. What do gear synchronizers do?
   a. They keep the output shaft in sync with the countershaft
   b. They synchronize the shift lever to the shift forks
   c. They match gear speed to output shaft speed
   d. They keep the transmission from going into two gears at once
2. A gear ratio of 3:1 means:
   a. The driven gear turns 3 times faster than the drive gear and has mechanical advantage
   b. The drive gear turns 3 times faster than the driven gear and has mechanical advantage
   c. The driven gear turns 3 times faster than the drive gear and does not have mechanical advantage
   d. The drive gear turns 3 times faster than the driven gear and does not have mechanical advantage
3. The path power flow takes through a transmission may include all the following parts EXCEPT:
   a. Transmission output shaft
   b. Counter shaft
   c. Transmission input shaft
   d. Shift fork
4. What is the difference between a transmission and a transaxle?
   a. Transmissions are used on front-wheel drive vehicles
   b. Transaxles have driveshafts with universal joints
   c. Transaxles integrate the transmission and the final drive into the same assembly
   d. Transmissions have larger bell housings than transaxles
5. What are snap rings commonly used for in a manual transmission?
   a. They hold gears in the proper position on the shafts
   b. They provide sliding metal-to-metal contact between components
   c. They mate with gears to achieve a specific gear ratio
   d. They are splined to transfer power to the drive wheels
6. How are manual transmissions/transaxles most often rated?
   a. Power measured in watts
   b. Speed measured in mph
   c. Torque measured in ft-lb
   d. Energy measured in BTUs
7. An electronically controlled transfer case needs all of these EXCEPT:
   a. Inputs (switches and sensors)
   b. Manual actuation (shift levers and manual locking hubs)
   c. Processing (control module)
   d. Outputs (motors and solenoids)
8. What assembly allows the drive wheels to turn at different speeds around corners?
   a. Transfer case
   b. Transmission
   c. Driveshaft
   d. Differential
9. Constant velocity (CV) joints are used on what type of axle?
   a. Independent suspension -- live
   b. Semi-floating
   c. Full floating
   d. Solid beam axle -- dead
10. All the following fluids are commonly used as a lubricant for manual transmissions EXCEPT:
    a. Gear lube
    b. Engine oil
    c. Automatic transmission fluid
    d. DOT 3 brake fluid
11. What do metal chunks in the manual transmission lubricant indicate?
    a. A normal condition
    b. Possible gear damage
    c. Clutch wear
    d. Pilot bearing damage

ASE Technician A/Technician B Style Questions

1. Technician A says that a differential allows drive wheels to turn at different speeds. Technician B says that differentials cause increased tire wear compared to solid axles. Who is correct?
   a. Technician A only
   b. Technician B only
   c. Both A and B
   d. Neither A nor B
2. Technician A says that an input gear with half as many teeth as the output gear has a ratio of 0.5:1. Technician B says that overdrive is when the driven gear is smaller than the drive gear. Who is correct?
   a. Technician A only
   b. Technician B only
   c. Both A and B
   d. Neither A nor B
3. Technician A says that the clutch connects and disconnects engine power to the drivetrain. Technician B says that the clutch allows the synchronizers to make smooth shifts. Who is correct?
   a. Technician A only
   b. Technician B only
   c. Both A and B
   d. Neither A nor B
4. Technician A says that axial loads are perpendicular to the shaft. Technician B says that radial loads are in line with the shaft. Who is correct?
   a. Technician A only
   b. Technician B only
   c. Both A and B
   d. Neither A nor B
5. Technician A says that a clutch allows a gradual application of engine power to the drivetrain. Technician B says that excessive slippage creates heat that can damage a clutch assembly. Who is correct?
   a. Technician A only
   b. Technician B only
   c. Both A and B
   d. Neither A nor B
6. Technician A says the gear members of a final drive are the ring gear and the pinion gear. Technician B says the pinion gear has more teeth than the ring gear. Who is correct?
   a. Technician A only
   b. Technician B only
   c. Both A and B
   d. Neither A nor B
7. Technician A says limited slip differentials apply power to both drive wheels even if one loses traction. Technician B says open differentials will only supply power to one wheel if it loses traction. Who is correct?
   a. Technician A only
   b. Technician B only
   c. Both A and B
   d. Neither A nor B
8. Technician A says that the engine needs to be idling to check the fluid level on a manual transmission. Technician B says that the fluid level should be at least 1" (25.4 mm) below the bottom of the fill hole. Who is correct?
   a. Technician A only
   b. Technician B only
   c. Both A and B
   d. Neither A nor B
9. Technician A says that some manual transmission fluid leakage is normal. Technician B says that manual transmissions have O-rings, gaskets, and seals that can potentially leak so a thorough visual inspection is routinely required. Who is correct?
   a. Technician A only
   b. Technician B only
   c. Both A and B
   d. Neither A nor B
10. Technician A says that manual transmission fluid is good for the life of the vehicle and only needs to be serviced if there is a leak. Technician B says to warm the transmission by driving it to help it drain better. Who is correct?
   a. Technician A only
   b. Technician B only
   c. Both A and B
   d. Neither A nor B

CHAPTER 22

The Clutch System

Learning Objectives

- **LO 22-01** Describe clutch principles.
- **LO 22-02** Describe the purpose and design of flywheels.
- **LO 22-03** Describe pressure plates and clutch discs.
- **LO 22-04** Describe throw-out bearing, clutch fork, and pilot bearing.
- **LO 22-05** Describe clutch operating mechanisms.
- **LO 22-06** Perform clutch maintenance.

ASE Education Foundation Tasks

See Appendix A to view the 2017 ASE Education Foundation Automobile Accreditation Task List Correlation Guide.

▶ Introduction

The clutch is a mechanical device located in the bell housing. It sits between the engine and the transmission (**FIGURE 22-1**). The clutch allows the driver to engage and disengage the engine from the transmission while operating the vehicle. The driver controls the clutch with his or her foot through the clutch pedal. The clutch has a series of components that help to make driving an interactive experience. It allows the driver to control the shifting of gears while driving. An automatic transmission does not use a manually operated clutch. This gives the driver less control over transmission operation other than to put it in or out of gear.

▶ Clutch Principles

LO 22-01 Describe clutch principles.

As mentioned, the clutch engages the engine to the transmission. But it needs to do so progressively. During application, it progressively transmits torque from the engine to the transmission. It also allows the transmission to be disconnected from the engine when shifting between gears.

Manual transmission clutches are typically dry clutches. Automatic transmissions use wet clutches, which run in a lubricating fluid (**FIGURE 22-2**). Dry clutches use the friction between the clutch surfaces to transmit torque. The amount of torque a clutch can transmit depends on the amount of friction between the components. In this case, between the clutch disc and the mating surfaces of the flywheel and pressure plate. Friction in a clutch is dependent on four variables:

- the coefficient of friction of the clutch facings,
- the diameter of the clutch,
- the number of clutch discs in the clutch assembly, and
- the total spring force clamping the parts together.

Increasing the friction of the clutch disc increases its torque-carrying ability. But it causes the clutch to grab. This makes it

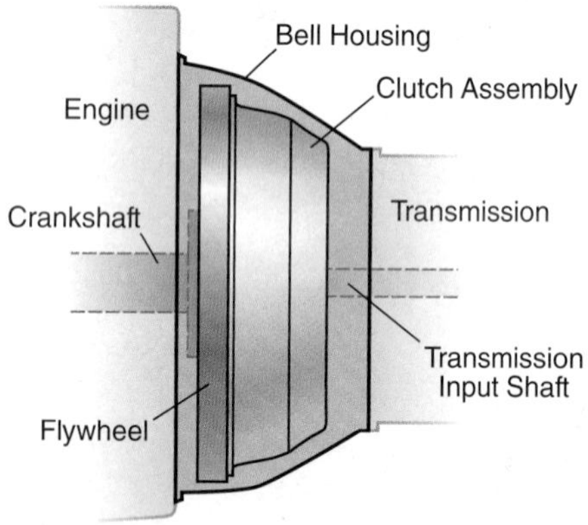

FIGURE 22-1 The clutch engages and disengages the engine from the transmission.

FIGURE 22-2 Types of clutches. **A.** Dry clutch from a manual transmission. **B.** Wet clutch from an auto transmission.

You Are the Automotive Technician

A customer brings a 2012 Ford Mustang into the shop for a clutch inspection. The clutch was slipping when it is in fourth gear, driving on the highway. You inform the customer that it would be beneficial to perform a test drive to properly diagnose this problem. The customer agrees, and you verify that the clutch slips when applying throttle while going up a hill. You drive back to the shop and inform the customer that you need to perform some additional diagnosis in the shop. Once you finish the diagnosis, you will let them know what you find.

1. What things should be done to confirm the clutch is worn out before removing the transmission?
2. Why is a testdrive important to help diagnose clutch system issues?
3. What are the main components to the clutch assembly?
4. What are the different clutch operating systems?

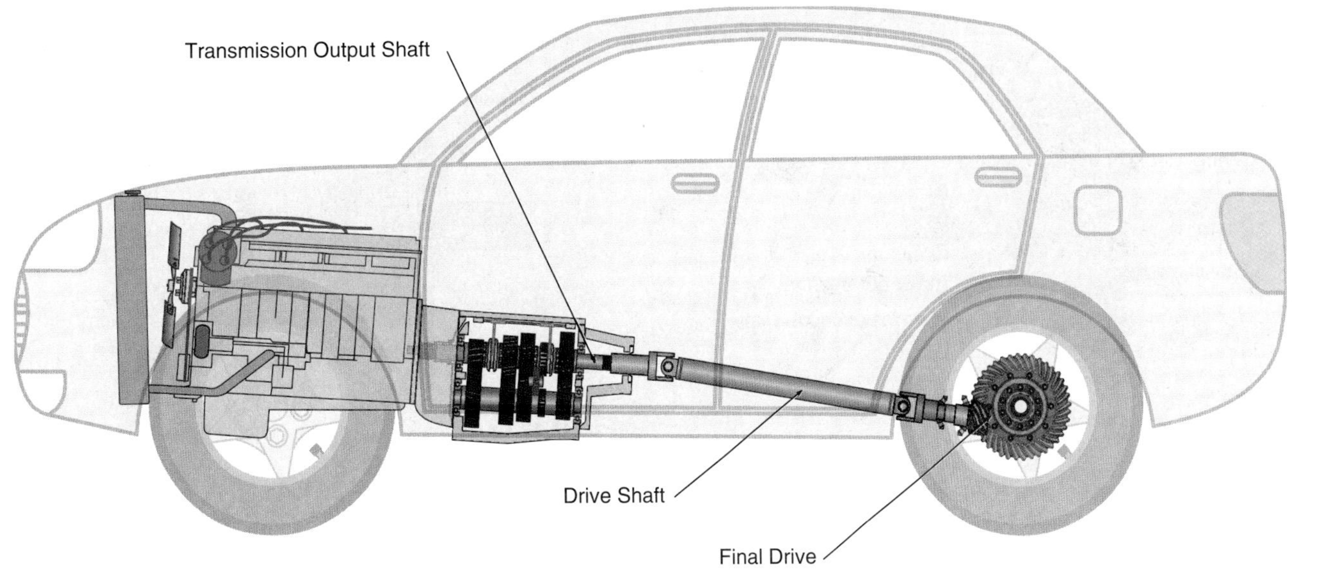

FIGURE 22-3 The transmission output shaft turns with the wheels because they are directly connected.

more difficult to move the vehicle from a stop. Increasing the diameter of the clutch gives it more leverage. This increases its torque capacity but takes up more area. Increasing the number of clutch discs increases the torque capacity. But the additional parts make it more complicated. Increasing the spring force clamping the parts together increases the torque capacity. But it takes more foot pressure to operate the clutch pedal.

Manufacturers have to balance all of these factors when designing a clutch for a particular application. This also comes into play if the customer modifies the engine to develop more horsepower. In this case, the clutch will likely need to be upgraded as well.

▶ TECHNICIAN TIP

Two or more clutch plates can be used to form a multi-plate clutch. Increasing the number of facings increases the torque capacity. This design is useful where a reduction in clutch diameter is desired. It is also an option when increasing the spring strength is undesirable. Increasing pedal effort increases driver discomfort.

Input and Output Shaft Speed

Vehicles are dependent on shifting between various gears at a variety of speeds. So, it is important to understand how the transmission input and output shaft speeds play a critical role in shifting. First, you need to understand that output shaft torque and speed can each be increased, but not at the same time. In low gear, the input shaft speed is relatively high while the output shaft speed is relatively low. This results in an increase of torque from the output shaft. However, the speed of the output shaft (and the speed of the vehicle) is relatively low. This setup makes for good acceleration from a stop or while driving at slow speeds.

As higher transmission gears are selected, the output shaft speed increases with each higher gear. However, the torque of the output shaft is lessened with each higher gear. This allows the vehicle to travel at higher speeds without over-revving the engine, but with decreased torque.

The second role that the input and output shaft speeds play is during the actual shift. When transitioning from one gear to another, the relative speeds of at least one shaft must change. This allows the gears to shift from one set to another. In order for a gear to be selected, at least one of the shafts must be able to turn freely for this speed change to happen.

The input shaft is able to connect to and disconnect from the engine's flywheel through the operation of the clutch. This allows the input shaft's speed to change during a shifting of the gears. The speed of the output shaft cannot be changed. This is because it is connected directly to the wheels through the drivetrain components (**FIGURE 22-3**). The faster the vehicle speed, the faster the rotation of the output shaft. This affects shifting because the speeds of both shafts must be such that the individual gears can be selected. This will be explored further when we discuss synchronizers.

Operation of Clutch Components

The main components of a clutch assembly are the **flywheel**, **clutch disc**, **pressure plate**, **throw-out bearing**, **clutch fork**, and **pilot bearing** (**FIGURE 22-4**). The flywheel is bolted to the rear of the crankshaft. The spring-loaded pressure plate is bolted to the flywheel. The clutch disc is clamped between the flywheel and pressure plate. It has two **friction facings** attached to a central hub. The hub is splined to engage the **transmission input shaft**.

With engine rotation, the flywheel, clutch disc, and pressure plate rotate together. The flywheel and pressure plate are the drive unit, and the clutch disc is the driven unit. Engine torque is transferred from the flywheel and pressure plate through the friction facings of the clutch disc. The center hub sends torque to the input shaft of the transmission.

Pushing the clutch pedal operates the clutch **release mechanism**. It does so by retracting the pressure plate against the force of its springs. This frees the friction disc from the clamping action of the pressure plate (**FIGURE 22-5**). Releasing the clutch pedal reapplies the clamping force. It reconnects the engine

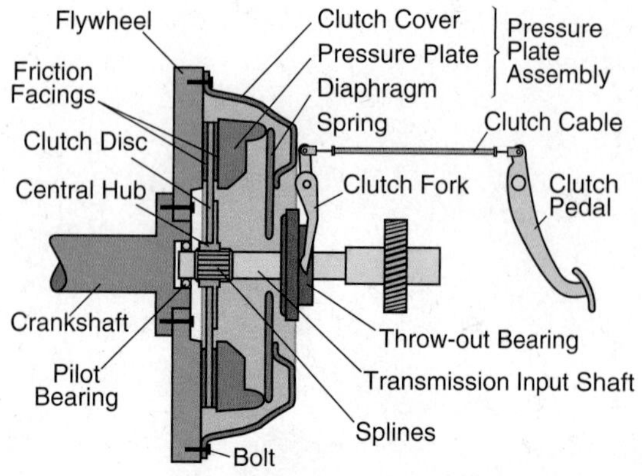

FIGURE 22-4 A standard light vehicle clutch.

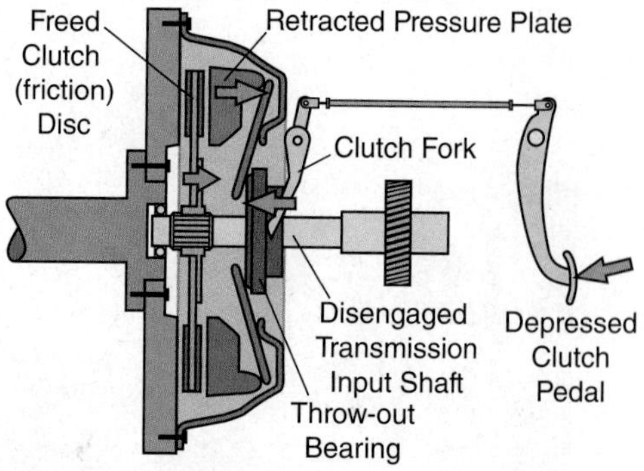

FIGURE 22-5 Depressing the clutch pedal retracts the pressure plate against the force of its springs or diaphragm and frees the friction disc from its clamping action.

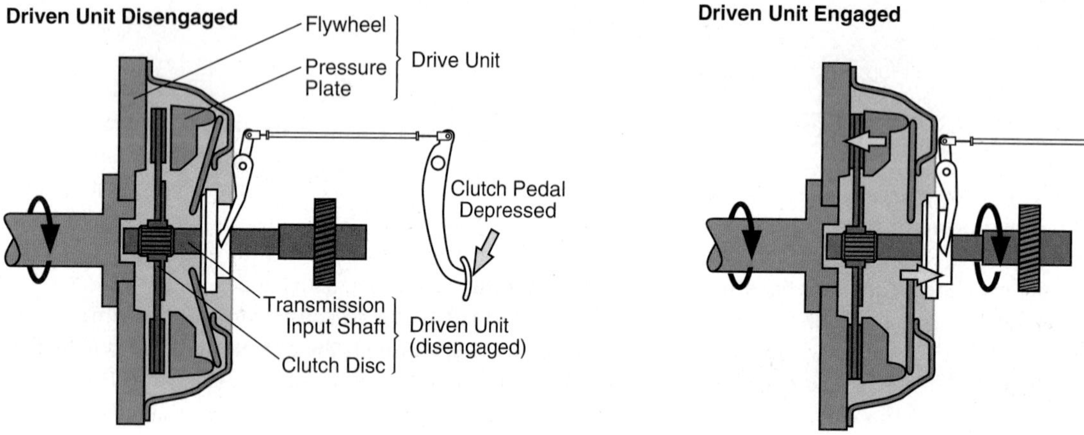

FIGURE 22-6 Releasing the clutch pedal reapplies the clamping force and reconnects the engine and transmission, firmly clamping them together to continue rotating as a unit.

and transmission by firmly clamping the clutch disc between the pressure plate and the flywheel. The clutch assembly now rotates as a unit (**FIGURE 22-6**).

▶ Clutch Components

LO 22-02 Describe the purpose and design of flywheels.

Flywheel

The main purpose of the flywheel is to smooth out the power pulses from the pistons during the power strokes. It also provides a friction surface for the clutch disc and a mounting surface for the pressure plate (**FIGURE 22-7**). The flywheel is quite heavy. It is usually made of cast iron so that it can store energy from each power pulse from the engine. It uses that energy to keep the crankshaft turning through the intake, compression, and exhaust strokes.

However, a heavy flywheel means that it slows the engine's acceleration. A lighter flywheel does not smooth out the power pulses as effectively, but it works well on a race car. Because it's lighter, it allows the engine to accelerate faster.

The flywheel is bolted to the rear of the crankshaft. It also incorporates the **starter ring gear**. The ring gear meshes with the starter motor drive gear to crank the engine over. In most cases, the ring gear is made of hardened steel and is press fit onto the outer edge of the flywheel.

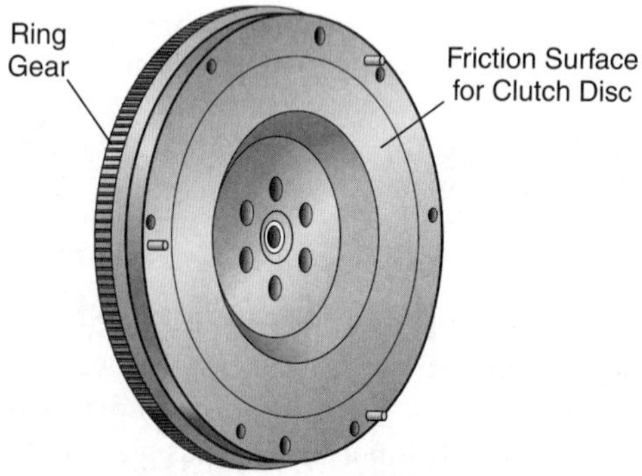

FIGURE 22-7 The flywheel.

▶ TECHNICIAN TIP

When replacing a starter motor, the teeth on the ring gear should be inspected thoroughly as well. If the teeth on the ring gear are damaged, they will likely destroy the matching teeth on the new starter motor.

Types of Flywheels

There are two main types of flywheels: one-piece flywheel and dual-mass flywheel. The one-piece flywheel is by far the most common on light vehicles. Its one-piece construction makes it simple, inexpensive, and reliable. It usually has a starter ring gear pressed onto its outer edge and a machined mating surface for the clutch disc.

The center of the flywheel has machined holes for bolts to mount it firmly to the flywheel. In some cases, the bolt pattern is equal, so it can be mounted in any position. In other cases, where the flywheel is used as the primary method of balancing the engine, it may have offset bolt holes. This makes it so it can be mounted in only one position on the crankshaft. Flywheels are generally designed for a particular application. So, do not try to use a flywheel from one type of engine on another type.

Some flywheels are of the stepped style (**FIGURE 22-8**). In this design, the friction surface of the flywheel is recessed. The outer diameter, against which the pressure plate is bolted, is raised. The depth of the step is critical for proper operation of the clutch assembly. If it is too deep, the clutch will slip or not engage. If it is too shallow, the clutch will be stiff and may not disengage. Both stepped and flat flywheels can be resurfaced if the wear or defects are minor.

> **TECHNICIAN TIP**
>
> Refinishing the flywheel moves the pressure plate toward the engine and away from the throw-out bearing. This increases clutch pedal free play. If too much material is removed from the flywheel surface, some release mechanisms will not be able to compensate for the loss. And the clutch will not fully release. Also, as the flywheel is machined thinner, the clutch center hub may contact the flywheel bolts. So even though most manufacturers do not list a minimum thickness for the flywheel, know that it can cause problems if it becomes too thin.

The other type of flywheel is the dual-mass flywheel. Its function is to absorb torsional crankshaft vibrations. Those are twisting forces created in opposite directions. A twisting force happens in one direction when a piston is on the compression stroke. It happens in the opposite direction on the power stroke. This vibration is magnified in diesel engines. This is due to their higher compression ratios than gasoline engines. By minimizing the torsional vibration, the dual-mass flywheel reduces potential damage to the transmission gear teeth.

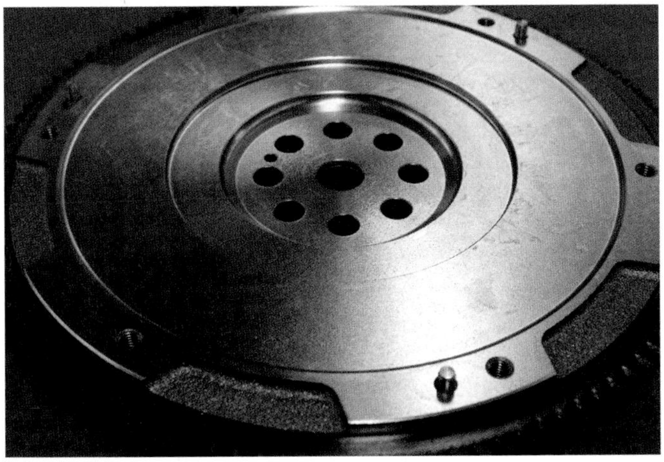

FIGURE 22-8 Stepped flywheels have a recessed friction surface for the clutch disc.

The dual-mass flywheel provides a small improvement in the engine's fuel economy. It does so by smoothing out the power pulses and focusing them in the direction of engine rotation. This also makes for smoother shifting.

There are two basic types of dual-mass flywheels. The first is composed of a primary and a secondary flywheel with a series of torsion springs and cushions. The second uses a planetary gear and torsional springs (**FIGURE 22-9**).

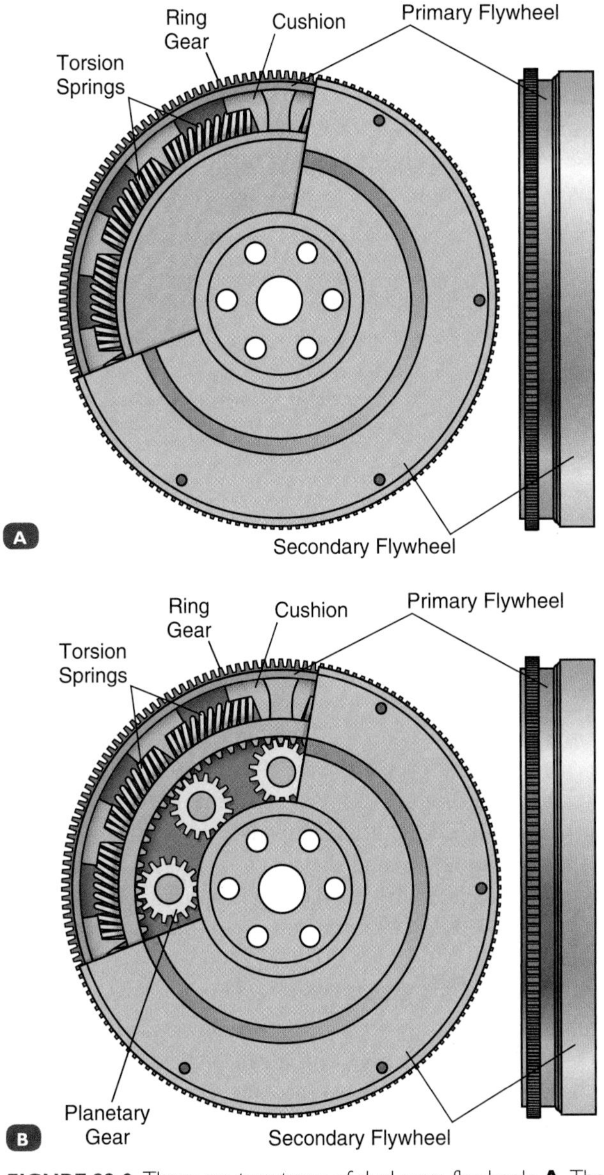

FIGURE 22-9 There are two types of dual-mass flywheels. **A.** The first is composed of a primary and a secondary flywheel with a series of torsion springs and cushions. **B.** The second uses a planetary gear and torsional springs.

In the first type of dual-mass flywheel, a friction ring is located between the inner and the outer flywheels. It allows the inner and the outer flywheels to slip. This feature is designed to minimize any damage to the transmission when torque loads exceed the rating of the transmission. The friction ring is the weak spot in the system. It can wear out if excessive engine torque loads are applied. This type of dual-mass flywheel also has a center support bearing. It carries the load between the inner and the outer flywheels. It is fitted with damper springs to absorb shocks.

The second type of dual-mass flywheel incorporates planetary gearing along with torsion springs. It is designed for engines with stronger vibrations at lower engine speeds. Because of this, the engine can have a lower idle speed. This reduces fuel consumption slightly. Some manufacturers use this style for their high-performance vehicles. It provides greater driving and shifting comfort. However, dual-mass flywheels are now being used in economy-type vehicles as well.

The torsional damper in a one-piece flywheel is located in the center hub of the clutch disc. Because of this it is very light duty. In a dual-mass flywheel, the torsional damper is located in the flywheel. This allows it to be much larger, and heavier duty. Because of this, it dampens engine torsional vibrations very effectively. Much more than is possible with standard clutch disc dampening technology.

▶ TECHNICIAN TIP

Dual-mass flywheels are designed to provide maximum isolation of the frequency below the engine's operating rpm, usually between 200 and 400 rpm. They are also very effective during engine startup and shutdown.

▶ Pressure Plates

LO 22-03 Describe pressure plates and clutch discs.

The pressure plate provides the clamping force in a clutch. It clamps the clutch disc between the pressure plate and the flywheel. This allows torque to be transmitted between those parts. The force is generated by one or more very strong springs. When the driver pushes on the clutch pedal, they push against the spring(s) in the pressure plate. Pushing the clutch pedal compresses the spring(s). This removes the clamping force from the clutch disc. Releasing the clutch pedal allows the spring(s) to apply their clamping force to the clutch disc again. Most automotive pressure plates are of either the diaphragm spring style or the coil spring style.

Diaphragm Pressure Plate

In light vehicles, the pressure plate is normally a diaphragm type and is serviced as an assembly. This means that it is not designed to be disassembled and repaired (**FIGURE 22-10**). A **diaphragm pressure plate** consists of:

- a pressed steel cover,
- a pressure plate with a machined flat surface,
- a number of spring steel drive straps, and
- the diaphragm spring.

The diaphragm spring is located inside the clutch cover on two **fulcrum rings**. It is held in place by a number of rivets passing through the diaphragm spring. The pressure plate is connected to the cover by the spring steel drive straps. The straps are riveted to the cover at one end, and to projecting lugs on the flat plate at the other. Retraction clips hold the flat plate in contact with the outer edge of the diaphragm spring.

During clutch operation, the throw-out bearing pushes on the diaphragm levers. The diaphragm spring pivots on the fulcrum rings. This pulls the outer edge of the diaphragm away from the flywheel. The retraction clips pull the pressure plate away from the clutch disc. Diaphragm clutches are known for their lighter feel when the clutch pedal is fully down. This makes them more comfortable to drive.

Coil Spring Pressure Plate

The **coil spring pressure plate** uses coil springs to create the clamping pressure. It uses release levers to release the clutch disc (**FIGURE 22-11**). Typically, three or four release levers are used, depending on the application. The release levers control

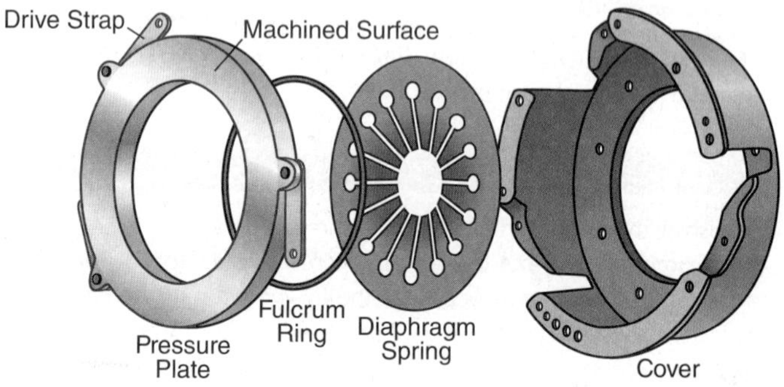

FIGURE 22-10 A diaphragm pressure plate consists of a pressed steel cover, a pressure plate with a machined flat surface, a number of spring steel drive straps, and the diaphragm spring.

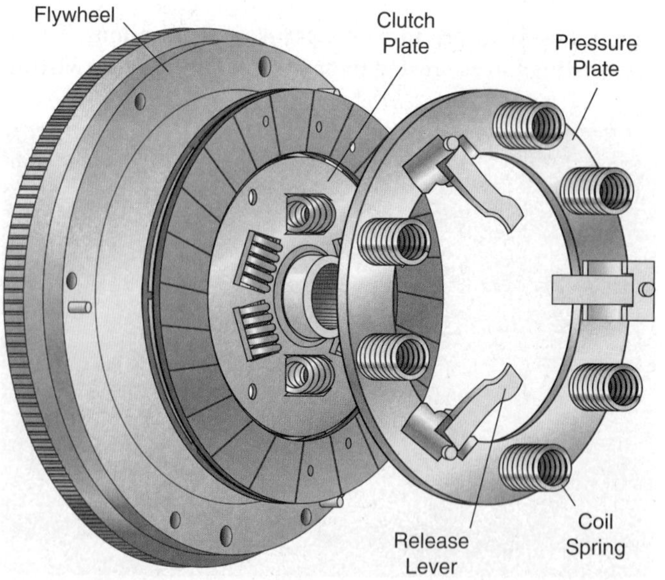

FIGURE 22-11 The coil spring pressure plate uses coil springs to create the clamping pressure and uses release levers to release the clutch disc.

the movement of the friction portion of the pressure plate. They pivot on the pedestals that are part of the pressure plate housing.

When the release bearing pushes the levers toward the flywheel, they pull the friction portion of the pressure plate back toward the driver. This releases the clamping force on the clutch disc. When the clutch pedal is released, the levers return to their rest position. The clamping force is restored to the clutch disc. It is now able to transmit torque, and move the vehicle forward.

An advantage of the coil spring pressure plate is that the more coil springs there are, the higher the torque capacity. A disadvantage is that if the clutch disc overheats, the springs can become weak. This reduces the clamping force. Reduced clamping force creates more slippage. More slippage creates more heat, which weakens the coil springs even further. Once a clutch starts slipping, it likely will have to be replaced very soon.

Another disadvantage is that the springs are less evenly spaced on many coil spring pressure plates. This can cause uneven wear on the friction surface of the pressure plate. And since there are fewer release levers, that can also uneven wear on the friction surface over time.

▶ TECHNICIAN TIP

The coil spring-type pressure plate generally requires more pedal pressure to hold the pedal down. This makes it less comfortable to drive. It is one reason why diaphragm pressure plate clutches are more desirable for light-duty passenger vehicles.

Clutch Disc

The clutch disc is also called a **driven center plate** or a friction disc (**FIGURE 22-12**). The clutch disc provides the friction material needed to transmit engine torque to the transmission. On the clutch disc, the friction facings are riveted to waved spring steel segments. The spring steel segments are themselves riveted to a steel disc. The central alloy-steel splined hub is separate from the steel disc.

Drive is transmitted from the steel disc to the hub through torsional coil springs or rubber blocks. This arrangement dampens torsional vibrations from the engine. It also absorbs shock loads imposed on the drive line by sudden or violent clutch engagement. A molded friction washer between the hub and the spring-retaining plate also acts as a damper.

The waved spring steel segments play an important role during clutch engagement. They allow a progressive application of the pressure plate clamping force to the clutch disc. They are located between the friction facings. The waved shape causes the facings to spread apart slightly when the clutch is disengaged. As the clutch pedal is released, they compress progressively. This results in a smoother engagement of the clutch when starting from a stop (**FIGURE 22-13**).

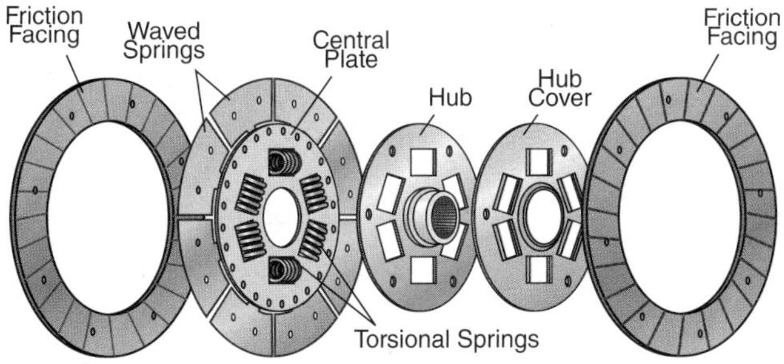

FIGURE 22-12 The clutch disc components.

▶ TECHNICIAN TIP

Some high-performance clutch discs do not use wavy springs between the clutch facings. This makes them more robust. But it also removes the progressive nature of the clutch action, making them much more "grabby."

Multiplate Clutches

Some vehicles such as heavy-duty diesel trucks and high-performance vehicles may use multiplate clutches. This allows them to handle the large amount of torque. Adding plates to a clutch unit increases its torque capacity. It does so without increasing the spring strength or clutch diameter (**FIGURE 22-14**). The increased torque capacity is due to the increased surface area of the friction materials. This spreads the torque load over a larger surface.

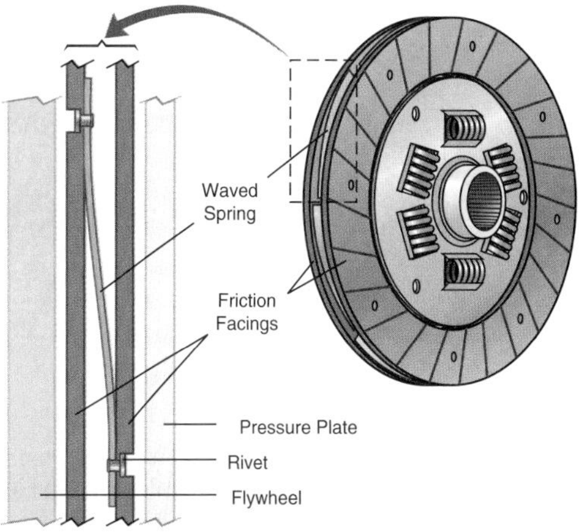

FIGURE 22-13 Waved springs allow progressive application of the clutch.

FIGURE 22-14 A multiplate clutch assembly.

A typical multiplate clutch assembly has two friction discs with friction material riveted to both sides of each disc. An internally splined hub on each disc mates with the splines on the transmission input shaft. These hubs can include a damper in normal automotive applications. In heavy-duty and high-performance applications, they are usually rigid with no damper.

The damper style has a series of springs to dampen the engine's power pulsations. It helps to prevent damage to the clutch disc. The rigid hub does not have a damper. It is susceptible to the power pulsations. When used in heavy truck applications, it is usually paired with a dual-mass flywheel to dampen power pulsations. On high-performance vehicles, the power pulsations are not of much concern.

A cast-iron separator plate fits between each disc. The separator plate is located on drive pins on the flywheel. The pins drive the separator plate with the flywheel and pressure plate. The separator plate provides a friction surface that mates to each of the two friction discs. The drive pins must be aligned properly for the separator plate to operate correctly. The friction discs and separator plate fit between the flywheel and the pressure plate. The pressure plate then provides a frictional clamping force on each of the mating surfaces.

When the clutch pedal is depressed, the release bearing acts on the pressure plate levers. They move the pressure plate away from the flywheel. This releases the clamping force on the facings and separator plate. When the clutch pedal is released, the spring tension forces the pressure plate, discs, and separator plate together. This clamps all the components together.

▶ Throw-Out Bearing and Clutch Fork

LO 22-04 Describe throw-out bearing, clutch fork, and pilot bearing.

The clutch throw-out bearing and clutch fork work together to compress the pressure plate springs. This occurs when the clutch pedal is pressed. Because the throw-out bearing pushes on the rotating pressure plate, it must be able to rotate as well. At the same time, the clutch fork does not rotate. Thus, the throw-out bearing must include a thrust bearing as part of its assembly.

Throw-out bearings usually use **thrust-type angular-contact ball bearings.** The bearing assembly is pressed onto a **carrier**. The carrier slides on the **front bearing retainer sleeve** that extends from the front of the transmission. This is considered a **push-type clutch** design. This is because the throw-out bearing pushes on the levers of the pressure plate. However, there are pull-type designs used on some applications. The bearing carrier is attached to the clutch fork with clips (**FIGURE 22-15**).

Moving the clutch release fork (clutch fork) brings the bearing thrust face into contact with the pressure plate levers. The bearing rotates with the pressure plate. The thrust-type angular contact ball bearing is packed with lubricant during manufacture. It requires no periodic maintenance during its service life, as long as it is not abused. This assumes that the clutch pedal free play is maintained. We will discuss this later.

The clutch fork is usually made of stamped steel or cast iron. It pivots either in the center, or at the end inside the bell housing. The pivot is generally screwed into the bell housing and is usually replaceable. The pivot should be inspected for wear and lubricated whenever the clutch is replaced. The release bearing engages in tabs in the clutch fork and is usually held in place by clips.

Not all release bearings are operated by a clutch fork. Some are operated directly by collar-style slave cylinders. These are sometimes called central or concentric slave cylinders because they are donut-shaped. This allows them to fit around the input shaft (**FIGURE 22-16**). They use hydraulic pressure to directly push the release bearing against the pressure plate fingers. Because of this, they do not use a clutch fork.

Pilot Bearing

The pilot bearing is essentially an alignment support bearing for the snout of the input shaft to ride on (**FIGURE 22-17**). Because it takes two bearings to support a rotating shaft, the pilot bearing is the front bearing for the input shaft. The other end of the input shaft is supported by the transmission input bearing. The pilot bearing can be a brass or bronze bushing, a needle-type bearing, or a roller-type bearing (**FIGURE 22-18**). Larger vehicles, such as trucks, may use a ball bearing type. Some of these bearings are installed in the cavity on the end of the crankshaft. Others may be placed or pressed into the center of the flywheel.

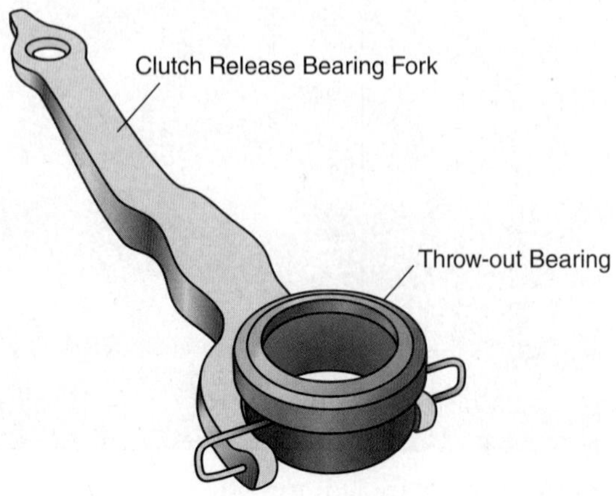

FIGURE 22-15 A clutch release bearing fork.

FIGURE 22-16 Collar-style slave cylinder operates the throw-out bearing directly, without a clutch fork.

To replace a pilot bearing, the transmission and clutch assembly must be removed. Some vehicles do not require a pilot bearing because of the construction of the input shaft. On this type of shaft, the input shaft is longer (typically front-wheel drive). So, it is supported on the front and back of the transaxle case by support bearings (**FIGURE 22-19**).

FIGURE 22-17 Pilot bearing and input shaft.

FIGURE 22-18 A. Brass or bronze bushing. **B.** Needle-type bearing. **C.** Roller-type bearing.

▶ Clutch Operating Mechanisms

LO 22-05 Describe clutch operating mechanisms.

Movement of the clutch pedal is transferred through an operating mechanism to the clutch assembly. This mechanism may be mechanically operated, or hydraulically operated. The two types of mechanical systems are the linkage style and the cable style.

Linkage-style systems use a system of links, rods, and levers. They transfer the operating force from the clutch pedal to the clutch fork. Cable-operated linkage uses a strong cable in a flexible housing to do the same thing. This style offers more flexibility and is easier to install in the factory.

Hydraulic systems use a clutch master cylinder and a slave cylinder to operate the clutch. The cylinders are connected by a steel, plastic, or reinforced rubber line. Brake fluid transfers the operating force from the master cylinder to the slave cylinder.

▶ TECHNICIAN TIP

Regardless of the operating mechanism, it is always a good idea for drivers to completely remove their foot from the clutch pedal. Do not rest your foot on the clutch pedal between shifts. This will help preserve the life of the throw-out bearing and friction facings.

Cable Mechanisms

Cable-operated clutch control systems are easily installed in the vehicle during manufacture. They also take up less overall engine compartment room. One end of the outer cable housing is fixed to the pedal support inside the vehicle. The other end is fixed to the transmission bell housing in the engine compartment.

The inner cable connects between the upper end of the clutch pedal and the clutch fork. The clutch fork operates the

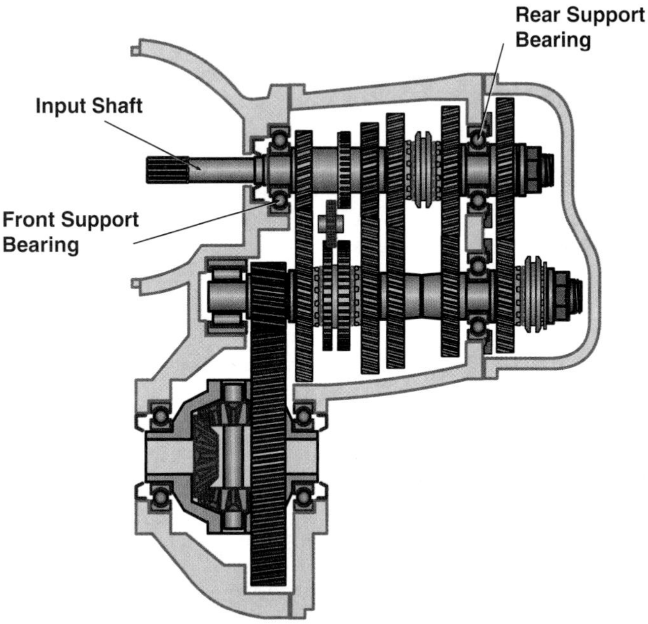

FIGURE 22-19 This vehicle doesn't use a pilot bearing because its two bearings are located in the transmission.

throw-out bearing. Depressing the clutch pedal transfers the movement through the cable to the clutch fork.

Most cable-operated clutches are designed in a way that they can be adjusted. Adjustment of the clutch cable provides for the specified amount of free play. Free play is the distance between the throw-out bearing and the pressure plate levers. It is measured at the clutch pedal when it is in the released position.

Free play prevents constant contact of the throw-out bearing with the pressure plate levers. It also prevents the subsequent rotation of the bearing when the engine is running. Some cable-operated clutches are adjusted manually. They usually have a threaded nut or collar that can be turned to obtain the proper free play (**FIGURE 22-20**). Some vehicles use a **quadrant ratchet**. It automatically adjusts the clutch pedal free play as needed (**FIGURE 22-21**). Many of these are adjusted just by lifting the clutch pedal with a toe.

Hydraulic Clutch Mechanisms

In hydraulic clutch release mechanisms, the clutch pedal acts on a master cylinder. It builds pressure in the system and sends fluid to the **slave cylinder**. A flexible hose connects the master cylinder to the slave cylinder. The slave cylinder can be mounted on (external type) or in (internal type) the transmission bell housing (**FIGURE 22-22**). The **slave cylinder** operates the clutch fork on an external type. Or it acts directly on the throw-out bearing on an internal type.

With the clutch pedal in the released position, the center valve in the master cylinder is clear of the inlet port. Fluid is free to flow between the reservoir and the master cylinder (**FIGURE 22-23**). This allows for expansion and contraction of the fluid as it heats and cools.

When the clutch pedal is initially depressed, the master cylinder piston moves forward. With slight movement of the piston, the

FIGURE 22-20 A cable-operated clutch with manual adjustment.

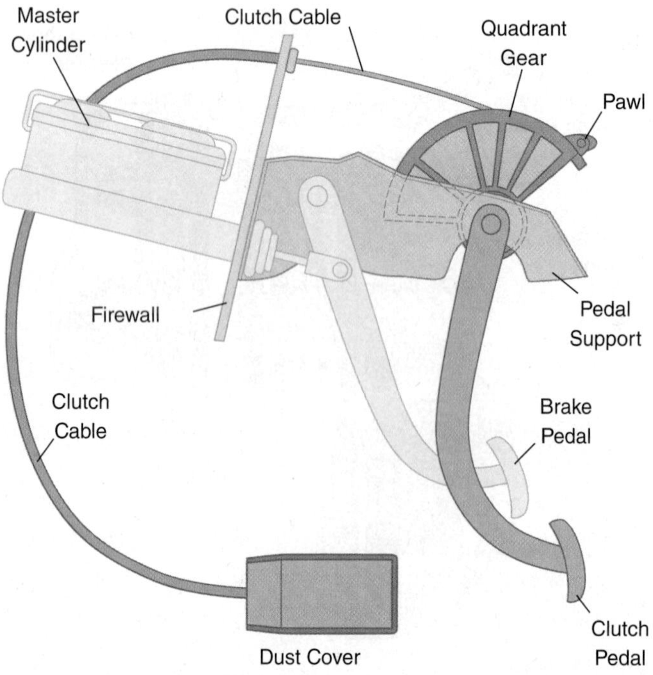

FIGURE 22-21 A cable-operated clutch with quadrant ratchet.

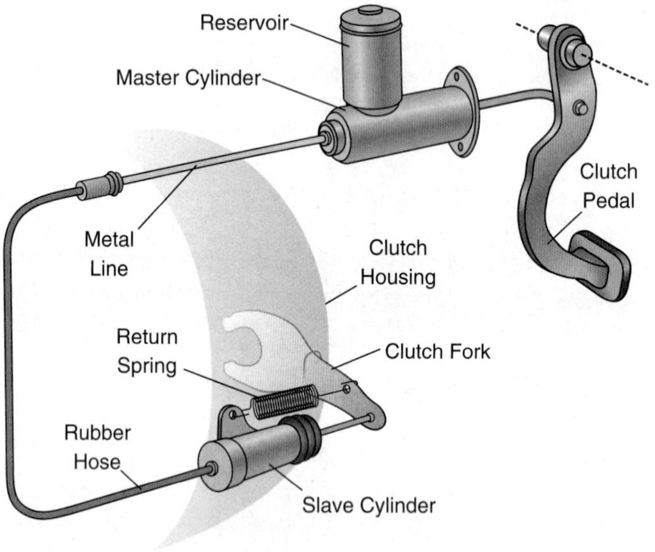

FIGURE 22-22 A hydraulic clutch control.

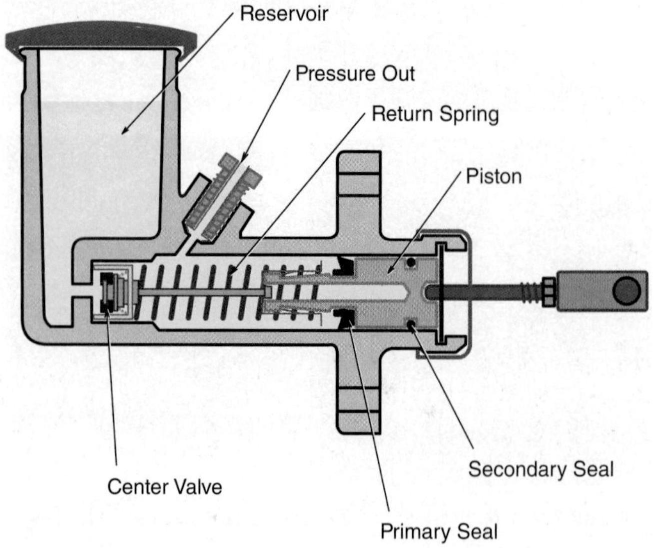

FIGURE 22-23 Cutaway view of a clutch master cylinder.

center valve closes off the inlet port from the reservoir. This traps fluid in the cylinder bore. Further piston movement displaces fluid through the outlet port of the master cylinder. Connecting lines carry the fluid to the slave cylinder. The slave cylinder piston then moves in accordance with the master cylinder piston.

The movement of the slave cylinder piston is transferred through a pushrod to the clutch fork. The clutch fork acts on the throw-out bearing. In other configurations, the slave cylinder pushes directly on the throw-out bearing (collar style). When the clutch pedal is released, the displaced fluid returns to the master cylinder. The center valve opens and allows excess fluid to return to the reservoir.

Most hydraulic clutches are self-adjusting. This is because the piston is allowed to return deeper into the slave cylinder as the clutch wears. Any excess hydraulic fluid vents to the master cylinder reservoir. This system automatically compensates for any clutch disc wear. Because of that, it makes it so the clutch, in most cases, does not have to be adjusted.

Linkage-Operated Systems

Older systems used a series of links and levers to operate the clutch. It used an equalizing mechanism called a bell crank (equalizer bar) that changed the direction of the force. It was attached to both the engine and the vehicle frame (**FIGURE 22-24**). The bell crank pivoted in plastic or nylon bushings which wore out over time.

Adjustments were made at the end of the link connected to the clutch fork. It consisted of a threaded rod and lock nuts. As the clutch plate wore, adjustments were necessary to maintain proper clutch pedal free play.

▶ TECHNICIAN TIP

You might think that as a clutch plate wears over time, the clutch free play would increase. In fact, it decreases. As the lining wears, the pressure plate levers move backward toward the throw-out bearing. This reduces the clearance between them resulting in less clutch pedal free play. If the clutch disc wears enough, all of the free play can be lost. This means that even with the driver's foot off the clutch pedal, the linkage is holding some pressure on the pressure plate. This reduces the pressure plate clamping force, and, if bad enough, can cause the clutch to slip. A slipping clutch quickly burns out and requires replacement.

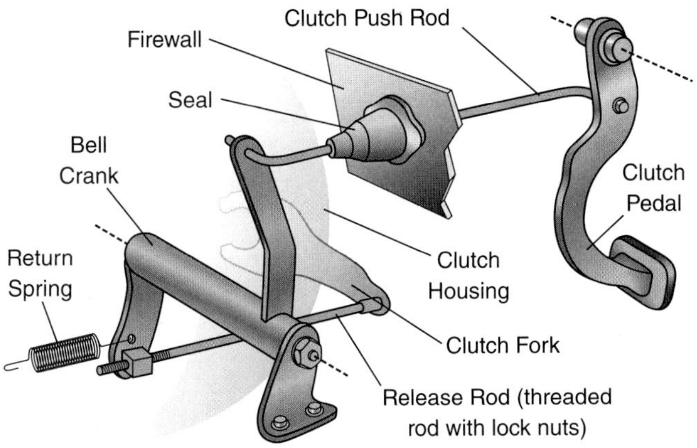

FIGURE 22-24 A linkage-operated system.

▶ Clutch Maintenance

LO 22-06 Perform clutch maintenance.

Clutch Safety and Hazards

Clutch maintenance can be hazardous if proper safety precautions are not taken. Wear appropriate clothing and eye protection while servicing all clutch components (**FIGURE 22-25**). Gaining access to work on clutches generally requires a lift or jack stands. Always make sure the vehicle is properly secured on lifts before performing any type of service. Always be mindful of spills, and clean the area as necessary to avoid any slips or falls.

Asbestos is a carcinogen. In fine particles, it can be breathed into the lungs, making it a health hazard. It was commonly present in older clutch disc linings. Modern technology has generally replaced asbestos. Newer clutch parts use organic compound resins with imbedded copper filings and some ceramic materials. Use equipment approved by OSHA to remove clutch dust and debris. And follow all local, state, and federal regulations for particulate disposal.

Never use compressed air to blow off clutch parts. It is possible that the dust does not contain asbestos. But even non-asbestos dust is hazardous and can damage lungs. Blowing the dust off parts puts it up in the air where it can be inhaled. OSHA-approved methods entail using a soap and water solution brushed onto the component to wet down and loosen the dust (**FIGURE 22-26**). The dust is then rinsed off and collected in a container. It can then be properly disposed of according to federal, state, and local laws.

Preventive Maintenance

There should be regular preventive maintenance performed on the clutch per manufacturer-specified intervals. Check and compare clutch pedal free play to specifications. Maintaining the specified clutch pedal free play is critical to maintaining clutch life. Follow the specified procedures in the service information.

The operation of the clutch pedal and return springs should be checked for binding and excessive movement. If a clutch switch is used on the clutch pedal, check that it is functional. Verify that

FIGURE 22-25 Protect yourself by wearing appropriate clothing and eye protection while servicing clutch components.

FIGURE 22-26 Brake and clutch dust wash station.

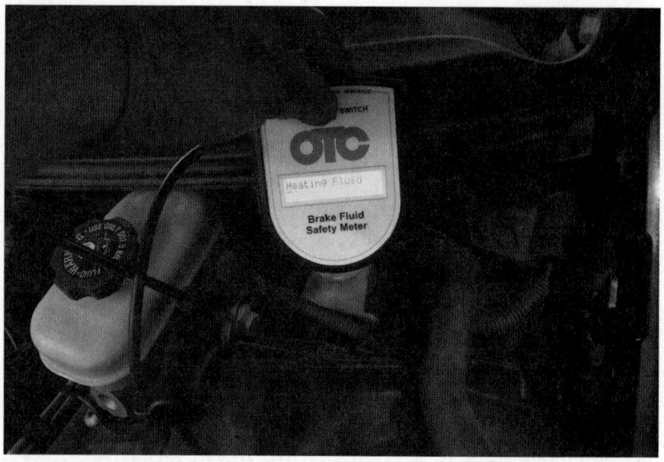

FIGURE 22-27 Checking brake fluid for excess moisture.

it prevents the engine from cranking over while the clutch pedal is released. Mechanical linkage should be inspected for damage. And all pivot points should be lubricated with the specified lubricant.

Hydraulic clutch fluid is typically brake fluid. It should be checked in a manner similar to checking brake fluid (**FIGURE 22-27**). Clutch fluid is often more neglected than even brake fluid. So, there is a good chance that the clutch fluid needs to be flushed if the vehicle is more than a few years old. Hydraulic clutch lines should be checked for leaks at the master cylinder and the slave cylinder. If accessible, pull the boot back on the slave cylinder and inspect it for signs of leakage. Check that all hydraulic lines are not kinked or heavily rusted. Make sure all hydraulic components are secure in their mountings.

Checking and Adjusting a Mechanical Clutch

As the clutch wears, the friction disc becomes thinner. This results in the pressure plate release levers moving closer to the release bearing. This causes the clutch linkage to lose its free play. Some clutches are self-compensating for wear, whereas others require checking and adjusting. You must refer to the manufacturer's service information to find the method of adjustment.

It is important to periodically check the clutch mechanism for proper operation and correct the free play. It is common to do so during every oil change. If the clutch pedal has too little free play, the throw-out bearing could remain in contact with the pressure plate levers. This prevents the pressure plate springs from applying full pressure. Incomplete clamping of the disc would occur, leading to premature failure of the clutch assembly. Also, the pedal will have to be released a long way before the clutch starts to engage.

If the free play is too large, the clutch pedal might not have enough travel to fully release the clutch. This causes the gears to clash (grind) when shifting. It also results in heavy synchronizer wear. With this condition, the car may creep forward even though the clutch pedal is fully applied. Or the clutch will start to engage right from the floor when releasing the pedal.

To check and adjust a mechanical-style clutch, follow the steps in **SKILL DRILL 22-1**.

SKILL DRILL 22-1 Checking and Adjusting a Mechanical-Style Clutch

1. Following the specified procedure, inspect the clutch linkage parts for damaged, worn, bent, or missing components. Look for signs of binding, looseness, and excessive wear. Operate the clutch pedal, and inspect all components under the dash.

2. Check the clutch linkage components under the hood for the same signs of wear or damage as the components under the dash.

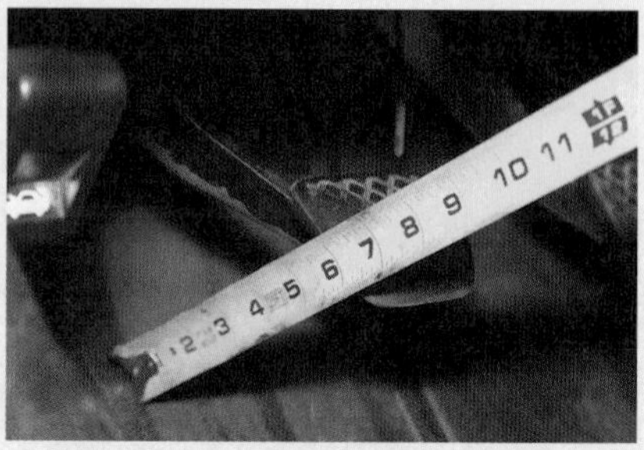

3. Measure the clutch pedal height. Compare your reading to the specifications, and determine any necessary actions to correct any fault.

SKILL DRILL 22-1 Checking and Adjusting a Mechanical-Style Clutch (Continued)

4. Measure the clutch pedal free play. Perform any adjustments as necessary, following the manufacturer's procedure.

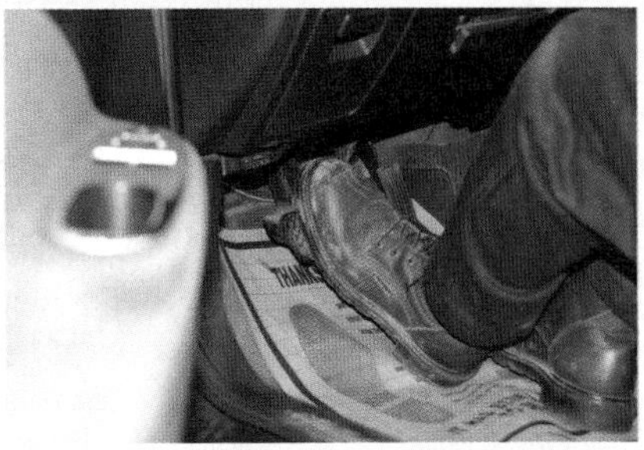

5. Start the vehicle and depress the clutch. The clutch should engage at the proper height and have the proper free play. Make a gear selection to ensure the gears do not clash going into mesh.

Checking, Adjusting, and Bleeding a Hydraulic Clutch

It is common to check the clutch hydraulic system during oil changes. If the hydraulic clutch system is improperly maintained, clutch operation could be compromised. This can lead to a shortened life of the clutch components. It can also lead to breakdowns. Also, improper or old fluid in the hydraulic system can cause master cylinder and slave cylinder damage and leaks.

To check and adjust a hydraulic clutch, follow the steps in **SKILL DRILL 22-2**.

▶ TECHNICIAN TIP

Make sure the floor mats or other obstructions will not affect the operation of the clutch pedal.

SKILL DRILL 22-2 Checking and Adjusting a Hydraulic Clutch

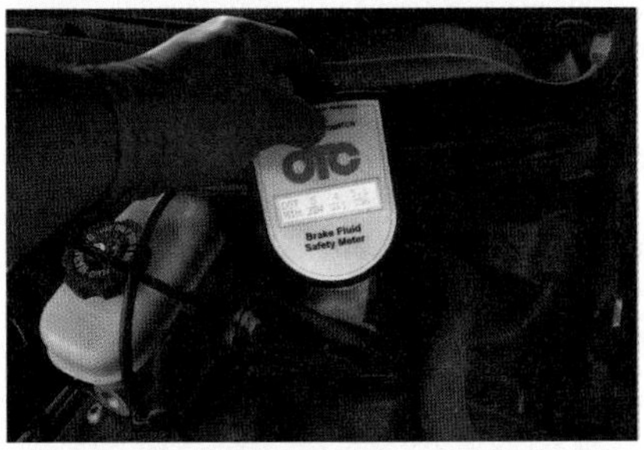

1. Inspect the clutch master cylinder for correct fluid level, and test the quality of the fluid.

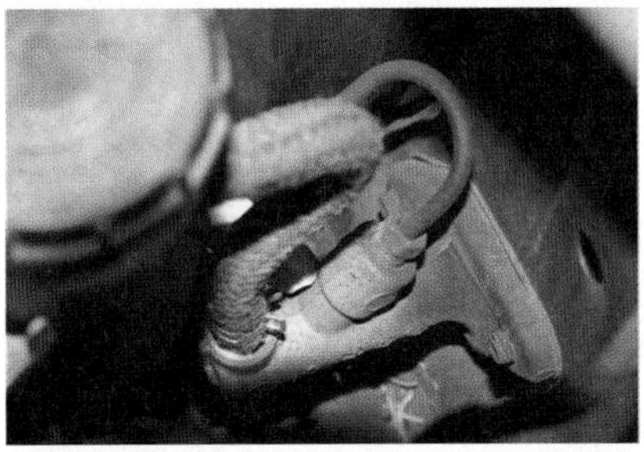

2. Inspect all line connections to the master cylinder.

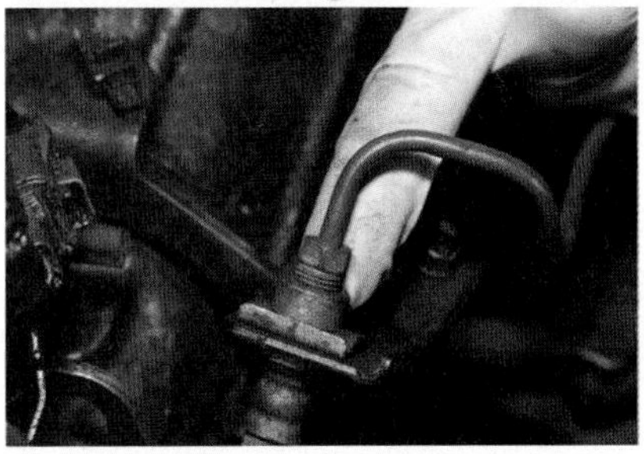

3. Check that all hydraulic lines are not kinked or leaking at their connections. This will require that the system be repaired and bled of any air. Check all rubber hoses for dry rot, bulges, or leaks. Make sure all hydraulic components are secure in their mountings.

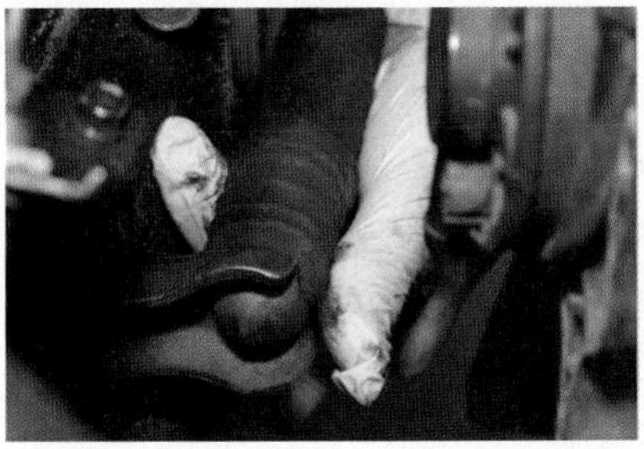

4. Check the boot on the slave cylinder for seepage, which may indicate a leaking slave cylinder piston seal.

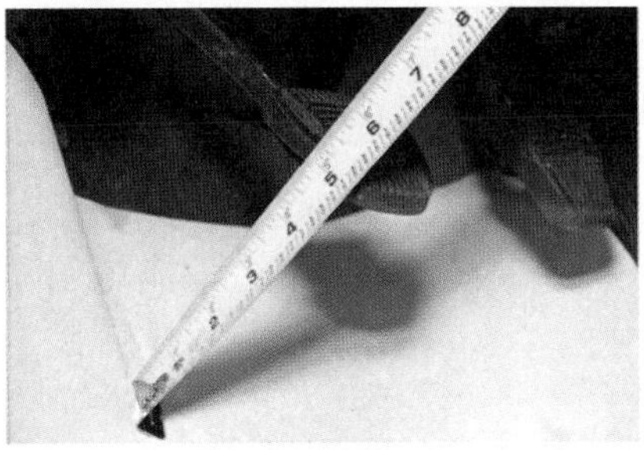

5. Check clutch pedal height. Measure clutch pedal free play, using a tape measure. Compare your readings to the specifications and determine any necessary actions to correct any fault.

Bleeding a Hydraulic Clutch System

In the case of a hydraulic clutch system failure, it may be necessary to bleed the air from the system. Bleeding is also needed whenever any hydraulic component is replaced. Or if the hydraulic fluid becomes unfit for use due to age or contamination. Not all systems are fitted with a bleeder screw. In this case, you may be able to bleed the system from one of the ends of the line: Either the end of the line entering the slave cylinder, or the end leaving the master cylinder. In some cases, the hydraulic clutch system is sealed from the factory. It is prefilled with fluid, and has no method of opening the system to bleed air from it.

Research the procedure and specifications for bleeding the hydraulic clutch system. There are three types of bleeding: gravity bleeding, manual bleeding, and pressure/vacuum bleeding. Determine the proper method of bleeding to use by consulting the service information.

Gravity bleeding uses gravity to push fluid and air from the master cylinder and lines out through the slave cylinder bleeder screw. In most vehicles, the clutch master cylinder is quite a bit higher than the slave cylinder. The weight of the fluid can therefore be used to supply the pressure to push fluid and air out of the system.

To bleed/flush a hydraulic clutch system using the gravity method, follow the steps in **SKILL DRILL 22-3**.

The manual bleeding method uses the clutch master cylinder to push fluid and air from the system. The procedure usually requires an assistant. The assistant holds the clutch pedal down while the other person opens the bleeder valve on the slave cylinder. This forces air and old fluid from the system.

You and the assistant need to coordinate actions. The assistant will have to hold the pedal down until you have the bleeder screw closed. Then, they allow the pedal to rise slowly once the bleeder screw is closed. This allows fluid to be drawn from the reservoir into the clutch master cylinder bore. Allowing the pedal to rise too fast can create a vacuum that draws air into the cylinder past the rubber seals.

The assistant should avoid pumping the pedal. This can result in a single air bubble breaking up into smaller bubbles which become distributed throughout the system. This will make it harder to bleed the system completely.

To bleed or flush a hydraulic clutch system using the manual method, follow the steps in **SKILL DRILL 22-4**.

▶ TECHNICIAN TIP

Be careful of the pressure that comes out of the slave cylinder when bleeding it. The fluid can easily splash or spray into your eyes. Also, brake fluid will soften paint. So, use fender covers when handling brake fluid. And clean up any spilled brake fluid with generous amounts of water.

SKILL DRILL 22-3 Bleeding/Flushing Hydraulic Clutch System Using the Gravity Method

1. If the fluid needs to be flushed, use a suction gun or old antifreeze tester to suck the fluid out of the clutch master cylinder reservoir.

2. Fill the reservoir with the specified fluid from a freshly opened bottle.

3. Open the bleeder screw on the slave cylinder.

4. Allow air and fluid to drain from the system into a container.

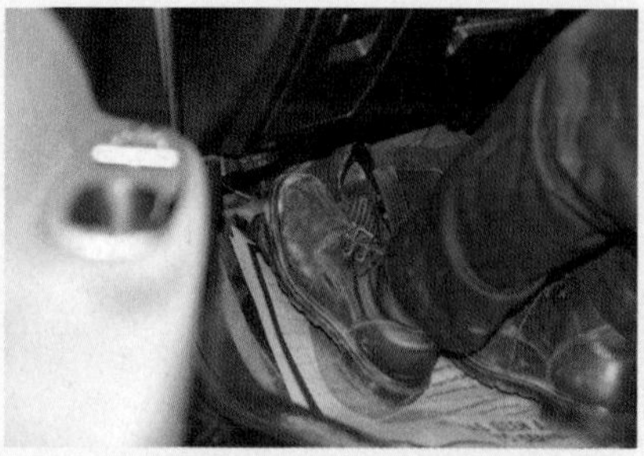

5. Keep the master cylinder filled. Once all air and old fluid are removed, close the bleeder screw, and operate the clutch pedal to check for normal operation.

6. After bleeding the clutch hydraulic system, fill the master cylinder to the correct level with the specified type of brake fluid.

SKILL DRILL 22-4 Bleeding or Flushing a Hydraulic Clutch System Using the Manual Method

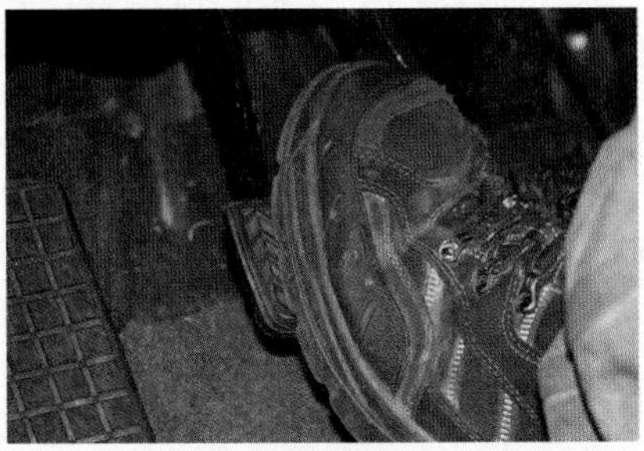

1. Remove all old fluid from the reservoir. Fill it with the specified fluid. Have an assistant depress the clutch pedal slowly.

2. Open the bleeder valve on the slave cylinder, and let fluid run out into a container. When all of the fluid stops flowing, close the bleeder valve, and slowly release the pedal. Repeat this process until all air and old fluid are removed from the system.

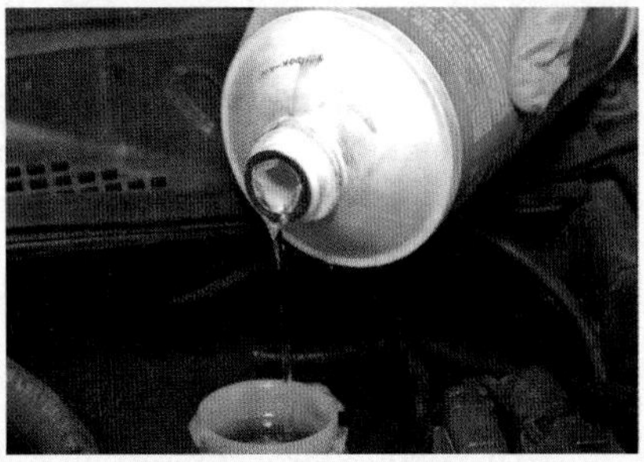

3. After bleeding the clutch hydraulic system, check for correct pedal feel, and fill the master cylinder to the correct level with the specified type of brake fluid.

The pressure or vacuum bleeding method is the most common method of bleeding a hydraulic clutch in a shop. It uses pressure or vacuum to push or pull fluid and air from the system. This method works well for systems that tend to trap air in the hydraulic system. The pressure or vacuum keeps the fluid moving continuously through the system. And that helps remove air bubbles that tend to get stuck in high spots. It does require special bleeding tools or equipment to create the pressure or vacuum.

To bleed a hydraulic clutch system using the pressure method, follow the steps in **SKILL DRILL 22-5**.

SKILL DRILL 22-5 Bleeding a Hydraulic Clutch System Using the Pressure or Vacuum Method

1. Remove all old fluid from the reservoir. Fill it with the specified fluid.

2. Hook up the pressure or vacuum bleeding tool to the vehicle with the correct adapters.

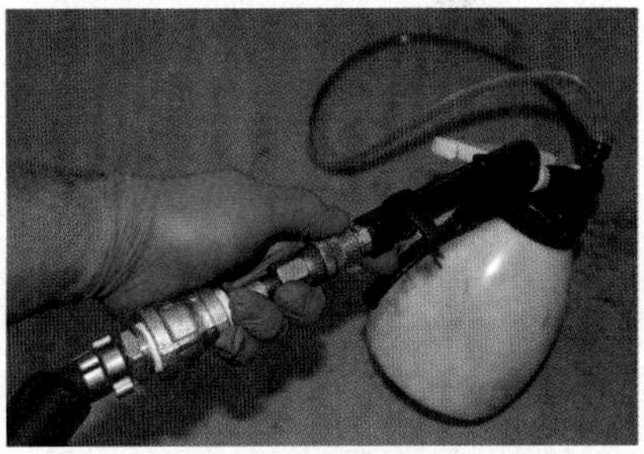

3. Apply pressure or vacuum to the system.

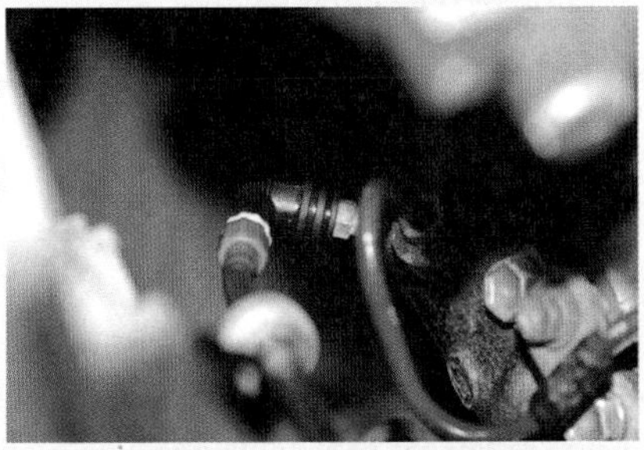

4. Open the bleeder screw, and allow the fluid and air to be purged from the system. Repeat this process as necessary.

5. After bleeding, check for correct pedal feel, and fill the clutch master cylinder to the correct level with the specified type of brake fluid.

▶ Wrap-Up

Ready for Review

- The amount of torque a clutch can transmit depends on the amount of friction between the clutch disc and the mating surfaces of the flywheel and pressure plate.
  - Clutch friction is dependent on the coefficient of friction of the clutch facings, the diameter of the clutch, the number of clutch discs in the clutch assembly, and the total spring force clamping the parts together.
- A flywheel smoothens out the power pulses from the pistons during the power strokes and provides a friction surface for the clutch disc and a mounting surface for the pressure plate.
  - One-piece flywheel; dual-mass flywheel
- A pressure plate clamps the clutch disc between the pressure plate and the flywheel and allows the transmission of torque. A clutch disc provides the friction material needed to transmit engine torque to the transmission.
- A clutch throw-out bearing and clutch fork work together to compress the pressure plate springs when the clutch pedal is pressed. A pilot bearing is essentially an alignment support bearing for the snout of the input shaft to ride on.
- Movement of the clutch pedal is transferred to the clutch assembly either mechanically or hydraulically. Mechanical systems can be of the linkage style and the cable style. Hydraulic systems use a clutch master cylinder and a slave cylinder to operate the clutch.
- Clutch maintenance should include checks for clutch pedal free play, hydraulic leaks, excess moisture in brake fluid, binding and excessive movement of clutch pedal and return springs, and mechanical linkages.

Key Terms

carrier Part of the throw-out bearing assembly. The part that the bearing is pressed on to.

clutch disc The center component of the clutch assembly, with friction material riveted on each side. Also called a clutch plate or friction disc.

clutch fork The part of the clutch linkage that operates the throw-out bearing.

coil spring pressure plate A type of pressure plate that uses coil springs to provide the clamping force.

diaphragm pressure plate A slightly conical, spring steel plate used to provide the clamping force for the clutch assembly.

driven center plate The friction disc that is held firmly against the flywheel by a pressure plate and that transfers power from the flywheel to the transmission input shaft.

front bearing retainer sleeve The part that the throw-out bearing assembly rides on. It usually retains the input shaft bearing on a manual transmission.

Flywheel A heavy, round metal disc attached to the end of the crankshaft to smooth out vibrations from the crankshaft assembly and provide one of the friction surfaces for a clutch disc used on a manual transmission.

fulcrum ring A steel ring that is used as a pivot point for the diaphragm spring in the pressure plate.

pilot bearing The bearing or bushing that supports the front of the transmission input shaft. It is mounted in the flywheel or the rear of the crankshaft.

pressure plate The friction surface of the clutch cover and the plate that squeezes the clutch disc against the flywheel.

push-type clutch A typical clutch system, used in modern vehicles, where the clutch fork pushes the release bearing forward to release the friction facing from the pressure plate.

quadrant ratchet The device used in some cable-operated clutches to provide self-adjustment as the clutch disc wears. Some quadrant ratchets adjust if you lift up on the clutch pedal.

release mechanisms Components that operate the clutch. Usually included are the throw-out bearing and the clutch fork. Some manufacturers include the operating system.

slave cylinder The component in a hydraulically operated clutch that converts hydraulic pressure to mechanical movement at the clutch fork.

starter ring gear Ring gear on a flywheel or flexplate that meshes with the starter drive gear.

throw-out bearing The part of the clutch release mechanism that imparts clutch pedal force to the rotating pressure plate levers.

thrust-type angular-contact ball bearing A type of bearing that uses a deep groove in the bearing races where the ball bearings ride; this design is for thrust conditions.

transmission input shaft The shaft that brings engine torque into the transmission.

Review Questions

1. Which of the following does NOT affect clutch torque carrying ability?
   - **a.** Diameter and number of clutch discs in the assembly
   - **b.** Coefficient of friction of the clutch facings
   - **c.** Length of the clutch pedal
   - **d.** Total spring force clamping parts together
2. When the clutch pedal is depressed, what is disengaged from the engine through the clutch?
   - **a.** The input shaft
   - **b.** The output shaft
   - **c.** The counter shaft
   - **d.** The shift shaft
3. All of the following are purposes of the flywheel in a manual transmission EXCEPT:
   - **a.** It smooths out power pulses from the cylinders firing
   - **b.** It provides a ring gear for the starter motor to turn

c. It provides the clamping force to the clutch using its diaphragm springs
d. It provides a mounting and friction surface for the clutch

4. Which component contains the springs that clamp the clutch to the flywheel when the clutch is engaged?
   a. The pressure plate
   b. The clutch plate
   c. The clutch disc
   d. The throw-out bearing
5. Which component contains the torsional springs that dampen the engine's power pulses and prevent damage to the clutch disc?
   a. The flex plate
   b. The clutch disc
   c. The pressure plate
   d. The pilot bearing
6. Which clutch disc component contains the splines that allow the clutch to ride on the input shaft?
   a. The wavy springs
   b. The friction facing
   c. The damper
   d. The hub
7. What supports the front of the input shaft on many manual transmission equipped vehicles?
   a. The throw-out bearing
   b. The hub bearing
   c. The pilot bearing
   d. The main bearing
8. How often should most hydraulic clutches be adjusted?
   a. Every oil change or 5,000 miles
   b. Once per year or 12,000 miles
   c. When the pedal freeplay is excessive
   d. Never, they are normally self-adjusting
9. If a technician is changing a clutch on a 30-year-old vehicle, what should they be especially concerned with on a vehicle that age?
   a. The vehicle may have old clutch linings containing asbestos
   b. The vehicle may have complicated procedures for clutch replacement
   c. The older vehicle may require special types of lubricants not available today
   d. There could be excessive fluid leaks which make tools slippery
10. Which of the following is NOT a common hydraulic clutch system bleeding procedure?
    a. Automated bleeding
    b. Manual bleeding
    c. Vacuum bleeding
    d. Gravity bleeding

ASE Technician A/Technician B Style Questions

1. A vehicle has been modified to increase its horsepower, and its clutch is slipping under heavy acceleration. Potential repairs are being discussed. Technician A states that a clutch assembly utilizing a higher coefficient of friction could be installed. Technician B states that a clutch assembly with stronger springs on the pressure plate could be installed. Who is correct?
   a. Technician A only
   b. Technician B only
   c. Both Technicians A and B
   d. Neither Technician A nor B
2. Two technicians are discussing the dry clutch system on a manual transmission. Technician A states that when depressing the clutch pedal, the clutch disc is squeezed harder and the clutch is applied. Technician B states that the clutch pedal is depressed (pushed) when changing gears, since it interrupts torque transfer from the engine to the transmission. Who is correct?
   a. Technician A only
   b. Technician B only
   c. Both Technicians A and B
   d. Neither Technician A nor B
3. Flywheels are being discussed. Technician A states that incorrect stepped flywheel machining can cause a clutch to slip. Technician B states that incorrect stepped flywheel machining can cause a clutch to drag, and not completely release. Who is correct?
   a. Technician A only
   b. Technician B only
   c. Both Technicians A and B
   d. Neither Technician A nor B
4. A vehicle is equipped with a dual-mass flywheel. Technician A states that its purpose is to increase the clutch torque holding capacity. Technician B states its purpose is to act as an engine dampener by absorbing the power pulses of the engine. Who is correct?
   a. Technician A only
   b. Technician B only
   c. Both Technicians A and B
   d. Neither Technician A nor B
5. A high-mileage vehicle comes into the shop with a slipping clutch. Technician A states that a worn clutch disc may be the cause. Technician B states that excessive clutch pedal free-play may be the cause? Who is correct?
   a. Technician A only
   b. Technician B only
   c. Both Technicians A and B
   d. Neither Technician A nor B

6. A vehicle comes into the shop with a slipping clutch. Technician A states that a worn throw-out bearing could be the cause. Technician B states that weak pressure plate springs could be the cause. Who is correct?
   a. Technician A only
   b. Technician B only
   c. Both Technicians A and B
   d. Neither Technician A nor B
7. A customer requires a clutch with a greater holding capacity since they installed twin turbochargers. They do not want to have a harder to depress clutch pedal. Technician A states that a twin disc clutch could be a good option. Technician B states that a pressure plate with stronger springs could be a good option. Who is correct?
   a. Technician A only
   b. Technician B only
   c. Both Technicians A and B
   d. Neither Technician A nor B
8. A noise is heard on a vehicle equipped with a manual transmission. The noise only occurs when the clutch pedal is completely depressed, and the noise goes away when the pedal is released. Technician A states that it could be a failed pilot bearing. Technician B states that to replace a pilot bearing, the transmission will have to be removed. Who is correct?
   a. Technician A only
   b. Technician B only
   c. Both Technicians A and B
   d. Neither Technician A nor B
9. Technicians are discussing a clutch release mechanism. Technician A states that a cable-operated clutch must maintain some amount of free play to ensure the throw-out bearing does not constantly press on the pressure plate. Technician B states that a cable type will automatically adjust the linkage between the pedal and the clutch master cylinder. Who is correct?
   a. Technician A only
   b. Technician B only
   c. Both Technicians A and B
   d. Neither Technician A nor B
10. A vehicle with a cable-operated clutch has not been regularly maintained. The customer complaint is grinding gears when shifting, and the vehicle creeps forward with the clutch pedal completely depressed and transmission in first gear. Technician A states that the clutch pedal free-play may be too tight. Technician A states that if this has been ongoing, the internal transmission synchronizer parts may be worn. Who is correct?
    a. Technician A only
    b. Technician B only
    c. Both Technicians A and B
    d. Neither Technician A nor B

CHAPTER 23

Driveshafts, Axles, and Final Drives

Learning Objectives

- **LO 23-01** Describe drivetrain layout.
- **LO 23-02** Inspect and service driveshafts.
- **LO 23-03** Describe the operation and maintenance of final drives and differentials.
- **LO 23-04** Service axles.
- **LO 23-05** Service U-joints and CV joints.

ASE Education Foundation Tasks

See Appendix A to view the 2017 ASE Education Foundation Automobile Accreditation Task List Correlation Guide.

▶ Drive Train Layout

LO 23-01 Describe drivetrain layout.

In previous chapters, we have covered how the engine's torque is delivered through the transmission. Now the torque must get from there to the appropriate wheels. That job is performed by the remainder of the drivetrain components. Those include driveshafts, transfer cases, final drives, differentials, and drive axles. Each of these assemblies is made up of subcomponents that allow them to perform their task.

Depending on the vehicle, the arrangement of these assemblies can vary. For example, most passenger vehicles are driven by two wheels. But most off-road vehicles are driven by all four wheels. This chapter discusses the various assemblies and components of the drivetrain. We will cover their layout, function, and, ultimately, how to maintain them.

Rear-Wheel Drive Layout

In a rear-wheel drive (RWD) vehicle, the engine and transmission only power the rear wheels. In a conventional RWD vehicle, the engine and transmission are mounted longitudinally. This means they are oriented front to back. Drive from the engine is transmitted to a rear axle assembly by a drive (propeller) shaft. Solid rear axle assemblies house the remaining components. This includes the final drive gears, differential gears, and axle shafts (**FIGURE 23-1**).

In vehicles with an independent rear suspension, the final drive unit is mounted on the chassis frame (**FIGURE 23-2**). Torque is transferred to each road wheel through external driveshafts (half-shafts). Each half-shaft has flexible joints on each end. Some RWD vehicles have rear- or mid-mounted engines. They use a transaxle and external half-shafts to drive the road wheels.

In solid rear axle applications, suspension action makes the final drive assembly rise and fall relative to the vehicle frame. This happens as a vehicle goes over bumps and various types of terrain. This movement changes the distance from the transmission to the final drive. It also changes the angle between the driveshaft and its connections. In addition, the pinion nose is forced up on acceleration by torque. And down when the brakes are applied on deceleration.

Despite these movements, the driveshaft must transfer the drive smoothly. Change in length is accommodated by a sliding coupling. It uses mating splines that allow the coupling to lengthen and shorten. When it is built into the front of the driveshaft, it is called a **slip yoke.**

The slip yoke has splines that mate to the splines on the transmission output shaft. The slip yoke is free to slide in and out on the output shaft. This allows the length to change as the

FIGURE 23-1 A typical RWD solid axle assembly with the final drive assembly enclosed in one housing.

FIGURE 23-2 Rear wheel drive–independent rear suspension

You Are the Automotive Technician

After a long, icy-cold Michigan winter, a customer visits your shop. They state that an unusual noise sometimes comes from the front of the vehicle. The vehicle is driven to and from work on Highway 94, which has multiple potholes due to the harsh winter weather. You complete a customer interview and take the 2013 Toyota Camry for a test drive. During the test drive, you notice that the vehicle makes a popping noise when turning corners. Your preliminary diagnosis is a worn constant velocity (CV) joint. But you will need to further inspect the vehicle in the service bay. The visual inspection of the underside of the vehicle confirms your preliminary diagnosis. The right front CV joint boot is torn and dripping water. There is also dirt mixed with the little bit of grease still inside the boot.

1. What is likely causing the customer's concern? And what will you recommend to correct it?
2. What are the three different drive axle arrangements used with an RWD solid axle? Which drive axle is most commonly used on medium to heavy trucks?
3. What is the difference between a final drive unit and the differential assembly?

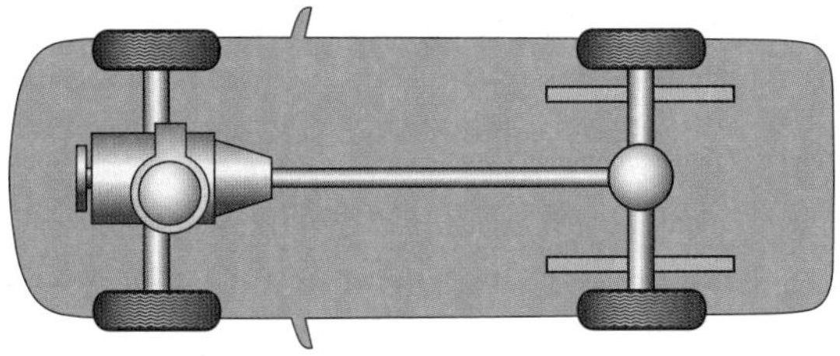

FIGURE 23-3 The centerline of the transmission is not in line with the axles.

FIGURE 23-4 The ring and pinion gear set allows power to be transferred 90 degrees from the power source.

suspension moves up and down. If the vehicle uses a **sliding spline driveshaft**, a two-piece driveshaft is joined in the middle with splines. The driveshaft can slide on itself to increase and decrease in length.

Universal joints (or simply **universal joints**(U-joints)) are fitted at the front and rear ends of the driveshafts. They allow for normal up-and-down suspension movement. U-joints need to be able to transmit full engine torque, so they are subject to wear. We will discuss these further in another section.

As shown in **FIGURE 23-3**, the centerline of the transmission is not in line with the axles. This means the power needs to be turned 90 degrees to get it to the rear wheels. A ring gear and pinion assembly is commonly used for this purpose (**FIGURE 23-4**). It also reduces the rotational speed of the axles relative to the driveshaft. This results in increased torque applied to the wheels.

Front-Wheel Drive Layout

In a front-wheel drive vehicle, the front wheels are driven by a transversely or longitudinally mounted engine. **Longitudinal** mounting means the front of the engine is facing the front of the vehicle. **Transverse** mounting means the front of the engine is facing one side of the vehicle. That is, the engine is facing one of the fenders. Most front-engine FWD vehicles use a transverse mounted engine coupled to a transaxle. Power is transmitted in line from the engine through the transmission and final drive. The front axles (half-shafts) (**FIGURE 23-5**) deliver the torque to the wheels.

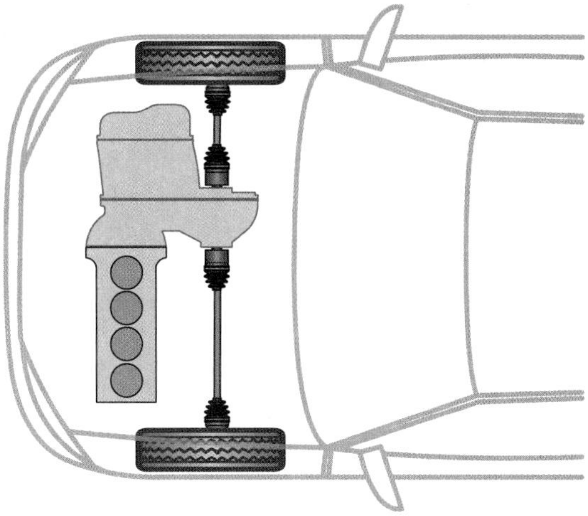

FIGURE 23-5 Powerflow in a transverse engine is in line with the engine, all the way to the wheels.

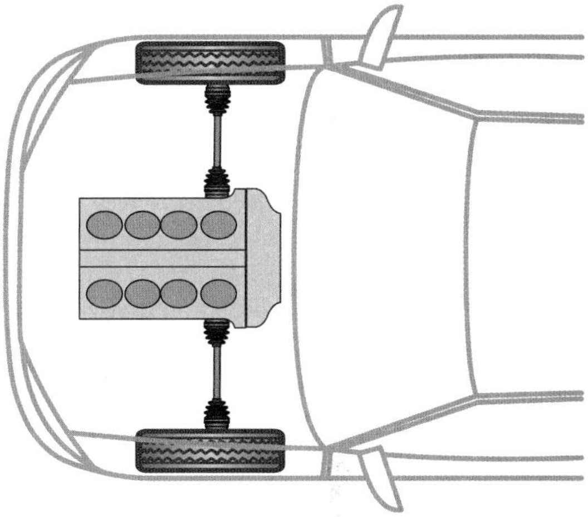

FIGURE 23-6 Powerflow in a longitudinally mounted engine must turn 90 degrees to power the wheels.

In transverse applications, the transaxle is mounted at the rear of the engine. A primary shaft engages with the splines of the clutch center plate. When a gear is selected, the drive is transferred to the secondary shaft. A pinion gear on the rear of the secondary shaft drives a helical ring gear. It is attached to the differential case. Drive is then transferred through the differential gears to each axle and then to each front wheel.

On a front-wheel drive (FWD) vehicle with a longitudinally mounted engine, the power must be turned 90 degrees to drive the axles (**FIGURE 23-6**). This is accomplished by the design of the transaxle. Power is transmitted through the clutch in a rearward direction. It is then typically transferred by a large chain or gears to the rest of the transaxle and final drive assembly. The final drive turns the power 90 degrees so half-shafts can transmit power to the wheels.

Four-Wheel Drive Layout

Vehicles with part-time four-wheel drive (4WD) are designed for optional off-road use. 4WD can be selected as needed for abnormal surfaces (mud, snow, ice, etc.). When driving on normal surfaces, the driver disconnects 4WD. Selection is made by a manual lever or by electronic control via a push button.

In 4WD applications, the engine and transmission are normally mounted longitudinally at the front. A transfer case is mounted to the rear of the transmission. The transfer case splits the power to both the front and the rear drive axles when 4WD is selected (**FIGURE 23-7**). Driveshafts are used to move power from the transfer case to the front and rear drive axles. The transfer case gives the driver control of four possible power transfer options:

1. The selection most often used is two-wheel drive (2WD), as most driving is done on clear non-slippery roads. 2WD sends power to the rear wheels only. The front axle is not being supplied with any power.
2. The next selection is typically neutral, which allows the vehicle to be towed. This can be if it is stuck in mud or snow, or towed down the highway with the engine off, such as behind a motor home. No power is being transmitted to either axle in this selection.
3. Another selection is 4WD *high range*, which transmits power to the front and rear drive axles at normal drive speed.
4. The last selection is 4WD *low range*. This position provides a gear reduction inside the transfer case. It results in higher torque to the front and rear drive axles. This position causes the vehicle to move slowly while the engine turns at a higher rpm. Thus, to avoid over-revving the engine, the low range should be selected only when driving at slow vehicle speeds.

TECHNICIAN TIP

A neutral position on the transfer case allows a power take-off (PTO) accessory to be powered without the vehicle having to move. Examples are a cable winch or a hydraulically driven dump bed. They are driven from a **PTO** gear mounted to the side of the transfer case.

All-Wheel Drive (AWD) Layout

AWD vehicles are widely used nowadays. They offer the convenience of having all four wheels driving the vehicle at most

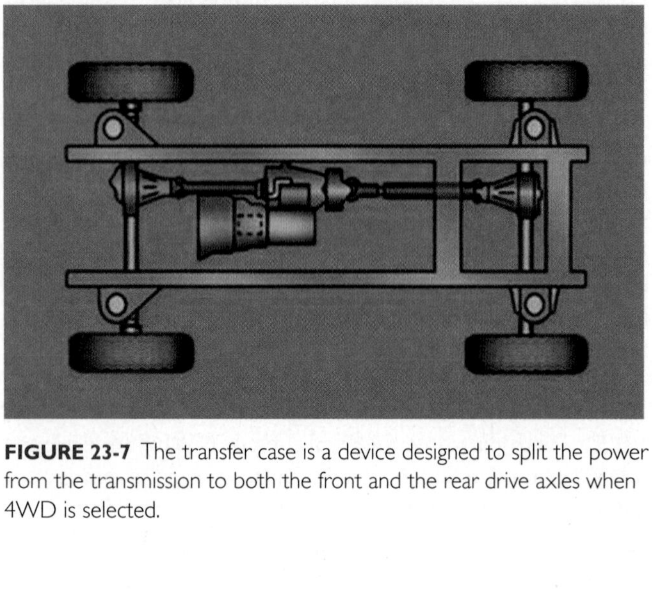

FIGURE 23-7 The transfer case is a device designed to split the power from the transmission to both the front and the rear drive axles when 4WD is selected.

times. This provides good traction when on-road (or occasionally off-road) conditions warrant it. Such vehicles are intended for lighter duty and are sometimes referred to as "boulevard SUVs."

In some AWD vehicles, only two of the wheels are normally powered, typically the front wheels. In this case, the RWD is applied automatically when needed to maintain traction. The driver is not normally aware of when this takes place. This provides peace of mind that it happens when needed. In other AWD vehicles, all four wheels are powered all of the time. But the torque may be split unevenly when driving under normal conditions. We cover that in more depth later in the chapter.

▶ Driveline Subassemblies and Components

LO 23-02 Inspect and service driveshafts.

The driveline subassemblies are made up of driveshafts, axle shafts, and half-shafts. Each shaft transfers torque from one component to the next. Configuration of the components varies depending on the type of drivetrain. For example, on most RWD vehicles, there is one driveshaft from the transmission to the final drive assembly. From there, two axle shafts deliver torque to the rear wheels.

On FWD vehicles, there are two half-shafts (axles)—one from each side of the transaxle to each of the front wheels. 4WD and AWD vehicles use combinations of driveshafts and half-shafts. The configuration depends on whether the axle shafts are independent or non-independent. Let's dig in.

Driveshafts—Rear-Wheel Drive

The driveshaft is a device that transfers torque from one component to another. Typically, this is from the transmission output shaft to the final drive assembly (**FIGURE 23-8**). The driveshaft is commonly made from metal tubing material. The driveshaft can be constructed from steel or aluminum. They come in various sizes and lengths depending upon the application.

The driveshaft can be a short one-piece assembly with a U-joint at each end. Or in some long-wheel base vehicles, such as a school bus, the driveshaft can be made up of as many as four separate segments. The driveshaft typically requires the use of

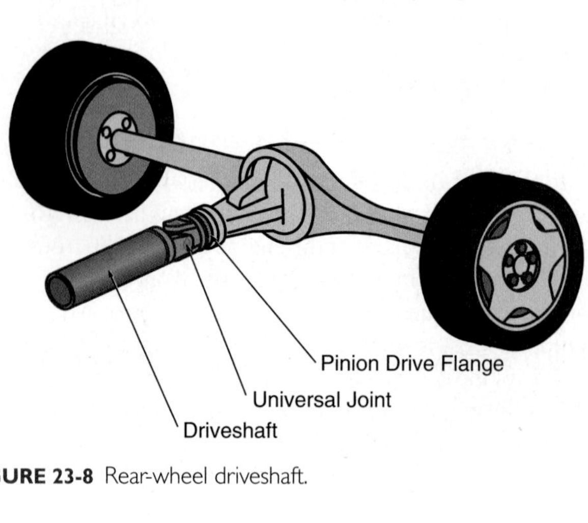

FIGURE 23-8 Rear-wheel driveshaft.

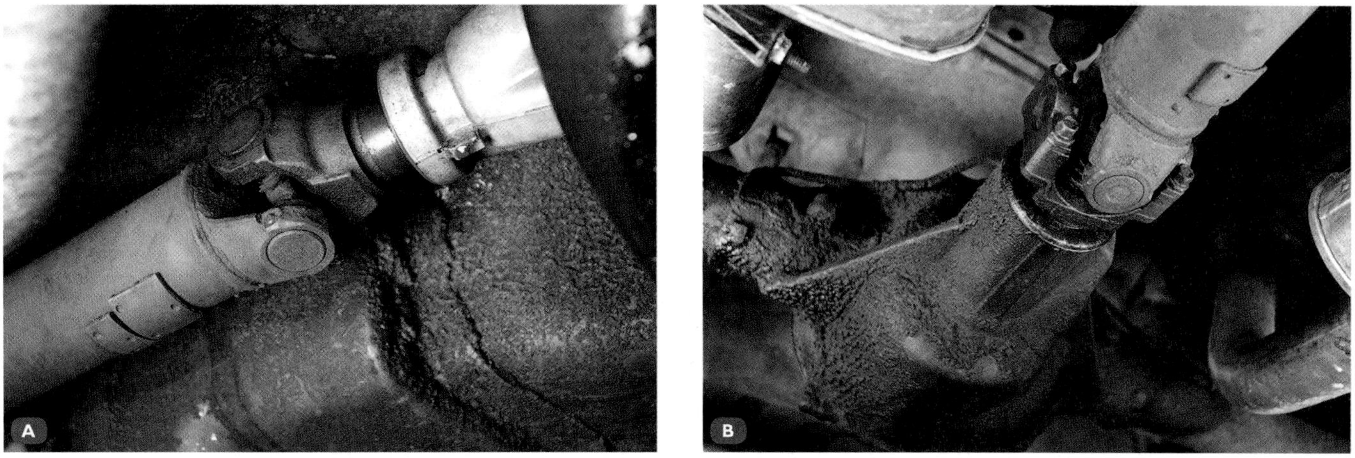

FIGURE 23-9 U-joint yokes are welded to each end of the driveshaft. **A.** Front. **B.** Rear.

joints at either end. They allow power to transfer smoothly as the driveshaft follows the movement of the suspension system.

U-joint yokes are typically welded to each end of the tube (**FIGURE 23-9**). The front U-joint connects to the driveshaft's internally splined slip yoke. It engages with splines on the transmission output shaft. The front slip yoke slides over the output shaft in the transmission. A bushing located inside the rear transmission housing supports the slip yoke. This design enables the slip yoke to move forward and backward as the shaft length changes with suspension movement.

The rear U-joint mates with a **companion flange** on the final drive pinion shaft (**FIGURE 23-10**). The companion flange is a splined flange that transmits power to the pinion gear. The companion flange can have recesses machined in it to accept the U-joint caps. If so, the U-joints are bolted to the companion flange with U bolts or straps. Some drive shafts have a flange similar to the companion flange. If so, the two flanges fit flat against one another and are bolted together. Unbolting the companion flange bolts is how the driveshaft is removed from the vehicle.

Some applications use a two-piece driveshaft that is able to flex in the middle. The carrier bearing supports the driveshafts where they meet. It consists of a non-friction bearing molded into a rubber mount (**FIGURE 23-11**). The carrier bearing is bolted to a mounting bracket on the vehicle frame. It supports the center of the driveshaft.

The maximum length of a single section of driveshaft tubing is approximately 72" (183 cm). Any longer and it would be prone to twisting of the tube from the torque output. Where longer driveshafts are needed, two or more sections may be used. The other reason to use the two-piece driveshaft is to minimize the angles that the U-joints have to operate at. The use of three or more U-joints breaks up any excessive angle between the transmission and the rear axle. Be sure to check the driveline angles when diagnosing this concern. Provision is normally made for adjusting the alignment of the two shafts. This is done by using shims under the carrier bearing.

When working with a two-piece driveshaft, it is critical that the yokes on both halves of the driveshaft be phased correctly. This means the pivot points of the U-joints are lined up in the same exact orientation from one shaft to the other. If the driveshaft halves are not assembled correctly, then vibration will result. With a one-piece driveshaft, the yokes are welded to the tube and cannot be moved. When disassembling a two-piece driveshaft, be sure to mark the position of each driveshaft half.

FIGURE 23-10 Companion flange on the final drive pinion shaft.

FIGURE 23-11 A carrier bearing supports the center of a two-piece driveshaft.

A marker or white paint mark will help to ensure the yokes line up during reassembly (**FIGURE 23-12**).

Checking Driveshaft Joints, Phasing, Angles, Balance, and Runout

Vehicles today are put through a tremendous variety of driving styles and terrains. Driver abuse can be a major cause of drivetrain wear. Each of the components can develop play as components wear. One of the most common places for wear is in the U-joints. As the U-joints wear, vibrations can occur, and major damage can happen if the joint were to break. Inspection of the driveline should be performed during oil changes, or other routine service.

U-joint phasing is important when multiple shafts are involved. If you are going to disassemble a multi-piece driveshaft, then it is critical to mark the parts in relation to each other. If the parts are not installed in the same position, problems will occur. First, if installed out-of-phase, vibration will occur even if the shaft is in balance. Second, if they are installed 180 degrees out of position, they will still be in phase, but not in balance. Along with balance and phasing, runout and concentricity should be checked. If any damage has occurred, then the damaged parts will need to be replaced.

Driveline angularity is the difference in the angles between the transmission output shaft and the final drive pinion shaft. It is measured in degrees. Ideally, both shafts would be horizontally and vertically parallel to each other (**FIGURE 23-13**). A difference in those angles can create a vibration in the system. Some manufacturers specify a maximum angle of 3 degrees between shafts. Beyond this, vibration and additional wear will occur. Proper angularity is designed into the system when the vehicle is made. This means it is usually only a problem if the vehicle has been modified or damaged.

An angle gauge should be used to measure these angles (**FIGURE 23-14**). If the angles are out of specifications, follow the manufacturer's procedure to bring the components into alignment. This typically involves using shims to change the angle of the carrier bearing. Or on some vehicles, shims are used to change the angle on the rear axle assembly.

To check driveshaft joints, phasing, angles, balance, and runout, follow the steps in **SKILL DRILL 23-1**.

Inspecting and Servicing Center Support Bearings

The center bearing is used on multi-piece drivelines. The bearing is usually supported by a rubber insert similar to a powertrain mount. It isolates noise and vibration from the chassis or body. The rubber can tear or deteriorate in the same way powertrain mounts do. The center bearing is subject to weather and the same abuses as a U-joint. Whenever an undervehicle service is performed, the center bearing should be inspected. If this bearing is equipped with a grease zerk, it should be lubricated whenever an oil change is performed. Without lubrication, the bearing will burn up. This can cause a vibration and noise concern to occur.

To inspect and service center support bearings, follow the steps in **SKILL DRILL 23-2**.

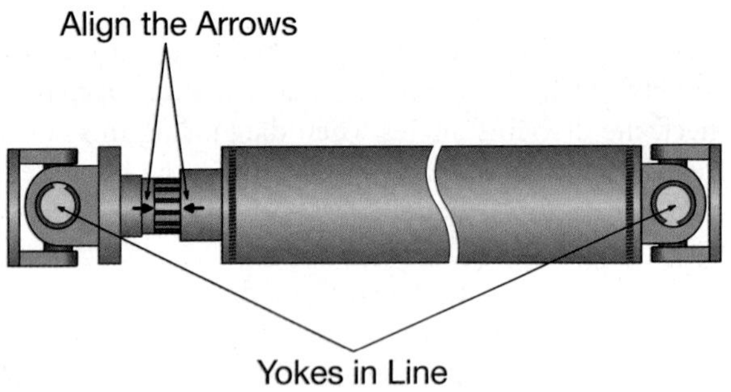

FIGURE 23-12 The driveshaft yoke alignment (phasing) is critical for smooth operation.

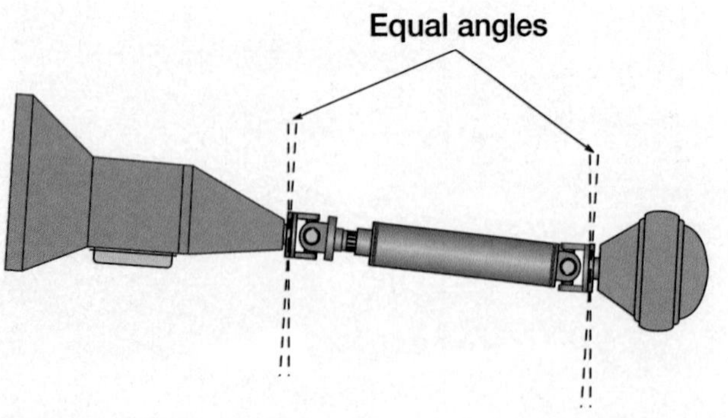

FIGURE 23-13 Driveline angularity is any difference in angle between the transmission output shaft and the final drive pinion shaft.

FIGURE 23-14 A driveline angle gauge.

SKILL DRILL 23-1 Checking Driveshaft Joints, Phasing, Angles, Balance, and Runout

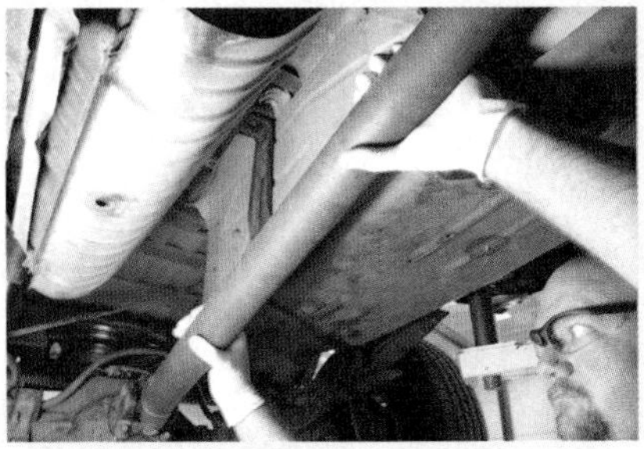

1. Raise the vehicle on a 2-post lift, making sure it is secure. Inspect the entire driveline by grasping the driveshaft and checking for abnormal play.

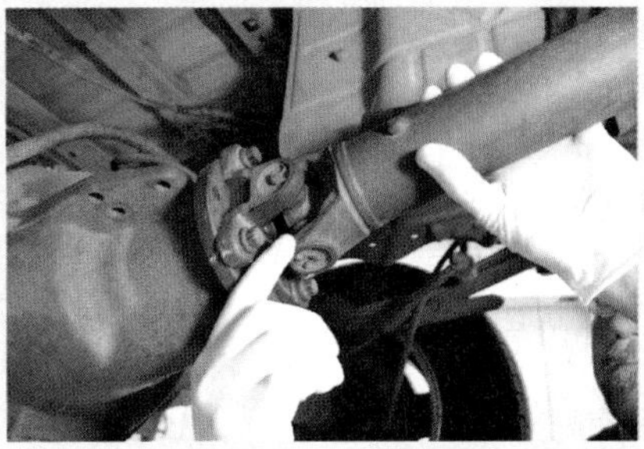

2. Move the shaft fore and aft and radially. Play indicates worn U-joints.

3. Check that the joints are in phase by checking the U-joint yoke on each end of the driveshaft with a level. Once finished, start the vehicle and safely run in drive at slow speed on the lift. Watch and listen to the joints. Any noise or off-centeredness indicates damaged U-joints.

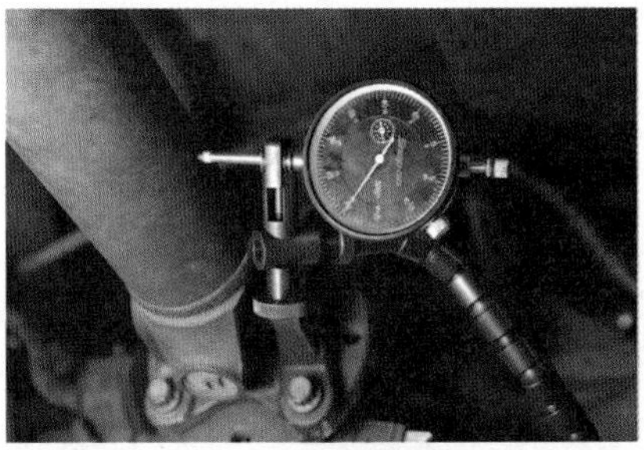

4. Slowly raise the speed and watch for balance issues on the driveshaft. If any balance issues are present, slow the speed so that the vehicle does not slip on the hoist. If all is well, stop the engine and measure the runout with a dial indicator, and check the driveshaft for concentricity.

5. Check the angularity using an angle gauge, and make any necessary adjustments. Shims may be required.

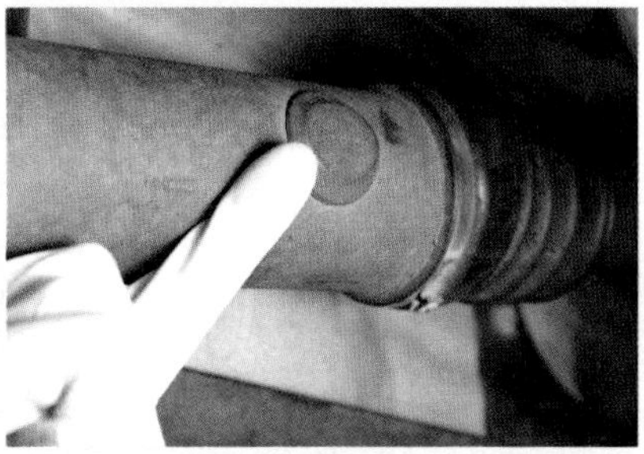

6. Check the driveshaft for any missing weights that are used to balance the shaft. Usually missing weights will be evident by the welds from the missing weights.

SKILL DRILL 23-2 Inspecting and Servicing Center Support Bearings

1. Safely raise the vehicle on an approved lift. Inspect the center bearing components for any major defects, such as looseness or noises.

2. Inspect the bearing mount rubber insert for dry rotting and cracking.

SKILL DRILL 23-2 Inspecting and Servicing Center Support Bearings (Continued)

3. If the bearing must be replaced, follow the manufacturer's procedures for proper installation of a new bearing. Typical bearing replacement may go as follows: mount the driveshaft in an approved vise.

4. Mark the shafts so they may be properly phased when put back together. Separate the two driveshafts.

5. Remove the U-shaped metal mounting bracket. Remove the rubber mount from around the center bearing.

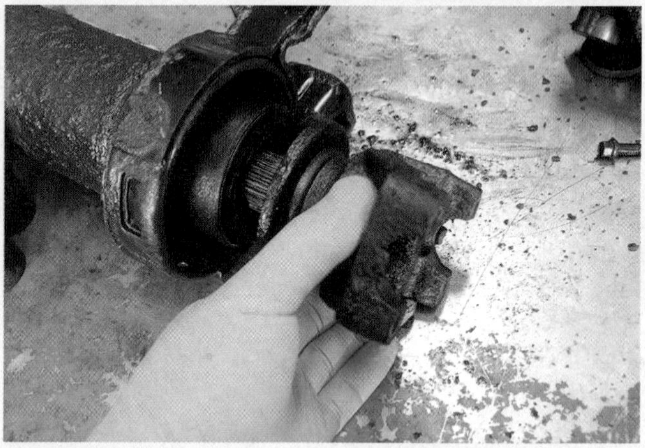

6. Remove any snap rings or circlips that may be holding the bearing in place.

7. Use an appropriate puller or press to remove the bearing from the driveshaft.

8. Check the splines for the slip yoke for any defects. Check the slip yoke on the mating shaft for wear and defects.

SKILL DRILL 23-2 Inspecting and Servicing Center Support Bearings (Continued)

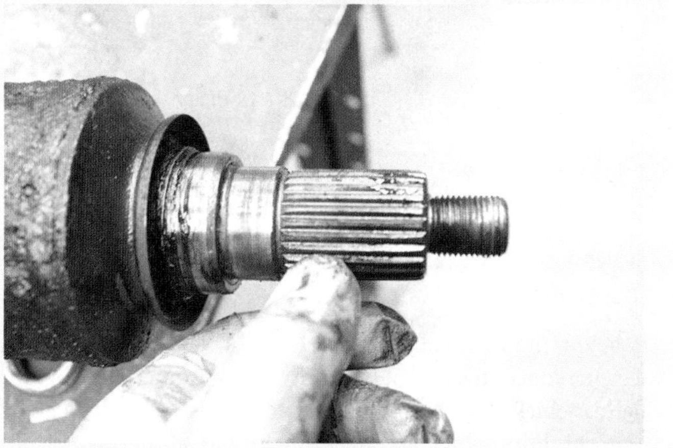

9. Press on a new bearing. Reinstall any necessary snap rings or circlips.

10. Install a new rubber mount around the new bearing. Reinstall the mounting bracket.

11. Put the two shafts back together, paying attention to driveshaft phasing.

12. Remove the driveshaft from the vise, and reinstall it into the vehicle.

▶ Final Drives/Differentials

LO 23-03 Describe the operation and maintenance of final drives and differentials.

The terms "final drive" and "differential" are confused by technicians. Specifically, the term differential gets used in two different ways. The first is the most technically accurate and refers to the components inside of the differential housing. They allow the axles to turn at different speeds when the vehicle is cornering or turning (**FIGURE 23-15**). These components make up what is called the differential assembly. Since the outside tire must travel farther around a corner than the inside tire, it must turn at a faster speed. The differential assembly allows this to happen. Otherwise, the tires would bind, skip, hop, and slide when going around a corner. This would then cause them to wear out quickly.

The second way "differential" is used is as another name for the final drive assembly. The final drive assembly provides the final gear reduction necessary for drivetrain operation. The final drive takes the power from the transmission and sends it to the wheels (**FIGURE 23-16**). The final drive also houses the "differential assembly."

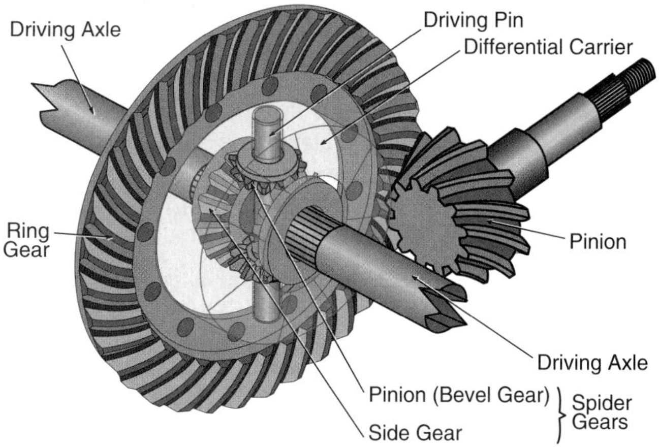

FIGURE 23-15 Differential gears are housed in the center of the ring gear.

Because the differential assembly is housed within the final drive, it is improperly called the differential. Thus, any time you hear the term "differential," ask whether the person means

FIGURE 23-16 Final drive assembly.

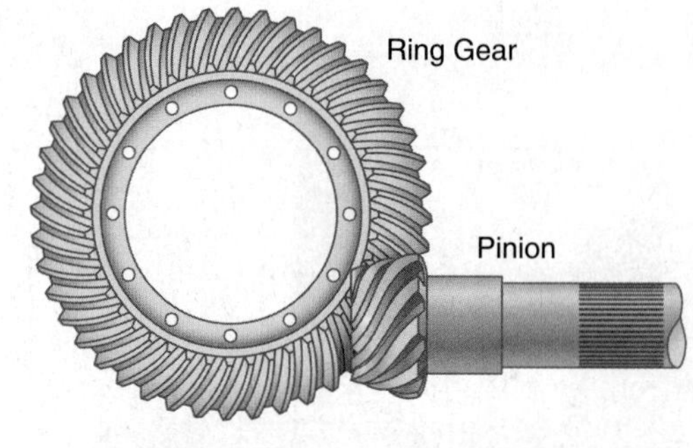

FIGURE 23-17 A hypoid gear arrangement.

the final drive assembly. Or if they mean only the differential assembly.

Final drives can be found in axles either at the front or rear of the vehicle. They can also be found in the transaxle of an FWD vehicle. The speed reduction gears in the final drive are called the ring and pinion gears. Various "rear axle" ratios are available to suit the demands of the vehicle use. Some "rear axles" allow for final drive ratios to be quickly changed, such as for racing on different track surfaces (dirt, paved, etc.) or track sizes.

Some final drive assemblies manufactured today incorporate the anti-lock brake system (ABS) wheel speed sensors. Speed sensors read wheel speed and can be part of the final drive gear set. If the vehicle is equipped with traction control, the traction control system can overcome the loss of traction. It does so by applying the brake on the wheel that is slipping. This applies more power to the wheel that is not slipping.

Differential assemblies come in two varieties. First there is the open differential, which means power is supplied to both wheels equally only when each tire maintains traction. On ice and snow, this is nearly impossible. If one wheel is stuck in the snow or ice, the other wheel can supply virtually no power to the other wheel. This is due to the nature of the open differential assembly. The second type is the limited-slip differential. It allows power to be delivered to both of the rear wheels. The one with traction can then drive the vehicle.

Preventive maintenance of the final drive assembly is important in preventing premature failure. The axles are filled with special lubricants. They promote long life and maintain quiet operation of the gears. Maintenance schedules list when each component requires maintenance or inspection.

Rear-Wheel Final Drive

In a conventional RWD vehicle, such as a pickup truck, a solid axle assembly is used. It incorporates the final drive gears, differential, and axle shafts in one housing. A ring gear and a pinion gear transfer power through 90 degrees. They also provide a final gear reduction to the driving road wheels. **Hypoid bevel gears** are normally used for this purpose. Hypoid gears are a special design of spiral bevel gears. The centerline of the pinion is below the centerline of the ring gear (**FIGURE 23-17**). This design reduces the height of the driveshaft tunnel, allowing for a flatter vehicle floor pan. The tooth shape also provides a greater area of tooth contact, and therefore greater strength.

In an RWD vehicle, the ring gear is bolted to the differential carrier. The carrier is supported by tapered roller side bearings. The bearings are retained by bearing caps and bolts. In some axle designs, threaded adjusting rings engage with threads in the housing. They push against the side bearings of the carrier. Because of this, they are used to set the carrier bearing preload and the backlash clearance between the ring and pinion. **Backlash** is the amount of movement between the teeth of the pinion gear and ring gear teeth. If you hold the ring gear and turn the pinion, there is a slight clearance back and forth, which is the backlash. On other axle designs, shims are used to adjust the side bearing preload and backlash.

Two smaller bevel gears, or pinions, are mounted on a driving pin that passes through the carrier (**FIGURE 23-18**). Two **side gears** mesh with the pinions and are in recesses in the differential carrier. The drive axles are splined to these side gears. With the ring gear bolted to the carrier, drive is transferred through the ring gear to the differential carrier pins. The pins carry the pinion gears with the carrier. The pinion gears carry the side gears with them, turning the axles and wheels. The pinion and side gears are also called spider gears.

When the vehicle travels in a straight line, the ring gear rotates the case. The driving pin and pinion gears rotate end-over-end, carrying the side gears with them. The side gears, which are splined to the axles, then turn the drive axles. There is no relative motion between the pinion gears and the side gears. Each side gear turns at the same speed.

As soon as the vehicle turns from a straight-ahead position, the inner wheel slows down. Its side gear turns more slowly than the differential case. The turning effort applied to the driving pin allows the pinion gears to rotate slowly on their pin. They walk around the inner side gear while still being turned end-over-end (**FIGURE 23-19**).

This rotation of the pinion gears makes the outer side gear and its road wheel speed up. The inner side gear and its road wheel slow down by an equivalent amount. The outer side gear then turns faster than the case. This provides an equal torque to each drive axle while allowing for their rotational speed difference.

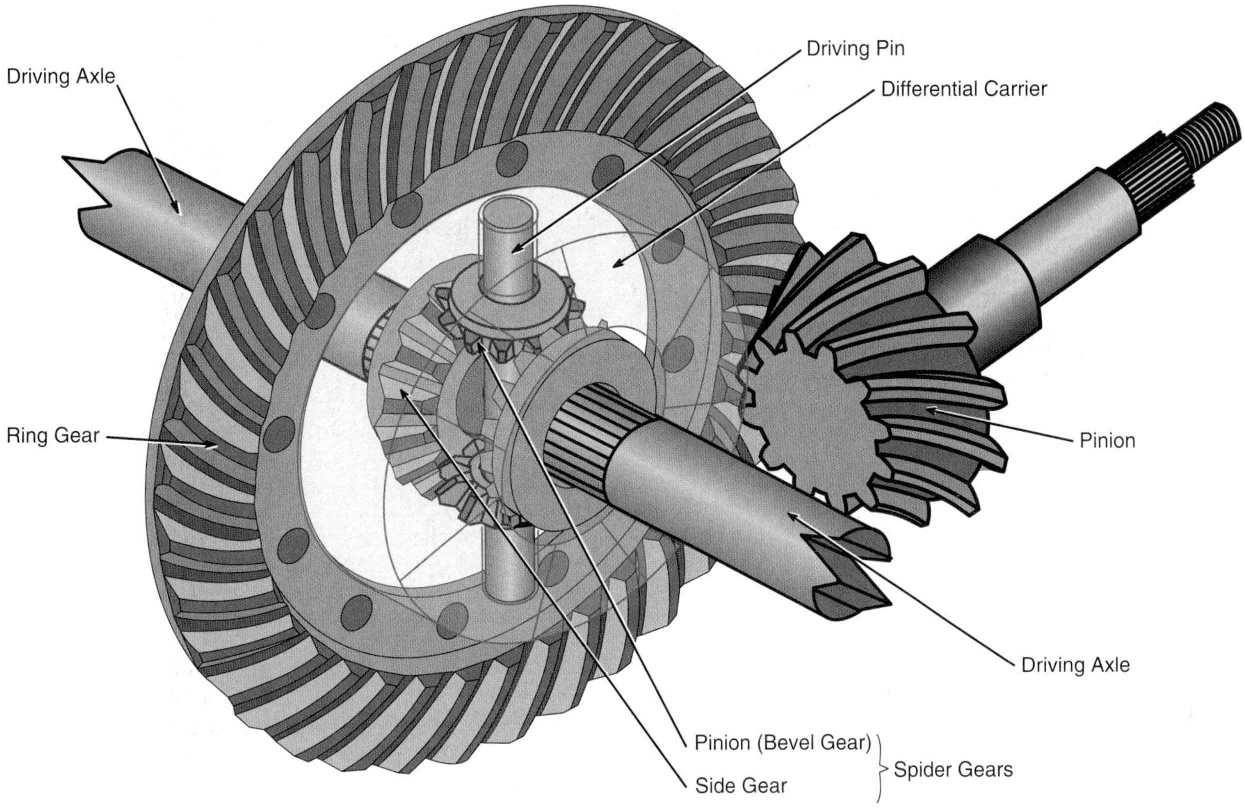
Driving Axle
Driving Pin
Differential Carrier
Ring Gear
Pinion
Driving Axle
Pinion (Bevel Gear)
Side Gear
Spider Gears

FIGURE 23-18 Differential pinion gear assembly.

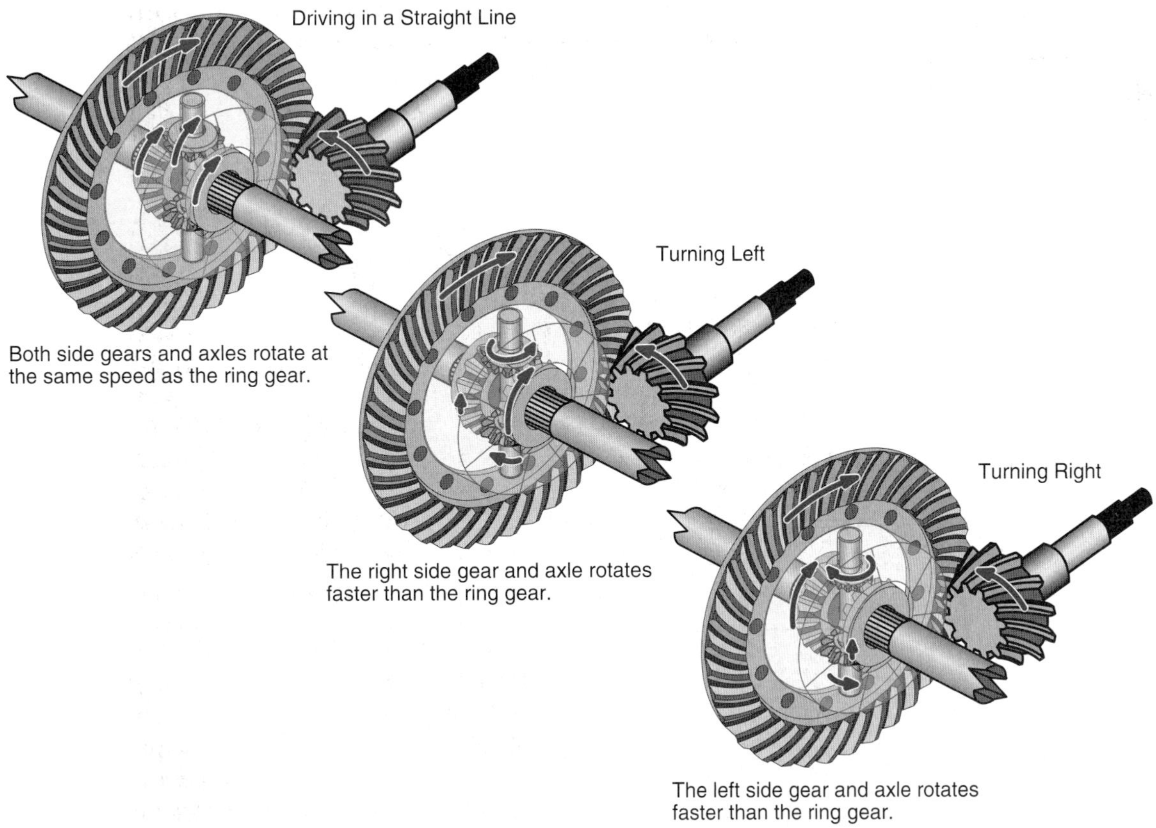
Driving in a Straight Line
Both side gears and axles rotate at the same speed as the ring gear.
Turning Left
The right side gear and axle rotates faster than the ring gear.
Turning Right
The left side gear and axle rotates faster than the ring gear.

FIGURE 23-19 The differential assembly allows the axles to rotate at different speeds during cornering.

Limited-Slip Differentials

Limited-slip differentials allow normal differential action under normal driving conditions. But when road conditions are slippery, the limited-slip differential reduces or prevents differential action. This makes it so that a wheel cannot spin freely. Drive is maintained to both wheels.

There are two main types of limited-slip differentials—clutch style and gear style. The clutch-style limited-slip differential uses a multi-plate clutch pack between each side gear and the differential case (**FIGURE 23-20**). Each clutch pack has two different sets of flat plates placed alternately in the pack. One set is plain steel, and the other is steel plate lined with friction material. One set of plates has internal splines that mate with splines on the side gear pressure ring. The other set of plates has driving lugs that locate in slots in the differential carrier.

The outside plate is dished, or cup shaped. This provides the initial tension on the clutch pack when the two halves of the carrier are assembled. Four differential pinions are mounted on two driving pins, at right angles to each other. They also mesh with the side gears. The pinion shafts are relieved, so as not to make contact at their intersection. The ends of the shafts have two flat surfaces forming wedge shapes, which fit into similar wedges in the carrier.

In straight-ahead driving, the driving force causes the pinion shafts to rotate end-over-end. This transfers the drive through the pinion gears and the side gears to the axles. There is no relative motion between the gears. However, the resistance at the road wheels forces the pinion shafts up the incline formed by the wedges in the case. As a result, the piston shafts are forced apart and the pinion gears exert a greater force on the side gears. This increases the force on the clutch packs. The force locks the side gears to the carrier, preventing any sudden spinning of either wheel. Under normal operating conditions, the driving torque is transmitted equally to each axle shaft and wheel. But when slippery patches are encountered, the ratio of torque delivered depends on the traction available at each road wheel.

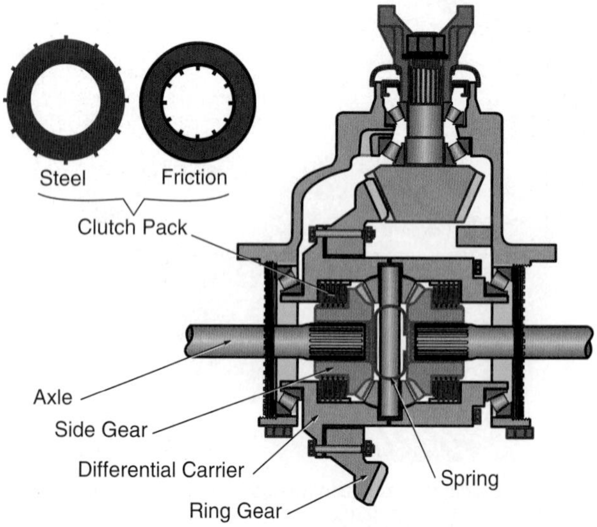

FIGURE 23-20 Clutch-style limited-slip differential.

The greatest amount of torque will be transmitted to the wheel with the most traction. When turning a corner, the limited-slip differential gives normal differential action. It permits the outer wheel to turn faster than the inner wheel. At the same time, the differential applies the major driving force to the inner wheel, improving stability and cornering.

Torsen Style

Some manufacturers use a gear-style limited-slip differential design (**FIGURE 23-21**). This design is used to improve vehicle stability and tire traction. These types go by various names: gear-based, torque-biased, or torque-sensing (Torsen). The heart of these limited-slip differentials is the parallel-axis helical gear set. The idea is to take torque from the wheel that is losing traction, and turn it over to the wheel with better traction. Torque application to the wheel with better traction begins because of the resistance between the sets of gears in mesh.

Helical-geared limited-slip differentials respond very quickly to changes in traction. They also do not bind from friction in turns like a clutch-style unit. The unit described here is the Torsen differential. It is a very popular limited-slip differential. Another manufacturer, Eaton, uses a similar design.

The side gears of the differential are cut in a worm gear configuration. The pinion gears are also cut with a worm gear cut. They are also splined to each other. When one wheel begins to slip under torque, the worm gears create a locking situation. This allows it to transfer power to the wheel with greater traction.

> ▶ TECHNICIAN TIP
>
> Oil changes are required less frequently with gear-style units. This is because there are no clutch discs to wear as the vehicle makes turns. Less clutch wear means less contamination of the gear oil.

FWD Differentials

The final drive and differential action are the same for FWD vehicles as they are for RWD differentials (**FIGURE 23-22**). The

FIGURE 23-21 A Torsen helical gear limited-slip differential.

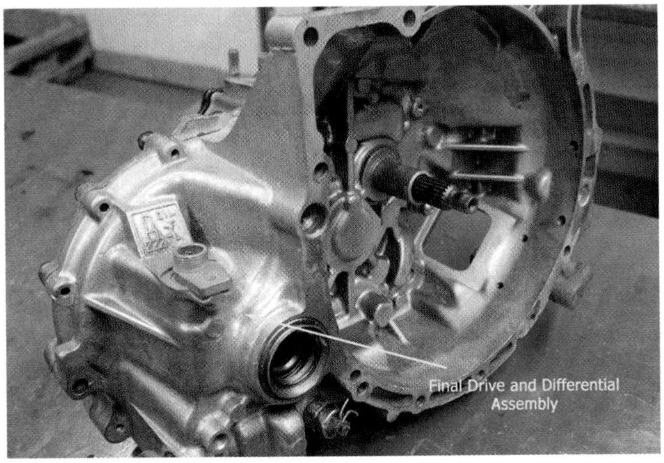

FIGURE 23-22 A typical FWD differential located in the transaxle case.

only real difference with an FWD vehicle is that the final drive and differential are located inside the transaxle. They are driven by the secondary shaft, which is connected to the pinion drive gear. Some automatic transaxles use a chain to drive the differential assembly.

Driveline and Axle Inspection and Repair

Inspecting Fluid Leakage

Fluid loss and leaks are a part of the aging process of a vehicle. Fluid leakage concerns are important and should be checked for on a regular basis. Axle seals, pinion seals, and gaskets used to seal the housing are the likely places to investigate. Seals are subject to wear that happens over time. They also harden or soften which reduces their sealing ability. Overfilling of the final drive will create a pressure buildup in the differential from foaming of the fluid. The breather may not be able to handle that and fluid can be pushed past the axle seal itself.

A lack of lubrication will cause the axle bearing to heat up and damage the seal. Brake heat coming from a dragging brake can also cause an axle seal to leak. A bent axle or axle bearing failure will speed up the failure of the seal. Visual inspection can identify some of these concerns and should be practiced on a regular basis.

To inspect fluid leakage, follow the steps in **SKILL DRILL 23-3**.

SKILL DRILL 23-3 Inspecting Fluid Leakage

1. Put the vehicle on an approved lift, and make sure it is secure. Visually inspect around the housing where the axle seats for any seepage.

2. Inspect the pinion flange for any seepage.

3. If necessary, remove the rear wheels and install a dial indicator on the axle flange to check for any distortion.

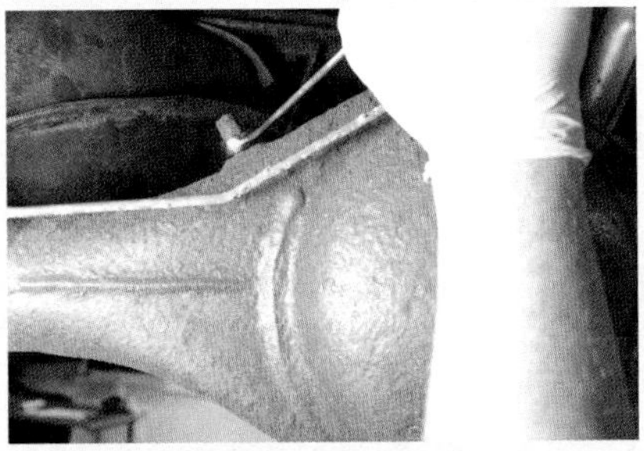

4. Check and clean the breather or vent for any obstructions that may cause a pressure buildup to occur.

5. Check the differential rear cover for a leaking gasket, if so equipped, and tighten if loose or replace as necessary.

6. Check the fluid level for lack of fluid or overfilling of the differential, as either one can indicate that a leak is present.

▶ Axles and Half-Shafts

LO 23-04 Service axles.

There are two types of axles used today—the dead axle and the live axle (drive axle). The dead axle does not provide any drive capabilities and is used on the rear of FWD vehicles and the front of RWD vehicles. It works the same as a trailer axle, allowing the wheels to spin freely and follow the drive axle. The purpose of live axle is to transfer torque to drive the wheels.

RWD Solid Axles

Drive axles used with an RWD solid axle assembly come in three varieties:

1. **Semi-floating axle**: On the semi-floating axle, the axle shafts are splined to the differential side gears. The outer bearing sits between the outer end of the axle shaft and the inside of the axle housing (**FIGURE 23-23**). The axles support the weight of the vehicle. This makes them subject to bending forces as the vehicle corners. If the vehicle with a semi-floating axle were to hit a curb and break the axle shaft, the wheel would come off the vehicle. Wheel studs are pressed into the axle flange at the end of this type of axle.
2. **Three-quarter floating axle**: On this type of axle, there is a single roller bearing between the hub and the outside of the axle housing (**FIGURE 23-24**). The axle flange is bolted to the housing and stabilizes the wheel during cornering. The vehicle's weight is supported by the hub and the bearing. The ¾-floating axle was used on older vehicles, such as Chrysler vehicles and pickup trucks.
3. **Full floating axle**: On the full floating axle, two tapered roller bearings sit between the hub and the outside of the axle housing (**FIGURE 23-25**). Because the tapered roller bearings are opposite to each other, they control sideways thrust very well. This arrangement also allows the weight of the vehicle to be carried by the hub and bearings, not on the axle itself. Since full floating axles can carry more weight, they are commonly used on medium- to heavy-duty vehicles. This type of axle floats between the axle side gears and the wheel hub. Torque is delivered to the wheel by the flange on the end of the axle that is bolted to the hub.

Some applications use conical wedges to keep the full floating axle centered on the hub. The flange requires a gasket or room temperature vulcanizing (RTV) sealer to seal between the axle flange and hub. The inner axle seal is part of the hub and can be serviced only by the removing the wheel and hub assembly.

▶ TECHNICIAN TIP

Because of their design, full floating axles can be removed without removing the wheels. Thus, the vehicle may not even need to be lifted off the ground.

Axle Flanges

In most cases, the flange on the end of the axle in either the semi-floating or the full floating axle is a molded part of the axle (**FIGURE 23-26**). The ¾-floating axle may or may not have a built-in flange as part of it. If there are wheel vibration concerns that are not related to the tires, it is good practice to check runout on the flange. If flange damage is diagnosed, replacement of the entire axle shaft will often be required.

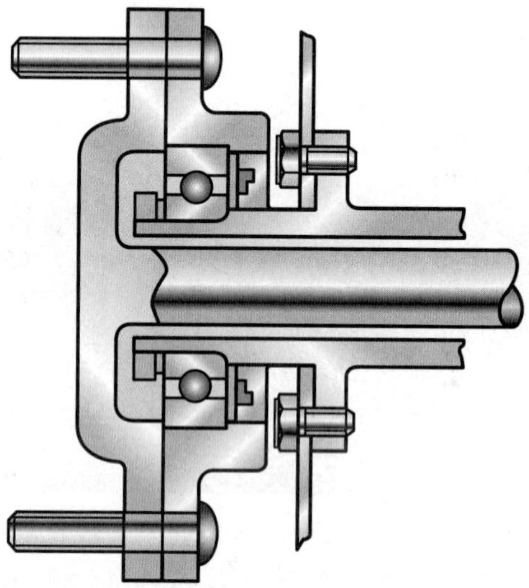

FIGURE 23-24 A ¾-floating axle—one bearing between the outside of the axle housing and the hub.

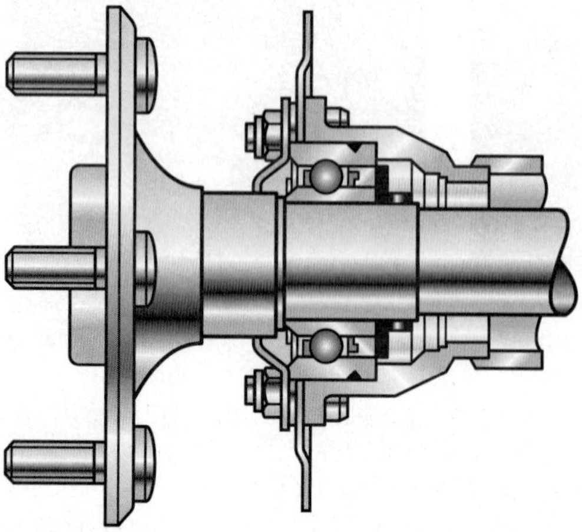

FIGURE 23-23 Semi-floating axle—bearing located between the axle and housing.

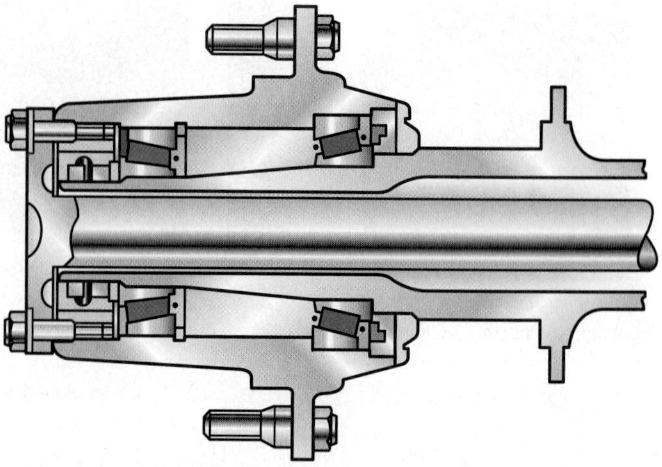

FIGURE 23-25 A full floating axle—two bearings between the axle housing and hub.

FIGURE 23-26 Axle flange with lug studs.

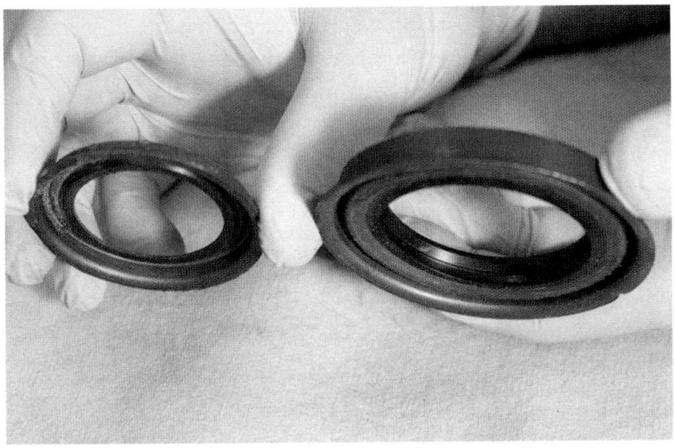

FIGURE 23-27 Single-lip and double-lip seals.

FIGURE 23-28 The speedy sleeve restores a damaged surface so that the seal has a clean smooth surface to ride against.

FIGURE 23-29 Typical half-shaft and CV joints.

Lug studs are commonly pressed into the axle flange. They are used to secure the wheel onto the axle. It is good practice to inspect the studs for damaged threads, stretching, or even breaking off. Lug studs that are damaged can usually be replaced.

Refer to the Disc Brake Systems chapter for the procedure to replace lug studs.

Axle Seals

Axle seals can be of the single-lip or double-lip design depending on the application (**FIGURE 23-27**). The purpose of axle seals is to contain the lubricating fluid in the final drive assembly. It also seals contaminants out of the sealed cavity. Too much oil inside the axle housing can cause the oil to be forced past the axle seals. Failure of any of the axle shaft bearings or differential bearings can cause leaking of the axle seals as well. Worn bearings allow the axle shafts to move. And the shaft movement will wear the seals quickly and create gaps that the fluid can leak out of.

Sometimes the seal surface on the axle becomes worn or nicked. An aftermarket repair kit called a **speedy sleeve** is available for some axles (**FIGURE 23-28**). The speedy sleeve is a thin metal sleeve that fits tightly over the seal surface of the axle. It provides a new, undamaged surface for the seal to ride against. The sleeve is thin enough that the same size, original seal can be used.

▶ TECHNICIAN TIP

A common issue you might find is the leaking of the axle grease onto the rear drum brakes. Typically, this will be found as part of a state inspection or during a routine brake system inspection. A customer may complain of the rear brakes grabbing. The sticking could be the result of axle gear oil creating a thin, sticky layer on the surface of the brake lining. This in turn makes the brake material grab the drum brake surface. This can cause the brakes to lockup, especially at slow speeds.

FWD/AWD Axles/Half-Shafts

In FWD vehicles and AWD vehicles, the axle shafts transfer the drive directly from the transaxle to the front wheels. FWD vehicles typically use axles called half-shafts. They have an inner and an outer CV joint; one joint on each end of each half-shaft (**FIGURE 23-29**).

An FWD transaxle typically does not place the final drive directly in the center of the vehicle. This is because the transmission is bolted to the end of the transversely mounted engine. Because the transmission is offset to one side, the half-shafts are usually of different lengths. The use of different-sized axle shafts can cause a problem known as torque steer. **Torque steer** is when

FIGURE 23-30 Intermediate shaft.

the vehicle pulls to one side during hard acceleration. When half-shafts are not of equal length, more torque is applied to the side with the short half-shaft. This creates the pulling condition.

In addition to the unequal axle length, the angles at the CV joints are different. This contributes to the pulling concern and can create vibration issues. To combat this condition, many manufacturers use an intermediate shaft (**FIGURE 23-30**). The intermediate shaft is a short section of shaft that typically has a bearing pressed onto it (similar to a carrier bearing). The intermediate shaft makes it so that both half-shafts are of the same length from left to right. In some cases, the manufacturer uses a longer half-shaft on one side. A rubber dynamic damper may be fitted to help absorb vibrations. However, this does not reduce torque steer issues.

Measuring Drive Axle Flange Runout and Shaft End Play

Any time wheel runout is located, it is good practice to measure the axle flange runout. It is possible that the wheel runout is caused by axle runout. It is also possible that there is runout in both the wheel and axle.

Once axles have been serviced or replaced, it is a good practice to again measure axle flange runout and axle end play. If all repairs have been done properly, runout and end play should be in specifications. Some procedures may differ due to the manufacturer's protocol. Always consult the service information to ensure proper procedures for runout and end play.

To measure drive axle flange runout and shaft end play, follow the steps in **SKILL DRILL 23-4**.

SKILL DRILL 23-4 Measuring Drive Axle Flange Runout and Shaft End Play

1. Remove the wheels and brake drum or rotor. Start by mounting a dial indicator on the backing plate, and set the dial gauge on the face of the axle flange. Zero out the indicator.

2. Slowly rotate the axle, and measure the runout between the high and low readings.

3. Compare with the manufacturer's specifications. If it does not match, replace the axle.

4. Once runout has been completed, zero out the dial indicator again.

SKILL DRILL 23-4 Measuring Drive Axle Flange Runout and Shaft End Play (Continued)

5. Following the service information procedure, pry the axle in and out, and obtain an end play reading.

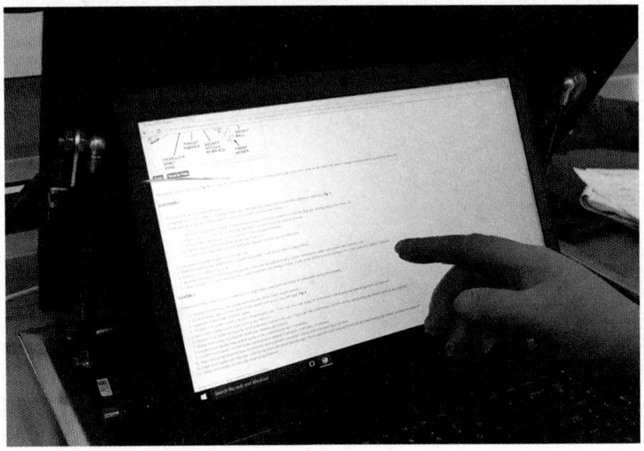

6. Compare to the manufacturer's specifications. If it does not match, the axle may have to be removed, and the axle groove should be inspected (integral axle type), or the bearings in the housing should be inspected or replaced (removable carrier type).

▶ Inspecting and Replacing Wheel Studs and Lug Nuts

Lug nuts and studs hold the drive wheels in place. Sometimes, improper torqueing of the lug nuts can cause the studs to fail or even lose a wheel when driving. Undertightening or overtightening can result in these conditions. Uneven tightening can also cause warping of brake rotors, which can cause brake pedal pulsations.

In some states, the law requires an inspection of the brakes that involves removal of the tires to gain access to the brake system. Lug nuts must be removed to do this. Inspection of the studs should be done in order to look for stretching or thread damage on these studs (**FIGURE 23-31**). If any defect or thread damage is present, the stud should be replaced. *Note:* Refer to the Servicing Disc Brakes chapter for the procedure to replace lug studs.

▶ TECHNICIAN TIP

Never use an impact wrench to torque lug nuts. Some shops allow the use of a torque stick followed by a torque wrench that is set to the manufacturer's torque specifications. Otherwise, lug and stud damage can occur. This can result in loss of the wheel and tire assembly while driving down the highway.

▶ Joints and Couplings

LO 23-05 Service U-joints and CV joints.

The driveshaft and half-shafts require a flexible joint at both ends to allow for changes in angle as the suspension travels up and down. There are two common types of joints used in this way—the U-joint and the CV joint. Both types of joints allow the shaft to transmit torque through a change of drive angle. The U-joint does so with an increase and decrease in velocity as the joint rotates every 90 degrees. A CV joint maintains the same velocity as it goes through its rotation.

Both types of joints can transfer torque through an angle. But the CV joint can do so at a much greater angle than the U-joint. This is why CV joints are used on front half-shafts where large steering angles must be accommodated.

FIGURE 23-31 Inspecting lug studs. **A.** Stretched lug stud. **B.** Cross-threaded lug stud.

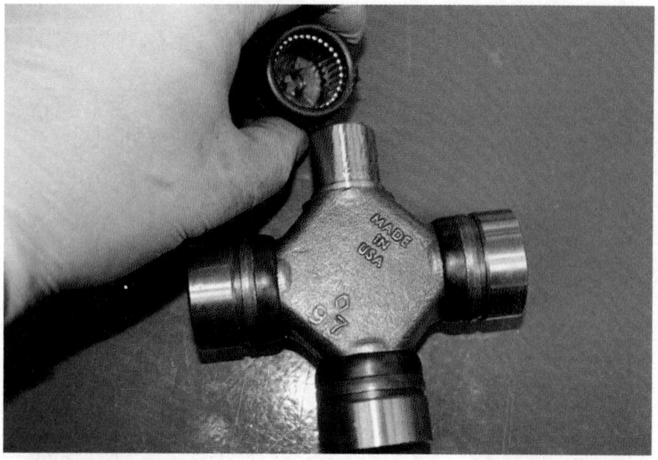

FIGURE 23-32 Cross-and-roller U-joint.

FIGURE 23-33 A typical double Cardan joint.

U-joints are typically used on driveshafts because they usually only operate at small angles. However, some manufacturers use CV joints to provide a smoother ride. CV joints are also used on some lifted off-road vehicles that have a lot of suspension travel. A large amount of torque is transmitted through the driveshaft and half-shafts. This makes the joints subject to wear over time, making regular inspection important.

Universal Joints

A **U-joint** is a cross-shaped flexible joint. There are caps that fit over the ends of the cross. Needle bearings fit between the ends of the cross and the caps, allowing the caps to rotate smoothly. The U-joint is considered a non-constant velocity joint.

The most common type of U-joint is a **Hooke's joint** (cross-and-roller joint) (**FIGURE 23-32**). The Hooke's joint consists of:

- a steel cross with four hardened bearing journals,
- hardened caps which locate the cross in the eyes of the final drive yoke, and
- needle rollers that fit between the journals on the cross and the hardened caps.

The cross swivels in the caps as the drive is transferred across the joint. However, the swiveling affects the speed transferred across the joint during rotation. In each revolution, the velocity of the yoke changes every 90 degrees due to the angles that the driveshafts are operating on. This change in the angular velocity increases as the angle that the joint operates on increases.

There are two planes of stop/start as the joint flexes. This is due to the cross shape of the joint. As the shaft rotates, it changes the velocity of the shaft up and down and then side to side. A speeding up and slowing down of the joint occurs. The sharper the angle of the joint, the more pronounced the velocity change becomes. The change in velocity can cause a vibration in the drivetrain.

The effect of the changing velocity can be minimized by aligning identical U-joints at each end of the drive shaft (phasing). This is done by having the yokes of the driveshaft in line with each other. An increase in the velocity of the front yoke is canceled out by a similar decrease in the velocity of the rear yoke in each revolution. In some designs, engineers fit a CV joint to the front end of the driveshaft. This allows for an increase of angles of the shafts without creating vibration.

A **double Cardan joint** greatly reduces the change in velocity of a single Cardan joint. It uses the second joint to cancel out the changes in velocity of the first joint. A double Cardan joint is considered by most technicians to be a CV joint under normal driveline angles. The double Cardan joint uses two Cardan joints housed in a short carrier (**FIGURE 23-33**). Each cross has either a centering ball or a socket that is connected to its back side. Each part joins with its mating component on the other cross. The ball-and-socket assembly keeps each Cardan joint operating at an equal angle. This effectively cancels out the change in velocity.

Replacing U-joints

When a U-joint is worn and exhibits play or binding, it needs to be replaced. But replacing a U-joint takes a bit of care to do it properly. First, the caps are retained in one of two ways. The first way is with C-clips, either on the inside or outside of the U-joint cap. The other method of retaining the caps is with an injected melted plastic that hardens and fills the groove when it cools. These are known as plastic-retained U-joints. When U-joints of this style are being removed, they usually need to be heated to soften the plastic.

U-joint caps are also press-fit into the yokes on the driveshaft. So, they will have to be pressed out of and into the yokes when being replaced. This can be done with a hydraulic press, a U-joint press, or a large vice. But any C-clips need to be removed before starting to press them apart. Also, use caution when pressing them apart as parts can be damaged, or shatter.

Many new U-joints come with a grease fitting so that the joint can be lubricated. The grease fitting screws into a threaded hole that is slightly offset to one side of the U-joint. In most cases, the U-joint needs to be installed so that the grease fitting will be toward the center of the driveshaft. Just don't install the grease fitting until the U-joint has been installed in the yoke. Otherwise, the grease fitting will get in the way of installing the U-joint. Also, don't grease the fitting until the U-joint is fully installed in both yokes. Otherwise, the grease will force the caps off the joints.

To properly replace a U-joint, follow the steps in **SKILL DRILL 23-5**.

SKILL DRILL 23-5 Replace a U-Joint

1. With the driveshaft removed, remove any C-clips that retain the caps. These can be on the outside or the inside.

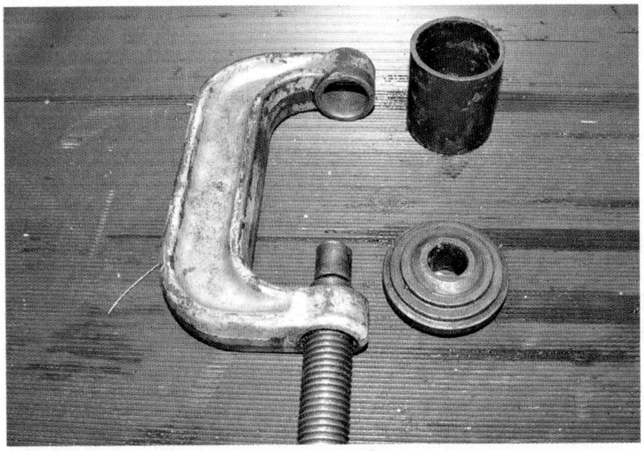

2. Select the correct adapters for the U-joint press and assemble the tool.

3. Place the tool over one of the caps and run the tool down so it is in contact with the cap.

4. Tighten the tool to force the cap on the opposite side out of the yoke. Be careful not to push the cap on this side to far. It will fall into the center of the yoke and make it hard to push back out.

SKILL DRILL 23-5 Replace a U-Joint (Continued)

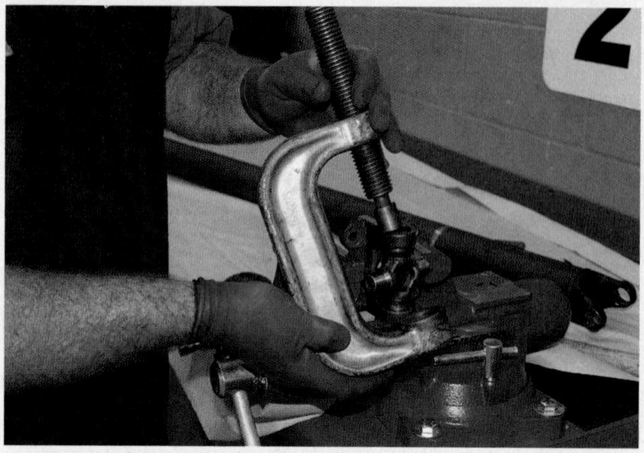

5. Remove the tool and install it on the other side. Press the cross back into the yoke to force the opposite cap out of the yoke.

6. Repeat steps 3–5 on the other two caps if they are captured in the other yoke.

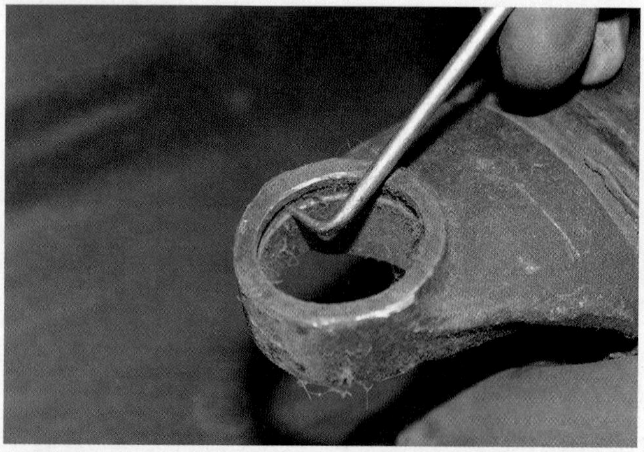

7. Clean the yoke and C-clip grooves with fine sand paper and a pick tool.

8. Carefully remove two opposing caps from the new cross. Make sure the needle bearings haven't fallen out of place.

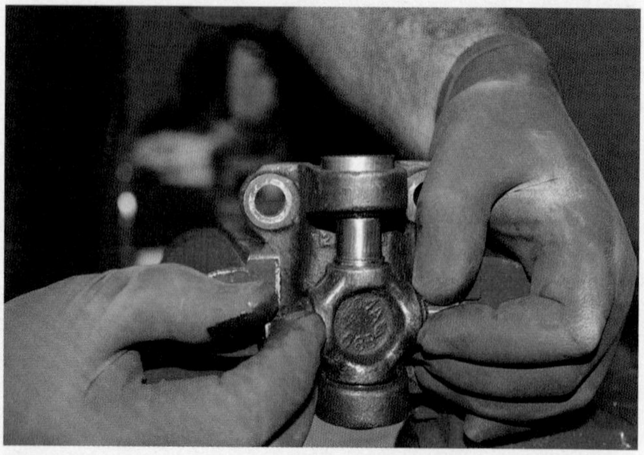

9. Orient the new cross in the yoke so that the grease fitting will be positioned correctly. This usually means that the fitting should face the center of the driveshaft. Make sure the fitting will have clearance once the U-joint if fully installed.

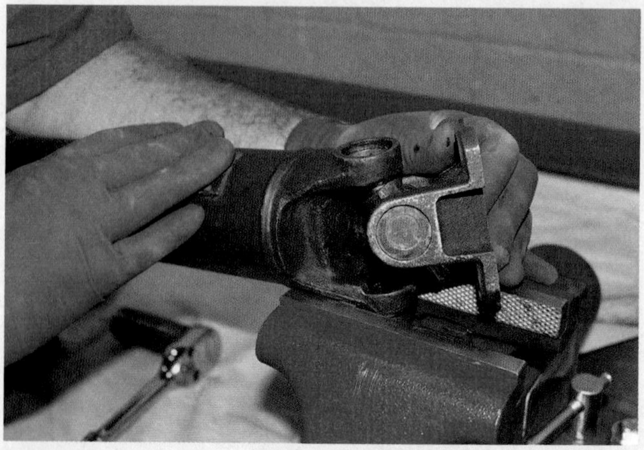

10. Carefully move the cross toward the top of one of the holes in the yoke. In this position, carefully place the bearing cap over the cross and into the hole in the yoke. Be careful to not knock over any of the needle bearings. Also, hold the cross as high in the hole as you can.

SKILL DRILL 23-5 Replace a U-Joint (Continued)

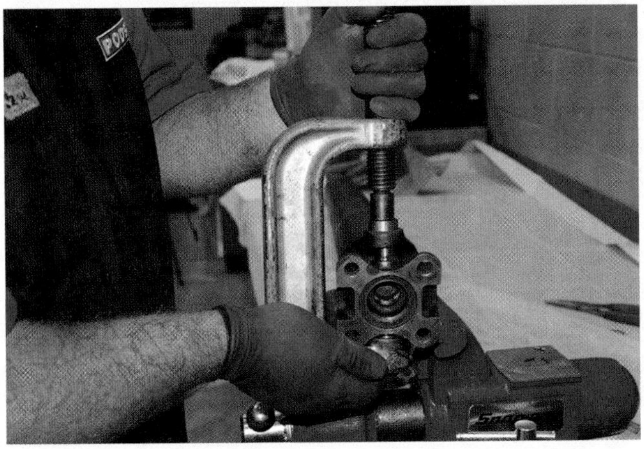

11. Use the tool to press the cap into the yoke. It may help to press it a small amount beyond normal so the cross will engage the needle bearings on the other cap, without falling out of the bearings on this cap.

12. Install the C-clip for the cap you just installed.

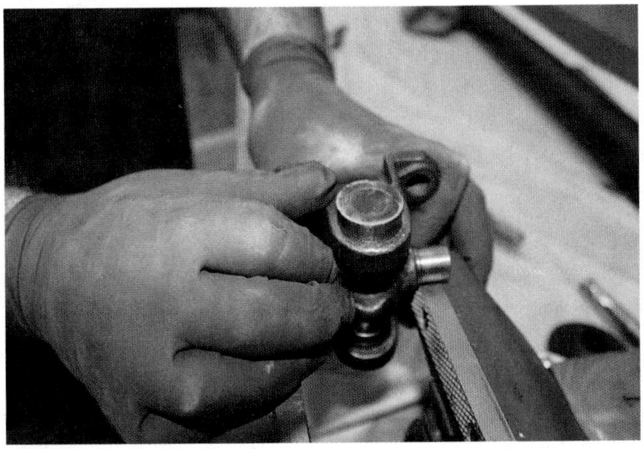

13. Hold the cross so it is still in the needle bearings on the cap you just installed, but raised up enough to engage the needle bearings on the cap you will install next.

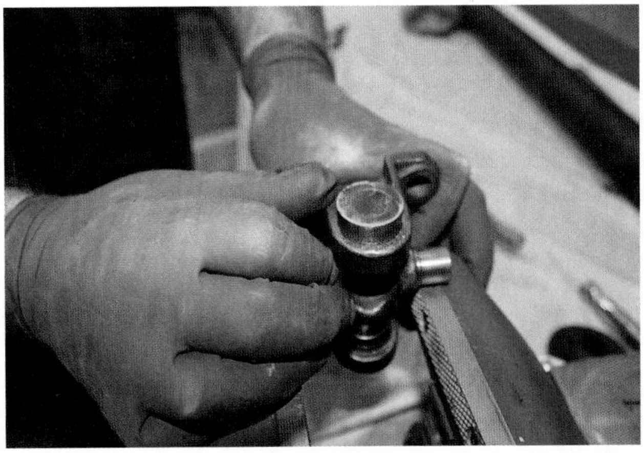

14. Push the cap in as far as you can by hand.

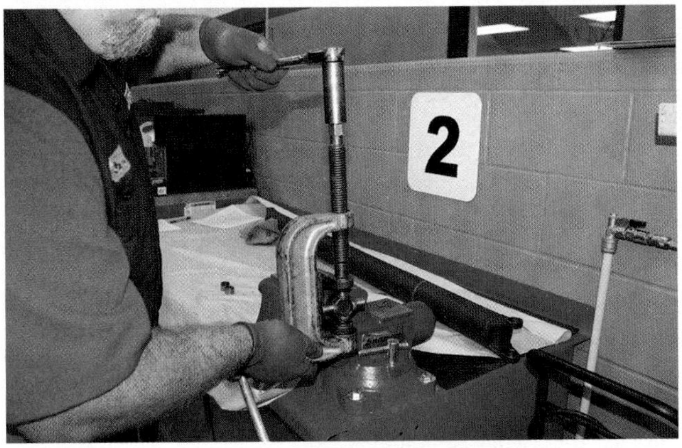

15. Use the tool to press the cap into the yoke. Press it in until the other cap is seated against the C-clip.

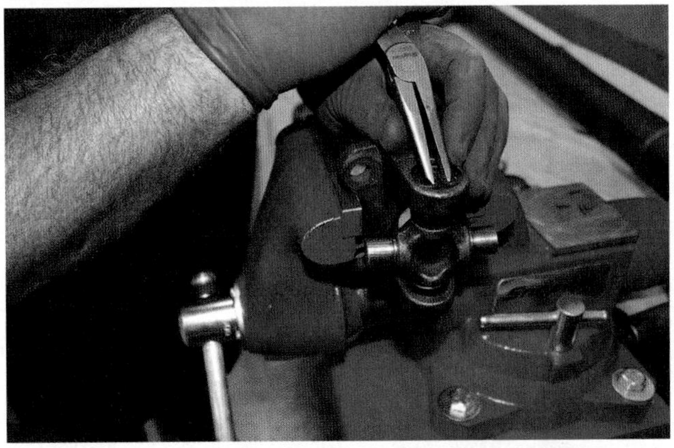

16. Install the C-clip for the cap you just installed. If both caps are fully installed, but you can't install the last C-clip, you may have knocked a needle bearing out of place. If so, you will have to push the cap/s back out and see if the bearings got damaged.

SKILL DRILL 23-5 Replace a U-Joint (Continued)

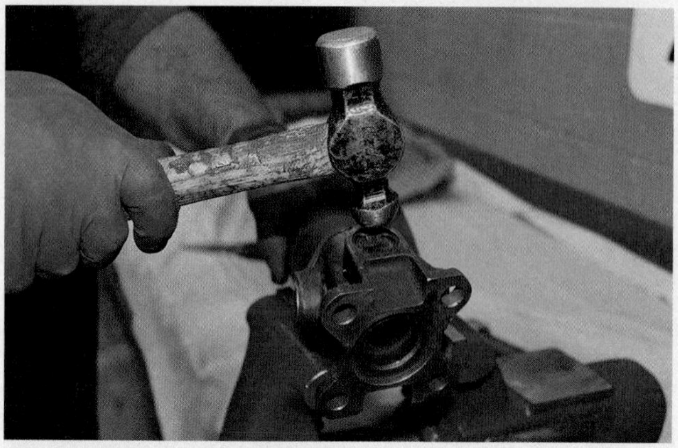

17. Once the caps are fully installed and the C-clips in their grooves, center the cross in the caps. Support the horizontal cross while giving a couple of firm raps of a hammer on the vertical yoke. Rotate the driveshaft and do this for each cap.

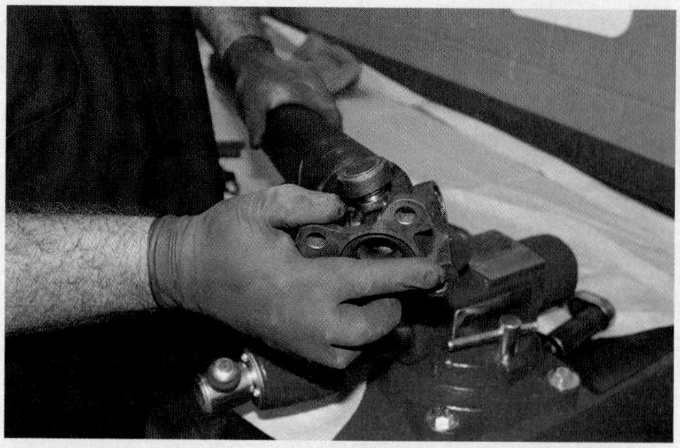

18. When finished, verify that the joint rotates freely. If all four caps are installed in yokes, install the grease fitting and lubricate the joint until grease barely comes past the rubber seals.

Constant-Velocity Joints

Vehicles with independent suspension use external drive axle shafts (half-shafts) to each road wheel. They typically use CV joints at each end of the half-shafts (**FIGURE 23-34**). Because of their construction, CV joints can operate at greater angles than U-joints. Also, CV joints allow for more torque transfer than U-joints due to their larger bearing surfaces.

To accommodate for changes in axle shaft length while driving, a sliding spline or a **plunge-type joint** is often used. It is typically the inner joint on the half-shaft. Plunge-type joints allow smooth power flow while allowing the joint to slide in and out. This effectively increases and decreases the length of the axle shaft during up-and-down suspension travel.

One type of plunge CV joint is the tulip or tripod joint. The **tulip/tripod joint** has three equally spaced fingers shaped like a star. On the ends of the star are three round-end caps. Needle bearings sit between the round-end caps and each finger. The round-end caps rotate on the needle bearings.

The outer housing of the CV joint has three straight grooves that run from side to side. The round-end caps slide in the grooves. This configuration allows in-and-out movement of the shaft while also allowing changes in the angle.

The **fixed-type joint** does not slide to allow for shaft lengthening or shortening. It simply allows for angle changes as the suspension moves. The fixed joint is typically used on the outboard side of the half-shaft. One type of fixed joint that you will service

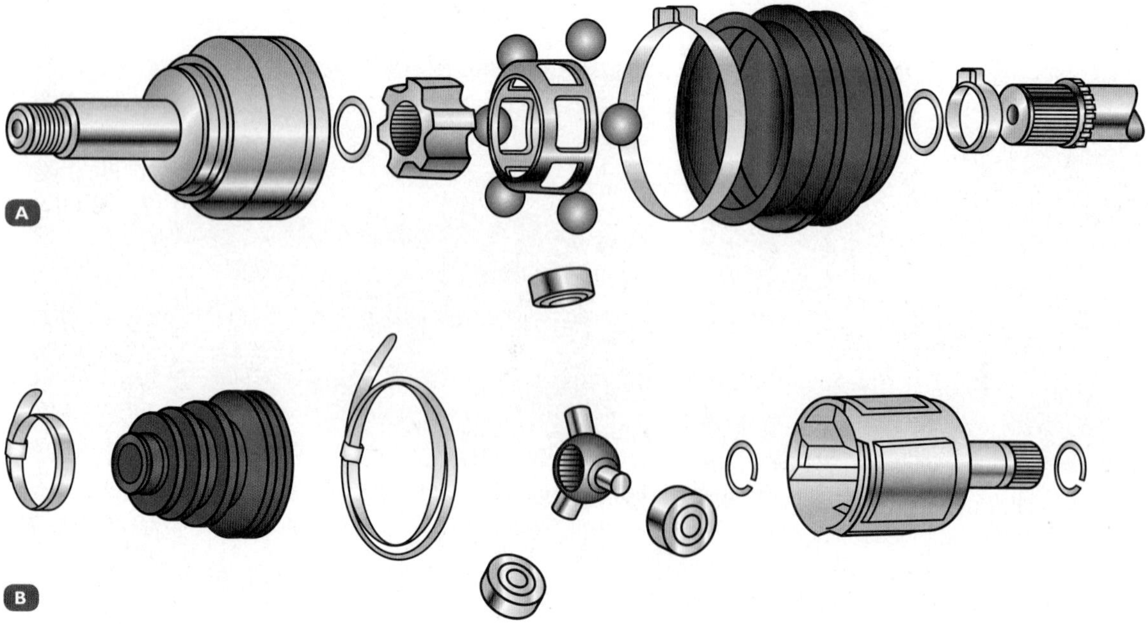

FIGURE 23-34 The CV joint comes in several forms. **A.** The Rzeppa joint. **B.** The tulip tripod joint. Both are commonly found on half-shafts.

as a technician is the **Rzeppa joint**. The Rzeppa joint has an inner race, six steel ball bearings, a bearing cage, and an outer race. The ball bearings, retained by a spherical cage, are carried in angular grooves in each race. These balls transfer the drive from one race to the other. The inner race is splined to the axle shaft while the outer race is splined to the wheel hub. The Rzeppa joint has been modified so that it will allow a limited amount of plunge capability.

CV Joint Lubrication and Inspection

Each CV joint is lubricated with grease. Special CV joint grease is used to keep the bearing surfaces moving freely. An accordion-type synthetic rubber boot, called a CV boot, is attached to the outside of the joint (**FIGURE 23-35**). It is designed to keep the grease inside the joint, and to keep dirt and water out.

FIGURE 23-35 CV boot.

The CV boot should be inspected for cracks and splits during routine oil changes. If a boot is leaking, traces of lubricant thrown on nearby components will be noted. In some cases, split-type boots can be used as replacements. Otherwise, the half-shaft will need to be removed to replace the defective boot. If any leakage is noted, the boot should be replaced before dirt and water can enter the joint, causing the joint to ultimately fail. If dirt or water has entered the joint, the joint should be replaced. This is typically done by replacing the entire half-shaft.

Although CV joints are larger than U-joints, they are still subject to wear.

SAFETY TIP

Some CV joint greases use lead to help cushion the high metal-to-metal contact loads within a CV joint. Always wash your hands thoroughly after working with any CV joint grease.

Diagnosing CV Joint Issues

CV joints have to withstand and transmit tremendous amounts of force when the vehicle is being driven. Grease helps to lubricate and cool the joint, which prevents or reduces wear. CV joints use rubber boots to contain the grease in the joint while still allowing the joint to pivot. Sometimes, the CV joint boots crack or become torn. The grease then runs out, and contaminants can get in. This combination of problems causes the joint to overheat and wear out, resulting in a failed CV joint that must be replaced. This is one reason that periodic inspection of the CV boots is important. If caught soon enough, the boot and grease can be replaced before the joint is damaged.

An indication of excessive wear is a clicking or popping noise. The noise occurs when the vehicle is driven with the wheels fully turned in one direction or the other. The noise is caused by the drive balls rolling into and out of the worn spots. Severe cases are indicated by noise at lesser steering angles. CV joints that make this noise have to be replaced; they cannot be repaired. A thorough test drive along with a visual inspection on a lift may be warranted in order to diagnose CV joint problems.

To diagnose a potential issue with a CV joint, talk to the customer to fully understand his or her concern. Verify the concern with a test drive or visual inspection, depending on the concern. You should then have enough information to make a diagnosis of the fault.

To diagnose CV joint issues, follow the steps in **SKILL DRILL 23-6**.

SKILL DRILL 23-6 Diagnosing CV Joint Issues

1. Thoroughly road test the vehicle to verify the customer complaint. Safely raise the vehicle on an approved lift, and make sure it is secure.

2. Visually inspect all four CV joints. Look for broken or ruptured boots. Look for cracked or dry-rotted boots, as they are subject to all types of weather.

SKILL DRILL 23-6 Diagnosing CV Joint Issues (Continued)

3. Manually move the axle shaft up and down, looking for unnecessary play or bad bearings. These conditions can cause a vibration coming from the side that is bad. They will also cause a clicking sound emanating from the side that is affected.

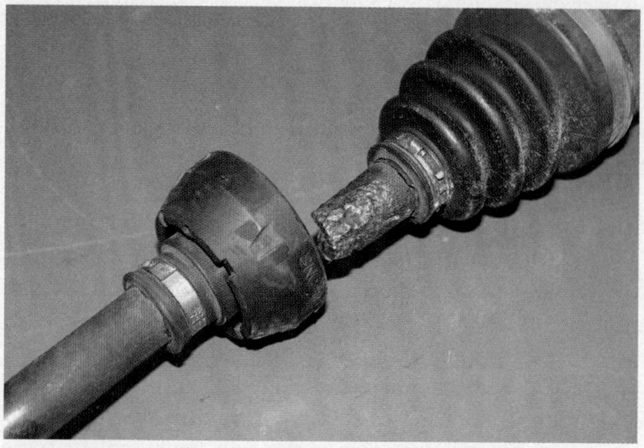

4. Look for any type of axle damage, such as a bent axle from a recent accident. Axle damage causes a vibration from the affected shaft as well.

Inspecting Half-Shaft Components

Service of CV joints depends on the type of failure that has occurred. For example, if the axle is clicking when turning sharply, then the joint or the entire half-shaft will need to be replaced. If the CV boot is torn, it can be replaced without replacing the entire joint or shaft. Some can be easy to replace, and some can be extremely hard. A thorough inspection will give you a good idea of whether to replace the half-shaft in its entirety. Or if you can just replace and service one CV joint or boot.

To inspect half-shaft components, follow the steps in **SKILL DRILL 23-7**.

> **TECHNICIAN TIP**
>
> Some CV joints do not require a retaining ring and just need to be driven off with a soft metal hammer, preferably a brass one.

SKILL DRILL 23-7 Inspecting Half-Shaft Components

1. Clamp the entire half-shaft into a soft-jawed vise, and make sure it is secure. Remove the retaining clamps from the CV boot. Slide the boot down the shaft, paying attention to the condition of the boot.

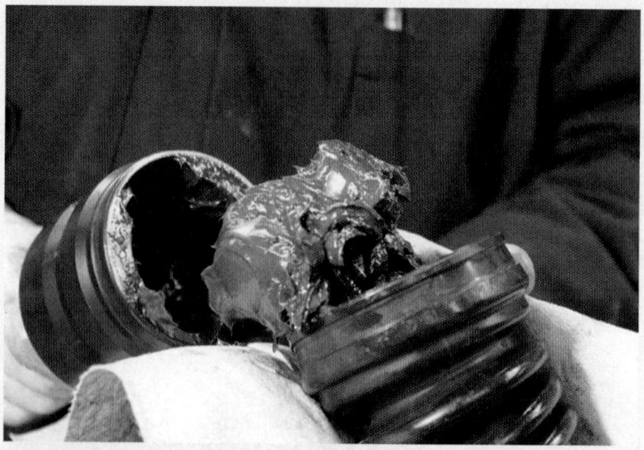

2. Wipe out as much grease as possible to be able to access the retaining ring from the CV joint itself. If there is a retaining ring present, remove the retaining ring with the appropriate tool. When the retaining ring is removed, remove the CV joint from the half-shaft, and inspect the splines on the end of the half-shaft. This also applies to the other end of the half-shaft.

SKILL DRILL 23-7 Inspecting Half-Shaft Components (Continued)

3. Inspect the old joint to gain an accurate assessment of the failure to prevent a reoccurrence of this failure. Reinstall the new CV joint onto the shaft splines as required.

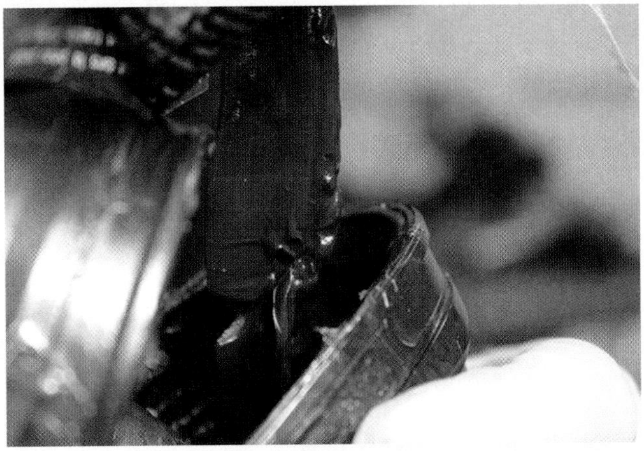

4. Apply the lubrication grease that comes with the new CV joint. Tighten the boot clamp to ensure that grease will not be lost. This applies to both sides of the half-shaft. Reinstall the half-shaft, following the manufacturer's guidelines.

▶ Wrap-Up

Ready for Review

- Different drivetrain layouts include: rear-wheel drive layout, front-wheel drive layout, four-wheel drive layout, and all-wheel drive (AWD) layout.
- In a driveshaft, the most common places for wear is in the U-joints. During multi-piece driveshaft disassembly, it is critical to mark the parts in relation to each other.
- A differential allows the axles to turn at different speeds when the vehicle is cornering or turning and the final drive assembly provides the final gear reduction necessary for drivetrain operation.
  - The differential assembly is housed within the final drive.
- Axle flange runout needs to be measured when wheel runout is located. Axle flange runout and axle end play should be measured when axles have been serviced or replaced.
- A u-joint must be replaced when it is worn and exhibits play or binding. The caps of u-joints are retained either with c-clips or with an injected melted plastic. The CV boot should be inspected for cracks and splits during routine oil changes. An indication of excessive wear in CV boots is a clicking or popping noise that occurs when the vehicle is driven with the wheels fully turned in one direction or the other.

Key Terms

backlash The required clearance between two meshing gears.

companion flange A splined flange attached to a vehicle component, such as a drive axle pinion shaft, that bolts to a flange yoke on a driveshaft.

double Cardan joint A type of joint that uses two Cardan joints housed in a short carrier and that reduces the change in velocity of a single Cardan joint by using the second joint to cancel out the changes in velocity of the first joint.

driveline angularity The angles at the universal joints.

fixed-type joint A joint that does not slide to allow for shaft lengthening or shortening; it simply allows for angle changes as the suspension moves.

Full floating axle An axle where the axle only carries the twisting force for the wheels. The axle does NOT carry the weight of the vehicle.

helical-geared limited slip differential A type of differential that responds very quickly to changes in traction and that does not bind from friction in turns or lose its effectiveness, because there are no clutches.

Hooke joint A joint with four trunnions and four bearing caps; also known as a Cardan joint or a universal joint.

Hypoid bevel gears A special design of spiral bevel gears where the centerline of the pinion is below the centerline of the ring gear

longitudinal The orientation of the engine in which the front of the engine is facing the front of the vehicle. It is most commonly found in rear-wheel-drive vehicles.

plunge-type joint The inner joint on the half-shaft that allows for changes in shaft length.

PTO Power take-off. Used on some trucks to power accessories such as a dump bed.

Rzeppa joint A type of fixed constant-velocity joint that has an inner race, six steel ball bearings, a bearing cage, and an outer race.

semi-floating axle An axle with the bearing placed between the axle and the axle housing; therefore, it carries the load of the vehicle on its outer end.

side gears A gear that is splined to the axle shaft and meshes with the spider gears, and that allows the axles to rotate at their own speeds when cornering and turning.

sliding spline driveshaft A two-piece driveshaft that is joined in the middle with splines. The driveshaft can slide on itself to increase or decrease in length.

slip yoke Part of a two-piece driveshaft that is splined and allows for a change in length of the shaft as the suspension compresses and rebounds.

speedy sleeve An aftermarket repair kit that consists of a thin metal sleeve that fits tightly over the seal surface of the axle, providing a new, undamaged surface for the seal to ride against.

three-quarter floating axle An axle on which there is only one wheel bearing that bears the weight of the vehicle, but the axle prevents the wheel from tipping inward or outward.

torque steer A condition in which the vehicle pulls to one side during hard acceleration. It can be the result of unequal axle lengths as designed, which cannot be repaired if it is by design.

transverse A term used to describe the side-to-side engine orientation when mounted in the engine compartment.

tulip/tripod joint A constant-velocity joint that has three equally spaced fingers shaped like a star. This configuration enables in-and-out movement of the shaft while also allowing flexing.

U-joint See *universal joint.*

universal joints A flexible cross-shaped joint used to transmit torque.

Review Questions

1. Differences in the distance and angles between the transmission and the final drive assembly are compensated for by all the following EXCEPT:
   a. Universal joints
   b. Slip yokes
   c. Sliding spline driveshafts
   d. Ring and pinion gearsets
2. What is meant by all-wheel drive?
   a. All-wheel drive is the same as 4-wheel drive
   b. All-wheel drive means all four wheels have power all the time
   c. All-wheel drive means the driver must lock the wheel hubs manually
   d. All-wheel drive means the vehicle must be in neutral to shift from high to low
3. All the following statements are true about universal joints EXCEPT:
   a. A universal joint connects the driveshaft to a companion flange on some RWD vehicles
   b. Some universal joints are held to the flange with U-bolts or straps
   c. Universal joints are considered constant velocity joints
   d. A universal joint is connected to the driveshaft through a slip yoke on some RWD vehicles
4. What is driveline angularity?
   a. It is the difference between the angle of the transmission output shaft and the final drive pinion shaft
   b. It is the difference between the angle of ring gear and the pinion gear
   c. It is the difference between the angle of the engine crank shaft and the clutch disk
   d. It is the difference between the angle of the chassis frame and the carrier bearing
5. Side gears and bevel pinion gears are referred to as:
   a. Rack gears
   b. Ring gears
   c. Sun gears
   d. Spider gears
6. Which drivetrain assemblies include final drive gears?
   a. Transaxles, front axles, and rear axles
   b. Transmissions, front axles, and rear axles
   c. Transaxles, transmissions, and rear axles
   d. Transmissions, transfer cases, and rear axles
7. What are the two common types of axles used today?
   a. Hemispherical and stratified
   b. Live and dead
   c. Internal and external
   d. Simpson and Ravigneaux
8. Which tool is axle flange runout measured with?
   a. A micrometer
   b. A vernier caliper
   c. A dial indicator
   d. Feeler gauges
9. What is the proper way to replace universal joints?
   a. Cut the joint out with a hole saw
   b. Hammer the old joint out and hammer the new one in
   c. Melt the joint caps with an acetylene torch
   d. Use a U-joint press to press the old joint out and the new joint in
10. What type of constant velocity joint allows a half-shaft to change length with suspension travel?
    a. A plunging joint
    b. A universal joint
    c. A flex joint
    d. A double Cardan joint

ASE Technician A/Technician B Style Questions

1. Technician A says that rear-wheel drive vehicles have a propeller(drive) shaft. Technician B says that some rear-wheel drive vehicles have transaxles. Who is correct?
   a. Technician A only
   b. Technician B only
   c. Both A and B
   d. Neither A nor B
2. Technician A says that a longitudinal mounted engine is most often found in rear-wheel drive and four-wheel drive vehicles. Technician B says front-wheel drive vehicles with transverse mounted engines need to turn the power flow 90 degrees to drive the wheels. Who is correct?
   a. Technician A only
   b. Technician B only
   c. Both A and B
   d. Neither A nor B
3. Technician A says that two-piece driveshafts use a carrier bearing. Technician B says that an out of phase driveshaft will cause a vibration. Who is correct?
   a. Technician A only
   b. Technician B only
   c. Both A and B
   d. Neither A nor B
4. Technician A says that driveline angularity can be checked with dial indicator. Technician B says that driveshaft phasing can be checked with a level. Who is correct?
   a. Technician A only
   b. Technician B only
   c. Both A and B
   d. Neither A nor B
5. Technician A says that backlash is the amount of movement between the ring gear and pinion gear. Technician B says that the axles are splined directly to the ring gear. Who is correct?
   a. Technician A only
   b. Technician B only
   c. Both A and B
   d. Neither A nor B
6. Technician A says that some limited-slip differentials use clutch packs with alternating friction and steel plates. Technician B says that some limited-slip differentials use a helical cut or worm gear sets. Who is correct?
   a. Technician A only
   b. Technician B only
   c. Both A and B
   d. Neither A nor B
7. Technician A says that the axle in full floating axles supports the vehicle's weight. Technician B says that semi-floating axles are commonly used on medium to heavy duty trucks. Who is correct?
   a. Technician A only
   b. Technician B only
   c. Both A and B
   d. Neither A nor B
8. Technician A says that unequal lug nut torque can cause brake pedal pulsations. Technician B says that overtorquing lug nuts with an impact gun is not possible. Who is correct?
   a. Technician A only
   b. Technician B only
   c. Both A and B
   d. Neither A nor B
9. Technician A says that Hooke's (cross-and-roller) universal joints are commonly used on driveshafts. Technician B says that constant velocity joints allow for greater operating angles to accommodate steering and independent suspension travel.
   a. Technician A only
   b. Technician B only
   c. Both A and B
   d. Neither A nor B
10. Technician A says that a clicking noise while turning sharply may be a worn outer constant velocity joint. Technician B says half-shaft replacement is required whenever a constant velocity boot is removed. Who is correct?
    a. Technician A only
    b. Technician B only
    c. Both A and B
    d. Neither A nor B

SECTION 5
Steering and Suspension

▶ CHAPTER 24 Wheels and Tires Theory

▶ CHAPTER 25 Servicing Wheels and Tires

▶ CHAPTER 26 Steering Systems Theory

▶ CHAPTER 27 Servicing Steering Systems

▶ CHAPTER 28 Suspension Systems Theory

▶ CHAPTER 29 Servicing Suspension Systems

▶ CHAPTER 30 Wheel Alignment

CHAPTER 24

Wheels and Tires Theory

Learning Objectives

- **LO 24-01** Describe tire and wheel physics.
- **LO 24-02** Describe wheel construction.
- **LO 24-03** Describe tire construction.
- **LO 24-04** Interpret tire markings.
- **LO 24-05** Describe tire safety features.

ASE Education Foundation Tasks

See Appendix A to view the 2017 ASE Education Foundation Automobile Accreditation Task List Correlation Guide.

▶ Introduction

The wheel and tire assembly is the only point of contact between the vehicle and the road surface. This contact plays a large role in determining how a vehicle handles and rides (**FIGURE 24-1**). The entire vehicle literally rests on the wheel and tire assemblies. So, today's newer, lightweight vehicles are more sensitive than ever to minor tire or wheel issues.

Poorly maintained wheels and tires decrease effective handling, fuel economy, and ride quality. They also increase the potential for accidents or breakdowns. Knowing tire and wheel theory will help you identify issues before they become unsafe.

▶ Tire and Wheel Physics

LO 24-01 Describe tire and wheel physics.

Tires and wheels work together to allow the vehicle to roll smoothly down the road. Tires by themselves have very little rigidity. They flex and deform whenever a force is put against them. On the other hand, wheels are very rigid and have no way of absorbing the unevenness of the road surface. So, driving would be annoying if they were used by themselves. Also, because wheels have a low coefficient of friction, traction would be almost nonexistent. But when you pair a tire with a wheel, you get the best of both worlds. First, the rigidity needed to provide directional control (from the wheel). And second, the flexibility and traction to grip the road surface while smoothing out the bumps (from the tire).

Tires on most passenger vehicles are called pneumatic tires. This is because they are filled with pressurized air, which gives them support. Some vehicles, such as forklifts, may use solid rubber tires (**FIGURE 24-2**). These tires are not susceptible to punctures. So they are used where puncture hazards are present such as garbage dumps or construction sites.

Early pneumatic tires were called tube-type tires because they used an inner tube inside the tire. It sealed the air in the tire/wheel assembly. Recent vehicles use tubeless tires. These tires are designed to seal tightly to the wheel in such a way that the pressurized air does not leak out.

The air pressure gives the tire its shape when the weight of the vehicle is sitting on it. Air pushes on the entire inside surface of the tire. It pushes it outward. Tires on many vehicles are inflated to about 32 pounds per square inch (psi) (220 kPa). This means that much air pressure pushing against every square inch of the inside of the tire. It would not be unreasonable to say there is approximately 1200 square inches of surface area inside a typical passenger car tire. So, 32 psi multiplied by 1200 square inches equals 38,400 lb (17,280 kg) of force trying to expand the tire.

So, why doesn't the tire expand like a balloon? This is because it has lots of strong reinforcing strands molded into it. They give it strength while still allowing it to be flexible. So the air inside the tire pushes outward on the tire, stretching the strands tightly,

FIGURE 24-1 Wheels and tires provide the connection between the vehicle and road surface.

FIGURE 24-2 Solid tires are used on machines where puncture hazards are present.

You Are the Automotive Technician

A customer brings her 2014 Mustang GT into your shop and needs to buy a new set of tires. She races the car regularly at the local road race track, so she wants to buy tires that will help her out there. She has several questions for you as she looks at the various options.

1. Would lower profile tires be helpful on a road race track? Why or why not?
2. Her tire placard on the driver's door shows 32 psi (~221 kPa), but the tire shows a maximum pressure of 40 psi (~226 kPa). Which pressure should she use?
3. Would bias-ply tires give her better traction than radial tires? Why or why not?
4. What types of tires could she use if she wanted to avoid having to change tires out on the track if she got a puncture?

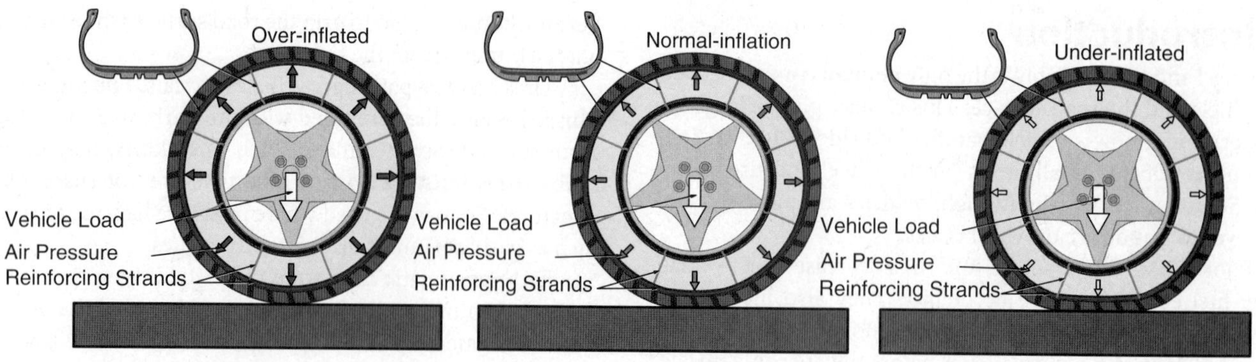

FIGURE 24-3 The strands in the tire are stretched tight by the air pressure inside the tire allowing it to support the weight of the vehicle. At the same time, the strands keep the tire from ballooning.

which prevents the tire from ballooning (**FIGURE 24-3**). Also, the relatively high pressure in the tire stiffens the tire so that it will support the weight of the vehicle. This design also explains the appearance of a tire that is substantially low on air pressure. The bottom will appear flat, and the sidewall relatively soft. The air pressure is not fully stretching the cords in the tire.

Tire Distortion

During cornering, centrifugal force acts on a vehicle to produce a **side force**. The side force is the pressure on the tire that pushes it toward one side of the rim as the vehicle makes a turn. In most instances, the friction between the tire and the road surface keeps the tire from sliding sideways. However, when the roads are slippery, or the vehicle is traveling too fast, the tire is unable to grip the road surface adequately. The side force overcomes tire to road friction. This results in the tire skidding sideways during a turn. Without the resistance created by friction, side force will cause the vehicle to continue in a straight line.

The tire provides this opposing force by being able to distort while still gripping the road (**FIGURE 24-4**). Because the tire's construction makes it elastic, it exerts a force, called **cornering force.** It allows the rubber to distort from its normal position. The tire's sideways distortion allows the vehicle to follow a path at an angle to the direction the road wheel is pointing (**FIGURE 24-5**). This is called the **slip angle.** As the cornering force increases, so does slip angle.

FIGURE 24-4 Tires distort during cornering, providing an opposing force that allows a vehicle to make a turn.

When a vehicle is being driven into a turn with a decreasing radius, both slip angle and cornering force increase. This continues until a point is reached where the tire slides. At this point, the only resistance comes from sliding **kinetic friction** across the road surface. The tire grips again only when the vehicle has slowed or the wheels are not turned so sharply. That is, when the side force is reduced to a level, the tire can withstand without skidding.

Both the front and rear tires develop a slip angle in a turn. The vehicle's path is determined by the steering of the front tires, and the slip angles of both the front and the rear tires. These slip angles depend on several factors:

- the weight distribution within the vehicle,
- the wheelbase,
- the tire track, and
- the overall length of the vehicle.

The weight distribution is affected by whether the engine is front, mid, or rear mounted. Another factor is whether the vehicle is front, rear, or all-wheel drive. Every vehicle has static weight distribution. This is the same whether the vehicle is at

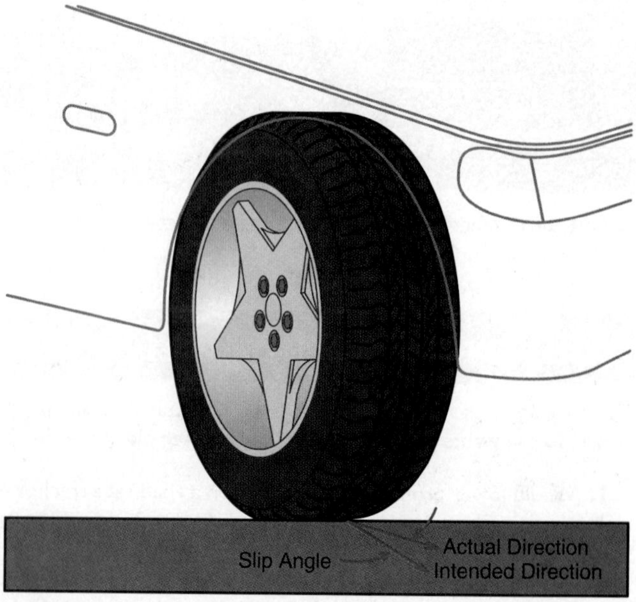

FIGURE 24-5 Slip angle is the difference in angle between the direction the tire is pointed and the direction the tire is going.

rest or traveling in a straight line at a steady speed. This distribution is changed side-to-side by centrifugal force when the vehicle is turning. It is changed front-to-back during acceleration or braking.

During cornering, centrifugal force puts more weight on the outside wheels. Acceleration puts more weight on the rear wheels. Deceleration or braking puts more weight on the front wheels. In a turn, centrifugal force tries to push the vehicle away from the corner. This is resisted by the cornering force of the tires. Thus, the tires' slip angles may not be equal, due to the cornering, deceleration, or acceleration forces acting on them.

When the front slip angles are larger than the rear slip angles, the vehicle is said to be in an **understeer** condition. This is referred to as the vehicle is "pushing" in the corners (**FIGURE 24-6**). Understeer results in the front of the vehicle being pushed toward the outside of the corner. At its worst, it causes the vehicle's front wheels to slide into the curb or off the track.

Oversteer is when the rear slip angle is larger than the front slip angle. It is referred to as the vehicle being "loose" in the corners (**FIGURE 24-7**). Oversteer tends to cause the rear of the vehicle to slide toward the outside of the corner. In this situation, the rear wheels lose traction. This typically results in the vehicle spinning out. When both the front and the rear tires have equal amounts of slip angle, the vehicle is said to have **neutral steer**. In this case, the front and rear of the vehicle tend to slide equally toward the outside of the corner as the slip angles increase.

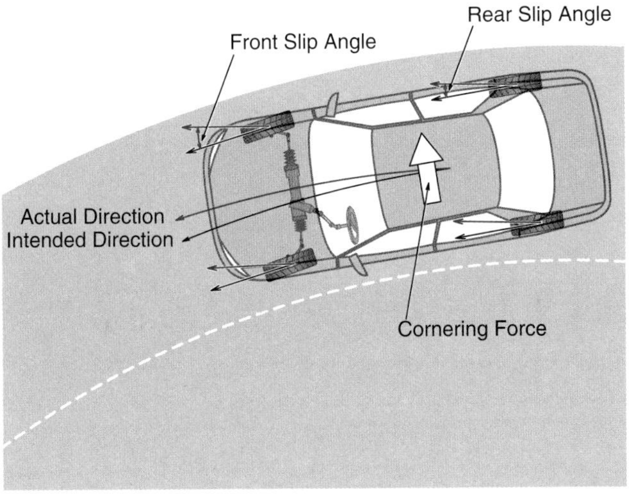

FIGURE 24-6 Understeer occurs when the front wheel slip angle is larger than the rear slip angle.

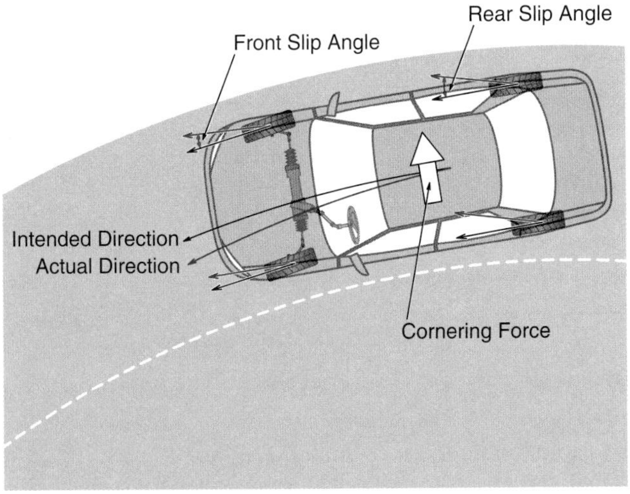

FIGURE 24-7 Oversteer occurs when the rear wheel slip angle is larger than the front slip angle.

Center of Gravity

The center of gravity is the balance point of the entire vehicle. Its actual position depends on the design of the vehicle and the location of its components (**FIGURE 24-8**). It is always located above the road surface and between the tires. When a vehicle is cornering, this is the point through which all centrifugal force is assumed to act. Its position is determined by the load carried by the front and rear wheels. That is, by how weight is distributed.

In a typical rear-wheel drive vehicle, the weight distribution is approximately 60% fore and 40% aft. Sixty percent of the weight is carried on the front wheels, 40% on the rear. This makes the center of gravity closer to the front than the rear. On a typical front-wheel drive vehicle, the weight distribution is approximately 75% fore and 25% aft. Lateral, or side-to-side, weight distribution can be expressed in the same way. It is affected by factors like the location of the fuel tank and battery.

> ▶ TECHNICIAN TIP
>
> The height of the center of gravity is determined by the height of the center of the mass above the road surface. Installing taller tires raises the vehicle's center of gravity, and vice versa.

▶ Wheels

LO 24-02 Describe wheel construction.

Wheel Construction

Wheels are usually made from pressed steel or cast aluminum alloy. They are lightweight, yet strong enough to withstand normal operational forces. Alloy wheels are popular because of their appearance and they are lighter than steel wheels. Aluminum is a better conductor of heat, so alloy wheels can dissipate heat from the brakes and tires more effectively than steel. Alloy wheels are often called mag or magnesium wheels. But wheels made of magnesium are rarely used on vehicles.

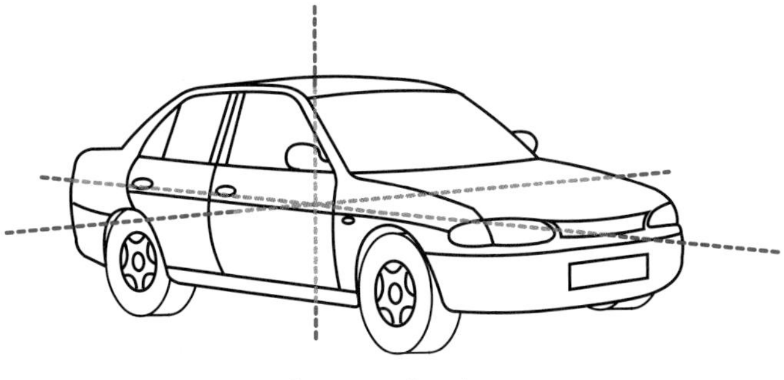

FIGURE 24-8 The vehicle's center of gravity.

FIGURE 24-9 Rim of the wheel.

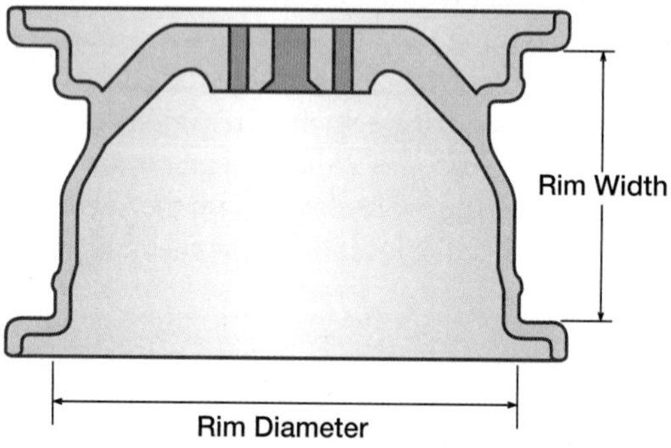

FIGURE 24-10 Rim width and diameter.

The terms "wheel" and "rim" are often used synonymously. Technically, the **rim** is a part of the wheel. It is better known as the rim flange. It is the outer circular lip of the wheel on which the inside edge of the tire is mounted (**FIGURE 24-9**). The purpose of the rim flange is to hold and seal the tire to the wheel.

Wheel width is the distance across the inside of the **rim flanges** at the **bead seat** (**FIGURE 24-10**). The bead seat is the edge of the rim flange that creates a seal between the tire bead and the wheel. The rim diameter is the distance across the center of the wheel from rim flange to rim flange. The width of the wheel and the diameter are traditionally stated in inches, although some are stated in millimeters.

▶ TECHNICIAN TIP

Wheels are stamped with a code. For example, a wheel designated "7 JJ by 14" would refer to a wheel measuring:

- 7" across the inside of the rim flanges;
- 14" in diameter from bead seat to bead seat;
- and with the flange profile conforming to a JJ code.

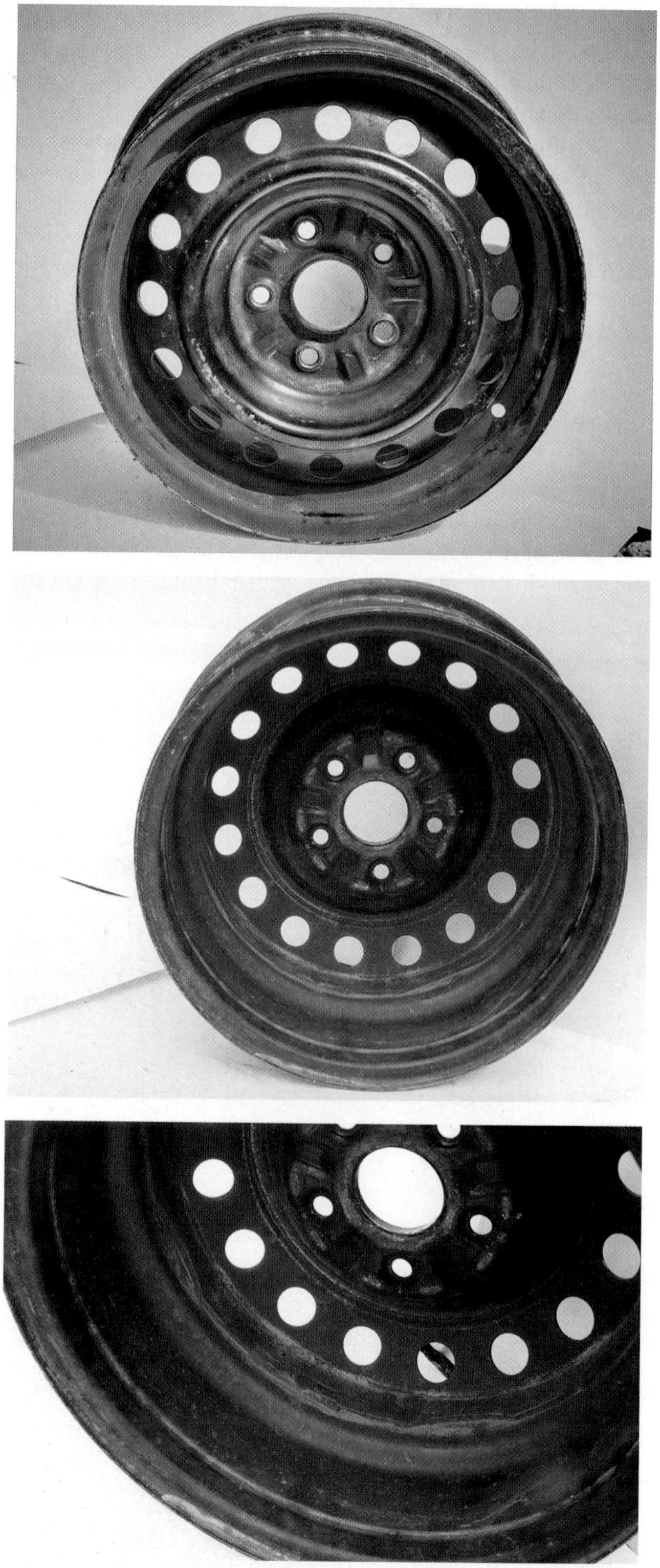

FIGURE 24-11 Parts of a steel wheel: wheel flange and rim.

Wheels can be made from two sections of pressed steel—a **wheel flange** or disc, and the rim (**FIGURE 24-11**). The wheel flange may be solid, but most manufacturers place holes in them to reduce the disc's weight. The holes also provide ventilation for cooling the brakes. The disc is either welded or riveted to the rim. The wheel flange is drilled for the wheel fasteners and bolted to the axle or axle flange. Additional types of wheels can be found in **TABLE 24-1**.

TABLE 24-1 Types of Wheels (Rims)

Type	Purpose
Steel wheels	Wheels that can be painted or chromed. Generally used on bottom-of-the-line vehicles.
One-piece alloy wheels	Wheels that are constructed in one piece, including the well, to facilitate the mounting and dismounting of the tire.
Multi-piece alloy wheel rims	Wheels that are constructed by welding together several separately molded pieces.
Custom wheels	Wheels that are built to suit the specific, unique needs and requirements of the customer.
Spinning wheels (spinners)	Wheels that are constructed with a piece that rotates independent from the wheel. High-speed roller bearings placed within the rim create the piece's movement.
Split rims	Rims that are constructed of a single long piece of metal formed into a circle, but whose ends are not welded together (a "split" ring). The split allows removal of the outside flange so the tire can be demounted. This type of rim is common on tractor trailers and other big-wheeled trucks.
Drop well wheels	Wheels constructed with a trough in the center of the rim that is smaller in diameter than the edge by the lip where the bead of the tire seals. The trough allows the tire to be mounted on the rim.
Safety rims	Rims designed to hold the tire bead in place on the rim in the event of inadequate tire pressure.

▶ TECHNICIAN TIP

The tire must be an exact fit on the rim to fulfill a number of functions:

- To ensure that the narrow contact area between the beads of the tire and the rim will seal the air in a tubeless tire
- To transfer all the forces between the tire and the wheel, without slipping or chafing
- To ensure that the friction between the tire and the rim prevents the tire from turning on the rim

Most passenger vehicle **wheels** are of the drop-center design. In drop-center wheels, part of the inner section sits lower than the sides (**FIGURE 24-12**). This design allows for tire removal and mounting. The drop-center rim is made in one piece and is permanently fastened to the wheel disc. When inflated, the tire is locked to the wheel.

Tires are locked to the rim in two main ways. The most common way uses safety ridges on the rim close to the bead seats (**FIGURE 24-13**). The tire expands over the safety ridge when it is installed. If the tire loses air pressure, the safety ridges help to hold the tire on the bead seat. Another way that tires are locked to the rim is done by tapering the rim and bead seat. They are tapered in such a way that the bead seat ramps toward the rim flange. As the tire is inflated, it gets pushed up the ramp, stretching it tightly on the rim.

▶ TECHNICIAN TIP

Most wheels have ventilation holes in the flange so that air can circulate to the brakes.

An important feature in a wheel is the location of the drop center in the wheel. The drop center can be closer to the front of the wheel, or to the rear of the wheel. This depends on the desired position of the drop center relative to the wheel flange. In most stock wheels, the drop center is closer to the front side of the wheel. In **deep dish wheels**, the drop center is closer to the rear of the wheel (**FIGURE 24-14**).

The drop center is crucial to removing the tire from the wheel. So, the side of the wheel that is closest to the drop center is the side the tire should be removed from. In fact, it is virtually impossible to remove the tire from the side of the wheel farthest from the drop center.

FIGURE 24-12 Drop-center rim.

FIGURE 24-13 Safety ridge on a wheel.

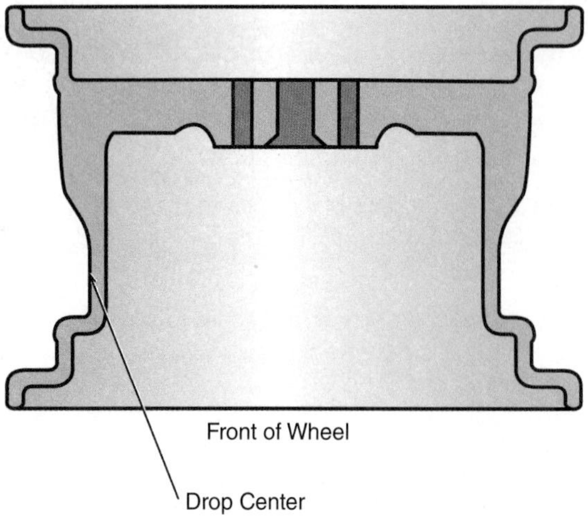

FIGURE 24-14 Deep dish wheel.

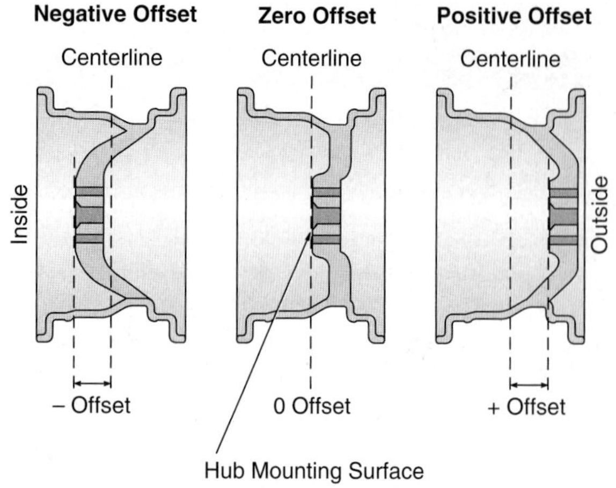

FIGURE 24-15 Zero, positive, and negative offset.

SAFETY TIP

While not as common, the wheel width and diameter can also be stated in millimeters. Metric wheels are not to be used with imperial tires. Nor are metric tires to be used on imperial wheels. Doing so could result in personal injury or a vehicle accident.

Applied Math

AM-1: Distance: The technician can measure distance using a variety of devices to determine conformance to the manufacturer's specifications and tolerances.

A technician may be called upon to measure distances with a variety of devices to determine conformance of components to manufacturer's tolerances and specifications. Some examples may include measuring:

- Tire tread with a depth gauge
- Brake pad thickness with a ruler
- Vehicle ride height with a tape measure
- Brake disc thickness with a vernier caliper
- Wheel runout with a dial gauge

Wheel Offset

The offset of a wheel is the distance from the wheel flange to the centerline of the wheel (**FIGURE 24-15**). Offset is important. It is typically used to bring the tire centerline into close alignment with the larger inner wheel bearing. It also reduces the bending load on the axle. This requires the inside of the wheel assembly to be shaped so that there is space for the brake assembly. This is especially true for disc brakes, which are typically larger in diameter than drum brakes.

The offset can be either zero, positive, or negative:

- **Zero offset**: The plane of the hub mounting surface is even with the centerline of the wheel. Vehicles manufactured in the mid-1970s through the 1980s were typically built with zero offset wheels.
- **Positive offset**: The plane of the hub mounting surface is shifted toward the outside or front side of the wheel. Positive offset wheels are generally found on front-wheel drive vehicles and the newer rear-wheel drive vehicles.
- **Negative offset**: The hub mounting surface is toward the brake side or back of the wheel's centerline. Older model vehicles with deep dish wheels often have wheels with negative offset. This is also the case with specialized high-performance vehicles.

Wheel Studs and Lug Nuts

Wheels are fastened to the hub by **wheel studs** and **lug nuts**. Wheel studs and lug nuts are highly stressed by loads from the weight of the vehicle and the forces generated by its motion. Wheel studs and nuts are made from heat-treated, high-grade alloy steel. The threads between the studs and the nuts are close fitting and accurately sized. All wheel nuts must be tightened to the specified torque in the proper sequence. Otherwise, the wheel could break free from the hub.

You must use a matching wheel retaining stud or nut with the wheel. There are several styles of lug nuts or lug studs (**FIGURE 24-16**). **Tapered seat lug nuts** have a tapered end that is placed toward the rim and fit into a matching taper in the rim to help center the wheel on the lug studs. **Flat seat lug nuts** use either an integrated washer or a separate washer. The lug nut is flat on the bottom which matches the wheel. Ball seat lug nuts have a rounded seat that has a matching surface on the wheel. As you can see, mixing and matching lug nuts and wheels would result in the tire not being held securely.

Prior to installing a wheel, it is critical that the mating surfaces of the wheel flange and the axle flange are clean. They must be free of dirt, mud, and rust. When installing the lug nuts or lug studs, make sure you install them in the correct orientation. The seat on the lug nut/stud must face the matching seat on the wheel. After the **lug nuts** have been installed, they must be torqued to the manufacturer's specification. Many shops request that the customer return and have the lug nut torque rechecked within a few days or about 100 miles.

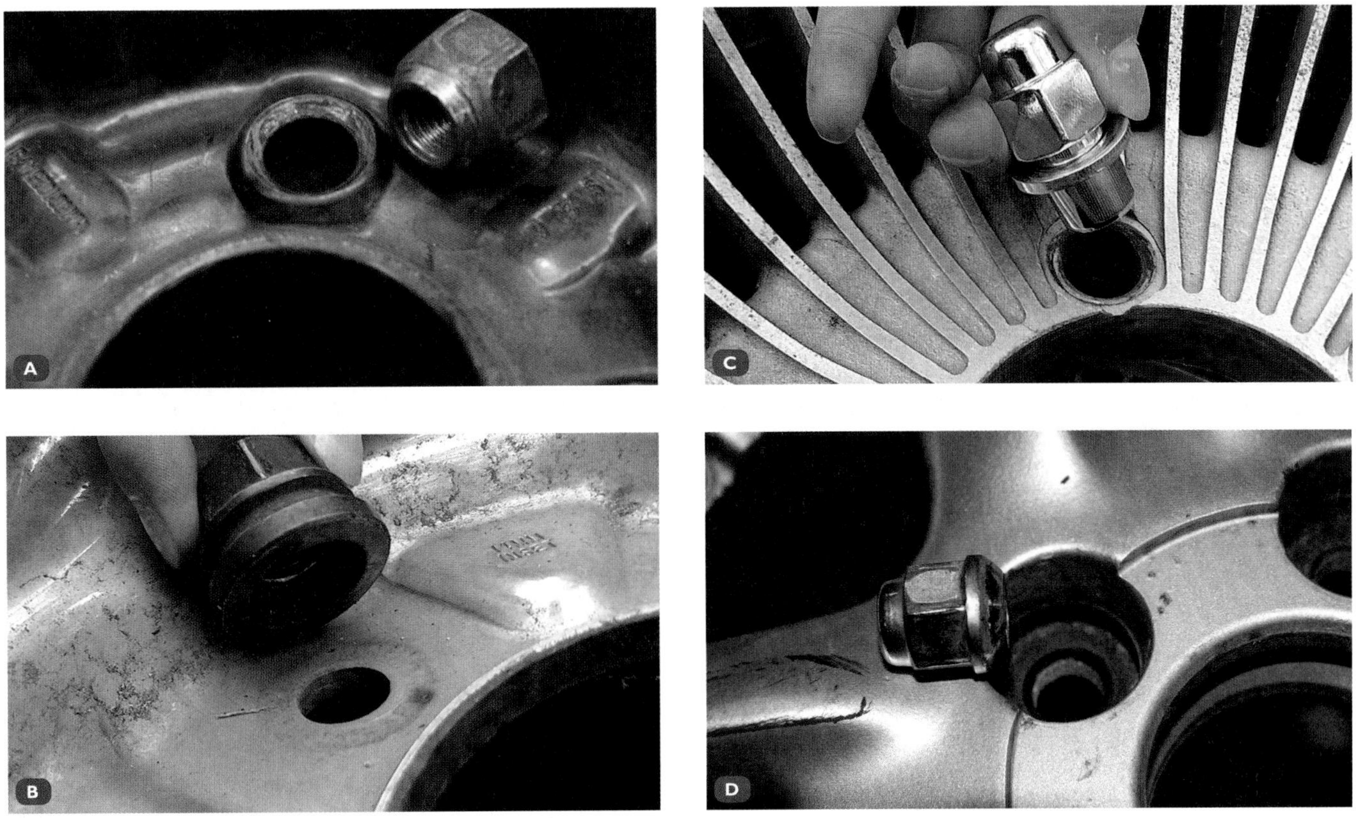

FIGURE 24-16 **A.** Tapered seat lug nut and wheel. **B.** Flat seat with integrated washer and wheel. **C.** Flat seat with separate washer and wheel. **D.** Ball seat.

If the lug nuts or studs are undertightened, then it is possible that they will loosen up while driving. This can cause the wheel could fall off. It also ruins the wheel and lug studs. If the nut or stud is overtightened, then there is a good chance that the studs will be stretched and weakened. They are then prone to break because of overstressing and fatigue fractures. Overtightening lug nuts can also cause the brake rotors to warp.

Most lug nuts and studs use right-hand threads, which means they tighten when turned clockwise. However, some manufacturers use lug nuts and studs with left-hand threads. In many cases, left-hand threads are used on the right side (passenger side) of the vehicle. Right-hand lug nuts and studs are used on the left side (driver side) of the vehicle. Manufacturers claim that left-hand lug nuts are less likely to back off when used on the right side of the vehicle. This may or may not be true. But properly torqued lug nuts do not generally loosen up, no matter if they are left-hand or right-hand.

Left-hand lug nuts can usually be identified by an "L" stamped on them or on the end of the lug stud (**FIGURE 24-17**). When you place the socket or lug wrench over the lug nut, make sure you check for the "L." If there is one, then you will need to turn the lug nut clockwise to loosen it.

FIGURE 24-17 Left-hand lug nut and stud identified by the "L" on the stud or nut.

▶ TECHNICIAN TIP

Don't be like the apprentice technician who was trying to get a vehicle ready for a brake job for the journeyman technician. When the journeyman came into work, the apprentice was just taking off the last wheel. The journeyman could hear the impact wrench hammering away and then the impact wrench "zing" to full speed. By the time he got to the apprentice, he was on the next to last lug nut. The journeyman stopped him. The apprentice said that all of the lug nuts were rusted on the studs on the right side of the car. And they all broke off when he tried to loosen them. The journeyman took the impact wrench, switched directions on it, and proceeded to easily remove the last two lug nuts. The apprentice was quite embarrassed, especially when the other technicians started calling him "Lefty."

If a lug nut does not start to loosen quickly when you are trying to remove it, always stop and try to turn it the other direction. It could be a left-hand lug nut, which you will have to turn clockwise to remove. And even if it is a right-hand lug nut, sometimes tightening it just a bit helps break loose any rust so that it comes off more easily.

The **wheel flange** usually has a machined hole in the center. It is used to accurately center the wheel on the axle flange. In some cases, the fit of the centering hole is snug enough that it can rust in place and stick to the axle flange. This requires penetrating oil, a hammer, or even heat from a torch to break it loose. Always seek the assistance of your supervisor if you encounter a wheel rusted to the axle flange.

Bolt Pattern

The bolt pattern refers to the number and spacing of the lug nuts and lug studs on the wheel and axle flange (**FIGURE 24-18**). Lug nuts/studs are most often evenly spaced around the wheel. But the spacing can be closer together, or further apart. The number of studs varies from vehicle to vehicle. For example, some smaller cars have three studs, some may have four studs, but most passenger cars have five. Pickup trucks and large SUVs can have as many as 6, 8, or 10 studs. The exact number and pattern of studs vary depending on the vehicle type and manufacturer design.

▶ TECHNICIAN TIP

Along with the bolt pattern configuration is **pitch circle diameter (PCD)**. PCD is the diameter of a circle drawn through the center of the wheel's bolt holes. It is a fixed measure set during manufacture that cannot be altered. PCD is measured in both inches and millimeters. It also indicates the number of studs or bolts on the wheel. One common configuration has four studs and a PCD of 100 mm, hence the size 4 × 100 designation. A gauge is typically used to determine the PCD on a wheel (**FIGURE 24-19**).

FIGURE 24-18 The bolt pattern refers to the number and spacing of lug nuts or wheel studs on the wheel.

Applied Math

AM-2: Distance: The technician can use standard and metric measurement instruments to determine correct sizes and distances.

In practice, both standard and metric measurement instruments must be used to determine correct sizes and distances. Specifications may be published in either scale. A common example includes measuring the diameter of wheels. They are almost universally measured in inches. Also, the PCD of wheel studs is usually quoted in millimeters.

▶ Tires

LO 24-03 Describe tire construction.

Tires are hollow, donut-shaped structures designed to provide traction to the wheel assembly. They also act as a cushion to absorb shock from road surfaces. The air in the tire supports the vehicle's mass. And the tire tread provides frictional contact with the road surface, so the vehicle can maneuver safely.

The tire itself is generally composed of the tread, plies or cords, sidewalls, inner liner, and bead (**FIGURE 24-20**).

- The tread is the exterior rubber portion of the tire that comes in contact with the road. It is configured in a variety of patterns based on the application of the tire, such as for driving in mud and snow.
- The plies or cords are the reinforcing material that gives the tread and sidewall their ability to hold their shape.
- The sidewalls are the lightly reinforced sides of the tire that provide lateral strength to the tire and prevent it from ballooning. They do not come into direct contact with the road.
- The inner liner is a covering of the casing material and seals the air in the tire.
- Bead strands are bands of steel wire coated in rubber that give the bead area the stiffness to hold the bead against the bead seat and to seal to the wheel.
- Tires can be inflated through a valve assembly located in the wheel.

FIGURE 24-19 A pitch circle diameter gauge being used on a wheel.

Modern tires are made from a range of materials. The rubber used is mostly synthetic, with carbon black added to increase strength and toughness. When used in the tread, this combination gives the tire a long life. Natural rubber is weaker than the synthetic version, so it is used mainly in sidewalls.

Cords of synthetic strands have high **tensile strength**. Because of their high strength, the cords resist stretching but are flexible under load. The cords are placed in parallel and impregnated with rubber to form sheets called plies or belts. Plies have high strength in one direction and are flexible in other directions. When cotton was used as a cord, the number of plies, or layers in a tire, was a measure of the tire's strength. Newer cord materials use fewer plies.

A modern steel-belted radial tire with a **six-ply rating** may have six plies in the tread, but just two plies in its sidewall. Having fewer plies makes the tire more flexible. Higher numbers of plies make a tire's response to bumps harsher, but the tire can withstand punctures much better.

The bead of the tire is made of a cord of high-tensile steel coated with rubber. The ends of the plies are wrapped around each bead, which is then wrapped in rubber to stop chafing of the plies. The length of the wire used for the bead determines the rim diameter of the tire.

Belts that reinforce the tread area of the tire are typically made of braided, high-tensile steel wire. But they can also be made of rayon or polyester. The inner liner of a tubeless tire is made of soft rubber. The inner liner must be flexible and airtight.

Tire Valve Stems, Cores, and Caps

The **valve stem** is a specially designed part that allows a tire to be inflated. The valve stem is a rubber or steel piece that attaches to the wheel. A **valve core** is threaded into the valve stem. The valve core is a **Schrader valve** (a spring-loaded, one-way valve) (**FIGURE 24-21**). It lets air into the tire when being inflated. It closes automatically to hold the air in.

A **valve stem cap** is used to keep debris out of the valve stem, which could cause the valve core to leak. The valve stem cap is removed to check air pressure and to inflate or deflate the tire. Some valve stem caps come with a built-in tool for tightening or loosening the valve core. The valve stem cap can also slow a leak from a valve core, but will not stop a leak completely. In newer vehicles, a green valve stem cap indicates that the tire is filled with nitrogen. To prevent dilution of the nitrogen, it should only be topped off with nitrogen.

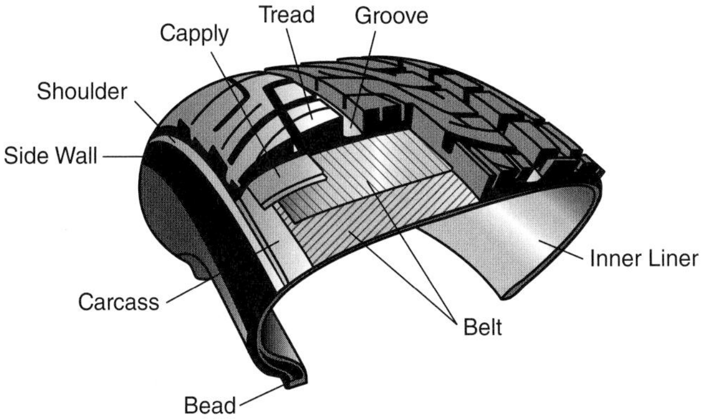

FIGURE 24-20 The tire itself is generally composed of the tread, sidewalls, inner liner, and bead.

Types of Tire Construction

Although there are two types of tire construction—**bias-ply** and **radial**—radial tires are by far the most common. In fact, one tire retailer estimates that 98% of the passenger vehicle tires sold today are radial tires.

Bias-Ply Tires

The bias-ply tire is the older form of tire. It is still in use on some trailers and off-road vehicles. This is primarily because of their more durable construction and slightly lower cost. Bias tire construction uses body ply cords extended diagonally from bead to bead. They run at 30- to 40-degree angles. Successive plies are laid at opposing angles. This results in a crisscross pattern onto which tread is applied (**FIGURE 24-22**).

Bias-ply design provides a strong, stable casing, but with relatively stiff sidewalls. However, during cornering, stiff sidewalls can distort the tread. This can partially lift the tread off the road surface. Friction between the road and the tire is reduced, which reduces traction.

> **▶ TECHNICIAN TIP**
>
> The bias-ply tire's stiff sidewalls can make tires run at a higher temperature. This is because, as the tire rotates, the body ply cords in the plies flex over one another, causing friction and heat. A bias-ply tire that overheats can wear prematurely. At the same time, stiffer sidewalls can resist punctures from sharp rocks or other objects better than a radial tire.

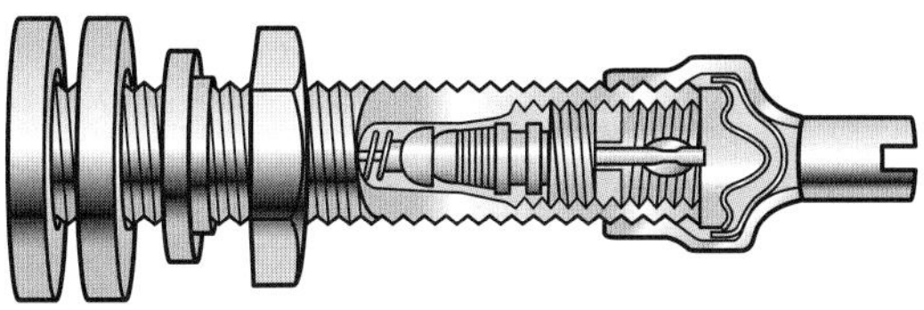

FIGURE 24-21 Components of a valve stem.

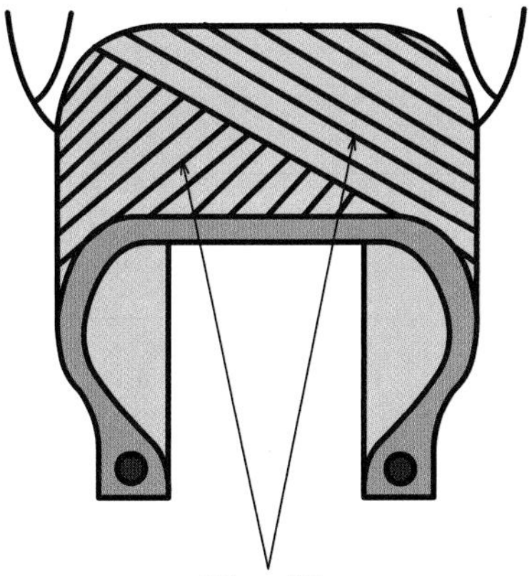

FIGURE 24-22 Bias ply tires. Notice the plies run in a crisscross pattern.

Radial Tires

Radial tire construction uses the same body ply cords, but they are laid across the tread extending from bead to bead. So, the cords end up parallel to each other at approximately 90 degrees to the centerline of the tread. Radial tires are extremely durable. They also maintain good traction with the road, providing better control of the vehicle. Most passenger cars now use radial tires, as do most four-wheel drive vehicles and heavy transport vehicles.

Radial ply tires have much more flexible sidewalls because of their construction (**FIGURE 24-23**). They use two or more layers of **casing plies**, which run radially from bead to bead. The casing plies give the tire shape and strength. They are loops of high-strength material (typically polyester) inlaid with rubber. They loop around the wires in the bead.

The sidewalls are more flexible because the casing plies do not cross over each other. However, a bracing layer of two or more high-strength belts is used to strengthen and stabilize the tire tread. The cords of the belts may be of fabric or steel. They are placed at 12 to 15 degrees to the circumference line of the tire. This design forms triangles where the belt cords cross over the radial cords. It gives the tire tread stability when accelerating, braking, and cornering.

The radial casing plies flex and deform in the sidewall area above the road contact patch. There are no heavy sidewall plies to distort, and flexing of the thin casing generates little heat. This allows a properly inflated radial tire to run cooler than a comparable bias-ply tire. Tread life is increased. Also, a radial tire has less rolling resistance, which increases fuel economy.

▶ TECHNICIAN TIP

The sidewalls of radial tires bulge where the tire meets the road. This makes it difficult to estimate the inflation pressure visually. Tire pressure needs to be checked with an accurate tire gauge. Using the correct inflation pressures extends the life of the tire and is vital for safety. Sidewalls of an underinflated tire flex too far, which causes the center section of the tread to be pushed up and away from the road surface. This causes wear at the shoulders of the tire. In an overinflated tire, the sidewalls are straightened, which pulls the edges of the tread away from the road and causes wear at the center of the tread.

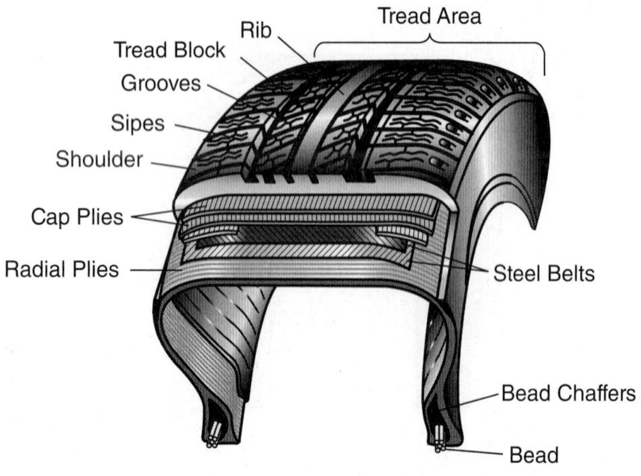

FIGURE 24-23 Radial tires. Notice that the plies from in a side-to-side pattern.

Tread Designs

Differing tread patterns give manufacturers the ability to design tires for special applications. They can be designed to work well in a variety of driving situations. These include rain, snow, mud, highway, or high-performance situations. Manufacturers spend a lot of money trying to design better tread patterns. And when they discover a good design, it can be very profitable. Tire tread patterns can be classified with these general characteristics (**FIGURE 24-24**):

- **Directional tread patterns** are designed to provide specific attributes during particular driving conditions. One example is driving in wet weather where there is a possibility of hydroplaning (sliding across water). The primary tread pattern is formed in such a manner as to provide maximum water removal. Thereby creating a surface with less water so the tire can adhere to it better. Specifically, these tread patterns actually pump water out from under the tire. This type of tire must only be mounted on the wheel so that it rotates in a particular direction so the tread will work properly. An arrow on the tire sidewall indicates the designated direction of forward travel.
- **Nondirectional tread patterns** are designed in such a way that the tire can be mounted for any direction of rotation. This allows the tire to be installed either way on the wheel. These tires are used for general applications. They are usually cheaper to manufacture than the specialized tires. They also allow tire stores to stock fewer tires.
- **Symmetric tread patterns** have the same tread pattern on both sides of the tire. A main feature of most symmetric tires is that they are nondirectional and thus can be fitted in either direction.
- **Asymmetric tread patterns** have a tread pattern that is different from one side of the tire to the other. They are designed to provide good grip when traveling straight and in turns. Asymmetric tires are labeled "outside" on the side of the tire that faces outward from the vehicle.
- **Directional and asymmetric tread patterns** are both directional and asymmetric. This means the tire is designed to rotate in only one direction and has one side that must face outward. This ensures that the tire performs as designed under operating conditions. These tires tend to be used on high-performance vehicles.

▶ Tire Markings

LO 24-04 Interpret tire markings.

Tires come in a variety of sizes and ratings. This allows them to accommodate the wide range of vehicles and driving situations. To help select the most appropriate tire for a particular application, tires use common sizing and rating systems. All tires

FIGURE 24-24 Tread designs: **A.** Directional tread pattern. **B.** Nondirectional tread pattern. **C.** Symmetric tread pattern. **D.** Asymmetric tread pattern. **E.** Directional and asymmetric tread pattern.

meeting legislative codes must have the following information clearly marked on the sidewall (**FIGURE 24-25**):

- Manufacturer or brand name
- International Organization for Standardization (ISO) tire class:
  - P—passenger
  - LT—light truck
  - C—commercial
  - T—temporary use as a spare wheel
  - No letter designation before the section width—either a European metric tire or an off-road tire
- Section width: measured in millimeters from sidewall to sidewall
- Aspect ratio: the height of the sidewall expressed as a percentage of the section width

FIGURE 24-25 Typical sidewall markings.

FIGURE 24-26 Maximum pressure that the tire is designed for. Always make sure this pressure is higher than the recommended pressure on the tire placard.

- Type of tire construction: R for radial; blank for bias-ply
- Wheel diameter: diameter of wheel from bead seat to bead seat, usually measured in inches
- Speed rating designation: the maximum speed for which a particular tire is rated. Vehicles should not be driven in excess of their speed ratings, with the exception of Z-rated tires. They are rated for speeds above 149 mph (240 kph). *Note:* Virtually all tires are rated for higher speeds than legal speed limits. Tire speed ratings do *not* give you legal cover for speeding! Speed ratings are as follows:
  - Q—up to 100 mph (161 kph)
  - R—up to 106 mph (171 kph)
  - S—up to 112 mph (180 kph)
  - T—up to 118 mph (190 kph)
  - U—up to 124 mph (200 kph)
  - H—up to 139 mph (224 kph)
  - V—up to 149 mph (240 kph)
  - W—up to 168 mph (270 kph)
  - Y—up to 186 mph (299 kph)
  - Z—149 mph (240 kph) and over
- Maximum air pressure: the maximum pressure that the tire can be inflated to (**FIGURE 24-26**). This does not indicate the vehicle manufacturer's recommended inflation pressure. Always refer to the vehicle's **tire placard** to determine proper inflation pressures.
- Load index: the maximum amount of weight that a tire can safely carry at the maximum rated tire pressure. When replacing tires, verify that the load index meets the vehicle manufacturer's weight rating for the front and rear axles.
- Uniform Tire Quality Grading (UTQG) system:
  - Tread wear grade: an approximation of how long a tire will last when compared with another tire from the same manufacturer. For example, a tire rated at 450 will last approximately three times as long as one rated 150.
  - Traction grade: a representation of a tire's wet traction characteristics. From highest to lowest: AA, A, B, and C.
  - Temperature grade: a representation of a tire's ability to resist and dissipate heat. From highest to lowest: A, B, and C.
- Department of Transportation compliance symbols and serial numbers, including date of manufacture code:
  - First two letters following the letters "DOT" identify the tire manufacturer and manufacturing plant.
  - The third and fourth letters are codes that indicate the tire's size.
  - The final three or four letters are codes indicating manufacturer-specified characteristics.
  - Week of manufacture: The first pair of digits represents the week of the year in which the tire was manufactured, starting in January.
  - Year of manufacture: The last two digits represent the year of manufacture; for example, "17" indicates a tire manufactured in 2017.

> **TECHNICIAN TIP**
>
> Radial tires are marked with the section width in millimeters, but with the rim diameter in inches.

Tire Sizes and Designations

The size of a tire must be appropriate for the vehicle application and intended use. The vehicle manufacturer designates the recommended size and load rating of the tires to be used on each particular vehicle. The bead diameter must match the wheel diameter. The section width must be suitable for use on the wheel. It also must be large enough to have a suitable load-carrying capacity for the vehicle. The overall tire size must allow sufficient clearance between the tire and the chassis and body components.

The section width of the tire is measured in millimeters, from sidewall to sidewall. The tire is installed on a standard wheel, inflated to its recommended pressure, and without any load on it. Then, the section width is measured by the tire manufacturer. Section width varies from manufacturer to manufacturer. So, do not assume tires with the same section width are all the same.

The **aspect ratio** of a tire is the ratio of its height to its width. It is usually given as a percentage. Information on tire aspect ratio is now included in the sidewall marking. The lower a tire's aspect ratio, the wider the tire is in relation to its height. An aspect ratio of 75 means the height of the sidewall is 75% as much as the section width. Generally, the higher the aspect ratio, the smoother the ride will be. But there will be more flex during cornering.

Low-profile tires have very short sidewalls and can be difficult to remove and install (**FIGURE 24-27**). They have an aspect ratio as low as 25. The low-profile tire improves cornering performance but sacrifices a smooth ride. There is also a greater danger of wheel and suspension damage when hitting potholes. Often, low-profile tires have a higher speed rating, as there is less centrifugal force trying to throw the tire apart.

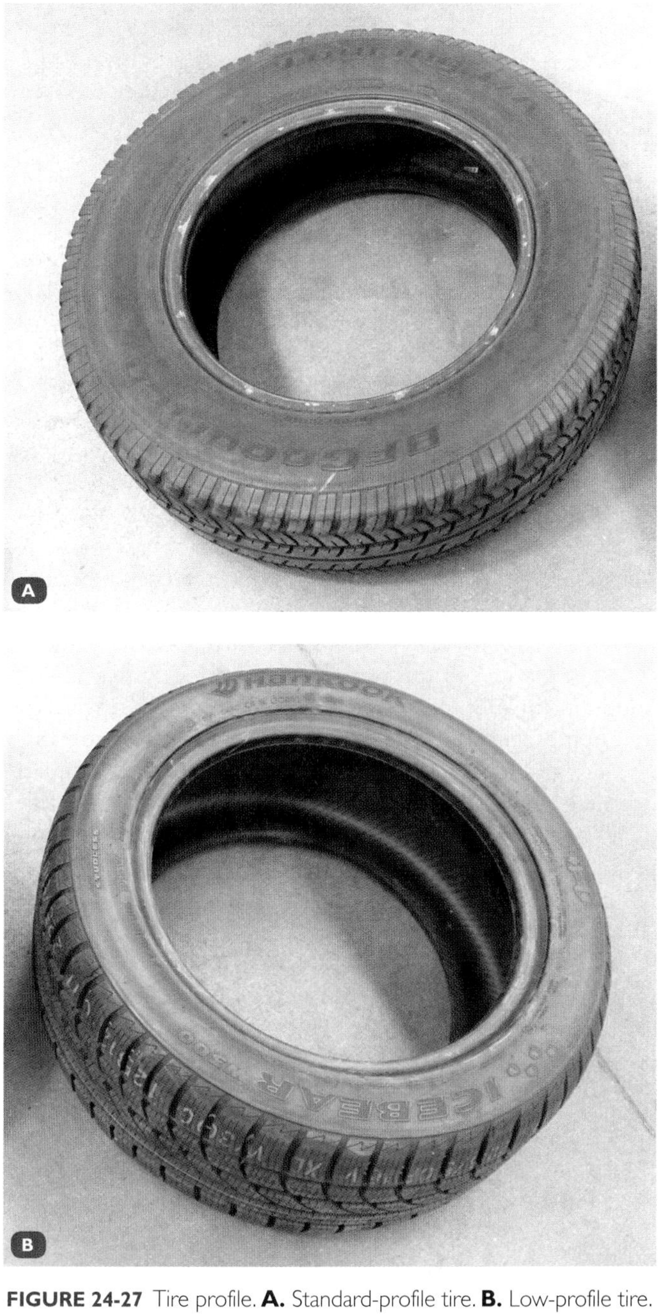

FIGURE 24-27 Tire profile. **A.** Standard-profile tire. **B.** Low-profile tire.

Tire Ratings for Tread Wear, Traction, and Temperature

One of the markings on the sidewall of a tire is a **Uniform Tire Quality Grading (UTQG)** rating (**FIGURE 24-28**). As discussed previously, the tire's UTQG rating provides information on three aspects of the tire's durability and operational characteristics. Those are: tread wear, traction, and temperature.

The **tread wear grade** comes from testing the tire in controlled conditions. The higher the number, the longer the life expectancy of the tread. This is when compared with another tire from the same manufacturer. No one vehicle will be subjected to exactly the same surfaces and same speeds as the controlled conditions. So, the number is only an indicator of expected tread life in normal conditions. The rating is based on a percentage of the projected wear life. For instance, when looking at two tires from the same manufacturer, one tire rated at 400 has a projected life of four times that of a tire rated at 100.

▶ TECHNICIAN TIP

There are many factors that influence wear. A few are: vehicle speed, road surface, climate, vehicle wheel alignment, and the driving characteristics of the driver. That is why the tread wear rating is only an indication of the anticipated tread life.

A **traction grade** is a letter-based indicator system. The rating is based on the tire's ability to stop a vehicle on wet concrete and asphalt in a straight-line situation. It does not indicate the tire's cornering ability. The tire traction indicators are rated from highest to lowest as AA, A, B, or C. It is important to note that the relevant rating does not indicate hydroplaning resistance; dry or snow traction capacity; or cornering capability in wet, dry, or snow conditions.

The **temperature grade** is a letter-based indicator system, performed by the tire manufacturer. The goal of the test is to determine how well a tire stands up to heat. It also measures how well the tire dissipates heat. The US government uses the UTQG criteria for tire temperature ratings. The tires are

FIGURE 24-28 Typical UTQG ratings on a tire sidewall.

graded from C (least tolerant of heat dissipation) to A (most tolerant of heat dissipation). Although a C-graded tire runs hotter, it is not necessarily unsafe. It is important to remember that these ratings are based on standardized test conditions. The tests do not reflect tires that are operated in overloaded, underinflated, and/or misaligned conditions. It should also be noted that one tire might be rated a low A and another a high B. So, the actual operating performance differences might be relatively small.

It is not uncommon for there to be differences in UTQG ratings within a given tire design. Sometimes, a particular vehicle manufacturer requires certain properties for the tires supplied to its vehicles. This can affect the ratings both positively and negatively. Sometimes there are differences between small sizes and large sizes in a given design. All of these aspects can affect the actual rating that is put on the sidewall.

Tire Date of Manufacture Coding

The **US Department of Transportation (DOT)** inspects everything and anything pertaining to transportation—including tires. As part of DOT regulations, there must be a tire manufacture date code stamped on the sidewall of every tire. In fact, it is illegal to sell a tire intended for use on a public road within the United States without a DOT stamp. Some manufacturers mold this code on only one sidewall. So, you might need to get under the vehicle and look at the inward-facing side of the tire. There, you will find a three- or four-digit code.

The tire date code indicates when the tire was manufactured. For safety concerns and as a rule of thumb, you should never use tires more than six years old. The rubber in tires degrades over time, irrespective of whether the tire is being used or not.

Reading the DOT code is relatively simple. A three-digit DOT code was used for tires manufactured before 2000. For example, "176" means that the tire was manufactured in the 17th week of the sixth year of the decade, say 1986.

For tires manufactured in the 1990s, there should be a small triangle after the DOT code. For example, a tire manufactured in the 17th week of 1996 might have the code "176Δ." After 2000, the code was switched to a four-digit code. For example, "30 13" means the tire was manufactured in the 30th week of 2013 (**FIGURE 24-29**).

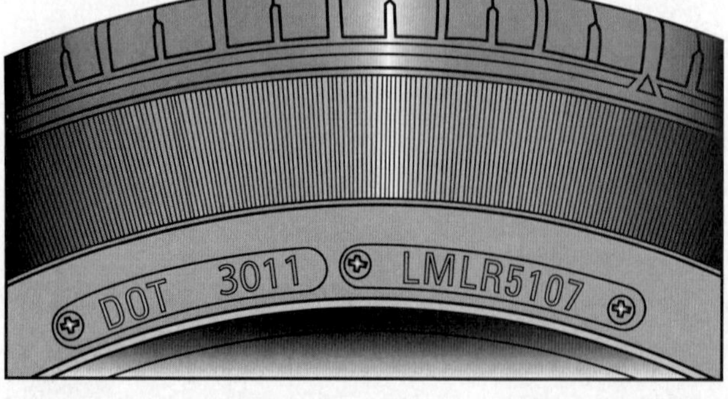

FIGURE 24-29 The DOT tire date manufacturing code is a four-digit code.

▶ Tire Safety Features

LO 24-05 Describe tire safety features.

Flat tires are a major cause of vehicle breakdowns. Generally speaking, vehicles cannot be driven with one or more flat tires. Attempting to do so will ruin a potentially repairable tire very quickly. Manufacturers address this situation in various ways. One of the best ways is to prevent the tire from becoming flat in the first place. The use of a **tire pressure monitoring system (TPMS)** warns the driver if one or more of the tires are low.

Another tactic that manufacturers use is run-flat tires. These tires are designed so that they can still be driven for a reasonable amount of time when the tire pressure is low or empty. Similar to the run-flat tires are self-sealing tires, which resist leaks and help maintain air pressure. The last safety feature involves the use of a spare tire. This is usually of the space-saving type, also called a temporary tire. It is designed to be used for reduced speeds and distances so that a driver can get the vehicle to a service facility. Each of these safety features is discussed further in the following sections.

Tire Pressure Monitoring Systems

Maintaining proper tire pressure is important for vehicle safety and performance. It helps in decreasing fuel consumption, reducing CO_2 emissions, and extending tire life. All tires lose inflation over time. With many modern vehicles having extended service intervals, tires can become very underinflated. Long periods of driving with low tire pressures can cause a lot of stress on the tire sidewalls. This results in increased operating temperatures that can lead to premature tire failure. Tires operating with low pressures can also affect the vehicle's handling and performance. In a worst-case scenario, underinflation can lead to a tire blowout or tread separation. As a result, the TREAD Act was passed. TREAD stands for The Transportation Recall Enhancement, Accountability, and Documentation. It called for all new passenger vehicles under 10,000 lb (4536 kg) to have a TPMS by October 1, 2007.

The automated tire pressure monitoring system (TPMS) monitors the tires for low air pressure. It alerts the driver when one or more tires are lower than (or in some cases, higher than) the designated thresholds. This alert can be an illuminated warning lamp or a chime (**FIGURE 24-30**). A TPMS can be fitted to all vehicles using conventional and run-flat tires. With some

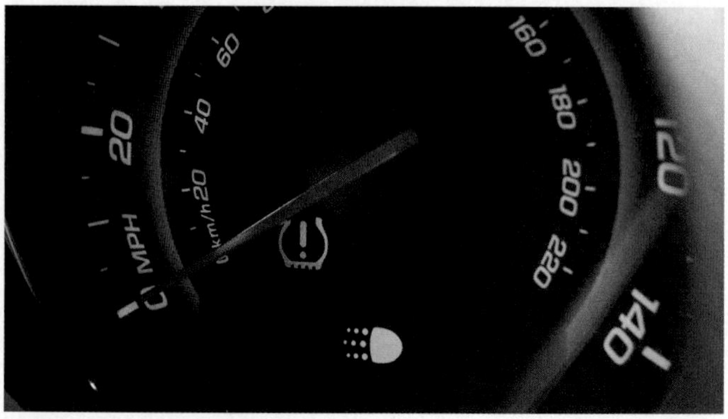

FIGURE 24-30 Low tire pressure warning on the instrument panel.

TPMS systems, drivers can monitor the tire pressures and temperatures from the driver's seat. This allows them to monitor the tires at all times. The TPMS is designed to ignore normal pressure variations caused by changes in ambient temperature.

There are two basic configurations used to monitor the vehicle's tire pressures: direct and indirect. A **direct TPMS** directly measures the tire pressure via a sensor that is installed inside each wheel. The sensor is able to respond to a drop in pressure of as little as 2 psi (14 kPa). Being that it is mounted inside the wheel helps protect it from damage. The sensor in each wheel is equipped with an antenna. It wirelessly relays the information to receivers located within the vehicle (**FIGURE 24-31**). The receivers send the signal to the control unit.

The control unit sends an appropriate signal to the driver's information circuit or, in some vehicles, the onboard computer. Then, an indicator is illuminated on the instrument panel to warn the driver of low tire pressure in a certain wheel. An audible and visual warning is usually used. The driver then knows to have the tires checked for leaks, and repaired. If no leak is found, then the tire can be reinflated to the specified pressure.

The sensors are powered by an internal battery that is designed to last between 5 and 10 years. The use of a **centrifugal switch** in the sensor allows the sensor to go to sleep when the vehicle stops, which extends battery life. When the battery goes dead, it has to be replaced. This usually means that the entire sensor must be replaced because the battery is typically sealed inside the sensor.

▶ TECHNICIAN TIP

In vehicles with a direct TPMS, the dashboard may show:

- the required tire pressure,
- the actual tire pressure,
- the tire pressure status, and
- the temperature of the tire.

In some vehicles, the driver can use the display control buttons to check the status of each tire.

A direct TPMS can be of two types: one-way communication or two-way communication. In one-way communication, the TPMS sensor can only transmit to the receiver. It cannot receive any information. In this type of system, the sensors usually use a centrifugal switch to turn them on and off to conserve battery energy. In a two-way communication TPMS, the sensor can receive as well as transmit signals. This allows the control unit to send signals to wake up or cause the sensor to sleep, thus extending the sensor's battery life. The two-way communication system is more complex and expensive than the one-way system.

The sensors are located inside the tire and fastened in some manner to the wheel. Many wheel sensors are integrated into a one-piece design along with the valve stem. Others are a separate unit screwed to the valve stem (**FIGURE 24-32**). A third type uses a band or strap that fastens the sensor in the drop center of the wheel. Band-style sensors are typically positioned opposite of the valve stem (**FIGURE 24-33**). When removing and replacing a tire on a wheel, you must be aware of the sensor's location. That way you can take steps to avoid damaging it.

▶ TECHNICIAN TIP

Most wheels with TPMS sensors use an aluminum valve stem as the antennae. Aluminum valve stems require a nickel-plated valve core to prevent corrosion.

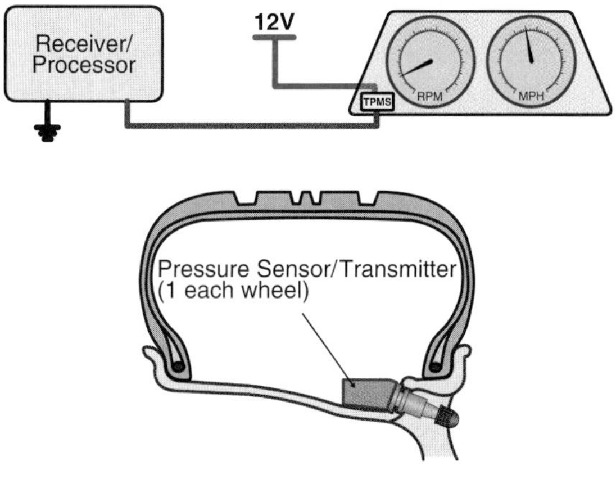

FIGURE 24-31 The tire pressure monitoring system (TPMS) using pressure sensors.

FIGURE 24-32 Typical pressure sensors used in TPMS.

FIGURE 24-33 Band-type TPMS sensor.

An **indirect TPMS** indirectly monitors tire air pressure. The most prevalent indirect TPMS in use today uses the anti-lock braking system (ABS) wheel speed sensors. The wheel speed sensors measure the rotational speed of each of the four wheels. A wheel that is rotating faster than the others indicates that the tire has lower pressure than the other tires (**FIGURE 24-34**). This is because a tire with low pressure has a smaller rolling radius. The smaller radius increases the speed of rotation. Because it is rolling faster than the other tires, the wheel speed sensor will send a slightly faster signal to the ABS control unit. The control unit monitors the changes in wheel speed. When a low tire pressure failure is detected, it activates the low tire pressure light and/or chime on the dashboard.

Run-Flat Tires

The main safety benefit of **run-flat technology** is that it allows a driver to maintain control of the vehicle if a tire suffers a rapid pressure loss. In addition, run-flat tires enable the driver to continue the journey within specified speed and distance limits. This is typically about 50 mph (80 kph) and at least 50 miles (80 km). This alleviates the need to replace the wheel on the side of the road or in an unsafe area. But as with any tire, the run-flat tire cannot continue to be driven if the sidewall has been compromised or blown out. In these instances, the tire needs to be changed. A TPMS is normally mandatory for all run-flat technology applications. A run-flat tire is designed to operate only for limited distances with no air pressure. So, the pressure needs to be monitored.

Tire manufacturers also claim that run-flat technology saves weight and space in the vehicle. It does this by eliminating the need to carry a spare tire. However, due to their construction, they are generally two to three times heavier than conventional tires. This adds unsprung weight (weight not held up by the vehicle's springs) to the vehicle. That affects handling, and increases fuel consumption.

Run-flat technology generally focuses on two goals: rigidity and heat resistance. The objective is for the tire to support the vehicle's weight when it is rotating with a total loss of air. The sidewall is constructed with reinforced rubber and is thicker than in conventional tires. This enables it to carry the vehicle's weight at zero pressure (**FIGURE 24-35**).

Because of the extra materials used in the construction of run-flat tires, they are also more expensive to purchase. Also, run-flat tires are usually harsher riding and noisier in operation. This can be a disadvantage in some applications, such as in luxury cars that are expected to have a quiet, smooth ride.

The bead shape of the run-flat tire is largely unchanged from conventional tires. This does enable them to be compatible with conventional rims. However, the bead wire is normally wider and reinforced. This ensures a more secure fit on a specialized rim, even at zero pressure. A special bead filler with low heat generation is used as part of the run-flat technology construction. This helps prevent excess heat buildup that can be generated with zero pressure in the tire.

The rim used with some run-flat technology tires is referred to as an **EH2 rim**. EH2 is an abbreviation for double-extended hump. These rims have a wider, or more extended, safety hump than a standard safety rim (**FIGURE 24-36**). This accommodates

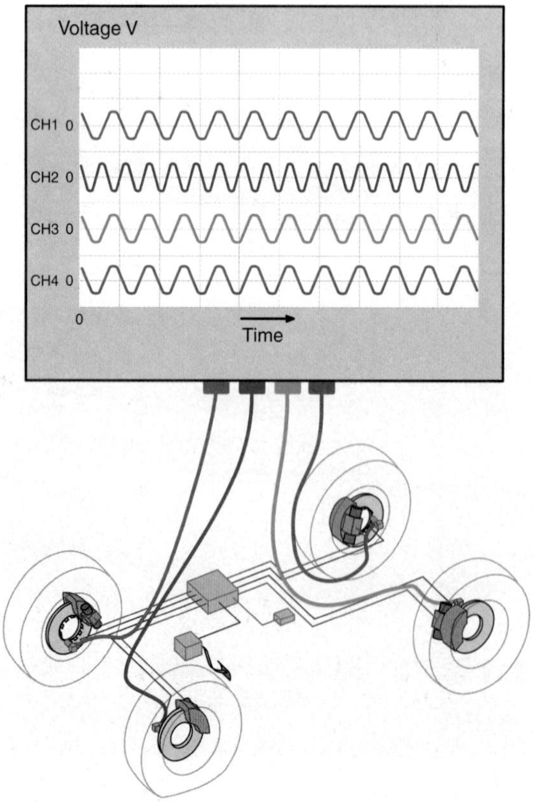

FIGURE 24-34 Indirect TPMS system showing a tire with low pressure.

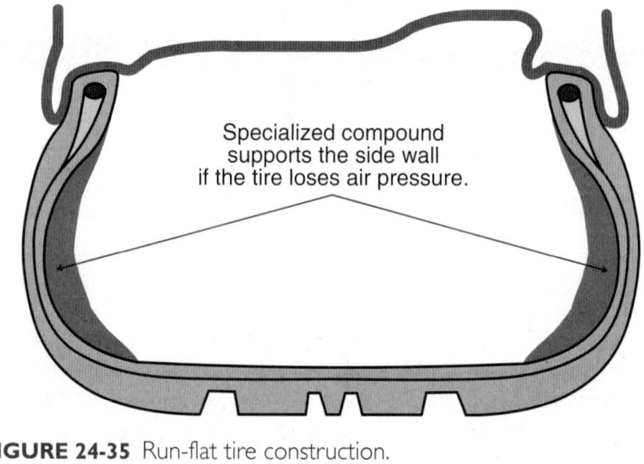

FIGURE 24-35 Run-flat tire construction.

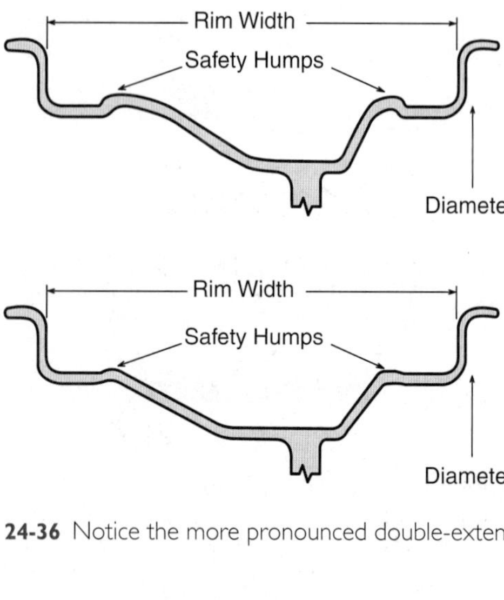

FIGURE 24-36 Notice the more pronounced double-extended safety humps.

the wider and more square-shaped tire bead better. It also does a better job of holding the tire against the rim flange when the pressure is low than a standard rim does.

▶ TECHNICIAN TIP

Tire manufacturers call their run-flat safety tire by various terminologies, but all have the same characteristics. Some run-flat tires are known as:

- Run-Flat Technology (RFT) tires
- **Extended Mobility Technology (EMT)** tires
- Continuous Mobility Technology (CMT) tires
- Zero-pressure (ZP) tires

These tire sidewalls can be six times thicker than the sidewalls of traditional tires. As a result, the manufacturers claim that run-flat tires can be driven in a deflated condition for an extended period of time. These tires are not indestructible. Major damage that slices the tire casing can still result in complete tire failure.

Self-Sealing Tires

Self-sealing tires were introduced recently. One self-sealing tire manufacturer states that its tire is designed to repair most small tread-area punctures instantly and permanently. Self-sealing tires feature standard tire construction. But they add a flexible and malleable lining inside the tire in the tread area (**FIGURE 24-37**). This lining serves as a puncture sealant that can permanently seal most punctures up to 3/16" (or 5 mm) in diameter.

Self-sealing tires first provide a seal around the object when the tire is punctured. If the object is removed, it then fills the hole in the tread. Note that not all punctures can be sealed in this way. Self-sealing tires can be repaired similar to a conventional tire, except the surface shouldn't be roughed up.

Since they are self-sealing, most drivers never even know that they just had a puncture. These tires can still leak air from between the bead and the rim or the valve stem. So, they still require a TPMS to detect any low-pressure conditions on 2008 and newer vehicles.

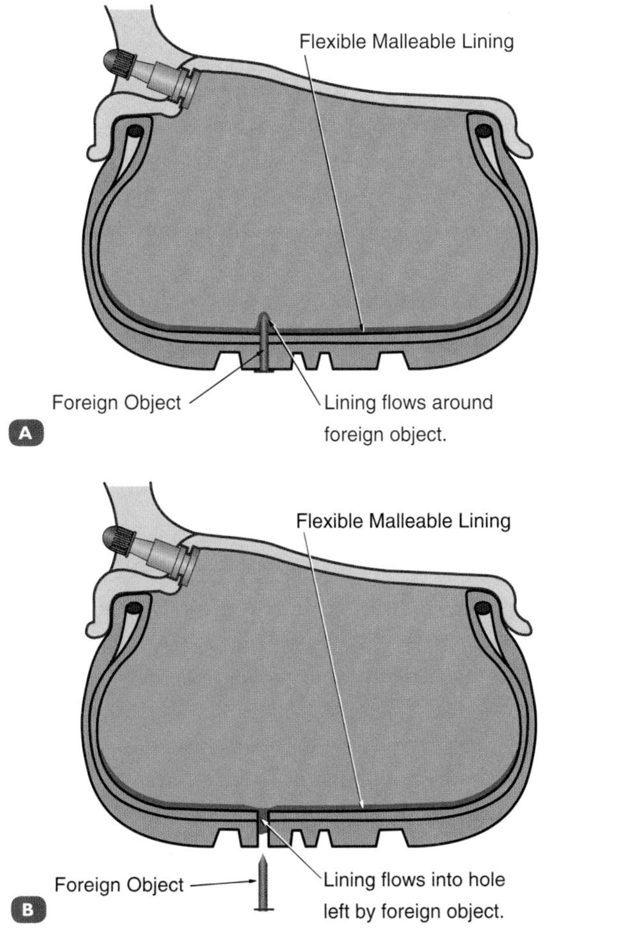

FIGURE 24-37 Self-sealing tire protects against small punctures.

Space-Saver Spare Tires

Space-saver spare tires are designed for emergency use only. They are designed to get the vehicle to a service center to have the regular tire repaired or replaced. Most manufacturers warn not to exceed 50 mph (80 kph) and 50 miles (80 km) on the space-saver spare tire. Otherwise, the tire may fail.

Some vehicles have miniature or collapsible space-saver spare tires as spares. These often need a specially charged canister for inflation when being installed (**FIGURE 24-38**). Other vehicles have small, temporary spare tires that have been inflated normally. But, because of their small size, they have a much higher pressure than normal road tires (**FIGURE 24-39**).

FIGURE 24-38 Collapsible space-saver spare and inflation canister.

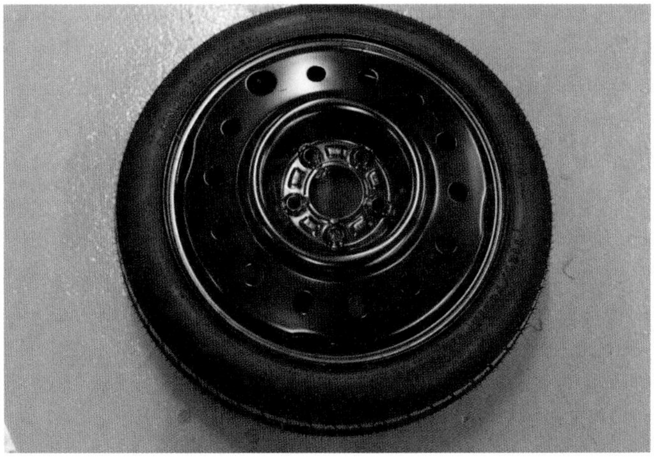

FIGURE 24-39 Typical temporary tire.

▶ Wrap-Up

Ready for Review

- Tires and wheel work together to provide the rigidity needed to provide directional control and the flexibility and traction to grip the road surface while smoothing out the bumps.
- Wheels are usually made from pressed steel or cast aluminum alloy. Rim is a part of the wheel and is the outer circular lip of the wheel on which the inside edge of the tire is mounted.
- Tires are hollow, donut-shaped structures that are constructed as bias ply or radial ply. Tire tread patterns can be classified as directional, nondirectional, Symmetric, Asymmetric, and Directional and asymmetric tread patterns.
- Tire markings should include: manufacturer or brand name, International Organization for Standardization (ISO) tire class, section width, aspect ratio, type of tire construction, wheel diameter, speed rating designation, maximum air pressure, load index, uniform tire quality grading (UTQG) system, and Department of Transportation compliance symbols and serial numbers.
- Various tire safety features used in a vehicle include tire pressure monitoring system, run-flat tires, self-sealing tires, and space-saver spare tires.

Key Terms

aspect ratio The ratio of sidewall height to section width of a tire.

asymmetric tread pattern A tread pattern that differs on each side and therefore is usually directional.

bead seat The part of the wheel that the tire seals against.

bias-ply A tire constructed in a latticed, crisscrossing structure, with alternate plies crossing over one another and laid with the cord angles in opposite directions.

casing plies Network of cords that give the tire shape and strength; also known as casing cords.

centrifugal switch A switch in the sensor of a TPMS that allows the sensor to go to sleep when the vehicle stops, which extends the TPMS's battery life.

cornering force The force between the tread and the road surface as a vehicle turns.

deep dish wheel A wheel with negative offset, which gives the outside of the wheel a deep dish appearance. Deep dish refers to the side of the wheel that is farthest from the drop center.

direct TPMS A type of automated tire pressure monitoring system that measures tire pressure and possibly temperature via a sensor installed inside each wheel.

directional and asymmetric tread pattern A tread pattern that is both directional and asymmetric, which means the tire is designed to rotate in only one direction and has one side that must face outward to ensure that the tire performs as designed under operating conditions.

directional tread pattern A tread pattern designed to pump water out from under the tire; each tire must be placed in a particular spot on the vehicle.

EH2 rim The specialized rim design that is used with some run-flat tires.

Extended Mobility Technology (EMT) Tires with thick sidewalls that allow the tire to be driven on even when it has no air pressure.

Flat seat lug nuts Lug nuts with a flat seat. The sides of the lug nut center the wheel to the axle flange.

indirect TPMS A type of automated tire pressure monitoring system that uses the antilock braking system of a vehicle to measure the difference in the rotational speed of the four wheels in order to determine tire pressure.

kinetic friction The friction between two surfaces that are sliding against each other.

lug nuts Fasteners that secure the wheel onto the wheel studs.

negative offset A condition in which the plane of the hub mounting surface is positioned toward the brake side or back of the wheel centerline.

neutral steer A condition in which both the front and the rear tires of a vehicle are experiencing the same slip angle.

nondirectional tread pattern A tread pattern that is nonspecific, allowing the tire to be placed on any wheel of a vehicle.

oversteer A condition when the front end steers more sharply than desired because the rear wheels lose adhesion during cornering. This causes the rear of the vehicle to swing out. The vehicle is said to be loose.

positive offset A condition in which the plane of the hub mounting surface is positioned toward the outside or front of the wheel centerline.

radial Type of tire where the casing plies run side-to-side from bead to bead.

rim The outer circular lip of the metal on which the inside edge of the tire is mounted.

rim flanges The outside edges of the wheel that help keep the tire from popping off the wheel.

run-flat technology A tire design that allows the vehicle to keep moving under driver control following a puncture or rapid loss of pressure.

Schrader valve A one-way valve used in a valve stem.

self-sealing tire A tire constructed with a flexible and malleable lining inside the tire around the inner tubeless membrane. The lining can seal small tread-area punctures instantly and permanently.

side force The pressure on the wheel that pushes it toward the outside or inside of the rim as the vehicle makes a turn.

six-ply rating Tire casing made up of six casing plies. Or a tire casing having the strength of six casing plies.

slip angle The angle between which the tire is pointing, and the vehicle is moving.

symmetric tread pattern A tread pattern with the same tread pattern on both sides of the tire; typically nondirectional.

Tapered seat lug nuts Lug nuts with a tapered seat. The tapered seat is used to center the wheel on the axle flange.

temperature grade a representation of a tire's ability to resist and dissipate heat. From highest to lowest: A, B, and C.

tensile strength The amount of force required before a material deforms or breaks when being pulled apart.

tire pressure monitoring system (TPMS) A federally mandated system to provide a means of reliable and continuous monitoring of vehicle tire pressure. It is designed to increase safety, decrease fuel consumption, and improve vehicle performance. A TPMS monitors the tires for low air pressure and alerts the driver when one or more tires are lower than (or in some cases, higher than) the designated thresholds. This alert can be an illuminated warning lamp or a chime.

traction grade A standardized grading system that indicates how well a tire will maintain contact with the road surface when wet.

tread wear grade The number imprinted on the sidewall of a tire by the manufacturer, as required by the National Highway Traffic Safety Administration (NHTSA), that indicates the tread life of a tire's tread.

understeer A condition in which the vehicle's front wheels are turned more sharply than the vehicle's actual direction because the front tires lose adhesion during cornering. The vehicle is said to be "pushing" in the corners.

Uniform Tire Quality Grading (UTQG) A standardized grading system, established by the National Highway Traffic Safety Administration (NHTSA), designed to provide tire buyers with a comparative measure of a tire's tread life, traction, and temperature characteristics.

US Department of Transportation (DOT) A federal agency that regulates transportation safety in the United States, including vehicles' wheels and tires. The DOT requires a code—a series of letters and numbers—to be stamped into the sidewall of every tire made for public use in the United States. These codes contain information such as the date of manufacture and the plant where the tire was manufactured.

valve core The one-way spring-loaded valve that screws into the valve stem; it allows air to be pumped into a tire and prevents it from flowing out.

valve stem Part which allows air to be added or removed from a tire.

valve stem cap A cap that fits tightly onto the valve stem to prevent debris from clogging it and acts as a secondary seal.

wheel flange The center portion of a wheel. Usually a formed disc which is welded to the rim.

wheel studs Threaded fasteners that are pressed into the wheel hub flange and used to bolt the wheel onto the vehicle.

Wheel width The distance between the bead seats on the wheel.

zero offset A condition in which the plane of the hub mounting surface is even with the centerline of the wheel.

Review Questions

1. Why are most car tires considered pneumatic?
   a. Because they mount on a rigid wheel assembly
   b. Because they contain many reinforcing strands
   c. Because they contact the road and divert water away
   d. Because they are filled with pressurized air
2. What is the typical weight distribution front to rear on a front-wheel drive, front engine vehicle?
   a. 50% front/50% rear
   b. 40% front/60% rear
   c. 75% front/25% rear
   d. 90% front/10% rear
3. What type of vehicle is most likely to have split rims?
   a. A tractor trailer
   b. A high-performance car
   c. An economy car
   d. A light duty truck
4. A customer requests installing aftermarket wheels that were purchased online, but without lug nuts. The original vehicle lug nuts are tapered seat lug nuts. The aftermarket wheels call for flat seat lug nuts. What should be done?
   a. The original lug nuts should be used and torqued to vehicle specification
   b. Use the original lug nuts but torque them 10% over specification
   c. Install flat seat lug nuts, torqued to specification, and recheck torque after 100 miles
   d. Leave the decision up to the customer, and do whatever they want done
5. Having fewer plies in the sidewall results in:
   a. A softer ride with better bump absorption
   b. More puncture resistance
   c. Better stiffness for high-speed turns
   d. A higher weight capacity
6. Which of the following could have been the date a tire with a DOT date code 0318 was made?
   a. March 1st 2018
   b. January 18th 2018
   c. January 3rd 2018
   d. March 15th 2018
7. If a vehicle is built with a top speed of 140 mph, what should its tires' minimum speed rating be?
   a. Q
   b. S
   c. V
   d. Z
8. What type of TPMS system uses the vehicle's wheel speed sensors to monitor the relative speed of each tire to determine the tire pressure?
   a. Direct
   b. Indirect
   c. Radial
   d. Relative

9. What allows direct-type TPMS sensors to go to sleep when the vehicle stops?
   a. A centrifugal switch
   b. The vehicle speed sensor
   c. The ABS wheel speed sensor
   d. The GPS system
10. What type of tires feature standard construction with a flexible and malleable lining inside the tread area for flat protection?
   a. Run flat tires
   b. Space saver tires
   c. Extended mobility tires
   d. Self-sealing tires

ASE Technician A/Technician B Style Questions

1. A vehicle that is raced on the weekends comes in for a handling concern. The customer states that when they are turning hard on the race track, the back of the vehicle tends to lose traction and they spin out. Technician A states this is known as oversteer. Technician B states this is because the rear slip angle is larger than the front slip angle. Who is correct?
   a. Technician A only
   b. Technician B only
   c. Both Technicians A and B
   d. Neither Technician A nor B
2. Advantages and disadvantages of bias ply tires and radial tires are being discussed. Technician A states that bias ply tires are more prone to side wall punctures. Technician B states that radial tires typically run cooler than bias ply. Who is correct?
   a. Technician A only
   b. Technician B only
   c. Both Technicians A and B
   d. Neither Technician A nor B
3. A vehicle comes into the shop. The left side lug nuts came off with ease, the right side lug nuts seem to be seized to the studs. Technician A states the lug nuts and studs could be very rusty and may require extra force to remove them. Technician B states the right side could be left-hand thread, in which case they must be turned clockwise to loosen. Who is correct?
   a. Technician A only
   b. Technician B only
   c. Both Technicians A and B
   d. Neither Technician A nor B
4. Aftermarket wheels are being discussed. Technician A states that many aftermarket wheels are known as "rims" since they have an extra ridge for performance. Technician B states that aftermarket wheels can be ordered with the customer's desired offset and backspacing. Who is correct?
   a. Technician A only
   b. Technician B only
   c. Both Technicians A and B
   d. Neither Technician A nor B
5. Tire speed ratings are being discussed. Technician A states a vehicle with Q-rated tires should not be driven over 100 mph on a race track. Technician B states a vehicle with high-perormance Z-rated tires means the vehicle may be legally driven at speeds slightly over the speed limit. Who is correct?
   a. Technician A only
   b. Technician B only
   c. Both Technicians A and B
   d. Neither Technician A nor B
6. Tire sizes are being discussed. Technician A states a tire with a higher aspect ratio is taller than one with a lower aspect ratio. Technician B states the first number in a tire size reading indicates its sidewall height. Who is correct?
   a. Technician A only
   b. Technician B only
   c. Both Technicians A and B
   d. Neither Technician A nor B
7. A customer is shopping for tires and wants advise from a technician on the best tire value. Technician A states the UTQG should be referenced since it will approximate tire performance and life. Technician B states the type of driving, speed, vehicle weight, and traction demands should be considered when choosing a tire. Who is correct?
   a. Technician A only
   b. Technician B only
   c. Both Technicians A and B
   d. Neither Technician A nor B
8. A customer wants tires for their Jeep, which they use exclusively for off-roading. They ask what tire changes could be made to give them more ground clearance. Technician A states a lower aspect ratio will give them more ground clearance. Technician B states a higher section width and the same aspect ratio will give them more ground clearance. Who is correct?
   a. Technician A only
   b. Technician B only
   c. Both Technicians A and B
   d. Neither Technician A nor B
9. A vehicle comes in with a TPMS lamp illuminated. Technician A states a sensor could have a dead internal battery. Technician B states some instrument panels in the vehicle will show the tire pressure, and it may just need air added. Who is correct?
   a. Technician A only
   b. Technician B only
   c. Both Technicians A and B
   d. Neither Technician A nor B
10. A customer is inquiring about run flat tires. Technician A states the purpose of run flat tires is to allow a customer to drive with a flat temporarily, until it can be replaced. Technician B states the run flat tire is kept in the trunk and is smaller with a higher pressure. Who is correct?
   a. Technician A only
   b. Technician B only
   c. Both Technicians A and B
   d. Neither Technician A nor B

CHAPTER 25

Servicing Wheels and Tires

Learning Objectives

- **LO 25-01** Describe tire maintenance preliminaries.
- **LO 25-02** Properly check and adjust tire pressure.
- **LO 25-03** Identify tire wear patterns and perform tire rotation.
- **LO 25-04** Perform tire balance.
- **LO 25-05** Dismount and mount a tire without TPMS.
- **LO 25-06** Dismount and remount a tire with TPMS.
- **LO 25-07** Replace tire valve stems.
- **LO 25-08** Perform TPMS service.
- **LO 25-09** Perform tire diagnosis and repair.
- **LO 25-10** Measure wheel, tire, axle flange, and hub runout.

ASE Education Foundation Tasks

See Appendix A to view the 2017 ASE Education Foundation Automobile Accreditation Task List Correlation Guide.

▶ Tire Maintenance Preliminaries

LO 25-01 Describe tire maintenance preliminaries.

Paying attention to the wheels and tires can also assist in identifying steering and suspension system problems. The wear patterns on a tire indicate wear or damage to particular vehicle components. Failing to inspect the wheel and tire assembly may result in missed faults. This can lead to vehicle breakdowns and more expensive repair costs.

A periodic inspection of the tires is necessary to ensure a long life (**FIGURE 25-1**). The tire pressure should be checked and adjusted on vehicles that do not have direct tire pressure monitoring system (TPMS). The tread area is inspected for cuts, flat spotting, or irregular wear patterns. The tread depth is inspected to ensure that it is above the built-in wear indicators. Tires that are at or below the wear indicators mean that the tire is at the end of its legal life and should be replaced. The sidewalls are inspected for cuts, gashes, and any signs of bulging.

At the same time, vehicle and tire manufacturers recommend that the tires be rotated during each oil change. Tire balance is performed when the tires are new. It is also performed any time there is a vibration concern that could be caused by tire balance faults.

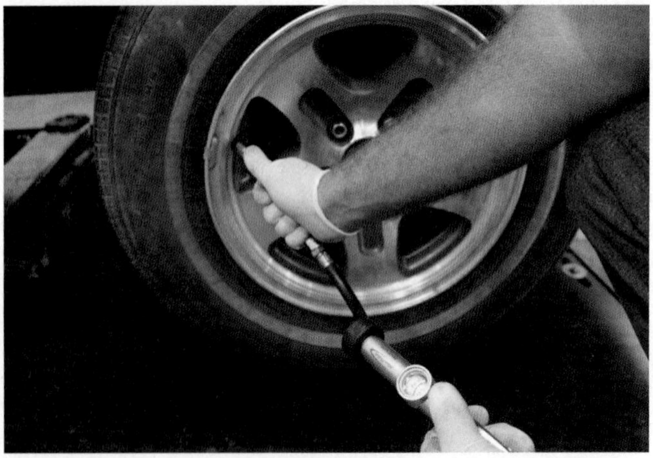

FIGURE 25-1 Whenever a vehicle comes into the shop, check the tires' air pressure and look for signs of wear.

Wheel alignment is another service that is performed on occasion. The alignment is normally checked and adjusted in cases such as:

- when there is abnormal tire wear,
- when steering/suspension parts are replaced,
- when the vehicle does not handle or drive correctly, and
- after an accident that could affect the alignment.

So, the goal of an alignment is to restore proper handling along with extending the life of the tires. Most tire shops recommend an alignment when new tires are purchased. This will help ensure that the tires don't wear out prematurely. This is usually a fairly easy sell, especially if the original tires wore out due to an alignment issue. We will cover wheel alignment in an upcoming chapter.

Tools

To service tires in an efficient manner, you must have the appropriate tools and equipment. The basic equipment list should consist of the items in (**FIGURE 25-2**):

- Tire pressure gauge—used to check the air pressure in tires.
- Tread depth gauge—used to measure the tire's tread depth.
- Valve stem tool—used to remove and install tubeless valve stems in rims.
- Valve core tool—used to remove and install valve cores in valve stems.
- TPMS torque tools—used to tighten TPMS sensor components.
- Tire-changing machine—capable of handling the range of tires that the shop stocks. In a shop that handles run-flat tires, the tire-changing machine must be designed to handle the more robust bead and sidewalls.
- Tire dunk tank—used to locate leaks in tires and rims.
- Tire spreader—used to spread the sidewalls of a tire for easier access during tire patching.
- Air tire buffer—used to lightly buff the inside surface of the tire as preparation for tire patching.
- Patch stitching tool—used to apply pressure to the patch when installing it.
- Tire inflation cage—used to contain the tire and rim during tire inflation in the event of a tire explosion.

You Are the Automotive Technician

A customer brings her 2010 Dodge Caravan to the dealership with the tire pressure light on. The customer tells you that the tires are filled with nitrogen, and on the drive to the dealership, the van pulled to the right. You bring the vehicle into the shop and notice that the right front tire looks low. Its pressure reads 22 psi. You inspect the tire and notice a small screw embedded near the center of the half-worn tread. You mark a big "X" across the tire (centered on the screw) to mark the hole. Air leaks out as you pull the screw out of the tire. You remove the wheel and check it for other leaks. There are none, so you prepare to repair the tire.

1. How can you identify whether tires have been inflated with nitrogen?
2. What is the proper method of repairing a tire?
3. What are some conditions when a tire should not be repaired?
4. What happens if you repair a high-performance speed-rated tire?
5. What must be done once the tire is repaired and back on the vehicle?

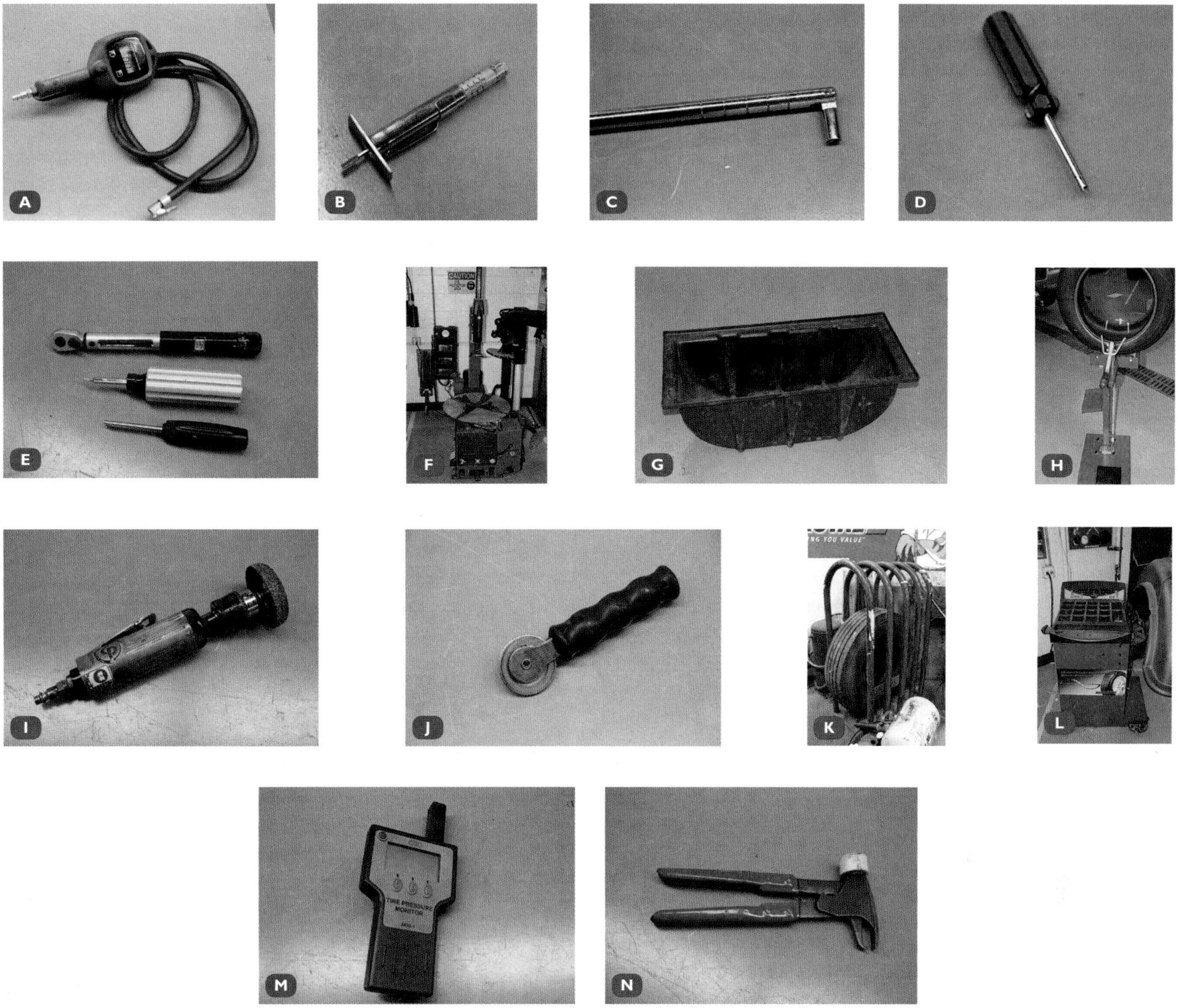

FIGURE 25-2 Typical tire repair tools and equipment.

- Wheel balancing machine—used to determine out-of-balance conditions on a tire so they can be balanced.
- Wheel weight hammer—used to remove and install wheel weights onto rims.
- TPMS reset tool—used to reset the TPMS (not required on all TPMS-equipped vehicles).

Common Tire and Wheel Issues

Common tire and wheel issues include:

- *Air loss:* The most common issue with tires is air loss. Tires normally lose a small amount of air over time and require periodic refilling. Punctures occur that cause leaks of various sizes. Valve stems and tire beads can also allow air to leak from the tire.
- *An out-of-balance tire or wheel:* The wheel and tires must be balanced, with the weight equally distributed throughout. When a tire rotates, any points of unequal weight will cause the tire to wobble. This places stress on the shocks, bearings, and wheel assembly. Additionally, the vehicle will typically vibrate at speeds of about 35 mph (56 kph) and above. It will have a rough ride, and the steering wheel may vibrate. A wheel balancer is used to identify and correct wheel imbalances.
- *Excessive loaded radial runout on the tire, wheel, and hub assembly:* With radial runout, the tire tread moves up and down. It is caused by incorrect manufacture of or damage to one of the components. This includes damage such as a broken belt in the tire. Correction involves replacing the out-of-round component.
- *Excessive lateral runout on the tire, wheel, and hub assembly:* With lateral runout, the tire tread moves side to side. It is caused by incorrect manufacture of or damage to one of the components. This includes damage, such as a bent rim or broken belt in the tire. The result is a wobble. When lateral runout occurs, the only method for fixing it is to replace the bent or improperly manufactured component.

- *Wheel trim imbalance (if fitted):* The term "wheel trim" refers to any pieces attached to a wheel that are not necessary for actual wheel function. This is typically a hubcap or trim ring. Imbalance or wear in trim pieces is corrected by removing or replacing the part.

Heavy pulling of a vehicle to either the left or the right while being driven is an indication of these possible problems:

- Mismatched tire sizes or pressures
- Tire with broken or misaligned belts
- Worn suspension or steering components
- A dragging front brake assembly
- Out-of-alignment wheels

Wheel alignment issues will be covered in the wheel alignment chapter. But it is good to visually inspect tires for the other conditions listed before recommending a wheel alignment. Let's dig into the tire maintenance procedures.

▶ Proper Tire Inflation

LO 25-02 Properly check and adjust tire pressure.

Tire inflation pressure is the amount of air pressure in the tire. It provides the tire with load-carrying capacity. It also affects the overall performance of the vehicle. Vehicle manufacturers determine the tire inflation pressure based on the vehicle's designed load limit and the tire size. The load limit is the greatest amount of weight a vehicle can safely carry.

The proper tire pressure for the vehicle is referred to as the recommended cold inflation pressure. It is measured in psi or kPa. You will find this information on the tire placard, expressed in psi or kPa (**FIGURE 25-3**). The tire placard is typically located on the A- or B-pillar door frame, in the glove compartment, or on the fuel filler flap.

Because tires are designed to be used on more than one type of vehicle, tire manufacturers list the maximum safe inflation pressure on the tire sidewall. This number is the highest amount of air pressure that should ever be put in the tire. The maximum tire pressure should never be exceeded. This is true even when the vehicle's tire placard indicates a higher pressure. As explained earlier, 32 psi of air pressure can generate over 30,000 lb of force on a tire.

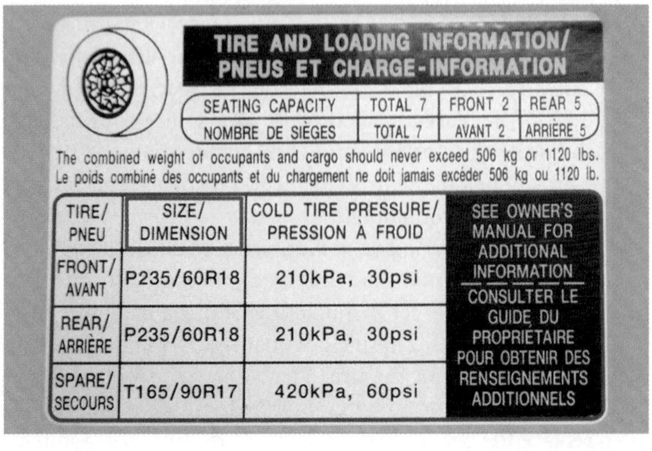

FIGURE 25-3 Typical tire placard.

If the maximum tire pressure is exceeded, even greater forces are generated and can cause the tire to explode violently. Too many young technicians have lost their life when they overinflated a tire and it blew up in their face. This is one of the more dangerous situations for inexperienced technicians. It is also why it is good to use a tire cage when inflating a tire (**FIGURE 25-4**). If the tire blows up, the explosion will affect your hearing. You also might be hit by some debris, but the cage is likely to contain most of the larger chunks of the tire.

To get an accurate pressure reading, the tires must be checked when cold. The term "cold" does not relate to the outside temperature. Rather, a cold tire is one that has not been driven on for at least three hours or less than 1 mile (1.6 km). While driving, tires get warmer because of the heat created by friction as the tire materials flex against each other. Flexing results in heat. The heated tire materials, in turn, heat the air within the tire. As the air within the tire warms, it expands, causing the air pressure to increase.

To obtain an accurate tire pressure reading, the tires must be cold. If you have a tire pressure compensation chart, you may be able to compensate for the extra pressure in warm tires. Some manufacturers supply this information, but others do not.

FIGURE 25-4 After installing a new tire, it is good to inflate it in a tire cage.

> ▶ TECHNICIAN TIP
>
> According to the U.S. Department of Energy, underinflated tires lower a vehicle's gas mileage. They claim it is reduced by 0.3% for every 1 psi drop in pressure in all four tires. So, running them 10 psi low reduces gas mileage approximately 3%. Underinflation wastes approximately 2 billion gallons of fuel each year in the United States alone.

Nitrogen Fill

Oxygen is harmful to rubber and other tire materials. The oxygen reacts with the rubber through oxidation. It also causes the rubber to lose its flexibility and sealing ability. This allows oxygen to permeate the rubber and degrade it further over time. Also, as the inner liner oxidizes, more air molecules can pass through it, causing an increased rate of pressure loss. In addition, moisture in shop air causes metal to rust or corrode. This can clog the valve stems, causing them to leak.

In recent years, some manufacturers have been filling their tires with pure or nearly pure nitrogen. Their hope is that the problems associated with oxygen-filled tires can be avoided (**FIGURE 25-5**). Nitrogen is an inert gas that does not react with the rubber compounds in the tire. Nitrogen generators also remove any moisture from the nitrogen gas. This reduces the corrosion effects on TPMS sensors and other metal parts of the wheels.

FIGURE 25-5 A nitrogen-filling station.

Advantages for Nitrogen Fills

Although both nitrogen and oxygen can permeate rubber, nitrogen does so at a much slower rate. It might take three months to lose 1 psi (14 kPa) with nitrogen, compared to just a month with normal air. In addition, nitrogen is far less reactive. It does not cause rust and corrosion on steel or aluminum, and it does not degrade rubber. Wheel surfaces stay smooth and clean, and rubber remains supple and resilient.

Because tires lose less pressure with nitrogen due to permeation, a tire's performance will be maintained over time. Reduced pressure loss increases the vehicle's fuel efficiency and tire life. It also helps prevent accidents. It does this by reducing the possibility of blowouts. It also maintains maximum tire traction with the road surface.

The air that we breathe and that is typically used to inflate tires consists of 78% nitrogen, 21% oxygen, and 1% other gases. When using pure nitrogen as an inflation gas, the composition of nitrogen increases from 78% to near 100%. To meet the standards for proper nitrogen fill means that the nitrogen level in the tire must be at 95% or higher. This typically requires two or three inflations and deflations to remove the oxygen. Although tire manufacturers generally support the use of nitrogen, they do not generally mandate its use.

Nitrogen-filled tires can be identified by the green valve stem caps that are placed on the valve stem (**FIGURE 25-6**). If you see a green valve stem, do not fill the tire with regular air, unless absolutely necessary. This would dilute the nitrogen.

Using a Tire Pressure Gauge

There are two main types of **tire pressure gauges**: fixed workshop gauges and portable pocket-size gauges (**FIGURE 25-7**). The three most popular types of pocket tire pressure gauges are the pencil type, the dial type, and the digital type. The pencil type looks similar to a pencil and contains a graduated sliding extension. The extension is forced out of the sleeve by air pressure when it is held against the tire valve. The dial type has a graduated gauge and movable needle. The digital type can look

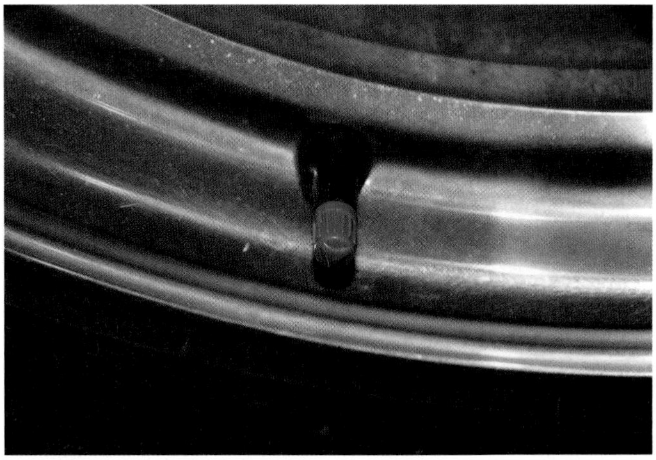

FIGURE 25-6 Nitrogen-filled tires can be identified by the green valve stem cap.

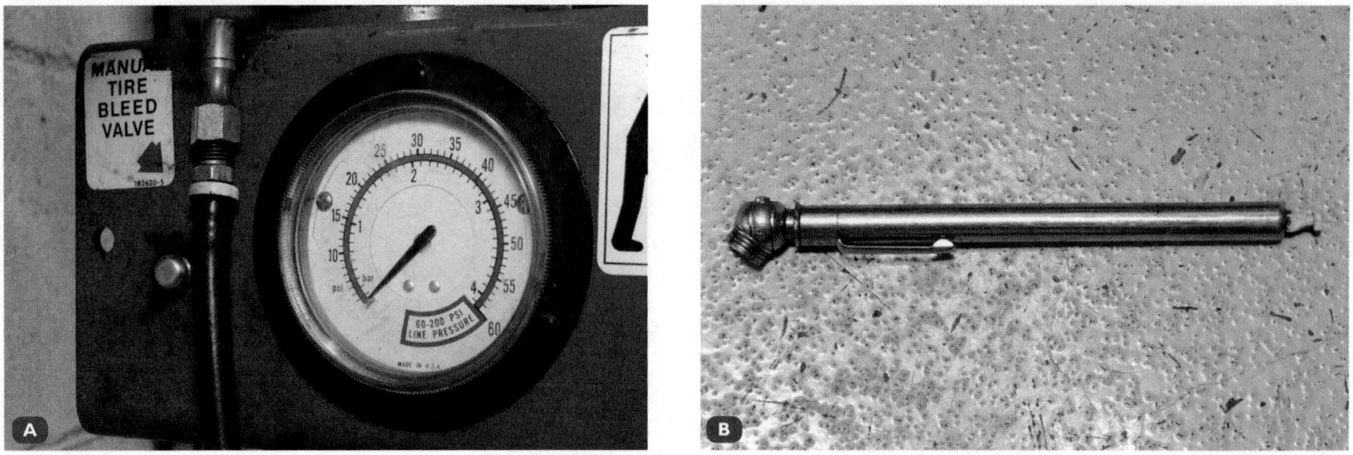

FIGURE 25-7 Tire pressure gauges. **A.** Fixed workshop gauge. **B.** Portable pocket-size gauge.

like any of the others, but it gives a digital reading of the pressure. It is generally the most accurate type of tire pressure gauge. Some digital pressure gauges can also read the temperature.

Each tire pressure gauge measures pressures in pounds per square inch (psi), kilopascals (kPa), or **bars**.

- One bar is equivalent to 14.5 psi or 100 kPa.
- One psi is equivalent to approximately 7 kPa. Some tire pressure gauges have scales for more than one unit of measurement.

The tire pressure will vary from vehicle to vehicle. It also depends on the intended use of the vehicle and driver preference. Recommended tire pressures are located on the vehicle's tire placard, usually located on the driver door pillar. The maximum tire pressure is stamped on the tire sidewall. Never inflate the tire above the maximum pressure listed on the sidewall. The tire may explode, or the wheel may give way and cause a blowout, which can easily be fatal.

▶ TECHNICIAN TIP

A common mistake when inflating tires is to use the pressure stamped on the sidewall instead of the one listed on the tire placard. The pressure branded on the sidewall is the maximum pressure that the tire is designed to withstand. It should *never*be exceeded, even when the placard lists a higher pressure! So you should always first check the placard. Then, verify that it is not higher than the maximum pressure stamped on the sidewall. If it is, someone installed underrated tires on the vehicle.

▶ TECHNICIAN TIP

Pocket-type tire pressure gauges are inexpensive. They are generally more accurate than those at service stations. Service station gauges are often damaged by weather, misuse, or being run over. If the same pocket-type tire pressure gauge is always used, then there will be no variation of readings.

To use a tire pressure gauge, follow the steps in **SKILL DRILL 25-1**.

▶ TECHNICIAN TIP

If you check the tire pressures after the vehicle has been driven and the tires are warm or hot, *do not* release any slightly excessive pressure. If you bleed the tire pressure down to the manufacturer's recommendation, it will be underinflated when the tire cools down. This could cause premature wear on the tires and handling issues with the vehicle. Most tire manufacturers recommend checking tire pressures before the vehicle has been driven more than 1 mile (1.6 km).

SKILL DRILL 25-1 Using a Tire Pressure Gauge

1. Remove the valve cap from the tire valve.

2. Make sure the graduated sleeve is seated into the gauge body.

3. Then push the tire gauge chuck firmly onto the head of the valve stem.

SKILL DRILL 25-1 Using a Tire Pressure Gauge (Continued)

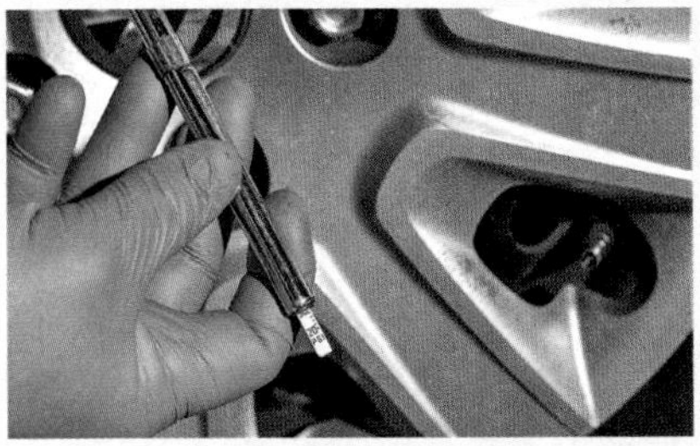

4. Remove the gauge and read the scale to determine the pressure.

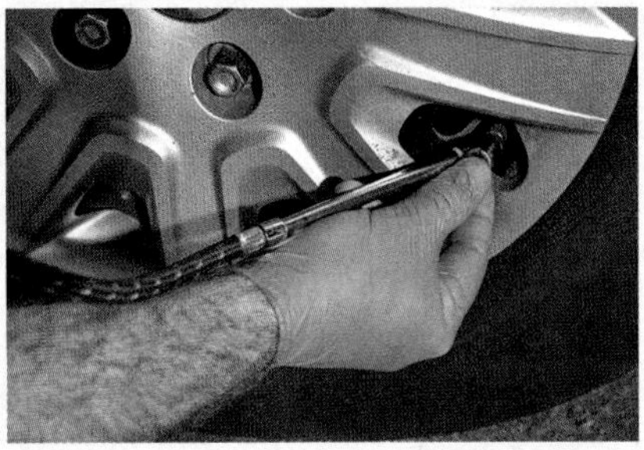

5. Attach the dial pressure gauge to the top of the valve stem. Adjust your hand pressure and angle so that no air escapes.

6. When the needle has jumped, remove the dial pressure gauge from the valve, and read the dial.

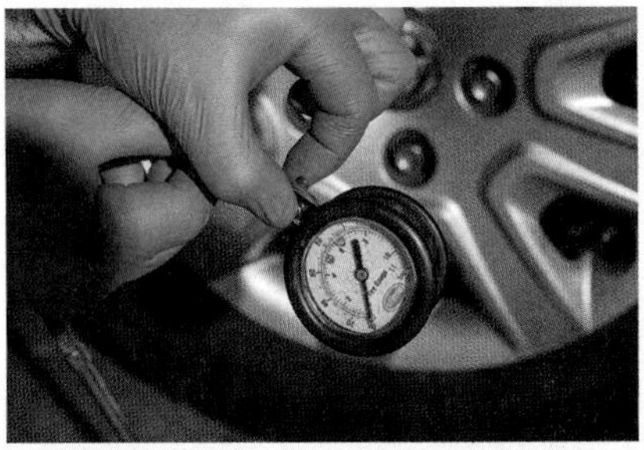

7. Reset the dial pressure gauge to zero by pressing the button on the neck of the dial.

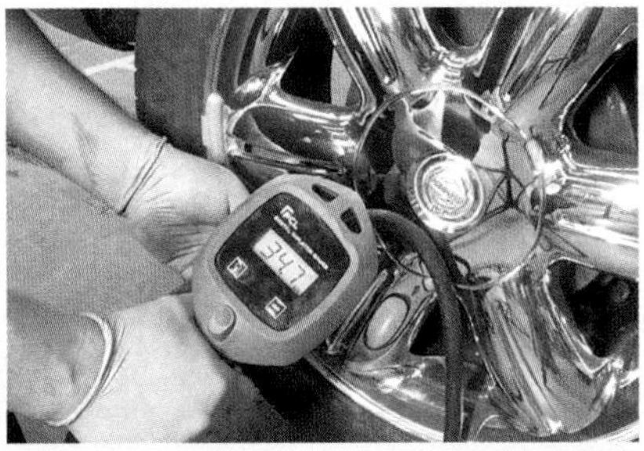

8. Attach the digital pressure gauge to the top of the valve stem. Adjust your hand pressure and angle so that no air escapes. Read the pressure displayed.

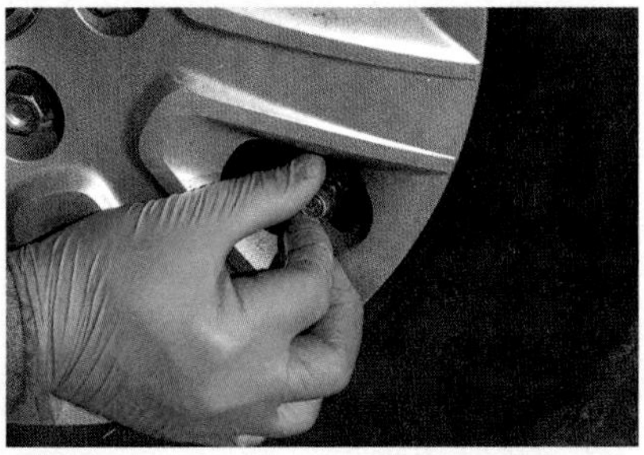

9. Repeat the procedure for all wheels. Remember to replace the valve cap on each wheel as you go.

Adjusting Tire Pressure

Even brand-new high-quality tires lose air over time, so tire pressure should be adjusted periodically. It is good practice to check tire pressure at least monthly to catch a leaking tire. Vehicles that are 2008 model year and newer are required to be equipped with a tire pressure monitoring system. This system is designed to monitor the vehicle for tires with low pressure. If a tire is outside of the specified limits, it will alert the driver of the problem. These vehicles may have a specific tire inflation and TPMS reset procedure required by the manufacturer. So, make sure you investigate that before inflating the tire. At the same time, tire pressures on vehicles with indirect TPMS can lose air slowly and evenly. So, don't assume that the tires are properly inflated just because it has a TPMS.

Many tires are filled with nitrogen instead of regular air. It is best to fill these tires only with nitrogen to avoid introducing oxygen into the tire. But regular compressed air can be used in an emergency. Nitrogen-filled tires can usually be identified by a green cap on the valve stem. Nitrogen-filled tires are best topped off at the manufacturer's dealership or a tire store.

When checking or adjusting tire pressure, check the recommended tire size and pressures on the tire placard. It is usually located on the driver's side door pillar. Compare the tire requirements to the tire information as molded into each tire. Don't be surprised if you find that a customer has installed mismatched tires. If the tire markings do not meet the specifications on the vehicle, inform your supervisor. Never allow a vehicle to be operated with tires that do not meet the vehicle manufacturer's specifications. Doing so can have serious safety repercussions.

> ▶ **TECHNICIAN TIP**
>
> The tire pressures should be checked when the tires are cold (~70°F [21°C]). On average, the pressure in a tire will increase or decrease by about 1 psi for each 10°F (or 12.5 kPa for each 2°C) the tire is above or below its normal operating temperature.

Always check the pressure when the tires are cold. Remove the cap from the valve stem on the first tire. Use a reliable tire gauge to check the air pressure in the tire. A pocket-type pencil, dial, or digital gauge is ideal for this purpose.

To adjust the tire pressure, follow the steps in **SKILL DRILL 25-2**.

SKILL DRILL 25-2 Adjusting Tire Pressure

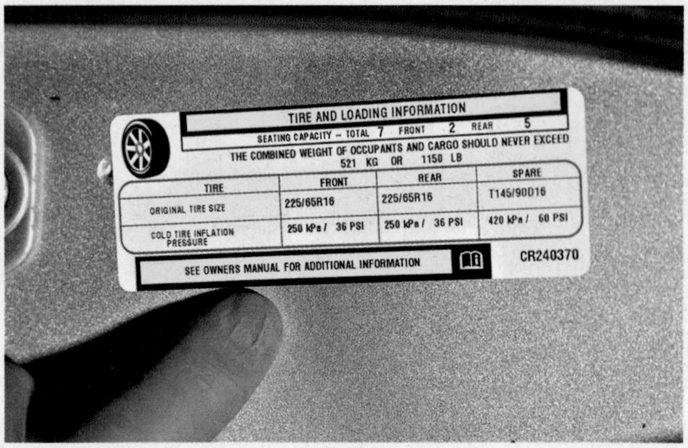

1. Park the vehicle so you can reach all four tires with the air hose. Check the tire placard for the specified pressure and load ratings.

2. Check the tire sidewall markings. Make sure the tires meet the pressure and load-carrying specifications. If not, inform your supervisor.

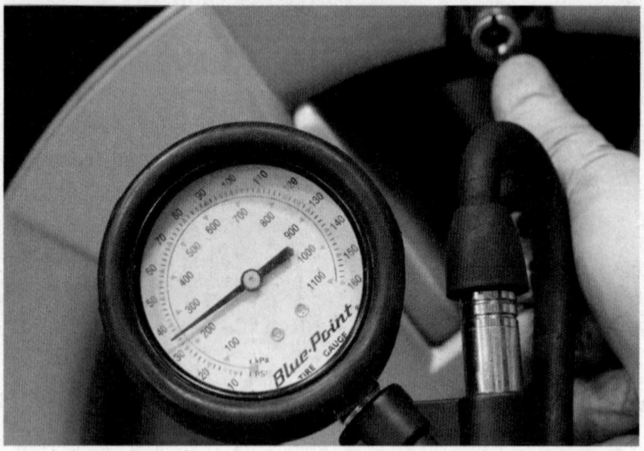

3. Remove the cap from the tire valve on the first tire. Use a reliable tire pressure gauge to check the air pressure in the tire.

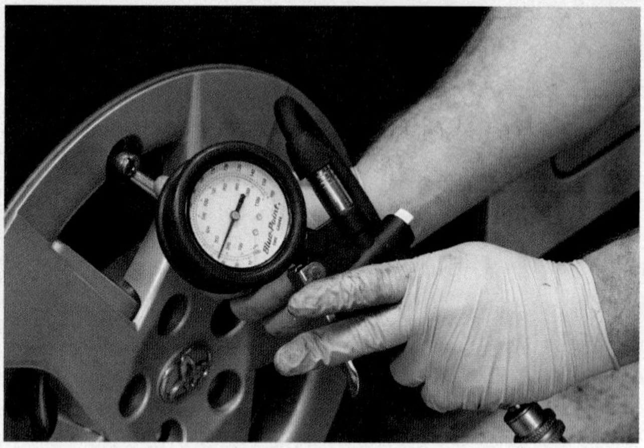

4. If you need to add air, use short bursts with the air hose so you do not overinflate the tire. Recheck the tire pressure after filling it, and replace the cap on the valve stem. Repeat the process for the other tires.

▶ Checking for Tire Wear Patterns

LO 25-03 Identify tire wear patterns and perform tire rotation.

Checking tires for improper wear patterns is generally performed as part of a tire rotation. It is also performed as part of a general vehicle inspection. Regardless of the exact pattern, all four tires should be inspected regularly to ensure they are wearing evenly. Irregular wear patterns are indicative of a problem. Common irregular wear patterns encountered include feathering, one-sided wear, cupping, center wear, and edge wear.

Feathering is observed as the tread ribs having a saw tooth pattern across the tread (**FIGURE 25-8**). This means it has a slightly rounded edge on one side of the ribs and a sharp edge on the other. This condition can be difficult to identify visually. A better option is to run a hand across the tire tread in both

FIGURE 25-8 Tires with feathering wear have a saw tooth wear pattern on the tread.

directions, feeling for the sharp edges. Feathering is most commonly caused by excessive toe-in or toe-out. If the tire's sharp edge is toward the outside of the tread ribs, then the tire is toed out. If there are sharp edges toward the inside of the tread ribs, then the tire is toed in.

One-sided wear refers to ribs on one side of the tire wearing out faster than those on the other side (**FIGURE 25-9**). This type of wear indicates that the wheels are not properly aligned. Cupping is the appearance of dips around the tread. It is usually most pronounced on just one side of the tire tread. But it can also be across the center of the tire. It occurs if the tire is allowed to bounce on the road. This creates uneven friction across the tire which cups the tread (**FIGURE 25-10**). It is usually caused by one or more suspension parts that are worn or bent.

Center wear is when the ribs in the middle of the tire wear faster than those on each side (**FIGURE 25-11**). It results from driving on overinflated tires. Edge wear occurs when the ribs on the outer edges of the tire wear out faster than those in the middle (the reverse of center wear). It indicates that the vehicle has been driven with the tires consistently underinflated. However, it can also be caused by regularly cornering the vehicle at excessive speeds.

Most tires have wear indicator bars incorporated into the tread pattern (**FIGURE 25-12**). Inspect the wear indicator bars. Tires should have at least 2/32" (2 mm) of tread remaining. The wear indicator bars are normally set at this depth. If the tread is worn down to that level or below, the tires are unserviceable and must be replaced.

If the vehicle is driven in wet conditions, many tire manufacturers recommend replacement at 4/32" (4 mm) tread. This helps to maintain traction in wet conditions and providing

FIGURE 25-9 Tires with one-sided wear have uneven wear across the tire tread.

FIGURE 25-10 Tires with cupped wear have dips around the tread.

FIGURE 25-11 **A.** Center wear. **B.** Edge wear.

FIGURE 25-12 Wear indicator bars on a tire.

SKILL DRILL 25-3 Checking for Tire Wear Patterns

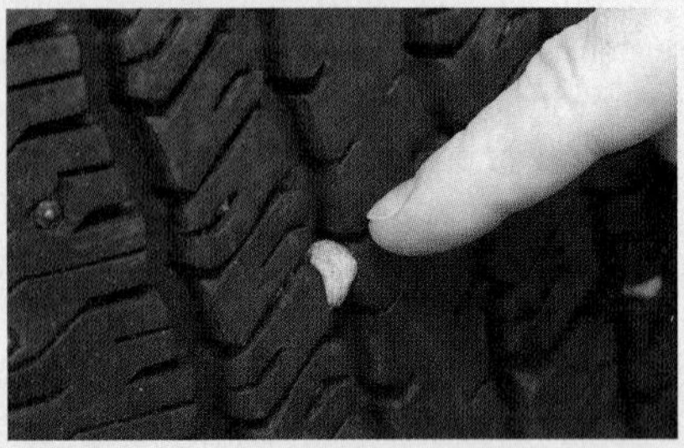

1. Inspect the tires for embedded objects in treads and remove them. If anything penetrates the tread, mark the hole with a tire crayon.

2. Look for signs of wear on all tires. Check the air pressure in the tires (see SKILL DRILL 37-2).

3. Check the tread wear depth. Inspect the wear indicator bars. Tires should have at least 2/32" (2 mm) of tread remaining. If the tread is worn down to that level or below, the tires are unserviceable and must be replaced.

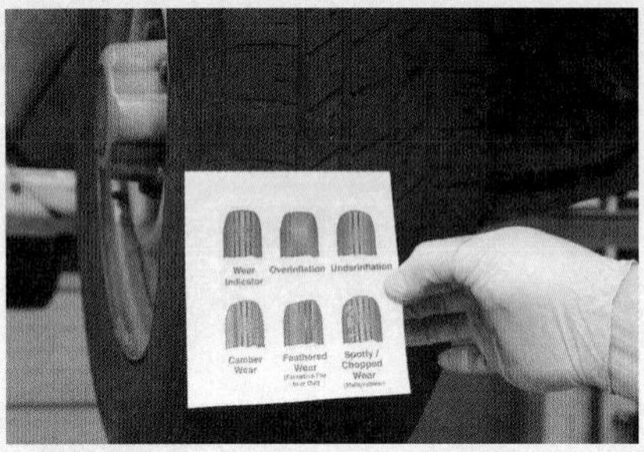

4. Check the tread wear patterns including the spare. Consult the service information to indicate the types of wear that have occurred.

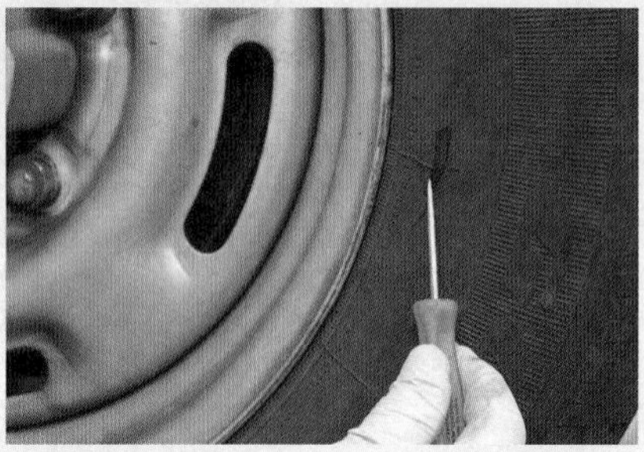

5. Inspect the sidewalls for signs of weather cracking and gouges from impacts with blunt objects. Carefully examine the tread area for separation. This is usually identified as bubbles under the tread area.

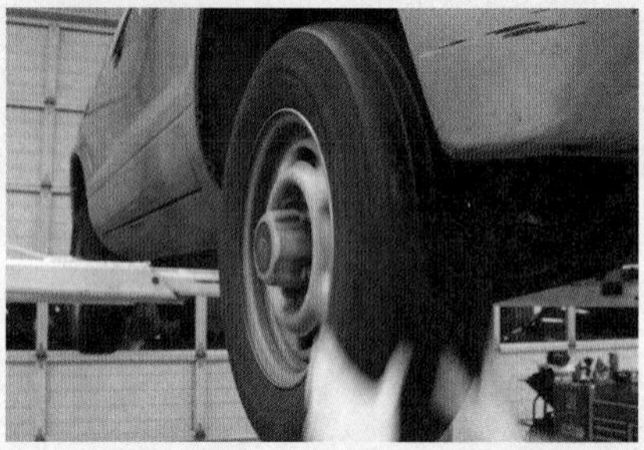

6. Spin the wheel and see if it is running true. If it is wobbling as it rotates, report it to your supervisor.

good stopping ability. If vehicles are driven in snowy or icy conditions, many tire manufacturers recommend replacement at 5/32" (6 mm). This maintains adequate traction under those conditions.

To check for tire wear patterns, follow the steps in **SKILL DRILL 25-3**.

Tire Rotation Pattern

Tire rotation service is the removal and relocation of each tire/wheel assembly on the vehicle. Each tire on the vehicle can wear differently. Regular tire rotation promotes uniform tread wear, extending the life of the set of tires. Each manufacturer designates the proper rotation sequence, depending on:

- whether the vehicle is front-wheel drive, rear-wheel drive, or all-wheel drive,
- the type of tires (e.g., directional), and
- whether a spare tire is involved in the rotation (**FIGURE 25-13**).

In general, a four-tire rotation with nondirectional tires can have one of three rotation patterns: forward-cross, rearward-cross, and X pattern. If the vehicle uses a full-sized spare tire, then the pattern should be one of two five-tire rotations. These are variations of the four-tire forward-cross or rearward-cross pattern.

Directional tires require keeping the tires on the same side of the vehicle. So they generally use a front-to-rear rotation pattern. If the directional tires are differently sized from front to rear, then the tires will have to be dismounted and remounted on the wheels from the other side of the vehicle. This keeps the tire rotating in the proper direction.

Many manufacturers recommend tire rotations at every oil change. This means that the tires generally should be rotated at intervals of approximately 5000 to 10,000 miles (8000 to 16,000 km). When the tires are removed, it is a good time to inspect the brake assembly. Measure the thickness of the brake lining and look for any leaks or damage to the brake assembly.

When reinstalling the lug nuts, make sure the correct side of the lug nut is facing the wheel. This is usually the tapered side that matches the taper in the wheel. Thread them on by hand at least two full turns. Do not put the nut or stud into the socket of an impact wrench and power them on directly. This practice can lead to the lug nuts or studs becoming cross-threaded. Once the lug nuts are started onto the studs, an impact wrench can be used to run the lug nuts down lightly. Do not tighten them with the impact wrench.

Use a torque wrench to tighten the lug nuts to the correct torque. Also, make sure to tighten them in the proper sequence

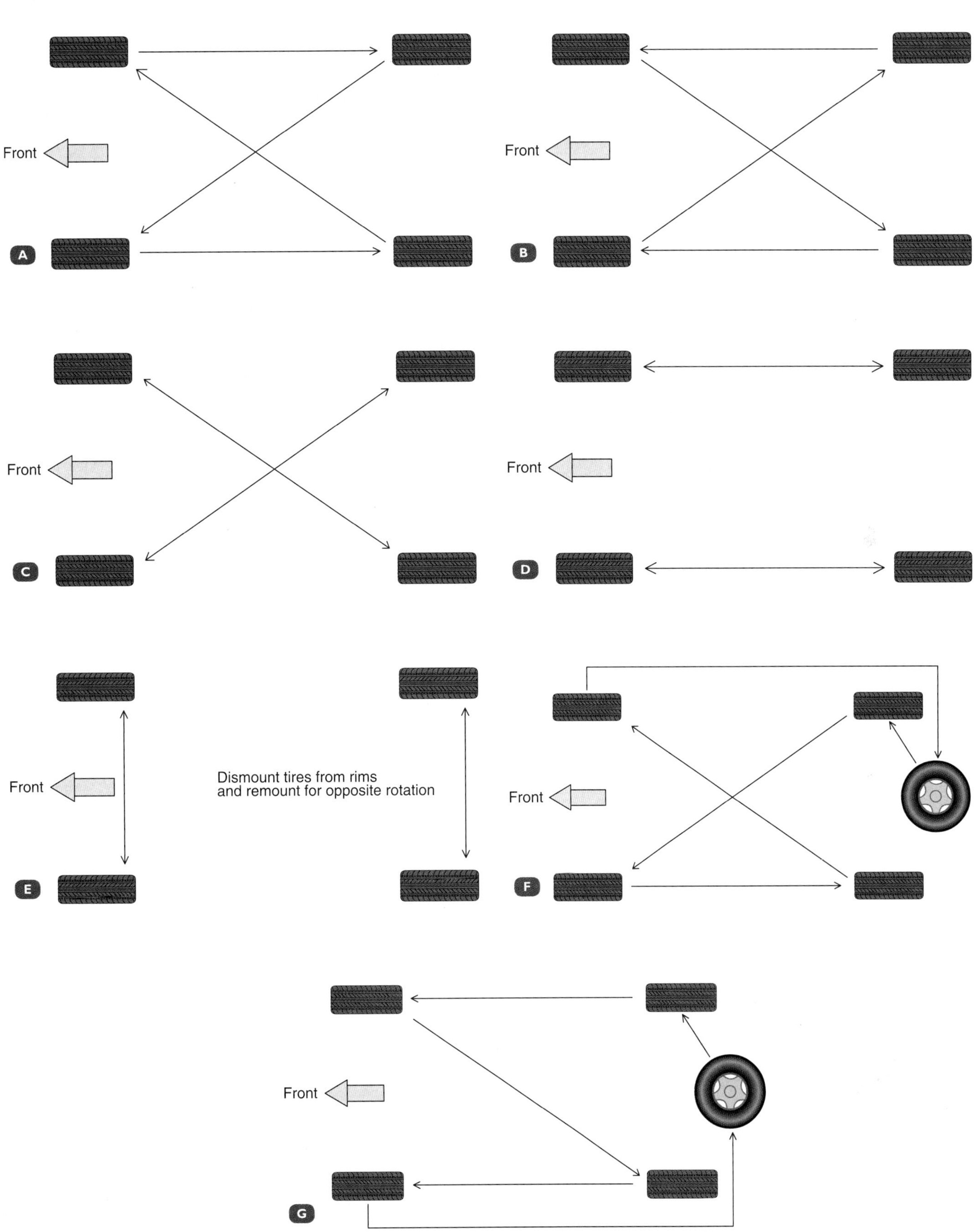

FIGURE 25-13 **A.** Forward-cross pattern. **B.** Rearward-cross pattern. **C.** X pattern. **D.** Front-to-rear pattern. **E.** Side-to-side pattern from the other side of the vehicle. **F.** Five-tire forward-cross pattern. **G.** Five-tire rearward-cross pattern.

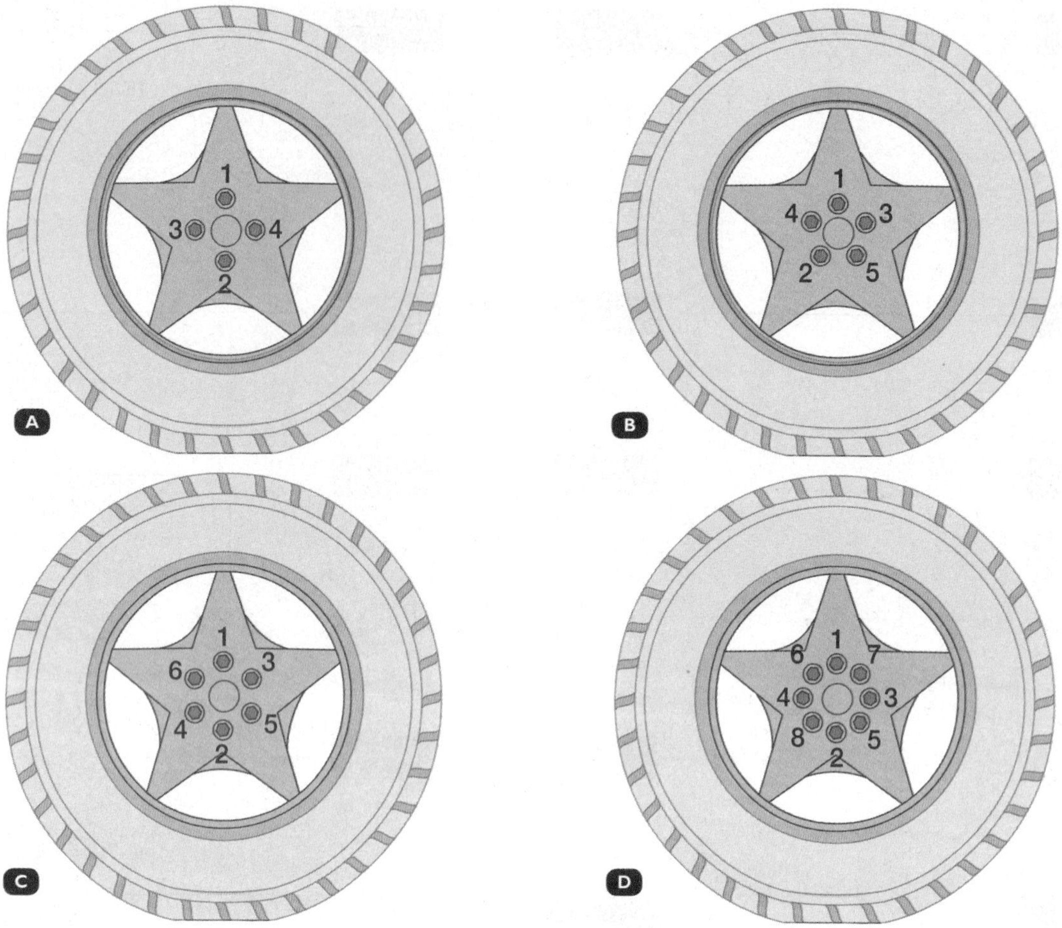

FIGURE 25-14 Typical lug nut torque patterns. **A.** 4-stud. **B.** 5-stud. **C.** 6-stud. **D.** 8-stud.

(**FIGURE 25-14**). This can be found in the service information. Typically, the lug nuts are tightened in a crisscross sequence on four stud wheels. On five-stud wheels, a star arrangement is used. Six- and eight-stud wheels have their own torque patterns. So refer to the service information.

Normally, the lug nuts should be tightened in two stages. First to 50% of the specified torque.Then a second time to 100% of the specified torque. Each stage uses the specified tightening sequence for the particular lug nut pattern.

▶ TECHNICIAN TIP

Lug nuts and lug bolts are designed with a specific grade (i.e., strength or amount of holding force) indicating a certain amount of stretch. Why do they stretch? Through proper torque, the bolt stretches just enough to allow the threads to tightly mate. This prevents them from working loose. Also, the torque listed for lug nuts is a dry torque, meaning that no lubricant should be used. Using a lubricant will cause the lug nuts to be over-tightened, which will likely result in failure.

As with all types of wheels, retorquing lug nuts is typically recommended between 25 and 100 miles (40 to 160 km) after the initial tightening. Always refer to the service information for proper factory specifications. They always take precedence over any general recommendations.

To rotate the tires, follow the steps in **SKILL DRILL 25-4**.

▶ Wheel Balance

LO 25-04 Perform tire balance.

Wheels and tires are mass-produced. This causes small (or sometimes big) imbalances to be manufactured into them. Also, they are not always perfectly round (concentric). But they can still be sold as long as they are within the tolerances specified by the manufacturer.

To complicate matters, tolerance stacking can happen. For example, the wheel hub may have an out-of-round tolerance of 0.005" (0.13 mm). The wheel may have a tolerance of 0.005" (0.13 mm). And the tire may have a tolerance of 0.025" (0.64 mm). If all of the tolerances stack up (one on top of another), then the wheel will experience balance and runout issues.

Wheels and tires that are out of balance generally produce an uncomfortable vibration. This results in premature wearing of suspension and steering components. It also causes uneven tire wear. An out-of-balance tire will cause a vibration at certain speeds, usually between 35 and 70 mph (56 to 113 kph). A tire is out of balance when one section of the tire is heavier, or stiffer, than the others. One ounce of imbalance on a front tire is enough to cause a noticeable vibration in the steering wheel at highway speeds.

SKILL DRILL 25-4 Rotating the Tires

1. Prepare the vehicle by removing any hubcaps or lug nut covers.

2. If using hand tools, break loose the lug or wheel nuts while some of the weight is still on the ground, and then raise the vehicle to a comfortable working position.

3. Remove the lug nuts and place them in a convenient place such as the arm of the hoist.

4. Remove the tire and wheel from the vehicle. Rotate the tires to the new specified position.

5. Reinstall the lug nuts by hand, at least two full turns, making sure the correct side of the lug nut is facing the wheel. Do not put the nut or stud into the socket of an impact wrench and power them on directly.

6. Use a torque wrench to tighten the lug nuts to the specified torque in the proper sequence.

There are two ways wheels can be balanced: off car and on car. Most shops use off-car balancers. However, if the brake rotor, drum, or hub is out of balance, then a balanced wheel cannot compensate for that. So, on-car balancers are sometimes used to balance the entire rotating assembly. The only problem with on-car balancing is that the wheel will need to be rebalanced when it is rotated to another position on the vehicle. This is one reason that most shops use off-car balancers (**FIGURE 25-15**).

Tires and wheels can be out of balance in three ways: static, dynamic, and road force. **Static imbalance** is when the imbalance is centered across the width of the tire (**FIGURE 25-16**). Static balancing does not take into consideration that the tire has width. Thus, static balancing can only address the imbalance as if it were centered at the width of the tire. Therefore, it can be effective only when the point of imbalance is actually centered within the width of the tire. Static imbalance tends to cause the tire to move purely up and down. When performing a static balance, the tire's imbalance is measured when the tire is stationary. Static balancing is no longer an acceptable method of balancing modern-dayvehicles.

Dynamic imbalance is when the imbalance is off-center of the tire width (**FIGURE 25-17**). Dynamic balancing is performed when the tire is rotating. This allows it to take into consideration imbalances that may be off-center. Dynamic imbalance can cause the tire to move side to side as well as up and down. Dynamic balancing is performed by placing specific amounts of weight on each side of the rim. The amount of weight provides the exact counterbalance needed. It can effectively account for an imbalance located anywhere within the volume of the tire/wheel assembly. Dynamic balancing is still common in most shops today.

It is possible that the tire and wheel can be perfectly in balance statically and dynamically. And yet, it can still experience a shimmy or shake that feels just like an out-of-balance tire. This problem is caused by **road force imbalance**. Road force imbalance occurs when the wheel or tire is not concentric. It can also occur if the tire's sidewall has uneven stiffness (**FIGURE 25-18**). As the tire rotates and contacts the surface of the road, any issues with concentricity or sidewall stiffness will push up and down on the vehicle. This causes a vibration similar to a tire that is out of balance. Road force balancing a tire takes into consideration all balance factors related to the wheel and tires. This gives the best ride quality out of all types of tire balancing.

When dynamically balancing a tire using an off-car balancer, the wheel and tire assembly is mounted on the balancer. The wheel is calibrated to the balancer. It is then spun up to speed so that the precise location and amount of the imbalance can be identified. The balancer then determines where and how much weight to add on each side of the wheel. The weights are either hammered into place, or stuck on with adhesive. The wheel is spun up again to ensure that the wheel is in balance.

If a road force balancer is used, then the tire is run up firmly against a roller, and any uneven forces are measured. If an uneven force is encountered, the machine will usually direct you to rotate the tire on the wheel. It will then spin the tire again to see if there are any concentricity issues with the wheel. It can also determine the best location for the tire on the rim to reduce road force fluctuations. After this process is completed, the wheel is reinstalled on the vehicle. And the lug nuts are torqued to the manufacturer's specification.

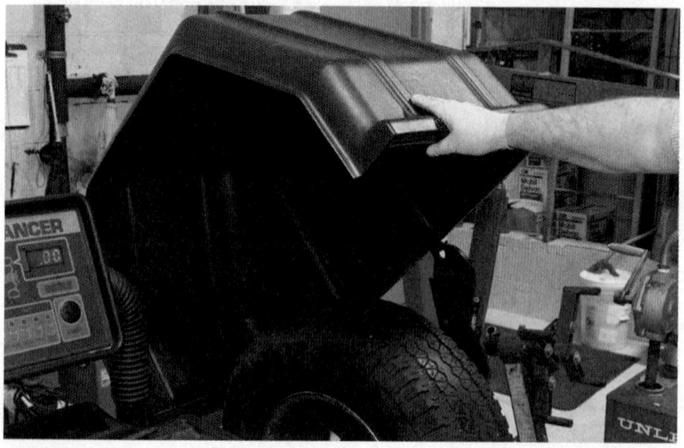

FIGURE 25-15 Off-car balancer.

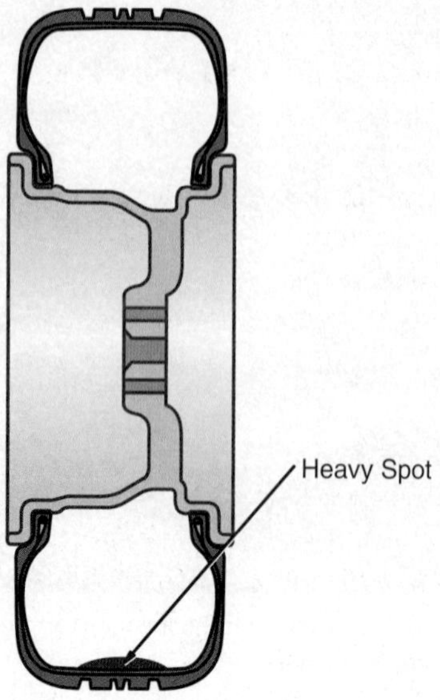

FIGURE 25-16 Static imbalance—the imbalance is centered across the width of the tire.

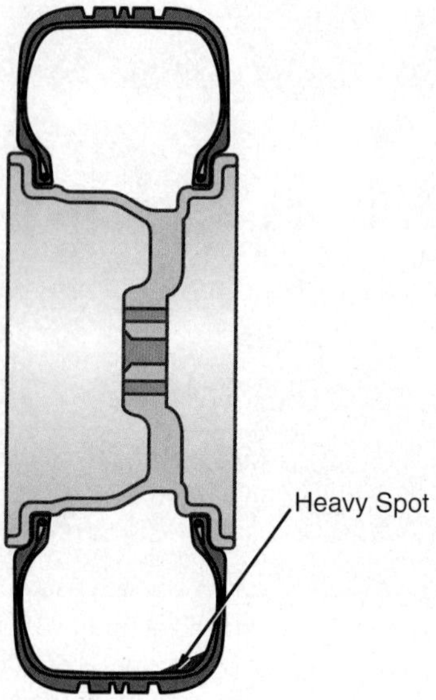

FIGURE 25-17 Dynamic imbalance—the imbalance is off-center of the tire's width.

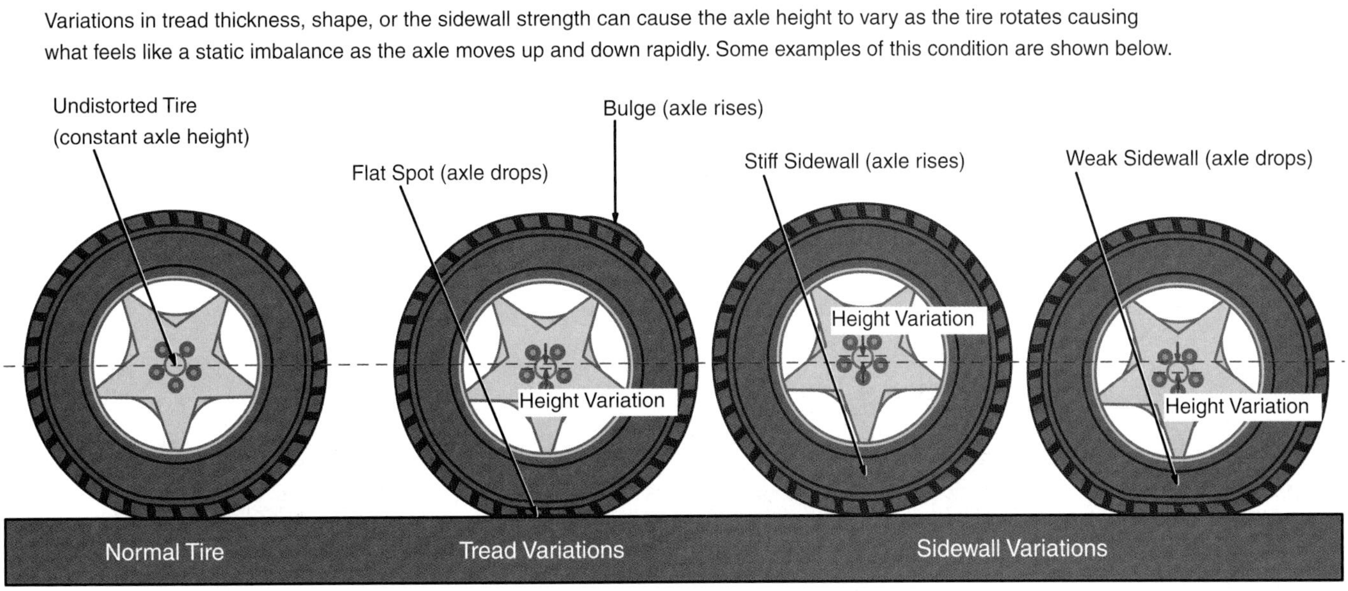
Variations in tread thickness, shape, or the sidewall strength can cause the axle height to vary as the tire rotates causing what feels like a static imbalance as the axle moves up and down rapidly. Some examples of this condition are shown below.
Undistorted Tire (constant axle height)
Flat Spot (axle drops)
Bulge (axle rises)
Stiff Sidewall (axle rises)
Weak Sidewall (axle drops)
Height Variation
Height Variation
Height Variation
Normal Tire
Tread Variations
Sidewall Variations

FIGURE 25-18 Road force imbalance—when the sidewall is either stiffer, or taller, in one place.

> ▶ **TECHNICIAN TIP**
>
> Not all shimmies and shakes are caused by tire balance issues. Sometimes belts inside the tire carcass break, creating a lump in the tire. Every time the lump hits the road, the tire pushes back. This really is just an extreme case of road force imbalance. But the tire will need to be replaced.

The majority of top-quality tires hold their balance reasonably well. This assumes that the vehicle is not driven over potholes. If the driver notices a vibration after a tire balance, it is possible that one of the balancing weights fell off. If the driver feels the vibration mostly in the steering wheel, the problem is most likely in a front wheel. If the vibration is mostly in the driver's seat, the problem is probably in one of the rear tires.

Applied Science

AS-1: Force, Balanced/Unbalanced: The technician can demonstrate an understanding of the role of balanced and unbalanced forces on linear or rotating vehicle assemblies.

When a tire is fitted to a wheel, two imperfectly weighted components are assembled together. They form a heavy, large-diameter rotational component. The chances of this assembly having perfect weight distribution are small. This confirms the importance of adding weight in specific areas to balance the assembly.

A car wheel rotates on a central axis, supported by bearings. When a wheel turning at high speed is out of balance, forces are created perpendicular to the rotational axis. This is the case either statically or dynamically. These forces are typically experienced as a vibration through the steering wheel if the front wheels are out of balance. However, a vibration may also be felt through the vehicle's body if the rear wheels are out of balance. When a wheel is in perfect balance, no forces exist to interrupt the smooth rotation around the central axis.

FIGURE 25-19 Wheel weights come in a variety of styles, materials, and coatings. So, make sure you use the proper one for the wheel you are working on.

Wheel weights come in a variety of styles to fit different rim configurations. This means that you will need to match the type of weight needed for a particular wheel (**FIGURE 25-19**). Weights come in 0.25-ounce or 5-gram increments. Although wheel weights have been primarily made of lead for decades, several states have outlawed the use of lead wheel weights. This is because of the potential environmental hazards of lead. In those states, steel weights coated with zinc or another protective layer are being used instead.

It is good practice to always use new wheel weights when balancing a tire. Do not reuse the old wheel weights. They are likely to be thrown from the wheel when driven at freeway speeds. If the weights you remove are made of lead, dispose of them properly.

Dynamic Balancing a Tire

A tire that is dynamically in balance is in balance when it is spinning as opposed to when it is stationary. Dynamic imbalance occurs when a spot on either the inside or the outside of the tire's centerline is heavy. This induces a side-to-side imbalance in the tire as it rotates. As a result, a vibration is created, which can cause the steering wheel to shimmy. Dynamic imbalance is usually a result of manufacturing variations. But it can be caused by a damaged tire or wheel. Obviously, dynamic imbalance reduces ride quality. It also tends to increase wear on the tires, and on steering and suspension system components. Dynamic balancing of the tires should be performed when new tires are installed on the vehicle. It is also performed any time that tire imbalance is suspected.

Dynamic balancers are capable of spinning the tire and then measuring the location of any dynamic imbalance. Some balancers are spun at low speed, but others are driven by the balancer at higher speeds. If the tire is spun by the balancer, embedded objects may fly off the tire, so it is important to wear safety glasses. If the wheel balancer is fitted with a safety hood, ensure that it is in place when the wheel is being rotated. This will further protect against flying objects.

All wheels require one of several specific designs of wheel weights. Incorrect weights can fly off when the vehicle is driven down the road. This can cause possible injury or vehicle damage. It is good practice to use new wheel weights when balancing a wheel for the same reason. If the vehicle has directional tires, ensure that they are reinstalled in their correct positions.

To balance a tire, follow the steps in **SKILL DRILL 25-5**.

SKILL DRILL 25-5 Balancing a Tire

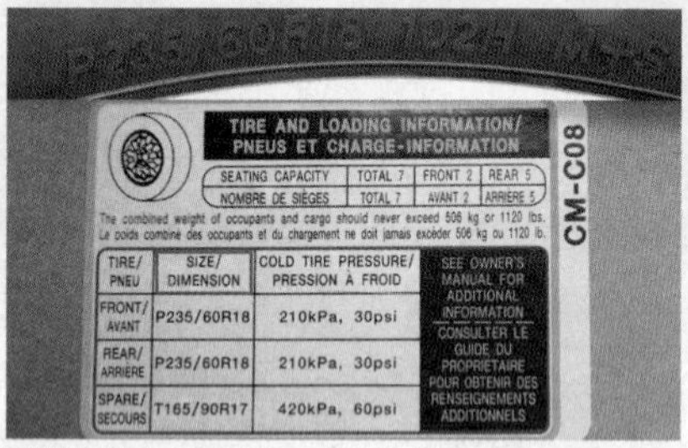

1. If using hand tools, prepare the vehicle by loosening the lug nuts. Then safely raise and support the vehicle. Check that the tires fitted to the wheels on the vehicle are the appropriate size and rating for the vehicle.

2. Mark the inside of the wheel or tire in relation to its location on the vehicle, and then remove it.

3. Check and adjust the tire pressure before balancing the tire. Mount the wheel and tire on the balancer, putting the inside part of the wheel toward the balancer in most cases. Secure the wheel by screwing the hub nut assembly on the balancer shaft.

4. Calibrate the balancer to the wheel by following the manufacturer's instruction manual. Input any data into the balancer's computer, if fitted. If no computer is fitted, set the balancer adjustments manually according to the instruction manual.

5. If equipped, lower the safety hood over the wheel. Spin the wheel.

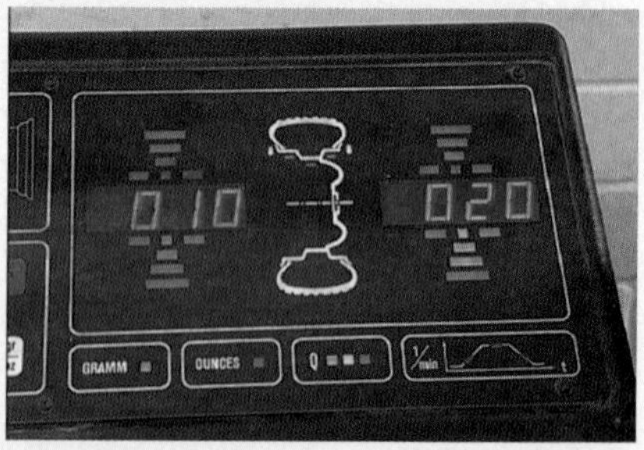

6. Read the balancer's analysis. If the wheel is out of balance, you should remove the old weights and recheck the balance of the wheel before adding new weights.

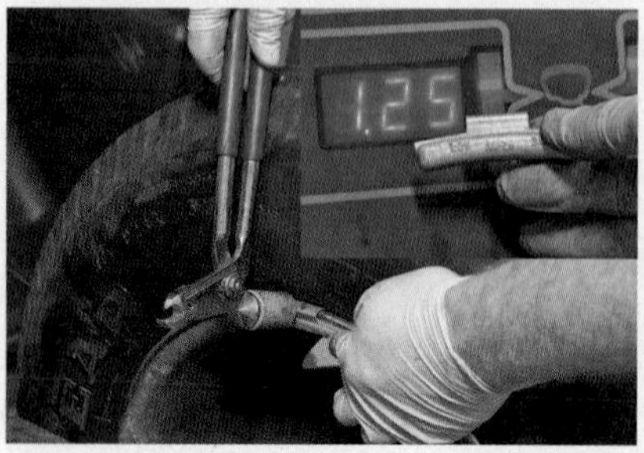

7. Install new weights as recommended by the machine's display.

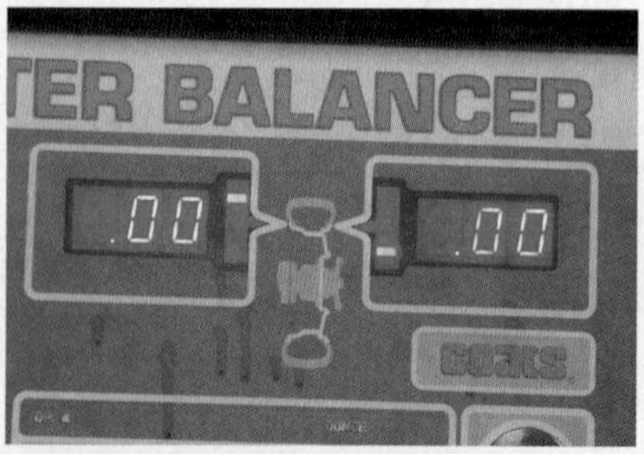

8. Respin the wheel to confirm that balance has been achieved. Repeat the process for the rest of the wheels and tires. Reinstall the wheels and tires to the vehicle.

▶ Dismounting a Tire

LO 25-05 Dismount and mount a tire without TPMS.

Tires generally have to be removed from a rim for only a few reasons. This includes replacing old tires with new tires, patching a leaky tire, and possibly switching between snow tires and regular tires. Because the bead of the tire seals against the wheel flange, repeatedly removing a tire from its wheel risks damaging the bead's sealing surface. So it is best to remove the tire only when absolutely necessary.

Dismounting a tire is usually performed on a tire machine, which is very powerful. So, extreme care must be exercised. Tire machines are strong enough to break the bead loose from the rim. They also hold the rim while the bead is forced over the flange during removal and installation. This means the tire machine has several operations that you must become familiar with. Different machines have different operating parameters. So, make sure you understand the operating procedure for the machine you are using.

The turntable jaws on the tire changer can usually hold the rim by grasping it from the outside or the inside. When mounting an alloy wheel, it is normal that the wheel be clamped from the outside. If the wheel is clamped from the outside on the tire machine, it is necessary to release the clamps before fully inflating the tire. Most steel rims are clamped from the inside. Always use the method that won't damage or scar the rim you are working on.

To dismount a tire without TPMS, follow the steps in **SKILL DRILL 25-6**.

SKILL DRILL 25-6 Dismounting a Tire Without TPMS

1. Before removing the tire, check to see if the wheel is equipped with a TPMS sensor. If it is, follow the manufacturer's procedures. Inspect the tread and sidewalls for any sharp cords sticking out that could injure you. If there is any damage, the tire should be discarded.

2. Check the wheel for any balance weights, and pry them off with the wheel weight tool.

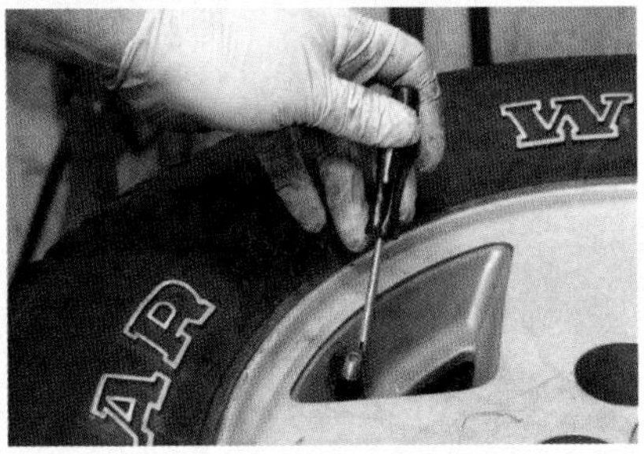

3. Locate the valve stem, unscrew the dust cap, and store it for later use. Using the valve core tool, unscrew the valve core carefully.

4. Once all the air has been removed from the tire, locate the wheel in the bead breaker, with the outside of the rim facing toward the blade. Locate the blade close to the edge of the rim while keeping your hands at a safe distance. Activate the bead breaker, which will force the tire bead away from the edge of the rim and over the safety ridge.

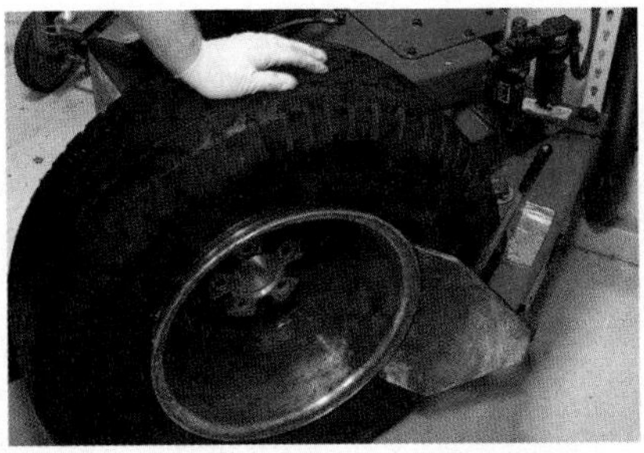

5. Release the blade, turn the wheel one-third to one-half of a turn, reposition the blade, and release this section of the tire as well. Release the blade. Roll the tire away from the machine, and reposition it with the inside of the rim facing toward the blade. Break the bead like the other side.

6. Set the wheel on the turntable with the shallow dish side of the wheel facing up. Position the wheel and tire assembly on the turntable, and activate the jaws so the wheel is centered. Activate the turntable to verify that the wheel is centered and securely held.

SKILL DRILL 25-6 Dismounting a Tire Without TPMS (Continued)

7. Lubricate the top bead with tire lubricant. Make sure you get the lube on the flat bead seat, not just the side of the bead.

8. Position the bead remover against the edge of the rim; if necessary, adjust it so it has the proper clearance between the rim and the roller.

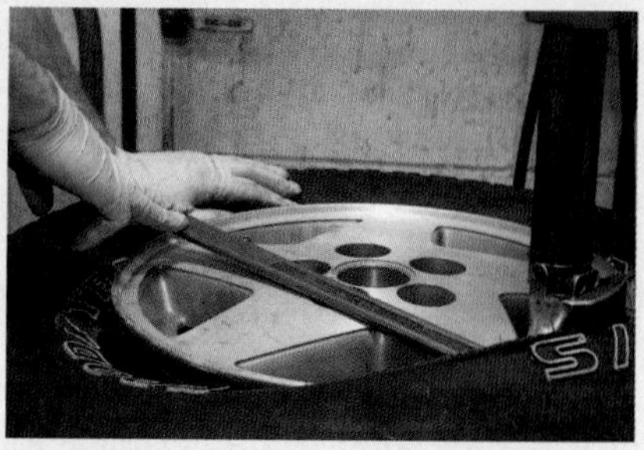

9. Use the tire lever to pry the top bead over the bead remover knuckle while pushing down on the sidewall on the opposite side of the tire.

10. Activate the turntable so the bead is guided off the rim. Once the bead is removed, stop the turntable, lift the tire slightly, and remove the tube, if fitted. Guide the lower bead into the drop center, and using the tire lever, pry the lower bead over the knuckle. Activate the turntable. The tire will come off the rim. If replacing the tire with a new one, remove the valve stem by either unscrewing it or using a valve stem tool. Discard the valve stem.

SAFETY TIP

For your safety, follow these tips:

- An inflated tire is a pressure vessel that must be treated with care and respect. Always fully deflate the tire before performing any repair tasks.
- Keep your hands clear of the bead breaker when it is operating. It applies great force to the tire that will cause you a severe injury if your hand is trapped.
- Always use correct lifting techniques when lifting a tire on and off the tire changer and vehicle.
- The tire tread comes in contact with many unknown substances that are transferred directly to your hands. For this reason, it is recommended that you wear protective gloves when handling tires. If you do not wear gloves, wash your hands after the tire change is completed and before eating.

Mounting a Tire

In modern-day vehicles, mounting a tire means more than just installing a properly sized tire onto the rim. Because tires do not come perfectly round and balanced, most tire manufacturers indicate the tire's highest point with a red dot. They also indicate the tire's lightest point with a yellow dot (**FIGURE 25-20**). The red dot should be lined up with the rim's lowest point, which is called **match mounting**. The yellow dot should be matched up with the rim's heaviest point, which is called **weight matching**. These points occur innately as part of the manufacturing process. It is impossible to create an absolutely perfect, even, round tire. Slight variations in density and shape of the tire are unavoidable.

Matching up the tire and rim using one of these methods will help avoid "tolerance stacking." This is when tolerances in mating parts are aligned such that the tolerances are added together, rather than canceling each other out. In the case of wheels and tires, this occurs when the tire's heavy or high points are aligned with the wheels' heavy or high points.

FIGURE 25-20 Red dot on tire indicates the highest point of the tire. Yellow dot indicates the lightest point on the tire.

In addition, the type of tread—directional, symmetric, and asymmetric—affects the direction in which tires are mounted on the wheels. It also affects which side of the vehicle they are installed on. Directional tires are designed to operate better in one direction. This requires that they be mounted accordingly. Symmetric tires can be operated in either direction, so they can be mounted in either direction. Asymmetric tires are generally side-specific, but they may also be directional. In this case, specific tires must be mounted on certain rims.

Do not fit a tire that is too wide or too narrow for the wheel. Check the tire manufacturer's recommendation for the correct range of wheel sizes for a particular tire. And never mix inch-based tires with metric-based wheels.

Inflating a tire for the first time is always dangerous. Some types of tires, such as split rims, must be inflated inside of a tire cage. Doing so contains the tire if it blows up while being inflated. Always follow the manufacturer's guidelines and shop policies when inflating a newly installed tire.

To mount a tire without TPMS, follow the steps in **SKILL DRILL 25-7.**

SKILL DRILL 25-7 Mounting a Tire Without TPMS

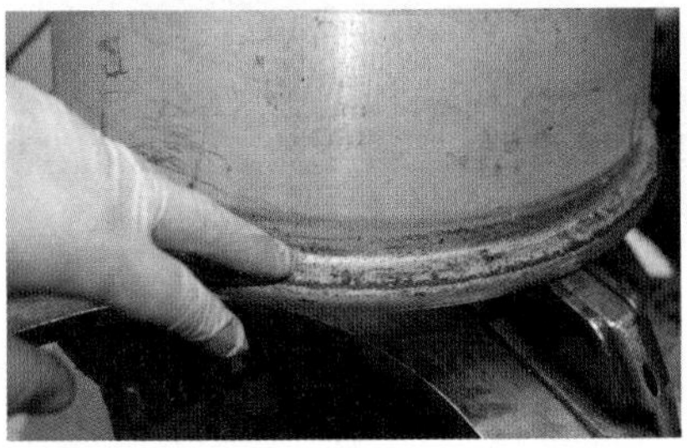

1. Mount the wheel to the tire machine. Examine the wheel, and remove any rust or dirt from the rim bead seat.

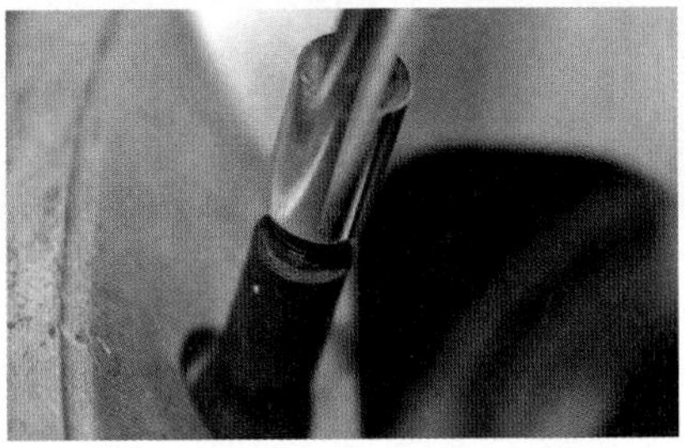

2. Select the correct type of tubeless valve stem, lube it with tire lube, and insert it through the hole in the rim from the inside. Using the valve stem tool, pull the stem through until its groove locates in the hole. Use the valve core tool to unscrew the valve core from the valve stem.

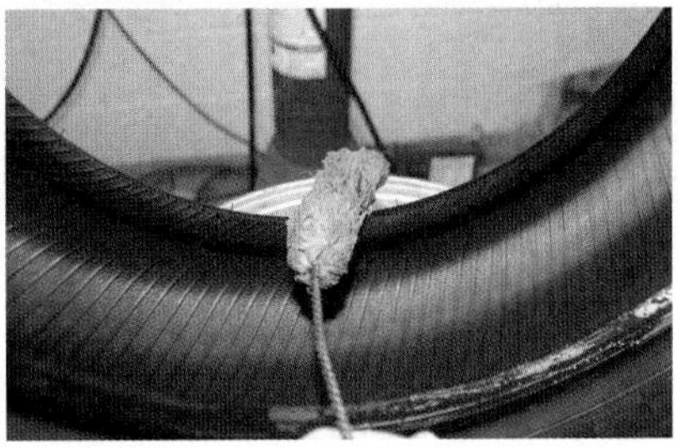

3. Apply some lubricant to the tire bead.

4. Position the tire on top of the rim so that a portion of the lower bead is positioned in the drop center while keeping the lower bead in the tire machine guide.

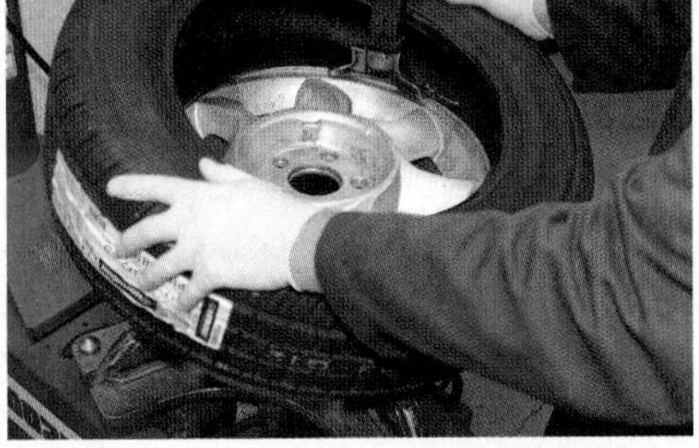

5. Activate the turntable and guide the lower tire bead onto the rim.

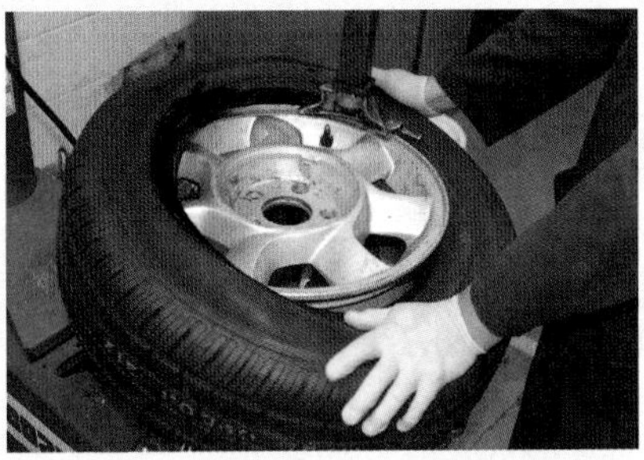

6. Once the lower bead is fitted, position the upper bead into the guide while holding the other side of the upper bead in the drop center.

SKILL DRILL 25-7 Mounting a Tire Without TPMS (Continued)

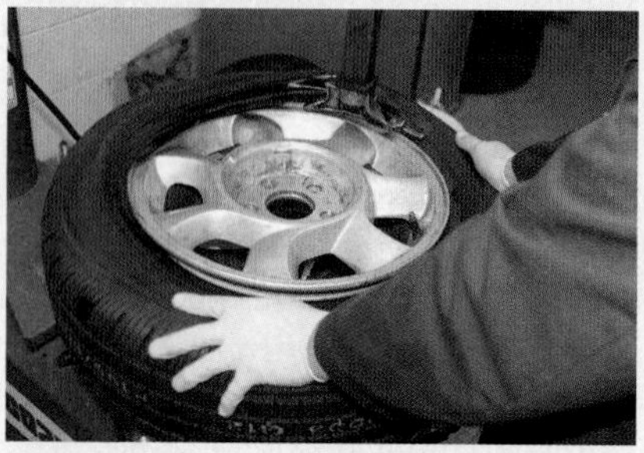

7. Activate the turntable and guide the tire onto the rim. As the turntable rotates, push the sidewall down, keeping your fingers clear of the rim, so that the tire bead is guided below the safety ridge into the drop center.

8. Attach the tire inflator chuck to the valve stem. Stand clear of the tire and inflate it, being careful to not exceed 30 psi (207 kPa) if both beads have not seated against the rim. If they have not seated by 30 psi, deflate the tire and inspect the rim and tire for damage. If they are okay, relube the tire and rim and reattempt to inflate the tire. If it still will not seat the beads by 30 psi, inform your supervisor.

9. Check the location of the bead indicator ridge to make sure the bead is fully seated. If the rim is clamped from the outside, it will be necessary to release the clamps so the tire can inflate fully.

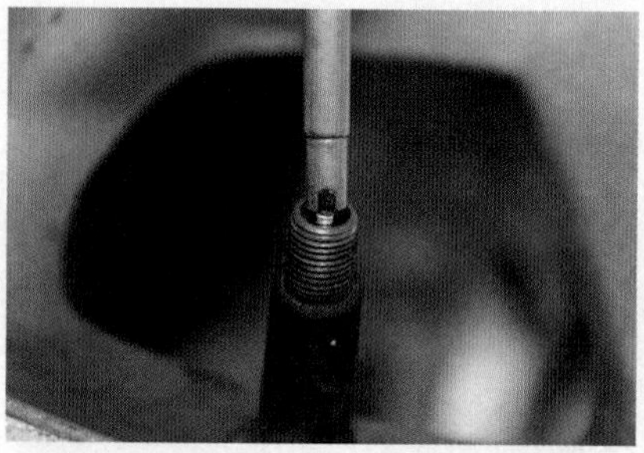

10. When the beads are properly seated, remove the tire inflator chuck, keeping your hands and face clear of the valve stem opening. Ice crystals can shoot out of the stem. Once the tire has completely deflated, screw the valve core into the valve stem, using the valve core tool.

11. Reattach the inflator, stand clear, and inflate the tire to the correct pressure as listed on the vehicle's tire placard. Be careful never to exceed the maximum tire pressure listed on the tire sidewall.

12. Use a soft brush to apply a small amount of soapy water to the bead and valve stem. If there are any air leaks, they will be indicated by bubbles.

▶ TECHNICIAN TIP

Always remove any balance weights from both sides of the rim before mounting it on the tire changer. If they are not removed, the bead remover could drag the weights around the rim, causing damage to the rim face. This is particularly important with alloy rims. The damage done by the balance weights is not repairable. This usually requires the rim to be replaced, costing the shop a lot of money.

SAFETY TIP

Overinflated tires can explode. Do not inflate the tire to a pressure greater than what is listed on the sidewall. The possibility of an explosion is why tire inflators have a spring-loaded (dead man's) trigger that does not lock into position. When the tire is being inflated, use a tire cage, if required. Or use an inflator that allows you to stand clear of the tire. Keep hands and body well away from the tire. When a tire explodes, the tire, rim, or components from the tire changer may cause serious injury or death to any person nearby.

▶ Dismounting, Inspecting, and Remounting a Tire on a Wheel Equipped with a TPMS Sensor

LO 25-06 Dismount and remount a tire with TPMS.

Remember that only direct reading TPMS systems have pressure sensors in the wheels. Indirect TPMS systems use the wheel speed sensors. Knowing the type of system the vehicle has will help you to not damage the sensor.

TPMS sensors need to be treated carefully when tires are being removed and reinstalled. Sometimes normal corrosion can damage the TPMS sensors. But more likely, a careless technician can damage the TPMS sensors by not removing the tire properly. If a sensor is damaged during a tire change, then the entire TPMS sensor unit will generally need to be replaced.

There are two general ways that TPMS sensors are mounted in the wheel. The first way involves it being attached to, or integrated with, the valve stem. So, care must be exercised when working around the valve stem area. It is easy to break the sensor with either the tire shovel, the tire bar, or the tire bead.

Begin with removing the air from the tire. If the TPMS sensors are integrated into a screw-in valve stem, it is usually safer to push the valve stem and senor into the tire. Do this by unscrewing the valve stem locknut. The valve stem and sensor can then be carefully pushed into the tire for safekeeping during disassembly. If this method is used, the tire can be dismounted normally. Just remember to remove the valve stem and sensor from the tire once the tire is off the rim.

When breaking the tire bead with the shovel, position the valve stem in the 6:00 position. If that doesn't break the bead all the way around, reposition the tire so the valve stem is in the 12:00 position. Then, use the shovel to break the remaining portion of the bead. Just be careful to not operate the shovel so the tire bead moves beyond the center line of the wheel. If it moves further than this, the sensor can be damaged.

The next step is for removing the tire from the rim. The location of the TPMS sensor will determine what needs to be done to protect the sensor. On vehicles with the TPMS sensor mounted to a rubber press-fit valve stem, the sensor is left in place during tire removal. To avoid damage, position the tire in the specified position on the tire machine before removal. This is typically with the valve stem located to the left (far side) of the tire machine's bead removal knuckle. The tire lever can then be used to carefully pry the top bead over the bead removal knuckle. The tire machine table can then be operated to remove the top bead from the rim.

Once the top bead is removed, reposition the wheel so the valve stem is again to the left (far side) of the bead removal knuckle. Use the tire lever to carefully pry the bottom bead over the bead removal knuckle. Operate the tire machine table to remove the bottom bead of the tire.

The other way TPMS sensors are attached to the wheel is by a band. It fits all the way around the drop center of the wheel. It holds the TPMS sensor firmly to the inside of the wheel. In many cases, the band-style TPMS sensor is positioned 180 degrees away from the valve stem. So, care must be exercised on the side opposite of the valve stem. This is done by positioning the valve stem at about 5:30 position relative to the bead removal knuckle. Use the tire lever to pry the top bead over the bead removal knuckle. Operate the machine table to remove the top bead of the tire.

Once the top bead is removed, reposition the rim so the valve stem is again at about the 5:30 position relative to the bead removal knuckle. Use the tire lever to carefully pry the bottom bead over the bead removal knuckle. Operate the tire machine table to remove the bottom bead of the tire.

To install a tire on a wheel with a screw-in TPMS valve stem, mount the tire normally. Make sure the sensor is properly mounted on the valve stem if it is the screw-in type. This includes tightening it to the proper torque. Place a new grommet or seal over the valve stem. Once the tire is installed on the wheel, push the top bead of the tire down and place the valve stem assembly up through the hole in the wheel. Place a new grommet or seal and lock nut over the valve stem. Tighten the lock nut to the specified torque.

To install a tire on a wheel with a press-fit valve stem and TPMS sensor, follow the instructions above. But lube up the side of the valve stem and pull it into the hole in the rim. Make sure it is fully seated in the rim.

To install a tire on a wheel with a band-style TPMS sensor, position the valve stem hole directly under the bead removal knuckle. Place the bottom bead of the tire in the knuckle. Hold the bottom bead of the tire in the drop center. Operate the tire machine table to install the bottom bead. Reposition the tire with the valve stem hole under the knuckle. Install the top bead like normal. If the tire spins on the rim, stop. Then, reposition the rim again, and try to install the top bead of the tire.

When mounting and dismounting tires with TPMS sensors, always follow the manufacturer's specified procedure. Failure to do so can ruin costly TPMS sensors. It may also delay finishing the job due to having to wait for replacement parts.

To dismount, inspect, and remount a tire on a wheel equipped with a TPMS sensor, follow the steps in **SKILL DRILL 25-8**.

SKILL DRILL 25-8 Dismounting, Inspecting, and Remounting a Tire with a TPMS Sensor

1. Following the specified procedure, remove the wheel from the vehicle, and deflate the tire by removing the valve stem lock nut if a screw-in stem. If a press-fit valve stem, remove the valve core.

2. Break the tire beads in the positions specified. Usually position the valve stem at 6:00 and 12:00 while you break the bead at 3:00 and 9:00.

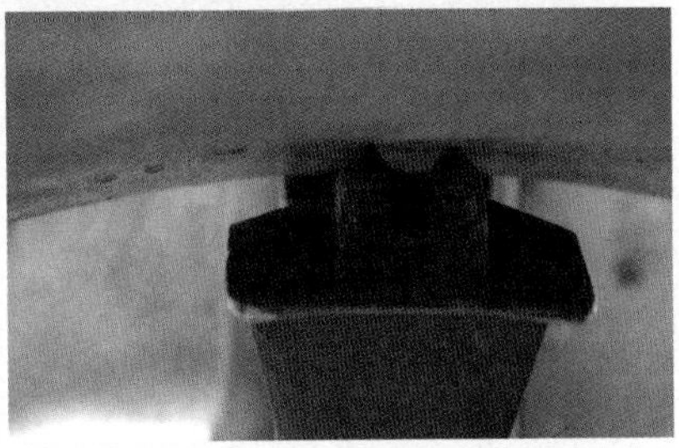

3. Position the tire assembly on the turntable with the shallow side of the wheel up, and engage the jaws to lock the wheel in place.

SKILL DRILL 25-8 Dismounting, Inspecting, and Remounting a Tire with a TPMS Sensor (Continued)

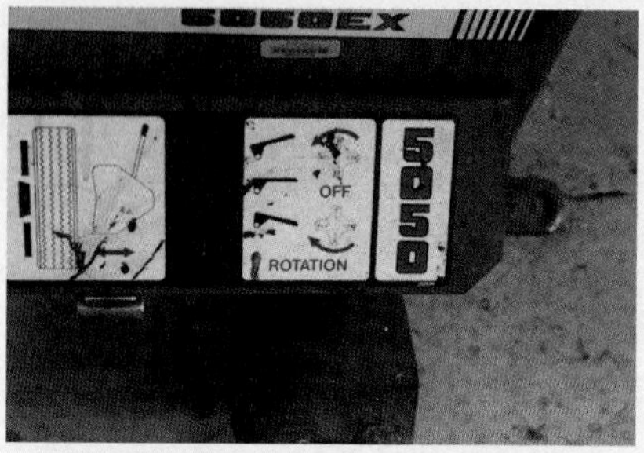

4. Activate the turntable to verify that the wheel is centered and securely held. Stop it in the specified position for removing the top bead.

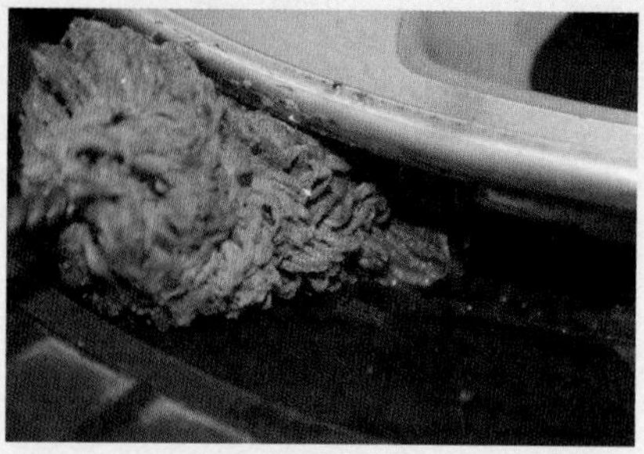

5. Lubricate both beads with tire lubricant. Make sure you get the lube on the flat bead seat, not just the side of the bead.

6. Position the bead remover against the edge of the rim. If necessary, adjust it so it has the proper clearance between the rim and the roller. Use the tire lever to pry the tire bead over the bead remover knuckle while pushing down on the sidewall on the opposite side of the tire.

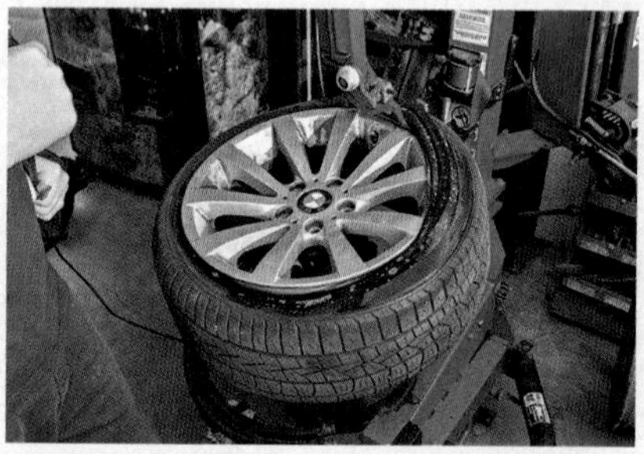

7. Activate the turntable while lifting up on the tire directly behind the remover knuckle to help work the top bead over the flange of the rim.

8. Rotate the turntable to the position specified for removing the lower bead.

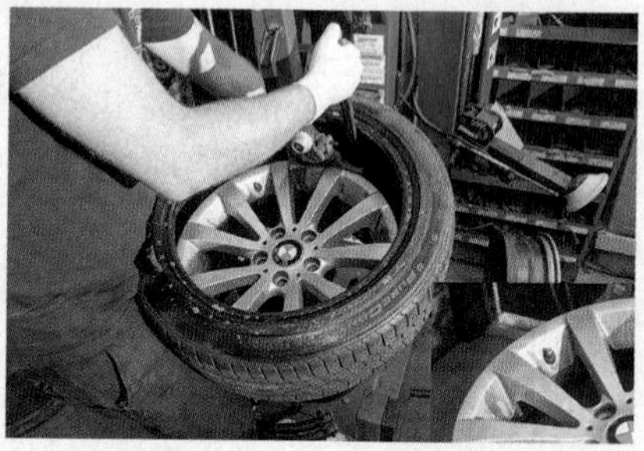

9. Carefully pry the lower bead over the knuckle while holding the other side of the tire up in the drop center.

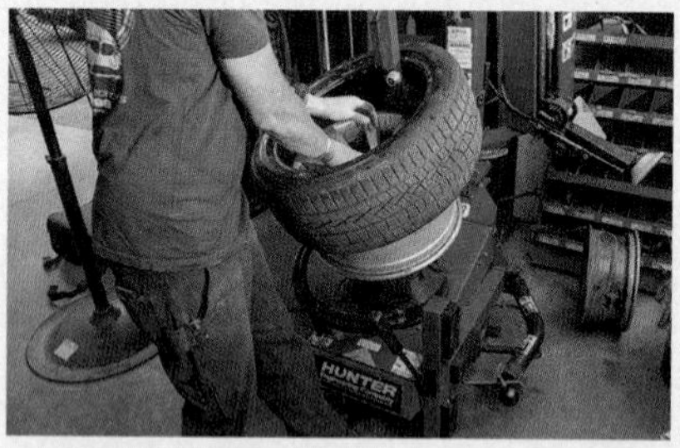

10. Activate the turntable while lifting up on the tire directly behind the remover knuckle to help work the bottom bead over the flange of the rim. Inspect the tire, rim, and TPMS sensor according to the manufacturer's procedure.

11. To install the tire, adequately lube both bead seats of the tire with tire lube. Activate the turntable to position the valve stem in the specified position for installing the lower bead.

12. Position the lower bead into the drop center while holding the lower bead in the tire machine shoe.

SKILL DRILL 25-8 Dismounting, Inspecting, and Remounting a Tire with a TPMS Sensor (Continued)

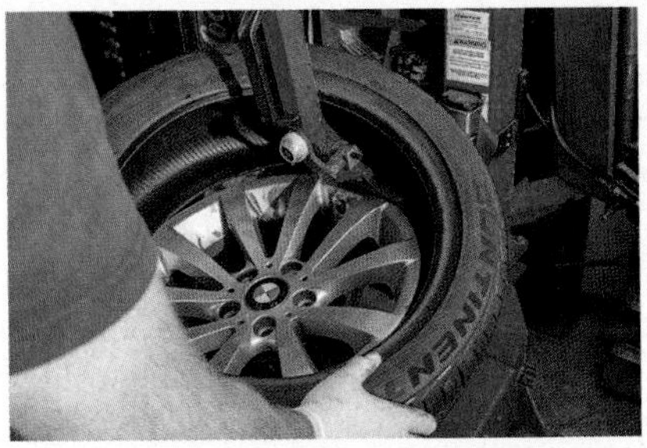

13. Activate the turntable while helping keep the lower bead in the drop center.

14. Activate the turntable to position the valve stem in the specified position for installing the upper bead.

15. Place the upper bead into the drop center while holding the upper bead in the tire machine shoe.

16. Activate the turntable while pushing the top bead into the drop center.

17. Install the TPMS valve stem

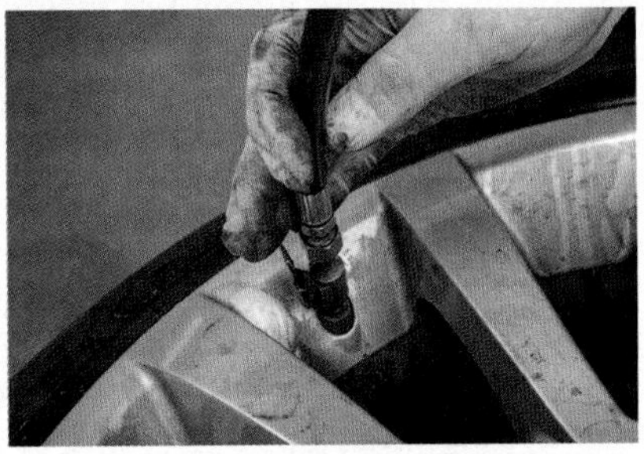

18. Inflate the tire as a non-TPMS wheel, and check for leaks.

▶ Replacing a Valve Stem

LO 25-07 Replace tire valve stems.

Valve stems come in a few styles. These include rubber press-fit valve stems, screw-in valve stems, and TPMS sensor-integrated valve stems (**FIGURE 25-21**). Rubber press-fit stems are normally replaced when new tires are mounted on the wheels. Screw-in valve stems are normally not replaced. But they may have rubber washers or O-rings that may require replacement when new tires are installed. TPMS-integrated valve stems may be made as part of the valve stem. Or they may screw onto, or snap into, the valve stem, which is replaceable.

It is important to know whether the TPMS sensor is connected to the valve stem. This is so you do not damage it during tire removal. Many TPMS valve stems have nickel-plated valve cores. They should not be replaced with brass valve cores. They should always be torqued with a valve core torque wrench. TPMS sensor replacement will be covered in another section.

Valve stems can also be damaged fairly easily. This can be caused by trauma from loose objects on the road. It can be from scraping into curbs or other low structures, or through improper tire changes. Luckily, valve stems are generally inexpensive and relatively easy to replace. To change the valve stem, the wheel must be removed from the vehicle. The tire is then deflated and the top bead broken loose. On some tires, the sidewall can be pushed down by hand to gain access to the valve stem. In other instances, the tire must be removed from the rim.

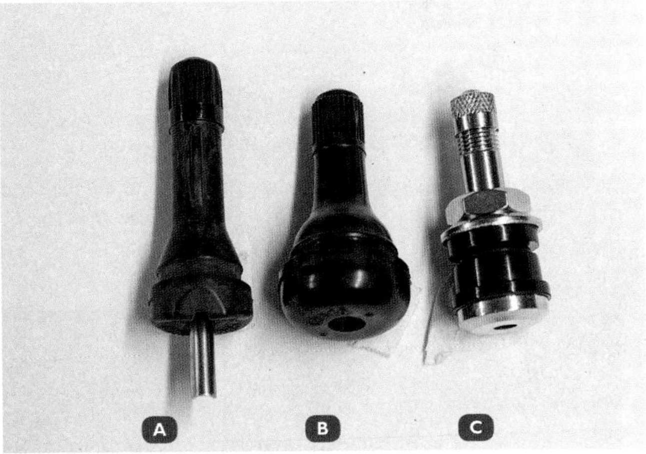

FIGURE 25-21 Types of valve stems. **A.** TPMS sensor instegrated. **B.** Rubber press-fit. **C.** Screw-in.

After removing the old rubber valve stem, a new one is put into place, with some lubricant around the base of the stem. The stem is then pulled into place using the proper installing tool. The tire is reassembled and inflated to the recommended pressure. Spray some soapy water around the valve stem and core to detect air leaks.

The other type of valve stem is the screw-in valve stem. It has a threaded shank with a locking nut and sealing washers. The wheel is removed from the vehicle and deflated, and the top bead is broken loose. The nut on the valve stem is then unscrewed, and the stem is removed from the inside of the wheel. The old sealing washers can be replaced with new ones. Then, the valve stem can be reinstalled in the wheel. In some cases, only the valve stem core requires replacement. In this case, the old core is unthreaded, and the new core is installed. Make sure to use a nickel-plated valve core if the valve stem holds the TPMS sensor. Also, tighten it to the specified torque. The tire pressure is then reinflated to the specified pressure.

To replace a rubber press-fit valve stem, follow the steps in **SKILL DRILL 25-9**.

To replace a screw-in valve stem, follow the steps in **SKILL DRILL 25-10**.

▶ Servicing TPMS sensors

LO 25-08 Perform TPMS service.

The TPMS sensor unit is a sealed component. Currently, TPMS sensors are powered by an internal battery which is designed to last 7–10 years. But they can last up to 12 years. At that time period, the battery or the entire sensor will need to be replaced. If it is one of the rare vehicles with a replaceable battery, the battery can be replaced (**FIGURE 25-22**). If the battery is not accessible, like in the vast majority of vehicles, then the entire sensor will need to be replaced. In fact, you will likely want to recommend replacing all four sensors. This is because they would all have similar life spans.

When replacing TPMS sensors, you will need to be aware of specific procedures. First, if it is a band-style TPMS sensor, you will need to install it in the proper location in the drop center of the rim. This is usually 180 degrees opposite of the tire valve stem (**FIGURE 25-23**). Mounting it in the wrong position will likely result in it being damaged when the tire is removed the next time.

When mounting a TPMS sensor that is screwed onto the tire's valve stem, use the new screw provided. Also, make sure you tighten it to the specified torque (**FIGURE 25-24**). This will

SKILL DRILL 25-9 Replacing a Rubber Press-Fit Valve Stem

1. Remove the wheel from the vehicle, deflate the tire, and break the top bead using the tire machine.

2. Screw the valve stem tool onto the old rubber press-fit stem.

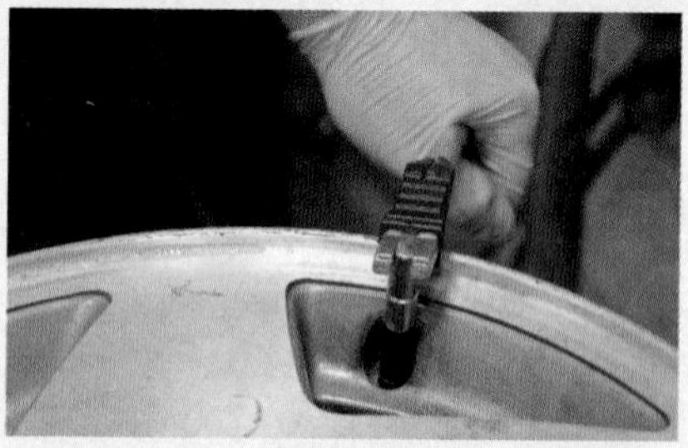

3. Pry the rubber press-fit valve stem from the wheel. Use one hand to hold onto the portion of the valve stem in case it breaks off.

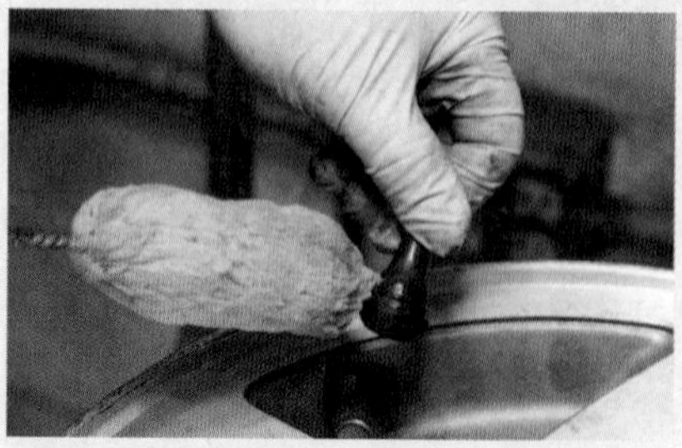

4. Clean and inspect the hole in the wheel. Clean any rust or corrosion with some sandpaper or other appropriate tools. Lubricate the new valve stem with tire lube.

5. Insert the valve stem into the hole in the wheel from the inside, remove the cap, and screw the valve stem tool onto the threaded end of the stem.

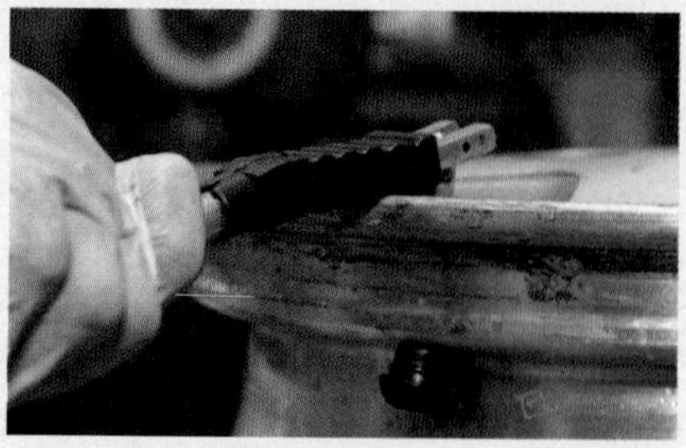

6. Use the handle of the valve stem tool as a lever to pull the retaining ridge through the hole in the wheel, and verify that the valve stem is properly installed.

SKILL DRILL 25-10 Replacing a Screw-In Valve Stem

1. Remove the wheel from the vehicle, deflate the tire, break the top bead using the tire machine, and mount the wheel on the machine.

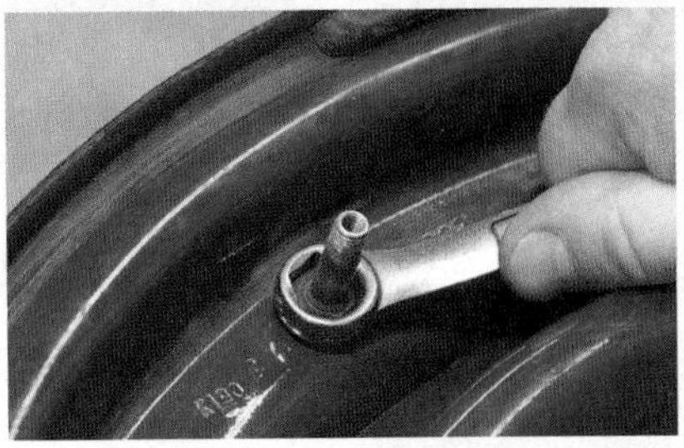

2. Unscrew the nut that holds the valve stem to the wheel. Remove the valve stem from the inside of the wheel.

3. Discard the old sealing washers, and replace them with new ones. Place the screw-in valve stem with one new sealing washer through the hole on the inside of the wheel.

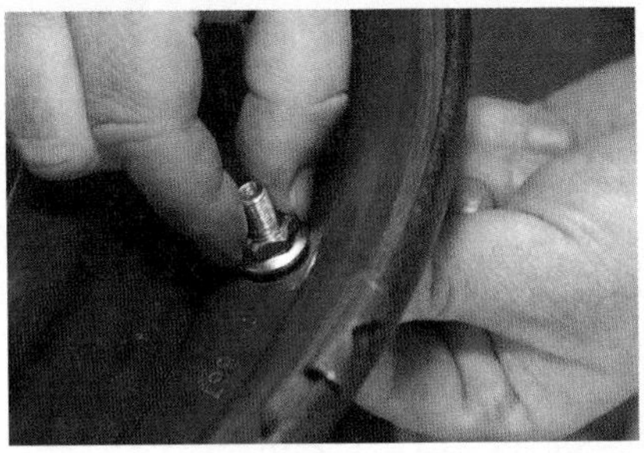

4. Place a new sealing washer over the valve stem, and thread the nut on by hand. Tighten the nut to the specified torque. Inflate the tire to the specified pressure.

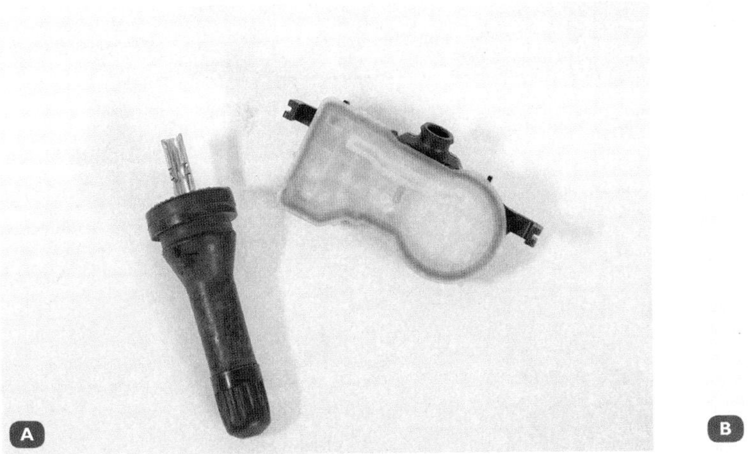

FIGURE 25-22 TPMS sensor with: **A.** Replaceable battery. **B.** Non-replaceable battery.

prevent it from loosening up down the road. It will also prevent it from being stripped out or broken off as it is tightened.

On systems where the TPMS sensor is bolted to, or molded into, the valve stem, you have additional steps to take. First, if it is a screw-in valve stem, you will need to use new grommets to seal the stem to the wheel. If it is a press-fit valve stem, replace the entire valve stem (**FIGURE 25-25**). Also, since the housing is made of aluminum, you must use a nickel-plated valve core (**FIGURE 25-26**). That will prevent galvanic reaction between the metals. Do NOT use a brass valve core. The valve core also needs to be tightened to the specified torque. This torque is a very small amount, measured in inch-pounds. Be careful,

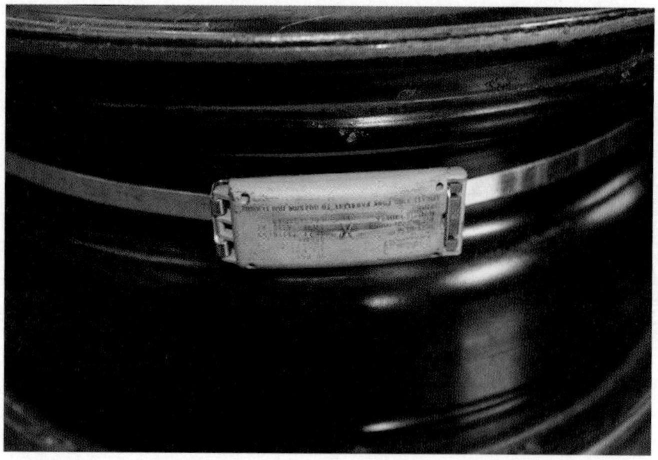

FIGURE 25-23 Most band-style TPMS sensors should be installed 180 degrees away from the valve stem.

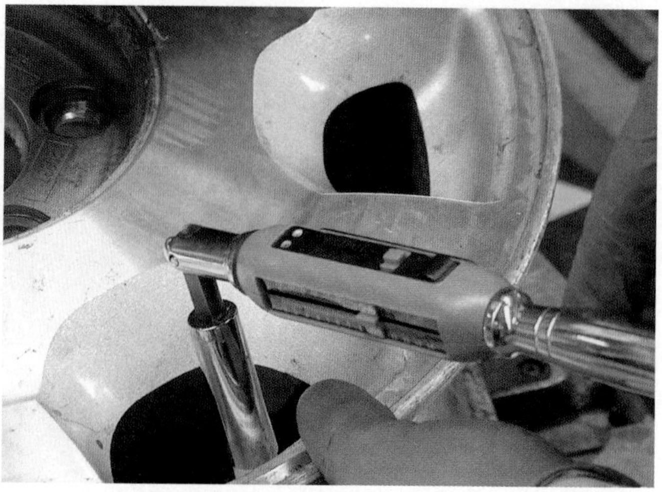

FIGURE 25-24 Always tighten the TPMS sensor screw to the specified torque.

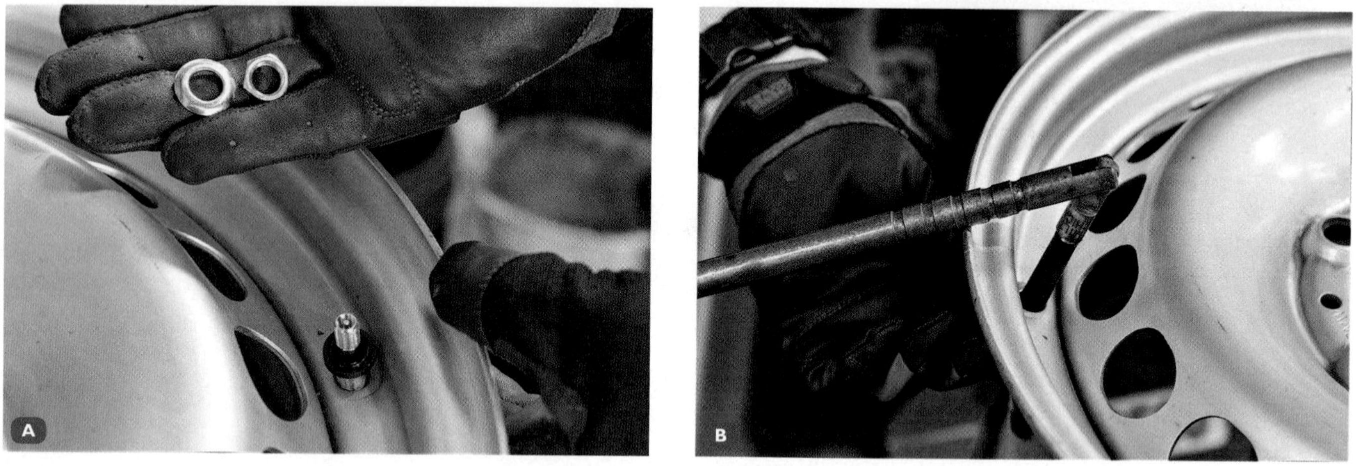

FIGURE 25-25 A. Replace the grommets on screw-in valve stems. **B.** Replace the entire valve stem on press-fit valve stems.

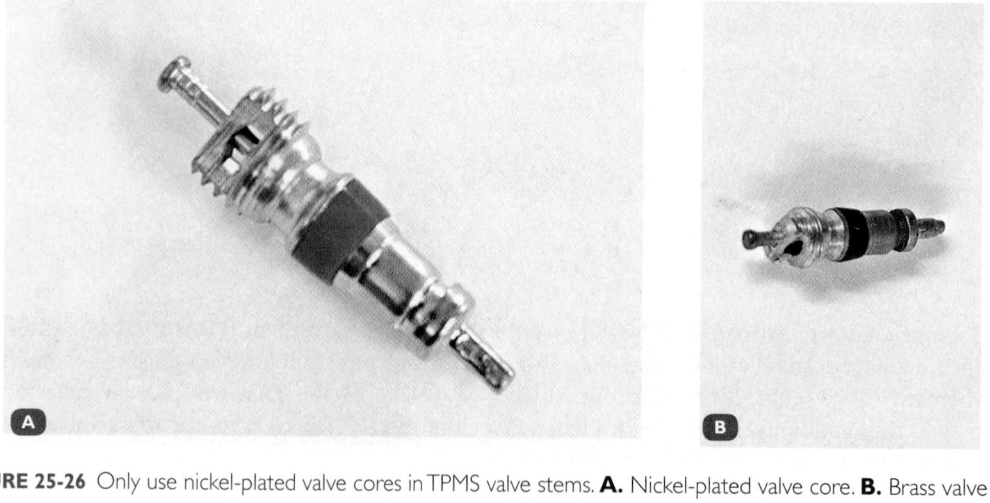

FIGURE 25-26 Only use nickel-plated valve cores in TPMS valve stems. **A.** Nickel-plated valve core. **B.** Brass valve core.

FIGURE 25-27 Use the specified tool to tighten the valve core to the specified torque.

as they are easily stripped or broken. Use the specified tool to tighten it (**FIGURE 25-27**).

It is also important to use either a plastic or nickel-plated valve stem cap (**FIGURE 25-28**). A chrome-plated cap will corrode to the stem. This makes it likely that the stem will be damaged while removing the cap. Also, be careful not to overtighten the valve stem cap. Doing so can break off the aluminum housing fairly easily.

Inspecting and Performing a Relearn on the TPMS

Inspection and relearning of the TPMS is required in several general situations. Inspection is needed whenever the tires have been dismounted from the wheels. In some cases, the TPMS sensor batteries need to be replaced and the sensor mounts need to be inspected. Or more likely, the sensors need to be replaced if they aren't equipped with replaceable batteries.

Diagnosis of system faults is required when the system detects a fault and turns on the warning light. A quick check of the system is made by turning the key to "run" and observing the tire pressure warning light or indicator. If the warning light stays on, this could be caused by one or more tires that are not at the proper pressure. This requires an inspection of the tires for leaks or incorrect pressures.

If the warning light is on, it could also indicate a fault in the system. A scan tool capable of communicating with the TPMS is used to interrogate the system. This allows you to determine if fault codes are present. If they are, then the service information will guide you through the diagnostic process.

When tires are rotated, the positions of the TPMS sensors move with the wheels. So, the TPMS needs to relearn the actual positions of the sensors after the tire rotation. This is called relearning. Some TPMS systems will automatically relearn after being driven for a period of time. In this case, the shop may want to test drive the vehicle to perform this relearn procedure. Or the customer may decide to perform the relearn by driving the vehicle themselves.

FIGURE 25-28 A. Always use either plastic or nickel-plated valve stem caps. **B.** Not chrome-plated caps.

A majority of vehicles will need to have a manual relearning procedure performed. This procedure can vary from manufacturer to manufacturer. But we will cover a couple of common ways it is performed. But no matter which type of manual relearning is performed, the tires first need to be set to the proper pressure. They also need to be mounted to their respective axles.

The magnet method uses a special magnet that fits over each of the valve stems, one at a time. The TPMS system is placed into relearn mode. Manufacturers do this in different ways. One way is by turning the ignition key on with the engine off. Then pressing both the lock and unlock buttons on the key fob. This puts the system into relearn mode. An initial chirp or honk is heard. Then, the magnet is placed over the left front valve stem. Another chirp/honk sounds when the TPMS system recognizes the signal from that wheel. The magnet is moved to the right front valve stem using the same process. After the chirp/honk, move to the right rear. And finally, the magnet is moved to the left rear. This should result in the relearn of all four TPMS sensors. Once complete, verify that the system is operating normally as indicated earlier.

Another way to perform the relearn procedure is with a TPMS scan tool. This method is needed if the vehicle's system can't retrieve the TPMS sensor IDs. It can also be used instead of the magnet method, if desired. The new sensor IDs can then be programmed directly into the vehicle's system. It is best to follow the instructions on the scan tool while performing this procedure.

Another way is the pressure-drop method. It starts by over-inflating the tires to a specific pressure. Then activating the relearn mode. Once in relearn mode, air is let out of the left front tire until the chirp/honk of the horn. Then move to each of the other tires, right front (RF), right rear (RR), and left rear (LR), and releasing air one at a time until the chirp. After the last tire, there will be two chirp/honks showing it is finished. The key is then turned off, and the tires inflated to their proper pressure.

To inspect, diagnose, and calibrate the TPMS, follow the steps in **SKILL DRILL 25-11**.

▶ Tire Diagnosis

LO 25-09 Perform tire diagnosis and repair.

Diagnosis of wheel and tire faults is a fairly common task for automotive technicians. Tires are subjected to much abuse compared with some of the other components on a vehicle. Following the strategy-based diagnostic process is key to identifying the cause of the fault. The first step is to verify the customer concern. It is best to make sure you have a good understanding of the concern. You may be able to get enough information from the repair order or the shop's customer concern questionnaire. If not, then discuss the concern further with either the service advisor or the customer.

Next, take the vehicle for a test drive if the vehicle is safe to do so. That way you can verify the concern and get clues to what is causing it, such as location and type of a noise or vibration. You will likely have to testdrive the vehicle in a variety of ways to glean as many clues as possible. Just don't violate any driving laws or company policies.

The next step is to research the possible faults. This includes researching TSBs and the service information. Use the clues gathered from the previous step to guide your research. Once the research is complete, you will need to use that to build an attack plan for diagnosing the fault. Your plan will have to take into account the following considerations:

- The most likely cause based on your test drive and research
- The test most likely to give you the most results in narrowing down the cause of the fault
- The order in which you will perform the tests based on ease and cost of performing
- How you will verify the cause once its source is found or suspected

SKILL DRILL 25-11 Inspecting, Diagnosing, and Calibrating the TPMS

1. Turn the ignition key to the run position and observe the TPMS warning light. If the light indicates low tire pressure at any wheel, check the pressure of that wheel with an accurate tire pressure gauge. If incorrect, adjust the pressure, using compressed air.

2. If the light indicates a system fault, connect an appropriate scan tool to the system and read any DTCs. Research DTCs in the appropriate service information, and follow the diagnostic steps listed to identify the cause of the fault.

3. Once diagnosis and repairs are complete, follow the specified relearn procedure. Here, a magnet is used to close the centrifugal switch in the sensor, identifying its location to the PCM.

Once you have a plan in place, it is time to perform focused testing to locate the root cause of the concern. This could be achieved by performing one or more of the tests that follow in this section. Each time you perform a test, you should know what you are expecting to find, and whether the part you are testing is the cause of the fault or not. This expectation comes from understanding the service information, as well as by the experience you are gaining from performing the maintenance tasks.

Once you have identified the root cause of the fault, get authorization to perform the repair. This could be replacement of damaged or worn parts or repairs such as balancing a tire, etc. Just make sure you perform the repair properly. There is very little room for errors when working on tires and wheels. Shortcuts or errors put passengers, others, and property at risk of severe injury.

Once the repair has been performed, verify that it corrected the fault. Initially, you can do this by performing the same test that identified the fault. If that checks out okay, then testdrive the vehicle to verify that it is operating properly. In some cases, you will have fixed the fault only to discover a secondary fault. For example, a vehicle may have a severe vibration caused by a tire with a broken belt. But when you testdrive it after the repair, the vehicle may pull to one side because of an alignment issue. The faulty tire may have been masking the alignment issue. This is another reason why vehicles should be testdriven after the repair. Next, we start to cover some of the tests and repairs for which you will need to gain experience.

Inspecting the Wheel Assembly for Air Loss

One of the most frustrating complaints from drivers is that they constantly have to pump their tires up as the result of an air loss. To efficiently check the suspect tire, it must be removed from the vehicle and aired up to its recommended pressure. A good method for checking where air loss is coming from is to immerse the wheel assembly in a container of water. The container must be large enough to immerse the tire either on its side, with the tire completely underwater. Or it can be upright, with approximately half the wheel assembly underwater. If there is a leak, it will be obvious by the discharge of air bubbles in the water (**FIGURE 25-29**). Mark the source of the air bubbles. Remove the wheel from the water tank, and carry out the appropriate repairs.

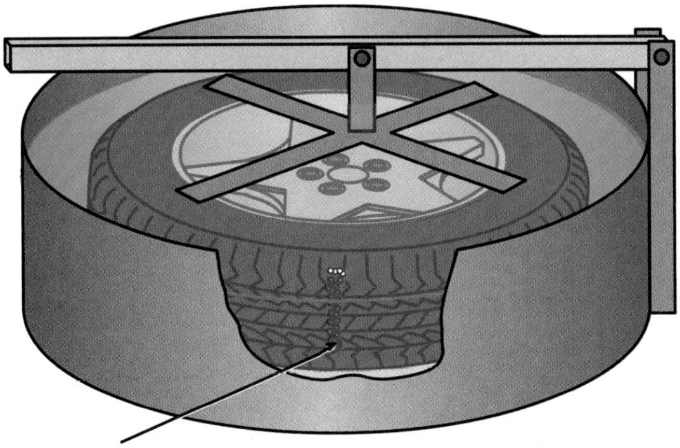
Air Bubbles show leak location, mark this spot for repair when the tire is removed from the tank.

FIGURE 25-29 A good method for identifying the location of an air leak is to immerse the tire assembly in water and watch for air bubbles.

To inspect the wheel assembly for air loss, using the dunk tank method, perform the steps in **SKILL DRILL 25-12**.

Another method for locating leaks can be carried out with a spray bottle of soapy water. You need to spray the soapy water around the valve stem and core. In addition, spray around the bead area. Also spray the entire tread area. If there is any air leakage, soapy air bubbles will indicate where the problem is. Mark the place on the tire where the air bubbles are coming from so that the repairs can be carried out.

SKILL DRILL 25-12 Inspecting the Wheel Assembly for Air Loss, Using the Dunk Tank Method

1. Remove the tire from the vehicle, and inflate it to the specified pressure.

2. Place the tire in the dunk tank and hold it below the surface of the water. Look for bubbles.

3. Flip the tire over and look for bubbles on the back side.

SKILL DRILL 25-13 Inspecting the Wheel Assembly for Air Loss, Using the Spray Bottle Method

1. Remove the tire from the vehicle, and inflate it to the specified pressure. Spray soapy water on the valve core and stem, and look for bubbles.

2. Spray soapy water on the tread, and look for bubbles.

3. Spray soapy water around the bead of the tire, and look for bubbles.

To inspect the wheel assembly for air loss, using the spray bottle method, perform the steps in **SKILL DRILL 25-13**.

Tire Repair

Driving even a short distance on a tire while it is severely underinflated can permanently damage it. Driving with low pressure overheats the tire, which weakens the sidewall belts. This quickly creates a dangerous, nonrepairable condition. The damage may not be visible from the outside. So, every tire requiring repairs must be removed from the wheel for inspection. This will allow its condition to be assessed and a determination of its reparability can be made.

The tire needs to be inspected externally for damage, first. This involves inspecting it for several conditions that would make it unrepairable. Those include:

- Worn down to, or beyond, the wear bars.
- Broken cords or belts as evidenced by bulges in the tread or sidewalls.
- Cords or belts sticking out of the tire.
- Any puncture outside of the center portion of the tread. Punctures in the shoulder and sidewall are NOT repairable.
- Punctures larger than 1/4" (6 mm)
- Any signs of serious internal damage, such as chafing of the sidewalls or cords sticking through the inner liner.
- Age of the tire. Most manufacturers recommend replacing tires that are 10 years old or more. So, you should not repair a tire that is at, or near, that age.

If the inspection shows no signs of nonrepairable failure, then the tire can be repaired. But the repairs must be made following the procedures recommended by the tire associations. This includes the Rubber Manufacturers Association. Among the criteria they require to perform a proper repair are:

- Repairs are limited to the tread area only.
- Puncture injury cannot be greater than 1/4" (6 mm) in diameter.
- Repairs must be performed by removing the tire from the rim/wheel assembly to perform a complete inspection to assess all damage that may be present.
- Repairs cannot overlap. This means that if there are two or more repairs required to fix the tire, then each repair patch must not overlap.
- A rubber stem, or plug, must be applied to fill the puncture injury, and a patch must be applied to seal the inner liner. A common repair unit is a one-piece plug patch with a stem and patch portion. A plug by itself is an unacceptable repair.

To patch a tire, you need several tools:

- A tire machine to remove the tire from the wheel.
- A buffer or scraper to clean the inside of the tire.
- Glue for attaching the plug patch.
- A tire plug patch.
- A stitching tool.

The tire plug patch is a superior product to the old-fashioned tire plug. In many states, it is the only legal way to repair a hole in a tire. It is more expensive. But is a much safer alternative for repairing a tire because it patches the inside of the tire in addition to filling the hole in the tire.

To patch a tire, follow the steps in **SKILL DRILL 25-14**.

▶ Measuring Wheel, Tire, Axle Flange, and Hub Runout

LO 25-10 Measure wheel, tire, axle flange, and hub runout.

Runout is the side-to-side or up-and-down variation in a part. It can occur in any of the wheel assembly components. In many cases, runout issues cannot be observed when the vehicle is stationary on the ground. Instead, runout problems can be felt as a vibration while the vehicle is being driven. In fact, it usually gets more noticeable as speed is increased. If the vibration is primarily observed in the steering column or hood, the runout is most likely in one or both of the front wheel assemblies. Vibration felt throughout the entire vehicle suggests that the problem is with one or both of the back wheels.

Runout can be defined as radial or lateral (**FIGURE 25-30**). Radial runout occurs when the component is out of round or off center. It is felt more as a vertical vibration. Lateral runout occurs when the component causes the wheels to jiggle side to side. It creates a horizontal vibration that feels like a shimmy. Lateral runout on front wheels is felt in the steering wheel even at slower speeds. Runout of the wheel, tire, axle, and hub are all measured in the same way. Either by using a dial indicator or a special runout gauge in contact with the surface of the component being measured.

To measure tire runout, follow the steps in **SKILL DRILL 25-15**.

To measure wheel runout, follow the steps in **SKILL DRILL 25-16**.

To measure axle flange or hub runout, follow the steps in **SKILL DRILL 25-17**.

SKILL DRILL 25-14 Patching a Tire

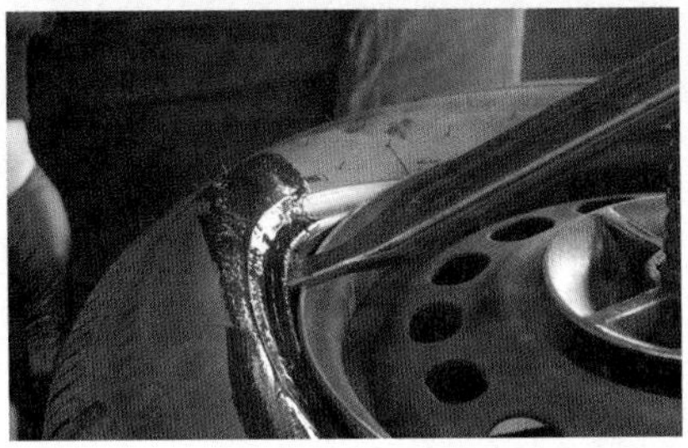

1. After marking the location of the air leak on the tread of the tire and the position of the tire and weights on the rim, remove the tire from the rim assembly.

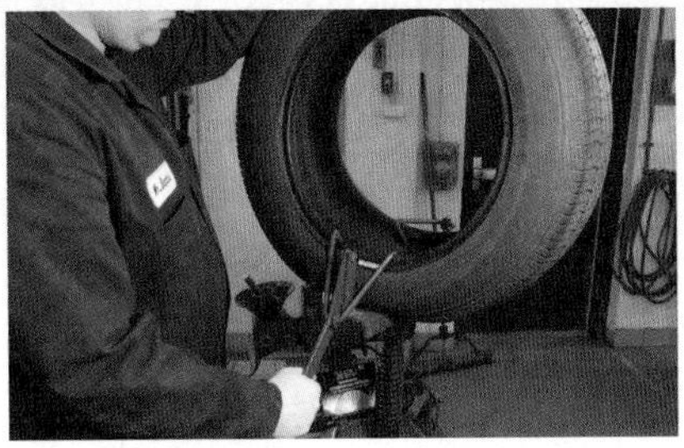

2. Mount the tire in a tire spreader so that the hole can be accessed from both the inside and the outside of the tire.

3. Use an air die grinder with the properly sized pointy bit that matches the plug patch, and drill into the tire where the leak was located.

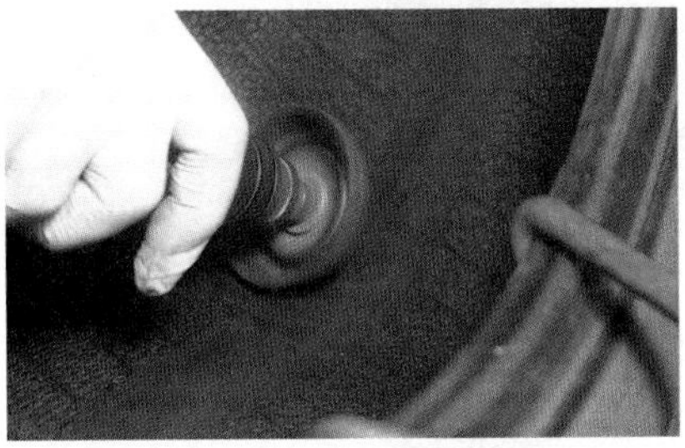

4. Use a tire buffer to smooth the area inside the tire around the hole. Smooth the area approximately 1/2" (13 mm) beyond the expected patch area.

5. After completing the buffing process, clean out all the accumulated debris with a vacuum.

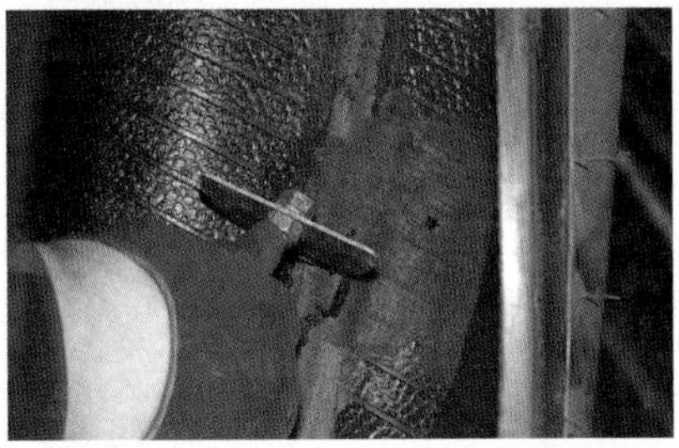

6. Liberally apply the liquid buffing solution to a clean rag, and scrub the area just buffed. Or use cleaner and a scraper to clean the area. Repeat this step once or twice, as needed.

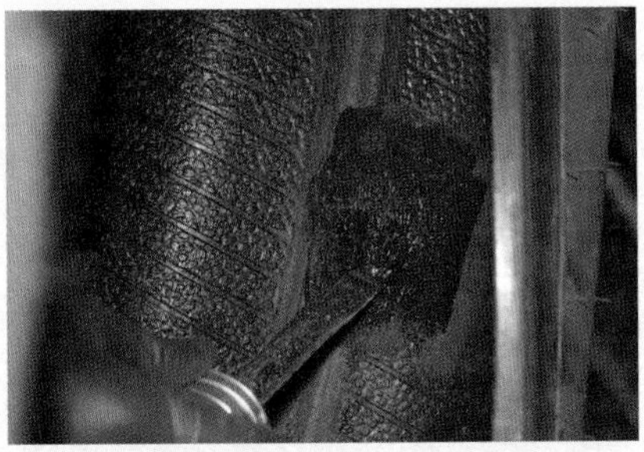

7. Apply vulcanizing cement evenly to the inner buffed surface of the tire. The cement needs to stand until it is relatively dry and is only tacky to touch.

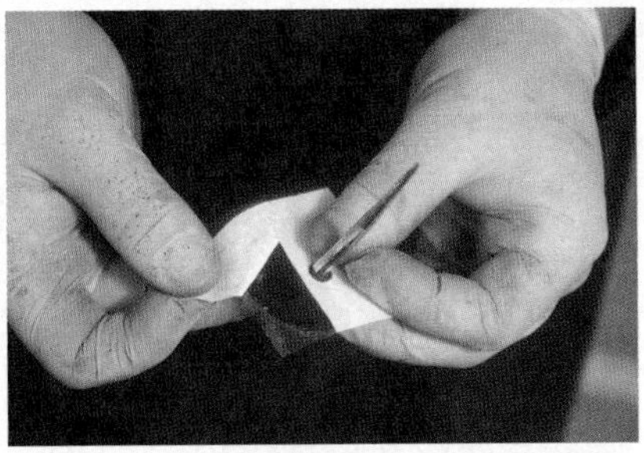

8. After selecting the appropriate tire patch, remove the plastic protective cover on the sticky side of the tire patch without getting your fingerprints on the sticky side.

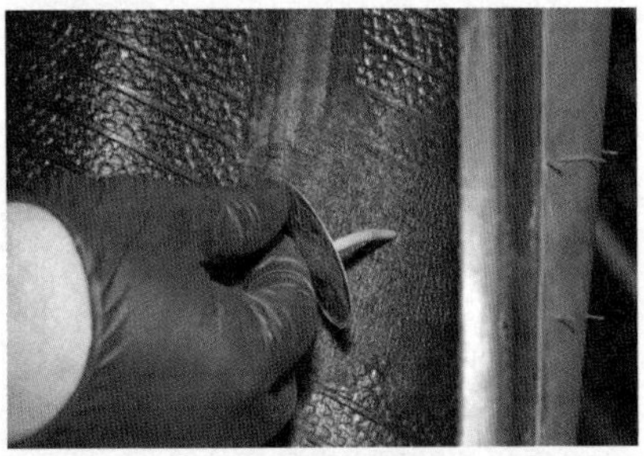

9. Take the pointed part of the plug-patch and push it through the inner side of the tire's hole that was roughed previously, pushing it through to the outside of the tire.

SKILL DRILL 25-14 Patching a Tire (Continued)

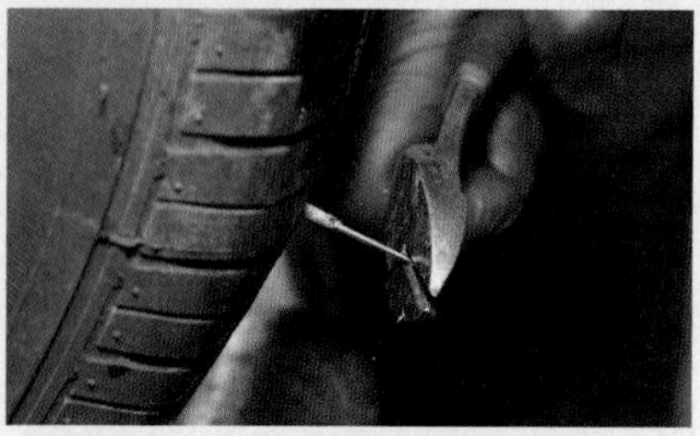

10. Using a pair of pliers, grip the stem of the plug-patch and pull it out so that the disc portion of the patch comes into contact with the cemented area. Pull this pointy part of the patch away from the tire's tread. The sticky side of the patch has now been tightly pressed onto the buffed surface.

11. Use a stitching tool and roll the inner side of the tire patch tight onto the inner surface of the tire. Start at the center of the patch and work outward. Cover the buffed area and newly applied patch with a rubber patch sealant. Trim the plug material even with or just above the tread, using a utility knife or other appropriate cutting tool.

12. After the rubber patch sealant has dried, the tire can be reassembled onto the wheel in its original position and inflated to the recommended pressure. Check the wheel assembly for balance before it is reinstalled on the vehicle.

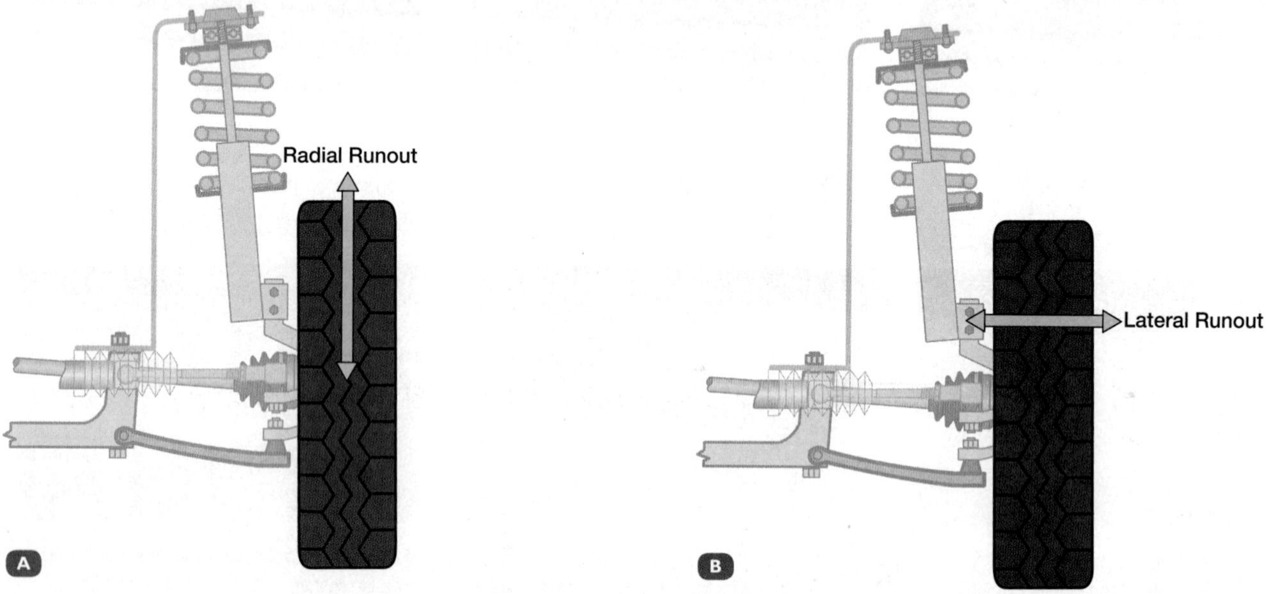

FIGURE 25-30 A. Radial runout. **B.** Lateral runout..

SKILL DRILL 25-15 Measuring Tire Runout

1. Research the procedure and specifications. Raise and support the vehicle on a hoist.

2. Select the runout gauge, attachment, and bracket that fit the tire. Mount the runout gauge on a firm surface to keep it still.

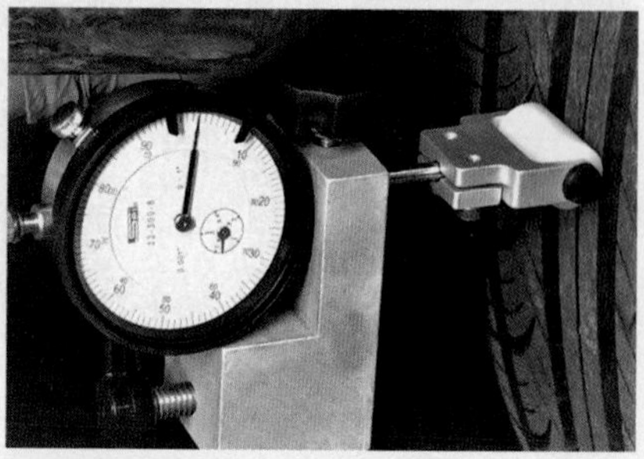

3. Adjust the runout gauge so it is 90 degrees to the tread of the tire. Rotate the tire to the low spot and zero the runout gauge.

SKILL DRILL 25-15 Measuring Tire Runout (Continued)

4. Carefully rotate the tire a couple of times while observing the runout gauge readings. If the pointer hovers around a single reading, the part has minimal runout, and the test is complete. If the gauge moves significantly, record the reading.

5. Compare these values to the manufacturer's specifications. If the deviation is greater than the specifications, the wheel and/or hub runout must be measured.

6. Reposition the runout gauge so it is 90 degrees to the sidewall of the tire. Perform steps 3 through 5 to obtain the lateral runout measurement for the tire, and compare to the vehicle's specifications.

SKILL DRILL 25-16 Measuring Wheel Runout

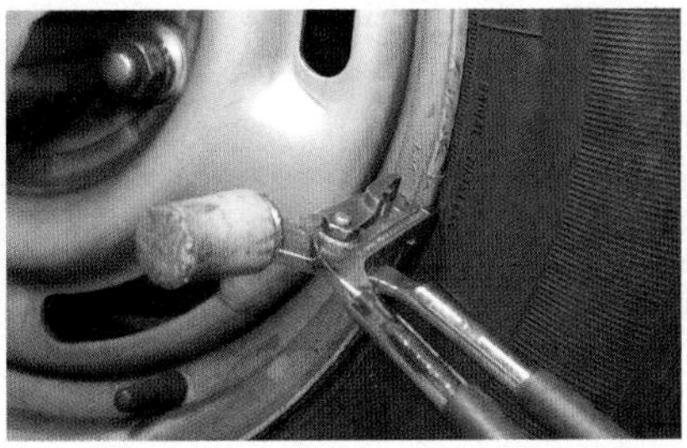

1. If equipped, remove the hubcap and use a wheel weight tool to remove any wheel weights.

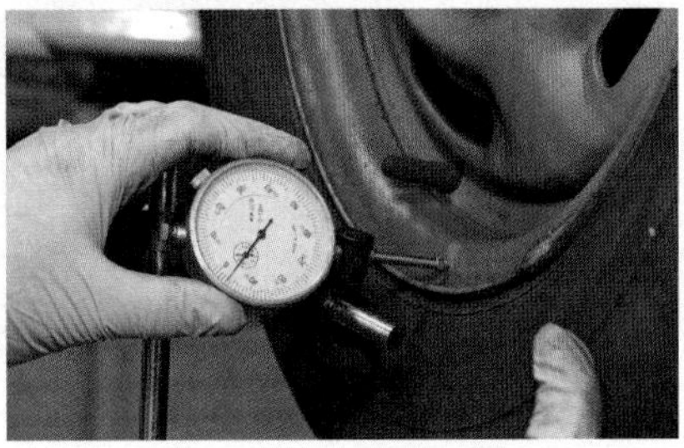

2. Select the runout gauge or dial indicator, attachment, and bracket that fit the wheel. Mount the dial indicator on a firm surface to keep it still. Adjust the dial indicator so the plunger is 90 degrees to the flat portion of the wheel where the wheel weights are installed.

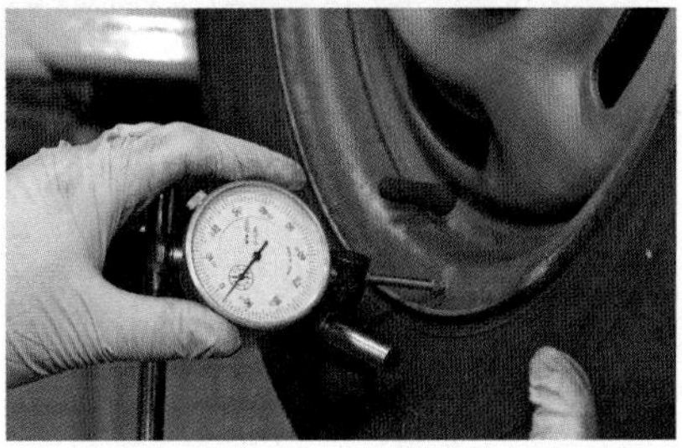

3. Press the dial indicator gently against the wheel, and rotate the wheel one full turn. Keep pressing until the plunger settles about halfway into the indicator. Verify that the plunger is still 90 degrees to the wheel, and lock the indicator assembly into position.

4. Rotate the wheel a couple of times while observing the dial readings. If the pointer hovers around a single graduation on the dial, the wheel has minimal runout or surface distortion and the test is complete. If the pointer moves significantly left and right, note the variations.

5. Find the point of maximum movement to the left and move the dial so that zero is over this point.

6. Continue to rotate the wheel. Find the point of maximum movement to the right and note the reading. Confirm this value by rotating the wheel to verify the zero point and high point. Compare these values with the manufacturer's specifications. If the deviation is greater than the specifications, the axle or hub runout must be measured.

SKILL DRILL 25-17 Measuring Axle Flange or Hub Runout

1. Prepare the vehicle by removing any hubcaps or lug nut covers. With the vehicle on the ground, loosen the lug or wheel nuts. Raise the vehicle, and remove the lug nuts and tire.

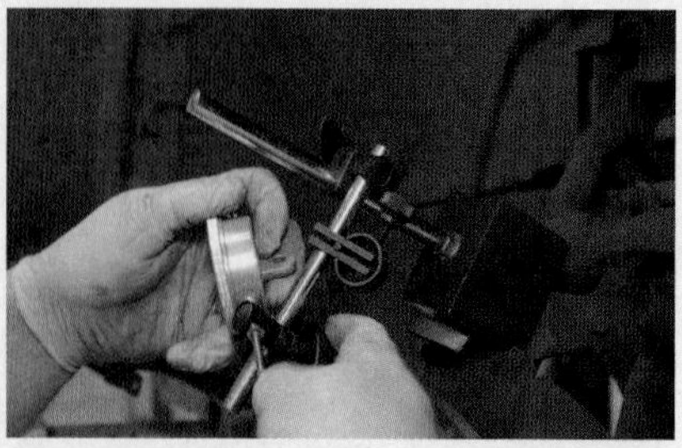

2. Select the dial indicator, attachment, and bracket that fit the axle flange or hub. Mount the dial indicator on a firm surface to keep it still.

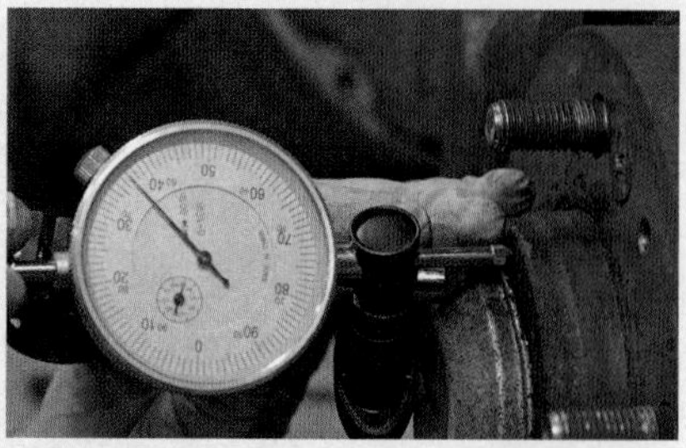

3. Adjust the plunger so it is 90 degrees to the axle flange or hub. Press the dial indicator gently against the flat surface of the axle flange or hub, and rotate it one full turn. Keep pressing until the plunger settles about halfway into the indicator. Verify that the plunger is still 90 degrees to the tire, and lock the indicator assembly into position. Rotate the axle flange or hub while observing the dial readings.

4. If the pointer hovers around a single graduation on the dial, the axle flange or hub has minimal runout or surface distortion, and the test is complete. If the pointer moves significantly left and right, note the variations. Find the point of maximum movement to the left, and move the dial so that zero is over this point.

5. Continue to rotate the axle flange or hub. Find the point of maximum movement to the right and note the reading. Confirm this value by rotating the axle several more times to verify the zero point and high point. Compare these values with the manufacturer's specifications. If the deviation is greater than the specifications, the axle or hub must be discarded. Reinstall the wheels and torque the lug nuts following the specified procedure and torque specifications.

▶ Wrap-Up

Ready for Review

- Common tire and wheel issues to look out for include: air loss, out of balance, excessive loaded radial runout, excessive lateral runout, and wheel trim imbalance.
- The recommended cold inflation pressure of a vehicle is available on the tire placard located on the A- or B-pillar door frame, in the glove compartment, or on the fuel filler flap.
- Common irregular tire wear patterns include feathering, one-sided wear, cupping, center wear, and edge wear. Tires can be rotated in the following patterns: forward-cross pattern, rearward-cross pattern, X pattern, front-to-rear pattern, side-to-side pattern from the other side of the vehicle, five-tire forward-cross pattern, and five-tire rearward-cross pattern.
- Dynamic balancing is performed when the tire is rotating and allows it to take into consideration imbalances that may be off-center. Dynamic balancing is performed by placing specific amounts of weight on each side of the rim as counterbalances.
- When mounting a tire, take care to perform match mounting and weight matching. Use extreme care when using a tire machine.
- TPMS sensor can be attached to, or integrated with, the valve stem. If the TPMS sensors are integrated into a screw-in valve stem, it is safer to push the valve stem and senor into the tire.
- Tire valve stems can be rubber press fit valve stems, screw-in valve stems, or TPMS sensor-integrated valve stems. To change the valve stem, the wheel must be removed from the vehicle.
- The TPMS sensor unit is a sealed component and the battery or the entire sensor will need to be replaced, if the battery is not accessible.

- Tire diagnosis can be performed information from the repair order, the shop's customer concern questionnaire, or from a test drive. Inspection for tire repair includes checking for several conditions: worn down to, or beyond, the wear bars; broken cords or belts; cords or belts sticking out of the tire; puncture outside of the center portion of the tread; punctures larger than 1 by 4 inch (6 mm); signs of serious internal damage; and tire age.
- Radial runout occurs when the component is out of round or off center and is felt more as a vertical vibration. Lateral runout occurs when the component causes the wheels to jiggle side to side and creates a horizontal vibration that feels like a shimmy.
  - Runout of the wheel, tire, axle, and hub are all measured using a dial indicator or a special runout gauge in contact with the surface of the component being measured.

Key Terms

bars A unit of measure of pressure. Short for barometric. One bar equals 14.7 psi.

dynamic imbalance A tire imbalance that causes the wheel assembly to turn inward and outward with each half revolution.

match mounting The process of matching up the tire's highest point with the rim's lowest point for the purpose of reducing the tire's radial runout.

road force imbalance Occurs when the wheel or tire is not concentric or when the tire's sidewall has uneven stiffness.

static imbalance Assumes the imbalance is centered across the width of the tire. A tire with static imbalance tends to vibrate vertically, with the heavy area slapping the road surface with each turn of the wheel.

tire inflation pressure Level of air in the tire that provides it with load-carrying capacity and affects overall vehicle performance.

tire pressure gauge A tool used to check the air pressure in tires.

weight matching The process of matching the tire's lightest point with the rim's heaviest point (generally at the valve stem) for the purpose of reducing the tire's radial imbalance.

Review Questions

1. What does the acronym TPMS stand for?
   a. Throttle Position Module Sensor
   b. Tire Pattern Monitor System
   c. Tire Pressure Monitor System
   d. Tire Pressure Modulating Sensor
2. What does the tire inflation number on the sidewall indicate?
   a. Average inflation pressure
   b. Maximum safe pressure
   c. Minimum safe pressure
   d. Proper pressure for tire regardless of vehicle
3. What is the most common tread wear on an underinflated tire?
   a. Feather wear
   b. Cupped wear
   c. Center wear
   d. Edge wear
4. What type of tire balancer provides the best ride quality?
   a. A static balancer
   b. A dynamic balancer
   c. A harmonic balancer
   d. A road force balancer
5. What is a directional tire?
   a. A tire that can be mounted in any of the four positions.
   b. A tire that can operate in either direction equally well.
   c. An electronic tire that indicates the directional heading of the vehicle.
   d. A tire designed to operate better in one direction.
6. To prevent TPMS sensor damage, the valve stem should be in the __________ position when breaking the tire bead with the shovel.
   a. 6:00
   b. 9:00
   c. 3:00
   d. 90 degree
7. What is used to find air leaks around the valve stem?
   a. A compression tester
   b. Soapy water
   c. Petroleum jelly
   d. A 5-gas analyzer
8. Why are nickel-plated valve cores recommended for valve stem-type TPMS sensors?
   a. To prevent a chemical reaction causing corrosion.
   b. To prevent a magnetic field that interferes with the sensor signal.
   c. To increase the cost of the TPMS sensors.
   d. To prevent environmental pollution.
9. What is a stitching tool used for when repairing a tire?
   a. To ream the puncture hole
   b. To buff the inside liner surface
   c. To thread the plug into the puncture hole
   d. To press the patch evenly into the inside liner
10. When using a dial indicator to measure hub runout, the runout measurement is:
    a. The sweep of the needle on the gauge as the hub is rotated.
    b. The average of four measurements around the hub.
    c. The sweep of the needle on the gauge as the hub is moved inboard and outboard.
    d. The average distance from the center of the hub to the ground as the hub is rotated.

ASE Technician A/Technician B Style Questions

1. Technician A says that an out-of-balance tire or wheel can cause a vibration. Technician B says that tires do not lose air pressure over time. Who is correct?
   a. Technician A only
   b. Technician B only
   c. Both A and B
   d. Neither A nor B
2. Technician A says that tire pressure should be checked after a test drive to warm up the tires. Technician B says that

digital pressure gauges are the least accurate type of tire pressure gauge. Who is correct?
   a. Technician A only
   b. Technician B only
   c. Both A and B
   d. Neither A nor B
3. Technician A says that directional tires should be rotated in a front-to-rear pattern. Technician B says to torque lug nuts in stages. Who is correct?
   a. Technician A only
   b. Technician B only
   c. Both A and B
   d. Neither A nor B
4. Technician A says that it is generally acceptable to reuse old wheel weights. Technician B says that tire balancers require routine calibration. Who is correct?
   a. Technician A only
   b. Technician B only
   c. Both A and B
   d. Neither A nor B
5. Technician A says to match-mount a tire on a wheel, the red dot should be lined up with the wheel's lowest point. Technician B says that match-mounting or weight-matching will improve the vehicle ride quality. Who is correct?
   a. Technician A only
   b. Technician B only
   c. Both A and B
   d. Neither A nor B
6. Technician A says that removing a valve stem-type TPMS sensor from the wheel before dismounting a tire will prevent bead-to-sensor contact damage. Technician B says that the band-type TPMS sensors are mounted outside of the drop center. Who is correct?
   a. Technician A only
   b. Technician B only
   c. Both A and B
   d. Neither A nor B
7. Technician A says that TPMS-integrated valve stems should be replaced when new tires are installed. Technician B says that valve stems can be replaced without removing the wheel from the vehicle. Who is correct?
   a. Technician A only
   b. Technician B only
   c. Both A and B
   d. Neither A nor B
8. Technician A recommends replacing all four TPMS sensors when one sensor's battery dies. Technician B says that band TPMS sensors should be installed next to the valve stem. Who is correct?
   a. Technician A only
   b. Technician B only
   c. Both A and B
   d. Neither A nor B
9. Technician A says that a puncture in a tire's shoulder or sidewall is not repairable. Technician B says that a plug patch repair from inside the tire is a safer repair than a tire plug repair from the outside of the tire. Who is correct?
   a. Technician A only
   b. Technician B only
   c. Both A and B
   d. Neither A nor B
10. Technician A says that excessive radial runout will cause a side-to-side (horizontal) vibration. Technician B says hub runout is measured with a runout gauge or dial indicator. Who is correct?
   a. Technician A only
   b. Technician B only
   c. Both A and B
   d. Neither A nor B

CHAPTER 26

Steering Systems Theory

Learning Objectives

- **LO 26-01** Describe steering system preliminaries.
- **LO 26-02** Describe steering geometry and rack and pinion layout.
- **LO 26-03** Describe parallelogram steering layout.
- **LO 26-04** Describe steering columns and their components.
- **LO 26-05** Describe rack and pinion steering boxes.
- **LO 26-06** Describe worm gear steering boxes.
- **LO 26-07** Describe hydraulic power steering system operation.
- **LO 26-08** Describe electric power steering system operation.
- **LO 26-09** Describe four-wheel steering operation.

ASE Education Foundation Tasks

See Appendix A to view the 2017 ASE Education Foundation Automobile Accreditation Task List Correlation Guide.

▶ Introduction

Every driver knows that turning the steering wheel steers the front wheels. Not as apparent are the many components that work together to make that happen. This chapter discusses the different steering systems and how they operate. We also cover the purpose and function of each part within them. This understanding will prepare you for maintenance and light repair of these systems. Maintaining a well-functioning steering system is necessary for safety and long vehicle life. But our journey starts with understanding how the steering system works.

▶ Steering System Overview

LO 26-01 Describe steering system preliminaries.

The **steering system** provides:

- control over the vehicle's direction of travel,
- good maneuverability for parking the vehicle,
- smooth recovery from turns as the driver releases the steering wheel, and
- minimal transmission of road shocks from the road surface through the steering wheel.

As vehicle technology has progressed, steering systems have gone through a number of refinements. Those refinements have enhanced vehicle safety, performance, and even fuel economy.

A basic steering system has four main assemblies:

- a steering column,
- a steering box,
- a steering linkage, and
- a steering Knuckles (**FIGURE 26-1**).

Added to the basic steering system is a power-assist system. It makes it easier for the driver to steer the vehicle. The power steering system can be either a hydraulic type or an electric type. In the case of electric power steering, the system can also be integrated with the electronic stability control system. This means that the powertrain control module (PCM) can take command of the steering to prevent loss of control of the vehicle. Electric power steering also makes four-wheel steering more of a feasible option.

The **steering column** transmits the driver's steering effort from the steering wheel to the steering box. The **steering box** converts the rotary motion of the steering wheel to the **linear motion** needed to pivot the wheels. The steering box also uses principles of gear reduction. This gives the driver mechanical advantage over the wheels, making it easier to steer them. The **steering linkage** transfers the linear steering effort to the wheels. It does so by connecting the steering box to the **steering arm** on each of the steering knuckles. The steering knuckles pivot on the ball joints, allowing the wheels to steer the vehicle.

There are two main types of steering systems used on vehicles today: rack and pinion and parallelogram. Each one has its strengths and weaknesses. But rack-and-pinion systems are the most commonly used. The rack-and-pinion system gets its name because a rotating pinion gear is used to move a flat, toothed rack (**FIGURE 26-2**). The rack is connected through pivoting socket ends and a tie rod directly to the steering arms. It is a simple, compact system with only a few moving and pivoting parts. It fits well into most engine compartments and takes up less space than other systems. This means it is less likely to get in the way of other components. And because of the orientation of the pinion gear to the rack, it gives a more precise steering feel. That makes it more responsive.

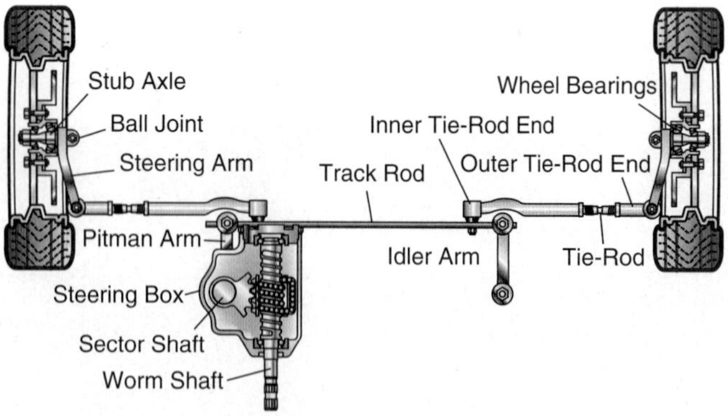

FIGURE 26-1 The components of a basic steering system.

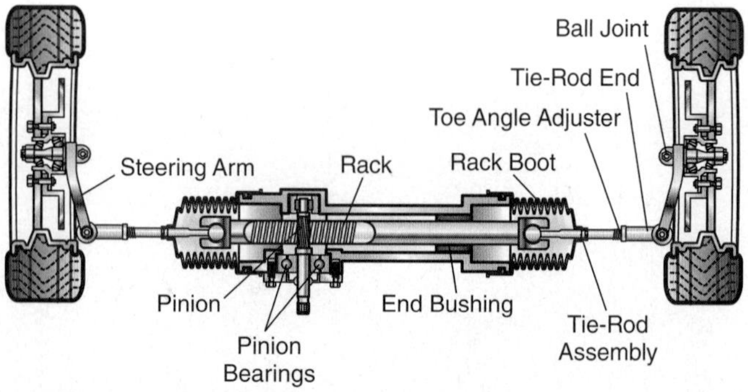

FIGURE 26-2 Rack-and-pinion steering system.

You Are the Automotive Technician

A regular customer who performs some of the maintenance on her car brings it to you for a wheel alignment. She has just replaced some steering components. She has been reading up on steering system theory and has some questions for you.

1. What is the Ackermann principle, and how does it affect the steering system?
2. What are the differences between rack-and-pinion and parallelogram steering systems?
3. What do the balls do in a recirculating ball steering gear?
4. What does the power steering switch do?

The **parallelogram steering system** gets its name because of the orientation of the steering linkage. The center link and axle, along with the pitman arm and idler arm, always move parallel to each other (**FIGURE 26-3**). Pivoting tie rods connect the parallelogram to the steering arms.

Parallelogram steering uses a **worm** gearbox. It provides mechanical advantage, along with changing the direction of rotation 90 degrees. The worm gear design reduces the road shock that is transmitted to the steering wheel. So, reduced road shock is a benefit of parallelogram designs. This is especially useful in off-road four-wheel drive vehicles, or non-sporty vehicles. A pitman arm and center link turn the rotary motion into lateral motion.

The **steering knuckles** are stout components. They firmly connect the wheels to the suspension and steering systems (**FIGURE 26-4**). Some styles provide a stub axle upon which the wheel bearings ride. Others use a hub-style wheel bearing assembly that is pressed or bolted onto the steering knuckle. The steering knuckle pivots on one or two ball joints, depending on the type of suspension. This will be covered in the Suspension Systems chapter. The steering arm transmits the steering force to the wheel assembly.

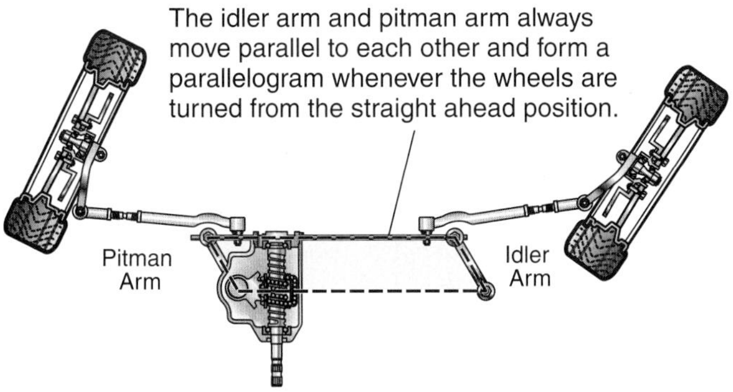
The idler arm and pitman arm always move parallel to each other and form a parallelogram whenever the wheels are turned from the straight ahead position.
Pitman Arm
Idler Arm

FIGURE 26-3 Parallelogram steering system.

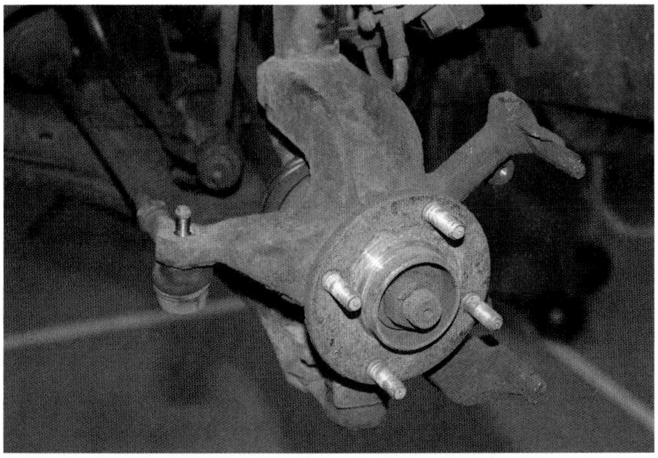

FIGURE 26-4 A typical steering knuckle that supports the wheel and brake assembly. It also is the connection point between the wheel and the steering and suspension system.

▶ Steering Geometry

LO 26-02 Describe steering geometry and rack and pinion layout.

The relationships between the steering system, the wheels, and the suspension system form what is called steering geometry. Steering geometry is a geometric arrangement of linkages in the steering system. They are designed to keep the wheels properly oriented through movement of the steering and suspension systems. As the wheels move up and down, the steering linkage swings vertically through an arc. The wheel would thus turn in and out as the vehicle goes over bumps, if it were not for steering geometry. In this case, the steering components go through a similar arc as the suspension components (**FIGURE 26-5**). This allows the wheel to track straight ahead, or in a consistent direction if it is in a turn.

Also, when rounding a corner, the inner and outer wheels must trace circles of different radii. Otherwise, the tires would be dragged across the road surface, which is called scrub. This is a challenge because no matter which way the vehicle turns, the inside wheel must always turn more sharply than the outside wheel (**FIGURE 26-6**). This is called toe-out on turns.

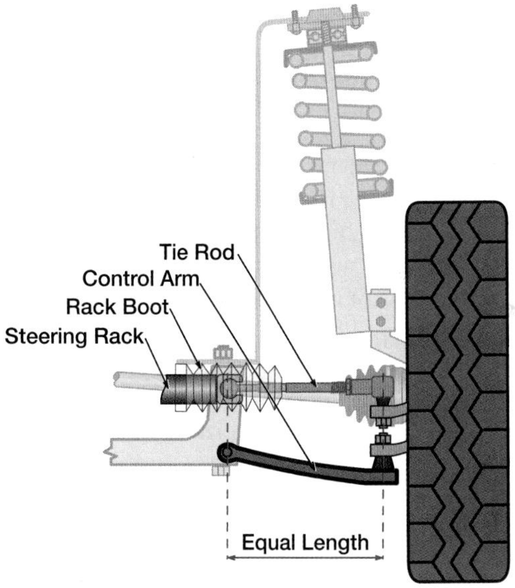
Tie Rod
Control Arm
Rack Boot
Steering Rack
Equal Length

FIGURE 26-5 The steering system and suspension system must operate through similar arcs so as not to affect the angle of the wheels when driven over bumps.

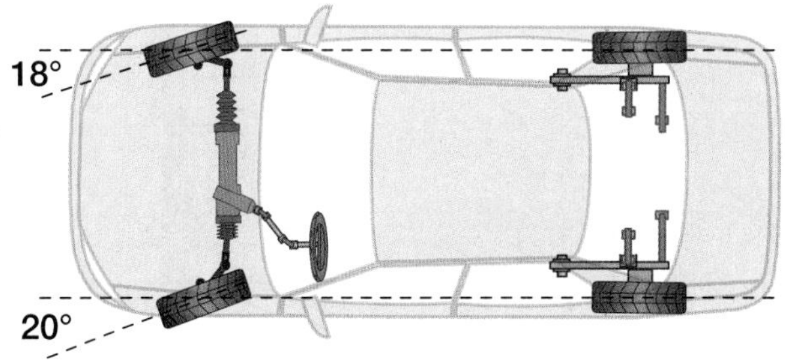
18°
20°

FIGURE 26-6 The inside tire must always turn sharper than the outside tire.

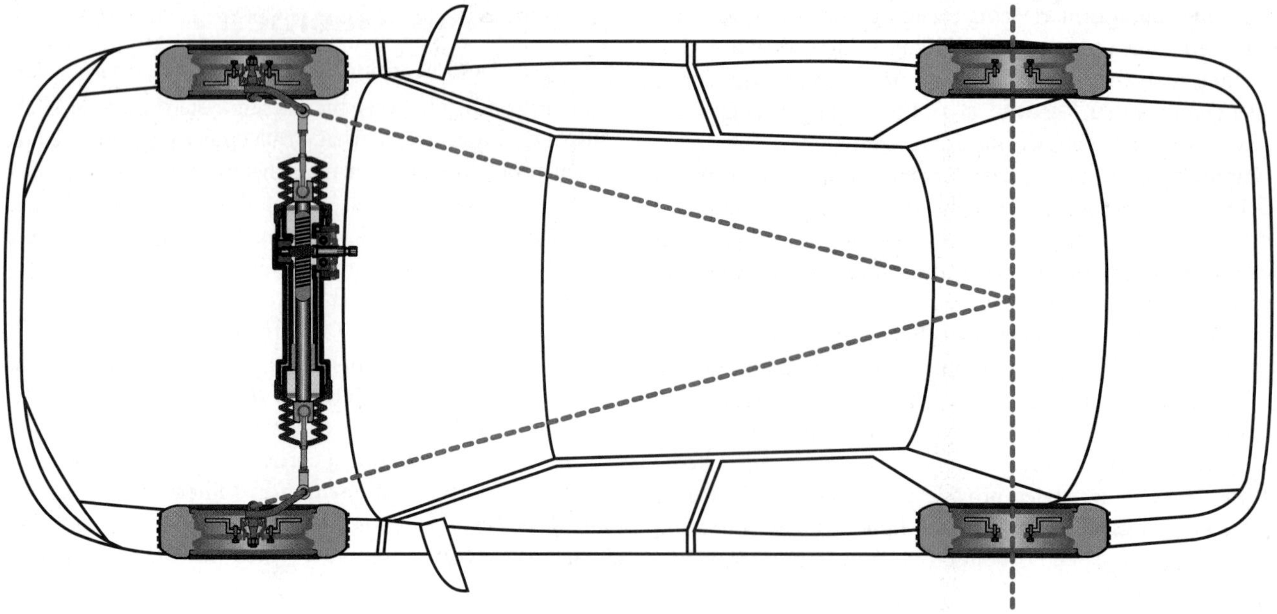

FIGURE 26-7 The Ackermann principle. Angling the steering arms inward allows the front wheels to travel around a common center point.

The Ackermann principle is named after the man who patented it in 1818. It provides the needed geometry when turning corners. The Ackermann principle affects the angles of the steering arms on the steering knuckles. They are angled toward the center of the vehicle. Imaginary lines drawn from the center of the steering knuckle pivot points, through the center of the outer tie-rod ends, intersect at the center of the rear axle (**FIGURE 26-7**). This angle is what allows the front wheels to navigate a corner around a common center point. Having a common center point minimizes tire scrub (**FIGURE 26-8**).

Steering and suspension geometry must be within the manufacturer's specifications for the vehicle to operate correctly. If it is not correct, it is usually because of worn or bent components. The technician will have to diagnose the faulty component. Then, replace it to restore proper operation. We will cover this in more detail in the Wheel Alignment chapter.

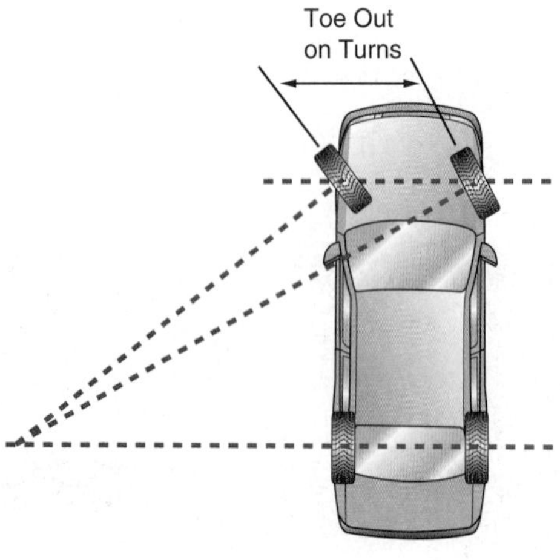

FIGURE 26-8 Wheel scrub is minimized when all wheels pivot around a common center.

Rack-and-Pinion Steering Linkage

The **rack-and-pinion steering system** is used on the majority of vehicles today. This is because of the limited under-hood space. It is also due to their compact and lightweight design (**FIGURE 26-9**). Steering response is very sharp because the rack directly operates the steering knuckle. There is also very little sliding and rotational resistance, which gives lighter operation. The primary components of the rack-and-pinion steering system are listed here:

- **Pinion**: A toothed helical cut gear that meshes with the rack (**FIGURE 26-10**). The pinion is connected to the steering column. As the driver turns the steering wheel, the forces are transferred to the pinion. Turning the pinion causes the rack to move in either direction. This is achieved by having the pinion in **constant mesh** with the rack.

FIGURE 26-9 A typical rack-and-pinion steering gear.

- **Rack**: A toothed, straight piece of metal that meshes with the pinion in the middle of the rack. It has tie rods on each end that fasten to the steering knuckle. The rack slides in the housing and is moved by the action of the **meshed pinion**. A spring-loaded or **adjustable bushing** is positioned opposite the pinion to control the components' meshing. This adjustable bushing allows the technician to adjust play out of the rack and pinion. The rack rides on a nylon bushing at the other end. The function of the rack is to transfer motion to the tie rods.
- **Inner tie rod or socket**: The inner tie rod has an inline ball-and-socket joint on one end of a shaft and is threaded on the other end of the shaft. The ball-and-socket joint threads onto one end of the rack (**FIGURE 26-11**). The other end of the inner tie rod threads into the outer tie-rod end. These two joints allow for suspension and steering angle movement. The tie rod is free to spin in the ball and socket when making adjustments in wheel alignment.
- **Outer tie-rod end**: Outer tie-rod end is attached between the inner tie-rod shaft and the steering arm (**FIGURE 26-12**). It transfers the movement of the rack to the steering arm. The outer tie-rod end is threaded onto the end of the inner tie-rod shaft. A jam nut locks the shaft and outer tie rod in place. The jam nut allows the length of the tie rod to be adjusted when making wheel alignment adjustments.
- **Rubber bellows**: The rubber bellows protect the inner joints from dirt and contaminants (**FIGURE 26-13**). In addition, they retain the grease lubricant inside the rack-and-pinion housing. There are two ends of the rack. Each side contains similar bellows.

▶ Parallelogram Steering Linkage

LO 26-03 Describe parallelogram steering layout.

The parallelogram steering system is used where ride comfort is more important than sporty handling. This is especially the case for larger vehicles. Parallelogram steering linkage is more complicated than the rack-and-pinion linkage. So, there are more wear points to know about and inspect. This is one reason

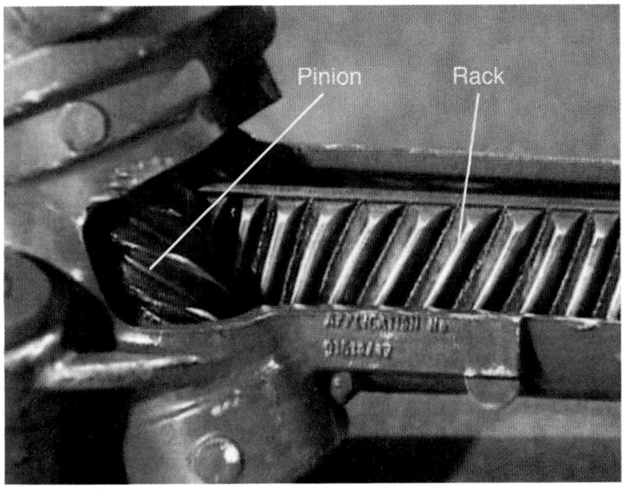

FIGURE 26-10 Pinion and rack.

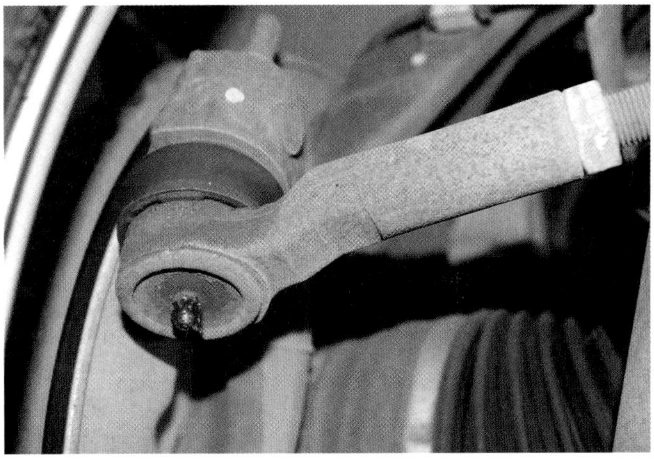

FIGURE 26-12 Outer tie-rod end.

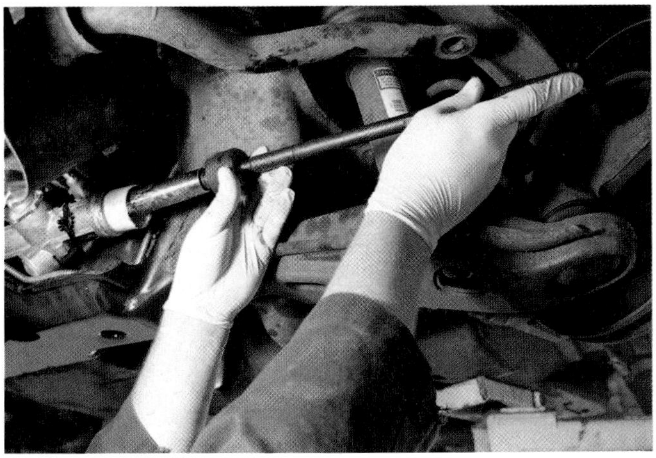

FIGURE 26-11 Inner tie-rod end or socket.

FIGURE 26-13 The rubber bellows seal the end of the rack-and-pinion assembly. Underneath is the inner tie-rod end socket.

why most vehicles today use rack-and-pinion steering. The primary components of the parallelogram steering system, from the gearbox, are as follows:

- **Pitman arm**: The pitman arm transfers movement from the steering box to the center link (**FIGURE 26-14**). It is attached to the steering box by a **spline** and nut. Splines are ridges on a shaft that mesh with grooves in a mating piece. They are used to transfer torque from one part to the other. As the driver turns the steering wheel, the steering box moves the steering linkages via the pitman arm. It moves either left or right, depending on the direction in which the steering wheel is turned.
- **Idler arm**: The idler arm is attached to the **chassis** (the frame of the vehicle). It is positioned parallel to the pitman arm (**FIGURE 26-15**). The idler arm assembly is the pivoting support for the steering linkage. The pivot is on one end and a ball socket is on the other end. Generally, an idler arm is attached between the opposite side of the center link from the pitman arm and the vehicle's frame. It holds the center link at the proper height so it can accurately relay the pitman arm's movement. Idler arms are generally more vulnerable to wear than pitman arms. This is because of the pivot function built into them.
- **Center link**: The center link connects the pitman arm to the idler arm (**FIGURE 26-16**). In this way, any movement in the pitman arm is directly applied to the idler arm. The center link may also be called a track rod or drag link.
- **Tie rod**: The tie rods connect the center link to the steering arms (**FIGURE 26-17**). Tie rods use a pivoting tie-rod end at each end. This allows the steering linkage to move only as needed.
- **Tie-rod end**: Tie-rod ends are attached to each end of each tie-rod shaft. The inner tie-rod ends attach each tie rod to the center link and serves as pivot points. The outer tie-rod ends attach each tie rod to the steering arm. Tie-rod ends pivot as the steering linkage is extended or retracted during turns. Tie rods and tie-rod ends are left- or right-hand threaded. When used with an adjustment sleeve, this allows the length of the tie rod to be adjusted.

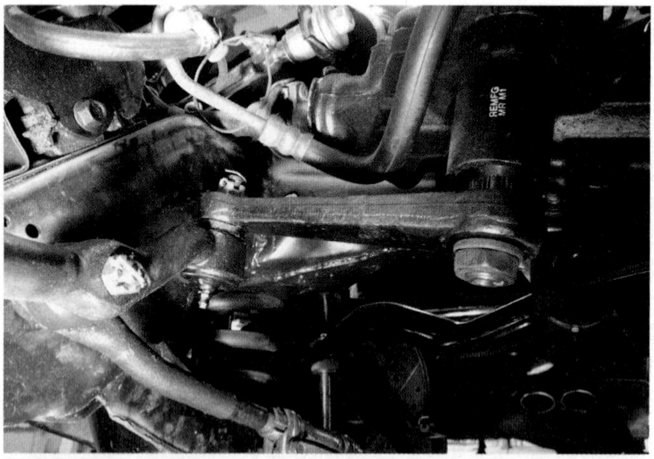

FIGURE 26-14 The pitman arm is bolted to the gearbox. It supports one end of the center link. It also transmits steering force to the center link.

FIGURE 26-16 The center link connects the pitman arm to the idler arm, so they both move together. Also, each end of the center link is connected to one of the tie rods.

FIGURE 26-15 The idler arm supports the other end of the center link and travels in parallel with the pitman arm.

FIGURE 26-17 Tie rods connect the center link to the steering arms. They have pivoting tie-rod ends at each end.

- **Adjustment sleeve**: The adjustment sleeve connects the tie rod to the tie-rod end (**FIGURE 26-18**). The adjustment sleeve is split lengthwise (either partially or completely). It has an internal thread that matches the external threads on the tie rods and tie-rod ends. It provides the adjustment point for adjusting the toe setting. It is similar to a turnbuckle. When a technician turns the sleeve, the tie rod lengthens or shortens. This is used to adjust the wheel alignment.

▶ TECHNICIAN TIP

Each type of steering system makes provision for adjustment of the **toe setting**. The toe setting is the symmetric angle that each wheel makes parallel with the length of the vehicle. The tie-rod ends are threaded to provide for the adjustment of the toe setting.

Each connection point in the steering linkage system has a flexible joint. This is typically a ball-and-socket construction (**FIGURE 26-19**). Flexible joints allow for steering and suspension movement. The tie-rods pivot at similar angles as the suspension components so as not to affect steering operation. Movement of the suspension that does affect steering is called bump steer.

Bump steer is the undesired condition produced when, upon hitting a bump, the vehicle darts to one side. This is due to the steering linkage being pushed or pulled as a result of the travel of the suspension. Typically, bump steer is due to a bent or wrong steering or suspension system component. This causes the angles between the steering and suspension system to not match. Going over a bump will then cause one or both wheels to change directions. If the design of the steering and suspension system components is correct, the suspension can move up and down without affecting vehicle steering.

In four-wheel drive vehicles with a **beam axle**, a single tie rod connects the steering arms on each wheel assembly. Movement of the pitman arm is transferred through the drag link to one of the wheels. The single tie rod transfers movement to the other wheel (**FIGURE 26-20**).

Four-wheel drive vehicles of this type often use a **steering damper** on the single tie rod. It resembles a shock absorber and operates on a similar principle. Specifically, it keeps the steering wheel from shaking when the driver hits a pothole or other road irregularity. The steering damper is mounted between the tie rod and either the rigid axle or the vehicle frame (**FIGURE 26-21**). Its purpose is to prevent shock forces being transmitted through the steering linkage and back to the steering wheel.

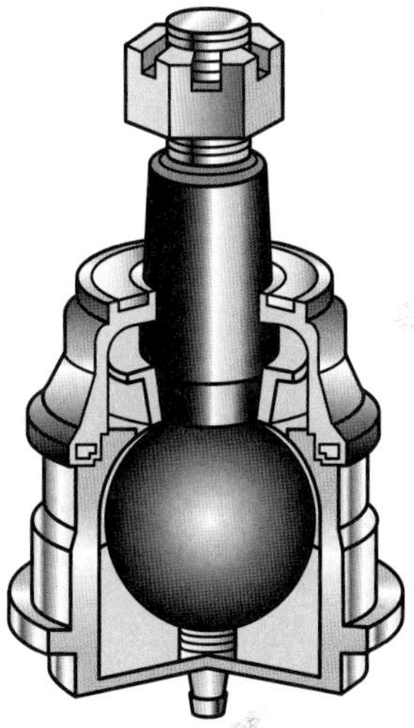

FIGURE 26-19 The flexible joints of the steering linkage are a ball-and-socket construction, which allows movement of the linkage with no play in the joint.

FIGURE 26-18 The adjustment sleeve is threaded to connect each tie rod to the tie-rod end. One end has right-hand threads, and the other end has left-hand threads for adjustment purposes.

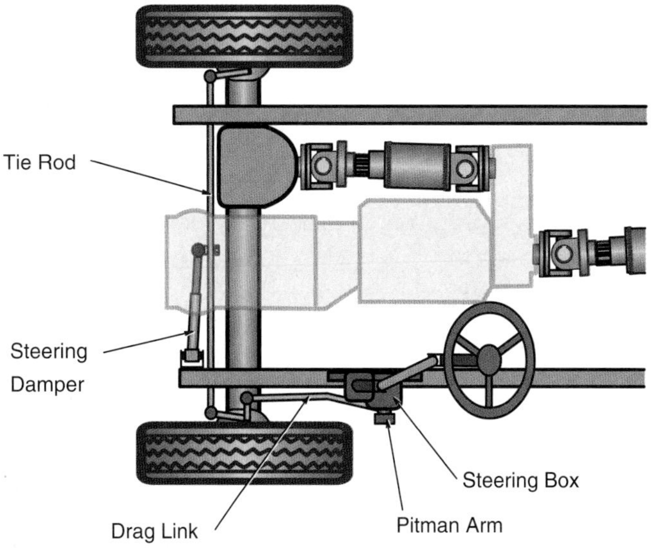

FIGURE 26-20 Steering linkage layout on a solid-axle four-wheel drive vehicle.

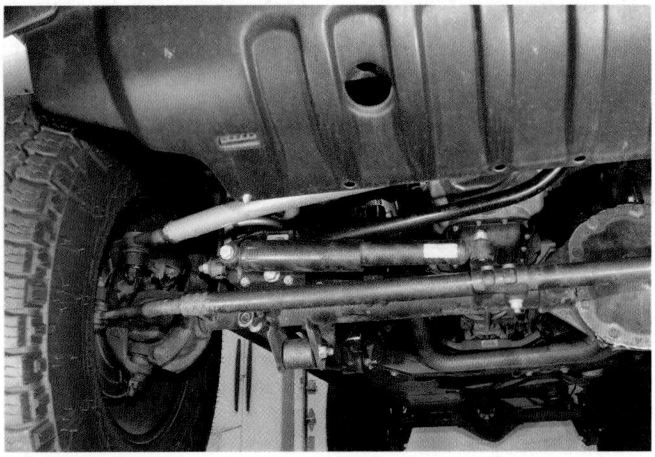

FIGURE 26-21 Steering damper mounted on a 4 × 4 vehicle

FIGURE 26-22 Collapsible steering column.

FIGURE 26-23 Steering column with an intermediate shaft for enhanced protection in an accident.

▶ Steering Columns

LO 26-04 Describe steering columns and their components.

Effort applied to the steering wheel is transferred down the steering column to a steering box. In early vehicles, the steering column was a straight shaft, running inside a hollow tube with bushings. The steering wheel was attached to one end, and the steering box to the other. In many frontal collisions, however, this design caused serious injury to the driver. The steering wheel would be forced back toward the driver's head and chest. And the sudden stop forced the driver into the steering wheel.

To reduce injuries, all steering columns are now fitted with collapsible sections that help protect the driver. During a collision, two forces are applied to the steering column. The first is the force of the steering box forcing the steering column toward the driver. Plastic shear pins allow the lower shaft to move over the upper shaft. The second force is the mass of the driver striking the steering wheel. This force breaks the brackets on the upper part of the column. The upper column is driven into the lower column (**FIGURE 26-22**).

Some steering columns use an **intermediate shaft**. It runs at an angle from the steering column to the steering gear. In a collision, the universal joints on the intermediate shaft allow the steering gear to fold under the steering column. This prevents the impact force from being transferred directly to the column (**FIGURE 26-23**). As added protection, most vehicles also integrate a driver's side airbag into the steering wheel. This system is discussed in greater detail in a bit.

The steering column is connected to the input shaft of the steering gear by a flexible joint (**FIGURE 26-24**). This flexible joint is needed because the steering column sits at an angle to the steering box. It also reduces the transmission of road shocks to the driver.

Some manufacturers fit sensors and an electric motor to the steering column (**FIGURE 26-25**). The sensors provide information to a steering control module. This can be used for electric power steering assist or for the electronic stability control system. The sensors are able to sense the rotational direction and speed exerted by the driver. The steering module is then able to control both the speed and the direction of the electric motor.

FIGURE 26-24 A flexible joint connects the steering column to the steering gear.

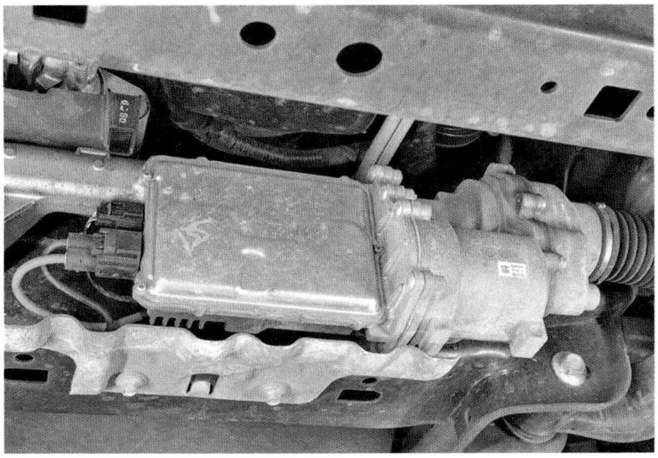

FIGURE 26-25 Steering column with electric motor and sensors.

FIGURE 26-26 **A.** Multifunction switch. **B.** Steering wheel entertainment controls.

Electric power assist can then be provided in either direction. This system will be discussed further in a later section.

The steering column generally accommodates for several controls:

- a horn button
- and an ignition switch and lock assembly, although more new vehicles are coming with a push-button start switch located in the dash.
- a multifunction switch (or separate switches). May include switches for lights, turn signal indicators, wipers, washer, and cruise control.
- The steering wheel itself may have integrated control switches for the entertainment system, cell phone control, and instrument panel display options (**FIGURE 26-26**).

Tilt/Telescoping Mechanism

Many manufacturers have added a tilting and/or telescoping function to their steering columns. These features allow drivers to control the steering wheel position. That way it can be set so that it best suits their needs. The tilting mechanism allows drivers to raise or lower the steering column position. The telescoping mechanism allows the driver to move the steering wheel closer or farther away. However, the amount of movement in any direction does have its limitations.

The tilt mechanism uses a heavy spring, a pivoting joint, and a ratchet mechanism (**FIGURE 26-27**). These added components make the steering column more complicated to repair. The telescoping mechanism includes a slip joint and locking mechanism. Both of these features tend to put additional stress on the wires in the steering column. This is caused by the increased movement allowed by the adjustment mechanisms. So, remember that when diagnosing steering column wiring faults.

Driver's Side Airbag

The driver's side airbag is designed to protect the driver in an accident. It provides a collapsing cushion that decelerates a driver's head and chest in a collision (**FIGRE 26-28**). One of the leading causes of vehicle fatalities is the rapid deceleration forces that occur within a body. The body's internal organs cannot handle the force of sudden deceleration. So, engineers have devised a variety of ways to slow deceleration during an accident. Airbags are one way to do that. The driver's side airbag prevents the head from contacting the steering wheel or dash.

Next to the seat belt, airbags are the biggest lifesaver in accidents. In most countries, airbag replacement is required as part of the collision repair process (**FIGURE 26-29**). If the airbag has been deployed in an accident, a new airbag and cover must be installed. This is required to meet legislative requirements in most states. At the same time, manufacturers may have further requirements. They may require specified components be inspected or replaced along with the airbags.

SAFETY TIP

Never lean over or in front of an armed airbag when working on the dash, instrument panel, or airbag system. Doing so can cause serious injury or death if it accidentally deploys. Always follow the manufacturer's procedure when disarming and arming the airbag.

Because of the critical nature of the airbag in an accident, it must always be connected electrically to its control module. This way it can be deployed when needed. Because the airbag is mounted in the steering wheel, a method of maintaining the electrical connection to the airbag is not easy. This is overcome by incorporating a **clock spring** into the steering column.

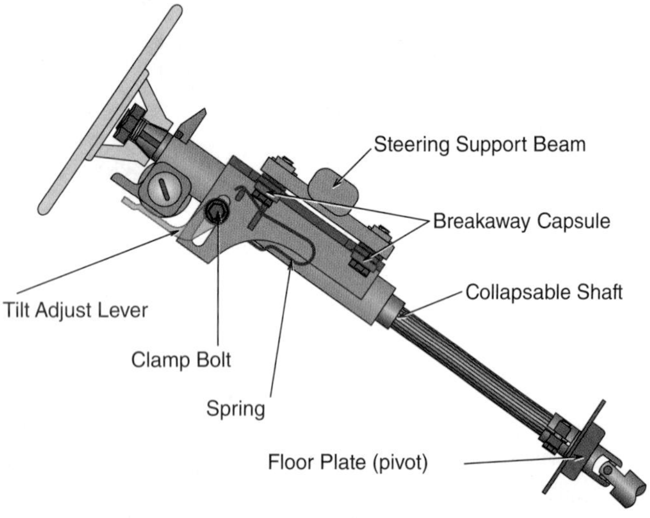

FIGURE 26-27 Typical tilt steering wheel mechanism.

FIGURE 26-28 The driver's side airbag being deployed in a crash.

The clock spring, which is also called a spiral cable, is a special rotary electrical connector (**FIGURE 26-30**). It is located between the steering wheel and the steering column. It maintains a constant electrical connection with the wiring while the steering wheel is being turned. The wires from the airbag connect through the clock spring. It coils and uncoils as the steering wheel is turned. This allows the airbag to maintain constant electrical contact.

When removing the airbag assembly from the steering wheel, the clock spring will need to be inspected. When the steering wheel is removed, the clock spring is easily visible. Inspect it to make sure that it and the clock spring wire are not damaged. Typically, if the electrical wiring inside the clock spring is damaged, the SRS light will turn on. This indicates that a code has been set. If the wiring has been damaged in any way, the clock spring must be replaced. It cannot be repaired.

▶ TECHNICIAN TIP

When removing the rack-and-pinion or steering gear, make sure the wheels are pointing straight ahead before disassembly. Also, either lock the steering column with the ignition key. Or use a steering wheel lock tool to prevent the clock spring from being moved off center. Make sure to also reassemble the parts in the straight-ahead position.

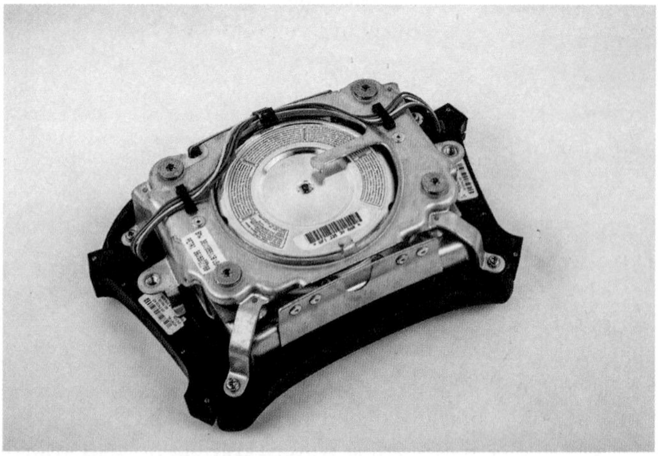

FIGURE 26-29 A replacement driver's side airbag. Never set it down this way!

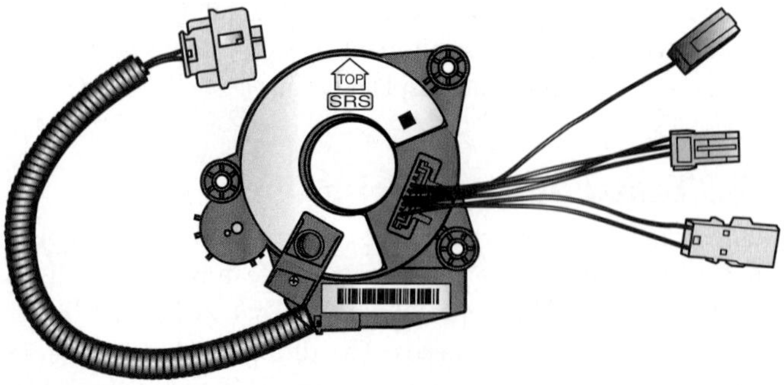

FIGURE 26-30 A typical clock spring arrangement.

▶ Steering Boxes

LO 26-05 Describe rack and pinion steering boxes.

There are a number of configurations of steering boxes (steering gears). But they are either manually operated or power assisted. Manual steering has no power source to assist the driver in turning the steering wheel. Whereas power steering can provide assist either hydraulically or electrically. If hydraulically assisted, a hydraulic pump is used. It can be driven by the engine or an electric motor. If electrically assisted, an electric motor directly assists the driver in turning the wheel.

Older vehicles were available with manual steering, as power steering was an expensive option. This is not the case now. Virtually all vehicles now come standard with a form of power steering.

The function of all steering boxes is the same, whether manual or power. They transfer the rotary motion of the steering wheel into the side-to-side motion needed to steer the wheels. They also provide gear reduction, which makes it easier to steer the wheels.

There are two basic types of steering boxes. Rack-and-pinion gearing is one type (**FIGURE 26-31**). The other type uses worm gearing and a sector shaft (**FIGURE 26-32**). As mentioned previously, both types have advantages and disadvantages. Some common variations of steering boxes are listed here:

- The rack-and-pinion gearbox
- The worm gearbox, consisting of
  - worm and sector,
  - worm and roller, or
  - worm and nut (commonly referred to as the recirculating ball).

Rack-and-Pinion Gearbox

The rack-and-pinion gearbox has a pinion gear that runs in mesh with the teeth on the rack. The pinion is turned by the steering column. The rack is connected to the tie rods. The meshing of the rack and pinion gives more direct operation. This enables the driver a better feel of the road through the steering wheel.

Turning the steering wheel rotates the pinion and moves the rack from side to side. On end take-off racks, ball sockets are mounted to the end of the rack. They provide flexibility during steering and suspension system movement (**FIGURE 26-33**). On center take-off racks, the tie rods connect to the center of the rack and pinion assembly (**FIGURE 26-34**). The center of the rack is what moves on this style of rack and pinion.

The teeth on both the pinion and the rack are helical gears. They enable smooth and quiet operation. The steering gears also provide mechanical advantage. The amount of mechanical

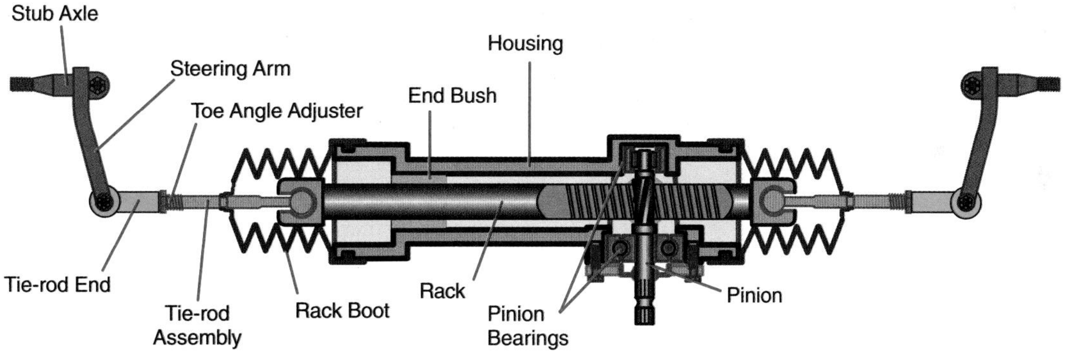

FIGURE 26-31 A rack-and-pinion steering gear.

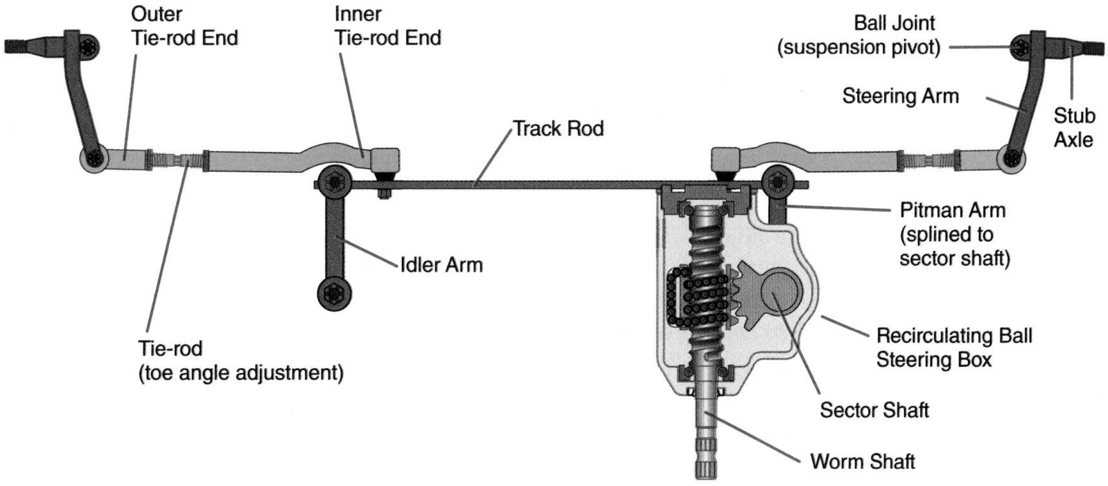

FIGURE 26-32 A worm steering gear.

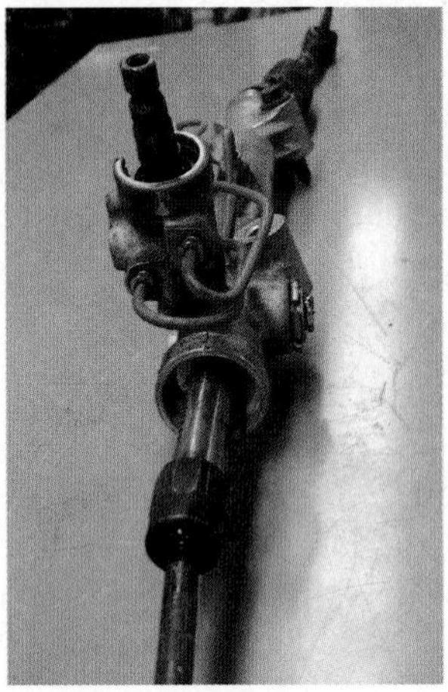

FIGURE 26-33 The end take-off rack uses ball sockets at each end of the rack.

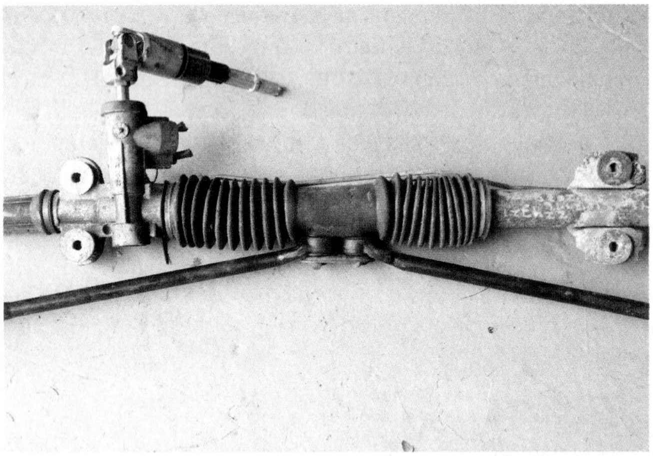

FIGURE 26-34 The center take-off rack uses ball sockets at the center of the rack.

advantage is determined by the steering **gear ratio**. This is the ratio, in degrees, between the turns of the steering wheel and the movement of the wheels. The value of this ratio depends on the size of the pinion. A small pinion gives easy steering. But it requires many turns of the steering wheel to turn the wheels from lock to lock. This is as far as the steering wheel can be turned from one side to the other. A large pinion means the number of turns of the steering wheel is reduced, but the steering wheel is harder to rotate.

In fixed-ratio rack-and-pinion systems, the ratio is the same for all positions of the wheels. It is a fixed ratio. This means that the rack-and-pinion system produces a single ratio from lock to lock. The steering ratio is calculated by turning the steering wheel one full turn and checking the number of degrees that the wheel moves. For example, if one turn of the steering wheel produces 20 degrees of wheel movement, then dividing 360 degrees of steering wheel rotation by 20 degrees of wheel movement results in an 18:1 ratio. Steering ratio is not the same on all vehicles. It is designed by engineers to provide a certain steering response for that particular vehicle.

In some applications, a variable-ratio gear is used. It allows a slow turn rate when the steering wheel is centered. But when turned

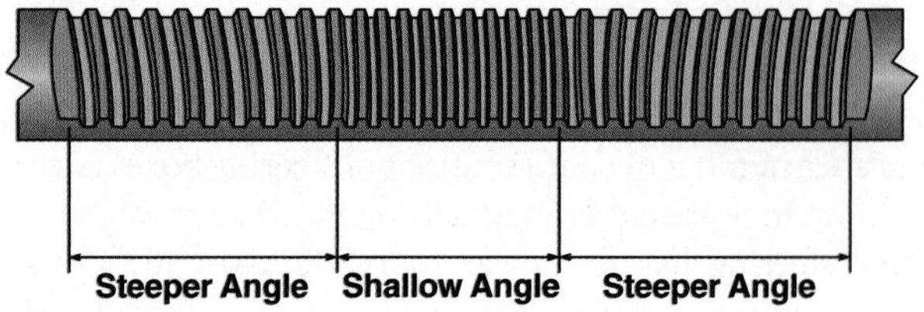

FIGURE 26-35 A variable-ratio rack showing the different angles on the teeth.

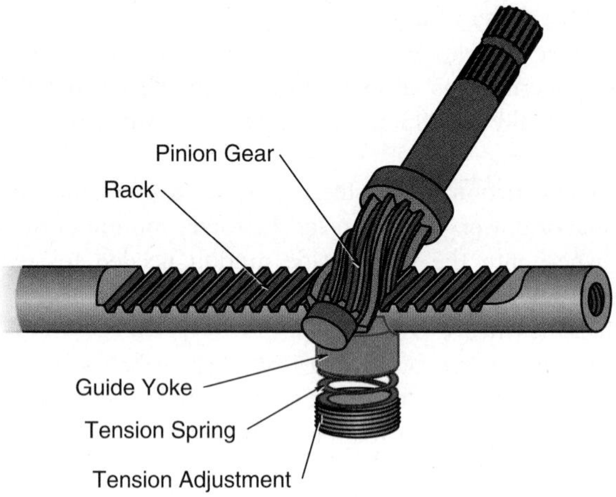

FIGURE 26-36 A spring-loaded rack guide yoke holds the rack against the pinion gear.

toward the steering lock, a faster rate is used. This feature provides very stable steering when traveling straight at high speeds.

The variable-ratio rack-and-pinion system uses a specially designed pinion gear tooth, which is more rounded. The rack has teeth that are narrowly spaced (at a shallow angle) in the center and further apart (at a steeper angle) toward the end (FIGURE 26-35). Closer (shallow angle) teeth provide slower movement of the rack in relation to the steering wheel.

On other vehicles, it is preferable that the steering be more responsive near the center point. This is ideal for maneuvering at slower speeds. It is accomplished by having widely spaced (at a steeper angle) teeth in the center of the rack, and closer spaced (at a shallow angle) teeth near the ends. This also helps prevent stiffer steering toward the full lock positions. In this position, the steering arm leverage is reduced.

A disadvantage of the rack-and-pinion steering system is that it typically is not manually adjustable. Over time, it does wear and develop backlash between the pinion teeth and the rack teeth. This is felt as play (looseness) in the steering wheel. In many cases, replacement of the rack and pinion is the only cure for excessive gear backlash.

The steering rack is supported at the pinion end by being sandwiched between the pinion and a **spring-loaded rack guide yoke** (FIGURE 26-36). It is sometimes called the rack bearing. Spring-loaded rack guide yokes are made of metal, nylon, or other durable material. They have a spring that pushes on the back side of the rack. This helps reduce the play between the rack and the pinion while still allowing for relative movement. There may be an adjuster plug that screws into the body

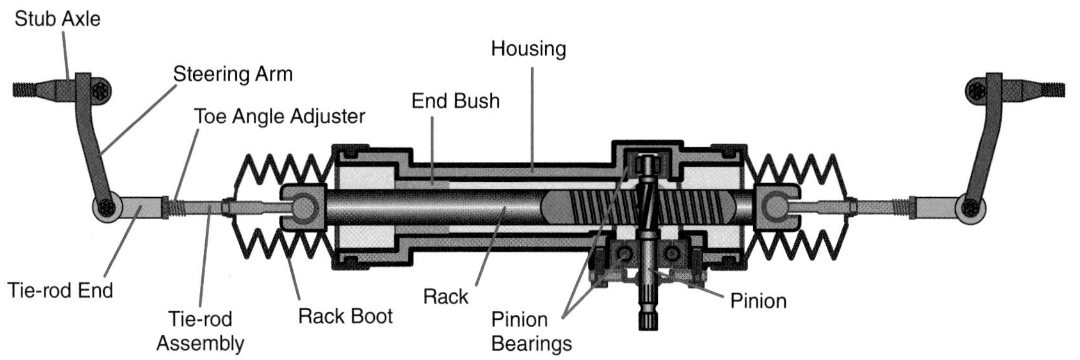

FIGURE 26-37 The pinion is supported by two bearings in the rack housing.

of the rack. It is there to put pressure on the rack guide yoke. This adjustment produces the correct mesh of the rack teeth to the pinion teeth. It also affects the amount of torque required to turn the pinion. If the setting is too tight, it will be harder to turn the pinion. This may result in a steering wheel that will not return properly. It is sometimes referred to as memory steer. Do not adjust the mesh of the rack-and-pinion gears unless directed by service information.

The pinion is supported by two bearings in the rack housing (see **FIGURE 26-37**). The bearings must be **preloaded** (lightly compressed). This holds the pinion in the correct position, relative to the rack. It also eliminates free play.

A rack-and-pinion steering box is normally lubricated by grease. Each end of the rack is protected from dirt and water by flexible synthetic rubber bellows. It is attached to the rack housing and to the tie rod. The rubber bellows extend and collapse as the rack moves. On some vehicles, the rubber bellows are interconnected by a tube. As the steering wheel is turned, air is transferred from the collapsing bellows side to the expanding bellows side. This process keeps the rubber bellows from collapsing.

With rack-and-pinion steering, the rack is directly threaded to the **tie-rod assembly**. The tie-rod assembly is attached to the steering knuckle. The tie-rod assembly consists of an inner and an outer tie-rod end that are threaded together.

Applied Math

AM-1: PROBABILITY: THE TECHNICIAN CAN RELATE PROBLEM SYMPTOMS TO THE PROBABILITY OF THE MALFUNCTION OF A SPECIFIC PART OR SYSTEM.

A common problem with steering systems, particularly in older vehicles, is excessive free play. Typically, the customer will describe the problem as the steering feeling "loose" or "sloppy." A road test will quickly verify the concern. There will be excessive movement of the steering wheel in the straight-ahead position, before the wheels even begin to turn.

With a pickup truck that has a parallelogram steering system, the excessive free play could be caused by wear in a number of points within the system. There is a higher probability of the fault being caused within the steering linkage than within the steering gearbox. As a result, components such as tie-rod ends, idler arms and bushings, and pitman arms and bushings should be checked and eliminated first, before inspecting the steering gearbox.

The inner tie rod is threaded onto the rack and has a ball-and-socket joint to allow swiveling movement. The other end of the inner tie rod is threaded to allow attachment of the outer tie-rod end, which provides the method used to adjust the toe angle. The steering arm, the **stub axle** knuckle, and the **stub-axle carrier** (the body of the stub axle knuckle) can be forged as one piece and referred to as a steering **knuckle**. They can also be made as separate units and assembled to form one piece (**FIGURE 26-38**).

FIGURE 26-38 Two-piece steering knuckle and steering arm.

▶ Worm Gearbox

LO 26-06 Describe worm gear steering boxes.

Worm gear steering boxes made the process of steering the front wheels an easier task for drivers of early automobiles. The worm gearbox uses two gears: a worm and a worm gear (also called a worm wheel) (**FIGURE 26-39**). The worm has teeth cut in a helical (spiral) shape and operates the same as a screw. In this case, the helix on the worm moves the worm wheel one tooth for each revolution of the worm. It converts the rotary motion of the steering wheel to the linear motion needed to control the wheels.

A worm gear also produces a large gear reduction. The gear ratio of the gearbox is a comparison of the angles of movement between the steering wheel and the wheel assembly. It is expressed as a ratio, for example, 18:1. The gear ratio of a worm gearbox increases output torque and reduces the effort the driver has to apply.

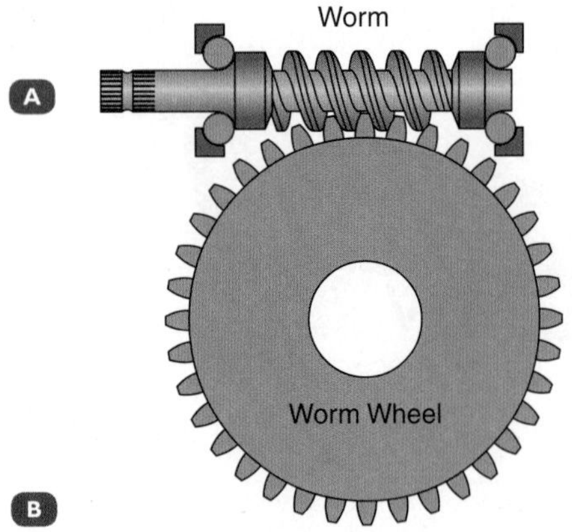

FIGURE 26-39 The parts of a worm gear assembly: **A.** Worm. **B.** Worm gear (or worm wheel).

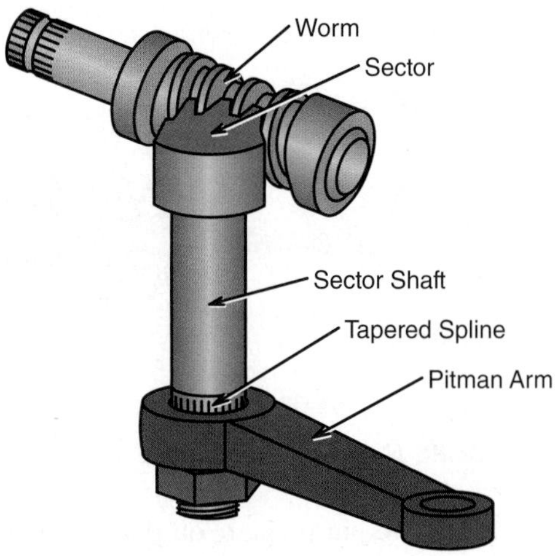

FIGURE 26-41 A worm-and-sector type of worm gear steering box.

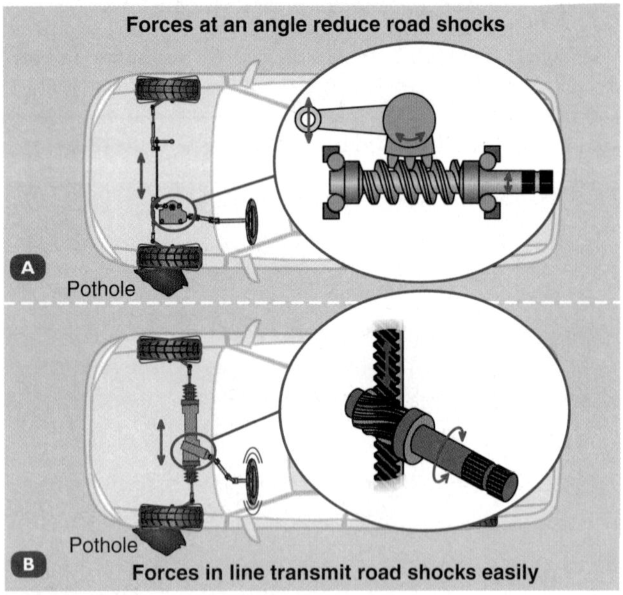

FIGURE 26-40 **A.** Road shock is minimized in a worm gear arrangement. **B.** Road shock is much larger in a rack-and-pinion gear.

FIGURE 26-42 The worm-and-roller steering gearbox.

Worm gears have another benefit. They do not transmit nearly as much road shock back to the driver as a rack-and-pinion gear assembly (**FIGURE 26-40**). When the worm is driven (by the steering wheel), the worm wheel moves easily. This is because each turn of the worm advances the worm wheel one tooth. But if the worm wheel is driven (such as by road shock), the worm wheel teeth butt up against the teeth on the worm. This prevents most of the force from being transmitted as rotary motion to the worm. This is because most of the force from the worm wheel is nearly perpendicular to the rotation of the worm gear.

Worm gear steering boxes are partially filled with a lubricant. The fluid lubricates the bearings and gears inside the steering gearbox. Seals and gaskets are used to contain the fluid, but they do wear out and leak. When inspecting steering gearboxes, always inspect them for fluid leaks. If found, inform your supervisor.

Several forms of the worm gearbox have been used throughout the history of the automobile. Three popular styles have included the:

- worm-and-sector,
- worm-and-roller, and
- recirculating ball gearboxes.

Worm-and-Sector Gearbox

The first worm gearbox was of the worm-and-sector style. The worm is meshed with a sector, or portion of a gear, mounted on its own shaft (called a sector shaft) (**FIGURE 26-41**). The outer end of the sector shaft has a spline that mates with an internal spline on the pitman arm. As the steering wheel rotates, the worm causes the sector to move through an arc. The sector shaft transfers the motion through the pitman arm to the steering linkage. This style of gearbox produces more friction between the gears than later worm gearboxes. Steering is more difficult because of this.

Worm-and-Roller Gearbox

The worm gear in a worm-and-roller steering box has an hourglass shape. It meshes with a double-ribbed roller. The roller is free to rotate on bearings which ride on a pin attached to the pitman shaft (**FIGURE 26-42**). As the driver turns the steering wheel, the worm rotates, and the roller rolls against the worm,

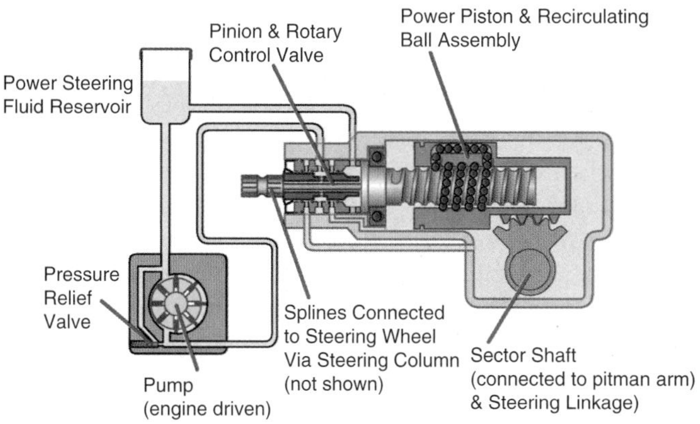

FIGURE 26-43 The recirculating ball steering system uses a worm gear that moves a ball nut by turning against ball bearings, reducing the force needed to turn the wheels.

FIGURE 26-44 A sector gear adjustment screw and lock nut maintains the proper engagement between the gears.

moving the roller in an arc. It follows the hourglass shape, and transfers the motion to the pitman shaft. The hourglass shape changes the steering ratio slightly as the roller nears each end.

The worm-and-roller gearbox was an improvement to the worm-and-sector style. It reduces the friction between the two gears, due to the roller. This makes turning the steering wheel easier.

Recirculating Ball Gearbox

The worm-and-nut steering gear is also known as a recirculating ball steering gear. The **recirculating ball steering box** contains a worm gear inside a ball nut. The ball nut has a very coarse, threaded hole, in it and gear teeth cut into its outside edge. The gear teeth engage teeth on the sector shaft to rotate the pitman arm. It was developed in 1940 and is an improved version of the worm-and-sector and worm-and-roller types of worm gear.

In the recirculating ball gearbox, both ends of the **worm shaft** are supported in the housing by angular bearings. They are preloaded to reduce lateral movement, known as end play. They also resist side thrust movement of the worm gear when it is under load.

A ball nut rides on the worm gear. Many steel balls roll between the spiral grooves of the worm and the inside of the nut (**FIGURE 26-43**). With rotation, the balls are rolled along the grooves, partly in the worm and partly in the nut. The balls form a low-friction internal thread. This causes the nut to thread up or down on the worm as it rotates. When the balls get to the end of the groove, they travel through **ball-return guides** to the other end of the nut. Ball-return guides are simply special passages through which the balls move.

▶ TECHNICIAN TIP

The steering box may be mounted so that the steering linkage is in front of (front steer) or behind (rear steer) the suspension cross member. When it is behind the cross member, it is protected by the cross member from possible damage. This position of the steering box also reduces the length of the steering column.

The sector gear and nut teeth are designed so that when the teeth are in the straight-ahead position, they have minimum clearance. This is called the high point. This design reduces free play when the steering wheel is straight.

The pitman shaft is supported by two caged needle roller bearings in the steering box housing. The sector teeth are angled. An adjustment mechanism on the steering housing cover allows for adjustment of the sector gear height. This adjustment provides proper engagement of the sector gear and nut teeth (**FIGURE 26-44**).

▶ TECHNICIAN TIP

In forward-control vehicles, the steering system is mounted in front of the engine and wheels. The steering box is mounted on the frame, with the pitman arm vertical (**FIGURE 26-45**). A **drag link** transfers movement of the pitman arm to a **relay lever**. The relay lever has two arms—one connected to the drag link and the other to the **track rod**. The track rod is supported on the other end by the idler arm. The longitudinal movement of the drag link pivots the relay lever. The movement is transferred through the track rod to the idler arm. And from there through the tie rods to the steering knuckles.

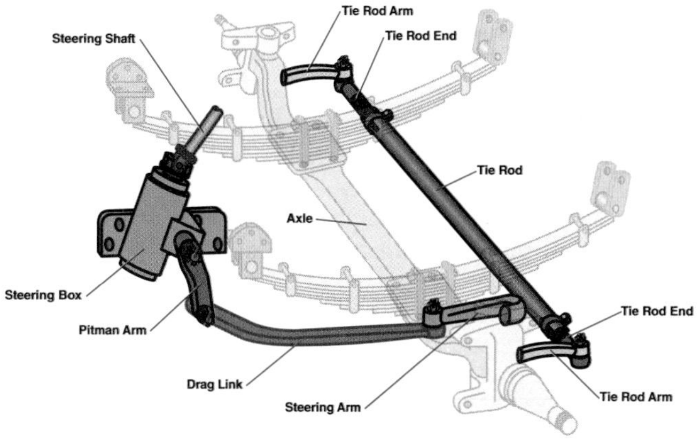

FIGURE 26-45 Forward control steering.

▶ Power Steering

LO 26-07 Describe hydraulic power steering system operation.

As vehicle design progressed, more steering effort became necessary. Front-wheel drive added weight to front end and wider low-profile tires are harder to steer. As a result, **power steering** was introduced. Power steering assists the driver in steering the wheels. The driver still needs to initiate the steering effort. But with just a little effort, the assist kicks in and helps. The power steering system is designed so that the vehicle can still be controlled even if the engine or the power steering system were to fail. If this were to happen, greater steering effort by the driver would be required.

There are three types of power steering:

1. Hydraulically assisted power steering
2. Electrically powered hydraulic steering
3. Fully electric power steering

Hydraulically Assisted Power Steering

Hydraulically assisted power steering uses pressurized hydraulic fluid to assist the driver in steering the wheels. This design is especially helpful at slower vehicle speeds, when steering wheel turning effort is much higher. An engine-driven hydraulic pump pressurizes the hydraulic fluid. The pressure operates the power unit at the steering gearbox, or rack-and-pinion assembly. Hydraulic pressure is delivered by the control valve and connecting hoses and pipes. A fluid reservoir holds a supply of power steering fluid. It can be mounted on the pump, or it can be remotely mounted.

With the engine running, fluid flows continuously through the system. It flows from the **power steering pump** to the steering gear control valve. The control valve sends it to the power unit and then back to the power steering pump. With the steering wheel in the neutral position, minimal pressure is needed. This means that minimal engine power is needed to operate the system in the neutral position. When additional assist is needed, the control valve redirects the fluid flow. The fluid that was freely returning to the pump is redirected to a power unit where it can provide assist. We will cover this in detail in a bit.

Some manufacturers task the power steering system with double duty. The pressure from power steering pumps has been used to power hydraulic brake boosters on some vehicles. It has also been used to power the radiator fan on a number of vehicles. When working on those systems, be aware how the power steering system is integrated into the other systems.

Power Steering Fluid and Hoses

Power steering fluid performs several functions. It must:

- withstand high temperatures and pressures.
- lubricate the pump and steering gear.
- preserve the system seals and pressure hoses.
- be able to flow freely at very cold temperatures.

Many manufacturers have specified certain types of automatic transmission fluid for their power steering systems over the years. Others call for specific formulations of a separate power steering fluid (**FIGURE 26-46**). Because there is typically only a quart or two of power steering fluid in the system, it gets worked pretty hard. The harder it is worked, the hotter it gets. The hotter it gets, the shorter its life. For this reason, some vehicles are equipped with a power steering fluid cooler (**FIGURE 26-47**). This is usually built into the high-pressure line. It helps cool the fluid before it gets to the steering gear.

Power steering fluid also gets contaminated with rubber and metallic particles from internal wear in the system. Because of this, some manufacturers install a replaceable filter in the system (**FIGURE 26-48**). Many manufacturers do not specify a power steering fluid change interval. But the fluid does degrade over time. And as mentioned, it becomes contaminated. Because

FIGURE 26-46 An example of one type of power steering fluid. Check the service information to determine the right kind for the vehicle.

FIGURE 26-47 A typical power steering fluid cooler mounted in the high-pressure line.

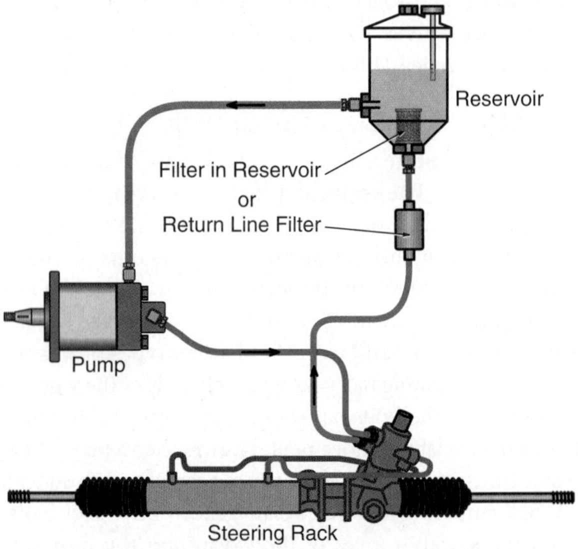

FIGURE 26-48 Some vehicles use a power steering system filter mounted either in the reservoir or in the return line.

of this, many technicians recommend flushing power steering fluid every 50,000 to 100,000 miles.

Power steering hoses are used to carry power steering fluid from the pump to the steering gear. This is the high-pressure hose and is usually made of flexible, high-pressure hose material (**FIGURE 26-49**). The return hose runs from the steering gear back to the pump reservoir. It carries fluid under much lower pressure. These hoses must also allow movement between the engine and chassis, so they can't be too stiff.

Over time, power steering hoses can become weak or damaged. If they leak, they can quickly cause the system to run dry. If the high-pressure hose leaks, it can spray hot power steering fluid all over the engine and exhaust. Power steering fluid is flammable, so leaks can cause extreme fire hazards. Because of this, they should be inspected for seepage or wear during each oil change. Also, most high-pressure power steering hoses use an O-ring to seal the end of each hose to the pump and steering gear (**FIGURE 26-50**). These seals can wear or leak, requiring replacement of the O-ring.

Steering Process

The hydraulic pressure is controlled by a rotary valve. It is located on the input shaft of the steering gear. Seals in the rotary valve prevent internal and external fluid leakage. When the steering wheel is turned, the rotary valve directs fluid to one side or the other of a piston. The piston is attached to the steering gear. Pressure then increases as required to provide steering assistance.

In a rack-and-pinion steering gear, the piston is formed centrally on the steering rack. The rack housing is the working cylinder (**FIGURE 26-51**). Pressure seals at each end of the cylinder isolate the **power section** from the rest of the rack and pinion.

FIGURE 26-49 High-pressure power steering hose.

FIGURE 26-50 A high-pressure power steering hose using an O-ring to seal the hose to the pump.

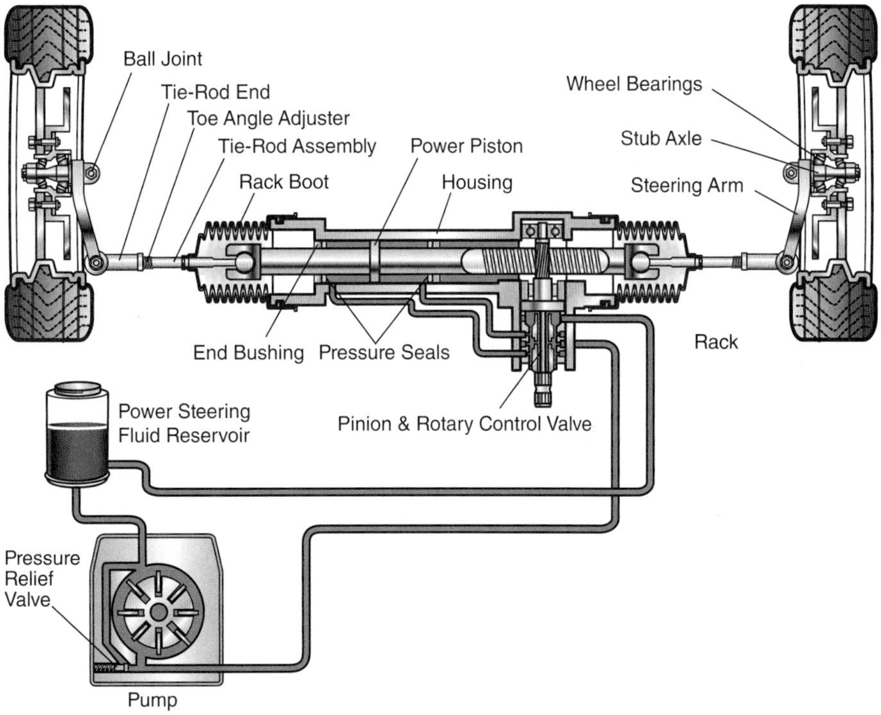

FIGURE 26-51 Power-assisted rack-and-pinion system.

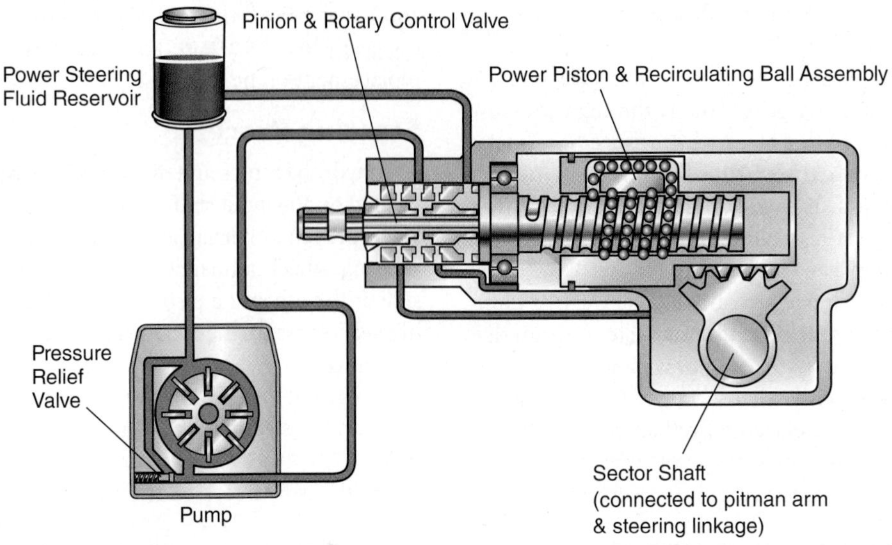

FIGURE 26-52 Power-assisted recirculating ball gearbox.

In a recirculating ball steering box, the power piston and seals slide in a cylinder in the housing (**FIGURE 26-52**). The power piston has an extension formed on one side, with teeth that engage on the sector shaft. Pressure applied to either side of the power piston produces a force. The force is transferred through the teeth, to help turn the sector shaft.

Connecting pipes transfer fluid from the rotary valve housing to one side of the piston at a time. This provides assistance, which acts directly on the steering gear. The rotary valve is located between the steering gear input shaft and the pinion gear, or worm. It consists of an inner member that is attached to the input shaft. A surrounding sleeve member is attached to the pinion gear or worm. It appears as a shaft within a shaft (**FIGURE 26-53**).

Turning the steering wheel makes both members rotate in the steering gear housing. But it is the slight rotary displacement of the inner member in the sleeve that directs the power steering fluid flow. This slight rotary displacement is allowed by a **torsion bar**, which is a spring-loaded steel rod. Its bottom end is connected to the pinion gear or worm and the outer member of the rotary valve. The top is connected to the input shaft and inner member of the rotary valve.

FIGURE 26-53 Power steering rotary valve and torsion bar.

When the steering wheel is turned, there is resistance from the front wheels at the road surface. This resistance is transmitted through the steering gear. The input shaft twists the torsion bar slightly. Because the inner member is also attached to the input shaft, this twisting provides a relative rotary displacement of the inner and outer members. It is this displacement that lets fluid flow through the valve to act on the piston at the steering gear. The input shaft can twist through only a small angle before it contacts a stop on the pinion gear or worm. This is needed as a fail-safe to provide manual steering when power assistance is not available.

With the steering in the neutral position, fluid flow is directed into and out of the valve assembly. The inner member lets fluid pass equally to both sides of the rack piston and return to the fluid reservoir. Equal, but low, pressure is applied to both sides of the power piston. No power assistance is provided.

When the steering wheel is turned, fluid is restricted from making a free return to the reservoir. It is now directed to the side that matches the turning action (**FIGURE 26-54**). At the same time, fluid on the opposite side is directed to the return circuit, back to the reservoir. Slight deflection of the valve gives a small amount of assistance. Assist becomes progressively greater as the torsion bar flexes further. The grooves of the inner member are precisely shaped. This allows them to meter the flow of fluid between the apply and release passageways.

Idle Speed Strategy

When steering a vehicle at slow speeds, the power steering demand will be high. But the engine speed will be low. The additional power steering pressure puts an added load on the engine. If the idle speed is not maintained during this condition, the engine could possibly stall.

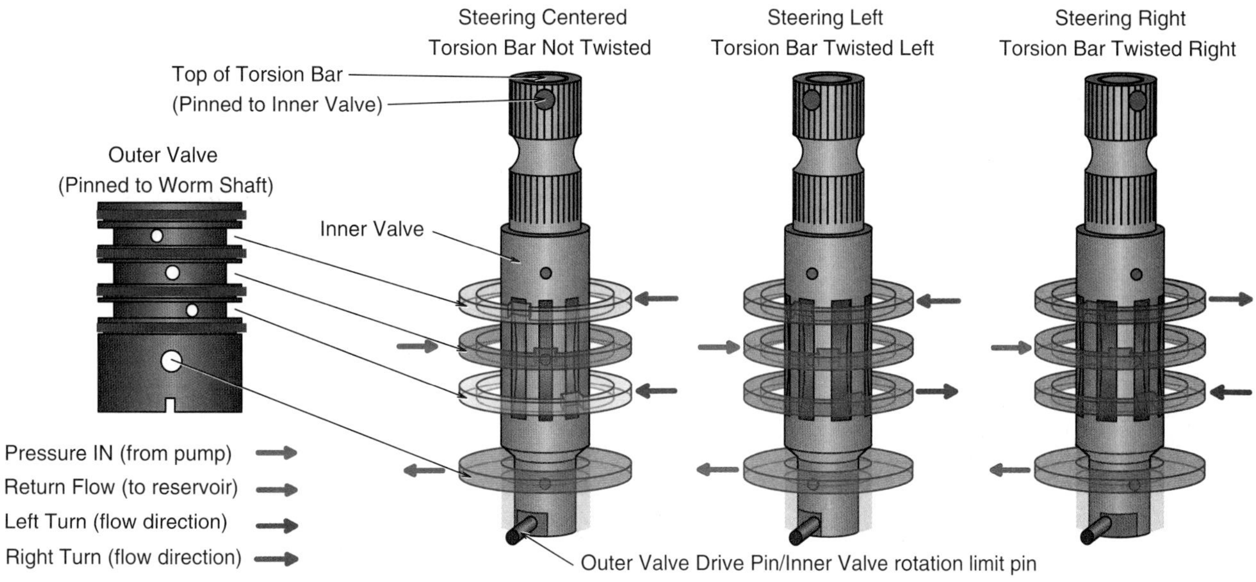

FIGURE 26-54 Slight flexing of the torsion bar allows fluid to be increased or decreased to either side of the power piston in the steering gear.

To prevent stalling, engineers program an idle speed strategy into the PCM. The PCM raises or maintains the idle speed when power steering pressure rises above a specified point. A power steering switch or sensor inputs power steering pressure information to the PCM. Based on this input and the actual engine idle speed, the PCM is able to raise or maintain the idle speed (**FIGURE 26-55**). It does so by increasing the amount of idle air flow. This is similar to opening the throttle slightly. When the power steering pressure is reduced, the PCM reduces the amount of idle air flow.

▶ Electric Power Steering

LO 26-08 Describe electric power steering system operation.

Electronics allow much more refined control in automotive steering systems. There are two types of electric power steering systems that we will discuss. One is pure electric, and the other is electrohydraulic. Both are alternatives to conventional hydraulic power steering systems. Electric power steering has several advantages over conventional hydraulic power steering systems:

- It is less expensive to run.
- It is easier to package and install.
- It is lighter and more compact.
- It also reacts faster to quick steering changes from the driver.

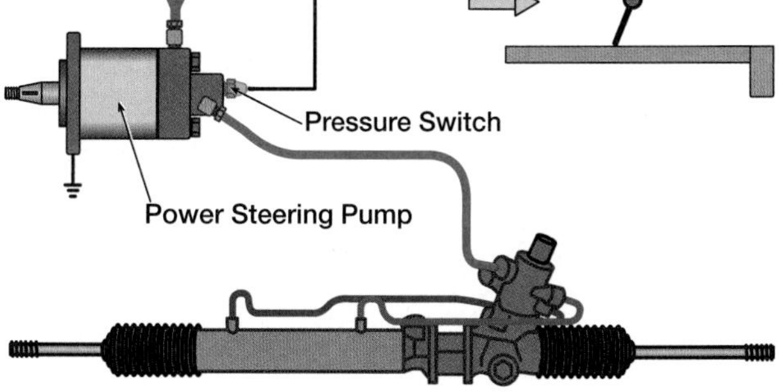

FIGURE 26-55 Idle speed control strategy. A power steering switch is activated by increased power steering system pressure. The PCM uses this signal to increase idle air flow to prevent the engine from stalling.

In **electrically powered hydraulic steering (EPHS)**, a brushless motor is used to drive the pump (**FIGURE 26-56**). It replaces the drive belts and pulleys that drive a conventional power steering pump. The pump is driven by an electric motor to reduce power drawn from the engine. Unlike an engine-driven pump, EPHS only operates on demand. Pump speed is regulated by an electronic controller to vary pump pressure and flow. This provides steering efforts tailored for different driving situations.

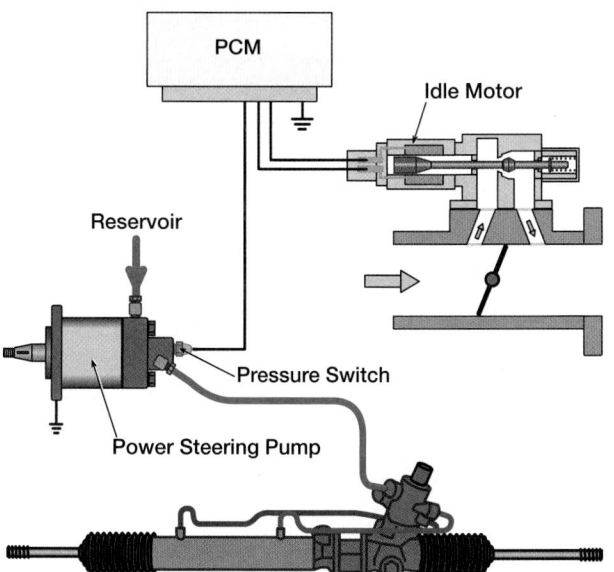

FIGURE 26-56 An electrically powered hydraulic steering pump and motor.

The pump can be run at low speed or shut off to provide energy savings during straight-ahead driving.

An EPHS system is said to use only 20% as much of the engine power used by a standard belt-driven pump. This improves fuel mileage substantially. The engine still contributes power to the steering system through electrical demand on the alternator. But it is greatly reduced from that of hydraulic power steering systems.

Electrically assisted steering (EAS) is a completely electrically powered power-assist system. It eliminates all hydraulic components and fluid. An electric motor replaces the hydraulic pump. EAS, or direct electric power steering, is a fully **electric power steering (EPS) system**. The reduction of components saves weight and reduces drag on the engine. This vastly improves fuel mileage. The EPS system is said to require only about 2% of the engine power that a belt-driven power steering pump uses. There is still a small amount of power required from the engine, which supplies the electrical demand. But it is greatly reduced.

An EPS steering system uses an electric motor to provide assist. It attaches either to the steering rack or to the steering column via a gear mechanism (**FIGURE 26-57**). It also incorporates a torque sensor. A microprocessor or electronic control unit control the steering dynamics and driver effort. Inputs include vehicle speed, steering wheel torque, angular position, and turning rate.

There are four primary types of electric power assist steering systems:

- Column-assist type: In this system, the **power assist unit**, controller, and **torque sensor** are attached to the steering column (**FIGURE 26-58**). The power assist unit is the electric motor. The controller is the electronic control unit. And the torque sensor measures the load on the steering wheel.
- Pinion-assist type: In this system, the power assist unit is attached to the steering gear pinion shaft. The power assist unit sits outside the vehicle passenger compartment. It allows assist torque to be increased greatly without raising interior compartment noise.
- Rack-assist type: In this system, the power assist unit is attached to the steering gear rack. It is located on the rack to allow for greater flexibility in the layout design (**FIGURE 26-59**).
- Direct-drive type: In this system, the rack-and-pinion and power assist unit form a single unit (**FIGURE 26-60**). The steering system is compact and fits easily into the engine compartment layout. Direct assistance to the rack enables low friction and **inertia** (resistance to a change in motion), which in turn gives an ideal steering feel.

In these systems, **active control** provides constant feedback from sensors to the **control unit**. The control unit processes

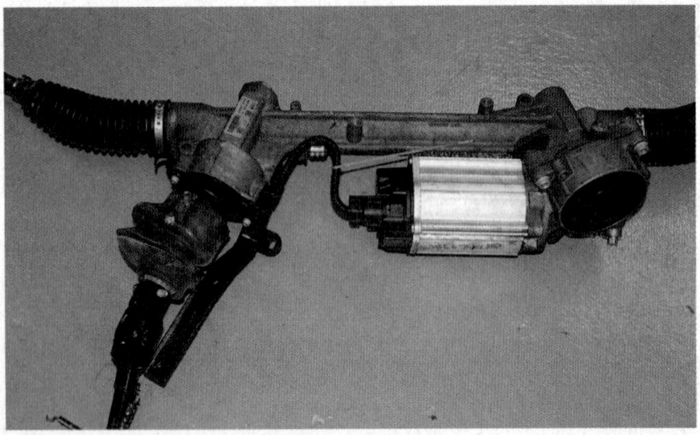

FIGURE 26-57 A typical electronic power steering rack-and-pinion assembly.

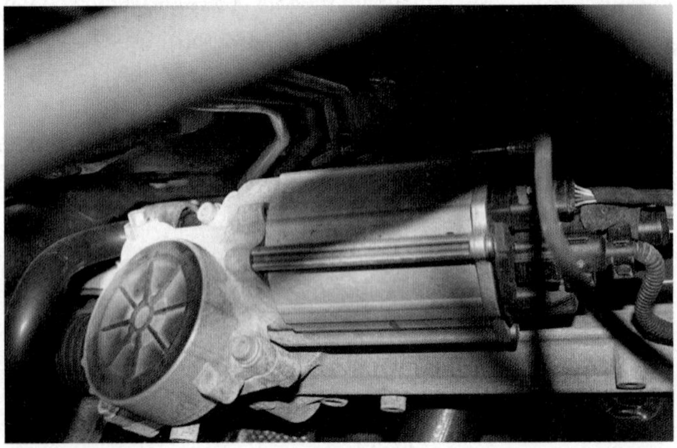

FIGURE 26-59 Rack-assist steering type.

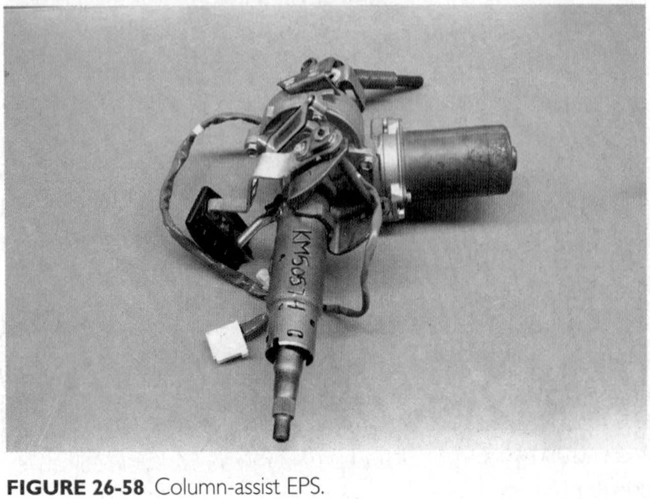

FIGURE 26-58 Column-assist EPS.

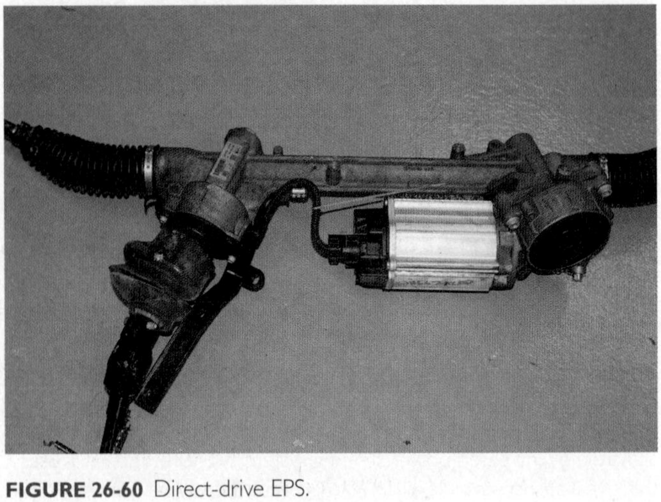

FIGURE 26-60 Direct-drive EPS.

data based on sophisticated computer algorithms. These features allow the steering system to react to the road, the weather, and even the type of driver. It can then provide assistance to the front or rear wheels, independent of direct driver input. For example, the electronic stability control system detects the start of an oversteer condition. The control unit can actually cause the EPS to steer the wheels in the opposite direction. It does so to prevent loss of control of the vehicle. Thus, an EPS system might steer the vehicle opposite of the driver's input if that input would lessen the control of the vehicle.

Basic EPS Operation

A steering sensor is located on the input shaft, where it is bolted to the gearbox housing. The **steering sensor** performs two functions. First, as a torque sensor, it converts steering torque input into voltage signals. The PCM monitors these signals. Second, as a rotation sensor, it converts the rotation speed and direction into voltage signals. An interfaced ECU converts the voltage signals from both sensors into signals that the PCM can process. Once processed, the PCM ultimately provides the proper output signal to the EPS motor.

The PCM also analyzes inputs from the vehicle's speed and wheel speed sensors. The sensor inputs are then compared to the **forces capability map data** stored in the PCM's memory. These map data are preprogrammed by the manufacturer. The PCM sends the appropriate command to the **power unit**. It supplies the electric motor with the necessary current to operate as commanded. The electric motor then pushes the rack to either the right or the left. The direction of rack movement depends on which way the current flows. Reversing the current flow reverses directional rotation of the electric motor. Increasing current to the electric motor increases the amount of power assist.

The EPS system has three operating modes:

- Normal control mode—provides left or right power assist in response to input from the torque and rotation sensor's inputs.
- Return control mode—assists steering return after completing a turn.
- Damper control mode—adjusts the amount of assist according to the vehicle speed to improve road feel and dampen kickback.

If the steering wheel is held in the full-lock position, steering assist reaches maximum. In this case, the power unit reduces current to the electric motor. It does so to prevent an overload situation. The power unit is also designed to protect the electric motor against voltage surges from a charging system fault.

The electronic steering control unit is capable of self-diagnosing faults. It does so by monitoring the system's inputs and outputs. It also monitors the amount of current flow the electric motor is drawing. If a problem occurs, the electronic steering control unit turns the system off. It does this by **actuating** (turning on) a fail-safe relay in the power unit. This eliminates all power assist, causing the system to revert to manual steering. An in-dash EPS warning light is also illuminated to alert the driver (**FIGURE 26-61**).

Higher Voltage Electrically Assisted Power Steering

Some electrically assisted power steering systems operate on higher voltage. They were developed for hybrid and electric vehicles (**FIGURE 26-62**). In hybrid vehicles, the engine is typically off during idling, deceleration, and braking. Thus, an EPS system was needed. The high-voltage battery provides all the power, with no reliance on engine or hydraulic power. The parts are similar to an electrically assisted power steering system.

These systems do not use the full battery voltage from the high-voltage battery pack. Instead, it is reduced to a lower voltage through a DC-to-DC converter. The converter basically just steps the voltage down. Some early hybrids simply used the 12-volt system that was used on gasoline vehicles. But a few models used voltage as high as 46 volts. Others used a mid-range voltage of 25–30 volts. With the development of more powerful electric motors, the manufacturers seem to be going back to 12-volt power steering motors. These are simpler systems to design and install on vehicles.

At the time, there were several advantages of using higher voltage to operate electric power steering motors. One was to create more power in the same-size motor. Smaller gauge wires could also be used in the electric steering system. This reduces weight in the vehicle. And these lighter weight components increase fuel economy slightly.

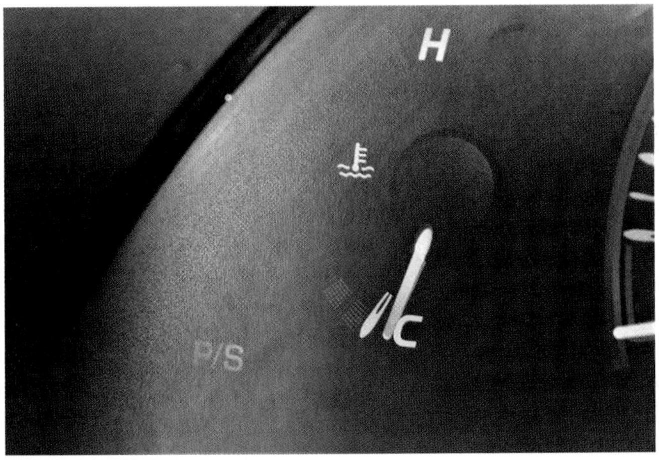

FIGURE 26-61 EPS light illuminated on the dash.

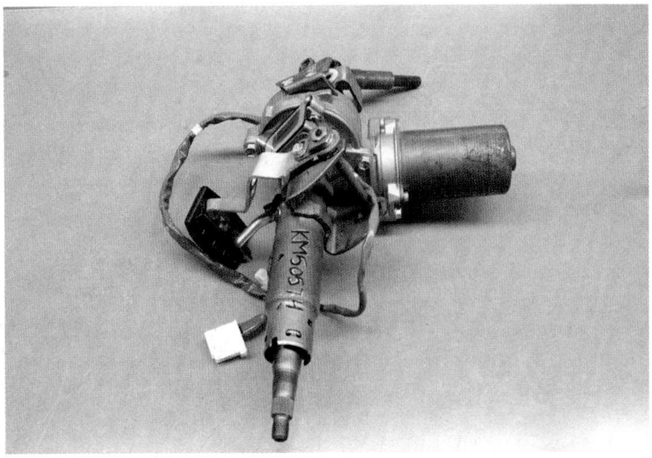

FIGURE 26-62 Hybrid vehicle EPS.

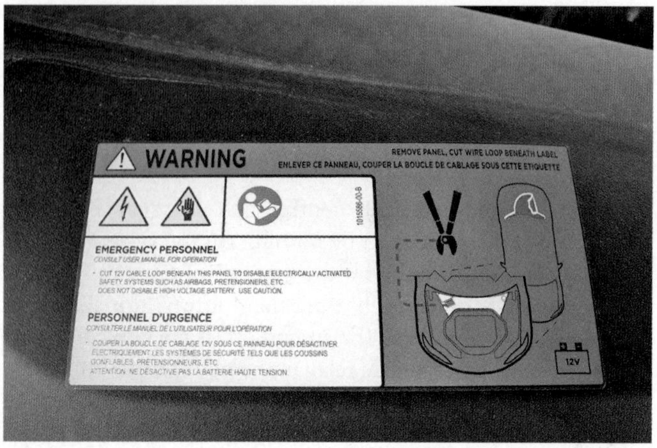

FIGURE 26-63 Even though EPS may not operate at the full high voltage, the high-voltage system will be activated.

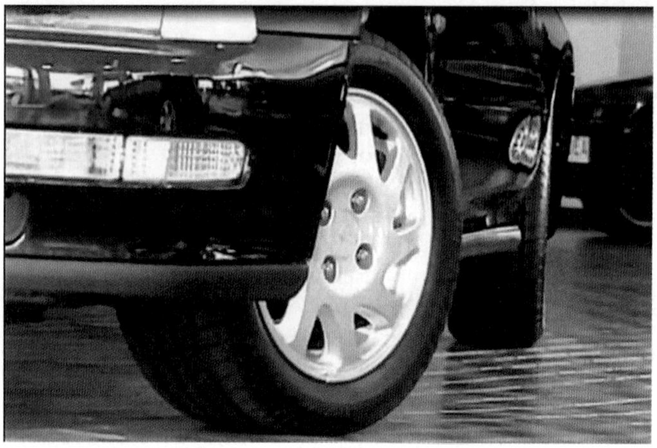

FIGURE 26-64 Four-wheel steering showing the rear wheels being steered in conjunction with the front wheels.

There is one very important thing to remember if you work on one of these higher voltage power steering systems. Voltages under 50 volts are usually safe against electric shock hazards. However, if the 46-volt system is live, so is the high-voltage system that you will be working around. And that requires special precautions:

- high-voltage gloves,
- a high-voltage meter,
- special insulated tools, and
- special training.

So, higher voltage electric power steering systems should only be serviced by highly trained individuals (**FIGURE 26-63**).

SAFETY TIP

On vehicles with high-voltage systems (50 volts or higher), you may be required to wear insulated gloves appropriate for the system voltage you are working on. Always make sure the gloves have been tested for electrical leaks within the required time frame. And always test the gloves before using them. Do this by rolling up the sleeve, trapping air inside, and verifying that there are no air leaks. If there are any air leaks, *do not* use them, as electricity could travel through the hole.

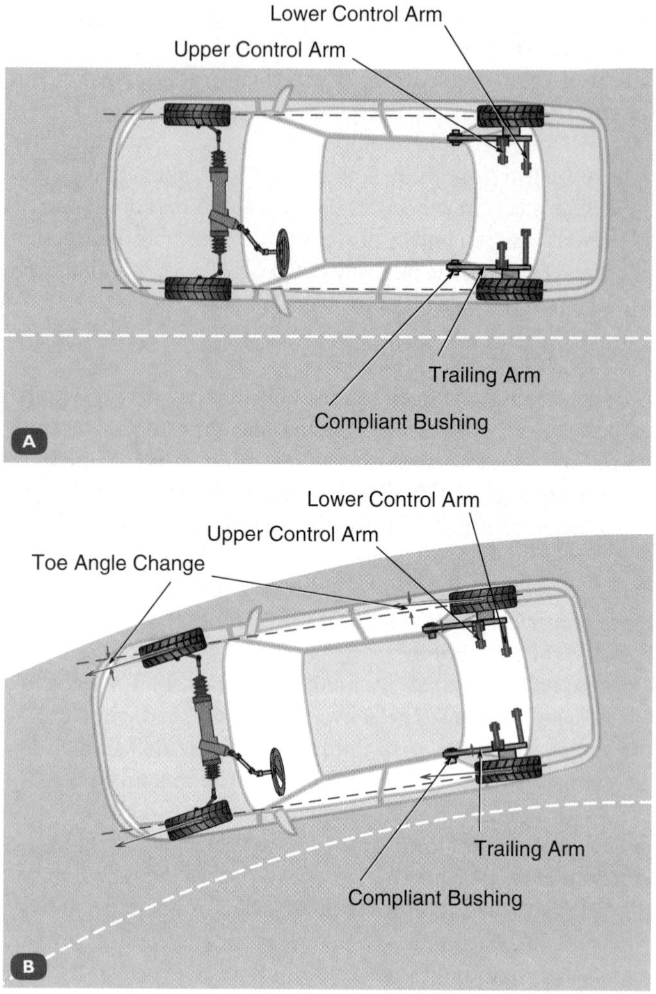

FIGURE 26-65 A passive four-wheel steering system uses compliant rubber bushings that allow a limited amount of rear-wheel steering during cornering.

▶ Four-Wheel Steering Systems

LO 26-09 Describe four-wheel steering operation.

Some vehicles have **four-wheel steering**. This means that the rear wheels can be steered independently of or in conjunction with the front wheels (**FIGURE 26-64**). The ability to steer the rear wheels improves high-speed handling. It also increases maneuverability during driving tasks such as U-turns and parallel parking.

There are two types of four-wheel steering systems—active and passive. Passive systems use compliant rubber bushings. They allow a limited amount of rear-wheel steering under load. This typically allows the wheels to angle slightly toward the inside of the corner, during turns (**FIGURE 26-65**). The passive system operates independently of the steering wheel and driver input. And this is why it is called a passive system.

Front wheels are controlled normally in active four-wheel steering systems. The rear wheels are usually steered through use of a computer and electric motors. In past designs, rear-wheel steering systems could be mechanically controlled. This would occur through a direct mechanical connection between the front and the rear steering boxes.

FIGURE 26-66 The rear actuator for an active four-wheel steering system turning the rear wheels.

In modern computer-controlled four-wheel steering, an actuator is used. It is similar to a front rack-and-pinion assembly. It attaches to the rear steering knuckles with tie-rod ends. The actuator turns the wheels when commanded by the steering control module (**FIGURE 26-66**). Originally, some actuators were powered hydraulically. Steering was controlled by the power steering pump and electronic control valves. Most systems today use a rack-and-pinion assembly driven by an electric motor. In most vehicles, the rear wheels can only be steered a small percentage compared with the front wheels.

Some manufacturers have adopted the standard that at low speeds, the rear wheels will turn in the opposite direction to the front wheels. This provides a substantially reduced turning radius (**FIGURE 26-67**). And at higher speeds, the rear wheels will turn in the same direction as the front wheels.

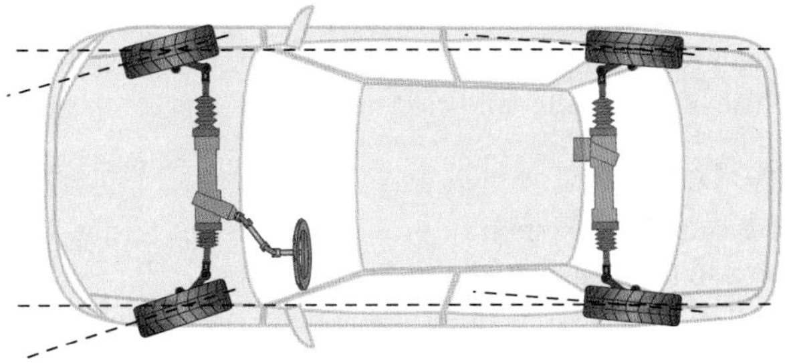

FIGURE 26-67 At slow speed, many four-wheel steering systems turn in the opposite direction to the front wheels. This reduces the turning radius.

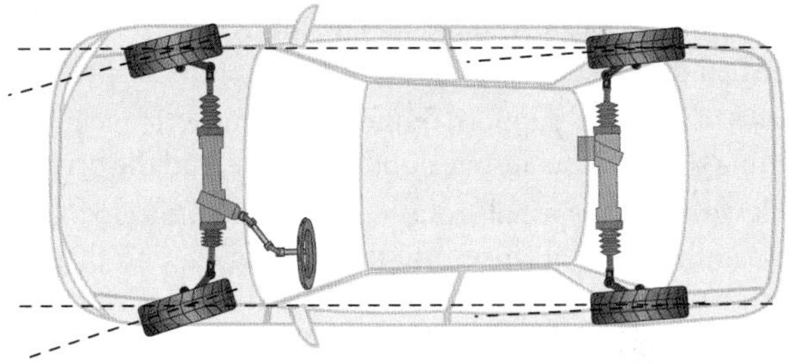

FIGURE 26-68 At high speed, many four-wheel steering systems turn in the same direction as the front wheels. This reduces the yaw forces.

In this way, smaller turning forces (yaw) are created during high-speed turning maneuvers (**FIGURE 26-68**). These standards provide better control of a vehicle at both high and low speeds. Other manufacturers use systems that disable rear steering at higher speeds. They only allow rear steering at slow speeds. This allows tighter cornering, such as when backing up a trailer.

▶ Wrap-Up

Ready for Review

- The steering system provides control over the vehicle's direction of travel and has a steering column, a steering box, a steering linkage, and a steering knuckle.
- The rack-and-pinion steering system makes use of the following components for a sharp steering response: pinion, rack, inner tie rod or socket, outer tie-rod end, and rubber bellows.
- Primary components of the parallelogram steering system, from the gearbox, are: pitman arm, idler arm, center link, tie rod, tie-rod end, and adjustment sleeve.
- The steering column is connected to the input shaft of the steering gear by a flexible joint as the steering column sits at an angle to the steering box.
  - An intermediate shaft runs at an angle from the steering column to the steering gear.
- The rack-and-pinion steering gearbox has a pinion gear that runs in mesh with the teeth on the rack. A fixed-ratio rack-and-pinion system uses the same ratio for all positions of the wheels. A variable-ratio gear allows a slow turn rate when the steering wheel is centered and a faster rate when turned toward the steering lock.
- Worm gear steering boxes use a worm and a worm gear to convert the rotary motion of the steering wheel to the linear motion needed to control the wheels.
  - Three popular styles of Worm gear steering boxes include: worm-and-sector, worm-and-roller, and recirculating ball gearboxes.
- Hydraulically assisted power steering uses pressurized hydraulic fluid from a power steering pump to assist the driver in steering the wheels.
- An electrically powered hydraulic steering (EPHS) makes use of a brushless motor to drive the pump and only operates on demand.
  - Electrically assisted steering (EAS) is a completely electrically powered power-assist system that eliminates all hydraulic components and fluid.

- Four-wheel steering systems can be passive systems that compliant rubber bushings or active systems that make use of a computer and electric motors.

Key Terms

active control A system of providing constant feedback from sensors in the vehicle to the control unit.

actuating The act of making something move or work.

adjustable bushing A brace or nylon part that pushes against the rack to adjust the mesh of the rack teeth to the pinion teeth.

beam axle Suspension system in which one set of wheels is connected laterally by a single beam or shaft.

bump steer The undesired condition produced when hitting a bump, where the vehicle darts to one side as the steering linkage is pushed or pulled as a result of the travel of the suspension.

chassis The main support frame in a vehicle. It includes the running gear, such as suspension, the engine, and the drivetrain.

clock spring A special rotary electrical connector located between the steering wheel and the steering column that maintains a constant electrical connection with the wiring system while the vehicle's steering wheel is being turned.

constant mesh A term used to describe two or more parts, such as gears, that are in constant contact with each other.

control unit Any device that controls another object, such as a computer.

drag link A steel or iron rod that transfers movement of the pitman arm to a relay lever.

electric power steering system (EPS) A steering system that uses an electric motor and sensors to provide feedback to the vehicle's computer systems in order to decrease steering effort.

electrically assisted steering (EAS) A power-assist system that uses an electric motor to replace the hydraulic pump to decrease steering effort.

electrically powered hydraulic steering (EPHS) A steering system that uses an electric motor to produce hydraulic assist for steering.

forces capability map data Data preprogrammed into the electronic control unit's memory by the manufacturer and used to determine how much power assistance is needed based on input from the vehicle's speed sensor and steering sensor.

four-wheel steering. Vehicle that has the ability to steer all four wheels.

gear ratio The relationship between two gears in mesh as a comparison to input versus output.

inertia The resistance to a change in motion.

intermediate shaft A steel rod positioned at an angle from the steering column to the steering gear that functions in transferring movement from one to the other.

knuckle The part that contains the wheel hub or spindle and attaches to the suspension components.

linear motion Movement in a straight line.

meshed pinion A pinion when it is mated with the rack.

parallelogram steering system A non–rack-and-pinion system that uses a series of parts, consisting of the pitman arm, idler arm, center link, and tie-rod assemblies, to relay movement from the steering gearbox to the wheel assembly.

power assist unit Assembly used to assist the driver in turning the steering wheel.

power section A chamber in the rack, where pressurized fluid acts upon pistons that assist in steering.

power steering pump A small hydraulic pump that provides assistance to the driver when turning the steering wheel.

power unit A belt- or gear-driven pump that produces hydraulic pressure for use in the steering box or rack.

preloaded A part that is already compressed from pressure.

rack-and-pinion steering system A steering system composed of a steering wheel, a main shaft, universal joints, and an intermediate shaft. When the steering wheel is turned, movement is transferred by the main shaft and intermediate shaft to the pinion.

recirculating ball steering box A worm gear steering box in which the worm rides on ball bearings

spline Ridges or teeth on a shaft that mesh with grooves in a mating piece and transfer torque to it, maintaining the angular correspondence between them.

spring-loaded rack guide yoke A spring-containing yoke that pushes on the back side of the rack to help reduce the play between the rack and the pinion while still allowing for relative movement.

steering arm An arm that extends from the steering knuckle. The tie rods connect to these arms in order to steer the wheels.

steering box The assembly which converts the rotary motion of the steering wheel into the linear motion of the steering linkage.

steering column Transmits the driver's steering effort from the steering wheel down to the steering box, usually made to collapse during a crash.

steering damper A device used to prevent shocks from irregular roads from being transmitted through the steering linkage and back to the steering wheel.

steering knuckle The knuckle located either between the lower control arm and MacPherson strut or upper control arm. It has either a spindle formed or bolted onto it, to which the wheel hub is attached, or it provides the location of the sealed front wheel bearing on front drive axles.

steering linkage Steel rods that connect the steering box to the steering arms on the steering knuckle.

steering sensor A torque sensor that converts steering torque input and direction into voltage signals for the power steering control module (PSCM) to monitor. Also a steering angle or position sensor, it converts the rotation speed and direction into voltage signals for the PSCM to monitor.

steering system A term used to describe all of the components and parts involved in steering a vehicle.

stub axle An axle used for one wheel.

stub-axle carrier The body of the stub-axle knuckle.

tie-rod assembly The part that fits between the rack and the steering arms and transfers the movement of the rack.

toe setting Setting of the toe-in or toe-out of the tires to the centerline of the vehicle.

torsion bar Spring steel rod used in power steering systems to allow relative movement between the steering wheel and steering gear. This allows the hydraulic control valve to direct power steering fluid pressure as needed.

torque sensor A device used to measure the load on the steering wheel.

track rod On forward-control vehicles, it connects the relay lever to the idler arm.

worm A gear with a helical, threaded shaft that is attached to the steering column and meshes with a worm wheel that transfers motion from the steering wheel to the steering linkage.

Worm gear steering A steering box consisting of a worm and worm gear.

worm shaft The protrusion of the worm gear that serves as the point of attachment to the steering column.

Review Questions

1. Which of the following is not a common steering system component?
   - a. Steering column
   - b. Steering knuckle
   - c. Steering box
   - d. Steering cylinder
2. Which part in the rack and pinion contains an inline ball and socket joint?
   - a. The Rack
   - b. The inner tie rod
   - c. The pinion
   - d. The rubber bellows
3. Which part in the parallelogram steering linkage connect the pitman arm to the idler arm?
   - a. The center link
   - b. The tie rod end
   - c. The steering damper
   - d. The adjustment sleeve
4. Why is the steering intermediate shaft equipped with a flexible joint?
   - a. To allow the angle of the wheels to change when they are steered.
   - b. To allow for drive by an electric motor on hybrid vehicles
   - c. Because the steering column electric motor needs insulation
   - d. Because the steering column sits at an angle to the box
5. What determines the turning rate on a rack and pinion?
   - a. The tension on the guide yoke or rack bearing
   - b. The gear ratio of the rack and pinion gears
   - c. The length of the steering intermediate shaft
   - d. The distance between the inner and outer tie rod
6. What is the purpose of the rolling balls on a recirculating ball steering box?
   - a. To take up wear on the worm shaft
   - b. To control the amount of play in the steering wheel
   - c. To form a low-friction internal thread
   - d. To stabilize the steering gear at high speeds
7. In a power steering system, which part controls the flow of fluid from side to side when the steering wheel is turned?
   - a. The power steering pump rotor
   - b. The power steering switch
   - c. The power steering cooler
   - d. The power steering rotary valve
8. In an electrically powered hydraulic steering (EPHS) system, where is the motor located?
   - a. Connected to the power steering pump
   - b. Connected to the rack and pinion assembly
   - c. Connected to the steering column
   - d. Connected to the steering intermediate shaft
9. In an electric power steering (EPS)system, what indicates steering wheel position and force to the ECU?
   - a. The steering sensor
   - b. The steering actuator
   - c. The electric motor
   - d. The EPS relay
10. In a passive four-wheel steering system, what allows the rear wheels to change angle around turns?
    - a. The rear rack and pinion
    - b. The rear steering actuator
    - c. The compliant bushings
    - d. The steering control module

ASE Technician A/Technician B Style Questions

1. Steering systems are being discussed. Technician A states that the steering knuckles typically pivot on ball joints. Technician B states that the steering arm converts rotary motion of the steering wheel to linear motion. Who is correct?
   - a. Technician A only
   - b. Technician B only
   - c. Both Technicians A and B
   - d. Neither Technician A nor B
2. Steering geometry is being discussed. Technician A states that in parallelogram steering system, the pitman arm and idler arm always move at right angles from one another. Technician B states that the steering arm transmits steering motion and force to the wheel assembly. Who is correct?
   - a. Technician A only
   - b. Technician B only
   - c. Both Technicians A and B
   - d. Neither Technician A nor B
3. Parallelogram steering linkage is being discussed. Technician A states that the pitman arm connects the steering box to the center link. Technician B states that the idler arm is splined to the steering box. Who is correct?
   - a. Technician A only
   - b. Technician B only
   - c. Both Technicians A and B
   - d. Neither Technician A nor B

4. Steering columns are being discussed. Technician A states that the steering column is equipped with a collapsible section to protect the driver in a collision. Technician B states that there may be various plastic shear pins and brackets which will break to allow extra movement in the case of a collision. Who is correct?
   a. Technician A only
   b. Technician B only
   c. Both Technicians A and B
   d. Neither Technician A nor B
5. A high-mileage vehicle comes into the shop with excessive steering wheel play (looseness) when driving. The play is found to be inside of the steering rack and pinion assembly. Technician A states that the play is most likely caused by wear and backlash between the rack-and-pinion gears. Technician B states that most rack and pinions can be easily adjusted manually to make up for worn gears. Who is correct?
   a. Technician A only
   b. Technician B only
   c. Both Technicians A and B
   d. Neither Technician A nor B
6. A worm steering gear box is being discussed. Technician A states that the steering gear box increases the driver's force through large gear reduction. Technician B states that compared to a rack and pinion, the worm gear box isolates the driver much more effectively from road shock. Who is correct?
   a. Technician A only
   b. Technician B only
   c. Both Technicians A and B
   d. Neither Technician A nor B
7. A vehicle with power steering is being discussed. Technician A states that a power steering reservoir is commonly mounted to the rack and pinion. Technician B states that the power steering pump is normally belt driven. Who is correct?
   a. Technician A only
   b. Technician B only
   c. Both Technicians A and B
   d. Neither Technician A nor B
8. A vehicle with power steering is being discussed. Technician A states that a pressure switch or sensor will normally cause the engine RPM to stay steady or increase slightly when turning hard at low speeds. Technician B states that the torsion bar in the rack and pinion is used to control fluid flow, which provides steering assistance. Who is correct?
   a. Technician A only
   b. Technician B only
   c. Both Technicians A and B
   d. Neither Technician A nor B
9. A vehicle with electronic power steering (EPS) comes into the shop. Technician A states that the system is designed with a failsafe so that it can still be steered in event of an electronic failure. Technician B states that the EPS system will likely be capable of some self-diagnosis and may set diagnostic trouble codes. Who is correct?
   a. Technician A only
   b. Technician B only
   c. Both Technicians A and B
   d. Neither Technician A nor B
10. A vehicle with four-wheel steering is being discussed. Technician A states that most newer model vehicles can be special ordered with rear wheel steering as an option. Technician B states that most modern rear wheel steering racks are operated with hydraulics. Who is correct?
   a. Technician A only
   b. Technician B only
   c. Both Technicians A and B
   d. Neither Technician A nor B

CHAPTER 27

Servicing Steering Systems

Learning Objectives

- **LO 27-01** Describe steering system service preliminaries.
- **LO 27-02** Describe the steering system diagnosis procedure.
- **LO 27-03** Perform power steering fluid maintenance.
- **LO 27-04** Perform rack-and-pinion service.
- **LO 27-05** Perform parallelogram steering linkage service.
- **LO 27-06** Inspect electric power steering and identify high-voltage electrical circuits.
- **LO 27-07** Disable SRS and service clock spring.

ASE Education Foundation Tasks

See Appendix A to view the 2017 ASE Education Foundation Automobile Accreditation Task List Correlation Guide.

▶ Introduction

LO 27-01 Describe steering system service preliminaries.

The steering system plays a critical role in controlling the vehicle. So, it is important that all scheduled maintenance and inspection tasks be performed as required. Also, any problems that are found must be diagnosed and repaired as soon as they appear. They can cause a loss of control of the vehicle, leading to an accident and injuries. So, pay close attention to the procedures listed in this chapter. And always follow the manufacturer's recommended procedures when repairing steering-related concerns. Manufacturers configure their vehicles in a variety of ways, so their procedures must be followed.

Steering systems are mostly mechanical. So, visual inspections and physical measurements make up most of the maintenance and diagnosis tasks. Mechanical joints and bushings, as well as motors, pumps, and switches, wear over time. This decreases their functionality and they eventually fail. Electrical faults are becoming more common. This is because there are more electronic power steering systems being produced. Sensors also are subject to wear, resulting in inaccurate readings and faulty reactions. The wires are subject to damage and breaks, as well as loose or corroded connections.

Consider the mechanical and electrical components when confronted with a steering system problem. Any problems in the suspension system can affect the steering system. So you need to have an understanding of both systems. Suspension systems are discussed more in the Suspension Systems chapters. For now, we focus primarily on steering system maintenance, diagnosis, and repair.

Tools and Equipment

There are two categories of tools used in diagnosing and repairing steering system problems. There are tools for mechanical diagnosis and repair. And there are tools for electrical diagnosis and repair. Tools for electrical diagnosis include the scan tool and the digital multimeter (DMM). Most manufacturers recommend a factory scan tool. It can perform tests and identify faults in the electrical part of steering systems. The scan tool can read almost every system on a vehicle. It provides valuable data to assist the technician in identifying problems. It is useful in assessing electronic steering systems for wiring and sensor malfunctions. Typically, electronic steering system faults set a diagnostic trouble code that the scan tool can retrieve.

ADMM is used along with the service information to diagnose electrical faults. The DMM is used to measure voltage and to check the continuity of electronic circuits. Breaks in the wiring of an electronic power steering (EPS) system are not uncommon. When this occurs, the system opens or shorts out. The DMM can be used to locate any opens or shorts.

The mechanical portion of the steering system needs specific tools and equipment. They include a power steering system pressure tester. Various measuring devices are used. These include dial indicators and belt tension gauges. The mechanical components have to be removed and replaced. This requires the special tools listed below (**FIGURE 27-1**). In addition, inch-pound and foot-pound torque wrenches, hammers, air hammers, and pry bars are also used.

The following tools are used in servicing the steering system:

- **Floor jacks and safety stands**: Jacks are used to lift a vehicle, and the safety stands provide stability while working on a raised vehicle.
- **Pry bar**: The pry bar is a lever used to apply pressure for testing purposes or to move various components.
- **Dial indicators**: Dial indicators are used to measure the runout or movement on different parts of the steering system, such as play in tie-rod ends.
- **Pitman arm puller**: This heavy-duty puller is made specially for removing the pressed-on pitman arm from the sector shaft.
- **Tie-rod end puller**: This tool is used to pull the tapered shaft on a tie-rod end from its mating steering component.
- **Tie-rod sleeve adjusting tool**: This tool has a tab designed to grab the slot in the sleeve and is used to turn the sleeve when adjusting the toe setting.
- **Pickle fork**: This U-shaped wedge is used for separating tie-rod ends and is operated by hammer or air hammer. A pickle fork will usually destroy the dust boot during the process, so it is not the best tool to use on tie rods that will be reused.
- **Inner tie-rod end tool**: This tool is used to loosen and tighten inner tie-rod ends.
- *Power steering system analyzer*: This tool includes a pressure and flow gauge set with various fittings to attach in line with the power steering pump. It is used to check volume of fluid flow, maximum pressure, and leaks internal to the steering gear.

You Are the Automotive Technician

A customer is driving on the highway and notices a burning oil smell and smoke. It appears to be coming from the front of his six-year-old vehicle with power rack-and-pinion steering. He is able to pull into a visitor's center and check under the vehicle's hood. He identifies a large amount of power steering fluid sprayed all over the engine compartment. And the power steering reservoir is nearly empty. He calls you for advice and wants to know if he can drive it to your shop 35 miles away.

1. What would you advise him to do, and why?
2. What are the most likely leak points?
3. While you are working on the vehicle, what other steering-related inspections would you perform?
4. If a vehicle with power steering comes into your shop with hard steering, what things would you check, and why?

- *Scan tool*: This tool is used to read codes and data from the vehicle's PCM when diagnosing the vehicle's computer-controlled systems.
- *DMM*: The DMM, which stands for digital multimeter, is used to measure voltage and ohms in a vehicle's electrical systems.
- *Serpentine belt gauge*: Used to check for excessive wear on serpentine belts.
- *Test light*: This is used to diagnose electrical problems.
- *Belt tension gauge*: This tool is designed to measure the amount of flex in a drive belt. Some manufacturers require the use of this tool to adjust proper belt tension.
- *Black light and dye kit*: This tool is used to pinpoint fluid leaks. Dye is added to the fluid of the leaking system, and the ultraviolet light makes the dye glow fluorescent.

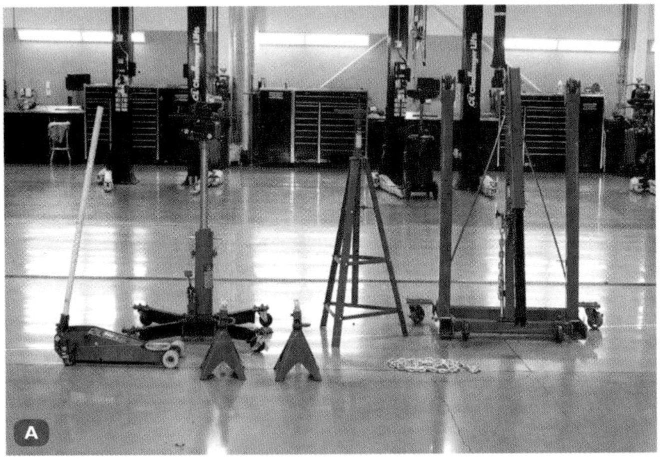

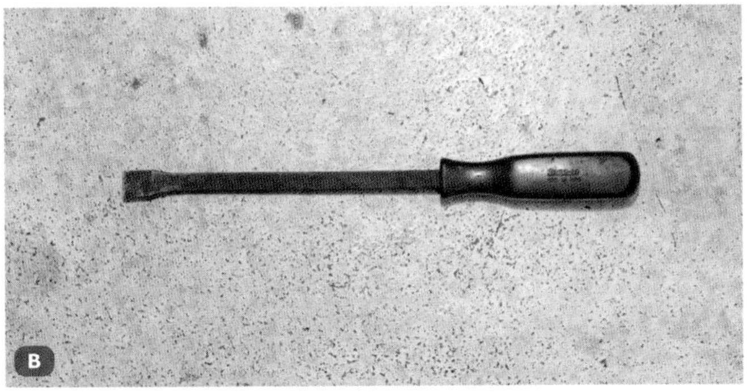

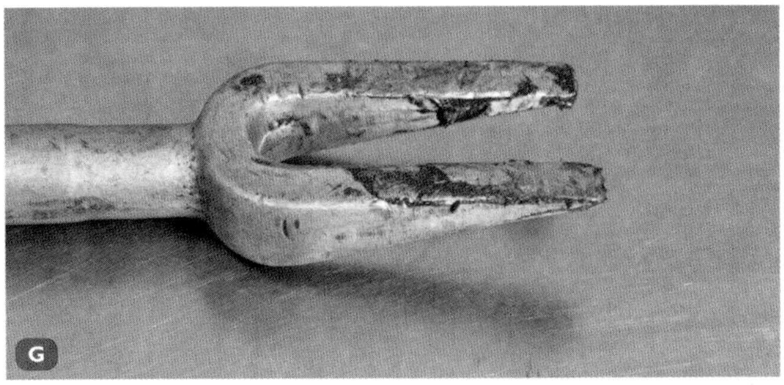

FIGURE 27-1 Steering system specialty tools. **A.** Floor jacks and safety stands **B.** Pry bar **C.** Dial indicators **D.** Pitman arm puller **E.** Tie-rod end puller **F.** Tie-rod sleeve adjusting tool **G.** Pickle fork

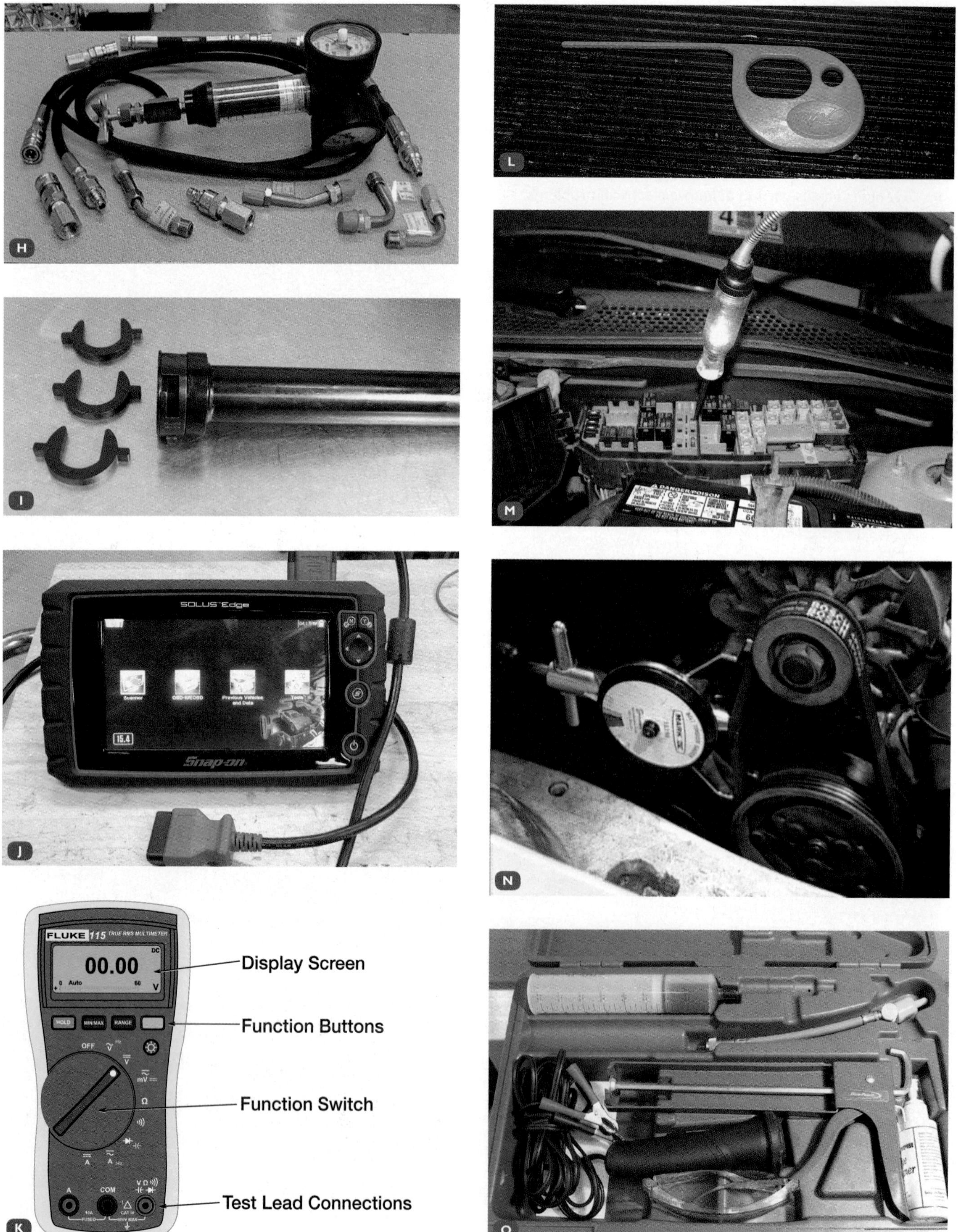

FIGURE 27-1 Steering system specialty tools. **H.** Inner tie-rod end tool **I.** Power steering system analyzer **J.** Scan tool **K.** DMM **L.** Serpentine belt gauge **M.** Test light **N.** Belt tension gauge **O.** Black light and dye kit

▶ Diagnosing Steering Systems

LO 27-02 Describe the steering system diagnosis procedure.

Steering systems can generate several different customer concerns. But there are main problems that arise. They are excessive play, unusual noises, vibrations/shimmies, and hard or inconsistent steering. The common culprits for these problems are wear, poor lubrication, or damaged parts. Play is the clearance in a part that is the result of wear in the part. In the steering system, play generally results from worn ball sockets and bushings. It can also be caused by too much clearance in the steering gearbox.

Hard or inconsistent steering is when it is difficult to turn the steering wheel. This can be either all the time or intermittently. There are many faults that can cause hard steering. From low tire pressure to overly tight adjustments in the steering gearbox. It can also be caused by power steering that is not working properly.

When diagnosing hard steering, it is important to thoroughly check the entire steering system. This way all of the faults causing the hard steering issue can be identified. The best way to check the steering system is by physically inspecting each of the components. This includes measuring play, looking for bent or damaged linkage, and testing the operation of the power steering system (**FIGURE 27-2**). We cover each of these tasks in this chapter.

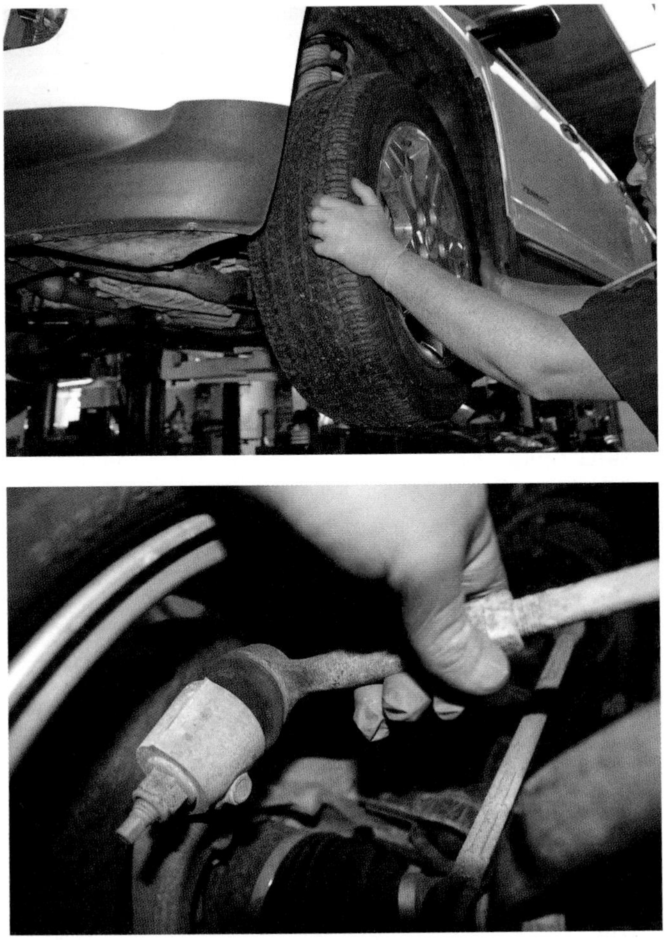

FIGURE 27-2 A technician inspecting a tie-rod end for excessive play.

Diagnosing Power Steering Fluid Leakage

Power steering fluid is critical to the proper functioning of the entire steering system. Any power steering fluid leak, no matter how small, is due for repair because of the fire hazard it presents. Because a leak can occur anywhere throughout the system, it is necessary to thoroughly inspect it all. One method for checking for leaks is to first clean all power steering fluid off the vehicle. Then run it for a little while. Turn off the vehicle and reexamine for new leaks. Any drops of fluid indicate a leak.

Another method for detecting leaks uses a special fluorescent dye. Following the dye manufacturer's instructions, the appropriate amount of dye is added to the power steering system. After running the vehicle for the recommended amount of time, turn the engine off. Inspect the power steering components with an ultraviolet light. The light will illuminate the fluorescent dye and make it easier to spot leaks. Always trace a leak to its source to identify what is causing the leak.

Diagnosing Power Steering Gear Issues (Non–Rack-and-Pinion)

Steering gear issues in a non–rack-and-pinion system can usually be one of the following:

- uneven effort needed for turning,
- looseness of steering,
- hard steering,
- unusual noises when steering, or
- leaks.

If the issue is related to the power steering system, such as leaks, noise, or hard steering, perform a visual inspection. Check the level and condition of the power steering fluid. Also, check the condition of the drive belt. If those are OK, you may need to perform a pressure test of the power steering system. That test is covered in our higher level course materials.

If the issue is related to the steering linkage, perform a visual inspection of the linkage, joints, and steering gear. With the vehicle raised and properly supported, have someone turn the steering wheel. Visually inspect the steering linkage as it is moved. Check for binding, excessive play, bent or binding components, and any other abnormal conditions.

To check for binding issues, you may need to separate the steering linkage joints. That way you can verify which components are binding. Binding can happen in the steering gearbox, linkage, ball joints, or steering column. For looseness concerns, do not forget to check the coupler between the steering shaft and the gearbox. These components are prone to wear, which creates play. Refer to the manufacturer's service information for diagnostic information for the concern you are working on.

Diagnosing Power Steering Gear Issues (Rack-and-Pinion)

Diagnosing a rack-and-pinion system is similar to a parallelogram system. It includes the same common complaints:

- Uneven effort needed for turning,
- looseness of steering,
- hard steering, and
- unusual noises or vibrations when steering.

The issue may be related to the power steering system. If so, perform the inspection and tests associated with power steering. If the issue is related to the steering linkage, perform a visual inspection of the linkage, joints, and steering gear. With the vehicle raised and properly supported, have someone turn the steering wheel. Visually inspect the steering linkage as it is moving. Check for binding, wear or loose connections, bent components, and debris. You also may need to disassemble components to further inspect them.

Diagnosing Steering Column Issues

Common steering column issues include unusual noises, looseness, and binding. These problems generally arise from either wear or debris lodged in the steering column. The process of diagnosing each of these problems is the same. The vehicle is raised and safely supported, and one person turns the steering wheel. The technician watches for looseness or binding and listens for noises. Once the location and cause are identified, research the service information to create a repair plan.

A large number of electrical switches and controls are mounted to the steering wheel. A device called a clock spring is used to transmit electrical signals across the rotating connection. The clock spring is a flexible ribbon that is coiled inside of a plastic holder. The clock spring must be centered when steering column work has been done. This allows the steering wheel to be turned from lock to lock without damaging the clock spring. Refer to service information on how to center the clock spring.

The typical clock spring will have four to five turns of movement before it runs out of travel (turn very gently or you will break the wires). The clock spring is normally marked to show the center position. This is where the clock spring should be positioned when the wheels are straight ahead.

If you do not know where the center point of the clock spring is, you will have to manually center it. Do this by following the procedure in the service information. This is always the safest way to center the clock spring. If there is no procedure listed, you can try this method. Carefully and gently turn the clock spring in both directions. Count the number of total turns between the end points. Then move back half that number of turns. The clock spring should now be centered. Also, be careful not to wind the clock spring in the opposite direction it was designed to be wound. That will greatly shorten its life.

The steering column of a vehicle with EPS or stability control is equipped with a torque sensor and steering angle sensor. If the steering column is being serviced, a steering angle sensor calibration is typically required. Refer to the service information for the correct method of recalibrating the steering angle sensor. Typically, a scan tool is required to perform this function.

▶ TECHNICIAN TIP

A technician who has removed the steering wheel must be extremely careful not to damage or impair the operation of any multifunction switch or the clock spring.

▶ Maintenance and Repair

LO 27-03 Perform power steering fluid maintenance.

Although the steering system is very critical to the safe operation of the vehicle, it does not normally need a lot of maintenance. In older vehicles, each of the steering and suspension joints must be lubricated with grease during each oil change. Newer vehicles use greased-for-life joints. So, most vehicles do not have grease zerks to lubricate the joints (**FIGURE 27-3**). However, always be on the lookout for them. Many replacement joints use zerk fittings to keep them lubricated. Another area of scheduled maintenance is replacement of the power steering belt. Also, the power steering fluid may need to be flushed.

To maintain a safe-operating vehicle, regular inspection of the steering components is essential. This includes inspecting the components and joints for damage, looseness, or leakage. It also involves operating the steering system. Look or feel for play in the components. If found, then measure the play and compare it with specifications. Also, listening can help identify binding in components. Always refer to the manufacturer's service information. It will list the inspection procedure and wear specifications.

Checking and Adjusting Power Steering Fluid

The power steering fluid transmits pressure throughout a vehicle's power steering system. It also provides lubrication to the moving parts. This means it is critical that the power steering fluid be inspected regularly for both level and contamination. The power steering fluid usually has both a hot and a cold indicator on the dipstick. This ensures an accurate reading of the fluid level in both conditions. Always refer to the vehicle's service information for the specified type of power steering fluid.

To check and adjust power steering fluid, follow the steps in **SKILL DRILL 27-1**.

FIGURE 27-3 A. A greased-for-life tie-rod end. **B.** Tie-rod end with a zerk fitting used to lubricate the joint.

SKILL DRILL 27-1 Checking and Adjusting Power Steering Fluid

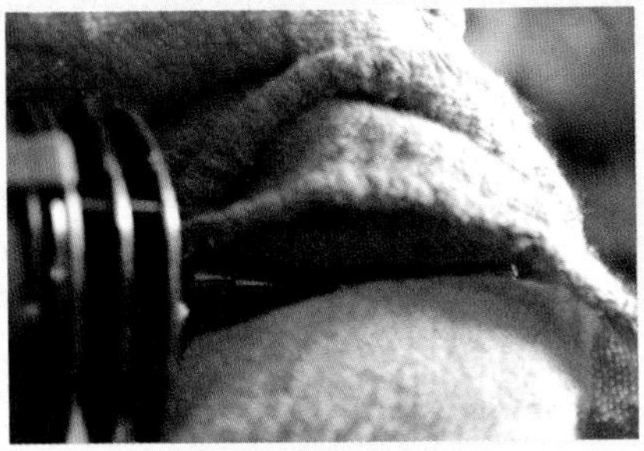

1. Locate the power steering reservoir. Clean around the cap if dirt is present. Remove the cap. Drip some fluid on a clean paper towel. Compare with the new fluid. If brown or black, recommend a fluid flush. Wipe the dipstick clean.

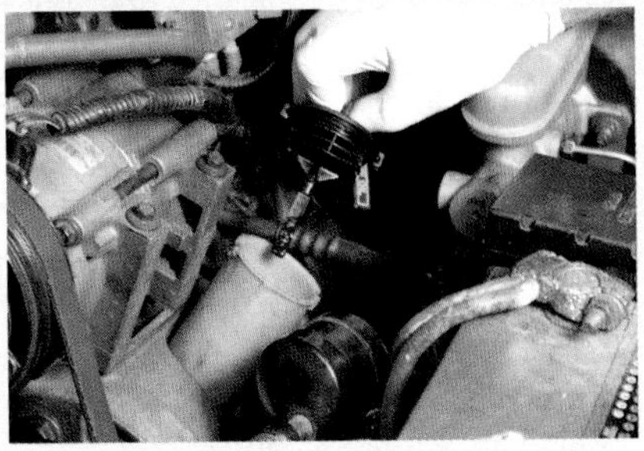

2. Reinstall the cap for a few seconds, then remove it again and check the fluid level on the dipstick. Verify that the level matches the temperature of the system—hot or cold.

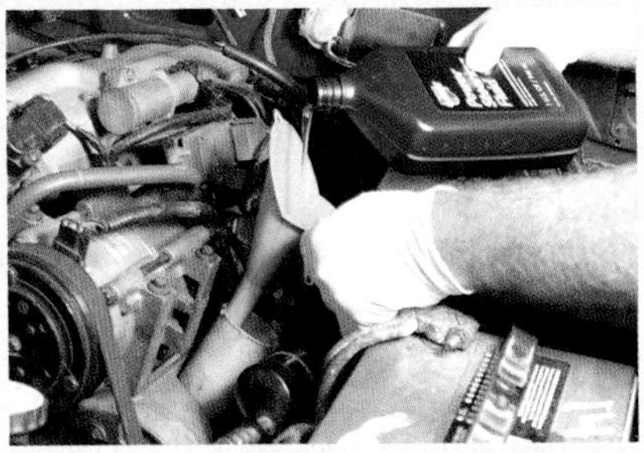

3. If the level is low, top if off with the specified fluid. If the vehicle has a plastic reservoir, check the marks to see if any fluid needs to be added.

Replacing Power Steering Pump Filter(s)

Some vehicles use a replaceable power steering pump filter. It is usually located in the fluid reservoir or in the return line from the steering gear (**FIGURE 27-4**). If it is in the reservoir, it may be a separate filter that sits in the bottom of the reservoir. If so, it can be carefully fished out of the reservoir and replaced. Other manufacturers build the filter into the reservoir such that the reservoir and filter are one unit. In this case, they are changed together.

If it is an inline filter, it is usually installed in the return line. This is because of the lower pressure. It also helps prevent contaminants from damaging the pump. It may be spliced into the hose between the steering gear and the pump. Or it could fit inside the return hose where the return hose connects onto the pump.

When changing the filter, you may want to consider flushing the power steering system first. This will allow removal of as much of the old fluid as possible before installing the new filter. Always check the service information to determine the scheduled service interval and the location of the filter, if equipped.

To replace the power steering pump filter(s), follow the steps in **SKILL DRILL 27-2**.

FIGURE 27-4 A. Inline power steering fluid filter. **B.** Reservoir style filter.

Flushing a Power Steering System

Flushing the power steering system is the process of removing all the old power steering fluid and replacing it with new. This is performed:

- whenever the manufacturer specifies a fluid change,
- the fluid appears contaminated or dirty,

SKILL DRILL 27-2 Replacing the Power Steering Pump Filter(s)

1. Research the vehicle you are working on to determine whether it has a power steering pump filter, what its change interval should be, and its location.
2. Following the specified procedure, remove the filter.
3. Inspect any hoses or fittings for cracks and damage.
4. Install the new filter, top off the reservoir with the correct fluid, and check for proper operation and leaks.

- a major part of the hydraulic steering system is replaced, or
- there is a serious mechanical problem involving the power steering pump or steering gear. In this case, metal shavings could be circulated through the system.

Only use the manufacturer's specified power steering fluid when flushing the system. When the service is complete, inspect the system for leaks. Turn the steering wheel to the full-lock position in both directions and back to the center. Then check for leaks. Also, ensure that none of the power steering hoses make contact with any other components.

Typically, a power steering system flush is performed with a flushing machine. The flushing machine is connected in line with the power steering system pump. It is used to remove all old dirty fluid and install new clean fluid without introduction of air. Note that air introduced into the power steering pump can be harmful to the pump. Since the pump is lubricated by the power steering fluid, scoring of the pump can result if the pump is run dry. There are several types of flushing machines, so follow the instructions that came with it. If one is not available, flushing will have to be performed as shown in Skill Drill 39-3. Be sure to follow the manufacturer's procedure for flushing power steering fluid.

To flush a power steering system, follow the steps in **SKILL DRILL 27-3**.

SKILL DRILL 27-3 Flushing a Power Steering System

1. Raise and support the vehicle, making sure the front wheels are off the ground. Using a suction gun, remove as much power steering fluid from the reservoir as possible.

2. Disconnect the power steering return hose from the reservoir, and stick the end in a suitable container.

3. Temporarily plug the return fitting on the reservoir with a snug-fitting rubber cap.

4. Add the recommended fluid into the reservoir until it is at the correct level.

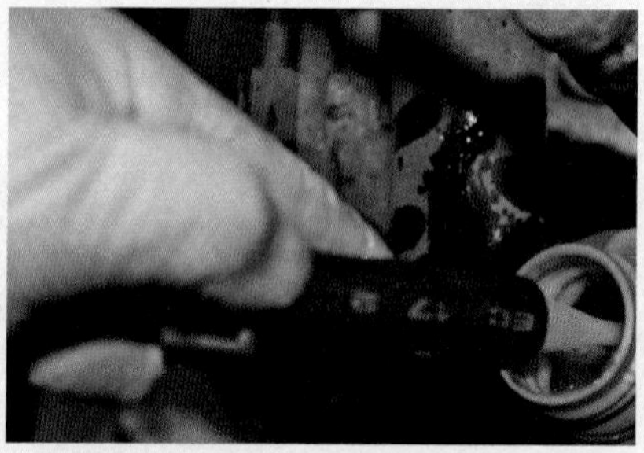

5. Start the engine, and have an assistant turn the steering wheel from lock to lock while keeping the reservoir at or near the full mark. When the fluid coming out of the return hose is clear, stop adding fluid. Keep running the engine until the level in the reservoir is just below the return line inlet. Turn off the engine.

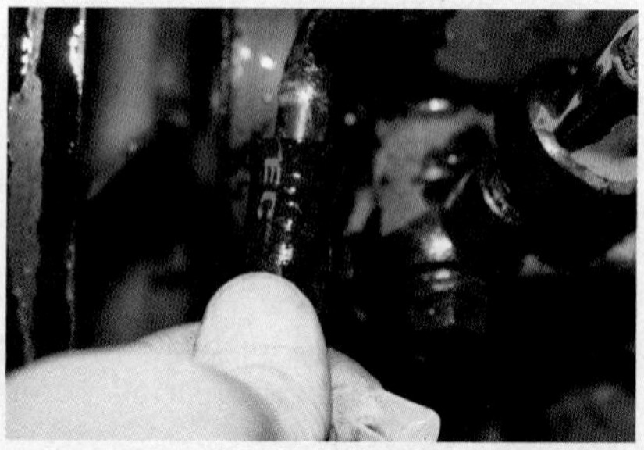

6. Reinstall the return hose on the reservoir, top off the fluid level, and start the engine. Turn the steering wheel lock to lock a few more times. Turn off the engine, and check the fluid level. Top off if necessary.

7. Lower the vehicle, start the engine, and turn the steering wheel from lock to lock. Check that the system is working correctly with the vehicle weight on the tires. Dispose of the waste power steering fluid in an environmentally approved manner.

▶ Perform Rack-and-Pinion Service

LO 27-04 Perform rack-and-pinion service.

Rack-and-pinion systems don't last forever. They wear out over time. Proper maintenance and good driving habits extend their life, but they still experience wear. When that occurs repairs and replacement parts are needed. We will cover some of the most common inspection, maintenance, and repair tasks that rack-and-pinion systems require.

Inspecting and Replacing Power Steering Hoses and Fittings

Any leakage of power steering fluid out of a power steering pressure hose can be catastrophic. Because it is under pressure, the fluid could spray all over the hot engine and exhaust system. The hot engine parts will cause the power steering fluid to smoke and billow out from under the hood. It can even catch on fire. So, it is important to inspect the condition of the hoses during oil changes. And don't forget that the ends of the hoses are often sealed with O-rings. These wear out, get brittle, and break, causing them to leak. Often the O-rings can be changed without having to change the hose.

Small leaks result in low fluid level in the power steering reservoir. The leaks also leave spots on the floor or parking space. If the fluid level gets low enough, it can cause a buzzing noise when driving. This noise is louder when steering the vehicle. The level of power assist may also fluctuate. If any of these symptoms are present, or to catch issues before they become a problem, regular inspection of the system is important. Always refer to the manufacturer's service information for any special precautions.

To inspect and replace power steering hoses and fittings, follow the steps in **SKILL DRILL 27-4**.

SKILL DRILL 27-4 Inspecting and Replacing Power Steering Hoses and Fittings

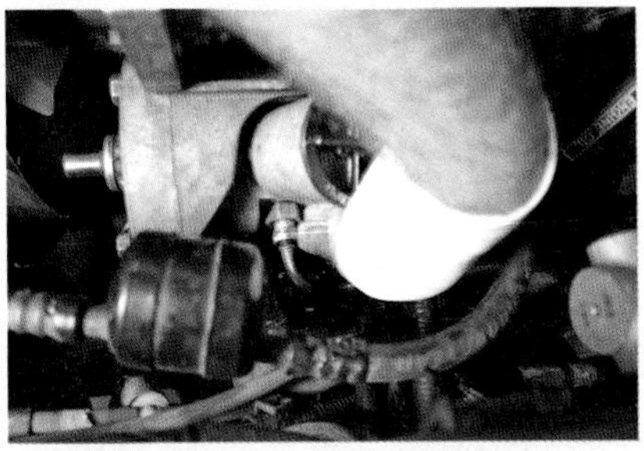

1. Safely raise and support the vehicle. Inspect each of the power steering hoses and fittings for leaks and damage.

2. Place a drain pan under the fitting to be disconnected. Use a flare nut wrench to loosen the fitting while you are holding the nut on the pump or steering gear with a wrench.

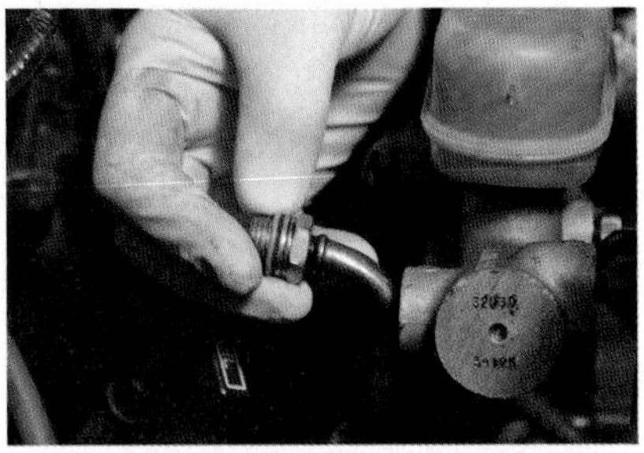

3. If replacing the O-ring, inspect the mating surfaces of the fitting for damage. If no damage is found, replace the O-ring and reinstall the fitting, being careful not to cross-thread or overtighten.

4. If replacing the hose, disconnect the fitting on the other end of the hose. Compare the old and new hoses to verify the correct replacement part.

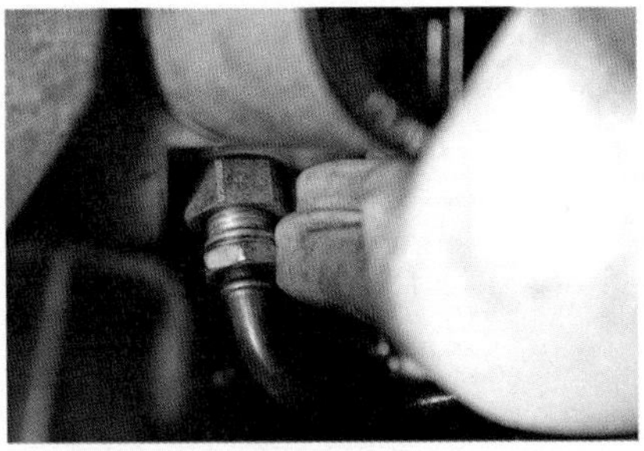

5. Carefully start the fitting on each end of the hose by hand, making sure they are not cross-threaded.

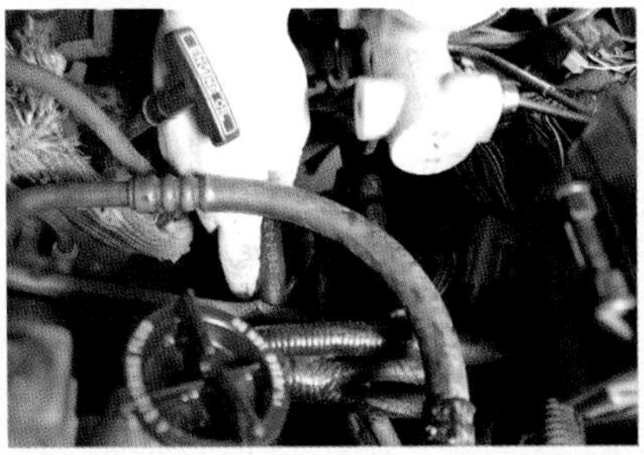

6. Check the routing of the new hoses, ensuring that they do not make contact with any components that could cause a failure. Then tighten the fittings to the proper torque.

7. Top off the reservoir with the specified fluid and start the engine. Turn the steering wheel from lock to lock a few times. Check for fluid leaks. Refill the fluid as necessary, and repeat the bleeding process as needed.

Inspecting Rack-and-Pinion Mounting Bushings and Brackets

During the normal vehicle use, the rack-and-pinion mounting bushings wear. This causes them to become compressed, brittle, or torn. Brackets may also be damaged or worn. If the vehicle has an oil or power steering fluid leak, the fluid may leak onto the mounting bushings. The oil tends to soften and degrade the bushings. When this happens, the driver typically complains of either loose steering or bumping noises when driving. In either scenario, a quick inspection of these bushings is warranted.

To inspect mounting bushings and brackets, follow the steps in **SKILL DRILL 27-5**.

Inspecting and Replacing Rack-and-Pinion Steering Gear Inner Tie-Rod Ends and Bellows Boots

Over time, the inner tie-rod ends wear and can cause excessive play in the steering linkage. Also, bellows boots may become torn, or dislodged from their seat. When bellows boots are torn, dirt or abrasives may enter the unit. This accelerates the wear of rack-and-pinion seals and bushings. Routine inspection can identify damage early, before additional problems arise.

When replacing the inner tie-rod end, a special tool will be needed to loosen and tighten it. Always use the specified tool. Otherwise, you will likely damage the rack-and-pinion assembly. Refer to the vehicle's service information when replacing the inner tie-rod ends and bellows boots.

To inspect and replace rack-and-pinion steering gear inner tie-rod ends and bellows boots, follow the steps in **SKILL DRILL 27-6**.

▶ Perform Parallelogram Steering Linkage Service

LO 27-05 Perform parallelogram steering linkage service.

As you know by now, parallelogram steering systems have a lot of links and pivot points. Each of those can become worn or damaged affecting vehicle safety. Because of this, we will cover the most common inspection, maintenance, and repair required of these systems.

SKILL DRILL 27-5 Inspecting Mounting Bushings and Brackets

1. Safely raise and support the vehicle, keeping the weight of the vehicle on the wheels, if possible. Inspect the bushings and brackets for any faults.

2. Try moving the rack-and-pinion assembly up and down by hand.

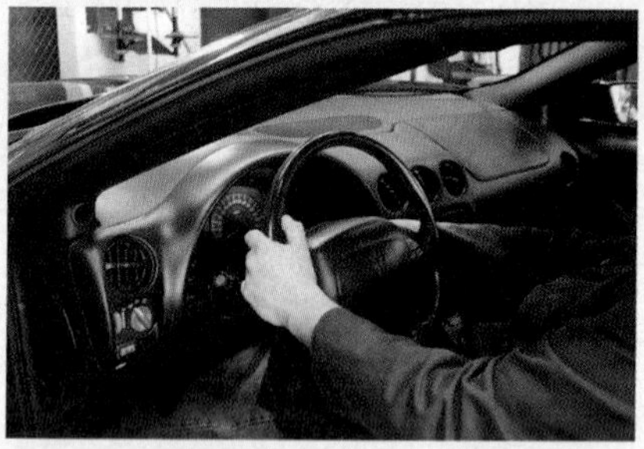

3. Have an assistant rock the steering wheel back and forth while you look for movement in the rack-and-pinion bushings. If the rack moves significantly, place a torque wrench on the mounting bolts, and tighten them to specification. If the rack still moves, then bushing replacement will be necessary.

4. Remove the bracket bolts.

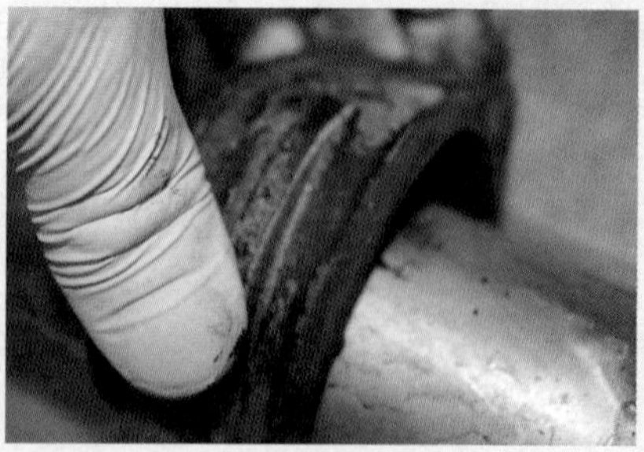

5. Remove the bushings from the rack.

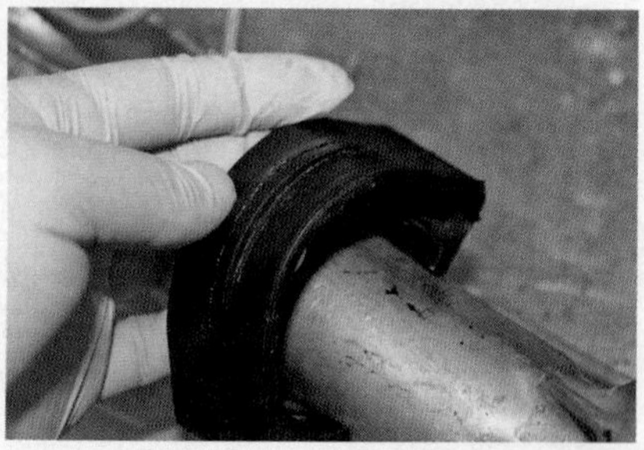

6. Install new bushings in reverse order of removal.

SKILL DRILL 27-6 Inspecting and Replacing Rack-and-Pinion Steering Gear Inner Tie-Rod Ends and Bellows Boots

1. Safely raise and support the vehicle. Inspect the rubber bellows for any signs of leaks, tears, or damage.

2. With the vehicle raised, have an assistant turn the steering wheel to one side and rock the steering wheel from side to side. On the side farthest out, squeeze the bellows until you make contact with the inner tie-rod joint. Feel for any play in the joint. Repeat this procedure for the other side. If play is found, replacement of the inner tie-rod ends will be necessary.

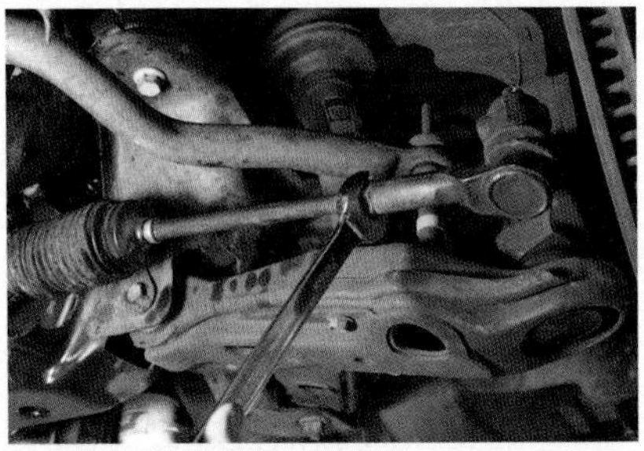

3. Remove the front wheel for the side being replaced, and loosen the locknut on the tie-rod end.

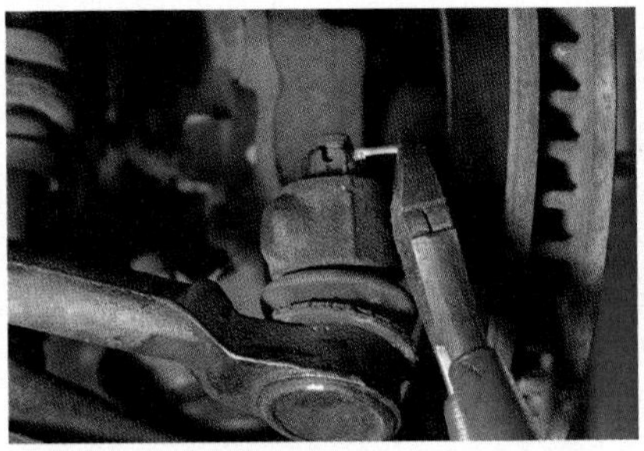

4. Remove the cotter pin and nut holding the outer tie-rod end to the steering arm. Separate the tie-rod end from the knuckle, using a tie-rod removal tool, the double-hammer method, or a pickle fork.

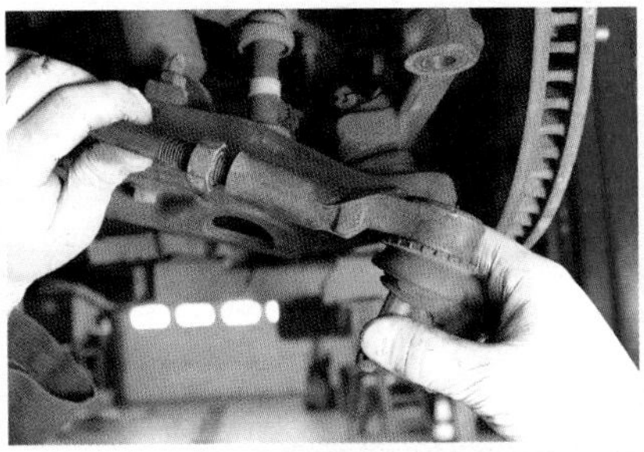

5. Count the number of turns to remove the outer tie-rod end from the threaded sleeve.

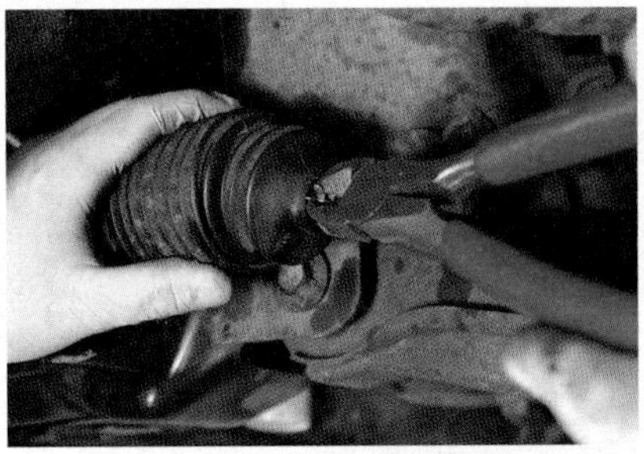

6. Remove the spring clamp from the bellows boot end to the inner tie-rod shaft. Remove the crimp clamp from the bellows boot to the rack-and-pinion housing. A new crimp clamp will be used on replacement of the boot.

7. Remove the bellows boot.

8. Using an inner tie-rod tool and the specified wrench to hold the rack, loosen the inner tie rod from the rack. Remove the inner tie-rod end.

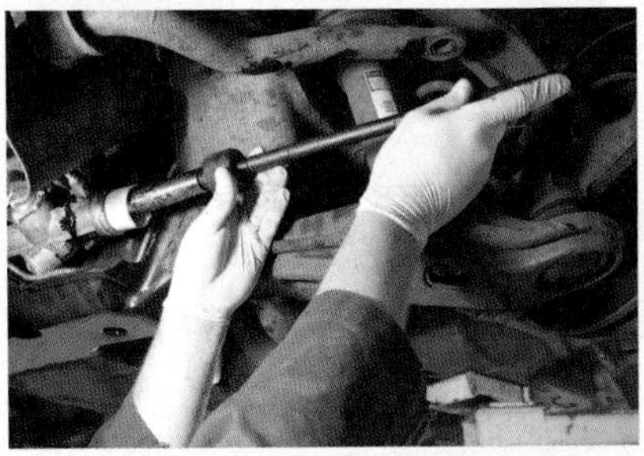

9. Install the new inner and outer tie rods in reverse order of removal, and verify that the play is gone. Perform alignment to reset toe after replacement is performed.

Inspecting and Replacing the Pitman Arm, Relay (Center Link/Intermediate), Rod, Idler Arm and Mountings, and Steering Linkage Damper

With any steering complaints, it is necessary to check the steering components for wear and damage. Slight problems in any of these components may result in significant steering problems and tire wear. Looseness in the idler arm bushings may cause excessive toe change on rough road surfaces. This can lead to a wandering condition and tire wear. Looseness in the tie-rod ends may be felt as loose steering. It is frequently mistaken as a steering gearbox problem. Because of these issues, each of the steering system joints needs to be inspected for excessive wear or damage. Also, note that replacement of any of these components will result in the need to perform a wheel alignment.

If the vehicle is equipped with a steering damper, a faulty one can cause a shimmy in the steering wheel. This usually occurs after hitting a bump in the road. On rare occasion, the damper can bind causing hard steering. Refer to the service information for the procedure and specifications for inspecting and replacing these components.

To inspect and replace the idler arm, follow the steps in **SKILL DRILL 27-7**.

To inspect and replace the pitman arm, follow the steps in **SKILL DRILL 27-8**.

To inspect and replace the center link (relay rod/intermediate rod), follow the steps in **SKILL DRILL 27-9**.

To inspect and replace the steering linkage damper, follow the steps in **SKILL DRILL 27-10**.

Inspecting and Replacing Tie-Rod Ends, Tie-Rod Sleeves, and Clamps

Tie rods make the final connection between the center link and the steering arms. The point of connection with the center link is considered the inner tie-rod end. The end that connects to the steering arm is considered the outer tie-rod end. Checking tie rods is important in identifying steering problems, because the ends are frequently worn or damaged.

SKILL DRILL 27-7 Inspecting and Replacing the Idler Arm

1. Safely raise and support the vehicle. Push the center link at the idler arm up and down, and watch the idler arm bushings for excessive movement.

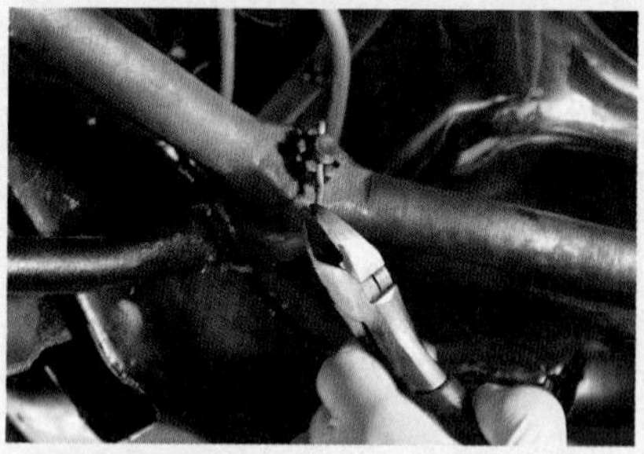

2. If the movement is out of specifications, the idler arm will have to be replaced. Remove the cotter pin, and loosen the nut connecting the idler arm taper stud to the center link.

3. Separate the idler arm taper stud, using the double-hammer method, a pickle fork, or the appropriate puller.

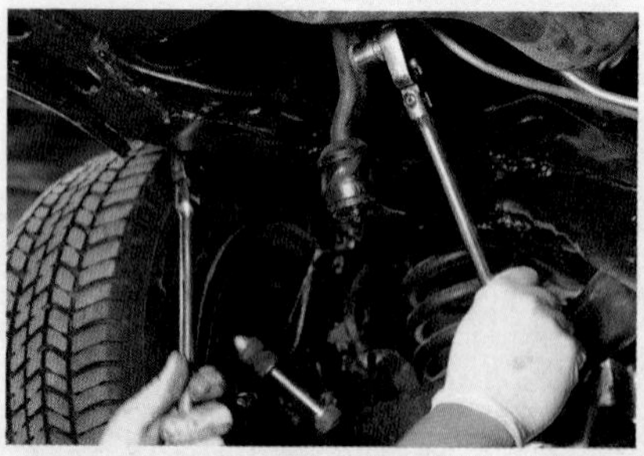

4. Remove the bolts holding the idler arm to the frame.

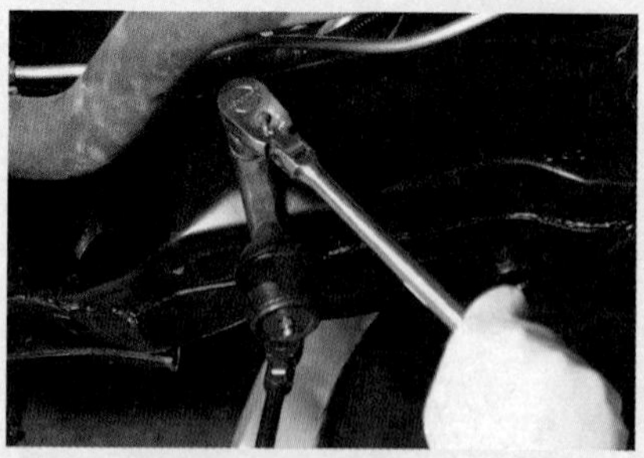

5. Install the new idler arm in reverse order of removal. Note that some idler arms must be installed with the steering centered. This will avoid inducing a twisting force on the steering linkage. Failure to do this can cause the vehicle to pull to one side.

SKILL DRILL 27-8 Inspecting and Replacing the Pitman Arm

1. Push and pull side to side on the front driver's side tire while watching for looseness in the pitman arm joint. If the movement is out of specifications, the joint will have to be replaced. This joint can be located on the pitman arm or the center link.

2. If the center link needs to be replaced, remove the cotter pin and nut holding the pitman arm taper stud to the center link.

3. If the pitman arm needs to be replaced, place an alignment mark on the pitman arm to the sector shaft, with white paint or a punch, to ensure correct positioning on reassembly.

4. Remove the pitman arm nut holding the arm to the sector shaft of the steering gear.

5. Using a pitman arm puller, pull the pitman arm free from the sector shaft.

6. Install the new pitman arm in reverse order of removal.

SKILL DRILL 27-9 Inspecting and Replacing the Center Link

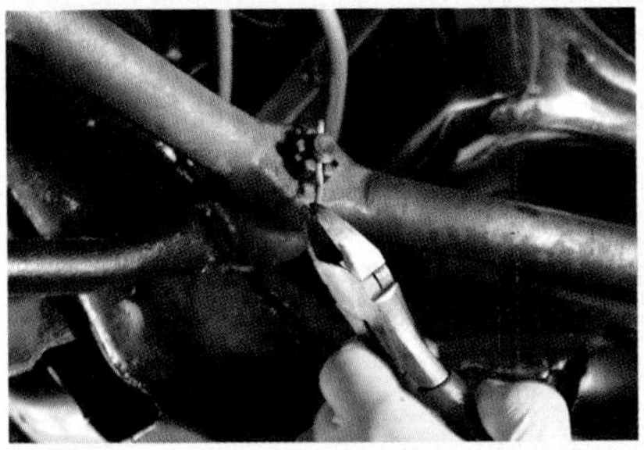

1. Push and pull the tire/wheel assembly from side to side, checking each of the center link joints for excess movement. If the movement is out of specifications, the joint(s) will have to be replaced. Remove the cotter pin and nuts from the taper studs of the pitman arm, idler arm, and inner tie-rod ends.

2. Separate the taper studs, using the double-hammer method, a pickle fork, or the approved puller. Remove the center link.

3. Install the new center link in reverse order of removal.

SKILL DRILL 27-10 Inspecting and Replacing the Steering Linkage Damper

1. Safely raise and support the vehicle. Look for fluid leaking out of the damper, a bent damper rod, and check that the mounting bushings are not worn.

2. Grab the wheels and turn them right and left. If the damper is working properly, there should be a fair amount of resistance when trying to turn the wheels quickly. If any of these conditions are found, replacement will be necessary. Remove the cotter pin and nut from the center link and frame attaching point, and remove the steering damper.

3. Install the new damper in reverse order of disassembly.

There are two basic types of tie-rod ends: spring-loaded and preloaded. A spring-loaded tie-rod end means that a fairly strong spring presses the ball into the socket. This accommodates for some wear over time. Because it is spring-loaded, it should never be tested by using pliers to compress it. It would fail even when new.

A preloaded tie-rod end has only enough internal clearance for the ball to pivot in the socket. The joint does not have a large spring and clearance, so it is not designed to be compressed. When inspecting tie-rod ends, there should be no up-and-down or side-to-side movement in the joint when moved by hand.

If removing a tie rod from an aluminum steering knuckle, do not use a pickle fork. It will damage the soft aluminum. Use the approved tie-rod end puller to separate the end from an aluminum knuckle or steering arm. *Note*: Replacement of tie-rod ends requires a wheel alignment to be performed; otherwise, rapid tire wear will occur. Be sure to follow the manufacturer's service information for inspecting and replacement of tie-rod ends.

To inspect and replace tie-rod ends, follow the steps in **SKILL DRILL 27-11**.

SKILL DRILL 27-11 Inspecting and Replacing Tie-Rod Ends

1. With the vehicle's weight on the tires, have an assistant gently rock the steering wheel between the 11 o'clock and the 1 o'clock positions. Note any side-to-side or up-and-down movement in the tie-rod ends. If the ball and socket is worn, replacement will be necessary. Loosen the tie-rod locknut found where the inner tie-rod end threads to the outer tie-rod end.

2. Remove the cotter pin and nut from the tapered stud of the tie-rod end. Separate it from the steering knuckle with the double-hammer method, a pickle fork, or the approved puller. Count the number of turns the tie-rod end rotates until it is backed out of the sleeve.

3. Install the new tie-rod end, using the number of turns counted in the previous step. This will get the toe angle close. Reinstall the new tie-rod in reverse order of removal. Perform a test drive to verify repair. Perform wheel alignment.

▶ Inspecting and Testing Electric Power-Assist Steering

LO 27-06 Inspect electric power steering and identify high-voltage electrical circuits.

On vehicles equipped with EPS, the system should be tested any time a driver complains of steering difficulties. This includes stiff, easy, intermittent, or lack of return to center concerns. The first step is to verify the concern. Start the vehicle and operate the steering wheel while observing the EPS warning lamp. If the lamp goes out as it should and the power steering feels normal, you may need to perform a test drive. This will allow you to check the operation of the steering system under driving conditions. The test drive should confirm the customer's concern. It will also give you valuable information regarding the fault.

If the EPS lamp is off, you will want to perform a visual inspection of the mechanical components. This includes tire pressure, tires, tie-rod ends, rack bushings, and steering column for excessive play or damage. If the EPS lamp indicates a fault, a scan tool will have to be hooked up to the vehicle's data link connector. The scanner accesses and retrieves data from the vehicle's onboard computer. This allows it to retrieve any stored diagnostic trouble codes (DTCs). The DTCs are used along with the manufacturer's flowchart to identify the exact problem. Consult the service information for the exact steps for diagnosis, repair, or replacement.

To inspect and test the electric power-assist steering, follow the steps in **SKILL DRILL 27-12**.

Identifying Hybrid Vehicle Power Steering System Electrical Circuits

Most full hybrid vehicles use electric power steering. In many cases, they operate at voltages higher than 12 volts. But usually not above 50 volts. At the same time, if the power steering system is operating, the hybrid vehicle's high-voltage circuit is also working. This can be upwards of 300 volts. This means that technicians could come into contact with the high-voltage circuitry even though they are working on the lower voltage EPS circuit. This means that technicians must be trained on the system before being allowed to work on it.

FIGURE 27-5 Manufacturers identify their high-voltage circuits with special colored wires. Here is one example using orange wire. Note that there is not a consistent color code used by all manufacturers.

Most manufacturers identify their high-voltage wires with specific colors. But there is not a uniform color designation across all manufacturers (**FIGURE 27-5**). Currently, orange, yellow, and blue are used by most manufacturers for circuit voltages higher than 12 volts. Always research the color of high-voltage wiring in the service information for any hybrid or electric vehicles prior to working on them. Also, never pierce the insulation on a high-voltage wire. It could cause serious bodily injury. It could also damage the expensive wiring harness.

SKILL DRILL 27-12 Inspecting and Testing the Electric Power-Assist Steering

1. Verify the concern by operating the steering wheel and observing the EPS warning lamp.

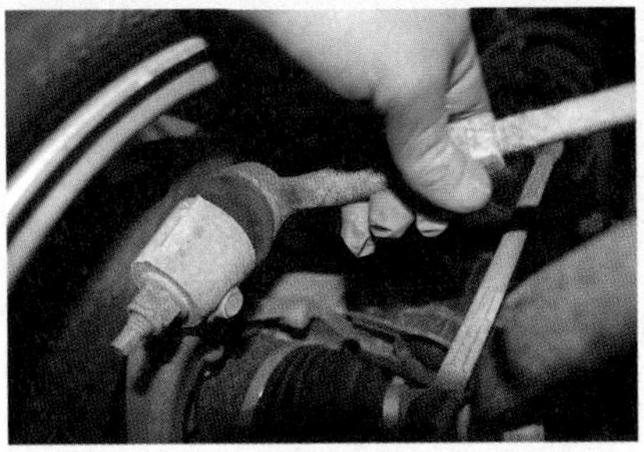

2. If the EPS lamp is off, inspect the steering system, including the tires.

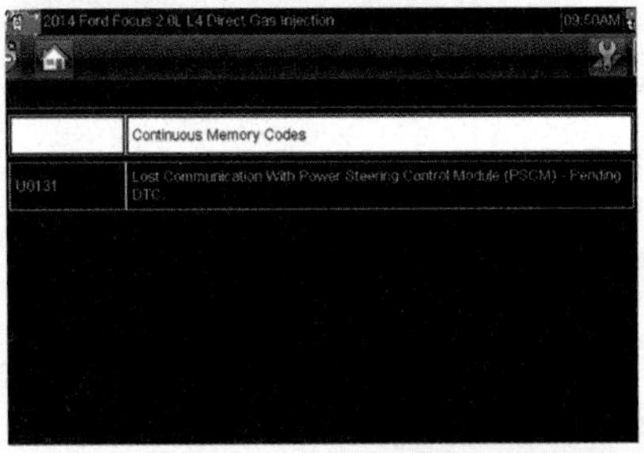

3. If the EPS lamp is on, connect a scan tool and read any EPS-related codes. Diagnose the fault by following the service procedure and using the strategy-based diagnostic process.

▶ Disabling and Enabling the SRS

LO 27-07 Disable SRS and service clock spring.

The driver's side airbag is located in the top cover of the steering wheel. The supplemental restraint system (SRS) must be disabled under the following circumstances:

- while working on or around the steering column,
- while working around any of the airbags or other pyrotechnic devices,
- while working around any of the sensors.

If not properly disabled, the airbag or other SRS devices could trigger accidentally. The large deployment force created by the ignited propellant can injure or kill the technician. It is important to know that most airbags are inflated by igniting a solid fuel similar to rocket fuel. Unintended deployment also creates the need to replace the airbag assembly and other required components. These components are expensive.

Generally, to disable an SRS, the correct SRS fuse must be located and removed. And then, the ignition switch is turned on and the SRS light is verified to remain on. Once the fuse is removed, the negative battery terminal is disconnected. The vehicle is allowed to sit for up to 15 minutes to discharge the capacitors. The capacitors store power to deploy the SRS devices in the event power is severed in an accident.

Note that a memory minder ***should not*** be used when working on the SRS system. Also, do not use a battery, voltmeter, ohmmeter, test light, or any other type of test equipment not specified in the service manual on any airbag, airbag squib, or airbag circuit. This will help to avoid accidental deployment of the airbag.

Some manufacturers have you disconnect the airbag connector. It is located under the dash near the steering column.

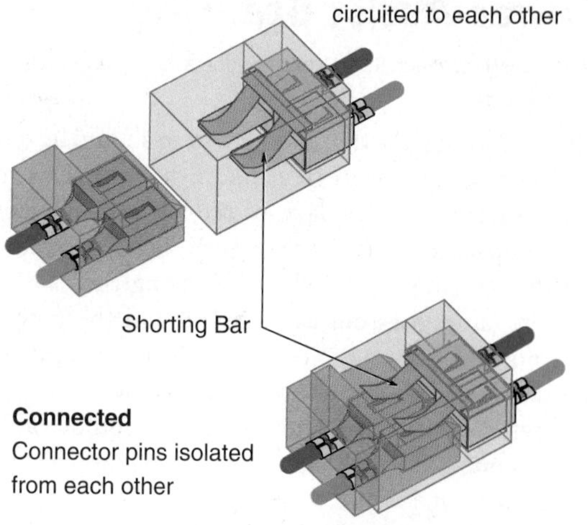

FIGURE 27-6 A shorting bar inside most airbag connectors shorts out the terminals when the connector is disconnected. This helps avoid accidental deployment when the airbag is removed from the vehicle.

Disconnecting this connector generally activates a shorting bar in the airbag side of the connector. The shorted connector makes it much harder for the airbag to accidentally deploy (**FIGURE 27-6**). With the SRS system disabled, you can carry out any needed diagnosis or repair on the steering column. Just remember to enable the SRS system once service is completed. Follow the vehicle's specific service information to properly disable and enable the SRS system.

To disable and enable the SRS, follow the steps in **SKILL DRILL 27-13**.

SKILL DRILL 27-13 Disabling and Enabling the SRS

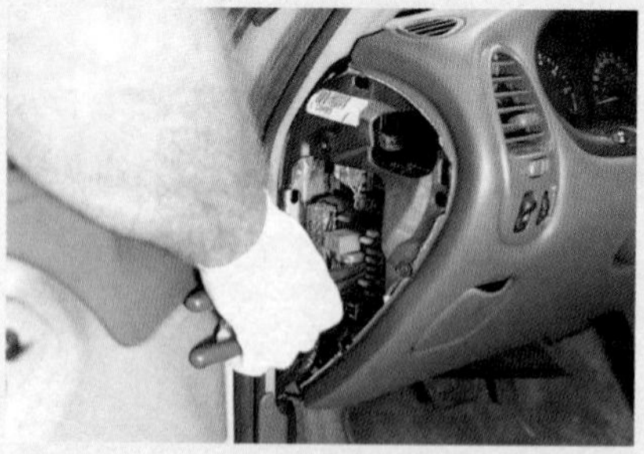

1. Find and remove the SRS fuse. Verify by turning the key on and observing that the SRS light remains lit for at least 30 seconds. If it goes out, you did not remove the correct fuse or all of the required fuses. Make sure the wheels are steered straight ahead. Turn the key off.

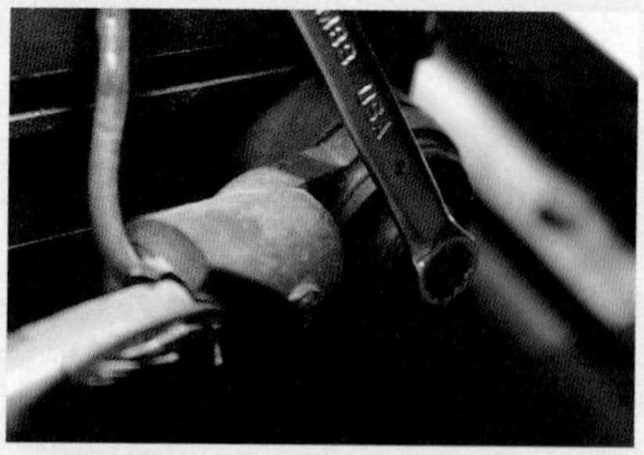

2. Remove the negative battery cable. Allow a minimum of 15 minutes to pass to let the SRS system capacitor's discharge. Note any radio presets or other memory features of the vehicle that will be erased when the battery is disconnected. *Do not use a memory minder or auxiliary power source!* If directed, disconnect the airbag connector.

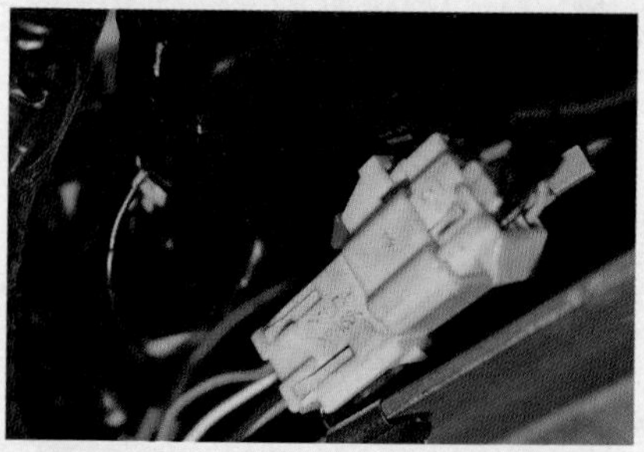

3. To enable the SRS system, verify that all SRS modules, components, and connectors are installed and connected properly.

SKILL DRILL 27-13 Disabling and Enabling the SRS (Continued)

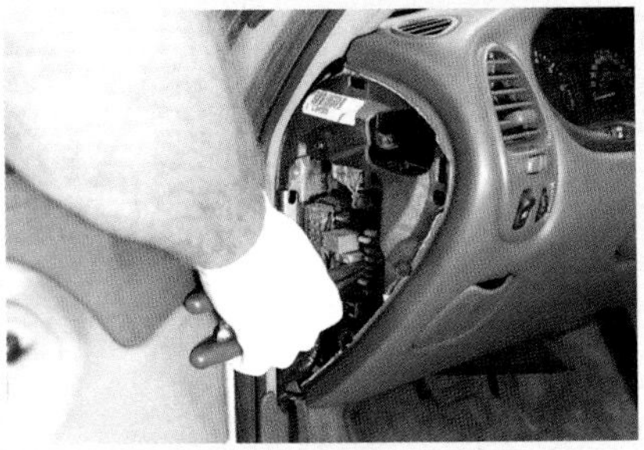

4. Reinstall the SRS fuse.

5. Reconnect the negative battery terminal, and tighten properly.

6. Without being in front of or reaching across the driver's side airbag, turn on the ignition switch and observe the SRS light. It should illuminate briefly and then go out and stay out. If so, the SRS system should be ready to be placed back into service.

SAFETY TIP

Many airbags are now of the two-stage variety. They can deploy with the proper amount of force for the accident. It can provide different levels of force depending on the severity of the accident, approximate weight of the occupant, etc. This means that even a deployed airbag is still potentially dangerous. So, treat it with caution during removal. It should be deployed soon after removal. Follow the manufacturer's specified procedure for deployment. This will render it safe for disposal.

Removing and Replacing the Steering Wheel and Center/Time the SRS Coil (Clock Spring)

The clock spring assists in supplying a constant electrical connection to a vehicle's driver's side airbag and horn. In some vehicles, it also supplies connections to the cruise control, message center controls, and entertainment controls. The clock spring coils and uncoils as needed during turning of the steering wheel. This ensures that the airbag is always ready to be deployed in the event of a collision.

Because of this winding and unwinding, the wires in the clock spring can break over time. This illuminates the SRS light and disables the SRS system. Because it is considered an important safety feature, it must be replaced if it is damaged in any way. Be sure to follow the manufacturer's service information to replace a clock spring. Failure to properly disable the airbag could result in serious injury or death. The below skill drill uses an example vehicle only. *Always* follow the service information for the vehicle you are working on.

To remove and replace the steering wheel and center/time the SRS coil, follow the steps in **SKILL DRILL 27-14**.

SKILL DRILL 27-14 Removing and Replacing the Steering Wheel and Centering/Timing the SRS Coil

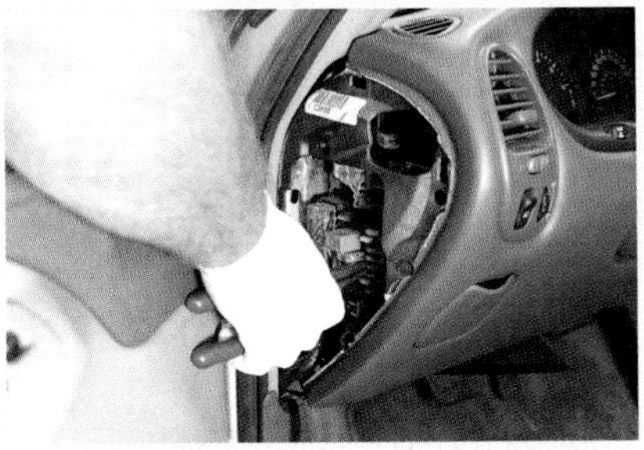

1. Find and remove the SRS fuse. Verify by turning the key on and observing that the SRS light remains lit for at least 30 seconds and does not go out. Make sure the wheels are straight ahead. Turn the key off.

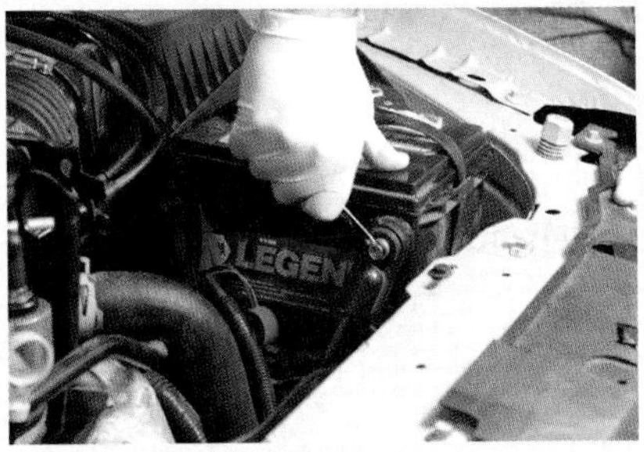

2. Remove the negative battery cable, and allow a minimum of 15 minutes to pass to let the SRS system's capacitors discharge. *Do not use a memory minder or auxiliary power source!*

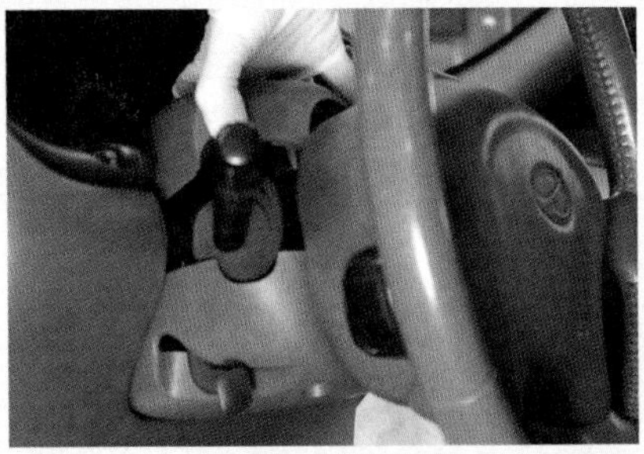

3. Locate the SRS connector at the bottom of the steering column and disconnect. Remove upper and lower trim panels.

SKILL DRILL 27-14 Removing and Replacing the Steering Wheel and Centering/Timing the SRS Coil (Continued)

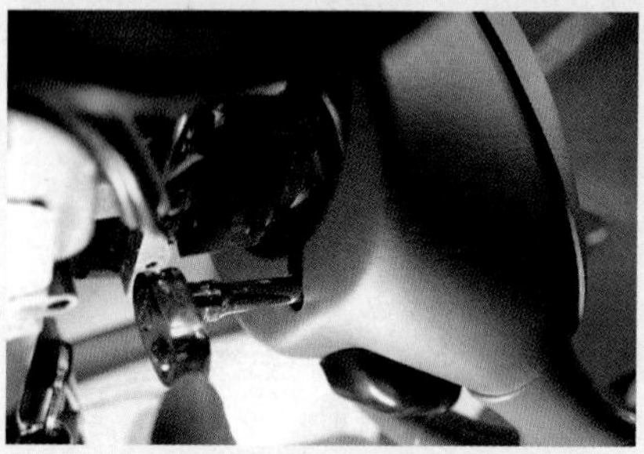

4. Locate the bolts or spring clips on the back of the steering wheel that hold the airbag to the steering wheel. Remove or release them.

5. Lift the airbag from the steering wheel, and disconnect the airbag connector to the steering wheel harness.

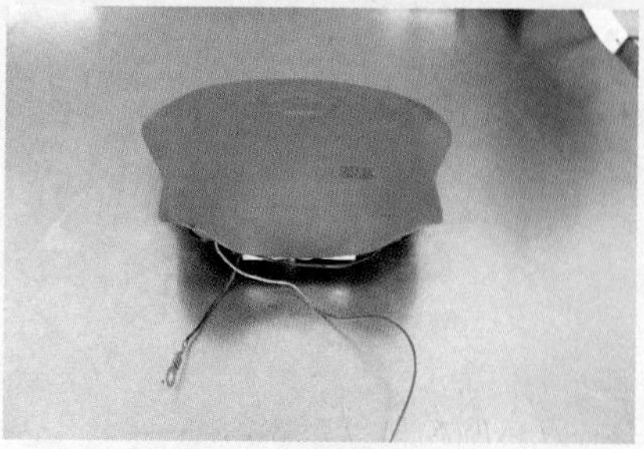

6. Sit the airbag on a bench, face up, in a safe place.

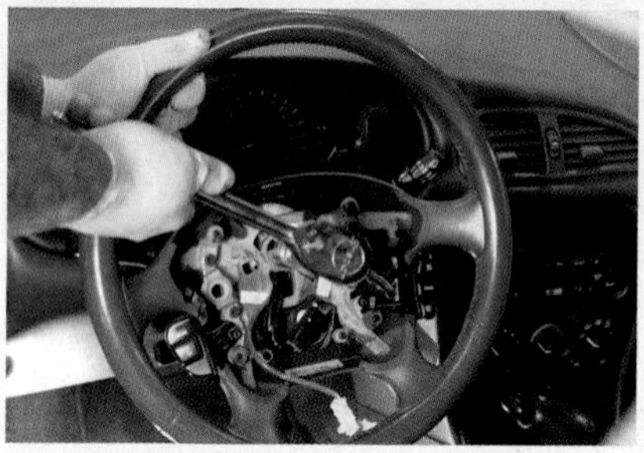

7. Remove the fasteners that hold the steering wheel on.

8. Remove the steering wheel with the manufacturer's recommended puller (this information can be found in the service information).

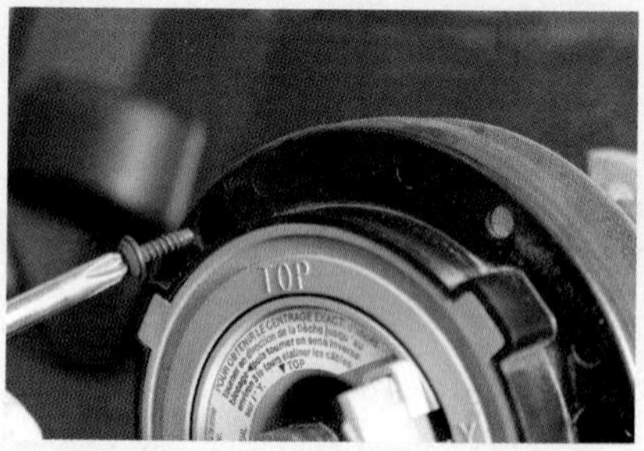

9. Remove the clock spring screws or snap ring, and lift the clock spring from the steering column shaft. Attach a thin wire to the clock spring connector (airbag connector was disconnected earlier) at the base of the steering column, and pull the clock spring and connector up through the steering column.

10. Gently pull the new clock spring harness down through the steering column.

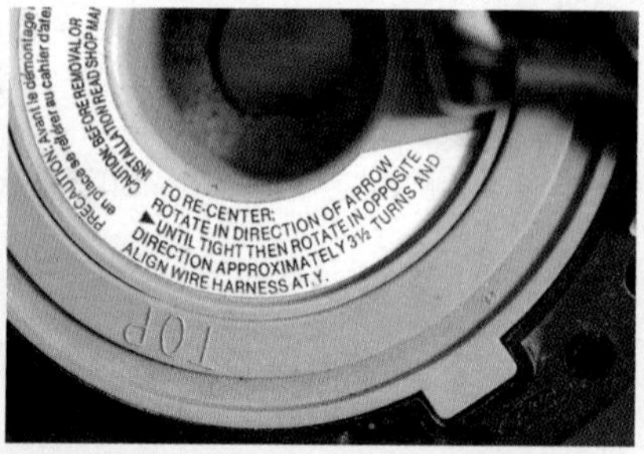

11. If the antirotation key is still installed, skip to step 13. If the antirotation key is not installed, center the clock spring, following the manufacturer's procedure. In this case, we gently rotate the inner rotor counterclockwise until it stops.

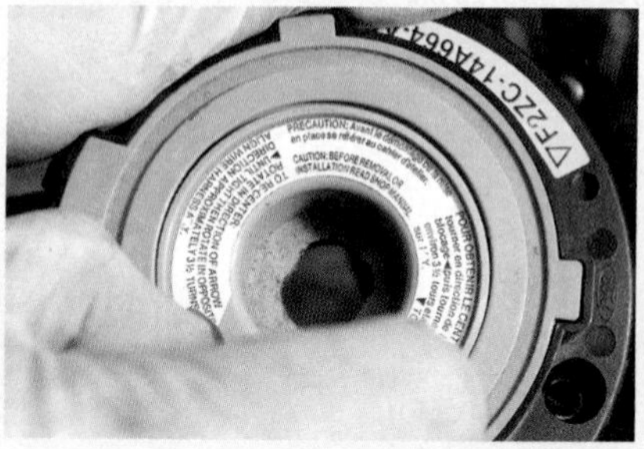

12. Rotate the inner rotor clockwise the required number of turns—in this case, four full turns. Verify that all of the marks are aligned in their specified positions.

13. Reinstall the clock spring assembly in the proper orientation, and secure it with the screws or snap ring. Reinstall components in the reverse order of removal. Ensure that torque specifications are followed. Also remove the antirotation key at the appropriate step.

▶ Wrap-Up

Ready for Review

- ▶ Visual inspections and physical measurements make up most of the maintenance and diagnosis tasks of steering systems. Tools for electrical diagnosis include the scan tool and the digital multimeter (DMM). Tools for mechanical diagnosis and repair include a power steering system pressure tester, various measuring devices, and special tools for component removal.
- ▶ Excessive play, unusual noises, vibrations/shimmies, and hard or inconsistent steering are commonly caused by wear, poor lubrication, or damaged parts. Physically inspecting each component is the best way to check the steering system and includes measuring play, looking for bent or damaged linkage, and testing the operation of the power steering system.
- ▶ The power steering fluid has both a hot and a cold indicator on the dipstick for an accurate reading of the fluid level in both conditions.
  - The power steering system should be flushed whenever the manufacturer specifies a fluid change, the fluid appears contaminated or dirty, a major part of the hydraulic steering system is replaced, or a serious mechanical problem involves the power steering pump or steering gear.
- ▶ Rack and pinion service should include inspection and replacement of power steering hoses and fittings, inspection of rack-and-pinion mounting bushings and brackets, and inspection and replacement of rack-and-pinion steering gear inner tie-rod ends and bellows boots.
- ▶ Parallelogram steering linkage service should include inspection and replacement of the pitman arm, relay (center link/intermediate), rod, idler arm and mountings, and steering linkage damper and inspection and replacement of tie-rod ends, tie-rod sleeves, and clamps.
- ▶ High-voltage circuits are wired with special colors. Always research the color of high-voltage wiring in the service information for any hybrid or electric vehicles prior to working on them.
  - If the EPS lamp indicates a fault, a scan tool can be used to access and retrieve data from the vehicle's onboard computer.
- ▶ The supplemental restraint system (SRS) must be disabled while working on or around: the steering column, the airbags, or any of the sensors.
  - A memory minder should not be used when working on the SRS system.

Key Terms

dial indicator A device for precision measurements used to measure small variations, such as end play, movement in a bearing, or run-out.

Floor jacks and safety stands Hydraulid jack used to lift the vehicle, and mechanical safety stands to support it once lifted.

pickle fork U-shaped wedge used for separating tie-rod ends that is operated by hammer or air hammer. This tool usually destroys the dust boot during the process, so it is not the best tool to use on tie-rods that will be reused.

pitman arm puller A heavy-duty puller made specially for removing the pressed-on pitman arm from the sector shaft.

pry bar A high-strength carbon steel rod with offsets for levering and prying. Also called a crowbar.

tie-rod end puller Used to pull the tapered shaft on a tie-rod end from its mating steering component.

tie-rod sleeve adjusting tool Has a tab designed to grab the slot in the sleeve and is used to turn the sleeve when adjusting the toe setting.

Review Questions

1. What tool is most often used to measure voltage and continuity on steering system components that are electrical or electronic?
   a. A dial indicator
   b. A power steering system analyzer
   c. A digital multimeter
   d. A scan tool
2. Steering wheel switches maintain electrical connection through what component?
   a. A torsion bar
   b. A clock spring
   c. A steering linkage
   d. An angle sensor
3. Common steering column issues include all the following EXCEPT:
   a. Unusual noises
   b. Tire chirp
   c. Looseness
   d. Binding
4. What is the primary benefit of using a power steering flush machine?
   a. To minimize the amount of air introduced into the system
   b. To push the filter out of the bottom of the reservoir
   c. To maximize the amount of air introduced into the system
   d. To backflush (reverse flow) the inline fluid filter
5. What service should be advised to the customer after a tie-rod end replacement?
   a. A power steering fluid flush
   b. A steering wheel replacement
   c. A wheel alignment
   d. A steering module reprogram
6. What is likely to happen if a bellows boot becomes torn?
   a. The vehicle will pull to the right.
   b. Road grime and debris will contaminate the internal rack-and-pinion seals.

c. The front tires will wear in the center of the tread.
d. The caster angle will be affected.

7. What tool is used to remove a pitman arm from the steering gear sector shaft?
   a. A pickle fork
   b. A crow-foot wrench
   c. A sector shaft separator
   d. A pitman arm puller
8. What type of tie-rod end accommodates for some wear over time?
   a. Preloaded
   b. Spring-loaded
   c. Center link
   d. Idler
9. What tool is typically used to begin diagnosis of an electric power steering system when the EPS lamp indicates a fault?
   a. A 12-volt test light
   b. A fused jumper wire
   c. A scan tool
   d. An oscilloscope
10. Why is disabling an SRS system important when working on the steering column?
   a. To keep the airbag or other SRS devices from triggering accidently.
   b. Disabling the SRS system deploys the airbags safely.
   c. To prevent the seat belt retractors from twisting the belts.
   d. To ensure the steering column stays indexed to the clock spring.

ASE Technician A/Technician B Style Questions

1. Technician A says that a pickle fork should be used to separate a tie rod that will be reinstalled later. Technician B says that a power steering system analyzer can measure power steering fluid flow and pressure. Who is correct?
   a. Technician A only
   b. Technician B only
   c. Both A and B
   d. Neither A nor B
2. Technician A says that florescent dye can be used to help identify the source of power steering fluid leaks. Technician B says that a binding steering system could be caused by a problem in the steering gearbox. Who is correct?
   a. Technician A only
   b. Technician B only
   c. Both A and B
   d. Neither A nor B
3. Technician A says that steering joints with zerk fittings should be lubricated regularly. Technician B says that some steering joints are greased-for-life, and do not need regular lubrication. Who is correct?
   a. Technician A only
   b. Technician B only
   c. Both A and B
   d. Neither A nor B
4. Technician A says that replacement power steering fluid is universal and can be used to top-off any power steering fluid type. Technician B says that power steering fluid contamination is checked with litmus paper. Who is correct?
   a. Technician A only
   b. Technician B only
   c. Both A and B
   d. Neither A nor B
5. Technician A says that power steering fluid leaks can be caused by faulty O-rings. Technician B says that fluid leakage onto rack-and-pinion bushings can cause bushing failure. Who is correct?
   a. Technician A only
   b. Technician B only
   c. Both A and B
   d. Neither A nor B
6. Technician A says that aluminum steering knuckles may require special pullers to remove tie-rod ends. Technician B says that some side-to-side movement in the tie-rod end joint is normal when rocking the steering back and forth with vehicle weight on the wheels. Who is correct?
   a. Technician A only
   b. Technician B only
   c. Both A and B
   d. Neither A nor B
7. Technician A says that a lack of steering return to center could be caused by electric power-assist. Technician B says that a common step in diagnosing an electric power-assist steering system is to retrieve codes. Who is correct?
   a. Technician A only
   b. Technician B only
   c. Both A and B
   d. Neither A nor B
8. Technician A says that special training is not needed to work on hybrid high-voltage EPS systems. Technician B says that manufacturers often identify their hybrid high-voltage circuits with special colored wiring harnesses. Who is correct?
   a. Technician A only
   b. Technician B only
   c. Both A and B
   d. Neither A nor B
9. Technician A says that it is safe to work on the SRS system immediately after disconnecting the negative battery cable. Technician B says that removed airbags should be placed on a workbench face up. Who is correct?
   a. Technician A only
   b. Technician B only
   c. Both A and B
   d. Neither A nor B
10. Technician A says that shorting bars inside airbag connectors prevent accidental deployment when the connector is disconnected. Technician B says that a memory minder should not be used while working on the SRS system. Who is correct?
   a. Technician A only
   b. Technician B only
   c. Both A and B
   d. Neither A nor B

CHAPTER 28

Suspension Systems Theory

Learning Objectives

- LO 28-01 Describe suspension system principles.
- LO 28-02 Describe suspension system spring components.
- LO 28-03 Describe fixed shock absorbers and struts.
- LO 28-04 Describe manually and automatically adjustable shocks.
- LO 28-05 Describe suspension system components.
- LO 28-06 Describe the main types of suspensions.
- LO 28-07 Describe the main types of front suspension systems.
- LO 28-08 Describe the main types of rear suspension systems.
- LO 28-09 Describe active and adaptive suspension systems.

ASE Education Foundation Tasks

See Appendix A to view the 2017 ASE Education Foundation Automobile Accreditation Task List Correlation Guide.

▶ Suspension System Principles

LO 28-01 Describe suspension system principles.

The suspension system of a vehicle is designed to keep the tires in contact with the road surface. This helps to maintain traction and control of the vehicle. At the same time, it provides a smooth ride for the passengers. It is composed of a network of springs, arms, struts, and shock absorbers that work together to achieve these purposes (**FIGURE 28-1**).

Wear from extended use is the main cause of suspension problems. However, impact from potholes and accidents may cause bent or broken components as well. There are several driver complaints that indicate problems in the suspension system. It can be a bouncy ride, a reduction in steering control, or unusual noises when going over bumps. Also, know that the suspension system interacts very closely with the steering system. Thus, part of the challenge in accurately diagnosing the vehicle is knowing how both systems operate. It will also help you perform regular inspection of each suspect component. Identifying wear before it causes severe problems is an important part of customer service.

Even if all the suspension and steering system components are in good working condition, the wheels also have to be aligned correctly. We will cover wheel alignment in a following chapter. This chapter will help you become familiar with the various types of suspension systems. It also covers their components, and how they operate. But as we go through this content, think about how each part controls wheel position and movement. That will help you understand the process of aligning the wheels when we get to it.

Sprung and Unsprung Weight

The **suspension system** reduces road shocks. Otherwise, they would be transferred through the chassis to the passengers. It also must keep the tires in contact with the road. The suspension system components must be strong. They have to withstand loads imposed by the vehicle's mass during cornering, accelerating, braking, and uneven road surfaces. This can make the parts quite heavy.

When a tire hits a speed bump, it creates a **reaction force**. This means that the tire will move in response to hitting the obstruction. The size of the reaction force generated depends on the **unsprung mass**, also called **unsprung weight**. Unsprung weight includes all of the parts of a vehicle that are not supported by the springs. This includes the wheels, tires, brakes, axles, and any steering and suspension parts not supported by the springs (**FIGURE 28-2**).

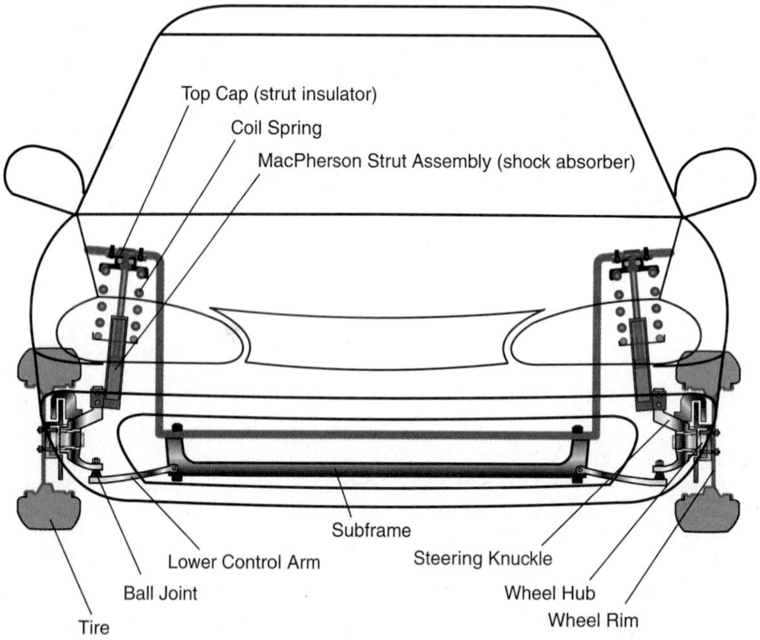

FIGURE 28-1 The suspension system.

You Are the Automotive Technician

A customer with a sporty import vehicle comes into your shop. She asks what she can do to make her vehicle handle better on the local autocross track. She says others have suggested that she reduce the sprung and unsprung weight of the vehicle, add a rear sway bar, and add some adjustable shock absorbers. How would you answer these questions she has?

1. What components are part of the unsprung weight of a vehicle? And what would be some ways to reduce it?
2. What components are part of the sprung weight of a vehicle? And what are some things she could do to reduce it on race day?
3. What are adjustable shocks, and how might they help her?
4. How would adding a rear sway bar affect the vehicle's handling?

Sprung weight is the body, drivetrain, and associated parts that are supported by the springs. The heavier the unsprung components, the greater the reaction force they will generate. When wheel assemblies encounter a speed bump, the reaction force may be large enough to make the tire keep moving upward after hitting the bump. This causes the tires to lose contact with the road surface. If the tires are not in contact with the road, they have no traction. They cannot control the vehicle's direction of travel, which is a dangerous situation. So, the heavier the unsprung components, the less control you have.

If the unsprung weight is very light, it will not generate as much upward reaction force. So it is easier for the spring to push it back down. This maintains contact with the road and therefore control of the vehicle. To get an idea of this concept, imagine how much more difficult it would be to stop a fast-moving bowling ball than a fast-moving beach ball. It's the same way with heavy unsprung components. Once in motion, they want to keep moving in the same direction.

Reaction forces also transfer a lot of momentum to the vehicle body. It moves upward as it absorbs the energy from the wheel assembly. This movement gets transferred to the occupants. So, lighter unsprung weight transmits less force to the body and makes for a more comfortable ride. As a general rule, unsprung weight should be kept as light as possible. But components need to be strong enough to handle the stresses under various driving conditions.

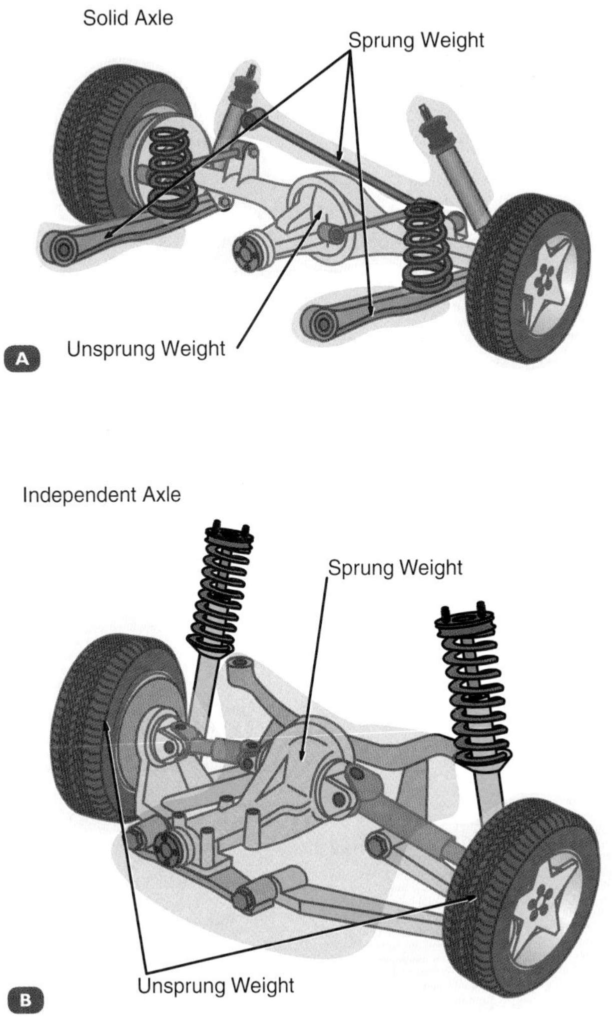

FIGURE 28-2 A. Unsprung weight includes all of the components between the spring and the ground. **B.** Sprung weight includes all of the components that are supported by the springs.

Suspension Function

Suspension systems must absorb the large road forces generated while driving on imperfect roads. This is so those forces won't be passed on to the occupants. And at the same time, the wheels and tires need to be held in the proper orientation for acceleration, braking, and steering. On top of all of that, the suspension system has to support the weight of the vehicle, occupants, and any additional load (**FIGURE 28-3**). To accomplish this, the

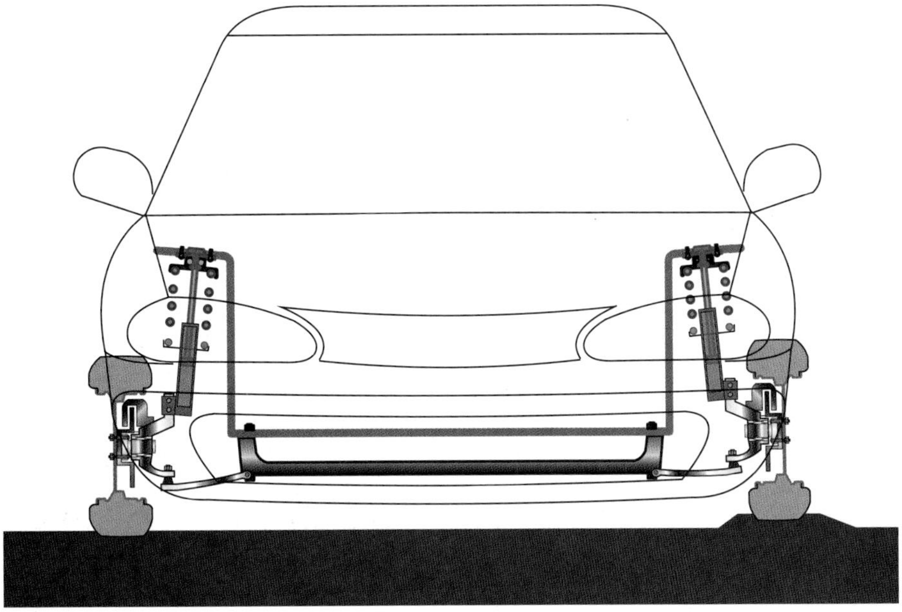

FIGURE 28-3 A suspension system subjected to a speed bump.

suspension system must be the flexible assembly between the wheels and the body. It has to be able to both flex and return to its original shape. This characteristic is called **elasticity**. Automotive suspension systems generally use the elastic properties of special metals formed into springs. They provide the elastic medium that supports the body.

There are three primary types of springs used in suspension systems: leaf springs, coil springs, and torsion bars (**FIGURE 28-4**). **Leaf springs** are located between the frame and the axle assemblies. They are typically semi-elliptical in shape.

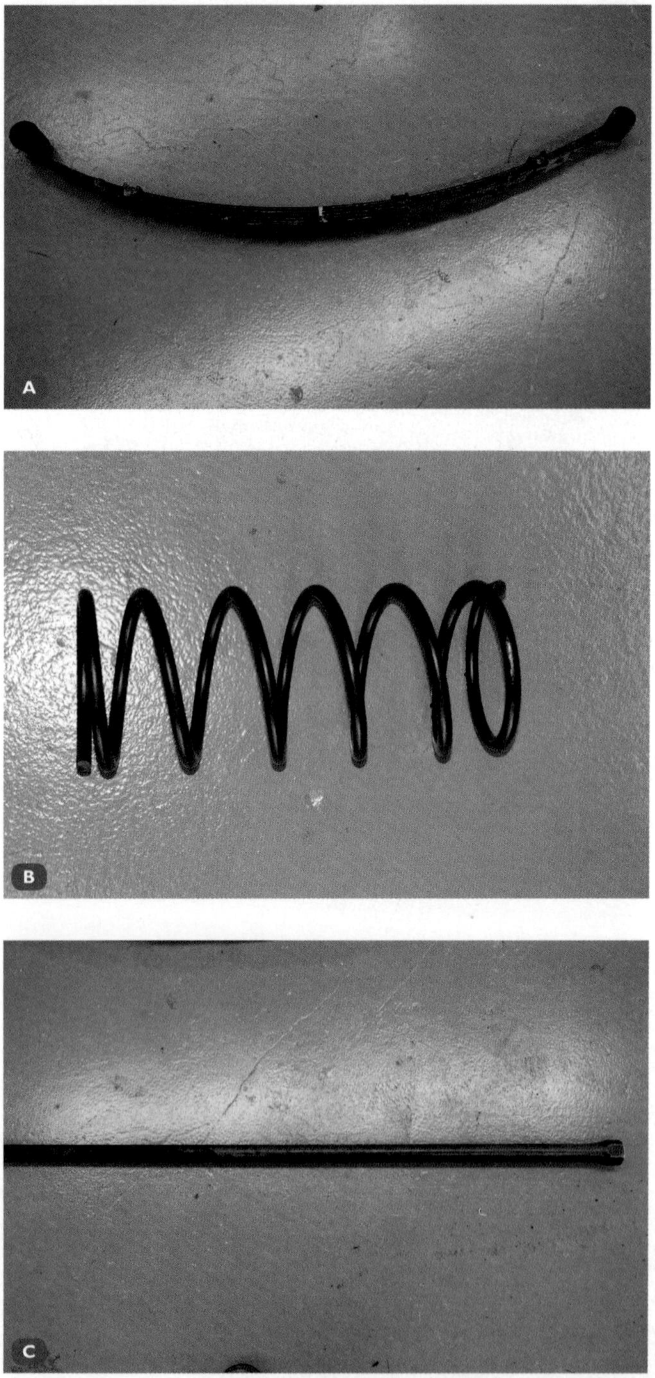

FIGURE 28-4 Three types of springs. **A.** Leaf spring. **B.** Coil spring. **C.** Torsion bar.

They absorb the **applied force** (pressure of the load) by flattening out under load. And they are often used at the rear end of a car or truck to help the vehicle carry large loads.

Coil springs are formed in a spiral from a single steel rod. They absorb the force of impact by twisting and compressing. They are commonly used on smaller vehicles to smooth the ride and improve handling.

Torsion bars are held rigid at one end and twist around their center as the other end is deflected. Torsion bars return to their original shape when the **deflecting force** is removed. They are typically used in the front of some pickup trucks because they handle better than leaf springs.

Nonmetallic materials, such as rubber, provide cushioning action. They are more commonly used as **stops** to limit extreme suspension movement. Stops are called jounce stops, bump stops, bumper stops, or rebound stops. They prevent parts from banging against each other, or the frame of the vehicle (**FIGURE 28-5**). They are typically a triangle or cone shaped to provide a collapsible cushion.

In some light vehicle applications, air is used for **ride height** control. Rubber bags filled with air are used to help support additional weight when placed at the rear of some vehicles. If placed at both ends of the vehicle, they are used to give additional ground clearance. This is predominately a function of SUVs for when off-road ground clearance can be a problem.

When springs are supporting the weight of the vehicle, they are in a partially compressed condition. When a wheel strikes a bump, the springs are compressed further. But they also impart an upward force on the mass of the vehicle. This causes it to move upward. Once past the bump, the spring starts to lengthen, pushing the wheel down but still pushing the body up. The spring tends to **overshoot**, or spring back, past its original length. And because the body is moving upward, there is not as much weight on the spring. So, this overshooting of the spring is magnified.

Once the body gets as high as it will go, the spring cannot hold it there, as it is higher than its normal compressed height. So the body falls, compressing the spring once again, producing oscillations in the spring and body (**FIGURE 28-6**). **Oscillations**

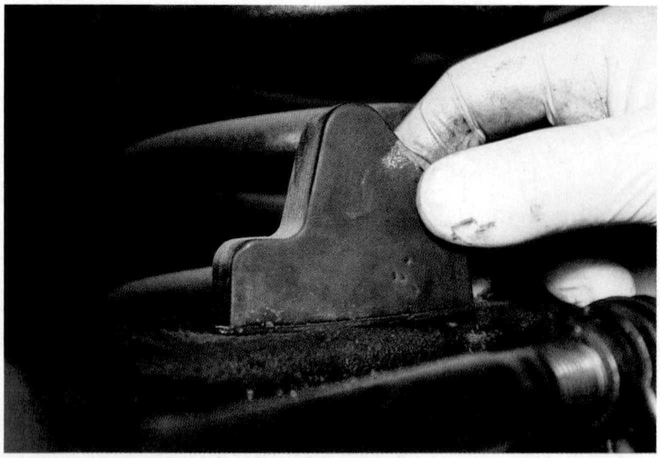

FIGURE 28-5 Rubber stops.

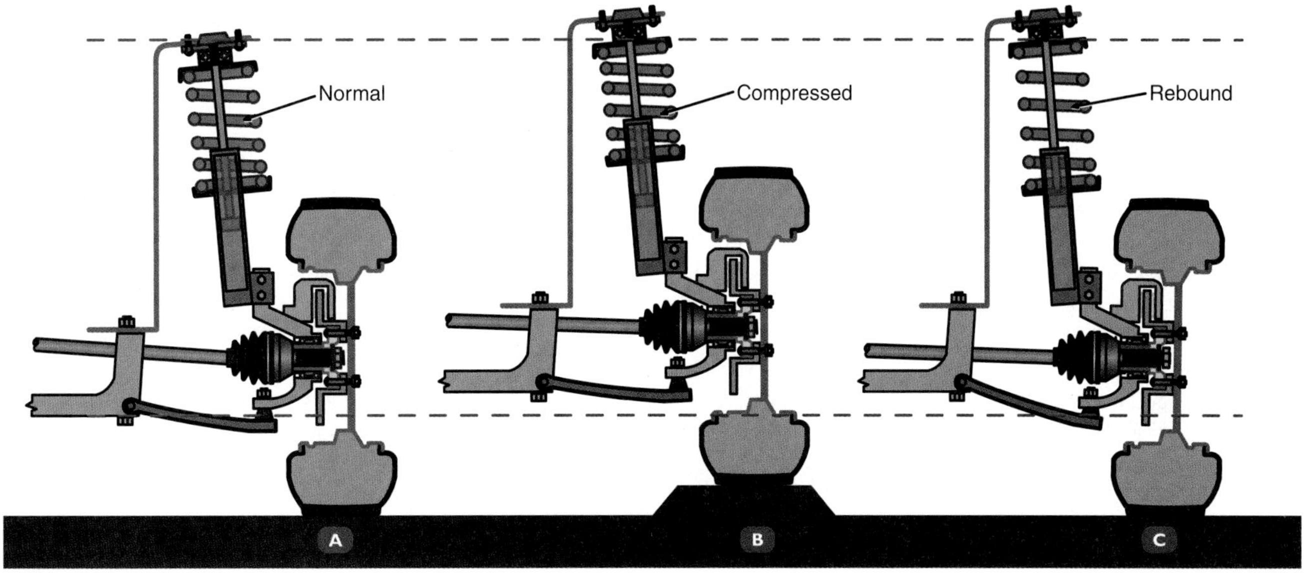

FIGURE 28-6 **A.** Spring at normal height. **B.** Spring compressed by hitting a bump and pushing the body upward. **C.** Spring overshoot on rebound.

means the spring compresses and rebounds over and over again. As a result, the vehicle bounces up and down, making the ride uncomfortable and unstable. It is what produces the forces that make the tires bounce off the ground.

Different materials have different levels of elasticity. Up to a certain point, they can be deformed and released, and they will return to their original condition. Beyond that point, they stay deformed. With some materials, returning to their original state too quickly can produce a bouncing effect. Preventing or reducing oscillations is called damping, more commonly referred to as dampening. It can occur in many different ways. The dampening material absorbs the energy from the oscillation. In a suspension system, a hydraulic shock absorber dampens oscillations in the spring. Rubber bushings further dampen road shock. These concepts are explored in more depth later in this chapter.

Controlling Forces in Suspension Systems

When a vehicle is in motion, driving thrust, braking torque, and cornering force exert pressure against the wheel units (**FIGURE 28-7**). **Driving thrust** is the force transferred from

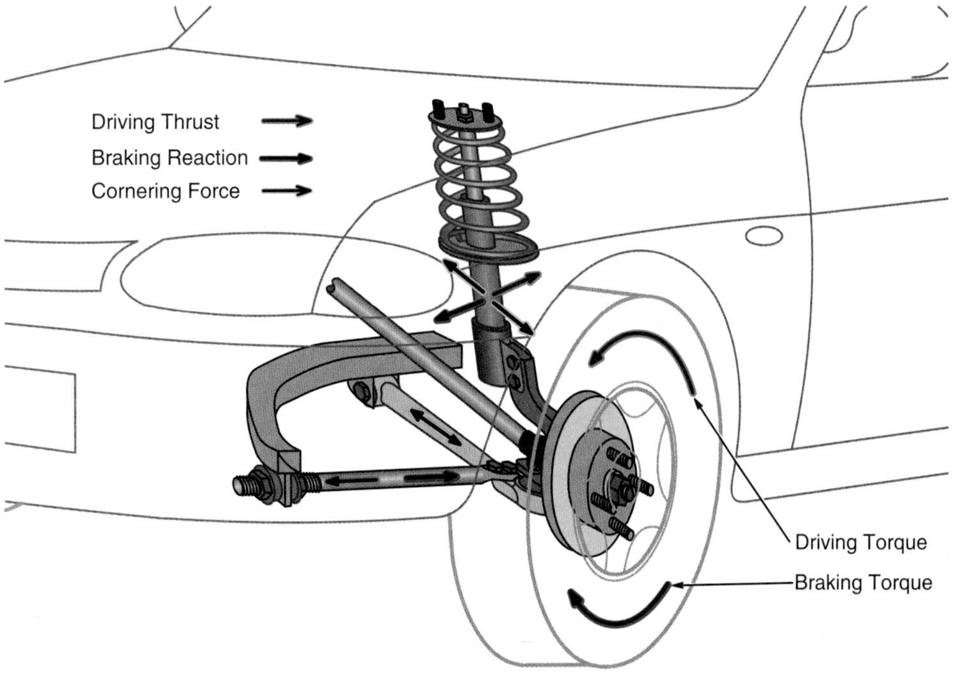

FIGURE 28-7 There are three common forces acting on suspension systems: driving thrust, braking torque, and cornering force.

the tire contact patch through the wheel. It places a twisting force on the suspension members that push the vehicle along the road. It also tends to try to move the wheel forward relative to the body.

Braking torque also places a twisting force on the suspension system. But it is in the opposite direction from driving thrust. It also tends to try to move the wheel backward relative to the body. **Cornering force** refers to the lateral movement of the suspension system during turning. It tries to push the wheel to one side or the other. It also tries to fold the wheel over.

These forces are transferred to the frame of the vehicle. But while they act, the wheel units must stay aligned with each other and with the frame. The wheels must be securely located longitudinally, laterally, and vertically. At the same time, they must have the freedom to move vertically to allow for suspension travel. They must also allow the wheels to pivot for steering.

The suspension system must absorb these forces while maintaining precise control of the wheels. Each of the suspension system components contributes to controlling these forces. At the same time, they keep the wheel units in proper alignment. But the forces do put stress on the joints and pivots, which tend to wear them out over time.

Yaw, Pitch, and Roll

The terms "yaw," "pitch," and "roll" describe the movement of a vehicle around three axes (**FIGURE 28-8**). The x-axis is the imaginary line drawn down the center of the vehicle from front to back. The y-axis is the imaginary line across the vehicle from left to right. The z-axis is the vertical line that runs through the center of the vehicle from top to bottom.

Roll is vehicular movement along its x-axis. It is the rolling motion you feel when making a sharp corner. It is generally what causes vehicle rollovers. **Pitch** is movement around the vehicle's y-axis. It is commonly felt during hard braking or fast acceleration. You can feel the front of the vehicle nose down or rise up slightly. **Yaw** is movement around the z-axis, felt when the vehicle deviates from its straight path. This occurs when the rear wheels slide out during drifting.

Movement around each axis must be controlled during all maneuvers of the vehicle for it to be safe. Safety systems, such as electronic stability control, are designed to keep vehicles within the safe limits of each axis. We will look at those systems in another chapter.

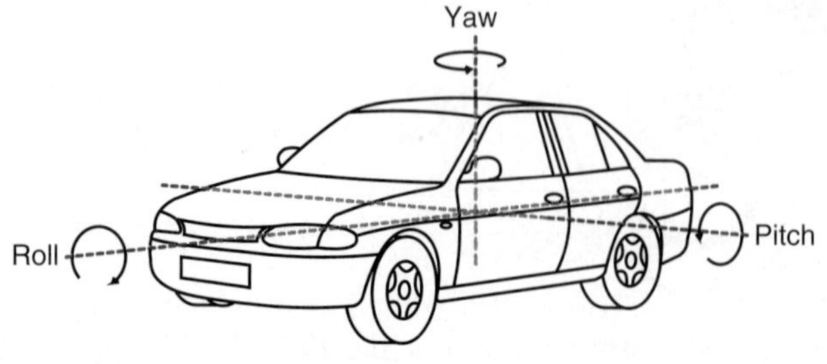

FIGURE 28-8 Yaw, pitch, and roll.

▶ Suspension System Components

LO 28-02 Describe suspension system spring components.

In order for the suspension system to perform its many functions, it is made up of many components. A basic suspension system consists of the following parts:

- **Springs**—the flexible components of the suspension. Basic types are leaf springs, coil springs, and torsion bars. Modern passenger vehicles usually use light coil springs.
- **Axles**—used to drive and/or support the wheels.
- **Shock absorbers**—dampen spring oscillations by forcing oil through small holes in a piston. The oil heats up as it absorbs the energy of the motion.
- **Control arms**—the primary load-bearing elements of a vehicle's suspension system. They are isolated from the chassis with rubber bushings and pivots that allow the up-and-down movement of the tire and wheel assembly.
- **Rods**—straight (or precisely formed) pieces of steel used to control motion within the vehicle's suspension system. They typically have pivoting or flexible mounts on each end.
- **Ball joints**—swivel connections mounted in the outer ends of the control arms.

FIGURE 28-9 shows the components of the suspension system.

Springs

Springs suspend the weight of the vehicle over the axles. They do so in such a manner that allows the wheels to follow unevenness in the road. Therefore, springs must be elastic. Most springs are made out of spring steel and sag over time, requiring replacement. Some manufacturers have used composite materials for their leaf springs. These are said to resist sagging better than spring steel, as well as substantially reduce weight. The most common spring configurations are coil springs, leaf springs, and torsion bars.

Coil Springs

Coil springs, also known as helical springs, are used on both the front and rear suspension. In many cases, they have replaced leaf springs in the rear suspension (**FIGURE 28-10**). A coil spring is made from a single length of special wire. It is heated and wound on a former in the shape of a coil to produce the required shape. The load-carrying ability of the spring depends on:

- the diameter of the wire,
- the overall diameter of the spring,
- its shape, and
- the spacing of the coils.

On a small passenger car, they are lighter and more flexible than springs on a heavier vehicle. The stiffer springs are designed to withstand heavier loads. They also make the ride much rougher.

The pitch of a spring is the distance from the center of one coil to the center of the adjacent coil. The coils may be evenly

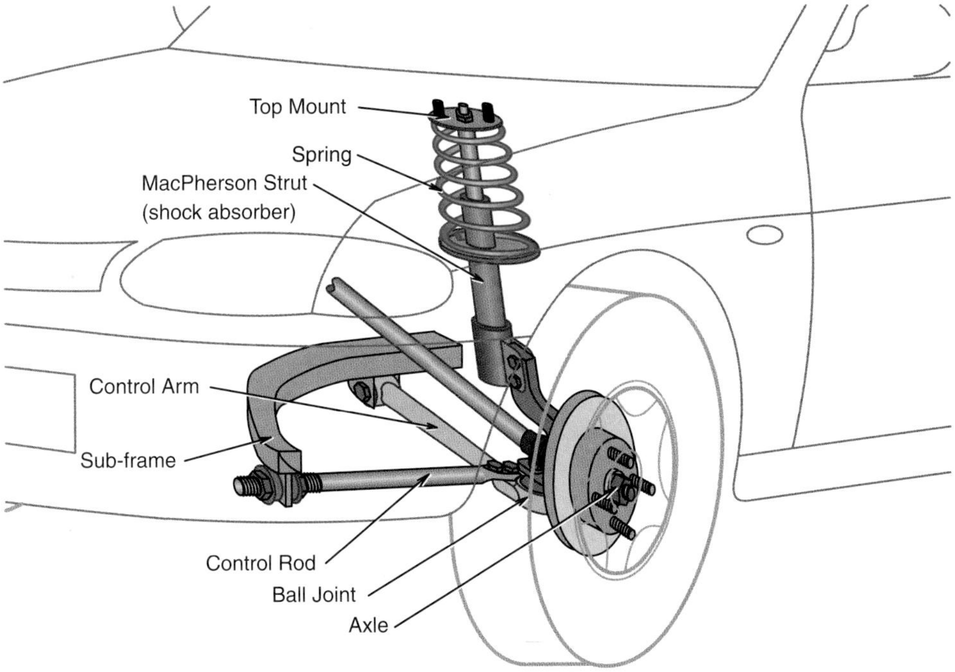

FIGURE 28-9 The components of a suspension system.

FIGURE 28-10 Coil springs, also known as helical springs, are used on the front suspension of most modern light vehicles.

FIGURE 28-11 A constant rate versus a progressive rate spring.

spaced, called **uniform pitch**, or unevenly spaced, called variable pitch. The wire can be the same thickness throughout, or it may taper toward the end of the spring. The spring itself may be cylindrical, barrel shaped, or conical.

Generally, a cylindrical spring with uniform wire diameter and uniform pitch has a constant rate of deflection. This means it takes the same amount of force to compress each coil. Its length reduces in direct proportion to the load applied.

When the pitch of a spring is varied, the deflection rate varies too (**FIGURE 28-11**). The spring is then said to have a **progressive rate of deflection**. It deflects easily under a light load, but its resistance increases as the load increases. This provides a softer ride when the vehicle is lightly loaded than if the vehicle has heavier constant-rate springs. Coil springs can look alike but have very different load ratings. They are often color coded for identification. They also normally use rubber pads at their seats to prevent the transmission of noise and vibration.

As a cylindrical coil compresses, it can become coil bound. This limits its travel and creates a harsh ride when heavily loaded. As conical and barrel-shaped springs compress, they have the ability to collapse into themselves (**FIGURE 28-12**). This creates a longer suspension travel for a given length of spring than for a cylindrical spring. It also means the spring is less likely to bottom out when heavily loaded. This type of spring is useful in the rear of some pickup trucks or vans.

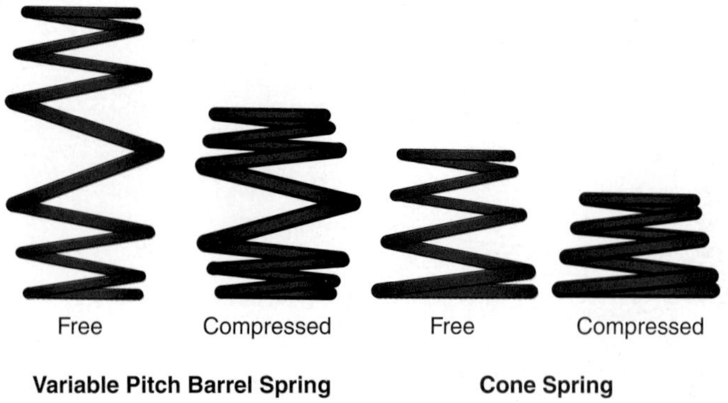

FIGURE 28-12 As conical and barrel-shaped springs compress, they have the ability to collapse into themselves.

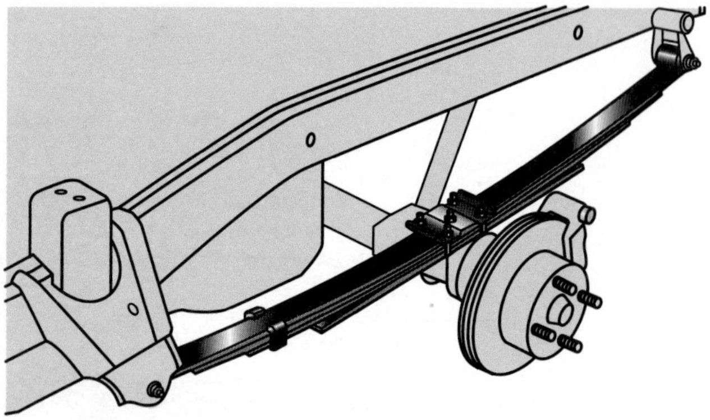

FIGURE 28-13 Leaf springs and components.

Leaf Springs

The leaf spring is one of the oldest forms of spring. It is usually used on rear-wheel drive vehicles and is mounted longitudinally. Leaf springs consist of one or more flat springs, commonly made of tempered steel (**FIGURE 28-13**). A number of leaves of different length are used to form a multi-leaf spring. Multi-leaf springs give the spring a progressive spring rate.

The multi-leaves are held together by a center bolt that passes through a hole in the center of each leaf. The center bolt is also used to locate the spring on the axle. The axle is then clamped to the multi-leaf spring by U-bolts. They wrap around the axle housing and through a spring plate on one side of the spring.

Rebound clips are metal straps wrapped at intervals around the leaf spring. They prevent excessive flexing of the main leaf during rebound. They also keep the end of the leaves in alignment. The longest leaf, called the main leaf, is rolled at both ends to form **spring eyes**. These eyes are used to mount the spring to the frame of the vehicle. Some multi-leaf springs have the ends of the second leaf rolled around the eyes of the main leaf as reinforcement. This leaf is called the **wrap leaf**.

The front of the multi-leaf spring is attached to a **rigid spring hanger** on the vehicle frame. This rigid hanger holds the spring, and ultimately the axle, in position with the frame. The rear of the multi-leaf is connected to the frame by a **swinging shackle**. It provides a link between the spring eye and a bracket on the frame. This swinging link is needed because the front of the spring is held rigidly to the frame. As the spring flattens out under load, the distance between the spring eyes increases. The swinging shackle allows for this lengthening and shortening.

Some multi-leaf springs have inserts between the leaves of plastic, nylon, or rubber. They act as insulators to reduce noise transfer and friction as the leaves move across each other. Some older vehicles completely enclose the leaf springs in grease for this same reason. The spring eyes are fitted with replaceable bushings. They are usually made with a flexible rubber section. But nylon and urethane bushings are also used. And in heavy-duty applications, bronze may be used.

Rubber insulating pads sit between the spring mounting pad and the spring. They also sit between the spring plate and the spring. Because they are rubber, they act as vibration insulators. Most springs are arched upward at the ends. But occasionally reverse-arch springs are used. This is found most often on lowered vehicles. No matter what, leaf springs support the weight of the vehicle while holding the rear axle in line with the frame.

FIGURE 28-14 A torsion bar is a long alloy-steel bar that is attached rigidly to the chassis or subframe at one end and to a control arm at the other.

Torsion Bars

A torsion bar is a long alloy-steel bar that is fixed rigidly to the chassis at one end, and to a control arm at the other end (**FIGURE 28-14**). The torsion bar is connected to the control arm in the unloaded condition (no pressure on the bar). As the control arm is raised, the torsion bar twists around its center. This places it under a **torsional load**. Torsional load is a twisting force. It is applied by anchoring one end of an object, and then applying a twisting force to the other end.

The torsion bar supports the vehicle load. Since it is anchored to the frame, it twists around its center to provide the springing action. Spring rate depends on the length of the torsion bar and its diameter. The shorter and thicker the torsion bar, the stiffer its spring rate. This is useful in heavier vehicles or vehicles that regularly haul loads.

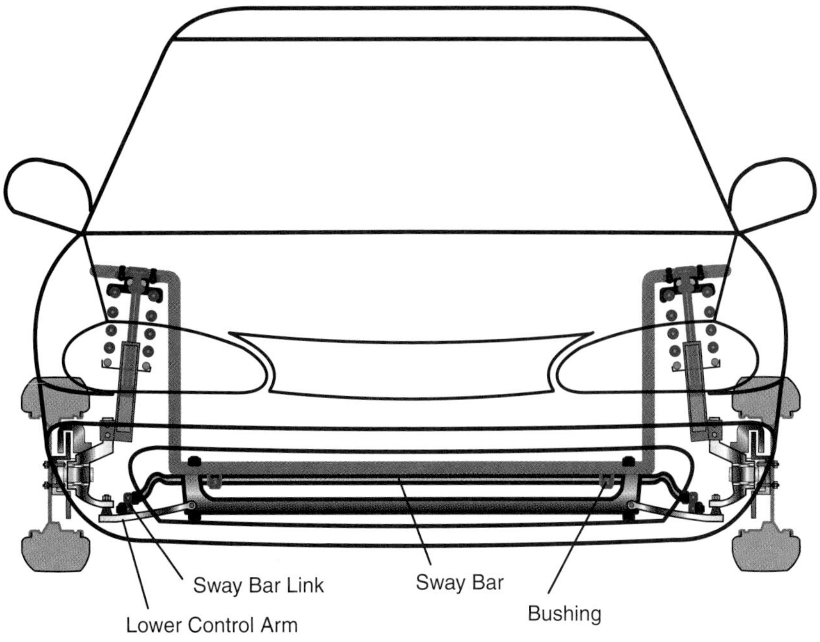

FIGURE 28-15 A sway bar reduces body roll when cornering.

Torsion bars can be used across the chassis frame in a **trailing arm suspension.** It can also be part of the connecting link between two-axle assemblies on a semi-rigid axle beam. After a lot of use, all springs can sag, including the torsion bar. This causes the ride height on one or both sides of the vehicle to be lower than specified. On many vehicles with torsion bars, the torsion bars can be adjusted. This can bring the vehicle back to its proper ride height. This is done by tightening the torsion bar adjuster bolt.

Sway Bars

A bar similar to the torsion bar is the **sway bar**, or antiroll bar. The sway bar is used in light vehicles as a stabilizer, or antiroll, bar. The center of the bar is connected to the chassis. Each end is connected to the lower control arm on each side of the suspension system (**FIGURE 28-15**). Sway bars are most common on the front suspension because it is most prone to sway. But they are also installed on the rear suspension in sporty vehicles.

When the vehicle is turning, centrifugal force acts on the body and tends to make it lean outward (roll). This tends to push up on the outward lower control arm, which lifts up on that end of the sway bar. The sway bar then transfers that twisting force to the other control arm. It tries to lift against that spring's downward pressure. Because the car is leaning, this helps keep the inside spring compressed while transferring its force to the other side of the suspension. Thus, the sway bar helps to resist this roll tendency.

In this way, the spring on the inside of the corner assists the spring on the outside of the corner. It allows inside spring to bear the additional load on the outside spring. This tends to make the vehicle corner more flatly. But because the sway bar can pivot in bushings on the frame, the sway bar just pivots when both wheels go over a speed bump. Also, stiffer sway bars can be installed in a vehicle to enhance its cornering ability.

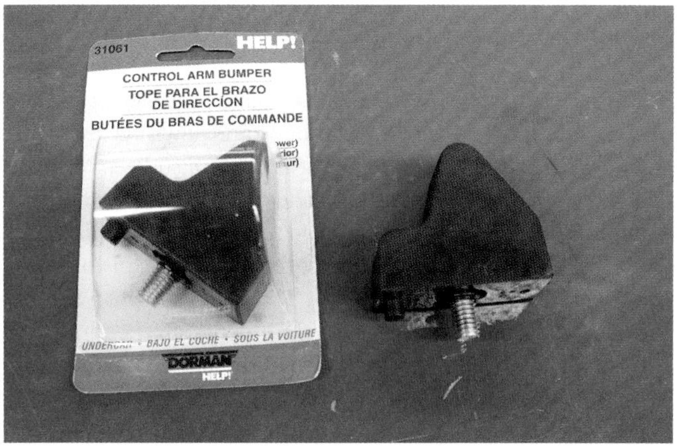

FIGURE 28-16 Type of rubber stop.

Rubber Stops

Rubber is used in most suspension systems as stops. If the suspension reaches its limit of travel, stops prevent direct metal-to-metal contact. This reduces jarring of the suspension components (**FIGURE 28-16**). The rubber stops protect the suspension parts when they bottom out. The stops also reduce the shock to the driver as the suspension bottoms out.

Shaped stops provide an auxiliary springing function. They gradually provide resistance when they make contact with the suspension. Rubber spring stops can be found on most vehicles. For vehicles with MacPherson **struts**, there is a cone-shaped rubber stop located on the strut below the strut bearing. Stops can also be found where the suspension components make contact with the frame.

▶ TECHNICIAN TIP

Rubber has a number of advantages. It doesn't need to be lubricated, it can be made into any shape, as required, and it is silent during use.

▶ Shock Absorbers and Struts

LO 28-03 Describe fixed shock absorbers and struts.

A shock absorber is a device designed to dampen spring oscillations. It absorbs shock loads while driving on irregular surfaces. The most widely used hydraulic shock absorber is the direct-acting telescopic type (**FIGURE 28-17**). It can be fitted to the suspension as a self-contained unit. It is referred to as a shock absorber. It can also be manufactured with a more heavy-duty housing. In that case, it is used to support the top of the steering knuckle as well as dampen spring oscillations. In this configuration, it is referred to as a strut. We will look at shock absorbers first.

Most **direct-acting telescopic shock absorbers** are of the twin-tube type. The bottom of the outer tube is attached to the suspension system. This tube contains an inner tube that provides a working cylinder for a piston. The piston slides up and down the cylinder. The piston is attached to a piston rod. The piston rod is connected to the vehicle's frame at its top end. A bushing in the top of the outer tube keeps the rod in alignment as it moves in and out of the shock absorber. It moves with **suspension action** (movement of the chassis up and down). A seal above the bushing prevents external oil leakage. It also keeps out dirt and moisture. A **shroud** (tube) made of steel, plastic, or rubber is typically placed over the shock rod to protect it from damage.

During compression (jounce), the rod and its piston move inward inside the shock absorber. This happens when the tire travels over a bump in the road. In extension (rebound), the rod and piston move outward inside of the shock absorber. This happens when the tire travels over a dip in the road. For dampening to be effective, resistance is needed in both directions. But the resistance may need to be in different proportions.

The resistance is provided by forcing oil past disc valves in the shock absorber's piston. Valves control the flow through the piston when it moves in one direction. Different valves control flow in the other direction. So, resistance on compression will vary as compared to rebound (**FIGURE 28-18**). Because of this, manufacturers design the desired ride characteristics into the shock absorber.

The valves provide control over the amount of force required to pass fluid through them at any given piston velocity. They can be made to open in stages, according to fluid pressure. This allows light resistance to motion when the piston moves slowly. And yet it provides heavy resistance when piston velocity is high.

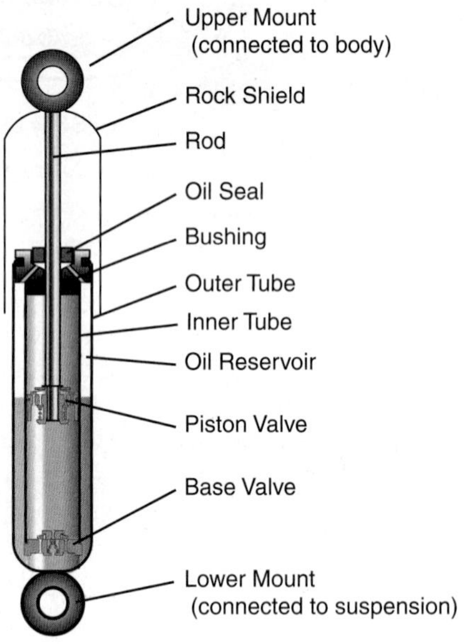

FIGURE 28-17 A shock absorber is a device designed to absorb shock loads caused from driving on irregular surfaces.

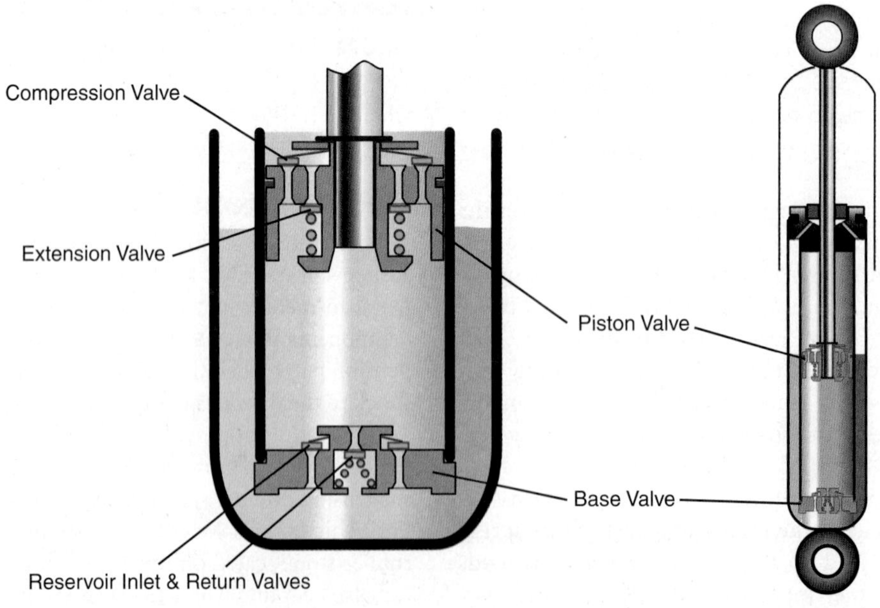

FIGURE 28-18 Shock absorber valves.

The rapid movement of the piston, continually forcing the oil backward and forward through the valves, causes the oil to heat up. The process of absorbing the energy of suspension movement heats the oil up. The heat is transferred through the outer tube to the outside air. However, the hotter the oil becomes, the greater its tendency to aerate. Aeration occurs because of the high velocity of the oil as it passes through the small passages in the valves. If the velocity is high enough, air dissolved in the oil creates small bubbles and forms foam. As we know from Pascal's law, fluid cannot be compressed, but air can. Air in the shock absorber results in a lack of dampening. This creates a soft, bouncy ride. It also allows the suspension members to contact the stops. A harsh ride is the result.

▶ TECHNICIAN TIP

A way to think about the basic function of a shock absorber is to think of a washer fastened to a rod through its center hole. A small hole is drilled through the flat of the washer so that it will let oil move from one side of the washer to the other. Some oil is added to a capped tube, and the rod and washer is inserted into the tube. Adding oil to the top, and capping the tube with a sealed cap, allows the rod to move in and out of the tube. But the oil does not leak.

Moving the rod up and down in the oil produces resistance, as oil will be forced through the small hole in the washer. Resistance is felt in both directions. If the size of the hole in the washer is increased, there will be less resistance. This is the basic operation of the shock absorber. Except in a shock absorber, one-way valves can be used to allow different rates of flow on compression and rebound.

Strut-Type Shock Absorbers

A **strut** functions exactly like a shock absorber. But it is much stronger because it is a structural part of the suspension system. It is integrated into MacPherson strut suspension systems. On a MacPherson strut system, the bottom of the strut connects solidly to the top of the steering knuckle. So, it must be much stronger than an ordinary shock absorber (**FIGURE 28-19**). The hydraulic shock absorber and the hydraulic strut provide their dampening action in the same way. They transfer oil under pressure through valves in the piston. The valves restrict the oil flow.

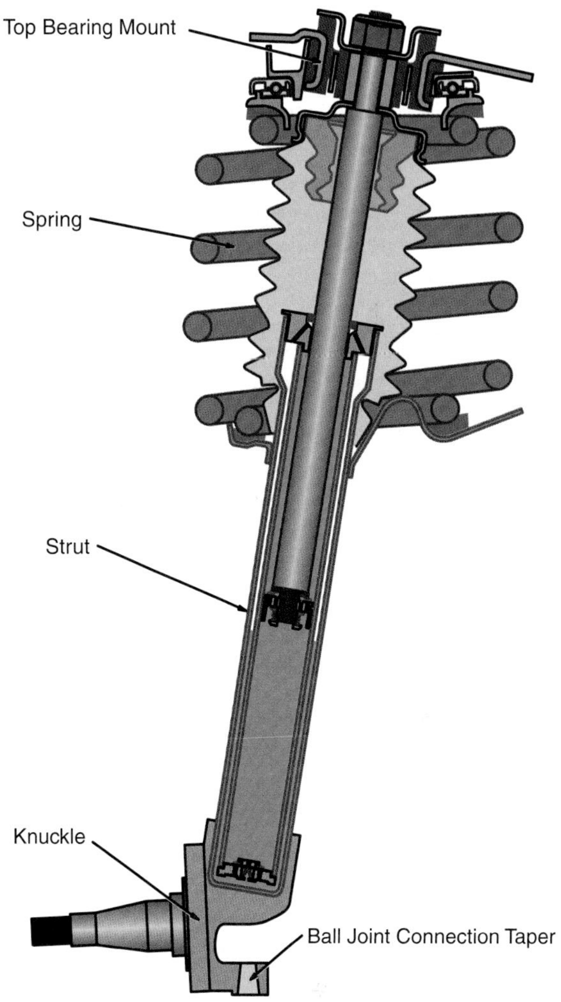

FIGURE 28-19 Cutaway view of a strut.

▶ TECHNICIAN TIP

Aerated oil has a certain amount of compressibility. So, it is unable to provide the dampening force previously achieved in the non-aerated condition. This is similar to a brake pedal that feels spongy when air is present in the brake's hydraulic system. The performance of the shock absorber is thus considerably reduced. This effect is called shock absorber dissolve.

Gas-Pressurized Shock Absorbers

Fluid fills the chambers above and below the piston of the hydraulic shock absorber. As the piston moves in the cylinder, valves control the movement of oil from one chamber to the other. In a gas shock, pressure on the oil is provided by nitrogen gas at the base of the cylinder. It acts on a free-floating separation piston that separates the gas from the oil (**FIGURE 28-20**).

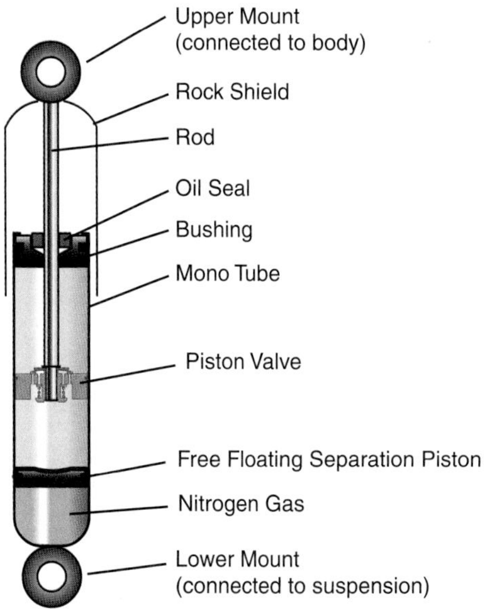

FIGURE 28-20 Gas-pressurized shock absorber.

On jounce, the piston moves downward. The movement of the piston rod displaces a quantity of oil equal to its volume. The separation piston is displaced accordingly, and gas pressure increases slightly. On rebound, the piston and rod move upward. Gas pressure reduces slightly as the separation piston makes up for the rod moving out of the shock. Pressure on the oil is maintained, even when the piston and rod are at the top of their stroke. The pressure applied to the oil keeps air bubbles (or aeration) from forming as easily. And this contributes to a more consistent job of dampening.

▶ Adjustable Shock Absorbers

LO 28-04 Describe manually and automatically adjustable shocks.

Manufacturers strive for their vehicles to have a comfortable ride. But they also want to maintain some level of sporty handling. Or they may want the ability to handle occasional heavy loads. This has led to development of a variety of shock absorbers with enhanced valving. As described, having valving with a few small passageways causes a high amount of resistance to jounce and rebound. This creates a sporty feel. At the same time, larger passageways provide less resistance, creating a smoother ride.

To be able to adjust the ride, manufacturers have machined more passageways into the valving. They then use manually or electrically opened and closed valves to control the ride. Opening and closing some of the valves softens and stiffens the dampening action. As far as extra loads, some shocks incorporate an expandable air bladder. It expands to support additional loads.

Load-Adjustable Shock Absorbers

When vehicles carry heavy loads, their suspension is compressed. This causes the rear of the vehicle to be lower than normal. If too much weight is added, steering can become lighter (less responsive). The alignment of the headlights becomes too high. And the length of suspension travel when going over bumps is reduced. This causes the suspension to bottom out more easily, which is uncomfortable for passengers.

To better handle extra loads, a **manually adjustable air spring** can be incorporated into the shock absorbers. This type of shock absorber is commonly referred to as an air shock (**FIGURE 28-21**). The **air spring** consists of a flexible rubber bladder, which seals the outside of the upper and lower halves of the shock absorber. When inflated, the bladder pushes the halves apart.

The shock absorber portion of an air shock is a standard hydraulic type. It provides normal dampening action. But when a heavy load is placed on the vehicle, the rubber air bladder can be pressurized with compressed air. This assists the vehicle's springs. By increasing the air pressure in the bladder, the ride height can be adjusted back to normal. Compressed air in the bladder can absorb smaller road shocks. It can also provide better ride characteristics than using stiffer springs instead.

The rubber air cylinder is connected to a filling valve by a flexible plastic hose. The filling valve looks similar to a tire valve stem. Air from an air compressor increases the pressure in the rubber cylinder. This allows the suspension to support more weight. The maximum air pressure setting must not be exceeded, as excessive air pressure can damage the shock absorber's air spring. When the load is removed, the extra air can be released through the filling valve, just as you would release air from a tire. This allows the suspension to return to its original height. However, a minimum air pressure must be maintained in the cylinder. This will prevent the rubber from chafing on itself during suspension action. Refer to the manufacturer's specifications for the air shock's minimum and maximum air pressure.

Another type of load-adjustable shock absorber uses a coil spring around the outside of the shock. It is called a coil-over shock. The spring is installed from the factory and under tension. It connects to both halves of the shock absorber. This tends to push the halves apart, assisting the regular springs in supporting the load (**FIGURE 28-22**).

Manual Adjustable-Rate Shock Absorbers

A **manual adjustable-rate shock absorber** has a manual, external damper rate adjustment. The number of valves and size of the passageways in the piston can then be manually selected. This varies the amount of restriction through the piston. It also varies the force needed to open the valves.

FIGURE 28-21 The air pressure can be adjusted in an air shock to help support heavier loads.

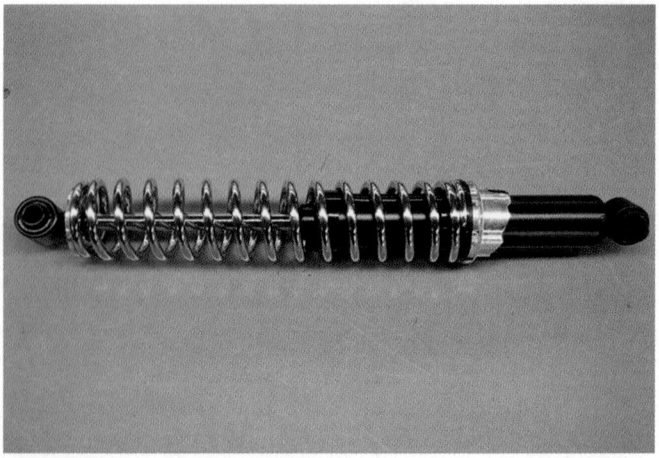

FIGURE 28-22 A coil-over shock acts as a helper spring in supporting a load.

When all of the orifices are open, a small dampening effect is applied to the oil. The spring force applied to the valve can also be reduced to allow the valve to open more easily. This means the oil can flow through the valves more easily. This gives a softer ride but can also allow more rolling and pitching of the body of the vehicle. Closing some of the orifices and increasing the spring force applied to the valves make it harder for fluid to flow through the piston. This increases the dampening effect of the shock absorber.

The method of changing the position of the valve varies. On some models, it is adjusted by turning a spindle located on the shock plunger rod. Turning the spindle moves the valves and changes the size of the orifices in the valve. Another style uses a selector knob near the bottom of the shock that can be adjusted at any time (**FIGURE 28-23**). On other models, when the shock absorber is extended to its maximum length, a pin is depressed, locking in an adjusting slide on the piston assembly. Twisting the two halves of the shock absorber changes the number of orifices and the spring force on the valves. The first two types of adjustable shock absorbers can be adjusted with the shock on the vehicle. The last type must be removed to change the dampening.

Electronic Adjustable-Rate Shock Absorbers

Some vehicles are equipped with electronic ride control systems. They provide driver-selected control of the ride quality. Typically, a selector is located inside the passenger compartment. It allows the driver to select between options such as sport, touring, or automatic. For example, if the driver selects sport, the ride will become firmer.

Electronic ride control can also be automatic, with no provision for driver control. In automatic ride control, the vehicle's electronic module chooses a softer or firmer ride. Sometimes, this is based on the speed of the vehicle or the type of driving that is being done.

In order to change dampening of the shocks, the shock must be adjustable. Some vehicles use a rotary stepper motor, called an actuator, to change the dampening. The shock works very similar to the manual adjustment shock that has the spindle that is turned.

In the electronically controlled shock, the actuator is mounted to the top of the shock. It turns the spindle as needed. The actuator changes the number of restrictions or orifices that the oil must pass through. When all orifices are open, oil can flow more easily through the passageways in the piston. Only a small dampening effect is applied to the oil. Closing some orifices makes it harder for fluid to flow through the piston. This increases the dampening effect of the shock absorber, providing a firmer ride. It is more suitable for higher speeds and faster cornering.

Another type of electronically adjustable shock is the solenoid-controlled shock. The solenoid is located in the side of the shock absorber or strut. It opens a passageway internal to the shock. Additional fluid is allowed to bypass the piston. If the solenoid is commanded closed, the fluid will not be able to bypass the piston. The result is more dampening. This type of shock is still controlled by the powertrain control module (PCM).

The newest style of electronically adjustable shock uses a special type of fluid. It's called **magneto-rheological fluid**. This fluid has the unique characteristic of changing viscosity when exposed to a magnetic field. General Motors has been using this fluid in vehicles equipped with MagneRide® suspension systems. The fluid is mixed with very small ferrous particles that react to magnetic fields. As the fluid is subjected to a magnetic field, the ferrous particles bind together. This effectively increases the viscosity of the fluid. The stronger the magnetic field, the thicker the fluid becomes.

The ability to increase and decrease the viscosity of the fluid eliminates the need to change orifice sizes. In this case, it is harder to pass a thicker fluid through a hole than a thinner one. Simply varying the magnetic field at the passageways causes varied dampening of the suspension (**FIGURE 28-24**). In fact, the viscosity can be changed in a millisecond or less. This allows active dampening of each individual shock absorber as the vehicle is being driven.

FIGURE 28-23 On some models, the position of the valve is adjusted by turning a spindle located on the shock plunger rod. This style uses a selector knob near the bottom of the shock that can be adjusted at any time.

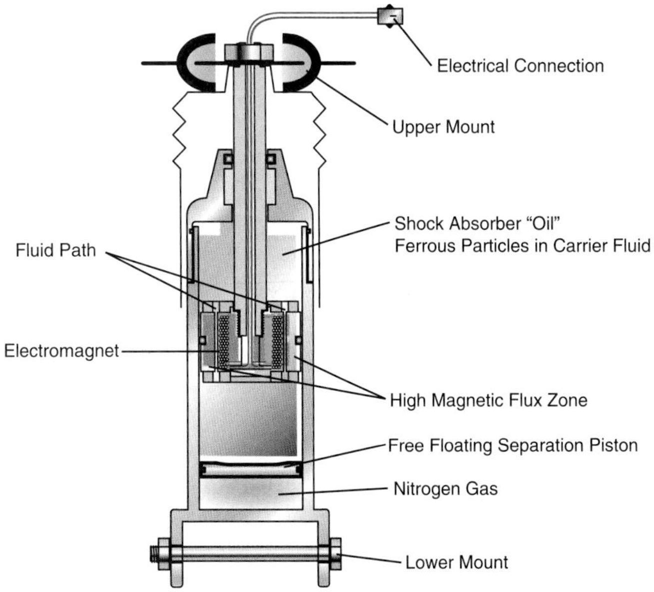

FIGURE 28-24 Electromagnetic shock absorber.

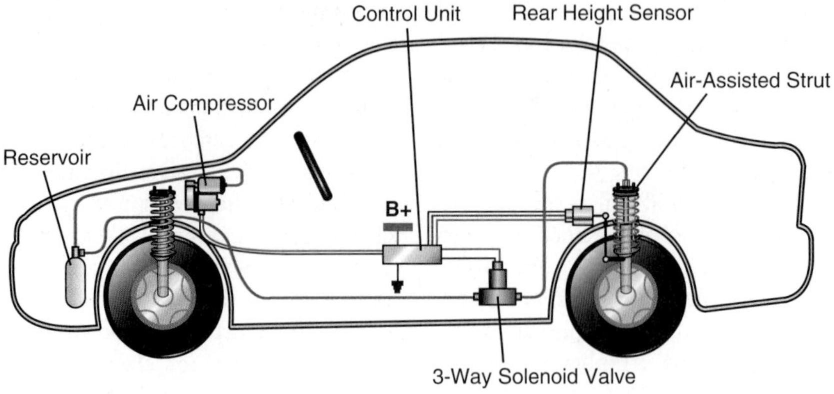

FIGURE 28-25 An automatic load-adjustable suspension system controls the vehicle ride height automatically, according to the load placed over the rear axle.

In this type of active dampening system, the PCM controls the strength of the magnetic fields in the shocks. It can stiffen the appropriate shocks in a continuously active manner as the vehicle is being driven. On the other hand, a particular shock can also be softened as a wheel goes over a bump. This can prevent much of the shock from being transmitted to the vehicle. Other manufacturers are beginning to use this technology because it offers a faster-acting type of dampening. This technology is also used in some motor mounts and other applications.

Automatic load-adjustable shock absorbers are also called **self-leveling**. This means they have a sensor that measures the ride height. It uses that information to adjust the self-leveling shocks. It makes adjustments according to the load placed over the rear axle. It has air-adjustable shock absorbers fitted to the rear suspension. An electrically driven compressor and air-dryer assembly supply the air. It also has a ride height sensor, a control unit, and associated wiring and tubing (**FIGURE 28-25**).

The ride height sensor is mounted to the cross member over the rear axle. A moveable link connects it to a rear suspension member. As the vehicle is loaded, the normal suspension springs are compressed. This lowers the height of the vehicle. When the ignition is switched on, the control unit senses the lowered ride height. It switches on the air compressor. Air is directed to the shock absorbers. This causes the airbag around them to expand and raise the suspension back to the normal ride height.

If the load is removed, the suspension springs expand, raising the height of the vehicle. The control unit senses the raised ride height. And air is exhausted from the shock absorbers. This causes the airbag to deflate and thereby lowers the suspension to the normal ride height.

During normal suspension operation, continual adjustment of vehicle ride height is prevented. A time delay in the control unit only allows the trim height to be adjusted when the PCM reads an out-of-trim signal for a short period of time. For example, 5 to 15 seconds. Thus, the system does not try to compensate for bumps in the road or weight transfer during braking.

The compressor run time or exhaust time is limited to a few minutes. Limiting the operational time prevents it from continuing to operate if the system develops an air leak or if an exhaust vent remains open. If a fault like this develops, most self-leveling systems will set a diagnostic code.

▶ Control Arms and Rods

LO 28-05 Describe suspension system components.

Control arms are components that serve as a primary load-bearing element of a vehicle's suspension system. Control arms can be formed in different shapes. The A-arm style is sometimes referred to as a **wishbone control arm**. The arm is a relatively flat triangular part that mounts to the frame or subframe at each leg of the A (**FIGURE 28-26**). These widely spaced mounting points prevent forward or backward movement of the steering knuckle. The mounting points typically use rubber bushings that allow the arm to pivot up and down. They also dampen or isolate the road shock and vibrations from the rest of the vehicle. The other end of the control arm has a ball joint. It connects the control arm to the steering knuckle assembly. This arrangement provides a smooth yet stable ride for the vehicle.

Another type of control arm uses only one contact point at the frame or body (**FIGURE 28-27**). With this style of control arm, another supporting piece, called a strut rod or radius rod, must be used. It keeps the control arm from pivoting forward and backward with changes in braking or acceleration. Other types of control arms are made from round tube material. These control arms may be called transverse arms or trailing arms.

FIGURE 28-26 Wishbone control arm.

Rods

Rods are typically straight (or precisely formed) pieces of steel. They are used to either transfer motion or prevent motion within a vehicle's suspension system. More specific names are given to rods based on their location or attachments. For example, a suspension system may use:

- tie rods,
- lateral rods,
- tension rods,
- control rods,
- Panhard rods (track bars),
- steering track rods, or
- strut rods (**FIGURE 28-28**).

Many of the rods use either a bushing or a joint on one or both ends. Each of these components is discussed elsewhere in this chapter. But it is important to remember that the exact number and types of rods within a vehicle vary greatly. It depends on how the vehicle was designed for its intended use.

Steering Knuckle

The steering knuckle, also known as a stub axle or spindle assembly, can be found in many variations. One type uses a forged piece containing the wheel hub or spindle and attaches to the suspension components. Other knuckles may provide a hole for the axle to pass through and the wheel hub to mount to (**FIGURE 28-29**). Some knuckles are cast iron. But newer vehicles may have cast aluminum steering knuckles. This reduces the unsprung weight as well as total vehicle weight.

Typically, the steering knuckle also has a steering arm. It is either cast as part of the knuckle or is a separate piece bolted to the knuckle. The steering arm serves to transmit the steering force to the steering knuckle when the driver turns the steering wheel. The steering knuckle pivots on a ball joint on the bottom and either a ball joint or a strut bearing on the top.

Ball Joints

Ball joints are swivel connections mounted in the outer ends of the control arms. They allow the steering knuckle to pivot. Ball joints are typically constructed of a ball and socket (**FIGURE 28-30**). Ball-and-socket joints are also used in most tie-rod ends. But the term "ball joints" is typically reserved for the primary joints that the steering knuckle pivots on. "Tie-rod ends" is the term used for the ball-and-socket joints on the steering linkage.

Ball joints allow the knuckle assembly to pivot as the control arms move up and down. They also allow the knuckle

FIGURE 28-27 Single-point control arm.

FIGURE 28-28 A strut rod is one type of rod used in steering and suspension systems.

FIGURE 28-29 A. Steering knuckle—front-wheel drive axle. **B.** Steering knuckle—dead vehicle.

FIGURE 28-30 Ball joint.

assembly to rotate for steering as well. The ball joint in most modern vehicles is a sealed, self-contained unit that is replaced as a unit when it is worn out.

The ball joint can be fastened to the control arm in several ways. In the past, a ball joint housing was threaded and then screwed into the control arm. This type of ball joint is no longer used, so it is only found on classic cars. The ball joint on modern vehicles is either press-fit into the control arm, or held by rivets or bolts. If the ball joint is pressed into the control arm, a special tool, called a ball joint press tool, is used to remove and install the joint.

A ball joint is made up of a pressed-steel housing fitted with **sintered** (bonded using pressure and heat) iron seats. The seats retain a hardened ball stud. Some ball joints use a Belleville spring to hold tension on the joint, which has to be compressed when tested.

Typically, a taper on the stud fits a mating taper on the steering knuckle. However, some use a straight stud with a crescent-shaped relief. A clamping bolt fits a hole in the knuckle. The crescent-shaped relief in the stud is aligned in the knuckle so the bolt can fit through the hole. The bolt then orients the stud and clamps it securely in the knuckle (**FIGURE 28-31**).

A rubber seal retains grease and keeps out dirt and water. Some ball joints have grease fittings (grease zerks) installed. They allow for periodic lubrication of the moving ball and stationary socket inside the ball joint (**FIGURE 28-32**). Grease fittings can sometimes be found on other suspension components. They need to be lubricated as part of a preventive maintenance program. Most light-duty vehicles manufactured today have maintenance-free suspension components. They do not provide access for lubrication. When inspecting the suspension, check to see whether grease fittings are present. They can be found on the suspension, steering, and drive line components.

Ball joints can be referred to as loaded or unloaded. A loaded ball joint supports the weight of the vehicle. An unloaded ball joint does not support any weight. It just holds the steering knuckle in position and is referred to as a follower ball joint. For example, a short/long-arm (SLA) suspension system has an upper and lower control arm. If the coil spring is located between the frame and the lower control arm, the lower ball joint is the loaded ball joint (**FIGURE 28-33**). This is because the force of the spring is placed against the lower

FIGURE 28-31 **A.** Tapered stud ball joint. **B.** Straight stud ball joint with crescent relief.

FIGURE 28-32 Some ball joints have grease fittings (grease zerks) installed, which allow for periodic lubrication of the moving ball and stationary socket inside the ball joint.

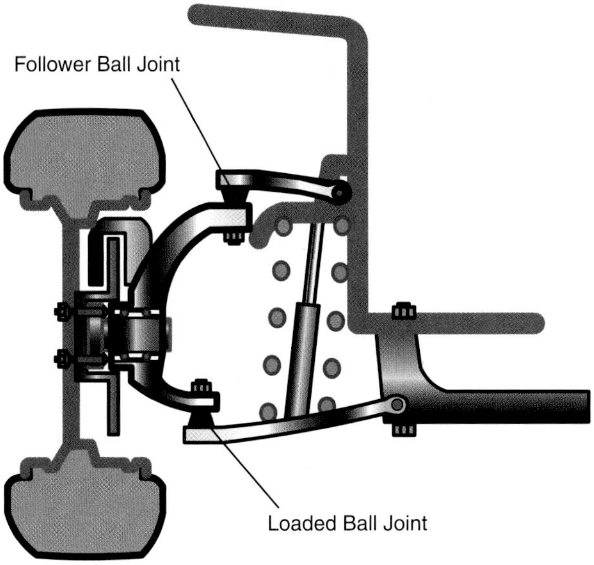

FIGURE 28-33 Loaded versus follower ball joints.

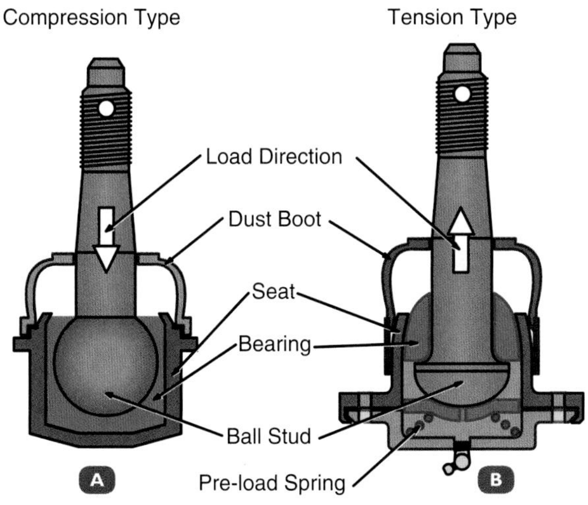

FIGURE 28-34 **A.** Ball joint made for primarily compression applications. **B.** Ball joint made for primarily tension applications.

control arm and ball joint. In this situation, the upper ball joint is the follower joint.

If the spring is located between the frame and the upper control arm, then the upper ball joint is the loaded joint. The lower ball joint is the follower joint. This distinction becomes important when testing the ball joints for play. This is because the technician has to unload the joint to accurately test for play.

Also, the ball joints can be designed to primarily carry either compression loads or tension loads (**FIGURE 28-34**). Compression forces tend to push the ball into the socket. So, those types of joints have most of their bearing surface and socket strength near the base of the ball. Tension ball joints are being pulled apart. So, they have most of their bearing surface and socket strength near the stud end of the ball. Never mix up the positioning of ball joints. Doing so will cause them to wear out quickly and fail.

FIGURE 28-35 Control arm bushing.

FIGURE 28-36 Strut rod bushings.

Bushings

Bushings act as pivot points and cushions at suspension fulcrum points. Control arm bushings and strut rod bushings are common examples. They allow for limited movement of the component while maintaining its alignment (**FIGURE 28-35**). They can be metallic or made of rubber, nylon, or urethane. Many rubber bushings have a metal inner and outer housing. The inner housing acts as a spacer, while also anchoring the rubber bushing. The outer housing allows the bushing to be more easily pressed into place.

Rubber bushings isolate noise and harshness. The rubber absorbs small impacts from the suspension without transmitting them directly to the driver. Rubber requires no lubrication. Rubber bushings can also be used on strut rods (**FIGURE 28-36**). Rubber-bonded bushings can be used to mount the steering rack to the vehicle frame.

FIGURE 28-37 Rubber-bonded bushing used in control arm bushings.

Spring shackle bushings can be molded to form two halves. Each half is fit into each side of the spring eye on the swinging shackle. With the spring loaded and the shackle plates tightened, the rubber is compressed in the spring eye. As the spring deflects, the rubber deflects without tearing.

Rubber-bonded bushings are normally used for the front eye of the spring at the fixed shackle point. They are also used in control arm applications. The rubber-bonded bushing has a steel outer housing and inner sleeve (**FIGURE 28-37**). The rubber medium is bonded between both to provide elasticity between them. The outer casing is normally pressed into place in the component. Relative movement between the casing and the inner sleeve causes the rubber to flex without tearing.

In control arm applications, particularly at the rear of a vehicle, the rubber bushing may be molded with a voided section (**FIGURE 28-38**). This is known as a **compliance bushing**. It allows the unit or component to comply with a controlled amount of movement in the direction of the void. This movement may be designed to allow compliance or deflection steer of the wheels when cornering. This influences the steering behavior of the vehicle. So, it is very important that the voided section be installed in its correct relative position. It is easy to forget how a bushing was positioned after it is removed. Marking the position with a paint mark is a good way to ensure that the bushing is installed correctly.

▶ Types of Suspension Systems

LO 28-06 Describe the main types of suspensions.

Manufacturers use various types of suspension systems. The design depends on the intended use of the vehicle, cost of manufacturing, and layout of the drive train. Each type comes in various configurations, which we examine in this chapter. Each type also has pros and cons that you need to be aware of. This is especially true when diagnosing and servicing suspension systems.

No matter they are of different types, the function of each suspension system is the same. They are designed to maintain the proper positioning of the wheels during all driving conditions. At the same time, they isolate the occupants from the road as much as possible. Suspension systems also need to accommodate systems that power, brake, and steer the wheels. Understanding the different suspension configurations gives you a good foundation for servicing them. We cover suspension system service in the next chapter.

Dead Axle/Live Axle

The terms **dead axle** and **live axle** refer to whether the axle is a driving axle or not (**FIGURE 28-39**). A dead axle simply holds the wheels in their proper orientation. A good example of a dead axle is a trailer (**FIGURE 28-40**). It does not power the wheels at all, but just holds them in place.

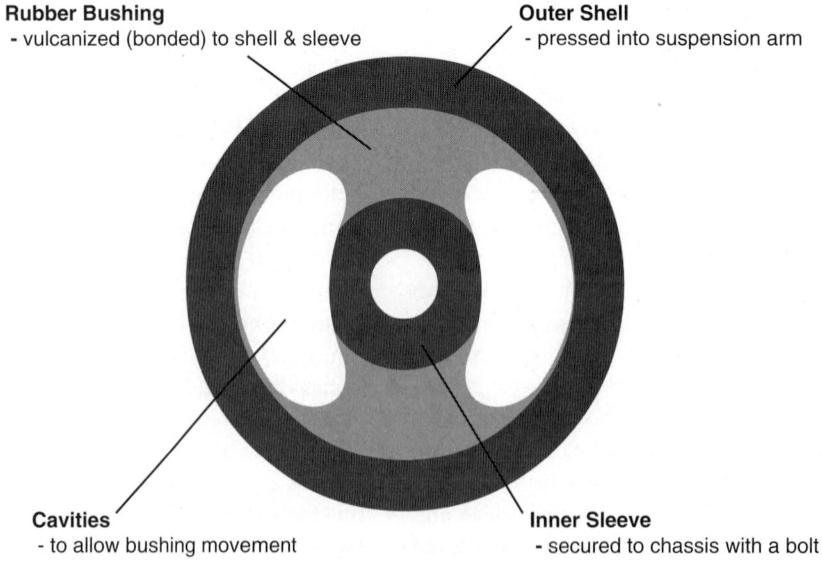

FIGURE 28-38 Compliance bushing.

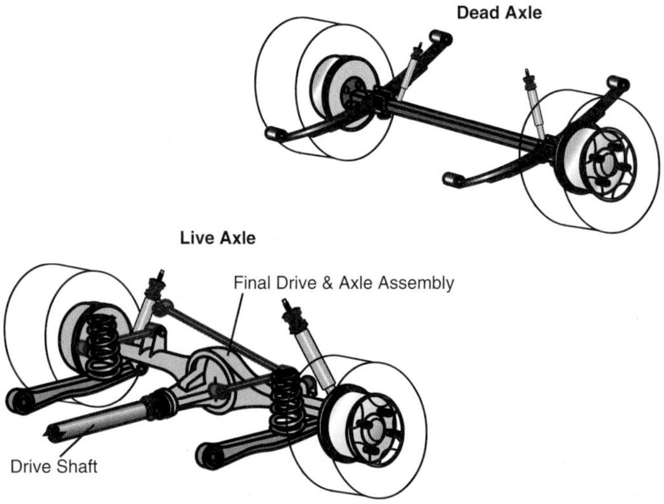

FIGURE 28-39 The terms "dead axle" and "live axle" refer to whether the axle is a driving axle or not.

FIGURE 28-40 Trailers are a good example of dead axles, where the axle simply holds the wheels in place.

A live axle not only holds the wheels in position but also drives them. Live axles are connected to the drive train in such a way that power can be transferred through the components in the live axle (**FIGURE 28-41**). Live axles can be part of either a solid axle suspension system or an independent suspension system. On front-wheel drive vehicles, a simple dead axle is used on the rear wheels. On a rear-wheel drive vehicle, the front axle is a dead axle. And on four-wheel drive or all-wheel drive vehicles, both the front and rear axles are live. Again, they can be of the independent or solid axle styles.

Solid Axle

The **solid axle** (beam axle) provides a simple means of mounting the hub and wheel assembly. Together with leaf or coil springs, it forms an effective, dependent suspension system (**FIGURE 28-42**). It is used in the rear suspension of many front-engine, rear-wheel drive vehicles. It is also used as the front suspension on many pickup trucks and full-size vans.

A solid axle forms a dependent suspension because the wheels on both sides of the axle are connected together. This means when one wheel goes over a bump, the other wheel tilts

FIGURE 28-41 A live rear axle not only holds the wheels in position, but it drives them as well.

FIGURE 28-42 A solid axle is a nonindependent suspension because the wheels on both sides of the axle are connected together.

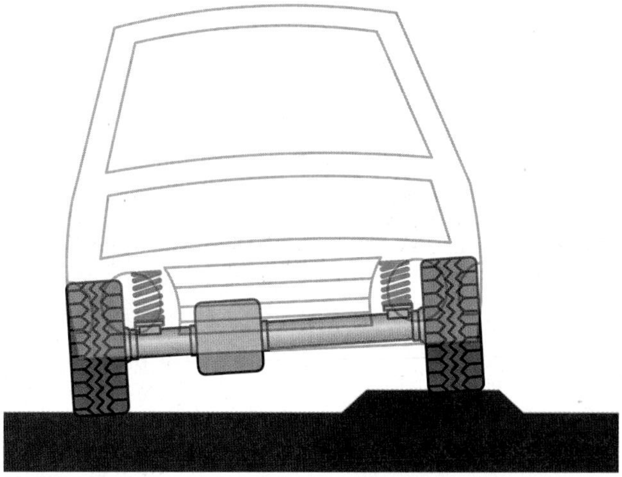

FIGURE 28-43 When one wheel on a solid axle goes over a bump, it affects the other wheel.

(**FIGURE 28-43**). This tilting reduces the tire-to-road contact patch, reducing friction. Thus, vehicles with solid axles are said to not handle as well as vehicles with independent suspensions. But they are inexpensive and good for hauling heavy loads.

Independent Suspension

An independent suspension allows the wheel on each side of the axle to move up and down independently of the other. In this arrangement, if one wheel hits a road irregularity, it will not upset the other wheel (**FIGURE 28-44**). One of the main benefits claimed for **independent suspension** is that unsprung mass can be kept low. This is because the heavy center axle housing is eliminated.

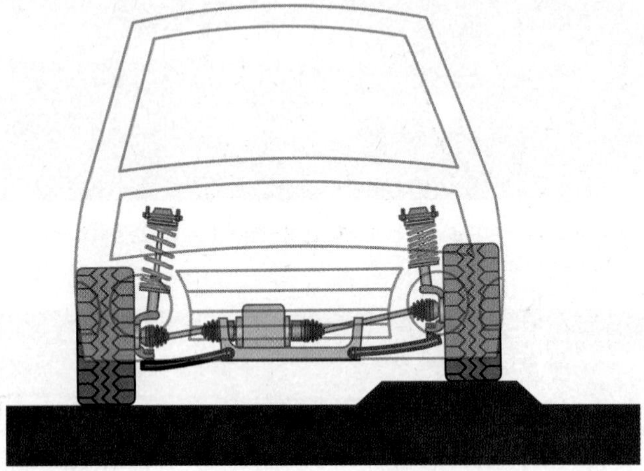

FIGURE 28-44 When one wheel on an independent suspension system goes over a bump, the other wheel is not affected.

FIGURE 28-45 The MacPherson strut suspension system. **A.** Front. **B.** Rear.

One of the simplest and most common independent suspension systems is the **MacPherson strut** (**FIGURE 28-45**). It can be used on either the front or the rear of the vehicle. It consists of a spring and heavy-duty shock absorber unit called a strut. On the front of the vehicle, the lower end of the strut is connected to the knuckle. The knuckle is located by a ball joint fitted to the end of the control arm. The strut's upper end is located in a strut tower formed in the unibody. It sits in a molded rubber mounting. If the strut is on the front, the upper mounting includes a bearing. The bearing allows the complete strut assembly to rotate with the steering.

A slightly more complicated, yet common type of independent suspension system is the SLA suspension system (**FIGURE 28-46**). It uses upper and lower control arms to control the movement of the knuckle. It typically uses either a coil spring or torsion bar to support the weight of the vehicle. This system is discussed further in the following section.

▶ Front Suspension

LO 28-07 Describe the main types of front suspension systems.

The front wheels of a vehicle can have the same type or a different type of suspension system than the rear wheels. The exact type of system can vary depending on the vehicle type, such as whether it is front-wheel or rear-wheel drive. There is one major difference between front and rear suspension systems. The front suspension must accommodate steering. Whereas most rear suspension systems don't. Another difference is that most front suspension systems incorporate a sway bar. The two main types of front suspension systems are the independent and the solid axle systems. Regardless of design, the primary function of the front suspension system is to keep the tires in constant contact with the road.

Strut Suspension

In strut suspension, the shock absorber is contained inside the strut (**FIGURE 28-47**). It is a direct-acting telescopic-type shock absorber. It is a type of hydraulic shock absorber. A coil spring is mounted over the strut. The strut assembly fits inside the suspension tower of the front wheel housing. The strut has an upper mounting point in the suspension tower.

FIGURE 28-46 The SLA suspension system.

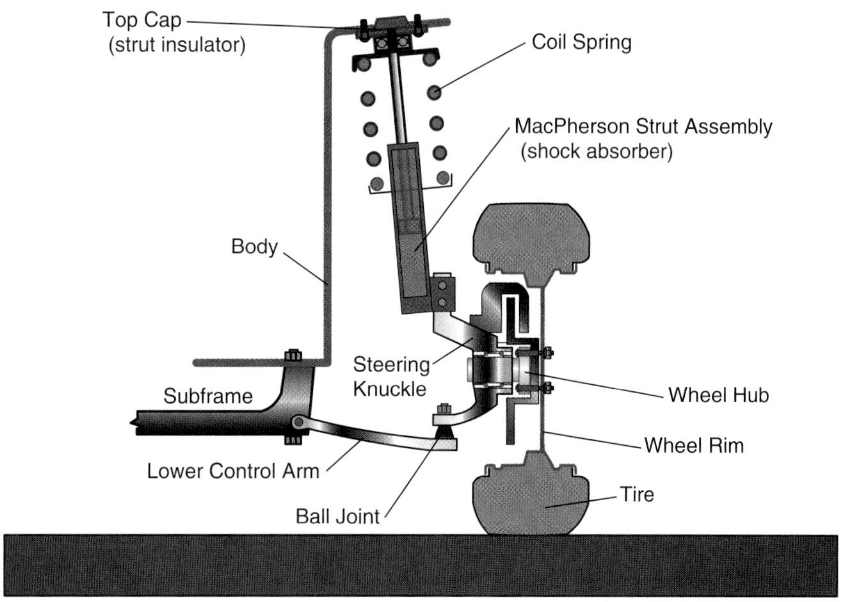

FIGURE 28-47 Strut suspension.

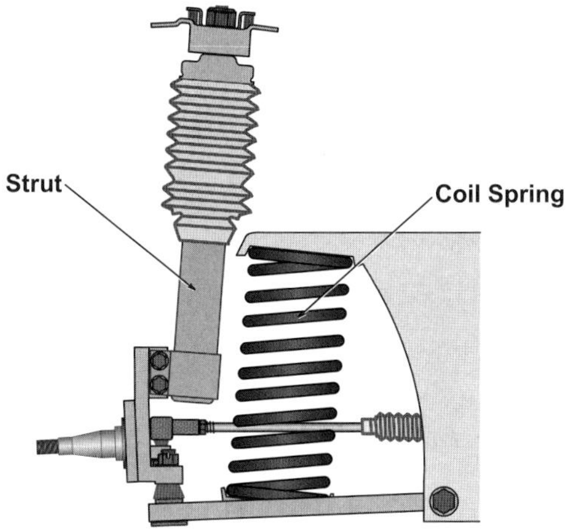

FIGURE 28-48 Modified strut suspension.

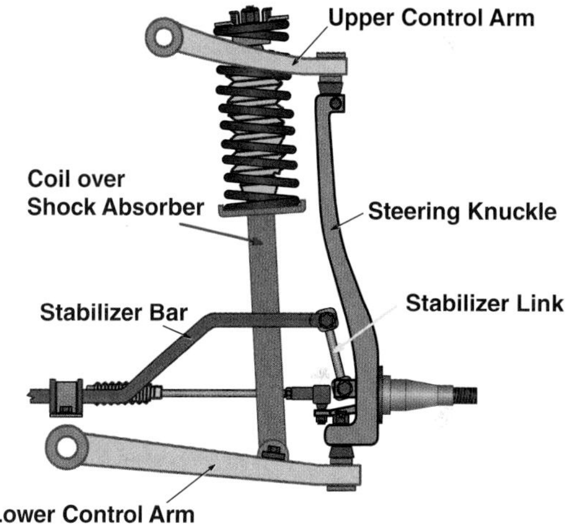

FIGURE 28-49 Double-wishbone modified strut.

When used on the front suspension, the strut's upper mounting is bearing mounted to allow for steering movement. The lower control arm is mounted (or held in place) to the frame or subframe by control arm bushings. The outer end of the control arm contains a ball joint connecting the steering knuckle to the control arm. The steering knuckle can then pivot on the ball joint on the bottom, and the strut bearing on the top. Note that the ball joint on a MacPherson strut is a follower joint, not a loaded joint. This is a compact, responsive, and efficient type of suspension system.

Modified Strut Suspension

The modified strut system (**FIGURE 28-48**) uses one control arm with the strut mounted in the same manner as a MacPherson strut. The main difference is that the coil spring is mounted on the lower control arm, which is usually an "A" arm. This allows for placement of the coil spring and controls fore-and-aft movements. It is used in the front on some vehicles and on the rear of others.

Double-Wishbone Modified Strut

The double-wishbone modified strut suspension uses a high-mount upper arm. **FIGURE 28-49** shows a coil over shock absorber (green arrow). The upper end of the stabilizer link (yellow arrow) attaches directly to the steering knuckle for a 1:1 motion ratio. A 1:1 ratio means they can call this a direct-acting stabilizer bar. Because of the additional leverage, it can be smaller and lighter. This makes it more responsive. The Cadillac CTS and ATS along with BMW use this type of suspension.

SLA Suspension

The **SLA suspension** gets its name from using two different-length control arms: one short upper control arm and one long lower control arm. The primary reason that the SLA suspension system was designed was to ensure correct wheel alignment as the vehicle corners. If the arms were the same length, then the wheel would stay perfectly straight as the suspension moved up and down over bumps. This is okay for straight ahead driving,

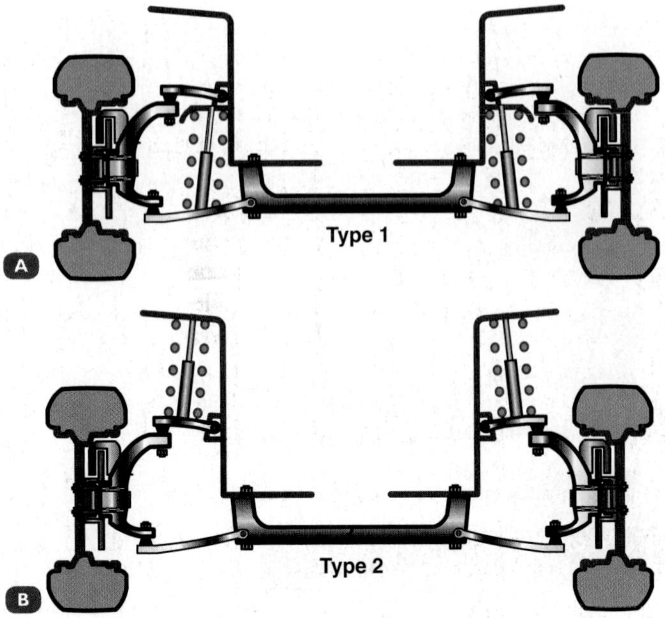

FIGURE 28-50 The coil spring can be mounted in two different ways: **A.** Between the frame and the lower control arm (type 1). **B.** Between the shock tower and the upper control arm (type 2).

but not when cornering. Having arms of the same length would result in incorrect positioning of the tire in relation to the road surface, as the body rolls. The body roll would tilt the top of the wheel outward (toward positive camber), reducing the tire-to-road contact patch. This would reduce traction.

The SLA system, with the short arm on top, tends to pull the top of the tire inward toward negative camber. This keeps the tire-to-road contact patch as large as possible. The shock absorber is located between the frame and the lower control arm. It is a direct-acting telescopic-type shock absorber. The coil spring or torsion bar can be mounted either between the frame and the lower control arm (type 1). Or between the shock tower (or frame with a torsion bar) and the upper control arm (type 2) (FIGURE 28-50).

Both control arms pivot on control arm bushings. These bushings twist on the control arm pins. The pins are bolted to the cross member or subframe of the vehicle. Rubber jounce and rebound stops are used to prevent direct metal-to-metal contact between the control arms and the frame. This occurs if the suspension reaches its maximum limit of travel.

The steering knuckle is mounted at the ends of the control arms by ball joints. They allow both up-and-down and steering movement of the wheel. The ball joints can be designed to be mounted so they are under compression or tension forces. Compression forces tend to push the ball into the socket. Whereas tension forces pull the ball away from the socket.

There are two arrangements of steering knuckles used on SLA suspensions: short knuckle and long knuckle (FIGURE 28-51). The short knuckle design locates the upper ball joint inside of the wheel. The long knuckle design moves the upper ball joint near the top of the tire. This means that the knuckle may even partially wrap around the tire. The long knuckle design affords the manufacturer a more ideal king pin geometry. It also gives

FIGURE 28-51 Steering knuckles. **A.** Long knuckle SLA. **B.** Short knuckle SLA.

better leverage against braking and cornering forces. King pin geometry is covered in the Wheel Alignment chapter.

Twin I-Beam Suspension

The twin I-beam suspension is a type of independent suspension. For each front wheel, it uses separate I-beams. Each I-beam pivots from the opposite side of the vehicle's frame or cross member (FIGURE 28-52). This system gives a wide radius that the wheel assembly swings through as the suspension compresses and rebounds. Most twin I-beam systems use coil springs to support the weight of the vehicle. This means each I-beam must be supported longitudinally, which is accomplished by use of a radius rod.

The radius rod connects to the I-beam from the rear. It is attached to the vehicle's frame with bushings. They allow the radius rod to pivot slightly as the suspension moves up and down.

▶ Rear Suspension

LO 28-08 Describe the main types of rear suspension systems.

The main function of the rear suspension system is to keep the rear tires in contact with the road. But they also must maintain their alignment with the front tires (FIGURE 28-53). However, rear-wheel drive or all-wheel drive rear suspension systems are a bit more complicated. They must also be engineered to transfer

FIGURE 28-52 Twin I-beam suspension system.

FIGURE 28-53 In front-wheel drive vehicles, the rear suspension system serves to keep the rear tires in contact with the road and aligned with the front tires.

engine torque to the rear wheels. And in vehicles with four-wheel steering, the rear suspension must allow the rear wheels to be steered. In this way, they are similar to the front wheels.

Rear-wheel suspension systems can be of either the independent or solid axle design. It is purely up to the preference of the designer. Most rear-wheel drive vehicles use a solid axle, dependent suspension system. On front-wheel drive vehicles, the rear suspension system is split between dependent and independent suspensions.

Rigid-Axle Leaf-Spring Suspension

A rigid-axle leaf-spring suspension can be used in both dead axles and live axles. The front of the leaf spring is attached to the chassis at the rigid spring hanger (**FIGURE 28-54**). The spring eyes typically use rubber bushings to connect with the vehicle's frame. The axle housing is rigid between each road wheel. Thus, any deflection to one side is transmitted to the other side. The rear of the leaf spring is attached to the swinging shackle. This allows for suspension movement. It allows the spring to extend or reduce in length, as the vehicle moves over uneven ground.

The top of the direct-acting shock absorber is attached to the chassis and to the spring pad at the bottom. The U-bolts attach the axle housing to the leaf spring. They have a clamping force that helps to keep the leaf spring together.

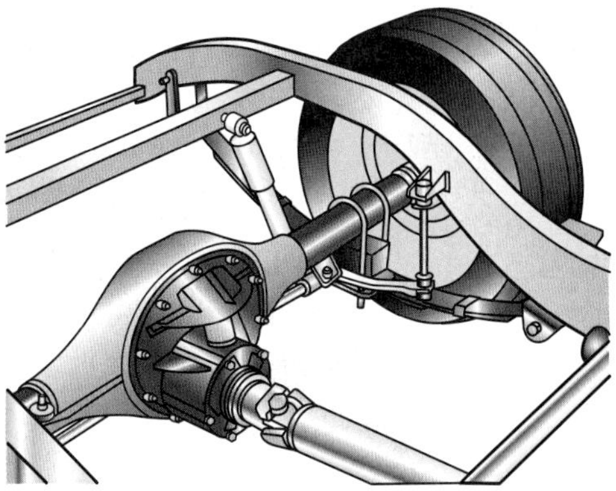

FIGURE 28-54 A rigid-axle leaf-spring suspension.

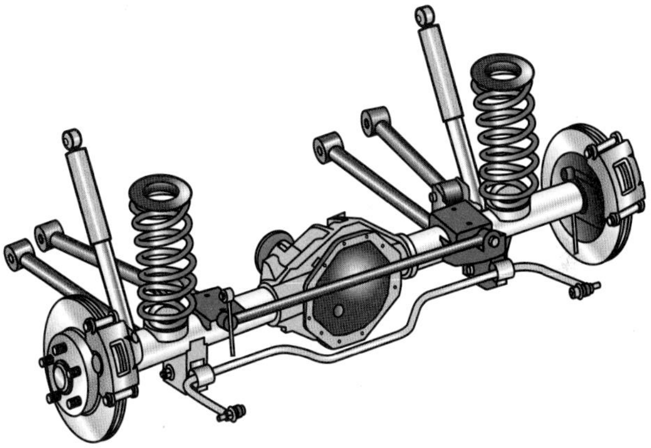

FIGURE 28-55 Rigid-axle coil-spring suspension.

Leaf springs hold the axle in position, both laterally and longitudinally. This eliminates the need for track bars. The leaf spring is usually made up of a number of leaves of different lengths. The top, or longest, leaf is normally referred to as the main leaf. Most leaf-spring suspensions rely only on the sideways stiffness of the leaf spring to keep the axle in position when turning a corner. This helps keep the axle from shifting sideways through turns.

Rigid-Axle Coil-Spring Suspension

In **rigid-axle coil-spring suspensions**, the coil spring is mounted between the axle housing and the vehicle body (**FIGURE 28-55**). One drawback of a coil spring is that it cannot provide any side-to-side or front-to-back stability to the axle. All it can do is suspend the body above the axle. This means that control rods must be used to control this potential axle movement. This occurs during braking, acceleration, and cornering. Manufacturers use a variety of configurations to address this concern.

The first style uses lower control arms near each coil spring that are parallel with the **centerline** of the vehicle. They connect between the frame and the axle. They maintain the longitudinal position of the axle. The upper control arms are angled toward the center of the vehicle. They counteract any lateral forces as well as twisting forces (**FIGURE 28-56**). Another style uses upper and lower control arms that all are parallel with the centerline of the vehicle. They control the twisting force of the axle but cannot control any lateral forces during cornering. One of two methods is used to control the lateral forces—a Panhard rod or a Watt's linkage.

A **Panhard rod**, also referred to as a track bar, sits parallel with the axle. One end connects to the frame of the vehicle, and the other end connects to the axle (**FIGURE 28-57**). The Panhard rod uses bushings on each end. This allows the joints to pivot as the suspension compresses and rebounds. The position of the axle is maintained laterally by the rod.

A **Watt's linkage** is a bit more complex but functions in a similar way (**FIGURE 28-58**). A lever that is able to pivot in the middle is mounted vertically to the rear axle near its center. The top of the lever is connected to a rod that is parallel to the axle. It is mounted high on one side of the frame. The bottom of the lever is connected to a similar rod. It is also parallel to the axle but mounted low on the other side of the frame. These two rods allow up-and-down movement of the axle but prevent lateral movement.

Rigid Dead Axle Suspension

A rigid dead axle is sometimes referred to as a beam axle. It can come in a variety of configurations. With **rigid dead axle suspension**, the longitudinal and lateral position of the axle must be maintained as in all axles. But because it is a dead axle, it typically has to withstand only braking forces.

One common rigid dead axle suspension uses lower trailing arms. They connect the vehicle frame to the axle. A strut assembly is used on top. The trailing arms maintain the longitudinal position of the axle (**FIGURE 28-59**). The springs and strut support the weight of the vehicle and assist in controlling any braking forces. The struts have an upper mounting point in the suspension tower. They are nonsteerable and therefore do not require an upper strut bearing. This type of suspension typically uses a Panhard rod to control lateral forces.

Another style, called a torsion beam axle, is a dead axle that uses a U-shaped axle beam with a torsion bar mounted inside it (**FIGURE 28-60**). A trailing arm on each side holds the axle longitudinally. A Panhard rod may be used to hold it laterally. And the strut and spring support the weight of the vehicle. The torsion bar provides a measure of resistance to twisting forces as one wheel goes over a bump.

Manufacturers use a variety of other types of rigid dead axles. But they all tend to use leaf springs, coil springs, or torsion bars, along with control arms, rods, and bushings. As you work out in the shop, be on the lookout for the different arrangements of these systems.

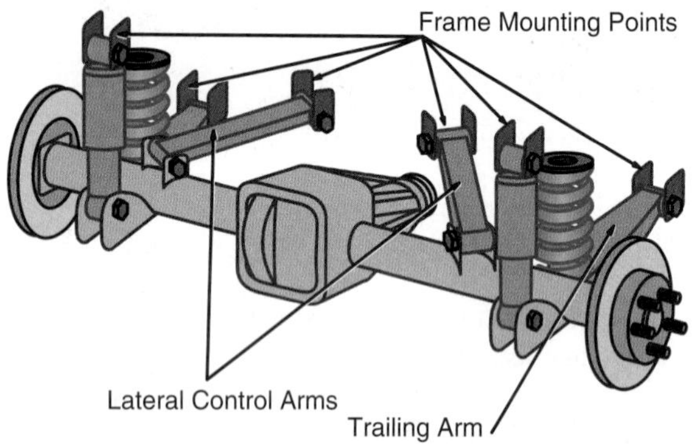

FIGURE 28-56 Lower control arms prevent any front–rear movement of the axle. Upper control arms prevent side-to-side and twisting movement of the axle.

FIGURE 28-57 A Panhard rod sits parallel with the axle.

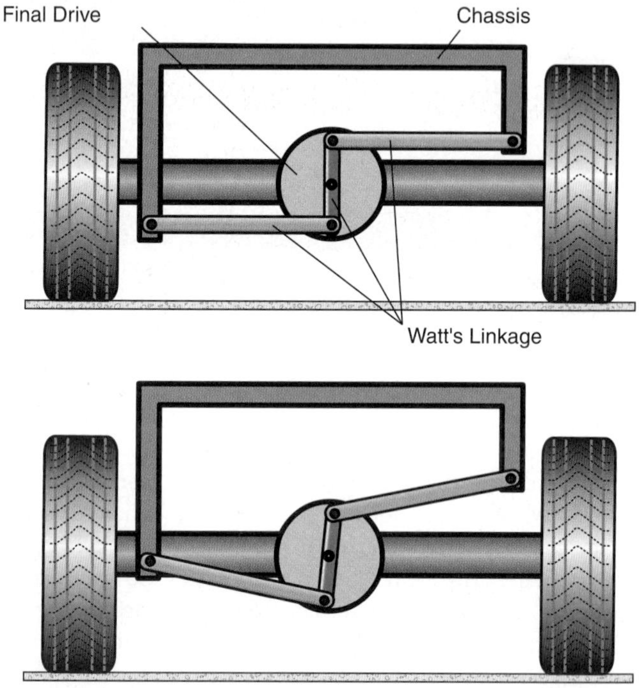

FIGURE 28-58 A Watt's linkage.

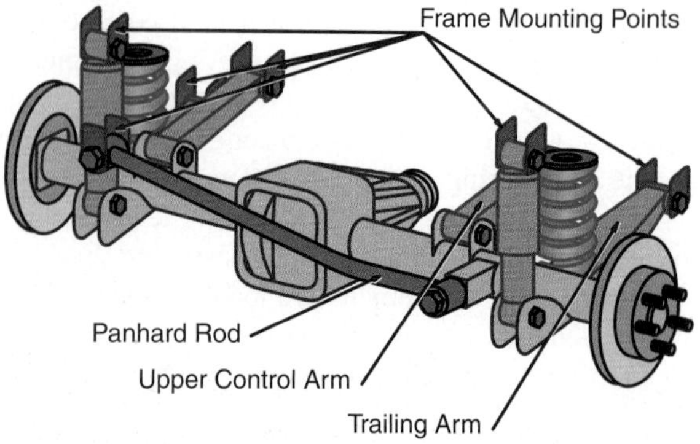

FIGURE 28-59 A rigid dead axle using a strut assembly, trailing arm, and Panhard rod to control axle movement.

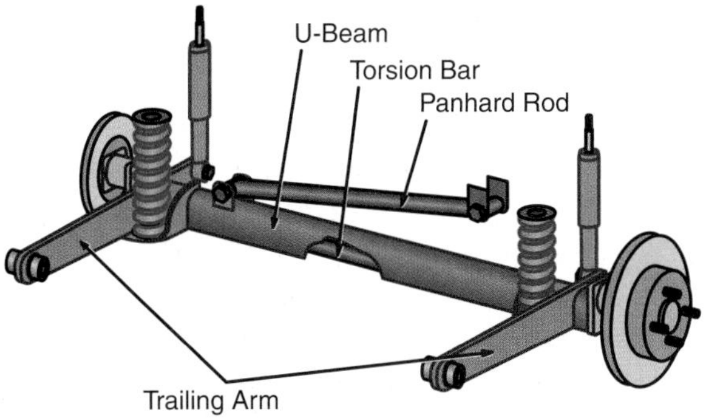

FIGURE 28-60 Torsion beam axle with internal torsion bar.

FIGURE 28-61 A typical MacPherson strut-type independent rear suspension (IRS).

Rear Independent Suspension (Dead Axle)

The kind of dead independent suspension used on the rear of a vehicle can be fairly simple. This is because the wheels typically do not steer or drive the vehicle. In this case, the suspension system only has to hold the wheels in the proper orientation while the vehicle is being driven. Because it is an independent suspension, each wheel is not connected to the other, so they can move separately.

A MacPherson strut system is commonly used at the rear. It can be similar to the front suspension system. It typically uses either a wishbone-shaped lower control arm or upper and lower control arms, along with a strut assembly (**FIGURE 28-61**). It can also be similar to Ford's Control Blade® system that uses a trailing arm with a lower control arm (**FIGURE 28-62**). A separate spring sits between the frame and control arm. A long travel shock absorber sits outside of the spring.

Rear-Wheel Drive Independent Suspension

In rear-wheel drive vehicles with independent rear suspension, the final drive unit is attached to the vehicle frame. This greatly reduces unsprung weight (**FIGURE 28-63**). Drive is transmitted to each wheel by **external driveshafts**. These shafts transfer power from the final drive to the wheels. Suspension is normally provided by coil springs. Each wheel unit is located by a combination of lateral and longitudinal control arms.

FIGURE 28-62 Ford's Control Blade IRS.

FIGURE 28-63 Live independent rear suspension. Note how the final drive is mounted directly to the vehicle frame.

The final drive assembly is normally bolted to the chassis. Because it must absorb the torque reaction, it must be securely fastened. Driveshafts use either conventional universal joints or constant-velocity joints. They allow flexing of the shaft while driving the wheels. Universal joints compensate for the up-and-down movement of the rear wheels. When universal joints are used, each driveshaft may have a **splined section called a slip joint**. It allows the shaft's length to change due to the suspension action (**FIGURE 28-64**).

On some vehicles equipped with an independent rear suspension, the half shafts themselves can be used as the upper link of the suspension. They provide the upper pivot point for the knuckle (**FIGURE 28-65**). In this case, the half shafts are fixed, meaning they do not extend or collapse. This makes the splined section unnecessary. So, the shafts can be made as a one-piece units.

The outer wheel bearing hub is held in position laterally and longitudinally at the bottom by a pivot on the end of the lower control arm. The half shaft holds the top of the hub in a vertical position. Because the hub pivots on both the lower control arm and the half shaft, the wheel can move up and down. The lower control arm has widely spaced pivots to provide stability. It is longer than the half shaft. This allows the camber to move slightly positive during cornering, just like an SLA suspension.

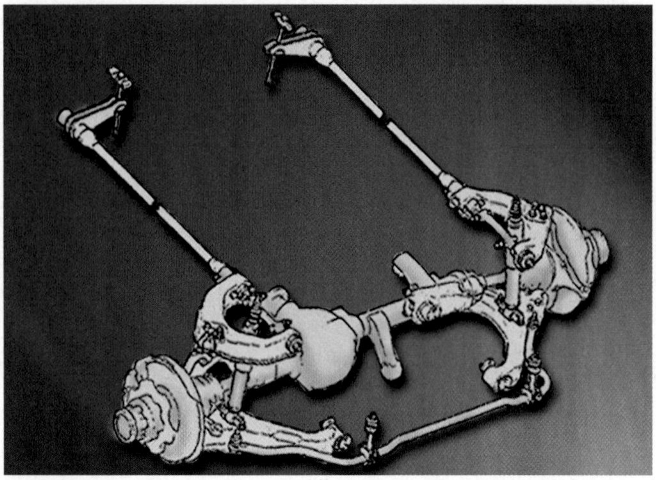

FIGURE 28-64 The independent rear suspension may have upper and lower control arms.

FIGURE 28-65 On some vehicles equipped with an independent rear suspension, the driveshafts or half shafts themselves can be used as the upper link of the suspension, providing the upper pivot point.

▶ TECHNICIAN TIP

External forces, such as curb impact or a collision, can damage control arms or linkages. This moves the wheel units from their correct position. The vehicle can then pull to one side, cause abnormal tire wear, and make the vehicle difficult to control.

▶ Active and Adaptive Suspension Systems

LO 28-09 Describe active and adaptive suspension systems.

There are two main categories of suspension system control. They are: active and adaptive (also called semi-active). An active suspension system acts independently of the driver. It fully controls the actions of each individual wheel while the vehicle is being driven. This means that it can do several things:

- An active suspension system can respond to irregularities in the road surface. It can raise or lower each wheel so it will follow the irregularities without causing much of a change in the vehicle's body height.
- It can automatically raise or lower the vehicle with changes in speed for aerodynamic or road condition reasons.
- It can prevent the vehicle from leaning in corners by raising the outside wheels when cornering. This maintains a flatter stance and keeps the tires perpendicular to the road surface, which enhances traction.
- It can also prevent the nose and tail of the vehicle from diving or rising during braking and accelerating maneuvers.
- At the same time, it can provide active dampening power to each individual shock absorber. This is based on the driver's preferred setting, road surface, and driving conditions.

Most active suspension systems are hydraulically actuated. This means that the servos are operated by high-pressure fluid from a hydraulic pump. The pressure is directed to or vented from the servos by a bank of electrically operated solenoid valves. Sensors monitor suspension height, vertical and horizontal acceleration, pitch, and body roll. This allows adjustments to be made almost instantaneously. Some systems can provide up to 3000 adjustments per second. So, you can see that this type of system can be very responsive. At the same time, these systems are very expensive and have large power requirements. Also, because of the amount of work they do, these systems tend to have high maintenance needs. This makes the costs prohibitive for most people. Therefore, few manufacturers include them in their vehicles.

Adaptive air suspension systems are less sophisticated than active systems. They provide electronically controlled dampening at all four shock absorbers. This means they are able to change the dampening power of the shocks when the road conditions change. At the same time, they can only change the dampening effect. They cannot actively raise or lower the wheel in response to road irregularities.

An adaptive air suspension system combines the ability to have sporty handling with a high level of ride comfort. Additionally, some systems include speed-dependent lowering of the body. This change in ride height means a low center of gravity. It results in significantly increased directional stability as well as increased aerodynamic efficiency. The ride height is adjusted by inflating or deflating air bags at each corner of the vehicle. Note that the adjustments in ride height are relatively slow to obtain. They don't happen in the same way that active suspension systems operate.

The information obtained from sensors is evaluated in the adaptive suspension's central control unit. These include sensors on the axles for ride height, and acceleration sensors on the body for vehicle pitch. The computer can adjust the dampening power of the individual shock absorbers. It can adjust within milliseconds, depending on driving situations. As long as minimal dampening forces are required, the damper settings remain soft. For instance, when driving straight ahead on good roads.

Specific adjustments to the dampening force can enhance either vehicle handling or occupant comfort. In some cases, adaptive dampening can reduce (not remove) rolling or pitching movements. This is the case when cornering, braking, or accelerating. Adaptive suspension systems are not as refined as active suspension systems. But, the differences between them are shrinking.

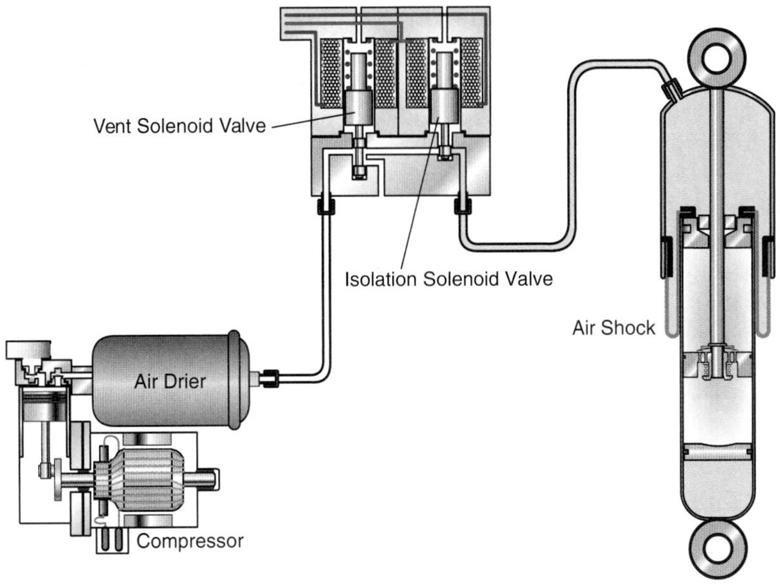

FIGURE 28-66 Adaptive air suspension.

Adaptive Air Suspension Operation

On some vehicles, when the ignition is switched on or when the vehicle's door is opened, the control system is activated. The height sensor monitors the distance between the vehicle's axle and its chassis. If the control system determines that the ride height is too low, it will command that air be added to the appropriate adaptive shock absorbers.

Height sensors monitor height and report it to the control unit. This includes when the vehicle is being loaded, unloaded, or lowered because of driver command or vehicle speed. The ECU compares this information to the stored reference values. The ECU activates the electric air compressor to provide compressed air for the system. To raise the vehicle, an isolation valve is opened. This allows compressed air to flow to the intended adaptive shock.

To lower the vehicle, an exhaust valve is opened. This vents pressure from the intended adaptive shock absorber (**FIGURE 28-66**). This also requires the isolation valve to close to maintain the required level, once reached. The adaptive shock absorber solenoid valves are subject to stringent leakage requirements. This is to maintain the vehicle's height, even when the system is not being operated.

Any dynamic air spring movement while the vehicle is in motion is ignored. It does not cause the control system to respond. The system can adjust the air pressure based on other parameters. These include vehicle speed, driver selection of a different mode, or any automatic adjustments that the control system determines are necessary.

▶ Wrap-Up

Ready for Review

- The suspension system of a vehicle reduces road shocks.
  - Unsprung weight includes all of the parts of a vehicle that are not supported by the springs. Sprung weight is the body, drivetrain, and associated parts that are supported by the springs.
- A basic suspension system consists: springs, axles, shock absorbers, control arms, rods, and ball joints. Most common spring configurations are coil springs, leaf springs, and torsion bars.
- A shock absorber is a device designed to dampen spring oscillations. It absorbs shock loads while driving on irregular surfaces. A strut is used to support the top of the steering knuckle as well as dampen spring oscillations.
- In adjustable valves, opening and closing some of the valves softens and stiffens the dampening action. Adjustable valves can be load-adjustable shock absorber, manual adjustable-rate shock absorber, and electronic adjustable-rate shock absorbers.
- Various components combine to form a suspension system and include: control arms and rods, steering knuckle, ball joints, and bushing.
- Manufacturers use various types of suspension systems, but the function of each suspension system is to maintain the proper positioning of the wheels during all driving conditions.
  - Different types of suspension system include dead axle/live axle, solid axle, and independent suspension.
- Regardless of design, the primary function of the front suspension system is to keep the tires in constant contact with the road. Types of front suspension system are: strut suspension, modified strut system, double-wishbone

modified strut suspension, SLA suspension, and twin I-beam suspension.

- Rear suspension system keep the rear tires in contact with the road and also maintain their alignment with the front tires. Rear-wheel suspension systems can be rigid-axle leaf-spring suspension, rigid-axle coil-spring suspension, rigid dead axle suspension, rear independent suspension, and rear-wheel drive independent suspension.
- An active suspension system acts independently of the driver and fully controls the actions of each individual wheel while the vehicle is being driven. Adaptive air suspension systems provide electronically controlled dampening at all four shock absorbers.

Key Terms

adaptive air suspension A suspension system that uses rubber bags or bladders filled with air to support the weight of the vehicle.

air spring A part that provides the springing action or auxiliary spring. It is typically used in air suspension systems or heavy truck applications.

air spring Consists of a flexible rubber bladder, which seals the outside of the upper and lower halves of the shock absorber. When inflated, the bladder pushes the halves apart.

applied force Pressure placed on something.

automatic load-adjustable shock absorber Typically, an air shock absorber used in an automatic load-sensing system that adjusts ride height (ground clearance) automatically, such as when additional weight is added to the vehicle; also called self-leveling.

axle A shaft connected to wheels that transmits the driving torque to the wheels.

ball joint A swivel connection mounted in the outer end of the front control arm. These swivels are typically constructed with a ball and socket to allow pivoting.

braking torque The torque acting to twist the axle housing around its center during braking.

Bushings A part that allows movement in a pivot point. It can also cushion suspension fulcrum points.

centerline A real or imaginary line drawn through the center of something.

coil spring Spring steel wire, heated and wound into a coil, that is used to support the weight of a vehicle.

compliance bushing A rubber bushing with a voided section molded in it that allows component movement under torque application. It is typically used in control arms on FWD vehicles to minimize torque steer issues.

control arm The primary load-bearing element of a vehicle's suspension system, commonly referred to as an A-arm or wishbone. These arms may be used as an upper and lower pivot point for the wheel assembly. They attach to the chassis with rubber bushings that allow up-and-down movement of the tire and wheel assembly.

cornering force The force between the tread and the road surface as a vehicle turns.

deflecting force A force that moves an object in a different direction or into a different shape.

direct-acting telescopic shock absorber A shock absorber designed to reduce spring oscillations.

driving thrust The force transferred from the tire contact patch through the axle housing and front half of the spring to the fixed shackle point that pushes the vehicle along the road.

elasticity The amount of stretch or give a material has.

external driveshaft A shaft used to transfer power from the transmission to the live axle.

independent suspension A system for allowing the up-and-down movement of one tire without affecting the other tire on that axle.

leaf spring A spring, made of one or more flat, tempered steel springs bracketed together, that is used in the suspension system to support the weight of the vehicle.

MacPherson strut A strut used on an independent suspension where the spring and shock are joined together; used on most front-wheel-drive vehicles.

magneto-rheological fluid A fluid that has the unique characteristic of changing viscosity when exposed to a magnetic field.

manual adjustable-rate shock absorber A shock absorber that allows manual adjustment of the dampening rate.

manually adjustable air spring A rubber air bag placed inside coil springs to increase the spring's load carrying ability. It is filled manually through a valve similar to a tire valve stem.

oscillation The fluctuation of an object between two states. With regard to suspension springs, it refers to the uncontrolled compression and decompression of the spring following overshoot.

overshoot The amount a spring extends (springs back) past its original length following compression.

panhard rod A metal rod that is mounted between the body or frame of the vehicle and the axle. It controls suspension movement side to side; also known as a transverse torque rod.

Pitch Movement around the vehicle's y-axis. Or on a spring, the distance from the center of one coil to the center of the adjacent coil.

progressive rate of deflection The change in deflection rate that occurs as the weight of the vehicle changes. The greater the weight, the lower the rate of deflection due to increased resistance.

reaction force A force that acts in the opposite direction to another force.

rebound clip A metal strap that is warped around the leaf spring to prevent excessive flexing of the main leaf during rebound.

ride height The amount of ground clearance a vehicle has, measured from a point on the body or frame, depending on the manufacturer; also known as ride height.

rigid dead axle suspension A type of dead axle suspension system that is nonindependent and uses a beam or solid axle.

rigid spring hanger The rigid part typically welded to the body or frame of the vehicle to which the front of the leaf spring is attached.

rigid-axle coil-spring suspension A dead axle that uses a coil spring.

roll Movement of a vehicle around its x-axis (the imaginary line down the center of the vehicle from front to back). It is commonly referred to as body roll or lean; when cornering, the body tries to move to the outside of the corner against the suspension.

Rods Straight (or precisely formed) pieces of steel used to control motion within the vehicle's suspension system.

rubber-bonded bushing A bushing that has a steel outer housing and inner sleeve with rubber inside; also known as a metalastic bushing.

self-leveling A vehicle with automatic load-adjustable shock absorbers.

shock absorber A device on a vehicle designed to absorb bumps and jolts caused from driving on irregular surfaces; it also dampens body movement.

shroud A steel or plastic cover placed over the shock rod.

sintered Solid part used as a bearing surface. It is made of small particles that are bonded together with heat and pressure.

SLA suspension See *short-/long-arm suspension.*

solid axle A single piece of steel that provides a simple means of mounting the hub and wheel units; also called beam axle or straight axle.

splined section A flat key made into a shaft to accommodate changes in shaft length due to movement in wheel camber with suspension action.

Springs Elastic component that supports the body of the vehicle. It allows the suspension system to flex with road irregularities.

spring eyes Rolled ends of some springs used to mount springs to the chassis.

spring shackle bushing A bushing that is positioned in the shackle to which the leaf spring mounts. Bushings allow the spring shackle to move as the leaf spring dimensions change over bumps.

stop A rubber part used to control the movement of control arms (suspension arms).

strut A rigid shock absorber assembly used on a MacPherson strut–type suspension.

suspension action Movement of the chassis up and down.

suspension system A vehicle system designed to isolate the vehicle body from road bumps and vibrations.

sway bar A part used in vehicles as a stabilizer, or antiroll, bar. It is connected to the chassis in the center, and each end is connected to one side of the suspension system. It is typically installed on the front, and sometimes the rear, suspension.

swinging shackle A shackle connected to the rear of the multileaf spring that allows the leaf spring to move downward when a load is placed on the rear of the vehicle.

torsional load A force that is applied by clamping one end of an object to another object that is then twisted.

trailing arm suspension A type of suspension system that uses upper and lower control arms.

uniform pitch A spring whose pitch (the distance from the center of one coil to the center of the adjacent coil) is the same distance throughout.

unsprung weight See *unsprung mass.*

Watt's linkage Another name for a rigid-axle coil-spring suspension that uses two bars similar to a Panhard rod and a pivot point on the axle, to keep the axle from moving in turns.

wishbone control arm Another term for an A-arm.

wrap leaf A spring containing spring eyes.

yaw Movement around the z-axis (vertical axis), felt when the vehicle deviates from its straight path, as when the rear wheels slide out during drifting.

Review Questions

1. All of the following are significant functions of a vehicle's suspension system EXCEPT?
   a. To keep tires in contact with the road surface
   b. To keep the vehicle body firmly mounted to the frame
   c. To maintain traction and control of the vehicle
   d. To provide a smooth ride for passengers
2. Which type of spring is often held together by a center bolt and attached to a rear axle by a u-bolt?
   a. Coil springs
   b. Torsion bars
   c. Sway bars
   d. Leaf springs
3. Which component is primarily responsible for dampening oscillations?
   a. The coil springs
   b. The torsion bars
   c. The sway bar
   d. The shock absorbers
4. On a MacPherson strut suspension design, what does the bottom of the strut connect to?
   a. The steering knuckle
   b. The ball joint
   c. The tie-rod end
   d. The wheel flange
5. In a load-adjustable shock absorber, why would an internal air spring be incorporated?
   a. To adjust the dampening action when going over bumps
   b. To support additional weight and adjust ride height
   c. To give it a stiff sporty feel on winding roads
   d. To make the steering more responsive to the driver
6. What is another common name for a steering knuckle?
   a. Tie rod
   b. Steering pinion
   c. Steering track rod
   d. Spindle assembly
7. When a solid rear axle simply supports the rear wheels without transmitting power to the wheels, what is it called?
   a. A live axle
   b. An independent axle
   c. A dead axle
   d. A drive axle
8. Which of the following is NOT a type of independent front suspension?
   a. Short long arm (SLA)
   b. Solid axle
   c. Double wishbone
   d. Twin I-beam

9. On a rear solid axle suspension, what part is mounted parallel to the axle, connecting the axle to the frame to control side-to-side axle movement?
   a. Leaf springs
   b. Coil springs
   c. Trailing arms
   d. Panhard rod
10. On an adaptive air suspension, how is a vehicle lowered?
   a. The air pump is cycled off and air travels back through the pump
   b. The air is directed from the rear shocks to the front
   c. An exhaust valve is opened until the desired height is achieved
   d. The isolation valve is opened, diverting air into the desired shock

ASE Technician A/Technician B Style Questions

1. A vehicle comes into the shop for a suspension inspection. Technician A states that the suspension should be inspected for wear and looseness from normal use. Technician B states that the suspension should be inspected for bent or damaged components due to impacts such as potholes or accidents. Who is correct?
   a. Technician A only
   b. Technician B only
   c. Both Technicians A and B
   d. Neither Technician A nor B
2. A vehicle with torsion bar suspension comes in to the shop with a low ride height. Technician A states that springs can wear and eventually start to sag. Technician B states that some torsion bar suspensions have an adjustment, which may return the vehicle back to its original ride height. Who is correct?
   a. Technician A only
   b. Technician B only
   c. Both Technicians A and B
   d. Neither Technician A nor B
3. A vehicle is being inspected and is found to have oil on the bottom of a shock absorber. Technician A states that this could indicate a leaking shock absorber which could affect ride quality. Technician B states that the oil level in the shock absorber should be checked and topped off if necessary. Who is correct?
   a. Technician A only
   b. Technician B only
   c. Both Technicians A and B
   d. Neither Technician A nor B
4. A high-performance sports car is being discussed. Technician A states that its adjustable shock absorbers may be controlled by externally mounted knobs that adjust fluid passage size. Technician B states that the adjustments may be electronic and controlled by a selector in the passenger compartment. Who is correct?
   a. Technician A only
   b. Technician B only
   c. Both Technicians A and B
   d. Neither Technician A nor B
5. Suspension components are being discussed. Technician A states that ball joints are also known as stub axles. Technician B states that ball joints may have a grease fitting which allows for periodic lubrication. Who is correct?
   a. Technician A only
   b. Technician B only
   c. Both Technicians A and B
   d. Neither Technician A nor B
6. Suspension types are being discussed. Technician A states that most rear-wheel-drive vehicles are equipped with a dead axle. Technician B states that most trailers are equipped with a live axle. Who is correct?
   a. Technician A only
   b. Technician B only
   c. Both Technicians A and B
   d. Neither Technician A nor B
7. Front suspension types are being discussed. Technician A states that a MacPherson strut design utilizes upper and lower control arms. Technician B states that an SLA type utilizes upper and lower ball joints. Who is correct?
   a. Technician A only
   b. Technician B only
   c. Both Technicians A and B
   d. Neither Technician A nor B
8. A vehicle with a double wishbone front suspension comes into the shop for maintenance. Technician A states that there are two lower control arms and no upper control arms. Technician B states that there is no upper ball joints, so only the lower ball joints should be inspected and serviced. Who is correct?
   a. Technician A only
   b. Technician B only
   c. Both Technicians A and B
   d. Neither Technician A nor B
9. Rear suspension systems are being discussed. Technician A states that some rear-wheel-drive vehicles are equipped with an independent rear suspension. Technician B states that some rear suspension systems are MacPherson strut design. Who is correct?
   a. Technician A only
   b. Technician B only
   c. Both Technicians A and B
   d. Neither Technician A nor B
10. Active and adaptive suspension systems are being discussed. Technician A states that active systems can respond to changes while driving. Technician B states most adaptive air suspension systems receive information from sensors mounted on the suspension. Who is correct?
   a. Technician A only
   b. Technician B only
   c. Both Technicians A and B
   d. Neither Technician A nor B

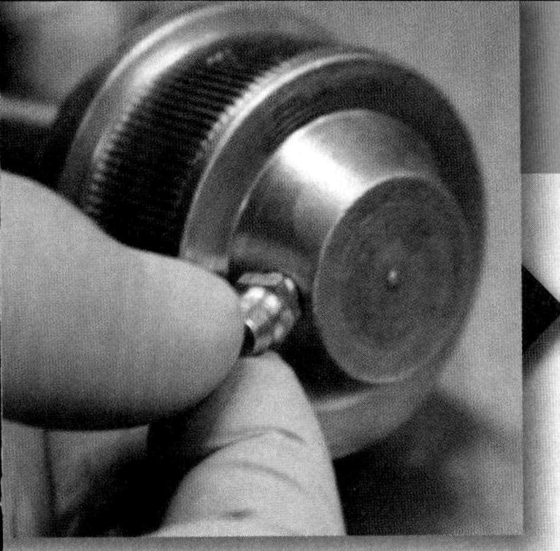

CHAPTER 29
Servicing Suspension Systems

Learning Objectives

- **LO 29-01** Describe suspension system service preliminaries.
- **LO 29-02** Describe suspension system diagnosis.
- **LO 29-03** Measure ride height and test shock absorbers.
- **LO 29-04** Unload a suspension and measure ball joint play.
- **LO 29-05** Replace stabilizer components and shock absorbers.
- **LO 29-06** Remove coil springs and steering knuckles.
- **LO 29-07** Remove control arms and ball joints.
- **LO 29-08** Install and lubricate SLA components.
- **LO 29-09** Inspect and service strut assembly.
- **LO 29-10** Inspect strut rods and bushings, leaf springs, and torsion bars.

ASE Education Foundation Tasks

See Appendix A to view the 2017 ASE Education Foundation Automobile Accreditation Task List Correlation Guide.

▶ Suspension System Service Preliminaries

LO 29-01 Describe supsension system service preliminaries.

This chapter covers the diagnosis, maintenance, and repair of modern suspension systems. In the previous chapter, you learned that the suspension system is primarily a mechanical system. Electric, electronic, and pneumatic devices may be added to it in some cases. But those devices enhance the function of the mechanical components; they do not replace them. You will usually be dealing with mechanical faults such as worn or damaged parts. Faults involving wear are often located during a visual inspection. They are then confirmed by feeling for play, or taking measurements.

Damaged components can typically be identified visually, but not always. In some cases, an alignment machine must be used to measure the steering and suspension angles. Those angles are then compared with specifications in order to identify damaged parts.

As you can see, you need a solid understanding of suspension system components so you can visually inspect them. This knowledge also comes in handy as you replace parts. Knowing how they are supposed to operate allows you to verify proper operation once you complete the repair.

Servicing Suspension Systems

After any suspension system work, the vehicle will likely need a wheel alignment. This will help ensure that the suspension system is adjusted properly. Wheel alignment is covered in the next chapter. Some instructors prefer to teach wheel alignment before servicing suspension systems. That way, students will understand the alignment angles before diving into the suspension system. So don't be afraid to read ahead and start to learn about the wheel alignment process. But for now, let's get started with an overview of some of the tools you will use during suspension system work.

Tools

Servicing suspension systems requires the use of special tools. Some were used in servicing steering systems. But most of the following are new tools:

- Dial indicators: Used to measure play in ball joints.
- Pry bars: Used to pry parts to check for play.
- Measuring tapes: Used to check ride height.
- Pitman arm puller: Used to pull the pitman arm from the sector shaft (**FIGURE 29-1A**).
- Tie-rod end puller: Should always be used on aluminum steering arms (**FIGURE 29-1B**).
- Pickle forks: Used to separate ball joints, but ruins the grease seal (**FIGURE 29-1C**).
- Electronic stethoscope: Used to locate unusual noise, vibration, or harshness (NVH); sometime called a chassis ear (**FIGURE 29-1D**).
- Coil spring compressor: Used to compress coil springs during removal and installation (**FIGURE 29-1E**).
- Scan tool: Used to check for codes in the vehicle's computer.
- Ball joint press tool: Used to remove and replace press-fit ball joints (**FIGURE 29-1F**).
- Air chisel: Air-operated hammer that can accept a variety of bits including chisels and punches (**FIGURE 29-1G**).
- Strut compressor: Used to compress MacPherson struts to remove the coil spring (**FIGURE 29-1H**).
- Strut servicing kit: Used for changing coil springs on struts (**FIGURE 29-1I**).
- Universal strut nut wrench kit: Used to disassemble a strut assembly.
- 24-mm strut rod socket: Used to remove a strut.
- Various lift devices (e.g., jack): Used to raise vehicle off of the ground and safety stands to hold it there. Devices include the air jacks found on the alignment lift.

▶ Diagnosis

LO 29-02 Describe suspension system diagnosis.

Problems in the suspension system generally result in driver complaints of a poor ride, poor handling, or noises. These may seem trivial. But they could indicate a serious and costly suspension problem. Left untreated, they could lead to additional issues or safety hazards. Diagnosis of suspension system problems follows the same strategy-based diagnostic process used on all vehicle systems. It's time we revisit those steps:

Step 1: Verifying the customer's concern
Step 2: Researching possible faults and gathering information
Step 3: Focused testing
Step 4: Performing the repair
Step 5: Verifying the repair

You Are the Automotive Technician

You work for County Fleet Services in the vehicle service department. One of the drivers wrote up a complaint sheet on a relatively new Dodge full-size delivery van. It is equipped with short-/long-arm suspension on front and leaf-spring suspension on the rear. The truck has a little more than 100,000 miles (160,000 km) on it. The complaint sheet lists several concerns. First, the operator admitted to hitting a curb with the right front wheel. Second, the vehicle pulls hard to the right when driven. Third, it makes an audible clunking noise in the front when going over bumps at around 10 mph (17 kph).

1. What are the potentially damaged components when the vehicle hits the curb?
2. What will you do to diagnose the pulling condition?
3. What are the possible causes of the clunking noise when hitting a bump?

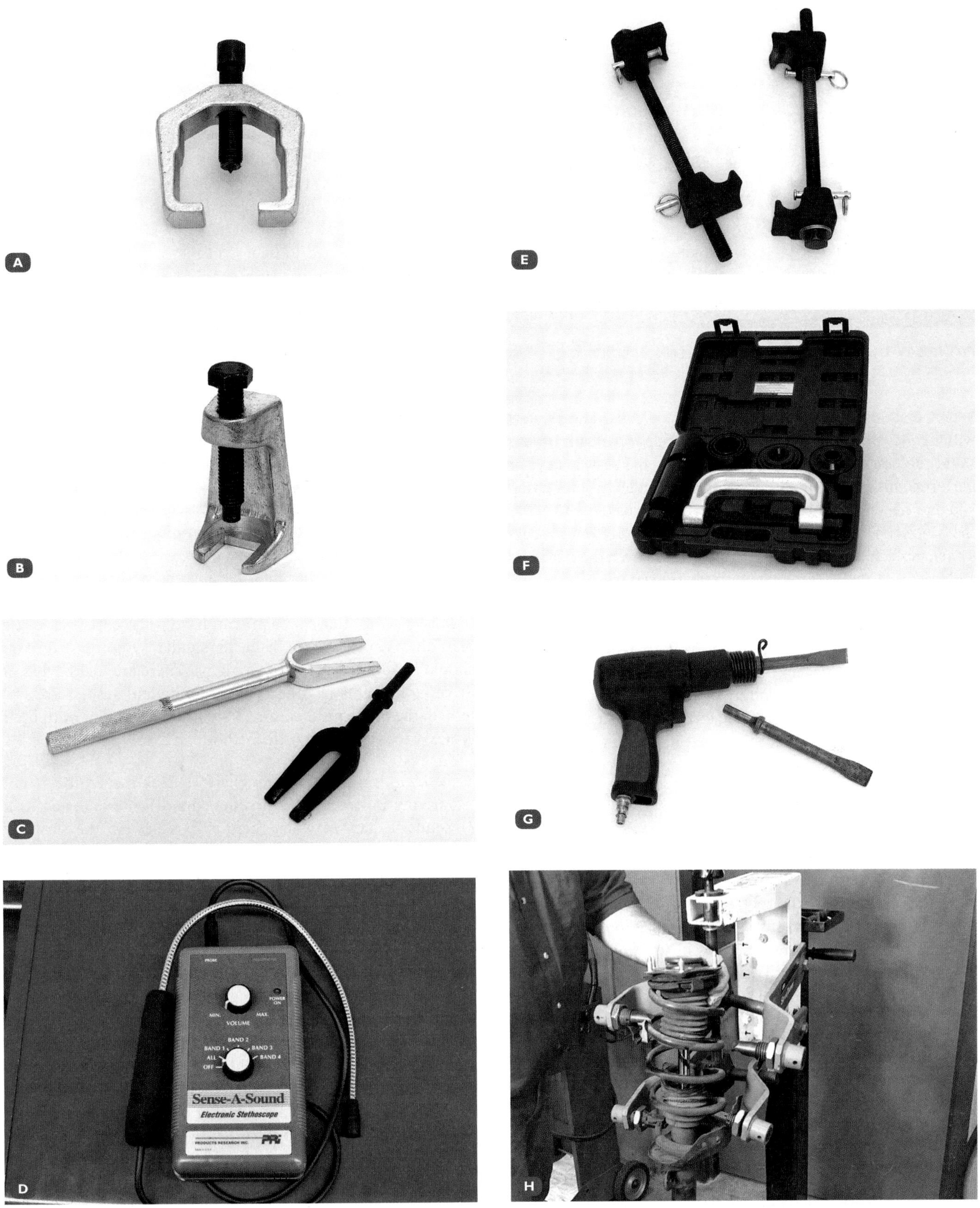

FIGURE 29-1 Suspension system tools. **A.** Pitman arm puller. **B.** Tie-rod end puller. **C.** Pickle forks. **D.** Electronic stethoscope. **E.** Coil spring compressor. **F.** Ball joint press. **G.** Air chisel. **H.** Strut compressor.

FIGURE 29-1 Suspension system tools. **I.** Universal strut nut kit.

When diagnosing suspension systems, you must always start with a good customer interview. This will help you to fully understand the concerns of the driver. Typically, the service advisor performs this and then writes up the repair order. In some cases, you may need to perform a test drive with the customer to verify the concern. Once you have gathered information from the customer, perform a test drive. With information from the customer and test drive, research the symptoms in the service information. Also, check to see if there are any related technical service bulletins (TSBs). If so, follow the information listed there.

If there are no related TSBs and you are familiar with the type of suspension system in the vehicle, take a minute to plan your testing strategy. This will involve using your understanding of the system and logic to decide what steps you should take initially. This usually includes a visual inspection of the tires, wheels, steering system, and suspension system. A general vehicle inspection can also point out related issues. If these initial inspections do not uncover the cause of the concern, you will need to dig deeper. For example, use an electronic stethoscope to identify any NVH faults. Or perform a wheel alignment to check the vehicle's wheel alignment.

Once you identify the root cause of the fault, present that information and the corrective actions needed to the customer. Obtain authorization to repair the vehicle. Now it is time to begin repairs. Follow the skill drills in this chapter along with the service information. If there is any difference between the skill drills in this book and the service information, always follow the service information. This is because it is directly applicable to that vehicle. Finally, once the repair is complete, test drive the vehicle. This will allow you to verify that the repair corrected the fault and that no other issues showed up. Doing this helps avoid comebacks. It also leaves a positive impression with the customer.

SAFETY TIP

Remember that steering and suspension concerns can result in injury or death to the driver and occupants. So, extreme attention to safety must be given while diagnosing suspension and steering concerns.

Common Issues

The most common problem in the suspension system is play or looseness of the parts. Many of the parts are designed with special connections that accommodate movement of the wheels. But excessive play is actually a bad thing. Excessive play magnifies the feel of road imperfections. It also makes the steering less responsive to steering wheel input. Excessive play is potentially very damaging to the connecting parts and tires.

Naturally, any amount of play in a fixed part is a big problem. It is frequently a result of fastener looseness or part failure. Remember, unwanted looseness in the suspension system can be extremely dangerous. So it should be corrected as soon as possible. In some cases, driving the vehicle should not be allowed until after the repair is performed.

Repair of excessive play typically means replacement of the loose part or parts. At the same time, some parts are designed to pivot and twist. So, a small amount of play may be within specifications. Always refer to the service information for proper testing procedures and specifications. That way you won't misdiagnose a fault. The upcoming skill drills are designed to work you through the common repair tasks, using common vehicles. But they don't replace the manufacturer's service information. Now let's look at some of the most common customer concerns related to the suspension system.

As discussed earlier, the steering and suspension systems work together to keep the wheels properly positioned. And because each of those systems can have worn parts that cause play, they can have the same or similar symptoms. So you should be familiar with both systems. There are many types of driver complaints relating to the steering and suspension system. Some of the most common are listed here along with their typical cause and diagnostic method:

Vehicle wander: This complaint means the vehicle is not driving in exactly the direction the driver is steering it. It tends to happen when there is looseness in the steering/suspension components. It can also be caused by the caster angle being wrong. A front-end alignment machine is the best tool for diagnosing caster angle problems. A pre-alignment inspection would find wear problems on joints, bushings, and components.

Pull: Pull is felt when the driver has to steer the vehicle one way just to keep it going straight. It can mean the air pressure in one or more of the tires is low. A simple tire pressure gauge is used to check the tire pressures. Pull can occur from a bad tire. Try switching the tires side to side. If the pull is now gone or has moved to the opposite side, then a tire is at fault. Pull can also happen when the camber or caster is out of adjustment from one side to the other by as little as half a degree. A vehicle will pull to the side with the most positive camber. If caster is the issue, the vehicle will pull to the side with the most negative caster. A four-wheel alignment is required to verify the alignment angles. Pulls can also result from power steering problems. Lifting the front tires off the ground and starting the engine can diagnose power steering problems. While doing this, watch for steering wheel movement. If the vehicle pulls only while braking, looseness in steering/suspension components and/or problems in the braking system may be the cause.

Bouncy ride/excessive body movement: This is typically a sign of worn shock absorbers/struts. If it only happens when turning corners, it could be a broken sway bar. Visually inspect the shocks/struts for leaks. Perform a shock absorber bounce test.

Shimmy/shake: This is typically a sign of tire imbalance or out-of-round condition. Checking tire balance and runout will identify either of these issues.

Hard steering: When the act of steering is difficult for the driver, there may be a problem in the power steering system. It can also be caused by too much positive caster, or binding in the steering system. A pressure gauge is used to check the power steering system, and a front-end alignment machine to check the caster. Hard steering could also be caused by something as simple as underinflated tires. A tire pressure gauge would be used to diagnose this. Check for binding in the steering system components by disconnecting components to verify the location of the binding.

Bump steer: Bump steer is when the vehicle steers itself upon hitting a bump. This is caused by incorrect angles in the steering linkage due to a component being bent. It can also be caused by an incorrectly positioned steering gear or rack and pinion. The steering linkage must follow the same arc as the suspension. Start with a visual inspection of the steering parts, and if necessary, check the alignment.

Torque steer: Torque steer is a pull that occurs during heavy acceleration. It can be the result of unequal axle lengths as designed, which cannot be repaired by the technician. Compare this vehicle to a known good vehicle to ensure it is a design issue. Abnormal torque steer can result from inner or outer tie-rod ends that are worn. Loose, worn, or broken engine or transmission mounts can also create this issue. Worn control arm bushings are another possibility. Installing a compliance bushing incorrectly with the void facing the wrong way can also create torque steer. Be sure to look for shifting components under load.

Steering return concerns: This condition is commonly referred to as memory steer. Memory steer can result from binding conditions in any pivot point in the steering or suspension system. This includes the ball joints, tie-rods, strut bearings, idler arm, binding steering gear, or rack and pinion. Or it can result from unbalanced power steering assist. It can also be caused by insufficient positive caster on the front wheels. Raise the vehicle wheels, and turn the steering while observing components. Disconnecting components one by one may be necessary to find binding in swivel points such as ball joints, strut bearings, or tie-rod ends. If these methods do not uncover the issue, check the caster with an alignment machine.

Noises: Noises can be tracked down through the use of a specialty tool called a chassis ear (**FIGURE 29-2**). This tool involves microphones that can be placed at various locations on the vehicle. A set of headphones is worn by the technician. The microphones can be selected one by one by the technician to pinpoint the noise.

Some of these tools can connect to a lab scope so that the vibrations/noises can be graphed on the screen. This allows the technician to measure the frequency, which can help identify the suspect component. *Operation of this tool during a road test should be performed by another person, and not the driver. Using the tool can be distracting to the driver.* Some noises can be identified and located while steering the vehicle when it is raised on a lift. Listen and feel for any noises and vibrations. Just be very careful around moving and hot components.

To diagnose a suspension noise that can be duplicated when not driving, bounce the vehicle up and down while listening for noise. If there are noises, confirm where the noise is coming from. Have an assistant bounce the vehicle while you check for the noise under the vehicle. You can listen by ear or with a stethoscope. Or you can feel by hand for vibrations or clunks. Listen around sway bar bushings, control arm bushings, springs, shock absorbers, component bolts, body panels, steering gear, the MacPherson strut mount, and ball joints. Inspect the component making the noise for damage, and replace if necessary. If the noise is found at a component bolt, check the bolt to see if it is loose. If so, retighten to the manufacturer's specifications.

FIGURE 29-2 Chassis ears can be used to help identify the location of suspension noises and vibrations.

Body Sway

To diagnose body sway, first test drive the vehicle to verify the customer's concern. Because you have to swerve from side to side to perform the test, drive the vehicle in a safe area. This would have little to no traffic, such as in a large empty parking lot. While driving the vehicle, move the steering wheel back and forth to make the vehicle swerve slightly. This will shift the vehicle weight from side to side. Observe how much body sway occurs.

It may be necessary to compare the sway of the vehicle to a like vehicle to ensure that the condition is not normal. If there is excessive sway, and the suspension is not electronically controlled, then check the vehicle's front and rear (if equipped) sway bar system. Inspect the sway bar bushings, brackets, and link bushings (**FIGURE 29-3**). Bushings should not be cracked and should have little give when pried. Replacement of faulty components is necessary.

If the vehicle is equipped with an electronically controlled suspension, it is possible that the system is not providing the active control needed to dampen body sway. In this case, the electronically controlled system will need to be checked for DTCs. Then, it can be diagnosed according to the service information. Electronic suspension diagnosis is beyond the scope of this book.

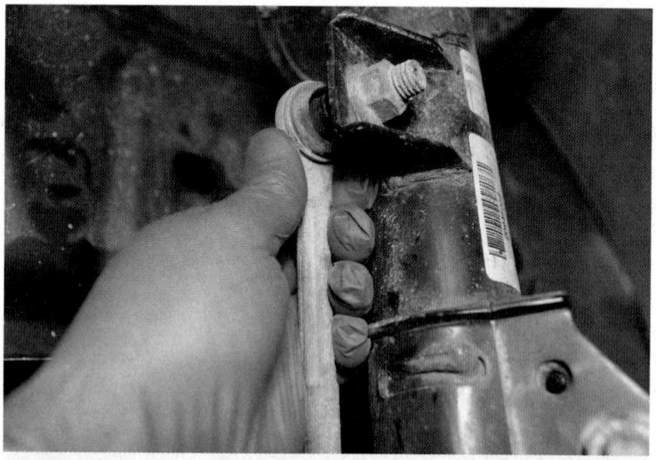

FIGURE 29-3 If the concern is excessive body sway, check the sway bar, bushing, brackets, and link bushings.

▶ Measure Ride Height

LO 29-03 Measure ride height and test shock absorbers.

The ride height of a vehicle can only be measured if it has matching tires that are properly inflated. There can also be no additional weight in the vehicle. Once these issues are taken care of, the ride height can be measured as specified by the service information. Most ride height specifications require the measurements to be within half of an inch side to side. If the ride height is out of specifications, it may be causing or contributing to the customer's concern.

Use a measuring tape from a fixed point on the vehicle's frame or body to the ground to check for improper ride height. Check for sagging vehicle springs, bent spring mounting points, a leaking gas-pressurized shock absorber, a faulty electronically controlled suspension system, or a bent frame or axle.

If the ride height issue is related to the coil or leaf spring, then replacement of the spring(s) may be necessary. If the vehicle is equipped with a torsion bar system, then you may be able to adjust the ride height. This is done by adjusting the torsion bar so that ride height is returned to specifications. Refer to the service information for proper adjustment of ride height. Remember that any time ride height is changed, a wheel alignment also must be performed.

To perform ride height diagnosis, follow the steps in **SKILL DRILL 29-1**.

Inspecting the Shock Absorbers

Shock absorbers and struts are located near each wheel and dampen body movement from bumps. Common reasons for testing shock absorbers are unusual tire wear, such as tires having a cupped appearance. Also, the driver may complain of a soft or bouncy ride. In some cases, a shock absorber can bind up, creating a very stiff ride.

If a vehicle has adjustable shock absorbers, make sure the shock absorber adjustments are the same for the left- and right-hand sides. Some shock absorbers contain pressurized gas,

SKILL DRILL 29-1 Performing Ride Height Diagnosis

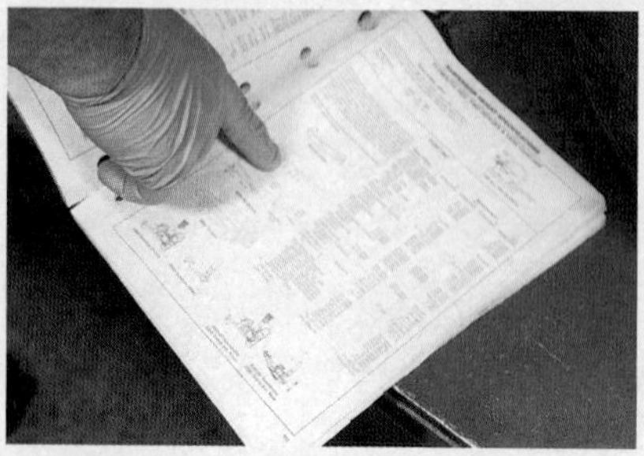

1. Refer to the manufacturer's service information for correct measurement points and specifications.

2. Check for properly sized, matching, and inflated tires. Correct any issues found.

3. Check the vehicle for any nonstandard loads in the trunk or luggage area. Remove them temporarily while measuring ride height.

SKILL DRILL 29-1 Performing Ride Height Diagnosis (Continued)

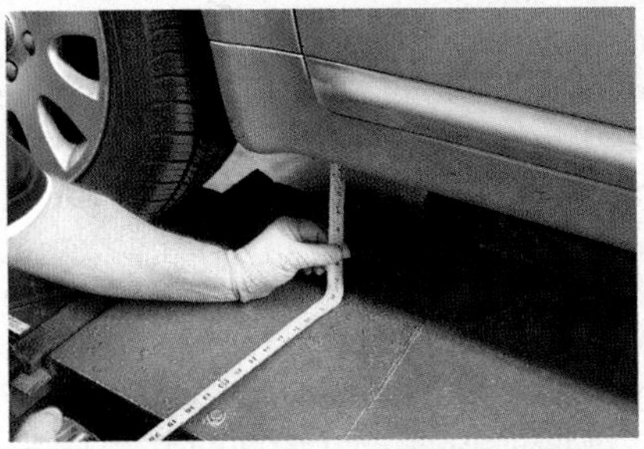

4. Measure from points specified, such as from frame to ground on all four corners of the vehicle. Compare measurements to specifications.

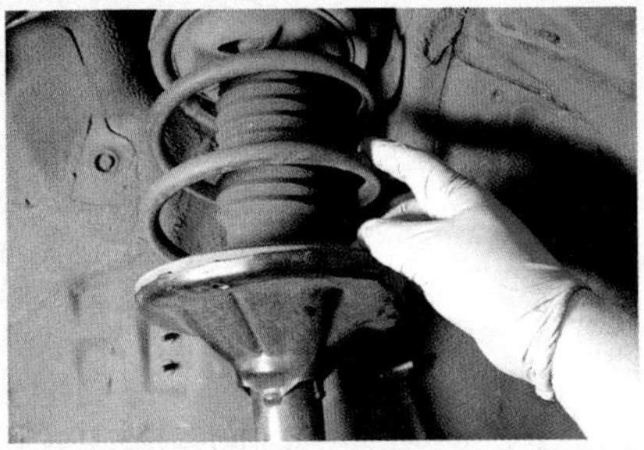

5. Inspect for bent components or a weak or broken spring if any measurements are not correct. If working with a torsion bar suspension, you may be able to adjust ride height to correct the condition.

SKILL DRILL 29-2 Checking Shock Absorbers

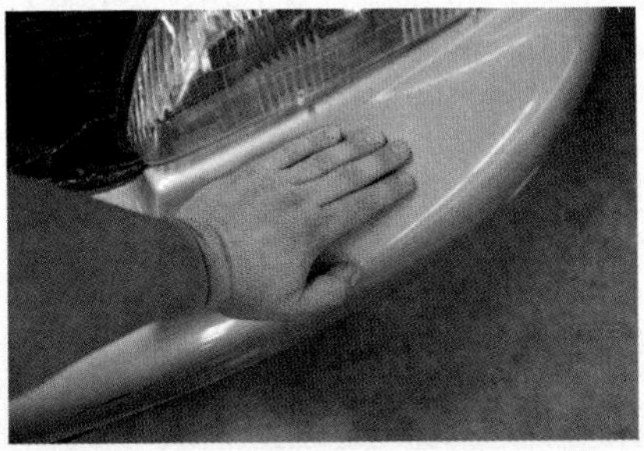

1. Place your weight on a bumper, and begin to bounce the vehicle until it reaches its maximum amount of travel produced by your weight. Stop bouncing at the bottom of the bounce. If the vehicle rebounds and compresses two or more times, replace the shock absorbers. If the shock absorbers are performing well, the vehicle will rebound once or one and a half times, then return to its original position.

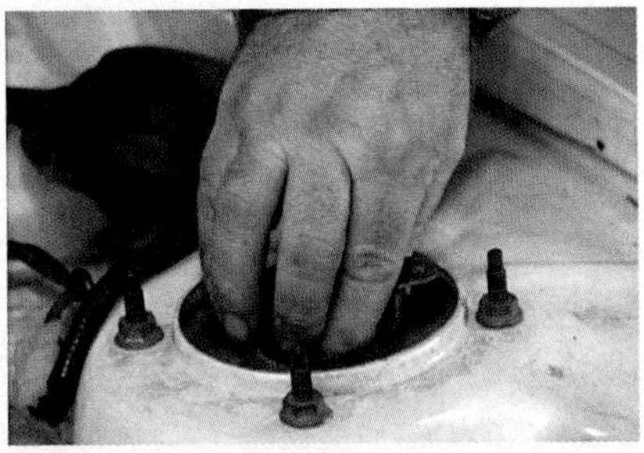

2. Pay particular attention to the top strut mounting during the bounce test. Place your hand on top of the mounting during the bounce test. Any noise or looseness in the mounting could indicate the need to replace the mount. While you are driving, the same test can be performed by stopping the vehicle suddenly from a very low speed. If the vehicle bounces up and down when coming to rest, you need to replace the shock absorbers.

3. Visually inspect the shock absorber mounting points for security and corrosion, and note any wet-looking patches on the sides of the shock absorbers. Slight dampness on the shock is typically normal. But a drip on the shock is not normal and indicates an excessive leak.

which can leak out. This causes an uneven ride height along with shock absorber performance issues.

Many of today's vehicles are equipped with a strut-type suspension instead of conventional shock absorbers. But testing either type of system involves the same procedure, a bounce test. While the vehicle is stationary, push up and down on a strong point at each corner of the vehicle several times. Make sure you don't push on the fenders as they can be dented. Watch how the vehicle responds after you release it. Typically, if you let go at the bottom, a good shock will allow the corner of the vehicle to rise and then settle back into position. On some vehicles with softer suspensions, it may allow the corner to rise, fall, and rise back into position. Any extra oscillations than that indicate worn shock absorbers.

Pay particular attention to the top strut mounting during the bounce test. Place your hand on top of the mount during the bounce test. Any noise or movement in the mounting could indicate the need to replace the strut mount. Also, have someone turn the steering wheel from lock to lock while feeling and listening to the top strut mount. If you do feel movement or hear noise, report it to your supervisor.

Visually inspect the shock absorber mounting points for security and corrosion. Note any wet-looking patches on the sides of the shock absorbers. A wet patch is a common indicator that the strut or shock absorber needs replacing because of a fluid leak. Slight dampness on the shock is typically normal. But drips on the shock are not normal and indicate an excessive leak.

To check shock absorbers, follow the steps in **SKILL DRILL 29-2**.

SAFETY TIP

To prevent personal injury, do not puncture or incinerate gas-charged shock absorbers.

▶ Unloading a Suspension to Measure Ball Joint Play

LO 29-04 Unload a suspension and measure ball joint play.

Play in the suspension system can be damaging to other components. It can also be a major safety hazard on the road. Testing for play requires the proper technique for the results to be accurate. For play to be measured, the joint must be unloaded. This means the joint cannot be under compression or tension forces. In the case of suspension ball joints, the joints cannot be supporting the weight of the vehicle when measuring the play.

The method of unloading the ball joints depends on the layout of the suspension. On a type 1 suspension, the coil spring or torsion bar pushes against the lower control arm. To unload the ball joint, a floor jack must be placed under the lower control arm and the wheel raised off the ground. The weight of the vehicle is thus supported through the spring to the control arm and then the jack. This leaves the ball joint only supporting the wheel, tire, spindle, and upper control arm. In this case, a pry bar can then be used to pry the wheel up and down while measuring the ball joint play. The upper ball joint can be tested by pushing in and out on the top of the tire. Any play in the joint is then measured and compared with specifications.

Some manufacturers specify that their type 1 suspensions be tested with the suspension system left hanging. The vehicle would be supported by the frame in this case. Manufacturers may also specify that only hand pressure be applied when checking for play in the joints. Check the service information before testing a particular vehicle.

On a type 2 suspension, the spring is on the top control arm. To unload the ball joint, it is best to fit a wooden block between the upper control arm and the frame. The block supports the control arm and spring in a position as close to normal ride height as possible. This method allows measurement of the play in the joint where maximum wear occurs. A pry bar can be used to pry the tire again so that play in the joint can be measured. The lower joint can be checked by rocking the bottom of the tire in and out.

In a MacPherson strut suspension, there is only a lower control arm. So, unloading the ball joint is performed by raising the vehicle by the frame and allowing the suspension to hang free. Place your hands at 12:00 and 6:00 on the tire. Rock the wheel in and out. Look for play in the lower ball joint. This approach tests the joint in a position it does not normally operate in. But if there is play, it will likely still be evident. Because it is a follower joint, most manufacturers say if it has any noticeable play, the joint must be replaced.

Be sure to test ball joints correctly. Refer to the manufacturer's instructions and specifications on how to test these joints. They are not all tested the same way and may require different tools.

To measure play, follow the steps in **SKILL DRILL 29-3**.

To measure play, follow the steps in **SKILL DRILL 29-4**.

To measure ball joint play in MacPherson Strut suspensions, follow the steps in **SKILL DRILL 29-5**.

Maintenance and Repair

▶ Removing, Inspecting, and Installing Stabilizer Components

LO 29-05 Replace stabilizer components and shock absorbers.

The stabilizer components (sway bar) help prevent body roll when cornering. The stabilizer bar itself rarely gives any trouble. But the rubber bushings or joints on the bar and links wear out. This usually results in increased body roll as well as a clunking noise in the suspension. The stabilizer components should be checked whenever there are handling concerns or suspension-related noises. Always refer to the service information for removing and inspecting the stabilizer components.

Note: We use the same vehicle to perform most of the SLA-related tasks. Also, we address each task as a sequence.

SKILL DRILL 29-3 Measuring Play: Loaded Lower Ball Joint

1. Place a floor or jack stand under the lower control arm.

2. Place a dial indicator on the lower control arm and vertically against the steering knuckle.

SKILL DRILL 29-3 Measuring Play: Loaded Lower Ball Joint (Continued)

3. Place a pry bar under the tire and pry it upward, watching the dial indicator reading as you pry and release. Record the total amount of movement in the joint and compare with the manufacturer's specifications.

4. Rock the tire in and out at the top, watching for any play in the upper ball joint, and compare with specifications.

SKILL DRILL 29-4 Measuring Play: Loaded Upper Ball Joint

1. Obtain a properly sized block of wood to fit between the upper control arm and the frame. Place the block of wood between the upper control arm and frame, so it is secure.

2. Place a dial indicator on the upper control arm and vertically against the steering knuckle.

3. Place a pry bar under the tire and pry it upward, watching the dial indicator reading as you pry and release. Record the total amount of movement in the joint, and compare with specifications.

4. Rock the bottom of the tire in and out, watching for any play in the lower ball joint, and compare with specifications.

This is as if you were doing the entire series of tasks, one after the other. So, the disassembly of each component assumes you have removed the previous components. Also, reassembly will be held off until all of the related tasks are completed.

To remove and inspect the stabilizer bar bushings and mount brackets, follow the steps in **SKILL DRILL 29-6**.

To remove and inspect the sway bar end links, follow the steps in **SKILL DRILL 29-7**.

SKILL DRILL 29-5 Measuring Play: MacPherson Strut

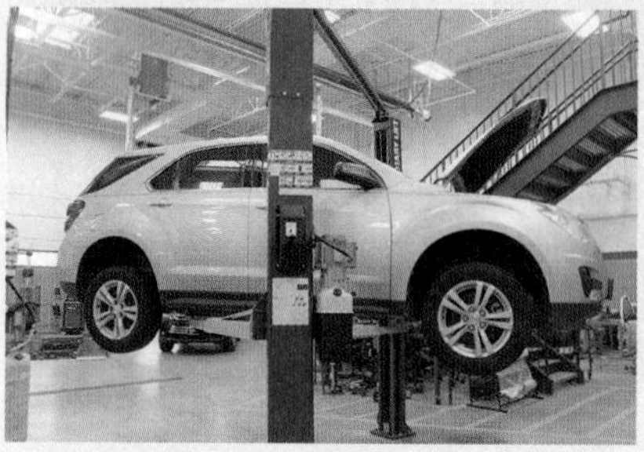

1. Raise and support the vehicle by the frame so that the suspension system hangs free.

2. Place hands on the tire at 12:00 and 6:00. Rock the tire in and out and feel for play in the lower ball joint.

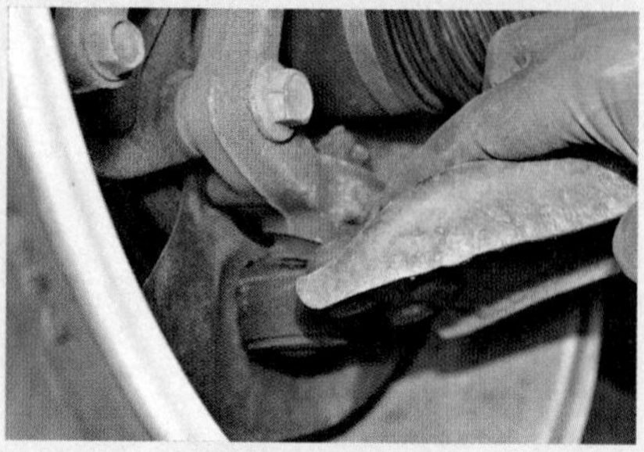

3. Watch the lower ball joint while rocking the tire. If there is any play, the joint likely needs to be replaced.

SKILL DRILL 29-6 Removing and Inspecting the Stabilizer Bar Bushings and Mount Brackets

1. Safely raise and support the vehicle. Remove the bolts holding the stabilizer bar bushing mount brackets, and inspect the brackets for cracks.

2. Remove the bushings by hand. Inspect the rubber in the bushings for cracks, brittleness, softness, or wear.

SKILL DRILL 29-7 Removing and Inspecting Sway Bar End Links

1. Safely raise and support the vehicle. Remove the nut holding the stabilizer bar link.

2. Remove the link by hand.

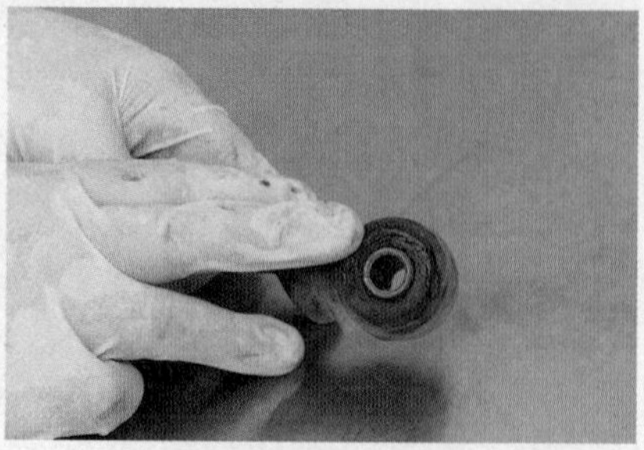

3. Inspect the links for damage. Inspect the rubber link grommets for cracks, softness, brittleness, or wear. Repeat with the other link.

Removing and Replacing Shock Absorbers

Worn shock absorbers cause the vehicle to ride poorly, especially on rough roads. When the tires encounter a bump in the road, a faulty shock absorber cannot dampen the spring oscillations. This causes a bouncy ride. The suspension continues to rebound and bounce. This action is then transferred to the vehicle frame and ultimately to the passengers.

> ▶ TECHNICIAN TIP
>
> Always replace shock absorbers in pairs so the suspension has the same characteristics for the left and right sides. Shock absorbers are rated for "bump" or "jounce," the rate at which they compress, and "rebound," the rate at which they expand.

Visually inspect the shock absorbers for any signs of oil leaking from the shaft seal. A slight oil film on the shock is considered normal. The mounting bushings must also be carefully inspected. Look for splits and deteriorated or missing rubber bushings. The mounting supports must be checked for damage and tightness.

Many shock absorbers come from the factory pressurized with gas. This reduces aeration of the fluid when operating. This pressure tends to expand the shock absorber, making it difficult to install. For this reason, pressurized shocks come compressed with a band holding them together. It is usually easiest to install one end or, in some cases, both ends of the shock before cutting the band.

> ▶ TECHNICIAN TIP
>
> Shock absorbers use rubber bushings to isolate them from the vehicle body. Always replace these bushings when replacing the shock absorbers.

On many suspension systems, the shocks provide the limit for full extension. This means the shocks may be holding the axle up when the shock is fully extended. In this case, removing the shock could cause the axle to slip, pinching fingers or causing the vehicle to shift on the hoist. It is always good practice to support the axle with stands while the shocks are being removed. If the top of the shock is held in by a stud and nut, you may have to use a wrench to hold the top of the stud while you unthread the nut.

If a shock absorber must be replaced, it is industry practice to replace them as pairs. This ensures the ride equilibrium of the vehicle. To replace a shock absorber, follow the steps in **SKILL DRILL 29-8**.

▶ Removing, Inspecting, and Installing SLA Suspension Coil Springs and Spring Insulators

LO 29-06 Remove coil springs and steering knuckles.

Coil springs absorb the road force by twisting, which compresses them. Whenever a driver complains of the way the vehicle sits or an issue related to the ride quality, the coil springs should be inspected. This should include measuring the vehicle's ride height.

Coil springs used in vehicles store very large amounts of energy. So, extreme caution is required when removing and installing them. Always use an adequately rated spring compressor that is in good working condition. Also, never stand in the direction the spring can fly, should something slip. This is generally in a position near where the tire sits, and beyond.

The spring can be mounted on either the lower control arm or the upper control arm. So, they must be supported in the specified way. Generally, the lower control arm will need to be supported with a floor jack. This will compress the spring, making installation of the spring compressor easier. Failure to follow service information can result in serious injury or death. In this skill drill, we work with the spring that is located on the lower control arm.

To remove and inspect SLA suspension system coil springs and spring insulators, follow the steps in **SKILL DRILL 29-9**.

SKILL DRILL 29-8 Replacing a Shock Absorber

1. Raise the vehicle on a lift, and support the axle/control arm with a jack stand. Remove the upper bolts holding the shocks in place.

2. Remove the lower bolts holding the shock in place.

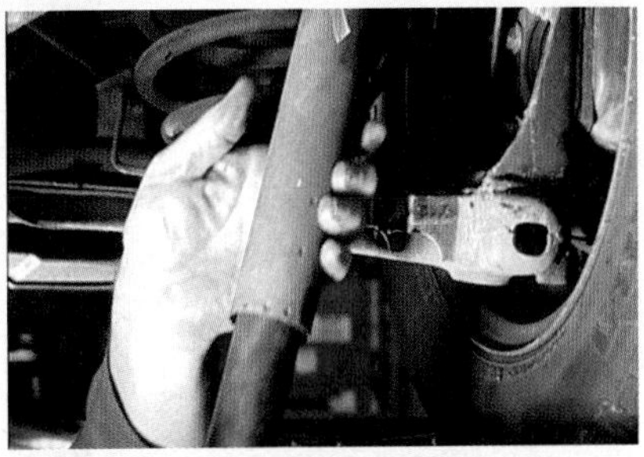

3. Pull the shock out by hand. Replace with new shock. Repeat on the other side.

SKILL DRILL 29-9 Removing and Inspecting SLA Suspension System Coil Springs and Spring Insulators

1. Perform the previous Skill Drills. Place a jacking device under the lower control arm that is being dismantled; maintain pressure on the lower control arm.

2. Place a spring compressor on the inside or outside of the coil spring to keep the pressure of the spring contained.

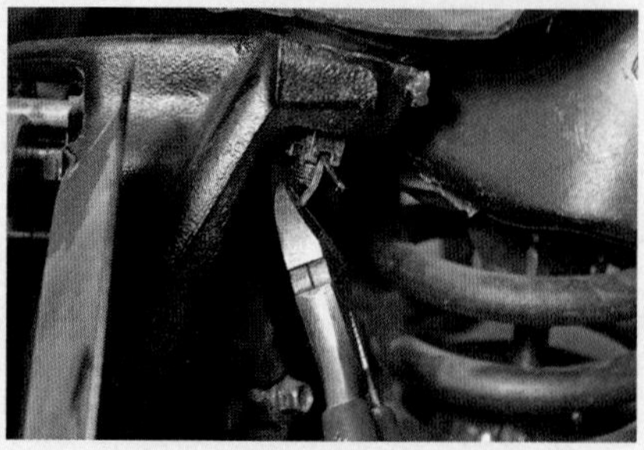

3. Remove the cotter pin, and loosen the nut from the upper ball joint.

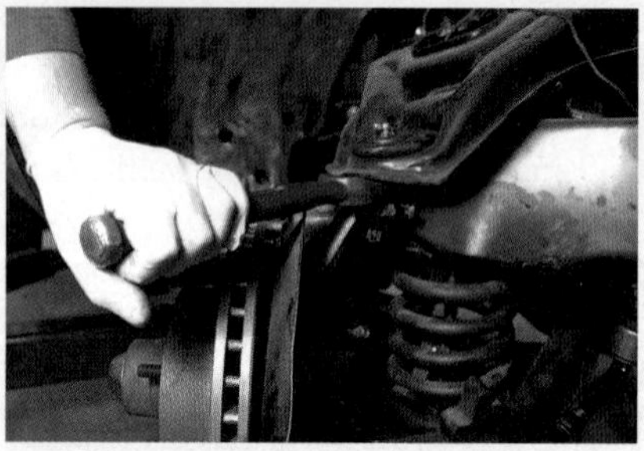

4. Using a ball joint separator or pickle fork, break loose the ball joint stud from the steering knuckle, and remove the nut.

5. Slowly and carefully lower the hydraulic jack supporting the lower control arm, to allow the coil spring to be removed. (Extreme caution must be used here so the spring compressor does not slip and injure you.)

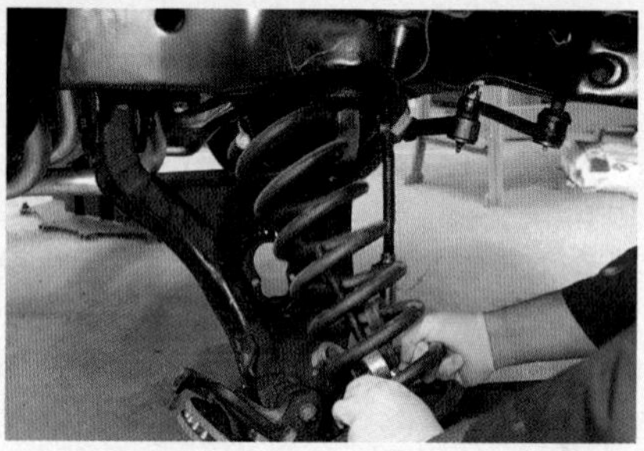

6. Release the spring compressor slowly. Clean and inspect the ball joint stud hole in the steering knuckle boss.

Removing, Inspecting, and Installing the Steering Knuckle Assembly

The steering knuckle serves as the pivot point for the wheels. Because they rotate on ball joints or tie-rod ends, they are not subject to much wear. But they can be damaged in an accident, requiring replacement. It is more likely that you will have to remove one in order to replace ball joints or control arms. This skill drill assumes a steering knuckle removal on the vehicle in the previous skill drill. Remember, coil springs are under extreme pressure! Follow the service information to correctly remove any steering knuckle.

To remove and inspect a steering knuckle assembly, follow the steps in **SKILL DRILL 29-10**.

SKILL DRILL 29-10 Removing and Inspecting a Steering Knuckle Assembly

1. Perform the previous skill drills. Disconnect the lower ball joint in the same manner as the upper ball joint.

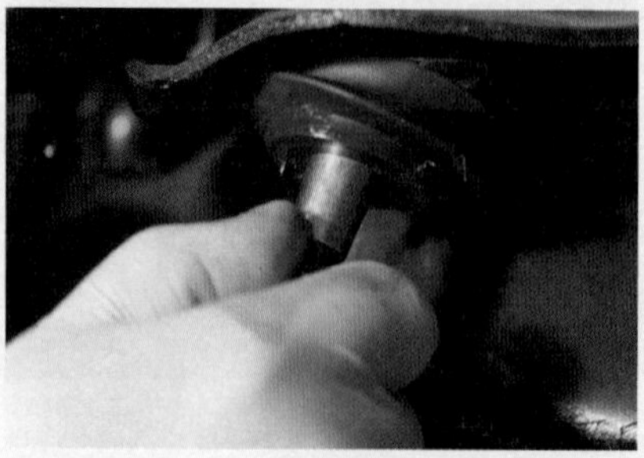

2. Remove the steering knuckle from the ball joint stud. Inspect the tapered portion of the stud for any bright, worn metal, which would indicate a worn steering knuckle boss.

3. Inspect the steering knuckle for any signs of wear or visible signs of bending or twisting.

▶ Removing and Inspecting Upper and Lower Control Arms and Components

LO 29-07 Remove control arms and ball joints.

Control arms themselves do not generally wear out, but the control arm bushings and ball joints do. So, you will likely have to remove them to service those components. However, some manufacturers do not sell bushings and ball joints separately. You may have to purchase the entire control arm and replace it as a unit. Also, control arms can become bent or damaged as the result of a collision or hitting a deep pothole. This means that control arms do need to be inspected for damage.

Finally, some customers lift or lower their vehicles. This involves modifying suspension parts. For example, switching out springs and control arms with new upgraded ones. This changes the ride height, alignment, and handling of the vehicle. Only very experienced and highly trained technicians should diagnose and repair these vehicles.

The vehicle used in this skill drill has a coil spring on the lower control arm. Not all vehicles are the same. Refer to the manufacturer's service information to correctly service the suspension system.

To remove, inspect, and install upper and lower control arms and components, follow the steps in **SKILL DRILL 29-11**.

SKILL DRILL 29-11 Removing, Inspecting, and Installing Upper and Lower Control Arms and Components

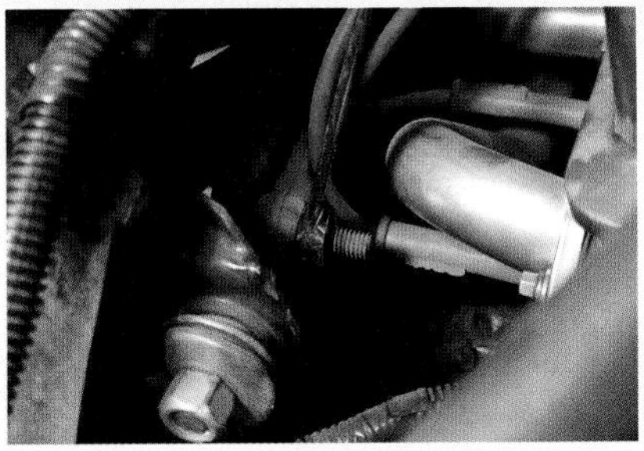

1. Perform the steps in previous skill drills. Loosen the bolts and remove the upper control arm.

2. Remove the lower control arm by removing the bolts holding the control arm in place. As you dismantle the system components, inspect the parts being removed to determine whether you need to order replacement parts.

3. Inspect the upper and lower control arm for wear or damage. If the control arm bushings show signs of excessive wear or deterioration, they must be replaced. Worn control arm bushings can be replaced with a hydraulic press or with an air chisel.

4. *Press method:* Place the control arm in a hydraulic press supported, so the bushing is level to the press head. Using the correct-sized adapter, press the bushing out.

5. Using the press, install the new bushing into the control arm. Be careful to start the bushing in straight, or you could bend the control arm. Press it in until the flange bottoms out.

6. *Air chisel method:* Install the control arm in a vise. Use an air chisel to work the bushing out. Be careful not to gouge the mating surface of the control arm.

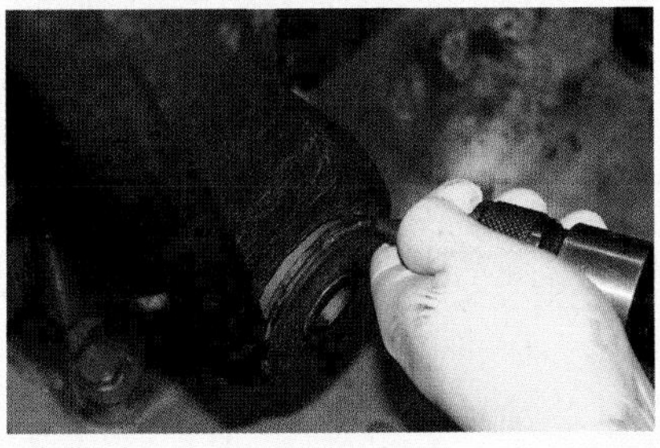

7. Using the air hammer, drive the new bushing into place, making sure it is fully seated.

▶ TECHNICIAN TIP

Remember that modifying a vehicle by raising or lowering the vehicle may be illegal in some states. It will also affect alignment angles.

Removing, Inspecting, and Installing Upper and Lower Ball Joints

The ball joint connects the upper and lower control arms to the steering knuckle. Over time, these parts wear and must be replaced. Removal of ball joints varies from vehicle to vehicle. Refer to the manufacturer's service information for the correct procedures. The below skill drill uses an SLA coil-spring suspension with two control arms. It is the more dangerous type that a technician services.

To remove, inspect, and install upper and lower ball joints, follow the steps in **SKILL DRILL 29-12**.

▶ Install SLA Suspension Components

LO 29-08 Install and lubricate SLA components.

Now that the individual components have been inspected and repaired, it is time to reinstall them. This requires attention to detail. All of the fasteners need to be properly torqued. New cotter pins need to be installed where required. And the spring needs to be handled with care. Make sure the spring is securely held by the spring compressor. They can crush fingers and bones if they slip. They can also pop out of place and shoot into you. Treat a compressed spring like a potential bomb.

Take care to route brake lines and hoses properly. It is easy to install them so they rub on a tire or suspension component. This can wear a hole in them over time, causing brake failure. And as always, follow the service information for the vehicle you are working on.

To reinstall SLA suspension components, follow the steps in **SKILL DRILL 29-13**.

Lubricating Steering and Suspension Systems

To function properly, the suspension and steering system must be lubricated. Lubrication keeps the parts from wearing on one another, which extends their life. On most modern vehicles, the joints are sealed and lubricated for life. This means there may be no grease fittings present. However, some vehicles do have grease fittings installed, either from the factory or on aftermarket parts. If grease fittings are present, then lubrication must be performed. Add enough grease to see the seal or rubber boot rise slightly. Under no circumstances should you overfill a lubricated joint with grease. Doing so can rupture the seal or rubber boot or bellows.

To lubricate a suspension system and a steering system, follow the steps in **SKILL DRILL 29-14**.

▶ TECHNICIAN TIP

Clean lubricating equipment very carefully. If you don't thoroughly clean the fitting or nozzle before pumping the grease into the fitting, dirt could be forced into the component. Any dirt entering a component will cause premature failure.

▶ Inspecting the Strut Cartridge or Assembly

LO 29-09 Inspect and service strut assembly.

Struts operate under harsh conditions. Because of this, they do wear out, cause customer concerns, and need to be inspected. Worn struts commonly cause cupped tire wear and ride comfort

SKILL DRILL 29-12 Removing, Inspecting, and Installing Upper and Lower Ball Joints

1. Remove the control arms as in the previous skill drills. Ball joints are held into the control arm either by a press fit or by rivets or bolts.

Press-fit style: Support the control arm with the correct adapter so there is room for the ball joint to be pressed out of it.

2. Press the joint out of the control arm.

3. Inspect the bore in the control arm to make sure it isn't cracked or damaged.

SKILL DRILL 29-12 Removing, Inspecting, and Installing Upper and Lower Ball Joints (Continued)

4. Turn the control arm over and support it, so the ball joint has room to be installed.

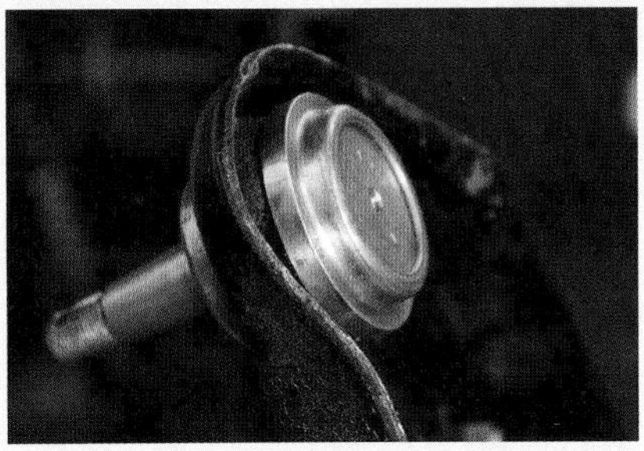

5. Press the ball joint into the control arm until it bottoms out.

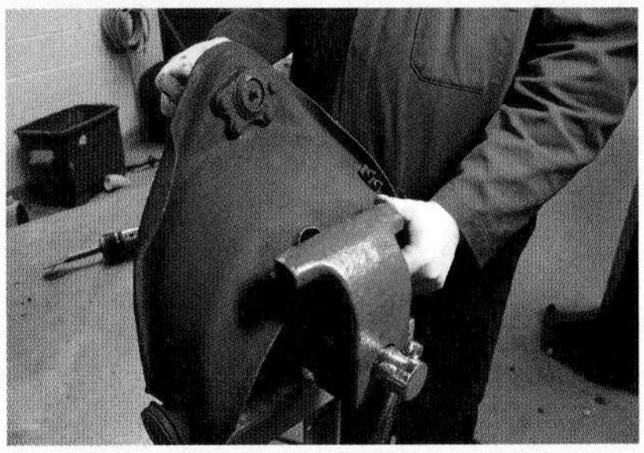

6. *Riveted style:* Mount the control arm in a vise with the rivet heads facing up or outward.

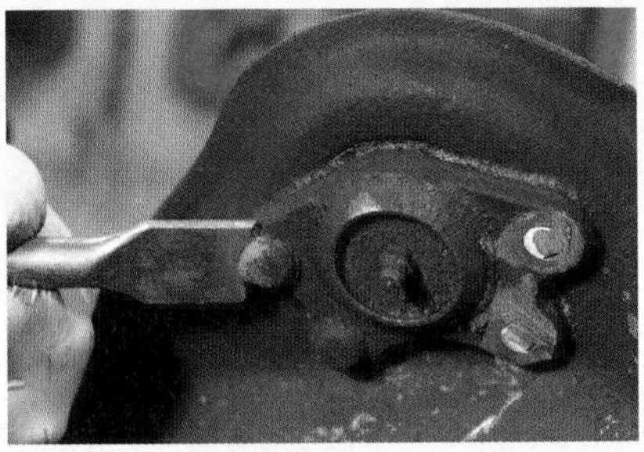

7. Use an air chisel to chisel off the rivet heads, or use a drill to drill approximately three-quarters the diameter of the rivet head. Carefully drill the head of the rivet off. Use a hammer and a punch to tap the rest of the rivet out of the ball joint and control arm.

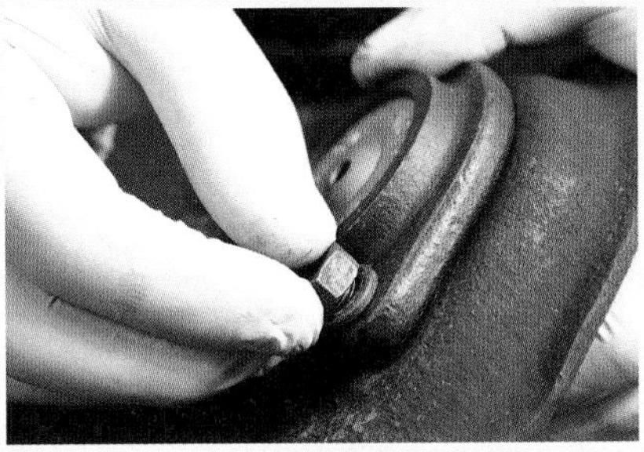

8. Inspect the holes, and remove any burrs with a file. Install the new ball joint with nuts and bolts, and tighten to the proper torque.

issues. They can also cause unusual noises, primarily when going over bumps or around corners.

Testing struts is similar to testing regular shock absorbers. Using your hand, push down on one corner of the front bumper continuously until the front of the vehicle is bouncing. When at the bottom, remove your hand and stop pushing. In most vehicles, the corner of the vehicle will rise and then settle into place. A vehicle with softer suspension may rise, fall, and rise to its settling point. If the vehicle bounces more than that, the strut needs to be replaced.

Most struts are one-piece units that include the shock absorber in the strut assembly. This type of strut is replaced as

SKILL DRILL 29-13 Reinstall SLA Suspension Components

1. Install the lower control arm in position, and wait until fully assembled to torque the bolts to specification.

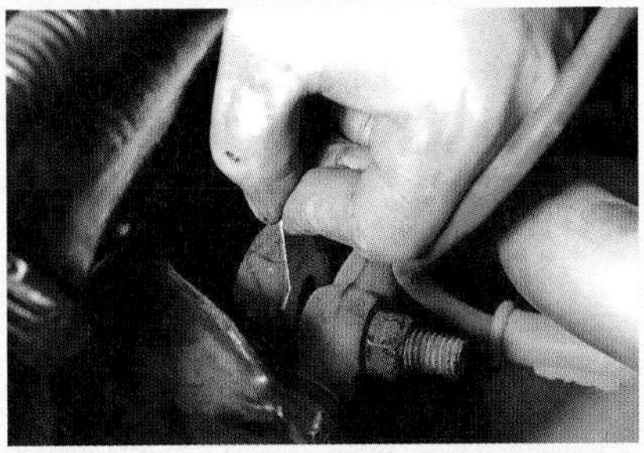

2. Install the upper control arm in position, and screw in the bolts finger tight. Insert shim packs, if equipped. Wait until fully assembled to torque the bolts to specifications.

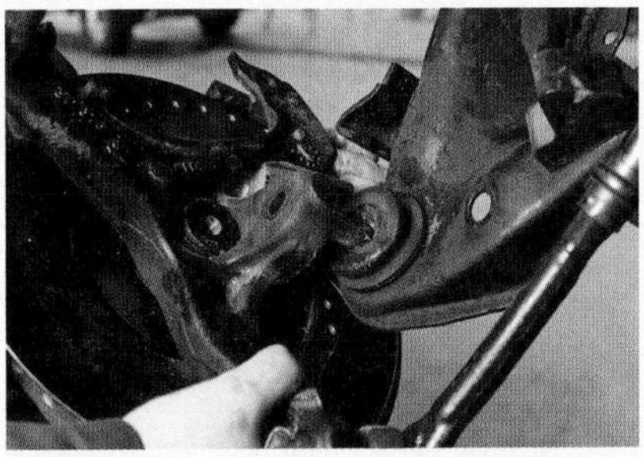

3. Place the steering knuckle in place on the lower ball joint stud. Finger-tighten the ball joint nut.

SKILL DRILL 29-13 Reinstall SLA Suspension Components (Continued)

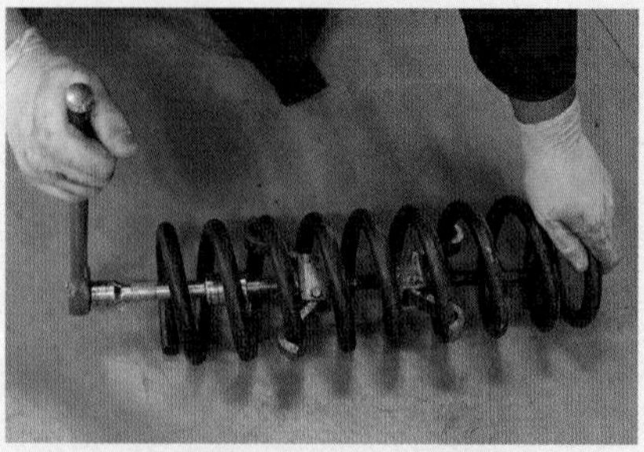

4. Install the spring compressor on the spring so that it can be installed in the vehicle between the lower control arm and the frame.

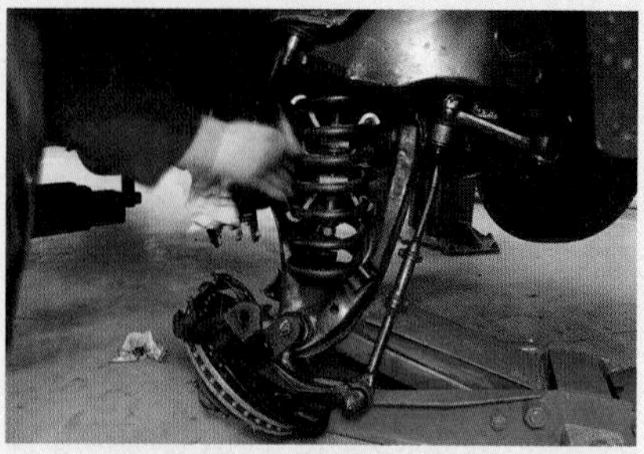

5. Position the spring on the lower control arm, and use a jack to hold it in place.

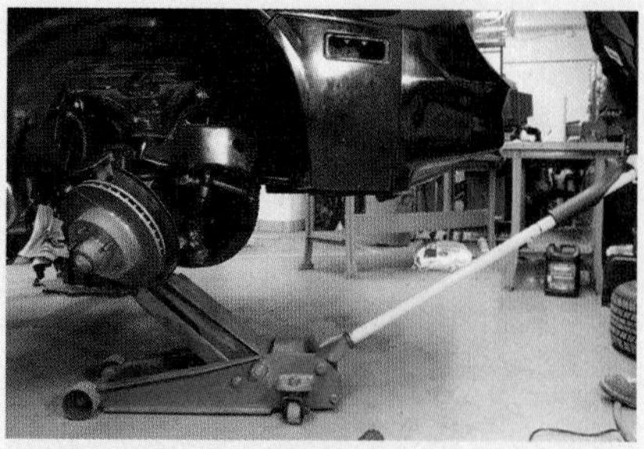

6. Continue to raise it while guiding the steering knuckle over the upper ball joint stud. Install the nut on the upper ball joint stud finger tight.

7. Use a cotter pin tool to turn the ball joint stud to align the cotter pin hole lengthwise to the vehicle so the cotter pin will be easy to install.

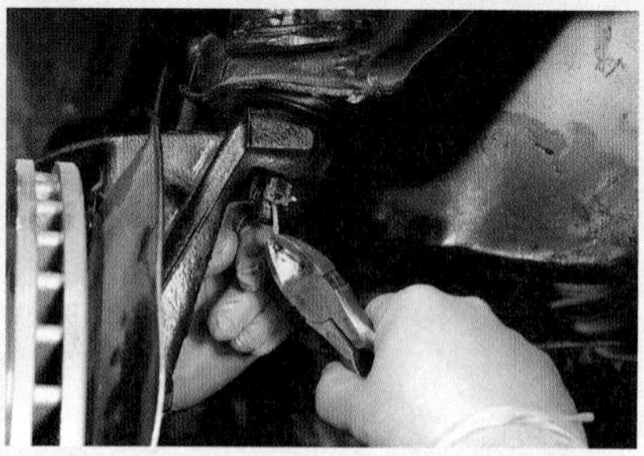

8. Torque both ball joints to specifications, and install the cotter pins.

9. Remove the spring compressor.

10. Install the new washers and bushings onto the new shock, if not already installed. Place the shock in the proper position by hand, and install the other bushings, if needed.

11. Install and properly torque the bolts or nuts to hold the shock in place using a torque wrench. Install the tie-rod end into the steering knuckle, torque the nut, and secure with a cotter pin.

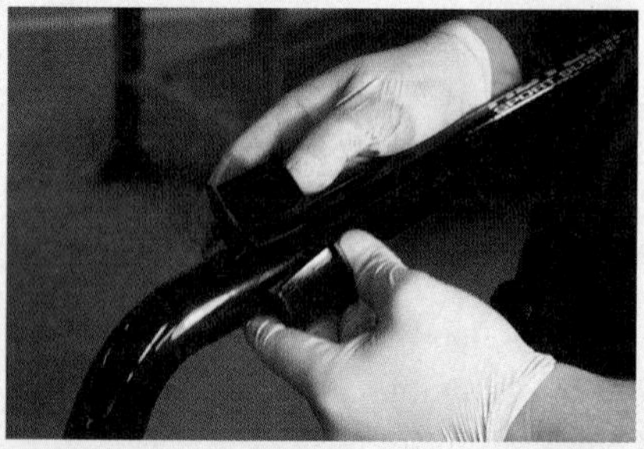

12. Install the new stabilizer bar bushings, placing the bushing around the bar.

13. Install bolts in the mounting bracket, and tighten to specifications.

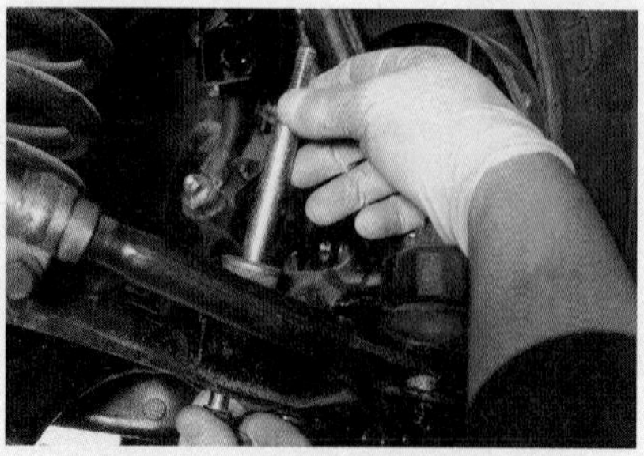

14. Install the stabilizer link to the control arm by assembling it onto the stabilizer bar and control arms with the spacer, washers, and rubber grommets installed in their proper places.

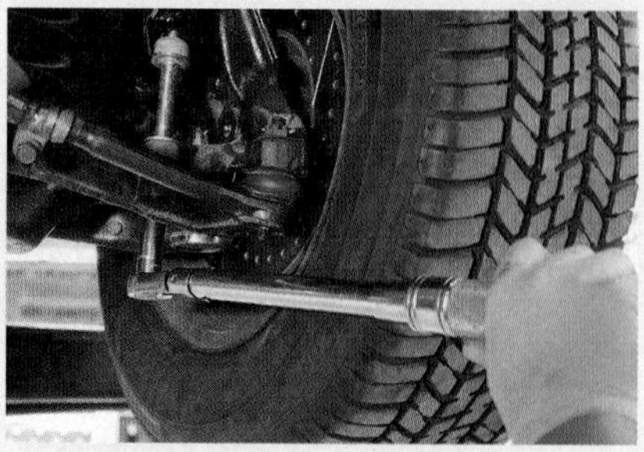

15. Install the nut on the threaded end of the link, and then tighten to the manufacturer's specifications.

SKILL DRILL 29-13 Reinstall SLA Suspension Components (Continued)

16. Reinstall the rotor assembly if removed.

17. Reinstall the brake caliper by placing the caliper over the rotor and tightening to the proper torque.

18. Jack up the lower control arm so the spring is compressed to its normal height, and properly torque the control arm bolts to specifications.

19. Install the wheel and snugly tighten the lug nuts. Lower the vehicle, and torque the wheel nuts to specifications.

a unit. Other vehicles have a strut housing with a replaceable strut insert cartridge. In this type of strut, some manufacturers fill the housing with a lightweight oil. This allows heat from dampening to be transferred to the outer housing more efficiently. This oil, along with the fluid inside the strut cartridge, can leak out as well. So be aware of that as you inspect the strut. To inspect the strut cartridge or assembly, follow the steps in **SKILL DRILL 29-15**.

On front wheels, the upper strut bearing wears out or can become damaged. So, it needs to be inspected periodically. This can be done by having someone turn the steering wheel with the vehicle on the ground. At the same time, feel and listen for roughness, popping, or looseness coming from the bearing. Also, safely raise the vehicle and inspect the mount for broken or damaged rubber bushings. To inspect the upper strut bearing mount, follow steps in **SKILL DRILL 29-16**.

SKILL DRILL 29-14 Lubricating a Suspension System and a Steering System

1. Check the service information to determine where the grease points are and the type of grease required. Also, look to see if any aftermarket grease fittings are installed. If so, grease them.

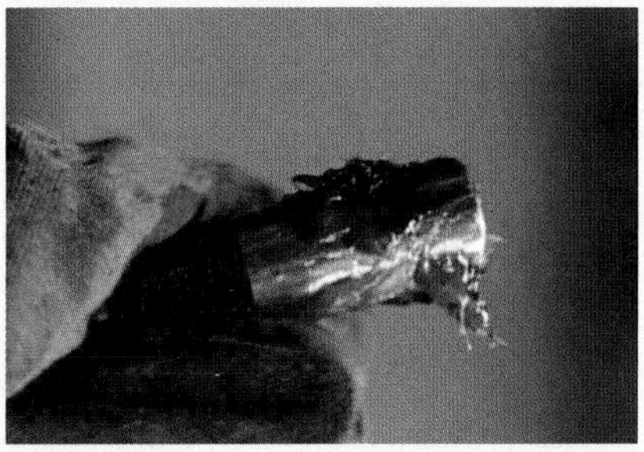

2. Clean each of the lubrication fittings and the grease gun nozzle by wiping all of them with a clean rag. You may need to remove a component's plugs and temporarily install a lubrication fitting. After the component has been lubricated, reinstall the original plug.

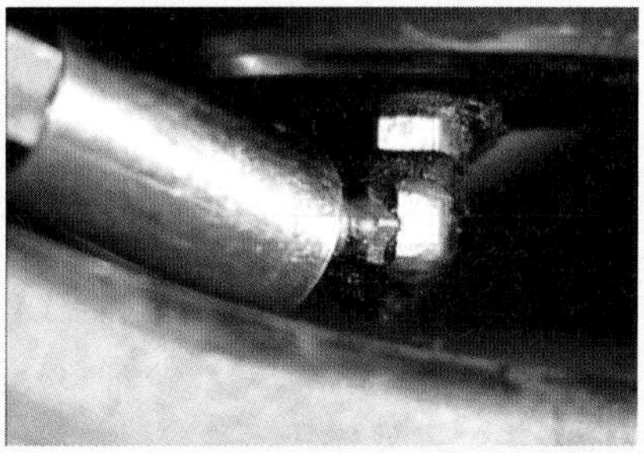

3. Push the grease gun nozzle fully over the fitting. It should snap into place. Add enough grease to see the seal or rubber boot rise slightly. Do not overfill a lubricated joint with grease.

SKILL DRILL 29-14 Lubricating a Suspension System and a Steering System (Continued)

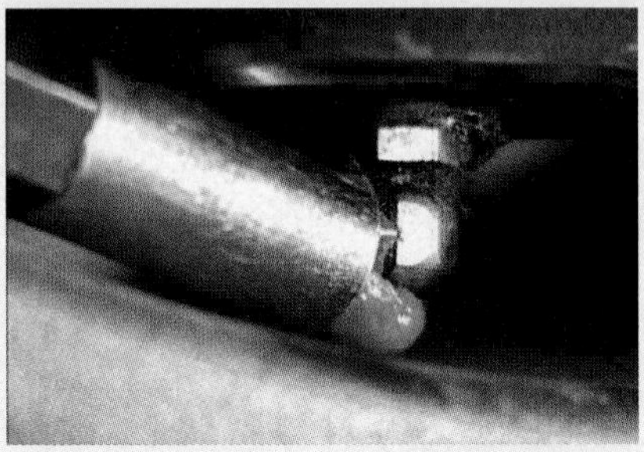

4. If the fitting is clean and will not take grease, remove the grease zerk, and check for blockage. If found, the fitting must be replaced with a new fitting of the same size and angle, and the joint relubricated.

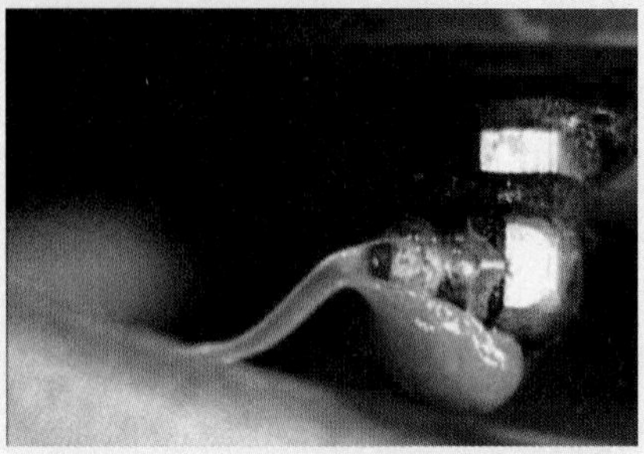

5. Remove the nozzle from the fitting and wipe away any excess grease from it. Repeat the procedure until all the appropriate joints have been lubricated.

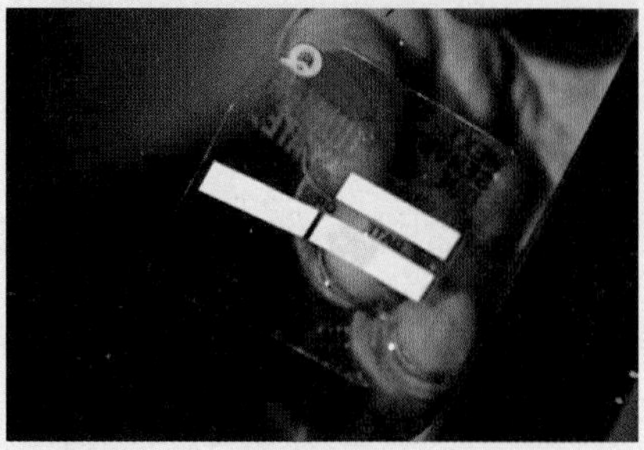

6. Attach a static cling sticker to the windshield, or reset the maintenance reminder system. Lower the vehicle and remove it from the lifting device.

SKILL DRILL 29-15 Inspecting the Strut Cartridge or Assembly

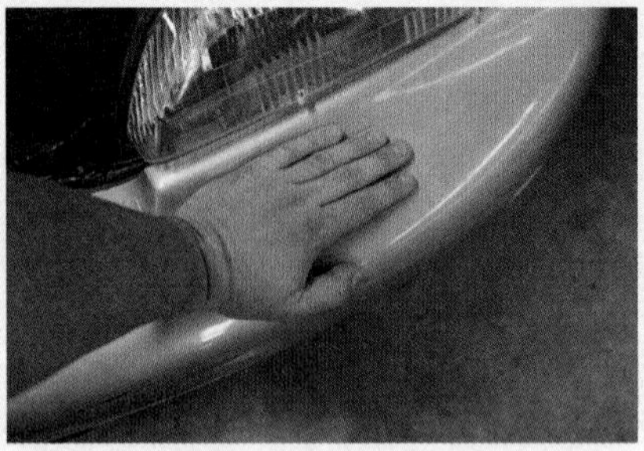

1. Bounce test each strut and check for lack of dampening, binding, or unusual noises.

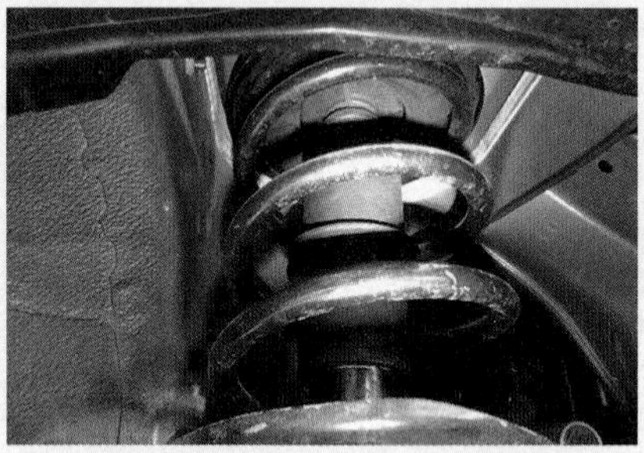

2. Inspect the strut assembly for damage.

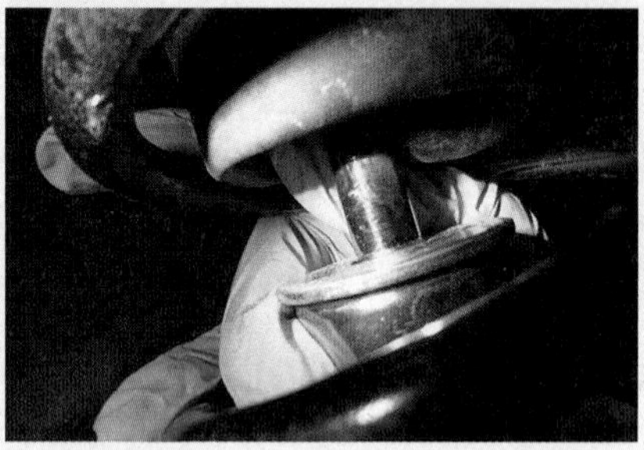

3. Inspect the top of the strut cartridge for leaks (a small amount of seepage is allowable for some vehicles).

SKILL DRILL 29-16 Inspecting the Front Strut Bearing and Mount

1. With the vehicle on the ground, inspect the upper strut bearing mount for damage or wear.

2. With your hand safely on the top of the strut mount (not in any holes), have an assistant turn the steering wheel. Feel for any roughness.

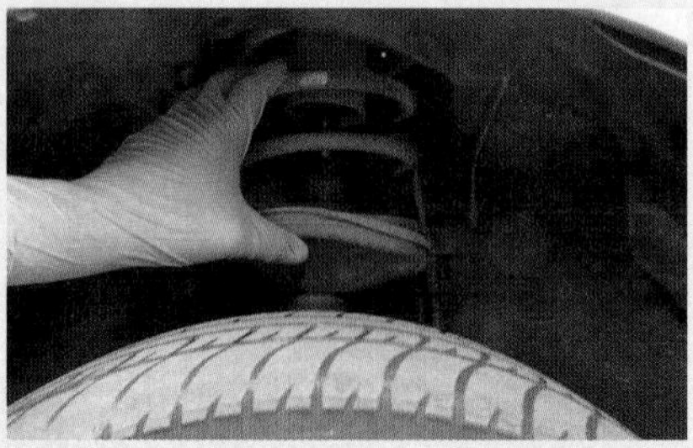

3. Lift the vehicle so that the weight is off the suspension. Inspect the upper strut mount for torn bushings or insulators. Also, inspect the spring and any insulators.

Removing, Inspecting, and Installing the Strut Cartridge or Assembly

Most vehicles have a one-piece strut that bolts into the knuckle on the lower end. The strut includes the shock absorber. The spring seat is located on the upper end of the strut. In a strut with an internal strut cartridge, some manufacturers fill the housing with a lightweight oil. This allows heat from dampening to be transferred to the outer housing more efficiently. Follow the service information regarding the type and amount of oil to use when replacing inserts.

Some parts suppliers are supplying full strut assemblies with the spring and upper strut bearing already installed. This means that the strut has all new parts and is quicker to install. The customer can get back on the road much faster this way. Be sure to follow the service information for the vehicle you are working on when replacing struts. The below Skill Drill uses a vehicle with a replaceable one-piece strut.

To remove the strut assembly, follow the steps in **SKILL DRILL 29-17**.

Removing, Inspecting, and Installing a Strut Coil Spring, Insulators, and the Upper Strut Bearing Mount

To remove, inspect, and install a strut coil spring, insulators, and the upper strut bearing mount, follow the steps in **SKILL DRILL 29-18**.

SAFETY TIP

Strut assemblies contain a coil spring that is held under compression. Always use a spring compressor to release the spring pressure before disassembly.

▶ Inspecting Strut Rods and Bushings

LO 29-10 Inspect strut rods and bushings, leaf springs, and torsion bars.

The strut rods and bushings hold the control arms in position longitudinally. The strut rod bushings wear and degrade over time, requiring replacement whenever they are loose. Since they are exposed, they are subject to damage. Bent strut rods are also fairly common. So inspect them for damage. On some vehicles, the strut rod is the adjustment point for caster. An alignment may have to be performed if the bushings get replaced. Refer to the manufacturer's service information to verify if this is needed.

To inspect strut rods and bushings, follow the steps in **SKILL DRILL 29-19**.

Inspecting Leaf Springs

The leaf springs need to be inspected whenever the vehicle's ride height does not meet the manufacturer's specifications. If found to be sagging, bent, or broken, they will need to be replaced.

SKILL DRILL 29-17 Removing the Strut Assembly

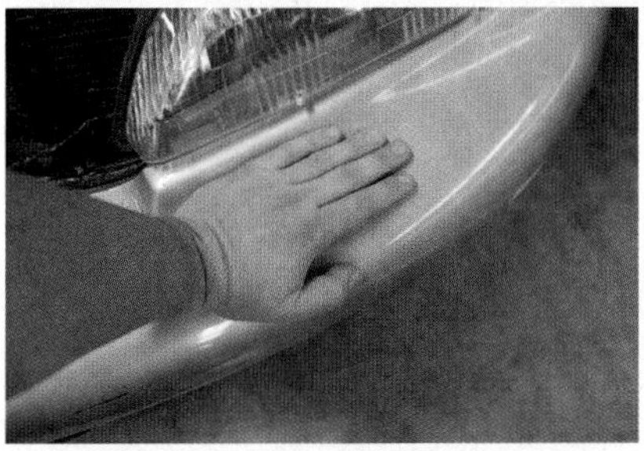

1. Bounce test each strut and check for lack of dampening, binding, or unusual noises.

2. Loosen the lug nuts one turn if removing by hand.

3. Safely raise and support the vehicle on a hoist or on jack stands.

4. Remove the lug nuts and then take off the wheels.

5. Remove the disc brake caliper, and wire it so it does not hang by its brake hose.

6. Remove the bolts holding the top of the strut to the strut tower (not the center strut bolt). Also remove the lower bolt/s and remove the strut assembly.

SKILL DRILL 29-18 Removing, Inspecting, and Installing a Strut Coil Spring, Insulators, and the Upper Strut Bearing Mount

1. Remove the strut from the vehicle, following the previous skill drill.

2. Place the strut into the strut compressor. Adjust the lower arms of the strut compressor so that the spring is supported properly.

3. Properly place the upper arms of the strut compressor onto the spring.

4. Slowly turn the handle of the strut compressor, and compress the spring. Ensure that the strut compressor arms are properly placed and are not sliding.

5. Once the spring is compressed enough that the strut is free from spring pressure, loosen and remove the center strut nut.

6. Remove the old strut from the spring.

7. Remove from the old strut any strut bellows, bumpers, spring seats that need to be transferred to the new strut. Install them on the new strut.

8. Install the new strut into the spring. Reinstall the upper strut mount in the correct position.

9. Install the upper washer and nut. Torque to the manufacturer's specification. Slowly release the spring compressor, and reinstall the strut into the vehicle.

Inspection is also necessary if noise is found to be coming from the leaf spring or bushings. If new leaf spring bushings will be needed, the use of proper press tools will ensure proper bushing replacement. Remember that when the leaf springs are removed, the axle will not be supported. The use of screw jacks to support the axle during removal will be necessary.

To inspect leaf springs, follow the steps in **SKILL DRILL 29-20**.

Inspecting Torsion Bar Suspension

The torsion bar provides the spring action in a torsion bar suspension system. In many cases, it is adjustable to allow for corrections in ride height. It has a large adjustment bolt in the torque arm. When adjusted, it changes the amount of pressure it places on the control arm to raise or lower the suspension. The torsion bar should be checked whenever the driver complains of suspension problems.

Torsion bars are typically stamped or marked with a left or right and must be installed in the proper side. If it has marks for left or right, then it is preset for twisting in that specific direction. It will not have the same spring force if twisted the opposite way.

When replacing torsion bars, if the torsion bars are not marked, then be sure to place a paint mark on them to identify left and right. They have been subjected to torsion in one direction for the life of the vehicle. Installing them backward can

SKILL DRILL 29-19 Inspecting Strut Rods and Bushings

1. Safely raise and support the vehicle on a hoist. Inspect the strut rods and bushings for wear and cracking.

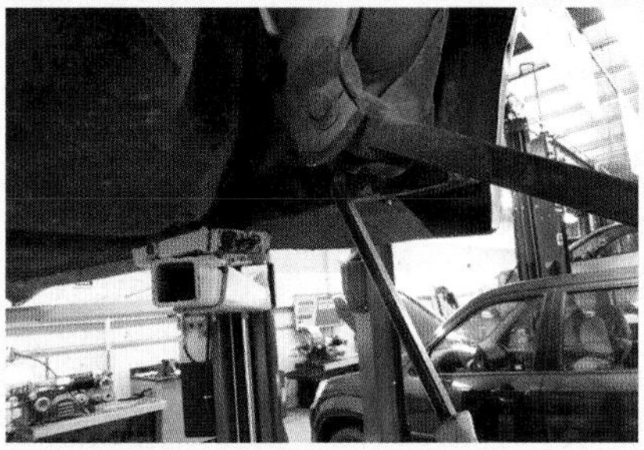

2. Pry the strut rod front to back and check for excessive looseness.

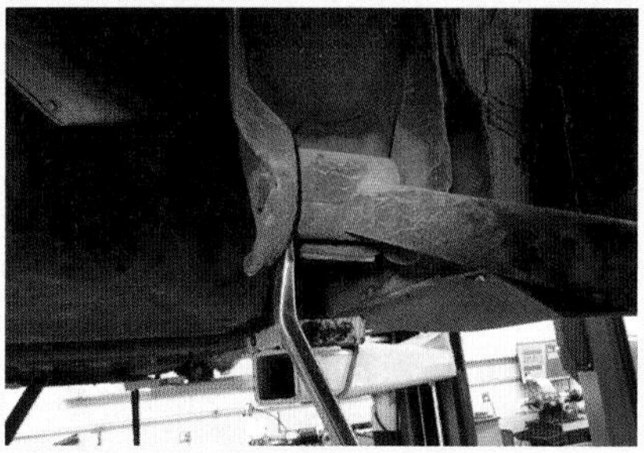

3. Pry the strut rod side to side and check for excessive looseness.

SKILL DRILL 29-20 Inspecting Leaf Springs

1. Safely raise and support the vehicle on a lift. Support the rear axle with tall screw jack stands. Check to see if any of the leaves are cracked or broken and that the noise deadening inserts are positioned correctly between the leaves.

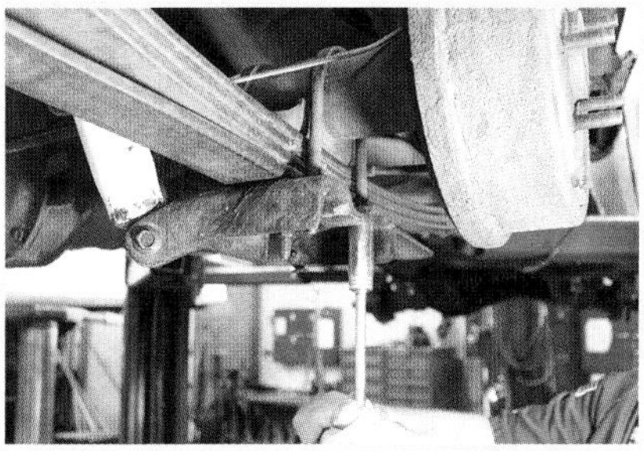

2. Test the security of the spring center bolt, and make sure the U-bolts are tight.

3. Check the condition of the bushings and the spring shackles by placing a lever between the frame and the eye of the spring and prying against the spring.

SKILL DRILL 29-21 Inspecting Torsion Bar Suspension

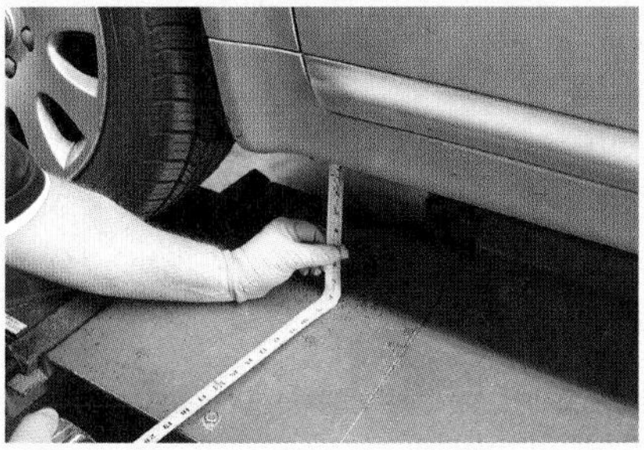

1. Measure the ride height and compare with specifications.

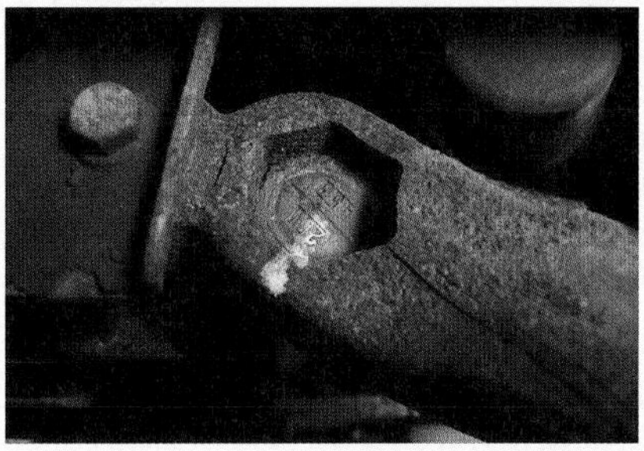

2. Safely raise and support the vehicle on a hoist. Inspect the torsion bars for damage and excessive rust.

3. If the ride height was out of specifications, readjust the torsion bars if possible. If still out of spec, inform your supervisor.

result in breakage or incorrect spring rate. The torsion bar may be found on either the lower or upper control arm on certain vehicles.

When inspecting torsion bars, look carefully for signs of excessive rust. As torsion bars rust, they weaken. Also watch for nicks in the torsion bar. Nicks create a weak spot that can result in breakage of the torsion bar. The above skill drill will not be the same for all vehicles. Follow the service information for the vehicle you are working on for proper operation of this task.

To inspect the torsion bar, follow the steps in **SKILL DRILL 29-21**.

▶ Wrap-Up

Ready for Review

- Servicing suspension systems requires the use of the following special tools: dial indicators, pry bars, measuring tapes, pitman arm puller, tie-rod end puller, pickle forks, electronic stethoscope, coil spring compressor, scan tool, ball joint press tool, air chisel, strut compressor, strut servicing kit, universal strut nut wrench kit, 24-mm strut rod socket, and various lift devices.
- Suspension system diagnosis follows a strategy-based diagnostic process that includes: verifying the customer's concern, researching possible faults and gathering information, focused testing, performing the repair, and verifying the repair.
- To measure the ride height of a vehicle, it should have matching tires that are properly inflated and no additional weight is present in the vehicle. Any time ride height is changed, a wheel alignment also must be performed.
  - Strut-type suspension or shock absorbers can be tested with a bounce test.
- The method of unloading the ball joints depends on the layout of the suspension.
  - On a type 1 suspension, a floor jack must be placed under the lower control arm and the wheel raised off the ground. On a type 2 suspension, it is best to fit a wooden block between the upper control arm and the frame.
- The stabilizer components should be checked whenever there are handling concerns or suspension related noises. If a shock absorber must be replaced, it must be replaced as pairs to ensure the ride equilibrium of the vehicle.
- An adequately rated spring compressor in good working condition must be used to remove and install coil springs. Failure to follow service information can result in serious injury or death.
  - The steering knuckle serves as the pivot point for the wheels and are not subject to much wear. But they can be damaged in an accident, requiring replacement.
- When control arm bushings and ball joints wear out, it may be necessary to replace the entire control arm as a unit. Refer to the manufacturer's service information for the correct procedures of removal of ball joints as the procedure varies from vehicle to vehicle.
- On most modern vehicles, the suspension and steering system joints are sealed and lubricated for life. If grease fittings are present, enough grease must be added to see the seal or rubber boot rise slightly.
- Testing struts is similar to testing regular shock absorbers using a bounce test. Struts that are one-piece units that include the shock absorber in the strut assembly are replaced as a unit.
- The strut rod bushings wear and degrade over time and require replacement whenever they are loose. Leaf springs need to be inspected whenever the vehicle's ride height does not meet the manufacturer's specifications. Screw jacks must be used to support the axle when leaf springs are removed. When inspecting torsion bars, look carefully for signs of excessive rust.

Key Terms

Bouncy ride/excessive body movement Up and down movement that isn't being controlled by the shock absorbers/struts.

bump steer The undesired condition produced when hitting a bump, where the vehicle darts to one side as the steering linkage is pushed or pulled as a result of the travel of the suspension.

hard steering When it is difficult for the driver to steer the vehicle. There may be a problem in the power steering system, too much positive caster, or binding in the steering system.

pull Felt when the driver feels the steering wheel wanting to go to one side. It can mean the air pressure in one or more of the front tires is low. A simple tire pressure gauge is used to check the tire pressures.

Shimmy/shake Vibration or wiggle in the steering or suspension system.

Steering return concerns Steering that catches, binds, or doesn't return properly.

torque steer A condition in which the vehicle pulls to one side during hard acceleration. It can be the result of unequal axle lengths as designed, which cannot be repaired if it is by design.

vehicle wander Occurs when the vehicle is not driving in exactly the direction the driver is steering it. This tends to happen when the caster angle is off or there is looseness in the steering/suspension components.

Review Questions

1. What tool is used to measure play in ball joints?
   a. Measuring tape
   b. Micrometer
   c. Dial indicator
   d. Stethoscope
2. What commonly causes bump steer?
   a. A tire out of balance
   b. Worn shock absorbers
   c. Insufficient positive caster
   d. Worn rack-and-pinion mounts
3. What can cause a vehicle's ride height to be out of specification?
   a. Weak leaf springs
   b. Misadjusted wheel alignment
   c. Weak valve springs
   d. Contaminated suspension bump stops
4. Where should the vehicle be supported when checking ball joint play on a MacPherson strut-type suspension?
   a. Under the lower control arm to compress the suspension.
   b. By the wheel to allow vehicle weight to compress the suspension.
   c. By the upper control arm to allow the suspension to hang free.
   d. Under the vehicle frame to allow the suspension to hang free.

5. Why should shock absorbers be replaced in pairs?
   a. To protect the coil springs from damage.
   b. To ensure ride equilibrium.
   c. To prevent premature ball joint wear.
   d. To compensate for road crown.
6. What component serves at the pivot for the wheels?
   a. Upper control arm
   b. Lower control arm
   c. Steering knuckle
   d. Torsion bar
7. How are rivetted ball joints removed properly?
   a. The rivets are melted out with an acetylene torch
   b. They are replaced with the control arm as an assembly.
   c. They are pressed out with a shop press.
   d. The rivet heads are chiseled off or drilled out.
8. Why are castellated nuts used to attach ball joint studs to the steering knuckle?
   a. They allow clearance for a cotter pin to keep the nut from backing off.
   b. They have a lock washer that keeps the nut from backing off.
   c. The provide the correct amount of thread contact to use red thread lock.
   d. They allow space for a welding rod to weld the nut and stud together.
9. Roughness at the top of the strut tower when an assistant turns the steering indicates:
   a. Loose ball joints
   b. Worn strut bearings
   c. Broken sway bar links
   d. Leaking rack-and-pinion gear
10. What alignment angle is set by adjusting the strut rod on some vehicles?
   a. Camber
   b. Caster
   c. Toe
   d. Included angle

ASE Technician A/Technician B Style Questions

1. Technician A says the final adjustment for the suspension system is performed during a wheel alignment. Technician B says a wheel alignment is often required after suspension components are replaced. Who is correct?
   a. Technician A only
   b. Technician B only
   c. Both A and B
   d. Neither A nor B
2. Technician A says worn sway bar linkages can cause a rattle over bumps. Technician B says some electronic suspension systems actively adjust to prevent body sway. Who is correct?
   a. Technician A only
   b. Technician B only
   c. Both A and B
   d. Neither A nor B
3. Technician A says shocks and struts should be measured for excessive runout. Technician B says shocks and struts should be inspected for leaking fluid. Who is correct?
   a. Technician A only
   b. Technician B only
   c. Both A and B
   d. Neither A nor B
4. Technician A says it is important to identify the layout of the suspension to accurately check ball joints for excessive play. Technician B says a pry bar can be used to pry the wheel up and down to measure ball joint play. Who is correct?
   a. Technician A only
   b. Technician B only
   c. Both A and B
   d. Neither A nor B
5. Technician A says worn sway bar bushings can cause a toe-in condition. Technician B says broken sway bar links can cause excessive body roll during cornering. Who is correct?
   a. Technician A only
   b. Technician B only
   c. Both A and B
   d. Neither A nor B
6. Technician A says extreme caution is required when removing and installing suspension springs. Technician B says to use wood spacers if the spring compressor doesn't compress the spring enough. Who is correct?
   a. Technician A only
   b. Technician B only
   c. Both A and B
   d. Neither a nor B
7. Technician A says control arms must be replaced if ball joints are worn on some vehicles. Technician B says ball joints can be pressed out of the control arms on some vehicles. Who is correct?
   a. Technician A only
   b. Technician B only
   c. Both A and B
   d. Neither A nor B
8. Technician A says to pump grease into the joint until you can see the grease coming out of the boot. Technician B says ball joints without grease fittings should be greased by piercing the boot with a grease needle. Who is correct?
   a. Technician A only
   b. Technician B only
   c. Both A and B
   d. Neither A nor B
9. Technician A says some struts are replaced as an assembly including the spring, mount, and cartridge. Technician B says struts should be checked for fluid leaks. Who is correct?
   a. Technician A only
   b. Technician B only
   c. Both A and B
   d. Neither A nor B
10. Technician A says torsion bars can be reinstalled on either side of the vehicle when removed. Technician B says leaf springs can be replaced without supporting the axle. Who is correct?
   a. Technician A only
   b. Technician B only
   c. Both A and B
   d. Neither A nor B

CHAPTER 30

Wheel Alignment

Learning Objectives

- **LO 30-01** Define camber, caster, and toe.
- **LO 30-02** Describe toe-out on turns and turning radius.
- **LO 30-03** Describe steering axis inclination, included angle, and scrub radius.
- **LO 30-04** Describe thrust angle, centerline, setback, and ride height.
- **LO 30-05** Describe types of wheel alignment.
- **LO 30-06** Describe wheel alignment preliminaries and adjustment methods.
- **LO 30-07** Prepare a vehicle for alignment.
- **LO 30-08** Adjust caster, camber, and toe.
- **LO 30-09** Measure secondary alignment angles.

ASE Education Foundation Tasks

See Appendix A to view the 2017 ASE Education Foundation Automobile Accreditation Task List Correlation Guide.

▶ Wheel Alignment Fundamentals

LO 30-01 Define camber, caster, and toe.

All wheels of a vehicle must be correctly positioned in relation to the vehicle and to one another. This will allow the vehicle to drive and steer properly (**FIGURE 30-1**). A driver should not need to keep manipulating the steering wheel when driving on straight, level roads. Similarly, little effort should be needed to turn the vehicle into curves. It should also return to the straight-ahead position once the curve has been negotiated.

Wheels are positioned on the suspension at certain angles to provide for easy driving of the vehicle. These angles, taken together, determine the vehicle's **wheel alignment**. The alignment of the wheels is maintained by the control arms, strut rods, tie rods, knuckles, and the vehicle frame. Alignments are normally performed for the following reasons:

- When handling issues related to alignment angles are found
- When tire wear shows any tire wearing angle issue
- When components are replaced that could affect the alignment
- Whenever new tires are being installed.

But before you can check alignment, you need to know what the alignment angles are and how they affect driving. Alignment angles are split between primary and secondary angles. Primary angles consist of camber, caster, and toe. They are more likely to be adjustable, but not always.

Secondary angles consist of:

- toe-out-on-turns,
- steering axis inclination,
- included angle,
- scrub radius,
- thrust angle,
- centerlines,
- setbacks, and
- ride height.

These angles are generally not readily adjustable. The exception is thrust line. It is adjustable on most independent rear suspensions. Secondary angles are used primarily for diagnostic purposes.

Finally, all of the steering and suspension system components must be within specifications. Any worn parts will not hold the wheels in their specified positions. So, aligning the wheels on a vehicle with worn parts is not effective. This is why a pre-alignment inspection is critical before performing a wheel alignment. Now, let's learn more about these alignment angles.

Camber

Camber is the side-to-side vertical tilt of the wheel. It is viewed from the front of the vehicle, and measured in degrees (**FIGURE 30-2**). A wheel where the top of the tire leans away from the center of the vehicle is said to have **positive camber**.

FIGURE 30-1 A technician checking the alignment on a vehicle.

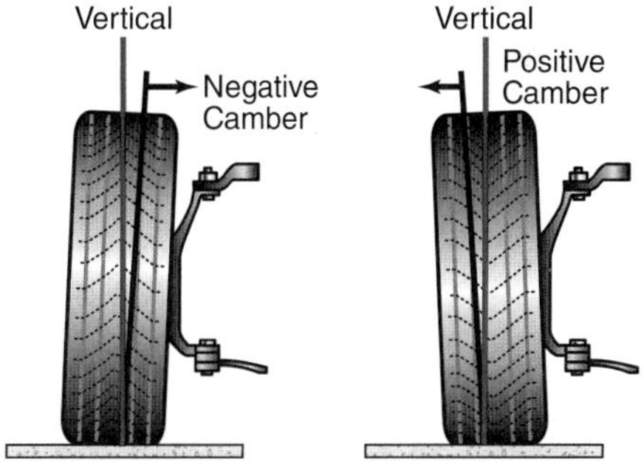

FIGURE 30-2 Camber.

You Are the Automotive Technician

A customer brings a four-year-old vehicle in for some new replacement tires because the ones on the vehicle are worn unevenly. You explain that for the tire warranty to be valid, you will have to perform a wheel alignment on the vehicle. And as part of the alignment, you need to perform a pre-alignment inspection. This will verify if the steering and suspension components are within specifications. You warn him that the wear patterns on the tires make you suspect that there are some worn parts that will have to be replaced. Please answer the customer's following questions.

1. How do worn steering and suspension parts wear out tires?
2. What components will you inspect during the pre-alignment inspection?
3. What are the three types of alignments, and when would you recommend each type?
4. What could cause a toe-out on turns measurement to be out of specifications?

A wheel that leans toward the center of the vehicle is said to have **negative camber**.

Camber tends to pull in the direction of the most positive camber. To picture why it pulls in this direction, think of a paper cup lying on its side. When it is rolled, it turns in the direction of the narrower end of the cup (**FIGURE 30-3**). Tires do the same thing in regard to their camber. Camber is affected by ride height and can be seen by looking at the tilt of the wheel in vehicles that are lowered incorrectly.

Camber used to be a heavy tire-wearing angle when tires were of the bias-ply type and had stiff sidewalls. Camber wear is now debatable, given the much more pliable sidewalls in radial tires. It is safe to say that tires can tolerate moderate amounts of improper camber much better than before. There are also studies that say most of what technicians diagnose as camber wear is really toe wear. Pay attention to this debate as it plays out.

On earlier vehicles with narrow, large-diameter tires, large camber angles were used. This was used to bring the centerline of road contact closer to the steering axis. Large camber angles also ensured the vehicle weight was carried by the large inner bearing. On modern vehicles, however, tires are much wider and generally smaller in diameter. Large camber angles would cause the tire to ride on the outer edges of the tires. The amount of camber is now reduced so that most vehicles in forward motion have what is called **zero camber**, or no tilt. This provides maximum tire patch contact with the road. It also increases tire life.

▶ TECHNICIAN TIP

Changes in running camber can be caused by many things. These include driving over road irregularities, load variations, and worn suspension components.

Caster

Caster is the forward or backward tilt of the steering axis from vertical when viewed from the side of the vehicle (**FIGURE 30-4**).

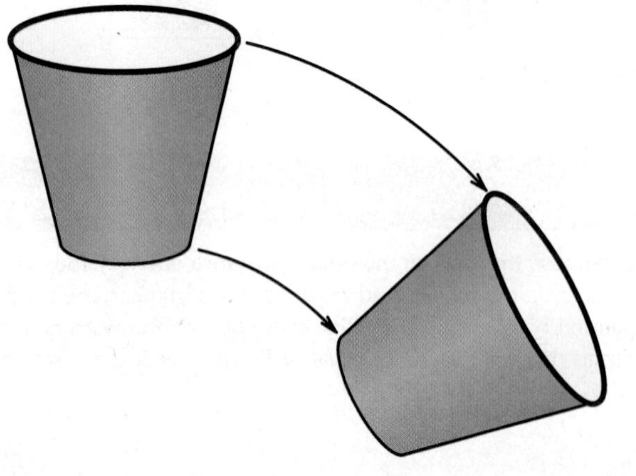

FIGURE 30-3 Camber pulls like a paper cup, toward the narrow end of the cup.

The steering axis is an imaginary line passing through the center of the ball joints on SLA suspensions. On MacPherson strut suspensions, the imaginary line passes through the center of the upper strut bearing and lower ball joint.

Caster is the tilt of the steering axis from vertical, measured in degrees. Backward tilt from the vertical is **positive caster**. Forward tilt is **negative caster**. When a vehicle has positive caster, the steering axis centerline meets the road surface ahead of the vertical centerline of the wheel. This puts the center of the tire contact point behind the steering axis centerline.

When the wheel is turned to the right, the tire contact point is moved to the left of the direction of travel (**FIGURE 30-5**). Conversely, when the wheel is turned to the left, the contact point is

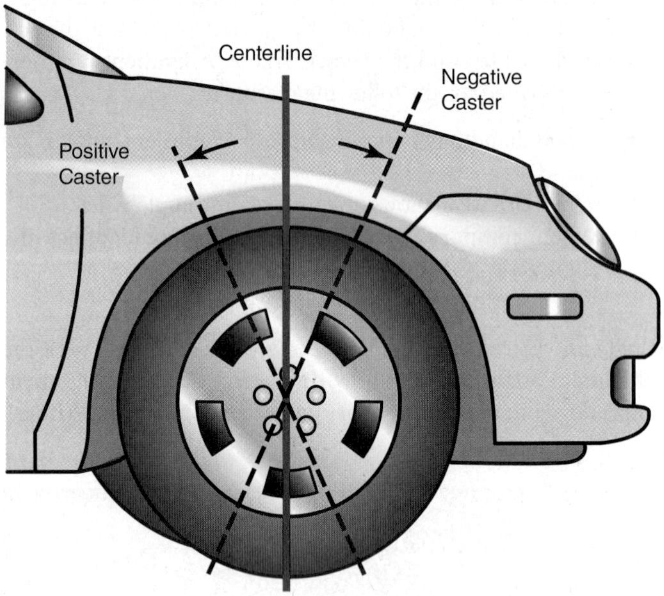

FIGURE 30-4 Caster.

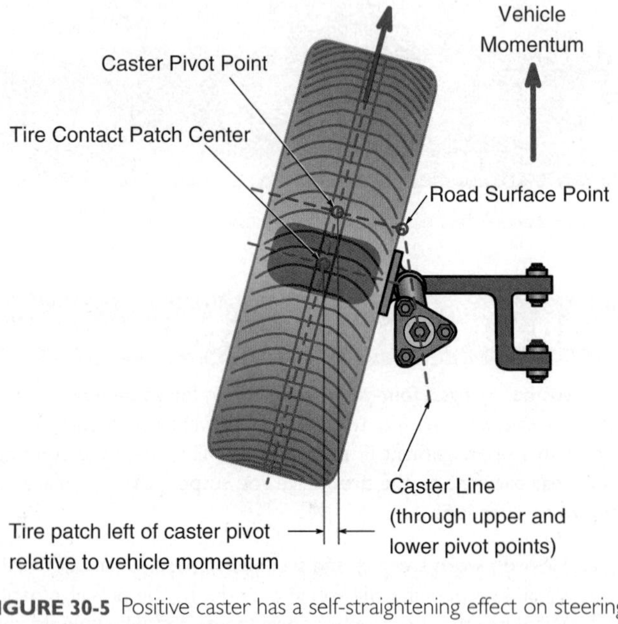

FIGURE 30-5 Positive caster has a self-straightening effect on steering when the wheels are turned.

moved to the right of the direction of travel. In forward motion, this generates a self-centering force on the tire. It helps return the wheels to the neutral position when the steering wheel is released.

Most vehicles have positive caster. This causes the tires to travel in a straight line with minimal driver input. However, as positive caster increases, more and more steering effort is needed to overcome the increased self-straightening force.

Positive caster also causes the spindle to tilt as the wheel is steered. This places more weight on the inside tire because of the body lifting that occurs with the inside tire. This lift can be seen by turning the steering from lock to lock while the vehicle is stationary.

Differences in caster cause a pull to the side with the most negative caster. Manufacturers generally specify the maximum difference in caster settings between the two wheels. This is called cross-caster. Many technicians will set the caster slightly less positive on the driver's side front wheel. This counteracts the natural pull of the road crown, which is toward the outside of the road.

▶ TECHNICIAN TIP

Some vehicles have, by design, a small amount of negative caster. This makes the steering light. Negative caster is created by placing the suspension system pivots so that they are tilted forward from the vertical line. Generally, such vehicles operate only at low speeds. Vehicles with negative caster can become unstable as speed increases. Caster is affected by ride height. On most modern vehicles, it is a nonadjustable angle without aftermarket components.

Toe-In and Toe-Out

Toe is the angle of the tires relative to one another when viewed from above (**FIGURE 30-6**). The condition in which the fronts of the wheels are closer together than the rears is called **toe-in**.

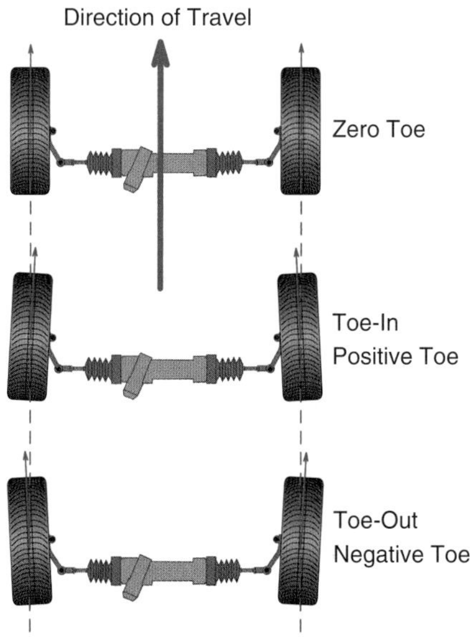

FIGURE 30-6 Toe is viewed from above the tire.

The condition in which the fronts of the wheels are farther apart than the rears is called **toe-out**. Some manufacturers use the terms "positive toe" for toe-in and "negative toe" for toe-out. Toe can be measured in inches, millimeters, or degrees.

The **static toe** setting is designed to compensate for slight wear in the system while the vehicle is being driven. The greater the wear, the more the wheels want to splay outward or inward while the vehicle is being driven. The static toe setting is designed to compensate for wear that is within specifications. So, it is designed into the vehicle based on several factors. Those are:

- front-wheel drive versus rear-wheel drive,
- front/rear brake system split versus diagonal brake system split, and
- specific desired handling characteristics.

Manufacturers specify the proper static toe such that the wheels will be parallel when the vehicle is in forward motion. This avoids scrubbing of the tires. Improper toe settings cause much faster tire wear than camber or caster.

▶ Toe-Out on Turns

LO 30-02 Describe toe-out on turns and turning radius.

Toe-out on turns is the relative toe setting of the front wheels as the vehicle turns. When a vehicle makes a turn, each wheel should rotate with true rolling motion that is free from tire scrub. True rolling motion is obtained only when each wheel is at 90 degrees to a common center point of the turn (**FIGURE 30-7**). This is found by drawing a line between the tire's horizontal centerline and the center point of the turn. Because the rear wheels are fixed, the center point of the turn will lie somewhere along a line extending from the rear axle. The actual point will depend on how far the steering wheel is turned from the straight-ahead position.

To provide true rolling motion, the inner wheel must be turned through a sharper angle than the outer wheel. This allows the inner wheel to have a smaller turning radius than

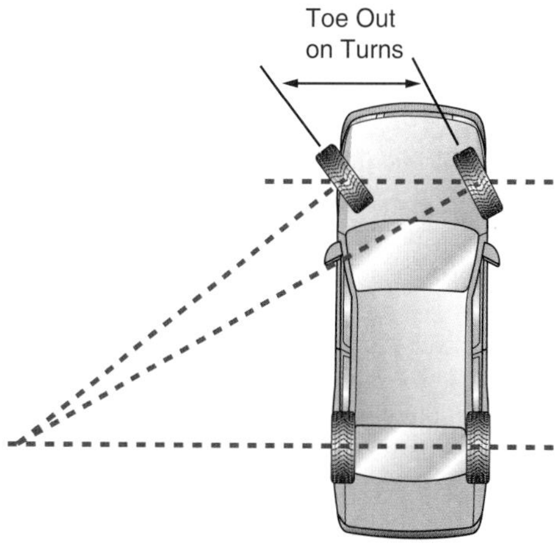

FIGURE 30-7 For toe-out turns to be correct, each wheel must be able to trace its own true arc when turning a corner.

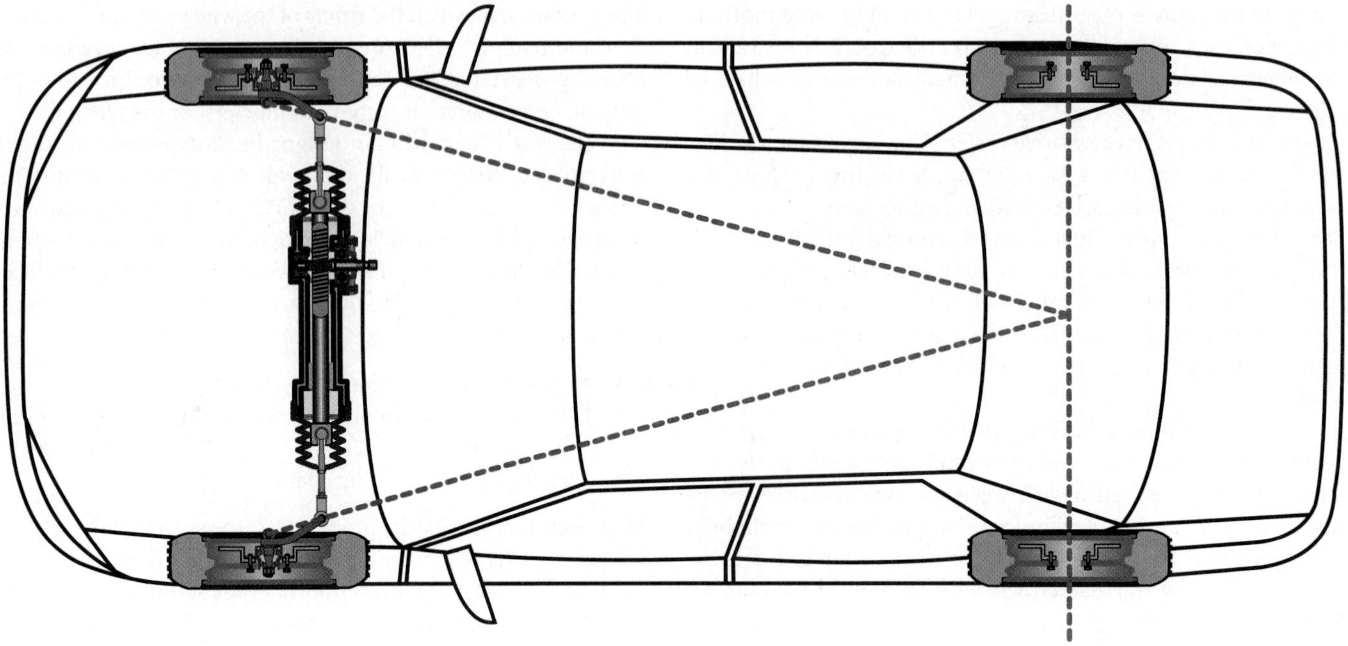

FIGURE 30-8 Ackermann angle.

the outer wheel. This correct positioning of the wheels when cornering is obtained by use of the **Ackermann principle**. This principle is related to the angle of the steering arms. They are set to the Ackermann angle. The Ackermann angle uses imaginary lines projected from the steering arms. They are at the Ackermann angle when the two lines meet at the center of the rear axle (**FIGURE 30-8**).

The angling of the steering arms forces the inner wheels to turn through a smaller circle when turned to that side. When the steering wheel is turned in the opposite direction, the wheels that were on the outside are now on the inside of the turn. They turn more sharply than the wheels that are now on the outside of the turn.

Applied Science

AS-1: Circular: The technician can demonstrate an understanding of circular motion as it relates to toe and camber on turns.

When you steer a vehicle through a turn, the outside front wheel has to travel a wider arc than the inside wheel. Therefore, the inside front wheel must steer at a sharper angle than the outside wheel. To provide true rolling motion, the inner wheel must be turned through a greater angle than the outer wheel. This allows the inner wheel to turn through a smaller turning radius than the outer wheel. With the tires on the turntable, toe-out on turns is measured by the turning angle gauges (turn plates) on the wheel alignment machine. The readings are measured electronically and displayed on the screen. Camber is the vertical angle of the wheels relative to the vehicle. This is best viewed from the front of the car. If the vehicle has negative camber, the tops of the wheels will be closer together than the bottom. During cornering, camber compensates for vehicle weight transfer and body roll.

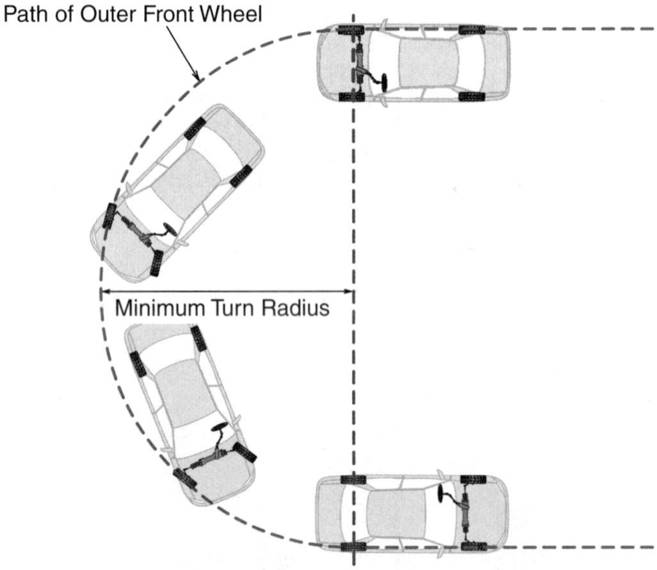

FIGURE 30-9 Turning radius is a measure of how small a circle the vehicle can turn in when the steering wheel is turned to the limit.

Toe-out on turns is a nonadjustable angle. If it does not meet the manufacturer's specifications, then you know that a steering component is bent. This is typically the steering arm. A common symptom when this is out of specifications is that the tires scrub when turning corners.

Turning Radius

Turning radius is a measure of how small a circle the vehicle can turn in when the steering wheel is turned to the limit (**FIGURE 30-9**). All vehicles have stops to limit how far the front wheels can turn. In some designs, these stops can be

FIGURE 30-10 Turning radius can be adjusted on some vehicles.

adjusted as part of a wheel alignment (**FIGURE 30-10**). If the stops are incorrectly adjusted, they could allow too sharp of a turning angle. This can bottom out the steering box which can damage it.

▶ Steering Axis Inclination

LO 30-03 Describe steering axis inclination, included angle, and scrub radius.

The axis around which the wheel assembly swivels as it turns to the right or left is called the steering axis. It is formed by drawing a line through the center of the upper and lower pivot points of the suspension assembly (**FIGURE 30-11**). Seen from the front of the vehicle, it is tilted inward at the top. The angle formed between this line and the vertical provides the **steering axis inclination (SAI)** angle. Because the SAI is not adjustable, if the camber angle is correct, then the SAI should also be correct. That is, it should match the manufacturer's specifications.

SAI acts with caster to provide a self-centering of the front wheels. When the wheels are in the straight-ahead position, the ends of the stub axles are almost horizontal. When the wheels turn to either side, the effect of SAI is to make the ends of the stub axle swing downward. But this tendency is prevented by the wheel and tire on the ground (**FIGURE 30-12**). The stub axle carrier then must move up, which raises the corner of the vehicle. When the steering wheel is released, the weight of the vehicle forces the stub carrier back down. This pushes the wheels back to a central position. When the wheels are turned the other way, the same thing happens on that side.

With a perfectly vertical steering axis, no self-centering would occur. The wheel would pivot on a radius, not under the center of the tire patch, but off to the inside of the tire. This would introduce a turning movement on the wheel whenever the tire hit a bump. It would also transmit more road shock back to the steering wheel. Steering would then be more difficult to control.

SAI also brings the pivot point close to the center of the tire contact patch at the road surface. For steering purposes, ideally, SAI intersects with the camber line (drawn through the center of the tire and the wheel) at the road surface. Any difference in distance between these two lines produces another suspension angle. This is called the scrub radius, which will be discussed shortly.

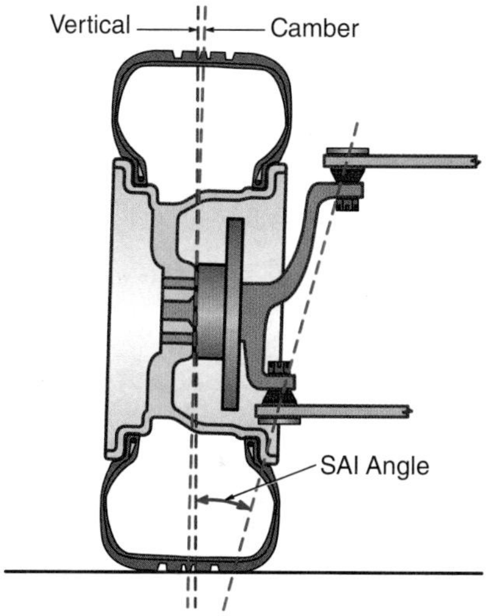

FIGURE 30-11 The axis around which the wheel assembly swivels as it turns to the right or left is called the steering axis. It is formed by drawing a line through the upper and lower pivot points of the suspension assembly.

Included Angle

The angle formed between the SAI and the camber line is called the **included angle** or diagnostic angle. It is found by adding the SAI angle and the camber angle together (**FIGURE 30-13**). If the camber angle is specified as negative, then it is subtracted from the SAI angle. When an angle is referred to as a diagnostic angle, it means the angle cannot be adjusted. But it is measured to determine if any parts are bent, such as a spindle, control arm, or steering arm.

Scrub Radius

Scrub radius is also known as steering offset and scrub geometry. It is the distance between two imaginary points on the road surface. One point is the centerline of the tire. The other point is where the SAI centerline contacts the road surface (**FIGURE 30-14**). If these two lines intersect at the center of the tire patch, then the vehicle is said to have zero offset, or **zero scrub radius**. If the tire's centerline is outside of the SAI line (intersect below the road surface), then it has positive offset, or **positive scrub radius**. If the tire's centerline is inside of the SAI line (intersect above the road surface), then it has negative offset, or **negative scrub radius**.

Scrub radius can be changed accidentally in a number of ways. Changing the diameter of the wheels and tires affects scrub radius. This is due to the increased distance the spindle is above the road. The difference in height changes where the SAI centerline intersects with the road (**FIGURE 30-15**).

Changing the distance that the centerline of the wheel sits relative to the vehicle body changes scrub radius. This is called wheel offset (**FIGURE 30-16**). Changing SAI and camber also affects scrub radius.

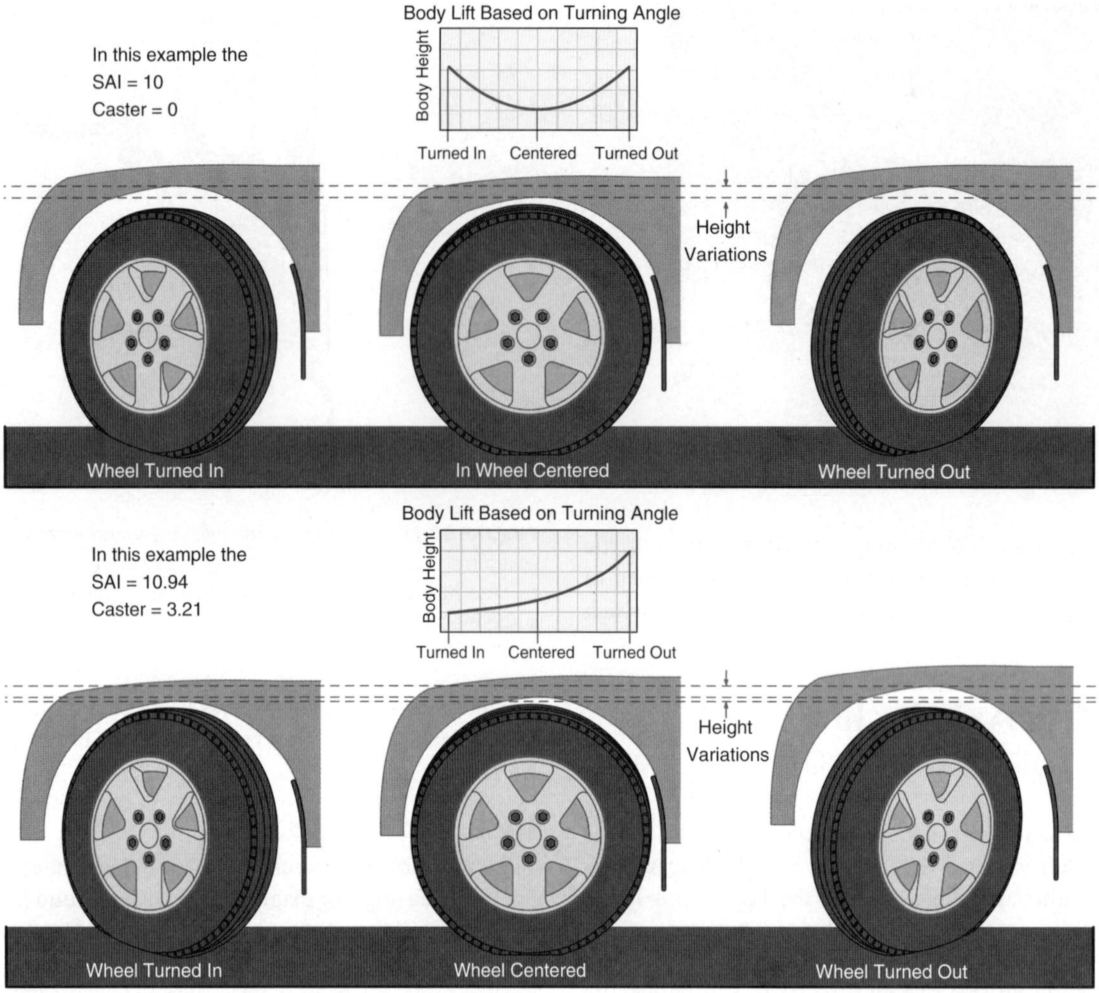

FIGURE 30-12 SAI and caster provide a self-centering effect of the front wheels. SAI tends to push each wheel down when turned either direction. Positive caster tends to push the inside wheel down when turned and lift the outside wheel.

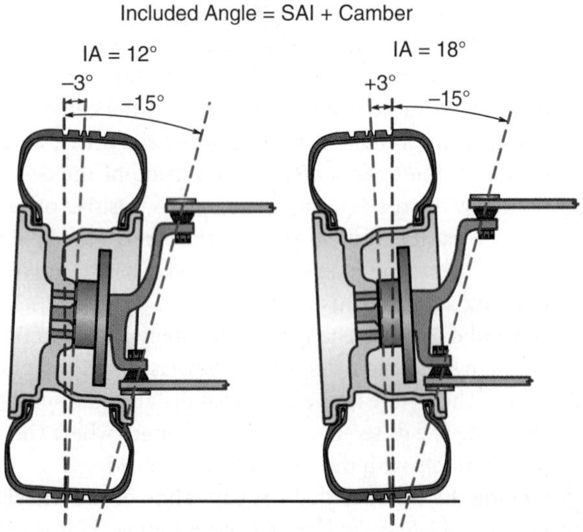

FIGURE 30-13 Included angle is found by adding the positive camber angle to the SAI angle or by subtracting negative camber from the SAI angle.

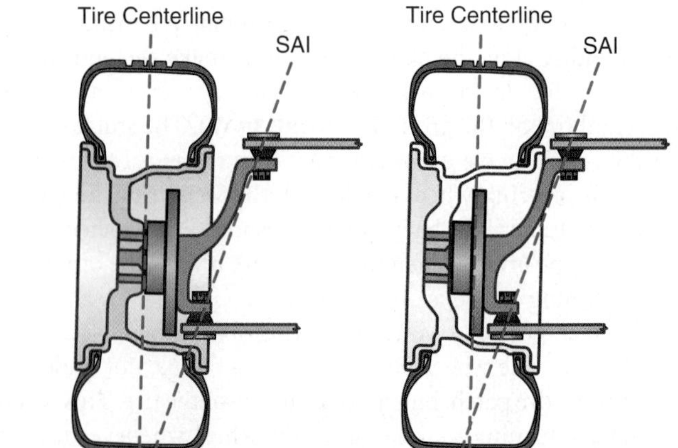

FIGURE 30-14 The scrub radius is the distance between two imaginary points on the road surface. One point is where the centerline of the tire contacts the road, and the other point is where the SAI centerline contacts the road.

On a rear-wheel drive vehicle with positive scrub radius, the vehicle's forward motion and the friction between the tire and the road tend to move the front wheels backward. This causes the wheels to toe-out. If it has negative scrub radius, the front wheels again tend to move back and the wheels now toe-in. On front-wheel drive vehicles, the opposite occurs. Positive scrub radius causes toe-in, and negative scrub causes toe-out.

Too much or too little scrub radius can produce unwanted effects. It can cause increased road shock when encountering irregularities in the road surface. Too much positive scrub radius causes the vehicle to dart to the side of greatest braking effort when the brakes are applied. Too much negative scrub radius causes the vehicle to veer away from the side of the greatest effort. How much it veers depends on the amount of the scrub radius. This is why vehicles with a diagonal-split brake system have negative scrub radius built into the steering geometry. If half of the brake system fails, then the vehicle will tend to pull up in a straighter line. This is due to pull from the negative scrub radius canceling out the pull from the braking action.

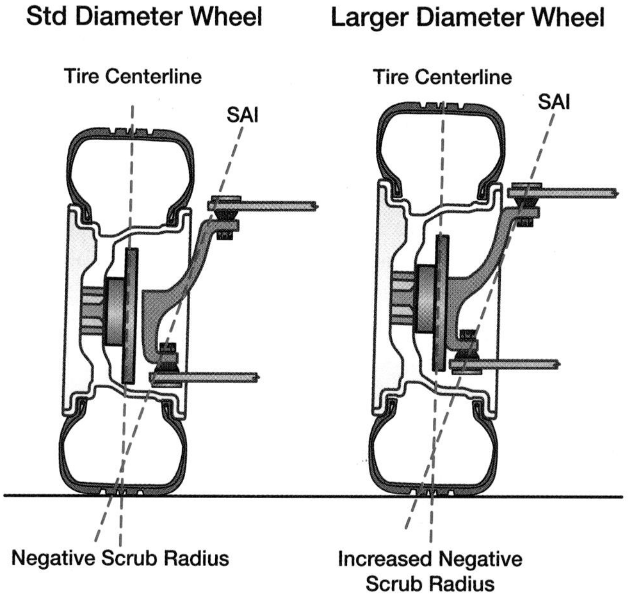

FIGURE 30-15 Increasing the diameter of the wheels and tires moves the scrub radius to more of a negative setting.

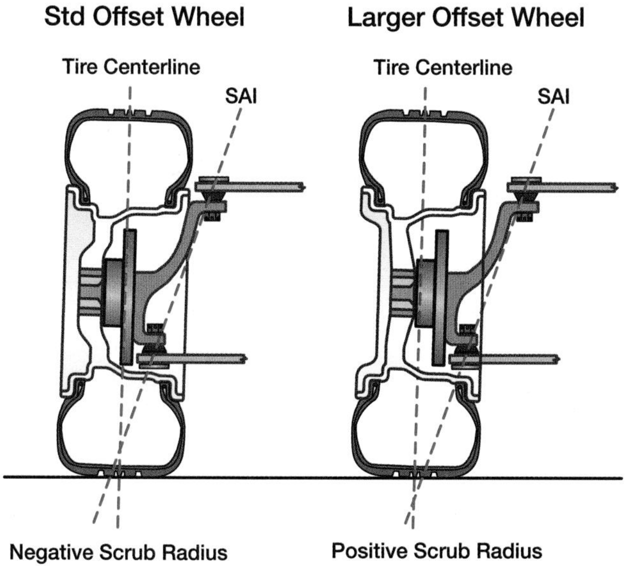

FIGURE 30-16 Changing wheel offset moves the centerline of the tire, which changes scrub radius.

▶ TECHNICIAN TIP

The offset of the wheel determines where the centerline of the tire meets the road surface. Because of this, it is important that the offset be maintained if wheels are being replaced. Changing the rim offset changes the scrub radius. It also changes the predictability of the vehicle handling, especially during brake failure. Also note that if tire size, ride height, or camber adjustment is changed, scrub radius also changes. This will affect vehicle handling, potentially creating a safety hazard. Understand that there is a liability risk when modifying a vehicle away from stock specifications.

▶ Thrust Angle, Centerlines, and Setback (Tracking)

LO 30-04 Describe thrust angle, centerline, setback, and ride height.

On a vehicle with independent rear suspension, performing a front-wheel-only alignment is considered an inadequate procedure. The rear tires also need to be aligned as well. If not, they will experience accelerated tire wear, and the vehicle will not drive properly. The rear wheels create a thrust line. The term **thrust line** refers to the direction in which the rear wheels are pointing (**FIGURE 30-17**). If the thrust line is not in line with the centerline of the vehicle, a **thrust angle is created. Thrust angle** refers to the angle between the centerline of the vehicle and the thrust line. **This tends to cause the vehicle to pull to the side opposite the thrust line.**

The thrust angle can be adjusted on vehicles with adjustable rear suspensions. On vehicles that do not have adjustable rear suspensions, a small amount of thrust angle can be compensated for. This is accomplished by aligning the front wheels to the rear wheels. No matter what, referencing the alignment of the front wheels to the rear wheels is very important.

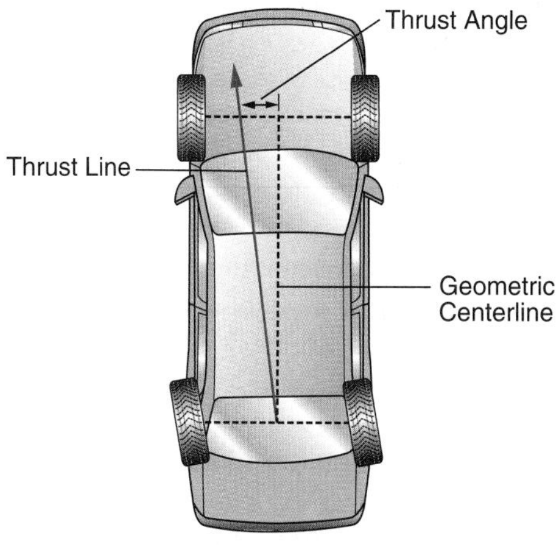

FIGURE 30-17 The thrust angle refers to the average angle of the rear wheels and its relationship to the vehicle's centerline.

Applied Math

AM-1: Visual Perception: The technician can visually perceive the geometric relationships of systems and subsystems requiring alignment.

Bill is an automotive technician trainee who is working with an experienced technician. Bill has an interest in front-end alignment. He has performed several alignments with the assistance of his mentor. A vehicle has just come into the shop for an alignment, and Bill has been given permission to do this job on his own. The experienced technician will do a final check to ensure that everything has been done properly.

Bill drives the vehicle onto the alignment rack, but before attaching the wheel sensors, he looks the car over. He positions himself at the front of the vehicle, sighting from the front tires toward the rear. Bill wants to see the relationship of the front tires to the rear. He observes that there is a small amount of the rear tread showing on each side. It appears to be an equal amount of tread on each side.

He begins the alignment by attaching the sensors on each wheel. The vehicle identification number (VIN) is scanned into the computerized system. The alignment system calculates the thrust angle and geometric centerline. The final result is that a slight toe adjustment on the rear wheels is needed. But everything else is within specifications. Bill's visual perception of the relationship of the components helped him to predict this reading before the alignment machine verified it.

Applied Math

AM-2: Trial and Error: The technician is able to solve problems by trying a suggested solution and observing the results.

A vehicle with a MacPherson strut design is in the shop for repairs. The technician who is assigned to this project has done similar jobs in the past. The repair order calls for two new front struts plus an alignment. In the past, the technician has not tried marking the position of the old strut. In this case, the technician is going to try a new method of carefully marking the old strut's position. This was suggested by another technician. The technician wants to observe the results of his procedure when the alignment is completed. The strategy is to produce a less involved alignment or perhaps eliminate the need for alignment adjustments to be necessary. The technician understands that this is only a trial-and-error procedure to be tested and could vary between vehicles. He was pleasantly surprised that the alignment was much closer to the specifications than in previous vehicles where he hadn't used this process.

Ideally, the thrust line and the vehicle's geometric centerline should line up closely. The centerline is drawn through points midway between each pair of wheels. However, the thrust line is not always as easy to determine. It is normally perpendicular to the rear-axle on solid-axle vehicles. In vehicles with an independent rear suspension, it is derived by splitting the toe angle of each of the rear wheels on the vehicle. For instance, the right rear wheel is toed in 6 degrees, and the left rear wheel is at 0 degrees. In this case, the thrust line will veer off 3 degrees to the left of the vehicle's centerline (**FIGURE 30-18**).

Ideally, the thrust line and centerline match. However, it is rare that they do. This is affected by the size of a vehicle, the tolerances during manufacture, operational stresses, and component wear. If the deviation is very small, then corrective action is normally unnecessary. However, a moderate deviation must be diagnosed and corrected. Under such conditions, the rear wheels steer the vehicle away from its centerline. The driver has to turn the steering wheel to one side to keep the vehicle going in a straight line.

Thrust angle problems may have several causes. Some examples include a broken center bolt on a leaf-spring suspension, a bent axle, a bent frame or unibody, worn or damaged suspension parts, or simply misadjusted rear toe. Look for these issues when the thrust angle is out of specifications.

Setback is a diagnostic measurement that is generally not adjustable. Setback is the distance one wheel is set back from the wheel on the opposite side of the axle. Setback is relative to imaginary lines running through the center of each wheel, perpendicular to the vehicle's centerline (**FIGURE 30-19**). Setback is best measured on an alignment machine (after the wheels are aligned). This is so it can be measured correctly from the centerline of the vehicle. But a preliminary check of setback can be made by measuring the distance of the wheelbase on both sides of the vehicle. And then comparing the measurements. If the vehicle has setback, then one wheelbase measurement will typically (but not always) be shorter than the other.

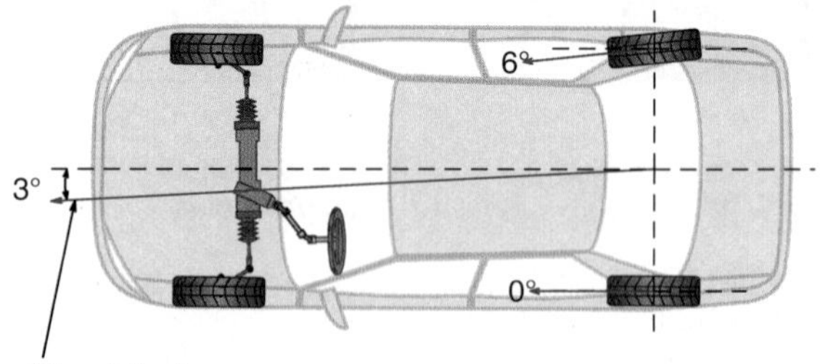

FIGURE 30-18 Rear wheel thrust line can be found by splitting the rear toe readings.

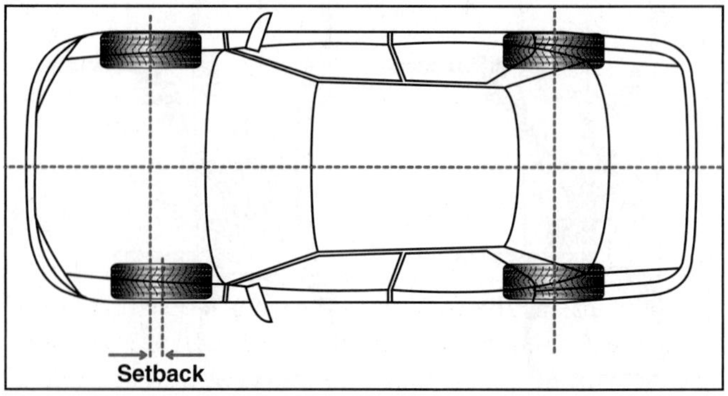

FIGURE 30-19 Setback refers to the distance one wheel is set back from the wheel on the opposite side of the axle. Setback is relative to lines running through the center of each wheel, perpendicular to the vehicle's centerline.

Setback can be incorrect due to damage from a collision. It can also be caused by a cradle that has been installed incorrectly or has shifted. Setback in the front wheels creates a pull condition to the side with the most setback. This is the wheel that is closest to the rear of the car. A setback issue in the rear will pull toward the side with the least setback.

Setback due to a cradle shift may also affect the camber or caster from side to side. On a vehicle that has no adjustment for camber or caster, check for a cradle shift. You can do this by looking for witness marks around the head of the bolt that holds the cradle to the body. If a partial shiny metal ring surrounds the bolt head, then the cradle has shifted. Loosen the cradle bolts and reposition to achieve proper alignment specifications. Usually, setback should be no more than a quarter of an inch from side to side.

▶ TECHNICIAN TIP

In extreme conditions of setback or thrust angle, the tracks the rear tires make are beside those of the front. This condition is known as dog-tracking or crabbing. It can cause diagonal tire wear patterns on the rear tires, as well as vehicle instability in some driving conditions. A vehicle that is dog-tracking will appear to be driving slightly sideways down the road.

Applied Math

AM-3: Parallel/Perpendicular: The technician can use measurement devices to determine the parallelism or perpendicularity of chassis, suspension, and other vehicle systems requiring the application of geometric alignment principles.

A vehicle was purchased at an auction and has been taken to an automotive repair shop for evaluation. The vehicle was set up on the alignment machine and alignment readings taken. Several of the angles were substantially out of specification. They include front-wheel setback, rear thrust angle, and SAI. These angles being out of specification indicate the potential of a bent frame or chassis. Upon further examination, the technician observes cracked undercoat on the subframe components. Also noted was a shifting of the engine cradle bolts. This verifies what the alignment angles are showing. The chassis is not parallel and perpendicular. This means that the wheels cannot be brought back into proper alignment since the chassis is out of alignment. The vehicle will need to be sent to a body shop for frame straightening.

Ride Height

Ride height, sometimes referred to as **trim height**, is the amount of distance between the ground and a specified part of the vehicle. Common points are the fender well, rocker panel, or frame (**FIGURE 30-20**). Ride height is measured with the vehicle unloaded. For cars, it is usually given with no cargo or passengers. Changes in ride height alter the position of the control arms. This can have an unfavorable effect on wheel alignment. Ride height that is not within specifications can be caused by improper tires or tire pressure; weak, sagging springs; or bent components such as a control arm, axle, or even a frame or unibody. Ride height should typically not vary by more than half an inch from side to side.

Ground clearance is similar to ride height. It is the distance from the ground to the lowest part of the chassis. This is typically a cross member, final drive, or oil pan. Ground clearance is especially important for off-road vehicles. Ground clearance can be increased by replacing stock components with aftermarket performance parts. But increasing ground clearance also increases the vehicle's center of gravity. This makes it more prone to rollover. It also affects scrub radius and potentially other wheel alignment angles. Thus, increasing ground clearance increases the risk of an accident. So, consider that carefully before modifying a vehicle in this way.

Applied Science

AS-1: Pneumatics: The technician can demonstrate an understanding of the forces and motions in pneumatic systems.

A vehicle is in the shop due to an inoperative automatic leveling control (ALC), which is part of an air suspension system. It maintains the correct rear suspension ride height. This height will be maintained even if a heavy weight is placed in the trunk. It is controlled by compressed air, which means it operates on pneumatics. In this case, a battery-powered air compressor is used to power the system. Nylon air lines connect from the compressor to the rear air shocks to lift the rear of the vehicle to the correct height. The vehicle height sensor triggers the air compressor to control the amount of run-time needed. When weight is removed from the trunk, the rear of the vehicle rises. At this point, the vehicle height sensor gives a release signal to the pressure release solenoid valve. It releases air from the shocks. The body of the vehicle returns to the correct height.

This type of system is usually found on luxury vehicles. The technician inspects the vehicle and discovers a broken airline. After replacing the line, the vehicle is tested by the addition of some heavy items in the trunk. The air compressor starts, and the rear of the vehicle rises to the correct ride height. When the weight is removed, the vehicle lowers to the proper height.

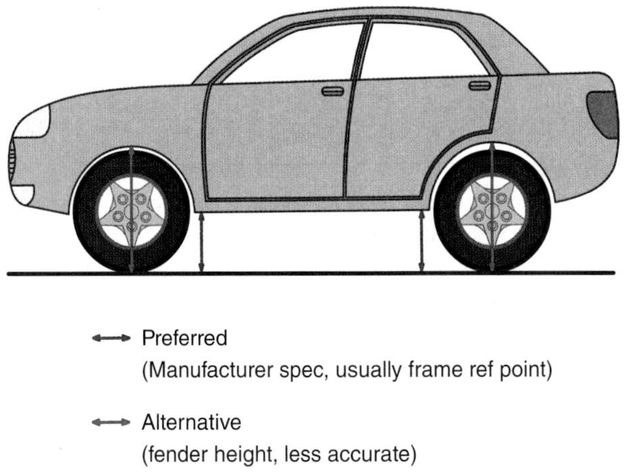

FIGURE 30-20 Ride height measurements.

▶ Performing a Wheel Alignment

LO 30-05 Describe types of wheel alignment.

Performing a wheel alignment means aligning the wheels to the specified angles. This starts with verifying that all of the suspension and steering system components are within specifications. Once that is verified, the wheel alignment angles are measured and compared with specifications. Then, any out of specification angles are brought back into specification. This is done by performing adjustments, or replacing damaged parts. So, let's dive in and see how to perform a wheel alignment.

Types of Wheel Alignment

Performing a wheel alignment requires the use of an alignment machine. In the past, simpler devices (such as using a measuring tape to set toe) were used to align the vehicle's wheels. Today's vehicles are more sensitive to the position of the wheels. So, using a tape measure to set toe is no longer acceptable. The technician uses a computerized alignment machine to ensure each of the alignment angles are correct.

The three basic types of wheel alignment are (1) front-end, two-wheel alignment; (2) thrust-angle alignment; and (3) four-wheel alignment. Two-wheel alignment is outdated and almost never performed on modern vehicles. But we cover it here briefly, so you will understand why it would be inappropriate for most vehicles.

In front-end, two-wheel alignment, the technician positions the vehicle on the alignment rack. Wheel sensors are then attached to the two front wheels. The sensors read the position of both front wheels and provide the measurements to the technician (**FIGURE 30-21**). The two-wheel alignment only compares the front wheel angles to each other. It does not look at rear wheel position, so it cannot take into account any thrust angle. The technician compares the measurements to the alignment specifications provided by the manufacturer. If they are out of specification, the technician adjusts the angles of the wheels until they match. If adjustments cannot correct the angles, parts need to be replaced, or shims used.

In a thrust-angle alignment, the technician attaches wheel sensors to all four wheels. The front wheels are compared with the angles of the rear wheels and adjusted to them (**FIGURE 30-22**). This alignment is typically performed on a vehicle with a solid rear axle where no adjustment is possible. If the thrust angle is out of specifications, the technician will need to diagnose what is causing the issue. This could be due to a shifted axle or collision damage. If the thrust angle is within specifications, the front wheels are adjusted to compensate for any slight rear thrust angle.

Applied Math

AM-4: Angles: The technician can use angle measurement equipment and techniques to determine any vehicle angle measurement variance from the manufacturer's specifications.

A front-wheel drive vehicle is in the shop for a routine tire rotation. The technician notices that one of the front tires is slightly worn on one side. He reports his findings to the service advisor, who contacts the vehicle owner. The customer authorizes a four-wheel alignment. The technician starts with a pre-alignment inspection, looking for any loose suspension components. No unusual problems are found. The technician is now ready to use angle measurement equipment and techniques to determine variance from the manufacturer's specifications.

An electronic sensor is attached to each of the four wheels. The technician scans the VIN. The alignment system automatically imports the correct specifications for the vehicle. In addition, the video screen gives a step-by-step approach to the procedure. The technician is shown the exact procedure for the vehicle being aligned. A color-coded system—red for out of specifications and green for within specifications—is also helpful.

Computer printouts with before and after adjustments are available to the customer. In this scenario, the alignment equipment indicates that a toe change is needed for the front wheels. As the technician makes the change, the display shows the results on the video screen. When the exact manufacturer's specifications are obtained, the technician locks down the jam nuts. This secures the setting.

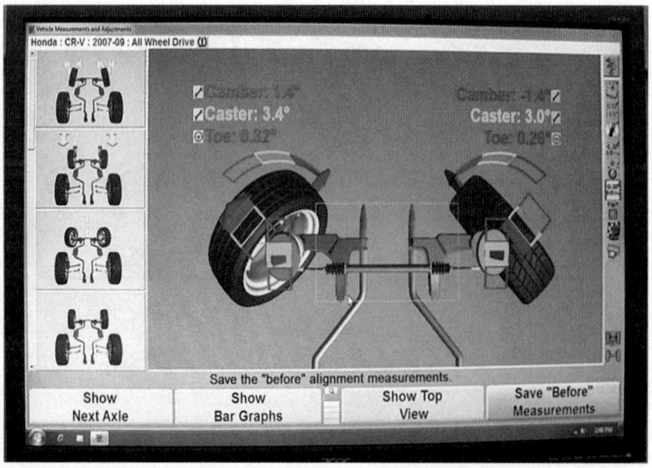

FIGURE 30-21 Readings for a two-wheel alignment.

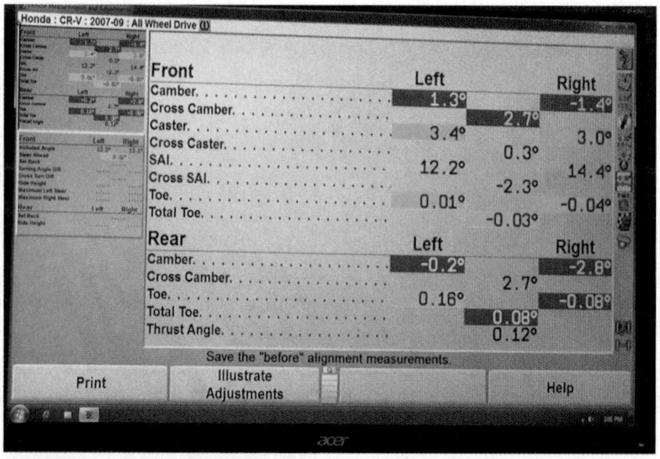

FIGURE 30-22 Readings for a thrust angle alignment.

In a four-wheel alignment, the technician positions the vehicle on the alignment rack. Wheel sensors are then attached to all four wheels. The rear wheels are adjusted first so that they conform to the vehicle's centerline. Then, the front wheels are adjusted to conform to the vehicle's centerline, and the position of the rear wheels (**FIGURE 30-23**). The four-wheel alignment provides the most accurate alignment of the wheels, but only if adjustment of the rear suspension is possible.

Remember that any time a worn steering or suspension component is removed, an alignment should be performed. For example, if the front struts (with or without adjustment slots) are being replaced, an alignment will have to be performed. This will ensure that the correct wheel alignment angles are restored. Any component that holds or allows movement of the wheels can affect the alignment, when replaced. So, it is good practice to check and align the wheels after these types of repairs.

▶ Wheel Alignment Preliminaries

LO 30-06 Describe wheel alignment preliminaries and adjustment methods.

Performing an alignment can be done for maintenance purposes. Or it can be done when tires or other steering and suspension components have been replaced. It can also be a helpful diagnostic step. For example, a driver is complaining of uneven tire wear, pulling, hard steering, or wandering conditions. The alignment machine helps you identify problems that might not be identifiable by a visual inspection. Bent or worn suspension components, or improper repairs of the steering or suspension system, are good examples.

Four-wheel alignments are the norm for current vehicles. But many technicians jokingly refer to a proper alignment as a five-wheel alignment. This is because the steering wheel also must be centered at the completion of a wheel alignment. Otherwise, the customer will be returning with a crooked steering wheel concern. Assuming you follow the proper procedures, the steering wheel itself will be centered when the alignment is complete.

But centering the steering wheel physically is not enough on most modern vehicles. A steering angle sensor tells the vehicle's stability control module where the steering wheel is pointed. It is possible for the steering wheel to be centered but for the sensor to show that it is off center. The steering angle sensor needs to be recalibrated if after the alignment the sensor does not match the steering wheel position. This process is performed with a scan tool once the alignment is within specifications and the steering wheel is physically centered (**FIGURE 30-24**).

Before performing the wheel alignment, you need first to verify the customer concern. Then, perform a pre-alignment inspection to ensure that the vehicle can be aligned. Ride height should be measured and compared with specifications. If any tire, wheel, suspension, or steering components are found to be worn or damaged, they must be replaced before aligning the wheels. Once any worn or damaged parts have been replaced, the technician should check the primary angles. These include caster (which is not adjustable on all vehicles), camber, and toe.

The secondary angles should also be inspected. They are the SAI, included angle, wheel setback, thrust angle, and toe-out on turns. The secondary angles are typically not adjustable. If out of specification, that normally indicates a suspension component or the vehicle frame is bent. One exception is when the vehicle has an adjustable rear independent suspension. In this case, thrust angle can typically be adjusted by adjusting the toe for each rear wheel, in or out. Typically, thrust angle is adjusted as close to the vehicle's centerline as possible.

Adjustment Methods

Adjusting wheel alignment angles must be done in this specific order: caster, camber, and then toe. Not all vehicles have adjustment for caster. If they do not, an aftermarket kit must be installed if caster has to be adjusted. Or the vehicle may have to go to a body shop for frame straightening. For vehicles equipped from the factory with provisions for adjustment, five common types of adjustment systems are used. These are: shim, eccentric bolts, slots, adjustable strut rod, and ball joint adjusting sleeve.

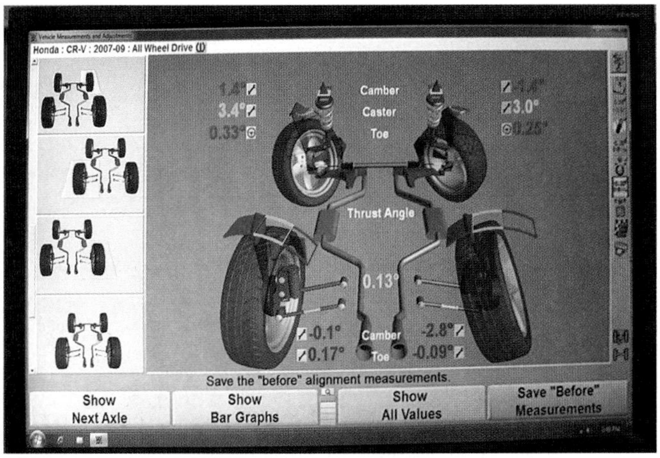

FIGURE 30-23 Readings for a four-wheel alignment.

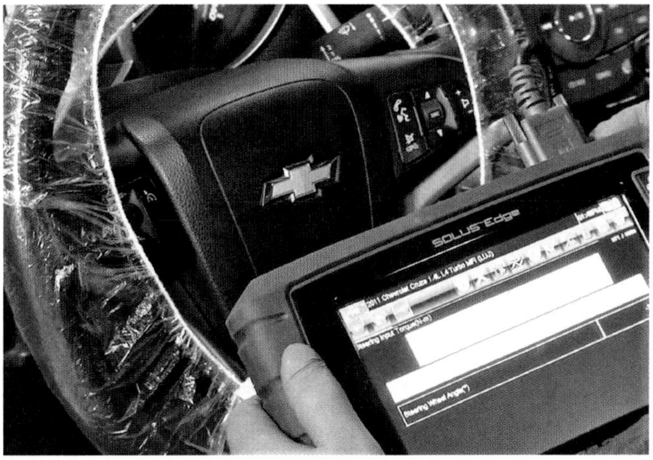

FIGURE 30-24 A scan tool being used to recalibrate a steering angle sensor.

The first type of adjustment is shim adjustment. Manufacturers have used shims on both the front and rear wheels over time. In this method, shims are added or removed from a shim pack. Shims are placed between the control arm pivot shaft and the vehicle frame at each of the two attachment bolts (**FIGURE 30-25**). There are two shim packs used—one at the front attaching bolt and one at the rear attaching bolt. Shims are added or removed to change the position of the pivot shaft. This ultimately positions the control arm and ball joint. Caster and camber angles can then be adjusted by adding or removing shims.

Applied Math

AM-5: Relationships: The technician verifies that the relationship of parallel lines and angles is in conformance with the manufacturer's specifications.

An experienced technician is assigned to assist a technician trainee with an alignment. After showing the trainee the basics of a pre-alignment inspection, the vehicle is pulled onto the alignment rack. The technician shows the trainee how to attach a target to each of the four wheels. The VIN number of the vehicle is scanned onto the alignment system. Manufacturer's data will be selected and available for alignment purposes.

The experienced technician explained that the alignment system is set up to guide the operator in a step-by-step procedure. On a video screen are instructions that are specific to the exact make and model of the vehicle. There are clear illustrations of all of the adjustments that are needed for each phase of the operation. Concerning the specifications for the vehicle being aligned, caster has a preferred setting of 3.33 degrees with a range of 2.33 to 4.33 degrees. Camber is 0 degrees preferred with a range of −1 to 1 degrees. Toe-in for front is 0.16" preferred, with a range of 0.11" minimum and 0.21" maximum. The technician and trainee then compare the alignment readings with the specifications to determine any needed adjustments.

Shim adjustment can affect both caster and camber. If the same number of shims is taken out of, or added to, each shim pack, the camber will be changed. If shims are taken from one shim pack and placed in the other, only the caster will be changed. Shims are available in various thicknesses and typically come in 1/64" (0.015 mm) up to 1/4" (0.250 mm). Typically, a shim change of 1/32" moves caster by 0.5 degrees. Camber moves around 0.3 degrees with the same shim change.

Shims are still used on some vehicles to adjust rear camber and toe. This type of shim is slightly wedge shaped and is placed behind the backing plate of the brake assembly. In some cases, it sits behind the hub of the wheel assembly. Its wedge shape moves the camber or toe to the correct position (**FIGURE 30-26**). These are aftermarket fixes and typically not used from the factory.

Another type of alignment adjustment for caster and/or camber is the eccentric bolt (**FIGURE 30-27**). The eccentric bolt has a slightly egg-shaped washer attached to one or both ends of the bolt. This egg-shaped washer pushes against ridges on the part being adjusted. This adjusts caster and/or camber.

If the vehicle has only one attachment point for the control arm to the body or frame, then it will use a strut rod to support the control arm. Some strut rods have an adjustment nut on the front and the rear of the attachment point to the control

FIGURE 30-25 A shim-type adjustment is probably the most time-consuming alignment adjustment. Moving shims from one attaching bolt to the other affects the caster setting, and changing shim pack size in both packs affects camber.

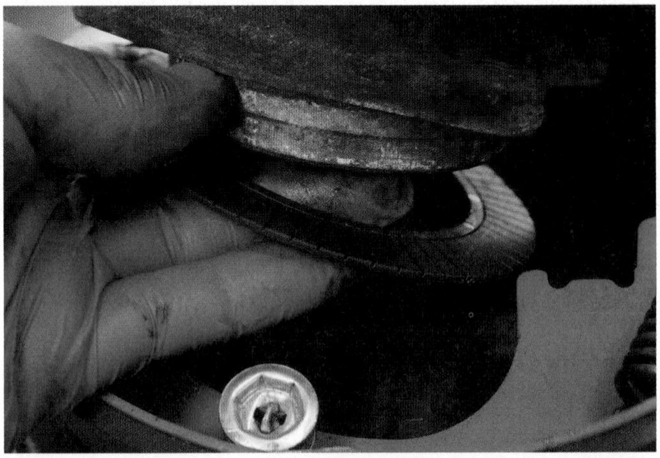

FIGURE 30-26 A typical shim used to adjust the alignment on a rear wheel.

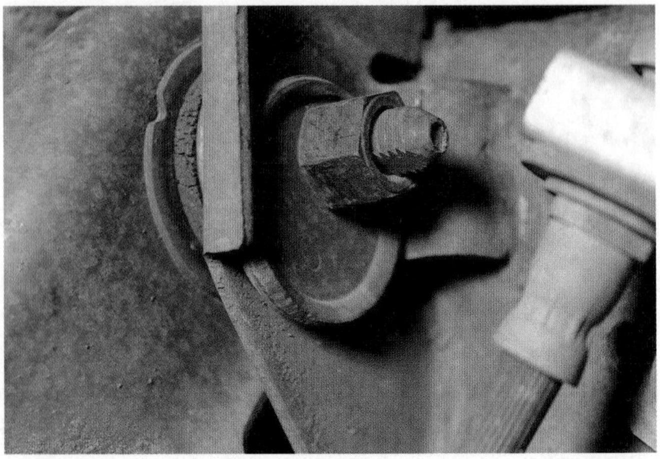

FIGURE 30-27 Eccentric bolt adjustment.

arm (**FIGURE 30-28**). If both nuts are moved, then the control arm will be pushed forward or pulled backward. This adjusts the caster setting.

The last method of adjusting caster and camber involves the ball joint adjusting sleeve (**FIGURE 30-29**). The adjusting sleeve is used on some four-wheel drive vehicles. It mounts into a solid axle or a twin I-beam axle. The ball joint tapered stud fits into the sleeve. The sleeve has a slightly offset hole so the ball joint tapered stud can fit into it. This sleeve can be rotated to change caster and camber slightly. When turning the adjusting sleeve, the ball joint stud is pushed either forward or backward, or in or out, changing caster or camber. The factory adjusting sleeve typically only gives a half degree of alignment change. If more adjustment is needed, an aftermarket adjusting sleeve is often installed.

Adjustment of toe is typically accomplished by lengthening or shortening the tie-rod assembly. Lengthening the tie-rod pushes the steering arm and changes the toe setting. Shortening moves the steering arm the opposite direction. The toe is changed in the other direction. On a vehicle with an adjusting sleeve, the technician must loosen the two clamp bolts that hold the adjusting sleeve tight. The sleeve can then be twisted in the proper direction for toe setting (**FIGURE 30-30**). Don't forget to tighten the clamp bolts when the toe is adjusted properly.

On a vehicle with a rack-and-pinion steering, the inner tie-rod typically threads into the outer tie-rod. No adjusting sleeve is used. To lengthen the tie-rod assembly on this vehicle, loosen the locknut on the outer tie-rod. Then, turn the inner tie-rod in or out to lengthen or shorten the tie-rod assembly (**FIGURE 30-31**). Tighten the locknut when adjustment is finalized.

▶ Tools

LO 30-07 Prepare a vehicle for alignment.

The tools commonly used in performing wheel alignments include (**FIGURE 30-32**):

- Wheel alignment machine: Used to measure and view alignment angles on a vehicle.
- Four-post alignment rack: Used to raise and level the vehicle off of the ground so you can align it.

FIGURE 30-28 The strut rod on some vehicles is the adjustment point for caster.

FIGURE 30-30 The tie-rod assembly comes in two forms and is adjusted by threading the assembly to lengthen or shorten.

FIGURE 30-29 The adjustable ball joint sleeve gives limited adjustment for caster and camber, and is often replaced with an aftermarket part with more adjustment potential.

FIGURE 30-31 The tie-rod on a rack-and-pinion steering system can be adjusted after loosening the locknut by turning the tie-rod either in or out.

- Turntables: Used on alignment racks so you can turn the front wheels to check caster angle.
- Wheel clamps: Used to hold alignment targets to the wheels.
- Specialty alignment tools: Used to align steering angles on a vehicle.
- Steering wheel holder: Used to keep steering wheel from turning when centering steering wheel during an alignment.
- Brake pedal depressor tool: Used to lock the brakes during caster sweep so wheels of vehicle cannot move.
- Steering angle reset tool or scan tool.

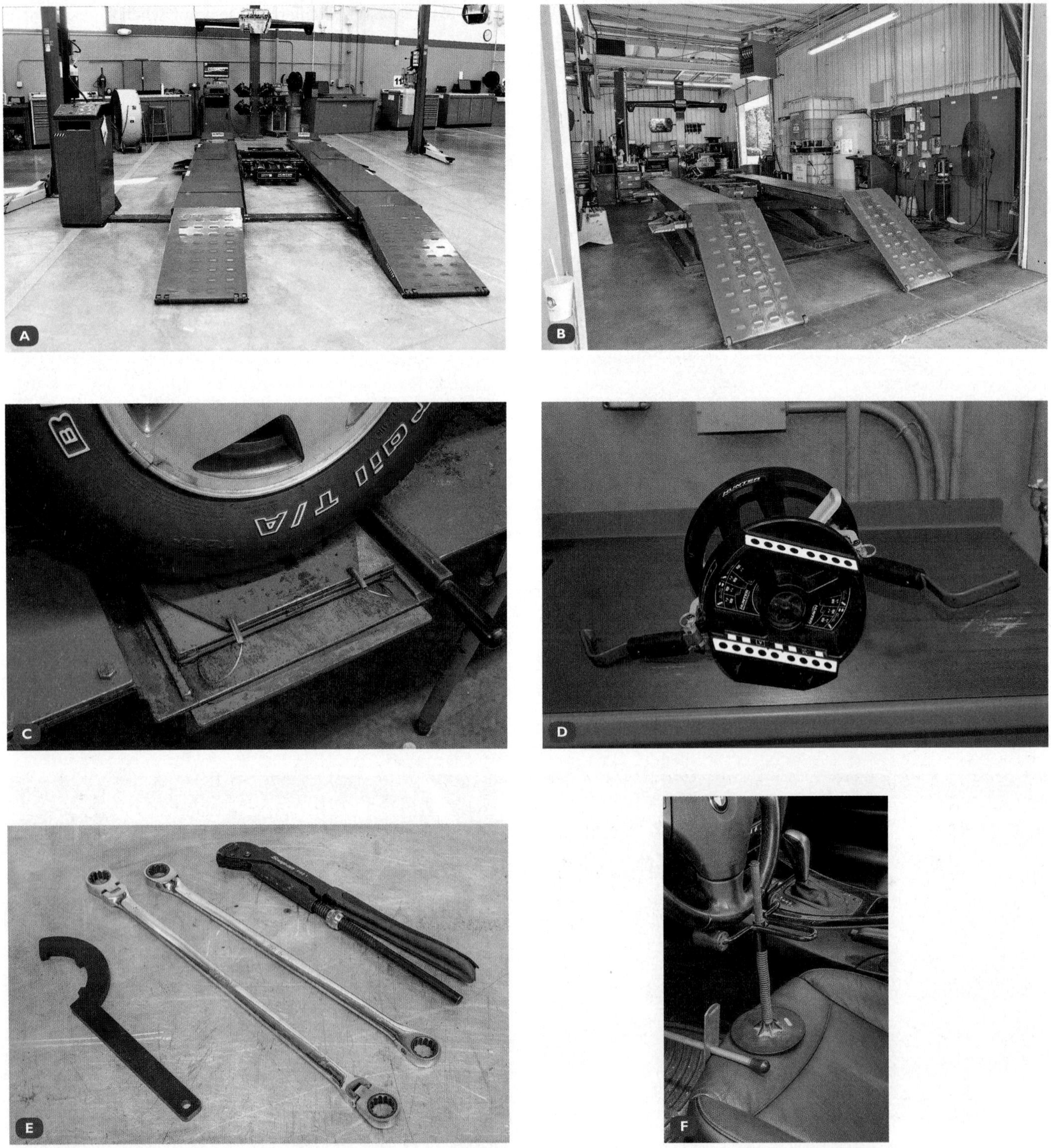

FIGURE 30-32 Common tools and equipment used on performing wheel alignments. **A.** Wheel alignment machine. **B.** Four-post alignment rack. **C.** Turntables. **D.** Wheel clamps/targets. **E.** Specialty alignment tools. **F.** Steering wheel holder.

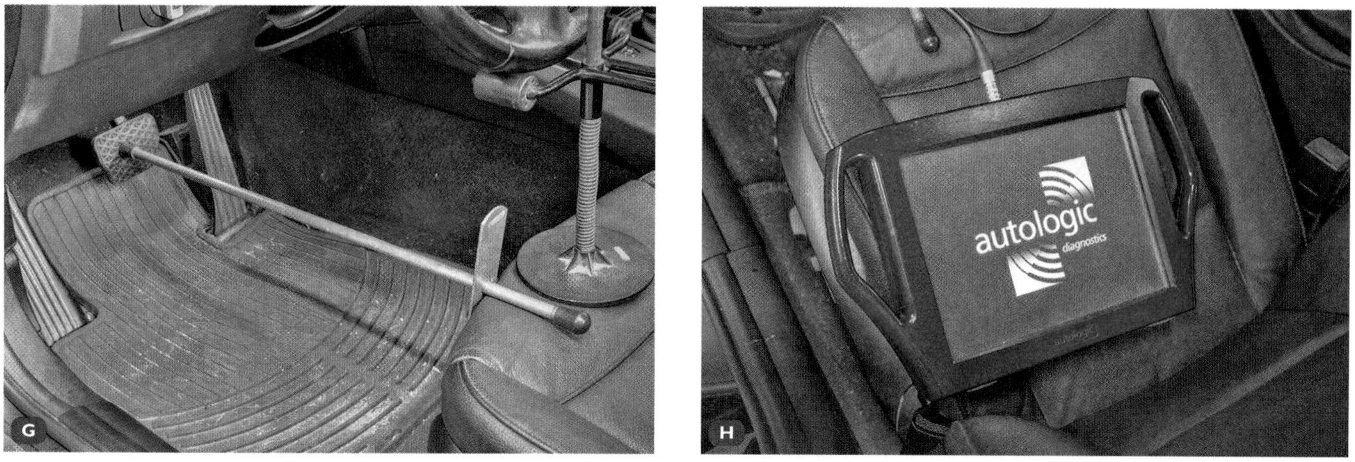

FIGURE 30-32 Common tools and equipment used on performing wheel alignments. **G.** Brake pedal depressor. **H.** Steering angle reset tool.

Performing a Pre-Alignment Inspection

A pre-alignment inspection is performed to ensure that the vehicle is ready for an alignment. Worn suspension or steering components, low tire pressure, and heavy objects in the trunk make it impossible to perform an accurate alignment. So, you need to identify any issues prior to aligning the wheels.

As with any diagnosis, verifying the customer's concern is the first step. Normally, it is best to test drive the vehicle prior to the pre-alignment inspection. This way you can listen carefully for any unusual noises. You can also note any improper driving issues related to the suspension and steering systems. Then, use a good pre-alignment inspection form to guide you through the inspection. These inspection tests were covered previously. Always document your findings for the customer.

To perform a pre-alignment inspection, follow the steps in **SKILL DRILL 30-1**.

SKILL DRILL 30-1 Preparing a Vehicle for a Wheel Alignment

1. Remove any heavy items from the trunk and passenger compartments.

2. Check the size and condition of all four tires. Adjust the air pressure to specifications.

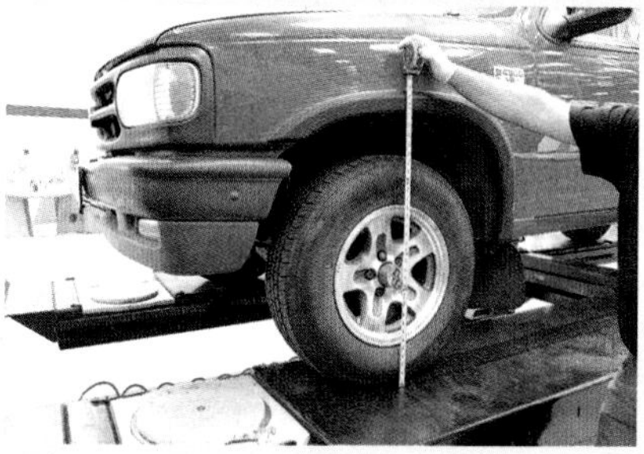

3. Measure the vehicle's ride height.

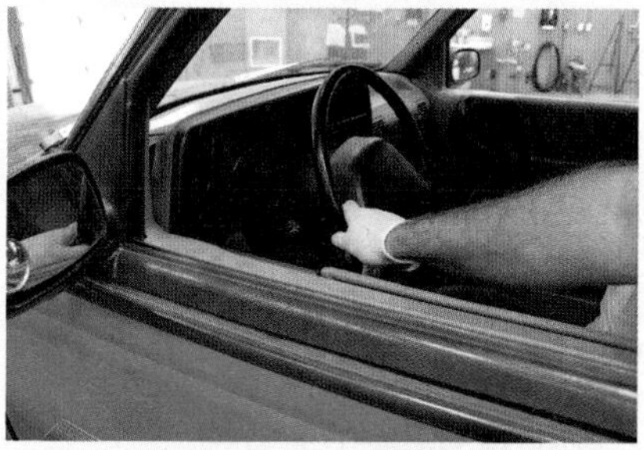

4. Check the play of the steering wheel. Correct any excess play before undertaking the wheel alignment.

5. Bounce each corner of the vehicle to check the correct functioning of the shock absorbers.

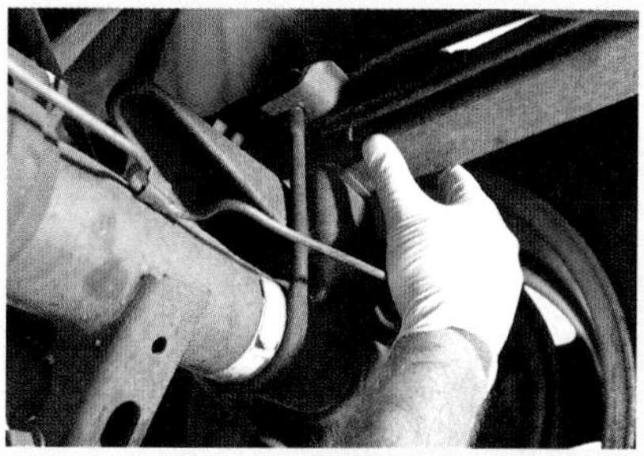

6. With the vehicle raised, inspect all suspension and steering components, including the wheel bearings. Repair or replace all damaged or worn suspension components.

SKILL DRILL 30-1 Preparing a Vehicle for a Wheel Alignment (Continued)

7. Position the vehicle on the wheel alignment ramp making sure the front tires are positioned correctly on the turntables.

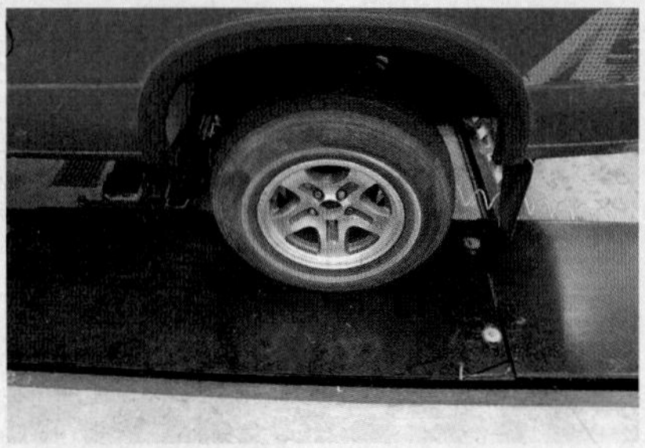

8. Position the rear wheels on the slip plates or rear turntables.

9. Attach the wheel units of the wheel alignment machine.

▶ Performing Four-Wheel Alignment

LO 30-08 Adjust caster, camber, and toe.

The alignment machine measures the angles of the wheels compared to one another. It also compares them with the centerline of the vehicle. These measurements are then compared with specifications to determine any needed adjustments.

The alignment machine's wheel adapters do not attach to the wheels perfectly straight. This means that the wheel adapter runout must be compensated for. Compensation ensures that the alignment machine can read the angles accurately. Alignment machines have the ability to compensate for wheel adapter runout. Each alignment machine has its own wheel adapter compensation process. So, make sure you follow the process for the machine in your shop.

Alignments always follow a specific order of adjustments. The rear wheels should be adjusted before the front wheels. The front wheels are then adjusted to the rear wheels. That way the vehicle can be made to track straight down the road. Once the wheels are aligned, you need to confirm that steering angle sensor is calibrated. The order is as follows:

1. rear caster,
2. camber, and
3. toe; then
4. front caster,
5. camber, and
6. toe (verify steering wheel is centered); then
7. calibrate steering angle sensor.

This sequence is for fully adjustable caster, camber, and toe on the front and rear of the vehicle. Performing the alignment adjustments in this way will prevent having to go back and readjust previous settings. This will save you time. It will also ensure that the alignment is accurate.

Not all vehicles have adjustable caster and camber. In this case, only toe can be set, unless aftermarket caster and camber kits are installed. This Skill Drill uses a vehicle equipped with two eccentric bolts attaching the lower control arm to the frame. This allows caster and camber to be adjusted. Follow the specified procedure in the service information, or let the alignment machine guide you. This Skill Drill is not designed to cover the procedure needed on all vehicles.

To perform a four-wheel alignment, follow the steps in **SKILL DRILL 30-2**.

SKILL DRILL 30-2 Performing Four-Wheel Alignment

1. Position the vehicle on the front-end rack. Raise the vehicle to a comfortable working level, and set the rack on its mechanical locks to provide a level surface.

2. Raise the vehicle with the air jacks on the alignment rack.

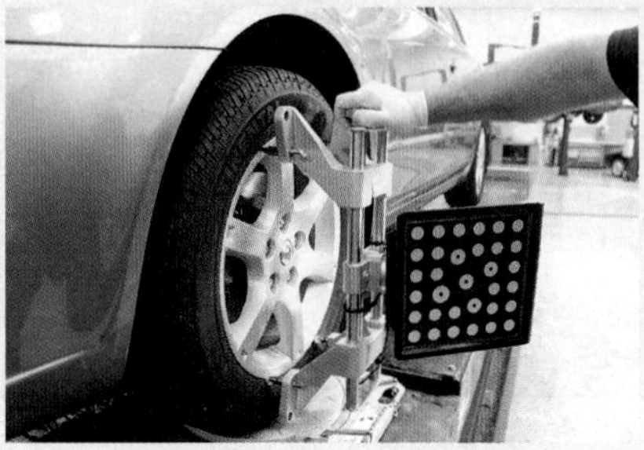

3. Attach sensors and compensate each one.

SKILL DRILL 30-2 Performing Four-Wheel Alignment (Continued)

4. Pull the lock pins from the slip plates and turntables.

5. Lower the vehicle as instructed by the machine. Install a brake pedal depressor.

6. Perform a caster sweep by selecting caster sweep on the machine and turning the wheel the number of degrees on the turntable as designated by the machine.

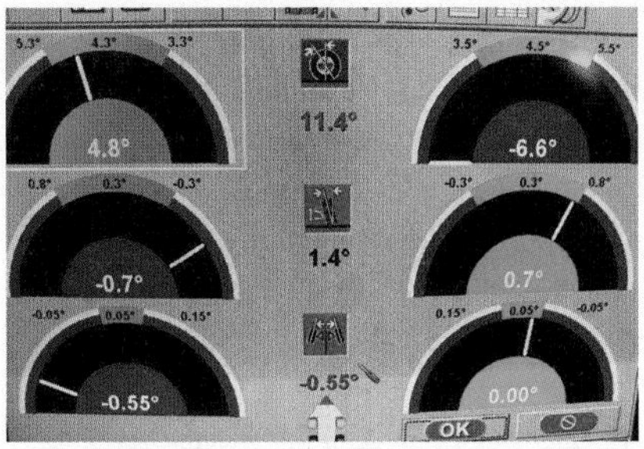

7. Take the alignment readings, and compare them with the vehicle manufacturer's specifications.

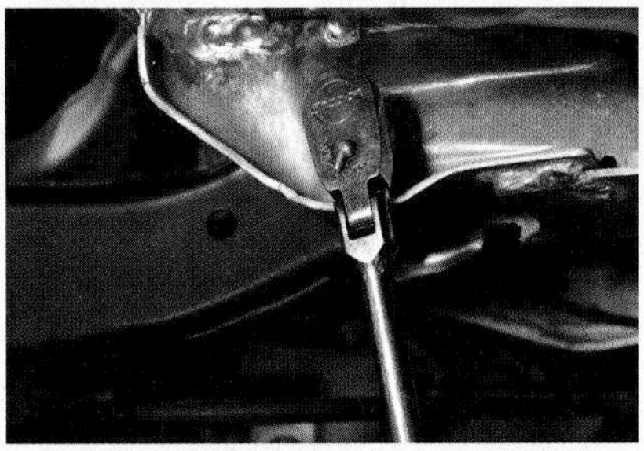

8. Prepare to adjust rear caster, camber, and toe, if possible, by loosening the eccentric bolts.

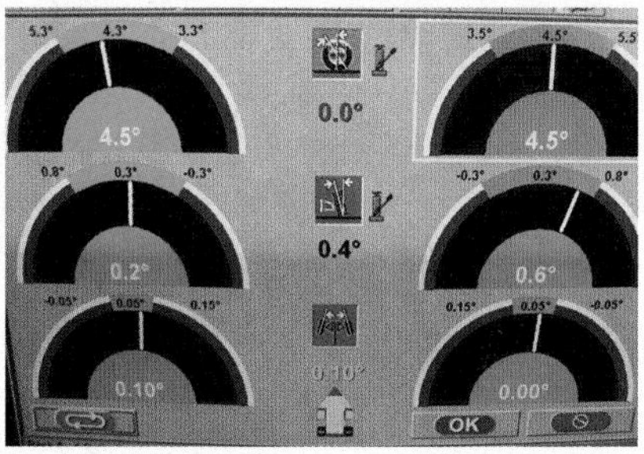

9. Adjust front caster and camber by turning the eccentric bolts attaching the control arm until alignment is within specifications.

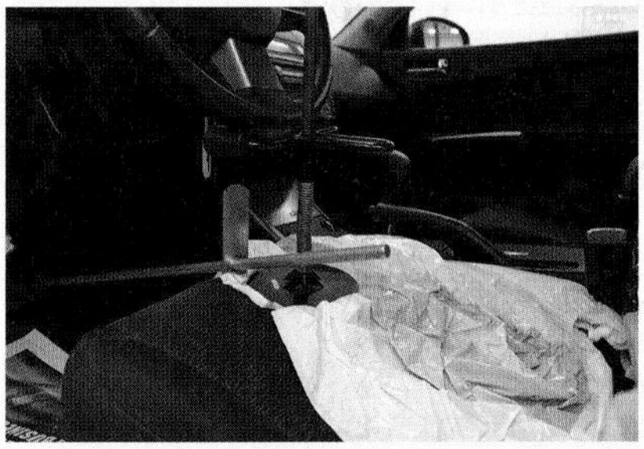

10. Install a steering wheel holder to center the steering wheel.

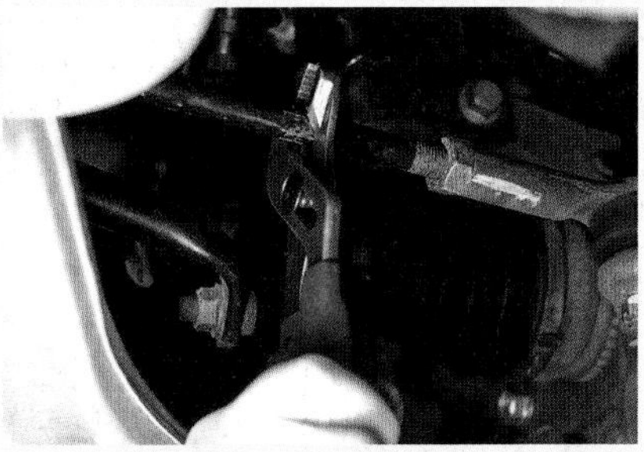

11. Adjust front toe by lengthening or shortening the tie-rod assemblies until toe is within specifications. Tighten the locknuts on the tie-rod assemblies.

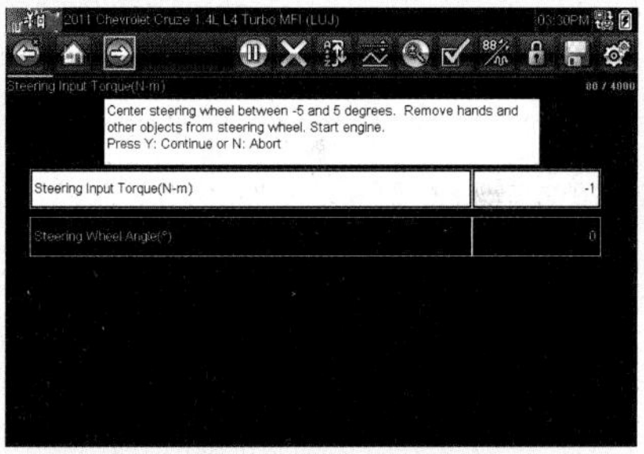

12. Verify that the steering wheel is centered, and calibrate the steering angle sensor. Test drive the vehicle to make sure the repair was successful.

▶ Checking Secondary Alignment Angles

LO 30-09 Measure secondary alignment angles.

The secondary alignment angles are typically diagnostic in nature. They help us understand if a component is bent or damaged. So, looking at them and verifying that they are within specifications will help ensure the vehicle drives properly. If any of the angles are out of specification, you will need to identify the cause. Likely, you will see evidence such as bent metal components. It can also show up visually when compared with the same part on the other side. We will cover each of these secondary alignment angles.

Checking Rear Wheel Thrust Angle

The rear thrust angle refers to the relationship between the rear wheels and the imaginary centerline of the vehicle. Rear thrust angle problems often result from an accident that bends the rear axle or axle mounting points. On independent rear suspension vehicles, incorrect thrust angle is often caused by incorrect rear toe. It can also be caused by component wear or damage.

Incorrect thrust angle will try to steer the rear of the vehicle in the direction the wheels are pointing. Think of a monster truck with rear steering, and the rear wheels turned. The rear of the vehicle will turn in the direction the rear wheels are pointing. The driver will have to turn the front wheels to make the vehicle crab-walk to drive in a straight line. As this vehicle

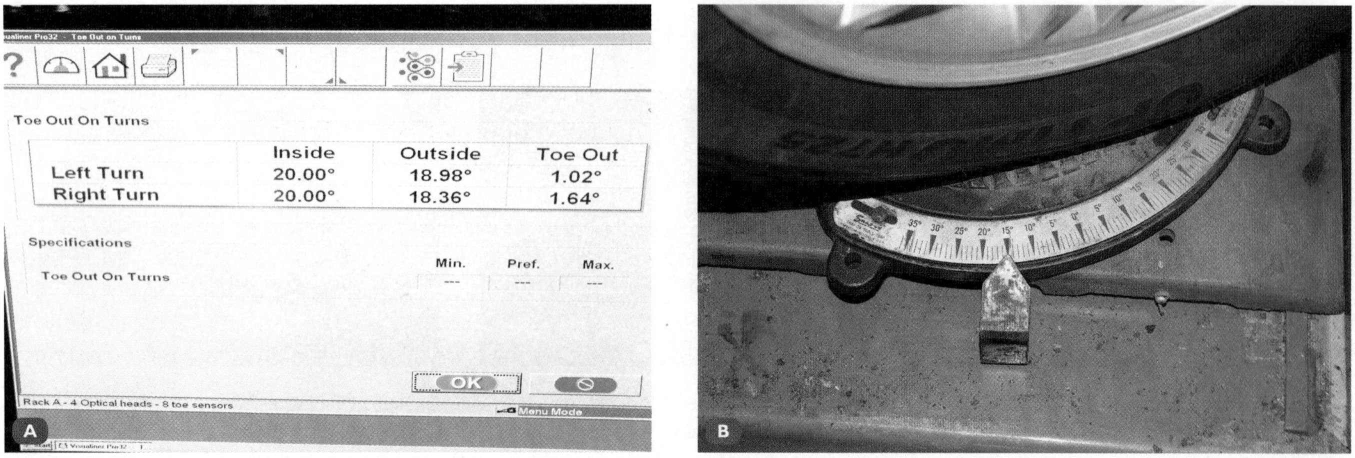

FIGURE 30-33 Displays of toe-out on turns: **A.** Alignment screen. **B.** Turn plates.

moves in a straight line, it will look like the body is sitting a bit sideways. Typically, the steering wheel will be off-center while driving. This is because the driver has to turn the front wheels to compensate for the rear wheels. The ideal thrust angle will be close to zero. But refer to the service information for the correct thrust angle for the vehicle you are working on.

To check the rear wheel thrust angle, follow the steps in **SKILL DRILL 30-3**.

Checking Toe-Out on Turns

When you steer a vehicle through a turn, the outside front wheel has to travel a wider arc than the inside wheel. For this reason, the inside front wheel must steer at a sharper angle than the outside wheel. This prevents the inside tire from dragging or sliding through a turn. It also provides true rolling motion while cornering. Toe-out on turns is sometimes referred to as TOOT, the Ackermann angle, or the track differential angle. Refer to service information to find the specifications for toe-out on turns.

With the tires on the turntable, toe-out on turns is measured electronically and displayed directly on the screen. It can also be read off the turning angle gauges (turn plates) (**FIGURE 30-33**). Refer to the manufacturer's specifications in the service information for the correct toe-out on turns' angles. Typically, a vehicle has an angle of 20 degrees on the inside tire and 18 degrees on the outside tire. This is not the same for every vehicle, so be sure to check specifications. Make sure the readings are at zero on each side when the wheels are straight ahead. Turn the steering wheel so that the inside wheel is at the specified angle. Then check the outside wheel angle and compare it with specifications.

If differences from specifications are found, then typically one or both steering arms are bent. It is also possible that the tie-rods are bent. Before checking this angle, be sure that caster, camber, and toe settings are within specifications. If those angles are incorrect, toe-out on turns will be incorrect.

Compare the maximum turning angle of each wheel as well. If there are differences from specifications, then check for bent or shifted components. This can be a rack that is shifted or a steering wheel that was not properly centered during a repair. In either case, the wheels will not turn as sharply in one direction as the other.

SKILL DRILL 30-3 Checking Rear Wheel Thrust Angle

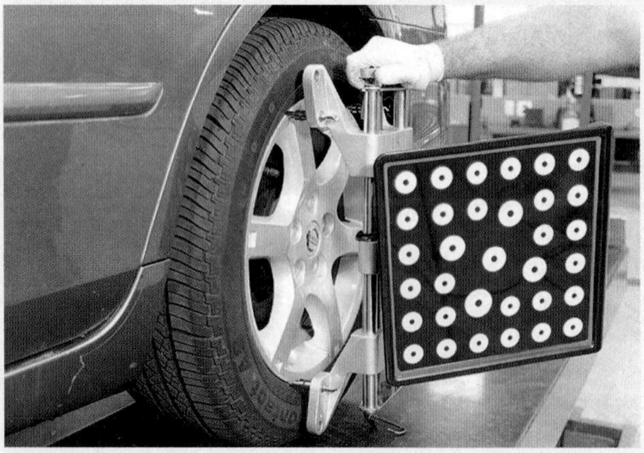

1. Position the vehicle on the alignment rack. Attach the wheel sensors on the vehicle to the locations specified by the sensor manufacturer, and compensate.

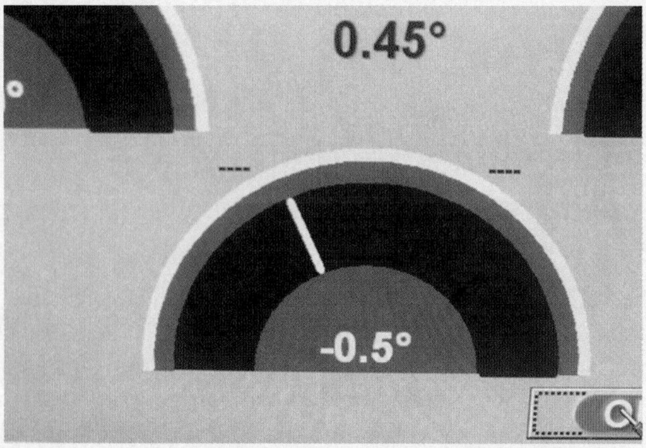

2. Thrust angle can be indicated on most alignment machines, although you may have to go to a special screen. Take the thrust angle reading, and compare it with the vehicle manufacturer's specifications.

Checking SAI and Included Angle

SAI is the angle formed by an imaginary line running through the upper and lower steering pivots relative to a vertical line. Included angle includes the camber reading with the SAI angle. They cannot be adjusted. So if they are out of specifications, something is bent. After an accident or body repair, a technician will often be asked to check SAI and included angle. This is to ensure they are within specifications.

Be sure that caster, camber, and toe settings are within specifications before checking SAI and included angle. Incorrect caster, camber, and toe angles will affect SAI and included angle. Generally, SAI and included angles should not vary more than half a degree from side-to-side. If SAI or included angle is incorrect, check for a bent strut, bent control arms, or a bent spindle or steering knuckle. If incorrect, SAI will commonly create an issue with the steering wheel not returning to center after cornering. To check SAI and included angle, follow the steps in **SKILL DRILL 30-4**.

Checking Front and/or Rear Cradle Alignment

Damage to the cradle of a vehicle can force the wheels out of alignment. This can possibly change caster, camber, toe, and setback readings. It can also create pull and tire wear issues. The cradle should always be perpendicular to the centerline of the chassis. It should also be centered side-to-side in the vehicle. It is possible that after an accident, the cradle mounting points have moved (bent). The vehicle will have to be sent to a frame shop to straighten the body before the alignment can be performed.

It is also important to ensure that the cradle is centered when removing and replacing it. Common examples are when replacing the engine or transmission. It is good practice to perform a wheel alignment after cradle removal and reinstallation. Refer to the manufacturer's service information on cradle adjustment.

To check cradle alignment, follow the steps in **SKILL DRILL 30-5**.

SKILL DRILL 30-4 Checking SAI and Included Angle

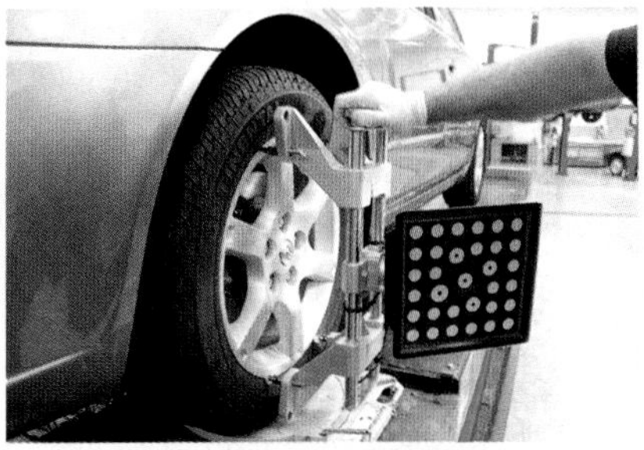

1. Position the vehicle on the alignment rack. Attach the wheel sensors on the vehicle to the locations specified by the sensor manufacturer, and compensate.

2. Follow the alignment machine instructions for taking the SAI measurements, and compare them with the vehicle manufacturer's specifications. Typically, the SAI reading will require a caster sweep to be performed. SAI is a nonadjustable angle. The angle helps to verify that suspension components are bent.

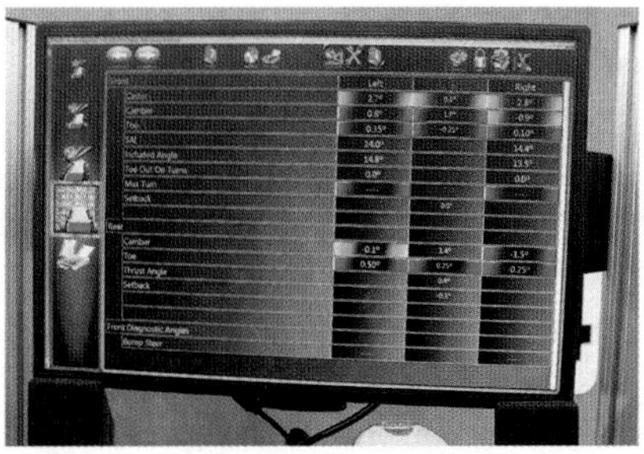

3. Calculate the included angle, if the alignment machine does not. Add the camber reading of each wheel to the SAI of each wheel, and compare with specifications. Remember that if camber is a negative number, you need to subtract the camber from the SAI to get the included angle.

SKILL DRILL 30-5 Checking Front and/or Rear Cradle Alignment

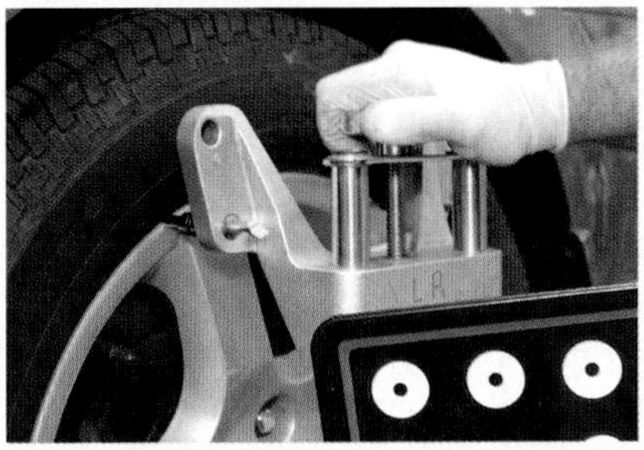

1. Position the vehicle on the alignment rack. Attach the wheel sensors on the vehicle, and compensate. Take the alignment readings and compare with the specifications. If camber and caster are incorrect and not adjustable, check for bent parts.

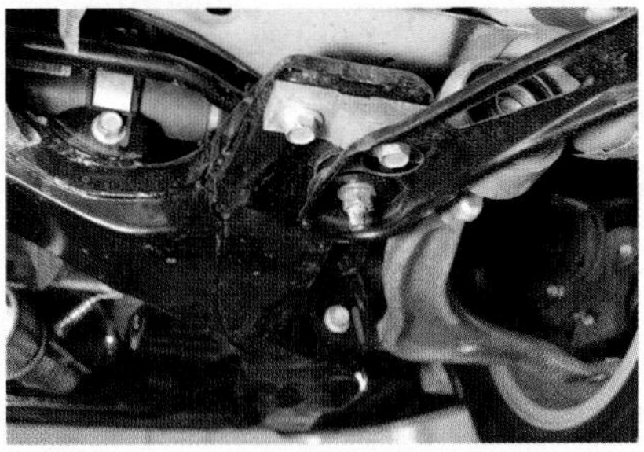

2. If the parts are not bent, check the positioning of the cradle. Loosen cradle bolts and shift in the necessary direction to correct alignment angles.

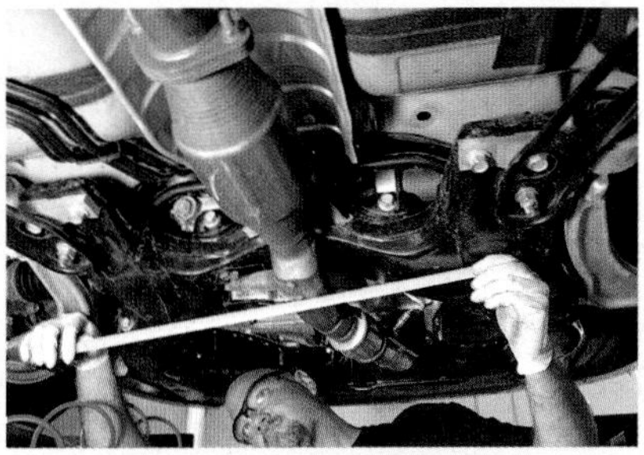

3. If adjustment is still not possible, check for a bent cradle or cradle mounting points by measuring from fixed points on one side compared with the same points on the other side. If they are different, the cradle needs adjusting.

▶ Wrap-Up

Ready for Review

- Camber is the side-to-side vertical tilt of the wheel; caster is the forward or backward tilt of the steering axis from vertical when viewed from the side of the vehicle; and toe is the angle of the tires relative to one another when viewed from above.
  - Caster and camber can be positive and negative; toe can be toe-in and toe-out
- Toe-out on turns is the relative toe setting of the front wheels as the vehicle turns. Turning radius is a measure of how small a circle the vehicle can turn in when the steering wheel is turned to the limit.
- Steering axis inclination (SAI) angle is not adjustable, if the camber angle is correct, then the SAI should also be correct. Included angle is found by adding the SAI angle and the camber angle together. Scrub radius is the distance between two imaginary points on the road surface and can be positive or negative.
- Thrust angle refers to the angle between the centerline of the vehicle and the thrust line and causes the vehicle to pull to the side opposite the thrust line. Setback is the distance one wheel is set back from the wheel on the opposite side of the axle. Ride height is the amount of distance between the ground and a specified part of the vehicle such as the fender well, rocker panel, or frame.
- Performing a wheel alignment requires the use of an alignment machine. Three basic types of wheel alignment are front-end, two-wheel alignment; thrust-angle alignment; and four-wheel alignment.
- Adjusting wheel alignment angles must be done in this specific order: caster, camber, and then toe.
- Commonly used tools in performing wheel alignments include: wheel alignment machine, four-post alignment rack, turntables, wheel clamps, specialty alignment tools, steering wheel holder, brake pedal depressor tool, and scan tool.
  - Four-wheel alignment follows the following order: rear caster, camber, and toe; then front caster, camber, and toe (verify steering wheel is centered); then calibrate steering angle sensor.
- Secondary alignment angles that need to be checked are rear wheel thrust angle, toe-out on turns, SAI and included angle, and front and/or rear cradle alignment.

Key Terms

Ackermann principle or angle The Ackermann principle angles the steering arms toward the center of the vehicle such that imaginary lines drawn from the center of the steering knuckle pivot points, through the center of the outer tie-rod ends, intersect at the center of the rear axle or at a point halfway between the two axles of a tandem drive.

camber The side-to-side vertical tilt of the wheel. It is viewed from the front of the vehicle and measured in degrees. Negative camber is when the top of the tire is closer to the center of the vehicle than the bottom of the tire.

caster The angle formed through the wheel pivot points when viewed from the side in comparison to a vertical line through the wheel.

included angle The angle of camber added or subtracted to the SAI angle. This is the angle of the steering knuckle pivot points in relation to the camber angle of the wheel; also referred to as a diagnostic angle.

negative camber Condition where the top of the tire leans toward the centerline of the vehicle.

negative caster Forward tilt of the steering knuckle pivot points from the vertical line.

negative scrub radius A condition in which the camber line is inside the steering axis centerline or when they intersect above the road surface.

positive camber Condition where the top of the tire leans away from the centerline of the vehicle.

positive caster Backward tilt of the steering knuckle pivot points from the vertical line.

positive scrub radius A condition in which the camber line is outside of the SAI line or where they intersect below the road surface.

scrub radius The distance between two imaginary lines—the camber line through the center of the tire, and the SAI line—or the point where they intersect above or below the surface of the road.

setback The distance one wheel is set back from the wheel on the opposite side of the axle.

static toe A setting, designed to compensate for slight wear in steering components, that may cause the wheels to turn slightly outward or inward while the vehicle is in motion.

steering axis inclination (SAI) The angle formed by an imaginary line running through the upper and lower steering pivots relative to vertical as viewed from the front.

thrust angle The angle formed between the perpendicular centerline of the rear axle in comparison to the centerline of the vehicle.

thrust line The imaginary line drawn perpendicular to the rear axle.

toe-in When the front of the wheels, as seen from above, are closer together than the rear of the wheels.

toe-out When the rear of the wheels, as seen from above, are closer together than the front of the wheels.

toe-out on turns (TOOT) The difference in turning angle of the inside tire in comparison to the outside tire. This angle difference allows the tires to roll through the corner rather than the inside tire dragging. Also referred to as Ackermann angle.

trim height The height from the ground to a specified part of the vehicle; also known as ride height or curb height.

turning radius A measure of how small a circle the outside front wheel (or the outside front corner of the vehicle body) can rotate around when the steering wheel is turned to the limit.

wheel alignment The practice of aligning the wheels of the vehicle to the centerline of the vehicle and to one another. It ensures that the vehicle will handle correctly and gives best tire wear.

zero camber A tire with no tilt, or zero camber angle.

zero scrub radius A condition in which the camber line through the center of the tire, intersects the SAI line at the road surface.

Review Questions

1. Which of the following is considered a secondary angle?
   a. Caster
   b. Toe
   c. Setback
   d. Camber
2. What alignment angle would cause the most rapid tire wear if not in specification?
   a. Caster
   b. Camber
   c. Toe
   d. Steering axis inclination
3. What is the term for how small a circle the vehicle can turn in with the steering wheel turned to the limit?
   a. Ackerman angle
   b. Maximum toe
   c. Circular diameter
   d. Turning radius
4. Included angle is the combination of camber plus _____.
   a. SAI
   b. Caster
   c. Scrub radius
   d. Total toe
5. What is the most likely cause for incorrect setback?
   a. Incorrect front toe
   b. Incorrect rear toe
   c. Cradle shift
   d. Sagging springs
6. What type of alignment is most likely to be performed on a rear wheel drive vehicle with a solid rear axle today?
   a. A 2-wheel alignment
   b. A thrust angle alignment
   c. A 4-wheel alignment
   d. A tape measure alignment
7. As part of a pre-alignment inspection, all of the following should be performed EXCEPT:
   a. Ride height measured or adjusted
   b. Loose steering parts replaced or repaired
   c. Bent suspension components replaced
   d. Previous alignment shims removed
8. When performing a four-wheel alignment. What is the last adjustment that should be made to the rear, before beginning adjustment in the front?
   a. Total caster
   b. Rear toe
   c. Rear caster
   d. Rear camber
9. What is the most likely secondary angle to cause a vehicle to "crab-walk"?
   a. Rear thrust angle
   b. Toe out on turns
   c. SAI
   d. Scrub radius
10. Generally, SAI and included angles should not vary more than how much side to side?
    a. A half degree
    b. One degree
    c. Three degrees
    d. Five degrees

ASE Technician A/Technician B Style Questions

1. A vehicle comes in for an alignment. Technician A notices the L/F wheel is tilted inward at the top and claims this is negative camber. Technician B observes excessive front tire wear and says it could be caused by incorrect toe. Who is correct?
   a. Technician A only
   b. Technician B only
   c. Both Technicians A and B
   d. Neither Technician A nor B
2. A vehicle comes in to the shop experiencing a pull to one side when driving. Technician A states the pull could be caused by a difference in caster from side to side. Technician B states an unequal camber from side to side could be the cause. Who is correct?
   a. Technician A only
   b. Technician B only
   c. Both Technicians A and B
   d. Neither Technician A nor B
3. Alignment angles are being discussed. Technician A states when turning, the inner wheel must be turned at a sharper angle than the outer wheel. Technician B states toe-out on turns is how quickly a vehicle can turn 180 degrees. Who is correct?
   a. Technician A only
   b. Technician B only
   c. Both Technicians A and B
   d. Neither Technician A nor B
4. Alignment angles are being discussed. Technician A states SAI is the axis around which a wheel assembly swivels as it turns. Technician B states SAI is formed by drawing a line

through the center of the upper and lower pivot points on the suspension. Who is correct?
   a. Technician A only
   b. Technician B only
   c. Both Technicians A and B
   d. Neither Technician A nor B
5. A vehicle is being inspected and aligned. Technician A states the vehicle thrust angle can be affected by bent parts or the result of impact. Technician B states that trim height cannot be affected by sagging springs. Who is correct?
   a. Technician A only
   b. Technician B only
   c. Both Technicians A and B
   d. Neither Technician A nor B
6. An alignment is being performed on a modern vehicle. Technician A states the vehicle should be aligned using the two-wheel alignment method. Technician B states the alignment should be performed after the worn or damaged components have been replaced. Who is correct?
   a. Technician A only
   b. Technician B only
   c. Both Technicians A and B
   d. Neither Technician A nor B
7. A customer is requesting an alignment be performed on their 2-year-old vehicle. Technician A states that an alignment should be performed as maintenance, or because of having steering or suspension parts replaced. Technician B states that the steering angle sensor may require recalibration after the alignment is performed. Who is correct?
   a. Technician A only
   b. Technician B only
   c. Both Technicians A and B
   d. Neither Technician A nor B

8. A four-wheel alignment is being performed on a modern vehicle. Technician A states the front toe should be set first. Technician B states the ride height should be adjusted last. Who is correct?
   a. Technician A only
   b. Technician B only
   c. Both Technicians A and B
   d. Neither Technician A nor B
9. An alignment is being performed. Technician A states that performing the alignment steps out of sequence can waste time and may cause inaccurate readings. Technician B states any heavy items should be removed from the trunk or cargo areas prior to alignment. Who is correct?
   a. Technician A only
   b. Technician B only
   c. Both Technicians A and B
   d. Neither Technician A nor B
10. A vehicle's alignment is being discussed. Technician A states caster, camber, and toe are the only relevant alignment angles. Technician B states secondary angles are used primarily for diagnosis. Who is correct?
   a. Technician A only
   b. Technician B only
   c. Both Technicians A and B
   d. Neither Technician A nor B

brembo

SECTION 6
Brakes

▶ CHAPTER 31 Principles of Braking

▶ CHAPTER 32 Hydraulics and Power Brakes Theory

▶ CHAPTER 33 Servicing Hydraulic Systems and Power Brakes

▶ CHAPTER 34 Disc Brake Systems Theory

▶ CHAPTER 35 Servicing Disc Brakes

▶ CHAPTER 36 Drum Brake Systems Theory

▶ CHAPTER 37 Servicing Drum Brakes

▶ CHAPTER 38 Wheel Bearings

▶ CHAPTER 39 Electronic Brake Control

CHAPTER 31

Principles of Braking

Learning Objectives

- **LO 31-01** Describe the history of brake development.
- **LO 31-02** Describe braking fundamentals.
- **LO 31-03** Describe the physics of braking.
- **LO 31-04** Describe friction, heat transfer, and brake fade.
- **LO 31-05** Describe rotational force, weight transfer, and levers.
- **LO 31-06** Describe the common types of automotive brakes.

ASE Education Foundation Tasks

See Appendix A to view the 2017 ASE Education Foundation Automobile Accreditation Task List Correlation Guide.

▶ Introduction

The brake system is one of the most critical systems on a vehicle. It allows the driver to slow or stop the vehicle as needed. In ideal situations, the driver will have enough time to anticipate the need to slow down well in advance of an event. This would allow the vehicle to slow down gradually. However, many situations require the quick use of a very efficient braking system to avoid an accident (**FIGURE 31-1**). In this chapter, we explore the history, theory, and operation of modern braking systems.

▶ The History of Brakes

LO 31-01 Describe the history of brake development.

Early automobiles evolved from horse-drawn buggies and used a similar scrub braking system. **Scrub brakes** are a simple mechanical system. They use leverage to force a friction block against one or more wheels (**FIGURE 31-2**). In a buggy, for example, the friction between the two surfaces transformed the energy of the moving buggy into heat energy. As heat was created in the friction materials, the buggy slowed down.

The scrub braking system was used for more than 2000 years with virtually no change. It worked reasonably well on dry wheel surfaces made of wood or steel. They became quickly outdated once rubber tires were developed around 1900. This is because the scrubbing action on the softer tires significantly decreased their life.

Faced with the need to replace the scrub braking system, designers had to consider other options. One option was the **band brake**. It used a metal band lined with friction material to clamp around the outside of a drum mounted to the axle or wheel (**FIGURE 31-3**). This system worked well in the forward direction. In the reverse direction, the band would try to unwind. The system was therefore impractical and abandoned after just a few short years.

FIGURE 31-1 Brakes must be fully functional in case of an emergency.

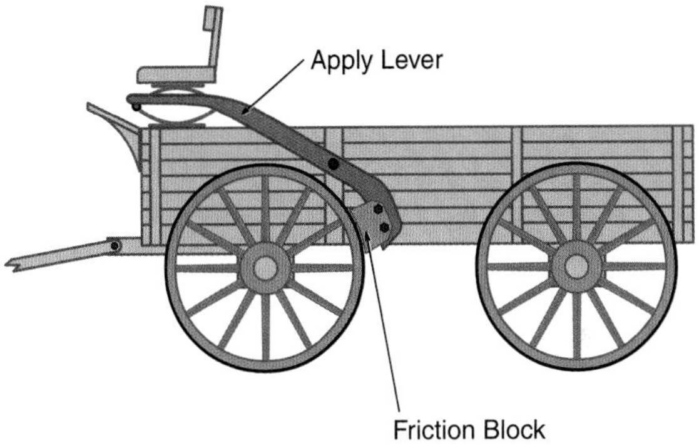

FIGURE 31-2 Scrub brakes operated directly on the wheel.

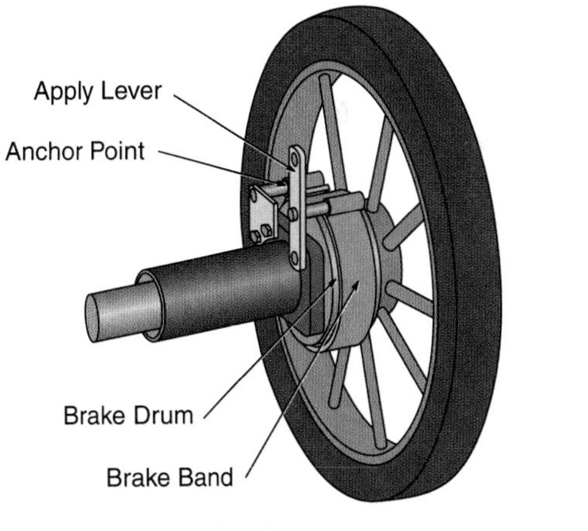

FIGURE 31-3 An old style band brake.

You Are the Automotive Technician

A customer comes into the shop with a brake concern on their 2012 Ford Taurus. They recently drove down a long mountain pass, using the brakes to stay within the speed limit. As they neared the bottom of the pass, the brakes felt like they weren't working, even though the engine was running and they were pushing very hard on the pedal. You recognize the smell of overheated brakes coming from the car. They ask you why the brakes didn't work correctly.

1. How will you explain what happened to the brakes?
2. What would you advise them to do if they are in that situation again?
3. Which types of brakes are more susceptible to this condition?
4. How are drum brakes and disc brakes different?

The next major development was the **drum brake**. It is similar to the drum brakes that are used today. The drum brake consists of two brake shoes that push against the inside of the brake drum. Early drum brake systems were mechanically operated by rods, links, and levers (**FIGURE 31-4**). This worked fine for applying a single brake unit.

As vehicle operating speeds increased, greater braking demands required brake units to be mounted to each wheel. With this mechanically operated system of rods, links, and levers, it was difficult to maintain equal braking forces at each wheel. It commonly caused the vehicle to veer dangerously to one side when braking. To prevent this, frequent adjustment of the brakes was required. Over time, the mechanical drum brake system gave way to the modern hydraulic drum brake system. This was due to its ability to automatically equalize braking forces at each wheel.

Disc brakes were originally developed in the early 1900s. But they did not find common use until the 1960s. The effectiveness of friction brakes depends on their ability to quickly dissipate heat into the atmosphere. This means that drum brakes with their friction materials on the inside were at a disadvantage. This led to the greater use of **disc brakes** on most vehicles (**FIGURE 31-5**). Disc brakes force brake pads against the outside of the brake rotor. Heat is created where atmospheric air can quickly remove the heat. This makes them more efficient under prolonged use.

Advanced Brake Systems

You might think, "If some brake force is good, then more would be better." But this is not the case. Applying too much brake force can cause the tire to lose traction and skid. If this happens, the driver can lose control of the vehicle. This problem led to the design of electronic brake control (EBC) systems. The first versions of EBC systems were the **antilock brake systems (ABS)**.

ABS systems have helped to reduce the number of vehicle accidents each year. They use a computer to monitor each wheel's speed during braking. That information is used to either hold, release, or apply hydraulic pressure to each wheel's brake unit. Wheel lockup is then prevented and maximum braking power is maintained, just short of brake lockup.

Additional safety demands led to the development of two major enhancements to ABS. The first was traction control systems (TCS) followed by electronic stability control (ESC). TCS systems help prevent tires from slipping during acceleration by first reducing engine torque. And, if necessary, it can apply brake pressure to any wheels that are slipping. ESC adds extra functionality to ABS and TCS. It helps prevent the tires from losing traction when the vehicle is being steered aggressively. It also works when evasive maneuvers are being undertaken.

In both TCS and ESC, the control unit can independently apply individual brake units. It can do this even though the driver is not stepping on the brake pedal. For more information on ABS brake systems, see the chapter on Electronic Brake Control.

Brake Assist and Brake-by-Wire Systems

EBC systems have been further enhanced. Additional programmed features such as **brake assist (BA)** gives greater control of the braking system to the computer. Since it can react more quickly and more deliberately than a driver, it is ideal in a panic situation.

An example of BA is the **brake-by-wire system**. A full brake-by-wire system does away with the hydraulic portion of the brake system. It replaces the hydraulic system with sensors, wires, an electronic control unit (ECU), and electrically actuated motors. The motors apply individual brake units at each wheel (**FIGURE 31-6**).

The driver applies foot pressure to a **brake pedal emulator**. This tells the computer how firmly the driver intends to brake. The control unit then sends control signals to the appropriate brake actuators. They generate the commanded clamping (brake) force and slow the vehicle. All of this is being monitored by sensors reporting data to the control unit, so the desired braking occurs.

FIGURE 31-4 An early drum brake system.

FIGURE 31-5 Disc brakes force the brake pads against the outside of the brake rotor.

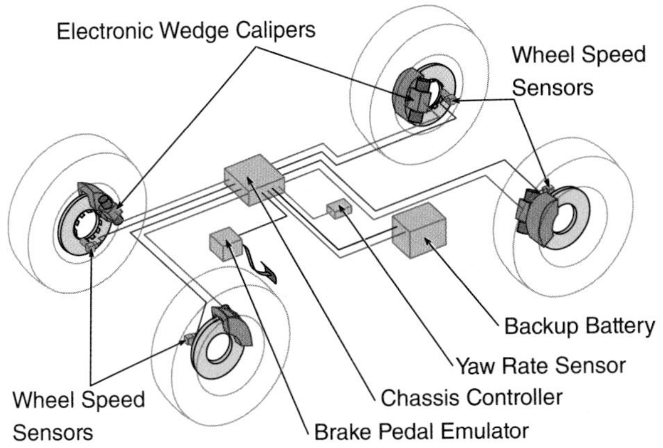

FIGURE 31-6 Schematic of a brake-by-wire system.

Giving greater control of the braking system to the computer increases driving safety. For example, it takes a certain amount of time and stopping distance to lift your foot from the accelerator pedal. It takes more time to step on the brake pedal. And even more time to apply the wheel brake units.

In a brake-by-wire system, that time and distance can be reduced. The computer can determine that the driver is in a potential panic stop. This is indicated by the quickly closing throttle plate. As the throttle pedal is being released quickly, the control system immediately applies the brakes lightly. This dries any moisture from the braking surfaces and takes up any clearance in the brake system. It also prepares the brakes to be fully applied by the control system if the driver steps on the brake pedal. The system can also prepare the brakes if the computer detects that a collision is about to happen.

▶ TECHNICIAN TIP

Many auto insurance companies give their customers a discount when the vehicle is equipped with brake safety features such as ABS systems. This is because they help avoid or minimize accidents.

Regenerative Brake Systems

Some vehicle manufacturers have hybrid vehicles with regenerative braking systems. This is due to increased government regulation geared toward improving fuel economy. Regenerative braking takes brake-by-wire technology to the next level. The regenerative braking system uses the electric motor as a generator. This is instead of applying friction brakes and losing energy as heat. The generator slows the vehicle by changing the vehicle's kinetic energy into electrical energy. The electricity generated controls the amount of stopping power (**FIGURE 31-7**). More stopping power means more electrical output created by the generator. This is limited to its highest rated output.

The regenerated electricity is stored in the vehicle's high-voltage battery. It can then be used later by the electric motor to drive the vehicle. This regeneration process makes the vehicle more fuel-efficient, especially in stop-and-go traffic.

Brake Fundamentals

▶ Service Brakes and Parking Brakes

LO 31-02 Describe braking fundamentals.

There are two brake systems on all vehicles—a service brake and a parking brake. The **service brake** is used for slowing or stopping the vehicle when it is in motion. It is operated by a foot pedal (**FIGURE 31-8**). Service brakes consist of drum and/or disc brakes. Some have disc brakes on the front wheels and drum brakes on the rear wheels. Others have disc brakes on all four wheels.

The **parking brake** is used for holding the vehicle in place when it is stationary. The parking brake is usually operated by hand. But some vehicles use a foot-activated pedal (**FIGURE 31-9**). It is not designed to be used when the vehicle is moving. That is also why the term "emergency brake" is not proper.

FIGURE 31-7 A vehicle display showing regeneration charging the high-voltage battery.

FIGURE 31-8 A vehicle service brake is used for slowing or stopping a moving vehicle.

Modern braking systems are hydraulically operated. They have two main sections: the brake units and the hydraulic system that applies them. The brake units are located at each wheel. The driver operates the brake pedal. Mechanical force is then applied to the pistons in the master cylinder. The pistons create hydraulic pressure in the master cylinder. The brake lines transfer the brake fluid throughout the hydraulic system. Equal pressure is maintained within the system. The hydraulic pressure pushes against the pistons in the cylinders at each wheel. The pistons force the friction material into contact with the braking surfaces (**FIGURE 31-10**).

Force is transmitted hydraulically through the fluid. For cylinders of the same size, the force on one is the same force created by the other. By using cylinders of different sizes, forces can be increased or reduced. This allows designers to obtain the desired braking force for each wheel (**FIGURE 31-11**). The cylinders force friction linings into contact with the braking surfaces. The resulting friction between the surfaces generates heat energy and slows the vehicle.

In disc brakes, pads are forced against the outside of a brake disc (**FIGURE 31-12**). In both systems, heat spreads into other parts and the atmosphere. This means that brake linings and drums, pads and rotors, and brake fluid must withstand high temperatures and high pressures. On modern vehicles, the basic brake system has some refinements. The two main ones are a power booster and EBC systems. These help the driver apply the brakes, prevent skidding, and maintain directional control of the vehicle. They do so under various driving situations (**FIGURE 31-13**).

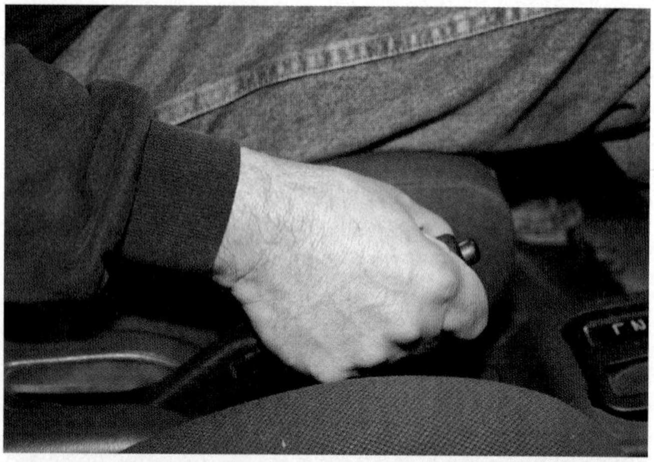

FIGURE 31-9 A vehicle parking brake is used to hold the vehicle in place when it is stationary.

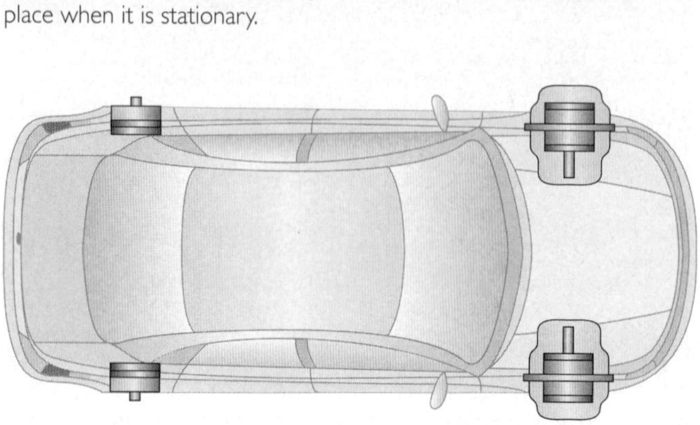

FIGURE 31-10 A braking system in a typical modern vehicle.

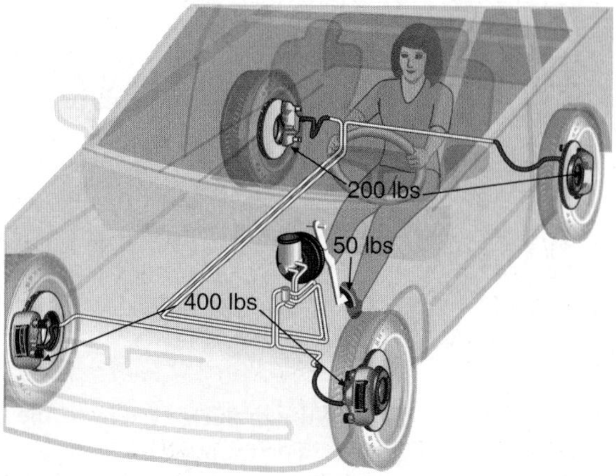

FIGURE 31-11 By using cylinders of different sizes, hydraulic forces can be increased or reduced, allowing designers to obtain the desired braking force for each wheel.

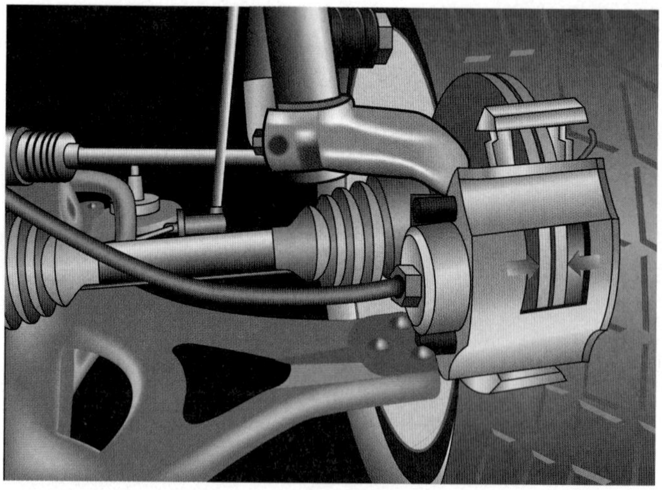

FIGURE 31-12 In disc brakes, pads are forced against the outside of a brake disc.

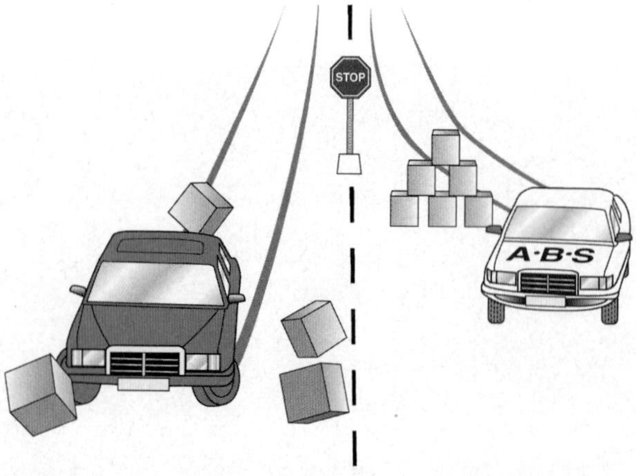

FIGURE 31-13 ABS brakes help prevent skidding and maintain directional control of the vehicle.

Factors That Affect Braking

A number of factors can influence vehicle braking. An effective braking system is designed so it will function in all of the following situations:

- **Road surface:** Generally, asphalt and concrete road surfaces allow for good braking while gravel surfaces or dirt roads do not.
- **Road conditions:** Roads that are wet, icy, or covered with loose gravel reduce the tires' traction and result in longer stopping distances. Extremely hot temperatures on asphalt roads can soften the asphalt, making it slippery.
- **Weight of the vehicle:** Heavier vehicles require more braking force to stop than lighter vehicles. Therefore, they usually have larger wheel brake units. Also, loading down a vehicle increases its stopping distance due to the vehicle's extra mass.
- **Load on the wheel during stopping:** Heavier loads increase the downward force on the wheels, thereby increasing tire traction (**FIGURE 31-14**).
- **Height of the vehicle:** Stopping power is exerted at the point where the tire and the road connect. The centerline of the vehicle's weight is above this tire-to-road contact point. The taller the vehicle, the greater the leverage on the contact point. During braking, this shifts some of the vehicle's weight from the rear wheels to the front wheels. Thus, controlling the vehicle in a panic situation becomes much more difficult.
- **How the vehicle is being driven:** Aggressive driving causes the tires to become hot and possibly overheated. This reduces the tire's ability to obtain maximum traction. Also, increased speed and aggressive handling force the brakes to work under extreme conditions.
- **The tires on the vehicle:** A tire's composition, tread style, tread condition, and inflation pressure all affect its traction (**FIGURE 31-15**). Manufacturers design tires with different qualities based on vehicle need. Tires are rated for their traction ability. Using the wrong tire will affect the vehicle's stopping power. For example, a tire with tread that is designed to channel water away from the tire-to-road contact point has greater traction when the road surface is wet than the tires used on drag race cars, which have a slick tread.

▶ Kinetic Energy

LO 31-03 Describe the physics of braking.

Kinetic energy is the energy of an object in motion. All moving objects have kinetic energy. Heavier objects have more kinetic energy than lighter objects moving at the same speed. If the weight doubles, the kinetic energy doubles.

FIGURE 31-14 Tires with differing loads. **A.** The empty pickup truck has low traction on its rear tires. **B.** A heavily loaded truck has higher traction on its rear wheels.

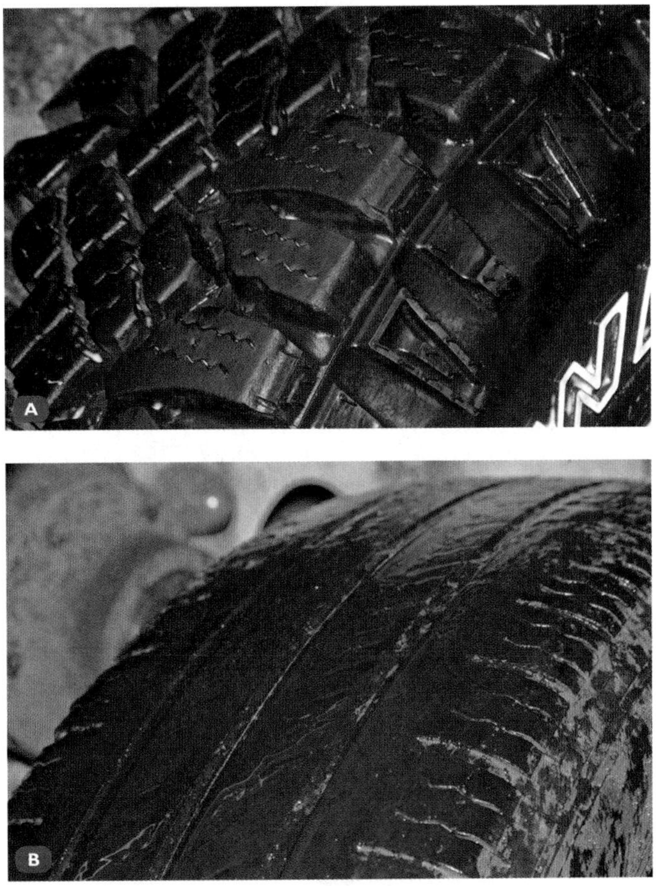

FIGURE 31-15 The condition of the tires affects braking performance. **A.** Tire in excellent condition. **B.** Worn-out tire.

Faster moving objects have more kinetic energy than slower moving objects of the same weight. Kinetic energy increases by the square of the speed. This means that if we double the speed of an object, the kinetic energy will increase by four times. If we triple the speed, the kinetic energy will increase by nine times. Thus, the heavier and faster an object is, the greater its kinetic energy (**FIGURE 31-16**).

During braking, the kinetic energy in the moving vehicle is converted to another form of energy. In most cases, it is converted into heat energy. This is what allows the vehicle to stop moving. Converting kinetic energy into heat energy is the function of the braking system.

Acceleration and Deceleration

Newton's first law of motion states that an object will stay at rest or uniform speed unless it is acted upon by an outside force. **Acceleration** refers to an increase in an object's speed. In an automobile, acceleration, or an increase in kinetic energy, is caused by the power from the engine. When the driver steps on the throttle pedal, the engine's power output is increased, and the vehicle accelerates (**FIGURE 31-17**). This acceleration requires a certain amount of energy. The heavier the vehicle, the more energy required to accelerate it to a given speed. A lighter vehicle requires less energy to accelerate. This is why race cars are stripped of all unnecessary weight.

Deceleration refers to a decrease in an object's speed. Remember that an outside force is needed for the speed of an object to change. So, we need an outside force to act upon the vehicle to cause it to decelerate. That force comes from the mass of the Earth. If you thought it came from the brakes, you would only be partially correct.

Imagine traveling at a high speed in a four-wheel drive vehicle and hitting a bit of a jump. Stepping on the brakes in mid-air to slow the vehicle wouldn't do you much good, would it? So the brakes only function when they connect the vehicle to the ground or roadway. In fact, that is what they do. They connect the vehicle to the ground through the rolling wheel and tire assembly. In doing so, they apply a varying amount of force from the ground to the vehicle. This is what causes the vehicle to decelerate.

The force of the brakes absorbs the kinetic energy of the vehicle as it brakes (**FIGURE 31-18**). The heavier the vehicle and the faster it is going, the more kinetic energy must be dissipated. Also, the harder the brakes must work.

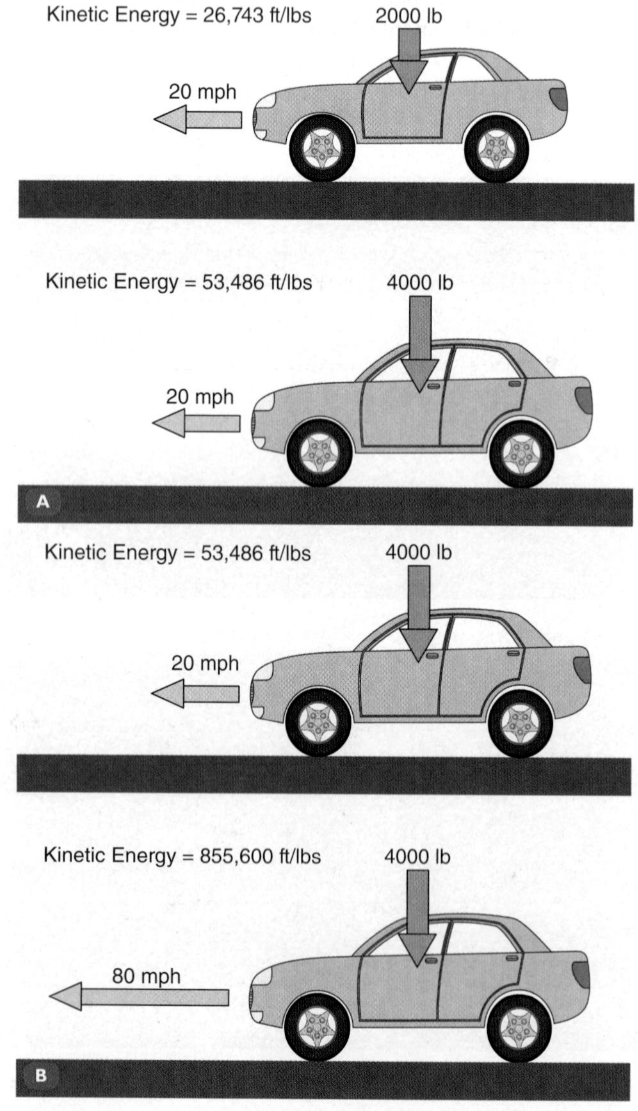

FIGURE 31-16 A. If weight doubles, kinetic energy doubles. **B.** If speed increases, kinetic energy increases by the square.

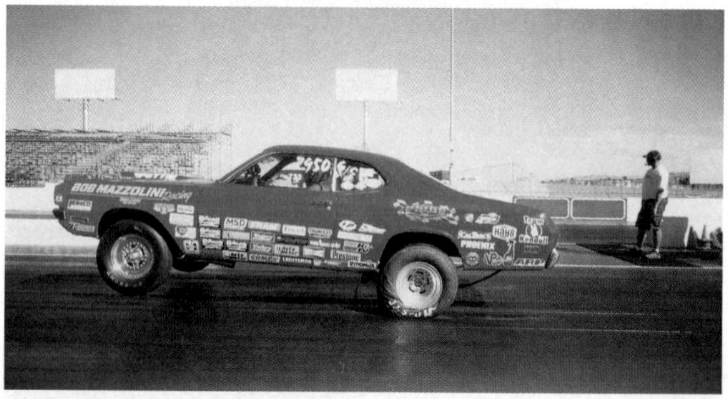

FIGURE 31-17 The engine's horsepower is accelerating this vehicle hard.

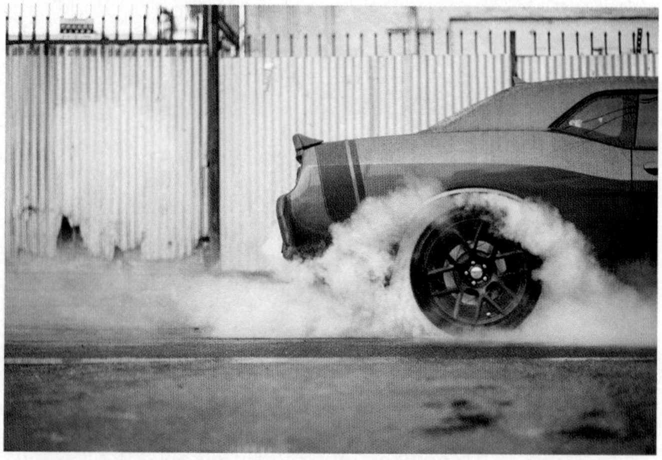

FIGURE 31-18 The vehicle's brakes are decelerating this vehicle hard.

Applied Science

AS-51: Acceleration/Deceleration: The technician can demonstrate an understanding of a vehicle's acceleration and deceleration as a function of vehicle weight and power.

Kinetic energy is the energy of an object in motion. All moving objects have kinetic energy. Heavier objects have more kinetic energy than lighter objects moving at the same speed. If the weight doubles, the kinetic energy doubles. Energy is needed to start a vehicle. Heat energy is generated in the engine via chemical energy (fuel). It is then converted via mechanical energy to kinetic energy, putting the vehicle in motion. Kinetic energy is converted to heat energy once again through the operation of the brakes. This heat energy is then dissipated in the surrounding air through the brake system, bringing the vehicle to rest.

Newton's first law of motion states that the greater the weight, the more energy is needed to accelerate and maintain speed. For example, it is easier for three people to push a two-door hatchback than an SUV. Also, the more the vehicle weighs, the more energy it takes to decelerate.

Energy Transformation

The law of **conservation of energy** states that energy cannot be created or destroyed. This means that the energy used to cause a vehicle to accelerate and decelerate must be transformed from one form of energy to another. Let's follow the cycle of energy transformation in a typical vehicle.

Gasoline or diesel fuels are potential energy in chemical form. A portion of the fuel's chemical energy is transformed within the engine, first into heat energy and then into mechanical energy. The mechanical energy is used to accelerate the vehicle. This converts the mechanical energy to kinetic energy. Once the vehicle is up to speed, the engine only needs to transform enough chemical energy into kinetic energy to keep it at speed. This involves overcoming wind resistance, climbing hills, and powering the vehicle's accessories. This is why most vehicles get better fuel economy while operating at steady speeds than in stop-and-go traffic. It takes a lot more energy to accelerate a vehicle than it does to maintain a particular speed.

Does deceleration require an energy transformation? Yes, it does. The kinetic energy has to be removed from the vehicle for it to decelerate. In other words, the kinetic energy must be transformed into another form of energy for the vehicle to slow down. In a standard vehicle, the braking system transforms the kinetic energy into heat energy (**FIGURE 31-19**). In essence, it takes the same amount of energy to slow a vehicle as it does to accelerate it. For safety's sake, we expect a vehicle to stop from a given speed faster than the time it took to accelerate to that speed. For this reason, the braking system can transform energy faster than the engine. Or in other words, the brake system can absorb energy at a faster rate than the engine can create it.

▶ Friction and Friction Brakes

LO 31-04 Describe friction, heat transfer, and brake fade.

Brakes transform kinetic energy to another form of energy. Standard brakes do this through the principle of friction. **Friction** is the resistance created by surfaces in contact. Static friction is resistance between nonmoving surfaces and is present in parking brakes. Kinetic friction is resistance between moving surfaces and is present in service brakes. Think of how rubbing sandpaper over a block of wood produces heat. In the same way, operating the brakes causes the moving friction surfaces to generate heat. This transformation of energy converts kinetic energy into heat energy and slows the vehicle.

The amount of friction between two surfaces in contact with each other is expressed as a factor, called the **coefficient of friction**. It can be found by comparing two forces. One is the amount of force pushing the two surfaces together. The other is the amount of resistive force generated between the two surfaces. For example, a stationary steel surface is pushed against a moving steel surface with 100 lb (45.36 kg) of force. It might generate 20 lb (9.07 kg) of resistive force. This is expressed as 20/100 (9.07/45.36), which equals a coefficient of friction of 0.20 (**FIGURE 31-20**). A stationary block of rubber that is pushed against a moving steel surface with 100 lb of force might generate 125 lb (56.7 kg) of friction. This would be 125/100 (56.7/45.36), which equals a coefficient of friction of 1.25.

▶ TECHNICIAN TIP

A higher coefficient of friction usually results in a faster wearing of the softer material, such as rubber. This is one reason why the brake lining is designed to be made of softer materials than the drum or rotor. In this case, the brake lining wears more than the drum or rotor.

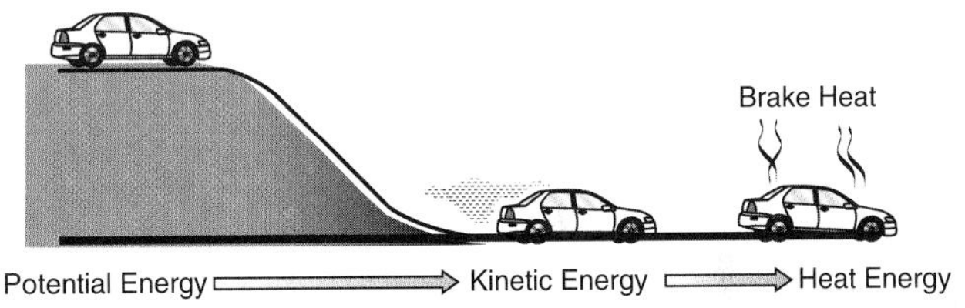

FIGURE 31-19 During deceleration, kinetic energy is transformed into heat energy.

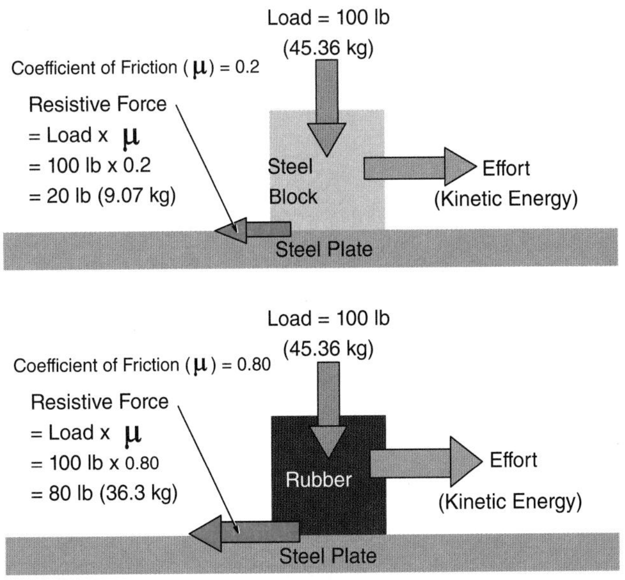

FIGURE 31-20 A. Coefficient of friction = 0.20. **B.** Coefficient of friction = 0.80.

Applied Science

AS-91: Friction: The technician can explain the role that friction plays in acceleration and deceleration.

Friction is very important to acceleration and deceleration. A vehicle's tires must maintain friction between the tire and the road in order to propel the car forward. If the tires are not in contact with the road, they will have nothing to push against. No friction is present and the car will not move forward.

Friction is also very important in deceleration, as the tire again has to maintain friction with the road. The braking force on the vehicle also uses friction in order to slow or decelerate the vehicle. As the brake pads press against the brake rotor, friction is created between the two surfaces. This friction creates heat and slows the vehicle.

Heat Transfer

A lot of kinetic energy must be converted to heat during the braking process. Heat transfer is critical to this process. Heat transfers from hot areas to cool areas. This process is used to continually transfer heat away from the friction materials. The brakes can then continue to perform their job of transforming the kinetic energy into heat energy.

Ultimately, most of the heat generated by the braking process radiates into the atmosphere. How it does this depends on the type of braking system (**FIGURE 31-21**). In drum brakes, the heat is created inside of the drum. It transfers through the drum to the outside surface where it radiates into the atmosphere. In disc brakes, the heat is created on the outer surfaces of the rotors. Air can more easily draw heat away from the rotor in this arrangement. Disc brakes also may have internal ventilation. This also helps to draw heat away from the outer surface of the rotor even faster.

Brake Fade

In automobiles, **brake fade** is the reduction in stopping power in the brake system. It is caused by a change in the brake system based on one of three factors. The first and most common is **heat fade**. Heat fade is caused by the buildup of heat in the braking surfaces. They get so hot they cannot create any additional heat. Because they cannot create any more heat, friction is reduced (**FIGURE 31-22**). Remember, the brakes must transform kinetic energy into heat energy to decelerate the vehicle. If heat energy cannot be generated, then the kinetic energy cannot be reduced. This means that the brakes will not be able to decelerate the vehicle.

Heat transfer is used to move heat away from the friction surfaces and allow them to continue generating heat. Once the temperature of the friction materials becomes so hot that they cannot generate any additional heat, the coefficient of friction drops. The brakes cannot generate stopping power until some of the heat dissipates.

▶ TECHNICIAN TIP

Excessive heat can cause warpage of disc brake rotors and brake drums. Brake fade and rotor warping can be reduced through proper braking techniques. When traveling on a long downgrade requiring braking, the driver should select a lower transmission gear. Periodic application of the brakes allows the brakes to cool between applications. Continuous light application of the brakes is sometimes referred to as riding the brakes. This can be particularly destructive in both wear and overheating of the brake components.

A driver experiences heat fade after using the brakes too much. This happens during high-performance driving or when going down a long, steep hill, particularly when towing. The brake pedal will be hard, but the braking effect "fades" away. The vehicle's rate of deceleration decreases. This is a dangerous condition and is why many long hills on freeways have truck escape ramps. The sand or other soft material slows a vehicle by absorbing the truck's kinetic energy in the soft material.

The second type of brake fade is called **water fade**. It is caused by water-soaked brake linings. The water acts like a lubricant between the friction surfaces. It lessens the coefficient

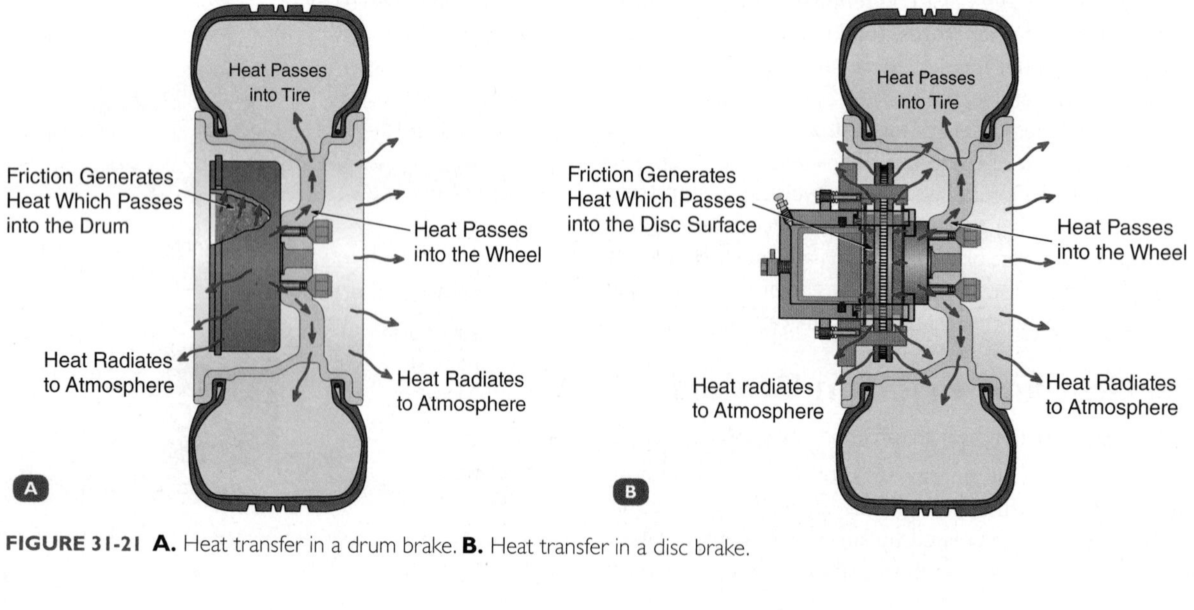

FIGURE 31-21 A. Heat transfer in a drum brake. **B.** Heat transfer in a disc brake.

of friction between the braking surfaces (**FIGURE 31-23**). This leads to a hard brake pedal but very little braking power. Once the water is removed from the friction surfaces through evaporation, the normal coefficient of friction is restored.

The third kind of brake fade is called **hydraulic fade**. It is caused by the brake fluid becoming so hot that it boils. Once it boils, it is no longer only a liquid. Part of it is now a vapor, which can be compressed. The brake fluid can no longer transfer force effectively to the wheel brake units. This prevents them from being applied firmly enough to create friction. Because the boiling fluid can be compressed, hydraulic fade can be recognized by the brake pedal becoming soft. It will also have increased pedal travel during brake usage (**FIGURE 31-24**).

▶ Rotational Force

LO 31-05 Describe rotational force, weight transfer, and levers.

When brakes are operated on a moving vehicle, a **rotational force** is generated. Friction between the brake components tends to twist the brake support in the direction of wheel rotation (**FIGURE 31-25**). The brake support is ultimately connected to the body of the vehicle. So, the body tends to rotate in the same direction. A good example of rotational force is when a motorcycle rider applies the front brake. If it is applied hard enough, the rear wheel is lifted completely off the ground. Suspension components must be able to withstand the rotational forces. But they can become worn and

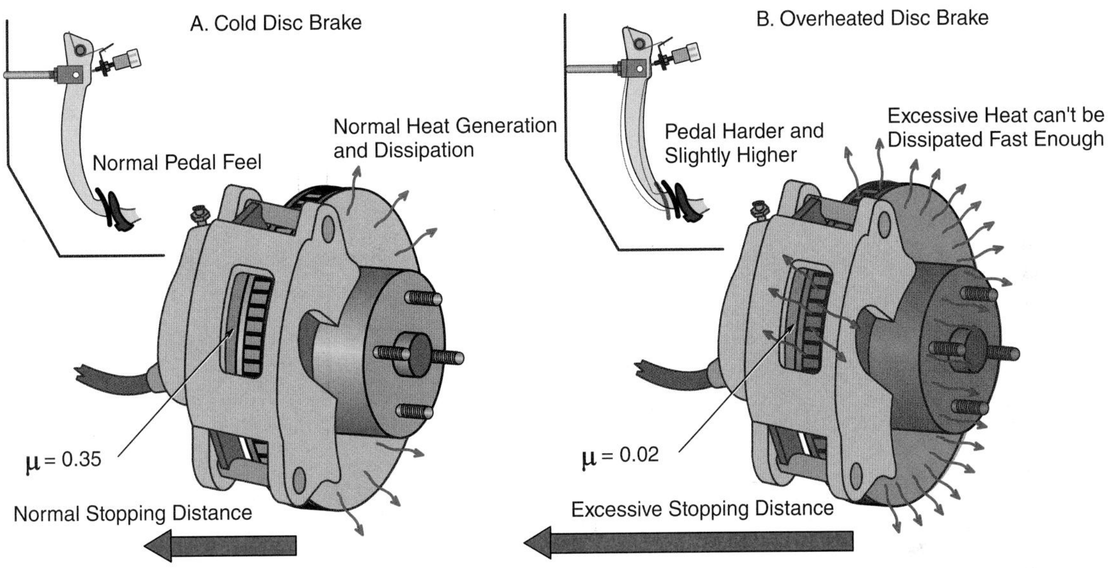

FIGURE 31-22 Heat fade—caused by the brake system reaching the temperature generated by the friction of the brake pads.

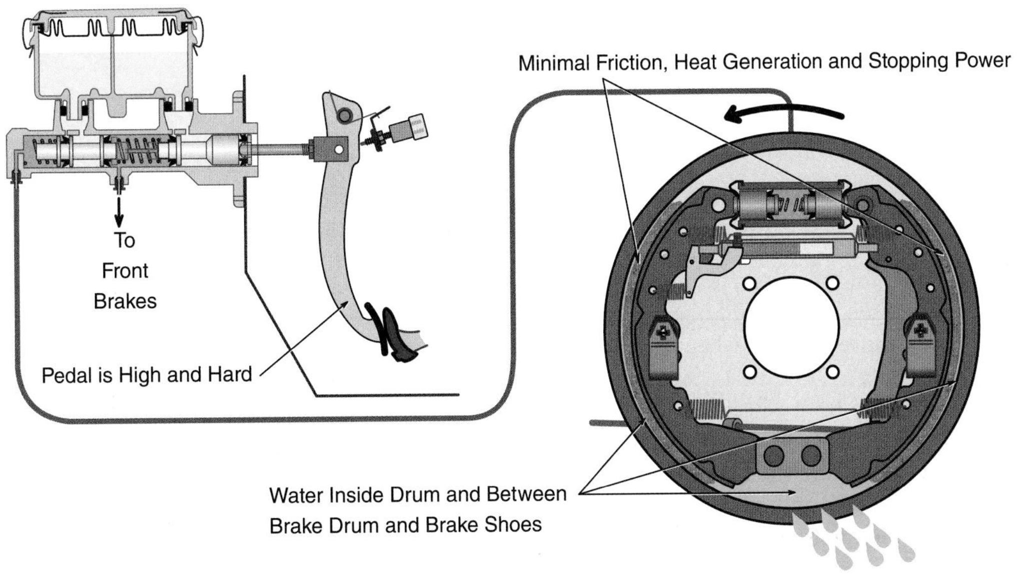

FIGURE 31-23 Water fade—water reduces the friction in brakes, causing a hard pedal with minimal stopping power until the brakes dry out.

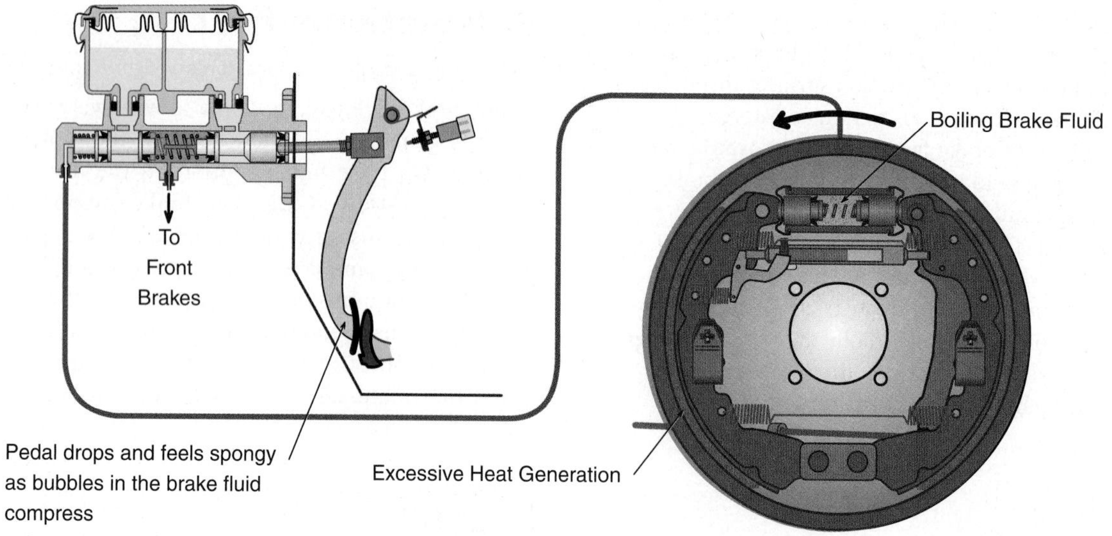

FIGURE 31-24 Hydraulic fade—when the brake fluid boils and can no longer transmit movement in the brake system to apply the brakes. The brake pedal is squishy and goes to the floor.

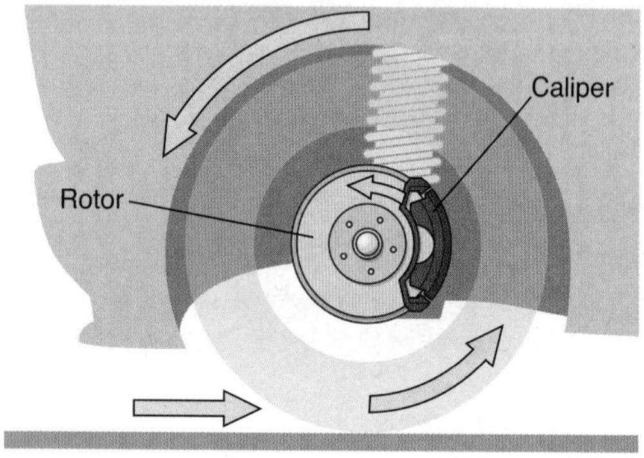

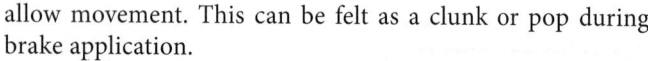

FIGURE 31-25 Braking rotational force occurs when the friction between the brake components twists the brake support in the direction of the wheel rotation.

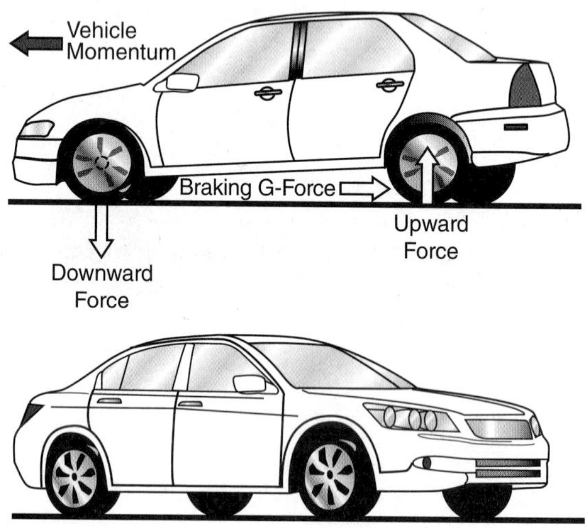

FIGURE 31-26 Weight transfer during braking.

allow movement. This can be felt as a clunk or pop during brake application.

Another result of rotational force is **weight transfer**. The rotational force tends to push the nose of the vehicle down and lift the rear of the vehicle. This transfers weight to the front wheels. Rotational force and weight transfer are also increased due to the centerline of the vehicle. The centerline of the vehicle is higher than the centerline of the axles. During braking, this tends to increase rotational force and weight transfer. Thus, the center of gravity tends to move forward when the brakes are applied firmly (**FIGURE 31-26**).

Weight transfer causes the front wheels to have increased traction. This allows them to bear more of the stopping load. It also causes the rear wheels to have less traction. This reduces the amount of braking load they can bear. Engineers take this into account when designing the brakes. Otherwise, the front wheels will not get enough stopping power, and the rear wheels will have too much. This results in rear wheel lockup and loss of control of the vehicle.

Rear wheel brake lockup is avoided by engineering the system properly. This includes the proper-sized master cylinder and wheel cylinders. It also uses valving that modifies the hydraulic pressure to the rear wheels under hard braking. This is covered in greater detail in the Hydraulics and Power Brakes chapter.

▶ TECHNICIAN TIP

Some owners raise their vehicles for better off-road clearance by installing a lift kit and/or large-diameter tires. Doing so raises the vehicle's center of gravity. It also increases the amount of weight transfer the vehicle experiences. This makes it more prone to rear wheel lockup and vehicle rollovers. It is important to only use lift kits engineered for the particular vehicle. Before installing such a kit, inform the driver of the effect that a higher center of gravity will have on vehicle operation.

Levers and Mechanical Advantage

Brake systems use levers and mechanical advantage to apply service and parking brakes. A simple example of a **lever** is a bar. The point around which a lever rotates and that supports the lever and the load is called the **fulcrum** (**FIGURE 31-27**). A lever allows the user to lift a large load over a small distance at one end. It does this by applying a small force over a greater distance from the other end. This is called mechanical advantage. The effort distance is from the fulcrum to the point the effort is applied. The load distance is from the fulcrum to the point the load is applied.

The effort required to move a load depends on the relative distance of the load and the effort from the fulcrum. The ratio of load and effort is called mechanical advantage. If the effort distance from the fulcrum is greater than the load distance, then the effort required will be less than the load being moved. If the load distance is greater than the effort distance, then the effort required is greater than the load being moved. This is known as a negative mechanical advantage, or **mechanical disadvantage**.

Using the right kind of lever in the right way allows a user to move larger loads with less effort. There are three basic types of levers:

1. Lever of the first order: The fulcrum is in the middle, between the load and the effort (**FIGURE 31-28**). Examples are a pry bar or a seesaw. The force applied in this situation is in the opposite direction of the load.
2. Lever of the second order: The load is in the middle, between the effort and the fulcrum (**FIGURE 31-29**). An example is a wheelbarrow. The force applied in this situation is in the direction of the load. Brake pedals are usually of the second order. They pivot at the top end (fulcrum). The foot pressure (effort) is applied to the bottom end. And the master cylinder (load) is applied between the two. Mechanical advantage is engineered into the brake pedal to provide the proper brake pedal application and feel (**FIGURE 31-30**).
3. Lever of the third order: The effort is in the middle, between the load and the fulcrum (**FIGURE 31-31**). An example is an oar when paddling a canoe. The hand holding the top of the oar is the fulcrum. The other hand holding the middle of the oar is providing the effort. And the water is the load. The force in this situation is in the direction of the load.

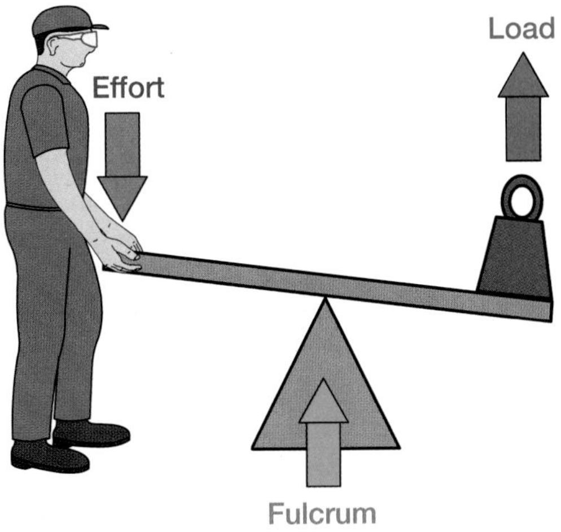

FIGURE 31-27 Parts of a lever.

FIGURE 31-28 Lever of the first order.

Applied Science

AS-12: Levers: The technician can explain how levers can be used to increase an applied force over distance.

The lever action is applied many times on a daily basis. The simplest example is the use of a wrench. If a bolt or nut is tight, then instead of using a wrench with a short handle, you can use a wrench with a longer handle. The longer handle improves the lever's mechanical advantage by increasing the ratio of the effort distance to the load distance. This provides more torque to undo the nut or bolt.

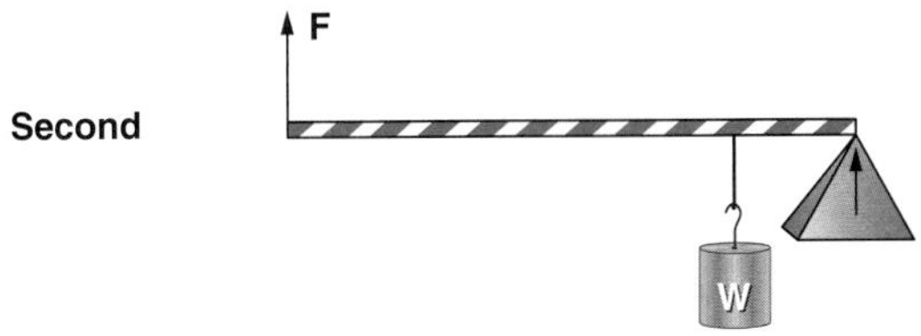

FIGURE 31-29 Lever of the second order.

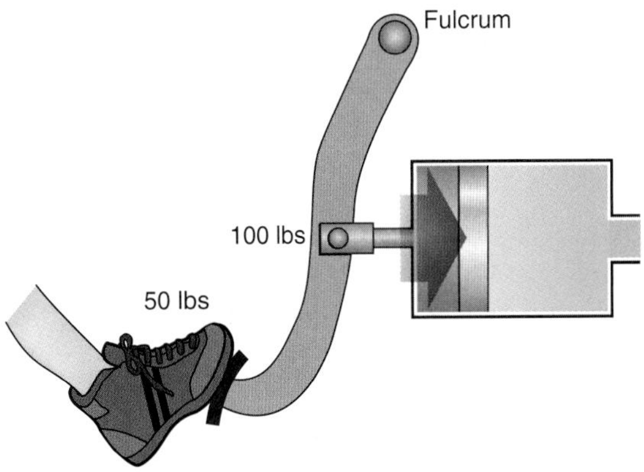

FIGURE 31-30 The brake pedal uses leverage to multiply the force applied to the master cylinder.

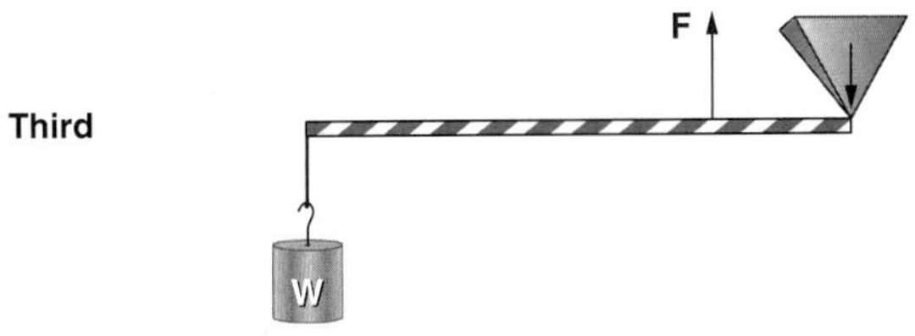

FIGURE 31-31 Lever of the third order.

▶ Types of Brake Systems

LO 31-06 Describe the common types of automotive brakes.

Now it is time to explore the different types of brakes used on various vehicles. Each vehicle application lends itself to a particular type of brake system. For example, conventional passenger vehicles use hydraulic brakes, which are fairly simple and reliable. They come in drum and disc configurations and are used on all four wheels.

Parking brakes are mechanically operated. They use mechanical force to apply two of the wheel brake units. The mechanical force can be generated by hand, foot, or electric motor. We discuss the various types of brakes in this section.

Hydraulic Brakes

Hydraulically operated friction brakes use two kinds of wheel brake units. Drum brakes have a drum attached to the wheel hub and rotate with the tire. Braking occurs by means of stationary brake shoes expanding against the inside of the drum. This creates friction and slows the vehicle.

Disc brakes have a disc brake rotor attached to the wheel hub and rotate with the tire. The braking occurs by means of stationary pads clamping against the outside of the rotor. This creates friction to slow the vehicle. On light vehicles, both of these systems are hydraulically operated. This means that they use hydraulic fluid to transfer the force from the driver. The driver operates the brake pedal, which operates the **master cylinder**. Hydraulic lines and hoses connect the master cylinder to the wheel brake units (**FIGURE 31-32**).

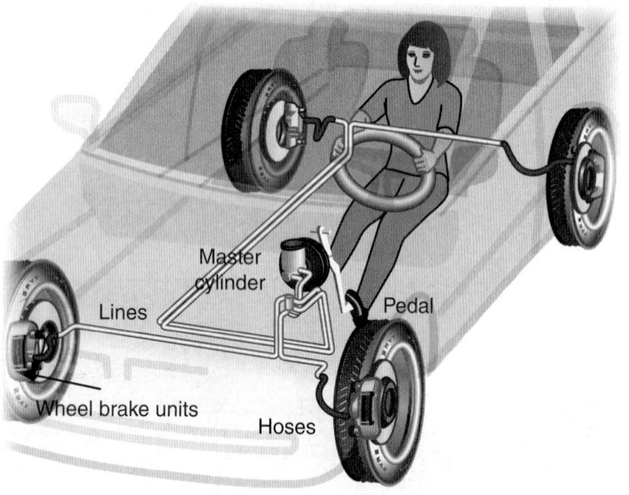

FIGURE 31-32 The hydraulic brake system.

Disc brakes require greater force to operate than drum brakes. It is why they usually include a power brake booster. The booster assists the driver by increasing the force applied to the master cylinder when the brake is operated (**FIGURE 31-33**). The additional force helps apply the brakes. But too much force can cause the wheels to skid.

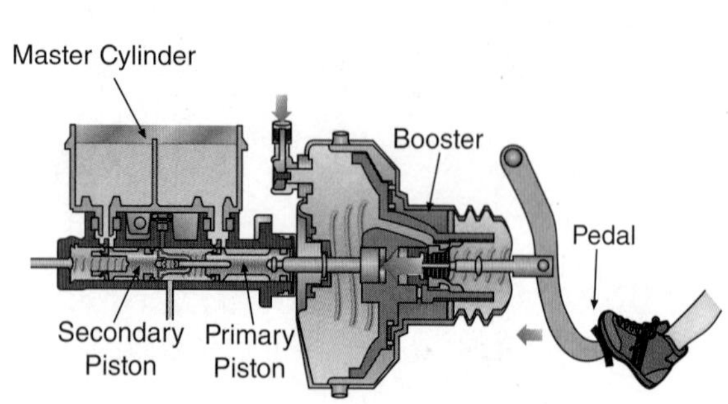

FIGURE 31-33 A power brake booster.

Modern drum and disc brake systems are regularly fitted with an ABS. It monitors the speed of each wheel and prevents wheel lockup or skidding. It doesn't matter how hard the brakes are applied or how slippery the road surface is. This allows the driver to better maintain directional control of the vehicle during hard braking. ABS also generally reduces stopping distances. The system consists of the following components:

- a brake pedal,
- power booster,
- master cylinder,
- wheel speed sensors,
- the ECU,
- and the hydraulic control unit, also called a hydraulic modulator.

Parking Brakes

As mentioned previously, all vehicles must be fitted with a service brake and a parking brake. The service brake is usually hydraulically operated and is used while the vehicle is being driven. It applies brake units at all four wheels. The parking brake is mechanically operated to hold the vehicle in place when it is parked. It applies the brake units on two wheels only.

There are several types of parking brakes. Some vehicles equipped with disc brakes incorporate a mechanically operated drum-style parking brake. This is commonly called a **top hat parking brake** design (**FIGURE 31-34**). It is mounted in the center of the rear disc brake rotors. Other vehicles use mechanical linkage to directly operate the disc-style service brake (**FIGURE 31-35**).

On drum brakes, a **drum-style parking brake** mechanically applies the brake shoes against the drum. The parking brake cable pulls on an actuating lever inside the brake drum assembly (**FIGURE 31-36**). The actuating lever is connected to the secondary brake shoe by a pin or tang. It actuates the primary shoe through use of a strut. Movement of the actuating lever forces both shoes against the drum.

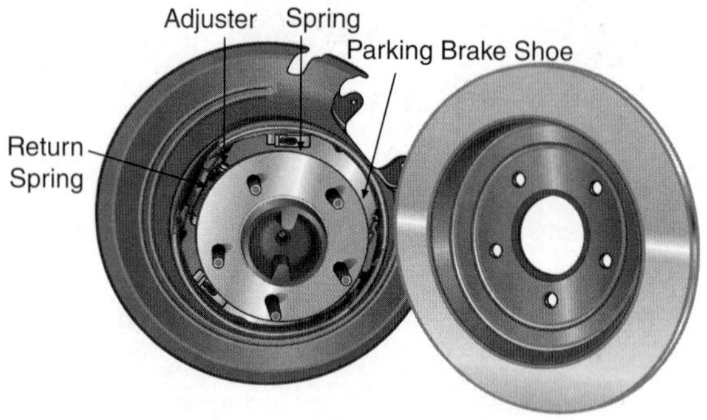

FIGURE 31-34 A top hat design parking brake.

Less common, older arrangements of parking brakes include a front wheel–mounted parking brake. It activates each front brake caliper. Another type of parking brake is the **transmission mounted parking brake**. It uses a small drum brake mounted between the transmission and drive shaft. It prevents the drive shaft from turning (**FIGURE 31-37**). This latter brake is sometimes called a transmission brake.

Parking Brake Cables

Parking brake cables transmit force from the parking brake actuating lever to the brake unit. Part of the cable is inside either a wound steel housing (**FIGURE 31-38**) or a plastic housing. The flexible housing allows it to be somewhat flexible, yet noncompressible. It is used to guide the cable and hold everything in place.

Other sections of the cable may be exposed. This can make them susceptible to damage by being caught on road hazards, especially on off-road vehicles. Also, because the cables are steel, they can rust and stick to the inside of the wound steel housing. This is especially true where deicing chemicals are used on roads. Periodic lubrication of the parking brake cables with the specified lubricant helps reduce this problem.

Parking Brake Apply Mechanisms

Parking brakes are commonly applied in two ways. The first is a hand-operated lever, usually mounted between the front seats. The lever has its fulcrum on the bottom end, and the cable is attached a few inches from the fulcrum. The driver pulls the handle on the end opposite of the fulcrum. This gives the driver substantial mechanical advantage. It also ensures that the parking brake is firmly applied (**FIGURE 31-39**).

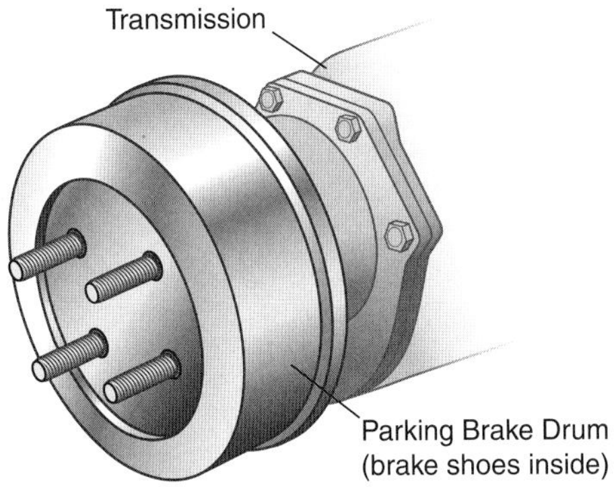

FIGURE 31-37 Parking brake—transmission style.

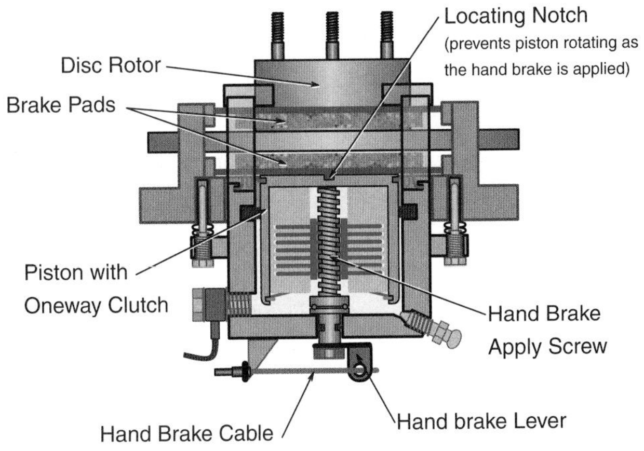

FIGURE 31-35 A disc brake system with an integral parking brake.

FIGURE 31-38 A parking brake cable.

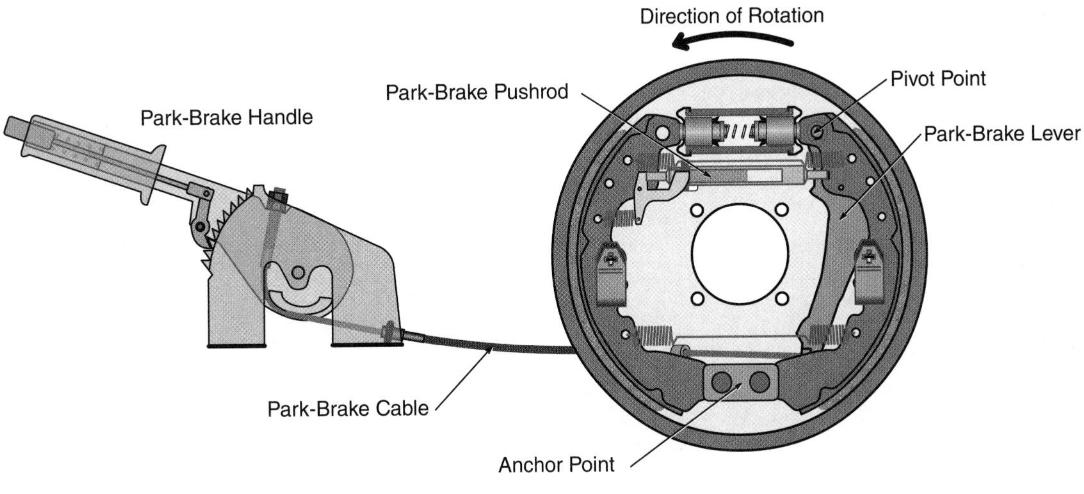

FIGURE 31-36 A drum-style parking brake.

The handle contains a ratcheting mechanism. It engages automatically when the driver pulls up on the parking brake lever. This mechanism holds the handle in the applied position and maintains tension on the parking brake cable. The driver can release the mechanism by lifting slightly up on the handle and pushing the release button. This retracts the ratcheting tab and allows the driver to lower the parking brake handle. Application pressure is now released on the parking brake.

The second type of operating mechanism works similarly, but is foot-operated. It also uses mechanical advantage to firmly apply the parking brake. The ratcheting mechanism holds it in place, once applied (**FIGURE 31-40**).

The pedal is usually mounted under the dash near the kick panel toward the outside of the car. Some foot-operated parking brake levers can be released by pulling a release handle. This retracts the ratcheting tang and allows the pedal to return to its rest position. Other pedals can be released by simply pushing the pedal a bit farther down. The ratchet automatically retracts and allows the pedal to return. The last release mechanism uses a vacuum or electric actuator to release the parking brake. It does so when the gear selector is moved out of the park position.

FIGURE 31-39 A hand-operated parking brake handle and ratcheting mechanism.

FIGURE 31-40 Foot-operated parking brake mechanism.

Parking Brake Adjustment

Parking brake systems incorporate some method of adjustment. However, it is not the same for all vehicles. The method of adjustment could be accomplished by adjustment nuts on the cable under the vehicle. The adjustment nuts can also be on the brake lever assembly. Or it can have adjustment nuts at the rear calipers (**FIGURE 31-41**). When adjusting parking brakes, the order of adjustment is as follows:

1. the parking brake cable, which should have slack in it;
2. the service brakes, which are adjusted according to the manufacturer's procedure; and
3. the parking brake.

This sequence ensures that both brake systems are adjusted properly.

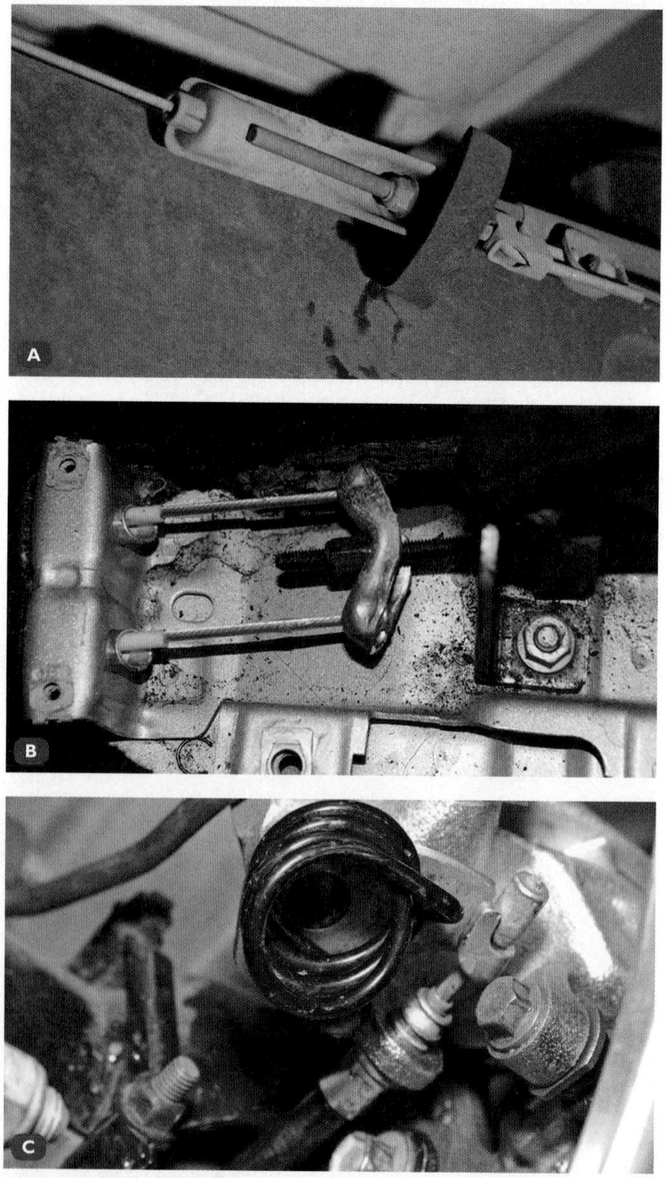

FIGURE 31-41 Methods of adjustment. **A.** Parking brake adjustment—under the car. **B.** Parking brake adjustment—brake lever. **C.** Parking brake adjustment—disc brake caliper.

▶ Wrap-Up

Ready for Review

- Early automobiles made use of scrub brakes. Scrub brakes were replaced by band brakes, which in turn were replaced by drum brakes. Disc brakes were developed in the early 1900s and gained wide use due to their ability to quickly dissipate heat into the atmosphere.
- The service brake is operated by a foot pedal and is used for slowing or stopping the vehicle when it is in motion. The parking brake is used for holding the vehicle in place when it is stationary.
    - Factors that affect braking include: road surface, road conditions, weight of the vehicle, load on the wheel during stopping, height of the vehicle, how the vehicle is being driven, and the tires on the vehicle.
- Kinetic energy increases by the square of the speed and if weight doubles, kinetic energy also doubles. During deceleration, kinetic energy is transformed into heat energy.
- Friction is the resistance created by surfaces in contact and can be static or kinetic. Kinetic energy must be converted to heat during the braking process and transferred away from the friction materials. Brake fade is the reduction in stopping power in the brake system and is caused by a change in the brake system due to heat fade, water fade, or hydraulic fade.
    - Static friction is resistance between nonmoving surfaces and is present in parking brakes. Kinetic friction is resistance between moving surfaces and is present in service brakes.
- Rotation force is generated when brakes are operated on a moving vehicle. Rotational force tends to push the nose of the vehicle down and lift the rear of the vehicle causing a transfer of weight to the front wheels. Basic types of levers are: lever of the first order, lever of the second order, and lever of the third order.
- Common types of automotive brakes include hydraulic brakes and parking brakes.
    - Hydraulic brakes come in drum and disc configurations and are used on all four wheels. Parking brakes are mechanically operated and use mechanical force to apply two of the wheel brake units.

Key Terms

acceleration An increase in a vehicle's speed.

antilock brake system (ABS) A safety measure for the braking system that uses a computer to monitor the speed of each wheel and control the hydraulic pressure to each wheel to prevent wheel lockup.

BA (brake assist) An enhanced safety system built in to some ABS systems that anticipates a panic stop and applies maximum braking force to slow the vehicle as quickly as possible.

band brake A type of brake that utilizes a steel band lined with friction material that wraps around a brake drum to slow the drum.

brake fade The reduction in stopping power caused by a change in the brake system such as overheating, water, or overheated brake fluid.

brake pedal emulator A brake pedal assembly used in electronically controlled braking systems to send the driver's braking intention to the computer; it mimics the feel of a standard brake pedal.

brake-by-wire system A braking system that uses no mechanical connection between the brake pedal and each brake unit. The system uses electrically actuated motors or a separate hydraulic system to apply brake force.

brakes A system made up of hydraulic and mechanical components designed to slow or stop a vehicle.

coefficient of friction (CoF) The amount of force required to move an object while in contact with another, divided by its weight.

conservation of energy A physical law that states that energy cannot be created or destroyed.

deceleration The process of decreasing a vehicle's speed.

disc brakes A type of brake system that forces stationary brake pads against the outside of a rotating brake rotor.

drum brakes A type of brake system that forces brake shoes against the inside of a brake drum.

drum-style parking brake A mechanically operated drum brake that can be set while the vehicle is not moving, to serve as a parking brake.

friction The relative resistance to motion between any two bodies in contact with each other.

fulcrum A half-round bearing that the rocker moves on as a bearing surface.

heat fade Brake fade caused by the buildup of heat in braking surfaces, which get so hot they cannot create any additional heat, leading to a loss of friction.

hydraulic fade Brake fade caused by boiling brake fluid; causes a spongy brake pedal.

kinetic energy The energy of an object in motion; it doubles with weight and increases by the square of the speed.

lever A simple machine that can allow a large object to be moved with less force.

master cylinder Converts the brake pedal force into hydraulic pressure, which is then transmitted via brake lines and hoses to one or more pistons at each wheel brake unit.

mechanical disadvantage When the load distance on a lever is greater than the effort distance, which means the effort required to move the load is greater than the load itself.

Newton's first law of motion A physical law that states that an object will stay at rest or uniform speed unless it is acted upon by an outside force.

parking brake A brake system used for holding the vehicle when it is stationary.

parking brake cable A mechanism used to transmit force from the parking brake actuating lever to the brake unit.

rotational force The force created by the rotating wheel when the brakes are applied; it causes the brake components to twist the brake support, and ultimately the vehicle, in the direction of wheel rotation.

scrub brakes A brake system that uses leverage to force a friction block against one or more wheels.

service brake A brake system that is operated while the vehicle is moving, in order to slow or stop the vehicle. Typically applied by foot.

top hat parking brake A drum brake that is located inside a disc brake rotor in order to act as a parking brake.

transmission mounted parking brake A drum brake that is mounted on the drive shaft, just after the transmission, to serve as a parking brake.

water fade Brake fade caused by water-soaked brake linings.

weight transfer Weight moving from one set of wheels to the other set of wheels during braking, acceleration, or cornering.

Review Questions

1. What was the primary reason for the development of hydraulic braking?
   a. To equalize braking forces at each wheel and prevent brake pull.
   b. To make brake pads and shoes last longer.
   c. To dissipate heat better preventing brake fade.
   d. To increase the amount of force required at the brake pedal.
2. All of these are factors that influence braking EXCEPT:
   a. Vehicle weight
   b. Tire condition
   c. Road condition
   d. TCS disabled
3. What system helps the driver apply the brakes by increasing the force applied to the master cylinder?
   a. Power booster
   b. Regenerative braking
   c. Brake-by-wire
   d. Band brakes
4. If a moving vehicle doubles its speed, its kinetic energy ______________.
   a. Doubles
   b. Triples
   c. Is squared
   d. Is multiplied by π
5. Why is heat transfer important to braking systems?
   a. Heat transfer to the brake fluid must be increased to keep it from boiling.
   b. Heat transfer must be increased to allow the brake linings to heat up properly.
   c. Heat transfer traps heat in the rotor to enhance braking performance.
   d. Heat transfer allows air to remove heat from the brakes to prevent heat fade.
6. What are the symptoms of hydraulic fade?
   a. Dragging rear brakes
   b. Grabbing front brakes
   c. Spongy and low pedal
   d. Steering wheel vibration
7. What force causes the rear of a vehicle to lift during braking?
   a. Clamping force
   b. Spring force
   c. Rotational force
   d. Centrifugal force
8. Weight transfer causes:
   a. Spongy pedal
   b. Low brake fluid level
   c. Increased traction
   d. Hydraulic fade
9. What are the two most common types of parking brake apply mechanisms?
   a. Hydraulic and pneumatic
   b. Hand lever and foot pedal
   c. Kinetic and potential
   d. Heat fade and water fade
10. How are park brakes commonly adjusted?
    a. Adjusting nuts on the cables
    b. Rotating an eccentric at the hand lever
    c. Applying the service brakes in reverse
    d. Topping off the master cylinder reservoir

ASE Technician A/Technician B Style Questions

1. Technician A says that an ABS prevents wheel lockup during hard braking. Technician B says that a TCS prevents tires from losing traction during acceleration. Who is correct?
   a. Technician A only
   b. Technician B only
   c. Both A and B
   d. Neither A nor B
2. Technician A says that brake-by-wire systems use braided steel cables instead of fluid to apply the brakes. Technician B says that regenerative braking uses friction to regenerate braking power. Who is correct?
   a. Technician A only
   b. Technician B only
   c. Both A and B
   d. Neither A nor B
3. Technician A says that parking brakes are designed to stop a vehicle in an emergency. Technician B says that service brakes are used to slow and stop a vehicle safely. Who is correct?
   a. Technician A only
   b. Technician B only
   c. Both A and B
   d. Neither A nor B

4. Technician A says that a vehicle changes chemical energy (in the form of gasoline or diesel) into kinetic energy (in the form of acceleration and speed). Technician B says that a vehicle's brake system changes kinetic energy (in the form of speed) into heat energy through friction. Who is correct?
   a. Technician A only
   b. Technician B only
   c. Both A and B
   d. Neither A nor B
5. Technician A says that a higher coefficient of friction typically results in less wear of the brake lining material. Technician B says that friction between the tire and road surface is sometimes referred to as traction. Who is correct?
   a. Technician A only
   b. Technician B only
   c. Both A and B
   d. Neither A nor B
6. Technician A says that water fade changes the coefficient of friction between the brake linings and drums until they dry out. Technician B says that hydraulic fade is caused by brake caliper seals leaking fluid onto the rotors. Who is correct?
   a. Technician A only
   b. Technician B only
   c. Both A and B
   d. Neither A nor B
7. Technician A says that the pivot point of a brake pedal lever can be referred to as the fulcrum. Technician B says that leverage is a form of mechanical advantage. Who is correct?
   a. Technician A only
   b. Technician B only
   c. Both A and B
   d. Neither A nor B
8. Technician A says that the size of the master cylinder pistons does not affect braking. Technician B says that valves may be used during hard braking to keep wheels from locking up. Who is correct?
   a. Technician A only
   b. Technician B only
   c. Both A and B
   d. Neither A nor B
9. Technician A says that disc brakes use friction linings attached to brake pads. Technician B says that disc brakes are mechanically operated. Who is correct?
   a. Technician A only
   b. Technician B only
   c. Both A and B
   d. Neither A nor B
10. Technician A says that drum brakes expand linings against a rotor to slow the vehicle. Technician B says that the power brake booster reduces the amount of force applied to the master cylinder. Who is correct?
    a. Technician A only
    b. Technician B only
    c. Both A and B
    d. Neither A nor B

CHAPTER 32
Hydraulics and Power Brakes Theory

Learning Objectives

- **LO 32-01** Describe hydraulic principles.
- **LO 32-02** Describe brake fluid types and characteristics.
- **LO 32-03** Describe master cylinder construction and operation.
- **LO 32-04** Describe quick take-up and ABS master cylinders.
- **LO 32-05** Describe brake pedal assemblies and divided hydraulic systems.
- **LO 32-06** Describe brake lines and hoses.
- **LO 32-07** Describe proportioning valves and their operation.
- **LO 32-08** Describe metering, pressure differential, and combination valves.
- **LO 32-09** Describe brake warning light and stop light operation.
- **LO 32-10** Describe the operation of vacuum brake boosters.
- **LO 32-11** Describe the operation of hydraulic brake boosters.

ASE Education Foundation Tasks

See Appendix A to view the 2017 ASE Education Foundation Automobile Accreditation Task List Correlation Guide.

▶ Introduction

In most modern vehicles, the wheel brake units are applied using the principles of hydraulics. This means the brakes are operated by the force transferred by noncompressible brake fluid (**FIGURE 32-1**). Pressing down on the brake pedal creates pressure in the brake fluid. Fluid transmits that pressure to the brake units. In turn, the brake units apply force to the friction materials. The friction materials transform the kinetic energy of the moving vehicle into heat energy. This causes the vehicle to decelerate.

To make it easier for the driver to apply the brakes, a power booster is fitted to the brake system. It increases the driver's brake pedal force to the master cylinder. This power booster can be operated by engine vacuum or through hydraulic pressure. Vacuum is typically supplied by the engine intake manifold or, in some cases, a vacuum pump. If it is operated by hydraulic pressure, that is usually generated by the power steering pump. Some vehicles use an electric-driven pump.

In this chapter, we explore the components of the hydraulic system, their purpose, and how they function. This will give you a solid foundation for understanding how to service them. This includes the maintenance, repair, and diagnosis process when working on the hydraulic system. Service tasks are presented in the next chapter.

▶ Principles of Hydraulics

LO 32-01 Describe hydraulic principles.

In the 1600s, Blaise Pascal observed the effects of pressure applied to a fluid in a closed system. **Pascal's law** states that pressure applied to a fluid in one part of a closed system will be transmitted without loss to all other areas of the system (**FIGURE 32-2**). This law is the principle behind hydraulic brakes.

Pressure created in the master cylinder is transmitted equally through the hydraulic braking system. It is transmitted without loss as long as the system remains closed and has no leaks. What happens to the pressure levels if there is a leak in the system? According to Pascal's law, a substantial leak prevents the pressure from building up. Therefore, the pressure within the system will be equally low. This means that the vehicle may lose some or all of its braking ability if a leak develops.

Pascal's law helps in diagnosing problems with the hydraulic braking system. For example, if the brake pedal is squishy (soft or spongy), there is a good chance that the hydraulic

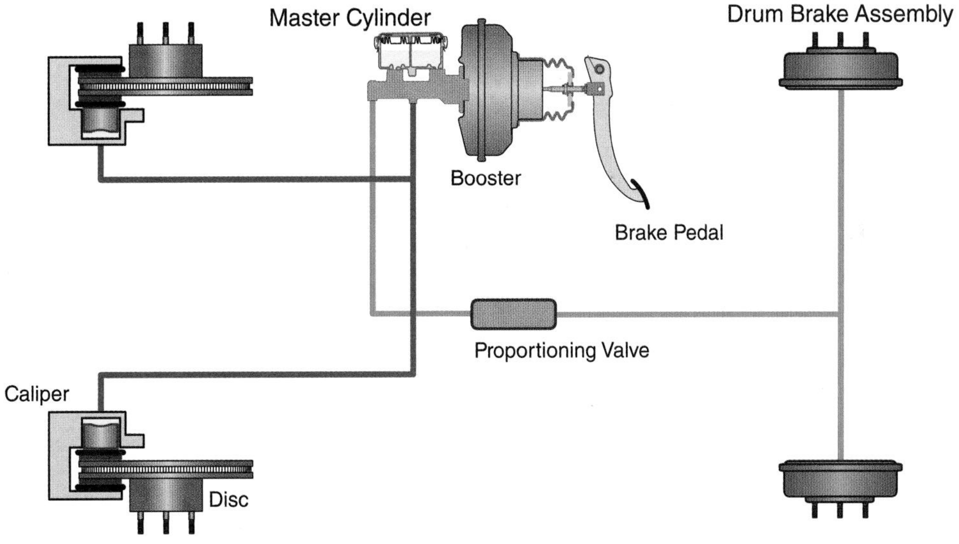

FIGURE 32-1 A schematic view of a hydraulic brake system.

You Are the Automotive Technician

A customer brings a 2011 Chevrolet Tahoe with 96,000 miles to the shop to have a brake concern addressed. The vehicle is regularly driven on the beach and has been pulling to the left, especially after braking. It has gotten worse over the last few weeks. You check the service history. It shows that the front brake pads were replaced, the rotors refinished, and the brake fluid flushed four years ago, at 54,000 miles. Your mentor technician wants to find out how much you know about braking systems before allowing you to start working on the vehicle. How will you answer his questions?

1. What is Pascal's law, and how can it be used to help you to diagnose this problem?
2. What are the different types of brake fluid, and when should they be used? When should they not be used?
3. What are the purposes of the proportioning, metering, and pressure differential valves?
4. What is the purpose of a tandem master cylinder, and how does it operate?
5. How does a vacuum brake booster operate?

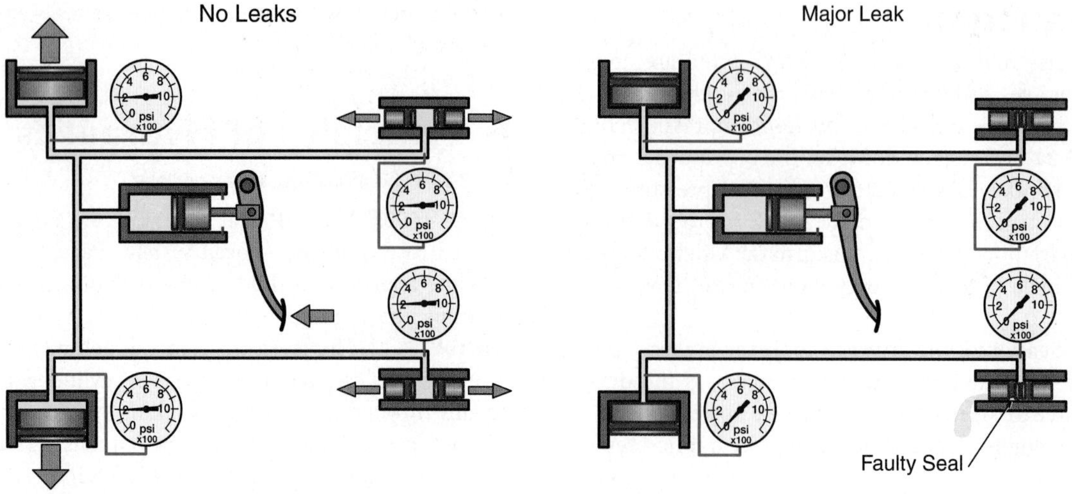

FIGURE 32-2 Pressure is transmitted without loss to all areas of a closed system.

braking system has air in it. This requires the system to be bled. If the brake pedal slowly sinks to the floor, there is likely a small leak in the system that must be found. If the vehicle pulls to one side, it could be that a brake hose is plugged up and is not transmitting pressure to one of the brake units. Knowing that pressure should be equal throughout the system helps you identify the cause, based on the reaction of the braking system. We delve more into diagnostics of the hydraulic system later.

Hydraulic Pressure and Force

Varying amounts of mechanical force can be extracted from a single amount of hydraulic pressure. Pressure is force per unit area (e.g., 50 lb psi, or 344.7 kPa). Therefore, the same pressure applied over different-sized surface areas will produce different levels of force (**FIGURE 32-3**). This principle allows engineers to design brakes to have a precise amount of braking force at each wheel. For example, the front wheels on some front-wheel drive vehicles can produce up to 80% of the vehicle's stopping power. This is because of the weight distribution and weight transfer. For these vehicles to brake well, more pressure must be applied to the front brake units than the rear brake units. This is accomplished through different diameter front and rear brake pistons. The larger brake pistons on the front wheels give greater braking power to the front wheels.

Input Force, Working Pressure, and Output Force

Figure 32-3 illustrates a hydraulic system that has cylinders of different diameters. When the brake pedal is pressed, the force against the master cylinder piston applies pressure to the brake fluid. This pressure is transmitted equally throughout the fluid. But each output piston develops a certain amount of output force depending on its diameter (surface area). The top cylinder is smaller than the master cylinder. This means the amount of output force it exerts will be less than the force applied to the master cylinder piston. The middle cylinder is the same size as the master cylinder, so the output force will be the same. The bottom cylinder is larger than the master cylinder, and so its output force will be greater. There are three variables to

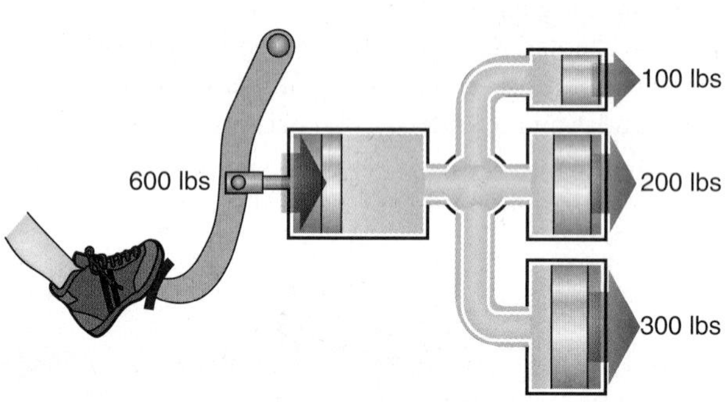

FIGURE 32-3 Engineers apply hydraulic principles to create varying amounts of mechanical force in hydraulic braking systems.

consider when discussing pressure and force in hydraulic systems (**FIGURE 32-4**):

- **Input force:** The force applied to the input piston is measured in pounds (lb), newtons (N), or kilograms (kg). For example, if 100 lb (45.36 kg) of force was applied to the input piston, this force would be labeled as 100 lb (45.36 kg).
- **Working pressure:** The working pressure of the hydraulic fluid is expressed as the amount of force per specified area. For example, 100 lb of force per square inch is labeled as 100 psi. It could also be expressed as 689.5 kilopascals (kPa). Note, 1 pascal = 1 newton per square meter, or 1 N/m^2. To find the working pressure, divide the input force by the area of the input piston. The example of 100 lb (45.36 kg) of force applied to a 1 sq in. piston creates 100 psi of working pressure. The same 100 lb of force applied to a 0.5 sq in. piston creates 200 psi of working pressure (100/0.5 = 200 psi). Conversely, the 100 lb of force applied to a 2 sq in. piston creates 50 psi of working pressure.
- **Output force:** Output force is exerted by the output piston and is expressed as pounds, newtons, or kilograms. Finding this measurement is fairly simple. Multiply the working pressure by the surface area of the output piston. For example, 200 psi of working pressure pushing on a 1 sq in.

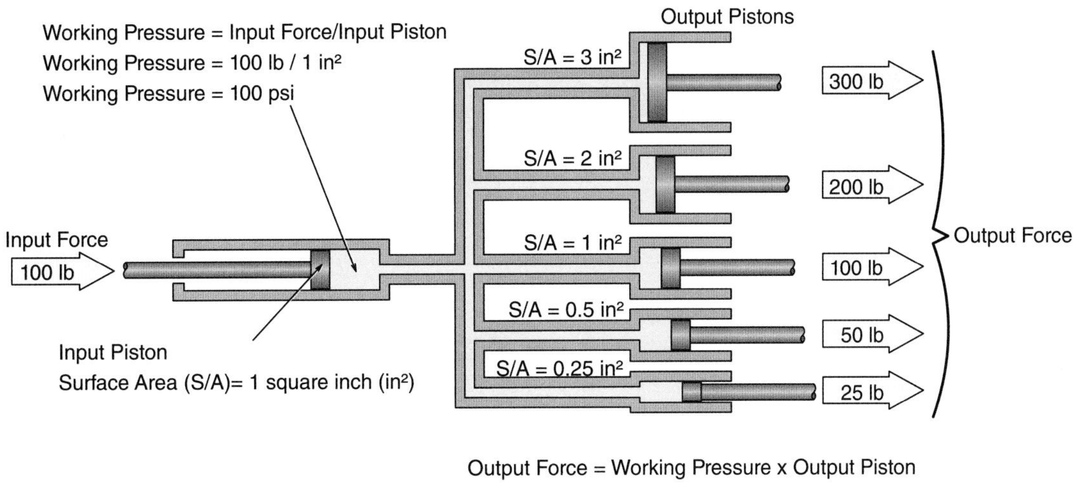

FIGURE 32-4 Working pressure and output force depend on the size of the pistons and the input force.

piston exerts 200 lb of force. The same 200 psi of working pressure acting on a 0.5 sq in. output piston creates 100 lb of output force. And if 200 psi of working pressure is applied to a 2 sq in. output piston, 400 lb of force will be created.

SAFETY TIP

It is critical that the hydraulic portion of the brake system not have any leaks or weak spots that could fail and cause leaks. Failure of the braking system could occur, putting the driver, passengers, and others in danger. Vehicles operated in corrosive environments, such as in areas where salt and certain deicers are used, are susceptible to brake line corrosion. Inspect all vehicles carefully for potential brake fluid leaks.

▶ TECHNICIAN TIP

Input force, output force, and working pressure are optimized during the design of the hydraulic system. This is based on a specific vehicle application. This is one reason why it is never acceptable to randomly substitute hydraulic components from another vehicle.

▶ Hydraulic Components

LO 32-02 Describe brake fluid types and characteristics.

The hydraulic system is made up of a number of components that work together to transmit the driver's effort to the brake units. These components must be able to withstand the force, pressure, and temperatures in the brake system. They must also be protected from the elements. This is because they are generally exposed to weather and road hazards. In this section, we explore each of the basic components of the hydraulic braking system.

Brake Fluid Types and Characteristics

Brake fluid is hydraulic fluid that has specific properties designed for mobile applications. It is used to transfer force, while under pressure, through hydraulic lines to the wheel braking units. Braking applications produce heat, so brake fluid must have a high boiling point. This allows it to remain effective under extreme temperatures. If brake fluid boils, it turns from a liquid to a vapor, which is compressible. This causes a spongy brake pedal and loss of braking ability. Brake fluid must also have a low freezing point so it will not freeze or thicken in cold conditions. If this were to happen, the force from the brake pedal would not be transferred to the wheel brake units.

Standard brake fluid is harmful to painted surfaces because it tends to soften paint. It must therefore be kept off all painted surfaces. Standard brake fluid is also **hygroscopic**, which means it absorbs water. It absorbs water from the atmosphere when it comes into contact with the air in the master cylinder reservoir. Over time, it can even absorb moisture through the flexible brake hoses.

Because water boils at a lower temperature than brake fluid, this will gradually reduce the brake fluid's boiling point. This makes the fluid more likely to boil and cause a hydraulic braking failure. Because of this, brake fluid must be flushed periodically to replace old contaminated fluid with new brake fluid.

▶ TECHNICIAN TIP

If the brake system needs to be bled, then the brake fluid should be tested. This allows you to see if it needs to be flushed out and replaced with new brake fluid.

Brake fluids are graded against compliance standards set by the US Department of Transportation (DOT) (**TABLE 32-1**). Brake fluids that meet these standards qualify for the DOT rating (**FIGURE 32-5**). They are considered to be quality brake fluids. Brake fluids are tested to ensure they meet the standards for the following:

- pH value
- Viscosity
- Resistance to oxidation
- Stability
- Boiling point

TABLE 32-1 DOT Ratings for Brake Fluid

Type	Materials	Minimum Dry Boiling Point	Minimum Wet Boiling Point	Additional Specifications
DOT 2	Castor oil based	Not specified	Not specified	• Outdated type of brake fluid that should not be used in any modern vehicles
DOT 3	Various glycol esters and ethers	401°F (205°C)	284°F (140°C)	• Good all-around brake fluid • Harms paint • Absorbs water
DOT 4	Various glycol esters and ethers	446°F (230°C)	311°F (155°C)	• Similar to DOT 3, but higher boiling point. • Compatible with DOT 3 and 5.1
DOT 5	Silicone based	500°F (260°C)	356°F (180°C)	• Not hygroscopic, will not absorb water • Less harmful to painted surfaces than glycol-based brake fluids • Provides better protection against corrosion • More suitable for use in wet driving conditions • *Not* to be used in any vehicle equipped with antilock brakes (ABS)
DOT 5.1	Contains polyalkylene glycol ether	500°F (260°C)	375°F (190.6°C)	• Suitable for ABS-equipped vehicles because of its high boiling point • More expensive than other brake fluids • Compatible with DOT 3 and 4

Source: United States Department of Transportation, Federal Motor Carrier Safety Administration: S5.1.2 Wet ERBP.

FIGURE 32-5 DOT 3, DOT 4, DOT 5, DOT 5.1 brake fluid.

SAFETY TIP

Because silicone-based fluid tends to **aerate** (tendency to create air bubbles) when forced at high pressure through small passages, it is *not* to be used in any vehicle equipped with ABS. The control valves in ABS brakes would cause DOT 5 brake fluid to aerate under an active ABS stop. This would lead to a spongy pedal and poor brake application.

▶ TECHNICIAN TIP

It is best not to mix brake fluids with different DOT ratings. Even if they have similar base composition.

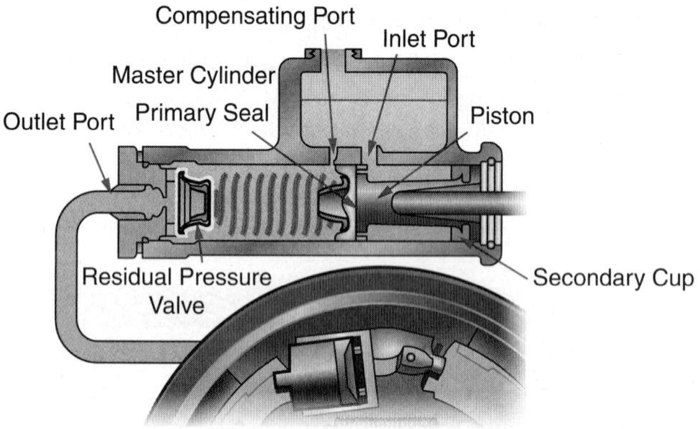

FIGURE 32-6 The master cylinder converts the driver's effort into hydraulic pressure.

▶ Master Cylinder

LO 32-03 Describe master cylinder construction and operation.

The master cylinder converts the brake pedal force into hydraulic pressure. The pressure is then used to operate the wheel brake units (**FIGURE 32-6**). Its mounting ears allow it to be mounted to a power booster or the firewall. The cylinder itself has threaded passageways to which the brake lines firmly connect. A brake fluid reservoir supplies the brake system with a reserve amount of brake fluid.

The master cylinder piston is operated by a pushrod from the power booster or the brake pedal. All vehicles 1968 and newer are required to use tandem master cylinders for safety purposes. But we will start our discussion with the less complicated single-piston master cylinder.

Single-Piston Master Cylinder

Single-piston master cylinders have one piston with two cups: a primary cup and a secondary cup (**FIGURE 32-7**). These cups are also known as seals because they keep the brake fluid from leaking past the piston. When force is applied to the piston by the pushrod, the **primary cup** seals the pressure in the cylinder. The **secondary cup** prevents loss of fluid past the rear end of the piston. An **outlet port** links the cylinder to the brake lines. An **inlet port** connects the reservoir with the space around the piston and between the piston cups. A **compensating port** connects the reservoir to the cylinder, just barely ahead of the primary cup.

With the brake pedal in the released position, the compensating port connects the brake system with the reservoir. The compensating port allows for expansion or contraction of the brake fluid ahead of the piston. This occurs as the brake fluid heats up and cools down. It can also compensate for brake fluid that does not return to the master cylinder. This is due to worn disc brake pads causing the caliper pistons to not return fully. This increases the brake fluid volume in the caliper. The compensating port allows brake fluid to move as needed when the brakes are released.

As the master cylinder piston moves forward, the compensating port is closed off. Brake fluid is then trapped ahead of the primary cup. Fluid can no longer return to the reservoir. Fluid trapped in the cylinder is then forced from the master cylinder outlet port into the brake lines.

When the brakes are released, the spring returns the master cylinder piston to its original position. When the piston fully returns against its stop, the primary cup uncovers the compensating port. Fluid ahead of the primary cup can now return to the reservoir as needed.

When the brake pedal is released quickly, a spring in the brake pedal pushes the piston back quickly. However, because of the restrictions in the hydraulic system, the brake fluid cannot return as quickly to the cylinder. This creates a low-pressure area ahead of the primary cup. As a result, air can be drawn into the system at the wheel cylinders on a drum brake system. To prevent this, small holes are drilled in the piston. This allows brake fluid from the reservoir to pass through the inlet port and past the edge of the primary cup. This prevents a vacuum from being created. The process of brake fluid moving past the primary seal is called **recuperation** (**FIGURE 32-8**).

On a drum brake system, brake fluid pressure needs to be held slightly above atmospheric pressure. This is accomplished by a valve called the **residual pressure valve**. Residual pressure helps to stop air from entering at the wheel cylinder cups when the brakes are not being applied. The residual pressure valve is located at the outlet end of the master cylinder on single-piston master cylinders. On tandem master cylinders, it is located under the tube seats where the brake lines connect.

Residual pressure valves are not used on disc brake circuits. This is because the caliper piston seal seals tightly between the piston and bore. Air cannot easily be drawn past the seal into the hydraulic system. Also, the residual pressure would keep the brake pads lightly applied against the brake rotors. This would cause brake drag, premature wear, and reduced fuel economy.

Tandem Master Cylinder

With a single-piston master cylinder, any fluid leak could mean the whole braking system fails. To reduce this risk, modern vehicles must have at least two separate hydraulic braking systems. Thus, the tandem master cylinder was developed. If one system fails, the other system can still provide a measure of braking ability. However, it will not be as effective. For example, the pressure can be 0 psi (0 kPa) in one circuit and normal in the other circuit. **Tandem master cylinders** combine two master cylinders within a common housing. They share a common cylinder bore (**FIGURE 32-9**).

Like two single-piston cylinders built end-to-end, a tandem cylinder has a primary piston and a secondary piston. The **primary piston** is in the rear of the cylinder. It is called primary because it is pushed directly by the pushrod. The **secondary piston** is in the front of the cylinder. The secondary piston has a rear-facing seal that seals fluid in the primary chamber. The rear-facing seal is what ultimately pushes the secondary piston during normal operation.

Each half of the cylinder has an inlet port, an outlet port, and a compensating port. There can be two separate reservoirs feeding each half of the cylinder. Or there can be just one

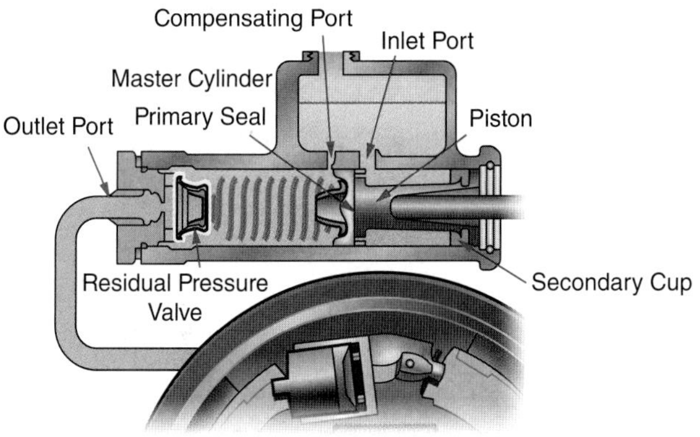

FIGURE 32-7 A single-piston master cylinder with primary and secondary cups.

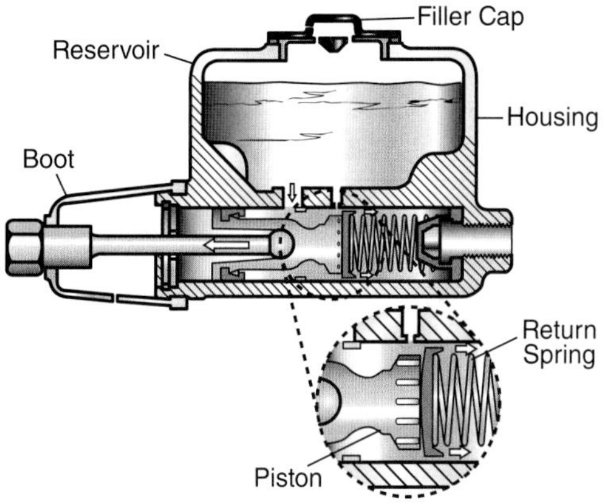

FIGURE 32-8 A single-piston master cylinder with small holes in the piston to allow for recuperation.

reservoir divided into separate sections. Dividing the reservoir prevents all of the fluid from draining out if there is a leak. This keeps a reserve amount of fluid for the working half of the master cylinder in case of a leak.

When the brakes are applied, the primary piston moves forward. It closes its compensating port. Continued movement of the piston causes fluid pressure in front of the primary piston to rise. The pressure acts upon the secondary piston, moving it forward. This closes its compensating port (**FIGURE 32-10**). Pressure now builds up equally in both circuits of the master cylinder. Both pistons continue moving forward. This displaces fluid into their separate circuits, and applies the brake units at each wheel.

Just like the single-piston master cylinder, the tandem master cylinder can develop a low-pressure area. It also occurs when the piston returns quickly. The tandem master cylinder overcomes this by using holes in the piston and grooves in the side of the primary cup. These primary cup grooves allow brake fluid to flow from the inlet port into the low-pressure area. This prevents low pressure from developing and air from entering the system.

What happens if there is a failure in the secondary circuit? The primary system pushes the secondary piston until it contacts the end of the cylinder bore (**FIGURE 32-11**). Once that happens, the primary circuit can start building pressure and operate its brake units. In this situation, it operates, but with increased pedal travel.

If the primary circuit fails, no pressure is generated to move the secondary piston. Thus, a rod attached to the front of the primary piston pushes the secondary piston directly. This allows the secondary piston to generate pressure to operate its brake units (**FIGURE 32-12**). This also results in a lower than normal brake pedal.

A differential pressure switch in the master cylinder or hydraulic system is activated if there is a leak in the system. It illuminates the brake warning light on the instrument panel. This alerts the driver of a loss of pressure between the two hydraulic circuits. We will cover this system later.

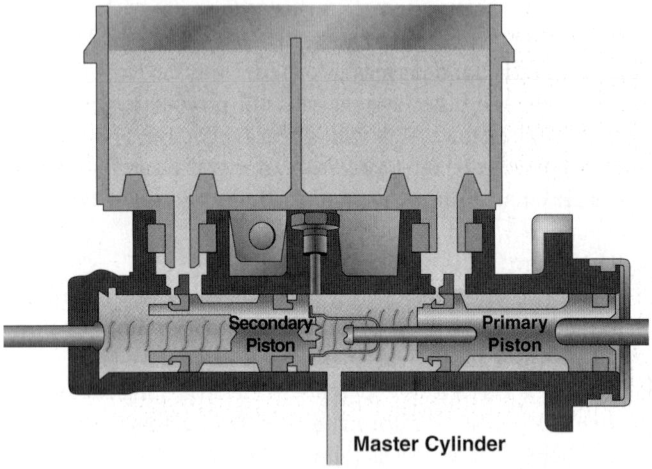

FIGURE 32-9 A tandem master cylinder showing common cylinder.

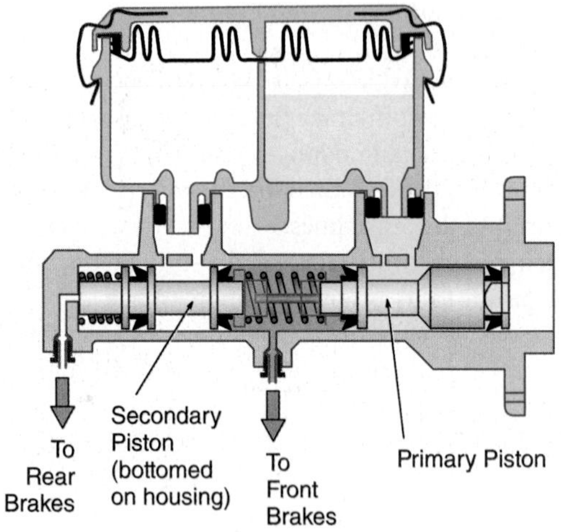

FIGURE 32-11 If there is a leak in the secondary circuit, the secondary piston travels to its stop, and then pressure builds in the primary circuit.

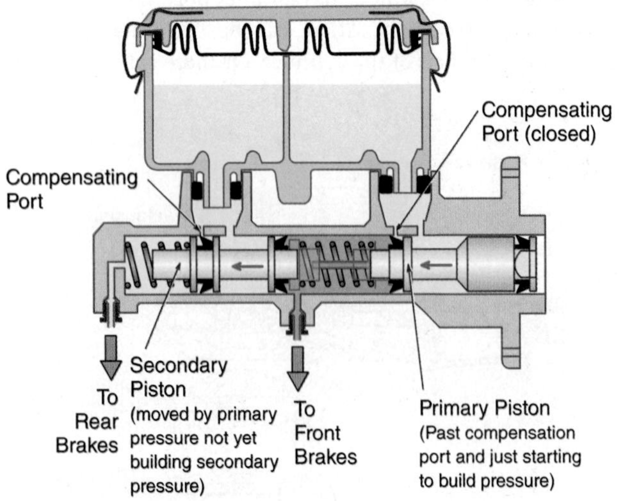

FIGURE 32-10 Moving the primary piston forward closes the compensating port and pushes the secondary piston forward, closing its compensating port.

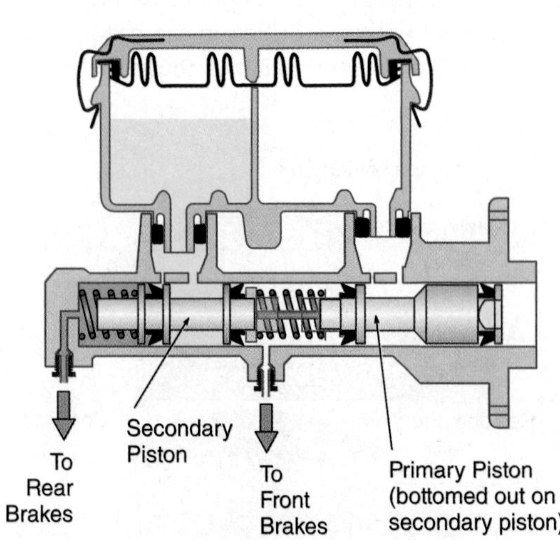

FIGURE 32-12 If there is a leak in the primary circuit, the primary piston travels until it contacts the secondary piston, moving the secondary piston and building pressure in the secondary circuit.

Reservoirs and Float Switches

Master cylinder reservoirs can be built into the master cylinder housing. They can also be a separate unit. Built-in reservoirs are made of the same material as the master cylinder, which is usually aluminum or cast iron. They are formed on top of the master cylinder (**FIGURE 32-13**). The reservoir cover on these master cylinders must be removed to inspect the brake fluid level. The cover uses a rubber diaphragm to help isolate the brake fluid from the air. These covers are usually held on by bail clips or tabs molded into the cover.

On two-piece master cylinders, the reservoirs are usually made of a see-through plastic material. They use grommets or O-rings to seal them to the master cylinder (**FIGURE 32-14**). Because they are see-through, it is usually unnecessary to remove the cover to check the brake fluid level. The covers can be screw-on caps or clip-on covers. The caps usually incorporate a diaphragm to minimize contact of the brake fluid with air. Some systems use a disc that floats on the brake fluid. This is also to minimize the surface area of the brake fluid in contact with the air.

To ensure that a leak in one brake circuit does not affect the other circuit, master cylinder reservoirs have two separate chambers. These reservoirs can be two totally separate chambers. Or they can use a divider in a common reservoir. The divider keeps the brake fluid level from falling below a minimum amount in the circuit that doesn't have a leak.

▶ TECHNICIAN TIP

Master cylinder reservoirs should always have air space at the top of the reservoir. This allows for the expansion of brake fluid as it heats up. So, never fill master cylinder reservoirs all the way to the top.

Most modern master cylinder reservoirs are equipped with a low brake fluid level float switch. It turns on the red brake warning light and/or sets a notification on the driver information system. The warning system can be activated by the float directly. Or a float with an embedded magnet may activate a switch when the float falls to a certain level (**FIGURE 32-15**).

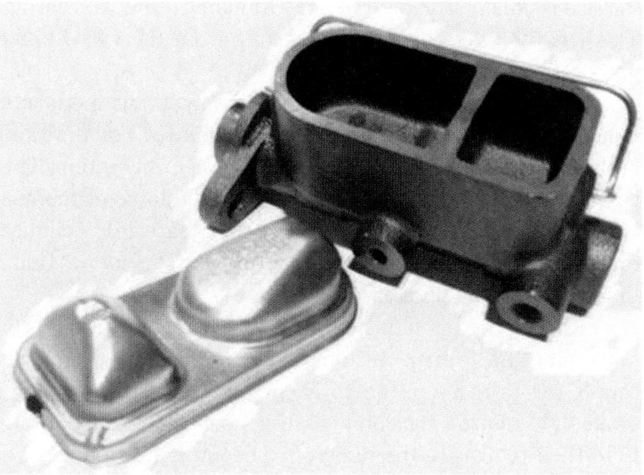

FIGURE 32-13 Master cylinder with built-in reservoir.

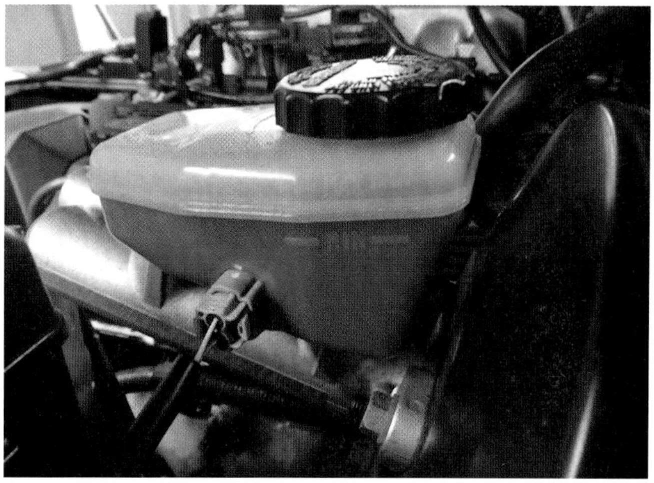

FIGURE 32-14 Two-piece master cylinder and reservoir.

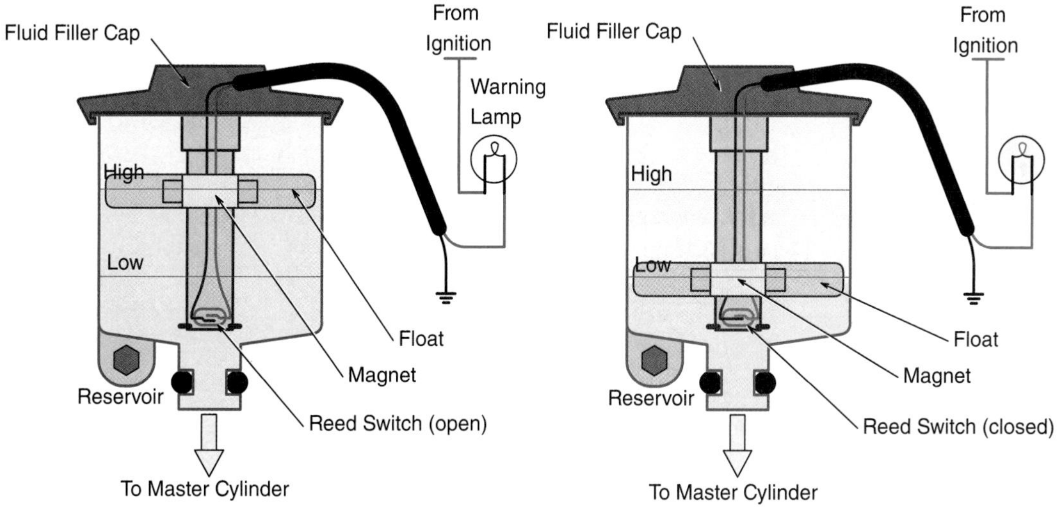

FIGURE 32-15 A magnetic switch is used to activate the brake warning light when the fluid level drops to a specified level.

The brake fluid level could be low for two reasons. The first is due to worn brake pad linings. The second could be due to a leak in the system. Both situations require further investigation. Adding brake fluid without further investigation into what is causing the low fluid condition could lead to a brake failure. This puts the driver and occupants in danger.

▶ Quick Take-up Master Cylinders

LO 32-04 Describe quick take-up and ABS master cylinders.

Quick take-up master cylinders are used on disc brake systems that are equipped with low-drag brake calipers. These calipers are designed to maintain a larger running clearance between the disc brake pads and rotor. If a standard master cylinder were used, the brake pedal would have to be pushed much farther down. Hence, the need for a quick take-up master cylinder.

Quick take-up master cylinders use a relatively large-diameter piston in the rear of the cylinder. It is able to push a large volume of fluid into the hydraulic system at low pressure (**FIGURE 32-16**). This moves the brake pads into contact with the rotor. Once the pressure rises above a predetermined point, a **quick take-up valve** opens. It bleeds off any extra pressure created by the large piston. At this point, a smaller diameter piston takes over and builds pressure within the hydraulic system. That pressure is used to apply the brakes normally.

ABS Master Cylinders

The ABS master cylinder is a tandem master cylinder used in divided systems. It also uses a primary piston and a secondary piston. It may also incorporate the quick take-up principles of operation. In some applications, the compensating port in the secondary chamber is removed. So, there is only an inlet port on the secondary chamber. The primary chamber still uses a compensating port and an inlet port.

The secondary piston incorporates a center valve. It controls the opening and closing of a supply port drilled into the piston. At rest, the supply port is open and connects the reservoir with the front brake circuits. This supply port replaces the compensating port in a normal master cylinder (**FIGURE 32-17**).

When the brake pedal is applied, the primary piston moves and closes its compensating port. Brake fluid pressure in the primary circuit rises slightly. The pressure moves the secondary piston forward, closing the center valve. The pressure builds in both circuits. If there is a leak in either circuit, the master cylinder only builds pressure in the working circuit.

When the brake pedal is released quickly, recuperation happens. When the primary piston is returned fully, any extra brake fluid from the wheel brake units flows through the compensating port. This prevents pressure from being trapped in front of the piston.

In the secondary circuit, the inlet port connects with the supply port drilling in the piston. Any difference in pressure lifts the center valve from its seat. Brake fluid enters the chamber ahead of the secondary seal. This prevents low pressure from developing. When the piston has returned to the "rest" position, the seal is pulled off its seat by the action of the link and spring. This lets the brake fluid still returning from the wheel brake units get back to the reservoir.

During some braking conditions, the hydraulic modulator must return brake fluid to the master cylinder. For the front brake circuits, brake fluid is returned to the front section. This forces the secondary piston back against the force of the rear brake pressure. So, both pistons are pushed rearward. If enough brake fluid returns, the center valve opens. This allows brake fluid to return to the reservoir.

If brake fluid is returned from the rear brake circuit, the secondary and primary pistons tend to be forced apart. This causes the primary piston to be driven rearward. If enough brake fluid returns, the compensating port is uncovered. Brake fluid then returns to the reservoir. The amount of brake fluid

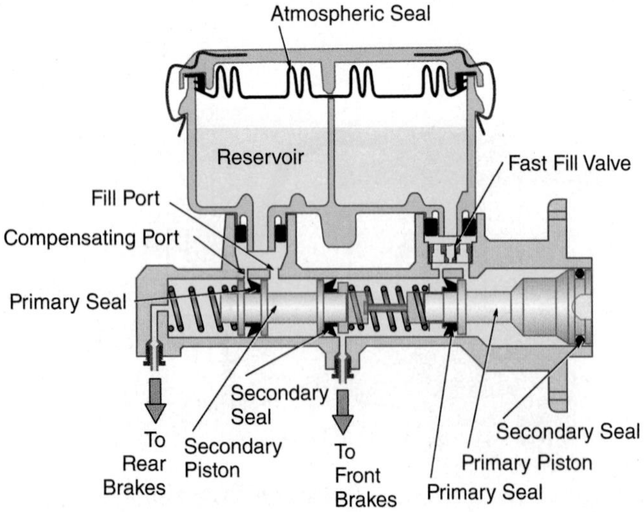

FIGURE 32-16 A quick take-up master cylinder with a larger diameter on the rear of the primary piston.

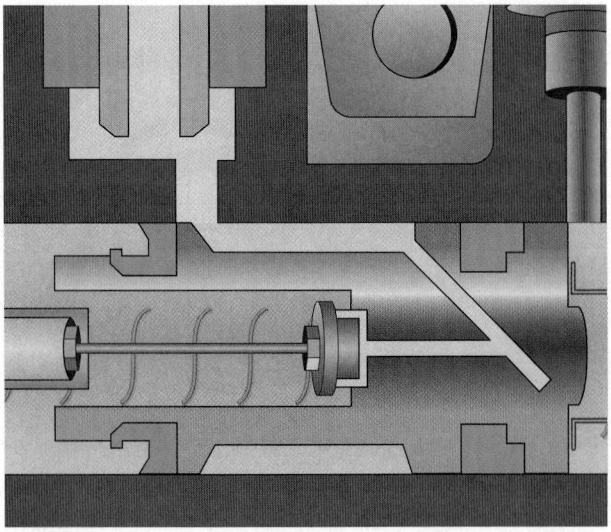

FIGURE 32-17 An ABS master cylinder.

that returns to the master cylinder is determined by the degree of ABS control. The driver may be aware of a rising brake pedal during this time.

▶ TECHNICIAN TIP

ABS pedal pulsation has been blamed for actually causing some accidents. The ABS system is normally only activated in a panic stop situation. Because of this, drivers who are unfamiliar with the pedal pulsation have been known to lift their foot off the brake pedal. And that can cause an accident. It is good for drivers to familiarize themselves with the feel of the ABS pedal pulsation before they need it. Do this by activating the ABS in a safe location, such as an abandoned parking lot. To address this issue, manufacturers have started to move toward electronic braking (brake-by-wire). This prevents hydraulic pulsations from being transmitted to the brake pedal.

▶ Brake Pedals

LO 32-05 Describe brake pedal assemblies and divided hydraulic systems.

The brake pedal uses leverage to multiply the effort from the driver's foot to the master cylinder. Different lever designs can be engineered to alter the brake pedal effort required of the driver. This is done by using different levels of mechanical advantage.

Brake pedals should be mounted securely and free from any excessive sideways movement. They need to be at a height and angle that will allow the driver to quickly move from the accelerator (throttle pedal) to the brake pedal. Brake pedals are covered with a rubber nonslip cover to maintain sure footing (**FIGURE 32-18**). These covers can become worn and lead to slippage. So, they need to be inspected periodically.

The brake pedal is usually suspended from a bracket between the dash panel and the firewall (**FIGURE 32-19**). It works as a force-multiplying lever. The pushrod transmits the brake pedal force either directly to the master cylinder or to the power booster. If the power assist fails, the brake pedal's leverage is designed to still generate a reasonable braking force. But with substantially increased foot pressure.

▶ TECHNICIAN TIP

Changes to how far the pedal travels or to its resistance can indicate brake problems. This can be whether it feels harder or softer than normal, or spongy. These can indicate faults in the power booster or air in the hydraulic system due to a leak.

▶ TECHNICIAN TIP

When ABS brakes are activated during heavy braking, the pulsations of the system can be felt by the driver, through the pedal. This is normal. However, a pulsating pedal during normal or light braking can indicate potential braking system problems. The most common problem is that a rotor has excessive thickness variation. But it could also possibly be warped. Either way, the rotor will need to be resurfaced or replaced.

Brake pedals must be free to return to their starting position when pressure is removed. This allows the master cylinder piston and pushrod to return to their undepressed position. The pedal is enabled to return to its starting position by a return spring (**FIGURE 32-20**). The spring action also causes the brake

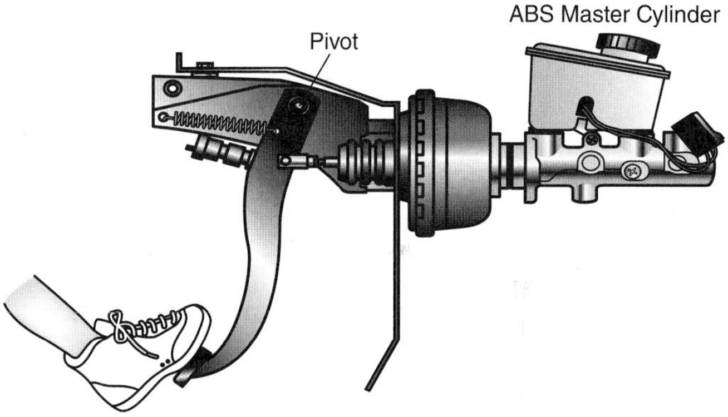

FIGURE 32-19 Brake pedal assembly.

FIGURE 32-18 Brake pedals have a nonslip cover that become worn over time.

FIGURE 32-20 Brake pedal return spring.

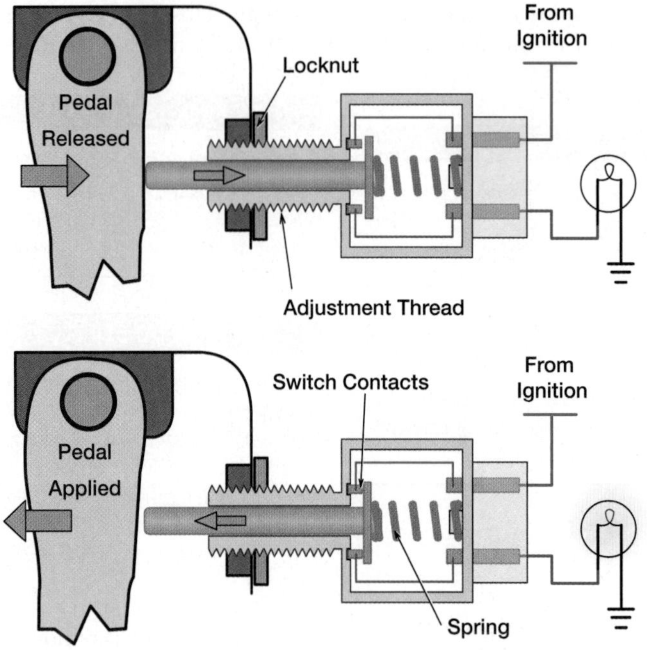

FIGURE 32-21 Brake light switch is turned off when the brake pedal is released.

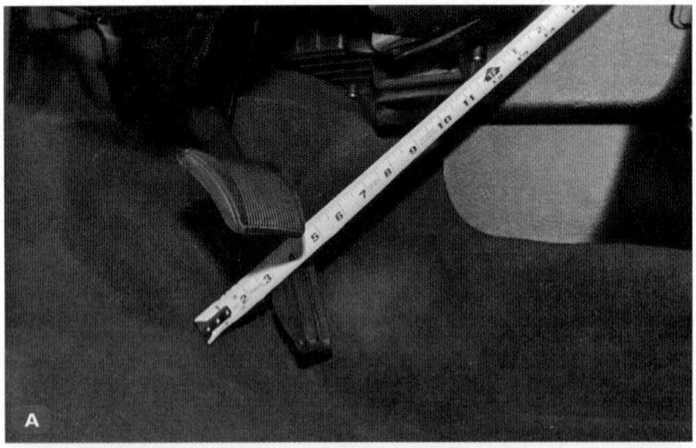

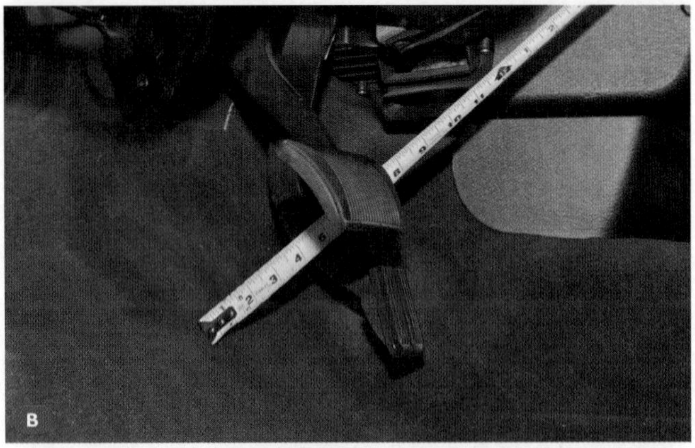

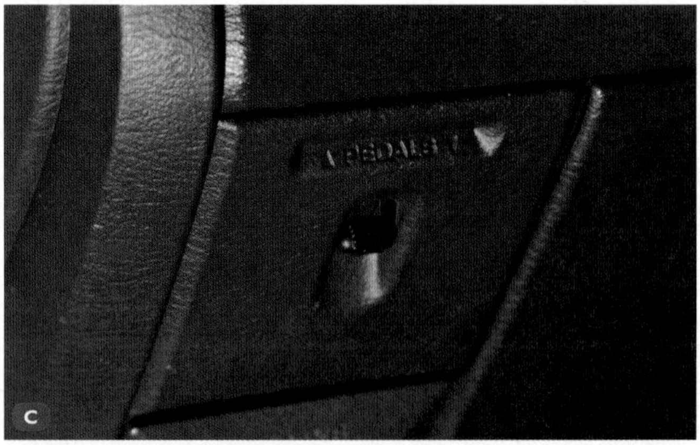

FIGURE 32-22 An adjustable pedal system. **A.** Low position. **B.** High position. **C.** Activating switch.

pedal to push the brake light switch open (**FIGURE 32-21**). This stops current flow to the brake lights.

When the brake pedal is applied, a lighter spring in the brake light switch causes the brake light switch contacts to close. This activates the brake lights. Brake light switches are adjustable on some vehicles. Misadjusted brake light switches can cause the brake lights to stay on when they should be off. It can also cause them to not come on when they are supposed to. Both situations can be corrected by adjusting the switch to the proper position. If the switch is nonadjustable, it will likely have to be replaced if it is not operating correctly.

Some vehicles come equipped with adjustable pedal assemblies. Such assemblies allow the driver to raise or lower the brake and throttle pedals for a better fit. These are usually adjusted by electrically driven motors. The motors are operated by a switch on the steering column or dash (**FIGURE 32-22**).

Types of Divided Hydraulic Systems

A wheel's braking ability depends on the load it is carrying. Therefore, the type of vehicle is a major factor in determining how its system should be divided (**FIGURE 32-23**). A front-engine,

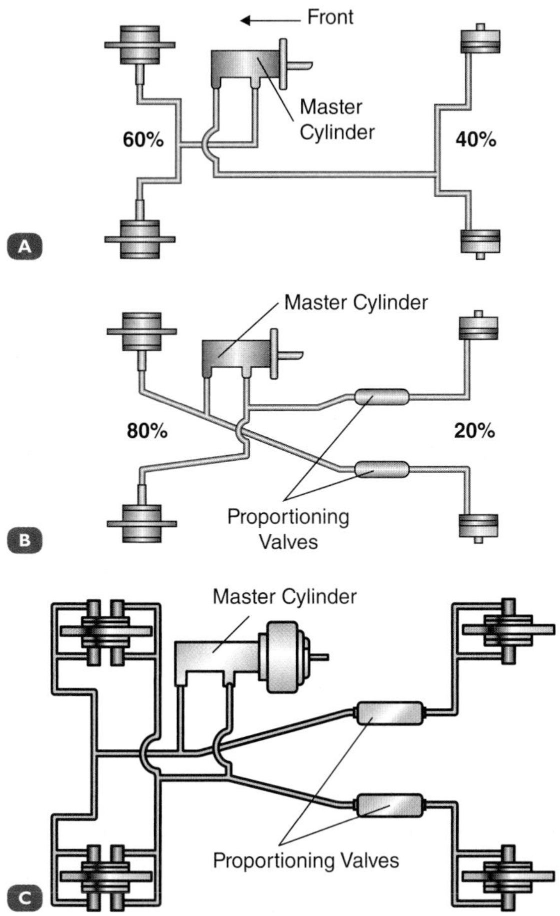

FIGURE 32-23 Divided hydraulic systems. **A.** Vertical, or front–rear, split hydraulic system. **B.** Diagonal, or X, pattern hydraulic system. **C.** L-split hydraulic system.

rear-wheel drive car has around 40% of its load on its rear wheels and 60% on its front wheels. Its braking system can therefore be divided in a vertical, or front–rear, split. This design puts the front wheels on a different system than the rear wheels. If half of the system fails—either the front or the rear—there is still enough separate braking capability left in the other half to stop the vehicle.

On a front-wheel drive vehicle, a load of about 20% on the rear wheels cannot provide enough braking force to adequately stop the vehicle. Therefore, front-engine, front-wheel drive vehicles use a braking system split in a diagonal, or X, pattern. The left-hand front brake unit is connected to the right-hand rear unit. And the left-hand rear unit is connected to the right-hand front unit. If one system fails, a 50% braking capability is available in the other system.

An alternative arrangement for front-engine, front-wheel drive vehicles is an L-split. The front disc brake units have four piston calipers. One inner and one outer piston on each front caliper connect to the right-hand rear brake unit. The other two pistons of each front caliper connect to the left-hand rear brake unit. As with the diagonal split system, if there is a failure of either half of the system, it still leaves 50% of the braking capability.

FIGURE 32-24 Steel brake line.

FIGURE 32-25 Flexible brake hose.

▶ TECHNICIAN TIP

When diagnosing and servicing brakes, it is helpful to know how the hydraulic system is divided. If the vehicle is pulling to one side due to a leak in half of the system, it will generally pull toward the side that is working in the front. Diagnosis and inspection of the other diagonal half usually leads to the leak. In the same way, if air is trapped in half of the hydraulic braking system, then **bleeding** air from the same diagonal half will make bleeding easier.

▶ Brake Lines

LO 32-06 Describe brake lines and hoses.

Brake lines and hoses carry brake fluid from the master cylinder to the brake units. They are basically the same on all brake systems and passenger vehicles. For most of their length, they are double-walled steel, coated to resist corrosion. They are attached to the body with clips or brackets to minimize damage (**FIGURE 32-24**). In some vehicles, the brake lines are inside the vehicle. This is to better protect them from corrosion and physical damage. Where the lines must move, flexible brake hoses allow for steering and suspension movement (**FIGURE 32-25**).

Brake Line Materials

The **brake lines** must be able to transmit considerable hydraulic pressure. Typically, this can be 1500 psi (10,342 kPa) or more during panic stops. They are made of seamless, double-walled steel. They should never be made from a softer, but less corrosive and easier-to-form, material such as copper. They also must conform to applicable standards such as those set by the Society of Automotive Engineers. Only brake lines that meet those standards can be used on vehicles.

If a brake line is damaged, it is common practice to replace the entire brake line with a factory replacement rather than repair it. At the same time, universal brake lines are available in a variety of lengths. They come factory flared with the correct fitting installed. They just need to be formed with the proper tubing bender to the specified shape. This allows them to follow the original brake line routing. Avoid kinks by only using the correct tubing bender. Kinked lines cannot be used or repaired.

Types of Brake Line Flares

Brake lines connect all of the various hydraulic braking system components. So, their ends must make leak-proof connections. This is accomplished by using the following types of flared lines and matching fittings:

- **Inverted double flare**: This type of flare is created by first flaring the end of the tube outward in a Y shape. Then, about half of the flared end is folded inside of itself (inverted). This leaves a double-thick section of brake line on the flared portion of the Y (**FIGURE 32-26**). The flared portion of the tube is clamped between the mating surfaces of the two fittings. This provides a secure, leak-proof connection when formed properly.
- **International Standards Organization (ISO) flare**: This type of flare is sometimes called a "bubble flare." The brake line is flared slightly out and then back in. This leaves the brake line "bubbled" near the end (**FIGURE 32-27**). The bubble is then clamped between two matching fittings.

SAFETY TIP

Never substitute a copper, aluminum, or non-approved line for the original. Doing so will lead to a brake failure, which you could be held liable for.

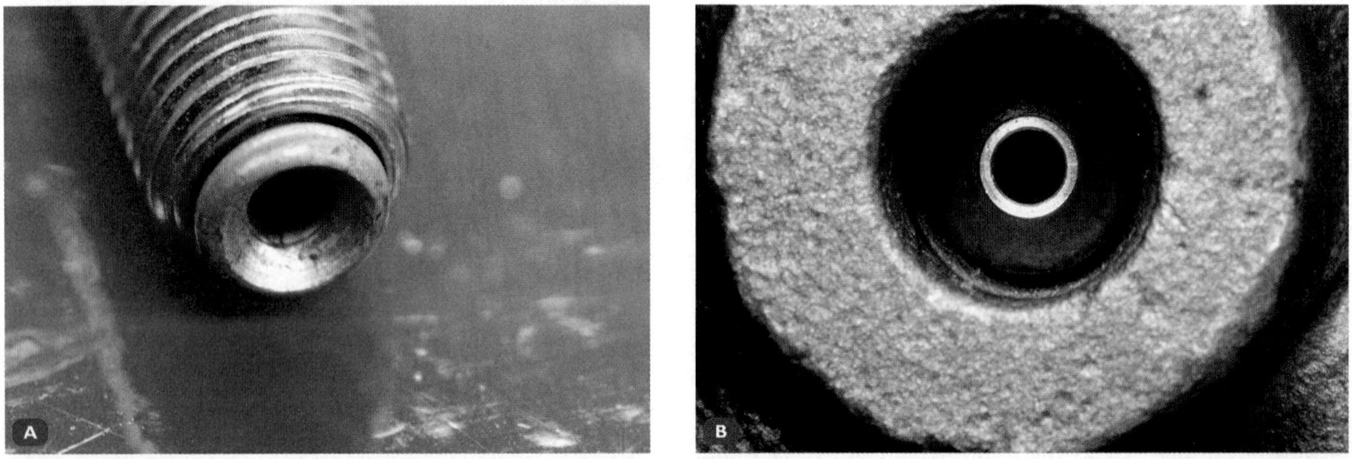

FIGURE 32-26 **A.** An inverted double-flared line. **B.** A matching fitting.

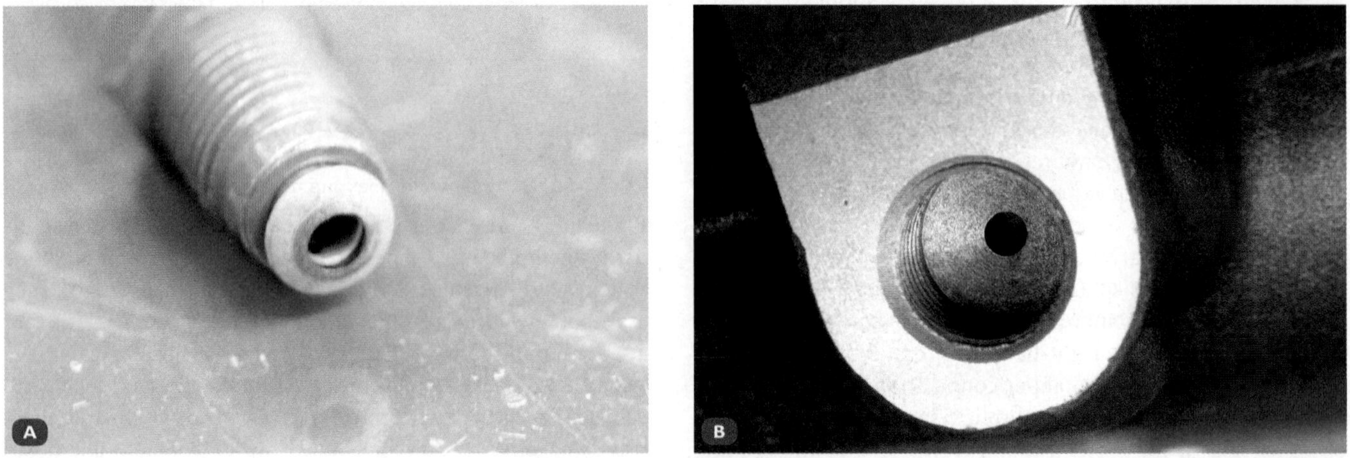

FIGURE 32-27 **A.** An ISO flared line. **B.** A matching fitting.

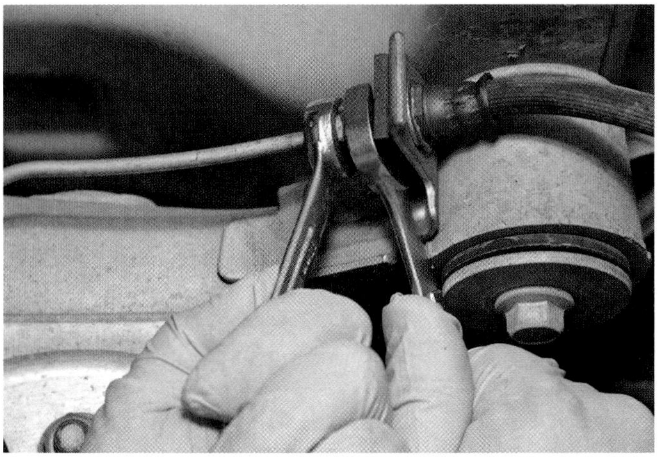

FIGURE 32-28 Double-wrench method.

> **▶ TECHNICIAN TIP**
>
> Many brake and fuel lines screw into an adapter. Always use a double-wrench method with flare nut wrenches when loosening and tightening brake and fuel lines. This technique will help prevent twisting of the steel lines (**FIGURE 32-28**). When using the double-wrench method, one wrench holds the adapter from moving. The other wrench turns the brake line fitting. Doing it in this way prevents the brake line from being kinked.

> **▶ TECHNICIAN TIP**
>
> Flared fittings form a seal by tightly compressing the brake line between the two halves of the fitting. No sealer is needed—nor should it be used—on these types of fittings.

Brake Hoses

A flexible section of the brake lines must be included between the body and suspension. This allows a flexible connection for steering and suspension movement. This is accomplished by using flexible hoses made of tough, reinforced tubing. These flexible **brake hoses** transmit the brake system hydraulic pressures to the wheel units (**FIGURE 32-29**). They also must be tough so as not to be damaged easily by road hazards.

When replacing brake hoses, always make sure they are of the proper length. If they are too short, they can be damaged by being stretched. If they are too long, they can contact moving components such as tires or a suspension member. This will weaken or wear a hole in the hose. This is especially common when a vehicle has been lifted or lowered.

Brake Hose Materials

Brake hoses are made of several layers of alternating materials. The inside is a liner that helps seal the brake fluid in and any moisture out. The liner is wrapped with two or more layers of flexible webbing. They are usually embedded in a synthetic rubber material. This provides reinforcement for the hose. These layers are covered with a tough flexible outer housing jacket designed to resist abrasion and damage (**FIGURE 32-30**).

FIGURE 32-29 A brake hose transmits hydraulic pressure to the moveable brake unit.

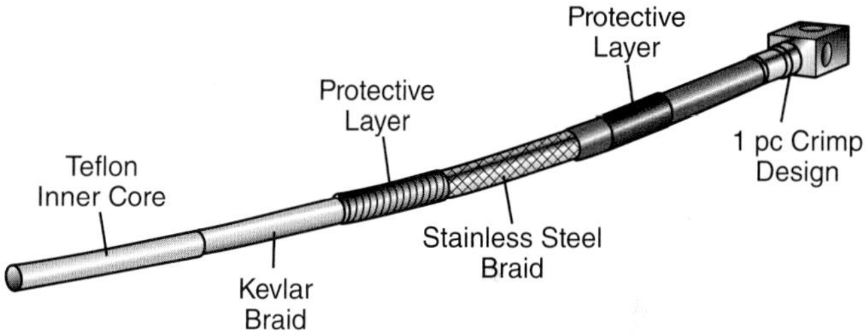

FIGURE 32-30 Flexible brake hose construction.

Although brake hoses are designed to be flexible, they should *never* be pinched, kinked, or bent tighter than a specified radius. Doing so reduces their life and can cause failure of the brake hose. Some technicians mistakenly use vise-grip pliers to crimp a brake hose while it is disconnected from the caliper or wheel cylinder. They do this to prevent brake fluid leaking out from the end of the brake hose. This practice can damage the brake hose and should not be used.

> **▶ TECHNICIAN TIP**
>
> *Never* hang a disconnected brake caliper by its flexible brake hose. Doing so can damage the brake hose. Always use a piece of wire or other material (e.g., cable or zip tie) to support the weight of the caliper assembly when it is removed.

Brake hoses should be inspected periodically for damage or defects (**FIGURE 32-31**). Some possible issues are:

- **Cracks:** The outer layers become brittle and crack over time. Replacement is required.

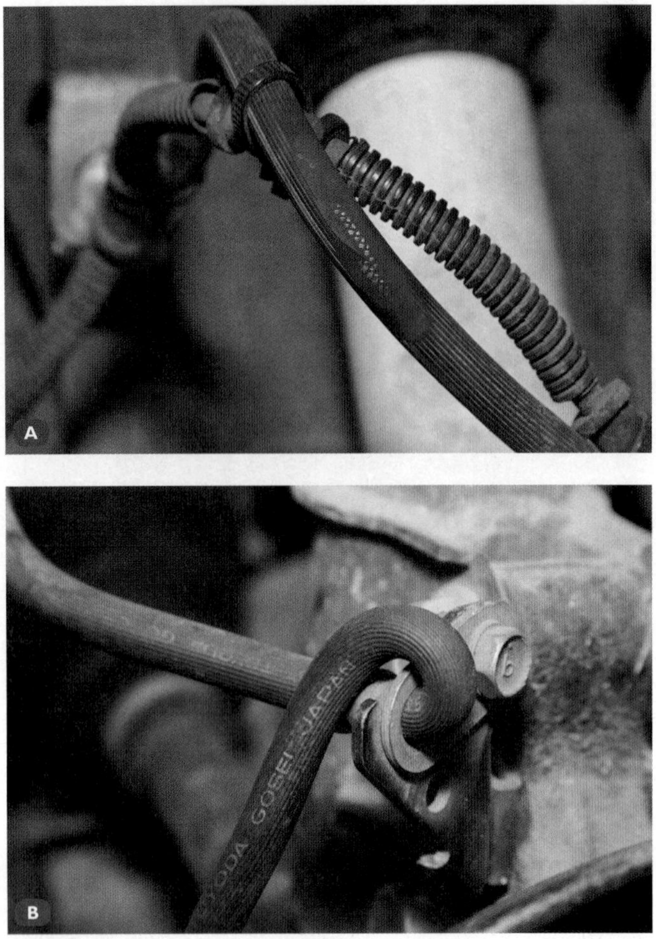

FIGURE 32-31 Possible brake hose issues. **A.** Brake hose with abrasion/wear. **B.** Brake hose with kinks.

- **Bulges:** The reinforcing layers become weak and break over time resulting in bulges in the outer cover. Replacement is required.
- **Abrasion or wear:** This usually happens because the brake hose was routed incorrectly or was too long or short for the application. This causes it to rub on a component, resulting, over time, in the abrasion. Replacement and rerouting are required.
- **Kinks:** Kinking usually happens when the brake hose has been pinched with vise grips or twisted on installation. Replacement is required.
- **Internal deterioration causing blockage of the passageway:** Replacement is required.

Sealing Washers and Fittings

Many brake hoses use banjo fittings to connect the hose to the wheel unit. These fittings comprise the banjo fitting, banjo bolt, and two copper or aluminum sealing washers (**FIGURE 32-32**). The banjo bolt, banjo fitting, and wheel unit usually have sealing ridges machined in them. These ridges dig into the softer sealing washers to ensure a leak-proof connection. The banjo bolt is hollow and allows the brake fluid flow through it. It clamps the banjo fitting to the wheel unit.

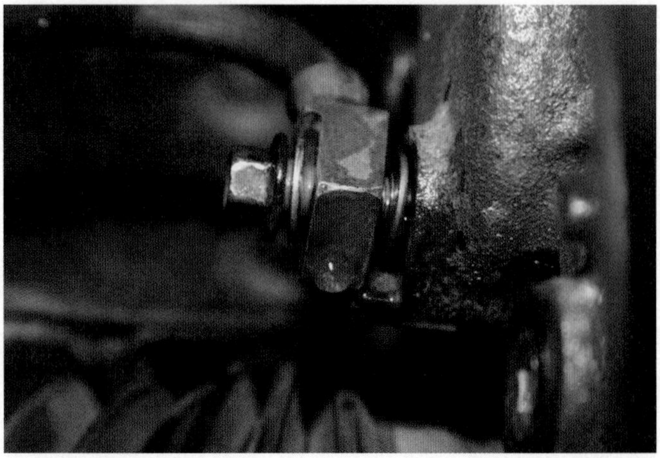

FIGURE 32-32 A loosened banjo fitting assembly.

The sealing washers fit on both sides of the banjo fitting. One sealing washer is between the head of the banjo bolt and the banjo fitting. And the other sealing washer is between the banjo fitting and the wheel unit. Always use the proper torque when tightening banjo bolts so that they are not twisted off or left loose and leak.

▶ TECHNICIAN TIP

Because the sealing washers are made of soft metal, they become crushed after use. It is good practice to replace them each time the banjo fittings are removed. Otherwise, leaks could occur.

▶ Hydraulic Braking System Control

LO 32-07 Describe proportioning valves and their operation.

The hydraulic braking system must be controlled accurately to maintain adequate control of the vehicle during braking. As we learned earlier, hydraulic pressure is equally applied throughout a sealed hydraulic braking system. We also learned that the hydraulic braking system can be designed to optimize the output force of each wheel brake unit. In a perfect scenario, that would work fine. But machines and roads are not perfect. The hydraulic pressure needs to be modified to accommodate different situations. The following components are used to modify or monitor the brake pressure:

- proportioning valves,
- metering valves,
- pressure differential valves, or
- antilock hydraulic control units

Let's see how each of them enhance the braking system.

Proportioning Valves

Proportioning valves reduce brake pressure to the rear wheels under certain circumstances. Proportioning valves can be pressure sensitive or load sensitive. The pressure-sensitive valve can be located in the master cylinder. However, it is often found in a separate unit in the rear brake circuit. The load-sensitive type is mounted on the body or on the axle, where it can respond to changes in the vehicle load.

The effectiveness of braking force is determined by tire-to-road friction. The greater the load on the tire, the greater the friction. The greater the friction, the greater the stopping ability. When a vehicle stops abruptly, a portion of the weight on the rear wheels transfers to the front wheels. This results in greater tire-to-road friction on the front tires and less on the rear. This is called **load transfer.** Weight is being transferred from the rear wheels to the front wheels.

If equal braking force is applied to the front and rear wheels, the smaller load in the rear can cause the rear wheels to lock up. Skidding tires on the surface of the road (kinetic friction) do not have as much friction as rolling tires. Stopping distance is then increased, which can lead to an accident. A proportioning valve reduces the pressure applied to the rear brakes under heavy braking. This helps to prevent rear wheel lockup by helping to maintain traction.

If the vehicle is equipped with ABS, it may not be equipped with a proportioning valve. The ABS system can make up for wheel slippage due to the changes in load transfer. This is covered in more detail in the Electronic Brake Control chapter.

Pressure-Sensitive Proportioning Valve Operation

The pressure-sensitive proportioning valve adjusts the braking force under heavy braking pressure. During normal braking, the poppet piston is in a relaxed position by a large pressure spring. The **poppet valve** holds against its retainer by a light return spring. Brake fluid passes through the pressure-sensitive proportioning valve to the rear brakes (**FIGURE 32-33**). In this condition, the rear brakes operate without any modification to the pressure.

During heavy braking, hydraulic pressure can reach the poppet valve's crack point. The pressure applied to the two different areas of the poppet piston creates unequal forces. The higher hydraulic pressure moves the poppet piston against the large pressure spring. This tends to close the pressure-sensitive proportioning valve. At a specified inlet pressure, the conical section of the pressure-sensitive proportioning valve seals against the seat. This holds the pressure steady to the rear brakes until there is a further change in inlet pressure (**FIGURE 32-34**).

As greater pedal force increases brake fluid pressure, it acts on the smaller end of the poppet piston. This combines with the force of the pressure spring to overcome the lower pressure now on the larger (output) end. As a result, the piston is forced back, opening the poppet valve and allowing pressure to rise to the rear brakes.

The increased pressure now acts on the larger end of the poppet piston. It again forces the piston forward. This closes the poppet valve and holds pressure steady. This repeated action causes a lowering of outlet pressure versus inlet pressure.

When the brake pedal is released, the pressure unseats the poppet valve. This lets the brake fluid return to the master cylinder. The pressure spring now returns the poppet piston to its relaxed position.

Some vehicles position the proportioning valve in a combination valve. Should the front brake system fail, the warning lamp spool moves forward. This takes the poppet valve with it (**FIGURE 32-35**). Pressure in the rear brakes rises, and the piston moves forward. However, it cannot close the poppet valve. So, full system pressure remains available for the rear brakes in this situation. Should the rear brake system fail, the warning lamp spool will move backward to activate the warning light.

A diagonally divided system needs one pressure-sensitive proportioning valve for each rear wheel. They are usually located in each rear brake line (**FIGURE 32-36**). It operates like the previous one. But it doesn't have the pressure differential warning light circuit.

Adjustable Proportioning Valves

Aftermarket adjustable proportioning valves are available for performance applications. They have a method of adjusting the crack point of the proportioning valve (**FIGURE 32-37**). This way they can be customized to a specific vehicle.

Adjustable proportioning valves are popular with kit car builders. The valves allow them to use components from a variety of vehicles that were not originally designed to work together. Adjustable proportioning valves are not recommended for most applications. This is due to the amount of trial and error necessary to set them properly.

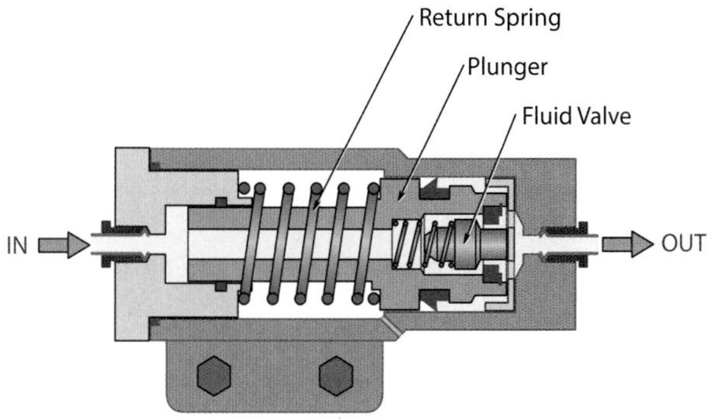

FIGURE 32-33 A pressure-sensitive proportioning valve in the open position.

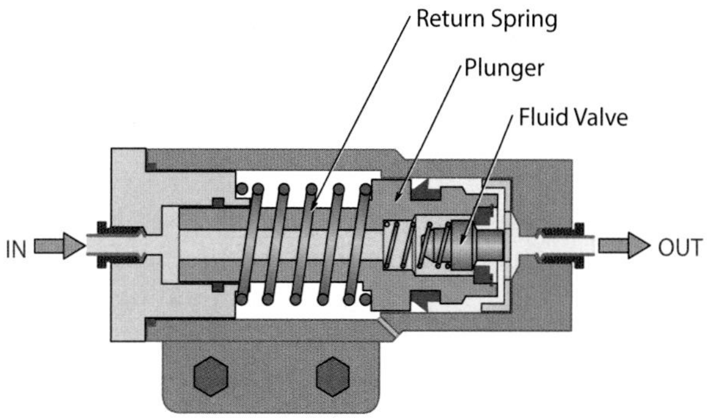

FIGURE 32-34 A pressure-sensitive proportioning valve in the closed position.

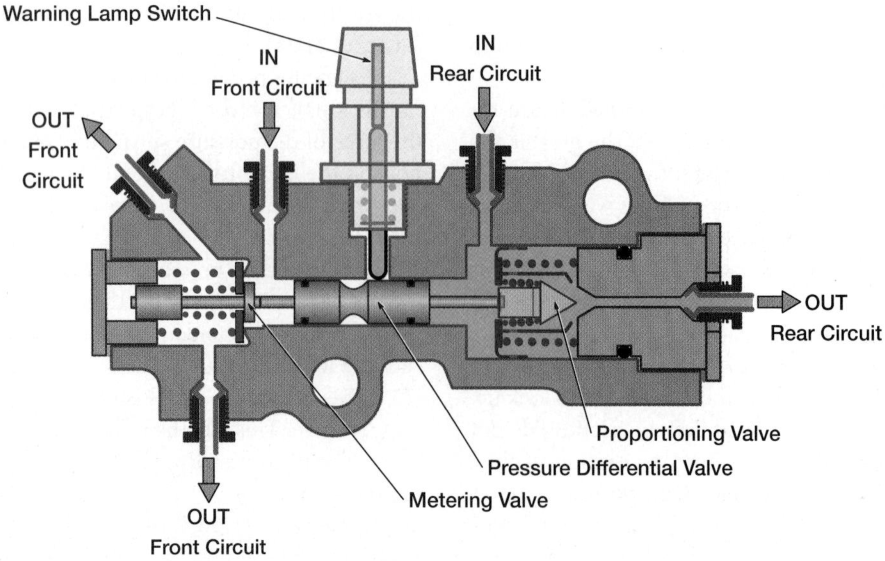

FIGURE 32-35 Proportioning valve action in a combination valve with a front brake circuit failure.

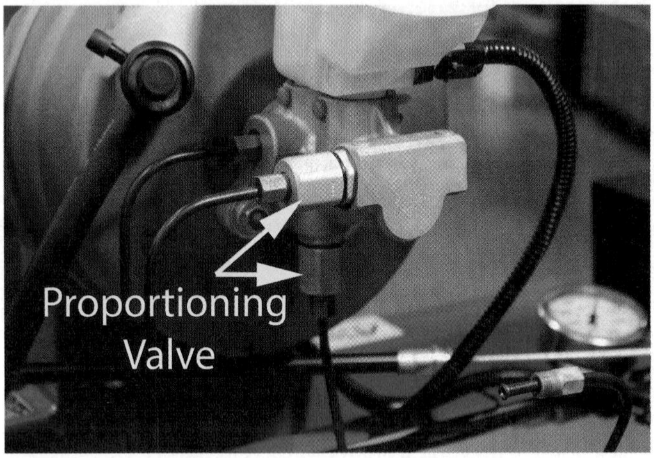

FIGURE 32-36 A pressure-sensitive proportioning valve on a diagonally split system.

FIGURE 32-37 Adjustable proportioning valve.

Load-Sensitive Proportioning Valve

The load-sensitive proportioning valve modifies brake pressure according to vehicle load. When the vehicle is lightly loaded, it further reduces pressure to the rear brakes. When it is heavily loaded, it allows higher rear brake pressure. The load-sensitive proportioning valve is usually located on the chassis. It has a lever that is connected to the rear axle (**FIGURE 32-38**).

As the vehicle is loaded, the chassis squats on the rear suspension. This moves the proportioning valve's lever. The load-sensitive proportioning valve then increases the rear brake pressure. The position of the lever changes the force on the load-sensitive proportioning valve. In this position, it further increases rear brake pressure.

A diagonally split system may have two load-sensitive proportioning valves. One valve for each rear brake unit. Each load-sensitive proportioning valve is mounted on the chassis, near the rear suspension.

▶ TECHNICIAN TIP

There are very few adjustable components in the hydraulic brake system. One notable exception is the load-sensitive proportioning valve. It is used on some pickup trucks and other load-carrying vehicles. They operate so that as load weight increases, more brake pressure is applied to the rear wheels as required.

In the case of a rear-wheel lockup on one of these vehicles, excessive pressure to the rear brakes may be suspected. Pressure gauges are fitted inline to the front and rear brakes. This is to check the operation and adjustment of load-sensing proportioning valves. After the hydraulic brake system is opened to install the gauges, any air is bled out. This is so accurate readings can be obtained.

The weight on the rear axle must be set according to charts or graphs published by the manufacturer. This will help determine the correct relationship between front and rear pressure at a given load. For example, look at the following graph. If a rear axle is loaded to 1984 lb (900 kg), and front brake pressure is raised to 1138 psi (7846 kPa), then rear brake pressure should fall within the range of 569–711 psi (3923–4902 kPa). The rear axle is then loaded to 3700 lb (1678 kg). If the rear

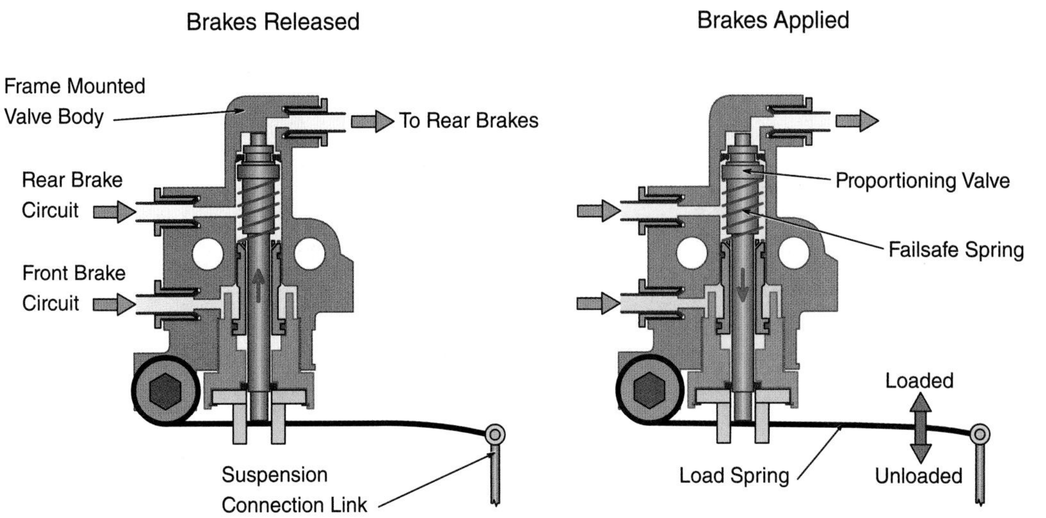

FIGURE 32-38 A load-sensitive proportioning valve adjust rear braking force based on changes in rear wheel load.

pressure is outside the specified range, adjustment of the linkage between the proportioning valve and the rear suspension may be required.

Rear Axle Load lb (kg)	Front Brake Pressure, psi (kPa)	Rear Brake Pressure, psi (kPa)
1984 (900)	1138 (7846)	569–711 (3923–4902)
3699 (1678)	1707 (11,769)	1323–1493 (9122–10,294)

Electronic Brake Proportioning

Many ABS-equipped vehicles integrate an electronic brake proportioning function in the system. It uses the hydraulic control unit to control pressure to the rear brake units. This function reduces pressure to the rear wheels under heavy braking. It operates like the proportioning valve. The pressure reduction is electronically controlled. This allows it to compensate for differences in traction by monitoring wheel slip. This is covered further in The Electronic Brake Control chapter.

▶ Metering Valves

LO 32-08 Describe metering, pressure differential, and combination valves.

Metering valves are used on vehicles with a disc/drum brake combination. Metering valves hold off the initial application of the front brakes (**FIGURE 32-39**). Drum brakes use springs to return the brake shoes to their rest position. It takes a certain amount of hydraulic pressure to overcome the tension of the return springs for the shoes to move. Disc brakes use the much smaller force of the square-cut O-ring to return the caliper piston to its rest position. Thus, very little hydraulic pressure is needed to move the brake pads into contact with the rotor. The metering valve is used in this situation to delay the application of the front drum brakes.

Vehicles typically handle better when the rear brakes engage before the front brakes. This helps keep the vehicle tracking straight while the brakes are initially applied. The metering valve keeps the front disc brakes from being applied until the rear drum brakes overcome the tension of the return springs. Vehicles with four-wheel disc brakes don't need metering valves. This is because the front and rear brakes apply at roughly the same time.

FIGURE 32-39 A metering valve.

Metering Valve Operation

The metering valve operates like a radiator cap. It has a spring that holds the metering valve closed until a specified pressure is reached (**FIGURE 32-40**). Pressure rises in the hydraulic braking system. Once it overcomes the rear brake return spring tension, the rear brakes start to apply. As pressure continues to rise, it overcomes the metering valve's crack point. Brake fluid flows to the front disc brake calipers and starts to apply the front brakes. The metering valve only has hydraulic pressure on the inlet side. So, any pressure above its crack point holds it open.

When the brake pedal is released, the metering valve is pushed closed by the spring. A fluid return valve opens up. It allows brake fluid to flow freely back to the master cylinder

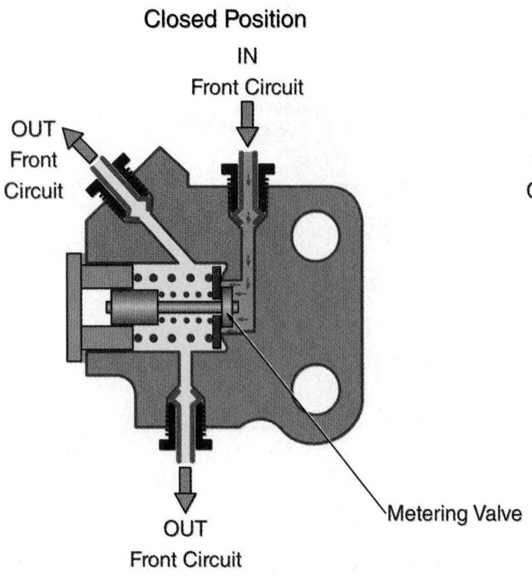

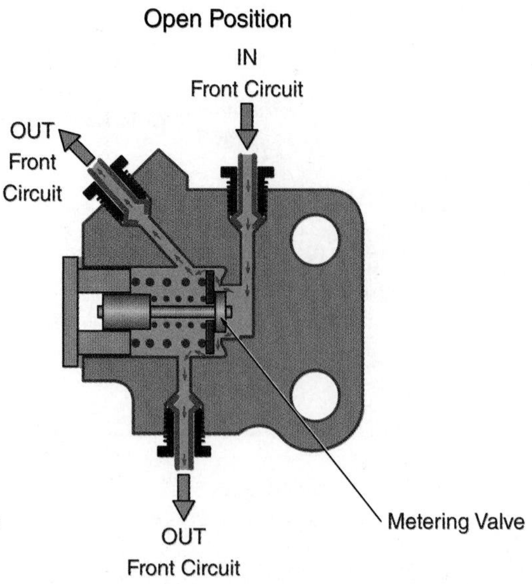

FIGURE 32-40 A metering valve holds pressure from applying the front brakes until a specified pressure is reached.

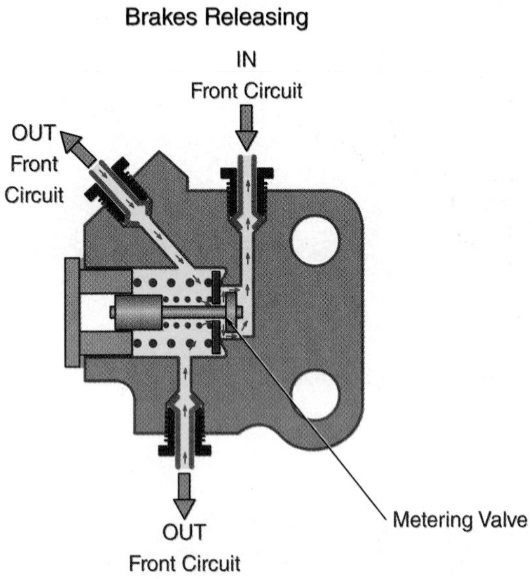

FIGURE 32-41 Brake fluid is free to flow through the fluid return valve back to the master cylinder when the brakes are released.

(**FIGURE 32-41**). During bleeding, if there is air in the lines, it may be difficult to raise the brake fluid pressure enough to open the metering valve. This makes it almost impossible to bleed any air out of the front part of the hydraulic braking system. Most metering valves are designed so that they can be manually held in the open position. This allows bleeding to take place. Also, many pressure bleeders operate at a pressure that is too low to open the metering valve. So, the valve has to be held open manually when bleeding with this method as well.

Vehicles equipped with a diagonally split system are generally front-wheel drive. They usually do not use metering valves for two reasons. First, the torque of the spinning engine during braking compensates for the earlier application of the front brake pads. This offsets not having a metering valve in the system. Second, because up to 80% of the braking occurs at the front wheels, they need to be applied as quickly as possible. So, a metering valve would delay application of the brakes.

SAFETY TIP

After bleeding the hydraulic braking system, always remember to remove the tool that is holding open the metering valve. Failure to do so could cause the vehicle to brake improperly.

Pressure Differential Valve

A **pressure differential valve** monitors any pressure difference between the two separate brake circuits. If there is a moderate leak anywhere in the system, it will illuminate the brake warning light on the instrument panel. Sometimes the light only flickers, or comes on when the brake pedal is pushed, indicating a small leak. The valve can be located in the master cylinder or in the combination valve.

▶ TECHNICIAN TIP

On a front-rear split system, there can also be a pressure difference if the rear drum brake shoes are grossly under-adjusted. When the brake pedal is pushed, the shoes do not make contact. The primary piston moves so far that it contacts the secondary piston, which starts to apply the front brakes. This raises the pressure in the secondary circuit higher than in the primary circuit. The pressure differential valve moves toward the primary circuit and turns the brake warning light on.

Pressure Differential Valve Operation

The pressure differential valve is connected between the two halves of the hydraulic braking system. Pressure is applied to

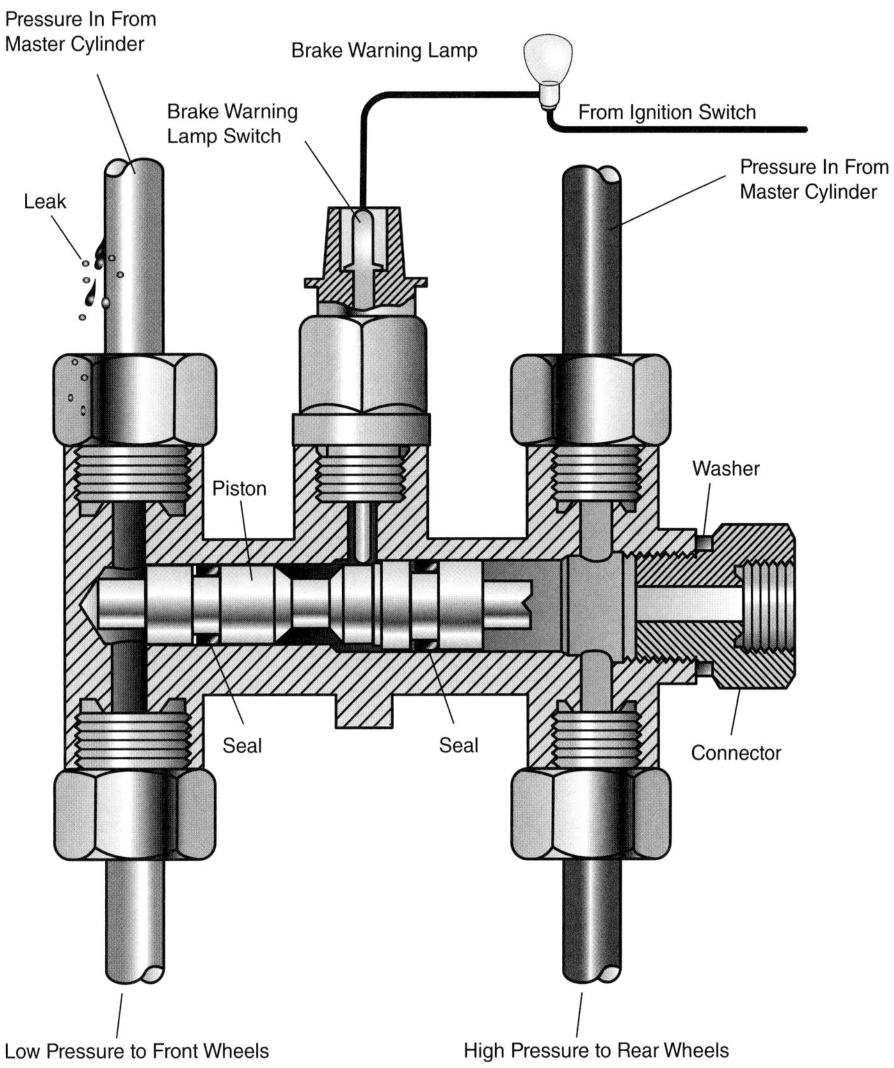

FIGURE 32-42 A pressure differential valve with a leak in the hydraulic braking system.

each end of the pressure differential valve. As long as the pressure stays the same in both circuits, the pressure differential valve remains centered. This keeps the light off. A moderate leak in the system lowers the pressure on that side of the circuit. The higher pressure in the non-leaking side pushes the pressure differential valve off center toward the side with the leak. The pressure differential switch then closes and illuminates the brake warning light. The light signals the driver that there is a serious leak in the hydraulic brake system (**FIGURE 32-42**).

During hydraulic braking system bleeding, the pressure differential valve may need to be centered. Most pressure differential valves have springs that help center them. Also, there is a small amount of clearance between the valve and bore. It allows a small amount of fluid to bypass the valve when the brakes are firmly applied. The spring then returns the pressure differential valve to center. On vehicles without this feature, first bleed the brakes. Then, you must bleed a small amount of brake fluid out of the opposite hydraulic circuit. This allows the pressure differential valve to move back to center. Follow the manufacturer's procedure.

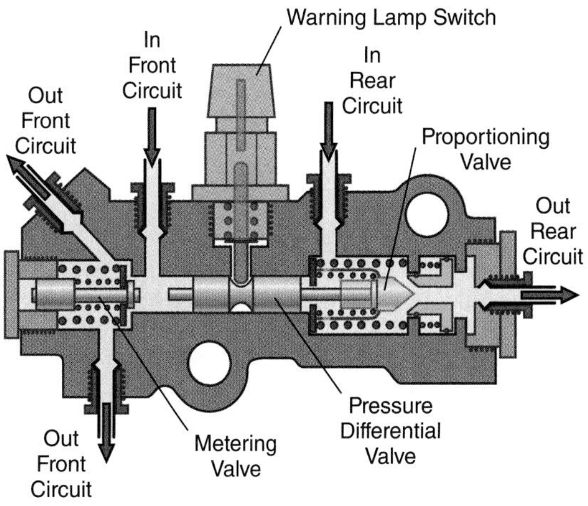

FIGURE 32-43 A combination valve.

Combination Valve

The combination valve combines the hydraulic brake system valves into one unit (**FIGURE 32-43**). They can include the pressure differential valve, metering valve, and proportioning valve.

Some combination valves combine only the pressure differential valve and proportioning valve(s). On others, the combination includes only the pressure differential valve and metering valve. Each valve operates and is contained in one unit. Combination valves are not serviceable. If they become faulty, replace them.

▶ Brake Warning Light and Stop Lights

LO 32-09 Describe brake warning light and stop light operation.

Red lights are used on vehicles as warning devices to warn drivers and others of specific conditions. When a red light comes on, the driver should take notice and respond appropriately. There are two general categories: the brake warning light and stop lights. The brake warning light is a single light located in the instrument panel. The stop lights are made up of at least three lights at the rear of the vehicle. Let's look at these systems.

Brake Warning Light

The brake warning light is located on the instrument panel. It is designed to warn the driver of a condition in the brake system that needs attention (**FIGURE 32-44**). Usually, the light can be illuminated by four causes (**FIGURE 32-45**):

- The first is when the parking brake is engaged. The light is turned on by the parking brake lever or pedal. This is to alert the driver that it is on so the driver will release it before driving.
- The second reason the brake warning light comes on is because the brake fluid level is too low in the master cylinder reservoir. This could be caused by a leak in the hydraulic braking system or worn disc brake pads.
- On vehicles without ABS, the third cause of the brake warning light coming on is unequal pressure in the hydraulic brake system. This causes the pressure differential valve to activate the brake warning light switch.
- The fourth cause of the brake warning light illuminating is called a "prove out" or "proofing" circuit. On most vehicles designed without controller area network bus (CANbus) system, turning the ignition switch to the crank position causes the warning light to illuminate. This is so the driver knows the bulb is good. On CANbus systems, the system turns the light on for a few seconds when the key is turned to the run position (**FIGURE 32-46**).

Stop Lights

Stop lights are also called brake lights. They are designed to warn others that the vehicle is braking. This information is critical to help avoid accidents. The regular stop lights are mounted on the rear of the vehicle and must conform to federal laws for brightness and location.

FIGURE 32-44 A brake warning light warns the driver of an issue in the braking system.

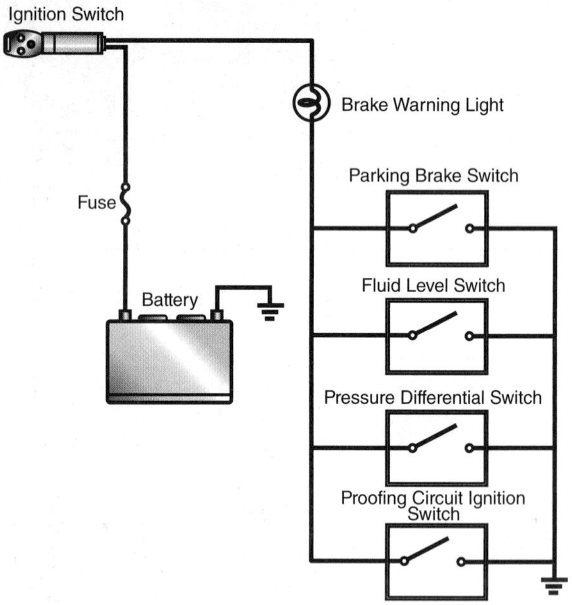

FIGURE 32-45 Brake warning light circuit.

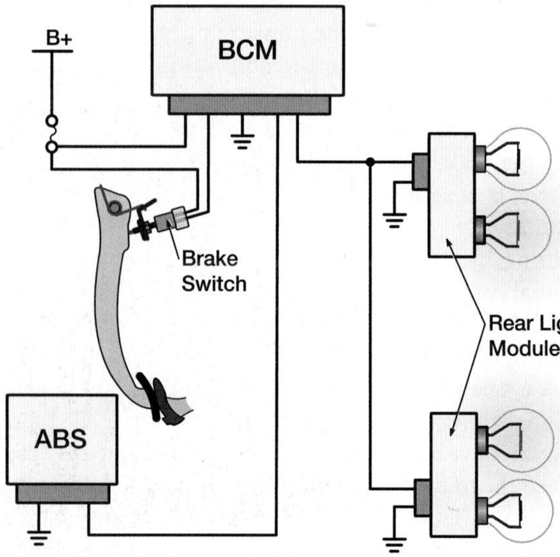

FIGURE 32-46 A CANbus brake warning light circuit.

Brake lights are activated by a normally closed stop light switch located on the brake pedal assembly. When the driver applies the brake, the brake pedal moves away from the stop light switch. A spring in the switch closes the contacts. This allows current to flow and illuminate the stop lights (**FIGURE 32-47**). Releasing the brakes causes the brake pedal to force the stop light switch open. This turns the stop lights off.

In 1986, North America mandated that all new passenger vehicles be equipped with a center high mount stop lamp (CHMSL) (**FIGURE 32-48**). Light trucks and vans were added in 1994. This lamp is located higher than the regular brake lights. Generally, it is near the centerline of the vehicle. It is designed to be in other drivers' line of sight while they are looking down the road to anticipate traffic hazards. This higher mounted lamp helps reduce rear-end collisions by being more visible to drivers. It does so by giving them more warning time for stopping.

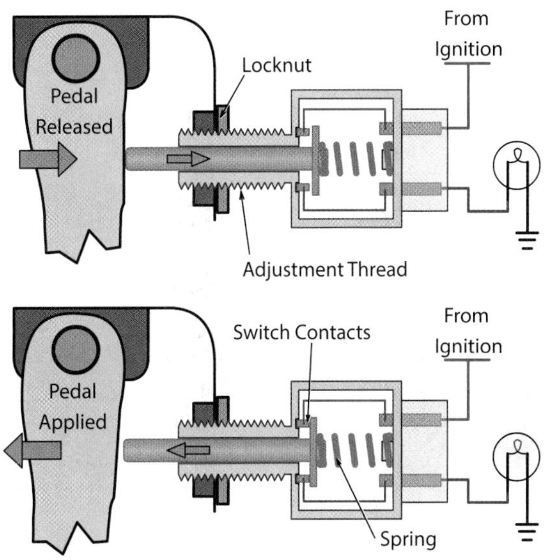

FIGURE 32-47 Brake light switch operation.

FIGURE 32-48 A center high mount stop lamp (CHMSL) mounted on a vehicle.

▶ Power Brakes

LO 32-10 Describe the operation of vacuum brake boosters.

A power brake booster assists the driver in applying the brakes. They use an external source of force to multiply the driver's pedal effort. This increases the force applied to the master cylinder pistons. They, in turn, increase the hydraulic pressure in the braking system. There are two main types of power brake units: vacuum-assist and hydraulic-assist (**FIGURE 32-49**). The vacuum-assist power booster is the most common.

Units on gasoline engines use the vacuum produced in the intake manifold to power the brake booster. This supplies approximately 20" of vacuum to the booster. Vehicles with diesel engines do not have manifold vacuum. So, they are fitted with an engine-driven vacuum pump.

The vacuum-assisted power booster operates between the brake pedal and the master cylinder. It uses the difference between engine vacuum and atmospheric pressure. This is what increases the force on the master cylinder pistons. The level of assistance depends on the pressure applied to the brake pedal. It also depends on the pressure difference between the vacuum side of the booster and the atmospheric pressure side.

The hydraulic-assisted power booster operates between the brake pedal and the master cylinder. It usually uses hydraulic

FIGURE 32-49 Power brake boosters. **A.** Vacuum-assist. **B.** Hydraulic-assist.

pressure from the power steering pump to increase the force on the master cylinder. The level of assistance this power booster gives depends on the pressure applied to the brake pedal. It also depends on the hydraulic pressure in the system.

Vacuum Booster

The most common types of vacuum boosters are the single diaphragm and the dual diaphragm. Just as the names imply, they have either one or two diaphragms. The diaphragms are used to extract force from atmospheric pressure. Using two diaphragms allows the diameter of the booster to be smaller. However, it is a bit longer than a single-diaphragm unit.

Vacuum boosters consist of the following (**FIGURE 32-50**):

- A housing, which encases all of the other parts
- One or two sealed diaphragms, which transmit the force to the master cylinder
- A diaphragm return spring, which pushes the diaphragm back when the brake pedal is released
- A control valve, which controls vacuum and atmospheric pressure to each chamber
- A pushrod and reaction disc, which operate in conjunction with the control valve
- A one-way check valve, which holds vacuum in the booster when it is higher than intake manifold vacuum
- Seals, which prevent air leaks

How much force can a vacuum booster create? Let's do the math, using a panic stop situation. The booster receives approximately 20" (51 cm) of vacuum from the intake manifold. That is equal to about 10 psi (69 kPa) of pressure because 2" (5.1 cm) of vacuum equals 1 psi (6.9 kPa). If the diaphragm has a circumference of 12" (31 cm), then it has an area of approximately 113 sq in. (729 cm^2). Therefore, 10 psi (69 kPa) × 113 square inches (729 cm^2) = 1130 lb (513 kg) of force.

That means that a 12" (31 cm) booster can add up to 1130 lb (513 kg) of force to whatever the driver's foot effort is. And that is after being multiplied through the leverage of the brake pedal. The total force is magnified further by the hydraulic system. This is why a vehicle weighing thousands of pounds can be stopped with minimal foot pressure.

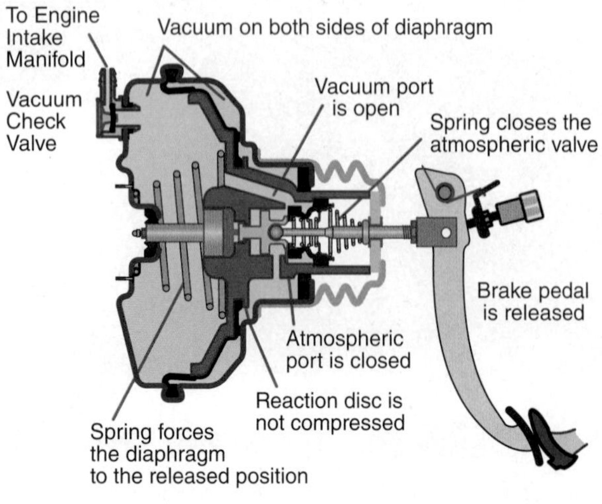

FIGURE 32-50 Vacuum brake booster in the released position.

Vacuum Booster Operation

When the driver steps on the brake pedal, it moves the brake pedal pushrod forward. The pushrod transmits movement through the power unit to the master cylinder. The master cylinder pistons apply the brakes. The pushrod also operates a control valve. This valve controls the flow of vacuum and atmospheric pressure to each side of the diaphragm. How it works depends on the position of the control valve.

A hose connects the intake manifold to a vacuum check valve on the power brake unit (**FIGURE 32-51**). With the engine running, the check valve allows air to be evacuated from the booster, but not to return. Vacuum in the intake manifold is used to evacuate the power unit. The valve holds vacuum in the booster in the case of an engine failure. In this case, it allows at least one full boosted brake application.

The booster chambers are separated by a flexible rubber diaphragm attached to the diaphragm plate (**FIGURE 32-52**). It

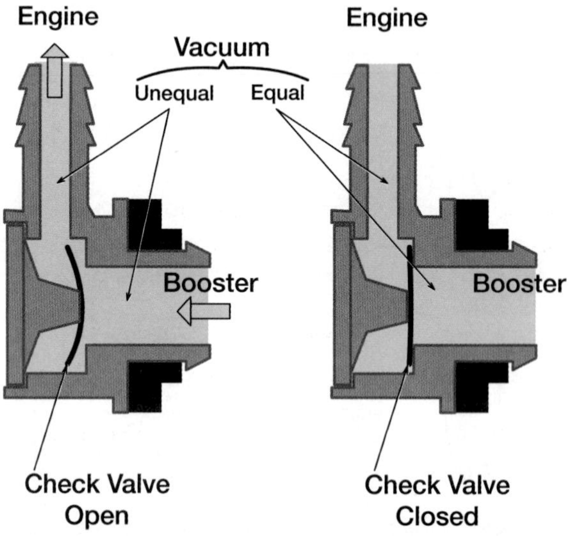

FIGURE 32-51 Power brake booster check valve. It holds vacuum in the booster if the engine dies.

FIGURE 32-52 Power brake booster diaphragm and return spring.

is held in the off position by a large-diaphragm return spring. The master cylinder pushrod and the control valve are centrally located on each side of the plate. The pushrod normally incorporates a system to adjust pushrod length. The adjustable length provides the proper play between the pushrod and the master cylinder piston. The length normally does not have to be adjusted. But if adjustment is required, use the proper tools and follow the service information completely.

As the brakes are applied, the pedal pushrod and plunger move forward in the diaphragm plate. This brings the vacuum valve into contact with the vacuum port seat. This closes the vacuum port, sealing off the passage connecting the two chambers. This holds the pressure and vacuum steady in each chamber. This is called the hold position.

Further movement of the pushrod and plunger moves the atmospheric valve away from the atmospheric port seat. Air at atmospheric pressure comes in through the air filter in the rear of the unit. From there, it enters the chamber behind the diaphragm. The difference in pressure on both sides of the diaphragm moves the diaphragm plate forward. It takes the master cylinder pushrod with it. This applies greater force to the master cylinder. This position is called the apply position. The vacuum valve is closed and the atmospheric valve is open (**FIGURE 32-53**).

Once the brake pedal stops moving, the atmospheric pressure continues to build up in the rear chamber. As hydraulic pressure rises, a counterforce acts through the master cylinder pushrod and the reaction disc. This counterforce allows the diaphragm plate and control valve to continue moving forward slightly. It moves until it causes the atmospheric valve to close off the atmospheric port. This stops the atmospheric pressure from entering the booster. The booster is again in the hold position. It is ready for increased or decreased pedal pressure from the driver.

During application, the reaction force against the valve plunger works against the driver to close the atmospheric port. With both the atmospheric and the vacuum ports closed, the power unit is in a hold position. It stays this way until increased pedal force reopens the atmospheric port, causing more boost.

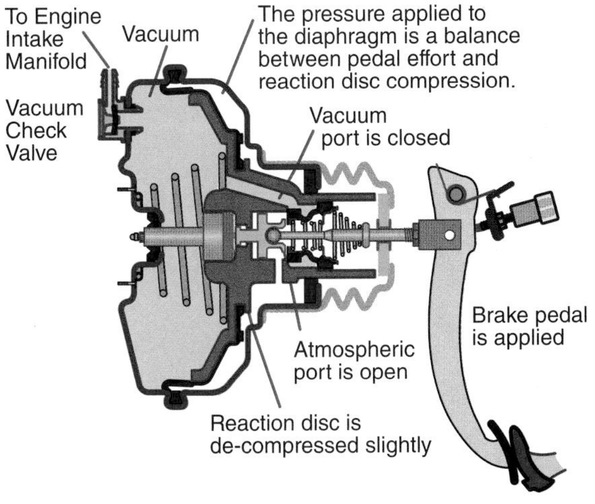

FIGURE 32-53 Vacuum brake booster in the apply position.

Or until a drop in pedal force reopens the vacuum port, reducing boost. When the force on the pedal is held constant, the valve always returns to the hold position.

When the brake pedal is released, the atmospheric valve is closed, and the vacuum valve opens. As a result, any atmospheric pressure is evacuated from the rear chamber. It travels through the front chamber. It leaves the booster through the check valve, and into the intake manifold. This reduces the atmospheric pressure pushing on the diaphragm plate. The return spring then pushes the diaphragm plate back to its off position. With the driver's foot off the brake pedal, the vacuum valve remains open. This ensures that there is equal vacuum on both sides of the diaphragm plate ready for the next application.

When the engine stops for any reason, no manifold vacuum is available to supply the booster. The vacuum remaining in the booster, held by the one-way check valve, will provide for at least one power-boosted brake application. After this, the brakes will still operate, but without power assistance. Much greater brake pedal effort will be required from the driver.

Dual-Diaphragm Boosters

Dual-diaphragm power boosters work on the same principle of operation as the single-diaphragm power booster. They are smaller in diameter to better fit in the limited space under the hood of some vehicles. However, two smaller diaphragms in tandem are used, one behind the other (**FIGURE 32-54**). This splits the surface area of the diaphragm between two diaphragm plates to provide the needed boost. There are separate vacuum and atmospheric chambers for each diaphragm plate. They use the same control valve concepts as a single-diaphragm booster. They apply, hold, and release in the same manner as a single-diaphragm booster.

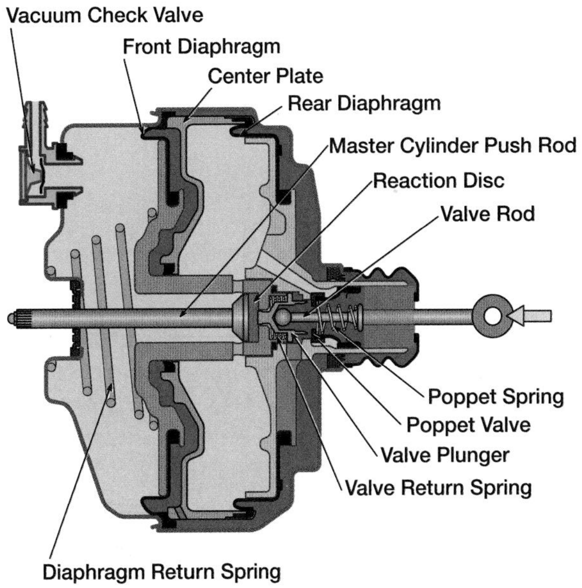

FIGURE 32-54 Dual diaphragm brake booster uses two diaphragms, one in front of the other, to create the required assistance.

▶ Hydraulic Brake Booster

LO 32-11 Describe the operation of hydraulic brake boosters.

Many vehicles are now equipped with hydraulically assisted brake boosters. The hydraulic booster system uses hydraulic pressure generated by the power steering pump. This is used rather than engine vacuum (**FIGURE 32-55**). This application is particularly suitable to vehicles with diesel engines, since they do not have manifold vacuum. Hydraulic brake boosters use power steering fluid to operate the booster. But the master cylinder portion of the system still uses brake fluid. Do not make the mistake of putting power steering fluid in the master cylinder reservoir. Also, don't put brake fluid in the power steering pump reservoir.

The hydraulic booster uses pressure from the power steering fluid that is circulating through the system. The hydraulic pressure is applied against the master cylinder actuating piston to provide boost. The hydraulic pressure generated by the power steering pump is stored in an accumulator. It is routed to the hydraulic booster unit. The booster applies mechanical force to the master cylinder when the brake pedal is applied (**FIGURE 32-56**).

The booster can generate pressures of 1200–2000 psi (8274–13,790 kPa) to activate the master cylinder. Hydraulic boosters can be completely separate components from the master cylinder. Or they can be integrated with the master cylinder. As a safety measure, part of the hydraulic booster system includes an accumulator.

Accumulators assist in maintaining a reserve of system pressure. Some accumulators are nitrogen pressurized. Others are spring loaded. The accumulator stores enough pressure to provide for a few full-power applications. If the engine stalls or the power steering belt breaks, accumulator pressure is used. Once the accumulator pressure is used up, there is no further brake booster function. This requires much higher pedal pressure to operate the brakes.

Electrohydraulic Brake Boosters

A less common brake booster system is electrohydraulic braking (EHB). **EHB** uses an electrically driven hydraulic pump (**FIGURE 32-57**). It is similar to the hydraulic booster system. It uses a high-pressure accumulator to store the required pressure to activate the master cylinder. But, it uses electric energy to "charge"

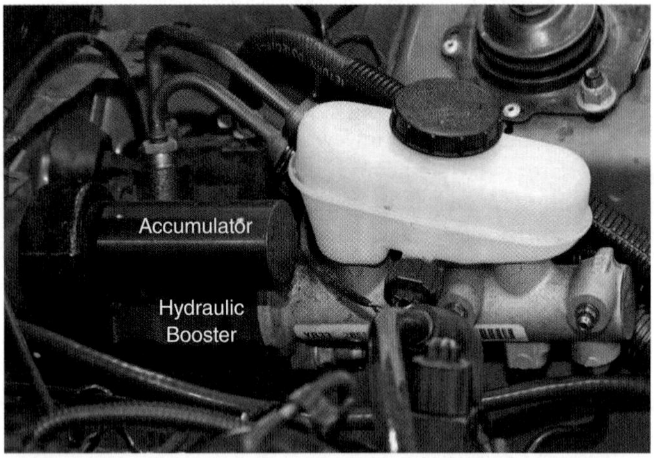

FIGURE 32-55 Hydraulic power brake booster.

FIGURE 32-57 An electrohydraulic power brake booster.

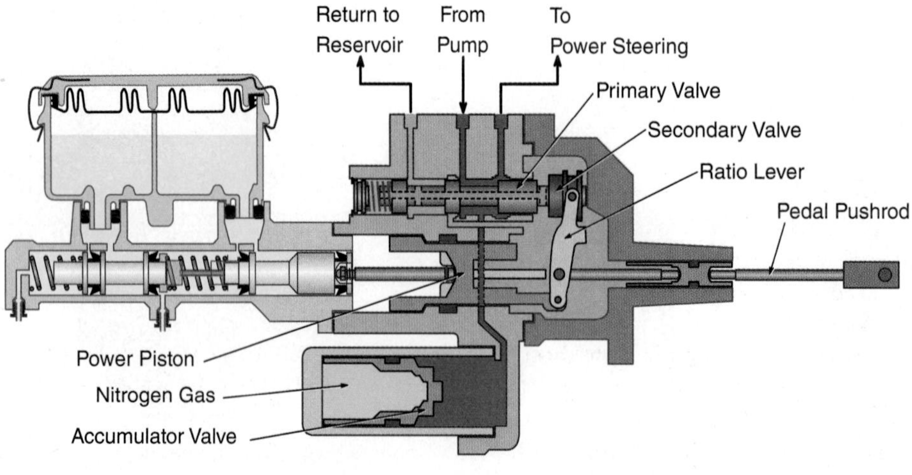

FIGURE 32-56 Cutaway view of a hydraulic power brake booster.

the accumulator. This means that less power is taken away from the engine during operation. Battery power is used only when needed, instead of continuously from the power steering pump. Also, there are fewer chances of problems with this system. This includes things such as a worn power steering pump, a slipping or broken pump drive belt, or leaky hose connections.

▶ Wrap-Up

Ready for Review

- Hydraulic brakes are operated by the force transferred by noncompressible brake fluid. According to Pascal's Law, pressure created in the master cylinder is transmitted equally through the hydraulic braking system.
  - Pressure and force in hydraulic systems are governed by input force, working pressure, and output force.
- Brake fluids are graded against compliance standards set by the US Department of Transportation (DOT) and are tested to ensure they meet the following standards: pH value, viscosity, resistance to oxidation, stability, and boiling point.
- The master cylinder converts the brake pedal force into hydraulic pressure. It has threaded passageways to which the brake lines firmly connect, a brake fluid reservoir supplies the brake system with a reserve amount of brake fluid, and is operated by a pushrod from the power booster or the brake pedal.
  - Single-piston master cylinder; tandem master cylinders
- Quick take-up master cylinders are designed to maintain a larger running clearance between the disc brake pads and rotor. The ABS master cylinder is a tandem master cylinder used in divided systems that also uses a primary piston and a secondary piston. The amount of brake fluid that returns to the master cylinder is determined by the degree of ABS control.
- Brake pedals should be mounted securely and free from any excessive sideways movement and need to be at a height and angle that will allow the driver to quickly move from the accelerator (throttle pedal) to the brake pedal. Different types of divided hydraulic systems include: vertical, or front–rear, split hydraulic system; diagonal, or X, pattern hydraulic system; and L-split hydraulic system.
- Brake lines and hoses carry brake fluid from the master cylinder to the brake units. They are double-walled steel and coated to resist corrosion. Brake hoses are made of several layers of alternating materials.
- Proportioning valves reduce brake pressure to the rear wheels under certain circumstances. They can be pressure sensitive or load sensitive. Aftermarket adjustable proportioning valves have a method of adjusting the crack point of the proportioning valve.
- Metering valves are used on vehicles with a disc/drum brake combination and hold off the initial application of the front brakes. Pressure differential valves monitor any pressure difference between the two separate brake circuits and illuminate the brake warning light on the instrument panel. Combination valve combines the hydraulic brake system valves into one unit and can include the pressure differential valve, metering valve, and proportioning valve.
- Brake warning light on the instrument panel is designed to warn the driver of a condition in the brake system that needs attention. The light can be illuminated under the following conditions: the parking brake is engaged, brake fluid level is too low, unequal pressure in the hydraulic brake system, or due to "prove out" or "proofing" circuit.
- Two main types of power brake units: vacuum-assist and hydraulic-assist.
  - The vacuum-assisted power booster uses the difference between engine vacuum and atmospheric pressure. The hydraulic-assisted power booster uses hydraulic pressure from the power steering pump to increase the force on the master cylinder.
- The hydraulic booster uses pressure from the power steering fluid that is circulating through the system to apply pressure against the master cylinder actuating piston to provide boost. The hydraulic pressure generated by the power steering pump is stored in an accumulator and is routed to the hydraulic booster unit.

Key Terms

aerate The tendency to create air bubbles in a fluid.

bleeding The process of removing air from a hydraulic braking system.

brake hose A flexible section of the brake lines between the body and suspension that allows for steering and suspension movement.

brake lines Made of seamless, double-walled steel and able to transmit over 1000 psi (6895 kPa) of hydraulic pressure through the hydraulic brake system.

compensating port Connects the brake fluid reservoir to the master cylinder bore when the piston is fully retracted, allowing for expansion and contraction of the brake fluid.

hygroscopic A property of a substance or liquid that causes it to attract and absorb moisture (water), as a sponge absorbs water. Brake fluid absorbs water out of the air; thus it is hygroscopic.

inlet port Connects the brake fluid reservoir to the master cylinder bore when the piston is fully retracted, allowing for expansion and contraction of the brake fluid.

International Standards Organization (ISO) flare method A method for joining brake lines, also called a bubble flare.

Created by flaring the line slightly out and then back in, leaving the line bubbled near the end.

inverted double flare A method for joining brake lines that forms a secure, leakproof connection.

load transfer Weight transfer from one set of wheels to the other set of wheels during braking, acceleration, or cornering.

metering valve A valve used on vehicles equipped with older rear drum/front disc brakes to delay application of the front disc brakes until the rear drum brakes are applied. Located in line with the front disc brakes.

outlet port The port leaving a cylinder or pump.

Pascal's law The law of physics that states that pressure applied to a fluid in one part of a closed system will be transmitted equally to all other areas of the system.

poppet valve A valve that controls the flow of brake fluid at usually preset pressures.

pressure differential valve A valve that monitors any pressure difference between the two separate hydraulic brake circuits; it usually contains a switch to turn on the brake warning light when there is a pressure difference.

primary cup A seal that holds pressure in the master cylinder when force is applied to the piston.

primary piston A brake piston in the master cylinder moved directly by the pushrod or the power booster; it generates hydraulic pressure to move the secondary piston.

proportioning valves Valves used mostly on older vehicles equipped with rear drum brakes to reduce rear wheel hydraulic brake pressure under hard braking or light loads. Located in line with the rear brakes.

quick take-up master cylinders Cylinders used on disc brake systems that are equipped with low-drag brake calipers to quickly move the brake pads into contact with the brake rotors.

quick take-up valve A valve used to release excess pressure from the larger piston in a quick take-up master cylinder once the brake pads have contacted the brake rotors.

recuperation Process by which brake fluid moves from the reservoir past the edges of the seal into the chamber in front of the piston. This prevents air from being drawn into the hydraulic system caused by low pressure when the brake pedal is released quickly.

residual pressure valve (residual check valve) In drum brake systems, a valve that maintains pressure in the wheel cylinders slightly above atmospheric pressure so that air does not enter the system through the seals in the wheel cylinders.

secondary cup A seal that prevents loss of fluid from the rear of each piston in the master cylinder.

secondary piston A piston that is moved by hydraulic pressure generated by the primary piston in the master cylinder.

single-piston master cylinder A master cylinder with a single piston that creates hydraulic pressure for all wheel units. If there is a leak in the system, there is a loss of pressure for all wheel units.

tandem master cylinder A master cylinder that has two pistons that operate separate braking circuits, so if a leak develops in one circuit, the other circuit can still operate.

Review Questions

1. Pascal's law is behind the hydraulic principals that is applied to brakes. What is Pascal's law?
   a. Fluid with pressure in part of a closed system will transmit higher pressureto other parts of the system.
   b. Fluid with pressure in a closed system will transmit with no loss to other parts of the system.
   c. Fluid with pressure in a closed system will transmit with some loss to other parts of the system.
   d. Fluid with pressure in part of a closed system will transmit with some loss to other parts of the system.
2. Which of the brake fluid types are silicone based?
   a. Dot 3
   b. Dot 5
   c. Dot 5.1
   d. Dot 4
3. The master cylinder uses hydraulic pressure to operate which units?
   a. Wheel brake
   b. Steering wheel
   c. Front wheels
   d. Input piston
4. Quick take-up master cylinders are used on what systems?
   a. Disc brake systems with ABS
   b. Disc brake systems with low drag calipers
   c. Disc/drum split systems
   d. Front/rear divided systems
5. What is the major factor(s) that determine(s) how a brake system is divided front to rear?
   a. Weight of the vehicle with passengers
   b. Year of the vehicle
   c. Height of the vehicle
   d. Weight distribution of the vehicle
6. What is the purpose of brake lines and hoses?
   a. To carry brake fluid from the master cylinder to the brake units
   b. To carry brake fluid from the power brake boosterto the master cylinder
   c. To modifythe brake fluid pressure to the brake units
   d. To protect the brake units from corrosion
7. How do pressure-sensitive proportioning valves affect the rear brakes?
   a. Increase brake pressure under light braking
   b. Increase brake pressure under heavy braking
   c. Decrease brake pressure under light braking
   d. Decrease brake pressure under heavy braking
8. What does the metering valve do?
   a. Delay the first application of the front brakes
   b. Monitor pressure differences between both brake circuits
   c. Limits the pressure to the rear brake units
   d. Centers the pressure differential valve

9. What is the purpose of the brake warning light?
   a. To notify the driver of high brake fluid level
   b. To notify the driver the brake system needs attention
   c. To notify others the vehicle is braking
   d. To notify the driver the vehicle is braking
10. How does the power brake assist the driver?
   a. It uses an outside source of force to increase the driver's pedal effort
   b. It recycles the brake fluid to increase the driver's pedal effort
   c. It uses a long lever to increase the driver's pedal effort
   d. It uses the pushrod length to increase the driver's pedal effort

ASE Technician A/Technician B Style Questions

1. A brake system is being discussed. Technician A states that a soft or spongy pedal is a good indication that there could be air in the hydraulic system. Technician B states that engineers use different input and output piston sizes to precisely control brake unit output force. Who is correct?
   a. Technician A only
   b. Technician B only
   c. Both Technicians A and B
   d. Neither Technician A nor B
2. A hydraulic brake system is being discussed. Technician A states that brake fluid must have a high boiling point since brakes create a lot of heat. Technician B states that hygroscopic means that brake fluid will not absorb water. Who is correct?
   a. Technician A only
   b. Technician B only
   c. Both Technicians A and B
   d. Neither Technician A nor B
3. A hydraulic brake system is being discussed. Technician A states that the brake master cylinder functions as an output piston. Technician B states that the single master cylinder has replaced the tandem master cylinder for increased safety. Who is correct?
   a. Technician A only
   b. Technician B only
   c. Both Technicians A and B
   d. Neither Technician A nor B
4. A vehicle comes in the shop with brake-related complaints. Technician A states that the brake system is normally split 50% front and 50% rear on front-wheel-drive vehicles. Technician B states that malfunctioning brake lights can be caused by a misadjusted brake light switch at the brake pedal. Who is correct?
   a. Technician A only
   b. Technician B only
   c. Both Technicians A and B
   d. Neither Technician A nor B
5. A hydraulic brake system is being discussed. Technician A states that a leaking brake line could be repaired with a new section of copper tubing. Technician B states that when fabricating a new brake line, a tubing bender should be used. Who is correct?
   a. Technician A only
   b. Technician B only
   c. Both Technicians A and B
   d. Neither Technician A nor B
6. A vehicle with a brake problem is being discussed; the rear wheels locks up when braking. Technician A states that the pressure-sensitive proportioning valve could be malfunctioning. Technician B states that the load-sensitive proportioning valve could require adjustment. Who is correct?
   a. Technician A only
   b. Technician B only
   c. Both Technicians A and B
   d. Neither Technician A nor B
7. A metering valve is being discussed. Technician A states that a defective metering valve could cause the rear brakes apply too quickly. Technician B states that a defective metering valve could cause grabby front brakes. Who is correct?
   a. Technician A only
   b. Technician B only
   c. Both Technicians A and B
   d. Neither Technician A nor B
8. A vehicle comes in to the shop with a red brake warning light illuminated. Technician A states that the parking brake could be applied or malfunctioning. Technician B states that there could be unequal pressure in the hydraulic brake system. Who is correct?
   a. Technician A only
   b. Technician B only
   c. Both Technicians A and B
   d. Neither Technician A nor B
9. A brake booster is being discussed. Technician A states that most diesel vehicles are equipped with brake boosters that utilize the engines intake manifold vacuum. Technician B states that the purpose of the brake booster diaphragm is to allow air to be evacuated from the booster, but not to return. Who is correct?
   a. Technician A only
   b. Technician B only
   c. Both Technicians A and B
   d. Neither Technician A nor B
10. A hydraulic brake booster is being discussed. Technician A states that a broken power steering belt could cause the brake booster to stop creating brake assist. Technician B states that some hydraulic brake boosters use an electric motor to create hydraulic pressure for the booster. Who is correct?
   a. Technician A only
   b. Technician B only
   c. Both Technicians A and B
   d. Neither Technician A nor B

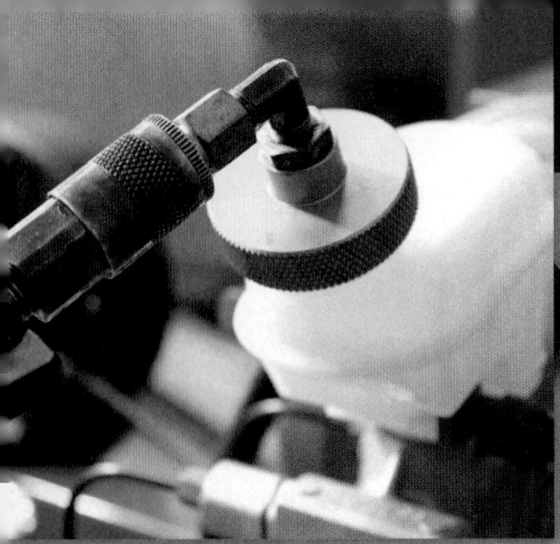

CHAPTER 33

Servicing Hydraulic Systems and Power Brakes

Learning Objectives

- **LO 33-01** Describe brake repair liability and brake tools.
- **LO 33-02** Inspect and test brake fluid.
- **LO 33-03** Bleed and flush brake systems.
- **LO 33-04** Measure brake pedal height, free play, and travel.
- **LO 33-05** Describe general brake hydraulic system diagnosis.
- **LO 33-06** Perform master cylinder service.
- **LO 33-07** Test vacuum-style brake boosters.
- **LO 33-08** Service brake lines and hoses.
- **LO 33-09** Test the operation of the brake warning lamp and stop lights.

ASE Education Foundation Tasks

See Appendix A to view the 2017 ASE Education Foundation Automobile Accreditation Task List Correlation Guide.

▶ Servicing Brake Hydraulic Systems

LO 33-01 Describe brake repair liability and brake tools.

Now that you know about the hydraulic brake components and basic hydraulic principles, we are ready to move on. It is time to cover the maintenance and light repair of the brake's hydraulic system (**FIGURE 33-1**). Although the hydraulic system is very dependable, it does require periodic maintenance. This begins with an initial visual inspection of the hydraulic components. It is followed up with checking the feel and operation of the brakes. Maintenance tasks typically revolve around checking and flushing the brake fluid. Diagnosis and repairs are in order if the maintenance checks indicate faults. Or if the owner brings the vehicle in with a brake-related concern. But before we cover these processes, we need to discuss brake repair liability issues.

Brake Repair Legal Standards and Technician Liability

Brake repair is right up with steering and suspension repair on the liability scale. Improperly repaired brakes can function reasonably well under normal driving situations. But they can fail during a panic situation, when brakes are needed the most. The likelihood of accidents, injury, or death goes up drastically in those situations. Shops and technicians have been successfully sued for improper brake repairs. Large cash settlements are a common result.

Technicians also risk being found criminally negligent (**FIGURE 33-2**). All it takes is for the court to determine that you have acted maliciously. Because of this, always follow the manufacturer's procedures when servicing brake systems. Also, research the technical service bulletins (TSBs) before performing any brake tasks. Never take shortcuts. It's not worth the risk.

Another issue that leads to liability is forgetting to tighten, or failing to properly tighten, components. When you are working on vehicles, it is easy for a distraction to interrupt your work. If that happens, then it is easy to think you already finished a part of the job when you really haven't. A good example is when you are torquing the lug nuts on the wheels. If you get interrupted while torquing the third wheel, it would be easy to come back and think you had torqued all of the wheels. And yet you left one untorqued and subject to falling off the vehicle down the road.

Good technicians reduce liability by creating processes to help ensure that steps of the job are not forgotten. This could be as simple as only installing the lug nuts when you are ready to immediately torque them (**FIGURE 33-3**). Another option if you can't torque them immediately is to only just start them on the lug studs (not running the lug nuts all the way down). This way it is obvious that they haven't been tightened or torqued. The same thing goes with most other parts. If you install them, torque them in place as soon as it is practical. Remember, safety first!

FIGURE 33-1 A technician flushing brake fluid with a vacuum bleeder tool.

FIGURE 33-2 A technician found criminally negligent for improper brake repair.

You Are the Automotive Technician

A customer brings in a four-wheel drive vehicle with a brake concern. The concern is that the brakes have to be applied very hard to stop the vehicle. The service history shows that the front brakes were replaced and the fluid flushed about five years ago. You test drive the vehicle and verify the hard brake pedal concern. During the visual inspection, you notice a stick that pushed the flexible brake hose against the right front tire, wearing the hose badly. The brake linings are in serviceable condition.

1. What faults could cause the hard brake pedal concern?
2. What tests would you do to confirm the cause of the hard pedal concern?
3. What fluid service is needed?
4. What do you need to do to repair the brake hose issue?

FIGURE 33-3 To reduce forgotten steps, create processes to help minimize missed steps.

SAFETY TIP

Asbestos is a naturally occurring mineral mined from the earth. Asbestos is a long, very thin fibrous crystal. When asbestos is disturbed, small needlelike fibers can break off and remain airborne, where people inhale them. Because these fibers are so small, they embed themselves deep within lung tissue, causing scarring. Repeated exposure can lead to asbestosis and lung cancer.

Asbestos has been removed from most brake and clutch materials. But it is still present in some replacement and old components. Therefore, you must treat all brake and clutch dust as if it contains asbestos. This is accomplished by wetting down the dust and washing it away. The best way is by using an aqueous brake wash station. Some states allow using an aerosol can of brake cleaning solution to carefully wash down the brake components.

Hydraulic System Maintenance

Tools

A number of tools make brake service much easier. Here are some of the most common:

- Brake bleeder wrenches are used to open and close bleeder screws. They come in a variety of configurations. They are designed to fit into the tight space where the bleeder screw is located. They are also usually designed with six sides. This decreases the likelihood of rounding off the bleeder screw hex head.
- Flare nut wrenches are used to loosen and tighten fittings on brake lines. Their open side slips over the brake line, and the wrench can still grab five or six points of the fitting.
- Vacuum brake bleeders can be operated by hand, air, or electricity. The hose from the vacuum bleeder is placed on the end of the bleeder screw, and the bleeder screw is opened. The vacuum is applied which pulls brake fluid from the hydraulic braking system through the bleeder screw. This fluid is captured in a container for later disposal.
- Pressure brake bleeders provide a reservoir of brake fluid under pressure. The pressure bleeder usually mounts to the master cylinder reservoir. It supplies a steady stream of clean brake fluid for bleeding or flushing. With the master cylinder pressurized, the bleeder screws on the wheel brake units can be opened one at a time. This bleeds or flushes them of old brake fluid and air.
- Brake fluid testers are used to test the boiling point/moisture content of brake fluid (**FIGURE 33-4**).

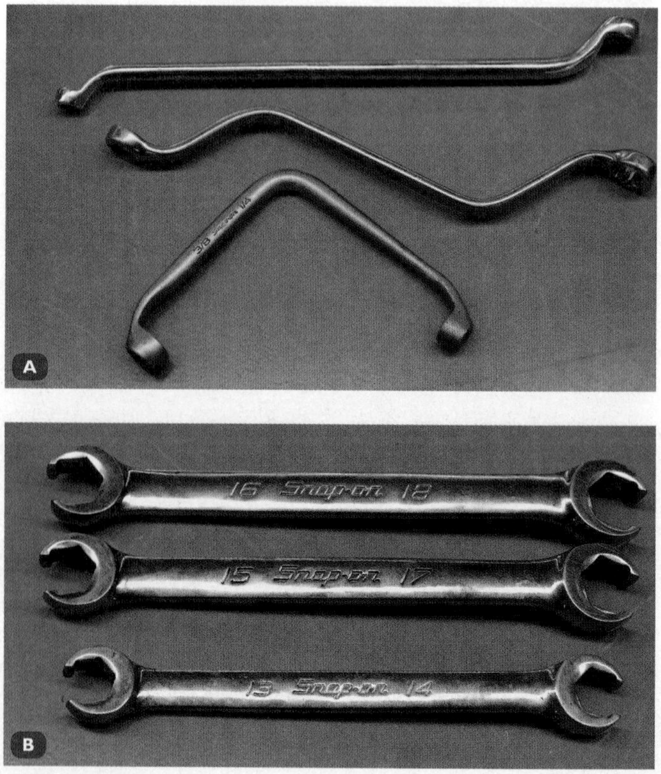

FIGURE 33-4 Common tools used to repair brakes: **A.** Brake bleeder wrenches. **B.** Flare nut wrenches.

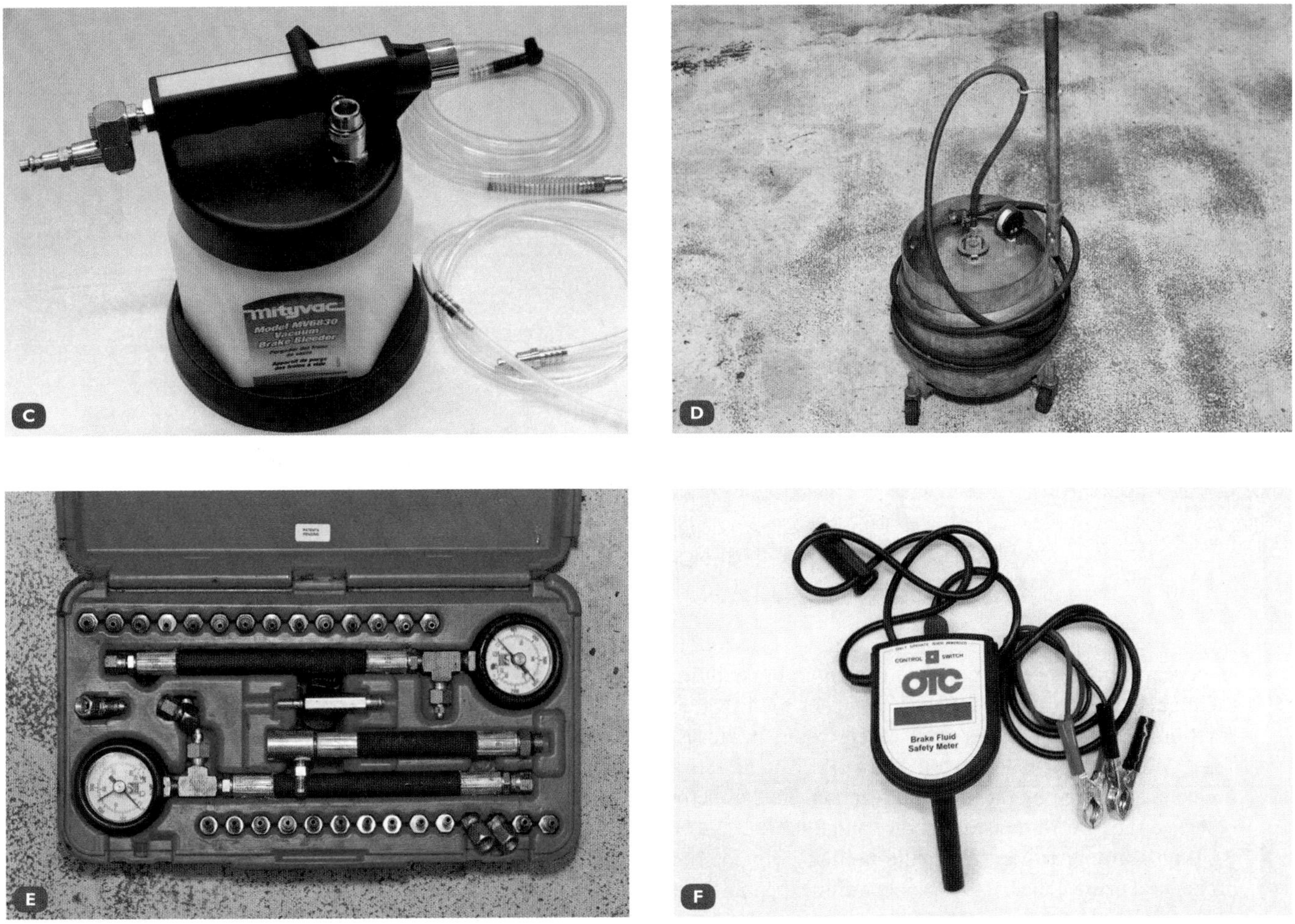

FIGURE 33-4 Common tools used to repair brakes: **C.** Vacuum brake bleeder. **D.** Pressure brake bleeder. **E.** Proportioning valve/metering valve gauge sets. **F.** Brake fluid tester.

▶ Brake Fluid Handling

LO 33-02 Inspect and test brake fluid.

Brake fluid is the lifeblood of the hydraulic braking system. So, you need to understand how to properly select and handle it. Failure to do so could cause damage to the hydraulic braking system. This can result in an unsafe situation for the driver and passengers. Because of this, brake fluid levels should be inspected during every oil change.

Brake fluid is considered a non-top-off fluid. This is because brake fluid does not get used up in the system. There are only two reasons that cause the brake fluid level to go down. The first reason is if the hydraulic system has a leak. In this case, the system will need to be inspected and the leak located. Once located, it can be repaired.

The second reason for a low brake fluid level is worn disc brake linings. In this case, the brake lining thickness will need to be measured and compared with specifications. If it is below the minimum specified thickness, the disc brakes will need to be serviced.

Normally, topping off the brake fluid should only be performed after the brakes have been repaired. Be careful not to spill brake fluid, as it can soften paint. If you do spill brake fluid on painted surfaces, immediately dilute it with fresh clean water. Do not rub the brake fluid with a cloth, as this could damage any softened paint.

To select, handle, store, and fill brake fluids to proper level, follow the steps in **SKILL DRILL 33-1**.

Brake Fluid Testing

Brake fluid replacement is a maintenance item for virtually all vehicles. Consult the service information to determine the correct intervals for brake fluid flushing. Some manufacturers neglect to specify a change interval. In this case, you will need to determine the proper interval. This is based on the environment the vehicle is driven in. In humid or wet climates, the brake fluid may need to be flushed every two years. In very dry climates, every three to four years might be appropriate.

There are several ways of determining if the brake fluid should be flushed:

- **Time/mileage:** The manufacturer may specify a time/mileage interval for flushing the brake fluid.
- **Digital multimeter (DMM)-galvanic reaction test:** The majority of today's braking systems use a combination of dissimilar metals. Manufacturers use aluminum in pistons and master cylinders, and steel in brake lines. When moisture mixes with brake fluid, a galvanic reaction (corrosion)

SKILL DRILL 33-1 Selecting, Handling, Storing, and Filling Brake Fluid

1. Research the specified type of brake fluid in the appropriate service information. Wipe around the master cylinder reservoir cover to prevent any dirt from entering the system. Remove the reservoir cover.

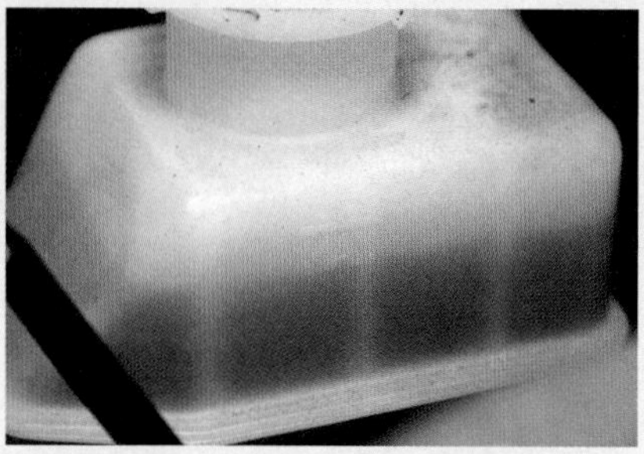

2. Check the fluid level in the reservoir. The fluid should be between the Full mark and Min mark on the side of the cylinder. Or within half an inch of the top of each chamber if there are no marks.

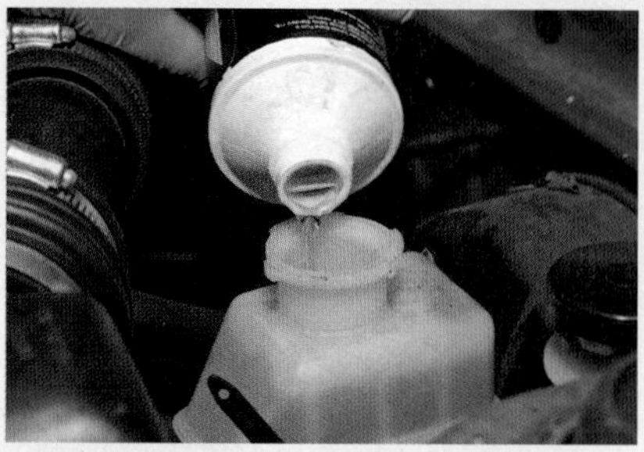

3. Only add the manufacturer's recommended brake fluid to bring the level to the full mark once all brake faults have been resolved. Replace the cover, and check that it is properly seated. Check for any leaks around the master cylinder. Dilute with fresh clean water if any brake fluid spilled.

can occur. The higher the moisture content in brake fluid, the higher the galvanic reaction. The higher the galvanic reaction, the greater the erosion or corrosion it causes. The DMM-galvanic reaction test uses a DMM to measure the voltage created by the galvanic reaction. This directly relates to the level of moisture in the fluid (**FIGURE 33-5**).

- **Boiling point test:** Measuring the boiling point of the brake fluid using a special tool can determine the moisture content of the brake fluid (**FIGURE 33-6**).
- **Test strip:** Measuring specific chemicals/metals in the brake fluid can indicate whether there is a chemical breakdown of the brake fluid. Moisture in the brake fluid leaches copper out of the steel brake lines. The more leached copper, the greater the amount of moisture. Some brake fluid test strips contain special color-changing pads. They change color based on the amount of copper in the brake fluid (**FIGURE 33-7**).

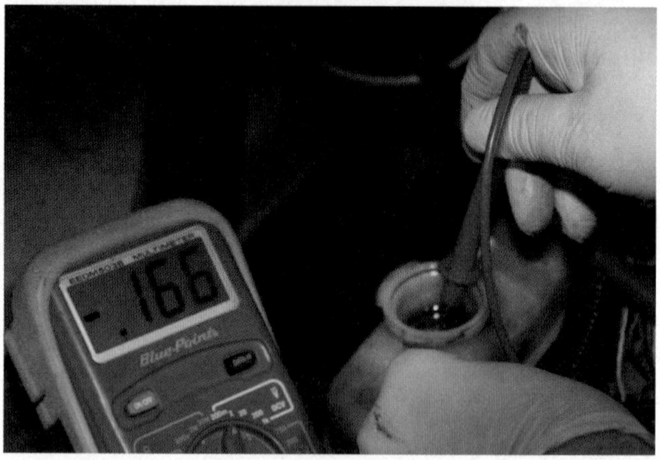

FIGURE 33-5 A DMM measures the voltage created by the galvanic reaction due to the level of moisture in the fluid.

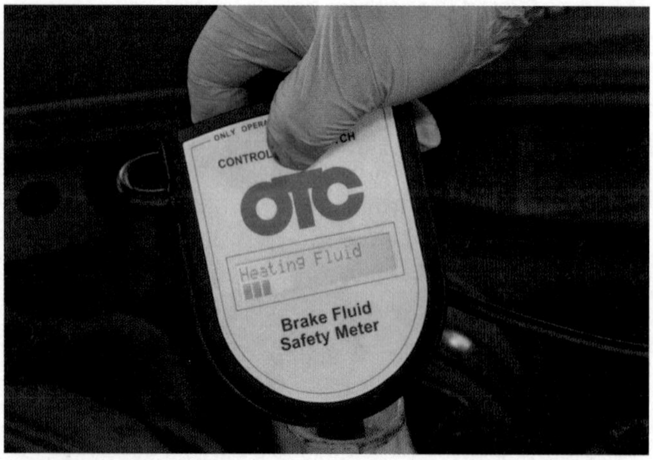

FIGURE 33-6 A brake fluid safety meter boils brake fluid to test for moisture.

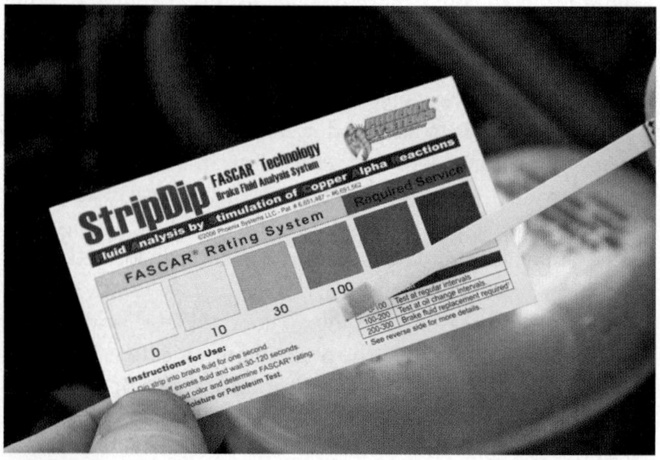

FIGURE 33-7 Brake fluid test strips react to moisture or chemicals in the brake fluid.

Brake fluid condition and level should be inspected at every oil change. Remember that the fluid should be replaced according to the manufacturer's service schedule. It should also be replaced if it fails one of the previously listed tests. For each of these tests, clean around the master cylinder cap before removing it. This helps prevent contaminants from entering the reservoir. Then, remove the master cylinder reservoir cap.

To perform a DMM-galvanic reaction test, follow the steps in **SKILL DRILL 33-2**.

To test brake fluid with a brake fluid tester, follow the steps in **SKILL DRILL 33-3**.

To test brake fluid with a brake fluid test strip, follow the steps in **SKILL DRILL 33-4**.

Applied Science

AS-1: Contamination: The technician can demonstrate an understanding of how a contaminated liquid can cause a chemical reaction that results in the deterioration of performance.

As moisture levels increase in brake fluid, the boiling point decreases. This is a result of the deterioration of the brake fluid's condition. In any liquid when boiling occurs, the liquid turns to vapor. Brakes work on the principle of liquid being incompressible. However, vapor is compressible. The result of boiling brake fluid in a brake system is a spongy pedal. This creates a loss of braking effort at the wheels, including the potential for complete brake failure. SAE field tests have documented that the average 1-year-old car has approximately 2% moisture in the brake fluid. This substantially reduces the boiling point of the brake fluid. Flushing brake fluid restores the boiling point.

SKILL DRILL 33-2 Performing a DMM-Galvanic Reaction Test

1. Set the DMM to DC volts. Insert the red lead in the "v/~" slot and the black lead in the "common" slot.

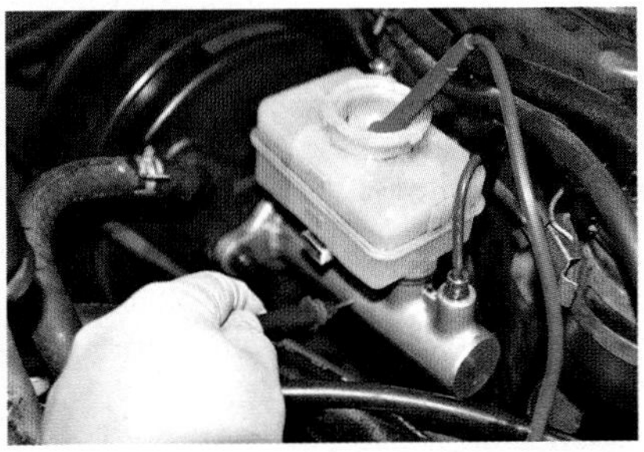

2. Place the red voltmeter probe in the reservoir brake fluid, and place the black lead on the metal housing of the master cylinder. Make sure to use an unpainted surface of the housing.

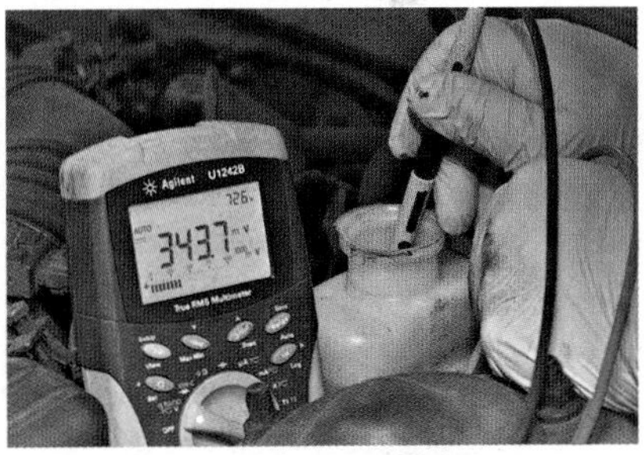

3. Compare the voltage reading you obtained to specifications. Less than 0.3 volt is okay. More than 0.3 volt means the fluid needs to be flushed.

SKILL DRILL 33-3 Testing Brake Fluid with a Brake Fluid Tester

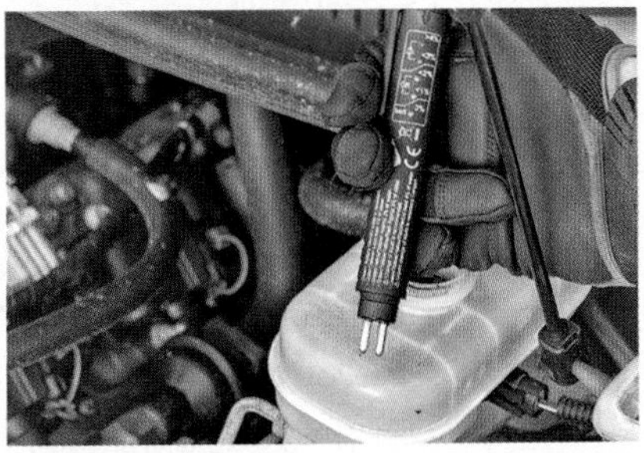

1. Read the directions for the tester you are using.

2. Place an amount of brake fluid from the master cylinder into the tester. Some testers are designed to be placed directly into the brake fluid. Test the fluid.

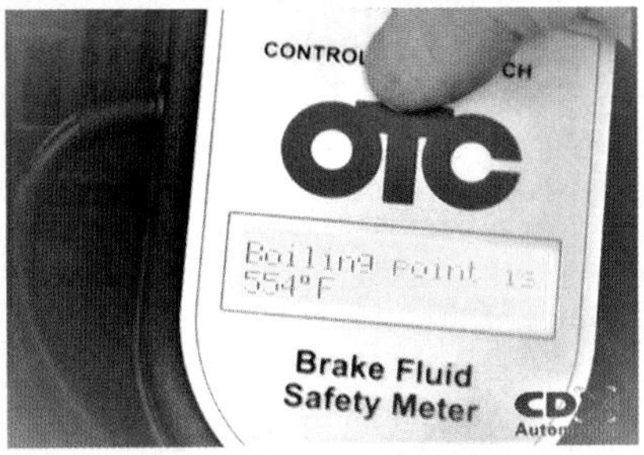

3. Compare the results you obtained to specifications. Most testers will tell you the boiling point or the percent of moisture in the brake fluid.

SKILL DRILL 33-4 Testing Brake Fluid with a Brake Fluid Test Strip

1. Read the directions for the brake fluid test strips you are using. Dip the test strip into the brake fluid for a specified amount of time (usually about one or two seconds).

2. Shake off excess fluid. An additional amount of specified time is allowed to pass before comparing the colored pad with the color chart.

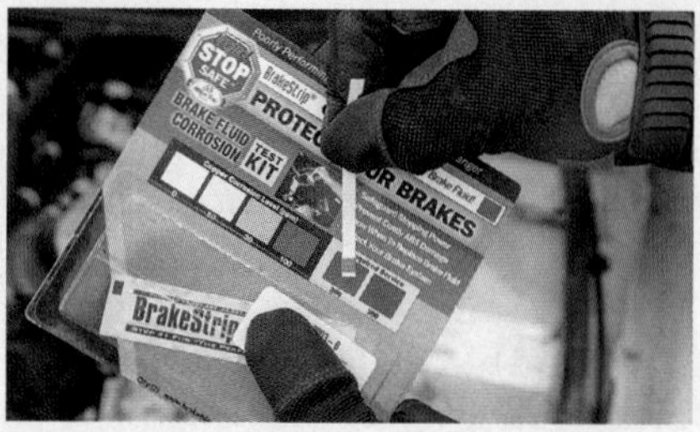

3. Compare the test strip to the chart to determine the level of contaminants or pH level.

▶ Bleeding Brake Systems

LO 33-03 Bleed and flush brake systems.

Fluids cannot be compressed. This is why fluids transmit pressure so well. Unlike fluids, gases are compressible. When pressure is applied against air (a gas), the air decreases in volume. This prevents the pressure from building as quickly. If air enters the brake system, brake pressure does not build up fully. The brakes feel spongy and are not fully functional. **Bleeding** the brakes is the process of removing any air from the system.

When bleeding or flushing brake systems, make sure the fluid you use is from a fresh bottle of brake fluid. Brake fluid absorbs water from the air when they are open. Always replace the cap immediately when using brake fluid from a bottle. If you do this, then the bottle will be considered fresh for up to a month after it was initially opened. But if it is allowed to sit open for even an hour or two, it is no longer considered fresh.

There are a number of different brake bleeding methods. The three most common are described here:

- **Manual bleeding**—Using a helper to manually operate the brake pedal while you open and close the bleeder screws on the wheel units. This allows the air and old brake fluid to be pushed out of the hydraulic braking system.
- **Vacuum bleeding**—Using a vacuum bleeder to pull the air and old brake fluid from the hydraulic braking system.
- **Pressure bleeding**—Using clean brake fluid under pressure to force the air and old brake fluid from the hydraulic braking system.

For all three methods, follow these general steps first. Remove as much of the old brake fluid from the reservoir as possible. A suction gun, old antifreeze tester, or turkey baster works well. Refill the master cylinder reservoir, using the specified fluid type. Then, determine which method of bleeding you will use. Also, research the service information to obtain the bleeding sequence for the vehicle. Typically, it is suggested to start bleeding the wheel farthest from the master cylinder. Then, move to the next closest, etc. So, a common bleeding sequence is: right rear (RR), left rear(LR), right front (RF), and left front (LF). Following the bleeding sequence will help you purge all of the air from the system.

The manual bleeding method requires the least amount of equipment and tools. But it requires more time if a large percentage of the brake fluid has to be changed. It also is less effective at removing trapped air in some systems. This is especially true on vehicles that have high spots in the brake lines. It's the same on vehicles that have a large vertical drop between the master cylinder and the wheel brake units. One example is a truck that has a lift kit installed. This method is best when only a small amount of brake fluid has to be bled. A good example is when the front brake calipers are replaced.

Using the manual bleeding method requires clear communication between you and the assistant. Otherwise, air will enter the system. Also, be sure that you *do not* bleed one wheel so much that the reservoir runs dry. This will allow air into the brake system. If this happens, it will be much harder to bleed the system. Since the air is compressible, it is harder to build pressure to bleed the system.

Another caution when manually bleeding brakes is to never push the brake pedal past its normal operating travel. Doing so can damage the master cylinder, especially on a master cylinder older than a few years. Let's look at why.

The master cylinder pistons normally only travel a short distance in the cylinder. This keeps the surface of the cylinder clean in the piston travel areas. The surface of the cylinder beyond piston travel builds up residue, and possible corrosion. When bleeding, forcing the master cylinder pistons beyond their normal travel pushes the seals into this area. The seals can become damaged by this. And it is likely that they will soon start leaking internally. This will be evidenced by a sinking brake pedal when applied. Customers will blame you for ruining their master cylinder, which now needs to be replaced. To avoid this, place one foot under the brake pedal while bleeding. This will prevent the pistons from traveling into the corroded area.

To perform the manual bleeding method, follow the steps in **SKILL DRILL 33-5**.

SKILL DRILL 33-5 Performing Manual Bleeding

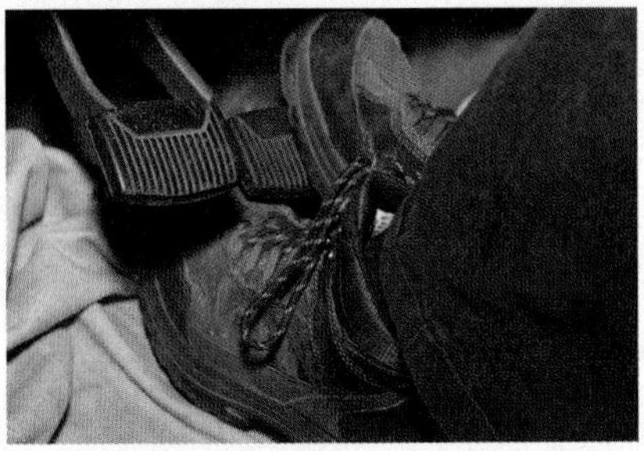

1. Ask an assistant to slowly push the brake pedal down gently, keeping their left foot underneath the pedal to limit full pedal travel.

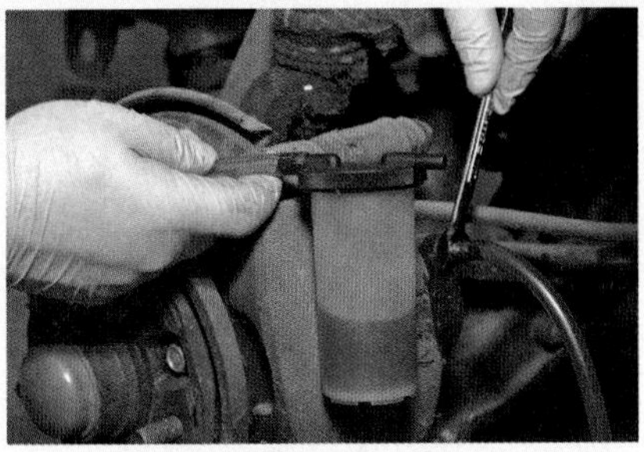

2. Install a clear hose on the farthest bleeder screw. Open the bleeder screw one-quarter to one-half turn. Observe any old brake fluid and air bubbles coming out. When the brake fluid stops, close the bleeder screw lightly. Have the assistant slowly release the brake pedal. Repeat these steps until there are no more air bubbles or old fluid coming out of the hose.

3. Close off the bleeder screw, and tighten it to the manufacturer's specifications. Check the level in the master cylinder reservoir, top it off, and reinstall the reservoir cap. Repeat this bleeding procedure for each of the brake units, moving closer to the master cylinder, one wheel at a time.

▶ TECHNICIAN TIP

Never run the master cylinder dry during bleeding. Doing so can put air into the ABS hydraulic control unit. This will make it much more difficult to bleed!

Vacuum bleeding uses an air-operated or hand-operated vacuum pump to bleed the brakes. This method is much faster than using the manual bleeding procedure. It is also easier to set up than the pressure bleeding system. Just be sure to watch how much fluid is collecting in the vacuum bottle, so you don't run the master cylinder dry. Also, remember that tandem master cylinders use divided reservoirs. The fluid level could be above half on one side and empty on the other side. So, watch the reservoir carefully. Top it off frequently to avoid getting air in the system.

Another issue when using a vacuum bleeder is that the tool draws air from around the threads of the bleeder screw while it is operating. These bubbles show up as bubbles in the fluid being removed from the brake system. When using a vacuum bleeder, ignore the bubbles and focus more on the color of the fluid. Once the fluid is coming out clear, all air should be out of the system.

To perform vacuum bleeding, follow the steps in **SKILL DRILL 33-6**.

The pressure bleeding method uses equipment that can provide a continuous supply of brake fluid. It provides brake fluid under pressure to the master cylinder reservoir. It may also include small-diameter hoses that attach to the bleeder screws. They collect all of the old brake fluid. This method takes a bit of work to connect the equipment to the vehicle. Also, you need to fill the reservoir with the specified brake fluid before installing the bleeder. This helps prevent creating bubbles in the reservoir. Because of

SKILL DRILL 33-6 Performing Vacuum Bleeding

1. Prepare the vacuum bleeder for use. Install the vacuum bleeder on the farthest bleeder screw. Open it one-quarter to one-half turn.

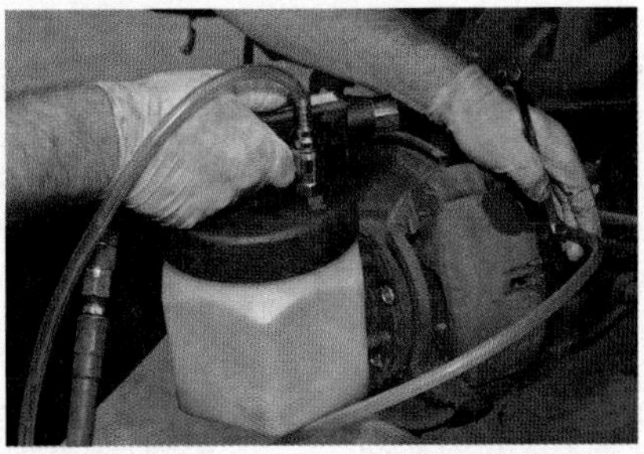

2. Operate the vacuum bleeder to pull brake fluid from the bleeder screw. Observe any old brake fluid and air bubbles coming out. Close off the bleeder screw, and tighten it to the manufacturer's specifications.

3. Check the level in the master cylinder reservoir, and top it off. Repeat this procedure for each of the brake units, moving closer to the master cylinder, one wheel at a time.

SKILL DRILL 33-7 Performing Pressure Bleeding

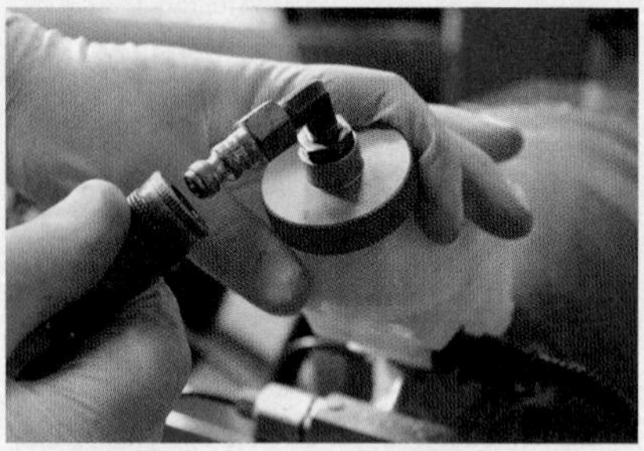

1. Prepare the pressure bleeder, and install it on the master cylinder reservoir.

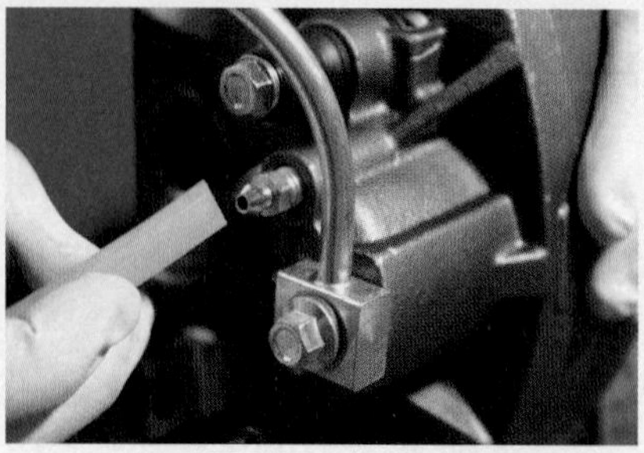

2. Install a clear hose on the farthest bleeder screw, and open it one-quarter to one-half turn. Observe any old brake fluid and air bubbles coming out.

3. Close off the bleeder screw when the brake fluid is clear and has no bubbles. Tighten it to the manufacturer's specifications. Repeat this procedure, moving closer to the master cylinder, one wheel at a time.

this extra work, it is best used when the hydraulic system needs a full flush, or on systems that have a tendency to trap air. To perform pressure bleeding, follow the steps in **SKILL DRILL 33-7**.

▶ TECHNICIAN TIP

One other method that you should know about is gravity bleeding. This method relies on the master cylinder being mounted higher than the wheel brake units. Gravity tends to pull the brake fluid down. So, gravity bleeding can occur just by opening one or more bleeder screws. If the height difference is great enough, the brake fluid will slowly drain out of the system. A technician can use this method to start the bleeding process while working on other wheel brake units. However, gravity bleeding can be a disadvantage. If a technician inadvertently leaves a hose disconnected or a bleeder screw open, *all* of the brake fluid can drain out of the system. This leaves it full of air and harder to bleed. Be very careful not to leave the hydraulic braking system open for too long.

After performing bleeding or flushing brakes, follow these steps:

1. Double-check that all bleeder screws are properly tightened.

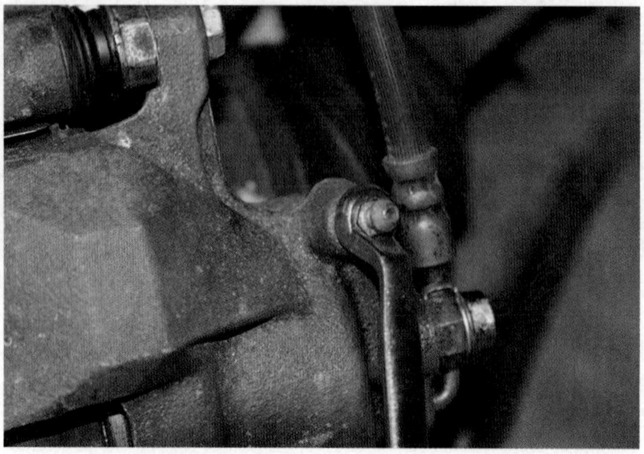

2. Replace all bleeder screw dust caps when finished.

3. Refill the master cylinder to the proper level, and reinstall the master cylinder reservoir cap.

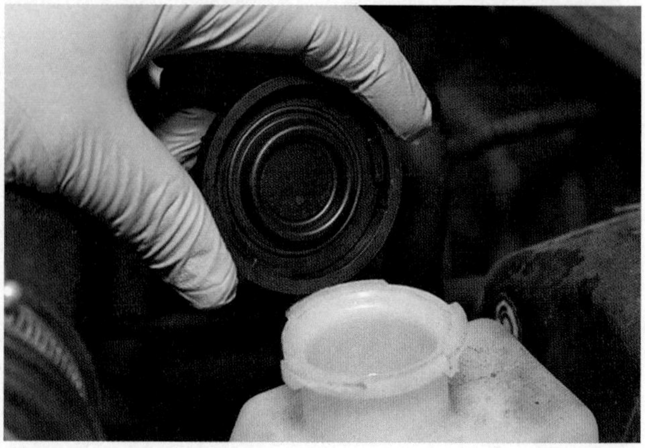

4. Start the vehicle, and check for proper brake pedal feel (it should be firm and high).

5. Verify that there are no leaks at each bleeder screw.

6. Dispose of any old brake fluid in an environmentally approved method.

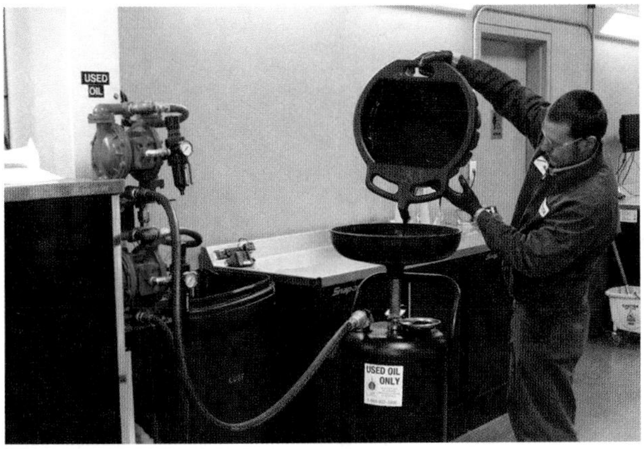

Flushing Brake Systems

Flushing brake fluid uses the same process as brake bleeding, just more thorough. It is designed not only to remove trapped air, but also to replace all of the old brake fluid with new brake fluid. Usually, clean brake fluid of the proper type is used for flushing the system. However, if the system has been contaminated with petroleum products, some manufacturers specify flushing it with clean alcohol. The alcohol helps dissolve any oil in addition to flushing the system. Once the alcohol flush is complete, dry compressed air can be used to blow the alcohol out of the lines. And, then the system can be flushed with clean brake fluid.

SAFETY TIP

Never use any petroleum or mineral-based products, such as gasoline or kerosene, to clean a hydraulic braking system or its components. They are not compatible with the seals used in the braking system. It will result in a failure of the hydraulic braking system and its components. This failure may result in improper braking, which may lead to injury or death.

▶ Inspecting the Brake Pedal

LO 33-04 Measure brake pedal height, free play, and travel.

Brake pedal height, free play, and travel are critical for proper brake operation. Having the proper brake pedal height helps to ensure that the brake pedal has enough starting height to fully apply the brakes. This is true even if one half of the hydraulic system is rendered useless by a leak. In other words, the pedal needs to have a certain amount of travel to be safe. This requires having the pedal start at the right height. Also, note that additional floor mats or other things can prevent full travel of the pedal. Make sure the area around the pedals is not obstructed.

Some vehicles have adjustments on the pedal assembly where pedal height can be adjusted. On most vehicles, it is non-adjustable. In this case, if the pedal height is wrong, it is possible that bushings are worn. Inspect them for wear, damage, or just plain missing. Another likely fault is that the brake linkage or supports are bent. A visual inspection will likely uncover this.

To measure brake pedal height, follow the steps in **SKILL DRILL 33-8**.

Free play is the amount of clearance between the joints in the brake pedal linkage and the master cylinder piston. To measure it, you can apply very light hand pressure to the brake pedal. Measure how far the pedal travels before you start to feel resistance. Moving the pedal up and down between the top and where resistance is felt is the free play (**FIGURE 33-8**).

If the free play is out of specifications, it is likely caused by a misadjusted brake pedal pushrod. However, it could also be caused by worn brake pedal bushings, or bent linkage. Research the service information to see if the brake pedal pushrod is adjustable.

To measure brake pedal free play, follow the steps in **SKILL DRILL 33-9**.

Some vehicles have specified brake pedal travel measurements. Other vehicles have specified reserve pedal measurements. Brake pedal travel is the distance the brake pedal travels from its rest position to its applied height. The

SKILL DRILL 33-8 Measuring Brake Pedal Height

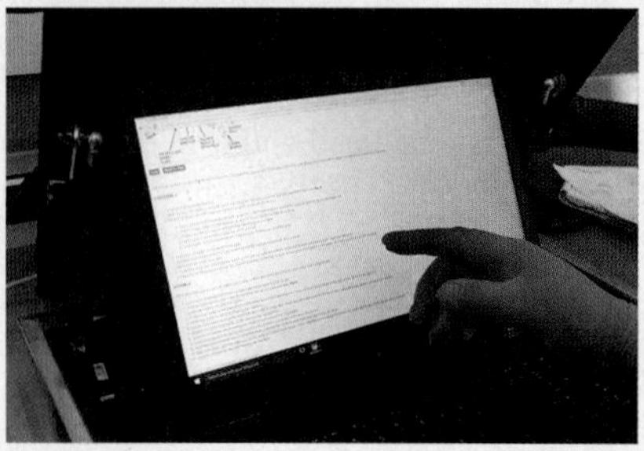

1. Research the procedure and specifications for measuring brake pedal height for the vehicle you are working on.

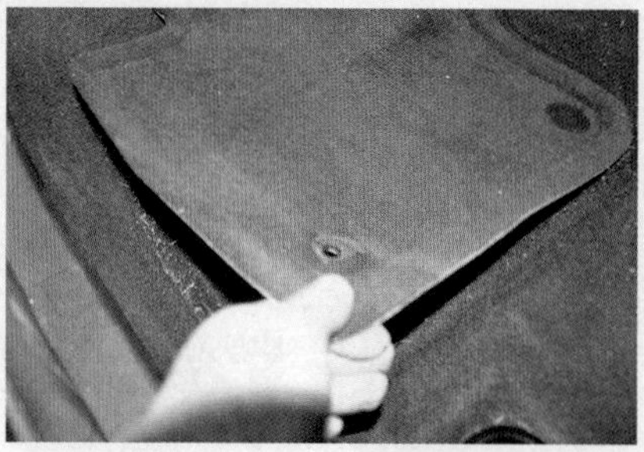

2. Remove any removable floor mats or anything lying on the floor near the brake pedal.

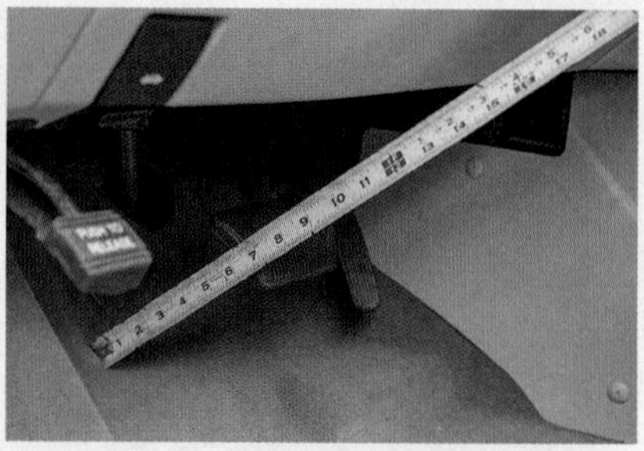

3. With the engine off, measure the brake pedal height between the two specified points, using a measuring stick. Compare this reading to the specifications, and determine any necessary actions.

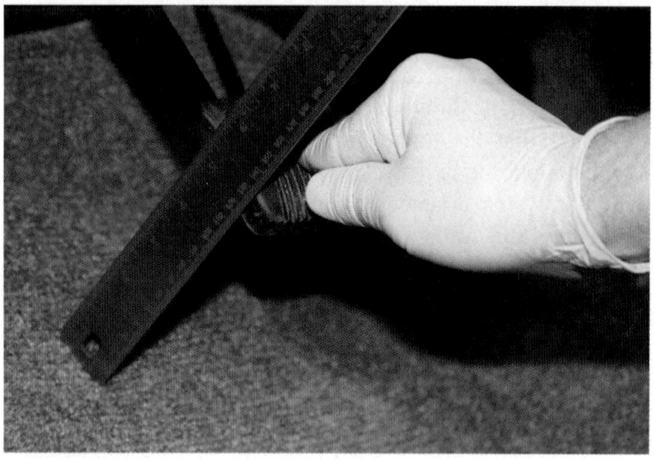

FIGURE 33-8 To measure free play, apply light hand pressure to the brake pedal. Measure how far the pedal travels before you feel resistance.

applied height might need to be at a specified pedal pressure. The travel is then measured by reading brake pedal height and subtracting the applied height from the floor (**FIGURE 33-9**). For example, the brake pedal height is 8" (203.2 mm). Travel takes it down to 5" (127 mm) off the floor. So, the travel is 3" (76.2 mm).

Reserve pedal is how much pedal travel remains when the brake pedal is applied. Again, the applied height might need to be at a specified pedal pressure. The measurement is from the floor to the height of the applied brake pedal. It represents how much reserve is left for the brake pedal to travel if needed (**FIGURE 33-10**).

If the reserve height or pedal travel measurement is out of specifications, it could be caused by a misadjusted brake pedal pushrod. It more likely indicates a hydraulic brake system fault. So, this is an important measurement to take.

To measure brake pedal travel, follow the steps in **SKILL DRILL 33-10.**

SKILL DRILL 33-9 Measuring Brake Pedal Free Play

1. Research the specified pedal free play measurement. Measure the brake pedal height.

2. Use light hand pressure to apply the brake pedal until all clearances are taken up. Measure the distance the brake pedal moved.

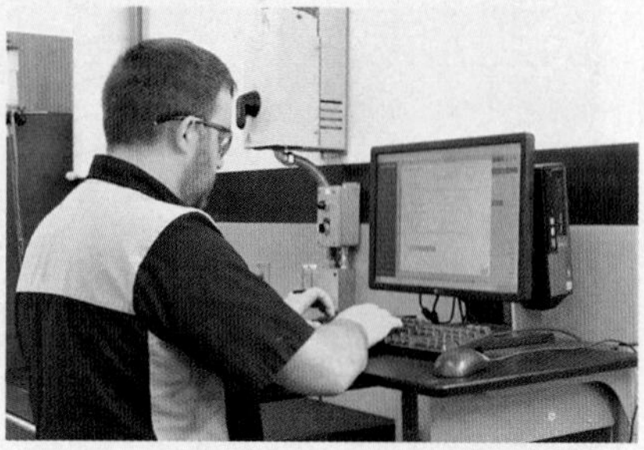

3. Compare this reading to the specifications, and determine any necessary actions.

FIGURE 33-9 Travel is measured by reading brake pedal height and subtracting the applied height from the floor.

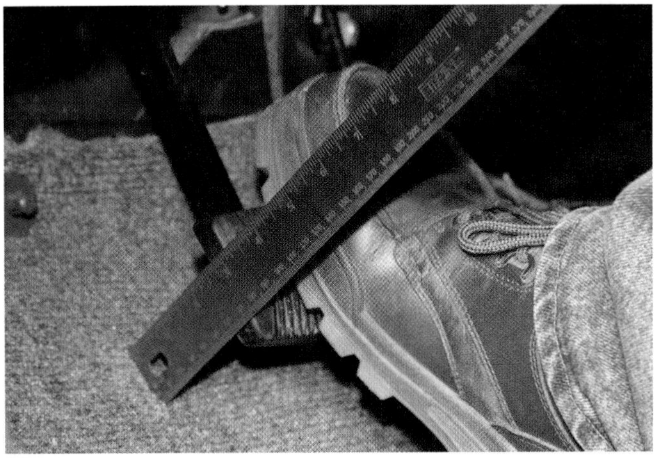

FIGURE 33-10 Reserve pedal is the measurement from the floor to the applied brake pedal height.

SKILL DRILL 33-10 Measuring Brake Pedal Travel

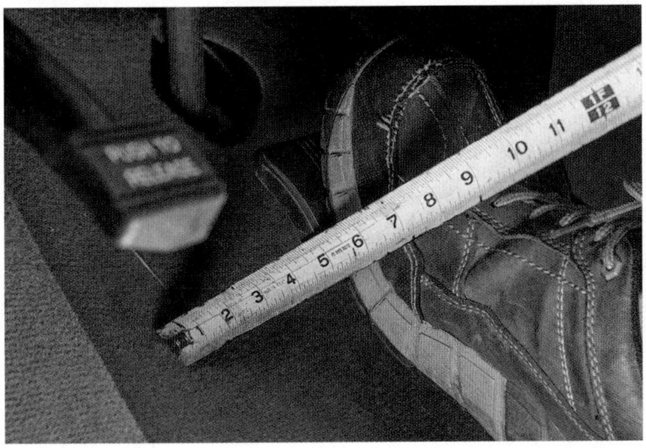

1. Research the specified procedure to measure the brake pedal travel or reserve pedal height.

2. Start the engine. This allows the booster to operate normally.

3. Apply the brake pedal with the specified force.

4. Measure the brake pedal travel or reserve height. Compare this reading with the specifications, and determine any necessary actions.

▶ Hydraulic System Component Diagnosis

LO 33-05 Describe general brake hydraulic system diagnosis.

Diagnosis of hydraulic braking system faults follows the strategy-based diagnostic process. It begins with verifying the concern. This means that you need a solid understanding of the customer concern. It is good practice to gather as much related information as possible about the concern. The more information you obtain, the better chance you have of diagnosing the issue correctly. Ask the customer questions such as these:

- Does the brake pedal sink under hard pressure? Or light pressure?
- Does the vehicle pull to the left when braking? Or the right?
- Does the wheel lock up under hard brake pressure? Or light pressure?

Once you understand how the vehicle is acting, try to determine what conditions are present with the concern. Ask questions such as these:

- Does it happen when it is wet? Dry? Or always?
- Does it happen when it is hot? Cold? Or always?
- Does it happen when you are driving at highway speeds? At stop-and-go speeds? Or always?
- Does it happen when you are carrying a heavy load? A light load? Or always?

If the vehicle is safe to drive, it is common practice to test-drive the vehicle to verify the customer concern. If it is an intermittent problem, or if it is hard to reproduce, you may need to take the customer along on the test drive. That way, they can point out what they are experiencing. While on the test drive, reproduce the conditions that the customer identified, and observe the concern. Do your best to determine the fault by safely operating the vehicle in a variety of ways.

As you gather information, compare what is happening in the vehicle with your knowledge of the hydraulic braking system. Use Pascal's law to help identify possible causes of the fault. For example, the vehicle pulls to the left when braking. Suspect that either a brake on the left side is grabbing or a brake on the right side has a lack of hydraulic pressure or reduced friction. If the fault is fairly substantial, the cause may be easy to identify. If the fault is not easily identified, then determine whether further test driving will be useful.

Once the test drive has given you as much information as possible, the next step is to research the manufacturer's TSBs. See whether any TSBs are related to the concern. If so, follow the information in the TSB to diagnose the vehicle. If there are no related TSBs, then research the manufacturer's diagnostic process in the service information.

Once you have researched the service information, it is time to develop a focused testing plan. Doing so may involve performing tests in the shop with a scan tool, pressure gauges, or a DMM. It may also require disassembly and visual inspection or measurements of the suspected component or system. If you have been thorough in your diagnostic process, this should lead you to the cause of the fault. If not, take what you learned from the test results and/or inspection. Then, go back to the service information and determine what the next best step is. Then perform it. Continue this process until you have identified the cause of the fault.

Once a cause of the fault is identified, consider whether it is the root cause. For example, the customer concern is that the brake pedal is spongy. You cannot stop your diagnosis when you find low brake fluid level in the master cylinder reservoir. Adding brake fluid and bleeding the brakes might restore proper function to the brake pedal for the time being. But you have to ask yourself, "Why was the brake fluid low?" It is likely that there is a leak in the hydraulic braking system that will cause the customer to return. In this case, the root cause might be a leaky wheel cylinder. Because the brake fluid was low, you should perform a visual inspection of the hydraulic braking system. This way you can verify if there are leaks, or not.

Once the repair is complete, you need to verify that the repair corrected the customer concern. This is usually done by retesting the component or system with the test that identified the fault. If it passes the test, then test drive the vehicle once again. This allows you to make sure the concern has been corrected and that no other issues are present.

Applied Science

AS-2: Hydraulics: The technician can explain how fluid pressure transmits force from one location to another.

Pascal's law states that incompressible liquids send pressure equally in all directions. This is the basic principle behind the operation of hydraulic braking systems on vehicles. When braking, brake pedal pressure is applied, via the master cylinder, to the brake fluid. The hydraulic fluid under pressure pushes against the caliper or wheel cylinder pistons. This forces friction material against the brake discs or drums to stop the movement of the vehicle.

For the sake of simplicity, let's think of a hydraulic brake system like a tube of toothpaste. Start with a full tube (no air). If you punch a hole in the tube and squeeze, toothpaste (a thick liquid) will be displaced from the hole. If you punch three holes through various parts of the tube and squeeze in only one place, toothpaste will be displaced from all three holes. This illustrates that force applied to a liquid in one area is transferred to all areas within a container.

▶ Inspecting the Master Cylinder

LO 33-06 Perform master cylinder service.

Inspecting the master cylinder for internal and external leaks is usually performed under the following conditions:

- the brake pedal sinks when the brakes are applied;
- the brake fluid is low in the reservoir;
- the brake warning light is on; or
- the brake pedal reserve height is too low.

Inspect the master cylinder for obvious signs of external leakage. Be sure to check all brake line fittings, sensor connections, reservoir seals, and the areas at the rear of the master cylinder near the power booster. Also, check the inside of the vacuum hose to the power booster for signs of brake fluid. Brake fluid would be present if it is leaking past the primary piston in the master cylinder and the front seal on the power booster.

When checking for a sinking brake pedal, start the vehicle, and apply the brake pedal, beginning with a very light pressure. The brake pedal should hold its position without sinking for at least 1 minute. If it continues to sink, the hydraulic brake system may have an external or internal leak. Further inspection of the system is needed. If the brake pedal holds steady, release the brake pedal, and then apply it a bit more firmly. Hold that pressure, and see if the brake pedal still holds steady. If it does, repeat this step, increasing the brake pedal pressure until firm pressure is applied.

External leaks can occur anywhere in the hydraulic system. They can be identified with a visual inspection of the system components as described above. If the pedal sinks and there is no external leak in the brake system, there is likely an internal leak in the master cylinder. There are two ways to verify this. The first test is the visual method, and the second is the block-off method. Both are explained in Skill Drill 33-11.

To check the master cylinder for internal or external leaks and proper operation, follow the steps in **SKILL DRILL 33-11**.

Master Cylinder Service and Bench Bleeding

Removing and bench bleeding the master cylinder are usually only performed when the master cylinder is being replaced. Bench bleeding makes bleeding the hydraulic system much easier. Removing all of the air from the master cylinder allows it to build normal pressure when bleeding the system. Failure to bench bleed the master cylinder results in very little brake fluid being bled out of each wheel cylinder or caliper. This increases the time it takes to bleed the brake system. Also, it will push a lot of air through the brake system. This can cause air to get stuck in the system.

To remove, bench bleed, and reinstall the master cylinder, follow the steps in **SKILL DRILL 33-12**.

Measuring and Adjusting Master Cylinder Pushrod Length

The master cylinder pushrod length is critical for proper brake operation. If it is too short, the driver will have to depress the brake pedal further than specified to operate the brakes. This reduces the amount of reserve pedal available in the case of a hydraulic leak. If the pushrod length is too long, it could prevent the master cylinder piston from returning far enough to

SKILL DRILL 33-11 Inspecting the Master Cylinder for Leakage

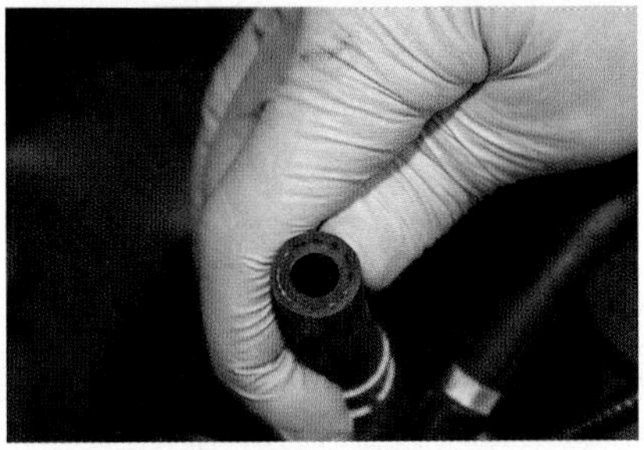

1. Check the brake fluid level to see if it is low. Inspect the master cylinder for obvious signs of external leakage. Also, check the inside of the vacuum hose to the power booster for signs of brake fluid. Start the vehicle, and apply the brake pedal, beginning with a very light pressure. The brake pedal should hold its position without sinking for at least 1 minute. Release the brake pedal, and apply it a bit more firmly. Hold that pressure, and see if the brake pedal holds steady. If it does, repeat this step with increasing pedal pressure until firm pressure is applied.

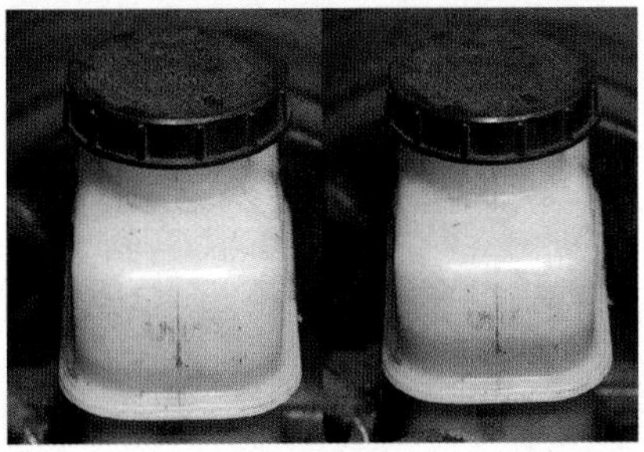

2. Inspect the master cylinder for internal leaks, using the visual method. Only perform this test if the brake pedal sinks.

- Remove the master cylinder reservoir cap.
- Have an assistant apply the brake pedal slowly and hold it steady.
- Watch the brake fluid in the reservoir. There should be an initial spurt of brake fluid from each of the two compensating ports as the brake pedal is first moved. If it is a continuous stream while the brake pedal is moving and the brake fluid level rises while holding the brake pedal down, then the master cylinder has an internal leak.

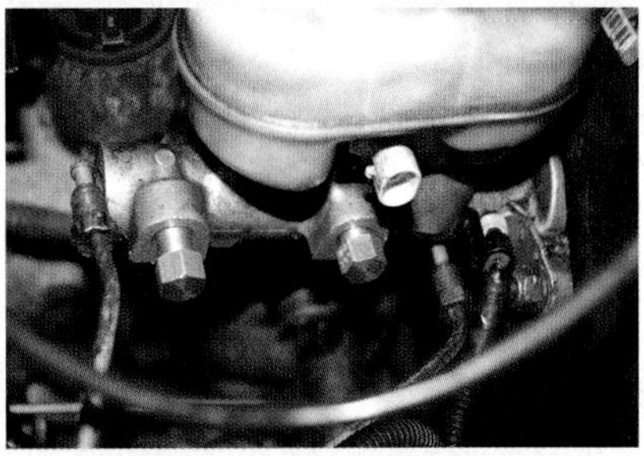

3. Inspect the master cylinder for internal leaks, using the blocked-off method. Only perform this test if the brake pedal sinks and the preceding inspection is not conclusive.

- Unscrew the brake line fittings from the master cylinder, and carefully move them out of the way.
- Install block-off fittings into the master cylinder.
- Bleed any air from each master cylinder port, and tighten the block-off fittings properly.
- Have an assistant apply the brake pedal slowly and hold it firmly.
- Verify that the fittings do not leak. If the brake pedal sinks, then the master cylinder has an internal leak.

SKILL DRILL 33-12 Performing Master Cylinder Service and Bench Bleeding

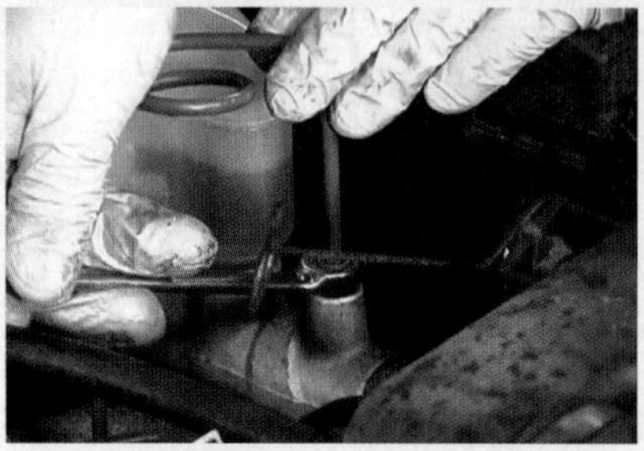

1. Compare the new master cylinder with the old one to verify that it is the correct replacement. Remove all of the old brake fluid from the master cylinder reservoir. Remove the master cylinder brake lines, using a flare wrench and, if necessary, the double wrench method.

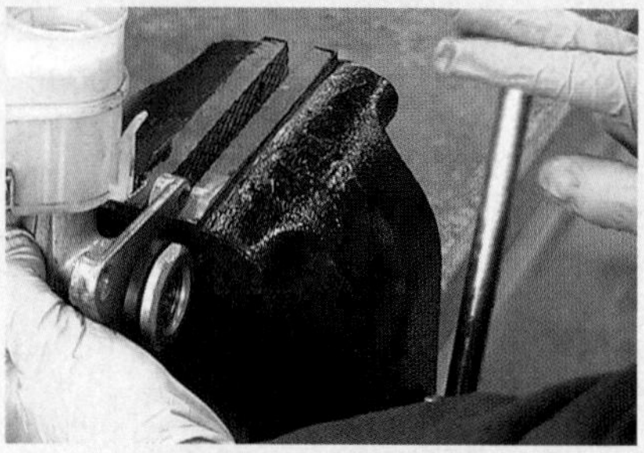

2. Remove the nuts holding the master cylinder to the power brake booster. Remove the master cylinder. Mount the master cylinder in a vise with the reservoir facing up. Now is a good time to perform the next task: measure and adjust pushrod length.

3. Install the bleeder lines into the master cylinder outlet ports. Position the ends of the lines deep within the master cylinder reservoir. Fill the reservoir about half-full with clean brake fluid. With an appropriate tool, slowly push the master cylinder piston into the bore. Allow the piston to return to its rest position. Repeat until all air bubbles have been removed from the master cylinder.

4. Place the master cylinder on the power booster, and install the nuts just far enough to hold it from falling off. Remove the bleeder lines.

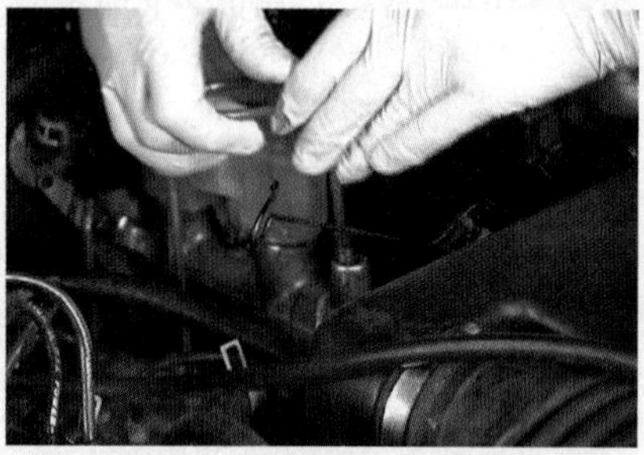

5. Carefully line up the brake lines. Start them using your fingers to thread them into the master cylinder outlet ports at least four or five threads from when they first catch. Do not use a wrench to start the fittings.

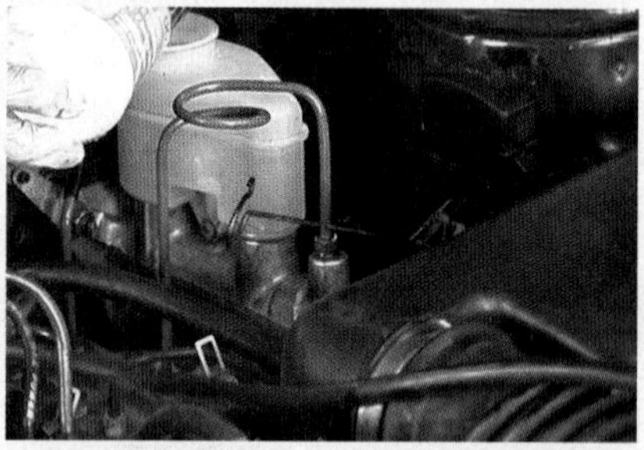

6. Tighten the brake line fittings using a flare nut or line wrench. Use the double wrench method if the master cylinder is fitted with an adapter. Top off the master cylinder, and bleed all wheel brake units to remove any remaining air. Start the engine, and check brake pedal feel to verify proper height and firmness. Visually inspect all fittings and bleeder screws for leaks.

uncover the compensating ports. This traps brake fluid pressure in front of the pistons and holds the brakes in the applied position.

As important as pushrod length is, it does not normally require adjustment, as it is locked in place. The situations that would call for adjusting it are as follows:

- Someone changed the adjustment setting;
- the brake pedal linkage has been repaired or adjusted; or
- the power booster is being replaced.

Just changing the master cylinder does not normally require pushrod length adjustment. Some manufacturers make gauges that are used to measure the pushrod length. To measure and adjust master cylinder pushrod length, follow the steps in **SKILL DRILL 33-13**.

▶ Diagnosing Power Brake Systems

LO 33-07 Test vacuum-style brake boosters.

All power brake systems should be inspected and tested whenever the customer complains that:

- the brakes are dragging,
- the brake pedal is harder to push than normal,
- the pedal height has changed, or
- if the engine operation changes more than a minimal amount when the brake pedal is applied.

Vacuum boosters should also be tested if the vehicle has an unlocated vacuum leak. Single-diaphragm and dual-diaphragm

SKILL DRILL 33-13 Measuring and Adjusting Master Cylinder Pushrod Length

1. Remove the master cylinder, following the specified procedure. Make sure the brake pedal is fully released and not bound up. Using the specified tool, measure the master cylinder pushrod length.

2. If the length is incorrect, loosen the locknut.

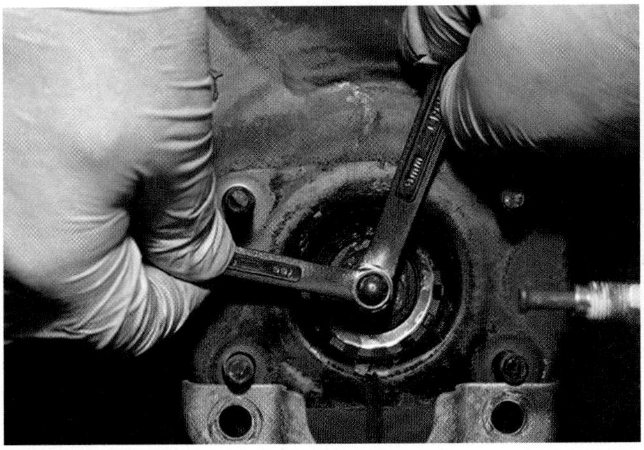

3. Adjust the pushrod to the proper length.

4. Retighten the locknut, and recheck the pushrod length. Reinstall the master cylinder and check pedal free play, pedal height, and reserve pedal.

vacuum brake boosters are diagnosed in the same manner. The following tests can be performed on both types of boosters:

- Brake pedal free travel: Test to see if there is the proper brake pedal linkage clearance.
- Performance/operation test: Test to see if the booster is operational.
- External leak test: Test for leaks to the atmosphere.
- Internal leak test: Test for leaks between the booster chambers.

Testing the Power Booster

Power booster testing starts with a brake pedal free travel test. It then follows up with a performance test to see if the booster is operating properly. If not, then an external leak test and internal leak test can determine why the booster isn't working properly.

The brake pedal free travel is critical for proper brake operation. The proper amount of free travel ensures that the master cylinder pistons return to their proper rest position. This uncovers the compensating ports. Insufficient free travel can cause the brakes to drag due to trapped fluid pressure in front of each piston. Brake fluid is unable to return to the reservoir through the blocked compensating ports. Excessive free travel is not good either. It reduces the amount of reserve pedal for braking in the event of a hydraulic brake system leak.

As important as brake pedal free travel is, it normally does not need to be adjusted because the components are locked in place. The situations that would call for adjusting it include the following:

- Someone changed the adjustment setting;
- the brake pedal linkage has been repaired or adjusted;

- the linkage has worn over time, leading to increased free travel; or
- the power booster is being replaced.

Just changing the master cylinder does not normally require brake pedal free travel adjustment. However, it is good practice to verify that it is within specifications. If adjustment is needed, the manufacturer usually incorporates a method of adjustment. One style uses a locking adjustment rod between the power booster and the brake pedal.

After verifying that the free travel is correct, you need to performance test the power booster. This test verifies that the booster is able to provide brake assist. To test pedal free travel and performance test the vacuum booster, follow the steps in **SKILL DRILL 33-14**.

Checking Vacuum Supply to Vacuum-Type Power Booster

The brake booster must have an adequate amount of vacuum to operate correctly. Insufficient vacuum requires the driver to increase foot pressure to activate the brakes. Excessive vacuum is not usually a problem, because the booster is designed to work using maximum engine vacuum. Many manufacturers specify a minimum of 16" of mercury (in. Hg; 406 mm Hg) of intake manifold vacuum. If the reading is insufficient, check for vacuum leaks or restrictions in the supply hose. If ok, check for an improperly tuned engine. To check vacuum supply to a vacuum-type power booster, follow the steps in **SKILL DRILL 33-15**.

SKILL DRILL 33-14 Testing Pedal Free Travel and Performance Testing the Vacuum Booster

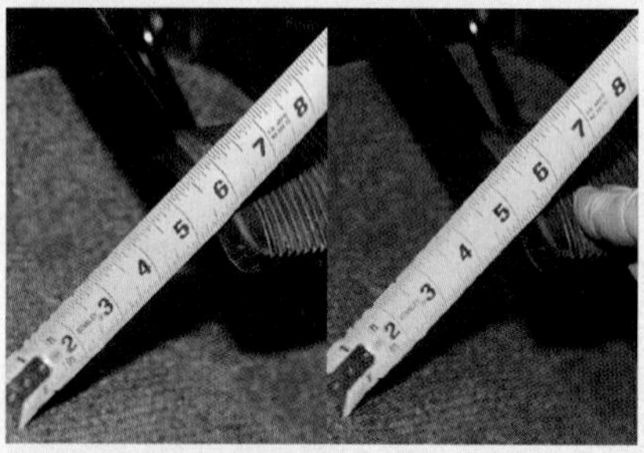

1. To test brake pedal free travel, start with the engine off. Depress the brake pedal several times to remove any vacuum or hydraulic pressure from the power booster. Measure the distance of the brake pedal free travel by depressing the brake pedal by hand until you just feel all of the slack taken up. Compare this measurement to specifications.

2. Performance test the booster by beginning with the vehicle engine off. Apply and release the brake pedal five or six times. This will bleed off any vacuum or hydraulic pressure in the power booster. Hold the brake pedal down with moderately firm pressure (20–30 lb [9.1–13.6 kg]).

3. Start the engine, and observe the brake pedal. On vacuum-assisted vehicles, if the pedal drops about an inch, the booster is providing boost. If not, the booster is not providing boost, and the following tests must be performed. On hydraulic-assisted vehicles, when starting the engine with your foot on the brake pedal, the pedal should either rise or fall about an inch (depending on the vehicle) if the booster is providing assist.

SKILL DRILL 33-15 Checking Vacuum Supply to Vacuum-Type Power Booster

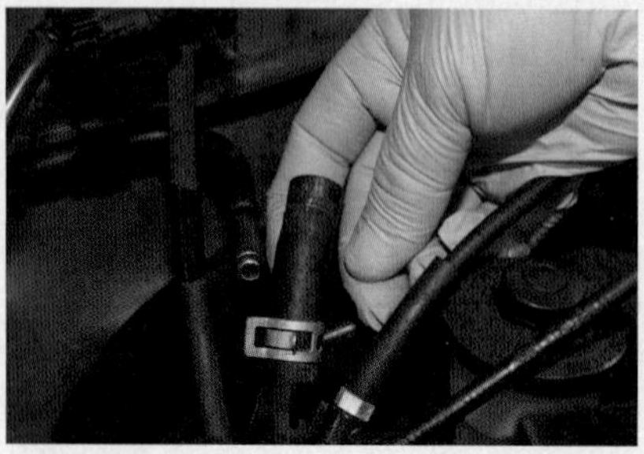

1. With the engine off, remove the inlet hose from the vacuum-type booster.

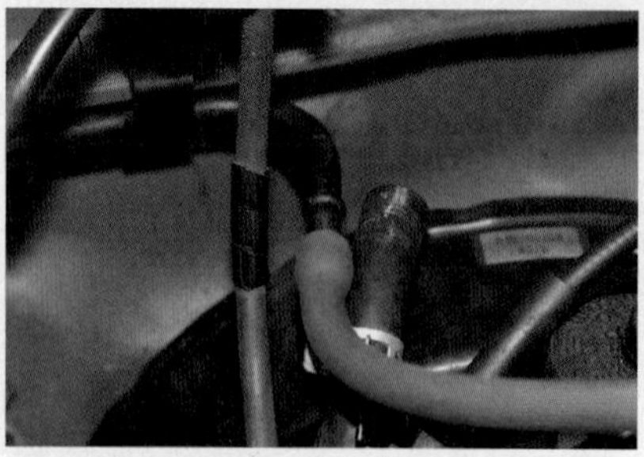

2. Connect a vacuum gauge to the vacuum supply end of the hose.

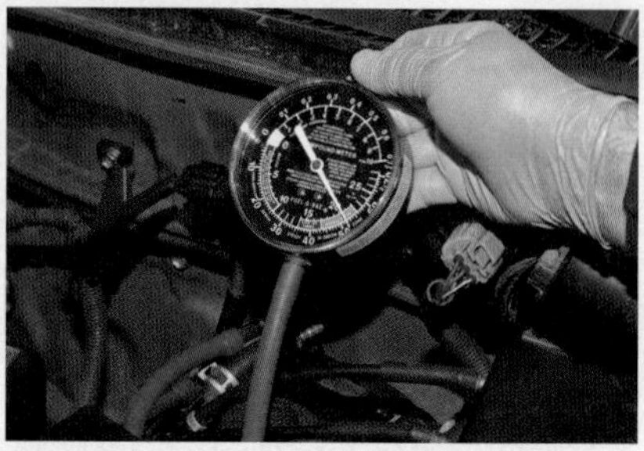

3. Start the engine, and read the vacuum supply available to the vacuum-type booster. Vacuum should be greater than 16" Hg (406 mm Hg) on most vehicles.

▶ TECHNICIAN TIP

Engines with performance camshafts typically have decreased amounts of manifold vacuum due to the camshaft profile. This lowered vacuum results in higher foot pressure required to stop the vehicle. An auxiliary vacuum pump may be needed to provide adequate braking. Also, vehicles operated at high altitude always have less vacuum. The general rule of thumb is that you will lose almost 1" Hg (25 mm Hg) of vacuum for every 1000' (305 m) of altitude gained.

Checking Vacuum-Type Power Booster Unit for Leaks and Inspecting the Check Valve

Vacuum leaks in the system may require increased driver foot pressure to apply the brakes. This is because the power booster operates off of vacuum. Because of this, a leaky power booster can affect the operation of the engine. The vacuum leak changes the air–fuel mixture.

Power boosters can leak internally or externally. Perform the external leak test before the internal leak test to avoid confusion in identifying the cause of the leak. To perform an external leak test:

- Start the engine, and allow it to run for at least 10 seconds.
- With your foot *off* the brake pedal, turn the engine off.
- Wait at least 10 minutes, and then apply the brake pedal with moderately firm pressure (20–30 lb[9.1–13.6 kg]). Note the feel and pedal reserve height.
- Apply the brake pedal a couple more times with the same moderately firm pressure. Each application should result in a higher and firmer brake pedal as the vacuum is released from the booster.

If it does operate properly, then there are no substantial external leaks in the system. If it does not hold vacuum, you need to inspect the booster for external vacuum leaks. The cause could be any of the following:

- The vacuum check valve and/or grommet
- The front or rear seals of the unit
- The atmospheric valve at the rear of the booster
- The case where the halves are crimped together
- A hole worn or rusted through the case

To check for the conditions listed above, use a stethoscope and listen for leaks around the outside of the booster. This includes the control valve at the rear of the booster, which is under the dash. Once you have identified that there are no external leaks, the second step is to perform an internal leak test:

- Begin by starting the engine and letting it idle.
- Apply the brake pedal with firm pressure (30–50 lb[13.6–22.7 kg]).
- Without moving your foot, shut off the engine, and observe the brake pedal for approximately 1 minute.

If the pedal stays steady, there are no internal leaks. If the brake pedal rises, there is an internal leak in the diaphragm, the control valve, or the check valve.

The third step is to perform a check valve operation test:

- Start the engine, and allow it to run for 10 seconds to evacuate the booster. Turn off the engine, wait at least 10 minutes, and then remove the check valve from the booster.

There should be a large rush of air into the booster. If there is, the check valve is holding a vacuum and is okay. If there is not, test the check valve by blowing air through it. Air should flow from the booster side of the check valve to the engine side only. If the check valve is okay and the tests indicate a leak, check for internal leaks.

To inspect the vacuum-type power booster unit for leaks and inspect the check valve for proper operation, follow the steps in **SKILL DRILL 33-16**.

▶ TECHNICIAN TIP

Holding the brake pedal in a steady manner should not affect the operation of the engine. If the engine runs rough when the brake pedal is held down or changes substantially when the brake pedal is released, it could indicate a vacuum leak in the booster. Perform the external and internal leak tests to identify any faults.

▶ Inspecting Brake Lines and Hoses

LO 33-08 Service brake lines and hoses.

Inspecting brake lines and brake hoses can show where maintenance is needed. This will help prevent vehicle breakdowns and unsafe vehicle operation. All manufacturers recommend regular inspection of these components. Be sure to take into account regional differences. Some regions are highly susceptible to corrosion from deicing chemicals. In this environment, steel lines would be prone to rust. Also, off-road vehicles may be more prone to dents, kinks, and cuts. Flexible brake hoses are susceptible to cracking, kinks, bulging, and abrasion. Inspect all lines and hoses carefully.

To inspect brake lines, brake hoses, and associated hardware, follow the steps in **SKILL DRILL 33-17**

Replacing Brake Lines, Brake Hoses, Fittings, and Supports

When brake lines and hoses have to be replaced, you will need to disconnect them at their fittings. Many times the brake line fitting is connected very tightly to a matching adapter. Since the brake line is pinched between the fitting and adapter, it will rotate with them. This will kink the line if the adapter turns, which is highly likely if you only use one wrench. In fact, it only takes about 1/16th of a turn of the adapter to kink the brake line beyond use.

The best way to remove fittings that are connected to adapters is with the double-wrench method. In this method, you place one flare nut wrench on the fitting, and one on the adapter. Hold the wrench that is on the adapter so it cannot turn. Then, twist the wrench that is on the fitting. This will break the

SKILL DRILL 33-16 Checking Vacuum-Type Power Booster Unit for Leaks and Inspecting the Check Valve

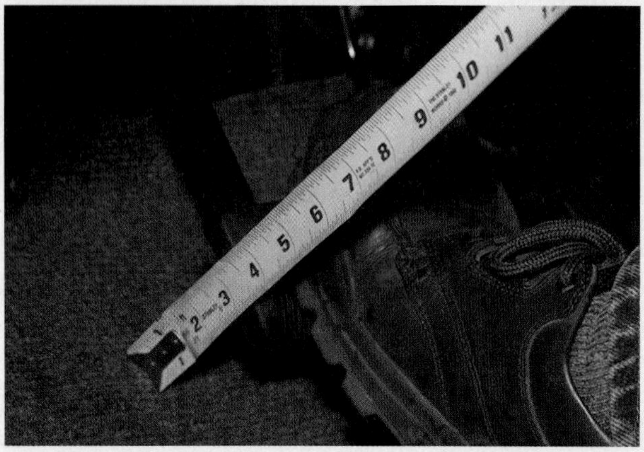

1. The first step is to perform an external leak test. Start the engine, and allow it to run for at least 10 seconds. With your foot *off* the brake pedal, turn the engine off. Wait at least 10 minutes, and then apply the brake pedal with moderately firm pressure (20–30 lb [9.1–13.6 kg]). Note the feel and pedal reserve height. Apply the brake pedal a couple more times with the same moderately firm pressure. Each application should result in a higher brake pedal as the vacuum is released from the booster.

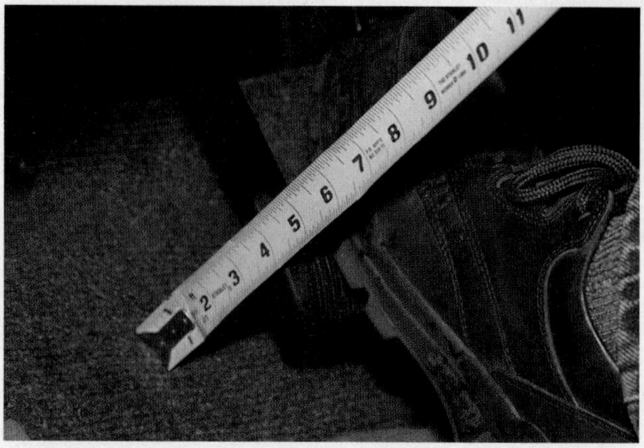

2. The second step is to perform an internal leak test. Begin by starting the engine and letting it idle. Apply the brake pedal with firm pressure (30–50 lb[13.6–22.7 kg]). Without moving your foot, shut off the engine, and observe the brake pedal for approximately 1 minute. If it stays steady, there are no internal leaks. If the brake pedal rises, there is an internal leak.

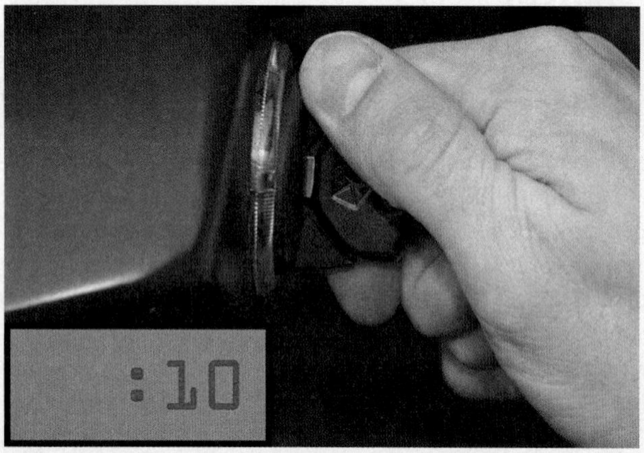

3. The third step is to perform a check valve operation test. Start the engine, and allow it to run for 10 seconds to evacuate the booster.

4. Turn off the engine, wait at least 10 minutes, and then remove the check valve from the booster. There should be a large rush of air into the booster if the check valve is holding a vacuum properly. If there is not, test the check valve by blowing through it.

fitting loose from the adapter. It also frees the brake line from being clamped between them. So, always use the double-wrench method when loosening brake line fittings.

To replace brake lines, brake hoses, fittings, and supports, follow the steps in **SKILL DRILL 33-18**.

Brake Line Flaring

To fabricate brake lines, decide whether you will be using the inverted double flare or ISO flare. First select new, double-walled steel brake line of the specified diameter. Make sure the end you are flaring is cut smooth and square, without dings or gouges. Clean up with a file, or recut the end with a tubing cutter. Obtain the flaring tool for the type of flare you are fabricating: double flare or ISO. Also, make sure you place any fittings required to be in place on the tubing before the tubing is flared.

To perform the inverted double flare method, follow the steps in **SKILL DRILL 33-19**.

To perform the **ISO flare method**, follow the steps in **SKILL DRILL 33-20**.

SKILL DRILL 33-17 Inspecting Brake Lines, Brake Hoses, and Associated Hardware

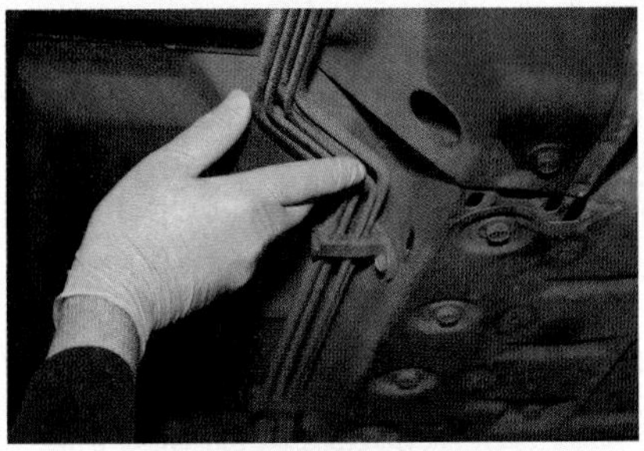

1. Safely raise the vehicle on a hoist. Trace all brake lines from the master cylinder to each wheel's brake assembly. Inspect the steel brake lines for leaks, dents, kinks, rust, and cracks.

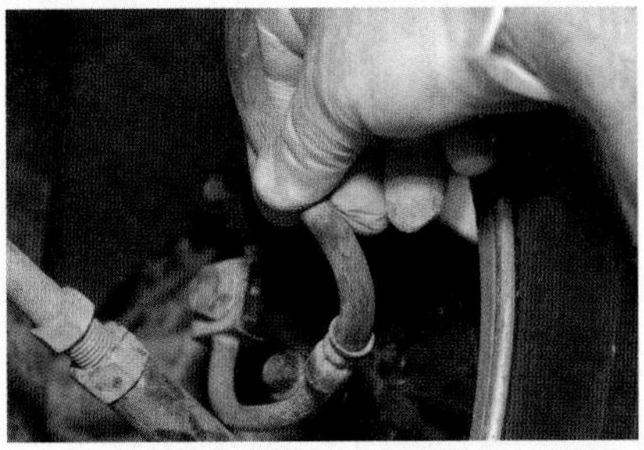

2. Inspect all flexible brake hoses for cracks, bulging, and wear.

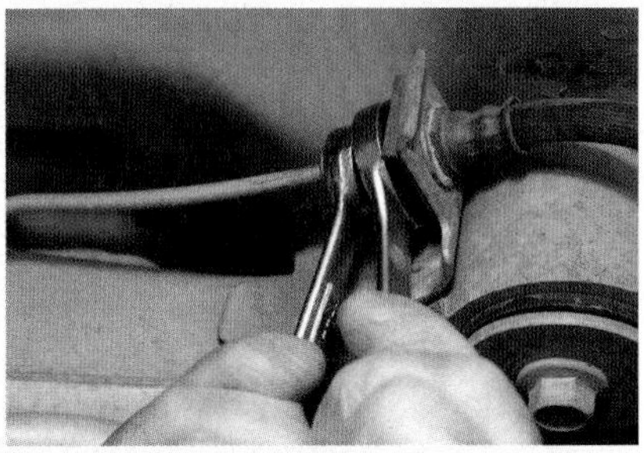

3. Check for any loose fittings and supports.

SKILL DRILL 33-18 Replacing Brake Lines, Hoses, Fittings, and Supports

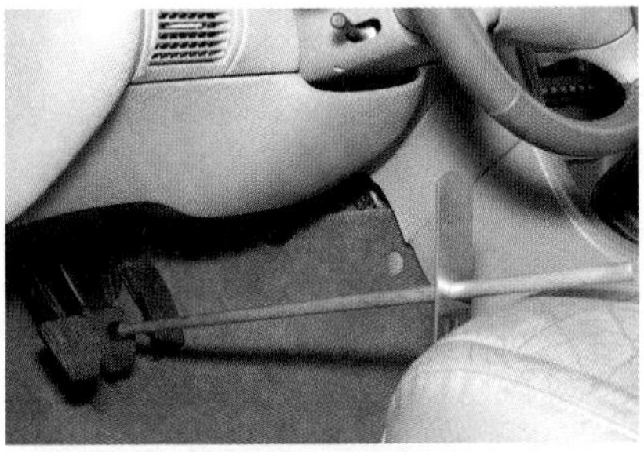

1. Depress the brake pedal with a brake pedal depressor to prevent brake fluid from draining out of the vehicle. Remove the stop light fuse to avoid draining the battery.

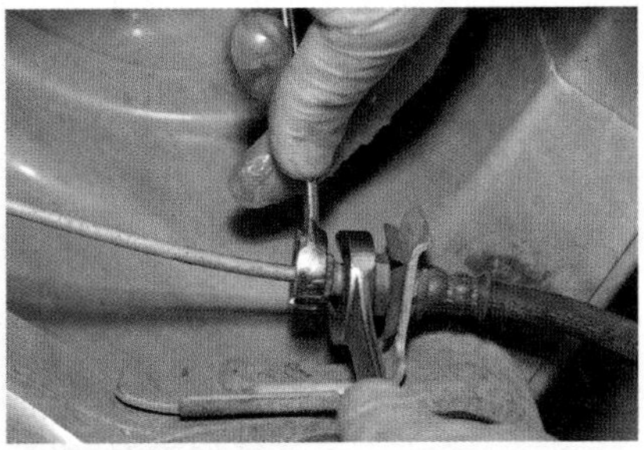

2. Safely raise and support the vehicle. Using flare nut wrenches and the double-wrench method, carefully remove any brake lines, hoses, fittings, and supports that are to be replaced. Inspect all components for damage and wear.

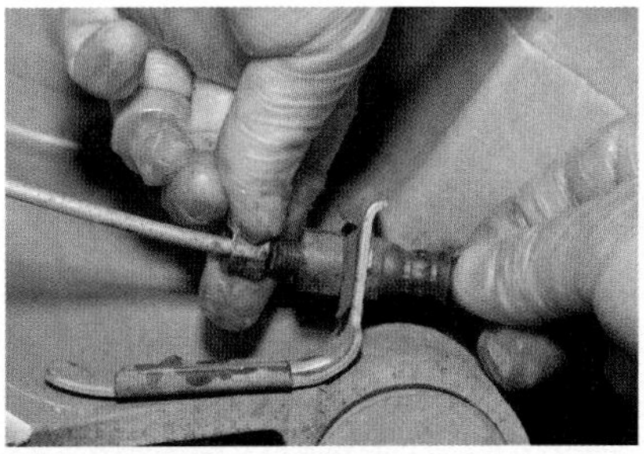

3. Carefully reassemble the removed components by using your fingers only. Once everything is assembled, tighten each fitting and support. Use flare nut wrenches and the double-wrench method. Bleed any trapped air from the system. Start the vehicle, and verify proper brake pedal height, firmness, and feel. Reinstall the brake light fuse.

SKILL DRILL 33-19 Fabricating Brake Lines, Using the Inverted Double Flare Method

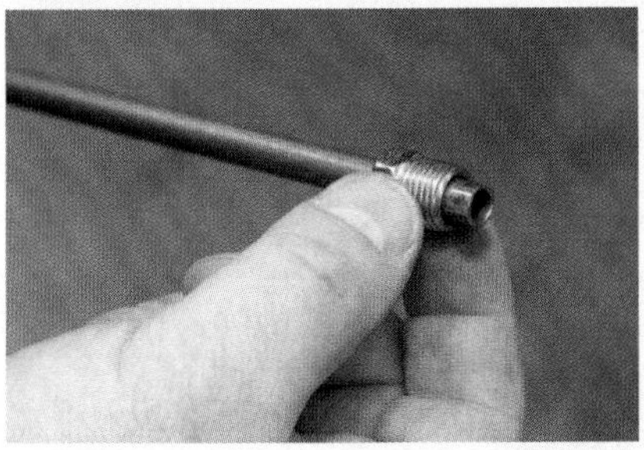

1. Choose the tube you will use to make the flare, and put the flare nut on the tube before creating the flare.

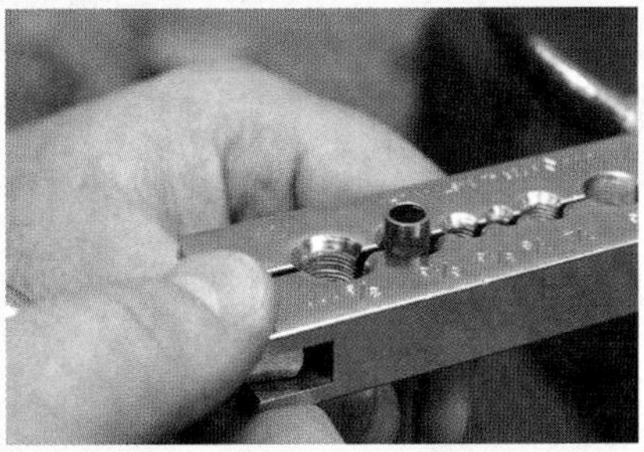

2. Use a bench vise to hold the brake line clamping tool. Match the size of the tube to the correct hole in the tubing clamp.

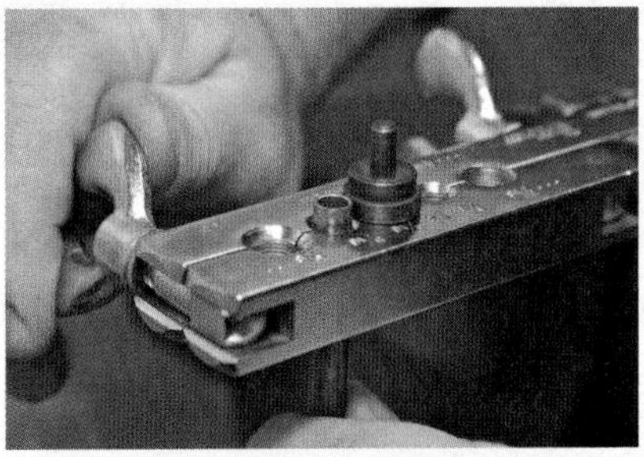

3. Select the proper sized button to match the tubing size. Use it to set the height of the tube in the clamp. Tighten down the clamp securely so the tube will not slip.

SKILL DRILL 33-19 Fabricating Brake Lines, Using the Inverted Double Flare Method (Continued)

4. Place the button in the end of the tube. Install the cone forming tool over the button. Turn the handle to make the bubble. Remove the forming tool and button from the tube.

5. Put the cone forming tool back on the clamping tool, and tighten the forming tool handle to create the double flare.

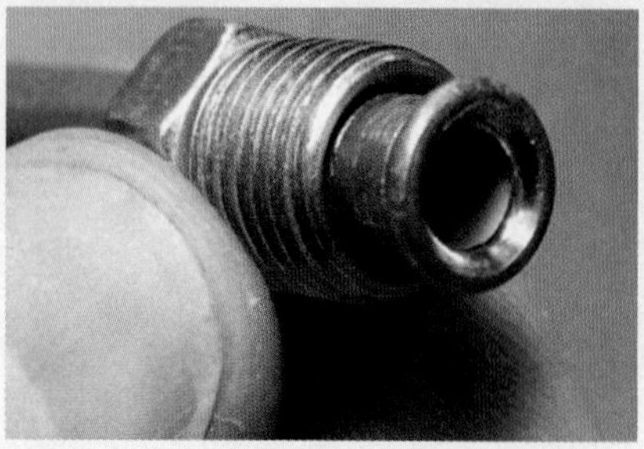

6. Remove the forming tool. Remove the tube from the clamping tool, and check the flare to ensure it is free of burrs. Also, check that it is formed correctly.

SKILL DRILL 33-20 Performing the ISO Flare Method

1. Install the proper fitting onto the brake line. Use a bench vise to hold the brake line clamping tool. Select the proper adapter to match the tubing size. Use it to set the height of the tube in the clamp. Tighten down the clamp securely.

2. Insert the adapter into the tubing, and install the flaring tool onto the clamp and over the adapter.

3. Tighten down the flaring tool until the adapter touches the clamp.

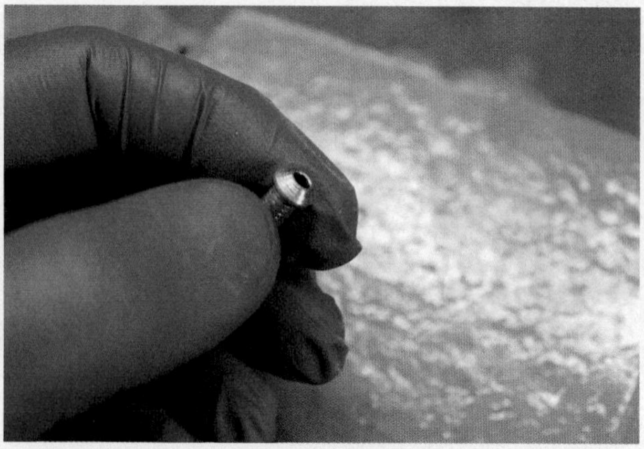

4. Remove the flaring tool, and inspect the flare to see if it has been formed correctly. Remove the flared line from the clamp, and give it a final inspection.

▶ Diagnosing the Brake Warning Lamp

LO 33-09 Test the operation of the brake warning lamp and stop lights.

A mechanical brake warning lamp system (non-CANbus) circuit is a simple circuit. It includes a lightbulb in series, with as many as four switches. The switches are connected in parallel with one another (**FIGURE 33-11**). Each switch can illuminate the warning lamp under the right conditions. The warning lamp should be off when the ignition key is in the run position if the parking brake is released and there are no faults in the hydraulic system. When the ignition is turned to the crank position, the warning lamp should illuminate. This is part of the bulb check function. If it doesn't come on in the crank position, the bulb is likely burnt out.

In the run position, the warning lamp can be illuminated only when one of the switches is closed. One switch is on the parking brake assembly. It illuminates the warning lamp when the parking brake is applied. This alerts the driver that the parking brake should be released before driving the vehicle. The second switch is located on the pressure differential assembly. It illuminates the warning lamp if there is a moderate-sized leak in the hydraulic system. The third switch, if the vehicle is so equipped, is located on the master cylinder reservoir. It illuminates the warning lamp if the fluid level falls below a certain point. The last switch, if the vehicle is so equipped, is located on the ignition switch. It acts as a bulb check feature, which illuminates the warning lamp when the ignition switch is in the crank position.

When diagnosing the brake warning lamp, it is good to start by verifying that the bulb is operational. Do this by using the circuit's bulb check feature if the vehicle is so equipped. To do so, turn the ignition key to the crank position. The brake warning lamp on the dash should illuminate on most vehicles.

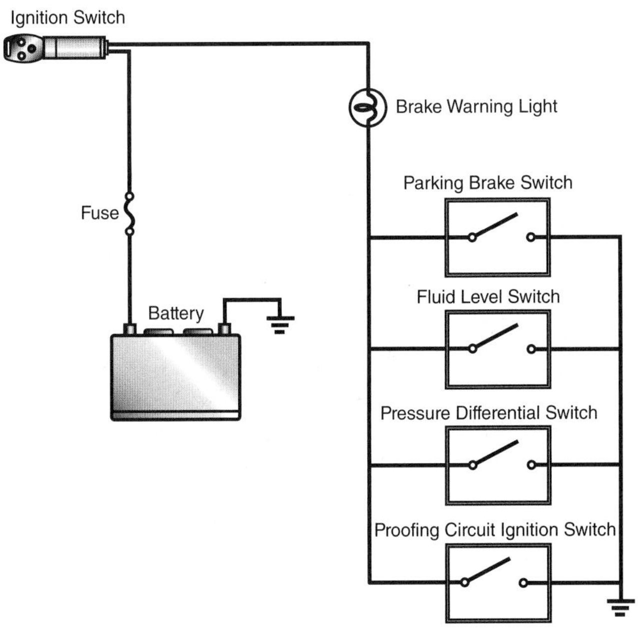

FIGURE 33-11 Brake warning light circuit.

If the bulb illuminates, then you know the bulb is good. You also know that it has power. Also, the ground circuit through the ignition switch is operating as it should.

If brake warning light does not illuminate, apply the parking brake, and recheck to see if the light is on. If not, check the fuse with a test light (**FIGURE 33-12**). If the fuse is okay, remove the bulb, and verify that it is not burned out. Measure its resistance with an ohmmeter and compare it with specifications or the resistance of a known good bulb.

If the brake warning lamp illuminated in the crank position, turn the ignition key to the run position. The brake warning lamp should be off. If it is, apply the parking brake. The warning lamp should illuminate. If it does not illuminate, you need to refer to the vehicle's wiring diagram to determine the best strategy to diagnose the system. If it is similar to the one in Figure 33-11, then locate the parking brake switch. Disconnect the wire harness connector from the switch, and jump it to ground with a jumper wire. With the ignition switch on, the brake warning light should now be illuminated. If it is, then the switch needs to be adjusted or replaced. If it is not, then there is a problem between the switch and the bulb. Use the same process for each of the other three switches.

If the brake warning lamp always stays on in the run position, you need to disconnect each of the four switches, one at a time. Check to see if the warning lamp goes off as you disconnect each one. If it does, then you need to test that switch. Determine whether it is adjusted properly or whether it is shorted.

If the light stayed on when the switches were disconnected, then there is likely a short in the wiring. It would be in one of the wires leading to the switches. This is the hardest part of testing the circuit. Thankfully, it doesn't happen very often.

In a **CANbus circuit**, the parking brake sensor and the low brake fluid sensors send status signals over the CANbus network (**FIGURE 33-13**). The appropriate module then signals the instrument panel module to illuminate the brake warning light. Diagnosis involves using a scan tool capable of interrogating the system. It also includes studying the wiring diagram. A DMM is used to perform specific tests. Refer to the Disc Brake System chapter for more information on parking brakes.

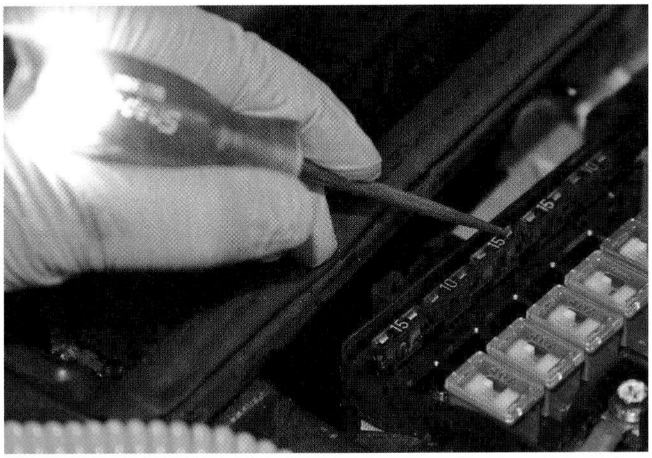

FIGURE 33-12 Checking a brake fuse with a test light.

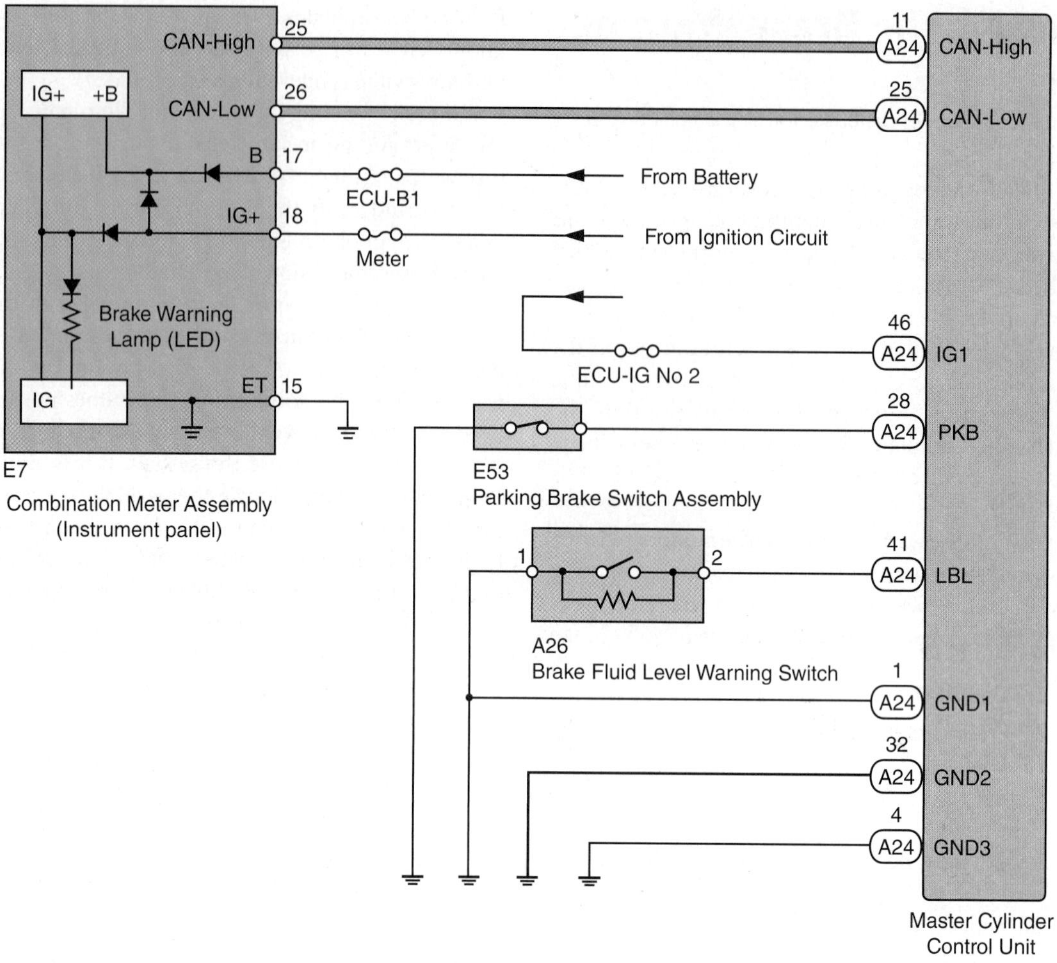

FIGURE 33-13 CANbus circuit for the brake warning light circuit.

Checking the Brake Warning Light System

To check a brake warning light system on a non-CANbus system, follow the steps in **SKILL DRILL 33-21**.

Checking the Parking Brake and Indicator Light System

To check the parking brake and indicator light system operation, follow the steps in **SKILL DRILL 33-22**.

SKILL DRILL 33-21 Checking the Brake Warning Light System in a Non-CANbus System

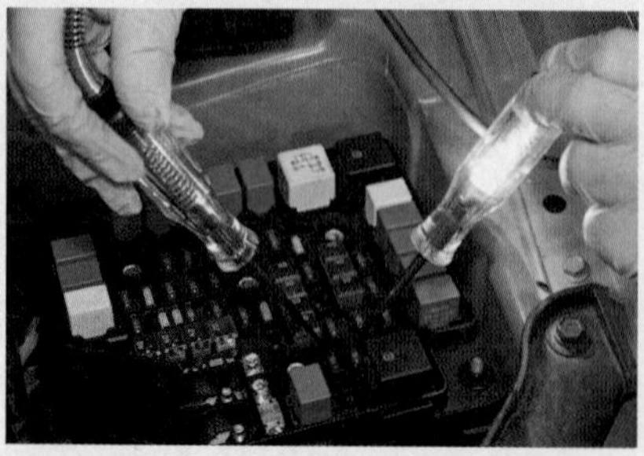

1. Perform a bulb check, and observe the brake warning light. If it is on, go to step 2. If it is off, check the fuse with a test light. If it is okay, check the warning lamp bulb with an ohmmeter.

2. Make sure the parking brake is released, and turn the ignition key to the run position. The brake warning lamp should be off. If not, go to step 6.

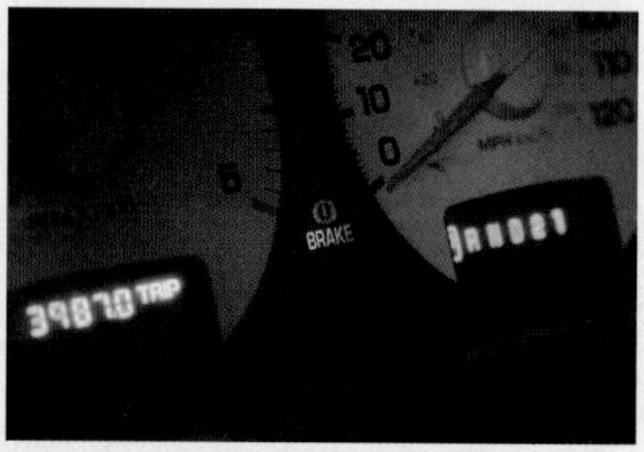

3. If the warning lamp is off, apply the parking brake. The lamp should illuminate. If it does, check the operation of the other switches. If they work fine, there are no faults present in the system.

SKILL DRILL 33-21 Checking the Brake Warning Light System in a Non-CANbus System (Continued)

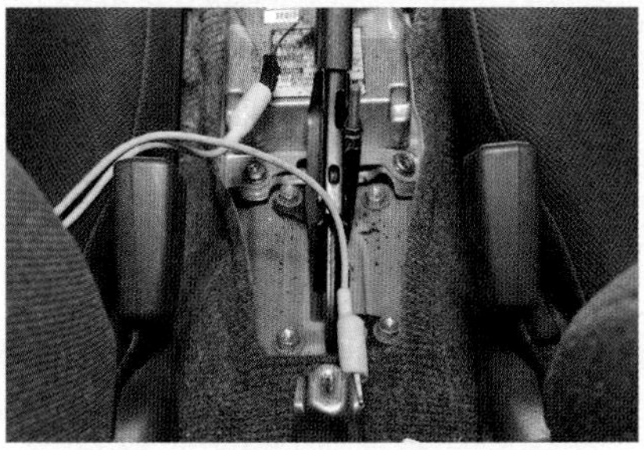

4. If the brake warning light is off when the parking brake is on, disconnect the wire from the parking brake switch, and ground the wire. The brake warning light should illuminate. If it does, test the parking brake switch with a DMM.

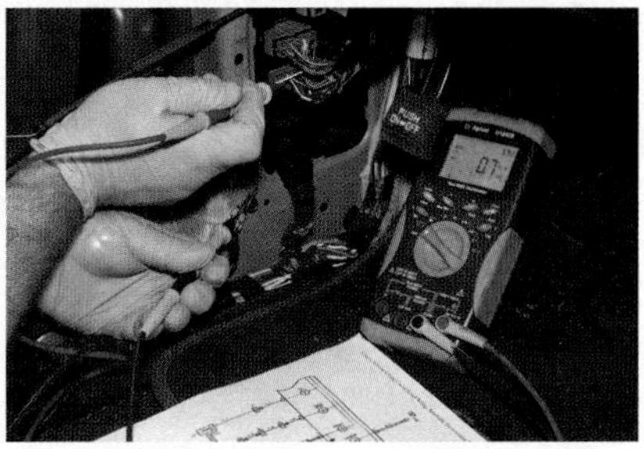

5. If the warning lamp does not illuminate with the parking brake wire grounded, suspect an open circuit between the parking brake wire and the warning lamp. Use a DMM and wiring diagram to diagnose the fault.

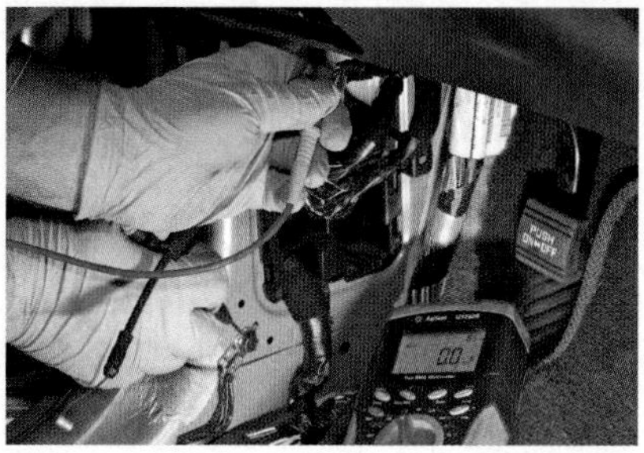

6. If the light stayed on in step 2, apply and release the parking brake several times. If this turns it off, clean or adjust the parking brake mechanism or switch. If it is still on, then disconnect each of the switches that activate it: parking brake, low fluid level, pressure differential, and ignition. If that turns the light off, then test that switch for misadjustment or a shorted condition. If the warning lamp is still on, suspect a short to ground in the wiring harness between the switches and the warning lamp bulb. Use a wiring diagram and DMM to identify the location of the fault.

SKILL DRILL 33-22 Checking Parking Brake and Indicator Light System Operation

1. Count the number of clicks it takes to apply the parking brake, and compare that with the manufacturer's specifications. If incorrect, follow the specified procedure to adjust it.
2. With the parking brake applied, turn the ignition switch to the run position. The red brake warning lamp should be illuminated.
3. Release the parking brake. The warning lamp should turn off. If it does, the parking brake warning light system is operating properly.
4. If the lamp does not turn off, disconnect the wire to the parking brake switch. If the light turns off, test the parking brake switch.
5. If it does not turn off, check the operation of the other warning lamp switches and circuit.

Diagnosing Stop Lights

The mechanical stop light system (non-CANbus) is a simple circuit. It consists of a normally closed switch in series with two to six brake lightbulbs. The bulbs are connected in parallel with each other (**FIGURE 33-14**). This circuit is also protected by a fuse. Some vehicles use separate fuses for the center high-mount stop lamp (CHMSL) and the normal stop lights. The switch is turned on and off by the movement of the brake pedal.

Diagnosis starts with operating the stop lights and observing their reaction. A wiring diagram shows how the manufacturer has wired the circuit. If the bulbs do not illuminate at all, check components common to all of the stop lights. This would be the fuse, stop light switch, and stop light ground circuit. If individual

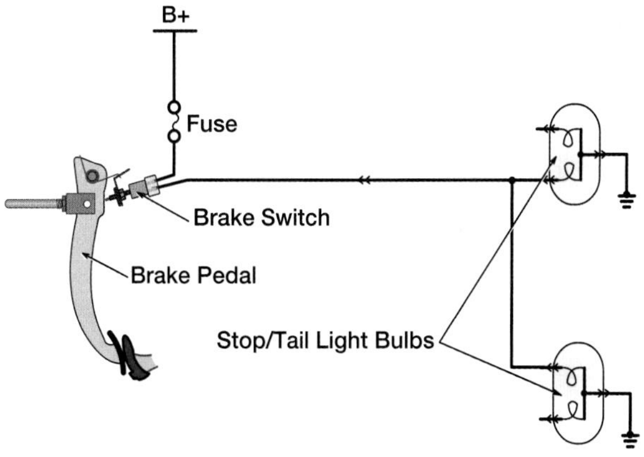

FIGURE 33-14 Typical non-CANbus stop light circuit.

lights do not illuminate, check those bulbs to see if they are burned out. If not, check to see if they have bad connections, terminals, or wires. A DMM works well for this.

Checking Operation of a Stop Light System

The stop lights are capable of operating at all times on most vehicles. This means the stop lights should come on whenever the brake pedal is pressed. It does not matter whether the key is on or off.

Checking the operation of the stop lights involves checking each bulb. If the bulbs are on even when the brake pedal is released, suspect:

- a misadjusted brake pedal,
- a defective brake switch, or
- a short to power in the circuit.

If none of the bulbs lights up when the brake pedal is depressed, suspect a fault that is common to all of the bulbs. This can be the fuse, brake switch, feed wire, or ground wire. If there is a problem with only some of the bulbs, suspect a fault that is common to only the bulbs that do not operate. This can be the bulbs or the individual wires. In any case, you will need to use the DMM to locate the open circuit, high resistance, or short circuit.

To check the operation of the brake stop light system and determine any necessary action, follow the steps in **SKILL DRILL 33-23**.

SKILL DRILL 33-23 Checking the Operation of the Brake Stop Light System

1. Verify that the stop lights are off when the brakes are released. If the lights are on, test the operation of the brake pedal and brake switch. Observe the stop lights when the brake pedal is applied. All lights, including the CHMSL, should be illuminated.

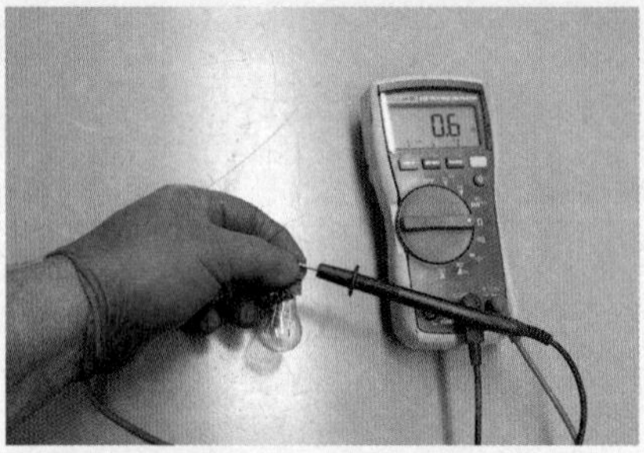

2. If some or all of the stop lamps do not illuminate, test them for an open or high resistance with a DMM.

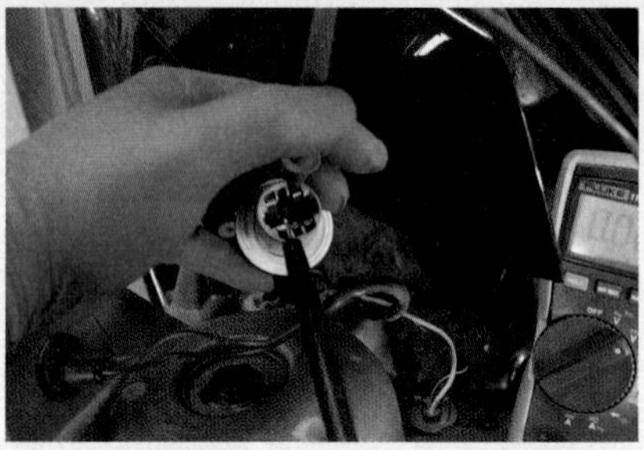

3. If the bulbs are okay, consult the wiring diagram to determine potential causes of the fault in the circuit. Use the wiring diagram and a DMM to identify the cause of the fault.

▶ Wrap-Up

Ready for Review

- A technician is liable for improperly repaired brakes as the likelihood of accidents, injury, or death increases. Another issue that leads to liability is forgetting to tighten, or failing to properly tighten, components.
  - Most common tools used for brake service include: brake bleeder wrenches, flare nut wrenches, vacuum brake bleeders, pressure brake bleeders, and brake fluid testers.
- Consult the service information to determine the correct intervals for brake fluid flushing. Brake fluid flush can be based on: time/mileage, digital multimeter (DMM)-galvanic reaction test, boiling point test, and test strip.
- Bleeding the brakes is the process of removing any air from the system. Air in a system prevents pressure buildup. Three most common brake bleeding methods include: manual bleeding, vacuum bleeding, and pressure bleeding.
- Brake pedal height should have a certain amount of travel to be safe. Brake pedal height can be measured between two specified points using a measuring stick. Moving the brake pedal up and down between the top and where resistance is felt is the free play. Brake pedal travel is the distance the brake pedal travels from its rest position to its applied height.
- Hydraulic braking system diagnosis follows the strategy-based diagnostic process that begins with verifying the concern and leads to the cause of fault. Once the repair is complete, verification of the repair is necessary.
- Master cylinder should be serviced if the brake pedal sinks when the brakes are applied; the brake fluid is low in the reservoir; the brake warning light is on; or the brake pedal reserve height is too low.

  - Master cylinder service includes bench bleeding and measuring and adjusting master cylinder pushrod length.
- Vacuum boosters should be tested if the brakes are dragging, the brake pedal is harder to push than normal, the pedal height has changed, if the engine operation changes more than a minimal amount when the brake pedal is applied, or if the vehicle has an unlocated vacuum leak.
  - Vacuum brake booster tests include: brake pedal free travel, performance/operation test, external leak test, and internal leak test.
- Regular inspection of brake lines and brake hoses will help prevent vehicle breakdowns and unsafe vehicle operation, especially in regions are highly susceptible to corrosion from deicing chemicals.
  - During brake lines and hoses replacement, they should be disconnected at their fittings with the double-wrench method.
- A brake warning light circuit consists of parking brake switch, fluid level switch, pressure differential switch, and proofing circuit ignition switch. These switches should be inspected when testing the operation of the brake warning lamp.

Key Terms

Boiling point test A test to determine the boiling point of brake fluid.

bleeding The process of removing air from a hydraulic braking system.

CANbus circuit A two-wire communication network that transmits status and command signals between control modules in a vehicle.

Digital multimeter (DMM)-galvanic reaction test A test using a DMM to measure the voltage created in worn out brake fluid.

free play The amount of movement between two mating parts.

ISO flare method A method for joining brake lines, also called a bubble flare. Created by flaring the line slightly out and then back in, leaving the line bubbled near the end.

manual bleeding A bleeding method where one person manually operates the brake pedal while the other person opens and closes the bleeder screws on the wheel brake units to allow the air and old brake fluid to be pushed out.

pressure bleeding A bleeding method that uses clean brake fluid under pressure from an auxiliary tool or piece of equipment to force the air and old brake fluid from the hydraulic braking system.

Test strip A chemically reactive strip that turns color when exposed to a reactive fluid.

Time/mileage A reference to the method of determining when a particular service should be performed.

vacuum bleeding Bleeding process that uses a vacuum bleeder to pull the air and old brake fluid from the system.

Review Questions

1. Brake lining material dust may be hazardous to breathe because it could contain ______________.
   a. Carbon dioxide
   b. Carbon monoxide
   c. Gasoline vapors
   d. Asbestos
2. When testing brake fluid, any voltage reading less than __________ is commonly acceptable.
   a. 0.3 Volts AC
   b. 3.0 Volts AC
   c. 0.3 Volts DC
   d. 3.0 Volts DC
3. Which method of brake fluid bleeding is ideal for when the fluid needs to be flushed out of a brake system?
   a. Manual bleeding
   b. Vacuum bleeding
   c. Pressure bleeding
   d. Free play bleeding
4. What could happen if the brake pedal is pushed past its normal travel during manual brake bleeding?
   a. The parking brake lever may stick
   b. The firewall may flex making it hard to bleed
   c. The master cylinder fluid lines may twist or kink
   d. The master cylinder piston seals my leak
5. If the pedal height is incorrect on a nonadjustable pedal assembly, what could be the cause?
   a. Leaking master cylinder cup seals
   b. Misadjusted drum brake shoes
   c. Air in the hydraulic system
   d. Worn pedal bushings
6. Some hydraulic brake concerns can be hard to identify. What should be done if you can't verify the problem?
   a. Tell the customer there is nothing wrong and give the vehicle back
   b. Search the Internet for a concern that sounds similar and diagnose that
   c. Replace each of the parts that might cause the customer concern
   d. Test drive the vehicle with the customer to gain a clear understanding of the concern
7. What does a master cylinder pushrod that is adjusted too long likely cause?
   a. Low brake pedal during brake application
   b. A spongy brake pedal feel
   c. Trapped pressure that applies the brakes when the pedal is released
   d. Stripped brake line fittings
8. A vacuum booster that leaks vacuum may cause all of the following EXCEPT:
   a. A lean running engine
   b. Incorrect push rod adjustment
   c. A rough idling engine
   d. Immediate loss of power brakes when the engine stalls

9. Why should the brake pedal be held in the depressed position when replacing brake lines or hoses?
   a. To prevent brake fluid drainage
   b. To prevent master cylinder return spring damage
   c. To keep the brake lights from discharging the battery
   d. To pressurize the new lines as soon as they are installed
10. Which service information document shows the path of electrical current flow and components involved in the stop light system?
   a. Parts and labor guide
   b. Removal and installation of the tail lamp assembly
   c. Stop lamp wiring diagram
   d. Stop lamp connector pin chart

ASE Technician A/Technician B Style Questions

1. Technician A says that technicians are not held legally responsible for improper brake repairs. Technician B says that it is important to develop work routines that eliminate the potential for errors. Who is correct?
   a. Technician A only
   b. Technician B only
   c. Both A and B
   d. Neither A nor B
2. Technician A says that brake fluid condition can be tested with a digital multimeter. Technician B says that brake fluid should be topped off whenever it is low. Who is correct?
   a. Technician A only
   b. Technician B only
   c. Both A and B
   d. Neither A nor B
3. Technician A says that manual bleeding a brake hydraulic system typically requires the help of an assistant. Technician B says that the order in which the brakes are bled is not important. Who is correct?
   a. Technician A only
   b. Technician B only
   c. Both A and B
   d. Neither A nor B
4. Technician A says that free play is measured by subtracting applied brake pedal height from the brake pedal height measurement at rest. Technician B says that if free play is out of specifications, it is likely a pushrod out of adjustment. Who is correct?
   a. Technician A only
   b. Technician B only
   c. Both A and B
   d. Neither A nor B
5. While discussing brake hydraulic system diagnosis. Technician A says that technical service bulletins (TSBs) should typically be researched before symptom diagnosis testing. Technician B says that once a fault is found, there may be an underlying root cause that requires further investigation. Who is correct?
   a. Technician A only
   b. Technician B only
   c. Both A and B
   d. Neither A nor B
6. Technician A says that with the engine running and the brakes applied, the brake pedal should hold its position. Technician B says that bench bleeding a master cylinder no longer saves time during replacement and is unnecessary. Who is correct?
   a. Technician A only
   b. Technician B only
   c. Both A and B
   d. Neither A nor B
7. Technician A says that while testing a vacuum booster, the pedal should not drop when the engine is started. Technician B says that the pressure supplied to the vacuum booster should be about 10 in Hg lower than engine manifold vacuum. Who is correct?
   a. Technician A only
   b. Technician B only
   c. Both A and B
   d. Neither A nor B
8. Technician A says that brake lines should be checked for abrasion and rust. Technician B says that open-end wrenches are just as good as flare nut wrenches when removing brake line fittings. Who is correct?
   a. Technician A only
   b. Technician B only
   c. Both A and B
   d. Neither A nor B
9. Technician A says that a brake fluid leak may cause the red brake warning light to illuminate. Technician B says that low brake fluid in the master cylinder reservoir may cause the red brake warning light to illuminate. Who is correct?
   a. Technician A only
   b. Technician B only
   c. Both A and B
   d. Neither A nor B
10. Technician A says that the normal stop lamps may be powered separately from the center high-mount stop lamp (CHMSL). Technician B says that a defective fluid level switch may cause the center high-mount stop lamp (CHMSL) to be inoperative. Who is correct?
   a. Technician A only
   b. Technician B only
   c. Both A and B
   d. Neither A nor B

CHAPTER 34

Disc Brake Systems Theory

Learning Objectives

- LO 34-01 Describe disc brake fundamentals.
- LO 34-02 Describe disc brake caliper operation.
- LO 34-03 Describe the brake pad assembly.
- LO 34-04 Describe brake rotor construction.
- LO 34-05 Describe parking brakes on disc brakes.

ASE Education Foundation Tasks

See Appendix A to view the 2017 ASE Education Foundation Automobile Accreditation Task List Correlation Guide.

▶ Disc Brake Fundamentals

LO 34-01 Describe disc brake fundamentals.

Disc brakes are so named because they create braking power by forcing friction pads against the sides of a rotating disc. This disc is also called a disc brake rotor, or rotor, and is bolted to the wheel (**FIGURE 34-1**). Disc brake calipers straddle the pads. Hydraulic pressure from the master cylinder causes the caliper to create a mechanical clamping action. This forces the brake pads onto the surface of the rotor, creating friction. Friction transforms the vehicle's kinetic energy into heat energy. As the vehicle's kinetic energy is transformed into heat energy, the vehicle's speed decreases.

Higher applied forces can be used in disc brakes than in drum brakes. This is because the rotor has to withstand only compressive forces. Drums have to withstand tension forces (**FIGURE 34-2**). Because heat is generated on the outside surfaces of the rotor, it can be easily transferred to the atmosphere. Because of these and other design features, disc brakes are effective at creating substantial braking power. And yet they use a fairly simple design that is relatively easy to service and repair.

▶ TECHNICIAN TIP

The purpose of the disc brake system is to provide an effective means to slow the vehicle under a variety of conditions in an acceptable distance and manner. The better a braking system can do this, the more likely the vehicle will avoid an accident.

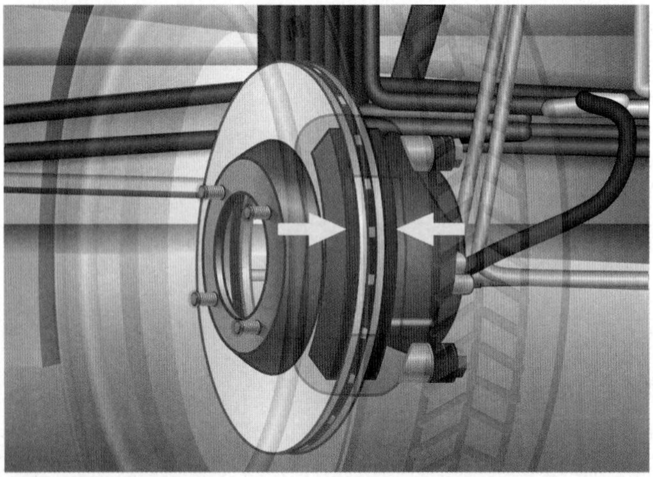

FIGURE 34-1 Disc brake operation.

Disc Brake System

Modern passenger vehicles are almost always equipped with disc brakes on at least the front two wheels. And many manufacturers are using them on all four wheels. The primary components of the disc brakes are as follows:

- Rotor
- Caliper
- Brake pads (**FIGURE 34-3**)

The **rotors** are the main rotating part of this brake system. They are durable and withstand the high temperatures that occur during braking. Most rotors are made from cast iron. Cast iron wears well and resists warpage. In high-performance vehicles, the rotors are made from composite materials such as ceramics or carbon fiber.

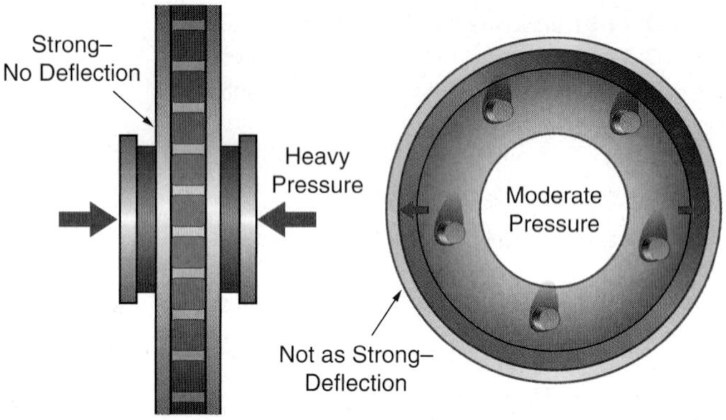

FIGURE 34-2 Disc versus drum brakes.

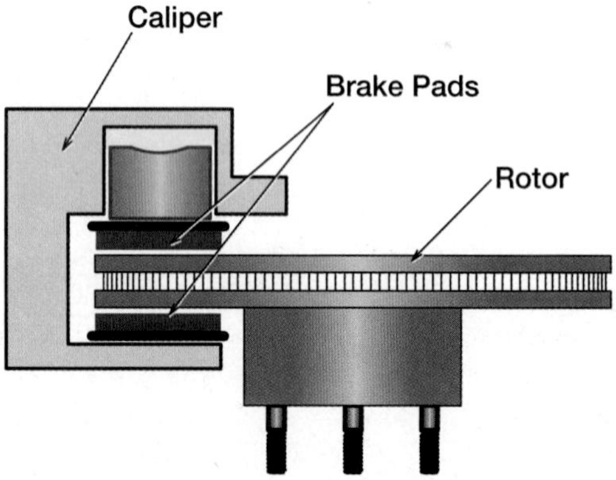

FIGURE 34-3 The disc brake system.

You Are the Automotive Technician

A customer comes into the service department. The customer concern is a pulsating brake pedal and high-pitched squealing during braking. The vehicle is a 2015 Town and Country minivan. It has 41,000 miles (69,202 km) on the odometer. After reviewing the vehicle's service history, you notice the vehicle has never had any brake service. You advise the customer that a thorough inspection will be needed to determine what is causing the issues. After visually inspecting the brakes and measuring the rotor's thickness and parallelism, you find the following. The front brake linings have worn down to the wear indicators, and the rotors are warped beyond specifications. You inform the customer of your findings and recommend that new front brake pads be installed and the rotors refinished. You also recommend a brake fluid flush, as the fluid is beyond its two-year life.

1. What conditions can cause a rotor to become warped?
2. What conditions would require replacement of the rotors rather than just refinishing them?
3. What are the relatively common maximum specifications for rotor runout and thickness variation?
4. How are floating calipers different than fixed calipers?

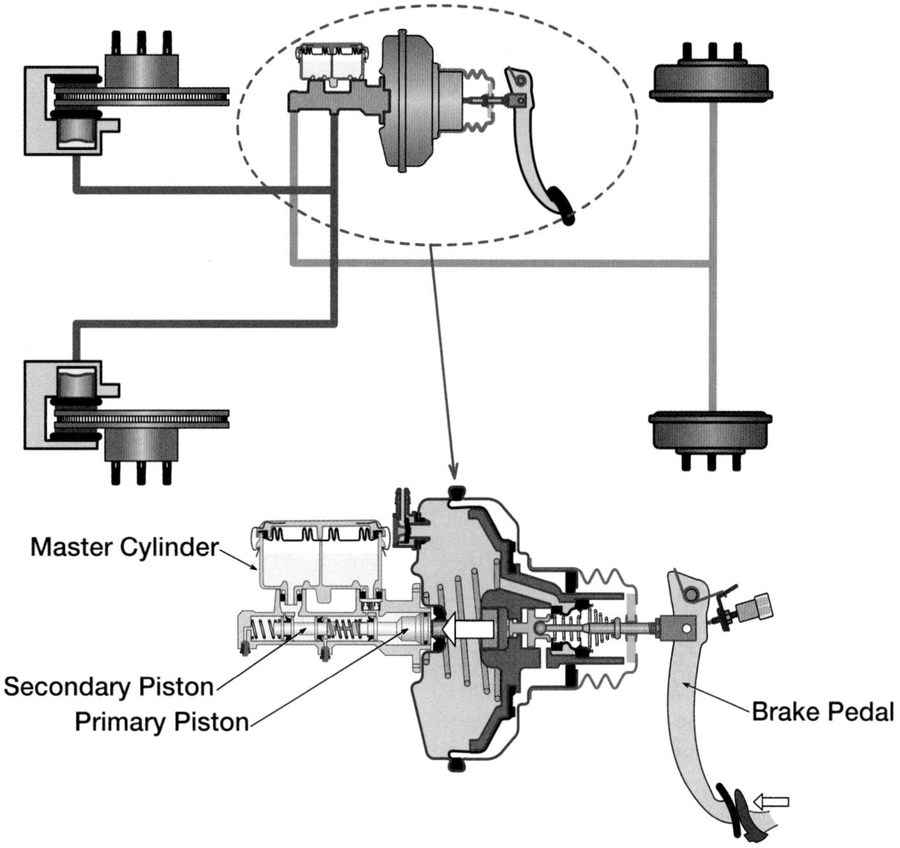

FIGURE 34-4 The master cylinder converts the pedal force into hydraulic pressure, which is then transmitted via brake lines and hoses to one or more pistons at each brake caliper.

The **caliper** straddles the rotor and houses the disc brake pads and an activating piston(s). The calipers use hydraulic pressure from the master cylinder to apply the brake pads. They are usually bolted to the steering knuckle or, in the case of a non-steering axle, to a suspension component. Calipers should be inspected at the same time as the brake pads.

The **disc brake pads** are located inside the caliper or caliper mounting bracket. The pads are forced against the rotor to slow or stop the vehicle. A disc brake pad consists of friction material bonded or riveted to a steel backing plate. With this design, the pads will wear out over time and must be replaced periodically.

Disc Brake Operation

Disc brakes can be used on all four wheels of a vehicle, or on the front wheels, with drum brakes on the rear. When the brake pedal is depressed, a **pushrod** transfers the force through a **brake booster** to a hydraulic master cylinder. The master cylinder converts the pedal force into hydraulic pressure. The pressure is then transmitted via brake lines and hoses to one or more pistons at each brake caliper (**FIGURE 34-4**). The pistons operate on friction pads to provide a clamping force on a rotor. The rotor is attached to the wheel hub. This clamping action is designed to slow or stop the rotation of the rotor and the wheel.

The rotors are free to rotate with the wheels, because of wheel bearings and the hubs that contain them. The hub can be part of the brake rotor, called a hub-style rotor. Or it could be separate from the rotor, called a hubless rotor. In this case, the rotor slips over the hub and is bolted to it by the wheel and lug nuts (**FIGURE 34-5**).

FIGURE 34-5 Hub-style and hubless rotors.

The brake caliper assembly is normally bolted to the steering knuckle or axle housing (**FIGURE 34-6**). In most cases, the brake is positioned as close as possible to the wheel, but there are exceptions. Some high-performance cars with **independent rear suspension (IRS)** use inboard disc brakes on the rear wheels. The calipers are mounted on or next to the final drive, which is directly mounted to the chassis. This

FIGURE 34-6 Outboard caliper.

configuration increases vehicle handling because it reduces the vehicle's unsprung weight. It does so by moving the final drive assembly and brakes from the axle to the chassis. Because the wheels and axles are now lighter, the vehicle's unsprung weight is much lower. This allows the springs to do a better job of keeping the wheels in contact with the ground.

Disc brake pads require a much higher application force to operate than drum brake shoes. This is because they are not self-energizing. The additional clamping pressure is created by increasing the diameter of the caliper pistons. Unfortunately, this means the brake pedal would have to travel farther to move the larger caliper pistons. This is not practical. Manufacturers have overcome this by equipping disc brake systems with a power booster.

Because of the high forces needed to apply a disc brake, using it as a parking brake is more challenging. Some manufacturers have chosen to design more complicated rear brake calipers. A few others have built an auxiliary drum-style parking brake assembly into the center of the rear brake rotors. This type of parking brake is referred to as a top hat design. We will cover this later.

Advantages and Disadvantages

Disc brakes have a number of advantages over drum brakes. They also have some disadvantages. In most cases, the advantages outweigh the disadvantages. One of the biggest advantages is that disc brakes can generate and transfer greater amounts of heat to the atmosphere. This is because most of the friction area of a rotor is exposed to air. So, cooling is far more rapid than for a drum brake. Faster cooling makes them better suited for high-performance driving or heavy-duty vehicles. This reduces the likelihood of brake fade.

Also, because of their shape, rotors tend to scrape off water more effectively. After being driven through water, disc brakes operate at peak performance almost immediately. Further, due to their design, disc brakes are self-adjusting. They do not need periodic maintenance or rely on a self-adjusting mechanism that is prone to sticking. Finally, in most cases, disc brakes are also easier to service than drum brakes.

Although disc brakes have many benefits over drum brakes, there are some disadvantages. The most clear disadvantage is that disc brakes are much more prone to making noise. Their design tends to create squeals and squeaks, which can be very annoying. Many a technician has spent time replacing perfectly functional brake pads because of excessive noise complaints.

Another issue is that the rotors warp more easily than drums. A warped rotor is when the rotor's friction surfaces have side-to-side runout. Runout above 0.003" (0.076 mm) causes brake pulsation. Warped rotors are caused by improperly torquing the lug nuts or overheating the brakes.

A related problem is excessive thickness variation of the rotor. Thickness variation is the difference in thickness of the parallel friction surfaces. Thickness variation as small as 0.0003" (0.0076 mm) can cause brake pedal pulsations. Resurfacing or replacement of the rotor is required. The last disadvantage is that disc brakes require a power booster. This is because disc brakes are not self-energizing, so they need higher clamping forces. This also makes it harder to use disc brakes as an effective parking brake.

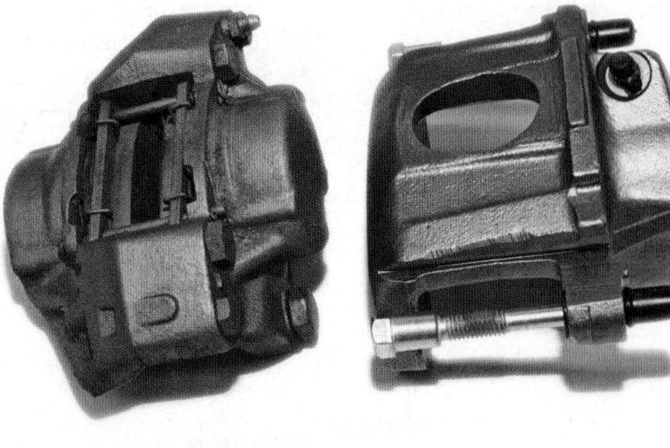

FIGURE 34-7 Fixed and floating calipers.

▶ TECHNICIAN TIP

As you see, disc brakes have more disadvantages than drum brakes. But the advantages are generally considered much more critical than the disadvantages. Thus, disc brakes are preferred over drum brakes in most applications.

▶ Disc Brake Calipers

LO 34-02 Describe disc brake caliper operation.

In most applications, the disc brake caliper assembly is bolted to the axle housing, or steering knuckle. It clamps the brake pads onto the rotors to slow the vehicle. There are two main types of calipers: **fixed calipers** and **sliding or floating calipers** (FIGURE 34-7). Sliding or floating calipers are the most common type used in passenger vehicles. This is because they are easier to build and are more compact.

All calipers are fitted with a **bleeder screw** at the top of the piston bore. It allows for the removal of air within the disc brake system. It is also used when performing routine brake fluid changes.

Fixed calipers are rigidly bolted in place. The caliper housing cannot move or slide. This makes their application of braking forces more precise than floating calipers. They commonly have one to four pistons on each side of the rotor (FIGURE 34-8). When the brakes are applied, hydraulic pressure forces the pistons on both sides of the caliper inward. This causes the brake pads to come in contact with the rotor (**FIGURE 34-9**). Once the pad-to-rotor clearance is taken up, the hydraulic pressure rises equally on each side of the caliper. Both brake pads are then applied equally.

Floating and sliding calipers have one or more pistons. But they are located only on one side of the caliper housing. This is usually on the inboard side of the rotor. Thus, the hydraulic force is generated on one side of the rotor. But the unique design applies equal braking force to both sides of the rotor. This distribution of force is possible because the caliper is mounted on pins, or slides. They allow the caliper housing to move (float or slide) from side to side, as necessary. This movement allows pistons on one side of the rotor to generate force on both sides of the rotor at the same time.

When the brakes are applied, hydraulic pressure forces the piston toward the rotor. This takes up any clearance between the inboard brake pad and the rotor. Then, because the caliper housing is free to move on the pins or slides, it gets pushed away from the rotor. This pulls the outboard brake pad into contact with the outboard side of the rotor. Once all clearance is taken up on the outboard brake pad, the clamping force increases equally on both brake pads (**FIGURE 34-10**).

Floating calipers are mounted in place by **guide pins** and bushings (**FIGURE 34-11**). The pins allow the caliper to move in and out as the brakes are operated and as the brake pads wear. Because the calipers move on the pins, the bushings must be lubricated. They are lubricated with high-temperature, waterproof disc brake caliper grease when they are serviced. This helps prevent them from binding or sticking. Inspecting, cleaning, and lubricating the pins, bushings, and dust boots are important steps in disc brake repair.

Sliding calipers have matching machined surfaces on the caliper and caliper mount. They allow the caliper to slide in the mount (**FIGURE 34-12**). They operate similar to the floating

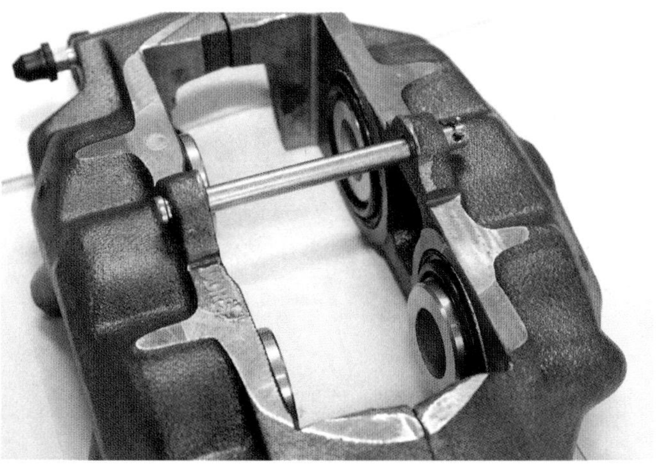

FIGURE 34-8 Fixed calipers with multiple pistons.

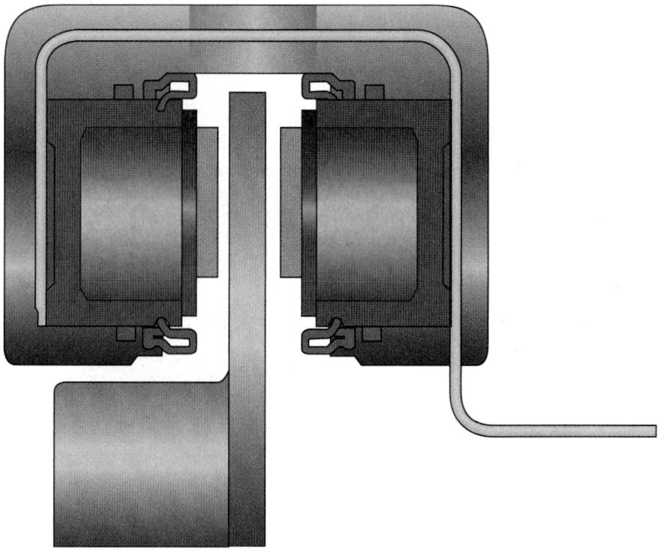

FIGURE 34-9 Fixed caliper application on solid rotor.

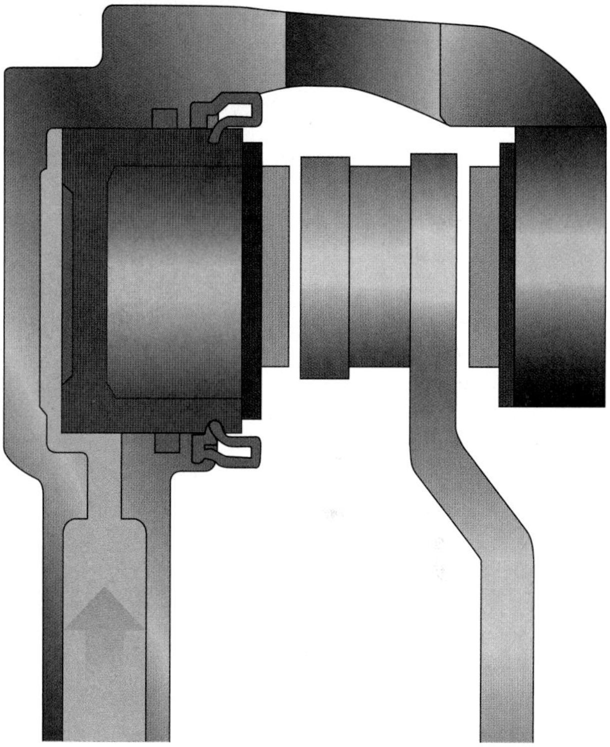

FIGURE 34-10 Sliding or floating caliper application on vented rotor.

FIGURE 34-11 Floating caliper and guide pins.

caliper. When the calipers are serviced, the mounting surfaces must be cleaned and lubricated. The same high-temperature, waterproof grease is used. Sliding calipers are held in place by clips or bolts that still allow them to slide. Sliding calipers do not slide as easily as floating calipers. So, most light-duty vehicles use the floating caliper design.

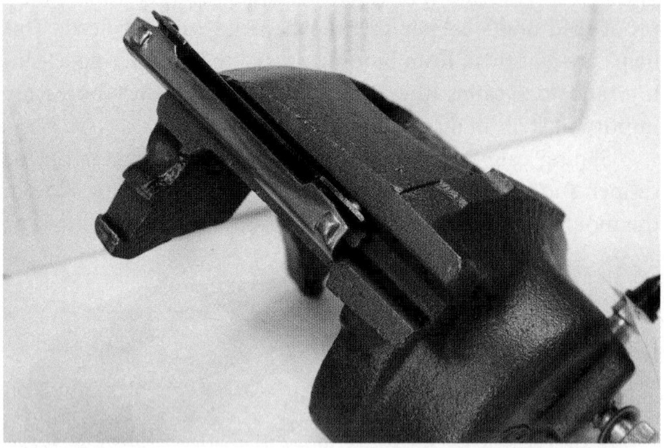

FIGURE 34-12 Sliding caliper.

O-Rings

In disc brake calipers, the piston is sealed by a stationary square section sealing ring. It is also called a **square-cut O-ring** (**FIGURE 34-13**). This O-ring has a square cross section and is fitted in a machined groove in the caliper. The O-ring is compressed between the piston and caliper housing. This creates a positive seal to keep the high-pressure brake fluid from leaking out. It also prevents air from being drawn into the system when the brake pedal is released quickly.

When the brakes are applied, the caliper piston moves outward. This movement slightly deforms the O-ring (**FIGURE 34-14**). When the brakes are released, the elasticity of the O-ring causes it to return to its original shape. This retracts the piston to provide a small running clearance between the rotor and pads. As the brake pads wear, the piston must move outward a bit farther than the sealing ring can stretch or flex. The sealing ring is designed to allow the piston to slide through it in this situation. This takes up the extra clearance, which makes disc brakes self-adjusting.

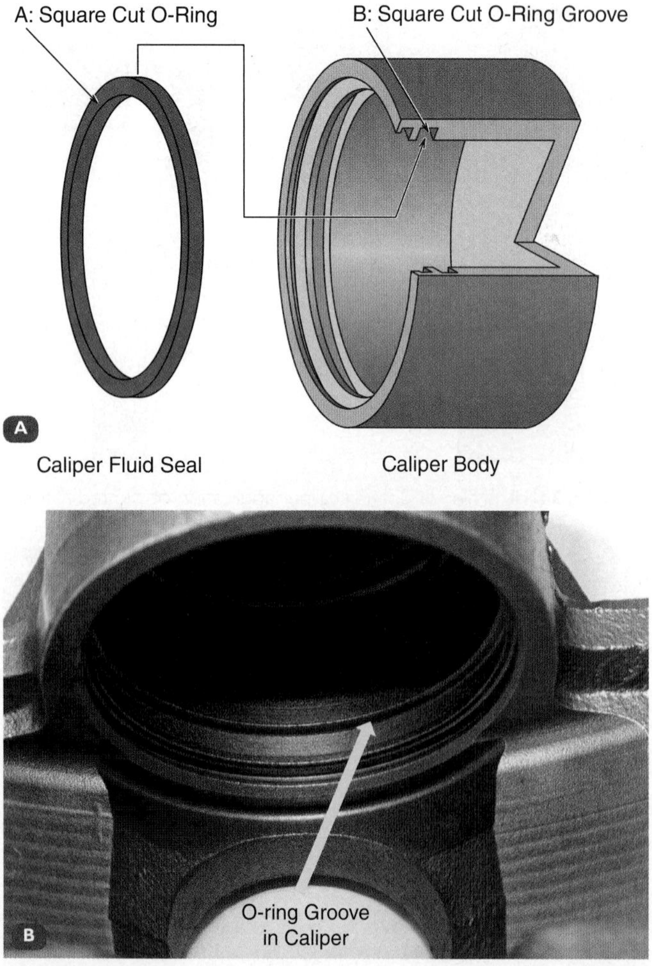

FIGURE 34-13 O-rings. **A.** Square-cut O-ring and O-ring cut to show square section. **B.** Square-cut O-ring groove in caliper.

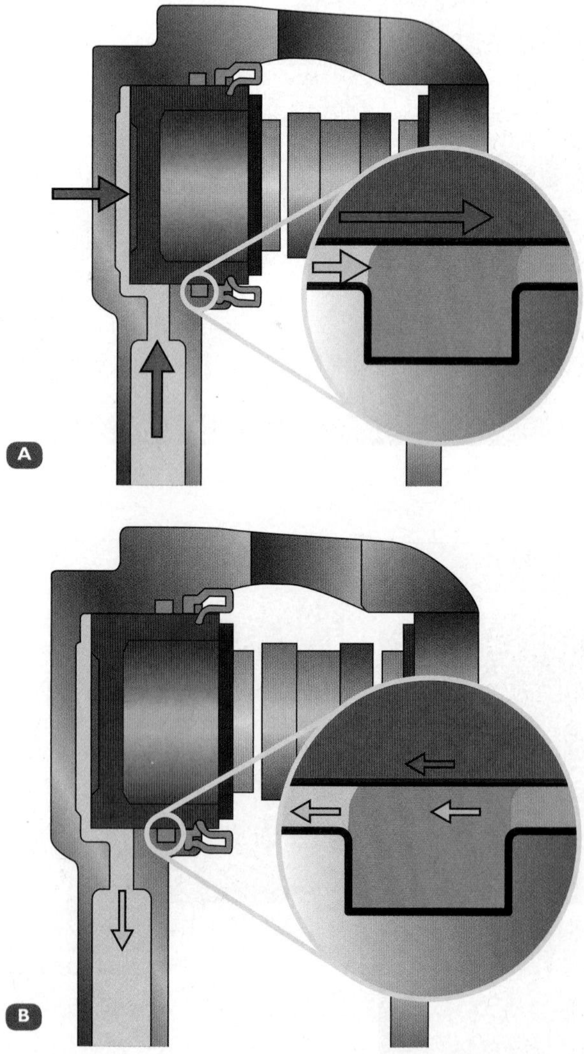

FIGURE 34-14 Square-cut O-ring. **A.** Square-cut O-ring during brake application. **B.** Square-cut O-ring during brake release.

▶ TECHNICIAN TIP

The force generated by the O-ring to retract the piston is fairly small. So, any corrosion or buildup on the piston or bore will cause the piston to stick and not retract. This holds the brakes in the applied position, causing brake drag, overheated brakes, and poor fuel economy. Technicians can identify this situation by test driving the vehicle. Then, an infrared temperature gun is used to measure the temperature of each brake rotor. The temperatures should be approximately the same on each side. If they are not, suspect a stuck or binding caliper on the hottest side.

Some calipers are known as **low-drag calipers.** They are designed to retract the brake pads farther, so they don't drag on the rotor. This gives a larger brake pad–to-rotor clearance. The calipers do this by retracting the pistons a little bit farther than normal. This is accomplished by modifying the sealing groove in the caliper. The outside of the sealing groove is machined at an angle toward the rotor (**FIGURE 34-15**). This allows the seal to flex a bit farther upon brake application. The O-ring then retracts the piston farther when the brakes are released. These systems use a "quick take-up" or "fast-fill" master cylinder to maintain adequate brake pedal reserve height.

The primary sealing surface is the outside diameter of the piston. It is critical that this surface be smooth and free of pitting or rust. Therefore, steel pistons are chrome plated. This gives the surface a hard, wear-resistant, and corrosion-resistant finish. Chrome can still rust, but it is much more corrosion resistant than steel.

Another way manufacturers have dealt with the corrosion issue is by making pistons out of a **phenolic resin**. Pool balls also are made from phenolic resin. It is very dense when it hardens and does not corrode or rust. This makes for a good sealing surface in brake systems. Although the phenolic pistons themselves do not corrode, the cast iron bore of the caliper can corrode. Rust can therefore cause a phenolic piston to seize in the bore (**FIGURE 34-16**).

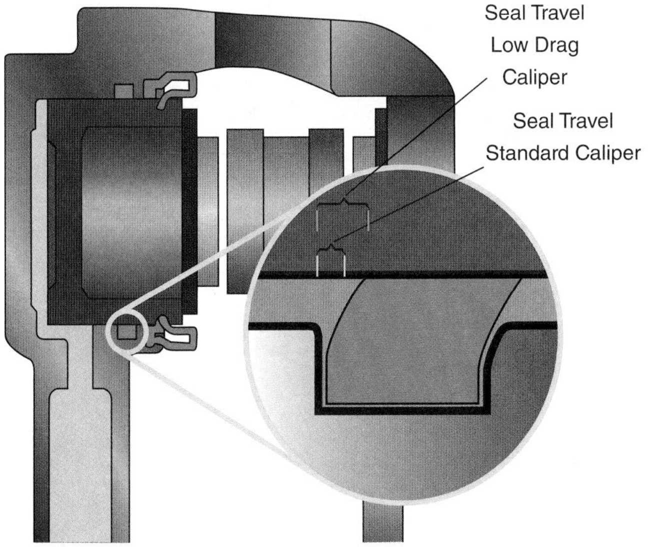

FIGURE 34-15 Low-drag caliper.

Phenolic pistons transfer heat more slowly than steel pistons (**FIGURE 34-17**). This is a good thing because heat transferred through the piston can cause the brake fluid to boil. Calipers with phenolic pistons are therefore less susceptible to brake failure. This is because they are less likely to cause the brake fluid to boil.

There is also a dust boot that seals the surfaces of the piston and caliper bore from outside dirt and moisture. This seal connects to both the piston and the caliper (**FIGURE 34-18**). It is

FIGURE 34-16 Corroded caliper piston bore.

FIGURE 34-17 Heat transfer. **A.** Phenolic piston (slow heat transfer). **B.** Steel piston (fast heat transfer).

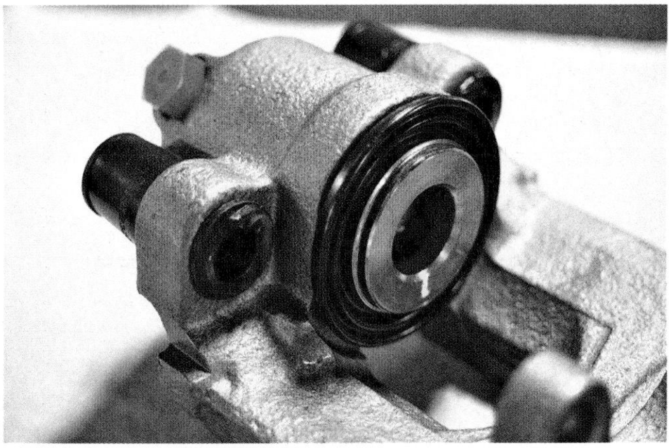

FIGURE 34-18 The caliper dust boot seals the piston to the caliper.

expandable to allow the piston to move outward as the brake pads wear. It also must be free from cuts and holes. Otherwise, the piston and bore could corrode. This can cause the piston to bind in the bore, causing brake drag.

▶ Disc Brake Pads and Friction Materials

LO 34-03 Describe the brake pad assembly.

Disc brake pads consist of friction material bonded or riveted onto a steel **backing plate** (**FIGURE 34-19**). **Bonded linings** are more common on light-duty vehicles. They are less expensive to build. Also, the bonding agent can fail under the very high temperatures of heavy-duty use. Because of this, **riveted linings** are used on heavier-duty or high-performance vehicles.

Metal rivets provide a mechanical connection to hold the lining to the backing plate. They are less susceptible to failure under high temperatures. At the same time, the rivets actually pinch some of the lining between the rivet head and the backing plate. This means the linings must be changed sooner than bonded linings. Otherwise, the rivet heads would contact the rotor and wear a groove in the face of the rotor.

The backing plate often has **lugs** that correctly position the pad in the caliper assembly. They help the brake pad maintain the proper position to the rotor (**FIGURE 34-20**). Disc brakes are usually designed so that the thickness of the pads can be checked easily once the wheel has been removed. Most disc brakes also are designed to allow the pads to be replaced with a minimum of disassembly.

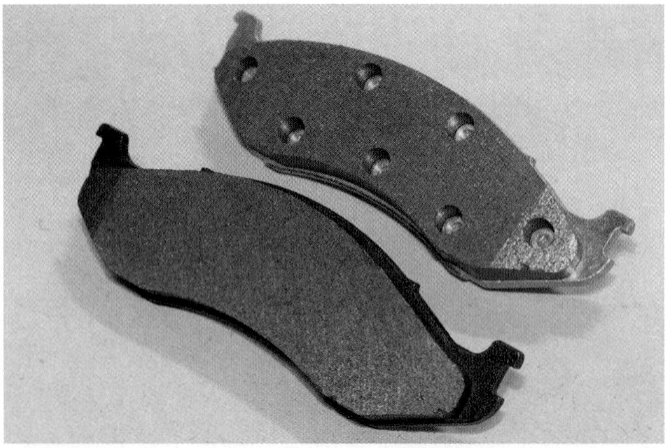

FIGURE 34-19 Bonded and riveted brake pads.

FIGURE 34-20 Brake pad locating lugs.

The composition of the friction material affects brake operation. Materials that provide good braking with low pedal pressures tend to lose efficiency when they get hot. This increases the stopping distance. They also tend to wear out sooner. Materials that maintain a stable friction coefficient over a wide temperature range generally require higher pedal pressures. They also have a tendency to put added wear on the disc brake rotor, reducing its useful life (**FIGURE 34-21**).

Brake Friction Materials

Friction is the force that acts to prevent two surfaces in contact from sliding against each other. The amount of friction between two surfaces is expressed as a factor. It is called the coefficient of friction. When friction occurs, the kinetic energy (motion) of the sliding surfaces is converted into thermal energy (heat). Some combinations of materials, such as a hockey puck on ice, have a very low coefficient of friction. There is very little friction between them and therefore almost no sliding resistance. Rubber tires against a dry, hard road surface have a high coefficient of friction. This means they tend to grip each other well.

Disc brake pads and drum brake linings are made from materials that have a moderate coefficient of friction (**TABLE 34-1**).

FIGURE 34-21 Brake rotor wear.

TABLE 34-1 Brake Lining Coefficient of Friction (Sliding)

Materials Involved	Coefficient of Friction-Dry Sliding
Rubber and concrete	0.6–0.85
Steel and cast iron	0.23
Copper and cast iron	0.29
Brass and cast iron	0.3
Leather and oak	0.52
Brake lining (FF rating)	0.35–0.45

They also must be able to absorb and disperse large amounts of heat without affecting the braking performance. As the heat in brake pads and linings builds up, the coefficient of friction is reduced. This means its stopping power is also reduced. When this happens, it is called brake fade. Minimizing fade is a major factor in the design of brakes and the development of friction materials.

Brake friction materials were historically made from asbestos compounds. Asbestos has excellent heat-resistant qualities. But it has proven to be hazardous. It is generally banned and is not normally used in vehicles. Nowadays, brake linings are manufactured from a variety of different materials, including the following (**FIGURE 34-22**):

- Non-asbestos organic (NAO) materials—organic materials such as Kevlar® and carbon
- Low-metallic NAO materials—small amounts of copper or steel and NOA materials
- Semi-metallic materials—a higher quantity of steel, copper, and/or brass
- Ceramic materials—ceramic fiber materials and possibly a small amount of copper

SAFETY TIP

Don't assume that a brake pad doesn't contain asbestos; it is still found in some applications.

The choice of brake lining compound depends on the application. Lighter passenger vehicles generate less brake heat than heavy or high-performance vehicles. Living in a very hilly region of the country puts added demands on the brake lining. The same goes for locations where there is a lot of stop-and-go traffic. The ideal brake composition for any given vehicle is a combination of weighted qualities, including the following:

- Stopping power
- Heat absorption and dispersion
- Resistance to fade
- Recovery speed from fade
- Wear rate
- Performance when wet
- Operating noise
- Price

For instance, owners of small economy vehicles tend to value longer pad life and minimal operating noise. They are not as concerned about resistance to fade in extreme conditions. Owners of high-performance cars, however, may consider fade resistance and stopping power to be most important. They may not worry so much about noise levels or wear rate.

The Society of Automotive Engineers (SAE) has adopted letter codes to rate the coefficient of friction of brake lining materials. The rating is written on the edge of the friction

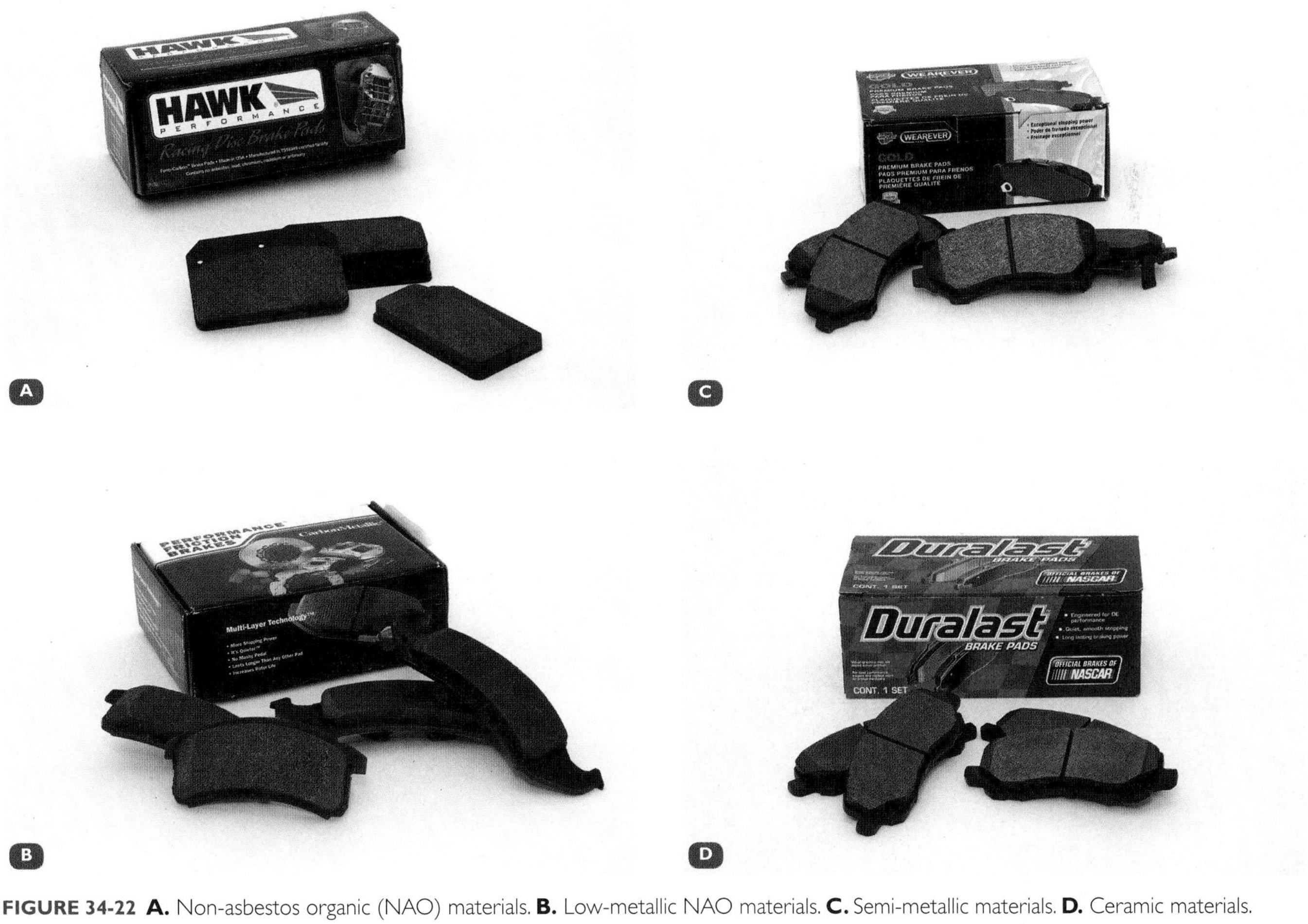

FIGURE 34-22 A. Non-asbestos organic (NAO) materials. **B.** Low-metallic NAO materials. **C.** Semi-metallic materials. **D.** Ceramic materials.

linings and is called the **edge code** (**FIGURE 34-23**). The lower the letter, the less friction the material has. This means that the brake pedal must be applied harder to achieve a given amount of stopping power. These code letters represent the following coefficients of friction:

- C: ≤0.15
- D: 0.15–0.25
- E: 0.25–0.35
- F: 0.35–0.45
- G: 0.045–0.55
- H: >0.55
- Z: Unclassified

The lining is tested both cool and hot. The rating is a two-letter designation such as "FF." The first letter is for the cool performance, and the second letter is for the hot performance. For example, FF has a cool coefficient of friction of 0.35–0.45 and the same coefficient of friction at the hot temperature. It also is very possible that the hot and cold ratings differ from each other. For example, if the rating is FE, the coefficient of friction reduces as the lining heats up.

The coefficient of friction range is wide for each letter designation. Linings with the same letter ratings may not have the same braking performance as each other. An EE-rated lining can have different characteristics from one manufacturer to another. Always use high-quality brake lining from reliable companies to help avoid brake issues.

Antinoise Measures

Disc brakes are more prone to annoying brake squealing than are drum brakes. Brake squealing is caused by vibrations set up between the brake pad and rotor. Manufacturers have addressed this problem in a number of ways:

1. Using softer linings with a higher coefficient of friction. These are less prone to noise than harder linings with a lower coefficient of friction.
2. Adding **brake pad shims and guides** to the brake pads, which help cushion the brake pad and absorb some of the vibration (**FIGURE 34-24**).
3. Using springs to tightly hold the pads in place to minimize vibration (**FIGURE 34-25**).
4. Contouring and grooving the lining material in a way that minimizes vibration. Grooving helps ventilate gases that build up at the surface of the friction material under heavy brake application. It can also help modify the harmonic vibration quality of the friction material to reduce brake squeal. On some pads, the depth of this grove is used as a wear indicator. When it is worn so it can no longer be seen, the pad should be replaced (**FIGURE 34-26**).
5. Incorporating **bendable tangs** on the brake pad backing plate that allows technicians to crimp the tangs so they are more firmly mounted in the caliper (**FIGURE 34-27**).

▶ TECHNICIAN TIP

Some manufacturers claim that refinishing their rotors removes too much material from the rotor. This can cause the brakes to squeal due to less mass. Less mass changes the harmonic vibration qualities of the rotor. They recommend replacing the rotors any time the rotors are worn enough to need resurfacing.

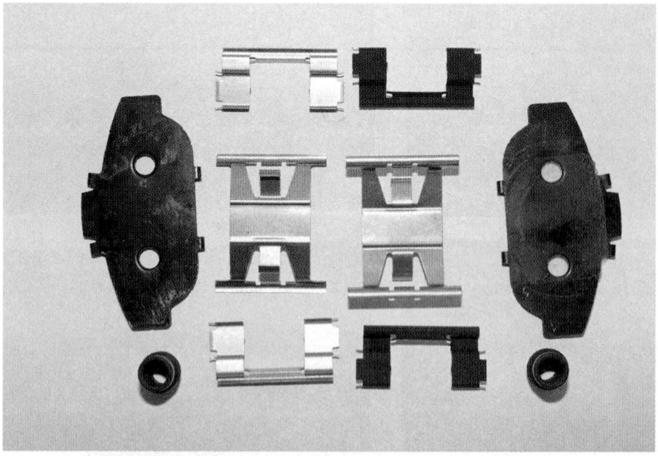

FIGURE 34-24 Brake pad shims and guides.

FIGURE 34-23 Brake lining edge code (FF).

FIGURE 34-25 Example of spring-loaded brake pad retainers.

FIGURE 34-26 Brake lining grooves and contouring.

Technicians also can apply noise-reducing compounds to the brake pads. One is a type of high-temperature liquid rubber compound that is applied to the back of the brake pad. When it cures, it stays flexible, absorbs brake pad vibrations, and helps reduce brake noise. Another compound is a specially designed liquid that is applied directly to the face of the lining material. This modifies the lining's coefficient of friction slightly, making it less likely that the lining will squeal. Make sure you apply the correct compound to the correct side of the brake pad.

Applied Science

AS-1: Sound: The technician can demonstrate an understanding of the role sound plays in identifying various problems in the vehicle.

Sound is a series of waves that travel through a gas, liquid, or solid and can often be heard by the human ear. Sound waves are created by vibrating objects, such as a guitar string. Moving objects that are in sliding contact with each other are highly likely to create sound. One example of this is fingernails dragging on a chalkboard. The fingernails vibrate on the surface of the chalkboard and create sound waves that can then be heard.

Technicians commonly use differences in sound to assist in diagnosing disc brake problems. For example, brake pads that are worn down to the metal backing plate make a deep grinding noise when the brakes are applied. Listening to hear which wheel or wheels the noise is coming from helps identify the source of the problem. In the same way, the brakes will make a high-pitched screeching noise if the brake pads are worn down to the scratcher. Many times this noise happens when the brakes are not being applied. One way to help determine which side the noise is coming from is by driving the vehicle next to a concrete wall or building. If the noise is on that side, it will get much louder when near the wall.

Wear Indicators

Some manufacturers provide a means of warning the driver that the brake pad linings are worn to their minimum limit. This helps ensure that the brake linings do not wear down to the point that they cannot properly perform their job anymore. Excessively thin brake linings tend to heat up more quickly than

FIGURE 34-27 Brake pad bendable tangs.

thicker linings. This can lead to premature brake fade. Not all manufacturers use a brake lining wear indicator. In those cases, it is especially important to inspect the brakes at regular intervals. This is commonly performed during tire rotations or oil changes.

Types

Some manufacturers use a mechanically operated wear indicator. This is achieved by a spring steel **scratcher** mounted to the brake pad (**FIGURE 34-28**). Part of the scratcher extends below the brake pad backing plate at the lining's minimum wear thickness. When the friction material wears down far enough, the scratcher contacts the surface of the rotor. It then makes a squealing noise similar to fingernails on a chalkboard. This distinctive noise means the brakes need service right away. When you replace the brake pads, make sure they come equipped with new scratchers. Also, make sure that they are set to the correct depth so they can function the next time the pads wear down.

▶ TECHNICIAN TIP

In many cases, the scratchers start to make noise when the brakes are not applied, and stop making noise when the brakes are applied. This is because applying the brakes tends to dampen the vibrations.

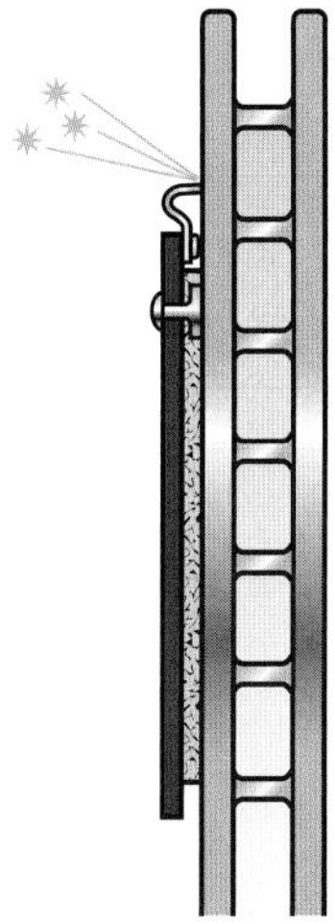

FIGURE 34-28 The scratcher brake wear indicator.

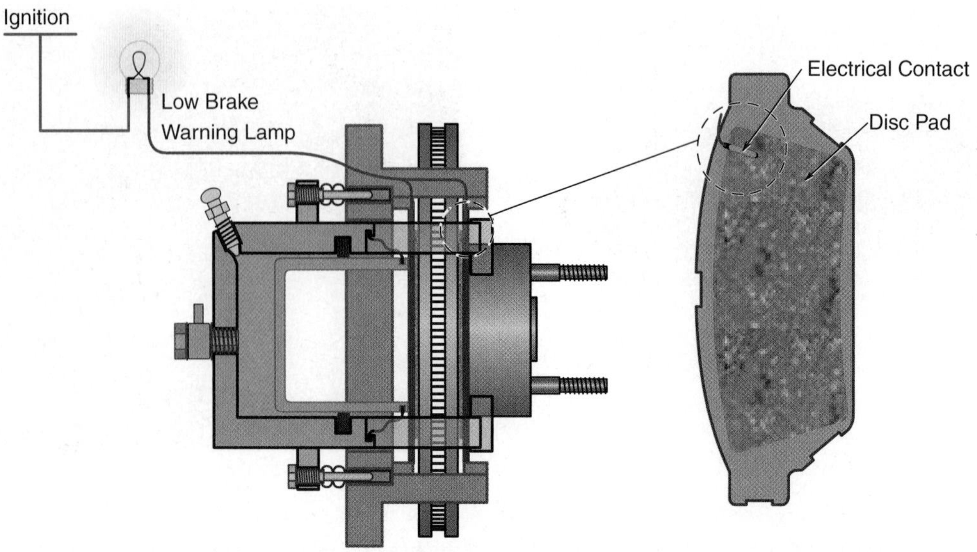

FIGURE 34-29 A brake pad wear indicator system illuminates a warning lamp on the dash when one or more brake pads wear down to a predetermined level.

Some manufacturers use a warning lamp or warning message on the dash to alert the driver that the lining is worn to its minimum thickness. These systems have an electrical contact installed on the brake pad. It is set at the point of the lining's minimum wear thickness. When the pad wears to this minimum thickness, the contact touches the rotor as the brakes are applied. This completes the circuit and activates a warning light or warning message. Again, this means that the disc brake pads are due for replacement (**FIGURE 34-29**). These contacts can be manufactured into the pad, or they can be clipped onto the pad. The contacts are normally replaced when the pads are replaced. Make sure that either the contacts come with the new pads or are ordered along with the pads.

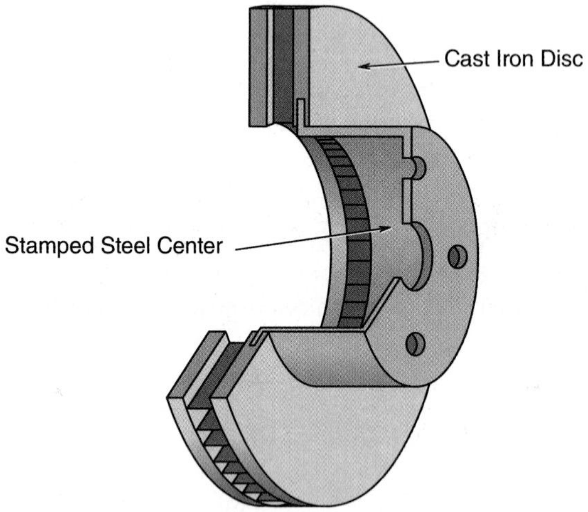

FIGURE 34-30 A composite rotor.

▶ Disc Brake Rotors

LO 34-04 Describe brake rotor construction.

The brake disc is also called a rotor. It is the main rotating component of the disc brake unit. The wheel and rotor rotate together. Because of this, some manufacturers integrate the ABS tone wheel into the rotor. The pads are forced onto the surface of the rotor with potentially thousands of pounds of force. So, the rotor must be strong and have a durable surface. Additionally, friction between the rotor and brake pads generates large amounts of heat. Rotors must be able to withstand the high temperatures generated. Cast iron is commonly used because it is both strong and withstands heat well.

To reduce weight, some manufacturers use a two-part rotor with a cast iron disc and a stamped steel center hat (**FIGURE 34-30**). This style of rotor is called a composite rotor. Some heavy-duty and/or high-performance vehicles have rotors made of reinforced carbon, carbon ceramic, or composite ceramic substances. This is to reduce weight and withstand much higher temperatures.

Because the rotor surfaces are squeezed between the two brake pads, any unevenness of the rotor surfaces causes the brakes to pulsate. This occurs as the thicker and thinner portions pass between the brake pads. The rotor surfaces must be parallel to each other to avoid this situation.

Rotors can fail in two ways: parallelism, which is also called thickness variation, and lateral runout (**FIGURE 34-31**). **Parallelism** is the most critical condition. If the rotor's thickness varies by as little as 0.0003" (0.0076 mm), the rotor tends to push the brake pads outward at any high spots. This tends to create more pressure on the brake pads and slows the vehicle down faster at that point. It also pushes up on the brake pedal as fluid is being forced back to the master cylinder. The result is a pulsation of the brake pedal and a surging of the vehicle while braking. This is usually more noticeable at lower braking speeds.

Lateral runout, also called warpage, is the side-to-side movement of the rotor surfaces as the rotor turns. A warped rotor can be within specifications for parallelism, but out of specification for lateral runout. Lateral runout tends to move the caliper pistons in the same direction as one another. In this situation, brake fluid is not pushed back to the master cylinder.

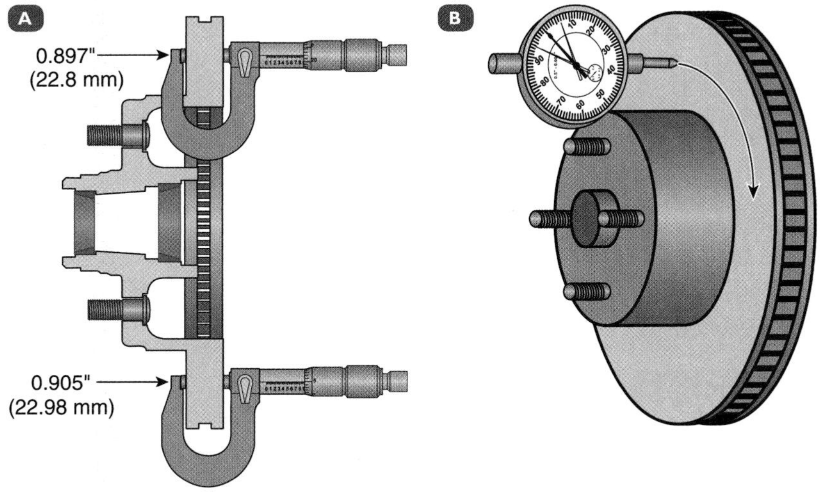

FIGURE 34-31 **A.** Example of disk thickness variation. **B.** Example of lateral runout.

However, the caliper tends to be moved side to side. This movement can cause the steering wheel to shimmy as the warped rotor follows the brake pads. Lateral runout that is greater than about 0.003" (0.076 mm) can cause this.

Also, runout causes the pads to rub on the high spots of the rotor as it turns freely. This causes uneven wear of the rotor. It can also cause pad material to be deposited on the rotor. Both of these conditions can create thickness variation concerns.

▶ TECHNICIAN TIP

Rotors can be warped by improperly torquing the lug nuts. Always use a properly calibrated torque wrench (or the proper torque stick if the shop policy allows) to torque the lug nuts to the manufacturer's specified torque. Also, torque them in the specified torque sequence.

FIGURE 34-32 A typical dust shield.

Disc brakes are equipped with a dust shield to help protect the rotor. Dust shields help keep dust, water, and other road debris away from the inside surface of the rotor (**FIGURE 34-32**). They also help direct air flow to the rotor to assist with heat transfer to the atmosphere. Dust shields are commonly made of plastic, but they can also be made of stamped sheet metal. Dust shields can become damaged during brake repair. So, always inspect them for proper clearance before installing the wheel assembly.

Applied Math

AM-1: Charts/Tables/Graphs: The technician can interpret charts, tables, and graphs to determine the manufacturer's specifications for a given system.

Algebra (9–12)

C3: Draw reasonable conclusions about a situation being modeled.

Representation

A2: Select, apply, and translate among mathematical representations to solve problems.

Technicians regularly apply math concepts to disc brake diagnosis and repair. For example, a vehicle has a brake pulsation problem. The technician measures the amount of rotor thickness variation and runout. Let's say the technician comes up with a thickness variation of 0.0025" (0.064 mm) and a runout of 0.0015" (0.038 mm) on one rotor. Next the technician looks up the manufacturer's specifications and finds that the maximum allowable thickness variation is 0.0005" (0.013 mm). The maximum allowable runout is 0.003" (0.076 mm). The technician determines that the thickness variation is excessive by 0.002" (0.051 mm) (0.0025" [0.064 mm] – 0.0005" [0.013 mm] = 0.002" [0.051 mm]). The runout is okay since it is under the maximum allowable specification. But even so, the rotor must be refinished or replaced to bring it back within specification. This is because the thickness variation is out of specification.

Solid and Ventilated Rotors

Rotors can be solid or ventilated (**FIGURE 34-33**). **Solid rotors** are less expensive and usually found on smaller vehicles. **Ventilated rotors** are used on heavier vehicles or high-performance vehicles. They improve heat transfer to the atmosphere. Passageways between the friction surfaces use centrifugal force to move air through the center of the rotor. Some ventilated rotors are directional. This means they are designed to force air through the rotor when turned in one direction only (**FIGURE 34-34**). If the rotor is rotated in the wrong direction, it will not move air properly and will overheat easier.

▶ TECHNICIAN TIP

If you hear an unusual scraping or grinding noise when you test drive a vehicle after servicing the brakes, check to see if the dust shield is rubbing on one of the rotors. If so, it can make a loud scraping or grinding noise. Because it is thin plastic or sheet metal, chances are good that it is either caught on something or bent. It can usually be put back into place or bent back into shape easily.

Disc brake rotors with holes or slots machined into their surface dissipate heat faster (**FIGURE 34-35**). They also help to remove water quickly from between the pad and rotor in wet driving conditions. The pads wipe across the holes or slots when the brakes are applied. This helps prevent the surface of the pad from becoming glazed. However, this scraping action reduces the overall life of the brake pad. So, these types of rotors are generally used in high-performance or heavy-duty vehicles.

Most disc brake rotors are stamped with the manufacturer's minimum thickness specification (**FIGURE 34-36**). This minimum thickness ensures an adequate amount of thermal mass for stopping power. When material is removed, there is not as much material to absorb heat. This causes the rotor to heat up faster (**FIGURE 34-37**). That means the brakes will experience brake fade sooner. Also, when the brake pads are worn out, and

FIGURE 34-33 Rotors. **A.** Standard solid rotors. **B.** Ventilated rotors.

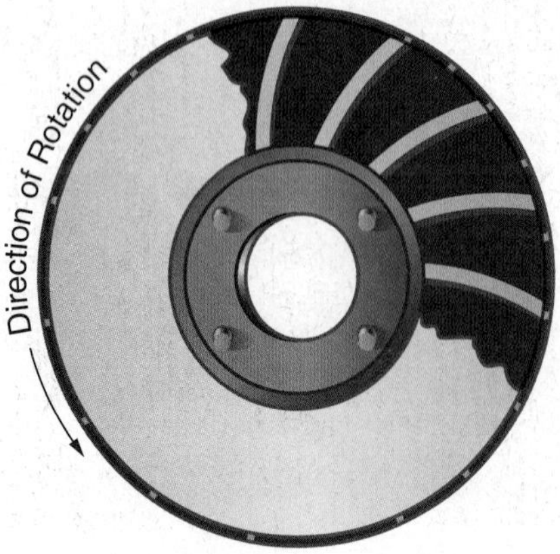

FIGURE 34-34 Directional ventilated rotor.

FIGURE 34-35 Slotted and drilled rotor.

FIGURE 34-36 Rotor with minimum thickness stamped on it.

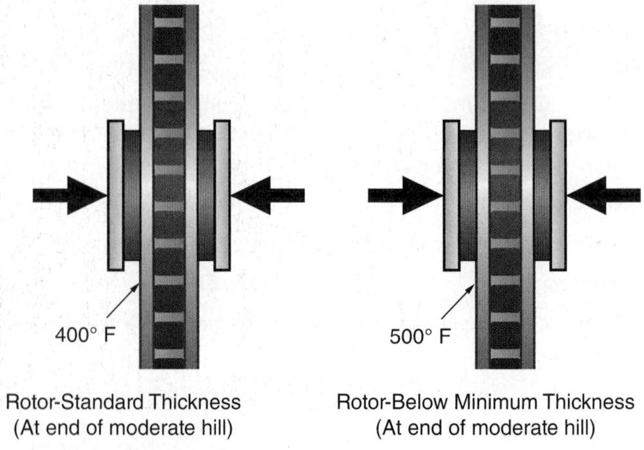

FIGURE 34-37 Rotor thickness and heat capacity.

if the rotor was substantially below the minimum thickness, the piston could be pushed out beyond the sealing ring. This would cause the brakes to lose hydraulic pressure and fail. Always make sure the rotors are above the manufacturer's minimum thickness before putting them back in service.

▶ Parking Brakes on Disc Brakes

LO 34-05 Describe parking brakes on disc brakes.

Parking brakes are designed to hold the vehicle stationary when parked. Manufacturers must meet minimum requirements when designing parking brakes. They must hold the vehicle for a given amount of time on a specified grade in both directions. The parking brake must also be activated separately from the service brakes. And the driver must be able to latch it in the applied position. Parking brakes can be foot operated or hand operated (**FIGURE 34-38**). They typically use a cable and ratcheting lever assembly.

Disc brakes require higher applied forces to operate. This makes them more difficult to use as parking brakes. However, manufacturers have overcome this challenge in a couple of ways.

Types of Parking Brakes

Parking brakes on disc brake units are primarily of two types: an integrated parking brake caliper and the top hat drum style. Alternatively, electric parking brakes are being used on some vehicles. The electric motor can pull on a conventional parking brake cable. On other types, the electric motor can be mounted on the caliper. From there, it directly operates the caliper piston to mechanically apply the brake pads.

Integrated Mechanical Parking Brake Calipers

The integrated parking brake mechanically forces the disc brake piston outward. This forces the brake pads to clamp the rotor when the parking brake is applied (**FIGURE 34-39**). A lever on the back side of the caliper is pulled by the parking brake cable. The lever turns a shaft that enters the rear of the caliper cylinder. The shaft uses a seal to prevent fluid leakage from the bore. The shaft has a coarse thread machined into it, which threads into a nut assembly inside the caliper piston. As the shaft is turned by the parking brake lever, the nut causes the piston to be forced outward. This applies the brakes. Releasing the parking brake cable allows a spring to unwind the shaft. This releases the pressure on the brake pads.

Top Hat Design Parking Brake

The top hat design gets its name from the shape of the rotor. The rotor has a deeper offset than normal, giving the appearance of a top hat. The offset portion allows room for a drum surface within the center of the rotor (**FIGURE 34-40**). Drum brake shoes

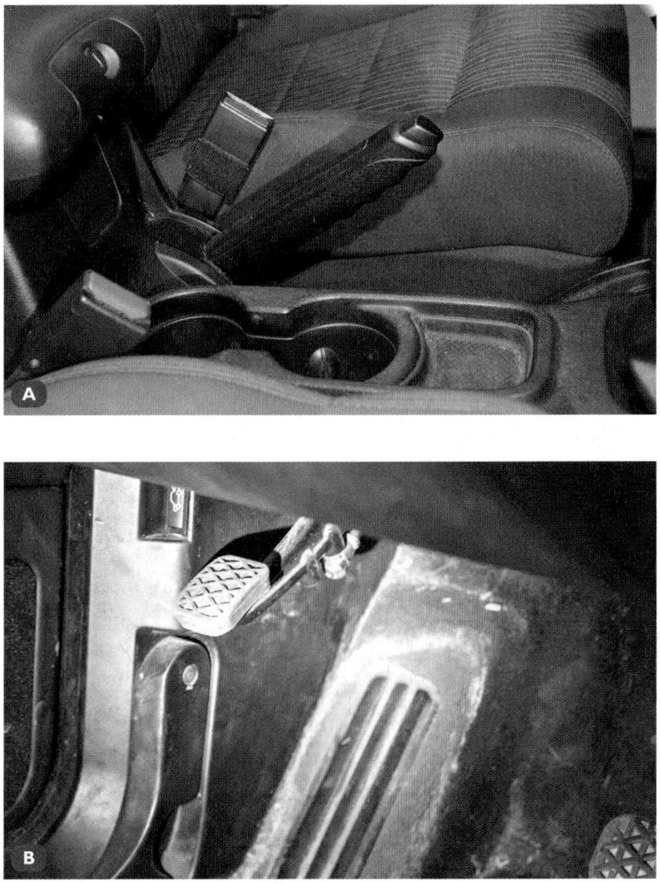

FIGURE 34-38 **A.** Hand-operated parking brake. **B.** Foot-operated parking brake.

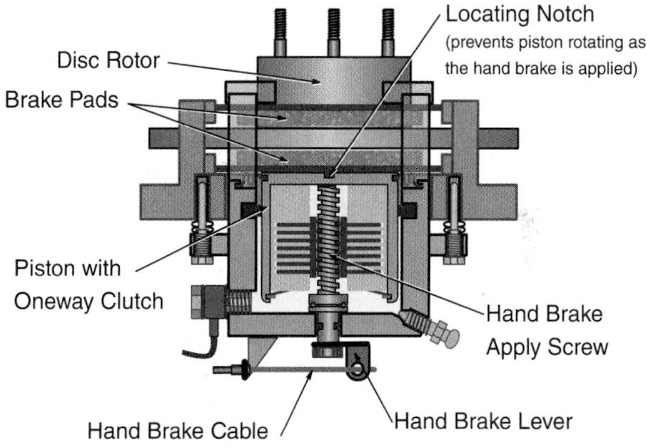

FIGURE 34-39 Integrated parking brake operation.

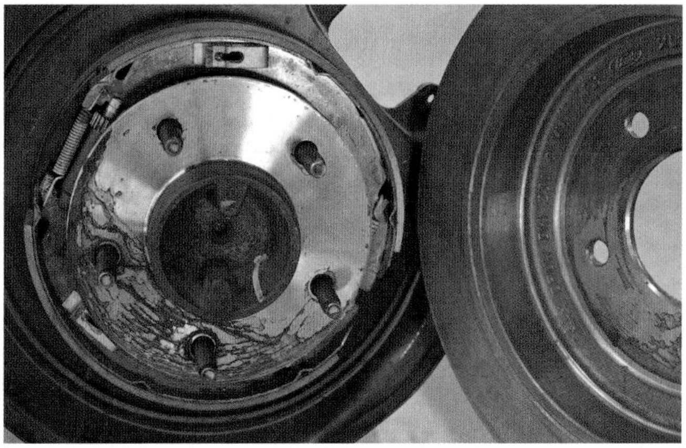

FIGURE 34-40 A top hat rotor and drum parking brake assembly.

are mechanically forced outward into contact with the inside of the brake drum, which locks the wheel. Releasing the parking brake allows the springs to retract the brake shoes from contact with the drum.

Electric Parking Brake

The electric parking brake uses an electric motor to apply the disc brake assemblies. The cable style uses an electric motor to pull standard parking brake cables. This operates standard integrated mechanical parking brake calipers.

The electrically integrated caliper style uses an electric motor mounted on the caliper. It directly applies the brakes. Some of these systems use a wedge and ramp to apply the piston (**FIGURE 34-41**). The piston then applies the brakes.

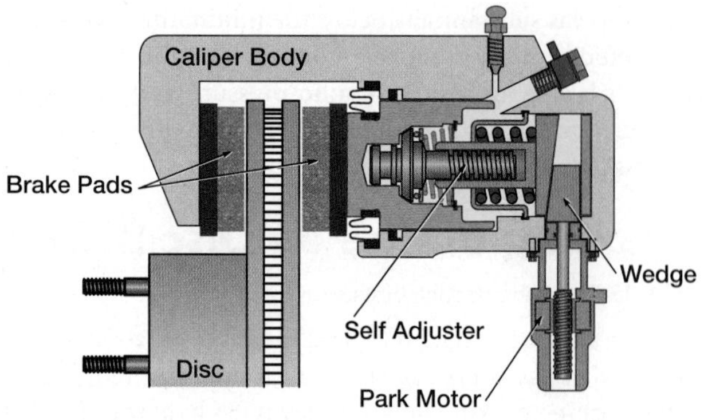

FIGURE 34-41 Electrically integrated parking brake caliper has a built-in motor that mechanically applies the brake pads.

Pushing a parking brake button on the dash causes the motor to either tension the cable or directly apply the parking brake. Electric parking brakes can also be integrated with the CANbus system. This system works with the parking brake to provide additional features beyond just holding the vehicle when it is parked. It can be used as a hill assist feature. In this case, it automatically holds the vehicle while it is stopped on a hill. This prevents it from rolling backward or forward. It can then be automatically released by the PCM when starting to move away from the stop.

Electric parking brakes may also work with the vehicle's proximity detector when backing up. If the system detects the vehicle getting too close to an object, the ECM can apply the electric parking brake. This can stop the vehicle to prevent it from striking the object.

▶ Wrap-Up

Ready for Review

- ▶ The design of disc brakes enables higher applied force and easy heat transfer to atmosphere. The primary components of the disc brakes are rotor, calipers, and brake pads.
- ▶ The disc brake caliper assembly is bolted to the axle housing, or steering knuckle and clamps the brake pads onto the rotors to slow the vehicle. Floating and sliding calipers have one or more pistons, but are located only on one side of the caliper housing.
- ▶ Disc brake pads consist of friction material bonded or riveted onto a steel backing plate. The backing plate has lugs that correctly position the pad in the caliper assembly and help the brake pad maintain the proper position to the rotor.
  - Brake linings materials include: non-asbestos organic, low-metallic NAO materials, semi-metallic materials, and ceramic materials.
- ▶ Brake rotor is the main rotating component of the disc brake unit and must be able to withstand the high temperatures generated. Rotors can be solid or ventilated.
  - Solid rotors are less expensive and usually found on smaller vehicles. Ventilated rotors are used on heavier vehicles or high-performance vehicles to improve heat transfer to the atmosphere.
- ▶ Parking brakes hold the vehicle for a given amount of time on a specified grade in both directions. They can be foot operated or hand operated. Parking brakes on disc brake units are primarily of two types: an integrated parking brake caliper and the top hat drum style.
  - An electric parking brake uses an electric motor to apply the disc brake assemblies.

Key Terms

backing plate A metal plate to which the brake lining is fixed.

bendable tangs Small tabs on the brake pad backing plate that are crimped on to the caliper, creating a secure fit and reducing noise.

bleeder screw A screw that allows air and brake fluid to be bled out of a hydraulic brake system when it is loosened, and seals the brake fluid in when it is tightened.

bonded linings Brake linings that are essentially glued to the brake pad backing plate; they are more common on light-duty vehicles.

brake booster A vacuum- or hydraulically operated device that increases the driver's braking effort.

brake pad shims and guides Small pieces of metal that cushion the brake pad and absorb some of the vibration, helping to cut down on unwanted noise.

caliper A hydraulic device that uses pressure from the master cylinder to apply the brake pads against the rotor.

disc brake pads Brake pads that consist of a friction material bonded or riveted to a steel backing plate; designed to wear out over time.

edge code A two-digit code printed on the edge of a friction lining that describes its coefficient of friction.

fixed caliper A type of brake caliper bolted firmly to the steering knuckle or axle housing, having at least one piston on each side of the rotor.

guide pins Pins that allow the caliper to move in and out as the brakes operate and as the brake pads wear.

independent rear suspension (IRS) A type of suspension system where each rear wheel is capable of moving independently of the other.

lateral runout The side-to-side movement of the rotor surfaces as the rotor turns; also called warpage.

low-drag caliper A caliper designed to maintain a larger brake pad–to-rotor clearance by retracting the pistons farther than normal.

lug A flange that is shaped to assist with aligning objects on other objects.

parallelism Both surfaces of the rotor should be perfectly parallel to each other so that brake pulsations do not occur. Also called thickness variation.

phenolic resin A very dense material used to create some brake pistons that is very resistant to corrosion and heat transfer.

pushrod A tubular rod that stands between the tappet and the rocker arm in an overhead valve engine; the pushrod transfers cam motion to the rocker arm. In the braking system, a mechanism used to transmit force from the brake pedal to the master cylinder.

riveted linings Brake linings riveted to the brake pad backing plate with metal rivets and used on heavier-duty or high-performance vehicles.

rotors The main rotating part of a disc brak system.

scratcher A thin, spring steel wear indicator that is fixed to the backing plate of the brake pad; it emits a high-pitched squeal when the brakes are applied if the brake pads have become too thin.

sliding or floating calipers A type of brake caliper that only has pistons on the inboard side of the rotor. The caliper is free to slide or float, thus pulling the outboard brake pad into the rotor when braking force is applied.

solid rotor A type of brake rotor made of solid metal, not ventilated.

square-cut O-ring An O-ring with a square cross section that is used to seal the pistons in disc brake calipers.

ventilated rotor A type of brake rotor with passages between the rotor surfaces that are used to improve heat transfer to the atmosphere.

Review Questions

1. Which of the following is a component in the disc brake system?
   a. Brake rotor
   b. Brake drum
   c. Self adjuster
   d. Wheel cylinder
2. How often are disc brakes adjusted by a technician?
   a. Once per year
   b. At every oil change
   c. Every 30,000 miles
   d. Never
3. All of the following are types of a disc brake caliper EXCEPT?
   a. Fixed
   b. Floating
   c. Master
   d. Low drag
4. How is a brake caliper piston sealed?
   a. A square cut O-ring
   b. A paper gasket
   c. A liquid RTV
   d. A brass ring
5. Some disc brake pads have their linings glued or bonded to them. Others are ______.
   a. Welded
   b. Riveted
   c. Magnetized
   d. Press fit
6. Which of the following is something that engineers consider when they design brake pads for a vehicle?
   a. Wheel cylinder size
   b. Piston pushback pressure
   c. Resistance to fade
   d. Recovery from boil
7. Why would a manufacturer use carbon ceramic brake rotors rather than cast iron brake rotors?
   a. Because it is less expensive to manufacture
   b. Because it is better for the environment
   c. Because it is lighter and withstands heat better
   d. Because it never wears out
8. What component is responsible for directing air toward the rotor and protecting from road debris?
   a. The dust shield
   b. The cooling fan
   c. The rotor hub
   d. The rotor slots
9. A vehicle equipped with rear disc brakes and a pedal-actuated rear parking brake is likely to be operated by a cable and ____.
   a. Cylinder
   b. Ratchet
   c. Spoke lock
   d. Parking pawl
10. What are parking brakes that use brake shoes inside of the rotor often called?
    a. Mini drum
    b. Lock-in barrel
    c. Top hat
    d. Combo shoes

ASE Technician A/Technician B Style Questions

1. A brake system is being discussed. Technician A states that disc brakes require more clamping pressure than drum brakes. Technician B states that disc brakes can transfer heat faster than drum brakes. Who is correct?
   a. Technician A only
   b. Technician B only

c. Both Technicians A and B
d. Neither Technician A nor B

2. A disc brake system being discussed. Technician A states that calipers straddle the rotor and house the brake pads. Technician B states that brake calipers expand the brake pads into the center of the rotor. Who is correct?
a. Technician A only
b. Technician B only
c. Both Technicians A and B
d. Neither Technician A nor B

3. Disc brake calipers are being discussed. Technician A states that the caliper bleeder is normally mounted at the bottom of the caliper. Technician B states that floating calipers are more precise than fixed since they are firmly bolted in place. Who is correct?
a. Technician A only
b. Technician B only
c. Both Technicians A and B
d. Neither Technician A nor B

4. A brake caliper is being discussed. Technician A states that floating caliper guide pins and bushings must be lubricated with high-temperature grease when serviced. Technician B states that the caliper piston O-ring retracts the piston. Who is correct?
a. Technician A only
b. Technician B only
c. Both Technicians A and B
d. Neither Technician A nor B

5. Brake pads are being discussed. Technician A states that brake pads should have the lowest coefficient of friction possible. Technician B states that brake pads are currently made of asbestos. Who is correct?
a. Technician A only
b. Technician B only
c. Both Technicians A and B
d. Neither Technician A nor B

6. A brake warning lamp is being discussed. Technician A states that the light is most likely caused by low brake fluid or a parking brake left on. Technician B states that some manufacturers have electrical contacts installed on the brake pad that will turn on this light when the brake pads are worn too low. Who is correct?
a. Technician A only
b. Technician B only
c. Both Technicians A and B
d. Neither Technician A nor B

7. Brake rotors are being discussed. Technician A states that brake rotors that are constructed of two materials are known as composite rotors. Technician B states that cast iron is commonly used in brake rotors because it is strong, and withstands heat well. Who is correct?
a. Technician A only
b. Technician B only
c. Both Technicians A and B
d. Neither Technician A nor B

8. A high-performance vehicle comes in to the shop. Technician A states that it is likely to have standard solid brake rotors. Technician B states that the rotors may have holes or slots for heat dissipation. Who is correct?
a. Technician A only
b. Technician B only
c. Both Technicians A and B
d. Neither Technician A nor B

9. A vehicle with disc brakes being discussed. Technician A states that the brake rotors often have a minimum thickness stamped on the rotor. Technician B states that if the brake rotors are too thick, they will not be able to withstand high levels of heat and could warp. Who is correct?
a. Technician A only
b. Technician B only
c. Both Technicians A and B
d. Neither Technician A nor B

10. A vehicle with an electronic parking brake is being discussed. Technician A states that the electronic parking brake may use an electric motor that pulls on the parking brake cables. Technician B states that the electronic parking brake may be integrated with the CANbus system which could allow it to be used by other systems. Who is correct?
a. Technician A only
b. Technician B only
c. Both Technicians A and B
d. Neither Technician A nor B

CHAPTER 35
Servicing Disc Brakes

Learning Objectives

- **LO 35-01** Describe disc brake diagnosis.
- **LO 35-02** Perform caliper and brake pad service.
- **LO 35-03** Measure and replace disc brake rotors.
- **LO 35-04** Refinish disc brake rotors.
- **LO 35-05** Inspect, replace, and torque lug nuts and studs.

ASE Education Foundation Tasks

See Appendix A to View the 2017 ASE Education Foundation Automobile Accreditation Task List Correlation Guide.

Servicing Disc Brakes

▶ Servicing Disc Brakes

LO 35-01 Describe disc brake diagnosis.

Disc brake diagnosis starts with understanding the customer's concern. Communicating directly with the customer is the best way to do that, but the customer is not always available. An experienced service advisor will gather the required information. So, read the service advisor's notes on the repair order carefully or speak with this person directly.

Once you understand the customer's concern, a test drive is usually needed to verify the accuracy of the concern. Depending on the concern, it may be as simple as stepping on the brake pedal and feeling the pedal sink to the floor. Or it could require a more detailed test drive to observe the fault the customer is describing. This is a good opportunity to test the brakes under a variety of conditions. Replicating the customer concern is important. That way you know you are addressing the situation that the customer is experiencing.

During the test drive, find a safe place to operate the brakes at a variety of speeds with a variety of brake pedal pressures. Especially, try to mimic the conditions the customer described. It may be necessary to go on a test drive with the customer driving. Allow them to operate the vehicle in the way that makes the problem evident. That way, they can point out the particular situation they are experiencing. Also, it is good to have the customer along in case the problem does not occur. In this case, they won't think you don't believe them, or that you are ignoring the issue.

If there is a concern related to the antilock brake system (ABS) that requires a test drive, extreme caution is required. Make sure that an accident doesn't happen. Because ABS operates only during extreme braking or poor traction conditions, you run the risk of being rear-ended. You also risk losing control of the vehicle if you apply the brakes hard enough to activate the ABS. So, make sure the vehicle is being tested away from all other traffic. Remember that if the yellow ABS warning lamp is illuminated, the ABS system is deactivated. In this case, do not try to activate ABS on a test drive.

If the yellow ABS lamp is off, the ABS system should be active and ready to activate. It is helpful when testing ABS to do so on a surface with limited traction, such as wet pavement or a dirt road. When you do apply the brakes firmly, ABS activation can typically be felt as pulsations in the brake pedal and possibly the steering wheel. The vehicle should brake quickly while maintaining steering control. In some cases, a "poor traction" lamp may illuminate, telling you that the ABS system had to activate. This warning lamp will usually turn off after several seconds. If the yellow ABS warning lamp is illuminated, it typically means that the vehicle's ECM has observed a fault in the ABS system. Be sure to check it for diagnostic trouble codes (DTCs).

Once the customer concern has been verified, research the concern in the service information and TSBs. Armed with this information, you should be ready to create a focused testing plan to begin testing the brake system. This could be as simple as:

- performing a visual inspection of the brake fluid level and condition,
- removing the wheels to disassemble and inspect the brake units, or
- measuring the thickness variation of the rotors.

It could also involve electrical diagnosis such as locating a short circuit in the brake pad warning indicator system.

Suspension and steering system faults can appear to be brake system faults. An example of this is a pulling condition while braking. If the strut rod bushings on the suspension are worn, then braking the vehicle will cause the wheel to move rearward. At the same time, that will cause the steering angle to change. The wheel will now point in a direction other than straight down the road. This imitates a brake pull. So, always inspect the suspension and steering systems when diagnosing concerns related to a brake "pull."

The braking system on a vehicle must be restored to its proper operation if one or more braking system faults are present. Diagnosis of any problem must identify all issues that would prevent the brakes from operating normally. Lawyers and technicians have been known to say, "He who touched the brake system last, owns it!" What this means is that if there is a problem in the braking system and you inspected it or worked on it, you are very likely liable for anything that went wrong with it. Brake

You Are the Automotive Technician

You are working on an 8-year-old vehicle with front disc brakes. The customer's concern is that the vehicle makes a loud noise, but only when braking. It has been getting worse over the last couple weeks. You test drive the vehicle and find that there is a heavy grinding noise coming from the right front wheel when braking. It also pulls to the right when braking. You inspect the vehicle and find that the brake fluid is a bit below the MIN line on the reservoir. The brake fluid is dark. The right front brake rotor is heavily scored on the inside. The inboard brake pad is metal-to-metal. The outboard pad has about 50% lining remaining.

1. What is causing the grinding noise?
2. What are the two possible options to correct the problem with the rotor?
3. What are the possible causes of one pad being down to metal and the other one at 50%?
4. What would you recommend to the customer regarding the calipers, and why?

system failures are more likely to lead to vehicle accidents than failures of most systems. This means that any diagnosis and subsequent repairs need to be thorough and complete.

Once you have identified the cause of the fault, determine the action that will correct the fault. This information can then be used for an estimate of repairs. It can then be given to the customer for authorization to perform the needed work. Once the repair is complete, retest the system to confirm that the concern has been fully addressed and no further issues are present.

▶ TECHNICIAN TIP

One way to identify whether a problem is coming from the front or rear brakes is through using the parking brake. Test drive the vehicle in a safe place at a relatively low speed and lightly apply the parking brake. If the condition is still present, the problem is with the rear brakes. You know this because the parking brakes are usually on the rear wheels. If the condition is not present, the problem is likely with the front brakes. This is not true on vehicles equipped with top-hat style parking brakes. This is because the parking brake is separate from the service brake.

Common Disc Brake Concerns

Braking concerns are generally easier to diagnose than most of the other systems on the vehicle. But there are a large variety of possible conditions. This means that you need a good understanding of disc brake and hydraulic theory. It also helps to use all of your senses to assist you in identifying the location of the fault. As a reminder, **TABLE 35-1** lists some of the common faults to consider for each concern.

TABLE 35-1 Causes of Common Disc Brake Concerns

Concern	Fault
Poor stopping	• Power booster not operating properly
	• Internal master cylinder leak; air in the hydraulic system
	• Metering valve or proportioning valve blocking fluid flow
	• Improperly adjusted drum brakes
	• Improper friction lining material
	• Contaminated linings
Noise	Friction lining material too hard
	• Wear indicator touching rotor (worn pads)
	• Lining worn down to metal
	• Worn caliper slides/guide pins
	• Component-specific noises
Vibration	• Improper friction lining material
	• Rotor surface finish not correct
	• Foreign object (mud, rocks, etc.) in rotor
	• Warped rotor/thickness variation excessive
	• ABS operating
Pulling	• Plugged or restricted brake hose
	• Stuck caliper piston
	• Seized caliper guide pins
	• Contaminated lining
	• Lining worn down to metal
	• Air in the hydraulic system
Grabbing	• Contaminated lining
	• Stuck caliper piston
	• Internal master cylinder leak
	• Misadjusted drum brakes
Dragging	• Stuck caliper piston
	• Seized caliper guide pins
	• Misadjusted master cylinder pushrod length
	• Binding brake pedal
	• Plugged restricted brake line or hose
Pulsation	• Warped rotors
	• Rotor parallelism
	• ABS operation

▶ Maintain and Repair Disc Brakes

LO 35-02 Perform caliper and brake pad service.

Disc brakes require periodic maintenance and repair as the vehicle ages. As stated previously, visual inspection of the braking system is part of the diagnostic procedure. It is also a critical part of maintenance and repair. A good visual inspection can turn up issues such as:

- lining that isn't wearing evenly,
- contaminated linings from a leaking brake caliper,
- a warped rotor, or
- even a binding caliper or caliper piston.

The following skill drills will lead you through the common disc brake maintenance and repair procedures.

Tools

The tools that are used to diagnose and repair brake systems include those shown in (**FIGURE 35-1**):

- **Brake lining thickness gauges**—Used to measure the thickness of the brake lining.
- **Brake wash station**—Used to clean drum and disc brake dust.
- **Caliper piston pliers**—Used to grip caliper pistons when removing them.
- **Disc brake rotor micrometer**—Used to measure the thickness and parallelism of a rotor.
- **Dial indicator**—Used to measure the lateral runout (side to side) of the rotor.
- **Parking brake cable pliers**—Used to install parking brake cables.
- **Caliper piston retracting tool**—Used to retract caliper pistons with integrated parking brakes.
- **C-clamp**—Used to push pistons back into the caliper bore on nonintegrated parking brakes.

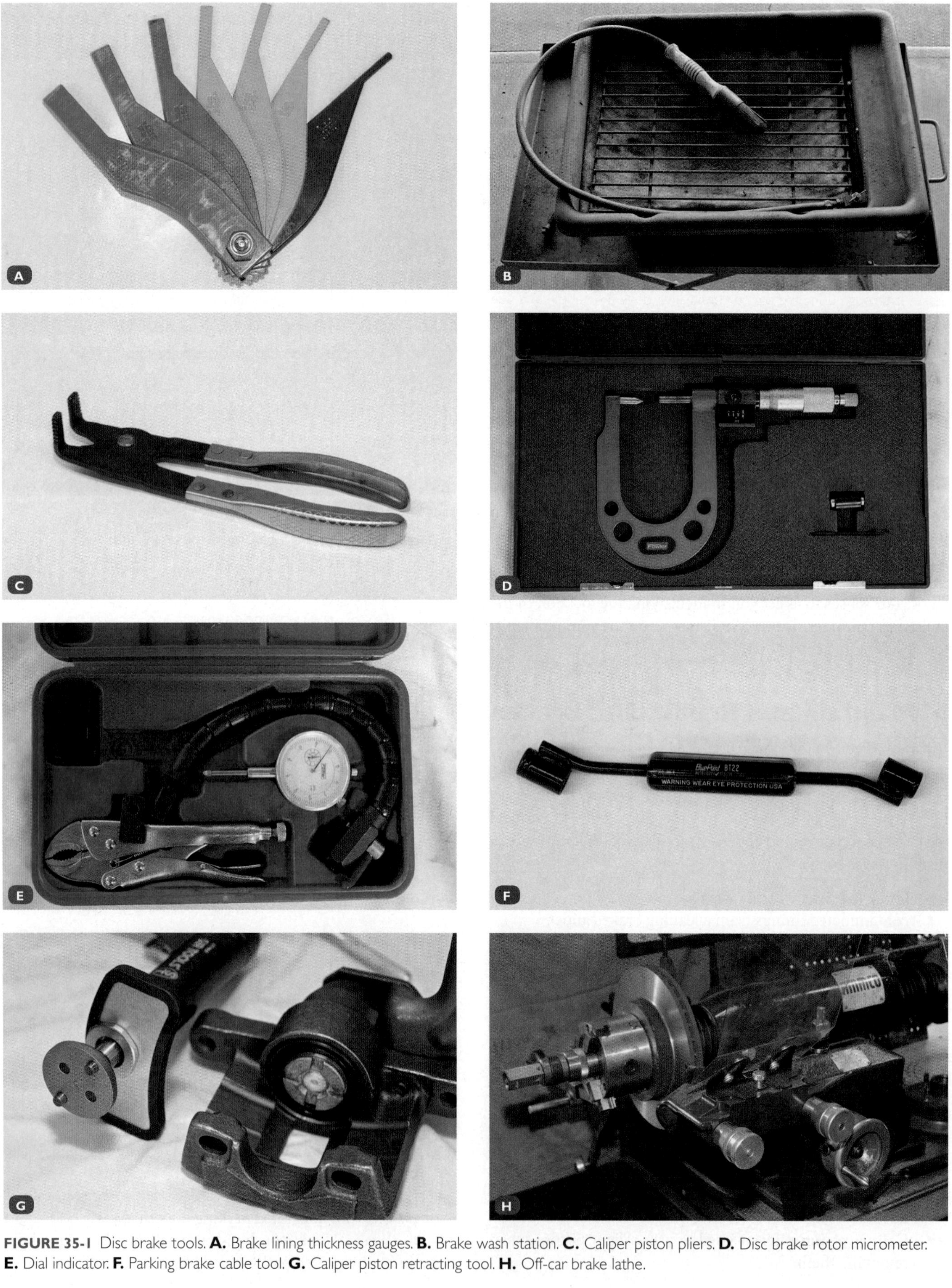

FIGURE 35-1 Disc brake tools. **A.** Brake lining thickness gauges. **B.** Brake wash station. **C.** Caliper piston pliers. **D.** Disc brake rotor micrometer. **E.** Dial indicator. **F.** Parking brake cable tool. **G.** Caliper piston retracting tool. **H.** Off-car brake lathe.

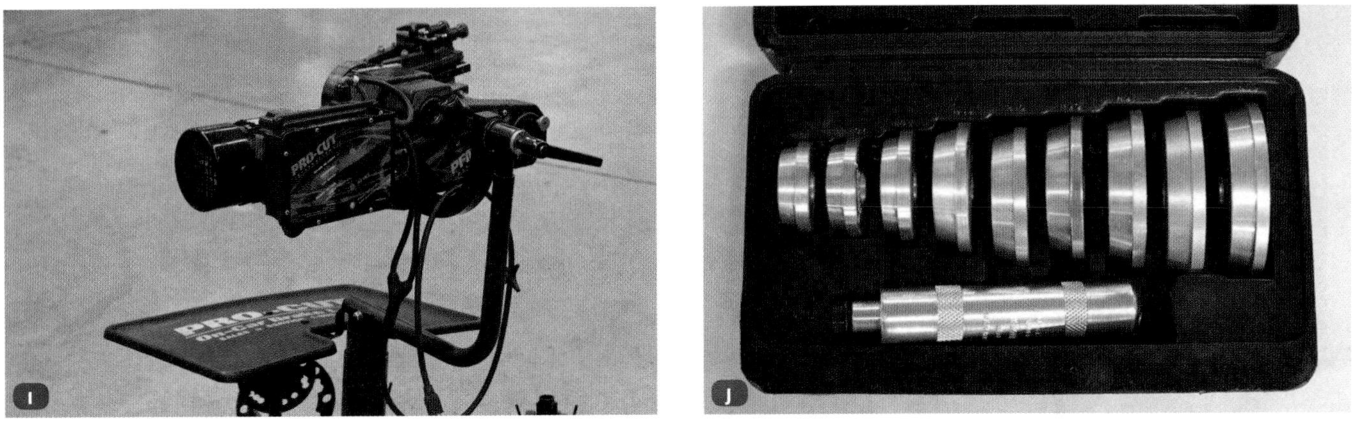

FIGURE 35-1 Disc brake tools. **I.** On-car brake lathe. **J.** Dust boot seal/bushing driver set.

- **Off-car brake lathe**—Used to machine drums and rotors that are off the vehicle.
- **On-car brake lathe**—Used to machine rotors that are on the vehicle.
- **Caliper dust boot seal driver set**—Consists of a driver and a variety of adapters used to install various sizes of dust boot seals.

Removing and Inspecting Calipers

Removing the caliper is necessary for replacing the brake pads on most vehicles. It also allows access for:

- machining of the rotors on the vehicle,
- removal of the rotors for replacement,
- machining of the rotors off the vehicle, and
- a thorough inspection of the caliper, pads, and rotor.

If the caliper is likely to be reinstalled on the vehicle, it is good practice to loosen the bleeder screws slightly and then retighten them. Doing so ensures that they are not seized in place. Failure to do this now could waste a lot of time later trying to repair a broken off bleeder screw. It is also good practice to flush the old brake fluid from the system at this time. That way, old brake fluid will not have to be bled through the new or newly rebuilt caliper. Also, leave the master cylinder reservoir level low. This will prevent overflowing the reservoir when pushing the caliper pistons into their bores.

Once the caliper is loose from its mount, push the caliper pistons back into their bores slightly. A small pry bar or C-clamp works well. That will usually force the piston back just far enough so that the caliper can be removed from the pads and rotor. To remove the caliper assembly, inspect for leaks and damage to the caliper housing, and determine any necessary actions, follow the steps in **SKILL DRILL 35-1**.

> **▶ TECHNICIAN TIP**
>
> Some technicians pinch off the flexible brake hoses with vise-grip pliers. This should be avoided because it crimps the hose, potentially damaging it internally and/or externally. It is a better idea to pull the brake fuse and use the brake pedal holding tool. That will block off the master cylinder compensating ports. Brake fluid is then prevented from leaking out of the system.

SKILL DRILL 35-1 Removing and Inspecting Calipers

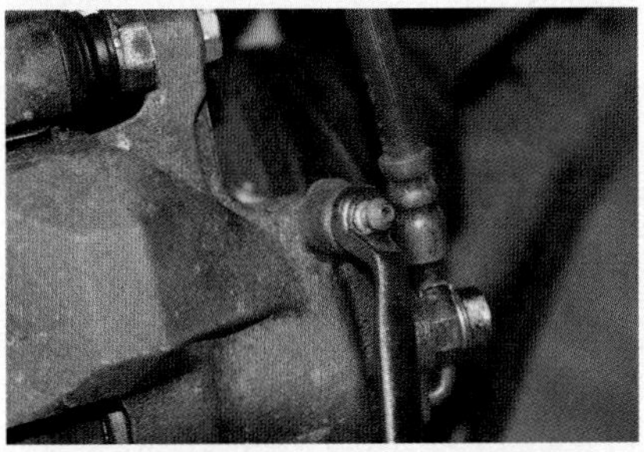

1. Research the procedure for removing the caliper in the appropriate service information. Loosen the bleeder screws slightly, and then retighten them.

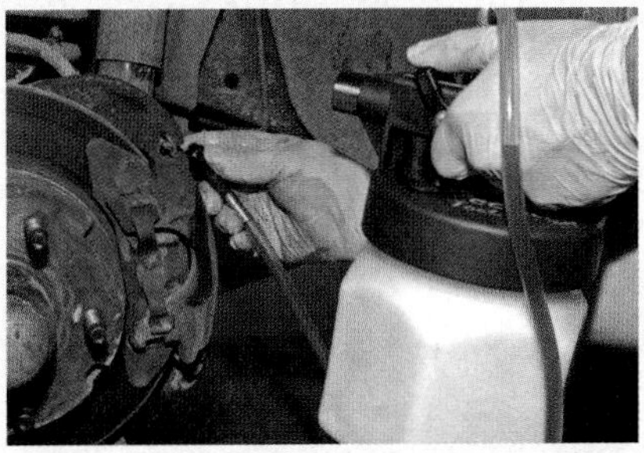

2. If the caliper is being rebuilt or a new caliper will be installed, it is good practice to flush the old brake fluid from the system at this time.

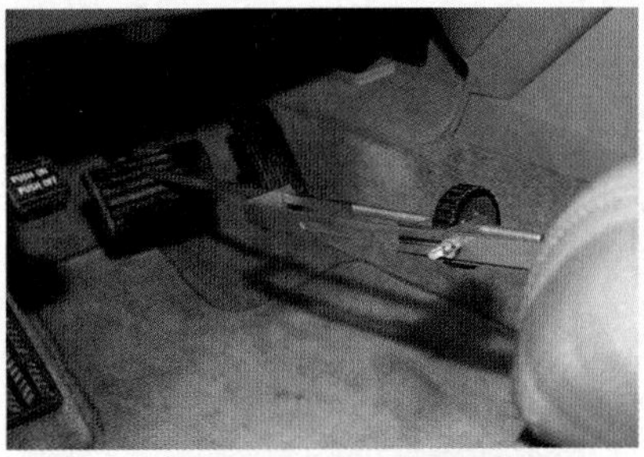

3. Use a brake pedal holding tool to slightly apply the brakes. This blocks off the compensating ports in the master cylinder to avoid excess fluid leakage.

SKILL DRILL 35-1 Removing and Inspecting Calipers (Continued)

4. Remove the brake line or hose from the caliper. Be careful not to lose any sealing rings.

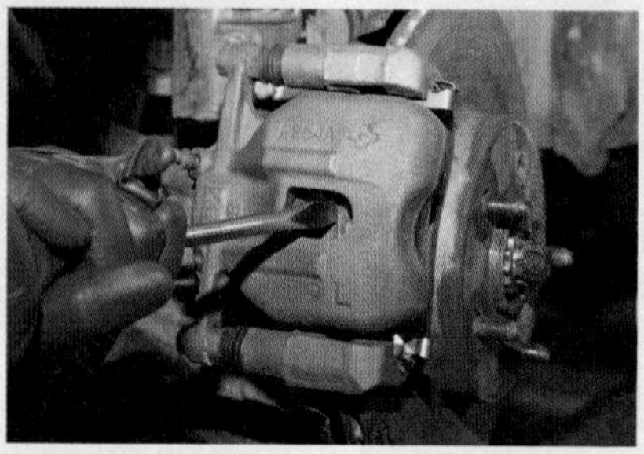

5. Push the caliper pistons back into their bores slightly. Although many technicians use a screwdriver, as shown, a pry bar or C-clamp is a safer choice.

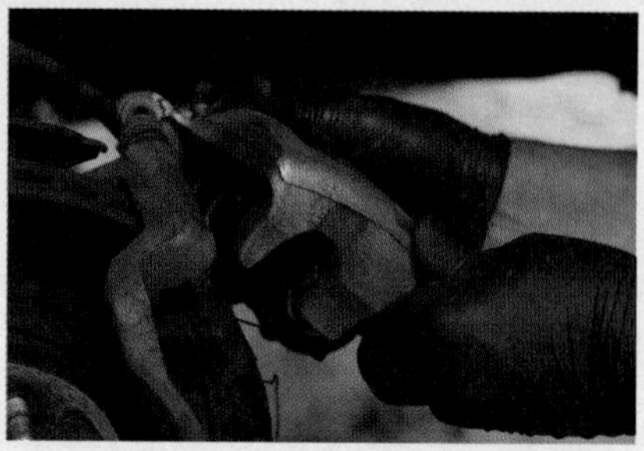

6. Remove the caliper assembly from its mountings.

7. Inspect the caliper, including the piston dust boot for leaks or damage. Determine any necessary actions.

Inspecting Caliper Mountings, Slides, and Pins

The caliper mountings and slides/pins are placed under heavy loads and forces. They also operate in harsh environments. Therefore, it is common that they experience wear over time. They can also corrode and bind up. Clean caliper mountings and slides/pins thoroughly, and inspect them closely for excessive wear.

To clean and inspect caliper mountings and slides/pins for operation, wear, and damage, and to determine any necessary actions, follow the steps in **SKILL DRILL 35-2**.

Inspecting Brake Pad Wear Indicators

The wear indicator system could be a scratcher type or a sensor type. It is also very possible that the vehicle does not incorporate any type of wear indicator system. Checking this system usually

SKILL DRILL 35-2 Inspecting Caliper Mountings, Slides, and Pins

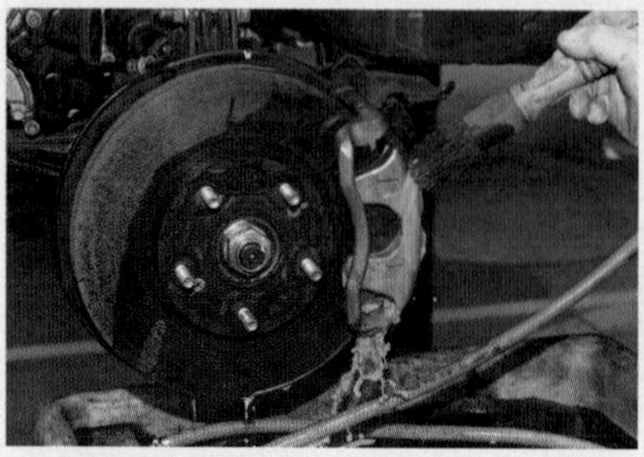

1. Clean the caliper mountings and slides/pins, using equipment/procedures for dealing with asbestos/hazardous dust.

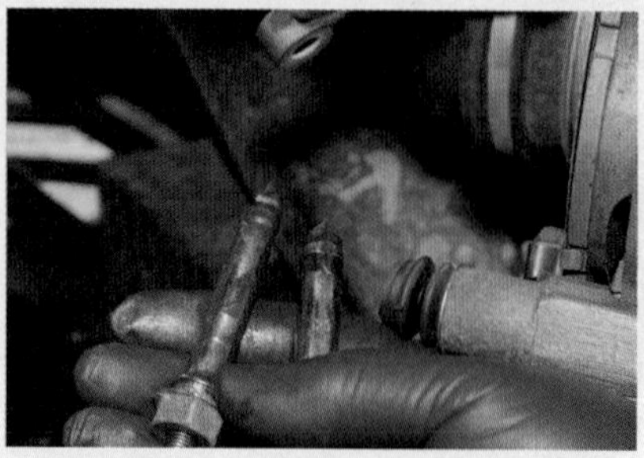

2. Once the dust is taken care of, you may need to use a brake-cleaning solvent to clean the components further.

3. Inspect the caliper mountings and slides/pins for wear and damage. Determine any necessary actions.

consists of verifying if the scratcher or sensor is contacting the brake rotor. If it is in contact, then the brake pads will need to be replaced. If not in contact, then the brake pad thickness will be measured. If inspection was due to a noise issue, then further inspection is necessary.

To test a brake pad warning system, the wiring harness may need to be tested. Some manufacturers require that the harness side of the brake pad sensor connector be grounded or jumped with a test lead. Then, verify that the brake pad warning lamp or warning message comes on. If not, then there is an open in the circuit that will need to be located.

To check the operation of the brake pad wear indicator system and determine any necessary actions, follow the steps in **SKILL DRILL 35-3**.

Checking Brake Pads

Brake pads are inspected during routine maintenance inspections. A simple visual inspection along with a measurement of the lining thickness is usually adequate. But if there are customer concerns with the brakes, it may be necessary to remove the pads for a more detailed inspection. Removing the brake pads is the most thorough way to inspect them. It will allow you to inspect not only the thickness of the lining but also the condition of the friction surface. This could show cracks in the lining or other defects that wouldn't otherwise be seen from the outside. It also allows for a more thorough inspection of the retaining hardware.

When removing pads, pay attention to the way the brake pads come off. Some disc brakes have different inboard and outboard pads. Slight differences (such as locating nubs) make it easy to install them incorrectly. To remove, inspect, and replace pads and retaining hardware, and to determine any necessary actions, follow the steps in **SKILL DRILL 35-4**.

> **▶ TECHNICIAN TIP**
>
> Because you have the rotor exposed, skip ahead to the sections on inspecting and servicing rotors. Complete the tasks listed there. This will save you time and effort. Pick up here once you are done with the rotor.

SKILL DRILL 35-3 Inspecting Brake Pads and Wear Indicators

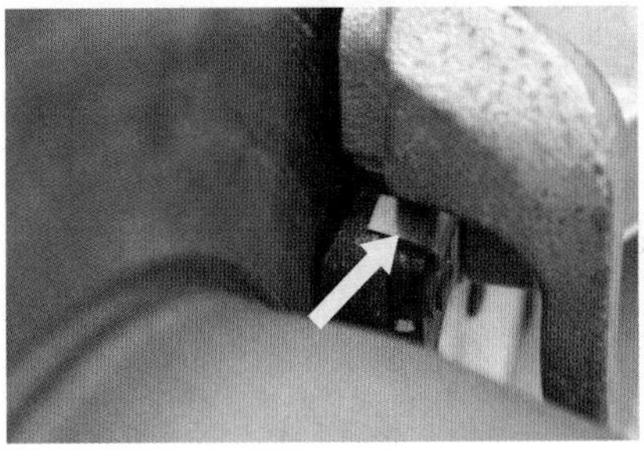

1. Determine the type of wear indicator system utilized. Check that it is not contacting or nearly contacting the rotor.

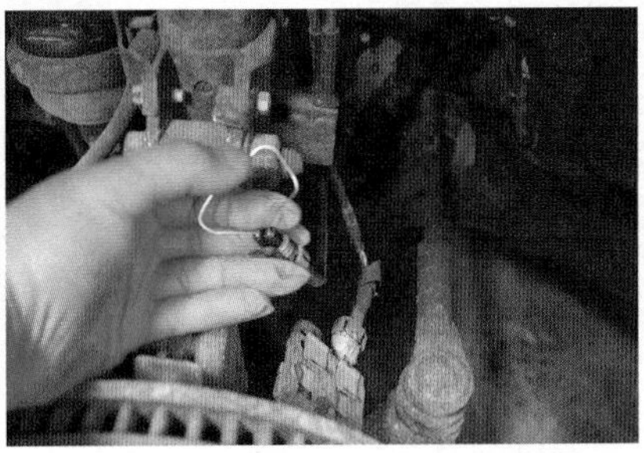

2. If the system is a sensor style, test the system. If the warning light is off, disconnect the brake pad connector and ground the harness side.

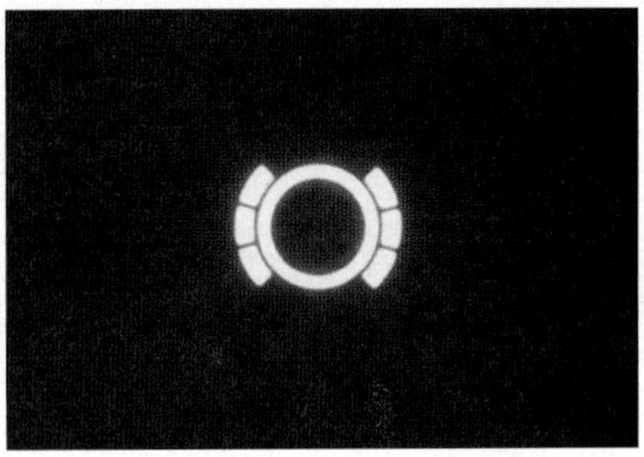

3. Check the brake pad warning light. It should be illuminated. Determine any necessary actions.

SKILL DRILL 35-4 Checking Brake Pads

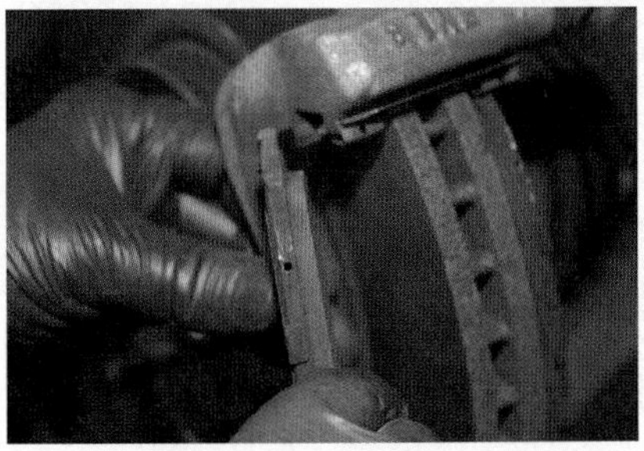

1. Remove the pads and retaining hardware.

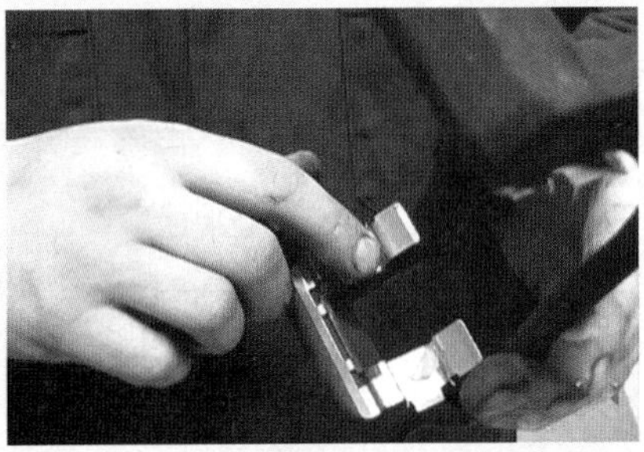

2. Inspect all pads, retaining hardware, and anti-noise shims for wear or damage.

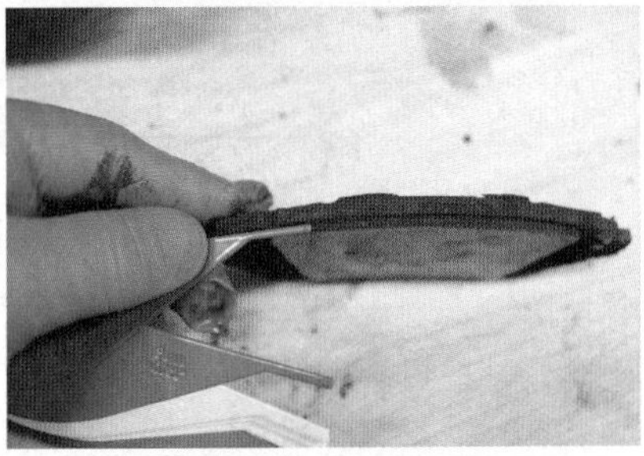

3. Measure the remaining brake pad thickness, and compare with specifications. Determine any necessary action(s).

▶ Disassembling Calipers

Calipers are disassembled and cleaned for a couple of reasons. The first is to diagnose a brake system concern related to one or more calipers. For example, if the vehicle has a brake pull, it could be caused by a sticking caliper piston. Disassembling the caliper allows you to verify whether that is the case. They also need to be disassembled and cleaned if they are going to be rebuilt. Some shops rebuild the calipers themselves, but most shops just replace them with rebuilt calipers if necessary. In fact, many shops use loaded calipers. They are new or rebuilt calipers that come with the brake pads preinstalled.

Disc brake calipers are usually side-specific. This means they are designed to be installed on a particular side of the vehicle. Failure to install them on the correct side results in the bleeder screws being on the bottom of the caliper bore. This prevents air from being bled from the caliper and results in a very spongy brake pedal.

When disassembling a caliper, compressed air is typically used to remove the piston. Be very careful, as the piston can shoot out with great force. It can be strong enough to break finger bones or pinch fingers off. Always use an approved wood or heavy cardboard cushion between the caliper piston and caliper housing. Keep your fingers away from the area.

Clean all of the caliper parts according to the specified procedure. Make sure the sealing ring groove is completely clean. This can be performed by scraping the groove with a variety of pick tools, such as dental picks. Just remember to wipe the groove out with a rag.

To disassemble and clean the caliper assembly; inspect parts for wear, rust, scoring, and damage; and replace the seal, boot, and damaged or worn parts, follow the steps in **SKILL DRILL 35-5**.

SKILL DRILL 35-5 Disassembling Calipers

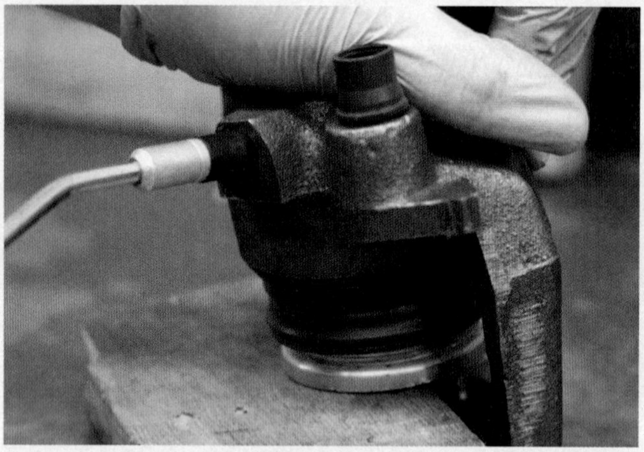

1. Disassemble the caliper, following the service manual procedure.

2. Clean all of the caliper parts, including the seal grooves, following the service manual procedure.

3. Inspect each of the parts for damage, rust, and wear. Also, check the caliper pin bores or bushings for wear or damage. Replace if they cannot be cleaned up.

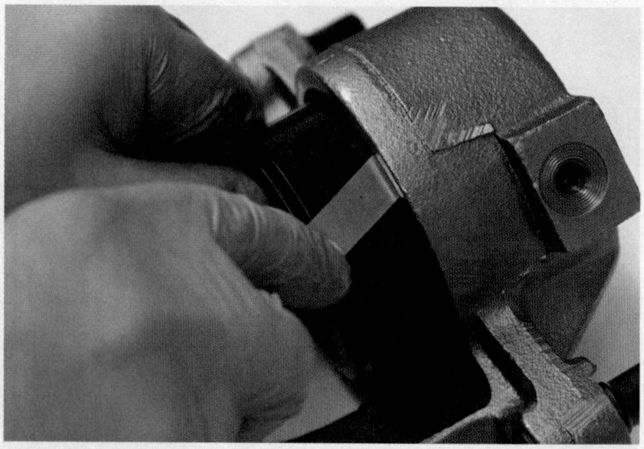

4. Measure the caliper bore-to-piston clearance with a feeler gauge, and compare with specifications. Determine any necessary actions.

▶ TECHNICIAN TIP

In some cases, the piston cannot be removed with compressed air because it is seized in the bore. If this happens, reinstall the calipers on the vehicle, without the pads. Bleed the brakes and use the brake pedal to force the stuck piston out of the caliper. You may have to block any non-seized pistons so they don't pop out. That way, just the seized piston gets pushed out.

Reassembling Caliper and Pad Assembly

Reassembling the caliper requires patience and attention to detail. Ensure that the sealing ring groove is spotless. Then, make sure that the O-ring gets seated fully in the groove. Apply clean brake fluid or approved caliper piston assembly lube on the piston and sealing ring prior to installing it. Be careful not to pinch, twist, or cut the sealing ring and dust boot.

Check the service information to see when the piston dust boot needs to be installed. Some calipers require the dust boot to be installed in the caliper before installing the piston. In this case, a special technique uses air pressure to "balloon" the dust boot over the piston. This takes a lot of experience, and it is easy to cut and bruise your fingers. Ask your supervisor to demonstrate this technique. On most calipers, the seal can be installed after the piston is installed. This is usually much easier to do.

When installing shims, springs, and clips, make sure you position them properly. Most of these parts can be installed in a variety of ways, many of which are wrong. The result will be rubbing, wear, noise, or spongy pedal. To avoid this issue, pay attention to how they came off. If you find yourself not remembering how a part fits, not all is lost. Many times, you can look at the wear patterns and determine the position the parts were originally in.

Once all brakes have been addembled and bled, seat the pads by applying the brake pedal several times. It is a good idea to place your left foot under the brake pedal while doing this. It prevents pushing the master cylinder pistons farther than normal into the master cylinder bore. This could dislodge sludge or cut the lips of the master cylinder primary seals.

Applying the brake forces the brake caliper pistons to adjust to the proper clearance. You may need to start the vehicle to enable the power booster to help you fully apply the brakes. This is especially true if the vehicle is equipped with integrated parking brake calipers. In that case, operating the parking brake while pumping the brake pedal can help it adjust.

To reassemble, lubricate, and reinstall the caliper, pads, and related hardware and the seat pads, and to inspect for leaks, follow the steps in **SKILL DRILL 35-6**.

SKILL DRILL 35-6 Reassembling Calipers

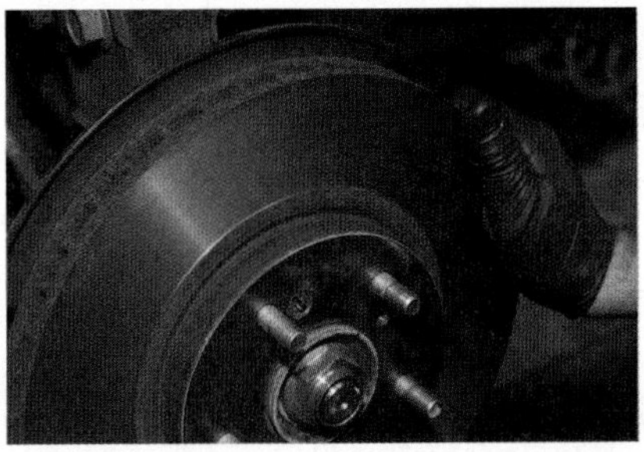

1. Make sure the rotor has been properly installed on the hub/spindle and that the hub surface is free of rust or dirt.

2. Install the piston seal, using brake fluid or brake assembly fluid.

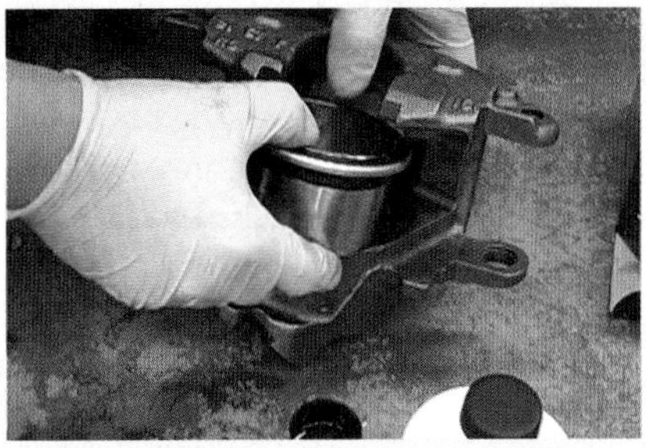

3. Install the piston by hand, being careful not to pinch, twist, or cut the sealing ring and dust boot.

4. Assemble the pads, hardware, and caliper on the caliper mountings, using the specified lubricant. Lubricate all moving parts.

5. Reinstall the brake line fittings, using two new copper washers (if the fittings are so equipped).

6. Tighten the brake line fitting and caliper bolts to the proper torque. Bleed the brakes, following the manufacturer's procedure.

SKILL DRILL 35-6 Reassembling Calipers (Continued)

7. Seat the pads by applying the brake pedal several times, not allowing the pedal to go all the way to the floor.

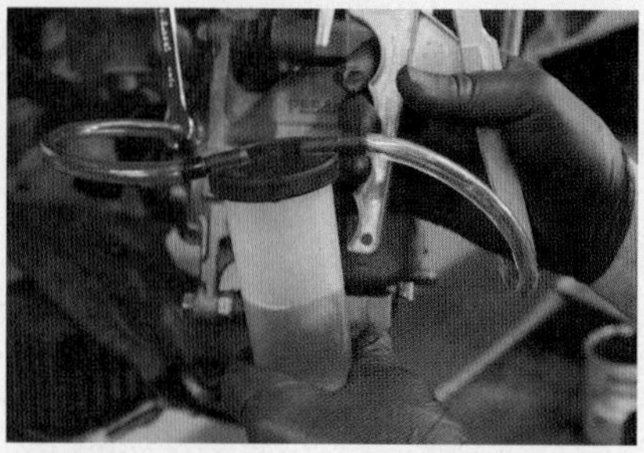

8. If the brake pedal is spongy, you need to bleed the brakes of any remaining trapped air in the system.

9. Inspect the system for any brake fluid leaks, no matter how small.

Retracting and Readjusting Pistons on an Integrated Parking Brake

Retracting the caliper piston on an integrated parking brake system is different than on a standard caliper. The integrated parking brake system uses a threaded shaft to force the piston outward from the caliper bore. Therefore, it cannot just be retracted with a C-clamp. The piston has to be screwed back in. It screws onto the threaded shaft which retracts it into the bore. This is accomplished by using a tool that mates to slots, grooves, or holes in the outer face of the piston. The tool is then turned by hand or wrench to screw the piston back into the bore.

One caution on this system is that in many cases, a short pin on the back of the inboard pad must fit into the slot, groove, or hole on the piston (**FIGURE 35-2**). This prevents the piston from turning when the parking brake is applied. So, when you retract the piston, the slot, groove, or hole must end up so that the pin on the back of the pad lines up with it. Failure to do so will cause the pin to sit on top of the piston, holding the pad away from the piston.

Several types of tools work for retracting the piston. One is a multisided cube with a variety of projections of varying configurations on each side. This cube fits on the end of a ratchet or extension. The tool must be held tightly against the piston so it does not slip (**FIGURE 35-3**). There are other more application-specific tools that use the caliper housing to keep the tool from slipping out of the holes in the piston. This style works best if you have access to one. It is shown in the following skill drill. Be careful not to tear the piston dust boot while twisting the piston in.

To retract the caliper piston on an integrated parking brake system, follow the steps in **SKILL DRILL 35-7**.

▶ Inspecting and Measuring Disc Brake Rotors

LO 35-03 Measure and replace disc brake rotors.

Rotors need to be inspected for conditions that would render them useless. A good visual inspection will identify obvious faults. One of the most common is excessive wear. This occurs when brake pads are allowed to wear down too far. In the case of bonded pads, if they wear down to the backing plate, they will damage the rotor very quickly. The steel backing plate grinds

FIGURE 35-2 Locating pin on the back of the inboard pad lines up with a slot in the caliper piston. This prevents the piston from twisting.

FIGURE 35-3 Using a cube-style caliper piston retracting tool.

SKILL DRILL 35-7 Retracting and Readjusting Pistons on an Integrated Parking Brake

1. Research the procedure for retracting the caliper piston. Select the proper adapter or tool to match the caliper piston.

2. Install the tool, and turn it in the direction that causes the piston to retract.

3. Continue turning until the piston is lightly seated at the bottom of its bore. If the inboard pad has a locating pin, make sure the slot in the piston lines up with the pin.

4. Make sure the dust boot is seated properly in its grooves.

away the cast iron of the rotor quickly (**FIGURE 35-4**). Many technicians have seen rotors that have worn down to the ventilation fins because of this. It doesn't take long to wear the rotor below the minimum thickness. So, if the pads wore down to the backing plate, the rotor almost always needs to be replaced.

If riveted pads wear down too far, the rivets will wear deep grooves in the rotor quickly. The heads of the rivets are higher than the brake pad backing plate. This means that the rivets will hit the rotor first. This creates deep grooves. Again, it doesn't take long before the grooves are below the minimum thickness.

Another issue that occurs is hot spots on the friction surface of the rotor. They are created when spots on the rotor get too hot. The cast iron hardens in those spots making them harder than the surrounding cast iron. So some people call them hard spots. Over time, the softer cast iron wears away, leaving a high

FIGURE 35-4 Worn rotor due to bonded pad worn down to the backing plate.

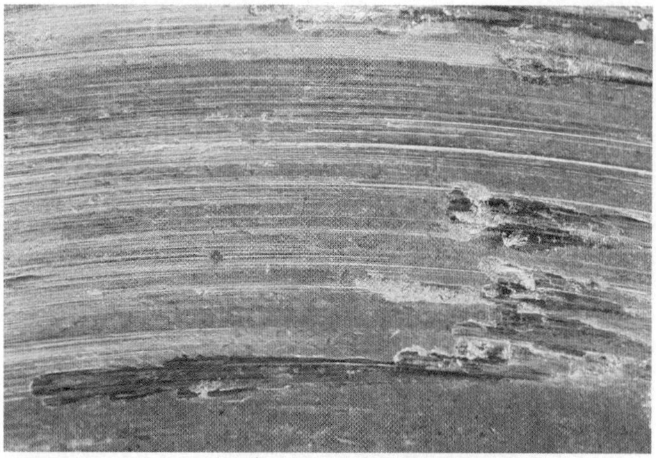

FIGURE 35-5 Rotor with hot spots.

spot where the hot spot is (**FIGURE 35-5**). This compounds the problem, since the high spot now experiences more friction, and it becomes hot quicker. You can feel hot spots, since they are raised up from the surface. And they are usually blue in color. Since hot spots are very hard, the rotor usually cannot be refinished.

For rotors to function properly, they also need to be within specifications. The three main measurements are minimum thickness, lateral runout, and thickness variation. Rotors that are too thin cannot handle as much heat as thicker rotors do. This means they experience brake fade sooner than thicker rotors do.

When measuring the rotor for minimum thickness, measure at the deepest groove or thinnest part of the rotor. If it is under the specified minimum thickness, it will have to be replaced. If it needs to be machined and it is above the minimum thickness, check to see how badly it is scored. Remember that removing 0.015" (0.38 mm) on each side of the rotor results in the thickness being reduced by 0.030" (0.76 mm). Many rotors only start with 0.060" (1.52 mm) of machinable material when they are new.

Excessive thickness variation causes brake pedal pulsation. The vehicle will also have a surging motion while coming to a stop. It may also cause the steering wheel to shimmy during braking. When measuring thickness variation, measure the thickness in four to eight places around the face of the rotor. Calculate the thickness variation by subtracting the smallest reading from the largest reading. Compare your answer with specifications.

Excessive lateral runout tends to cause the steering wheel to shimmy when braking. Since the rotor faces are parallel, they do not usually cause brake pedal pulsations. But over time, it can cause an excessive thickness variation problem. This is due to the high spots of the rotor continuously hitting the brake pad while driving down the road. This constant rubbing wears the high spot, leading to excessive thickness variation.

To clean, inspect, and measure rotor thickness, lateral runout, and thickness variation and to determine any necessary action(s), follow the steps in **SKILL DRILL 35-8**.

Removing and Reinstalling Rotors

Removing the rotor is required when the rotor needs to be replaced. This is usually because it is under the specified minimum thickness or would be after machining. It also would need to be removed to be refinished on an off-car brake lathe. Also, hubless rotors that are being machined with an on-car brake lathe need to be removed so they can be cleaned. Rust and dirt accumulate between the rotor and the hub which can cause the rotor to not turn true.

The hub-style rotor has the wheel bearing hub cast into it. This style generally requires disassembly of the wheel bearing assembly to remove the rotor from the vehicle. The hubless style uses a wheel bearing hub separate from the rotor. The rotor is held onto the hub by the wheel studs and lug nuts. Some hubs

SKILL DRILL 35-8 Inspecting and Measuring Disc Brake Rotors

1. Research the procedure and specifications for inspecting the rotor. If you have not already done so, remove the caliper assembly, brake pads, and any hardware. Clean the rotor with approved asbestos removal equipment.

2. Inspect the rotor for hot spots or scoring, cracks, and damage.

3. Measure the rotor thickness at the deepest groove or thinnest part of the rotor and compare with specifications.

SKILL DRILL 35-8 Inspecting and Measuring Disc Brake Rotors (Continued)

4. Measure the thickness of the rotor in four to eight places around the face of the rotor. Calculate the maximum thickness variation, and compare with specifications.

5. Set up a dial indicator to measure lateral runout. Rotate the rotor and find the lowest spot on the rotor; then zero the dial indicator.

6. Slowly rotate the rotor to find the highest spot on the rotor. Read the dial indicator showing maximum runout.

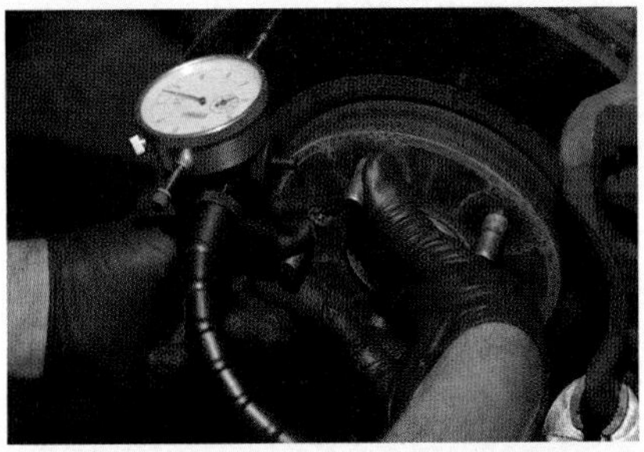

7. Keep turning the rotor to make sure the dial indicator does not read below zero. If it does, re-zero the dial caliper on the lowest spot. Keep turning the rotor to find the highest spot, and reread the dial indicator. Compare all of your readings with the specifications. Determine if the rotor is fit for service, is machinable, or needs to be replaced.

also use small screws to hold the rotor on the hub. Removing hubless rotors is generally easier than removing hub-style rotors. However, some manufacturers design their rotors to unbolt from the rear side of the bearing hub. In these applications, the wheel bearing hub must be removed before the rotor can be removed from the hub.

When removing hubless rotors, mark the rotor so that it can be reinstalled in the same position. A permanent marker, crayon, or center punch can be used to do this. When removing a hub-style rotor, you need to remove the wheel bearings. First, remove the wheel bearing locking mechanism (cotter pin, locknut, or peened washer). Remove the wheel bearing adjusting nut, thrust washer, and outer bearing. The rear bearing must be removed for service or replacement of the rotor. Use the following procedure to remove the inner bearing and grease seal.

- Reinstall the adjusting nut onto the spindle about five turns.
- Grasp the rotor on the top and bottom, and push it toward the center of the vehicle.
- Hold slight downward pressure as you firmly pull the rotor toward yourself. This should cause the adjusting nut to catch the wheel bearing and to pull it and the seal out of the rear of the hub.
- The grease seal will have to be replaced with a new one.
- Now would be a good time to perform any other rotor-related tasks, such as inspecting and measuring a rotor, refinishing a rotor, and servicing wheel bearings. Perform those tasks, and return here to reinstall the rotor.
- If a new rotor is being installed, make sure to clean off any anticorrosion coating it was shipped with. Follow the rotor manufacturer's procedure to remove this coating.

To remove and reinstall the rotor, follow the steps in **SKILL DRILL 35-9**.

SKILL DRILL 35-9 Removing and Reinstalling Rotor

1. Research the procedure for removing and reinstalling the brake rotor. If you have not already done so, remove the caliper assembly, brake pads, and any hardware. If the caliper mount straddles the rotor, remove it.

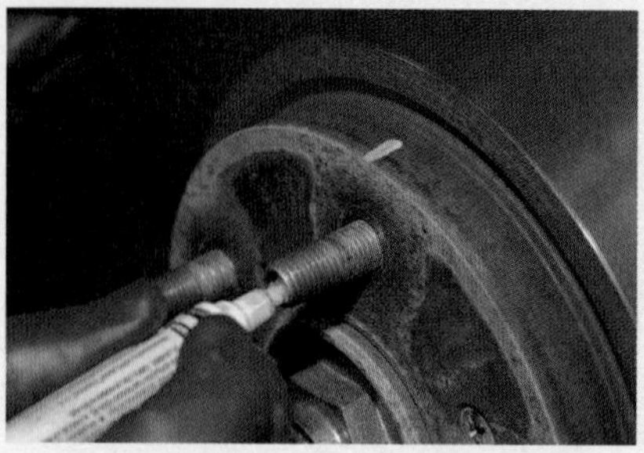

2. To remove the hubless-style rotor, mark the rotor for proper reinstallation.

3. If there are screws or speed nuts holding the rotor to the hub, remove them.

4. Remove the rotor from the hub.

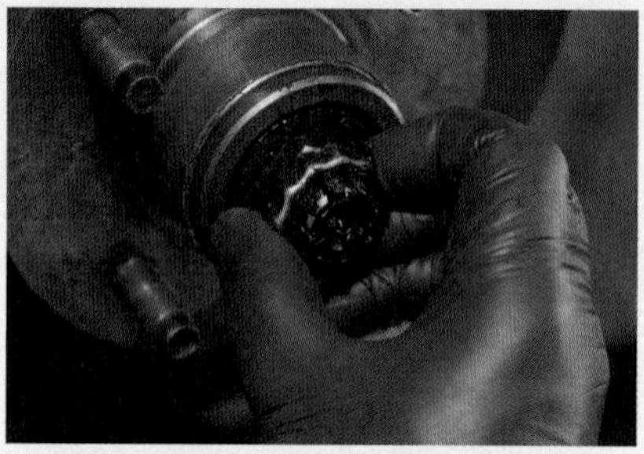

5. To remove the hub-style rotor, remove the wheel bearing locking mechanism.

6. Remove the wheel bearing adjusting nut, thrust washer, and outer bearing.

7. Follow the steps of the inner bearing and grease seal removal procedure if the rear bearing must be removed.

8. To reinstall a hubless-style rotor, clean all mounting surfaces on the hub and rotor, and remove any burrs.

9. Slip the rotor over the wheel studs.

10. If the rotor uses small screws to hold the rotor to the hub, reinstall those and tighten to the proper torque.

11. Spin the rotor to ensure it spins true and does not contact any other components such as the dust shield.

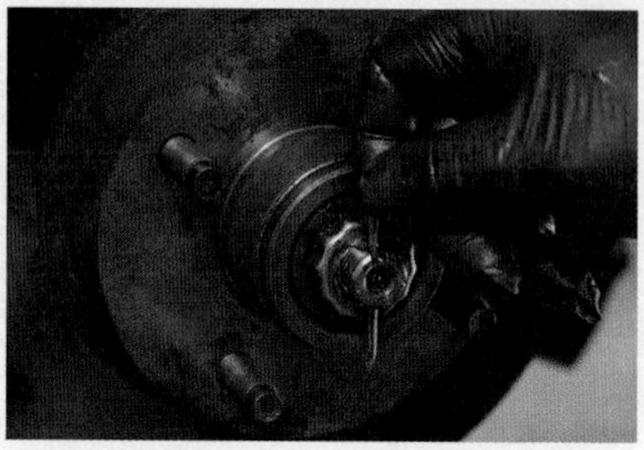

12. To reinstall a hub-style rotor, see the wheel bearing service Skill Drills in the Wheel Bearing chapter. Be sure to install the locking device. Spin the rotor and listen for contact.

▶ Refinishing Rotors on Vehicle

LO 35-04 Refinish disc brake rotors.

Rotors need to be refinished when they have excessive runout, thickness variation, or grooving. A brake lathe refinishes the rotor surfaces by removing metal and truing the surfaces. If the grooving or surface defects are too great, refinishing the rotor may make it too thin. The rotor thickness should always be remeasured once the refinishing is complete. This is to ensure it is above the manufacturer's minimum thickness. Rotors under the minimum thickness need to be replaced.

Rotors can be refinished while on the vehicle or off the vehicle. On-vehicle refinishing is preferred by most manufacturers (those who allow refinishing of their rotors). This is because it minimizes runout issues between the hub and rotor. When the rotor is machined on the vehicle, it is refinished true to the hub. This minimizes any lateral runout issues. Because the rotor is still on the vehicle, the brake lathe drives the rotor during the refinishing process. This requires using the proper adapter and mounting it to the lug studs.

Before refinishing can begin, you need to verify that mating surfaces of the hubless rotor and the hub are free from any dirt, rust, and debris. On hub-style rotors, you need to adjust the wheel bearings so that there isn't any end play. Just remember to readjust them after refinishing is complete. When making a cut, set the cutting bits to the proper cutting depth for machining. This is usually between 0.004" and 0.015". Too little removal tends to overheat the tip of the cutting bit. Too much can overload the lathe, as well as make the rotors much thinner.

Many newer machines use an elliptical motion to give a nondirectional finish. This makes it so a finish cut or separate nondirectional finish is not needed when using these machines. If using an older machine, you may need to give the rotor a nondirectional finish with sandpaper or a special tool.

To refinish a rotor while it is on the vehicle and to measure final rotor thickness, follow the steps in **SKILL DRILL 35-10**.

Refinishing Rotors Off Vehicle

Refinishing a rotor while it is off the vehicle is a bit different than on-vehicle refinishing. The major difference is in the setup. Hub-style and hubless-style rotors each require their own way of being mounted on the brake lathe. Most hub-style rotors use the **bearing races** to center and drive the rotor on the lathe spindle. Thus, bearing adapters of the proper size have to be selected and used. The spindle nut then clamps the rotor onto the spindle through these bearing adapters and races.

Hubless rotors are centered in one of two ways. The first way uses a spring-loaded centering cone. It aligns the rotor's centering hole with the spindle. Clam shell clamps are then used on each side of the rotor to clamp it to the lathe spindle. Composite hubless rotors use a special adapter. It drives the rotor from the center hole while clamping it firmly between solid plates. This type of adapter can be used on standard hubless rotors as well.

Once the rotor is mounted and centered, you need to prevent chatter. This is done in one of two ways. The best way is to use an anti-chatter belt. It is a rubberized belt that wraps tightly around the outer edge of the rotor. It absorbs any vibration in the rotor during the machining process. They work well on ventilated rotors.

If you are working on a solid rotor, the belt is likely too wide to be used. In this case, you will need to use the two anti-chatter pucks on the back of the machine. Install them firmly against the surface of the rotor, opposite to one another. That way, they will ride against the rotor during machining, absorbing vibrations.

Since most brake lathes can refinish both drums and rotors, it is easy to move the wrong lever. Doing so can cause the lathe to machine a deep groove into the rotor or drum. This will most likely ruin the drum or rotor. So, make sure you know what each of the controls do before operating the lathe.

SKILL DRILL 35-10 Refinishing Hubless Rotors on Vehicle

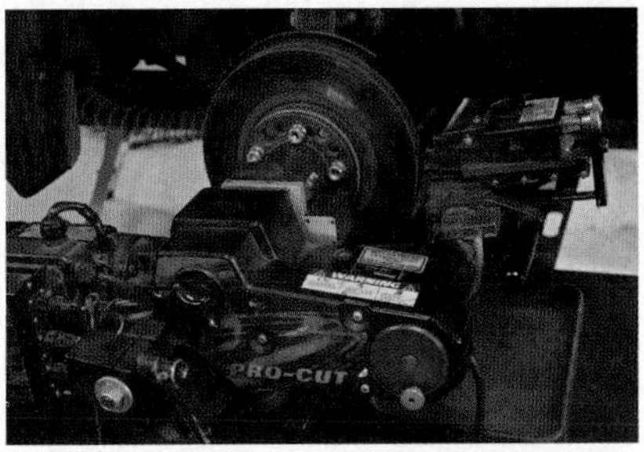

1. Research the brake lathe manufacturer's procedure for properly refinishing the rotor. Mount the on-car brake lathe to the rotor after cleaning the rust and dirt from between the rotor and hub or adjusting the wheel bearing so there is no end play.

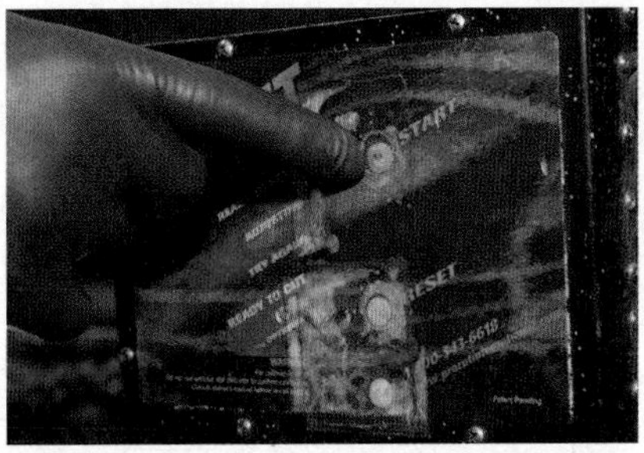

2. Perform the runout calibration on the brake lathe. Some brake lathes require manual compensation, whereas other machines can perform this automatically.

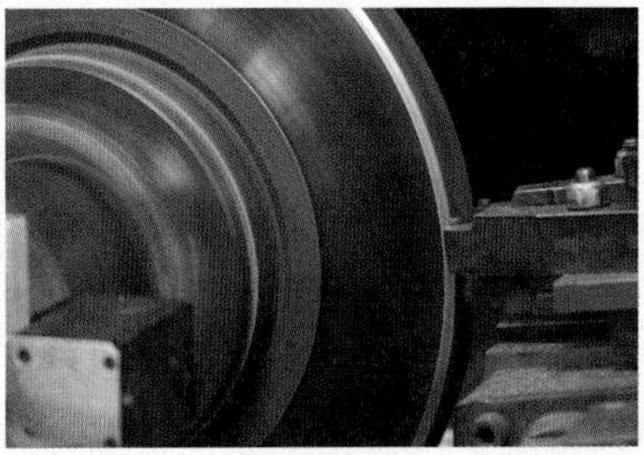

3. Adjust the cutting bits, and cut off any lip at the edge of the rotor.

SKILL DRILL 35-10 Refinishing Hubless Rotors on Vehicle (Continued)

4. Make sure the cutting bits will not contact the rotor face, and move the cutting head toward the inner diameter of the rotor face. Set the cutting bits to the proper cutting depth for machining.

5. Install the antichatter device, if specified.

6. Engage the automatic feed, and watch for proper machining action. If necessary, repeat this step until all damaged surface areas have been removed on both sides of the rotor.

7. If necessary, perform a finish cut on the rotor.

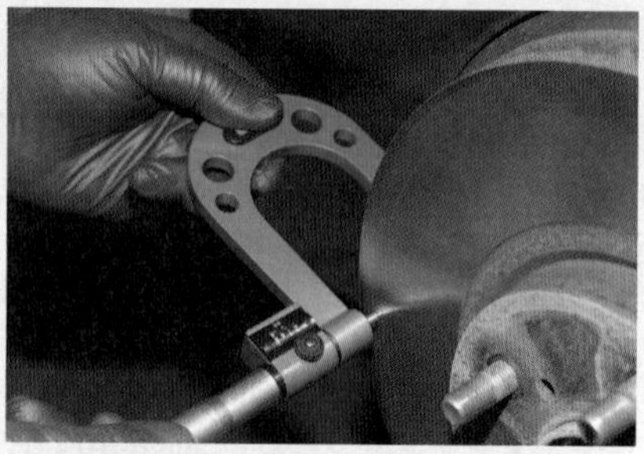

8. Remeasure the rotor thickness to determine if the rotor is above minimum thickness specifications. Readjust the wheel bearings if necessary.

Also, do not remove the drum or rotor from the brake lathe before determining that it is finished. This may mean bringing your instructor to the brake lathe to verify it is finished. Do not remove it and take it to your instructor, unless directed to do so.

To refinish a rotor while it is off the vehicle and measure final rotor thickness, follow the steps in **SKILL DRILL 35-11**.

▶ Inspecting and Replacing Wheel Studs

LO 35-05 Inspect, replace, and torque lug nuts and studs.

Wheel studs need to be replaced when they become damaged. Typically, this means that they have become stretched, broken off, or cross-threaded due to improper installation. Wheel studs can be damaged by being overtightened and stretched. Look for a necked-down or thinned-out section of the stud. This is most likely to happen within the threaded area of the stud. Stretched studs must be replaced.

Studs can also have their threads damaged by cross-threading or seizing of the lug nut on the stud. It is best to replace the stud and nut if this occurs. Finally, wheel studs may even break off if they are overtightened beyond their stretch point. It is always a good idea to consider replacing all of the studs on a wheel (and maybe the ones on the other wheels also) if one stud is broken off. It is likely that the others have been weakened as well.

Some assemblies are designed to allow for the removal and replacement of the wheel studs while the flange is still installed on the vehicle. The manufacturer may have provided a recessed spot in the knuckle or hub where the studs have enough clearance to be removed. Thus, the flange has to be positioned in that particular position. Other vehicles do not have enough clearance on the back side for the stud to be removed. In these cases, the flange must be removed from the vehicle.

There are two primary methods of replacing lug studs: the **drawing-in method** and the **hydraulic press method**. The drawing-in method uses the lug nut to draw the wheel stud into the flange. It is accomplished by inserting the new stud into the wheel stud hole in the flange. Then, a special bearing tool or enough heavy-duty washers are placed over the stud. This allows the lug nut to draw the wheel stud into the flange when tightened. The lug nut is placed on the stud, flat side in, and tightened.

Make sure that as the stud is pulled in, the lug nut does not run out of threads on the stud. If it does, remove the nut and

SKILL DRILL 35-11 Refinishing Rotors Off Vehicle

1. Research the brake lathe manufacturer's procedure for properly refinishing the rotor. Clean nicks, burrs, or debris from the mounting surfaces of the rotor, including the centering hole.

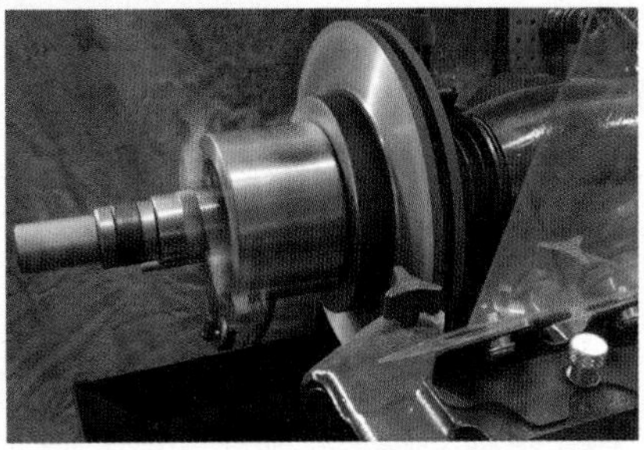

2. Mount the rotor. Check that the rotor is running true on the lathe.

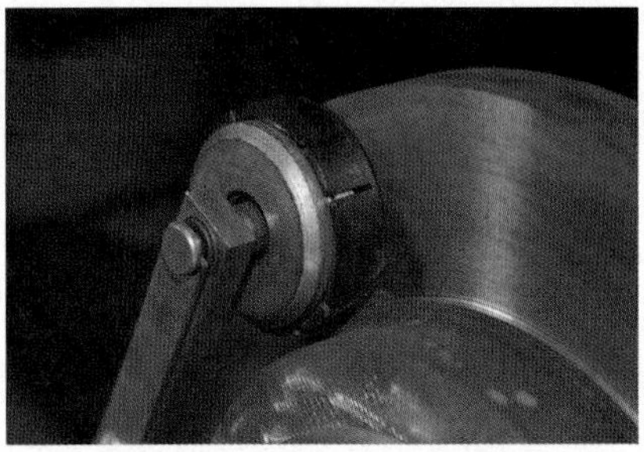

3. Install the antichatter band or antichatter pucks on the rotor.

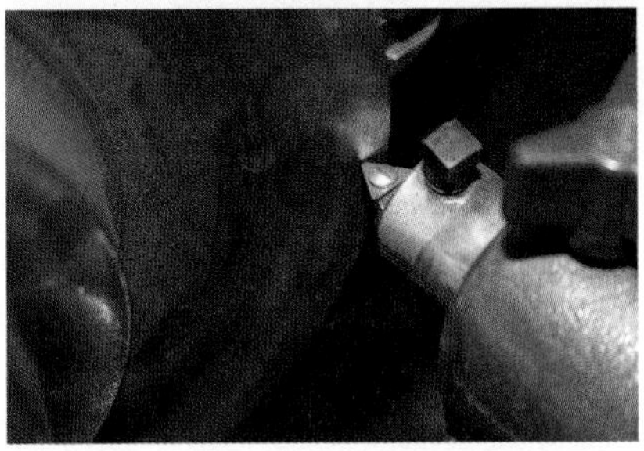

4. Position the cutting head about one-quarter the way in from the outer diameter of the rotor. Turn on the lathe.

5. Set the cutting bits to the proper cutting depth for machining the ridge.

6. By hand, move the cutting head outward toward the ridge. Slowly remove the ridge.

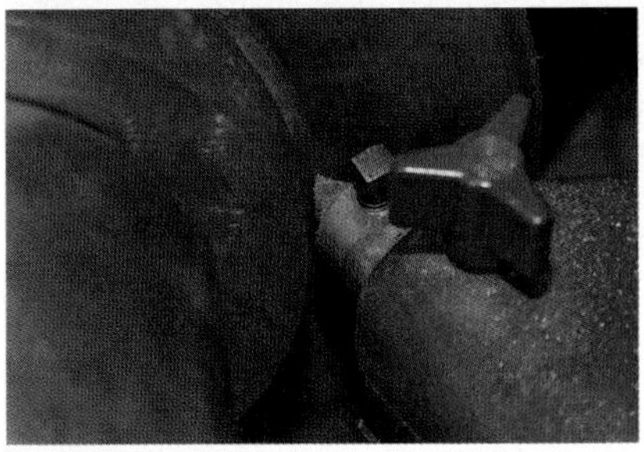

7. Once the ridge is removed, run the cutting head all the way into the inner face of the rotor.

8. Set the cutting bits to the proper cutting depth for machining the rotor face.

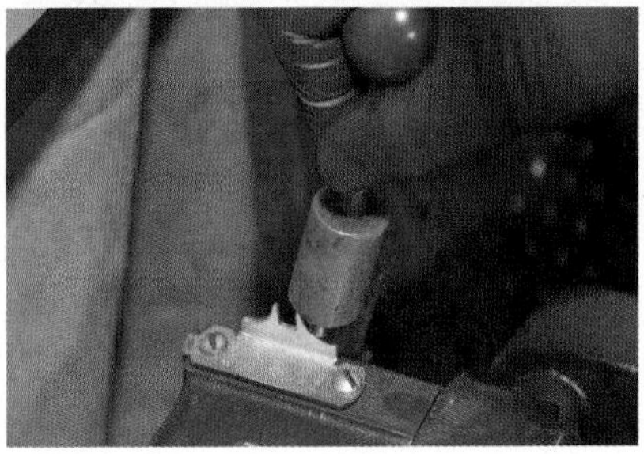

9. Engage the automatic feed on "fast" cut, and watch for proper machining action. Some single-cut machines only have one cutting speed. If necessary, repeat this step until all damaged surface areas have been removed on both sides of the rotor.

10. If necessary, perform a finish cut on the rotor. This is usually done on "slow" speed.

11. Use sandpaper or a drill with a sanding pad to give the rotor faces a nondirectional finish.

12. Remeasure the rotor thickness to determine if the rotor is above minimum thickness specifications. Wash the machined rotor in a hot, soapy water solution or parts washing cabinet to remove any metal particles, and dry.

add washers. Keep tightening the lug nut until the wheel stud bottoms out in the flange. Verify that the head of the stud is fully seated in the flange.

The hydraulic press method uses a press to force the wheel stud into the flange until it bottoms out. This method requires the flange to be removed from the vehicle. Make sure the flange is positioned and supported properly on the press table. Sometimes, it is best to use a short piece of pipe (a bit longer than the wheel stud) to support the flange while the stud is being pressed in. Once the stud is installed, verify that it is fully seated in the flange.

To inspect and replace wheel studs, follow the steps in **SKILL DRILL 35-12**.

SKILL DRILL 35-12 Inspecting and Replacing Wheel Studs

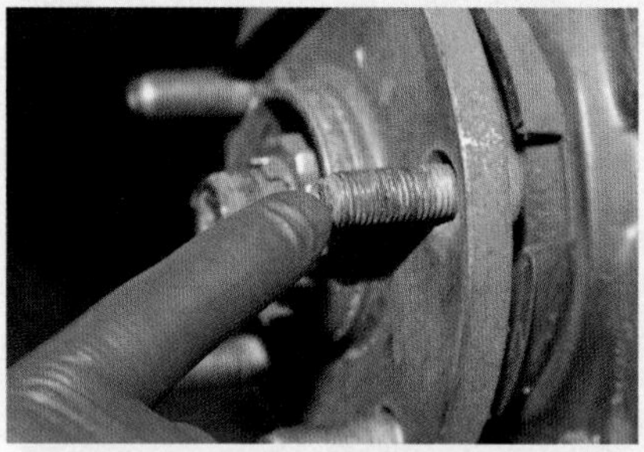

1. Inspect the wheel studs and lug nuts for damage. Look for signs of stretched studs, cross-threaded lug nuts, and broken-off studs. Determine whether the stud can be removed with the flange on or off the vehicle.

2. To perform the drawing-in method, position the flange so the stud has clearance on the back side to be removed.

3. Remove any damaged studs with a hammer. Be careful not to damage any of the surfaces on the flange and hub, including the wheel speed sensor and tone ring.

4. Insert the stud in the hole in the flange, and rotate it so that all of the flutes on the stud line up with the notches in the flange.

5. Place the special bearing tool or enough heavy-duty washers over the stud to prevent the lug nut from bottoming out on the threads.

6. Place the lug nut onto the stud, flat side in. Tighten it until the stud bottoms out in the flange. Inspect the threads on the stud and lug nut to make sure they did not get damaged.

7. To perform the hydraulic press method, remove the hub and/or wheel flange from the vehicle following the manufacturer's procedure.

8. Use the press to push out any damaged lug studs.

9. Insert the new lug stud into the wheel flange hole. Line up the flutes on the stud with the notches in the flange.

SKILL DRILL 35-12 Inspecting and Replacing Wheel Studs (Continued)

10. Support the flange so the press is pushing the stud straight into the flange.

11. Push the stud in until it bottoms out in the flange.

12. To complete the drawing-in method and the hydraulic press method, verify that each stud is fully seated in the flange.

Installing Wheels, Torquing Lug Nuts, and Making Final Checks

This step, although fairly simple, can result in problems if it is not done properly. Overtightening the lug nuts can cause the wheel studs to break. This can happen either immediately or, worse, after the vehicle has been driven for a period of time. Overtightening also can cause warpage of the rotors. You will then need to refinish or replace them.

Undertightening can lead to loosening of the lug nuts and result in the wheel working its way off the vehicle. This can cause the driver to lose control of the vehicle and potentially result in an accident. Lug nuts should be tightened to the proper torque, in the specified sequence. All manufacturers specify the sequences for each of their vehicles (**FIGURE 35-6**). The torque pattern is usually in some form of a star or cross.

Be careful which way you install the lug nuts. Many wheels use a tapered hole that matches the tapered end of the lug nut. This centers the wheel on the wheel flange. Other wheels use a flat surface that matches flat surfaces on the lug nuts. But no matter what, the lug nut surface *must* match the mating surface of the wheel. Always check that these surfaces match.

When installing lug nuts, the weight should be off the vehicle. This means that the wheel is off the ground or is barely touching the ground. The lug nuts should easily center the wheel on the hub. This is especially important on aluminum wheels that use a flat lug nut seating surface. These lug nuts can dig into the sides of the lug nut holes and cause the wheel to not center properly. Once these lug nuts are against the aluminum wheel, work the wheel onto the lug nut shafts. Then, you can torque them down properly.

FIGURE 35-6 Common torque sequence for lug nuts.

▶ TECHNICIAN TIP

There is some controversy regarding the use of anti-seize on wheel studs and lug nuts. Because the purpose of anti-seize is to prevent the components from sticking, CDX errs on the side of not using those products, for fear of the lug nuts coming loose. Also, the torque given for the lug nuts is dry (no lubricant). So, torquing them to specifications would lead to overtightening. At the same time, some areas of the country are prone to excessive rust. It is understandable why technicians there would want to put a very light amount of anti-seize only on the threads of the wheel stud. They are careful not to get any on the contact seat of the wheel and lug nut. Doing so prevents rust from building up on the threads. However, the lug nut torque would have to be reduced due to the lubricant. CDX prefers to install lug nuts dry and avoid this issue altogether. Many shops have their customers sign on the repair order that they will return to the shop, so the lug nut torque can be rechecked after 50–500 miles of driving. This is to make sure the lug nuts have not loosened.

Once the lug nuts are torqued properly it is time to check the brake pedal feel. If you installed a brake pedal depressor tool, now is the time to put it away. Also, verify that the brake fuse is reinstalled, if removed. Pump the brake pedal several times. This adjusts the disc brake pistons in the calipers. It also prevents the brake pedal from going to the floor when you go to move the vehicle out of the shop. Many accidents have occurred because of this. While you are at it, operate the parking brake to make sure it is adjusted properly.

Next, check for any leaks. Look at each wheel brake assembly to make sure that brake fluid is not leaking out. Also verify that the dust caps are reinstalled on the bleeder screws. If there are no leaks, check the brake fluid level in the master cylinder reservoir. It should be at the Full/Max level. If not, top it up. Just make sure the master cylinder reservoir cap is properly installed.

It is now time for a test drive. Proper brake operation can then be verified. You will also be able to burnish the new brake pads and rotor surfaces. Burnishing is also called bedding in. It is the process of transferring pad material onto the rotor evenly. It also cooks off the resins that are used to bind the friction material together in the pad. Burnishing results in long, quiet brake life, which creates satisfied customers.

For burnishing to happen properly, the rotor and pad material should be heated slowly and evenly. This is done by making a specified number of stops from a specified speed, with the appropriate wait times between stops. In some cases, brake lining manufacturers want you to perform a series of stops at light-to-moderate brake application. Other manufacturers specify a series of moderate-to-heavy stops. Always check the manufacturer's procedure for burnishing the brakes once the brake job is complete. This process will help avoid the dreaded disc brake squeal.

To install the wheel, torque lug nuts, and make final checks and adjustments, follow the steps in **SKILL DRILL 35-13**.

SKILL DRILL 35-13 Installing Wheels, Torquing Lug Nuts, and Making Final Checks

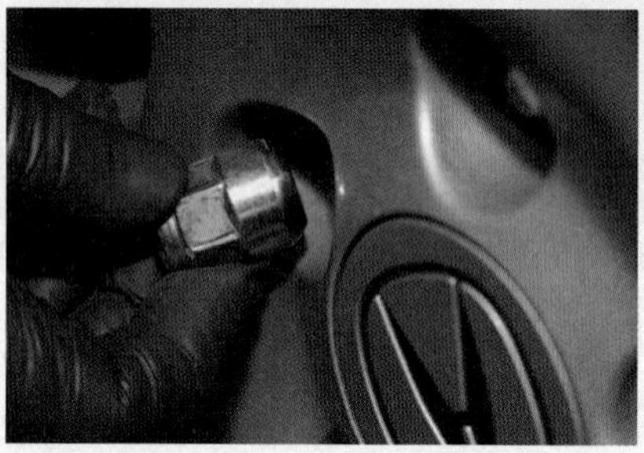

1. Start the lug nuts on the wheel studs, being careful to match up the surfaces.

2. Carefully run all of the lug nuts down so they are seated in the wheel.

3. Lower the vehicle so the tires are partially on the ground to keep them from turning while tightening the lug nuts.

4. Use a torque wrench to tighten each lug nut to the proper torque in the proper sequence.

5. Once all of the lug nuts have been torqued, go around them again, this time in a circular pattern to ensure that you did not miss any in the previous pattern.

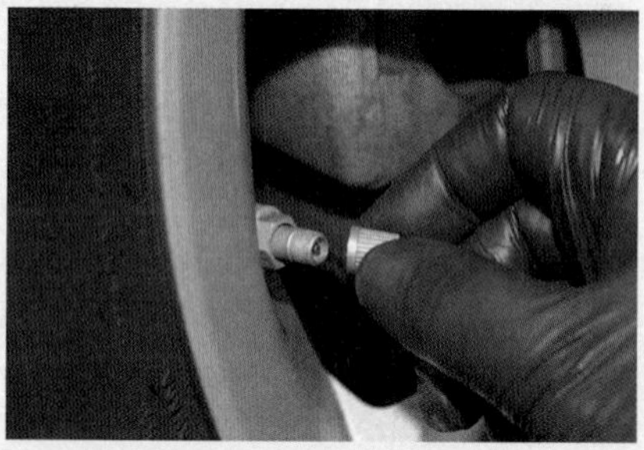

6. If the vehicle was equipped with hubcaps and valve stem caps, reinstall them.

7. Check the brake fluid level in the master cylinder reservoir. Start the vehicle, and check the brake pedal for proper feel and height. Check the parking brake for proper operation. Also, inspect the system for any brake fluid leaks and loose or missing fasteners.

▶ Wrap-Up

Ready for Review

- Disk brake test can include: visual inspection of brake fluid level and condition, removal of the wheels for brake units for inspection, or measurement of the thickness variation of the rotors.
- Caliper and brake pad service should include inspection of caliper mountings, slides, and pins; inspection of brake pad wear indicators; brake pad checks; caliper disassembly and cleaning; and retracting and readjusting pistons on an integrated parking brake.
- Three main measurements for disc brake rotors are minimum thickness, lateral runout, and thickness variation. Excessive thickness variation causes brake pedal pulsation and excessive lateral runout tends to cause the steering wheel to shimmy when braking.
- Disc brake rotors need to be refinished when they have excessive runout, thickness variation, or grooving. Rotors can be refinished while on the vehicle or off the vehicle.
  - The major difference between on the vehicle or off the vehicle refinishing is in the setup.
- Inspection of the wheel studs and lug nuts for damage should include checks for signs of stretched studs, cross-threaded lug nuts, and broken-off studs. Primary methods of replacing lug studs are the drawing-in method and the hydraulic press method.
  - The lug nut surface must match the mating surface of the wheel. When installing lug nuts, the weight should be off the vehicle.

Key Terms

bearing races Hardened metal surfaces that roller or ball bearings fit into when a bearing is properly assembled.

brake lining thickness gauge A tool used to measure the thickness of the brake lining.

brake wash station A piece of equipment designed to safely clean brake dust from drum and disc brake components.

caliper dust boot seal driver set A set of drivers used to install metal-backed caliper dust boot seals.

caliper piston pliers A tool used to grip caliper pistons while removing them.

caliper piston retracting tool A tool used to retract caliper pistons on integrated parking brake systems.

C-clamp A clamp shaped like the letter C; it comes in various sizes and can clamp various items.

dial indicator A device for precision measurements used to measure small variations, such as end play, movement in a bearing, or run-out.

disc brake rotor micrometer A specially designed micrometer used to measure the thickness of a rotor.

drawing-in method A method for replacing wheel studs that uses the lug nut to draw the wheel stud into the hub or flange.

hydraulic press method A method for replacing wheel studs that uses a press to force the wheel stud into the flange until it bottoms out.

off-car brake lathe A tool used to machine (refinish) drums and rotors after they have been removed from the vehicle.

on-car brake lathe A tool used to machine (refinish) rotors while they are still attached to the vehicle.

parking brake cable pliers A tool used to install parking brake cables.

wheel studs Threaded fasteners that are pressed into the wheel hub flange and used to bolt the wheel onto the vehicle.

Review Questions

1. What can cause brake vibration and pulsations that are considered normal?
   a. ABS operation
   b. Rotor parallelism
   c. Rotor surface finish
   d. Strut rod bushing play
2. What concern is often the first indication that brake pads are worn below specification?
   a. Grabbing brakes
   b. Pulling brakes
   c. Dragging brakes
   d. Noisy (squealing) brakes
3. What will likely happen if brake calipers are installed on the wrong side of the vehicle?
   a. The ABS will activate at inappropriate times.
   b. The brakes will grind constantly.
   c. The brakes will not bleed properly.
   d. The caliper slide pins will bend.
4. What is the proper procedure to retract the pistons on most integrated parking brake calipers?
   a. Compress the piston with a C-clamp being careful not to damage the piston.
   b. Rotate the piston using a special tool that engages the slots or holes on the face.
   c. Release parking brake cable tension and push the piston in by hand.
   d. Cycle the ignition three times with the parking brake applied.
5. What are the three main rotor measurements?
   a. Lateral runout, end play, and thickness variation
   b. Thickness variation, minimum thickness, and maximum diameter
   c. Force variation, maximum diameter, and lateral runout
   d. Minimum thickness, lateral runout, and thickness variation
6. If a rotor is 0.040" thicker than the minimum thickness, how much metal can commonly be machined off each side before it should be discarded?
   a. 0.040"
   b. 0.020"

c. 0.010"
d. 0.005"

7. When refinishing rotors on the vehicle why should the cutting depth not be less than 0.004"?
   a. Too shallow of a cut will overheat the tip of the cutting bit.
   b. A shallow cut will leave a directional finish.
   c. Cutting less than 0.004" will cause the rotors to chatter.
   d. It will cause hot spots in the rotors.
8. What may happen if the wrong lever is moved during rotor machining?
   a. Excessive lateral runout will warp the rotor and bend the lathe arbor shaft.
   b. The lathe will rotate backwards causing the rotor to be machined backwards.
   c. The lathe will flash a warning light and a buzzer will sound.
   d. The lathe will feed in the wrong direction cutting a deep groove in the rotor surface.
9. What is burnishing?
   a. Centering the wheel on the hub with tapered lug nuts.
   b. Torqueing the lug nuts so the rotors do not warp.
   c. The transfer of lining material to the rotor evenly.
   d. Performing four ABS stops to heat the rotors rapidly.
10. What is used to draw in the wheel stud using the drawing-in method?
   a. A wheel stud puller
   b. Vise grip pliers
   c. A brass hammer
   d. A lug nut and washers

ASE Technician A/Technician B Style Questions

1. Technician A says that steering and suspension problems may be misdiagnosed as brake issues. Technician B says that professional technicians can be personally held liable for the quality of their brake service and repairs. Who is correct?
   a. Technician A only
   b. Technician B only
   c. Both A and B
   d. Neither A nor B
2. Technician A says that disc brake vibrations are commonly caused by restricted brake lines or hoses. Technician B says that grabbing brakes could be caused by fluid leaking onto the linings. Who is correct?
   a. Technician A only
   b. Technician B only
   c. Both A and B
   d. Neither A nor B
3. Technician A says that a dial indicator is used to measure the thickness of the brake pad linings. Technician B says that a bench brake lathe is used to machine drums and rotors. Who is correct?
   a. Technician A only
   b. Technician B only
   c. Both A and B
   d. Neither A nor B
4. Technician A says that on some vehicles the inboard pad is different from the outboard pad. Technician B says that when performing disc brake procedures, it is okay to let the calipers hang by the hoses. Who is correct?
   a. Technician A only
   b. Technician B only
   c. Both A and B
   d. Neither A nor B
5. Technician A says that rotors should be marked when removed from the hub so that it can be reinstalled in the same position. Technician B says that hub-style rotors may need a new grease seal installed before reinstalling the rotor. Who is correct?
   a. Technician A only
   b. Technician B only
   c. Both A and B
   d. Neither A nor B
6. Technician A says that excessive thickness variation typically causes metal-to-metal contact. Technician B says that excessive lateral runout tends to cause the steering wheel to shimmy. Who is correct?
   a. Technician A only
   b. Technician B only
   c. Both A and B
   d. Neither A nor B
7. Technician A says that most manufacturers who allow rotors to be refinished prefer off-car machining in order to maximize runout tolerances. Technician B says that when a rotor is machined on the vehicle, it is refinished true to the hub. Who is correct?
   a. Technician A only
   b. Technician B only
   c. Both A and B
   d. Neither A nor B
8. Technician A says that hubless rotors can be centered with an adapter that drives the rotor from the center hole. Technician B says that anti-chatter belts absorb vibration during the machining process to provide a smooth finish. Who is correct?
   a. Technician A only
   b. Technician B only
   c. Both A and B
   d. Neither A nor B
9. Technician A says that over-torqued lug nuts can cause wheel studs to break. Technician B says that the hub and flange must be replaced if wheel studs break on most vehicles. Who is correct?
   a. Technician A only
   b. Technician B only
   c. Both A and B
   d. Neither A nor B
10. Technician A says that pumping the brake pedal after brake work is important to adjust the caliper pistons and avoid a potential accident. Technician B says to be careful not to lose the bleeder screw dust caps because they must be reinstalled. Who is correct?
   a. Technician A only
   b. Technician B only
   c. Both A and B
   d. Neither A nor B

CHAPTER 36

Drum Brake Systems Theory

Learning Objectives

- **LO 36-01** Describe drum brake fundamentals.
- **LO 36-02** Describe the types of drum brake systems.
- **LO 36-03** Describe brake drums and backing plates.
- **LO 36-04** Describe wheel cylinders.
- **LO 36-05** Describe brake shoes and lining.
- **LO 36-06** Describe drum brake springs.
- **LO 36-07** Describe drum brake self adjusters and parking brake operation.

ASE Education Foundation Tasks

See Appendix A to view the 2017 ASE Education Foundation Automobile Accreditation Task List Correlation Guide.

▶ Drum Brake Fundamentals

LO 36-01 Describe drum brake fundamentals.

Drum brakes get their name from the rotating drum-shaped component called a brake drum (or simply drum). Drums are bolted to the vehicle's axle flange by the lug nuts. This means the wheels and drums rotate together. Hydraulic wheel cylinders force the brake shoes against the inside of the drums. The brake shoes have friction lining which converts the vehicle's kinetic energy into heat energy. Heat is created on the inside the drums. It has to transfer through the drum material before it can be transferred to the atmosphere. Situations such as heavy loads or long downhill grades result in heavy brake use. This can lead to overheating of the lining, drums, or brake fluid, resulting in a loss of braking called brake fade.

To inspect the condition of the brake lining and drum surface, the drums must be removed (**FIGURE 36-1**). Drum brakes have more parts and components, so repairing them can be a bit more complicated than disc brakes. However, drum brakes are generally less expensive to manufacture. They are also easier to adapt a parking brake too.

Drum Brake System Overview

Although many vehicles use disc brakes on all four wheels, drum brakes are still used on vehicles. They can be found on the rear wheels of vehicles with disc/drum brakes, and on all four wheels of older vehicles. Drum brakes are designed to match the braking requirements of various vehicles. Heavier duty vehicles like pickup trucks use larger-diameter brake drums and shoes. They can also use wider brake shoes and drum surfaces. These factors allow the brakes to create, absorb, and transfer a large amount of heat energy. The main components of the drum brake system are as follows (**FIGURE 36-2**):

- **Brake drum:** The brake drum fits over the brake linings and forms the braking surface for the brake linings. It is usually made from cast iron and machined so the inside surface rotates true.
- **Backing plate:** The backing plate is made from stamped steel and is bolted to the steering or suspension components. It supports the wheel cylinder(s), brake shoes, and hardware.
- **Wheel cylinder:** The wheel cylinder is attached to the backing plate. The wheel cylinder pistons push the brake shoes into contact with the brake drum.
- **Brake shoes:** The brake shoe consists of the steel shoe and the brake lining friction material. The brake shoes are held against the backing plate by hold-down springs and clips.
- **Springs and clips:** Return springs retract the brake shoes when the brakes are released. Other springs work with the self-adjuster and parking brake linkage operation.
- **Automatic brake self-adjuster:** The automatic self-adjuster automatically adjusts the brakes. It maintains a specified amount of running clearance between the shoes and drum. It operates in one of two ways—either when using the brakes while backing up or as part of applying the parking brake. It also makes periodic brake adjustment unnecessary.

FIGURE 36-1 Drum brake with the drum removed.

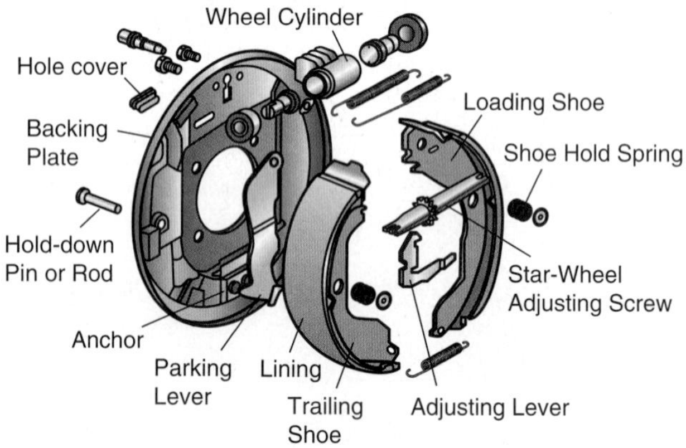

FIGURE 36-2 The main components of a drum brake system.

You Are the Automotive Technician

You are working in San Francisco, California, on your company's fleet vehicles, which are about six years old. Today you are performing a rear brake inspection on a light-duty truck. The driver has indicated that the parking brake is not able to completely hold the vehicle on some of the hills he has to park on. You operate the parking brake and notice that it can be pushed all the way to the floor before it is completely tight. This means you will have to visually inspect the rear brake assemblies.

1. How is the parking brake similar to, and different from, the rear drum brakes?
2. What are the possible faults in the service brakes (drum brake style) that could cause the parking brake to be out of adjustment?
3. How are the primary brake shoe and secondary brake shoe different?
4. How are servo-action and non-servo brakes different from each other?

- **Parking brake mechanism:** The parking brake linkage mechanically operates the brake shoes (service brakes). It holds the vehicle stationary when the driver applies the parking brake.

Drum Brake Operation

In drum brake systems, when the brake pedal is depressed, a pushrod transfers the force to the master cylinder. The master cylinder converts the brake pedal force into hydraulic pressure. The pressure is then transmitted without loss via the brake lines and hoses to the wheel cylinders.

There are one or two wheel cylinders at each drum brake assembly.

The pistons within each wheel cylinder are forced outward by the hydraulic pressure. They apply force to both brake shoes, forcing them into each rotating drum (**FIGURE 36-3**). The brake shoes are anchored to the backing plate. This prevents them from rotating freely with the drum. The friction generated between the moving surfaces slows the rotation of the drum and wheel.

Each drum brake has two brake shoes with an attached friction material called a lining. These shoes expand against the inside surface of a brake drum and slow the wheel. The harder the linings are forced against the brake drum, the greater the braking force applied. They can be expanded mechanically or hydraulically.

Drum brake systems have to be adjusted to allow for wear of the lining. As the lining wears, the brake shoes must be pushed farther outward to contact the drum. Over time, this wear causes only the top portion of the linings to contact the drum. This reduces the surface area of the lining that can dissipate the created heat. Also, if the brakes are not adjusted, the brake pedal reserve height will be too low to be safe. Because of this, drum brake systems must incorporate an automatic adjuster. It keeps the brakes properly adjusted.

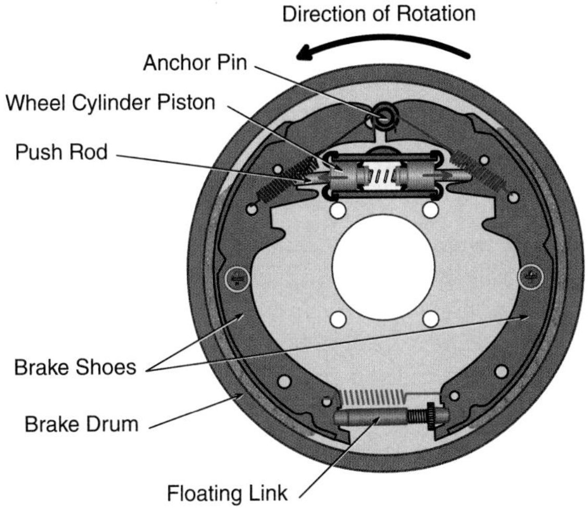

FIGURE 36-3 The pistons within each wheel cylinder are forced outward by the hydraulic pressure and apply force to the brake shoes, forcing them into each rotating drum.

▶ TECHNICIAN TIP

Servicing the wheel bearings is a common part of a brake job if the vehicle has serviceable wheel bearings. If the vehicle uses non-serviceable bearings, the bearings should be checked during a brake job. If they are worn out, they need to be replaced. See the Wheel Bearings chapter for more information on wheel bearings and service.

Self-Energizing and Servo Action

Drum brakes are **self-energizing**. This means they can increase the force with which they are applied. When brake shoes come into contact with the moving drum, the friction tends to carry them in the direction the drum is rotating. Because the brake shoes are inside the drum and anchored at one end, this has a wedging effect on the brake shoe. This wedging effect assists the driver in applying the brakes. The driver then does not have to push so hard on the brake pedal (**FIGURE 36-4**).

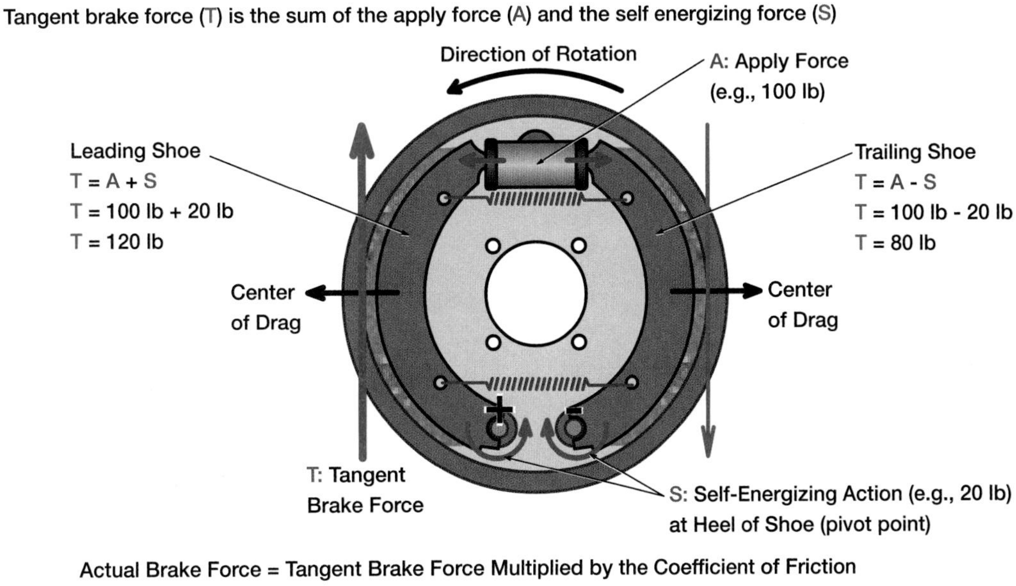

FIGURE 36-4 Drum brake shoes are self-energizing.

The positioning of the brake shoes determines whether a brake shoe is self-energizing. Brake shoes are designed in a leading or trailing manner. **Leading shoes** are installed so the direction they are applied is the same as the forward rotation of the drum. Leading shoes are self-energizing. **Trailing shoes** are installed so the direction they are applied is opposite to the forward rotation of the drum. Trailing shoes are not self-energizing. In fact, they tend to have reduced energization. This means they are not nearly as efficient at developing braking force as leading shoes are.

Servo action as related to brakes means that one brake shoe, when activated, applies an increased activating force to the other brake shoe. This is in proportion to the initial activating force. It further enhances the self-energizing feature of some drum brakes. We cover this topic in greater depth later in this chapter.

▶ Types of Drum Brake Systems

LO 36-02 Describe the types of drum brake systems.

There are three main types of drum brake systems:

- twin leading shoe,
- leading/trailing shoe (also called single leading shoe), and
- duo-servo.

Each type uses similar drum brake components but functions a bit differently. All three types are self-energizing in at least one direction. The first two types are non-servo brakes. The duo-servo drum brake uses servo action in both directions. Each of the three types have pros and cons. Designers use the type that best fits the vehicle application.

Twin Leading Shoe Drum Brake Systems

The **twin leading shoe drum brake system** is the least common type in modern automotive use. This system was once popular on front wheels. It is very efficient at braking in the forward direction. Because vehicles travel much faster forward than in reverse, this matched the braking needs well. The large forward stopping power this system generated allowed it to operate without a power brake booster.

Twin leading shoe drum brake systems use two single-piston wheel cylinders. This type of wheel cylinder is also called a single-acting wheel cylinder. One cylinder is near the top of the backing plate and the other one is near the bottom (**FIGURE 36-5**). Each wheel cylinder activates one of the brake shoes. The brake shoes are anchored at the closed end of the opposite wheel cylinder.

It is called a twin leading shoe drum brake system because both shoes are arranged in a leading shoe (self-energizing) configuration. This arrangement gives very good stopping power in the forward direction. When applied in the reverse direction, the braking force is only about 30% as efficient. This type of drum brake system was usually accompanied by one of the other types of brakes on the rear wheels. The rear brake would then be used as a parking brake. The twin leading shoe drum brake system is very well suited for motorcycles. This is because they are driven mostly in the forward direction and rarely in reverse.

Leading/Trailing Shoe Drum Brake Systems

The **leading/trailing shoe drum brake system** is very common on the rear wheels of front-wheel drive vehicles. This is because of their equal braking forces in both the forward and reverse directions. They use a single wheel cylinder with two pistons. This type is also called a double-acting wheel cylinder. The wheel cylinder is usually mounted near the top of the backing plate (**FIGURE 36-6**). Each piston operates one of the brake shoes, and each shoe is anchored at the bottom of the backing plate.

This arrangement makes one shoe a leading shoe and the other a trailing shoe. In the forward direction, the front piston forces the front brake shoe into the drum. So, it acts like a leading shoe. The rear piston pushes the rear brake shoe into contact with the drum. But it acts as a trailing shoe. So, it does not create as much braking power.

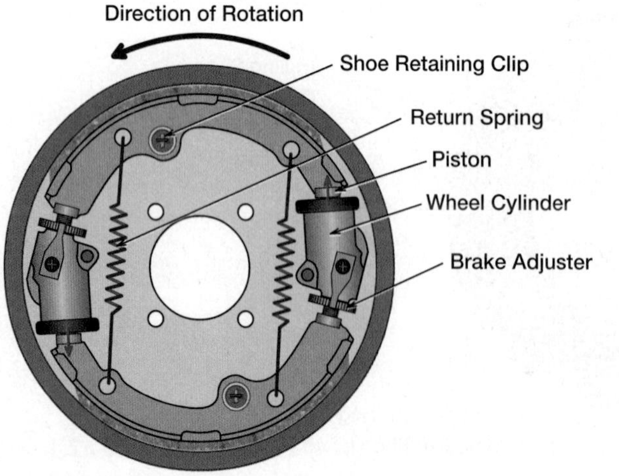

FIGURE 36-5 A twin leading shoe drum brake works very well in the forward direction.

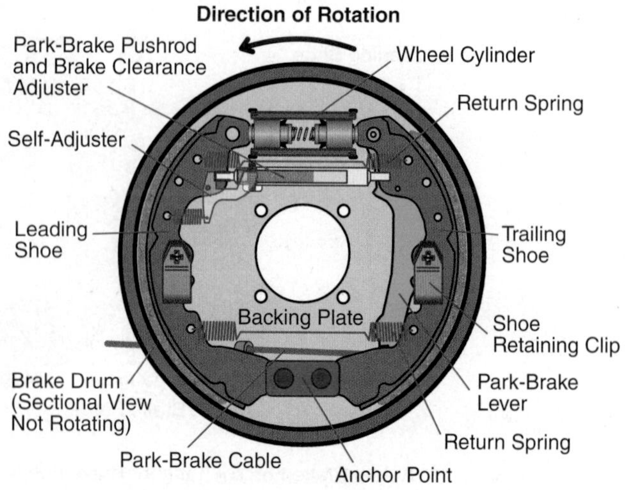

FIGURE 36-6 The leading/trailing shoe drum brake system.

When the car is in reverse or facing uphill and the brakes are applied, the rear shoe becomes a leading shoe and the front shoe becomes a trailing shoe. The leading/trailing shoe drum brake system works equally well in both directions. It is also important to note that it does not provide maximum braking in either direction. It tends to produce a lesser but equal amount of force in both directions. This makes it ideal for the rear wheels of front-wheel drive vehicles. This is because approximately 70%–80% of the braking power needed occurs at the front wheels. Only 20%–30% is needed at the rear wheels.

Duo-Servo Drum Brake Systems

Duo-servo drum brake systems get their name from using the servo action in both the forward and reverse directions. Like the leading/trailing system, the system uses a single wheel cylinder with two pistons. It is also called a double-acting wheel cylinder. It is usually mounted near the top of the backing plate (**FIGURE 36-7**). The bottom of each brake shoe is not anchored to the backing plate. They are connected by an adjustable, floating link. This configuration allows the bottom of the brake shoes to move in the direction of the drum. What keeps the shoes from just spinning around with the drum? There is an anchor pin at the top of the backing plate above the wheel cylinder. It prevents each shoe from rotating past that point. The shoes can move away from the anchor pin, but they are stopped by it when they rotate toward it.

When the brakes are applied, the front piston overcomes the weaker front return spring tension. This moves the forward shoe into contact with the drum. The friction causes it to rotate with the drum. This applies a small force through the bottom connecting link. It is applied (servo action) to the bottom of the rear shoe, pushing it into contact with the rotating drum. The top of the rear shoe is thus carried into the anchor pin at the top of the backing plate. This causes both shoes to stop rotating. The brakes generate a small amount of braking force at this point.

As the driver applies more force to the brake pedal, the forward piston in the wheel cylinder pushes the front shoe harder into the rotating drum. This causes the front shoe to apply more force to the rear shoe. This forces the rear shoe harder into the drum as the anchor pin prevents it from rotating. Because hydraulic pressure is the same on both pistons in the wheel cylinder, the rear piston also tends to push the rear shoe outward into the drum. This helps apply it, although not in a completely complimentary direction.

Because the front shoe multiplies the force to the rear shoe, the rear shoe does more of the braking work. If the linings were the same length on the front and rear, then the rear shoe would wear out much faster. This is why manufacturers put longer lining on the rear shoe than on the front shoe. They might also use linings with different coefficients of friction for each of the shoes. This allows them to get the desired braking load between the two shoes. It is important, therefore, to install the correct shoe in the correct position. Failure to do so will cause the brake linings to wear unevenly as well as work improperly.

The reason this type is called a duo-servo brake system is because it uses servo action in both directions. When braking in reverse, the rear shoe multiplies the force applying the front shoe. This type of brake works well when braking in both directions. However, it is designed to work best in the forward direction.

▶ Drum Brake Components

LO 36-03 Describe brake drums and backing plates.

Drum brakes are made up of a variety of parts. They all work together to:

- apply the brake shoes,
- release the brake shoes, and
- adjust the brake shoes

We will cover each of the brake systems components and the roles they play. This will provide a strong foundation for understanding how to service brake systems.

Brake Drums

Brake drums provide the rotating friction surface that the brake lining contacts. They are usually made from cast iron. Although, some manufacturers have used aluminum drums in the past. Cast iron is able to withstand high temperatures, absorb a lot of heat, and maintain its shape. To enhance the cooling of the brake drum, some manufacturers add cooling fins to the outside of the brake drum (**FIGURE 36-8**).

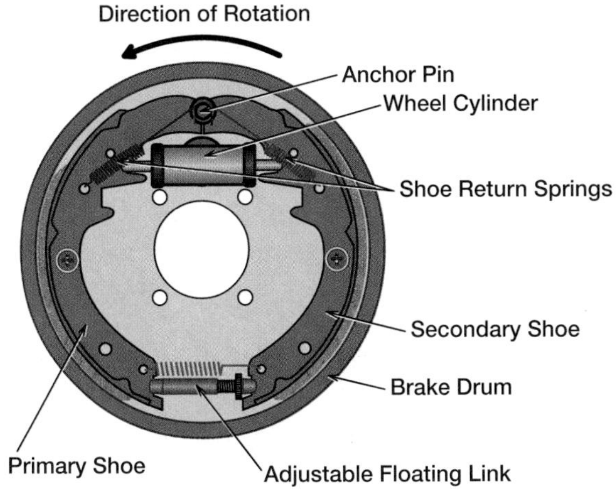

FIGURE 36-7 A duo-servo drum brake system.

FIGURE 36-8 Brake drums. **A.** Without cooling fins. **B.** With cooling fins.

Brake drums are machined to a specified diameter by the manufacturer. This is called its standard diameter. Manufacturers specify the maximum allowable inside diameter a brake drum can be worn to, or machined to. This diameter is usually stamped or cast on the outside of the brake drum (**FIGURE 36-9**). This specification is commonly 0.060" (1.524 mm) over the standard diameter on many brake drums. But it can be as small as 0.030" (0.762 mm) over standard. Or it can be as high as 0.090" (2.286 mm) over standard on some passenger vehicles. Always check the manufacturer's specifications. When refinishing drums, remember that taking 0.015" (0.381 mm) off one side of the brake drum surface also removes 0.015" (0.381 mm) from the other side. This will increase the diameter of the drum a total of 0.030" (0.762 mm).

Brake drums that are heavily grooved or warped likely must be replaced. Even if they are not grooved, they have to be measured. This is because they may have been refinished one or more times before, making them oversized. Never put an oversized brake drum back in service because it does not have as much mass to absorb brake heat. It also is not as strong as a drum that is within specifications.

FIGURE 36-9 Brake drum with maximum diameter stamped/cast into it.

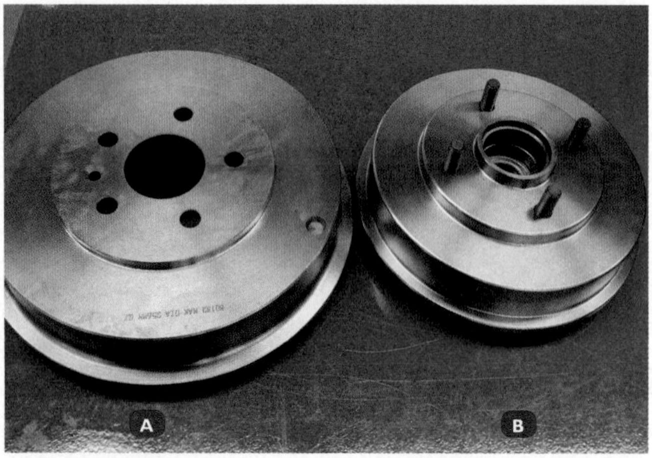

FIGURE 36-10 A. Hubless drum. **B.** Hub-style drum.

Types of Brake Drums

Like rotors, brake drums can have an integrated hub, called a hub-style drum, or a separate hub called a hubless-style drum (**FIGURE 36-10**). Hubless drums are the most common in passenger vehicles. This is because of the high percentage of sealed wheel bearings being used. The wheel bearings in this case do not have to be packed with grease periodically. So, they do not have to be removed unless they wear out. The brake drums slip over the lug studs on the wheel flange and are held on by the wheel and lug nuts. Hubless drums are less expensive to replace when they no longer meet the manufacturer's specifications.

Hub-style drums have a one-piece integrated hub/drum assembly. The wheel bearings are housed in the hub and are usually serviceable. If they are serviceable, then it is standard procedure to service them during a brake job. This includes packing the wheel bearings with the specified grease and replacing the grease seals. More steps are required to remove hub-style drums from the vehicle than hubless drums. They are also more expensive to replace.

Backing Plate

All of the brake unit components except the brake drum are mounted on the backing plate. The backing plate is bolted to the vehicle axle housing or knuckle. The backing plate is usually pressed into a very specific configuration from heavy-gauge steel (**FIGURE 36-11**). It uses a labyrinth seal on its outer edge to help keep out dirt and water spray. The labyrinth seal is made up of a raised edge on the outer surface of the backing plate. It fits into a groove or recess in the brake drum (**FIGURE 36-12**). This makes it so there isn't a direct path for water spray and dirt to enter the drum.

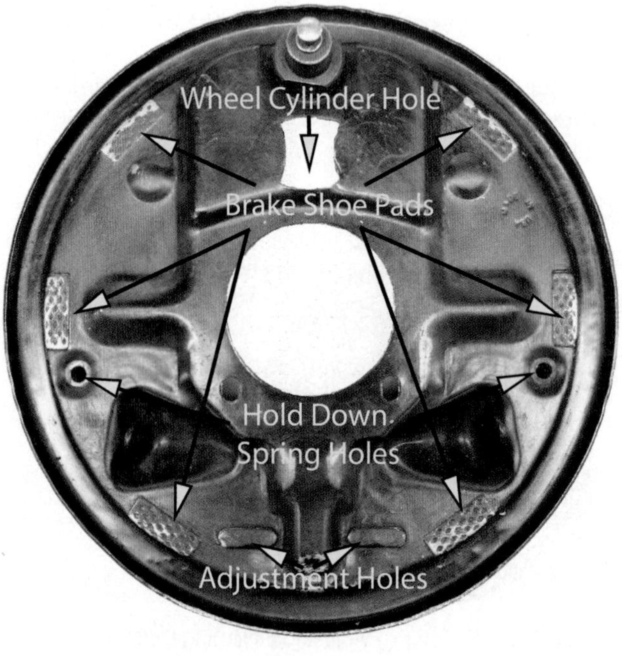

FIGURE 36-11 The backing plate.

The backing plate also has holes stamped in it. These are for the wheel cylinder, hold-down pins, and the parking brake cable. Some vehicles have one or two openings to allow for manual adjustment of the brake shoes. These adjustment holes are plugged with rubber grommets that must be reinstalled after adjusting the brakes.

One of the most important components of the backing plate is the **anchor pin** (or anchor block). The anchor pin must be able to take all of the braking force when the brakes are applied. This means it must be strong and firmly attached to the backing plate (**FIGURE 36-13**). The anchor pin may also hold the shoe guide, return springs, and self-adjuster cable on duo-servo-style brakes.

The inside surface of the backing plate has flat or raised brake shoe contact pads stamped into it. The brake shoes are held against the brake shoe contact pads by spring pressure. The brake shoes are free to move side to side on the contact pads. Over time, the brake shoes can wear a groove in the surface of the contact pad (**FIGURE 36-14**).

During brake shoe replacement, these contact pads have to be cleaned and inspected. If there is no wear, then a very light coat of white lithium grease can be applied to the contact pads. This lessens any wear during the life of the new brake shoes. If there is light wear, this can be cleaned up with a file or small handheld grinder. If the grooves are too deep, the backing plate will have to be replaced.

▶ Wheel Cylinders

LO 36-04 Describe wheel cylinders.

The wheel cylinder is located inside the brake drum. It is either bolted or firmly clipped to the backing plate (**FIGURE 36-15**). It converts hydraulic pressure from the master cylinder into mechanical force. The force pushes the brake linings against the inside of the brake drum.

Wheel cylinders usually contain:

- a cylinder housing,
- one or two pistons,
- a lip seal for each piston,
- a spring and expander set,

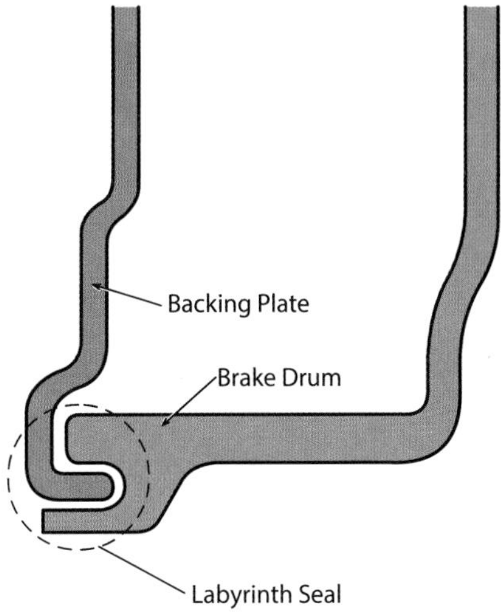

FIGURE 36-12 A labyrinth seal prevents a direct path for water spray and dirt to enter the brake drum.

FIGURE 36-14 Backing plate with worn brake shoe contact pads.

FIGURE 36-13 Two styles of anchor pins mount on backing plates.

FIGURE 36-15 The wheel cylinder mounted on the backing plate.

- a dust boot for each open end of the cylinder,
- a pushrod for each piston, and
- a bleeder screw (**FIGURE 36-16**).

Wheel cylinder housings are usually made of cast iron or aluminum alloy. This allows them to operate under high pressures and temperatures. The **cylinder bore**, or inside diameter of the cylinder, is created by drawing a properly sized ball bearing through the bore. This technique gives the surface a hard, smooth finish. It is also why most manufacturers recommend replacing, rather than honing, the wheel cylinder if it has pits or corrosion. Cylinder bores on aluminum wheel cylinders are usually anodized to help resist corrosion. They also should not be honed, as this would remove the protective finish. Some cylinder bores are sleeved with stainless steel to be longer wearing and more resistant to corrosion.

No matter the type of material used, the cylinder and pistons are manufactured to a precise diameter. This provides the proper amount of clearance between them. It also maintains the proper tension of the piston seal to the cylinder bore. This helps hold pressure in and air out. These are other reasons why honing the cylinder bore is discouraged.

Contamination, particularly from water, causes pitting and rusting of the inner surface of the wheel cylinder. Such damage can result in leakage of brake fluid from the cylinder. It can also cause the pistons to stick in the bore.

The sealing surface in a wheel cylinder is the inside surface of the cylinder bore. This surface must be clean, smooth, and free of pitting. The wheel cylinder cups seal against the surface of the cylinder bore. Hydraulic pressure forces the lip of the seal into the surface of the cylinder bore even harder when the brakes are applied.

The sealing cups are made of materials that are compatible with the type of brake fluid specified for the particular vehicle. Using the wrong seal compound or adding the wrong brake fluid in the brake system can cause seal damage and brake failure. Always verify that the brake fluid and components are compatible with each other.

Wheel cylinder pistons are usually made of anodized aluminum. They have a built-in mating surface for the pushrods or brake shoes to engage. Most pistons use a flat surface that supports the piston seal, but some pistons use a seal that fits within a machined groove at its inner end. Piston-to-cylinder clearance is critical for proper operation. So, verify that the cylinder and piston are not worn beyond the specified clearance.

Wheel cylinders may be fitted with a spreader and a light expansion spring. They help to keep the lips of the seal in contact with the cylinder bore during times of low pressure. Low pressure occurs during retraction and while at rest (**FIGURE 36-17**). This helps keep air from being drawn into the cylinder. A flexible dust boot fits over the open ends of the cylinder and allows for piston movement. At the same time, it helps keep brake dust and moisture away from the inside of the cylinder and piston.

Wheel cylinders are fitted with a bleeder screw to allow for bleeding of air and old brake fluid from the hydraulic brake system (**FIGURE 36-18**). The bleeder screw is a hollow screw. It has a taper on the end that mates with a matching tapered seat in the wheel cylinder. These tapered seats seal when the bleeder screw is closed. The bleeder screw has been cross-drilled into the center hole just above the taper.

When the bleeder screw is loosened slightly, the tapered seat opens. Brake fluid enters the cross-drilled passage and into the center hole. It then flows out the end of the bleeder screw. Tightening the bleeder screw closes off the tapered seat and holds pressure in the wheel cylinder. Bleeder screws have a small rubber dust cap that fits over the exposed end of the screw. This keeps water, dirt, and debris from entering the center hole and plugging it up or rusting it in place. Always remember to replace these caps when you finish bleeding the brakes.

▶ TECHNICIAN TIP

During a brake inspection, it is good practice to carefully peel back the dust boot from the wheel cylinder to see if there is brake fluid behind the dust boot. If there is, the wheel cylinder is starting to leak and should be replaced.

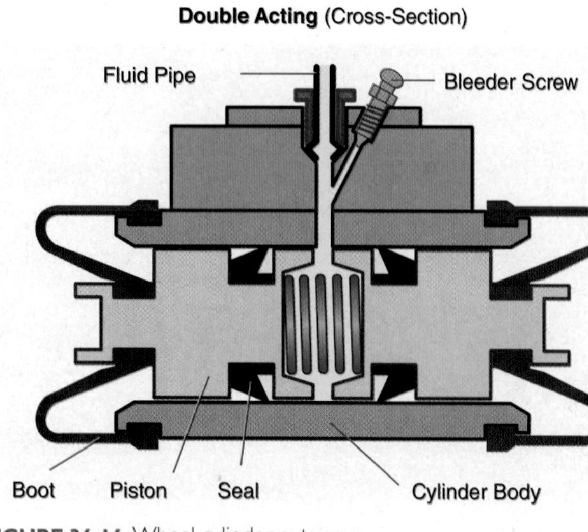

FIGURE 36-16 Wheel cylinder cutaway.

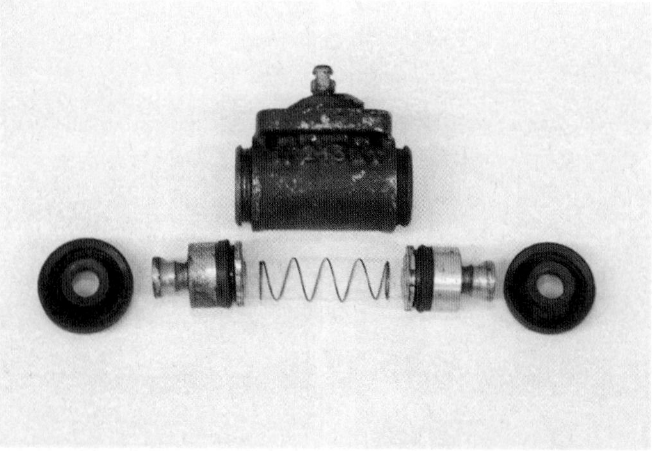

FIGURE 36-17 Wheel cylinder components.

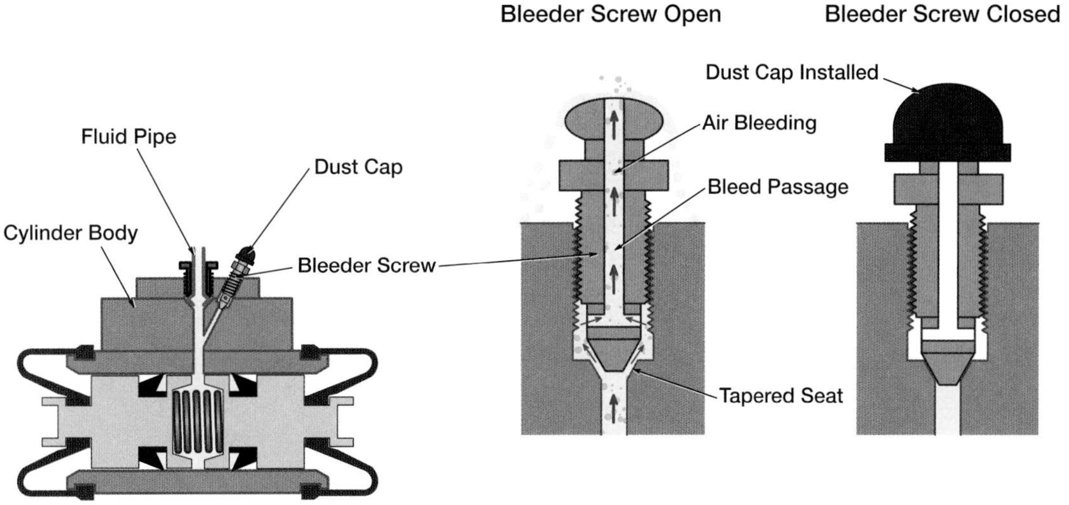

FIGURE 36-18 Bleeder screw allows air and fluid to be bled from the system.

Types of Wheel Cylinders

Wheel cylinders come in different configurations. They are either single acting or double acting. Single-acting cylinders use a single piston. This means that the force is generated in one direction only. Double-acting cylinders use two pistons opposite to each other. This means that the force acts in two different directions (**FIGURE 36-19**).

Most modern vehicles use double-acting wheel cylinders. This is because they are simpler to design, install, and bleed. Double-acting wheel cylinders use a common cylinder with a piston and lip seal in each end. There is usually a coil spring with expanders on each end, positioned between the lip seals. The expander helps to hold the seal lips against the cylinder bore when there is little or no hydraulic pressure.

Single-acting wheel cylinders are used on some non-servo drum brakes. Each wheel cylinder has only one piston, so the cylinder bore is closed off at the opposite end. There are two of these cylinders on each wheel assembly, one for each brake shoe. The wheel cylinder is very similar to the double-acting cylinder. It is made of the same materials and has an aluminum piston, a lip seal, a spring and expander, a dust boot, a pushrod, and a bleeder screw.

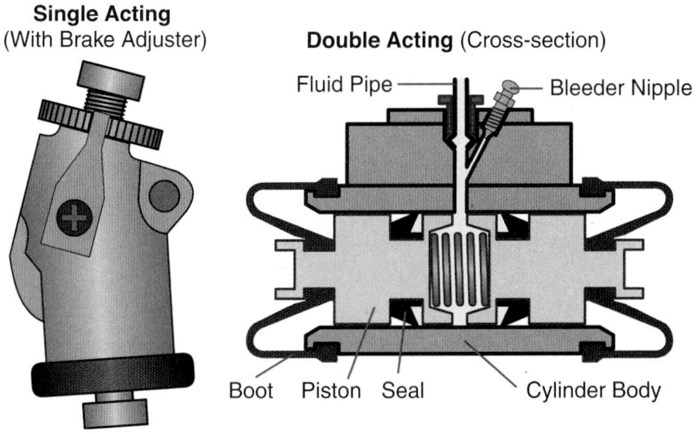

FIGURE 36-19 Types of wheel cylinders.

▶ Brake Shoes and Linings

LO 36-05 Describe brake shoes and lining.

The drum brake system uses metal brake shoes. They have holes, slots, and tabs for springs and hardware to attach to. The metal brake shoes have friction material called linings attached to them. Linings can be riveted but are more often bonded to the brake shoes (**FIGURE 36-20**). The brake shoe is technically just the metal portion. But technicians generally use the term "brake shoe" to mean the metal shoe and lining assembly.

The composition of the lining material affects brake operation. Materials that provide good braking with low pedal pressures tend to lose efficiency when they get hot (**FIGURE 36-21**). This means the stopping distance will be increased. They also tend to wear out more quickly. Materials that maintain a stable friction coefficient over a wide temperature range generally require higher pedal pressures. They also tend to put added wear on the brake drum friction surface, reducing its useful life.

The Society of Automotive Engineers has adopted codes to rate the coefficient of friction of brake lining materials. The rating is written on the edge of the friction linings and is called

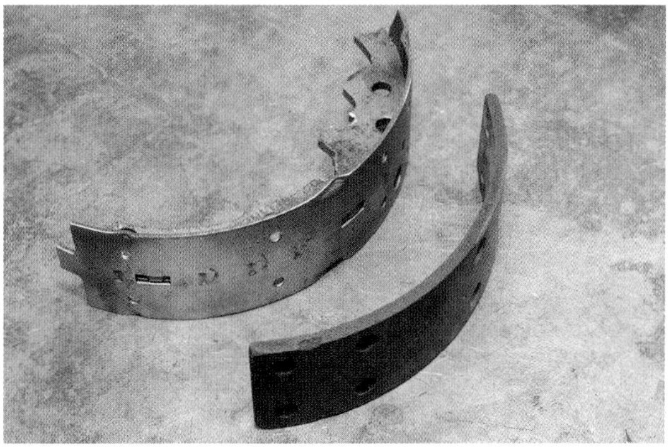

FIGURE 36-20 The drum brake shoe and lining (Lining removed for clarity).

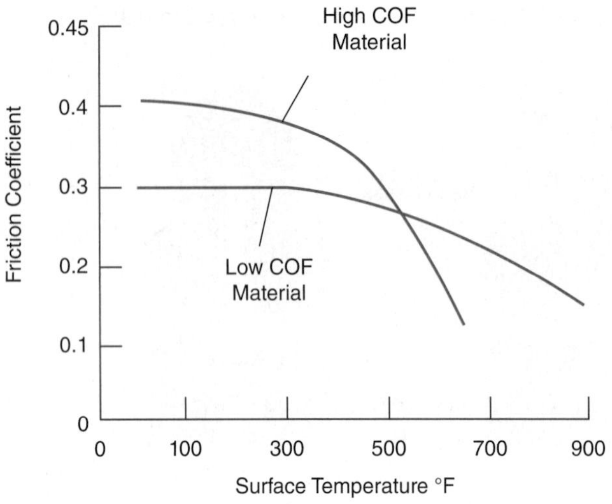

FIGURE 36-21 The brake lining coefficient of friction is affected by temperature.

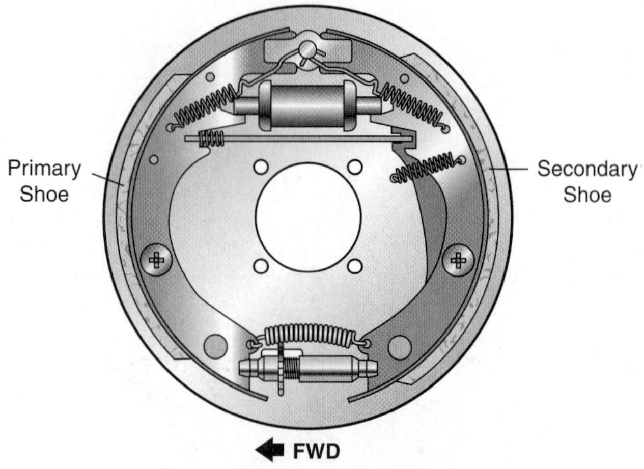

FIGURE 36-22 Primary and secondary brake shoes.

the edge code. The Disc Brake System chapter discusses these ratings in detail. Drum brakes are usually designed so that the condition of the lining can only be checked once the drum has been removed.

Brake friction materials were historically made from asbestos compounds. Asbestos has excellent heat-resistant qualities. Asbestos has since been proven to be hazardous. It is generally banned and is not normally used in vehicles. Today, brake linings are manufactured from a variety of different materials. They include the following:

- Non-asbestos organic (NAO) materials—organic materials such as Kevlar® and carbon
- Low-metallic NAO materials—small amounts of copper or steel and NOA materials
- Semi-metallic materials—a higher quantity of steel, copper, and/or brass
- Ceramic materials—ceramic fiber materials and possibly a small amount of copper

▶ TECHNICIAN TIP

As the heat in brake pads and linings builds up, the coefficient of friction capability of the material—and consequently its stopping power—is reduced. This reduction is called brake fade. Minimizing or overcoming brake fade is a major factor in the design of brakes and the development of brake friction materials.

Primary and Secondary Brake Shoes

The terms "primary" and "secondary" refer to the brake shoes in a duo-servo brake system. The primary shoe goes toward the front of the vehicle. The secondary shoe goes toward the rear of the vehicle (**FIGURE 36-22**). In most cases, the primary and secondary metal shoes are the same. But the linings installed on the brake shoes are of different length.

Because the primary shoe applies the secondary shoe, the secondary shoe is responsible for doing most of the braking work. Therefore, the primary shoe lining is shorter in length, and the secondary shoe lining is longer. The primary lining may also have a different coefficient of friction than the secondary lining. New shoes are normally labeled with "pri" (for primary) or "sec" (for secondary) on the edge of the lining. Verify that you are installing the lining in the proper position.

▶ TECHNICIAN TIP

New technicians tend to make a couple of rookie mistakes when it comes to duo-servo brake installation. The first relates to the way the new brake shoes are packaged in their box. The manufacturer generally puts the brake shoes in the box in like pairs. What this means is that both primary shoes are in the bottom of the box. This means that both secondary shoes are in the top of the box (or vice versa). When students remove the top two shoes from the box and compare them to each other, they find that the shoes match. They shouldn't if they are for a duo-servo system. The students assume the matching shoes belong on the same side of the vehicle and install them accordingly (**FIGURE 36-23**). They then do the same with the other two brake shoes (which also match each other) on the other side of the vehicle. So, both primary shoes are now installed on one side of the vehicle, and both secondary shoes are installed on the other side.

FIGURE 36-23 A common installation error: The primary shoes are on one side of the vehicle, and the secondary shoes are on the other side.

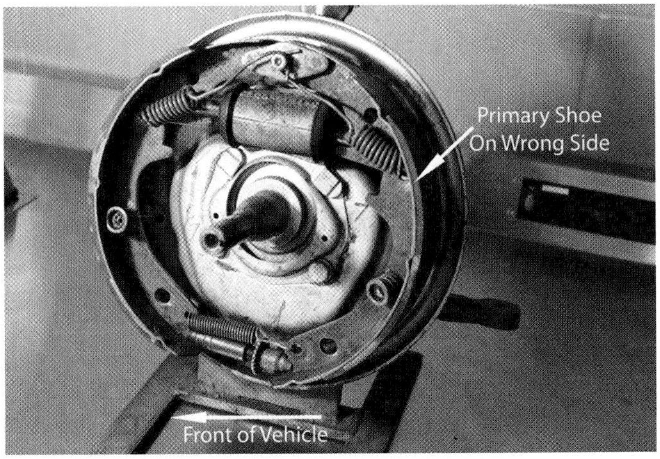

FIGURE 36-24 Another common installation error: The primary and secondary shoes are swapped from side to side.

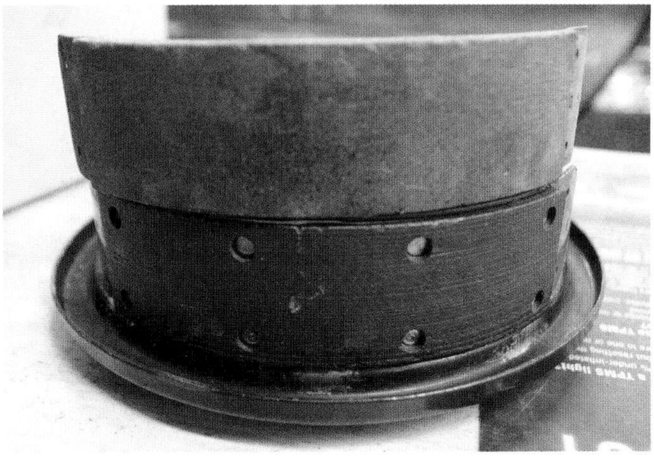

FIGURE 36-25 **A.** Bonded brake shoe lining. **B.** Riveted brake shoe lining.

The second mistake involves installing the shoes in the wrong position on the backing plate. When students install the brake shoes on the first assembly, they may look carefully at the assembly and determine (correctly) that the primary shoe goes on the forward side of the backing plate. For the driver's side brake assembly, this is the left side of the backing plate. They then determine that the secondary shoe goes on the right side of the backing plate. When they move to the passenger side of the vehicle, they believe they have already determined that the primary shoe goes on the left side of the backing plate just like the other side. Unfortunately, that is the *rear* side of the backing plate. The secondary shoe should be installed in that position. So, the brake shoes are installed correctly on one side of the vehicle. But they are installed incorrectly on the other side (**FIGURE 36-24**). Watch for these mistakes as you perform brake inspections and service.

Riveted and Bonded Friction Materials

Lining on brake shoes is much thinner than on disc brake pads. This is due to the much greater surface area of the brake shoe lining. Drum brake shoes consist of friction material or lining bonded or riveted onto a steel shoe (**FIGURE 36-25**). Bonded brake linings are more common on light-duty vehicles because they are less expensive to build. Also, lighter vehicles do not subject the brake linings to as much heat. So, the bonding agent is not as likely to fail as it would on heavier vehicles. Bonded linings are glued to the metal brake shoe under high pressure and temperature. This ensures that the bonding is as strong as possible.

Riveted linings are used on heavier duty or high-performance vehicles. Metal rivets are usually made of copper or aluminum. They provide a mechanical connection to hold the brake lining to the shoe. Rivets are less susceptible to failure under high temperatures. At the same time, the rivets actually pinch some of the lining between the rivet head and the brake shoe. The brake linings cannot wear down as much before the rivets contact the drum. This means that riveted linings must be changed sooner than bonded linings. This will prevent the rivet heads from contacting the drum and wearing a groove in the friction surface.

FIGURE 36-26 Brake dust.

Drum Brake Noises

Drum brakes are not prone to squealing like disc brakes are, but that does not mean they never make noise. One of the most common sounds that drum brakes make is a groaning noise. It can be caused by excessive brake dust in the drum. The dust causes the brake shoes to skip and catch, which sounds like a groan. Removing the drum and cleaning the excess brake dust or replacing the brake shoes due to wear usually resolves this issue (**FIGURE 36-26**).

Another noise that drum brakes can make is a grinding noise. If the friction lining wears all the way down to the metal shoe, a metal-on-metal grinding noise will result (**FIGURE 36-27**). This noise can only be resolved by replacing the brake shoes and refinishing or replacing the drum. Keeping drum brakes from making unwanted noises is pretty simple. Keep them clean and inspect the brake lining thickness periodically. Replace them when they get to the specified minimum thickness.

The last common noise that drum brakes can make is a clicking noise. This can be caused by two situations. First, it could be that the brake shoes have worn grooves in the contact pads on the backing plate. They cause the clicking noise when the shoe moves into and out of the groove.

FIGURE 36-27 Brake shoes worn down to the metal.

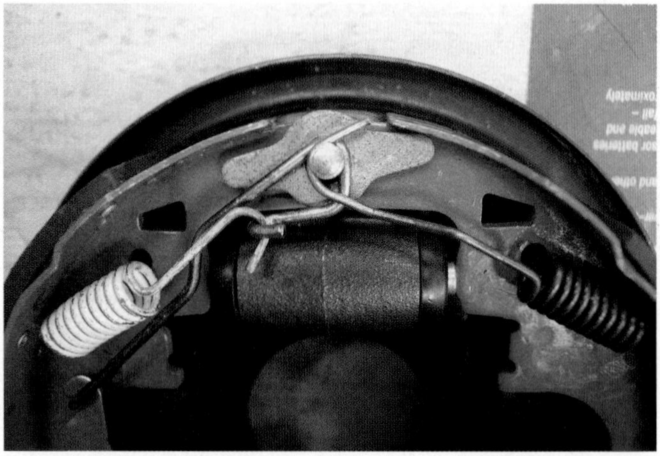

FIGURE 36-29 Return springs pull the shoes back to their rest position.

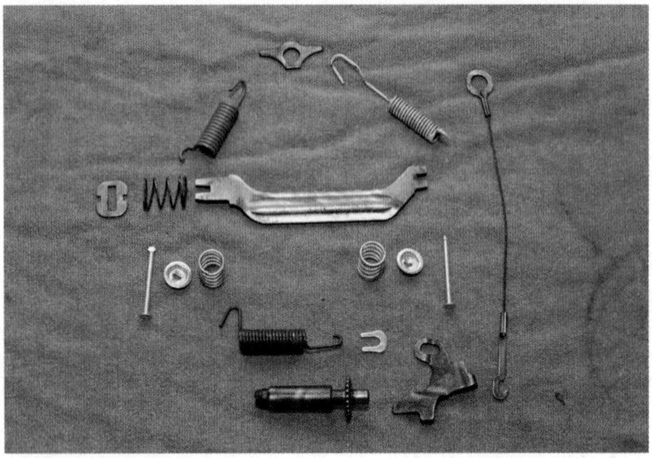

FIGURE 36-28 Examples of brake springs and hardware.

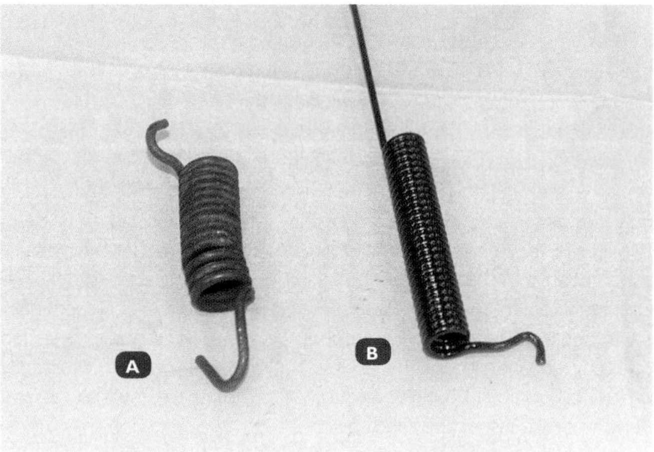

FIGURE 36-30 **A.** Worn return spring with stretched coils. **B.** New return spring with tight coils.

It could also be that the brake drum's surface finish cut is too rough and acts like the threads on a screw. When the lining contacts the surface of the drum, the lining follows the threaded finish of the drum. The linings thread themselves away from the backing plate. When the brakes are released, the hold-down springs snap the brake shoes back against the backing plate.

▶ Drum Brake Springs

LO 36-06 Describe drum brake springs.

Drum brakes use a variety of springs and hardware to control the action of the brake shoes (**FIGURE 36-28**). Each type of brake uses its own arrangement of these components. Pay close attention to how they are installed before removing them. It is good practice to only disassemble and reassemble the brake components on one side of the vehicle at a time. That way you can refer to the other side if you forget how something came apart. Generally speaking, there are several types of springs used in drum brakes. They are: return, hold-down, and specialty.

Return Springs

Return springs retract the brake shoes when the driver releases the brake pedal. Each return spring either connects to one brake shoe and the backing plate anchor pin or directly to the other brake shoe. In both cases, they pull the brake shoes back into their rest position (**FIGURE 36-29**). Return springs can have one or more coil springs in the length of the spring. Or it can be a large U-shaped spring.

Return springs are generally quite stiff, making them a challenge to install. Brake pliers can make installation easier. Return springs may look the same front to rear within a particular brake assembly. But be aware that they can have different amounts of strength. On duo-servo brakes, the primary return spring is weaker than the secondary return spring. This allows the primary shoe to be applied before the secondary shoe.

The coils on return springs should be tightly wound together. This means that the coils should be touching each other, with no space between them (**FIGURE 36-30**). Any space indicates that the return spring has been stretched and must be replaced. Return springs are normally replaced in sets. If

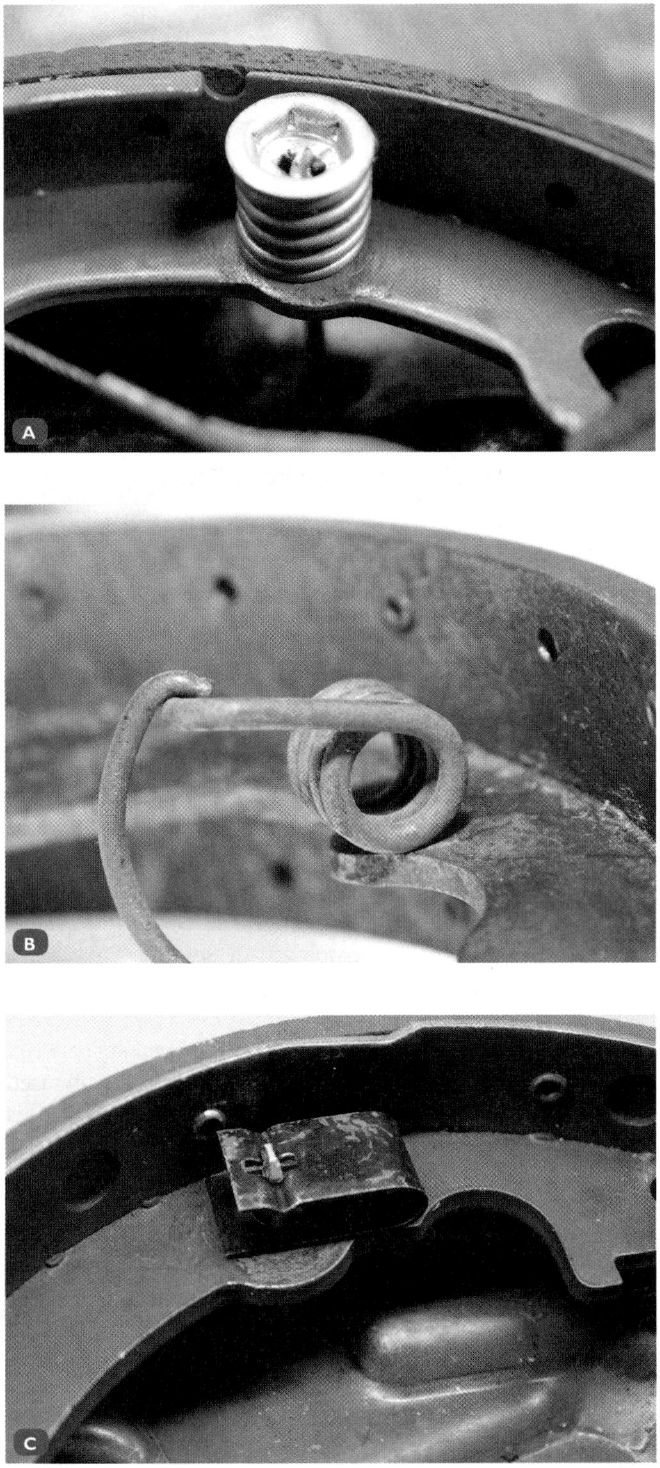

FIGURE 36-31 Hold-down springs. **A.** Coil spring style. **B.** Sideways coil style. **C.** U-shaped clip style.

one spring has to be replaced, they all should be replaced to avoid mismatched components. Also, it is good practice to replace the return springs and hardware during every brake job. This is because of the high temperatures and the large number of apply and return cycles that drum brake springs must endure.

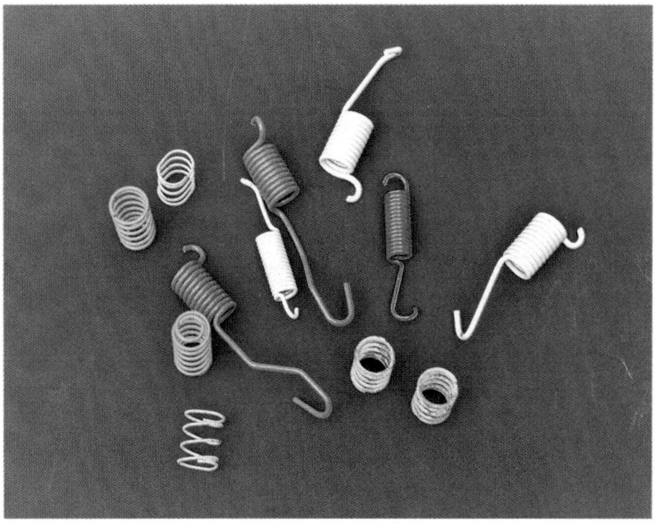

FIGURE 36-32 Brake specialty springs.

Hold-Down Springs

Hold-down springs do just what their name suggests. They hold the brake shoes against the backing plate (**FIGURE 36-31**). They can be coil springs in combination with a spring retainer and a pin, sometimes referred to as a brake nail. The pin extends through the backing plate, brake shoe, and center of the spring. Or they can be coil springs lying on their side and connected to the pin beside them. They can also use U-shaped spring steel clips in combination with a pin that extends through the backing plate and brake shoe.

Specialty Springs

Specialty springs are used to return links and levers on the parking brake system or the self-adjuster mechanism. Specialty springs can be of all different shapes and sizes (**FIGURE 36-32**). They can be used to push or pull components into proper position. Be sure to install each specialty spring in the correct position so that each of the brake systems works as intended. If in doubt, check the other wheel brake unit or the service information to see how it is assembled.

▶ Self-Adjusters

LO 36-07 Describe drum brake self adjusters and parking brake operation.

Brake shoe lining wears over time. This increases the clearance between the brake lining and drum. This causes the wheel cylinder pistons to travel farther before the brake shoes contact the drums. It also causes the brake pedal to travel farther before activating the brakes. Reserve pedal height is then reduced. And this leaves the driver vulnerable if a hydraulic failure occurs.

Since 1968, manufacturers needed to incorporate a self-adjusting device into their drum brakes. This helped to safeguard against these dangers. The self-adjuster must be capable of maintaining proper shoe-to-drum clearance. Manufacturers have come up with a couple ways to meet the rule.

Types of Self-Adjusters

The first type of self-adjuster is used on most duo-servo brakes. It has an adjustable, threaded star wheel assembly holding the bottoms of the brake shoes apart. The link has three main parts:

- the adjuster screw,
- the non-threaded barrel (and possible thrust washer), and
- the star wheel, which is threaded on one end and smooth on the other.

The adjuster screw and barrel have slots that fit to tabs on the brake shoes. This prevents them from turning. The star wheel's threaded end screws into the adjustment screw. The smooth end of the star wheel fits into the non-threaded barrel (**FIGURE 36-33**). As the star wheel is turned, the thread causes the self-adjuster to lengthen. This takes up excess brake shoe clearance, as needed. The question is what turns the star wheel?

A movable self-adjuster link is held against the star wheel (**FIGURE 36-34**). The link is moved up and down by the action of applying the brakes while the vehicle is backing up. If the link moves far enough, it will catch another tooth of the star wheel. The spring action will turn the star wheel one tooth, adjusting the brakes shoes outward just a bit. This process continues until the excess brake shoe clearance is taken up. At this point, the self-adjuster link cannot move far enough to catch another tooth on the star wheel. As the brake linings wear, the clearance increases. This allows the movement of the link to increase to the point that it catches another tooth. This continues for the life of the brake lining.

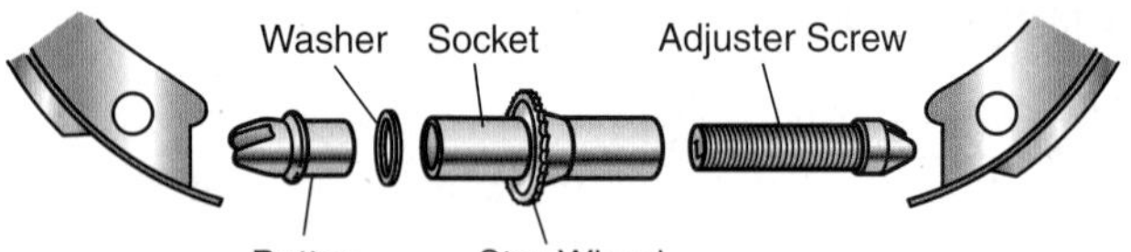

FIGURE 36-33 A star wheel assembly.

▶ TECHNICIAN TIP

Some drivers never use their brakes when backing up. Instead, they back up slowly. They put the transmission in first gear and use their clutch or automatic transmission to stop the vehicle. They start going forward at the same time. Backing without braking prevents the self-adjusters from adjusting the drum brakes on vehicles equipped with this type of self-adjuster. Educating customers on how the self-adjuster works will help them keep their brakes adjusted as well as get longer life out of the linings.

▶ TECHNICIAN TIP

Be careful to not switch the self-adjusters from one side of the vehicle to the other. If this happens, the self-adjuster will retract the adjustment. This will cause the brake shoe clearance to increase as the brakes adjust. The customer will report that the brake pedal keeps getting lower and lower. But it was fine when he/she picked up the vehicle.

The self-adjuster link is moved by a cable or rod, which is connected to the anchor pin at the top of the backing plate. The movement of the secondary brake shoe, when the brakes are applied while backing up, causes the self-adjuster link to be pulled by the cable or rod. The position of the link and cable or rod is critical to the operation of the self-adjuster assembly. The link should be level with the star wheel and close to its centerline. Also, the star wheel assembly is side-specific. This means that the threads on each star wheel are opposite to each other. So they turn in opposite directions.

A similar style of self-adjuster works off movement between one brake shoe and the parking brake strut. It is usually used on non-servo brakes. The star wheel linkage is threaded. It is

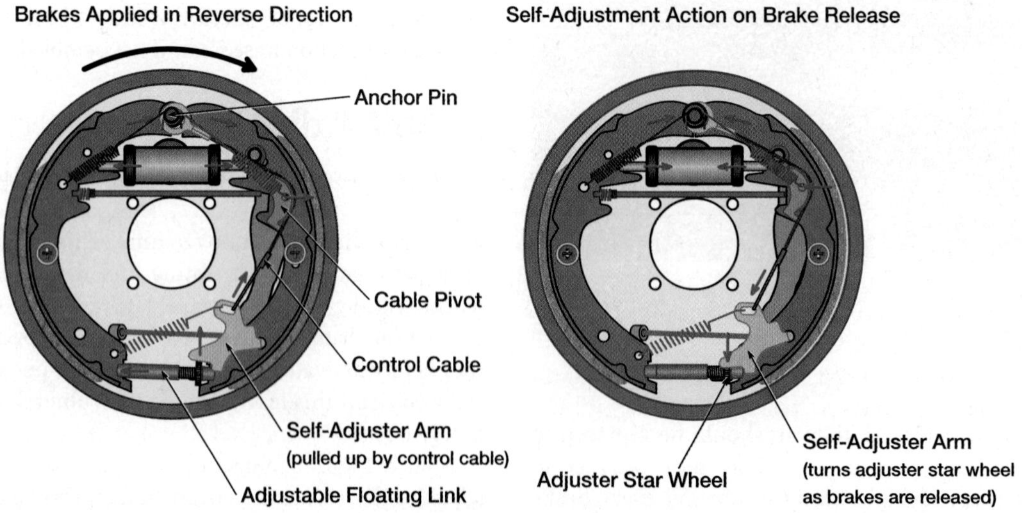

FIGURE 36-34 A self-adjuster on a duo-servo-style brake.

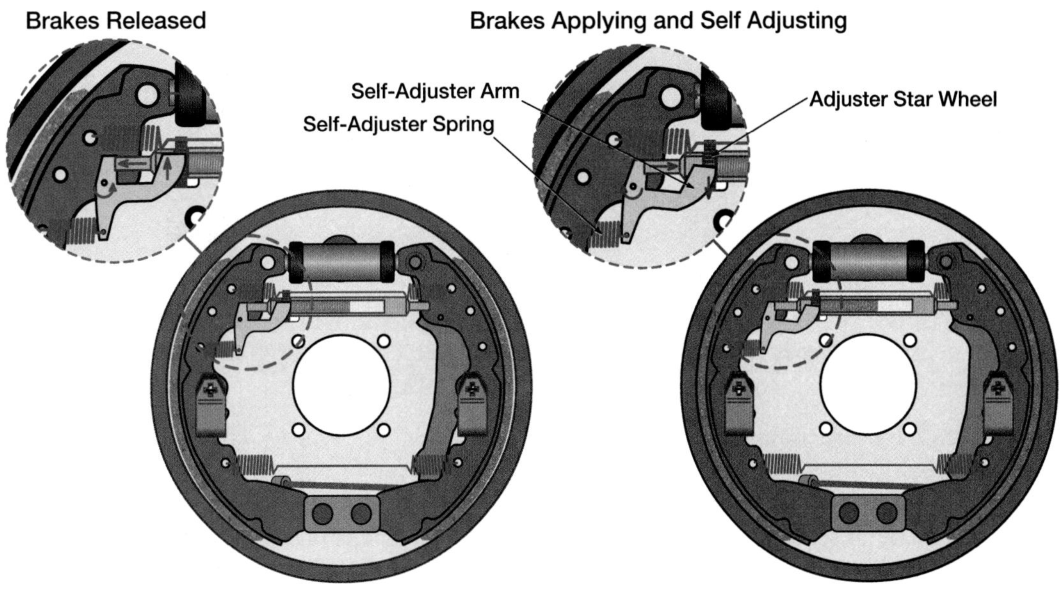

FIGURE 36-35 Alternate style of automatic star wheel brake adjuster.

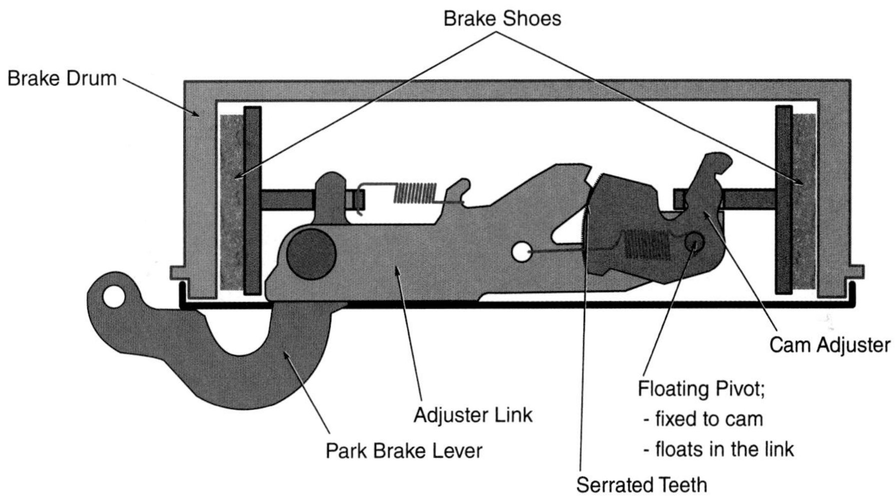

FIGURE 36-36 A ratchet-style adjuster adjusts if excess brake clearance is present when the service brake is applied.

usually mounted between the brake shoes just below the wheel cylinder (**FIGURE 36-35**). It acts as the parking brake strut. There is a pivoting self-adjuster lever attached to one brake shoe.

When the service brake is applied, a spring pulls the pivoting lever down. The pivoting lever is pushed up by the end of the parking brake strut when the brake pedal is released. If the lever travels far enough when the service brake is applied, it will catch the next tooth on the star wheel. As the brake pedal is released, the end of the strut pivots the self-adjuster link upward against spring pressure. This turns the star wheel linkage slightly and lengthens the parking brake strut slightly. Brake shoe clearance is decreased slightly. This process continues as the brake pedal is applied and released until the excess clearance is taken up. The self-adjuster lever will then be unable to catch another tooth on the star wheel. At least until the clearance opens a bit more as the brake linings continue to wear.

Some manufacturers use a ratcheting-style self-adjuster, sometimes called a cam-style adjuster (**FIGURE 36-36**). This type uses two toothed pieces held in contact with each other by spring pressure. They can slide over each other in one direction but hold in the other direction. The two pieces allow excessive shoe-to-drum clearance to be taken up when the brake pedal is applied. They then prevent the shoes from fully returning to their rest position when the brakes are released.

As the service brakes are applied, the shoes move apart. If they travel far enough, the toothed pieces slide over each other until the shoes contact the drum. When the brakes are released, the components allow a small amount of inward

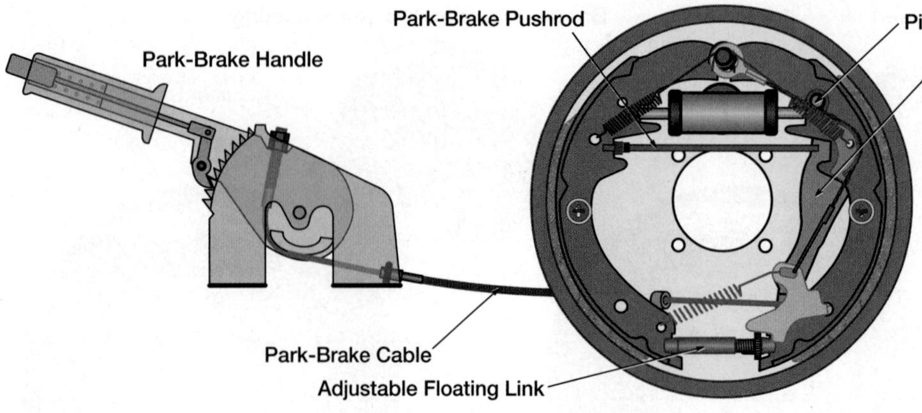

FIGURE 36-37 A parking brake assembly for a drum brake.

brake shoe movement to occur. This creates the proper shoe-to-drum clearance. This style of self-adjuster generally can be adjusted in one brake pedal application. The amount of time spent pre-adjusting the brake shoes is substantially reduced.

Parking Brake Systems

Drum parking brake systems mechanically apply the regular service brake shoes. Because drum brakes are self-energizing, it is easier to generate the force needed to apply them mechanically than it is to apply disc brakes. The parking brake cable attaches to the bottom of the parking brake actuating lever in the drum brake assembly. The other end of the lever is attached to the top end of one of the brake shoes. A strut rod runs from near the top of the actuating lever to the other brake shoe (**FIGURE 36-37**).

When the cable is pulled, the lever pivots on the strut rod. This pushes the top of the brake shoe rearward and the strut rod forward. Because the strut rod is connected to the front shoe, it is pushed forward. This results in the brake shoes being forced apart. They are then held in firm contact with the drum, holding the vehicle stationary. Releasing the cable allows the brake retracting springs to pull the brake shoes away from the drums.

▶ Wrap-Up

Ready for Review

- Brake drums are bolted to the vehicle's axle flange by the lug nuts. The main components of the drum brake system are: brake drum, backing plate, wheel cylinder, brake shoe, springs and clips, automatic brake self-adjuster, and parking brake mechanism.
- Three main types of drum brake systems: twin leading shoe, leading/trailing shoe, and duo-servo. All three types are self-energizing in at least one direction.
- Drum brakes function to apply the brake shoes, release the brake shoes, and adjust the brake shoes. The backing plate houses all of the brake unit components except the brake drum.
  - One of the most important components of the backing plate is the anchor pin.
- The wheel cylinder converts hydraulic pressure from the master cylinder into mechanical force that pushes the brake linings against the inside of the brake drum. Wheel cylinders usually contain: a cylinder housing, one or two pistons, a lip seal for each piston, a spring and expander set, a dust boot for each open end of the cylinder, a pushrod for each piston, and a bleeder screw.
- The drum brake system makes use of metal brake shoes that have friction material called linings attached to them. The composition of the lining material affects brake operation.
  - Brake linings materials include: non-asbestos organic, low-metallic NAO materials, semi-metallic materials, and ceramic materials.
- The action of the brake shoes is controlled with the help of return, hold-down, or specialty springs. Each type of brake uses its own arrangement of these components.
- A self-adjuster must be capable of maintaining proper shoe-to-drum clearance. Types of self-adjuster: adjustable, threaded star wheel assembly; automatic star wheel brake adjuster; ratchet-style adjuster.
  - It is easier to generate the force needed to apply drum parking brake systems mechanically than it is to apply disc brakes.

Key Terms

anchor pin A component of the backing plate that takes all of the braking force from the brake shoes.

automatic brake self-adjuster A system on drum brakes that automatically adjusts the brakes to maintain a specified amount of running clearance between the shoes and drum.

backing plate A metal plate to which the brake lining is fixed.

brake drum A short, wide, hollow cylinder that is capped on one end and bolted to a vehicle's wheel; it has an inner friction surface that the brake shoe is forced against.

brake shoe The arched metal shoes and friction material in a drum brake system.

cylinder bore The diameter of the hole that the piston moves in; it is one of the main components that is included in the size of the engine.

duo-servo drum brake system A system that uses servo action in both the forward and reverse direction.

hold-down spring tool A tool used for removing and installing hold-down springs.

leading shoes Brake shoes that are installed so that they are applied in the same direction as the forward rotation of the drum and thus are self-energizing.

leading/trailing shoe drum brake system Type of brake shoe arrangement where one shoe is positioned in a leading manner, and the other shoe in a trailing manner.

parking brake mechanism A mechanism that operates the brake shoes or pads to hold the vehicle stationary when the parking brake is applied.

return springs Springs that retract the brake shoes to their released position.

self-energizing The property of drum brakes that assists the driver in applying the brakes; when brake shoes come into contact with the moving drum, the friction tends to wedge the shoes against the drum, thus increasing the braking force.

Servo action A drum brake design where one brake shoe, when activated, applies an increased activating force to the other brake shoe, in proportion to the initial activating force; further enhances the self-energizing feature of some drum brakes.

specialty springs Springs used to return links and levers on the parking brake system or the self-adjuster mechanism.

springs and clips Various devices that hold the brake shoes in place or return them to their proper place.

trailing shoes Brake shoes installed so that they are applied in the opposite direction to the forward rotation of the brake drum; not self-energizing and less efficient at developing braking force.

twin leading shoe drum brake system Brake shoe arrangement in which both brake shoes are self-energizing in the forward direction.

wheel cylinder A hydraulic cylinder with one or two pistons, seals, dust boots, and a bleeder screw that pushes the brake shoes into contact with the brake drum to slow or stop the vehicle.

Review Questions

1. Which of the following is true about drum brakes?
   a. They are generally expensive to manufacture
   b. They are more resistant to overheating
   c. They are easier to adapt a parking brake to
   d. They don't require removal of the drum to inspect linings
2. What is the purpose of the wheel cylinder?
   a. To support the brake shoes and hardware
   b. To self-adjust the brakes to compensate for wear
   c. To form the braking surface for the brake linings
   d. To push the brake shoes into contact with the drums
3. What type of drum system is most likely to be found on the rear of a modern front-wheel drive vehicle?
   a. Twin leading shoe
   b. Leading/trailing shoe
   c. Duo-servo
   d. Non-anchored
4. Which type of drum brake design is very efficient at braking in forward, but the least effective in reverse?
   a. Duo-servo
   b. Leading/trailing shoe
   c. Double-Anchor
   d. Twin leading shoe
5. A drum with a minimum thickness specification of 10.060" is measured at 10.070". Which of the following is most true?
   a. The drum can be machined another 0.010"
   b. The drum should be inspected and machined
   c. The drum can be reused if there are no grooves
   d. The drum must be replaced
6. How is a wheel cylinder bleeder screw sealed?
   a. The piston seals
   b. The bleeder o-ring
   c. The tapered seat
   d. The bleed passage lip seal
7. In a duo-servo drum brake system, in what location does the primary shoe belong?
   a. Toward the front of the vehicle
   b. On the driver's side of the vehicle
   c. On the passenger side of the vehicle
   d. Toward the rear of the vehicle
8. What is the most likely condition that would cause brake shoes to click?
   a. The brake shoes are worn to metal on metal
   b. There is excessive dust in the brake drum
   c. The brake shoes have worn grooves in the backing plate
   d. The brake shoes are nearing minimum thickness
9. All of the following are examples of typical drum brake springs EXCEPT?
   a. Return
   b. Apply
   c. Specialty
   d. Hold-down
10. When were manufacturers required to integrate self-adjusting devices into their drum brakes?
    a. 1968
    b. 1977
    c. 1986
    d. 2000

ASE Technician A/Technician B Style Questions

1. Drum brakes are being discussed. Technician A states that long downhill grades or heavy use can overheat drums. Technician B states that since drum brakes have more material, they are more resistant to overheating than disc brakes. Who is correct?
   a. Technician A only
   b. Technician B only
   c. Both Technicians A and B
   d. Neither Technician A nor B
2. A brake repair is being discussed. Technician A states that drum brakes have less moving parts and are easier to replace than disc brakes. Technician B states that drum brakes that are misadjusted will have a low brake pedal height. Who is correct?
   a. Technician A only
   b. Technician B only
   c. Both Technicians A and B
   d. Neither Technician A nor B
3. Rear drum brake types are being discussed. Technician A states that a duo-servo system benefits from servo action in forward only. Technician B states that the leading/trailing shoe design produces maximum braking in forward only, so it is commonly found on heavy duty trucks. Who is correct?
   a. Technician A only
   b. Technician B only
   c. Both Technicians A and B
   d. Neither Technician A nor B
4. Drum brakes are being discussed. Technician A states that direct road spray to the brake shoes is prevented by a labyrinth seal in-between the drum and backing plate. Technician B states that the shoe to backing plate contact pad should be lightly greased during a brake service. Who is correct?
   a. Technician A only
   b. Technician B only
   c. Both Technicians A and B
   d. Neither Technician A nor B
5. A wheel cylinder is being discussed. Technician A states that the wheel cylinder can be bolted or firmly clipped to the backing plate. Technician B states that most wheel cylinders are equipped with 1 piston and 2 bleeders. Who is correct?
   a. Technician A only
   b. Technician B only
   c. Both Technicians A and B
   d. Neither Technician A nor B
6. A wheel cylinder is being discussed. Technician A states that if the cylinder bore surface in a wheel cylinder is pitted, it should be replaced rather than honed. Technician B states that the purpose of a dust boot is to keep brake dust and moisture away from the piston and cylinder. Who is correct?
   a. Technician A only
   b. Technician B only
   c. Both Technicians A and B
   d. Neither Technician A nor B
7. Brake shoes are being discussed. Technician A states that some brake shoes have lining riveted to the backing and others are bolted. Technician B states that it is a common mistake for new technicians to install both primary shoes on one side of the vehicle, and both secondary shoes on the other side. Who is correct?
   a. Technician A only
   b. Technician B only
   c. Both Technicians A and B
   d. Neither Technician A nor B
8. Drum brake springs being discussed. Technician A states that it is a good idea to leave one side of the vehicle together for reference while the first side is finished since the spring locations can be confusing. Technician B states that the return springs are generally soft and easy to install by hand. Who is correct?
   a. Technician A only
   b. Technician B only
   c. Both Technicians A and B
   d. Neither Technician A nor B
9. A vehicle with drum rear brakes is being discussed. Technician A states that the self-adjusting mechanism often works to take up slack when stopping in reverse or by applying the parking brake. Technician B states that self-adjusters lengthen over time as the brake shoe linings wear. Who is correct?
   a. Technician A only
   b. Technician B only
   c. Both Technicians A and B
   d. Neither Technician A nor B
10. A parking brake is being discussed. Technician A states that the parking brake typically applies the rear drum brakes hydraulically. Technician B states that drum brakes require less force to apply as a parking brake than disc brakes because of their self-energizing action. Who is correct?
    a. Technician A only
    b. Technician B only
    c. Both Technicians A and B
    d. Neither Technician A nor B

CHAPTER 37
Servicing Drum Brakes

Learning Objectives

- **LO 37-01** Describe drum brake diagnosis.
- **LO 37-02** Inspect and measure brake drums.
- **LO 37-03** Refinish brake drums.
- **LO 37-04** Service brake shoes and hardware.
- **LO 37-05** Service wheel cylinders.
- **LO 37-06** Install drums, wheels, and perform final checks.

ASE Education Foundation Tasks

See Appendix A to view the 2017 ASE Education Foundation Automobile Accreditation Task List Correlation Guide.

Servicing Drum Brakes

▶ Servicing Drum Brakes

LO 37-01 Describe drum brake diagnosis.

Drum brake diagnosis starts with understanding the customer's concern. Communicating directly with the customer is the best way to do that, but the customer is not always available. An experienced service advisor will gather the required information. So, read the service advisor's notes on the repair order carefully or speak with them directly. Once you understand the customer's concern, a test drive is usually needed to verify the accuracy of the concern. Depending on the concern, it may be as simple as stepping on the brake pedal and feeling the pedal sink to the floor. Or it could require a more detailed test drive to observe the fault the customer is describing. This is a good opportunity to test the brakes under a variety of conditions. Replicating the customer concern is important. That way you know you are addressing the situation that the customer is experiencing.

During the test drive, find a safe place to operate the brakes. You need to test them at a variety of speeds and with a variety of brake pedal pressures. Especially try to mimic the conditions the customer described. If you are having difficulty replicating the concern, it may be necessary to go on a test drive with the customer driving. Allow him/her to operate the vehicle in the way that makes the problem evident. It also allows them to point out the particular situation they are experiencing. Also, it is good to have the customer along in case the problem does not occur. In this case, the customer will understand why you can't diagnose the problem.

Test driving a vehicle with an ABS issue requires extreme caution. Make sure that an accident doesn't happen. ABS operates only during extreme braking or poor traction conditions. So, you run the risk of being rear-ended or losing control of the vehicle if you apply the brakes hard enough to activate the ABS. Always make sure the vehicle is being test driven away from all other traffic. Remember that if the yellow ABS warning lamp is illuminated, the ABS system is likely to be deactivated. In this situation, trying to activate ABS on a test drive requires even more caution.

If the yellow ABS lamp is off, the ABS system should be active and ready to activate. It is helpful when testing ABS to do so on a surface with limited traction, such as wet pavement or a dirt road. When you do apply the brakes firmly, ABS activation can often be felt as pulsations in the brake pedal and possibly the steering wheel. The vehicle should brake quickly while maintaining steering control. In some cases, a "poor traction" lamp may illuminate, telling you that the ABS system was activated. This warning lamp usually turns off after several seconds of driving. If the yellow ABS warning lamp illuminates, it typically means that the vehicle's ECM has recorded a fault in the ABS system. A DTC is likely set that you should retrieve with a scan tool.

Once the customer concern has been verified, research the concern in the service information and TSBs. Armed with this information, you should be ready to create a plan to begin focused testing of the brake system. This could be as simple as:

- performing a visual inspection of the brake fluid level and condition.
- removing the wheels and drums to inspect the brake units.
- measuring the diameter and runout of the drums.

It could also involve electrical diagnosis such as locating a short circuit in the brake warning indicator system.

Suspension and steering system faults can appear to be brake system faults. An example of this is a pulling condition while braking. If the control arm bushings are worn, then braking the vehicle will cause the wheel to pivot outward. This makes the wheel point in a direction other than straight down the road, imitating a brake pull. So, always inspect the suspension and steering systems when diagnosing concerns related to a brake "pull."

The vehicle's braking system must always be restored to its proper operation when working on it. This also means that any diagnosis must identify all of the issues that would prevent the brakes from operating normally. Lawyers and technicians have been known to say, "He who touched the brake system last,

You Are the Automotive Technician

A vehicle owner has indicated that all of a sudden, the brake pedal is lower than normal. It also doesn't stop as well as it should. You step on the brake pedal and find that it is indeed very low. Because of the safety concerns, you skip the test drive and take it straight to the shop for a visual inspection. There are no related TSBs. The service information indicates that the vehicle is equipped with duo-servo brakes. You inspect the vehicle and find that the brake fluid is empty in one half of the reservoir. The fluid in the other side is below the MIN line and very dark. Brake fluid is found dripping out of the right rear brake drum. And the brake shoes are saturated with brake fluid.

1. What can cause the reservoir to be empty in only one half?
2. What service should be performed based on the dark brake fluid?
3. Where is it likely the brake fluid came from that was on the brake shoes?
4. Can the fluid-soaked brake shoes be reused? Why or why not?

owns it!" What this means is that if there is a problem in the braking system and you inspected it or worked on it, you are very likely liable for anything that went wrong with it. Brake system failures are more likely to lead to vehicle accidents than failures of most other systems. So, any diagnosis and subsequent repairs need to be thorough and complete.

Once you have identified the cause of the fault, determine the action that will correct the fault. This information can then be used for preparing an estimate of repairs. This estimate can then be given to the customer for authorization to perform the repairs. Once the repair is complete, retest the system to confirm that the concern has been fully addressed and no further issues are present.

> ▶ TECHNICIAN TIP
>
> It is common to have more than one fault present in a brake system when you diagnose it. For example, worn brake shoes, a leaky wheel cylinder, and a leaky seal in the master cylinder. It is good practice to perform a thorough inspection of the brake system whenever one condition is present. This way you can determine if any others are present at the same time. This prevents you from having to go back to the customer to get permission to perform the additional work. Since this is after he/she agreed to the initial repair, the customer has reason to doubt your competence or integrity.

> ▶ TECHNICIAN TIP
>
> One way to identify whether a problem is coming from the front or rear brakes is by using the parking brake. But the vehicle must have a manually applied parking brake, not an electronic parking brake. Test drive the vehicle in a safe place at a relatively low speed and lightly apply the parking brake. If the condition is still present, the problem is with the rear brakes. You know this because the parking brakes are usually only on the rear wheels. If the condition is not present, the problem is likely with the front brakes. This is not true on vehicles equipped with top-hat style parking brakes. This is because the parking brake is separate from the service brake.

▶ Maintain and Repair Drum Brakes

LO 37-02 Inspect and measure brake drums.

Drum brakes require periodic maintenance and repair as the vehicle ages. As stated previously, visual inspection of the braking system is part of the diagnostic procedure. It is also a critical part of maintenance and repair. A good visual inspection can turn up issues such as:

- lining that isn't wearing evenly,
- contaminated linings from a leaking wheel cylinder,
- a warped drum, or
- even broken springs and hardware.

The following skill drills will lead you through the common drum brake maintenance and repair procedures.

Tools

Like most systems on the vehicle, the brake system requires special tools when working on them. They make the job safer and easier to perform. Always use the proper tool for the job at hand. Also, do not exceed the capabilities of any tools. Doing so can lead to personal injury. The tools used to diagnose repair drum brake systems include (**FIGURE 37-1**):

- **Brake wash station**—Used to clean drum and disc brake dust.
- **Brake spring pliers**—Used to remove and install return springs.
- **Hold-down spring tool**—Used to remove and install hold-down springs.
- **Drum brake micrometer**—Used to measure inside brake drum diameter.
- **Brake shoe adjustment gauge**—Used to pre-adjust brake shoes before installing the drum.
- **Brake spoon**—Used to adjust brake shoes when the drum is installed.
- **Wheel cylinder piston clamp**—Used to hold the pistons in the wheel cylinder while the brake shoes are removed.

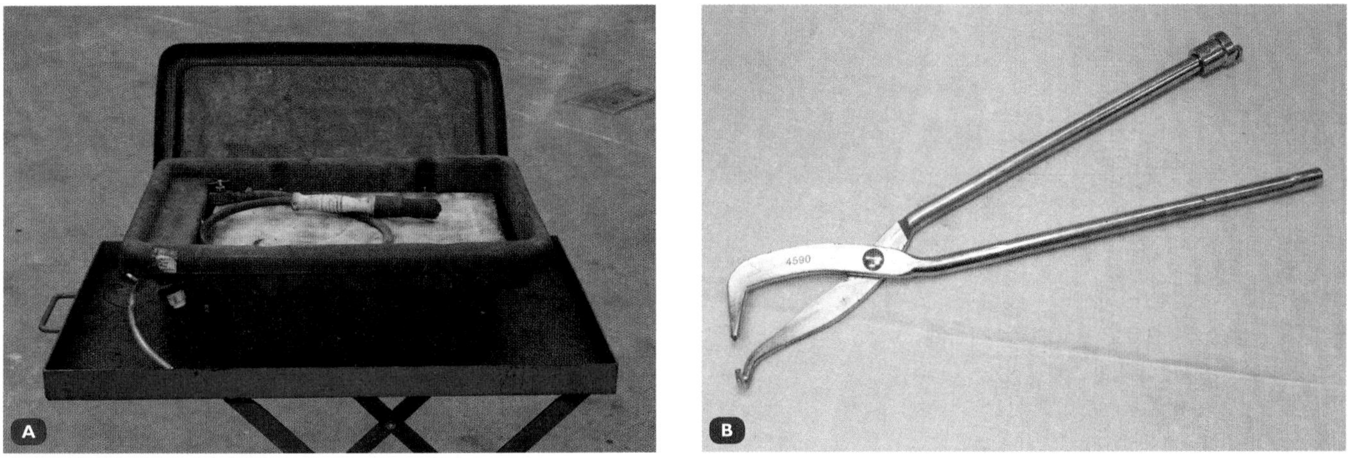

FIGURE 37-1 Drum brake tools. **A.** Brake wash station. **B.** Brake spring pliers.

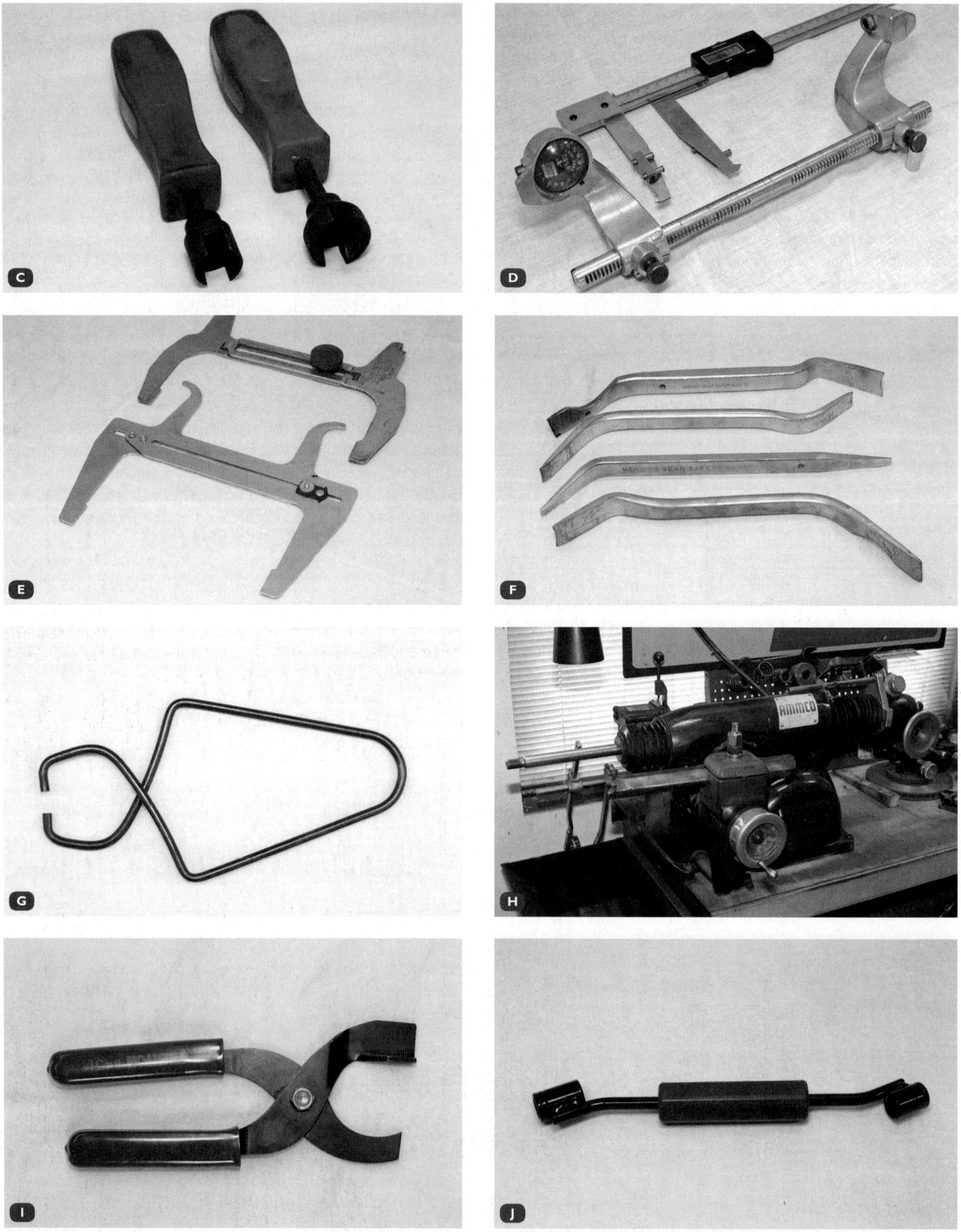

FIGURE 37-1 Drum brake tools. **C.** Hold-down spring tool. **D.** Drum brake micrometer. **E.** Brake shoe adjustment gauge. **F.** Brake spoon. **G.** Wheel cylinder piston clamp. **H.** Off-car brake lathe. **I.** Parking brake cable pliers. **J.** Parking brake cable removal tool.

- **Off-car brake lathe**—Used to machine drums and rotors that are off the vehicle.
- **Parking brake cable pliers**—Used to install parking brake cables on the parking brake lever.
- **Parking brake cable removal tool**—Used to remove the parking brake cable from the backing plate.

Removing, Cleaning, Inspecting, and Measuring Brake Drums

Brake drums need to be removed for a variety of reasons. For example, inspecting the thickness of the brake linings, checking for wheel cylinder leaks, and packing wheel bearings. While removed, the brake drums should be inspected visually for any damage. They should also be measured with a drum micrometer. This will allow you to verify that they are smaller than the specified maximum diameter. If not, they should be replaced.

Before removing the brake drum, you need to determine if it is a hub style, or hubless (slip-off) style. Most light-duty vehicles use hubless drums. You can usually tell which style of brake drum you are working on by looking at the holes for the lug studs. If there is clearance between the stud and drum, it is likely a hubless brake drum. You can also look at the clearance between the center hole of the brake drum and the hub. A hubless brake drum has a parting line at the centering hole. A hub-style brake drum appears to be one solid piece.

You can take several steps to ease the drum removal process. First, back off the brake shoe adjustment before removing the drum. The shoes and drum are in close proximity to each other. If you don't back off the adjustment, they can bind during the removal process. This is especially true if there is a ridge on the edge of the drum or in case the brake shoes are heavily grooved.

Second, because of the tight clearances between the wheel flange and brake drum, the surfaces can rust together. You may have to use a medium or large ball-peen hammer to hammer on the drum between the lug studs. Make sure you do not hit the lug studs because you will damage the threads. Some drums can be rusted on very solidly. Sometimes, a torch may be needed to heat up the drum to break the rust bond. Some drum brakes have two threaded holes in the drum that allow bolts to be installed. By tightening the bolts, the drum is forced off the wheel flange.

In case the drum still can't be removed, cut the heads off the hold-down pins. This can be done with a diagonal cutter from the back side of the backing plate. This releases the brake shoes from the backing plate and allows the shoes to come off with the drum. If you are having difficulty breaking the drum loose, seek the assistance of your supervisor.

Once the drum is removed, inspect it for damage. Any deep grooves, cracks, hot spots, or heat checking means the drum needs to be replaced. If there is no visible damage, clean it using a brake wash station. Then measure its diameter. Also, calculate the amount of out-of-round it has.

To measure a brake drum, a brake drum micrometer or special Vernier tool is used. The drum micrometer is an interesting tool that needs to be set to the drum size before it can be used. You do this by adjusting each of the arms to the desired setting on the rails. Each arm should be set to the same reading. For example, on a 10-inch (254 mm) drum, each arm should be set to the 10-inch mark. The dial indicator will then read the amount the measurement is above 10 inches (254 mm).

When placing the drum micrometer in the drum, the dial indicator side goes first. To find the maximum diameter, hold the pivoting end firmly against the drum. Then, move the dial indicator end side-to-side to find the largest reading. Compare this with specifications. Remember that on drums, the measurement must be smaller than the specified maximum diameter.

To find the amount the drum is out-of-round, move the tool about 1/3 of the way around the drum. Take another measurement and record the reading. Then move the tool another 1/3 of the way around the drum again. Take another reading and record it. Subtract the smallest number from the largest number to get the out-of-round measurement. Compare that with specifications.

To remove, clean, inspect, and measure brake drums and determine any further actions, follow the steps in **SKILL DRILL 37-1**.

SKILL DRILL 37-1 Removing, Cleaning, Inspecting, and Measuring Brake Drums

1. To perform this procedure on a hubless-style drum, first make matching marks on the drum and wheel flange for reinstallation in the correct position. A permanent marker, crayon, or center punch can be used.

2. If there are screws or speed nuts holding the drum to the wheel flange, remove them following the specified procedure. The speed nuts can be discarded and are not needed upon reassembly. The screws will be reused.

3. Remove the drum from the wheel flange.

SKILL DRILL 37-1 Removing, Cleaning, Inspecting, and Measuring Brake Drums (Continued)

4. To perform this procedure on a hub-style drum, remove the wheel bearing locking mechanism (cotter pin, locknut, or peened washer).

5. Remove the wheel bearing adjusting nut, thrust washer, and outer bearing.

6. If the inner bearing needs to be removed for service or replacement of the drum, use the following procedure to remove it and the grease seal.

- Reinstall the adjusting nut onto the spindle about five turns.
- Grasp the drum/hub assembly on the top and bottom. Push it toward the center of the vehicle. Hold with slight downward pressure as you firmly pull the drum/hub assembly toward yourself. This should cause the adjusting nut to catch the wheel bearing and pull it and the seal out of the hub.

7. Clean the drum with approved asbestos removal equipment. Inspect the drum for hard spots/hot spots, scoring, cracks, and damage.

8. Measure the drum diameter at the deepest groove or most worn part of the drum, and compare with specifications. If it is over the size limit, it will have to be replaced. If it has to be machined and is below the maximum diameter, check to see how badly it is scored.

9. Find the smallest diameter of the drum, and write down the reading. Find the largest diameter of the drum, and write down that reading. Calculate the difference between the two readings, (out-of-round), and compare it with specifications. Determine any necessary actions.

▶ TECHNICIAN TIP

When backing off the self-adjuster, remember that you will have to hold the self-adjuster link away from the star wheel so that the star wheel can be retracted.

▶ Refinishing Brake Drums

LO 37-03 Refinish brake drums.

Brake drums must be refinished when they have excessive grooving or are out-of-round. A **brake lathe** refinishes the drum friction surface. It does so by removing metal and making it perfectly round with the proper finish. If the grooving or surface defects are too great, the drum may require the removal of too much metal to satisfactorily refinish the surface. The drum diameter should always be remeasured once the refinishing is complete. This will help to ensure that it is under the manufacturer's maximum diameter. Never put a brake drum that is over the maximum size back in service. It does not have the ability to absorb as much heat and will experience brake fade much sooner.

Hub-style and hubless-style drums are mounted on the brake lathe differently. Most hub-style drums use the bearing races to center and drive the drum on the lathe spindle. Bearing adapters of the proper size need to be selected and fit to the bearing races. The spindle nut then clamps the drum onto the spindle using the bearing adapters and races.

Hubless drums can be mounted in two ways. The first way uses the composite rotor adapter. The second way uses two clamshells and a centering cone. Both methods mount in a similar manner as on a disc brake rotor. The clamshell system uses a spring-loaded centering cone to align the drum's centering hole with the spindle. The clamshell clamps the drum to the lathe spindle.

Once the drum is mounted to the brake lathe, check to make sure it is running true. If so, install the anti-chatter belt. This is a rubber belt that wraps snugly around the outside of the drum. It absorbs vibrations from the machining process. Follow the brake lathe manufacturer's instructions for refinishing the brake drum.

▶ TECHNICIAN TIP

Most brake lathes are able to machine both drums and rotors. So care must be taken while machining a drum. Avoid an inadvertent adjustment that renders the drum scrap metal with only a quarter turn of the hand wheel. This means you need to be very familiar with brake lathe operation.

To refinish a brake drum and measure final drum diameter, follow **SKILL DRILL 37-2**.

SKILL DRILL 37-2 Refinishing Brake Drums

1. Clean any nicks, burrs, or rust from the mounting surfaces of the drum, including the centering hole, if used.

2. Mount the drum on the brake lathe.

3. Visually check to see that the drum is running true on the lathe and is not wobbling. If it wobbles, turn off the lathe and recheck for condition that causes it to wobble.

4. Install the anti-chatter band on the drum. It will prevent the drum from vibrating during the machining process, which would create a rough finish and dull the cutting bit.

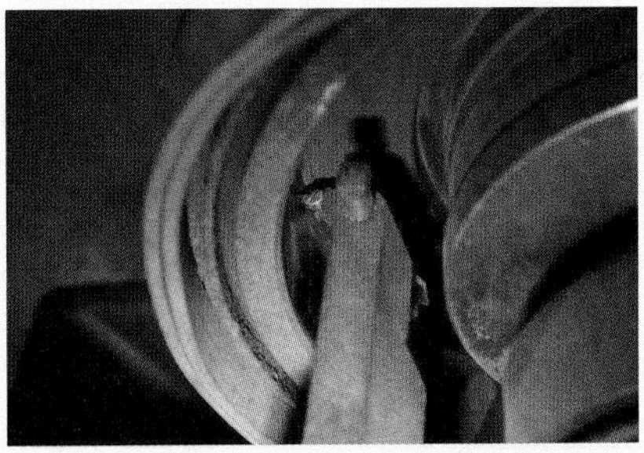

5. Set the position of the cutting tool so that the brake drum is close to the brake lathe when the cutting bit is in the far corner of the drum.

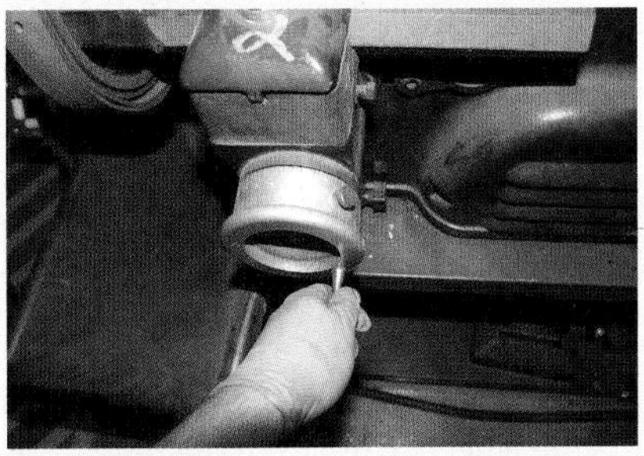

6. Make sure the cutting bits will not contact the face of the drum, and move the cutting head about 0.5" (12.7 mm) in from the outside of the drum.

SKILL DRILL 37-2 Refinishing Brake Drums (Continued)

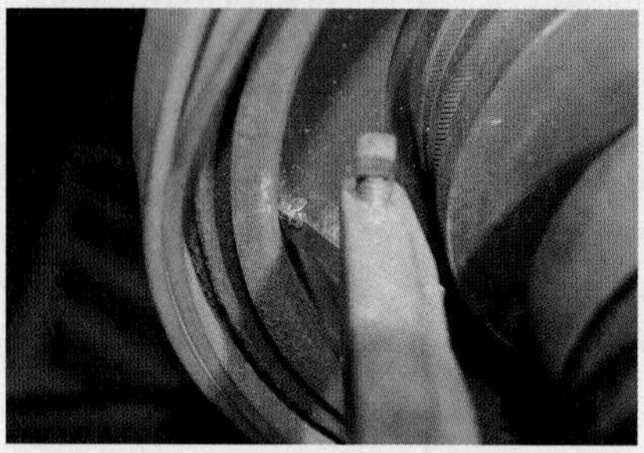

7. Turn on the brake lathe, and set the depth of the cutting tool so it just touches the surface of the drum. Rotate the brake lathe's handwheel so that the drum moves outward and the cutting bit contacts the ridge. Keep turning slowly to remove the ridge.

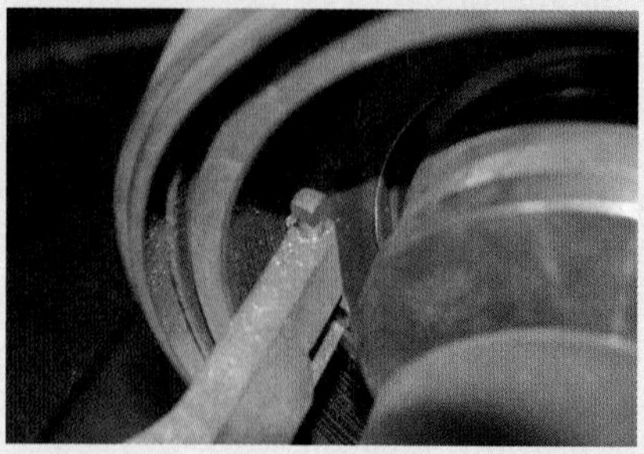

8. Once the ridge is removed, run the drum all the way in so the cutting bit is in the inner corner of the drum. Set the cutting bit to the proper depth for machining the surface of the drum, and lock it in place.

9. Engage the automatic spindle feed, and set it to the proper speed; lock it in place, and watch for proper machining action. Some single-cut machines only have one speed.

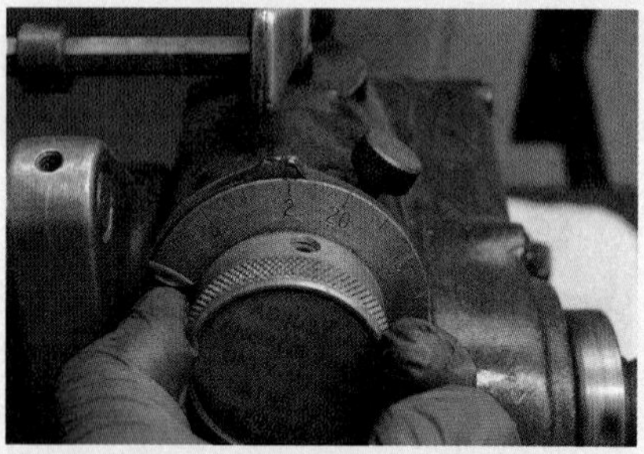

10. If necessary, repeat this step until the worn surface areas have been removed all the way around the surface of the drum. If the brake lathe is not a single-cut machine, perform a finish cut on the drum. This cut is usually done on a slower spindle feed speed.

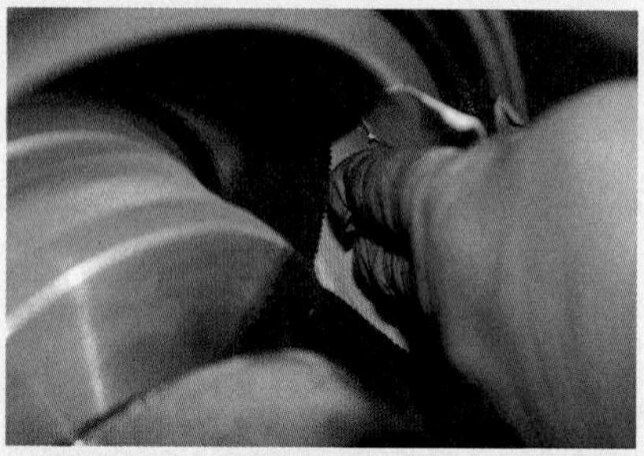

11. Move the drum well away from the cutting bit, and use sandpaper to give the drum surface a nondirectional finish.

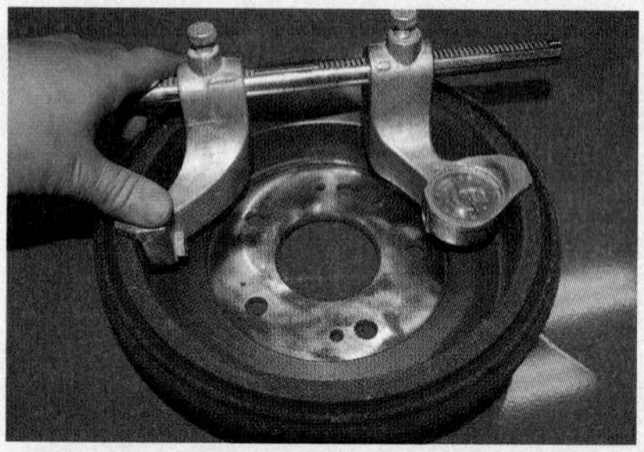

12. Remeasure the drum diameter to determine whether the drum is above the maximum-diameter specifications. If so, discard the drum.

▶ Removing, Cleaning, and Inspecting Brake Shoes and Hardware

LO 37-04 Service brake shoes and hardware.

Drum brakes need to be disassembled for a variety of reasons such as:

- when the brake lining has worn beyond specifications;
- the wheel cylinders are faulty;
- the self-adjuster is stuck or damaged; or
- the axle seal has failed and leaked gear oil or wheel bearing grease onto the brake lining.

Drum brake shoes are generally more complicated to change than disc brake pads. So, care and attention needs to be taken when working on them. They are held together by a combination of springs, links, levers, guides, and retainers. It is easy to install these items incorrectly. It is good practice to carefully examine each brake assembly before disassembling. That way you can check to see if it was assembled correctly previously. It will also help you remember how to reassemble it. You may also want to take a picture that you can refer to during reassembly.

Brake shoes and springs are always replaced in axle sets. If one brake assembly has linings contaminated with grease or brake fluid, the lining on both wheel brake units will have to be replaced. The same goes with springs. The other parts are left to the discretion of the technician and any shop policies.

Regarding star wheel assemblies, there is some controversy as to whether the threads should be lubricated. The argument on one side is that if you do not lubricate the threads, then they will likely rust and freeze up. The other side argues that any lubricant on the threads will attract brake dust and gum up the threads. There is some truth to both arguments. CDX would suggest using a light coating of lubricant on the threads in locations where the star wheel is likely to come in contact with salt, mud, or water. This is because rust and corrosion are a greater hazard than dust gumming up the threads. In areas where the assembly is unlikely to come in contact with those conditions, use no lubricant on the threads. But always follow the manufacturer's recommended procedure, if specified.

▶ TECHNICIAN TIP

It can be hard to remember where all of the parts fit inside of a drum brake assembly. It is good practice to disassemble only one side at a time. This leaves the other side as a reference for when you reassemble the first side. Or you can take a couple of reference pictures before disassembly.

To remove, clean, inspect, and reassemble a duo-servo brake, follow the steps in **SKILL DRILL 37-3**.

To disassemble, clean, inspect, and reassemble a non-servo brake, follow the steps in **SKILL DRILL 37-4**.

SKILL DRILL 37-3 Removing, Cleaning, Inspecting, and Reassembling a Duo-Servo Brake

Note: In the appropriate service information, research and follow the procedure for disassembling the brake assembly. Manufacturers use many drum brake configurations, so the following steps are general in nature and should not be substituted for the manufacturer's procedure.

1. Clean the brake shoes, hardware, and backing plates using equipment and procedures for dealing with asbestos and/or dust.

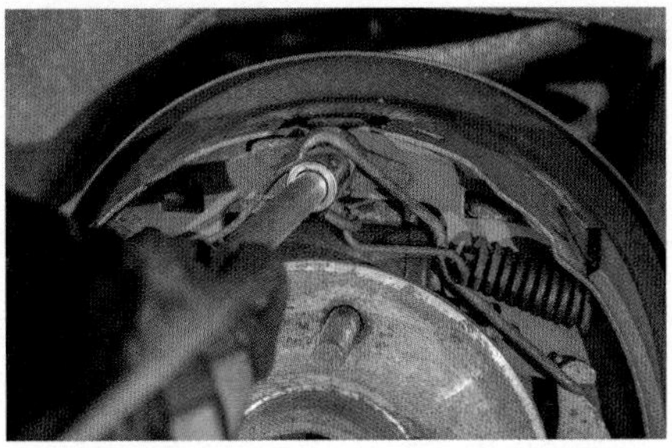

2. To disassemble a duo-servo brake, first remove the return springs, cable guide (if installed), and shoe guide. Set everything aside in the order it will go back on. Remove the parking brake strut and spring.

3. Remove the primary shoe hold-down spring, retainer, and pin.

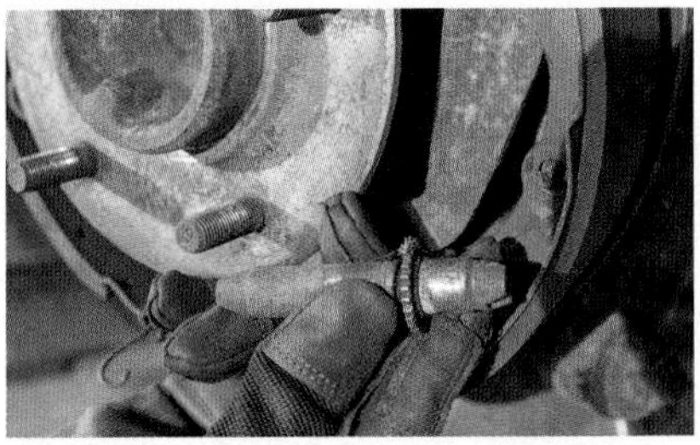

4. Remove the self-adjuster spring and star wheel assembly and primary shoe.

5. Remove the secondary hold-down spring, retainer, pin, and secondary shoe.

6. Disassemble the parking brake lever from the brake shoe and hardware from the backing plate, being careful not to lose any parts and remembering how they go back together.

7. Finally, clean and inspect all parts according to the manufacturer's procedure.

8. To reassemble a duo-servo brake, first reassemble the parking brake lever on the brake shoe and parking.

9. Reassemble the star wheel assembly, lubricate the floating end, and set aside. Also, lubricate the contact pads on the backing plate.

SKILL DRILL 37-3 Removing, Cleaning, Inspecting, and Reassembling a Duo-Servo Brake (Continued)

10. Install both shoes to the backing plate with the hold-down spring assemblies.

11. Install the shoe guide and self-adjuster cable over the anchor pin.

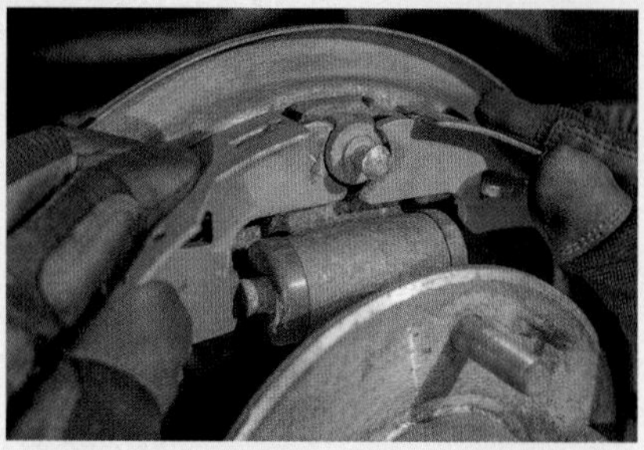

12. Install the cable guide and return spring in the secondary shoe. Also, align the wheel cylinder pushrod in the shoe.

13. Position the secondary shoe in place and use brake spring pliers to install the return spring over the anchor pin.

14. Install the parking brake strut rod onto the secondary shoe, and pull the primary shoe engaged with the parking brake strut rod.

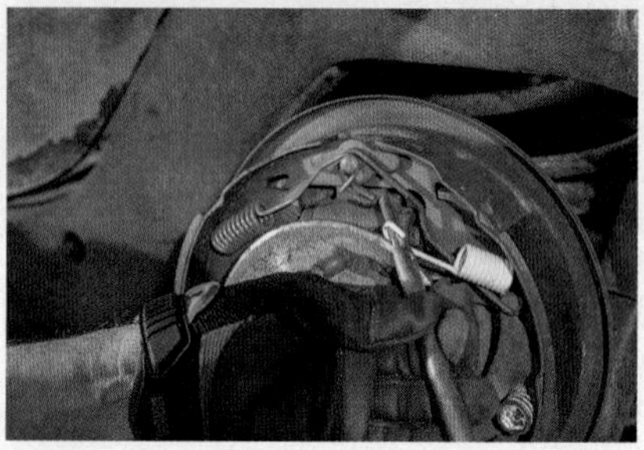

15. Install the return spring in the primary shoe and use brake spring pliers to stretch the return spring over the anchor pin.

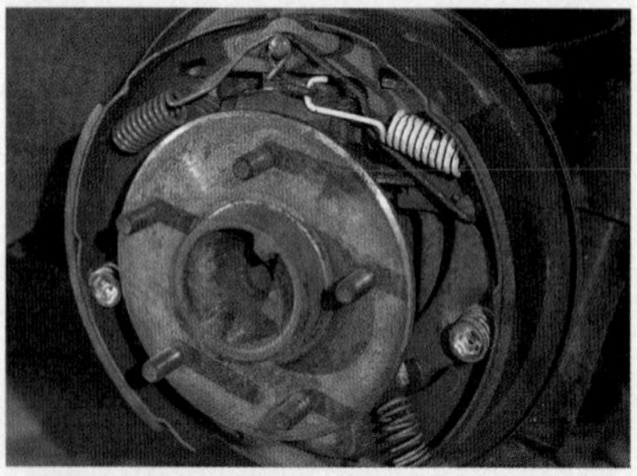

16. Install the self-adjuster link, cable, and spring into position.

17. Install the star wheel between the bottoms of the two shoes.

18. Finally, check the fit of all springs, clips, and levers. Operate the self-adjuster cable or lever.

▶ Removing, Inspecting, and Installing Wheel Cylinders

LO 37-05 Service wheel cylinders.

If a wheel cylinder is leaking or binding, there is a good chance that the cylinder bore is corroded and pitted. This requires removal and replacement of the wheel cylinder. Wheel cylinders also must be removed if the backing plate is being replaced. Very few shops rebuild wheel cylinders anymore. If you are reusing the wheel cylinder, make sure the bleeder screw can be opened. Open it before you remove the wheel cylinder. If it breaks off, you will need to replace the wheel cylinder. Because bleeder screws are hollow, they break off easily. Once broken off, bleeding is virtually impossible.

Some wheel cylinders are bolted to the backing plate. Others use a spring clip to retain the wheel cylinder to the backing plate. And others use a large C-clip. Use the correct tool to remove the bolts or clips.

When reinstalling a wheel cylinder, it is very important that you don't cross-thread the brake line. This is VERY easy to do. The best way to avoid this is by installing the brake line before installing the mounting bolts or clips. In this way, you can move

SKILL DRILL 37-4 Dissembling, Cleaning, Inspecting, and Reassembling a Non-Servo Brake

Note: In the appropriate service information, research and follow the procedure for disassembling the brake assembly. Manufacturers use many drum brake configurations, so the following steps are general in nature and should not be substituted for the manufacturer's procedure.

1. Clean the brake shoes, hardware, and backing plates, using equipment and procedures for dealing with asbestos and/or dust.

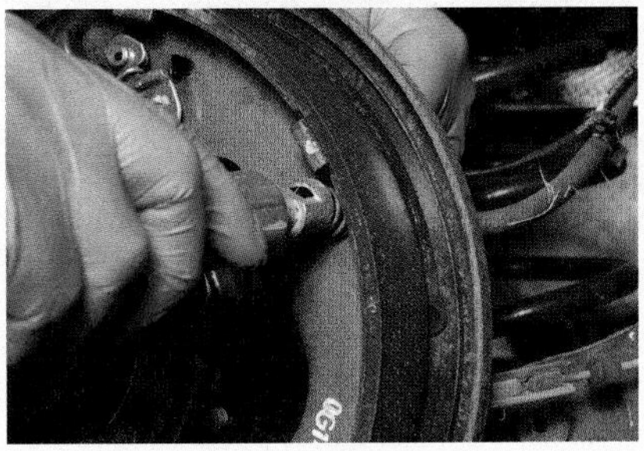

2. To disassemble a non-servo brake, first remove the hold-down springs, retainers, and pins.

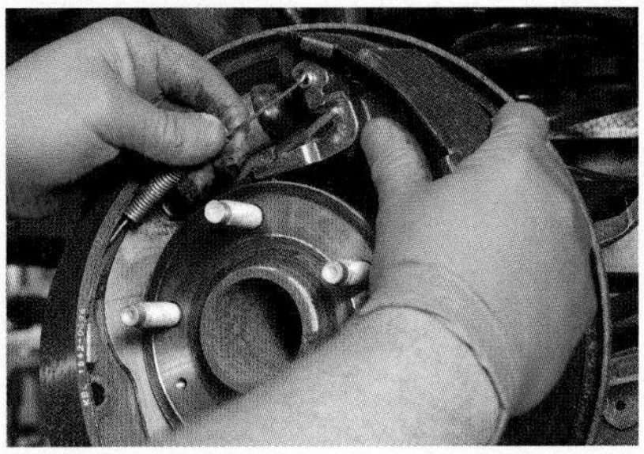

3. Spread the shoes apart and remove the parking brake strut and self-adjuster components.

4. Remove the return springs.

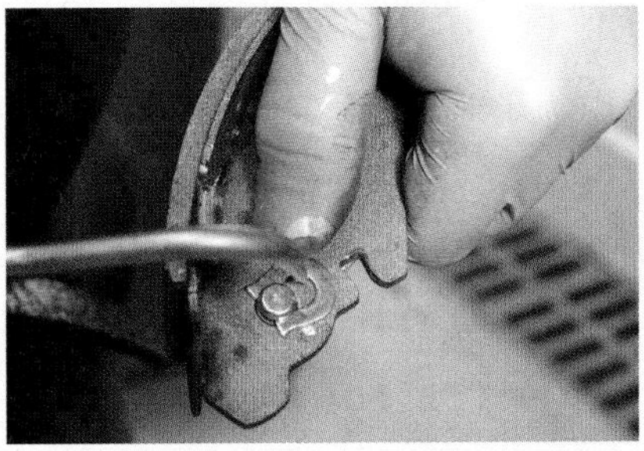

5. Disassemble the parking brake lever from the brake shoe and hardware from the backing plate, being careful not to lose any parts and remembering how they go back together.

6. Clean and inspect all parts according to the manufacturer's procedure.

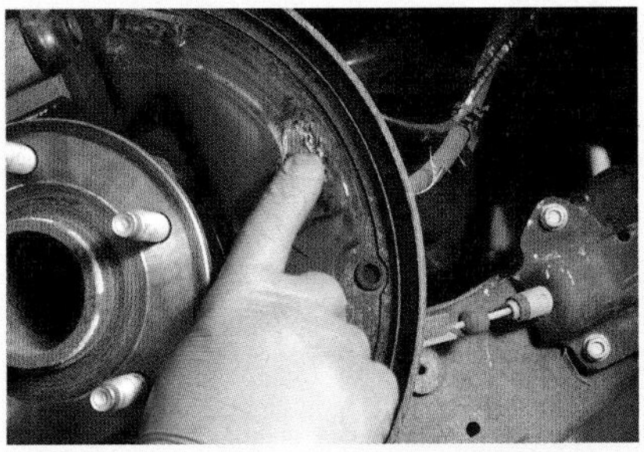

7. To reassemble a non-servo brake, first assemble and lube the self-adjuster/parking brake strut assembly. Lube the backing plate pads.

8. Place one shoe on the backing plate and install the hold-down spring and pin.

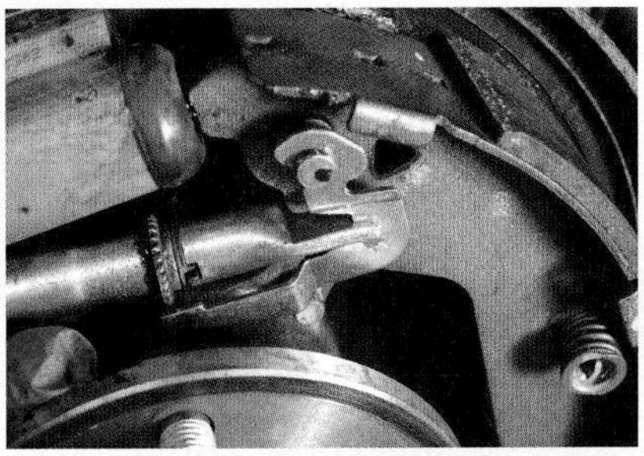

9. Place the self-adjuster/parking brake strut on the installed brake shoe.

SKILL DRILL 37-4 Dissembling, Cleaning, Inspecting, and Reassembling a Non-Servo Brake (Continued)

10. Place the retracting springs on both shoes, and fit the loose shoe to the backing plate, being sure to line up the wheel cylinder pushrods, the self-adjuster, and the parking brake mechanism.

11. Install the hold-down spring and pin.

12. Finally, check the fit of all springs, clips, and levers.

the wheel cylinder so that the threads line up with the brake line fitting. Make sure you can screw the brake line in by finger at least three complete turns. Once this is done, you can install the mounting bolts or clips and tighten them down. Then, use a flare wrench to tighten the brake line fitting.

> **TECHNICIAN TIP**
>
> If one wheel cylinder is leaking, how long until the other wheel cylinder starts leaking? And if it does leak, will it contaminate the new brake lining? If so, the new brake shoes will need to be replaced again. Take into consideration the age and condition of the components when you inspect the system. It is usually good practice to recommend replacement of both wheel cylinders if one is bad.

To remove, inspect, and install wheel cylinders, follow the steps in **SKILL DRILL 37-5**.

▶ Pre-Adjusting Brakes and Installing Drums

LO 37-06 Install drums, wheels, and perform final checks.

Pre-adjusting the brake shoes and parking brake is a routine step in a drum brake job. It saves time because it is faster to adjust the brakes with the drum off than it is when the drum is installed. The pre-adjustment sequence is important. Start by making sure the parking brake is fully released. Then, look at the shoes to make sure the parking brake adjustment is not holding the brake shoes in the applied position. You can see this by verifying that the brake shoes are up against their stops on the top and bottom of each shoe. If they are not firmly against them, loosen the parking brake adjustment. Then adjust the service brakes.

To adjust the service brake, you will extend the adjusting mechanism. To know how far to adjust it, use a brake shoe

SKILL DRILL 37-5 Removing, Inspecting, and Installing Wheel Cylinders

1. Peel back the dust boots, and check for brake fluid behind them. Determine any necessary actions.

2. Use a flare nut or line wrench to unscrew the brake line from the wheel cylinder.

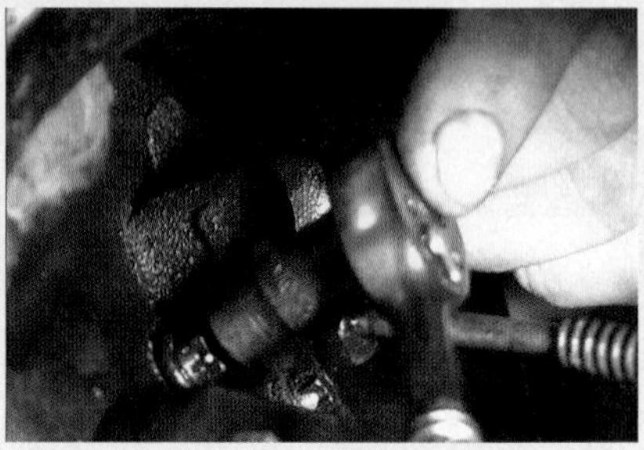

3. Remove wheel cylinder mounting bolts and wheel cylinder from the backing plate.

SKILL DRILL 37-5 Removing, Inspecting, and Installing Wheel Cylinders (Continued)

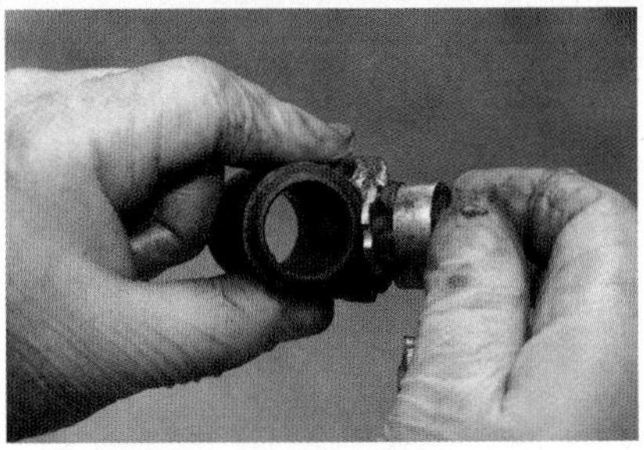

4. Disassemble the wheel cylinder, and inspect each part. Determine any necessary actions.

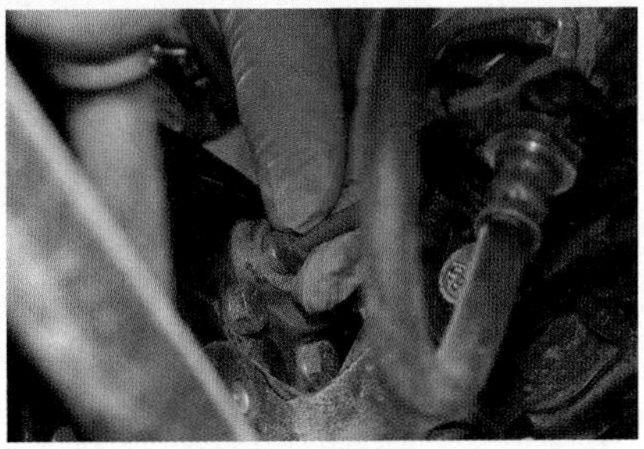

5. If the cylinder can be reused, rebuild it with new seals and dust boots. In most cases, replace it with a new wheel cylinder. Reinstall the brake line by hand.

6. Install and tighten any mounting screws. After that, tighten the brake line with a flare nut or line wrench.

adjustment gauge. Place the gauge inside the brake drum. Expand it so it is in the widest part of the drum. Then lock it in place. Take it to the brake assembly. Place it over the widest part of the brake shoes. Extend the adjusting mechanism until the brake shoes just touch the gauge. Once they are adjusted, install the drums and perform a final adjustment of the brake following the service information. Once the service brakes are adjusted, adjust the parking brake.

To pre-adjust brake shoes and the parking brake and to install brake drums, follow the steps in **SKILL DRILL 37-6**.

SKILL DRILL 37-6 Pre-Adjusting Brakes and Installing Drums

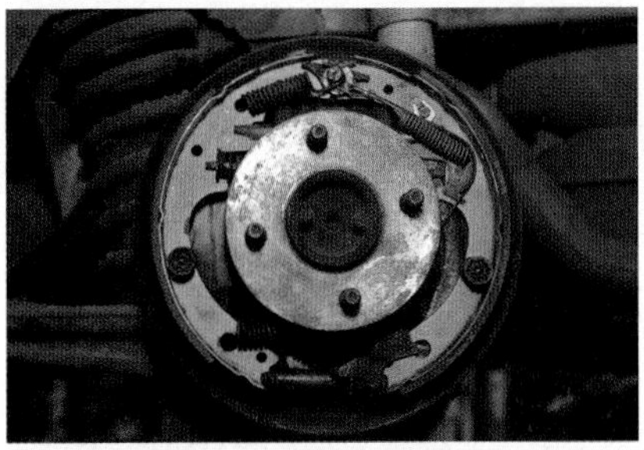

1. Make sure the brake shoes are fully up against their stops and centered on the backing plate.

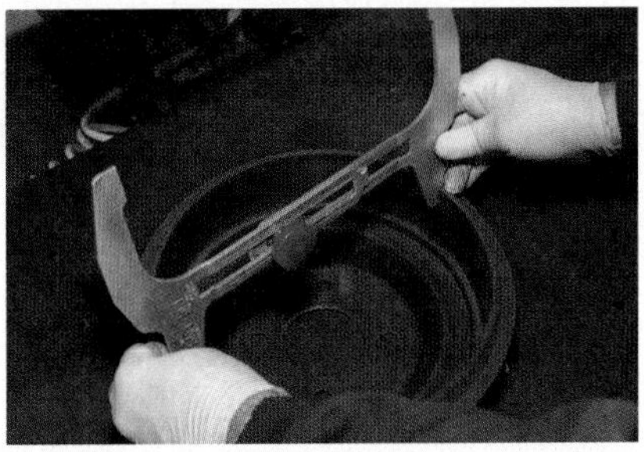

2. Set the pre-adjustment gauge to the drum diameter, and lock it in place.

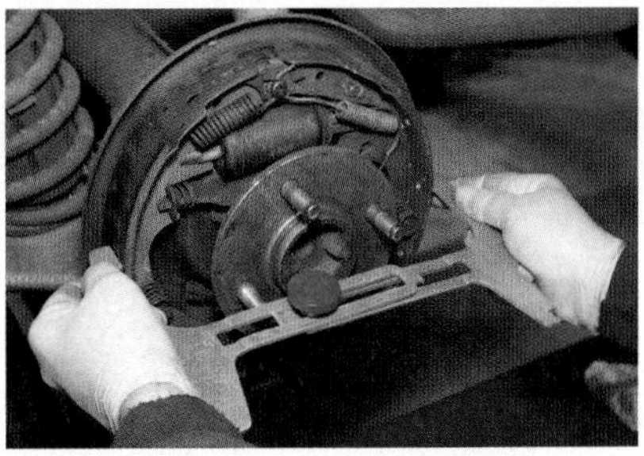

3. Place the pre-adjustment gauge over the center of the brake shoes.

4. Adjust the star wheel until the centers of the brake shoes just contact the pre-adjustment gauge.

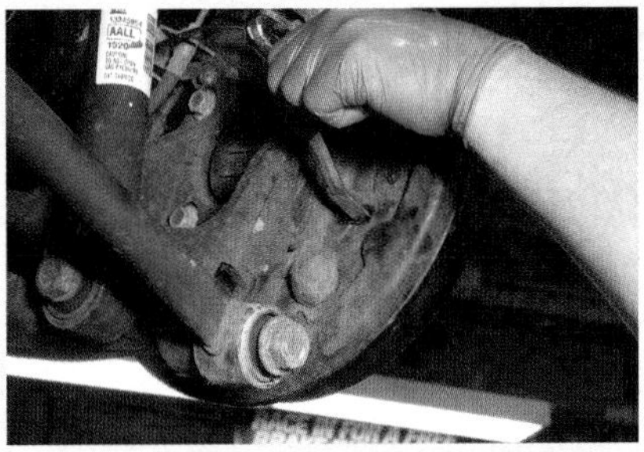

5. Test install the brake drum to verify the drum fits. Adjust the shoes as necessary.

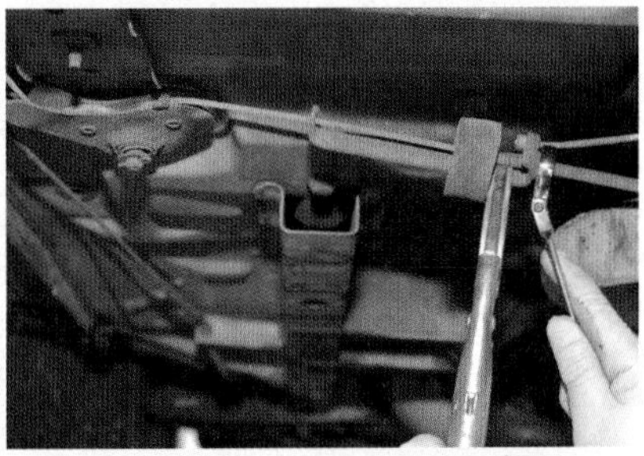

6. Adjust the parking brake according to the manufacturer's procedure. If the drum has serviceable wheel bearings, repack, install, adjust, and secure them (see Wheel Bearing chapter).

Installing Wheels, Torquing Lug Nuts, and Making Final Checks

This procedure, although fairly simple, can result in problems if not performed properly. Overtightening the lug nuts can cause the wheel studs to break. This can happen either immediately or, worse, after the vehicle has been driven for a period of time. If this happens, there is a strong risk of the wheel falling off.

Undertightening can lead to loosening of the lug nuts. It can also result in the wheel working its way off the vehicle. This can cause the driver to lose control of the vehicle and will potentially result in an accident. Lug nuts should be tightened to the proper torque, in the specified sequence. All manufacturers specify these details for each of their vehicles. The torque pattern is usually in some form of either a star or cross. Once you torque the lug nuts in the specified pattern, go around it again one-by-one in a circle. That way you will know that you didn't miss any when torquing them.

Be careful which way you install the lug nuts. Many wheels use a tapered hole that matches the tapered end of the lug nut. The taper centers the wheel on the wheel flange. Other wheels use a flat surface that matches flat surfaces on the lug nuts. No matter what, the lug nut surface *must* match the mating surface of the wheel. Always check that these surfaces match.

When installing lug nuts, the weight should be off the vehicle. The lug nuts should easily center the wheel on the hub. Proper centering is especially important on aluminum wheels that use a flat lug nut sealing surface. The lug nuts can dig into the sides of the lug nut holes and cause the wheel not to center correctly. Once the lug nuts are up against the wheel, work the wheel up onto the lug nut shafts. Then you can torque them down properly.

▶ TECHNICIAN TIP

There is some controversy regarding the use of a lubricant or antiseize on wheel studs and lug nuts. Because the purpose of a lubricant or antiseize is to prevent the components from sticking, CDX errs on the side of not using those products on lug nuts. This is due to fear of the lug nuts coming loose. Also, the torque given for the lug nuts is dry (no lubricant), so torquing them to specification would lead to overtightening. At the same time, in areas of the country prone to rust, it is understandable why some want to put a very light amount of antiseize only on the threads of the wheel stud. (Not on the contact seat of the wheel and lug nut.) Doing so would prevent rust from building up on the threads. However, the lug nut torque would need to be reduced due to the lubricant. CDX prefers to install lug nuts dry and avoid this issue altogether. Even then, many shops have their customers sign on the repair order that they will return to the shop so the lug nut torque can be rechecked after 50–100 miles of driving. This is to make sure the wheels have not loosened. Antiseize on lug nuts would make this problem worse.

To install the wheel, torque lug nuts, and make final checks and adjustments, follow the steps in **SKILL DRILL 37-7**.

SKILL DRILL 37-7 Installing Wheels, Torquing Lug Nuts, and Making Final Checks

1. Start the lug nuts on the wheel studs by hand. Carefully run them down so they are seated in the wheel. Lower the vehicle so the tires are partially on the ground. Use a torque wrench to tighten each lug nut to the proper torque in the proper sequence. Once all of the lug nuts have been torqued, go around them again, this time in a circular pattern.

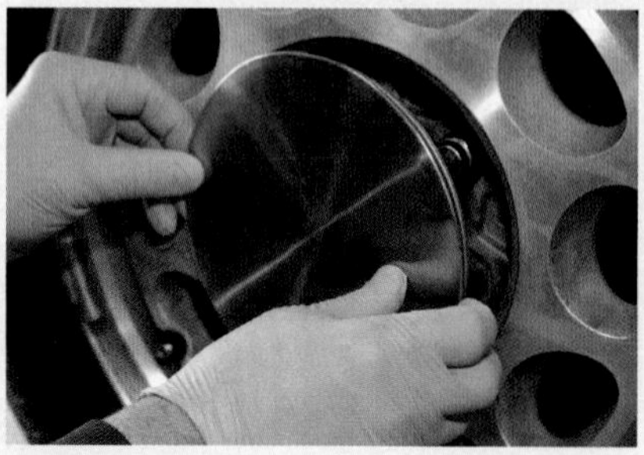

2. Reinstall hubcaps and valve stem caps.

3. Check the brake fluid level in the master cylinder reservoir. Start the vehicle and check the brake pedal for proper feel and height. Check the parking brake for proper operation. Inspect the system for any brake fluid leaks, missing bleeder screw dust caps, and loose or missing fasteners.

▶ Wrap-Up

Ready for Review

- ▶ Replicating the customer concern is an important part of drum brake diagnosis. Focused testing of the brake system should include: a visual inspection of the brake fluid level and condition, brake unit inspection, and measurement of drum diameter and runout.
- ▶ The tools used to diagnose repair drum brake systems include: brake wash station, brake spring pliers, hold-down spring tool, drum brake micrometer, brake shoe adjustment gauge, brake spoon, wheel cylinder piston clamp, off-car brake lathe, parking brake cable pliers, and parking brake cable removal tool.
  - A brake drum micrometer or special Vernier tool is used to measure a brake drum.
- ▶ Brake drums must be refinished when they have excessive grooving or are out-of-round. A brake lathe is used to refinishes the drum friction surface by removing metal and making it perfectly round with the proper finish.
- ▶ Drum brake shoes are held together by a combination of springs, links, levers, guides, and retainers. Taking a picture during disassembly can help during reassembly. Brake shoes and springs are always replaced in axle sets.
- ▶ A leaking or binding wheel cylinder needs to be replaced. When reinstalling a wheel cylinder, it is very important not to cross-thread the brake line. The best way to avoid this is by installing the brake line before installing the mounting bolts or clips.
- ▶ Pre-adjusting the brake shoes and parking brake saves time as it is faster to adjust the brakes with the drum off than it is when the drum is installed. Overtightening the lug nuts can cause the wheel studs to break, while undertightening can lead to loosening of the lug nuts. Lug nuts should be tightened to the proper torque, in the specified sequence.

Key Terms

brake lathe A tool used to refinish the drum surface by removing a small amount of metal and returning it to a concentric, nondirectional finish.

brake shoe adjustment gauge An adjustable tool used to pre-adjust the brake shoes to the diameter of the brake drum.

brake spoon A tool used to adjust the brake lining–to-drum clearance when the drum is installed on the vehicle.

brake spring pliers A tool used for removing and installing brake return springs.

brake wash station A piece of equipment designed to safely clean brake dust from drum and disc brake components.

drum brake micrometer A tool used for measuring the inside diameter of the brake drum.

hold-down spring tool A tool used for removing and installing hold-down springs.

off-car brake lathe A tool used to machine (refinish) drums and rotors after they have been removed from the vehicle.

parking brake cable pliers A tool used to install parking brake cables.

parking brake cable removal tool A tool used to compress the spring steel fingers of the parking brake cable so that the cable can be removed from the backing plate.

wheel cylinder piston clamp A tool that prevents the pistons from being pushed out of the wheel cylinders while the brake shoes are being replaced.

Review Questions

1. What is the best way a technician can protect themselves from legal action while performing brake service work?
   a. By closely following the manufacturer's service procedures.
   b. By following the process their grandfather used on his classic vehicle.
   c. By mimicking someone's homemade Internet video.
   d. By following the advice of an unproven "expert" in an Internet forum.
2. What indicator light may illuminate when the ABS system activates?
   a. The check engine light
   b. The four-wheel low indicator light
   c. The limited slip indicator light
   d. The poor traction indicator light
3. What tool is used to measure brake drum diameter?
   a. A depth micrometer
   b. A drum micrometer
   c. A tape measure
   d. Feeler gauges
4. While removing hubless drums using a hammer, where should the drum be hit?
   a. Between the lug studs
   b. On the lip of the drum at the backing plate
   c. On the hub bearing race
   d. On the end of the lug studs
5. Why should a drum that is over the maximum diameter not to be put back in service?
   a. It will cause the wheel to spin faster and may activate the ABS.
   b. The parking brake may not allow the brake to release.
   c. It may not absorb enough heat causing premature brake fade.
   d. The drum surface finish will be rough causing a noise.
6. Why should sandpaper be used on the drum surface after machining?
   a. To keep the drum from overheating.
   b. To give the drum surface a nondirectional finish.
   c. To clean the surface of metal filings.
   d. To keep the drum from rusting to the hub flange.

7. Why is it advisable to disassemble and reassemble drum brakes one side at a time?
   a. It allows the side not being disassembled to be used as a reference during reassembly.
   b. Because the secondary shoe is different from the primary shoe.
   c. To keep the master cylinder reservoir from overflowing.
   d. To ensure that all of the parts get replaced.
8. When are wheel cylinders typically replaced?
   a. When they are leaking, binding, or have a broken bleeder screw.
   b. When there is a brake pedal pulsation concern.
   c. When the brake drums are replaced.
   d. When the star wheel adjuster binds causing the brakes to stop adjusting.
9. Which type of wrench should be used to tighten the brake line to the wheel cylinder.
   a. Box-end
   b. Open-end
   c. Click-type torque
   d. Flare-nut
10. What special tool is required to perform a drum brake pre-adjustment?
   a. A drum micrometer
   b. Brake spring pliers
   c. A brake shoe adjustment gauge
   d. A wheel cylinder piston clamp

ASE Technician A/Technician B Style Questions

1. Technician A says that if the yellow ABS light is illuminated, the ABS will likely not work. Technician B says that testing ABS brake operation in traffic can be dangerous. Who is correct?
   a. Technician A only
   b. Technician B only
   c. Both A and B
   d. Neither A nor B
2. Technician A says that some vehicles have hub-style drums so that the wheel bearings must be removed to remove the drum. Technician B says that because of the tight clearances between the drum and wheel flange, rusting between these surfaces seldom occurs. Who is correct?
   a. Technician A only
   b. Technician B only
   c. Both A and B
   d. Neither A nor B
3. Technician A says that in order to find drum out-of-round reading, you should take three diameter measurements, add them together and divide by three. Technician B says that drums with cracks, deep groves, and hotspots can almost always be machined and reused. Who is correct?
   a. Technician A only
   b. Technician B only
   c. Both A and B
   d. Neither A nor B
4. Technician A says that when machining drums, a cutting depth set at 0.008" will increase the diameter by 0.016". Technician B says that on a multiple speed lathe, the final cut should be set to the fastest feed speed. Who is correct?
   a. Technician A only
   b. Technician B only
   c. Both A and B
   d. Neither A nor B
5. Technician A says that hub-style drums are mounted to the lathe with bearing adapters that center the drum using the wheel bearing races. Technician B says that hubless drums may be mounted with a spring-loaded cone and two clamshell adapters. Who is correct?
   a. Technician A only
   b. Technician B only
   c. Both A and B
   d. Neither A nor B
6. Technician A says that most drum brake assemblies are easier to service than disc brake assemblies. Technician B says that drum brake linings and springs can be replaced on one side only when contaminated. Who is correct?
   a. Technician A only
   b. Technician B only
   c. Both A and B
   d. Neither A nor B
7. Technician A says that brake spring pliers are safer to use when installing brake springs than needle nose pliers. Technician B says that a small amount of lubricant should be applied to the backing plate contact pads. Who is correct?
   a. Technician A only
   b. Technician B only
   c. Both A and B
   d. Neither A nor B
8. Technician A says that some wheel cylinders are bolted to the backing plate. Technician B says that wheel cylinders are rebuilt more often than replaced. Who is correct?
   a. Technician A only
   b. Technician B only
   c. Both A and B
   d. Neither A nor B
9. Technician A says that the parking brake should be applied while pre-adjusting the brake shoes. Technician B says that the brake shoes should be against their stops at the top and bottom during the pre-adjustment. Who is correct?
   a. Technician A only
   b. Technician B only
   c. Both A and B
   d. Neither A nor B
10. Technician A says that tapered lug nuts should be installed with the tapered side towards the wheel. Technician B says that under-torqueing lug nuts often causes drums to warp out-of-round. Who is correct?
   a. Technician A only
   b. Technician B only
   c. Both A and B
   d. Neither A nor B

CHAPTER 38
Wheel Bearings

Learning Objectives

- **LO 38-01** Describe wheel bearing fundamentals.
- **LO 38-02** Describe wheel bearing types.
- **LO 38-03** Describe grease seals and lubricants.
- **LO 38-04** Describe wheel bearing arrangements for rear drive axles.
- **LO 38-05** Describe wheel bearing diagnosis and failure analysis.
- **LO 38-06** Perform maintenance tasks on serviceable wheel bearings.
- **LO 38-07** Remove and reinstall sealed wheel bearings.

ASE Education Foundation Tasks

See Appendix A to view the 2017 ASE Education Foundation Automobile Accreditation Task List Correlation Guide.

▶ Wheel Bearing Theory

LO 38-01 Describe wheel bearing fundamentals.

In this chapter, you will learn about wheel bearings and the role they play in the vehicle. Because wheel bearings can fail, you will learn how to diagnose common issues. You also learn how to properly maintain and, if necessary, replace today's wheel bearings. This will prepare you to discuss wheel bearing issues with customers and recommend repair options.

Wheel bearings are a commonly overlooked component. All wheel bearings used to be of the serviceable type. They required periodic maintenance every 24,000–30,000 miles (38,624–48,280 km). This consisted of disassembling, cleaning, inspecting, lubricating, reassembling, and adjusting them (**FIGURE 38-1**).

With the introduction of sealed wheel bearings, periodic service is not needed. Many manufacturers switched to sealed wheel bearings on most of their light-duty vehicles. As a result, many technicians do not give the wheel bearings as much consideration as they once did. However, both types of wheel bearings can fail. And vehicles that still use serviceable wheel bearings still need maintenance. So, it is important that you become familiar with the various types of wheel bearings.

FIGURE 38-1 Technician servicing a serviceable wheel bearing.

Wheel Bearings Overview

Wheel bearings allow the wheels to roll with a minimum of friction. At the same time, they still maintain accurate wheel positioning under all driving conditions. Most wheel bearing assemblies include the following components:

- an outer race,
- an inner race,
- roller bearings or ball bearings, and
- a bearing cage to hold the rollers or balls in place (**FIGURE 38-2**).

The rollers or balls are made of hardened metal and roll between the two races. The races are also made of hardened metal. They are carefully formed to match the contour of the rollers or balls. This allows all of the components roll easily against each other.

Each race either fits firmly within a housing or on a shaft. In many situations, the races are held in place by an **interference fit** with the housing or shaft. This means they must be pressed into place with a high amount of force. In other words, the races are designed so that when they are installed, they should not rotate in or on their respective components. The rollers or balls are specifically designed to provide the rolling function. Because the races, rollers, and balls are made of hardened metals, they are designed to resist wear and damage. However, overloading the vehicle can put the wheel bearings under a greater load than they are rated for, which can cause them to fail. In the same way, using improper or insufficient lubricant can cause them to fail. Wheel bearing failure is discussed in detail later in this chapter.

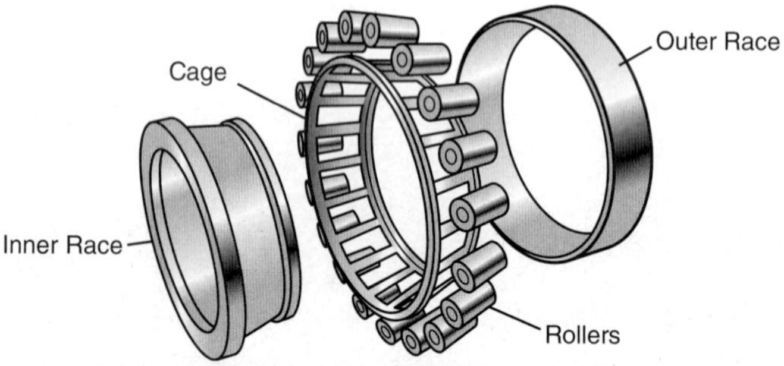

FIGURE 38-2 Components of a typical wheel bearing.

You Are the Automotive Technician

A customer pulls a 2008 Ford F250 pickup truck into your shop. They complain that the vehicle is making a howling sound when driving more than about 20 mph. The noise seems to gradually increase as the speed of the vehicle increases. They are concerned that the vehicle might break down and leave them stranded. You ask where the noise is coming from, and they explain it is coming from the center or rear of the vehicle. You notice that the truck is equipped with a solid live rear axle. The tires are in good shape, with a fairly smooth highway tread on them.

1. On a test drive, how can you determine whether the noise is coming from a wheel bearing or a transmission bearing?
2. What wheel bearing arrangements can be used on a solid live rear axle, and how do they differ?
3. How are rear wheel bearings lubricated on each type of solid live rear axle?
4. How is a sealed bearing different from a serviceable bearing?

There are two categories of wheel bearings: **serviceable bearings** and **sealed bearings** (**FIGURE 38-3**). Serviceable bearings are designed so they can be disassembled and serviced. Sealed bearings are not serviceable.

Serviceable bearings need occasional service. This includes many things, such as disassembling, cleaning, inspecting, and repacking them with the specified lubricant. They also need to be adjusted and retained once installed. Sealed bearings are designed so they cannot be disassembled or adjusted. Sealed bearings are manufactured with the proper clearance and filled with the specified lubricant from the factory. They are designed to last the life of the vehicle. But bearings do wear out or fail, which then requires replacement. We discuss sealed bearing failure and replacement in detail later in the chapter.

▶ TECHNICIAN TIP

Roller and ball bearing assemblies are called **antifriction bearings**. This is because the components are in rolling contact with one another. Therefore, they have minimal friction. A sleeve-type bearing, or bushing, such as a clutch pilot bushing, is called a **friction bearing**. These components are in sliding contact with one another. Antifriction bearings operate much more freely than friction bearings.

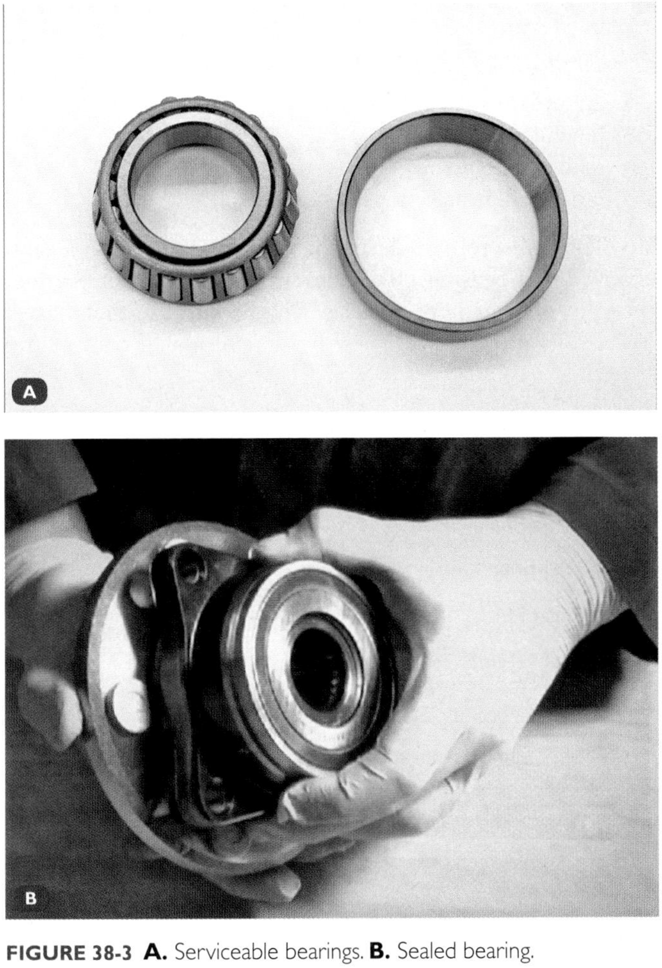

FIGURE 38-3 **A.** Serviceable bearings. **B.** Sealed bearing.

▶ Wheel Bearing Types

LO 38-02 Describe wheel bearing types.

Wheel bearings commonly are of the following types:

- cylindrical roller bearing,
- tapered roller bearing,
- ball bearing,
- double-row ball bearings, and
- and double-row tapered roller bearings.

Each one is designed for a particular application. For example, roller bearings support the load over a larger surface area. This allows them to carry heavier loads than ball bearings. Manufacturers determine the type of wheel bearings they will use based on the particular application. You need to be familiar with each type of wheel bearing. You will then be able to successfully service a variety of vehicles.

Cylindrical Roller Bearings

Cylindrical roller bearing assemblies use rollers that are cylindrical in shape. This means that the races are parallel to each other with the rollers between them (**FIGURE 38-4**). Cylindrical bearings are used where they are not subject to side loads. They are commonly used in rear-wheel drive vehicles. Cylindrical bearings require the use of thrust bearings. They prevent the axle from moving side to side. In this arrangement, all side-to-side movement is controlled within the differential assembly. So, cylindrical roller bearing assemblies solely support the weight of the vehicle.

In many instances, cylindrical roller bearing assemblies use the surface of the axle as the inner bearing race. In this case, the cylindrical roller bearings ride directly on the axle shaft. If the bearing assembly fails, the axle shaft will most likely be damaged. It will have to be replaced along with the bearing assembly. However, some bearing manufacturers have designed replacement cylindrical roller bearing assemblies. They ride farther out on the axle shaft than the original bearing. So, the axle may not have to be replaced in this situation.

Cylindrical roller bearing assemblies are made so they have the proper running clearance between the rollers and races. This means that they don't require adjustment. But like all bearings discussed in this chapter, they do require lubrication. Lubrication cushions and cools the rollers and races while operating.

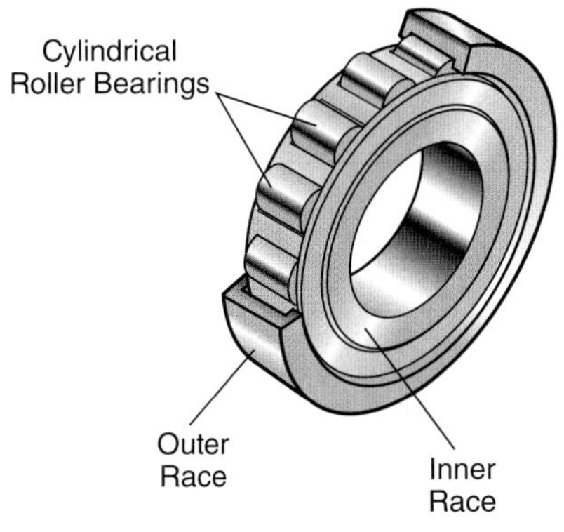

FIGURE 38-4 Cylindrical roller bearing assembly.

Tapered Roller Bearings

Tapered roller bearings have races and rollers that are tapered. They are tapered in such a manner that all of the angles meet together at a common point (**FIGURE 38-5**). This design allows the tapered rollers to freely roll between the angled inner and outer races. The tapered rollers are contained in a bearing cage. It holds the tapered bearings to the inner race as a unit. The inner race is called the cone, and the outer race is called the cup. Together, the cone and cup make up a tapered roller bearing assembly (**FIGURE 38-6**).

Tapered roller bearing assemblies are commonly used where heavy loads and side loads (thrust) must be supported. Wheel bearing side load conditions occur when the vehicle is cornering. This pushes the bottom of the outside wheel inward and the bottom of the inside wheel outward. The wheels and bearings are then put under a side load condition. In this situation, cylindrical roller bearing assemblies would allow the axle to slide sideways. This means they cannot control the side load (thrust) condition. However, tapered roller bearing assemblies can.

The components are on an angle to the centerline of (and not parallel to) the axle shaft. They can then control side movement (thrust) in one direction. Because of this, tapered roller bearing assemblies are generally used in opposing pairs. That way, they can control side movement (thrust) in both an inward and outward directions. When used in pairs, they are generally referred to as inner (or inboard) and outer (or outboard) bearings (**FIGURE 38-7**). The inner bearing assembly supports most of the vehicle weight. Because of this, the inner bearing assembly is typically larger than the outer bearing assembly.

Some manufacturers have designed double-row tapered roller bearing assemblies. They combine two opposing tapered roller bearing assemblies. They are housed in a sealed unit. This design provides excellent side thrust–carrying capacity. It also has excellent load-carrying capacity. For more information, see the Sealed Wheel Bearings section.

Tapered roller bearing assemblies use tapered components. So, the wheel bearing assembly must be adjusted to have the proper **running clearance** between the tapered rollers and races when the bearing assemblies are installed. The running clearance is the amount of space between components during operation. It can be a problem if the tapered roller bearing clearance is too tight. The components will bind and overheat due to increased pressure and because the rollers squeeze out too much grease. This makes the lubricating film too thin. There is also a problem if the tapered roller bearing is adjusted with too much clearance. Excessive side-to-side and up-and-down movement will occur. This can cause the components to hammer against each other. This will damage the surfaces of the tapered roller bearings and races.

Ball Bearings

Ball bearing assemblies consist of:

- an inner race,
- an outer race,
- ball bearings, and
- and a ball bearing cage (**FIGURE 38-8**).

The balls roll in deep channels in the races. Deeply grooved ball bearing assemblies are used as wheel bearings on a lot of light-duty vehicles. This is because they have a lower rolling resistance. There is a much smaller contact area

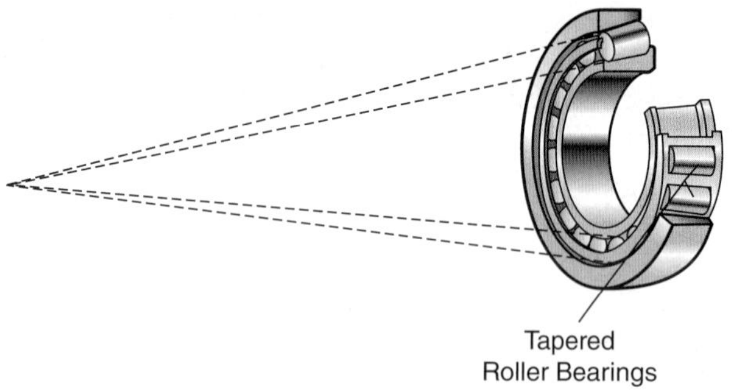

FIGURE 38-5 With tapered roller bearings, the common axis of bearings and races provides minimal rolling resistance.

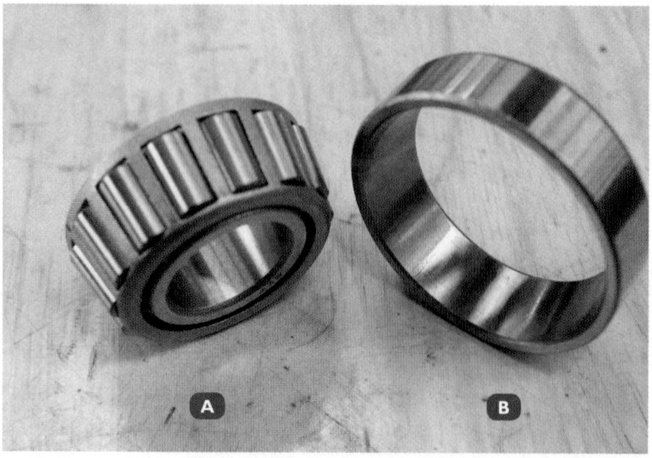

FIGURE 38-6 **A.** Cone. **B.** Cup.

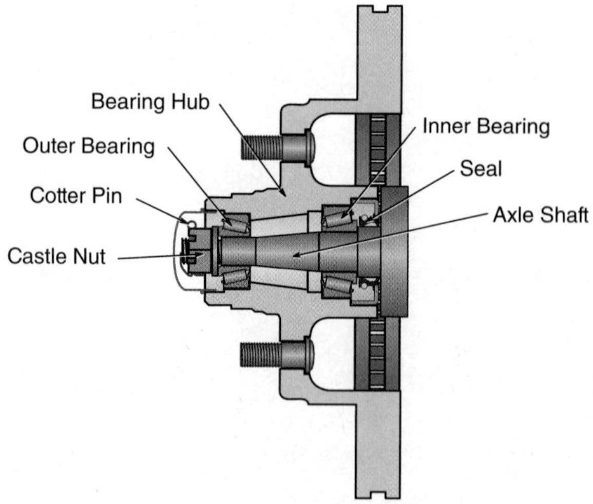

FIGURE 38-7 Tapered roller bearings used to control thrust in both directions.

between the balls and races on a ball bearing assembly. This keeps them from being used on larger vehicles, which experience higher loads.

Side loads are controlled by the balls rolling against the sides of the channels. There is much less surface area between the balls and the sides of the channels compared with the tapered roller bearing. So, ball bearing assemblies are limited in how much side load they can handle. At the same time, the small surface area allows them to roll freer than roller bearings. They, therefore, help manufacturers decrease drag on a vehicle, which increases fuel efficiency.

Ball bearing assemblies are also available as a **double-row ball bearing** (**FIGURE 38-9**). The double-row configuration gives the ball bearing assembly twice the surface contact area. This allows it to control greater amounts of loads and side loads than a single-row bearing. Most wheel bearings using a ball bearing assembly are the double-row ball bearing. They are usually used in automotive light vehicle applications.

The outer race is usually a one-piece unit, whereas the inner race is usually two separate pieces. The inner races are manufactured to butt up against each other. When the assembly is torqued in place, they create the correct running clearance.

Sealed Wheel Bearings

Some vehicles use sealed wheel bearings. These bearing assemblies are designed and manufactured as completely sealed units. They can be single row or double row, depending on the vehicle application. They can be made up of ball, cylindrical roller, or tapered roller types of bearings. The bearing assemblies are prefilled with lubricant. Integrated grease seals contain the lubricant (**FIGURE 38-10**).

Sealed wheel bearings are manufactured with the proper running clearance. So they do not need to be adjusted. This saves the vehicle manufacturer money by decreasing the time it takes to install them during vehicle assembly. It also makes for a more reliable and consistent installation process. This is because every unit comes preset at the proper clearance.

Sealed wheel bearings are also beneficial to vehicle owners. They are designed to last the life of the vehicle without periodic maintenance. This saves owners money.

Another application for sealed wheel bearings is the **unitized wheel bearing hub** (**FIGURE 38-11**). These usually include either a double-row ball bearing assembly or a double-row tapered roller bearing assembly. It is installed in a housing that is bolted onto the knuckle or axle housing. In many cases, it also has the wheel flange pressed into the inner race of the wheel bearing. In this way, the old bearing assembly can be unbolted and a new one bolted

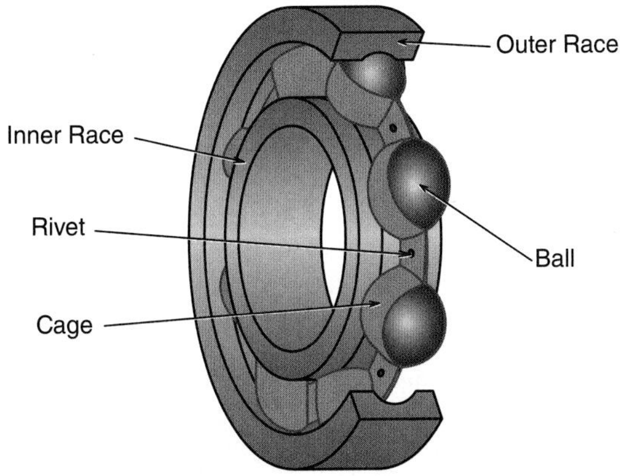

FIGURE 38-8 A typical ball bearing assembly.

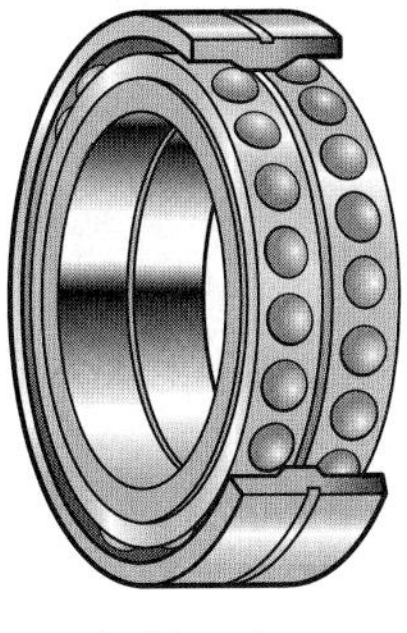

FIGURE 38-9 A double-row ball bearing assembly.

FIGURE 38-10 Sealed bearing.

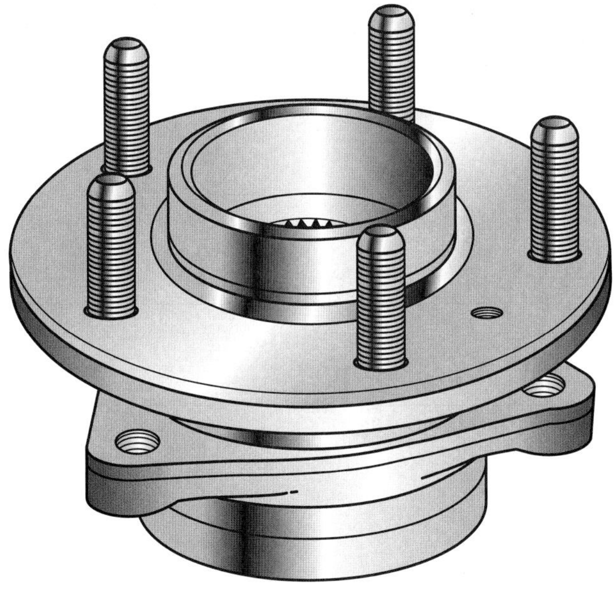

FIGURE 38-11 A unitized wheel bearing hub assembly.

in its place with minimal labor and no adjustments needed. This is virtually a zero maintenance system until it wears out. It is then replaced as a unit. On some vehicles, the ABS sensor is integrated into the unitized wheel bearing assembly.

▶ Grease Seals and Axle Seals

LO 38-03 Describe grease seals and lubricants.

All wheel bearings rely on some sort of seal to keep the lubricant in and contaminants out. Some wheel bearings have the seal built right into the bearing assembly. This is the case with sealed bearings. Other wheel bearings rely on a completely separate **grease seal**. In some applications, the grease seal is a press fit into the axle housing, which is stationary. It seals against the axle shaft, which rotates. In other applications, the grease seal is press fit into the wheel hub, which rotates. It seals against the spindle, which does not rotate.

Most axle seals consist of a stamped sheet metal case and flexible sealing lip with an internal **garter spring** (**FIGURE 38-12**). The metal seal case is designed to be a press fit when installed in the housing. The outside surface of the metal case usually comes precoated with a thin layer of sealer compound (**FIGURE 38-13**). It seals minor surface imperfections between the seal case and the housing.

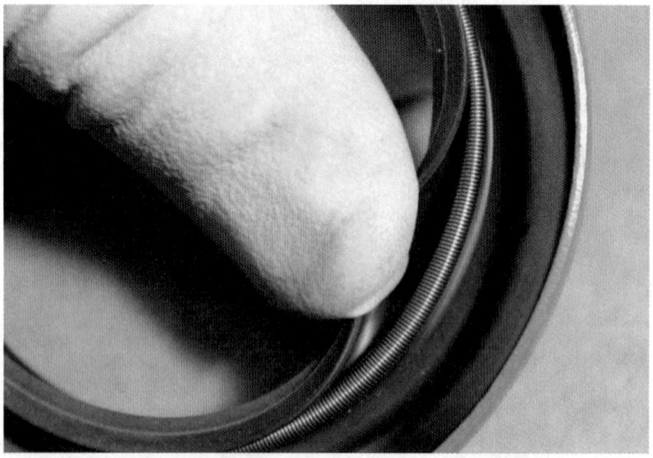

FIGURE 38-12 Components of a typical seal.

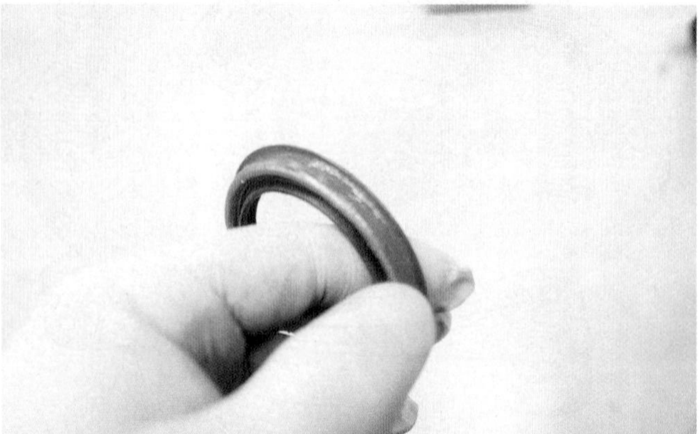

FIGURE 38-13 Seals come coated with a thin layer of sealer compound.

The sealing lip is carefully designed so that it will seal the specified lubricant. Many seals use a garter spring to help hold the lips of the seal in contact with the shaft it is sealing. The garter spring helps maintain an adequate seal if the parts are slightly out of alignment. It also allows for a small amount of runout or clearance between the seal and shaft.

Because the garter spring holds the seal against the rotating component, the seal can wear a groove in that component. Always inspect the sealing surface of the rotating component to verify that it doesn't have excessive wear. In some cases, you can purchase a very thin metal repair sleeve (**FIGURE 38-14**). It fits snugly over the sealing surface of the shaft. The seal now has a new surface to ride against. If a repair sleeve is not available, the grooved component may have to be replaced if the wear is great enough.

Seals come in a variety of configurations, so it is critical to use the specified seal for the application you are servicing. The sealing lip can also be made from a variety of materials. Always purchase seals from a reputable manufacturer so you can be confident they will do their job and last a long time. When you install a seal, always remember to pre-lube the sealing lip with a small amount of oil or grease. This will prevent it from overheating during initial use. Overheating could lead to premature failure of the seal.

▶ TECHNICIAN TIP

Grease seals are designed to be used only one time. If you have to remove a seal for any reason, replace it with a new one. Failure to do so will likely result in a leaky seal. Also, if you damage a seal during installation, replace it with a new one. Always use care when installing seals, and use the correct installation tool.

Lubrication

All wheel bearings require lubrication to extend their useful life. There are two common types of lubricants used—**gear lube** and bearing grease (**FIGURE 38-15**). Each application specifies one or

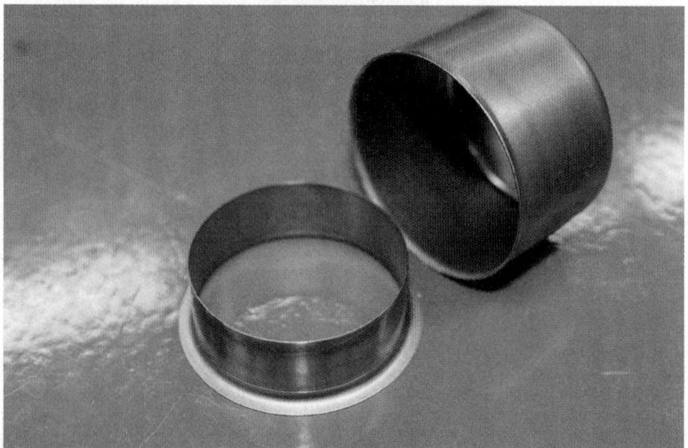

FIGURE 38-14 Grooves in some axles can be repaired with a special thin metal sleeve.

FIGURE 38-15 **A.** Gear Lube. **B.** Bearing Grease.

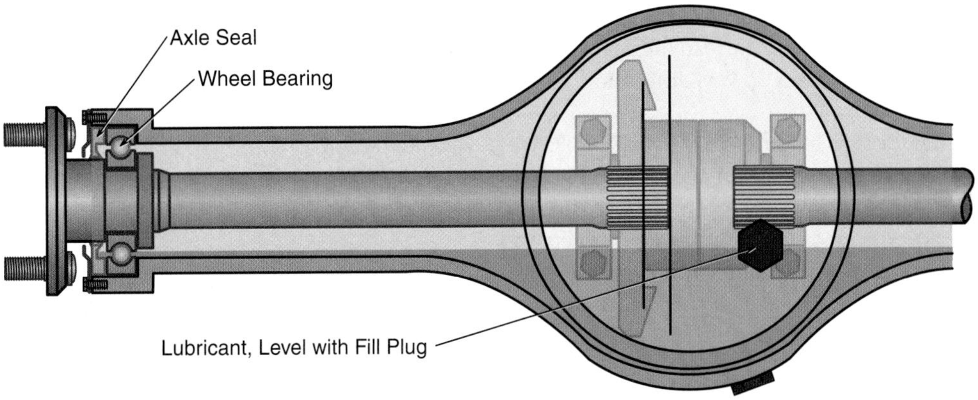

FIGURE 38-16 Gear lube at the proper level lubricates some types of axle bearings.

the other. They cannot be substituted. This is due to the different housing and seal designs each lubricant demands.

Gear lube is somewhat thicker than engine oil. Bearing grease is much thicker than gear lube. Bearing assemblies that use gear lube run in a housing partially filled with gear lube. The gear lube is thin enough to flow around and between the rollers, lubricating them. It also helps prevent overheating of the wheel bearing assemblies.

In many rear-wheel drive vehicles, the wheel bearing assemblies are open to the axle housing, which is partially filled with gear lube. In this case, the gear lube lubricates the wheel bearings along with the final drive assembly (**FIGURE 38-16**). Maintenance of this system involves draining the old gear lube and refilling the system with the specified new gear lube. In most cases, this can be performed without disassembling the wheel bearings. A **drain plug** or bolt is used for draining the system. Filling the system is made through the **fill plug** (**FIGURE 38-17**).

A fill plug is usually threaded. It can be removed to allow the level of a fluid to be checked and filled. Some manufacturers have chosen to use a rubber fill plug that snaps into the fill hole. The level of gear lube should normally be within 0.25" (6.35 mm) of the bottom of the fill plug hole.

FIGURE 38-17 Drain and fill points for rear axle assembly.

▶ TECHNICIAN TIP

If the level of gear lube is lower than it should be, suspect a leaking axle shaft grease seal. This can usually be verified by looking at the inside of each tire and the back side of the brake backing plate. If gear lube is present, the grease seal is leaking and needs to be replaced. Also, check the wheel bearing to make sure it is not faulty. To do so, follow the bearing diagnosis procedure listed later in this chapter.

Other bearings are lubricated with bearing grease. Since bearing grease is very thick, it doesn't flow very well. This means that the bearings need to be packed with grease between the rollers (**FIGURE 38-18**). That way the grease can lubricate the rollers and races. Bearings packed with grease still need grease seals. They keep the grease contained when it heats up and starts to flow a bit. We will discuss packing of wheel bearings later in the chapter.

In order to properly maintain wheel bearings, you need to understand the different characteristics of gear lube and bearing grease. This helps ensure that you select the proper lubricant for the task you are performing. Gear lube is classified by the Society of Automotive Engineers (SAE) according to its viscosity. It is classified by the American Petroleum Institute (API) according to its service grade. **Viscosity** refers to the thickness of the gear lube; the higher the number, the thicker the gear lube. Vehicle manufacturers specify a certain viscosity of gear lube based on two main factors. The first is the climate the vehicle is operated in. The other is based on the load the bearings are carrying. So, understanding viscosity is very important when servicing vehicles.

Standard viscosities for gear lube are 70W, 75W, 80W, 85W, 90, and 140. The "W" stands for "winter," or the gear lube's cold temperature viscosity. The non-W ratings are the viscosity of the gear lube at a predetermined hot temperature. Similar to engine oil, gear lube is available in multi-viscosity configurations. Some examples include 75W-90, 80W-90, and 85-140. Unlike engine oil, the temperature for the "W" rating varies according to the standard being met. For example, 70W has a maximum allowable temperature for its viscosity of –67°F (–55°C), and 85W is –10°F (–23°C). Manufacturers have gone to great lengths to specify the proper gear lube for their vehicles. Make sure you follow their recommendations when choosing the gear lube for a particular vehicle.

Current ratings of the API service grade are GL-4 and GL-5. Generally, the higher the number, the better the lubricant. GL-4 is intended for use with bevel-type gears operating under moderate speeds and loads. These are common in many manual transmissions.

GL-5 has about twice as much extreme pressure additive as GL-4. So, it is intended for gears that operate under high-speed/low-speed, high-torque, and shock-load conditions. This is common in most final drives that use hypoid-type gears. It can also be used in some manual transmissions. Always check the manufacturer's specifications to determine the proper gear lube for the application you are working on.

Most serviceable wheel bearings require **grease** as their lubricant. Grease is made of a base oil, plus a thickening agent. Specific additives are used to meet the requirements of the application. **Lithium soap** is a common thickening agent in automotive grease (**FIGURE 38-19**). Other greases use calcium or **molybdenum thickening agents** (**FIGURE 38-20**). Some add small amounts of copper and/or lead to enhance the grease's ability to withstand extreme pressures.

Automotive wheel bearing grease is a thickened lubricant, designated as a plastic solid. This means that it is thick enough at room temperature to maintain its shape if left undisturbed. It is also thin enough to be squeezed into and out of small spaces. Its consistency is similar to a glob of gel toothpaste.

FIGURE 38-18 Packing a bearing with grease by hand.

FIGURE 38-19 Lithium soap grease.

FIGURE 38-20 Molybdenum grease.

The thickness of grease is graded by the **National Lubricating Grease Institute (NLGI)** (**TABLE 38-1**). Because it does not flow at room temperature, the bearings must be packed with grease. This is performed when the bearings are installed. It is also performed as maintenance according to the manufacturer's maintenance schedule. This is typically every 4 years, or whenever the brakes are replaced.

▶ Wheel Bearing Arrangements for Rear Drive Axles

LO 38-04 Describe wheel bearing arrangements for rear drive axles.

Rear drive axles come in three different designations: full-floating, semi-floating, and ¾ floating. Each designation refers to how the axle and wheel are supported by the wheel bearings. It is also important for you to understand the differences so that you will be able to service each style properly.

In a full floating axle arrangement, the axle only carries a twisting force. The weight of the vehicle is fully carried by a pair of tapered roller bearing assemblies. The bearings ride between the hub and axle tube (**FIGURE 38-21**). The axle does not carry any of the vehicle load because the wheel is bolted directly to the bearing hub. The hub also controls side thrust. Full floating axles handle heavy loads better than the other styles. This makes them ideal for use in heavy-duty applications such as many one-ton pickups, trucks, and vans.

In a semi-floating axle, the wheel flange is part of the axle. The axle is supported by a single bearing assembly near the flange end of the axle (**FIGURE 38-22**). The bearing assembly is made up of either the ball bearing type or the cylindrical roller bearing type. The bearing rides between the axle and the inside of the axle tube. This puts all of the weight on the axle flange, which transfers the weight to the wheel bearing assembly. In this case, both the axle and bearing assembly carry the full weight of the vehicle. The axle also provides the twisting force for the wheel. This arrangement is generally considered the lightest duty of the three types of axle designations.

In a ¾-floating axle design, there is a single bearing assembly between the outside of the axle tube and the hub (**FIGURE 38-23**). It is typically made up of either the ball bearing type or the cylindrical roller bearing type. The axle has a wheel flange that bolts to the hub and provides lateral support for the

TABLE 38-1 NLGI Rating System

NLGI Number	Consistency
00	Semifluid
0	Very soft
1	Semisoft
2	Semifirm
3	Soft—common wheel bearing grease
4	Firm
5	Very firm

Courtesy of the National Lubricating Grease Institute.

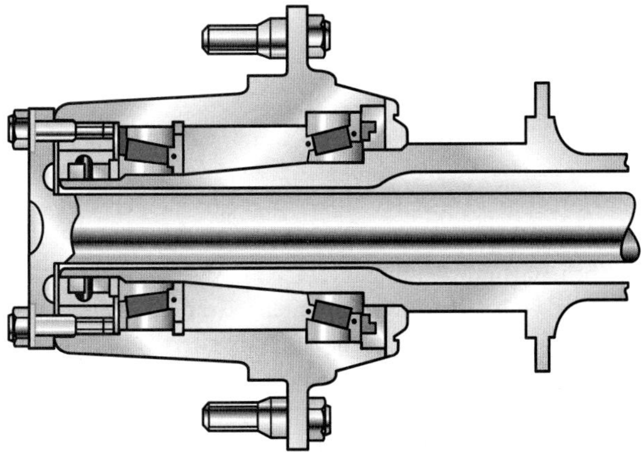

FIGURE 38-21 Full floating axle—two tapered roller bearings between the axle housing and hub.

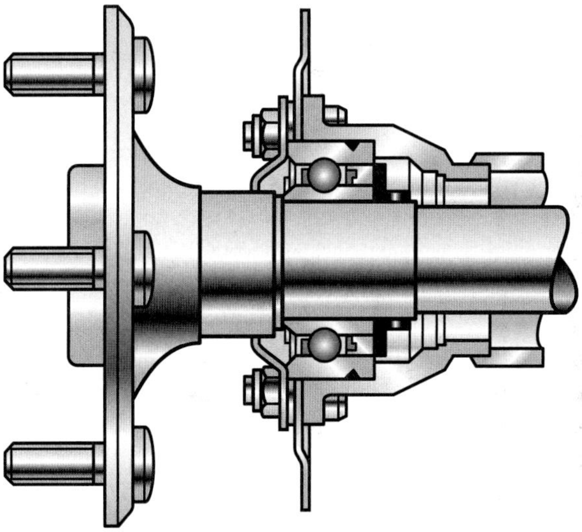

FIGURE 38-22 Semi-floating axle—bearing located between the axle and housing.

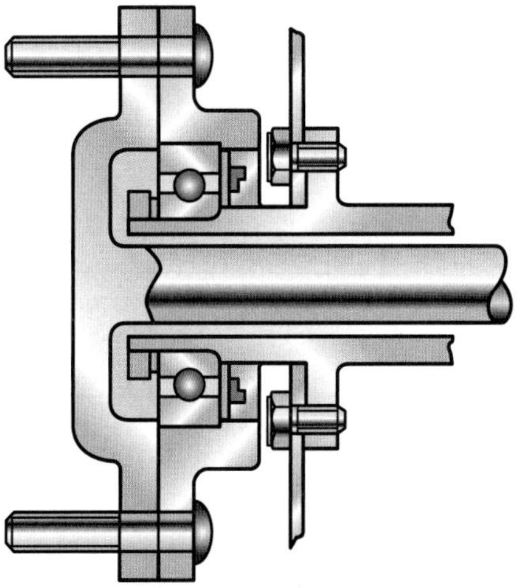

FIGURE 38-23 Three-quarter floating axle—one bearing between the outside of the axle housing and the hub.

hub and wheel. The bearing assembly supports the weight of the vehicle. The axle also provides the twisting force for the wheel. This arrangement is generally considered heavier duty than the semi-floating axle. But it is lighter duty than the full floating axle.

▶ Diagnosis

LO 38-05 Describe wheel bearing diagnosis and failure analysis.

Wheel bearings can be damaged from excessive loads such as:

- overloading the vehicle,
- shock loads such as hitting a large pot hole,
- improper adjustment, or
- just plain wear over time.

The wheel bearings must hold up under difficult circumstances. Faulty bearings should be suspected whenever an unusual sound comes from the wheel areas when the vehicle is being driven. Common wheel bearing noises include rumbling, whirring, and howling. The noise usually can be heard once the vehicle gets up to 15–20 mph (24.1–32.2 kph). It usually gets louder as the vehicle speeds up.

Loose or worn wheel bearings can also cause the vehicle to wander, shimmy, or vibrate. For these concerns, it is best to lift the wheels off the ground and check the wheel bearings for looseness. Grab the tire at the 6 o'clock and 12 o'clock positions and lightly wiggle it back and forth. Watch the inside of the wheel to verify that the play is coming from the wheel bearings, and not the ball joints. Then, grab the wheel at the 3 o'clock and 9 o'clock positions, and again lightly wiggle the wheel back and forth. Watch to verify that the play is coming from the wheel bearings, and not the steering linkage.

If there is play in the wheel bearings and the vehicle uses sealed bearings, they will need to be replaced. If the vehicle uses serviceable bearings, the maximum allowable play is approximately 0.010" (0.254 mm). If greater than that, they will need to be serviced. This includes disassembling, cleaning, and inspecting them for wear. If they are in good condition, they will need to be repacked, reinstalled, and readjusted.

One way to isolate a wheel bearing noise from a transmission noise is to drive the vehicle at the speed at which it is making the noise. Then shift into a higher or lower transmission gear while maintaining the same speed. If the noise speeds up or slows down, then it is a transmission-related issue. If it stays relatively the same, accelerate, coast, and decelerate the vehicle. If the noise changes, suspect the differential or universal joints. If the noise stays relatively the same, it is most likely related to the wheel bearings.

To determine which side the bearing noise is coming from, drive the vehicle at the speed at which it is making the noise. Then, lightly rock the car side to side, using the steering wheel. If the noise gets louder when the car is steered right, then it is usually the left side with the bad bearing, and vice versa. Only perform this test in a safe place such as an abandoned parking lot. Another challenge is distinguishing a wheel bearing noise or a tire noise. The best approach to this problem is to drive over different road surfaces, such as asphalt and concrete. If the noise changes, it is likely a tire problem. If the noise does not change, it is likely to be a faulty wheel bearing.

If you cannot determine the source of the noise on a test drive, you might be able to do so by placing the vehicle on a hoist and spinning the wheels. If the vehicle has a MacPherson strut suspension, raise and support the vehicle. Then, spin the wheel by hand while holding on to the coil spring with the other hand. The spring tends to magnify the wheel bearing roughness, which can be felt with some practice.

If the suspect wheel is a drive wheel, under close guidance of your supervisor, have an assistant drive the vehicle in gear while on the hoist. Listen to the wheel bearings with a stethoscope. Because the vehicle is running on the hoist, this can be a hazardous situation. It must only be performed under the close guidance of your supervisor.

Applied Science

AS-1: Vibrations/Waves: The technician can demonstrate an understanding of the types and causes of vibrations caused by out-of-balance or excessively worn systems.

Sound is very important in diagnosing problems in a motor vehicle. Sound is a series of waves that travel through a gas, liquid, or solid, many of which can be heard by the human ear. For example, the solid rumble strips on the sides of the highway make a continuous noise or vibration when driven over by a motor vehicle.

Sound can help you diagnose a worn wheel bearing that needs replacing. As a wheel bearing becomes worn, the case hardening on the bearing race begins to wear away. This is due to constant friction between the bearing race and the bearing rollers. As the case hardening starts to wear, the bearing race develops low spots in the race. As the bearing rollers start to roll over the high and low spots in the bearing race, a noise occurs.

To diagnose the noise of a worn bearing in a motor vehicle, you can hoist the vehicle, spin the wheel, and listen for a noise. Or you can test drive the vehicle and gently turn the steering wheel from side to side. As the vehicle moves from side to side, more weight is placed on each side of the bearing due to load shifting. You will hear a louder noise as you turn away from the side with the faulty bearing.

Failure Analysis

Wheel bearings can be inspected and determinations made about why they failed. Identifying what caused the problem helps ensure that the same thing does not happen to the new wheel bearing. Failure analysis starts with removing the faulty wheel bearing and cleaning it. Once it is clean, visually inspect each of the wheel bearing components. Compare your findings with the manufacturer's failure analysis chart (**FIGURE 38-24**). This information should lead you to what caused the wheel bearing failure.

Once the cause of the wheel bearing failure has been determined, take measures to ensure that the failure does not repeat

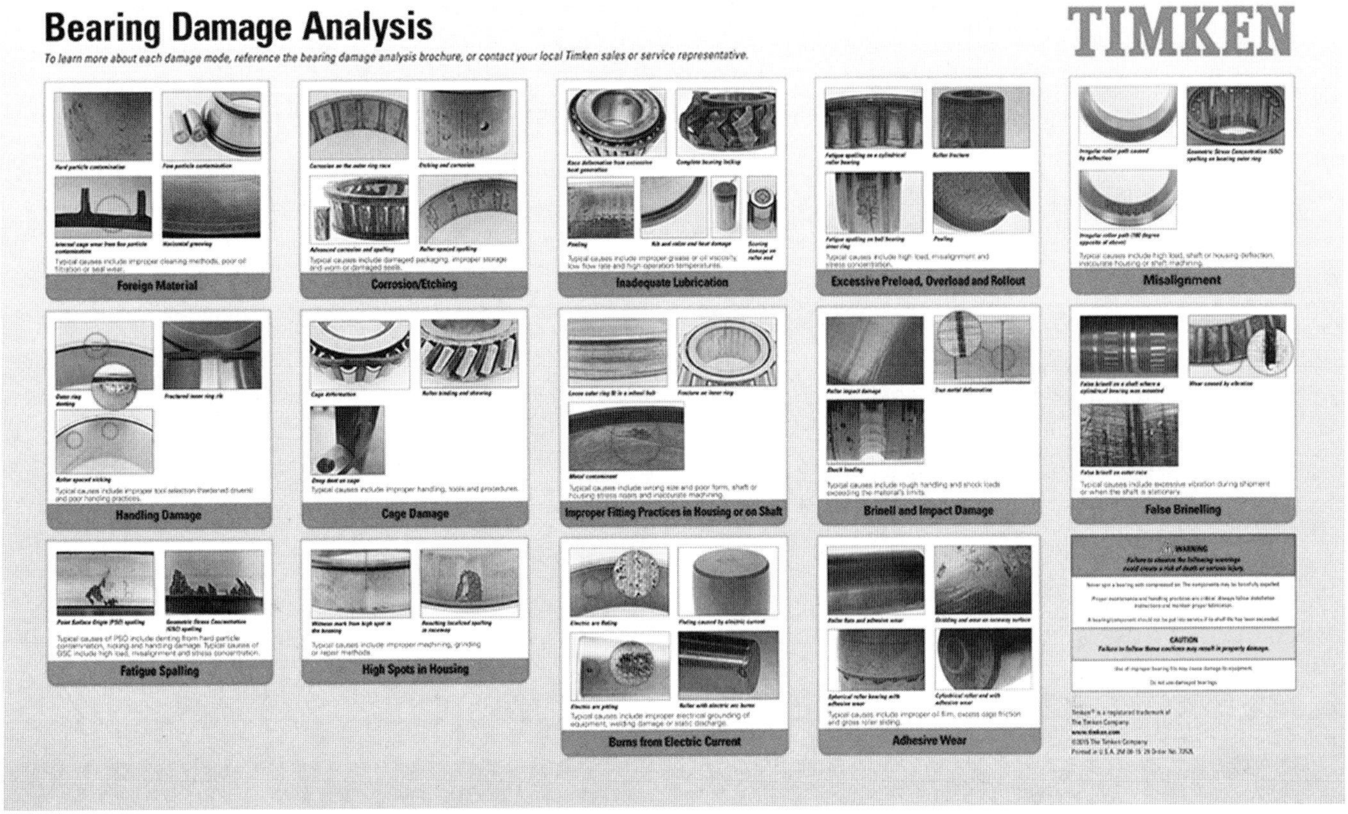

FIGURE 38-24 Wheel bearing failure chart.

itself. For example, if the bearing and race surfaces show signs of rust or corrosion, inspect the sealing surfaces of the hub. Check to see if water is getting past the dust cover or grease seal. Also, if there are metal shavings in the wheel hub and grease, a thorough cleaning of the wheel hub will be required. All of the metal shavings must be removed or they will ruin the new bearing assembly.

Some wheel bearing failures involve spun races. Normally, races are press fit into their housing. Races should not spin in either the machined bore of the wheel hub or on the surface of the spindle/axle. If they do, it can lead to wear of the wheel hub or spindle/axle. If this occurs, replacement of these components is required. Be sure to inspect the wheel hub and spindle/axle closely for any damage every time the wheel bearings are serviced.

FIGURE 38-25 Cotter pin used to retain a wheel bearing adjustment nut.

▶ Maintenance and Repair

LO 38-06 Perform maintenance tasks on serviceable wheel bearings.

Sealed wheel bearings are used on a higher percentage of vehicles. They need no maintenance. But serviceable wheel bearings do need periodic maintenance. Maintenance consists of disassembling, cleaning, inspecting, repacking, installing, and adjusting the bearings. This is commonly performed during brake shoe/pad replacement. Service intervals are also specified by the vehicle manufacturer. They usually specify service at 24,000–30,000 miles (38,624–48,280 km).

A normal part of servicing wheel bearings includes the replacement of the old grease seal with a new one. The **cotter pin** (if used) is also replaced. The cotter pin is a soft metal pin that can be bent into shape and is used to retain the bearing adjusting nut (**FIGURE 38-25**). Ensure that the correct replacement parts and grease are available before starting the job.

Tools

Here is a list of common tools used to maintain and repair wheel bearings (**FIGURE 38-26**):

- Bearing packer
- Seal puller
- Wheel bearing race installer/seal installer set
- Wheel bearing **locknut** sockets
- Cotter pin removal tool
- Dust cap pliers

Wheel Bearing Adjustment

All wheel bearings need the proper end play or preload to operate correctly. **End play**, in the context of wheel bearings, refers to the amount of inward and outward movement of the hub. This is due to the clearance within the bearing assembly.

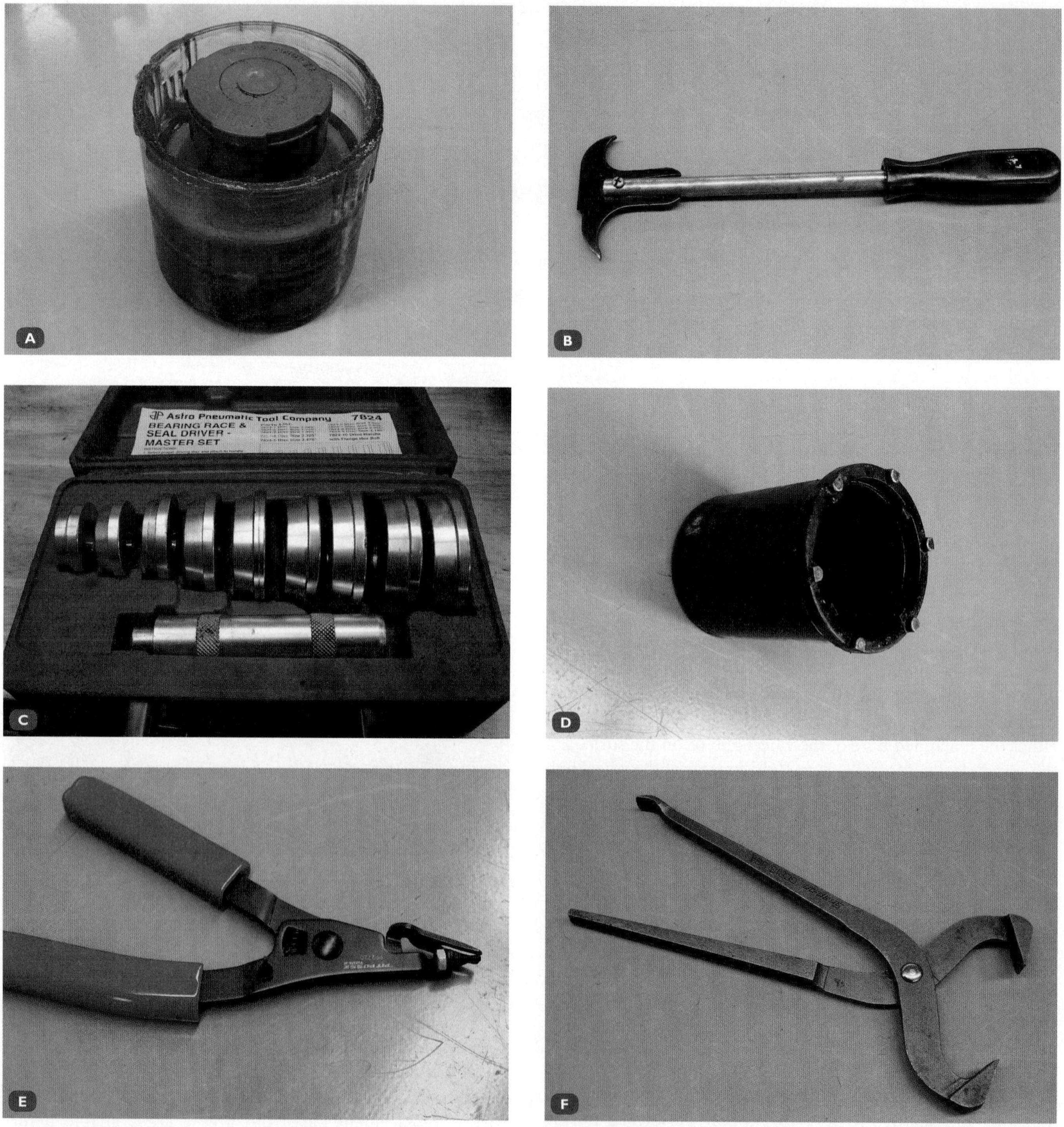

FIGURE 38-26 A. Bearing packer. **B.** Seal puller. **C.** Wheel bearing race/seal installer set. **D.** Wheel bearing locknut sockets. **E.** Cotter pin removal tool. **F.** Dust cap pliers.

Preload refers to a specified amount of pressure forcing the bearing components together.

Sealed bearings come from the manufacturer with the proper clearance machined into them. These wheel bearings are designed so that when the components are tightened together, the races butt up against each other in such a way that the proper clearance is maintained. For these wheel bearings, it is only critical that the retaining bolts are torqued to the proper specification. This is usually quite high and can typically exceed 200 ft-lb (271.1 N·m). Be sure to check the manufacturer's torque specifications for the application you are working on.

On adjustable wheel bearings, the proper clearance must be set using the **adjusting nut**. In this case, the adjusting nut is initially tightened to about 20 ft-lb (27.1 N·m). This squeezes out any grease between the races and rollers. It is then loosened one-sixth to one-quarter turn. Then, it is only tightened lightly (usually about 15–25 in-lb [1.69–2.82 N·m]). This provides a small amount of clearance or preload between the rollers and races. The adjusting nut is then locked in place by a locking mechanism so that it cannot loosen.

The locking mechanism is important to prevent the wheel from falling off. This could cause an accident. The most common locking mechanism uses a **keyed washer** (hardened), adjusting nut, **lock cage**, and cotter pin (**FIGURE 38-27**). On four-wheel drive vehicles, the locking mechanism usually includes several things. These include a keyed washer (hardened), adjusting nut, **keyed lock washer** (or tang washer), and locknut (**FIGURE 38-28**).

Repacking and Adjusting Wheel Bearings

Serviceable wheel bearings should be serviced periodically:

- according to the manufacturer's scheduled maintenance chart,
- whenever brake work is being performed, or
- if a faulty wheel bearing is suspected.

When servicing wheel bearings, it is critical that the proper grease is used. Also, avoid mixing different types of grease. Do this by thoroughly cleaning all old grease from the wheel bearings. If packing a bearing by hand, it is not good enough to just wipe grease around the bearing. You need to force the grease to fill the gaps between the rollers.

Another option for packing bearings is to use a bearing packer. It will also force the old grease out of the wheel bearing (**FIGURE 38-29**). Just make sure it has the specified type of grease. Also, make sure the grease is not contaminated with dirt or debris.

When reinstalling wheel bearings, it is critical to follow the manufacturer's adjustment procedure. Always use new grease seals and cotter pins (if used). Doing so will prevent grease leaks. It will also ensure that the adjusting nut does not back off, causing an unsafe driving situation.

To remove, clean, inspect, repack, install, and adjust wheel bearings, follow the steps in Skill Drills 38-1 through 38-7. To remove, clean, and inspect the wheel bearings, follow the steps in **SKILL DRILL 38-1**.

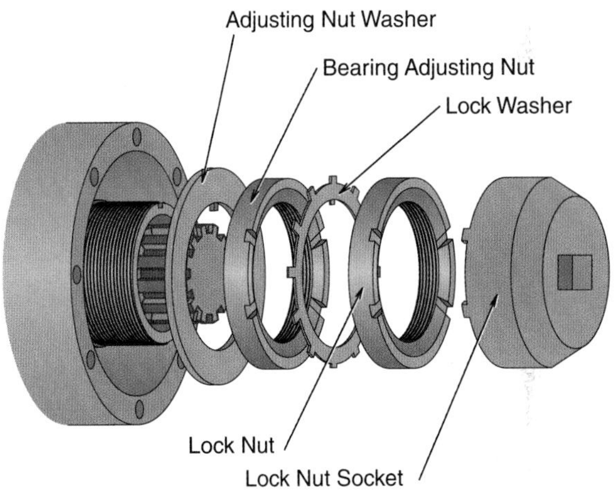

FIGURE 38-28 Typical locknut-style wheel bearing locking mechanism.

FIGURE 38-27 Typical cotter pin–style wheel bearing locking mechanism.

FIGURE 38-29 A bearing packer.

SKILL DRILL 38-1 Removing, Cleaning, and Inspecting Wheel Bearings

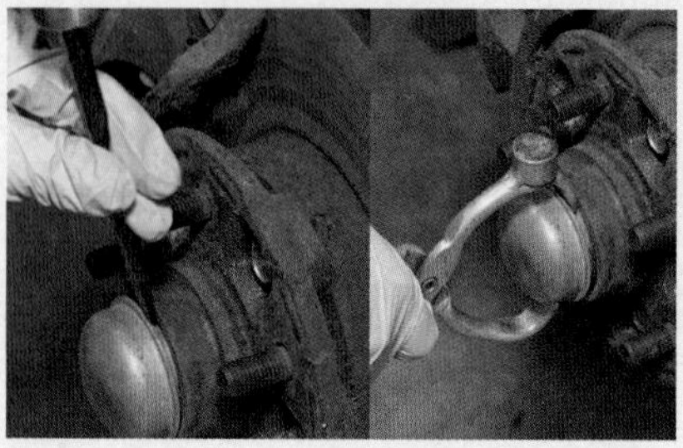

1. Remove the wheel bearing dust cap with dust cap pliers or a narrow cold chisel and hammer.

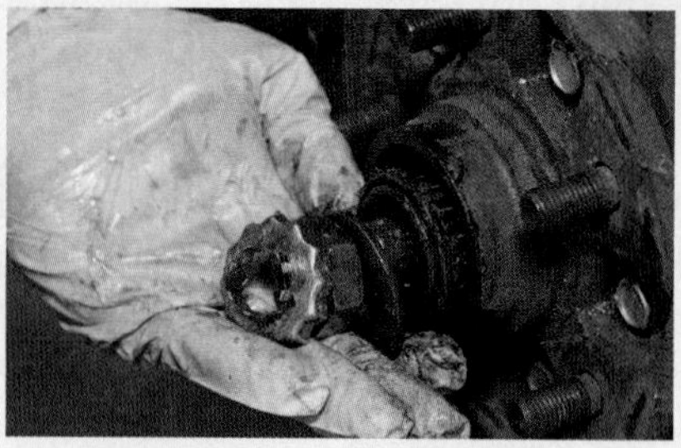

2. Remove the locking mechanism, adjusting nut, keyed washer, and outer bearing. Reinstall the adjusting nut approximately five turns back onto the spindle.

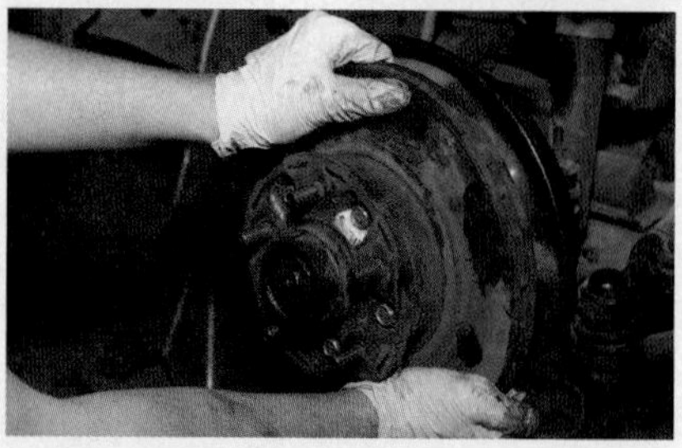

3. Grasp the drum/rotor at the 1 o'clock and 7 o'clock positions or 11 o'clock and 5 o'clock positions. While holding downward pressure, quickly pull the drum/rotor toward you. The adjusting nut should catch the inner bearing race and pop the grease seal and bearing out of the hub, leaving them sitting on the spindle.

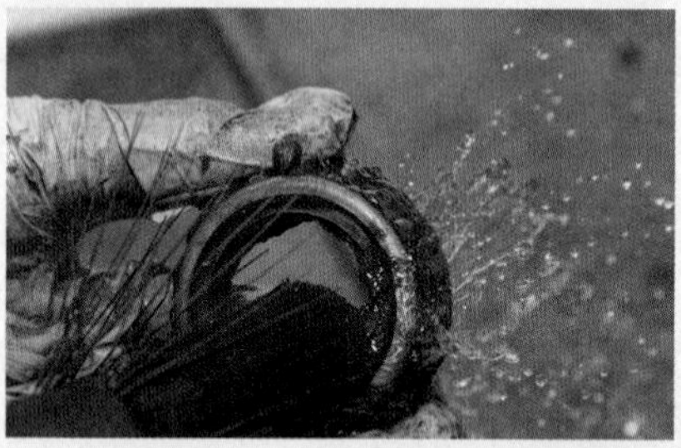

4. Wipe any old grease off of the wheel bearings, races, and spindle with a rag and give them a quick visual inspection. Consult the bearing diagnosis chart to identify any faults.

5. If the wheel bearings are in serviceable condition, completely clean the wheel bearings, races, and hub. If using solvent to clean any of the components, make sure there is no solvent-contaminated grease left on or in the parts.

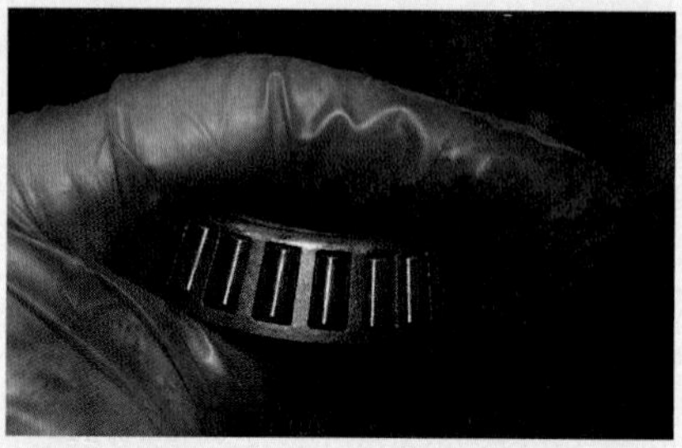

6. Give the parts a final inspection, and consult the bearing diagnosis chart if there are any signs of damage. Using the specified grease, pack both wheel bearings, being careful to keep dirt and debris out of the grease (see Skill Drills 38-2 and 38-3).

To pack grease by hand, follow the steps in **SKILL DRILL 38-2**.

SKILL DRILL 38-2 Packing Grease by Hand

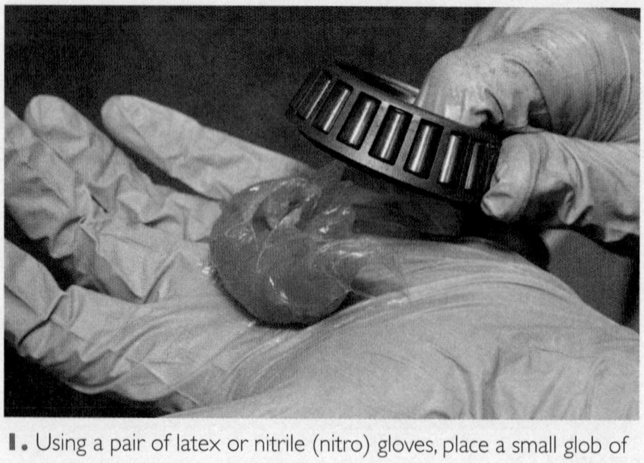

1. Using a pair of latex or nitrile (nitro) gloves, place a small glob of grease in the palm of your nondominant hand. Place the index finger of your other hand through the bearing center hole, with the larger diameter facing down.

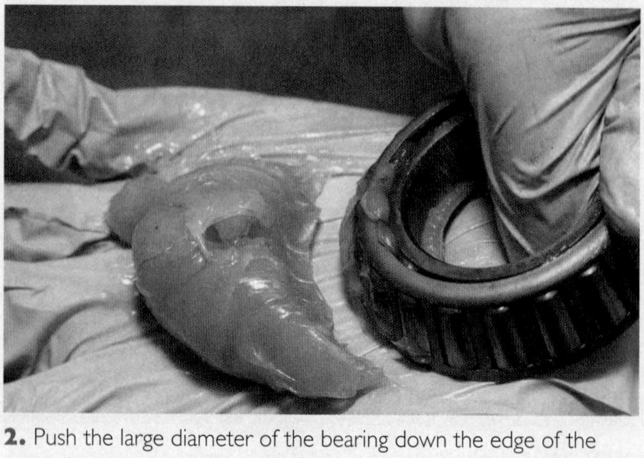

2. Push the large diameter of the bearing down the edge of the grease into your palm. This should force grease into the space between the bearings and race. Continue this process until grease comes out of the top of the bearing.

SKILL DRILL 38-2 Packing Grease by Hand (Continued)

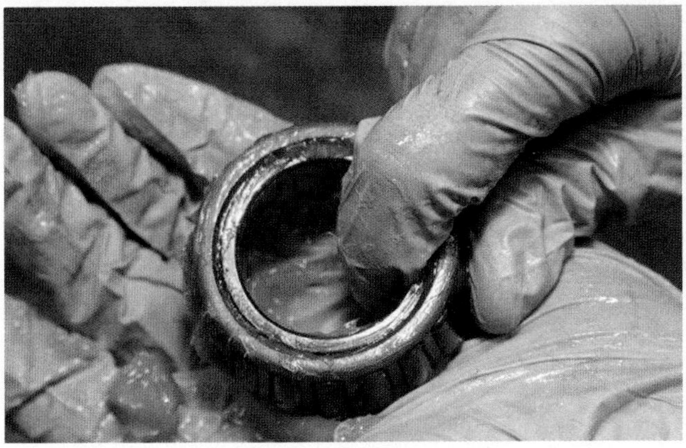

3. Carefully turn the bearing as a unit to a new space, and keep forcing grease between the bearings. Do this until all of the spaces are full.

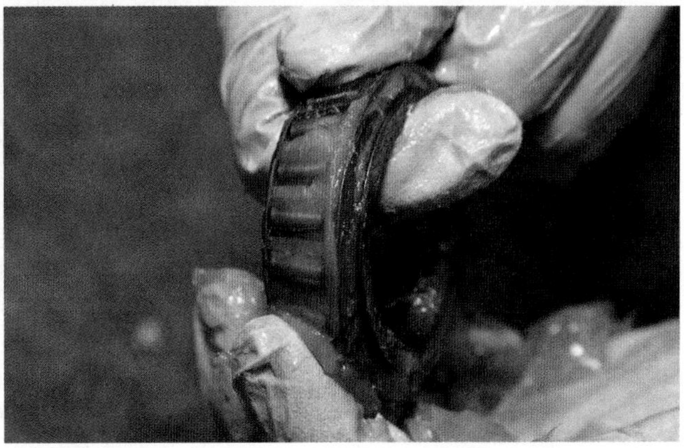

4. Smear some grease around the outside of the bearing. Repeat this process on the other bearing.

To pack grease with a bearing packer, follow the steps in **SKILL DRILL 38-3**.

SKILL DRILL 38-3 Packing Grease with a Bearing Packer

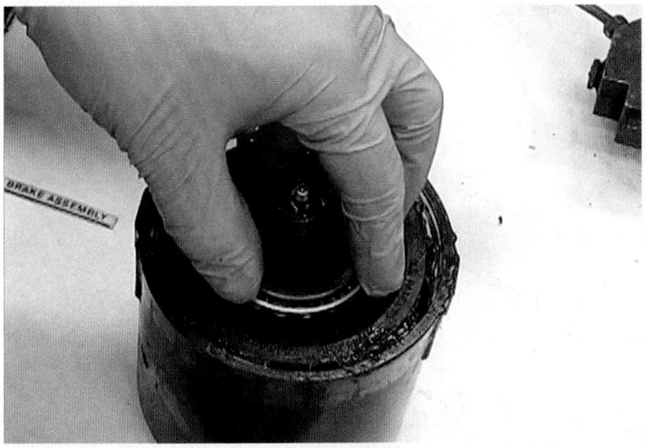

1. Make sure the bearing packer has the proper grease and that it is uncontaminated. Place the wheel bearing with the narrow side down.

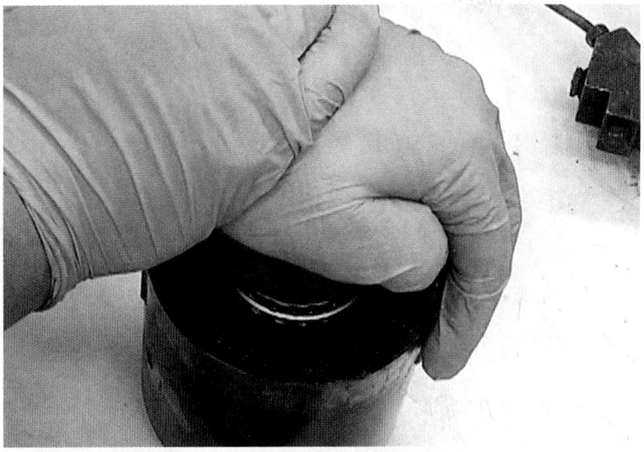

2. Place the packer cone on the top of the wheel bearing. Pack the wheel bearing, following the packer's instructions.

3. Remove the wheel bearing. Wipe off any old grease.

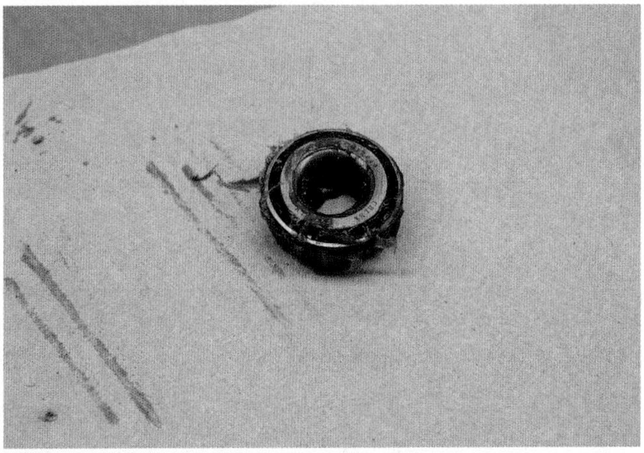

4. Smear some new grease around the outside of the wheel bearing. Repeat this process on the other wheel bearing. Be sure to set a packed wheel bearing down on a clean surface.

SKILL DRILL 38-4 Installing Wheel Bearings

1. Place a small amount of extra grease in the center of the hub. Do not fill it completely. Place the inner wheel bearing in its race, narrow side toward the race.

2. Using a seal installer or hammer, install the new grease seal, being careful not to damage it. Place a small amount of grease on the lip of the seal to provide it with initial lubrication.

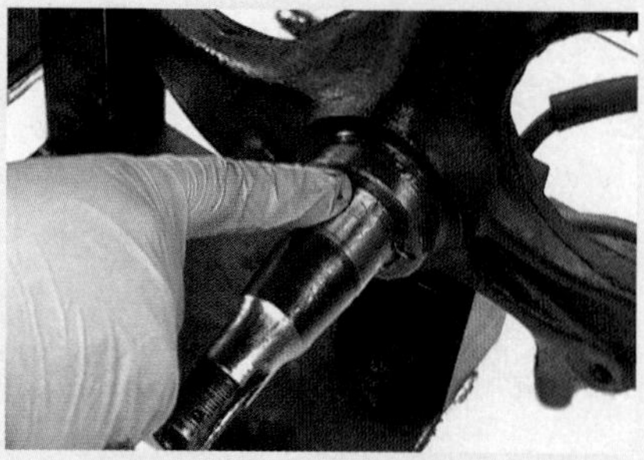

3. Make sure the spindle is clean, including the mating surface for the seal.

4. Without getting grease on the drum/rotor, carefully install it on the spindle, making sure the inner bearing fully seats against the spindle flange.

5. Install the outer bearing on the spindle and into the race.

6. Install the keyed washer and adjusting nut on the spindle, and tighten until finger tight.

7. Tighten the adjusting nut to the specified seating torque (usually about 20 ft-lb [27.1 N·m]) while turning the drum/rotor. This squeezes the excess grease out from between the wheel bearings and races while seating the bearings. *Do not leave the bearing this tight!*

8. Loosen the adjusting nut approximately one-sixth to one-quarter turn without turning the drum/rotor.

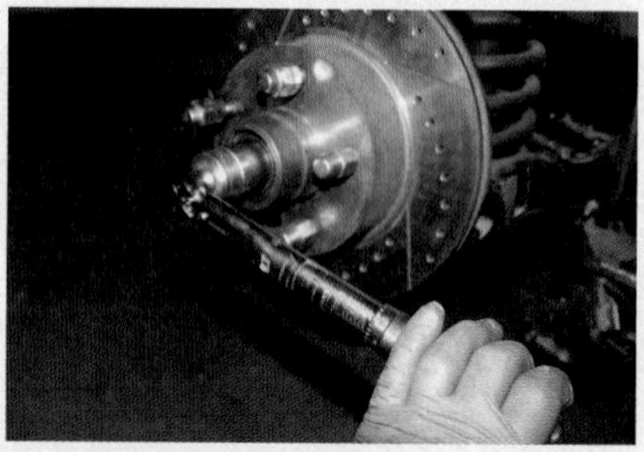

9. Tighten the adjusting nut to the specified preload torque. This is usually about 15–25 in.-lb (1.69–2.82 N·m).

To install wheel bearings, follow the steps in **SKILL DRILL 38-4**.

▶ TECHNICIAN TIP

Some technicians simulate the preload torque by placing a 12" crescent wrench on the nut. Then use only the hanging weight of the crescent wrench when it is parallel to the ground. Do not let it drop into position. Just lower the handle to where it stops turning the nut. This should be when the handle is approximately level. If not, reposition the crescent wrench on the nut, and allow it to lower until it stops parallel.

To install the locking mechanism, follow the steps in **SKILL DRILL 38-5**.

SKILL DRILL 38-5 Installing the Locking Mechanism

1. If the locking mechanism is a cotter pin, insert the new cotter pin through the castellated nut or locking cage, and the spindle. The short leg of the cotter pin should be against the castellated nut, and the long leg should be toward you. With the cotter pin fully engaged in the notch, bend the outer leg toward you and up over the end of the spindle. Cut it off just beyond the spindle. Also, cut the short leg off at the nut or cage. Make sure the cotter pin will not hit the inside of the dust cap.

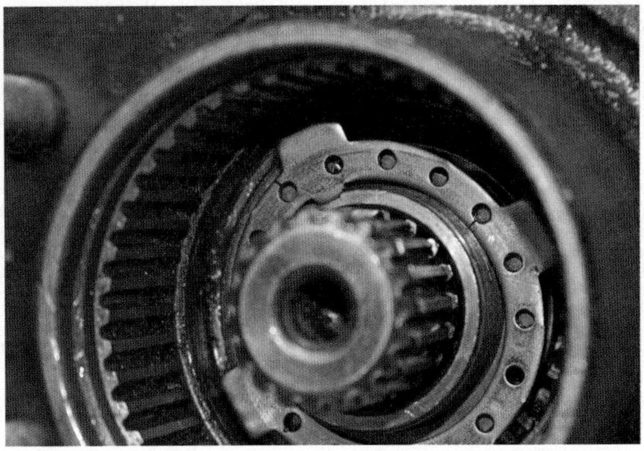

2. If it is a pin and hole style, or a bendable tang locking style, then place the washer against the adjusting nut (with the pin lined up, if that style) and thread the locking nut up against it.

3. Tighten the locknut to the specified torque. This is usually a substantial torque of 50 ft-lb (67.79 N·m) or way more. If a bendable tang style, bend the appropriate tang out toward you and against the flat side of the locking nut with a small pry bar to lock the adjustment in place.

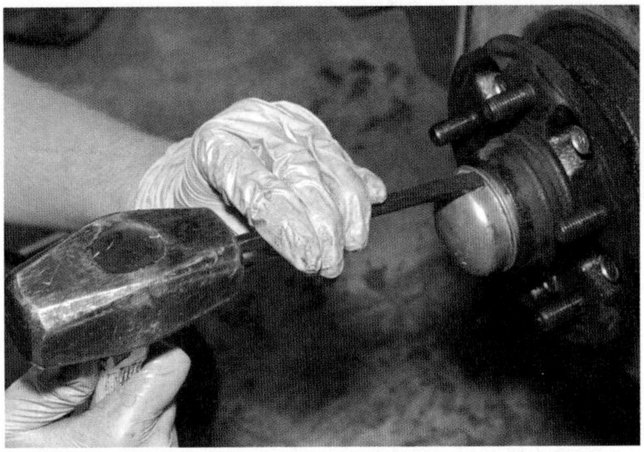

4. Install the dust cap, being sure it is fully seated in the hub. Make sure the drum/rotor turns freely without binding or making any unusual noises.

Replacing Wheel Bearings and Races

Wheel bearings and races have to be replaced only when they are damaged. If one part of the wheel bearing is damaged, all parts must be replaced. On serviceable bearings, the inner race, roller bearings, and bearing cage are one unit. They generally slip off of the spindle. The outer race is usually press fit into the hub. It has to be driven or pressed out and a new one press fit back in. It is critical that the seat in the hub be spotlessly clean. This means it must have no burrs, or the bearing race will not seat properly and the bearing will fail prematurely.

To replace a wheel bearing and race, follow the steps in **SKILL DRILL 38-6**.

SKILL DRILL 38-6 Replacing Wheel Bearings and Races

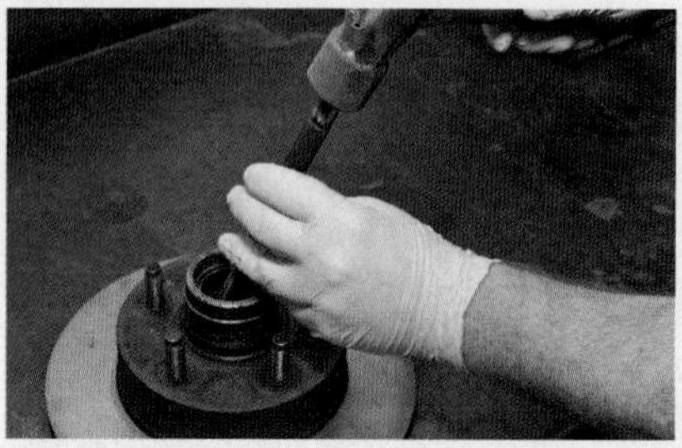

1. With the wheel bearings removed from the wheel hub, clean and inspect the bearing and race for damage. Determine which bearing and race have to be replaced. Using a hydraulic press, or a hammer and punch from the opposite side of the hub, carefully force the race from the hub. Keep it as straight as possible while removing it.

2. Remove any burrs with a fine file or Dremel™, and remove any debris from the seat. Lightly lubricate the outside surface of the new race, and set it thick side down in the hub.

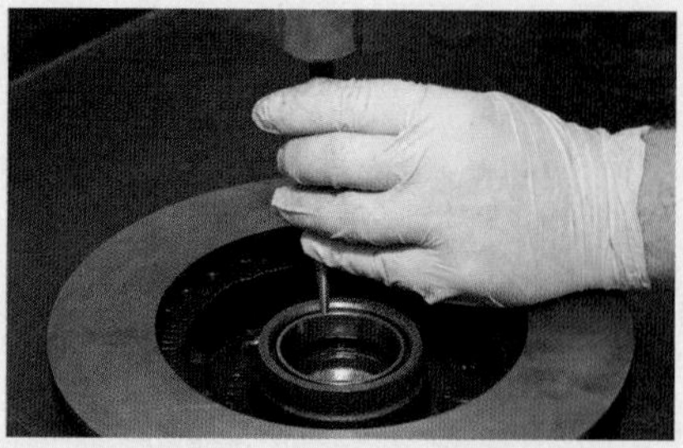

3. Using a hydraulic press or a hammer and bearing race installer, carefully drive the race until it is fully seated in the hub. When you are using a hammer and punch, a distinct sharp metallic sound should be produced when it seats. Inspect the race to verify that it is fully seated. Also, check for any damage caused by installation. If everything is good, pack the new bearing and install it according to Skill Drills 38-2 through 38-5.

FIGURE 38-30 Sealed wheel bearing.

FIGURE 38-31 Unitized wheel bearing hub assembly.

▶ Removing and Reinstalling Sealed Wheel Bearings

LO 38-07 Remove and reinstall sealed wheel bearings.

Sealed wheel bearings come in two configurations. The first is a replaceable sealed bearing only (**FIGURE 38-30**). On most front wheels, this wheel bearing is pressed between the hub and wheel flange. It is the more difficult of the two to replace. The second configuration consists of a unitized wheel bearing hub assembly (**FIGURE 38-31**). It includes a sealed wheel bearing, a removable wheel bearing hub, and possibly the wheel flange. In most cases, this type can be unbolted from the suspension system and a new one bolted in its place, and it is ready to go. This is the most common arrangement on recent vehicles.

The replaceable bearing style needs to be pressed apart with a hydraulic press. A special sealed bearing removal/installing tool can also be used on some vehicles. If using the hydraulic press method, the steering knuckle will have to be removed from the vehicle. This is so it can be placed on the hydraulic press. If the special sealed bearing tool is used, most bearings can be removed while the steering knuckle is still installed on the vehicle. This can save the technician a fair amount of time. To remove and reinstall a sealed wheel bearing assembly using the unitized wheel bearing hub style, follow the steps in **SKILL DRILL 38-7**.

SKILL DRILL 38-7 Removing and Reinstalling Sealed Wheel Bearings Using the Unitized Wheel Bearing Hub Style

1. Loosen the axle hub nut, if equipped, while the tire is still on the ground. Remove the wheel and brake assembly, following the specified procedure. Also disconnect the ABS connector and/or sensor if mounted to the hub.

2. If the wheel you are working on is a drive wheel, remove the axle hub nut, and tap the drive axle loose with a dead blow hammer.

3. Unbolt and remove the hub assembly from the steering knuckle. Clean the knuckle assembly, and check the hub seat for nicks, burrs, or other damage.

4. Carefully compare the new hub to the old one. Fit the new hub assembly (over the axle shaft, if equipped) to the knuckle, making sure it is fully seated in place. Torque the mounting bolts to the specified torque.

5. Reassemble the brake assembly and ABS sensor, if removed, following the specified procedure. Install the wheel, and torque the lug nuts. Be sure the correct ends of the lug nuts are facing the wheel.

6. Install the drive axle nut, if equipped. Use a new hub nut if called for by the manufacturer, and torque to specifications.

▶ Wrap-Up

Ready for Review

- Wheel bearings allow the wheels to roll with a minimum of friction and maintain accurate wheel positioning under all driving conditions. Most wheel bearing assemblies include an outer race, an inner race, roller bearings or ball bearings, and a bearing cage.
- Common type of wheel bearing include cylindrical roller bearing, tapered roller bearing, ball bearing, double-row ball bearings, and double-row tapered roller bearings. Each bearing is designed for a particular application.
- Grease seal can be press fit into the axle housing or the wheel hub. Two common types of lubricants are gear lube and bearing grease. Gear lube is somewhat thicker than engine oil and bearing grease is much thicker than gear lube.
- Rear drive axles can be designated as full-floating, semi-floating, and quarter floating.
  - In a full floating axle arrangement, the axle only carries a twisting force; in a semi-floating axle, the wheel flange is part of the axle; and in a quarter floating axle design, a single bearing assembly is located between the outside of the axle tube and the hub.
- Common wheel bearing noises include rumbling, whirring, and howling. Loose or worn wheel bearings can also cause the vehicle to wander, shimmy, or vibrate. Failure analysis of wheel bearing starts with removal of the faulty wheel bearing, visual inspection of the wheel bearing components, and comparison of the findings with the manufacturer's failure analysis chart.

- Serviceable wheel bearings should be serviced periodically: according to the manufacturer's scheduled maintenance chart, whenever brake work is being performed, or if a faulty wheel bearing is suspected.
  - Proper grease should be used when servicing wheel bearings.
- Sealed wheel bearings come in two configurations: the first is a replaceable sealed bearing only and the second configuration consists of a unitized wheel bearing hub assembly that includes a sealed wheel bearing, a removable wheel bearing hub, and possibly the wheel flange. A special sealed bearing removal/installing tool needs to be used to remove a replaceable bearing.

Key Terms

adjusting nut The nut used to adjust the end play or preload of a wheel bearing.

antifriction bearing A bearing that uses rolling elements to reduce friction.

cotter pin A single-use, soft, metal pin that can be bent into shape and is used to retain bearing adjusting nuts.

cylindrical roller bearing assembly A type of wheel bearing with races and rollers that are cylindrical in shape and roll between inner and outer races, which are parallel to each other.

double-row ball bearing assembly A single ball bearing assembly using two rows of ball bearings riding in two channels in the races.

drain plug Typically a threaded plug used to remove fluid from a system.

End play Fore and aft movement between mating parts.

fill plug Usually a threaded plug that can be removed to allow the level of a fluid to be checked and filled. This could also be a rubber snap fit plug.

garter spring A coiled spring that is fitted to the inside of the sealing lip of many seals, used to hold the lip in contact with the shaft.

grease A lubricating liquid thickened to make it suitable for use with many wheel bearings.

grease seal A component that is designed to keep grease from leaking out and contaminants from leaking in.

interference fit A condition in which two parts are held together by friction because the outside diameter of the inner component is slightly larger than the inside diameter of the outer component.

keyed lock washer The washer that fits between the adjusting nut and the locknut; the face of the washer is drilled with a series of holes that mate to a short pin from the adjusting nut, locking it to the spindle.

keyed washer The washer that fits between the adjusting nut and the wheel bearing and that has the center hole keyed to fit a slot on the spindle or axle tube.

lithium soap A thickening agent for grease to give it the proper consistency.

lock cage The stamped sheet metal cap that fits over the bearing adjustment nut and is secured by a cotter pin going through it and the spindle/axle.

locknut The nut that holds the adjusting nut from turning; usually tightened much tighter than the adjusting nut.

molybdenum thickening agent A compound used in some greases to give them the needed consistency.

National Lubricating Grease Institute (NLGI) An organization that grades the thickness of automotive and industrial grease.

preload Further pressure applied to bearing-supported parts after all the free play is taken up.

running clearance The amount of space between wheel bearing components while in operation.

sealed bearings Wheel bearings that are assembled by the manufacturer, with the proper lubrication, and sealed for life; cannot normally be disassembled.

serviceable bearings Wheel bearings that can be disassembled, cleaned, inspected, packed, reinstalled, and adjusted.

tapered roller bearing A type of wheel bearing with races and rollers that are tapered in such a manner that all of the tapered angles meet at a common point, which allows them to roll freely and yet control thrust.

unitized wheel bearing hub An assembly consisting of the hub, wheel bearing(s), and possibly the wheel flange, which is preassembled and ready to be installed on a vehicle.

viscosity The measurement of how easily a liquid flows; the most common organization that rates lubricating fluids is SAE.

Review Questions

1. How often did older serviceable type wheel bearings need to be serviced?
   a. Once per year
   b. Every other oil change
   c. 24,000–30,000 miles
   d. 50,000–60,000 miles
2. Which of the following is a common wheel bearing component?
   a. The grease fitting
   b. The roller cage
   c. The lower race
   d. The center race
3. Which of the following is true about cylindrical roller bearings?
   a. They require adjustment as periodic maintenance
   b. They support large thrust (side to side) loads
   c. They do not require any type of lubrication
   d. They spread the force of the load across the large roller surface
4. Which type of bearing must be used in pairs to support large side thrust loads?
   a. Ball bearings
   b. Tapered roller bearings
   c. Cylindrical roller bearings
   d. Sealed bearings

5. What is the purpose of a garter spring in a grease seal?
   a. To hold the seal firmly into the housing
   b. To keep tension on the bearings of the component
   c. To hold the seal tight against the rotating component
   d. To expand the outer diameter of the seal to fit the housing
6. All of the following are common types of lubricants for axle and wheel bearings EXCEPT?
   a. GL-5 Gear lube
   b. Bearing grease
   c. 5W-30 oil
   d. Lithium soap grease
7. If an axle only carries a twisting force, therefore not carrying any vehicle weight, what type of rear axle is it?
   a. A full floating
   b. A semi-floating
   c. A ¾ floating
   d. A ½ floating
8. Why is it important to inspect a failed wheel bearing while referencing the bearing failure analysis chart?
   a. To understand which direction to install the new part
   b. To submit a failure report to the bearing manufacturer
   c. To identify the cause so the same thing does not happen again
   d. To help determine if the bearing type should be changed or modified
9. When servicing wheel bearings, what should be done with the grease seal?
   a. The original seal should be carefully pressed or driven in
   b. The original seal should be cleaned and inspected for reuse
   c. The original seal should be coated with a sealer before installation
   d. The original seal should be discarded and replaced with a new one
10. What is true about most unitized wheel bearing and hub assemblies?
   a. They are normally changed by unbolting and bolting a replacement in
   b. They often require use of the hydraulic press or special removal/installation tool
   c. They are the most difficult to replace because the unit is so large
   d. They require preload adjustment after being greased and reassembled

ASE Technician A/Technician B Style Questions

1. Wheel bearings are being discussed. Technician A states that serviceable wheel bearings typically need to be adjusted on reassembly. Technician B states that sealed wheel bearings do not require periodic maintenance. Who is correct?
   a. Technician A only
   b. Technician B only
   c. Both Technicians A and B
   d. Neither Technician A nor B
2. Wheel bearings are being discussed. Technician A states that many bearing races are interference fit into housings or onto shafts. Technician B states that bearing races are commonly made of soft brass. Who is correct?
   a. Technician A only
   b. Technician B only
   c. Both Technicians A and B
   d. Neither Technician A nor B
3. Wheel bearing types are being discussed. Technician A states that cylindrical wheel bearings support large axial (side to side) loads. Technician B states that opposing tapered roller bearings support large axial (side to side) loads. Who is correct?
   a. Technician A only
   b. Technician B only
   c. Both Technicians A and B
   d. Neither Technician A nor B
4. Seals are being discussed. Technician A states that lip seals should be pre-lubed with a small amount of oil or grease to prevent premature failure. Technician B states that grease seals should not be reused and often require use of a special tool to install correctly. Who is correct?
   a. Technician A only
   b. Technician B only
   c. Both Technicians A and B
   d. Neither Technician A nor B
5. Rear axle bearings are being discussed. Technician A states that full floating axles use a pair of tapered roller bearings in a hub to carry the vehicle weight. Technician B states that full floating axles cannot carry as much weight as semi-floating axles. Who is correct?
   a. Technician A only
   b. Technician B only
   c. Both Technicians A and B
   d. Neither Technician A nor B
6. Wheel bearings are being discussed. Technician A states that loose wheel bearings can be identified by wiggling a tire back and forth at 6 and 12 o'clock while on a lift. Technician B states that a wheel bearing noise will vary with vehicle speed while a transmission bearing noise will vary with engine RPM and gear range. Who is correct?
   a. Technician A only
   b. Technician B only
   c. Both Technicians A and B
   d. Neither Technician A nor B
7. Wheel bearings are being discussed. Technician A states that a defective wheel bearing does not require close inspection since it cannot be repaired, only replaced. Technician B states that failed wheel bearings can make noises such as rumbling, whirring, or howling. Who is correct?
   a. Technician A only
   b. Technician B only
   c. Both Technicians A and B
   d. Neither Technician A nor B

8. Wheel bearing replacement is being discussed. Technician A states that the cotter pin is one-time use and needs to be replaced each time it is removed. Technician B states that the specified grease must be used when installing new wheel bearings. Who is correct?
   a. Technician A only
   b. Technician B only
   c. Both Technicians A and B
   d. Neither Technician A nor B
9. Serviceable wheel bearings are being discussed. Technician A states that after cleaning the old grease from bearings, it is not critical that all solvent be removed before installing new grease. Technician B states that the race needs to be replaced when a new bearing is installed. Who is correct?
   a. Technician A only
   b. Technician B only
   c. Both Technicians A and B
   d. Neither Technician A nor B
10. Sealed wheel bearing replacement is being discussed. Technician A states that some sealed wheel bearings require removal of the knuckle and set up on a hydraulic press to remove and install new bearings. Technician B states that some sealed bearings can be replaced on the car using a special tool. Who is correct?
    a. Technician A only
    b. Technician B only
    c. Both Technicians A and B
    d. Neither Technician A nor B

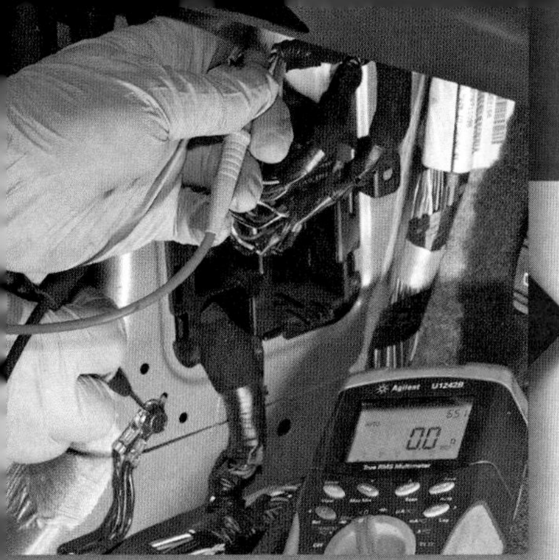

CHAPTER 39
Electronic Brake Control

Learning Objectives

- **LO 39-01** Describe the evolution of electronic brake control systems.
- **LO 39-02** Describe ABS system operation.
- **LO 39-03** Describe ABS master cylinder and HCU operation.
- **LO 39-04** Describe ABS wheel speed sensor and brake switch operation.
- **LO 39-05** Describe ABS EBCM operation.
- **LO 39-06** Describe Traction Control System operation.
- **LO 39-07** Describe Electronic Stability Control System operation.

ASE Education Foundation Tasks

See Appendix A to view the 2017 ASE Education Foundation Automobile Accreditation Task List Correlation Guide.

▶ Evolution of Electronic Brake Control Systems

LO 39-01 Describe the evolution of electronic brake control systems.

Electronic brake control (EBC) systems have greatly increased the safety of vehicles over the years. This is due to integrating computer-controlled hydraulics into the braking system (**FIGURE 39-1**). Standard hydraulic brake systems have limitations on how effectively they can stop a vehicle. The driver can only input braking force to the system through the brake pedal. It applies hydraulic pressure at predetermined ratios to the front and rear brakes. In a panic situation, the driver is unable to apply the exact amount of force needed to maintain maximum braking. With too little force, the vehicle does not stop as quickly. With too much force, the tires skid, making the vehicle's stopping distance even longer. At the same time, if the front wheels skid, the driver loses the ability to steer the vehicle. If the rear wheels skid, the car could spin out and possibly roll over.

Even if the driver could apply the perfect amount of braking force, there is no way to accommodate different amounts of traction at each wheel. When tires on one side of the vehicle are on dry pavement, they have good traction. If the other tires are on wet pavement, they have less traction. Braking the vehicle under these conditions in a panic stop could lead to a loss of control.

FIGURE 39-1 Brakes must be fully functional in case of an emergency.

In the quest for increased safety, manufacturers developed a series of EBC systems. The first-generation EBC system was the antilock brake system (ABS). It was designed to prevent wheels from locking up under braking conditions. This helps the driver maintain steering control of the vehicle. Maintaining steering control allows drivers to be better able to avoid collisions. ABS also shortens most panic stop distances by preventing wheel lock up.

ABS was a great start. But manufacturers found that they could get additional functionality by modifying the system slightly. They added a few components and modified the computer's software. This allowed them to provide traction control capability to the vehicle. This helps the driver maintain control of the vehicle while accelerating. But, ABS works only when braking the vehicle.

With some more refinements, they were able to provide electronic stability control. This added a few more components and advancements in the software. Electronic stability control helps prevent loss of driving control, especially vehicle rollovers. And the enhancements keep coming. Some systems now provide engine braking control, trailer sway control, and crash avoidance braking.

Basic Operation of Electronic Brake Control Systems

ABS systems use a computer that monitors the speed of each wheel as the brakes are applied. If one or more wheels begin to lock up, the computer sends electrical signals to **solenoid valves**. They momentarily hold or release hydraulic pressure to that wheel until it speeds up and starts rolling again. Once it starts rolling, the computer allows hydraulic pressure to be applied to that wheel again, slowing it down. This process is repeated very rapidly as the vehicle is brought to a stop.

Because the tires remain in rolling contact with the road surface, the vehicle can be steered. This allows the driver to maintain directional control (steerability) (**FIGURE 39-2**). These actions are completely dependent on the driver applying pressure to the brake pedal.

The basic ABS system does a good job of managing the braking effort of the driver in a panic stop situation. But it is

You Are the Automotive Technician

A longtime customer brings a 2012 Ford Explorer into the shop to get the brakes inspected. He claims that the brake pedal pulsates when brakes are applied. You ask if any recent work has been done on the vehicle. He explains that new tires were put on recently. He also explains that after leaving the shop, a car pulled out in front of him, and he had to lock up the brakes to avoid hitting it. He was startled by very heavy brake pedal pulsations. Ever since then, the brakes have a small pulsation that seems to be getting worse. He is wondering if there is a problem with the antilock brakes.

1. Is there a problem with the ABS system? What will you say to the customer?
2. What do you suspect is causing the pedal pulsations, and how did it occur?
3. How would you diagnose a problem when the ABS warning lamp is illuminated?
4. In what two primary ways does traction control reduce wheel slip?

limited to using the hydraulic pressure the driver exerts on the system. This means that the standard ABS system by itself cannot increase the hydraulic pressure in the brake system. Nor can it apply hydraulic pressure separate from the driver. As long as the driver is exerting firm pressure on the brake pedal, ABS can work fully.

The second-generation EBC system was the **traction control system (TCS)**. It added a high-pressure pump and a few **isolation valves** to the basic ABS system. Manufacturers found that they could now assist the driver in minimizing wheel slip while the vehicle is being accelerated. This is especially effective on slippery road surfaces such as gravel, snow, and ice.

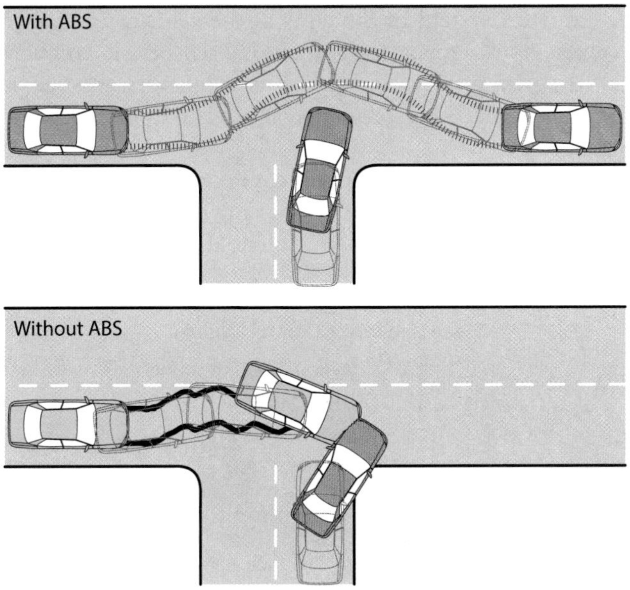

FIGURE 39-2 ABS helps the driver avoid accidents in a panic stop situation by maintaining steering control of the vehicle.

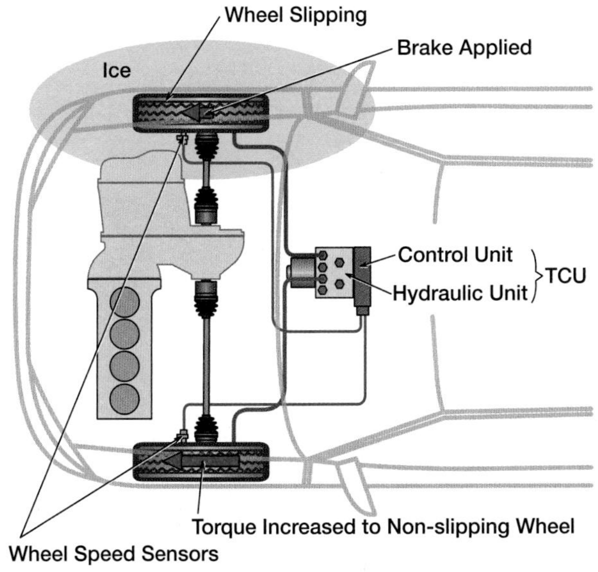

FIGURE 39-3 If a wheel is slipping, the TCS system applies the brake pressure to the slipping wheel. This causes torque to be sent to the wheel with more traction.

In most vehicles, the vehicle's traction is only as good as the tire with the least traction. So if one tire is on a patch of ice, the vehicle may not have enough traction to move. It may ultimately become stuck. The TCS system applies brake pressure to the slipping tire. This causes more of the engine's torque to be transmitted to the wheel or wheels with the most traction (**FIGURE 39-3**). If necessary, the TCS system can also request that the engine's powertrain control module reduce the power output of the engine. This further enhances traction. These actions are controlled by the PCM and do not require any input from the driver.

The next-generation EBC system was the **electronic stability control (ESC) system**. ESC takes the ABS and TCS systems one step further. It uses two more sensors. One sensor signals information regarding the driver's directional intent (**steering wheel position sensor**). The other sensor signals information regarding the vehicle's actual direction (**yaw sensor**). The EBC module (EBCM) can then detect the start of an **understeer, oversteer**, or potential rollover condition.

Understeer and oversteer are conditions that happen when a vehicle is traveling too fast for a particular corner. During understeer, the vehicle's front wheels are turned more sharply than the vehicle's path (**FIGURE 39-4**). The front tires

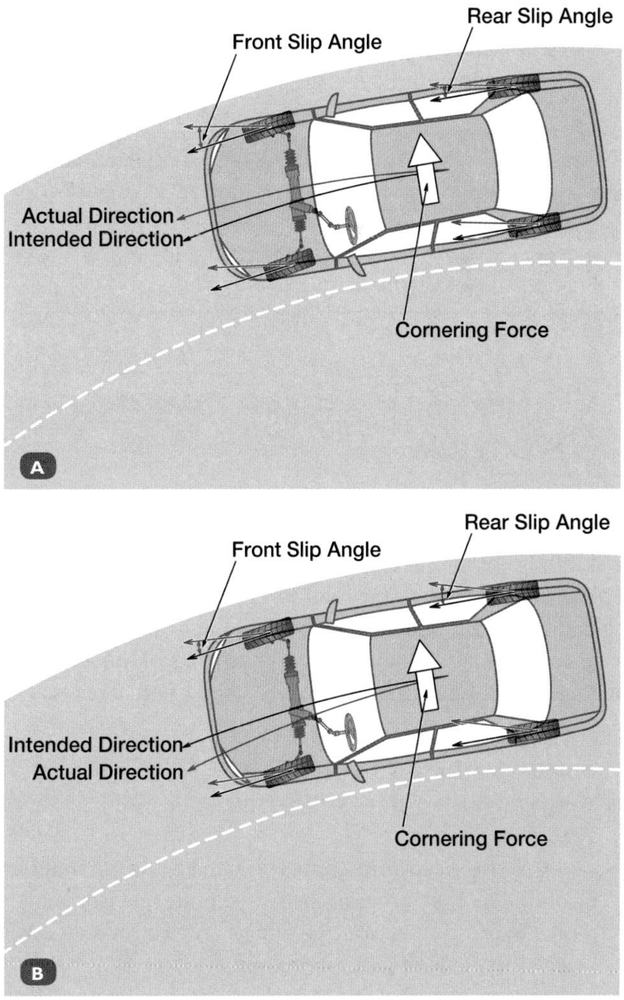

FIGURE 39-4 **A.** Understeer condition. **B.** Oversteer condition.

are actually sliding sideways toward the outside of the corner. The greater the understeer, the more the tires slide. Understeer is also referred to as "push," as in "the vehicle is pushing in the corners."

Oversteer is just the opposite. It occurs when the vehicle is turning more sharply than the front wheels are being steered. This happens when the rear tires are sliding sideways toward the outside of the corner. Oversteer is also referred to as "loose," as in "the vehicle is getting loose in the corners." The rear tires lose traction while cornering during oversteer. Most passenger vehicles are designed to have a bit of understeer. Understeer is easier for a driver to recover from than oversteer.

The EBCM uses information provided by the sensors of the ESC system along with the wheel speed sensors. It monitors the stability of the vehicle. If necessary, the EBCM can command individual brake units to be applied. It can also command a decrease in engine torque. For example, a vehicle is traveling too fast around a right-hand corner. The front wheels start to lose traction (understeer). The stability control system can apply the right rear brake. This helps to pivot the vehicle around the right rear tire. The vehicle is assisted in turning, and at the same time slowing the vehicle slightly. If additional measures are needed, additional brakes can be applied and the engine torque reduced. The control system performs these functions automatically. It does so without any driver input, other than steering the vehicle in the desired direction.

▶ TECHNICIAN TIP

It is important for customers to know that ABS, TCS, and ESC are not guarantees of avoiding an accident. These systems are designed to help drivers who are driving in a responsible manner to avoid an accident. Drivers can easily exceed the ability of these systems.

▶ Antilock Braking System Overview

LO 39-02 Describe ABS system operation.

The ABS system is designed to prevent wheels from locking or skidding. In fact, it doesn't matter how hard the brakes are applied or how slippery the road surface is. This is so that steering control of the vehicle can be maintained, and the stopping distance shortened. The primary components of the ABS braking system are shown in (**FIGURE 39-5**) and listed here:

- **ABS master cylinder**: Creates hydraulic pressure for each of the two hydraulic brake circuits.
- **Power booster**: Boosts driver brake pedal force on the master cylinder.
- EBCM or electronic control unit (ECU). An onboard computer. It is programmed to monitor sensor data. It then sends output control signals to electronic solenoid valves in the hydraulic control unit.
- **Hydraulic control unit (HCU)** or modulator: Contains electric solenoid valves controlled by the EBCM. It modifies hydraulic pressure in each hydraulic circuit (**FIGURE 39-6**). Most systems also contain an **accumulator** to store brake fluid under pressure.
- **Wheel speed sensor**: A device that monitors wheel speed and sends that signal to the EBCM.
- **Brake switch**: An on/off switch mounted at the brake pedal that informs the EBCM whether the driver is applying the brakes.

The EBCM may be located inside the vehicle or mounted near the HCU. It could be integrated into the HCU. In many cases, it is a separate module from the powertrain control module. However, it may be part of the vehicle's **body control module (BCM)**. The BCM is the computer that controls the electrical system in the body of the vehicle. The EBCM receives input signals from the ABS sensors. It compares that data with information stored in its memory and decides what actions are necessary. It then sends output commands to the HCU.

The HCU or modulator is connected in line with the brake lines. It sits between the master cylinder and the wheel brake units. It houses electric solenoid valves that control

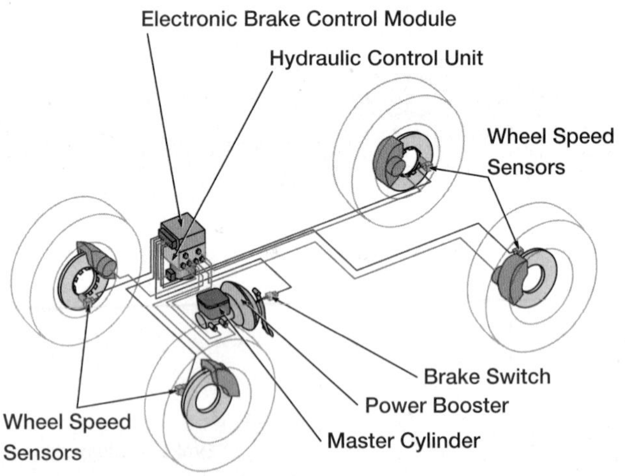

FIGURE 39-5 Typical ABS system.

FIGURE 39-6 A hydraulic control unit (HCU).

the flow of brake fluid to each wheel. The HCU receives operating signals from the EBCM to control the brakes during ABS conditions.

The power booster and master cylinder are mounted on the firewall. In most current applications, these components operate similarly to their non-ABS counterparts. Although, some manufacturers use a portless master cylinder. Brake fluid can return to the master cylinder reservoir easier in this design than a master cylinder fitted with a compensating port.

When the brakes are operating without ABS action, the brake pressure is controlled by the driver's foot pressure. However, the foot pressure is assisted by the power booster. In other words, the ABS system only affects brake pressure when one or more wheels are starting to skid.

Wheel speed sensors consist of two main parts. The first part is a toothed tone wheel (or tone ring). It rotates with the road wheels. The second part is a pickup assembly. It works with the tone ring to generate a speed signal that varies with the speed of the wheel. The wheel speed sensor is located near the wheel hub in many applications (**FIGURE 39-7**).

There are two main types of wheel speed sensors. The first is a **variable reluctance sensor** (magnetic induction). This type generates an analog AC sine wave signal (**FIGURE 39-8**). They can also be of the magneto-resistive or Hall effect type. This type generates a digital square wave signal. These signals can be used by the EBCM to determine the speed of each wheel. We cover the operation of these sensors in much greater depth in the ABS Components section.

Antilock Braking System Operation

When the ignition switch is turned on, the ABS controller illuminates the yellow ABS warning lamp. At that time, it also performs an automatic self-check of the system. If the system check passes, the controller will extinguish the warning lamp. This indicates to the driver that the ABS system is functional. Some ABS systems perform an additional self-check once the vehicle reaches approximately 3–5 mph (4.8–8 kph). Failures in the ABS system cause the controller to illuminate the ABS warning light in the instrument panel. If the light is illuminated, the ABS system normally shuts down and doesn't operate.

As the wheels start to turn, the wheel speed sensors generate small electrical signals and send them to the EBCM. When the brakes are applied, the wheels' rotational speed slows down. As the speed changes, the signal sent to the EBCM changes in like manner. If the control unit detects that a wheel is slowing too quickly and starting to lock, it takes corrective action. An output signal is sent to the appropriate solenoid valve in the HCU. This modifies the hydraulic pressure to the affected wheel brake unit. Let's explore how that works.

Principles of ABS Braking

Braking force and the tendency of the wheels to lock up are affected by a combination of factors such as:

- the friction of the road surface;
- the type, condition, and loading of each tire; and
- the difference between the vehicle speed and the speed of the wheels.

It should be noted that maximum traction happens with approximately 10%–20% tire slip. Thus, maximum braking traction occurs when the wheels are rotating 10%–20% slower than the vehicle speed. At the same time, traction falls off quickly above approximately 20% wheel slip. This is why ABS is so effective. It allows just enough slip to keep the tires at close to their maximum traction. It does so by rapidly modulating the hydraulic pressure in the vehicle's brake system.

During normal braking, the rotational speed of each wheel falls equally. So, no ABS intervention is needed. In this condition, the EBCM does not energize the solenoid valves in the hydraulic unit. The master cylinder's hydraulic pressure is applied to the wheel brake units. The ABS is not involved. However, even though the ABS is passive during normal braking, the EBCM constantly monitors the speed of each wheel. It looks for any wheel that begins to decelerate more rapidly than any of the other wheels.

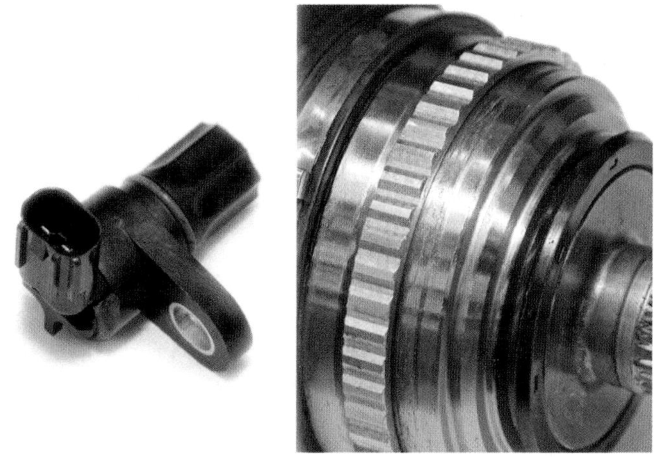

FIGURE 39-7 A wheel speed sensor and tone wheel.

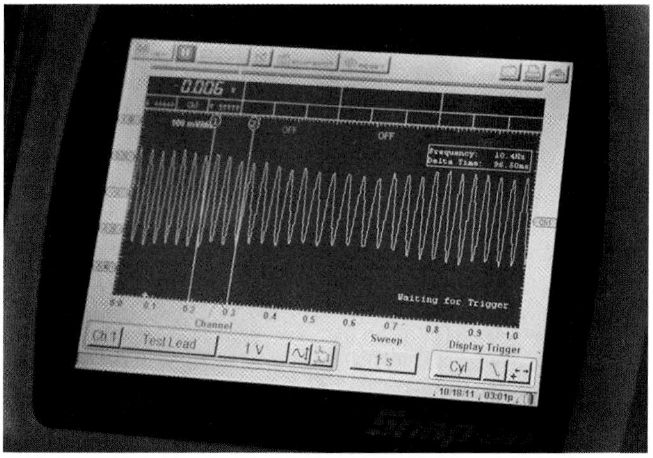

FIGURE 39-8 An oscilloscope pattern from a variable reluctance wheel speed sensor.

If one wheel speed sensor signals more severe wheel deceleration, it means the wheel is beginning to skid. The EBCM sends current to the appropriate solenoid valve (**FIGURE 39-9**). The first level of valve action isolates that brake circuit from the master cylinder. This prevents the braking pressure at that wheel from rising, and holds it there. If the wheel is still decelerating too rapidly, the EBCM commands the appropriate solenoid valve to release braking pressure. The solenoid valve opens a passage from the brake circuit on that wheel. This releases the hydraulic pressure to that brake unit. Brake fluid is released back to the master cylinder.

If the wheel speed sensors indicate that the wheel is rolling again, the EBCM de-energizes the solenoid valves. This lets the hydraulic pressure to be applied to that brake unit again. This cycle repeats itself at up to about 16 times per second. It is normal in an ABS system for the valves in the HCU to keep changing position as they modulate the brake pressure that is being applied. These changes in valve position cause rapid hydraulic pulsations. On some vehicles, this can be felt by the driver through the brake pedal. The solenoid valves also make a fairly loud clicking noise as they cycle on and off.

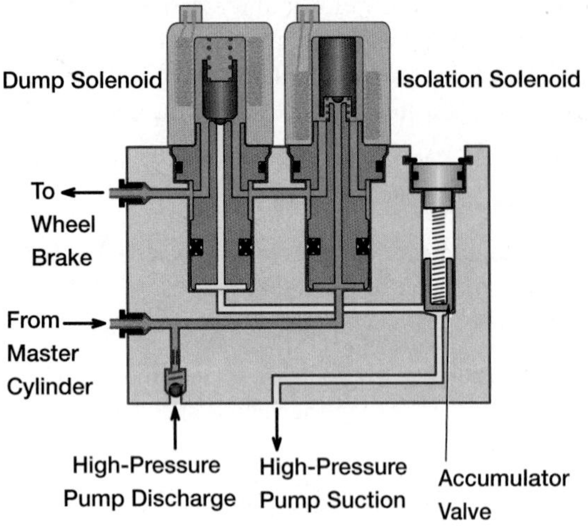

FIGURE 39-9 Hydraulic Control Unit solenoid valve arrangement for modulating ABS hydraulic pressure.

> **▶ TECHNICIAN TIP**
>
> Drivers should be taught to expect ABS brake pedal pulsation when in a panic stop. Some drivers who have never experienced this actually let up on the brake pedal in a panic stop. This is because of the rapid pulsations and accompanying noise. When in a panic stop, drivers should push hard on the brake pedal and not let up until the vehicle is stopped or out of danger. Brake-by-wire systems are not subject to this brake pedal pulsation situation. This is because the brake pedal is not part of the hydraulic system.

ABS Components

▶ ABS Master Cylinder

LO 39-03 Describe ABS master cylinder and HCU operation.

ABS master cylinders come in two major configurations: integral and nonintegral (**FIGURE 39-10**). **Integral ABS systems** are mostly found on older vehicles. They combine the tandem master cylinder, HCU, and power booster in one unit. The power booster consists of a high-pressure electric pump and accumulator. It operates the integrated master cylinder. Brake fluid passes from the master cylinder portion of the assembly to the HCU portion. There, the pressures are modified by the computer-controlled solenoid valves.

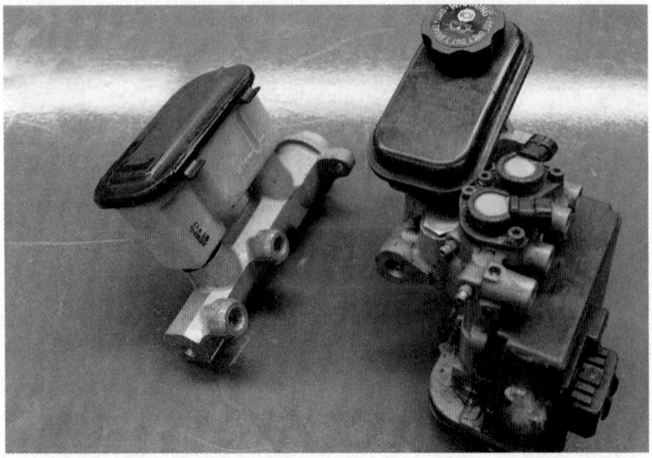

FIGURE 39-10 Nonintegral and integral master cylinder assemblies.

Nonintegral ABS systems use a fairly standard tandem master cylinder and a typical vacuum or hydraulic power booster. The booster assists the driver in applying force to the master cylinder. The master cylinder sends fluid under pressure to the HCU, which is a separate assembly. It is installed in line between the master cylinder and the wheel brake units. The HCU houses the computer-controlled solenoid valves.

Purpose and Operation of the ABS Master Cylinder

Nonintegral ABS master cylinders are usually identical to non-ABS master cylinders. They both use primary and secondary pistons in a common housing with a **common bore**. Some of these master cylinders utilize a portless ABS master cylinder design. They do not use a compensating port on the secondary circuit. Instead, the secondary piston incorporates a center valve (**FIGURE 39-11**). It controls the opening and closing of a supply port in the piston. At rest, the supply port is open and connects the reservoir with the front brake circuit. The primary piston still uses an inlet port and a compensating port. Therefore, the portless design is only used on the secondary circuit.

When the brake is applied, the primary piston moves and closes its compensating port. Fluid pressure in the primary circuit rises. It acts with the primary piston spring to move the secondary piston forward, closing the center valve. Pressure builds in both circuits, which applies the brakes in both circuits.

If braking conditions are such that the hydraulic control unit must return brake fluid to the master cylinder. Then, for the front brake circuits, brake fluid is returned to the front section. This forces the secondary piston back against the force of the primary piston spring and the rear brake pressure. In this case, both pistons move rearward. If enough brake fluid returns, the center valve in the secondary piston opens. Brake fluid is then allowed to return to the master cylinder reservoir.

If brake fluid is returned from the rear brake circuit, the secondary and primary pistons tend to be forced apart. This generally moves the primary piston rearward. If it travels far enough, brake fluid will return to the reservoir through the compensating port. The amount of brake fluid that returns to the master cylinder is determined by the degree of antilock braking control.

With as many as 16 ABS control cycles per second, the rapid changes in hydraulic pressure cause brake fluid pulsations to be sent back to the master cylinder. These pulsations can be felt by the driver at the brake pedal.

HCU–Control Valve Operation

In a standard ABS system, the HCU houses electrically operated hydraulic control valves (**FIGURE 39-12**). They are also called solenoid valves. They control brake pressure to specific wheel brake circuits. Each separate hydraulic circuit within the HCU has one or two solenoid valves. They provide three operating conditions: apply, hold, and release (**FIGURE 39-13**). During the apply mode, the solenoid valves allow brake fluid to freely flow through the HCU to the specific wheel brake unit. In this case, the driver is in full control of the brakes, through the master cylinder.

In the hold mode, a solenoid valve "isolates" the master cylinder from the brake circuit. This prevents brake pressure from building any further. The brake pressure to the wheel is held at that level. In the release mode, a solenoid valve releases (dumps) the brake circuit pressure to the wheel. The wheel is then allowed to start rolling again. The solenoid valve opens

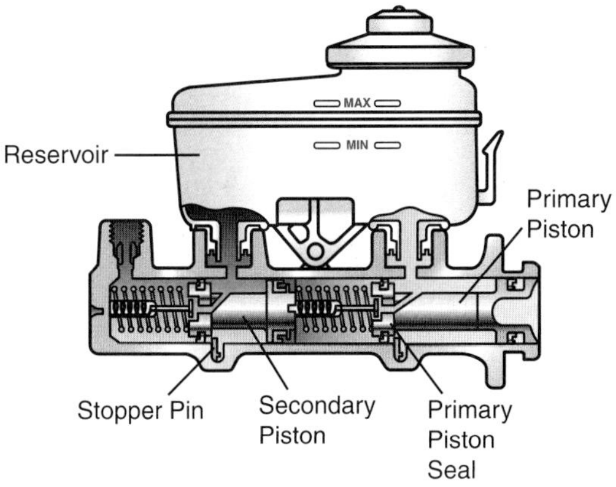

FIGURE 39-11 Portless ABS master cylinder.

FIGURE 39-12 Hydraulic Control Unit.

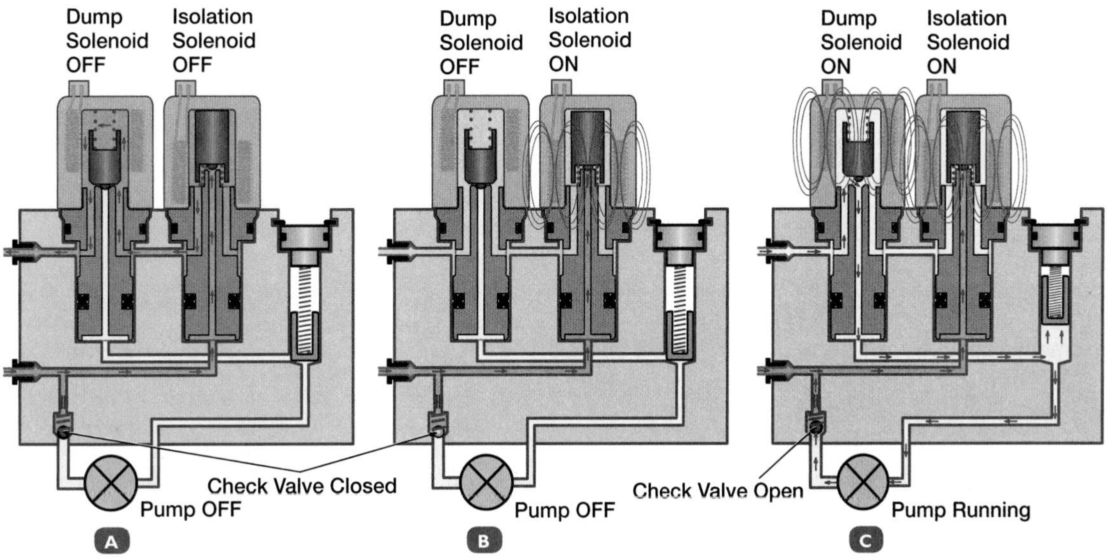

FIGURE 39-13 **A.** Apply. **B.** Hold. **C.** Release.

a passage back to the accumulator. Brake fluid is stored in the accumulator until it can be returned by an electric pump to the master cylinder reservoir. Notice that the brake pressure is held or dumped even though the driver is applying the brake pedal.

▶ TECHNICIAN TIP

Many, but not all, HCUs are sealed units and cannot be serviced. If you are working on a vehicle with a sealed HCU and it is faulty, it will have to be replaced. This can be quite costly.

Operation of the Hydraulic Control Unit

The ABS control module (or EBCM) sends commands in the form of electrical signals to the HCU. The HCU executes the commands, using one or two solenoid valves for each hydraulic circuit. The control valves are located between the master cylinder and the wheel brake units. They can apply, hold, or release hydraulic pressure going to the brake units.

In a normal non-ABS braking scenario, brake pedal force is transmitted to the master cylinder. It pressurizes the brake fluid which is then sent to the HCU. Fluid flows through the non-energized open isolation valves to the brake units at the wheel. The pressurized hydraulic fluid flows freely through the HCU to the brake units at each wheel.

When the control unit detects a wheel is locking up, it sends a command current to the isolation solenoid valve for that brake circuit. This current causes the solenoid valve to close, isolating the brake circuit from the master cylinder. That holds the hydraulic pressure between the solenoid valve and the brake circuit constant. This is regardless of whether the master cylinder hydraulic pressure rises or falls.

If excessive wheel deceleration continues, the control module commands the dump valve for that brake circuit to open. This reduces the braking pressure by opening a passage from the brake circuit to the accumulator. A pump in the HCU sends brake fluid back to the master cylinder. One or both pistons are pushed rearward in the bore and fluid is vented to the reservoir.

If the wheel starts to roll again, the EBCM signals the dump valve to close and the isolation valve to open. The hydraulic pressure from the master cylinder is again allowed to apply the brakes. The wheel is again slowed. This process continues until either the vehicle comes to a stop, or the driver lifts their foot from the brake pedal. In most standard ABS systems, the hydraulic pressure in the brake circuits can never rise above the master cylinder pressure.

Types of Hydraulic Control Units

There are a number of HCUs that vehicle manufacturers use. But they generally fall into a few categories. The first category relates to how many channels the system has. A **channel** generally means the number of wheel speed sensor circuits and hydraulic circuits a system has (**FIGURE 39-14**).

A single-channel system uses a single wheel speed sensor. It is typically located in the differential. It also uses one hydraulic control circuit to control both rear wheels. Since it has only one sensor, the ABS system has nothing to compare the deceleration to. So, maximum deceleration is programmed into the control unit. If the deceleration of the differential exceeds the maximum deceleration, then the HCU will be activated.

A two-channel system uses two separate speed sensors and hydraulic control circuits. It uses one wheel speed sensor and one hydraulic control circuit for each rear wheel. The two hydraulic control circuits apply brake pressure separately to the rear wheels.

A three-channel system controls both the front and rear wheels. It is configured so that each front wheel has its own wheel speed sensor and hydraulic control circuit. The rear brakes use a single-wheel speed sensor with a single hydraulic control circuit.

A four-channel system uses separate wheel speed sensors and hydraulic control circuits for each of the four wheels. Since these systems have more than one sensor, the control unit not only watches for wheel deceleration that is faster than the programmed maximum. It also watches for individual wheels decelerating faster than others. This gives a more robust ABS control.

Another difference among types of HCUs is the number of solenoid valves per hydraulic control circuit. Some HCU units use a single, three-position solenoid valve per circuit. Others use dual, two-position valves per hydraulic circuit (**FIGURE 39-15**).

The first position of the single, three-position valve allows brake fluid to flow through the apply port while blocking the release port. The second position blocks the apply port and the release port. The third position blocks the apply port and opens the release port. Thus, the single, three-position valve provides all three conditions: apply, hold, and release.

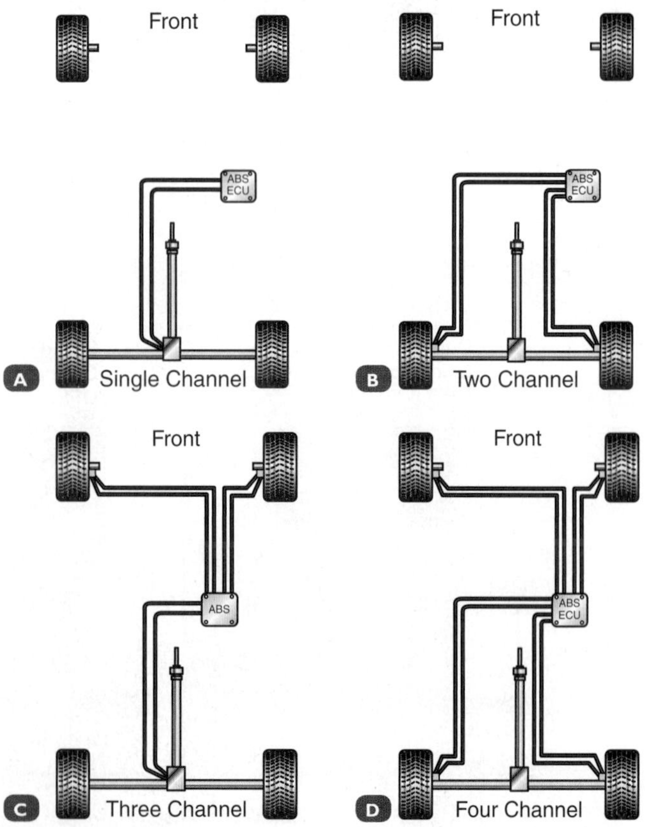

FIGURE 39-14 The four types of ABS channels. **A.** Single-channel system. **B.** Two-channel system. **C.** Three-channel system. **D.** Four-channel system.

The dual, two-position valve style of HCU uses one solenoid valve to open and close the apply port. This is commonly called the isolation valve. When this valve is not energized, the apply port is open. The second solenoid valve opens and closes the release port. When this valve is not energized, the release port is blocked. The EBCM operates each of these valves independently to obtain apply, hold, and release functions. Because there are twice as many solenoid valves, the EBCM is more complicated and costly to build. This is because each valve needs its own electrical control circuit. Therefore, EBCMs and HCUs cannot be randomly interchanged.

Another difference between HCUs is the type of accumulator used—low pressure or high pressure. **Low-pressure accumulators** hold brake fluid in a spring-loaded chamber when it is released by the dump valves (**FIGURE 39-16**). The hydraulic pressure remains fairly low in this type of accumulator. This is

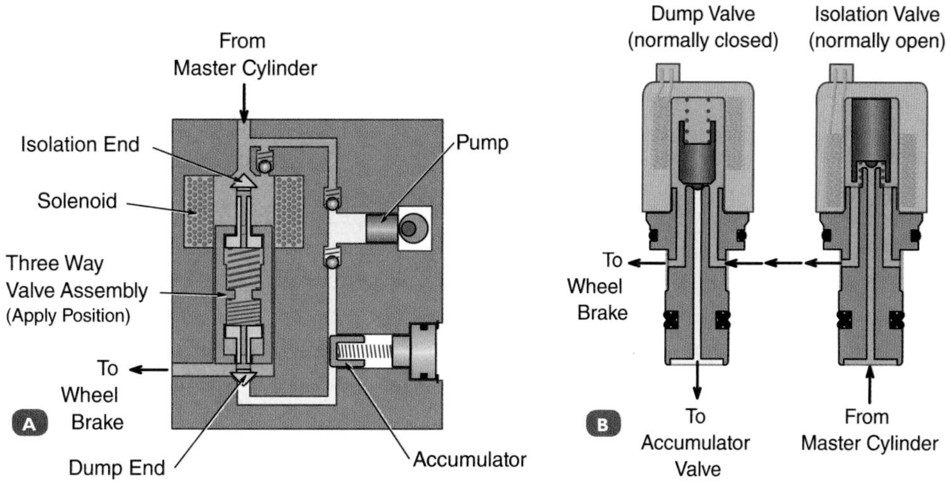

FIGURE 39-15 A. Single three-position valve. **B.** Dual, two-position valves.

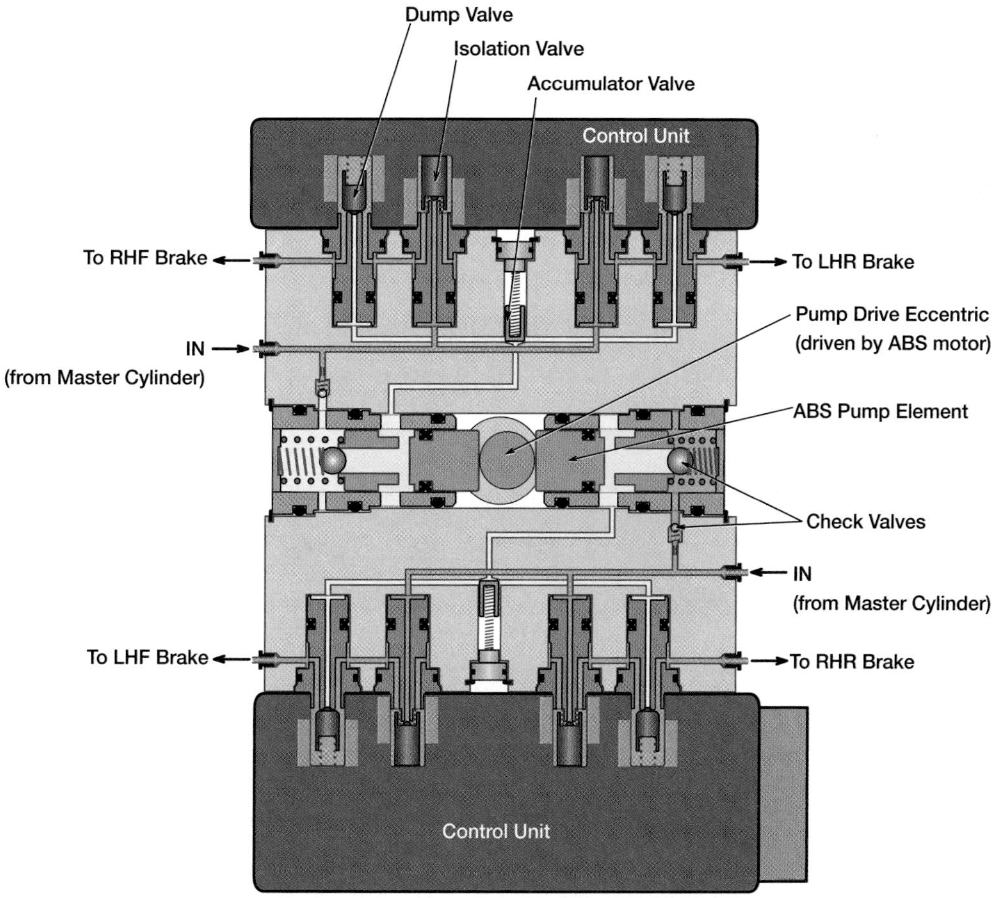

FIGURE 39-16 Low-pressure accumulator holds brake fluid until it can get returned to the master cylinder reservoir.

because an electric pump returns the released brake fluid to the master cylinder. It does so once the brake fluid reaches a certain level in the accumulator. When the electric pump turns on, the fluid returning to the master cylinder pushes the brake pedal toward the driver's foot. This causes the brake pedal to rise. This can be confusing to drivers who aren't expecting the brake pedal to be pushed back toward them.

High-pressure accumulators are used to store brake fluid under high pressure for one of two purposes. The first purpose is when it is used as a power booster for applying the integrated master cylinder. The second purpose is when it is used to independently apply the wheel brake units when the EBCM commands it.

When used as a power booster, pressure in the accumulator is maintained by a high-pressure electric pump. The pump is activated by a pressure switch and relay when the hydraulic pressure falls below a certain point. When the pressure reaches the upper pressure limit, the pressure switch opens and deactivates the pump. The hydraulic pressure is then used to boost the driver's foot pressure on the master cylinder when the driver depresses the brake pedal. If the high-pressure pump fails for any reason, the accumulator holds enough pressure to apply the brakes 10 to 20 times before it is used up. If that occurs, the brakes will still operate but will need much higher foot pressure.

The accumulator used to supply brake pressure to the HCU also uses a high-pressure pump, pressure switch, and relay. This system maintains an operating pressure of approximately 1200–2700 psi (8274–18,616 kPa), depending on the system (**FIGURE 39-17**).

SAFETY TIP

Be careful to follow the manufacturer's procedures when working on EBC systems. High-pressure brake fluid stored in the accumulator is dangerous.

The high-pressure pump pushes the brake fluid against a high-pressure nitrogen chamber. This holds pressure on the brake fluid. The hydraulic pressure is used to independently apply the brakes during a TCS or ESC event. This will be covered further in the Electronic Stability Control section. If the high-pressure pump fails and the hydraulic pressure falls below the pump's specified "on" pressure, the EBCM will disable the ABS system. It will also illuminate the yellow warning lamp, alerting the driver to an ABS system fault.

▶ Wheel Speed Sensors

LO 39-04 Describe ABS wheel speed sensor and brake switch operation.

Wheel speed sensors create electrical signals based on the rotational speed of each wheel they monitor. Wheel speed sensors do so by using principles of electromagnetism to generate an analog or digital electrical signal. This signal is read by the

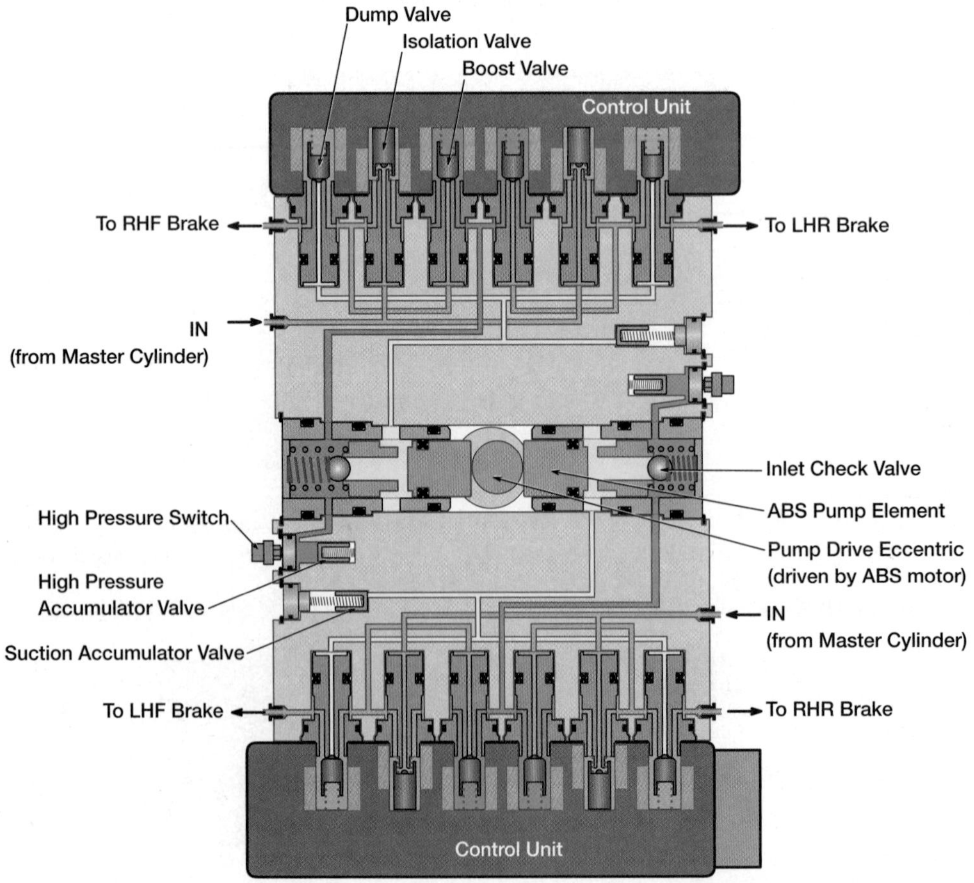

FIGURE 39-17 High-pressure accumulator.

EBCM to determine the speed of each wheel. It also monitors the rate of deceleration of each wheel. This information is used to determine if a wheel is starting to lock up and skid.

A wheel sensor assembly consists of a toothed tone wheel (or tone ring) and stationary pickup assembly. The tone wheel rotates with the wheel. The pickup assembly is attached to the hub or axle housing. The **pickup assembly** and **tone wheel** do not touch each other. A small gap, called an **air gap**, must be maintained at the specified clearance. Because there is no mechanical connection, there is virtually no wear on these parts. That is, unless a foreign object gets between them.

Types of Wheel Speed Sensors

The three most common types of wheel speed sensors are:

- variable reluctance (magnetic induction style),
- **magneto-resistive**, and
- **Hall effect** styles.

The variable reluctance type is simpler and usually less expensive for manufacturers to use. This style is sometimes called a passive system. This is because it needs no outside power to function, so it is self-contained. Magnetic induction occurs when the teeth on the tone wheel pass the sensor, creating an analog AC voltage signal.

As each tooth of the tone wheel approaches the pickup, the magnetic field creates a small voltage. The voltage pushes current flow in one direction inside the pickup assembly. As each tooth leaves the pickup assembly, voltage is generated that pushes current flow in the opposite direction. This process creates a full-cycle sine wave for each tooth on the tone wheel (**FIGURE 39-18**). The faster the wheel is turned, the faster the sine wave rises and falls. The speed at which the sine wave rises and falls is referred to as frequency. Frequency is measured in hertz, where one hertz equals one full-cycle sine wave per second.

The height of the sine wave, called its amplitude, also tends to change with speed. At very slow vehicle speeds, when the vehicle is just creeping along, the amplitude of the sine wave is very low. As the speed increases, so does the amplitude, along with the frequency. This AC signal is sent to the ECU, where it is processed. Once processed, it is then compared with the AC signals from the other wheels to determine wheel lockup.

Most variable reluctance wheel speed sensors are two-wire sensors. The two wires complete the circuit to and from the ECU. The variable reluctance sensor assembly consists of a coil of wire wound around a permanent magnet. Each end of the coil is connected to one of the wheel speed sensor terminals, which connect directly into the EBCM (**FIGURE 39-19**).

Because this type of sensor operates on principles of magnetism, the air gap between the toothed tone wheel and pickup coil is critical. If the air gap is too small, the parts could contact each other, damaging them. If the air gap is too large, the sensor output signal to the ECU could be too weak. This could trigger a code or cause the sensor to work intermittently.

One drawback to the variable reluctance sensor is that it does not function effectively below vehicle speeds of around 5 mph (8 kph). This is because it depends on the speed of movement of the tone wheel to create a signal. In other words, the amplitude of the sine wave it creates at slow speeds is not high enough for the EBCM to read it. This can prevent the ABS from

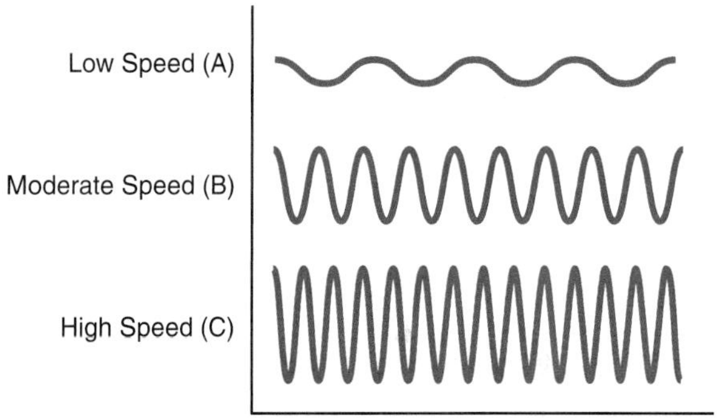

FIGURE 39-18 Wheel speed sensor sine wave. **A.** Signal during low vehicle speed. **B.** Signal during moderate vehicle speed. **C.** Signal during high vehicle speed.

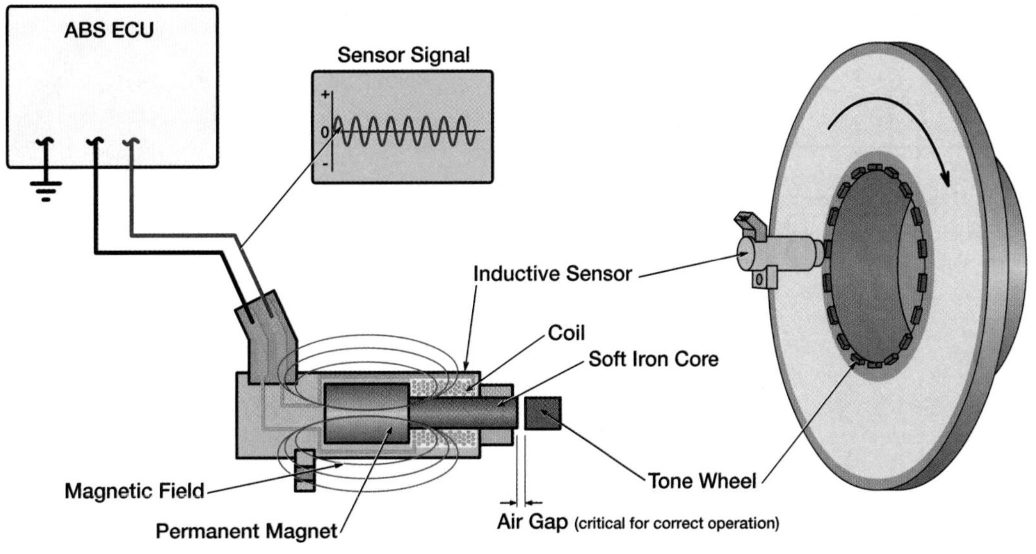

FIGURE 39-19 Variable reluctance wheel speed sensor assembly.

functioning during the last part of a braking event. The lack of ABS functionality at that speed could lengthen the stopping distance significantly. This is especially true on a very slippery road surface such as ice.

The magneto-resistive and Hall effect sensor systems are called active systems. This is because they require an outside power source to operate. If the sensor loses power or ground, it cannot generate an output signal. The power wire originates from the EBCM. It normally supplies the speed sensors with a reference voltage of between 5 and 12 volts, depending on the manufacturer. This helps ensure that the sensor is not affected by changes in the vehicle's electrical system voltage. A signal wire transmits the output signal from the sensor to the EBCM. The magneto-resistive and Hall effect sensors can be a three-wire arrangement, with the third wire being a dedicated ground. On some vehicles, a two-wire arrangement is used with ground being provided by the chassis.

The magneto-resistive and Hall effect wheel speed sensors operate similarly to all Hall effect sensors. A reference voltage and ground are supplied to the sensor assembly. An internal circuitry causes a small current to flow across the semiconductor bridge/Hall material (**FIGURE 39-20**). If the bridge/Hall material is exposed to a magnetic force, the magnetism forces the current to flow to one side of the bridge/Hall material. This produces a small difference in voltage across the sides of the bridge/Hall material (**FIGURE 39-21**). Voltage is then amplified and processed into a digital "on" signal (circuit is pulled to ground) and sent to the EBCM.

As the magnetic field is removed, the small signal voltage across the bridge/Hall material falls to 0V. The signal sent to the EBCM will be a digital "off" signal (reference voltage). As the magnetic field is alternately applied and removed, the sensor will send a digital square wave on/off signal. Because the magnetic field does not have to be moving for the bridge/Hall effect voltage to be created, the sensor works all the way down to 0 mph (0 kph). This allows the ABS to continue functioning until the vehicle comes to a virtual full stop.

▶ TECHNICIAN TIP

Testing wheel speed sensors is dependent on knowing which kind of sensor the vehicle uses. Do not assume all two-wire sensors are of the variable reluctance style. Turn the ignition switch to the Run position and the wheels stationary. Use your digital multimeter (DMM) to properly back-probe both sensor wires for voltage. If neither wire has voltage, suspect a variable reluctance sensor. If one of the two wires has a reference voltage, you are likely dealing with a magneto-resistive or Hall effect sensor.

Brake Switch

In addition to activating the rear brake lights, the brake switch sends an electrical input signal to the EBCM. This tells the EBCM when the driver is applying the brakes. If the brakes are being applied, the EBCM will activate the appropriate solenoid valves if the wheels are starting to lock up. If the brake switch indicates that the brakes are not being applied and the wheel speed sensors are showing unequal speeds, the EBCM on some vehicles illuminates a low-traction warning lamp. This is to alert the driver to the low-traction condition.

The brake switch is a normally closed switch. This means that if the switch is not affected by any outside force, electrical current will flow through it. The brake pedal pushes the brake switch open when the brake pedal is released (**FIGURE 39-22**). As soon as the driver steps on the brake pedal, the spring in the brake switch closes the contacts. This sends an electrical current (signal) to activate the brake lights. This electrical signal is also sent to the EBCM, signaling it that the driver is applying the brakes. In some systems, the brake switch is only used to signal the BCM. It then sends current to illuminate the brake lights.

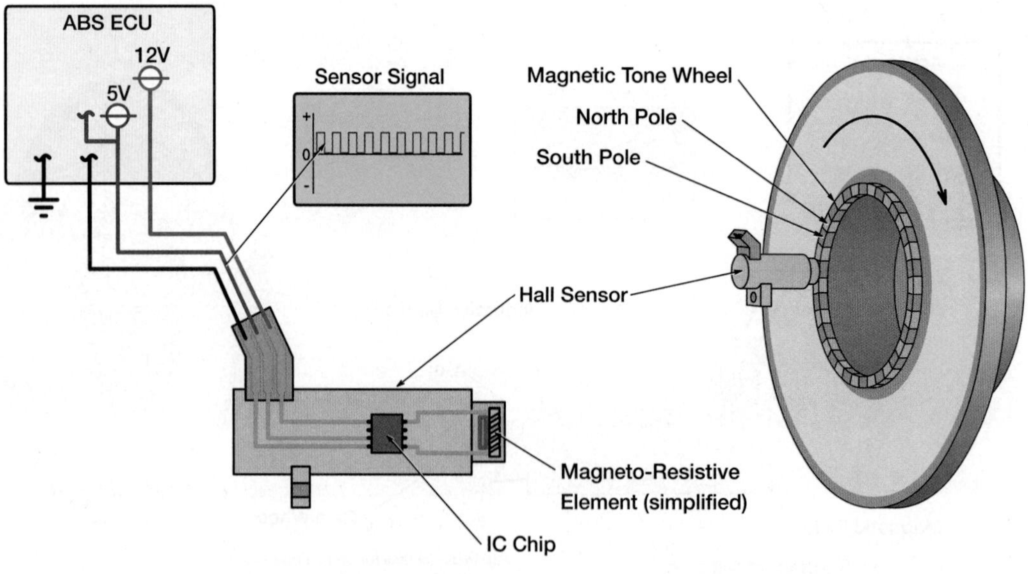

FIGURE 39-20 Hall effect wheel speed sensor assembly.

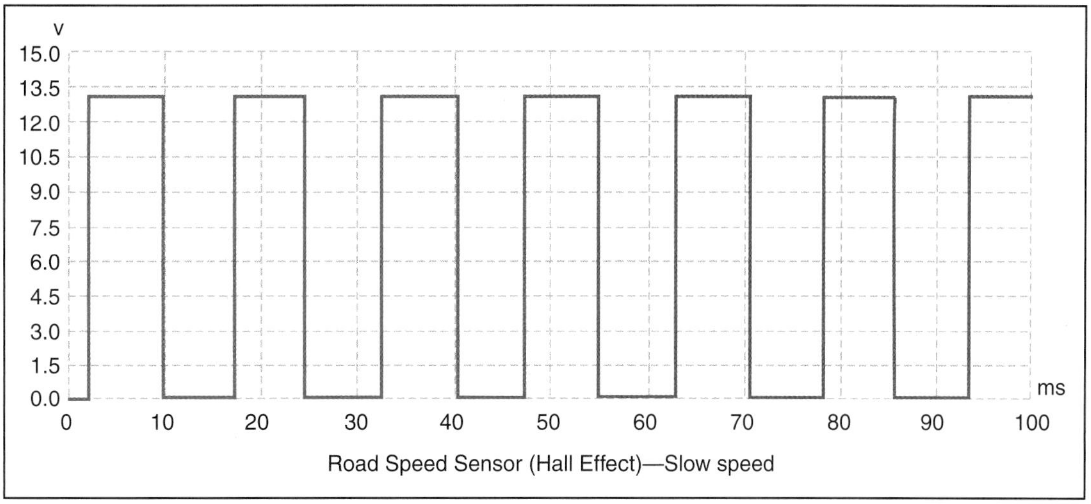

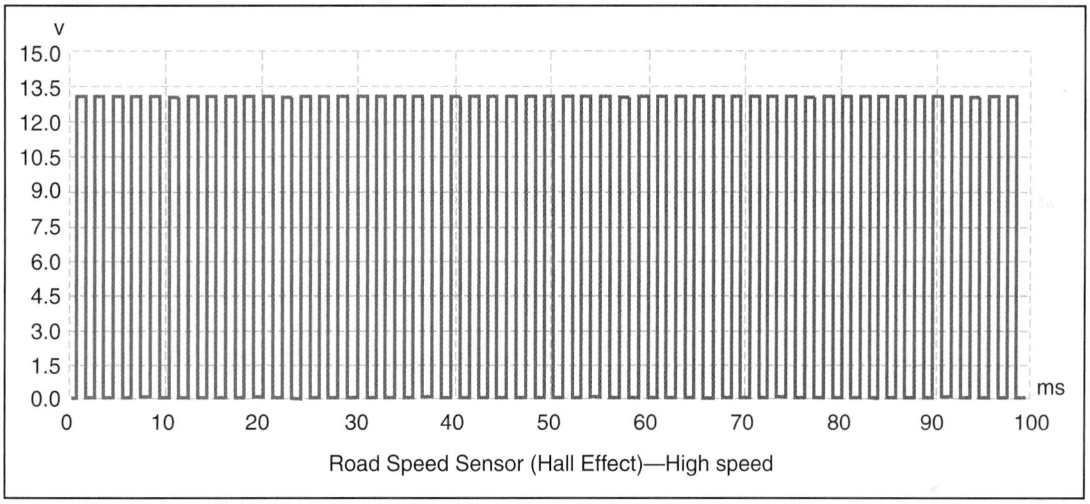

FIGURE 39-21 Hall effect operation.

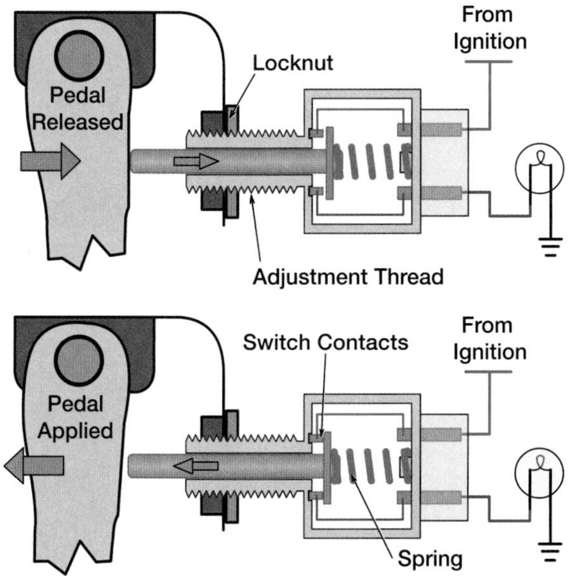

FIGURE 39-22 The brake light switch is an important input for the ABS. The switch is held open by the released brake pedal.

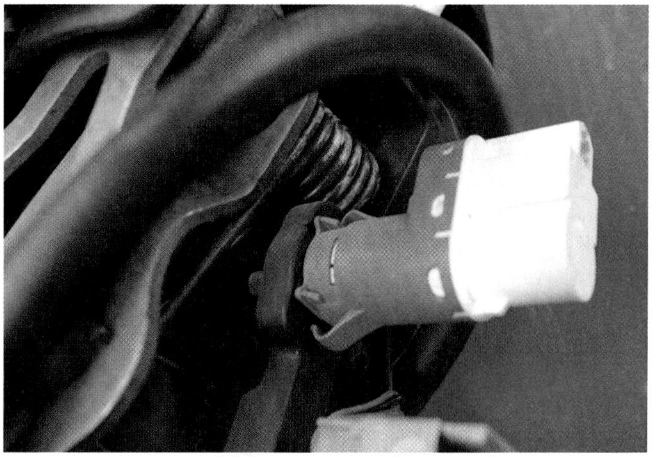

FIGURE 39-23 Typical brake pedal position sensor.

More advanced EBC systems may use a brake pedal position sensor. It indicates how far and fast the brake pedal is being pushed (**FIGURE 39-23**). It sends a variable signal based on the application of the brakes. The EBCM uses this signal to determine brake pedal travel and speed. This gives the

EBCM additional information about the type of braking that is being performed. It then allows the EBCM to modify the ABS intervention.

▶ ABS Electronic Brake Control Module (EBCM)

LO 39-05 Describe ABS EBCM operation.

The EBCM is made up of (**FIGURE 39-24**):

- electronic signal processor,
- an electronic data processor,
- computer memory, and
- and output drivers to control the output devices such as the electric solenoid valves.

The EBCM is programmed from the manufacturer to make brake control decisions. The decisions are based on sensor input data, which are compared to the data maps stored in its memory. These maps are designed to account for all of the reasonable braking conditions that the vehicle could experience. The EBCM then sends the appropriate output commands to the controlled devices. The EBCM continuously monitors the sensor data for any indication that one or more wheels are about to lock up.

The EBCM receives signals from several sources (**FIGURE 39-25**). A switch at the brake pedal provides a brake on/off condition or, on some vehicles, a brake pedal position signal. An input from the ignition switch signals that the driver has turned the ignition on. Some control units monitor the battery voltage. They use the rise in charging system voltage to indicate that the engine is actually running. The **vehicle speed sensor** reports the speed of the vehicle. Each of these input signals is used by the EBCM to monitor the vehicle and driver actions. It then decides which ABS actions are necessary to prevent a full skid condition.

Some ABS control modules have additional functionality designed into them. One example is electronic brake proportioning. This feature does away with the mechanical proportioning valve. It duplicates that action electronically. It uses the ABS valves to reduce rear brake hydraulic pressure under moderate brake pedal application. The EBCM restricts pressure to the rear wheels based on how hard the brake pedal is being applied. In this case, the EBCM does not wait until a wheel sensor reports that one or both rear wheels are locking up. Instead, it reduces rear brake pressure slightly to help prevent lockup. It occurs during moderate and heavy braking. This is needed because of weight being transferred away from the rear wheels to the front wheels.

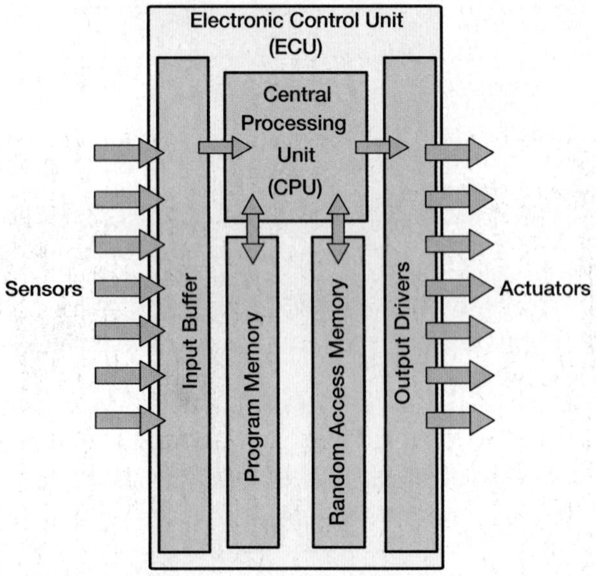

FIGURE 39-24 EBCM operation.

The EBCM performs an automatic system self-check on the ABS system every time the key is turned to the Run position. The EBCM turns on the warning lamp as part of a bulb check. If the EBCM detects a fault in the system, the ABS warning lamp will remain on. Most, but not all, systems store the fault in the EBCM memory. The faults are stored as diagnostic trouble codes (DTCs). They are retrieved by technicians when diagnosing ABS system faults.

Some older ABS systems provide **blink codes**, also known as flash codes, through the ABS warning lamp. This is activated when a specific terminal is grounded or two specific terminals are shorted together. If a code is stored in the EBCM memory, the EBCM will blink the ABS warning lamp in a manner that indicates a particular trouble code. For example, a code 12 would be one blink followed by a short pause, then two rapid blinks followed by a long pause. Each code is usually displayed three times before the next code is displayed. Once all codes have been displayed, the codes start at the beginning again.

Most ABS systems require a scan tool to read the **fault codes**. It connects to the EBCM or powertrain control module data link connector. Fault codes indicate which circuit is experiencing a fault, such as an open left front wheel speed sensor circuit (**FIGURE 39-26**). The fault codes also can indicate if there is a condition the EBCM determines is out-of-acceptable tolerances. An example being wheel speeds that do not match within the specified tolerance. The cause could be as simple as having a tire of the wrong size installed on the vehicle. It could also be properly sized tires that are not inflated to the same pressure. Or it could be a faulty wheel speed sensor. Once the codes have been retrieved, technicians use service information to diagnose and locate the cause of the fault.

▶ TECHNICIAN TIP

It is important for the customer to know that the ABS system is disabled when the ABS yellow warning lamp is on. The brakes will work normally, but without ABS function.

▶ Traction Control System (TCS) Overview

LO 39-06 Describe Traction Control System operation.

A basic ABS system can prevent skidding by holding or releasing individual brake circuit pressure. But it has no ability to apply the brakes apart from the driver-created hydraulic pressure. This system works fine as long as the tire slippage is a result of the driver applying the brakes. However,

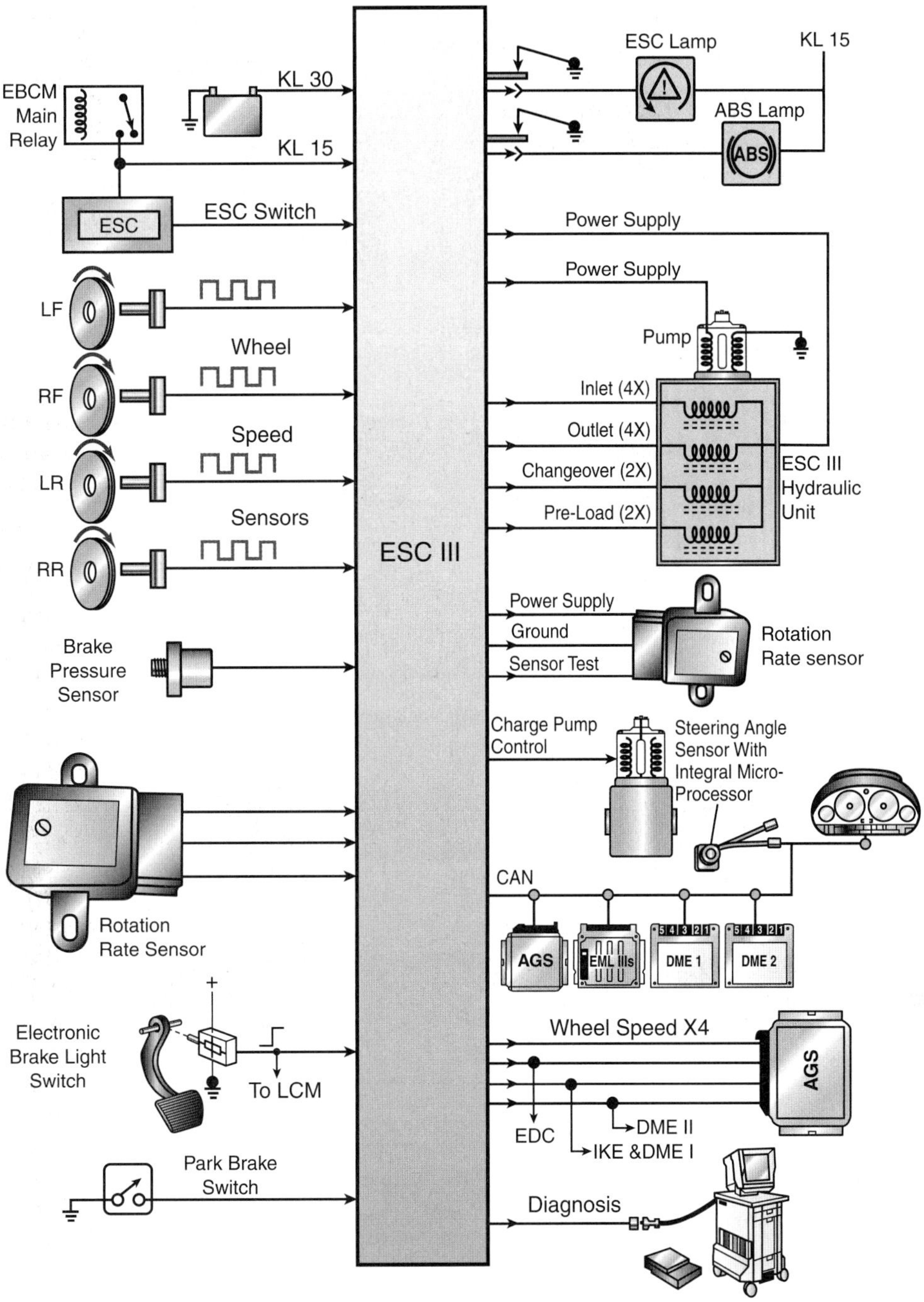

FIGURE 39-25 EBCM circuit.

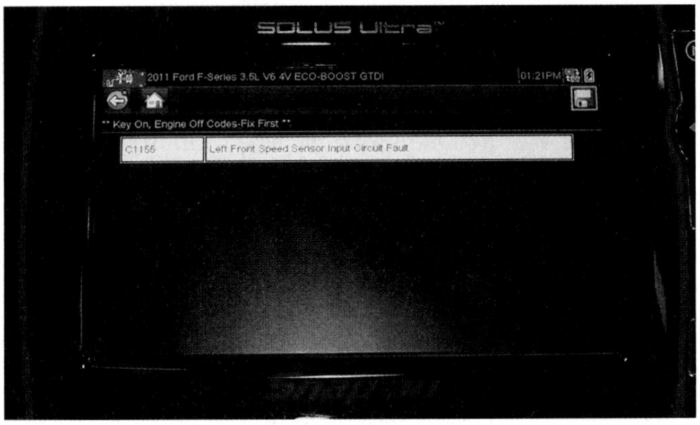

FIGURE 39-26 Scan tool showing an open LF wheel speed sensor.

tires also slip because the engine torque accelerating them exceeds their traction with the road. In this scenario, they can slip, spin, or break loose, causing a loss of control of the vehicle. The TCS system was developed to prevent the drive wheels from slipping while the vehicle is being accelerated. It is active up to a manufacturer-specified speed. Above that speed, traction control is deactivated by the EBCM. This is because further acceleration is unlikely to cause the wheels to lose traction.

To obtain traction control capabilities, manufacturers have added a few design features to the basic ABS system. One of which is the high-pressure pump and accumulator that was discussed earlier. This pressure is used to activate brake units on the drive wheels independently of the driver.

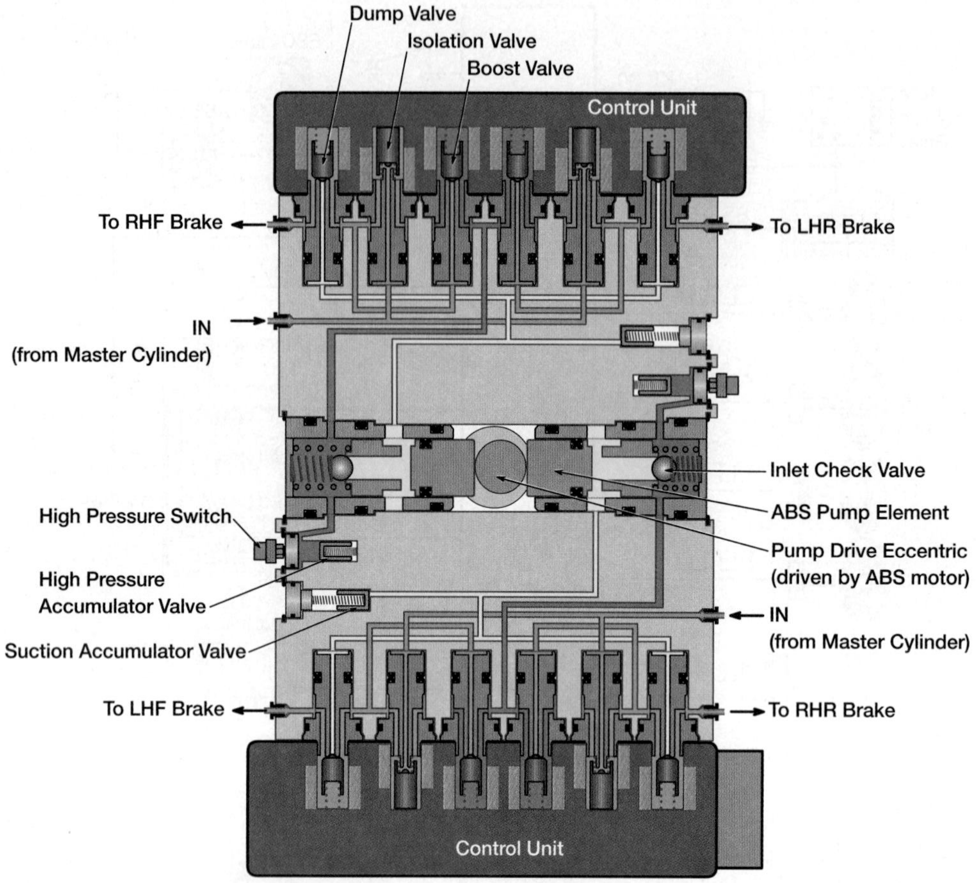

FIGURE 39-27 An HCU with boost valves.

The sensors are the same as in the ABS system. But the ability to apply the individual drive wheel brakes is needed. Thus, two to four extra solenoid valves, called **boost valves**, are added to the HCU (**FIGURE 39-27**). These boost valves direct hydraulic pressure from the accumulator to the ABS solenoid valves. This is so that individual wheel brake units can be applied independently. Additional programming is added to the EBCM. It is used to control the high-pressure pump and extra HCU valves. It also decides when each of them needs to be activated.

Operation of the TCS

When the TCS is active, the EBCM monitors the speed of the individual drive and non-drive wheels. It also monitors the vehicle speed from the vehicle speed sensor. If the driven wheels are accelerating at different speeds from each other or the non-driven wheels, the EBCM can identify which wheel is slipping.

If one or more wheels are slipping, the EBCM will take action to reduce the torque to the appropriate wheels. It does so in steps. First, it applies the brake to any wheels that are slipping. It does this by activating the isolation valve to close off the supply port from the master cylinder. It then activates the boost valve to pressurize the brake circuit on the spinning wheel to slow it down. If that is not enough to prevent the slippage, the EBCM will request reduced power from the engine. This can be accomplished in several ways:

- by reducing the throttle plate opening,
- shutting down one or more fuel injectors,
- reducing the engine timing, or
- selecting a higher gear in the transmission.

Once the wheel speeds return to proper parameters, the EBCM will return the TCS system to normal. It will also continue to monitor the wheels for slippage.

Some TCS systems can be temporarily deactivated by a TCS function switch located on the dash or center console (**FIGURE 39-28**). If the driver deactivates TCS, the system will not intervene during wheel slip. Drivers will disable the TCS for a variety of reasons. They might be climbing a long hill on a rough gravel road. This would continuously activate traction control, overheating the brakes. Or they might want to show off by "roasting" the tires. Or they might want to experience driving without traction control on a racetrack. The TCS will automatically default back to "On" during the next ignition switch cycle. In most cases, if the TCS is deactivated, the ABS system will still be active.

▶ TECHNICIAN TIP

Manufacturers use various strategies in their TCS systems, and not all of them apply the brakes as a first step. Some of them reduce engine power first. Even so, EBCMs today operate very fast, so there may only be a few milliseconds between each action.

▶ Electronic Stability Control (ESC) Overview

LO 39-07 Describe Electronic Stability Control System operation.

ABS does a good job of preventing wheels from locking up under hard braking or poor traction conditions. It also allows the driver to maintain directional control of the vehicle. TCS also does a good job of maintaining traction when the vehicle is accelerated in a relatively straight line. However, drivers can lose directional control of the vehicle. This can occur:

- while driving aggressively,
- taking emergency steering actions,
- or if there are sudden changes in the traction of the road surface while in a turn.

These situations can cause the vehicle to understeer (push) or oversteer (loose, or fishtail). It can also cause vehicles with a high center of gravity, such as an SUV, to roll over (**FIGURE 39-29**). All of these situations can lead to serious accidents.

If any of these situations start to happen, the ESC system can activate individual wheel brake units as necessary. This can help keep the driver from losing control of the vehicle. ESC utilizes the ABS and TCS systems, but it does so with a few enhancements. This allows it to more actively interface with the vehicle's operation in maintaining directional stability while the vehicle is being steered. A US Insurance Institute for Highway Safety 2006 study found that approximately 10,000 fatal accidents in the United States could be avoided each year. How? If all vehicles were equipped with ESC. This finding led the Department of Transportation to require that all vehicles of less than 10,000 lb (4536 kg) gross vehicle weight, and manufactured after September 1, 2011, be equipped with an ESC system that meets their minimum specifications.

The ESC system integrates a:

- yaw sensor,
- **steering angle sensor**, and
- sometimes a **roll-rate sensor** into the basic ABS and TCS systems (**FIGURE 39-30**).

It also adds new programming parameters into the EBCM to monitor the vehicle's stability. Additional output command capabilities apply individual non–drive wheel brake units independent of the driver.

The yaw sensor measures the amount of directional rotation of the vehicle on its vertical axis. In other words, it tells the

FIGURE 39-28 A switch for deactivating the traction control system.

FIGURE 39-29 Electronic stability control helps the driver maintain control of the vehicle when driving.

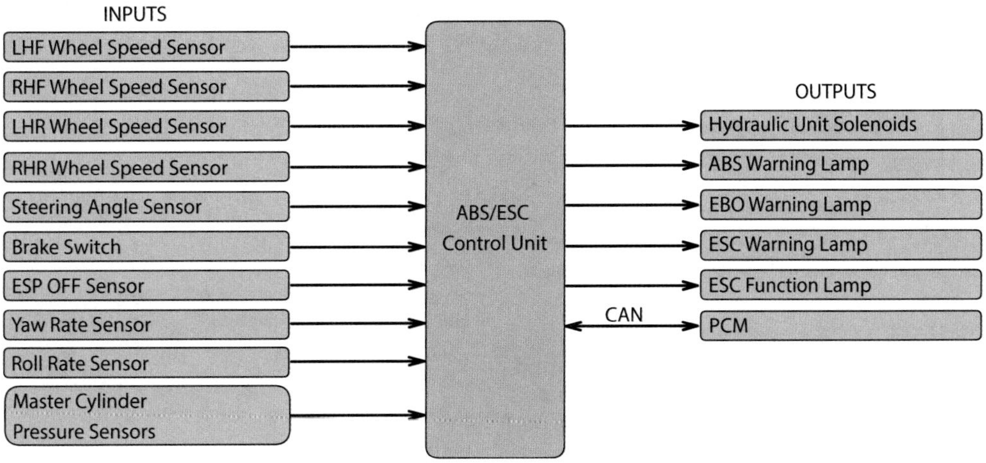

FIGURE 39-30 Typical ESC schematic.

EBCM the rate at which the vehicle is turning. The steering angle sensor tells the computer what the driver's directional intent is. If equipped, the roll-rate sensor tells the computer the rate of roll and the amount of roll that the vehicle is experiencing. The EBCM continuously monitors these signals and compares them with preprogrammed scenarios. It decides which, if any, brake units need to be applied, and if engine torque needs to be reduced to keep the vehicle stable. This process is a good example of the computer feedback loop: Input → Control Logic Process → Output.

SAFETY TIP

Most standard passenger vehicles are designed with a bias toward understeer. It is generally agreed that understeer is easier for the average driver to recover from. However, many performance vehicles are designed with a slight bias toward oversteer. This can be managed by an experienced driver while driving aggressively.

▶ TECHNICIAN TIP

The yaw sensor operates similarly to a Wii or other video game controllers. Its internal circuitry senses movement and sends a signal which is directly related to the movement it senses.

Electronic Stability Control (ESC) Operation

If the ESC system is activated, the ECBM monitors:

- the yaw sensor signal,
- the steering angle sensor signal,
- the roll-rate sensor signal, as well as
- the wheel speed sensor signals.

If the vehicle is beginning to understeer, oversteer, or roll, the EBCM detects it in the signal values. It then applies up to three wheel brake units to help bring the vehicle back within proper stability parameters. If that does not stop the stability issue, the EBCM will request a reduction in engine power. It does this through the powertrain control module. This helps slow the vehicle further.

On most vehicles, the EBCM can warn the driver before a loss of control. It does so by illuminating a warning lamp on the dash or sounding a beeper when the ESC system has detected the start of a skid. This way the driver will be informed that they are on the verge of losing control of the vehicle.

On most vehicles, the ESC system defaults to "on" so it is always active. Some vehicles have a switch on the dash or center console to temporarily deactivate the system. This can be useful when driving in mud or sand when traction is virtually nonexistent and the ESC system cannot function effectively. Even though some ESC systems can be turned off, they may still reactivate the ESC system under certain situations. Common examples are: driving above a specified speed, or if a spin is detected while the brakes are being applied.

Some ESC systems incorporate a switch that allows the driver to select one or more varying levels of assist from the ESC system. Selections may be, "touring," "track," or "sport"

FIGURE 39-31 Switch to select the level of ESC assist desired.

(**FIGURE 39-31**). This option allows the driver to experience differing levels of wheel slip than when ESC is fully activated. The driver is able to push the vehicle closer to the edge of control. At the same time, the ESC system is available as a backup, but with limited assistance. When driving on a racetrack, for example, the driver may want full control of the vehicle instead of being limited by the ESC system.

In the continuous search for new bells and whistles to impress customers and enhance safety, manufacturers have designed other features into ESC systems, such as these:

- Hill assist: Holds the brake pressure until the throttle is depressed and the vehicle starts to move forward.
- All-wheel drive traction control: Applies brake pressure as needed to any of the four individual wheels that may be slipping to maintain power to the wheels with the most traction.
- Engine braking control: Increases the engine torque if the ESC system detects wheel slippage during deceleration.
- Panic stop assist: Detects a driver's rapid throttle release and lightly applies the brakes to dry the rotors and prepare the brakes for a panic stop.
- Accident avoidance: Works in conjunction with adaptive cruise control to monitor objects in front of the vehicle. If the ESC system detects an imminent collision, it can apply the brakes or boost the brake pressure above driver pressure.
- Hill descent control: Works in conjunction with the ESC system to control the speed of the vehicle when going down loose, rough, or slippery slopes.
- Trailer sway control: Detects trailer sway and uses the ESC system to keep it under control.
- Optimized hydraulic braking: Monitors brake pressure in each brake circuit. Increases brake pressure above boosted pressure if necessary.

▶ TECHNICIAN TIP

Many ESC-equipped vehicles monitor signals from other sensors as well to help prevent a loss of control of the vehicle. These sensors include the throttle position sensor, vehicle speed sensor, and brake pedal position sensor. When diagnosing an ESC system fault, research the sensors monitored by the EBCM.

▶ Wrap-Up

Ready for Review

- The first-generation EBC system was the antilock brake system (ABS). It was designed to prevent wheels from locking up under braking conditions. Modification of ABS allowed manufacturers to provide traction control capability to the vehicle. Electronic stability control helps prevent loss of driving control, especially vehicle rollovers.
- ABS system makes use of the following components to prevent wheel lockup: ABS master cylinder, power booster, EBCM, hydraulic control unit, wheel speed sensor, and brake switch.
- Integral ABS systems combine the tandem master cylinder, HCU, and power booster in one unit. It operates the integrated master cylinder and brake fluid passes from the master cylinder portion of the assembly to the HCU portion. Nonintegral ABS systems use a fairly standard tandem master cylinder and a typical vacuum or hydraulic power booster that assists the driver in applying force to the master cylinder.
  - The HCU houses electrically operated hydraulic control valves that control brake pressure to specific wheel brake circuits.
- Wheel speed sensors use principles of electromagnetism to generate an analog or digital electrical signal that is read by the EBCM to determine the speed of each wheel. Three most common types of wheel speed sensors are: variable reluctance, magneto-resistive, and Hall effect styles. The brake switch is a normally closed switch that is used to activate the rear brake lights and send an electrical input signal to the EBCM.
- The EBCM is programmed to make brake control decisions based on sensor input data, which are compared to the data maps stored in its memory. The EBCM is made up of electronic signal processor, an electronic data processor, computer memory, and output drivers.
- In a TCS, the EBCM will take action to reduce the torque to the appropriate wheels by applying the brake to the slipping wheels and activating the boost valve to pressurize the brake circuit on the spinning wheel to slow it down.
  - The EBCM can also reduce power from the engine by reducing the throttle plate opening, shutting down one or more fuel injectors, reducing the engine timing, or selecting a higher gear in the transmission.
- ESC utilizes the ABS and TCS systems to more actively interface with the vehicle's operation in maintaining directional stability while the vehicle is being steered. If the ESC system is activated, the ECBM monitors: the yaw sensor signal, the steering angle sensor signal, the roll-rate sensor signal as well as the wheel speed sensor signals.

Key Terms

ABS master cylinder Master cylinder used in anti-lock brake systems.

accumulator A device placed between the evaporator and the compressor to collect liquid refrigerant and prevent it from entering the compressor.

air gap The space or clearance between two components, such as the space between the tone wheel and the pickup coil in a wheel speed sensor.

blink code A method of providing fault code data for a specific system, which involves counting the number of flashes from a warning lamp and observing longer pauses between the light blinks.

body control module (BCM) An onboard computer that controls many vehicle functions, including the vehicle interior and exterior lighting, horn, door locks, power seats, and windows.

boost valve A valve located in the HCU that is controlled by the EBCM; it allows brake fluid under high pressure to flow into the HCU hydraulic circuits, to apply the brakes when commanded.

brake switch The electrical switch that is activated by the brake pedal; it turns on the brake lights and signals the EBCM that the brakes are being applied.

channel The number of wheel speed sensor circuits and hydraulic circuits the EBCM monitors and controls.

common bore When a single cylinder is used for two pistons. A tandem master cylinder would be an example of two pistons in one bore.

electronic brake control (EBC) system A hydraulic brake system that has integrated electronic components for the purpose of closely controlling hydraulic pressure in the brake system.

electronic stability control (ESC) system A computer-controlled system added to ABS and TCS to assist the driver in maintaining vehicle stability while steering.

fault codes An alphanumeric code system used to identify potential problems in a vehicle system.

Hall effect An electrical effect where electrons tend to flow on one side of a special material when exposed to a magnetic field, causing a difference in voltage across the special material. When the magnetic field is removed, the electrons flow normally, and there is no difference of voltage across the special material. This effect can be used to determine the position or speed of an object.

high-pressure accumulator A storage container designed to contain high-pressure liquids such as brake fluid.

hydraulic control unit (HCU) An assembly that houses electrically operated solenoid valves used in electronic braking systems; also called a modulator.

integral ABS system A brake system in which the master cylinder, power booster, and HCU are all combined in a common unit.

isolation valve The valve in the HCU that either allows or blocks brake fluid that comes from the master cylinder from entering the HCU hydraulic circuit.

magneto-resistive Type of wheel speed sensor that uses an effect similar to a Hall effect sensor to create a digital signal.

non-integral ABS systems A brake system in which the master cylinder, power booster, and HCU are all separate units.

oversteer A condition when the front end steers more sharply than desired because the rear wheels lose adhesion during cornering. This causes the rear of the vehicle to swing out. The vehicle is said to be loose.

pickup assembly A component with a wire coil wrapped around a ferrous metal core; it is used to generate an electrical signal when a magnetic field passes through it.

Power booster Device used to increase the brake pedal force on the master cylinder. Can be vacuum or hydraulic types.

roll-rate sensor A sensor that measures the amount of roll around the vehicle's horizontal axis that a vehicle is experiencing.

solenoid valve A type of electromechanically operated valve that uses an electric current to control fluid flow.

steering angle sensor A sensor that measures the amount of turning a driver desires. This information is used by the ESC system to know the driver's directional intent.

steering wheel position sensor (SWPS) A sensor that signals to the EBCM both the position of the steering wheel and the speed at which it is being turned.

tone wheel The part of the wheel speed sensor that has ribs and valleys used to create an electrical signal inside of the pickup assembly.

traction control system (TCS) A computer-controlled system added to ABS to help prevent loss of traction while the vehicle is accelerating.

understeer A condition in which the vehicle's front wheels are turned more sharply than the vehicle's actual direction because the front tires lose adhesion during cornering. The vehicle is said to be "pushing" in the corners.

variable reluctance sensor A sensor that uses the principle of magnetic induction to create its signal. It is used to measure rotational speed, including wheel speed, machine speed, engine speed, and camshaft and crankshaft position.

vehicle speed sensor (VSS) A sensor used by the PCM to measure vehicle speed. It is often located in the transmission extension housing. The output signal may be analog or digital.

wheel speed sensor A device that creates an analog or digital signal according to the speed of the wheel.

yaw sensor A sensor that measures the amount a vehicle is turning around its vertical axis. This information is used by the ESC system to know how much a vehicle is turning.

Review Questions

1. ABS (Antilock Braking System), TCS (Traction Control System), and ESC (Electronic Stability Control) are all subsystems of:
   a. PCS (Powertrain Control System)
   b. ERC (Electronic Ride Control)
   c. EBC (Electronic Brake Control)
   d. ASD (Automatic Shutdown)
2. Electronic Stability Control is designed to control:
   a. Tire traction during acceleration
   b. Understeer and oversteer
   c. Limited slip differentials
   d. The wheel speed sensor signal stability
3. Maximum braking occurs when the wheels are rotating _________ slower than the vehicle speed.
   a. 0%–5%
   b. 10%–20%
   c. 85%–95%
   d. 100%
4. What is a nonintegral ABS System?
   a. The electronic brake control module and hydraulic control unit are joined with the brake master cylinder into one unit.
   b. The electronic brake control module and hydraulic control unit are separate from the brake master cylinder.
   c. The ABS system is controlled by the body control module and does not have an electronic brake control unit.
   d. The hydraulic control unit is controlled by the brake master cylinder and does not have an electronic brake control module.
5. An ABS system that can control each wheel independently is considered a ______________system.
   a. Two-channel
   b. Three-channel
   c. Four-channel
   d. Single-channel
6. Which ABS system uses wheel speed sensors that require an outside power source to operate called?
   a. A passive system
   b. An active system
   c. A dominant system
   d. A submissive system
7. Which of these is a common additional function of the ABS control module?
   a. Brake lining wear indication
   b. Brake shoe adjustment
   c. Park brake bulb check
   d. Electronic brake proportioning
8. Where does the hydraulic pressure for traction control system originate?
   a. From a brake fluid pump and reservoir called the accumulator.
   b. From the driver applying the brakes at the brake pedal.
   c. From centrifugal force of the brake rotors or drums.
   d. From the pressure heat causes when the fluid in the calipers and wheel cylinders expands.

9. Traction control systems commonly use all of these to reduce engine power EXCEPT:
   a. Reducing engine timing
   b. Shutting down fuel injectors
   c. Moving the gear select lever to neutral
   d. Reducing the throttle plate opening
10. Which sensor tells the traction control module the driver's intended direction?
   a. Steering angle sensor
   b. Roll-rate sensor
   c. Yaw sensor
   d. Wheel speed sensor

ASE Technician A/Technician B Style Questions

1. Technician A says that the ABS system rapidly cycles the brakes between hold, release, and apply during panic stops. Technician B says that first-generation ABS systems create braking pressure without driver input. Who is correct?
   a. Technician A only
   b. Technician B only
   c. Both A and B
   d. Neither A nor B
2. Technician A says that Hall effect wheel speed sensors create an analog AC signal that changes frequency with wheel speed. Technician B says that variable reluctance wheel speed sensors create a digital DC signal that changes amplitude with vehicle speed. Who is correct?
   a. Technician A only
   b. Technician B only
   c. Both A and B
   d. Neither A nor B
3. Technician A says that the ABS hold, release, and apply cycle can repeat itself up to about 16 times per second. Technician B says that the ABS cycle can be felt in the brake pedal on some vehicles. Who is correct?
   a. Technician A only
   b. Technician B only
   c. Both A and B
   d. Neither A nor B
4. Technician A says that in hold mode, the hold solenoid valve is open and fluid pressure builds to "hold" the brakes. Technician B says that in release mode, fluid pressure is released from the brake caliper and pumped back to the master cylinder. Who is correct?
   a. Technician A only
   b. Technician B only
   c. Both A and B
   d. Neither A nor B
5. Technician A says that a variable reluctance wheel speed sensor signal may be incorrect below vehicle speeds of 5 mph. Technician B says that the distance from the sensor to the tone wheel (air gap) will not affect the sensor signal. Who is correct?
   a. Technician A only
   b. Technician B only
   c. Both A and B
   d. Neither A nor B
6. Technician A says that the brake pedal pushes the brake switch open when the brake pedal is applied. Technician B says that the EBCM uses the brake switch signal to determine whether to activate the ABS or not. Who is correct?
   a. Technician A only
   b. Technician B only
   c. Both A and B
   d. Neither A nor B
7. Technician A says that the EBCM (electronic brake control module) monitors the wheel speed sensors for 30 seconds, once per drive cycle. Technician B says that the EBCM has self-diagnostic capabilities and can set fault codes. Who is correct?
   a. Technician A only
   b. Technician B only
   c. Both A and B
   d. Neither A nor B
8. Technician A says that the traction control system applies the brakes to all four wheels when slippage is detected. Technician B says that traction control systems do not set trouble codes. Who is correct?
   a. Technician A only
   b. Technician B only
   c. Both A and B
   d. Neither A nor B
9. Technician A says that ESC (electronic stability control) applies the brakes at individual wheels to control vehicle rotation around its vertical axis. Technician B says that most cars and light trucks built after 2011 should have ESC. Who is correct?
   a. Technician A only
   b. Technician B only
   c. Both A and B
   d. Neither A nor B
10. Technician A says that some ESC (electronic stability control) systems can be deactivated. Technician B says that some ESC systems can detect trailer sway. Who is correct?
   a. Technician A only
   b. Technician B only
   c. Both A and B
   d. Neither A nor B

REC
HOLD
ACA/DCA
MIN/MAX
REL
ON
OFF

SECTION 7

Electric

▶ CHAPTER 40 Principles of Electrical Systems

▶ CHAPTER 41 Electrical Components and Repair

▶ CHAPTER 42 Meter Usage and Circuit Diagnosis

▶ CHAPTER 43 Battery Systems

▶ CHAPTER 44 Starting and Charging Systems

▶ CHAPTER 45 Lighting Systems

▶ CHAPTER 46 Body Electrical System

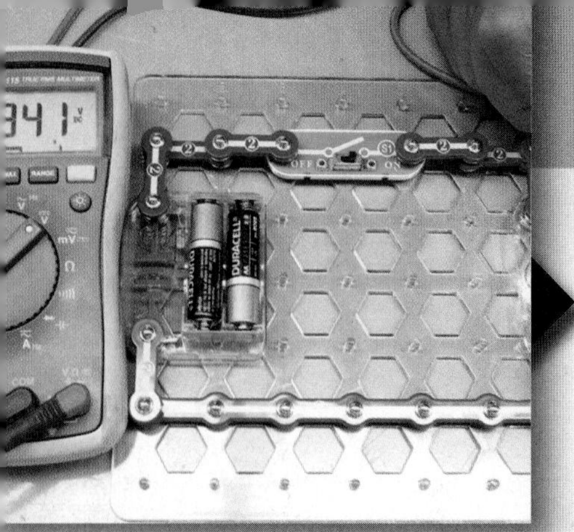

CHAPTER 40

Principles of Electrical Systems

Learning Objectives

- **LO 40-01** Describe the importance of learning electrical theory.
- **LO 40-02** Explain conductor, insulator, and semiconductor materials.
- **LO 40-03** Describe the process of electron movement in a simple circuit.
- **LO 40-04** Explain Volts, Amps, Ohms, Power, and Ground.
- **LO 40-05** Describe the sources of electricity.
- **LO 40-06** Describe the effects of electricity.
- **LO 40-07** Use Ohms Law to calculate values.
- **LO 40-08** Use Watts Law to calculate values.
- **LO 40-09** Describe series circuits and use its laws to calculate values.
- **LO 40-10** Describe parallel and series-parallel circuits and calculate values.
- **LO 40-11** Describe DC and AC and Kirchhoff's Current Law.
- **LO 40-12** Explain how to use electrical concepts to solve problems.

ASE Education Foundation Tasks

See Appendix A to view the 2017 ASE Education Foundation Automobile Accreditation Task List Correlation Guide.

▶ Importance of Learning Electrical Theory

LO 40-01 Describe the importance of learning electrical theory.

Electrical complexity of vehicles has continued to increase. At the same time, hydrocarbons have become scarcer and more expensive. So, efficiency will continue to drive the need for advanced electrical and electronic systems. We can see this trend through the increasing popularity of hybrid vehicles. Manufacturers are also investing in future technologies. Electric vehicles and fuel cell technology (**FIGURE 40-1**) are only a few examples.

Technicians need to have a solid understanding of electrical principles. They need to be knowledgeable in order to service today's vehicles. This includes electrical terminology, the behavior of electricity, and circuit theory. Your success depends largely on your ability to apply these electrical principles. You also need to understand how they relate to the operation of almost every system in the vehicle. This makes the use of electrical principles in the repair of today's vehicles critical for technicians.

▶ TECHNICIAN TIP

Many students find that going through electrical theory one time isn't enough to fully understand it. Successful students study a portion of the content and then go back and review it. This may take a few times so that they can build a strong understanding of the concepts. Electricity is understandable, but you have to apply yourself. At the same time, becoming a successful technician will be much easier with a sound knowledge of this topic. Please consider this.

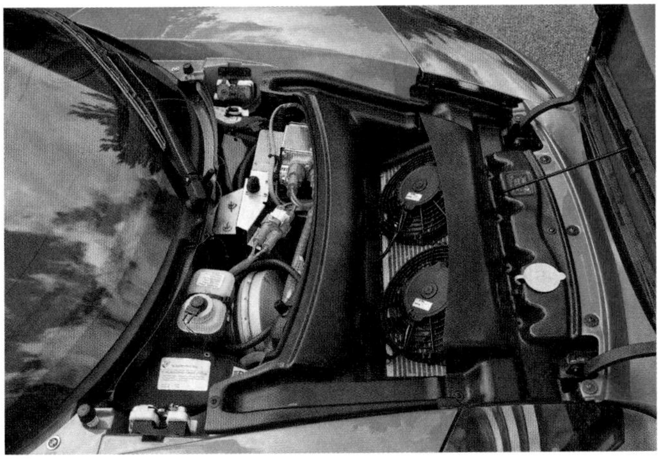

FIGURE 40-1 Manufacturers are relying on electricity and electronics more and more to meet stringent government and customer requirements.

Electrical Fundamentals

Understanding the behavior of electricity can be more difficult than understanding other concepts. For all intents and purposes, electricity cannot be seen. You have to use your imagination and visualize what it is doing. At the same time, electricity is governed by the laws of physics. So learning how electricity behaves can be approached in a logical manner, as with any other science.

In this chapter, we explore the various types of circuits and how electricity behaves within each type. We will also explore the various ways that electricity can be created. That will help you understand how sensors operate. We will also explore each of the common ways that electricity can be put to use. This is used in the output devices to carry out the computer's commands. Understanding all of this will lead to a strong electrical foundation.

To get started, it is helpful to know that electricity is made up of tangible objects. We cannot normally see them with our eyes because these particles are so small. But we can imagine them in our minds. In fact, it may be helpful to think of electricity as nothing more than the movement of specific particles from one point to another. Imagine a line of marbles rolling through a tube, or drops of water through a pipe (**FIGURE 40-2**). The moving marbles or drops of water can perform work if they are directed against another object with force. In the same way, electricity can perform work. But it must be directed at objects that can extract energy from the moving particles. Common examples are lights and electric motors.

So, what makes the marbles move? That is where some of electricity's magic comes in. Remember back when you were in grade school? Unlike charges attract, and like charges repel (**FIGURE 40-3**). These attracting and repelling charges are what cause the electrical particles to move and perform work. As we continue, just remember that electricity is the movement of these particles from one place to another. The next question is, where do these charges come from?

You Are the Automotive Technician

You are supervising an intern at the shop where you work. She has been a great help to you, increasing your productivity by doing the tasks she has learned so far. She will be starting electrical training next semester. She has been told it can be difficult to learn, so she wants to get a head start on it. She asks you the following questions. How would you respond?

1. What exactly are volts, ohms, and amps?
2. What do "continuity," "open," and "short" mean?
3. What is Ohm's law?
4. What is the difference between a series circuit and a parallel circuit?

FIGURE 40-2 Electricity is like marbles moving through a tube, whose movement can do work.

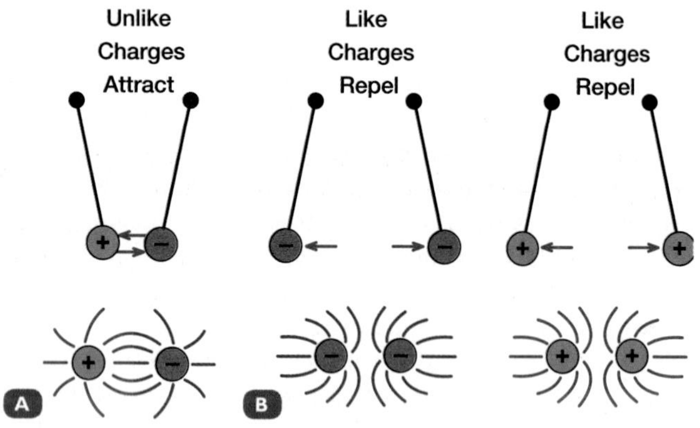

FIGURE 40-3 **A.** Unlike charges attract. **B.** Like charges repel.

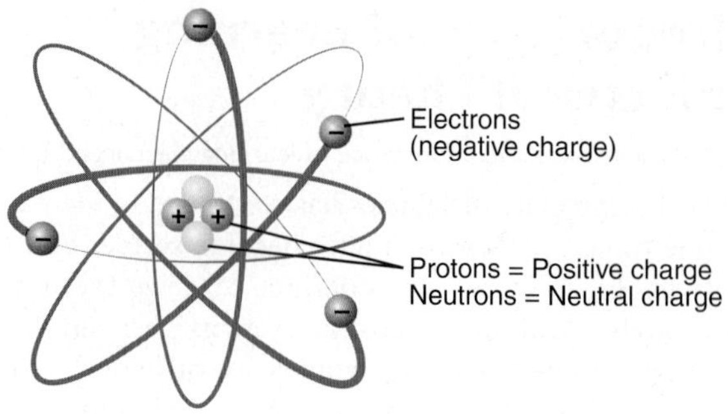

FIGURE 40-4 Parts of an atom.

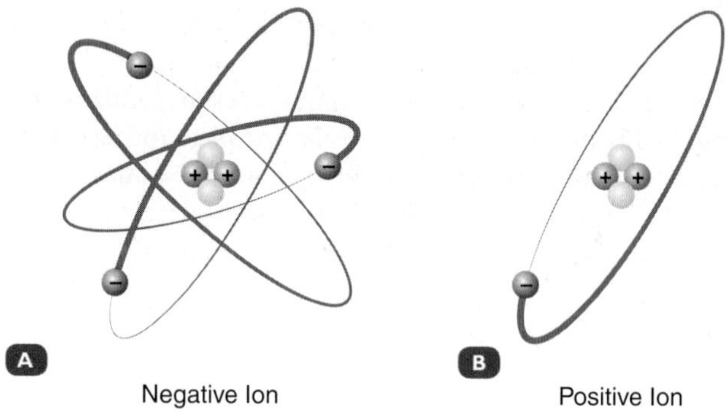

FIGURE 40-5 Ions. **A.** Negative ion. **B.** Positive ion.

▶ Electrical Fundamentals

Basic Electricity

LO 40-02 Explain conductor, insulator, and semiconductor materials.

All questions about the nature of electricity lead to the composition of matter. All matter is made up of atoms (FIGURE 40-4). Atoms, in turn, are made up of positively, negatively, and neutrally charged particles. The nucleus has at least one positively charged proton and, in most atoms, at least one neutron that has no charge (neutral). Moving around the nucleus are one or more negatively charged electrons.

Electrons travel in different rings, or shells, around the nucleus. Each ring, or shell, can contain a specific maximum number of electrons. Any additional electrons must fit into the next higher ring, or shell. With equal numbers of protons and electrons, the charges within an atom cancel each other out. This leaves the atom in a balanced state, with no overall charge. In this state, the electrons and protons are content to stay in the atom just as they are.

Not all atoms are balanced. This means that some atoms are unbalanced. An atom with more electrons than protons has an overall negative charge. Because of this, it is called a negative ion (FIGURE 40-5). "Ion" simply means the atom has an imbalance of electrical charges due to the gain or loss of electrons. A negative ion is not balanced, since it has an overall negative charge. Because each of the electrons has a negative charge, they repel one another. The repelling force wants to push one of the electrons away from this atom.

A deficiency of electrons gives the atom an overall positive charge. This atom is called a positive ion, because it is also not balanced. In this case, it is exerting an attracting force on electrons. It tries to pull an electron from another ion or atom.

If a negative ion and positive ion are close enough, electrons can move from one ion to another ion. The negative charge of the negative ion exerts a repelling force on its extra electron. This tends to push the electron away from itself. At the same time, the positive ion exerts an attracting force on the extra electron. This tends to pull the electron from the negative ion. The transfer of the electron balances both atoms out in this case. The flow of electrons from atom to atom (ion to ion) is called current flow. It is the basic concept of electricity (FIGURE 40-6).

Not all atoms can give up or accept electrons easily. Materials that can do so easily are called conductors. Those materials that cannot give up electrons easily are called insulators (FIGURE 40-7A). The explanation of what makes a good conductor or a good insulator is quite complex. It is found in the theories of quantum mechanics, which address the arrangement and behavior of electrons around the nuclei of atoms.

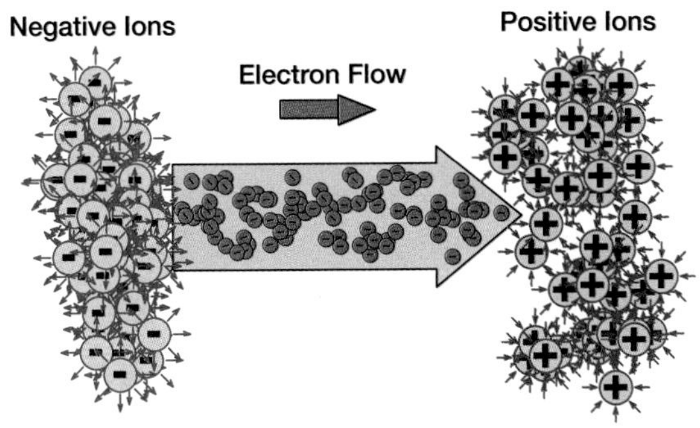

FIGURE 40-6 Electrons being pushed away from the negative side and pulled to the positive side. This movement is current flow.

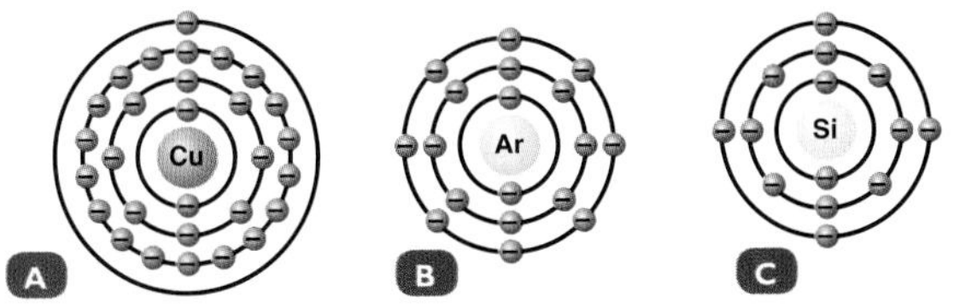

FIGURE 40-7 A. Conductor. **B.** Insulator. **C.** Semiconductor.

To simplify matters, it is safe to say that in some materials, there are electrons that are charge carriers. They are called **free electrons**. They are located on the outer ring of the atom, called the valence ring. These electrons are only loosely held by the nucleus. They have more freedom to move from one atom to another when an electrical potential (pressure) is applied. In fact, atoms with fewer electrons in the valence ring are the best **conductors**. Having one electron in the valence ring is the best conductive material. This is because the single electron in the valence ring by itself is held the most loosely by the nucleus.

Materials made up of atoms with one to three electrons in the valence ring are considered conductors. The more atoms with free electrons a particular material has, the better the material can conduct electrons. Metals typically have lots of free electrons because of the atoms' structure. They are therefore good conductors.

Every substance, even air, will conduct an electrical current if enough electrical pressure (voltage) is applied to it. But the word "conductor" is normally used for materials that allow electrons to flow with little resistance. Most metals are good conductors. The most common conductor used in automobiles is copper. It is used in virtually all of the wiring that connects automotive components together. The more electrons a conductor must carry, the heavier the gauge or thickness the wire needs to be.

Materials that do not conduct electrons easily are called **insulators**. Most plastics are good insulators. The plastic covering on a wire is a good example of this. The ceramic portion of a spark plug is also a good insulator. In insulators, electrons in the valence ring are bound much more tightly to the nucleus. A good insulator does not allow current flow because it has no free electrons. And the electrons it does have cannot move freely.

Therefore, an insulator resists the movement of electrons when an electrical potential is applied. Insulators are made up of atoms that have five to eight electrons in the valence ring. The greater the number of electrons in the valence ring, the harder it is for them to break away. So, these atoms are better insulators (**FIGURE 40-7B**).

Semiconductors are materials that conduct electricity easier than insulators. But they don't conduct electricity as well as conductors. Semiconductor materials such as silicon are crucial in electronics. They are used to make electronic components, such as transistors and microchips. The semiconductor material can switch from being a conductor to an insulator and back again. And it can happen very quickly. And they do it without mechanical means.

Atoms that have four electrons in the valence ring are considered semiconductors (**FIGURE 40-7C**). Note that the semiconductor material has precisely four electrons. So, it is only one electron away from becoming either an insulator or a conductor. If an electron is added, it becomes an insulator. If an electron is removed, it becomes a conductor. Thus, the semiconductor material can be used as a switch. The semiconductor material can be controlled. Electrons will either be allowed to flow though the semiconductor material or stopped by it. All we have to do is add electrons to, or subtract them from, the semiconductor material. We will explore this further in a later topic.

Applied Science

AS-1: Conductors: The technician can explain the difference between an electrical conductor and an insulator.

The electrical conductivity of a material refers to the freedom of the electrons within the atoms of the material to move around. Materials with high electron mobility are known as conductors while materials with low electron mobility are called insulators. High electron mobility allows electrons to flow when a voltage is applied across two points of the material.

Most familiar conductors are metallic. However, there are nonmetallic conductors such as graphite and salt solutions. Silver is the best conductor, but copper is most commonly used for electrical wiring. This is due to its many desirable properties, including high tensile strength and ductility.

An insulator is a material that resists the flow of an electrical charge. Glass and paper are both very good insulators. But polymers and plastics are more commonly used to insulate wiring. The purpose of an insulator in electrical applications is to support or separate conductors. It does so without allowing current to flow through the insulator.

▶ Movement of Free Electrons

LO 40-03 Describe the process of electron movement in a simple circuit.

Free electrons are necessary for electrical current flow. But for the free electrons to move easily, they need two things. The first is a complete pathway, called a circuit. And the second is a force that makes them move. The electrical force from a battery can cause electrons to move. Here is how. Like charges repel, so the negative electrons repel each other and are forced from the negative terminal of the battery. Unlike charges attract, so the electrons are attracted toward the positive protons in the positive side of the battery. In this case, free electrons flow from the negative terminal to the positive terminal (**FIGURE 40-8**).

This flow of electrons is called current flow, or amperage. It is measured in amps. Note that the current is flowing in one direction only, a condition called direct current (DC). Most, but not all, circuits in passenger vehicles operate on DC. The larger the charge between the negative and the positive terminal, the more strongly the positive terminal attracts and the negative terminal repels the free electrons. This is called **electromotive force** and is typically referred to as voltage. So, you could say that **voltage** is the force that motivates electrons to move through a circuit. The unit of measurement for voltage is volts. The greater the voltage, the stronger it pushes on the electrons. So, higher voltage causes higher amperage.

> ▶ TECHNICIAN TIP
>
> Electrical current is the flow of electrons. So, it is natural to say that current flows in the direction in which the free electrons move. As discussed—from negative to positive. This is called the **electron theory**. However, it was previously thought that the natural way for them to flow was from positive to negative. This is called the **conventional theory**. This was before the discovery that electrons are negatively charged. Most wiring diagrams are written from the conventional theory perspective. Whereas, electronic circuits are typically designed and operate on the electron theory perspective. Thus, both concepts are still in use. In fact, a third theory exists. It closely mirrors the conventional theory. The **hole theory** states that although negative electrons do move from negative to positive, holes move from positive to negative as electrons move from atom to atom. Holes (current) flow from positive to negative, and are sometimes called "positive holes." What's most important to remember is that voltage causes current to flow through conductive paths (resistance). *From this point on, we use the conventional and hole theories when explaining the electrical concepts, unless discussing batteries, or otherwise noted.*

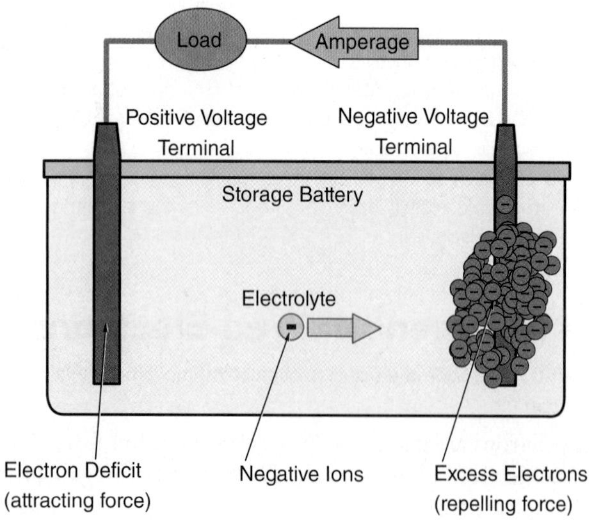

FIGURE 40-8 An excess of electrons on the negative side of the battery pushes electrons toward the positive side of the battery. The positive side of the battery pulls on the electrons as well. The force is called voltage. The current flow is called amperage.

Resistance to Current Flow

Also affecting the current flow in a circuit is **electrical resistance.** It is measured in **ohms (Ω)**. Resistance slows down the movement of electrons in a circuit. All materials have some resistance—even good conductors. There are four factors that determine the amount of electrical resistance:

1. **Type of material**: This refers to how many free electrons are in a material.
2. **Length of the conductor**: As length increases, so does resistance.
3. **Diameter of the conductor**: The larger the conductor, the greater the amount of current it can carry.
4. **Temperature of the conductor**: The higher the temperature, the harder it is for free electrons to pass through and the higher the electrical resistance.

As mentioned, all materials have some resistance to current flow. But a **resistor** is a component designed to extract energy from the current as it flows through the resistor. A typical resistor has a set resistance, usually marked or coded on its surface. Electrical resistance is the electrical equivalent of friction in the mechanical world. It is the degree to which a material opposes, or resists, the passage of current flow. Current flowing through resistance creates heat.

Resistance is measured in ohms. Under most conditions (except temperature change), the resistance of an object is a constant. It does not depend on the amount of voltage or current passing through it.

Electrical Circuits

Electrical circuits are designed to perform electrical work in a controlled manner. They can be compared to a small city:

- The roads are like wires;
- the stoplights are like switches;
- businesses are like electrical devices where work happens;
- and cars are like electrons that deliver the workers to the workplace.

Electrical circuits can be very basic. Minimally they consist of a power supply, a fuse, a switch, a component that performs work, and wires connecting them all together (**FIGURE 40-9**).

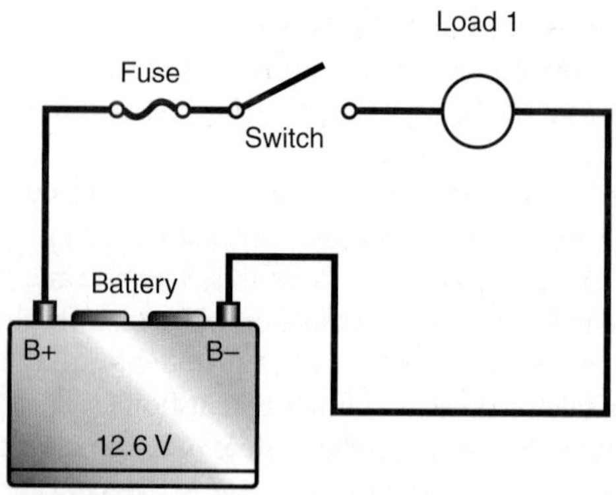

FIGURE 40-9 Simple circuit.

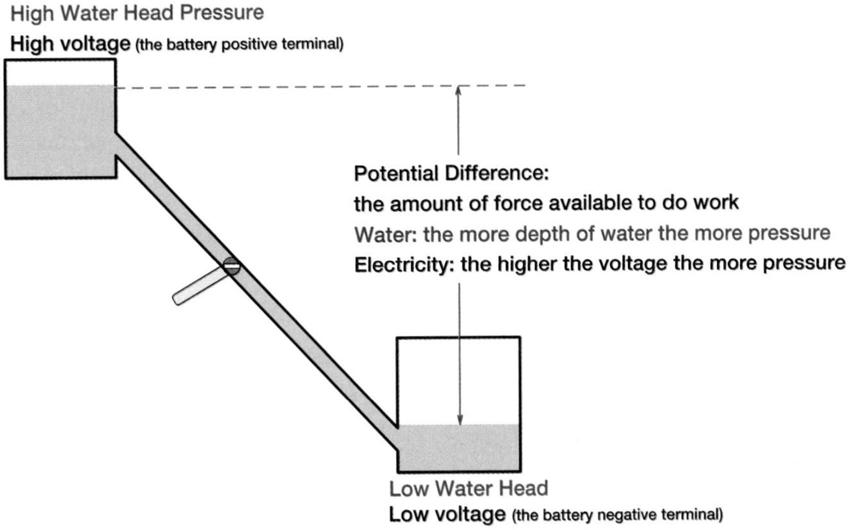
High Water Head Pressure
High voltage (the battery positive terminal)
Potential Difference:
the amount of force available to do work
Water: the more depth of water the more pressure
Electricity: the higher the voltage the more pressure
Low Water Head
Low voltage (the battery negative terminal)

FIGURE 40-10 Pressure (volts) in electrical systems is similar to pressure in a water system. The pressure wants to even out.

In this circuit, the battery is the power source. It creates a potential difference across its terminals, measured in volts. Voltage pushes a flow of electrons (current flow), measured in amps, through the circuit. Current is allowed to flow in the circuit when the switch is in the closed position. The current flows through the fuse into the circuit wires, through the switch and to the lamp. Current flowing through the resistance of the lamp filament causes it to glow, producing light. The current leaves the lamp filament and returns back to the battery. This completes the circuit. When the switch is moved to the open position, the current path is broken. Current flow stops, turning the lamp filament off.

A circuit can be much more complex than the one just described. But even then, most circuits contain a power source, circuit protection device, control mechanism, load, and connecting wires. Don't worry, we will help you understand how electricity behaves. Once you know that, you will be able to understand more complicated circuits.

▶ Volts, Amps, and Ohms

LO 40-04 Explain Volts, Amps, Ohms, Power, and Ground.

Volts, amps, and ohms are three basic units of electrical measurement. Voltage is the potential or difference in electrical pressure between two points in a circuit. It is measured in **volts**. For example, the voltage of a typical car battery is 12 volts. This is the potential difference, or electrical pressure, between the positive and the negative battery terminals. It can be measured with a voltmeter or **multimeter** set to read voltage.

It might be easier to understand if you think of voltage just like the water pressure that exists in the bottom of a full tank of water (**FIGURE 40-10**). The weight of the water creates pressure. The fuller the tank, the greater the pressure. To measure voltage, a voltmeter is used. It measures the difference in voltage (electrical pressure) between two points in a circuit. One meter lead is used at each of the two points in the circuit.

The ampere, or **amp**, is the unit used to describe how much current flow or how many electrons are flowing past a given point in 1 second. This occurs when work is being performed, such as when a lamp is operating. An amp is equal to 6.28 billion billion electrons past a given point in 1 second. Yes, billion billion is correct. To say it another way, think of a pile of 1 billion electrons. One amp would equal the number of electrons in 6.28 billion of those piles traveling past a given point in a circuit in 1 second (**FIGURE 40-11**). That is impressive! And just to stretch your thinking, a starter motor may draw about 200 amps.

Finished counting all those electrons yet?
1,876,001...1,876,002 Oh not yet boss.
Darn forgot where I was up to.
Do I really have to count all 6.28 billion piles..... can I please just count one of the piles of 1 billion electrons instead.............
CDX

FIGURE 40-11 One amp equals 6.28 billion piles of electrons, each pile containing a billion electrons.

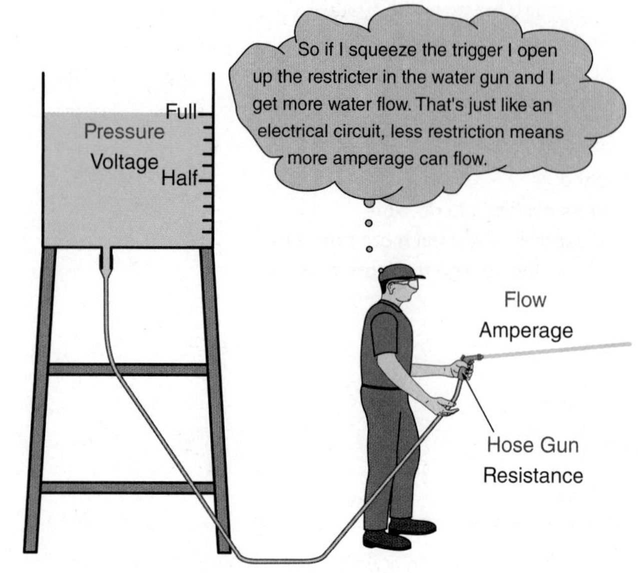

FIGURE 40-12 Current flow (amps) in an electrical circuit is like water flowing through a pipe.

Amperage, or current flow, can be thought of as a faucet being turned on and water flowing. Each drop of water is like an electron (**FIGURE 40-12**). Current flow is measured in amps by placing an **ammeter** into the circuit so that the current flows through the meter.

The ohm is the unit used to describe the amount of electrical resistance in a circuit or component. The higher the resistance, the less current (amps) that will flow in the circuit for any particular voltage. The lower the resistance, the higher the current that will flow in the circuit. Using the water analogy, if you kink a hose with the water running, less water will come out of the hose for a given pressure (**FIGURE 40-13**). If you kink the hose more, more resistance will be added, and even less water will flow through the hose. Lessening the kink lowers the resistance and allows more water to flow. Resistance in a simple electrical circuit works the same way.

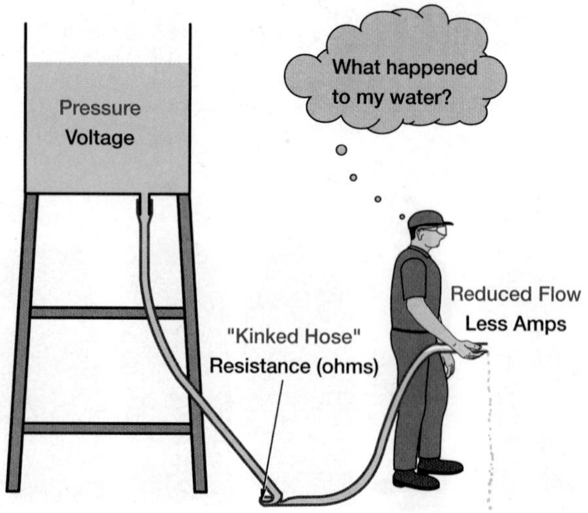

FIGURE 40-13 Resistance (ohms) in an electrical circuit is like a restriction in a water hose.

An ohmmeter is used to measure the amount of resistance in a component or circuit. The component or wire must be disconnected from the rest of the circuit. The ohmmeter pushes a small amount of current through the part being tested. The amount of resistance in the component changes the amount of current that the ohmmeter can push through the component. The more current the ohmmeter can push through the component, the lower the resistance will read on the ohmmeter.

Power (Source or Feed) and Ground

"Power" and "ground" are terms to describe the beginning and end of a circuit. Power, source, and feed signify the supply side (beginning) of the circuit. This is where the electricity originates. Ground signifies the return side of the circuit (**FIGURE 40-14**). In conventional theory, the supply side is the positive side of the circuit, and the return side is the negative side of the circuit. We will be using this theory throughout the rest of this text. In a vehicle, the positive battery post is considered the source. The power or feed side of a circuit refers to the wires and components that originate at the positive post of the battery. The power side ends at the load.

Ground is a term used by technicians to indicate the portion of the circuit that returns current flow to the negative post of the battery. The ground side of the circuit starts at the negative post of the battery and ends at the load. The term "ground" is also used to mean a direct connection with the negative side of the circuit. For example, "the switch is grounded." Also, if you are told to ground something, it means that object has to be connected to the negative side of the circuit.

Many vehicles connect the chassis, body, and engine block to the negative battery terminal. This means most of the metal components on the vehicle are grounded. As a result, many manufacturers use the chassis as the return path for current flow. This cuts down the amount of wire needed in the vehicle. At the same time, many computer circuits use dedicated ground wires from their sensors back to the computer. This helps ensure that the electrical signal is accurate and not affected by any stray electrical signals on the ground circuit.

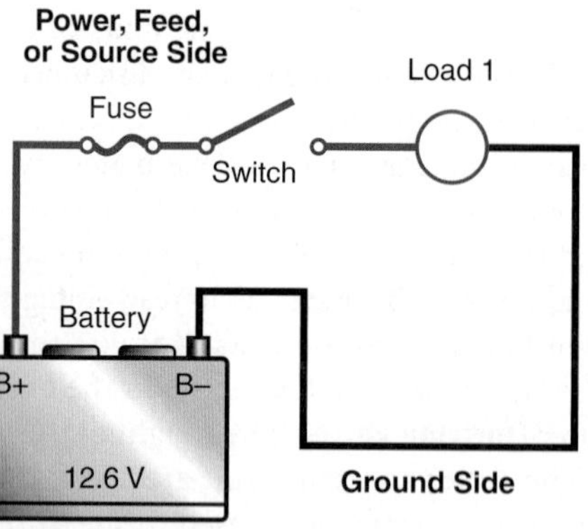

FIGURE 40-14 Power and ground.

Applied Science

AS-2: Ground: The technician can demonstrate an understanding of the problems associated with having an electrical circuit inadequately grounded.

Ground problems can present some of the most challenging diagnostic issues. Let's illustrate why ground points are so important on motor vehicles. In a simple circuit, we might have a battery, a fuse, and a load, all connected by wire. In automotive applications, manufacturers take advantage of the conductivity of metal bodies. They connect the ground side of electrical components to the vehicle's body or chassis. The negative side of the battery is also connected to the body. So, instead of electrical current returning to ground via wiring, it is conducted through the body.

If we have a component with no ground connection, we have an open circuit and no current flow. Loose or dirty ground connections create a point of high resistance in the circuit. This results in excessive voltage drop and lower current flow. Lamps illuminate dimly, electric motors do not work or not at full speed, and running problems in computer-controlled systems. In some circuits, issues may be encountered where electricity finds alternative paths to ground. You may sometimes see vehicles with brake lights flashing along with, or opposite to, turn signal lights. This is commonly caused by a poor ground in the taillight circuit.

▶ Sources of Electricity

LO 40-05 Describe the sources of electricity.

Electricity is a unique source of energy. It can be transformed into a variety of other kinds of energy. Common examples include thermal energy, light energy, chemical energy, and mechanical energy. It can also be created easily in a variety of ways. Think of other types of energy, such as coal and oil. They are much harder to create, generally relying on natural processes over a long period of time. This section explores some of the many ways that electricity can be produced.

An easily observable source of electricity is the static electricity produced in thunderstorms. Other ways that electricity can be produced include:

- the movement of a conductor through a magnetic field,
- the application of pressure to a special type of crystal,
- the conversion of sunlight by solar cells,
- and chemical reactions.

Regardless of how it is produced, electricity is always the movement of electrons in a circuit. Let's see how we can get those electrons moving so they can do work for us.

Electrostatic Energy

Static electricity can be induced by rubbing two insulators together. During this process, one material loses electrons to the other. The insulator losing electrons becomes positively charged. The other insulator gains electrons and becomes negatively charged (**FIGURE 40-15**).

When these two highly charged surfaces are brought close enough together, electrons leap across the gap. This causes a spark and cancels out the charge imbalance. This electron transfer can be experienced as an electric shock. It applies to any charged surfaces where the imbalance in charge is great enough to make the electrons leap the gap. The spark can be dangerous. When it is near fuel vapor at a gas station, it can even cause an explosion.

SAFETY TIP

Many fires have been caused by the operator during the refueling process. The operator gets back into the vehicle to check on a child or to put a credit card away. The operator slides back out across the seat of the vehicle (without touching the metal of the vehicle). Then, he/she grabs the refueling nozzle, at which time a spark ignites the fumes escaping from the filling tank. So, always touch the metal of the vehicle or fuel pump before touching the refueling nozzle. Also, fires have resulted from customers sliding their plastic gas cans out of the back of a plastic-lined pickup bed. Here the spark was created when the nozzle touched the can or when the pour spout touched the gas tank.

Thermoelectric Energy

If two different metals are joined and heated, a small electrical current can be generated. For a temperature rise of around 392°F (200°C), the potential difference created is about 9 millivolts. The point that is heated is called a **hot junction**. The whole unit is called a **thermocouple** (**FIGURE 40-16**).

When developing engines, manufacturers use thermocouples. They measure the high temperatures of components such as spark plugs and exhaust systems. Thermocouples are also used to measure exhaust gas temperature on race cars. The best air–fuel ratio can then be determined so that the engine parts will not overheat. Pilots in general aviation use thermocouples to measure the exhaust gas temperature. This allows them to adjust the fuel mixture as the airplane changes altitude.

Electrochemical Energy

When two dissimilar metals are immersed in an acidic liquid called an **electrolyte**, the chemicals breakdown into charged

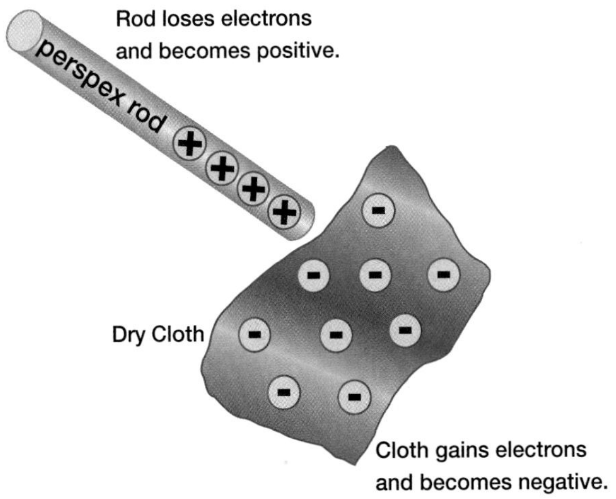

FIGURE 40-15 Rubbing two insulators together pulls electrons from one surface to the other, creating an electrical charge.

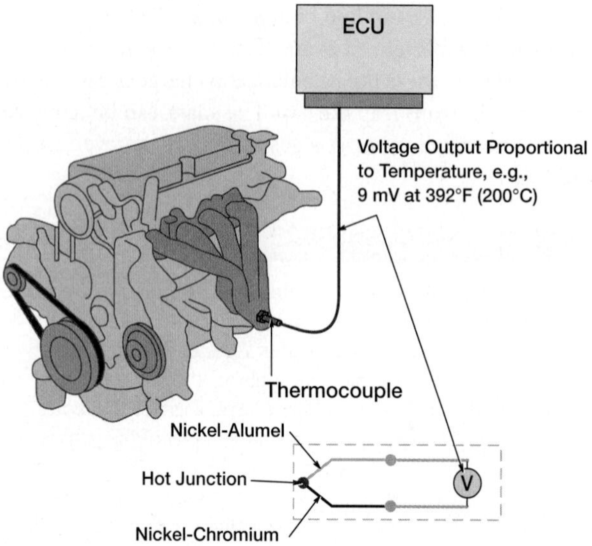

FIGURE 40-16 A thermocouple creates a small electrical current flow that can be used to indicate temperature, for example, that of exhaust gases.

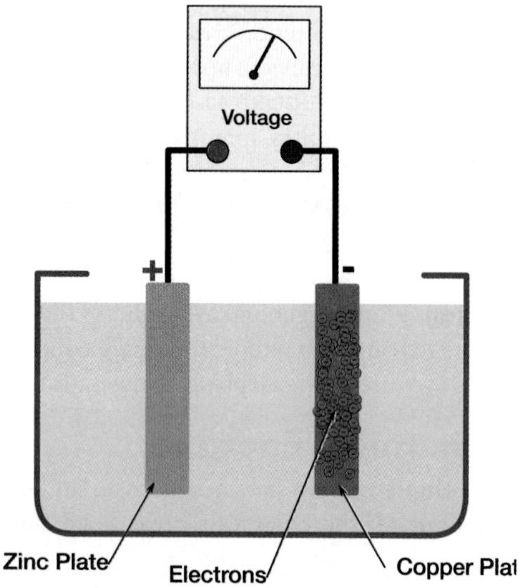

FIGURE 40-17 Electrolysis chemically separates electrons to one plate, creating an electrical charge.

particles, called **ions**. As you learned earlier, ions are unbalanced atoms. They allow electricity to flow (**FIGURE 40-17**). The process is called **electrolysis**. This principle is applied in the standard lead-acid battery used in most vehicles. It is covered more thoroughly in the Batteries chapter.

Photovoltaic Energy

Solar cells convert sunlight directly into electricity. They are used in a wide variety of applications to create electricity. They are made of semiconducting materials similar to those used in computer chips. When these materials absorb sunlight, the photons knock electrons loose from their atoms. This causes the electrons to flow through the material. And this produces

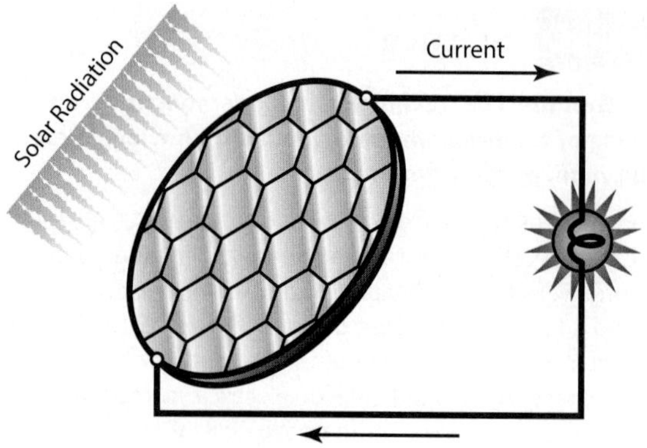

FIGURE 40-18 Photovoltaic effect.

electricity. This process is called the **photovoltaic (PV) effect** (**FIGURE 40-18**). Sun load sensors use this principle to determine how much sunlight is hitting the vehicle. This is used to control the heating, ventilation, and air-conditioning (HVAC) system. It is also used to sense the amount of daylight to operate automatically dimming headlights.

▶ TECHNICIAN TIP

The term "photovoltaic" comes from the Greek word *photos*, meaning light, and the name of the Italian physicist Volta, after whom the volt (and consequently voltage) is named. Photovoltaic means literally "of light and electricity."

Piezoelectric Energy

An electrical potential occurs when mechanical stress is placed on certain crystals (**FIGURE 40-19**). An example would be quartz. The process is also reversible. A potential electrical difference across the crystal physically distorts the crystal. The best-known simple application of **piezoelectric energy** is the sparking lighter. They ignite gas in grills, barbecues, or cooktops. But the piezoelectric principle is also used in automobiles. Knock sensors, pressure sensors, and some electronic fuel injectors are a few examples.

Electromagnetic Induction

Electromagnetic induction is the creation of an electrical voltage across a wire within a changing magnetic field. When a conductor cuts across a magnetic field, current flows in the conductor. It flows one way when the conductor cuts the field in one direction. It then reverses as it cuts the field in the opposite direction. Thus, it creates alternating current, or AC.

Moving a wire inside a magnetic field produces a current flow. Similarly, moving a magnet inside a stationary coil of wire produces the same effect. For example, a magnet can be rotated next to a winding. Current flow is then produced (**FIGURE 40-20**). As the magnet rotates, the ammeter deflects according to the direction of current flow. For every half revolution, current flow reverses. Increasing the speed of the magnet increases the amount of electrical energy produced. Electromagnetic induction is used in alternators, ignition coils, and some sensors on the vehicle.

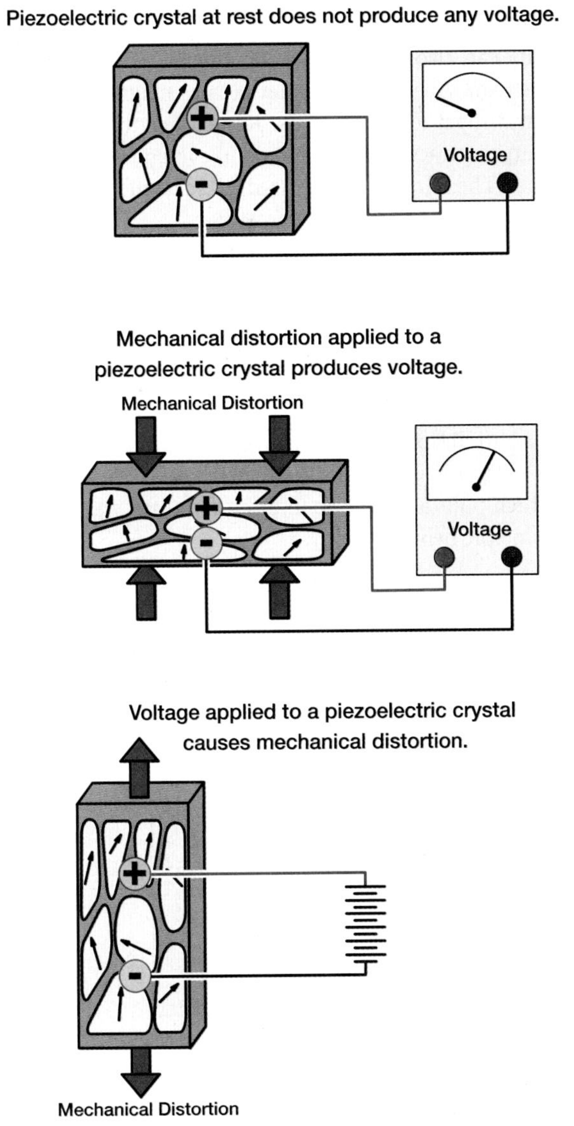
Piezoelectric crystal at rest does not produce any voltage.
Voltage
Mechanical distortion applied to a piezoelectric crystal produces voltage.
Mechanical Distortion
Voltage
Voltage applied to a piezoelectric crystal causes mechanical distortion.
Mechanical Distortion

FIGURE 40-19 Piezoelectric energy.

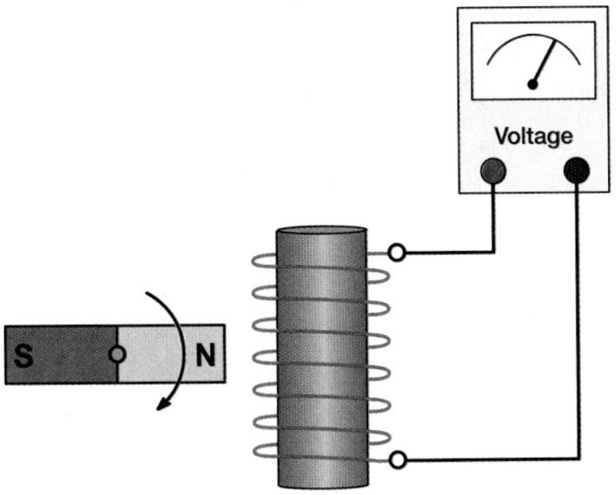
Voltage
S
N

FIGURE 40-20 Electromagnetic induction using a magnet rotating next to a winding.

To have induction, you must have three things:

- a winding,
- a magnet,
- and relative movement (i.e., movement of one past, or through, the other).

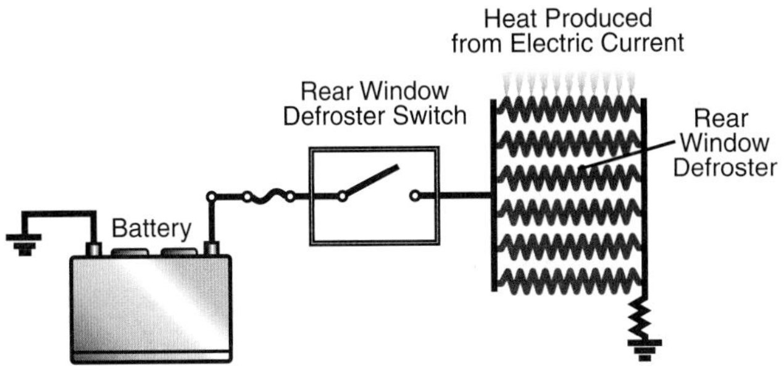
Heat Produced from Electric Current
Rear Window Defroster Switch
Rear Window Defroster
Battery

FIGURE 40-21 Heating effect of electricity.

The amount of induction is dependent upon:

- the strength of the magnetic field,
- the number of windings,
- the speed of the movement,
- and the relative distance between the field and the winding.

Applied Science

AS-3: Generators: The technician can explain how the movement of a conductor in a magnetic field generates electricity.

When a wire is moved through a magnetic field, an electric current is generated within the wire. This phenomenon is known as electromagnetic induction. At the same time, the conductor can remain stationary while the magnetic field is moved in relation to it. The generation of a voltage depends directly on the movement of either the wire or the magnetic field. Without relative movement, there is no voltage generated.

Electromagnetic induction works by exposing the electrons within the conductor to a magnetic force. This force acts perpendicular to both their motion and the magnetic field. The electrons are then forced to move through the conductor, creating a charge.

▶ Effects of Electricity

LO 40-06 Describe the effects of electricity.

Whenever electricity flows, it produces effects. For example, light, heat, chemical reactions, and magnetism are all effects of the flow of electricity. Sometimes those effects are the primary reason for the activity. For example, magnetic effects are used to generate electricity. At other times, the effects are just by-products of the activity. For example, current flowing in the conductors of a generator produces heat as a by-product and is wasted energy. No transformation of energy from one form to another is 100% efficient. There is always waste, which usually shows up as heat. For example, a typical incandescent lightbulb creates light. But only about 10% of the electricity is converted to light. About 90% is wasted as heat.

Heating Effects

Heat is not always bad. Sometimes we need to create heat, such as in circuit breakers and rear window defoggers (**FIGURE 40-21**). As current flows through resistors or components with resistance, heat is created. Headlights transform the energy into intense heat that then makes the bulb filament glow white-hot.

Since they glow white-hot, they produce light. Electricity can also create heat to defog a window. In some circuit breakers, excess current flow creates heat that pops the breaker open. This stops current flow. It then provides circuit protection from excessive current flow.

Chemical Effects

In a previous section, we said the electricity can be created by chemical reactions. This is the way a lead-acid battery creates electricity to start the engine. But electricity can also create chemical reactions. This also happens in lead-acid batteries. When a battery is discharged, the charging system reverses the current flow in the battery. This forces the sulfuric acid out of the lead plates where it recombines with water to increase the strength of the electrolyte. This will be covered in much greater depth in the Battery chapter.

Light Effects

The production of light can also be an effect of electricity. An example of this is the **light-emitting diode (LED)**. The **LED** is a semiconductor diode that creates light. It does so by emitting photons from its PN junction when a current flows through it (**FIGURE 40-22**).

At one time, LEDs were available only in red, green, and yellow tints. More recently, they have become available in many different colors. They have also grown in light intensity. They are now replacing traditional bulbs as lighting sources. They produce as much or more light with less heat loss, using much less energy. LEDs are also being used as headlights on some vehicles. It is likely that as LED technology improves and prices come down, they will be installed on more new vehicles.

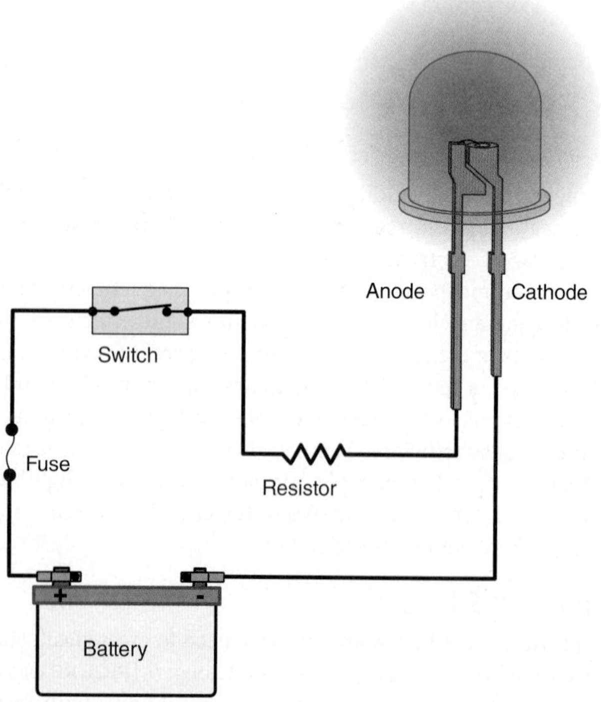

FIGURE 40-22 Light created by an LED.

Electromagnetic Effects

Electricity can also create magnetic effects, referred to as electromagnetism. An **electromagnet** creates magnetic forces that attract and repel ferrous metals. Remember that unlike magnetic charges attract each other. And like magnetic charges repel each other. These forces can be used to create mechanical movement, such as in a relay or electric motor.

In electrical systems, **magnetism** occurs when a current passes through a conductor. A magnetic field is created around the conductor. If the wire is wound into a coil, the magnetic fields combine to create a stronger and denser magnetic field. The field has a north and a south pole, just like a permanent magnet. Turn the current flow off, and the magnetic field collapses and disappears. By turning current to the coil on and off, this magnetic effect can be turned into mechanical movement which pulls a switch open or closed (**FIGURE 40-23**). This is the principle behind a **relay**.

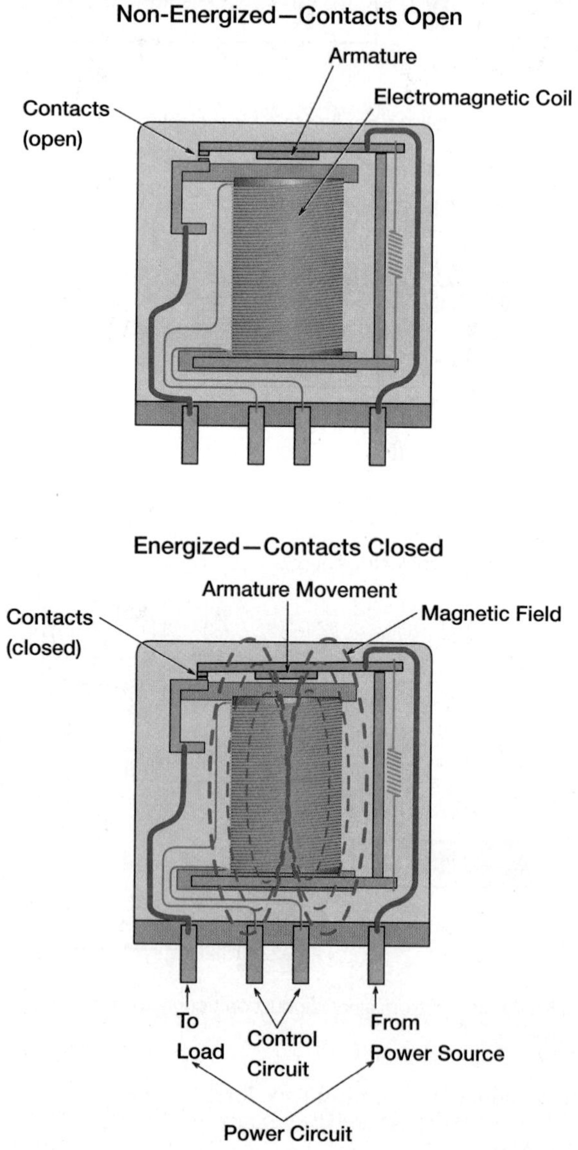

FIGURE 40-23 A relay uses electromagnetic force to open and close a switch.

Electromagnets are made by winding a conductor wire, many hundreds or thousands of times, around a soft iron core. Passing a current through the coil creates magnetism. Increasing the number of turns of wire on the coil increases the strength of the magnetic field. Also, increasing the current flow through the coil increases the strength of the magnet.

The metal core uniformly aligns the magnetic fields. This strengthens the magnetic effect. The size of the electromagnet is determined by its application. A fuel injector contains a small electromagnetic coil using very fine wire. But a starter motor uses heavier wire in larger coils. Also, reversing the current flow through the winding reverses the north and south poles of the electromagnet.

Applied Science

AS-4: Electromagnetism: The technician can explain the relationship between current in a conductor and strength of the magnetic field.

Current flowing through a conductor creates a magnetic field around the conductor. Increasing the current increases the strength of the magnetic field. Also, winding the conductor around a soft iron core concentrates the magnetic field. This increases its strength. The greater the number of turns on the winding, the stronger the magnetic field. Electromagnets use this principle. The strength of the electromagnet is controlled by controlling the current. More current means a stronger magnetic field and vice versa.

▶ Ohm's Law

LO 40-07 Use Ohms Law to calculate values.

Ohm's law helps us to understand the relationship between volts, amps, and ohms. Understanding how they are related to each other will make electrical diagnosis much easier. Ohm's law shows us that if one of the three units changes, then at least one of the other two must also change. But they always have to balance out mathematically.

Here's how Ohm's law works. It tells us that it takes 1 volt to push 1 amp through 1 ohm of resistance. That gives us the relationship among the three units. Volts and resistance are physical things. Voltage is the electrical pressure in a circuit. Resistance is the physical restriction in the circuit. Amperage is the amount of electrons moved. This means that amperage is the result of both the voltage and resistance.

If voltage or resistance changes, then amperage will also change. For example, if voltage stays the same and resistance doubles, half as much amperage can be pushed through the resistance (**FIGURE 40-24**). Conversely, if resistance stays the same but voltage doubles, then the greater force pushes twice as much current through the resistance. Or, if voltage is cut in half and the resistance stays the same, then amperage will be cut in half. Knowing how amperage responds to changes in voltage and resistance will help you with diagnosis. It will help you know what tests to perform to identify the location of circuit faults. We discuss this further in the chapter on meters.

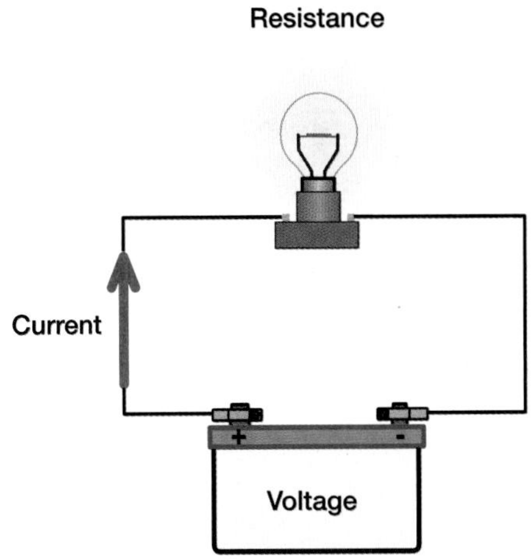

FIGURE 40-24 Current flow is the result of voltage and resistance.

Ohm's Law and Ohm's Law Calculations

As we saw, Ohm's law is a relationship between volts, amps, and ohms that must always balance out. That means that if we know any two of the values, then we can calculate the third. In calculating Ohm's law, R stands for resistance, V for voltage, and A for amps. Depending on which value you wish to solve for, apply one of the following three formulas:

- $A = V/R$
- $V = A \times R$
- $R = V \div A$

▶ TECHNICIAN TIP

Note that some sources use E for voltage and I for current. Also, we want to highlight the fact that amperage is *always* the result (product) of voltage and resistance. If the amperage in a circuit is not correct (the device is not operating correctly), it is because either the voltage or resistance is wrong. Remembering this will help you diagnose electrical problems.

Using the Ohm's law circle will help you remember which math operation to use (**FIGURE 40-25**). All you have to do is place your finger over the value you are looking for. If you place your finger on the top value (volts), then you would multiply amps by resistance. If you place your finger on one of the side values, then you would divide volts by the other value. This means two things. First, the values always have to balance. And second, as long as you know any two values, the third can be calculated.

Let's look at an example. The battery voltage can be measured. Let's use 12 volts. The value of the resistor, 4 ohms, is on its casing. Current then equals voltage (12 volts) divided by resistance (4 ohms). We can quickly calculate that there should

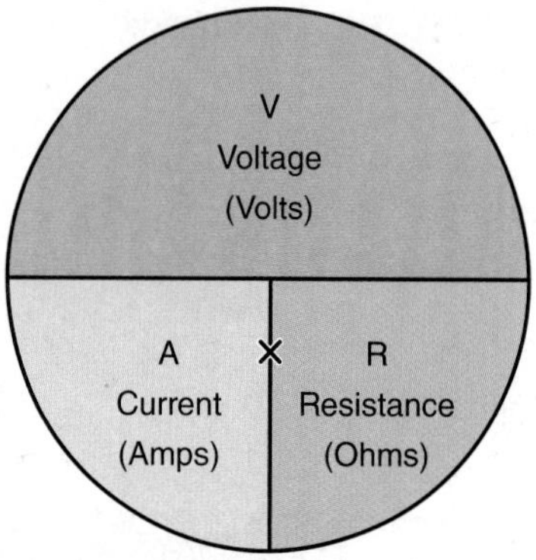

FIGURE 40-25 Ohm's law circle.

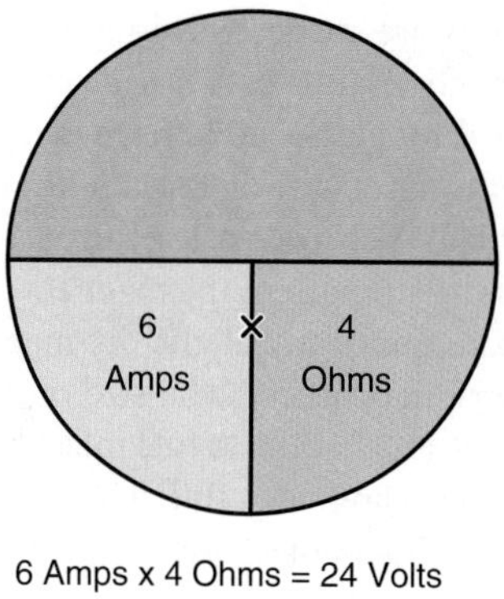

FIGURE 40-27 To find the value of V, multiply A by R, or 6 amps by 4 ohms.

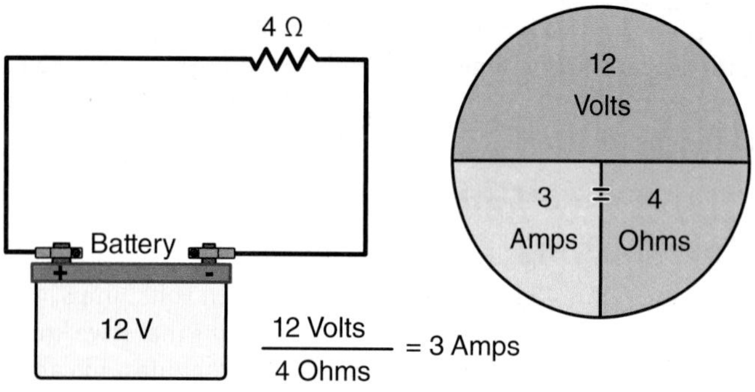

FIGURE 40-26 Calculating current flow in a circuit.

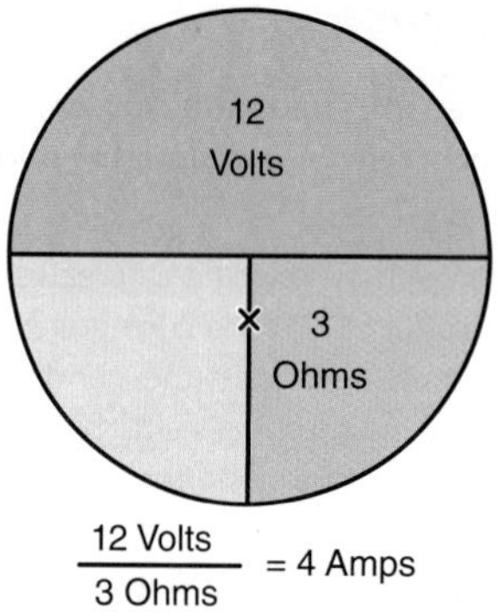

FIGURE 40-28 To find the value of A, divide V by R, or in this case 12 volts divided by 3 ohms.

be 3 amps of current flowing through every point in the circuit (**FIGURE 40-26**).

Using the formulas of Ohm's law gives an accurate method of determining expected values in an electrical circuit. We can then compare these expected values with measured values to determine if an electrical fault exists. If a fault exists, we can use our understanding of Ohm's law to find where it is located. To assist with these problems, use the Ohm's law circle.

- If the value of V and R are known, then to find A, V is divided by R. Place your thumb over A and the circle shows you this formula.
- Similarly, if V and A are known, then R can be found by dividing V by A.
- If A and R, are known, then V is found by multiplying A by R.

In the circuit diagram in (**FIGURE 40-27**), the value of the applied voltage is not known, but the amperage and resistance are. To find the value of V, multiply A by R, or 6 amps by 4 ohms. The answer is 24 volts.

In the circuit diagram in (**FIGURE 40-28**), the value of the current flow is unknown. To find the value of A, divide V by R, or in this case 12 volts divided by 3 ohms. The value of A is 4 amps.

Applied Science

AS-5: Ohm's Law: The technician can demonstrate an understanding of and explain the use of Ohm's law in verifying circuit parameters (resistance, voltage, amperage).

Ohm's law quantifies the relationship between amperage, voltage, and resistance in any given circuit. It provides a means to calculate any one of these parameters from the values of the other two. Ohm's law is applied most easily through the use of a circle diagram like the one shown here:

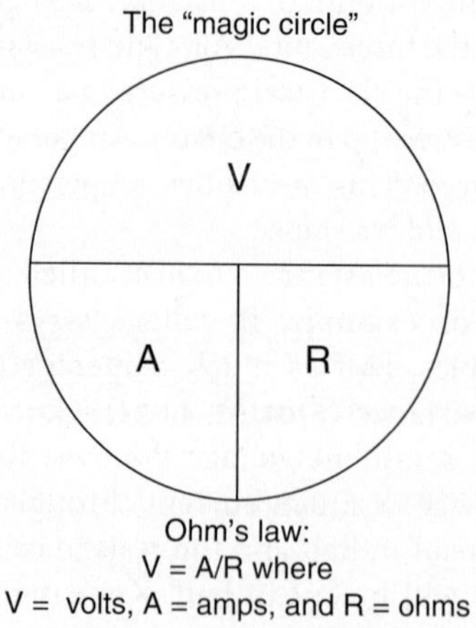

These diagrams are used by placing your finger over the parameter you are trying to calculate. Then, read the remainder of the diagram to identify the calculation you need to perform. For example, if you need to work out the amperage flowing in the circuit, block off A. The calculation is V ÷ R, or voltage divided by resistance. For a 12-volt automotive circuit with 20 ohms of resistance, your calculation is 12 ÷ 20, to arrive at 0.6 amp.

Applied Math

AM-1: Formulas: Using formulas, the technician can predict the outcome(s) under different variations.

Ohm's law, a mathematical formula, can be applied to predict the outcome of changes made to electrical systems. Say we have a 12-volt automotive electrical circuit to drive a bulb in an interior light. A customer has complained that the interior light in his car is not bright enough. They requested that a higher wattage bulb be fitted. (*Caution*: We are only using this as an example. Additional wattage creates additional heat. This can cause the fuse to blow or, in an extreme case, a fire. It is recommended that you never modify a vehicle from the manufacturer's original condition.) The resistance of bulbs varies according to their wattage. Generally, the higher the wattage, the lower the resistance.

The original 5-watt bulb fitted to the vehicle had a resistance of 30 ohms. Through Ohm's law, we know that current flow equals voltage divided by resistance, or in this case 0.4 amp. The new 10-watt bulb to be fitted has half the resistance, 15 ohms. Current flow in the circuit is now calculated as 12 volts divided by 15 ohms, which is 0.8 amp, which provides twice the wattage.

▶ Electrical Power and the Power Equation

LO 40-08 Use Watts Law to calculate values.

Energy is the potential to do work. However, work is done only when energy is released. A disconnected battery is not doing work. But it has the potential to do work and is therefore a source of energy. The difference in electron supply at the battery terminals creates electrical force. It is sometimes called the potential difference or voltage. In the case of a standard charged automotive battery, it has a potential difference of approximately 12 volts. Tapping this potential means turning the battery's chemical energy into electrical energy.

Turning one form of energy into another is called energy transformation. The amount of energy transformed is the amount of work done. When a person's legs turn the pedals of a bicycle, chemical energy (from oxygen and food) is being turned into mechanical energy. A motorcycle engine turns chemical energy first into thermal energy. The thermal energy is then turned into mechanical energy. In each case, **work** is being done. But there is a difference with the motorcycle. It does the work more quickly than the bicycle, delivering more mechanical energy faster. That difference is called **power**.

Power is the rate at which work is performed. It is also known as the rate of transforming energy. In an electrical circuit, power refers to the rate at which electrical energy is transformed into another kind of energy.

The unit of **electrical power** is the **watt**. One watt of electrical power is produced when 1 volt is used while pushing 1 amp of current through a resistance. From this comes the power equation:

- P (the power in watts), equals A (the current in amps), multiplied by V (the voltage in volts). It looks like $P = V \times A$

This calculation is applied similarly to Ohm's law and is typically represented as a triangle (**FIGURE 40-29**).

When current flows through a resistor, the resistor becomes hotter as it converts electrical energy into heat energy. The greater the voltage used by, or the amperage pushed through, the resistor, the greater the wattage. A lightbulb uses a certain amount of electrical power. If a 12-volt circuit with a single light has a current flow of 5 amps, then applying the formula will yield:

- P (watts) = A × V
- P (watts) = 5 A × 12 V
- P (watts) = 60
- The power consumed by the circuit is 60 watts, and the bulb will carry a rating of 60 watts.

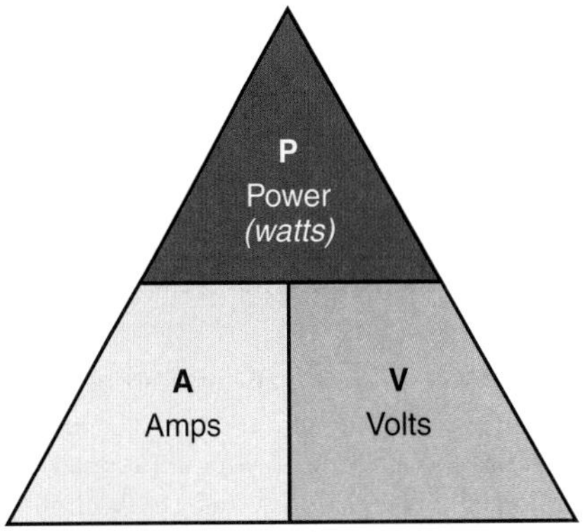

FIGURE 40-29 Watt's law triangle.

Applied Science

AS-6: Resistance: The technician can demonstrate an understanding of the relationship of resistance to heat, voltage drop, and circuit parameters.

The electrical resistance of a component is the component's opposition to the flow of current. Any resistance in a circuit creates a voltage drop across the resistive component, be it a resistor, a bulb, an electric motor, or merely a wire. The amount of voltage drop is proportional to the amount of resistance. Resistors work by dissipating energy in the form of heat. Therefore, for a given amount of current flowing in a circuit, the higher the resistance of a component, the more heat it will produce.

Let's look at an example using a rear window defogger. The defogger elements form a resistor in the circuit. In this case, heat is produced by the resistor grid to defog the window. The resistance of each grid line creates a steady voltage drop across the length of the grid line. Current flowing through the grid line then creates heat.

Broken grid lines cannot conduct current flow. So, there is no progressive voltage drop, nor heat generated in that grid line.

Battery voltage (12 volts) is applied across the element when you switch on the defogger. If you were to measure the voltage at the center of one of the grid lines, you should expect to see a voltage of approximately 6 volts. If the grid is burned open on the ground side, then the voltage would read about 12 volts in the middle. This is because there cannot be a voltage drop if there is no current flow. If the grid was burned open on the power side, the voltage would be near 0 volts. This is because the current flow cannot get past the open circuit to the middle of the grid line. Moving the voltmeter toward the side with the break can pinpoint the exact location of the broken grid line. That way, repairs can be made.

If a circuit is powered by a 12-volt battery with a load using 20 amps, using the power equation (P = A × V), we can determine that the load is using 240 watts of power. For reference purposes, 746 watts equal 1 horsepower. So, 240W is about 1/3 horsepower.

It is also possible to simplify and transpose the power equation. If power = voltage × amps, then:

- Voltage equals power divided by amps: V = P ÷ A
- Amps equals power divided by voltage: A = P ÷ V

Applied Math

AM-2: Algebraic Expressions: The technician can use Ohm's law and the power law to determine circuit parameters that are out of tolerance.

The power law, or Watt's law, is a mathematical equation. It describes the relationship of voltage, current, and power in electrical circuits. Power is defined as the amount of work done in a given period of time. Understanding this equation is useful in calculating characteristics within circuits. The power law is written as: Power (watts) = Voltage (volts) × Current (amps).

An unknown parameter can be calculated from any two known parameters. Like Ohm's law, the power law is commonly represented in a triangle diagram, as shown here:

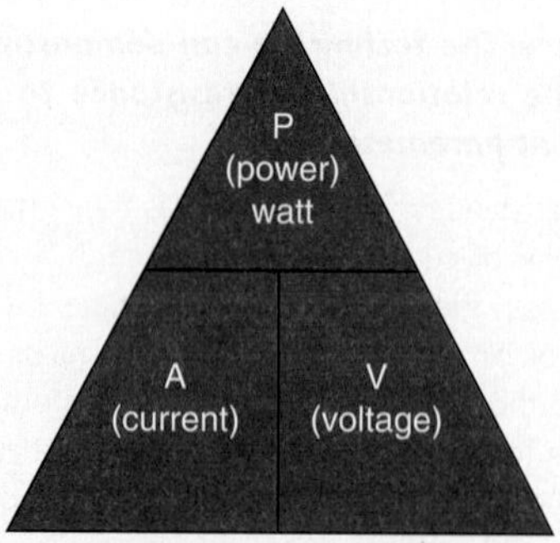

For example, we have an electrical circuit driving a power window motor in a passenger car. The vehicle operates on a 12-volt electrical system. The wattage of the window motor is 22 watts. To calculate the current flow in the circuit, we divide the power (or the motor's wattage) by the voltage, or P/V. In this case, the amperage would be 22/12, or 1.8 amps.

You can see that increasing the amps from 5A to 20A increased the watts in the same proportion, from 60W to 240W. The same thing would happen if the voltage were increased.

This formula can be applied to any circuit where the voltage and current flow are known. However, if the values of voltage or current flow are not known, then Ohm's law can be used to determine the missing value.

▶ Series Circuits

LO 40-09 Describe series circuits and use its laws to calculate values.

A series circuit is the simplest type of electrical circuit. In a series circuit, there is only one path for current to flow. All of the current flows to each component in turn. It also means that all the electrons flow at the same rate (amps) throughout all parts of the circuit. Amperage is equal everywhere within a series circuit.

In a series circuit, if there is more than one resistance in the circuit, those resistances are connected one after the other. Thus, the resistances add up in series. The total resistance in a series circuit is the sum of all of the individual resistances. For example, imagine a circuit with three resistors in series, each having 4 ohms of resistance. Total resistance of the three 4-ohm resistors is 4 + 4 + 4 =12 ohms, since resistance adds up in a series circuit.

However, the voltage drops from a potential difference of 12 volts as it leaves the battery to virtually no voltage at all as it returns to the battery. At each point in the circuit where current flows through a resistance, a drop in voltage occurs. This is called a **voltage drop**. Voltage drops are good when they occur inside of an intended load. They are bad when they occur where they are not wanted; outside of intended loads.

In **FIGURE 40-30**, after the first resistor, voltage has dropped from 12 to 8 volts. After the second resistor, it is down to 4 volts. After the third resistor, it is 0 volts. So, all of the source voltage has been used up by the end of the circuit.

Ohm's law can be used in series circuits to calculate voltage, resistance, and current. Any one of these can be calculated as long as the values of the other two are known.

The series circuit laws listed here provide a summary of how electricity behaves in a series circuit. A series circuit is

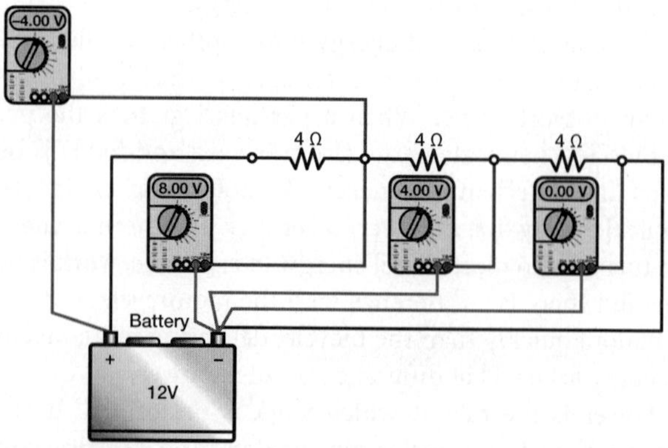

FIGURE 40-30 Voltage drop in a series circuit.

defined as a circuit with one or more loads, but with only one path for current to flow. Here are the laws for a series circuit:

- Current flow stays the same in a series circuit. Current flow is the same in all parts of the circuit. It doesn't add up or drop. The pressure pushing the amperage is what gets used up.
- Voltage drops (gets used up) as current goes through resistance(s) in series. The source voltage is equal to the sum of the individual voltage drops in the circuit (Kirchhoff's voltage law).
- Resistance adds up in series. Total circuit resistance is equal to the sum of the individual resistances; for example, $R_T = R_1 + R_2 + R_3$, and so on.

We can use those laws along with Ohm's law to help us understand the behavior of electricity. For example, in a circuit with three 2-ohm bulbs in series and a 12-volt battery, we can figure out what is happening electrically in the circuit (**FIGURE 40-31**). First, we have to calculate the total resistance in the circuit, which is 6 ohms. We know the circuit has 12 volts, so 12V (total voltage) ÷ 6Ω (total resistance) = 2A (total amps).

Because current flow stays the same in a series circuit, then each of the three bulbs has 2 amps flowing through it. And because each bulb has 2 ohms of resistance, we can calculate that the voltage used by each one (voltage drop) is 2A × 2Ω = 4-volt drop across each bulb.

We can do it again with slightly different resistances. So, this example has three 8-ohm bulbs in series and a 12-volt battery (**FIGURE 40-32**). Total resistance is 24 ohms. Total voltage is 12V. So, total amps are 0.5 amp.

> **TECHNICIAN TIP**
>
> There is one important thing to know about current flow in a circuit. It doesn't flow like dominoes, with a little bit of delay between each domino. Current flow is more like the close-fitting marbles inside a tube that we described earlier. As you push on a marble at one end, the marble at the other end falls out. Current flow works the same way, except the effect of current flow travels at the speed of light. Thus, if you add an electron at one end of a wire that is 186,000 miles (300,000 km) long, an electron at the other end of that wire will be pushed out 1 second later. That is why we say that current flow stays the same throughout a series circuit. As one electron is moved, all of the others move with it.

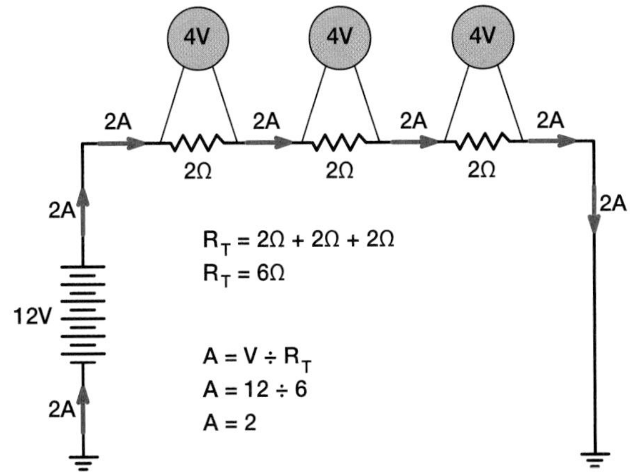

FIGURE 40-31 Using the series circuit laws and Ohm's law to understand how electricity behaves in this circuit.

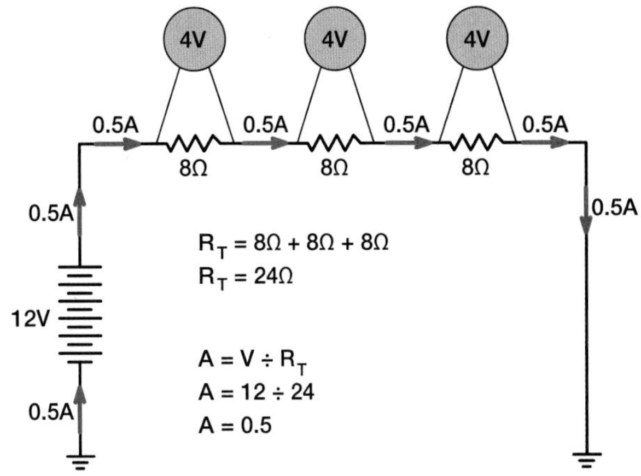

FIGURE 40-32 Notice that increasing the resistance of each of the three bulbs decreased the current flow. But didn't change the individual voltage drops.

> **Applied Math**
>
> ***AM-3: Mentally: The technician can determine the proper mathematical operation (addition, subtraction, multiplication, or division) and mentally arrive at the solution.***
>
> Determining the behavior of electricity in electrical circuits uses different mathematical operations. Some examples may include:
>
> - adding the resistance of individual components in a series circuit to determine the total resistance of the circuit,
> - dividing the voltage by the resistance to determine the current flow in a parallel circuit,
> - or multiplying the amperage of a circuit by its resistance to determine the voltage a device is using.

Because current flow stays the same in a series circuit, each of the three bulbs has 0.5 amp flowing through it. And because each bulb has 8 ohms of resistance, we can then calculate the voltage used by each one (voltage drop) is 0.5A × 8Ω = 4-volt drop across each bulb.

▶ Parallel Circuits

LO 40-10 Describe parallel and series-parallel circuits and calculate values.

In a series circuit, components are connected like links in a chain. If any link opens, current to all of the components stops flowing. In a parallel circuit, there is more than one path for current to flow (**FIGURE 40-33**). If any connection or component fails in one branch of a parallel circuit, current continues to flow normally through the remaining branches. This is one reason why parallel circuits are used in automotive headlight and taillight systems. If one lamp fails, current continues to flow through the other lamps in parallel. In a series circuit, all would go out, which could be disastrous.

Electricity always behaves according to the laws of physics. But when it operates in a parallel circuit, there are some additional laws you need to know. The parallel circuit laws

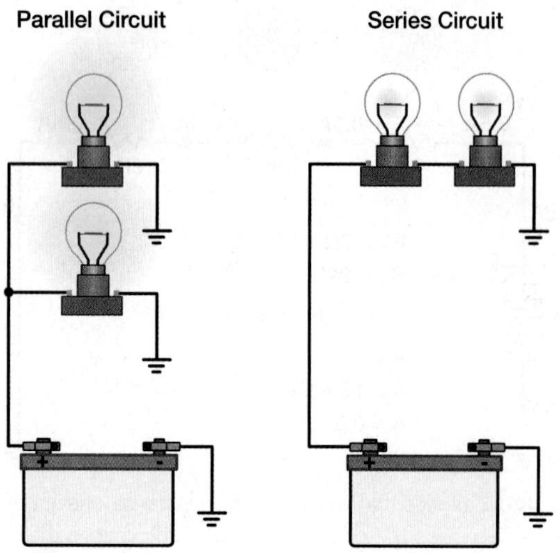

FIGURE 40-33 Typical parallel circuit compared to a series circuit.

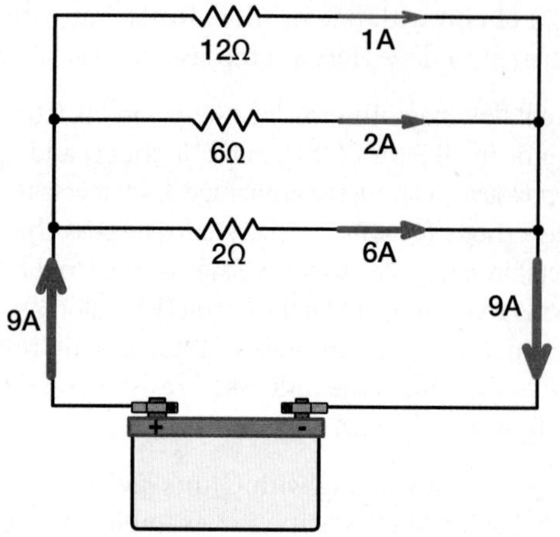

FIGURE 40-34 Parallel circuit with unequal resistances.

listed here should be applied when working with parallel circuits:

- The voltage is the same at the input of all branches of a parallel circuit. The voltage drop is the same across branches in parallel.
- The total current in a parallel circuit equals the sum of the current flowing in each branch of the circuit. Current flow adds up in parallel.
- The total resistance of a parallel circuit decreases as more branches are added. Total parallel circuit resistance will always be less than the branch with the lowest resistance.

Let's look a bit more closely at how those laws work. A feature of a properly working parallel circuit is that the voltage drop across each parallel branch is the same. No matter how many branches are added, or removed, the voltage drop across them will be the same. Another feature of a parallel circuit is that the current flowing in each branch is determined by the resistance and voltage drop of that branch. So, current flow can vary in each branch, they don't have to be the same.

In a parallel circuit where the resistors in each branch are the same, the current flowing in each branch is also the same. The sum of their individual current flows is equal to the total current flowing in the entire parallel circuit. When the resistances are not equal, the current divides according to the resistance of each branch. But the total current flow is still the sum of the current flowing in each branch (**FIGURE 40-34**).

Parallel Circuit Resistance

Resistance in a parallel circuit is not as easily calculated as it is in a series circuit. This is because as branches are added, another path for current to flow to ground is added. Adding extra paths reduces the circuit's total resistance to current flow.

For example, you have a 12-volt parallel circuit with three branches, and each branch has a 12-ohm resistor that allows 1 amp of current flow. So, the total current flow would be 3 amps. If you then add a fourth parallel branch with a 12-ohm resistor to the circuit, the total current increases from 3 amps to 4 amps. This is because in a parallel circuit, adding more branches provides more pathways for current to flow. This decreases the overall resistance to current flow. And thus, current flow increases. This is like having a freeway with two lanes and bumper-to-bumper traffic. If you add a third lane, the resistance to traffic flow decreases, and cars can move more easily.

Here is another visual example: You have a bucket full of water with three holes in the bottom. If you add another hole, the total flow out of the bucket would speed up. This means that resistance to water flow decreases even though another path with resistance is added. To calculate resistance, use one of the two formulas in (**FIGURE 40-35**). Use the ohm value of each resistor in the place of R_1, R_2, etc.

For 2 resistor in parallel

$$R_T = \frac{R1 \times R2}{R1 + R2}$$

For more than 2 resistor

$$R_T = \frac{1}{\frac{1}{R1} + \frac{1}{R2} + \frac{1}{R3}}$$

For example if R1 = 2Ω and R2 = 4Ω and R3 = 6Ω

$$R_T = \frac{2\Omega \times 4\Omega}{2\Omega + 4\Omega}$$

$$R_T = \frac{8\Omega}{6\Omega}$$

$$R_T = 1.33\Omega$$

$$R_T = \frac{1}{\frac{1}{R1} + \frac{1}{R2} + \frac{1}{R3}}$$

$$R_T = \frac{1}{\frac{1}{2\Omega} + \frac{1}{4\Omega} + \frac{1}{6\Omega}}$$

$$R_T = \frac{1}{0.5\Omega + 0.25\Omega + 0.16\Omega}$$

$$R_T = \frac{1}{0.91\Omega}$$

$$R_T = 1.09\Omega$$

FIGURE 40-35 Calculating total resistance in parallel circuits.

Applied Science

AS-7: Parallel/Series Circuits: The technician can explain current flow and voltage in series and parallel circuits.

The terms "series" and "parallel" describe different ways in which components in a circuit are connected. In a series circuit, all components are connected along a single path. Electricity can flow only one way, so the current flow within the circuit is the same at all points. The voltage in the system changes at different points due to voltage drops at various resistors.

In a parallel circuit, components are connected so that the circuit divides into two or more paths. All components are connected to the same voltage supply, so the same voltage is applied to each component. Current flow in different branches changes depending on the resistance in each branch. Therefore, different branches of the circuit may experience differing amperage. But only if their resistance values differ.

Series-Parallel Circuits

When electrical components are wired together one after another, with only one path for current to flow, they are said to be wired "in series." When they are wired together side by side, so there is more than one path for current to flow, and they are said to be wired "in parallel." A **series-parallel circuit** is made of both a series circuit and a parallel circuit. The series circuit portion can be before or after the parallel portion of the circuit. Series-parallel circuits can be analyzed using the same laws that apply to separate series or parallel circuits. You just have to apply the series laws to the series portion of the circuit and the parallel laws to the parallel portion.

A circuit for dash lights is an example of a series-parallel circuit. In this circuit, a variable resistor is connected in series with a number of dash lights. The dash lights are connected in parallel to each other (**FIGURE 40-36**). When the variable resistor is turned, the resistance changes. As the resistance increases, the voltage drop across the variable resistor increases. This decreases the voltage and current flow to the bulbs. This makes the dash lights dimmer. When the variable resistance is lowered, the voltage drop across it is reduced, and the current flow in the circuit increases. The increased voltage and current flow makes the dash lights brighter. Because the dash lights are in parallel, each of them receives the same amount of voltage. Since the total circuit current flow is divided equally between them, they are equally as bright as one another.

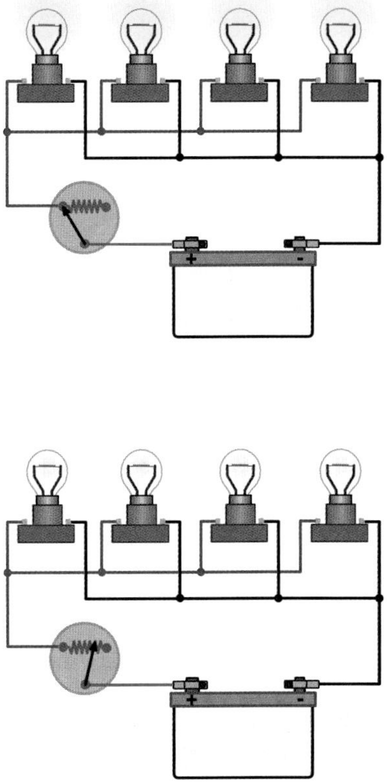

FIGURE 40-36 A dash light circuit is a good example of a series-parallel circuit.

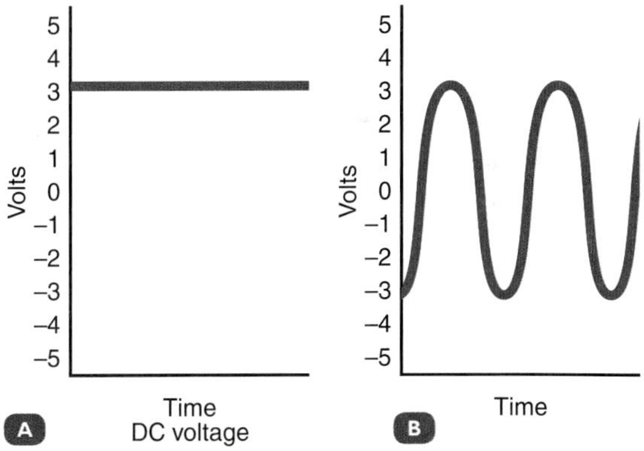

FIGURE 40-37 Waveforms. **A.** Direct current. **B.** Alternating current.

▶ Direct Current and Alternating Current

LO 40-11 Describe DC and AC and Kirchhoff's Current Law.

Electrons must flow in a circuit for work or action to be undertaken. For example, the action of a lamp glowing brightly is caused by the flow of electrons through the filament. The electrons heat the filament up which causes it to glow. There are two fundamental types of **current flow: DC** and **AC** (**FIGURE 40-37**). DC is produced by a battery. The battery maintains the same positive and negative **polarity**. Therefore, the current flows in one direction only. The characteristics of DC are the fixed polarity of the applied voltage and the flow of charges in only one direction. It is possible to have varying DC. However, the charges always flow in one direction, and the applied voltage polarity remains the same.

AC is the type of current in your home electricity supply. The alternating current repeatedly reverses or alternates its polarity. Thus, the current flow moves back and forth within a circuit. AC is produced in what is called a **sine wave**. It operates on a cycle, gradually building to a maximum current flow in one direction (positive value). Then gradually reducing to zero. It then gradually builds to a maximum current flow in the other direction (negative value). And finally gradually reduces back to zero current (**FIGURE 40-38**). In most cases, this cycle can occur many times a second. For example, the AC flow in the house supply has a cycle rate of 60 times per second in the US and Canada. Other

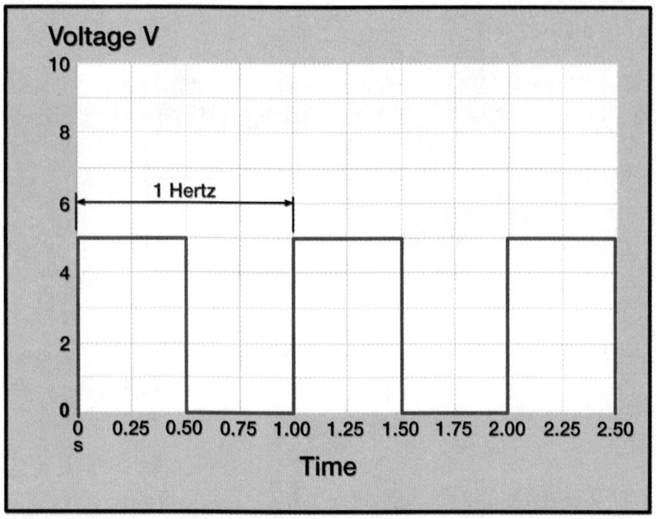

FIGURE 40-38 One cycle per second = 1 Hertz.

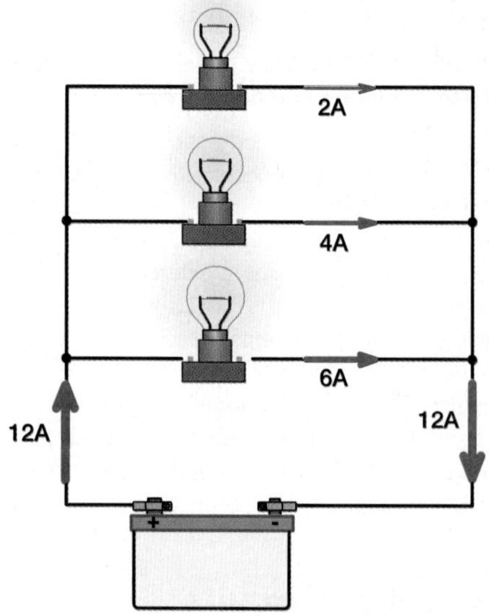

FIGURE 40-39 Kirchhoff's current law: Current entering a junction is equal to the sum of current flowing out of the junction.

parts of the world operate on 50 cycles per second. **Hertz** is the measurement of frequency. It indicates the number of cycles per second. So, Hertz simply means "cycles per second."

Applied Science

AS-8: AC/DC: The technician can explain the difference between direct and alternating current.

Direct current (DC) and alternating current (AC) are two forms of electricity. They are produced differently and have different uses. DC electricity is the simpler form. It starts in one place and then flows in the same direction to its destination. AC electricity flows in one direction for a period of time, then changes direction over and over again.

In modern automotive applications, AC electricity is generated by the alternator using electromagnets. The AC electricity is converted to DC, or rectified, by diodes in the alternator. It is then supplied to the vehicle's electrical system or used to charge the battery.

AC is used in vehicles to a lesser extent than DC. Alternators use it to generate electrical power. The AC is first transformed into DC before it leaves the alternator. It can then be effectively put to use in the DC electrical system.

AC is used in the electric motors on most hybrid vehicles. Those motors generally require high amounts of electrical power. Because AC is more efficient than DC, AC is more advantageous for that application. The AC is created by a sophisticated electronic inverter. In general, electrical components are designed to work on either AC or DC, but not both. For example, a DC motor will not effectively work on AC, and vice versa.

Kirchhoff's Current Law

Kirchhoff's current law describes a fundamental electrical principle. It is used by technicians in understanding how parallel circuits work. Many technicians understand the principle behind the law without necessarily associating it with Kirchhoff. Simply stated, the law says that current entering any junction is equal to the sum of current flowing out of the junction. For example, a junction has three conductors. Ten amps of current is flowing in through conductor A. Current flows out by the other two conductors, B and C (**FIGURE 40-39**). According to Kirchhoff's current law, a total of 10 amps flow out of conductors B and C. We do not know how many amps flow out of B or C, but the total must be 10 amps. Kirchhoff's law is used with parallel circuit laws. It will allow you to be able to better understand the behavior of electricity in parallel circuits.

▶ Using Basic Electrical Concepts to Solve Problems

LO 40-12 Explain how to use electrical concepts to solve problems.

If you understand the preceding concepts, then you are well on your way to diagnosing electrical problems. Here is a question: If the amps in a circuit are lower than they should be, what are the two possible causes? If you said that either the voltage is too low or the resistance is too high, you would be correct. If the available voltage to the device is low (charging system not working), then the full amount of current will not flow. In the same way, if the resistance is too high (dirty, loose connection), then current will also be too low.

Second question: If the amps in the circuit are higher than they should be, what are the two possible causes? If you said that the voltage is too high or the resistance is too low, you would be correct. If the voltage is too high (charging system is putting out too much voltage), then more current flow will be pushed through the resistor. If the resistance is too low (short circuit), then too much current will flow.

As you can see, if you keep your eye on the current flow, you will have a good clue as to what type of fault you are looking for. In many cases, that doesn't require using a meter. You can usually see it with your eye. For example, a light is dim. In this case, you know that the current flow is too low because work is not being performed like it should be. And that means there is either not enough available voltage or too much resistance. Quickly testing the available voltage at the device will indicate whether you need to perform a voltage drop test on both sides of the circuit. Measuring the resistance of the electrical device will indicate if a high resistance or an open circuit condition is present in the component. Comparing your readings with the specifications will indicate if the resistance is correct or not.

Applied Math

AM-4: Equivalent Form: The technician can write or rewrite an algebraic equation to solve for any unknown variables.

Ohm's law is commonly used to calculate the value of an unknown variable from known values in electrical circuits. The equation is commonly stated as voltage (V) equals current flow (A) multiplied by resistance (R), or V = AR. To calculate current flow or resistance, instead of voltage, the equation must be rewritten. To apply the rules of basic algebra, anything we do to one side of the equation, we must also do to the other side. To calculate current flow, we need the A by itself on one side of the equation. So, we divide both sides by R to arrive at A = V/R. To calculate resistance, the R must stand alone. So, we divide both sides of the equation by A to arrive at R = V/A. Ohm's law is commonly represented in a circle format. This makes the correct algebraic equation very simple to find.

Applied Science

AS-9: Electricity: The technician can demonstrate an understanding of and explain the properties of electricity as they relate to lighting, engine management, and other electrical systems in the vehicle.

Some sensors generate their own voltage. Examples include:

- zirconia oxygen sensors;
- analog crankshaft or camshaft position sensors, which generate AC voltage;
- and piezoelectric knock sensors.

Voltages generated from these sensors are monitored by the powertrain control module (PCM). They are used to determine engine-operating parameters.

Some computer-controlled devices can simply be turned on and off. A cooling fan may be turned on when the coolant reaches a temperature of 219°F (104°C), for example. It can then be turned back off when the coolant temperature falls below 201°F (94°C). Pulse-width modulation provides a means for a computer to control a device with variable operating parameters. Fuel injectors, for example, need to be open for differing amounts of time, depending on the fueling requirements of the engine. An ECU can control injector opening time by sending a pulsed signal with shorter or longer duration.

▶ Wrap-Up

Ready for Review

- ▶ A solid understanding of electrical terminology, the behavior of electricity, and circuit theory is required to service today's vehicles. Electricity is the movement of particles due to like and unlike charges from one place to another.
- ▶ Materials that allow electrons to flow with little resistance are called conductors, materials that do not conduct electrons easily are called insulators, and materials that conduct electricity easier than insulators but not as well as conductors are semiconductors.
- ▶ Easy movement of electrons requires a circuit and a force that makes them move. Free electrons flow from the negative terminal to the positive terminal and this flow of electrons is called current flow.
- ▶ Voltage is the potential or difference in electrical pressure between two points in a circuit; ampere is the unit used to describe how many electrons are flowing past a given point in 1 second, ohms is the unit used to describe the amount of electrical resistance in a circuit or component, power signifies the supply side of the circuit, and ground indicates the portion of the circuit that returns current flow to the negative post of the battery.
  - Voltage is measured in volts, current flow is measured in amps, and resistance is measured in ohms.
- ▶ Electricity can be transformed into a variety of other kinds of energy. Some of the sources of electricity include: electrostatic energy, thermoelectric energy, electrochemical energy, photovoltaic energy, piezoelectric energy, and electromagnetic induction.
- ▶ Electricity flow produces various effects such as heating effects, chemical effects, light effects, and electromagnetic effects.
- ▶ Ohm's law is a relationship between volts, amps, and ohms. Three formulas: $A = V/R$, $V = A \times R$, and $R = V \div A$. Ohm's law circle can also be used to determine the math operation to be used for the relationship between volts, amps, and ohms
- ▶ Power equation states that P (the power in watts), equals A (the current in amps), multiplied by V (the voltage in volts). This calculation is applied similarly to Ohm's law and is typically represented as a triangle.
- ▶ In a series circuit, there is only one path for current to flow and all of the current flows to each component in turn. Laws for a series circuit: current flow stays the same in a series circuit, voltage drops (gets used up) as current goes through resistance(s) in series, and resistance adds up in series.
  - Total circuit resistance is equal to the sum of the individual resistances.

- Laws of parallel circuits: the voltage drop is the same across branches in parallel, the total current in a parallel circuit equals the sum of the current flowing in each branch of the circuit, and the total resistance of a parallel circuit decreases as more branches are added.
  - A series-parallel circuit is made of both a series circuit and a parallel circuit. Apply the series laws to the series portion of the circuit and the parallel laws to the parallel portion.
- DC is produced by a battery and AC is produced in a sine wave. AC is used in vehicles to a lesser extent than DC. Kirchhoff's current law says that current entering any junction is equal to the sum of current flowing out of the junction.
- Low amps in a circuit can be due low voltage or high resistance. High amps in a circuit can be due high voltage or low resistance. Measuring the resistance of an electrical device will indicate if a high resistance or an open circuit condition is present.

Key Terms

AC See *alternating current.*

amp An abbreviation for amperes, the unit for current measurement.

ammeter A device used to measure current flow.

current flow The flow of electrons, typically within a circuit or component.

conductors A material that allows electricity to flow through it easily. It is made up of atoms with one to three valance ring electrons.

conventional theory. The theory that electrons flow from positive to negative.

DC See *direct current.*

electrical power A measurement of the rate at which electricity is consumed or created.

electrical resistance A material's property that reduces voltage and amperage in an electrical current.

electrolysis The process of pulling metals apart by using electricity or by creating electricity through the use of chemicals and dissimilar metals.

electromagnet A conductor wound in a coil that produces a magnetic field when current flows through it.

electromagnetic induction The production of an electrical current in a conductor when it moves through a magnetic field or a magnetic field moves past it.

electromotive force An electrical pressure or voltage.

electron theory The theory that electrons, being negatively charged, repel other electrons and are attracted to positively charged objects; thus electrons flow from negative to positive.

energy The ability to do work.

free electron An electron located on the outer ring, called the valence ring, that is only loosely held by the nucleus and that is free to move from one atom to another when an electrical potential (pressure) is applied.

ground The return path for electrical current in a vehicle chassis, other metal of the vehicle, or dedicated wire.

hole theory The theory that as electrons flow from negative to positive, holes flow from positive to negative.

hot junction The heating point of a thermocouple.

insulator A material that holds electrons tightly and prevents electron movement.

ion An atom that has fewer electrons than protons (positive) or that has more electrons than protons (negative).

Kirchhoff's current law An electrical law stating that the sum of the current flowing into a junction is the same as the current flowing out of the junction.

LED See *light-emitting diode.*

light-emitting diode (LED) Diodes that produces light when current flows across the P–N junction.

magnetism The force that attracts or repels magnetic charges; the property of a material to respond to a magnetic field.

multimeter A test instrument used to measure volts, ohms, and amps. A digital multimeter may also be called a digital volt-ohmmeter (DVOM).

ohm The unit for measuring electrical resistance.

Ohm's law A law that defines the relationship among current, resistance, and voltage.

piezoelectric energy A type of electricity in which a material such as a quartz crystal produces voltage when mechanical pressure sistorts it.

photovoltaic (PV) effect The conversion of sunlight into electricity.

polarity The state of charge, positive or negative.

power The rate at which work is done; electrical power is measured in watts.

relay An electromechanical switching device whereby the magnetism from a coil winding acts on a lever that switches a set of contacts.

resistor A component designed to have a fixed resistance.

semiconductor A device, usually made from silicon, that has been doped with boron and phosphorus to create two or more distinct layers. The layers can be joined to create components like diodes and transistors. A semiconductor has four valence electrons, and the doping creates a layer that has more positive charge carriers and a layer with more negative charge carriers. Current and voltage are used to manipulate the charge to provide an insulating or conduction function in a semiconductor.

series-parallel circuit A circuit that has both a series and a parallel circuit combined into one circuit.

sine wave The shape of an AC waveform as it changes from positive to negative, graphed as a function of time.

thermocouple A temperature-sensing component that consists of two dissimilar metals that produce voltage proportional to temperature.

volt The unit used to measure potential difference or electrical pressure.

voltage The electrical pressure that causes current to flow in a circuit.

voltage drop The amount of potential difference between two points in a circuit.

watt The unit for measuring electrical power.

work The result of force creating movement; or the transformation of energy from one type to another.

Review Questions

1. Understanding the behavior of electricity can be more difficult than understanding other concepts, primarily because it cannot be _______.
   a. Controlled
   b. Seen
   c. Stopped
   d. Felt
2. Materials that are composed of atoms that easily give up and accept electrons are known as ________.
   a. Conductors
   b. Insulators
   c. Resistive
   d. Ions
3. What happens to the resistance of a conductor as its length increases?
   a. It increases
   b. It decreases
   c. It stays the same
   d. It depends on the material
4. What is the term for electrical flow?
   a. Voltage
   b. Amperage
   c. Resistance
   d. Conductance
5. What best describes electromagnetic induction?
   a. Electricity that is created inside of special cells using sunlight
   b. Electricity that is created when a crystal is distorted or under mechanical stress
   c. Electricity that is formed by two dissimilar metals submerged in electrolyte
   d. Electricity that is created when a wire passes through a magnetic field
6. A circuit has 12 volts, and 4 ohms of resistance. What is the amperage according to the ohms law calculation?
   a. 12 amps
   b. 8 amps
   c. 3 amps
   d. 48 amp
7. If a load is using 10 amps at 12 volts, how many watts of power is this according to the watts law calculation?
   a. 22 watts
   b. 100 watts
   c. 120 watts
   d. 180 watts
8. If a series circuit has 2 loads, each being 8 ohms, what is the combined resistance of the loads?
   a. 4 ohms
   b. 8 ohms
   c. 16 ohms
   d. 64 ohms
9. If a circuit has 2 loads in parallel, each being 4 ohms, what is the resistance of the loads in the circuit?
   a. .5 ohms
   b. 2 ohms
   c. 8 ohms
   d. 16 ohms
10. If a lightbulb is too dim, what is the most likely cause?
   a. Circuit voltage too high
   b. Circuit resistance too low
   c. Circuit amperage too high
   d. Circuit resistance too high

ASE Technician A/Technician B Style Questions

1. Electrical fundamentals are being discussed. Technician A states that amperage is a measure of electrical pressure. Technician B states that resistance is measured in ohms. Who is correct?
   a. Technician A only
   b. Technician B only
   c. Both Technicians A and B
   d. Neither Technician A nor B
2. The laws of electricity are being discussed. Technician A states that as a conductor heats up, its resistance increases. Technician B states that resistors are designed to extract energy from the current flow as it passes through the resistor. Who is correct?
   a. Technician A only
   b. Technician B only
   c. Both Technicians A and B
   d. Neither Technician A nor B
3. Measuring electricity and electrical components are being discussed. Technician A states that a volt meter is used to measure current flow. Technician B states that an ohm meter is used to measure resistance. Who is correct?
   a. Technician A only
   b. Technician B only
   c. Both Technicians A and B
   d. Neither Technician A nor B
4. Electricity being applied to modern vehicles is being discussed. Technician A states that some vehicles use photovoltaic technology to sense daylight. Technician B states that electromagnetic induction is rarely used on modern vehicles. Who is correct?
   a. Technician A only
   b. Technician B only
   c. Both Technicians A and B
   d. Neither Technician A nor B

5. LEDs are being discussed. Technician A states that LEDs create much more heat compared to traditional bulbs. Technician B states that LEDs are replacing traditional bulbs in more and more new vehicles. Who is correct?
   a. Technician A only
   b. Technician B only
   c. Both Technicians A and B
   d. Neither Technician A nor B
6. An electrical circuit and Ohms law are being discussed. Technician A states that if voltage increases, and resistance stays the same, amperage will decrease. Technician B states that if resistance decreases and voltage stays the same, amperage will decrease. Who is correct?
   a. Technician A only
   b. Technician B only
   c. Both Technicians A and B
   d. Neither Technician A nor B
7. An electrical circuit and Watts law are being disused. Technician A states that if amperage increases and voltage stays the same, the power will increase. Technician B states that if voltage goes up, and amperage goes up, the power will increase. Who is correct?
   a. Technician A only
   b. Technician B only
   c. Both Technicians A and B
   d. Neither Technician A nor B
8. An electrical circuit is being discussed. Technician A states that if one load is added in series, the total resistance decreases. Technician B states that the voltage applied to a series circuit will be shared between each of the loads. Who is correct?
   a. Technician A only
   b. Technician B only
   c. Both Technicians A and B
   d. Neither Technician A nor B
9. Electrical current is being discussed. Technician A states that AC is normally stored in a battery. Technician B states that hertz is the number of times per second an AC waveform cycles between positive and negative. Who is correct?
   a. Technician A only
   b. Technician B only
   c. Both Technicians A and B
   d. Neither Technician A nor B
10. An electrical circuit is being discussed. Technician A states that if resistance is added, the current will decrease. Technician B states that if voltage is decreased, amperage would increase. Who is correct?
   a. Technician A only
   b. Technician B only
   c. Both Technicians A and B
   d. Neither Technician A nor B

CHAPTER 41

Electrical Components and Repair

Learning Objectives

- **LO 41-01** Describe Electrical Switches.
- **LO 41-02** Describe circuit protection devices.
- **LO 41-03** Describe the operation of relays and solenoids.
- **LO 41-04** Describe the basic operation of motors and transformers.
- **LO 41-05** Describe the common types of resistors.
- **LO 41-06** Describe wire.
- **LO 41-07** Describe wire harnesses.
- **LO 41-08** Use wiring diagrams to trace circuits.
- **LO 41-09** Replace wire terminals.
- **LO 41-10** Perform solder repairs.

ASE Education Foundation Tasks

See Appendix A to view the 2017 ASE Education Foundation Automobile Accreditation Task List Correlation Guide.

▶ Electrical Component Preliminaries

LO 41-01 Describe Electrical Switches.

Electrical components are used to modify or manage the flow of current in a circuit. Some examples include switches, fuses, circuit breakers, resistors, capacitors, and relays. Each component performs a specific task within the circuit. Electrical components also have terminals for connecting into electrical circuits. The terminals are typically snapped into plastic connectors. For testing purposes, connectors are often numbered or marked to identify the terminals (**FIGURE 41-1**). Wiring diagrams or schematics are labeled with these markings.

Some components are polarity-sensitive. This means they must be connected into the circuit with the correct polarity. For example, some capacitors, and most semiconductor components, are polarity-sensitive. In polarity-sensitive components, the polarity is marked on at least one connection (**FIGURE 41-2**). This helps to ensure that they are connected into the circuit correctly.

Electrical components are connected together by wires in wire harnesses. They carry electrical signals in virtually every system on the vehicle. If you know how these components and circuits work, it will help you diagnose and repair them.

Electrical Components

Switches

A **switch** is an electrical device used to turn the current on and off in a circuit. When turned off, switches open the circuit, stopping current flow. When turned on, they close the circuit, allowing current to flow. For example, switch off—light goes out. Switch on—light comes on.

There are many different types and configurations of switches. They include toggle switches, push-button switches, and specialty switches. Two examples of specialty switches are for turn signals and windshield wipers (**FIGURE 41-3**). The most basic switches are two-terminal toggle switches. They simply turn on or off, such as a transmission overdrive switch. Momentary switches are spring loaded to one position (on or off) and can be pushed to the other position. A horn switch would be a good example of a momentary switch.

More complex switches have many terminals and contacts inside them. They are able to switch a number of circuits at the same time. The **turn signal switch** is an example of a more complex switch. It has three positions: center for off, moved down for the left turn signals, moved up for the right turn signals.

Circuit or schematic diagrams are often called wiring diagrams. They typically show switches, their contacts, and the surrounding circuits. This is so that technicians can identify

FIGURE 41-1 Connectors are commonly marked with numbers so that you can identify the terminal.

FIGURE 41-2 Polarity-sensitive devices usually have the polarity marked for at least one of the terminals.

You Are the Automotive Technician

You are preparing to be a guest speaker at the local automotive technician training school. You specialize in electrical diagnosis and repair. Because of this, the instructor would like you to explain some of the things the students need to know to do your job. You need to prepare for the following questions.

1. What does a relay do, and how does it work?
2. What does an ignition coil do, and how does it work?
3. How do you go about tracing wires on a wiring diagram to help you understand how the circuit operates?
4. How do you perform a solder repair on a broken wire in a wire harness?

how they operate in a circuit (**FIGURE 41-4**). The terminals on the switch are often numbered or lettered so they can be identified. This helps when connecting them into their circuits. Wiring diagrams also show the terminal identification of the mating connectors.

▶ Circuit Protection Devices

LO 41-02 Describe circuit protection devices.

Fuses and **circuit breakers** are designed to protect electrical circuits. They do so by opening the circuit if the current flow is excessive. The most common kinds of circuit protection devices are:

- fuses,
- fusible links,
- circuit breakers,
- and **positive temperature coefficient** (**PTC**) thermistor protection devices.

Fuses and circuit breakers are rated in amps, and their ratings are usually marked on them. Fusible links are typically rated by their wire size.

Usually, a fuse contains a precisely shaped metal strip. It is designed to overheat and melt when subjected to a specified excessive level of current flow. This permanently breaks the circuit and stops the excessive current flow. Potential damage to the wiring harness and more valuable components is likely avoided.

Fuses come in a variety of configurations, from cylindrical glass cartridge fuses to plastic blade fuses (**FIGURE 41-5**). Blade fuses are the most commonly used on recent vehicles. They also come in a variety of amp ratings (**TABLE 41-1**). Blade fuses also come in a variety of physical sizes including micro, mini, regular, and maxi. Fuses are typically housed in fuse boxes. Fuse boxes are located around the vehicle, typically under the hood and/or dash (**FIGURE 41-6**). But they can also be found in the trunk of some vehicles. Fuse boxes typically include relays and sometimes diodes as well.

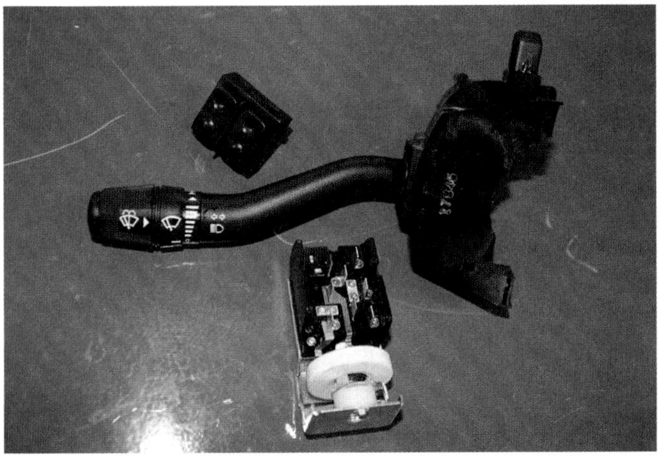

FIGURE 41-3 Typical automotive switches.

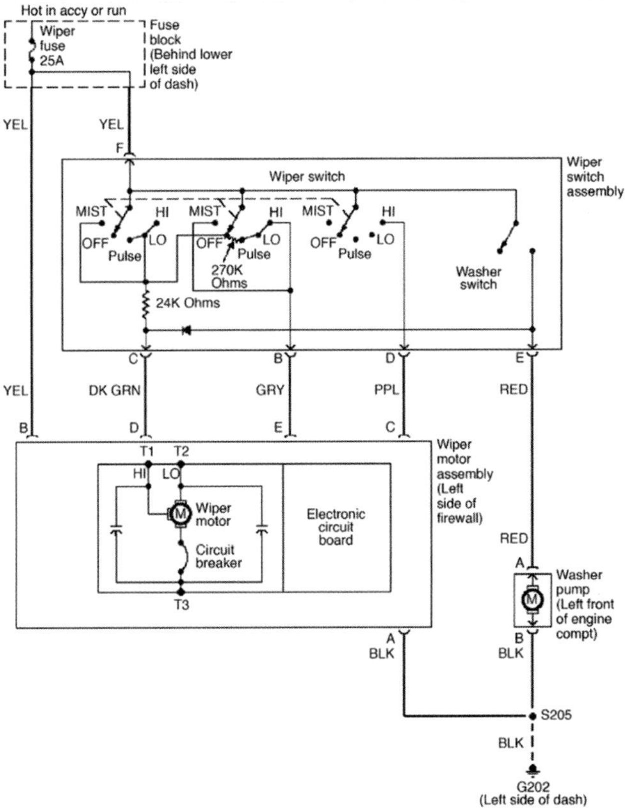

FIGURE 41-4 Typical wiring diagram showing how switches are wired into the circuit.

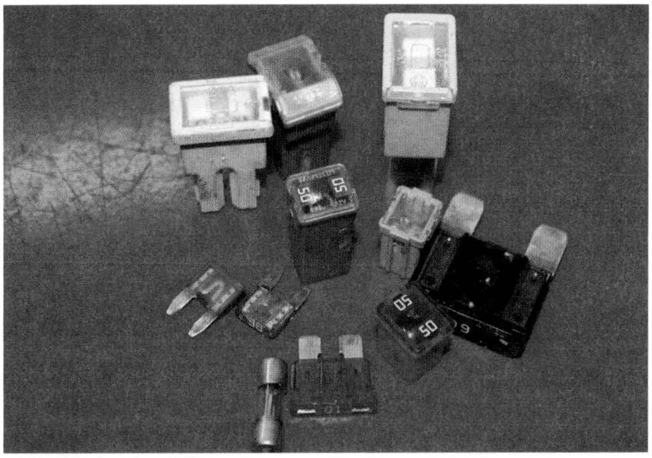

FIGURE 41-5 Various fuses.

TABLE 41-1 Blade Fuse Color Codes for Common Fuse Sizes

Color	Amp Rating
Tan	5 A
Brown	7.5 A
Red	10 A
Blue	15 A
Yellow	20 A
Clear	25 A
Green	30 A

A fusible link is made of a short length (usually 6" [15 cm] or less) of smaller diameter wire. The wire has a lower melting point than standard wire, and insulation that is fire-resistant. Fusible links are typically placed near the battery. This helps protect the wiring harness between the battery and any fuse boxes. In most cases, they are used to carry higher current flows than fuses. They also typically feed power to more than one circuit (**FIGURE 41-7**).

Fusible links are fairly durable and do not fail very often. When they do fail, it is usually due to a substantial short circuit in the system. Another common cause is when the fusible link wire is abused by excessive flexing or pulling on it. Some newer vehicles use maxi-fuses instead of fusible links. These are large blade-type fuses (**FIGURE 41-8**).

Circuit breakers are different from fuses and fusible links in two ways. First, they are not destroyed by excess current. And second, they can be reset, either automatically or manually. In a circuit breaker, a bimetallic strip heats up and bends. This opens a set of contacts and breaks the circuit when current flow becomes excessive (**FIGURE 41-9**). In most types, as the strip cools, it returns to its original shape. The contacts then close,

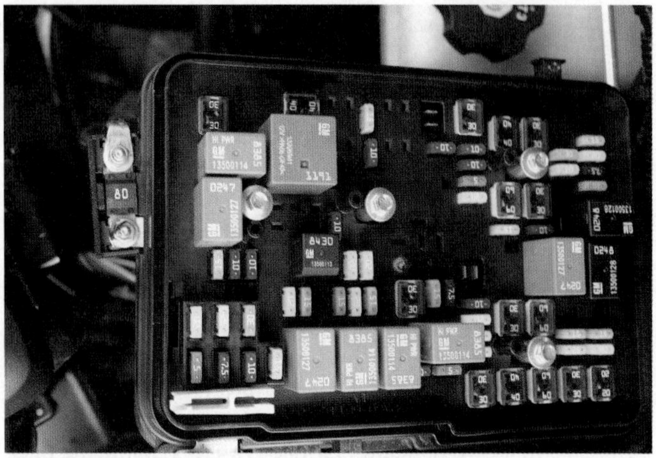

FIGURE 41-6 Typical underhood fuse box.

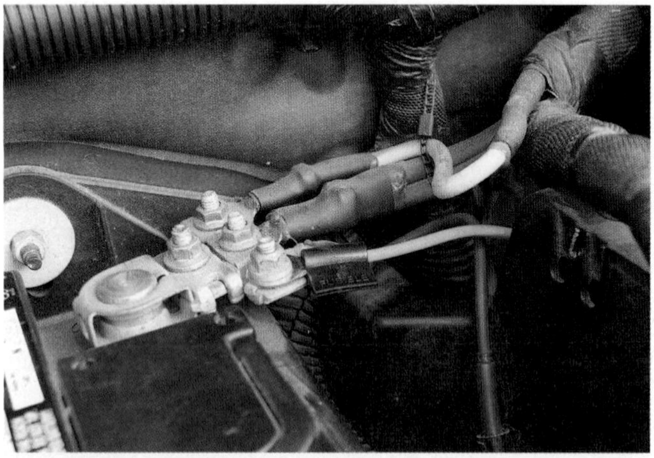

FIGURE 41-7 Fusible links are typically placed near the battery and carry the current needed to power an individual circuit or a range of circuits.

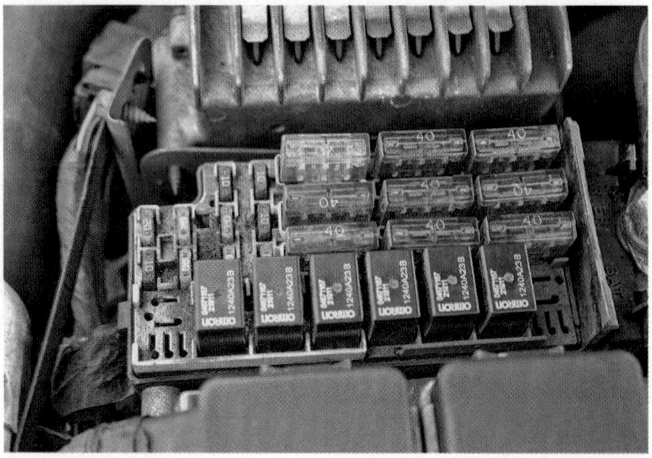

FIGURE 41-8 Maxi-fuses are used in place of fusible links.

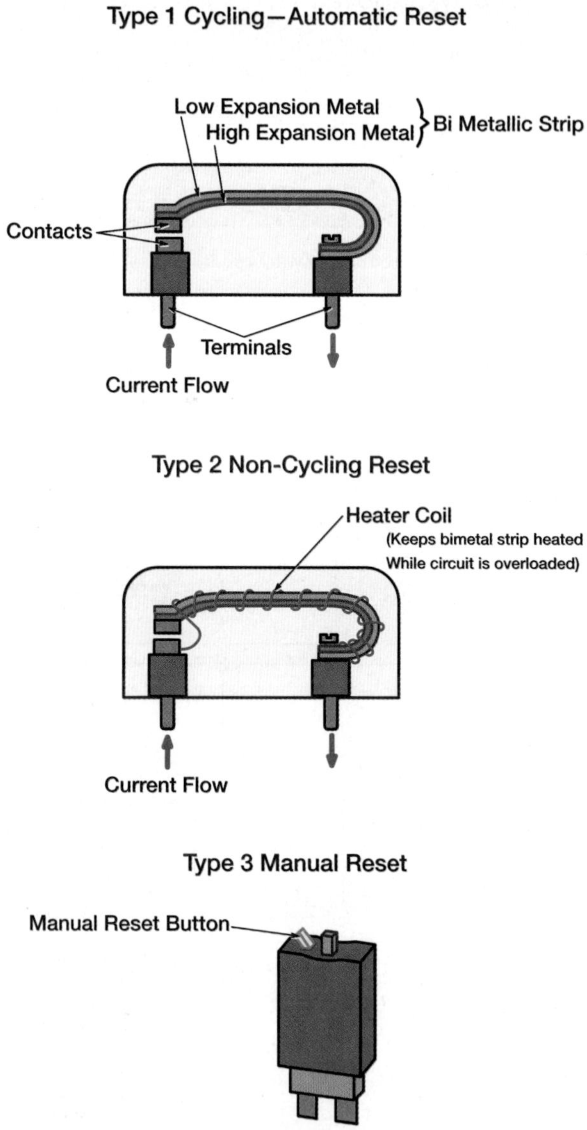

FIGURE 41-9 A circuit breaker opens the circuit if the current goes too high. It can usually be reset either automatically or manually.

completing the circuit once more. These are called self-resetting circuit breakers. Manual breakers must be reset by hand. This usually involves flipping a lever or inserting a small rod to reset the bimetal spring once it cools down.

PTC thermistors are also used as circuit protection devices. They have very low resistance at room temperature. But as the temperature increases, so does their resistance. If too much current starts to flow through a PTC, the small voltage drop creates heat in the PTC. The increased heat produces increased resistance. This further increases the voltage drop. This cycle continues quickly until the PTC reaches its maximum resistance. This effectively shuts off most of the current flow to the protected device. PTCs generally reset once power is removed and they are allowed to cool. They are typically integrated into components such as power window motors and door locks.

Applied Science

AS-69: Fuse: The technician can explain the role of a fuse or fusible link as a protective device in an electrical or electronic circuit.

A fuse, or fusible link, is a form of overcurrent protection device. It consists of a conductive metal strip that melts when excessive current flows through it. When the fuse "blows," the circuit is broken, and no current flows in the remainder of the circuit. Fuses are installed for two reasons. First, they can prevent excessive current flow from damaging more expensive components in the circuit. And second, they can prevent overheating within wiring and components, which could potentially cause a fire.

Flasher Can/Control

The **flasher can** is the control mechanism for the turn signal lights on the vehicle. As the name suggests, it flashes or turns the turn signal lights on and off. It does so at a regular rate to indicate the driver's intent to change the vehicle's direction. While flasher cans are mechanical devices, flasher controls are generally electronic devices (**FIGURE 41-10**). Both types perform the same job. But the electronic version is more reliable and consistent.

Flasher cans operate like an automatically resetting circuit breaker. This means that they use a bimetallic strip to open and close the switch contacts. This opening and closing gives them a distinctive clicking sound. It is used along with the turn signal indicator lights on the dash to tell the driver when the turn signals are on. This is helpful in remembering to turn them off when needed.

Electronic flasher controls control the on/off function electronically. This means they can be designed to operate over a wider range of current flow. This makes them ideal for trailer towing. In many cases, the hazard lamps are also operated by a flasher can or flasher control. In newer vehicles, the flasher operation is activated by the body control module. It sends signals over the communication network to the taillight module. The module actually turns the turn signal on and off.

▶ Relays

LO 41-03 Describe the operation of relays and solenoids.

Relays are switches that are turned on and off by a small electrical current. They are ideal for using a small current to control a larger current. An example is the horn circuit. A small current flow in the steering column can turn on a larger current flow to the horn. The small current activates the relay, which sends the larger current to the horn. Thus, larger wires are needed only up front, where they can be shorter. Smaller wires can be used to run up the steering column.

Many electrical components found in a motor vehicle are controlled by relays. Electronic control units (ECUs) use relays to control components such as fuel pumps, headlights, and the cooling fan. All of these circuits carry large electrical loads.

The relay is made up of an electromagnet, a set of switch contacts, terminals, and the case (**FIGURE 41-11**). The electromagnet is a winding of fine metal-insulated wire wrapped around an iron core. Each end of the winding is connected to one of the relay terminals. There are usually three contacts—two fixed and one movable. The movable contact is fixed to a spring-loaded armature blade located between the fixed contacts. It is held against one of the fixed contacts, which is called the **normally closed** (NC) contact. When the relay coil is activated, the electromagnet pulls the movable armature blade and contact away from the NC contact. It is pulled against the **normally open** (NO) contact. This sends power out to the controlled device.

A **solid-state relay** acts like a mechanical relay. But it does not have any moving parts (**FIGURE 41-12**). This means that

FIGURE 41-10 Flasher control and flasher can.

FIGURE 41-11 Inside a typical relay.

FIGURE 41-12 Inside a typical electronic relay.

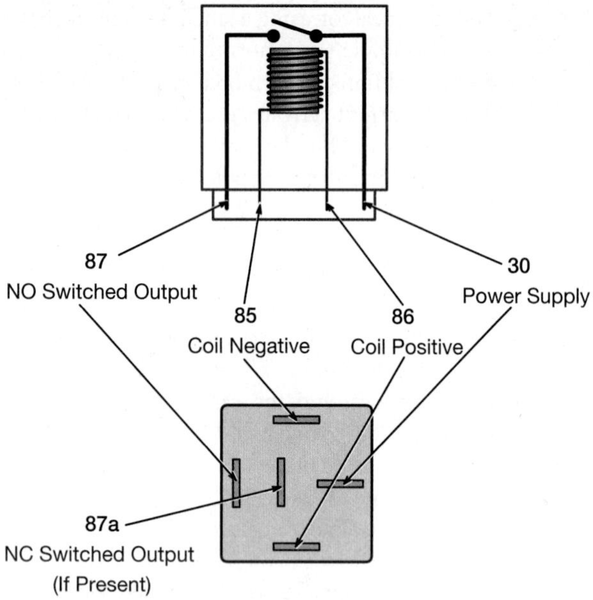

FIGURE 41-13 Relay schematic.

electronic relays do not use mechanical switches. They use transistorized circuitry to turn the circuit on and off. They also do not make any sound, unless they were specifically designed with components to do that.

Automotive relay terminals use one of two standard labeling systems: 85, 86, 30, 87a, and 87 or 1, 2, 3, 4, 5 (**FIGURE 41-13**).

- 85 or 1 = One end of the relay winding
- 86 or 2 = The other end of the relay winding
- 30 or 3 = Common-movable switch contact
- 87a or 4 = NC fixed contact to the common
- 87 or 5 = NO fixed contact to the common

When electromagnetic relays are de-energized, the collapse of the magnetic field induces a large voltage spike in the relay coil windings. If unchecked, this voltage can be transmitted back into the circuit. This can damage electronic components. Some relays deal with this danger by placing either a suppression diode or a resistor in parallel with the winding (**FIGURE 41-14**). Doing so reduces the voltage spike. The voltage is shunted from the output side of the coil back to the input side of the coil. In

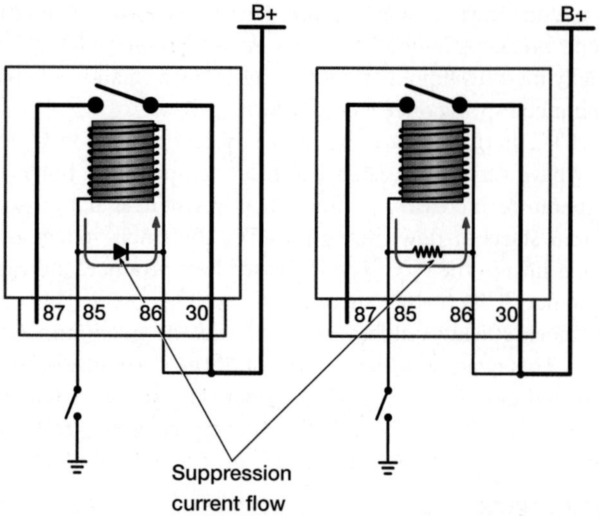

FIGURE 41-14 Spike-protected relays.

doing so, it is dissipated in the loop. This design prevents the voltage from spiking so high when the relay is de-energized. When replacing a relay that has spike protection, make sure to use a new relay that is specified for the application.

Solenoids

A **solenoid** is an electromechanical device. It converts electrical energy into mechanical linear (back-and-forth) movement. Solenoids can be used to pull or push. In a simple solenoid, insulated wire is wound many times around a hollow tube. A sliding mild steel core is made to fit inside the tube. When the winding is energized, the resulting magnetic field attracts the mild steel. It is drawn into the tube and produces linear motion. Solenoids can be very strong and produce a lot of mechanical force. Solenoid design may incorporate return springs, multiple windings, electrical contacts, and mechanical connections.

Fuel injectors and starter motor solenoids are two of the many solenoid-type components used in a motor vehicle (**FIGURE 41-15**). The operation of a solenoid is similar to a relay. But where a relay uses a magnetic field to close an electrical circuit, a solenoid uses a magnetic field to create lateral movement. In the case of a starter solenoid, it may also close heavy electrical contacts.

The movable metal core used in an electromagnet is referred to as an **armature**. It is spring loaded so that it is positioned partially outside the electromagnetic coil. It is free to move in and out. When the coil is energized, the magnetic field draws the armature into the center of the coil. If the armature is attached to a lever or plunger, it will move as well. Stopping current flow causes the electromagnet to de-energize. The spring then pushes the armature back out.

Another device that uses a solenoid is a vehicle horn. When the armature is drawn in by the electromagnetic coil, it opens a set of electrical contacts so that current flow through the coil is stopped. This stoppage causes the armature to move out again, closing the contacts, drawing the armature back in. This process occurs at very high speeds. The vibration caused by the rapid movement is transferred to a diaphragm, and the familiar horn sound is produced (**FIGURE 41-16**). Refer to the Starting and Charging Systems chapter for a detailed description of the starter solenoid operation.

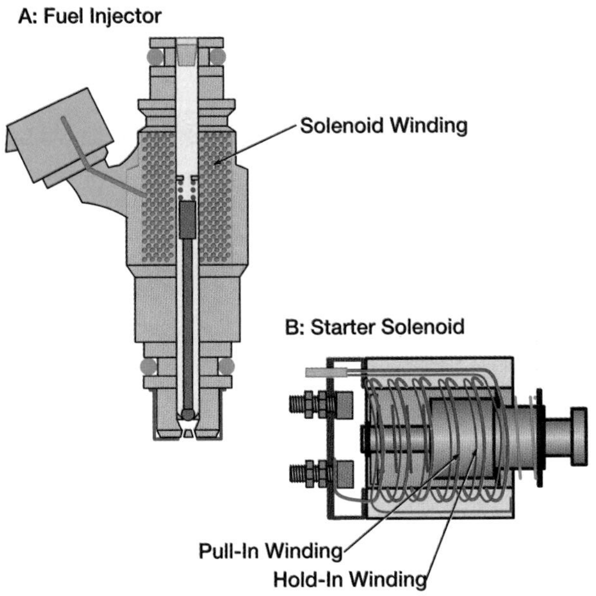

FIGURE 41-15 Examples of solenoids: **A.** Fuel injector. **B.** Starter motor solenoid.

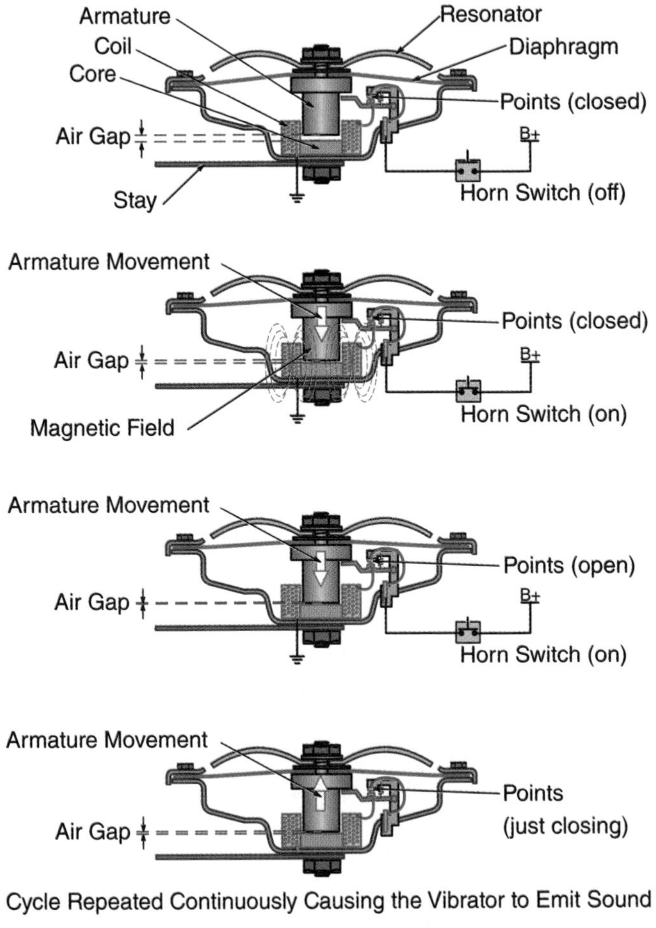

FIGURE 41-16 A horn is a type of solenoid designed to create noise as its output.

▶ Motors

LO 41-04 Describe the basic operation of motors and transformers.

Although solenoids use magnetic fields to create lateral movement, electric motors use magnetic fields to create rotary movement. Motors consist of two main components: the field and the armature. The field contains either electromagnets or permanent magnets. In either type of field magnet, the magnetic field is constant. The armature contains electromagnetic coils. Electromagnets can be turned on and off. This means that the magnetic field can be turned on and off in the armature. The interaction between the two magnetic fields causes the armature to rotate.

When turned on, a magnetic field is created in one of the armature coils. That magnetic field is attracted to the magnetic field in one field (**FIGURE 41-17**). It is also repelled by the magnetic field in an adjacent field. This causes the armature to rotate a small amount. Once the armature rotates far enough, the current is switched from one armature coil to the next. This creates the next magnetic field that is attracted and repelled by the field.

The armature has a number of individual windings. Each winding has two ends. Each of the ends is connected to a copper bar. The bars make up a segmented component of the armature called a **commutator**. A set of carbon brushes transfer electricity to the bars on the **commutator**. The commutator and brushes act as switches to control the current flow through the windings of the armature (**FIGURE 41-18**). The brushes allow the

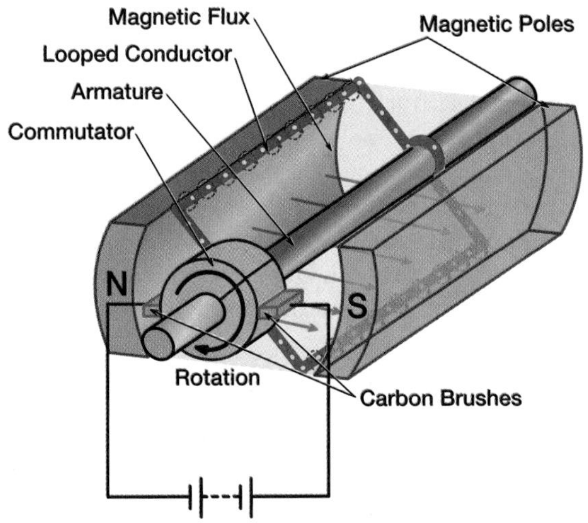

FIGURE 41-17 Interaction of the magnetic fields causes the motor to turn.

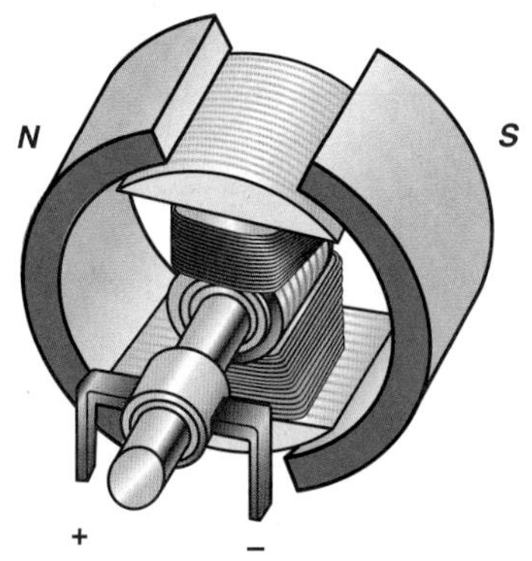

FIGURE 41-18 Simplified electric motor diagram.

electrical connection to occur even when the armature is spinning. Refer to the Starting and Charging Systems chapter for more information on motors.

Ignition Coils and Transformers

Ignition coils and transformers both operate under the principles of electromagnetic induction. They use electromagnetism to produce electricity rather than mechanical movement. An ignition coil can be described as a **step-up transformer**. This is because the output can be raised much higher than the input voltage. For example, an input voltage of 12 volts can be raised to 60,000 volts (60 kV) or more volts in an ignition coil.

Two sets of coil windings are used. One coil winding is referred to as a **primary winding**. It is wound around a second winding called, the **secondary winding** (**FIGURE 41-19**). The primary coil typically has 200 to 300 turns of light-gauge wire. The secondary winding has approximately 30,000 to 60,000 turns of very fine wire.

When current is passed through the primary winding, the magnetic field builds. It surrounds both windings. When the current is turned off, the magnetic field collapses. It collapses with enough speed to induce high voltage in the primary winding (self-induction). It also induces very high voltage in the secondary winding (mutual induction). This is because voltage is induced into each of the thousands of windings of the secondary coil. This voltage is strong enough to overcome the infinite resistance of the spark plug gap. It pushes current across the gap, causing a spark to ignite the air–fuel mixture in the cylinder.

The **transformer action** causes heat to be produced. In the past, the internal coils were immersed in cooling oil, allowing the heat to be conducted to the case. Modern ignition coils do not use oil (**FIGURE 41-20**). They are usually constructed using a heat-conducting hard resin. Cooling happens through their location on a heat sink or in a stream of air. The use of computer controls has allowed the time that current flows through the primary windings to be minimized. This reduces heat and electrical loads while still providing enough spark to ignite the air–fuel mixture.

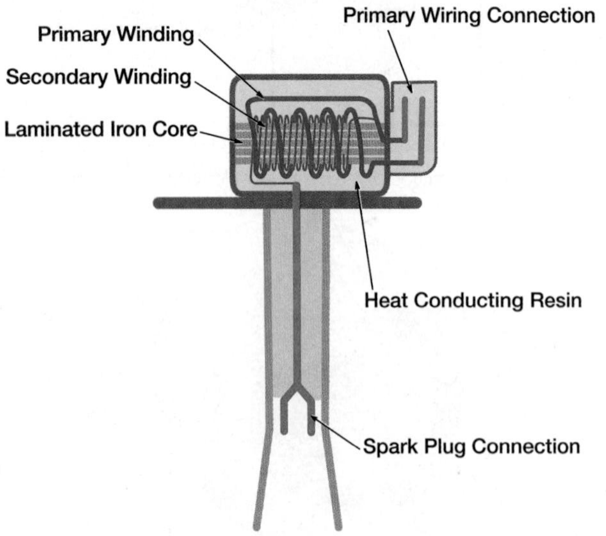

FIGURE 41-19 An ignition coil uses induction to step up the voltage from one coil to much higher voltage in the other coil.

Step-down transformers operate under the same operating principles. The only difference is that the secondary coil has fewer turns than the primary. This provides a lower induced output. These transformers are used on power poles to lower the voltage to your house or school. They are also used on low-voltage devices in your home plugged into 110-volt outlets. Good examples are cell phone chargers.

▶ TECHNICIAN TIP

You might expect a transformer to be a great way to boost the amount of electrical power that is transformed, but it isn't. The amount of power after the transformation is relatively the same as before the transformation. For example, if we raise the voltage from 12 volts to 120 volts, the amperage will decrease from, say, 10 amps to 1 amp. Thus, the wattage stays the same:

12 volts × 10 amps = 120 watts
120 volts × 1 amp = 120 watts

Also, remember that transformers are not 100% efficient, so some of the power is lost as heat.

▶ Resistors

LO 41-05 Describe the common types of resistors.

Resistors are electrical components that resist current running through them. Putting a resistor in a circuit reduces the amperage in the circuit. It also causes a voltage drop across the resistor. Resistors are commonly used to control the voltage and amperage

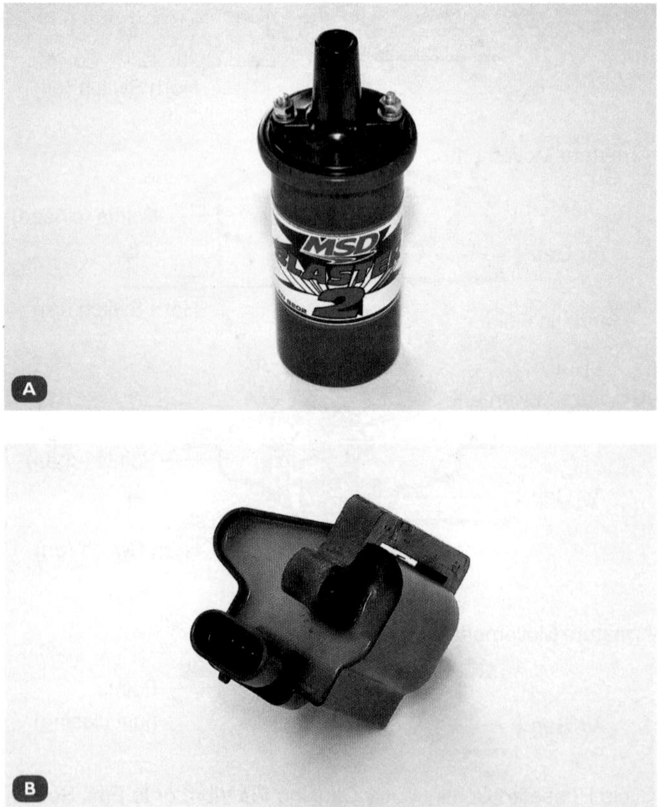

FIGURE 41-20 A. Oil-cooled ignition coil. **B.** Heat-conducting resin coil.

going to various components. Resistors take many forms. Most high-wattage resistors contain a coil of high-resistance wire wound around a ceramic form. This helps dissipate heat.

Fixed Resistors

Fixed resistors are generally cylindrical in shape. They have metal leads projecting from each end, which is why they are called axial resistors. A series of colored stripes are printed on these types of resistors. They indicate the resistance and tolerance levels of the resistor (**FIGURE 41-21**). Fixed resistors can be manufactured as very tiny devices without leads. In this case, they can be placed into integrated circuits along with other miniaturized components.

Resistor Ratings

Resistance is measured in ohms, represented by the Greek letter omega (Ω). Resistors are rated in ohms as well. The amount of resistance indicates how strongly the resistor will oppose any current flowing through them. Because resistors work by converting some of the electrical energy passing through them into heat, they also have a power rating. Only the resistance value is marked. The resistor's power rating is determined by its size.

Regardless of their power rating, resistors are small, so identification by numbers is impractical. To identify their value, many resistors are typically marked with four or five colored bands. Each color represents a numeric value. The color bands are set close to each other and read from left to right. The last band, or tolerance band, is spaced farther apart and shows the resistance is within a certain percentage of that is listed (**FIGURE 41-22**).

Variable Resistors

Resistors found on circuit boards are normally fixed in value. Some resistors found in the vehicle are variable. The value of **variable resistors** can be changed by movement of a slide or with temperature change. The three types of variable resistors are rheostats, potentiometers, and thermistors. Variable resistors can be linear. This means their resistance value varies proportionally across their range. They can also be nonlinear, meaning the resistance change is not proportional across their range.

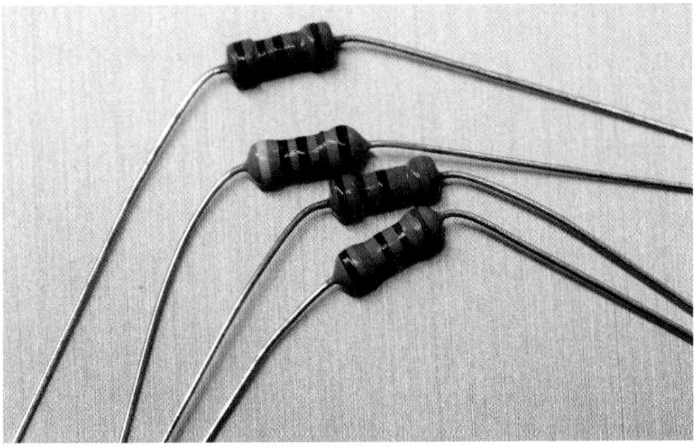

FIGURE 41-21 Typical fixed resistors.

Rheostats

A **rheostat** is a mechanical variable resistor with two connections. It consists of a resistance wire wrapped in a loose coil connected to the supply at one end only. A moveable wiper moves over the wire manually (**FIGURE 41-23**). It sends current out to the controlled device. When the wiper is close to the beginning of the coil, the total resistance value is very small. As the wiper is positioned closer to the end of the coil, the resistance value increases. Rheostats are commonly used in dash light dimmer circuits and some fuel gauge sender units. They work well in altering the current flow and voltage in a circuit.

Potentiometers

Potentiometers are variable resistors with three connections, two fixed and one moveable. They act as voltage dividers and as such alter the voltage in a circuit. A resistance wire is

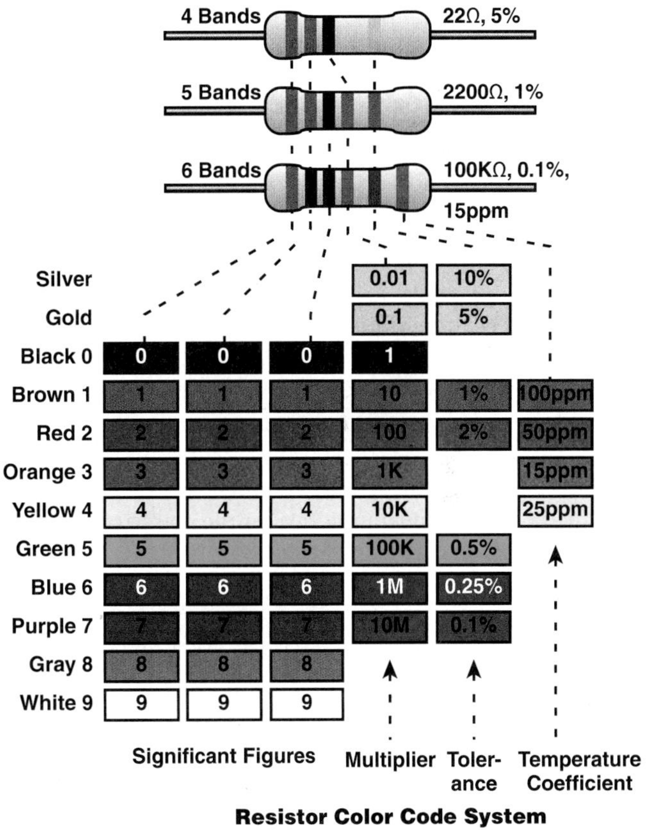

FIGURE 41-22 Resistor color codes.

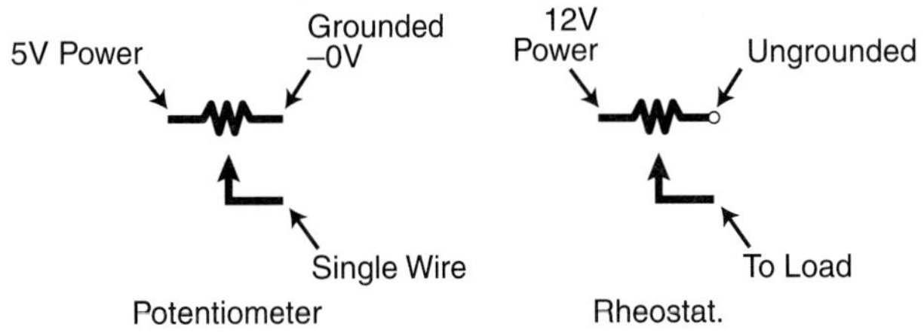

FIGURE 41-23 Variable resistors.

FIGURE 41-24 Potentiometer.

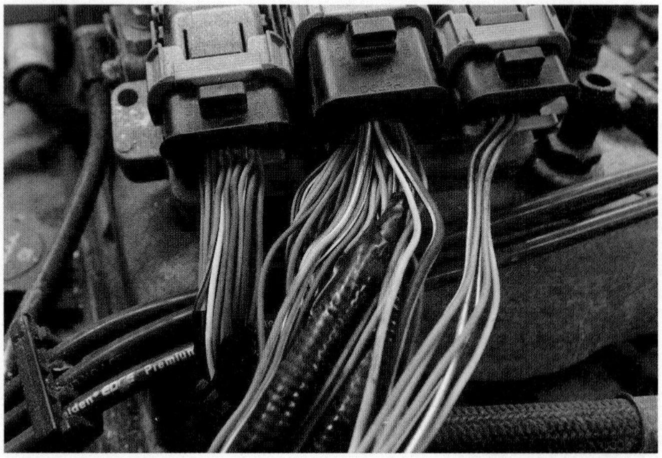

FIGURE 41-26 Wires carry information and commands throughout the vehicle.

wrapped between two fixed connections. One fixed connection is attached to the electrical supply, the other to a ground. The third connection is moved across the coil by a wiper, similar to a rheostat (**FIGURE 41-24**). The variable voltage output is taken from this point. Throttle position sensors are potentiometers.

Thermistors

We have already discussed **thermistors.** Their resistance values change with temperature. There are two types. One is the **negative temperature coefficient** (NTC) thermistor. The other is the **PTC** thermistor (**FIGURE 41-25**). As the temperature increases, the resistance value decreases for NTC thermistors. On PTC thermistors, as the temperature decreases, the resistance increases. NTC thermistors are the most common. They are commonly used in temperature sensors. Two examples are coolant temperature sensors and air temperature sensors. They then provide temperature data for the powertrain control module (PCM).

▶ Wires

LO 41-06 Describe wire.

Wires and wiring harnesses are the arteries of the vehicle's electrical system. As such, they need to be kept in good condition, free of any damage or corrosion. They carry the electrical power and signals through the vehicle. This is how virtually every system on a vehicle is controlled (**FIGURE 41-26**).

As technology in vehicles has increased, so too has the number of wires installed on vehicles. Wires are still the main signal carriers in a vehicle. However, wireless communication is used in some vehicle security, entertainment, and tire pressure monitoring systems. To help protect wires and keep them organized, they are bundled together in a wiring harness. Several wiring harnesses are located throughout the vehicle.

Types of Wires

Electrical wires are used to conduct current around the vehicle. Wire can also be referred to as cable, which typically refers to large-diameter wire. Automotive **wire** is commonly a multi-stranded copper core. It is wrapped with seamless plastic insulation (**FIGURE 41-27**). Copper is typically used as it offers low electrical resistance. It also remains flexible even after years of use. The insulation is designed to protect the wire. It also prevents leakage of the current flow so that it can get to its intended destination.

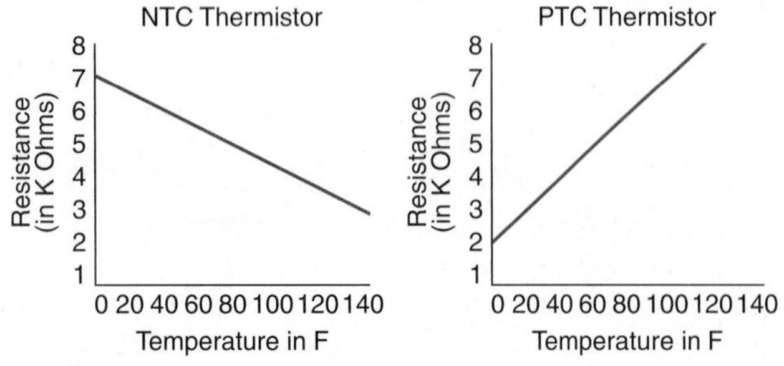

FIGURE 41-25 Thermistors. **A.** NTC thermistor resistance. **B.** PTC thermistor resistance.

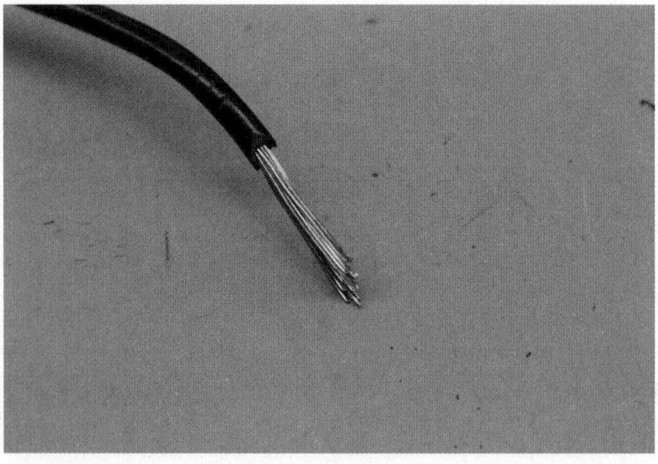

FIGURE 41-27 Multi-strand wire (stripped).

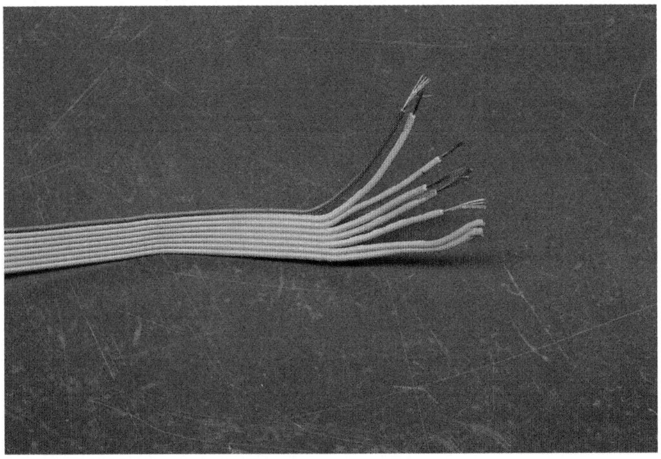

FIGURE 41-28 Ribbon cable (stripped).

Ribbon cable is a series of wires that are formed side by side and joined along the wire insulation. They are flat like a ribbon (**FIGURE 41-28**). Ribbon cable works well when several wires run from one component to another. The ribbon design groups them so they can be routed neatly and easily. Ribbon cable is often found inside computers and other electronic components. It is used for connecting between printed circuits or between printed circuits and other components.

Some wires, especially signal wires and communication wires, are shielded. This helps to prevent electromagnetic interference (EMI), also referred to as "noise." We will discuss this further in a later section.

Wire Sizes

Wire size is very important for the correct operation of electrical circuits. Selecting a wire gauge that is too small for an application causes an excessive voltage drop. In extreme cases, the wire will get hot enough to melt the insulation. This can short wires together causing all kinds of problems. Selecting a wire gauge that is too large increases the cost, weight, and size of wiring harness.

The diameter of a wire affects its resistance and therefore how much current it can carry. Even good conductors have a slight amount of resistance. The resistance of a wire is determined by its:

- length,
- diameter,
- construction material,
- and temperature.

The longer the wire and the smaller the diameter, the higher the resistance. The shorter the wire and the larger the diameter, the lower the resistance.

Two scales are used to measure the sizes of wires. One is the metric wire gauge, which measures the cross-sectional area of the conductor in square millimeters (**TABLE 41-2**). The second is the **American wire gauge (AWG)**. The AWG system uses a rating number. The larger the rating number, the smaller the wire and the lower its current-carrying capability (**FIGURE 41-29**). Most countries use the metric scale. American manufacturers are split, with some using AWG and others using the metric scale.

Vehicles use a variety of wire sizes depending on the requirements of each particular circuit. Manufacturers and standards bodies use wire size charts to specify how much current each wire gauge can safely carry. The correct wire size for an application can be looked up on a wire size chart. You just need to know the amperage of the circuit and the length of the wire. For example, a particular12-volt circuit is designed for a maximum current flow of 10 amps. The wire is approximately 15' (6.1 m) long. Using the AWG table as a reference, the correct wire gauge to use is 12 AWG (**TABLE 41-3**). But be careful of the chart you use, as many of them allow up to a 10% voltage drop over the length of the wire. This is significantly more voltage drop than is allowed in most automotive circuits.

TABLE 41-2 Metric Wire Size and American Wire Gauge (AWG) Comparison

Metric Wire Sizes (Square mm)	AWG Wire Sizes (Gauge)
0.22	24
0.35	22
0.5	20
0.8	18
1.0	16
2.0	14
3.0	12
5.0	10
8.0	8
13.0	6
19.0	4
32.0	2

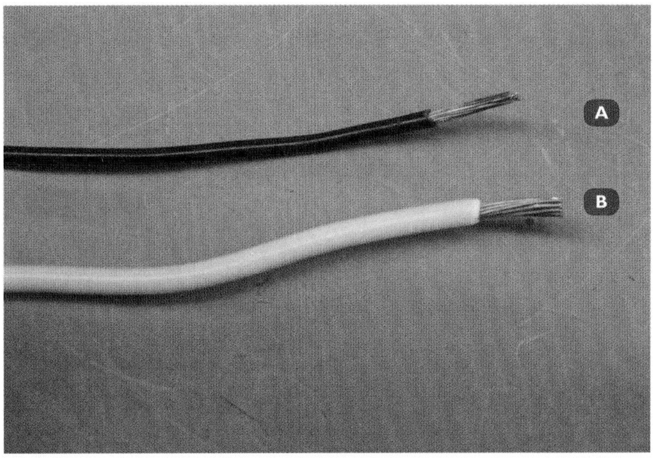

FIGURE 41-29 In the AWG system, the larger the wire number size, the smaller the diameter of the wire. **A.** 16 gauge. **B.** 12 gauge.

TABLE 41-3 AWG Wire Sizes Based on Amperage and Wire Length*

	Wire Length from Battery to Load						
Circuit Amps	2'	5'	7.5'	10'	15'	20'	25'
2	20	20	20	18	18	18	16
5	18	18	18	18	16	14	14
8	18	16	16	14	14	12	12
10	16	16	16	14	12	12	10
12	16	16	14	14	12	12	10
15	16	16	14	12	10	10	8
18	16	14	12	12	10	8	8
20	14	14	12	10	10	8	8
25	14	12	12	10	8	8	6
30	12	12	10	10	8	6	6

* Chart is based on a maximum 0.4-volt drop per wire size. Shorter distances are less than a 0.4-volt drop.

Two different methods describe the conductor size within these standards. A wire may be described in metric size as 5.0. This indicates that it is a single strand that has a cross-sectional area of 5.0 millimeters squared (mm^2). It can also be expressed as 10/0.5. This indicates that there are 10 strands of wire, each with a cross-sectional area of 0.5 mm^2. The same system can be applied to the AWG rating using the gauge size.

▶ Wiring Harnesses

LO 41-07 Describe wire harnesses.

Wiring harnesses are also known as wiring looms or cable harnesses. They are used throughout the vehicle to group two or more wires together. They are typically placed within a sheath of tubing or insulating tape (**FIGURE 41-30**). Often, harnesses on modern vehicles contain many wires. The wires terminate at crimped terminals inserted into connector or harness plugs.

There are usually a number of harnesses within the vehicle. They interconnect with various connector plugs as required to form the wiring system of the vehicle. Wiring harnesses run around the engine bay, through the dash and interior cabin, and to the rear of the vehicle. They are attached to the vehicle with harness fasteners such as body clips or wire ties. Rubber sealing grommets are used when the harness passes through the metal bodywork.

Terminals and Connectors

Terminals are installed on the ends of wires to provide low-resistance termination. They allow electricity to be conducted from the end of one wire to the end of another wire. In many cases, they allow the wires to be disconnected and reconnected. They come in many different types and sizes to suit various wire sizes and termination requirements (**FIGURE 41-31**). For example, there are:

- **push-on spade terminals,**
- **eye ring terminals** to accommodate screws,

FIGURE 41-30 Typical wiring harness.

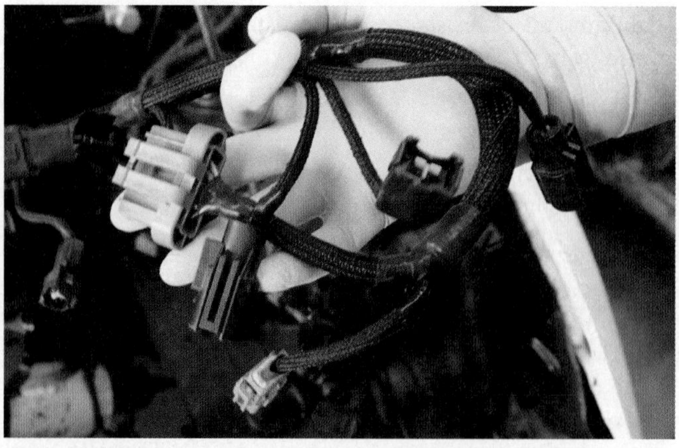

FIGURE 41-31 Terminals and connectors are installed to the ends of wires to provide low-resistance termination of the wires.

- **butt connectors**,
- and **male and female terminals** that are designed to be separated and reconnected.

Most terminals are of the crimp type. These require the use of special tools to crimp the terminal to the end of the wire. Terminals can be insulated or non-insulated.

Some **solder-type terminals** are still in use. They require the use of electric or gas soldering irons, flux, and solder to make the connection. When soldering wiring, always use a rosin or rosin-core solder. Never use an acid-core solder. The acid can cause corrosion and high resistance over time.

Terminals can be installed as a single terminal on a wire. They can also be grouped together in a wiring harness with a connector housing. These are also called **wiring harness connectors. Connector** housings have male and female sides. They are usually shaped so that they can be connected in only one way (**FIGURE 41-32**). They often incorporate a locking mechanism so the connector cannot accidentally work loose. Many of these connectors are weatherproofed to keep moisture out. Special tools are needed to remove the terminals from the connector housing.

Shielding

In certain locations within a vehicle, strong EMI is present. This subjects wiring harnesses to unwanted electromagnetic induction. This interference is referred to as electrical noise or EMI noise. To prevent noise, many vehicles use **shielded wiring harnesses**. The type of shielding used can be one of three forms: twisted pair, Mylar tape, or drain lines.

Twisted Pair

Twisted pair uses two wires delivering signals between common components. The wires are uniformly twisted through the entire length of the harness. The harness ends at a terminating resistor (**FIGURE 41-33**). The twisted wires, along with the terminating resistor, help cancel any noise that occurs in the wires. This reduces the loss of data in the transmitted signals.

The controlled area network bus, or CANbus, in a modern vehicle may use one or more twisted pairs to connect all the vehicle control units. This creates a common data line(s) to share information over. Refer to the On-Board Diagnostics chapter for more information on the CANbus system.

Mylar Tape

Mylar tape is an electrically conductive material. It is wrapped around a wiring harness. This can be either inside or outside the outer harness layer (**FIGURE 41-34**). Any noise that attempts to reach the wires inside the shield is absorbed by the Mylar. The Mylar conducts the noise to ground via a ground connection. The shielding is important to prevent electrical noise penetrating into the electrical wiring. If the harness is exposed, the Mylar will have to be rewrapped so that noise cannot penetrate into the harness.

Drain Lines

A **drain line** is a non-insulated wire that is wrapped within a wiring harness (**FIGURE 41-35**). The drain wire is connected to ground at the harness source end. It conducts any noise to ground, negating the noise effect. If the drain wire is cut, it will be inoperative. So, it is important that the wire not be cut or left disconnected.

Applied Science

AS-78: Magnetic Fields/Forces: The technician can explain the effect of magnetic fields on unshielded circuits in control modules.

Shielding is applied to electrical circuits to prevent unwanted electromagnetic induction. Errant voltages may be generated in unshielded circuits due to EMI. Some control modules work with high sensitivity and very low voltages. This is particularly the case with sensor inputs. In these cases, even a small induced voltage, known as "noise," can have serious effects on the running of the vehicle. It can also affect the functionality of safety systems. The control unit may effectively respond to the noise as if it is actually a signal from a sensor.

FIGURE 41-32 Male and female wire connector.

FIGURE 41-33 A twisted pair used to carry signals between components, and also act together to cancel any electrical noise.

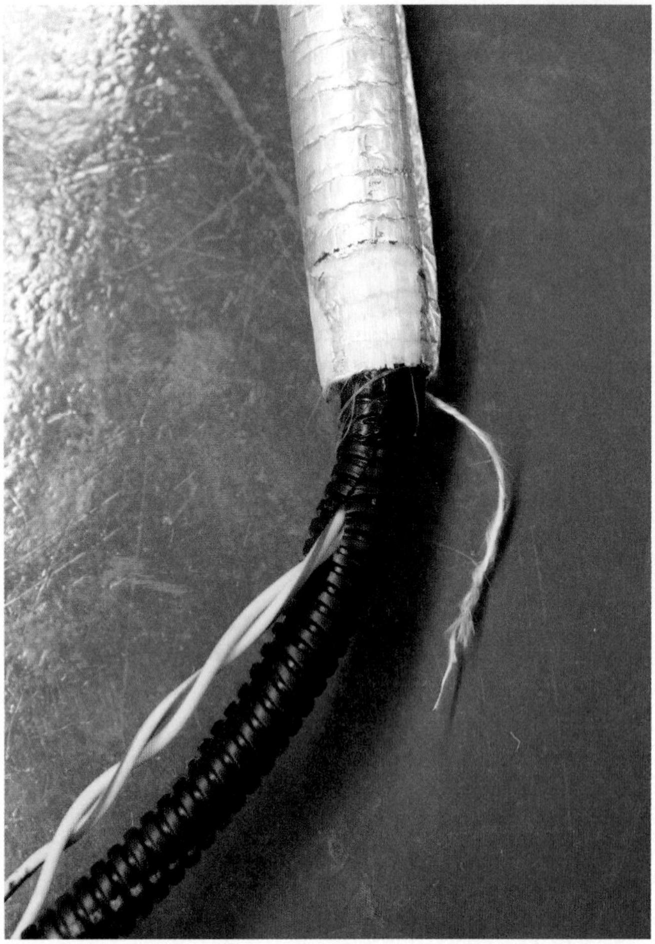

FIGURE 41-34 Mylar tape is used around wiring harnesses to shield them from electrical noise.

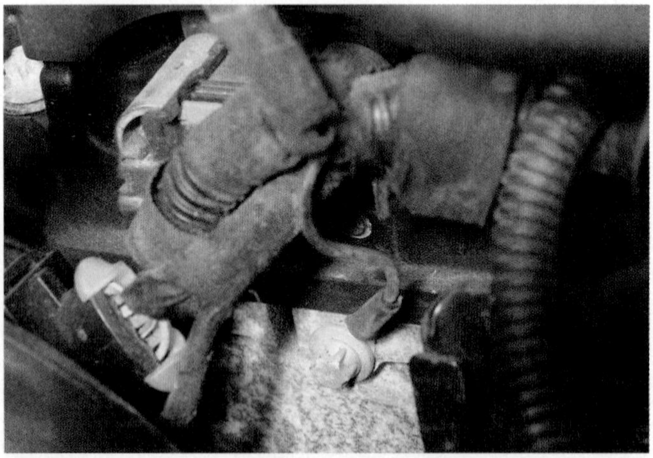

FIGURE 41-35 A drain line inside a wire harness, used to drain electrical noise to ground.

▶ Wiring Diagram Fundamentals

LO 41-08 Use wiring diagrams to trace circuits.

Wiring diagrams are also known as electrical schematics or electrical diagrams. They use symbols to represent electrical components. They are like a map of all of the electrical components and their connections (**FIGURE 41-36**).

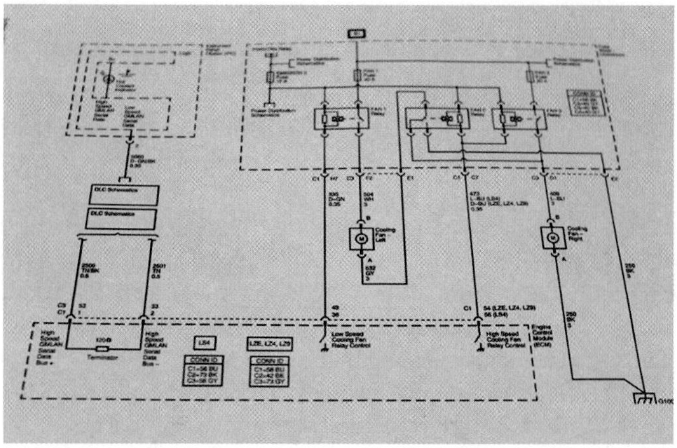

FIGURE 41-36 Typical wiring diagram.

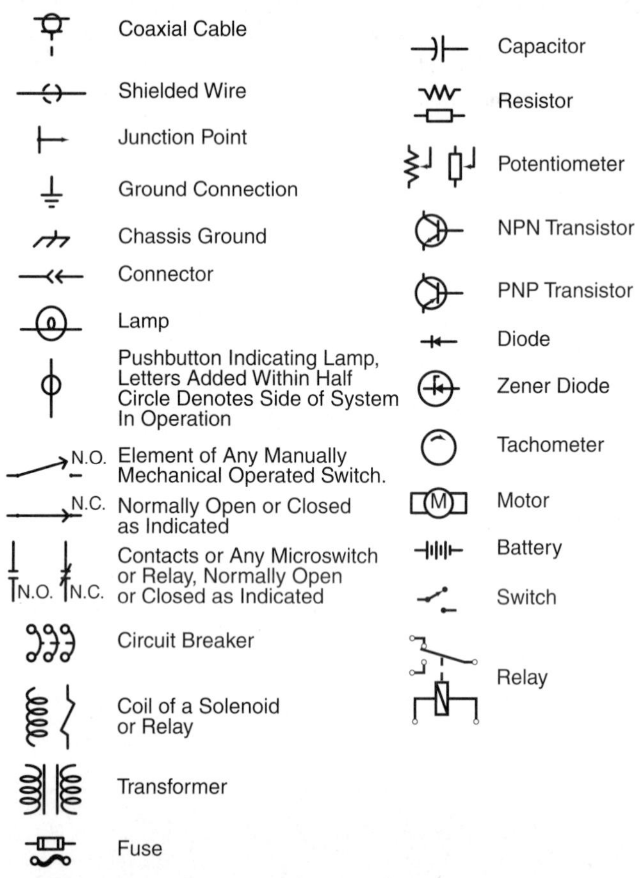

FIGURE 41-37 Every electrical device and component has a corresponding electrical symbol.

The wiring on modern vehicles is very complex. There are many wires and interconnected components. A single wiring diagram of the whole vehicle would be very difficult to use. To make it easier, the wiring diagrams are split up into systems and subsystems. This reduces the complexity on each page. For example, there may be a wiring diagram for the starter system and a separate one for the charging system. There are other diagrams for the engine, transmission, antilock brakes, headlights, and so on.

To assist in understanding the wiring diagrams, manufacturers supply keys for the diagrams. One of which is a list of the component symbols and their names (**FIGURE 41-37**). Other

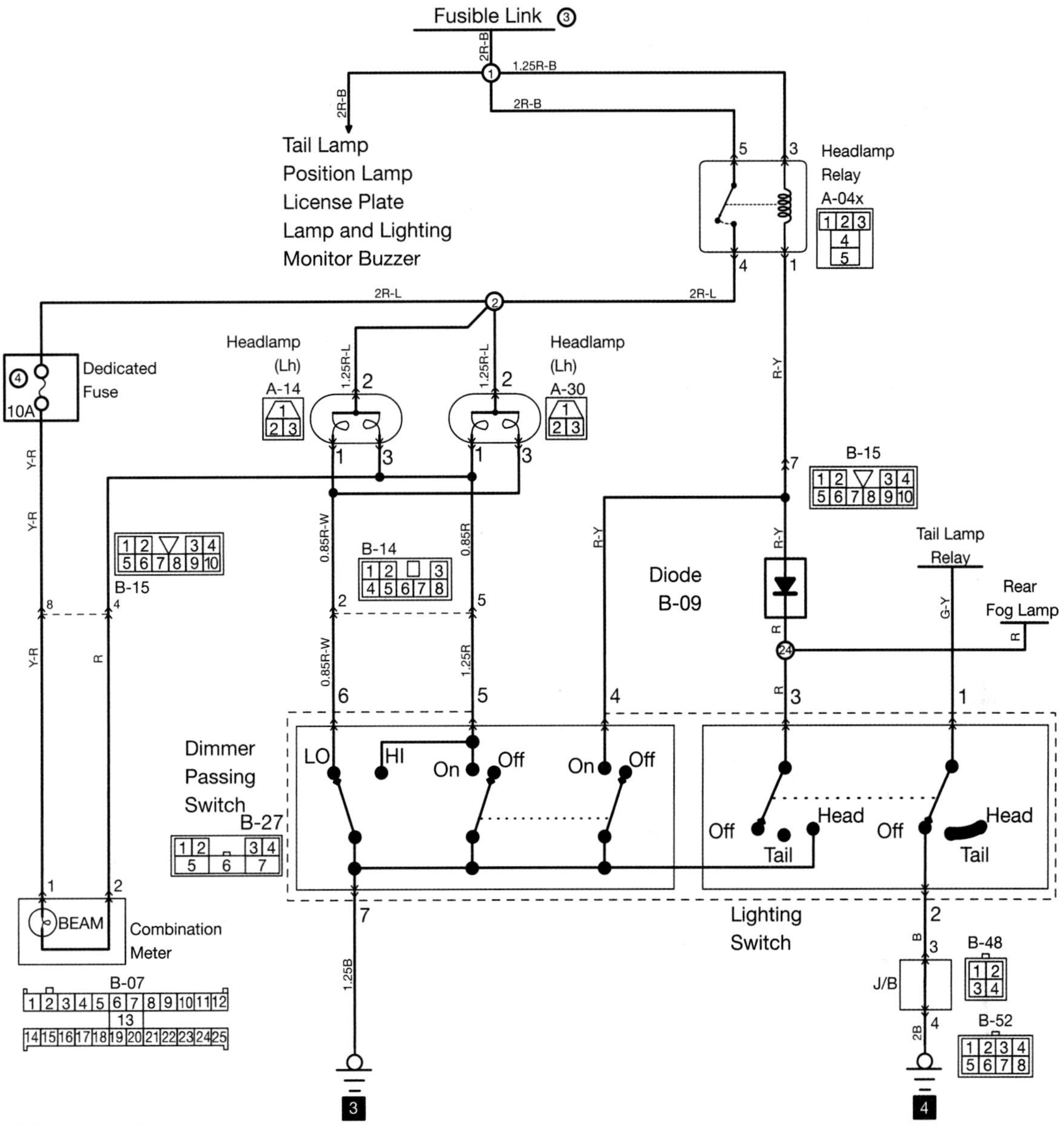

Wire Color Code

B: Black LG: Light Green G: Green L: Blue W: White Y: Yellow SB: Sky Blue
BR: Brown O: Orange GR: Gray R: Red P: Pink V: Violet

FIGURE 41-38 Typical key to wire identification codes.

keys refer to the color codes of the wires, wire size, harness connector identification, and pin numbers (**FIGURE 41-38**). Many of the symbols are standardized and used universally by manufacturers. However, in some cases, variations may exist.

In many cases, wires have two colors. The first color is the solid color. The second color is the stripe, also called a tracer. So, "BLK/WHT" would mean the wire uses a solid color of black, with a white tracer (stripe) (**FIGURE 41-39**).

In many cases, wiring diagrams are set up with power on the top of the diagram and ground near the bottom. This makes it easier to follow the flow of electricity in the diagram (**FIGURE 41-40**). Armed with all of this information, the technician can read the wiring diagram. This is done by identifying circuits on the diagram as they relate to the actual circuits on the vehicle.

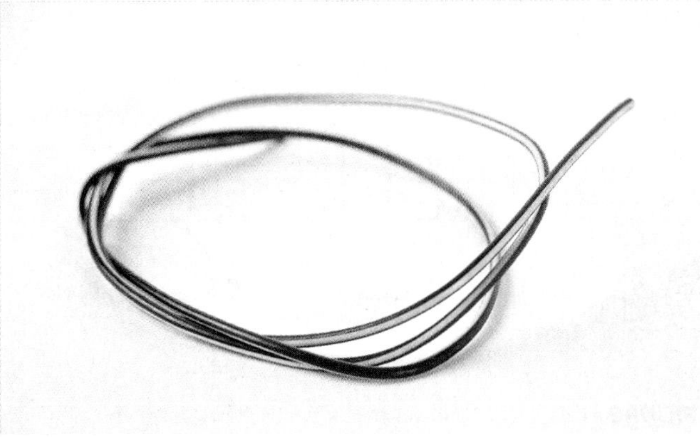

FIGURE 41-39 Black wire with a white tracer.

Wiring diagrams are a critical tool when diagnosing electrical circuit faults. In the strategy-based diagnostic process, it falls under step 2—Research. Wiring diagrams allow you to understand how the circuit was designed. And by tracing the circuits, as you will learn next, you will be able to determine how electricity flows through the circuits. That information is used along with your understanding of Ohm's law and the circuit laws. They allow you to predict what voltages should be present at the various connection points.

Knowing the expected voltages then allows you to measure the actual voltages present on the vehicle. Any substantial difference between the actual voltage and predicted voltage will point you in the direction of the fault. For example, the customer concern is that the starter motor runs very slowly. The wiring diagram shows no resistance between the battery positive terminal and the starter motor. You predict that there should be battery voltage at the starter motor while cranking (**FIGURE 41-41**). You check the voltage at that point and find that the actual voltage is only 4 volts. You then check voltage at upstream points in the circuit (solenoid, positive cable, etc.). Keep checking until you locate the high-resistance fault. These concepts are covered in much more depth in the next chapter.

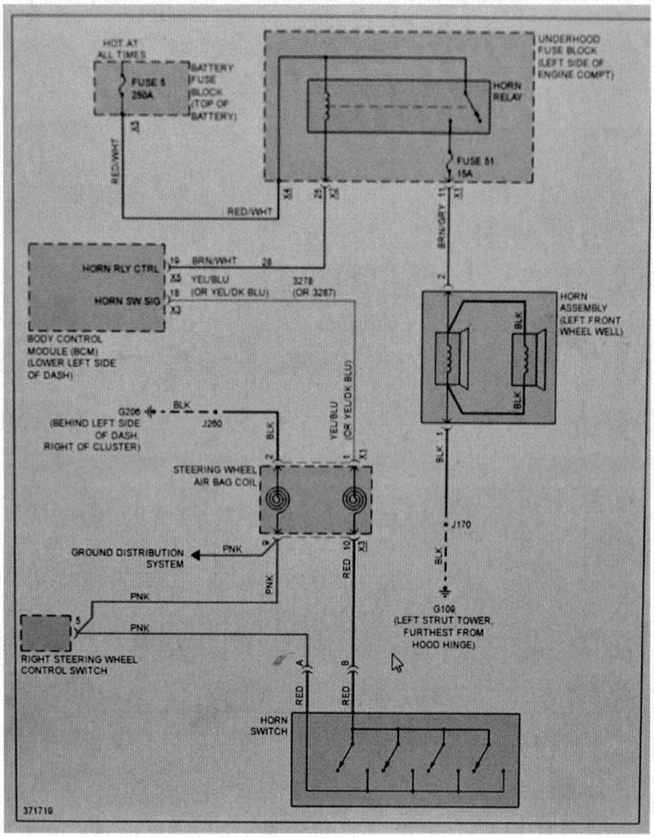

FIGURE 41-40 Wiring diagram with power at the top and ground near the bottom.

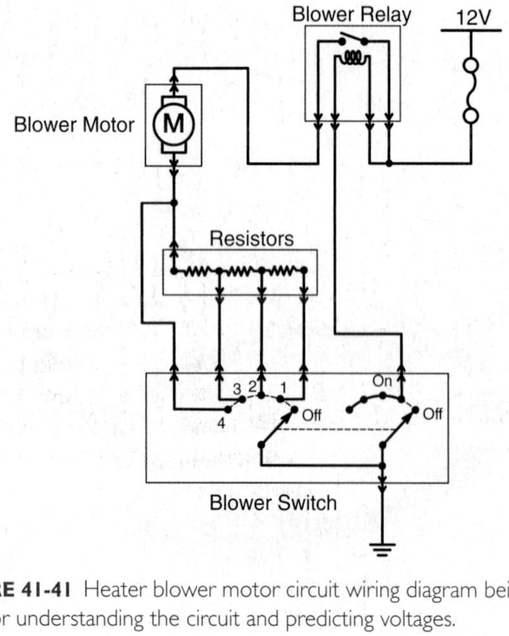

FIGURE 41-41 Heater blower motor circuit wiring diagram being used for understanding the circuit and predicting voltages.

Using Wiring Diagrams

Vehicle wiring diagrams are also known as schematics. They may be available in various formats. Some are paper-based manuals, while others are online resources. They are produced by manufacturers and some aftermarket publishing companies. Repair information is, more and more often, found via the Internet, using subscription services. You need an understanding of the symbols, abbreviations, and connector coding to use wiring diagrams. These are usually found on the diagram or in information pages.

Reading a wiring diagram is like reading a road map. There are a lot of interconnected circuits, wires, and components to decipher. Learning to read wiring diagrams takes a bit of time and experience. But knowing that circuits usually consist of a power source, circuit protection device, a switch, a load, and a ground is a good start. Jorge Menchu of AESWave has been promoting a novel approach. His process uses colored crayons or highlighters to trace out wires on the diagram. This helps you understand how a particular circuit operates (**FIGURE 41-42**). The following is a paraphrased version of that process.

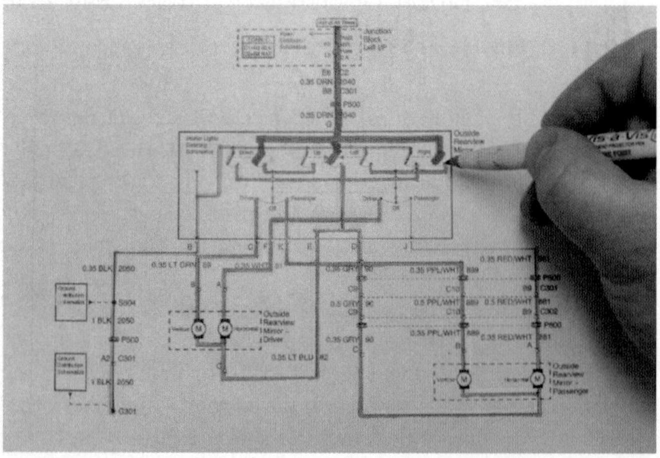

FIGURE 41-42 Color coding of the wiring diagram helps to understand the circuit.

Begin by printing out a copy of the wiring diagram for the circuit being diagnosed. You then identify each wire in the circuit and color it according to the following designations:

- Color all of the wires green that are directly connected to "ground."
- Color all of the wires red that are "hot" at all times.
- Color all of the wires orange that are "switched to power."
- Color all of the wires yellow that are "switched to ground."
- If there are any wires that reverse polarity, such as power window motor wires, mark those with side-by-side orange and yellow lines.
- Finally, color any variable wires, such as signal wires, blue.

Coloring the wires on the wiring diagram in this way does several things. First, it forces you to determine what each wire in the diagram does. This helps you get a total picture of the circuit. Second, it helps to organize your thoughts. This makes it easier to understand how electricity flows through the circuit. Third, it helps to keep you from losing your place or forgetting what a particular wire does. And fourth, it can give you confidence that you have properly diagnosed the problem. This is because you know why the circuit is not working properly and exactly where the problem is located.

To use wiring diagrams to diagnose electrical circuits, follow the steps in **SKILL DRILL 41-1**.

SKILL DRILL 41-1 Using Wiring Diagrams to Diagnose Electrical Circuits

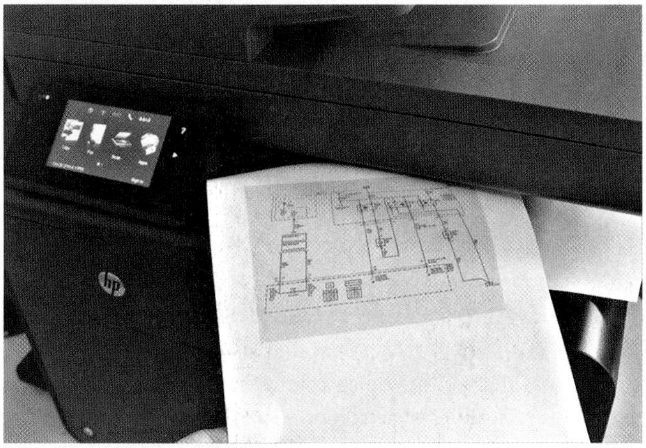

1. Identify the correct wiring diagram for the vehicle and system circuit being repaired, and print a copy.

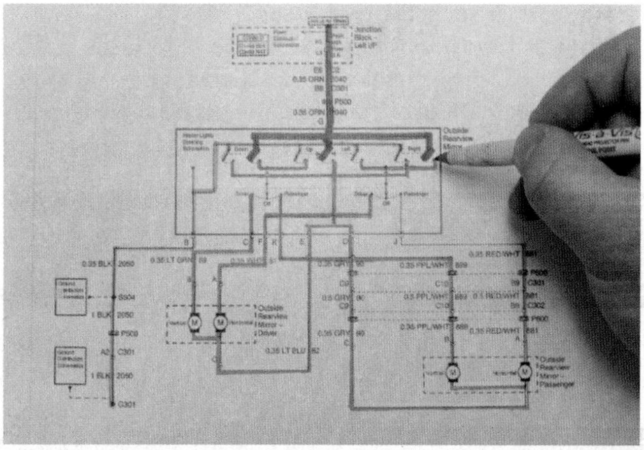

2. Color each wire (using the color coding system listed in the text) on the wiring diagram for the circuit that requires diagnosis. Note components, wire coding, and harness connectors.

3. Determine circuit test points and their location on the wiring diagram. Find the same test point on the vehicle, and perform the appropriate electrical test.

4. Depending on the results of the test, continue to use the wiring diagram to guide you in performing additional tests on the circuit until the fault has been located.

▶ Wire Maintenance and Repair

LO 41-09 Replace wire terminals.

Wires are generally trouble-free and long-lasting. But they can be damaged. Generally speaking, any issues with wiring are more likely to be with the terminals than with the wires themselves (**FIGURE 41-43**). Terminals can corrode, lose their tension, or push back up inside the connector. Each of these leads to poor connections and voltage drops. If you suspect a problem with a wire, first inspect the ends. If no problems are found, look for mechanical damage to the wire or wiring harness itself.

When a wire is damaged, it is usually due to one of several conditions. One possibility is that the wire has been physically broken. This happens due to excessive flexing or stress. Another issue is when a wire gets pinched between components. This happens when a wire falls between the engine and the transaxle when replacing a transmission. The pinched wire can cause either a short circuit or an open circuit.

Wires can also be misrouted so that they lie on a hot surface such as the exhaust manifold. The insulation melts and causes the wires to short. In all of these cases, the problem can typically be spotted visually. One problem that may be harder to spot is if a circuit shorts out and melts one or more wires together within a harness. In this case, you may have to open up the wiring harness and inspect the wires. Next, we cover how to repair wires, terminals, and connectors.

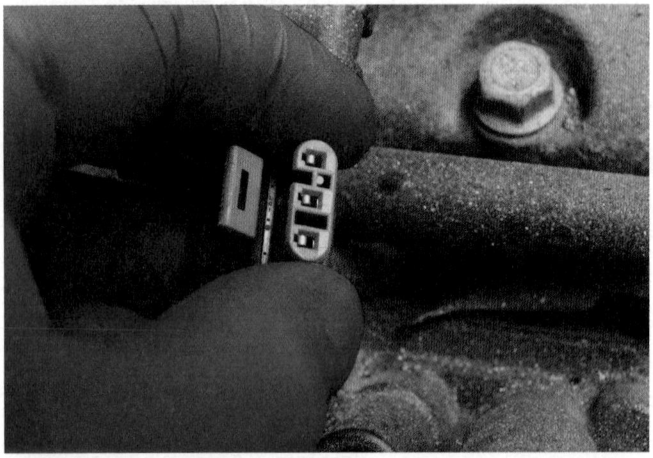

FIGURE 41-43 Inspecting wire terminals for faults.

Stripping Wire Insulation

An insulating layer of plastic covers electrical wire used in automotive wiring harnesses. When electrical wire is joined to other wires or connected to a terminal, the insulation has to be removed. Wire stripping tools come in various configurations, but they all perform the same task. The type of tool you use or purchase will depend on personal preference. A good pair of wire strippers removes the insulation without damaging the wire strands. Never use a knife or other type of sharp tool to cut away the insulation. It often cuts away some of the strands of wire as well. This is known as ringing the wire, which effectively reduces the current-carrying capacity of the wire.

Only remove as much insulation as is necessary to do the job. Insufficient bare wire may not achieve a good connection. And excessive bare wire may expose the wire to a potential short circuit with other circuits or to ground. Also, removing more than 0.5" (12.7 mm) of insulation at a time can stretch and damage the wire core.

To strip wire insulation, follow the steps in **SKILL DRILL 41-2**.

Installing a Solderless Terminal

If a wire itself needs to be repaired, it should be soldered back together. You should not use solderless terminals to reconnect the wires. Factories use solderless terminals throughout the vehicle, primarily at connectors. Solderless terminals are quick to install. They are also effective at conducting electricity across joints that are designed to be disconnected. Solderless terminals require a clean, tight connection. It is important to make sure the wire and the connection are clean before attaching any terminals.

You should use connections that match the size of the wire. Many solderless connectors are color coded for the size

SKILL DRILL 41-2 Stripping Wire Insulation

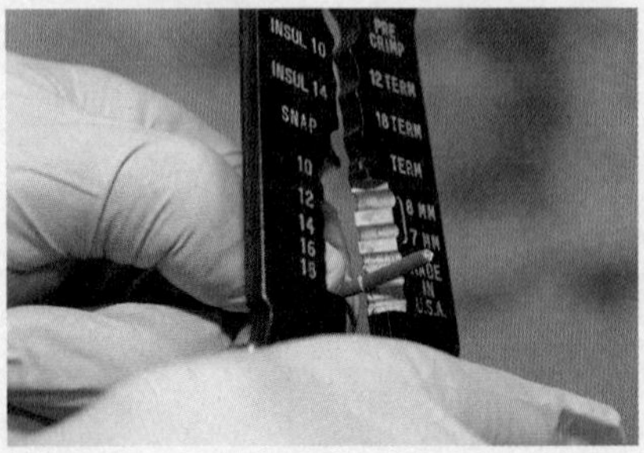

1. Choose the correct stripping tool.

2. Select the hole that matches the diameter of the wire to be stripped. Place the wire in the hole, and close the jaws firmly around it to cut the insulation.

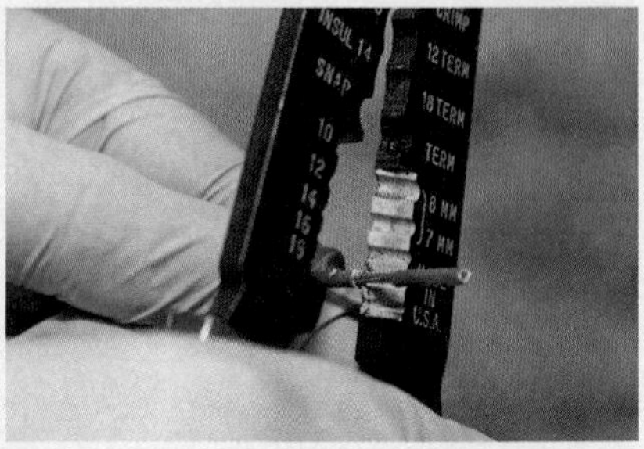

3. Remove the insulation. To keep the strands together, give them a light twist.

of wire they are designed to work with (**FIGURE 41-44**). Some examples include yellow 12–10 AWG, blue 16–14 AWG, and red 22–18 AWG.

Use the correct wire stripper to strip only as much insulation off as needed to allow the wire to fully engage the terminal. To keep the wires together after stripping them, give them a slight twist. Do not twist the wire too much; otherwise, you risk a poor wire-to-terminal connection. Use the correct crimping tool for the connection, as many manufacturers use special crimping tools. Using the wrong type of tool will cause the connection to have a poor grip on the wire. It may also damage the terminal.

FIGURE 41-44 Various solderless connectors.

To install a universal solderless terminal, follow the steps in **SKILL DRILL 41-3**.

▶ Soldering Wires and Terminals

LO 41-10 Perform solder repairs.

Soldering involves joining wires using molten metal that solidifies when it cools. Solder used in automotive electrical applications is an alloy. It is typically made up of 60% tin and 40% lead. Solder needs to change from a solid state into liquid easily as it is heated. It must also return to its solid state quickly as it is cooled.

Solder is available as solid or flux cored. Flux is needed to prevent the metals from being oxidized when they are heated. Solid solder requires an external flux to be applied in the soldering process. Flux-cored solder has a bead of flux within the center of the solder.

Flux can have either an acid base or a rosin base (**FIGURE 41-45**). Acid flux is designed to be used on nonelectrical metal joints such as radiators. Acid flux must be removed

SKILL DRILL 41-3 Installing a Solderless Terminal

1. Make sure you have the correct size of terminal for the wire to be terminated and that the terminal has the correct volt/amp rating. Remove an appropriate amount of the protective insulation from the wire.

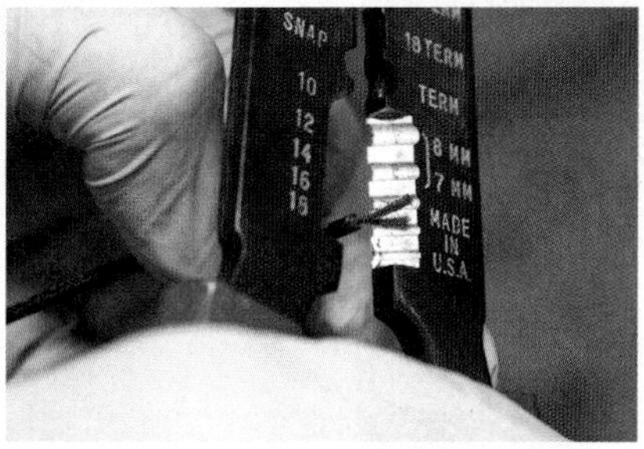

2. Lightly twist the wire strands in their normal direction, and place the terminal onto the wire.

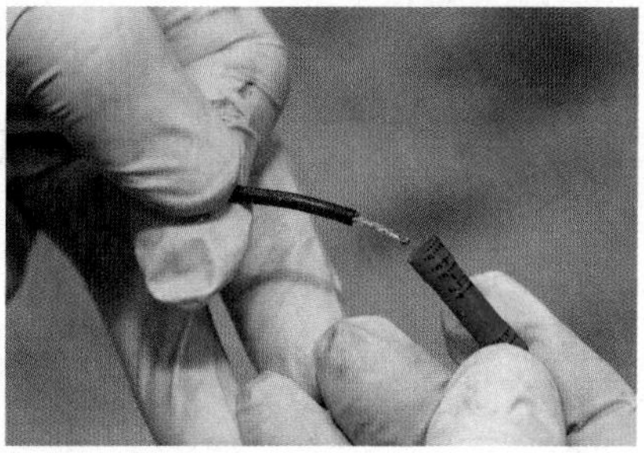

3. Use a proper crimping tool for the terminal you are crimping. Do *not* use pliers, as they have a tendency to cut through the connection. Select the proper anvil.

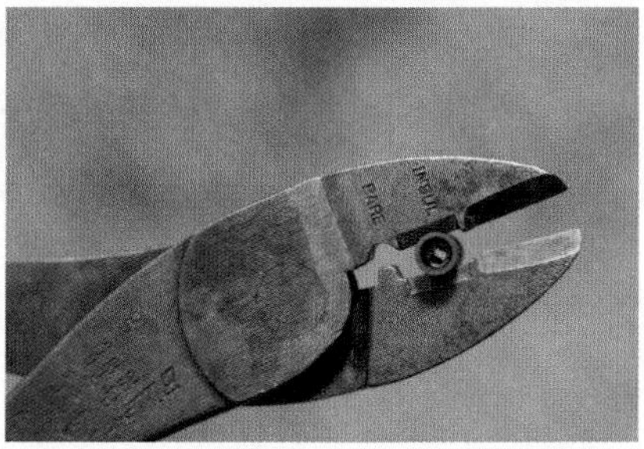

4. Crimp the core section first. Use firm pressure so that a good electrical contact will be made, but do not use excessive force, as this can bend the pin or terminal.

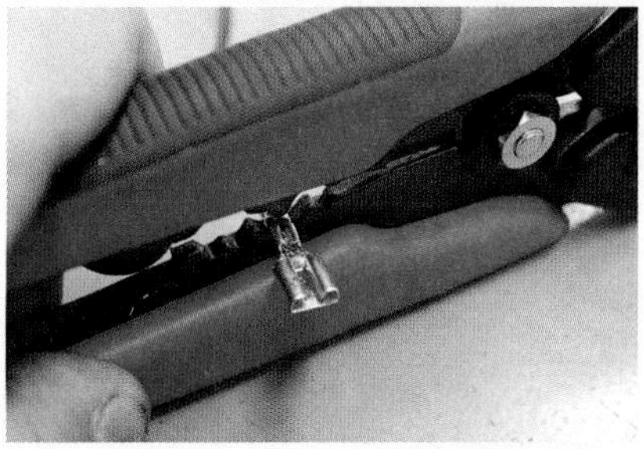

5. If crimping an insulated terminal, lightly crimp the insulation tabs so that they hold the insulation firmly.

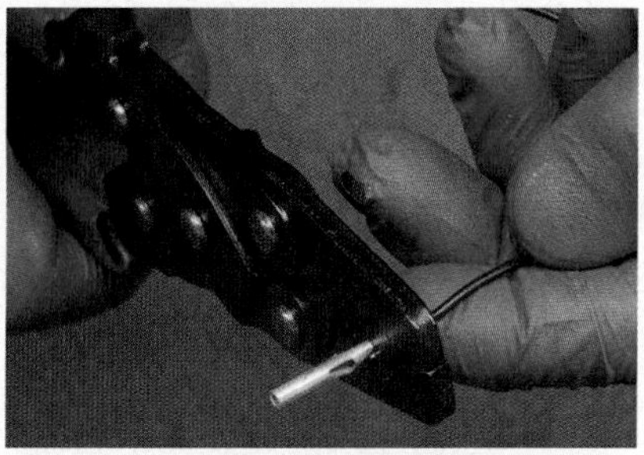

6. If crimping a factory terminal, use the proper tool and follow the instructions.

after the soldering process so that the joint does not corrode. Rosin flux is used on electrical connections. This is because it is much less likely to corrode the metals than acid flux. Acid flux and rosin flux also come in paste form that can be brushed onto the joint if using solid-core solder.

Solder is applied with a hot soldering iron. The soldering iron is heated electrically or by an external source such as a butane torch (**FIGURE 41-46**). The soldering iron tip absorbs heat that is then applied to the wires to be joined. Once they are hot enough, solder can be melted between the wires. It solidifies as it cools, "gluing" the wires together.

For a connection to be successful, the soldering iron has to be clean and "tinned." Cleaning may be as simple as heating the tip and wiping it on a damp cloth. Or, with the soldering iron cold, you may need to use a file to remove oxidized metal (**FIGURE 41-47**). It can also be reshaped with the file so it will effectively transfer heat to the wires. The tinning process assists in transferring heat to the wire by leaving a small amount of molten solder on the tip (**FIGURE 41-48**). This increases the surface area where the tip contacts the wires. To tin the soldering iron, the tip is heated, and a small amount of solder is applied to the tip. Excess solder is removed with a cloth rag.

The soldering iron tip is heated until it can melt solder. It is then applied to the wire so heat is transferred to the wire. The solder is then applied to the wire opposite the soldering iron. Once the wire is up to soldering temperature, it will melt the solder and pull the solder into the strands of wire. This produces a strong, effective joint. Do not apply too much heat to the wire. Otherwise, two things will happen. First, the solder will be drawn too far up the strands of wire. This makes a very long, nonflexible joint that is subject to breaking. The second problem is that the insulation may overheat and melt.

Once the joint is soldered, it needs to be protected. This is best done with heat shrink tubing. Heat shrink tubing shrinks when heat is applied to it with a heat gun (**FIGURE 41-49**). Most types of heat shrink tubing are hollow. Some types contain a sealer, which, when heated with the heat gun, melts into and seals the joint. If there is no heat shrink tubing available, then it is possible to seal and protect the splice with electrical insulating tape.

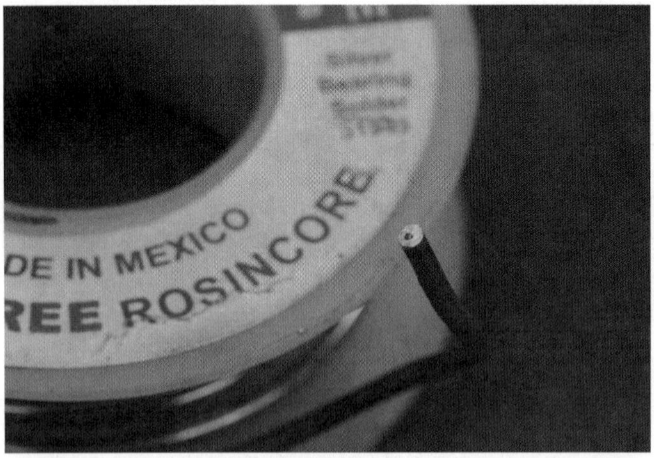

FIGURE 41-45 Rosin core solder is used for soldering electrical components and wires.

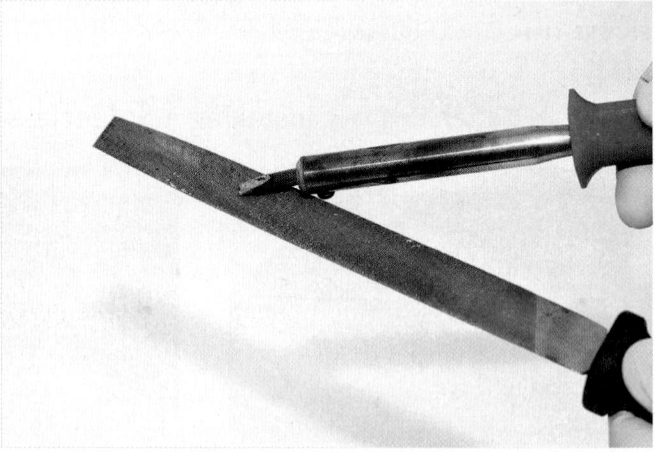

FIGURE 41-47 Cleaning a soldering iron tip with a file.

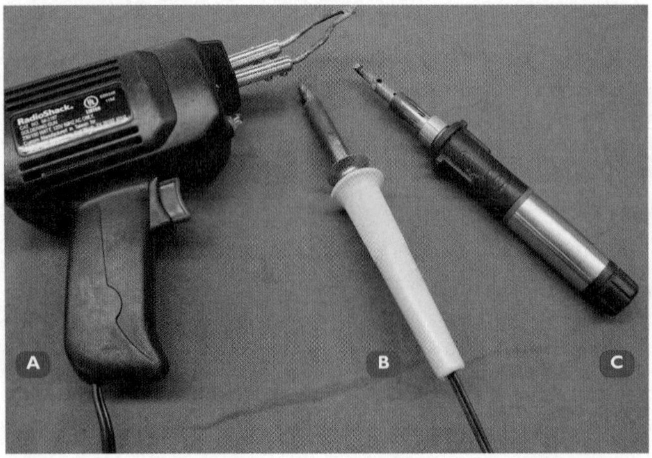

FIGURE 41-46 A. Electric soldering gun. **B.** Pencil-type soldering iron. **C.** Butane soldering iron.

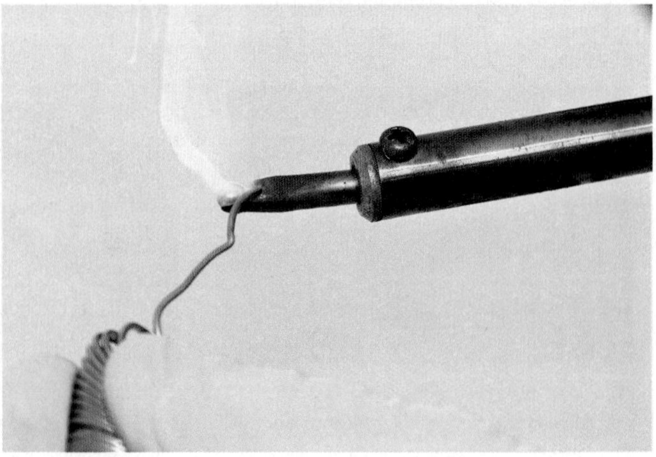

FIGURE 41-48 Tinning a soldering iron.

FIGURE 41-49 Heat shrink tubing shrinks when heated.

SAFETY TIP

Although soldering is generally thought of as a simple process, it can be very dangerous. The solder, soldering iron, and wires are very hot and can cause severe burns. Be careful what you grab or where you set hot items. Molten solder can be flicked by a springy wire up into your eyes. So, always wear safety glasses or goggles.

▶ TECHNICIAN TIP

One mistake students make is trying to apply the solder directly to the tip of the soldering iron while the iron is heating up the wires. This does melt the solder, but it is likely that the wire is not hot enough for the solder to stick to it. Instead, the solder just globs on top of the wires, leading to what is called a cold joint. A cold joint has high resistance, and the wires are likely to break loose from the solder. One sign that the solder joint is good is that you can clearly see the outline of the wires on the surface of the solder all the way around the joint.

When using a soldering iron, you must be careful not to burn yourself or any part of the vehicle you are working on. The tip of the soldering iron must be hot enough to melt metal solder. So, make sure it is in a safe position and not touching anything while it is heating up. To solder wires and connectors, follow the steps in **SKILL DRILL 41-4**.

SKILL DRILL 41-4 Soldering Wires and Connectors

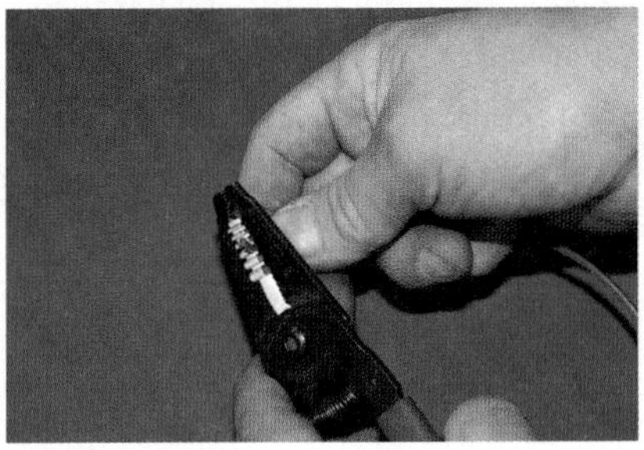

1. Safely position the soldering iron while it is heating up. While the soldering iron is heating, use wire strippers to remove an appropriate amount of insulation from the wires.

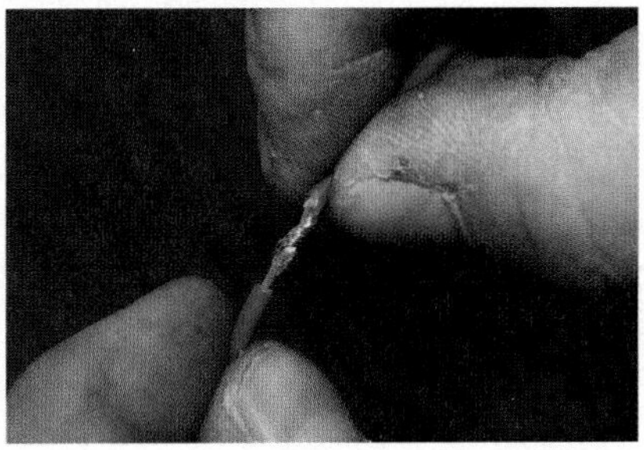

2. Place heat shrink tubing down one of the wires. Twist the wires together to make a good mechanical connection between them.

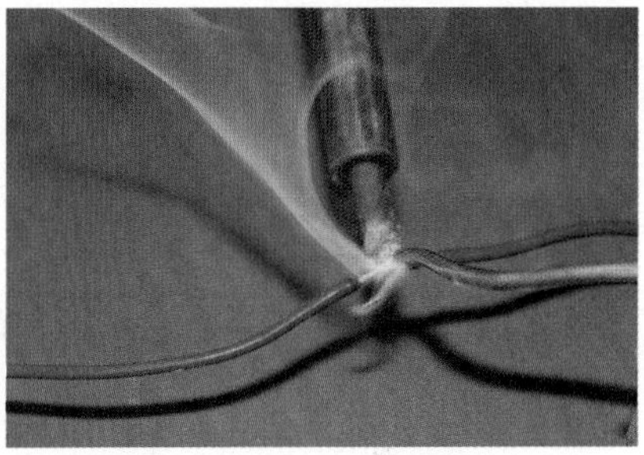

3. Tin the soldering iron tip, and gently heat up the wires while placing the solder opposite of the soldering iron. Allow the solder to be drawn into the joint.

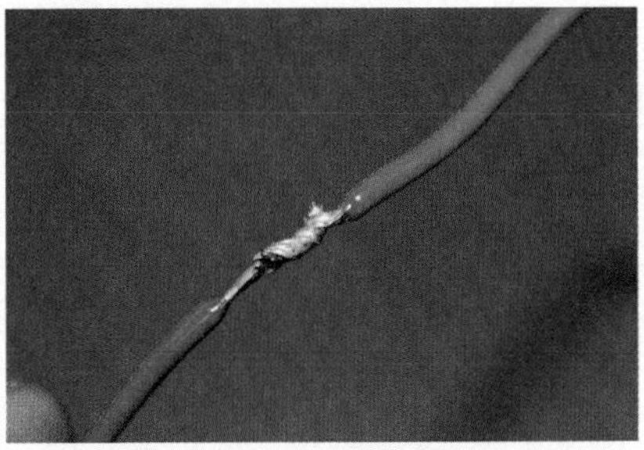

4. Pictured is a good solder joint where the solder has been drawn in.

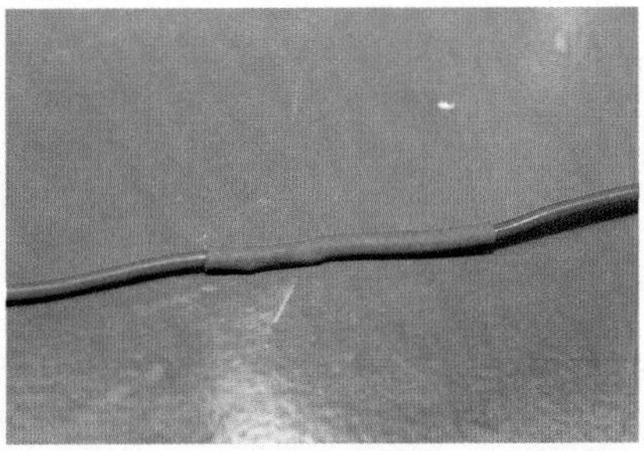

5. Once the electrical connection has been made and it has cooled, slide the heat shrink tube over the joint. Use a heat gun to shrink the tubing around the joint.

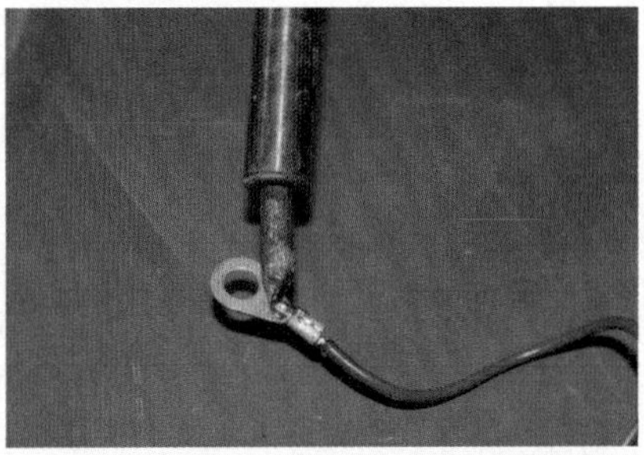

6. To solder a wire to a terminal connector, it is best to crimp it in place as before and use the solder to "glue" the joint together. Place the heated iron onto the terminal. Apply the solder to the end of the crimped wire tabs. Some solder will be pulled between the terminal and the wire. Cover the terminal with heat shrink tubing.

Wrap-Up

Ready for Review

- An electrical switch is a device used to turn the current on and off in a circuit. Different types and configurations of switches include toggle switches, push-button switches, and specialty switches. Complex switches have many terminals and contacts inside them and are able to switch a number of circuits at the same time.
- Fuses and circuit breakers are designed to protect electrical circuits by opening the circuit if the current flow is excessive. Common kinds of circuit protection devices are: fuses, fusible links, circuit breakers, and positive temperature coefficient (PTC) thermistor protection devices.
  - Fuses and circuit breakers are rated in amps and fusible links are typically rated by their wire size.
- Relays are switches that are turned on and off by a small electrical current. They are made up of an electromagnet, a set of switch contacts, terminals, and the case. A solenoid converts electrical energy into mechanical linear movement.
  - A relay uses a magnetic field to close an electrical circuit, while a solenoid uses a magnetic field to create lateral movement.
- Electric motors consist of two main components: the field and the armature and use magnetic fields to create rotary movement. The armature of a motor rotates due to the interaction of the magnetic fields. Ignition coils and transformers use electromagnetism to produce electricity rather than mechanical movement. A transformer makes use of primary winding, secondary winding, and an iron core.
- Resistors are used to control the voltage and amperage going to various components. Different forms of resistor are fixed resistors, variable resistors, rheostats, potentiometers, and thermistors.
  - The resistor's power rating is determined by its size and are typically marked with four or five colored bands with each color representing a numeric value.
- Automotive wire is commonly a multi-stranded copper core that is wrapped with seamless plastic insulation. The resistance of a wire is determined by its: length, diameter, construction material, and temperature.
  - Metric wire gauge measures the cross-sectional area of the conductor in square millimeters. The American wire gauge (AWG) uses a rating number, the larger the rating number, the smaller the wire and the lower its current-carrying capability.
- Wiring harnesses group two or more wires together and are typically placed within a sheath of tubing or insulating tape. Wiring harnesses run around the engine bay, through the dash and interior cabin, and to the rear of the vehicle.
- Wiring diagrams use symbols to represent electrical components and provide a map of all of the electrical components and their connections. Wiring diagrams of a vehicle are split up into systems and subsystems to reduce complexity. Color coding of the wiring diagram helps to understand the circuit.
- Issues with wiring are more likely to be with the terminals than with the wires themselves. Wire stripping tools can be used to remove the insulation without damaging the wire strands. A wire that requires repair should be soldered back together.
- Solid solder requires an external flux to be applied in the soldering process. Flux-cored solder has a bead of flux within the center of the solder. For a connection to be successful, the soldering iron has to be clean and "tinned." Soldered joint should be protected with heat shrink tubing.

Key Terms

American Wire Gauge (AWG) Standardized wire gauge used in North America. The higher the number, the smaller the wire is, and the lower the current-carrying capacity.

armature The rotating wire coils in motors and generators. It is also the moving part of a solenoid or relay, and the pole piece in a permanent magnet generator.

butt connector A crimp or solder joint that creates a permanent connection.

circuit breaker A device that trips and opens a circuit, preventing excessive current flow in a circuit. It can be reset to allow for reuse.

circuit or schematic diagram A pictorial representation or road map of the wiring and electrical components.

commutator A device made on armatures of electric generators and motors to control the direction of current flow in the armature windings.

Connector The plastic housing on the end of a wiring harness that holds the wire teminals in place. It can also refer to a type of wire terminal that connects wires together or to a common point such as a bolt.

drain line A wire included in a harness, with one end grounded to reduce interference or noise being induced into the harness.

eye ring terminal A type of crimp or solder terminal that has an enclosed eyelet to connect the terminal with a bolt or screw.

fixed resistor A resistor that has a fixed value.

flasher can A mechanical device that switches the vehicle's turn signal and hazard flasher bulbs on and off.

Fuses A circuit protection device with a conductive metal strip that melts when excessive current flows through it.

male and female terminal A crimp or solder terminal on which the male and female ends join to create a removable low-resistance connection.

Mylar tape Polyester film that may be metalized and incorporated into a wiring harness to provide electrical shielding.

NC (normally closed) An electrical contact that is closed in the at-rest position.

negative temperature coefficient (NTC) A characteristic of materials whereby resistance decreases as temperature increases.

NO (normally open) An electrical contact that is open in the at-rest position.

polymeric positive temperature coefficient (PPTC) device A thermistor-like electronic device used to protect against circuit overloads. Also called resettable fuse.

potentiometer A three-terminal resistive device with one terminal connected to the input of the resistor, one terminal connected to the output of the resistor, and the third terminal connected to a movable wiper arm that moves up and down the resistor, creating a varying voltage signal. Also called a pot, or variable resistor.

primary winding The coil of wire in the low-voltage circuit that creates the magnetic field in a step-up transformer.

push-on spade terminal A disconnectable type of crimp or solder terminal used to terminate electrical wires.

rheostat A variable resistor constructed of a fixed input terminal and a variable output terminal, which vary current flow by passing current through a long resistive tightly coiled wire.

ribbon cable A type of flat harness in which cables are insulated from one another but joined together side by side.

secondary winding The coil of wire in which high voltage is induced in a step-up transformer.

shielded wiring harness A wiring harness that has shielding built into it to protect it from induced electrical interference.

solder-type terminal A terminal that requires soldering to fasten the terminal to the cable or wire, instead of being crimped.

solenoid An electromagnet with a moving iron core that is used to cause mechanical motion.

solid-state relay A relay that performs the function of a mechanical relay but uses only electronic components.

step-down transformer A component that converts high-voltage, low-current AC power from a wall outlet (or engine) to a lower-voltage, higher-current AC or DC output.

step-up transformer A transformer used to increase the voltage from a lower input voltage to a higher output, such as an ignition coil.

switch An electrical device with contacts that turns current flow on and off.

terminal Metal connectors that are attached to wire ends. They are used to create electrical connections that can be disconnected and reconnected.

thermistor A variable resistor that changes its resistance based on temperature. Most thermistors have a negative temperature coefficient, meaning that their resistance decreases as temperature increases. Commonly used to measure coolant, oil, fuel, and air temperatures.

transformer action The transfer of electrical energy from one coil to another through induction in a transformer.

turn signal switch A switch that turns the left and right turn signal lights on and off.

twisted pair Two conductors that are twisted together to reduce electrical interference.

variable resistor A component that has a mechanism for varying resistance.

wire A conductor usually made of multistranded copper with an external insulated coating; used to transmit electricity within circuits.

wiring diagram A schematic drawing and symbol representation of the wiring and components; also called an electrical schematic.

wiring harness The network of wires, connectors, and terminals, pre-formed into bundles, that carry current within electrical circuits.

wiring harness connector A plug that contains multiple terminals with male and female ends.

Review Questions

1. A switch is used to:
   a. protect a circuit.
   b. allow current to flow only one way.
   c. reduce the current flow.
   d. turn current flow on and off.
2. Fuses prevent circuit damage by:
   a. stopping excessive current flow.
   b. reducing wiring length.
   c. limiting voltage increases.
   d. decreasing circuit resistance.
3. By using a small current in the relay winding, the contacts in the relay controls:
   a. a larger current.
   b. a smaller current.
   c. the exact same amount of current.
   d. the amperage at which the fuse blows.
4. The two main components of motors are the armature and:
   a. The field
   b. The commutator
   c. The brushes
   d. The Magnets
5. Which of these is NOT a type of resistor?
   a. Stationary resistor.
   b. Thermistor.
   c. Variable resistor.
   d. Fixed resistor.
6. If a wire is described as being metric size of 8.0 it is:
   a. 8.0 mm (square) cross sectional area.
   b. 8.0 cm (square) cross sectional area.
   c. 8 inch (square) cross sectional area.
   d. 0.8 mm (square) cross sectional area.

7. Learning to read wiring diagrams takes a bit of time and experience, but knowing that circuits usually consist of _______ and a ground is a good start.
   a. A power source
   b. A switch
   c. A load
   d. All of the above
8. The type of solder that is safe for electrical wires and incorporates flux in the core of the solder is referred to as:
   a. rosin cored solder.
   b. acid cored solder.
   c. silver solder.
   d. tinning solder.
9. Shielded wiring harnesses primarily help prevent:
   a. wires from chaffing
   b. unwanted short circuits between wires
   c. wires from becoming corroded
   d. unwanted electromagnetic induction.
10. When stripping wire insulation:
   a. use a knife or razor blade
   b. strip off at least 0.5" beyond the joint or terminal
   c. strip off no more than 0.5" at a time.
   d. use a propane torch to soften the insulation first

ASE Technician A/Technician B Style Questions

1. Two technicians are discussing solder repair. Technician A says that solder should be applied to the soldering iron tip while soldering. Technician B says that the solder should be applied to the wire joint while soldering. Who is correct?
   a. Technician A
   b. Technician B
   c. Both Technician A and B
   d. Neither Technician A nor B
2. Two technicians are discussing wiring diagrams. Technician A says that wire colors are listed. Technician B says that connectors are listed. Who is correct?
   a. Technician A
   b. Technician B
   c. Both Technicians A and B
   d. Neither Technician A nor B
3. Technician A says that 18-gauge AWG wire can carry more current flow that 12 gauge AWG wire. Technician B says that metric wire is sized by its cross-sectional area. Who is correct?
   a. Technician A
   b. Technician B
   c. Both Technicians A and B
   d. Neither Technician A nor B
4. Two technicians are discussing circuit protection devices. Technician A says that fusible links are short sections of special conductor material, used to protect wires such as between the battery and fuse box. Technician B says that automotive circuit breakers are generally designed to automatically reset if they are tripped. Who is correct?
   a. Technician A
   b. Technician B
   c. Both Technician A and B
   d. Neither Technician A nor B
5. Tech A says that some relays are equipped with a suppression diode in parallel with the winding. Tech B says that some relays are equipped with a resistor in parallel with the winding. Who is correct?
   a. Tech A
   b. Tech B
   c. Both A and B
   d. Neither A nor B
6. Tech A says that a twisted pair is two wires, twisted together, that deliver signals between common components. Tech B says that in many cases, wiring diagrams are set up with power on the top of the diagram and ground on the bottom. Who is correct?
   a. Tech A
   b. Tech B
   c. Both A and B
   d. Neither A nor B
7. Tech A says that solderless terminals are universal in size meaning that any terminal can be used with any size wire. Tech B says that many manufacturers use special crimping tools. Who is correct?
   a. Tech A
   b. Tech B
   c. Both A and B
   d. Neither A nor B
8. Two techs are discussing an electric motor. Tech A says that a set of carbon brushes transfer electricity to the bars on the commutator. Tech B says that the armature has one long winding. Who is correct?
   a. Tech A
   b. Tech B
   c. Both A and B
   d. Neither A nor B
9. Technician A says relays are turned on and off by a small amount of current. Technician B says many relays have both normally closed (NC) and normally open (NO) contacts. Who is correct?
   a. Technician A
   b. Technician B
   c. Both Technician A and B
   d. Neither Technician A nor B
10. Tech A says that most switches have numbered or lettered terminals. Tech B says that momentary switches can only be used once, then they need to be replaced. Who is correct?
   a. Technician A
   b. Technician B
   c. Both Technician A and B
   d. Neither Technician A nor B

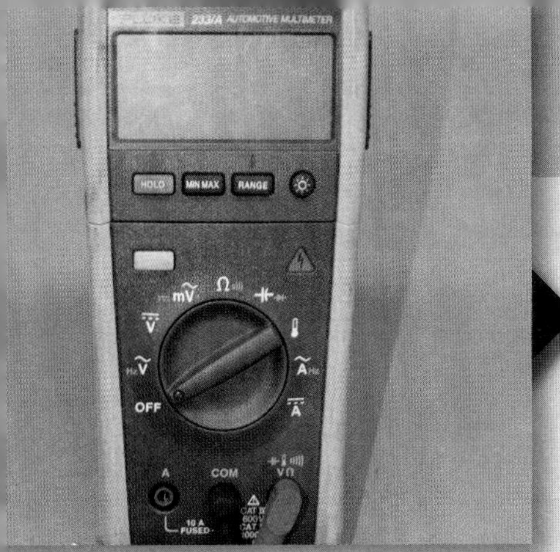

CHAPTER 42

Meter Usage and Circuit Diagnosis

Learning Objectives

- **LO 42-01** Describe basic meter info.
- **LO 42-02** Describe basic meter layout and ranges.
- **LO 42-03** Describe special meter settings probing techniques.
- **LO 42-04** Describe how to measure volts, amps, and ohms.
- **LO 42-05** Perform available voltage and voltage drop measurements.
- **LO 42-06** Perform resistance measurements.
- **LO 42-07** Perform current measurements.
- **LO 42-08** Perform series circuit measurements.
- **LO 42-09** Perform parallel circuit measurements.
- **LO 42-10** Perform series-parallel circuit measurements.
- **LO 42-11** Perform measurements on variable resistors.
- **LO 42-12** Describe electrical circuit testing.
- **LO 42-13** Perform voltage and voltage drop measurements.
- **LO 42-14** Locate opens, shorts, grounds, and high resistance.
- **LO 42-15** Test circuits with a test light and fused jumper wire.
- **LO 42-16** Test circuit protection devices, switches, and relays.

ASE Education Foundation Tasks

See Appendix A to view the 2017 ASE Education Foundation Automobile Accreditation Task List Correlation Guide.

▶ Introduction to Multimeters

LO 42-01 Describe basic meter info.

Digital multimeters (DMMs) are tools for taking electrical measurements. They are frequently used to diagnose electrical faults (**FIGURE 42-1**). Like many diagnostic tools, practice is required to understand how they are used. This includes knowing how to properly connect them into electrical circuits. Once a reading is obtained, it has to be interpreted and applied to what is happening in the circuit. The reading is used in conjunction with knowledge of electrical theory to diagnose the cause of the fault. In this chapter, we explain how to set up and use DMMs for measuring voltage, amperage, and resistance. We will also help you to understand how to use those readings to identify electrical faults.

This chapter also includes a number of exercises. They relate circuit examples and DMM readings to Ohm's law calculations. This will help expand your understanding of electrical behavior. Knowing how to properly use DMMs, as well as how to interpret and apply their readings, will allow you to diagnose electrical faults. This will make you very valuable to your employer.

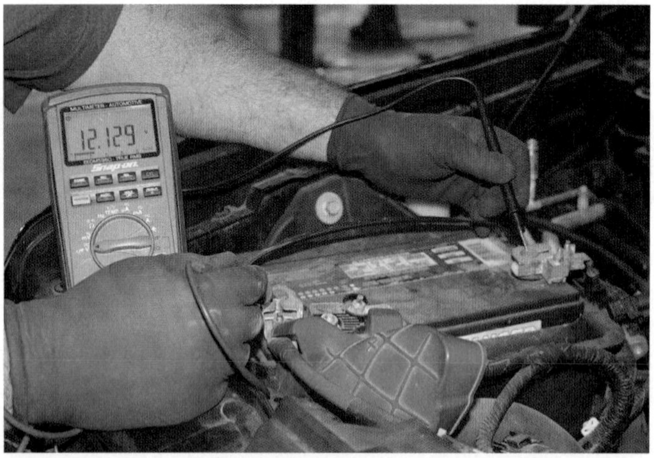

FIGURE 42-1 Digital multimeters (DMMs) allow technicians to look inside electrical circuits when diagnosing problems.

Digital Multimeters

A DMM or a digital volt-ohmmeter (DVOM) is a versatile and useful piece of test equipment (**FIGURE 42-2**). They are called digital meters because they give a numerical reading on a digital display. An analog meter, by comparison, uses a needle that hovers over a series of scales. This requires the technician to determine the numerical value of the reading. DMMs are easier to read, which means that a technician is less likely to get the wrong reading. The DMM tends to be

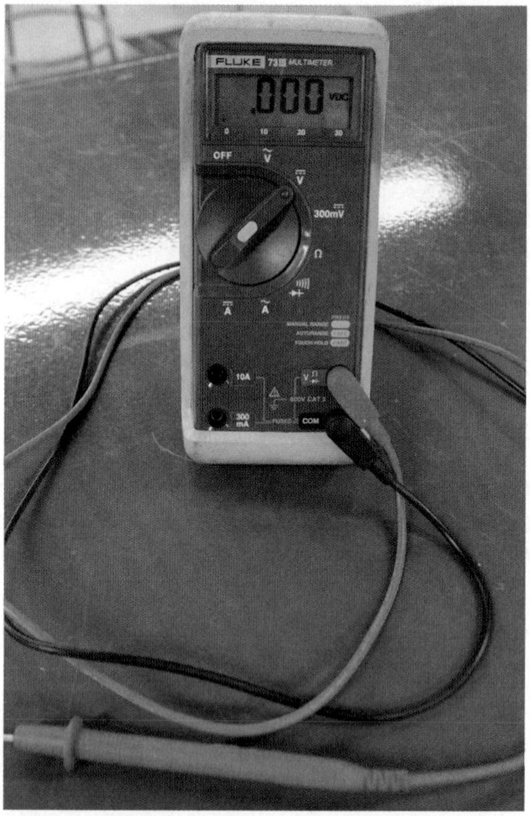

FIGURE 42-2 A digital multimeter (DMM) is a versatile and useful piece of test equipment.

You Are the Automotive Technician

Your supervisor requested that you train the shop apprentice on how to use a DMM. You will be using the DMM to teach the apprentice how to take simple measurements. That will come in handy when doing vehicle inspection tasks. Some examples are measuring battery and charging system voltage, testing a fuse to see if it is blown, and performing a voltage drop on a simple circuit. You demonstrate how to set up the meter for each of these tests. Once competent in setting up the meter, you demonstrate how to perform each type of reading. The following is the list of questions your supervisor has asked you to review with the new employee after the hands-on training:

1. What are the three most common measurements taken by a DMM, and how is the meter connected to the circuit for each one?
2. What are the steps in setting up a meter to take a reading?
3. What is "available voltage," and how is it measured?
4. Describe an unwanted voltage drop and what causes it.
5. Describe the process of measuring voltage drop.
6. What two conditions can cause the current flow in a circuit to be too high?

the first test tool selected for electrical diagnosis and repairs. Basic DMMs can measure:

- alternating current (AC) and direct current (DC) voltage,
- AC and DC amperage,
- and resistance.

Most modern DMMs can also measure frequency and temperature. They may also have a dedicated diode test capability.

DMMs come in a variety of layouts and qualities. You need to get used to the meters in your shop so you will know their capabilities and how to use them. Most DMMs of average quality or better are "fused". This means that one or more "fast-blow" fuses are included inside the DMM (**FIGURE 42-3**). If the amperage is too high, the fuse will blow, protecting the meter. If the meter is unfused, it will not be protected. It could be damaged if used incorrectly when measuring amperage.

DMMs and test leads also should have a CAT rating listed on the front amperage (**FIGURE 42-4**). CAT is short for "category." Each level, or CAT, is designed to work safely on higher powered electrical systems (**TABLE 42-1**). CAT ratings were not

TABLE 42-1 Meter CAT Ratings

Overvoltage Category	Short Description	Examples
CAT I	Electronics	Low-power electronic equipment such as copiers, etc.
CAT II	Single-phase plug-in tools and equipment	Portable tools, appliances, etc.
CAT III	Three-phase fixed equipment and single-phase commercial lighting	Equipment in fixed installations (this is the minimum rating required for hybrid vehicles)
CAT IV	Three-phase utility connection, any outdoor wires	Main power wires from the utility company; outside wire for lighting (this rating is for heavier duty meters and leads for hybrid vehicles)

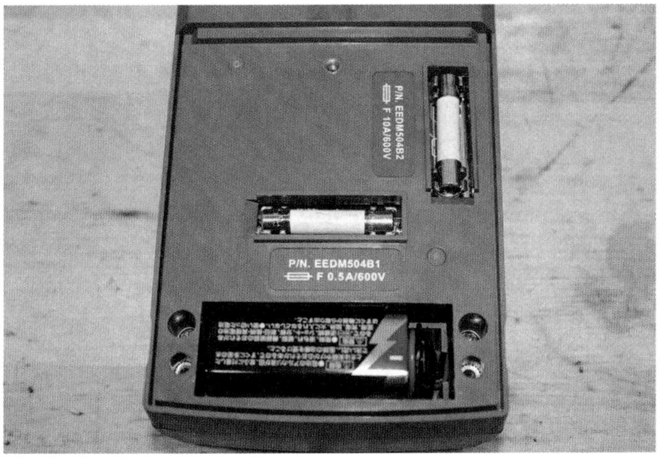

FIGURE 42-3 Meter fuses installed in a meter.

FIGURE 42-4 Meters must meet the CAT rating for the voltage you are working on.

designed initially for automotive meters because most vehicles use low voltage. But an increasing number of hybrid and electric vehicles are on the road. They operate on very high voltages. This means that CAT ratings are becoming important for automotive technicians to know.

Hybrid vehicles typically require meters and test leads rated as CAT III or CAT IV. Another thing to keep in mind if you are working on high-voltage systems is that you need to wear a pair of certified and tested rubber-insulated gloves. Most likely they will need to be worn with leather protectors over them. Always use the proper CAT-rated meter and leads along with the proper personal protective equipment when working on high-voltage systems.

▶ Digital Multimeter Purpose

LO 42-02 Describe basic meter layout and ranges.

DMMs are used to take many different electrical circuit measurements. They are one of the first tools used when conducting electrical repair or diagnosis work. As a voltmeter, the DMM can measure electrical voltage within circuits. For example, measuring the voltage available at a fuse, switch, or lamp. In this way, voltage problems can be identified.

The DMM can also measure the resistance of a component, connector, or cable. For example, measuring the resistance of an ignition coil. The resistance can be compared with specifications. And any resistance issues will be discovered.

DMMs can also measure current flow in circuits. For example, the amount of current flow through an air conditioning compressor clutch can be measured. The current flow can be compared with specifications. Any deviation will indicate an issue with either resistance or voltage.

Clearly, a DMM is a very versatile tool. This explains why it is the most commonly used electrical diagnostic tool. In the next several sections, we further explore DMMs and how to use them.

Digital Multimeter Layout and Accessories

There are two main components of a DMM: the main instrument body, and the test leads that connect the DMM to the circuit. The DMM main instrument body has:

- a function switch to choose the type of electrical measurement to be taken,
- a digital display to report the readings,
- and slots to connect test leads (**FIGURE 42-5**).

Test leads are used to connect the DMM to the circuit being tested and come in pairs: one red, the other black. Basic leads have a probe on one end for making the connection with the electrical circuit being tested. On the other end is a terminal for plugging into the slots of the DMM. A wide variety of test leads and adapters are available to make it easier to use the DMM. For example, alligator clips enable hands-free connection of the leads. Accessories such as temperature probes and inductive current clamps connect to the input slots of the DMM. They convert temperature or current flow into a voltage that can be measured by the DMM (**FIGURE 42-6**).

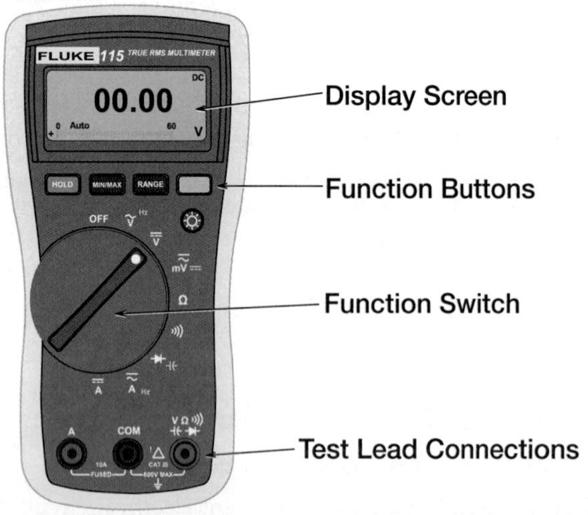

FIGURE 42-5 The body contains the display screen, function switch, function buttons, and slots to connect the leads.

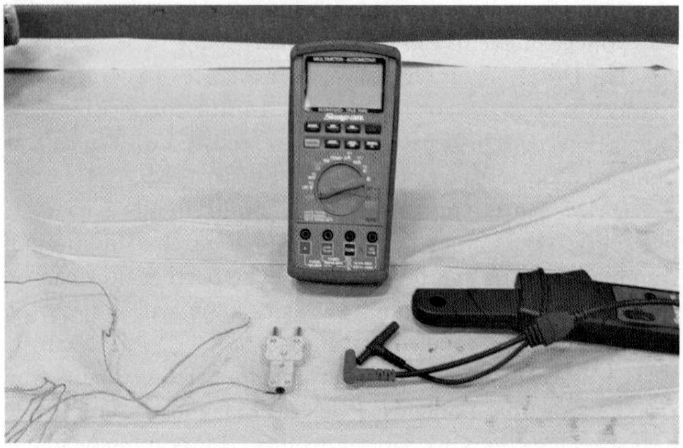

FIGURE 42-6 Accessories such as temperature probes and inductive current clamps are available for DMMs, extending the meter's usefulness.

Ranges and Scales

For measurements, DMMs read in a wide range. They read from very small quantities, such as 1/10,000 of a unit, up to very large quantities in the range of millions of units. It is not possible for DMMs to effectively and accurately measure such a wide range in only a single range or scale. This means DMMs must have multiple ranges or scales.

Before we can talk about those ranges, we need to understand that the DMM screen can only display four or five digits. This means that symbols must be used to substitute for some of the digits. (**TABLE 42-2**) shows the common symbols, their prefix, and the factor they represent. You will have to place the appropriate electrical symbol—V, A, or Ω—behind the factor symbol. For example, 2168 mV would be the same as 2168 millivolts. It could also be called 2.168 volts, as there are 1000 millivolts in 1 volt. Either designation is correct. The challenge is taking the meter reading and making sense of it, which takes practice.

Once you understand the symbols and what value each represents, you are ready to decide which range to set the meter to. (**TABLE 42-3**) lists a typical set of DMM ranges. However, there is no single range or scale value used by all DMM manufacturers. The resolution indicates the smallest value that can be measured on that range. Note that the resolution is different for each range. To achieve the most accurate reading, always select the lowest range possible for the value being measured. For example, if you are measuring 12 volts, you should select the 60-volt range. This is because 6 volts would be too low, and 600 volts would be less accurate. At the same time, if you are unsure of the value being tested, always start on the highest range. Then, turn it down to the lowest acceptable range based on that reading.

TABLE 42-2 DMM Values

Factor	Prefix	Symbol
1,000,000	Mega	M
1000	Kilo	K
1	No prefix	
0.001	Milli	M
0.000001	Micro	μ

TABLE 42-3 DMM Ranges

Function	Range	Resolution
mV DC	0–600.0 mV	0.1 mV
V DC	0–6.000 V	0.001 V
	0–60.00 V	0.01 V
	0–600.0 V	0.1 V
	0–1000 V	1 V
Ohms	0–600.0 Ω	0.1 Ω
	0–6.000 kΩ	0.0001 kΩ
	0–60.00 kΩ	0.01 kΩ
	0–600.0 kΩ	0.0001 MΩ
	0–40.00 MΩ	0.01 MΩ

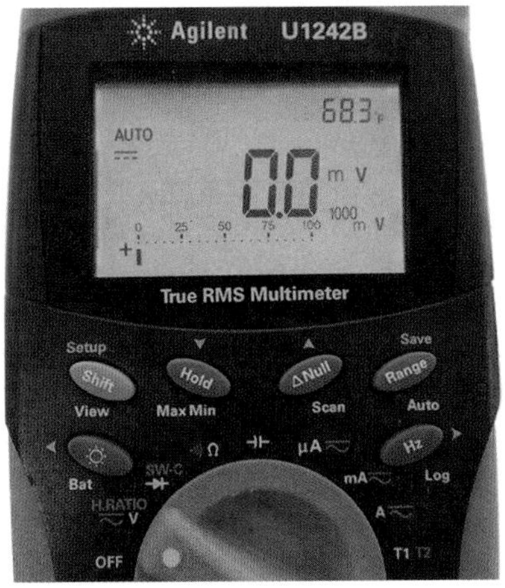

FIGURE 42-7 Meter set to auto range and reading mV.

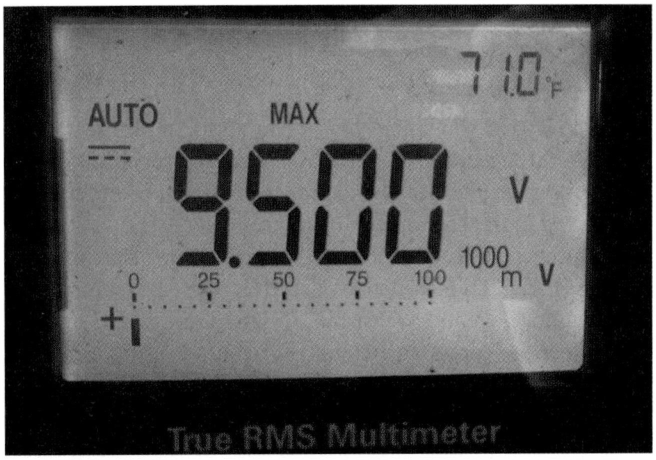

FIGURE 42-8 Meter showing MAX reading.

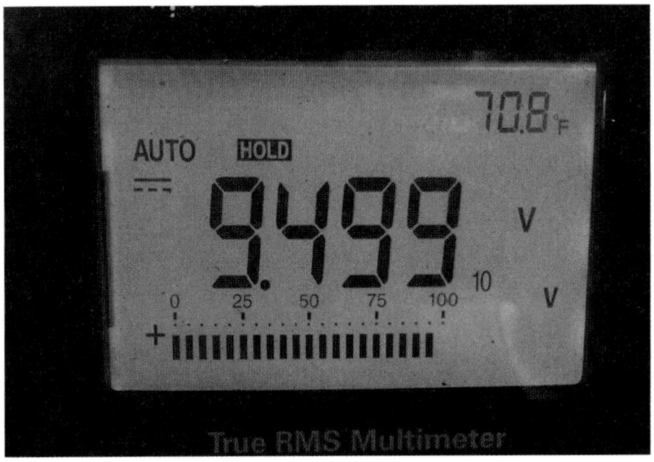

FIGURE 42-9 Meter showing HOLD reading.

Most modern DMMs have an automatic ranging capability, shown as "auto" on the screen. In fact, most DMMs default to "auto" when they are turned on. They also maintain the ability to be used in a manually selected range, but that must be selected separately. When used in the "auto" range, the DMM selects the best range for the value being measured. This is so that the technician does not have to be concerned with manually setting the range (**FIGURE 42-7**).

But be careful! The meter does not give you flashing light warnings that it has changed ranges. So, it is extremely easy to miss that. Many a technician has been led down the wrong diagnostic path by thinking the 12.6 on the meter was volts when in fact the meter had auto-ranged to millivolts. So, instead of having full power, the circuit being tested had almost *no* power. To prevent this mistake, many instructors require their students to use only manual ranges when using their meter.

▶ Min/Max and Hold Setting

LO 42-03 Describe special meter settings probing techniques.

Special settings are incorporated into the design of many DMMs. They make it easier to take certain measurements. In the **min/max setting**, the DMM records in memory the maximum and minimum reading obtained during the test (**FIGURE 42-8**). The min/max setting can be used to measure the throttle position sensor (TPS) voltage. The meter is connected to the TPS signal wire and the ignition turned on. Then the throttle is slowly opened all the way. The DMM captures the minimum and maximum TPS voltage, which can indicate a fault in the TPS.

A limitation in the use of the DMM is the sample rate, which is the speed at which the DMM can sample the voltage. The DMM does not continuously sample the voltage. Rather, it checks the voltage at regular intervals, or at a sample rate. This can occur very quickly—for example, every 100 milliseconds. However, it does mean that if a transient voltage occurs between samples, it will not be recorded by the DMM. Where quicker sample rates are required, other tools such as oscilloscopes can be used.

The **hold function** allows the display to be frozen. The hold button is pushed after a reading is taken. The display will hold the value on the display until the function or the DMM is turned off (**FIGURE 42-9**). A variation of the hold function is the auto-hold function found on some DMMs. When the auto-hold function is turned on, you can then take a measurement. The auto-hold feature will hold the display until the function or DMM is turned off. This function can be useful when taking measurements in difficult locations, such as underneath a dash. In this case, you may not be able to watch the meter display while making the meter connections.

Setting Up a DMM

To set up a DMM to take accurate measurements, you need to take several steps. You need to know whether you will be measuring resistance, voltage, or amperage. You should also know the reading that you are expecting. If very high voltages are to be measured, it is important to make sure the DMM and leads match the appropriate CAT rating.

All of this information determines the way in which you set up the DMM. This includes the connections you need to make on the DMM and the range you select. Resistance measurements on wires should be performed with the circuit unpowered. If measuring the resistance of components, they should be removed from the circuit.

The following steps describe how to set up a DMM:

1. Know what you are testing—volts, amps, or ohms.
2. Know the value you expect to be reading (specification).
3. Select the meter, leads, and probes with the appropriate CAT rating to suit the measuring task.
4. Connect the leads to the DMM in the proper slots.
5. Use the function switch to select the type of measurement to be undertaken (e.g., resistance, volts, or amps; DC or AC).
6. Select the correct meter range if you are using a manual ranging meter.
7. Prepare the circuit. Power it up if doing a voltage drop or amperage measurement. Power it down or disconnect it if taking a resistance measurement.
8. Connect the leads to the circuit being tested.
9. Read the meter display; remember to look at the factor symbol, and apply that to the reading.

Test Leads: Common and Probing

Many people incorrectly label the red lead as positive and the black lead as negative. However, if you look at your meter near the test lead slots, you will not see a "+" or a "–" anywhere. What you will see is "A" (typically 10 A), "mA," "common," and "V/Ω" (**FIGURE 42-10**). "Common" just means that the slot is common to all of the functions of the meter. In other words, this lead does not need to be moved when different functions of the meter are typically accessed. On the other hand, the red lead does have to move, depending on what function of the meter is being used. That is why it is labeled with the V/Ω symbol, and not "+."

If you find this distinction questionable, consider the following. Later on, when various electrical signals are measured at the same time on an oscilloscope, more than just the red lead is needed. In fact, we also typically use a yellow, a blue, and green test leads. In all of these situations, the test lead (no matter the color) acts as a probe into the circuit. So, rather than referring to the red lead as the positive lead, it is more accurate to refer to it as the probing lead for the DMM. We can then introduce probing leads of other colors when we use an oscilloscope, red being one of them.

There is one other important fact to note. And that is: the meter screen will always read what the probing lead is touching as compared with what the common lead is touching. For example, the common lead is touching the battery's negative post. And the probing lead is touching the positive post. In this case, the meter screen will likely display a "+" before the reading (**FIGURE 42-11**). That means that the probing lead is touching something more positive than the common lead.

Now reverse the leads by putting the red lead on the battery's negative post. And the common lead on the positive post. The meter screen will display a "–" before the number. That means that the probing lead is touching something more negative than the common lead (**FIGURE 42-12**). When you understand this

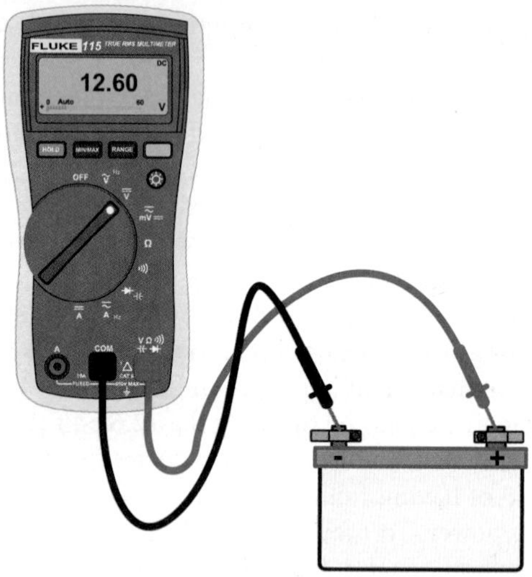

FIGURE 42-11 The screen reads what the red lead is measuring as compared with the black lead. If the black lead is on the negative battery post and the red lead is on the positive battery post, the screen will likely, but not always have a "+" in front of the reading.

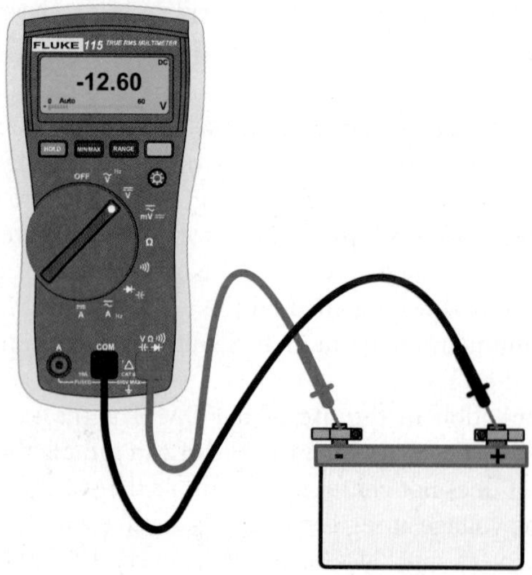

FIGURE 42-12 In this situation, the probing lead is connected to something more negative than the common lead. This is indicated by the "–" in front of the reading.

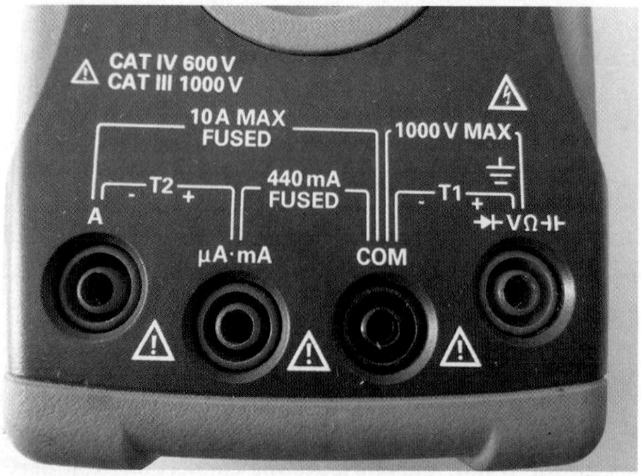

FIGURE 42-10 Slots for meter leads.

concept, it will help in keeping you from jumping to the conclusion that "the meter leads are hooked up backward." You will then be ready to start diagnosing all kinds of strange electrical problems. This is especially true with ground issues and charging system issues.

Probes and Probing Techniques

Gaining access to test points in a circuit requires a variety of probes and **probing techniques**. Some examples of probes are alligator clips, fine-pin probes, and insulation piercing clips (**FIGURE 42-13**). Make sure you know the voltage limits of the probes you use. High-voltage measurements require special probes that are designed for that purpose.

The standard probe leads that are supplied with a DMM are basic straight metal probes. They are useful for making quick measurements in circuits. But they do require the use of both hands to hold them in place. Leads with alligator clips come in various sizes. They allow the DMM leads to be clipped onto the circuit and held in place by themselves. This frees up your hands for other tasks. These clips are particularly useful for connecting to larger terminals, such as battery terminals.

Not all terminals are exposed. Using standard probes in this situation will not work. Back-probing is used to probe the backside of a connector to make contact with the terminal. To perform this task, very fine pins are used. The pins are designed to slip into the back of connectors and provide contact without causing damage (**FIGURE 42-14**).

Insulation-piercing probes are also available but should rarely be used. They have sharp fine pins that pierce a wire's insulation to create a connection with the wire. If used, remember to always seal any holes that the probe makes. This will help prevent any corrosion. Use liquid insulation or a similar product to reinsulate. Do not use room temperature vulcanizing (RTV) silicone. It attracts moisture as it cures, potentially causing the wire to corrode. Because it may result in damage to the insulation or conductor, many vehicle manufacturers do *not* allow this type of probe.

▶ Measuring Volts, Ohms, and Amps

LO 42-04 Describe how to measure volts, amps, and ohms.

The most common measurements taken with DMMs are voltage, resistance, and amperage. To take voltage measurements, the probing lead (red) is connected to the volts/ohms, or V/Ω, slot. The common lead (black) is connected to the common, or COM, slot. An appropriate range or auto range is selected on either AC or DC voltage, depending on the voltage to be measured.

The probing lead is typically connected to the positive side of the circuit being tested. The common lead is typically connected to the negative post of the battery, or a good ground. Connecting the voltmeter in this way is used to check the voltage available at various places in the circuit. This test is called the available voltage test (**FIGURE 42-15**). Use the red lead to probe connections in the circuit you are testing. Watch your screen. If the voltage is not what you were expecting, you will need to perform further tests.

If the "+" or "–" is not what you were expecting, check the leads to verify they are connected the way you intended. If you still get an unexpected reading, stop and analyze the situation. Ask yourself, what could cause the meter to read that way? Then brainstorm the options.

▶ TECHNICIAN TIP

A technician recently posted an electrical problem on a technical forum. He said that he had hooked up a voltmeter with the black (common) lead on the negative battery terminal. He also connected the red (probing) lead to the vehicle engine ground, with the engine running. The meter read a negative number. He asked the forum if he had the meter leads hooked up backward. He received several comments saying that yes, he had hooked them up backward. However, those technicians were not correct. His probing lead was registering a reading that was more negative than the common lead.

But what could be more negative than the negative post of the battery? When the engine is running, the alternator can be more negative than the negative battery post. So, what his DMM was trying to tell

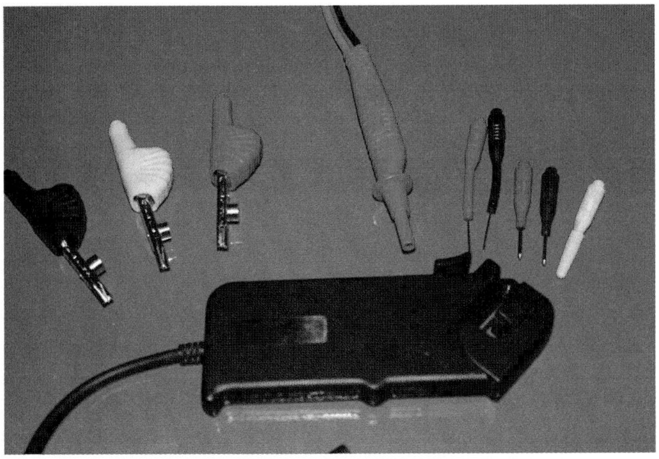

FIGURE 42-13 There are many different leads and probes you can use, depending on the circuit being tested.

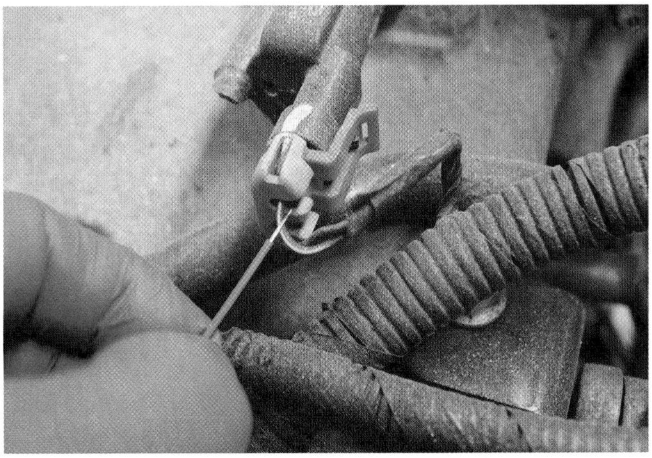

FIGURE 42-14 Back-probing a connector to take a voltage reading.

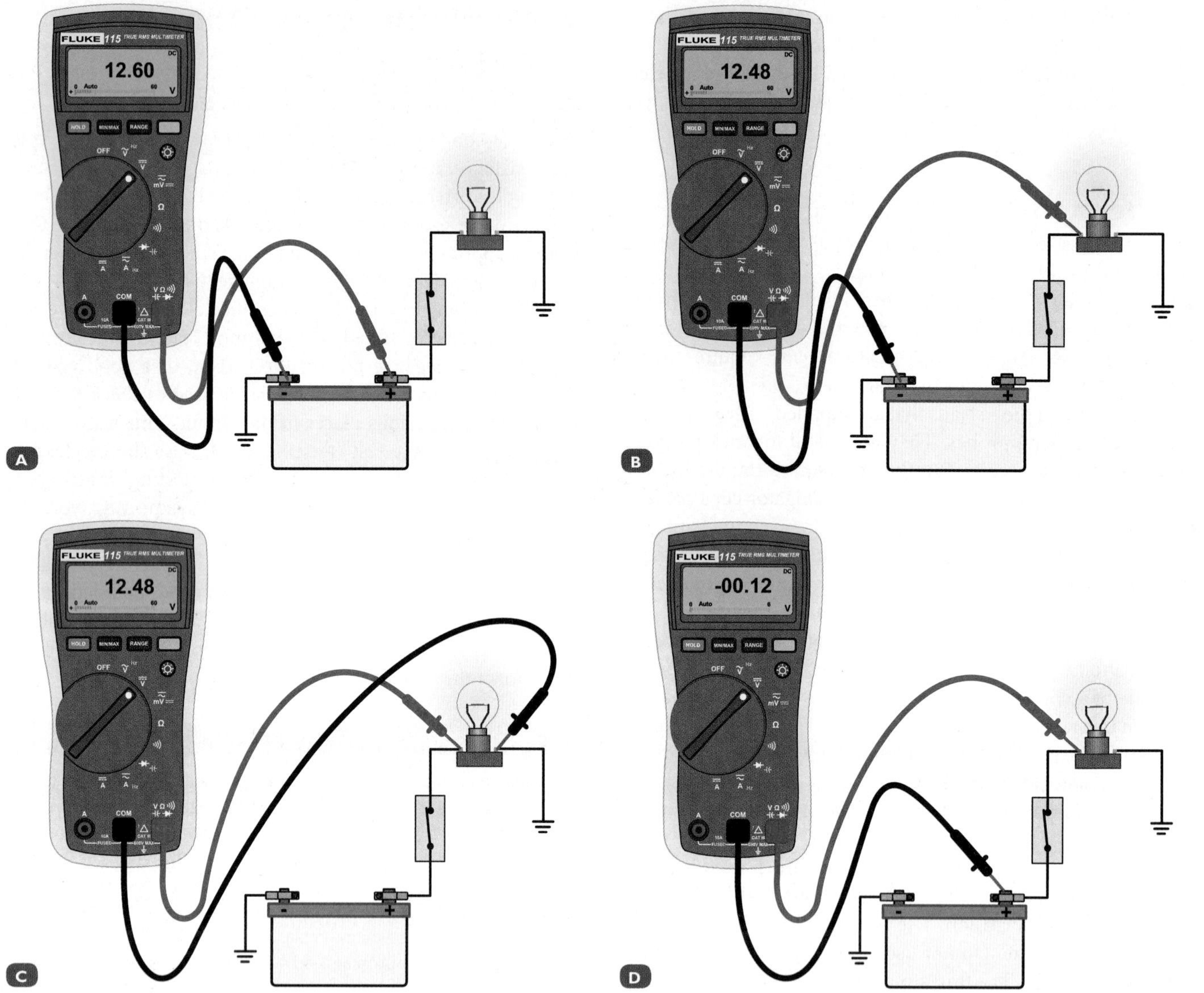

FIGURE 42-15 When connecting a voltmeter in the circuit, the voltmeter leads can be placed anywhere. The meter will then give you the difference in voltage between those points. Note that **A** and **B** are available voltage tests. **C** and **D** are voltage drop tests.

him was that there was a voltage drop between the negative battery post and the alternator frame. If he had understood that the probing lead was not lying to him and that it was reporting exactly what it was touching compared with what the common lead was touching, then, he could have started down the path to diagnosing what it indicated. In this case, he should have been looking for a dirty ground connection between the negative battery post and the engine block.

The amps (A) or milliamps (mA) flowing through a circuit can be measured with an ammeter. Most DMMs can measure milliamps or 10 to 20 amps directly through the meter. The correct range needs to be selected, along with AC or DC. The red probe is connected to the A or mA slot (always start high and work your way down if necessary). The black lead is connected to the COM slot.

To measure current, the DMM is connected in series with the circuit. This means that one wire in the circuit has to be disconnected. The meter leads connect to each disconnected end of the circuit. This is so that the meter is in series with the load (**FIGURE 42-16**). The probing lead is connected closest to the positive terminal of the power supply or battery.

▶ TECHNICIAN TIP

There is one big caution to remember when you finish using an ammeter. *Always* move the red test lead from the A slot to the V/Ω slot immediately after finishing. If you leave the red lead in the A slot and then change the selector to measure volts, there is still a straight path through the A slot to the common slot. This means the DMM will create a short circuit when hooked up to measure voltage. This will blow the fuse in the meter as well as potentially damage the circuit.

If larger amperage must be measured, then current clamps can be connected to the DMM. Depending on their range, they

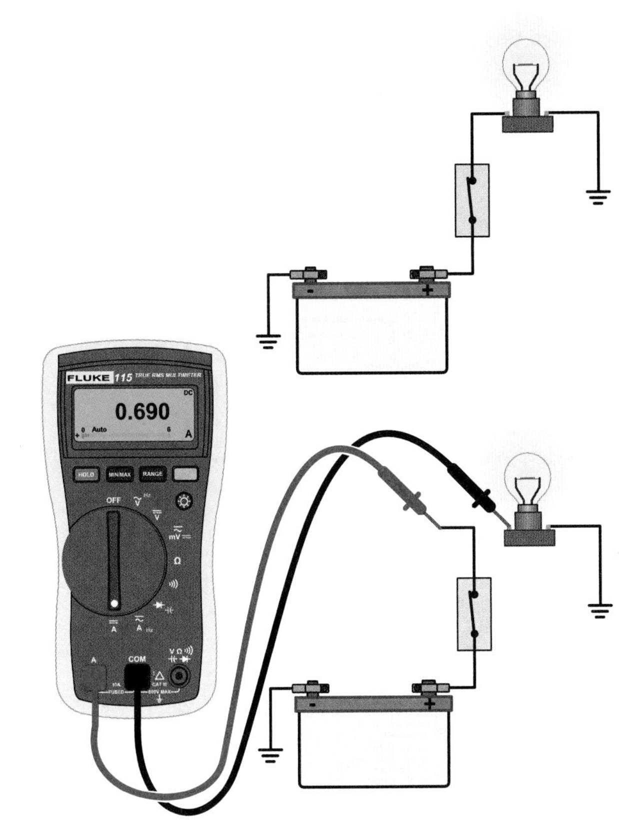

FIGURE 42-16 When measuring amps, one wire in series with the load has to be disconnected, and the ammeter leads connected between the disconnected ends.

can measure high currents. Some can measure starter motor current draw of 400 amps or more. Current clamps are available in a variety of current-measuring ranges.

The current clamp fastens around the conductor. It measures the strength of the magnetic field produced from current flowing through the conductor. It then outputs a voltage that the DMM reads as voltage. This is directly related to the amount of current flowing in amps.

When using current clamps, the DMM is set to read volts. Current clamps also have the advantage that they clamp around the conductor. So, the circuit does not need to be broken apart as you would have to do with a standard ammeter (**FIGURE 42-17**).

When measuring resistance in a circuit, always make sure the power is disconnected. In order to read resistance, batteries inside the DMM supply the circuit with power. If power is not removed from the circuit being tested, it disrupts the measurement. It can also provide a false reading or potentially damage the DMM.

To take resistance measurements, the red (probing) lead is connected to the V/Ω slot. The black (common) lead is connected to the COM slot. You then need to select an appropriate range or auto range. To accurately measure the resistance of a component, you should remove or isolate the component from the circuit. Doing so eliminates the possibility of any parallel circuit resistance affecting the measurement. The red probing lead is connected to the input side of the component being tested. The black common lead is connected to the output side of the component (**FIGURE 42-18**).

FIGURE 42-17 A current clamp being used with a DMM. Note that the current clamp fits around the wire and is not connected in series.

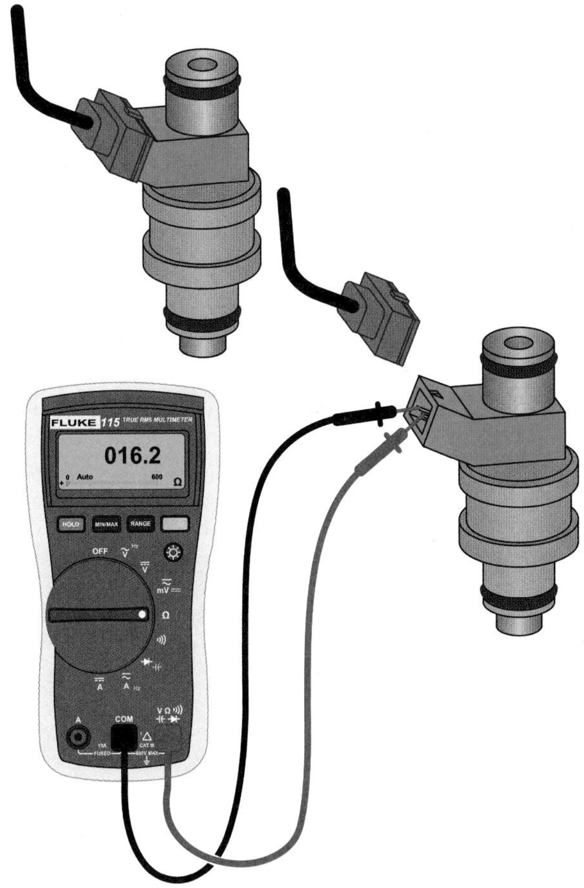

FIGURE 42-18 When measuring resistance, always isolate the component from the circuit, and place the meter leads across the input and output terminals.

▶ TECHNICIAN TIP

DMMs come in many forms. Always follow the specific manufacturer's instructions in the use of the DMM. Serious damage either to the DMM and/or to the electrical circuit could result.

Applied Math

AM-47: Specified Symbols: The technician can use conventional symbols (E for voltage, etc.) to solve problems using formulas such as Ohm's law, E = IR.

Voltage, current, and resistance have a specific relationship to one another. This concept forms the basis for electrical diagnosis.

George Ohm discovered that it takes 1 volt to push 1 amp through 1 ohm of resistance. Ohm's law is E = I × R, or Voltage = Amps × Resistance. If you know two of the three values for an electrical circuit, you can find the missing one. To find current flow, the formula is volts divided by resistance. To find voltage, the formula is amps multiplied by resistance. To find resistance, the formula is volts divided by amps.

In this example, we have an electrical circuit that has a resistance of 6 ohms with a 12-volt supply. We want to find the amperage. Our formula will be A = V/R:

A = V/R
A = 12 volts/6 ohms
A = 2 amps, which is our current flow.

▶ Voltage Exercises

LO 42-05 Perform available voltage and voltage drop measurements.

Voltage Ranges

Typically, a DMM has both an auto range and a manual range capability. The way in which you select auto range and manual range varies depending on the DMM. Different DMMs have different range settings. For example, one DMM's setting could be 6 V, 60 V, and 600 V, and another's 4 V, 40 V, and 400 V. **FIGURE 42-19** shows a circuit with two resistors in a series with a 12-volt DC supply. Various DMM ranges can be compared by measuring the voltage at the input and output of each resistor. The sample DMM we used has the following ranges: 600.0 mV, 6.000 V, 60.00 V, 600.0 V, and 1000 V.

Key Learnings: Using the lowest range that will still read gives the most accurate reading. Also, note that the voltage drops every time current goes through a resistor.

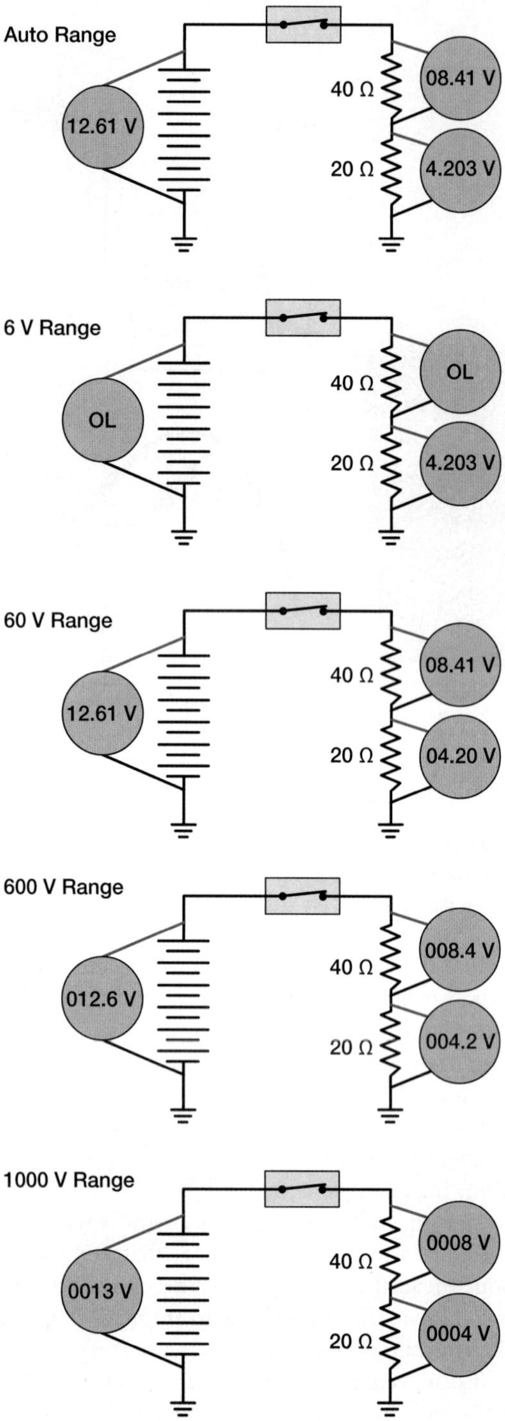

FIGURE 42-19 Measuring voltage using various ranges, in a circuit with two unequal resistors connected in series with a 12-volt DC supply.

Measuring Voltage

FIGURE 42-20 shows a series circuit of two resistors with a 12-volt battery and switch. In this example, the voltage will be measured in various parts of the circuit with the switch in the open position. Notice what happens to the voltage in the circuit. No voltage got past the open switch, so the loads (resistances) didn't get any current flow. This is because there was no voltage after the switch. Open circuits prevent current from flowing.

Key Learnings: With the switch open, no current can flow anywhere in the circuit. With no current flow, there is no voltage used by the resistors. The only place that has a voltage drop is from the input to the output of the open switch.

FIGURE 42-21 shows the same series circuit, but with the switch in the closed position. Notice that voltage passed through the closed switch to the first resistor. But not as much voltage made it to the second resistor. Remember that when current flows through a resistor, it uses up voltage. In fact, if we subtract the amount of voltage present after a resistor from the amount before the resistor, we can see how much voltage the resistor

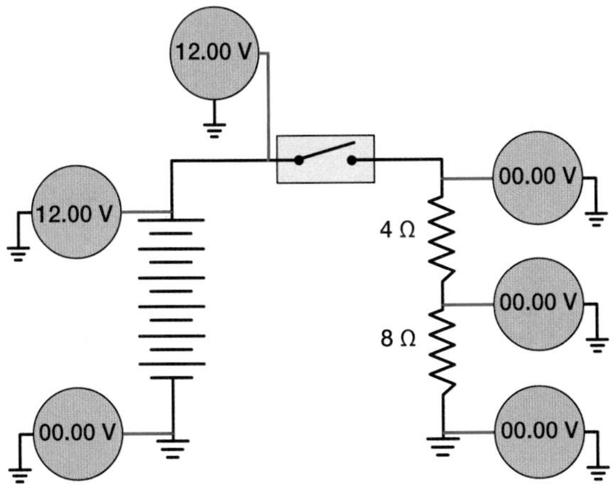

FIGURE 42-20 Measuring voltage at each load in a circuit with two unequal resistors connected in series with a 12-volt battery and with the switch open.

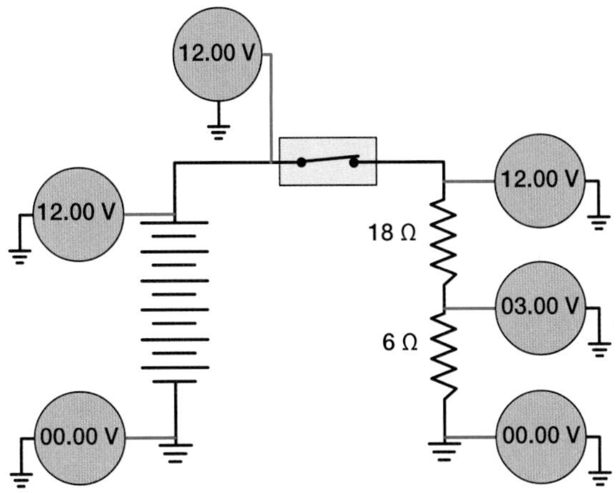

FIGURE 42-22 Another example of measuring voltage in a circuit with two resistors of unequal resistance connected in series.

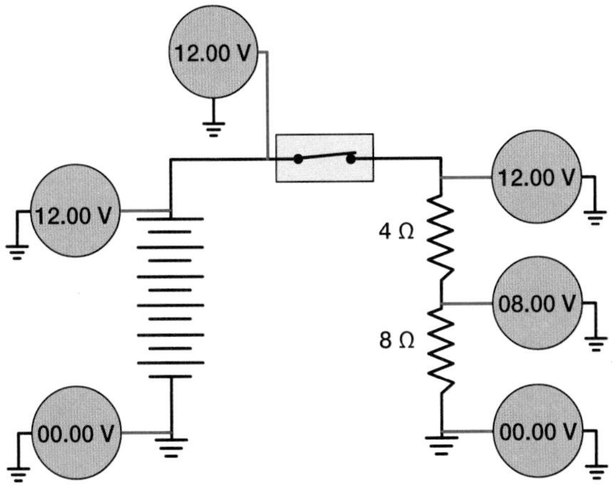

FIGURE 42-21 Measuring voltage in a circuit with two unequal resistors connected in series, with a 12-volt battery and the switch closed.

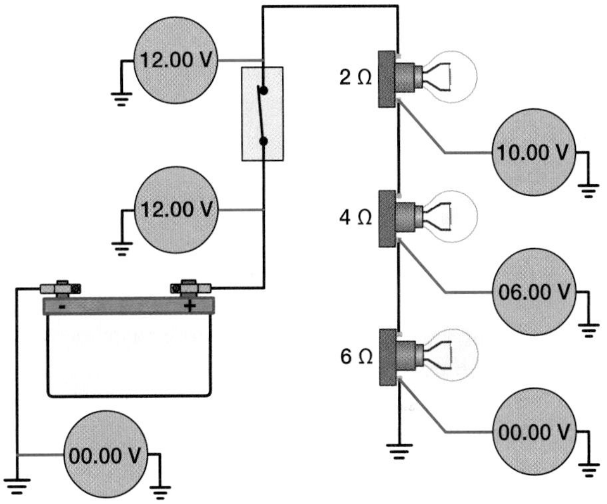

FIGURE 42-23 Notice how the voltage splits up when we have three resistors rather than just two.

used. In this case, the first resistor had 12 V at the input and 8 V at the output. That means the first resistor used 4 V. The second resistor had 8 V at the input and 0 V at the output. That means the second resistor used 8V. Also, note that after the second resistor, all of the voltage has been used up.

Because R_1 and R_2 are of different resistances, each will have a voltage drop proportional to its resistance. In a series circuit, the higher resistance has a greater proportion of the voltage drop. Notice that R_2 has twice the amount of resistance that R_1 has. Thus, it uses up twice the amount of voltage pushing the current through R_2 as it does through R_1.

Key Learnings: Current is now flowing through the entire circuit. Here are important points:

- All of the voltage gets used up by the end of the circuit.
- Each load uses some of the voltage.
- The higher the resistance, the more voltage used.

If we change the resistance of each resistor, the voltage used by the resistors also changes. In this case, both resistances are higher than in the previous example (**FIGURE 42-22**). Notice how the voltage used by each resistance changes. But it is still in proportion to the size of each resistance. Also, notice how putting the larger resistance first means the first resistor now uses the most voltage.

Key Learnings: The resistor with the highest resistance always uses the most voltage in a series circuit. But the sum of the voltage drops still equals source voltage.

What happens to voltage in a circuit when we change the number of resistors? In this example, we add one more resistor so there is a total of three resistors in series (**FIGURE 42-23**). Notice how the voltage changes. But once again the voltage used by each resistor changes in proportion to the other's.

Key Learnings: Adding a third resistor splits the voltage up three ways instead of two. The highest-resistance resistor still uses the most voltage. And the sum of all three voltage drops still equals source voltage.

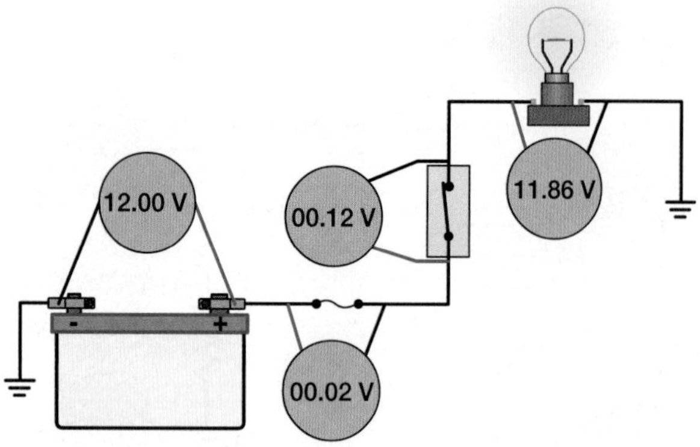

FIGURE 42-24 Typical voltage drops in a properly functioning circuit.

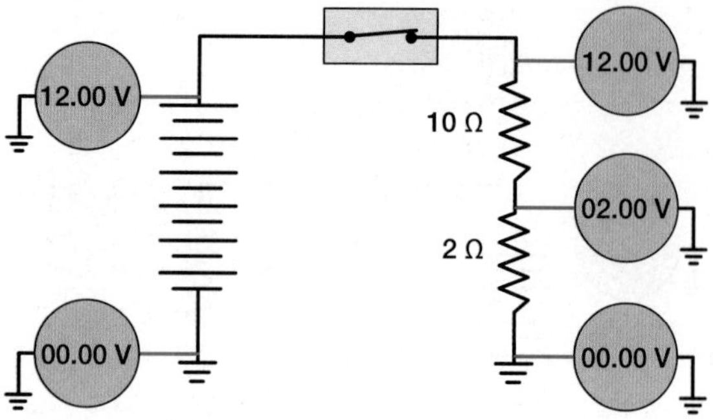

FIGURE 42-25 Available voltage testing.

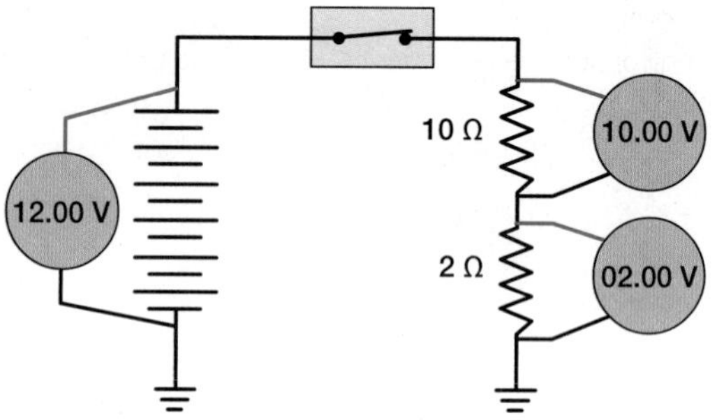

FIGURE 42-26 Voltage drop testing.

Voltage Drop

Voltage drop is the difference in voltage between two points in a circuit. Voltage drop occurs when current flows through a resistance. The voltage coming out of the resistance is less than the voltage going into the resistance. The higher the resistance, the higher the voltage drop, which also lowers the current flow.

In a properly operating circuit, the battery voltage is typically used up in the load (**FIGURE 42-24**). Most loads are designed to operate on full battery voltage. So, any voltage drops before or after the load reduce the available voltage at the load. This means there is less voltage to be used by the load. This reduces the power used by the load. You will remember that electrical power (W) = V × A. Voltage drops tend to reduce both the voltage and amperage available to the load. Therefore, power used by the load is also reduced, making it operate less effectively.

Voltage drops become a problem when they become excessive and occur in parts of the circuit other than the load. In reality, a small amount of resistance exists in each cable, connector, and switch within the circuit. A problem arises when the voltage drops become excessive in these components. This is because they starve the load.

Measuring Voltage Drop

In the earlier voltage exercises, we measured the voltage available at each side of the resistor. We then used math to calculate how much voltage each resistor used. To take these measurements, we placed the common lead on a good ground. The red probing lead was placed in the circuit to find out how much voltage was available at that point. Using the meter this way is called doing an **available voltage test.** In other words, we measured the voltage available both before and after each resistor (**FIGURE 42-25**).

Remember that a voltmeter measures the difference in voltage between the black lead and the red lead. Because of this, we can use the meter to measure voltage drop directly. This can be done by placing one voltmeter lead on the input of a component and the other lead on the output. The meter will read the difference in voltage between the two (**FIGURE 42-26**). This is how we perform a **voltage drop test**.

▶ Resistance Exercises

LO 42-06 Perform resistance measurements.

In this section, the exercises are designed to explain the use of the DMM in measuring resistance. Resistance measurements are used to check components or circuits against the manufacturer's specifications. For example, the resistance of sensors should be within the specified range. Examples are given to demonstrate measuring resistance. It is important to understand that Ohm's law tells us that current flow is inversely proportional to resistance. The higher the resistance, the less current will flow. The reverse is also true. The lower the resistance, the higher the current flow.

Measuring Resistance

In this exercise, a total circuit resistance measurement is taken. For resistance measurements, you need to select "auto range Ω" or the correct "manual range" on the DMM. Then, connect the red lead to V/Ω slot and the black lead to COM slot. If using "manual range," select an appropriate range. Start at the highest range and work your way down. Resistance measurements should only be taken with power disconnected. And ideally, the component should be disconnected from the circuit (**FIGURE 42-27**).

What happens to total circuit resistance when we add another resistor in series with the first one? Our series circuit law for resistance tells us that the resistances add up. So let's see if that happens (**FIGURE 42-28**).

Let's add one more resistor of equal resistance in series with the other two resistors in the circuit. Notice that the total circuit resistance just continues to add up. This is just like the series circuit law tells us (**FIGURE 42-29**).

Let's try one last experiment. What happens to total circuit resistance if we have three unequal-resistance resistors in series with each other? Note that the total resistance is just the sum of the individual resistances. This is just like the series law tells us (**FIGURE 42-30**).

▶ Current Exercises

LO 42-07 Perform current measurements.

In this section, the exercises are designed to explain the use of the DMM when taking DC current measurements. Undertaking the exercises will improve your understanding of Ohm's law and current measurements. Examples are given to demonstrate measuring current. We also explore the magnetic fields that are produced around a conductor when current flows. It is important to understand that current is the same in all parts of a properly working series circuit.

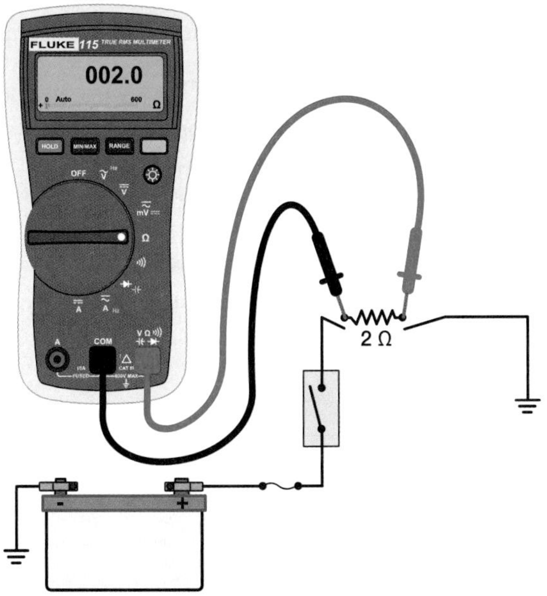

FIGURE 42-27 Measuring total resistance in a circuit with a single resistor.

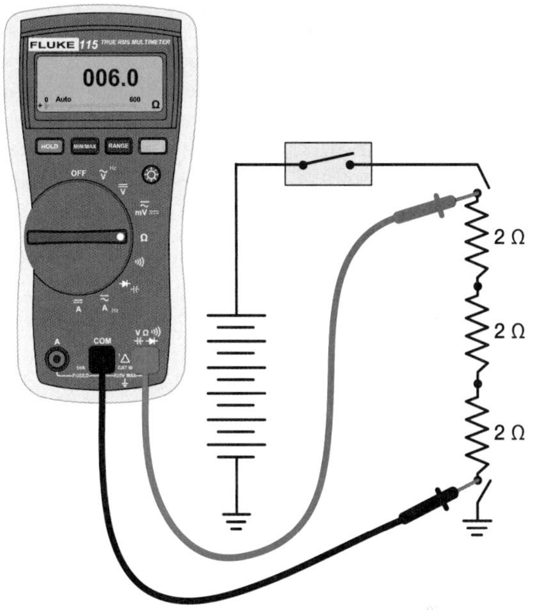

FIGURE 42-29 Measuring total resistance in a circuit with three equal-resistance resistors.

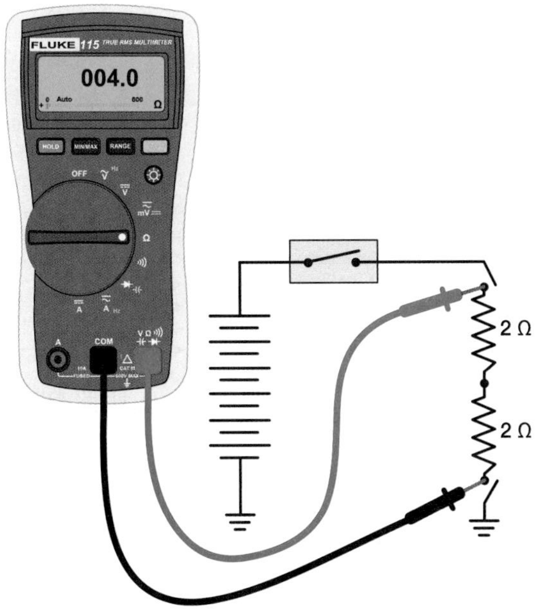

FIGURE 42-28 Measuring total resistance in a circuit with two equal-resistance resistors.

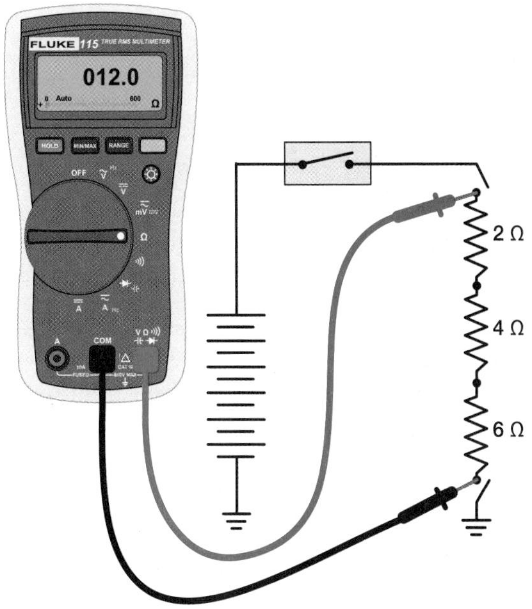

FIGURE 42-30 Measuring total resistance in a circuit with three unequal-resistance resistors.

Applied Math

AM-3: Mentally: The technician can mentally add two or more numbers to determine conformance with the manufacturer's specifications.

AM-6: Mentally: The technician can mentally subtract decimal and whole numbers to arrive at a difference for comparison with the manufacturer's specifications.

In this scenario, a technician is using a DMM to test the resistance of an ignition coil pack for a V-6 engine. The coil pack is suspected to be faulty. This is due to a failed power balance test in which the cylinder was not properly contributing to the performance of the engine. Because coil designs are different, manufacturers' testing procedures vary. With the key off and the battery lead to the coil disconnected, the ohmmeter function of the DMM is used to check resistance. The technician measures the resistance of the primary and secondary windings. In this case, the specifications for the primary windings are 0.3–1.0 ohm. Before taking this reading, the technician checks the resistance in the leads of his DMM. There is 0.2 ohm of resistance. This will need to be subtracted (mentally) from the resistance reading. Using service information, the technician places the leads of the DMM in proper slots of the component. A reading of 0.8 ohm is obtained. He mentally subtracts 0.2 ohm for the resistance in the meter leads, for a corrected reading of 0.6 ohm. This is within specifications for the primary windings.

The next step is the reading for the secondary windings of the coil pack. The manufacturer's specifications are 8000–9000 ohms. The technician places the meter leads as shown in the service information and obtains a reading of 6200 ohms. Mentally, the technician subtracts this from the minimum specification and records the answer on the repair order. The coil pack is 1800 ohms below minimum specifications. The technician will install a known good component (coil pack) and test to see how the engine performs.

Always remember that an ammeter must be connected in series within the circuit. That means that the circuit must be broken into two. And each end of the ammeter should be connected to one of the two broken ends. This method ensures that all of the current flowing through the circuit flows through the ammeter (**FIGURE 42-31**).

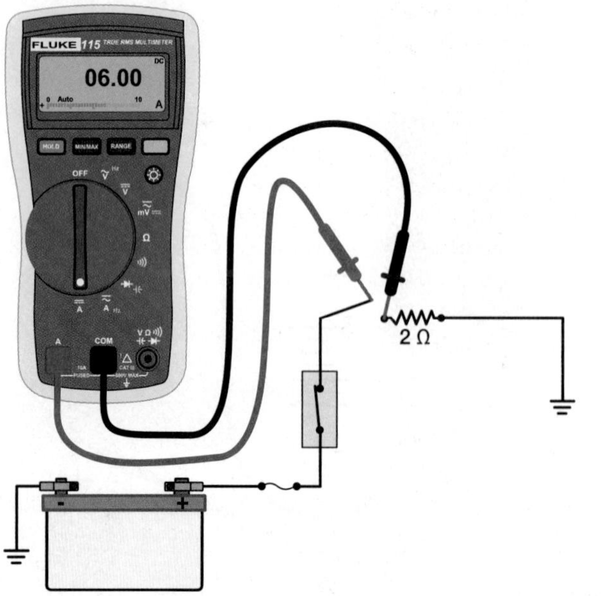

FIGURE 42-31 Measuring current flow in a circuit with a single resistor.

The series circuit law for current flow tells us that current flow stays the same throughout all parts of the circuit. Watch what happens when we place a second resistor of equal resistance in the series circuit. The current flow stays the same at all parts of the circuit. But also note that the current went down from the previous example. If you said that is because the resistance went up, you would be correct (**FIGURE 42-32**).

What happens to current flow if we have a circuit with unequal-resistance resistors? According to our series circuit laws, the current flow should stay the same throughout the circuit. Let's see if that is true (**FIGURE 42-33**).

Current and Magnetic Fields

In this example, a relay controlled by a switch is used to switch the current through a lightbulb. The compass is used to demonstrate that a magnetic field is produced around the relay winding

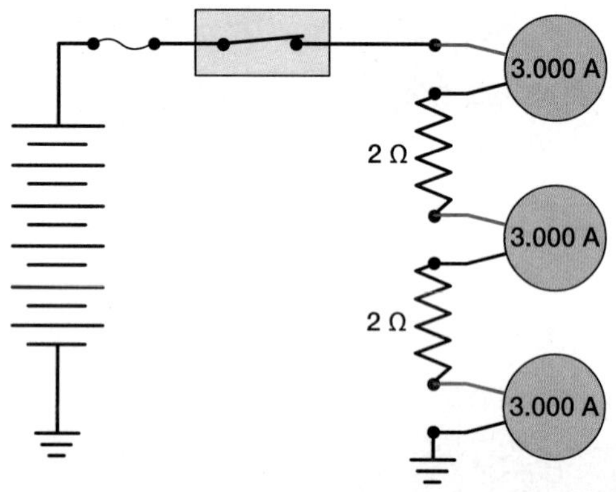

FIGURE 42-32 Measuring current flow in a circuit with two equal-resistance resistors in series.

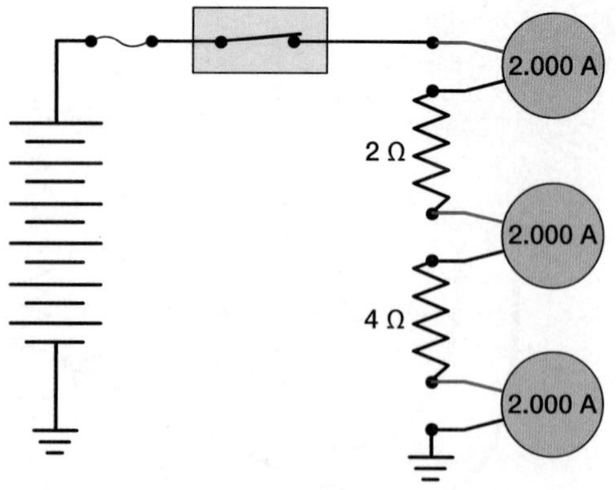

FIGURE 42-33 Measuring current flow in a circuit with two unequal-resistance resistors in series.

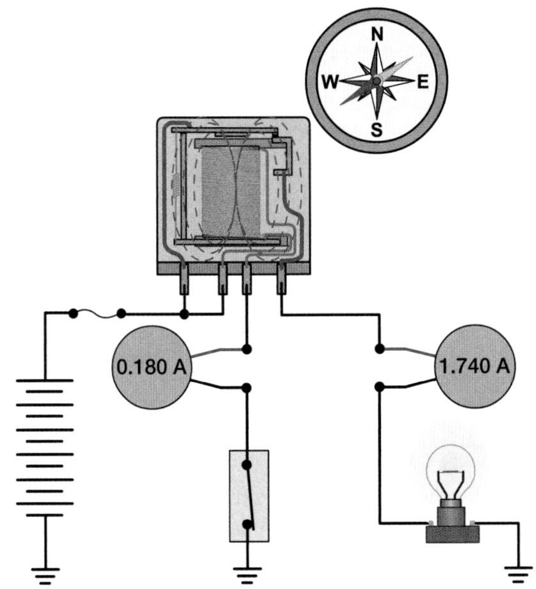

FIGURE 42-34 Measuring amperage in a circuit with a relay controlled by a switch and a single lightbulb with a 12-volt DC supply.

when the current flows through it. To conduct this experiment, set the DMM to measure "DC amps." Connect the red lead to the A slot and the black lead to the COM slot. If using a manual-range DMM, select an appropriate range.

FIGURE 42-34 shows a circuit with a relay controlled by a switch and a single lightbulb with a 12-volt DC supply. Notice that the control current (relay winding) is much smaller than the lightbulb current. Finally, the compass is used to show that when energized, the relay winding produces a magnetic field. This is one way to check whether the relay winding is operating without removing the relay to test it.

Applied Math

AM-9: Mentally: The technician can mentally divide decimal and whole numbers to determine conformance with the manufacturer's specifications.

In this scenario, the technician is using a DMM to measure cooling fan current flow in amps. The circuit resistance in ohms will be determined, using Ohm's law.

The first step is to locate proper service information for the vehicle. It should cover the following items: fuse block diagram and cooling fan circuit diagram. Also, locate the specified current flow for cooling fan motor. Next, the cooling fan relay is removed. The DMM is set up to measure current by placing the red lead in the 10A slot on the meter. The black lead should be placed in the COM slot. In this testing procedure, the DMM is connected in series with the red lead in socket 30 and the black lead in socket 87. With the cooling fan motor in operation, the technician observes a reading of 3 amps. The specifications for this vehicle call for a range of 2.3–4.6 amps to operate the cooling fan motor.

To determine the resistance of this circuit, we use the formula, R = V/A. This is a form of Ohm's law (Voltage = Amps × Resistance); R = 12 volts/3 amps = 4 ohms.

▶ Perform Series Circuit Measurements

LO 42-08 Perform series circuit measurements.

These series circuit exercises show the use of the DMM for measuring voltage, resistance, and amperage in a series circuit. This is critical for understanding how to diagnose electrical faults. If current flow is wrong, it is wrong because either the voltage is wrong or the resistance is wrong. So, knowing how to measure volts, amps, and resistance allows you to gather data about the circuit. This will be critical for determining whether voltage or resistance is causing the fault. The examples given here demonstrate how current flow is affected by voltage and resistance in a series circuit.

It is important to remember that current flow stays the same in all parts of a series circuit. Also, the sum of the voltage drops across individual resistances in a series circuit is equal to the supply voltage. The exercises also examine how changing resistance in a series circuit affects the current flow and voltage drops.

One *very* important point should be noted in the following exercises. We are using lightbulbs to represent the loads. This is so that you can see the relative brightness of each bulb to represent the work being done in the circuit. One thing to know about lightbulbs is that the resistance of the filament increases greatly when they light up. This is due to the heat causing it to glow. We are *not* taking that increase in resistance into consideration in these exercises.

So, if you were to measure the current flow in a 12-volt circuit with a 6-ohm bulb (cold resistance), it wouldn't have 2 amps flowing through it. It would probably have about 0.25 amp due to the increase in resistance from the filament being white hot. This has a minimal effect on the exercises listed below. But we want you to be fully aware of this situation. And that is, the measured resistance of a bulb will be much lower than the calculated resistance of the bulb when lit. This just demonstrates the effect that heat has on resistance. It does not affect the concepts presented below.

Series Circuit Exercise 1

In this exercise, voltage and voltage drop measurements are taken in a series circuit. To measure voltage drop, set the DMM on the voltage range. Select "auto range volts DC" or the correct "manual range" on the DMM. Then, connect the black lead to the COM slot and the red lead to the V/Ω slot. You can measure voltage drop across components, connectors, or cables. But the current has to be flowing to get accurate measurements.

Remember, when checking voltage, the leads can be placed in either direction. Just remember which way you placed them so you understand what the reading means. Also, when measuring current flow, always break the circuit open and insert the ammeter in series with the load in the circuit. (**FIGURE 42-35**) shows a typical circuit with a battery supply, fuse, switch, and lightbulb.

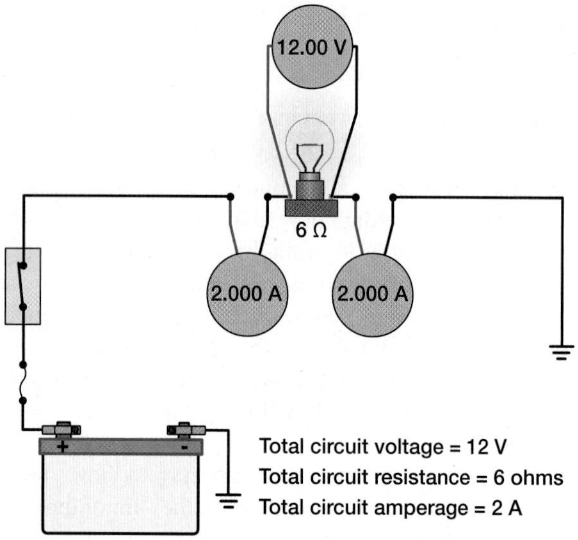

FIGURE 42-35 Electrical behavior in a typical series circuit with a battery supply, fuse, switch, and lightbulb.

Key Learnings: Note that the lightbulb operates at full brightness. Also, note the following:

- All of the voltage is used up as current flows through the circuit.
- Virtually all of the voltage is used up in the load on a properly operating circuit.
- The current flow stays the same throughout the circuit.
- Because there is only one resistance in the circuit, that is also the total circuit resistance.

Series Circuit Exercise 2

In this exercise, we add a second equal-resistance bulb in series with the first bulb. Let's look at how this affects resistance, current flow, and voltage drop in a series circuit (**FIGURE 42-36**).

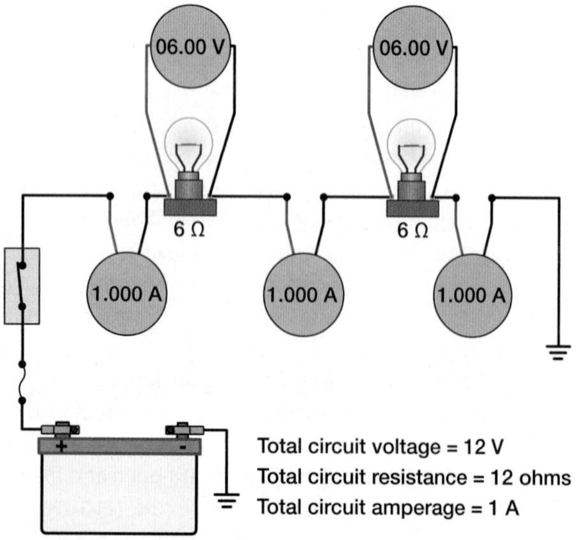

FIGURE 42-36 Electrical behavior in a circuit with two lightbulbs of equal value connected in series.

Key Learnings: Note that having two equal-resistance bulbs means the bulbs are equally bright, but much dimmer than with only one bulb. Also, note the following:

- The voltage split up evenly (voltage drop) between the two bulbs due to their equal-resistance value.
- The total resistance doubled, which reduced total current flow in the circuit by half.
- But the current flow stayed at that lower level at every point in the circuit.
- If one bulb burns out, it creates an open circuit. This means that no current flow will flow through the circuit, including the other bulb.

Series Circuit Exercise 3

In this exercise, we have two unequal-resistance bulbs in series. This changes how electricity behaves in the circuit. But that behavior can be predicted by applying the appropriate Ohm's laws and series circuit laws. Let's see how this new arrangement operates (**FIGURE 42-37**).

Key Learnings: Note that having two unequal bulbs in series makes the bulbs operate differently from each other. Also, note the following:

- The voltage split up unevenly (voltage drop) between the two bulbs due to their unequal-resistance value.
- The sum of the voltage drops still equals source voltage.
- The first bulb is the brightest. This is because it has the larger voltage drop due to its higher resistance.
- The total current flow stayed the same throughout the entire circuit, even though the resistances are unequal.
- Total resistance is the sum of the individual resistances.

Series Circuit Exercise 4

In this exercise, we use the same two bulbs as in the last exercise. But this time we reverse their order. In the previous exercise,

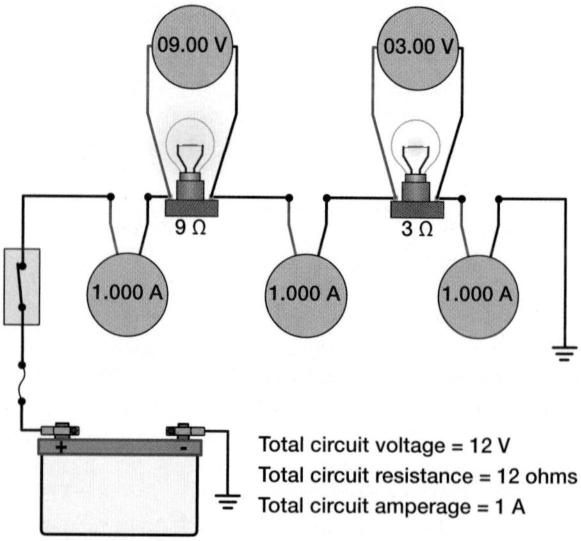

FIGURE 42-37 Electrical behavior in a circuit with two unequal lightbulbs connected in series.

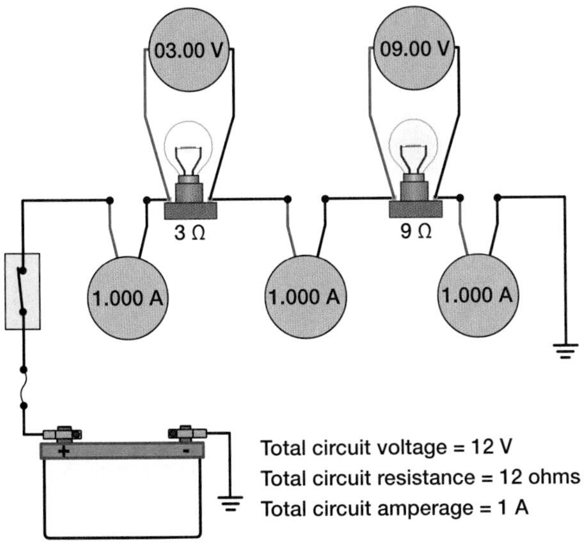

FIGURE 42-38 Electrical behavior in a series circuit with the two unequal-resistance lightbulbs reversed.

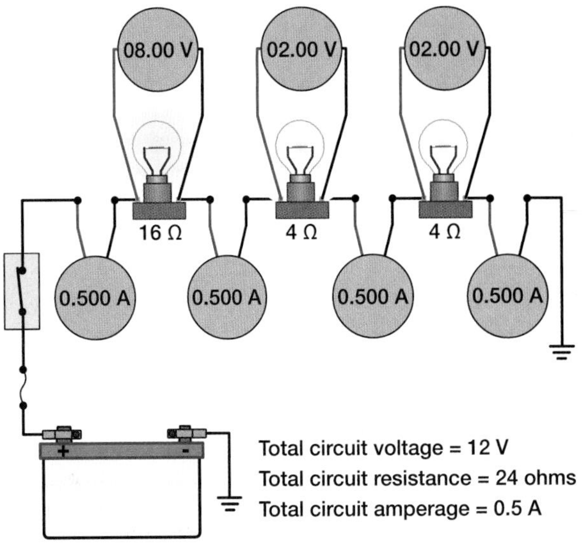

FIGURE 42-39 Electrical behavior in a circuit with three unequal-resistance lightbulbs in series.

the first bulb (the small one) was relatively bright, whereas the second bulb (the larger one) was relatively dim. Which bulb do you think will be bright this time? The larger first bulb? Or the smaller second bulb? Let's take a look to see which one uses the most voltage (**FIGURE 42-38**).

Key Learnings: Note that swapping the position of the bulbs in the circuit didn't change the operation of the bulbs. They still operate just like they did in the previous exercise. Note the following:

- The brightest bulb is still the highest resistance bulb, even though it is last in the circuit.
- It uses the most voltage, making it brighter.
- Current flow stays the same throughout the circuit, even though the resistances of each bulb are not the same.
- The sum of the voltage drops still equals source voltage.
- The resistance still adds up in a series circuit.

Series Circuit Exercise 5

In this exercise, we place three unequal-resistance lightbulbs in series to see how that affects the behavior of electricity. Again, apply Ohm's law and the series circuit laws to predict the meter readings and calculations shown in (**FIGURE 42-39**).

Key Learnings: Note that there are now three bulbs in the series instead of two. Adding the extra bulb in this manner affected the circuit in the following ways:

- The voltage splits up three ways. But the sum of the individual voltage drops still equals source voltage.
- The sum of the individual resistances adds up to total circuit resistance. So, adding more bulbs increases the resistance.
- The higher total resistance reduced the total current flow from the previous circuits.
- The current flow stays the same at all points in the circuit.

▶ Perform Parallel Circuit Measurements

LO 42-09 Perform parallel circuit measurements.

In this section, the exercises show the use of the DMM for measuring volts, amps, and ohms in a parallel circuit. Parallel circuits are commonly used in the vehicle's electrical system, especially for lights. Understanding the relationships between voltage, amperage, and resistance in parallel circuits helps you to diagnose electrical faults. Examples are given to demonstrate how to measure volts, amps, and ohms. This will help show how current flows and voltage drops in parallel circuits. It is important to remember the laws for a parallel circuit:

- resistance goes down when more parallel paths are added;
- current flow from individual legs adds up in parallel;
- and voltage stays the same at all common parallel circuit inputs.

The following exercises help to reinforce the understanding of these laws.

One of the main differences in a parallel circuit compared with a series circuit is that resistance goes down as more parallel loads are added. This means that the total parallel resistance is always lower than the smallest parallel resistance. So, we can't just add up the resistance values to find total parallel circuit resistance. We need a new way of determining that. There are actually two ways. The first is the easiest, but it only works if there are only two resistors in parallel. It can't calculate the resistance for more than two resistances. Here, it is using R_1 = 6 ohms and R_2 = 3 ohms as an example:

$$R_T = \frac{R_1 \times R_2}{R_1 + R_2} = \frac{6 \times 3}{6 + 3} = \frac{18}{9} = \frac{2}{1}, \text{ so } R_T = 2 \text{ ohms}$$

So, the total resistance for a 6-ohm resistor in parallel with a 3-ohm resistor is 2 ohms.

The second way of calculating parallel resistance is a bit more involved. But it can accommodate any number of parallel resistances. You just have to add more values to the right-hand side of the equation.

$$R_T = \frac{1}{\frac{1}{R_1}+\frac{1}{R_2}+\frac{1}{R_3}}$$

$$R_T = \frac{1}{\frac{1}{2}+\frac{1}{4}+\frac{1}{2}}$$

$$= \frac{1}{\frac{5}{4}}$$

$$= \frac{1}{1.25}; R_T = .8\Omega$$

Parallel Circuits—Exercise 1

In this exercise, we explore how adding an equal-resistance bulb in parallel with an existing bulb affects resistance, current flow, and voltage drops. **FIGURE 42-40** shows two circuits—one series circuit and one parallel circuit. We will explore the behavior of electricity in a parallel circuit. The extra resistor in parallel causes an increase in circuit current flow. This also decreases total circuit resistance.

Key Learnings: Adding an equal-resistance bulb in parallel allows both bulbs to be lit to full brightness. Also, note the following:

- Adding another bulb in parallel provides another path for current flow. This reduces the total circuit resistance, which increases total circuit amperage.
- Total resistance of the bulbs in parallel is lower than the resistance in the lowest resistance bulb, just as the parallel circuit law for resistance claimed.
- Each bulb uses the full 12 volts, so each bulb is lit to full brightness.
- If one bulb burns out, the other bulb will still operate because each bulb has its own power and ground.

Parallel Circuits—Exercise 2

In this exercise, we see how unequal resistance bulbs affect voltage, resistance, and current flow in a parallel circuit. **FIGURE 42-41** shows two unequal-resistance bulbs in parallel. Notice how the unequal resistance changes the current flow in each leg of the circuit.

Key Learnings: Notice that the current flow in each leg of the parallel circuit is proportional to its resistance. Each leg acts independently in a parallel circuit. Also, note the following:

- Total circuit resistance is still lower than the lowest resistance.
- Voltage stayed the same at both parallel inputs.
- Total circuit current flow is found by adding up the current flow for each leg.

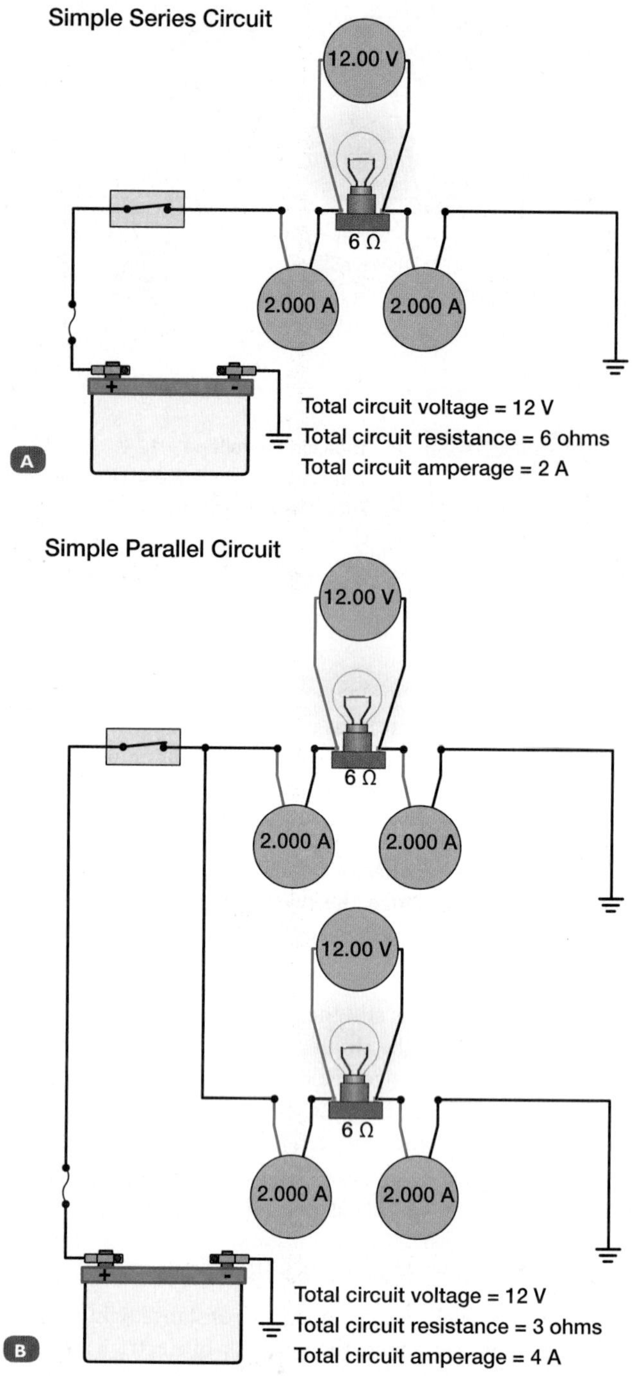

FIGURE 42-40 A. Electrical behavior in a series circuit. **B.** Electrical values in a parallel circuit with equal-resistance bulbs.

- From the above, we can determine that the voltage drop across one parallel branch is the same for all parallel branches, regardless of their resistance.

Parallel Circuits—Exercise 3

In this exercise, we add one more bulb in parallel with the existing two bulbs, which are also in parallel. (**FIGURE 42-42**) has three unequal resistors in parallel. Notice how the electrical values change in this exercise.

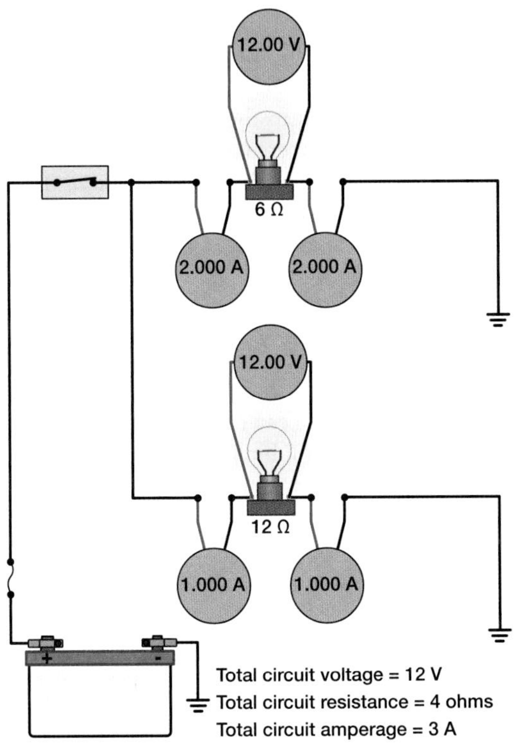

FIGURE 42-41 Electrical behavior in a parallel circuit with unequal-resistance bulbs.

Key Learnings: Adding another bulb in parallel made for three bright lights instead of two. Also, note the following:

- Total circuit resistance has gone down further and is still lower than the lowest individual resistance.
- Total circuit current flow has increased with the new parallel bulb.
- Each bulb is using full source voltage.

SAFETY TIP

Notice how the current in the circuit increases with each additional load. This is also what happens with a power strip. Each time an additional item is plugged in, the current increases. Overloading the circuit may cause the protection device to trip or in extreme cases could cause a fire. Never overload an electrical circuit, whether on a vehicle or in the shop.

Applied Math

AM-45: <, >, =, e.g.: The technician can interpret symbols to determine conformance with the manufacturer's specifications.

In this scenario, we look at three of the most common symbols used in technical manuals.

<, >: These symbols are lesser than (<) and greater than (>). An example of their use is describing the specification for the maximum limit for an AC voltage from an alternator (<0.5 volt AC).

=: This symbol is a mathematical symbol used to indicate equality. An example is the formula for Ohm's law, which is E = I × R.

e.g.: This means "for example." It comes from the Latin expression *exempli gratia*, or "for the sake of an example." An example of this is "The technician uses a precision measuring instrument (e.g., micrometer) to measure the part."

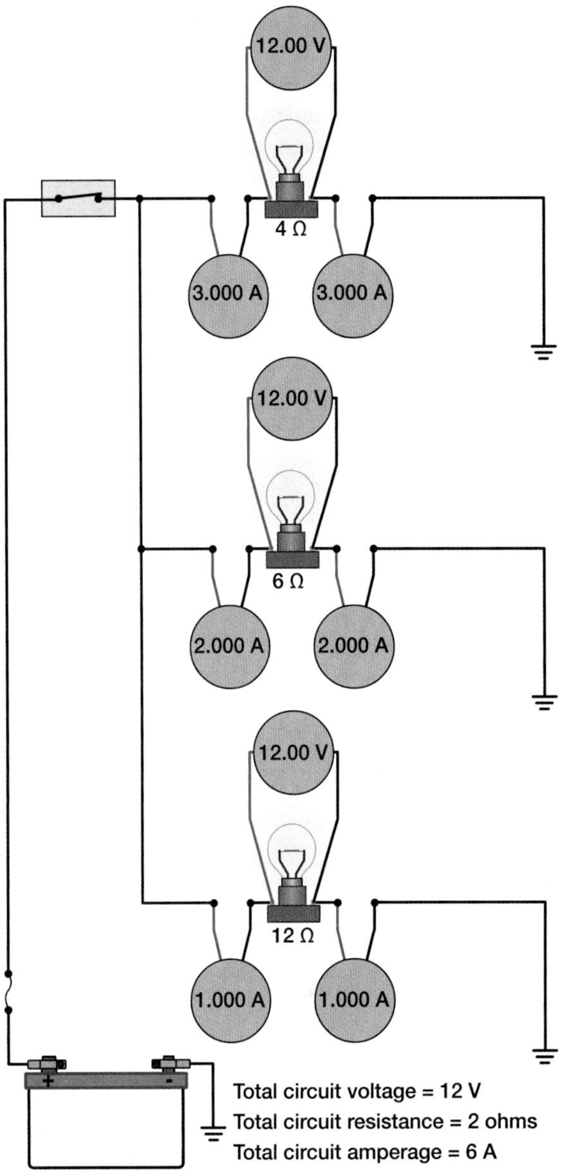

FIGURE 42-42 Electrical behavior in a parallel circuit with three unequal-resistance bulbs.

▶ Perform Series-Parallel Circuit Measurements

LO 42-10 Perform series-parallel circuit measurements.

In this section, the exercises are designed to demonstrate the use of the DMM in measuring volts and amps in a series-parallel circuit. Series-parallel circuits are found in vehicles, such as in dash light dimmer circuits. However, series-parallel circuits are not as common as parallel circuits.

Typically, a series-parallel circuit happens when unwanted resistance shows up in series with a parallel circuit. For example, consider if the brake light switch contacts become worn out. They can create a voltage drop in series with the brake lights. This will cause all the brake lights to be dimmer than they should be.

It is important to understand how series-parallel circuits work. The relationships between current flow, voltage drops, and resistance will help you diagnose these types of electrical faults. Examples are given to show how to measure voltage and current in a series-parallel circuit. They will show how current flow and voltage drop are affected by resistance in a series-parallel circuit.

It is important to understand that to calculate current flow and voltage drop, the total resistance of the circuit must be known. That means that you will need to boil the entire circuit down to a series circuit equivalent. Then, you can work out what is happening in each separate part of the circuit. You will see how that is done in the experiments.

As always, the voltage drop will be the same across each parallel branch. Also, the sum of the current flow in each branch is equal to the total parallel circuit current flow. And the series circuit laws apply to the series portion of the circuit. These exercises also examine how the addition of resistors in series with a parallel circuit affects the circuit current flow, voltage drops, and circuit resistance.

Series Parallel—Exercise I

In this exercise, voltage, resistance, and current measurements are taken from the series-parallel circuit formed by resistors R_1, R_2, and R_3. This will help you observe the behavior of electricity in this type of circuit (**FIGURE 42-43**). To measure voltage drop, the DMM must be set on the voltage range. Select "auto range volts DC" on the DMM. Then, connect the black lead to the COM slot and the red lead to the V/Ω slot. Voltage drop can be measured across components, connectors, or cables as long as current is flowing in the circuit. The red lead of the DMM is normally connected to the positive side of the component.

To measure current, the DMM must be set to read DC amps. Connect the red lead to the A slot and the black lead to the COM slot. If using a manual-range DMM, select an appropriate range. Please note that calculations may have been rounded in some instances.

Key Learnings: We used meters to measure the electrical values in the circuit. But you can calculate them as well by using the circuit laws along with Ohm's law. To calculate the values, you need to first effectively convert them to a series circuit. You do this by finding the total resistance of the parallel circuit. Because the parallel portion is in series with the series circuit, the parallel resistance can be added to the series resistance. This will give you the total circuit resistance.

Then, you can calculate the total current flow by dividing the total circuit voltage by the total circuit resistance. We know the

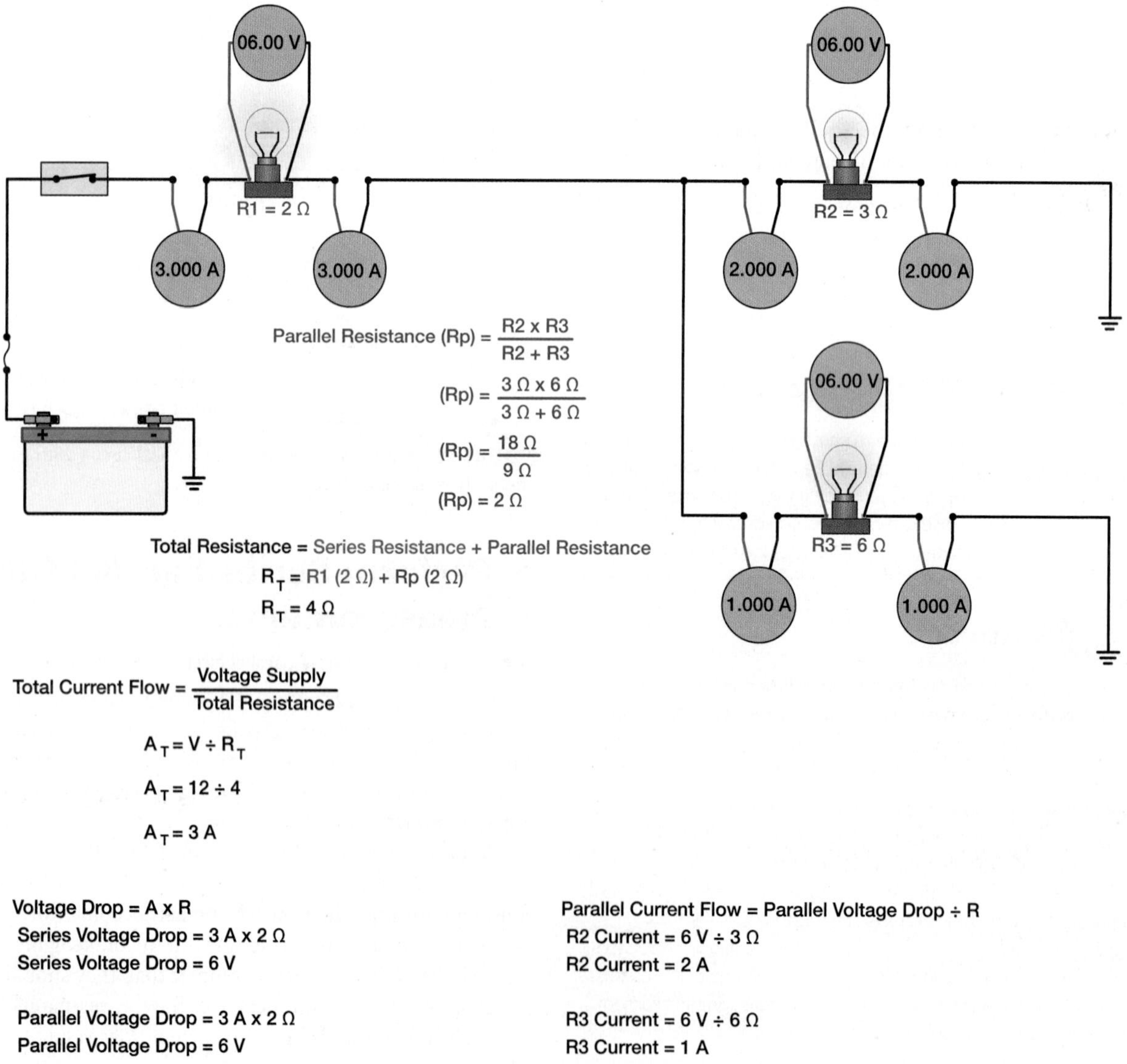
06.00 V
06.00 V
R1 = 2 Ω
R2 = 3 Ω
3.000 A
3.000 A
2.000 A
2.000 A
Parallel Resistance (Rp) = R2 x R3 / R2 + R3
(Rp) = 3 Ω x 6 Ω / 3 Ω + 6 Ω
(Rp) = 18 Ω / 9 Ω
(Rp) = 2 Ω
06.00 V
R3 = 6 Ω
1.000 A
1.000 A
Total Resistance = Series Resistance + Parallel Resistance
R_T = R1 (2 Ω) + Rp (2 Ω)
R_T = 4 Ω
Total Current Flow = Voltage Supply / Total Resistance
A_T = V ÷ R_T
A_T = 12 ÷ 4
A_T = 3 A
Voltage Drop = A x R
Series Voltage Drop = 3 A x 2 Ω
Series Voltage Drop = 6 V
Parallel Voltage Drop = 3 A x 2 Ω
Parallel Voltage Drop = 6 V
Parallel Current Flow = Parallel Voltage Drop ÷ R
R2 Current = 6 V ÷ 3 Ω
R2 Current = 2 A
R3 Current = 6 V ÷ 6 Ω
R3 Current = 1 A

FIGURE 42-43 A series-parallel circuit formed by bulbs R_1, R_2, and R_3.

total circuit current flows through the series portion of the circuit. So, we can then calculate the voltage drop of the series circuit by multiplying the current flowing through R_1 times the resistance of (R_1). We can then subtract the voltage drop of R_1 from source voltage to get the voltage applied to the parallel circuit.

We know that each branch of the parallel circuit gets the full voltage of what we just calculated. And since each branch is grounded, each branch drops that full applied voltage. Then, we can calculate the current flow through each parallel resistor by dividing the parallel circuit voltage drop by each parallel resistance (R_2 and R_3). Also, note these findings:

- The total circuit current flows through the series bulb.
- The total source voltage is shared between the series and parallel portions of the circuit.
- Each of the parallel bulbs receives the full parallel circuit voltage.
- The total circuit amperage is split up between each branch of the parallel portion of the circuit according to the resistance of each bulb.
- The total resistance of the parallel circuit is lower than any single parallel branch resistance.

Series Parallel Exercise 2

In this exercise, we have added another bulb in the parallel portion of the circuit. Voltage and current measurements will be taken from the series-parallel circuit formed by resistors R_1, R_2, R_3, and R_4. You will then see how the electricity behaves differently from the last circuit (**FIGURE 42-44**).

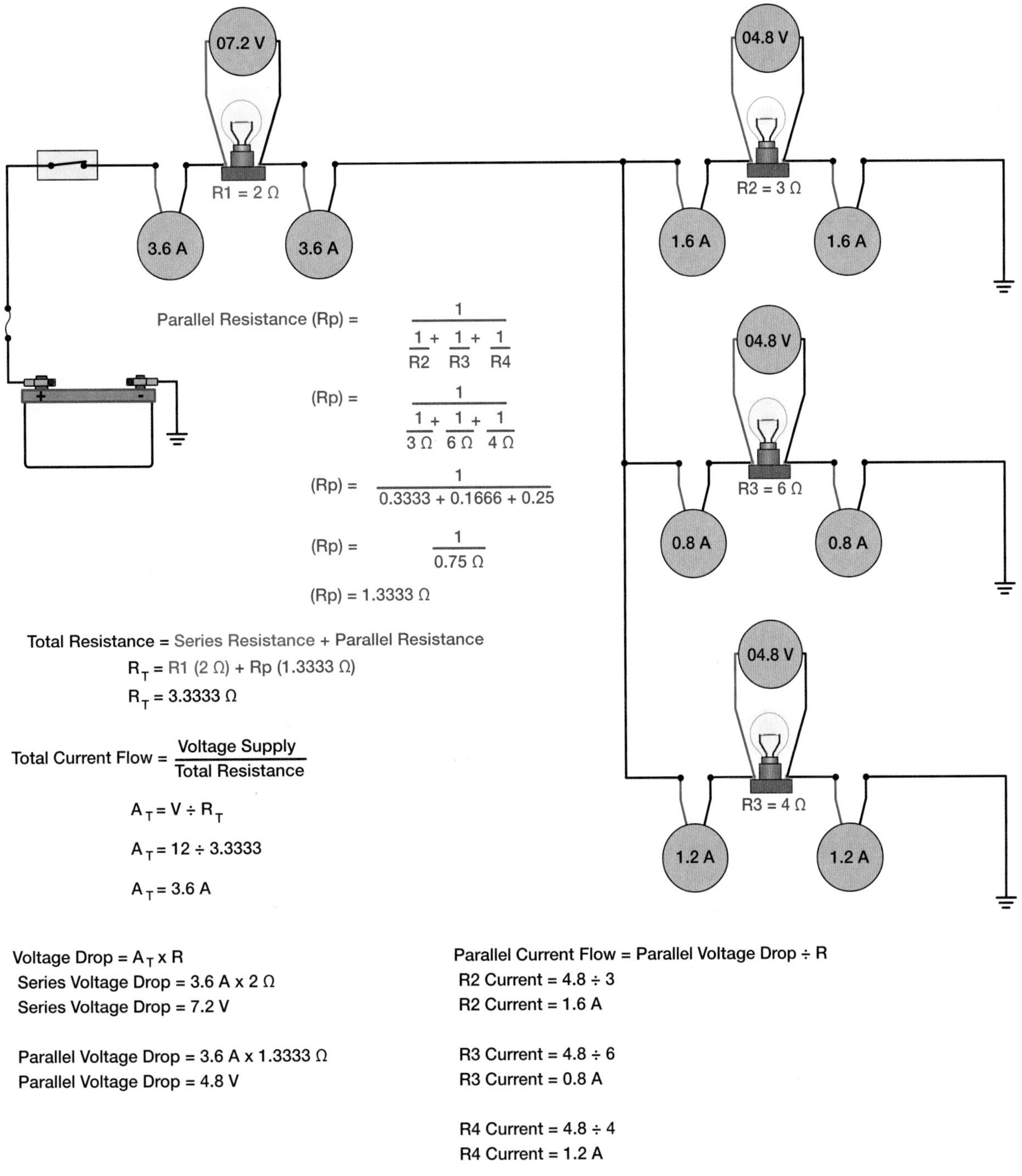

FIGURE 42-44 A series-parallel circuit formed by bulbs R_1, R_2, R_3, and R_4.

Key Learnings: Adding the additional bulb in the parallel portion of the circuit reduced the total resistance of the parallel circuit. This also lowered the resistance of the entire circuit. The decreased resistance increased the total circuit current flow. The increased current flow through R_1 increased the voltage used (voltage drop) of (R_1). This increased voltage drop reduced the voltage to the parallel portion of the circuit. Also, the parallel branches had to share the total parallel current flow with the additional branch. So, each branch received slightly less current flow than the previous exercise. Also, note these findings:

- Adding the additional bulb in parallel reduced the total circuit resistance slightly.
- The reduced resistance allowed a bit more total circuit current to flow.
- The increased current flow increased the voltage drop across R_1, making it a bit brighter.
- The parallel bulbs got dimmer for two reasons. First, the voltage they received was less. And second, each branch got less current flow. Because Power is V × A, and the bulbs used less of each of these, they operated on lower power.

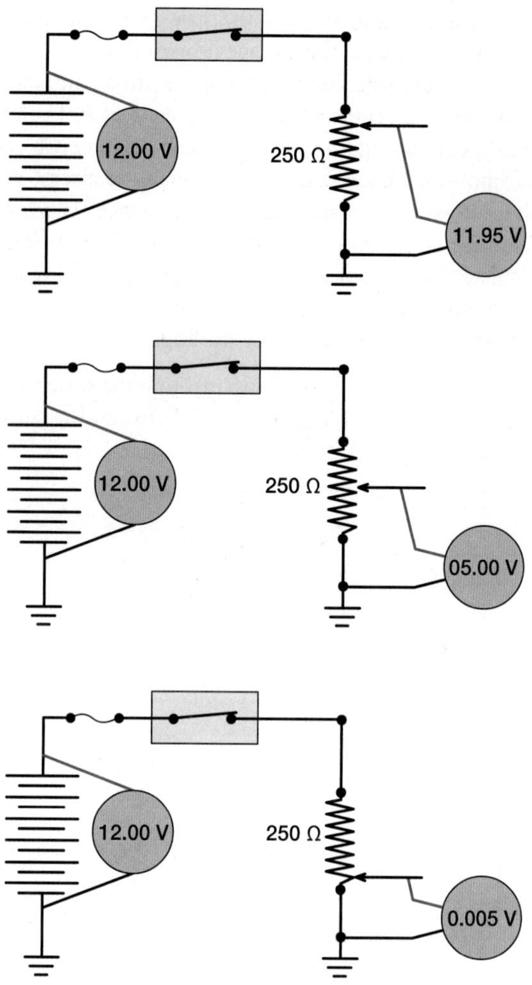

FIGURE 42-45 Measuring voltage in circuits with a 250 Ω variable resistor and a 12-volt DC supply.

▶ Variable Resistors

LO 42-11 Perform measurements on variable resistors.

In this section, the exercises show the use of the DMM in measuring voltage and amperage in a circuit with a variable resistor. This will help you understand how voltage, resistance, and current change as the potentiometer is adjusted. Examples are given to demonstrate measuring voltage and current. It is important to understand that as the position of the wiper is changed, so does the voltage and current flow.

Variable Resistors Exercise

In this exercise, a variable resistor is used as a potentiometer, or voltage divider. For voltage measurements, you need to select "auto range volts DC" on the DMM. Then, connect the red lead to the V/Ω slot and the black lead to the COM slot. For current measurements, select "auto range milliamps DC" on the DMM. Connect the red lead to the A slot and the black lead to the COM slot. If using a manual-range DMM, you will need to select an appropriate range.

FIGURE 42-45 shows circuits with a 250 Ω variable resistor in the circuit. It is used as a voltage divider with a 12-volt DC supply. In FIGURE 42-46A, the wiper of the variable resistor is set so that maximum voltage will occur. In FIGURE 42-46C, the variable resistor is set so that minimum voltage can occur. The variable resistor is continuously variable between these points. The voltage varies depending on the position of the wiper (FIGURE 42-43C). When the wiper in FIGURE 42-46 is at the upper or lower limits, in practice, there may be a small amount of the variable resistor left on the outer edges. So, the voltage may not quite reach the full 12 volts or 0 volts.

Key Learnings: A variable resistor can provide an infinitely variable voltage reading from ground to input voltage, depending on the position of the wiper. This makes it ideal for use as a position sensor.

▶ Electrical Circuit Testing

LO 42-12 Describe electrical circuit testing.

Electrical circuit testing begins with understanding circuit types and how electricity behaves within them. Add to that the ability to use meters to measure the values of voltage, amperage, and resistance. Also critical is understanding how to read wiring diagrams so you will know how the circuits are constructed. Once you know all of this, you will be well on your way to diagnosing electrical faults successfully. Those are the concepts we will explore in this section. Feel free to refer back to the previous sections in this chapter. They will help you remember how electricity behaves as well as how meters are hooked up for specific measurements. Let's kick this off by seeing how Ohm's law can help us predict the behavior of electricity.

Using Ohm's Law to Diagnose Circuits

Ohm's law can be used in two ways to help diagnose electrical circuit faults. The first is by using it to perform the math to predict and verify measurements. The second way is by using the relationships between volts, ohms, and amps that Ohm's law demonstrates. These relationships will help guide you through the diagnostic process.

When using Ohm's law in the first way, you calculate electrical quantities in a circuit. That is valuable in cross-checking actual measured results within the circuit. For example, if the resistance and voltage of a circuit are known, then the theoretical current can be calculated using Ohm's law. The calculated result can then be compared with the specified or measured amperage. This will allow you to determine if the circuit is functioning correctly.

Technicians often do a quick calculation, sometimes just in their head. This gives them an approximate value of an electrical quantity before they take actual measurements. Doing so allows them to anticipate what they will be measuring and to set the DMM to the correct range.

Always remember that a calculation may only give an approximate value. In actual circuits, variations exist in components. This causes differences between calculated values and actual measurements.

Using Ohm's law the second way helps you understand the relationships between volts, amps, and ohms. For example, if voltage stays the same but resistance decreases, amperage must increase (**FIGURE 42-46**). In the case of a short circuit, the resistance decreases and the amperage increases. This can potentially blow the fuse. In the opposite scenario, where resistance increases, current flow decreases (**FIGURE 42-47**). This is the case when a corroded or loose connection introduces excessive resistance into the circuit. It also results in less electrical power (volts and amps) to operate the intended load.

What does Ohm's law tell us to expect when the voltage changes? If voltage decreases and the resistance stays the same, then amperage will decrease (**FIGURE 42-48**). This results in less power being able to operate the load. If the voltage increases and the resistance stays the same, then amperage will increase (**FIGURE 42-49**). If amperage and voltage both increase, then the

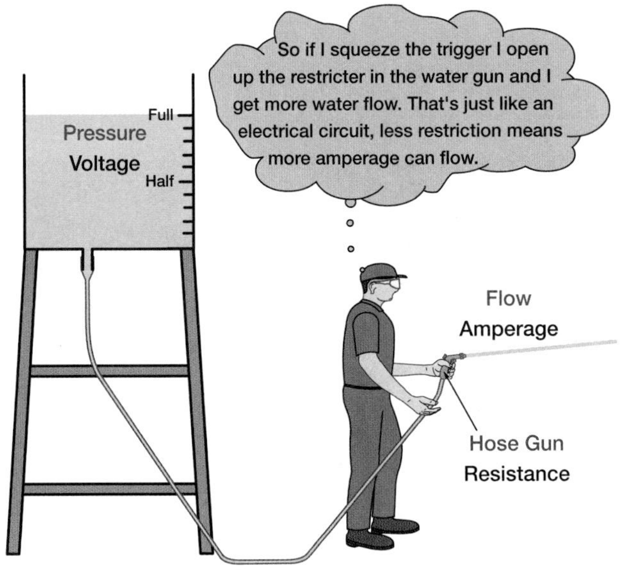

FIGURE 42-46 If voltage stays the same but resistance decreases, amperage must increase.

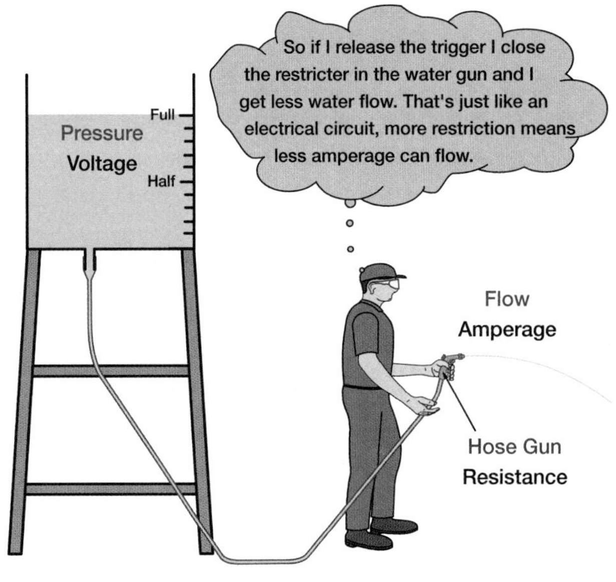

FIGURE 42-47 If voltage stays the same but resistance increases, amperage must decrease.

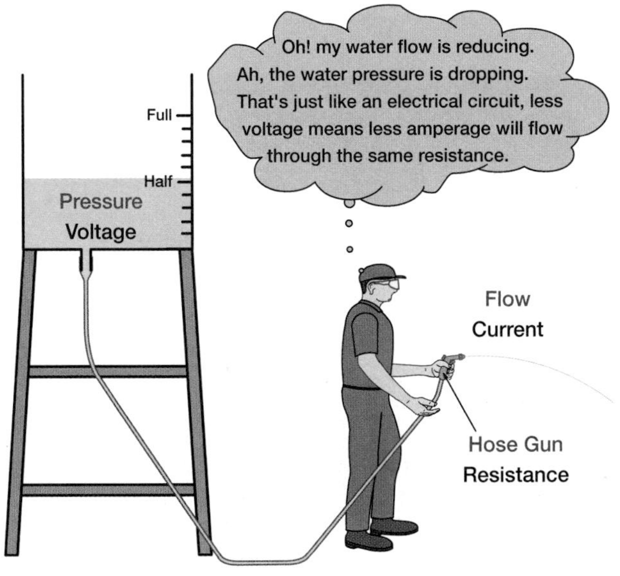

FIGURE 42-48 If voltage decreases but resistance stays the same, amperage must decrease.

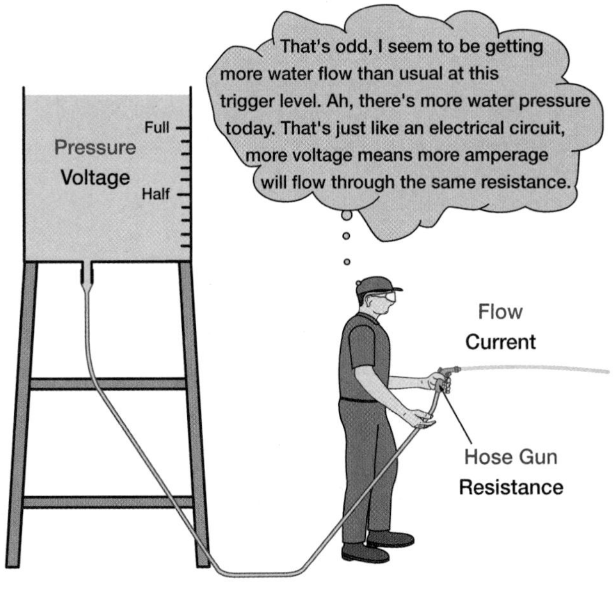

FIGURE 42-49 If voltage increases but resistance stays the same, amperage must increase.

electrical power operating the load is also increased. This condition can shorten the life of, or even burn out, the load.

So, how do amperage changes affect volts and resistance? That is a good question. But if you think about it, amperage is a result of, or product of, the voltage and resistance. Amperage does not exist without both voltage and resistance. There must be a means of pushing the amperage (voltage) and a path for the amperage to flow (resistance). If you ask yourself, "What is the amperage doing in a circuit?" the answer will always be that it is doing whatever the voltage and resistance allow it to do. If the amperage is low, then you know that one of two conditions is present—either the voltage is low or the resistance is high (**FIGURE 42-50**). If the amperage is high, then either the voltage is high or the resistance is low (**FIGURE 42-51**). Understanding the relationships between volts, amps, and ohms helps you know what test you need to perform next during diagnosis.

Because amperage is a product of voltage and resistance, it is a good idea to keep your eye on the amperage. What this means is that if you have a circuit fault, you can generally see what the amperage is doing; it is either high or low. If you truly cannot "see" what the current is doing in the load, like in an internal solenoid, you will have to measure it and compare it with specifications. In most cases, you can see it, and you can do so without measuring it. For example, you can "see" that the fuse blew (**FIGURE 42-52**): It blew because the current was too high. You can "see" that the light is dim. It is dim because the current is too low (most common scenario) because either the voltage is low or the resistance is high. On the flip side, if your eye determines that the current is high, then either the voltage is high or the resistance is low.

For example, let's say that the left front headlight is dim, but the right one is bright (**FIGURE 42-53**). Your eye determines that

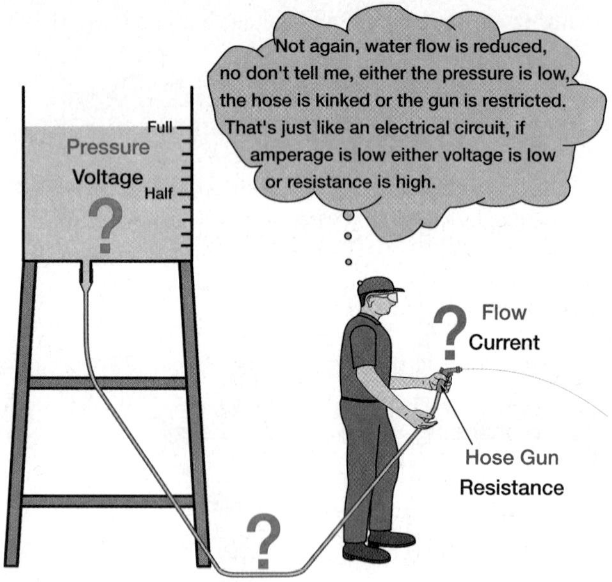

FIGURE 42-50 If amperage is low, then either the voltage is low or the resistance is high.

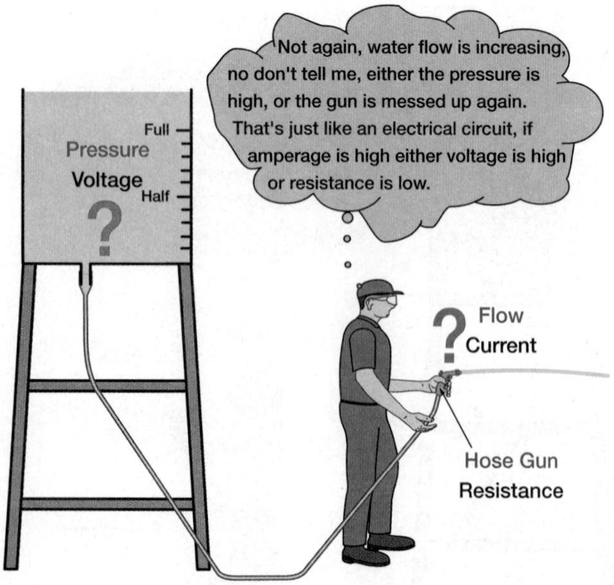

FIGURE 42-51 If amperage is high, then either the voltage is high or the resistance is low.

FIGURE 42-52 You can "see" that the fuse is blown: It most likely blew because of high current flow.

FIGURE 42-53 Our eye can determine that the left front headlight is dim due to low current flow.

the current in the left front circuit is low. Thus, either the voltage is low or the resistance is high. Now, you only have to test for those two things. Because the right side is bright, the battery voltage is okay and not causing the dim light. But we want to know battery voltage so we can use it for comparison purposes in the next step.

Use a voltmeter to measure the voltage at the battery with the lights on, and write it down (**FIGURE 42-54**). If it is low, determine why, such as a faulty charging system. Next, check the voltage drop across the input and output terminals of the headlight with the circuit on (**FIGURE 42-55**). The voltage drop should be well within 1.0 volt of the battery voltage. If not, then switch to looking for the high resistance in the circuit. You do this by first measuring the voltage drop on each side of the headlight circuit back to the battery (**FIGURE 42-56**). If there is an excessive voltage drop on one or both sides, continue to take voltage drop readings. Do so by moving the leads closer together in the circuit (**FIGURE 42-57**). If the voltage drop across the headlight is well within 1.0 volt of battery voltage, the problem is most likely the headlight itself. You may be able to check the resistance of the headlight filament and compare it with a known good bulb. Or you may have to measure the current flow through the bulb and compare that with specifications. Remember that its resistance increases greatly due to heat when it is illuminated.

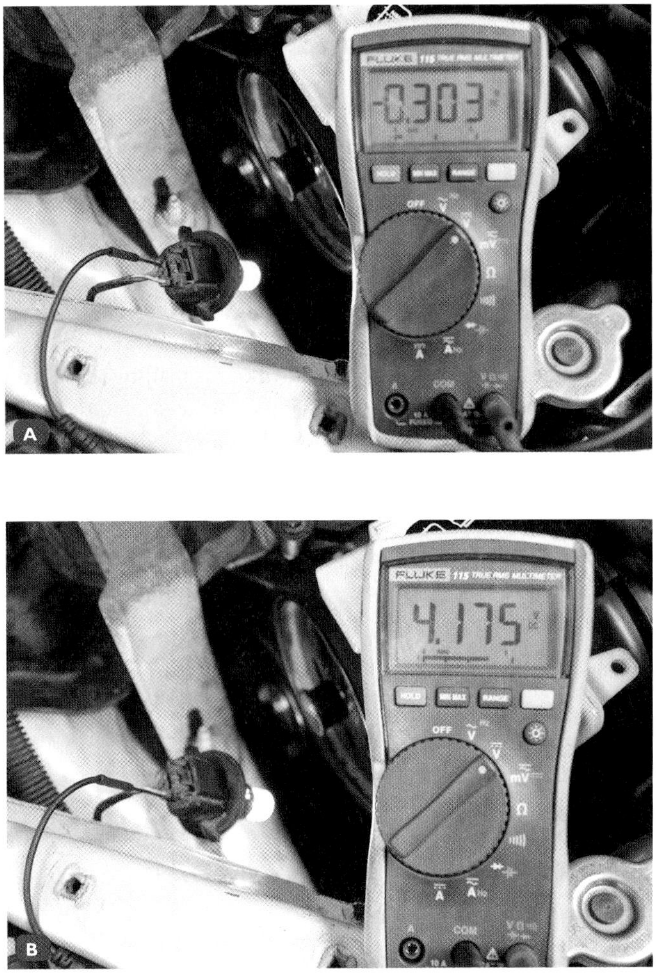

FIGURE 42-56 Measure the voltage drop on each side of the headlight back to the battery. **A**. Power side shows an acceptable voltage drop. **B**. Ground side measurement indicates a fault.

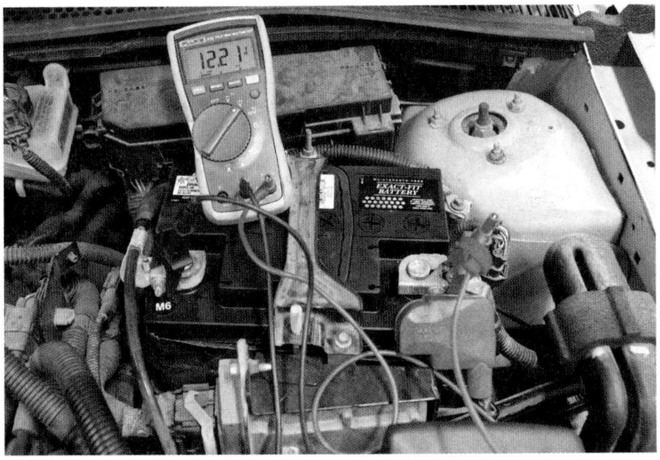

FIGURE 42-54 Measure battery voltage with the headlights on.

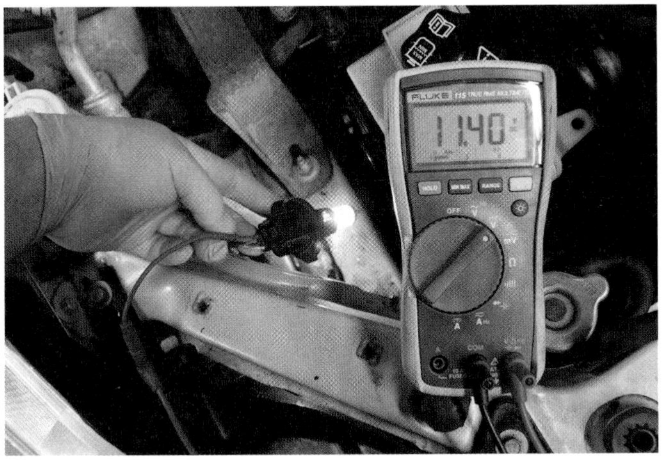

FIGURE 42-55 Measure the available voltage at the dim headlight with the light on. It should be well within 1.0 volt of battery voltage.

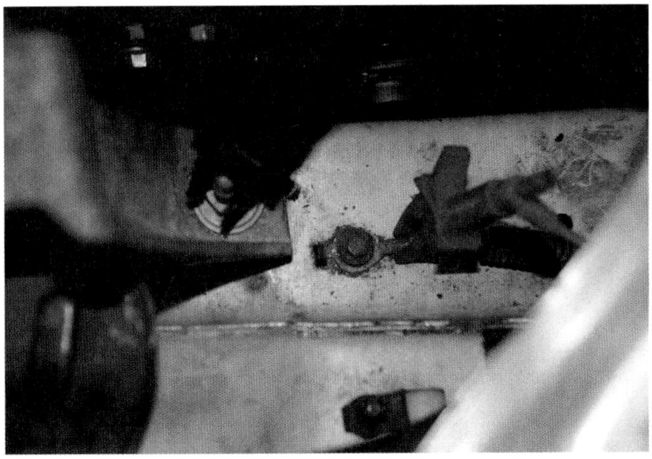

FIGURE 42-57 Here, most of the excessive voltage drop is on the dirty and loose ground connection.

▶ TECHNICIAN TIP

I once ran into a vehicle that had very dim headlights on both sides of the vehicle. They were dim on both the low beams and high beams, so I suspected a low-voltage or high-resistance problem in the wiring harness. But when I opened the hood, there was a lot of light from the headlights shining into the engine compartment. A closer look showed that the reflective surface inside the headlights had flaked off. It was all sitting on the bottom of the headlights. So, most of the light from the filament was shining backward into the engine compartment, and not being reflected forward. Replacing the headlights fixed the problem.

If the current flow appears too high in a circuit, then it may have too much voltage or too little resistance. Measuring the battery voltage is an easy way to check for too much voltage. If the voltage is okay, then the resistance must be too low. Resistance of loads can be checked using an ohmmeter and comparing the reading with specifications. An ohms reading that is too low indicates that it is shorted. A high ohms reading indicates a high-resistance fault. The ohmmeter can also be used to check the wire harness for any short circuit conditions as well. We cover this in much more detail later in this chapter.

▶ Using a DMM to Measure Voltage and Voltage Drop

LO 42-13 Perform voltage and voltage drop measurements.

The electrical system is becoming increasingly complex on modern vehicles. Measuring voltages with a DMM is a very common task when diagnosing electrical faults. Ensure that you do not exceed the maximum allowable voltage or current for the CAT rating of the DMM. If you are measuring high voltages, wear appropriate personal protective equipment. These include items such as high-voltage safety gloves, long-sleeved shirts and pants, and protective eyewear. Remove any personal jewelry or items that may cause an accidental short circuit. For most measurements, set the DMM to auto range for ease of use. Select DMM leads and probe ends to match the task at hand. For example, if you need to take a measurement but require both hands to be free, use probe ends with alligator clips.

When using a voltmeter to measure voltage, you have a couple of options. One is the available voltage test. This typically involves placing the common lead on a good ground. The red probing lead can be placed anywhere in the circuit to find out how much voltage is available at that point (**FIGURE 42-58**). This

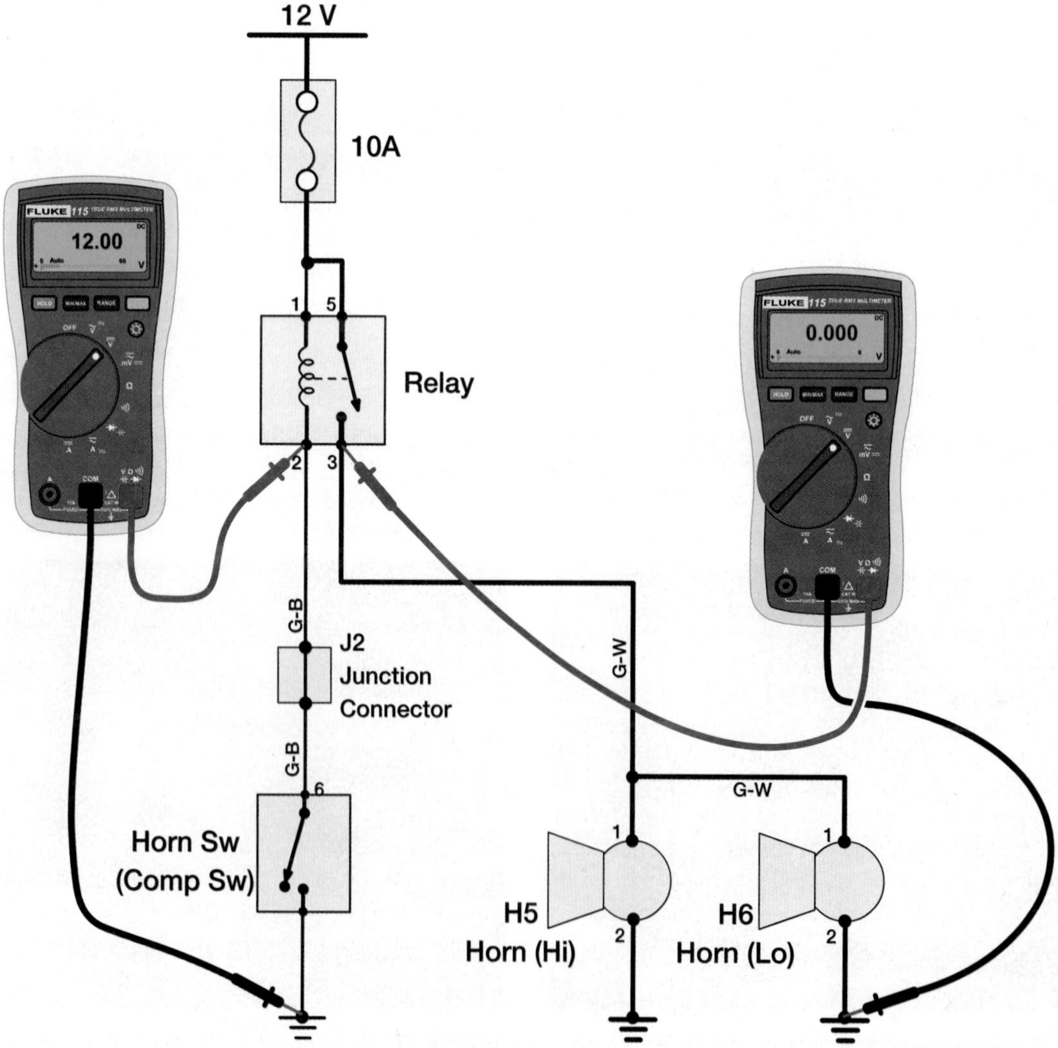

FIGURE 42-58 An available voltage test is performed by placing the black lead on a good ground and the red probe in the circuit where you want to know how much voltage is available.

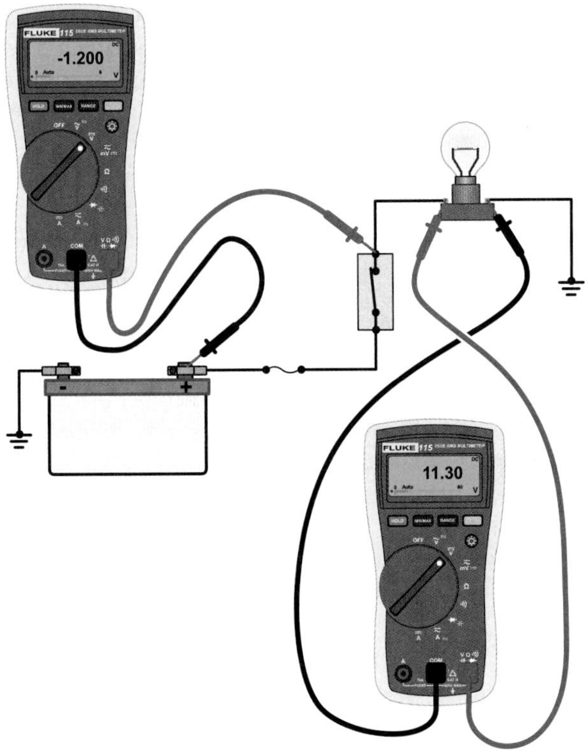

FIGURE 42-59 Excessive voltage drops in an improperly functioning circuit.

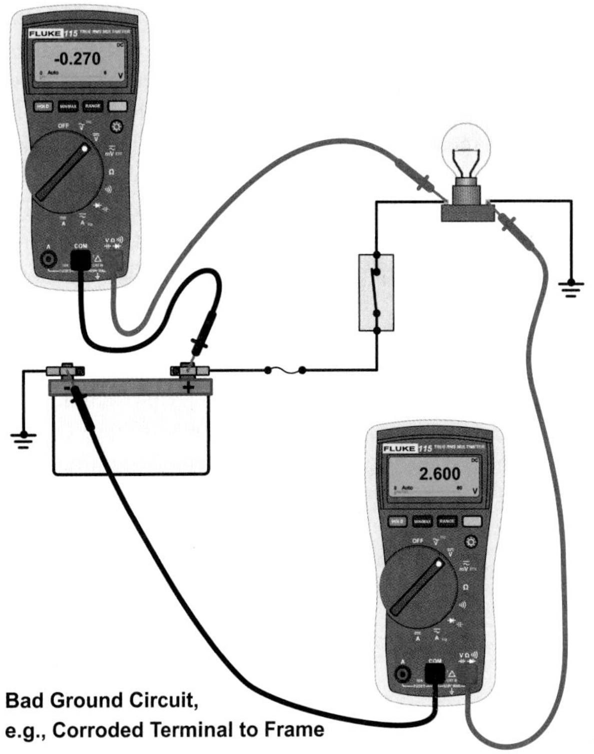

FIGURE 42-60 Measuring voltage drop on each side of the circuit.

gives you a reading of how much more voltage is at the probing lead than is at the common lead. Although that is helpful, it only gives you an indication of voltage. It does not tell you how much voltage we started with. Nor does it tell you how much voltage was lost getting to the test point. To do that, you need to do one of two things. Either measure the available voltage at the battery and compare that voltage reading with voltage at the component. Or perform a voltage drop test.

A voltage drop occurs when current flows through a resistance. The higher the resistance, the higher the voltage drop. We could say that a voltage drop test measures the excessive resistance in a circuit. When testing for a voltage drop, the circuit must be turned on. That way, the circuit will have current flowing, thereby making voltage drops evident. Just to be clear, in a complete circuit, no current flow means an open circuit. So, the only voltage drop will be across the open in the circuit, and it will be a voltage drop of full battery voltage. Otherwise, if current is flowing, then there will be voltage drops at all resistances in the circuit. This includes wanted and unwanted resistances.

To test for unwanted voltage drop, measure the voltage across each part of the circuit. This includes the conductors, switches, and connectors. Compare the readings with specifications (**FIGURE 42-59**). In general, in a 12-volt system, the individual voltage drops across an individual wire, connection, or switch should be less than 0.2 volt. Additionally, the total voltage drop across each side of the whole circuit (from battery + to load, or from load to battery –) should not exceed 0.5 volt on 12-volt circuits, or 1.0 volt on 24-volt circuits.

You could measure the voltage drop on each connection and wire on each side of the circuit. You would then have to add them together to see if they are more than 0.5 volt. It is faster if you measure the total voltage drop on each side of the circuit first. This allows you to identify which side of the circuit has the excessive voltage drop (**FIGURE 42-60**).

For the positive side, you can do this easily by placing the black voltmeter lead on the Bat + terminal and the red lead on the input of the load. Make sure the circuit is activated. If the voltage reading is 0.5 volt or lower, that side of the circuit is fine. If it is larger than –0.5 volt, then you need to check each individual connections and wires in that side of the circuit.

The negative side of the circuit can be checked by placing the black voltmeter lead on the Bat – terminal, and the red lead on the output side of the load. Again, the reading should be less than 0.5 volt. If the voltage drop on each side of the circuit is okay, and the load is not working properly, suspect that the load has a fault. We will show how to check loads later in this chapter.

Performing a Voltage Drop Test

To measure voltage drop, the DMM needs to be set to the voltage position. You need to set the function switch to either "auto range volts DC" or the correct manual range. Then, connect the black lead to COM slot and the red (probing) lead to the V/Ω slot.

For this example, the customer concern is that the horn is not very loud. You verify the concern by activating the horn and hearing that it is not nearly as loud as it should be. You print out the wiring diagram for the horn circuit to familiarize yourself with it. You measure the source voltage at the battery terminals with the horn operating, and it reads 12.48 volts. You then connect the meter across the horn input and output terminals to measure the voltage drop of the horn. Activating the horn shows a voltage drop across the horn of 8.28 volts (**FIGURE 42-61**). Because the circuit started with 12.48 volts, but the horn is only using 8.28 volts, you know that there is another voltage drop of 4.2 volts. It is split between the feed side and ground side of the circuit. But we do not yet know how much voltage drop is on each side.

You decide to start on the feed side of the horn circuit. So, you connect the black lead to the positive terminal of the battery and the red lead to the input wire of the horn (the wire connected to the horn). When you activate the horn, the voltmeter reads –4.2 volts. This means that there is a voltage drop of –4.2 volts in the feed side of the circuit. In this case, the "–" means "less than." This means that the voltage is 4.2 volts less at the input of the horn (red lead) than it is at the positive battery post (black lead) (**FIGURE 42-62**). Because the voltage drop is more than 0.5 volt, this is an excessive voltage drop in that side of the circuit. The test leads have to be moved, wire by wire, closer together in the circuit until the point of the voltage drop is located. In this case, there is a –4.03-volt drop across the relay contacts.

You could make the same measurement with the DMM leads reversed. If you place the red lead on the positive battery post and the black lead on the input of the horn, the meter would then read 4.2 volts. In this case, it shows positive. This is because the red lead is on the positive post of the battery, which is 4.2 volts higher than the horn input where the black lead is connected (**FIGURE 42-63**). As you can see, voltmeter leads can be hooked up in a couple of ways. Just remember that the meter always reads what the red lead is touching as compared with what the black lead is touching.

Don't forget that excessive voltage drops can happen on the ground side as well. For example, a corroded or bad chassis ground can cause a voltage drop. It will reduce the voltage and current available to loads as well. **FIGURE 42-64** shows a simple

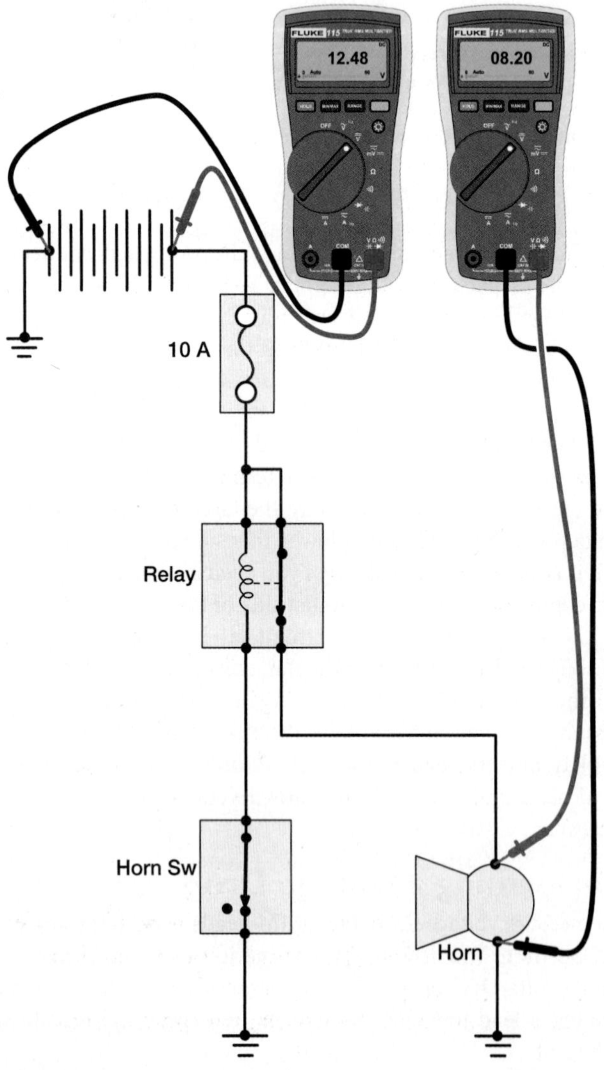

FIGURE 42-61 Comparing available voltage at the battery to the available voltage at the horn shows that the horn is not operating on full battery voltage.

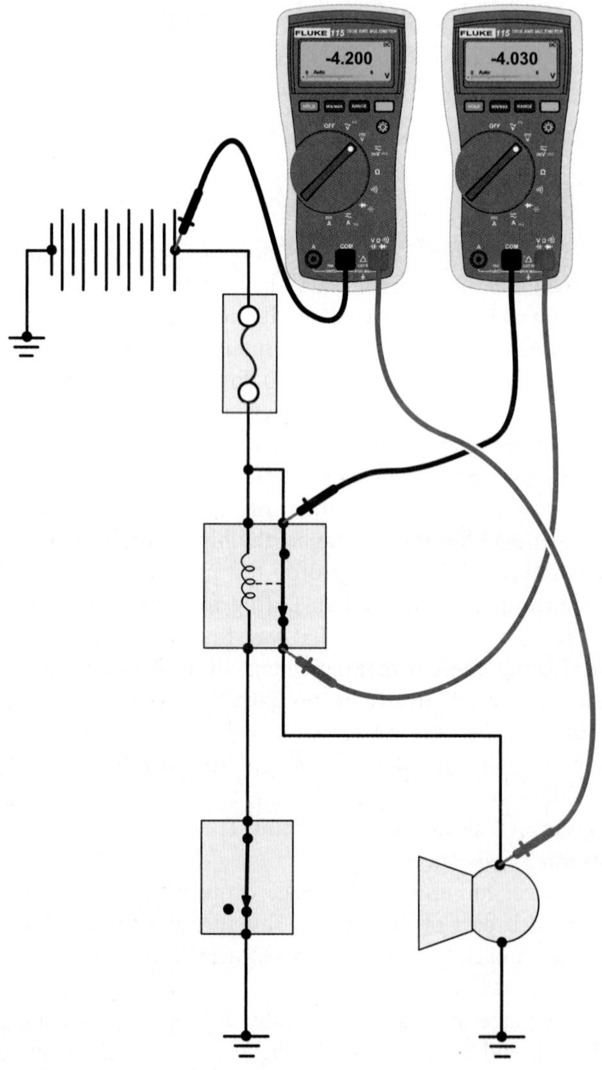

FIGURE 42-62 Excessive voltage drop on the feed side of the circuit due to worn horn relay contacts.

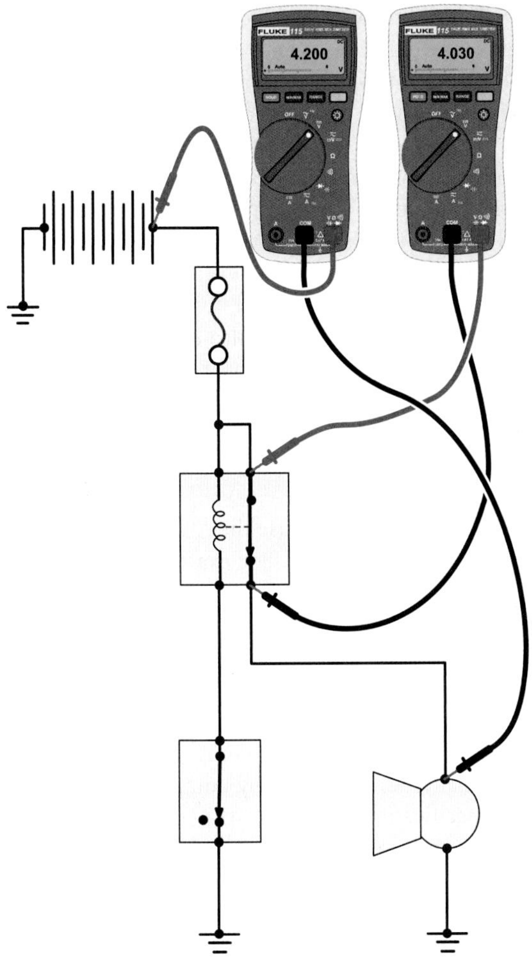

FIGURE 42-63 Reversing the meter leads in the circuit shows 4.2 more volts at the battery + terminal than at the input of the horn.

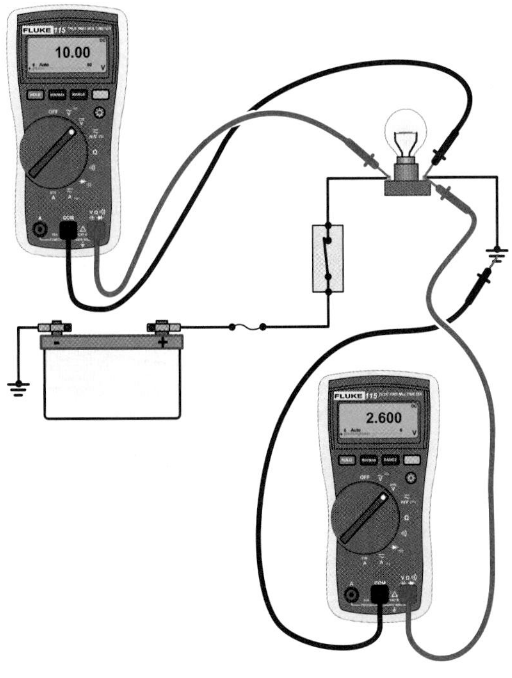

FIGURE 42-64 Voltage drops can occur on the negative side of the circuit. Don't neglect checking the ground side when loads do not operate correctly.

circuit with a bulb connected via a switch across a 12-volt circuit. In this circuit, a corroded ground connection has caused a resistance that is dropping 2.6 volts across it. The excessive ground circuit voltage drop is in series with the bulb. Because of this, it reduces the voltage drop across the bulb. This in turn causes poor illumination.

Applied Math

AM-12: Mentally: The technician can mentally multiply numbers that include decimal numbers to determine conformance with the manufacturer's specifications.

In this scenario, a technician is conducting a voltage drop test on the positive side of the fuel pump circuit. As described in the text, a voltage drop occurs when current flows through a resistance. The higher the resistance, the higher the voltage drops. So, we could say that this test checks for excessive resistance in a circuit. When testing for a voltage drop, always have the circuit operating. That way, the circuit will have current flowing, so any voltage drops will be evident.

The technician begins by placing the DMM selector on the 20-volt DC scale. The black lead is connected to the battery positive terminal. The red lead is connected to the input terminal of the fuel pump. When the fuel pump is activated, the meter reads 1.20 volts. This is in excess of the 0.5 volts maximum allowed. Pinpoint tests will need to be used to identify the point of high resistance.

The technician starts at the battery. The checks are made step by step all the way to the fuel pump using the direct method of voltage drop testing. This means that both test leads are placed on the same side of the circuit. The tests include from the battery post to the cable clamp. Then, from the cable clamp to the end of the cable at the relay. And then across the relay contacts. Then from the relay to the fuel pump. At each of the test points, an assistant turns the ignition key on to load the circuit. Each of the four connection points has a reading of 0.30 volt. The technician mentally multiplies those decimal numbers to obtain a total of 1.20 volts (4×0.30 volt $=$ 1.20 volts). In this way, the technician knows that he has found all of the voltage drops on that side of the circuit.

▶ Locating Opens, Shorts, Grounds, and High Resistance

LO 42-14 Locate Opens, Shorts, Grounds, and High Resistance.

DMMs, test lamps, and simulated loads tend to be the tools used most often for locating opens, shorts, grounds, and high-resistance faults. The Principles of Electrical Systems chapter has more information on opens, shorts, grounds, and high-resistance faults. As review, an **open circuit** is a break in the electrical circuit. It is where either the power or ground has been interrupted. A systematic check of the circuit is required. This is done by first performing a voltage drop check on each side of the affected circuit to determine which side is open (**FIGURE 42-65**). An open circuit causes a voltage drop equal to the source voltage. Once the voltage drop is isolated to one side of the circuit, voltage drop testing can continue on that side. Just work the leads closer together in steps.

Also, use your understanding of electrical systems to consider the most likely places for the open circuit. High-probability items are a blown fuse or a faulty switch. And don't forget that the load could also be open. If the voltage drop test on each side of the circuit is within specifications, use an ohmmeter to check whether the load is open. Some loads such as diodes cannot be tested with a standard ohmmeter. In this case, follow the manufacturer's diagnostic procedure.

High resistance refers to a circuit where there is unintended resistance. This causes the circuit to not perform properly. It can be caused by a number of faults. These include corroded or loose harness connectors, incorrectly sized cable for the circuit current flow, incorrectly fitted terminals, and poorly soldered joints. The high resistance causes an unintended voltage drop in the circuit when the current flows. This drop reduces the amount of voltage that can be used by the load. The high-resistance fault also reduces the current flow in the circuit. The reduction in voltage and current reduces the amount of electrical power to load (Power = Voltage × Amperage). Because of this, the load will not perform as designed. Unwanted high resistance can best be located by conducting a voltage drop test. Do so on the power and ground circuits just like you did for an open circuit. The only difference is that the voltage drop will be less than battery voltage (**FIGURE 42-66**).

If the high resistance is within the load, such as a relay coil, the resistance can be checked with an ohmmeter. Then, compare the reading with specifications. Some devices, such as fuel injectors or ignition coils, may need further testing using an oscilloscope. In this way, the waveform can be evaluated. The waveform can indicate issues that an ohmmeter cannot identify as easily. Oscilloscopes are beyond the scope of this book.

Shorts, or **short circuits**, can occur anywhere in the circuit. They can be difficult to locate, especially if they are intermittent. A short is a circuit fault in which current travels along an accidental or unintended route. It can be thought of as a shorter path for current to flow. The short may occur within the load, such as shorted relay windings. It can also be in the wiring, where a wire is shorted to ground or to supply voltage. A short typically causes lower than normal circuit resistance. The low-resistance fault causes an abnormally high current flow in the circuit. This may cause the circuit protection devices, such as fuses or circuit breakers, to open the circuit.

A short to power may cause the circuit to remain live even after the switch is turned off. For example, a short between a wire with constant power and a wire switched by the ignition switch. This would cause the circuit controlled by the ignition switch to remain on even after the switch is turned off. Just remember that shorts can be caused by faulty components or damaged wiring.

Shorts that happen within components, such as a relay coil, can usually best be tested by comparing the ohm's reading with specifications (**FIGURE 42-67**). Shorts that occur in

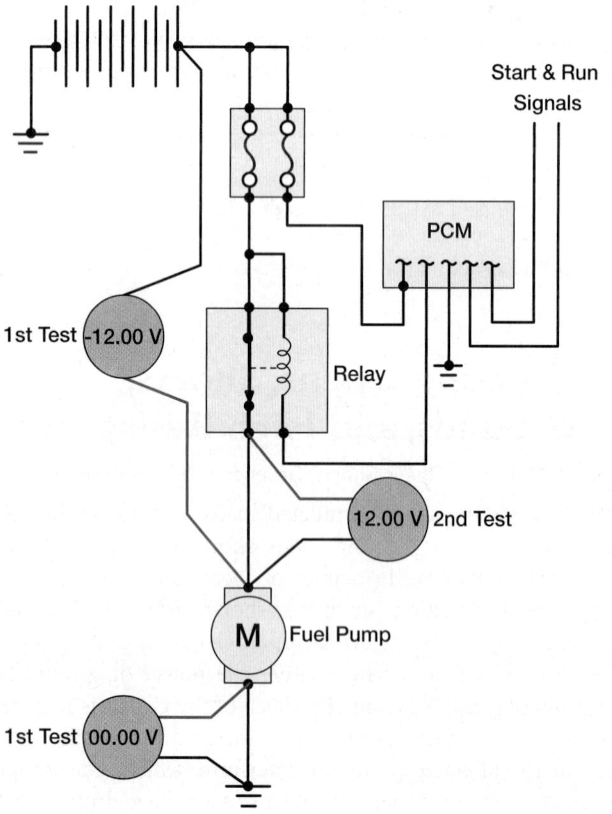

FIGURE 42-65 Locating an open circuit fault starts with a voltage drop test on each side of the circuit, then isolation tests by moving the voltmeter leads together.

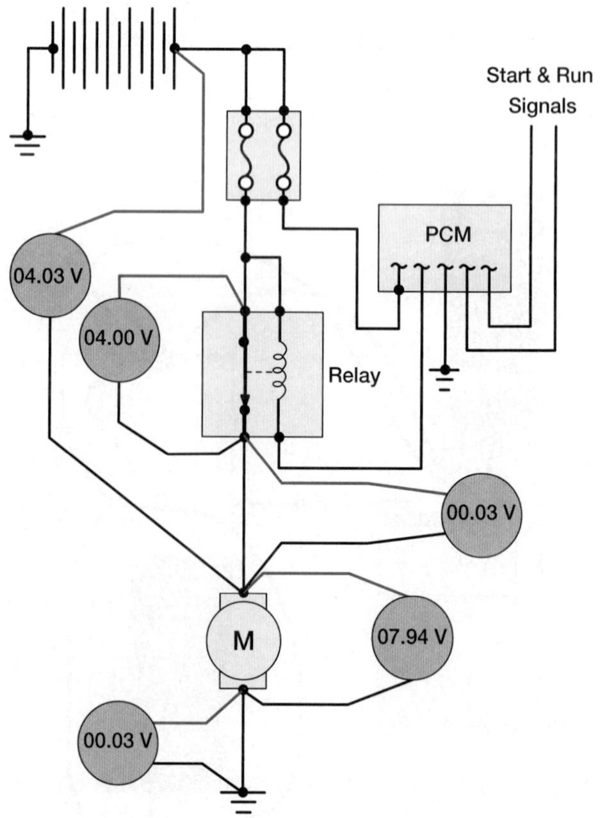

FIGURE 42-66 Example of a high-resistance fault causing an excessive voltage drop.

wire harnesses are usually best tested by disconnecting each end of the affected harness. Then, use an ohmmeter to test for unwanted continuity between the various wires. A reading on the ohmmeter when connected to two separate wires indicates a short circuit between them. A true short between wires is indicated by a very low ohm reading, typically around 1 ohm or less.

Short to ground refers to a situation where a point in the circuit is unintentionally connected to ground. This can happen on the feed side of the circuit or the ground side. If on the feed side, the current flow will be very high and likely blow a fuse. If on the ground side, then the load will likely run anytime the circuit has power. An initial test can be conducted by disconnecting the fuse and the loads on that fuse's circuit. Then, checking for continuity between the output of the fuse and ground (**FIGURE 42-68**). If there is continuity, you will need to find a connector between the fuse and load. Disconnect it to see which side of the harness has continuity to ground. Keep tracing it until you locate the fault.

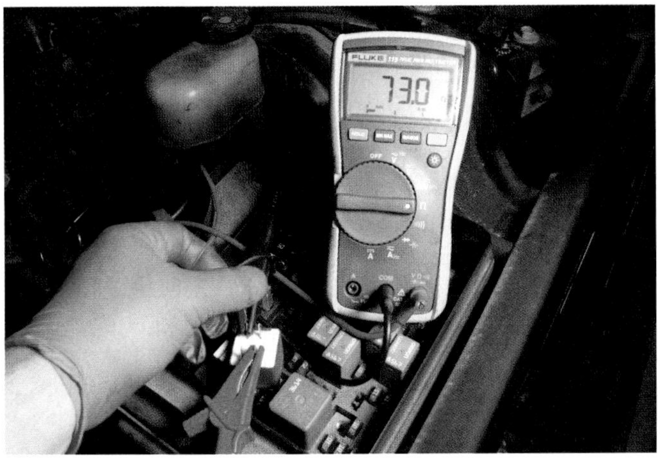

FIGURE 42-67 Shorts within components are usually best tested with an ohmmeter and compared with specifications.

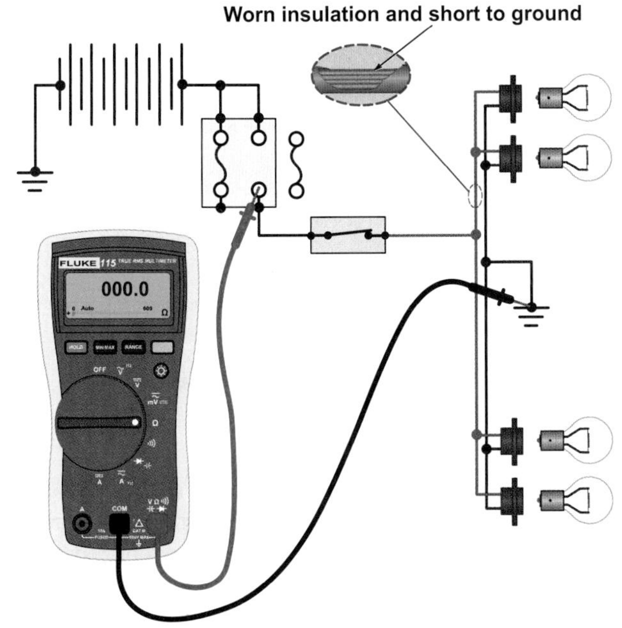

FIGURE 42-68 To find a short to ground, disconnect the loads on the circuit, and check for continuity between the output of the fuse and ground.

If, after disconnecting the loads, there was no continuity between the harness and ground, you need to check the resistance of each load. This will verify if any loads are shorted internally. For example, if testing the blower motor circuit, first disconnect the blower motor. Check for continuity between the output of the fuse and ground. If the short is still in place, then the wiring between the fuse and the load must be at fault. To further narrow down the site of the short to ground, inspect the wiring harness. Look for obvious signs of damage.

Another test can be conducted to locate shorts to ground. Do this by connecting a test lamp or buzzer in place of the fuse (**FIGURE 42-69**). Current will flow through the test lamp or buzzer and find a ground through the short. Parts of the circuit can then be disconnected along the wiring harness to narrow down the location of the short.

Specialized short-circuit detection tools are also available. They work by sending a signal through the wiring harness. A receiving device is then moved along the wire loom and indicates when a short is located. This type of device can be very useful in situations where it is difficult to access the wiring, such as within large wire looms or under vehicle trim.

Short to power refers to a condition where power from one circuit leaks into another circuit. A short-to-power situation usually causes strange electrical issues. In some cases, one or more circuits operate when they should not. In the case of sensor wires, the short to power causes incorrect signals to be sent to the computer. It may then make very wrong decisions based on the faulty signal (**FIGURE 42-70**). In this case, the engine, transmission, or other computer-controlled component can react strangely. Shorts to power are diagnosed first with a voltmeter to check for the unwanted voltage. Next, an ohmmeter is used to isolate the problem in the wire harness.

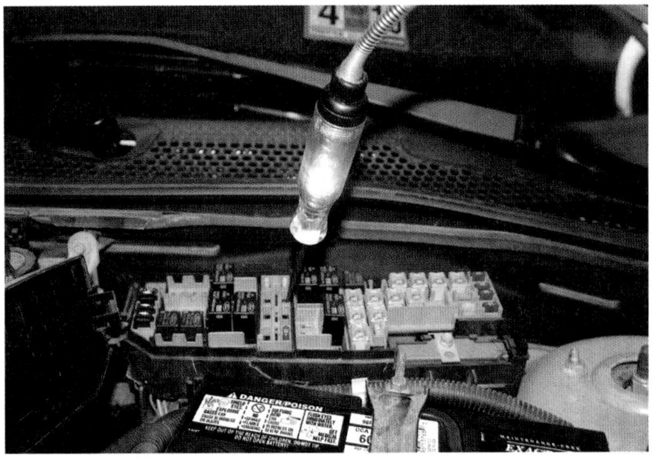

FIGURE 42-69 A test light being used in place of a fuse to indicate current is flowing through the circuit. Disconnect wires until the light goes off.

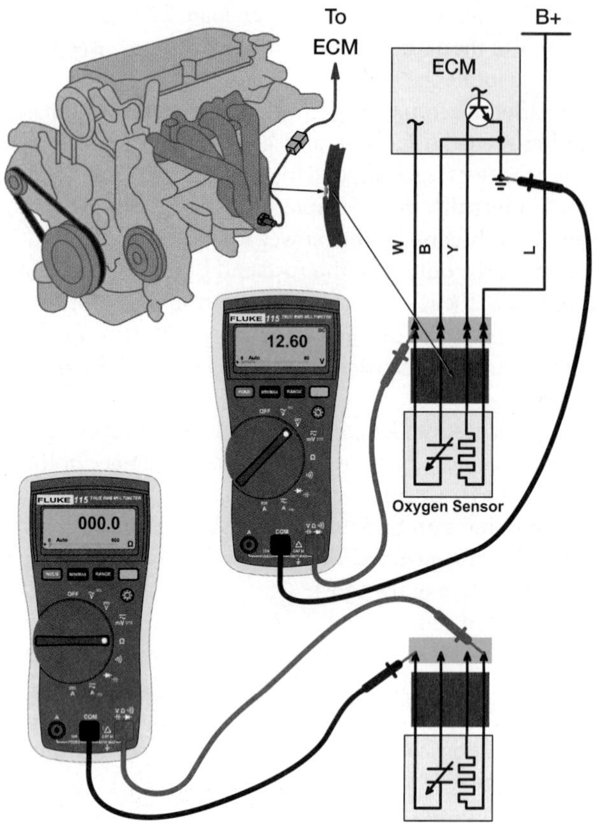

FIGURE 42-70 Short to power: The oxygen sensor heater wire melted to the signal wire, sending battery voltage to the PCM. This can be verified with a voltmeter; then, an ohmmeter is used to show continuity between the wires.

▶ Checking Circuits with a Test Light

LO 42-15 Test circuits with a test light and fused jumper wire.

Non-powered test lamps are useful in determining whether electrical power is present in a part of a circuit. But you should always first test the test light on a known good power and ground before using it to test a circuit. It is possible that the bulb in the test light is burned out. You want to know that before performing any tests.

For the test light to light, it needs both power and ground. If the test light illuminates, you know you have both of those. You also know that the test light bulb is good. If the light does not illuminate, the circuit is missing one or both of those elements. Or the test light is faulty. So, always double check the test light.

Test lights are great for performing simple tests such as testing fuses. The test light alligator clip can be quickly grounded and the probe end touched to each end of the suspect fuse. If both ends light, the fuse itself is good (but the fuse box terminal could be loose). If only one side of the fuse lights the test lamp, then the fuse is blown. The circuit will then have to be diagnosed to find out why it blew.

To avoid damaging the test light, make sure the circuit voltage you are testing does not exceed the test light's rating. Most test lights are rated for 6- or 12-volt systems. Using the light in a 24-volt system will blow the bulb. You should *not* use a test light to test SRS (supplemental restraint systems). An unintended deployment of the airbags could result. This is a very dangerous and costly mistake. Also, using a test light on a computer circuit designed for very small amounts of current flow can damage the electronics inside the modules.

To check circuits with a test light, follow the steps in **SKILL DRILL 42-1**.

Checking Circuits with Fused Jumper Leads

Jumper leads can be used in a number of ways to assist in checking circuits. They can be purchased in a range of sizes, lengths, and fittings, or connectors. Or they can be built by the technician.

Jumper leads have a number of uses. They are used to extend test leads so that circuit readings to be undertaken with a DMM. They are also used to jump across terminals on fuses, relays, and other components. This is a quick way to see if a circuit will operate if you bypass part of it. In this way, jumper leads provide an alternate power or ground source for components being tested.

Regardless of their application, it is important that the circuit remains protected by a fuse of the correct size. To determine the correct size of fuse for any particular application, refer to the manufacturer's amperage specifications for the components being tested. Also, know that the fuse protects the jumper lead. It may also provide some protection to the circuit being tested. But if power is supplied to the wrong part, it could still be over-powered causing it to be damaged.

To check circuits with fused jumper leads, follow the steps in **SKILL DRILL 42-2**.

SAFETY TIP

Be very careful how you hook up any type of jumper leads, fused or unfused. If you hook them up to the wrong branch of a circuit, especially electronic circuitry, damage can be extensive. There is the old "magic smoke" saying: "Electrical and electronic components work off of the principle of magic smoke. Once the magic smoke is allowed to escape from the component, the component will never function again." Don't use jumper leads in a way that would let the "magic smoke" out of the circuit.

Inspecting and Testing Circuit Protection Devices

Protection devices are designed to prevent excessive current from flowing in the circuit. Protection devices like fuses and fusible links are sacrificial. This means that if excessive current flows, they will blow or burn open and must be replaced. Circuit breakers can be reset. Once they trip, they either reset automatically or require a manual reset by pushing a button or moving a lever. Fuses, fusible links, and circuit breakers are available in various ratings, types, and sizes. They must always be replaced with the same rating and type.

SKILL DRILL 42-1 Checking a Circuit with a Test Light

1. Connect the end of the light with the clip on it to the negative battery terminal. Touch the probe end of the test light to the positive battery terminal. The light should come on.

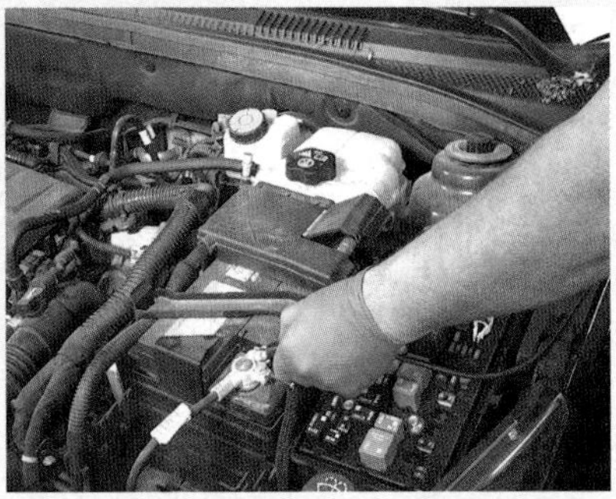

2. Connect the clip to any known good ground. A typical known good ground is any unpainted metal surface on the vehicle that is directly attached to the battery ground return system.

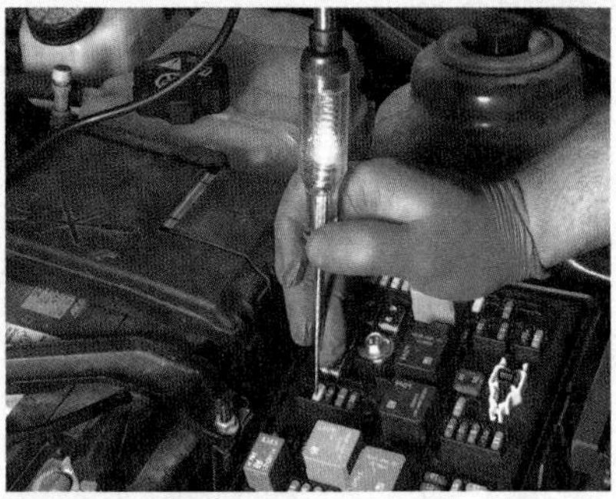

3. Place the probe on the terminal to be tested. If voltage is present, the light will come on.

SKILL DRILL 42-2 Checking Circuits with Fused Jumper Leads

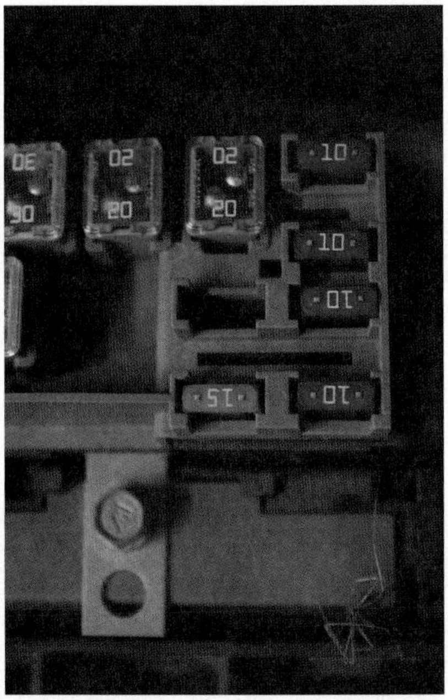

1. Identify the circuit to be checked, and determine the fuse rating for the circuit.

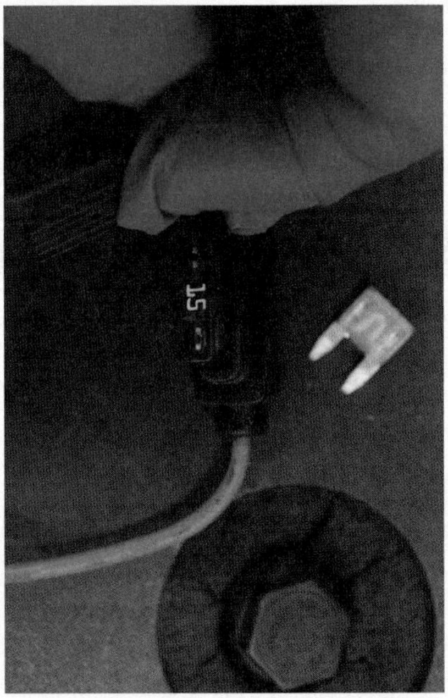

2. Select the appropriate jumper lead, install the correct fuse in it, and connect one end to the battery positive terminal.

3. Touch the jumper lead quickly to the horn input terminal. Never jump across a load. Doing so bypasses circuit resistance, causing excessive current in the circuit and damaging it.

In most vehicles, protection devices are situated in the power or feed side of the circuit. A blown or faulty fuse can be tested using a DMM or test lamp. A good fuse has virtually the same voltage on both sides. A blown fuse typically has battery voltage on one side and 0 volts on the other side (**FIGURE 42-71**). If there are 0 volts on both sides of the fuse, either the ignition may have to be turned on. Or a fusible link or maxi-fuse supplying the fuse box may be burned open. Fusible links can be checked for voltage as well. Voltage on both ends means it is good. Voltage only on one end means it is open. Just be sure to test it with a load on the circuit.

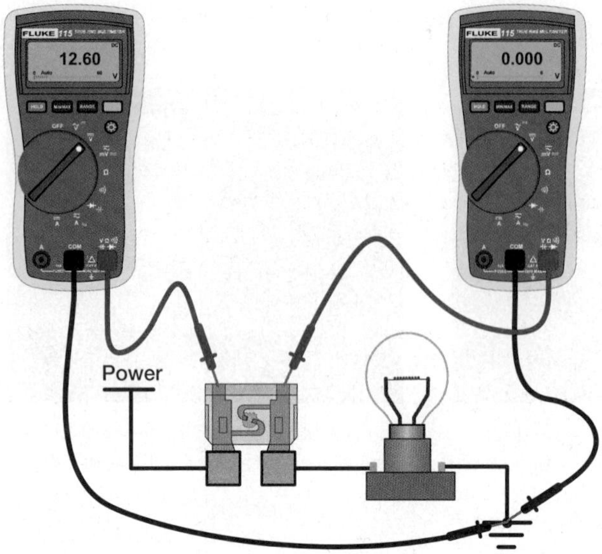

FIGURE 42-71 A blown fuse has battery voltage on one side and 0 volts on the other side.

Fuses can typically be visually inspected (**FIGURE 42-72**). This may require the removal of the fuse from the fuse holder. The fusible element should be intact and, if measured by an ohmmeter, should have a very low resistance. The contacts on both the fuse and the fuse holder should be clean and free of corrosion. They should both fit snugly together.

FIGURE 42-72 Burned open fuse.

▶ TECHNICIAN TIP

It is fine to condemn fuses if they are obviously blown, but if they appear intact, do not rely on your eyes. Over time, fuses heat slightly and cool. This heating and cooling process can cause the fuse to become brittle and crack. The crack can be very fine, almost invisible. And yet it will not conduct electricity, leading your diagnosis astray. So check fuses with a test light or voltmeter.

To inspect and test circuit protection devices, follow the steps in **SKILL DRILL42-3**.

SKILL DRILL 42-3 Inspecting and Testing Circuit Protection Devices

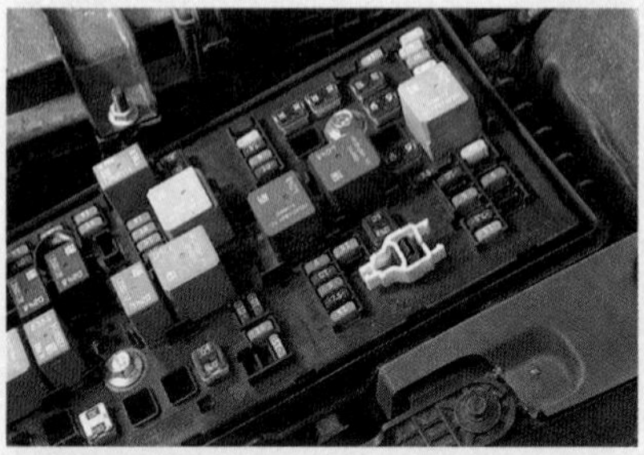

1. Identify the protection device to be inspected and tested.

2. Conduct a visual inspection.

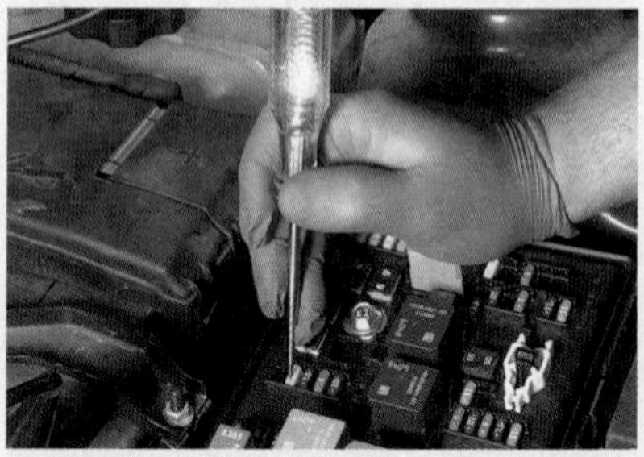

3. Set up a DMM to read volts, or use a test lamp.

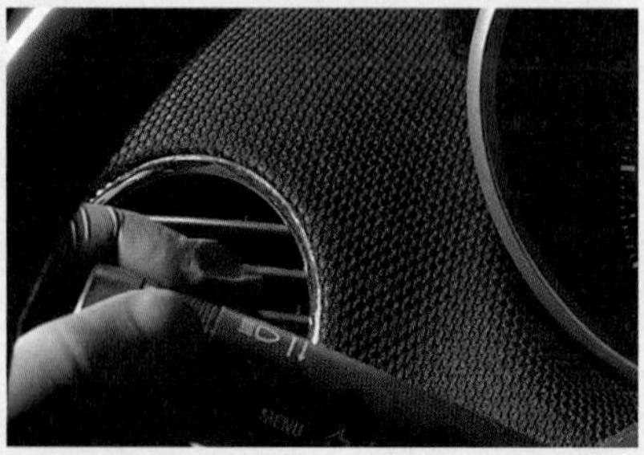

4. Energize the affected circuit, if necessary.

5. Test for voltage on both sides of the circuit protection device.

6. Determine and perform any necessary actions.

▶ Inspecting and Testing Switches, Connectors, Relays, Solenoid Solid-State Devices, and Wires

LO 42-16 Test circuit protection devices, switches, and relays.

Manufacturers produce wiring diagrams and diagnostic flowcharts. They are used to guide the technician through a diagnostic sequence. Use the customer concern to direct you to the specific diagnostic sequence. To do this, gather as much information as possible before beginning. Start with the customer concern and then verify it by operating the circuit. All of this allows you to determine a testing sequence for diagnosing the fault.

Inspection of electrical devices and wires usually starts with a visual inspection. That is usually followed up with electrical testing. The visual inspection looks for breakage, corrosion, or deformity. It includes examination of the insulation for any worn or melted spots.

In the case of switches, solenoid contacts, and relay contacts, electrical testing is necessary. For example, a voltage drop test is used to see if they have excessive resistance (**FIGURE 42-73**). Performing a voltage drop test on these components will determine if they are faulty. Voltage drop testing is performed by measuring the voltage from the input of the switch to the output while operating the circuit. Voltage drops above 0.2 volts indicate high resistance on most switches. Note that some starter solenoids allow a slightly larger voltage drop across the solenoid contacts. The same goes for battery cables.

Relays are made up of two separate components: the winding and the contacts. The contacts are tested like a normal switch. The circuit is operated so the relay is energized. Then, a voltage drop test is performed across the contacts. The voltage drop should be less than 0.2 volts.

The winding can be tested with an ohmmeter. But it needs to be disconnected from the circuit. Once it is disconnected, the resistance can be measured from the input to the output of the winding. Compare the reading with specifications to determine if the winding is ok.

If the relay is a spike-protected relay that has an internal diode, special testing is required. They make expensive relay testing tools that can be used. Another way is to use a fresh 9-volt battery. Place the two battery terminals across the winding terminals of the relay. In one direction of the battery, the relay will click. In the other direction of the battery, it should not click. If it clicks both ways, the diode if open. If it doesn't click in either direction, either the diode is shorted, or the winding is open. In either case, the relay will need to be replaced. Make sure to replace it with a specified replacement.

Some solenoids can be disassembled and visually inspected. In this case, the solenoid end cap may be removed and the contacts visually inspected (**FIGURE 42-74**). Typically, if there is an excessive voltage drop across the contacts, the contacts will be pitted and burned. Measuring resistance also comes into play when a shorted relay or solenoid winding is suspected.

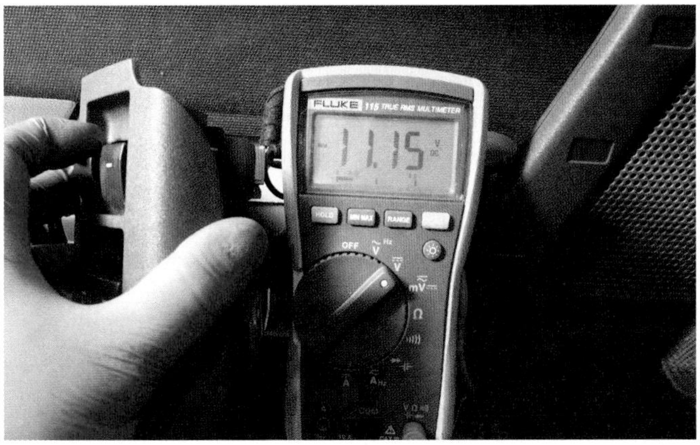

FIGURE 42-73 Switches can be tested for excessive resistance with a voltage drop test.

FIGURE 42-74 Some solenoids can be disassembled and the contacts inspected or replaced.

▶ Wrap-Up

Ready for Review

- ▶ Basic DMMs can measure: alternating current (AC) and direct current (DC) voltage, AC and DC amperage, and resistance. DMMs and test leads should have a CAT rating listed on the front amperage.
- ▶ The DMM main instrument body has: a function switch to choose the type of electrical measurement to be taken, a digital display to report the readings, and slots to connect test leads. Most modern DMMs default to automatic ranging capability to select the best range for the value being measured.
  - To achieve the most accurate reading, always select the lowest range possible for the value being measured.

- In the min/max setting, the DMM records in memory the maximum and minimum reading obtained during the test. The hold function allows the display to be frozen until the function or the DMM is turned off.
- To measure voltage, the probing lead (red) is connected to the volts/ohms, or V/Ω, slot and the common lead (black) is connected to the common, or COM, slot. To measure amps, the red probe is connected to the A or mA slot and the black lead is connected to the COM slot. To measure current, the DMM is connected in series with the circuit. To measure resistance, the red lead is connected to the V/Ω slot and the black lead is connected to the COM slot.
- In an available voltage test, the common lead is placed on a good ground and the red lead is placed in the circuit to determine the voltage available at that point. In voltage drop test, one voltmeter lead is placed on the input of a component and the other lead is placed on the output to read the voltage difference between the two.
- To measure resistance, auto range Ω or manual range is selected in the DMM, the red lead is connected to the V/Ω slot, and the black lead is connected to COM slot.
  - Resistance measurements should only be taken with power disconnected.
- To measure current, set the DMM to measure "DC amps" and connect the red lead to the A slot and the black lead to the COM slot. An ammeter must be connected in series within the circuit when measuring current.
- To measure voltage drop in a series circuit, set the DMM on the voltage range, select "auto range volts DC" or the correct "manual range" on the DMM, and connect the black lead to the COM slot and the red lead to the V/Ω slot.
- In a parallel circuit: resistance goes down when more parallel paths are added; current flow from individual legs adds up in parallel; and voltage stays the same at all common parallel circuit inputs.
- To measure voltage drop in a series-parallel circuit, set "auto range volts DC" on the DMM, connect the black lead to the COM slot, and connect the red lead to the V/Ω slot.
- For voltage measurements of a variable resistor, select "auto range volts DC" on the DMM, connect the red lead to the V/Ω slot, and connect the black lead to the COM slot. For current measurements of a variable resistor, select "auto range milliamps DC" on the DMM, connect the red lead to the A slot, and connect the black lead to the COM slot.
- To diagnose electrical circuit faults, ohm's law can be used to perform the math to predict and verify measurements or the relationships between volts, ohms, and amps can be used.
- Available voltage test can be used when measuring voltages with a DMM. To measure voltage drop, set the DMM to either "auto range volts DC" or the correct manual range, connect the black lead to COM slot, and connect the red lead to the V/Ω slot.
- To check for an open circuit, perform a voltage drop check on each side of the affected circuit. Once the voltage drop is isolated to one side of the circuit, voltage drop testing can continue on that side. To check for shorts, an ohmmeter can be used to test for unwanted continuity between various wires. An initial test for short to ground can be conducted by disconnecting the fuse and the loads on that fuse's circuit and checking for continuity between the output of the fuse and ground. Shorts to power are diagnosed with a voltmeter to check for the unwanted voltage and with an ohmmeter to isolate the problem in the wire harness.
- Test lights are great for performing simple tests such as testing fuses. A test light should not be used to test SRS (supplemental restraint systems). Jumper leads can be used to extend test leads so that circuit readings to be undertaken with a DMM and also to jump across terminals on fuses, relays, and other components.
- A blown or faulty fuse can be tested using a DMM or test lamp. A voltage drop test can be used to test switches, solenoid contacts, and relay contacts.

Key Terms

available voltage test Measurement of voltage at various points in a circuit, with the black meter lead on ground and the red lead probing the circuit.

hold function A setting on a DMM to store the present reading.

min/max setting A setting on a DVOM to display the maximum and minimum readings.

probing technique The way in which test probes are connected to a circuit.

short circuit A condition in which the current flows along an unintended route; also called a short.

short to ground Fault conditions in a circuit where the circuit is unintentionally contacting a grounded component or wire. This may result in a short, in the case of a power wire, or it could cause a circuit to stay live, as in the case of a switched ground circuit.

short to power A condition in which current flows from one circuit into another.

voltage drop test Measurement of the difference in voltage between two points in a circuit: the black lead on the end point being tested and the red lead on the beginning point.

Review Questions

1. What is the minimum DMM rating for hybrid vehicles?
   a. CAT I
   b. CAT II
   c. CAT III
   d. CAT IV
2. A technician is setting up their DMM to take a voltage reading. Where should the black lead be installed in the meter?
   a. In the A slot
   b. In the V/Ω slot

c. In the mA slot
d. In the COM slot

3. All of the following are true about voltage in a functional series circuit EXCEPT:
a. All of the voltage gets used up by the end of the circuit
b. Each of the loads use up the same amount of voltage
c. Each load uses some of the voltage
d. The load with higher resistance uses more voltage

4. When checking resistance what should the readings be compared to?
a. Manufacturer's specifications
b. The owner's manual readings
c. The meter's guide book
d. Watt's Law

5. Which of the following is true about a series circuit?
a. The current varies at different parts of the circuit
b. The resistance of every load is the same
c. An open in one load still allows the others to work
d. Nearly all of the voltage should be used up in the load

6. Which of the following is true about a parallel circuit?
a. Circuit resistance goes up as more loads are added
b. Current flow in all legs are the same
c. Each load has access to the same voltage input
d. If one load is open, none of the other loads will work

7. Which of the following is true about a series-parallel circuit?
a. The total source voltage is shared between the series and parallel loads
b. The series loads and parallel loads have access to the same source voltage input
c. Adding a load in the series portion will reduce the total circuit resistance
d. Adding a load in the parallel portion will increase the total circuit resistance

8. When performing an available voltage test on a circuit, which of the following should be performed first:
a. Place the red lead before the load, and the black lead after the load
b. Place the red lead before the load and the black lead on a good ground
c. Measure the source voltage of the battery using your red and black leads
d. Check the condition of the ground using voltage drop to the negative battery terminal

9. When testing a circuit with a test light, what must be done first?
a. Power on the test light by pressing its ON button
b. Turn off power to the circuit before testing
c. Set the test light to the correct voltage range
d. Test the light with a known good power and ground

10. What type of circuit protection device can be reset?
a. A fuse
b. A fusible link
c. A relay
d. A circuit breaker

ASE Technician A/Technician B Style Questions

1. A voltage reading is being taken using a DMM. The screen reads 4997 mv. Technician A states that this is 0.4997 volts. Technician B states that this reading is equivalent to 4.997 volts. Who is correct?
a. Technician A only
b. Technician B only
c. Both Technicians A and B
d. Neither Technician A nor B

2. A throttle position sensor is being checked with a DMM. Technician A states that the DMM sample rate may be too slow, and an oscilloscope would likely have a faster sample rate. Technician B states that the min/max function can be used to assist the technician in finding an electrical fault in the sensor. Who is correct?
a. Technician A only
b. Technician B only
c. Both Technicians A and B
d. Neither Technician A nor B

3. Two technicians are discussing measuring resistance. Technician A states that power should be applied to the circuit when checking resistance. Technician B states that when checking resistance on manual range, start at the highest range and work your way down. Who is correct?
a. Technician A only
b. Technician B only
c. Both Technicians A and B
d. Neither Technician A nor B

4. Two technicians are discussing measuring current in a circuit. Technician A states that the circuit must be opened and the DMM inserted in series. Technician B states that the circuit must be powered off to measure current. Who is correct?
a. Technician A only
b. Technician B only
c. Both Technicians A and B
d. Neither Technician A nor B

5. Variable resistors are being discussed. Technician A states that variable resistors can be used to indicate position. Technician B states that a potentiometer is a type of variable resistor. Who is correct?
a. Technician A only
b. Technician B only
c. Both Technicians A and B
d. Neither Technician A nor B

6. An electrical circuit is being discussed. Technician A states that if voltage stays the same, but resistance increases, current must increase. Technician B states that if voltage decreases, and resistance stays the same, current will increase. Who is correct?
a. Technician A only
b. Technician B only
c. Both Technicians A and B
d. Neither Technician A nor B

7. A voltage drop test is being discussed. Technician A states that the black lead should be placed on the negative terminal of the battery, and the red lead at the input of the load. Technician B states that the red lead should be placed at the input of the load, and the black lead at the positive battery terminal. Who is correct?
   a. Technician A only
   b. Technician B only
   c. Both Technicians A and B
   d. Neither Technician A nor B
8. A short to ground is being discussed. Technician A states that if the short to ground is before the load, it will cause excessive current. Technician B states that a short to ground after the load will cause a fuse to blow. Who is correct?
   a. Technician A only
   b. Technician B only
   c. Both Technicians A and B
   d. Neither Technician A nor B
9. Two technicians are discussing testing a circuit with a test light. Technician A states that if a test light illuminates, it is receiving power and ground. Technician B states that test lights should never be used with supplemental restraint systems. Who is correct?
   a. Technician A only
   b. Technician B only
   c. Both Technicians A and B
   d. Neither Technician A nor B
10. A relay is being tested. Technician A states that the relay winding can be tested with an ohmmeter while it is powered on. Technician B states that some relays use a diode for circuit protection and they require different tests than conventional relays. Who is correct?
   a. Technician A only
   b. Technician B only
   c. Both Technicians A and B
   d. Neither Technician A nor B

CHAPTER 43

Battery Systems

Learning Objectives

- LO 43-01 Describe basic battery construction and operation.
- LO 43-02 Describe basic types of batteries.
- LO 43-03 Describe battery configurations, terminals, and cables.
- LO 43-04 Describe battery ratings and the charge–discharge cycle.
- LO 43-05 Describe conditions that shorten/lengthen the life of a battery.
- LO 43-06 Describe the purpose and types of battery maintenance.
- LO 43-07 Inspect, clean, fill, and replace the battery and cables.
- LO 43-08 Perform battery charging and jump-starting.
- LO 43-09 Perform battery state of charge and specific gravity tests.
- LO 43-10 Perform battery capacity tests.
- LO 43-11 Maintain and restore electronic memories.
- LO 43-12 Measure parasitic draw.

ASE Education Foundation Tasks

See Appendix A to view the 2017 ASE Education Foundation Automobile Accreditation Task List Correlation Guide.

▶ Introduction to the Battery

LO 43-01 Describe basic battery construction and operation.

The electrical system on modern vehicles is becoming more complex. Electrical power controls almost all aspects of the vehicle. It goes beyond the engine and transmission. It includes brakes, suspension, navigation, entertainment, and much more. Virtually every system on modern vehicles relies on electrical power.

Fundamental to the automobile is the 12-volt storage battery. It provides the initial electrical power for the engine starting system. Once the engine is started, all of the vehicle's electrical systems rely on the charging system. It provides sufficient power output and precise voltage control for the vehicle's onboard circuits. Let's start with the battery. It provides the necessary standby (and reserve) power to get things going.

FIGURE 43-1 Typical automotive storage battery installed in a vehicle.

The Battery

As you learned in the Principles of Electrical Systems chapter, electricity is a very flexible and useful energy. It can be used easily in a variety of ways. But one drawback to electricity is that it cannot be stored easily in its electrical form for later use. This means that electricity must be stored in another form of energy and reconverted to electricity when needed. This is where batteries enter the discussion.

Batteries were developed in the early 1800s. Since that time many varieties and designs have been developed. The battery is part of everyday life and is widely used in modern electrical and electronic devices (**FIGURE 43-1**). Batteries store electricity in chemical form. This is possible because electricity causes a chemical reaction within the battery. In other words, the electrical energy is transformed into chemical energy. The chemical reactions change the composition of the chemicals. They are then stored until the electrical energy is needed. When electricity is needed, a chemical reaction occurs. This transforms the chemical energy back into electrical energy.

How does a battery transform electrical to chemical energy and vice versa? It all starts with the components of a battery. A battery consists of:

- two dissimilar metals,
- an insulator material separating the metals,
- and an electrolyte, which is an electrically conductive solution (**FIGURE 43-2**).

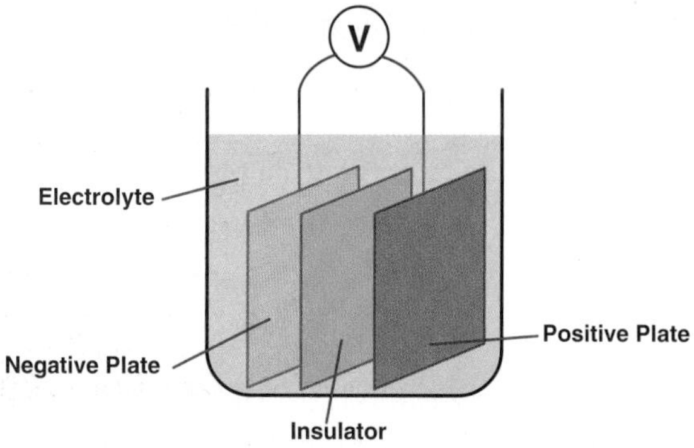

FIGURE 43-2 Components of a simple battery.

The output of the battery depends on the materials used. The traditional automotive battery type is the lead-acid battery. It is available in many different shapes, sizes, and designs. This allows them to meet the requirements for various applications. For example, there are two main types of batteries: starting and deep cycle (**FIGURE 43-3**). The battery used for starting a vehicle's engine is different from a deep cycle battery used in a golf

You Are the Automotive Technician

A seven-year-old vehicle has just been towed into the shop with a no-crank condition. The driver informs you that he forgot to turn the lights off while in a store for about 30 minutes. When he came out, the car cranked very slowly for a second or two and then stopped. He is surprised that it died so quickly. After writing up the repair order and looking up the service history, you inspect the vehicle visually. You notice that the battery is the original one, and the terminals are quite dirty. After using a battery terminal tool to clean the battery terminals, you test the battery's capacity. You find that it has degraded from its original 650 CCAs to 185 CCAs. With a booster battery connected, the vehicle starts and runs normally.

1. What test is used to measure the battery's CCA capacity?
2. Which parts need to be replaced and why?
3. How would you maintain any adaptive memories while disconnecting the battery?
4. What is parasitic draw, and how is it measured?

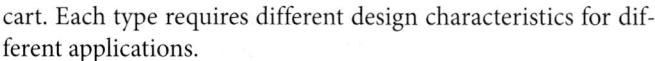

FIGURE 43-3 **A.** Starting battery. **B.** Deep cycle battery.

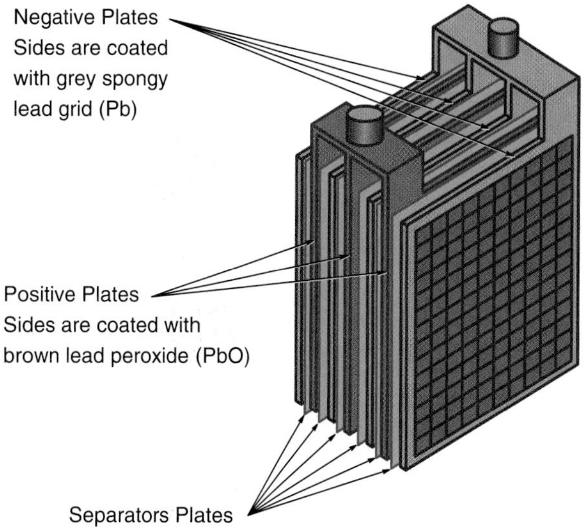

FIGURE 43-4 Typical plate arrangement in a wet cell battery.

cart. Each type requires different design characteristics for different applications.

Starting-type batteries provide high current draws for short periods of time. Whereas deep cycle batteries withstand deep discharge much better. This allows them to supply smaller, continuous loads over longer periods of time. The design may change. But the fundamental components and their operation remain the same.

Lead-Acid Flooded Cell Batteries

The wet cell lead-acid battery is the main storage device in automotive use. It is called a flooded cell battery because the lead plates are immersed in a water–acid electrolyte solution. An automotive battery can supply very high discharge currents while maintaining a high voltage. This is useful when starting a cold engine. It gives a high power output for its compact size, and it is rechargeable.

The standard 12-volt car battery consists of six cells connected in series. Each cell has a nominal 2.1 volts. This gives a total of 12.6 volts for a fully charged "12-volt" battery. Each cell contains two sets of electrodes (called plates). One set is made of sponge lead (Pb) and the other set is lead dioxide (PbO_2). They are immersed in an electrolyte solution of diluted sulfuric acid (H_2SO_4) and water (about 36% sulfuric acid and 64% water). As the battery discharges, the sulfuric acid is absorbed into the lead plates. This slowly turns both of the plates into lead sulfate. At the same time, the strength of the electrolyte becomes less acidic as the acid is absorbed into the plates. Recharging the battery reverses this process.

Inside automotive storage batteries, the main storage part is the wet cell. The nominal 2.1 volts of each cell does not depend on the size of the cell. However, its current capacity does depend on size. The surface area of the plates in a cell determines the cell's current capacity. In a lead-acid battery, positively and negatively charged plates are assembled so that they are alternated. Inside of a cell, all of the positive plates are connected to each other in parallel. The negative plates are also connected to each other in parallel (**FIGURE 43-4**). The more plates, the greater the current capacity of the cell.

Because the plates are arranged alternately, they have to be close to each other, but not touching. If they touch, an internal short circuit in the cell will occur. This is typically what has happened in a battery with a "dead cell." The positive and negative plates have shorted together and therefore discharge that cell completely. Normally, the plates are kept from touching each other by separators, usually made of plastic.

The battery's six 2.1-volt cells are connected in series to form the battery. This gives the battery a nominal voltage of 12.6 volts. The cells are sealed from each other and filled with dilute sulfuric acid. The battery case is usually made of plastic or hard rubber. With the cells in series, one end of the battery is connected to the negative post. The other end is connected to the positive post (**FIGURE 43-5**).

▶ Low-Maintenance and Maintenance-Free Batteries and Cells

LO 43-02 Describe basic types of batteries.

Many types of batteries are available for vehicles. They have many design variations. Some batteries are designed for starting. Some are for deep-cycle usage. And some are for low-maintenance or maintenance-free applications. Deep-cycle batteries are made with heavier lead plates that tolerate deep discharging better than starting batteries. But they are heavier and bulkier than starting batteries and have a lower output per pound.

Low-maintenance batteries require little, if any, topping off the water in the electrolyte. The plates and venting system are designed so that they do not normally gas. This means they

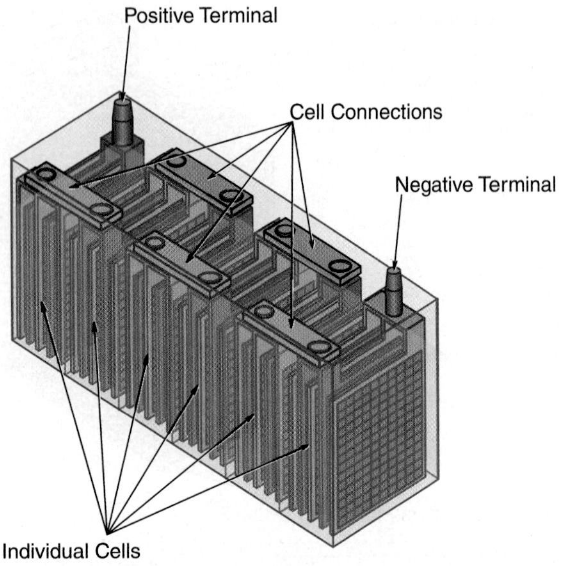

FIGURE 43-5 Interconnections between all six cells in a battery, showing the most negative and positive points of the battery.

do not normally release water vapor to the atmosphere. Low-maintenance batteries still have removable caps. That is so the electrolyte can be checked and topped off if necessary.

▶ TECHNICIAN TIP

Because lead is relatively soft, when it is used in the lead plates of a battery, it is prone to bending and stretching. Many years ago, antimony was added to the plates to strengthen them. But this caused the cells to gas off water, which caused the electrolyte level to fall over time, requiring periodic topping off. Calcium is now added to one or both plates to reduce water usage.

Maintenance-free batteries are also called valve-regulated lead-acid batteries. They are fully sealed, but equipped with a pressure relief valve. They do not require the electrolyte to be topped off. In some cases, they use a gel-type electrolyte instead of a liquid.

Absorbed glass mat batteries have the electrolyte absorbed within a mat of fine glass fibers. The plates in this type of fully sealed battery can be made flat or wound in a cylindrical cell (**FIGURE 43-6**). The electrolyte in absorbed glass mat batteries is a gel, which does not spill. This type of battery is especially handy for rough handling or tipping. In fact, this type of battery can even be mounted on its side and will still perform well. Thus, the absorbed glass mat battery is especially suited for off-road and racing vehicles.

A sealed or maintenance-free battery typically has no removable cell covers. This means that you cannot adjust or test the fluid levels inside. However, some of these batteries do have a visual indicator. It is called a single-cell hydrometer float. It provides information on the status of the charge and condition of one of the battery cells (**FIGURE 43-7**). Each manufacturer provides details of these visual indicators. Refer to these when performing an inspection.

FIGURE 43-6 AGM battery.

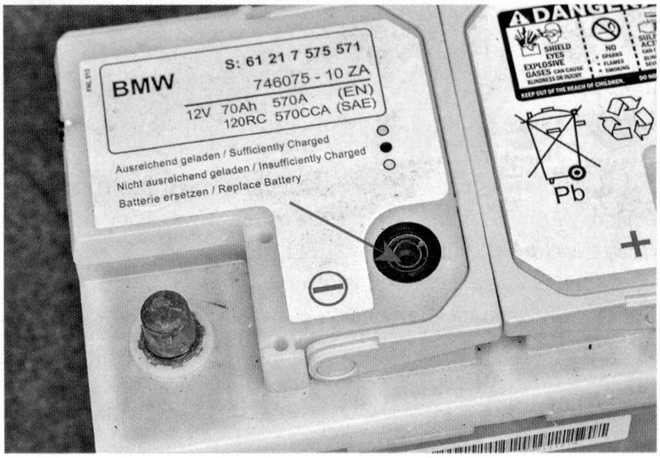

FIGURE 43-7 A status indicator is built into some batteries.

Advanced Batteries

New batteries and cells continue to be developed. Over time, they become more efficient and higher density. Consumer electronic devices have been driving the development of new battery technologies. Examples include cell phones and tablet computers. They may be nickel-cadmium (Ni-Cd), nickel-metal hydride (Ni-MH), and lithium ion (Li-ion). Each of these batteries is types of rechargeable, lead-acid cell batteries.

Nickel-cadmium batteries contain older technology. Nickel-metal hydride batteries have replaced them. Both have a cell voltage of 1.2 volts. The nickel-metal hydride battery can have two to three times the energy though. This made nickel-metal hydride batteries especially useful for drive motor applications. They were used in early hybrid electric, plug-in hybrid, and battery electric vehicles. They also tended not to have the memory effect that plagued the nickel-cadmium batteries of the past. The memory effect required nickel-cadmium batteries to be completely discharged between charge cycles. This helped ensure that the greatest performance of the battery was maintained.

The lithium-ion battery is a newer type of rechargeable cell. It is now used in many consumer electronic devices, such as cell phones and tablet computers. Such applications paved the way for the use of lithium-ion batteries in hybrid-electric or battery-electric vehicles. These vehicles use what is known as the rechargeable energy storage system (RESS). A lithium-ion battery has one of the highest energy density ratios of common batteries in production today. This high energy density means the battery can store more energy than other comparable batteries. This is a real advantage for vehicle applications. They also have a low self-discharge rate. This means they can sit for long periods without discharging.

The actual cell voltage depends on the final materials used to make the cell. The typical cell voltage for a lithium-ion battery is 3.6 volts. In contrast, cell voltage of a nickel-cadmium or nickel-metal hydride battery is 1.2 volts. And a typical lead-acid battery cell is 2.1 volts. Like all batteries, the lithium-ion cell has an anode, a cathode, and electrolyte. When discharging, the electrons are removed from the anode and added to the cathode. When the cell is charging, the reverse occurs.

Lithium-ion batteries may suffer from **thermal runaway** and cell rupture if overheated or overcharged. In extreme cases, thermal runaway may result in an explosion. Extreme care should be taken when handling or charging lithium-ion batteries. Always follow the manufacturer's recommendations.

In hybrid or electric vehicle applications, many small individual cells are used. A number of these cells are connected in series to each other. Then, these are connected in parallel with other series arrangements. This forms a battery pack that delivers the power requirements of the vehicle (**FIGURE 43-8**). Many RESS battery packs develop voltages in excess of 200 volts. And that means it is extremely important to follow the high-voltage safety precautions.

Advantages of lithium-ion batteries include:

- High energy density: There is more power per pound.
- Low self-discharge: Self-discharge is typically less than half of that of nickel-cadmium.
- Low maintenance: No periodic discharge is required.
- No memory.
- Low internal resistance: They are good for high current requirements.

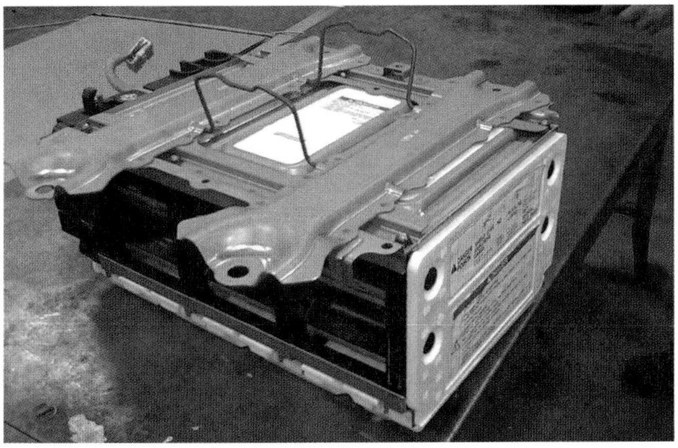

FIGURE 43-8 A typical hybrid vehicle battery stack is made of many small cells.

Disadvantages of lithium-ion batteries include:

- Need for circuit protection to ensure current and voltage are within safe limits
- Sensitivity to high temperatures
- Increased cost of manufacturing (however, these costs are being reduced as research improves the technology)
- Potential for damage if completely discharged

It should be noted that battery research is extremely active right now. It is driven by the search for evermore energy-dense renewable energy storage systems. As new developments find their way to market, expect to see even more variations of lithium-type batteries. There are likely to be discoveries of other compounds that work even better. No matter what, battery development will be critical to the future of electric vehicles.

▶ Battery Configurations

LO 43-03 Describe battery configurations, terminals, and cables.

Batteries have different size and configuration requirements. They must fit the space allotted for the battery and its location in the vehicle (**FIGURE 43-9**). Physical attributes of the automotive battery include:

FIGURE 43-9 Vehicle manufacturers locate the battery in a variety of places in the vehicle. **A.** Underhood placement. **B.** Under-seat placement.

- the size of the battery case,
- the location of the battery terminals (top or side mounted and left and right),
- and the size or type of battery terminal (round-tapered posts, screw-on terminals, or both) (**FIGURE 43-10**).

Battery sizes are designated by the Battery Council International (BCI). BCI designates the group number assigned to each configuration (**FIGURE 43-11**). Individual groups are specified by dimension in length, width, and height. Note that a physically larger battery does not necessarily mean a higher electrical capacity. So, always check the battery's ratings in addition to the BCI group size. Other designations relate to the **battery terminal configuration**. This refers to the placing of the positive and negative terminals on the battery.

Battery Cables and Terminals

Battery cables and terminals are designed to carry high discharge currents. This primarily occurs during cranking of the engine. Battery cable terminals are usually made of solid lead or zinc-plated brass. There are a couple of common designs. The most common being a cone design that provides a large surface contact area. A nut and bolt tighten the terminal onto the battery post. The tapered cone allows easy removal of the clamp when the bolt is loosened. The larger positive and smaller negative terminals are slightly different in size. This helps to ensure correct connection of the cables when servicing the battery (**FIGURE 43-12A**).

Another type of battery terminal is the side terminal. It gets its name because the battery connection is on the side of the battery. The terminal is a flat circle with a center bolt. The bolt holds the terminal tightly to the lead battery connection (see **FIGURE 43-12B**). Side-terminal batteries seem to avoid some, but

FIGURE 43-10 Battery posts. **A.** Side terminal. **B.** Top post.

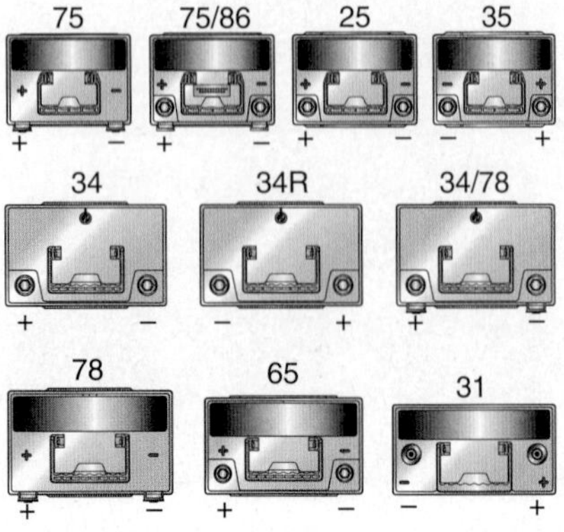

FIGURE 43-11 Typical types of layouts that use a lettering system for identification purposes.

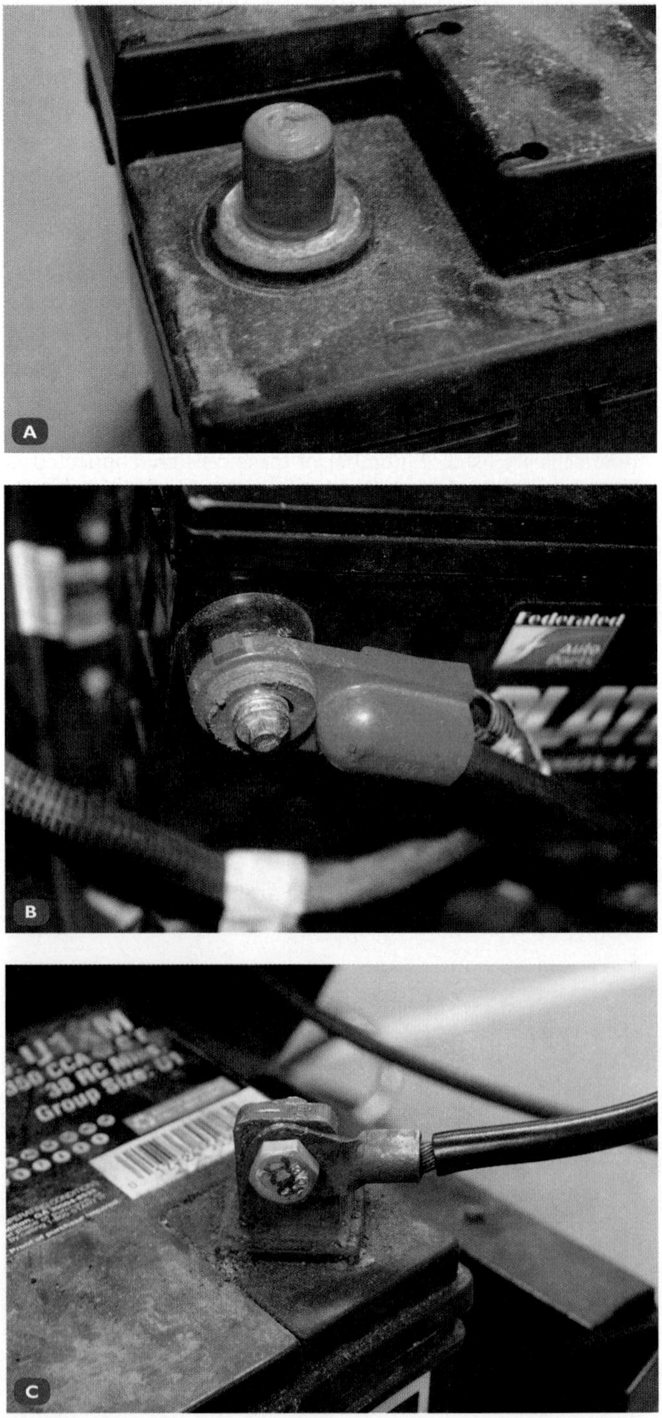

FIGURE 43-12 **A.** Tapered-post battery terminal. **B.** Side-post cable terminal. **C.** Lug terminal and cable.

not all, of the oxidation issues of a top-post battery. One shortcoming of a side-terminal battery is that it is harder to get a good connection when using jumper cables to jump-start the vehicle.

One less common battery terminal is a flat terminal (lug) with a hole through the center. The post on the battery sticks up vertically. It has at least one flat side and a matching hole through the post. The battery cable end butts up against the flat post. A bolt inserted through the battery post and terminal holds them firmly together (see **FIGURE 43-12C**).

> **▶ TECHNICIAN TIP**
>
> Installing a battery backward into a vehicle destroys expensive onboard electronics. Unfortunately, some technicians and do-it-yourself vehicle owners have experienced this. Always be careful to connect the cables to the correct polarity. And don't assume that a red cable means positive or a black cable means negative. Someone may have previously changed them with the wrong color.

Battery cables are usually made of many fine strands of copper wire. They are bound tightly together and insulated to make a cable with high current capacity. A number of cable sizes are available to handle various current capacities. For example, a small-displacement spark-ignition engine requires far less current to crank than a high-compression diesel engine. So, make sure you use the proper sized cable when replacing them (**FIGURE 43-13**).

Battery terminals are usually crimped or soldered onto the battery cables. This ensures a strong, low-resistance connection. Often, heat-shrink tubing with sealing adhesive is used over the joint (**FIGURE 43-14**). This helps keep it clean and protected from corrosion. Battery cables should also be protected from chafing or damage. And terminals should be kept clean and free of corrosion.

> **▶ TECHNICIAN TIP**
>
> Corrosion from the battery can creep into and destroy a battery cable. It can get beneath the insulation where it may not be visible. A cranking voltage drop test on the battery cable is the most reliable way to detect such high resistance caused by corrosion.

FIGURE 43-13 A. Battery cable for a small displacement engine. **B.** Battery cable for a large displacement engine.

> **▶ TECHNICIAN TIP**
>
> When buying battery cables or jumper cables, be careful to verify the wire size. Some wire manufacturers place very thick insulation over a small wire size. This makes the cable look bigger than it is. Remember, it is the size of the wire that determines current flow, not the size of the insulation.

▶ Battery Ratings

LO 43-04 Describe battery ratings and the charge–discharge cycle.

The **electrical capacity** that a typical lead-acid battery can store is determined primarily by the total surface area of the plates. But it is also due to some extent by the thickness of the plates. The more plate surface area there is, the higher the electrical capacity of the battery. Automotive battery plates tend to be manufactured in a standard size. So, a common method of increasing surface area is to increase the number of plates per cell. For example, a 12-volt, 11-plate cell has a higher capacity than a 12-volt, 9-plate cell.

There are several methods used to rate automotive battery capacity. The three most common are **cold cranking amps (CCAs)**, **cranking amps (CAs)**, and **reserve capacity (RC)**. The ratings are usually marked on automotive batteries. Technicians use these ratings when testing battery performance. They also use the ratings when picking batteries for particular vehicles (**FIGURE 43-15**).

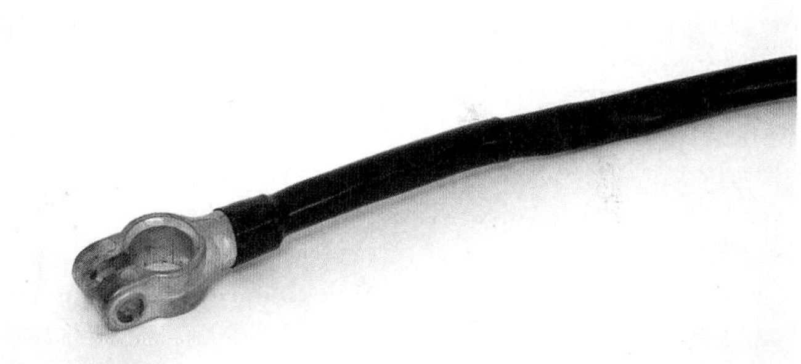

FIGURE 43-14 Battery cable with sealed end.

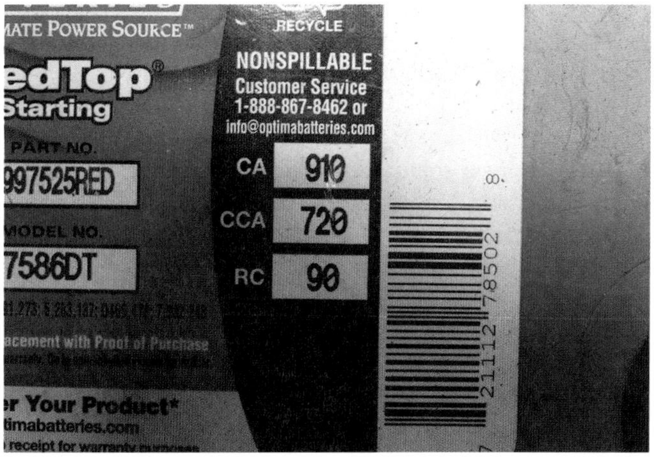

FIGURE 43-15 Battery ratings.

- CCA measures the load in amps that a battery can deliver for 30 seconds while maintaining a voltage of 1.2 volts per cell (7.2 volts for a 12-volt battery) or higher at 0°F (−18°C).
- CA measures the same thing, but at a higher temperature −32°F (0°C). This can cause confusion because a 500 CCA battery has about 20% more capacity than a 500 CA battery. So, it is important to keep the ratings straight.
- RC is the time in minutes that a new fully charged battery at 80°F (26.7°C) will supply a constant load of 25 amps without its voltage dropping below 10.5 volts for a 12-volt battery. This rating approximates the amount of time that a vehicle can be driven before the battery dies, if the charging system fails completely.

The CCA, CA, and RC are typically marked on automotive batteries, and technicians use these ratings when testing battery performance and when selecting batteries for particular applications (FIGURE 43-15).

Battery Discharging and Charging Cycle

As the battery creates current flow to operate electrical devices, it is being discharged (**FIGURE 43-16**). As it discharges, the sulfuric acid in the electrolyte joins with lead dioxide to form lead sulfate. The oxygen from the plate joins the hydrogen from the electrolyte to form water. Lead sulfate also forms at the negative plate. This occurs as sponge lead reacts with sulfate from the electrolyte.

In a discharged lead-acid cell, the active material of both plates becomes lead sulfate. This means they are no longer dissimilar metals. At the same time, the electrolyte becomes mostly water. This is because the acid leaves the electrolyte and is absorbed into the plates. The result is a very weak sulfuric acid solution in the electrolyte. This, along with the plate material becoming similar, creates a weak electrical output of the battery. And ultimately, once the battery is discharged, it is considered "dead."

When being charged, electrical pressure (voltage) is higher than that of the battery's total cell voltage. This pushes electricity back into the battery, reversing the chemical process (**FIGURE 43-17**). The charging device acts like an electron pump. It forces electrons to move from the positive plates back to the negative plates in the battery. At the negative plates, sulfate is forced out of the plate and back into the electrolyte. This changes the negative plates back into sponge lead. It also creates a stronger solution of sulfuric acid in the electrolyte. At the same time, lead dioxide is formed at the positive plates. This occurs as the sulfuric acid is forced back into the electrolyte. This process restores the cell's electrical potential or voltage.

The charging process increases the amount of acid in the electrolyte. The electrolyte then becomes stronger. When further charging no longer makes the electrolyte stronger, charging is complete. Connecting a lead-acid battery to a load reverses the process.

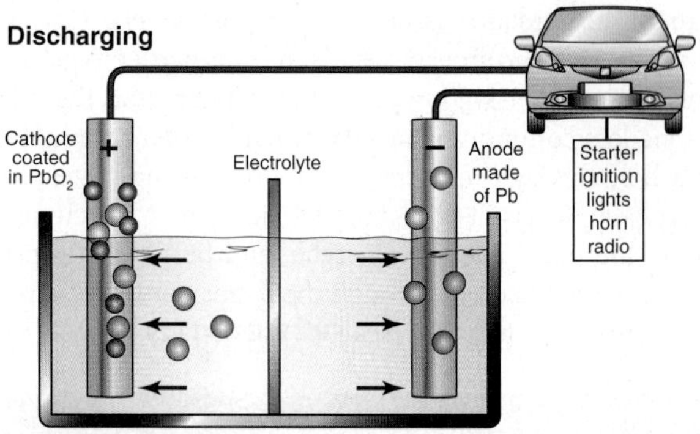

FIGURE 43-16 Discharging cycle.

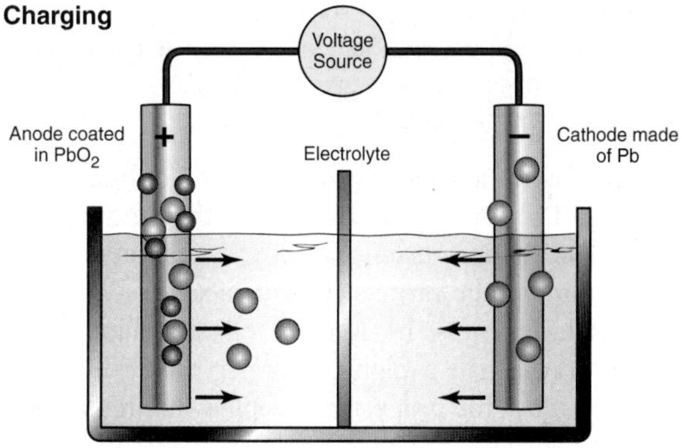

FIGURE 43-17 Charging cycle.

Applied Math

AM-10: Whole Numbers: The technician can multiply whole numbers to determine differences for comparison with the manufacturer's specifications.

An automotive battery is composed of six individual cell compartments within a case. The individual cells are joined by cell connectors in series with one another. Each cell has an open circuit voltage of approximately 2.1 volts at 80°F (26.7°C). When fully charged, a 12-volt automotive battery has an actual open circuit voltage of 12.6 volts at 80°F (26.7°C). For easy calculations with whole numbers, we have 2 volts per cell × 6 cells = 12 volts, which is why we refer to it as a 12-volt battery.

▶ Battery Life

LO 43-05 Describe conditions that shorten/lengthen the life of a battery.

Batteries do not last forever. Over time and use, the plates start to lose effectiveness. Lead paste falls off the plates and collects in the bottom of the battery case. This type of damage is especially likely to develop in a battery subjected to high vibration and shock. A good example of this would happen in an off-road vehicle. As plate material becomes deposited in the bottom of the battery case, it eventually bridges the plates, shorting them

Applied Science

AS-54: Batteries: The technician can demonstrate an understanding of the electrochemical reactions that occur in wet cell and dry cell batteries.

A vehicle that has a no-crank condition is towed into a dealership. The technician assigned to the job has completed his diagnosis of the problem. It is determined that the vehicle has a faulty battery. An apprentice technician has also discovered that the hold-down clamp on the battery is missing.

At break time, the apprentice asks the technician why he thinks the battery failed. The technician explains his thoughts by going back to the basics of battery construction and the electrochemical reactions that occur during various events. The battery case can be thought of as a container. Inside this container are six separate compartments, or cells, which are the energy-producing units. Each cell has a number of plates soaked in an electrolyte solution of about 64% water and 36% sulfuric acid. The plates are separated into the anode, or negative, side. And the other is the cathode, or positive, side.

As stated in our text, connecting a lead-acid battery to a load causes chemical changes as the battery discharges. At the positive plate, sulfate from the electrolyte joins with lead to form lead sulfate. And oxygen from the plate joins the hydrogen from the electrolyte to form water. Lead sulfate also forms at the negative plate. This occurs as sponge lead forms with sulfate from the electrolyte. Overall, the percentage of acid in the electrolyte falls and the percentage of water rises. This reduces the strength of the electrolyte. As the cell discharges, the plates develop the same composition, which reduces the potential of the cell. Recharging the battery again restores the difference between its sets of plates.

The technician explains his theory of the failure of the battery, attributing it to the missing hold-down clamp. With the battery not being securely fastened, plate material breaks off and drops down into the bottom of the battery case. This eventually causes the plates to short out. He goes on to explain that all batteries will eventually wear out. But excessive vibration likely shortened the life of this battery.

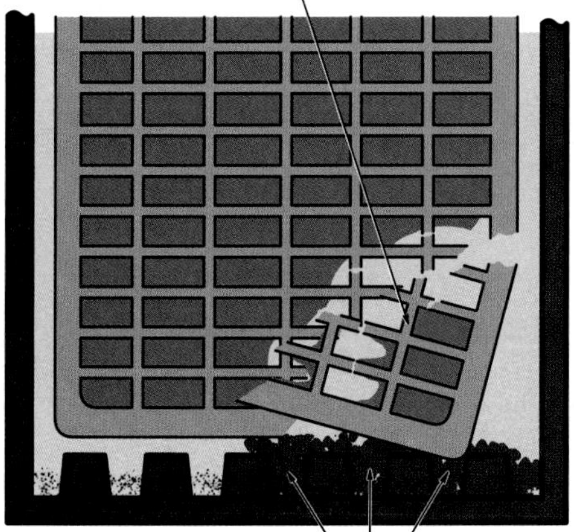

FIGURE 43-18 Over time, lead from the plates can flake off and build up high enough to bridge the gap between the plates, shorting out the cell.

out (**FIGURE 43-18**). This renders that cell useless. This lowers the battery's open circuit voltage and reduces the battery's capacity.

A battery may also lose its capacity as it becomes sulfated. This means that the surfaces of the plates harden due to the acid sitting in the plates for extended times. This makes it more difficult for the acid to be absorbed deeper into the plate, creating less current flow. As a result, the battery can no longer readily accept a charge. Nor can it adequately discharge under load.

Leaving a battery on a slow charger overnight and then finding that it fails a load test in the morning is an indication of a sulfated battery. Another indicator is when a conductance test shows that the battery's capacity in CCAs is significantly lower than specified.

Sulfated batteries cannot create the necessary current flow and should be replaced and recycled. Leaving a battery less than 80% charged greatly speeds up the process of sulfation. The further discharged, the quicker the battery sulfates. Keeping a battery fully charged reduces the problem of sulfation.

Applied Science

AS-55: Acids/Bases: The technician can identify the effects of the pH of a solution on various systems.

A technician has replaced a heater core in a vehicle, a task that required many hours of labor due to its location. The technician filled the cooling system with the type of antifreeze recommended by the manufacturer, with a 50/50 mix of distilled water. After running the engine to check for leaks, the technician used a coolant test strip to verify the proper pH level of the coolant. The test strip indicated a pH level of 10, which is in the correct range of 9.8 to 10.2 pH. Any reading below 9.0 indicates that the coolant may cause damage to the water pump, radiator, heater core, or other components. This condition is known as electrolysis.

An acid is a substance that produces hydrogen ions. A base is a substance that produces hydroxide ions. The pH scale is used to tell how acidic or base a substance is. A handheld pH meter is a precision device used to measure the pH of liquids. This instrument is more accurate than pH test strips.

In addition to pH testing of the cooling system, brake fluid can be also checked with test strips or a pH meter. The safe range for brake fluid ranges from 7.0 to 11.5 on the pH scale. Battery electrolyte specific gravity can also be tested with a hydrometer or a refractometer. When a battery is discharging, the specific gravity of the electrolyte is dropping. When a battery is being charged, the situation is reversed. Electric current is forced into the battery by the vehicle's alternator or a battery charger. The specific gravity of the battery goes up as a result.

Battery Life Factors

Batteries operate under a range of conditions that affect their performance and life. Batteries should be kept clean, dry, and fully charged. They should have the correct level of electrolyte. They should be kept at a moderate temperature (e.g., 77°F [25°C]). And they should be well secured. A battery's lifetime is shortened by the following conditions:

- Being fully discharged or having deep discharge cycles
- Remaining overcharged or undercharged
- Experiencing high discharge rates for extended periods
- Experiencing excessive vibration
- Being exposed to extremes of temperature
- Having dirt or moisture on the case
- Developing corrosion

As you can see, poor maintenance and operating conditions reduce the battery's operating life.

▶ TECHNICIAN TIP

Back in the day when many people worked on their own cars, a very handy do-it-yourselfer bought for his vehicle a top-shelf battery from the local parts store. After a couple of weeks, the customer brought the battery back, saying it had a dead cell. The store employee checked it out, and sure enough, it had a dead cell, so they gave him a new battery. A few weeks later, the customer returned with the same problem. The store employee apologized profusely and gave him another battery. Once again, the customer was back within the month with another bad battery.

By this time, the store owner figured something had to be going on, so he went to the customer's home to check it out. When he got there, he saw a beautiful '57 Chevy with a tilt front end. One quick look told him everything he needed to know. When the tilt front end was installed, the battery tray had to be removed. The customer found a nice new spot to mount the battery tray. He had bolted it right to the top of the upper control arm for the suspension. So, every bump in the road bounced the battery and literally shook it to death. Needless to say, the customer ended up paying for a few ruined batteries.

▶ Battery Maintenance

LO 43-06 Describe the purpose and types of battery maintenance.

Batteries need regular maintenance. This includes inspection, cleaning, testing, and charging when discharged. Batteries should be checked during scheduled oil changes. They should also be inspected in the fall, before cold weather sets in. Winter weather makes the battery work harder. So, any needed maintenance should be performed before then.

When inspecting the battery, check that the battery electrolyte level is within the markings on the case. Or look in each cell to ensure the plates are well covered (**FIGURE 43-19**). Make sure the exterior case is dry and free of dirt. Dirt on top of the battery can actually cause premature self-discharge of the battery. This is because current "leaks" across the path of dirt and grime. The grime becomes conductive and drains the battery over time.

To verify whether the surface of the battery has to be cleaned, use a digital multimeter (DMM). Set the meter to volts to measure the voltage on the top surface of the battery. You can do this by placing the black lead on the negative battery post. Then, drag the red lead around the top of the battery. Measure the voltage present across the surface (**FIGURE 43-20**). Any

FIGURE 43-19 The electrolyte level should be **A.** between the lines on the battery case or **B.** near or touching the bottom of the fill hole.

FIGURE 43-20 Use a DMM to measure the voltage on the top of the battery. Any voltage higher than about 0.2 volt means the battery should be cleaned.

voltage over about 0.2 volt means the battery should be cleaned. A mixture of baking soda and water works well for this. Just make sure not to get any of that mixture down inside of the battery. Doing so will tend to neutralize the electrolyte, damaging the battery.

Keeping the battery and terminals clean is one of the best maintenance tasks for the money. The lead in the battery posts and terminals oxidizes over time. Unfortunately, lead oxide is an insulator. Therefore, if the lead surfaces of the post and terminal oxidize, the insulator effect prevents current flow. This can cause the vehicle not to crank over, stranding the driver. Lead oxide can be identified by its dark gray or black color (**FIGURE 43-21**).

The only effective way of removing lead oxide is with a special battery terminal scraper or wire brush. This removes the oxidized lead and restores the lead to its shiny silver color. Once the battery terminal is reinstalled and tightened, the terminals should be coated with a battery oxidation inhibitor. This will help slow down the oxidation process. Note that baking soda and water will NOT remove oxidation. Oxidation must be scraped off.

Batteries sometimes have to be recharged when lights are left on, or when the vehicle has not been driven for several weeks. Performing a battery state-of-charge (SOC) test is a good indicator of whether the battery has to be charged. A hydrometer or refractometer is used for this test. Another battery maintenance task is performing a conductance test. It measures the capacity of the battery. This can be used to determine how much life is left in the battery. These tests are explained in the following topics.

FIGURE 43-21 A. Heavily oxidized battery post. **B.** Clean battery post.

SAFETY TIP

To ensure safety when working on batteries, take the following precautions:

- Make sure the hood is secured with a hood prop rod.
- Wear the appropriate personal protection equipment (PPE) before starting the job. Remember that batteries contain acid. If this acid is splashed in your eyes, it can blind you or seriously damage your eyes.
- Follow the manufacturer's personal safety instructions to prevent personal injury or damage to the vehicle you are working on.

Battery Service Precautions

- When servicing batteries, always ensure that you have the right PPE (e.g., safety eyewear, gloves, and shop clothing).
- Never wear any jewelry, such as neck chains, watches, or rings, when working on or near batteries. Jewelry may provide an accidental short-circuit path for high currents.
- If the battery is being, or has been recently, charged, ensure that the space around the battery is well ventilated.
- Avoid making any sparks, which may ignite the **gassing** hydrogen–air mixture. Never create a low-resistance connection or short across the battery terminals.
- Always remove the negative or ground terminal first when disconnecting battery cables. This is to reduce the possibility of a wrench creating an accidental short to ground because you are removing the negative terminal first. If the wrench connects from the negative battery terminal to ground, nothing will happen. And once the negative terminal is off the battery, if the wrench touches ground while on the positive post, there is not a complete circuit back to the negative battery terminal. So, no spark will occur.

SAFETY TIP

The author has been in the vicinity of two batteries when they blew up. One was from across the shop and sounded like a bomb going off. A student had been welding a van door when a spark from the welding ignited the hydrogen gases around the battery. Fortunately, the battery was on the opposite side of the door from where the student was working. The battery case was blown apart and acid was thrown 5 to 10 feet in all directions. Luckily, no students were within that range. Always keep sparks away from batteries.

Battery Recycling

Batteries contain many environmentally damaging chemicals and metals. If they find their way into a landfill, the acid and metals can contaminate the soil and waterways. Correct disposal of batteries by recycling them is good for the environment. Also, the precious metals can be reclaimed for reuse. Many municipalities require battery recycling. To ensure this, they levy a "core charge" on every new automotive battery sold. The core charge is refunded if an old battery is brought in and exchanged for the new one. This process helps prevent batteries from being discarded in the trash or left lying around. Check local laws and regulations to ensure that batteries are disposed of properly.

▶ Inspecting, Cleaning, Filling, and Replacing the Battery and Cables

LO 43-07 Inspect, clean, fill, and replace the battery and cables.

As noted earlier, batteries last longer if they are properly maintained. In fact, one of the most common causes of vehicle no-starts is dirty or corroded battery cables. Inspecting, cleaning, and filling batteries (if not maintenance free) are common maintenance tasks. These tasks should be performed every six months to one year on top-post batteries. On side-post batteries, they should be performed every one to two years.

Automotive batteries can look lighter than they really are. Always lift them with care, and get help if needed. When reconnecting the battery, be sure you do not connect the battery with reverse polarity. Doing so will send current in the reverse direction through the electrical system. The reverse current flow will likely damage many of the electronic control units (ECUs) in the vehicle. So, what started out as a simple maintenance task becomes an expensive diagnose and repair job. In this case, the shop has to perform it at no charge to the customer.

SAFETY TIP

When disconnecting the battery terminals, always remove the negative terminal completely first. And always install the negative terminal only after the positive terminal is fully tightened. Here is why. Once the negative terminal is removed, if the wrench touches the chassis when it is on the positive terminal, no spark will be created. This is because there is no path back to the negative terminal. You still need to be careful that the wrench does not bridge straight from the positive terminal to the negative terminal. As that would cause a huge spark.

To inspect, clean, fill, or replace the battery, battery cables, clamps, connectors, and hold-downs, follow the steps in **SKILL DRILL 43-1**.

SKILL DRILL 43-1 Inspecting, Cleaning, Filling, and Replacing the Battery and Cables

1. Measure the voltage on the top of the battery with a DMM. Place the black lead on the negative post. Move the red lead across the top of the battery to find the highest reading.

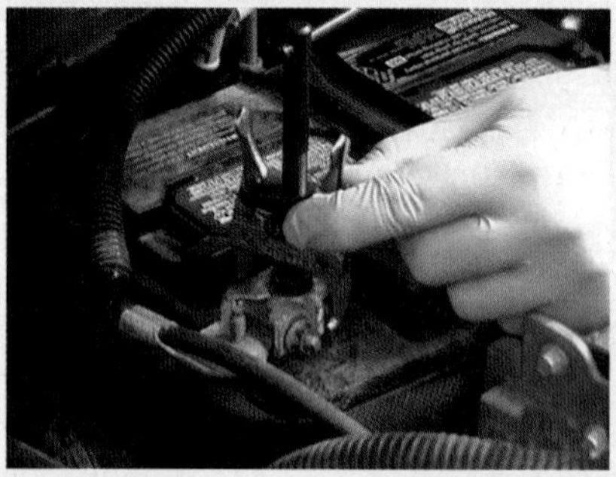

2. Remove the cable clamp from the negative terminal first. Then, remove the positive terminal. Bend the cables back out of the way so that they cannot fall back and touch the battery terminals accidentally.

3. Remove the battery hold-downs or other hardware securing the battery.

4. Keeping it upright, remove the battery from its tray. Place it on a clean work surface. Inspect the battery for damage.

5. Carefully clean the battery case and the battery tray.

6. Clean the battery posts with a battery terminal tool. Clean the cable terminals with the same battery terminal tool. Examine the battery cables for fraying or corrosion.

SKILL DRILL 43-1 Inspecting, Cleaning, Filling, and Replacing the Battery and Cables (Continued)

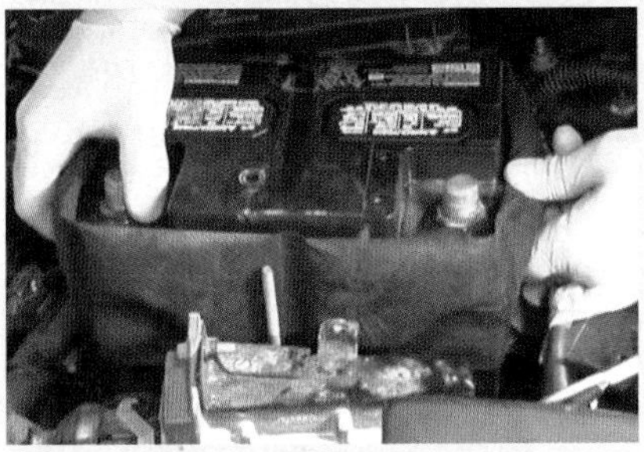

7. Reinstall the cleaned and serviced battery. Replace the hold-downs, and make sure the battery is securely held in position. If installing a new battery, ensure that it meets the original manufacturer's specifications.

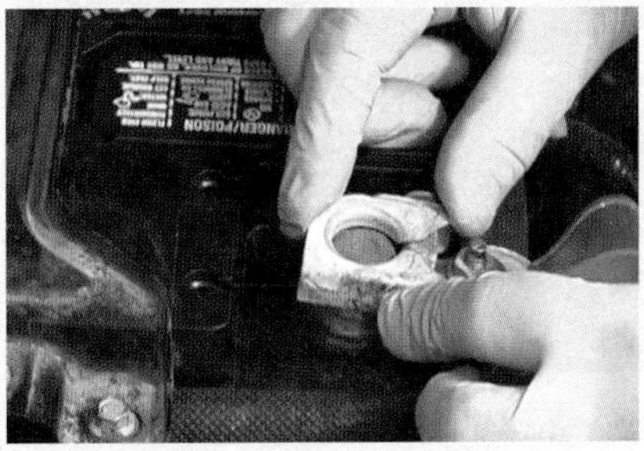

8. Reconnect the positive battery terminal and tighten it in place. Once the positive terminal is finished, reconnect the negative terminal and tighten it.

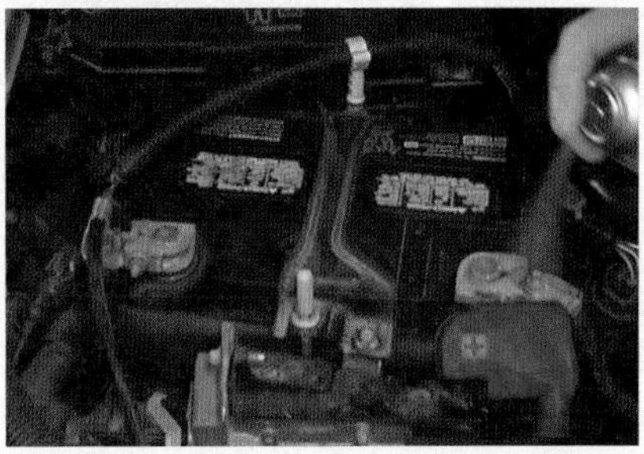

9. Coat the terminal connections with anticorrosive paste or spray to keep oxygen from the terminal connections. Test that you have a good electrical connection by starting the vehicle.

Applied Science

AS-74: Activity of Metals: The technician can explain the conductivity problems in a circuit when connectors corrode due to electrochemical reactions.

A dealership technician has a do-it-yourself neighbor who enjoys working on his own vehicles. The neighbor has questions from time to time for the professional technician. The do-it-yourselfer was restoring a vehicle in his spare time. He had just purchased a new battery and a new starter. But the engine still would not crank over. The technician demonstrated how to do a voltage drop test using a DMM. The problem was in a battery cable connection at the starter solenoid. When the cable end was removed from the solenoid, corrosion was found. After removing the corrosion from both surfaces and reattaching the cable, the engine cranked over properly. The neighbor realized that he did not actually need a new battery and starter.

This is an example of a conductivity problem when connectors corrode due to electrochemical reactions. All metals on the vehicle can rust. Battery terminal corrosion is one of the main concerns regarding vehicle maintenance. The terminal corrosion can be removed with a wire brush. External corrosion on battery terminals is very easily spotted. But internal corrosion between the battery post and cable clamp is hidden. It is necessary to take the connection apart and clean each surface with a battery terminal cleaning brush. Connections with corrosion problems increase resistance at the battery terminals. This reduces the voltage to the electrical system.

▶ Charging the Battery

LO 43-08 Perform battery charging and jump-starting.

Vehicle batteries may become discharged and require charging. This occurs if:

- the lights or other accessories were left on.
- the vehicle has been sitting idle without starting for more than a couple of weeks.
- the charging system is faulty.
- the vehicle is driven for very short trips.
- the battery surface is dirty and drains the battery over time.

Also, before testing a battery, it should ideally be fully charged. If charging is needed, you have to determine the charge rate for the battery. But remember that slow charging is less stressful on a battery than fast charging.

- To determine the ideal charging rate: CCAs divided by 70. Example: 500 CCA = 7 A
- To determine the max charging rate: CCAs divided by 40. Example: 500 CCA = 12.5 A

A slow charger usually charges at a rate of less than 5 amperes. A fast charger charges at a much higher ampere rate, depending on the original battery's state of charge. So, it should only be used under constant observation. To fully charge a battery can take more than 20 hours, depending on the state of charge.

Some manufacturers recommend removing the negative battery terminal while charging a battery. This is to reduce the risk of damaging the vehicle's electronics. If the battery has to be disconnected during charging, it may lose pre-sets stored in memory. So, remember to verify if the vehicle's adaptive memory will need to be maintained. If the battery terminals are heavily corroded, they may need to be cleaned before charging the battery. Also, monitor the battery voltage during charging.

- *Never exceed 15.5 volts* when charging a 12-volt flooded cell battery.
- *Never exceed 14.8 volts* when charging a 12-volt AGM battery.
- *Never exceed 14.3 volts* when charging a 12-volt gel cell battery.

After charging the battery, it is good practice to clean the battery posts, cable terminals, and battery case.

To charge a battery, follow the steps in **SKILL DRILL 43-2**.

SKILL DRILL 43-2 Charging a Battery

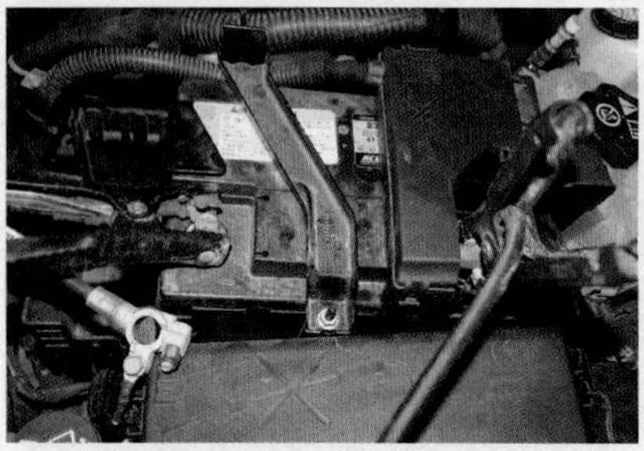

1. Ensure that the battery has not been frozen. Install a memory saver into the DLC, and disconnect the negative terminal. Verify that the charger is unplugged and off. Connect the red lead to the positive terminal and the black lead to the negative terminal.

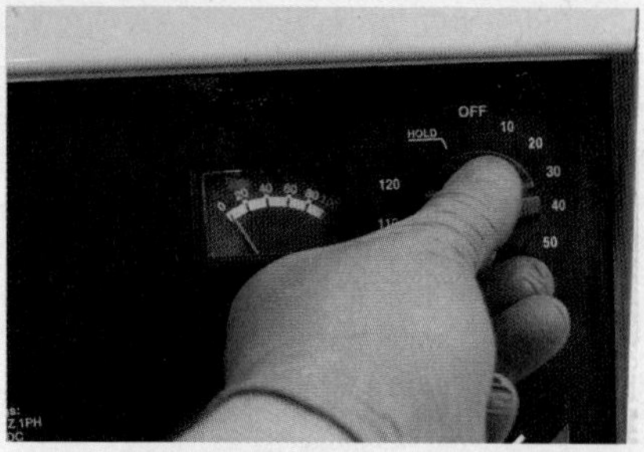

2. Calculate the ideal and maximum charge rates. Plug in the charger, and turn it to the appropriate charge rate. Monitor the battery voltage while charging.

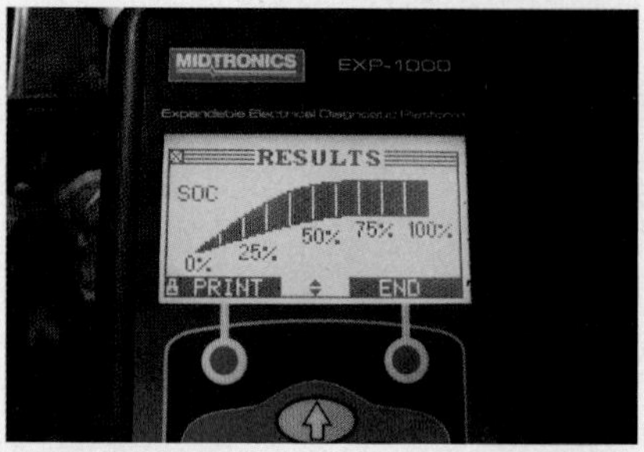

3. Once the battery is charged, turn the charger off. Remove the surface charge by turning on the headlights for three to five minutes. Then let the battery sit for another 2–3 minutes. Using a conductance tester, load tester, DMM, refractometer, or hydrometer, test the battery's state of charge and capacity.

Applied Science

AS-73: Electrochemical Reactions: The technician can demonstrate an understanding of the ion transfer process that occurs in an automotive battery.

A vehicle is in the shop for routine maintenance, and a general inspection is being performed. Jim, the technician, observes that the battery appears to be old. The code on the battery indicates it was manufactured six years ago.

At lunch time, the technician discusses the topic of batteries with another technician. He has the opinion that an automotive battery should be replaced every five years to prevent possible starting problems. Jim does not agree as he believes there are too many variables to make this determination. This leads to a discussion of how a lead-acid battery works.

As stated in our text, in a discharged lead-acid cell, the active material of both plates becomes lead sulfate. At the same time, the electrolyte becomes mostly water as the acid is absorbed into the plates. The result is a very weak sulfuric acid solution in the electrolyte. When being charged, the battery is connected to a DC electrical supply. The electrical pressure (voltage) from the charger is higher than that of the battery's total cell voltage. The charging device acts like an electron pump. It forces electrons to move from the positive plates to the negative plates in the battery.

At the negative plates, sulfate is discharged back into the electrolyte. This changes the chemical composition of the plates into sponge lead. It also creates a stronger solution of sulfuric acid in the electrolyte. At the same time, lead dioxide is formed at the positive plates. All of this helps to restore the cell's electrical potential or voltage.

The charging process increases the amount of acid in the electrolyte, making it stronger. When further charging no longer makes the electrolyte stronger, charging is complete.

Jump-Starting the Vehicle

Jump-starting a vehicle is the process of using one vehicle to start another vehicle. The vehicle with a charged battery provides electrical energy to start the vehicle that has a discharged battery. Starting a vehicle requires a high amount of electrical energy. So, jump-starting puts stresses on both vehicles. When the discharged vehicle is being cranked, the battery voltage tends to fall very low. This is because the battery is already discharged. This causes the alternator on the running vehicle to put out its maximum current, which puts it under heavy load. But as soon as the jumped vehicle stops cranking, the voltage shoots up quickly. This can cause a potentially high voltage spike. It can be large enough to damage electronic components in either vehicle. To help reduce this risk, it is good practice to turn on the headlights on the charged battery while cranking the dead vehicle. The headlights will help absorb any voltage spike. The same voltage spike can happen when the jumper cables are being disconnected.

Another risk is overheating and damaging the dead vehicle's alternator. With all of the electrical draws on the vehicle, today's alternators work harder than ever. Adding the job of fully recharging a discharged battery on top of the regular electrical loads can push an alternator into the danger zone. So, keeping as many electrical loads off after a vehicle has been jump-started will reduce the load on the alternator.

There is a risk of damaging the electronics and alternator. So, many manufacturers prohibit jump-starting on their vehicles. And some tow companies are now refusing to jump-start newer vehicles. They now recommend either externally slow charging the battery or replacing the battery with a charged battery.

If you do decide to jump-start a vehicle, always read the owner's manual for both vehicles. And then be sure to follow their jump-starting guidelines. Also, never attempt to jump-start a frozen battery.

It is usually best to let the running vehicle charge the battery on the other vehicle for 5 to 10 minutes before trying to start the vehicle. Once the dead vehicle's engine has started, let both vehicles come to an idle for a moment. Do not turn off the vehicle with the discharged battery. It needs an extended amount of time to recharge. Before you disconnect the service battery from the discharged battery, it is good practice to place a load on the charged battery. Do this by turning on an accessory such as the headlights. This helps absorb any sudden rise in voltage that may occur as the load on the alternator is suddenly decreased. Another method of reducing the risk of damage to sensitive electronic devices is by using jumper leads that have a built-in surge protector.

An option that is less risky when jump-starting a vehicle is the use of a jump box. A jump box does not involve a charging system. This greatly lessens the risk of voltage spikes. This device contains a relatively high-capacity battery. It has short cables and spring-loaded clamps that can be connected directly to the dead battery. It provides a substantial additional boost of electrical energy to start the vehicle.

SAFETY TIP

When connecting jumper cables, a spark almost always occurs on the last connection you make. That is why it is critical that you make the last connection on the engine block. This keeps it away from the battery. A spark also occurs when you disconnect the first jumper cable connection. So, that should also be the black lead's connection at the engine block.

To jump-start a vehicle, follow the steps in **SKILL DRILL 43-3**.

▶ TECHNICIAN TIP

Alternators are not generally designed to charge a dead battery while running several accessories. If a battery must be jump-started, it is always best to recharge the battery using a battery charger. That way, the alternator will not have to work so hard. Some technicians say it takes only 15 minutes to burn up an alternator when charging a dead battery. And if you must drive the vehicle after being jump-started, turn off as many accessories as possible. This will lessen the load on the alternator.

SKILL DRILL 43-3 Jump-Starting a Vehicle

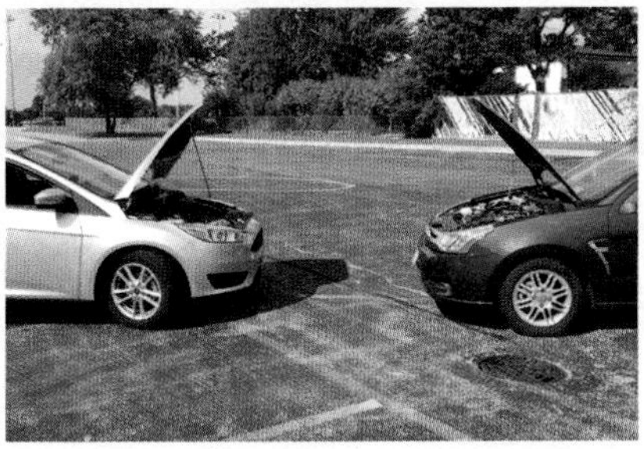

1. Position the charged battery close enough to the discharged battery that it is within comfortable range of your jumper cables. If the charged battery is in another vehicle, make sure the two vehicles are not touching.

2. First, connect the red jumper lead to the positive terminal of the discharged battery in the vehicle you are trying to start.

3. Next, connect the other end of this lead to the positive terminal of the charged battery or the remote terminal.

4. Then, connect the black jumper lead to the negative terminal of the charged battery or the battery remote terminal.

5. Connect the other end of the negative lead to a good ground on the engine block of the vehicle with the discharged battery. Make it as far away as possible from the battery.

6. Do not connect the lead to the negative terminal of the discharged battery itself. Doing so may cause a dangerous spark. Also, do not connect the negative lead to the body or chassis as the ground wire from the body back to the negative battery terminal is usually too small to carry current needed for jump-starting the vehicle.

SKILL DRILL 43-3 Jump-Starting a Vehicle (Continued)

7. Try to start the vehicle with the discharged battery. If the booster battery does not have enough charge or the jumper cables are too small in diameter to do this, start the engine in the booster vehicle. Allow it to partially charge the discharged battery for several minutes. Turn on the headlights on the booster vehicle to reduce the possibility of a voltage spike damaging electronic equipment, and try starting the discharged vehicle again.

8. Disconnect the leads in the reverse order of connecting them.

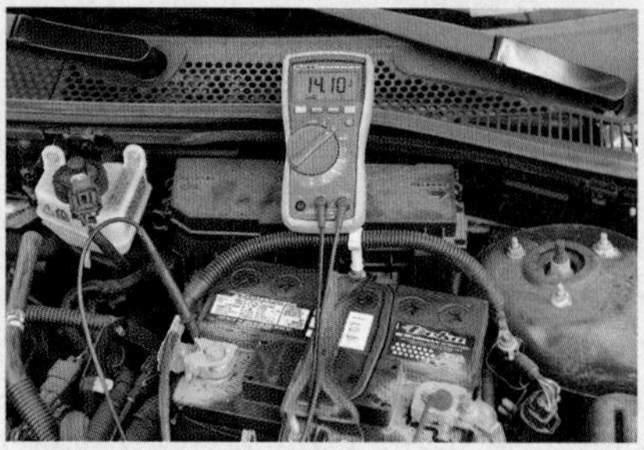

9. If the charging system is working correctly and the battery is in good condition, the battery will be recharged while the engine is running. However, it could end up overheating and damaging the alternator.

▶ Testing Battery State of Charge and Specific Gravity

LO 43-09 Perform battery state of charge and specific gravity tests.

The conductance test is currently the preferred test for determining battery condition. But other tests still can be used to give us some helpful information. One of those tests is the SOC test. SOC testing indicates how charged or discharged a battery is, not how much capacity it has. The degree of the battery's charge is handy to know when testing starting and charging system issues. It is also helpful in just about any other electrical issue as well.

There are two tests for determining the battery's state of charge: the specific gravity test and the open circuit voltage test. The specific gravity test measures the electrolyte's specific gravity. This indicates the acid content, and so the state of charge. Although specific gravity testing is the most accurate, most batteries are of the maintenance-free type. So, they may not provide access to the electrolyte in the cells for specific gravity testing. Older batteries with removable caps may be tested for specific gravity.

Recall that the acid concentration drops as the battery becomes discharged. This lowers the electrolyte's specific gravity. The higher the specific gravity, the higher the percentage of acid in the electrolyte. This corresponds to a high battery state of charge (**TABLE 43-1**). When measuring specific gravity with a hydrometer, know that temperature can affect the reading. The hydrometer's reading, unlike that of the refractometer, must be corrected for electrolyte temperature. If so, refer to the temperature correction chart for electrolyte above and below 80°F (26.7°C). The correction will need to be added to or subtracted from the base reading. *Note:* If there is a difference in specific gravity reading of 0.050 or more between the cells, the battery should be replaced.

TABLE 43-1 Specific Gravity and the Corresponding State of Charge

State of Charge	Specific Gravity @ 80°F (26.7°C)
100%	1.265 or higher
75%	1.225
50%	1.190
25%	1.155
0%	1.120

▶ TECHNICIAN TIP

When performing a SOC test, keep these tips in mind:

- When filling a battery that is not fully charged, never fill it to the top of the full line, as charging the battery raises the electrolyte level.
- Small amounts of electrolyte in the hydrometer may leak out and damage the vehicle's paint or your clothing.
- Do not inadvertently remove electrolyte from one cell or add it to another cell when testing; doing so causes incorrect readings.

When drawing electrolyte into the hydrometer, draw in enough so that the float is floating. It should not be sitting on the bottom or hitting the top. If the electrolyte level is too low to do this, then you will have to add distilled water. Then, the battery will have to be fully charged to mix the water and acid.

A very quick and reasonably accurate indicator of battery state of charge is the open circuit voltage test. This test uses a DMM to accurately measure the voltage of a battery. The battery must not have been charged for at least four hours. Or, if recently charged, the surface charge has to be removed. It can be removed by turning on the headlights for 3–5 minutes. Once the surface charge is removed,

Applied Science

AS-56: Density/Specific Gravity: The technician can explain the role of specific gravity in determining the condition of the system.

A technician is assigned to perform a battery SOC test on an automotive battery with removable filler caps. This procedure is done using a battery hydrometer. The technician would normally use an electronic battery conductance tester. But because the conductance tester was dropped, it is out for repairs.

Observing the safety precautions, the technician uses a hydrometer to draw some electrolyte from each cell. The results are recorded on a sheet of paper. As stated in our text, a very low overall reading of 1.150 or below indicates a low state of charge. A high overall reading of about 1.280 indicates a high state of charge. The reading from each cell should be the same. If one or more cells are 0.050 or higher from the rest, it indicates a fault in the battery. See the temperature corrections chart to adjust for temperature variations that are above or below 80°F (26.7°C) for the battery electrolyte temperature.

TABLE 43-2 State of Charge as Indicated by Voltage Reading

Voltage	Percentage of Charge
12.6 or greater	100
12.4–12.6	75–100
12.2–12.4	50–75
12.0–12.2	25–50
11.7–12.0	>0–25
0.0–11.7	0 (no charge)

the headlights can be turned off. The battery should be allowed to sit for 2–3 minutes, and then the open circuit voltage can be measured. The voltage reading will then roughly indicate the battery's state of charge. For example, 12.6 volts or above indicates a fully charged battery. 12.4 volts indicates a charge of approximately 75%. And 12.0 indicates a battery with very little charge (**TABLE 43-2**).

To perform a battery SOC test, follow the steps in **SKILL DRILL 43-4**.

Applied Math

AM-5: Decimals: The technician can subtract decimal numbers to determine conformance with the manufacturer's specifications.

A technician is performing a battery state-of-charge test, using a hydrometer. This procedure can be used on automotive batteries that have removable filler caps. The tube of the hydrometer is inserted into one of the battery cells. The technician gently draws electrolyte into the hydrometer. The float indicator rises and floats freely. The specific gravity can be obtained by reading the indicator at eye level. When doing this procedure, it is important to consider the hazards of battery electrolyte, which contains sulfuric acid. Eye protection and rubber gloves are needed for personal protection. After reading the hydrometer, the electrolyte is returned to the cell from which it was withdrawn. The same procedure is repeated for the other five cells.

Manufacturer's specifications state that a fully charged battery should show a specific gravity of approximately 1.260 with electrolyte temperature at 80°F (26.7°C). The results of the readings of each cell are compared to a chart that shows specific gravity as compared to state of charge. If the electrolyte temperature is above or below 80°F (26.7°C), corrections must be made by one of several methods. In this case, the technician consulted a temperature correction chart.

The technician observes that at 40°F (4.4°C), the correction factor is to subtract 0.016 points. The reading for cell 1 is 1.240. To compensate for the huge temperature difference, corrections are needed. In this example, we have 1.240 minus (–) 0.016 = 1.224 for the adjusted specific gravity of one of the battery cells. The next step would be to look at the readings for the other five cells of the battery.

SKILL DRILL 43-4 Testing the Battery's State of Charge

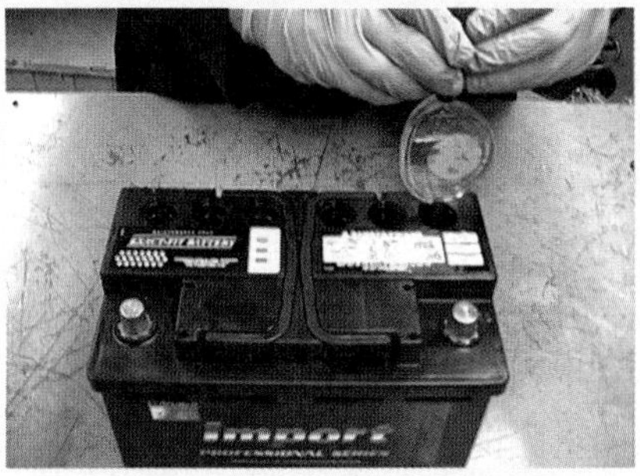

1. Test the specific gravity of each of the cells by using a battery hydrometer. Draw some of the electrolyte into the tester, and read the scale. Compensate for the temperature of the electrolyte if necessary.

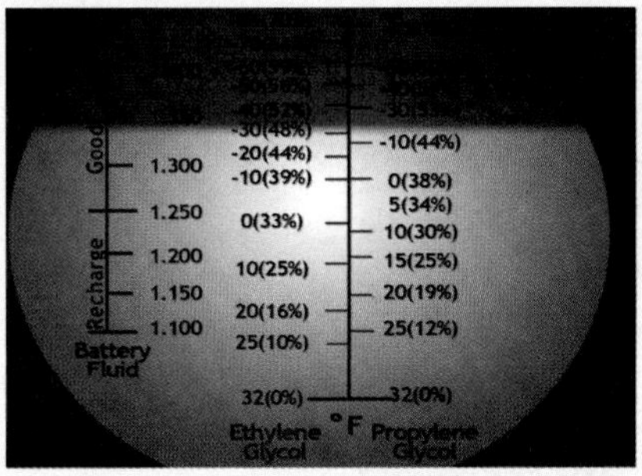

2. To test the specific gravity with a refractometer, place a drop or two of electrolyte on the specimen window, and lower the cover plate. Look into the eyepiece with the refractometer under a bright light. Read the scale for battery acid. The point where the dark area meets the light area is the reading.

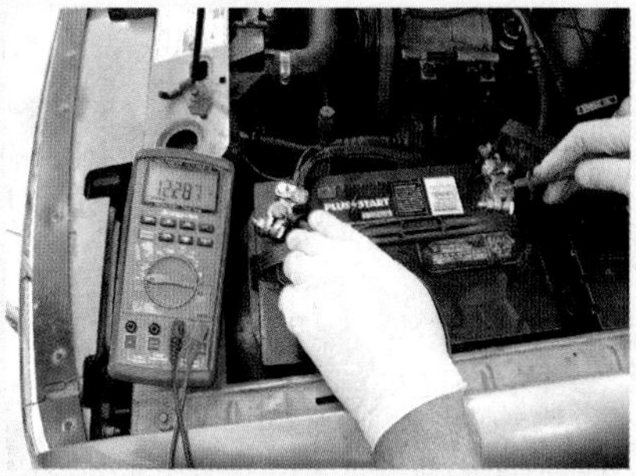

3. To conduct an open circuit voltage test, select the "volts DC" position on your DMM. Attach the probes to the battery terminals (red to positive, black to negative). Compare the reading to those given in (Table 43-2).

▶ Testing Battery Capacity Preliminaries

LO 43-10 Perform battery capacity tests.

When in doubt of how well a battery can meet the demands placed upon it, its capacity should be tested. But before testing begins, you need to make sure that the battery meets the capacity rating for the vehicle. Look up the service information, and compare it with the rating tag on the battery. Using a battery with too little capacity will overstress the battery and the starting system. Using a battery with substantially too much capacity overloads the battery tray. This is because the larger battery weighs more. Additionally, the hold-downs are not likely to fit it. It also may put a larger load on the charging system. So, always verify the battery is correct for the application.

Conductance Testing for Capacity

Once the application has been verified, you can go about testing the battery. The use of electronic battery testers has by and large replaced the need for hydrometer and other invasive types of battery testing. In fact, many manufacturers now require the use of a conductance test for warranty purposes (**FIGURE 43-22**). No longer will they accept the results of a high-amperage load test.

Many conductance testers have integrated printers. They allow for easy printout of the test results. That way the printout can go with the warranty paperwork.

The conductance tester sends low-frequency signals into the battery. This allows it to determine the battery's ability to conduct current. The greater the ability to conduct current, the higher the CCA capacity of the battery. Because batteries deteriorate over time, their CCA capacity also reduces. The conductance tester is able to determine a battery's CCA capacity by measuring its conductance. This is because there is a near-linear comparison between the two. Dirty battery terminals can affect the conductivity. So, the terminals may have to be cleaned before taking the reading.

FIGURE 43-22 Conductance tester being used to measure a battery's existing capacity.

Conductance testing takes only a minute or two to complete. This provides a good way to show customers the condition of their battery. It also shows how that condition degrades over time. Once the conductance test shows that the battery no longer meets specifications, then customers will have confidence that the battery is at the end of its useful life. If they want to avoid having the battery fail and leave them stranded, the battery will need to be replaced. But they will have had the opportunity to prepare for that, by watching the battery capacity decrease over time.

Battery Load Testing

The load test has been used for years to test a battery's capacity and internal condition. But some manufacturers have been saying that their batteries should *not* be load tested. Instead, they require their batteries to be conductance tested. They claim that load testing can damage the battery. Therefore, always check the service information before performing a load test.

As the name suggests, the load test subjects the battery to a high rate of discharge. The voltage is then measured at the end of a set time to see how well the battery creates that current flow. In other words, can the battery maintain a high rate of discharge for a specified time? And is the voltage still relatively high? If so, then you know the battery has good capacity. And therefore, it is in relatively good condition. Conversely, if the voltage falls off fairly quickly, the battery's capacity is low.

It is kind of like two people running a mile sprint. The one who does it in 5 minutes is likely in pretty good shape. The one who takes 20 minutes is likely in poor shape. It is similar with batteries. The faster the rate at which a battery can create current flow, the higher its voltage. This indicates a high capacity.

Another way to think of a load test is that we are testing the battery's ability to produce the high starting current to crank an engine over. But it also needs to maintain enough voltage to operate the ignition and electronic control systems. Plus, it has to do it a number of times, if needed.

You will remember that CCAs reflect the load in amps that a battery can deliver for 30 seconds while maintaining a voltage of 7.2 volts or higher at 0°F (−18°C). Because vehicle and battery manufacturers specify the CCA rating for every vehicle and battery, that rating is used to calculate the load placed on the battery when load testing. The battery can be either in or out of the vehicle. But it must be at, or near, a full state of charge for the test to be accurate. The electrolyte temperature should be approximately 70°F (21°C) for the most accurate results. This is because a cold battery cannot produce current flow as efficiently. So, it will show a false fail result. When load testing a battery, use the following testing parameters:

Test load = Half the CCA of the battery (verify it is sized correctly for the vehicle)

Load test time = 15 seconds

Results: Pass = 9.6 volts or higher; Fail = less than 9.6 volts

If the battery fails the load test, one further test is required before condemning the battery. The battery needs to be slow charged until it is fully charged (can take up to 20 hours). Then, repeat the load test. If it still fails, the battery is sulfated and has to be replaced. If it passes, test the vehicle's charging system to see if it is the cause of the discharged battery.

▶ TECHNICIAN TIP

Technicians used to perform a three-minute charge test on batteries if they failed the load test. But battery and vehicle manufacturers are discouraging use of this test because of its potential of damaging the battery or vehicle. Thus, we do not discuss this test method further.

To load test a battery, follow the steps in **SKILL DRILL 43-5**.

Applied Math

AM-4: Whole Numbers: The technician can divide whole numbers to determine differences for comparison with the manufacturer's specifications.

AM-7: Whole Numbers: The technician can divide whole numbers to determine differences for comparison with the manufacturer's specifications.

Battery load testing is a procedure that uses a simulated starter current draw on a battery. The results of this test can be compared with manufacturer's specifications to determine the condition of the battery. There are a number of steps to complete regarding a battery load test using a carbon pile unit. One of the steps is to use one-half CCA rating as the load to be applied to the battery. CCA stands for "cold cranking amps," which is one of the battery's primary ratings. In this example, we have a 770 CCA battery, and the load to be applied is one-half of that amount. To determine this, we can divide by two to give us the 385 CCA load to apply. Another way to describe this would be: 770 CCA/2 = 385 CCA (test load).

In this example, we are dividing whole numbers. Division is the opposite of multiplication.

▶ Identifying Modules That Lose Their Initialization During Battery Removal

LO 43-11 Maintain and restore electronic memories.

Many electronic modules in vehicles require a small amount of power to maintain their **keep alive memory (KAM)**. Some examples are radio stations and other driver-specific presets. When the battery is disconnected, the memory of these presets is usually lost. The powertrain control module (PCM) may also lose its adaptive learning data. This means that the vehicle will have to relearn this information during a period of driving that could take several days. These situations can be annoying.

For other systems, such as the security system, loss of memory may prevent the vehicle from being restarted. It may also prevent the radio from being used. At the very least, it may require the dealer to be contacted for vehicle-specific codes to reinitialize the vehicle systems. Check the manufacturers' and owners' information to determine which systems will be affected by the power loss. You need to identify any system that requires security or initialization codes to be reentered. Then, ensure the procedures and equipment are available to reinitialize systems or modules. Do this before disconnecting the battery.

In some cases, it may be possible to use a memory saver to maintain the vehicle's memory while the battery is disconnected. Many technicians use an external 12-volt DC power supply connected to the data link control with a suitable cable (**FIGURE 43-23**). Consider that many cigarette lighter sockets are always powered on. This allows power to be supplied through them while the battery cables are disconnected. Remember that providing power back into the circuit makes the system susceptible to short circuits. This happens if you ground any of the powered wires or terminals, including the battery-positive terminal.

SKILL DRILL 43-5 Load Testing a Battery

1. With the tester controls off and the load control turned to the Off position, connect the tester leads to the battery. Place the inductive amps clamp around either the black or the red tester cables in the correct orientation.

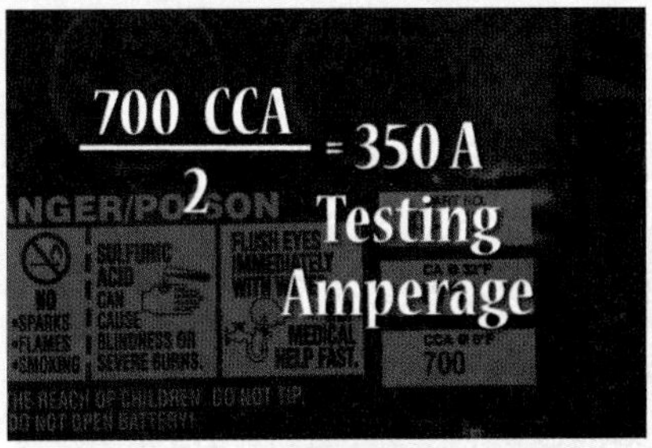

2. Verify that the temperature of the battery is within the testing parameters. Use an infrared temperature gun to measure the temperature of the side of the battery. If you are using an automatic load tester, enter the battery's CCA and select "Test" or "Start." If you are using a manual load tester, calculate the test load, which is usually half of the CCA.

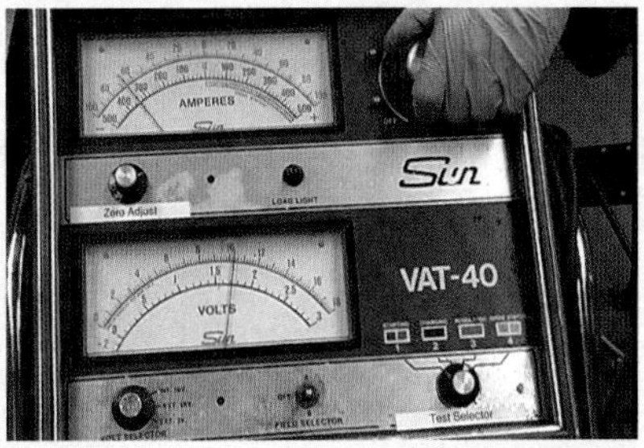

3. Maintain this load for 15 seconds while watching the voltmeter. Read the voltmeter, and immediately turn the control knob off. At room temperature, the voltage should be 9.6 volts or higher at the end of the 15-second draw. If the battery is colder than room temperature, look up the compensated minimum voltage. Determine any necessary action.

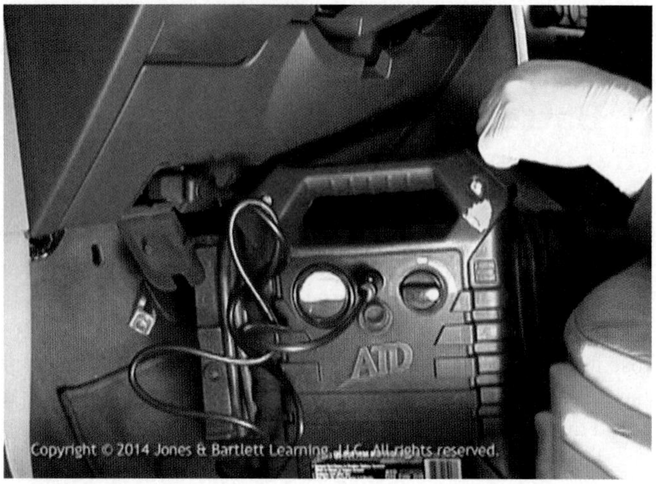

FIGURE 43-23 Memory saver and cable.

To identify electronic modules, security systems, radios, and other accessories that require reinitialization or code entry following battery disconnect, follow the steps in **SKILL DRILL 43-6.**

To maintain or restore electronic memory functions, follow the steps in **SKILL DRILL 43-7.**

▶ Measuring Parasitic Draw

LO 43-12 Measure parasitic draw.

All modern vehicles have a small amount of current draw when the ignition is turned off. This is called **parasitic draw.** Parasitic current draw is used by the vehicle's KAM circuits. KAM systems maintain memory functions and monitor systems. For example, the vehicle's theft-deterrent system is part of this parasitic current draw. The total parasitic draw should be a relatively small amount of current. This is because

SKILL DRILL 43-6 Identifying Accessories That Require Reinitialization

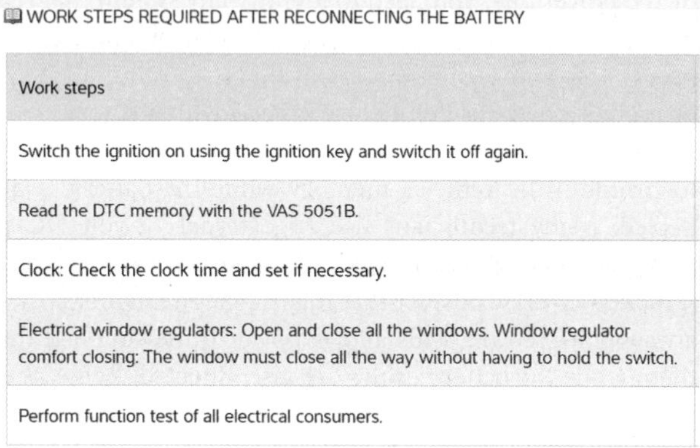

WORK STEPS REQUIRED AFTER RECONNECTING THE BATTERY

Work steps
Switch the ignition on using the ignition key and switch it off again.
Read the DTC memory with the VAS 5051B.
Clock: Check the clock time and set if necessary.
Electrical window regulators: Open and close all the windows. Window regulator comfort closing: The window must close all the way without having to hold the switch.
Perform function test of all electrical consumers.

1. In the appropriate service information or owner's manual, find the section that lists the information on the electronic modules, security systems, radios, and other accessories.

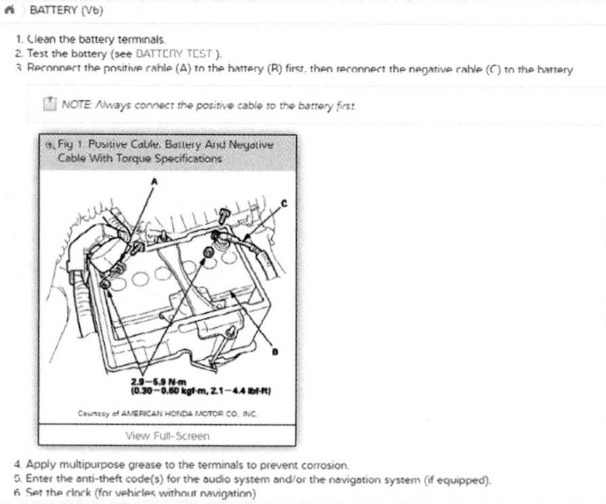

BATTERY (Vb)

1. Clean the battery terminals.
2. Test the battery (see BATTERY TEST).
3. Reconnect the positive cable (A) to the battery (B) first, then reconnect the negative cable (C) to the battery

NOTE: Always connect the positive cable to the battery first.

Fig 1. Positive Cable, Battery And Negative Cable With Torque Specifications

4. Apply multipurpose grease to the terminals to prevent corrosion.
5. Enter the anti-theft code(s) for the audio system and/or the navigation system (if equipped).
6. Set the clock (for vehicles without navigation)

2. From the service information, list the systems and modules that may require initialization.

SKILL DRILL 43-7 Maintaining or Restoring Electronic Memory Functions

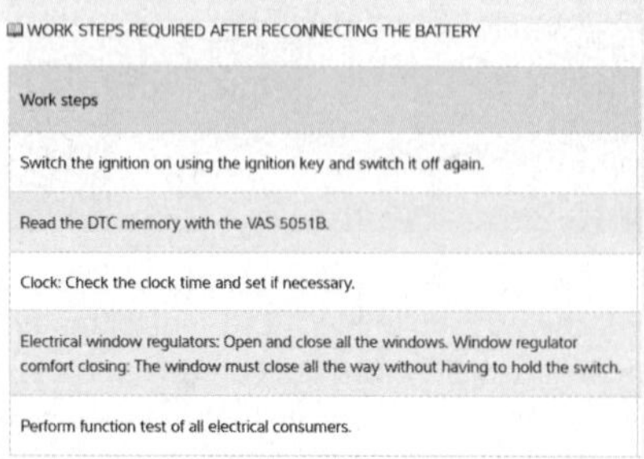

WORK STEPS REQUIRED AFTER RECONNECTING THE BATTERY

Work steps
Switch the ignition on using the ignition key and switch it off again.
Read the DTC memory with the VAS 5051B.
Clock: Check the clock time and set if necessary.
Electrical window regulators: Open and close all the windows. Window regulator comfort closing: The window must close all the way without having to hold the switch.
Perform function test of all electrical consumers.

1. Identify which modules, if any, require reinitialization or code entry when the battery is disconnected, following SKILL DRILL 43-6.

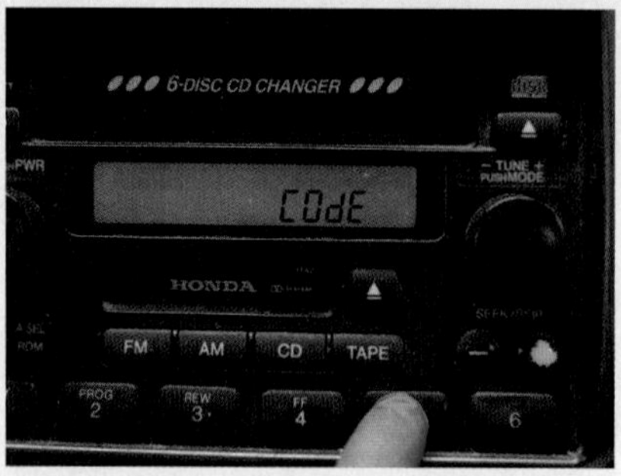

2. Identify the correct procedure and any needed tools, and verify that initialization codes are available.

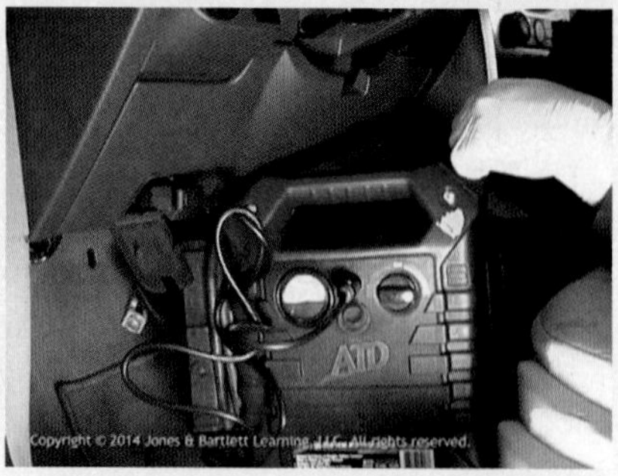

3. If maintaining memory function, install a memory minder prior to the vehicle battery being disconnected. If reinitializing an electronic module, use the correct codes to reinitialize the modules if required.

excessive parasitic draw will discharge the battery over a short amount of time.

Parasitic draw does not necessarily immediately drop to its lowest level the instant the ignition is turned off. Typically, when the ignition is switched off, the vehicle's engine stops. Then, the vehicle systems begin to shut down, which is called "going to sleep" or "entering sleep mode." Vehicle systems will turn off at different intervals. For example, some vehicles may take up to a couple of hours before the last system goes to sleep.

Consult the manufacturer's service information to determine the maximum allowable parasitic current draw. At the same time, determine the amount of time you need to wait after the ignition is turned off. If you are lucky, the service information will list the time it takes the modules to go to sleep. Parasitic draw that is higher than specified discharges the battery prematurely. This is hard on the battery, and it can potentially leave the passengers stranded.

Parasitic draw can be measured in several ways. The most common is the process of using an ammeter capable of measuring milliamps. The ammeter is inserted in series between one of the battery posts and the battery terminal. Note that the modules may reset during this process. So, you may have to wait for the modules to go back to sleep. If excessive parasitic draw is measured, disconnect fuses or systems one at a time while monitoring parasitic current draw. This will help you determine the systems causing excessive draw.

Disconnecting the battery can be avoided if a sensitive low-current (i.e., milliamps) clamp is available. The low-amp **current clamp** measures the magnetic field generated by a very small current flow through a wire or cable. You can measure parasitic draw by placing the low-amp current clamp around the negative battery cable. If excessive parasitic draw is measured, disconnect fuses or systems one at a time. Monitor the parasitic current draw to determine the systems causing the excessive draw. Is the parasitic draw still high after removing all of the fuses? If so, suspect an unfused circuit. Some examples are the alternator diodes, the ignition circuit, or control module feed wires.

To measure parasitic draw with a standard parasitic load test, follow the steps in **SKILL DRILL 43-8**.

The last way to measure a parasitic draw is a bit controversial. But it is well worth trying out as it can save a lot of time. And it requires no special tools other than a DMM. It is named the Chesney parasitic load test after its creator, Sean Chesney. Instead of using an ammeter to measure the draw, an ohmmeter is used, which sounds wrong—but stay with me.

Before doing anything, set the ohmmeter to ohms (the lowest scale), touch the meter leads together, and read the screen. The reading is the resistance of the meter leads and is called the "delta" value. This is the meter's true zero when used with those leads. Typically, an ohmmeter will read about 0.1 ohm when the leads are touched together. Remember this number for later. Some meters have a delta feature that recalibrates the ohmmeter to zero when the leads are placed together and the delta button is pressed. If your meter has this delta feature, you can use it so that you will not have to remember the delta reading.

Next, with the battery terminals still connected to the battery, place the black lead on the negative post of the battery. Then, place the red lead first on an unpainted surface of the alternator housing. Read the ohmmeter, and subtract the delta value from the reading. This reading corresponds to the relative parasitic draw on the system, which we describe next. Then, take the same reading, but with the red lead on a good body ground, and remember the reading.

SKILL DRILL 43-8 Measuring Parasitic Draw—Standard Test

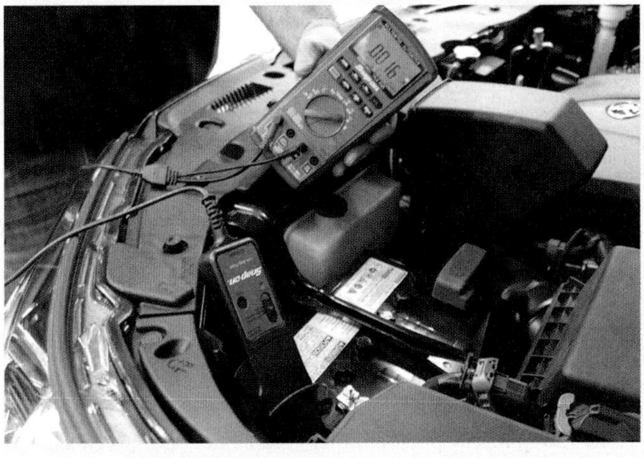

1. Research the parasitic draw specifications in the service information. Connect the low-current clamp around the negative battery cable. Measure the parasitic draw. Compare the parasitic draw to specifications.

2. Disconnect the circuit fuses one at a time to determine which circuit has the excessive parasitic current draw. Determine any necessary actions.

SKILL DRILL 43-9 Measuring Parasitic Draw—Chesney Parasitic Load Test

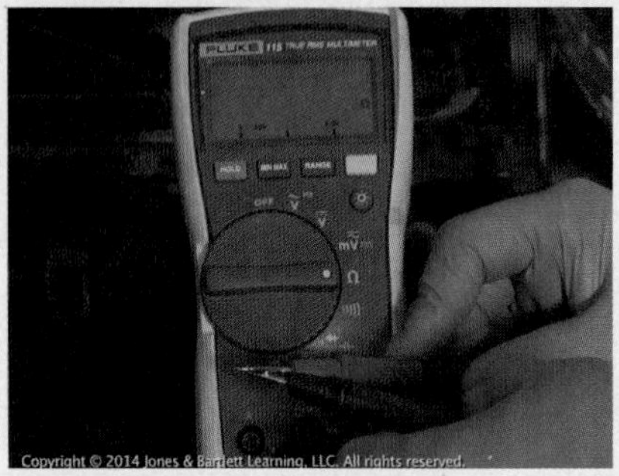

1. Set the DMM to read ohms (lowest scale if available). Connect the leads together and read the meter screen. This is the meter's delta reading.

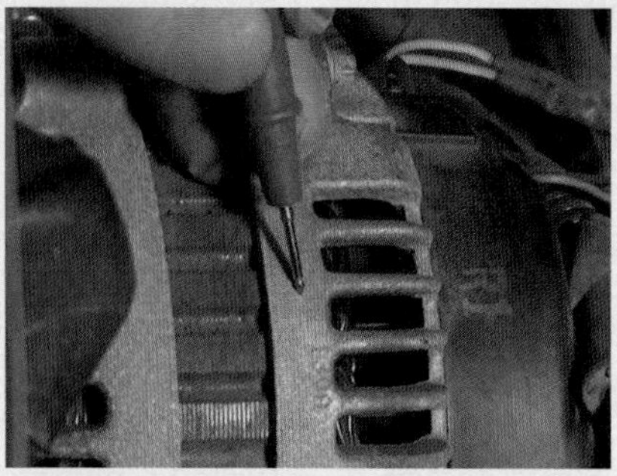

2. Place the black meter lead on the negative battery post and the red lead first on the alternator case and then on a good body ground.

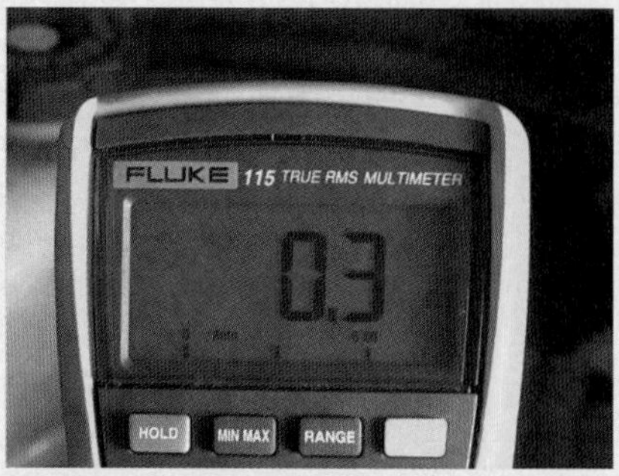

3. Read the meter in both places, and compare the reading to the Chesney parasitic load test- ratio graph. If there is excessive parasitic load, pull fuses one at a time to identify the faulty circuit.

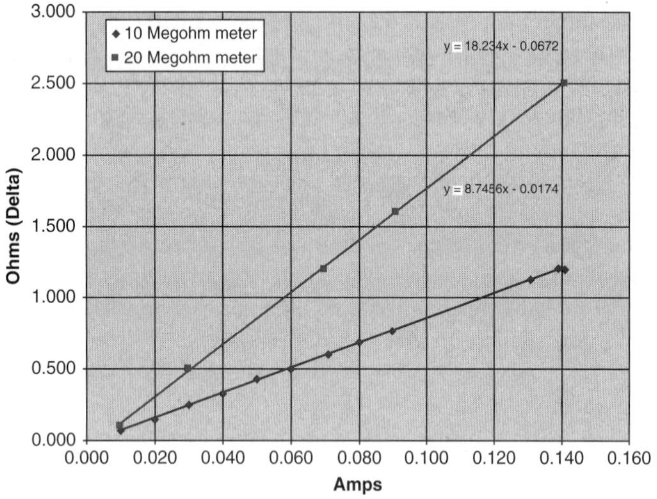

FIGURE 43-24 Chesney parasitic load test-ratio graph.

Through testing, Chesney found that a draw of about 35 milliamps equaled an ohm reading of about 0.3 ohm (above the delta value). This is on a DMM with 10 megohms of impedance. On a DMM with 20 megohms of impedance, it read about 0.6 ohm (**FIGURE 43-24**). Anything above those readings indicates an excessive parasitic draw. Locate the excessive draw as explained above.

You may be skeptical of this method, as was I. But don't discount it. Give it a try on any vehicle. Connect the meter as above. Then, simulate a parasitic draw by opening the driver's door. This will illuminate the dome light. Watch the ohmmeter. It went up, right? Close the door. As soon as the light went off, the ohmmeter reading went back down, right? Amazing, isn't it?

To measure parasitic draw with a Chesney parasitic load test, follow the steps in **SKILL DRILL 43-9**.

▶ Wrap-Up

Ready for Review

- A battery consists of: two dissimilar metals, an insulator material separating the metals, and an electrolyte. Starting-type batteries provide high current draws for short periods of time. Deep cycle batteries withstand deep discharge much better.
- Different types of batteries available for vehicles include: deep-cycle batteries, low-maintenance batteries, Maintenance-free batteries, Absorbed glass mat batteries, and lithium-ion batteries.
- Battery sizes are designated by the Battery Council International (BCI) based on dimension in length, width, and height. Battery terminal configuration refers to the placing of the positive and negative terminals on the battery. Common designs of battery cable and terminals: tapered-post battery terminal, side-post cable terminal, and lug terminal and cable.
- Methods used to rate automotive battery capacity include: cold cranking amps (CCAs), cranking amps (CAs), and reserve capacity (RC). These ratings can be used for testing battery performance and selecting batteries for particular applications.
- Conditions that shorten/lengthen the life of a battery: being fully discharged or having deep discharge cycles,

remaining overcharged or undercharged, experiencing high discharge rates for extended periods, experiencing excessive vibration, being exposed to extremes of temperature, having dirt or moisture on the case, or developing corrosion.

- Regular maintenance of battery includes inspection, cleaning, testing, and charging when discharged. The electrolyte level of a battery should be between the lines on the battery case or near or touching the bottom of the fill hole. Lead oxide on battery terminals can be removed with a special battery terminal scraper or wire brush.
- Do not connect the battery with reverse polarity when reconnecting it. When disconnecting the battery terminals, always remove the negative terminal completely first and always install the negative terminal only after the positive terminal is fully tightened.
- Formula to determine the ideal charging rate of a battery is CCAs divided by 70 and the formula to determine the max charging rate is CCAs divided by 40. Slow charging is less stressful on a battery than fast charging.
- When jump stating a vehicle, it is good practice to turn on the headlights on the charged battery while cranking the dead vehicle to absorb any voltage spike. A jump box can be used to lessen the risk of voltage spikes when jump-starting a vehicle.
- The specific gravity test or the open circuit voltage test can be used for determining the battery's state of charge.
  - The specific gravity test measures the electrolyte's specific gravity.
- A conductance tester sends low-frequency signals into the battery to determine the battery's ability to conduct current. The greater the ability to conduct current, the higher the CCA capacity of the battery. Load test subjects the battery to a high rate of discharge and the voltage is then measured at the end of a set time.
- A memory saver can be used to maintain the vehicle's memory while the battery is disconnected. An external 12-volt DC power supply connected to the data link control with a suitable cable can be used as a memory saver.
- Parasitic draw can be measured by inserting an ammeter in series between one of the battery posts and the battery terminal, by placing the low-amp current clamp around the negative battery cable, or with Chesney parasitic load test.

Key Terms

absorbed glass mat Batteries that have the electrolyte absorbed within a mat of fine glass fibers.

battery terminal configuration The placement of positive and negative battery terminals.

cold cranking amps (CCA) The load in amps that a battery can deliver for 30 seconds while maintaining a voltage of 1.2 volts per cell (7.2 volts for a 12-volt battery) or higher at 0°F (−18°C).

cranking amps (CA) A standard similar to CCA, but that measures the battery's function at a higher temperature (32°F [0°C]).

current clamp A device that clamps around a conductor to measure current flow. It is often used in conjunction with a digital volt-ohmmeter (DVOM).

electrical capacity The ability of a circuit or component to carry electrical loads.

gassing When gas escapes the battery; caused by overcharging or rapid charging a battery.

keep alive memory (KAM) Memory that is retained by the ECM when the key is off.

parasitic draw The unwanted current draw that occurs once the vehicle has been turned off and the systems have gone to sleep.

reserve capacity (RC) Refers to the length of time, measured in minutes, that a new, fully charged 12-volt battery discharges under a specified load of 25 amps at 80°F (26.6°C) before battery cell voltage drops below 1.75 volts per cell (10.5 volts for a 12-volt battery).

thermal runaway During thermal runaway, the high heat of the failing cell will propagate to neighboring cells, causing them to become thermally unstable as well. When lithium-ion batteries enter the thermal runaway, extreme overheating—and in some cases, fire—can be expected. Thermal runway is also referred to as venting the flame.

Review Questions

1. The electrolyte in an automotive lead acid battery is:
   a. dilute sulfuric acid.
   b. hydrochloric acid.
   c. sulfur dioxide.
   d. nitric acid.
2. Absorbed glass mat batteries have the electrolyte absorbed within a mat of fine glass fibers, and are a type of:
   a. Low-maintenance battery.
   b. Unsealed battery.
   c. Maintenance-free battery.
   d. Single-cell hydrometer battery.
3. The most common battery cable terminal is a ________ that provides a large surface contact area with the ability to tighten the terminal onto the battery post using a nut and bolt.
   a. Side terminal design
   b. Cone design
   c. Flat terminal design
   d. Back terminal design
4. CCA measures the load in amps that a battery can deliver for ________ while maintaining a voltage of ___ volts per cell.
   a. 60 seconds; 2.1
   b. 30 seconds; 2.1
   c. 60 seconds; 1.2
   d. 30 seconds; 1.2

5. Clean the battery with a mixture of ________, but make sure not to get any of that mixture down inside of the battery, as it will tend to neutralize the electrolyte, damaging the battery.
   a. Baking soda and water
   b. Soap and water
   c. Salt and water
   d. Alcohol and water
6. What is the correct sequence for disconnecting battery terminals?
   a. Disconnect the negative terminal and then the positive terminal
   b. Disconnect whichever terminal is the easiest first
   c. Disconnect whichever terminal is hardest first
   d. Disconnect the positive terminal and then the negative terminal
7. When a vehicle has been shut off, it can have a:
   a. thermal runaway.
   b. parasitic draw.
   c. paralysis draw.
   d. heat runaway.
8. When performing an open circuit voltage test to determine the battery state of charge, what should a fully charged battery read?
   a. 12.0 volts
   b. 12.2 volts
   c. 12.6 volts
   d. 13.5 volts
9. When performing a battery load test, what should the load be set to?
   a. 150 amps
   b. 300 amps
   c. Two times the cold cranking amps (CCA)
   d. Half the cold cranking amps (CCA)
10. When using a DMM to measure parasitic draw, connect the meter leads:
   a. between one of the battery posts and terminal on the battery cable
   b. from one battery post to the other
   c. from the positive battery post to a good engine ground
   d. across the high current contacts on the starter solenoid

ASE Technician A/Technician B Style Questions

1. Technician A says that a battery stores electrical energy in chemical form. Technician B says that the chemical reactions change the composition of the chemicals, which then are stored until the electrical energy is needed. Who is correct?
   a. Technician A
   b. Technician B
   c. Both Technicians A and B
   d. Neither Technician A nor B
2. Technician A says that when disconnecting the battery, the negative terminal should be disconnected first. Technician B says that when disconnecting the battery terminals, always remove the positive terminal completely first. Who is correct?
   a. Technician A
   b. Technician B
   c. Both Technicians A and B
   d. Neither Technician A nor B
3. Two technicians are discussing the specific gravity test. Technician A says that if the electrolyte level is too low, then you will have to add distilled water, and the battery will have to be fully charged to mix the water and acid. Technician B says that the battery should be topped up with a mixture of acid and water. Who is correct?
   a. Technician A
   b. Technician B
   c. Both Technicians A and B
   d. Neither Technician A or B
4. Two technicians are discussing battery load testing. Technician A says that if the battery fails the load test, it is bad and should be replaced. Technician B says if it fails the load test, it should be fully charged and the test repeated. Who is correct?
   a. Technician A
   b. Technician B
   c. Both Technicians A and B
   d. Neither Technician A nor B
5. Two technicians are discussing measuring parasitic draw. Technician A says that the parasitic draw is measured with an ammeter. Technician B says that the parasitic draw is measured with a voltmeter. Who is correct?
   a. Technician A
   b. Technician B
   c. Both Technicians A and B
   d. Neither Technician A nor B
6. Technician A says that batteries should be charged as fast as possible. Technician B says that the ideal charging rate is the CCA divided by 70. Who is correct?
   a. Technician A
   b. Technician B
   c. Both Technician A and B
   d. Neither Technician A nor B
7. Technician A says that lead oxide acts as an insulator on the battery posts and has to be scraped away. Technician B says that current can leak across the dirt on the surface of the battery. Who is correct?
   a. Technician A
   b. Technician B
   c. Both Technician A and B
   d. Neither Technician A nor B

8. Technician A says that checking the specific gravity will indicate the battery's cold cranking amps. Technician B says that a battery load test should be performed when the battery is heavily discharged. Who is correct?
   a. Technician A
   b. Technician B
   c. Both Technician A and B
   d. Neither Technician A nor B
9. Technician A says that a 12-volt battery has six cells. Technician B says that the more plates a cell in a battery has, the more voltage it creates. Who is correct?
   a. Technician A
   b. Technician B
   c. Both Technician A and B
   d. Neither Technician A nor B
10. Technician A says that the specific gravity of the electrolyte can be checked with a hydrometer. Technician B says that the specific gravity of the electrolyte can be checked with a refractometer. Who is correct?
   a. Technician A
   b. Technician B
   c. Both Technician A and B
   d. Neither Technician A nor B

CHAPTER 44

Starting and Charging Systems

Learning Objectives

- **LO 44-01** Describe starting system fundamentals.
- **LO 44-02** Describe starter motor construction.
- **LO 44-03** Describe starter motor engagement.
- **LO 44-04** Describe armature and starter drive operation.
- **LO 44-05** Describe solenoid operation.
- **LO 44-06** Describe starter control circuit operation.
- **LO 44-07** Describe solenoid operation.
- **LO 44-08** Test starter high-current circuit voltage drop.
- **LO 44-09** Test starter control circuit voltage drop.
- **LO 44-10** Test starter relays and solenoids.
- **LO 44-11** Remove and install a starter.
- **LO 44-12** Describe idle stop–start stop system operation.
- **LO 44-13** Describe charging system operation.
- **LO 44-14** Describe the rotor, slip ring, and brushes.
- **LO 44-15** Describe the stator, end frames, fan, and pulley.
- **LO 44-16** Describe rectification.
- **LO 44-17** Describe voltage regulation.
- **LO 44-18** Perform a charging system output test.
- **LO 44-19** Perform charging system circuit voltage and voltage drop tests.
- **LO 44-20** Replace alternator.

ASE Education Foundation Tasks

See Appendix A to view the 2017 ASE Education Foundation Automobile Accreditation Task List Correlation Guide.

▶ Introduction to Starting Systems

LO 44-01 Describe starting system fundamentals.

The starting system provides a method of rotating (cranking) the vehicle's internal combustion engine (ICE). This is so the combustion cycle can begin. In early vehicles, this was done by the use of a hand-crank (**FIGURE 44-1**). Modern vehicles use an electric starter motor. It draws its electrical power from the vehicle's battery (**FIGURE 44-2**). The starter is designed to only work for short periods of time. But it must crank the engine at sufficient speed in order for it to start. Modern starting systems are very effective provided that they, and the battery, are well maintained.

FIGURE 44-1 Early engines were hand-cranked to start them.

Engine Starting (Cranking) Systems

The starting/cranking system consists of two electrical circuits. One circuit is high-amperage circuit and the other is low-amperage circuit. The high-amperage circuit powers the starter for cranking the engine over. The low-amperage circuit is used to control the high-amperage circuit. The high-amperage circuit consists of the following:

- battery,
- high-amperage side of the solenoid,
- and starter motor assembly.

The control circuit consists of the following:

- battery,
- ignition switch,
- safety switch (clutch switch or neutral safety switch),
- and the low-amperage side of the solenoid.

On PCM-activated starting systems, there are also the:

- PCM (along with all of its sensors)
- and a relay.

The onboard computer (PCM) may incorporate a security/antitheft system. If so, it determines when and if the cranking circuit will function.

During the cranking process, two actions occur. First, the pinion of the starter motor engages with the flywheel ring

FIGURE 44-2 Modern electric starter bolted to the transmission.

You Are the Automotive Technician

The owner of an eight-year-old Ford F-150 4X4 has the vehicle towed to your shop because it won't start. The customer tells you that they went into a store, and when they came out, it wouldn't restart. They agree to let you diagnose the problem while they wait in the customer lounge. You verify the problem by trying to crank the engine over. The starter motor clicks once, and the starter current is very low.

1. What are the most likely faults that would cause an engine not to crank over?
2. What starting system components are likely *not* causing this problem?
3. What tests will you perform to diagnose the problem, and what results will you be expecting?
4. After repairing the starting concern, how would you verify that the charging system is working properly?

gear. Then, the starter motor rotates to turn over, or crank, the engine. The starter motor is an electric motor mounted on the engine block or transmission. It is typically powered by the 12-volt storage battery. However, some hybrid vehicles use the high-voltage battery to operate the starter motor.

Starters are designed to have high turning effort (torque) at low speeds. This requires a high amount of current flow so the starter cables are the heaviest in the vehicle. The starter motor rotates the engine flywheel and crankshaft from a resting position. It keeps them rotating until the engine fires and runs on its own.

Starter Motor Principles

The starter motor converts electrical energy to mechanical energy for the purpose of cranking the engine over. There are three sections to the typical starter:

- the electric motor,
- the drive mechanism,
- and the solenoid (**FIGURE 44-3**).

The starter motor is mounted on the transmission or cylinder block. It is positioned to engage a ring gear. The ring gear is positioned around the outside edge of the flywheel, flexplate, or torque converter.

Starting is usually initiated by the operator activating a starter switch built into the ignition lock assembly. Some vehicles use a start button instead. A relatively small current then flows through a neutral safety switch or clutch switch to a starter relay. The relay sends a larger current to operate the starter solenoid, which is typically mounted atop the starter motor.

The solenoid plunger moves the drive pinion gear into engagement with the ring gear. It also closes a set of heavy-duty contacts. This allows a very large current to flow from the battery to the starter motor. The starter motor rotates the armature and drive pinion gear, causing the crankshaft to rotate. When the engine starts and is able to run on its own, the operator usually releases the key.

The solenoid spring withdraws the pinion gear from the ring gear. It also disconnects the heavy-duty contacts which stops current flow to the starter motor. On many modern vehicles, the PCM signals the relay to continue the cranking process until the vehicle starts. In the case that the engine does not start, the starting operation will time out after a preset amount of time.

Starter Types—Direct-Drive/Gear Reduction

Starter motors can be designed to drive the pinion gear in one of two ways. They can be either direct-drive or gear reduction. In the direct-drive system, the starter drive is mounted directly on one end of the armature shaft. The starter drive transfers the rotating force of the armature directly to the engine flywheel (**FIGURE 44-4**). In this arrangement, the only gear reduction is the reduction between the pinion gear and the ring gear.

Gear reduction starters use an extra gear between the armature and the starter drive mechanism. They have a reduction of about 4:1. The gear reduction allows the starter to spin at a higher speed with lower current (**FIGURE 44-5**). It also enables the starter to be downsized and yet create a higher torque output. Two types of gearing systems are normally used: spur gears or planetary gears. Spur gears require the armature to be offset via a gear housing that holds the starter drive. A gear reduction system using planetary gears does not require an offset housing. The planetary gears are housed in the drive-end housing in line with the starter drive.

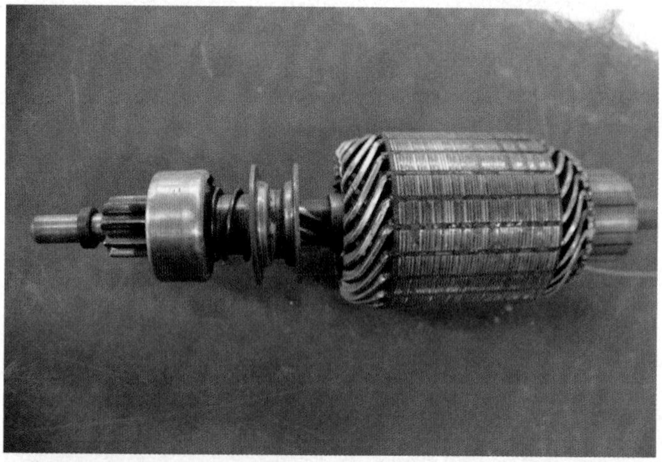

FIGURE 44-4 Armature and starter drive from a direct-drive starter.

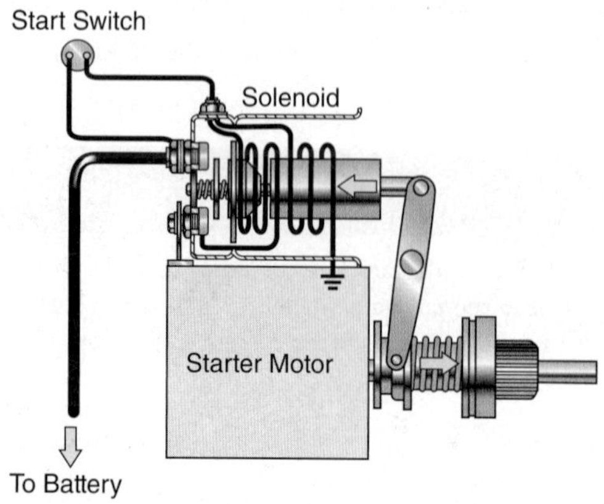

FIGURE 44-3 Typical starter motor, solenoid, and starter drive.

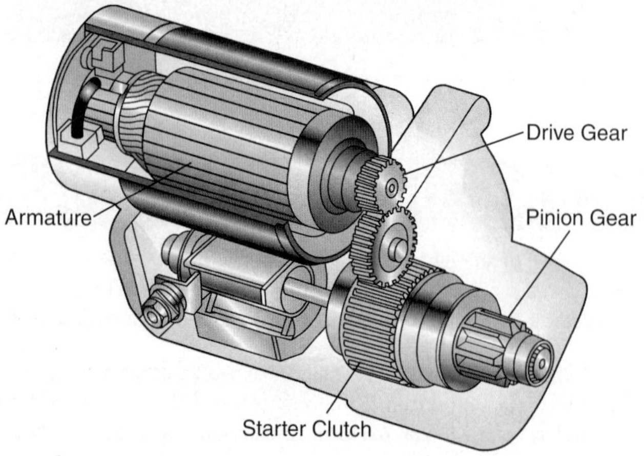

FIGURE 44-5 Gear reduction starter.

▶ Starter Motor Construction

LO 44-02 Describe starter motor construction.

A starter motor normally consists of the following components:

- field coils or large permanent magnets,
- an armature,
- a commutator,
- brushes,
- a drive pinion with an overrunning clutch,
- and a drive pinion engagement solenoid and shift fork (**FIGURE 44-6**).

The armature is the revolving component of the DC motor. The armature shaft is supported at each end by bushings or bearings pressed into end frames. The bushings locate the armature centrally in the outer casing (i.e., the "barrel") of the motor. They hold the armature between the field coils or permanent magnets. Direct drive starters are equipped with a brake washer. It is used to slow the armature when the starter is disengaged after the engine starts.

The commutator end frame carries copper-impregnated carbon brushes. They conduct current through the armature when it is being rotated in operation. The brushes are mounted in brush holders. The brushes are kept in contact with the commutator by tensioned spiral springs (**FIGURE 44-7**).

Half of the brushes are connected directly to the end frame and ground the armature windings. The other brushes are insulated from the end frame. They connect to the positive battery terminal via the starter solenoid. This connection is direct from the solenoid in the case of a permanent magnet starter. In a series-wound motor, the positive connection goes through the field poles first (**FIGURE 44-8**).

Starter Magnet Types

Starter motors use two magnet types: electromagnetic and permanent magnet (**FIGURE 44-9**). Electromagnetic fields are created as current flows through heavy copper windings. The windings are wound around iron pole shoes to concentrate the magnetic field. The pole shoes are then fastened to the starter case/barrel.

Permanent magnets are located similarly. But they do not need electricity and therefore occupy less space. On both styles, the case is made of iron. It serves to further concentrate the magnetic field produced by the field magnets.

Starter motors with electromagnetic field windings are typically series-wound motors. This means that the field windings are in series with the armature (**FIGURE 44-10**). Because the resistance of the field and armature windings is low, the current flow

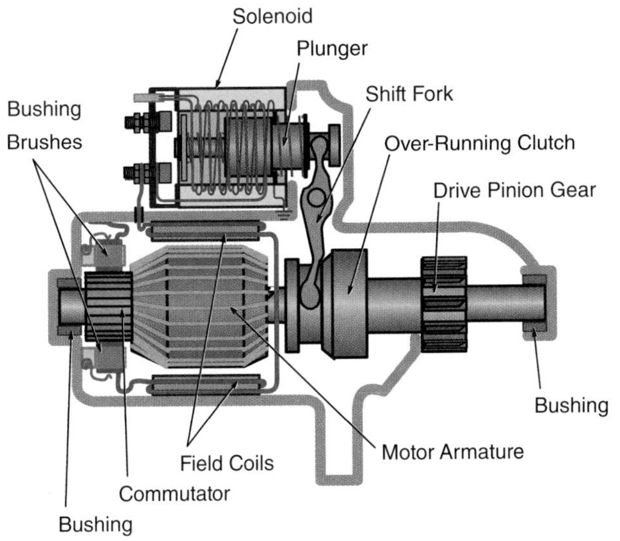

FIGURE 44-6 Cutaway view of a starter motor.

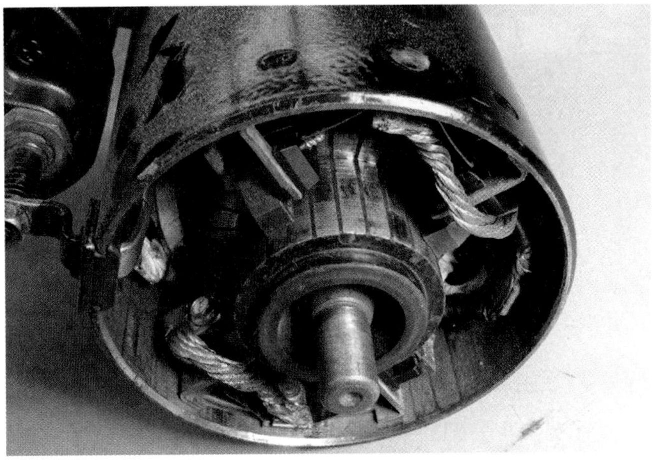

FIGURE 44-7 The brushes transfer electricity to the rotating commutator.

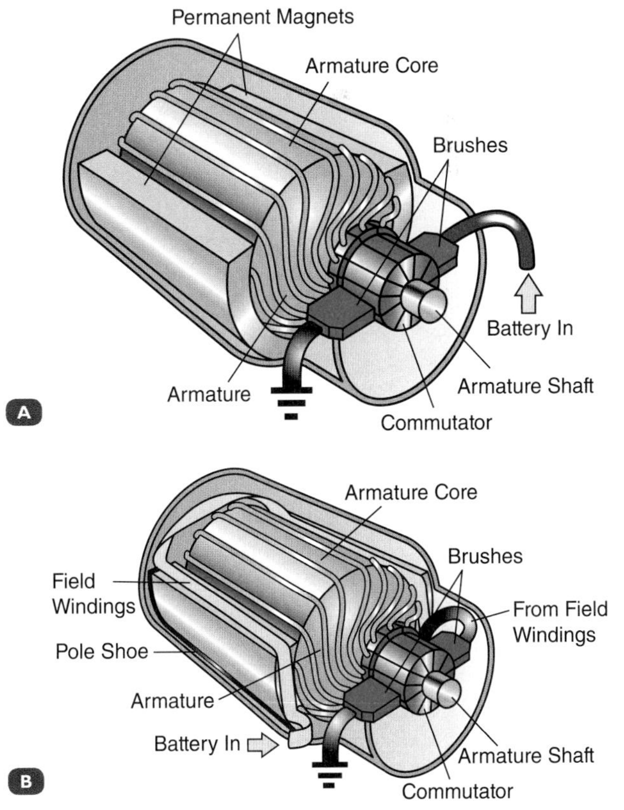

FIGURE 44-8 Electrical schematic of the power flow. **A.** Permanent magnet starter. **B.** Series-wound starter.

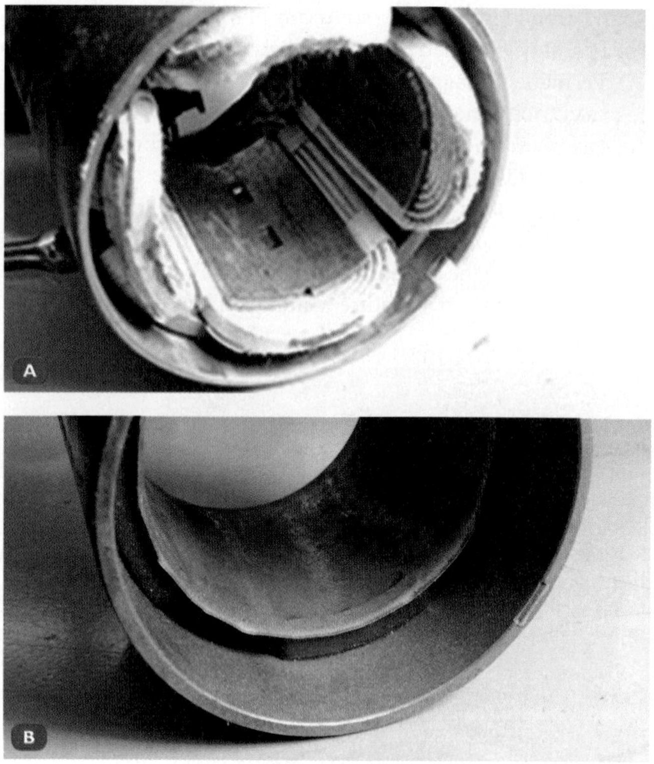

FIGURE 44-9 **A.** Electromagnetic fields. **B.** Permanent magnetic fields.

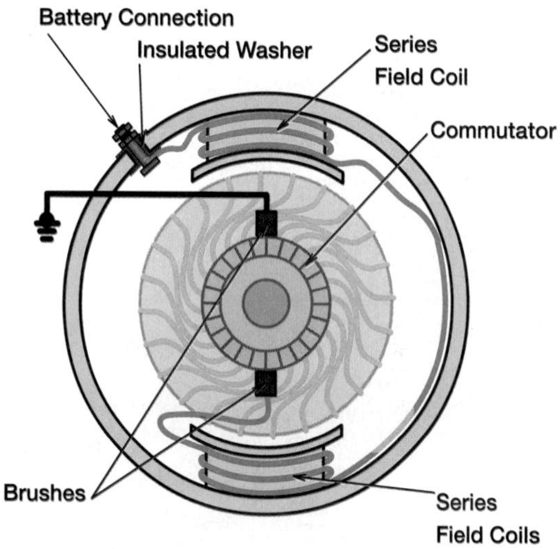

FIGURE 44-10 Series wound motor.

is high when the motor starts under load. This generates a strong magnetic field that produces high torque at low speeds. This high initial torque drops sharply as the motor speed increases. This is because of the **counter-electromotive force** (CEMF).

As the armature spins in the magnetic field, voltage is generated in the armature windings. This is how CEMF is created. The CEMF increases with armature speed. It opposes current flowing through the starter motor. Both current flow and torque output are reduced. The faster the motor turns, the less current it draws and the less torque it develops. For example, on a typical V6 engine with standard compression, the initial surge through the starter is usually over 400 amps. But as the armature starts

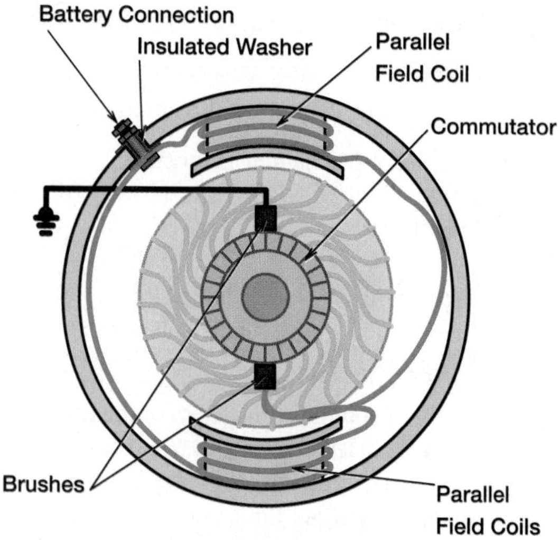

FIGURE 44-11 Series-parallel-wound motor.

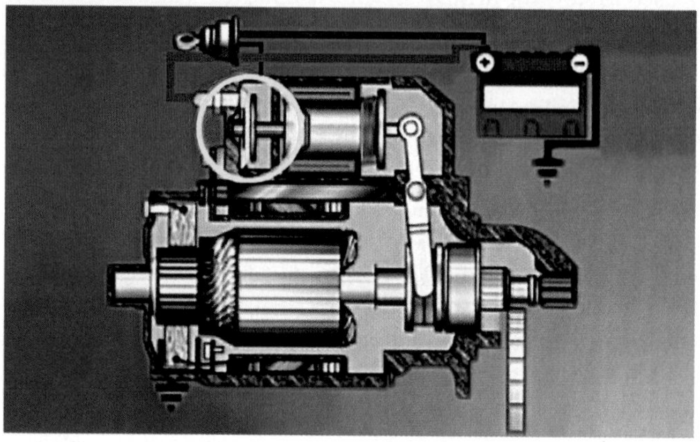

FIGURE 44-12 Cutaway view of starter showing actuating assembly in the engaged position.

to spin, the CEMF opposes the current flow. This reduces the current flow to around 120–150 amps (average) within about a second. This initial surge and subsequent drop of amperage can be observed on an oscilloscope when testing the starter motor.

Some series-wound motors have parallel-wired field windings (**FIGURE 44-11**). But they are still wired in series with the armature. These starters are referred to as series-parallel-wound starter motors. By connecting the field windings in this way, more current can flow in the circuit. This results in an overall increase in torque output.

▶ Starter Motor Engagement

LO 44-03 Describe starter motor engagement.

Engagement is initiated by operation of the ignition switch. A starter-mounted solenoid is activated. The solenoid plunger is attached to a pinion shift lever and operating fork. Solenoid operation moves the operating fork. This causes the pinion to engage with the ring gear. It also causes the plunger contacts to bridge the main starter terminals (**FIGURE 44-12**).

The fork is located in a guide ring on the pinion drive. The pinion drive is coupled to the pinion gear via a roller-type overrunning clutch. It is designed to transmit drive in one direction only. It freewheels in the opposite direction.

The pinion drive is mounted on a slight helix that is machined onto the armature shaft. It forms a very coarse thread. This arrangement allows the pinion drive to rotate slightly when it is moved toward the ring gear. This feature works together with a chamfer on the leading edge of the ring gear and pinion gear teeth. Together, the rotation of the pinion and chamfer of the teeth help in meshing the gears. However, if the pinion gear teeth butt against the ring gear teeth and engagement is prevented, the guide ring continues its movement. It slides over the sleeve of the drive and compresses a meshing spring. This allows the solenoid plunger contacts to bridge the main terminals, and the armature begins to turn.

Slight armature rotation and the force from the meshing spring push the pinion teeth into mesh with the ring gear. The meshing spring forces the pinion into the ring gear until the pinion contacts a stop ring on the armature shaft. This prevents further axial movement. The starter drive is locked to the shaft via the helix. The one-way clutch drives the pinion gear and transfers the armature rotation to the ring gear.

The pinion has only a small number of teeth compared with the ring gear, usually around 17:1. This means that the armature will rotate 17 times for each revolution of the flywheel (**FIGURE 44-13**). This gear reduction multiplies the torque from the starter motor 17 times. This allows a relatively small electric motor to turn the much larger ICE. If a gear reduction starter is used, then the torque is multiplied further, giving more cranking power from the same size starter motor.

As soon as the engine starts, it may easily run at 1000 revolutions per minute (rpm) or more. If still engaged, the engine would then turn the starter pinion gear about 17 times faster, or 17,000 rpm. Turning that fast would destroy the armature. At this instant, the free-wheeling of the overrunning clutch prevents the armature from turning too fast.

The pinion remains meshed as long as the engaging lever is held in the engaged position. Releasing the starter switch allows the solenoid plunger return spring to disengage the pinion gear from the ring gear. This returns the engaging lever, starter drive, and pinion gear to their original position (**FIGURE 44-14**).

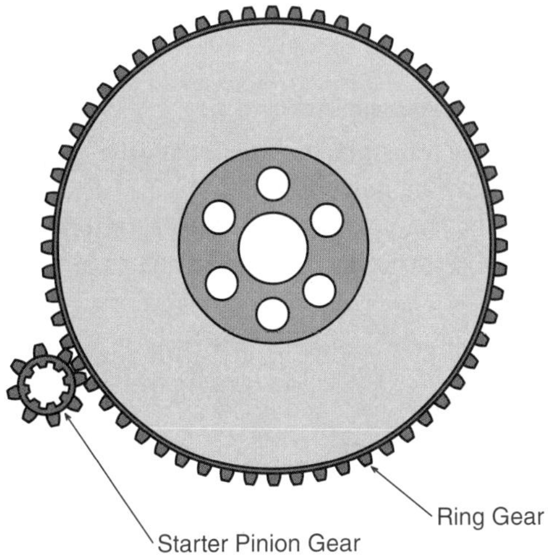

FIGURE 44-13 The starter gear is much smaller than the flywheel ring gear, typically about 17:1.

▶ Armature Windings and Commutator

LO 44-04 Describe armature and starter drive operation.

When current flows in a conductor, an electromagnetic field is generated around it. If the conductor is placed so that it cuts across a stationary magnetic field, the conductor will be forced out of the stationary field. This occurs when the lines of force are distorted by the electromagnetic field around the conductor. They try to return to a straight-line condition. Reversing the direction of current flow in the conductor will cause the conductor to move in the opposite direction. This is known as the motor effect. It is greatest when the current-carrying conductor and the stationary magnetic field are at right angles to each other.

A conductor loop that can freely rotate within the magnetic field is the most efficient motor design. In this position, when current flows through the loop, the stationary magnetic field is distorted and the lines of force try to straighten. This forces one side of the loop up and the other side of the loop down, thus turning the loop (**FIGURE 44-15**). It causes the loop to rotate

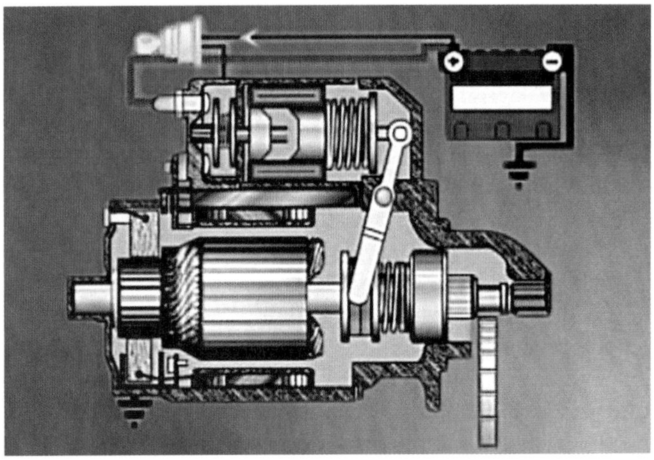

FIGURE 44-14 Cutaway view of starter showing actuating assembly in the released position.

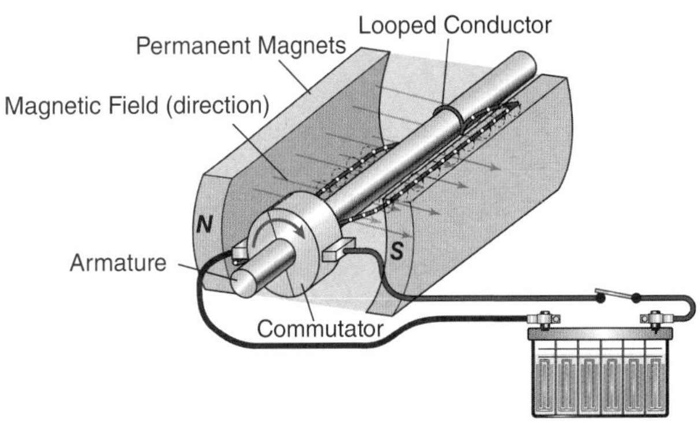

FIGURE 44-15 Simple single-loop motor and electromagnetic fields—with commutator and brushes.

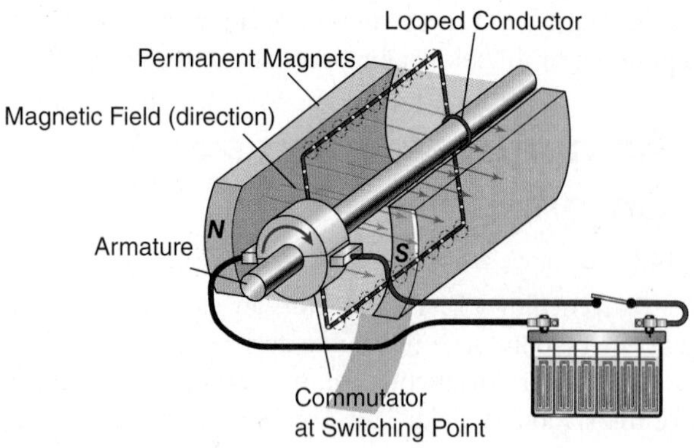

FIGURE 44-16 Simple single-loop motor and electromagnetic fields at the switching point of the commutator.

until it is at 90 degrees to the magnetic field. To continue rotation, the direction of current flow in the conductor must be reversed at this static neutral point.

A commutator is used to continually reverse the current flow in the loop (**FIGURE 44-16**). This keeps the armature rotating. For example, a commutator consists of two semicircular segments that are connected to the two ends of the loop. The commutator segments are insulated from each other. Carbon-impregnated brushes provide a sliding connection to the commutator. This completes the circuit and allows current to flow through the loop.

Rotation begins with both sides of the conductor loop cutting the stationary field. The loop passes the point where the field is no longer being cut. Momentum of rotation carries the loop and the commutator segments so that the brushes make contact again. This maintains current flow in the same direction in each side of the loop relative to the stationary field.

This process maintains a consistent direction of rotation of the loop. In order to achieve a uniform motion and torque output, the number of loops must be increased. The additional loops smooth out the rotational forces. A starter motor armature has a large number of conductor loops. Therefore, there are many segments on the commutator (**FIGURE 44-17**).

Applied Science

AS-75: Electromagnetism: The technician can explain the relationship between current in a conductor and strength of the magnetic field.

Electromagnetism can be defined as the physical relationship between a magnet and electricity. Michael Faraday is best known for his discoveries of electromagnetism back in 1831. His biggest breakthrough was his invention of the electric motor. A magnetic field is created when current flows through the field winding. The armature spins as a result of the magnetic field. To control the output of the motor, more or less current is supplied to the motor. This strengthens or weakens the magnetic field.

In the case of a charging system, we can increase the output of the alternator by increasing the rotor's magnetic field. This is accomplished by increasing the current flow through the rotor.

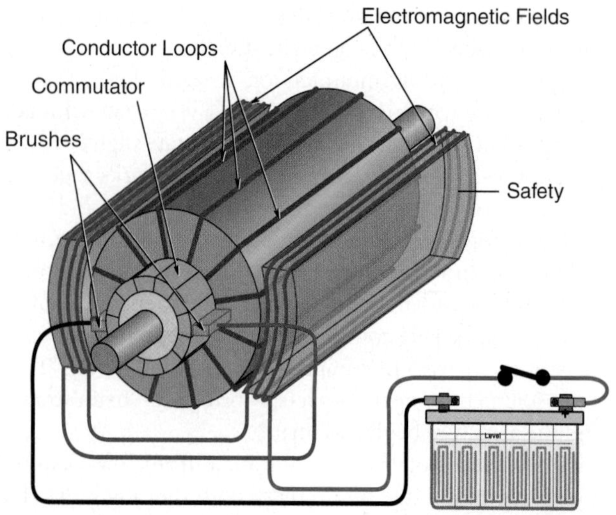

FIGURE 44-17 Simple multiloop motor and electromagnetic fields—with commutator and brushes.

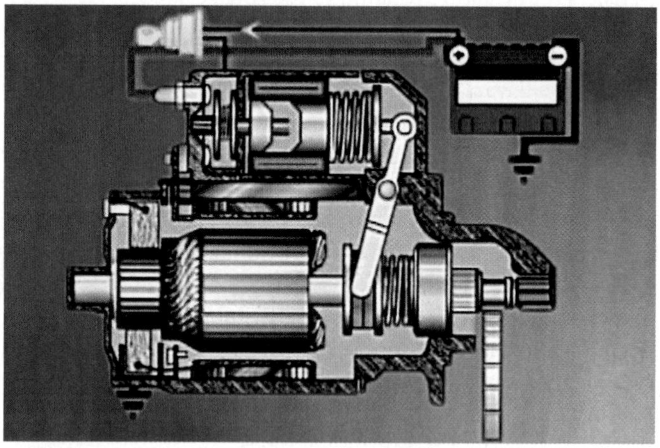

FIGURE 44-18 Cutaway view of a starter drive.

▶ Starter Drives and the Ring Gear

LO 44-05 Describe solenoid operation.

The starter drive transmits the rotational drive from the starter armature to the engine via the ring gear. The ring gear is mounted on the engine flywheel, flexplate, or torque converter. The starter drive is composed of a (**FIGURE 44-18**):

- pinion gear,
- an internal spline that mates with the slightly curved external spline on the armature shaft,
- an overrunning clutch,
- and a meshing spring.

The pinion gear is small in comparison to the ring gear. This means the starter turns many times faster than the engine ring gear. It also gives a large amount of mechanical advantage to the starter motor, allowing it to crank over the much larger ICE.

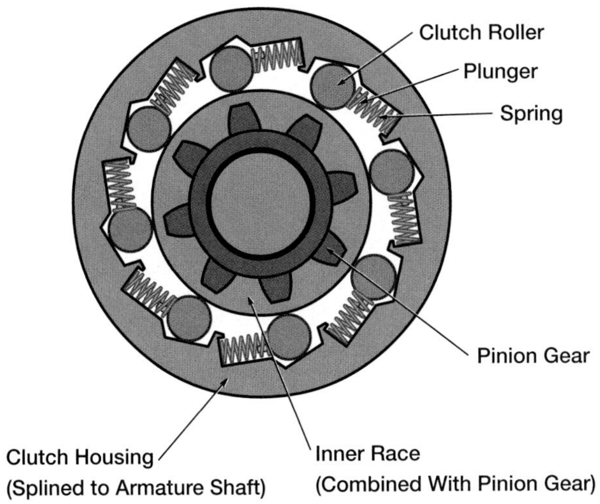

FIGURE 44-19 Starter drive one-way clutch.

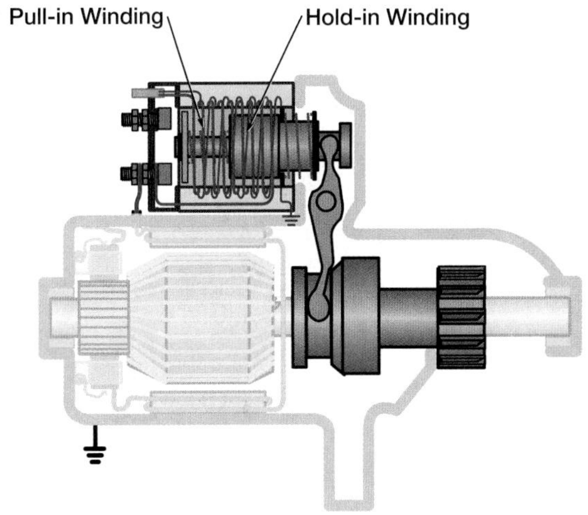

FIGURE 44-20 The solenoid uses two electrical windings: a hold-in winding and a pull-in winding.

The overrunning clutch drives the pinion gear in one direction while allowing it to freewheel in the opposite direction. The overrunning clutch uses roller bearings housed between an inner shell and an outer shell. The roller bearings ride in tapered ramps built into the outer shell (**FIGURE 44-19**). Springs push the rollers toward the tapered ends of the ramps. In the forward direction, the rollers roll slightly between the tapered ramps. This pinches them between the inner and outer shells, locking the assembly and driving the pinion gear.

When the engine starts, it wants to spin the pinion in the opposite direction. This causes the rollers to roll up the inclined ramps against spring pressure. The rollers unlock and allow the pinion gear to freewheel.

The overrunning clutch is an important part of the vehicle. It prevents the starter motor from being driven by the engine once the engine starts. If that happened, the armature would spin faster than it could handle. Without the overrunning clutch, the starter motor would overspeed, causing significant damage.

▶ Solenoid Operation

LO 44-06 Describe starter control circuit operation.

The solenoid on the starter motor performs two main functions. It switches the high current flow required by the starter motor on and off. It also engages the starter drive with the ring gear. The solenoid is typically a cylindrical device mounted on the starter motor (**FIGURE 44-20**). It is constructed with two electrical windings. One is a **pull-in winding** and the other is a **hold-in winding**, which will be explained in a moment.

One end of the solenoid has a moving soft iron plunger. It is connected to a lever that moves the starter drive. The other end has an insulated cap with electrical connections to the solenoid windings. It also houses a set of high-current contacts and a movable copper disc. The movable disc completes the cranking circuit when the solenoid plunger is drawn forward (**FIGURE 44-21**). Once the cranking circuit is completed, current flows from the battery to the starter fields and armature.

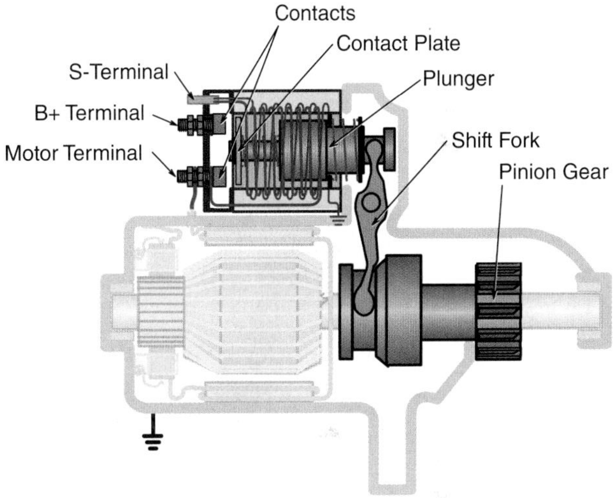

FIGURE 44-21 Solenoid starter contacts and starter drive linkage.

The starter control circuit activates the solenoid winding to draw the plunger forward. The control circuit can be activated directly by the ignition switch on some vehicles. On other vehicles, the PCM activates it. When the control circuit is activated, it supplies power to the two windings in the solenoid. The inputs of both windings are connected to the S-terminal (control circuit) on the solenoid.

The pull-in winding draws a higher current and creates a stronger magnetic field than the hold-in winding. Both windings are wound in the same direction in the solenoid housing. The output of the pull-in winding is connected to the main starter input terminal. This provides ground to the pull-in winding until the solenoid contacts close. The output of the hold-in winding is connected to ground on the starter casing. This provides ground to the hold-in winding at all times.

When the starter is activated, current passes through both starter windings. The magnetic fields from both windings work together. They attract the solenoid plunger toward the main starter terminals in the solenoid cap (**FIGURE 44-22**). Plunger

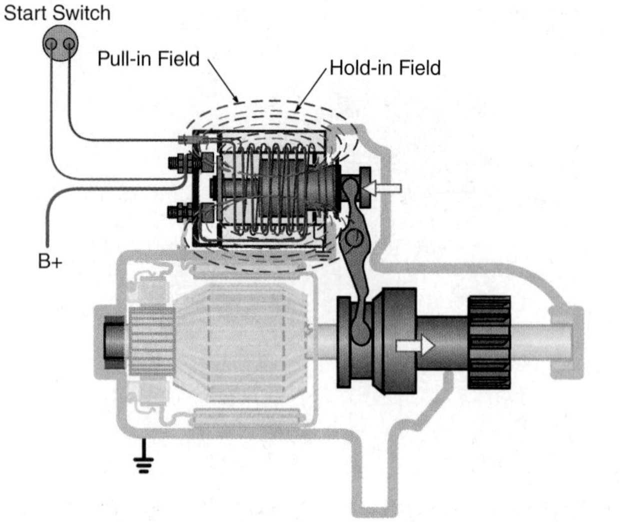

FIGURE 44-22 Both windings energized, creating maximum magnetism, and solenoid plunger starting to move toward the cap.

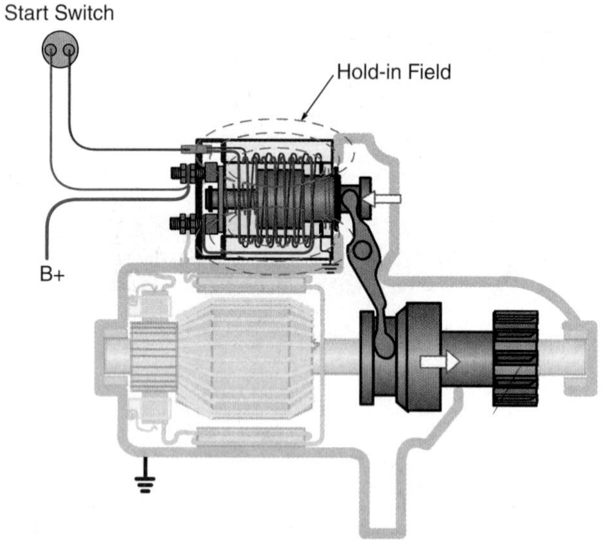

FIGURE 44-23 The solenoid plunger bridging the contacts and causing the motor to turn and shorting-to-power the output of the pull-in winding, stopping current flow through the pull-in winding, and stopping its magnetic field.

movement also operates the shift fork lever which engages the drive pinion with the ring gear.

The plunger also contacts a switching pin. It transfers the motion through a contact spring, closing the main solenoid terminals. This allows a large current to flow from the battery through the starter motor windings. The armature and pinion rotate, turning the engine crankshaft.

However, closing the contacts does another very important task. It shorts-to-power the output wire of the pull-in winding. This results in two sequential actions. First, the short-to-power means that now the pull-in winding has battery voltage applied to both the input and the output of the winding. This stops the current flow through the pull-in winding. Because there is no current flow, the magnetic field in the pull-in winding collapses.

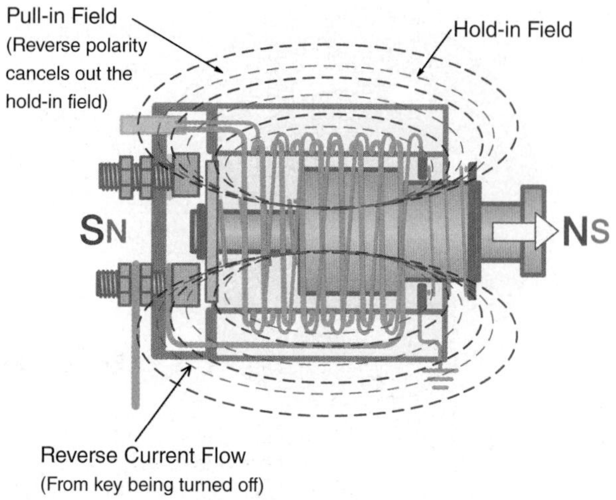

FIGURE 44-24 Current flow goes backward through the pull-in winding when the ignition key is turned from the crank position. This creates an opposite magnetic field, which cancels out the magnetic field from the hold-in winding.

But the hold-in winding still has power from the control circuit. So, it continues to hold the plunger in place while the starter cranks. This holds the moving contact against the main starter terminals (**FIGURE 44-23**). The starter continues to crank the engine over.

Once the engine starts, the control circuit is deactivated. The current stops flowing through the control circuit supplying the hold-in and pull-in windings. This is when the second action of the short-to-power condition of the pull-in winding comes into play. Current flows backward through the pull-in winding and then forward through the hold-in winding.

Notice that the current flows in the two windings opposite to each other. This creates magnetic fields that are opposite to each other (**FIGURE 44-24**). This tends to cancel both magnetic fields out. Because of this, the plunger is retracted by the return spring. This opens the main solenoid contacts and disconnects power to the starter motor. It also disconnects the power flowing to the pull-in and hold-in windings. The starter motor stops cranking the engine. As the plunger returns, it also retracts the pinion gear to its rest position.

▶ Starter Control Circuit

LO 44-07 Describe solenoid operation.

The starter control circuit provides a means of operating the starter motor only within certain parameters. Some of these parameters are when:

- the transmission is in Park,
- the clutch is depressed,
- the brake pedal is applied,
- or the proper ignition key is being used.

These requirements help prevent accidentally starting the vehicle in gear (**FIGURE 44-25**). They also help prevent the vehicle from being stolen.

For many years, manufacturers have placed switches in series with the starter solenoid windings. They prevent the

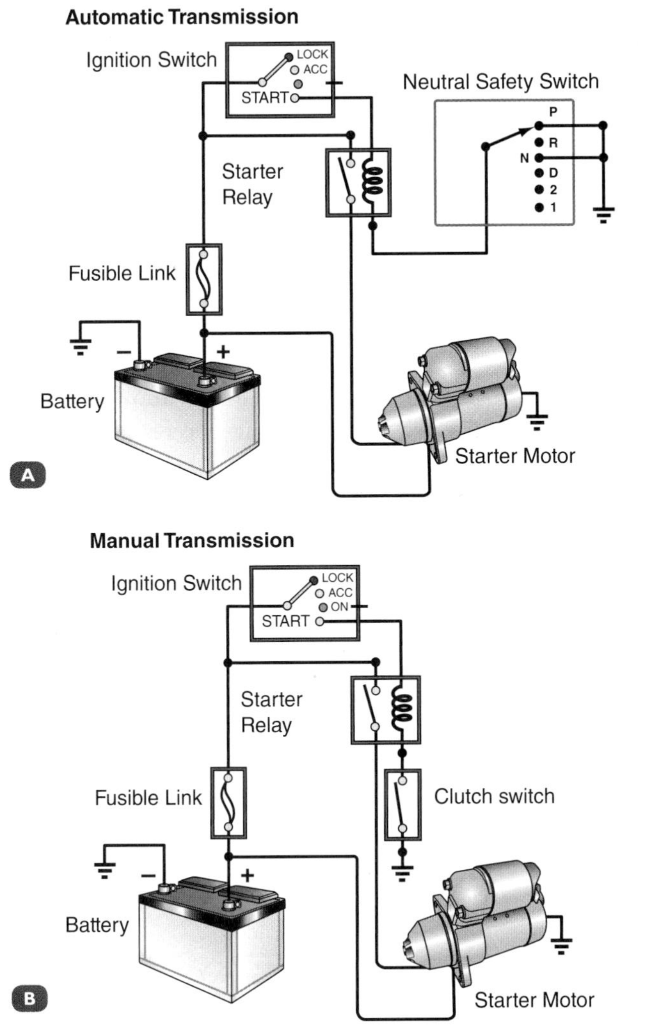

FIGURE 44-25 Basic starter control circuit. **A.** Neutral safety switch circuit. **B.** Clutch switch circuit.

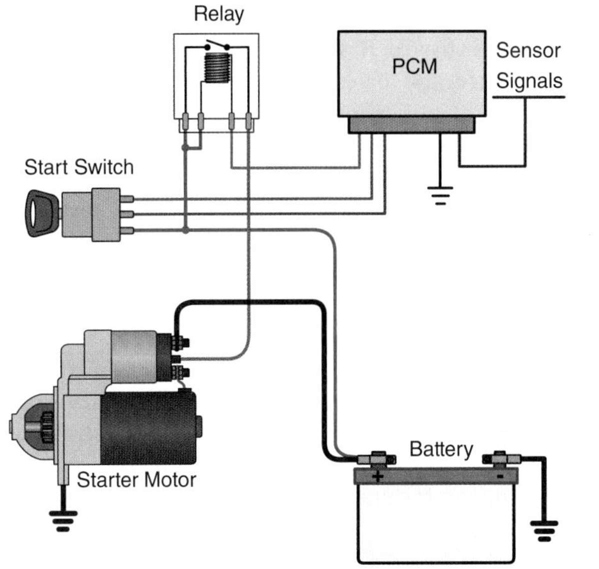

FIGURE 44-26 PCM-controlled starter circuit.

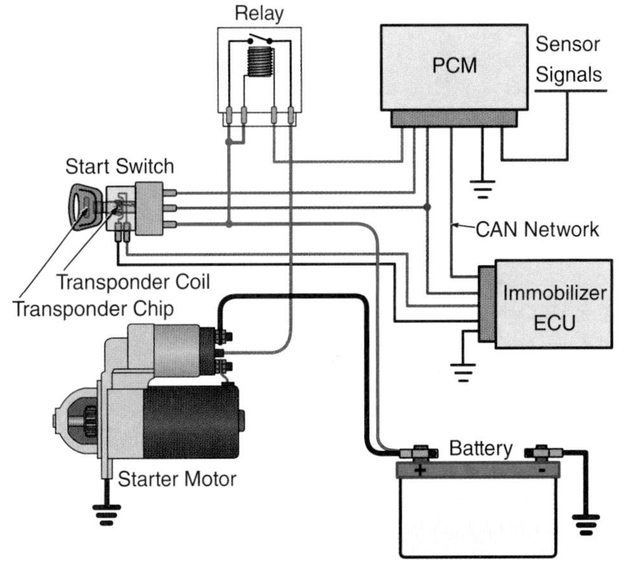

FIGURE 44-27 A typical vehicle immobilization system.

starter from being activated unless each of the switches is closed. If the vehicle is equipped with an automatic transmission, then a neutral safety switch is incorporated into the shifter linkage. This switch is only closed when the transmission is in Park or Neutral. If the vehicle is equipped with a standard transmission, then a clutch switch is installed. It closes when the clutch pedal is pressed to the floor.

Newer vehicles with their computer controls can monitor the same information, and more. The starter can be prevented until all of the required parameters are met. Once they are met, the PCM either activates a starter relay, or the solenoid directly (**FIGURE 44-26**). With PCM-controlled starters, the starter control circuit becomes part of the vehicle theft-deterrent system. It can disable the starter to prevent the vehicle from being started and stolen. We cover this in the next section.

Vehicle Immobilization Systems

There are many different names given to vehicle immobilization systems. Each manufacturer produces its own version, each with subtle differences. Vehicle immobilizers are computer-managed security systems. They disable the vehicle starter and engine systems. They do so by using an electronic system to uniquely identify each vehicle key by a security code system. Some keys have a built-in electronic circuit board used to store the code. This system makes it very difficult to start the vehicle with anything other than a correctly coded vehicle key. It also means that the key not only needs to be cut to fit the lock. But it must also be coded electronically to match the vehicle. Key coding is usually done with a scan tool with the correct software and a pass-through device. This enables the body control module (BCM) to be programmed from the Internet.

Another type of immobilizer system uses a static code programmed into circuitry built into the key. The circuit is powered by electromagnetic induction from a small coil surrounding the key lock. Once powered up, the circuit in the key sends the static binary code wirelessly to the receiving coil. From there, it is sent to the immobilizer module and compared with the registered code (**FIGURE 44-27**). Newer systems use an immobilizer control unit that generates rolling codes. These codes change each time the system is activated. In this case, the circuitry in the key works with the control unit to determine the rolling codes.

The key identification system has two states of operation: mobilized and immobilized (or secure). When mobilized, the vehicle and engine components are allowed to operate normally. While in the immobilized or secure state, the key identification system is activated. This prevents the engine from starting and/or cranking. The system is set to immobilize when the engine is switched off and the key removed. Usually the system will incorporate a flashing warning lamp on the dash to identify that it is immobilized. To mobilize the system, the key needs to be inserted into the ignition switch and turned to the on position. Or in the case of an electronic smart key, the key needs to be within certain distances of the sensors. If so, the operator can step on the brake pedal, and push the start button.

Some General Motors vehicles have used a key with a built-in resistor (**FIGURE 44-28**). When the key with the correct value of resistance is inserted into the ignition switch, the immobilizer module enables the starter control circuit to operate.

Keyless Starting/Remote Starting

Immobilizer systems now use keyless starting. The vehicle has a start button on the dash and does not require the key to be inserted into an ignition switch. In this type of wireless system, the start button will start the vehicle only if the key is in the proximity of the vehicle. For example, if it is in the driver's pocket and if the driver is stepping on the brake pedal. The vehicle detects the key wirelessly and mobilizes the system so the engine can start if the button is pressed (**FIGURE 44-29**).

A further variation of keyless starting is where the vehicle can be started remotely (e.g., inside the house). A start button is located on the key fob (**FIGURE 44-30**). Pressing the button starts the vehicle remotely. This is handy in hot or cold climates when warming or cooling the vehicle is desired prior to entering the vehicle. The vehicle will typically run for a predetermined amount of time. After that, it will shut off.

▶ Starter Draw Testing

LO 44-08 Test starter high-current circuit voltage drop.

Starter motors can be tested in two ways: on vehicle or off vehicle. The on-vehicle test is usually called a starter draw test. The off-vehicle test is called a starter no-load test. Many manufacturers will provide specifications for one or both of these tests.

Ideally, the starter motor should be tested under load. This gives the best indication of any starter issues. When performing the test, it must be performed with a fully charged battery. And the battery capacity must be correct for the vehicle.

Starter current draw is at its highest when the engine is just starting to first rotate. As the starter motor and engine cranking speed increase, the current draw decreases. It quickly stabilizes once the engine reaches full cranking speed. It is at this point that the amperage is read and then compared with specifications.

There is a variety of starter test equipment used to test starter draw. Each device will operate slightly differently. But

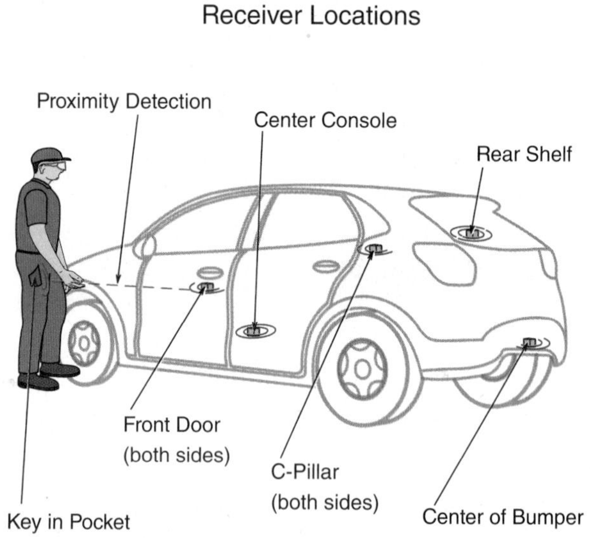

FIGURE 44-29 The vehicle has a number of receivers that detect the key wirelessly, allowing the driver to start the engine with the key in their pocket or purse.

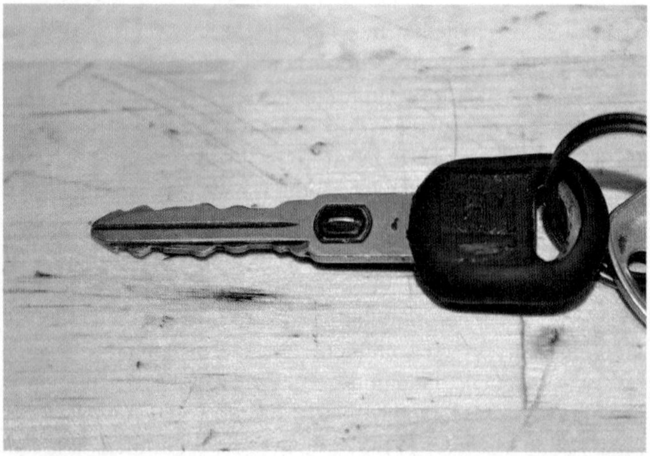

FIGURE 44-28 Key with built-in resistor as part of a theft-deterrent system.

FIGURE 44-30 Key fob with remote start button.

all of them should have an inductive high-current ammeter to measure the cranking current flow. The inductive ammeter is easier to use than a standard ammeter. This is because it does not require any battery cables to be removed. It can be quickly clamped around the main starter cable. The tester should also have a voltmeter to measure the cranking voltage.

When conducting the test, the engine must be disabled so it will crank but not start. This is done in one of the four ways listed in the skill drill. The current flow and voltage will be measured during cranking and compared with specifications.

To test the starter draw, follow the steps in **SKILL DRILL 44-1**.

SKILL DRILL 44-1 Testing the Starter Draw

1. Research the specifications for the starter draw test. Prepare the starter tester by setting it up to measure starter current. Connect the red lead to the positive terminal of the battery and the black lead to the negative terminal of the battery.

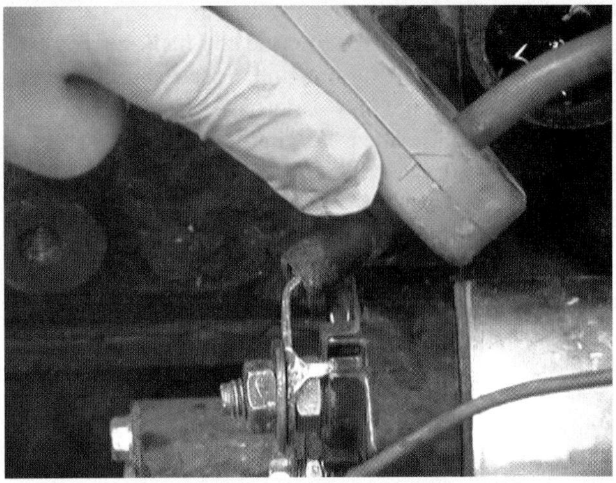

2. Connect the amps clamp around either the positive or the negative battery lead in the correct orientation. Make sure all of the appropriate wires are inside the clamp and the clamp is completely closed.

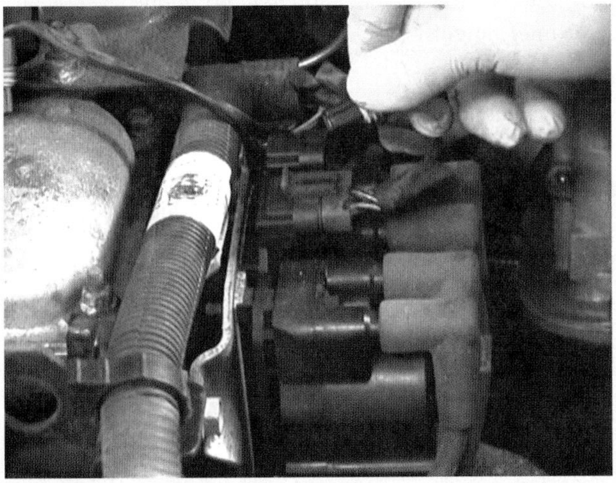

3. Disable the engine from starting by one of the following methods:

- Clear flood mode: This mode is programmed by some manufacturers on electronic fuel-injected vehicles. It can be activated by holding the throttle down to the floor before turning the key. If the engine starts, lift your foot off the throttle and try another method.
- Pull the fuel pump relay and run the engine until it dies.
- Disconnect the fuel injectors or ignition coils.
- Disconnect the spark plug wires.

4. With the engine disabled, crank the engine and read the amps and volts as soon as the amps stabilize. Compare the readings with specifications and determine any necessary actions.

▶ Testing Starter High-Current Circuit Voltage Drop

LO 44-09 Test starter control circuit voltage drop.

The electrical circuit of the starter motor consists of a high-current circuit and a control circuit. The control circuit activates the solenoid. It can be either PCM-controlled or non-PCM-controlled. Voltage drop can occur across both the high-current and control circuits. However, the high-current circuit is more susceptible to voltage drop. This is due to the much larger amount of current flowing in the circuit. We will look at testing the control circuit voltage drop in the next topic.

When testing the voltage drop on the high-current side, for the measurement to be meaningful, the starter must be activated. The starter does not necessarily have to spin, but the solenoid must at least click when the ignition switch is engaged. If the solenoid does not click, then you will need to test the control side of the starter circuit.

A digital multimeter (DMM) is used to measure voltage drop across all parts of the circuit. A voltmeter with a minimum/maximum range setting is very useful when measuring voltage drop. It will record and hold the maximum and minimum voltage drop that occurs for a particular operation cycle. Refer to the Meter Usage and Circuit Diagnosis chapter for more information on DMMs.

Voltage drop is tested while the circuit is under load. Check the service information for safe methods of disabling the engine so it will not start. The DMM is connected in parallel across the component or section of the circuit that is to be tested. Usually, the most efficient method is to test large sections of the circuit first. If required, narrow down the test to individual components. This will allow you to identify precisely where excessive voltage drop is located. For example, connect the black probe to the positive battery terminal. Then, connect the red probe to the heavy gauge input wire leading into the starter motor. Operate the starter with the engine disabled. Record the voltage drop across the positive side of the circuit while the starter motor is energized. Many manufacturers will specify the maximum allowable voltage drop. But a rule of thumb is no more than 0.5 volts (500 millivolts) for a 12-volt circuit.

▶ TECHNICIAN TIP

On most vehicles, the starter cable is connected to the input of the solenoid. The output of the solenoid is connected to the input of the starter motor. The heavy contacts are located between the input and the output terminals of the solenoid. Where possible, it is best to measure the voltage drop from the positive battery post to the starter motor input (not the solenoid input). That way you are measuring any voltage drop across the solenoid contacts. These contacts are a high-probability failure point.

The same test should also be performed on the negative side of the circuit. To do so, connect the black meter lead to the negative battery terminal. The red meter lead connects to the starter housing. Crank the engine and read the voltmeter. Compare the reading with specifications.

To test starter circuit voltage drop, follow the steps in **SKILL DRILL 44-2.**

▶ TECHNICIAN TIP

A faulty battery will affect voltage drop tests, so always ensure that the battery is fully charged and in good condition before performing starter tests.

Inspecting and Testing the Starter Control Circuit

The starter control circuit activates the starter solenoid, which activates the starter motor. If there is a problem in the starter control circuit, the vehicle will likely not crank over at all, or maybe intermittently. The control circuit is made up of the:

- battery,
- fusible link,
- ignition switch,
- neutral safety switch (automatic transmission vehicles),
- clutch switch (manual transmission vehicles),
- starter relay,
- and solenoid windings.

Applied Math

AM-13: Add/Subtract/Divide/Multiply: The technician can estimate the results of basic arithmetic operations and can accurately round numbers up or down.

The following scenario will give us a review of the application of basic arithmetic operations related to starter current draw.

An automobile with a V8 engine has a manufacturer's specification for starter current draw of 180 amps. The specification allows for a tolerance of plus or minus 10%. The technician must determine whether the starter circuit will fall within manufacturer's specifications.

Let's determine what those numbers will be in terms of amps. 180 × 0.90 will give us the minus 10% factor. 180 × 0.90 = 162 amps. Now, for the plus 10% factor. 180 × 1.10 = 198 will give us the plus 10% factor. Now we know that the specifications for starter draw are 162–198 amps, with 180 amps considered the ideal current draw. We could round these off to 160 and 200.

When the starter current draw is measured, it reads 310 amps. Our specification was198 amps maximum. If we subtract 198 from 310, we have 112 amps over the specification, indicating a faulty starter motor or partially seized engine.

To accurately round numbers up or down, here are some tips:

- You can round numbers to give a rough idea of an amount or quantity.
- 862 bolts in stock are closer to 900 than to 800, so you could round to 900.
- A population of 76,310 could be rounded to 76,000.

SKILL DRILL 44-2 Testing Starter Circuit Voltage Drop

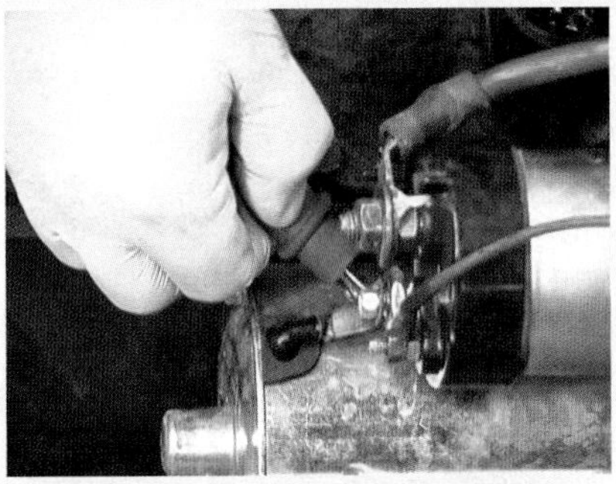

1. Set the DMM to volts. Connect the black lead to the positive battery post and the red lead to the input of the starter (not the input of the solenoid unless that is the only accessible terminal).

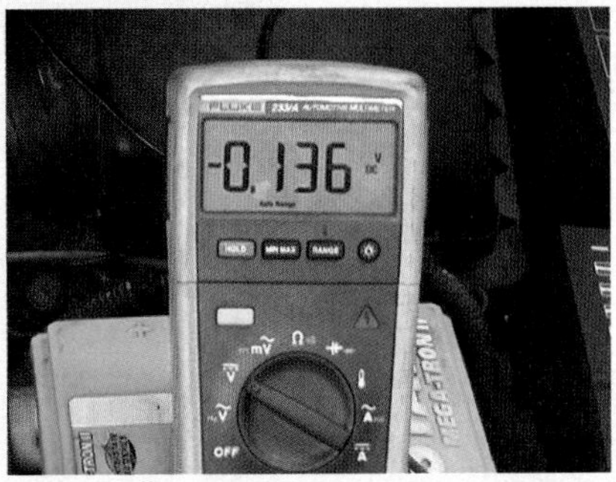

2. Disable the engine from starting. Crank the engine and read the maximum voltage drop for the positive side of the circuit.

3. Connect the black lead to the negative battery post and the red lead to the starter housing. Crank the engine and read the voltage drop. If the voltage drop is more than 0.5 volts on either side of the circuit, use the voltmeter and wiring diagram to isolate the voltage drop. Determine any necessary actions.

If the starter is controlled by the PCM, then you must be aware of all of the related circuits. This includes the immobilizer circuit and the PCM itself.

Before performing any tests, you should know and confirm the customer's concern. The manufacturer's wiring diagram should also be consulted. Use it and the service information to determine the circuit operation. It will also allow you to identify all of the components in the starter control circuit.

Once an understanding of the customer concern and circuit operation is obtained, it is time to test. Start by placing the DMM's black lead on the battery positive terminal. Place the red lead on the solenoid's input terminal (control circuit terminal). Measure the voltage drop with the key in the crank position. At that point, assuming a fault in the control circuit is present, the voltage drop reading at the control circuit will be more than about 0.5 volts. If it is, you will need to perform individual voltage drop tests on the power side of the control circuit. This will allow you to determine which part(s) of the circuit the voltage drop is located in.

If the voltage drop is less than 0.5 volts, then measure the voltage drop on the starter ground circuit. If the voltage drop is excessive, perform individual voltage drops on the ground leg. If both the control circuit power and the ground circuit voltage drops are within specifications, the resistance of the solenoid pull-in and hold-in windings will need to be measured. If out of specifications, the solenoid or starter motor (and solenoid) will need to be replaced.

To inspect and test the starter control circuit, follow the steps in **SKILL DRILL 44-3**.

SKILL DRILL 44-3 Testing Starter Circuit Voltage Drop

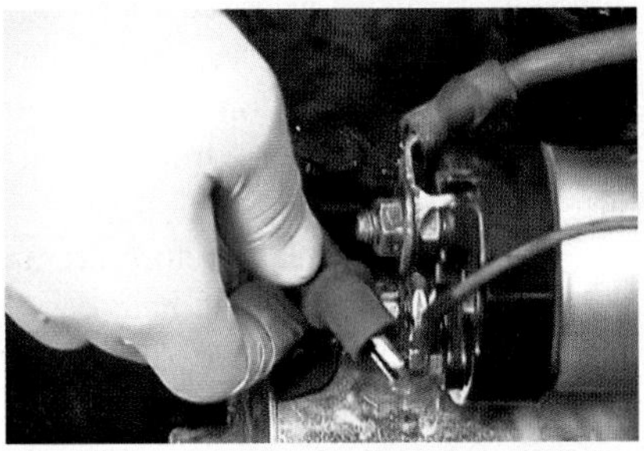

1. Use a DMM to measure the voltage drop on the positive side of the control circuit. Connect the black lead to the positive battery terminal and the red lead to the control circuit terminal on the solenoid.

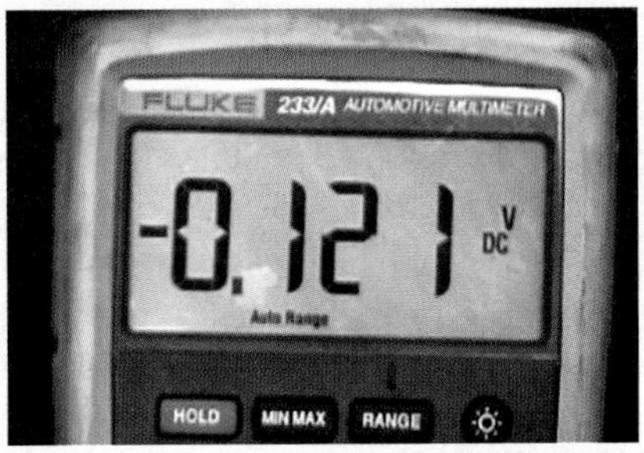

2. Crank the engine over and read the meter.

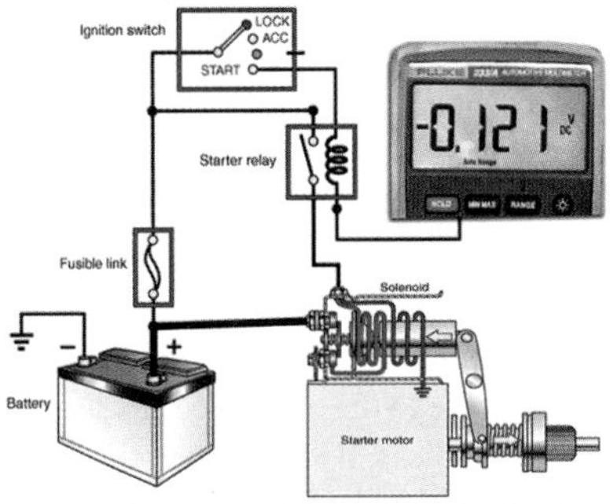

3. Measure the voltage drop on the ground side of the circuit (starter housing to battery negative terminal). If the voltage drop is more than 0.5 volts on either side, use a wiring diagram to determine where to measure the individual voltage drops on that side of the circuit.

▶ Inspecting and Testing Relays and Solenoids

LO 44-10 Test starter relays and solenoids.

The starting system typically contains solenoids and relays that activate the control circuit. The solenoid is mounted on the starter motor, while the starter circuit relay is usually in or near the main fuse box.

Consider a couple of things before performing any tests. You need to ensure that the vehicle battery is charged and in good condition. The manufacturer's wiring diagrams should be checked. Use them and the service information to determine the circuit operation, identification, and location of all components in the starter circuit.

Relays must be tested in two or three ways depending on the relay. The simplest test is to measure the resistance of the relay winding. If it is out of specifications, the relay will need to be replaced. If it is OK, the contacts will need to be tested for an excessive voltage drop. The best way to do this is by using an adapter that fits between the relay and the relay socket (**FIGURE 44-31**). This will allow the normal circuit current flow to flow through the contacts so that a voltage drop measurement can be taken. Any excessive voltage drop across the relay contacts will require the replacement of the relay.

The last test is used only on relays with a suppression diode in parallel with the relay winding. Connect a reasonably fresh 9-volt battery across the relay winding terminals in one direction. Then, switch polarity by turning the battery around. If the diode is good, the relay should click in one direction and not in the other (**FIGURE 44-32**). If it clicks in both directions, the diode is open. If it does not click in either direction, the relay winding is open or the diode is shorted. Both of these conditions require replacement of the relay.

Solenoids can be difficult to test on the vehicle due to poor access. The tests will usually be limited to voltage and voltage drop tests on the main contacts. For other tests, such as pull-in and hold-in winding tests, the starter motor will usually need to be removed. Care should be taken when testing relays and solenoids. Be careful to ensure that cables are not shorted to ground or the engine is not accidentally cranked over.

The first test to perform is a voltage drop test across the solenoid contacts. Place the red lead on the solenoid B+ input. Place the black lead on the solenoid B+ output. The voltage drop should be less than 0.5 volts. If not, replace the starter assembly.

Testing of the solenoid windings requires partial disassembly of the solenoid. Therefore, it is usually best to disconnect the control circuit connector from the solenoid. Use a jumper wire to apply battery voltage to the control circuit terminal on the solenoid. See if the solenoid clicks. If it does, then there is likely a fault in the control circuit wiring. If the solenoid still does NOT click (and the ground circuit is good), then the solenoid windings or starter brushes are likely worn. Note that sometimes tapping on the starter while the key is turned to the crank position will free up the brushes enough that the pull-in winding can operate. If the solenoid still does not work, then the solenoid is likely faulty and will need to be replaced.

To inspect and test relays and solenoids, follow the steps in **SKILL DRILL 44-4**.

FIGURE 44-31 Relay test adapter is used to test relays.

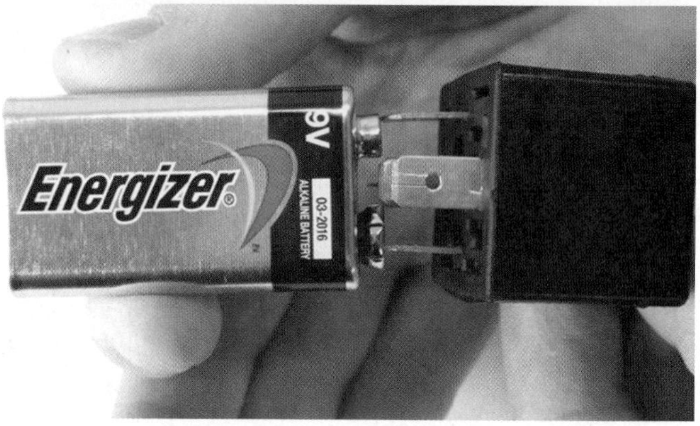

FIGURE 44-32 Testing a relay with a 9-volt battery.

▶ Removing and Installing a Starter

LO 44-11 Remove and install a starter.

Starter motors are usually located close to the flywheel end of the engine. They can be in difficult-to-reach locations. Some engine components or covers may need to be removed to gain access. In most cases, the starter can be accessed more easily from underneath the vehicle. On some vehicles, the starter is located under the intake manifold. So the top of the engine will need to be disassembled. Check the service information to determine where the starter is located.

Always disconnect the negative battery lead before attempting to remove or install a starter motor. This will help prevent sparks from accidental short circuits. And be sure to use a memory minder if vehicle memory needs to be maintained.

Remember that the starter is heavy, so support it well while removing and installing it. Also, look for shims that fit between the starter and engine. They are used on some vehicles to adjust the clearance between the starter drive gear and ring gear. Reuse any shims unless the starter instructions direct you otherwise. Be careful to properly tighten all bolts and nuts when installing the starter. This includes the mounting bolts as well as the connections for the wires and cables.

To remove and install a starter in a vehicle, follow the steps in **SKILL DRILL 44-5**.

SKILL DRILL 44-4 Inspecting and Testing Relays and Solenoids

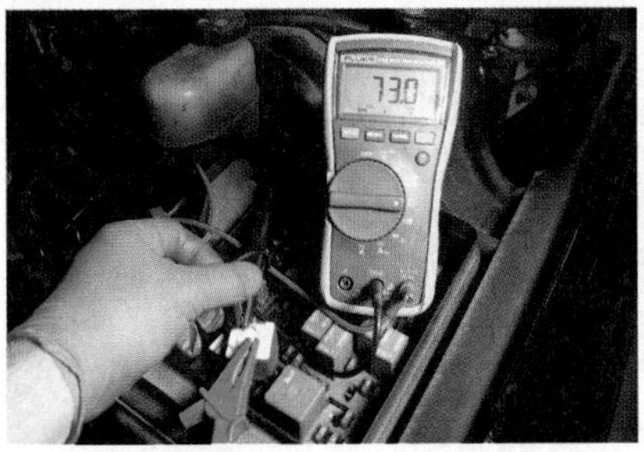

1. To test a relay, measure the resistance of the relay winding and compare with specifications. If out of specifications, replace the relay.

2. Use a relay adapter to mount the relay on top of the relay socket so you can check the control circuit wiring and perform voltage drop tests on the contacts. Activate the relay while measuring the voltage across the relay winding. If it is near battery voltage, the control circuit wiring is okay.

3. Measure the voltage across the contacts with the relay NOT activated. This should read near battery voltage if both sides of the switched circuit are OK. If not, perform voltage drop tests on each side of the switch circuit. Activate the relay while measuring the voltage drop across the contacts. If it is more than 0.2 volts, the relay will need to be replaced.

4. To test a starter solenoid, measure the voltage drop across the solenoid contact terminals with the key in the crank position. If more than 0.5 volts, replace the solenoid or starter assembly.

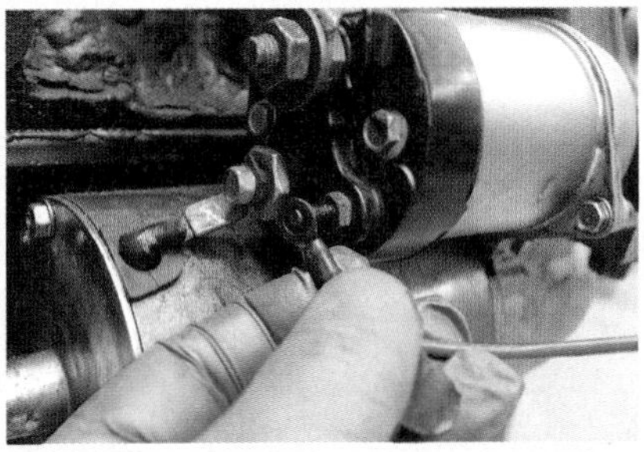

5. If the solenoid does NOT click with the key in the crank position, remove the electrical connection for the control circuit at the solenoid.

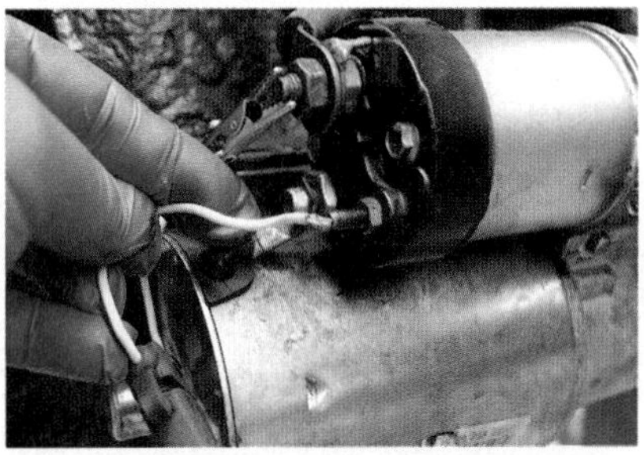

6. Use a jumper wire to apply battery voltage to the control circuit terminal on the solenoid and see if the solenoid clicks. If it clicks, check the control circuit. If it doesn't click, replace the starter and solenoid.

SKILL DRILL 44-5 Removing and Installing a Starter

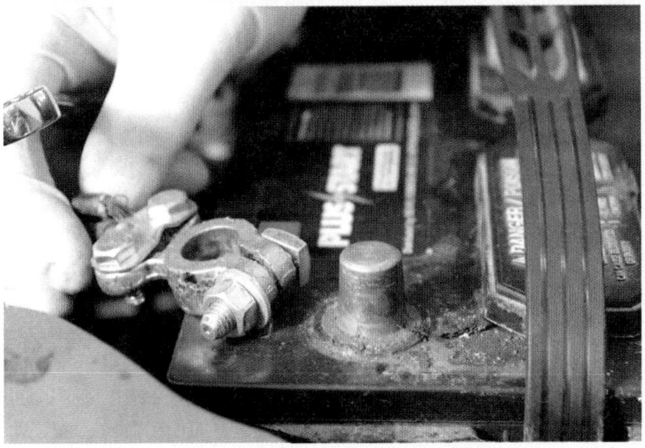

1. Disconnect the negative terminal of the battery after determining whether a memory minder is required.

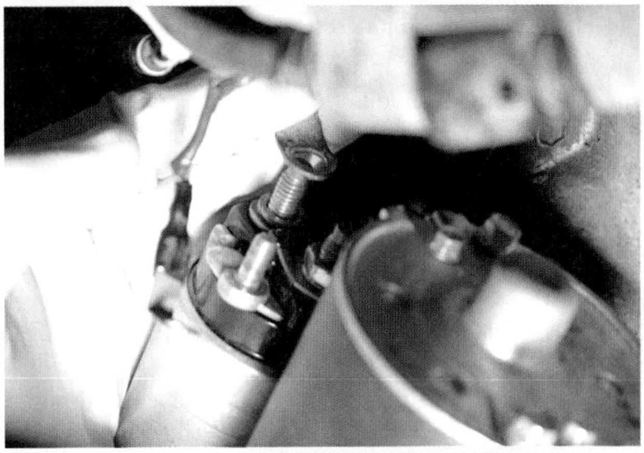

2. Remove any engine covers or components required to gain access to the starter. Remove starter motor electrical connections, noting how the wires were routed. In some vehicles, the wires cannot be accessed until the starter is removed.

SKILL DRILL 44-5 Removing and Installing a Starter (Continued)

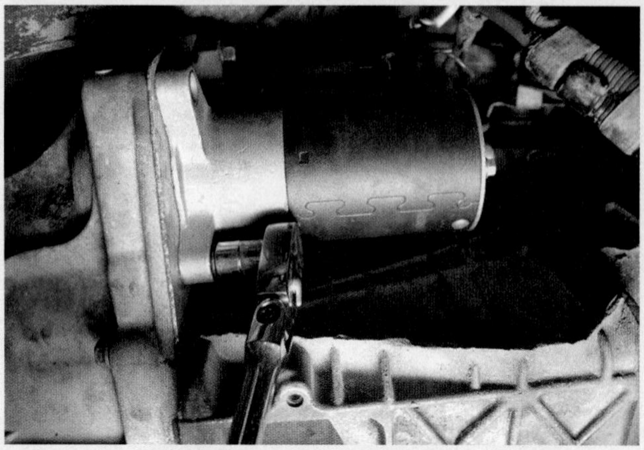

3. Loosen the starter motor mounting bolts, and remove them while holding the starter so it does not fall.

4. Remove the starter motor, being careful to catch any shims that might be between the starter and the block. Reinstall the starter motor by reversing these steps. Verify its proper operation.

Applied Math

AM-2: Decimals: The technician can add decimal numbers to determine conformance with the manufacturer's specifications.

In this example, we are working with decimal numbers. Decimals are numbers that are expressed using a decimal point.

A starter has been rebuilt, and the technician wants to check the pinion clearance. A feeler gauge is used to determine this clearance. Manufacturer's specifications for this clearance are from 0.010" to 0.140", with 0.070" considered the midpoint. The technician's feeler gauge set only goes to 0.045" which fits too loosely in the gap. He places a 0.025" blade next to the 0.045" blade, which, together, fit in the gap with just the right tension. The selected gauges are 0.025" plus 0.045", to equal a total of 0.070".

▶ Idle–Stop/Start–Stop Systems

LO 44-12 Describe idle stop-start stop system operation.

Starting systems have not changed a lot over the years. But one recent change is the addition of idle–stop/start–stop functionality. This system automatically shuts the engine off at times, typically when the vehicle is stopped. It then restarts the engine when needed. This significantly reduces CO_2 emissions and fuel consumption. For the most part, existing components are used. However, in some cases, they may be upgraded to handle the greater frequency of use. We will look at the three most common methodologies used in idle–stop systems, also called start–stop systems.

The simplest system uses unmodified components already on the vehicle. It is on full hybrid vehicles that use a flywheel mounted motor/generator, or a geared motor/generator. This motor/generator is much more powerful than a regular starter motor. So it spins the engine over faster, which is ideal for idle–stop systems. In this case, the hybrid control module activates the motor/generator to spin the engine over when needed. This provides a very smooth, quick start of the engine. The motor/generator in this case can also be used for regeneration when braking the vehicle. This enhances fuel economy even more.

The second type of system uses an upgraded starter motor. It is more powerful and robust than a regular starter. It is computer controlled so it can be activated to start the engine on demand. With its higher power, it can spin the engine over faster, making the engine start quickly. Some vehicles equipped this way use an integrated starter/alternator instead of just a starter. In this case, the alternator function may also be used for regenerative braking. This recaptures some of the wasted braking energy.

The third type of system is unique since it does not rely solely on a starter motor. Instead, it ignites a combustible mixture in an appropriate cylinder. This aids in causing the engine to spin. It can only do this if the engine is equipped with a gas direct injection system. Through crankshaft and camshaft position sensors, the PCM knows where each cylinder is at in the firing order as the engine is shutting down.

To make sure the pistons stop in the same position every time, some systems manipulate the throttle plate to increase/decrease engine vacuum against pistons. They may also manipulate the alternator output to increase/decrease the load on the crankshaft. These functions cause the engine to stop at the same position every time. The PCM then knows which cylinder is on the power stroke. It injects fuel into that cylinder, and then causes the ignition system to generate a spark in that cylinder. The burning air–fuel mixture pushes the piston down the cylinder. This effectively starts to crank the engine over and fire other cylinders. At the same time as the first cylinder is being ignited, the starter motor is also engaged. Both processes work in tandem to crank the engine over faster than could be done separately. Ultimately, this causes the engine to start faster. In this arrangement, the starter motor is also used for cranking the engine when it is cold.

No matter how the engine is cranked over, the process for stopping and starting the engine is similar. When the engine is up to operating temperature (including the catalytic converter), and the vehicle is being driven to a stop, the PCM shuts off the fuel injectors. This can happen on some vehicles as it is braking to a stop. But many vehicles wait until the vehicle has actually stopped for a few seconds. While the driver holds the brake pedal down, the engine stays off.

On a non-hybrid vehicle, as soon as the brake pedal is released, the PCM commands the engine to be started. The appropriate modules take over to crank and start the engine. On hybrid vehicles, when the driver releases the brake pedal, the engine may not need to be started. In this case, the electric traction motor may power the vehicle away from the stop. The PCM will determine when the engine needs to be started, which may be down the road a ways.

Some standard transmission vehicles also incorporate stop/start functionality. One manufacturer controls the system in this way. As the vehicle is approaching a stop, the driver depresses the clutch pedal and moves the gear shift to neutral, and then releases the clutch. The engine goes into idle stop if the vehicle is either below a minimum speed, or stopped. The engine stays stopped until the driver depresses the clutch. This signals the system to restart, allowing the driver to select a gear and drive away from the stop.

▶ Charging Systems

LO 44-13 Describe charging system operation.

Modern vehicles are more dependent on electronic and electrical systems than ever. It is also a trend that will continue into the future. These systems require a constant and reliable supply of electrical power. This is supplied by the charging system. Alternators are sometimes called AC generators, or just generators. They create the needed electrical energy whenever the engine is operating (**FIGURE 44-33**).

The terms "generator" and "alternator" are often used to describe the same thing. They describe the electrical generating component. Strictly speaking, the DC generator has not been used on most vehicles since the 1960s. Thanks to solid-state electronics and circuitry, alternators have taken over. This is due to their superior operating characteristics, which include the following:

- Greater wattage output at lower rpm
- Higher rpm
- Smaller physical size and weight for a given output
- Greater reliability and longer service life

Both DC generators and alternators produce electricity. They both do so by relative movement between conductors and a magnetic field. This induces an electrical potential or voltage within the conductors. There is a key difference between an alternator and a DC generator. It is concerning which component generates electricity. In the DC generator, the conductors that generate power rotate as part of the armature. They rotate within a magnetic field created by the stationary pole shoes (**FIGURE 44-34**).

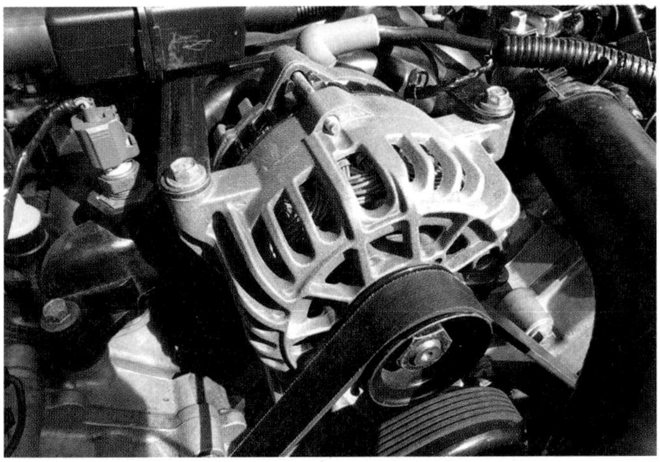

FIGURE 44-33 Typical alternator installed on engine.

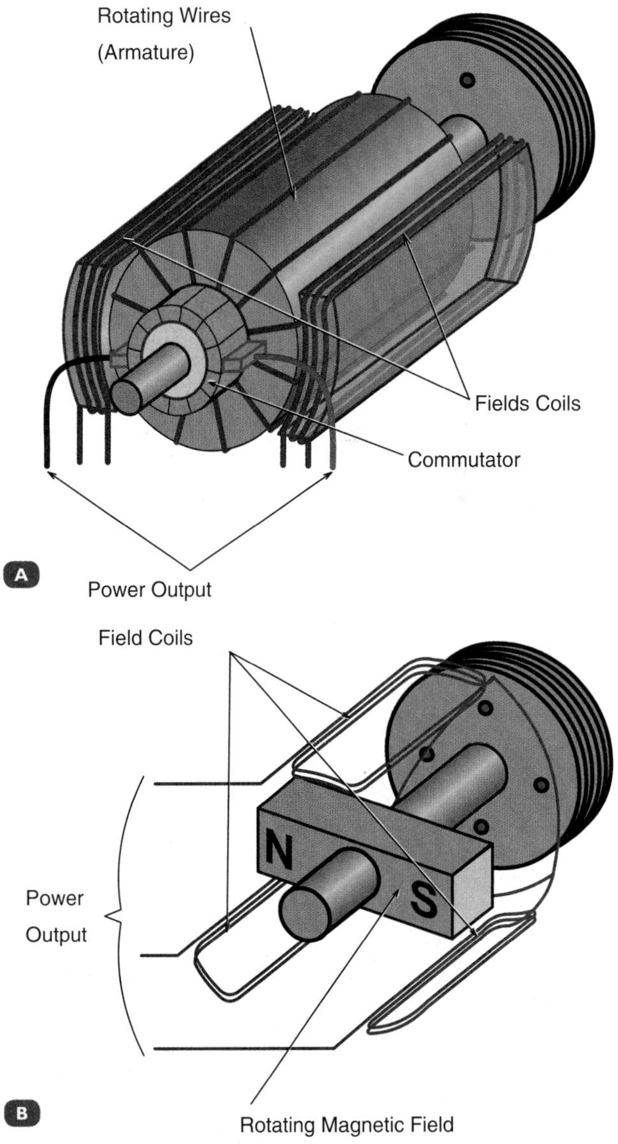

FIGURE 44-34 **A.** DC generators use rotating wires in a stationary magnetic field. **B.** Alternators use a rotating magnetic field in stationary wires.

In the alternator, the magnetic field is created by the rotor. The rotor rotates within the stationary stator windings. The rotating magnetic field generates electricity in the stationary stator windings. In both cases, there is relative movement between the magnetic field and the conductors.

The main parts of the charging system include the:

- battery,
- alternator,
- voltage regulator (internal or external),
- charge warning light,
- and wiring that completes the circuits.

The battery stores electrical energy in chemical form. Because of its storage capacity, it acts as an electrical dampening device. It dampens variations in voltage or voltage spikes. It also provides the electrical energy for cranking the engine. Once the engine is running, the alternator converts some of the engine's mechanical energy into electrical energy. This is used to operate all the electrical components on the vehicle. The alternator also charges the battery to replace the energy used to start the engine.

The voltage regulator circuit maintains optimal battery state of charge. It does so by sensing battery voltage continuously. It then controls the output of the alternator to maintain that voltage.

Older vehicles have separate regulators attached on the firewall (**FIGURE 44-35**). Newer charging systems included regulators that were located inside the alternator. After that, the PCM is used to control the charging system. It can control alternator output based on several parameters. Some examples are the electrical load, and engine load and rpm. Also included are the alternator capability, battery temperature, and fuel economy.

FIGURE 44-35 External voltage regulator.

Electromagnetic Induction in Alternators

The alternator converts mechanical energy into electrical energy by electromagnetic induction. In a simplified version, a bar magnet rotates in an iron yoke, which concentrates the magnetic field. A coil of wire is wound around the stem of the yoke. As the magnet turns, voltage is induced in the coil, producing a current flow (**FIGURE 44-36**). When the north pole is to the right and south is to the left, voltage is induced in the coil. The voltage produces current flow in one direction. As the magnet rotates, the positions of the poles reverse. This reverses the polarity of the voltage as well. And as a result, so does the direction of current flow. Current that changes direction in this way is called alternating current, or AC. In this example, the change in direction occurs twice for every complete revolution of the magnet.

The value of the electromotive force (EMF) or voltage potential induced by an AC generator depends on four factors:

- The strength of the magnetic field—increasing the strength of the magnetic field increases the value (voltage output) of the induced EMF
- The speed at which the magnet rotates
- The relative distance between the magnet and conductors
- The number of turns of wire on the stationary coil

A single-phase AC generator has only one stationary coil. Because of this, it creates a single sine wave (**FIGURE 44-37**). In a typical alternator, three or possibly four separate coils of wire, or phase windings, are common. The windings are arranged so that when the magnet is rotated, it generates a three-phase (or four-phase) output. The phases are equally spaced in time. This results in a phase shift of one phase every 120 degrees, in the case of a three-phase alternator (**FIGURE 44-38**).

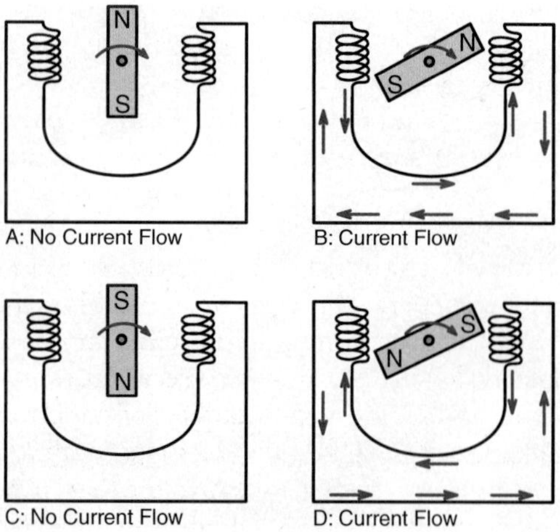

FIGURE 44-36 Electromagnetic induction.

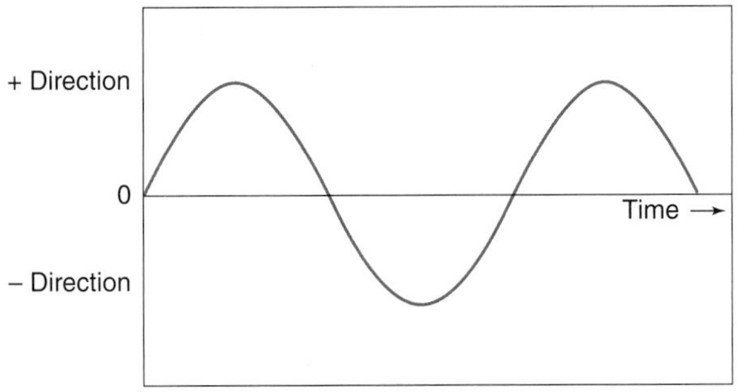

FIGURE 44-37 Single-phase AC signal.

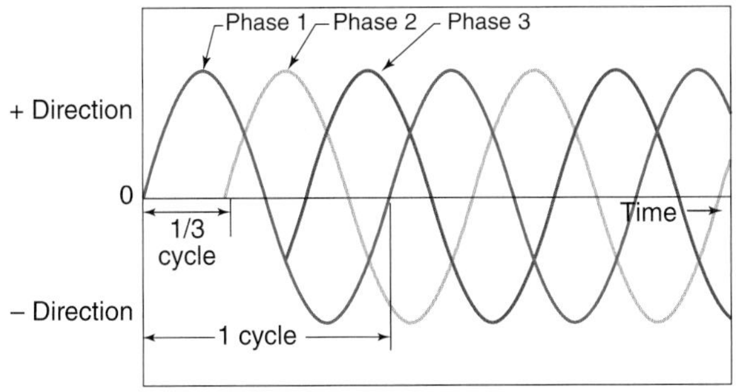

FIGURE 44-38 Three-phase AC signal.

▶ Alternator Component Overview

LO 44-14 Describe the rotor, slip ring, and brushes.

The alternator consists of:

- a stationary winding assembly called the stator,
- a rotating electromagnet called the rotor with a slip ring,
- a brush assembly,
- a rectifier assembly,
- two end frames,
- and a cooling fan and drive pulley (**FIGURE 44-39**).

A voltage regulator monitors battery voltage and varies current flow through the rotor field circuit. This varies the strength of the rotating magnetic field. Field current varies as required to perhaps 5 or more amps. It controls the output of the alternator up to about 150 amps for a typical automobile. The voltage regulator's job then is to control the output of the alternator. It does so by maintaining the system voltage within specified limits. Regardless of the make or model, the fundamental construction and components of alternators are very similar.

Rotor

The rotor is an electromagnet that rotates freely in the alternator. The rotor is supported on each end by ball bearings. It consists of a coil of insulated wire wound around an iron core and pressed onto a steel shaft. An iron pole piece is fastened on each side of the coil assembly so that the projections or claws (which are bent) interlace. The ends of the rotor coil winding are connected to insulated slip rings mounted on the shaft. The spring-loaded brushes maintain contact with the slip rings at all times. This allows current to flow into and out of the rotor winding.

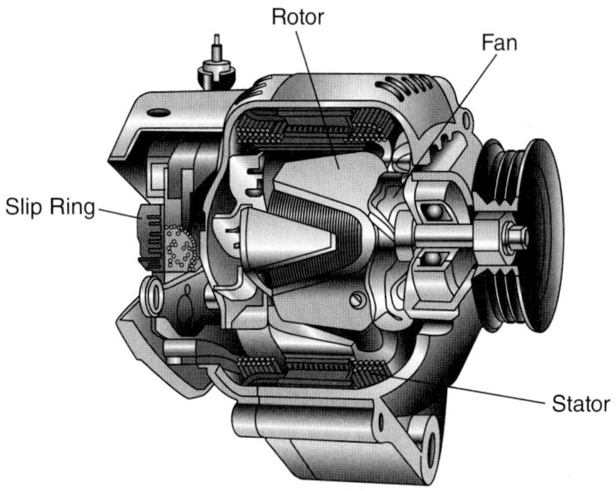

FIGURE 44-39 The alternator.

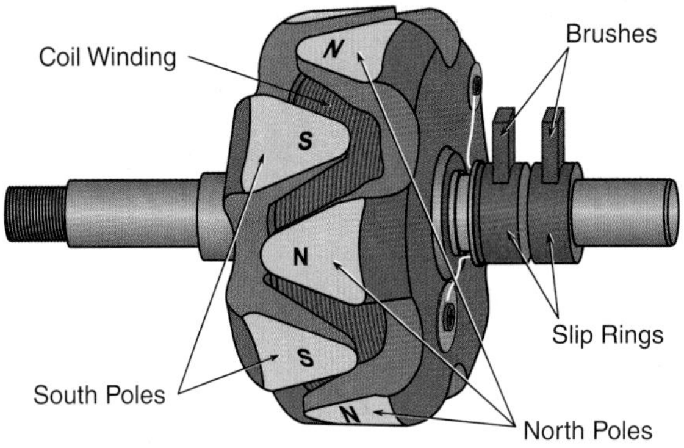

FIGURE 44-40 The rotor has projections that create alternating north and south magnetic fields.

When current is passed through the slip rings and the coil winding, it establishes strong north and south poles at the ends of the iron core and the shaft (**FIGURE 44-40**). The projections then take on the same polarity as the end of the shaft on which they are mounted. This forms pairs of north and south poles that are alternately spaced around the rotor circumference. The rotor usually has 8–12 poles, which are tapered to reduce noise and create a smooth AC sine wave output.

When the rotor is fully energized, the magnetic field is at its strongest. It also induces maximum current flow in the stator. When the rotor is fully energized, it requires a fair amount of mechanical energy to rotate the rotor inside of the stator. In fact, a high-current alternator can use more than 5 horsepower from the crankshaft to operate. Thus, any sizable electrical load affects fuel economy. Also, a fan is either pressed or slipped onto the rotor shaft. It assists in cooling the rotor, stator windings, and rectifier assembly.

Slip Rings and Brush Assembly

Slip rings and brushes aid in making an electrical connection to the rotating rotor assembly (**FIGURE 44-41**). The slip rings are normally copper bands. They are molded onto an insulating material, and then pressed onto the steel shaft of the rotor. Each end of the rotor winding is connected to one of the copper bands. As the rotor rotates, the brushes maintain a constant connection with each end of the winding. Brushes are made of a combination of copper and carbon. They are carried in brush holders mounted in the end frame of the alternator. Springs help maintain contact with the slip rings. Brushes do wear out over time.

Brushless Alternators

Because alternator brushes are a wear item, they can cause breakdowns when they wear out. Because of this, some manufacturers have designed brushless alternators. Brushless alternators induce current flow into the rotor. They do this through one stationary field winding in the housing and a separate armature on the rotor (**FIGURE 44-42**).

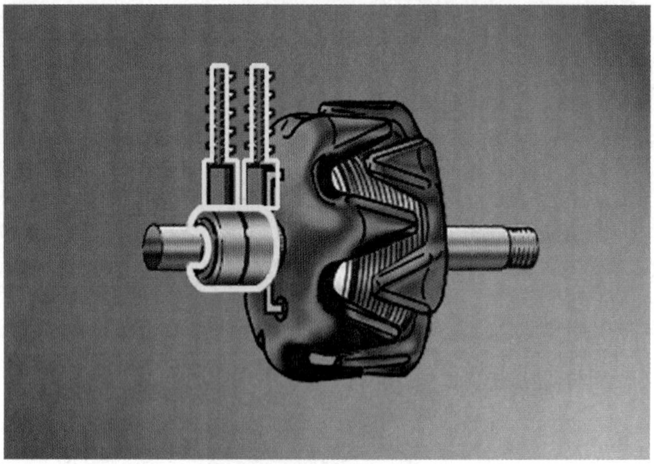

FIGURE 44-41 Slip rings and brushes aid in making an electrical connection to the rotating rotor winding.

The induced exciter current flow that is created in the rotor armature is AC. A rectifier on the rotor turns the AC into DC. This DC is supplied to the main rotor winding, where it is used to create the rotor's electromagnetic force. The main current in the rotor is controlled by the strength of the exciter field.

The greater the current flow in the exciter field, the greater the exciter magnetic field. The greater this magnetic field, the greater the current flow in the exciter armature. The greater this current flow, the greater the current flow in the main rotor winding. The greater the rotor current flow, the greater the strength of the rotor's magnetic field. The greater this magnetic field, the greater the alternator output.

▶ Stator

LO 44-15 Describe the stator, end frames, fan, and pulley.

The stator consists of a cylindrical, laminated iron core. It carries the three (or four)-phase windings in slots on the inside (**FIGURE 44-43**). The windings are insulated from each other and also from the iron core. They form a large number of conductor loops, which are each subjected to the rotating magnetic fields. The stator is mounted between the two end housings. It holds the stator windings stationary so that the rotating magnetic fields can permeate the windings.

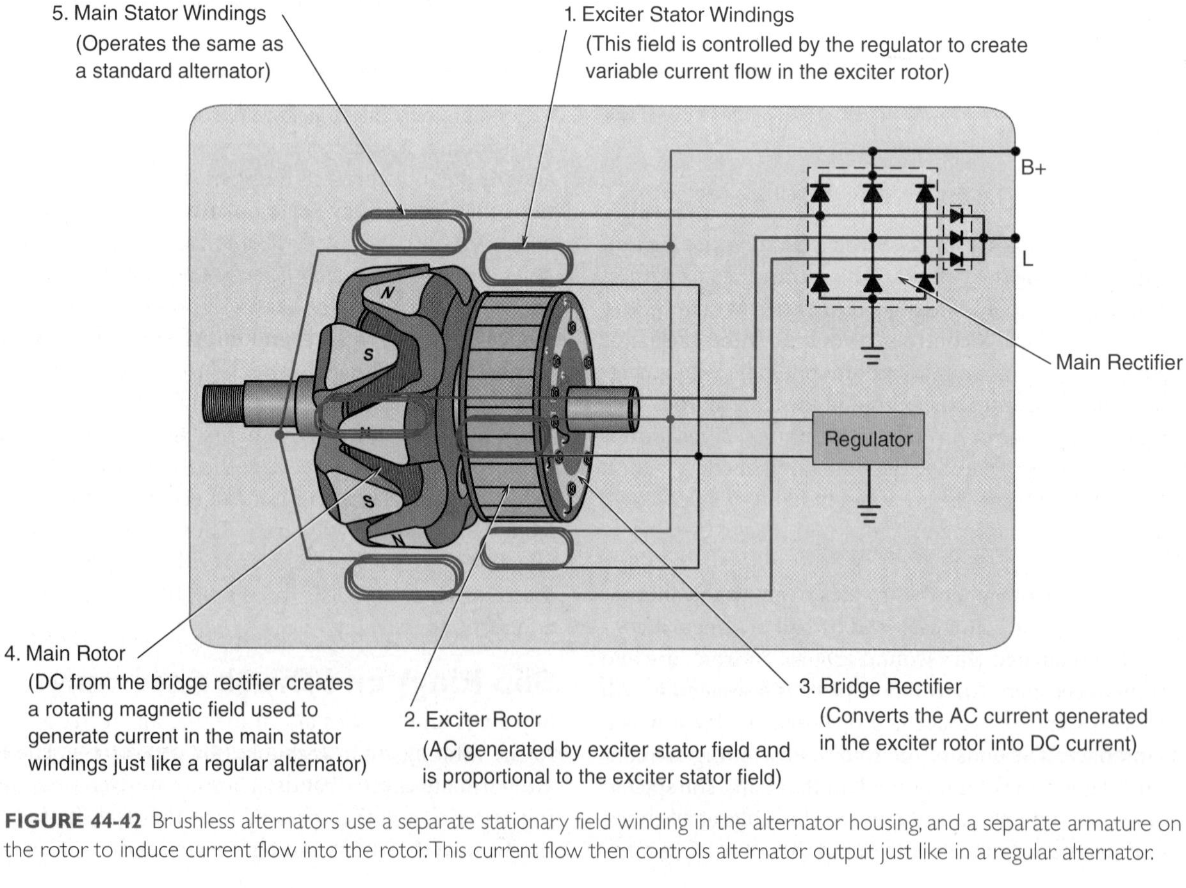
5. Main Stator Windings
(Operates the same as a standard alternator)
1. Exciter Stator Windings
(This field is controlled by the regulator to create variable current flow in the exciter rotor)
B+
L
N
S
N
S
N
Main Rectifier
Regulator
4. Main Rotor
(DC from the bridge rectifier creates a rotating magnetic field used to generate current in the main stator windings just like a regular alternator)
2. Exciter Rotor
(AC generated by exciter stator field and is proportional to the exciter stator field)
3. Bridge Rectifier
(Converts the AC current generated in the exciter rotor into DC current)

FIGURE 44-42 Brushless alternators use a separate stationary field winding in the alternator housing, and a separate armature on the rotor to induce current flow into the rotor. This current flow then controls alternator output just like in a regular alternator.

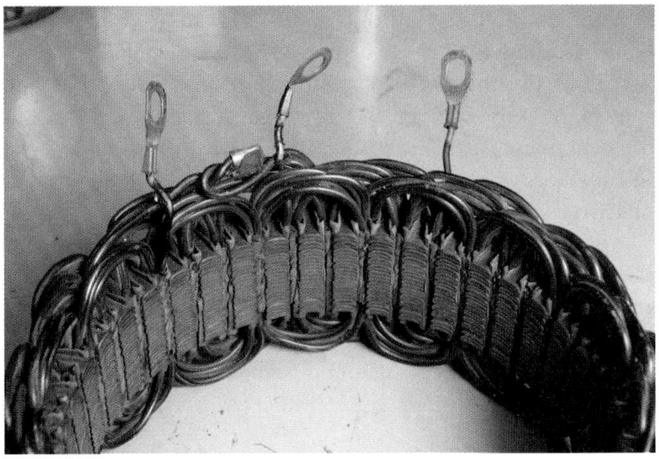

FIGURE 44-43 The stator consists of a cylindrical, laminated iron core, which carries the three-phase windings in slots on the inside.

FIGURE 44-44 End frames and bearings.

FIGURE 44-45 Alternator fan and pulley.

FIGURE 44-46 Typical overrunning alternator decoupler (OAD).

Alternator End Frames and Bearings

The alternator housings are typically constructed from aluminum. They have vents within the frames to provide for a large amount of airflow to assist in dissipating heat (**FIGURE 44-44**). The housings accept the bearing assemblies, which support the rotor at the drive and slip ring ends. A pulley that is driven by a belt is mounted at the drive end of the alternator. Some housings also accept some or all of the diodes, which are pressed into holes in the housing.

Alternator Cooling Fan and Pulley

The alternator's cooling fan is a powerful centrifugal type of fan (**FIGURE 44-45**). It is mounted on the rotor shaft and may be an integral part of the drive pulley or part of the rotor. It is essential to maintaining a cooling stream of air over the diodes and stator. Due to the short length of the rotor, the cooling air needs to be spiraled in. This creates a long cooling path for the air. Temperatures are then maintained within the manufacturer's specifications. To achieve this spiraling effect, the cooling fins on the plate have different openings—small and large. To get the maximum cooling effect, the alternator fan must be driven in the correct direction. Refer to the service information to determine the correct direction if replacing the fan.

A new feature that has been added to alternators over the past number of years is the overrunning alternator pulley (OAP). Another type is called an overrunning alternator decoupler (OAD) (**FIGURE 44-46**). They may also be known as alternator decoupling pulleys (ADPs). These devices are in place of standard pulleys on alternators. They lock in one direction to drive the alternator while the engine is accelerating or at cruise. They freewheel when the engine is decelerated. The following functions are provided:

- Reduce belt noise and vibration
- Reduce stress placed on the tensioner and belt
- Extend belt and tensioner life
- Improve fuel economy

ADPs are prone to wear just like the tensioner and belt, so they need to be inspected and replaced when faulty. It is also good practice to replace the ADP for maintenance purposes when the tensioner and belt are replaced.

To diagnose an ADP, a couple of tests can be performed. First, you can run the engine at approximately 2000 rpm. Then, turn the engine off while listening for a clicking, popping, or buzzing noise coming from the ADP. The second test involves using special tools to turn the ADP (with the belt still on) in both directions. It should turn smoothly in the freewheeling direction with less than the maximum specified torque. Also, it should *not* turn in the other direction. If any faults are found with the ADP, replace it along with the belt and tensioner. Make sure you replace the ADP with the specified replacement part. They are *not* interchangeable.

▶ Rectification

LO 44-16 Describe rectification.

Rectification is a process of converting AC into DC. DC is required by the battery and nearly all of the automobile systems. To change AC to DC, automotive alternators use a rectifier assembly. It consists of diodes in a specific configuration. Remember that a diode allows current to flow in one direction but blocks the flow of current in the other direction.

A so-called three-phase "bridge" rectifier has a minimum of six diodes (three positive and three negative). It rectifies the AC output of the stator windings to DC. A diode bridge gets its name from two diodes in series, bridged with a wire.

As the rotor rotates, the north and south magnetic fields permeate the stator windings. Each time the polarity changes, the current in the stator windings changes direction. No matter what direction the current is flowing in the stator windings, the diodes in the rectifier only allow current to flow into the rectifier and out of the rectifier in one direction (DC). For example, when winding A comes under the influence of the magnetic field, a voltage is induced in that stator winding. This generates current flow in one direction (**FIGURE 44-47**).

When the rotor's magnet rotates further, windings B and then C come under the influence of the magnetic field. The current path is then through those windings(**FIGURE 44-48**). As the rotor moves through its various positions, individual phase currents change in magnitude and polarity. But the output current remains in one direction only. This is because the individual phase windings are set 120 degrees apart (three-phase system).

Thus, two things are happening. First, the three phases are split so that as the voltage in one phase is falling, the voltage in the next phase is still rising. As the second phase is falling, the third phase is rising. And as the third phase is falling, the first phase is rising again (**FIGURE 44-49**).

Second, by redirecting the current so that it is flowing in the positive direction, the diodes effectively flip all of the negative current flow activity. The activity from below the neutral point of the graph is flipped to the positive side. This smoothes out the power flow even further (**FIGURE 44-50**).

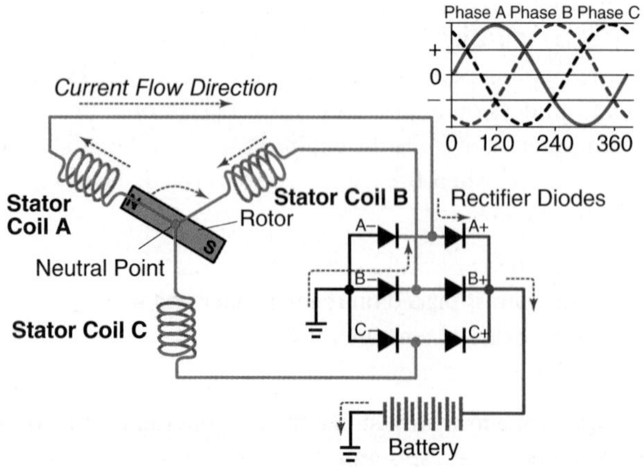

FIGURE 44-47 Current flow through a single phase in the forward direction.

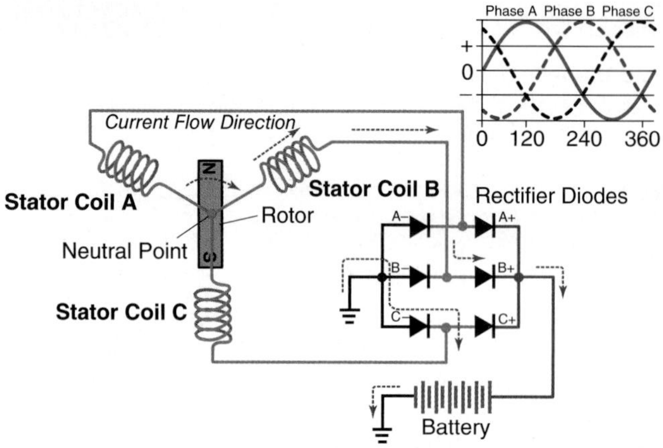

FIGURE 44-48 Current flow through a single phase in the reverse direction.

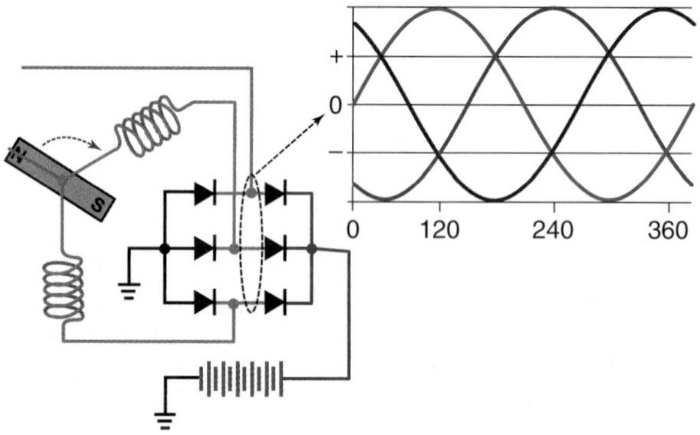

FIGURE 44-49 Three phases—not rectified.

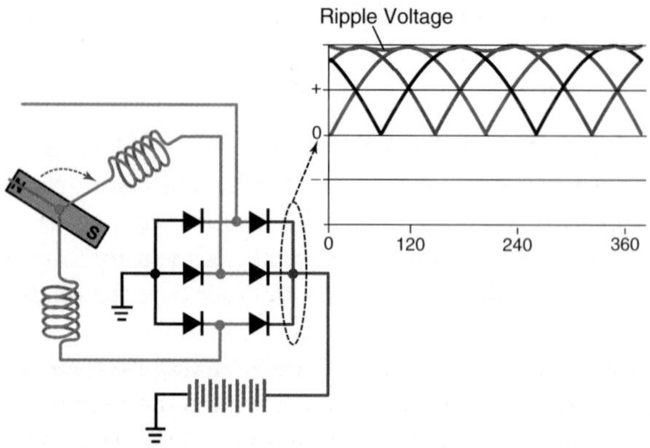

FIGURE 44-50 Three phases—rectified.

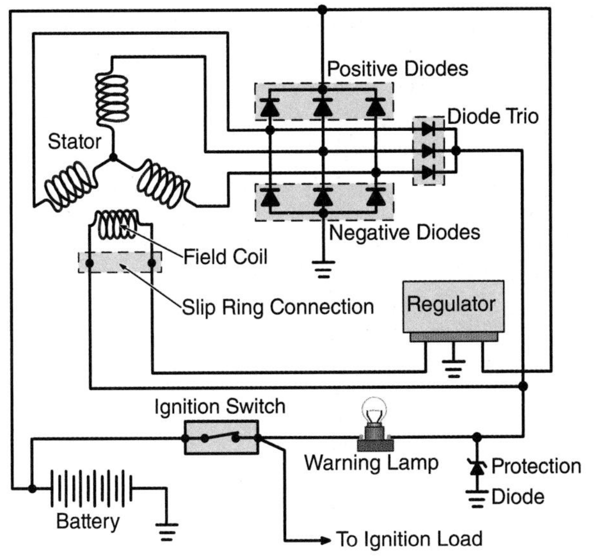

FIGURE 44-51 Positive and negative diode schematic, including diode trio.

The top of the waveform is called the alternator ripple. The ripple should be consistent across each winding. If so, this indicates that the stator windings and diodes are each creating consistent current flow and voltage. Inconsistent ripple indicates a high-resistance fault in either the diodes or the stator windings. Study the illustrations carefully. You need to understand the role the diode bridge plays in providing the relatively smooth DC output.

Rectifier Assembly

The diodes for rectification (converting AC to DC) are mounted on heat sinks. This assists in dissipating the heat generated in the diodes. Each diode drops the voltage approximately 0.7 volt and thus tends to create heat. The diodes must be properly cooled to avoid premature failure. Three or more diodes are mounted on each heat sink. One heat sink has the positive diodes, and the other has the negative diodes (**FIGURE 44-51**). The positive diode heat sink is insulated from the frame. It is connected through the output terminal to the positive battery terminal. The negative diode heat sink is connected to the frame. This provides the return circuit, via the negative battery terminal, to be completed. Some alternators use a diode trio assembly to supply power to the field windings once the alternator is charging. Each of these diodes is connected to one end of a stator winding.

▶ Voltage Regulation

LO 44-17 Describe voltage regulation.

The alternator's output is determined by the amount of current flowing through the rotor. The greater the current flow through the rotor, the stronger the magnetic field. The stronger the magnetic field, the stronger the alternator output. The weaker the magnetic field, the weaker the alternator output. The voltage regulator monitors battery voltage. It adjusts the current flowing through the rotor appropriately. This controls the strength of the electromagnet.

When voltage output is low, the regulator allows more current to flow through the rotor field winding. This increased current flow strengthens the magnetic field. This increases the output of the alternator. As the output voltage increases to the maximum regulated voltage, the voltage regulator reduces the current flow through the rotor. This reduces the strength of the magnetic field and alternator output.

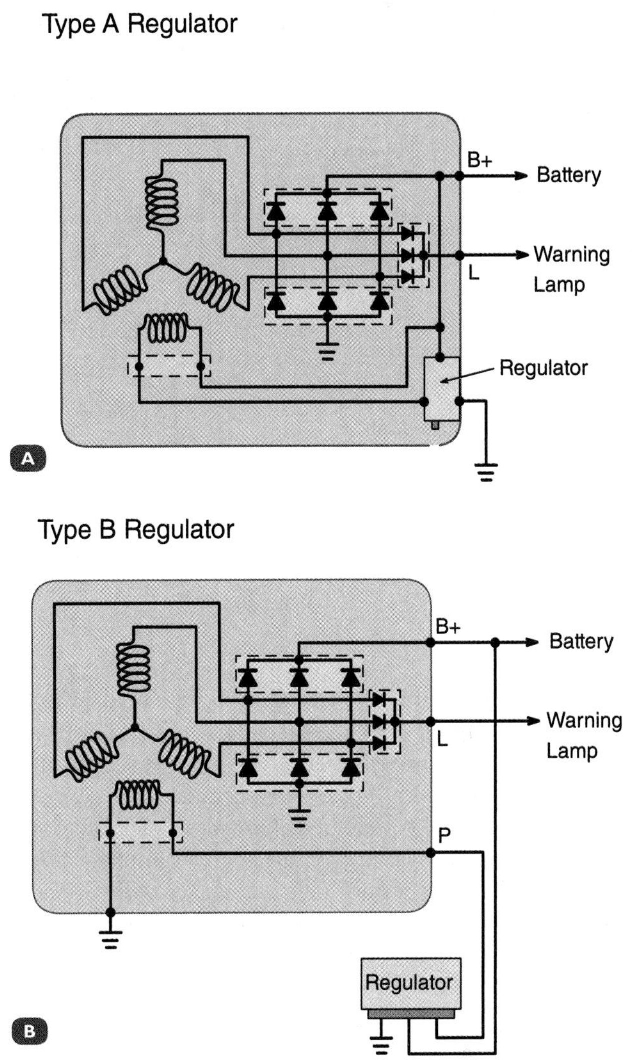

FIGURE 44-52 A. Regulator circuit—A Type. **B.** Regulator circuit—B type.

Regulator switching takes place in milliseconds. So, the voltage output is fairly constant. This process occurs fast. Fast enough to maintain consistent voltage even as loads switch on or off. Today's field circuits are electronically controlled by a pulse-width-modulated signal. This provides for even smoother regulation and quicker control.

A-Type or B-Type Regulating Circuits

The field current may be controlled on the positive side of the rotor windings (B type). It may also be controlled or on the ground side (A type) (**FIGURE 44-52**).

- In an A-type regulating circuit, alternator B+ output is fed directly to the rotor. Voltage regulation is done on the ground side of the field.
- In a B-type circuit, the voltage regulator is on the positive side of the field. The ground is constant.

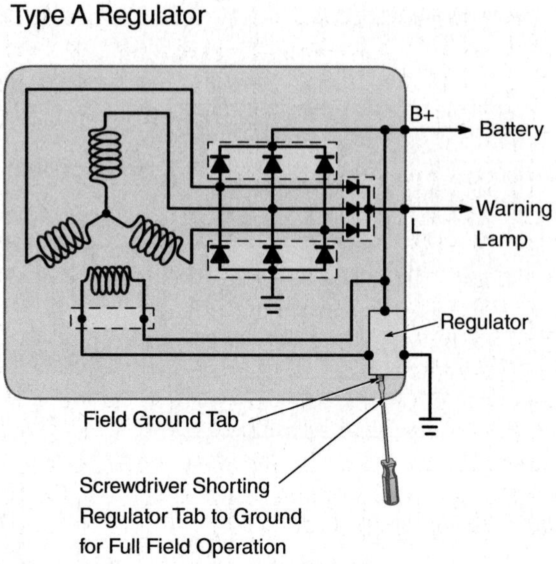

FIGURE 44-53 Full-fielding an A-type regulator circuit.

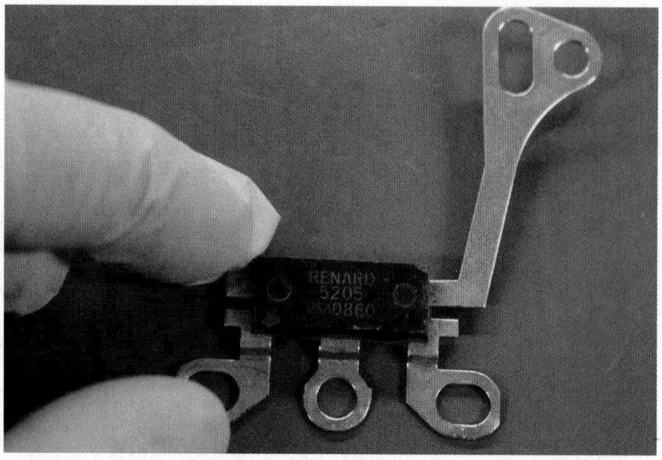

FIGURE 44-54 Diode trio supplies power to the rotor in some alternators.

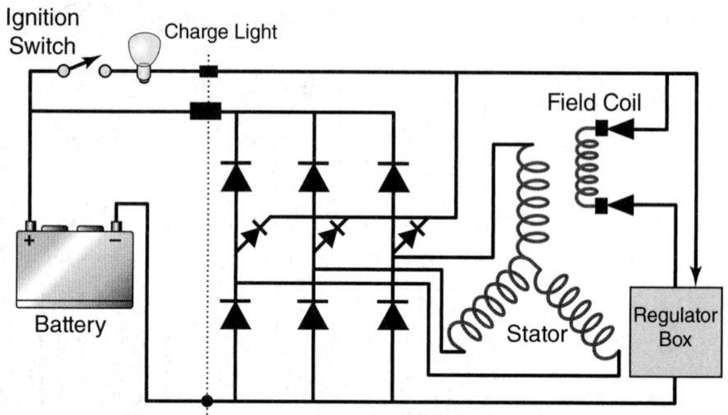

FIGURE 44-55 Current flow from the charge indicator lamp to start the alternator charging.

Knowing the difference is necessary. You will need to know this to properly bypass a voltage regulator (full-fielding) when diagnosing some charging systems. For example, many alternators are designed with A-type circuits and a grounding tab on the regulator. The grounding tab provides a method of full-fielding the charging system with the engine running. The tab is grounded by placing a small metal rod or screwdriver on the metal tab. And then shorting it out against the alternator frame (**FIGURE 44-53**). This grounds the A-type field, causing full current to flow through the rotor, creating full alternator output. *Caution:* This procedure puts the charging system in an unregulated condition. This can cause the system voltage to increase high enough to damage electronic components. Only perform this test at idle, and only for a few seconds.

If the charging system is a B type, then battery power has to be fed to the field. This is usually accomplished by first disconnecting the regulator connector. Power is then applied with a jumper lead to the correct terminal. Make sure you do not mix up the types of charging systems or supply power or ground to the wrong place. Doing so can damage the electronic components.

Rotor Circuit Control

Current is provided to the rotor by means of the copper slip rings on the rotor and brushes. The rotor can be supplied with power in a variety of ways. On older vehicles, rotors on many vehicles were supplied power by means of three extra diodes, called a diode trio. They are connected to the bridge rectifier circuit (**FIGURE 44-54**). These extra diodes are known as field diodes or exciter diodes. In this case, the alternator is said to be self-exciting. However, this self-excitation can only occur when the alternator is producing an output.

When the ignition is on, current flows from the positive battery terminal through the ignition switch. It then flows through the charge indicator lamp to the L terminal of the alternator (**FIGURE 44-55**). The circuit is completed through the slip rings and rotor field winding. Then through the voltage regulator to ground on the vehicle frame.

The small amount of current flowing in the circuit illuminates the indicator lamp. It also provides the initial excitation of the field winding. This slightly magnetizes the rotor's iron

Applied Science

AS-72: Voltage: The technician can demonstrate an understanding of and explain system voltage generation, uses, and characteristics.

Voltage generation relies on an engine-driven alternator. Consider the principle of electromagnetic induction. Alternators generate alternating current (AC) through the use of a moving electromagnet. AC electricity must be converted to direct current (DC) before being used in automotive systems. The AC electricity is "rectified," or converted to DC, by being passed through a bridge of diodes. Diodes are small solid-state devices that allow current flow in only one direction. The resulting DC electricity can power various vehicle systems. It can also charge the vehicle's battery. Generators produce DC electricity. They were once used in automotive applications, but then replaced by alternators. This happened when small, inexpensive diodes became available. Alternators are more practical, simpler to produce, and need less-intensive maintenance.

claw pole shoes. The weak magnetic field gets the generation process started.

When the rotor is driven (turned by the pulley) from the engine crankshaft, the rotating magnetic field induces a voltage in the stator phase windings. That is then fed to the bridge rectifier to create DC. From there, DC goes to the B+ alternator output terminal. Stator output is also fed to the exciter diodes so that DC current can flow to the field circuit. This strengthens the magnetic field, and the output voltage rises quickly.

The voltage regulator takes control of the field circuit. It limits field current to maintain a preset regulated output voltage of approximately 14 volts. As the voltage on each side of the charge indicator lamp is now equal (between the lamp and the diode trio), there is no current flow through the lamp. This turns the lamp off. The alternator is now charging. Because the output voltage at the B+ alternator terminal is greater than that of the battery, current flows to the battery. This begins the recharging process.

Voltage Regulator

The voltage regulator in modern vehicles is a solid-state electronic device. It may be installed inside the alternator. For older vehicles, it may be mounted on the fender well as a separate component. The internal regulator is a replacement part and is not usually serviceable (**FIGURE 44-56**). The regulator's electronic circuit senses the battery voltage and switches the rotor circuit on and off rapidly. This maintains a constant voltage output up to the alternator's maximum output current.

Modern vehicles control the alternator output voltage from within the PCM or BCM. The control principle is the same. Output voltage is controlled by switching the rotor's field circuit on and off. But, the PCM makes voltage adjustments based on a larger number of inputs. Some of these are as follows:

- engine battery, and ambient temperature,
- engine cranking,
- engine load,
- desired regeneration during deceleration,
- and electrical load.

FIGURE 44-56 Internal regulator mounted inside the alternator.

In most cases, an alternator that is PCM controlled also uses an internal regulator. The PCM works in conjunction with the regulator to control the output of the alternator. In some cases, the PCM communicates the desired charge rate to the regulator. The regulator then adjusts the current flow through the rotor appropriately. The regulator then reports the alternator load and any faults to the PCM. If any faults develop, the PCM will store one or more diagnostic trouble codes. It may also illuminate the malfunction indicator light on the instrument cluster.

▶ Charging System Output Test

LO 44-18 Perform a charging system output test.

Vehicle charging systems are voltage regulated. This means that the charging system will try to maintain a set output voltage from the alternator. As electrical current load increases in the vehicle systems, voltage starts to drop. The voltage regulator senses this voltage drop and increases the current output of the alternator. This in turn increases system voltage to try to maintain the correct voltage in the system.

The testing of an alternator output initially involves testing the system's regulated voltage. This is done using a voltmeter. Regulated voltage is the voltage at which the regulator is limiting the alternator output too. This happens once the battery is relatively charged, as evidenced by the greatly reduced current output. Regulated voltage should be between the manufacturer's specified minimum and maximum regulated voltage. If it is incorrect, verify that there are no voltage drops on the alternator and regulator. If no voltage drops are found, the regulator is likely faulty and will have to be replaced.

Once the regulated voltage is confirmed, it's time to check the charging system output. This is done by using an external electrical load. A carbon pile is commonly used to reduce the battery voltage. This tricks the regulator into full-fielding the alternator, making it produce maximum output. This output is read using an inductive ammeter. It should then be compared with the manufacturer's rated output specifications. An alternator that puts out within 10% of its rated output is okay. Less than that indicates a fault. This could be a faulty regulator, faulty alternator, or excessive voltage drop(s) on the alternator output, ground, or control circuits.

▶ TECHNICIAN TIP

Back in the day of DC generators, it was acceptable to test the charging system by removing one of the battery terminals from the battery post while the engine was running. If the engine kept running, the charging system was working. If the engine died, the charging system was not working. On alternator-equipped vehicles, doing this test is very risky, as any voltage spike caused from disconnecting the battery terminal can destroy any of the electronics in the vehicle. Thus, this test is no longer valid on virtually all vehicles on the road today.

To perform a charging system output test, follow the steps in **SKILL DRILL 44-6**.

SKILL DRILL 44-6 Performing a Charging System Output Test.

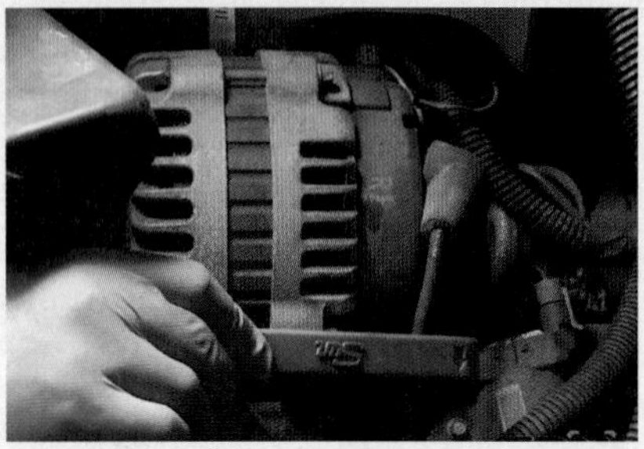

1. Connect a charging system tester to the battery with the red lead to the positive post, the black lead to the negative post, and the amps clamp around the alternator output wire.

2. Start the engine, turn off all accessories, and measure the regulated voltage at around 1500 rpm. The regulated voltage is the highest voltage the system achieves once the battery is relatively charged, as evidenced by the ammeter reading less than about 15–20 amps. Typical regulated voltage specifications are wider than they used to be because of the ability of the PCM to adjust the output voltage for a wide range of conditions. Always check the specifications.

3. Operate the engine at about 1500 rpm, and either manually or automatically load down the battery just enough to obtain the maximum amperage output without pulling battery voltage below 12.0 volts. This reading should be compared against the alternator's rated output. Normally, readings more than 10% out of specifications indicate a problem.

Applied Math

AM-1: Whole Numbers: The technician can add whole numbers to determine measurement conformance with the manufacturer's specifications.

An alternator is being tested to determine if it meets manufacturer's specifications. If an alternator is damaged due to a blown diode or similar problem, it is usually out of specifications by a wide margin. For this type of alternator, the output specification is 95 amperes. The technician tests the alternator and it puts out 65 amps. The service material states a good alternator will provide an output that is within 15 amps of its rated value. The technician adds 15 amps to the original 65 amps, for a total of 80 amps. This is below the specification of 95 amps for this type of alternator. In this example, we are working with whole numbers. If a number has a negative sign, a decimal point, or a part that's a fraction, it is not considered a whole number.

▶ Testing Charging System Circuit Voltage Drop

LO 44-19 Perform charging system circuit voltage and voltage drop tests.

An excessive voltage drop in the charging system output or ground circuit tends to cause one of two problems: (1) The battery will not be fully charged because, although the alternator is creating the specified voltage, the voltage drop is reducing the amount of voltage to the battery, or (2) the battery is fully charged, but the alternator is working at a higher voltage to do so, potentially overheating it. Which of the two issues is occurring depends on where the voltage is sensed. If it is sensed at the alternator, then the battery will generally be undercharged. If the voltage is sensed at the battery, then the alternator will work at the higher voltage. Knowing the system will help you diagnose voltage drop issues in the output and ground circuits of the charging system.

The external alternator output circuit consists of the wires and fuse. Or, fusible link between the positive battery post and the alternator output terminal. It also includes the ground wire back to the battery from the engine and chassis. Voltage drop may occur anywhere in the output current circuit and ground circuit. But is especially common at the terminals and connectors because of the high charging system current flowing through them. Even a small amount of resistance can cause significant voltage drop when conducting high current flow.

As with testing for voltage drop at the starter circuit, a DMM is used to measure voltage drop across all parts of the circuit. A voltmeter with a minimum/maximum range setting is very useful when measuring voltage drop. It will record and hold the maximum voltage drop that occurs for a particular operation cycle. Voltage drop tests are only valid when the circuit is under load. This is because voltage drops can only occur when circuits are energized. The greater the flow, the greater the voltage drop, if resistance is present. Therefore, when testing for voltage drop, always perform the test when the circuit is being operated.

To measure for voltage drop, the DMM is connected in parallel across the component, cable, or connection that is to be tested. Usually, it is most efficient to first measure the total voltage drop on both sides of the circuit. If required, narrow down the test to individual components to identify precisely where excessive voltage drop is located. For example, connect the black probe to the output side of the alternator. Connect the red probe to the positive post of the battery. Then, operate the charging system under a heavy load by turning on as many electrical

SKILL DRILL 44-7 Testing Charging Circuit Voltage Drop

1. Set the DMM up to measure voltage, and select min/max if available. Connect the red probe of the DMM to the output terminal of the alternator and the black probe on the positive post of the battery.

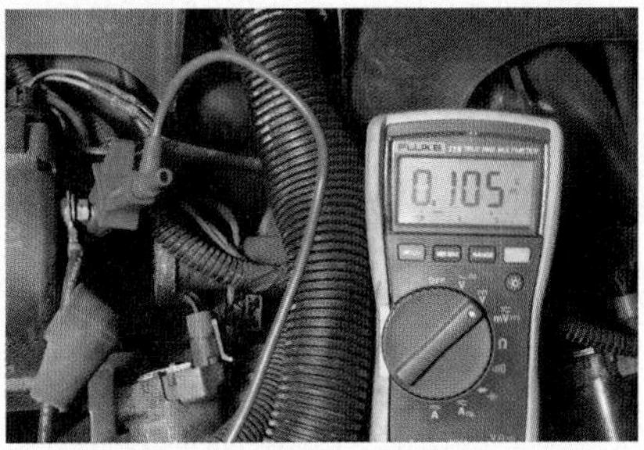

2. Start the engine and turn on as many electrical loads as possible, or use an external load bank to load the battery. Read the maximum voltage drop for the output circuit.

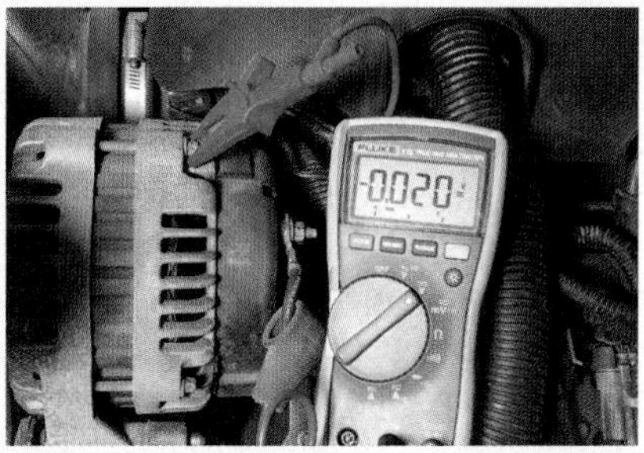

3. Move the leads to measure the voltage drop on the ground circuit by placing the red probe on the alternator case and the black probe on the negative terminal of the battery. Read the maximum voltage drop for the ground circuit. Determine any necessary actions.

items as possible. You can also use a load tester to load down the battery. Record the voltage drop across the output side of the circuit while the alternator is fully charging. Then, perform the same test on the ground side of the circuit. Perform the test from the negative battery terminal to the frame of the alternator. Many manufacturers specify the maximum allowable voltage drop. But a rule of thumb is no more than 0.5 volt (500 millivolts) for each side of a 12-volt circuit.

To perform a charging circuit voltage drop test, follow the steps in **SKILL DRILL 44-7**.

Applied Science

AS-64: Motors: The technician can demonstrate an understanding of the role of the generator/alternator in maintaining battery and system voltage.

A vehicle is in the shop for an illuminated battery light on the instrument panel. The repair order states that the customer had to jump-start the vehicle in order to drive it to the shop. An apprentice technician has been given the job of diagnosing this vehicle with the help of an experienced technician, if required. The apprentice starts by verifying the customer's concerns and finds the vehicle once again requires a jump-start. When the engine starts, the battery light on the dash remains on. Checking under the hood, the apprentice finds what appears to be a new battery. Reading further on the repair order, the customer states that he replaced the battery two days ago. The apprentice connects his DMM and obtains a reading of 10.2 volts. It is obvious to the apprentice that the alternator is not charging at all. The experienced technician performs a voltage drop test on the charging system. All wires are good, indicating that the alternator needs to be replaced.

As stated in our text, the alternator converts mechanical energy into electrical energy by electromagnetic induction. In a simple version, a bar magnet rotates in an iron yoke, which concentrates the magnetic field. A coil of wire is wound around the stem of the yoke. As the magnet turns, voltage is induced in the coil, producing a current flow. A typical charging voltage will be from 13.8 to 14.5 volts DC, which keeps the battery fully charged at 12.6 volts at 70°F (21°C).

Diagnosing Undercharge and Overcharge

Undercharging and overcharging are both bad for a vehicle. Undercharging leads to electrical systems that do not function fully. Also, the battery may not fully charge, leading to an early death due to sulfation. Overcharging can lead to short life of bulbs and other electrical devices. At the same time, overcharging the battery can increase gassing and loss of water from the electrolyte. Maintaining a proper charge is critical to long life and proper operation of the electrical system.

The voltage regulator keeps a constant voltage output from the alternator. It does so up to the maximum rated output. Always check the manufacturer's specifications for alternator regulated voltages. More and more manufacturers are using PCM-controlled charging systems. This means the voltages can have a wider range than non-PCM-controlled systems.

As a rule of thumb, the regulated alternator output voltage should be between 13.8 and 14.5 volts. This is for a typical non-PCM-controlled 12-volt charging system. If the voltage measured is outside of the manufacturer's specifications, further testing will have to be performed to determine the cause.

For this task, we assume that the charging system is charging (not a no-charge condition) and that it is either not charging fully or is overcharging. If the system is overcharging, it can almost always be tracked to either a faulty voltage regulator or a faulty regulator ground. Both of which can cause the alternator to overcharge.

An undercharge condition has many more potential causes such as:

- a loose drive belt,
- voltage drop in the charging system wiring,
- faulty regulator,
- worn brushes in the alternator,
- high resistance in the rotor,
- open or shorted diodes,
- open or shorted stator windings,
- or even a shorted cell on a battery.

Always check for the easiest and most common faults first. For example, check that the drive belts are not slipping, particularly under load. Also verify that the battery does not have a shorted cell and that its capacity is adequate.

If nothing obvious can be found wrong, then you will need to dig deeper. One way to do this is to use an oscilloscope to check the alternator ripple. The ripple indicates faulty stator windings or diodes. This would require that the alternator be replaced. If the ripple is okay and the voltage output is low, then testing for voltage drops on the output and ground circuit is a good step. If voltage drops are found, then they will have to be repaired and the system retested. If there are no voltage drops, then the fault is likely to be located inside of the alternator. A faulty regulator, worn brushes, or a faulty rotor would also require replacement of the alternator.

▶ Replacing an Alternator

LO 44-20 Replace alternator.

Alternators have to be replaced whenever they are electrically or mechanically faulty. Electrical faults include:

- no-charge,
- undercharge,
- or overcharge conditions.

Mechanical faults include worn bearings, or other internal or external mechanical damage. When replacing an alternator, always disconnect the negative terminal of the battery. When the battery is connected, battery voltage is always present at the output terminal of the alternator. If the battery negative terminal is not removed from the battery, a wrench could ground out the output terminal of the alternator. If this happened, a very large spark could occur. So always disconnect the negative battery terminal before removing the alternator. Also, check the service information about maintaining the adaptive memory, if necessary.

When replacing an alternator, check the belts, pulleys, and tensioners to make sure they are in good shape. If not, replace them as well. Don't forget to check belt tension. Even on a vehicle with an automatic tensioner. They get weak over time, and may be causing the charging system issue.

Even though most vehicle electrical systems are described as being 12 volt, they typically operate at between 13.8 and 14.5 volts. If the system is not operating in this range, it could be caused by several problems. The drive belt may have become loose. Or excessive voltage drops in the charging or voltage-sensing circuit may be the cause. If these components check out, the alternator, external voltage regulator, internal regulator, or even the built-in voltage-regulating circuit in the ECM/BCM may be faulty. Make sure you have identified the cause of the fault before removing the alternator.

To replace an alternator, follow the steps in **SKILL DRILL 44-8**.

SKILL DRILL 44-8 Replacing an Alternator

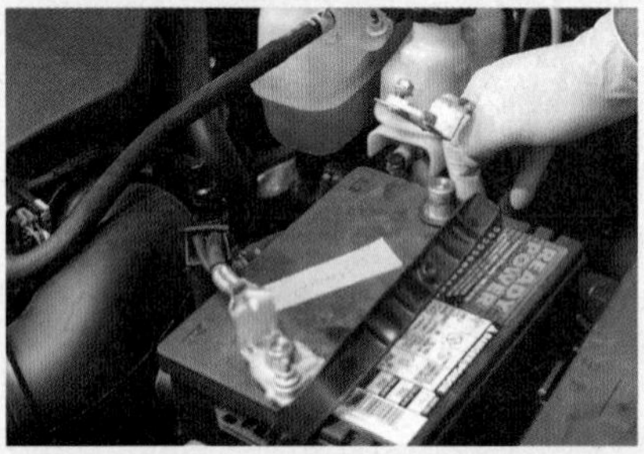

1. Install fender covers. Verify any memory issues, and remove the negative terminal of the battery.

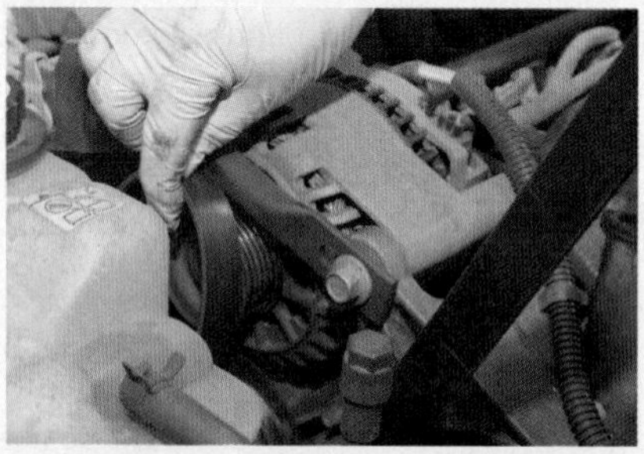

2. Loosen the drive belt and remove it from the alternator pulley. Check the condition of the belt to see if it is still serviceable.

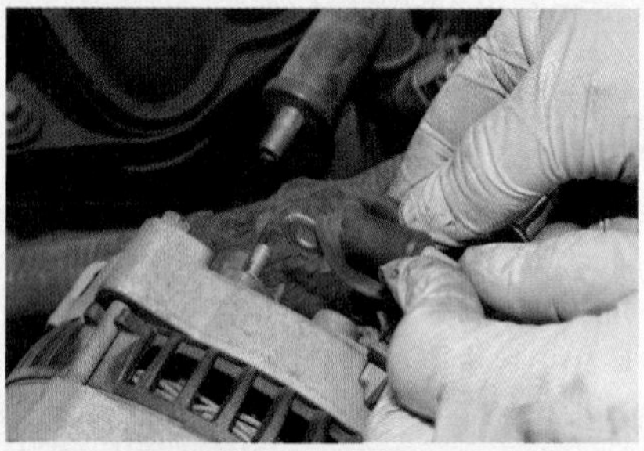

3. Locate the electrical connections at the rear of the alternator, and note their positions. Loosen any securing fasteners or covers, and remove terminals one at a time.

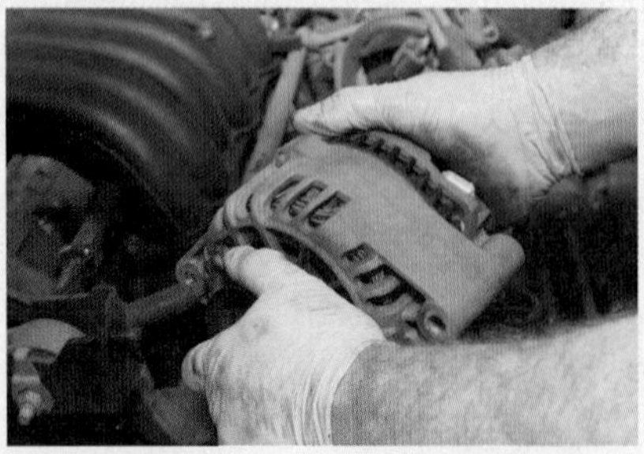

4. Loosen the securing fasteners that hold the alternator to its mounting bracket(s), making sure the alternator is supported. Remove the alternator.

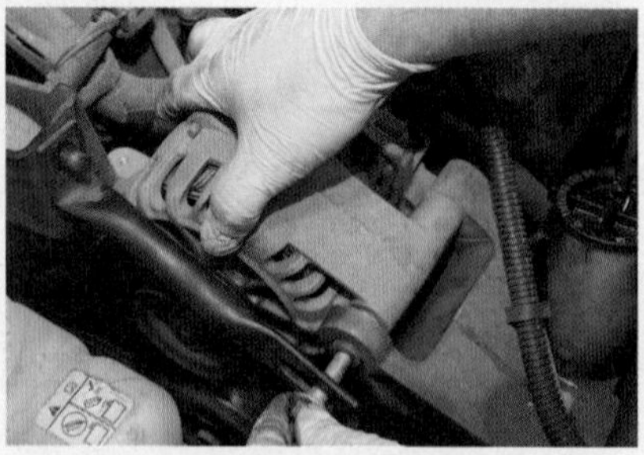

5. Reinstall the alternator. Situate the alternator in the mounting bracket(s) and, while still supporting the alternator, loosely start the securing fasteners that hold the alternator to its mounting bracket(s).

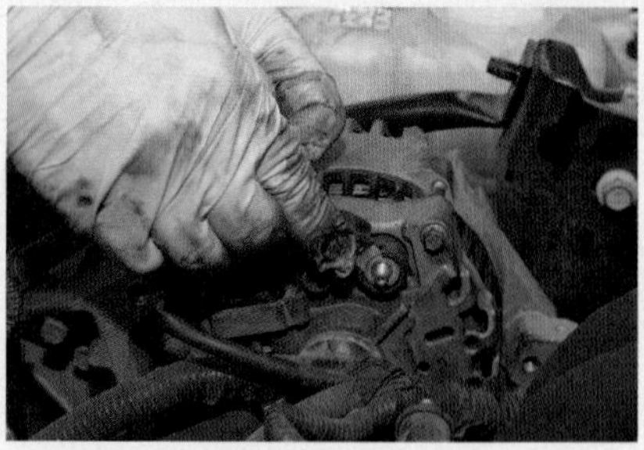

6. Reinstall the electrical wires to their correct terminals, referring to the manufacturer's information. Check the security of any fastening devices.

SKILL DRILL 44-8 Replacing an Alternator (Continued)

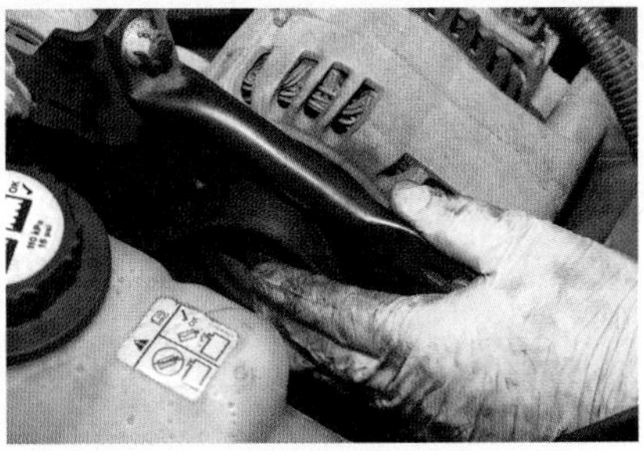

7. Install the drive belt over the alternator drive pulley and, using the correct tools, adjust the belt to the correct tension.

8. Reattach to the negative post of the battery. Make sure the fastener is tight, and replace any battery terminal covers.

9. Turn the ignition to the On position, and make sure the charge light on the instrument panel illuminates. Start the engine: the charge light should go off. Measure the regulated voltage and maximum alternator amperage output. Remove the fender covers, and return any tools used to their correct place.

▶ Wrap-Up

Ready for Review

- The starting/cranking system consists of high-amperage circuit and low-amperage circuit. The high-amperage circuit powers the starter for cranking the engine over and the low-amperage circuit is used to control the high-amperage circuit.
- A starter motor consists of field coils or large permanent magnets, an armature, a commutator, brushes, a drive pinion with an overrunning clutch, and a drive pinion engagement solenoid and shift fork.
    - Starter motors use two magnet types: electromagnetic and permanent magnet.
- Starter motor engagement is initiated by operation of the ignition switch that allows the pinion drive to rotate slightly when it is moved toward the ring gear. A gear reduction between the pinion and ring gear multiplies the torque from the starter motor to crank the engine.
- The starter drive transmits the rotational drive from the starter armature to the engine via the ring gear that is mounted on the engine flywheel, flexplate, or torque converter. The starter drive is composed of a pinion gear, an internal spline that mates with the slightly curved external spline on the armature shaft, an overrunning clutch, and a meshing spring.
- The solenoid on a starter motor switches the high current flow required by the starter motor on and off and also engages the starter drive with the ring gear. It is constructed with a pull-in winding and a hold-in winding.
- Starter control circuit can be operated only when the transmission is in Park, the clutch is depressed, the brake pedal is applied, or the proper ignition key is being used. In PCM-controlled starters, the starter control circuit is a part of the vehicle theft-deterrent system.
- The on-vehicle test of a starter motor is called a starter draw test and the off-vehicle test is called a starter no-load test. Starter current draw is at its highest when the engine is just starting to first rotate. Starter test equipment should have an inductive high-current ammeter to measure the cranking current flow.
- When testing the voltage drop on the high-current side of a starter motor, the starter must be activated. Voltage drop should be tested while the circuit is under load with the DMM connected in parallel across the component or section of the circuit that is to be tested.
- To test starter control circuit voltage drop, place the DMM's black lead on the battery positive terminal, place the red lead on the solenoid's input terminal, and measure the voltage drop with the key in the crank position.
- The simplest method to test relays is to measure the resistance of the relay winding. Solenoids tests will usually be limited to voltage and voltage drop tests on the main contacts.
    - A voltage drop test should be performed across the solenoid contacts by placing the red lead on the solenoid B+ input and the black lead on the solenoid B+ output.
- In most cases, the starter can be accessed more easily from underneath the vehicle. Disconnect the negative battery lead before attempting to remove or install a starter motor. A starter should be well supported before its removal or installation.
- Three most common methodologies used in idle–stop systems are a flywheel mounted motor/generator, an upgraded starter motor, and a system that ignites a combustible mixture in an appropriate cylinder.

- DC generators and alternators produce electricity by relative movement between conductors and a magnetic field. This induces an electrical potential or voltage within the conductors. The main parts of the charging system include the: battery, alternator, voltage regulator, charge warning light, and wiring.
- The rotor is an electromagnet that rotates freely in the alternator and is supported on each end by ball bearings. Slip rings and brushes aid in making an electrical connection to the rotating rotor assembly.
  - Slip rings are normally copper bands that are molded onto an insulating material, and then pressed onto the steel shaft of the rotor.
- The stator consists of a cylindrical, laminated iron core which carries the three (or four)-phase windings in slots on the inside. Alternator housings house the bearing assemblies that support the rotor at the drive and slip ring ends. The alternator's cooling fan is mounted on the rotor shaft and may be an integral part of the drive pulley or part of the rotor.
- Alternators use a rectifier assembly that consists of diodes in a specific configuration to convert AC into DC. The diodes for rectification are mounted on heat sinks to dissipate the heat generated in the diodes.
- Voltage regulator monitors battery voltage and adjusts the current flowing through the rotor to control the strength of the electromagnet. The voltage regulator in modern vehicles is a solid-state electronic device with an electronic circuit that senses the battery voltage and switches the rotor circuit on and off rapidly.
- A charging system output test initially involves testing the system's regulated voltage with a voltmeter. Regulated voltage should be between the manufacturer's specified minimum and maximum regulated voltage.
- A DMM is used to measure voltage drop across all parts of the charging system circuit. To measure for voltage drop across charging system circuit, the DMM is connected in parallel across the component, cable, or connection that is to be tested.
  - Voltage drop tests are only valid when the circuit is under load.
- Electrical faults of an alternator include: no-charge, undercharge, or overcharge conditions. Mechanical faults include worn bearings, or other internal or external mechanical damage. When replacing an alternator, the negative terminal of the battery should be disconnected.

Key Terms

counter-electromotive force (CEMF) An electromagnetic force produced by the spinning magnetic field of the armature, which induces current in the opposite direction of battery current through the motor.

hold-in winding The winding that is responsible for holding the solenoid in the "on" position; typically draws less current than the pull-in winding.

pull-in winding A high-current winding found in starter solenoids that pulls the solenoid plunger into the activated position.

rectification A process of converting alternating current (AC) into direct current (DC).

Review Questions

1. Which of the following is part of the starter control circuit?
   a. The high amperage side of the solenoid
   b. The starter motor assembly
   c. The ignition switch
   d. The flywheel
2. Starter motor magnet types are typically permanent magnet, and _____.
   a. Electromagnet
   b. Semi-permanent magnet
   c. Temporary magnet
   d. Part-time magnet
3. Which part of the starter assembly is responsible for moving the drive pinion gear into engagement with the ring gear?
   a. The electric motor
   b. The solenoid plunger
   c. The armature windings
   d. The one-way clutch
4. Which of the following is true when performing a starter draw test?
   a. The engine should be allowed to start up during the test
   b. The starter should be off-vehicle and placed in a test stand
   c. The engine should be disabled so it will crank but not start
   d. The spark plugs should be removed from the engine
5. All of the following are ways to test a starter relay EXCEPT:
   a. Measure resistance of the coil windings
   b. Check voltage drop across the contacts
   c. Use a 9 V battery on the winding terminals
   d. Reverse the relay position in the fuse box
6. How should a technician prevent sparks or short circuits when replacing a starter?
   a. Disconnect the negative battery terminal
   b. Remove the starter fuse
   c. Remove the starter relay
   d. Take the key out of the ignition
7. Which charging system component is responsible for maintaining the optimal battery state of charge?
   a. The stator windings
   b. The charge warning light
   c. The voltage regulator
   d. The slip rings and brushes
8. Which component is an electromagnet that rotates freely inside an alternator?
   a. The brush assembly
   b. The slip ring
   c. The rectifier
   d. The rotor

9. If a technician uses a screwdriver to short out the shorting tab on the back of an alternator, what should happen?
   a. The circuit breaker should trip to protect the circuit
   b. The alternator should charge at full output
   c. The engine should stall out and not restart
   d. The charging light should turn on in the instrument panel
10. Which of the following is most likely to cause an overcharging condition?
   a. A loose or slipping drive belt
   b. Worn brushes in the alternator
   c. Voltage drop in the alternator output circuit
   d. A faulty voltage regulator

ASE Technician A/Technician B Style Questions

1. A starter motor is being discussed. Technician A states that if a conductor is placed so it cuts across a magnetic field, the conductor will be forced out of the magnetic field. Technician B states that the commutator reverses the current flow through the armature as it spins. Which Who is correct?
   a. Technician A only
   b. Technician B only
   c. Both Technicians A and B
   d. Neither Technician A nor B
2. A starter solenoid is being discussed. Technician A states that the starter solenoid switches the high current flow to the starter motor on and off. Technician B states that the solenoid is controlled by the PCM on some vehicles. Who is correct?
   a. Technician A only
   b. Technician B only
   c. Both Technicians A and B
   d. Neither Technician A nor B
3. A starter control circuit is being discussed. Technician A states that the neutral safety switch is used to prevent the engine from being started in gear. Technician B states that automatic transmission equipped vehicles use a clutch pedal switch as part of the starter control circuit. Who is correct?
   a. Technician A only
   b. Technician B only
   c. Both Technicians A and B
   d. Neither Technician A nor B
4. Testing high current starter circuit voltage drop is being discussed. Technician A states that the battery should be disconnected for this test. Technician B states that typically the starter cable should not drop more than 0.5 volts during cranking. Who is correct?
   a. Technician A only
   b. Technician B only
   c. Both Technicians A and B
   d. Neither Technician A nor B
5. Testing the starter control circuit is being discussed. Technician A states that the voltage tests on the circuit should be performed with the key in the crank position. Technician B states that a problem in the starter control circuit often causes the starter to crank slowly. Who is correct?
   a. Technician A only
   b. Technician B only
   c. Both Technicians A and B
   d. Neither Technician A nor B
6. The idle stop/start system is being discussed. Technician A states that the purpose of start/stop systems is to make the starter last longer. Technician B states that some start/stop systems don't use a starter motor to crank the engine over. Who is correct?
   a. Technician A only
   b. Technician B only
   c. Both Technicians A and B
   d. Neither Technician A nor B
7. An alternator is being discussed. Technician A states that the stator is the rotating part connected to the pulley. Technician B states that some alternators have an overrunning pulley that can be turned on or off by the PCM. Who is correct?
   a. Technician A only
   b. Technician B only
   c. Both Technicians A and B
   d. Neither Technician A nor B
8. Alternators are being discussed. Technician A states that the purpose of a rectifier is to convert AC to DC. Technician B states that a rectifier consists of a total of 3 diodes. Who is correct?
   a. Technician A only
   b. Technician B only
   c. Both Technicians A and B
   d. Neither Technician A nor B
9. A charging system output test is being performed. Technician A states that the alternator output involves measuring the system voltage with a volt meter. Technician B states that a carbon pile tester can be used to trick the alternator into charging at full output. Who is correct?
   a. Technician A only
   b. Technician B only
   c. Both Technicians A and B
   d. Neither Technician A nor B
10. An alternator is being replaced. Technician A states that the battery negative terminal must be disconnected to prevent a possible short circuit when removing the output terminal. Technician B states that belt tension needs to be checked as part of replacement. Who is correct?
   a. Technician A only
   b. Technician B only
   c. Both Technicians A and B
   d. Neither Technician A nor B

CHAPTER 45

Lighting Systems

Learning Objectives

- LO 45-01 Describe the purpose of the lighting system.
- LO 45-02 Describe the types of lights.
- LO 45-03 Describe light bulb configurations.
- LO 45-04 Describe park, tail, marker, and license lights.
- LO 45-05 Describe driving, fog, and cornering lights.
- LO 45-06 Describe brake and backup lights.
- LO 45-07 Describe turn signal and hazard lights.
- LO 45-08 Describe headlights and head light systems.
- LO 45-09 Describe lighting system testing and precautions.
- LO 45-10 Perform peripheral lighting service.
- LO 45-11 Perform headlight service.

ASE Education Foundation Tasks

See Appendix A to view the 2017 ASE Education Foundation Automobile Accreditation Task List Correlation Guide.

▶ Lighting System Introduction

LO 45-01 Describe the purpose of the lighting system.

Well-designed vehicle lighting systems enhance vehicle safety. They do so by increasing the driver's visibility while operating the vehicle. They also signal the driver's intent to those around the vehicle (**FIGURE 45-1**). Lighting systems are also used inside the vehicle. They indicate messages to the driver and provide convenience to any occupants. Manufacturers continue to improve their lighting systems as new technologies are developed. Modern lighting systems include the increased use of:

- light-emitting diode (LED),
- electronic body control units to manage lighting,
- and high-intensity discharge (HID) lamps and xenon lighting. These are much brighter than the traditional sealed-beam or halogen units.

Many vehicles also have systems that warn you if lamps are not working properly.

FIGURE 45-1 Lighting systems not only increase the driver's visibility, they also signal the driver's intent to others.

Lighting Systems

Lighting systems improve the driver's visibility at night. They also make a vehicle visible to other road users. The headlight switch activates taillights, park lights, and headlights. They allow the driver to see the road ahead. A beam selector switch allows the driver to change the beams from high to low, or vice versa.

Brake lights operate when the brake pedal is depressed. On some vehicles, they operate when the brake control module automatically applies the brakes. Red or amber turn signals alert other drivers of a change in direction. They are mounted so they can be seen from the front, the rear, and sometimes the sides of the automobile. The emergency flasher system operates both front and rear turn signals at the same time as a warning to others. Other circuits operate courtesy, or convenience, lights, backup (reverse) lights, fog lights, and fault indicators. We will cover each of these lighting systems later in this chapter.

▶ Types of Lamps

LO 45-02 Describe the types of lights.

Modern vehicles use many different kinds and sizes of lamps. These are known in some places as lightbulbs or light globes. There are several lamp types available. These include:

- standard incandescent lamps,
- halogen lamps,
- vacuum tube fluorescent (VTF) lighting,
- HID xenon gas systems,
- LEDs, and more.

Conventional incandescent lamps are being replaced in many applications by these other more efficient types of lights.

You Are the Automotive Technician

A customer brings a 2014 Dodge Avenger into the dealership for a 60,000-mile service checkup. You take the vehicle back to the service bay. Part of the 60,000-mile service is to check the lighting and peripheral systems. Your coworker operates each of the lights while you check the front lights and then the rear lights. You notice that the center high-mount stop light (CHMSL) on the trunk lid is not functioning, but the brake lights are working normally. The customer agrees to allow you to diagnose it.

After referring to the wiring diagram, you see that the CHMSL is on the same fuse and brake switch as the brake lights. So those components do not need to be investigated further. Based on your reading of the wiring diagram, you determine that the problem is either:

- between the brake switch and CHMSL or
- between the CHMSL and the ground, or
- the CHMSL bulbs or unit are faulty.

You test the power and ground at the CHMSL terminals with the brake pedal applied. You find 12.2 volts available. This indicates that the CHMSL is faulty. It is a sealed LED-style assembly, so it will have to be replaced. The customer approves the repair; you install the new CHMSL and verify that the light is working properly. You review the work and invoice with the customer, thank them for their business, and set them up in the automated email system. This will send a reminder for the next scheduled maintenance visit.

1. Why didn't you have to check the fuse or brake switch?
2. What did the 12.2-volt reading at the CHMSL terminals tell you?
3. How do wiring diagrams save time?
4. How would you locate the fault if the voltage drop across the CHMSL was 8.3 volts?

Incandescent and Halogen

Incandescent lamps consist of one or more tungsten filaments. The filaments heat up to approximately 5000°F (2760°C) and glow white-hot (**FIGURE 45-2**). The filament material does not burn because there is no oxygen in the bulb. Most of the oxygen has been replaced by inert gases that stop combustion from occurring. The power in watts consumed is often marked on the lamp. Wattage (work performed) is found by multiplying the voltage used by the lamp by the current flowing through it. The higher the wattage, the more light that can be created when compared with a similar type of bulb. Incandescent bulbs are inefficient, converting only about 10% of the electricity to visible light.

Halogen lamps are another type of incandescent lamp. They are filled with a halogen gas such as bromine or iodine (**FIGURE 45-3**). These lamps have a much longer life and are generally brighter than standard incandescent bulbs. They also produce more light per unit of power consumed. However, they become very hot in use. They are manufactured from highly heat-resistant materials. The bulbs must be handled carefully because they can be damaged even by fingerprint residue.

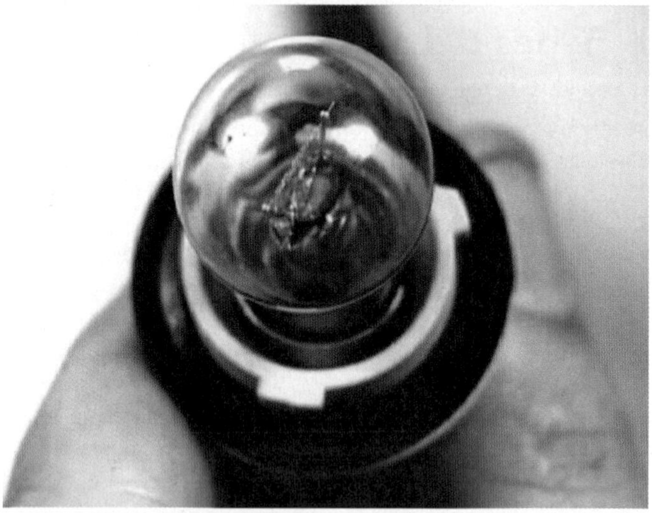

FIGURE 45-2 Incandescent bulb with single filament.

Light-Emitting Diode

Light-emitting diodes (LEDs) are not new in automotive applications. Examples include warning indicators and alphanumeric displays. More recent developments have seen the production of a wider range of LED colors. They are also becoming much brighter than previous types. It is now possible to get LEDs that emit bright red, green, blue, yellow, and clear or white light. This has made it possible to use LEDs for many new applications, such as more general lighting applications. For example, LEDs are now often used for stop lights, turn signals, and interior lighting (**FIGURE 45-4**).

One of the advantages of LEDs is that they turn on instantly. This is particularly useful in brake lights. In fact, LEDs can reduce the brake light response time by two-tenths of a second. This translates to an extra 16' (4.9 meters) of stopping distance for vehicles traveling at highway speeds. LEDs also have better visibility in inclement weather. They operate at cooler temperatures, consume less energy, and are much smaller. And they can last up to 100 times longer, reducing the cost of repair. LEDs can be specifically designed for use in LED lighting units. They can also be used as LED replacement bulbs for more traditional bulb sockets.

When used in automotive lighting, several LEDs may be required to provide a specified amount of light. To do so, they are usually connected in groups called series strings. A number of series strings are then connected in parallel. In this way, they can give off the required amount of light.

A typical LED has a voltage drop of 1.2 to 3.5 volts across it depending upon the color. LEDs work best when the voltage to them and the current flow through them remain constant at a preset level. There are two main ways to achieve this. The first is via a resistor connected in series with the LED. The second

FIGURE 45-3 Halogen bulb.

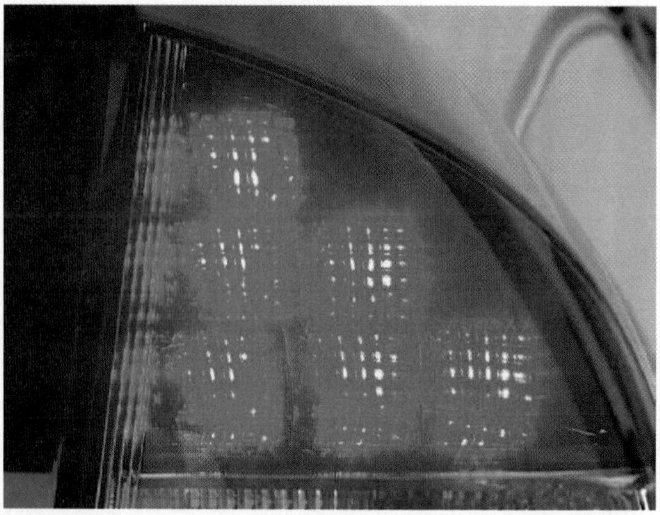

FIGURE 45-4 LED lights.

and more preferred way is through the use of a voltage regulation circuit.

Some LED lights are multivoltage, which means they can work on both 12- and 24-volt systems. These lights are normally used in aftermarket products, which can be installed in a wide range of vehicles.

Vacuum Tube Fluorescent

Vacuum tube fluorescent (**VTF**) lighting is also called a vacuum fluorescent display (VFD). They are used for instrumentation displays on instrument panel clusters. This type of lighting emits a very bright light with high contrast. They can display various colors (**FIGURE 45-5**). VTFs usually display:

- bar graphs,
- seven-segment numerals,
- multi-segment alphanumeric characters,
- or a dot-matrix pattern.

VTF displays include different kinds of alphanumeric characters and symbols to alert drivers of various conditions.

High-Intensity Discharge

High-intensity discharge (**HID**) headlamps produce light with an electric arc. This is instead of a glowing filament (**FIGURE 45-6**). The high intensity of the arc comes from metallic salts that are vaporized within an arc chamber. HIDs produce more light for a given level of power consumption than ordinary tungsten or halogen bulbs. Automotive HID lamps are commonly called xenon headlamps. However, they are actually metal halide lamps that contain xenon gas. The light from HID headlamps exhibits a distinct bluish tint. This is compared with the yellow-white color of tungsten-filament headlamps.

HID headlamp bulbs do not run on low-voltage direct current. They require a **ballast** with an internal or an external igniter. It is either integrated into the bulb or included as a separate unit, called a ballast. The ballast increases the voltage substantially. It also controls the current to the bulb.

HID headlamps typically produce between 2800 and 3500 lumens of light. They do this while using between 35 and 38 watts of electricity. Comparatively, halogen headlamp bulbs produce between 700 and 2100 lumens of light. They use between 40 and 72 watts of electricity. As you can see, HID headlight systems offer substantially greater light output than halogen bulbs. If the higher output HID light source is used in a well-engineered headlamp lens, the driver gets more usable light. Studies indicate that drivers react sooner to roadway obstacles when using HID headlamps. Therefore, good HID headlamps contribute to driving safety.

The contrary argument is that HID headlamps can negatively impact the vision of oncoming traffic. The increased intensity makes it easy to blind oncoming traffic. This is especially true if the headlights are aimed improperly. Additionally, there are scientific studies showing the impact of HID headlamps. The light from HID headlamps is 40% more glaring than the light from standard headlamps. These issues pose increased risks of a head-on collisions with blinded drivers.

Some countries mandate that HID headlamps may only be installed on certain vehicles. The vehicles (except motorcycles) must then be equipped with lens-cleaning systems (**FIGURE 45-7**). This reduces glare. They also must have automatic self-leveling

FIGURE 45-6 HID headlamp assembly.

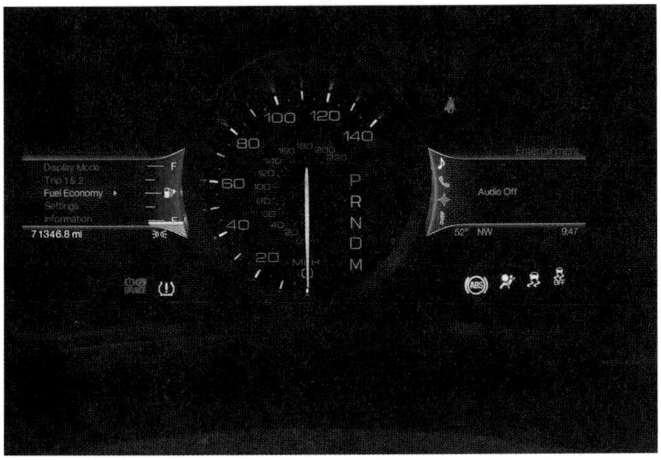

FIGURE 45-5 Typical vacuum fluorescent display.

FIGURE 45-7 HID headlight with lens cleaning system.

systems, which prevents dazzling oncoming traffic. These systems are usually not present on non-HID equipped vehicles. If a halogen headlamp is retrofitted with an HID bulb, excessive amounts of glare may be produced.

Another disadvantage of HID headlamps is that they are much more expensive to produce, install, and repair. However, some of this cost is offset by the longer lifespan of the HID burner.

▶ Lamp/Lightbulb Configurations

LO 45-03 Describe light bulb configurations.

All lamps or lightbulbs have letters and numbers on them (**FIGURE 45-8**). They typically indicate the part number, operating voltage, and power consumed. For instance, in a bulb marked 12V/21W, the filament will consume 21 watts of power when 12 volts is applied across the filament. The wattage is not necessarily an indication of light output. But it can generally be assumed that the higher the wattage, the greater the light output. This is at least true when comparing the same types of lights with each other.

Lamps and lightbulbs come in a variety of configurations to fit the various applications within a vehicle. One designation is how many filaments the bulb has. Single-filament bulbs are common for use as courtesy lights, dash lights, and warning lights. Dual-filament bulbs have two filaments of different wattage (**FIGURE 45-9**). One filament emits a small amount of light. The second filament emits more light. These bulbs work well as a combination taillight and brake light. The taillight emits less light and the brake light emits more light.

Headlights can also be dual-filament bulbs. In some cases, the low-beam filament is lower wattage than the high-beam filament, but not always. In a headlight, the filaments are positioned to give a different profile of light. Low beams emit light closer to the vehicle and angled slightly toward the side of the road. Whereas high beams tend to focus farther down the road and straight ahead.

Another feature that differs among lights is the type of base on the lamp. In other words, what type of socket it is retained in. Bayonet-style bulbs have been around for a long time. They get their name from the two retaining pins on the side of the base (**FIGURE 45-10**). The pins follow slots in the side and bottom of

FIGURE 45-8 Bulbs have identifying numbers on them.

FIGURE 45-9 Dual-filament bulb.

FIGURE 45-10 Bayonet-style bulbs.

the socket. At the bottom, the slots turn sideways into a small pocket. The pins are retained in the pocket by the spring-loaded base in the bottom of the socket. This pushes the bulb upward and locks it in place. This design resists vibration very well.

Removal of the bulb requires carefully pushing the bulb in, rotating it slightly counterclockwise, and pulling it out. One or two electrical contacts are built into the bottom of the bulb's base. If the bulb is a dual-filament bulb, it will have two contacts on the base. Also, the pins will be of unequal height. This is so the bulb contacts will be registered properly with the contacts in the socket.

SAFETY TIP

Bayonet-style glass bulbs use very thin glass. Many a technician has suffered severe cuts from them. This occurs when pushing the bulb in while removing or installing it. If the glass bulb shatters, it can cut your fingers down to the bone. Always use a doubled-up shop rag when replacing these bulbs.

Many newer bulbs use a wedge base either made from the glass bulb itself or with a built-in plastic base (**FIGURE 45-11**). The bulbs are pushed straight into the socket, and tension from the socket retains the bulb. The electrical contact on the glass-based bulbs is made by wires extending from the base of the bulb and bent over opposite sides of the wedge.

Some dome lights use festoon lights, which have a base on each end of a cylindrical lightbulb (**FIGURE 45-12**). Each end of the filament is connected to one of the bases. Generally, the bases fit in spring steel contacts, which hold the lightbulb securely in place.

▶ Types of Lighting Systems

LO 45-04 Describe park, tail, marker, and license lights.

There are many different lighting systems or circuits. Each system is designed to perform its specific roles. For example, turn indicators, stop lights, taillights, courtesy lamps, and headlamps all perform different roles. We will explore each of these systems below.

Regulations govern the lighting systems. They regulate many things. These include lamp locations, color, and brightness. Regulations make sure there is consistency in the use of lighting on vehicles. This reduces confusion and improves safety while driving. Also, lighting systems that are well maintained improve road safety for all drivers. Check lighting regulations before changing the vehicle's lighting systems.

Park/Tail/Marker/License Lights

Park, tail, and marker lights are all low-intensity or low-wattage bulbs. They are used to mark the outline or width of the vehicle. Park and tail lamps tend to be installed close to the corners of the vehicle. Park lamps are placed at the front of the vehicle, and taillights in the rear. In some cases, park lights are incorporated into the headlight assembly. Park lamps are yellow or white in color.

Park lights are used at night when the vehicle is parked on the side of the road. Park lights operate when the light switch is moved to the park light position. For safety reasons, park lights and taillights continue to operate when the light switch is moved to the headlight position. The bulbs are connected in parallel with each other. This allows the rest of the bulbs in the circuit to operate if one bulb burns out.

Tail lamps are red and located at the rear of the vehicle (**FIGURE 45-13**). They are not as bright as stop lights, but they are usually installed in a cluster assembly with them. Government regulations control the height of the lamps and their brightness. Tail lamps are connected in parallel with each other. Tail and park lamps may use separate fuses. If one circuit fails, the other will continue to operate.

Marker lights are used to mark the sides of some vehicles. They are often located down the sides of the vehicle or trailer. They can also be located on the front and rear fenders. On newer vehicles, they are sometimes placed on side-view mirrors. Other

FIGURE 45-11 Wedge-style bulbs.

FIGURE 45-12 Festoon-style bulbs.

FIGURE 45-13 Tail lamps are red in color like brake lights, but not nearly as bright.

FIGURE 45-15 License lights are designed to illuminate the license plate, without seeing the bulb.

FIGURE 45-14 Marker lights on an SUV.

times, they are placed between the front and the rear doors on large SUVs or pickups (**FIGURE 45-14**).

Red marker lamps face toward the rear. Yellow marker lamps face toward the front of the vehicle. These lights are designed to work when the park lights or headlights are selected. They sometimes operate as turn signals, so a driver can warn others of the intent to switch lanes or turn a corner.

License plate lamps produce a white light. They are designed to illuminate the lettering on the license plate at night. They are positioned so that the bulb itself cannot be seen from the rear (**FIGURE 45-15**). The bulbs are connected in parallel to each other and the taillights. So they operate whenever the taillights are on.

▶ TECHNICIAN TIP

In many vehicles produced before the mid-1960s, front park lights turn off when the headlights are on. The thinking was that if the headlights are on, then the park lights were unnecessary. The problem arose when a headlight failed. Only one light illuminated on the front of the vehicle, mimicking a motorcycle. This gave the impression of the vehicle being narrower than it is. Designing the park lights to stay on with the headlights helped to prevent this unsafe situation.

On computer-controlled lighting systems, the park lights and taillights are controlled by the body control module (BCM). The lights are supplied with power and ground through a networked light controller. The controller is connected to a network/bus system (discussed later in this chapter) by the twisted pair of communication wires. The park light switch is hardwired to a module. When the park light switch is activated, the module sends a park light request out on the network. The appropriate controllers then pick up the message. Power or ground is then sent to the appropriate lightbulb filaments in order to illuminate the park lights.

▶ Driving Lights

LO 45-05 Describe driving, fog, and cornering lights.

Vehicle headlight systems include the use of driving lights (**FIGURE 45-16**). They provide high-intensity light over long distances. They do this better than standard headlight systems. Vehicle regulations specify the positioning and lens configuration of driving lights. Local regulations must be followed when adding or changing driving lights.

Many types of driving lights are available. They come in different sizes, shapes, lens pattern, and bulb wattage. In some instances, a single driving light can be installed to suit particular applications. But driving lights are normally installed in pairs.

Most driving lights use quartz halogen bulbs in the 55- to 120-watt range. The quality of the reflector is extremely important in driving lights to get the best performance. Driving lights are wired so that they operate only when the high beam is operating. When the headlights are switched from high beam to low beam, the driving lights turn off. This ensures that oncoming traffic is not blinded by excessive light.

FIGURE 45-16 Vehicle with factory driving lights.

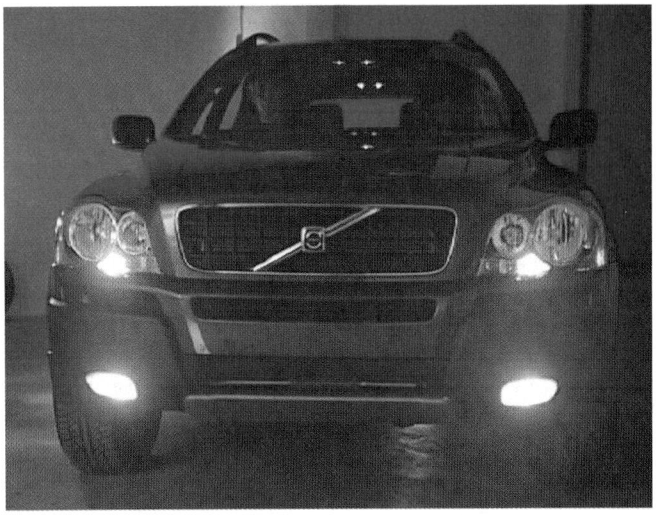

FIGURE 45-17 Vehicle with factory fog lights.

Although many performance vehicles come equipped with driving lights, they can be added to almost any vehicle. If they are, a relay and circuit breaker should be used for circuit protection reasons. And always follow local regulations.

Fog Lights

Fog lights are used with other vehicle lighting in poor weather such as thick fog, driving rain, or blowing snow. Fog is made up of water droplets suspended in the air. Therefore, it can reflect light from the headlights back into the driver's eyes at night. In such conditions, fog lights can help drivers see farther ahead at reasonable speeds. They are used with park lights and low-beam headlights, but not with high beams.

Most older fog lights have yellow-colored reflectors. However, more recently, white fog lights have become more widely used. This is because yellow lenses reduce fog light brilliance by about 30%. Fog lights typically use quartz halogen bulbs and are available in different shapes and sizes. Fog lights are usually mounted lower than headlights to avoid reflection. They tend to be aimed straightforward and low (**FIGURE 45-17**). Fog light lenses have a sharp cutoff pattern. Most of the light projected remains below the driver's eye level.

Fog lights are typically wired with a relay and circuit breaker. The method of connection of fog lights depends on local regulations. They may be wired to work only with park lights and to turn off when headlights are used. Or they may be wired to work when low beams are used. The BCM normally controls the function of the fog lights if they are installed as original equipment.

Cornering Lights

To improve visibility during night driving, some vehicles have cornering lights (**FIGURE 45-18**). Cornering lights are white lights usually installed into the bumper or fender. They are designed to provide side lighting when the vehicle is turning corners. The additional lighting helps the driver to see the curb and any obstacles that may not be illuminated by the headlights. Cornering lights turn on only when the headlights and turn signal switches are both on. They turn off automatically when the turn signal cancels. Even though they come on with the turn signals, their light is steady. That is, they don't blink.

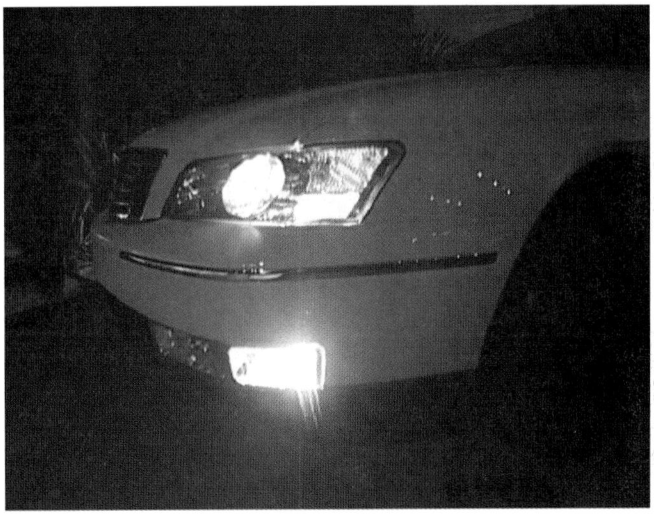

FIGURE 45-18 Cornering lights.

▶ Brake Lights and CHMSL

LO 45-06 Describe brake and backup lights.

Brake lights, which may also be called stop lights, are red lights mounted to the rear of the vehicle. They are usually incorporated in the taillight cluster. Many vehicles by law now have a higher additional third brake light. It is mounted on top of the trunk lid or near the rear window. This light is called CHMSL, or "chimsul" (**FIGURE 45-19**). The brake lights are activated whenever the driver operates the foot brake to slow or to stop the vehicle. They may also come on when a control module automatically applies the brakes.

FIGURE 45-19 Center high-mount stop light (CHMSL).

FIGURE 45-20 Typical backup lights.

Some vehicles may use the same bulbs for the brake lights and the turn signals. If so, then the turn signal flashes the brake light when needed. We will discuss this further in the turn signal section. Other vehicles use separate bulbs for the turn signals and brake lights. Today, no matter which type of brake and turn signal arrangement, both are computer-controlled by the BCM. This occurs when the computer sees an input from either the brake pedal switch or turn signal switch.

Backup Lights

The backup lights are also called reverse lights. They are white lights mounted at the rear of a vehicle (**FIGURE 45-20**). They provide the driver with vision behind the vehicle at night. They also alert other drivers to the fact that the vehicle is in reverse. They come on any time the key is in the run position and the gear selector is in reverse. Modern vehicles use network/bus systems and the BCM to command the backup lights to come (**FIGURE 45-21**).

Backup lights are the only white lights visible on the rear of the vehicle. Because they operate only in reverse, other drivers can tell that the driver is backing up. This is why it is illegal to have broken tail/brake/turn signal lenses in the rear of the vehicle. White light would be visible, confusing other drivers.

▶ Turn Signal Lights

LO 45-07 Describe turn signal and hazard lights.

Turn signal indicators are located on the extreme corners of the vehicle. They are usually amber in the front and can be either red or amber in the rear. A column-mounted switch, operated by the driver, directs a pulsing current to the indicator lights on one side of the vehicle or the other. These pulsing lights warn other road users of the driver's intended change of direction.

Once activated, they continue until the switch is canceled. This is done either by the operator or by a canceling mechanism in the switch. The canceling mechanism returns the switch to

Applied Science

AS-33: Refraction: The technician can demonstrate an understanding of refraction as it occurs in systems that employ fiber optics.

A vehicle is in the shop for repairs to the lighting system. The repair order lists the customer concern as the console-mounted shift indicator is not working. The driver is unable to determine which gear is selected at night.

Wendy is the apprentice technician who was given the assignment of solving this concern. The first step is to verify the customer concern. She turns on the headlights and sees no light while moving the shift lever. Wendy looks up the service information on a shop computer. Expecting to find an illustration showing the location of the bulb, she finds the procedure for removal of the illumination control. She then begins the disassembly procedure to get to the components involved. Rather than a conventional light socket and bulb, Wendy finds a fiber-optic lighting assembly. It provides illumination for the shifter indicator.

Fiber optics components have been used by Mercedes-Benz, BMW, Daimler, Audi, Porsche, Volvo, and Cadillac. Some of the applications include dashboard lights, interior lights, taillights, and entertainment systems.

Refraction is defined as the change in direction of light due to a change in speed as it passes from one medium to another. A reference for the path of light is described with respect to the normal path of light. If light enters a new medium and is slowed down, it bends to the normal path. If it speeds up, it bends away from the normal path.

Aftermarket lighting accessories are available. They include fiber-optic taillights and reverse lights for certain vehicles. These lights are advertised to be considerably brighter than stock units. They also light up faster and run cooler. Due to the use of fiber rather than lightbulbs, fiber-optic lights are said to be vibration-proof.

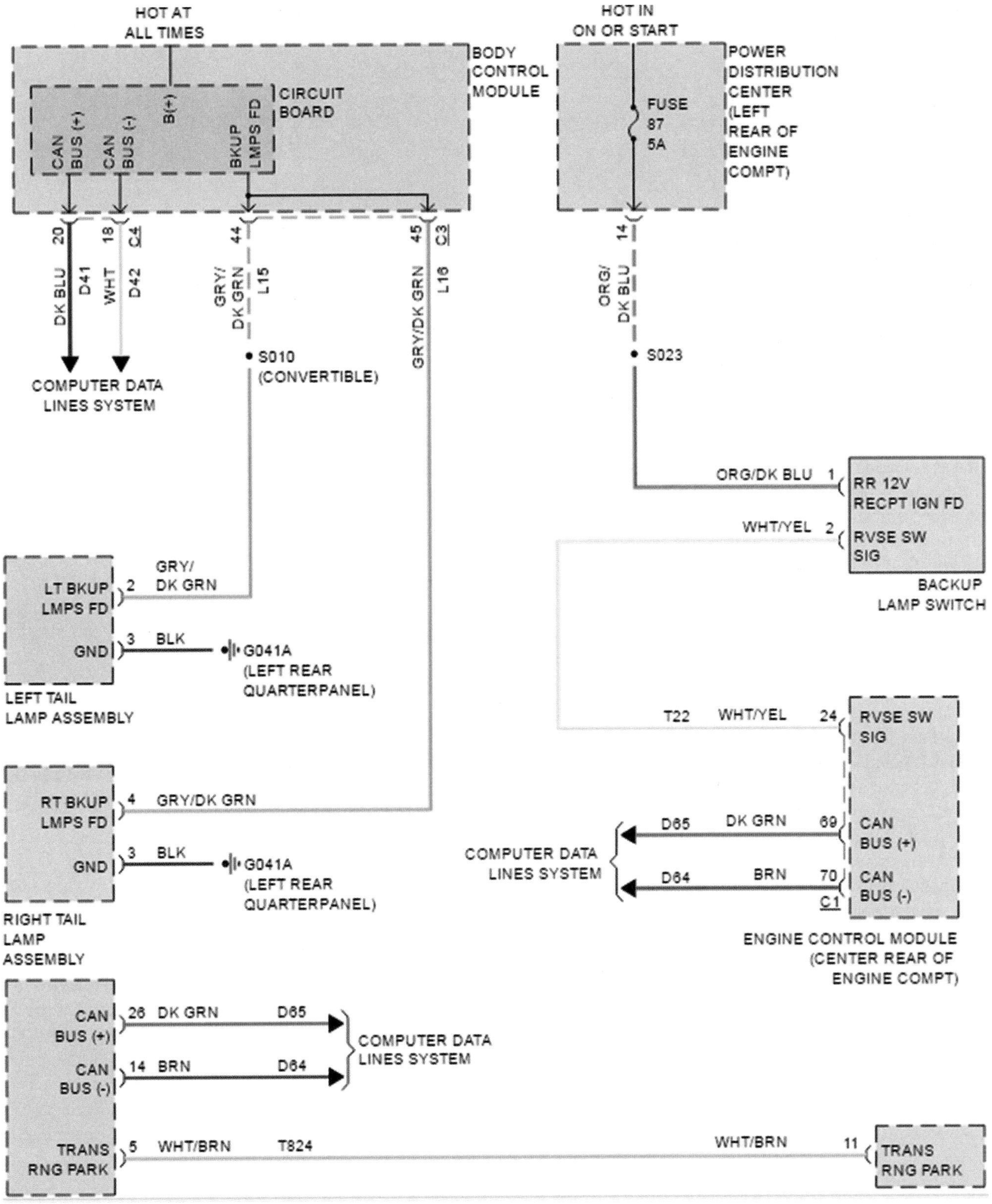

FIGURE 45-21 Typical wiring diagram for a CAN/Bus backup light circuit.

its Off position after a turn has been completed and the steering wheel is returned to the straight-ahead position.

The pulsing current is created by a flasher unit. It uses a timing circuit to pulse the current flowing out of the flasher unit 60 to 120 times per minute. This pulsing current is directed through the indicator switch to the indicator lights at the front and rear of the vehicle. This causes the lamps to flash on and off. An indicator light on the instrument cluster also blinks in

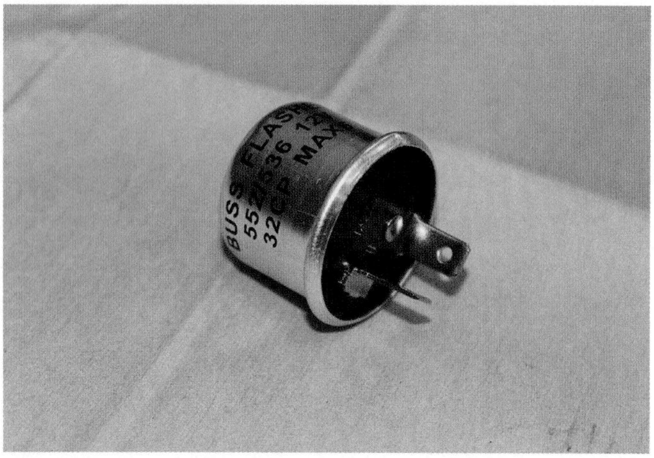

FIGURE 45-22 Typical flasher can.

FIGURE 45-23 Turn signals—Imported style flashes the individual yellow light.

sync with the turn signals. The operation of the flasher unit also produces a clicking sound. It audibly informs the driver that the indicators are in operation.

Older vehicles used a thermo-mechanical flasher unit (**FIGURE 45-22**). It relied on heat from the current flow to cause the flasher unit to work. It is very important to use bulbs of the proper wattage on all types of flasher units. Otherwise, the speed of the flash may be incorrect if incorrect bulbs are used.

These indicators can also be computer controlled. The BCM commands the appropriate turn indicators to come on as it sees an input from the turn signal switch. The BCM causes them to flash it at the proper rate. The computer turns off the turn signals when the steering angle sensor signals the steering wheel is being centered. It can also cancel the turn indicators if the vehicle is driven for a programmed amount of time or distance without the steering wheel being turned. Often a chime will alert the driver if the turn indicator has been left on for too long. On many computer-controlled turn signals, the turn signals will not work at all when the computer senses the wrong amperage flow in the circuit.

Turn Signal Lights—Domestic and Import Systems

All turn signal lamps flash, but there are variations in layout and design. This is most noticeable when comparing some domestic and imported vehicles. Imported vehicles tend to have separate amber-colored turn signal lamps on both the front and the rear of the vehicle (**FIGURE 45-23**). Some domestic vehicles use the rear brake lamps as turn signals. They flash the brake lamp on one side to indicate the turn (**FIGURE 45-24**).

In such cases, current for the brake lights must flow first from the brake switch through the turn signal switch. It then flows to the individual brake lights. This design tends to make following the current path more complicated (**FIGURE 45-25**). This is because there are two inputs on the turn signal switch. One for the turn signals and one for the brake lights. So the turn signal switch is much more complicated on a domestic-style turn signal system.

FIGURE45-24 Turn signals—Domestic style flashes the brake light.

Hazard Warning Lights

All modern vehicles are equipped with hazard warning lights. This circuit connects into the turn indicator lights. It pulses all exterior turn indicator lights and both indicator lights on the instrument panel (**FIGURE 45-26**). Hazard lights can warn other road users that a hazardous condition exists. Or that the vehicle is standing or parked in a dangerous position on the side of the road. The hazard lights also use a flasher unit that can be a separate unit or the same as that used for the turn signals. The BCM may control the hazard lights. It can do so under two conditions. The first is when it sees an input from the hazard switch, indicating the driver activated it. And second, when a signal from the restraints control module indicates a vehicle crash.

▶ Headlights

LO 45-08 Describe headlights and head light systems.

Headlights are built into the front of a vehicle. They illuminate the road ahead of the vehicle when driving at night or in conditions of reduced visibility. In headlights, most vehicles require both a high beam and a low beam. The beams are created by

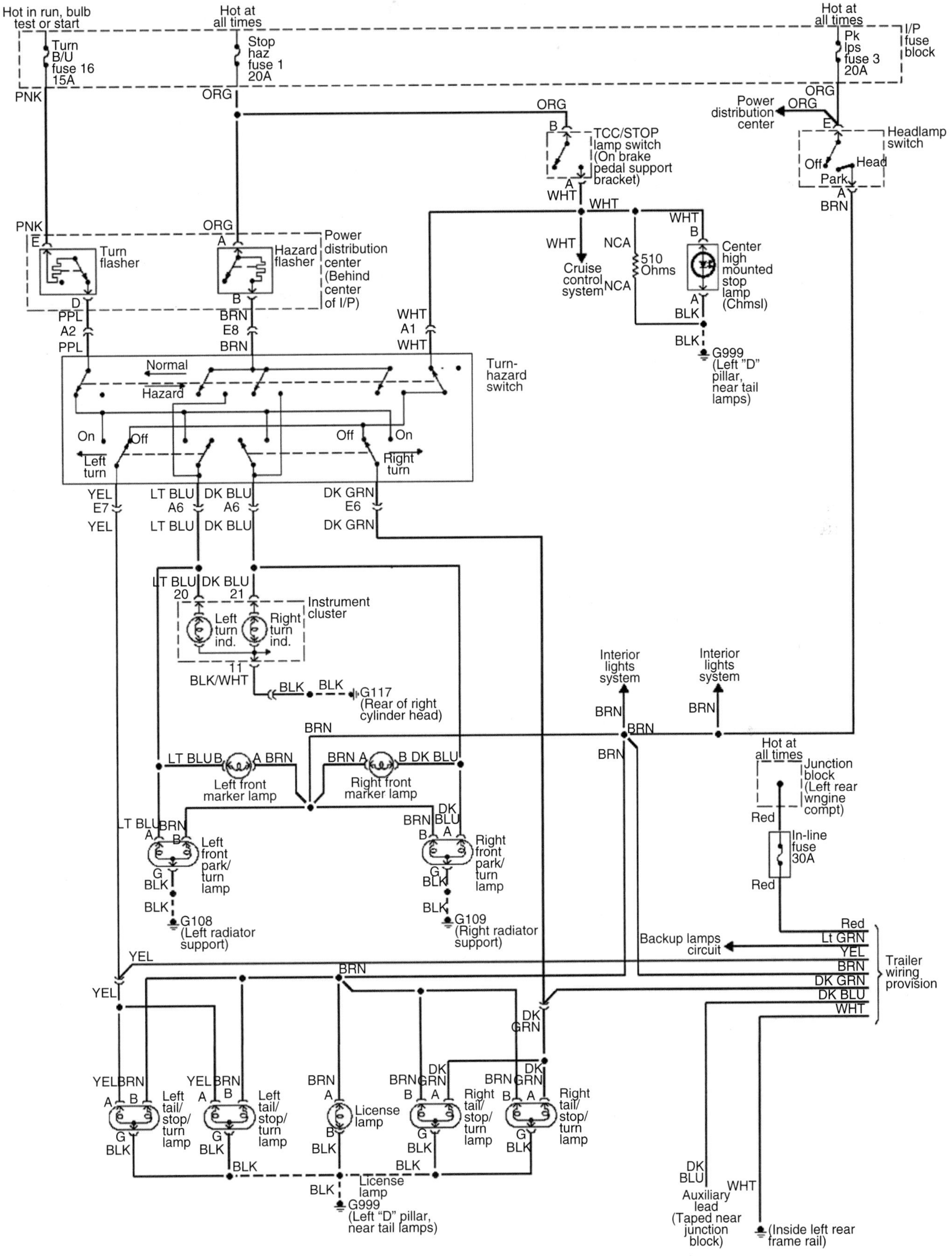

FIGURE 45-25 Typical wiring diagram showing domestic-type turn signals.

FIGURE 45-26 Import-style turn signal circuit with integrated hazard warning lights.

FIGURE 45-28 Low-beam filament position in relation to the reflector.

FIGURE 45-27 High- and low-beam filament position in relation to the reflector.

separate filaments. They can be included either in one lightbulb or placed in separate bulbs.

These filaments must be positioned correctly in relation to the highly polished reflector. This is called focusing and is carried out when designing the light assembly for the vehicle. The high-beam filament is precisely positioned. It projects the maximum amount of light forward and parallel to the road (**FIGURE 45-27**). This light is then shaped by the lens, which is made up of many small glass or plastic prisms fused together. These prisms bend the light horizontally and vertically to achieve the desired light pattern.

The low-beam filament is often placed above and slightly to one side of the high-beam filament. This produces a beam of light that is projected slightly downward and toward the curb (**FIGURE 45-28**). With this arrangement, the high-beam filament produces the most concentrated light output. Whereas the low-beam filament gives a downward and dispersed beam that is less likely to blind oncoming drivers.

A beam selector switch selects between low beam and high beam. The switch is typically located on the steering column. However, older vehicles had the switch located on the floor (**FIGURE 45-29**). It was operated by foot. In many cases, the beam selector switch operates relays that send power to the appropriate lights. On body-controlled lighting, the beam selector switch sends a signal on the network. The appropriate module reacts to the signal and commands the appropriate lights to turn on or off.

Types of Headlights

A sealed-beam headlight has a highly polished aluminized glass reflector that is fused to the optically designed lens. It is a completely sealed unit that has the filaments accurately positioned in relation to the reflector. Most older vehicles used one of the following:

- two large 7" (178 mm) round lamps,
- four small 5.25" (133 mm) round lamps,
- two large rectangular lamps,
- four small rectangular lamps.

Regardless of size, when a filament fails in a sealed-beam light, the whole sealed unit must be replaced.

A semi-sealed beam headlight uses a replaceable bulb with a prefocus collar. The collar locates the bulb in the headlight. It also controls the correct positioning of the filaments to the reflector and lens (**FIGURE 45-30**). Some replaceable headlight bulbs have a partial shield below the low-beam filament. This shield prevents light from the filament from striking the lower part of the reflector. If it did, it would be reflected higher than the midpoint of the lamp. The shield provides the primary shape of the low beam.

The final shaping of the beam is carried out by small cylindrical prisms in the headlight lens. They provide a low beam that is asymmetrical. The asymmetrical lens pattern causes light to be thrown upward at a 15-degree angle on the curb side. This helps to illuminate objects, persons, or animals close to the road.

An alternative to a reflector-type lighting system is a projection-type headlight system. This type of headlight often has a smaller front lens. However, it produces a high-intensity

FIGURE 45-29 A. High-beam switch as part of the turn signal switch. **B.** Foot-operated beam selector switch.

forward beam. It uses a lens system, rather than the traditional reflector system, to project the light forward (**FIGURE 45-31**). A projector-style light can use a standard incandescent bulb or, more commonly, an HID light.

HID lights use light from an electric arc rather than heating up a filament until it glows (**FIGURE 45-32**). High voltage is applied to tungsten electrodes. Xenon gas inside the bulb is then ionized. This creates an electrical path between the electrodes, which lowers the resistance of the gap. As the temperature rises, metallic salts are vaporized. They provide a stable arc, emitting much light.

To initially jump the gap, high voltage is needed. A ballast and igniter raise the vehicle's low-voltage direct current (DC) to as much as 25,000 volts alternating current (AC). Once the light reaches full operation, not as much voltage is needed. So voltage is maintained between approximately 40 and 85 volts

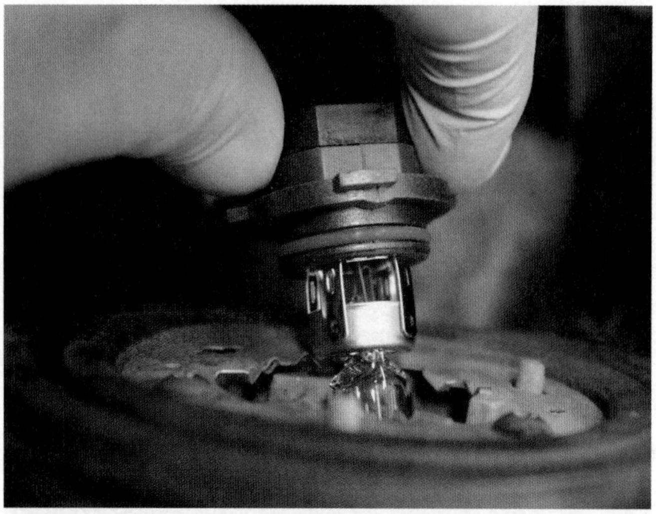

FIGURE 45-30 Replaceable halogen bulb.

FIGURE 45-32 HID headlight assembly.

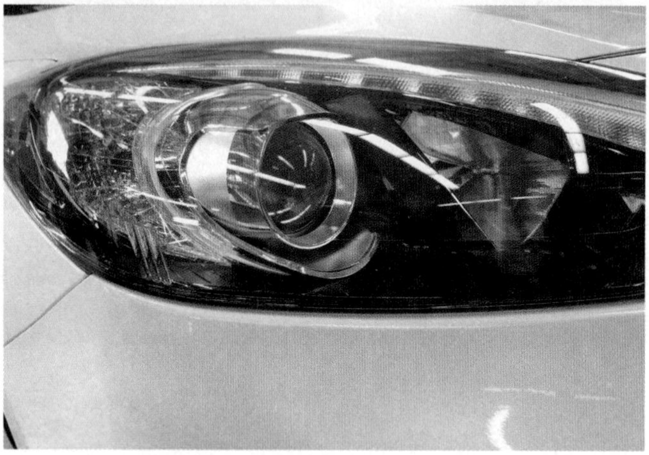

FIGURE 45-31 Projector bulb assembly.

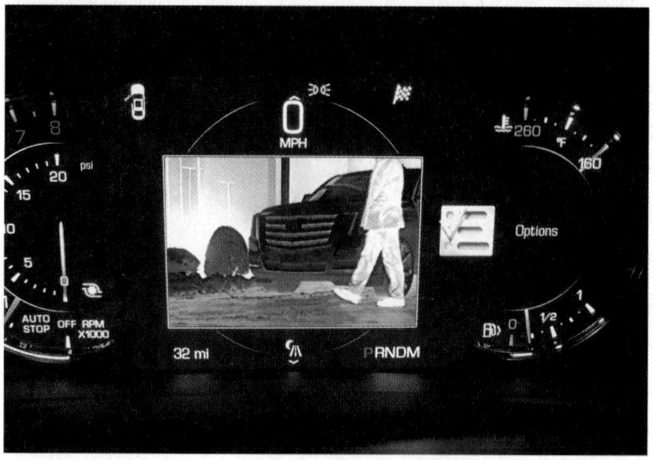

FIGURE 45-33 Night vision enhances visibility under dark or poor weather conditions.

AC, depending on the system. HID lights give off a brighter and bluer light than halogen bulbs. They do so with less electrical energy, so they help fuel efficiency slightly. They last two to four times as long as a halogen bulb but are quite a bit more expensive. Also, they take up more room in the engine compartment.

Some vehicles use LED lights as headlights. They are not currently as bright as HID lights but are at least as bright as halogen lights. LED lights are known for low power consumption. But for LEDs to create enough light to function as headlights, they must consume almost as much energy as HID lights. Still, LED technology continues to improve. One drawback of current LED headlights is that high temperatures degrade or damage them. Thus, heat sinks and cooling measures are needed, which add complexity, cost, and space.

Night vision is another relatively new technology. It enhances a driver's visual perception in dark or poor weather conditions. There are two types of night vision systems: active and passive. Active systems use an infrared light generator that projects infrared light in front of and to the side of the roadway ahead. A special camera picks up the reflected infrared radiation. The image is then displayed either on the windshield, using a heads-up display, or on an LCD screen on the dash or navigation system (**FIGURE 45-33**).

Passive night vision systems use a heat-sensing camera (thermal imaging). It picks up thermal radiation emitted by objects. This system does not have an infrared light source on the vehicle. The captured thermal image is then displayed either on a heads-up display or on an LCD screen. One benefit of the thermal system is that it can be programmed to recognize pedestrians and animals. It can then either place an outline around them on the display or flash a warning symbol.

Daytime Running Lights

Daytime running lights (DRLs) are an additional safety feature designed to improve the vehicle's visibility to other drivers. They use existing lights that turn on when the vehicle is running and turn off when the engine stops. The system operates the headlights in the front, which are typically operated at about a 60% power level. They provide light without excessively decreasing bulb life or using full electrical power. The system operates the taillights on the rear.

DRLs are mandatory on modern vehicles licensed in Canada and some other countries. In the United States, DRLs are permitted, but not required. Their use is somewhat controversial. First, if they are too bright, they can cause daytime glare. Second, they tend to mask the visibility of turn signals, making it harder for other drivers to determine a vehicle's intent. Overall, they have not been proven to reduce accidents or increase safety. They also require energy to operate. This reduces fuel economy and increases carbon dioxide emissions.

▶ Lighting Circuit Testing and Service

LO 45-09 Describe lighting system testing and precautions.

Each lighting circuit has particular operating characteristics. They are based on its purpose and the system design. Knowing how a circuit is designed to operate and how it interacts with other related circuits is critical. It will help you to better know how to go about diagnosing the system. Manufacturers design the circuits to operate in specific ways. They then create schematics or wiring diagrams, which are a diagrammatic layout of the entire circuit. Learning how to read wiring diagrams takes time and practice. Understanding wiring diagram information can then be used to diagnose circuit faults.

Lighting System Wiring Diagrams

The layouts of electrical circuits and their components are shown as diagrams. The diagrams are made up of symbols and connecting lines. Being able to read a wiring diagram is probably the most important skill when diagnosing an electrical fault. For detailed explanation of wiring diagrams, see the Electrical Components and Wiring Diagrams chapter.

Not all wiring diagrams use the same symbols or the same numbering system. It is helpful to refer to the service information for specific details on how to read their own wiring diagrams. Some of the common symbols and what they represent are shown in **TABLE 45-1**.

HID Safety Precautions

HID lamps produce a very bright white light. Manufacturers use various designs of HID lamps. Regardless of the design, they generally require a very high-voltage spark of up to 25,000 volts to start the arc. They tend to have a high operating voltage (e.g., 40–85 volts AC) to maintain light. To generate the voltages required to operate, a transformer and electronic circuitry called a ballast are used.

Several safety precautions should be taken when working on HID systems. There is a risk of electrocution, burns, or shock from the high voltages generated by the HID system. If diagnosing the HID system, be very careful when working on the system when it is live. You should wear safety glasses, high-voltage safety gloves, and safety boots. You should ensure that the vehicle, engine compartment, and ground under the vehicle are dry. Do not touch the ballast while it is operating. It will often generate a lot of heat. Persons with active electronic implants, such as heart pacemakers, should not work on HID headlamps. If changing out the bulb, make sure the headlights are turned off. The manufacturer may specify that the battery be disconnected when replacing the bulb. If so, follow the manufacturer's instructions.

There is also a risk of injury caused by exposure to ultraviolet light produced by the HID lamp if the lamp is operated outside of its housing. Once ignited, the pressure inside an HID bulb can build up to a very high pressure (around 220 pounds per square inch[1517 kilopascals]). This is due to the high operating temperature (about 1500°F [816°C]). This pressure creates a potential explosion hazard. So do not attempt to power an HID bulb outside of the headlamp assembly to test it. Nor should you operate it near flammable gases or liquids. Also, the bulb must be maintained in a horizontal position when it is on. Otherwise, it may overheat and fail.

HID headlamps use various heavy metals in their construction. Therefore, it is important to always dispose of the bulbs in an environmentally friendly way. Avoid breaking bulbs, as there is also a risk of poisoning caused by inhalation or skin contact of heavy metal vapors and toxic salts.

▶ Courtesy Lights

LO 45-10 Perform peripheral lighting service.

Courtesy lights or lamps are used to provide ambient lighting in the cabin (**FIGURE 45-34**). They are usually low-intensity or low-wattage bulbs. They can be overhead, in the doors, glove compartment, and the trunk. Door and latch switches control the various lamps for the trunk and glove compartment. Courtesy lights are usually controlled by the vehicle body computer. It monitors inputs from the ignition and door switches which are either in the handle, latch, or door pillar. Courtesy lights may be designed with timing circuits. They allow for the cabin lamps to stay on for a short period after all the doors are shut or the vehicle is locked.

Checking Lighting and Peripheral Systems

When checking lighting and peripheral systems, be sure to work in a systematic manner to avoid missing a faulty bulb or other component. A vehicle may have warning lights that activate only if that circuit is in use. You may need to turn that circuit on

TABLE 45-1 Common Symbols

Battery and fuse symbols

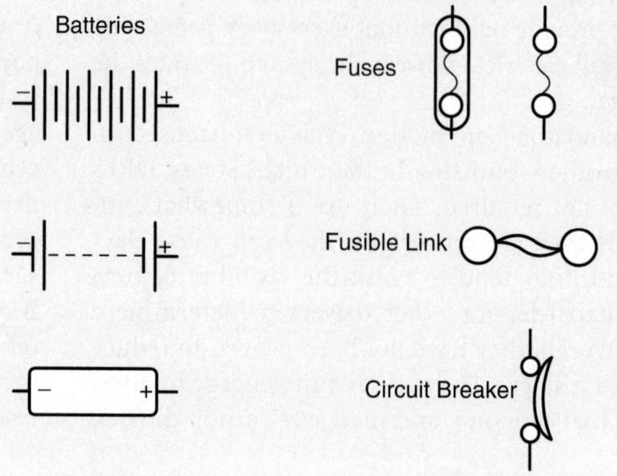

Ground and connector symbols

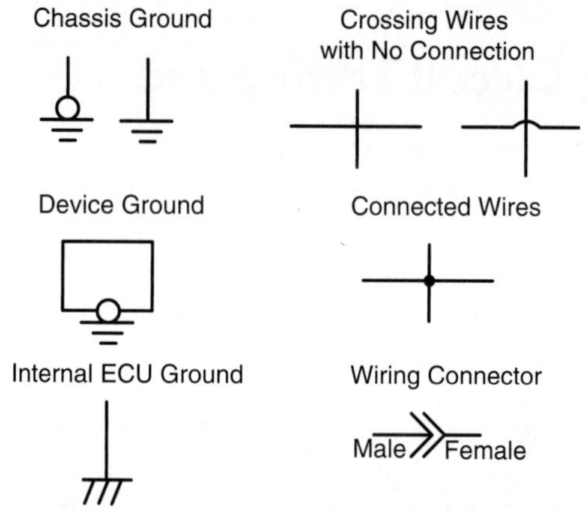

Switch symbols

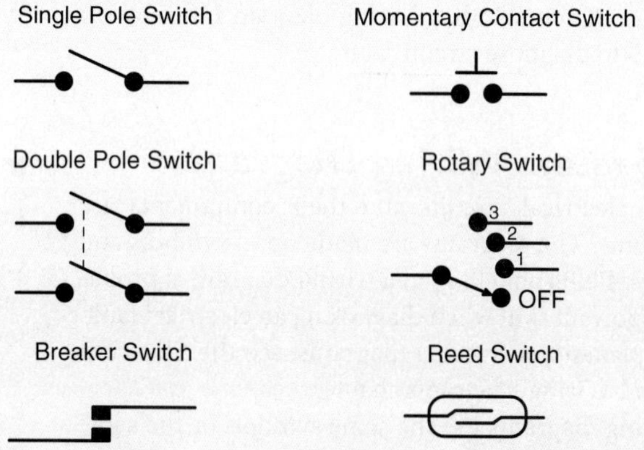

Resistors, coils, and relay symbols

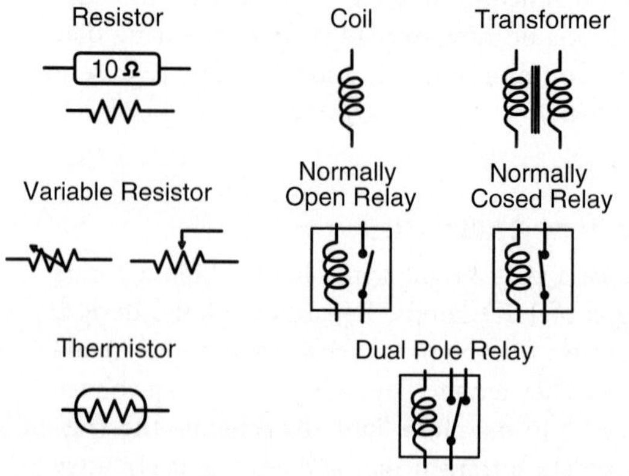

TABLE 45-1 Common Symbols (*Continued*)

Semiconductor symbols	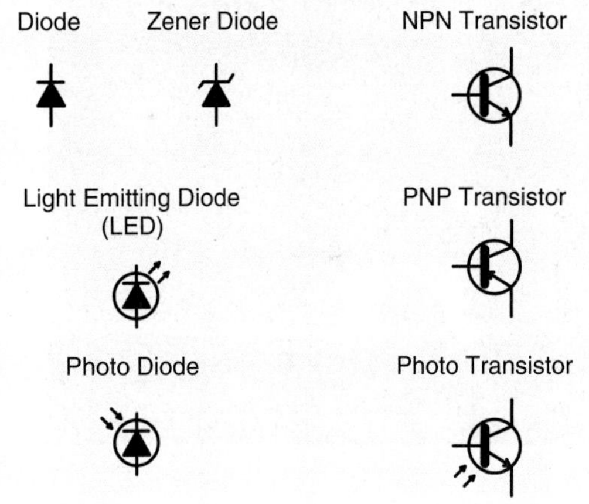Diode Zener Diode NPN Transistor Light Emitting Diode (LED) PNP Transistor Photo Diode Photo Transistor
Capacitor and device symbols	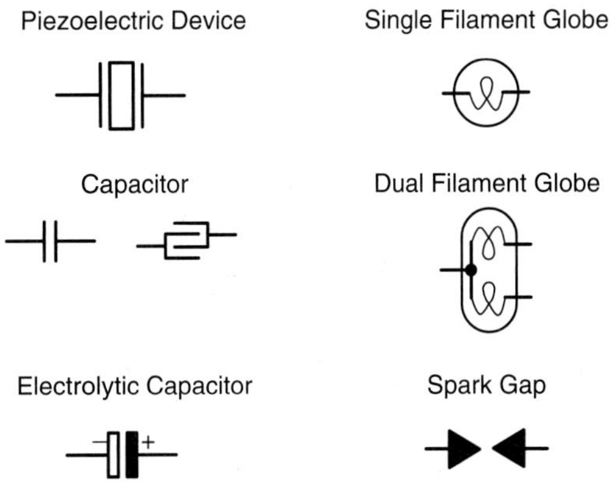Piezoelectric Device Single Filament Globe Capacitor Dual Filament Globe Electrolytic Capacitor + Spark Gap
Motors, generators, and solenoid symbols	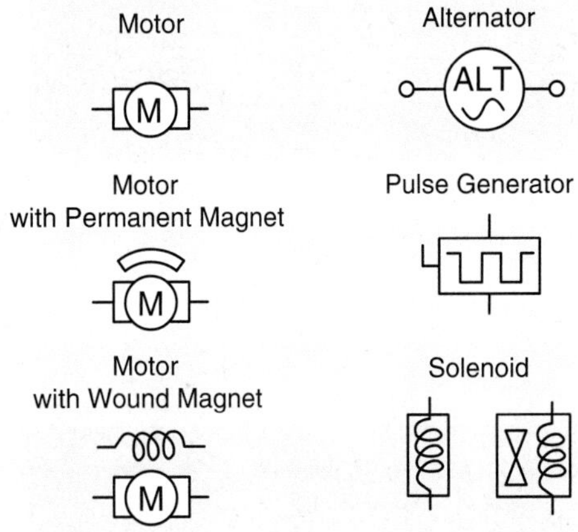
Gauges and warning device symbols	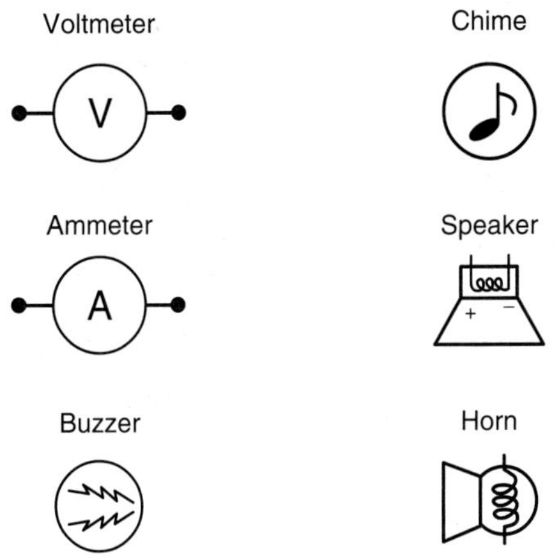

FIGURE 45-34 Courtesy lights provide ambient light where needed.

to see the warning lights. If you are unsure of where these are, ask your supervisor.

To check lighting and peripheral systems, follow the steps in **SKILL DRILL 45-1**.

Inspecting and Changing an Exterior Lightbulb

Lightbulbs have a limited life span and burn out occasionally. Inspecting the lighting system's operation periodically will help identify any lightbulb issues. It is good practice to do this during oil changes. If none of the bulbs in a particular circuit are working, there may be a bigger electrical problem to resolve. This will require diagnosis.

SKILL DRILL 45-1 Checking Lighting and Peripheral Systems

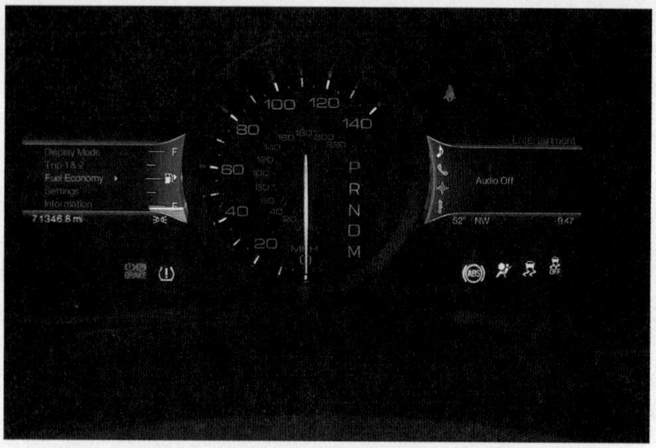

1. In a darkened area, turn on or activate the ignition. The dash warning lights should be displayed. Start the engine. If any warning light stays on after the engine is started, it could indicate a problem in one of the vehicle's safety or mechanical systems. If you are unsure about what any of the warning lights mean, consult the owner's manual.

2. Have someone stand behind the vehicle to report any problems. Turn the ignition to run. Switch on the park lights, taillights, marker lights, and license plate lights. Do the same for left and right turn indicator lights. Depress the brake pedal to make sure the brake lights work and have the proper intensity. Also, place the transmission in reverse and check the backup lights.

3. With someone in front of the vehicle, make sure the high and low headlight beams, park lights, turn signals, driving lights, fog lights, and cornering lights are all working properly.

4. With the interior light switch in the correct position, open the driver's side door to make sure the interior lights work. If any of these lights do not operate, you may need to replace a bulb or diagnose the fault.

Many tail, brake, and signal lightbulbs have more than one filament inside them. These bulbs normally have offset pins to ensure proper orientation of the electrical contacts in the socket. Be sure to look carefully at the bulb you are replacing to make sure you do not try to force the bulb in the wrong way. Also, make sure you don't replace a dual-filament bulb with a single-filament bulb. Doing so can short both lighting circuits together, causing the lights to work very strangely. Some bulbs have a colored glass globe that enables them to be used with a clear lens. If you replace a bulb of this type, make sure you replace it with one of the same color.

SAFETY TIP

The glass in a lightbulb is thin, and if it breaks, it is extremely sharp. Because bayonet-style lights need to be pushed in to be released, and they can become corroded in place, it may take a lot of force to push them in. Placed under such high pressure, the glass can break and severely cut the skin and even the tendons in fingers. If a bulb does not come out easily, stop, use a folded rag, and try to remove it again. If it shatters, the rag will help to protect your fingers.

To check and change an exterior lightbulb, follow the steps in **SKILL DRILL 45-2**.

▶ Checking and Changing a Headlight Bulb

LO 45-11 Perform headlight service.

There are many types of headlight bulbs available. Always make sure you replace a bulb with one of exactly the same type. Sealed-beam units require that the whole light be replaced when one filament has failed. If the reflector in the sealed-beam light shows signs of degradation, it also indicates that you must change the unit. If both lights operate but are not bright when switched on, start the engine to see if this solves the problem; the battery may be in a poor state of charge. When replacing a halogen bulb, avoid touching it with your fingers, which can leave a residue from your fingers on the outer surface. This residue can cause the bulb to crack or shatter and burn out after a short time of operation. If you inadvertently touch the bulb, clean it with alcohol and a lint-free cloth. Do not use gasoline or paraffin to clean the bulb.

To check and change a headlight bulb, follow the steps in **SKILL DRILL 45-3**.

▶ TECHNICIAN TIP

A customer, with his vehicle having sealed-beam headlights, once complained of poor headlight performance. When the technician turned on the lights, they both worked but didn't seem to create much light. As the technician opened the hood to investigate further, he noticed something strange. The headlights were shining light out the back of the lights, illuminating the engine compartment. He inspected the lights and found that the metallic reflector in the sealed beams had pretty much all flaked away. This allowed most of the light to exit the rear of the bulb, instead of being reflected forward. Replacing the sealed-beam headlights with new ones fixed the customer's concern.

Aiming Headlights

Although the principle of aiming headlights is the same in the majority of cases, the legal rules can differ from region to region. Be sure to check the requirements for your location. If you are unsure of what these are, ask your supervisor. Some manufacturers may suggest that the headlights be aimed on high beam. Others should be aimed on low beam. And in cases of separate high and low beams, both beams may need to be aimed independently. The manufacturer may also suggest that a load be placed in the vehicle. This is to simulate the ride height of the vehicle when it is traveling down the road. Headlights are typically aligned both

SKILL DRILL 45-2 Checking and Changing an Exterior Lightbulb

1. Remove the cover to expose the bulb. If the bulb is pin mounted, gently grip the bulb and push it inward. Turn the bulb slightly counterclockwise, and remove it from the bulb holder. Some bulbs pull straight out.

2. Inspect the bulb holder to make sure there is no corrosion. If there is, clean it with a bulb socket wire brush or emery cloth.

3. Insert the new bulb into the bulb holder, depress it fully, turn it slightly clockwise, and release it. Test it by switching it on and off. Then replace the cover, and test it again.

SKILL DRILL 45-3 Checking and Changing a Headlight Bulb

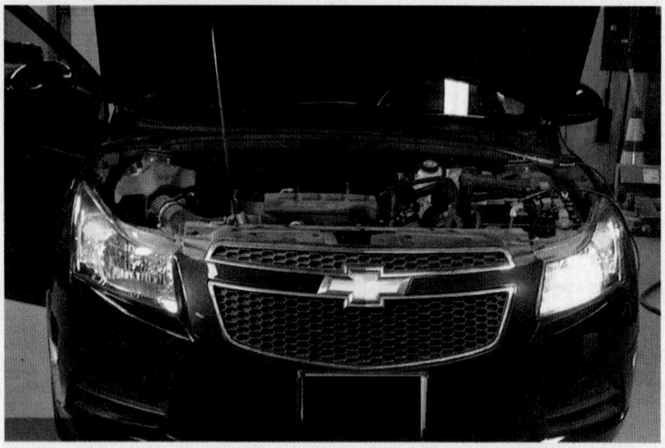

1. Switch the headlights on at low beam, then to high beam. Check that the high-beam indicator is operating. If one of the lights does not operate or is dim, that headlight will have to be diagnosed and potentially replaced.

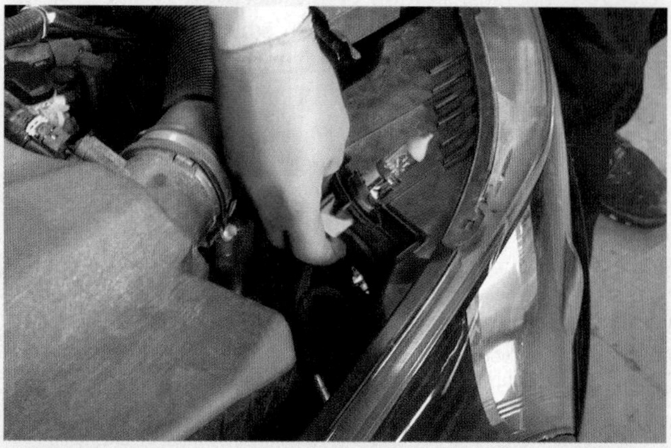

2. Test the vehicle headlights. Obtain the replacement lamp for the vehicle. Unplug the electrical connector at the back of the lamp unit.

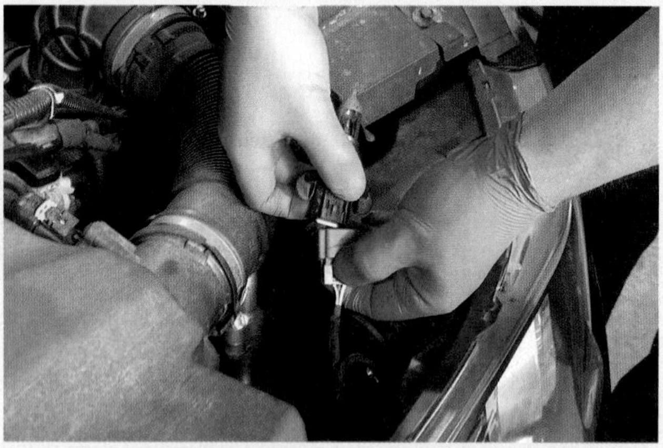

3. Remove the old bulb, and replace it with the new one. Handle the new bulb only by its base or, if supplied, by the card cover.

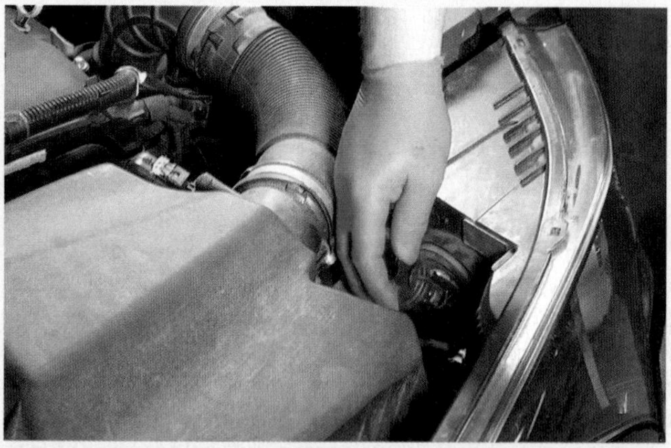

4. Replace the unit and the retaining ring or bulb assembly, and then plug in the connector. Switch on the lights again to confirm that they are both operating correctly.

vertically and horizontally. This illuminates the road as much as possible without blinding oncoming traffic. Some vehicles have headlights that adjust automatically. Do not try to adjust those unless you are diagnosing a fault in the system.

When aiming or aligning headlights, there are a couple of possible methods to utilize. The first, while no longer commonly used, involves an on-vehicle headlight aligner. This method was used extensively on sealed-beam headlights. The aligners are calibrated to the floor at the point where the front and rear wheels contact the concrete. Then, the aligners are installed on the headlights with a built-in suction cup, which holds them in place. The aligners use a set of mirrors that can be calibrated to local regulations so that the headlights can be aligned properly.

Off-vehicle headlight aligners have become popular. This was after the introduction of aerodynamic headlights using removable halogen bulbs. These aligners sit in front of the vehicle after calibration for the floor slope and orientation with the vehicle. With the headlights on, adjustments can be made by following the instructions on the aligner. Refer to the manufacturer's information for specific information about headlight aiming.

To aim headlights, follow the steps in **SKILL DRILL 45-4**.

SKILL DRILL 45-4 Aiming Headlights

1. Make sure the tires are inflated properly, the wheels point straight ahead, and there is no extra weight in the vehicle. Position the vehicle correctly in relation to the headlamp aligner unit, following the equipment manufacturer's instructions. Calibrate the aligner for any floor slope and for the vehicle being tested.

2. On the types of aligners that require the headlights to be on during alignment, turn the headlights on to a low-beam setting. The center of the illuminating beams should be in the lower right quadrants of the chart or wall markings or as specified by the manufacturer.

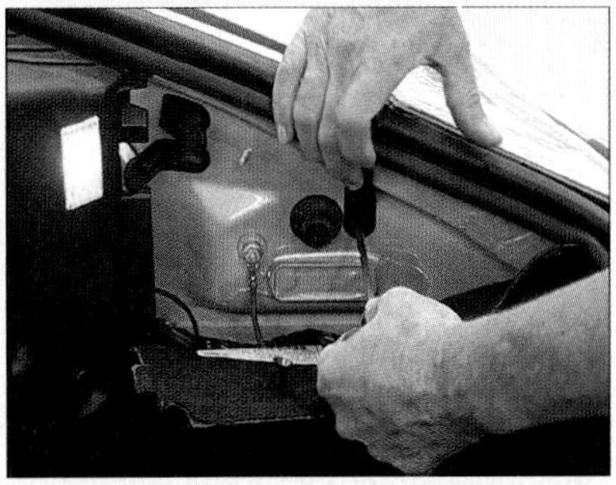

3. The high beam should be centered, falling on the intersections of the horizontal and vertical marks or as specified by the manufacturer. If necessary, turn the adjustment screws on the headlight so the lights point to the correct places or bubbles on the levels are centered, depending on the type of aligner equipment you are using.

Headlight Brightness

Headlight brightness is critical to safe driving. If too dim, the road ahead will not be adequately illuminated. If too bright, the life of the headlight will be shortened, making it prone to early burnout. In most cases, the headlights are more likely to be too dim than too bright. Bulbs that are too bright can indicate that the charging system voltage is too high. Or someone may have replaced the headlights with an aftermarket set, which may be illegal. Checking bulb numbers will verify this condition.

If the headlights are too dim, there can be several causes. The most common one is high resistance in the light circuit. This can be checked by measuring the voltage drop on both the power side of the bulb and the ground side. The voltage drop should be less than 0.5 volt on each side. If an excessive voltage drop is indicated, use a wiring diagram to research the circuit. Once you know how it is wired, use the DMM to isolate the voltage drop by using the meter to measure each segment of the circuit. When a high resistance is located, perform the needed repair, whether that is a wire repair or a connector, switch, or relay replacement. If the voltage drop is less than 0.5 volt on each side, suspect the lightbulb is wearing out. One way to verify this is with a light intensity meter. This meter can measure the amount of light energy the lamp is producing, so you can compare that with specifications (**FIGURE 45-35**). If the light is being supplied with the specified voltage, but it is not creating enough light, the bulb will have to be replaced.

FIGURE 45-35 A light meter being used to measure light intensity.

Wrap-Up

Ready for Review

- Lighting systems improve the driver's visibility at night. The headlight switch activates taillights, park lights, and headlights to allow the driver to see the road ahead. Brake lights operate when the brake pedal is depressed. Red or amber turn signals alert other drivers of a change in direction.
- Several lamp types are used in modern vehicles and they include: standard incandescent lamps, halogen lamps, vacuum tube fluorescent (VTF) lighting, HID xenon gas systems, LEDs, and more.
- Lamps and lightbulbs come in a variety of configurations to fit the various applications within a vehicle. The different configurations are: single-filament bulbs, dual-filament bulbs, bayonet-style bulbs, wedge-style bulbs, and festoon-style bulbs.
- Park, tail, and marker lights are all low-intensity or low-wattage bulbs that are used to mark the outline or width of the vehicle. These lights reduce confusion and improves safety while driving.
- Driving lights provide high-intensity light over long distances. They come in different sizes, shapes, lens pattern, and bulb wattage. Fog lights are used with other vehicle lighting in poor weather such as thick fog, driving rain, or blowing snow. Cornering lights are designed to provide side lighting when the vehicle is turning corners.
    - Most driving lights use quartz halogen bulbs in the 55- to 120-watt range, fog lights typically use quartz halogen bulbs and are available in different shapes and sizes, and cornering lights are white lights usually installed into the bumper or fender.
- Brake lights are activated whenever the driver operates the foot brake to slow or to stop the vehicle. The backup lights are white lights mounted at the rear of a vehicle that provide the driver with vision behind the vehicle at night. Modern vehicles use network/bus systems and the BCM to control the backup lights.
- Turn signal indicators are located on the extreme corners of the vehicle, are usually amber in the front and can be either red or amber in the rear, and are operated by a column-mounted switch. The circuit for hazard warning lights connects into the turn indicator lights and pulses all exterior turn indicator lights and both indicator lights on the instrument panel.
- Headlights require both a high beam and a low beam to illuminate the road ahead of the vehicle. A beam selector switch typically located on the steering column selects between low beam and high beam. Different types of headlight systems include: reflector-type lighting system and projection-type headlight system.
- Information from wiring diagram can be used to diagnose lighting system circuit faults. Safety glasses, high-voltage safety gloves, and safety boots should be worn when working on HID systems. Persons with active electronic implants, such as heart pacemakers, should not work on HID headlamps. Bulbs in HID headlamps must be disposed in an environmentally friendly way.
- A systematic approach should be followed when inspecting lighting and peripheral systems to avoid missing a faulty bulb or other component. Take care not to force the bulb in the wrong way during replacement. A dual-filament bulb should not be replaced with a single-filament bulb.
- When replacing a halogen bulb, avoid touching it with your fingers. Gasoline or paraffin should not be used to clean a halogen bulb. Check local regulations for the purpose of aiming headlights. A light intensity meter can be used to measure the amount of light energy produced by a lamp.

Key Terms

ballast A device that increases lighting voltage substantially and controls the current to the bulb.

halogen lamp A type of incandescent lamp that is filled with a halogen gas such as bromine or iodine.

high-intensity discharge (HID) A type of lighting that produces light with an electric arc rather than a glowing filament.

incandescent lamp The traditional bulb that uses a heated filament to produce light.

vacuum tube fluorescent (VTF) A type of lighting used for instrumentation displays on vehicle instrument panel clusters. This type of lighting emits a very bright light with high contrast and can display in various colors; also called vacuum fluorescent display (VFD).

Review Questions

1. _______ headlamps produce light with an electric arc rather than a glowing filament.
    a. Vacuum tube fluorescent
    b. High-intensity discharge
    c. Incandescent and halogen
    d. Light-emitting diode
2. All of the following turn on with the taillights, EXCEPT:
    a. backup lights.
    b. license plate lights.
    c. park lights.
    d. side markers.
3. Which of the following is NOT a part of the reverse light circuit?
    a. Brake light switch.
    b. Reverse light switch.
    c. Ignition switch.
    d. Vehicle battery.

4. Red or amber turn signals alert other drivers of a change in direction and are mounted so they can always be seen from the _______ of the automobile.
   a. Front
   b. Rear
   c. Sides
   d. All of the answers
5. Many vehicles use halogen light bulbs. What must you avoid when handling halogen bulbs?
   a. Touching the glass.
   b. Touching the metal.
   c. Touching the halogen gas.
   d. Touching the terminal.
6. Which lights are wired in parallel with the taillights and operate whenever the taillights are switched on?
   a. Headlights
   b. Turn signal lights
   c. Backup lights
   d. License plate lights
7. All of the following statements with respect to the function of headlights are true EXCEPT:
   a. They illuminate the road ahead.
   b. They help drivers at the time of reduced visibility.
   c. They provide two beams, high and low, to serve different purposes.
   d. They are connected in series with each other.
8. When aiming headlights:
   a. make sure the wheels are pointed 20 degrees to the right
   b. make sure the vehicle ride height is correct
   c. adjust the tire pressure after aligning the headlights
   d. place 100 pounds of weight in the trunk
9. Which of the following lamps produce more lumens with a bluish tinge for the given wattage when compared with all other lamps?
   a. Incandescent lamps
   b. Halogen lamps
   c. High-intensity discharge lamps
   d. VTF lamps
10. Which type of light bulb have a base on each end of a cylindrical bulb?
   a. Bayonette style
   b. Festoon style
   c. Wedge style
   d. Dual-filament style

ASE Technician A/Technician B Style Questions

1. Technician A says light-emitting diodes (LEDs) have better visibility in inclement weather, operate at cooler temperatures, consume less energy, are much smaller, and can last up to 100 times longer than traditional bulbs. Technician B says LEDs can reduce the braking light response time by two-tenths of a second. Who is correct?
   a. Technician A
   b. Technician B
   c. Both Technicians A and B
   d. Neither Technician A nor B
2. Two technicians are discussing brake light bulbs. Technician A says that a two-filament bulb uses the second filament as a back-up if the first filament burns out. Technician B says that two-filament bulbs have different wattage filaments. Who is correct?
   a. Technician A
   b. Technician B
   c. Both Technicians A and B
   d. Neither Technician A nor B
3. Technician A says that all modern vehicles by law must be equipped with a center high mount stop lamp (CHMSL). Technician B says the CHMSL is usually mounted on top of the trunk lid or in the rear window of a vehicle. Who is correct?
   a. Technician A
   b. Technician B
   c. Both Technicians A and B
   d. Neither Technician A nor B
4. Technician A says hazard warning lights use a flasher unit that can be a separate unit or the same as that used for the turn signals. Technician B says daytime running lights are used to warn other road users that a hazardous condition exists or that the vehicle is standing or parked in a dangerous position on the side of the road. Who is correct?
   a. Technician A
   b. Technician B
   c. Both Technicians A and B
   d. Neither Technician A nor B
5. Technician A says high-intensity discharge (HID) headlamps produce light with an electric arc rather than a glowing filament. Technician B says HID lamps are commonly called xenon headlamps. Who is correct?
   a. Technician A
   b. Technician B
   c. Both Technicians A and B
   d. Neither Technician A nor B
6. Technician A says park, tail, and marker lights are all high-intensity or high-wattage bulbs. Technician B says license plate illumination lamps are connected in series with the taillights. Who is correct?
   a. Technician A
   b. Technician B
   c. Both Technicians A and B
   d. Neither Technician A nor B
7. Technician A says cornering lights are typically used in poor weather such as thick fog, driving rain, or blowing snow. Technician B says fog lights are red lights usually installed into the bumper or fender and are designed to provide side lighting when the vehicle is turning. Who is correct?
   a. Technician A
   b. Technician B
   c. Both Technicians A and B
   d. Neither Technician A nor B

8. Tech A says that some brake lights get power from the brake switch through the turn signal switch. Tech B says many turn signals use amber lights. Who is correct?
   a. Tech A
   b. Tech B
   c. Both A and B
   d. Neither A nor B
9. Tech A says that incandescent bulbs resist vibration well. Tech B says that HID headlamps require a ballast to raise the voltage for the light. Who is correct?
   a. Tech A
   b. Tech B
   c. Both A and B
   d. Neither A nor B
10. Tech A says that a light intensity meter is used to measure the brightness of headlights. Tech B says that if a bulb is dim, you should perform a voltage drop test on the power and ground side of the bulb. Who is correct?
    a. Tech A
    b. Tech B
    c. Both A and B
    d. Neither A nor B

CHAPTER 46

Body Electrical System

Learning Objectives

- **LO 46-01** Describe the reasons for vehicle networks.
- **LO 46-02** Describe the types of vehicle networks.
- **LO 46-03** Use scan tool to check for module communication and software updates.
- **LO 46-04** Test electric motor circuits.
- **LO 46-05** Test power door locks and remove door panels.
- **LO 46-06** Describe the operation of keyless entry/remote start systems.
- **LO 46-07** Test horn systems.
- **LO 46-08** Test wiper and washer systems.
- **LO 46-09** Describe supplemental restraint systems and how to disable–enable them.

ASE Education Foundation Tasks

See Appendix A to view the 2017 ASE Education Foundation Automobile Accreditation Task List Correlation Guide.

▶ Vehicle Networks

LO 46-01 Describe the reasons for vehicle networks.

Electrical systems on modern vehicles are becoming increasingly complex. A range of electronic and accessory systems have been added to the base vehicle. Global positioning systems, entertainment systems, security systems, electric seats, and heated glass are just a few examples. Most of these systems are controlled by electronic modules. The modules are controlled by a body computer system. Increasingly, the modules are interconnected with each other through vehicle data networks. The networks transfer data and commands to each of the modules. Technicians require an in-depth knowledge of these systems. This includes knowing how the network functions. We will start with some of the basic principles. That way, you will gain a knowledge base upon which you can build your further studies.

Networking and Multiplexing

Even the most basic modern vehicles include many electronically controlled systems. If each electronic system had its own ECU, harness, and sensors, the weight of the added components would negate some of the efficiency it provides. A vehicle's many electronic systems could require miles of insulated wiring, splices, and terminals. It would also add cost to the vehicle.

One solution is to reduce the amount of wiring required in the vehicle. This can be done by integrating sensor information into a common wiring harness, called a bus. The bus combines individual systems, wherever possible, into a **multiplexed** serial communications network. An added advantage is that with less wire and fewer connections, there is less chance of dirty connections causing faults. Of course, if a network fails, it can shut down the entire network and with it the controlled devices. Also, troubleshooting a networked system requires additional training and tools. This is compared to a simple digital multimeter (DMM).

With networking, there are one or more twisted pairs of wires with terminating resistors at the end of each pair. They carry digital information throughout the entire vehicle. In many cases, there is a medium- or low-speed bus for nonsafety systems. For safety systems, such as antilock brakes, supplemental restraint systems, and many of the engine controls, a high-speed bus is used. All components "listen" to the signals carried on the network. But they only act when requested to do so. Such a system is often referred to as a controller area network bus (CANbus) (**FIGURE 46-1**). The thin twisted pair of wires connects all the onboard control modules to each other. Thus, the powertrain control module (PCM), **body control module (BCM)**, electronic brake control module, transmission control module, and many others are networked together. They communicate, much like "party lines" on old-fashioned telephone systems. A certain ring tone identified a specific phone. But any phone could hear the conversation or respond.

Output devices that respond to commands on the network are referred to as nodes. All nodes hear all data on the network.

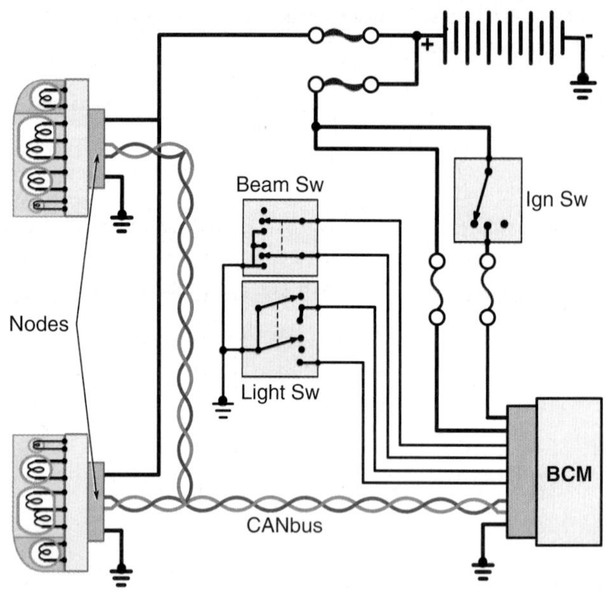

FIGURE 46-1 Typical CANbus diagram.

You Are the Automotive Technician

A 2014 minivan has a power window that is not working properly. The customer informs you that the "one-touch" function works sometimes, but not at others. After performing a VIN check for service history, you look up technical service bulletins (TSBs). You notice one related to the "one-touch" feature on this vehicle. The issue may be related to a binding window track. Plus there is a software update to help address the issue down the road.

The first step is to verify whether there is an overcurrent condition on the window motor. You connect the scan tool and see a code indicating an overcurrent shutdown on the driver's window. Next, you clean out the exposed window track with a clean shop rag. You then spray silicone lubricant into the window track and operate the window several times. Because the window is moving smoothly, the software can be updated. You follow the flash reprogramming steps detailed in the manufacturer's guide. The new software allows a slightly higher current flow in the circuit. This will prevent the computer from disabling the feature. Once the update is complete, you verify that the "one-touch" feature is functioning properly.

1. What are the various types of automotive networks?
2. What is a CANbus system?
3. How can software updates correct problems in the vehicle?
4. What are the precautions before updating software on a vehicle?

But they only respond when called upon to perform some action or diagnostic function.

The advantage of a multiplex network is that it reduces the number of dedicated wires for each function. This results in a reduction in the number of wires in the wiring harness. It also reduces system weight while improving reliability, serviceability, and installation.

Sensor data multiplexed on the network are prioritized. For example, information such as vehicle speed and engine temperature is shared on the network. It is used for several systems such as the instrument panel cluster and the PCM. They monitor the data for their own purposes. This design helps to reduce the number of sensors needed.

Some information being shared takes priority over others. For example, a message related to an impending collision would take precedence over a lamp-out message. In such a case, the vehicle must take certain actions. In this case, the system prioritizes that information.

Networking of data also allows greater vehicle content flexibility. This is because functions can be added or modified through software changes. Other control units, such as those related to trailer towing, can be added to the system as required. This is done by simply connecting them into the network and updating the software to integrate it into the network. You can think of this process as being similar to adding a printer to your home computer system. Simply plug it in, configure it, and start using it.

A diagnostic scan tool is connected to the CANbus network. It can both monitor and extract operational information to assist in diagnosis and fault finding. Factory scan tools and many aftermarket scan tools are bidirectional. In addition to receiving data, they can be used to command engine and vehicle components to operate. This is a very useful function when diagnosing faults in networked systems.

▶ Vehicle Communications Networks

LO 46-02 Describe the types of vehicle networks.

The additional features and accessories have increased the need for interaction between the vehicle's systems and components. For example, the following systems must know the speed of the vehicle:

- the instrument panel for the speedometer,
- the transmission for gear selection,
- the radio for speed-sensitive volume control, and
- so on (**FIGURE 46-2**).

In the same way, most of the other systems need common sensor information as well. If separate wires were needed to connect all of the sensors and actuators together, it would require a lot of wires.

Manufacturers solved this issue by using a multiplexed system. Information is transmitted across one or two wires that run to all of the control modules. The control modules interpret the information and act on it, if needed. They control the operation of actuators or other electrical/electronic devices. The modules are also supplied with power and ground. This allows the modules to operate, as well as supply power and ground to the devices they control. This makes for a much simpler wiring harness with fewer wires.

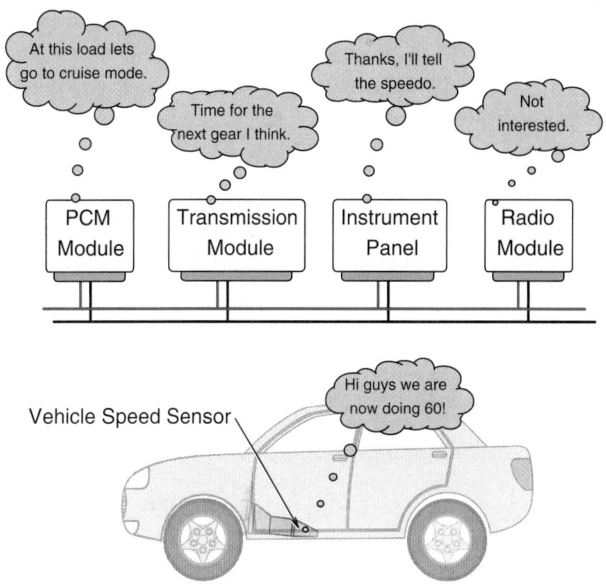

FIGURE 46-2 A networked system allows data to be sent over the system to all modules. The ones that need to know that information act on it, and the rest ignore it.

There are several types of networks: LIN, LAN, CAN, CAN FD, FlexRay, MOST, and Ethernet-based systems. Each type has specific benefits and limitations. Many vehicles incorporate more than one type of network. This is due to the varying requirements of the different systems. For example, systems like power mirrors and power seats do not need high-speed networks to operate effectively. However, fuel injector control and airbag deployment need networks with much higher speed. Also, usually, the faster the network, the more expensive it is. So, it makes sense to use lower speed/lower cost networks to handle the less-data-intensive systems. The higher speed/higher cost networks can be used for fuel injector control and airbag deployment systems. This does mean that the different networks must be able to share information where necessary.

Local Interconnect Network (LIN)

We look at each one briefly, starting with LIN, which stands for local interconnect network. And, like most other networks, is used to send serial data between modules, called nodes. In the LIN system, there is one master node and up to 15 slave nodes. They communicate over a single wire at speeds up to 20 kbits/s (**FIGURE 46-3**). This network works well for body-controlled systems such as wipers, mirrors, and lights.

Controller Area Network (CAN)

This network can accommodate up to 100 nodes. Many high-end vehicles typically have up to about 70 nodes. Each node is a master node capable of receiving, transmitting, and processing the signals independently.

The nodes communicate over a two-wire twisted pair, which reduces electrical interference (**FIGURE 46-4**). One wire designated as CAN–H (CAN–High, or CAN+) carries the high voltage signal. The other wire is CAN–L (CAN–Low, or CAN–).

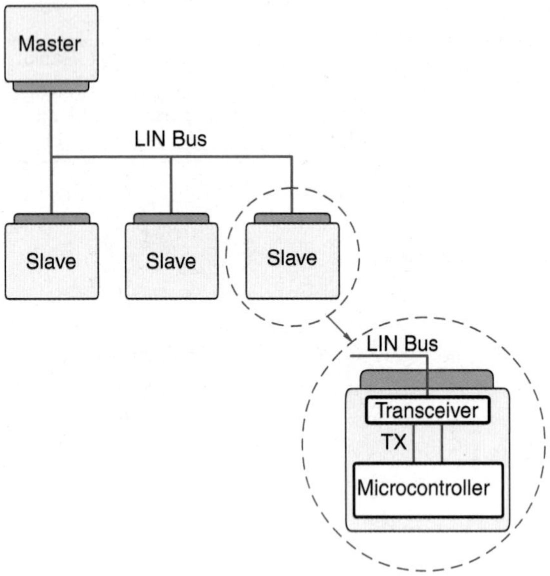

FIGURE 46-3 A typical LIN network using a single-data communication wire.

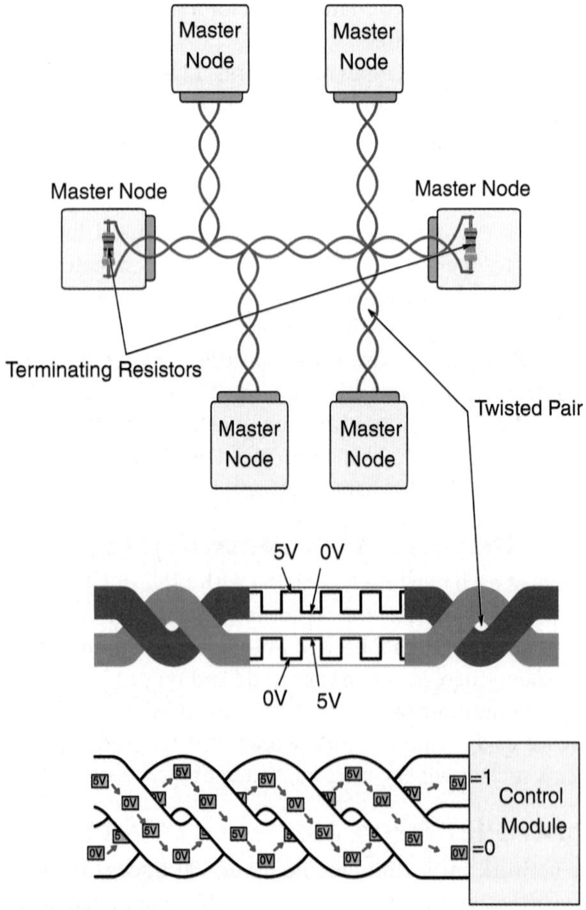

FIGURE 46-4 A typical CAN system with several nodes, each of which is a master node.

It carries the mirrored low-voltage signal. This pairing of wires with opposite voltage signals is used to verify the accuracy of the signal being sent by comparing the individual signals to one another.

The high-speed portion of the system sends 8 bits per frame and operates at speeds of up to 1 Mbit/s. And the low-speed portion (called fault-tolerant CAN) operates between 40 and 125 kbits/s. This dual-speed system works well for high-speed dependent network systems along with lower speed systems. This cuts down on the number of sensors required if two separate systems were used.

Configurations and Terminal Resistors

The CANbus-H and CANbus-L system have terminating resistors connected to the data lines to form a loop through the resistors. This is regardless of how the network is connected. The value of resistors can vary by manufacturer. However, high-speed CANbus systems tend to use dual 120-ohm resistors. This gives a total of 60 ohms per pair because they are connected in parallel, but usually at opposite ends of the bus (**FIGURE 46-5**).

The resistors can be connected externally in the wiring harness or connectors, but they can also be integrated internally into modules. The data lines are also twisted together with a specific twist rotation per length. This twisting, in combination with the terminating resistors, reduces the amount of interference on the data lines. The CANbus system has a **data link connector (DLC)** that provides a connection point into the data network. The DLC is also used to connect scan tools into the network. It can also be used to conduct voltage and waveform tests on the CANbus system.

Controller Area Network with Flexible Data-Rate (CAN FD)

This is an enhanced version of CAN. It operates on up to 64 data bytes per frame, compared with 8 bytes per frame, and at a speed of up to 5 Mbits/s. Because of the increased speed and data bit size, the network can handle data traffic more effectively.

FlexRay Networks

FlexRay networks operate at higher speeds of up to 10 Mbits/s. They use either one twisted pair of wires (single channel), or two twisted pairs of wires (dual channel). This increases the

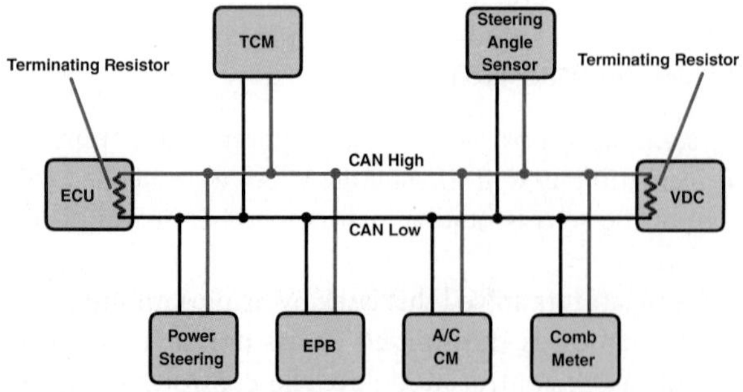

FIGURE 46-5 A network diagram with terminating resistors.

fault tolerance of the network (**FIGURE 46-6**). Because this network has a high transfer rate, it works well in data-intensive systems such as engine management.

Media-Oriented Systems Transport Networks

Media-Oriented Systems Transport (MOST) networks are designed to transmit high-speed audio, voice, video, and data signals. There can be up to 64 MOST devices. The devices are plug-and-play type, and hence can be added to the system easily, even after the vehicle is in service. There are several versions of MOST systems, each having increased bandwidth over the previous version (25, 50, and 150 Mbit/s). MOST systems send data signals in one of two ways. The first is optically, using plastic optic fiber. The second is electrically, using electrical conductors (**FIGURE 46-7**).

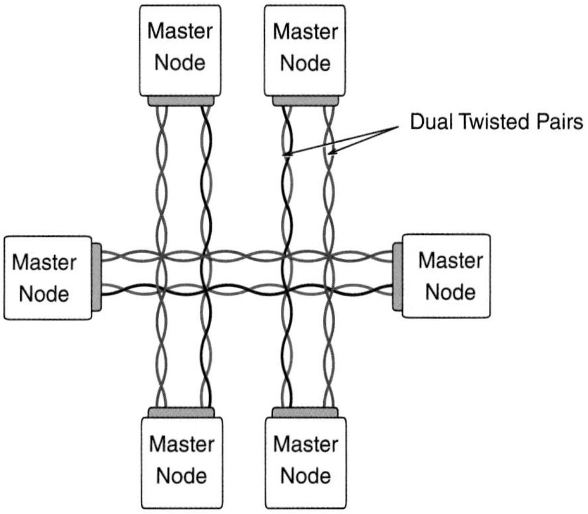

FIGURE 46-6 A typical FlexRay network with a dual-channel system.

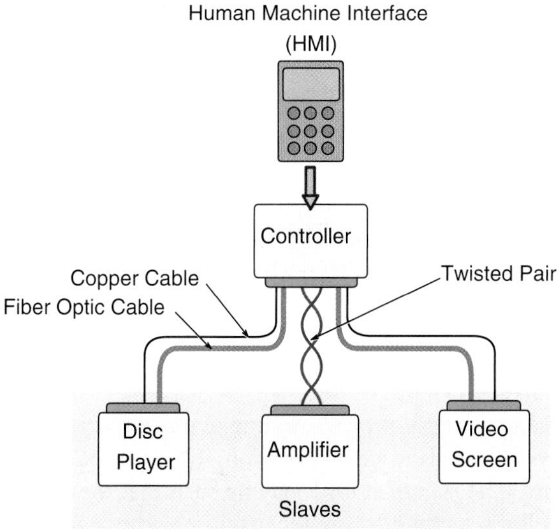

FIGURE 46-7 A typical MOST network with both a fiber optic and electric communication systems.

Ethernet Network Systems

Ethernet network systems have been used in the past to connect into the vehicle's onboard diagnostics system. It has also been used to perform software upgrades when accessing vehicle network systems. Now they are just being implemented as in-vehicle network systems (**FIGURE 46-8**). Ethernet currently operates at 10 and 100 Mbit/s, with work underway to enhance it to 1 Gbit/s. This makes it ideal for video and entertainment systems. It is also good for connecting to handheld electronic devices such as cell phones and tablet computers. Also, it is being investigated for use in 360-degree camera systems used in drive-by-wire systems. Currently, Ethernet systems use four data wires, but there is an industry push to adopt a single twisted-pair wire for Ethernet networks in automotive applications. This would cut down in weight, wire harness size, and cost.

The network configurations are how the network connects to each module. They vary with each manufacturer and type of network. Some use a simple series configuration. Others use variations of series networks and parallel network connections with descriptive names, such as looped series, star parallel, or bussed parallel (**FIGURE 46-9**).

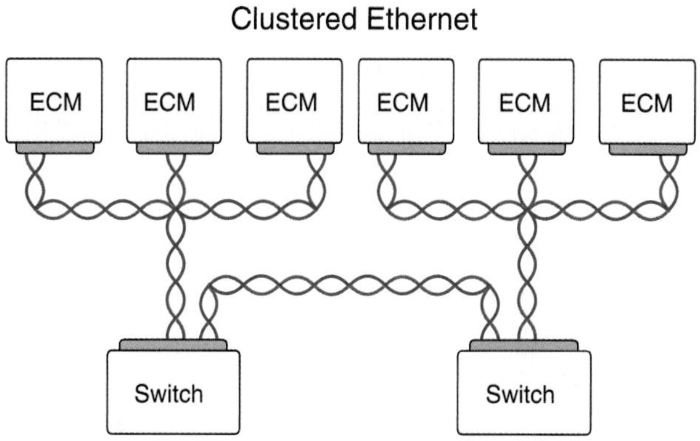

FIGURE 46-8 Ethernet networks are being used as in-vehicle networks.

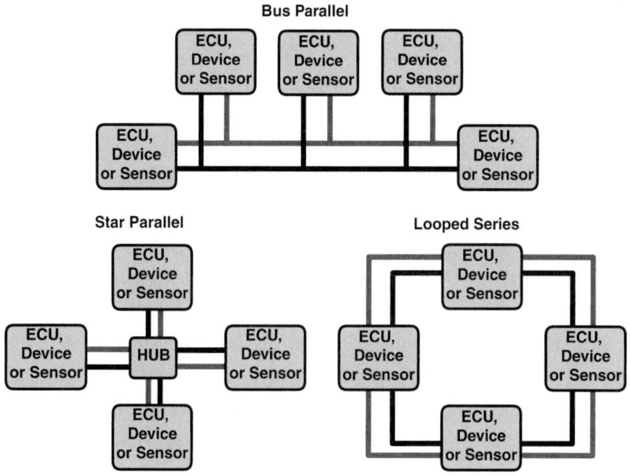

FIGURE 46-9 Network configurations.

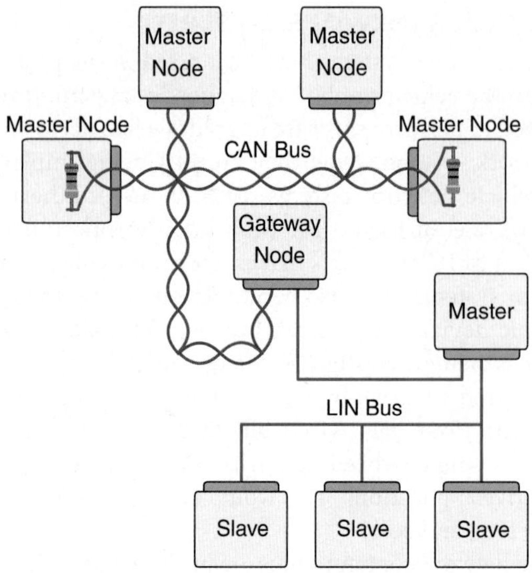

FIGURE 46-10 A gateway node connects two types of networks together so that information can be shared on both.

Most vehicles are equipped with more than one type of network because slower networks are less costly and faster networks needed for critical systems are more expensive. Networks are usually connected together so that sensor information can be shared. This means that duplicate sensors for each network are not needed. But because the networks operate differently, there has to be an interface between them so the data can be transferred properly. This function is provided by gateway modules (nodes) (**FIGURE 46-10**). A vehicle will have one gateway module between each different type of network.

▶ Using a Scan Tool

LO 46-03 Use scan tool to check for module communication and software updates.

Many different types and brands of scan tools are available. The vehicle and system coverage varies with each scan tool. Some scan tools require hardware keys to be inserted for each different type of vehicle. Others may have provisions for vehicle manufacturer and model to be selected using software. Always ensure that you understand how to connect and set up the specific scan tool for the vehicle you are working on.

Using the scan tool may require the vehicle's ignition system to be on for an extended period of time without the engine running. The electrical load may be such that the vehicle's battery will discharge. A battery support unit may need to be connected to ensure the vehicle's battery does not discharge. Battery support units are high-current output power supplies with a steady voltage supply. They are used to support the battery when the engine is not running. Do not use a standard battery charger to perform this role. Many chargers do not have a steady direct current (DC) voltage supply. This may cause interference on the network communication circuits or damage components.

Checking for Module Communication Errors Using a Scan Tool

Modern vehicles have multiple modules to control the various vehicle systems. The number of modules and the purpose of each vary among manufacturers and even vehicle models. Modules communicate using the network on the vehicle. If there are communication errors on the network, some or all of the modules will not be able to operate. Also, one or more U codes will typically be set, indicating a fault in the communication network. If the network is active or partially active, a scan tool may be able to help you diagnose the fault.

One of the best things you can do is to connect a scan tool to the network DLC. Then, check the scan tool to see if there are any modules that are missing or not communicating. Typically, if a device is not showing as present, you need to physically locate that module. Then check it for power and ground. If those are correct, use the scan tool to command the module to operate the device. For example, use a scan tool to *actively test* the output component to see if it operates. If it operates correctly, then that means there is nothing wrong on the output side of the circuit. If it doesn't operate, check the communication wire on the module to see if it is receiving the signal.

Also, use a scan tool to check the data list for a "state of change" in signal such as a switch while turning the circuit on and off. If the data correspond to the state of change signal, then the input circuit is working properly.

When diagnosing the network system, if

- the power and ground are sufficient,
- the signal wire(s) is communicating the proper signal, and
- the module is not responding,

chances are good that the module is faulty and has to be replaced.

To check for module communication (including CANbus systems) errors using a scan tool, follow the steps in **SKILL DRILL 46-1**.

Performing Software Transfers, Updates, or Flash Reprogramming on Electronic Modules

There are many things you may need to diagnose and repair vehicles. They include software transfers, updates, or flash reprogramming of the vehicle's electronic modules. Manufacturers put out software updates for vehicle modules to enhance vehicle operation. The manufacturer-supplied updates are usually available over the Internet. Software updates are usually obtained through a subscription service. Flash reprogramming tools are available to transfer the software update. They connect to the vehicle's DLC.

Using the flash program tool may mean the vehicle's ignition system needs to be on for a long time without the engine running. The electrical load may be such that the vehicle battery will discharge. If this happens, it may cause a fault in the reprogramming. A battery support unit needs to be connected to ensure that the vehicle's battery does not discharge. Battery

SKILL DRILL 46-1 Checking for Module Communication Errors Using a Scan Tool

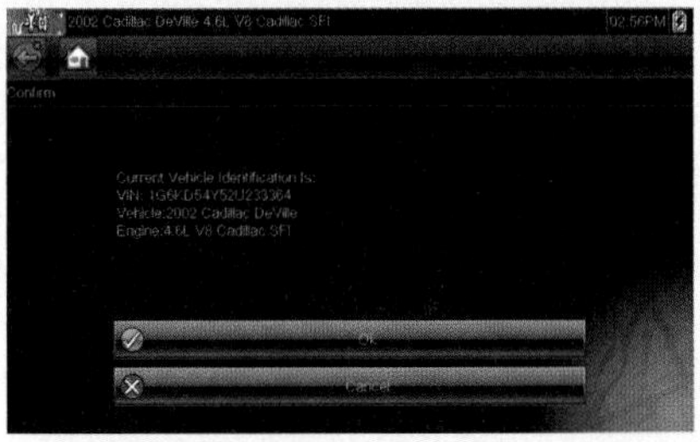

1. After researching the correct procedure in the manufacturer's information, locate the DLC, and connect the scan tool. Power on the scan tool and turn the ignition on. Establish scan tool communications with the vehicle.

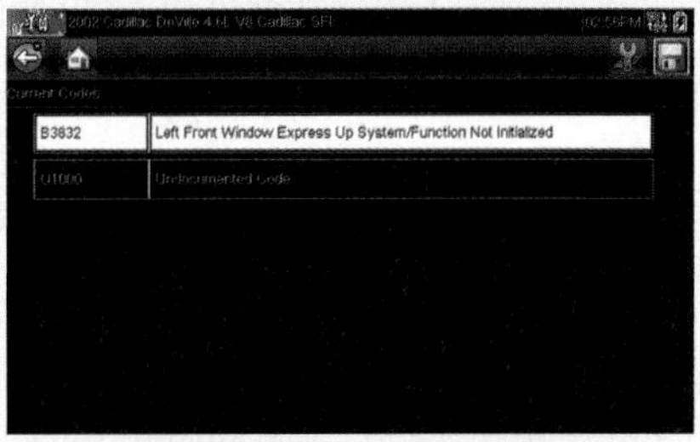

2. Check fault codes using the scan tool, and follow the diagnostic procedure in the service information. Check to see if any modules are inactive, or not listed as active, that should be.

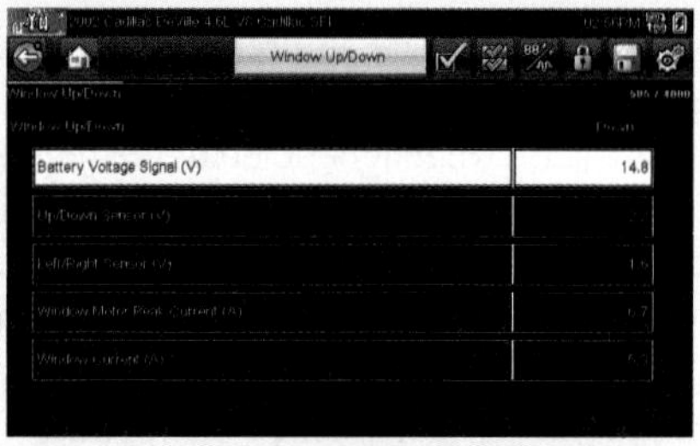

3. Command the module to take the appropriate action, and observe the response. If you do not receive the proper response, refer to the service information for the diagnostic procedure.

support units are high-current output power supplies with a steady voltage supply. They are used to support the battery when the engine is not running. Do not use a standard battery charger to perform this role, as many do not have a steady DC voltage supply. This may cause interference or damage components resulting in damage to the vehicle. Never shut the key off or lose power when flashing a module, as damage can occur.

To perform software transfers, software updates, or flash reprogramming on electronic modules, follow the steps in **SKILL DRILL 46-2**.

SKILL DRILL 46-2 Performing Software Transfers, Software Updates, or Flash Reprogramming on Modules

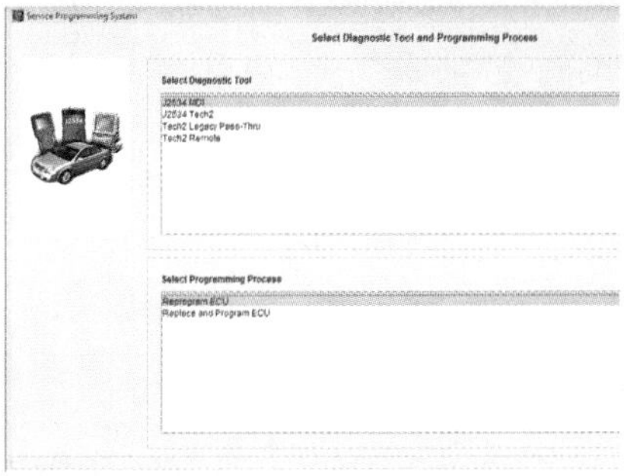

1. Obtain the latest vehicle software program updates from the manufacturer, and load the updates into the flash programmer. Connect a battery support unit to the vehicle.

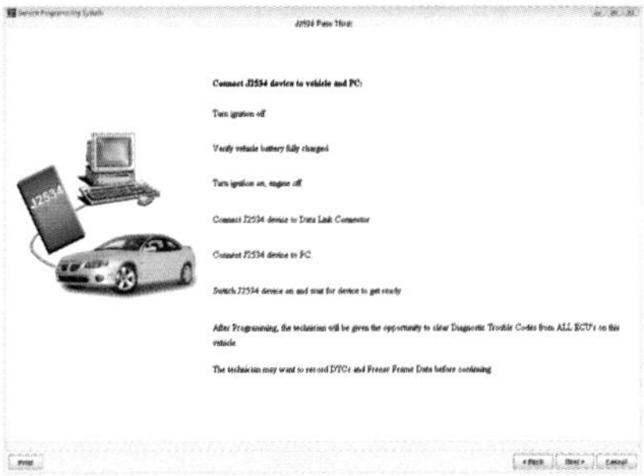

2. Locate the DLC. Connect and power on the flash program tool. Turn the ignition on.

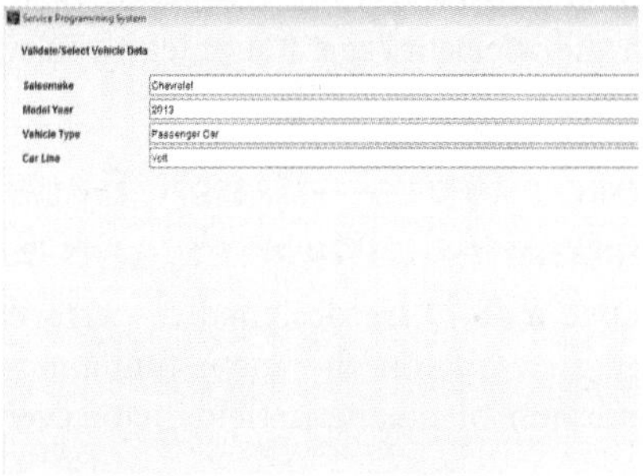

3. Establish flash program tool communication with the vehicle.

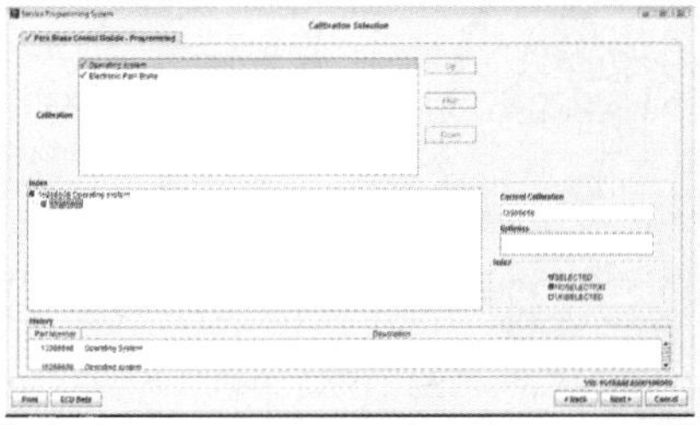

4. Identify the vehicle modules to be updated, and use the flash program tool to perform software transfers, software updates, or flash reprogramming.

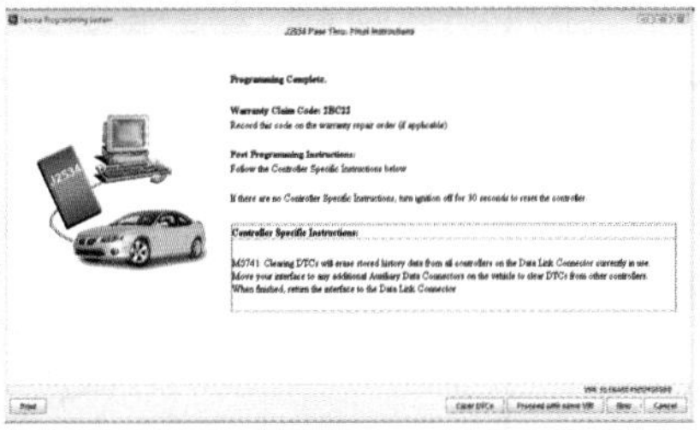

5. Disconnect the flash programmer, and check the vehicle's functionality.

▶ Electric Accessory Motors

LO 46-04 Test electric motor circuits.

Electric motors rotate as a result of the interaction of magnetic fields. That is, magnetic poles repel each other, and opposite poles attract. Two magnetic fields are required for motor action. One is in the casing and the other is in the rotating armature. The magnetic field in the casing can be generated by permanent magnets which are always magnetic. Or they can be generated by electromagnets made up of many windings of fine copper wire wound in loops called field coils. The armature's magnetic field is generated in loops of wire that form the armature windings. Motor action occurs through the interaction of the magnetic fields between the armature and field coils. This interaction causes a rotational force to act on the armature, creating the turning motion (**FIGURE 46-11**).

The rotary motion can be used to drive a variety of devices. Many times, the speed of the motor must be geared down so that more torque can be obtained. This is the case with windshield wipers and many actuators. Most of the time, the gears are located inside the motor housing. If one or more gears go bad (e.g., stripped teeth), then the motor may still spin. But the output gear or lever will not move as designed. Thus, listening for motor rotation can help you diagnose the problem.

For example, if the windshield wipers do not move when activated, yet you can hear the motor running, you know that the electrical circuitry is working. But the movement is not getting from the motor to the wipers. This example could be an internal gear problem in the motor, or it could be a problem with the wiper linkage. But at least you will have a good head start on diagnosing the fault.

Motor Types—Brush, Brushless, Permanent Magnet, and Stepper Type

Electric motors are designed to perform specific functions. They all work on the same fundamental principle of the interaction of magnetic fields. However, their construction varies. One of the most common types of DC electric motors is the **permanent magnet type**. In this type of motor, the magnetic field in the casing is produced by permanent magnets. While the armature has an electromagnetic field generated by passing electrical current through loops or windings (**FIGURE 46-12**). Power is supplied to the armature loops by brushes riding on the commutator. They continually change the current flow in the armature loops as it rotates. This changes the polarity of the magnetic field in the armature. The armature magnetic field interacts with the permanent magnets in the casing.

A **brushless DC motor**, as the name suggests, does not have any brushes. It is sometimes called an "electronically commutated motor." In this type of motor, an electronic control module replaces the brushes and commutator. A brushless motor has permanent magnets rotating around a fixed armature or stator (**FIGURE 46-13**). Having a fixed stator eliminates the need to have brushes. The current flow still needs to be

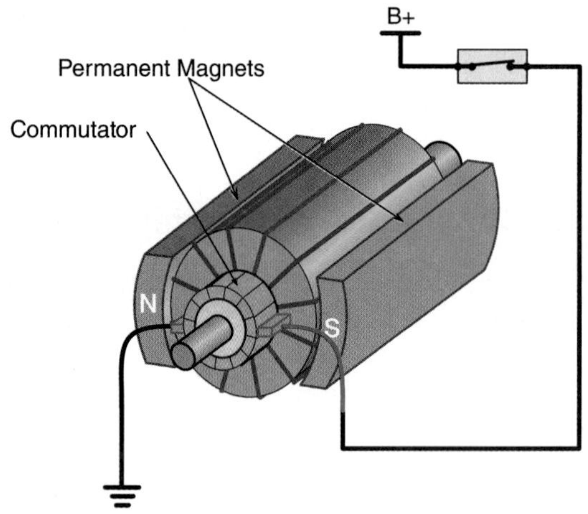

FIGURE 46-12 A typical permanent magnet electric motor.

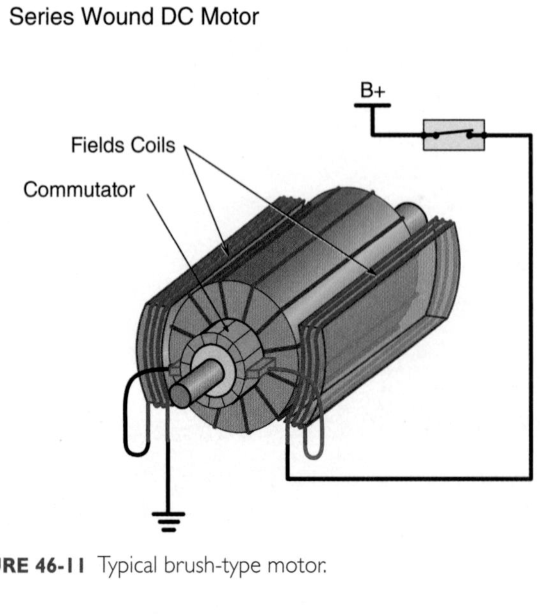

FIGURE 46-11 Typical brush-type motor.

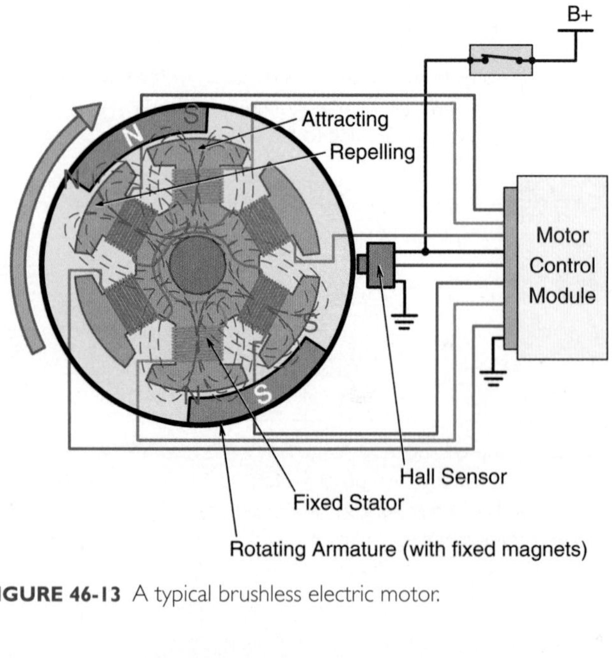

FIGURE 46-13 A typical brushless electric motor.

continually alternated in the stationary stator to produce motor action. This is achieved by an electronic control module. A brushless DC motor has several advantages over brush motors; these include:

- improved efficiency,
- improved reliability, and
- less electrical interference created (this is because there are no sparks created by brushes making and breaking contact on the commutator).

A **stepper motor** is a type of brushless motor with a key difference. It is designed to rotate in fixed steps, each step being a set number of degrees. The stepper motor uses a rotor with teeth of alternating polarity, and a number of wound coils (typically four in automotive use). Each pair of coils is controlled by an electronic microcontroller. Each time the controller energizes a pair of coils for a short period of time, the motor armature turns a set number of degrees. This causes two teeth line up with that set of coils and stop. The teeth to the immediate left and right are halfway between alignment with the other set of coils (**FIGURE 46-14**). Energizing each coil in rapid succession creates rotational movement.

Stepper motors can rotate a short distance and then stop or go in the reverse direction for a set number of degrees. This is because each coil is controlled individually through the microcontroller. Stepper motors are increasingly used in vehicles. For example, the electronic throttle body uses a stepper motor to open and close the throttle plate one step at a time. It is controlled by the PCM.

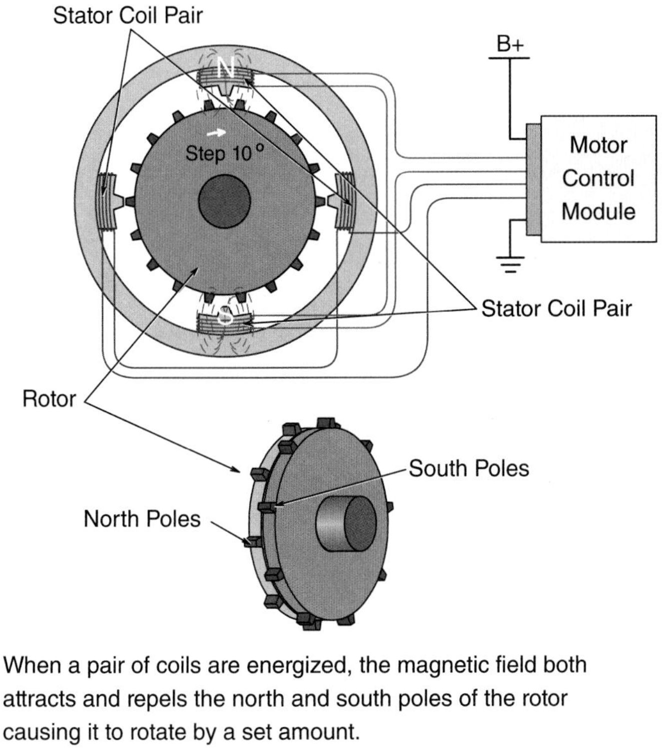

FIGURE 46-14 Typical stepper motor.

Blower Motor and Circuits

The **blower motor** in a vehicle is usually the permanent magnet type. It rotates a fan which moves air over the air-conditioning evaporator and heater core. The blower motor circuit incorporates a speed control for the motor. Speed can be controlled by switching between a number of resistors connected in series which changes the amount of power to the motor. Or it can be a more compact but complex electronic speed control module that operates electronically.

The blower motor circuit consists of the power supply, fuse, on/off switch, speed control, relay, and the motor itself (**FIGURE 46-15**). Consider in particular the vehicles equipped with air conditioning. In them, the blower motor control circuit is incorporated into the control unit.

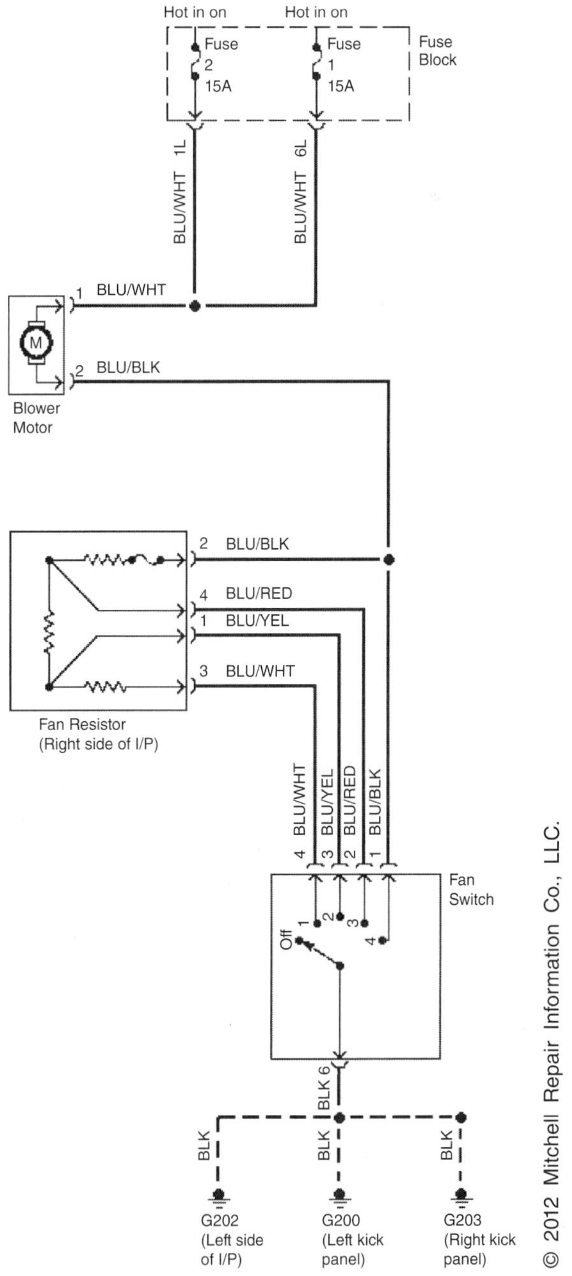

FIGURE 46-15 A blower motor circuit (non-computer control).

Cooling Fans and Circuits

Electric cooling fans for the radiator are becoming more common. This is particularly true for transversely mounted engines. They are also used as supplementary cooling fans on vehicles equipped with air conditioning. Cooling fans in modern vehicles are usually switched by relays. The relays are controlled by an electronic control unit. They are switched based on information from the coolant temperature sensor. Control units can be quite complex and can incorporate variable speed and on/off control. The control circuits are designed to keep the engine temperature at its most efficient temperature.

The basic electric cooling fan circuit contains the battery or power supply, fuse, relay, fan control circuits, and fans (**FIGURE 46-16**). The actual circuit and its complexity will vary by manufacturer and vehicle model. Examine the circuit diagram or schematic for the fan system before attempting any diagnosis.

Testing Electric Motors

Electric motors wear out and fail. Their brushes, bearings, and windings go bad. But they can also be affected by faulty electrical circuits and control devices. DMMs will be used to measure voltages and test fuses, relays, switches, and power and ground connections. You will also need to visually inspect the external portion of the motor and any drive mechanism for damage. These issues can be the cause of no motor operation.

The DC motors may be located in doors or in locations that require the removal of vehicle trim components. These pieces are prone to breakage, so disassemble them carefully. Permanent magnets within DC motors are brittle and may be damaged if the outer casing is struck with a hammer or other hard objects.

Diagnosis and repair activities require you to work with motors, gears, and mechanical levers. These may create a crush injury hazard so keep fingers away from mechanisms. Always wear the correct protective eyewear and clothing. And use the appropriate safety equipment. This includes fender covers, seat protectors, and floor mat protectors.

To test the electric motor, follow the steps in **SKILL DRILL 46-3**.

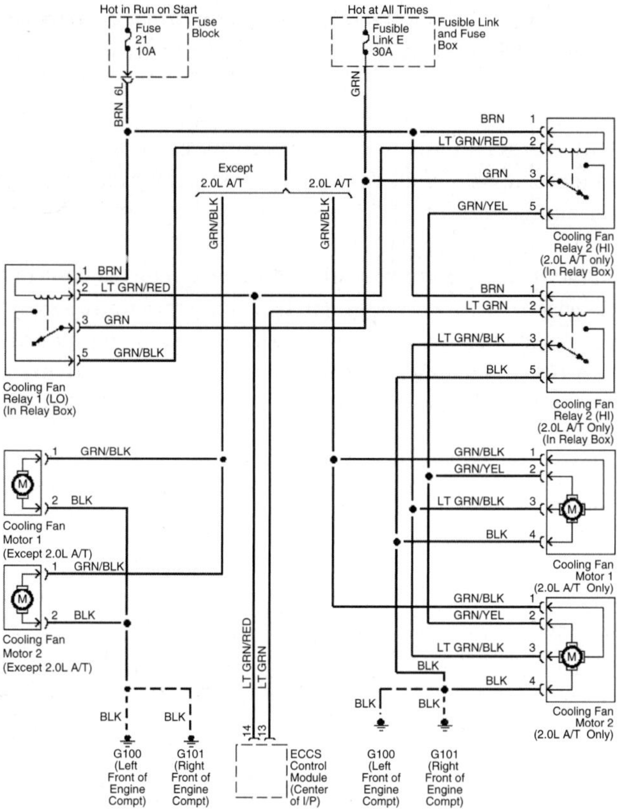

FIGURE 46-16 An electric cooling fan circuit.

SKILL DRILL 46-3 Testing the Electric Motor

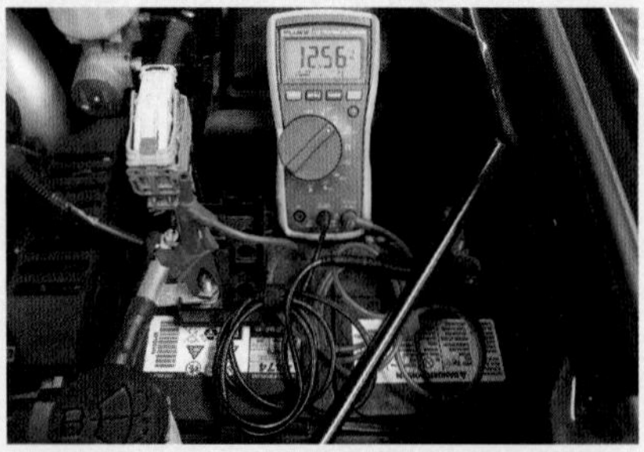

1. Research circuit operation and circuit diagrams for the motor-driven accessory requiring repair, from the manufacturer's information. Measure and record the voltage at the battery.

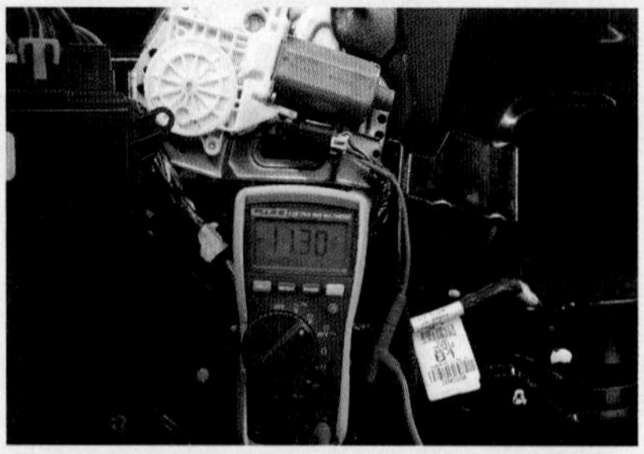

2. Use a DMM to measure the power and ground at the motor while it is being operated. If the voltage reading is more than 1.0 volt lower than battery voltage, perform voltage drop tests on each side of the circuit to determine which side has the fault.

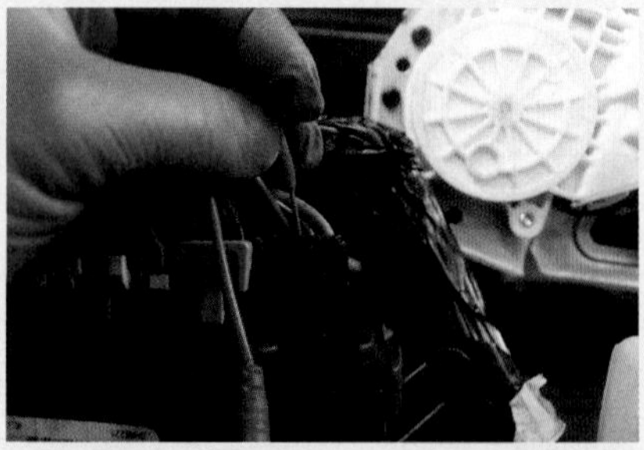

3. Once the faulty side has been identified, measure the voltage drop on each section of the circuit to locate the high resistance. If the total power and ground voltage to the motor were within 1.0 volt of battery voltage, there is a fault in the motor. Determine any necessary actions.

Applied Science

AS-66: Electricity/Measurement: The technician can demonstrate an understanding of the correct procedure to measure the electrical parameters of voltage, current, and resistance.

A vehicle is in the shop for multiple electrical problems. The systems affected are the power seats, power windows, power door locks, and interior lighting. This problem was initially intermittent, but at this point the affected systems are not functioning.

In diagnosing the vehicle, the technician measures the voltage across one of the power seat motors. The voltage is several volts below the battery voltage when the seat switch is operated. The technician notices that these circuits share a common ground connection. So a voltage drop test on the ground side is needed. The technician attaches a long jumper lead to the negative battery post and the black test lead of the DMM. The DMM is set to read DC volts, and the red lead is connected to the power seat motor ground terminal. The seat is operated, and the voltage drop measured. A good ground should have a reading of 0.50 volt or less. The technician finds a very high voltage drop reading indicating the fault is on the ground side. Using the service information, the technician locates the four ground connections secured by a single bolt behind the driver's side kick panel. The terminals are loose and corroded. Emery cloth is used to clean each ground connector on both sides. Then, the frame is cleaned, and the bolt securely tightened. The DMM reading is now 0.07 volt, and all circuits are working properly.

AS-67: Ammeter/Voltmeter: The technician can demonstrate an understanding of how to correctly measure electrical current and voltage in a circuit.

To measure electrical current with the DMM, it is helpful to know that current is the volume of electrons flowing in a circuit. Amps and milliamps are a measure of the volume of electrons. To measure current below 10 amps, place the meter in series with the load. If the current is expected to be above 10 amps, an inductive pickup must be used.

Turn off the power to the circuit, and determine where the test leads should be attached. Put the black test lead into the COM slot. Plug the red test lead into the 10 amps or 300 milliamps slot, and select DC amps. Open the circuit at a connector and connect the test probe's tips to each connector so current will flow through the meter in series. Turn on the circuit and read the meter.

To measure voltage with a DMM, it is helpful to know that voltage is the electrical pressure that causes current to flow. Voltage is the difference in electrical potential between two points in a circuit. To make a voltage measurement, the meter leads must be connected across the load or component being measured. To measure battery voltage, place the red lead on the positive post of the battery. Place the black lead on the negative post of the battery. A fully charged automotive battery should be approximately 12.66 volts at 70°F (21°C).

▶ Power Door Locks

LO 46-05 Test power door locks and remove door panels.

Power or electric door locks are a common option in modern vehicles and are installed in each door. They are often integrated with the vehicle's security system and remote locking systems. They are often called central locking systems. Many vehicles also have trunk locks incorporated into the system. The driver's side door is usually a master door lock. This means that when it is locked or unlocked with the key (or the remote fob), the other locks follow suit.

Power door locks use an electric actuator to move the mechanical door lock mechanism. There are two main types of actuators: a DC motor actuator and a solenoid type. The DC reversible permanent magnet motor type is enclosed in a single casing with a number of gears for speed reduction. The gears operate a rack-and-pinion gear set. It converts the rotational motion into linear motion to operate the lock (**FIGURE 46-17**). A centrifugal clutch built into the motor main gear allows the door locks to be operated manually.

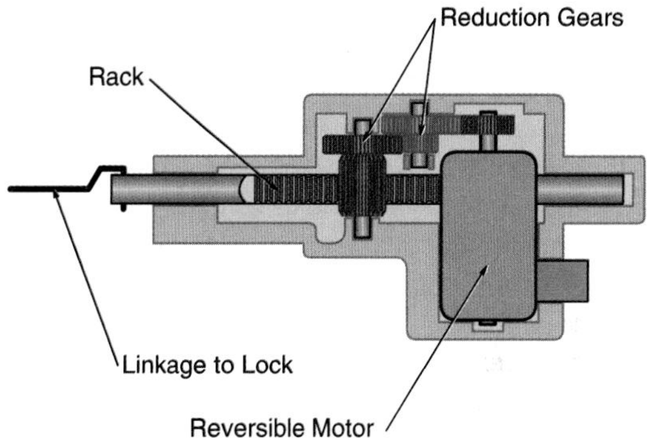

FIGURE 46-17 Power door lock actuator—DC motor type.

In a solenoid-type actuator, there is a permanent magnet plunger with a stationary coil wound around the outside. The north and south poles on the plunger react with the magnetic field created in the coil. This produces linear movement of the plunger (**FIGURE 46-18**). The plunger's direction of movement can be reversed by controlling the direction of current flow through the coil. Changing directions of the current flow changes the north and south magnetic fields that are created. The fields either attract or repel the plunger. The connecting rod of the plunger is connected to the door locking mechanism. The electric door lock circuit consists of the power supply, fuses

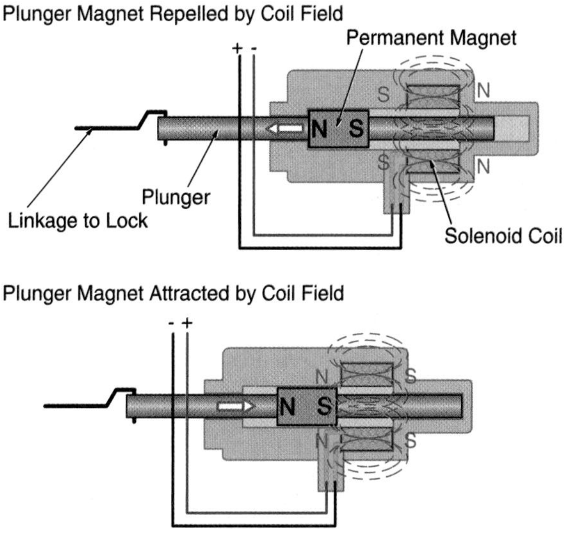

FIGURE 46-18 Power door lock actuator—solenoid type.

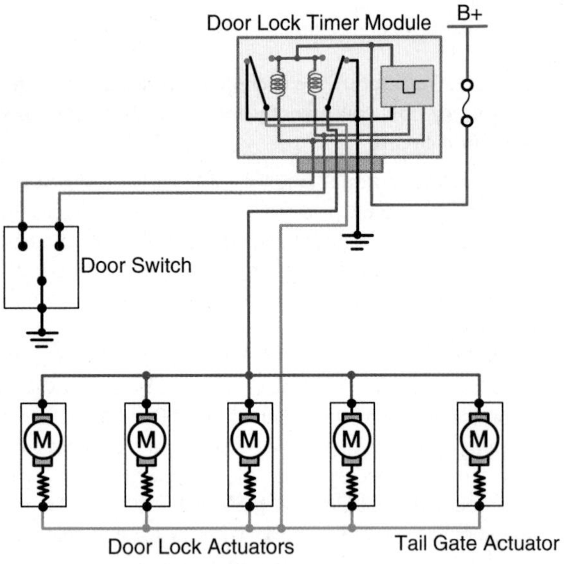

FIGURE 46-19 Typical power door lock circuit.

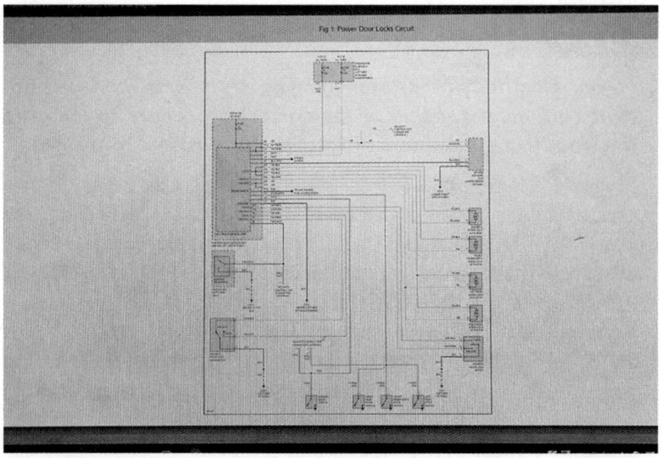

FIGURE 46-20 Typical non-CANbus power door lock wiring diagram.

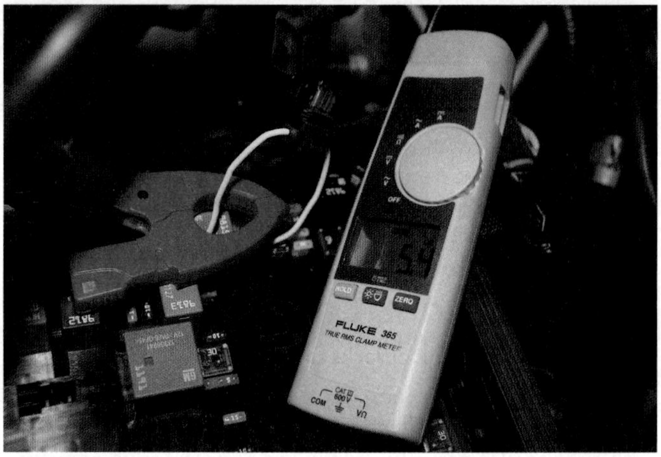

FIGURE 46-21 Using an inductive ammeter around a jumper lead placed in the fuse holder allows you to measure the current flowing when the door lock switches are being activated.

or circuit breakers, the actuator (motor or solenoid), relays, switches, and wiring. Relays may be used to control the motors or solenoids (**FIGURE 46-19**).

Testing Electric Locks

Door lock systems can fail for a variety of reasons, from a dead battery in the remote fob to worn-out door lock switches and actuators. If the system is a CANbus system, the network or control module could be faulty. When starting a diagnosis on the electric lock system, it is good to verify what works and what doesn't. For example, if the system works from the lock button on the driver's door, but not on the passenger door, you probably want to focus on the passenger door switch.

Once you have an idea of how the system is functioning, it is a good idea to consult the system's wiring diagram (**FIGURE 46-20**). This will help you to understand how the circuit is supposed to work. Then, you can build your diagnostic strategy from there. At that point, it is probably time to break out the DMM, automotive test light, or scan tool. They will help you trace how electricity is flowing (or not flowing) in the circuit.

It is important to measure current flow through circuits when switches are activated. This can be done by having an inductive ammeter connected around a jumper lead placed in the power door lock fuse holder (**FIGURE 46-21**). You will be looking for opens, shorts, high resistance, and improper signals in the electrical circuit. Also look for mechanical faults in the motors, solenoids, and linkage.

Removing the Door Panel

Removing door panels may require the removal of the arm rest, door lever, window crank (if equipped), and any switch panels mounted on the door. In most cases, the door panel itself will be attached to the door using clips. They will need to either be pried out of their holes, or slid off of their stops. Some manufacturers use a few screws to provide extra security. To pry the door panel from the door, it is best to use a door panel tool to prevent damage to the door or paint finishes. Special tools may be required to release the clip that holds the manual window handles in place, if equipped. Preserve the inner lining for reuse by carefully peeling it back from the door frame. Carefully examine the door panel for screws and fastenings that may need to be removed before attempting to pry off the door panel.

To remove the door panel, follow the steps in **SKILL DRILL 46-4**.

▶ Electric Lock and Keyless Entry Systems

LO 46-06 Describe the operation of keyless entry/remote start systems.

Keyless entry systems and the engine immobilizer or security systems are separate systems. However, they are usually linked so that the security system is enabled or disabled as the vehicle is locked and unlocked. Keyless entry is also known as remote key locking. Additional features such as alarms and warning devices may be incorporated into the system. The system is usually

SKILL DRILL 46-4 Removing the Door Panel

1. Research removal and reinstallation of the door panel from the manufacturer's service information. Remove fixtures such as the arm rest.

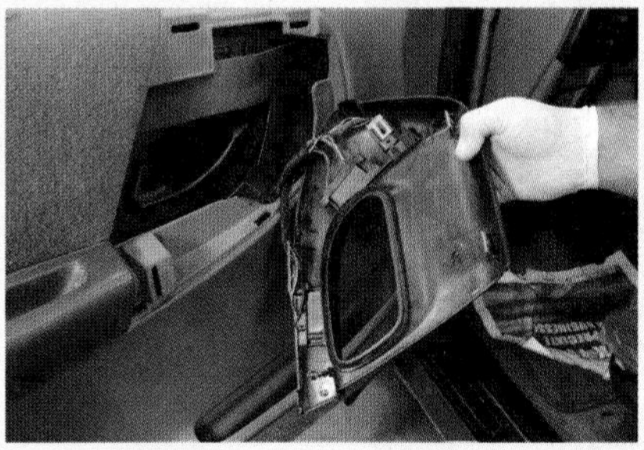

2. Remove the switch panel.

3. Remove the cables, and carefully pry the panel from the frame.

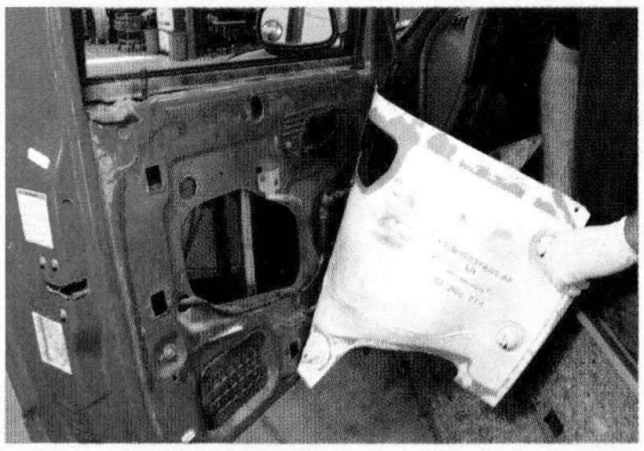

4. Peel back the inner liner from the frame, preserving it for reuse.

5. Reinstall the door panel using the reverse procedure while ensuring all electrical connections are reinstalled.

FIGURE 46-22 Wireless key fob with buttons.

controlled by a wireless fob. It is used to lock and unlock the vehicle, provided it is near enough (**FIGURE 46-22**).

Some vehicles have a remote start button on the key fob. It allows the vehicle's engine to be started remotely when the button is pushed. This is very useful in extremely cold or hot weather.

On networked vehicles, many systems are programmed to perform actions automatically. Some examples are:

- locking the vehicle when it is put into gear (automatic transmission),
- locking the vehicle when it reaches a certain miles per hour (mph), and
- unlocking it when put back into park.

The driver can program most of these systems to enable or disable these features. Some vehicles have stepped unlocking. This allows one push of the remote fob to unlock the driver's door only and two pushes to unlock all doors. This feature can be programmed on some vehicles as well.

Types of Systems

Manufacturers continue to develop various systems for controlling vehicle access and starting operations. And yet, each has its own subtle differences. Despite the differences, the systems tend to fall into two common types: **remote keyless entry (RKE)** and **passive keyless entry (PKE)**. The RKE system uses a fob transmitter that is usually attached to the key or key ring. The fob has a number of buttons or keys for remotely locking and unlocking the doors or trunk. It may also have a panic button that can activate the horn when pushed.

The PKE system uses a fob transmitter but does not require any action by the user. The vehicle senses if the fob is within range of the vehicle. If so, it automatically unlocks the door when the person touches the handle, without the need to press any buttons (**FIGURE 46-23**). It can also lock the doors automatically once the driver walks away from the vehicle.

FIGURE 46-23 Passive keyless entry fob automatically communicates with the vehicle. Notice that the remote key is extended.

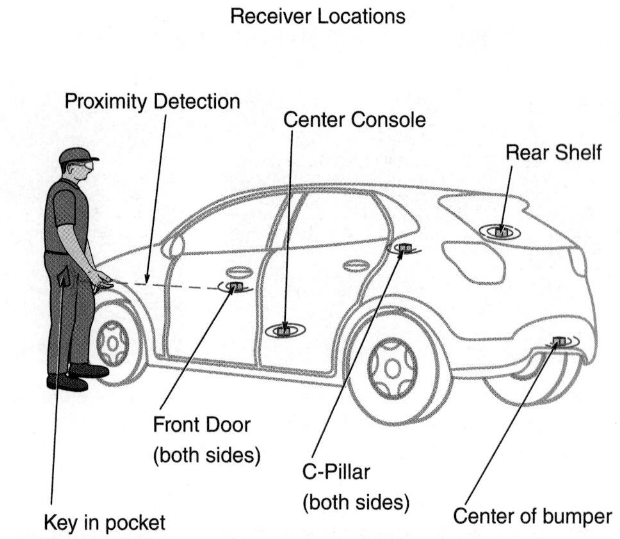

FIGURE 46-24 Typical layout for a passive keyless entry system.

Principles of Operation

The RKE system consists of a fob transmitter, and a receiver inside the vehicle. When a key or button is pressed, the fob transmits a coded signal that is received by the vehicle's receiver, provided it is within range. The BCM recognizes the key's code and activates the vehicle's electric door locks.

The PKE system consists of an electronic key that is carried on the person's body. A series of receivers determine where the electronic key is located once it is in close proximity to the vehicle (**FIGURE 46-24**). Communication is then established between the vehicle's system and the key. If the key is authenticated, the vehicle unlocks the door when the closest door handle is touched, without the need for any buttons to be pressed. The same authentication system can be used to control the vehicle's remote start system, security system, and engine immobilizer.

▶ Horn Systems

LO 46-07 Test horn systems.

The vehicle horn is a sound-generating safety device to warn others of a vehicle's presence. Legislation requires vehicles to have a working horn and in most cases also prescribes the horn's required sound level. Single horns may be used, but often horns are installed in pairs. One horn has a high tone and the other with a low tone. This produces a harmonious sound and gives redundancy if one horn fails.

Horn, Relay, Switch, and Clock Spring

The sound of a horn is produced by the vibration of a metal diaphragm. The diaphragm is operated by an electromagnet switched by a set of contacts. The diaphragm is connected to an armature that moves inside an electromagnet (**FIGURE 46-25**). When the electromagnet is on, the diaphragm is pulled by the armature toward the electromagnet. As this occurs, a trip ring opens the contacts, allowing the diaphragm to spring back to its original position. This cycle happens many times a second, producing sound through the rapid vibration of the diaphragm.

The horn switch is usually mounted in the steering wheel. It requires a method to maintain an electrical connection to the circuit as the steering wheel rotates. On vehicles with a driver's side airbag, the clock spring, also called a spiral cable, performs this function. In vehicles without a driver's side airbag, a slip ring and brush assembly is used. The clock spring is also used to carry the electrical signals for the driver airbag circuit. It may also be used for the sound system or cruise control circuits if they are built into the steering wheel.

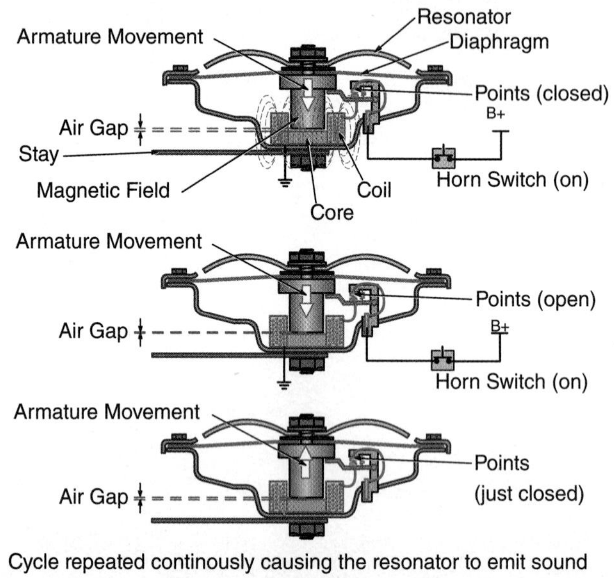

FIGURE 46-25 The inside components of a horn.

The clock spring is a rotating device with flexible ribbon cable. The clock spring rotates as the steering wheel rotates (**FIGURE 46-26**). It is designed to turn two or three times in each direction from the center point. It cannot rotate endlessly in a single direction. This means that it must be centered (timed) to match the position of the steering system when it is installed. Otherwise, it would reach the end of its travel before the steering wheel reaches its stop.

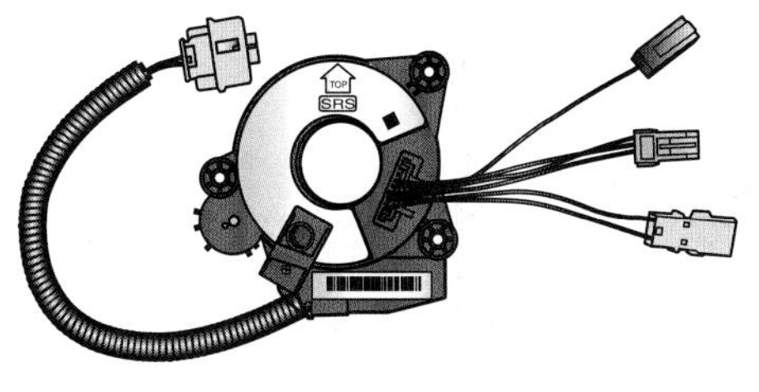

FIGURE 46-26 A clock spring.

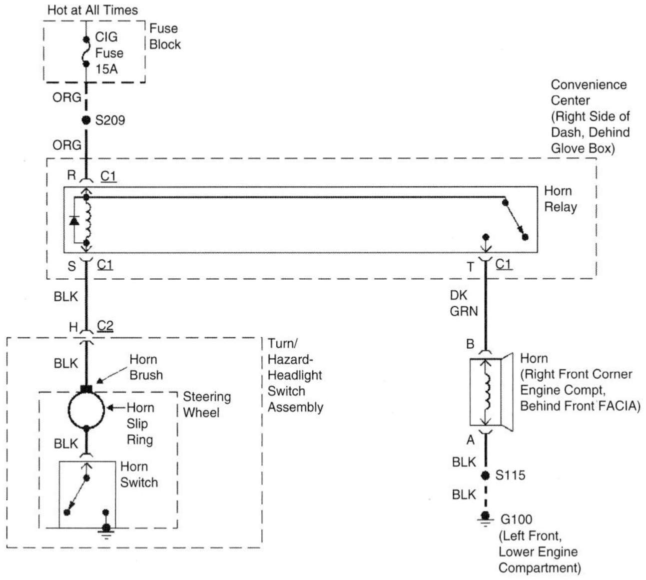

© 2012 Mitchell Repair Information Co., LLC.

FIGURE 46-27 A typical horn circuit.

The horn circuit is made up of the battery power supply, a fuse or circuit breaker, a relay, the horn switch, and the clock spring. The horn switch is usually part of the grounding circuit. The switch grounds the relay winding, which in turn supplies power through the relay contacts to operate the horn (**FIGURE 46-27**). The horns on many newer vehicles can also be operated by pressing the panic button on a fob. They may chirp when locking or unlocking, and honk when a breach of security has happened. In most cases, these horns are operated by a body computer when the correct inputs arrive.

Testing the Horn System

The horn circuit can fail in many ways. Any of the components in the circuit can go bad, as can the wiring. Using a DMM to measure voltages and test voltage drops on fuses, relays, switches, and power and ground connections will guide you when testing the circuit. The horn switch is usually mounted within the steering wheel and is usually connected through the steering column by a clock spring. The horn switch usually grounds the horn relay winding, which in turn sends power to the horn.

When testing the horn, it may be easiest to first test to see if power is getting to the horn when the horn is activated. It is not uncommon for horns to go bad, especially if they are rarely used or are overused. If power is getting to the horn, check the voltage drop on the ground side. If the voltage drop is fine, the horn is bad. In this case, tap the horn while it is energized. It is possible the contacts inside the horn are dirty or rusty. Tapping on the horn may free them up. If so, you will know that the rest of the circuit is operational. Even so, you will probably want to replace the horn as it is likely to go bad again soon.

If the horn does not have power when activated, it is good to go to the horn relay to perform the next tests. At the relay, you have access to each of the legs in the circuit:

- power from the horn fuse,
- the output terminal to the horn,
- power from the fuse feeding the winding, and
- the horn switch terminal.

Using the DMM and a jumper lead, you can check each of these legs. Remove the relay and using the DMM, check for power from each fuse. If present, use the jumper lead from the horn fuse input to the horn output (at the relay base) and see if the horn honks. If it does not honk, check the wiring between the relay and the horn. If it does honk, use a DMM to check the continuity of the horn switch leg with the horn switch depressed. There should be continuity between the horn switch terminal on the relay block and ground. If there is, then the relay is faulty.

▶ TECHNICIAN TIP

It is common for technicians who suspect a faulty relay to swap a known good relay for the horn relay. If this fixes the problem, then the horn relay is faulty. If you do this, make sure you use a matching relay. Not all relays are the same.

To diagnose the horn system, follow the steps in **SKILL DRILL 46-5**.

▶ Wiper/Washer System

LO 46-08 Test wiper and washer systems.

The wiper and washer system is an important vehicle safety system. The wipers ensure that the driver has a clear line of sight through the windshield. They do so by removing any excess moisture from the glass. The wipers usually have a number of speeds and intermittent operation for varying rainfall conditions. The washer system provides a cleaning spray. When used in conjunction with the wipers, it helps to clean road grime from the windshield. Some vehicles use a wiper and washer for the rear glass, and even some headlights as well (**FIGURE 46-28**).

Wiper and Delay Circuits

The wiper motor usually has at least a high and low speed and a time delay or intermittent operation. Intermittent operation is controlled by a timer circuit. It can be a discrete electronic timer. Or it can be controlled by the vehicle body computer. Incorporated into the circuit and activated by the motor gear is a parking switch. It ensures that the motor stops when the wiper blades are in the park position.

The DC motor is dynamically braked to make sure the motor stops instantly, without overrun, in the park position. To

SKILL DRILL 46-5 Testing the Horn System

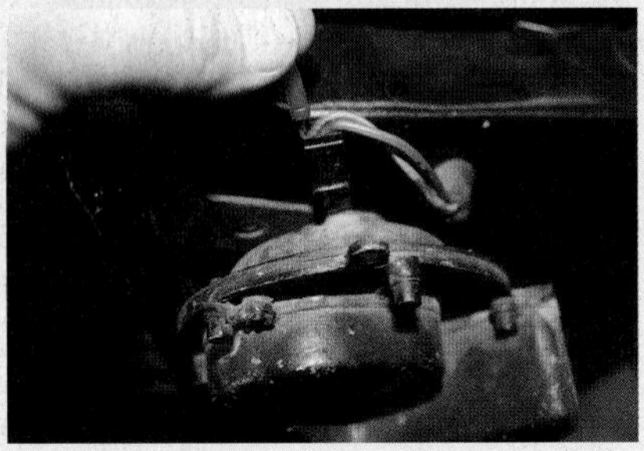

1. Confirm that the horn does not operate. Research circuit operation and circuit diagrams for the horn from the service information. Check for power and ground at horn when the horn switch is operated. If power and ground are present, the horn is faulty and needs to be replaced.

2. If there is no power to the horn, remove the relay, and check for battery voltage at terminal 30 (3). If battery voltage is not present, check the horn fuse and the rest of the feed circuit.

3. If battery voltage is present, jump terminal 30 (3) to terminal 87 (5). The horn should honk. If it doesn't, check for an open wire between terminal 87 (5) and the horn with a DMM.

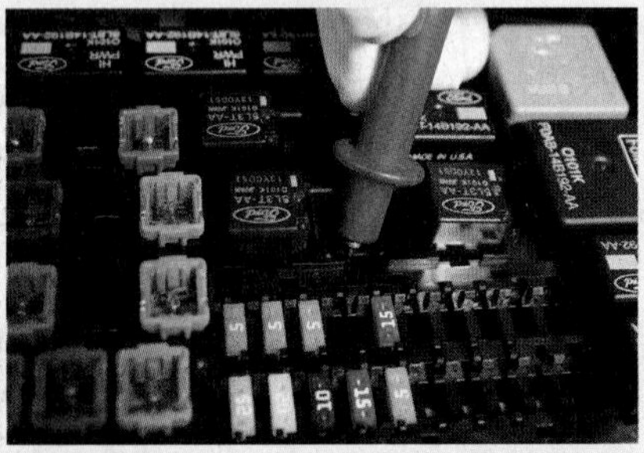

4. If the horn now honks, check that either terminal 85 (1) or 86 (2) has battery voltage. If 85 has battery voltage, then 86 is controlled by the horn switch, or vice versa.

5. Connect a DMM between the switch leg terminal and ground. Operate the horn switch; resistance should decrease to less than an ohm. If it doesn't, search out the high resistance or open circuit back to the horn switch. Suspect either the clock spring or the horn switch itself.

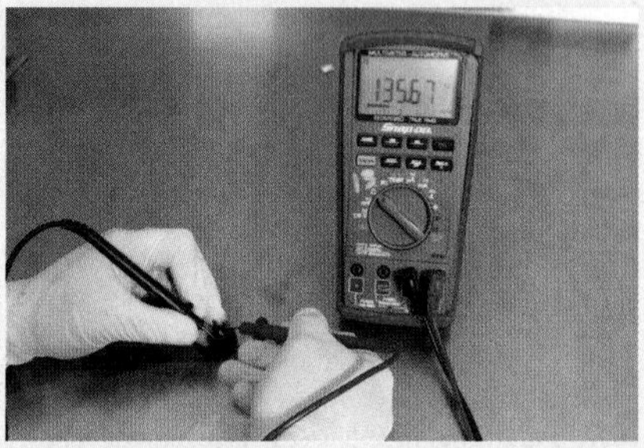

6. If the switch leg is good, measure the resistance between terminals 85 (1) and 86 (2) on the horn relay itself. The resistance should be between about 40 and 100 ohms. If not, replace the relay.

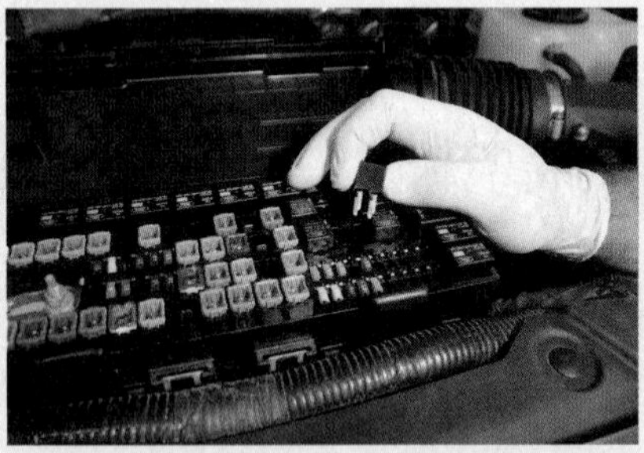

7. If the resistance is good, either measure the voltage drop across the switch contacts of the relay 30 (3) and 87 (5) or substitute a known good relay.

provide high and low speed, the wiper motor is equipped with three brushes. Two brushes are placed 180 degrees apart, as they are in a standard motor. An additional or third brush is located off-center (**FIGURE 46-29**). When current is switched from the low-speed brush to the high-speed brush, it provides higher rotational speed of the armature. This is due to the different interaction of the magnetic fields created by both brushes.

The wiper circuit consists of the battery, a fuse or circuit breaker, the ignition switch, the wiper switch, an intermittent timer and/or body computer, and the motor assembly with park switch (**FIGURE 46-30**). The park switch is built inside the motor and gearbox housing. It switches the power feed to the low-speed brush when the wiper switch is in the Off position and the wiper blades move toward the park position (**FIGURE 46-31**).

FIGURE 46-28 Some headlights use a wiper and washer to keep them clean.

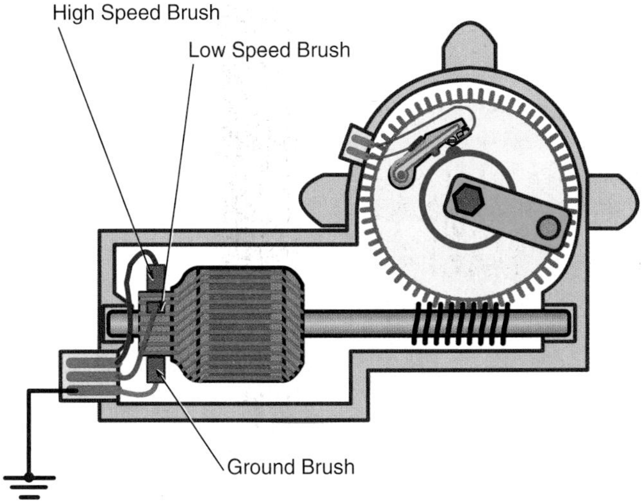

FIGURE 46-29 Many wiper motors use three brushes to get both a low and high speed.

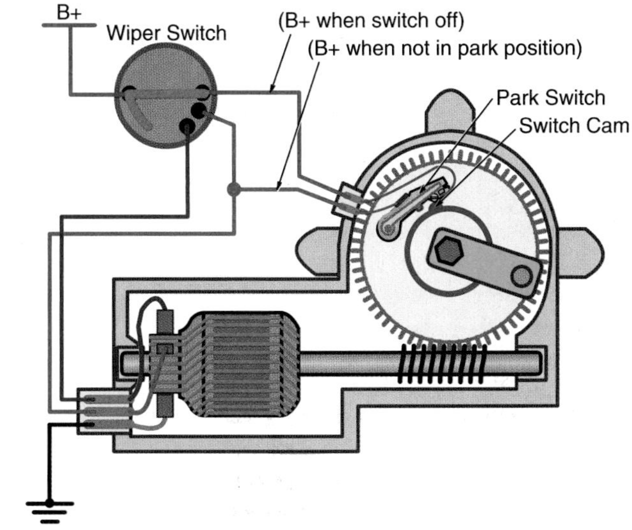

FIGURE 46-30 A wiper circuit diagram.

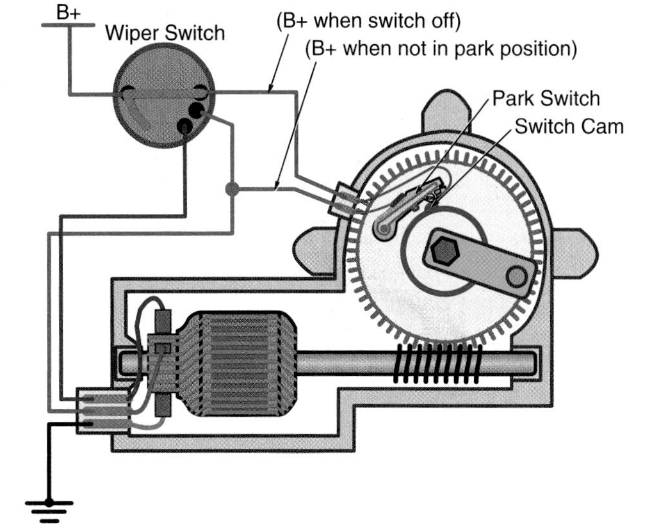

FIGURE 46-31 The park switch allows the wipers to be powered until they reach the park position, and then they stop.

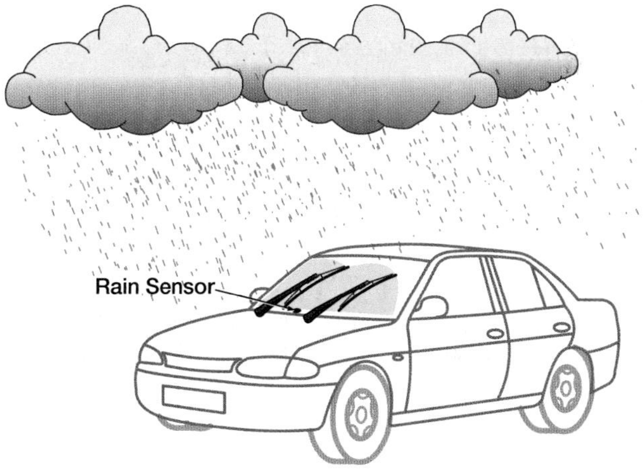

FIGURE 46-32 Rain sensor.

Once in the park position, the park switch opens the circuit to the motor. Dynamic braking of the motor occurs, bringing the motor to a quick stop. Intermittent operation is achieved by pulsing the low-speed circuit with a temporary power, causing the motor to move off the Park position. With the power pulse removed, the motor continues on a single wipe of the windshield through the park switch until it again reaches the park position.

Wiper circuits can also be ground-switched circuits. For example, the ignition switch provides a power feed to the motor. The path is through a fuse or circuit breaker, and the wiper switch, depending on the switch position. The high- or low-speed brush is then switched to ground. Some vehicles use a rain sensor mounted in or near the windshield (**FIGURE 46-32**). It signals the body computer when water is on the windshield. The body computer then operates the wipers accordingly.

Testing the Wiper System

The windshield wipers can fail in a number of ways. First, they may not work at all, such as when a fuse or motor is bad. Second, they may work on only some speeds, such as when the switch or one of the brushes in the motor is faulty. Third, they may not work on the intermittent speed, such as when the delay module is faulty. Fourth, they may not park, as when the park contacts in the motor are faulty or the wiper switch is bad. Each of these faults can be diagnosed by using a DMM and referring to the wiring diagram to locate the fault. Another fault with wiper systems is that the linkage wears out, which can prevent one or both of the wiper arms from operating, or operating properly. To diagnose this, you have to inspect the linkage for worn or disconnected joints.

This activity requires you to work with motors, gears, and mechanical levers, which may create a crush injury hazard. So keep fingers away from mechanisms. Also realize that if you remove the wiper blades and allow the wiper arms to slam back in place, they can break the windshield. Always place a folded-up fender cover or heavy cardboard on the windshield when removing wiper blades.

Testing the Washer System

Windshield washer systems are made up of:

- a washer fluid bottle to hold the washer fluid,
- a DC electric pump,
- nozzles to spray the glass,
- tubing to carry the washer fluid from the pump to the nozzles, and
- the electric circuitry.

The DC motor on the pump is a permanent magnet motor. It has power supplied to one brush and ground supplied by the other brush. The switch for the washer is usually located on the same switch assembly as the wipers. Operating the washer switch usually also activates the wipers for at least one cycle. However, some systems allow for multiple wipe cycles before parking the wiper blades.

Both the electrical circuit and the pump discharge circuit need to be tested for correct operation. The electrical circuit consists of the power supply, fuse, switch, and DC pump motor, and can be tested with a test lamp or DMM. Test for improper voltage and voltage drops in the circuit. If the voltage and ground to the pump are good, then test the pump for continuity. Performing these tests will identify any electrical faults.

The pump or washer fluid discharge circuit is made up of the pump, the washer fluid supply or tank, the tubes that carry the washer fluid to the washer nozzles, and the nozzles. You may find that blowing air through the discharge circuit and nozzles cleans out any debris that may be plugging the system. Just be careful that you don't overpressurize the system and cause an inaccessible hose to be blown off of a fitting. Always be sure to use the specified washer fluid at the correct concentration. In cold climates, improper concentrations of washer fluid can freeze, breaking the bottle. Also, ensure an adequate quantity of washer fluid is in the tank before conducting any tests.

To test the washer system, follow the steps in **SKILL DRILL 46-6**.

SKILL DRILL 46-6 Testing the Washer System

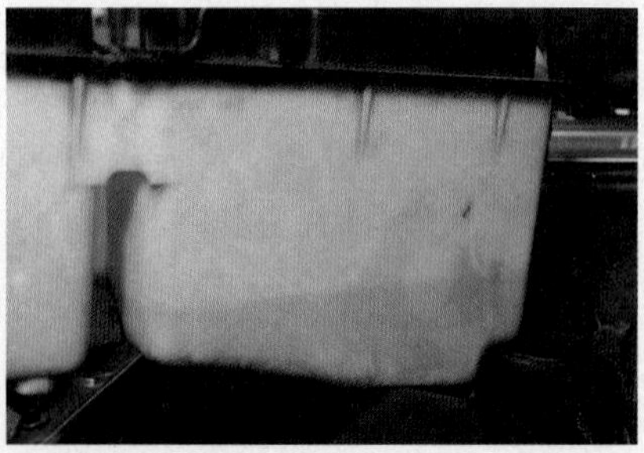

1. Check for the correct solution level in tanks, and listen for washer pump operation.

2. If an electrical fault is indicated by lack of a washer motor sound, check for battery voltage and ground at the pump when it is activated.

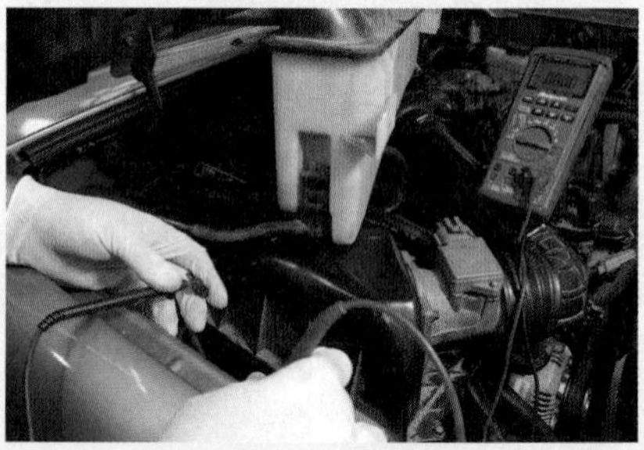

3. If the power or ground circuit is faulty, perform voltage drop tests on the faulty side until the fault is located.

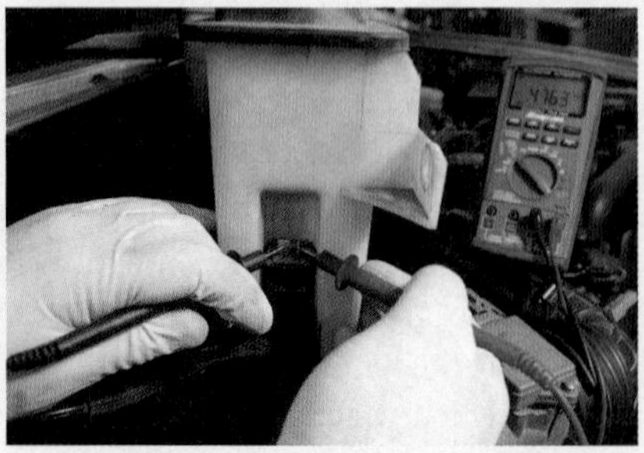

4. If the power and ground circuits are good, measure the resistance of the pump motor, and compare with specifications.

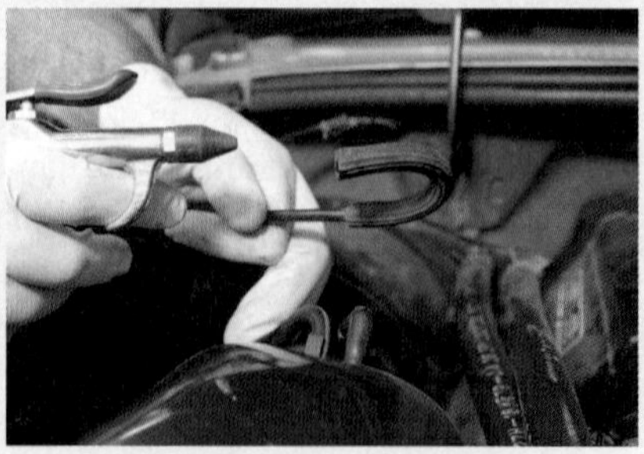

5. If the pump spins when activated but does not pump washer fluid, check the hoses and nozzles for obstructions. Blow out with compressed air if necessary.

▶ Supplemental Restraint Systems

LO 46-09 Describe supplemental restraint systems and how to disable–enable them.

The **supplemental restraint system (SRS)** is used to describe passenger safety devices. These include airbags (including side curtains and knee bags) and seat belt pretensioners (**FIGURE 46-33**) installed on vehicles. Manufacturers continue to develop and improve SRS technology to make vehicles safer. When first introduced, a single airbag was usually installed on the driver's side. Now, it is not unusual for modern vehicles to have more than six airbags. They may have these along with side-impact protection, known as side curtains (**FIGURE 46-34**). One of the newest additions is the pedestrian airbag. These devices are explained in the following sections.

Purpose and Operation of the SRS

Vehicle safety systems are designed to protect occupants during accidents. They can be classified as primary (passive) safety systems and secondary (active) safety systems. Primary safety systems are ready to operate in any accident. They include bumper bars, body panels, seat belts, crumple zones, and collapsible steering columns. A secondary safety system has to be activated to work. It is only necessary in severe accidents (**FIGURE 46-35**). The two most common types of secondary systems are airbags and seat belt pretensioners.

Seat belts secure and restrain the occupant within the seat and vehicle cabin. In minor collisions, they perform their task well. In a more severe impact, inertia causes the occupant to move forward with greater force. This increases the possibility of injury caused by the restraining force exerted by the seat belt. Or in extreme situations, from the occupant striking interior components. In these situations, airbags provide additional protection for the occupants.

In a vehicle equipped with airbags, they deploy during a severe collision, offering a greater degree of protection from injury. Airbags provide cushioning against the effects of inertia. The bag deploys toward the occupant's approaching body, inflated rapidly by pressurized nitrogen gas. Typically, inflation takes no longer than about three-hundredths (0.03) of a second. The airbag is not a nice soft pillow. It provides a strong

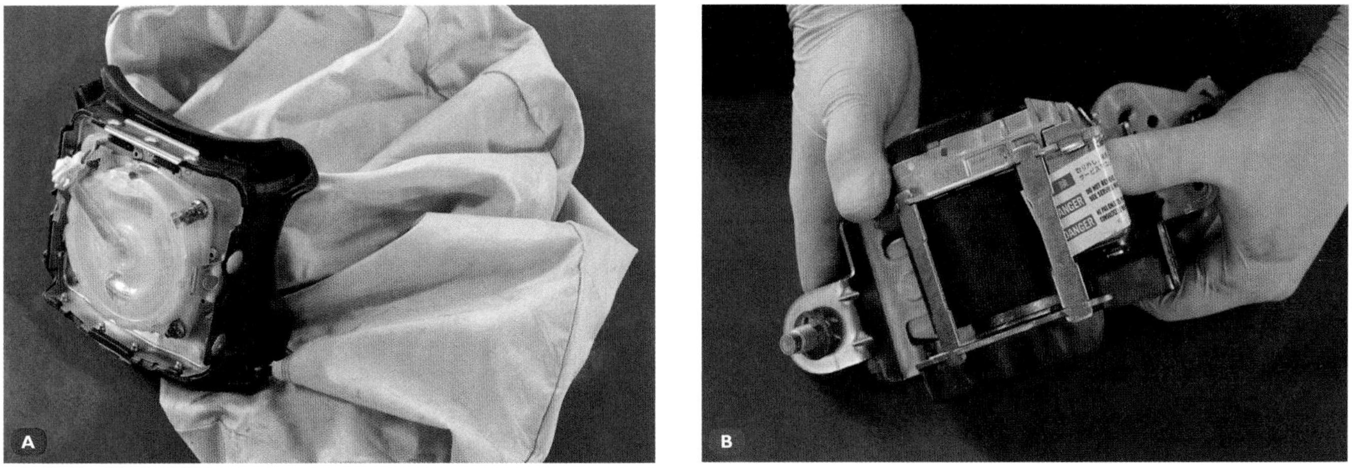

FIGURE 46-33 **A.** Airbag. **B.** Seat belt pretensioner.

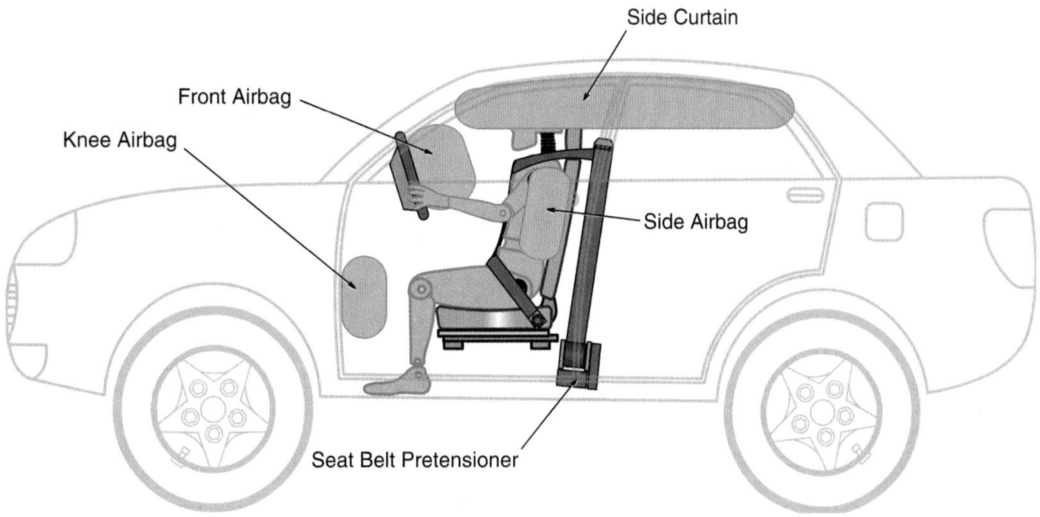

FIGURE 46-34 Typical SRS injury reduction devices.

counterforce to react against the inertia of the occupants. It is not designed to be comfortable. It is designed to minimize injury. Immediately after absorbing the momentum, the airbag deflates, having done its job.

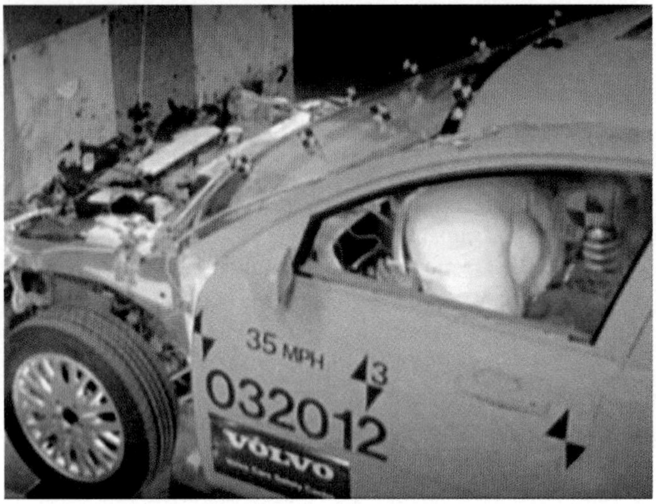

FIGURE 46-35 SRS in action during a collision.

Airbags and Pretensioners

Airbags are usually described as being "supplemental," but in some countries, wearing seat belts is not mandatory. In this case, airbags become the primary restraint mechanism. So they must be able to trigger at lower speeds and be larger in volume. There are a number of different types of airbags whose size and location are determined by the type of protection they offer (**FIGURE 46-36**). However, just because a vehicle has airbags doesn't mean its passengers don't need to wear seat belts. Seat belts are a primary safety system, meaning they are very important and should be worn. Airbags are designed to supplement seat belts, not replace them.

The most common location for an airbag is in the center of the steering wheel. Here, it protects the driver from frontal impacts. Airbags are also commonly installed on the passenger side of dashes for the same reason. Side-impact airbags are located in the sides of front seats to protect the occupants from side impacts. Curtain side airbags are located in the side edge of roof linings to protect the occupant's head from side impacts. Knee airbags are located under the dash and protect the lower legs. The airbag assembly consists of a nylon bag, **squib**, igniter, gas generator, and airbag-triggering mechanism (**FIGURE 46-37**).

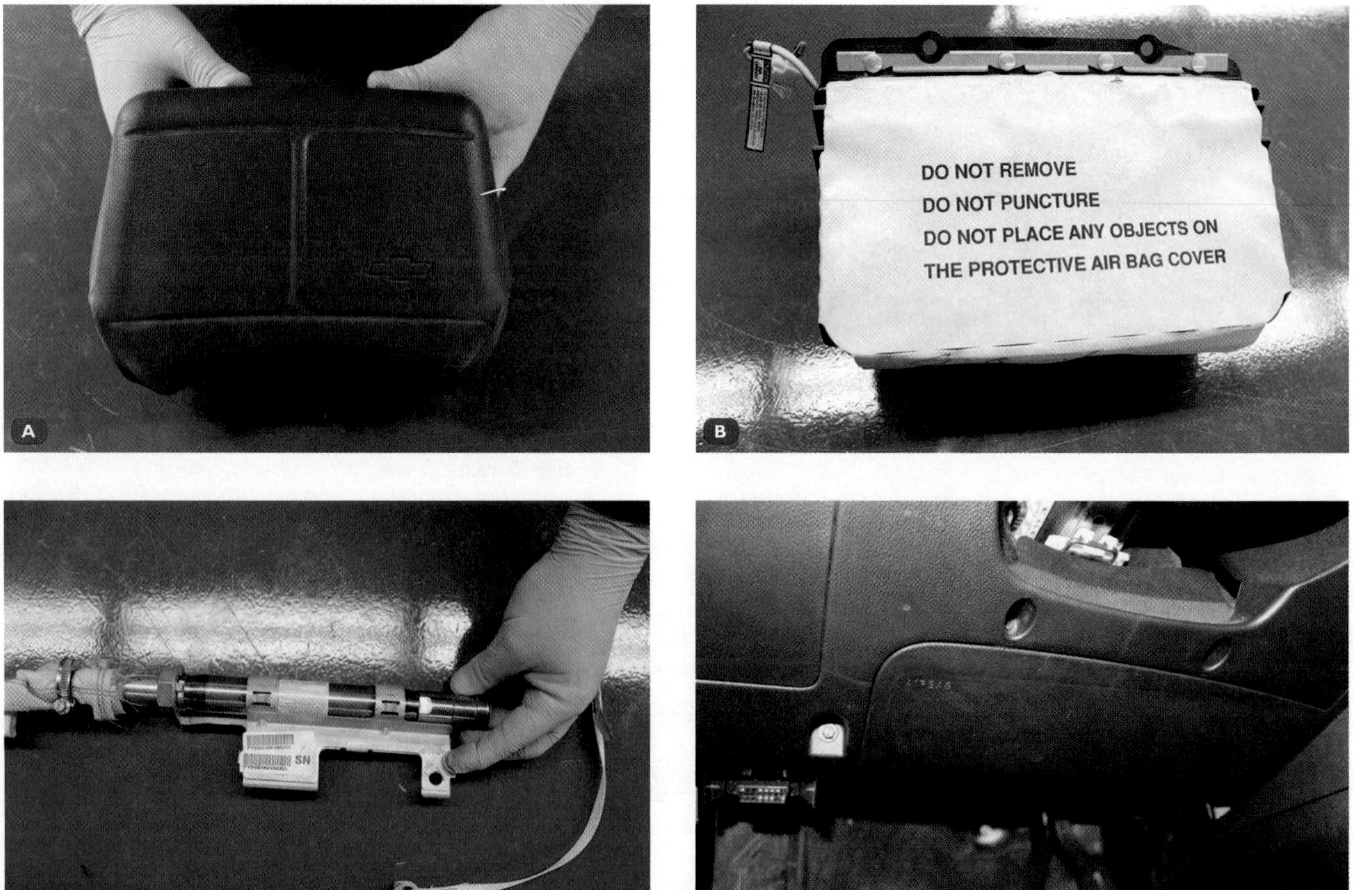

FIGURE 46-36 **A.** Driver's side airbag. **B.** Passenger's side airbag. **C.** Side airbag. **D.** Knee airbag.

There are two types of airbag-triggering mechanisms: electrical and mechanical. Most airbags are triggered electrically. A small electrical current is delivered from an SRS control module. Mechanically activated systems use inertia to move a triggering pin. The airbag deploys due to simultaneous actions occurring within the squib, the igniter, and the gas generator. This is regardless of the type of triggering mechanism. These three parts are located in a metal housing attached to the back of the airbag assembly.

When the control module determines that the airbag should be deployed, the electrical current triggers the squib. The heat generated causes the igniter to burn, which in turn ignites the gas generator. High-pressure nitrogen gas is produced, and the airbag rapidly inflates. When the airbag assembly is mounted, it sits behind a pad, which may have a fracture line cast into the inner face. The force of the gas generated when the airbag deploys causes the cover pad to rupture, allowing the bag to fully inflate (**FIGURE 46-38**).

SAFETY TIP

Airbags, when they deploy, inflate toward the occupant at more than 100 mph (160 kph). Because of this force and speed, they have been known to break wrists and even cause death when not used properly. Always keep the seat positioned so you are at approximately arm's length from the steering wheel. That is, the steering wheel should be at least 10" to 12" (25 to 50 cm) from your chest. Also make sure your hands are in the 9 o'clock and 3 o'clock positions. The old "10 and 2" position can lead to your hands being thrown back toward you or outward (remember that 100+ mph), both of which can injure you.

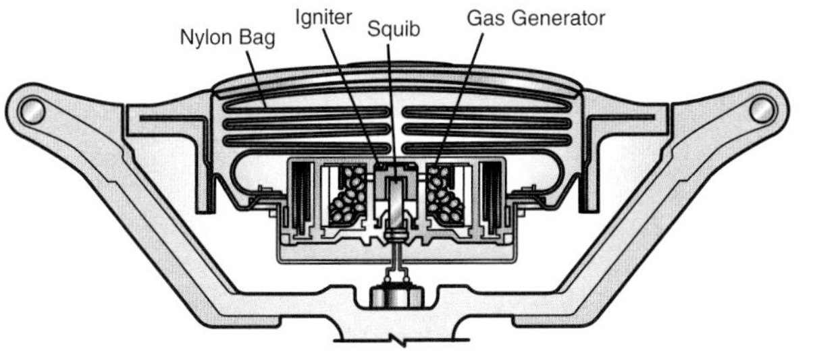

FIGURE 46-37 Parts of an airbag assembly.

FIGURE 46-38 You can see the fracture line on this airbag, where it will rupture if deployed.

Mechanically deployed airbags do not have any electrical circuitry. The squib is ignited with a firing pin. Under severe deceleration, inertia causes a steel ball to release a firing pin into the squib. Once the squib has been triggered, the deployment process is identical to electrically triggered airbags. The airbag is fully inflated within three-hundredths (0.03) of a second, cushioning the occupant as his/her body moves forward.

The airbag is made from nylon and is folded into the front face of the assembly. Older airbags were coated in cornflower or talcum powder, which acted as a lubricant for the fabric during deployment. Newer ones use either more slippery fabric or silicone to allow the airbag to inflate without binding. Relatively large holes are usually located in the rear face of the airbag. They allow the nitrogen gas to escape once the airbag has deployed (**FIGURE 46-39**).

A new type of airbag recently released is called a pedestrian airbag. It is located under the rear edge of the hood, typically in the cowl area. When a collision with a pedestrian is imminent, the hood retracts slightly, and the airbag inflates to cover a portion of the windshield (**FIGURE 46-40**). This helps to protect pedestrians from direct impact with the windshield.

Seat Belt Pretensioners

Seat belt pretensioners are used to tighten the seat belt in a severe frontal accident. There are three types of pretensioners: electric, mechanical, and pyrotechnic. Each is described in the text below. The most common type, sometimes called a ballistic pretensioner, deploys in an actual accident. It relies on an explosive charge that

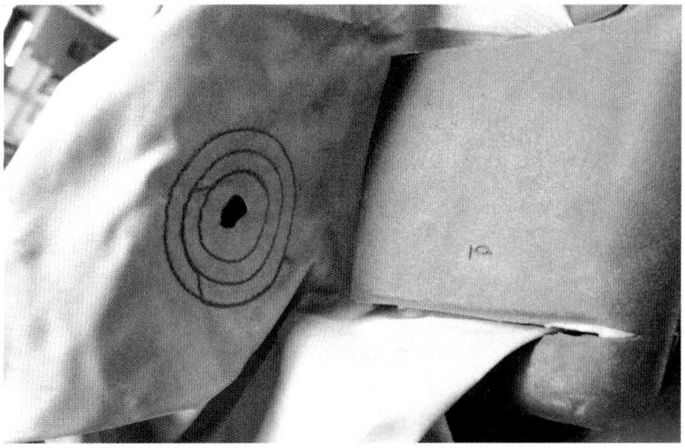

FIGURE 46-39 Large vent holes for deflating the airbag quickly after deployment.

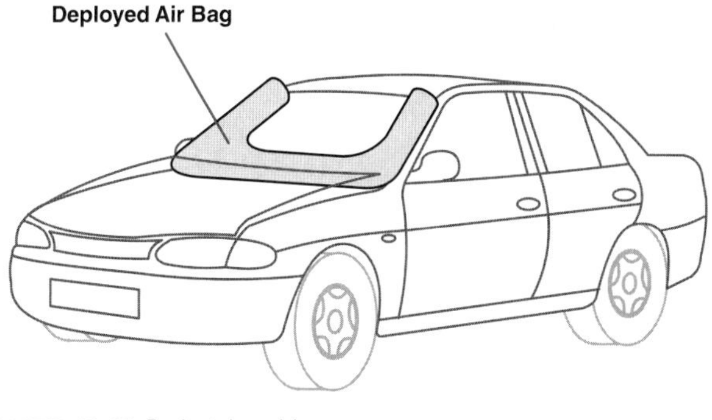

FIGURE 46-40 Pedestrian airbag.

is detonated electronically by an actuator within the seat belt tensioning mechanism (**FIGURE 46-41**). This explosion moves a piston that pulls on a steel cable, causing the belt to tighten by approximately 4" (100 mm). The design allows for the belt to tension before the occupant has moved forward in the seat.

Mechanical systems rely on inertia to move a sensing mass. During the beginning phase of an accident, the movement of the sensing mass releases a spring to pull on a cable, thus tightening the belt. Once the pretensioner has triggered, a ratchet prevents the seat belt from loosening. When the seat belt is removed from the buckle, it cannot be reinserted, and the assembly will have to be replaced.

A newer style of seat belt pretensioner uses an electric motor to pull in the slack on the seat belts (**FIGURE 46-42**). This type is reusable, meaning it can be activated repeatedly. Because of this, it can be deployed much earlier in a potential accident. On some vehicles, the system will tug on the seat belt to get the driver's attention, along with preparing for a collision.

Rip stitching is used on seat belts, in conjunction with an airbag and seat belt pretensioners. During a collision, the pretensioners initially pull the seat belt tight. However, the stitching gradually tears, allowing the occupant to move forward into the airbag at a controlled rate.

For safety reasons, these belts must be replaced once they have had their stitching ripped. Manufacturers generally fit warning labels within the fold to indicate the belt is to be replaced when the label is revealed (**FIGURE 46-43**). Ripping the stitching exposes the warning label.

Sensors, Control Module, and Circuitry

Crash sensors can be installed in various positions throughout the vehicle. Their location depends upon the direction of deceleration they are designed to detect. Some manufacturers place the sensors within the PCM. Others are located behind the front bumper, headlights, and dash (**FIGURE 46-44**).

Side-impact sensors are located in the doorsills or B-pillar. They inform the SRS control module of a side impact. The module then decides whether to deploy the left- or right-side airbags. The sensors indicate when a predetermined deceleration rate

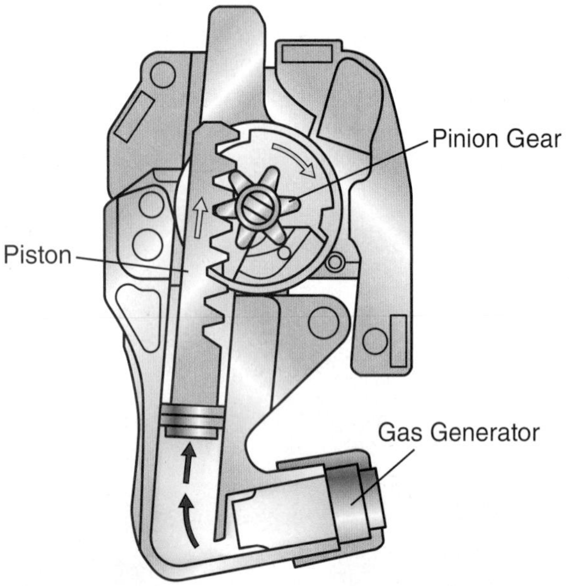

FIGURE 46-41 Seat belt pretensioner.

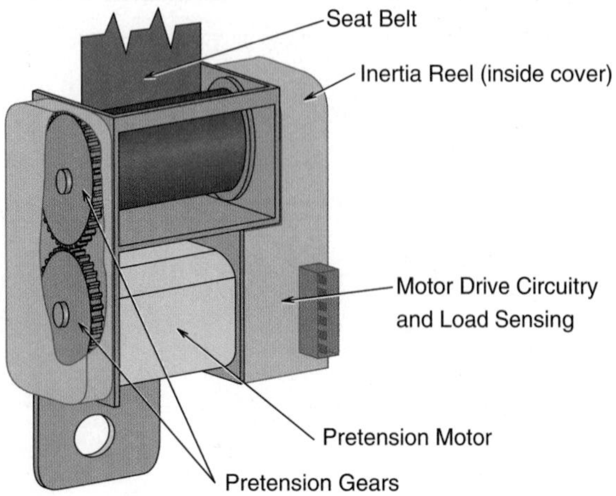

FIGURE 46-42 An electric motor–style seat belt pretensioner assembly.

FIGURE 46-43 Rip stitching rips away at a controlled rate so the deceleration rate of the occupant is not as severe.

FIGURE 46-44 Typical crash sensor bolted to the front radiator support on the vehicle.

has been exceeded. When it is from the appropriate direction, the control module deploys the relevant airbags.

If the collision is from the front, the driver and passenger airbags will deploy. If the collision is from the side, the module determines whether to deploy the seat-mounted airbag or the curtain airbags for one side of the vehicle. With more refined designs, the passenger airbag deploys only if there is an occupant in the seat. Deployment can also depend on the weight of the occupant and whether the passenger airbag switch, if equipped, is turned on.

Some systems include a safing sensor mounted within the SRS control module (**FIGURE 46-45**). This is to prevent incorrect and unnecessary deployment. The SRS control module will only pass current through the squib if both the safing sensor and a crash sensor show, at the same time, that a predetermined deceleration rate has been exceeded.

Capacitors within the SRS control module are used to store electricity. They act as a backup power supply if the vehicle has its battery destroyed or disconnected in an accident. The capacitors supply the electricity required to keep the SRS system operational for a short time. If a fault is detected in the system, the SRS warning light is illuminated and stays on.

Some seat-mounted side-impact airbags also operate without electricity. When the side of a vehicle is crushed inward, a detonator mounted on the lower outside edge of the seat is activated. Pyrotechnic tubes connect the detonator to the airbags, which in turn ignite the squib. Many vehicles use two-stage side-impact bags. This design provides protection to an occupant's upper torso over a more extended time.

If any of the SRS-related devices are deployed for any reason, most manufacturers require all of the deployed devices, the crash sensors, and the control module to be replaced. This can be quite expensive. It is especially true if you accidentally deployed one or more airbags while working on the vehicle.

Disabling and Enabling the SRS

The driver's side airbag is located in the top cover of the steering wheel. The SRS system must be disabled while working on or around the steering column. The system must also be disabled when working on any of the other airbags or other pyrotechnic devices, and any of the sensors. If not properly disabled, the airbag or other SRS devices could trigger accidentally. This can injure or kill the technician because of the large deployment force created by the ignited propellant. It is important to know that most airbags are inflated by igniting a solid fuel similar to rocket fuel. Unintended deployment also creates the need to replace the airbag assembly and other required parts. This can be very expensive.

Generally, to disable an SRS, the correct SRS fuse must be located and removed. And then the ignition switch is turned on and the SRS light is verified to remain on. Once the fuse is removed, the negative battery terminal is disconnected. The vehicle is allowed to sit for up to 15 minutes to discharge the capacitors. Note that a memory minder ***should not*** be used when working on, or around, the SRS system. Also, do not use a battery, voltmeter, ohmmeter, test light, or any other type of test equipment not specified in the service manual on any SRS circuit. This will help to avoid accidental deployment of the airbag, which could result in injury or death.

Some manufacturers have you disconnect the airbag connector. It is located under the dash near the steering column. Disconnecting this connector generally activates a shorting bar in the airbag side of the connector. The shorted connector makes it much harder for the airbag to accidentally deploy (**FIGURE 46-46**). With the SRS system disabled, you can carry out any needed diagnosis or repair on the steering column. Just remember to enable the SRS system once service is completed. Follow the vehicle's specific service information to properly disable and enable the SRS system.

To disable and enable the SRS, follow the steps in **SKILL DRILL 46-7**.

SAFETY TIP

Many airbags are now of the two-stage variety. They can deploy one or both stages to deliver the proper amount of force. It depends on the severity of the accident, approximate weight of the occupant, etc. This means that even a deployed airbag is still potentially dangerous as it may have one stage undeployed. So it needs to be treated with caution during removal. It should be deployed soon after removal, following the manufacturer's specified procedure, which will render it safe for disposal.

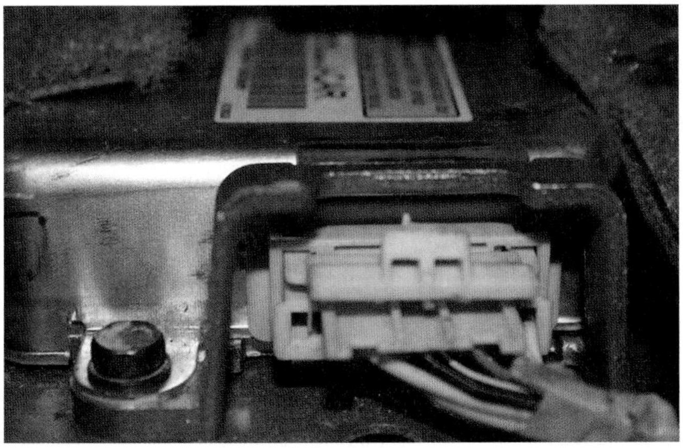

FIGURE 46-45 The safing sensor is typically located inside the SRS control module.

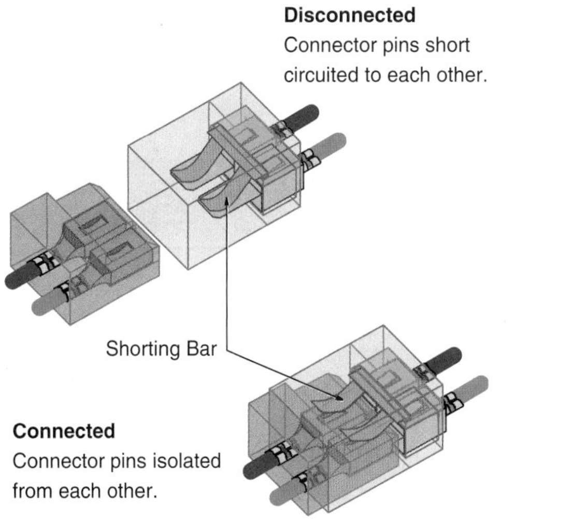
Disconnected
Connector pins short circuited to each other.
Shorting Bar
Connected
Connector pins isolated from each other.

FIGURE 46-46 A shorting bar inside most air bag connectors shorts out the terminals when the connector is disconnected. This helps avoid accidental deployment when the air bag is removed from the vehicle.

SKILL DRILL 46-7 Disabling and Enabling the SRS

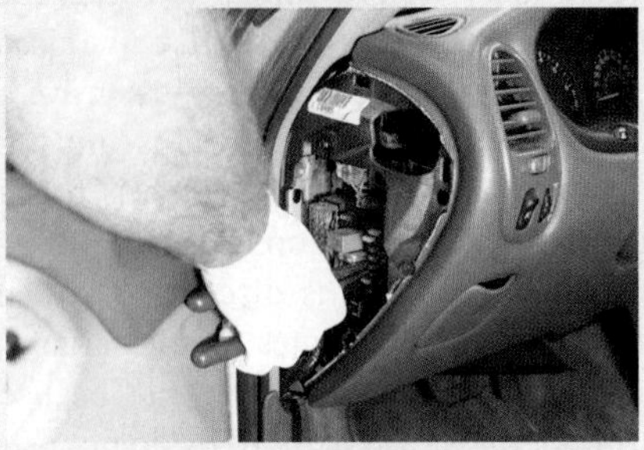

1. Find and remove the SRS fuse. Verify by turning the key on and observing that the SRS light remains lit for at least 30 seconds. If it goes out, you did not remove the correct fuse or all of the required fuses. Make sure the wheels are steered straight ahead. Turn the key off.

2. Remove the negative battery cable, and allow a minimum of 15 minutes to pass to let the SRS system capacitors discharge. Note any radio presets or other memory features of the vehicle that will be erased when the battery is disconnected. *Do not use a memory minder or auxiliary power source!*

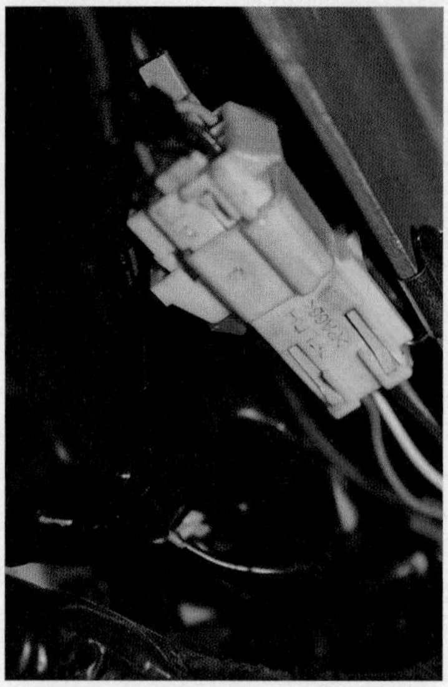

3. To enable the SRS system, verify that all SRS modules, components, and connectors are installed and connected properly.

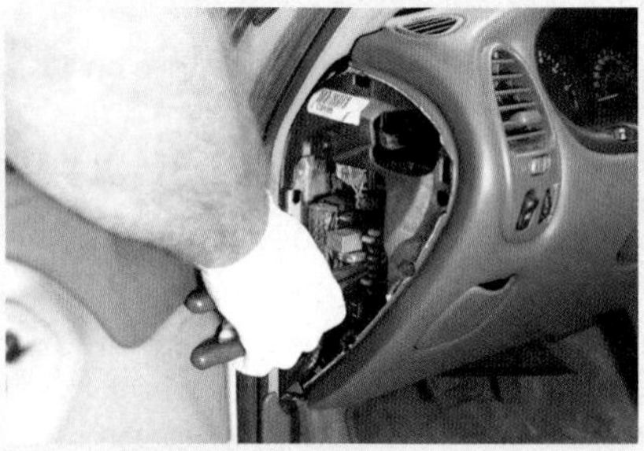

4. Reinstall the SRS fuse.

5. Reconnect the negative battery terminal, and tighten properly.

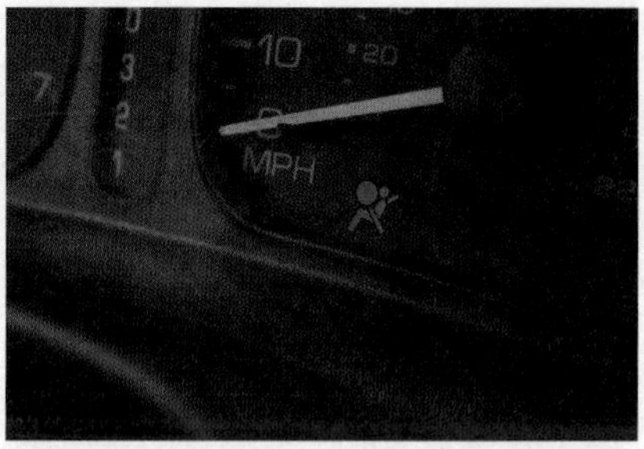

6. Without being in front of or reaching across the driver's side airbag, turn on the ignition switch and observe the SRS light. It should illuminate briefly and then go out and stay out. If so, the SRS system should be ready to be placed back into service.

▶ Wrap-Up

Ready for Review

- Integrating sensor information into a common wiring harness reduces the number of dedicated wires for each function. It also reduces system weight while improving reliability, serviceability, and installation. Networking of data also allows greater vehicle content flexibility.
- In a multiplexed system, information is transmitted across one or two wires that run to all of the control modules. Different types of vehicle networks include: LIN, LAN, CAN, CANFD, FlexRay, MOST, and Ethernet-based systems.
- A scan tool can be connected to the network DLC to check if there are any missing modules or not communicating, it can be used to actively test the output component to see if it operates, and it can be used to check the data list for a "state of change" in signal. Flash reprogramming tools can be used to transfer the software update of the vehicle's electronic modules to the vehicle's DLC.

- DMMs can be used to measure voltages and test fuses, relays, switches, and power and ground connections of electric motors. Circuit operation and circuit diagrams for motor-driven accessory requiring repair can be accessed from the manufacturer's information.
- Door lock systems can fail for a variety of reasons, isolating the general location of fault can help in diagnosis. Current flow through circuits can be measured by an inductive ammeter connected around a jumper lead placed in the power door lock fuse holder. Removal of door panels may require the removal of the arm rest, door lever, window crank (if equipped), and any switch panels mounted on the door.
- Remote keyless entry system consists of a fob transmitter, and a receiver inside the vehicle. When a key or button is pressed, the fob transmits a coded signal that is received by the vehicle's receiver, provided it is within range. The passive keyless entry system consists of an electronic key that is carried on the person's body. A series of receivers determine where the electronic key is located once it is in close proximity to the vehicle.
- First test when testing a horn is to see if power is getting to the horn when the horn is activated. If the horn does not have power when activated, the following legs in the circuit need to be tested: power from the horn fuse, the output terminal to the horn, power from the fuse feeding the winding, and the horn switch terminal.
- Windshield wipers can fail due to a bad fuse or motor, faulty motor brush, faulty delay module, or a faulty wiper switch. A DMM can be used to diagnose the faults and located by referring to the wiring diagram. Both the electrical circuit and the pump discharge circuit of a windshield washer system need to be tested for correct operation.
- To disable an SRS, the correct SRS fuse must be located and removed, and then the ignition switch must be turned on to verify that the SRS light remains on. The negative battery cable should be removed and a minimum of 15 minutes should be allowed to let the SRS system capacitors discharge. To enable the SRS system, the SRS fuse should be reinstalled and the negative battery terminal should be reconnected.

Key Terms

blower motor An electric motor, usually the permanent magnet type, that moves air over the air-conditioning evaporator and heater core.

body control module (BCM) An onboard computer that controls many vehicle functions, including the vehicle interior and exterior lighting, horn, door locks, power seats, and windows.

brushless DC motor An electric motor that does not have any brushes and is sometimes called an electronically commutated motor. In this type of motor, an electronic control module replaces the brushes and commutator.

data link connector (DLC) The under-dash connector through which the scan tool communicates to the vehicle's computers; it displays the readings from the various sensors and can retrieve trouble codes, freeze-frame data, and system monitor data.

multiplexed The carrying of multiple signals on one wiring circuit. Digital signals are muptiplexed on a dedicated network; also referred to by the abbreviation MUX.

permanent magnet type An electrica motor in which the nagnetic field in the casing is produced by permanent magnets, and the armature has an electromagnetic field generated by passing electrical current through loops or winding, thereby producking the motor action.

PKE (passive keyless entry) An active system that senses the proximity of a fob and locks or unlocks the vehicle.

remote keyless entry (RKE) A system that remotely unlocks and locks the vehicle without the use of a traditional key.

squib Used to ignite a device, like the propellant in an airbag module. While the device does not explode, it does create a very aggressive chemical reaction that creates the heat to expand and deploy the airbag from within its storage packaging.

stepper motor A specialized DC motor that has a rotor that is operated by a series of coils that surround the rotor. The rotor is stepped, or moved incrementally, by pulsing the coils in sequence, causing the rotor to move in a specific direction and amount of rotation. The coils can be pulsed in either direction, so the rotor can move clockwise or counterclockwise. These motors are often used to move a component a very specific amount for precise control of the related output system, like the doors in heating and air conditioning duct work.

supplemental restraint system (SRS) A passenger safety system, such as airbags and seat belt pretensioners.

Review Questions

1. Using networking and multiplexing has many advantages, all of the following are advantages of networking EXCEPT:
   a. They require less dedicated wires, terminals, and connections
   b. They require no special training or advanced diagnostic tools
   c. They allow greater vehicle content flexibility and modules can be added
   d. They save weight and improve serviceability and installation
2. What does a high-speed CANbus network use to reduce interference on data lines?
   a. A single highly insulated wire between modules
   b. A pair of wires surrounded in its own metal conduit
   c. A high-performing voltage filter on the bus
   d. A pair of twisted wires and 2 terminating resistors
3. If a communication error occurs between modules on the network, what will the first letter in the diagnostic trouble code (DTC) be?
   a. P
   b. B
   c. C
   d. U

4. A brushless DC motor eliminates the need for the brushes and the ______
   a. Armature
   b. Commutator
   c. Windings
   d. Bushings
5. How do power door locks that use a DC motor convert the spinning motion of the motor to linear motion of the lock?
   a. A rack-and-pinion gearset
   b. The solenoid magnetic field
   c. A centrifugal clutch
   d. The control module
6. What is the main difference between an RKE and a PKE keyless entry system?
   a. The RKE system has more buttons on the fob
   b. The PKE system does not require a fob battery
   c. The PKE system does not require input from the user
   d. The RKE system does not require the user to press buttons
7. On a modern vehicle equipped with a driver side airbag, what component allows the air bag to maintain its electrical connection as the steering wheel is turned?
   a. The slip ring and brush
   b. The horn fuse and relay
   c. The wireless contacts
   d. The clock spring
8. Which component is responsible for making sure the wipers stop at the correct point, the bottom of the windshield?
   a. The wiper control switch
   b. The park switch
   c. The rain sensor
   d. The DC electric pump
9. What material is an air bag typically made of?
   a. Rubber
   b. Plastic
   c. Cotton
   d. Nylon
10. Why is it critical to disable an SRS system before performing any related service?
   a. Servicing may set DTCs, causing a warning light
   b. Not disabling will damage the module permanently
   c. An accidental airbag deployment can kill a technician
   d. Accidental deployment will require recharging the airbag gas

ASE Technician A/Technician B Style Questions

1. A network system in a modern vehicle is being discussed. Technician A states that the network system allows fewer wires to be used and saves weight. Technician B states that if a network component malfunctions, it may completely shut down the network. Who is correct?
   a. Technician A only
   b. Technician B only
   c. Both Technicians A and B
   d. Neither Technician A nor B
2. Different network communication types are being discussed. Technician A states that LIN networks are the fastest type today. Technician B states that CAN networks can accommodate up to 100 communicating modules. Who is correct?
   a. Technician A only
   b. Technician B only
   c. Both Technicians A and B
   d. Neither Technician A nor B
3. A malfunctioning network system is being discussed. Technician A states that a scan tool can be used to see which modules are communicating on the network. Technician B states that if a module is not communicating on the network, it needs to be replaced. Who is correct?
   a. Technician A only
   b. Technician B only
   c. Both Technicians A and B
   d. Neither Technician A nor B
4. A malfunctioning windshield wiper system is being discussed. The wipers do not move. Technician A states that if the motor can be heard running, the problem is likely mechanical rather than electrical. Technician B states that if the motor cannot be heard, the problem could be the motor, or a problem in the electrical circuit or control device. Who is correct?
   a. Technician A only
   b. Technician B only
   c. Both Technicians A and B
   d. Neither Technician A nor B
5. Power door locks are being discussed. Technician A states that a centrifugal clutch allows motor-type power door locks to be manually operated. Technician B states that if power door locks work from the passenger door switch, but not from the driver door switch, diagnosis should focus on the passenger door. Who is correct?
   a. Technician A only
   b. Technician B only
   c. Both Technicians A and B
   d. Neither Technician A nor B
6. Keyless entry systems are being discussed. Technician A states that a PKE system will often unlock when a key comes close enough to the vehicle, even if a button was not pressed. Technician B states that on some vehicle key fobs, there may be a button to remotely start the engine. Who is correct?
   a. Technician A only
   b. Technician B only
   c. Both Technicians A and B
   d. Neither Technician A nor B
7. Vehicle horns are being discussed. Technician A states that typical vehicle horns make their sound when an electric motor spins and air is blown out of the horn rapidly. Technician B states that the vehicle's horn switch on the steering wheel has a wireless connection to the control module. Who is correct?
   a. Technician A only
   b. Technician B only

c. Both Technicians A and B
d. Neither Technician A nor B

8. Wiper and washer systems are being discussed. Technician A states that before testing a washer system, be sure there is adequate liquid in the washer tank. Technician B states that some vehicles use a rain sensor to indicate water on the windshield to the body computer. Who is correct?
   a. Technician A only
   b. Technician B only
   c. Both Technicians A and B
   d. Neither Technician A nor B
9. SRS systems are being discussed. Technician A states that the purpose of SRS is to supplement the seat belt in preventing passenger injury. Technician B states that inertia created by rapid deceleration is what triggers air bag deployment. Who is correct?
   a. Technician A only
   b. Technician B only
   c. Both Technicians A and B
   d. Neither Technician A nor B
10. SRS systems are being discussed. Technician A states that most air bags are activated by an electrical signal from the SRS control module. Technician B states that air bags take about 1.3 seconds to completely inflate. Who is correct?
   a. Technician A only
   b. Technician B only
   c. Both Technicians A and B
   d. Neither Technician A nor B

150
10
00
5
50

SECTION 8
Heating and Air Conditioning

▶ CHAPTER 47 Principles of Heating and Air-Conditioning Systems

CHAPTER 47

Principles of Heating and Air-Conditioning Systems

Learning Objectives

- **LO 47-01** Describe the history of air conditioning and the required licensure.
- **LO 47-02** Explain the physics that allows air conditioning.
- **LO 47-03** Explain the qualities of refrigerant and the refrigerant cycle.
- **LO 47-04** Explain the main air-conditioning components and their operation.
- **LO 47-05** Explain the types of refrigerant and refrigerant oils.
- **LO 47-06** Describe the heating and ventilation system operation.
- **LO 47-07** Describe the operation of the defroster and blower motor operation.
- **LO 47-08** Performance test and check HVAC system.

ASE Education Foundation Tasks

See Appendix A to view the 2017 ASE Education Foundation Automobile Accreditation Task List Correlation Guide.

▶ HVAC Introduction

LO 47-01 Describe the history of air conditioning and the required licensure.

This chapter explains the processes involved in controlling the environment inside the vehicle's passenger compartment. The condition of the air influences a driver's performance and concentration. So it promotes both safety and comfort. The heating, ventilation, and air-conditioning system is called the HVAC system for short. It is designed to maintain a comfortable temperature in the vehicle cabin. It also provides fresh, filtered, air that is conditioned to the driver's preference. And it does so on a constant basis (**FIGURE 47-1**). The supplied air is either heated, cooled, or both. This depends on the outside temperature and the desired cabin temperature.

The HVAC system in most passenger vehicles uses the heat created by normal engine operation to raise the temperature in the passenger compartment. Refrigerant is used to cool the air in the passenger compartment. To ensure that the air brought into the vehicle is free from dust and debris, many vehicles pass the air through a cabin air filter.

Without proper maintenance, the HVAC system can degrade. This results in a less pleasant ride for the passengers, particularly during the summer and winter months. It is critical that automotive technicians be familiar with the theory of operation of the system and its components. This is so that you can maintain, diagnose, and repair it when necessary.

FIGURE 47-1 The HVAC system provides fresh, filtered, and conditioned air to the passenger compartment.

History of Automotive Heating and Cooling

There is some debate about the origins of systems for heating the passenger compartment of vehicles. What has been confirmed is the use of gas lamps in the early 1900s for heat. Passengers would bring a small gas lamp into the compartment for heat. But these were not connected to the vehicle in any way. Around 1917–1920, some vehicles transferred exhaust gas through pipes into the compartment to provide heat. In the late 1920s, some vehicles had simple blowers that simply funneled air from around the engine into the cabin. By the 1930s, the first true heating systems using engine coolant were developed. They were similar to what is still used today.

Automotive air-conditioning has a long history. It begins with a discovery by Michael Faraday. In 1820, Faraday discovered how to create cool air. He passed air over an evaporating coil of compressed ammonia. This was a great advance in comfort. However, ammonia is a caustic agent and hazardous to humans. It causes skin burns, lung damage, and possibly asphyxiation. In 1902, Willis Carrier invented and patented the first electric air conditioner. A great advance was made in 1928 when a research division of General Motors produced **chlorofluorocarbons (CFCs)**. This is a chemical compound containing chlorine, fluorine, and carbon. CFCs made it possible for air-conditioning systems to be used by consumers with no apparent harm, unlike the ammonia systems.

In 1939, Packard introduced the first automotive air-conditioning system. It was an add-on luxury unit (**FIGURE 47-2**). The add-on was offered for $274, which in today's money would be more than $5000. Automotive air-conditioning is now standard on most production vehicles because of customer demand. This is not only for the comfort it provides. It is also for the increased fuel economy that comes from increased aerodynamic efficiency. Having the windows rolled up while driving at moderate and high speeds reduces wind drag.

▶ TECHNICIAN TIP

Almost all manufacturers rely on the air conditioner's ability to remove moisture from the air to meet the defrost time standard set by the U.S. Department of Transportation.

You Are the Automotive Technician

A customer brings his vehicle to the auto shop for a routine maintenance check. He will be driving his vehicle from New York to Florida this fall and won't be returning until the following spring. He is worried about how his vehicle's HVAC system will perform during his long trip and during his stay in sunny Florida. He isn't familiar with auto shops in Florida, so he wants you to inspect the system and suggest any repairs it may need.

1. What are the two types of automotive air-conditioning systems?
2. Refrigerants are another important part of the HVAC system. Why is identifying refrigerants an important part of repairing air-conditioning systems?
3. What license is required for technicians to work with refrigerants?
4. What are the four main components in an air-conditioning system?

FIGURE 47-2 In 1939, Packard introduced the first automotive air-conditioning system as an add-on luxury unit.

HVAC Regulation

The **Environmental Protection Agency (EPA) regulates** automotive air-conditioning. Section 609 of the **Clean Air Act (CAA)** contains all the motor vehicle air-conditioning (MVAC) requirements and laws. Federal law regulates air emissions from stationary and mobile sources. This law authorizes the EPA to establish National Ambient Air Quality Standards (NAAQS). This is meant to protect public health and public welfare. One way it does so is by regulating emissions of hazardous air pollutants.

This act sets the regulations for shops that do air-conditioning work. It mandates:

- that each shop maintain a current government-issued license,
- what certifications are required for technicians doing MVAC service,
- the specific refrigerants that can be safely used for each application,
- how refrigerants must be stored and reclaimed, and
- the specific approved equipment technicians must use.

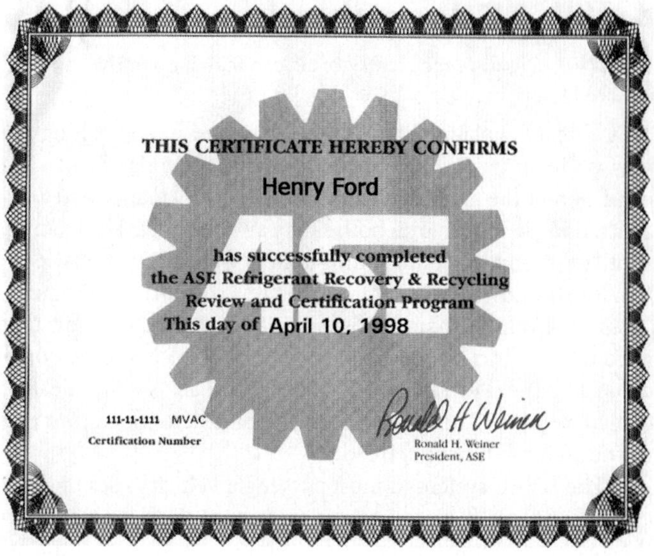

THIS CERTIFICATE HEREBY CONFIRMS
Henry Ford
has successfully completed
the ASE Refrigerant Recovery & Recycling
Review and Certification Program
This day of April 10, 1998
111-11-1111 MVAC
Certification Number
Ronald H. Weiner
President, ASE

FIGURE 47-3 EPA 609 license.

Licensure

Automotive air-conditioning service technicians need to have a special license (**FIGURE 47-3**). It is granted after passing the 609 test. This is a self-study course. It can be found online through companies such as Automotive Service Excellence (ASE), Mobile Air-conditioning Society (MACS), and Mainstream Engineering. Passing the test results in a lifelong license. Because the license never expires, the technician is responsible for keeping up to date with all new regulations. The 609 license also allows the technician to purchase a few select refrigerants, such as R-12, R-134a, and HFC 1234. The 609 license does not mean that the technician knows how to repair automotive air-conditioning systems. It just means that the technician knows the laws and agrees to comply with all regulations.

> ▶ TECHNICIAN TIP
>
> The equipment used to recover refrigerants from MVAC systems is also covered in the 609 MVAC laws. The technician is responsible for using properly certified equipment when servicing MVAC systems. The current fine for not complying with the 609 laws is $37,500 per day per incident. This fine is applicable to the person or entity not in compliance with the 609 regulations.

▶ HVAC Principles

LO 47-02 Explain the physics that allows air conditioning.

Consider the function of the air-conditioning system. What is it meant to do? It reduces the temperature and humidity inside the passenger compartment to a comfortable level (**FIGURE 47-4**). The HVAC system achieves this by removing excess heat from the passenger compartment. It does so through a series of thermal and chemical transformations. In other words, heat is removed from the passenger compartment and transferred by the air-conditioning system to the outside air.

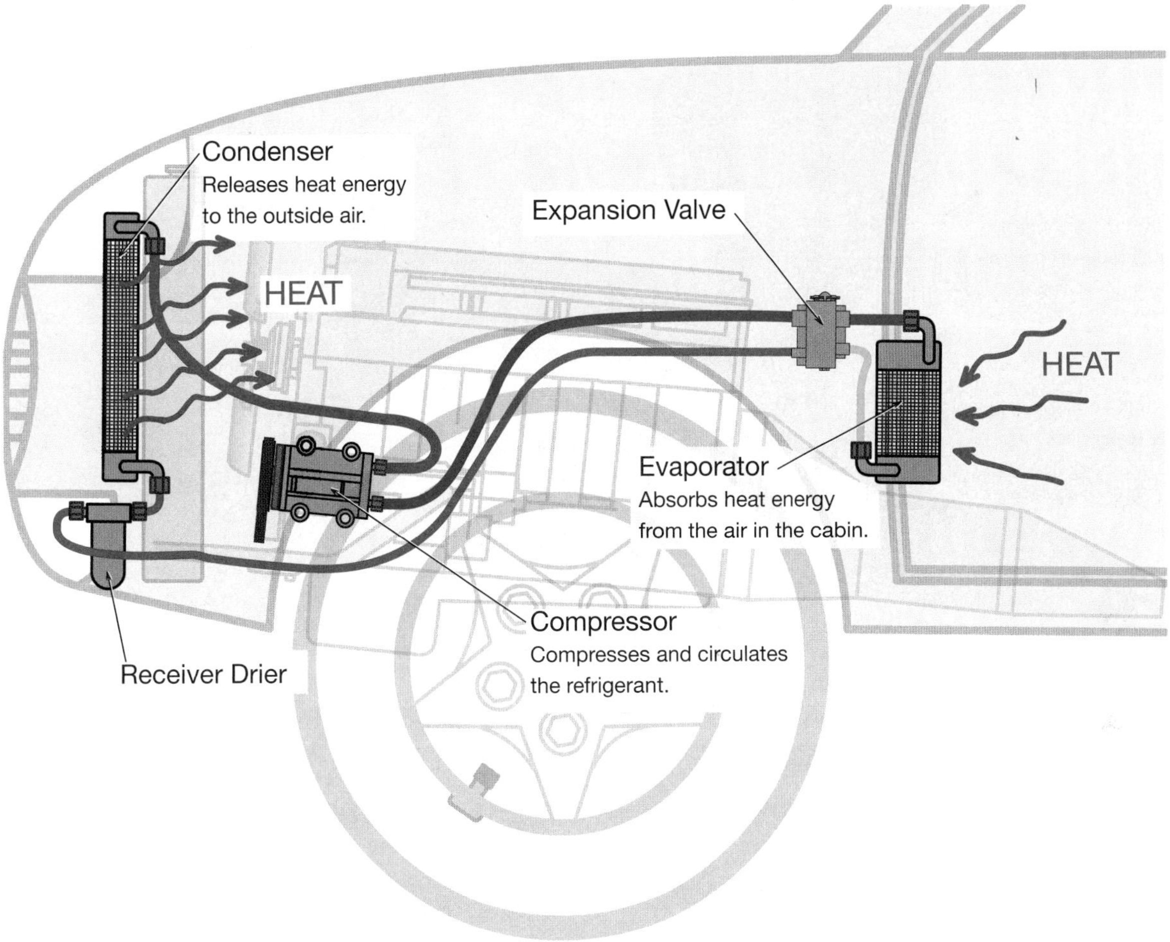

FIGURE 47-4 The HVAC system reduces the temperature inside the passenger compartment by removing excess heat through a series of thermal and chemical transformations.

The HVAC system is governed by the principles of physics just like all of the other systems on a vehicle. Understanding the physics will help you to know when a system is operating properly or not. One of the simplest concepts concerns heat transfer:

- Heat always transfers from hot objects to cold objects (**FIGURE 47-5**).

Another concept concerns density:

- Hot liquids and gases are less dense and tend to rise (**FIGURE 47-6**).
- Cool liquids and gases tend to be pulled to the bottom by gravity.

Another simple concept is the relationship between temperature and pressure:

- When pressure is raised on a gas, its temperature also rises.
- When pressure is lowered, so does its temperature.

These and other principles will be explored in greater depth in the following sections.

FIGURE 47-5 Heat always travels from hot places to cold places.

FIGURE 47-6 Hot things tend to rise, and cold things tend to fall.

Applied Science

AS-84: Relative Humidity: The technician can demonstrate an understanding of and discuss relative humidity in terms of its effect on automotive heating and air-conditioning systems.

Relative humidity is essentially a measure of how moist the air is. Or in more detail, it's the amount of water vapor in the air expressed as a percentage of the amount that the air can "hold" at a given temperature. The warmer air is, the more moisture it is able to hold.

In a car air-conditioning system, warm air is forced through a cold (below dew point) evaporator core. This condenses the water vapor in the air into its liquid form and drains it away. This is much like the way water condenses on the outside of a cold soda can. Thus, water vapor in the incoming air is removed, and the relative humidity in the vehicle's cabin is lowered. On humid days, more water vapor is in the air. So you may notice larger than usual amounts of water draining from the evaporator onto the ground underneath the vehicle.

Although you may expect heating the incoming air to have the opposite effect, in reality this is not the case. Cool air is passed through a hot heater core (generally heated by coolant from the engine's cooling system). This heats the air up, which increases the water-carrying capacity of the air. But because no water is being added, heating up the air lowers its relative humidity in the cabin.

Cold and Hot

The conditions "cold" and "hot" are more of a feeling than a science. Cold is simply a relative term for the absence of heat. It can be regarded as a condition that exists after heat has been removed. Heat energy is in all matter, from gases (air), to liquids (water), to solids (metal). One might think that 32°F (0°C) is cold, but more heat energy is present there than at −19°F (−28°C). Heat energy cannot be created or destroyed. It can only be transferred from one object to another. Heat always travels from a warmer object to a cooler object.

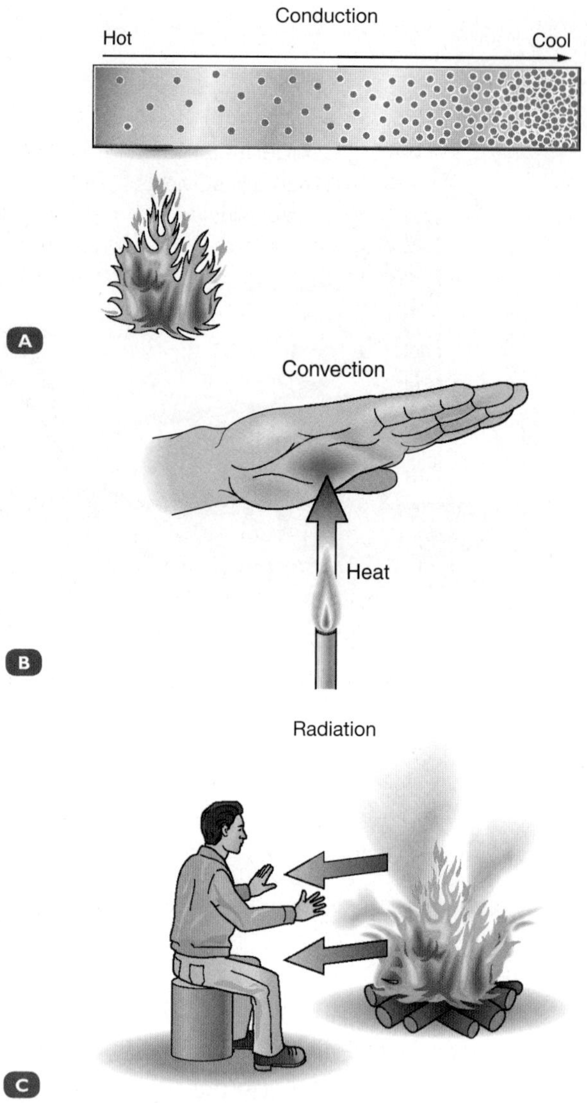

FIGURE 47-7 Heat transfer. **A.** Conduction. **B.** Convection. **C.** Radiation.

Consider this age-old question: Does ice make your hand cold, or does your hand warm up the ice? The answer is that your hand is warming up the ice, and the extreme transfer of heat energy makes your hand feel cold. The warming of the ice is known as **heat transfer**. Heat transfer can take place by one of three methods: conduction, convection, or radiation (**FIGURE 47-7**). The HVAC system relies on all three types of heat transfer. **Conduction** is the process of transferring heat through matter by the movement of heat energy through solids from one particle to another. For example, if you hold a lighter up to a steel pipe, the heat from the flame will travel through the steel, eventually heating the entire pipe.

Heat transfer by **convection** is the circulatory movement that occurs in a gas or fluid with areas of differing temperatures.

This is due to the variation of the density and the action of gravity. Consider a container of water being heated on a stove. If you put a drop of food coloring in one part of the water and start to heat the water, you can readily see the circulation of the water within the container. As the water heats up, it expands and becomes lighter than the surrounding water. This causes a column of water to rise to the top. The cooler water then falls to the bottom. Thus, a cycle occurs of warmer water rising to the top, pushing cooler water to the bottom. The convection currents have the same effect of a small pump pushing the water around in a set pathway.

Radiation is the transfer of heat through the emission of energy in the form of invisible waves. The sun radiates energy to the Earth over the vast miles of outer space. The electromagnetic radiation readily travels through the vacuum of space. When this energy is absorbed, it becomes converted to heat. That is, the heat you feel as the sun touches your body on a warm day. Of course, the direction in which the radiation travels can be changed or redirected. An example is when a sheet of shiny aluminum foil reflects the sun's rays and bounces the energy in another direction.

Let's see how heat transfer is used in cooling the passenger compartment. Radiation of heat from the rays of the sun penetrates the windows and heats up the interior components and air. The driver turns on the air-conditioning function of the HVAC system. Convection currents from the heated air in the passenger compartment are pushed through an evaporator core filled with refrigerant. Heat is then conducted through the metal coils of the evaporator. The metal coils of the evaporator then transfer the heat by conduction to the refrigerant. This causes convection currents to transfer heat through the refrigerant. The refrigerant is then circulated by the compressor and carried to the condenser. There, conduction and convection dissipate the heat to the surrounding air, and ultimately to the atmosphere.

Heat Energy

Heat energy is measured in **British thermal units (Btus)**. The Btu is a standard unit of measurement of heat quantity. It describes the amount of heat it takes to move temperature up or down in degrees. One pound of water at **atmospheric pressure** and at a temperature of 32°F (0°C) would require 1 Btu of energy added to increase the temperature 1 degree Fahrenheit. A change in the temperature of 1 pound of water from 32°F (the freezing point of water) to 212°F (100°C) (the boiling point of water) would require 180 Btus of heat energy.

Water cannot get any hotter than 212°F at atmospheric pressure. The temperature will not rise any further. The molecules have reached their maximum speed of movement at atmospheric pressure. Temperature is increased only by moving the molecules faster. If the water is exposed to any more energy, it will start to change its state of matter from a liquid to a gas: steam. Turning a pound of water at 212°F at atmospheric pressure into steam requires 970 Btu. The heat required to change the water at its maximum temperature from a liquid to a gas is called **latent heat of evaporation** (**FIGURE 47-8**).

This latent heat is called "hidden heat" because it cannot be felt or measured with a thermometer. Heat that can be felt

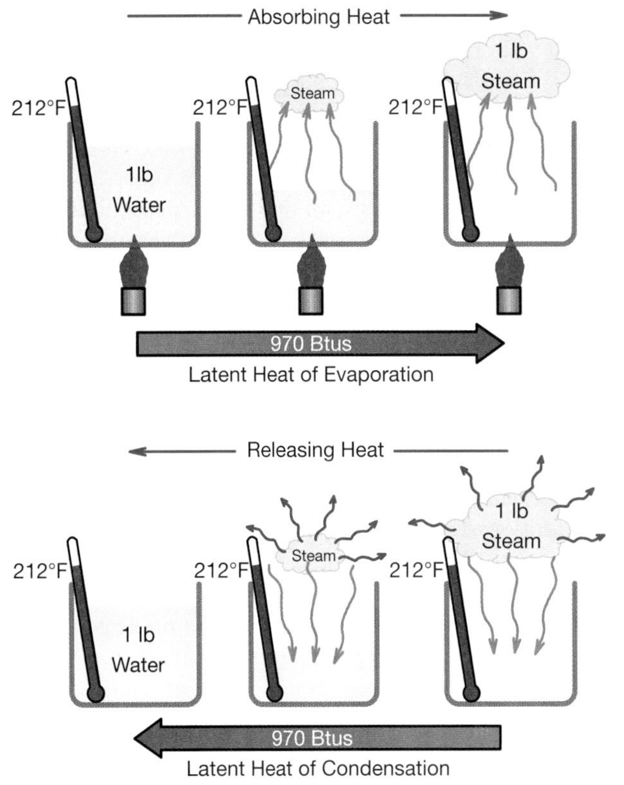

FIGURE 47-8 It takes 970 Btus of heat to turn 1 pound of water into steam, and vice versa.

and measured with a thermometer is called sensible heat. The amount of heat given up by steam to turn back into a liquid is the same 970 Btu and is called **latent heat of condensation**.

When you take the same pound of water from a liquid to a solid, it freezes. To do this, heat must be removed from the liquid to make ice. At 32°F, the molecules in the water cannot move any slower without changing state. At 32°F, the **latent heat of freezing** begins. It takes the removal of 140 Btu from 1 pound of 32°F water to change its state into ice.

The changing of state in liquids and gases, referred to as vaporizing and condensing, is the basis of air-conditioning. Because it takes more heat to evaporate a liquid than to freeze a liquid, air-conditioning uses the process of evaporating a liquid to absorb a large amount of heat energy. By evaporating liquids, the air conditioner can remove the heat that causes vaporization from the cabin air. This makes the passengers feel cooler. It is the same principle by which sweat cools human beings. Heat is absorbed by the evaporating sweat droplets on the skin, thereby removing heat from the body and cooling it. Similarly, by evaporating liquids, the air conditioner can remove large amounts of heat from the air and make the occupant feel cooler.

▶ Refrigerant Principles

LO 47-03 Explain the qualities of refrigerant and the refrigerant cycle.

Manipulating nature's laws of **vaporization** and **condensation** for liquids and gases is how the air conditioner works. Consider the cycle of vaporization and condensation. The air conditioner must have a fluid that vaporizes and condenses at the right temperature and pressure. It must also be able to change its state in the continuous cycle of vaporization and condensation without breaking down. The fluid used for these purposes is called **refrigerant**.

The refrigerant must have a boiling point well below freezing. This is so that its boiling and condensing points can be changed by changing pressure. The super-low boiling point of most refrigerants is an important factor. It allows the system to control the vaporization and condensation points of the refrigerant at normal ambient temperatures. A commonly used refrigerant is R-134a (also known as tetrafluoroethane). It is a hydrofluorocarbon. It contains none of the ozone-depleting chlorine found in previous refrigerants.

With refrigerants, maintaining vaporization and condensing points at normal ambient temperatures is simply done by raising or lowering the pressure (**FIGURE 47-9**). At atmospheric pressure, R-134a boils at about −15°F (−26°C). So in most normal conditions, it is in its gaseous state. Adding a small amount of pressure to the refrigerant causes the boiling point of the refrigerant to rise. Indeed, a low pressure (approximately 15 psi [103 kPa]) on a refrigerant will force its boiling point to just above freezing (32°F [0°C]). Likewise, to allow the R-134a to condense back into a liquid, its boiling point can be increased well above the ambient temperature by further increasing its pressure. Forcing the refrigerant's boiling point so that it is well above ambient temperatures allows it to be cooled by ambient air. Then, it will condense back into a liquid.

With the refrigerant boiling below the temperature of the air in the passenger compartment, the refrigerant will forcefully absorb heat from the passenger compartment's air. It does so by using the heat from the air to boil the refrigerant. R-134a under low pressure (15–25 psi [103–172 kPa]) boils at about 32°F (0°C). Warm air from the passenger compartment is circulated across the evaporator containing R-134a, causing it to boil. As R-134a boils, it absorbs the heat. The heat is then pulled along with the moving R-134a away from the evaporator. The process continually repeats as the fan circulates the air, and the compressor continually circulates refrigerant. This process results in a continual supply of cool air to the cabin. The air returning to the passenger compartment will feel cool because heat was removed from the air. The state of the refrigerant changes from liquid to gas while also cooling the air.

If the temperature of the refrigerant falls below the freezing point of water during this process, the air conditioner must shut off. The air passing over the evaporating refrigerant is full of moisture. So the extreme heat removal causes any moisture in the air to condense on the cold surface of the evaporator (**FIGURE 47-10**). The condensed water could freeze on the outside of the evaporator and block the airflow that is needed to vaporize the refrigerant. The air must keep flowing over the evaporator. This renews the heat source which keeps the refrigerant evaporating (boiling) efficiently without the water freezing. The system is designed to prevent freezing if it is operating correctly.

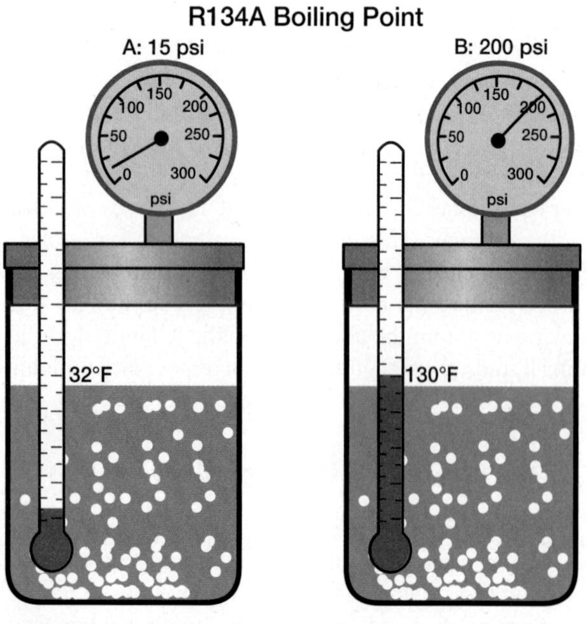

FIGURE 47-9 Boiling points of R-134a. **A.** At 15 psi (103 kPa). **B.** At 200 psi.

Applied Science

AS-23: Phases/States: The technician can explain in detail the three states of matter.

Solids, liquids, and gases are the three most common states of matter on Earth. The distinction between them is based on characteristics relating to the shape and volume of the piece of matter in question.

Solid matter retains a fixed volume and shape. There are many examples of solid matter used in motor vehicles. Some examples include cast iron, steel, and aluminum used to build engines, and the fabric and plastic materials used in interiors. Liquid matter retains a fixed volume, but the shape adapts to the shape of the container it is held within. There are many types of liquids used in motor vehicles. In the case of gases, the matter does not retain a fixed volume. It expands to fill whatever volume is available. Air-conditioning refrigerant is an interesting example of a compound. It takes the form of both a liquid and a gas in different sections of the air-conditioning system.

FIGURE 47-10 Water vapor condenses on cold surfaces.

Applied Science

AS-28: Conduction/Convection: The technician is able to explain the concept of heat transfer in terms of conduction, radiation, and convection in automotive systems.

Heat can be transferred via three methods: conduction, convection, and radiation. Put simply, conduction is the transfer of heat through solid matter. Convection is the transfer of heat through liquid or gas relying on the movement of currents. And radiation is the transfer of heat energy through space by means of electromagnetic waves.

All three of these methods of transfer come into play when trying to keep the inside of a car cool. The heat you feel in the cabin originates from the sun. The sun's heat travels by radiation. You feel it coming through the glass windows into the vehicle cabin. Heat is transferred through the cabin via convection, with the hot glass panels heating the air inside and causing it to move in currents. When turned on, the air-conditioning system moves the hot air through the cold evaporator core. The heat is moved to the evaporator core via convection and dissipated through the metal evaporator core by conduction.

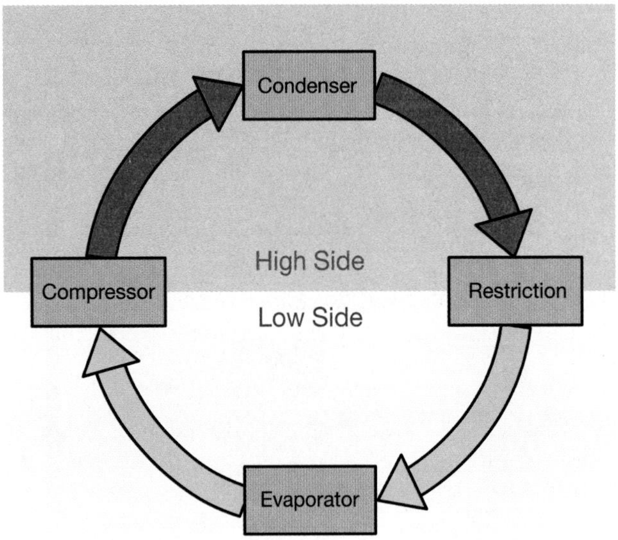

FIGURE 47-11 Low-side components and high-side components.

Differential Gas Pressure (Refrigerant Cycle)

Every vehicle's air conditioner has a low-pressure side. It is also called the low side or suction side (approximately 20–40 psi [138–276 kPa]). They also have a high-pressure side, called the high side or discharge side (approximately 170–230 psi [1172–1586 kPa]). The low-pressure side is designed to allow easy transformation of the refrigerant from a liquid to a gas (evaporation). The high-pressure side enables the transfer of heat out of the refrigerant, so it changes back from a gas to liquid (condensation) (**FIGURE 47-11**).

Starting at the compressor, a low-pressure gas is compressed into a high-pressure gas. It is pumped into the condenser located in front of the vehicle radiator. The gaseous refrigerant is cooled by the air passing through the condenser. It leaves the condenser as a high-pressure liquid. The high pressure is required to allow the change from gas to liquid. From there, the liquid travels through a small orifice (restriction) that reduces the pressure of the refrigerant (a fixed orifice tube or thermal expansion valve).

After the restriction, the low side of the system begins. The low-pressure liquid leaves the metering device and travels into the evaporator in the passenger compartment. As the low-pressure liquid travels through the evaporator, a fan blows the warm to hot cabin air across the fins. This adds heat to the refrigerant until it boils. As it boils, it absorbs heat from the air. The liquid refrigerant is then transformed back into a gas. The gaseous refrigerant then travels back to the compressor to start the cycle over.

Applied Science

AS-29: Expansion/Contraction: The technician is able to demonstrate an understanding of the expansion and contraction of system parts as a result of heat generated during the use of the system.

When matter is heated, it expands, or becomes larger in size. Conversely, when cooled, it contracts, or becomes smaller. This assumes that the change in temperature does not cause the matter to change state. Thermal expansion (or TX) valves, found in many air-conditioning systems, operate via these principles. The function of a TX valve is to regulate the flow of liquid refrigerant from the evaporator back to the compressor. Thereby, it also regulates the temperature of the refrigerant.

The valve is actuated against spring pressure by a temperature-sensing bulb. It is filled with a gas similar to the refrigerant used in the air-conditioning system. As the refrigerant temperature at the evaporator outlet increases, the pressure in the bulb increases. This is due to expansion of the gas which causes the TX valve to open. When the evaporator outlet temperature drops, the pressure in the bulb decreases. The gas contracts and allows spring pressure to close the valve.

▶ Air-Conditioning Components and Operation

LO 47-04 Explain the main air-conditioning components and their operation.

Each HVAC system uses refrigerant and four major components. Each component has a specific job and purpose. The components are the compressor, condenser, restriction, and evaporator (**FIGURE 47-12**). The air-conditioning cycle starts with the **compressor**. The compressor changes a low-pressure refrigerant gas to a high-pressure gas. The compressor also provides the needed refrigerant movement in the system. Without movement, the system could not exchange the heat from the condenser and the evaporator. Moving refrigerant also allows for the pressure drop to occur in the restriction. Without movement, the restriction could not drop the pressure. Refrigerant enters the compressor as a low-pressure gas from the evaporator. The compressor's function is to increase the pressure of the gaseous refrigerant. It then pushes the refrigerant into the condenser.

Forcing a gas to become a liquid is achieved through condensing. So the next component after the compressor in the air-conditioning system is the **condenser**. Hot, high-pressure

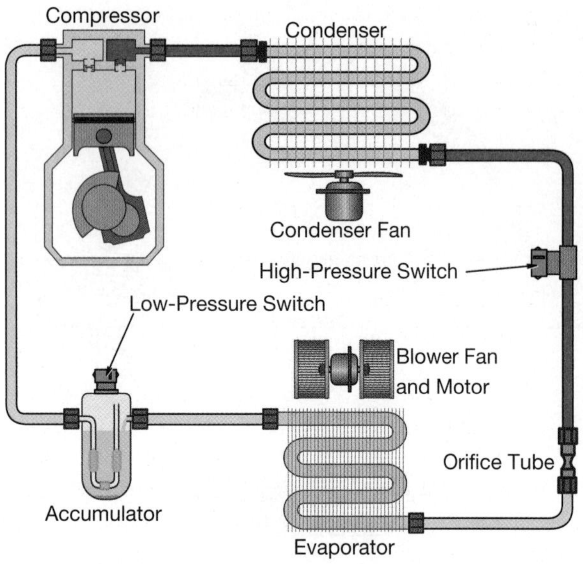

FIGURE 47-12 The components of an air-conditioning unit.

gas enters the condenser from the compressor. The gas flows through a series of coils in the condenser, and ambient air passes over the outside of the finned coils. The engine's radiator fan and/or dedicated condenser fans move ambient air across the condenser coils and fins. The cooler outside air removes the heat from the hot, high-pressure gas. The gas begins to condense into a liquid. Emerging from the outlet of the condenser is a relatively warm, high-pressure liquid. The condenser transforms gas to liquid on the high side of the system.

The air conditioner is a **closed-loop system**, meaning that nothing else enters or exits the system. The refrigerant just continues to cycle through the system, changing pressure and state to complete the task of cooling the cabin. Before the liquid can be evaporated back into a gas for the compressor, the pressure has to drop. This is so the boiling point can be lowered below the passenger compartment air temperature. Vaporization is then used to remove the heat from the passenger compartment air. Lowering the pressure occurs by flowing a liquid through a **restriction**.

The restriction lowers the pressure by using the Bernoulli principle. Daniel Bernoulli stated that if the flow of a fluid remains constant, the energies of the fluid remain constant. Changing one of the two energies (up or down) affects the other energy oppositely. The two energies in fluid are velocity and pressure. These energies react opposite to each other. When velocity increases, the pressure must drop if the flow remains constant. The restriction causes a significant increase in refrigerant velocity. This is because the compressor is constantly moving fluid in the closed-loop system.

The fluid flowing through the restriction is a liquid. Liquids are considered incompressible (Pascal's law). Forcing a noncompressible fluid through a small orifice causes the velocity in the restriction to rise. If the velocity of the liquid rises, the pressure of the liquid must drop. This is according to the Bernoulli principle. Raising the velocity and dropping the pressure is how the restriction transforms the high-pressure liquid refrigerant to a low-pressure liquid refrigerant. The reduced pressure reduces the boiling point of the refrigerant, ready for the evaporator to vaporize into a gas.

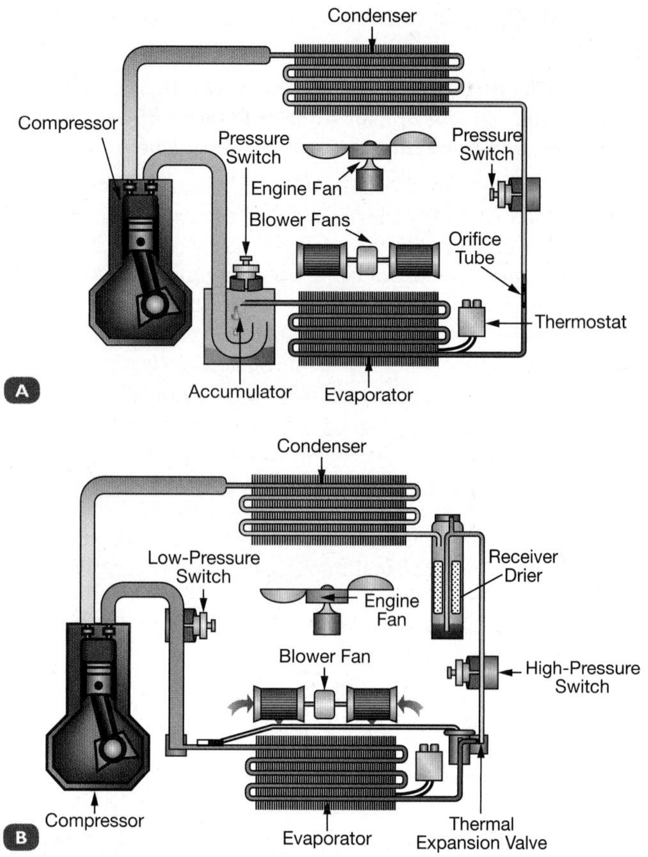

FIGURE 47-13 Types of automotive air-conditioning systems. **A.** Fixed orifice tube system. **B.** Thermal expansion valve system.

The next component in the air-conditioning system is the **evaporator**. The evaporator transforms low-pressure liquid to a low-pressure gas. The transformation from liquid to gas begins when the air from the passenger compartment passes over the evaporator coils and fins. The pressure on the refrigerant is typically held between approximately 20 and 40 psi (138 and 276 kPa). The pressure on the low side of the system reduces the refrigerant's boiling point to a bit over 33°F (1°C). The liquid refrigerant absorbs heat from any source available as it boils. This heat comes from the warm air flowing through the evaporator core.

The vaporization of the refrigerant is forced to happen just above the freezing point of water so that the evaporator core will not freeze. By moving air across the evaporator coils, the convection currents pass over the evaporator coils. This conducts heat to the refrigerant through the tubes of the evaporator. The conduction of heat through the coils then provides the needed energy to boil (evaporate) the liquid refrigerant into a gas. It also strips the air of much of its heat energy and moisture. This makes the air flowing from the ductwork cool and crisp.

Types of Automotive Air-Conditioning Systems

There are two definitive types of air-conditioning systems. The first is the **fixed orifice tube system**. The other is the **thermal expansion valve (TXV) system** (**FIGURE 47-13**). The difference is that the TXV is adjustable based on the temperature of the outlet pipe from the evaporator. Whereas the fixed orifice tube provides a nonadjustable passage for the refrigerant to pass through. Each type of air-conditioning system contains one subcomponent that is vital to the operation and efficiency of that type of system. The type of restriction being used will determine which subcomponent is required. Fixed orifice tube systems use an accumulator to ensure that a pure gas is delivered to the compressor. TXV systems use a receiver filter drier to ensure that a pure liquid is delivered to the restriction.

The fixed orifice tube system has the accumulator on the low-side line between the evaporator and the compressor. This system cannot adapt to temperature changes in the cabin. So, it is likely that during cooler weather or as the cabin temperature decreases, the refrigerant in the evaporator will not fully boil. This means that some liquid refrigerant will still be present as it heads for the compressor. Liquid cannot be compressed. Its presence in the compressor would cause damage from trying to compress a noncompressible substance.

The accumulator's job is to prevent liquid from getting into the compressor and causing damage. Liquid and gaseous refrigerant are both able to enter the accumulator. In the accumulator, the liquid falls to the bottom, and the gas rises to the top (**FIGURE 47-14**). The outlet of the accumulator only draws from the top. Therefore, it only allows gaseous refrigerant to reach the compressor. The liquid in the bottom is heated by the gas passing through it and heat from the engine compartment. As it boils, it becomes a gas.

The TXV system has a receiver drier in the high-side line between the condenser and the TXV. This system can adapt to the temperature changes in the evaporator. So, unlike the fixed orifice system, it does not need the safety device for the compressor. The receiver drier also lets in liquid and gaseous refrigerant, but it actually draws the liquid from the bottom on the outlet side (**FIGURE 47-15**). The receiver stores the accumulated liquid refrigerant in case of a sudden change of temperature in the cabin. If the cabin is cool because the air-conditioning has been on for a while, and then the customer opens a window or door, hot air rushes in. This raises the outlet temperature of the evaporator and causes the TXV to rapidly open. This rapid opening causes a rush of the extra liquid refrigerant from the receiver drier to the evaporator. The receiver drier holds a reserve of liquid refrigerant. This prevents vapor from being drawn in and making cooling inefficient.

▶ Types of Refrigerant

LO 47-05 Explain the types of refrigerant and refrigerant oils.

Proper air-conditioning operation needs many things. It needs all the components to work well. This includes the proper refrigerant type and amount. Otherwise, the air conditioner will not cool the passenger compartment. Each component requires that the refrigerant level be correct. And that the previous component delivers the refrigerant to it in the correct state. The evaporator requires a low-pressure liquid. The compressor requires a low-pressure gas. The condenser requires a high-pressure gas. And the restriction requires a high-pressure liquid.

There are many types of refrigerants available. But automotive applications generally use one of the three most common: R-12, R-134a, or HFO-1234yf (**FIGURE 47-16**). These refrigerants are not compatible and must not be mixed together. We explore each of these refrigerants next.

Dichlorodifluoromethane (R-12) is a member of the chlorofluorocarbon, or CFC, family of gases. It was the first common refrigerant to be used in automotive air conditioners. It is nonflammable, nontoxic, and stable at all temperatures. It does not react with aluminum, steel, or copper; and is soluble in mineral oils. So mineral oil is used to lubricate the compressor. R-12 has a boiling point of −21.8°F (−29.9°C). The first part of the word refers to chlorine, a primary component of CFCs that is banned by the U.S. government in refrigerants. This is because of its harmful effects on the atmosphere. In the early 1990s, R-12 was found to be one of the leading causes of the depletion of the ozone layer. Soon after, it was outlawed for new vehicle installation. Although you will not find R-12 in today's production vehicles, you may still have customers with older vehicles using R-12.

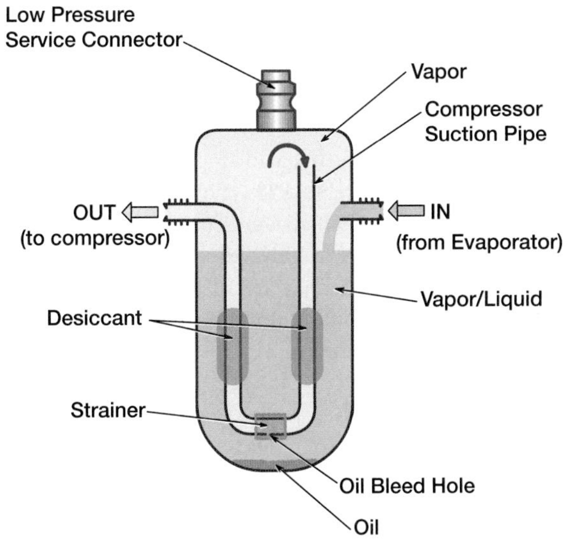

FIGURE 47-14 An accumulator is used on fixed orifice tube systems. It prevents liquids from entering the compressor.

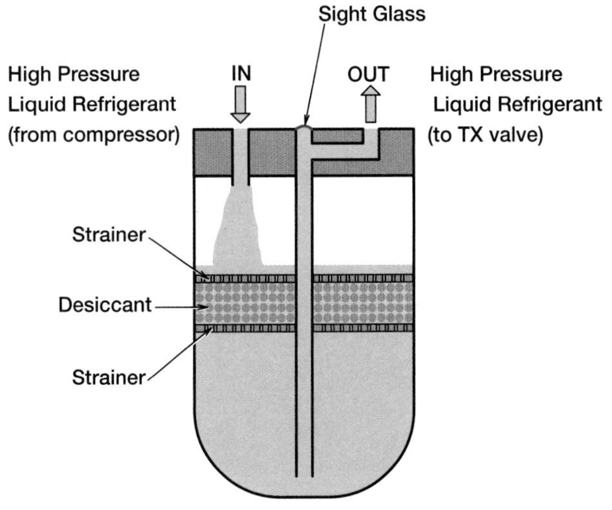

FIGURE 47-15 A receiver drier is used on TXV systems. It sends liquid to the restriction.

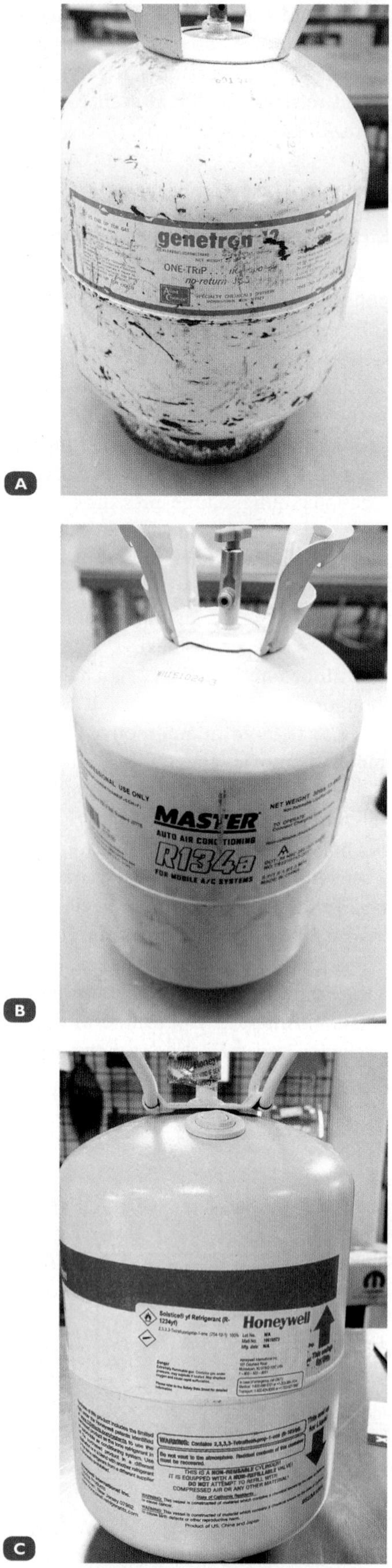

FIGURE 47-16 Types of refrigerant. **A.** R-12. **B.** R-134a. **C.** HFO-1234yf.

Tetrafluoroethane (R-134a) was the replacement for R-12. It is a hydrofluorocarbon, or HFC. R-134a has a boiling point of −15.3°F (−26.3°C), which is nearly the same as R-12's boiling point. R-134a can only be substituted for R-12 (retrofitted) by following a specific procedure. R-134a uses a polyalkalene glycol (PAG) oil to lubricate the air-conditioning components. R-12, the less common refrigerant, uses mineral oil. One reason it is so important to always use the correct refrigerant is that PAG oil mixed with mineral oil creates a hazardous gas that can corrode the air-conditioning system. The mixture is also considered a contaminant and must be handled as hazardous waste. To help prevent the wrong lubricant or refrigerant being installed during servicing, the service ports on air-conditioning systems have been changed. The service equipment for an R-12 system cannot be connected to an R-134a system. Common PAG oils are PAG 46, PAG 100, PAG 133, and PAG 150.

R-134a is nonflammable. It is accepted by the automotive world to be safe for automotive air conditioners. It is a liquid and boils well below normal room temperatures. When it vaporizes, it absorbs tremendous amounts of heat. Unfortunately, recent discoveries show that R-134a is a major contributor to greenhouse gases, so it is being phased out. Tetrafluoropropene (HFO-1234yf) is now being used in new vehicles.

▶ TECHNICIAN TIP

It is important to identify refrigerants as a part of repairing air-conditioning systems. Mixing refrigerants or refrigerant oils could cause problems. These problems may damage the system and the machines that fix air conditioners. So always identify the type of refrigerant in the vehicle before recovering a system. This helps prevent the accidental contamination of the refrigerant in your recovering machine.

HFO-1234yf is the latest refrigerant for mobile air-conditioning systems. It has a greatly reduced global warming potential (GWP) of 4. This is compared with a GWP of 1430 for R-134a refrigerant. Its atmospheric life-time is only 11 days, compared with 13 years for R-134a. It has a boiling point of −20.2°F (−29°C). It is slightly flammable. But the safety risk has been determined to be significantly less than the risk associated with gasoline. Also, HFO-1234yf is more costly to produce than R-134a. So that will impact customers as their systems need servicing. HFO-1234yf uses a PAG-type oil with different additives for belt-driven compressors. Polyolester (POE) oil is used in electrically driven compressors. HFO-1234yf uses service ports that are different than those used with R-12 or R-134a. So contamination issues can be minimized (**FIGURE 47-17**).

Refrigerant Oils

The oil used in refrigeration systems must be compatible with the refrigerant used. The old refrigerant R-12 used mineral oil. R-134a uses a PAG oil. And either PAG or POE oil is recommended for use with HFO-1234yf depending on the system (**FIGURE 47-18**).

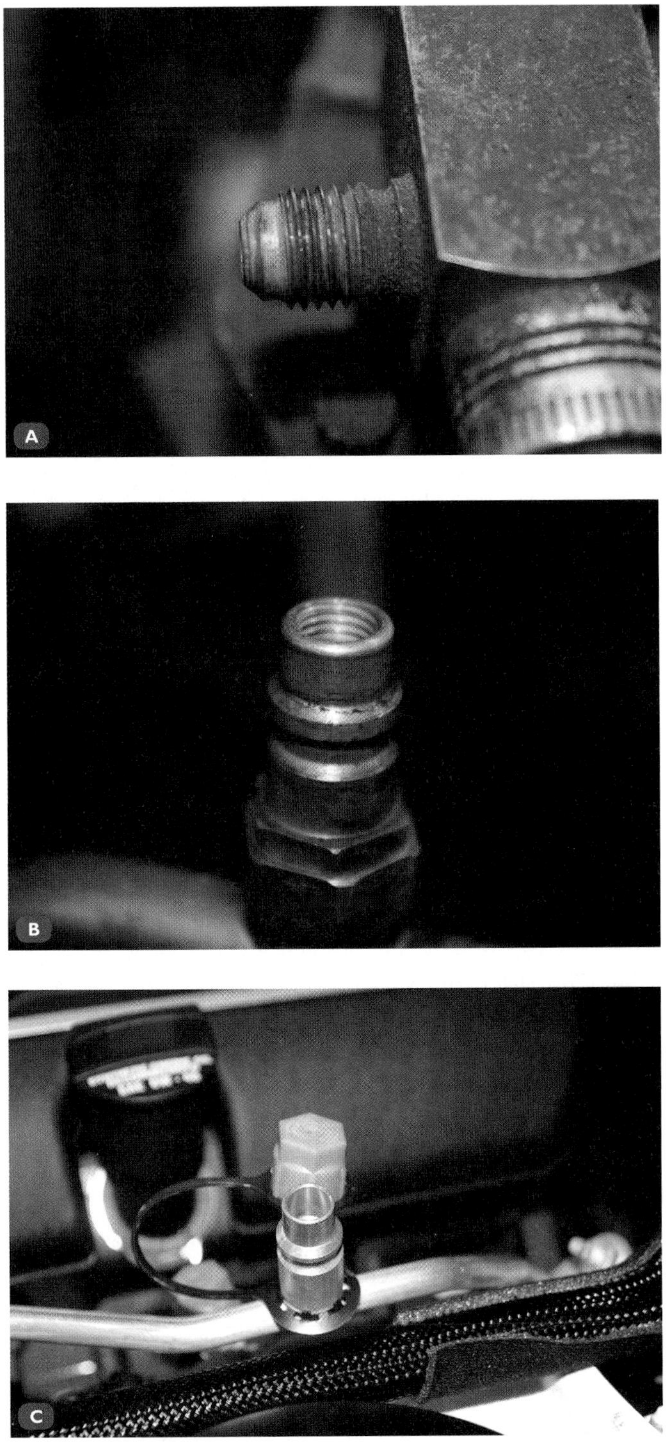

FIGURE 47-17 Service ports. **A.** R-12. **B.** R-134a. **C.** HFO-1234yf.

FIGURE 47-18 Refrigerant oils. **A.** Mineral oil. **B.** PAG oil. **C.** POE oil.

The oil is necessary to keep moving parts in the compressor lubricated. It is also used to protect the gaskets and seals while helping them seal. It must be able to move throughout the system without foaming and be compatible with the pressure changes taking place. The oil is picked up and carried throughout the system in the refrigerant. So some oil will be found in all the major air-conditioning system components (compressor, evaporator, receiver drier or accumulator, and condenser). The oil in the compressor is just as important as oil in an engine. Without it, the compressor will overheat and destroy itself.

Mineral oil is clear to light yellow, and PAG oil is usually a light blue color. As the oil picks up dirt or becomes contaminated, it will turn brown or black, depending on the level of contamination.

POE oil can sometimes be used with R-134a. It can also be recommended when retrofitting R-134a in R-12 systems. This is because POE oils are not as reactive to small traces of mineral oil residue. Both POE and PAG oils absorb moisture. So be sure to cap the oil bottle whenever you are not pouring it.

▶ Heating and Ventilation System Overview

LO 47-06 Describe the heating and ventilation system operation.

Many of the components used to warm the passenger cabin of a vehicle also function in removing heat from the engine. The key shared components are the radiator, thermostat, water pump, and upper and lower radiator hoses. As the engine heats up from the process of burning fuel to create power, the coolant in the engine is also heated. It reaches temperatures of 180–220°F (82–104°C) under normal conditions. In hot climates or under heavy loads, it can reach 235°F (113°C) or more.

The heat from the hot coolant opens the thermostat, and the water pump pushes the coolant into the radiator. Airflow from the vehicle moving and/or from the cooling fans passes through the radiator. Heat is transferred to the air, warming the air and reducing the temperature of the coolant. The coolant is then cycled back through the engine to restart the process. Heated coolant is also pumped to a heater core. The blower motor causes air to flow through the heater core and is warmed. The heated air is directed into the passenger compartment.

The pressure inside of the air-conditioning system is manipulated to control the boiling point of the refrigerant. Similarly, the cooling system is kept under pressure to prevent the coolant from boiling at the normal boiling point of water, which is 212°F (100°C). The cap holds the pressure at approximately 8–20 psi (55–138 kPa) depending on the manufacturer's design. Raising the pressure 1 psi (7 kPa) raises the boiling point of water approximately 3°F. So, water in a cooling system with a 15 psi (103 kPa) radiator cap will not boil until 45°F above 212°F, or 257°F (125°C). And if the water is mixed with antifreeze, the boiling point will be even higher.

SAFETY TIP

In the same way that liquid refrigerant boils when its pressure is lowered, so too does coolant. If an engine is overheating, never remove the radiator pressure cap. If you do, and the coolant is above 212°F (100°C), then the coolant will instantly turn into steam. It will shoot out of the radiator like a geyser, potentially burning you badly. Let all engines cool before opening the radiator pressure cap.

Applied Math

AM-40: Proportion: The technician can solve problems that determine the proportion of variables of a solution and determine if that proportion is within the manufacturer's specifications.

Engine coolant is composed of a mixture of antifreeze. This is commonly ethylene glycol or propylene glycol, and water. Different operating temperature ranges may require different ratios of antifreeze to water. This is to ensure adequate cooling system performance. A common mixture ratio is 50% antifreeze to 50% water. But other ratios may be specified by antifreeze manufacturers for extreme temperature ranges.

Coolant concentration can be checked with an antifreeze hydrometer, which indicates the proportion of antifreeze in the system. Coolant concentration should be checked during vehicle servicing. Topping up cooling systems with water is a common practice, but the proportion of antifreeze in the system is reduced. If measurement indicates that the proportion of antifreeze in the system is low, some coolant must be drained from the system. It can then be refilled with antifreeze to the correct proportion. Too much antifreeze in the system can cause poor cooling. If this occurs, some coolant must be drained from the system. It can then be refilled with water to the correct proportion.

In the heating system, two hoses run from the engine to a heater core (small radiator). The heater core is located inside the cabin in the plastic box with the air-conditioning evaporator. As air is moved through the fins of the heater core, hot air is removed and sent out of the vents to heat the passenger cabin. The heater core is made from tubes that have thin metal fins connected to them. This increases the surface area tremendously (**FIGURE 47-19**). Because the tubes are also made of relatively thin metal, they can corrode from the inside. Pinhole leaks can develop from old acidic antifreeze. Changing a heater core can be a time-consuming and costly job in many of today's vehicles. So regular cooling system maintenance can pay off for a customer.

To adjust the amount of heat delivered to the cabin, some heating systems use a control valve on the inlet of the heater

FIGURE 47-19 Typical heater core.

core. It controls the flow of hot coolant through the heater core. The valve can be operated in three ways:

- mechanically operated by a cable,
- vacuum operated by a vacuum diaphragm, or
- electrically operated by an electric motor.

Other vehicles use a blend door to control how much airflow goes through the heater core. It is positioned based on the driver's request, using the temperature settings on the dash. Blend doors can also be operated mechanically, by vacuum, or by electric motors. Most vehicles either use a heater control valve or a blend door to control the amount of heat delivered to the passenger compartment (**FIGURE 47-20**).

To improve driver and passenger safety, a supply of fresh air is necessary to maintain comfort and reduce fatigue. Fresh air is drawn from outside, usually through air ducts at the base of the windshield, and directed to the passenger compartment. The air is then delivered through a system of ducts and air doors. They are positioned to direct air to particular points inside the passenger compartment.

Some of the heat generated by the burning fuel in the engine is removed by the cooling system. The hot coolant is pumped through a heater core usually located in the vehicle cabin. As cool air blows across the heater core, the air is heated by the warm coolant. The coolant is then pumped back through the engine to pick up heat from the engine and the cycle continues. In the dash is a control for the blower fan speed. The blower fan moves air through the system. In most vehicles, all air flows through the evaporator on its way to the heater core (**FIGURE 47-21**).

If cold air is desired, then the air-conditioning system is activated to cool the evaporator. If the cool or cold setting is selected, then the blend door will divert the air around the heater core. Or the heater control valve will prevent coolant from flowing through it. If warm or hot air is desired, then the air conditioner will be off. And either the blend door will divert the air through the heater core. Or the heater control valve will allow hot coolant to flow through the heater core.

If defrost is desired, then both the air conditioner and the heater will operate. The air first goes through the evaporator. There, most of the moisture is pulled from the air. The air is then directed through the heater core where it is heated. This causes hot, dry air to be blown onto the windshield, which defrosts it quickly.

Another control, called a mode switch or selector, regulates the location of airflow. Several doors are designed to direct the air to various places. Common places are the feet (floor ducts), head (vent ducts), windshield (defrosting ducts), or a combination of locations, such as the feet and windshield (**FIGURE 47-22**).

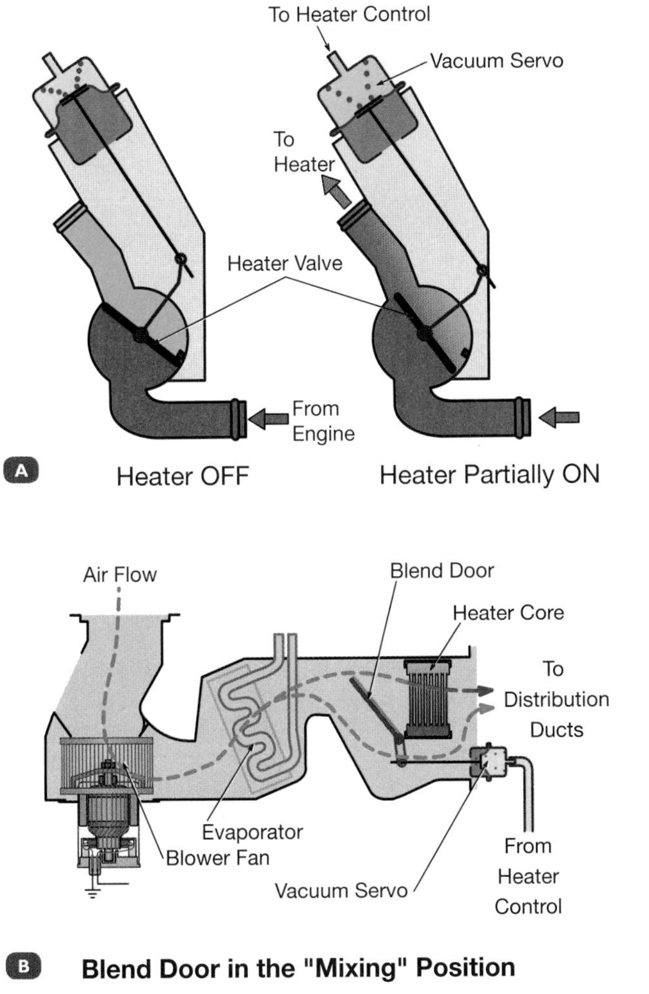

FIGURE 47-20 Most vehicles either use a heater control valve or blend door to control the amount of heat delivered to the passenger compartment. **A.** Typical heater control valve. **B.** Typical blend door.

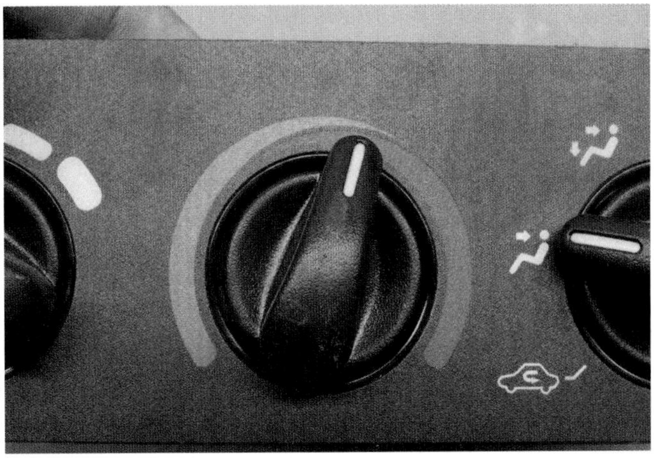

FIGURE 47-21 A typical temperature control on an HVAC system.

FIGURE 47-22 A typical mode switch on an HVAC system.

This control also has a vent setting that allows fresh air in, and then directs it to the vents.

Vehicles with air-conditioning can have a recirculation switch or lever, or they can have a "max" or "maximum" setting. This setting closes the door to the outside air. The air-conditioning system then only works with the air already present in the cabin. Using only the air within the cabin produces better cooling. This is because the air-conditioning system will be cooling air that has already been through the air-conditioning system, rather than hot, muggy air from outside the vehicle. Recirculation can also be used on some vehicles to help keep the moisture out when using the defroster.

▶ Defroster

LO 47-07 Describe the operation of the defroster and blower motor operation.

When the air conditioner is allowed to operate during the defrost mode, the air is dried from the moisture removal occurring as air passes over the evaporator. This occurs before passing the air over the heater core. The result of air passing over both cores is dry, heated air. The hot, dry air from the defrost duct hits the windshield, and the moisture from the glass is quickly drawn into the air. The defrost time of passenger vehicles has decreased considerably. This is because of the air conditioner being switched on during the defrost cycle. Having the air-conditioning operate during defrost also keeps refrigerant and oil circulating through the AC components and hoses during the winter months. This is when the AC is not likely to be used. This keeps the system active and lengthens its life.

Applied Science

AS-9: Scientific Methods: The technician develops a theory relative to the cause of the problem based on the information provided, and then tests the hypothesis to determine the solution.

A vehicle is brought to the shop with an air-conditioning concern. The customer is concerned that all air from the ventilation system is being directed to the windshield. Turning the mode control to either the face or feet positions has no effect.

The vehicle in question uses a vacuum-operated system to control blend-door positions in the HVAC system. The technician forms a theory that the system may have a vacuum leak. Based on the fact that, in the event of no vacuum, the system defaults to defroster operation. This ensures the driver retains visibility and is able to safely operate the vehicle.

The technician first confirms that the HVAC vacuum line is connected to the engine's intake manifold. Then, forms a hypothesis that the vacuum leak is occurring within the HVAC system located inside the vehicle's cabin. He is able to confirm this hypothesis by connecting a vacuum pump to the HVAC line in the engine bay and confirming that the system does not hold vacuum. Individual components and sections of vacuum line in the cabin are then individually tested to ensure they hold vacuum. A common failure in this scenario is an internal vacuum leak in the mode control assembly.

Blower Motors

Blower motors are electric motors that determine the rate of airflow out of the vents. Attached to the spinning shaft of the motor is what is called the "squirrel cage." It is a plastic cage with fins to pull or push air (**FIGURE 47-23**). The speed is controlled by the driver (except in electronically controlled systems) through the fan switch and a series of resistors to limit the fan speed. When placed after the evaporator, they pull the air through the evaporator and heater core in the box. Placing the blower before the evaporator pushes air through the evaporator and heater core in the box.

Blower Motor Resistor Packs/Speed Controls

The resistor pack or resistor block is a common blower motor control (**FIGURE 47-24**). It uses resistors in series to regulate the speed of the fan (**FIGURE 47-25**). The more resistance there is in the circuit, the less current available to spin the fan. The lowest speed (speed 1) passes the current through several resistors. The highest speed uses no resistors, so the motor has full current flow. The amount of resistors in a pack depends on the number of different speed settings.

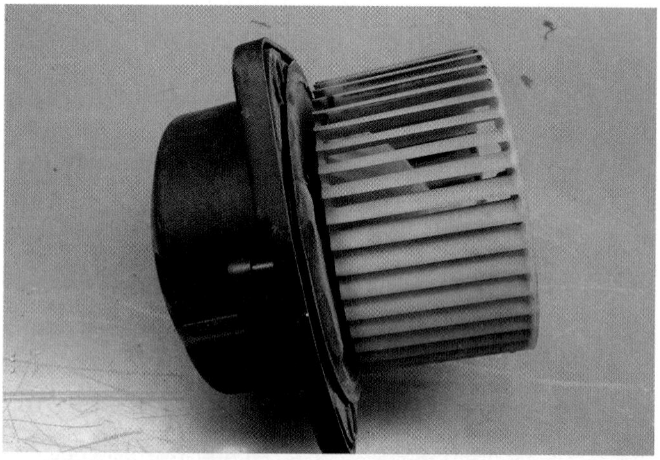

FIGURE 47-23 Blower motor and squirrel cage fan.

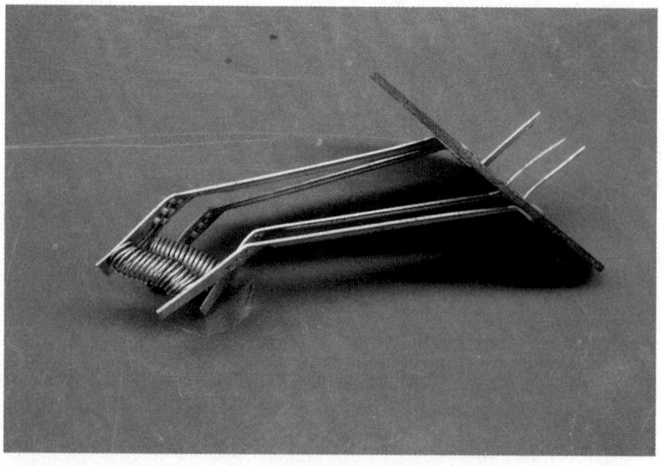

FIGURE 47-24 A typical blower motor resistor pack.

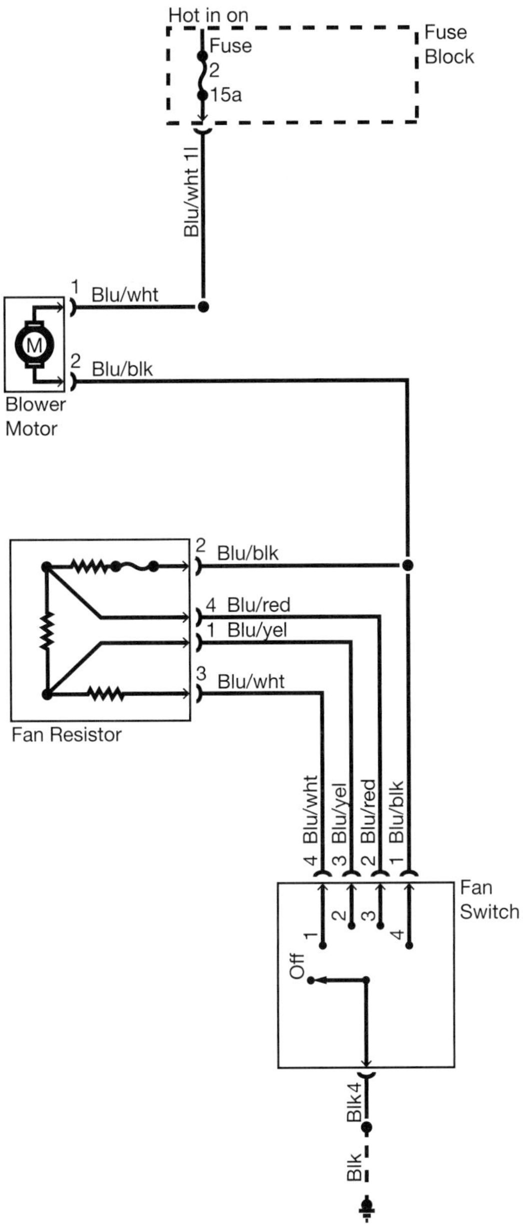

FIGURE 47-25 Blower motor resistor and switch schematic.

On vehicles where the customer just sets a desired temperature, the fan is controlled by AC control unit. It uses **pulse width modulation (PWM)** to control fan speed. Normally, this is done by controlling "on" time compared with "off" time on the ground side of the circuit. The longer the on time, the faster the fan will spin. This is covered in the Electrical chapter.

▶ Performance Testing

LO 47-08 Performance test and check HVAC system.

The performance test and visual inspection of the air-conditioning system are the first step in the diagnostic process. *Performance testing* is the term used to describe the standard air-conditioning testing process. You need to look at all of the components and compare how the air-conditioning unit is functioning versus how it is designed to function. When testing the air-conditioning system, make sure to have all of the controls at the maximum settings. The heater fan should be set on the highest speed. The heater control knob should be on the coldest setting. The airflow should be set to the dash vents. Also, it is good practice to use an external fan to supply air to the condenser to simulate driving conditions. Failure to do this typically causes the system to fail the performance test.

The engine's rpm should be at 1200. This is the ideal and average rpm for the majority of air-conditioning systems on vehicles. Before testing the air-conditioning system, let the system stabilize for a few minutes. It will allow the refrigerant to equalize and the temperatures to reach the proper level.

To performance test an air-conditioning system, follow the steps in **SKILL DRILL 47-1**.

▶ TECHNICIAN TIP

The one rule for diagnosis of an air-conditioning system is "when in doubt, suck it out." This means if you cannot find the problem while diagnosing, remove all refrigerant and start over from a base charge. Starting with a known base charge removes any question as to whether the refrigerant is too low or too high or whether there is any moisture causing a problem in the system.

SKILL DRILL 47-1 Performance Testing an Air-Conditioning System

1. Turn on the vehicle. Place a fan in front of the vehicle to simulate the airflow that occurs when driving.

2. Close all windows. Turn the air conditioner to its maximum cold setting.

3. Raise the engine rpm to 1200–2000. Check the vent temperature using a thermometer. Compare the temperature recorded to the diagnostic chart in the service information.

▶ TECHNICIAN TIP

Performance testing the air-conditioning system must be done before and after the repair. Doing so helps ensure that you identify any faults before starting the repairs. Also, it verifies that the faults are completely repaired, and nothing new showed up.

Abnormal Noises

The compressor is the most common source of abnormal noises arising from the air conditioner. If the compressor fails internally, it may make a knocking noise. Restrictions are another cause of odd noises. Restrictions rattle the air-conditioning pipes, especially if the restriction comes and goes. Abnormal noises can occur while the air-conditioning system is operating. Listen for noises that may occur when the compressor clutch is engaged. If the noise disappears when the compressor clutch disengages, then the noise is from the air-conditioning system. Be sure to listen for noises under the hood as well as in the vehicle's passenger compartment. Turn the compressor clutch on and off, and move the controls on the dash, listening to determine whether the noises appear and disappear.

Applied Math

AM–33: Temperature: The technician can use standard and metric temperature measurement instruments to measure system temperature and determine conformance to metric specifications.

What do you think the most common air-conditioning concern is? It's the air-conditioning system failing to effectively cool the cabin of the vehicle. Technicians need to determine whether the system is operating effectively before proceeding with diagnosis. This is done by measuring the vent temperature as the air enters the vehicle cabin.

Once the vent temperature is measured, it must be compared with the manufacturer's specifications. Then, it can be determined if the system is operating effectively. The specifications may be published in metric units. Then, a metric thermometer should be used. Or readings must be converted from Fahrenheit to Centigrade. This will help determine if the measured temperature conforms to specifications.

▶ TECHNICIAN TIP

Here are some general rules of thumb when diagnosing an AC system:

- The duct temperature should be no greater than 17°F (−8.3°C) above the evaporator temperature. If so, check the heater control valve or blend door.
- The optimum vent temperature is from 38°F to 48°F (3.3–8.8°C), depending on outside temperature and humidity.
- The air-conditioning clutch cycle should not allow the duct temperature to have greater than a 5° swing from high to low. This is while staying within the range of 38°F to 48°F. If the swing is out of the 5° range, the thermal switch or the pressure cycle switch may have to be replaced, if the vehicle is so equipped.

Inspecting the Condenser for Airflow Restrictions

The condenser is normally inspected when high-side pressures are too high and the system is not cooling well. Road debris, leaves, and animal fur and feathers are common culprits, but anything that sits in front of the condenser can cause this issue. Also, air dams play an important role in directing airflow through the condenser. Make sure they are installed properly and not damaged. This inspection should also be performed with any air-conditioning tune-up or inspection, such as those offered by many shops before summer. Use a flashlight on the back of the condenser. Look in from the outside of the vehicle for the light coming through the condenser (**FIGURE 47-26**). Move the light across the entire condenser. If the light is not showing, the condenser should be cleaned.

When cleaning the condenser for debris, shop air or a water hose may be used to remove the debris, but be careful not to fold over the fins. If the shop air and a water hose do not remove the restriction, the condenser must be removed and professionally cleaned.

▶ TECHNICIAN TIP

There are factory adjustments on both the thermal cycle switch and the pressure cycle switch. The adjustments are for achieving the best performance at the factory. Trying to adjust either of the switches because of an incorrect temperature swing may mask the faulty switch for a short time. If these switches are incorrectly functioning, they should be replaced.

Inspecting the Evaporator Housing Water Drain

As the air-conditioning system is running, the evaporator sweats water throughout the day. The water collected has to be drained from the evaporator housing. A drain hose is connected to the

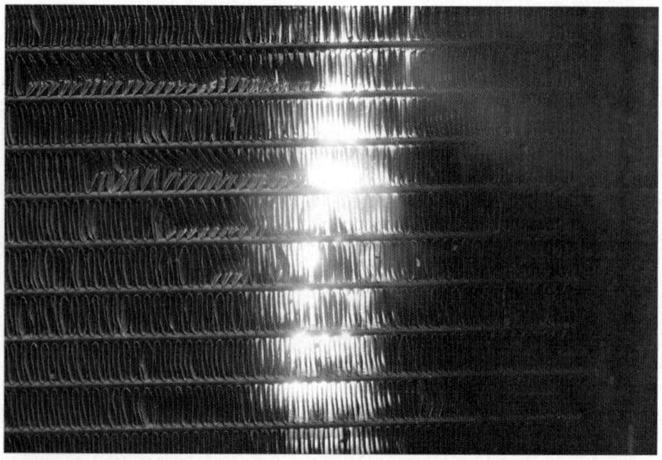

FIGURE 47-26 Using a flashlight to see if the condenser fins are restricted.

SKILL DRILL 47-2 Inspecting the Evaporator Housing Water Drain

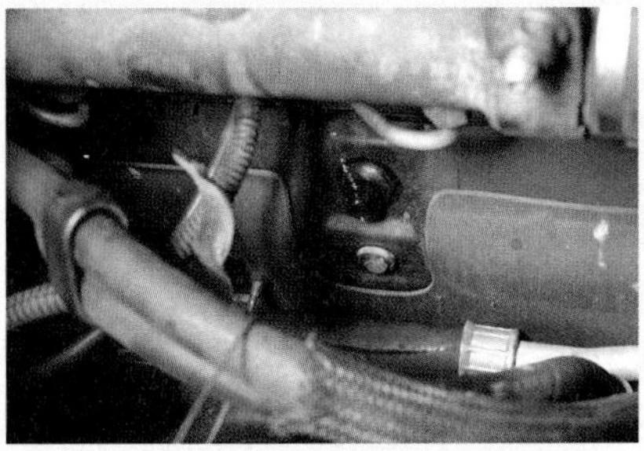

1. Determine that the drain tube is clogged by allowing the air-conditioning system to run while observing the drain tube for water drops. In a plugged drain, no water or very few drops are found.

2. Carefully use a rod to clear the inside of the hose, if possible.

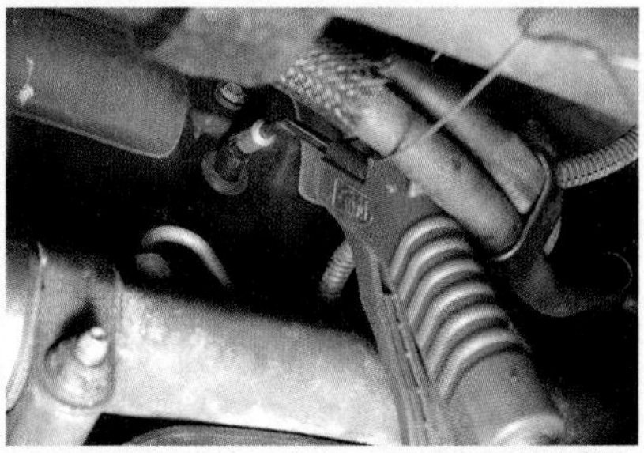

3. If that doesn't work, try a quick blast of low-pressure shop air and an air nozzle up into the drain tube. Foul smelling water will come out of the drain tube when the clog is removed; stand back.

evaporator housing and exits through the vehicle's firewall. Checking that the drain is not plugged is a common diagnostic procedure. If it is clogged, try to clear it with a rod or a blast of low-pressure compressed air. Once the water has completely drained, the task has been performed. Advise the customer to periodically check for a water puddle under the vehicle; if one is not present, the clog may have reoccurred, and removal of the air box may be necessary to open the air box and clean out any remaining debris.

To inspect the evaporator housing water drain, follow the steps in **SKILL DRILL 47-2**.

Eliminating Air-Conditioning System Odors

Air-conditioning systems are prone to unpleasant odors. Because warm, moist air is passed through the cold evaporator, most of the moisture condenses on the surface of the evaporator and drips to the bottom of the air box. This moisture then sits in the air box and ducts where it is dark, which is an ideal place for mold and bacteria to grow. The mold and bacteria produce odors that are directed at the passengers when the fan is turned on. The ducts also provide an appealing space that mice and other small animals use for building their nests. The nests themselves can give off odors, and sometimes one or more rodents die, giving off a very foul odor.

Identifying and eliminating the cause of the odor usually makes customers very appreciative. If you find that an odor is stronger in one vent position than in the others, the smell is likely coming from that duct. If the odor is in all positions, the problem is likely in the heater/evaporator housing. Keep testing until you can pinpoint the source of the odor. Check for odors as you try each of the following: selecting "fresh" air and turning the fan on high, selecting the dash vent, selecting each vent setting, turning the heater on high, and turning the air conditioner on.

If the drain tube becomes clogged with leaves or debris over time, the water will not be able to drain; it will stagnate and begin growing bacteria, creating an unpleasant odor. The solution to this concern is to unclog the drain tube. If odors continue to be a problem and the drain tube is open, the use of an anti-odor kit may be necessary. These kits require gaining access to the evaporator, either through a specified vent, complete removal of the evaporator, removal of the fan, removal of the resistor pack, or by drilling a small hole into the air box in a precise location. When access is gained to the evaporator, a spray chemical cleaner is used on the evaporator fins and allowed to dry.

A follow-up chemical coating is then applied, which keeps bacteria from growing on the fins of the evaporator. If a hole was drilled into the air box according to the directions of the kit, a repair kit is usually included to repair the hole. Sometimes this will be a very strong piece of adhesive tape or a plug that will attach to the hole. Drilling a hole into the air box is not the preferred method and can result in air leaks and noise. Follow the directions of the kit to ensure a quality repair.

To eliminate air-conditioning system odors in the case of a clogged drain, follow the steps in **SKILL DRILL 47-3**.

Cabin Air Filter

Many manufacturers have added cabin air filters to the inlet side of the heating/air-conditioning box where fresh air enters. This filter is designed to catch dust and outside contaminants so they do not get blown into the cabin. Many are paper, just like the air filter for the engine. As such, they should be checked regularly and replaced when they begin to clog. Otherwise, they will cause the system to be inefficient due to poor airflow. Some manufacturers are using activated charcoal cabin filters. They help trap odors and airborne pollutants such as carbon monoxide and oxides of nitrogen.

SKILL DRILL 47-3 Eliminating Air-Conditioning System Odors

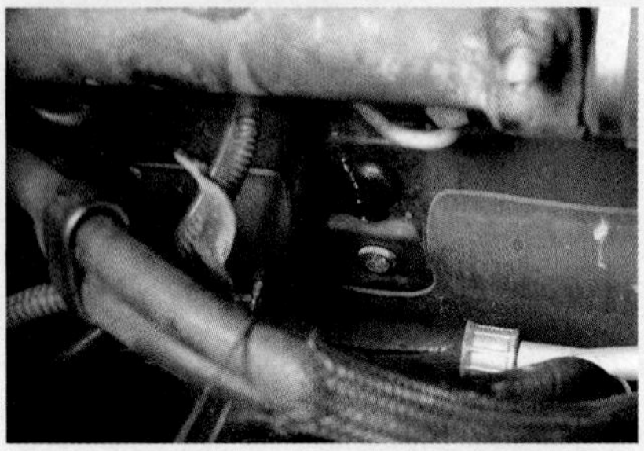

1. Verify that the drain is not plugged. Open all of the doors on the vehicle to allow it to air out a bit.

2. Turn the blower fan on medium speed. Operate the system in all zones to see if the smell is stronger in one position than the others. If the smell is equally strong in all positions, inspect the heater/evaporator housing.

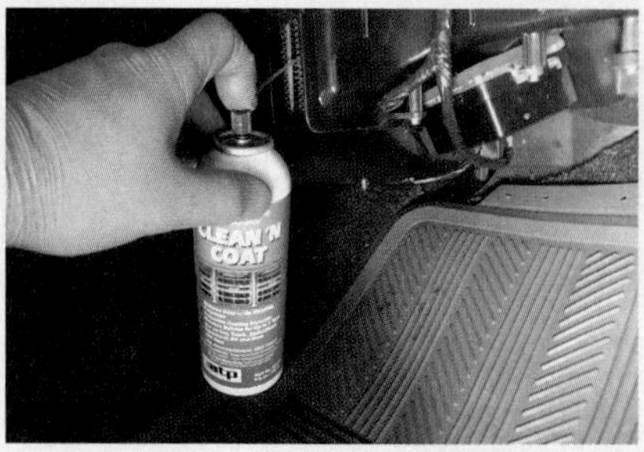

3. If the odor is caused by mold or bacteria buildup, use an anti-odor kit to clean and kill any buildup.

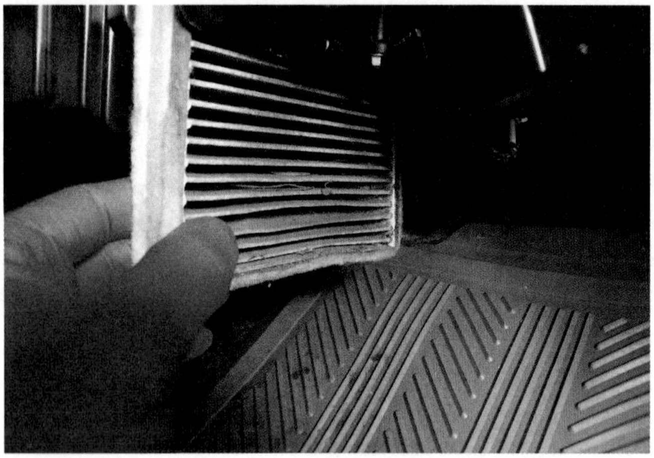

FIGURE 47-27 A typical location of a cabin filter.

The filter is housed in the air box and can be accessed from one of a variety of positions, depending on the vehicle. The access may be from under the hood near the firewall, under the windshield, or behind the glove box (**FIGURE 47-27**). It is usually fairly easy to remove and replace once you find the access cover. This filter should be inspected during every service and replaced according to the manufacturer's specified interval, typically once a year or every 12,000 to 15,000 miles (19,000 to 24,000 km). When inspecting the cabin air filter, use the same guidelines as for an engine air filter. Hold it up to a light, and look at it to see if the light shines through or if it is blocked from being dirty. Also check it for any cracks, tears, or deformities that would cause it to be ineffective (**FIGURE 47-28**).

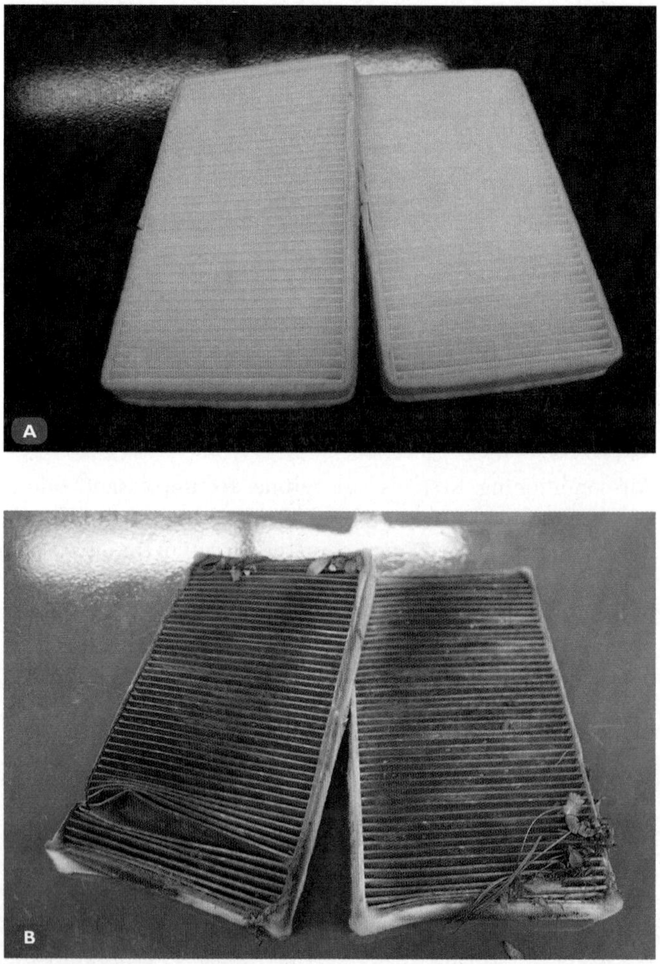

FIGURE 47-28 Cabin filters. **A.** Clean. **B.** Dirty.

▶ Wrap-Up

Ready for Review

- In 1902, Willis Carrier invented and patented the first electric air conditioner. In 1928, a research division of General Motors produced chlorofluorocarbons (CFCs) that made it possible for air-conditioning systems to be used by consumers with no apparent harm.
  - Automotive air-conditioning service technicians need to have a special license is granted after passing the 609 test. The 609 license also allows the technician to purchase a few select refrigerants, such as R-12, R-134a, and HFC 1234.
- The HVAC system reduces the temperature inside the passenger compartment by removing excess heat through a series of thermal and chemical transformations. Heat transfer can occur through conduction, convection, or radiation.
- The refrigerant must have a boiling point well below freezing. Maintaining vaporization and condensing points of refrigerant at normal ambient temperatures is done by raising or lowering the pressure.
- Four major components of HVAC systems are compressor, condenser, restriction, and evaporator. The air conditioner is a closed-loop system in which the refrigerant just continues to cycle through the system, changing pressure and state to complete the task of cooling the cabin.
  - Two definitive types of air-conditioning systems are the fixed orifice tube system and the thermal expansion valve (TXV) system.
- Different types of refrigerants used in automotive applications include: R-12, R-134a, or HFO-1234yf. The oil used in refrigeration systems must be compatible with the refrigerant used. Different types of oil are mineral oil, PAG oil, and POE oil.
- In the heating system, as air is moved through the fins of the heater core, hot air is removed and sent out of the vents to heat the passenger cabin. A control valve on the inlet of the heater is used to adjust the amount of heat delivered to the cabin.
- In the defrost mode, air passes over the heater core and produces dry, heated air. Hot, dry air from the defrost duct hits the windshield, and the moisture from the glass is quickly drawn into the air. Blower motors determine the rate of airflow out of the vents. Placing the blower before the evaporator pushes air through the evaporator and heater core in the box.
- Performance testing of HVAC systems includes checks for abnormal noise, condenser inspection for airflow restrictions, evaporator housing water drain inspection, elimination of air-conditioning system odors in the case of a clogged drain, cabin air filter inspection.

Key Terms

atmospheric pressure (atm) The pressure of the air surrounding everything, caused by gravity and the weight of air. The higher the altitude from sea level, the lower the atmospheric pressure.

British thermal unit (Btu) A measure of heat energy. It takes 1 Btu to raise the temperature of 1 pound of water 1°F.

compressor A belt- or electrically driven device designed to increase refrigerant pressure and cause refrigerant to travel through the air-conditioning system.

chlorofluorocarbon (CFC) A chlorine-based fluorocarbon compound.

Clean Air Act (CAA) A policy signed into law in 1990 that sets standards for air pollution to eliminate ozone-depleting elements.

closed-loop system A refrigerant system where nothing else enters or exits the system.

condensation Change of state from a vapor to a liquid, such as the moisture that collects on a cool surface.

condenser A component of the HVAC system that transfers heat from the system to the atmosphere.

conduction The process of transferring heat through matter by the movement of heat energy through solids from one particle to another.

convection The process of transferring heat by the circulatory movement that occurs in a gas or fluid as areas of differing temperatures exchange places due to variations in density and the action of gravity.

dichlorodifluoromethane An inert, colorless gas that can be used as a refrigerant. It is stored in white containers.

Environmental Protection Agency (EPA) A US federal government agency that deals with issues related to environmental safety.

evaporator The cold surface of the air-conditioning system that absorbs heat from a cab or vehicle.

fixed orifice tube system A system with a fixed orifice tube that uses an accumulator between the evaporator and the compressor.

heat transfer The flow of heat from a hotter part to a cooler part; it can occur in solids, liquids, or gases.

latent heat of condensation The amount of heat removal necessary to change the state from a gas to a liquid without changing the actual gauge temperature.

latent heat of evaporation The amount of heat required to change the state from a liquid to a gas without raising the actual gauge temperature.

latent heat of freezing The amount of heat removal required to change the state from a liquid to a solid without changing the actual gauge temperature.

pulse width modulation (PWM) A digital on/off electrical signal. PWM is a very precise control method for an output device on a varying frequency. Usually, only fuel injectors are operated in this format, as the software programming and related circuits are very complex. Most other components are duty-cycled on a fixed frequency.

radiation The movement of energy through space, such as the movement of energy from the sun to the earth.

refrigerant The name given to a chemical compound designed to meet the needs of the refrigeration system.

restriction A blockage that partially stops or slows the flow of a material such as refrigerant.

tetrafluoroethane An inert, colorless gas that can be used as a refrigerant. It is stored in light blue containers.

thermal expansion valve (TXV) system A system with a valve designed to sense evaporator outlet temperature and vary the inlet orifice size accordingly.

vaporization Change of state from a liquid to a gas.

Review Questions

1. What year was the first automotive air-conditioning system offered on a new vehicle?
   a. 1931
   b. 1939
   c. 1958
   d. 1976
2. Which certification is a technician required to hold in order to legally handle refrigerant in a vehicle?
   a. 609
   b. R-22
   c. ASE A7
   d. MVAC
3. Which type of heat transfer occurs when cold air surrounds your body and cools it?
   a. Conduction
   b. Convention
   c. Convection
   d. Radiation
4. At atmospheric pressure, what temperature does R-134a boil at?
   a. −15°F
   b. 32°F
   c. 100°C
   d. 212°F
5. In refrigeration, how can we force refrigerant to evaporate?
   a. By compressing it
   b. By speeding it up
   c. By slowing it down
   d. By reducing its pressure
6. In an air-conditioning system, which component receives refrigerant as a low-pressure liquid and changes it to a low-pressure gas, and as a result, absorbs surrounding heat?
   a. The condenser
   b. The compressor
   c. The evaporator
   d. The accumulator
7. In an air-conditioning system, which device changes the high-pressure gas refrigerant into a high-pressure liquid after giving off its stored heat?
   a. The evaporator
   b. The condenser
   c. The accumulator
   d. The orifice
8. R-12 refrigerant was found to be a leading cause of ozone layer depletion; what is the latest refrigerant used by manufacturers that has the least effect on the ozone layer?
   a. Freon
   b. R-134a
   c. Tetraflouroethane
   d. HFO-1234yf
9. In the HVAC system, which component heats the surrounding air for passenger comfort when the system is set to a warm temperature?
   a. The thermostat
   b. The radiator
   c. The heater core
   d. The evaporator
10. When performance testing an AC system on the max cool setting, how would a technician know the acceptable temperature range?
    a. It should be between 39°F and 45°F on a hot day
    b. They should compare it to another vehicle temperature that day
    c. They should ask the customer what the temperature was
    d. They should refer to the diagnostic chart in service information

ASE Technician A/Technician B Style Questions

1. Refrigerant handling certification is being discussed. Technician A states that a technician can obtain the 609 certification from ASE. Technician B states that a technician can obtain the 609 certification from MACS. Who is correct?
   a. Technician A only
   b. Technician B only
   c. Both Technicians A and B
   d. Neither Technician A nor B
2. Air-conditioning principles are being discussed. Technician A states that heat always travels from hot to cold. Technician B states that convection is heat that travels through solids. Who is correct?
   a. Technician A only
   b. Technician B only
   c. Both Technicians A and B
   d. Neither Technician A nor B
3. Air-conditioning fundamentals are being discussed. Technician A states that the sun heating your body through a windshield on a sunny day is called radiant heat. Technician B

states that heat that cannot be felt or measured is called latent heat. Who is correct?
  a. Technician A only
  b. Technician B only
  c. Both Technicians A and B
  d. Neither Technician A nor B
4. Principles of AC are being discussed. Technician A states that if R-134a is pressurized to 15 psi, it will boil at just over 32°F. Technician B states that as R-134a boils in the AC system, it gives off heat to air surrounding it. Who is correct?
  a. Technician A only
  b. Technician B only
  c. Both Technicians A and B
  d. Neither Technician A nor B
5. The states of refrigerant are being discussed. Technician A states that when a refrigerant changes from a gas to a liquid, it absorbs heat. Technician B states that when a refrigerant changes from a liquid to a gas, it gives off heat. Who is correct?
  a. Technician A only
  b. Technician B only
  c. Both Technicians A and B
  d. Neither Technician A nor B
6. Air-conditioning components are being discussed. Technician A states that a restriction device can be a fixed orifice or a thermal expansion valve. Technician B states that the purpose of the compressor is to circulate refrigerant, and with the help of a restriction, control the refrigerant pressure. Who is correct?
  a. Technician A only
  b. Technician B only
  c. Both Technicians A and B
  d. Neither Technician A nor B
7. Refrigerant types are being discussed. Technician A states that R-134a is being phased out because it was recently found to be a major contributor to greenhouse gases. Technician B states that HFO-1234yf is non-flammable. Who is correct?
  a. Technician A only
  b. Technician B only
  c. Both Technicians A and B
  d. Neither Technician A nor B
8. An HVAC system is being discussed. Technician A states that on most vehicles, the MAX cold setting will pull air from the outside air to provide maximum cooling. Technician B states that on some vehicles, a heater control valve blocks coolant from entering the heater core unless the heater is turned on. Who is correct?
  a. Technician A only
  b. Technician B only
  c. Both Technicians A and B
  d. Neither Technician A nor B
9. An HVAC system is being discussed. Technician A states that on some vehicles, the AC may be turned on automatically when defrosting the windshield since it dries the air. Technician B states that many vehicles use a blower motor resistor pack to control blower motor speeds. Who is correct?
  a. Technician A only
  b. Technician B only
  c. Both Technicians A and B
  d. Neither Technician A nor B
10. An HVAC system is being discussed. Technician A states that water dripping from the center of the vehicle on a humid day with the AC on may be normal. Technician B states that some manufacturers are using charcoal cabin filters which trap odors and airborne pollutants before they enter the cabin. Who is correct?
  a. Technician A only
  b. Technician B only
  c. Both Technicians A and B
  d. Neither Technician A nor B

SECTION 9
Engine Performance

- ▶ CHAPTER 48 Ignition Systems
- ▶ CHAPTER 49 Gasoline Fuel Systems
- ▶ CHAPTER 50 Engine Management System
- ▶ CHAPTER 51 On-Board Diagnostics
- ▶ CHAPTER 52 Induction and Exhaust
- ▶ CHAPTER 53 Emission Control
- ▶ CHAPTER 54 Alternative Fuel Systems

CHAPTER 48
Ignition Systems

Learning Objectives

- **LO 48-01** Describe ignition system preliminaries.
- **LO 48-02** Describe the operation of the primary and secondary ignition systems.
- **LO 48-03** Explain required voltage and available voltage.
- **LO 48-04** Explain spark timing.
- **LO 48-05** Explain common ignition components.
- **LO 48-06** Describe spark plugs.
- **LO 48-07** Describe contact breaker point ignition systems.
- **LO 48-08** Describe mechanical spark timing systems.
- **LO 48-09** Describe electronic ignition systems.
- **LO 48-10** Describe distributorless ignition systems.
- **LO 48-11** Perform ignition system maintenance.

ASE Education Foundation Tasks

See Appendix A to view the 2017 ASE Education Foundation Automobile Accreditation Task List Correlation Guide.

▶ Ignition System Introduction

LO 48-01 Describe ignition system preliminaries.

The air–fuel mixture inside each cylinder must be ignited for it to release its energy. When it does, it creates the pressure that forces the piston down the cylinder on the power stroke (**FIGURE 48-1**). Gasoline engines use the heat of a high-voltage spark to ignite the mixture. The purpose of the ignition system is to create the high-voltage spark and deliver it at the right time to each cylinder.

The ignition system consists of a primary (low-voltage) circuit and a secondary (high-voltage) circuit. The primary circuit activates the ignition coil. The coil changes the low voltage of the vehicle battery into the high voltage needed to jump across the spark plug electrodes. The secondary circuit transmits the high voltage from the coil, or coils, to the spark plug at each cylinder. Although there are several types of ignition systems, the following components are common to all of them. They include:

- the spark plugs,
- the ignition coil, and
- a device for triggering the ignition coil.

For an engine to run smoothly and efficiently, the high-voltage spark must jump across the spark plug electrode at the right time. This typically occurs as the piston approaches top dead center (TDC) of the **compression stroke**. The ignition system must also be able to advance or retard the timing of the spark. It does so based on engine conditions such as load, speed, and driver input. This chapter explains the principles and operation of modern ignition systems. It also covers the basics of diagnosis, maintenance, and repair.

Ignition Principles

When the driver turns the key to the start position (or presses the electronic start button), power is sent to the primary winding of the ignition coil. As the engine is cranked, a switching circuit turns the primary ignition coil circuit on and off. Each coil amplifies the battery's low voltage and high current signal into a very high voltage and very low current spark. This high voltage is delivered from each ignition coil to the spark plug in each cylinder (**FIGURE 48-2**). As the high voltage pushes current across the spark plug air gap, the air–fuel mixture is ignited. This causes cylinder pressure to increase greatly, which in turn pushes the piston down the cylinder.

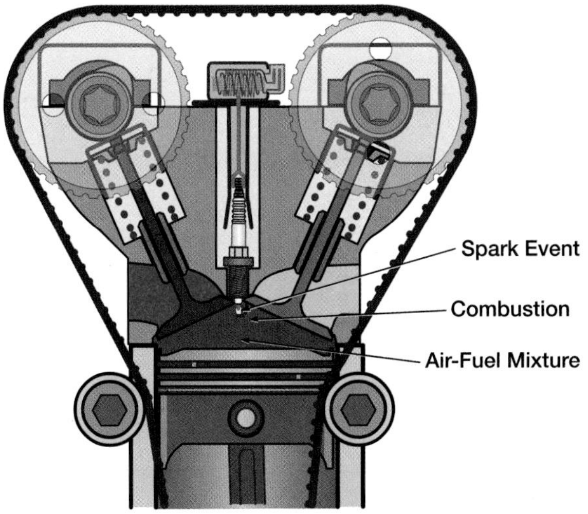

FIGURE 48-1 The purpose of the ignition system is to create a spark to ignite the air–fuel mixture in the combustion chamber.

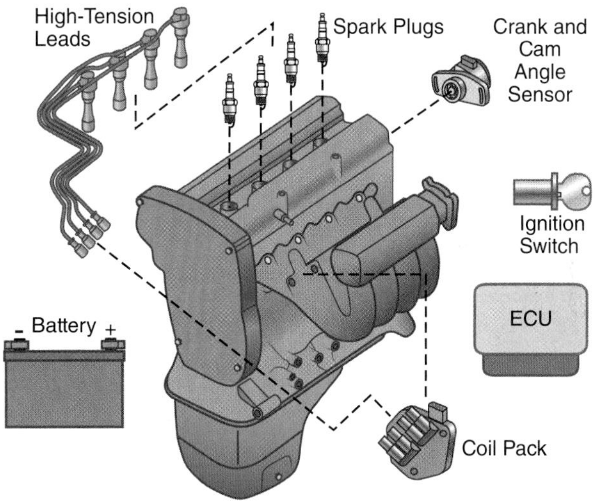

FIGURE 48-2 An ignition system circuit.

You Are the Automotive Technician

Today a customer visits your shop for a scheduled 60,000-mile service on a 2013 vehicle. The service information says to replace the spark plugs and inspect the secondary ignition system. It consists of the ignition coils and spark plug boots. When you twist the coils to pull them off the plugs, several of the boots tear, requiring replacement. After the boots are off, you use compressed air (while wearing safety glasses) to blow out any debris from around the spark plugs. You remove the plugs one at a time and inspect them. The deposits are light and show that they are burning correctly—neither too hot nor too cold. The gaps are worn, as expected. You inspect the ignition coils. The terminals and insulation look good, and you will be replacing all of the boots.

1. How would you prevent the boots from sticking to the spark plugs the next time?
2. How would you test to see if the ignition system was working on a vehicle that cranked but didn't start?
3. If this customer drove only short trips of a mile or two, would you recommend a spark plug with a hotter heat range or colder heat range? Why or why not?
4. How can a spark plug boot cause a cylinder to misfire?

The ignition system has undergone changes in technology over the decades. This has allowed engines to meet requirements for dependability, reduced maintenance expectations, and strict emission standards. The original system was the **contact breaker point ignition system**. It was a mechanical system with a switch, called **contact breaker points**. They opened and closed as the engine was running (**FIGURE 48-3**). The points turned the primary ignition circuit on and off. This created high voltage in the secondary circuit that was distributed to each spark plug in the firing order.

The first advancement was the replacement of the mechanical switch. An electronic switching device was developed (**FIGURE 48-4**). This device had no moving components, meaning minimal wear on the switching device. So it required very little, if any, maintenance. This ignition system was called an **electronic ignition system-distributor type**. It still used a single coil and distributor to dispense the spark to the various cylinders.

The next advancement was eliminating the distributor by using dedicated ignition coils. It used one coil for each pair of cylinders. This system was called a **waste spark ignition system** (**FIGURE 48-5**). A four-cylinder engine would have two ignition coils. A six-cylinder engine would have three coils; and so on. Eliminating the distributor meant doing away with the last mechanical component of the ignition system. This again resulted in increased reliability and reduced maintenance issues.

The latest development has been to give each cylinder its own ignition coil. This system is called a **direct ignition system** or **coil-on-plug** (COP) ignition system (**FIGURE 48-6**). This system puts the ignition coil directly on top of the spark plug. So it eliminates the spark plug wires, which are subject to leakage, damage, and wear.

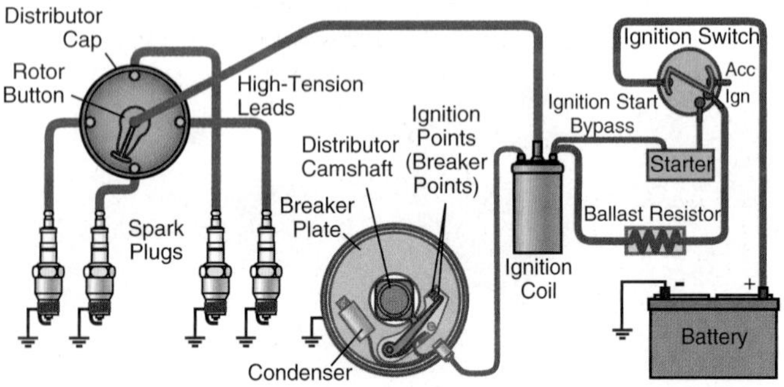

FIGURE 48-3 Breaker point ignition system.

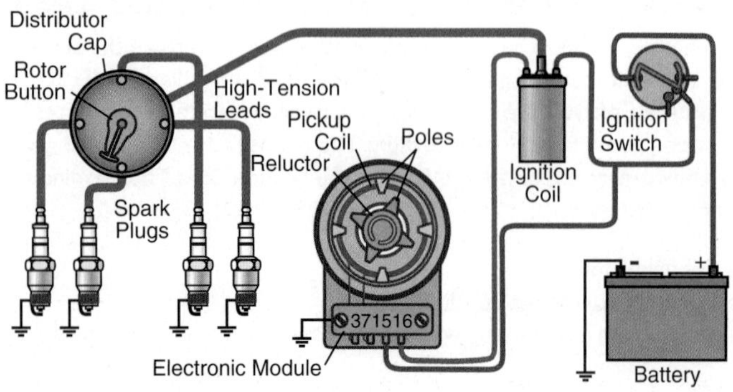

FIGURE 48-4 Electronic ignition system-distributor type.

▶ Primary and Secondary Circuits

LO 48-02 Describe the operation of the primary and secondary ignition systems.

The ignition system uses an **induction coil**. It converts low-voltage and high-current flow into very high-voltage and very low-current flow. The low-voltage side is called the **primary circuit**. The high-voltage side is called the **secondary circuit**. The battery supplies the low-voltage and high-current flow to power the primary circuit. The induction coil steps up the low voltage in the primary circuit to the high voltage in the secondary circuit. A problem in the primary circuit will affect the output of the secondary circuit. Technicians need to understand both circuits when identifying faults in the ignition system (**TABLE 48-1**). Notice how the types of primary circuit components stay the same over each type of system. And yet the types of secondary circuit components reduce in number as newer systems are introduced.

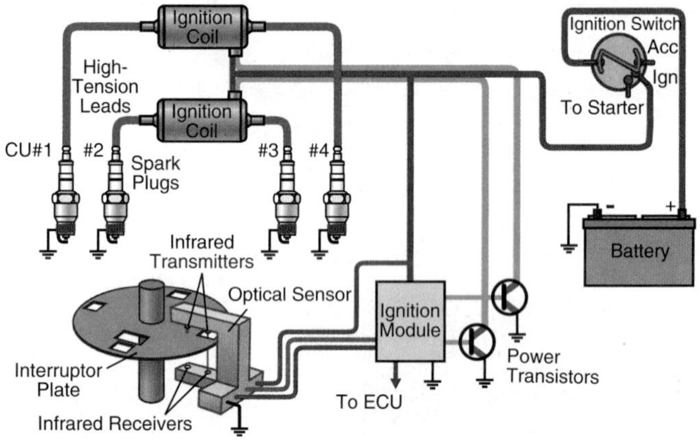

FIGURE 48-5 A waste spark ignition system.

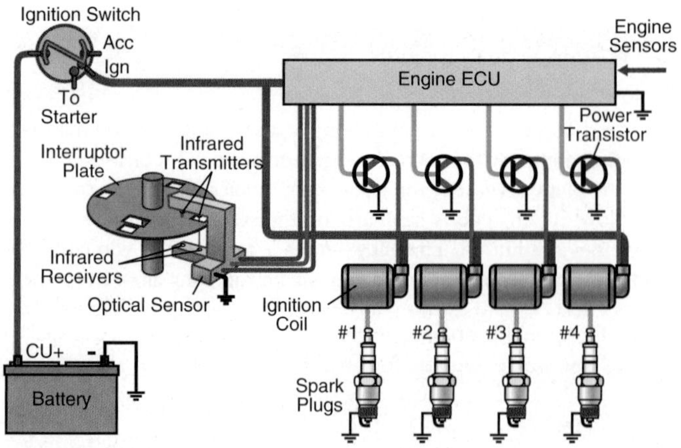

FIGURE 48-6 A coil-on-plug ignition system.

TABLE 48-1 Primary and Secondary Circuit Components

Primary Circuit Components	Secondary Circuit Components
Contact Breaker System	
Contact breaker system	Ignition coil—secondary winding
Ignition switch	Coil wire
Ballast resistor	Distributor cap
Ignition coil—primary winding	Rotor
Capacitor	Spark plug wires
Contact breaker points	Spark plugs
Electronic Ignition—Distributor-Style System	
Primary Circuit Components	**Secondary Circuit Components**
Battery	Ignition coil—secondary winding
Ignition switch	Coil wire
Ignition coil—primary winding	Distributor cap
Ignition module	Rotor
Triggering device	Spark plug wires
	Spark plugs
Electronic Ignition—Distributorless-Style System	
Primary Circuit Components	**Secondary Circuit Components**
Battery	Ignition coils—secondary windings
Ignition switch	Spark plug wires
Ignition coils—primary windings	Spark plugs
Ignition module	
Triggering device	
Direct Ignition System	
Primary Circuit Components	**Secondary Circuit Components**
Battery	Ignition coils—secondary windings
Ignition switch	Spark plugs
Ignition coils—primary windings	
Ignition module	
Triggering device	

Faraday's Law

Most automotive ignition systems use ignition coils that operate on the principles of an induction coil. Automotive induction coils step up the nominal battery voltage of 12 volts to the voltage needed to bridge the gap across the spark plug electrodes. This can be up to 100,000 volts. These induction coils operate according to Faraday's law.

Faraday's law states that relative movement between a conductor and a magnetic field allows four ways by which voltage can be induced in a conductor:

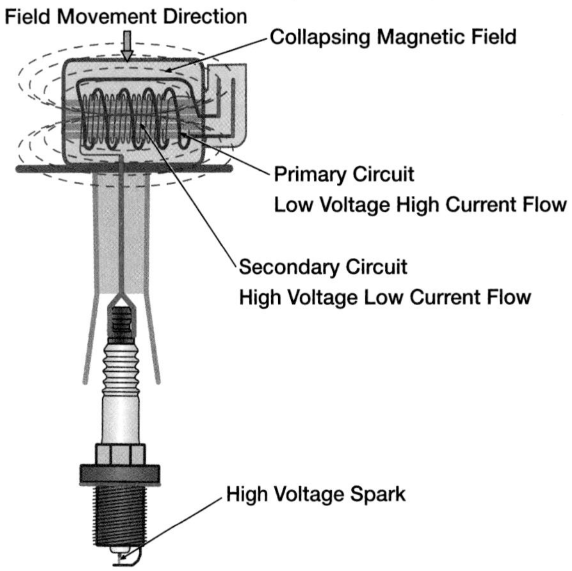

FIGURE 48-7 The ignition coil induces high voltage, but low current flow in the secondary windings of the coil.

1. Moving a magnet so that the magnetic lines of force cut across a conductor, as in an alternator.
2. Moving a conductor so that it cuts across the stationary magnetic field, as in a generator.
3. Starting, stopping, or changing the rate of current flow in a conductor. This causes the conductor to induce an electromagnetic field into itself and occurs in the primary windings of an ignition coil. This process is called self-induction.
4. Starting, stopping, or changing the rate of current flow in a conductor that is positioned close to a second conductor. This is called mutual induction. It is used to induce high voltage in the secondary winding of the ignition coil.

When any of these methods are used to induce voltage in a conductor, the value of that voltage depends on the following conditions:

- The density, or strength, of the magnetic field—the stronger the field, the greater the induced voltage
- The number of turns of the windings in the coil—the more turns, the greater the induced voltage
- The speed at which the lines of force are cut—the greater the speed, the greater the induced voltage.

In the induction coil, the secondary winding has many thousands of turns of fine enameled copper wire. The primary winding has a few hundred turns of relatively heavy wire (**FIGURE 48-7**). It is positioned around the outside of the secondary winding. A soft iron core is positioned centrally to concentrate the magnetic field. Current flow through the primary winding establishes a magnetic field around the windings. The higher the current flow, the stronger the field.

Sudden interruption of the primary current effectively stops the current flow. The magnetic field collapses into the iron core. This returns its stored energy to the coil by cutting across the coil's primary and secondary windings. This produces a

Applied Science

AS-76: Coil: The technician can explain how a coil can increase the battery voltage needed to fire a spark plug.

An ignition coil is a step-up transformer. It increases the voltage from the battery to a high voltage necessary to fire the spark plugs and ignite the air–fuel mixture in the cylinders. Coils operate via the concept of electromagnetic induction. It dictates that a moving magnetic field (or a change in a stationery magnetic field in the case of an ignition coil) can induce a current in a wire exposed to the field.

The internal makeup of a coil is basically an iron core with two windings of wire, both wound around the core. The secondary winding is wound with considerably more turns than the primary winding. Typically, it is wound at a ratio of about 100:1.

In operation, battery voltage is applied to the primary winding, creating a magnetic field. This process is called saturation. The secondary winding is exposed to this field. To create a spark, the primary circuit is interrupted, or turned off, by a powertrain control module, ignition module, or breaker points. This interruption causes the magnetic field to collapse very rapidly. A high voltage is then induced into the primary and secondary windings. The increase in voltage is provided by the difference in the number of turns between the windings. Current in the secondary winding is directed to the spark plugs, where it creates a spark as it jumps across the air gap.

self-induced voltage in the primary winding and a mutually induced voltage in the secondary winding.

The maximum value of the secondary voltage is partly determined by the ratio of the number of turns between the two windings. In this case, it is approximately 100 to 1. It is also determined by the value of the self-induced voltage in the primary winding. In this case, it is about 300 volts. If the coil is 100% efficient, the maximum voltage available from the secondary winding would be 300 volts multiplied by 100, or 30,000 volts.

The value of the self-induced voltage in the primary winding is also influenced by the rate of collapse of the magnetic field. This is determined by how quickly the current flow is stopped through the coil. So it is critical that the primary current be switched off as quickly as possible. All ignition systems make provisions to ensure that this occurs. This subject is covered in greater detail later in this chapter.

▶ Required Voltage Versus Available Voltage

LO 48-03 Explain required voltage and available voltage.

Understanding the terms "required voltage" and "available voltage" will help you diagnose ignition systems. **Required voltage** is the amount of voltage required to initially jump the spark plug gap. Once the voltage reaches the point where current is flowing in the secondary circuit, the required voltage drops to a much lower level. It remains just sufficient enough to sustain current flow and the spark. This gives more opportunity for the air–fuel mixture to ignite during the duration of the spark.

The other factor is **available voltage**. It is the maximum amount of voltage available to jump the spark plug gap if the gaps were infinite. In other words, available voltage is the maximum amount of voltage that the ignition coil can put out. It is very important that the available voltage is always higher than the required voltage (**FIGURE 48-8**). If the available voltage falls below required voltage, there will not be enough voltage to push current across the spark plug gap. So no spark will be created. Likewise, if required voltage is ever higher than the available voltage, the spark will not be able to jump the gap. Because the spark is required to ignite the air–fuel mixture, higher required voltage or lower available voltage will prevent the spark from occurring. The cylinder will then not fire. This is called a misfire.

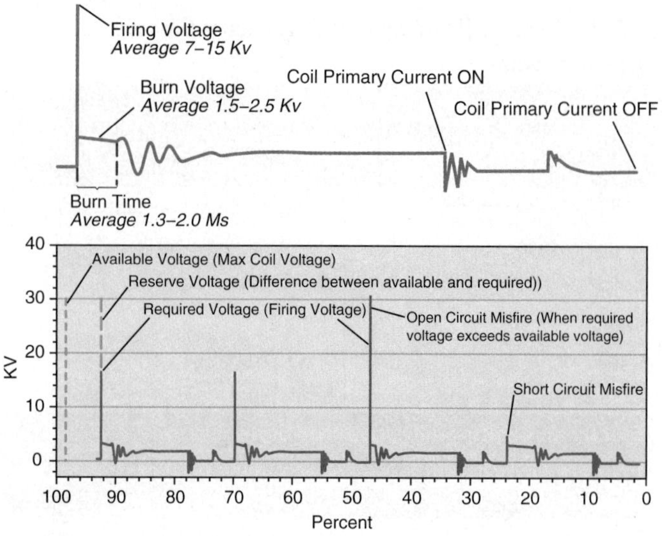

FIGURE 48-8 Available voltage, required voltage, reserve voltage, and misfire.

As a vehicle is driven over time, the required voltage will typically increase while the available voltage decreases. The required voltage increases primarily because of the growing gap of the spark plug as it wears over time. However, it can also increase because of any open circuits in the secondary side of the circuit. Examples include a broken rotor tip or an open spark plug wire.

Manufacturers design their ignition systems to have an amount of reserve voltage above the required voltage to prevent misfire. But if the vehicle's ignition system is not maintained, this reserve voltage is reduced. Engine misfire will begin to occur. This is especially noticeable when accelerating or climbing a hill. Inspecting all of the components and replacing any that are worn will restore the available voltage to what it should be. It will also help reduce the required voltage to within specifications. This helps restore the ignition system to its proper operation.

▶ TECHNICIAN TIP

Required voltage increases because of:

- spark plug gaps that are worn or rounded,
- compression pressures that are higher than normal, or
- lean mixtures.

Available voltage is affected by faults in the primary or secondary circuits. Shorted coil windings or high resistance are examples.

▶ Spark Timing

LO 48-04 Explain spark timing.

The timing of the spark is critical to the smooth and efficient operation of the engine. For any given engine speed and load, the correct **spark timing** varies. It varies according to a number of factors, including:

- Air–fuel ratio
- Detected knock
- Engine speed
- Engine load
- Engine temperature
- Air temperature
- Transmission gear selected
- Throttle position

At higher engine speeds, the time available to burn the air–fuel mixture decreases. But the time required to completely burn the mixture remains essentially the same (**FIGURE 48-9**). Therefore, the spark must occur sooner in the cycle (advanced). This is to give the air–fuel mixture enough time to burn and create maximum cylinder pressure soon after TDC.

At low engine revolutions per minute (rpm), when the throttle is opened, the engine load increases. This speeds up the burn rate of the air–fuel mixture. The spark does not have to be advanced as much because the engine is not spinning very fast. Thus, there is plenty of time for the air–fuel mixture to burn before the piston moves very far in the cycle.

As engine speed increases, there is increasingly less time for the air–fuel mixture to be ignited and for the maximum pressure to develop. Therefore, the ignition point must be advanced earlier in the cycle. That is, it must start earlier in relation to the piston's position during the compression stroke. This adjustment must occur automatically in relation to engine speed and engine load.

If the timing is advanced too far, then the engine begins to knock, which can destroy an engine. So the timing needs to be precisely controlled. If it is equipped with a knock sensor located on the engine block, the sensor sends a signal to the **powertrain control module (PCM)**. The PCM may also be called an **engine control module (ECM)** or an electronic control unit (ECU). The PCM then retards the timing to decrease or eliminate the knock. At lower engine temperatures, the fuel does not atomize as quickly. So the spark must be advanced more than when the engine is fully warmed up.

In early vehicles that use a distributor, base spark timing is set at idle speeds. It is set by positioning the distributor body in relation to its rotating cam. The timing is almost always indexed to the **number-one cylinder**. The contact breaker points are operated by each cam lobe to provide the same timing point for succeeding cylinders in the firing order (**FIGURE 48-10**). This initial setting, which in most, but not all, cases occurs before TDC. This allows time for maximum pressure in the cylinder to develop, just as the piston is starting to descend on the power stroke.

Today's vehicles use electronically triggered ignition systems. The timing is engineered into the design of the engine and is not adjustable. The timing components are the crankshaft position sensor and the camshaft position sensor. These sensors are fixed in position on the engine and are triggered by the crankshaft or flywheel (**FIGURE 48-11**) on most engines, or

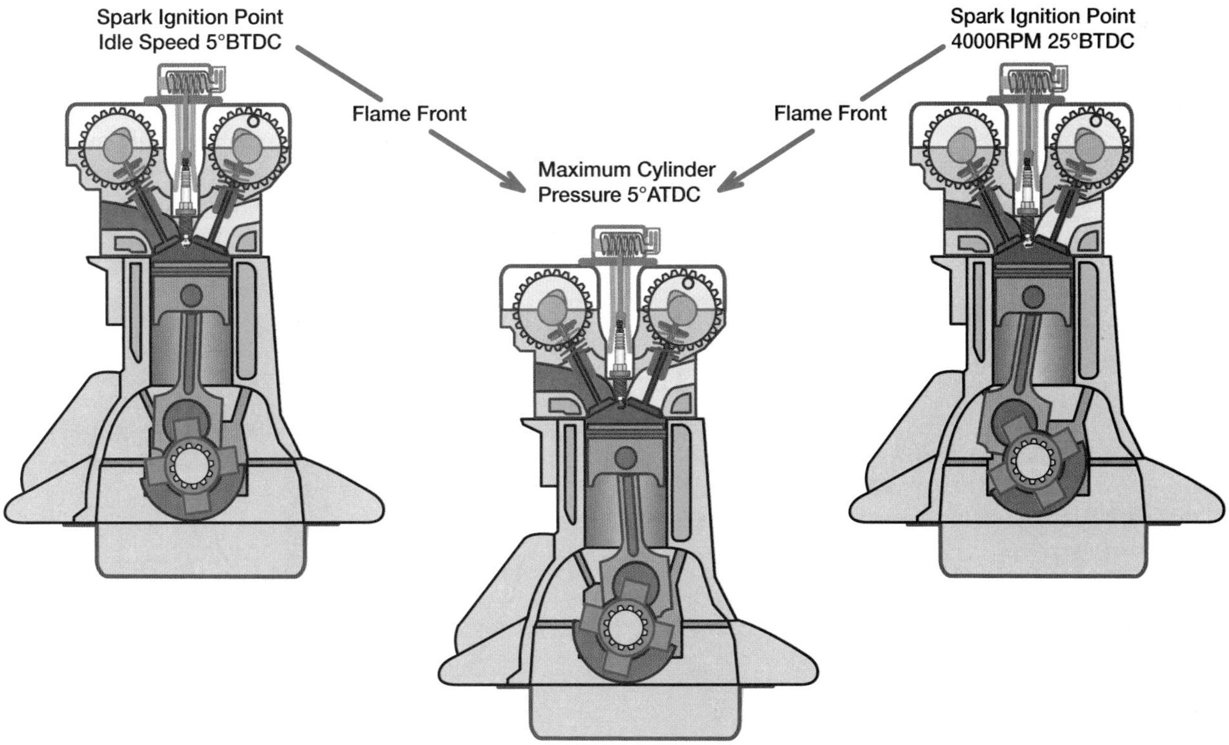

FIGURE 48-9 At higher engine speeds, the spark needs to occur earlier in the cycle so that maximum pressure will develop shortly after TDC.

within the distributor (if used). The initial timing specification is programmed into the PCM software.

▶ Components Common to All Ignition Systems

LO 48-05 Explain common ignition components.

Most ignition systems include a battery, ignition switch, ignition coil(s), high-tension leads (also called spark plug wires), and spark plugs. The battery is the same one used to start the vehicle and provide electrical power to all electrical loads when the engine is not running. The ignition switch is used to turn the electrical power to the ignition system on and off. The ignition coil is needed to amplify the battery voltage into very high voltage.

The high-tension leads transmit the high voltage from the coil(s) to the spark plugs. The spark plugs are threaded into the top of each cylinder. They provide a place for the spark to jump within the combustion chamber.

FIGURE 48-10 Breaker points and cam determine timing for each cylinder.

FIGURE 48-11 A crankshaft position sensor.

Ignition Switch

The **ignition switch** has more functions than simply starting and stopping the engine. It provides a way to disconnect most electrical accessories from the battery with an ignition key. The common positions of an ignition switch include the following (**FIGURE 48-12**):

- **Lock**: The key can only be removed from the lock position. When this occurs, all nonessential electrical circuits are disabled, and the steering column lock is enabled. If equipped, the engine immobilizer and theft-deterrent system are normally activated at this time.
- **Off**: Turning the key from the lock to the Off position unlocks the steering column. It does not enable any electrical systems or disable the engine immobilizer or theft-deterrent system.
- **Accessory**: This position allows power to be supplied to the vehicle entertainment system and the blower fan. Some vehicles also supply the wipers, electric windows, and sunroof. The features enabled by the accessory position are mainly for passenger convenience. Prolonged use of any of them without the engine running will drain the battery.
- **On/Run**: When the switch is turned to the On position, most warning lamps on the instrument panel should illuminate. This is to test the operation of the lamps. On vehicles that are not equipped with engine immobilizers, this position also activates the accessories and ignition system. Vehicles that are equipped with engine immobilizers do not normally activate these systems until the key is turned to the start position.
- **Start/Crank**: The start position activates the starter motor relay and/or solenoid. This enables the engine to crank and start the engine. Vehicles without engine immobilizers can start immediately. The required electrical systems will be activated when the key was turned to the On position. In vehicles equipped with engine immobilizers, a number of essential electrical systems are not normally enabled until the key is turned to the start position. When this occurs, communication between other control units determines

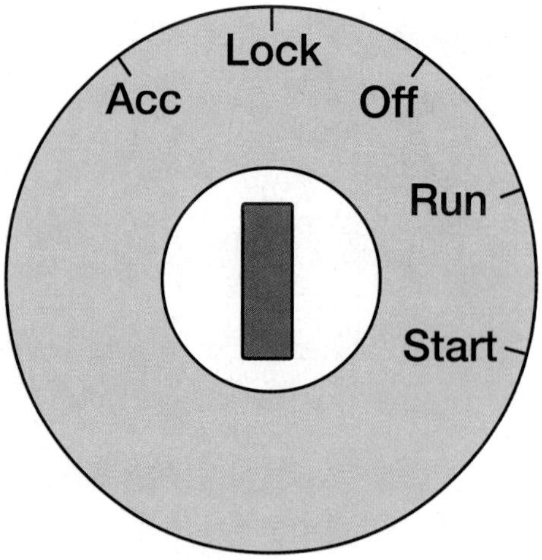

FIGURE 48-12 Ignition switch positions.

whether to allow the engine to start or not. They deactivate the immobilizer and activate the ignition, fuel, and charging systems. This will allow it to start. There may be a slight delay while this communication is occurring.

The key itself can consist of two parts. The first is a mechanical type that works in the key barrel and turns to unlock the steering column. It also moves the switch through the various stages of Off, Accessories, On, and Start. The second portion consists of a transponder that is part of an immobilizer system. It disables the ignition, fuel, and starter systems. A transponder device with the correct code and within close range of a receiver in the vehicle enables the engine to start. This system was covered in more depth in a previous chapter.

In many automatic transmission vehicles, the ignition switch also has a transmission shift interlock device connected to it. In these vehicles, the gear selector must be moved into the Park position before the key can be removed from the lock. In the same way, the transmission shift lever cannot be moved out of Park until two things happen. The key must be turned to the On/Run position and the brake pedal depressed.

Ignition Coil

Ignition coils are basically step-up transformers. They amplify the battery's low voltage to the very high voltage needed for the current to jump the spark plug gap (**FIGURE 48-13**). There are two sets of windings in an ignition coil: the primary winding and the secondary winding.

When battery voltage pushes current through the primary windings, a magnetic field is created. When the primary circuit opens, the collapsing magnetic field induces a voltage of more than 20,000 volts in the secondary winding. This voltage pushes current through the secondary circuit to the spark plug electrodes. The high voltage pushes current across the gap. This creates a spark needed to ignite the air–fuel mixture.

A standard ignition coil has a rod-shaped laminated iron core. It is located centrally by an insulator at its base. The secondary winding, with 15,000 to 30,000 turns of very thin **enameled copper wire**, is wound around the core. It is insulated from the core by layers of treated insulated paper. The primary winding, with a few hundred turns of much heavier copper wire, is wound on the outside of the secondary winding. A shield of soft iron surrounds the outer windings. The complete assembly is inserted into a one-piece steel or aluminum container or encased in an epoxy housing. On older coils, the container is then filled with special oil, which provides good electrical insulation. It also permits rapid heat dissipation. Epoxy-encased coils are cooled by the surrounding air.

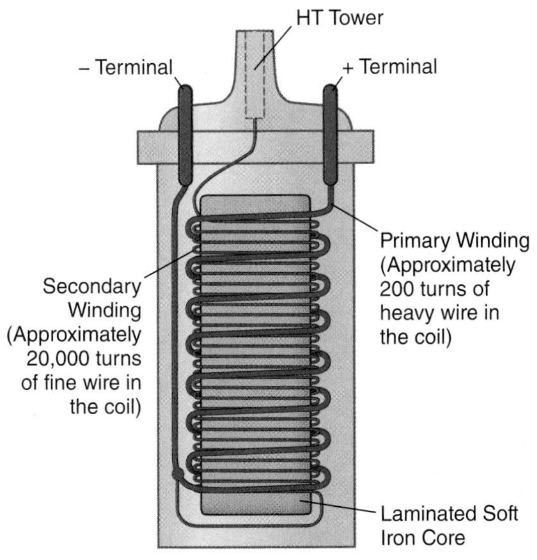

FIGURE 48-13 Cutaway of a basic ignition coil.

Applied Science

AS-65: Transformers: The technician can explain the ignition coil transformer's role in generating the high voltage required to fire a spark plug.

The 12-volt battery voltage used in light vehicles is insufficient to jump spark plug gaps. Much higher voltage is needed to fire the plugs and initiate combustion. This is especially true considering the high pressures and turbulent conditions inside the cylinders. Ignition systems allow a modest battery voltage to generate the very high voltages required to fire the spark plugs.

The ignition coil is a step-up transformer. The function of a transformer is to transfer voltage from one circuit to another. Either as a higher voltage (as in this case) or as a lower voltage (referred to as a step-down transformer).

Depending on the ignition system, the 12 volts applied to the coil may be amplified to between 20,000 and 100,000 volts. It does so via a process called electromagnetic induction.

The coil has two terminals, positive and negative, for external connection to the primary circuit. The ends of the primary winding are connected internally to each of these terminals. Provisions are also made for connecting the high-tension coil lead to the coil at a heavy insulated center terminal.

One end of the secondary winding is connected to this center terminal. The other end is connected either to one end of the primary winding or in some applications to a separate ground connection. On waste spark systems, each end of the secondary winding is attached to a spark plug wire (two plugs per coil). COP and coil-near-plug (CNP) systems usually have one end of the secondary winding grounded. The other end is connected to the high-tension terminal of the spark plug.

High-Tension Leads

High-tension leads are also known as HT leads or spark plug wires. They connect the secondary ignition components together. They connect the coil to the distributor cap and the distributor cap to the spark plugs. The high-tension leads conduct the high voltage generated in the secondary ignition circuit when each ignition pulse occurs.

Today's electronic ignition systems have to endure higher voltage. This is because of the increased spark plug gaps and leaner air–fuel mixtures. The insulation material on the high-tension leads is now much thicker than in earlier models. The core of the high-tension lead is typically made of carbon-impregnated linen or fiberglass (**FIGURE 48-14**). It has a specific resistance to current flow. This helps reduce the radio frequency interference (RFI) emitted from the high-tension leads. A crimped terminal at each end provides for connection of the components.

If a distributor is used, typically a coil wire transmits the high voltage from the ignition coil to the center terminal of the distributor cap. On systems where the coil is mounted inside the distributor, there is no coil wire. The high voltage is transmitted through a terminal in the distributor cap. On systems that do not use a distributor, there is no coil wire. Instead, the high-tension leads connect the terminals of coil packs directly to the terminals of the spark plugs.

▶ TECHNICIAN TIP

With the high voltage capacity of today's ignition systems, routing of the spark plug wires is of utmost importance. As the current is conducted down the leads, a magnetic field around the lead is created. If the leads are not spaced far enough apart and run parallel to each other, then an **induced voltage** can be generated in the other wires. In some instances, this could lead to premature firing of the adjacent spark plug. Severe engine damage can result when the engine has two cylinders fire well before one of them is up on the compression stroke. Make sure the plug wires are routed and secured according to the manufacturer's specifications.

▶ Spark Plug

LO 48-06 Describe spark plugs.

The **spark plug** consists of a plated metal shell with a ceramic insulator. An electrode extends through the center of the insulator (**FIGURE 48-15**). Threads on the metal shell allow it to be screwed into the cylinder head. A short side electrode is attached to one side and bent toward the center electrode. It provides a ground path for the spark. Some spark plugs may have as many as four side electrodes. The electrodes are made from a special alloy wire with a manufacturer-specified gap between them. Modern spark plugs also include an internal resistor to suppress voltage spikes when the spark plug is energized. The reduced voltage spikes reduce RFI.

Spark plugs are identified by three different features:

- Thread size or diameter
- Reach or length of the thread
- Heat range or operating temperature

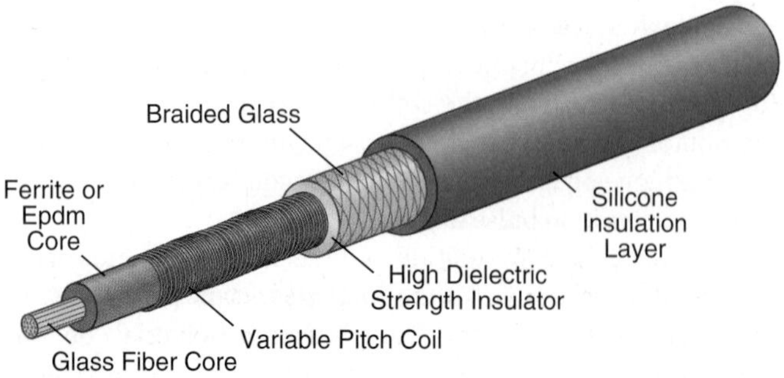

FIGURE 48-14 The parts of a high-tension lead.

Spark Plug Design

The metal case of a spark plug removes heat from the insulator and passes it on to the cylinder head. It also provides structural strength to withstand the torquing force applied when tightening the plug into place. The case also acts as the ground for the current passing through the electrodes.

The insulator covering the center electrode is usually made from an aluminum oxide ceramic. It has a high tolerance to heat and electrical voltage. The ribbing design, the composition of the insulator material, and the length of the insulator all determine the plug's heat range.

The spark plug seals the combustion chamber when installed. The seal can be of two types. The first is a hollow metal washer. It seals by being partially crushed between the flat surface of the head and the plug just above the threads. The seal can also be provided by a carefully machined taper on the spark plug body. It mates with a matching machined taper in the spark plug hole in the cylinder head (**FIGURE 48-16**).

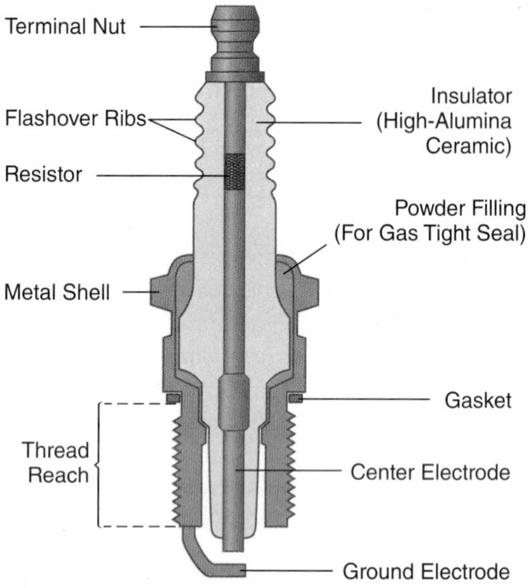

FIGURE 48-15 Spark plug.

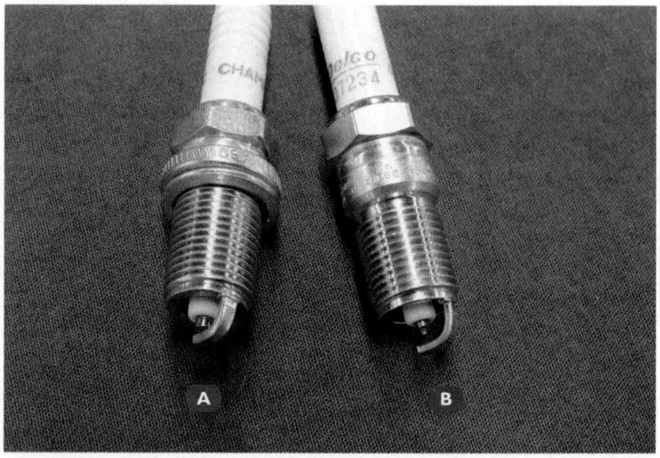

FIGURE 48-16 **A.** Metal gasket seal. **B.** Tapered seat design.

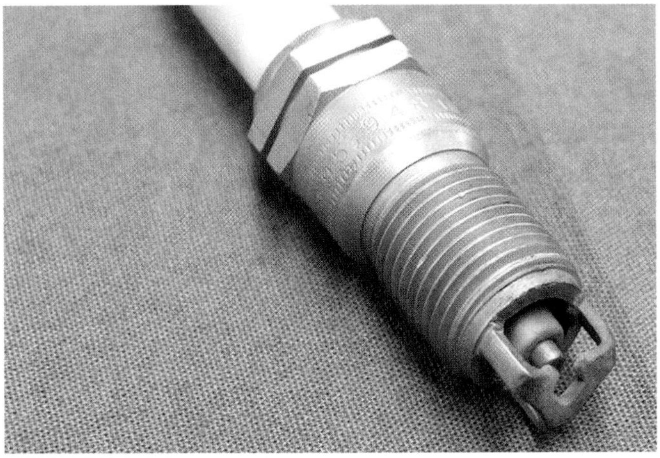

FIGURE 48-17 Some spark plugs have multiple side electrodes.

A terminal at the outer end of the spark plug connects it electrically to the ignition system. Most ignition leads clip on the spark plug. In a few applications, they may be held in place by a threaded nut. The side electrode runs very hot and is usually made of nickel steel or another high-temperature metal. It is welded to the side of the metal spark plug case. Some spark plug designs have multiple side electrodes. They create the gap between the side of the center electrode and the side electrodes (**FIGURE 48-17**).

Spark plug condition affects required voltage. Some of the factors that increase the required voltage are:

- An excessive gap raises the amount of voltage that is required to jump the gap.
- The electrons tend to stream easily from sharp edges on the electrodes. As these edges erode and become less sharp, it becomes more difficult for the electrons to jump the air gap. This causes an increase in the required voltage.
- Generally it takes less voltage for a spark to jump from a hot surface to a cooler surface. The **center electrode** is the hottest part of the spark plug. So many older ignition systems were designed to jump from the center electrode to the side electrode. However, with today's high-energy ignition systems, there is enough available voltage to overcome this issue. So some systems are designed so that the spark jumps from the ground electrode to the center electrode.

It used to be necessary to maintain spark plugs by sandblasting built-up deposits from the internal surfaces and filing the electrodes sharp again. Now, low-erosion materials such as platinum and iridium have become available. They last a lot longer (up to 100,000 miles [160,000 km]) and are replaced rather than refurbished.

Be careful when installing plugs on some waste spark systems. Some of these systems use polarity-sensitive spark plugs. They have platinum only on one electrode. They are designed for the spark to jump from either the center electrode or from the ground electrode. These types of spark plugs have to be installed in the correct spark plug hole. That way, the spark will jump the correct direction across the plug gap (**FIGURE 48-18**). Manufacturer's service information provides this information.

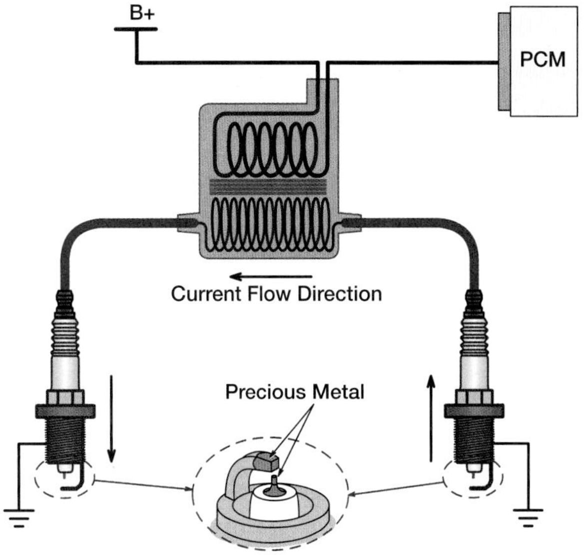

FIGURE 48-18 Polarity-sensitive spark plugs must be installed in the proper orientation so that the spark jumps the intended direction across the air gap.

▶ TECHNICIAN TIP

The same spark plug can sometimes be used in different engines with different gap settings, or sometimes gaps can be damaged during shipment. Therefore, the gap should be checked with an accurate spark plug gauge. If necessary, the gap should be adjusted according to the specifications when the spark plug is first installed. The gap is critical to engine performance. A narrow gap may give a spark too weak to ignite the air–fuel mixture effectively. A gap too wide may fail to produce any spark at all. Either way, the wrong gap is likely to produce a drivability concern. One exception to the gapping rule is iridium spark plugs. The gap should not be adjusted, as the iridium is brittle and can snap off. Verify that their gap meets specifications. If the gap is wrong, they will have to be replaced.

Spark Plug Size

Spark plug size refers to the diameter of the threads of the spark plug. Nearly all of today's light-duty vehicles use either a 14-mm or an 18-mm spark plug. However, some manufacturers use 16-mm "high-thread" spark plugs or even 12-mm long-reach spark plugs (**FIGURE 48-19**). Most manufacturers have been switching to smaller plugs to allow for larger or more valves in the cylinder head. Both the Society of Automotive Engineers (SAE) and the International Organization for Standardization (ISO) have a set of standards for spark plugs. They cover length, hex size, thread diameter, and thread pitch.

Spark Plug Reach

Spark plug reach is the distance from the seat of the spark plug to the end of the spark plug threads. The purpose of spark plug reach is to ensure that the spark plug electrodes are in the most efficient position for combustion within the cylinder. If the reach is too long, the piston may strike the spark plug while moving up. If the reach is too short, the spark may occur

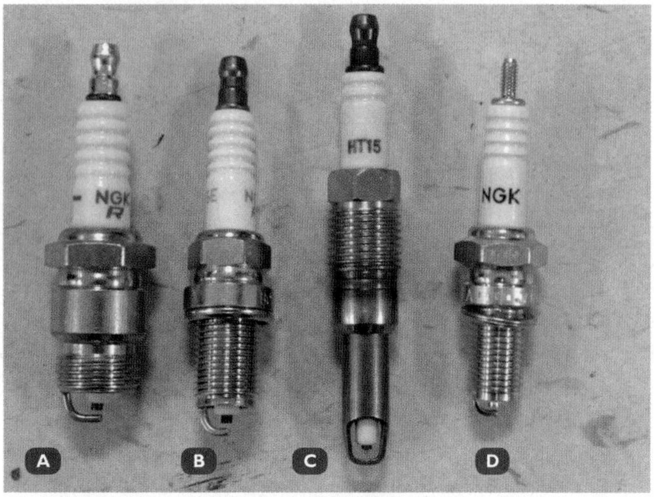

FIGURE 48-19 Common spark plug sizes. **A.** 18 mm. **B.** 14 mm. **C.** 16-mm "high thread." **D.** 12-mm long reach.

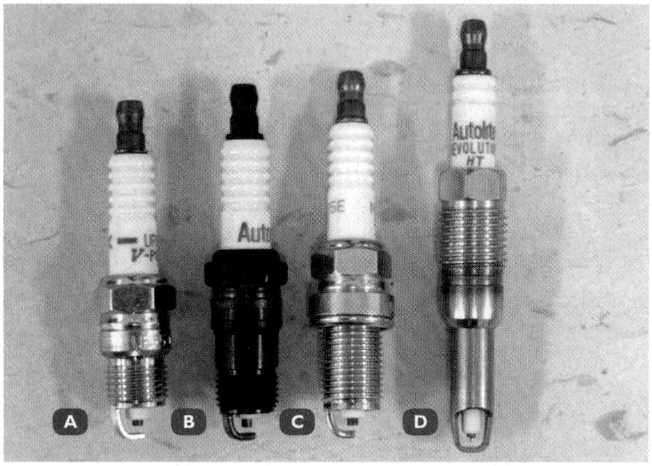

FIGURE 48-20 Spark plug reach. **A.** Short. **B.** Long (half thread). **C.** Long (full thread). **D.** High thread.

inside of the threaded spark plug hole, resulting in a misfire. Some spark plugs are only threaded on the bottom half or, in the case of "high-thread" plugs, the top half of their reach. Always install the specified spark plug for the engine you are working on (**FIGURE 48-20**).

Heat Range

The operating temperature of a spark plug refers to the temperature at the sparking tip of the spark plug inside a running engine. This is referred to as the **heat range**. Spark plugs should operate between average temperatures of 746°F and 1460°F, or 400°C and 800°C. The temperature that a spark plug will reach depends on the distance the heat must travel from the insulator on the firing end to reach the outer shell of the plug. There, the heat enters the cylinder head and the water jacket (**FIGURE 48-21**). If the heat path is long, the spark plug will retain more heat. It will therefore run at a higher temperature than one with a short heat path.

If a spark plug is too cold, deposits may form on the insulator. This can provide a path for the spark to travel, leading to misfire. If it is too hot, the heat can ignite the air–fuel mixture itself, causing preignition. Using a spark plug with the proper heat range is critical for the operation of the engine.

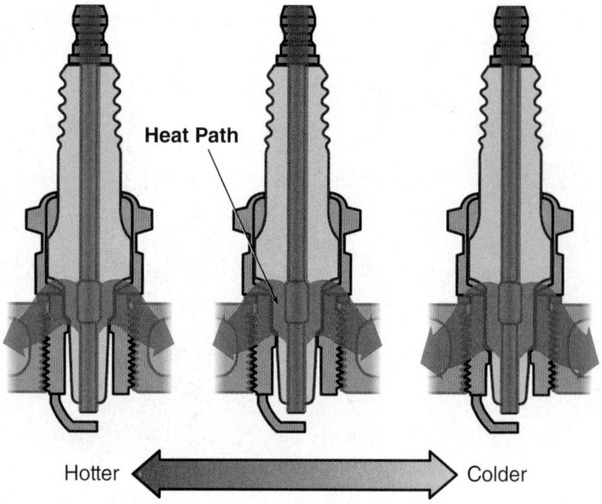

FIGURE 48-21 Heat range of spark plugs.

▶ TECHNICIAN TIP

Spark plug cooling is affected by spark plug torque. Almost all of the heat from the spark plug must be transferred through the threads of the spark plug to the cylinder head. Properly torquing of the spark plugs keeps the threads in contact with the head. This allows heat to transfer more quickly. The plug is able to operate at its designed temperature, instead of overheating and causing preignition.

▶ Types of Ignition Systems

LO 48-07 Describe contact breaker point ignition systems.

Several types of ignition systems have been used over the years. Contact breaker point ignition systems were used on earlier vehicles. They relied on mechanical devices to create the spark. On later vehicles, the mechanical contact breaker was eliminated. An electronic means of controlling the primary circuit was used. Modern vehicles no longer use a distributor. Instead, they send the high voltage from the coil packs or individual coils directly to the spark plugs.

The main difference among these systems is in the way the primary circuits are controlled. The primary circuit needs to be switched on and off to create and collapse the primary magnetic field. This switching is done with breaker points in older vehicles. In more modern vehicles, it is done by electronic switches.

Contact Breaker Point Ignition Systems

Contact breaker point ignition systems were used on early-model vehicles. They provided a mechanical means of quickly connecting and disconnecting the primary side of the ignition coil to ground. The contact breaker is a mechanically operated electrical switch. It is fixed to the distributor base plate and opened and closed by the distributor cam. The distributor cam rotates with the engine (**FIGURE 48-22**). This process builds up and collapses the magnetic field in the ignition coil.

The contact points are opened by **cam lobes** on the otherwise round distributor shaft. They are closed by a spring. The number of lobes is equal to the number of cylinders. This allows the contact breaker points open the circuit and activate the ignition coil at the end of each cylinder's compression stroke.

In order for the spark plugs to fire at the correct moment, the distributor must be installed in the right position. Most manufacturers specify that the number-one piston be at TDC of the compression stroke. The ignition **rotor** must then be aligned with the number-one terminal on the distributor cap.

Contact breaker point ignition systems provide a simple means of establishing and interrupting the current flowing in the primary ignition circuit. A basic system consists of the following:

- The battery—provides a source of energy.
- The ignition switch—provides the driver control over system operation.
- An ignition coil—provides step-up transformer action.
- Contact breaker points—opened and closed by lobes on the distributor cam as the engine rotates. This makes and breaks the primary circuit at the correct time in the ignition cycle (**FIGURE 48-23**).
- A capacitor—also called a condenser, assists in the rapid collapse of the ignition coil's magnetic field. Any voltage surge across the contacts will charge the capacitor, rather than cause damaging arcing.

FIGURE 48-22 Typical contact breaker point distributor.

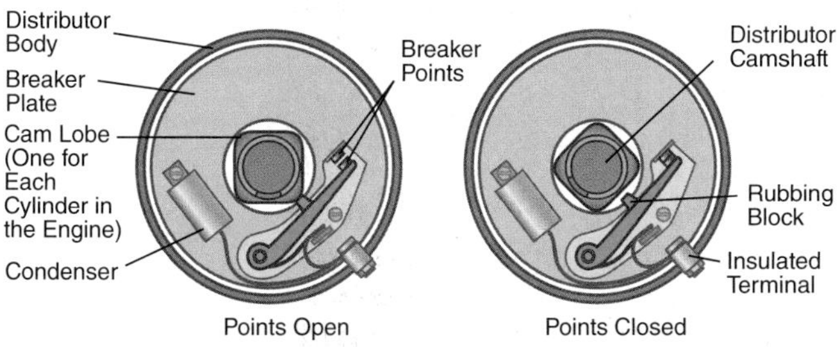

FIGURE 48-23 Contact breaker point operation.

- A distributor—rotates at half the speed of the crankshaft. It houses the contact breaker points and distributes the high voltage from the ignition coil to the spark plugs in the correct firing order.
- Connecting wires and leads—suitable for conducting the current flowing in the ignition system, at the appropriate voltage level.

Opening and closing the contact breaker switches the primary current off and on. When the contacts begin to separate, the primary current wants to continue to flow. This produces an arc across the contacts. The capacitor absorbs this surge of **inductive current** by providing a very short-term alternative path, in parallel with the opening contacts. The capacitor charges to the peak value of the primary winding voltage almost instantaneously. By the time this has occurred, the gap between the contacts is too wide for a spark to jump between them. This abrupt interruption of the primary circuit assists in the rapid collapse of the magnetic field. And that increases the value of the voltage induced in the primary and secondary windings.

As soon as the secondary voltage reaches a value great enough to bridge the gap across the spark plug, a spark occurs. The voltage in the primary circuit falls to a value below that of the charged capacitor (**FIGURE 48-24**). The capacitor then discharges back and forth between itself and the primary winding of the coil. When the points close, the capacitor discharges completely.

The spark at the spark plug needs to last for about a minimum of 1 millisecond (0.001 second). This provides enough heat and time to ignite the air–fuel mixture. The time the current is flowing across the electrodes is called the spark duration. Extra-long spark duration causes excessive wear on the spark plug electrodes. Too short, and it may not ignite the air–fuel mixture.

The condenser is made up of two plates constructed from narrow strips of aluminum foil. They are insulated from each other by a special waxed paper called a dielectric. The plates and insulating paper are rolled up tightly together and sealed in a metal can by crimping the end over onto a gasket (**FIGURE 48-25**). A spring in the base forces the plates and insulation against the gasket to keep out moisture. One plate is connected to the capacitor case and, through its retaining screw, to ground. The other plate is connected to the external connecting lead. When the capacitor is installed, any voltage surge across the contacts will charge the capacitor rather than erode the contacts on the

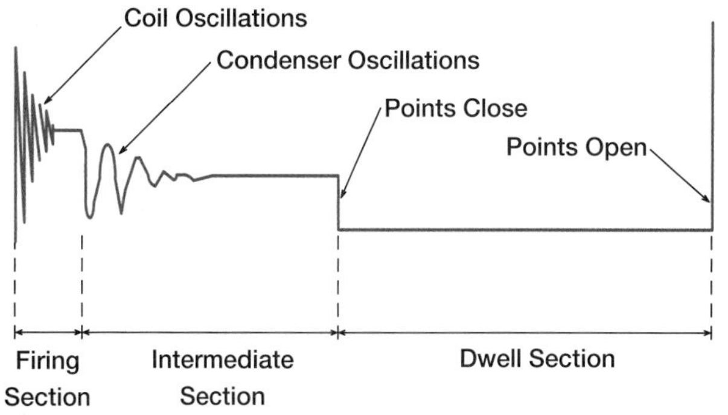

FIGURE 48-24 Primary circuit voltage on an oscilloscope.

breaker points. The condenser is typically mounted to either the inside or the outside of the distributor.

> TECHNICIAN TIP

The condenser is a critical component of the primary circuit in point-type ignition systems. Without it, the points arc as they are opened. This arcing causes a slow stopping of the primary current flow. This, in turn, causes a slow collapsing of the magnetic field in the ignition coil. And this causes low secondary voltage output of the coil. If you have weak spark output on a point-type ignition system, do not be like a former student and overlook the very simple and inexpensive condenser. This student ended up replacing the spark plugs, spark plug wires, cap, rotor, points, ignition coil, and carburetor before he asked for help. The wise old technician, after listening to his story, pulled out a used condenser. He connected it to the negative side of the coil and ground with a couple of test leads. Then he asked the student to try starting the engine. The student was amazed when it started right up and ran like new.

> TECHNICIAN TIP

The secondary spark occurs when the points just start to open, not when they close. It is the rapid collapse of the magnetic field that induces the high voltage in the secondary windings. Knowing that the spark occurs just as the points open helps a technician static-time an engine. This is needed after an engine rebuild or when the distributor has been removed.

Dwell Angle

For a four-stroke, four-cylinder engine running at 2000 rpm, 4000 sparks must be supplied every minute. The time available to make and break the primary circuit each time is very short. As engine speed rises, the time available is even shorter. Therefore, the length of time the current flows through the primary winding must be sufficient to create the necessary magnetic field. If not, then there will be insufficient voltage for the spark.

The condition in which the magnetic field builds to its maximum strength is called coil saturation. In contact breaker systems, it is during the brief time that the contacts are closed. This is when the primary current flows and the magnetic field builds.

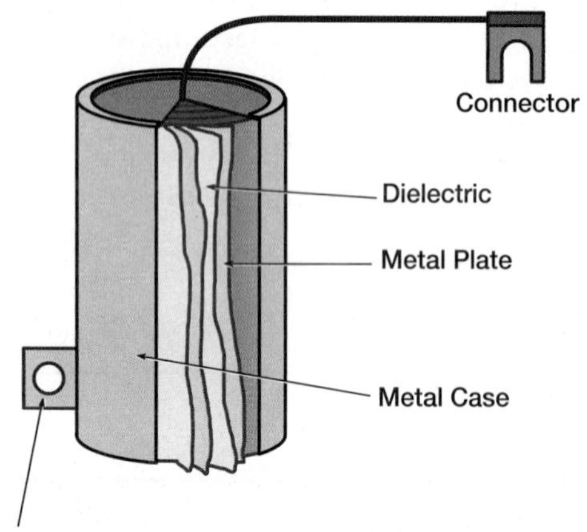

FIGURE 48-25 Typical condenser construction.

Current flows through the coil primary windings until the next eccentric cam pushes one contact point away from the other. The amount of time, in degrees of distributor rotation, that the contacts are closed is called the **dwell angle** (**FIGURE 48-26**).

The setting of the contact breaker gap influences the dwell angle. On most systems, once it has been set, the dwell angle remains fixed regardless of engine speed. A large gap gives a small dwell angle, as the contact points are closed for a shorter time. A small gap gives a large dwell angle because the contact points are closed for a longer time. The manufacturer's recommended gap provides the specified dwell angle for each application. This assumes there is no wear on the distributor cam. The dwell angle should be adjusted to specifications. Then, the point gap should be checked to identify any cam lobe wear. The proper gap also ensures there is sufficient spacing between the contacts to prevent arcing.

> TECHNICIAN TIP

As the rubbing block wears, the point gap decreases and the dwell increases. Setting the dwell to the lower end of the specifications allows the dwell to stay within specifications longer. Also, for every degree the dwell increases as the rubbing block wears, the ignition timing retards about one degree. This reduces fuel mileage and power. As a result, engines need to be "tuned up" every 10,000 to 15,000 miles (16,000 to 24,000 km). A new set of points and condenser would be installed, the dwell angle set, and the ignition timing adjusted.

Ballast Resistor and Bypass Circuit

A **ballast resistor** limits the amount of current flowing in the ignition primary circuit (**FIGURE 48-27**). It is inserted in series between the ignition switch and the positive terminal of the ignition coil. It is usually located near the ignition coil, where it can dissipate its heat into the air.

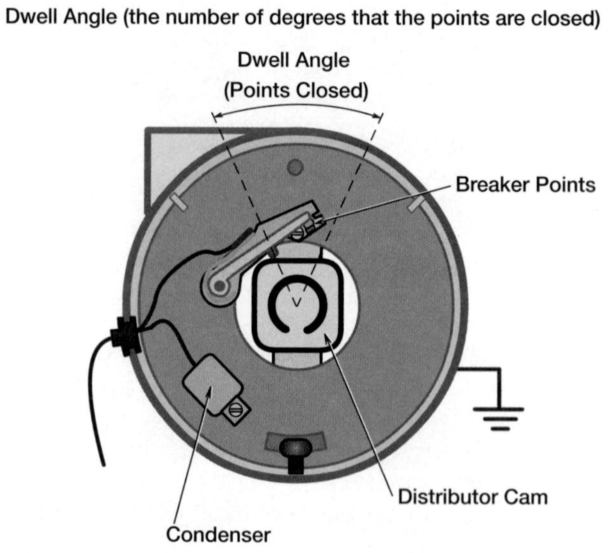

FIGURE 48-26 Dwell is the time that the points are closed and the magnetic field is building in the ignition coil.

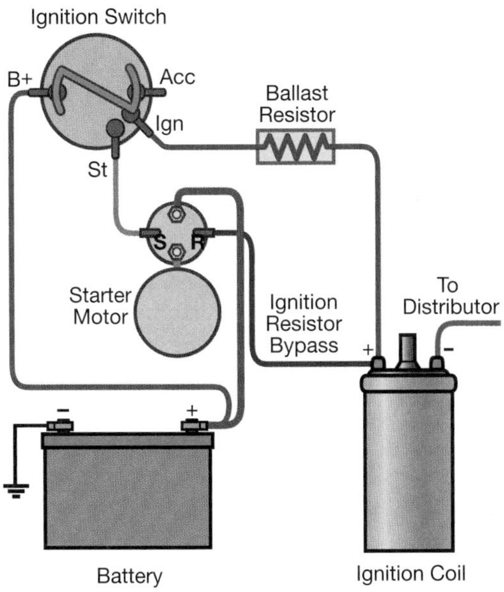

FIGURE 48-27 A ballast resistor in an ignition circuit.

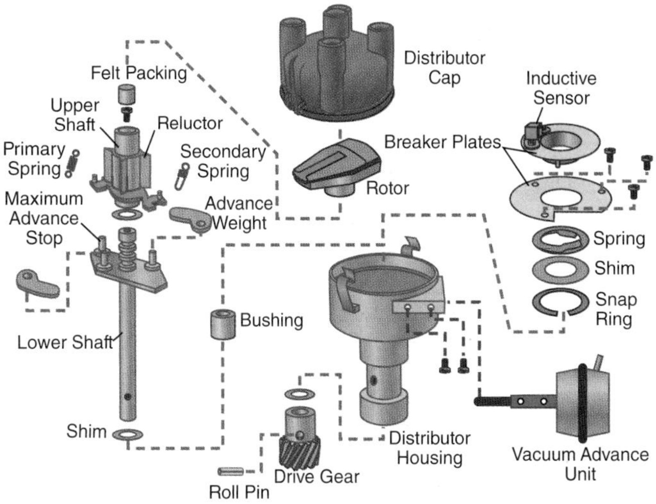

FIGURE 48-28 Distributor.

FIGURE 48-29 Typical distributor with the coil mounted on top of the distributor cap.

When the ignition key is turned to the start position and the starter begins to crank the engine, a heavy load is placed on the battery. This load causes the battery voltage to drop to about 10 volts. Ignition systems must be able to fire the air–fuel mixture on this reduced voltage. Yet, when the engine is running, ignition systems must be able to run on approximately 14 volts. This is because the charging system is charging the battery.

The ballast resistor and bypass circuit are key to providing the correct voltage in a point-type ignition system. When the voltage is low because the engine is cranking over, a resistor bypass circuit bypasses the resistor. Full cranking voltage is provided to the coil. When the ignition switch is turned back to the Run position, the resistor bypass circuit is disabled. Current has to flow through the resistor, dropping its voltage from 14 volts to approximately 9–10 volts. More modern solid-state electronic ignition systems do not need a ballast resistor. This is because they have much more reserve voltage. It stays above the required voltage even when primary voltage drops during cranking.

Distributors

The main function of the **distributor** is to distribute the spark to the spark plugs. It must do so in the correct sequence and at the correct time in the engine cycle (**FIGURE 48-28**). The distributor includes a distributor cap, rotor, switching device, and a shaft with cam lobes. The distributor also houses the vacuum and mechanical (centrifugal) timing **advance mechanisms** on older engines. These are used to advance the timing of the spark under certain driving conditions.

The **distributor cap** covers the end of the distributor to protect the components inside. It is held onto the distributor by clips or screws. It also provides a connection point between the rotor and the spark plug leads. The rotor is a high-voltage rotating switch. It transfers the secondary voltage from the center terminal of the distributor to the side terminals. The distributor distributes the high-voltage pulses to the individual spark plug wires in the correct sequence.

In some applications, the ignition coil is mounted inside of the distributor. The cap covers the coil and provides an electrical path from the coil's output terminal to the center of the rotor. In other applications, the ignition coil is bolted to the top of the distributor cap and covered with a plastic shield (**FIGURE 48-29**). In this instance, the center of the coil sits right above the rotor. A hole in the cap allows a spring-loaded button to transfer the high voltage directly to the center of the rotor.

An insulated **rotor arm** with a brass or steel electrode is keyed to the shaft directly above the cam. It rotates within a molded insulated distributor cap. An internal spring-loaded carbon brush conducts each high-voltage spark from the central terminal of the cap to the center of the rotor as it turns.

The high-voltage spark is timed to occur when the rotor electrode is aligned with a fixed electrode inside the distributor cap. The high voltage bridges the small gap between the electrodes. This drives current through the ignition terminal for that cylinder. It bridges the gap at the spark plug.

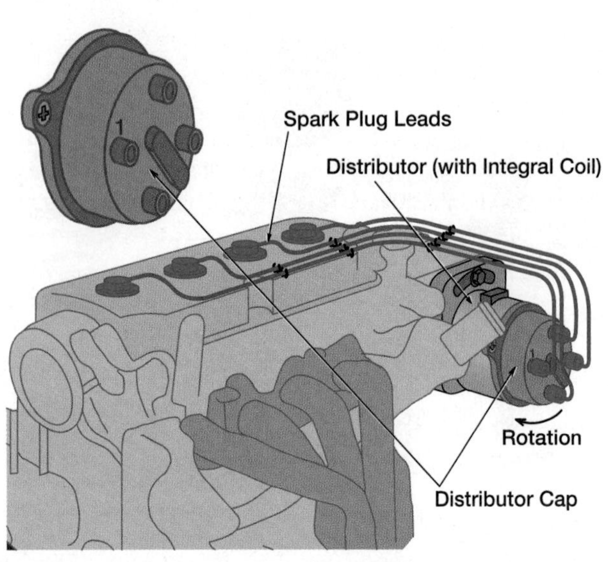

FIGURE 48-30 The spark plug wires are arranged in the firing order on the distributor cap.

As the rotor turns inside the distributor cap, the high-voltage electrical sparks take place successively for each cylinder according to the firing order. For example, an in-line four-cylinder engine typically has a firing order of 1–3–4–2. If the distributor shaft were designed to spin clockwise, the high-tension leads would be arranged clockwise. They would be placed every 90 degrees in the firing order (**FIGURE 48-30**).

▶ Engine Spark Timing

LO 48-08 Describe mechanical spark timing systems.

The ignition timing refers to the point at which the spark plug "fires" near the end of the compression stroke (typically 5–10 degrees before top dead center (BTDC). On older vehicles with distributors, the base timing can be adjusted. That is done by rotating the position of the distributor housing in relation to the engine. On newer engines, the timing is not adjustable. This is because the computer controls timing based on all of the sensor information.

Engine manufacturers include timing marks on many of their engines. This is so that the technician can check and adjust the engine timing if it has changed. There are two types of timing on gasoline engines: valve timing and ignition timing. Valve timing refers to the position of the valves in relation to the piston stroke. During the four-stroke cycle, the valves must open and close at the proper times. Engine manufacturers place marks on the gears, sprockets, or pulleys and the corresponding surfaces of the block and head. This allows the valve timing to be properly set whenever the valve timing components are being replaced.

▶ TECHNICIAN TIP

Some newer engines still use a distributor for distributing the high-voltage spark. But they use a crankshaft position sensor to control the timing of the spark. Turning the distributor on this application interferes with the rotor to cap alignment. It does not actually change the timing. So just because it has a distributor does not mean that adjusting it will change the spark timing.

Centrifugal Advance Units

The **centrifugal advance mechanism** (also called mechanical advance) controls ignition timing in relation to engine speed. It is located within the distributor and can be above or below the contact points' base plate. The distributor shaft has two sections that rotate on the same axis. The inner shaft is driven by the camshaft. The outer shaft has the cam lobes that control the opening time of contact points. Both parts are joined by the centrifugal advance mechanism (**FIGURE 48-31**). Its function is to allow the outer shaft to rotate forward, within limits, in relation to the inner shaft, proportional to engine speed. This forward rotation advances the timing.

The inner shaft has two flyweights attached. Each pivot at opposite ends and are controlled by a spring. As the inner shaft is turned, they are thrown outward by the effect of centrifugal force. The faster the shaft turns, the more they move outward. Slowing the shaft speed reduces the amount of centrifugal force, and spring force pulls them back in.

A slot in each weight is keyed to the outer shaft. As the weights move outward, they turn the outer shaft forward in relation to the inner shaft. This movement has the effect of advancing the opening of the contact points in relation to engine speed.

The relationship between speed and advance is determined by two factors. The first is how heavy the weights are. The

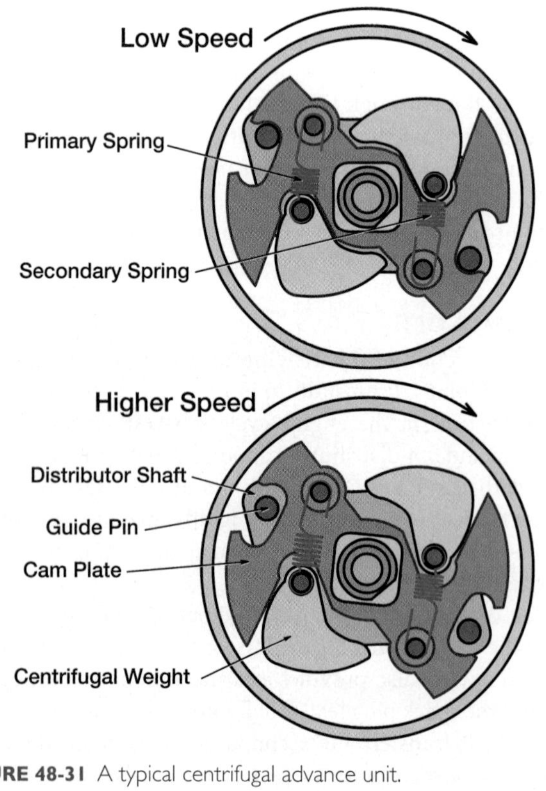

FIGURE 48-31 A typical centrifugal advance unit.

second is by the tension of the springs. The maximum advance is determined by limiting the movement of the weights with slots and pins, or bushings.

Vacuum Advance Units

The vacuum advance mechanism controls **ignition advance** in relation to engine load. At light load, an engine is more efficient when the timing is advanced slightly. Under heavy load, the timing must be retarded slightly to avoid "pinging," or spark knock. The function of ignition advance is to improve fuel economy and, in doing so, reduces exhaust emissions.

Vacuum is routed from the carburetor to a spring-loaded diaphragm housing attached to the distributor. The diaphragm is attached to a pull rod that, when operating, acts on the **distributor base plate**. The pull rod turns the base plate opposite to the direction of distributor shaft rotation (**FIGURE 48-32**). This movement results in the opening time of the contact points being advanced during light engine loads. During heavy engine loads, the timing is retarded.

Manufacturers designed their vacuum advance to operate on one of three types of vacuum: ported, manifold, or venturi (**FIGURE 48-33**). On older vehicles, ported vacuum is the most common. It is obtained through a small slit in the throttle bore located just above the throttle plate. This slot has no vacuum at idle, creating no vacuum advance at idle. As the throttle is opened, the vacuum to the slit increases, causing some vacuum advance, which tapers off at wide-open throttle.

Manifold vacuum is obtained through a small hole in the throttle bore just below the throttle plate. It provides strong vacuum at idle and less vacuum as the throttle is opened. Venturi vacuum is obtained from a port that is connected to the smallest diameter of the venturi. This system has no vacuum at idle, but as airflow increases through the venturi, the vacuum increases. The vacuum in one of these systems acts on the vacuum diaphragm. It causes the timing to advance as vacuum increases, or retard as vacuum decreases. The vacuum responds differently in each type of vacuum. Therefore, the manufacturer specifies the base timing and the advance curve profile for the engine.

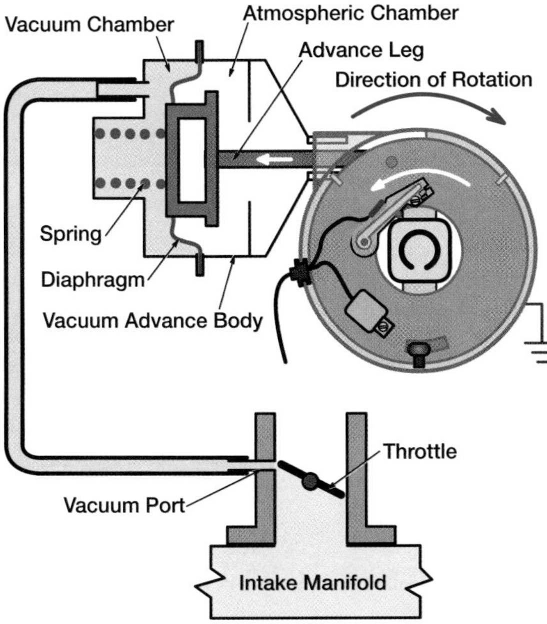

FIGURE 48-32 Typical vacuum advance unit, which advances the ignition timing under light engine loads.

▶ Electronic Ignition Systems

LO 48-09 Describe electronic ignition systems.

In **electronic ignition systems**, the contact breaker points are eliminated. They are replaced by both an electronic triggering system and an electronic switching device (**FIGURE 48-34**). The triggering system is called a pickup assembly. It generates an

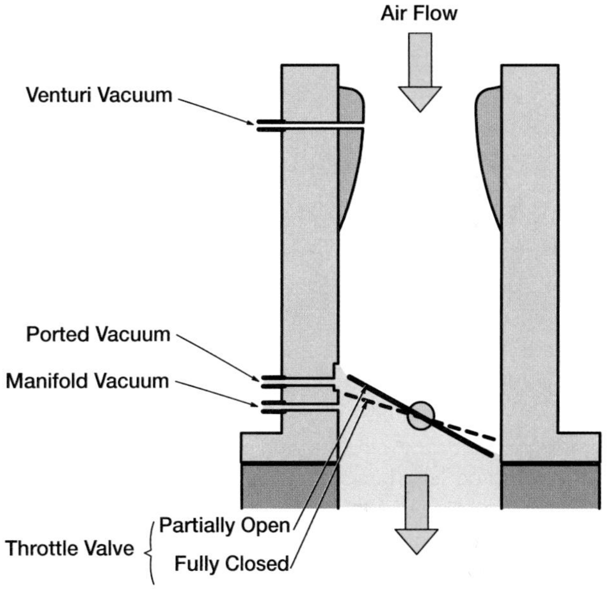

FIGURE 48-33 Vacuum sources: ported, manifold, and venturi.

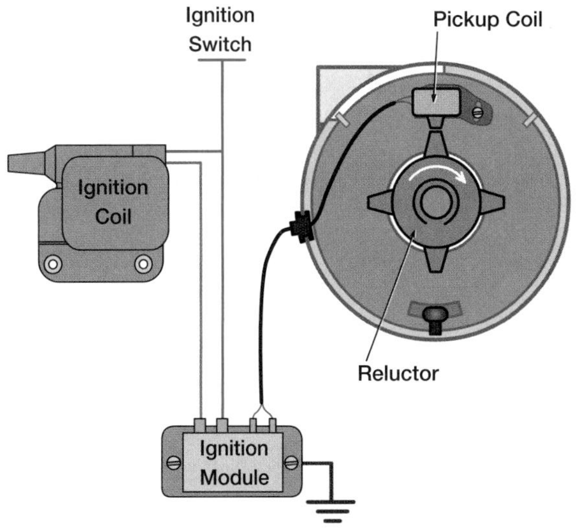

FIGURE 48-34 Electronic ignition systems eliminated the contact breaker points and condenser and replaced them with a pickup assembly and ignition module.

electronic signal that turns the electronic switching device on and off. The switching device is called an ignition module. It uses a power transistor to turn the primary circuit on and off. This causes the magnetic field to build up and collapse in the primary coil winding. It occurs in the same way that the point ignition system does.

In general, a pickup assembly has two main components. One component is stationary. And the other component spins with the distributor shaft. When the two components interact as designed, a signal is sent to the ignition module. This is accomplished in a few ways. It can be through the use of either an inductive pickup assembly, a **Hall-effect sensor**, or an **optical sensor**. Early electronic ignition systems included an ignition module and a triggering device. They were coupled with a standard distributor.

Many of the components in an electronic ignition system are essentially the same as those in a contact breaker point ignition system. The following sections describe only those components that are not part of a contact breaker point ignition system.

Components of Distributor-Type Electronic Ignition Systems

Ignition Modules

Ignition modules are used in virtually every type of ignition system except the contact breaker point system. It uses the information from the triggering device to turn the primary circuit on and off. As technology progressed, processing power was added to the module. This is so that the dwell time could be modified. Thus, the dwell time could be shortened during low engine rpm and lengthened as rpm increased. This flexibility helps to prevent overheating of the ignition coil and transistor during low speed. It also maintained adequate time for coil saturation during all engine speeds. Further refinement saw current-limiting circuitry added to the ignition module. This allowed for the use of low-impedance ignition coils.

In the 1980s, manufacturers started to use their PCMs to work along with the ignition module. The module typically provided for base timing while the engine was being started. Once the engine was running, it then switched over and allowed the PCM to control ignition timing through the ignition module.

In early electronic-ignition vehicles, the ignition module was usually located in one of three places. These were: in the distributor, on the distributor, or away from the distributor. If it was away, it was on the firewall or inner fender (**FIGURE 48-35**). On many waste spark systems, the ignition module is located either under the ignition coils or within the coil assemblies. On newer vehicles, the ignition module functions are controlled in one of two ways. Either by the internal circuitry of the PCM or circuitry housed in the ignition coils themselves. In many cases, this eliminates the need for an external ignition module.

Induction-Type Pickup Assemblies

Induction-type systems are electronic systems that use a magnetic pulse generator to generate an AC signal. They are also called variable reluctor sensors (VR sensors). They come in a

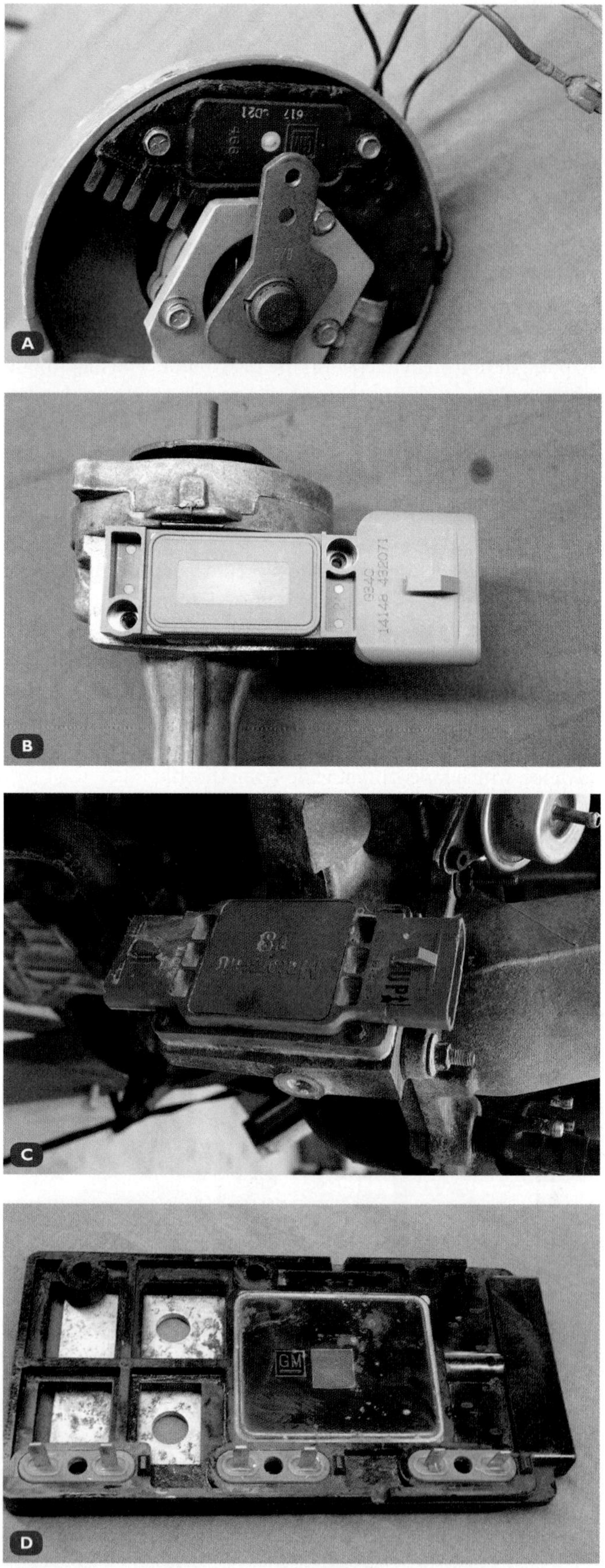

FIGURE 48-35 Externally mounted ignition modules. **A.** In the distributor. **B.** On the side of the distributor. **C.** On the firewall or fender. **D.** Under the coils.

variety of configurations. But all of them have a **stator** (also called a pickup coil) mounted on the distributor body. They also have a rotor unit (also called a **reluctor** or trigger wheel) attached to the distributor shaft (**FIGURE 48-36**).

The reluctor has one tooth for each cylinder. As it spins, the teeth interact with the stator to trigger the ignition module. The stator may have only one projection with a stationary coil of fine enameled wire wound around it. Or it may have a circular permanent magnet with a number of projections or teeth corresponding to the number of engine cylinders. A stationary coil of fine enameled copper wire is wound on a plastic reel and positioned inside or around the magnet.

As the **stationary winding** is influenced by the magnetic field, a voltage is induced in the winding each time the magnetic field changes. As the teeth approach, the strength of the magnetic field increases. This induces a voltage and current flow in the winding. The polarity of the voltage is said to be positive as it produces a current flow in a certain direction. When the teeth are in alignment, the magnetic field is at its strongest. But at that point, it is not changing. Voltage and current now fall to zero (**FIGURE 48-37**).

As the teeth move away, the strength of the magnetic field changes again. Once more, voltage and current are induced in the winding. This time, current flow is in the opposite direction, and the polarity is now said to be negative. Because polarity changes every time the teeth approach and leave the stator teeth, the voltage produced is an AC voltage. Current flow also alternates (**FIGURE 48-38**).

Hall-Effect Sensors and Operation

Another type of electronic ignition system is the Hall-effect type. In Hall-effect systems, a Hall-effect generator is located inside the distributor. It is also called a Hall-effect sensor or switch. It signals the ignition module to turn the primary circuit on and off. Hall-effect generators operate by using a potential difference, or voltage, created when a current-carrying conductor is exposed to a magnetic field. If a magnetic field is applied at right angles to the direction of current flow in a conductor, the lines of magnetic force permeate the conductor. Electrons flowing in the conductor are deflected to one side of the conductor. This deflection creates a potential difference or voltage across the conductor (**FIGURE 48-39**). The stronger the magnetic field, the higher the voltage. This dynamic is called the Hall-effect voltage. If the magnetic field is alternately shielded and exposed, it can be used to trigger a switching device. Hall devices are made of semiconductor material. They respond quickly when shielded from or exposed to a magnetic field.

In a distributor, the Hall-effect generator and its **integrated circuit** are located on one leg of a U-shaped assembly. That assembly is mounted on the distributor base plate (**FIGURE 48-40**). An integrated circuit is a semiconductor chip that contains miniature versions of various electrical components within one housing. A permanent magnet is located on

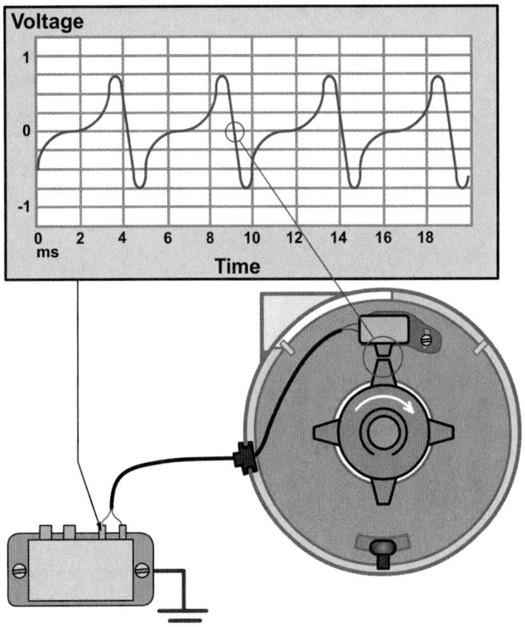

FIGURE 48-37 Inductive pickup coil and the signal it generates.

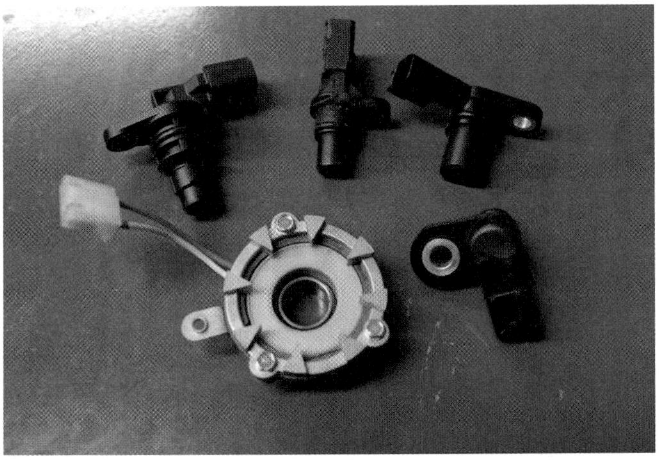

FIGURE 48-36 Types of inductive ignition pickup coils.

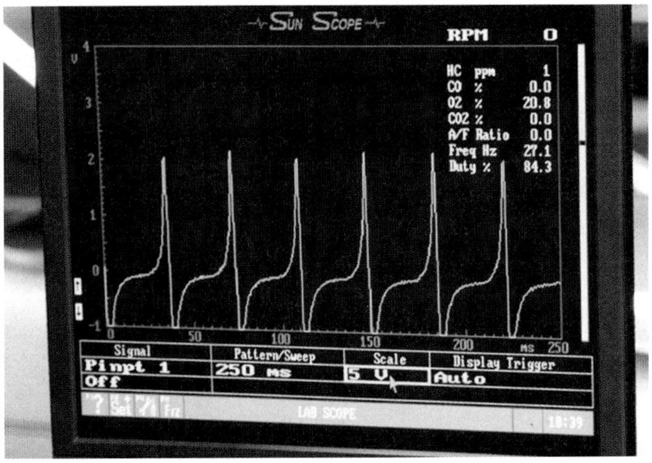

FIGURE 48-38 Typical ignition pickup coil scope pattern.

the other leg, and an air gap is formed between them. An **interrupter ring**, which has the same number of blades and windows as engine cylinders, is rotated by the distributor shaft. It rotates the blades through the air gap. The purpose of an interrupter ring is to systematically block and expose the magnetic field. The ring is shaped like a very shallow cup with slits, or windows, cut into it at evenly spaced intervals. It is made of a ferrous metal.

When a window is aligned with the assembly, the magnetic field is at its strongest. Its lines of magnetic force permeate the Hall generator material. The Hall generator material creates a voltage that is used to switch the primary circuit to ground. When the interrupter ring rotates so that the blade aligns with the assembly, the magnetic field is shielded from the generator. The Hall material does not create a signal voltage. Thus, the primary circuit is not switched to ground. With continuous rotation, the blades repeatedly move in and out of the air gap, and the signal voltage turns on and off repeatedly. This can be used to control the operation of the ignition coil primary circuit.

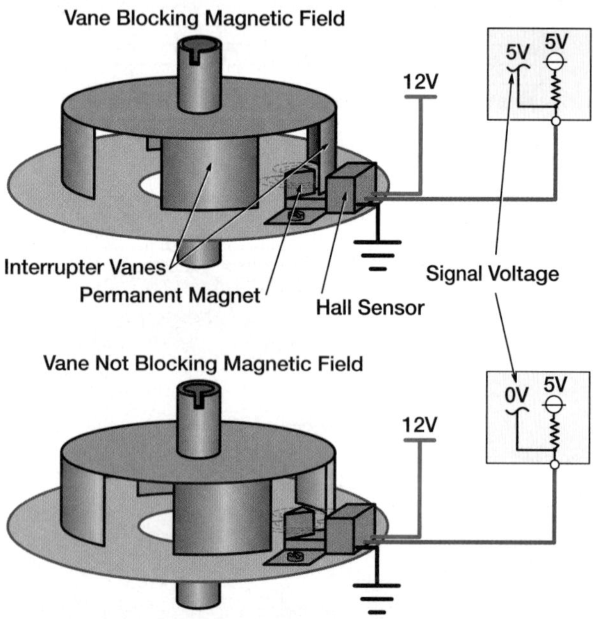

FIGURE 48-39 Hall-effect sensor operation.

Optical-Type Sensors

Another type of electronic ignition system uses a light-emitting diode (LED) and a phototransistor. Together, they create an optical sensor. The optical sensor is located inside the distributor and can be used to sense the position of the crankshaft. It then sends an appropriate voltage signal to the ignition module or PCM. A signal rotor plate is attached to the distributor shaft (**FIGURE 48-41**). Although various designs are used, a typical rotor plate has 360 slits at 1-degree intervals. Inboard of these slits are four slits on a four-cylinder engine, six slits on a six-cylinder engine, and so on. One slit is larger than the others to identify the position of the number-one cylinder to the PCM.

As the rotor plate turns, it passes between an LED above the rotor plate and a phototransistor below the plate (**FIGURE 48-42**). When provided with a suitable voltage, LEDs transmit a fine beam of light. Phototransistors receive this light and use it to make a voltage output signal. When a slit is in alignment, the light beam passes through it, and a signal is transmitted to the control unit. When the slit is out of alignment, the light beam is interrupted, and the signal falls to zero.

The control unit uses the signals from the 1-degree outer slits to gauge engine rpm. They are also used to determine **crank angle position** in 1-degree increments. It uses the signals from the inner slits to gauge the piston position, with the signature slit identifying the number-one piston. The signals from both sets of diodes are converted to on/off pulses in the sensor's internal circuitry.

The control unit determines ignition timing. It does so after computing data from the various input sensors. It computes it against the optimum settings for each operating condition recorded in its memory. It then switches the primary circuit on and off by controlling the operation of a power transistor in the ignition module.

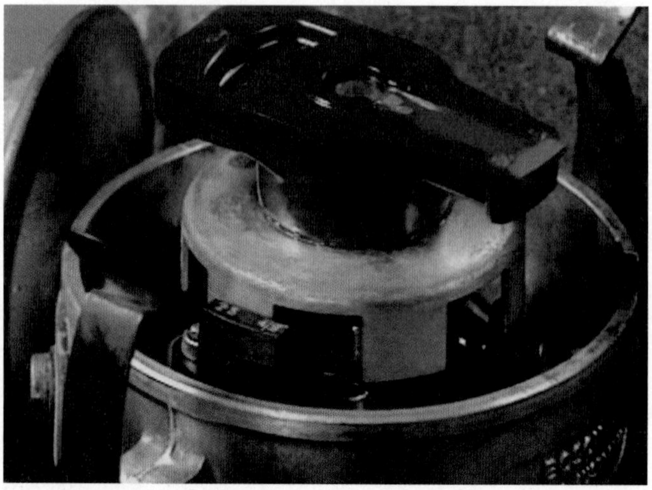
FIGURE 48-40 Hall-effect setup in a distributor.

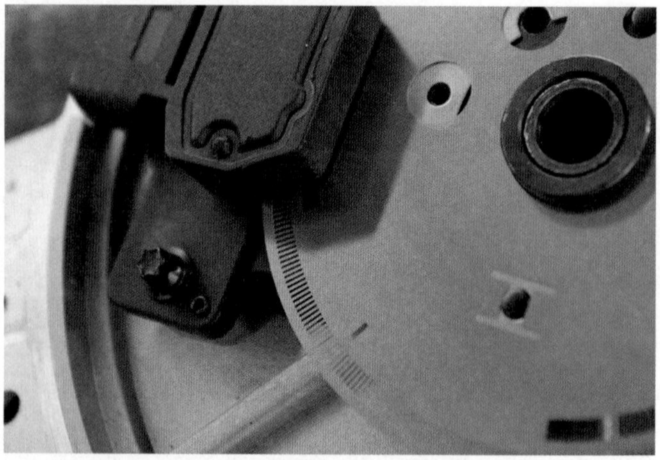
FIGURE 48-41 Optical sensor ignition system.

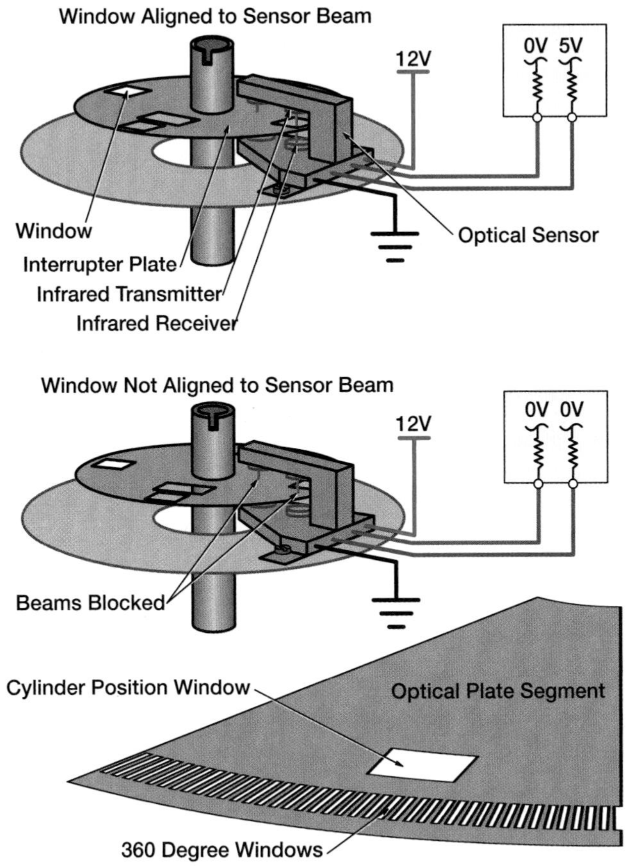

FIGURE 48-42 A typical optical sensor assembly.

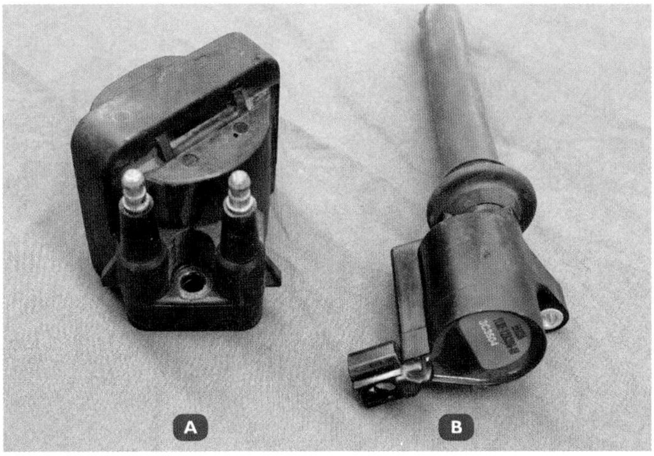

FIGURE 48-43 **A.** Waste spark ignition coil. **B.** Coil-on-plug ignition coil.

> ▶ TECHNICIAN TIP
>
> The optical sensor is very precise due to the small slits that the light passes through. However, because the slits are so small, they can easily become blocked by dirt, debris, or even oil. Make sure the slits in the rotor plate stay clean when you are working around it.

▶ Components of Distributorless-Type Systems

LO 48-10 Describe distributorless ignition systems.

In **distributorless ignition systems**, the distributor is eliminated. It is replaced by several ignition coils. What is the advantage of the distributorless ignition system? It is the elimination of all the mechanical parts included in a distributor assembly. This includes bushings, bearings, the **breaker plate** assembly, and advance mechanisms. The result is that distributorless ignition systems have eliminated much of the required maintenance. Now, items such as a distributor cap, rotor, contact points, and condenser are no longer used.

There are two main types of distributorless ignition systems: waste spark, and COP (**FIGURE 48-43**). In waste spark systems, there is one ignition coil for each pair of companion cylinders. For COP systems, there is one ignition coil for each cylinder. Signals from a **crankshaft position sensor** and, if equipped, a **camshaft position sensor** are used by the PCM to control the on/off signal to the primary windings of the coils. Both of these sensors are either Hall-effect or induction-type sensors. They let the PCM know the positions of the crankshaft and the camshaft. Not all engines use a camshaft position sensor. On waste spark ignition engines, the PCM only needs to know when the number-one cylinder is at TDC (compression or exhaust doesn't matter). It can then determine the proper spark timing for the other cylinders.

Crank and Cam Sensors

Crankshaft and camshaft position sensors are used in distributorless ignition systems. They are used for calculating engine speed and determining piston position. The crankshaft position sensor can be mounted at the front, middle, or back of the engine. It functions like a pulse generator in a distributor. The reluctor or trigger wheel is mounted to the flywheel, the crankshaft, or the crankshaft pulley.

The camshaft position sensor monitors the position of the camshaft. It lets the PCM know which cylinder is approaching its power stroke. The two sensors work together to control the firing of the spark plugs. This includes advancing and retarding the timing (**FIGURE 48-44**). They are also used for fuel injector control. These sensors provide more accurate readings than contact breaker points. There are fewer mechanical parts to wear as well, which affect accuracy.

As an example, in one system, the identification of each pair of cylinders and a signal for their triggering is provided by a dual crank sensor. It is fixed to the engine timing cover. This contains two Hall-effect switches sharing a central magnet. The magnet forms two air gaps between them. The crank sensor contains two **concentric interrupter rings**, which look like one shallow cup placed inside a larger shallow cup. Both rings have a number of blades and windows on their outer edges. They are mounted on the rear of the crankshaft balancer and rotate through the air gaps (**FIGURE 48-45**). The voltage signals provided allow the ignition module to identify which pair of companion cylinders is ready for ignition.

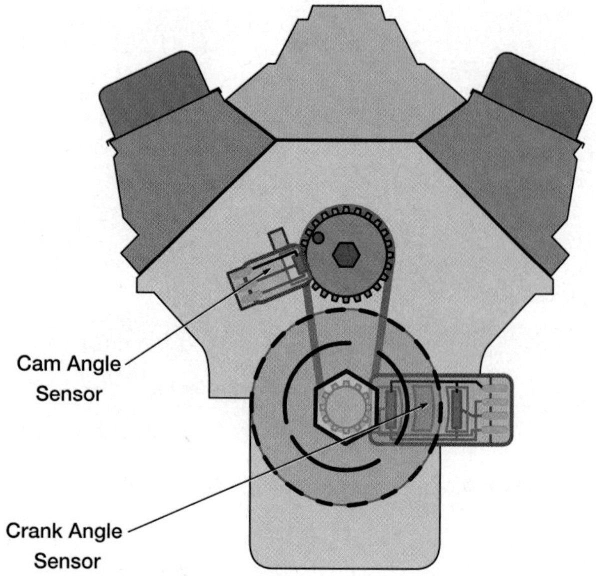

FIGURE 48-44 The crankshaft position sensor and the camshaft position sensor work together to provide timing signals that the PCM uses to fire the spark plugs and control variable valve timing.

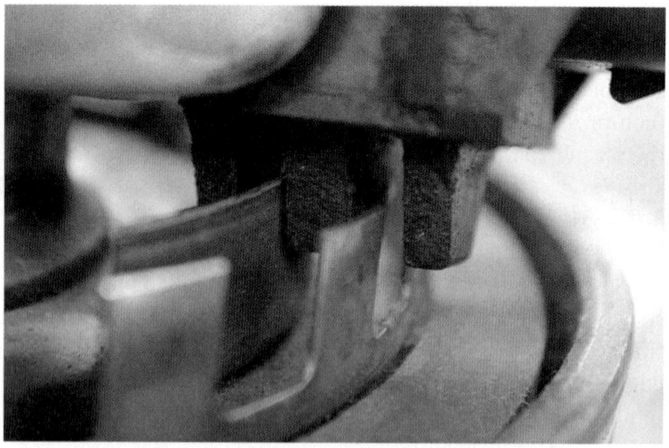

FIGURE 48-45 The interrupter rings on an engine's crankshaft pulley.

SAFETY TIP

Crankshaft and camshaft sensors are typically located around the front of the engine. They are near the belts and other moving parts. Always remove the key from the ignition before working around the moving parts of the engine.

Waste Spark Ignition System

In a waste spark system, each ignition coil serves two cylinders. Each end of the secondary winding is attached by a high-tension lead to a spark plug. These two plugs are on companion cylinders. That is, cylinders where the pistons reach TDC at the same time, but on opposite strokes (**FIGURE 48-46**). The cylinder on the compression stroke is said to be the **event cylinder**. This is because it is the one getting ready to produce power. The cylinder on the exhaust stroke is the **waste cylinder**, as the spark has no effect on engine operation.

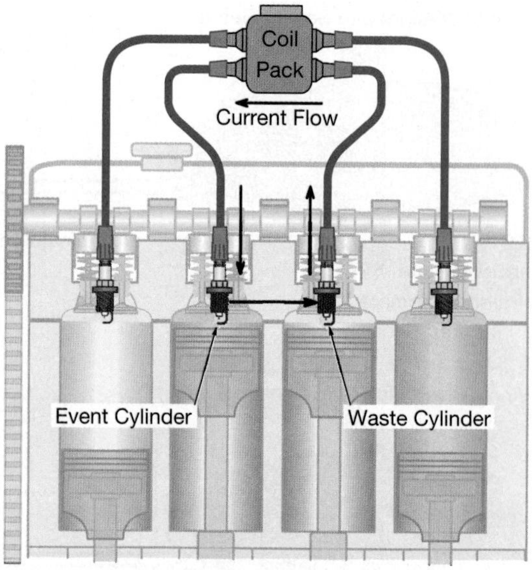

FIGURE 48-46 A waste spark system fires two spark plugs at a time: the event cylinder and the waste cylinder.

When the high voltage is induced in the secondary winding, the secondary circuit is completed by:

- current flowing through the high-tension lead to the center electrode on one spark plug
- current bridges the gap and creates a spark in that cylinder
- current travels through the cylinder head,
- current bridges the spark plug gap in the companion cylinder,
- current flows back through that high-tension lead to its starting point at the ignition coil.

The cylinder on the compression stroke with its charge of fuel and air is "fired" by its spark. This drives the piston down on the power stroke. The spark at the plug of the cylinder on the exhaust stroke simply serves to complete the circuit, and is "wasted."

When the crankshaft rotates through one revolution, the roles of the cylinders are reversed. The waste cylinder becomes the event cylinder. Firing of both spark plugs again takes place, and this cylinder now drives its piston down on the power stroke. This same process occurs on each pair of companion cylinders as they approach the TDC position. The primary circuit in each coil must therefore trigger at the correct time in each crankshaft revolution. Timing is controlled by the PCM based on all of the engine sensor data.

Because each coil feeds two cylinders, there are three ignition coils for a six-cylinder engine. Each coil has its own primary and secondary windings. The coils can be combined to form one coil pack. Or they can be three separate coils (**FIGURE 48-47**). An eight-cylinder engine has four ignition coils.

Coil-on-Plug Ignition System

Most current production gasoline engines use a single ignition coil for each cylinder (**FIGURE 48-48**). There are two common layouts: COP and CNP. The COP system positions each

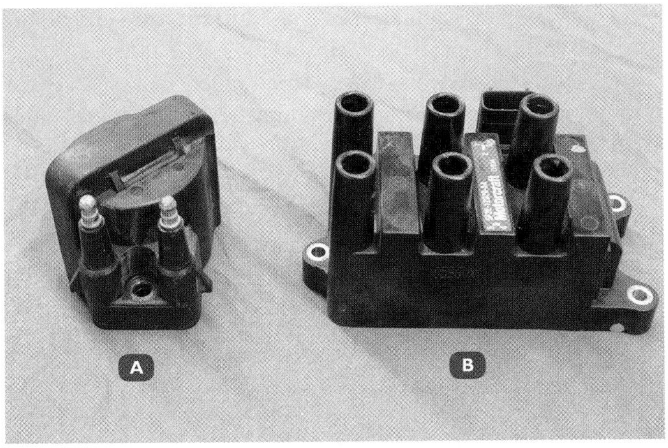

FIGURE 48-47 Typical waste spark ignition coils. **A.** Individual ignition coil. **B.** Coil pack.

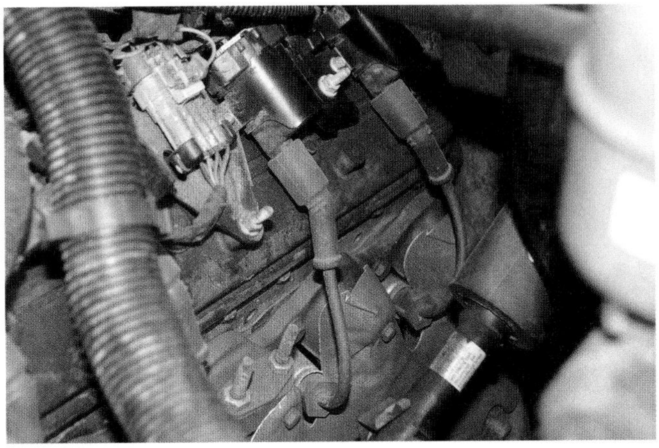

FIGURE 48-49 Typical coil-near-plug ignition system with the spark plugs in the side of the head and the coils nearby.

FIGURE 48-48 Typical coil-on-plug ignition system with the coils partially removed.

coil directly on top of each spark plug. In fact, the coil has a replaceable, integrated spark plug boot on the end of the coil. The CNP system positions the coils a short distance away from the spark plugs. This is usually due to the spark plugs being in the side of the cylinder head instead of the top. This system uses short spark plug wires to deliver the spark to the spark plugs (**FIGURE 48-49**).

In many ways, these systems are the simplest ignition system in terms of components. They do away with most of the secondary components. They also use existing engine management components to replace some of the dedicated primary circuit components in older ignition systems (**FIGURE 48-50**). In fact, in most of these systems, the PCM acts as an ignition module for the primary circuit of each coil. This controls the saturation time of each coil and the time of each spark.

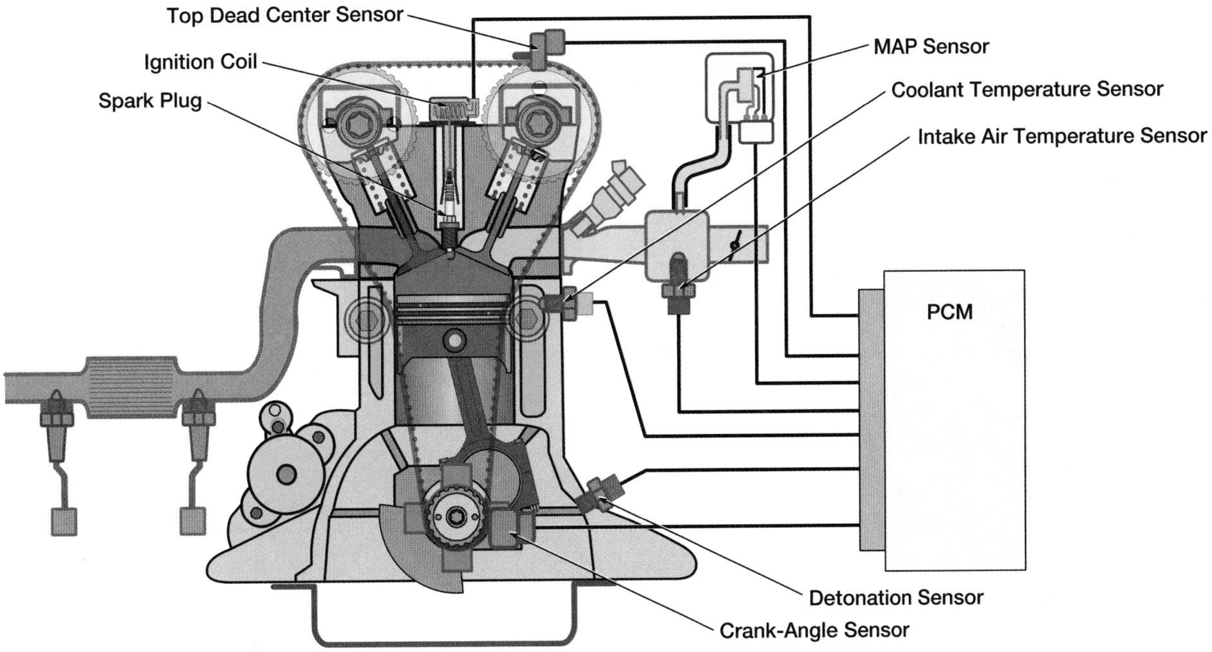

FIGURE 48-50 Schematic of a typical COP ignition system on a four-cylinder engine. Notice that there are many fewer secondary components than a distributor-style ignition system.

COP and CNP systems have several benefits over other ignition systems:

- Better high rpm performance. Each coil only sparks once per cylinder, per cycle, meaning they have to supply a lot fewer sparks. This allows a greater amount of time for coil saturation between sparks, especially at high rpm.
- Ability to create higher voltage spark output without over-stressing the coils.
- Fewer parts required, which reduces cost, weight, failures, and maintenance.
- Better control of emissions due to a hotter spark.

COP systems come in two physical configurations. The coils can be separate with each coil mounted on top of one spark plug as mentioned. Or they can be molded into an insulated cassette that is mounted directly over the top of the spark plugs (**FIGURE 48-51**). The individual COP coil systems have become more common on recent vehicles.

▶ Ignition System Maintenance

LO 48-11 Perform ignition system maintenance.

Maintenance Tools

The ignition system is capable of putting out up to 100,000 volts. So the proper tools and safety practices are necessary when working around the ignition system. Here are several tools you will need:

- A special spark plug socket is needed to remove and reinstall the spark plugs. These sockets are deeper than regular sockets, come primarily in three sizes ($^{9}/_{16}$", $^{5}/_{8}$", and $^{13}/_{16}$"), and have a rubber insert. The rubber insert makes removal and installation less likely to damage the spark plug. This insert allows the spark plug socket to grip the spark plug when the spark plug is being installed in a deep well. It also holds the spark plug centrally within the socket. This prevents the spark plug from being cocked sideways inside the socket, which can break the ceramic insulator.
- A spark plug gapping tool is used to measure and adjust the spark plug gap before installation (**FIGURE 48-52**).
- A high-tension insulated wire puller makes it easier to remove stuck high-tension wires from spark plugs (**FIGURE 48-53**).
- A tool for removing broken-off high-thread spark plugs (**FIGURE 48-54**).

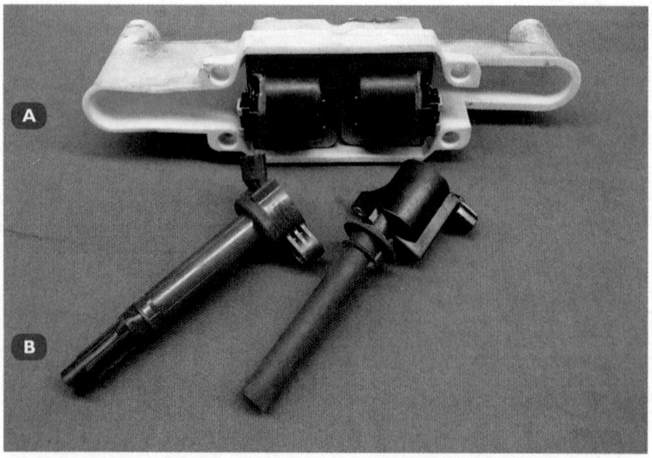

FIGURE 48-51 **A.** Cartridge style coil-on-plug assembly. **B.** Individual coil-on-plug assemblies.

FIGURE 48-52 Spark plug gapping tool.

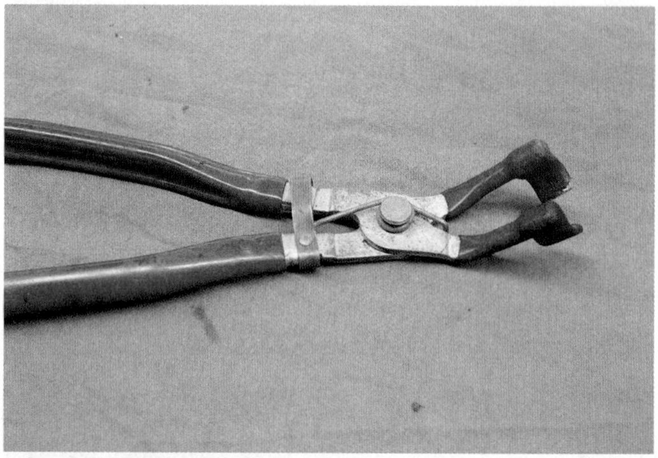

FIGURE 48-53 High-tension insulated wire puller.

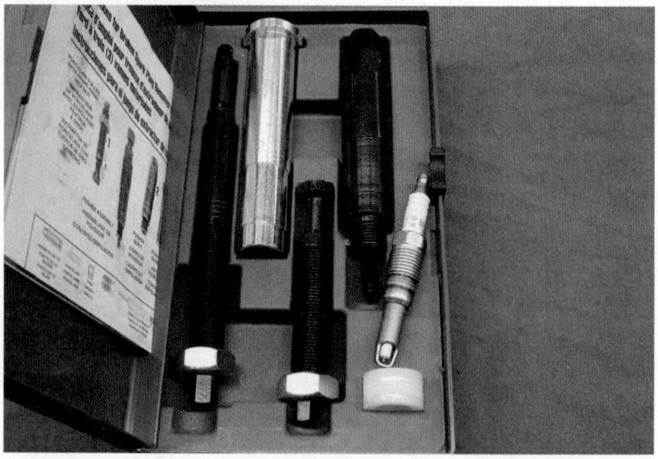

FIGURE 48-54 Tool for removing broken-off high-thread spark plugs.

Using Service Information

Service information is critical for servicing every system on the car. This includes the ignition system. This information is usually accessed in an online format. It can be from a subscription service. Or, it can be from the manufacturer-supplied service information system. There are three main areas of service information you need to access. These include specifications, technical service bulletins (TSBs), and diagnostic/repair information. These topics are covered in the Vehicle Service Information and Diagnostic Process chapter.

When working on ignition systems, you need to research the TSBs. Do this whenever you start a diagnosis. You will also need to research the specifications when servicing or diagnosing the system. And you will need to research the diagnostic, testing, and repair procedures when you are performing any ignition system work.

To use service information, follow the steps in **SKILL DRILL 48-1**.

Removing and Replacing Secondary Ignition Components

As the vehicle ages, the components of the secondary ignition system will wear. Wear parts need to be replaced every so often. These things include the spark plugs, high-tension leads (spark plug wires) or coil boots, and the distributor cap and rotor. This will maintain vehicle performance and meet emission standards. The main components to be replaced on today's vehicles are the spark plugs and the high-tension leads (if equipped). On COP systems, the coil boots may need replacement. Hard starting, misfiring under load, or high tailpipe emissions may indicate a need for service. We explore the process of removing, inspecting, and replacing these secondary ignition components next.

Servicing the Spark Plug

According to many maintenance schedules, the spark plugs should be replaced every 30,000 to 60,000 miles (50,000 to 80,000 km). Some newer vehicles use high-performance spark plugs that are made with platinum, iridium, or gold palladium. They do not require replacement until the engine reaches 100,000 miles (160,000 km) (**FIGURE 48-55**).

When service is needed, avoid attaching the spark plug wires to the wrong cylinder. This is best done by replacing the spark plugs one at a time. Note that some manufacturers recommend that the spark plugs only be replaced on a cold engine. This is to avoid damaging the threads in the cylinder head. Check the service procedure before attempting to replace the spark plugs on a hot engine. Some manufacturers have very specific procedures for removing their spark plugs. The procedure should be followed exactly, or broken-off plugs can occur. Even when following exact procedures, some applications still result in broken-off spark plugs. You will then have to use the proper tools and follow the correct procedures for removal of the broken plug.

It is quicker to check the spark plug gap on all of the new spark plugs at one time. Most spark plugs come pre-gapped. But they should be checked to make sure they were not damaged during shipping. If the gap of the spark plug is correct, there will be a slight drag between the gap gauge and the spark plug electrodes. If the gap is too big, the outer electrode should be bent down slightly until it is correct. If the gap is set too

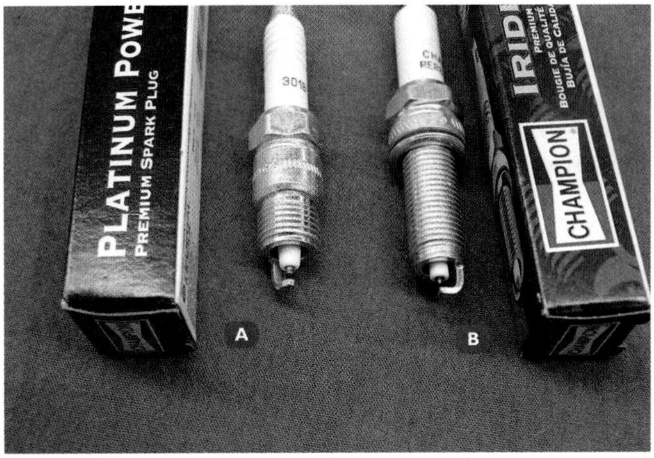

FIGURE 48-55 A. Platinum spark plug. **B.** Iridium spark plug.

SKILL DRILL 48-1 Using Service Information

1. Obtain the website address for the service information system used by your repair shop. There may be a user name and password for the business as a whole or each technician may have individual login information.
2. If not using an online service, obtain the service manual for the vehicle you are servicing.
3. Using the year, make, and model of the vehicle, locate the section that covers the ignition system. This section is usually located within the "Engine Performance" or "Engine Electrical" section. It may also be within the "Engine Controls" section.
4. Locate and record the specifications for the pickup sensors, spark plug wires, ignition coil (primary and secondary windings), or any other related component. The specifications will be in ohms or volts, or sometimes in frequency (Hz).
5. Print out or write down the specifications for use at the vehicle while you are testing.
6. Research the TSBs to see if there are any updated procedures, parts, or related symptoms that can guide you through your task.

SKILL DRILL 48-2 Replacing the Spark Plugs

1. Check the electrode gap of each spark plug by finding the correct-sized gauge on the gapping tool. Attempt to place it between the center and the side electrode.

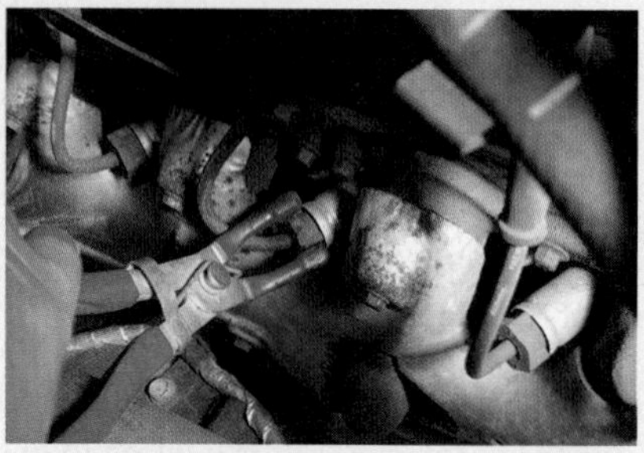

2. To gain access to a spark plug, first remove the spark plug wire by gripping its boot by hand. Or use a high-tension wire puller to gently pull it off the spark plug.

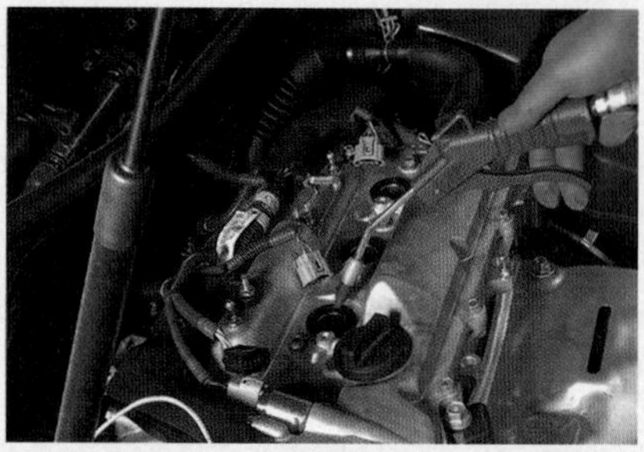

3. Clean out the spark plug pocket with compressed air.

4. Remove the spark plug, using the correct-sized spark plug socket.

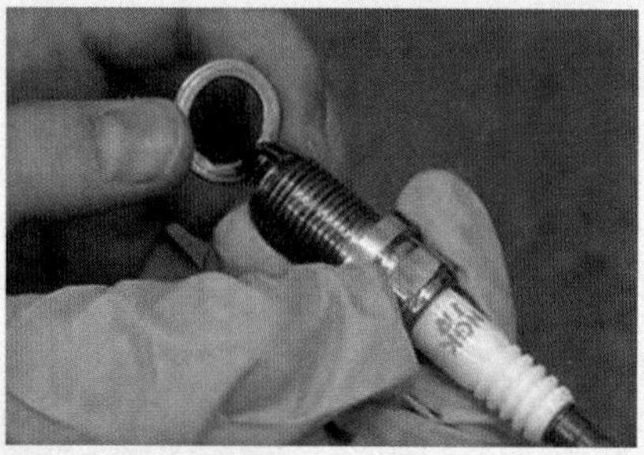

5. If the new plugs are equipped with a gasket, ensure they are installed properly. Install the spark plugs by hand to avoid cross-threading. Use a torque wrench to properly tighten the spark plug.

6. Once the spark plug is installed, reinstall the spark plug wire by pushing it on to the terminal of the spark plug until it snaps into place.

small, bend the outer electrode away from the center electrode until the gapping tool fits snuggly between the two electrodes. Some spark plugs, such as those made with iridium, are not to be adjusted. If the gap is wrong, they are to be replaced, as they break very easily.

To replace the spark plugs, follow the steps in **SKILL DRILL 48-2**.

Replacing the Spark Plug Wires or COP Boots

The spark plug wires, high-tension leads, or COP boots should be replaced periodically. There is not usually a recommended replacement interval stated in the owner's maintenance schedules for spark plug wires. Many shops recommend replacement at about the 100,000-mile (160,000 km) range. When removing spark plug wires, twist the boots to break them loose from the spark plug. Also, make sure you remove them by pulling on the boot and not the wire. Otherwise, the terminal can pull off the end of the wire. Spark plug boot pliers can help you reach down to grab the boot in arrangements where the boots are obstructed.

When replacing a set of plug wires, replacing one spark plug wire at a time will help prevent you from getting the wires and cylinders mixed up. It is also good to lay out the wires shortest to longest so that you can match up the wires if they are not numbered. When reinstalling the plug wires, some spark plug boots should have a small amount of high-temperature dielectric grease smeared on the inside. This helps prevent the boots from sticking to the spark plug porcelain. In some cases, you need to burp any trapped air from the spark plug wire boots with a small, dull screwdriver or thin wire. This is so that the trapped air does not push the boots back off after they are installed.

To replace the spark plug wires, follow the steps in **SKILL DRILL 48-3**.

SKILL DRILL 48-3 Replacing the Spark Plug Wires

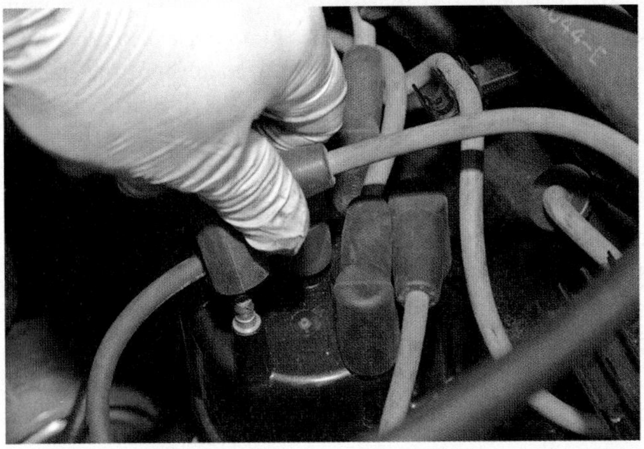

1. Remove one end of the wire (or boot) from the distributor cap, coil pack, or COP coil. Remove the other end from the spark plug with a high-tension wire puller by twisting the boot on the spark plug while gently pulling it off.

2. If dielectric grease is specified for the spark plug boots, use a cotton swab to spread a small amount around the inside of the boot.

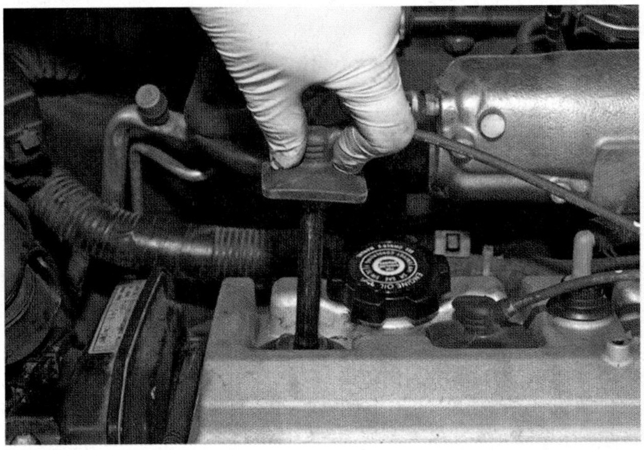

3. Install the new spark plug wire by pushing one end onto the spark plug and the other onto the distributor cap, coil pack, or COP coil. The spark plug wires should be routed in their factory positions to prevent damage.

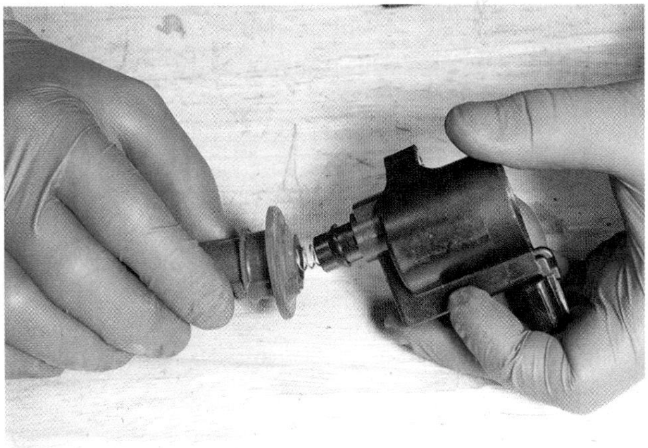

4. If replacing COP boots, carefully work the boot over the secondary terminal until it is fully seated.

▶ Wrap-Up

Ready for Review

- The ignition system consists of a primary circuit that activates the ignition coil and a secondary circuit that transmits the high voltage from the coil, or coils, to the spark plug at each cylinder. Different types of ignition system include: contact breaker point ignition system, electronic ignition system-distributor type, waste spark ignition system, and direct ignition system.
- The induction coil of an ignition system converts low-voltage and high-current flow into very high-voltage and very low-current flow. In the induction coil, the secondary winding has many thousands of turns of fine enameled copper wire and the primary winding has a few hundred turns of relatively heavy wire.
- Required voltage is the amount of voltage required to initially jump the spark plug gap. Available voltage is the maximum amount of voltage available to jump the spark plug gap if the gaps were infinite.
- Spark timing of an engine varies according to the following factors: air–fuel ratio, detected knock, engine

speed, engine load, engine temperature, air temperature, transmission gear selected, and throttle position.
- Ignition systems include a battery, ignition switch, ignition coil(s), high-tension leads (also called spark plug wires), and spark plugs.
- Modern spark plugs also include an internal resistor to suppress voltage spikes when the spark plug is energized. Spark plugs are identified by thread size or diameter, reach or length of the thread, and heat range or operating temperature.
- Contact breaker point ignition systems provide a simple means of establishing and interrupting the current flowing in the primary ignition circuit. It consists of the battery, the ignition switch, an ignition coil, contact breaker points, a capacitor, a distributor, and connecting wires and leads.
- The centrifugal advance mechanism controls ignition timing in relation to engine speed. The vacuum advance mechanism controls ignition advance in relation to engine load.
- Electronic ignition systems make use of an electronic triggering system and an electronic switching device. Different types of electronic ignition systems include: distributor-type electronic ignition systems, induction-type pickup assemblies, hall-effect systems, and an electronic ignition system that uses a light-emitting diode (LED) and a phototransistor.
- The distributor in a distributorless ignition system is replaced by several ignition coils. Two main types of distributorless ignition systems include: waste spark, and COP.
    - Waste spark systems use one ignition coil for each pair of companion cylinders and COP systems use one ignition coil for each cylinder.
- Tools required to work safely on an ignition system include: special spark plug socket, spark plug gapping tool, high-tension insulated wire puller, and a tool for removing broken-off high-thread spark plugs. The main components to be replaced on today's vehicles are the spark plugs and the high-tension leads.

Key Terms

advance mechanism A device used to trigger an earlier spark based on engine conditions.

available voltage The maximum amount of voltage that the induction coil secondary is capable of putting out.

ballast resistor Used to limit the amount of current flowing in the ignition primary circuit.

breaker plate The movable plate that the breaker points are mounted on that pivots as the vacuum advance pulls on it.

cam lobes Raised areas or protrusions on an otherwise round shaft.

camshaft position sensor A sensor mounted near the camshaft and used to send camshaft and valve position information to the PCM.

center electrode The electrode located in the center of a spark plug. It is the hottest part of the spark plug.

centrifugal advance mechanism An ignition timing device, located above or beneath the distributor base plate, that rotates with the distributor cam and is used to advance the spark. As engine speed rises, the flyweights on the advance mechanism are thrown outward by centrifugal force. Because the distributor cam is able to rotate on the distributor shaft, the weights act against their springs and move the distributor cam forward.

compression stroke The stroke of the piston during which air and fuel is being compressed into a small area prior to ignition.

concentric interrupter rings Two interrupter rings that have the same center.

contact breaker point ignition system A type of ignition system that uses a mechanical means of turning the primary circuit on and off.

contact breaker points A mechanically operated electrical switch that is fixed to the distributor base plate and opened and closed by the distributor cam with the rotation of the engine. The contacts normally form a self-contained unit, fixed to the base plate by a retaining screw engaged in a slot in the fixed contact.

crank angle position The position of the crankshaft, measured in degrees.

crankshaft position (CKP) sensor A sensor used by the PCM to monitor engine speed. It can be one of three types of sensors—Hall effect, magnetic pickup, or optical.

direct ignition system May refer to a waste spark ignition or a coil-on-plug ignition system, in which the coils are directly attached to the spark plugs.

distributor The part of an ignition system that distributes the spark to the spark plugs in the correct sequence and at the correct time. It includes a distributor cap, rotor, shaft, and usually a switching device.

distributor base plate A round metal plate near the top of the distributor that is attached to a distributor housing; also called a breaker plate.

distributor cap The top portion of a distributor, used to make a connection between the spinning rotor and the high-tension leads.

distributorless ignition system An ignition system that does not include a distributor. It uses signals from the crankshaft position sensor and the camshaft position sensor sent to the PCM to determine when to send a signal to the ignition module.

dwell angle The amount of time that the primary circuit is energized, measured in degrees of distributor rotation.

electronic ignition system An ignition system that uses a nonmechanical (electronic) method of triggering the ignition coil's primary circuit.

electronic ignition system—distributor type An ignition system that uses a distributor but replaces the contact points with an electronic triggering device and control module.

enameled copper wire Wire that uses a thin layer of enamel as insulating material. The thinness of the insulation allows the wire to be closely wound in a coil, creating a dense magnetic field when current flows through it.

engine control module (ECM) A computer that controls the ignition and fuel control and emissions control systems on an engine; also called the electronic control unit (ECU) or powertrain control module (PCM).

event cylinder The cylinder that is on compression and ready for the spark to ignite the air-fuel mixture on a waste spark ignition system.

Hall-effect sensor A sensor commonly used to measure the rotational speed of a shaft; they have the advantage of producing a digital signal square waveform and have strong signal strength at low shaft rotational speeds.

heat range The rating of a spark plug's operating temperature.

high-tension leads The heavy insulated wires used to connect the distributor cap terminals to the spark plugs, and the ignition coil to the distributor cap; on waste spark systems, they connect the coils to the spark plugs.

ignition advance The means of causing the spark to occur earlier within the compression stroke for better performance and fuel economy during changing engine conditions.

ignition coil A device used to amplify an input voltage into the much higher voltage needed to jump the electrodes of a spark plug.

ignition module An electronic component that electronically controls the ignition coil or coils.

ignition switch A switch operated by a key or start/stop button and used to turn a vehicle's electrical and ignition system on or off.

induced voltage The creation of voltage in a conductor by movement of a magnetic field that is near that conductor.

induction coil An electrical transformer that uses magnetic fields to produce high-voltage pulses from low-voltage direct current.

induction-type system A type of ignition system that uses a magnetic pulse generator to trigger the spark.

inductive current The current that has been created across a conductor by moving it through a magnetic field.

integrated circuit (IC) A complete circuit that is formed at a micro level. Processor chips include thousands of integrated circuits.

interrupter ring A ferrous metal ring shaped like a very shallow cup with slits or windows cut into it at evenly spaced intervals. The ring has the same number of blades and windows as engine cylinders and is rotated by the engine moving the blades through an air gap. The purpose of an interrupter ring is to systematically block, and expose, the magnetic field in a Hall-effect sensor in order to turn the primary ignition circuit on and off.

number-one cylinder Typically the cylinder located farthest forward on the engine. It is the first cylinder in the firing order.

optical sensor A sensor that generates a voltage when excited by a beam of light.

powertrain control module (PCM) A computer that controls the ignition, fuel, and emissions control systems on an engine; also called the electronic control unit (ECU) or engine control module (ECM).

primary circuit The low-voltage circuit that turns the coil on and off.

reluctor A rotating, toothed wheel that changes the reluctance of a material to conduct magnetic lines of force.

required voltage The amount of voltage needed to push current across the electrodes of a spark plug located in the combustion chamber.

rotor A high-voltage rotating switch that transfers voltage from the distributor cap's center terminal to the outer terminals.

rotor arm The portion of the rotor that extends toward, but not touching, the outer distributor cap terminals.

secondary circuit The part of an ignition system that operates on higher voltage and delivers the necessary high voltage to the spark plugs.

spark plug A device that provides a gap for the high-voltage spark to occur in each cylinder.

spark plug reach The length of the spark plug from the seat to the end of the threads.

spark timing The point at which a spark occurs at the spark plug relative to the position of the piston.

stationary winding An extended length of wire wrapped into a circle. These windings are fixed, as opposed to some types, which are meant to spin.

stator Portion of an electronic ignition system that is mounted to the base of the distributor. It has a circular permanent magnet with a number of projections or teeth corresponding to the number of engine cylinders, and a stationary coil of fine enameled copper wire wound on a plastic reel and positioned inside the magnet.

waste cylinder The cylinder in a waste spark ignition system that receives a spark near the top of its exhaust stroke.

waste spark ignition system An ignition system in which each ignition coil served two cylinders, with ach end of the secondary winding attacted by a high-tension lead to a spark plug. The spark is used to ignite the air-fuel mixture in one cylinder and has not effect on the other cylinder.

Review Questions

1. Which of the following is the most recent type of ignition system?
   a. The direct ignition system
   b. The waste spark system
   c. The contact point system
   d. The electronic ignition distributer system
2. Which component is responsible for increasing the voltage from about 12 V to as high as 100,000 V?
   a. The spark plug
   b. The contact points
   c. The distributor
   d. The ignition coil

3. In an ignition system, what is "required voltage"?
   a. The minimum voltage to power the coil on
   b. The available voltage to the ignition coil from the fuse
   c. The minimum voltage to initially jump the spark plug gap
   d. The voltage required for the points to open and close
4. All of the following have a large effect on ignition timing EXCEPT:
   a. Engine load
   b. Fuel temperature
   c. Engine speed
   d. Engine temperature
5. On a vehicle with a distributor, what delivers the spark to the spark plugs?
   a. High-tension leads
   b. The coil boot
   c. Primary wiring
   d. The ignition coil
6. What is the purpose of the side electrode on a spark plug?
   a. To prevent the spark plug from loosening
   b. To insulate the high voltage from the head
   c. To provide a path to ground for the spark
   d. To provide structural strength
7. In a contact breaker point ignition system, what is the purpose of the condenser?
   a. To keep the sparks close together
   b. To control when the spark occurs
   c. To send the spark to the correct cylinder
   d. To prevent arcing in the contacts
8. What is the typical base ignition timing on older vehicles with adjustable timing?
   a. 0 degrees, or right at TDC
   b. 5–10 degrees before TDC
   c. 5–10 degrees after TDC
   d. 15–20 degrees before TDC
9. In an electronic ignition system using an optical sensor, what is used to sense the position of the crankshaft?
   a. A breaker point contact and condenser
   b. A Hall effect sensor with chip
   c. An inductive pick-up and reluctor
   d. An LED, phototransistor, and plate with slits
10. In a distributorless ignition system, what indicates engine position and speed?
   a. The throttle position sensor
   b. The optical sensor and its spinning plate
   c. The crankshaft and camshaft sensors
   d. The spinning breaker plate

ASE Technician A/Technician B Style Questions

1. Ignition systems are being discussed. Technician A states that primary circuits are very high voltage. Technician B states that secondary circuits are low voltage. Who is correct?
   a. Technician A only
   b. Technician B only
   c. Both Technicians A and B
   d. Neither Technician A nor B
2. An ignition coil is being discussed. Technician A states that the available voltage is the minimum voltage required to jump the spark plug gap. Technician B states that manufacturers design their ignition systems to keep available voltage higher than required voltage at all times. Who is correct?
   a. Technician A only
   b. Technician B only
   c. Both Technicians A and B
   d. Neither Technician A nor B
3. Ignition timing is being discussed. Technician A states that ignition timing was adjustable on older vehicles with distributors. Technician B states that new vehicles with direct ignition do not have adjustable ignition timing, it is done by the PCM. Who is correct?
   a. Technician A only
   b. Technician B only
   c. Both Technicians A and B
   d. Neither Technician A nor B
4. Ignition systems are being discussed. Technician A states that coil-on-plug systems have two high-tension leads going to two spark plugs. Technician B states that on a distributor type, the high voltage is transmitted through a terminal on the distributor cap. Who is correct?
   a. Technician A only
   b. Technician B only
   c. Both Technicians A and B
   d. Neither Technician A nor B
5. A spark plug is being discussed. Technician A states that raising the spark plug gap increases the voltage required to jump the gap. Technician B states that spark plug reach is the distance the spark is able to jump across the electrode. Who is correct?
   a. Technician A only
   b. Technician B only
   c. Both Technicians A and B
   d. Neither Technician A nor B
6. A distributor-style ignition system is being discussed. Technician A states that if the coil fails on this type, it will cause one cylinder to misfire. Technician B states that on this type, the coil can be mounted inside of the distributor cap. Who is correct?
   a. Technician A only
   b. Technician B only
   c. Both Technicians A and B
   d. Neither Technician A nor B
7. Ignition timing is being discussed. Technician A states that some vehicles use vacuum devices to advance ignition timing based on engine load. Technician B states that some vehicles use centrifugal force and weights to advance ignition timing based on increased engine RPM. Who is correct?
   a. Technician A only
   b. Technician B only

c. Both Technicians A and B
d. Neither Technician A nor B

8. Technician A states that spark plug pockets in the cylinder head should be blown out after removing the spark plugs. Technician B states that spark plugs can overheat if they are not tightened to the specified torque. Who is correct?
   a. Technician A only
   b. Technician B only
   c. Both Technicians A and B
   d. Neither Technician A nor B
9. A distributorless ignition system is being discussed. Technician A states that a waste spark system requires more frequent maintenance when compared to distributor types. Technician B states that coil-on-plug systems have very few secondary ignition components. Who is correct?
   a. Technician A only
   b. Technician B only
   c. Both Technicians A and B
   d. Neither Technician A nor B
10. Technician A states that an insulated wire puller is useful to remove stuck high-tension wires. Technician B states that on a modern vehicle, spark plugs should be replaced every year, or every 12,000 miles. Who is correct?
    a. Technician A only
    b. Technician B only
    c. Both Technicians A and B
    d. Neither Technician A nor B

CHAPTER 49
Gasoline Fuel Systems

Learning Objectives

- **LO 49-01** Describe the types of fuel systems.
- **LO 49-02** Describe gasoline fuel characteristics.
- **LO 49-03** Describe normal and abnormal combustion.
- **LO 49-04** Describe the fuel–air requirements for internal combustion.
- **LO 49-05** Describe the fuel tank components.
- **LO 49-06** Describe the fuel pump components.
- **LO 49-07** Describe fuel lines, fuel filters, and fuel rails.
- **LO 49-08** Describe how fuel pressure is regulated.
- **LO 49-09** Describe fuel injectors and their operation.
- **LO 49-10** Describe throttle body and multi-point fuel injection.
- **LO 49-11** Describe the operation of gas direct injection.
- **LO 49-12** Describe the drawbacks of GDI systems.
- **LO 49-13** Describe basic carburation operation.
- **LO 49-14** Describe the circuits on a carburetor.
- **LO 49-15** Describe the barrels on a carburetor.
- **LO 49-16** Remove and replace a fuel filter.
- **LO 49-17** Test fuel pressure and volume.
- **LO 49-18** Check for fuel contaminants and alcohol content.
- **LO 49-19** Test fuel injectors.

ASE Education Foundation Tasks

See Appendix A to view the 2017 ASE Education Foundation Automobile Accreditation Task List Correlation Guide.

▶ Introduction

LO 49-01 Describe the types of fuel systems.

Today's gasoline **fuel systems** must meter a precise amount of fuel into the engine. And it must do so under a wide range of operating conditions. The most important job of the fuel system is optimizing engine performance. This should happen while keeping fuel consumption and emissions to a minimum. Although carburetors were commonly used on vehicles up until about 30 years ago, they didn't perform this job adequately (**FIGURE 49-1**). Their limitations prompted the development of fuel injection, starting with mechanically operated systems.

Mechanically operated fuel injection systems sprayed a continuous flow of fuel through the fuel injectors. It had a fuel injector at each individual intake port. These systems were sometimes called "bug sprayers" because they put out a continuous fine mist. The amount of fuel sprayed was controlled mechanically by an airflow-operated fuel distributor. The more air that entered the engine, the more fuel that was sprayed.

Mechanical fuel injection still could not meet the increasingly stringent emission standards. So **electronic fuel injection (EFI)** was introduced. Fuel delivery could then be controlled electronically. This meant that the injectors could be turned on and off very precisely to provide continuous adjustments to the air–fuel ratio. Most early EFI systems used **throttle body injection (TBI)**. It had either one or two injectors that were mounted in a throttle body. These systems looked like a simplified carburetor (**FIGURE 49-2**).

As technology progressed, **multipoint fuel injection (MPFI or just PFI)** became more common. It consists of one injector installed in each runner of the intake manifold (**FIGURE 49-3**). This design provided more equal distribution of fuel among the cylinders. Fuel economy is improved and emissions are reduced further.

But even that wasn't the end of the line for new innovations. **Gasoline direct injection (GDI)** places the fuel injectors in the cylinder head near the spark plugs. They spray fuel directly into the combustion chamber (**FIGURE 49-4**). This technology allows the engine to operate on much leaner mixtures during certain operating conditions.

To make EFI function correctly, an electronic control unit (ECU) is used. It is typically called a PCM. It determines the proper quantity of fuel to be injected. For the PCM to make the

FIGURE 49-1 Carburetor fuel system.

FIGURE 49-2 Throttle body fuel injection.

You Are the Automotive Technician

A new customer brings a late-model vehicle into your shop, complaining that it hasn't been "running right" for the last few weeks. It was taken to another shop, where the technician said it needed a tune-up and changed the spark plugs, oxygen sensors, fuel filter, and air filter. It improved at first, but then it started getting worse. The malfunction indicator lamp (MIL) came on shortly after getting it out of the shop. The vehicle seems to be low on power as well, and yesterday it didn't start. When it started today, the customer decided to bring it in. You explain that you will need to diagnose the problem. The diagnostic charge is authorized. You scan the powertrain control module (PCM) and find a P0171 (system too lean bank 1), and a P0174 (system too lean bank 2). Because the oxygen sensors are new and look like they are original equipment manufacturer (OEM) replacements, you decide to perform a fuel pump pressure and volume test. You will also perform an alcohol content test on a small sample of the fuel after the volume test.

1. Why do you suspect a fuel pump problem in this situation?
2. If the fuel pump passes the pressure test, should you perform the volume test? Why or why not?
3. How could too much alcohol in the gasoline cause the engine to run too lean?
4. How can a scan tool and test drive be used to find a low fuel pressure and volume condition?

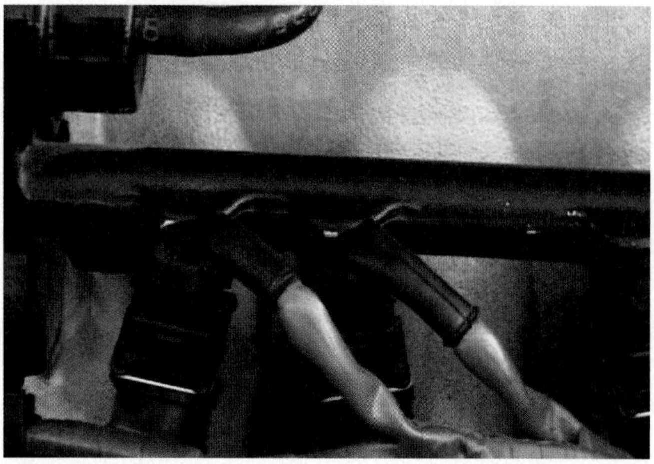

FIGURE 49-3 Port fuel injection.

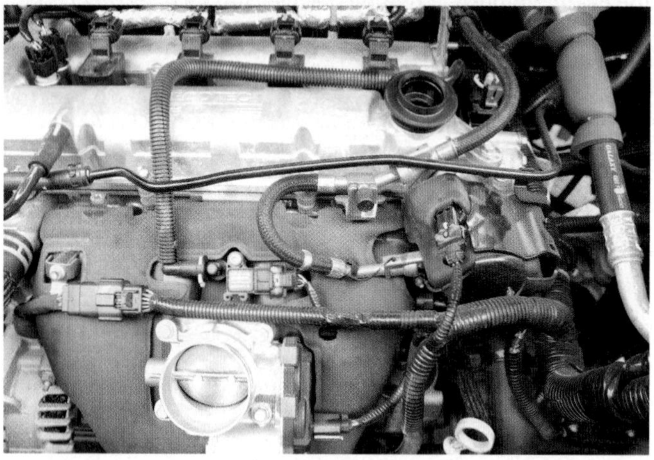

FIGURE 49-4 Gasoline direct injection.

proper determination, information from a variety of sensors is needed. The wide variety of information allows the fuel injection system to operate very precisely.

It is important to have a basic understanding of how these systems operate. This understanding will give you a head start on properly inspecting and maintaining the fuel system for your customers. This will help keep their vehicles operating at peak efficiency and performance.

▶ Gasoline Fuel System Principles

The purpose of a fuel system is to provide the ideal air–fuel mixture for the operating conditions of the engine. Liquid fuel will not burn. Fuel has to be vaporized, turning from a liquid to a gas. It then has to be mixed with the proper amount of air. It takes time and temperature for fuel to vaporize fully. The smaller the liquid droplets, the faster they can be vaporized. The process of making the droplets small is called atomization. A fuel system needs to atomize the fuel as small as possible so that it can be vaporized before it is ignited.

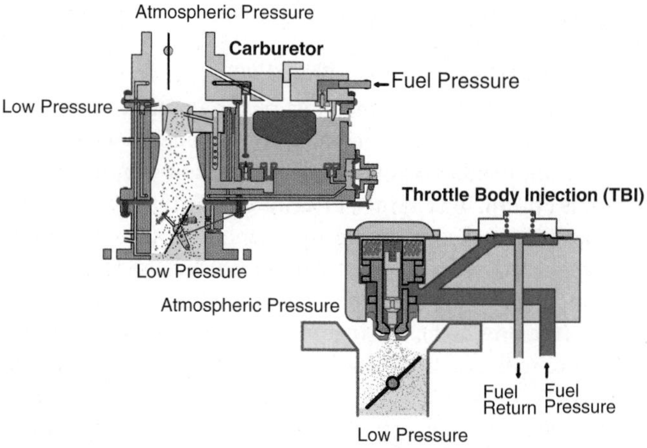

FIGURE 49-5 The main difference between carburetion and fuel injection is that carburetion works off of pressures below atmospheric pressure (vacuum). Fuel injection works off of pressures substantially above atmospheric pressure.

As explained earlier, older cars and trucks used a carburetor to atomize the fuel and mix it with air. Modern gasoline-powered vehicles use EFI systems. They atomize the fuel into much smaller particles. This allows the fuel to vaporize more easily, giving better efficiency and performance over carburetors.

Why can't carburetors atomize fuel as efficiently as fuel injection systems? It is because the principles of operation are different. Pressure differential is central to all fuel systems. The greater the pressure differential, the easier it is to atomize the fuel. A carburetor works off pressures below atmospheric pressure (vacuum). This involves a relatively low pressure differential.

Fuel injection, in contrast, works off pressures above atmospheric pressure. This generally involves a much higher pressure differential (**FIGURE 49-5**). With a very small pressure difference, as in a carburetor, fuel passages must be much larger. This means the fuel entering the airstream in a carburetor is in relatively large droplets. However, fuel that is sprayed out of a very small hole at high pressure creates much smaller droplets. This occurs in fuel injection systems.

Modern fuel injection systems have three subsystems:

- **Fuel supply system:** This system provides pressurized, filtered gasoline to the fuel injectors or carburetor (in older vehicles). The fuel supply system draws in gasoline from the gas tank (fuel cell). It delivers fuel under pressure to a fuel metering device. Today's vehicles typically use an in-tank electric fuel pump. Older vehicles used a mechanical fuel pump. It was typically mounted on the engine and driven by the camshaft.
- **Air supply system:** The air supply system is also called the induction system. It provides clean, filtered air for combustion in the engine. An air filter is a paper filter or **element** that filters the incoming dirty/dusty air. The element is usually housed in a plastic or metal air cleaner housing. It is located in the engine compartment. Engineers have made numerous configurations to prevent water and dirt

intrusion. In addition, air intake sounds have been carefully analyzed. And intake systems have been engineered to give the optimum vehicle performance while providing smooth, quiet operation.

- **Fuel metering system**: This system constantly meters and adjusts the amount of fuel that the engine is burning. The most common types of EFI fuel metering are MPFI and gas direct injection. Although specific systems vary, many of the systems have similar parts. For example, vehicles with these EFI systems have sensors, fuel injectors, and a PCM.

▶ Gasoline Fuel

LO 49-02 Describe gasoline fuel characteristics.

Gasoline fuel is derived from crude oil. Crude oil is taken out of the ground as a liquid mixture of highly flammable compounds. They include hydrogen and carbon, called hydrocarbons, together with impurities. Crude oil is then processed into many fuel and lubricant products at an oil refinery. A fractional distillation process is used. The crude oil is heated in the base of a tower and allowed to condense at different temperatures (levels) of the tower. More volatile compounds rise to the top of the tower, where they condense. Less volatile compounds condense lower in the tower (**FIGURE 49-6**).

Gasoline is very volatile. This means that it mixes easily with air to form a combustible gas or vapor. The more effectively liquid gasoline is changed into vapor, the more efficiently it burns in the engine. Thus, high volatility is desirable. However, gasoline vapor allowed to mix with air in the open is highly explosive. Therefore, it can be very dangerous, so gasoline must be handled with care. High volatility also can create excessive hydrocarbon emissions. If these are vented to the atmosphere, they contribute to air pollution and smog. If liquid gasoline is heated, it vaporizes. If it vaporizes in the fuel pump, vapor is pumped through the system instead of liquid. This situation is called **vapor lock**. It causes the engine to run very poorly, and to the extent that the engine may even die. Vapor lock was fairly common on carbureted engines. It rarely happens on fuel-injected engines.

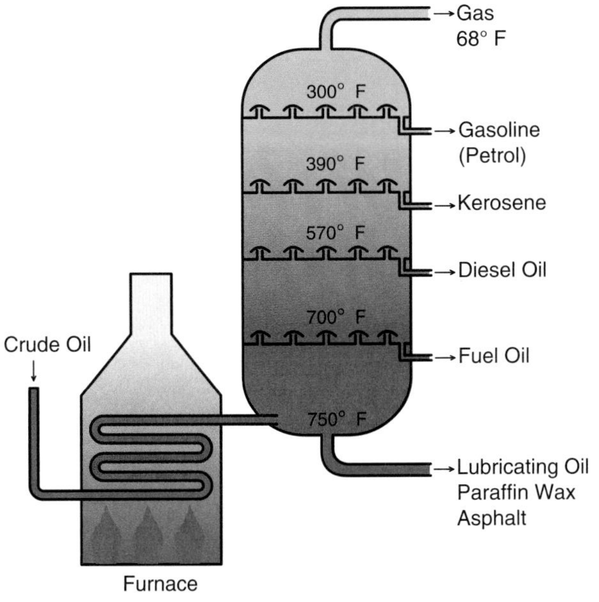

FIGURE 49-6 In processing gasoline from crude oil, more volatile compounds rise to the top of the tower while less volatile compounds stay low in the tower.

Gasoline is mainly a mixture of paraffins, naphthenes, aromatics, and olefins. Gasoline also contains some other organic compounds and contaminants such as sulfur. Some contaminants can cause corrosion, so they must be removed. Tight regulations in some countries limit the allowed proportion of aromatics, olefins, and sulfur in gasoline.

Gasoline in its raw processed form is not suitable for use in vehicle engines. It must be enhanced with different additives. Some examples are detergents, octane boosters, and oxygenates such as ethanol. Detergents help to keep the fuel system clean. This especially applies to the fuel injector nozzles and intake valves. Octane boosters make it harder to ignite the gasoline. This makes it less prone to be ignited before the spark occurs. And oxygenates add oxygen to the fuel. This helps reduce carbon monoxide pollutants by slightly leaning out the air–fuel mixture (**FIGURE 49-7**).

When a mixture of gasoline (petrol) and air is compressed inside an engine cylinder, it heats up. If the compression of the engine is high enough, the air–fuel mixture may spontaneously ignite before the spark occurs. This is called preignition, also known as **knocking**.

Gasoline fuel can be modified during processing so that it is less prone to spontaneously ignite. This is done by including additives such as octane boosters. This resistance to ignition is known as its **octane rating**. The less easily the fuel ignites, the higher the octane rating. Higher compression engines are more susceptible to engine knock. So they require fuels with a higher octane rating.

A gasoline's octane rating is measured by the producer. Two different methods are used to measure the octane rating of a fuel—the Research Octane Number (RON) and the Motor Octane Number (MON). Depending on the composition of the fuel, the MON of a modern gasoline will be about 8 to 10 points

FIGURE 49-7 Typical oxygenated fuel sticker.

lower than the RON. However, there is no direct link between RON and MON. But the MON is a better measure of how the fuel behaves when under load.

In most countries, the RON rating is the one that is usually displayed on the pump at filling stations. In the United States and Canada, and some other countries, the displayed fuel rating is an average of the RON and the MON rating: (RON + MON)/2 (**FIGURE 49-8**). This is also called the Anti-knock Index. Consequently, the rating number for identical fuels will on average be about 4 to 5 points higher in Europe than in the United States.

There is a popular belief that fuels with higher octane ratings will improve performance in vehicles that are designed to run on low octane fuels. This is largely a myth, although some premium fuels do have higher energy ratings. Higher powered engines usually have a higher compression ratio. Higher compression ratios generally require more expensive higher octane fuels. This is to prevent preignition and detonation. Higher octane does not in itself mean higher energy output. A fuel designed for a high-compression engine will not deliver any more power in a lower compression engine. In fact, it will likely be lower. Engines perform best when used with the octane rating recommended by the engine manufacturer.

Gasoline's octane is boosted with additives. Prior to the introduction of catalytic converters, tetraethyl lead was added to boost octane. Lead was also used to lubricate valve faces and seats, which slowed down wear. But lead also coats the catalytic converter and oxygen sensor, rendering them inoperable. So leaded fuel was phased out for general vehicle use, and unleaded fuel took its place. The so-called unleaded gasoline may contain small amounts of lead, but maximum levels are tightly controlled. Octane is now boosted with additives of ethanol, aromatic hydrocarbons, and other chemicals.

▶ Controlling Fuel Burn

LO 49-03 Describe normal and abnormal combustion.

For gasoline to burn properly, it must be mixed with the right amount of air. The air–fuel ratio where all of the fuel and all of the oxygen in the air is burned is about 14.7 to 1 by mass. By volume, it is about 11,000 to 1 (**FIGURE 49-9**). This means that it takes a lot of air to burn a small amount of gasoline.

A lean air–fuel mixture has more air in proportion to the amount of fuel. A slightly lean mixture gives good fuel economy and low exhaust emissions. This is suitable for cruising conditions. A mixture that is too lean can make an engine run rough and overheat. A rich air–fuel mixture has less air in proportion to the amount of fuel (**FIGURE 49-10**).

A slightly rich mixture can produce more power. But the extra fuel it uses increases fuel consumption and emissions. A mixture that is slightly rich causes incomplete burning. This reduces efficiency and increases carbon monoxide emissions significantly. If the mixture is extremely rich, it can foul spark plugs and increase carbon deposits in the combustion chamber. Controlling the air–fuel ratio so it is at the proper level for each operating condition is driving changes in fuel system technology.

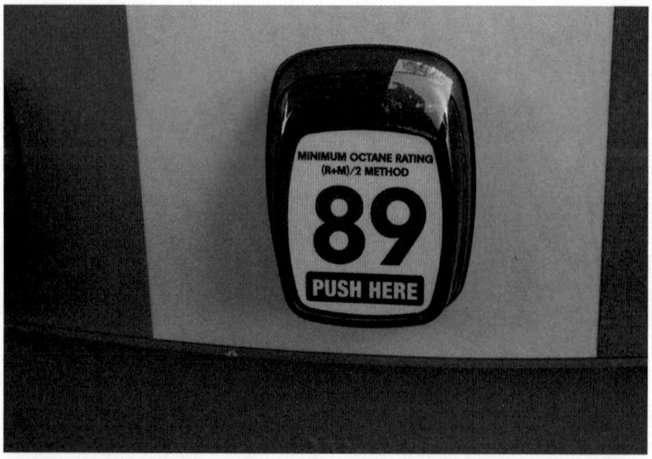

FIGURE 49-8 Typical Sticker on a gas pump listing its Anti-knock Index.

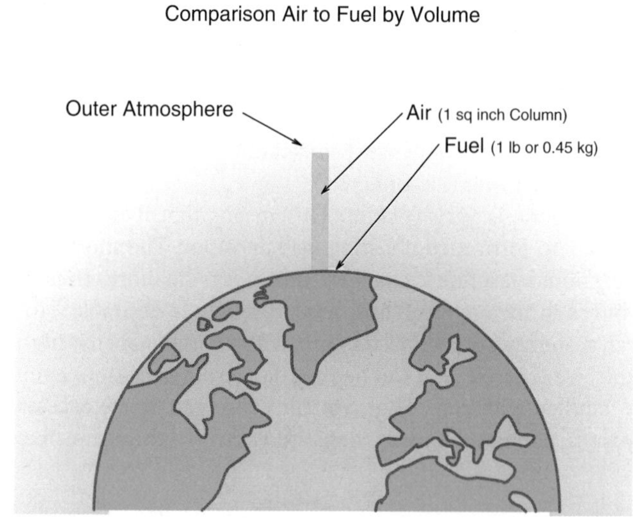

FIGURE 49-9 Air–fuel ratio where all of the fuel and all of the oxygen is burned is about 14.7:1 by mass and 11,000:1 by volume.

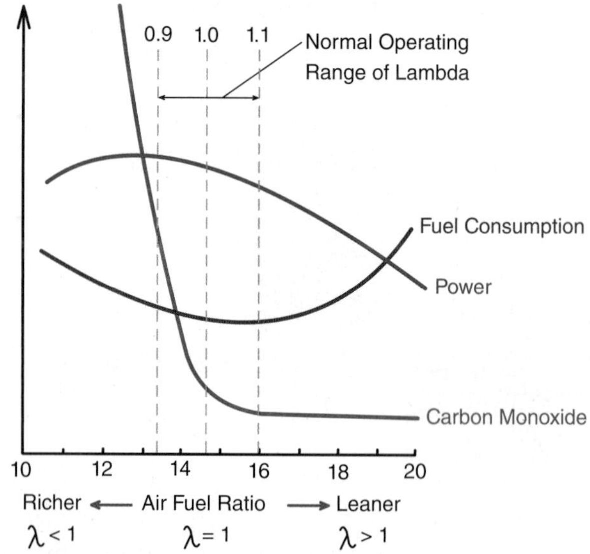

FIGURE 49-10 Rich and lean air–fuel ratios.

Combustion

In normal combustion, the spark jumps across the spark plug gap. A small ball of flame forms around the tip of the plug (**FIGURE 49-11**). The piston finishes compressing the mixture. The flame spreads faster and moves evenly to halfway through the mixture as the piston reaches top dead center. The flame picks up more speed, and then shoots out to consume the rest of the mixture. Combustion ends with the piston a short way down the cylinder. Ideally, this would completely burn all of the fuel that entered the cylinder. And thus, extract the maximum thermal energy from the gasoline. As you will see, this process can go wrong.

Detonation/Preignition

Detonation is a violent collision of flame fronts in the cylinder, caused by uncontrolled combustion. It occurs after the spark plug has fired. The sudden rise in pressure can cause the remainder of the mixture to ignite spontaneously. This creates a second flame front that collides with the first front, making a knocking sound (**FIGURE 49-12**). Sustained detonations can raise temperatures enough to melt holes in the tops of pistons.

Preignition occurs before normal combustion. Something in the combustion chamber heats up enough to ignite the mixture before the spark plug fires. This is often caused by a hot piece of carbon. Detonation and preignition can cause severe damage and must be avoided. Many engine management systems use knock sensors to detect detonation. The PCM then retards the ignition timing within certain limits to try to stop it. Knock sensors are discussed in greater detail in the next chapter.

An engine that keeps running after it is switched off is said to be running-on or **dieseling**. That is because, as in a diesel engine, the fuel is ignited just from heat in the cylinder, not from a spark from the plug. This situation may even cause an engine to run backwards for a brief time when it is being turned off. Dieseling can be caused by:

- a high idling speed,
- an overheated engine,
- too many carbon deposits in the chamber,
- use of a gasoline with an octane rating that is too low, or
- a leaky fuel injector.

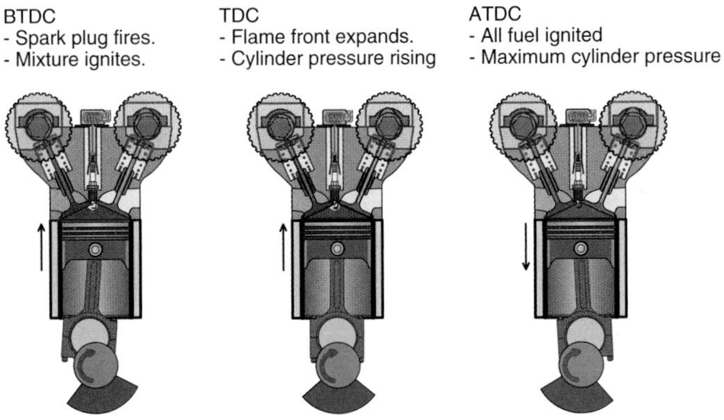

FIGURE 49-11 In normal combustion, the spark plug ignites the mixture, and a small ball of flame forms around the tip of the plug.

On carbureted engines, it can be prevented by a special valve in the carburetor idle circuit. The valve cuts off fuel when the ignition is turned off. Or a solenoid can be used to close the **throttle** to a below-idle position. Fuel-injected vehicles do not typically experience dieseling because the PCM shuts off the injector(s) during shutdown. One exception is if there is a leaking injector when the ignition is turned off. Then the engine could diesel.

▶ Fuel/Air Requirements for Internal Combustion

LO 49-04 Describe the fuel–air requirements for internal combustion.

There are three main fuel and air requirements for internal combustion. They are:

- Fuel—Gasoline of the correct octane, and not degraded. Also must be at the correct ratio.
- Air—An oxygen-rich gas. It also must be at the correct ratio.
- Pressure differential—Used to get fuel and air into the combustion chamber.

▶ Fuel

Already discussed in detail previously in this chapter.

Air

Air is one of the essential components of the internal combustion engine. The components of air are a mixture of gases and small particles, which vary in composition. There are four primary components of the Earth's atmosphere: nitrogen, oxygen, argon, and carbon dioxide (**FIGURE 49-13**). Nitrogen, the largest component of air, is considered inert. This means that it does not burn under normal temperatures. So it typically just passes through the engine.

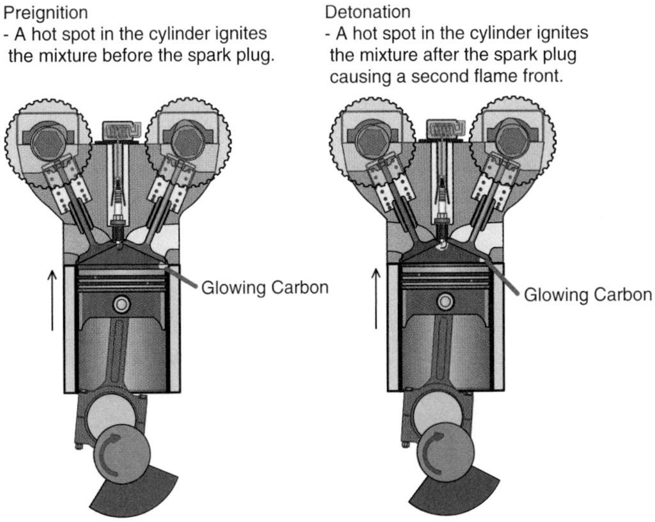

Preignition
- A hot spot in the cylinder ignites the mixture before the spark plug.
Detonation
- A hot spot in the cylinder ignites the mixture after the spark plug causing a second flame front.
Glowing Carbon
Glowing Carbon

FIGURE 49-12 Preignition occurs before normal ignition. Detonation occurs after normal ignition.

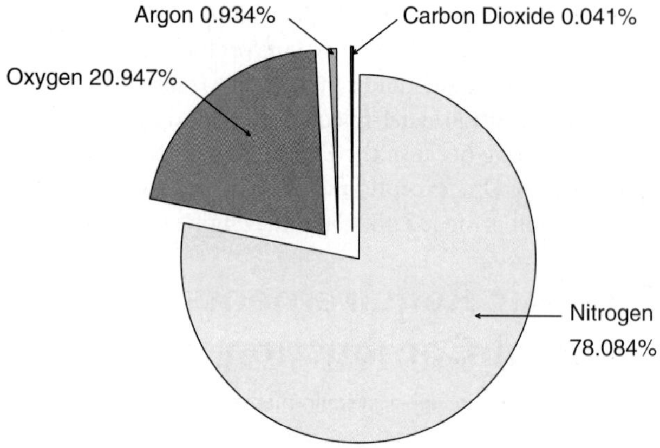

FIGURE 49-13 The four primary components of the Earth's atmosphere are nitrogen, oxygen, argon, and carbon dioxide.

Oxygen is a highly reactive gas that promotes combustion of all types of fuels. It is the burnable part of air that mixes with fuel to power the engine. Argon and carbon dioxide are trace elements in air. Neither burn so they also pass through the engine. These components are discussed further in the Emission Systems chapter.

Stoichiometric Ratio

The term **stoichiometric ratio** is represented by the Greek letter lambda (λ). It describes the chemically correct air–fuel ratio necessary to achieve complete combustion of the fuel and air. In other words, it is when all of the oxygen completely combines with the fuel, leaving no molecules of either remaining. All of the oxygen and fuel have been chemically combined through complete combustion.

For gasoline fuel, the stoichiometric air–fuel ratio is 14.7 parts air to 1 part fuel by mass, not volume. By volume, that equals 11,000 gallons (or liters) of air to every 1 gallon (or liter) of fuel—a ratio of 11,000:1. So if the air–fuel mixture is at the stoichiometric air–fuel ratio of 14.7 to 1, then the lambda value is 1 ($\lambda = 1$).

If an air–fuel mixture has a higher figure, say a lambda value of 1.05, there is more air in proportion to the fuel than 14.7 to 1. In fact, it is about 15.5:1, and the mixture is said to be slightly lean. A mixture with a lower lambda value, say 0.95, has proportionately less air than fuel. In this case, approximately 14.0:1, and the mixture is said to be slightly rich.

The exhaust gas oxygen sensor (often written as "O_2" sensor) is also called the lambda sensor. It is used to indicate the amount of oxygen in the exhaust. That way, the PCM can maintain a lambda reading that oscillates just above and below 1. The upstream oxygen sensor is usually installed in the exhaust manifold. A high percentage of oxygen may mean there isn't enough fuel to burn up all the oxygen. The mixture is too lean, and lambda is greater than 1.

The oxygen sensor delivers this information to the PCM. It adjusts the mixture to richen it up slightly. Similarly, a low percentage of oxygen in the exhaust may mean there isn't enough oxygen to burn all of the fuel, and lambda is less than 1. The PCM would then adjust the mixture to lean it out slightly.

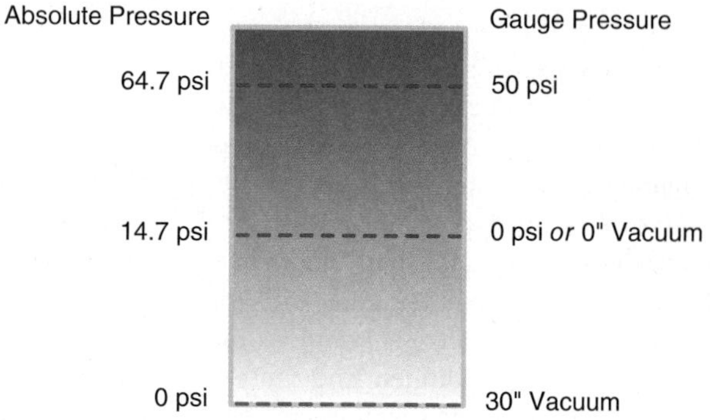

FIGURE 49-14 Absolute pressure versus gauge pressure.

Different fuels have a different stoichiometric ratio. For instance, methanol's air–fuel ratio is 6.4:1. Ethanol's air–fuel ratio is 9:1. Again, these are measured by mass, not volume.

The density of air is its mass per unit volume. Thus, a volume of air at high density has a higher mass than the same volume at low density. And if there is a larger mass of air, it will contain proportionally more oxygen. The density of air in the atmosphere changes at different temperatures and altitudes. That means the air that enters an engine at different locations could have very different amounts of oxygen. The amount of oxygen in air directly affects how well it supports combustion. So knowing the density of the air is very important in determining an air–fuel ratio for an engine.

Pressure and Vacuum

Pressure and negative pressure, or **vacuum**, are terms used daily in the automotive industry. Manufacturers recommend a pressure to which tires should be inflated. And manifold vacuum is used to operate a power brake booster. Gases exert pressure on all bodies they make contact with. This applies also to the air in the Earth's atmosphere. Air has mass, and as a result it exerts pressure, called atmospheric pressure, not only on the Earth's surface, but also on all objects on the Earth's surface.

Atmospheric pressure reduces with altitude. At sea level, it is calculated as 14.7 pounds per square inch (psi), or 101 kilopascals (kPa). But if this is so, why does a pressure gauge read zero when it is not in use? This is because the gauge indicates only pressure above atmospheric pressure. This reading is called gauge pressure (**FIGURE 49-14**). If you needed to know absolute pressure, an absolute pressure gauge would be required, and it would read 14.7 psi at sea level. Absolute pressure equals gauge pressure plus atmospheric pressure. Readings on a tire gauge are based on gauge pressure, not absolute pressure. They would need 14.7 psi added to them (at sea level) for atmospheric pressure. In automotive use, it is most common for pressures to be measured in gauge pressure.

▶ TECHNICIAN TIP

When an engine is supercharged, air is pumped at higher pressure into the intake manifold. Atmospheric pressure is 14.7 psi at sea level. So if a supercharger provides 7 psi of boost, it would increase the engine's horsepower about 50%. At 15 psi of boost, the engine's horsepower would be increased about 100%.

FIGURE 49-15 Engine vacuum. **A.** At idle. **B.** Partial throttle.

What if pressure below atmospheric pressure is being measured? A pressure gauge with its zero reading (gauge pressure) cannot measure those lower pressures. Therefore, pressure below atmospheric pressure is called "vacuum," and a vacuum gauge is normally used to measure it. An example of vacuum is intake manifold vacuum.

In most gasoline engines, the position of the throttle plate controls the volume of air entering the manifold. At **idle** speed, the pistons draw air away from the manifold at a faster rate than it can pass the throttle plate into the manifold. This creates a high vacuum, or low absolute pressure. At wide-open throttle, depending on load, the vacuum is much less. Pressure in the manifold rises closer to atmospheric pressure (**FIGURE 49-15**).

A vacuum gauge can be calibrated in inches of mercury in a scale reading from 0" to 30". Or it can use millimeters of mercury, on a scale from 0 to 760 mm. The scale is derived from the fact that atmospheric pressure at sea level supports a column of mercury approximately 30" (760 mm) high. This glass tube, closed at one end, is filled with mercury and then inverted in a bowl of mercury. The space above the mercury is a vacuum. Atmospheric pressure pushes on the exposed surface of the mercury and supports the column (**FIGURE 49-16**). At higher altitude, its height falls, which indicates that atmospheric pressure at that altitude is less. Thus, this scale can be used to indicate the amount of

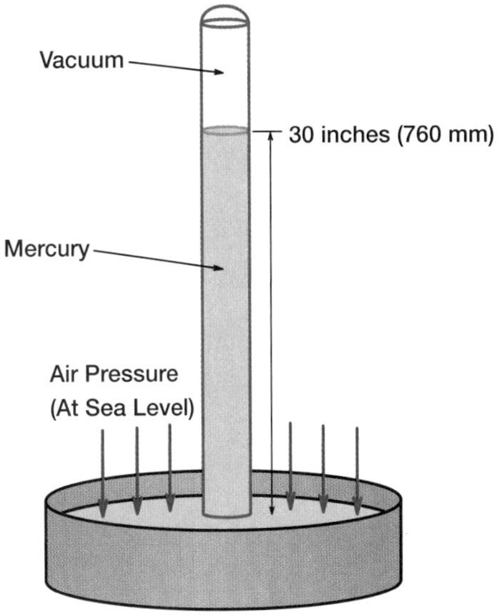

FIGURE 49-16 Atmospheric pressure at sea level supports a column of mercury approximately 30 inches (760 mm) high.

vacuum that exists below atmospheric pressure. At idle speeds, a vacuum gauge connected to the intake manifold will indicate a reading of approximately 15" to 21" (308–530 mm) of mercury, depending on the altitude.

Above sea level, air pressure is reduced. This means that it contains less oxygen. To maintain the correct air–fuel ratio, the amount of fuel delivered to the engine must be reduced to match the amount of oxygen present. Some engine management systems measure changes in atmospheric pressure with a barometric pressure sensor. This information is used along with the oxygen sensor information to determine the proper amount of fuel to inject.

▶ Fuel Delivery System Components

LO 49-05 Describe the fuel tank components.

The fuel supply system consists of several components that all play key roles in delivering gasoline to the fuel injection system. We will look much more in depth at each of the following:

- Fuel tank—stores fuel for delivery to the engine as needed.
- Fuel filler neck—allows fuel to be added to the fuel tank.
- Gas cap—prevents leaks into and out of the filler neck.
- Evaporative emission control system (EVAP)—temporarily stores fuel vapors until they can be burned in the engine.
- Fuel pump relay—sends power to operate the fuel pump on some vehicles.
- Fuel pump—pumps fuel at the proper pressure and volume to the injectors.
- Fuel tank sending unit—provides a method of signaling the level of fuel in the fuel tank to the fuel gauge on the instrument panel.
- Fuel filter—filters the fuel going to the fuel injectors to strain out particles of dirt and debris.

- Fuel lines—provide a route for fuel to flow from the tank to the injectors. Also typically provides a location for a fuel filter.
- Fuel rail—a hollow tube that delivers fuel to the injectors. May have a pressure tap and pressure regulator mounted on it.
- Fuel pressure regulator—the device that maintains the proper pressure in the delivery system. Can be mechanical or electronic.
- Fuel injector(s)—electromechanical solenoid valves that control the flow of fuel into the intake manifold or combustion chamber.

Fuel Tank

The fuel tank (or gas tank, as it is sometimes called) is the primary reservoir of the onboard fuel supply. The typical fuel tank of modern vehicles consists of a gas cap, filler neck, fuel, fuel pump, and **fuel gauge sending unit** (**FIGURE 49-17**). Its primary function is to safely hold an adequate supply of gasoline for prolonged engine operation.

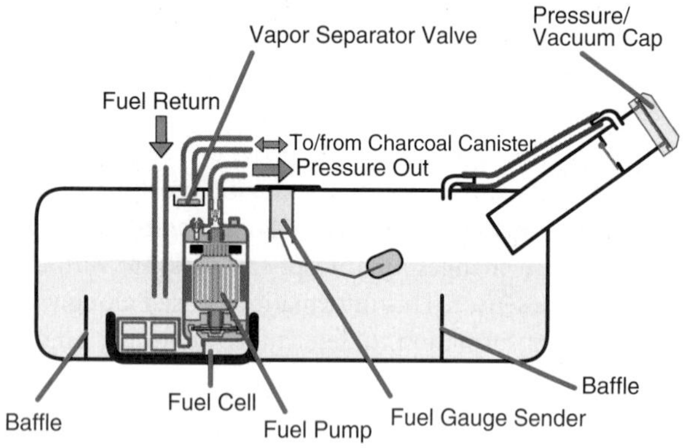

FIGURE 49-17 The typical fuel tank of modern vehicles consists of a gas cap, filler neck, fuel, fuel pump, and gauge sending unit.

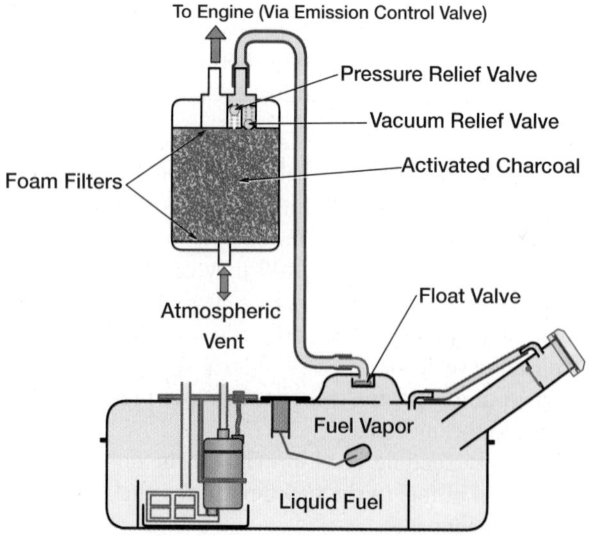

FIGURE 49-18 The fuel tank is vented through the evaporative emissions system.

Where the tank is mounted depends on space and styling. Safety demands that it be positioned well away from heated components. It must also be located outside the passenger compartment. Tanks are made either of tinned sheet steel that has been pressed into shape or of nonmetallic materials. Aluminum or steel is used on commercial vehicles. The metal tank is usually in two parts, joined by a continuous weld around the flanges where the parts fit together. Baffles make the tanks more rigid. They also prevent the sloshing of fuel and ensure that fuel is available at the pickup tube.

Because fuel expands and contracts as temperature rises and falls, fuel tanks are vented to let them breathe. Modern emission controls prevent tanks from being vented directly to the atmosphere. They use evaporative control systems. Vapor from the fuel tank is vented through a charcoal canister. Fuel vapors are then stored until they are burned in the engine (**FIGURE 49-18**). A vapor or vent line with a check valve connects the space above the liquid fuel with the canister. This valve opens above a specified pressure and lets through vapor, but not liquid.

Liquid fuel closes the check valve and blocks the line, which stops it from reaching the charcoal. Some systems have a small container, called a liquid–vapor separator, above the fuel tank. It also prevents liquid fuel from reaching the charcoal canister.

Fuel Filler Neck

The fuel filler is where fuel enters the tank. The **fuel filler neck** is a pipe that extends above the fuel tank (**FIGURE 49-19**). On vehicles with catalytic converters, the filler neck is designed to prevent leaded fuel being added. Its diameter is smaller than those on older leaded fuel vehicles. A trapdoor inside the filler is opened only by the nozzle of the unleaded gasoline spout.

The location of the filler neck depends on the design of the vehicle and location of the tank. The filler neck can incorporate the use of a blowback ball valve. It helps prevent fuel from blowing back out the filler neck during fill-ups. It also deters gas theft.

FIGURE 49-19 The fuel filler neck.

SAFETY TIP

Never use a hose and your mouth to siphon gasoline from a vehicle. It is easy to swallow or inhale the gasoline, which is poisonous. If gas must be removed from a vehicle, always use a gas caddy with a suction pump.

Gas Cap

Modern vehicles are required by the Environmental Protection Agency (EPA) to have a non-vented gas cap. This non-vented cap prevents fuel vapors from being directly vented to the atmosphere. Vented caps waste gasoline and contribute to air pollution. To help ensure that the cap is tightened properly, many of these caps use a ratchet system. Drivers tighten the cap until it ratchets (**FIGURE 49-20**).

Newer Ford vehicles are using an Easy Fuel capless system (**FIGURE 49-21**). There is no gas cap to remove when refueling the vehicle. When the properly sized nozzle is placed on the top of the cap, latches release. They open the valve and allow the nozzle to be fully inserted. This cap also reduces the amount of gasoline vapors that escape to the atmosphere when compared with a standard cap.

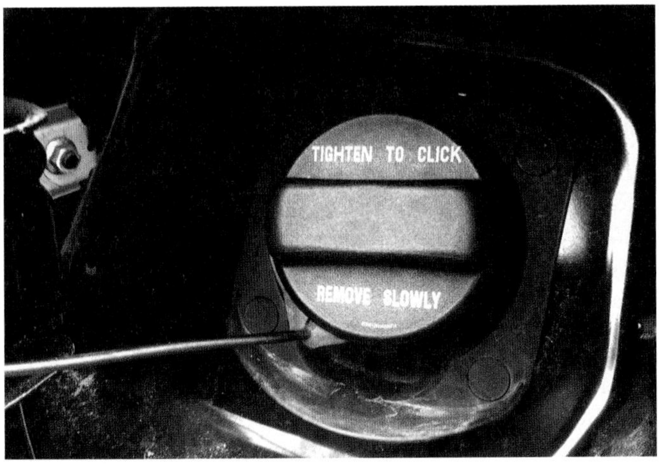

FIGURE 49-20 Many gas caps need to be tightened until they click.

FIGURE 49-21 New capless system.

▶ TECHNICIAN TIP

A loose or leaky gas cap can cause the MIL to illuminate on the dash. This occurs when the evaporative emission system monitor runs and the leak is detected. Always check the tightness of the cap if an evaporative emission system leak code is present. If the cap is tight, you may want to test the cap for leaks. Do so on a cap tester before digging into the evaporative emission system to locate the leak.

▶ Fuel Pump

LO 49-06 Describe the fuel pump components.

Most fuel-injected vehicles use one or two electric pumps to supply the fuel system with pressurized fuel. The pump is typically located inside the fuel tank. On some vehicles, it is mounted on the frame. If located in the tank, the pump is the submersible type. Electric fuel pumps can be the low-pressure type or the high-pressure type. It depends on the fuel system used on the vehicle.

The fuel pump is electrically operated and electronically controlled. It is driven by a permanent-magnet electric motor. It is a sealed unit integral with the pump (**FIGURE 49-22**). Fuel flows through the pump and around the electric motor when it is running. There is never an ignitable mixture (of air and fuel) inside the pump housing. There is only fuel, so there is virtually no danger of explosion. The pump is designed to deliver more fuel than the maximum requirement of the engine. Pressure is controlled in one of two ways: Either by a fuel pressure regulator that maintains fuel pressure between the pump and the injectors, or by controlling the speed of the pump to maintain the correct pressure.

For a short time after an engine is switched off, the engine temperature keeps rising. This can vaporize fuel in the fuel lines. A pressure check valve in the pump maintains the fuel pressure in the system after engine shutdown. It prevents the fuel from boiling (vaporizing) as it absorbs heat from the engine (heat soak). This pressure diminishes after about 20 to 30 minutes. But it ensures the effective hot-starting capability of the engine.

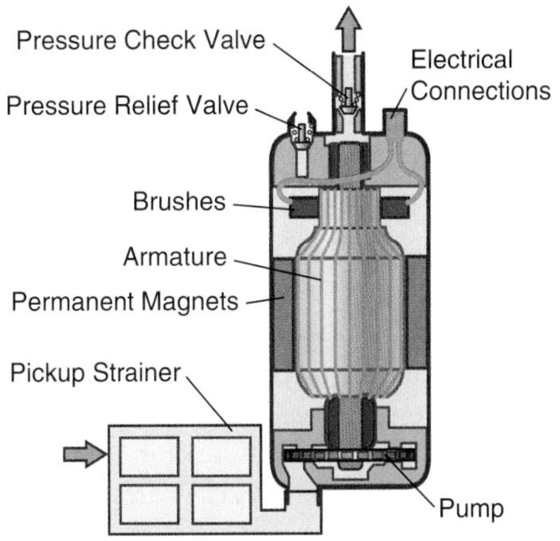

FIGURE 49-22 The fuel pump is driven by a permanent-magnet electric motor, a sealed unit integral with the pump.

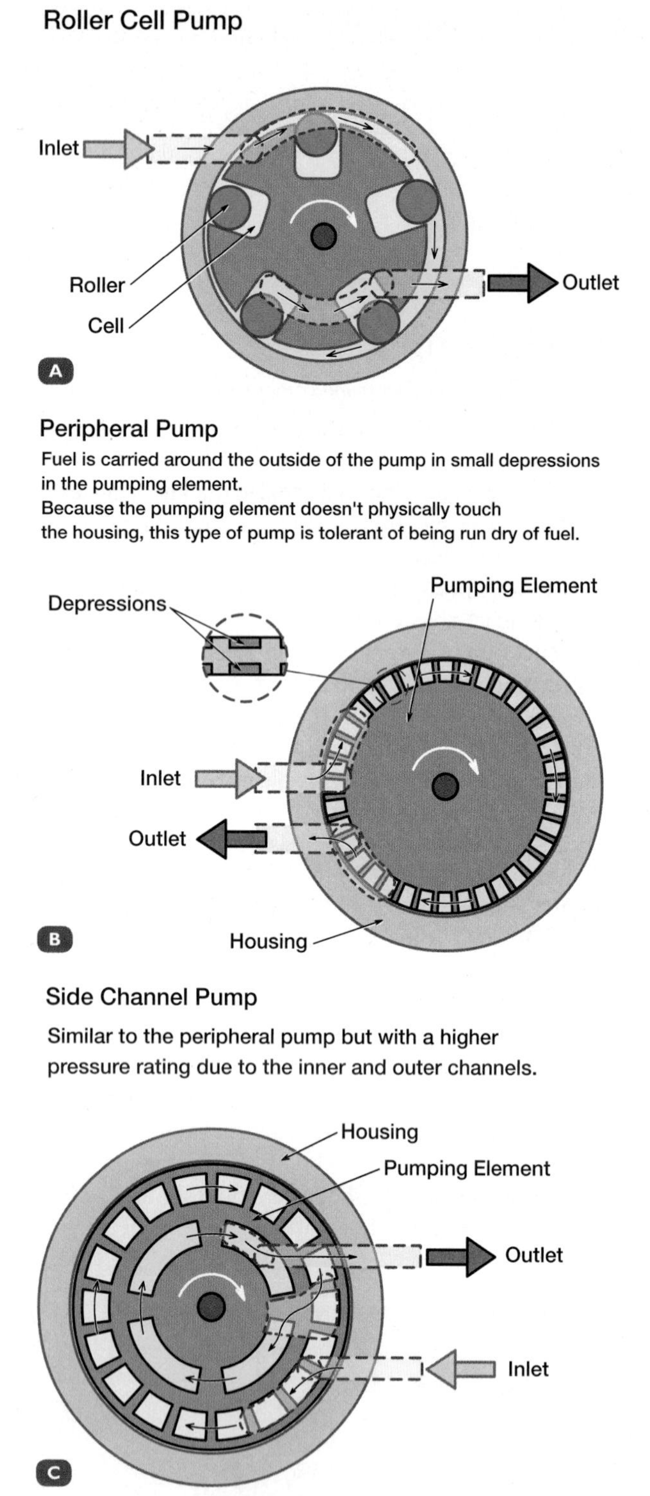
Roller Cell Pump
Inlet
Roller
Cell
Outlet
A
Peripheral Pump
Fuel is carried around the outside of the pump in small depressions in the pumping element.
Because the pumping element doesn't physically touch the housing, this type of pump is tolerant of being run dry of fuel.
Depressions
Pumping Element
Inlet
Outlet
B
Housing
Side Channel Pump
Similar to the peripheral pump but with a higher pressure rating due to the inner and outer channels.
Housing
Pumping Element
Outlet
Inlet
C

FIGURE 49-23 Types of pumps used in fuel pump assemblies.

Electric fuel pumps use various types of pump chambers. One of which is the roller cell (**FIGURE 49-23**). In this style, rollers float in channels in an offset rotor. As the rotor turns, the rollers are thrown outward. Fuel is drawn in as the space between the rollers expands. Fuel is then forced out as the space decreases. Other types are the peripheral type and the side channel pump.

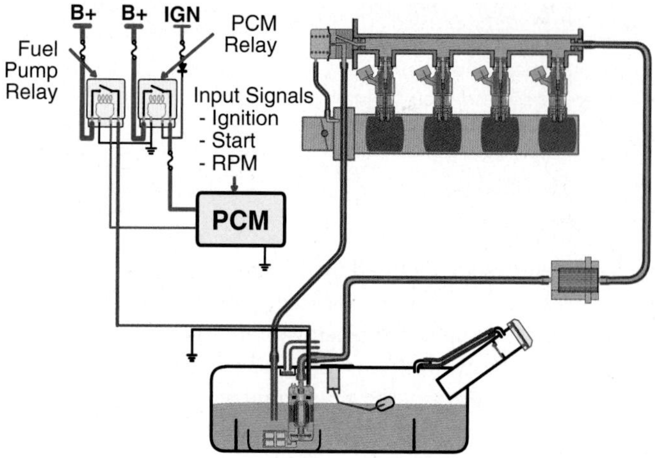
B+
B+
IGN
PCM Relay
Fuel Pump Relay
Input Signals
- Ignition
- Start
- RPM
PCM

FIGURE 49-24 Fuel pump circuit.

▶ Fuel Pump Relay

Fuel pumps can draw a considerable amount of current while they run. To carry that high current load, modern vehicles use a **fuel pump relay**. The relay is an electromagnetically operated switch. It activates the fuel pump when the:

- ignition is turned on for priming the fuel system,
- engine is cranked, and
- engine is running.

As low-amperage current is passed through the winding in the relay, a magnetic field is created. It pulls the contacts together, controlling the larger current flow going to the fuel pump (**FIGURE 49-24**). The current operating the fuel pump relay winding is typically controlled by the PCM. In most cases, the PCM grounds the winding when it wants the fuel pump to operate.

▶ Fuel Tank Sending Unit

For obvious reasons, it is important for the driver to know how much fuel the vehicle has in the tank. The primary job of the **sending unit** is to convert the level of fuel into an electrical signal. The signal is sent to either the gas gauge located in the instrument panel, or to the BCM, which then controls the gas gauge.

The sending unit is a variable resistor that is attached to a float mechanism (**FIGURE 49-25**). As the fuel level changes, the float rises and falls. As it does, the resistance changes, which in turn, changes the amount of current sent to the fuel gauge. This current is calibrated to directly indicate an accurate amount of fuel in the tank.

On today's EFI vehicles, the sending unit is housed with the fuel pump. In many cases, the assembly also includes a replaceable or nonreplaceable fuel filter. The entire assembly has three jobs (**FIGURE 49-26**):

- pick up fuel from the bottom of the tank through the strainer,
- pressurize the fuel by way of the fuel pump,
- filter the fuel, and
- create a fuel level signal to send to the fuel gauge.

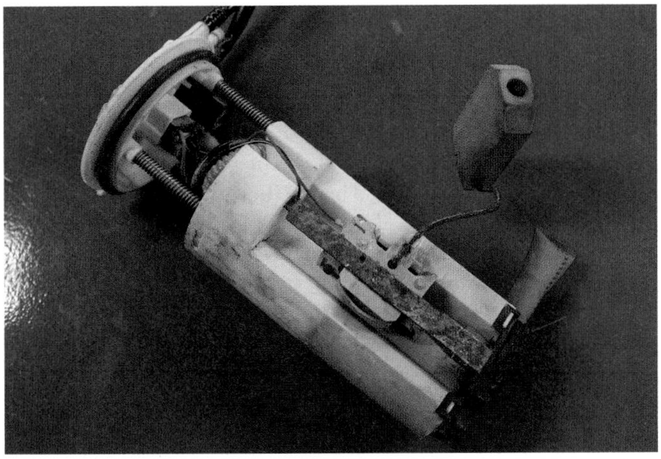

FIGURE 49-25 The sending unit is a variable resistor that is attached to a float mechanism.

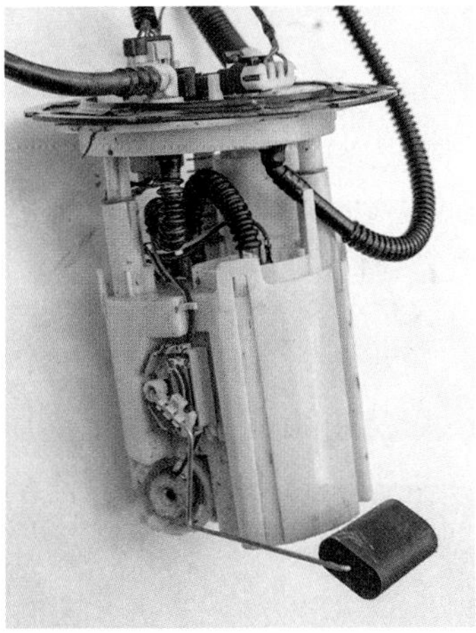

FIGURE 49-26 Fuel pump assembly with strainer, pump, filter, and sending unit.

FIGURE 49-27 A typical fuel sock installed on the bottom of a fuel pump assembly.

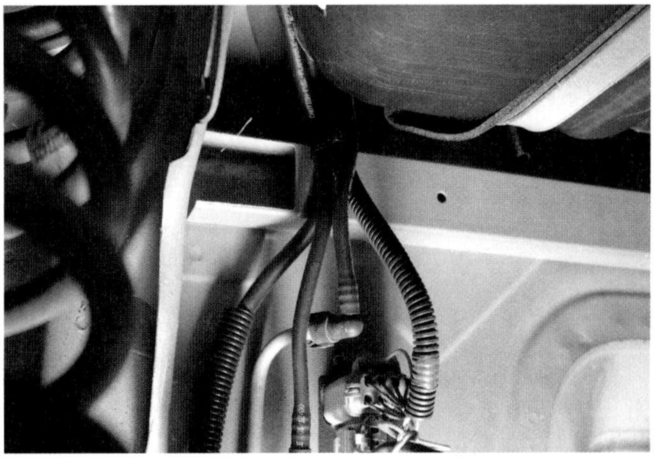

FIGURE 49-28 Fuel and evaporative emission lines.

These components are meant to last at least the life of the warranty.

Filter Sock

Fuel contamination can be extremely detrimental to the fuel supply system as well as the engine itself. To this end, a **filter sock** is the first line of defense. It is incorporated into the end of the fuel pickup tube. The sock typically consists of a fine mesh. It prevents most small particles from being drawn into the fuel pump and sent through the rest of the fuel system (**FIGURE 49-27**). It typically has to be replaced when the fuel pump is replaced. Otherwise, the manufacturer won't warranty the fuel pump. They can also become plugged up due to varnish, water, or dirty fuel.

▶ Fuel Lines

LO 49-07 Describe fuel lines, fuel filters, and fuel rails.

Fuel lines are usually made of metal tubing or synthetic materials. A fuel supply line carries fuel from the tank to the engine. A return line may also be provided to allow excess fuel to return to the tank. Returning the excess fuel helps prevent the formation of vapor that can occur in the fuel supply line during hot conditions.

Other lines that run along with the fuel lines include evaporative emission system lines. They connect the fuel tank, charcoal canister, and intake manifold (**FIGURE 49-28**). These lines can be made of steel or plastic. Steel lines can rust out in parts of the country that use salt on the roads. Both types of lines can be damaged by improperly lifting or supporting a vehicle with a jack, jack stand, or hoist.

▶ Fuel Filter

The **fuel filter** typically consists of a pleated paper filter housed in a sealed container. Its primary function is to prevent contaminants from reaching the injectors. The paper filter can also prevent small amounts of water from getting past it. This is because the paper absorbs the water, causing the fibers in the filter to swell. It is then harder for liquid to get through it.

Most fuel filters are considered maintenance items. This means that they need to be replaced at specified intervals. A fuel filter that restricts fuel flow due to being plugged up puts extra wear on the fuel pump. So don't neglect replacing the filter when specified.

Most fuel filters are directional, meaning they must be installed in the proper direction. Usually an arrow on the filter indicates the direction of flow. Fuel filters can typically be found in two places. One is on the frame rail or near the tank (**FIGURE 49-29**). The other is in the tank as part of the fuel pump assembly (**FIGURE 49-30**).

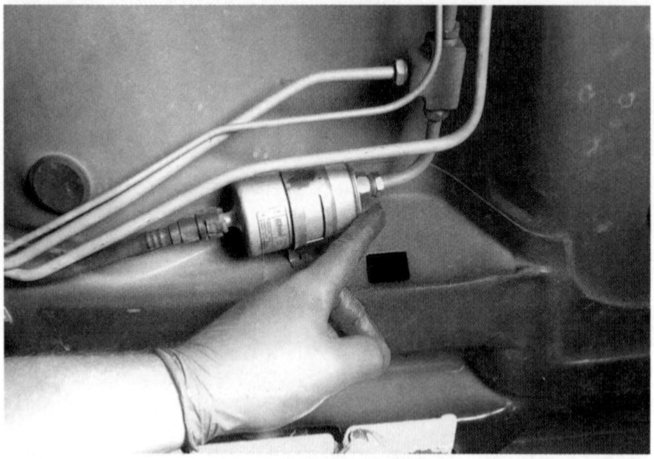

FIGURE 49-29 Typical inline fuel filter.

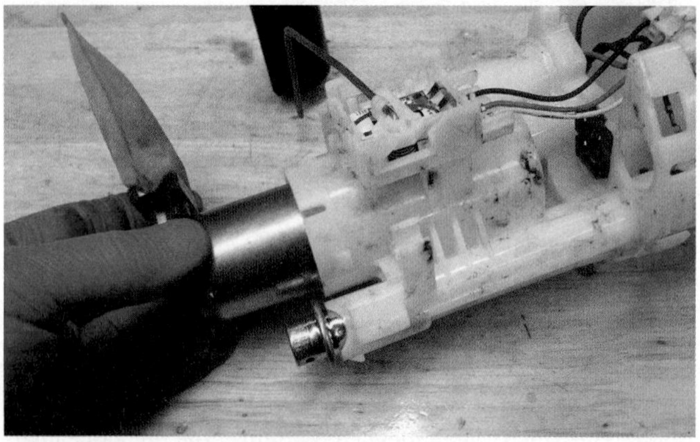

FIGURE 49-30 In-tank fuel filter.

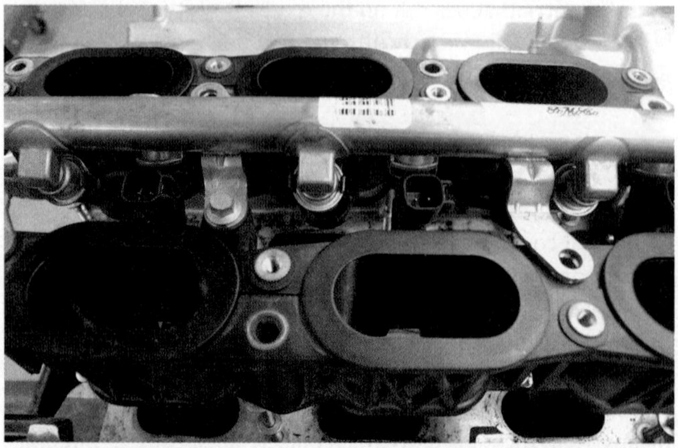

FIGURE 49-31 Typical fuel rail that sits on top of the fuel injectors.

▶ Fuel Rail

A **fuel rail** is a special manifold designed to provide a reservoir of pressurized fuel for the fuel injectors. The fuel rail typically sits on top of the fuel injectors (**FIGURE 49-31**). The gasoline is fed from the fuel tank by way of the fuel lines. The fuel, under pressure of course, is supplied by the fuel pump. The fuel rail can also serve as a mounting place for the fuel pressure regulator.

The couplings where the fuel lines meet the fuel rail are usually sealed by:

- flared ends,
- quick-connect, high-pressure O-rings, or
- banjo fittings (**FIGURE 49-32**).

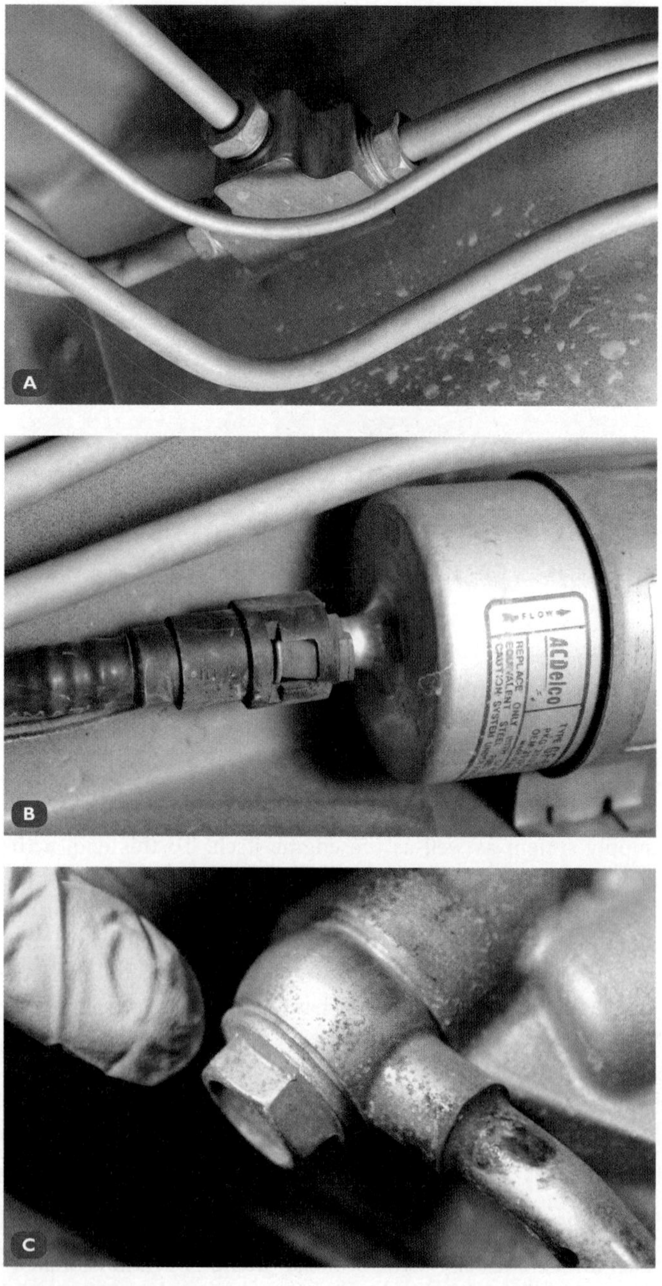

FIGURE 49-32 Types of fuel line connections. **A.** Flare. **B.** Quick couplings. **C.** Banjo fitting.

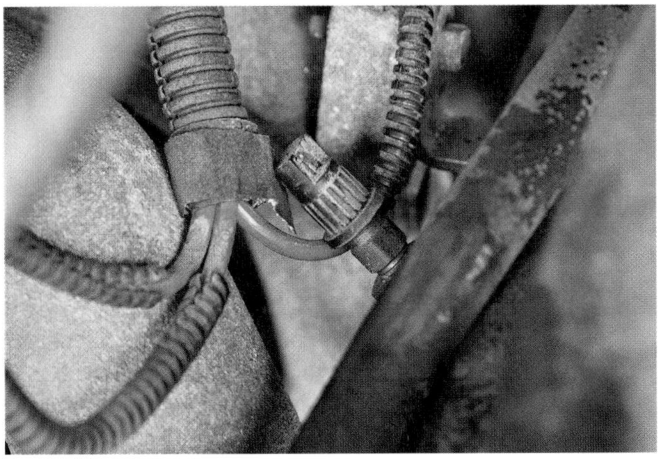

FIGURE 49-33 Fuel pressure test port.

These methods ensure a sealed, leak-proof joint between the parts. Many, but not all, fuel rails provide a means for connecting a fuel pressure tester to the rail. This design is called a Schrader valve (**FIGURE 49-33**). It is a one-way valve that holds pressure in the system while the pressure tester is connected and disconnected to the fuel rail. Schrader valves usually are covered by a screw-on cap with a seal inside. This prevents leaks in the event the Schrader valve goes bad.

▶ Fuel Pressure Regulation

LO 49-08 Describe how fuel pressure is regulated.

Fuel pressure must be regulated in some manner. On port fuel-injected engines, the fuel pressure has to be maintained at a certain amount above manifold pressure. Because manifold pressure changes with engine load, fuel pressure also has to change with it. Fuel pressure is managed either by a mechanical pressure regulator, or by controlling the speed of the fuel pump. If the vehicle uses a mechanical pressure regulator, it is commonly mounted on the fuel rail. Some vehicles have it located in the fuel tank.

The EFI pressure regulation system can be either a circulation type or a returnless type. In a circulation system:

- fuel is drawn from the tank by a fuel pump,
- fuel is delivered to solenoid-operated injection valves, called **injectors**,
- fuel pressure at the injector(s) is maintained by a fuel pressure regulator, and
- excess fuel flows back to the tank through a return line.

In most returnless systems, fuel is delivered to the fuel injectors, the same as that in circulation types. But the pressure is controlled by the speed of the pump. Also, the pump only delivers the amount of fuel needed by the injectors. This means that there is no need for a return line back to the tank.

A **fuel pressure regulator** mounted on the fuel rail incorporates a diaphragm-operated valve (**FIGURE 49-34**). The fuel pressure regulator inlet is connected to the fuel rail, and an outlet lets fuel return to the tank. Between them is a diaphragm-operated valve and pressure spring. The position of the valve determines how much fuel is returned to the tank. Movement of the diaphragm opens and closes the valve.

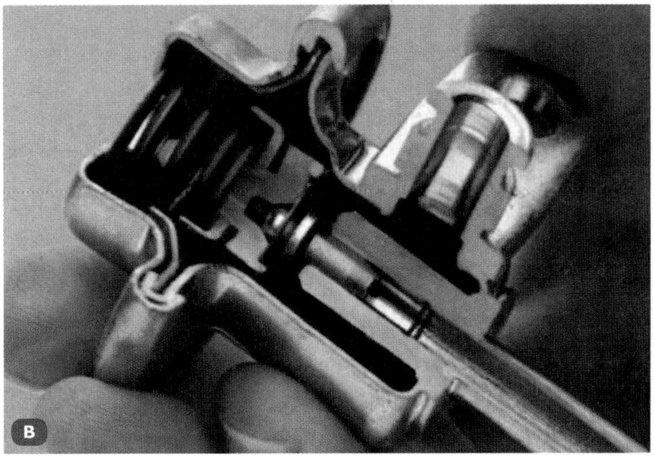

FIGURE 49-34 A fuel pressure regulator mounted on the fuel rail incorporates a diaphragm-operated valve.

Fuel pressure builds against the diaphragm. Once it hits the preset pressure, the valve opens slightly and bleeds off pressure. This action reduces the pressure on the diaphragm which closes the valve. With the valve closed, the pressure rises again until the valve opens once again. Thus, the strength of the pressure spring determines fuel pressure in the fuel rail. In this way, it keeps it at a fixed value.

The pressure in the intake manifold varies a lot with changes in engine speed and load. So, for any injection duration, if fuel is held at constant pressure, then as manifold pressure varies, so does the amount of fuel delivered. That means fuel pressure must be continuously adjusted to maintain a constant pressure drop across the injector. Remember that the pressure drop we mean is between the fuel pressure and manifold vacuum.

A constant pressure drop across the injector is maintained by exposing the spring side of the diaphragm to intake manifold vacuum. The spring-loaded vacuum chamber is connected by a manifold vacuum line to the intake manifold. The vacuum pulls against spring pressure to modify the pressure at which the valve opens. This changes the pressure at which fuel is allowed to return to the fuel tank.

As the throttle is opened, the manifold vacuum reduces. That, in turn, allows more spring pressure to push on the diaphragm. It then requires more fuel pressure to open the valve, providing higher fuel pressure in the fuel rail (**FIGURE 49-35**). This maintains a consistent pressure drop across the fuel injector as the throttle is operated by the driver.

When manifold vacuum is high (low absolute pressure), as it is at idling, fuel pressure is low. As manifold vacuum decreases (high absolute pressure), toward open throttle, fuel pressure rises. Because the injectors are all subjected to the same pressure drop, they all inject an equal amount of fuel for a given "on" time (pulse width). The quantity of fuel delivered is thus controlled accurately by the pulse width of the injector.

Note that manifold pressure is not needed to assist the spring against the diaphragm in TBI systems. This is because injection occurs above the throttle plate, at atmospheric pressure. Thus, fuel pressure is determined solely by the force of the regulator spring acting on the diaphragm. When the engine is running, the circulation of fuel ensures that cool fuel is delivered at all times. This helps prevent the formation of vapor in the lines.

The pump control circuit controls the pump. It operates the fuel pump for only a few seconds when the ignition is switched on, but the engine is not running or cranking. This is used for priming the system. The pump turns off after that if it doesn't see the engine running above a certain rpm. The pump control circuit allows the pump to operate during cranking and when the engine is running above the specified minimum rpm.

Newer vehicles have been using **returnless fuel injection systems** for the past 15 years or so. Manufacturers turned to using returnless fuel systems to reduce evaporative emissions. On a return-type system, the fuel returning to the tank is hot from engine heat. Hot fuel will readily vaporize. This makes the evaporative system work harder, requiring additional fuel vapor storage capacity. Because no hot fuel is returned to the tank in a returnless system, the fuel in the tank stays relatively cool. This minimizes vaporization.

There are two types of returnless systems. One uses a mechanical pressure regulator in the fuel tank. The in-tank regulator uses a spring-loaded pressure regulator similar to the type mounted to the fuel rail. But it does not change with manifold pressure and therefore does not use a vacuum hose. Excess fuel pressure is simply vented into the tank (**FIGURE 49-36**).

The second type of system controls the speed of the fuel pump electronically to modify fuel pressure. The PCM sends a square wave (digital) signal to control the speed of the fuel pump (**FIGURE 49-37**). The PCM monitors the fuel pressure through a fuel pressure sensor mounted on the fuel rail. The PCM controls the fuel pump based on engine speed, load, and other factors. The faster it turns, the higher the pressure and flow. The slower it turns, the lower the pressure and flow.

▶ Fuel Injectors

LO 49-09 Describe fuel injectors and their operation.

Most modern fuel injectors are simply spring-loaded, electric-solenoid spray nozzles (**FIGURE 49-38**). They incorporate a filter screen in the inlet. They are usually sealed with O-rings to the fuel rail and intake manifold. The nozzle can be one of several types, such as rotating disc style, pintle style, and ball-valve style. Each type has its own pros and cons.

The job of a fuel injector is to spray the proper amount of gasoline in the proper pattern. This occurs either in the intake ports, directly into the combustion chamber, or into a pre-chamber in the combustion chamber. The fuel injectors are positioned between the fuel rail and the intake manifold. Or in the case of a GDI fuel system, between the fuel rail and the combustion chamber. They spray in response to signals from the PCM (**FIGURE 49-39**).

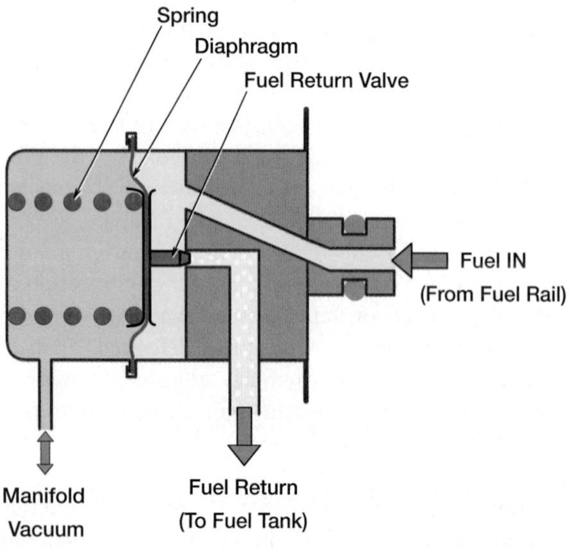

FIGURE 49-35 Typical fuel pressure regulator that uses intake manifold vacuum to maintain a constant pressure drop across the fuel injector.

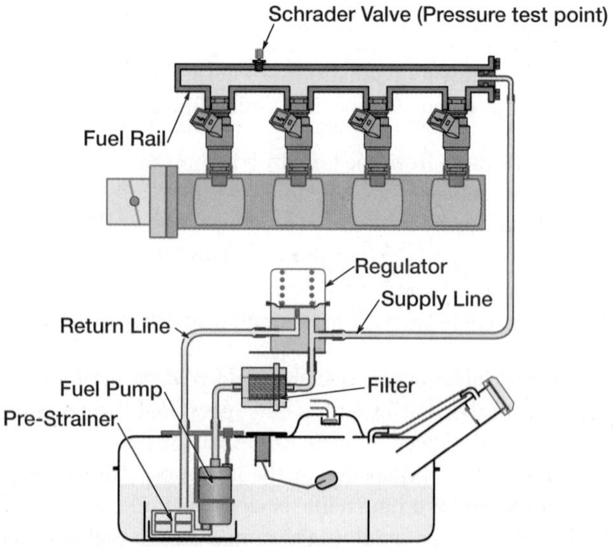

FIGURE 49-36 A mechanical returnless fuel system. Note that the pressure regulator can be either inside or outside the tank.

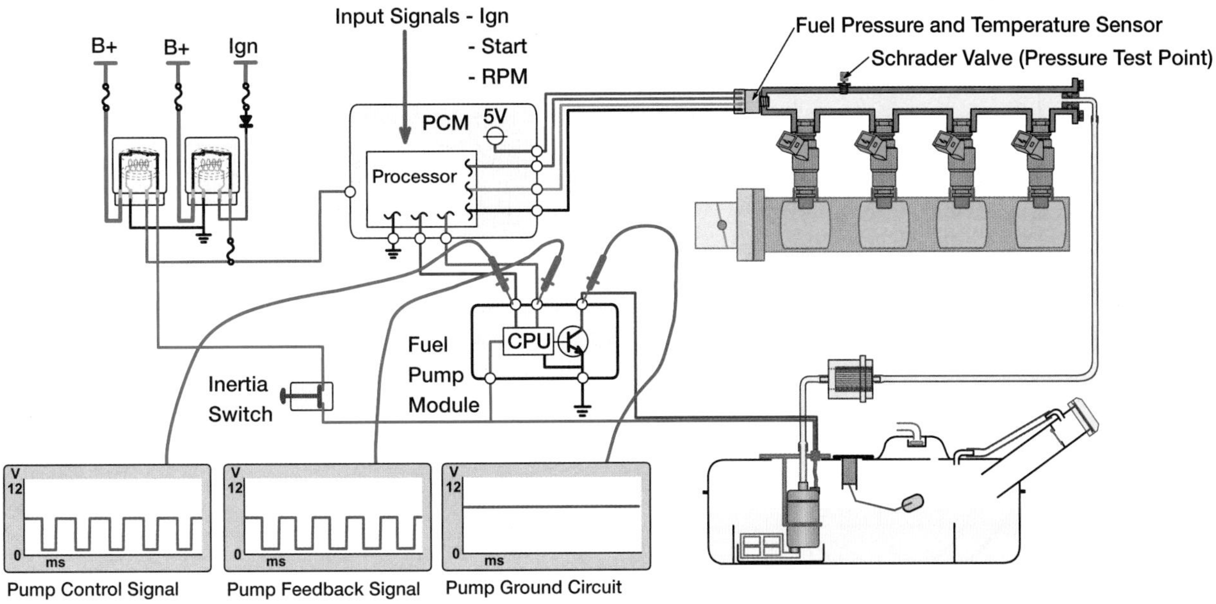

FIGURE 49-37 Electronically controlled fuel pressure.

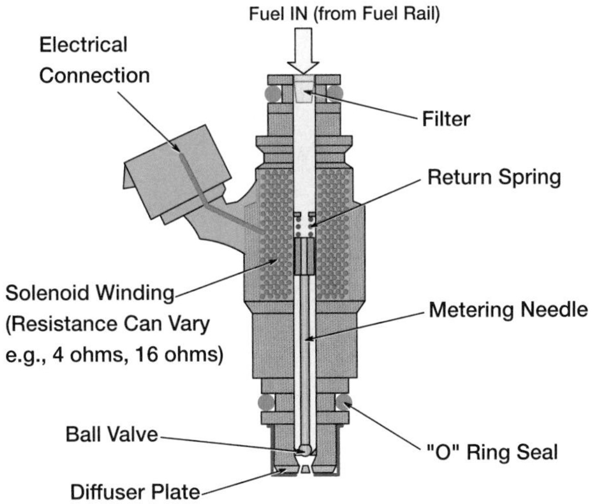

FIGURE 49-38 The inside components of a port fuel injector.

FIGURE 49-39 The modern fuel injector sprays a fine mist of fuel when activated.

As you know, fuel is pressurized and waiting in the fuel rail. When the PCM grounds the electrical circuit from the fuel injector, current flows through the windings of the injector (**FIGURE 49-40**). A magnetic field is created which opens the pintle valve. When it is time to close the pintle valve, the PCM opens the circuit to the fuel injector which stops the current flow. An internal spring closes the pintle valve, and it waits for the next "on" command. This happens repeatedly hundreds of times per minute at each fuel injector.

The fuel injectors are designed and built to very exacting tolerances. This gives them a high degree of fuel delivery precision and a long service life. The response time to lift the injector needle to the fully open position is about 1 millisecond. If battery voltage is low, this response time takes longer, and the cylinders receive less fuel. As the voltage falls, the PCM can compensate for this slower opening time by extending the pulse width of the injector(s).

The injectors are sealed into the manifold by O-rings that prevent air entering at that point. O-rings are also commonly used to seal the injector to the fuel rail. The O-rings, together with plastic caps on the injector nozzles, also act as a barrier to heat being transferred to the injector body (**FIGURE 49-41**).

▶ Types of EFI Systems

LO 49-10 Describe Throttle Body and Multi-point Fuel Injection.

There are three basic EFI systems:

- TBI, also called single-point injection,
- MPFI, and
- GDI, or simply direct injection (**FIGURE 49-42**).

The specific characteristics of each of these designs will be explained further. But they all operate on similar principles. Fuel is supplied to the injectors at the specified pressure. The PCM

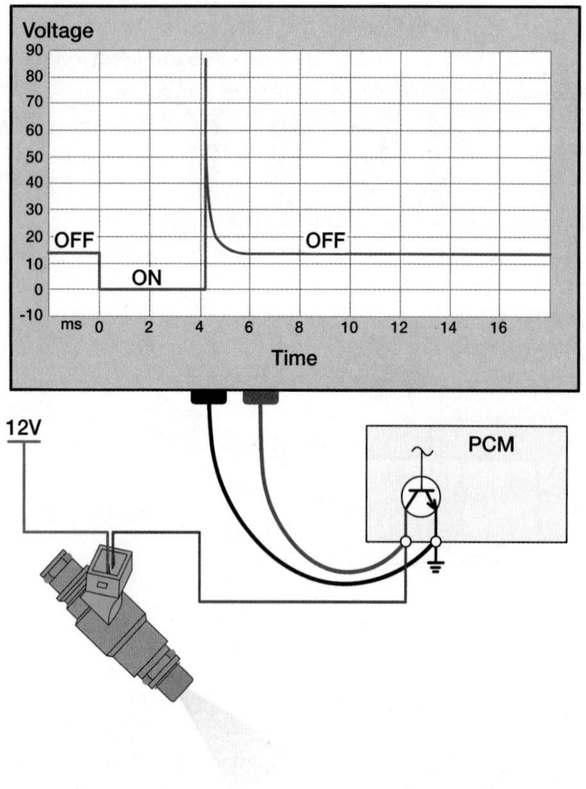

FIGURE 49-40 Most fuel injectors are controlled on the ground side. So the PCM grounds the injector to turn it on.

FIGURE 49-41 Typical O-rings and plastic cap act as heat barriers on a fuel injector.

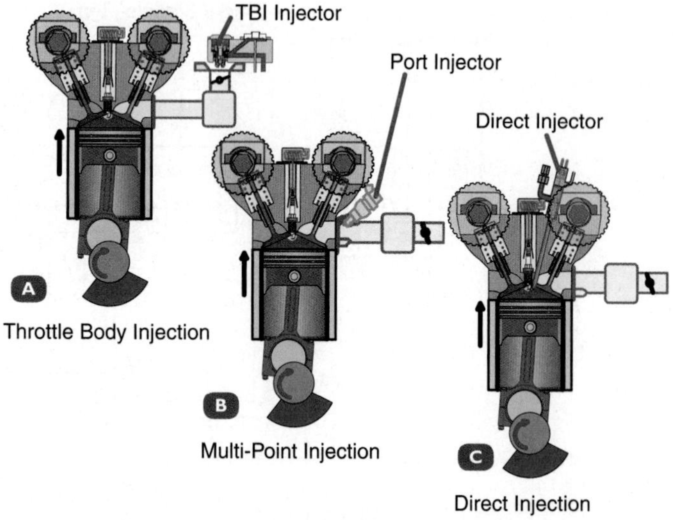

FIGURE 49-42 The three basic systems. **A.** Throttle body injection. **B.** Multipoint fuel injection. **C.** Direct injection.

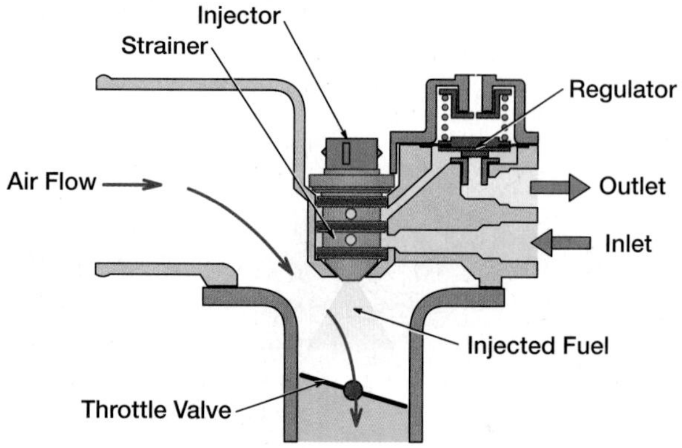

FIGURE 49-43 Typical TBI system.

sends an electric signal to each injector to cause it to open for a certain amount of time (pulse width). And fuel is injected into the intake manifold or combustion chamber.

Throttle Body Injection Systems

TBI is a system with one or two fuel injectors located centrally on the intake manifold, right above the throttle plates (**FIGURE 49-43**). It is also known as single-point injection or central-point injection. Fuel is sprayed into the top center of the throttle body and then mixed with the incoming filtered air. This air–fuel mixture is then delivered to each of the engine's cylinders relatively evenly. TBI is the simplest type of EFI system. It requires only one or two injectors. This allows the PCM to be a simpler, less powerful design. And because the fuel is sprayed above the throttle plates, it is at atmospheric pressure. So the pressure drop across the injector is always the same. Thus, fuel pressure does not have to change with throttle opening or engine load.

In TBI systems, the central injector is normally triggered on every ignition pulse. However, if there are two injectors, alternate triggering may be used. At idling speeds, the frequency may be less to provide finer control. TBI is the predecessor to modern-day multipoint (or multiport) fuel injection.

In TBI systems, fuel pressure and flow are provided by an electric fuel pump. It is typically located on the frame rail but can also be located in the fuel tank. Most TBI systems operate on fairly low fuel pressure, typically between about 15 and 20 psi (103 and 138 kPa).

Multipoint Fuel Injection Systems

In MPFI systems, a fuel injector is used for each cylinder. One injector is located in the intake manifold near each intake valve. It sprays fuel toward the valve. Each injector is connected to the fuel rail, which supplies the injector with fuel under pressure (**FIGURE 49-44**). Each injector also has an electrical connector that provides it with power and ground. Most electronic fuel injectors are supplied with constant battery voltage when the ignition key is in the Run or Crank position. The PCM then switches the ground side of the injector circuit on and off. When the PCM switches the circuit on, the fuel injector opens and sprays fuel. When the PCM switches the circuit off, the spring inside the injector closes the injector, and fuel stops spraying.

In MPFI systems, fuel pressure and flow are provided by an electric fuel pump. It is most likely located in the fuel tank. The system typically operates on higher fuel pressure than TBI, of approximately 35–70 psi (241–482 kPa).

Simultaneous Fuel Injection

In MPFI, the injectors can be triggered in various ways. The simplest way is to trigger them at the same time or in groups. This is called simultaneous injection, grouped injection, or banked injection. This arrangement operates the injectors twice per cycle—once each crankshaft revolution. Each time, half the fuel is delivered for the cycle. In a six-cylinder engine, groups of three injectors are triggered every third ignition pulse (**FIGURE 49-45**). Grouping the injectors allows for a less powerful processor in the PCM and fewer drivers inside. It also means fewer terminals and wires coming out from the PCM. This is because there is only one control wire for each group of injectors.

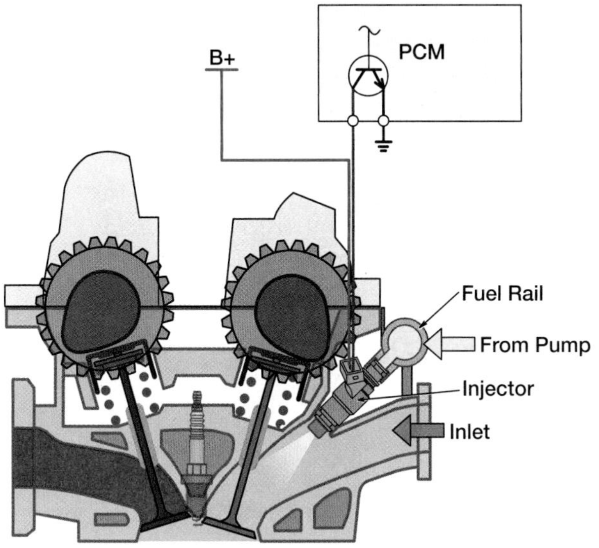

FIGURE 49-44 Typical MPFI system.

Sequential Fuel Injection

Sequential injection means injection occurs in the sequence of the firing order. The injectors spray fuel following the firing order of the engine. Each injector typically opens only once in each cycle to deliver the fuel needed. Because each injector is controlled separately, the PCM has to have individual drivers to turn each injector on and off. This requires more computing power to manage each injector circuit. It also requires more wires back to the PCM, as each injector requires its own circuit (**FIGURE 49-46**). Because each injector fires only once per cycle, the timing of the injection pulse is important. Therefore, the position of both the number-one cylinder and the camshaft must be known. This also adds to the complexity of the system over a TBI or simultaneous-injection system.

▶ Gasoline Direct Injection Systems

LO 49-11 Describe the operation of Gas Direct Injection.

There are continued pressures to reduce emissions and increase fuel economy. So, manufacturers have looked to new technologies in fuel injection. Just as TBI replaced carburetion, GDI is the natural successor to **indirect fuel injection**. In fact, approximately half of all new vehicles sold in the United States are equipped with GDI engines. So what makes it different?

Indirect fuel injection atomizes fuel at or near the intake valve. GDI systems take their design cues from diesel technology by spraying the fuel directly into the cylinder (**FIGURE 49-47**). Because of this, direct injection engines can run effectively on extremely lean fuel mixtures—mixtures much leaner than stoichiometric (14.7 parts fuel to 1 part air). In fact, their fuel mixtures can be as lean as about 65 to 1! Also, because the fuel can be injected near the top of the compression stroke, detonation can be minimized. This means that compression ratios can be higher, which also increases efficiency.

The fuel injector for each cylinder is located in the cylinder head. Fuel is directly sprayed into the combustion chamber as a

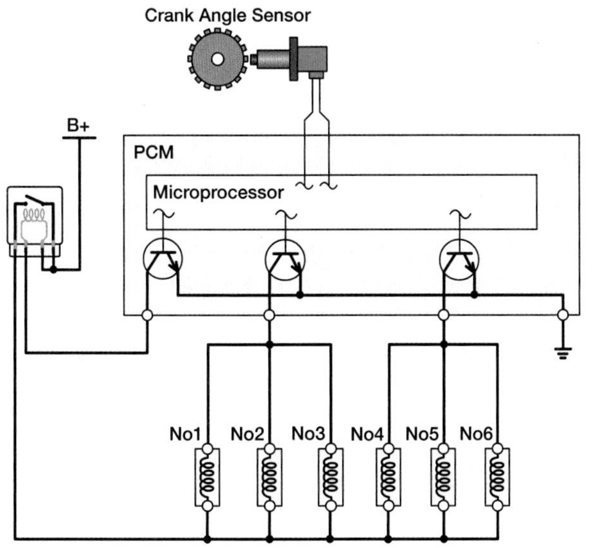

FIGURE 49-45 Typical simultaneous fuel injection circuit showing each group of injectors on a six-cylinder engine.

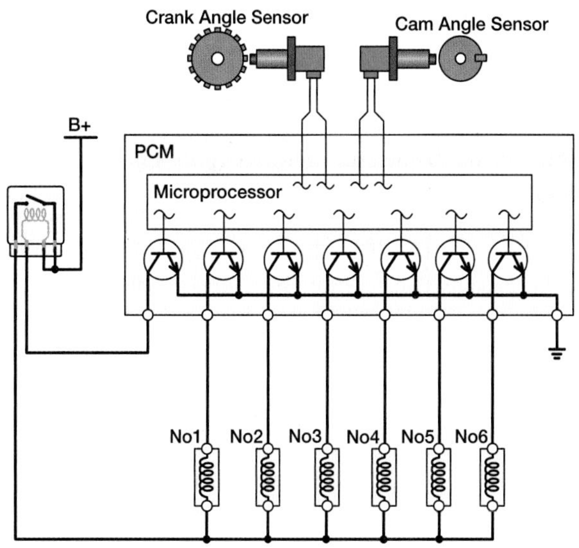

FIGURE 49-46 A typical sequential fuel injection schematic showing how each injector is controlled by the PCM.

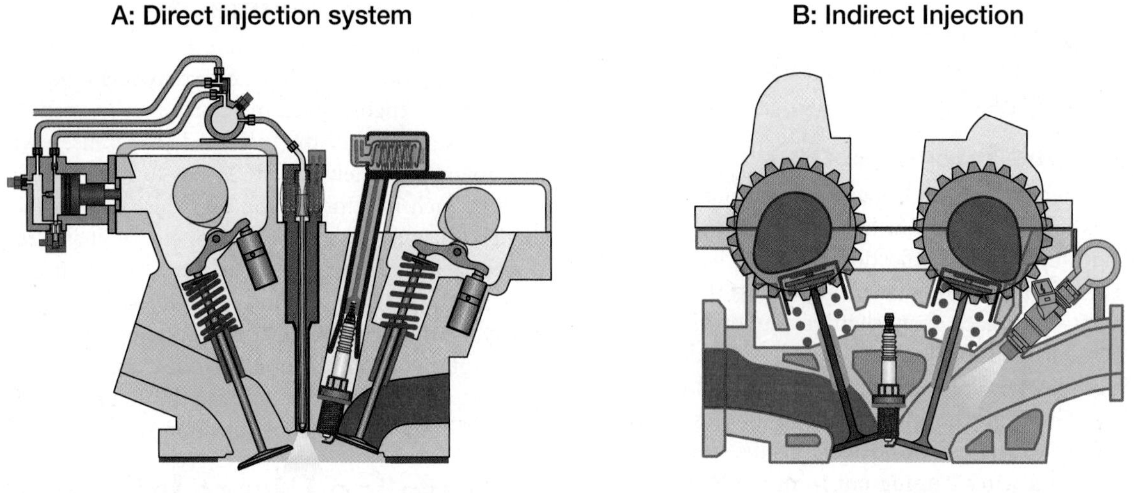

FIGURE 49-47 Injection systems. **A.** Direct injection system. **B.** Indirect injection system.

very highly atomized mist. It is sprayed at the precise time it is needed, depending on the operating conditions of the engine. There are four typical modes of operation, plus variations of these:

- stratified charge,
- stoichiometric,
- full power, and
- catalyst heating.

The PCM selects the mode best suited for the operating conditions.

The **stratified charge** mode is designed to be used during light throttle, cruise conditions. It provides the best fuel economy of all of the modes. This is because it runs so much leaner than the stoichiometric ratio. The reason for this is because the injector can inject fuel near the end of the compression stroke, when the piston is near the spark plug. The pistons have a specially designed pocket that helps concentrate the fuel in a small space near the spark plug. Various pocket designs are in use. This makes it so the fuel particles can be close enough together that they all burn (**FIGURE 49-48**).

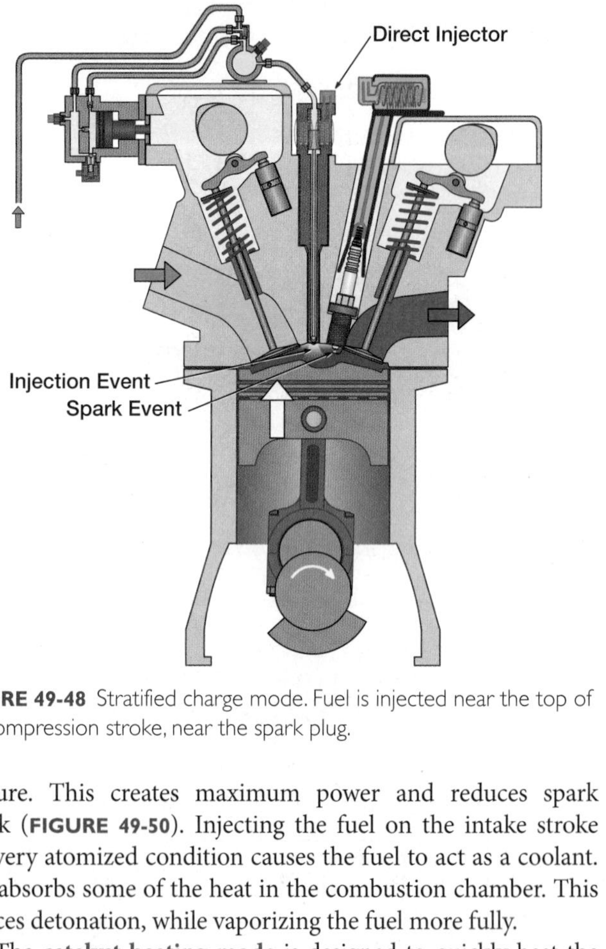

FIGURE 49-48 Stratified charge mode. Fuel is injected near the top of the compression stroke, near the spark plug.

The surrounding area outside this pocket is primarily air and recirculated exhaust gases. They provide a barrier so the flame and heat stay further away from the cylinder walls. Heat loss is reduced in the combustion chamber, which also reduces the load on the cooling system. This is called a stratified charge because a combustible volume of fuel is mixed with air in just a small area. It is not thoroughly dispersed among the entire amount of air in the combustion chamber. This design greatly increases fuel economy.

The stoichiometric mode is designed for moderate engine load conditions. It is perfect when a bit more power is needed than for light cruising. During this mode, just enough fuel is injected during the intake stroke to create a stoichiometric air–fuel ratio. Injecting on the intake stroke gives more time for the additional fuel to mix more fully with the air. Because the fuel is injected on the intake stroke, it is able to disperse more evenly throughout the combustion chamber. It is not just contained by the pocket in the piston (**FIGURE 49-49**). This is called a **homogeneous mixture** and results in good power and a very clean burn.

The **full power mode** is designed for heavy engine load conditions, when full power is needed. During this mode, a slightly rich mixture is injected during the intake stroke in a homogenous mixture. This creates maximum power and reduces spark knock (**FIGURE 49-50**). Injecting the fuel on the intake stroke in a very atomized condition causes the fuel to act as a coolant. Fuel absorbs some of the heat in the combustion chamber. This reduces detonation, while vaporizing the fuel more fully.

The **catalyst heating mode** is designed to quickly heat the catalytic converter. It injects fuel in the stratified mode near the top of the compression stroke. It also injects a small amount of fuel during the power stroke. This delayed fuel continues to burn for a longer period of time and is used to heat up the catalytic converter.

So far GDI engines still use spark plugs. Using spark plugs allows the GDI system to accommodate the different operating modes listed above. For example, if ignition of the air–fuel mixture depended on the heat of compression, then fuel couldn't be added during the intake stroke. This is because it would ignite prematurely on the compression stroke, making the engine

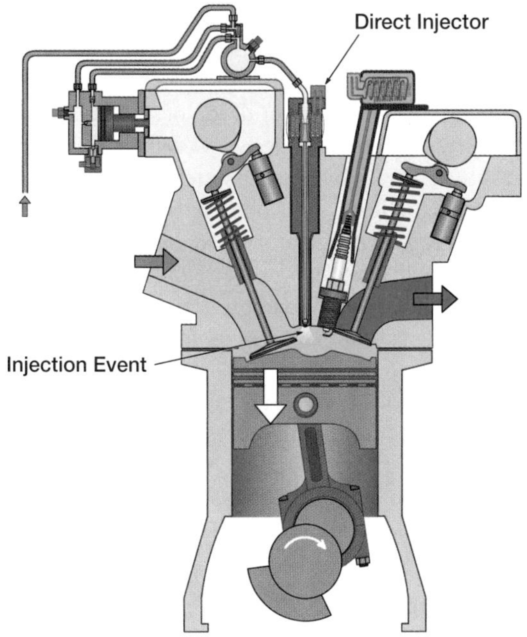

FIGURE 49-49 Stoichiometric mode. Fuel is injected during the intake stroke, fully mixing the air and fuel.

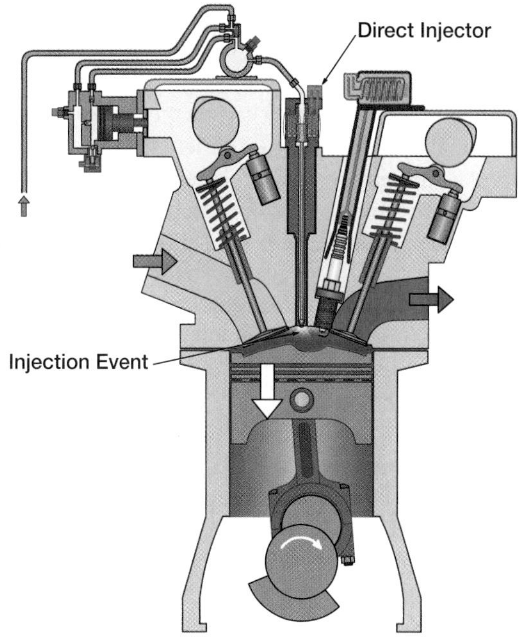

FIGURE 49-51 A multi-pulse injection event.

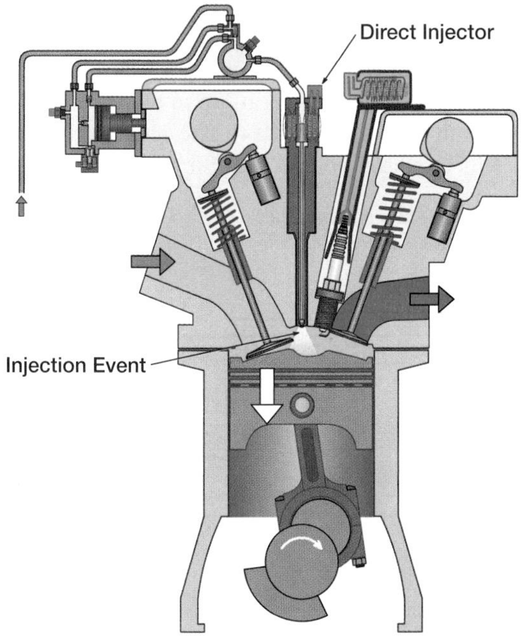

FIGURE 49-50 Full power mode. A slightly rich mixture is injected on the intake stroke. The additional fuel helps cool the charge to reduce detonation.

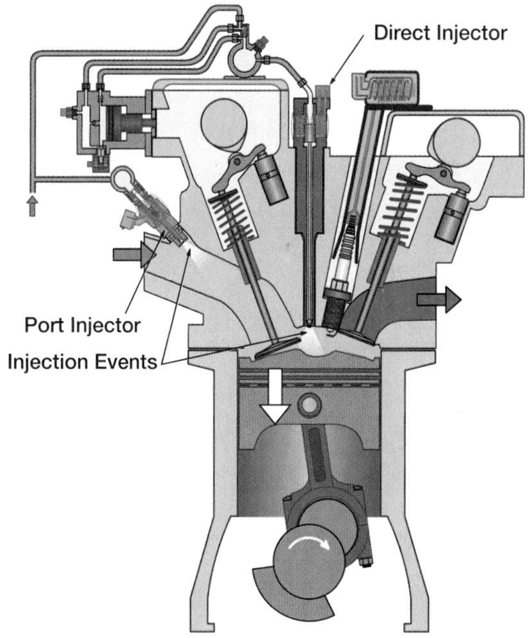

FIGURE 49-52 Some manufacturers use both a direct injection system and an indirect injection system.

detonate and destroy itself. Using a spark plug allows the ignition timing to be carefully controlled and adjusted as needed.

Speaking of timing, many GDI fuel systems can provide more than one injection event per cycle. For example, a small amount of fuel may be injected in stratified charge mode at the top of the compression stroke. Then, there is a small second injection event following ignition. It provides a small amount of additional power while still maintaining a lean mixture (**FIGURE 49-51**). If the second injection event is delayed, then it provides the ability to quickly heat the catalytic converter.

Another example of multi-pulse injection events is when a moderately lean fuel mixture is injected during the intake stroke. Then, a small amount of fuel is injected near the top of the compression stroke to act as a primer charge. This results in a more complete burn of a moderately lean mixture. Some manufacturers accomplish this by using two sets of injectors. One is the direct injection fuel injectors in the cylinder head. And the second set is indirect fuel injectors in the intake manifold (**FIGURE 49-52**). The indirect fuel injectors operate when fuel is needed for moderate and heavy power demands. They inject fuel just like a sequential EFI system. The direct fuel injectors operate under light loads or in tandem when full power is needed.

GDI fuel systems can be of the low-pressure variety (only inject during the intake stroke or early in the compression stroke). Or they can be of a high-pressure variety (can inject during the intake stroke, the compression stroke, and potentially the power stroke). Low-pressure systems operate on a fuel pump system, similar to an MPFI system, with about 35–70 psi (241–482 kPa). High-pressure systems generally have a low-pressure pump, like the one just described. It feeds a high-pressure pump that raises the fuel pressure up to as much as 3000 psi (20,684 kPa).

Most high-pressure GDI systems include the following components (**FIGURE 49-53**):

- Low-pressure pump—An in-tank electric pump used to supply fuel to the high-pressure pump at the correct pressure and volume. The pressure is typically controlled by the PCM.
- Low-pressure sensor—Used to monitor the fuel pressure in the low-pressure system. The PCM uses this reading to control the output of the low-pressure fuel pump to maintain the proper pressure.
- High-pressure pump—A mechanical pump capable of creating fuel pressure of up to approximately 3000 psi (20,684 kPa) in some vehicles. It is typically operated off an eccentric with two or three cam lobes on the camshaft. A roller rides on the lobes and moves a plunger in a bore. This creates the high pressure needed in GDI systems. The pressure is maintained in the fuel rail, so it is available to all injectors equally.
- High-pressure sensor—Used to monitor the fuel pressure in the high-pressure system. The PCM uses this reading to control the output of the fuel pressure regulator valve, based on the operating condition of the engine.
- Fuel pressure regulator valve—Used to control the fuel pressure in the high-pressure system. This valve is typically electromagnetically controlled by signals from the PCM. The fuel pressure regulator valve uses a check valve in the outlet. It holds the high pressure in the fuel rail when the engine is shut down. It is not unusual for the high-pressure side to hold substantial pressure for up to two hours.
- High-pressure fuel injectors—These injectors have to accurately control very high fuel pressures. Also, the nozzles are subject to the high pressures and temperatures of combustion. These injectors are usually electromagnetically operated solenoid valves. Some of them operate on voltage higher than 12 volts.

▶ GDI Drawbacks

LO 49-12 Describe the drawbacks of GDI systems.

GDI systems have a lot of positive qualities such as increased fuel economy, reduced emissions, and more power. But there are some negative points as well. GDI systems are typically more expensive, complicated, and noisy than standard fuel injection systems. One of the most common drawbacks that technicians run into is their tendency for carbon to build up on the intake valves and runners (**FIGURE 49-54**). This reduces airflow into the engine, which reduces volumetric efficiency and engine power.

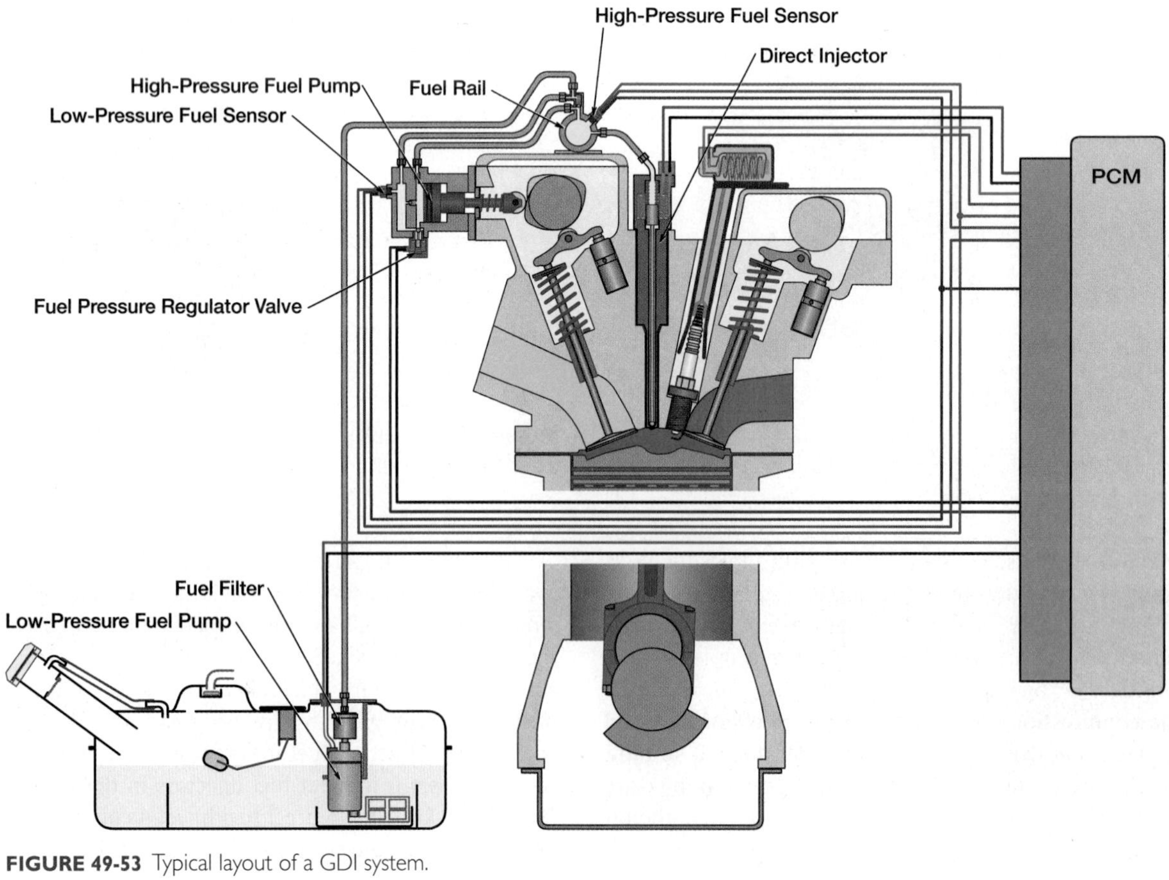

FIGURE 49-53 Typical layout of a GDI system.

Because of this tendency, it is critical to use gasoline with adequate detergents. One way to do this is to use Top Tier Detergent Gasoline™ fuel. It has much more deposit-reducing detergent than the minimum specified by the EPA. It has also been tested to demonstrate its ability to reduce the buildup of carbon deposits. Another method of reducing carbon buildup is to use a high-quality motor oil specified for the vehicle. Then, make sure to change the oil at the recommended intervals. This will prevent oil vapor from creating additional carbon deposits.

If customers don't take the preventive actions above, and the valves carbon up, there are several ways to remove the carbon buildup. But none of them are quick, easy, and without risk. Here are the options:

- Remove the cylinder head, disassemble it, and clean it. This is absolutely the least risky way to remove the carbon, especially if the buildup is very heavy. But it is also likely to be the most expensive method. Unless, that is, you factor in that the other methods have various levels of catastrophic risk to the engine.
- Remove the intake manifold and use a crushed walnut shell blaster to blast away the carbon. The walnut shells are soft enough not to damage the base metal while blasting away the carbon. There is a risk of walnut shells and debris getting where it shouldn't be and causing problems down the road. But that is minimized if this procedure is carried out carefully.
- Remove the intake manifold, and use a chemical cleaner to soften and dissolve the carbon so it can be removed. This is a fairly safe method as long as all of the cleaner and loosened carbon is removed. There are two primary dangers. First, if there are hard carbon deposits that don't get removed, they can pass into the combustion chamber and get wedged between the top of the piston and cylinder head. This can create an engine knock and possibly damage engine bearings, pistons, or connecting rods. The second way is if chemical cleaner gets caught in an inaccessible cavity and is drawn into the cylinder during start-up. This can cause hydrolocking of the cylinder, which can cause the same damage as just listed.
- Induction cleaning is the least invasive because minimal engine disassembly is required. It is a good option on engines that only have soft carbon buildup, not hard carbon deposits. This is likely to be the case on engines with less than about 30,000 miles (48,280 km). This method has the same dangers as with the previous option. That is, the possibility that there are unknown hard carbon deposits that will be loosened up and damage the engine. Or chemical cleaner can pool in cavities and cause hydrolocking. Another problem with this method is on vehicles equipped with turbochargers. The induction cleaning chemicals can overheat the turbo and even the catalytic converter when the vehicle is driven while the chemical is working.

Another drawback to GDI systems is the condition called low-speed preignition. It causes severe pressure waves in the combustion chamber. They can catastrophically damage components in just a few revolutions of the engine. It is thought to be caused by very small, poor-quality oil droplets that work their way past the piston rings into the combustion chamber. Once there, they ignite during the compression stroke (preignition). They then ignite the air–fuel mixture, causing severe pressure waves as the piston is still moving up. This condition can be reduced by using quality oil that is specified by the manufacturer. In fact, Toyota mandates a specific oil to reduce this condition in some of their GDI vehicles.

▶ Carbureted Fuel Systems

LO 49-13 Describe basic carburation operation.

We are covering carburetion lightly. This is because some states still require the teaching of carburetors. Also, some instructors like to relate the mechanical systems of the carburetor to the EFI components that do the same jobs. This section explains how carburetors operate in case you ever have to work on one. If you have no interest in carburetors, skip this section.

Initially, manufacturers used carburetors and mechanical fuel pumps to deliver fuel to the engine. The carburetor was usually fed gasoline by an engine-driven mechanical fuel pump. Traditionally, mechanical fuel pumps were a diaphragm-type pump mounted to the engine block. They were driven directly by the camshaft. Whenever the engine was running, the pump would draw fuel from the tank and deliver it to the carburetor with a low amount of fuel pressure. This delivery of fuel was kept in the carburetor's float bowl.

The fuel level in the float bowl was maintained by a needle and seat (**FIGURE 49-55**). When the float lowered enough,

FIGURE 49-54 GDI engines are prone to carbon buildup on the intake valves and runners.

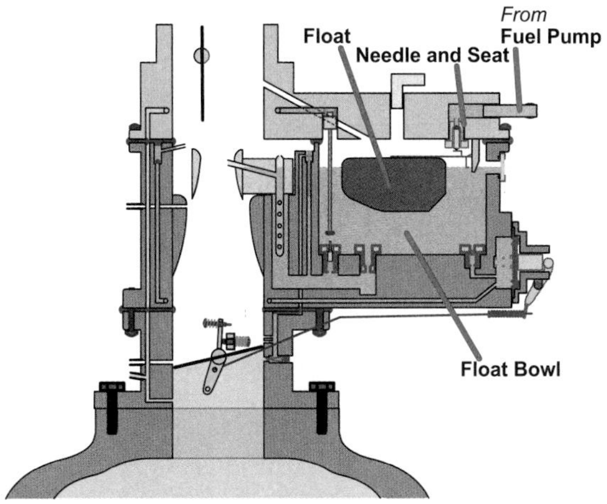

FIGURE 49-55 Float bowl and float system.

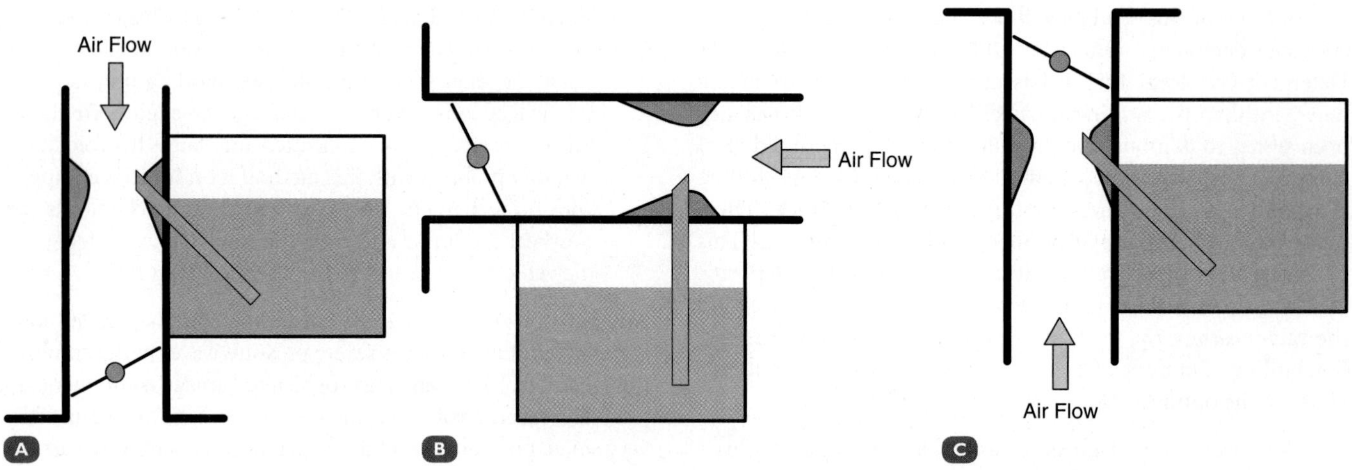

FIGURE 49-56 Types of carburetors. **A.** Downdraft. **B.** Side draft. **C.** Updraft.

fuel passed through the needle and seat. Once the bowl was full again, the needle was closed off against the seat. In actual steady speed operation, the float dropped just enough to allow the same amount of fuel in that was being drawn out.

Carburetors, when tuned properly, delivered fuel effectively. But they need to be properly maintained over time. If not, they would cause drivability and emissions problems. As electronics came into play, carburetors were phased out. Manufacturers opted for more effective, efficient, and reliable EFI systems.

The carburetor supplies the engine with the approximately correct air–fuel mixture for all conditions of operation.

- It atomizes the fuel and mixes it with air.
- It controls the delivery of this approximately correct mixture to the intake system.

Carburetors come in different designs (**FIGURE 49-56**). Most carburetors are the downdraft design. But there are also side-draft carburetors and updraft carburetors. These terms refer to the direction air flows through the carburetor. Each type has a float bowl where a float and a needle valve control the fuel level. The air horn and **venturi** are located in the top of the barrel of the carburetor. A throttle valve controls airflow through the venturi and is linked to the throttle pedal.

As the piston moves through its intake stroke, it creates a low-pressure area. As a result, air from the atmosphere flows through the venturi (**FIGURE 49-57**). The venturi is narrower than the rest of the barrel, and it is shaped to make the air speed up as it passes through. A similar effect occurs around the wings of aircraft. The shape of the wing section speeds up the airflow over the top of the wing. This creates a low-pressure area, lower than the atmospheric pressure, below. The result is an upward force that provides lift for the aircraft.

The shape of the venturi is designed to apply the same principle, known as the Bernoulli effect. It creates a low-pressure area where the end of the nozzle protrudes into the airflow. Atmospheric pressure in the float bowl is greater than the pressure on the end of the nozzle. This pressure difference forces fuel from the nozzle. The fuel mixes with the passing air. This breaks the fuel into droplets, which is part of the process of being atomized.

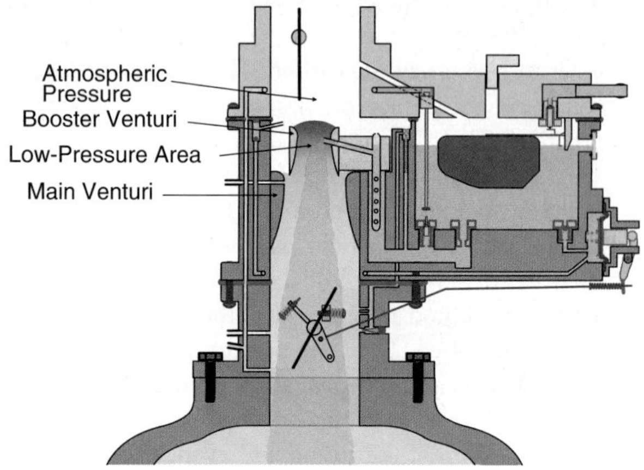

FIGURE 49-57 As air is drawn through the venturi, its speed increases and its pressure drops, creating a low-pressure area.

Depressing the accelerator increases air speed through the carburetor, thus lowering air pressure at the nozzle. Pressure on the fuel in the float bowl stays constant. So more fuel is forced into the venturi to mix with the increased air. This keeps the air–fuel ratio roughly constant for a range of throttle openings. The throttle valve also controls flow of mixture into the engine. Opening it allows more mixture to be delivered, which increases engine power and speed. Closing it has the opposite effect.

▶ Carburetor Operation

The carburetor is a mechanical device that delivers the approximately correct mixture of air and fuel to each of the cylinders. It is bolted to the intake manifold. While the engine is running, the intake stroke of each piston creates a low-pressure area (vacuum) in the intake manifold. The vacuum created in a mechanically sound carbureted engine at sea level is typically between 18" and 21" (457 and 533 mm) of mercury. With atmospheric pressure at sea level being 14.7 psi (101 kPa), a moderate pressure differential results. That differential in pressure allows the clean filtered air to enter the carburetor. It is also used to meter out the corresponding amount of fuel over varying and constantly changing conditions. Ideally, the carburetor controls the air–fuel mixture as precisely as possible.

▶ Carburetor Circuits

LO 49-14 Describe the circuits on a carburetor.

The basic carburetor components are housed inside a metal casting called the carburetor body. It serves as the mounting point for the various carburetor system components. The components are grouped together to create carburetor circuits. Each of the circuits is designed to provide a certain function to the carburetor. The six circuits are the float, idle, main metering, power, accelerator pump, and choke.

▶ Float Circuit

The **float chamber** holds a quantity of ready-to-use fuel at atmospheric pressure. Its supply is refilled by a float-driven valve. As the fuel level drops, the float drops too. It opens an inlet valve, which allows the fuel pump to deliver more fuel to the float chamber. The float rises with the replenished fuel level, closing off the inlet valve.

To allow atmospheric pressure to act on the fuel, the float bowl must not be sealed. It must be vented in one of the three ways.

- to the atmosphere (unbalanced carburetor)
- to the air horn above the venturi (balanced carburetor), or the **charcoal canister** (evaporative emission carburetor) (**FIGURE 49-58**).

If the float level is too low, more airflow through the venturi will be required to pull out the fuel. This causes a lean air–fuel ratio. Consequently, too high a float level will cause the mixture to be too rich. Float adjustment is important when rebuilding a carburetor. Flooding a carburetor also produces rich mixtures. Flooding can be caused by a worn needle and seat, or by dirt trapped between the needle and seat. This prevents the fuel from being shut off. So the level in the float bowl continues to rise. Fuel then dribbles from the nozzle with little or no venturi action.

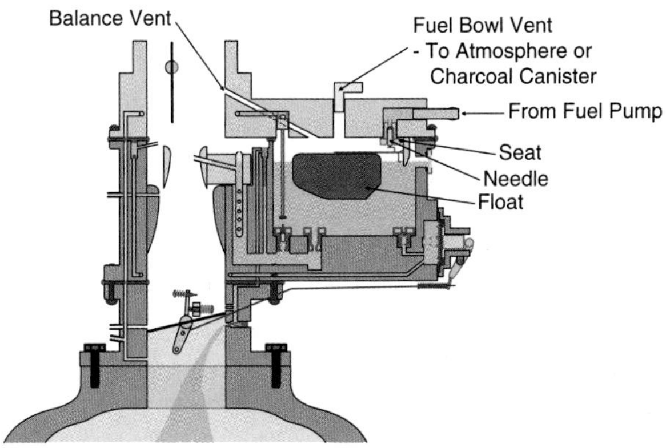

FIGURE 49-58 Float bowl, float, and needle and seat.

▶ Idle and Off-Idle Circuits

When the throttle valve is closed or nearly closed, manifold vacuum is created behind the throttle. It is sufficient to pull a small amount of fuel and air through small openings located after the butterfly valve (**FIGURE 49-59**). This is called the idle circuit. It enables the engine to keep running when there is not enough air speed through the venturi to create a vacuum. As the throttle valve opens slightly, the manifold vacuum is reduced. So additional small openings are uncovered to compensate for this. This design is the "off-idle" circuit.

▶ Main Metering Circuit

The main metering circuit comes into action above fast idle. This is when airflow through the venturi increases. A main **metering jet** in the float bowl meters fuel passing into the discharge nozzle (**FIGURE 49-60**). How much fuel leaves the nozzle depends on the pressure difference created by the airflow through the venturi. As the throttle opens, airflow increases and speeds up. More and more fuel is drawn from the discharge nozzle. However, the mass of air does not increase in proportion

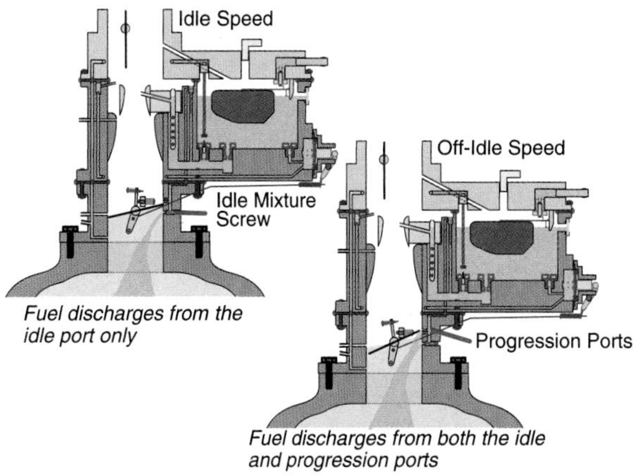

FIGURE 49-59 Idle and off-idle circuit.

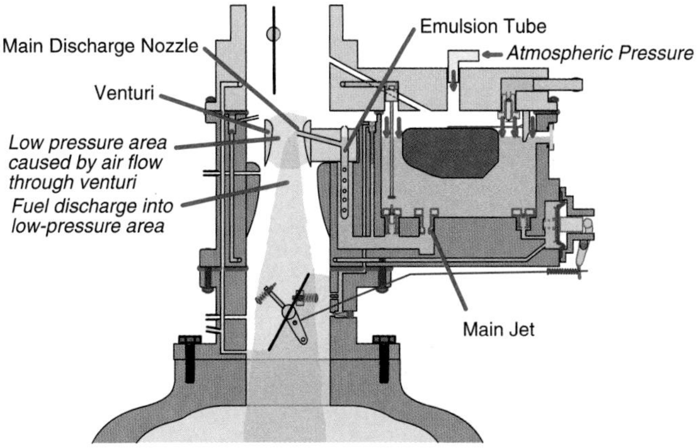

FIGURE 49-60 A main metering jet in the float bowl meters fuel passing into the discharge nozzle.

with the speed. The result is that high speeds can produce a mixture that is too rich. To correct this, more air can be added. This is called compensation by air correction.

As the throttle opens and engine speed increases, the fuel level in the jet well falls. This exposes air bleed holes in the discharge tube. Air can now mix with the fuel and prevent the mixture from becoming too rich. As the throttle opens farther, the fuel level falls further, exposing more air-holes. More air bleeds in to maintain the correct mixture. Main metering fuel flow can typically be adjusted by replacing the removable jets with jets having larger or smaller orifices.

▶ Power Circuit

The size of the main jet is selected to provide the best mixture for economy under cruising conditions. When the throttle is open wide for maximum power, a richer mixture is required. The extra fuel is provided by a power valve. It uses a vacuum piston and rod opening it as it is needed. At low speeds, intake manifold vacuum is transferred through a passage to the vacuum piston. This holds the piston up and keeps the power valve closed. With the throttle valve fully open for full engine power, the vacuum in the intake manifold falls. A spring pushes down the vacuum piston and rod to open the power valve. Additional fuel flows through the power valve to enter the fuel well and add to the fuel from the main jet. This provides the extra fuel needed to enrich the mixture for full power.

Some carburetors use metering rods instead of a vacuum piston. The metering rods are pulled down into the main jets at idle and cruise to restrict the fuel flow. When manifold vacuum drops under heavy load, springs push the metering rod(s) up. This increases the opening size of the main jet(s). Other carburetors use a diaphragm-type power valve that opens an additional passage when vacuum drops under load.

▶ Accelerator Pump Circuit

Extra fuel is also needed for accelerating. Suddenly opening the throttle increases the airflow. But fuel cannot flow from the discharge nozzle quickly enough to match it. An extra squirt of fuel is needed, which is where the **accelerator pump circuit** comes into play. Depressing the pedal compresses a duration spring. It exerts a force on the plunger of a small plunger pump (**FIGURE 49-61**). This pressurizes fuel below the plunger and closes off the inlet valve. Fuel flows past a check valve and out a discharge nozzle above the venturi. The duration spring extends the time for delivering the fuel.

Releasing the pedal lets the linkage move the plunger upward. The check valve closes, and the inlet valve opens to let fuel refill the pump chamber from the float bowl. This primes it for the next shot of fuel. Thus, whenever the throttle is opened, the **accelerator pump** discharges a small amount of fuel into the throat of the carburetor. So constantly working the throttle pedal when driving down the road wastes fuel.

▶ The Choke

Fuel ignites less readily when cold. If the engine is also cold, then some fuel vapor can condense out of the air–fuel mixture onto the intake manifold and cylinder walls. This loss of fuel makes the combustible mixture leaner. To compensate for this, a valve known as the **choke** restricts the flow of air at the entrance to the air horn. This lowers the pressure at the venturi and off-idle circuits. In this way, additional fuel is sucked into the incoming air. It does it through all the fuel circuits—idle, off-idle, and main—at the same time.

The choke can be controlled manually by a cable that operates the valve. However, most are controlled automatically. In this way, the valve is closed when the engine is cold and opens progressively as the engine warms up (**FIGURE 49-62**).

When the engine is warm, the fuel drawn into the manifold vaporizes readily. The engine can be started without the aid of a choke. The choke should operate as briefly as possible. Overusing it produces rich mixtures. That causes extreme amounts of exhaust pollution. It also increases fuel consumption. Some later-model carburetors that used a cable-operated choke also used a spring-loaded choke release. It turned the choke off after a set amount of time.

▶ Carburetor Barrels

LO 49-15 Describe the barrels on a carburetor.

Carburetors can have one, two, three, or four barrels (**FIGURE 49-63**). Extra barrels improve performance, particularly at high speeds. Letting more air and fuel enter the cylinders increases power. However, a carburetor that is too large reduces engine

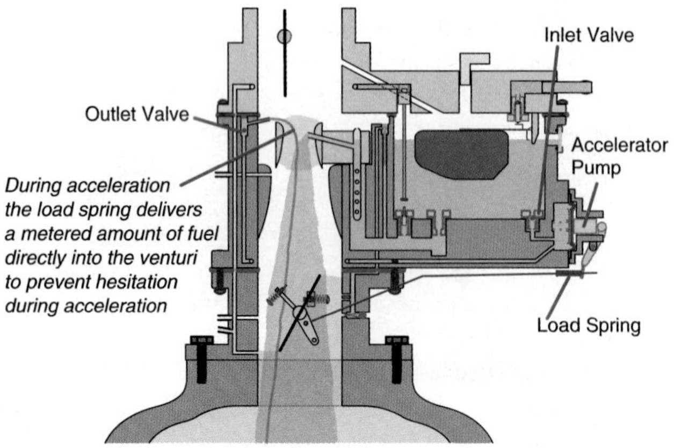

FIGURE 49-61 Accelerator pump circuit.

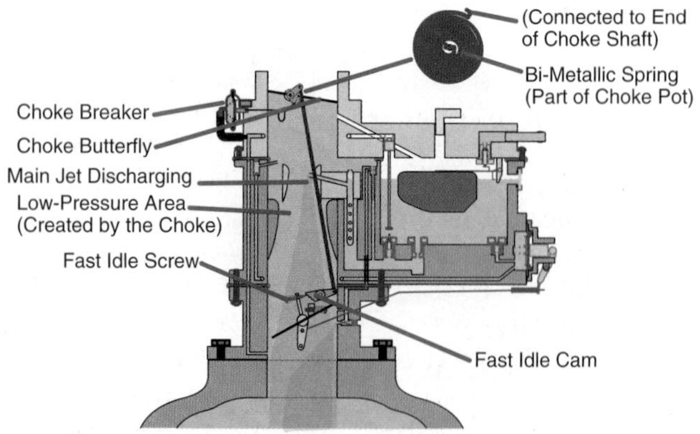

FIGURE 49-62 The choke.

FIGURE 49-63 **A.** One-barrel. **B.** Two-barrel. **C.** Four-barrel.

power. This is because not enough airflow can be maintained through the venturi to draw fuel out of the float bowl.

Two-barrel carburetors have two venturis and two outlets to the intake manifold. They come in two basic designs. One has a common float chamber, with each barrel having a complete set of all other circuits. The throttles open simultaneously. The float chamber may be straddled by two connected floats that almost surround the air passage. This design leaves the float chamber unaffected by cornering, climbing, accelerating, or braking.

In the other basic design, called a progressive carburetor, throttles open in two stages (**FIGURE 49-64**). This design combines the two barrels to act as a single carburetor. The two stages combine good low-speed operation of a single-barrel design with the extra airflow of two barrels.

FIGURE 49-64 A progressive carburetor.

FIGURE 49-65 Mechanically controlled secondaries.

One of the two barrels of the carburetor has all the circuits needed to supply mixtures for the whole range of operation. It is called the primary side. The other barrel is called the secondary side. It supplies extra mixture, but only at high speed or full throttle. It normally has a main metering system and a nonadjustable idling system. The primary side has a choke for cold starting.

When the engine is being started, the throttle on the secondary side is already closed. So a choke is not needed. From idle to medium speeds, only the primary throttle is open. When engine speed rises to where additional breathing capacity is needed, the secondary throttle opens. It admits more air–fuel mixture. By the time the primary throttle is wide open, so is the secondary throttle.

Opening the throttles can be controlled mechanically (**FIGURE 49-65**). Or it can be by a vacuum unit that is connected by a pull rod to a lever on the secondary shaft (**FIGURE 49-66**). When air flows past ports in the venturis, it produces low-pressure areas. A hose transmits this low pressure to the diaphragm chamber. This low pressure acts on the diaphragm and opens the secondary throttle.

Large-capacity V8 engines may use a four-barrel carburetor with four venturis. These carburetors are typically of a two-stage design. This is effectively two, two-stage carburetors combined.

FIGURE 49-66 Vacuum-operated secondaries.

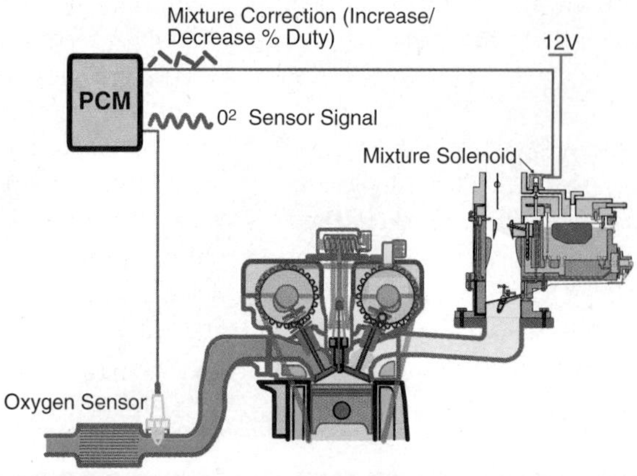

FIGURE 49-67 A computer-controlled carburetor normally uses an electronically controlled solenoid valve.

Some carburetors have a central molded plastic fuel bowl incorporated into the design. The result is lowered fuel temperatures with more precise fuel metering of the air–fuel ratios.

▶ Computer-Controlled Carburetors

A **computer-controlled carburetor** normally uses an electronically controlled solenoid valve. It is called a mixture control solenoid and responds to the PCM commands (**FIGURE 49-67**). This type of carburetor system uses various sensors in the exhaust and engine to monitor operating conditions. In turn, these sensors send that information to the PCM. According to the sensor data, the PCM then knows if the mixture is rich or lean. The PCM adjusts the air–fuel mixture accordingly. The computer constantly sends commands to the mixture-control solenoid. So the solenoid constantly opens and closes the air and fuel passages in the carburetor.

▶ Maintenance and Repair

LO 49-16 Remove and replace a fuel filter.

Replacing a Fuel Filter

Fuel filters are a maintenance item and need to be replaced according to the manufacturer's specified replacement schedule. They also need to be replaced anytime restricted flow is encountered. Fuel filters can be located in a variety of places, depending on the vehicle. Common locations are:

- under the vehicle in the fuel line,
- under the hood in the engine compartment, and
- inside the fuel tank as part of the fuel pump assembly.

Fuel pressure is maintained in all electric fuel pump systems, even when the engine is not running. So it is important to release the pressure before opening the system. This helps prevent fuel from spraying everywhere.

There are several ways to relieve the pressure in the fuel system before removing the fuel lines. One way is to remove the fuel pump relay or the fuel pump fuse, and run the engine until it dies. Another method is to connect a fuel pump pressure gauge (that has a bleed valve) to the fuel rail test point. Then release the excess pressure from the system into a gas can. On GDI-equipped engines, you may need to use a scan tool to release the fuel pressure. Just remember that fuel pressure on the high side of a GDI system can be up to 3000 psi (20,684 kPa). So following the manufacturer's specified procedure is critical for safety purposes.

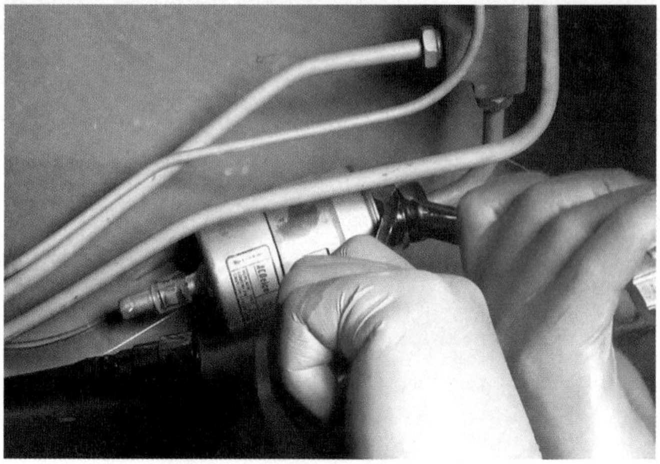

FIGURE 49-68 When removing filters with flared fittings or banjo fittings, use the double-wrench method to loosen and tighten them.

Once the static pressure is released in the fuel system, vent the pressure in the gas tank by removing the gas cap. Doing so reduces the amount of gas that dribbles out of the lines when removing the filter. Always refer to the service information for the specified procedure for replacing the fuel filter for the vehicle you are working on.

Some vehicles use metal or plastic lines that bolt onto the filter. They are usually equipped with either a flared fitting or a banjo bolt. For either type, you need to use the double wrench method to unbolt the line. Otherwise, you may twist or kink the metal or plastic line. The double-wrench method uses two wrenches. One wrench is placed on the filter nut and the other on the flare nut or banjo bolt (**FIGURE 49-68**). Use this wrench to prevent the filter from twisting while applying force to the other wrench. Often you can place the wrench handles so they are at a slight angle to each other. Then squeeze them together to break the fittings loose.

Many newer vehicles use quick couplers to retain the fuel lines on the fuel filter. These couplers make it very quick and easy for the factory to install the filter. This is because the fuel filter lines only have to be pushed into or onto the filter inlet and outlet. The quick coupler may use a plastic clip or a coiled spring to retain the line on the filter. Only use the proper release tool when removing the line. Otherwise, you could damage the fitting or line. Also, most quick coupler systems use O-rings to seal the fuel line to the filter. Make sure the O-rings are in good shape, or replace them with new ones.

Some fuel lines are connected to the fuel tank and the filter with flexible hoses rather than metal lines. Check their condition to determine whether it is necessary to replace the hoses and clamps along with the filter. Some replacement filters come with these items; when they are supplied, you should always use them. If they are not supplied, make sure to use the proper type of new fuel line and suitable clamps.

There are different types of clamps for flexible fuel lines, including spring type, worm type, and rolled edge. You will need to use the appropriate tool when installing new clamps on the hoses.

Some manufacturers have been installing the fuel filter in the fuel tank along with the fuel pump assembly (**FIGURE 49-69**). The only way to get to the pump and filter is through the top of the fuel tank. Sometimes manufacturers provide a removable access cover under backseat or trunk that allows you to get to the top of the tank. Other times the tank has to be removed to gain access. Once the covers are removed, the fuel pump assembly can be carefully removed. On some vehicles, the filter can be replaced separately. On others, the fuel filter is changed as part of the fuel pump assembly. In this case, it is not normally a maintenance item.

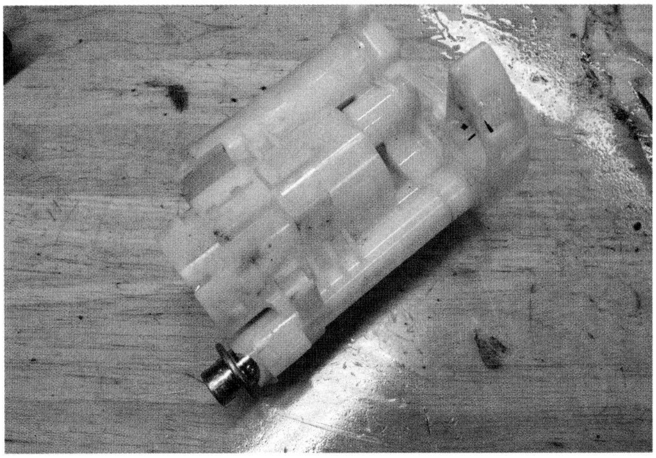

FIGURE 49-69 Typical fuel filter assembly located in the fuel tank.

SAFETY TIP

Gasoline fuel vapor is explosive and highly flammable. Be careful not to spill any fuel onto a hot engine component, where it could evaporate, ignite, and start a fire. Also, take care not to cause any sparks while you are changing a fuel filter. Collect the gasoline waste in a metal container, and dispose of it in an environmentally prescribed way. Always wear the appropriate personal protection equipment before starting the job.

To replace a fuel filter, follow the steps in **SKILL DRILL 49-1**.

SKILL DRILL 49-1 Replacing a Fuel Filter

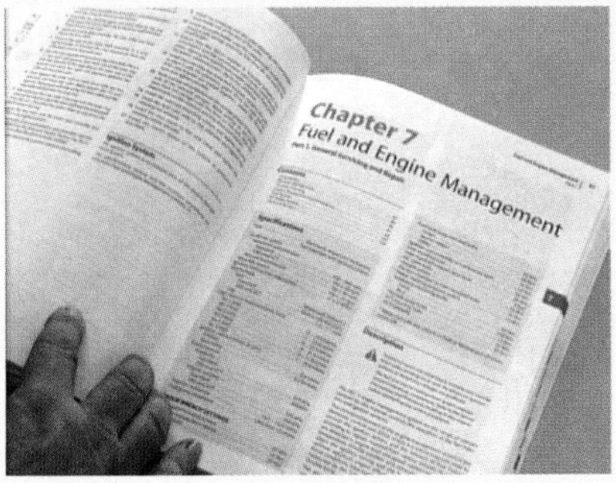

1. Refer to the service information to identify the location, type of fuel filter, and the procedure for removing and replacing it. If it is equipped with an electric fuel pump, release the pressure according to the service information.

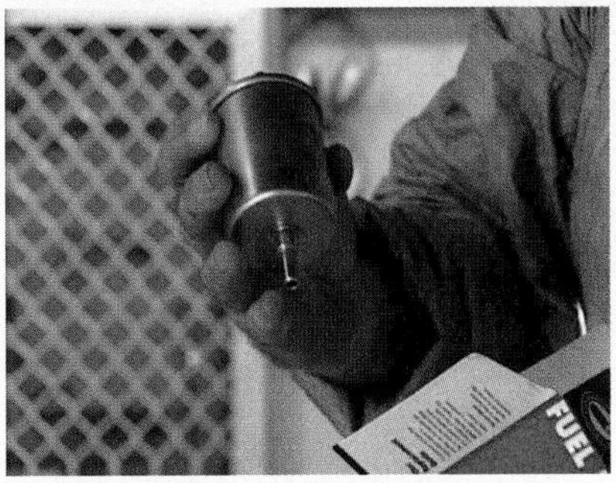

2. Obtain the correct replacement filter and components. Loosen the bracket holding the filter in place, if equipped. Follow the steps below according to the type of filter you are replacing.

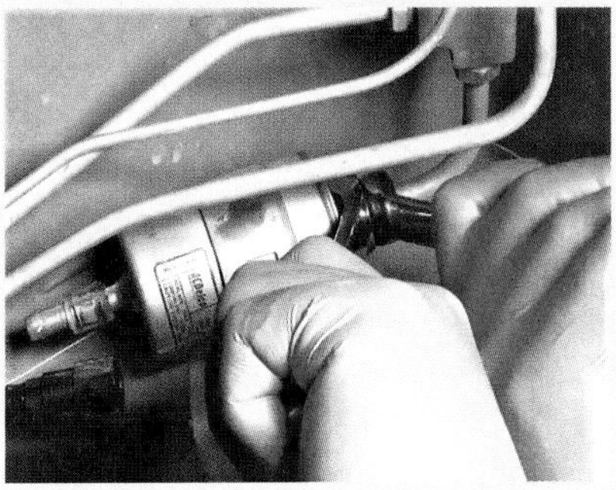

3. On flared fitting types, disconnect the fuel line on the engine side of the filter, using the double-wrench method. If necessary, drain any excess fuel into a fuel-proof container.

SKILL DRILL 49-1 Replacing a Fuel Filter (Continued)

4. Some low-pressure types use clamps to seal the connections.

5. Reinstall the filter, making sure you have the filter facing in the right direction, with the flow indicator arrow pointing toward the engine. Tighten the fittings on both ends, using the double-wrench method.

6. For a quick disconnect filter: Using the correct tool, release the quick disconnect connectors from the outlet end of the filter, catching any leaking fuel in a fuel-proof container.

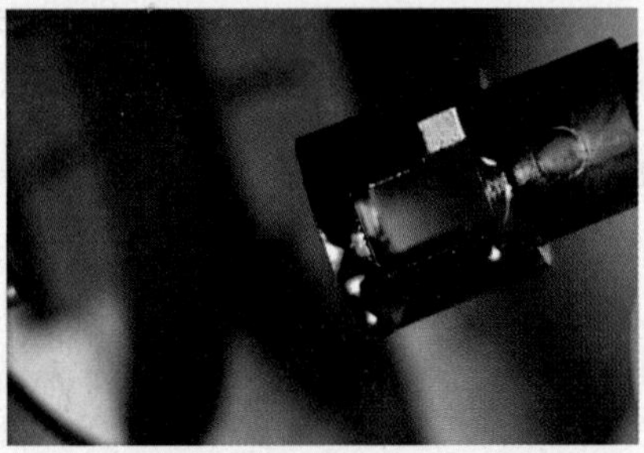

7. Release the quick disconnect connectors from the inlet end of the filter, and remove the filter from the lines. Inspect or replace any O-rings.

8. Reinstall the filter, with the flow indicator arrow pointing toward the engine, and fully engage the lines, making sure they are secure. On vehicles with the filter in the tank, remove the fuel pump assembly.

9. On vehicles with the filter in the tank, remove the fuel pump assembly.

10. Carefully remove the in-tank fuel filter from the pump assembly, and replace it with a new one, if it is a replaceable type.

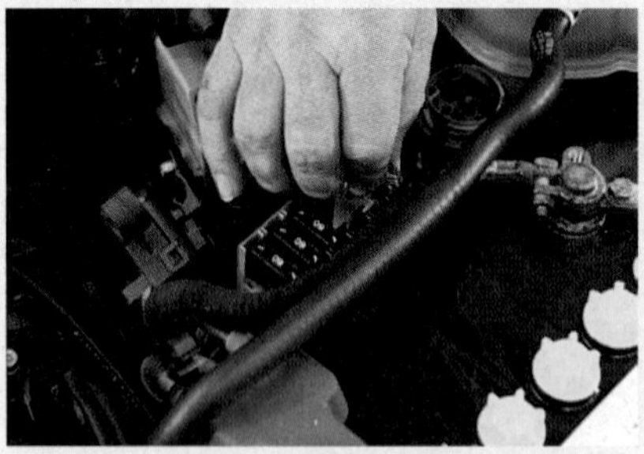

11. Wipe any residual fuel off with a clean shop rag. It is helpful to write the date and mileage on the filter. Remember to replace the fuel pump fuse, if removed.

12. Turn the key to the On position for a few seconds, but *do not start the engine*, and then turn it back to off. Repeat the process two more times, checking the filter connections for leaks. If no leaks are found, start the vehicle, letting it run for two to three minutes before shutting it off. Recheck the filter connections for leaks.

▶ Inspecting and Testing Fuel Pumps

LO 49-17 Test fuel pressure and volume.

Fuel pumps have to be tested if a vehicle experiences low engine power or if the vehicle will not start due to a fuel-related issue. Some shops test a fuel pump's performance, looking to see if it is nearing failure. There are several ways to test a fuel pump:

- pressure/volume test,
- scan tool data stream test, and
- lab scope inductive current flow test.

The first two will be described in this section.

The pressure/volume test measures the fuel pressure being delivered to the fuel rail along with the volume of the fuel pump. Many, but not all, fuel-injected engines provide a test port that a fuel pressure tester can connect to. Once connected, the pressure can be measured with either just the key on or with the engine running. Manufacturers give pressure specifications to compare the pressure readings with.

If the pressure is low, and the system uses a return line from the pressure regulator on the fuel rail, you may need to pinch off the return line. This will allow you to see if the pressure goes up due to a faulty pressure regulator. If the pressure is within specifications, turn the engine off, but watch the pressure gauge. It should stay steady. If it drops, there is a leak in the system, either at the:

- pressure regulator,
- the check valve in the fuel pump, or
- even a fuel injector.

The pressure may be within specifications and hold pressure, but the fuel pump volume may be low. Specifications are given in volume per amount of time, such as 1 pint in 30 seconds. It can also be given in pounds per amount of time. For example, 28 lb per hour, which usually has to be converted to volume per amount of time.

The volume is measured with the engine idling. Most EFI fuel pressure gauges have a valve that allows fuel to be taken from the rail without killing the engine. The fuel can then be caught in a calibrated container. If the fuel flow is low, check to see if the fuel filter is plugged and restricting the flow. If the filter is okay, you may have to remove the fuel pump from the fuel tank and inspect the fuel sock at the bottom of the fuel pump to see if it is plugged. If all of the filters are not restricted but the pump volume is low, the available voltage at the pump will have to be measured. If the voltage is good, the pump will need to be replaced.

One of the best ways to verify that there is a fuel delivery fault is to use a scan tool and test drive. If possible, set the scan tool to record:

- rpm,
- vehicle speed sensor (VSS),
- front oxygen sensors,
- short-term fuel trim,
- long-term fuel trim, and
- if equipped, a fuel pressure sensor

Then go on a test drive. In a safe place and manner, briefly drive the vehicle under heavy load and at higher rpm. After returning to the shop, look at the recording. If you see a low fuel pressure, or a continuous lean oxygen signal with an associated large increase in fuel trim, a fuel delivery problem is indicated. This problem could be as simple as a restricted fuel filter. Or it could be something more severe such as a faulty pump or plugged fuel injectors.

To inspect and test fuel pumps, follow the steps in **SKILL DRILL 49-2**.

SKILL DRILL 49-2 Inspecting and Testing Fuel Pumps

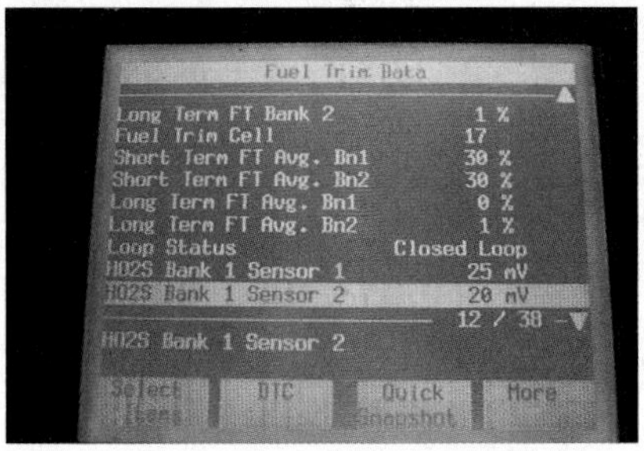

1. Connect a scan tool to the data link connector (DLC) and set up to record rpm, VSS, front oxygen sensors, both fuel trims, and fuel pressure, if equipped. Test drive the vehicle in a safe place under heavy load and at higher rpm. After returning to the shop, look at your recording. If you see a low fuel pressure or a lean oxygen signal with a large increase in fuel trim, the vehicle has a fuel delivery problem.

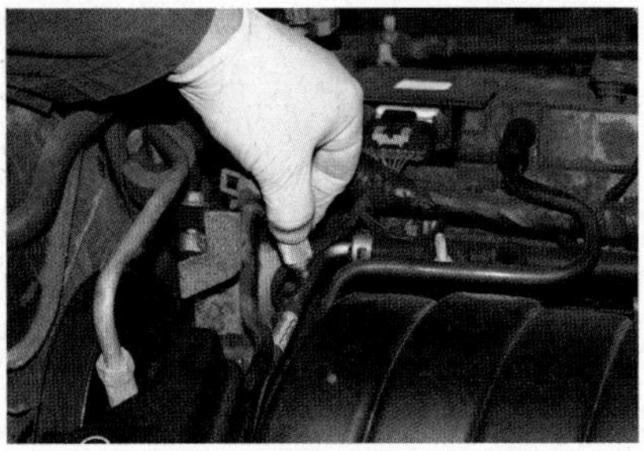

2. After researching the procedure for testing the fuel pump in the service information, install a fuel pump gauge on the fuel rail test port.

3. Turn the key to the Run position and measure the pressure. If none, test the fuel pump's electrical circuit.

SKILL DRILL 49-2 Inspecting and Testing Fuel Pumps (Continued)

4. Start the engine, measure the pressure, and compare with specifications. With the engine running and the end of the fuel line from the fuel pressure gauge in a 1-quart plastic bottle, open the valve on the gauge and stop the fuel flow after 15 seconds.

5. Measure the amount of fuel delivered in that time, and compare with specifications.

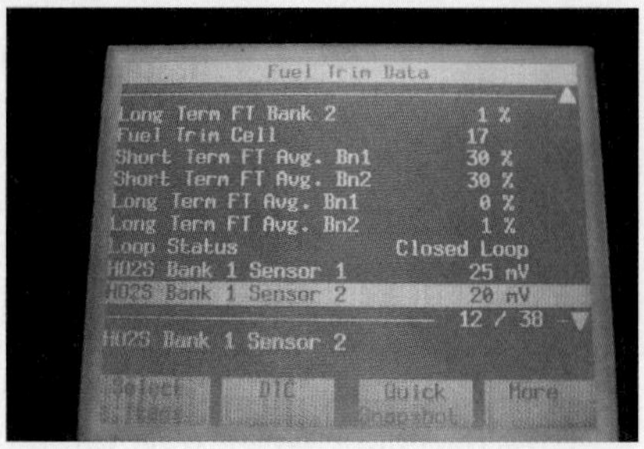

6. After repair, reconnect the scan tool and test drive the vehicle to verify the repair by monitoring the oxygen sensors and fuel trim.

Applied Math

AM-18: Standard/Metric: The technician can measure/test with tools designed for standard or metric measurements, and then convert the measurement to the system used by the manufacturer for specifications and tolerances.

A vehicle is in the shop for engine performance concerns. The technician needs to check the fuel pressure. As he looks in his toolbox for the gauge, he remembers that his friend has not returned his fuel pressure gauge. He has another one, but it is calibrated only in the metric unit bar. He connects this pressure gauge to the Schrader valve and starts the engine. With the engine running at idle, the gauge shows a reading of three bars. The technician uses the conversion factor of 1 bar = 14.503 psi multiplied by 3 to obtain the result of 43.5 psi. The service information states that the vehicle should have 44 psi, plus or minus 3 psi. This confirms that the fuel pressure is acceptable, and the technician will now conduct a fuel volume test.

▶ Checking Fuel for Contaminants and Quality

LO 49-18 Check for fuel contaminants and alcohol content.

It is easy to forget that fuel by itself can cause drivability issues if it becomes contaminated or old. It can also be the wrong fuel for the vehicle. In fact, with more people owning diesel-powered vehicles, you may encounter vehicles filled with the wrong fuel. It is not uncommon for a diesel vehicle to be filled with gasoline, or vice versa. Also, higher blends of alcohol in gasoline make it easy to fill up a standard gasoline vehicle with E-85. It is made up of 85% alcohol and 15% gasoline. A flex fuel vehicle will operate correctly on that blend, but a non-flex fuel vehicle will run very lean. This is because alcohol requires a much richer mixture to run properly. Anytime a vehicle is not running properly (and other common causes have not identified the issue), you need to check the fuel for contaminants and quality.

Short of sending fuel out to be tested by a lab, there are some simpler tests that can be done. All require getting a sample from the fuel supply. This is best taken from the fuel rail. That way you will be testing the fuel that the engine is running on. You will want to collect a cup or two of fuel in a clear container, preferably not glass, which breaks easily. Allow the fuel to settle, and then look at it. Is it the right color? Is it cloudy? Does it have a separation line due to being contaminated with water? Does it smell right? If any of those conditions are present, you will need to drain the fuel system, flush it out, and replace the fuel filters.

If the fuel passes the visual inspection, it is time to perform an alcohol content test. For this test, you will use a tall 100-mL graduated cylinder. A mixture of 10 mL of clean water and 90 mL of gasoline is carefully agitated for 30 seconds. The mixture is allowed to settle for a minute or two. Any alcohol in the fuel will absorb the water. This raises the level above the original 10 mL. Each milliliter above 10 equals the percentage of alcohol in the gasoline. Most vehicles can tolerate 10% alcohol without causing any drivability issues. And flex fuel vehicles can typically operate normally on 85% alcohol.

To check fuel for contaminants and quality, follow the steps in **SKILL DRILL 49-3**.

▶ TECHNICIAN TIP

This author was recently at a Ford dealer catching up on new technologies. The service manager related a story about a four-month-old diesel truck that the customer inadvertently filled with gasoline and drove for a short distance. In this situation, to maintain the factory warranty, all of the high-pressure fuel system components had to be replaced at a cost of about $15,000. This was a very expensive mistake that the customer was responsible for. Be careful to use the correct fuel.

SKILL DRILL 49-3 Checking Fuel for Contaminants and Quality

1. Collect a quantity of fuel from the vehicle's fuel rail in a clear plastic fuel container. Let it settle, and check for contaminants, cloudiness, or improper odor.

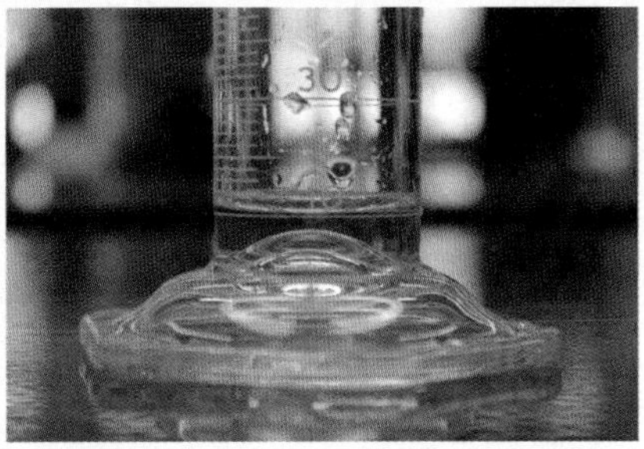

2. Pour 10 mL of water into the 100 mL graduated test tube.

3. Add 90 mL of gasoline, bringing the total volume to 100 mL.

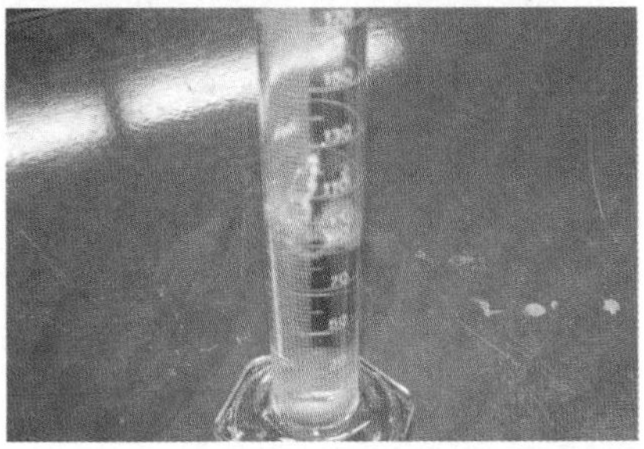

4. Cap the test tube tightly. Slowly agitate the fuel–water mixture for 30 seconds. If there is any alcohol in the fuel, this motion will allow the water to be absorbed by the alcohol.

5. Allow the mixture to settle. Observe the level of the water in the bottom of the test tube. Anything higher than the initial 10 mL is the amount of alcohol in the fuel. List your observations, and determine any necessary actions.

6. Carefully pour off the fuel in the test tube back into the fuel container. Make sure no water leaves the test tube. Properly dispose of the remaining water–fuel mixture.

▶ Inspecting and Testing Fuel Injectors (Non-GDI)

LO 49-19 Test fuel injectors.

Fuel injectors can fail electrically or mechanically. They can also become plugged, or deposits can build up on the nozzle, affecting their spray pattern. If the injector fails completely, it will cause the cylinder to misfire. If the injectors have a poor spray pattern due to deposits, the engine may experience general roughness or random misfires.

Before testing an injector, it should be identified as potentially causing a drivability issue. Other faults can cause one or more cylinders not to work correctly, just like a fuel injector problem. Some examples are faults in the ignition system, poor compression, vacuum leaks, and other engine mechanical issues.

Use all of your diagnostic skills and test equipment to narrow the fault down to the injectors. For example, you may need to retrieve the codes, read the data stream, or access mode 6 data. These will help you narrow down the area of the fault. If you suspect a fuel injector is causing the fault, you can quickly measure the resistance of the injector winding. Then compare the reading with specifications. Any improper resistance will require replacement of the injector.

If the injector resistance is OK, check the on/off signal from the PCM. This is done with the use of noid lights or lab scope. Noid lights are easy to use. Plug the correct one into the wiring harness for the fuel injector. Then crank or start the engine. The noid light should blink indicating the control circuit is working. If no blinking, inspect the power and ground sides of the control circuit.

If the resistance and control circuit are good, you can perform an injector pressure drop test. This will determine if the injector is restricted with deposits. This test uses a fuel pressure gauge and injector pulsing tool. The fuel pressure gauge is connected to the test port in the fuel rail. The pulsing tool is connected to the battery and the suspect injector. The pulse tool pulses the injector a certain number of times for a certain amount of time. The idea is that you will pressurize the fuel rail by turning the key to the On position which charges the fuel rail. Then start the pulse tool, which pulses the injector the set amount of times. Read the pressure on the fuel pressure gauge. Compare this reading with the pressure readings after doing the same test on the other injectors.

If the flow is the same between the injectors, the pressure at the end of each test should be within a psi or two of each other. If not, the one with the higher pressure at the end of the test is not allowing as much fuel to flow as the others. This can indicate deposit buildup on the nozzle or a restricted filter screen on the injector. Also, note that all of the injectors could be restricted and giving relatively even readings. And finally, just because the pressure drops are within specifications, the spray pattern can still be wrong. This would prevent good atomization of the fuel which can cause misfires. In this case, the fuel injectors may need to be cleaned or replaced.

To inspect and test fuel injectors, first research the fuel injector testing procedure and specifications in the appropriate service information. Inspect the fuel injectors for leaks and damage. Then, follow the service information to test the fuel injectors. If the information does not provide a method for testing the injectors, follow the steps in **SKILL DRILL 49-4**.

SKILL DRILL 49-4 Inspecting and Testing Fuel Injectors

1. Measure the resistance of the suspect fuel injector and compare with specifications. If the resistance is not in specification, replace the injector.

2. Install the proper noid light in the fuel injector wire harness. Crank or start the engine and observe the noid light. Blinking light indicates proper control circuit operation. No blinking indicates a fault on the power or ground side of the injector.

3. Install the fuel pressure gauge on the fuel rail. Connect the injector pulsing tool to one fuel injector, according to the tool maker's instructions. Set it for the appropriate number and time of pulses.

4. Pressurize the fuel rail by turning on the ignition switch for a few seconds. Then turn the ignition switch off. Record the fuel pressure.

5. Activate the injector pulsing tool for the appropriate amount of time. Watch the pressure gauge, and record the pressure after the injector has been pulsed.

6. Repeat this test on each fuel injector, pressurizing the fuel rail each time before activating the injector pulsing tool. Record the pressure readings before and after each test. Compare all readings with specifications and to one another. Determine any necessary actions.

▶ Wrap-Up

Ready for Review

- Different types of fuel systems used to meter a precise amount of fuel into the engine are mechanically operated fuel injection systems, electronic fuel injection, throttle body injection, multipoint fuel injection, and gasoline direct injection.
- Gasoline fuel is derived from crude oil and is mainly a mixture of paraffins, naphthenes, aromatics, and olefins. Gasoline must be enhanced with additives such as detergents, octane boosters, and oxygenates to make it suitable for use in vehicle engines.
- For gasoline to burn properly, the air–fuel ratio should be about 14.7 to 1 by mass. A slightly lean mixture gives good fuel economy and low exhaust emissions and a slightly rich mixture can produce more power, but the extra fuel it uses increases fuel consumption and emissions.
- Three main fuel and air requirements for internal combustion are fuel, air, and pressure differential.
- Fuel supply system makes use of the following components to deliver gasoline to the fuel injection system: fuel tank, fuel filler neck, gas cap, evaporative emission control system, fuel pump relay, fuel pump, fuel tank sending unit, fuel filter, fuel lines, fuel rail, fuel pressure regulator, and fuel injector.
- The fuel pump is electrically operated and electronically controlled and is driven by a permanent-magnet electric motor. Types of pumps used in fuel pump assemblies are peripheral pump, side channel pump, and roller cell pump.
- Fuel lines are usually made of metal tubing or synthetic materials and carry fuel from the tank to the engine. The primary function of fuel filter is to prevent contaminants from reaching the injectors. A fuel rail is a special manifold designed to provide a reservoir of pressurized fuel for the fuel injectors.
- Fuel pressure is controlled by a fuel pressure regulator or by controlling the speed of the pump. The EFI pressure regulation system can be either a circulation type or a returnless type.
- Fuel injectors spray the proper amount of gasoline in the proper pattern in the intake ports, directly into the combustion chamber, or into a pre-chamber in the combustion chamber. The nozzle of an injector can be one of several types, such as rotating disc style, pintle style, and ball-valve style.
- Three basic EFI systems are: TBI, MPFI, and GDI. In TBI systems, fuel is sprayed into the top center of the throttle body and then mixed with the incoming filtered air. In MPFI systems, a fuel injector is used for each cylinder. Each injector is connected to the fuel rail, which supplies the injector with fuel under pressure.
- GDI systems spray the fuel directly into the cylinder and can run effectively on extremely lean fuel mixtures. Four typical modes of GDI operation are: stratified charge, stoichiometric, full power, and catalyst heating.
  - High-pressure GDI systems include the following components: low-pressure pump, low-pressure sensor, high-pressure pump, high-pressure sensor, fuel pressure regulator valve, and high-pressure fuel injectors.
- Drawback of GDI systems include: carbon build up on the intake valves and runners, low-speed preignition, expensive, complicated, and noisy operation than standard fuel injection systems.
- The carburetor supplies the engine with the approximately correct air–fuel mixture for all conditions of operation, atomizes the fuel and mixes it with air, and controls the delivery of the mixture. Carburetors design can have a downdraft design, side-draft design, or updraft design.
- The components of a carburetor are grouped together to create carburetor circuits. The six circuits are the float, idle, main metering, power, accelerator pump, and choke.
- Carburetors can have one, two, three, or four barrels to improve performance, particularly at high speeds. Throttle opening can be controlled mechanically or by a vacuum unit.
- Fuel filters can be located under the vehicle in the fuel line, under the hood in the engine compartment, or inside the fuel tank. Refer to the service information for the specified procedure for replacing the fuel filter.
- Fuel pump can be tested using a pressure/volume test, scan tool data stream test, or lab scope inductive current flow test. The pressure/volume test measures the fuel pressure being delivered to the fuel rail along with the volume of the fuel pump. A fuel delivery fault can be verified with a scan tool and a test drive.
- Check the fuel for contaminants and quality if a vehicle is not running properly and other common causes have not identified the issue. A visual inspection and an alcohol content test can be used to easily test the fuel.
- Retrieving codes, reading the data stream, or accessing mode 6 data can help narrow down the area of the fault when testing a fuel injector. An injector pressure drop test can be used to determine if the injector is restricted with deposits.

Key Terms

accelerator pump A small pump, usually located inside the carburetor, that sprays an extra amount of fuel into the carburetor air horn during acceleration.

accelerator pump circuit The carburetor circuit involved with heavy acceleration, directly connected to the accelerator pump. Also used to prime the engine during cold starting.

air supply system Equipment in a motor vehicle that delivers air to the engine.

catalyst heating mode An operating mode in GDI systems where fuel is injected near the top of the compression stroke; the fuel is ignited; and then an additional amount of fuel is injected during the power stroke to create heat later in the cycle so that the catalytic convert can be warmed up faster.

charcoal canister A device used to trap the fuel vapors. The fuel vapors adhere to the charcoal until the engine is started, and engine vacuum can be used to draw (purge) the vapors into the engine so that they can be burned along with the air-fuel mixture.

choke A device that provides a rich air-fuel mixture until the engine warms up by restricting the flow of air at the entrance to the carburetor, before the venturi.

computer-controlled carburetor A carburetor that changes and sets air-fuel ratio based on commands from a PCM that uses memory and input from sensors.

detonation The condition in which the remaining fuel charge fires or burns too rapidly after the initial combustion of the air-fuel mixture. It is audible through the combustion chamber walls as a knocking noise.

dieseling A condition in which the engine continues to run after the ignition key is turned off. Also referred to as run-on.

electronic fuel injection (EFI) An injection system in which fuel delivery is controlled electronically, allowing continuous adjustments to the air-fuel ratio.

element The replaceable portion of a filter, such as an air filter element or oil filter element.

filter sock The first line of defense in the fuel supply system. The sock typically consists of a fine mesh, which prevents most small particles from being drawn into the fuel pump and sent through the rest of the fuel system.

float chamber A chamber that holds a quantity of fuel at atmospheric pressure, ready for use.

fuel filter A device that removes impurities from the fuel before they reach to carburetor or injection system. Filters may be made of metal or plastic screen; paper, or gauze.

fuel filler neck The upper end of the fuel filler tube leading down to the fuel tank, which accepts the fuel hose nozzle at the gas station pump.

fuel gauge sending unit A device used for creating a signal to control a fuel gauge.

fuel metering system Equipment in a motor vehicle that delivers the proper amount of fuel to each cylinder.

fuel pressure regulator A system that controls the pressure of fuel entering the injectors.

fuel pump relay A relay to turn on or off the high-amperage circuit of the fuel pump.

fuel rail Tubing that connects several injectors to the main fuel line.

fuel supply system Equipment in a motor vehicle that delivers fuel to the engine.

fuel system Equipment in a motor vehicle that delivers fuel to the engine.

full power mode An operating condition in a GDI fuel system where the fuel is injected on the intake stroke in a slightly rich condition and allowed to fully mix with the air in the cylinder. This creates maximum power and reduces detonation.

gasoline A volatile, flammable liquid mixture of hydrocarbons, obtained from crude oil and used as fuel for internal combustion engines.

gasoline direct injection (GDI) A fuel injection system in which fuel is sprayed directly into the combustion chamber.

homogeneous mixture An air-fuel mixture evenly dispersed throughout the combustion chamber.

idle The speed at which an engine runs without any throttle applied.

indirect fuel injection Any fuel injection that is not sprayed directly into the combustion chamber.

injector A valve that is controlled by a solenoid or spring pressure to inject fuel into the engine.

knocking A noise heard when the air-fuel mixture spontaneously ignites before the spark plug is fired at the optimum ignition moment.

metering jet A calibrated orifice in a carburetor for fuel to flow through. Often it is replaceable for performance or economy desires.

multipoint fuel injection (MPFI) An injection system in which fuel is injected into the intake ports just upstream of each cylinder's intake valve, rather than at a central point within an intake manifold. Also called multiport injection.

octane rating A standard measure of the performance of a motor or aviation fuel. The higher the octane number, the more heat the fuel can withstand before self-igniting.

Pressure The force per unit area applied to the surface of an object.

Pressure The result of resistance to fluid flow.

returnless fuel injection system A type of injection system in which no hot fuel is returned to the tank, thus keeping the fuel in the tank relatively cool and minimizing vaporization.

sending unit The device that changes a temperature, fluid level, or pressure into a varying resistance. This varying resistance is in series with the related instrument cluster gauge, and the resulting voltage drop across the sending unit is what causes the needle on the gauge to indicate the related temperature, fluid level, or pressure.

stoichiometric ratio The optimum ratio of air to fuel for combustion—14.7 parts air to 1 part fuel by weight.

stratified charge A layering of the air-fuel mixture in the combustion chamber. Used when the fuel injector places fuel near the spark plug, but not the surrounding space.

throttle A device used to produce acceleration by controlling the air-fuel mixture.

throttle body injection (TBI) A fuel injection system that uses one or more fuel injectors mounted above or in the throttle body itself; also called single-point injection.

vacuum A pressure in an enclsed area that is lower than atmospheric pressure.

vapor lock A situation in which vapor forms in the fuel line, and the bubbles of vapor block the flow of fuel and stop the engine.

venturi A restriction (narrowed area) in the air horn.

Review Questions

1. Which type of fuel injection system uses a throttle body with one or two injectors mounted in it?
   a. MPFI
   b. PFI
   c. TBI
   d. GDI
2. What is the term for fuel that evaporates in the fuel pump and is then pumped through the lines, causing the engine to run poorly or even stall?
   a. Fuel lock
   b. Vapor lock
   c. Air bound
   d. Pre-ignition
3. What is a fuel's resistance to ignition called?
   a. Dilution
   b. Resistance
   c. Lead
   d. Octane
4. When an engine experiences two flame fronts colliding violently within its combustion chamber after the spark has occurred, it is called ______.
   a. Detonation
   b. Pre-ignition
   c. Flame out
   d. Dieseling
5. The perfect air-to-gasoline ratio of 14.7:1 for complete combustion in an engine is known as:
   a. Rich ratio
   b. Lean ratio
   c. Stoichiometric ratio
   d. Median ratio
6. All of the following are functions of a fuel pump assembly mounted in the fuel tank EXCEPT:
   a. To power the fuel pump relay
   b. To pressurize the fuel
   c. To strain the fuel
   d. To provide a fuel level signal
7. On most fuel injectors, what force closes the injector after opening to spray?
   a. The solenoid reversal
   b. The metering needle weight
   c. The stepper motor
   d. The return spring
8. On a 4-cylinder engine with a typical multipoint fuel injection system, how many fuel injectors would it have and where would they be mounted?
   a. Four, all in the throttle body
   b. Four, one near each intake valve
   c. Four, one near each spark plug
   d. Four, one in each cylinder
9. Despite the many benefits of a GDI fuel system, one drawback is carbon build-up. What should a customer do to help prevent the carbon build-up?
   a. Replace the spark plugs and filters every year
   b. Use the specified high-quality engine oil
   c. Use top-tier detergent gasoline fuel
   d. Add fuel injector cleaner at every fill-up
10. When replacing a fuel filter, if it is not found in a fuel line, where is it likely to be located?
   a. Inside of the throttle body
   b. Beside the fuel injectors
   c. Underneath the engine cover
   d. Inside of the fuel tank

ASE Technician A/Technician B Style Questions

1. Air–fuel ratio is being discussed. Technician A states that a gasoline ratio with excess air is known as rich. Technician B states that no matter the fuel type, 14.1 air to 1 fuel is ideal. Who is correct?
   a. Technician A only
   b. Technician B only
   c. Both Technicians A and B
   d. Neither Technician A nor B
2. Fuel delivery components are being discussed. Technician A states that the fuel filler neck allows fuel to be added to the tank. Technician B states that the fuel tank sending unit temporarily stores fuel vapors until they can be burned in the engine. Who is correct?
   a. Technician A only
   b. Technician B only
   c. Both Technicians A and B
   d. Neither Technician A nor B
3. Fuel pumps and filters are being discussed. Technician A states that a fuel pump gets its power directly from the PCM. Technician B states that the typical replaceable fuel filter is a metal screen on the fuel pump assembly. Who is correct?
   a. Technician A only
   b. Technician B only
   c. Both Technicians A and B
   d. Neither Technician A nor B
4. Fuel injectors are being discussed. Technician A states that fuel injectors need to spray the proper quantity of fuel. Technician B states that injectors need to spray fuel in the correct pattern. Who is correct?
   a. Technician A only
   b. Technician B only
   c. Both Technicians A and B
   d. Neither Technician A nor B
5. Fuel injection system types and their operation are being discussed. Technician A states that the sequential fuel injection type requires a driver for each injector in the PCM. Technician B states that in a simultaneous fuel injection system, the PCM will signal the injectors in

groups which require less wiring and more simple circuitry. Who is correct?
   - **a.** Technician A only
   - **b.** Technician B only
   - **c.** Both Technicians A and B
   - **d.** Neither Technician A nor B

6. Fuel injection systems are being discussed. Technician A states that most current fuel systems are returnless, which reduces evaporative emissions. Technician B states that some fuel pumps vary speed using the PCM to regulate pressure. Who is correct?
   - **a.** Technician A only
   - **b.** Technician B only
   - **c.** Both Technicians A and B
   - **d.** Neither Technician A nor B

7. GDI fuel systems are being discussed. Technician A states that an in-tank electric fuel pump can supply pressures of 3000 psi. Technician B states that the GDI low-pressure pump is normally driven by the engine camshaft. Who is correct?
   - **a.** Technician A only
   - **b.** Technician B only
   - **c.** Both Technicians A and B
   - **d.** Neither Technician A nor B

8. Fuel filter replacement is being discussed. Technician A states that the first step in fuel filter replacement is to drain the fuel tank completely. Technician B states that a scan tool may be required to relieve fuel pressure before service on GDI engines. Who is correct?
   - **a.** Technician A only
   - **b.** Technician B only
   - **c.** Both Technicians A and B
   - **d.** Neither Technician A nor B

9. A fuel pump test is being discussed. Technician A states that fuel pumps may warrant testing if an engine experiences low power or won't start. Technician B states that one of the first steps in testing a fuel pump is to remove it from the tank and inspect the fuel sock to see if it is plugged. Who is correct?
   - **a.** Technician A only
   - **b.** Technician B only
   - **c.** Both Technicians A and B
   - **d.** Neither Technician A nor B

10. Testing electronic fuel injectors is being discussed. Technician A states that fuel injector windings can be tested electrically by checking resistance. Technician B states that injectors can be tested mechanically, and one way is to perform a pressure drop test. Who is correct?
   - **a.** Technician A only
   - **b.** Technician B only
   - **c.** Both Technicians A and B
   - **d.** Neither Technician A nor B

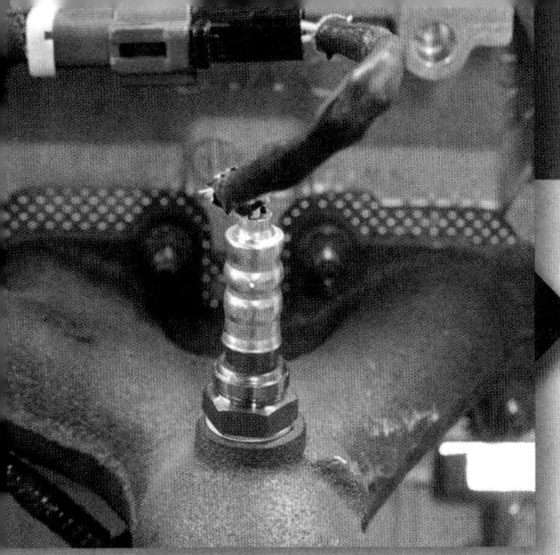

CHAPTER 50
Engine Management System

Learning Objectives

- **LO 50-01** Explain analog and digital signals.
- **LO 50-02** Explain potentiomer-based sensor operation.
- **LO 50-03** Explain thermistor-based sensor operation.
- **LO 50-04** Explain position sensor and speed sensor operation.
- **LO 50-05** Explain oxygen sensor operation.
- **LO 50-06** Explain how airflow is measured.
- **LO 50-07** Explain how air pressure is measured.
- **LO 50-08** Explain engine knock and how it is detected.
- **LO 50-09** Explain how switches are used in engine management.
- **LO 50-10** Explain the sections in a PCM.
- **LO 50-11** Describe controlled devices in engine management systems.
- **LO 50-12** Explain feedback and looping.
- **LO 50-13** Explain short- and long-term fuel trim and fuel shut off mode.

ASE Education Foundation Tasks

See Appendix A to view the 2017 ASE Education Foundation Automobile Accreditation Task List Correlation Guide.

▶ Introduction

LO 50-01 Explain Analog and Digital signals.

Previously, in the Motive Power Types chapter, we discussed the inefficiency of gasoline engines from the 1970s and before. These engines were about 25% efficient. This means that only 25% of the available energy in gasoline was used to power the vehicle down the road. The other 75% was wasted. Wasting large amounts of gasoline became unacceptable. This was because gasoline became more expensive and environmental concerns heightened. A movement began to improve the efficiency of the gasoline engine. This led to the addition of a whole range of technology and systems to address this issue.

Engine management systems were born out of a need for greater efficiency and emission control. So they have grown in scope and complexity over time. Here, we won't cover the systems as they evolved, but we look at current engine management systems. For the most part, older systems used many of the same components.

The engine management system controls the operation of the engine and all of its related systems. But we will focus primarily on the fuel and ignition system portions of the engine management system. We cover the diagnostics portion of the engine management system in the On-Board Diagnostics chapter. And we cover the emission controls portion of engine management in the Emissions Control chapter. For each of these topics, the engine management system has components in one of three separate groups:

- Sensors monitoring any condition that the powertrain control module (PCM) needs.
- The PCM receives sensor information, processes that information, and compares it with stored data. It then sends output commands to the controlled devices. This is called input–process–output.
- Controlled devices are also called output devices or actuators. They implement the commands sent by the PCM.

Today's engine management systems are much more powerful than previous systems. They also use many additional sensors and controlled devices. This allows for maintaining a high degree of control over the engine operation. But before we discuss each of the main components in the system, we need to explore the various types of electrical signals utilized in the system.

▶ Digital and Analog Signals

There are two main types of electrical signals, analog and digital. Analog signals are continuously variable. They typically change in strength over time. Conversely, a digital signal is a direct on/off signal with no in-between transition (**FIGURE 50-1**). Both types of signals are useful and can be used as either sensor signals or output signals.

Analog signals show how something changes over time. Or they can cause something to operate in a variable manner (**FIGURE 50-2**). For example, a throttle position sensor (TPS) typically sends an analog signal to the PCM. The signal voltage generated at closed throttle is typically around 1 volt. The signal voltage rises as the throttle is opened, reaching a maximum reading of almost 5 volts. Dashlight control is a good example of an analog output signal. As the dash light dimmer switch is rotated, it turns a rheostat. The rheostat sends a proportional output signal to the lights, causing them to dim or brighten.

Digital signals can be used to show whether something is on or off, or to command something on or off (**FIGURE 50-3**). A good example of a digital signal is the brake on/off signal. When the driver steps on the brake pedal, the brake on/off

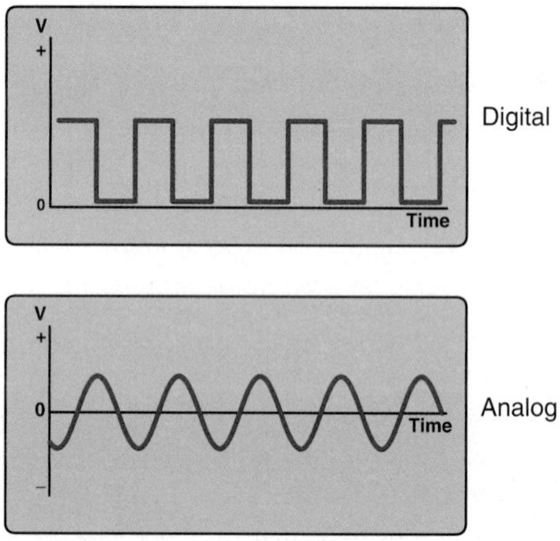

FIGURE 50-1 Typical analog and digital waveforms.

You Are the Automotive Technician

The intern at your shop is preparing for an upcoming Skills USA technical competition. It will include test questions about engine management systems and related terminology. Each competitor received the following questions to study prior to the competition. You are asked to give an overview of each concept to help get started.

1. What is the difference between digital and analog signals?
2. How do potentiometer-based sensors operate?
3. How do PTC and NTC thermistors operate?
4. How do inductive and Hall-effect sensors operate?
5. How does a wide-band oxygen sensor operate differently than a stoichiometric sensor?
6. What are the distinct sections of a PCM, and what do they do?
7. How are relays, solenoids, modules, and motors used to control engine operation?

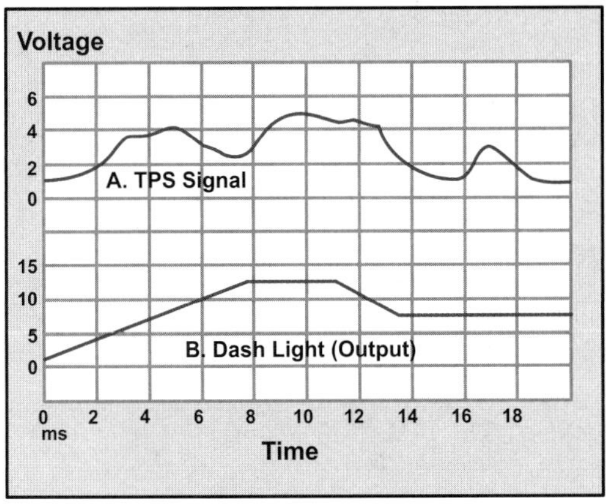

FIGURE 50-2 Analog signals. **A.** TPS sensor signal (analog). **B.** Dash light control output (analog).

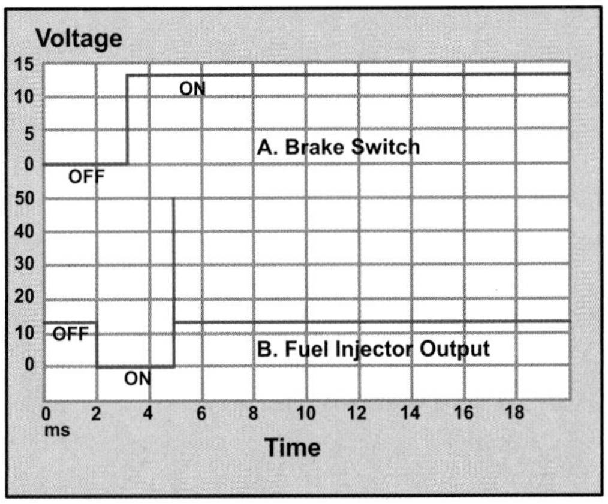

FIGURE 50-3 Digital signals. **A.** Brake on/off switch (digital). **B.** Fuel injector output (digital).

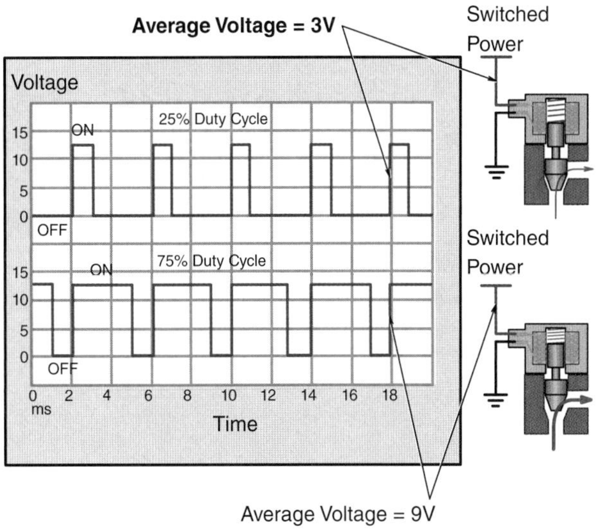

FIGURE 50-4 A digital signal can be used to create a varying output signal, as in this pulse-width-modulated **exhaust gas recirculation (EGR)** valve.

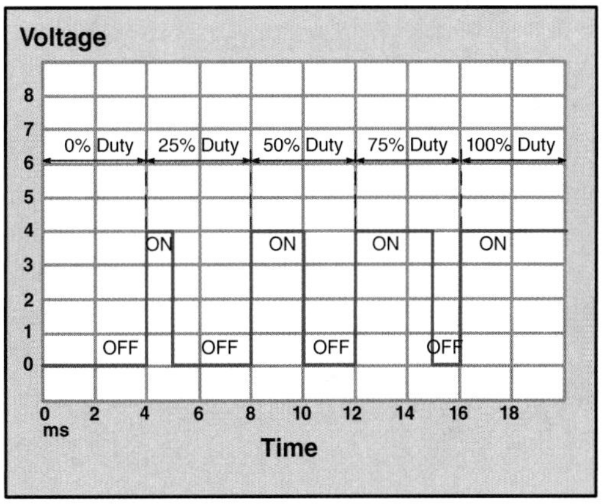

FIGURE 50-5 Duty cycle is a type of digital signal that uses pulse width modulation (PWM) to control the strength of the output signal generated. Duty cycle can generally be controlled from 0% to 100% on-time.

switch sends a full voltage signal to the PCM. When the brake pedal is released, the signal drops instantly to 0 volts. Digital signals can also be sent by the PCM as an output signal to turn devices on or off. For example, the PCM grounds the negative side of a fuel injector to turn it on. This allows it to inject fuel into the engine for a certain amount of time. When the PCM opens the ground, the injector turns off.

When cycled very quickly, digital signals can be used to create a varying output control signal. This results in a variable control similar to that provided by an analog signal (**FIGURE 50-4**). The PCM creates an alternating on/off signal many times per second. The percentage of time that the signal is off, versus on, can be varied by the PCM. This is called duty cycle (**FIGURE 50-5**). It allows a varying output signal to be sent to the controlled device.

In a varying analog signal, variable resistance is used to control the strength of the output signal. The electrical energy used by the resistance is wasted as heat. When a digital duty cycle signal is used, the circuit alternates between full on and full off. This gives a lower average current flow and voltage to the load. The percent of on-time determines the strength of the output control signal (average current flow and voltage). Therefore, there is no energy that is wasted as resistance. The increased energy efficiency of digital technology is being used more and more to control and monitor all of the systems in a vehicle. This includes the engine management system.

Another type of signal used to transmit data is frequency. Frequency is measured in cycles per second. A cycle is the distance between waves or wavelengths. A higher number of cycles per second indicates a higher frequency (**FIGURE 50-6**). Both analog and digital signals have frequency because both signals cycle between a high and low state.

Sound waves are a good way to illustrate frequency. Sound travels through the air by producing pressure waves—areas of high pressure and areas of low pressure. The rate at which these waves are created is called **frequency**. The higher the frequency, the higher the pitch of the sound. The range of human hearing is approximately 20 to 20,000 cycles per second.

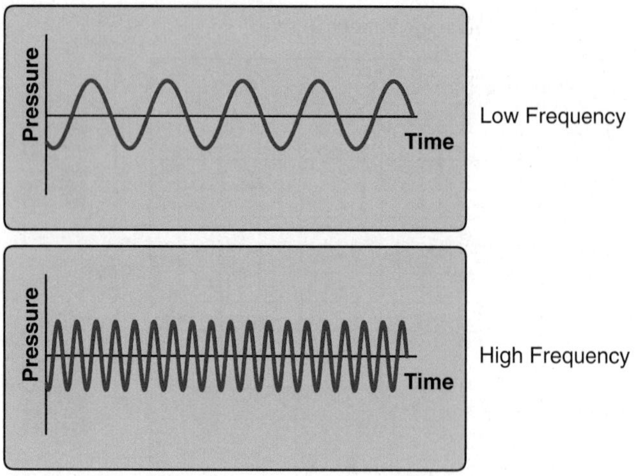

FIGURE 50-6 The higher the frequency, the greater the number of cycles per second.

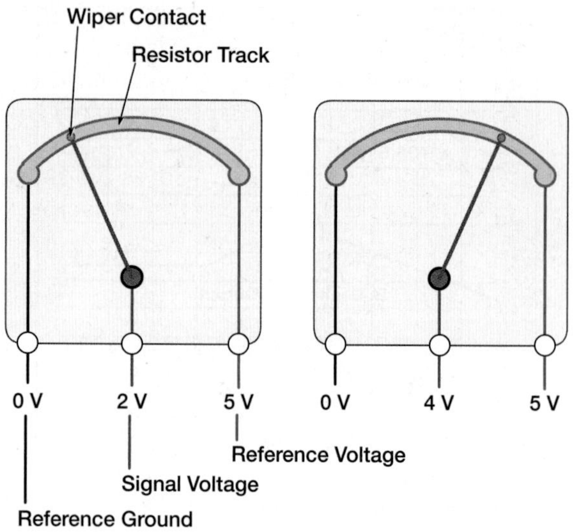

FIGURE 50-7 A potentiometer has a fixed resistor that drops the voltage evenly from input to output. The wiper contacts the resistor and sends the voltage out at that point on the resistor as a voltage signal.

An engine produces sounds across a wide range of frequencies. The mufflers and resonators in the engine exhaust system reduce these sounds to an acceptable level. In the combustion chamber, burning the air–fuel mixture creates sound. Proper burning creates a moderate level of sound. Detonation creates a much louder and sharper level of sound. A sensor can be used to monitor the frequency and volume of sound created in the combustion chamber. This can indicate whether the fuel is being burned properly. The PCM can then make adjustments when necessary. Frequency is also used in wheel speed sensors to indicate wheel speed and deceleration.

▶ Engine Management Sensors

LO 50-02 Explain potentiomer-based sensor operation.

As mentioned earlier, the engine management system is made up of three main parts. The first of which are input sensors. Each sensor monitors specific conditions. It then creates a signal that corresponds to the conditions. The signal is then sent to the PCM, where it is evaluated and ultimately used to make decisions. The most common sensors used in recent engine management systems include the following:

- Throttle position sensor (TPS)
- Accelerator pedal position (APP) sensor
- Engine coolant temperature (ECT or CTS) sensor
- Intake air temperature (IAT) sensor
- Crankshaft position (CKP) sensor
- Camshaft position (CMP) sensor
- Oxygen sensor (before and after catalytic converts) (O_2 or HO_2S) sensor
- Manifold absolute pressure (MAP) sensor
- Barometric pressure (BARO) sensor
- Mass airflow (MAF) sensor
- Air-conditioning compressor clutch signal
- Brake on/off (BOO) switch
- Knock (KS) sensor
- Vehicle speed sensor (VSS)
- Fuel pressure sensors
- Ignition pickup assembly

▶ Potentiometers

Some sensors use a **potentiometer** to create a variable voltage signal that corresponds to a specific condition. A potentiometer is a resistor with a mechanically variable wiper. The wiper contacts the resistor along its length. This allows it to access an increasing or decreasing level of voltage as the wiper moves (**FIGURE 50-7**). The resistor in engine management sensors is generally a thin film that is straight or circular in construction. It has three electrical connecting points. One each at each end of the resistor, and a third attached to a center sliding contact, called a wiper. The wiper is designed to move across the resistor. A reference voltage (typically less than battery voltage; 5 V is the most common) is applied to the resistor. This causes a steady current flow through the resistor, creating a consistent voltage drop. As the wiper contact moves across the resistor, it picks up the voltage at that point. The voltage is sent as a signal referencing that position.

▶ Throttle Position Sensor

The TPS gathers information on throttle position. It allows the control unit to make adjustments according to operating conditions. It is located on the throttle body and is operated by rotation of the throttle shaft.

A potentiometer-type sensor monitors throttle position over its full range. One end has a 5-volt reference voltage from the control unit (**FIGURE 50-8**). The other end is connected to the control unit ground. A third wire runs from a sliding contact in the TPS to the input circuits of the control unit. The sensor is a variable resistor. As the angle of the throttle valve changes, so does the voltage signal from the third wire.

At closed throttle, the reading is usually below 1 volt. As the throttle valve opens, the voltage signal rises. At wide-open throttle, the signal is about 4.5 volts. Ongoing monitoring of throttle position provides accurate data for the control unit.

To test a TPS, measure the voltage on the signal return wire with the vehicle in key on, engine off (KOEO) mode. Using the throttle pedal, slowly open the throttle all the way, and release it. The voltage should rise steadily and return to the closed-throttle level. While a DMM will catch large TPS faults, an oscilloscope may be needed for smaller ones.

▶ Accelerator Pedal Position Sensor

APP sensors are used on vehicles with an electronic throttle control. The APP sends signals about the position, direction, and speed of the throttle pedal. The PCM uses this information to determine driver intent, related to engine power needed. The APP is made up of two TPSs that work opposite to each other. One sensor moves from low voltage to high voltage. The other one moves from high voltage to low voltage (**FIGURE 50-9**). This creates a redundant circuit of opposite readings. The PCM can then quickly identify a problem with one or both of the APP sensor signals.

▶ Thermistors

LO 50-03 Explain thermistor-based sensor operation.

A resistor that changes its resistance with changes in temperature is called a **thermistor**. There are two types of thermistors. One is a positive temperature coefficient (PTC) type. The other is a negative temperature coefficient (NTC) type (**FIGURE 50-10**). In PTC thermistors, the resistance goes up with the temperature. In NTC thermistors, the resistance goes down as the temperature goes up.

Thermistors are used in various temperature-related sensors on modern vehicles. Coolant temperature sensors, fuel temperature sensors, and ambient temperature sensors are a few types (**FIGURE 50-11**). For example, the coolant temperature sensor is typically an NTC thermistor. As the engine temperature warms up, the internal resistance decreases. This resistance

FIGURE 50-8 A potentiometer-type sensor monitors throttle position over its full range. One end has a 5-volt reference voltage from the control unit.

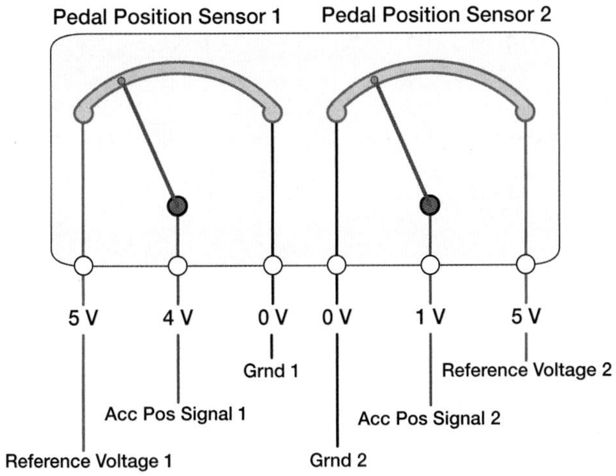

FIGURE 50-9 Typical APP sensor operation.

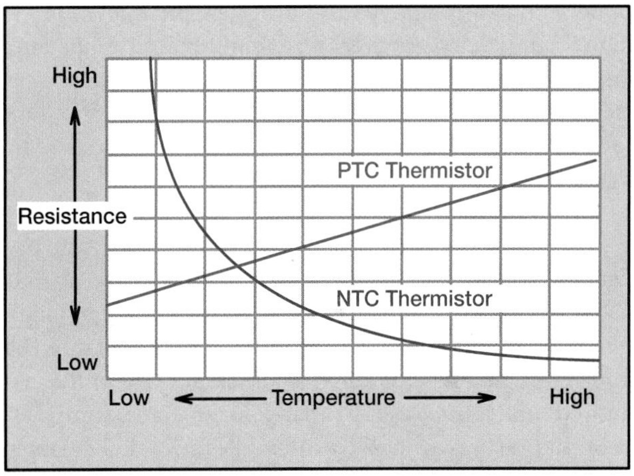

FIGURE 50-10 Typical resistance changes of PTC and NTC thermistors at different temperatures.

FIGURE 50-11 A typical thermistor used as a coolant temperature sensor.

change causes changes in a voltage signal used by the PCM. It decreases the injector on-time as necessary to keep the engine from running too rich as it warms up. As a result, the engine operates with the proper air–fuel ratio at various operating temperatures.

Engine Temperature Sensors

To maintain the air–fuel ratio within an optimum range, the control unit must account for coolant temperature and air temperature. Extra fuel is needed when the engine is cold. This is because the air is colder and therefore denser. It also does not vaporize liquid fuel as well.

The engine coolant temperature (ECT) sensor is immersed in coolant in the cylinder head, block, or intake manifold. It has a hollow threaded tube that has a resistor sealed inside it (**FIGURE 50-12**). This resistor is made of a thermistor semiconductor material. Its electrical resistance falls as temperature rises. The signal from the ECT is used by the PCM to control the air–fuel mixture of the engine throughout its operating temperature.

Enrichment of the air–fuel mixture occurs during engine cranking, then reduces as the engine warms up. The control unit monitors coolant temperature during engine operation. This design ensures that the engine runs smoothly at all temperatures. Some manufacturers use a cylinder head temperature sensor instead of an ECT to signal engine temperature. This sensor screws into the metal casting of the cylinder head to sense head temperature.

Air temperature is one factor that changes the density of the air. So air temperature has to be monitored by the PCM to better judge how much fuel needs to be injected. An air temperature sensor does just that. If the air temperature sensor is installed in the airflow sensor, it is called an **intake air temperature (IAT) sensor** (**FIGURE 50-13**). When it is installed in one of the intake manifold runners, it is called a manifold air temperature (MAT) sensor. In both cases, it relays information about air temperature and therefore density. The control unit can then vary the injector pulse width accordingly.

FIGURE 50-12 The engine coolant temperature (ECT) sensor is immersed in coolant in the cylinder head, block, or intake manifold.

▶ Crankshaft Position Sensor

LO 50-04 Explain position sensor and speed sensor operation.

Crank angle sensing sends information on the speed and position of the crankshaft to the PCM. It is used for the control of ignition timing and injection sequencing. The control unit can then trigger the ignition and injection to suit virtually all operating conditions. The **crankshaft position (CKP) sensor** may be mounted externally on the crankcase housing. In some applications, it may be mounted inside the housing of the ignition distributor.

There are different kinds of CKP sensors. **Inductive-type sensors** typically sense the movement of a toothed disc on the crank pulley (reluctor) (**FIGURE 50-14**). On some applications, they sense the ring gear teeth on the flywheel. These sensors do not make physical contact. The sensor is mounted on the crankcase housing. It consists of a stator with a central permanent magnet, and a soft iron core surrounded by an induction winding. The housing around all of these components is insulated from the iron core and windings. The iron core interacts magnetically with the teeth on the reluctor.

FIGURE 50-13 Intake air temperature (IAT) sensor.

FIGURE 50-14 An inductive-type sensor and toothed wheel.

The stator is positioned so that it has a very small clearance, or air gap, between the end of the soft iron core and the reluctor teeth. As the reluctor rotates, the teeth approach and leave the stator. This changes the air gap, which changes the strength of the magnetic field. Changing the magnetic field changes the voltage and current induced in the circuit.

As the tooth approaches the stator, the strength of the magnetic field increases. This induces a voltage, and current flow, in the winding. The polarity of the voltage is said to be positive, as it produces a current flow in a certain direction. When the tooth aligns with the stator, the magnetic field is at its strongest, but at that point it is not changing. Voltage and current flow fall to zero.

As the tooth moves away from the stator, the strength of the magnetic field changes again. Once again voltage and current are induced in the winding. But this time, current flow is in the opposite direction. The polarity of the voltage is now said to be negative.

Because polarity changes every time a tooth approaches and leaves the stator, the alternating voltage produces an alternating current (AC). It is the frequency of this alternating voltage that is used by the control unit to calculate engine rpm.

Crankshaft position is detected by a separate inductive sensor. It sends a signal to the control unit when a tooth or bore passes, and generates one pulse per revolution. It signals the control unit that the number-one piston is, for example, 80 degrees before top dead center. Ignition timing is then decided according to the operating conditions. It is then triggered to occur a certain number of degrees from that point.

When only one sensor is used, the reluctor is shaped to provide information on both crankshaft speed and position. The reluctor has a number of equally spaced teeth around its circumference, but one or two teeth are omitted. The frequency of the pulse from each tooth gives engine speed in rpm. But on each revolution, the pulse alters as the gap from the one or two missing teeth passes the sensor. This, again, gives the position of the number-one piston.

A Hall-effect CKP sensor measures the position of the crankshaft, as well as engine speed (rpm). The sensor is in very close proximity to a metal disc that is typically attached to the crankshaft or camshaft. Around the circumference of the disc, there are evenly spaced windows and shutters in one or two rows (**FIGURE 50-15**). The sensor is stationary and is typically mounted to the block. It has a magnet on one side and a sensor on the other side. The windows and shutters alternately allow and block the magnetic field from the sensor as the crankshaft rotates.

The sensor is made up of Hall-effect material. It reacts to the varying magnetic field and creates an on/off digital signal. This is what gives the PCM a measurement of the engine rpm. This sensor may have two rows of windows and shutters. If so, one is used for an rpm signal. The other is used for the crankshaft position signal. It determines ignition module and injector timing.

▶ Camshaft Position Sensor

The **camshaft position (CMP) sensor** (or cam sensor) sends constant data to the computer. It lets the computer know when the number-one piston is on the compression stroke. From there, the position of other pistons on their power strokes can be determined as well. It also identifies the position of the camshaft and valves on variable valve-timing engines. It does this by constantly reading a point on the camshaft or on the distributor (**FIGURE 50-16**). The CMP sensor works in conjunction with the CKP and a knock sensor. It provides information the computer needs to keep the ignition timing adjusted for all load conditions. These work together so the computer can reduce pinging and knocking.

▶ Ignition Pickup-Style Position Sensor

This type of ignition pickup is primarily a Ford design. Known as PIP, which stands for profile ignition pickup. It is a Hall-effect switch that provides camshaft position data to the ignition module and Ford ECC-IV processor (**FIGURE 50-17**). The pickup assembly is located inside of the distributor, which rotates at camshaft speed. Because the distributor is driven by the camshaft, the processor can determine crankshaft position.

FIGURE 50-15 A Hall-effect CKP sensor measures engine speed, represented in rpm of the crankshaft.

FIGURE 50-16 The camshaft position (CMP) sensor sends data to the computer to let it know which cylinder is on its power stroke as well as the position of the camshaft and valve.

FIGURE 50-17 A typical Ford EEC-IV ignition pickup assembly.

FIGURE 50-18 A typical VSS magnetic pickup sensor.

This data allows the Ford ECC-IV system to make adjustments to the ignition timing and fuel injection. One point to note. As the timing chain wears, the looseness of the chain affects the sensor signal. It is delayed from the actual crankshaft position.

▶ Vehicle Speed Sensor (VSS)

The vehicle speed sensor (VSS) is a detection device that sends the vehicle speed information (i.e., how fast the vehicle is traveling) to the PCM. The VSS is sometimes located in the instrument cluster as part of the speedometer. However, most manufacturers mount these sensors on the transmission or transfer case.

Most VSSs are a type of sensor called a magnetic pickup, or magnetic reluctance, sensor. This is similar to a magnetic CKP sensor. The sensor assembly consists of an iron core wrapped with fine copper wiring. There is also a reluctor wheel, which has raised "teeth" located around the circumference of the wheel. **FIGURE 50-18** shows the parts of a magnetic pickup sensor.

The reluctor wheel is attached to the output shaft of the transmission and spins with the output shaft. As the teeth of the reluctor wheel get closer to the iron core, a positive voltage is produced. As the tooth moves away, the voltage becomes negative. This process repeats over and over again as the reluctor wheel spins, creating an AC voltage. This signal is sent to the PCM, where the vehicle speed can be determined from the signal frequency. The magnetic pickup sensor produces its own voltage while the reluctor wheel is spun. These sensors do not need a separate wire to supply them with a reference voltage.

Another type of VSS is called a **reed switch**. There is a thin, movable blade switch inside the sensor. A magnet is installed in a rotating part, called the rotor, on the output shaft (**FIGURE 50-19**). As the output shaft spins, the rotor spins. When the magnet comes near the reed switch, the switch closes, allowing current to flow through the switch. When the magnet moves away from the switch, the switch opens, breaking the current flow. This opening and closing of the switch creates a DC square wave pattern that the PCM can use to determine vehicle speed.

Reed switches have to be supplied with electricity to operate. The amount of voltage supplied to the sensor varies with

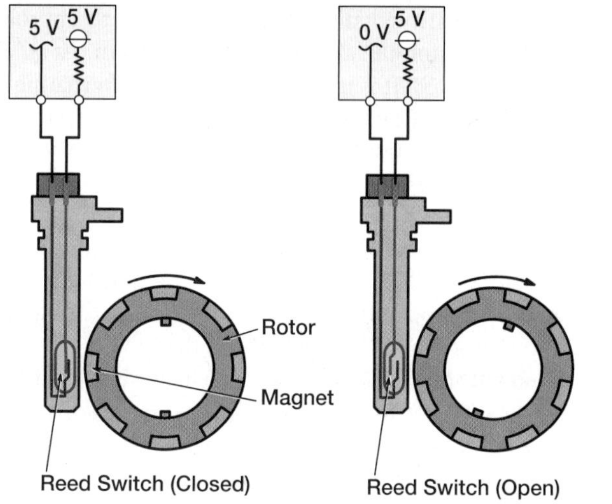

FIGURE 50-19 A reed switch VSS. As the magnet in the rotor passes the reed switch, the switch closes, allowing current to flow.

different vehicle manufacturers. Some of these switches apply a ground to a wire coming from the PCM. Others simply allow a voltage to pass from a voltage source, through the reed switch, to the PCM.

▶ Oxygen Sensor

LO 50-05 Explain oxygen sensor operation.

Oxygen sensors are positioned in the exhaust system (**FIGURE 50-20**). They provide the engine management PCM with an electrical signal that relates to the amount of oxygen in the exhaust gas. The PCM uses this and other information to determine the correct amount of fuel to be injected into the engine.

Originally, automotive oxygen sensors were stoichiometric sensors. Although these are still in use today, they indicate only whether the air–fuel ratio is rich (deficient oxygen) or lean (excess oxygen). They do not indicate how rich or how lean. Their output signal changes almost vertically on either side of an air–fuel ratio of stoichiometric (14.7:1). Above 0.45 volt indicates rich; below 0.45 volt indicates lean (**FIGURE 50-21**).

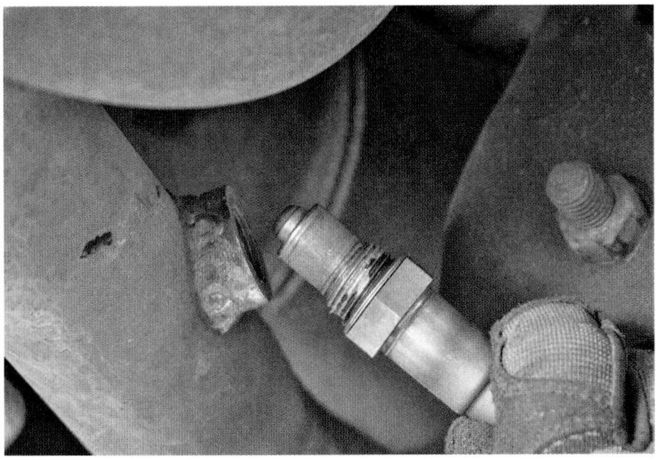

FIGURE 50-20 Typical oxygen sensor mounted in the exhaust system.

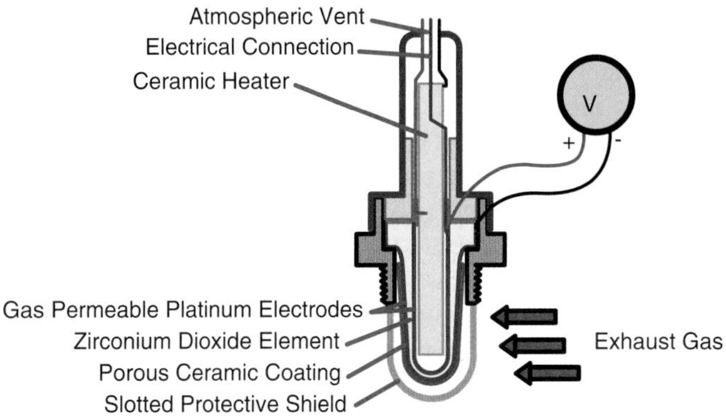

FIGURE 50-22 Oxygen sensors, when at temperature, create an electrical signal based on the difference in oxygen in the exhaust stream compared to ambient air.

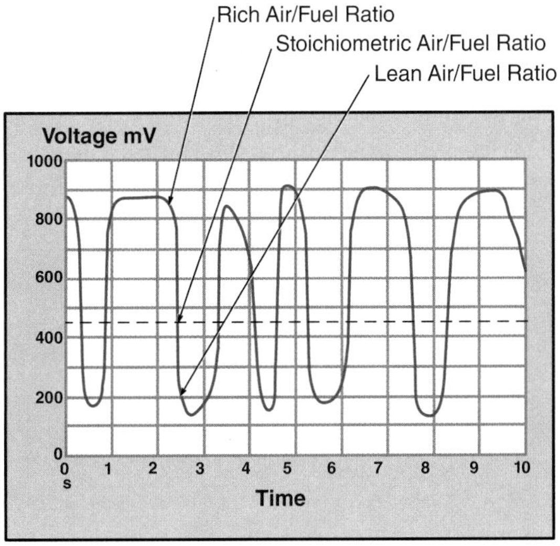

FIGURE 50-21 Many oxygen sensors typically have a switching point of 0.45 volt, indicating a stoichiometric air–fuel ratio.

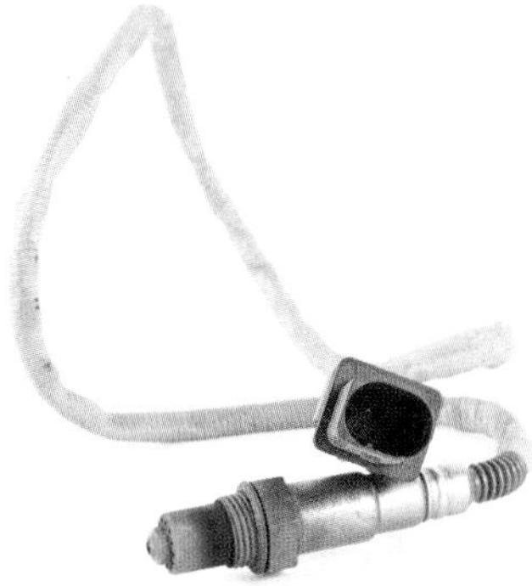

FIGURE 50-23 Wide-band oxygen sensor.

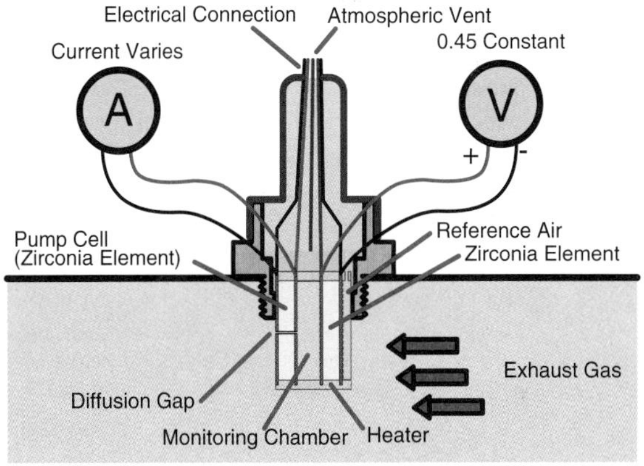

FIGURE 50-24 On wide-band oxygen sensors, the PCM sensor uses a solid-state pump to add or remove oxygen from the exhaust gas chamber.

It is the "Nernst cell" inside the sensor that produces this signal voltage. The Nernst cell operates by comparing the amount of oxygen in the exhaust gas with oxygen levels in the outside air (**FIGURE 50-22**). To operate, this cell has to be hot, approximately 600°F (315°C) or more. Exhaust gas can heat the sensor. However, when an internal heater is added, the sensor becomes operational more quickly. It is called a heated oxygen sensor (HO_2S). This is most relevant when the sensor is positioned in cooler parts of the exhaust system, away from the exhaust manifold.

As emission standards become tighter for both gasoline and diesel engines, a more precise signal is required. In these systems, the broadband oxygen sensor (or wide-band oxygen sensor) is used (**FIGURE 50-23**). It informs the PCM of a range of air–fuel ratios from about 9:1 up to atmospheric air. These systems are ideal for gasoline or diesel engines, particularly with the extremely lean-running gas direct injection (GDI) engines.

The wide-band oxygen sensor is far more sophisticated in operation than the earlier sensors. However, it still uses a Nernst cell. But there is a difference. The exhaust gas oxygen levels are referenced to a sealed chamber of air within the sensor and not outside air. It is from this chamber that the Nernst cell samples exhaust gas. This sensor uses a solid-state pump to add or remove oxygen from the exhaust gas chamber (**FIGURE 50-24**).

The computer controls the current flowing through the oxygen pump. This is so that the Nernst cell output voltage is maintained at that of a stoichiometric ratio. Current flowing in one direction through the pump adds oxygen. Current flowing in the opposite direction removes oxygen. The value and direction of current required to do this represents the level of oxygen in the exhaust gas. Reading this current allows the PCM to determine the amount of oxygen in the exhaust. This allows the PCM to precisely monitor the air–fuel ratio between approximately 9:1 and atmospheric air. The PCM can then more accurately control the amount of fuel injected to maintain low emission levels.

These sensors have an electrical heating element that heats the sensor quickly from a cold start. Typically, they warm up in less than 10 seconds. This is a shorter time than that taken by the older sensors. This faster heating is achieved because the sensors have far less material within the sensing element. Current through the heater is controlled by the PCM. This allows the correct operating temperature to be continuously maintained.

All 1996 and newer vehicles have an oxygen sensor before and after the catalytic converter. This is to monitor the converter operation. For a catalytic converter to change exhaust gases correctly, it must be capable of storing and releasing oxygen from the catalyst. When this happens, the amount of oxygen entering the converter will differ from that leaving. The PCM uses these two signals, along with response time, to determine if the catalytic converter is functioning properly. If a malfunction is detected for a predetermined period, the MIL will be illuminated. Many manufacturers also use the rear oxygen sensor to fine-tune the fuel trim. If the PCM determines that there is a catalyst-damaging event occurring, most systems will flash the MIL light to warn the driver.

Applied Science

AS-10: Standard/Metric: The technician can convert measurements taken in the standard or metric system to specifications stated in either system.

An apprentice technician has been assigned to replace an intake manifold and fuel rail on a vehicle. An experienced technician is working with the apprentice on certain key points. The technicians search the manufacturer's service information. They locate and discuss the proper torque specifications which are in standard and metric units. The experienced technician explains that if there is only one specification, conversion factors can be used to determine the other torque value.

The common English term "foot-pound" (ft-lb) is sometimes interchangeably used with the preferred SAE term "pound-foot" (lb-ft). To convert 1 lb-ft to the metric unit of Newton meters (N·m), we multiply the lb-ft value by 1.355 N·m. For example, 10 lb-ft = 13.55 N·m. To convert 1 N·m to lb-ft, we multiply the N·m value by 0.737 lb-ft. For example, 10 N·m = 7.37 lb-ft.

Conversion tables are available. A number of interactive calculators are available online to convert various units from standard to metric and metric to standard.

▶ Air Supply

LO 50-06 Explain how airflow is measured.

In MPFI and GDI systems, the air required for the combustion of the fuel travels first through the air filter. It then travels through the throttle valve, and into the common manifold, or plenum chamber. From there, individual intake runners, or pipes, branch off to each cylinder. Ideally, all of these runners are of equal length (**FIGURE 50-25**). This allows for a more uniform distribution of the air and fuel delivered to the cylinders. The design of the intake system determines how large an air mass can be drawn into a cylinder at any given engine rpm.

With unobstructed passages to each cylinder, the cylinder fills with air as efficiently as possible. The breathing of the engine, or its volumetric efficiency, is improved. As more air is forced into the cylinder, the air–fuel mixture becomes denser. This increases the pressure on the piston during the power stroke, as well as engine power output.

The temperature of the air influences density of the air–fuel mixture. Cold air is denser than hot air, so it has a greater mass in any given volume. Thus, on most intake systems, the air entry is located away from engine heat. Because the manifold carries air only, it does not have to be heated by coolant. Filtered air arrives at the intake port as cool and dense as possible, ready for mixing with the fuel from the injector. Refer to the Induction and Exhaust chapter for more information on intake manifolds.

Air Measurement

The amount of air entering the engine must be measured or calculated. This is so that the amount of fuel injected forms a suitable mixture for the engine operating conditions at that time. The volume of air can be determined in two ways—speed density calculation or direct measurement.

The speed density method calculates the volume of air entering the engine, which varies according to the engine speed and the load. If calculations are used, they are based on the following:

- engine rpm as measured by the CKP sensor,
- engine load as measured by the MAP sensor,
- intake air temperature,
- throttle position, and
- coolant temperature.

The airflow is then calculated using preprogrammed fuel maps. The oxygen sensor is used to provide feedback about the richness or leanness of the fuel that has been injected.

Mass Airflow Sensor

A mass airflow (MAF) sensor directly measures the mass of filtered air entering the engine. It measures in grams per second (gps). The MAF sensor is typically installed in the air intake

FIGURE 50-25 Individual intake runners branch off to each cylinder.

tube leading to the throttle body assembly. However, some MAF sensors are built into the throttle body assembly and others are mounted on the air cleaner assembly.

All air entering the intake flows through the MAF sensor. It then flows through a sealed, flexible tube or duct to the throttle body, after which it enters the engine. Older systems used a vane-type airflow meter. These meters have a spring-loaded vane, or flap, that deflects according to how much air enters (**FIGURE 50-26**). The electrical signal to the PCM varies with this deflection as contacts attached to the vane move across a potentiometer.

A newer style uses a hot wire to measure the mass of air entering the engine. The sensor uses a platinum wire placed in the airstream within the MAF sensor housing (**FIGURE 50-27**). A constant voltage drop is maintained across the wire. As the wire heats up, its resistance rises, and current flow drops. As the engine is started and air is drawn across the wire, the air cools the wire. This lowers its resistance and increases the current flow. The more air that flows across it, the higher the current flow. Also, air temperature and density affect the amount of heat removed from the wire. The cooler and denser the air, the more it cools, and the greater the current flow. This cooling effect is directly related to the mass of air passing through the airflow meter. So the airflow meter can send a signal to the PCM that indicates the mass of the air flowing through it.

Some hot-wire MAF sensors have a built-in cleaning feature. The PCM applies a higher current flow across the hot wire for a second or two to burn off any residue or contaminants. This typically occurs just after engine shutdown. Some hot-wire sensors require periodic manual cleaning. Clean these components carefully, using a cleaner recommended for MAF sensor cleaning. Note that some manufacturers specifically say not to clean their MAF sensors. Instead, they recommend replacing them if they are not reading correctly. Be careful to check the manufacturer's recommendation regarding this. And always follow the manufacturer's directions precisely if cleaning is allowed.

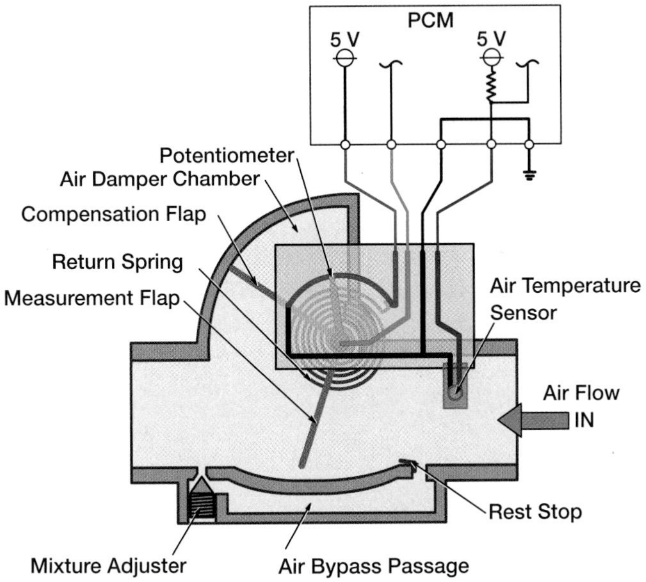

FIGURE 50-26 Typical vane-type airflow meter.

The Karman vortex airflow sensor measures airflow disruptions. There are three types of vortex sensors. But all of them use an air regulator to first smooth out or reduce any turbulence in the airflow entering the sensor. A triangular vortex column then disrupts a portion of the airflow creating vortices or whirlpools in the airflow (**FIGURE 50-28**). These "whirlpools" are then measured by one of three different methods:

1. Optical type—The air vortices cause a mirror to oscillate and interrupt an infrared signal.

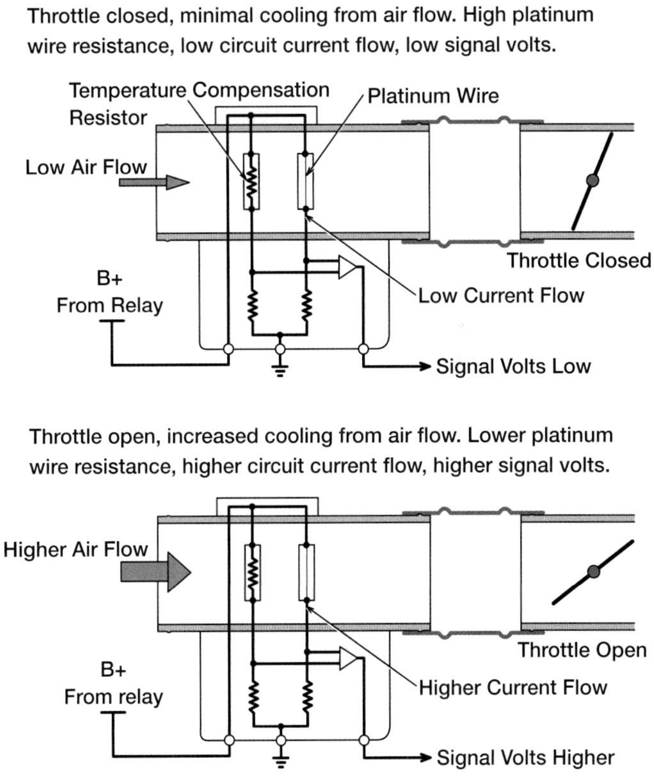

FIGURE 50-27 A hot-wire mass airflow sensor uses the cooling effect of air passing across the heated wire to change the current flow. The change in current flow indicates the mass of the air entering the engine.

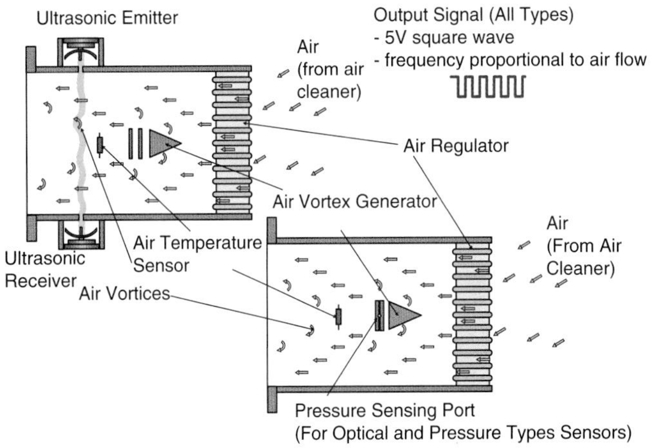

FIGURE 50-28 A triangular, vortex-generating rod disturbs the airflow and causes whirlpools of air to form in the body of the sensor.

2. Ultrasonic type—The whirlpools disrupt an ultrasonic beam that is transmitted through the air stream.
3. Pressure type—The pressure waves caused by the vortex column are measured by what is essentially a pressure sensor, similar to a MAP sensor. This is now the most common type.

Regardless of the method used to measure the whirlpools, all of these sensor types produce a 5-volt digital signal. The frequency of the signal is proportional to the airflow through the sensor.

The intake system must be airtight, with no leaks downstream of the airflow meter. Otherwise, inaccurate signals will be sent to the PCM. Any air that leaks into the system without going through the airflow meter is called false air, or unmetered air. If air bypasses the airflow meter, it tends to lean out the mixture. This is because the PCM is not accounting for the extra air. Always be on the lookout for cracked or unplugged vacuum hoses that may contribute to this problem.

▶ Manifold Absolute Pressure Sensor

LO 50-07 Explain how air pressure is measured.

Changes in engine speed and load cause changes in intake manifold pressure. A MAP sensor measures these pressure changes and converts them into an electrical signal. The signal may be an output voltage or a frequency. By monitoring output signals, the PCM senses manifold pressure. It uses this information to calculate the basic fuel requirement.

The MAP sensor can use a piezoelectric crystal. If there is a change in the pressure exerted on this crystal, the resistance changes. This alters its output signal. Some manufacturers also use the changes in manifold pressure to determine whether the **EGR valve** opened or not. This is a perfect job for the MAP sensor.

The sensor is either connected to the intake manifold by a small-diameter, flexible tube. Or it is screwed directly into the intake manifold. The control unit typically sends a 5-volt reference signal to the sensor. As manifold pressure changes, so does the electrical resistance of the sensor. This, in turn, produces a change in the output voltage (**FIGURE 50-29**). During idling, manifold pressure is low (high vacuum), which produces a comparatively low MAP output voltage. With a wide-open throttle, manifold pressure is higher. It is closer to atmospheric pressure (low vacuum), so the output voltage is higher.

Another type of MAP sensor that has been used by some manufacturers is one that sends a varying frequency to the PCM. As the manifold pressure changes, the sensor sends a corresponding frequency signal to the PCM. Frequency can be read on a digital multimeter that has a frequency function.

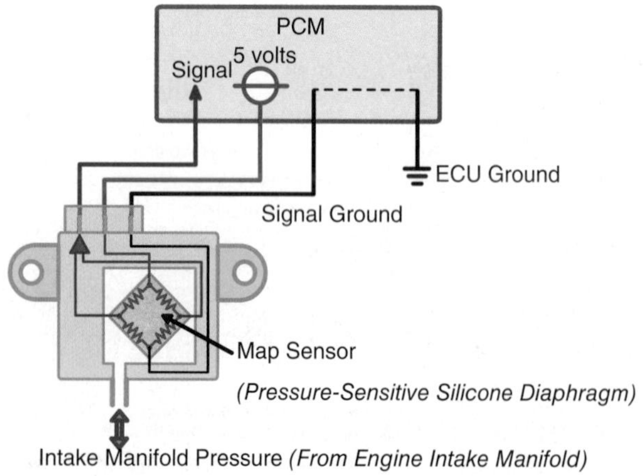

FIGURE 50-29 As manifold pressure changes, so does the electrical resistance of the sensor, which in turn produces a change in the output voltage.

▶ Barometric Pressure Sensor

Barometric pressure is the pressure on the air around us. It is caused by gravity pulling the air towards the Earth. Barometric pressure changes primarily with temperature and altitude. Barometric pressure affects the mass of air in a given volume. The higher the barometric pressure, the more air molecules in the space. The more air molecules there are, the more oxygen in the space. The more oxygen in the space, the more fuel is needed for proper combustion.

The barometric pressures at sea level and, for example, in Denver, Colorado are extremely different. This is because Denver, which is called the mile-high city, sits at a fairly high altitude. So the air in Denver is at a much lower barometric pressure than a city by the ocean. This means that the air in Denver has less oxygen. In fact, it has about 10% less since barometric pressure decreases by about 2% per 1000 feet of elevation. And yet, the vehicle must function properly at both locations or anywhere in-between.

A **barometric pressure (BARO) sensor** is very similar to a MAP sensor. But it specifically measures barometric pressure (**FIGURE 50-30**). It is used to modify the base airflow calculation. The BARO sensor works in conjunction with the MAP sensor to provide constant pressure difference information to the PCM. This information is critical for proper air–fuel mixture as well as ignition timing.

Older BARO sensors were typically mounted on the firewall or along the fender in the engine bay. Most late-model vehicles actually use the MAP sensor to take a barometric pressure reading before the vehicle is started. So they don't have a separate BARO sensor.

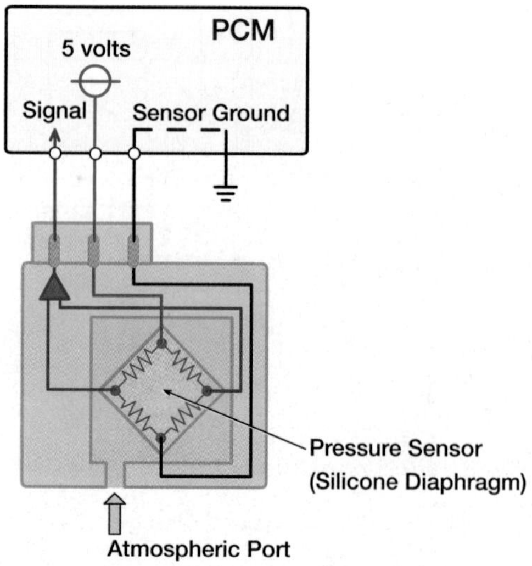

FIGURE 50-30 Illustrated view of a BARO sensor.

▶ Fuel Pressure Sensor

Most vehicles now use a fuel pressure sensor mounted on the fuel rail (**FIGURE 50-31**). This sensor allows the PCM to monitor the fuel pressure in the fuel rail. The PCM can then control the speed of the fuel pump to maintain the correct pressure in the system. GDI systems typically use two fuel pressure sensors. One sensor is used in the low-pressure system. And a separate sensor is used in the high-pressure system.

Applied Science

AS-83: Barometric Pressure: The technician can demonstrate an understanding of the relationship of barometric pressure to engine performance.

Atmospheric pressure is a measure of the pressure exerted by the weight of the atmosphere. This is also called barometric pressure because barometers are used to measure it. The term "barometer" comes from the Greek words for "weight" and "measure." A mercury barometer compares atmospheric pressure with a column of mercury and measures the results in inches or millimeters of mercury.

A barometric pressure (BARO) sensor measures barometric pressure. It is crucial to most fuel injections systems. The barometric pressure varies by geographic elevation. The vehicle must function properly in any given location. Accurate pressure readings are critical for proper air–fuel mixture. It is also important for ignition timing.

The principle of the operation of barometric sensors is based upon the flexing of a silicon chip. The output voltage signal varies based upon the amount that the chip flexes due to barometric pressure. The PCM uses this information along with other data to determine the proper adjustments that should be made to the fuel system as well as other systems.

▶ Knock Sensor

LO 50-08 Explain engine knock and how it is detected.

Engine knock occurs in the combustion chamber when there is an unwanted spike in pressure, caused by preignition or detonation. Some vehicles are equipped with a **knock sensor** (**FIGURE 50-32**). It monitors the noise that is created by the pressure spike. This noise, or "knock," can be created by an excessive load on the engine, ignition timing that is too advanced, or an overheated engine.

The PCM can adjust the ignition timing to help reduce the knocking. In addition, adjustments to the ignition timing and fuel delivery can be made if the vehicle begins to overheat. Following is a brief description of what occurs during excessive engine load and during engine overheating:

- Excessive load on the engine
  1. Ignition starts.
  2. The expanding gases create a pressure wave designed to push the piston down.
  3. The forces opposing piston movement are too high.
  4. The piston accelerates slowly and maintains a small volume above the piston.
  5. The unburnt mixture is compressed by the advancing pressure wave.
  6. This fuel self-ignites due to an increase in temperature as the mixture is compressed and creates its own flame front.
  7. The two advancing flame fronts create a huge spike in pressure, creating engine knock.
  8. This knock has enough energy to badly damage pistons, rings, spark plugs, and rod bearings.
- Overheated engine—faulty thermostat
  1. Ignition starts.
  2. The expanding gases create a pressure wave designed to push the piston down.
  3. The temperature of the unburnt fuel is too high due to the overheated engine.
  4. The unburnt mixture is compressed further by the advancing pressure wave.
  5. This fuel self-ignites due to an increase in pressure/temperature and creates its own flame front.
  6. The two advancing flame fronts create a huge spike in pressure, causing engine knock.

One way of overcoming engine knock is to allow the fuel to be ignited later in the compression stroke. Doing so requires

FIGURE 50-31 Fuel pressure sensor mounted on the fuel rail.

FIGURE 50-32 Knock sensor mounted on an engine.

having less ignition advance. If the fuel is ignited later, then the pressures above the piston will be less during the early stages of the power stroke. This reduces engine knock.

▶ TECHNICIAN TIP

Another way of preventing engine knock is to use fuel with a higher octane rating. This makes the fuel harder to ignite and slower to burn, both of which reduce engine knock.

The function of the knock sensor is to produce an electrical signal that corresponds to engine knock. The sensor is screwed into the engine block, where it is influenced by all engine vibrations (**FIGURE 50-33**). Using a piezoelectric crystal, the sensor produces a signal voltage proportional to the vibrations applied to it. The PCM interprets the strength, frequency, and timing of the signal to determine if knock has occurred.

If knock is occurring, the PCM will retard the timing a few degrees and watch the knock sensor signal. It does this to determine whether the knock is still present. If so, the PCM will continue to retard the timing in steps, within its limits, until knock is eliminated. If knock is not occurring, the PCM will advance the timing a degree or two and watch the knock sensor signal. Thus, the timing is kept at the optimum point for the driving conditions. V-configured engines often have two knock sensors installed, known as twin knock sensors. This design allows the PCM to have control over knock on separate banks.

Knock sensor technology has improved. But, it cannot always know the difference between noises. For example, between an engine knock and a loose air-conditioning compressor bracket that rattles. In this situation, the rattling air-conditioning compressor bracket causes a noise that the knock sensor picks up. But because the "knock" in this case does not go away as the timing is retarded, the PCM gradually reduces ignition advance until it reaches its limits. This causes low power, poor fuel economy, and possible engine overheating. Monitoring the knock sensor signal and the amount of ignition timing retard on a scan tool will help you identify this issue.

FIGURE 50-33 The sensor is screwed into the engine block, where it is influenced by all engine vibrations.

▶ Switches

LO 50-09 Explain how switches are used in engine management.

Switches provide a digital on/off signal to the PCM. This means they can be used to indicate when devices and circuits are activated that require compensation by the PCM. For example, the brake on/off switch indicates to the PCM that the driver is applying the brakes. This information is useful for reducing or stopping fuel injection pulses, deactivating cruise control, and so on.

While at idle, engine speed can be adversely affected by fairly minor changes in load. For example, loads from transmissions, power steering pumps, and air-conditioning compressors. For optimum engine control, it is necessary for the PCM to be informed of these loads before the engine speed drops. Switches do that (**FIGURE 50-34**). They inform the PCM when the power steering pump pressure is high. Or that a drive gear in an automatic transmission is selected. Or that the air-conditioning compressor is being engaged. When the PCM receives the signal from the switch, it can immediately take action. It can start to increase the engine's power to meet the need of the additional load. The result is excellent PCM control of idle speed as loads are placed on the engine.

Applied Math

AM-15: Mean/Median/Mode: The technician can calculate the average (mean) of several measurements to determine any variance from the manufacturer's specifications.

A new vehicle was purchased that was advertised to have a fuel economy rating of 24 mpg city driving and 30 mpg highway driving. The owner decided to calculate the average (mean) of several trips to see if the advertised ratings were correct. The plan was to check both city and highway driving separately, using several tests for each category.

The city mileage test consisted of three separate fill-ups, with miles recorded from the odometer. The first fill-up required 14.1 gallons of fuel for 330 miles of driving, equaling 23.4 mpg. The second fill-up required 13.6 gallons of fuel for 338 miles, equaling 24.9 mpg. The third fill-up required 12.9 gallons for 286 miles, equaling 22.2 mpg. To calculate the average (mean), he added 23.4 mpg + 24.8 mpg + 22.2 mpg and divided by 3. The result is 23.5 mpg city driving, which is very close to the advertised 24 mpg.

The highway driving test consisted of two fill-ups on a road trip. The first fill-up required 9.5 gallons for 260 miles, equaling 27.4 mpg. The second fill-up required 10.1 gallons for 305 miles, equaling 30.2 mpg. To determine the mean, he added 27.4 mpg + 30.2 mpg and divided by 2. The result is 28.8 mpg, which is slightly less than the advertised 30 mpg for highway driving.

▶ Powertrain Control Module

LO 50-10 Explain the sections in a PCM.

The PCM is like the brain of the engine. It gathers information, processes that information based on stored data, makes decisions, and commands components to operate in specified ways. It has the ability to learn from its actions and to use that information to make better decisions in the future. It can also perform self-diagnostics on virtually all of the components in the system, including itself. In basic terms, the PCM is composed of four distinct parts (**FIGURE 50-35**):

- A memory and storage section
- A sensor input/output signal processing section
- A data processing section
- An output drivers section

The memory and storage section contains the software for the computer. They also serve as a place to store diagnostic trouble codes (DTCs), freeze-frame data, and learned data. Software is loaded into the memory, which is the basis for processing data. The software contains all the information needed for the processor to interpret the sensor information. Then, it can make decisions based on the programmed information. It also contains the instructions for performing self-diagnostic tests on most of the circuits and the PCM itself.

Often, you can update the software in the PCM's memory by uploading software updates. This is called a reflash process. By using software that can be updated, manufacturers can design fixes for many of the issues that may arise once a vehicle goes into operation. A software update is much easier, quicker, and less expensive for resolving common customer or drivability issues than replacing parts.

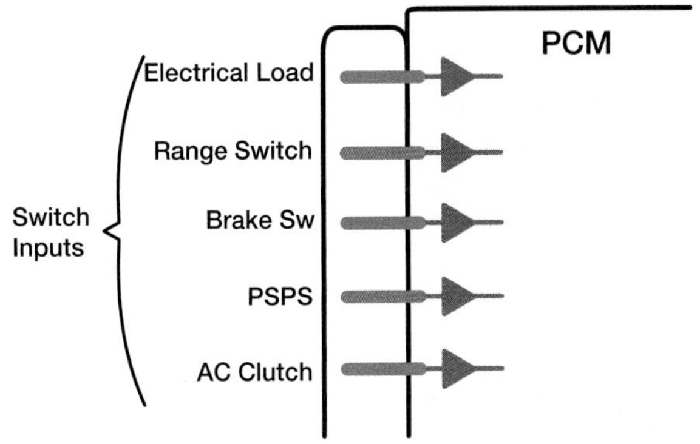

FIGURE 50-34 Switches signal the PCM when certain actions take place.

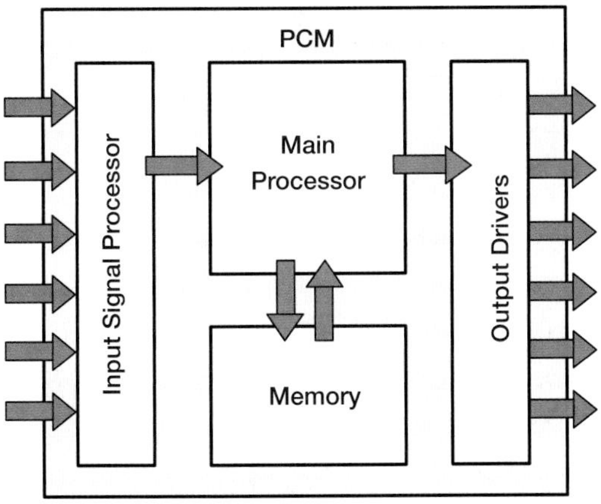

FIGURE 50-35 The sections of a PCM.

The input signal processing section handles sending the proper reference voltage to many of the sensors. It also receives the sensor signals. Then, it processes them into information that the main computer processor can use (**FIGURE 50-36**).

The data processor and drivers are hardware related. So they make up the physical portion of the PCM. The processor function receives the data from the input signal processing section and compares it with data maps in the memory. Once the sensor data have been analyzed, the processor looks up the appropriate data map. It then makes a decision about what actions to take (**FIGURE 50-37**).

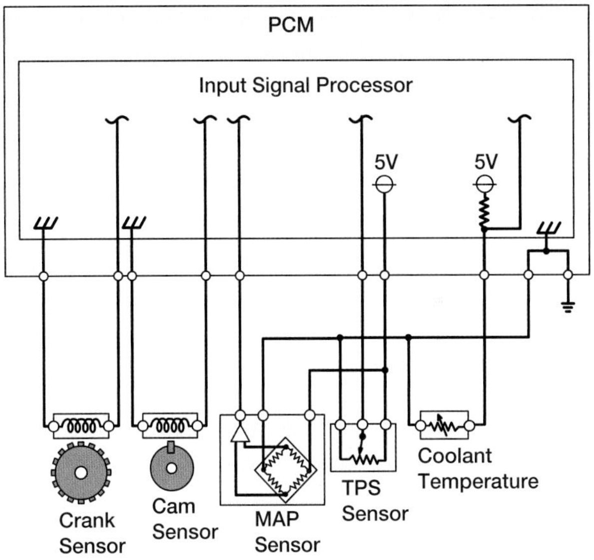

FIGURE 50-36 The input signal processing section sends reference voltage to many of the sensors. It also processes the signals returning from the sensors so the data processor can use the data.

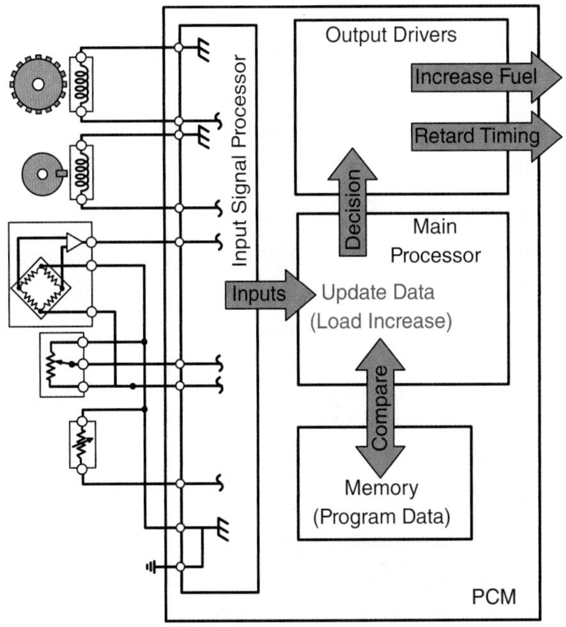

FIGURE 50-37 The data processor evaluates the sensor data and compares it with data maps and decides what actions to take. It then commands the output devices to send output signals to the actuators.

The processor then sends the appropriate commands to each of the output drivers. They are electronic switching or control devices. They control each of the actuators in the engine management system. The processor also commands the diagnostic tests to be run on the system and evaluates the results compared with the information stored in memory. The output drivers typically send either simple on/off signals or pulse-width-modulated signals to the actuators. The type of signal sent depends on the actuator being controlled (**FIGURE 50-38**).

The PCM also communicates with other electronic control modules on the vehicle. This is because the control modules share data from each other's sensors through a communication network. Some of those control modules will be on the same network. Others will be on other networks, because most late-model vehicles use more than one communication network. This is outlined in the chapter on Body Electrical System. In the case of multiple communication networks, the PCM sends a command over the network system. The appropriate module carries out that command when the message is received. The module has its own output drivers that control the actuator.

The PCM is a gateway node that is connected to each of the other networks. A gateway node translates the information from, or to, the other networks so it can be shared between networks. Here is an example of how it would work. Suppose the driver begins to drive the vehicle forward. The PCM monitors the VSS signal for engine management purposes. The body control module (BCM) is programmed to lock the doors at a predetermined speed. The BCM monitors the VSS signal coming across the network from the PCM. At the predetermined speed, the BCM sends the command to lock the doors on its network. Each door control module receives the message and commands its own output driver to power the circuits to lock the door. The doors then lock at the predetermined vehicle speed. A confirmation may be sent back across the network that the message was received and the action taken (**FIGURE 50-39**).

▶ Controlled Devices

LO 50-11 Describe controlled devices in engine management systems.

Remember, sensors monitor conditions and send information to the PCM. Actuators such as relays, solenoids, modules, and motors do the work. When a specific criterion is met, or the vehicle driver gives a certain command, an input signal is sent to the PCM for processing. Then, an output signal is sent to the corresponding controlled device (relay/solenoid/module/motor) to make an action happen. Next, we look at the most common engine management output devices related to the fuel and ignition systems.

▶ Relays

A **relay** is often used as an actuator to control other actuators when the computer must control a high-current load. The computer typically controls the relay ground circuit. When a high-current job is requested or commanded, the computer simply completes the ground circuit. The relay coil field pulls the mechanical contacts closed. This allows the high current levels to flow to the appropriate load. Therefore, relays can be used to control many different devices. Some examples are fuel pumps, starter motor solenoids, and AC compressor clutches.

Most fuel pumps are operated by a fuel pump relay. The relay controls power to the fuel pump and is typically controlled by the PCM (**FIGURE 50-40**). On a mechanical pump, it would quit pumping if the vehicle were involved in an accident that resulted in the engine dying. But an electric fuel pump will continue

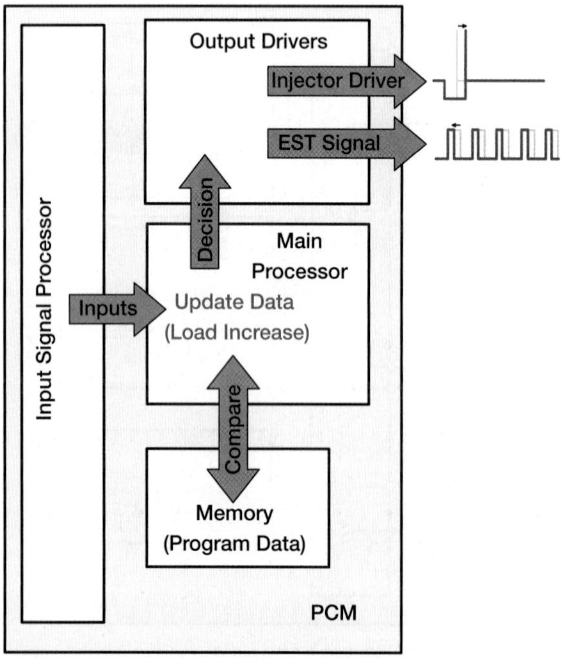

FIGURE 50-38 The output drivers send either on/off signals, pulse-width-modulated signals, or timing signals to the actuators.

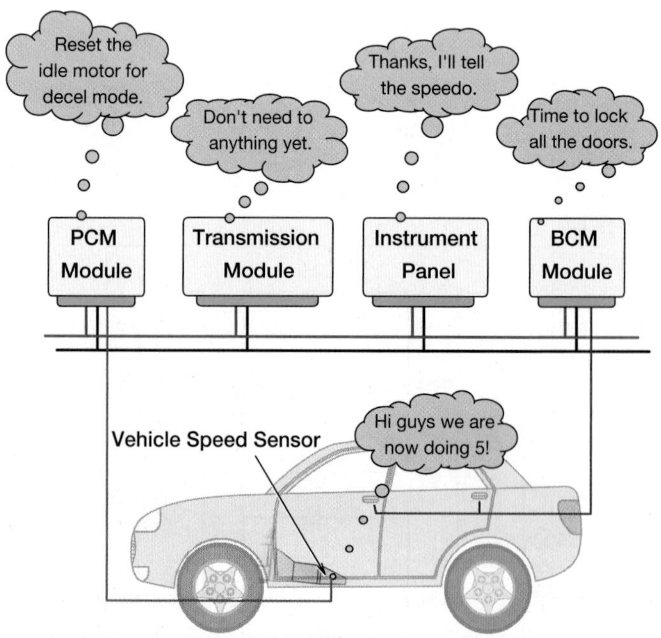

FIGURE 50-39 The PCM shares sensor data on the vehicle networks so that other control modules have access to those data they need.

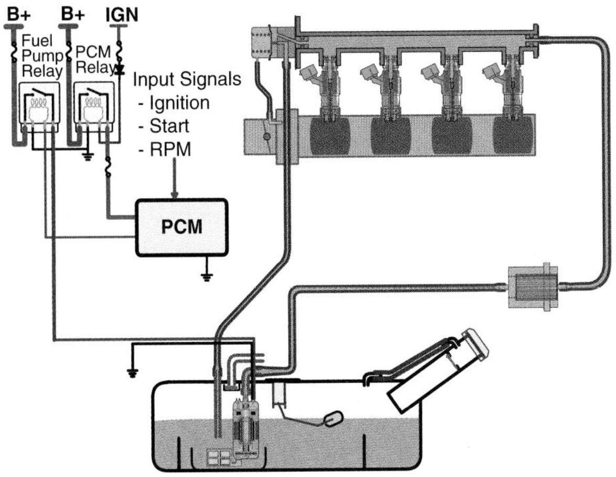

FIGURE 50-40 Relays are one type of controlled device activated by the PCM. Relays then power all kinds of other actuators.

FIGURE 50-41 A typical inertia switch with reset button on top.

running as long as electricity is flowing to it. For safety's sake, the PCM will shut down the fuel pump relay if it detects that the engine died. It determines this by watching the CKP sensor. It keeps the fuel pump running as long as the CKP shows that the engine is running above about 350 rpm. This way, the fuel pump can be shut down in the event of an accident. The PCM can also operate the fuel pump for a few seconds if the ignition switch is turned on. It does this to prime the system. The PCM can also command the fuel pump on while the vehicle is being cranked.

Some vehicles use an inertia switch to turn the fuel pump off in the event of an accident. The inertia switch is held closed by a spring-loaded trigger. A loose weight is free to move within the area of the trigger. If the vehicle experiences enough of a jolt in any direction, the weight will dislodge the trigger, opening the switch. Once the switch opens, it has to be reset manually (**FIGURE 50-41**).

▶ Solenoids

Solenoids are another type of controlled device used to carry out commands from the PCM. A solenoid is an electromechanical device similar to a relay. But instead of moving a switch magnetically, it moves a valve or linkage. When solenoids are used to open and close valves, they can control hydraulic or pneumatic

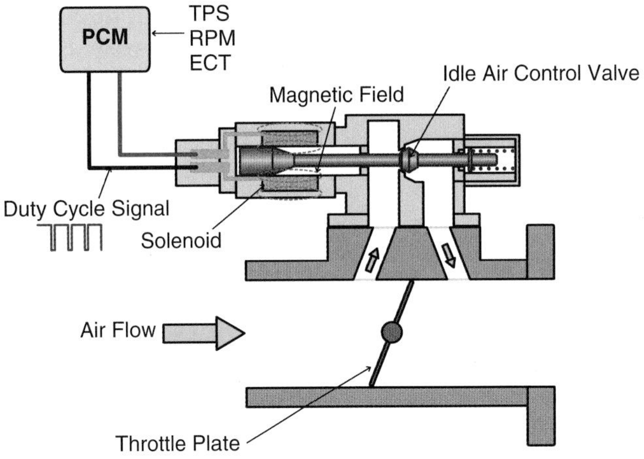

FIGURE 50-42 The solenoid-type air control valve acts on signals from the control unit to bypass a measured airflow around the throttle plate.

circuits. A few of the common solenoids that are controlled by the PCM are listed here:

- Fuel injectors
- Fuel pressure control valves in pressure regulators
- Idle air control valves
- Vacuum control valves
- Vent valves
- Oil pressure control valves (variable cam drives, etc.)
- EGR valves

Solenoids can be controlled on either the power side or the ground side of the circuit. In most cases, solenoids like vacuum control valves are designed to withstand continuous duty cycle. This means they can be on for long periods of time. In this case, they would receive typical on/off signals from the PCM. Other solenoids, like fuel injectors, are designed to be cycled on and off continuously at a very rapid rate. In this case, they receive a pulse-width-modulated signal from the PCM, as discussed earlier.

The solenoid-type air control valve acts on signals from the PCM to allow a measured amount of airflow to bypass the throttle plate (**FIGURE 50-42**). The position of the valve depends on how much current the control unit applies to the solenoid. Maximum current flow opens the valve fully to give maximum airflow. This is generally the starting position. Thereafter, the tapered valve is positioned by pulsing the solenoid winding circuit at a predetermined frequency.

The amount of valve opening, and therefore the amount of airflow, depends on the "on" time of the pulse. A long "on" pulse with a short "off" pulse will produce a high average voltage, and therefore a large opening of the valve. A short "on" pulse with a long "off" pulse will produce a low average voltage, and therefore a small valve opening and an increase in airflow. This "on" and "off" time is called the duty cycle, and it is generally expressed as a percentage.

▶ Control Modules

Most electronic modules on vehicles have their own switching circuit. It is built inside the module that controls any associated actuators. Modules need a timing signal to know when to switch the output circuit on and off. In this case, the **ignition control module (ICM)** controls the ignition coil's primary circuit. When the ICM receives a timing signal from the PCM, it then uses that timing signal to know when to switch the ignition coil primary circuit on and off.

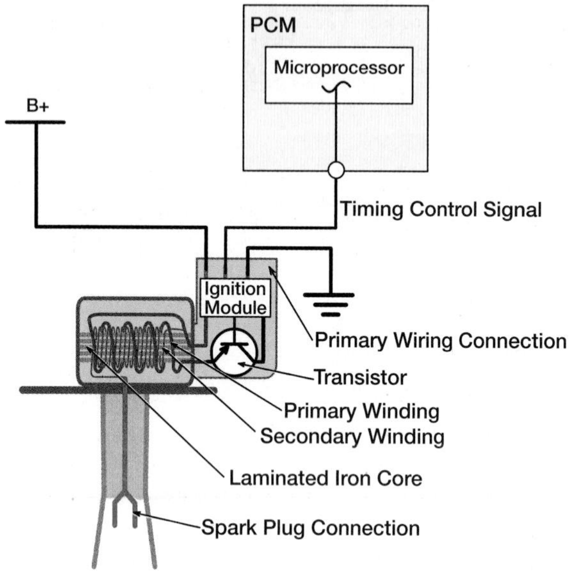

FIGURE 50-43 A coil-on-plug ignition module built into the coil.

There are a wide range of configurations of PCM-controlled ignition control, from the simple to the complex. Examples of simple ignition modules are many coil-on-plug modules built directly into the ignition coils. Most of them are simple switching devices that make no adjustments to spark plug timing. They only receive on/off timing signals from the PCM, and they turn the primary circuit on and off based on those commands (**FIGURE 50-43**). All of the timing calculations are handled by the PCM in this case. And because the PCM has access to all of the engine-related sensor data, it is probably the best-suited component to manage ignition system timing.

Some ignition modules have more functionality than coil-on-plug ignition modules. They receive signals from the CMP sensor, the CKP sensor, and the PCM (**FIGURE 50-44**). This type of module is capable of processing the CMP and CKP signals to determine crankshaft position and rpm. In this case, the module is able to control timing, within limits, across a broad rpm range without the timing signal from the PCM. These types of modules typically completely control the ignition timing during engine start-up. They then switch over to PCM-controlled timing once the engine is running. Because of this, if the PCM timing signal is ever lost, the ignition module can provide enough control over the ignition system to enable the vehicle to be driven to a shop for service. As long as the PCM timing signal is present, its calculations are based on a lot more sensor

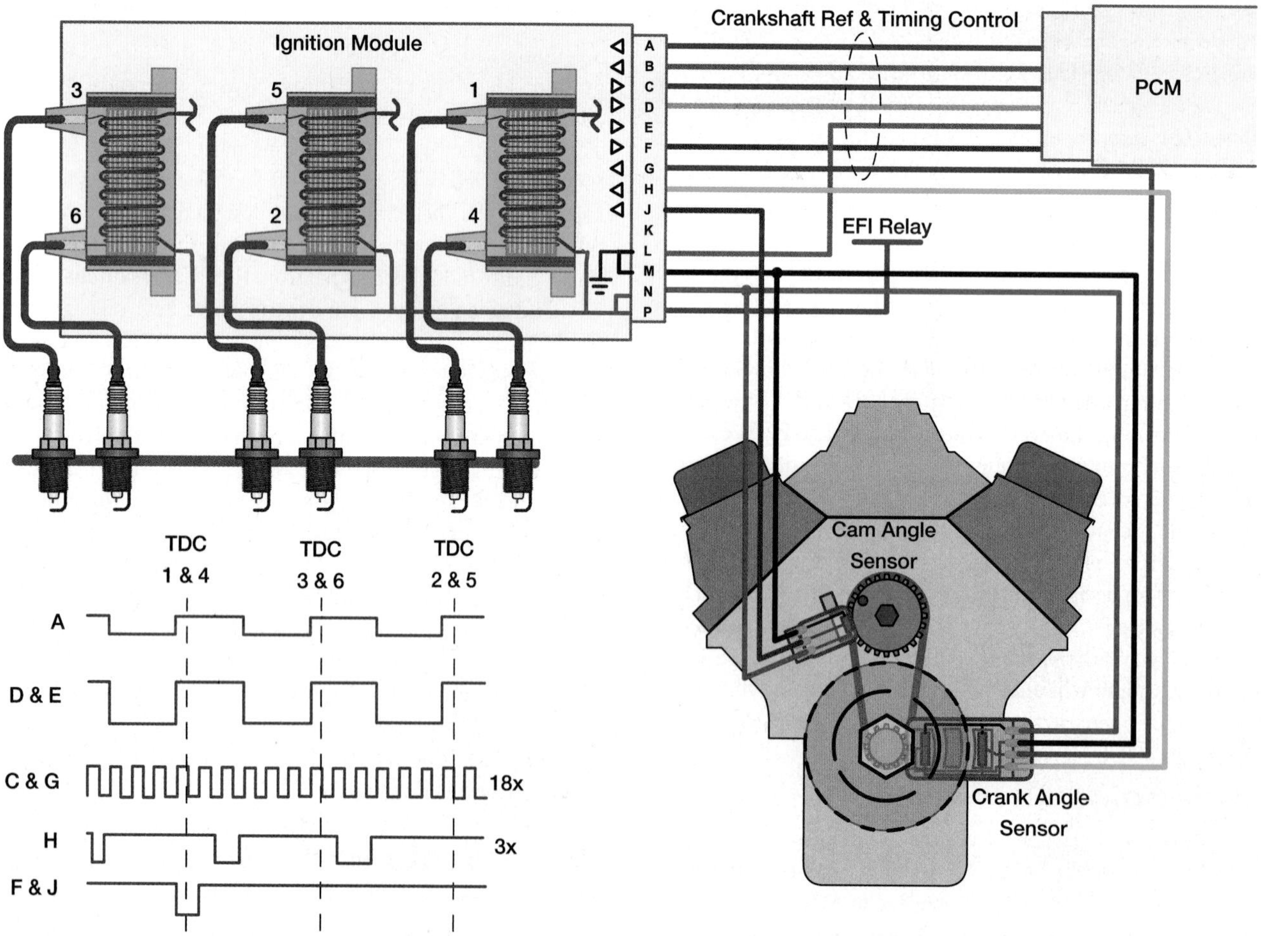

FIGURE 50-44 Typical waste spark ignition module circuit.

data, so its timing decisions are the most valid. These types of ICMs are usually mounted under or near the ignition coils on a waste spark system. If the engine has a distributor-type system, they may be mounted on/in the distributor.

▶ Electric Motor Actuators

Another type of actuator is an electric motor. Electric motors operate many devices on a vehicle from starters to the power windows to the throttle plates. Any type of motor can be controlled by the PCM; it just needs the right type of signal to operate. For example, a power window uses a typical permanent magnet DC motor, and therefore, to move the window up and down, it only needs an on/off signal that can be reversed. A stepper motor needs two to four alternately reversing signals, which moves the motor one step per signal.

An idle air control valve that uses a stepper motor has a tapered pintle valve that is opened and closed using a screw and nut assembly (**FIGURE 50-45**). The nut is part of a centrally located, permanent magnet armature that engages with the screw on the pintle. Rotating the armature extends or retracts the pintle, which closes or opens the air passage. The further the valve is open, the more air that bypasses the throttle plate, and vice versa. Rotation occurs by switching current flow in two coils. The pintle turns in forward or reverse steps, numbered from zero up to 255, with the bypass air passage fully closed for maximum airflow. This large number of steps allows the pintle position to be finely controlled.

When the idle speed is controlled by the position of the throttle plate, a direct current motor may be used as a variable throttle stop. The controlling circuit can let current flow through the motor in either direction, to extend or retract the linear actuator.

▶ Electronically Controlled Throttle

An **electronically controlled throttle** is often referred to as drive-by-wire or throttle-by-wire. It is the replacement device for the long-standing use of a throttle cable (**FIGURE 50-46**). This system uses a pedal position sensor to calculate the driver's desired throttle position. The command is sent to the PCM. If the conditions are right, the PCM sends a command to a throttle servomotor to open or close the throttle plates. The computer can adjust cruise consistency, acceleration, and deceleration. This is done in order to improve fuel economy, reduce emissions, and prevent sudden changes in speed that could affect the life of the drivetrain components.

Some late-model, high-end vehicles may have a driver-controlled switch. It provides a throttle response choice between "touring" and "sport." Touring has a slower and more conservative throttle response, which increases fuel economy. The sport setting provides a much quicker and more aggressive throttle response. This makes performance a priority over fuel economy.

▶ Feedback and Looping

LO 50-12 Explain feedback and looping.

For efficient operation of the catalytic converter, the air–fuel mixture must be maintained close to the stoichiometric ratio. This is 14.7 parts air to 1 part fuel (or leaner during certain operating parameters). Efficiency of combustion can be monitored by measuring the percentage of oxygen in the exhaust gas. A high percentage may mean the mixture entering the cylinder is too lean (not enough fuel).

The oxygen sensor in the exhaust stream tells the PCM how much oxygen is in the exhaust gas. The PCM then varies injector opening time to achieve the correct air–fuel ratio. Monitoring the data over time also allows the base fuel settings to be updated in the PCM memory as components age. This is called **adaptive learning**. The PCM memorizes its fuel settings and their results for different operating conditions and stores them for future use. If a fault occurs in a component or part of a system, a fault code will be stored in memory. Fault codes can be retrieved by connecting a scan tool to the data link connector. Information on the fault can then be analyzed.

Although most injectors operate at the nominal battery voltage of 12 volts, many sensors have a reference voltage of

FIGURE 50-45 In the stepper motor type of idle control, the tapered pintle is positioned using a screw and nut assembly.

FIGURE 50-46 An electronically controlled throttle is the replacement device for a throttle cable.

5 volts. This voltage is maintained by a voltage regulator in the control unit. Reference voltages provide for more accurate signals, as fluctuations due to changes in battery voltage do not occur at that reduced voltage.

During engine operation, three main **pollutants** are created. These include oxides of nitrogen, carbon monoxide, and hydrocarbons. These three pollutants enter the catalytic converter. How efficiently they are converted depends on several things. One is composition of the exhaust gases, and that depends on the air–fuel mixture sent into the cylinder for combustion. The more ideal the mixture, the fewer pollutants produced.

If the air–fuel ratio supplied to the engine is too rich, then the nitrogen oxides are converted efficiently, but the carbon monoxide and the hydrocarbons are not. If the air–fuel mixture supplied to the engine is too lean, the opposite occurs. Carbon monoxide and hydrocarbons in the exhaust gas are converted efficiently, but the nitrogen oxides are not.

Highly efficient conversions of all three pollutants occur only in a narrow range of air–fuel ratios. This range, the stoichiometric point, occurs around the ideal air–fuel ratio of 14.7 to 1 (**FIGURE 50-47**).

If the mixture ratio falls outside this range, the efficiency of conversion of these pollutants will rapidly decrease. Because of this strict requirement, vehicles with a three-way catalytic converter have a feedback system, called looping. It is made up of the fuel injectors, PCM, and oxygen sensors. When the feedback loop is active, it is called closed loop. When the feedback loop is not active, it is called **open loop**. During closed loop, the PCM adjusts the air–fuel ratio based on the readings from the oxygen sensors, also known as a lambda sensors. This is called a feedback loop because the PCM constantly adjusts the injector "on" time while monitoring the oxygen sensor reading. In this way, it determines whether the mixture it sent was rich or lean.

For example, if the oxygen sensor indicates a rich condition, the PCM will command a slightly shorter pulse width of the injector. The oxygen sensor should then indicate a lean condition. If so, the PCM will command a slightly longer pulse width of the injector to enrichen it slightly. In stoichiometric mode, the oxygen sensor should toggle back and forth between slightly lean and slightly rich conditions. This is ideal for most catalytic converters (**FIGURE 50-48**).

When an engine is first started, it starts in open loop. This means that the PCM pretty much ignores the oxygen sensor signals. It determines injector pulse width primarily on the speed-density or MAF system, plus information from the other sensors. During normal operation, open loop may also occur during extended idle times, or under maximum power. It can also occur if there is a fault that causes the air–fuel ratio to be excessively rich or too lean for long periods.

For the system to be in closed loop, several conditions must be met, but they are different for each manufacturer. Generally, the engine temperature must be above a certain point. The oxygen sensor must be active above a certain voltage. And a minimum amount of time must have passed since the vehicle was started. Until those conditions occur, the vehicle operates in open loop.

When the oxygen sensor reaches its operating temperature, around 600°F (315°C), it sends an output voltage to the control unit to signal whether the mixture is richer or leaner than a lambda value of 1.00. That is, the air–fuel ratio is richer or leaner than 14.7 to 1. When the mixture deviates from this point, the output voltage changes sharply above or below its switching point (approximately 0.45 volt) (**FIGURE 50-49**).

The oxygen sensor voltage signal changes sharply when the air–fuel mixture changes from lean to rich. If the control unit sees the voltage as high, then the quantity of fuel injected is reduced. Similarly, if the mixture changes from rich to lean and the voltage is low, more fuel is injected to enrich the mixture.

The control unit then adjusts the pulse width of the injector. This ensures the most efficient operation of the catalytic converter. During engine operation, this adjusting is continuous, and almost instantaneous, trying to maintain an air–fuel ratio for lambda very near 1.

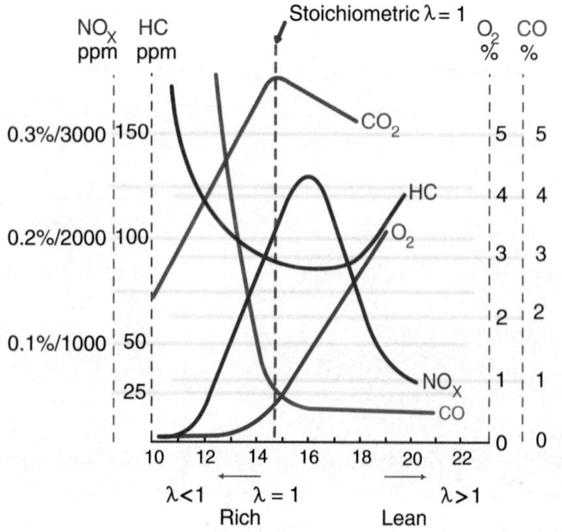

FIGURE 50-47 The stoichiometric point.

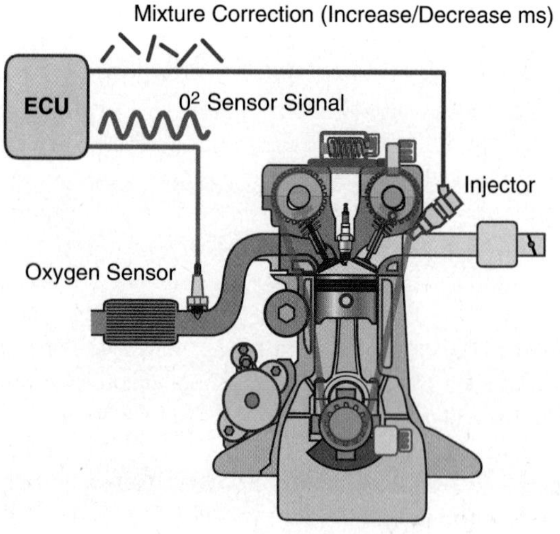

FIGURE 50-48 Closed loop means the control unit is using feedback from the oxygen sensor to alter the injection pulse width.

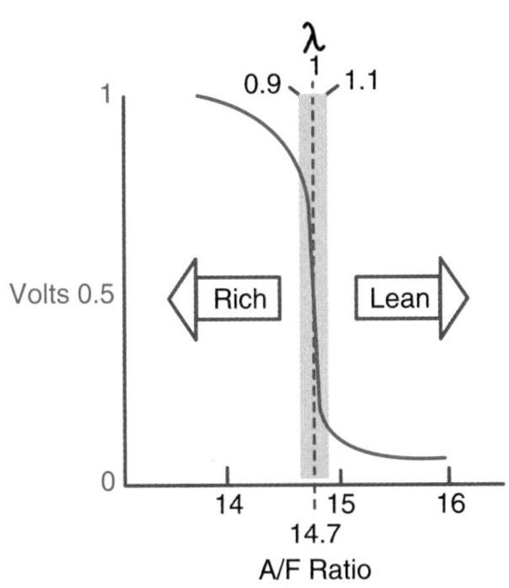

FIGURE 50-49 When the mixture deviates from the stoichiometric point, the output voltage changes sharply above or below its switching point of approximately 0.45 volt.

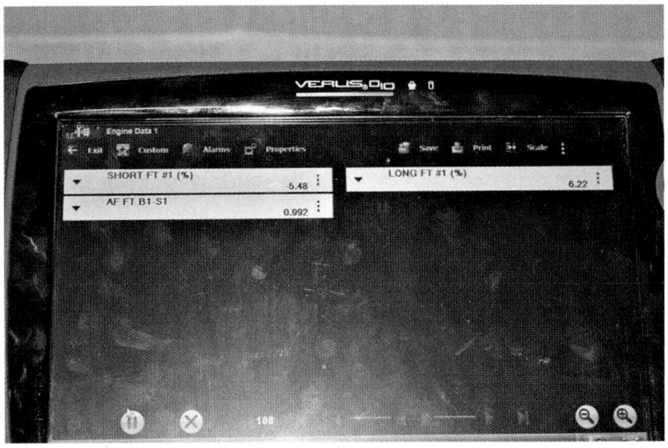

FIGURE 50-50 Scan tool showing short-term and long-term fuel trim.

▶ Short- and Long-Term Fuel Trim

LO 50-13 Explain short- and long-term fuel trim and fuel shut off mode.

The computer is programmed with the expected injector pulse width for nearly all operating conditions. But circumstances such as engine and fuel system wear, or differences in fuel, cause the programmed settings to result in an improper mixture. This is indicated by the oxygen sensors. The oxygen sensors are the computer's window into the exhaust stream. So, they are used to confirm the accuracy of the computer's pulse width signal. If the mixture is off, then the PCM will add or subtract fuel from the base amount. The difference between the stored pulse width and the actual pulse width required to keep the mixture at the correct ratio is called fuel trim. This means that the PCM is adjusting the amount of fuel injected, plus or minus, from the programmed amount. It is also part of the PCM's adaptive learning.

The PCM watches the average of the enriching and leaning over short periods of time (less than a second). It calculates the percentage changed from the calculated value. It then displays it as a positive (adding fuel) or negative (subtracting fuel) percentage. This is called short-term fuel trim (STFT). The PCM aims to keep the STFT at its midpoint so that fuel trim adjustments can respond quickly. If it sees the STFT start to creep away from the midpoint, the PCM will increase or decrease its base pulse-width setting (long-term fuel trim (LTFT)) to bring the STFT back to center. It will then record the updated base pulse width and store it in the PCM's memory. It will also store the difference between the original programmed pulse width and current pulse width, as a LTFT percentage, either positive or negative. Note that the STFT is not normally stored in memory when the key is turned off.

Both STFT and LTFT can be displayed on a scan tool connected to the PCM (**FIGURE 50-50**). STFT and LTFT readings are useful when diagnosing drivability concerns and give us a window into how the engine management system is operating. As you can imagine, if readings for both LTFT and STFT are near zero, then it is likely that the system is operating close to its design. If you see the LTFT is more than single digits positive, it means the PCM thinks the vehicle is running too lean and is substantially enriching the mixture. This can indicate a:

- plugged fuel filter,
- worn fuel pump,
- restricted fuel injectors,
- vacuum leak,
- exhaust manifold leak, or
- oxygen sensors that are weak and reading too low a voltage.

If the LTFT is more than single digits negative, it means the PCM thinks it is running too rich and is leaning it out. This could be caused by:

- a leaky fuel injector,
- a leaky pressure regulator,
- a purge valve that is stuck open, or
- gasoline-contaminated oil in the oil pan.

It could also be that the oxygen sensors that are malfunctioning by sending too high a voltage (this is less common).

▶ Fuel Shutoff Mode and Clear Flood Mode

Another operating mode is called the **fuel shutoff mode**. This mode turns the fuel injectors off when the vehicle is coasting above a certain rpm and speed with the throttle closed. Rather than inject fuel into the cylinders during coasting, which would be wasted, the PCM shuts the fuel injectors off. This continues until either the rpm drops below a certain level or the throttle is opened.

Clear flood mode is like fuel shutoff mode, but for a different reason. Consider a carbureted vehicle. If the engine is flooded with fuel, the driver would hold the throttle plate open while cranking the engine over to clear the flooded condition.

Some manufacturers have built that functionality into their fuel-injected vehicles. Clear flood mode is activated if the PCM sees the TPS indicating a wide-open throttle when the ignition is first turned on or cranked. If this occurs, the PCM will keep the injectors turned off to clear a flooded engine condition.

Technicians use the clear flood mode to fool the engine into cranking without starting. This is commonly done during a cranking sound diagnosis test. Holding the throttle to the floor, and then turning the key to crank the engine over, allows the engine to crank, but not start. This allows the technician to evaluate the cranking sound. And they can do this without having to disable the ignition system or fuel system manually. Just be aware that not all electronically fuel-injected vehicles are equipped with this functionality.

▶ Wrap-Up

Ready for Review

- ▶ Analog signals are continuously variable and typically change in strength over time. Analog signals can be used to show how something changes over time. A digital signal is a direct on/off signal with no in-between transition. Digital signals can be used to show whether a component is on or off, or to command a component on or off.
- ▶ A potentiometer is a resistor with a mechanically variable wiper that contacts the resistor along its length. As the wiper contact moves across the resistor, it picks up the voltage at that point and the voltage is sent as a signal referencing that position.
- ▶ A thermistor changes its resistance with a change in temperature. In PTC thermistors, the resistance goes up with the temperature. In NTC thermistors, the resistance goes down as the temperature goes up.
- ▶ Inductive-type CKP sensors typically sense the movement of a toothed disc on the crank pulley. The sensor is mounted on the crankcase housing and consists of a stator with a central permanent magnet, and a soft iron core surrounded by an induction winding. Hall-effect CKP sensor measures the position of the crankshaft, as well as engine speed. The sensor reacts to varying magnetic field and creates an on/off digital signal.
- ▶ Oxygen sensors, when at temperature, create an electrical signal based on the difference in oxygen in the exhaust stream compared to ambient air. A wide-band oxygen sensor informs the PCM of a range of air–fuel ratios from about 9:1 up to atmospheric air.
- ▶ Speed density calculation for air flow is based on engine rpm as measured by the CKP sensor, engine load as measured by the MAP sensor, intake air temperature, throttle position, and coolant temperature. A mass airflow (MAF) sensor directly measures the mass of filtered air entering the engine.
- ▶ A MAP sensor measures pressure changes in an engine, converts them into an electrical signal, and sends a varying frequency to the PCM. A barometric pressure sensor measures barometric pressure that is used to modify the base airflow calculation.
- ▶ Engine knock occurs in the combustion chamber due to an unwanted spike in pressure, caused by preignition or detonation. A knock sensor monitors the noise that is created by the pressure spike. A fuel with a higher octane rating can also be used to prevent engine knock.
- ▶ The PCM is composed of four distinct parts: a memory and storage section, a sensor input/output signal processing section, a data processing section, and an output drivers section.
- ▶ Relays, solenoids, modules, and motors act as actuators based on the inputs from a PCM. A relay is used as an actuator to control other actuators when the computer must control a high-current load. A solenoid is an electromechanical device used to moves a valve or linkage.
- ▶ Vehicles with a three-way catalytic converter have a feedback system that is made up of the fuel injectors, PCM, and oxygen sensors. During closed loop, the PCM adjusts the air–fuel ratio based on the readings from the oxygen sensors. Several conditions must be met for the system to be in closed loop.
- ▶ Fuel trim is the difference between the stored pulse width and the actual pulse width required to keep the mixture at the correct ratio. STFT and LTFT readings can be used to determine the operation of engine management system. Fuel shutoff mode turns the fuel injectors off when the vehicle is coasting above a certain rpm and speed with the throttle closed.

Key Terms

adaptive learning Software that can learn and change strategy based on different factors.

barometric pressure (BARO) sensor A sensor that measures atmospheric pressure.

camshaft position (CMP) sensor A detection device that signals to the PCM the rotational position of the camshaft.

crankshaft position (CKP) sensor A sensor used by the PCM to monitor engine speed. It can be one of three types of sensors—Hall effect, magnetic pickup, or optical.

EGR (exhaust gas recirculation) system A system that recirculates a portion of burned gases back into the combustion chamber to displace air and fuel and cool combustion temperatures.

EGR valve A valve that controls recirculation of a portion of burned gases back into the combustion chamber to displace air and fuel and cool combustion temperatures.

electronically controlled throttle A system that uses electronic, instead of mechanical, signals to control the throttle. Sometimes called drive-by-wire.

frequency The number of events or cycles that occur in a period, usually 1 second.

fuel shutoff mode A safety precaution by which fuel is shut off when certain conditions are met during a vehicle crash.

ICM (ignition control module) A general control unit of some electronic ignition systems, usually with current and dwell angle control; driver and output stage; and, in some cases, electronic spark timing functions.

inductive-type sensor A sensor mounted on the crankcase housing that is used to sense the movement of the ring gear teeth on the flywheel, or a toothed disc on the crank pulley.

intake air temperature (IAT) sensor A sensor that measures the temperature of the incoming air through the air filtration system.

knock sensor An engine sensor that detects pre-ignition, detonation, and knocking.

open loop When the computer system feedback loop is not active.

Oxygen sensors An exhaust sensor used to measure the amount of oxygen in the exhaust gases produced by the engine.

pollutant A potential threat to human health or the environment, resulting from excessive amounts of chemicals and waste.

potentiometer A three-terminal resistive device with one terminal connected to the input of the resistor, one terminal connected to the output of the resistor, and the third terminal connected to a movable wiper arm that moves up and down the resistor, creating a varying voltage signal. Also called a pot, or variable resistor.

reed switch A type of speed sensor that uses a magnetic field to open and close a movable set of contacts. It is used with a rotating magnet to measure rpm of a shaft and send the signal to the PCM.

relay An electromechanical switching device whereby the magnetism from a coil winding acts on a lever that switches a set of contacts.

thermistor A variable resistor that changes its resistance based on temperature. Most thermistors have a negative temperature coefficient, meaning that their resistance decreases as temperature increases. Commonly used to measure coolant, oil, fuel, and air temperatures.

Review Questions

1. Which sensor is responsible for measuring the amount of air entering an engine?
   a. MAP
   b. MAF
   c. HO2S
   d. BARO
2. Which of the following is true regarding inductive-type crankshaft position sensors?
   a. The sensor is spun and indicates speed by centrifugal force
   b. The sensor is stationary and contacts a spinning reluctor
   c. The sensor's iron core reacts magnetically with teeth on the spinning reluctor
   d. The sensor is stationary, and it puts out DC voltage as the rotor passes
3. What is the typical voltage for a stoichiometric ratio as indicated by an oxygen sensor
   a. 2.5 volts
   b. 1.0 volts
   c. .45 volts
   d. 6.0 volts
4. What is false air?
   a. Air that is not pure oxygen as seen by the O_2 sensor
   b. Air that has not been filtered through the airbox
   c. Air that has not been measured by the MAF sensor
   d. Air that is not detected by the MAP sensor
5. When interpreting a MAP sensor reading, what would indicate high manifold pressure (low vacuum)?
   a. High voltage
   b. Low voltage
   c. High current
   d. Low current
6. A knock sensor indicating knock is likely to be a result of all of the following EXCEPT:
   a. Excessive engine load
   b. Over-advanced ignition timing
   c. An overheated engine
   d. A faulty spark plug
7. Which of the following is most likely to be indicated by a switch rather than a sensor?
   a. Engine coolant temperature
   b. Intake manifold pressure
   c. Brake pedal position
   d. Throttle position
8. What is updating a PCM's software often called?
   a. Reflashing
   b. Unzipping
   c. Renewing
   d. Clearing
9. What is being ignored when a vehicle is operating in open loop?
   a. The fuel injectors
   b. The TPS
   c. The ECT
   d. The O_2 sensors
10. Which of the following is most likely to cause an excessively negative long-term fuel trim condition?
    a. A leaking fuel injector
    b. Restricted fuel injectors
    c. A vacuum leak
    d. Low fuel pressure

ASE Technician A/Technician B Style Questions

1. Engine sensors are being discussed. Technician A states that analog sensors make a square wave style waveform. Technician B states that digital sensors make a rounded or variable-type waveform. Who is correct?
   a. Technician A only
   b. Technician B only
   c. Both Technicians A and B
   d. Neither Technician A nor B
2. Thermistors are being discussed. Technician A states that thermistors are used commonly for position sensors in vehicles. Technician B states that as a negative temperature coefficient (NTC) thermistor heats up, its resistance decreases. Who is correct?
   a. Technician A only
   b. Technician B only
   c. Both Technicians A and B
   d. Neither Technician A nor B
3. A crankshaft position sensor is being discussed. Technician A states that it relies on a small air gap between it and the spinning reluctor. Technician B states that most crankshaft position sensors are optical in nature and use LEDs and a phototransistor. Who is correct?
   a. Technician A only
   b. Technician B only
   c. Both Technicians A and B
   d. Neither Technician A nor B
4. Oxygen sensors are being discussed. Technician A states that early oxygen sensors could measure precisely how rich or how lean an engine was running. Technician B states that wideband oxygen sensors can only tell that the engine is rich or lean, but not how much. Who is correct?
   a. Technician A only
   b. Technician B only
   c. Both Technicians A and B
   d. Neither Technician A nor B
5. Airflow measurement is being discussed. Technician A states that the speed density method relies on several engine sensors and is then calculated based on a preprogrammed fuel map. Technician B states that in types with a MAF, there can be no leaks in the intake tube or manifold for the sensor reading to be accurate. Who is correct?
   a. Technician A only
   b. Technician B only
   c. Both Technicians A and B
   d. Neither Technician A nor B
6. A knock sensor is being discussed. Technician A states that knock sensors are normally mounted to the chassis or firewall. Technician B states that most knock sensors use a piezoelectric crystal. Who is correct?
   a. Technician A only
   b. Technician B only
   c. Both Technicians A and B
   d. Neither Technician A nor B
7. Pressure sensors are being discussed. Technician A states that fuel pressure sensors are often mounted inside the fuel tank. Technician B states that BARO sensors provide readings to the PCM to control the air–fuel mixture. Who is correct?
   a. Technician A only
   b. Technician B only
   c. Both Technicians A and B
   d. Neither Technician A nor B
8. PCMs are being discussed. Technician A states that the PCM acts as the brain of the engine management system by monitoring inputs and commanding outputs. Technician B states that the PCM refers to a data map to make decisions. Who is correct?
   a. Technician A only
   b. Technician B only
   c. Both Technicians A and B
   d. Neither Technician A nor B
9. PCM actuators are being discussed. Technician A states that the throttle position sensor is an example of an actuator. Technician B states that a fuel injector is an example of an actuator. Who is correct?
   a. Technician A only
   b. Technician B only
   c. Both Technicians A and B
   d. Neither Technician A nor B
10. Fuel trim is being discussed. Technician A states that negative fuel trim can be caused by a vacuum leak. Technician B states that positive fuel trim can be caused by a restricted fuel filter. Who is correct?
    a. Technician A only
    b. Technician B only
    c. Both Technicians A and B
    d. Neither Technician A nor B

CHAPTER 51

On-Board Diagnostics

Learning Objectives

- **LO 51-01** Describe why onboard diagnostic systems are needed.
- **LO 51-02** Describe OBD I and OBDII.
- **LO 51-03** Describe Diagnostic Trouble Codes.
- **LO 51-04** Describe MIL operation and Freeze Frame Data.
- **LO 51-05** Describe the purpose of drive cycles and system readiness monitors.
- **LO 51-06** Describe the purpose of a scan tool.
- **LO 51-07** Retrieve DTCs, Monitor Status, and Freeze Frame Data.

ASE Education Foundation Tasks

See Appendix A to view the 2017 ASE Education Foundation Automobile Accreditation Task List Correlation Guide.

▶ Introduction

LO 51-01 Describe why onboard diagnostic systems are needed.

The automobile we drive today has certainly evolved significantly. This is especially true in the recent past. Properly operating vehicles are running cleaner than ever. And they put out only a small amount of pollution compared with vehicles of 40 years ago. Yet despite all the innovations, **emissions** from automobiles can still harm the environment, as well as our health, if something fails. In fact, one badly operating new vehicle can put out as much emissions as hundreds of properly running vehicles combined. Because of this, the engine management system needs to be in proper operating condition. The onboard diagnostic system monitors the engine management system for faults. Any faults it identifies should be diagnosed and repaired in a timely manner.

▶ Reasons for Onboard Diagnostic Systems

In the United States, Congress has passed federal emission regulations, starting with the Clean Air Act in 1963. This program was expanded in 1967 to help curb emissions from automobiles. Major amendments to the law, requiring regulatory controls for air pollution, were enacted in 1970, 1977, and 1990.

There are parts of the country where the airshed is in noncompliance with clean air laws. There, vehicle emission testing and repair remedies are a big part of day-to-day business. Consider when a vehicle's malfunction indicator lamp (MIL)—formerly called a "check engine light"—is illuminated. It means the onboard diagnostic system has identified a fault that is likely causing it not to comply with clean air regulations (**FIGURE 51-1**). In the interest of public health, keeping cars running clean is a mandate. Low tailpipe emissions also mean a vehicle is running efficiently. This also helps to conserve fuel, in turn reducing carbon dioxide emissions. So the onboard diagnostic system is critical to meeting these requirements.

Pollutants

Vehicles emit pollutants to the atmosphere in two major ways: evaporative emissions and tailpipe emissions. Evaporative emissions are known as **volatile organic compounds (VOCs)**. VOCs may be fuel or oil vapors emitted primarily from the:

- fuel tank,
- fuel lines, or
- engine crankcase.

In earlier vehicles, engine draft tubes spewed VOCs from the crankcase onto the road. This was evidenced by dark center-of-lane deposits on uphill stretches of our highways. Likewise, hot summer days once meant the smell of gasoline fumes around parked cars—but not today!

The tailpipe emits a variety of pollutants, including the following:

- Carbon monoxide (CO)
- Hydrocarbons (HCs), or VOCs
- Particulate matter (PM)
- Carbon dioxide (CO_2)
- Sulfur dioxide (SO_2)
- Oxides of nitrogen—nitric oxide (NO) and nitrogen dioxide (NO_2)

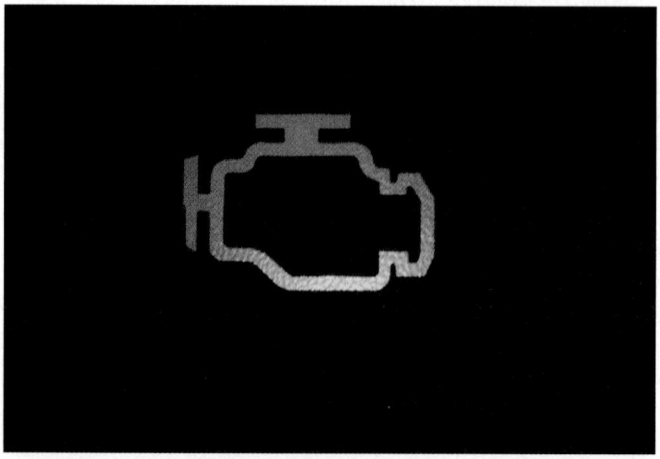

FIGURE 51-1 The malfunction indicator lamp (MIL) indicates to the driver that the powertrain control module (PCM) has identified a fault in the engine management system.

You Are the Automotive Technician

A new customer, Jim, brings his vehicle into your shop for an illuminated MIL. He says it has been on for about a week. He wants to know if you have one of "those computers" that will tell you what is wrong with his vehicle. You tell him, as you chuckle, that you wish there was a tool you could buy that would do that. But you do have several scan tools and the training to know how to use them. With that and some good technical service information, you can diagnose what is causing the MIL to be on, so it can be fixed right the first time. Jim says that makes sense, as one of his coworkers had a similar problem. The shop he went to told him that the vehicle needed a particular part, but that part didn't fix the problem. The coworker ended up buying several parts over a couple of visits, and the MIL always came back on. Jim agrees to let you diagnose the problem.

1. What does the MIL being illuminated indicate?
2. What is "freeze-frame" data? And how can it help you diagnose the problem?
3. What do monitors that report "Ready" and "Not Ready" indicate?
4. What is the "data stream," and how is it used during diagnosis?

Carbon monoxide is highly toxic, as discussed in the Personal Safety chapter. Oxides of nitrogen and hydrocarbons react together in the presence of sunlight to create ground-level ozone. It is considered a health hazard. Ground-level ozone especially affects children, the elderly, and those with respiratory problems. On "ozone alert days," such people are advised to stay indoors. And the operation of VOC-spewing lawn and garden equipment is discouraged.

All of these pollutants are hazardous to one degree or another. Most of them, or their control systems, are monitored by the onboard diagnostic system. The onboard diagnostic system performs self-tests on the control systems at engine start-up and continues to monitor them once underway. If the PCM determines there is a fault, several actions occur:

- the driver is alerted via the MIL,
- a snapshot of operating conditions is saved as freeze-frame data for technicians to examine later, and
- if needed, alternate operating strategies are initiated to protect the vehicle and the environment.

Many strategies have been implemented to clean up vehicle emissions. In more recent years, closed-loop electronic systems have assumed the major responsibility. The following is an abbreviated timeline of the many strategies devised for curtailing vehicle emissions in the United States since the 1960s:

- 1961—First positive crankcase ventilation (PCV) systems required in California
- 1964—Sealed crankcase systems replace crankcase "draft tubes"
- 1968—Controls for crankcase VOCs, tailpipe CO, and hydrocarbons (PCV, AIR, etc.)
- 1969—Commencement of limiting oxides of nitrogen emissions by retarding ignition and valve timing
- 1971—Introduction of evaporative emission (EVAP) control systems (charcoal canisters)
- 1973—Exhaust gas recirculation (EGR) systems used to control oxides of nitrogen
- 1975—Unleaded gasoline phase-in
- 1975—Oxidizing catalytic converters introduced
- 1981—Closed-loop systems and three-way catalysts introduced; first-generation onboard diagnostics (OBD I) enacted nationwide (OBD I started with California vehicles)
- 1996—Second-generation onboard diagnostics (OBD II) enacted nationwide
- 1998—Commencement of vehicle refueling emission controls
- 2003—Widespread adoption of controller area network (CAN) communication systems (originated in 1998 with Robert Bosch–equipped vehicles like Volkswagens)
- 2006—The FlexRay network was first added to a production vehicle. It is faster and more reliable than CAN.
- 2009—The EPA decided that greenhouse gasses can be regulated under the Clean Air Act
- 2010—The EPA set standards for 2012–2016 light-duty vehicles to reduce greenhouse gas emissions.
- 2012—The EPA and NHTSA extend the greenhouse gas requirements for 2017–2025 light-duty trucks and medium-duty passenger vehicles.
- 2015—The EPA and NHTSA propose greenhouse gas emissions and fuel economy standards for 2018–2027 medium- and heavy-duty vehicles.

Let's look further into these onboard diagnostic (OBD) systems. We will explore further how they are designed to meet the EPA standards, and how they function.

▶ Onboard Diagnostic Systems

LO 51-02 Describe OBD I and OBDII.

OBD systems are a great resource. They help the technician work more efficiently. They also remove some guesswork. This is important when dealing with today's sophisticated vehicles. Onboard computers are sometimes referred to as **modules**. They monitor the systems and components. And they inform the vehicle operator when a fault happens. The systems can also provide information about the fault. Diagnosis starts by using diagnostic equipment to access the information in the onboard computers. Together, they help to identify problems in engine management systems and components.

In the United States, two main generations of OBD systems have been used. The first generation of onboard diagnostic (**OBD I**) systems operated under manufacturer standards. It started with California vehicles and become nationwide in 1981. The second generation of onboard diagnostic (**OBD II**) systems operates under standards set by the Society of Automotive Engineers (SAE). They conform to U.S. Environmental Protection Agency (EPA) regulations. Some manufacturers started using OBD II with a few of their models in 1994 and 1995, in order to become familiar with the systems. But for the 1996 model year, OBD II was standardized and initiated nationwide.

OBD I systems mainly monitored hard electrical circuit malfunctions. So these systems were typically limited to well under 100 possible faults. OBD II is an enhanced diagnostic system. It identifies faults that may affect the vehicle's emission output and drivability. This system can accommodate up to several thousand faults. OBD II constantly monitors and manages the engine management system as it controls the engine and vehicle's emission systems. This in turn enables much tighter emission standards to be enforced. Cutoff points are far lower than those previously possible.

OBD I allowed each vehicle manufacturer to have its own:

- OBD system,
- with its own diagnostic connector (**FIGURE 51-2**),
- its own trouble codes,
- its own component names, and
- its own scan tool.

This made diagnosing and repairing vehicles much more difficult for technicians. The OBD II requirements standardized all of this. It provided a certain degree of uniformity.

FIGURE 51-2 OBD I diagnostic connectors. **A.** Ford. **B.** Chrysler. **C.** GM. **D.** Toyota.

All vehicle manufacturers must adhere to certain standards regarding the:

- diagnostic connector,
- fault code descriptions, and
- nomenclature (names of parts).

This means that any so-called "generic" OBD II faults and background data can be accessed and read by **aftermarket** scan tools. It can also be read with original equipment manufacturer (OEM) scan tools. This greatly reduced the cost and complexity of repairing these vehicles.

Both OEM and aftermarket OBD II scan tools serve to access OBD information via the data link connector (DLC). The DLC enables the scanner to access data stored in the vehicle's various computers (**FIGURE 51-3**). The DLC itself is a 16-pin connector with a common (SAE J1962) size, shape, and pin layout (**FIGURE 51-4**). Regardless of vehicle make and model, its location is also now fairly standardized. It is typically within 2' (0.61 m) of the steering wheel below the driver's side instrument panel.

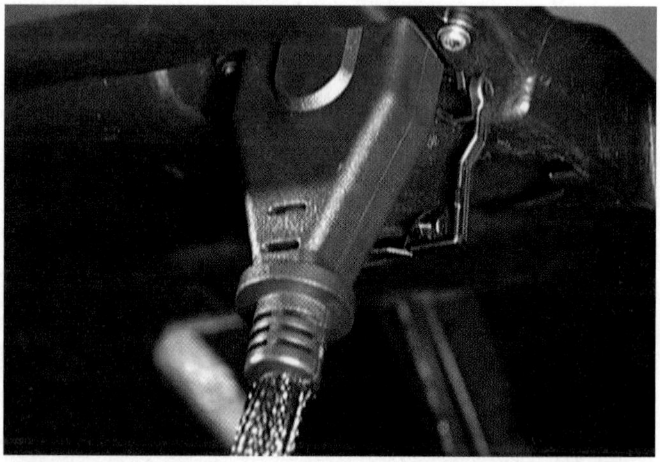

FIGURE 51-3 The data link connector (DLC) enables the scanner to access data stored in the vehicle's various computers.

▶ OBD Terminology

The automotive technician needs to know and understand the terms, or "lingo." These terms are used in the diagnosis and repair of OBD systems. Before the SAE published the appropriate "recommended practices," OEMs used nonstandardized nomenclature. The SAE has helped to standardize automotive terms used by engineers and technicians alike. A complete listing of OBD I and OBD II terms is available from the SAE (**SAE J1930**). Some aftermarket and manufacturer service information also list the OBD terms.

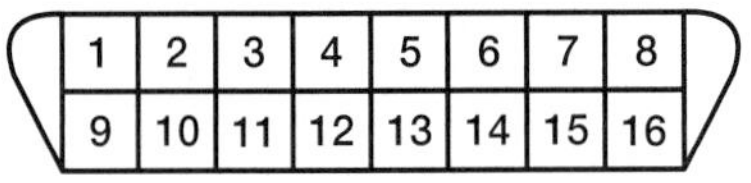

1 Manufacturer's discretion
2 Bus + Line SAE J1850
3 Manufacturer's discretion
4 Ground
5 Ground
6 Manufacturer's discretion
7 K Line, ISO 9141
8 Manufacturer's discretion
9 Manufacturer's discretion
10 Bus – Line, SAE J1850
11 Manufacturer's discretion
12 Manufacturer's discretion
13 Manufacturer's discretion
14 Manufacturer's discretion
15 L Line, ISO 9141
16 Power supply

FIGURE 51-4 Pin layout for a data link connector (DLC).

FIGURE 51-5 The malfunction indicator lamp (MIL).

Data Stream

One of the advantages of OBD II is that the sensor data can be accessed with a scan tool. That means that you can use the scan tool to access the information from virtually any sensor in the system, through the PCM. The scan tool displays the interpreted sensor readings that the computer is receiving. This means that you can examine all of the sensor data while hunting for clues to the cause of the fault.

For example, the customer complains that the engine temperature gauge reads way colder than it used to. You can connect the scan tool and access the coolant temperature sensor data. If it shows full engine operating temperature of 195°F (90.5°C), then it is likely that the temperature gauge or sending unit is faulty. You can then focus your further testing there. If the coolant temperature sensor showed the engine operating temperature at 153°F (66.7°C), then it is likely that the engine thermostat is stuck open.

▶ Diagnostic Trouble Codes

LO 51-03 Describe Diagnostic Trouble Codes.

Diagnostic trouble codes (DTCs) indicate the component or circuit in which a fault has been detected. **Codes** are "set" (stored) in the memory of one or more of the vehicle's PCMs once a fault is detected. An emission-related fault is generated when values being monitored indicate the emissions would be 1.5 times the EPA Federal Test Procedure (FTP). For example, an open in the engine's coolant temperature circuit will cause the fuel system to run rich. This increases the emissions so a DTC will be set. Also, if two sensors send conflicting data, a fault is recorded, and the MIL is illuminated.

Some DTCs set due to implausibility and conditions that are out of parameters. So it is not always a case of a failed component or wiring. Understanding how the affected system operates is very important when troubleshooting the OBD II system. If you don't know how it works, how can you fix it? Locating, reading, and understanding the strategy used by the manufacturer to generate the DTC (enabling criteria) is of utmost importance.

Emission-related DTCs are the same (generic DTCs) across all vehicle makes and models. It's the same with the SAE-recommended names used to describe components and systems. Generic DTCs—those that are in any way emission related—are now the same from one vehicle to the next. Refer to **SAE J2012** for more information on interpreting generic DTCs.

Most manufacturers go beyond the generic OBD II requirements. They program their own proprietary functionality into the PCM. This can include manufacturer (non-generic) codes, sensor information, and bidirectional tests. Refer to OEM service information for accessing and understanding OEM-specific (non-generic) system codes. This is also important for performing manufacturer specific tests. The manufacturer side of the OBD II system can be preferable to the generic OBD II. This is because it has more options and information. Many scan tools can access the manufacturer side of the system. That is generally a good option. And manufacturer scan tools will give you even more options and control over the system.

If a vehicle system experiences a fault, the MIL located in the instrument cluster will likely illuminate (**FIGURE 51-5**). The MIL alerts the driver that there is a problem in the engine management system. Even a loose gas cap can trigger a DTC and illuminate the MIL (or a special gas cap warning light) under certain conditions. Sometimes, it takes two consecutive drive cycles, or "trips," to illuminate the MIL. If the catalytic converter is at risk, such as from a continuous misfire, the MIL will flash. This warns the driver to stop the vehicle as soon as safely possible and turn off the engine. Serious damage to the catalytic converter can be avoided in this way.

Once the MIL is on, it will remain on until one of two things occurs. The first is if the technician clears the codes. The second is if the fault has not been detected during three consecutive trips when the **monitor** is run. OBD systems store DTCs in the computer's so-called keep alive memory (KAM). On older vehicles, the codes remain in memory until power is disconnected. For newer vehicles, DTCs are saved even if the vehicle's battery is disconnected. DTC information remains stored in the respective control module's long-term memory. This is regardless of whether a "hard" (continuous) fault or an intermittent fault has set the code.

DTCs are stored according to their status. **Current DTC** means that the test failed during the last one to three drive cycles (**FIGURE 51-6**). There is a good probability that this is a hard fault, meaning that it is not intermittent. **Pending codes** are codes that have not been validated by failing a second consecutive test.

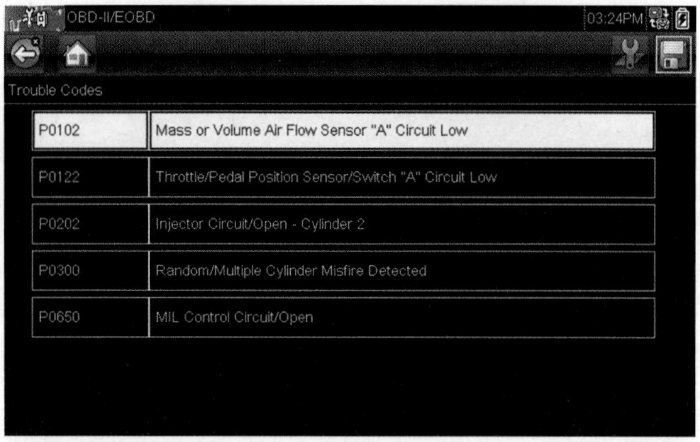

FIGURE 51-6 A scan tool displaying current codes.

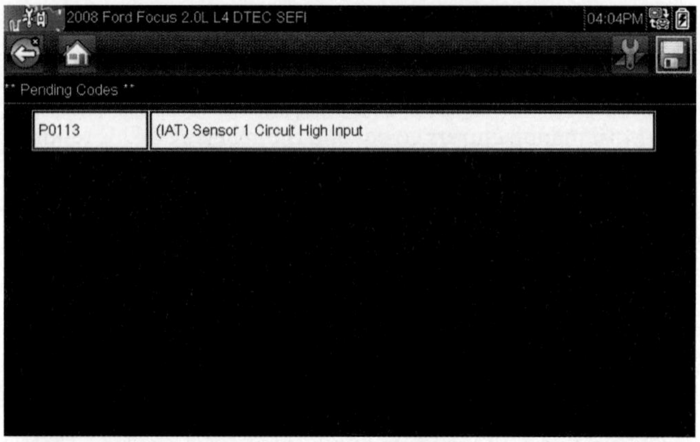

FIGURE 51-7 A scan tool displaying pending codes.

Or they can be a current code that was downgraded because it is intermittent (**FIGURE 51-7**). **History codes** are pending or current codes that have happened in the past but haven't been cleared. This could be either manually with a scan tool, or automatically after approximately 40 warm-up cycles without the code reoccurring (**FIGURE 51-8**). Warm-up cycle is a term used to determine when automatic clearing of DTCs occurs.

A **warm-up cycle** is when the engine:

- is started and run until the engine operating temperature reaches 160°F (71.1°C)
- has increased in temperature at least 40°F (22°C), and
- the engine is turned off again.

A DTC may be generated by any of the numerous computer modules on board. DTCs serve not only to inform the driver of a fault but also to enable the automotive technician to determine where the fault has occurred. DTCs are used in conjunction with diagnostic flowcharts found in the manufacturer's service information. Together, they assist technicians in determining the likely cause of a failure.

OBD II codes are made up of a series of letters and numbers that are grouped together. They identify which system, component, or circuit is at fault (**FIGURE 51-9**). The first character of the code is a letter. It identifies whether the fault is located within the:

- powertrain (P),
- body (B),

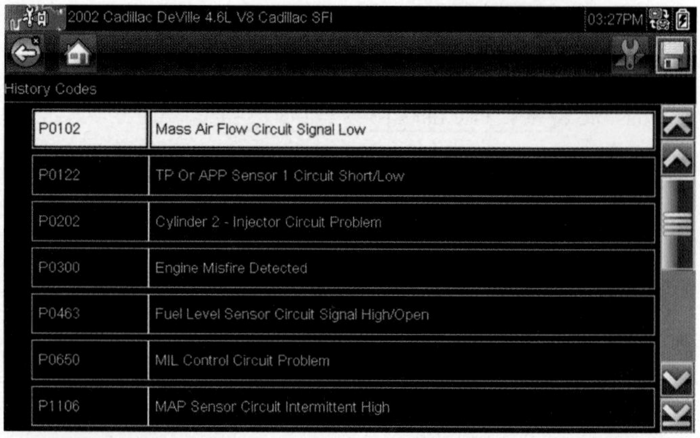

FIGURE 51-8 A scan tool displaying history codes.

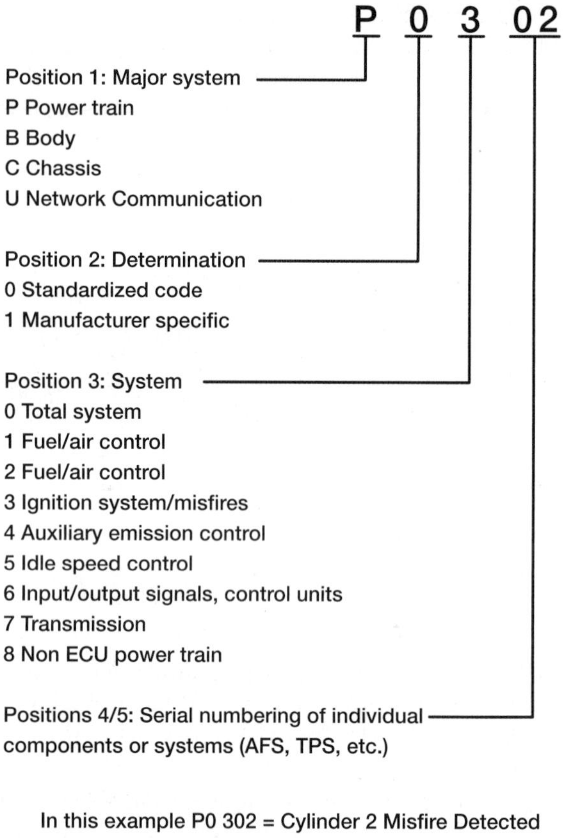
P 0 3 02
Position 1: Major system
P Power train
B Body
C Chassis
U Network Communication
Position 2: Determination
0 Standardized code
1 Manufacturer specific
Position 3: System
0 Total system
1 Fuel/air control
2 Fuel/air control
3 Ignition system/misfires
4 Auxiliary emission control
5 Idle speed control
6 Input/output signals, control units
7 Transmission
8 Non ECU power train
Positions 4/5: Serial numbering of individual components or systems (AFS, TPS, etc.)
In this example P0 302 = Cylinder 2 Misfire Detected

FIGURE 51-9 Decoding a diagnostic trouble code (DTC).

- chassis controller (C), or
- communication system (U).

The second digit indicates SAE-controlled (generic) or manufacturer-specific (OEM)- DTC.

- A "0" or "2" indicates SAE-controlled (generic) code.
- A "1" indicates manufacturer-specific (OEM) code.
- A "3" indicates both SAE-controlled (generic) and manufacturer-specific (OEM) code.

The third digit indicates the powertrain subgroup DTC and usually from "0 to 9" as listed in Figure 51-9.

The fourth/fifth digit indicates the area or component involved in the fault. Refer to DTC charts (on the Internet or in service information) for a complete listing of the hundreds of possible codes. The list grows constantly as onboard systems become even more sophisticated. A good example is the additional codes required for hybrid-electric vehicles.

▶ Malfunction Indicator Lamp (MIL) Operation

LO 51-04 Describe MIL operation and Freeze Frame Data.

Under OBD II requirements, the MIL can come on or flash depending on a number of factors. Even when a fault is detected, the MIL and the storing of a DTC will only occur under the following conditions:

- On a one-trip DTC, the MIL will illuminate when the DTC and freeze-frame data are stored.
- On a two-trip DTC, if the same fault is detected in two consecutive drive cycles, the MIL is illuminated, and a DTC and freeze-frame data are stored.
- If a catalyst-damaging fault occurs, the MIL will flash when the fault is present.
- If a fault occurs that triggers limp home mode, the warning lamp may be illuminated, and a DTC will be stored even if the fault only occurs in one drive cycle.

The engine warning lamp will go out if no faults are detected in three consecutive drive cycles. The DTC will, however, remain as a history code. To turn off the MIL manually, use a scan tool to erase the codes and freeze-frame data.

Freeze-Frame Data

Since OBD II came about in 1996, technicians have had the ability to access **freeze-frame data**. That data is automatically recorded in the vehicle's PCM when a vehicle DTC is stored. Freeze-frame data are a list of sensor data (parameter ID [PID]) that are associated with the specific DTC. Typically, freeze-frame data can only be saved for one DTC, generally the highest priority DTC. If a more serious fault supersedes another previously less serious recorded fault, freeze-frame data are updated. DTCs are rated highest to lowest priority in this order.

- DTCs related to misfires
- one-trip DTCs
- two-trip DTCs
- nonemissions-related DTCs.

Difficult-to-find intermittent faults can thus be diagnosed by carefully reviewing the freeze-frame data. Look there for clues to the location of the fault. For example, a vehicle set a misfire DTC while crossing a rough set of railroad tracks, but at no other time. Check the freeze frame for any out-of-range data. If you see data that is out of range for a particular circuit, you will have a good place to start looking for the fault. One possibility is that crossing the train tracks causes the TPS sensor to glitch. This affects fuel injector ontime which messes up the air–fuel ratio causing the misfire. Seeing the TPS reading out of range compared with the other sensor readings is a huge clue. You would then check the sensor and sensor terminals for issues. Modern scan tools can retrieve freeze-frame data and display them for analysis after the fact (**FIGURE 51-10**).

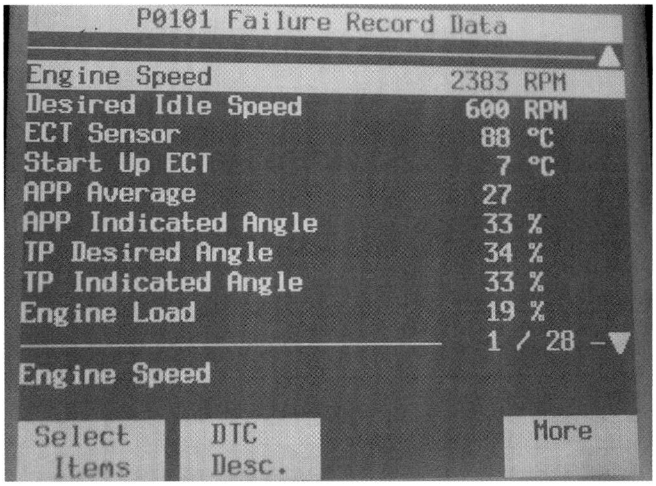

FIGURE 51-10 A scan tool can retrieve and display freeze-frame data.

▶ Drive Cycles

LO 51-05 Describe the purpose of drive cycles and system readiness monitors.

A **drive cycle** is designed to operate the vehicle in a specific manner. It needs to be operated so that it meets the **enabling criteria** to cause the PCM to perform the readiness monitor tests. This typically includes the following events:

- a vehicle starts,
- warms up,
- is accelerated,
- cruises,
- slows down,
- accelerates once more,
- decelerates,
- stops, and
- cools down.

The EPA has established a "drive cycle" for emission testing of various kinds of vehicles. The light-duty vehicle (240-second) drive cycle is the basis for how OBD II readiness monitors (tests) are run (**FIGURE 51-11**). A complete drive cycle should include running the monitors on all systems. They should be completed in under 15 minutes. Sometimes, certain monitors cannot be run during a drive cycle. For example, if enabling criteria for running that particular monitor are not met. An EVAP monitor will not run if the fuel tank is less than one-quarter full or more than three-quarters full. In some cases, particular monitors cannot run until after other monitors have been run and passed. EVAP monitoring will also not run if the engine coolant temperature monitor has not passed. State-run emission inspection stations in the Unites States will sometimes allow a vehicle to pass an emission inspection, even if one or two monitors have not run.

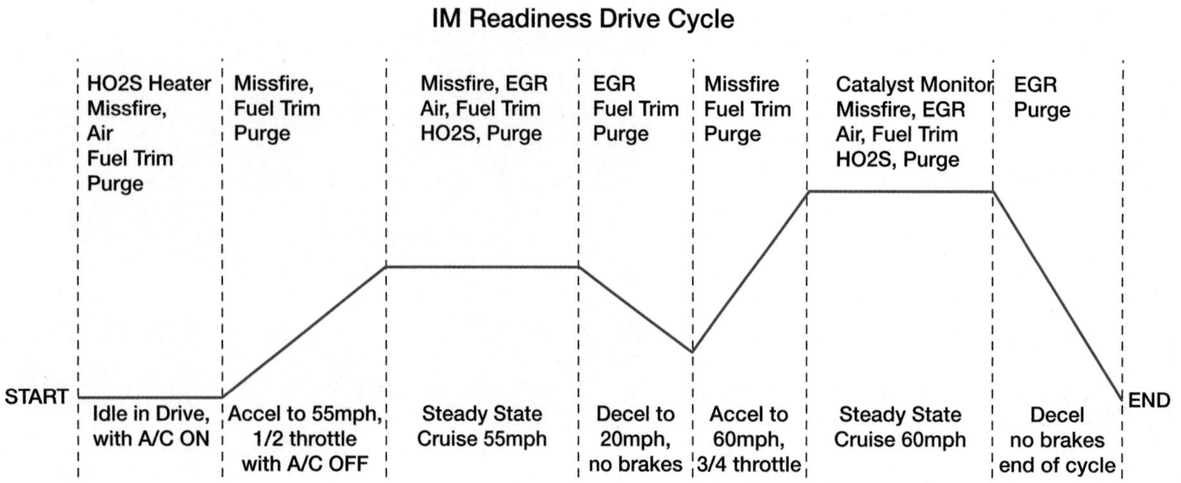

FIGURE 51-11 240-second drive cycle.

▶ System Readiness Monitors

A monitor is a computer's test of its system components and circuits. It checks for malfunctions that would cause emissions to increase 1½ times more than FTP regulations allowed. OBD II standards state that a vehicle's PCM must monitor the parts and systems with two priorities in mind. They are high priority (**continuous monitors**) and low priority (**noncontinuous monitors**) (**FIGURE 51-12**). High priority faults that require continuous monitoring are listed here:

- Misfire monitor—It continuously monitors either crankshaft speed or combustion chamber feedback that shows misfiring cylinders. If the monitor detects misfires between the lower and upper threshold, a "pending DTC" will be set. Freeze-frame data will also be stored. If the misfire reoccurs during the next consecutive trip, a "current DTC" will be set and the MIL illuminated. If the monitor detects misfires above the upper threshold, the MIL will blink immediately and a "current DTC" will be set. It typically doesn't require a second trip. This may also cause the engine management system to default to open loop status to better control fuel trim.
- Fuel system monitor—It monitors the amount of long-term fuel trim (LTFT) correction needed to keep the air–fuel mixture correct. Above or below the specified fuel trim threshold, a "pending DTC" will be set. It typically takes two consecutive trips to turn the MIL on and convert this to a "current DTC."
- Continuous component monitor—It continuously monitors the engine management sensors and actuators for faults. Failure on the first trip, during which the failure causes the emissions to exceed the specified limits, results in a DTC set, freeze-frame data stored, and MIL illuminated. Failure on the first trip during which the failure does not cause an emission failure sets a pending DTC. It requires a second consecutive failure to turn the MIL on.

Note that continuous monitors do not have a monitor readiness status indicator because they run continuously.

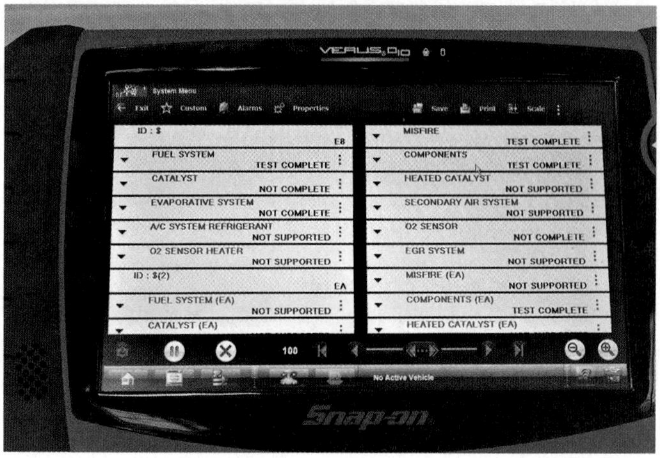

FIGURE 51-12 Typical status of monitors on a scan tool.

Lower priority faults are monitored with noncontinuous monitors such as these:

- Catalytic converter monitor—It monitors the efficiency of the catalytic converter.
- Heated catalyst monitor—It monitors the condition of the catalyst heaters.
- Oxygen sensor monitor—It monitors the activity and response time of the oxygen sensors.
- Oxygen sensor heater monitor—It monitors the condition of the oxygen sensor heaters.
- Evaporative system monitor—It monitors the system for leaks as well as the ability to purge the canister of vapors.
- EGR system monitor—It monitors the ability of the EGR to allow exhaust gases to recirculate. May also monitor variable valve timing on engines that use that in place of an EGR valve system.
- Secondary air system monitor—It monitors the secondary air system, if equipped, to verify that air is directed where specified.

In such cases, noncontinuous monitors are run only once during each engine warm-up cycle. Or even less often. It depends on certain requirements. This may include ambient temperatures or fuel level. Because of this, they have readiness monitor status indicators that indicate whether the readiness monitor has run.

Each time the engine is started and the vehicle is driven according to the enabling criteria for a readiness monitor to run, the PCM runs them one at a time. They verify whether the emission system it monitors is functioning correctly. If the system completes the monitor readiness test, then the monitor status will be logged as "Ready" in the PCM's memory. This doesn't mean that the monitored system doesn't have a fault—but only that the test completed. Until the monitor runs and completes the system readiness test, the status will be listed as "Not Ready" or "Not Completed." On one-trip DTCs, if a fault is detected, the MIL is illuminated after the first fail. On two-trip DTCs, it will illuminate after the second consecutive fail. Both scenarios indicate that the vehicle needs attention.

Some vehicles can indicate whether the readiness monitors have run and are "ready" without using a scan tool. On those vehicles, this mode can be activated by turning the ignition to On (not Crank) and watching the MIL. The MIL should come on and stay on steady if all of the readiness monitors have passed. If after 15–20 seconds the MIL blinks, it indicates that not all of the readiness monitors have passed.

Applied Science

AS-18: Time: The technician uses direct and indirect methods to measure time and compare the results with the manufacturer's specifications.

A technician is troubleshooting a vehicle. It has an intermittent problem with hesitation on acceleration. The technician has started the evaluation by conducting a visual inspection. After this, a scan tool is used to check for codes but there are none. Fuel pressure and volume are checked, and both are within the manufacturer's specifications. The technician checks for service bulletins for this symptom but has not found any leads in this area.

Another technician suggests confirming that the injector pulse width meets the manufacturer's specifications. He has had a similar vehicle that had problems in this area. The technician checks the service manual and finds that the injector pulse width should be 2.5 to 2.7 milliseconds at engine idle. The manual also gives other pulse-width specifications for other engine speeds. Using a scan tool, the technician is able to verify that the pulse width is within the manufacturer's specifications at all published engine speeds. The scan tool data is on a direct-time basis as real-time engine data is being used.

Pulse width is the amount of ontime measured in milliseconds that a fuel injector delivers fuel to the cylinder. The injector pulse width depends on the input signals supplied to the PCM from its various engine sensors. The pulse width expands on acceleration and contracts under lighter loads.

▶ Scan Tools

LO 51-06 Describe the purpose of a scan tool.

A scan tool (scanner) is a device able to electronically communicate with and extract data from the vehicle's onboard computers. Onboard computer modules include the PCM, electronic brake control module (EBCM), body control module (BCM), transmission control module (TCM), and perhaps numerous others. Simple scan tools from the 1980s could read and erase fault codes and little more. As onboard systems became more complex, so did professional-grade scan tools used in the service bay (**FIGURE 51-13**). Today, adapters are available that work with smartphones. The dongle plugs into the DLC and wirelessly communicates with the smartphone. The smartphone has an app, which causes the phone to work as a scan tool (**FIGURE 51-14**). These devices are available for $25–$100.

Cost and complexity increase with the level of bells and whistles desired in a scan tool. Automotive technicians today are finding that they must use faster and more accurate diagnostic instruments. These include graphing scan tools. They can see hidden faults in components or system waveforms. Scan tools and digital storage oscilloscopes or PC oscilloscopes are now used to display engine compression and vacuum readings, and so forth.

Scan tools with bidirectional capability, called **bidirectional scanners**, are used to send commands to the vehicle's

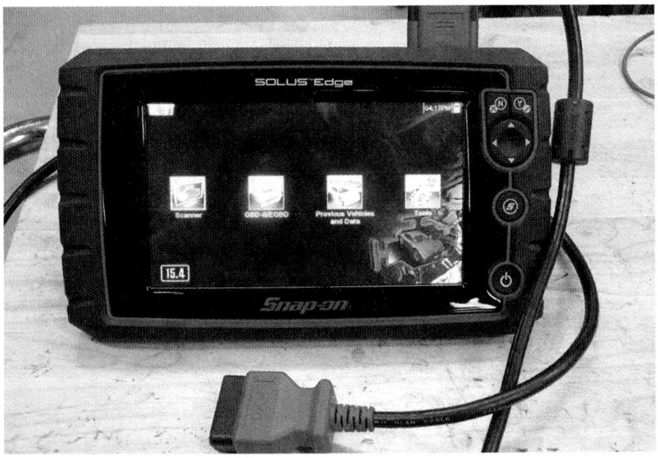

FIGURE 51-13 Typical scan tool used in a shop.

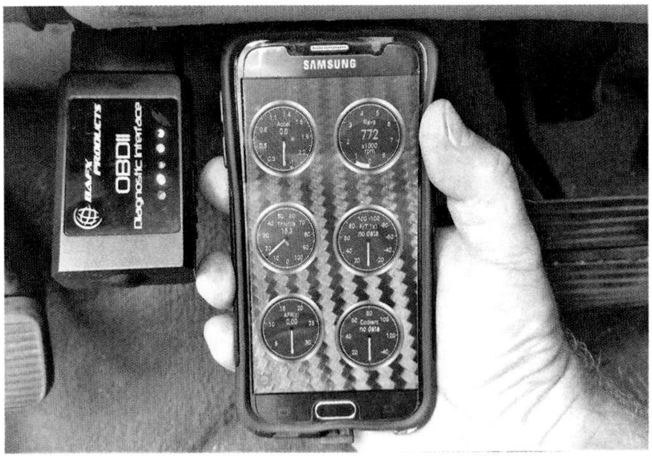

FIGURE 51-14 Typical smartphone dongle which turns the smartphone into a scan tool.

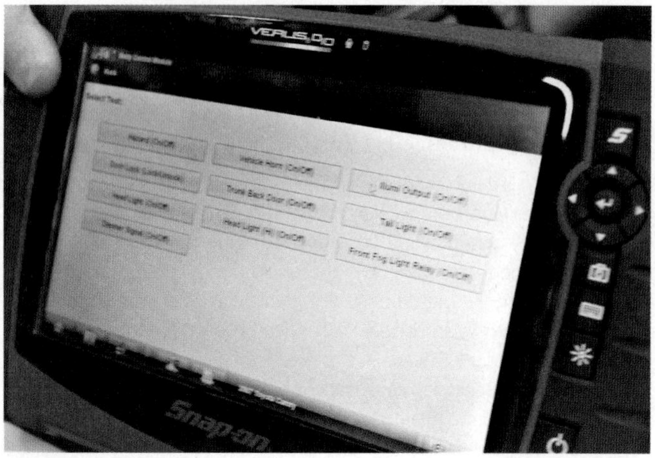

FIGURE 51-15 Bidirectional scanners allow you to activate some components on the vehicle through the scanner.

PCM. The PCM will then cause various components and systems to operate for testing purposes. For example, using the scan tool to command the radiator fan to operate gives you lots of information (**FIGURE 51-15**). First, if the fan does run, then you know that the fan is good. Power/ground to the fan is also good. And the PCM is able to cause it to operate. This would then cast suspicion on a faulty coolant temperature sensor or a thermostat stuck open. Both of these could be the cause of the fan not running.

If the fan did not operate, then you would know that the PCM is *not* able to activate it. So checking power, ground, wires, and the fan itself would be in order. Scan tools are used along with digital storage oscilloscopes, portable five-gas **emission analyzers**, and other diagnostic equipment for much more effective time-saving diagnostic routines.

The scan tool is not a "magic bullet," however. A fault code found does not necessarily point directly to the problem. For example, a P0171 Fuel System Too Lean code could be caused by a defective oxygen sensor. But it could also be caused by an underreporting mass airflow sensor, a vacuum leak, use of the wrong fuel, and so on. So further testing is often required.

Training and experience are important in effectively using modern information-gathering tools. It is also important to understand the following:

- principles of combustion theory
- internal combustion engine operation, and
- emission systems. These are all essential ingredients for success.

Engines, transmissions, and braking systems are still, by nature, mechanical devices. Electronics can control and monitor them. But they cannot overcome major physical defects. Consider the following example. Worn valve guides can cause excessive oil consumption. This may go undiagnosed because a code is not set.

After the scanner is used to read any codes, a visual inspection of the vehicle should be performed. This should happen before any electronic troubleshooting begins. Simple faults like a cracked or loose vacuum hose, or even a loose gas cap, may easily be found and corrected, whether a code is set or not.

Indeed, the field is demanding. It requires extensive training and also a sizeable investment. Opportunities abound for the technician who understands the theory and masters the use of diagnostic tools. Used together, you will be able to efficiently analyze, diagnose, and solve problems on today's advanced engine systems.

▶ Retrieving and Recording DTCs, OBD Monitor Status, and Freeze-Frame Data

LO 51-07 Retrieve DTCs, Monitor Status, and Freeze Frame Data.

Modern vehicles are required to meet strict environmental emission regulations. This continues over the life of their operation. Because of this, vehicles have sophisticated electronic control units and sensors. They monitor and adjust the engine, fuel, ignition, and emission systems. They also provide diagnostic capability to meet the required standards.

There are many integral steps in diagnosing faults within vehicle systems. These include retrieving and recording DTCs, monitor status, and freeze-frame data. The technician must be able to do so in many situations. This includes anytime a vehicle illuminates the MIL, sets a DTC, fails the system monitors, or fails an emission test.

Every vehicle sold is required to meet the requirements of the FTP, which is designed to simulate actual driving conditions. The OBD II system constantly tests and analyzes each system's performance over the drive cycle. This is achieved through a set of PCM programs called monitors, which test each system's operation. The monitor sets a ready or not-ready status based on each system test.

The PCM will set and store DTCs if a fault is detected. Freeze-frame data are also stored. It provides a snapshot of particular data before, during, and after the time of the fault being detected. Never delete or reset the DTCs until all testing has been completed. This is because erasing the codes also erases the freeze-frame data and the status of the monitors. And even then, it may be better not to erase the codes after repair. It may be better to let the PCM turn the MIL off. This way, all of the monitors will not have to rerun to change their status to "ready." If you do need to erase the codes before diagnosing the fault, make sure that any codes and freeze-frame data are recorded and safely stored.

To retrieve and record DTCs, OBD monitor status, and freeze-frame data, follow the steps in **SKILL DRILL 51-1**.

SKILL DRILL 51-1 Retrieving and Recording DTCs, OBD Monitor Status, and Freeze-Frame Data

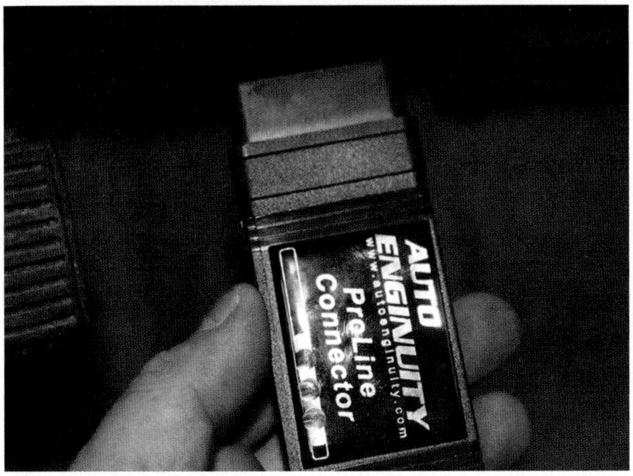

1. Select the scan tool to provide the best coverage for the type and make of vehicle. Locate the DLC and connect the scan tool.

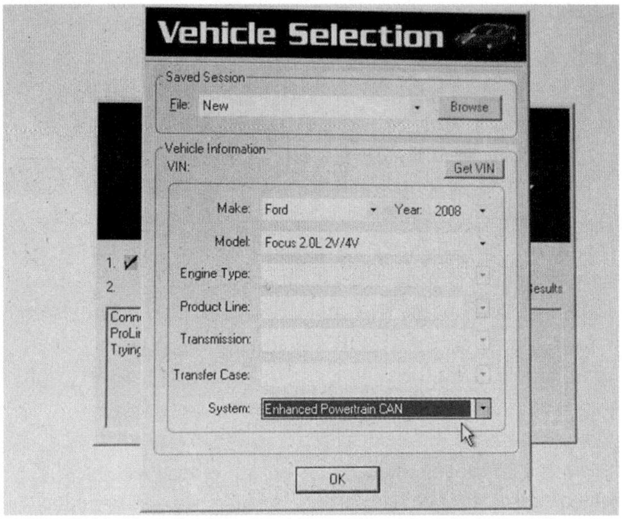

2. Power on the scan tool, and turn the ignition on. Establish scan tool communications with the vehicle.

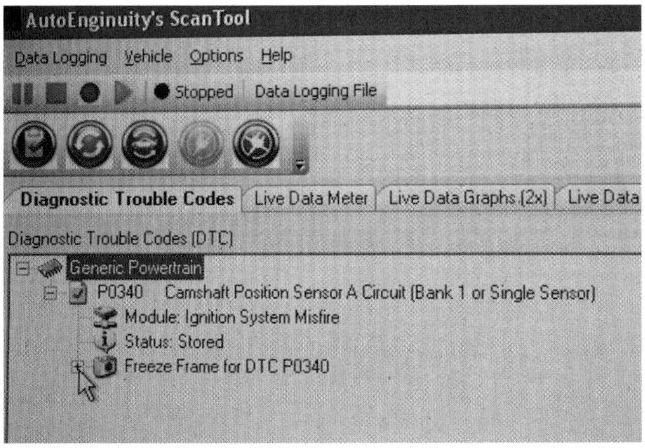

3. Retrieve and record the DTCs.

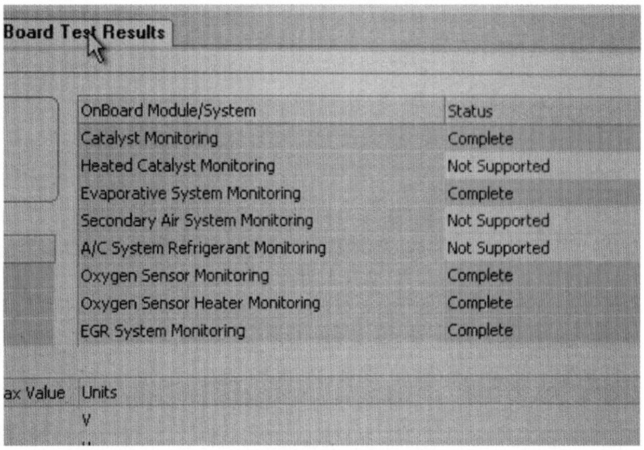

4. Retrieve and record OBD monitor status.

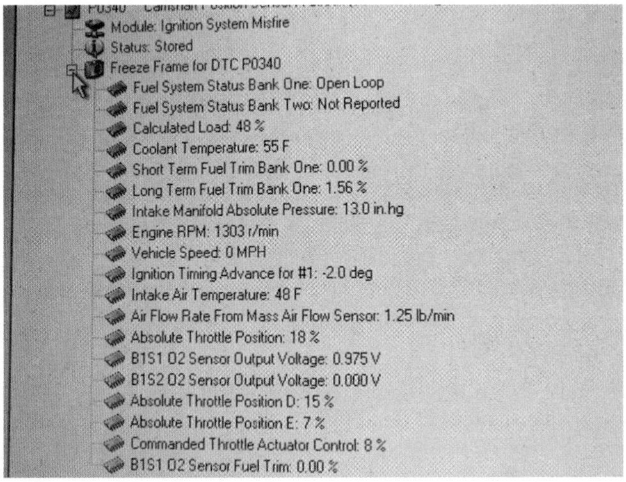

5. Retrieve and record freeze-frame data applicable to DTCs and monitors.

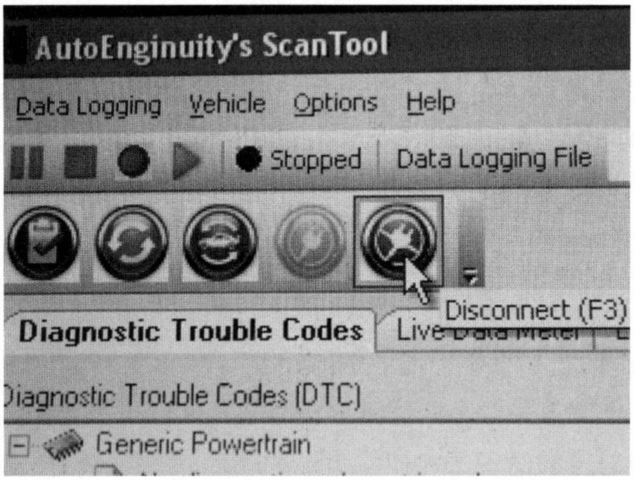

6. Power off the scan tool, turn the ignition off, and disconnect the scan tool.

▶ Wrap-Up

Ready for Review

- Onboard diagnostic system monitors the engine management system for faults and help reduce tail pipe emissions to conserve fuel and reduce carbon dioxide emissions. The system helps reduce volatile organic compounds from fuel tank, fuel lines, and engine crankcase.
- OBD I systems mainly monitored hard electrical circuit malfunctions and were typically limited to well under 100 possible faults. OBD II is an enhanced diagnostic system that identifies faults that may affect the vehicle's emission output and drivability. OBD II constantly monitors and manages the engine management system as it controls the engine and vehicle's emission systems.
- Diagnostic trouble codes are set in the memory of one or more of the vehicle's PCMs once a fault is detected. Emission-related DTCs are the same across all vehicle makes and models.
  - Current DTC means that the test failed during the last one to three drive cycles. Pending codes are codes that have not been validated by failing a second test consecutive test. History codes are pending or current codes that have happened in the past but haven't been cleared.
- MIL can come on or flash depending on a number of factors: on a one-trip DTC, the MIL will illuminate when the DTC and freeze-frame data are stored; on a two-trip DTC, if the same fault is detected in two consecutive drive cycles, the MIL is illuminated; if a catalyst-damaging fault occurs, the MIL will flash; and if a fault occurs that triggers limp home mode, the warning lamp may be illuminated.
- A drive cycle is designed to operate the vehicle in a specific manner. It needs to be operated so that it meets the enabling criteria to cause the PCM to perform the readiness monitor tests. High priority faults that require continuous monitoring are: misfire monitor, fuel system monitor, and continuous component monitor.
  - Lower priority faults are monitored with noncontinuous monitors.
- A scan tool is used to electronically communicate with and extract data from the vehicle's onboard computers such as the PCM, EBCM, BCM, and TCM. Bidirectional scanners are used to send commands to the vehicle's PCM.
- A scan tool can be used to retrieve and record the DTCs, monitor status, and freeze-frame data. Do not delete or reset the DTCs until all testing has been completed.

Key Terms

aftermarket A company other than the original manufacturer that produces equipment or provides services.

bidirectional scanners Scanners used to cause various components and systems to operate for test purposes.

Codes Another name for a diagnostic trouble code

current DTC Condition indicating that the test failed during the last ignition cycle (for two-trip codes, on the previous cycle).

drive cycle A series of prescribed automobile operating conditions during which emissions testing is performed.

emission analyzer A service bay or lab device used for detecting/measuring vehicle tailpipe emissions.

emissions Tailpipe and volatile organic compound pollutants emitted by the automobile.

enabling criteria The operation conditions required before a monitor is allowed to run.

freeze-frame data Refers to snapshots that are automatically stored in a vehicle's powertrain control module (PCM) when a fault occurs (only available on model year 1996 and newer).

History codes A fault code that has occurred but is not current. It is saved in the PCM's memory for 40 warm up cycles.

modules An electronic computer or circuit board that controls specific functions.

monitor An OBD II test run to ensure that a specific component or system is working properly.

noncontinuous monitor A monitor that runs only once per drive cycle.

OBD I The first generation of onboard diagnostic systems that originated for California vehicles.

OBD II The second generation of onboard diagnostic systems, which have been in effect for all US vehicles since 1996.

pending codes Codes that have not been validated by failing a second test.

SAE J1930 An SAE standard for across-the-board standardization of parts and systems nomenclature.

SAE J2012 An SAE standard for across-the-board identification of generic DTCs.

volatile organic compound (VOC) The hydrocarbons in petroleum products that contribute to combustion.

warm-up cycle When the engine is started and run until the engine operating temperature reaches 160°F (71.1°C), has increased in temperature at least 40°F (22°C), and the engine is turned off again.

Review Questions

1. When was the PCV systems first used in California?
   a. 1931
   b. 1961
   c. 1982
   d. 1996
2. Which feature allows sensor data readings to be accessed with a scan tool?
   a. The OBD window
   b. The DTC chart
   c. The data stream
   d. The sensor page

3. If an emissions related sensor or component is monitored by the OBD II system and indicates emissions would exceed 1.5 times the standard limit, what happens?
   a. The vehicle shuts off for 24 hours or until the new part is installed
   b. The government receives a notification and customer gets a letter
   c. The OBD system isolates the system, lights the MIL, and begins self-repair
   d. A DTC is set, information stored, and the MIL is illuminated
4. A vehicle previously set a DTC and the MIL was illuminated. The vehicle was repaired and driven for a couple of weeks, but the codes had not been cleared; what type of DTC is it?
   a. Previous code
   b. History code
   c. Pending code
   d. Current code
5. What is a DTC that needs to be detected in two consecutive drive cycles to become current called?
   a. A double DTC
   b. A catalyst-damaging DTC
   c. A two-trip DTC
   d. A double redundant DTC
6. Typically, what happens if freeze-frame data is set for a DTC, then a higher priority DTC is set?
   a. Freeze-frame data is stored for both, and every DTC that sets afterward
   b. Freeze-frame data stays for the first, and the second doesn't record
   c. The first freeze-frame is transferred to the backup drive and the new data is in the PCM
   d. Freeze-frame data is erased for the first, and the second is stored in its place
7. All of the following are part of the drive cycle EXCEPT:
   a. Vehicle warms up
   b. Vehicle is accelerated
   c. Vehicle is steered
   d. Vehicle slows down
8. Which of the following is an example of a continuous monitor?
   a. Misfire monitor
   b. Catalytic converter monitor
   c. Oxygen sensor heater monitor
   d. Evaporative system monitor
9. A scan tool that is capable of receiving information, but also sending commands to the vehicle, is known as:
   a. An EBCM
   b. Bidirectional
   c. A scanner
   d. An oscilloscope
10. When diagnosing with a scan tool, when should the DTCs be cleared?
   a. Right away to start diagnosis with a completely blank slate
   b. After writing down the DTCs on a piece of paper
   c. After looking up the DTC service information and printing it out
   d. After the diagnosis is complete or freeze-frame data is saved

ASE Technician A/Technician B Style Questions

1. Onboard diagnostics and emissions systems are being discussed. Technician A states that the three-way catalytic converter was introduced in 1981. Technician B states that the first generation of OBD I was introduced in 1996. Who is correct?
   a. Technician A only
   b. Technician B only
   c. Both Technicians A and B
   d. Neither Technician A nor B
2. Onboard Diagnostics are being discussed. Technician A states that prior to OBD II, onboard diagnostics operation and scan tool connections were different from one manufacturer to the next. Technician B states that when OBD II was launched in 1996, it was required to meet SAE standards, as well as US EPA regulations, which gave them some level of uniformity. Who is correct?
   a. Technician A only
   b. Technician B only
   c. Both Technicians A and B
   d. Neither Technician A nor B
3. A vehicle with a current P0128 DTC is being discussed. Technician A states that the P means it is related to the powertrain system. Technician B states that the 0 means it is manufacturer specific. Who is correct?
   a. Technician A only
   b. Technician B only
   c. Both Technicians A and B
   d. Neither Technician A nor B
4. DTCs are being discussed. Technician A states that a pending DTC is set when the failure is damaging the catalytic converter. Technician B states that DTCs can be cleared by the technician, or automatically by passing the test that set the DTC a specific number of times consecutively. Who is correct?
   a. Technician A only
   b. Technician B only
   c. Both Technicians A and B
   d. Neither Technician A nor B
5. DTCs and the MIL are being discussed. Technician A states that when a one-trip DTC sets, the MIL is illuminated immediately. Technician B states that if a catalyst damaging fault occurs, it will turn the MIL red rather than yellow. Who is correct?
   a. Technician A only
   b. Technician B only
   c. Both Technicians A and B
   d. Neither Technician A nor B

6. DTCs and freeze-frame data are being discussed. Technician A states that when a DTC is set, various related PIDs are stored in a freeze-frame to aid in diagnosis. Technician B states that DTCs related to misfires are the lowest priority DTC. Which Who is correct?
   a. Technician A only
   b. Technician B only
   c. Both Technicians A and B
   d. Neither Technician A nor B

7. A drive cycle is being discussed. Technician A states that the drive cycle should typically be able to be completed in 15 minutes if all conditions are met. Technician B states that an EVAP monitor will not run if the fuel level is not within specified range. Who is correct?
   a. Technician A only
   b. Technician B only
   c. Both Technicians A and B
   d. Neither Technician A nor B

8. System readiness monitors are being discussed. Technician A states that some monitors are monitored continuously while the engine is running. Technician B states that some monitors may only run once per drive cycle, or even less. Who is correct?
   a. Technician A only
   b. Technician B only
   c. Both Technicians A and B
   d. Neither Technician A nor B

9. Scan tools are being discussed. Technician A states that some scan tools can communicate with various onboard computers including engine, brakes, and transmission. Technician B states that a modern, professional scan tool can only read and erase DTCs. Who is correct?
   a. Technician A only
   b. Technician B only
   c. Both Technicians A and B
   d. Neither Technician A nor B

10. Using a scan tool for diagnosis is being discussed. Technician A states that clearing the DTCs will leave the readiness monitor status unchanged. Technician B states that erasing DTCs will also erase the freeze-frame data. Who is correct?
   a. Technician A only
   b. Technician B only
   c. Both Technicians A and B
   d. Neither Technician A nor B

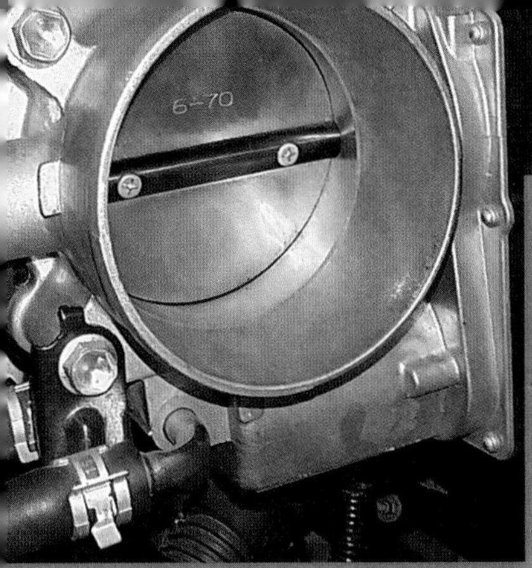

CHAPTER 52

Induction and Exhaust

Learning Objectives

- **LO 52-01** Describe the intake system.
- **LO 52-02** Describe the air cleaner assembly.
- **LO 52-03** Describe heated air intake systems.
- **LO 52-04** Describe differences in intake manifold designs.
- **LO 52-05** Describe volumetric efficiency and forced induction.
- **LO 52-06** Describe turbocharger and intercooler operation.
- **LO 52-07** Describe the exhaust system and exhaust pipe.
- **LO 52-08** Describe catalytic converters and connecting pipes.
- **LO 52-09** Describe mufflers and resonators.
- **LO 52-10** Inspect air filters, housings, and duct work.
- **LO 52-11** Inspect the intake system for leaks.
- **LO 52-12** Inspect integrity of exhaust system components.

ASE Education Foundation Tasks

See Appendix A to view the 2017 ASE Education Foundation Automobile Accreditation Task List Correlation Guide.

▶ Introduction

LO 52-01 Describe the intake system.

The intake and exhaust systems are critical parts in the operation of the internal combustion engine. You can think of it as "breathing." The air enters the engine through the intake system and exits out the exhaust system. The intake system ensures that clean air is supplied to the engine. The air is then mixed with fuel and burned in the combustion chamber. Thermal expansion of the burning gases pushes the piston down the cylinder. In fact, the more air an engine can take in, the more power it can make.

Clean air is essential for a proper, long-lasting engine. If dirt were to get between the close-moving parts, it would cause premature wear of the engine. In fact, it only takes a tablespoon or two of dirt entering the engine through the intake system to ruin an engine.

The intake system must provide a sealed passageway to the combustion chambers. This ensures that no contaminants leak into the system. It also means that air is not allowed to bypass the airflow sensor, which could create a drivability problem. On most vehicles, the intake system also controls the amount of air entering the engine by use of a throttle plate. This is how engine speed and power are controlled.

The exhaust system provides a path for the burned exhaust gases to exit the engine and travel safely out the rear of the vehicle. In doing so, it provides a method of reducing the noise from the power pulses. It also includes components that help reduce the harmful emissions in the exhaust stream.

Getting rid of exhaust gases is just as important as getting air into the engine. If exhaust gases cannot leave the engine easily, then they will back up in the system. This reduces the amount of air that will be able to enter the engine, reducing power output. An efficient, free-flowing exhaust system assists the engine in creating maximum power with minimal emissions. In this chapter, we explore the operation, testing, and light repair of the components in both the intake and the exhaust systems.

▶ The Intake System

The intake system is designed to control the flow of air delivered to the combustion chamber. In many fuel-injected and carbureted engines, fuel is mixed in the intake system. This is so that air and fuel are being delivered together. In diesel and gasoline direct-injection engines, fuel is injected directly into the combustion chamber. There it is mixed with air. Thus, only air (plus some emissions gasses) flows through the intake system.

For any fuel to burn efficiently, no matter where it is injected, it must be vaporized and fully mixed with the proper amount of air. Vaporization begins when the fuel system atomizes the fuel by breaking it up into very small particles. This occurs by using high pressure to spray the fuel into the charge of air. These small particles make it much easier for the fuel to vaporize. Heat and low pressure also may be used to vaporize the atomized fuel.

Intake System Components

The primary components of the automotive intake system are the:

- intake manifold,
- throttle body, and
- the air induction system.

The intake manifold is bolted to the cylinder head. Its construction and design depend on the engine for which it is created. The intake manifold directs airflow into each cylinder (**FIGURE 52-1**). When restricted by the throttle plate, the intake manifold provides a source of vacuum for systems such as the power brake system.

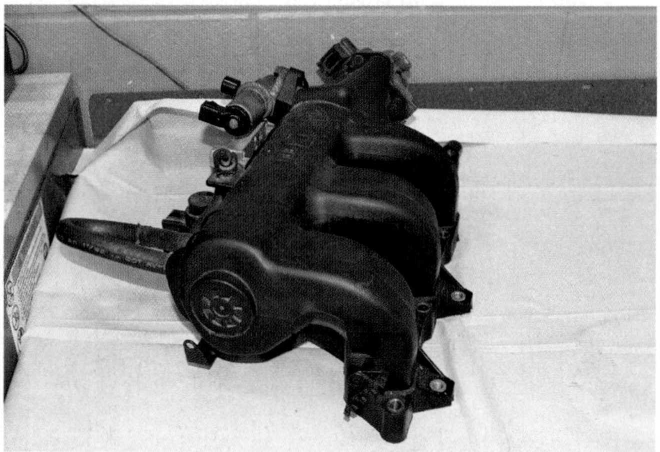

FIGURE 52-1 The intake manifold.

You Are the Automotive Technician

A customer is concerned that their 12-year-old Ford Mustang doesn't idle as well as it should. Also, that the malfunction indicator lamp (MIL) is illuminated. You scan it for codes and notice that it has a P0131 (Circuit Low Voltage B1S1), a P0171 (Fuel Trim Lean B1), and a P0300 (Engine Misfire) detected. As an apprentice technician, you refer to the service information to look up the process for diagnosing the codes. You notice that there could be several things that cause those codes. They include a vacuum leak, weak oxygen sensor, and dirty fuel injectors. You decide to start by looking for a vacuum leak because you hear what sounds like a leak when the vehicle is running.

1. What are the different ways of locating a vacuum leak?
2. Where are the possible places that a vacuum leak can occur?
3. How does a vacuum leak affect a vehicle equipped with a mass airflow sensor?
4. How does a vacuum leak affect the oxygen sensor reading?

No matter the type of gasoline fuel system, the intake manifold creates a mounting place for a throttle body assembly. In throttle body injection, one or two fuel injectors are mounted in the top of the throttle body. They spray fuel down into the intake manifold. In a port fuel–injected engine, the injectors are mounted in the intake manifold near the intake valves (**FIGURE 52-2**). In gas direct injection, the injectors are mounted in the cylinder heads.

The throttle body controls airflow with a butterfly valve or valves. They are also called throttle valves. The throttle valves are opened and closed in one of two ways. Mechanically controlled systems use a throttle cable. On vehicles that use an electronic throttle control, an electric motor opens and closes the throttle plates (**FIGURE 52-3**). Attached to the throttle body is a throttle position sensor. It provides throttle position information to the engine computer. The throttle body also includes vacuum ports for the operation of vacuum-controlled devices. Common examples are the evaporative emission system and the power brake booster.

In front of the throttle body are the following components:

- an air cleaner and housing,
- solid and flexible-duct tubing,
- connectors, and
- sometimes a mass airflow (MAF) sensor (**FIGURE 52-4**).

The inlet opening of the induction system may be located in various positions under the hood. It depends mostly on the available space that the automotive engineer had to work with.

▶ Air Cleaner

LO 52-02 Describe the air cleaner assembly.

The air cleaner assembly filters the incoming air. In past designs, the air cleaner housing was made of stamped metal and housed a round air filter. Most air cleaners are now made of

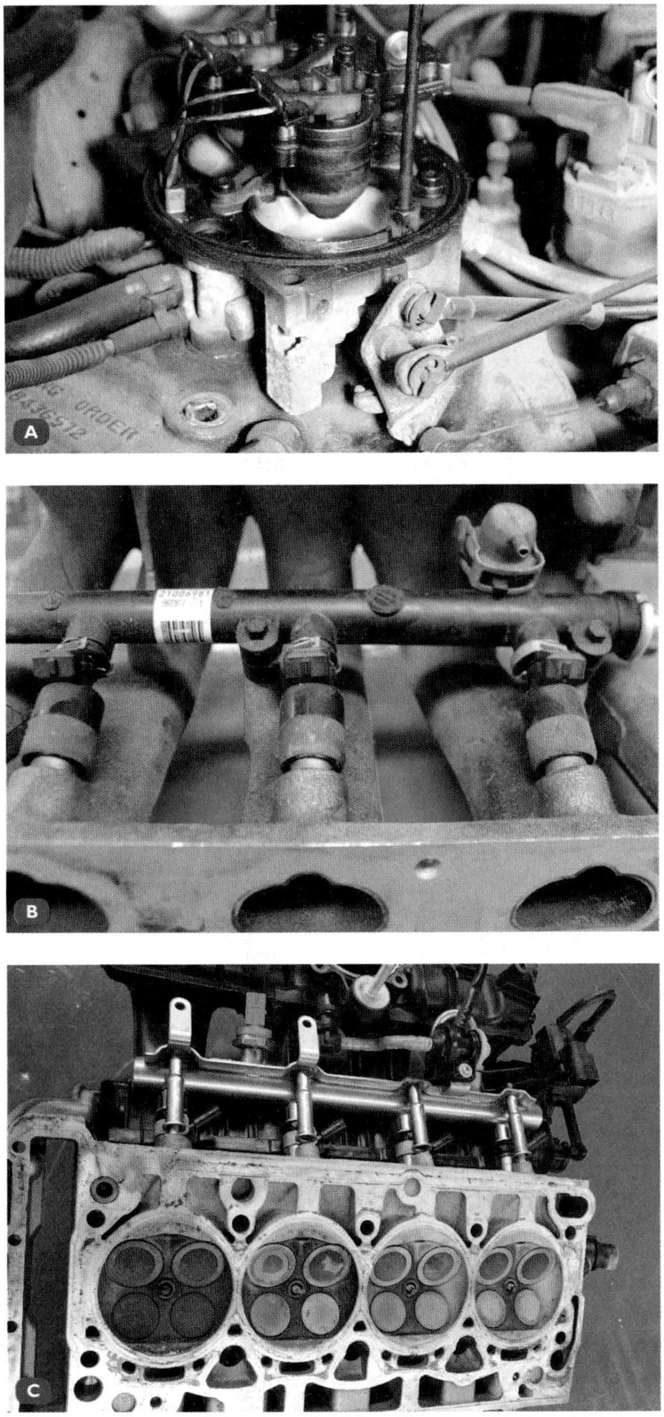

FIGURE 52-2 A. Throttle body injection. **B.** Port fuel injection. **C.** Gas direct injection.

FIGURE 52-3 The throttle plates can be: **A.** Cable operated. **B.** Electric motor operated.

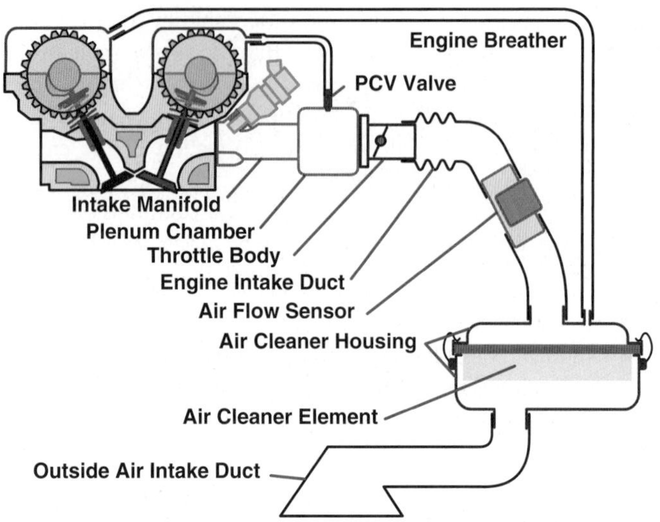

FIGURE 52-4 The air induction system directs air from the inlet through an air filtration device and on to the intake manifold.

plastic and can vary in shape from square to round. The air filter element may be manufactured from pleated paper or from oil-impregnated cloth or felt. In much older vehicles, it was manufactured in an oil bath configuration (**FIGURE 52-5**).

Another function of the air cleaner assembly is to muffle the noise of the intake pulses and the incoming air. The air cleaner can also act as a flame arrester. If a gasoline engine backfires, the air cleaner can contain the flame within the intake system. In many vehicles, the air inlet is mounted where it can obtain cool, clean air. But the location depends on the available space and the hood design.

A lot of air passes through the intake system into the engine. In a gasoline engine, the air–fuel mixture, by weight, is about 14.7 parts air to 1 part fuel. By volume, that is about 10,000 times more air than fuel. When an engine consumes 10 gallons of gas, the air filter will have filtered 100,000 gallons of air. The air–fuel mixture enters the engine, so the air must be clean. Any abrasives that enter the engine can quickly cause wear and damage.

▶ TECHNICIAN TIP

One way to tell if a pleated paper air filter has to be changed is by holding it up to a light and seeing how much light passes through it. Just because it is a bit dirty on the outside doesn't mean it has to be replaced. In fact, a lightly plugged filter prevents particles from going through the filter better than a brand new filter. But if it is too plugged, it will restrict airflow to the engine. If you can see a lot of light through it, it is reusable. If it mostly restricts the light, it should be replaced.

Many electronically fuel-injected systems have an airflow sensor between the air cleaner and the throttle body. It accurately measures all of the air entering the engine. It is essential that there be no air leaks after the airflow sensor, as leaks will upset the air–fuel mixture. It is interesting to note that a MAF sensor can measure air entering the engine down to tenths of a gram per second.

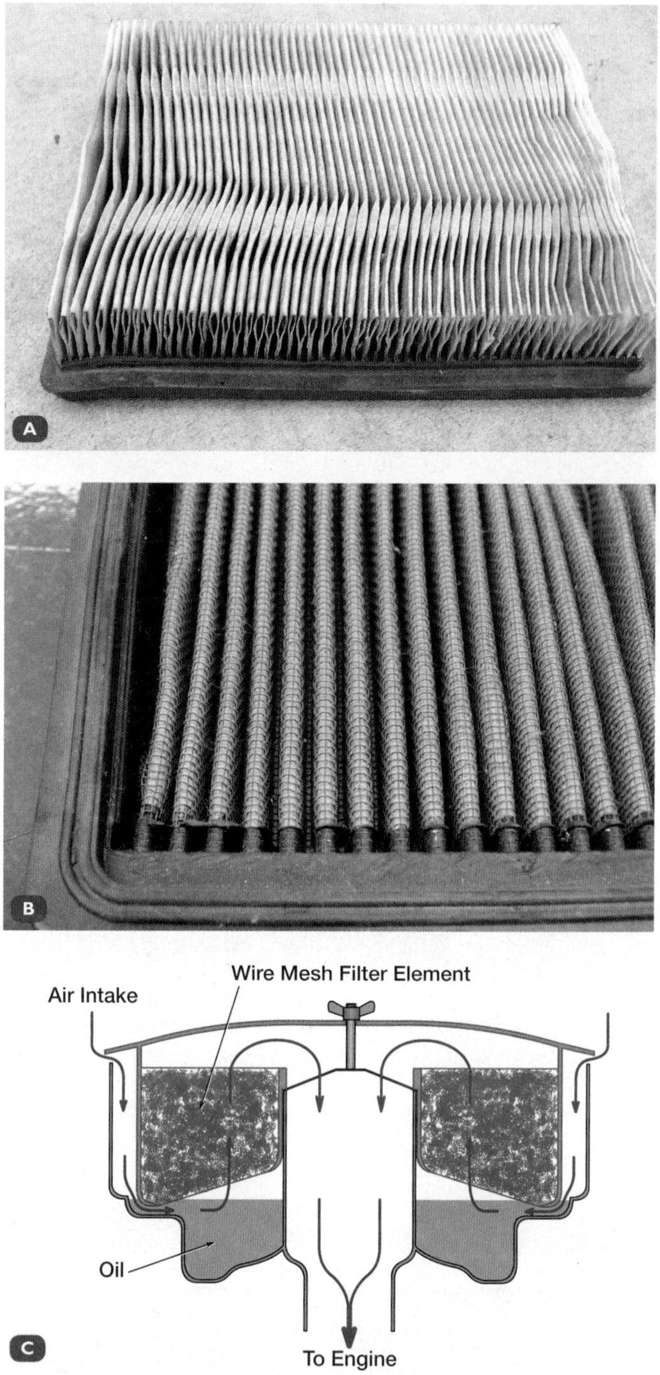

FIGURE 52-5 Types of air filter systems. **A.** Pleated paper. **B.** Oil-impregnated cloth or felt. **C.** Oil bath air cleaner.

On most heavy-duty diesel engines and a few gasoline engines, the air cleaner assembly uses an air filter indicator. It identifies whether the filter needs to be serviced (**FIGURE 52-6**). The indicator typically has a red band that will show if the filter is restricted. Some indicators lock in place, even when the engine is stopped. Others indicate filter condition only when the engine is running. The indicator is mounted between the air cleaner and the engine. When the air filter creates enough of a restriction, the vacuum produced in the intake tube causes the indicator to display the warning.

Some heavy-duty air cleaners also incorporate a cyclone-type pre-cleaner. It is usually mounted directly onto the air cleaner unit (**FIGURE 52-7**). The cyclone system uses angled vanes that give the incoming air a swirling motion. Centrifugal force throws the heavier dirt particles outward. They collect in a bowl at the bottom of the cleaner, where they can be removed manually. An efficient cyclone pre-cleaner can remove up to 90% of particles before they reach the main filter element, greatly extending its life.

One type of air cleaner system is the long-life filtration system that is currently used on the Ford Focus (**FIGURE 52-8**). This air cleaner assembly is a nonserviceable unit that is designed to last 150,000 miles (240,000 km) or more. It is designed to also capture any hydrocarbons being released from the throttle body when the engine is off. The hydrocarbons are then pulled back into the engine when the engine is running. The filter element is made of specially designed foam. This filter can be serviced only by replacement of the entire assembly. It is quite expensive, typically costing several hundred dollars.

▶ Ducting

The ducting is what connects the ridged air cleaner to the throttle body of the intake manifold. Ducting is typically made of hardened plastic. It uses flexible rubber couplings to absorb engine movement. It may have a MAF sensor in line and be retained with worm-style clamps to seal the duct to the sensor.

Some ducting also includes air dampening, sometimes referred to as a Helmholtz resonator. A **Helmholtz resonator** is a container that is sealed and specially shaped to cancel noise created by pressure waves. The pressure wave bounces off the walls of the resonator and collides with the incoming waves, cancelling the noise. On the vehicle, this resonator is a tube that is connected to the air ducting. It is simply a sealed chamber that is open where it is connected to the air duct (**FIGURE 52-9**). The pressure waves created when the engine pulls air into the air duct produce a loud suction sound. Use of the resonator minimizes the noise produced in the air intake system.

FIGURE 52-6 Air cleaner with filter service indicator.

FIGURE 52-7 Cyclone-type pre-cleaner.

FIGURE 52-8 Typical long-life air filter that is not serviceable.

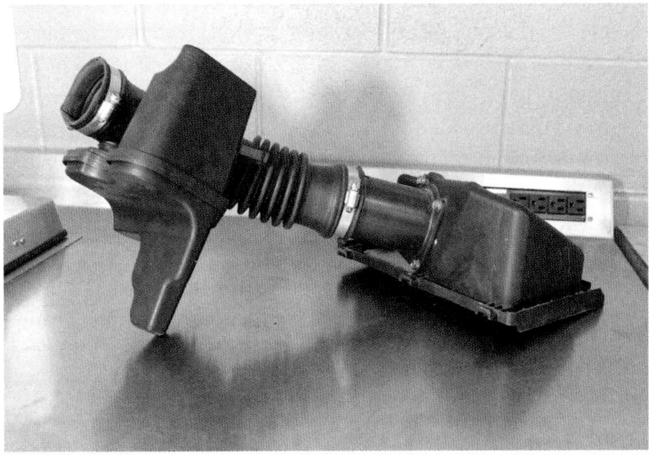

FIGURE 52-9 The air duct on some vehicles contains a resonator to cancel air noise as air flows through the intake.

Applied Science

AS-38: Amplification: The technician can explain to a customer how sound can be amplified in a vehicle due to resonant cavities and other physical characteristics of the vehicle.

The noise level of a vehicle plays an important role in the overall satisfaction rating by the owner. A quiet vehicle is expected by the majority of vehicle owners. However, some owners prefer a more natural sounding engine. Air intake and exhaust systems can be manufactured and modified for a variety of different sounds.

Back in 1863, physicist Herman Von Helmholtz did many studies regarding air-dampening devices. He discovered a resonator that uses the principle of sound waves colliding, resulting in the concept of cancelling noise. The resonator can be used in the induction system. The benefit is the ability to muffle airflow noise. The Helmholtz resonator is basically a simple device consisting of a cavity with one or more short narrow tubes. For each application, the device must be precisely tuned. Concerning the air induction system, the resonator is installed between the air filter and the engine inlet. The design of the Helmholtz resonator creates a sound frequency that cancels out some of the intake air noise. On some vehicles, there is a vacuum-activated valve that disables the effects of the resonator at a certain rpm.

AS-39: Carriers/Insulators: The technician can demonstrate an understanding of how sound generated in one place can be carried to other parts of the body or engine through metal and materials.

Sound travels through steel approximately 17 times faster than through the air. This is because the molecules in steel are closer together. Sound is produced when an object vibrates. Vibrations can pass from molecule to molecule quickly in materials such as steel.

When a vehicle has a bad wheel bearing that is producing a rumbling sound, it is often difficult to tell which of the four wheels the noise is coming from. In this case, you might try driving the vehicle alongside a building or a concrete block wall. Sound waves travel in all directions from their source. Some of them will strike the wall, and the noise will be reflected back toward the vehicle. A greater proportion of the sound waves will strike the wall when it is near the side of the vehicle with the faulty wheel bearing. This is because more sound waves will be reflected toward the vehicle. Thus, if the noise is louder when the passenger side is near the wall, the technician will know the faulty bearing is on the passenger side.

A mechanic's stethoscope can also be used to locate sounds on vehicles. The tip of the metal probe can be used to trace the sound to the source of the problem. There are now electronic stethoscopes that can be used to locate sounds on vehicles.

▶ Intake Air Heating

LO 52-03 Describe heated air intake systems.

Heating of the intake manifold is required in a carbureted or throttle body–injected engine. It provides greater vaporization of the fuel and ensures that fuel does not collect on the cold walls, creating a lean mixture. Port fuel–injected or gas direct injection (GDI) intake manifolds do not normally have to be heated. This is because the manifold does not carry fuel.

There are several methods that manufacturers have used to heat intake manifolds, including:

- hot engine coolant,
- hot exhaust gases, and
- electric heaters.

Flowing hot engine coolant through passageways in the intake manifold heats the manifold (**FIGURE 52-10**). However, it takes awhile for the coolant to become hot. So this is not the fastest way to heat the intake manifold. In this type of system, coolant continuously circulates through the manifold, even when the engine is fully warmed up.

On V-type engines, exhaust gases can also be directed through passageways in the intake manifold during engine warm-up. The exhaust gases are hot immediately. They therefore heat the manifold up more quickly than coolant does. A heat riser valve is located at the end of one of the exhaust manifolds. It blocks most of the exhaust gases from exiting through that exhaust pipe. The exhaust gases are then directed through a passageway in the cylinder head. Then they travel through a chamber in the bottom of the intake manifold which is warmed by the heat. The gases then travel through the passageway in the other cylinder head. Finally, they exit into the exhaust manifold on the opposite side of the engine.

Once the engine is warm, the heat riser opens the valve in the exhaust manifold. Exhaust gases can now flow normally out of the exhaust pipe. If the heat riser valve sticks closed, the engine will run hotter than normal. It will also likely be low on power.

Electric heaters have also been used to preheat the air. A preheat grid is typically placed between the intake manifold and the throttle body (**FIGURE 52-11**). During cold engine operation, current flows through the heating grid, warming the air passing through it. Once the engine reaches a certain temperature, the current flow to the grid is turned off, and the engine operates normally. If the heater stays on too long, the heater grid can melt down. This drops particles down the intake manifold, which can damage the engine.

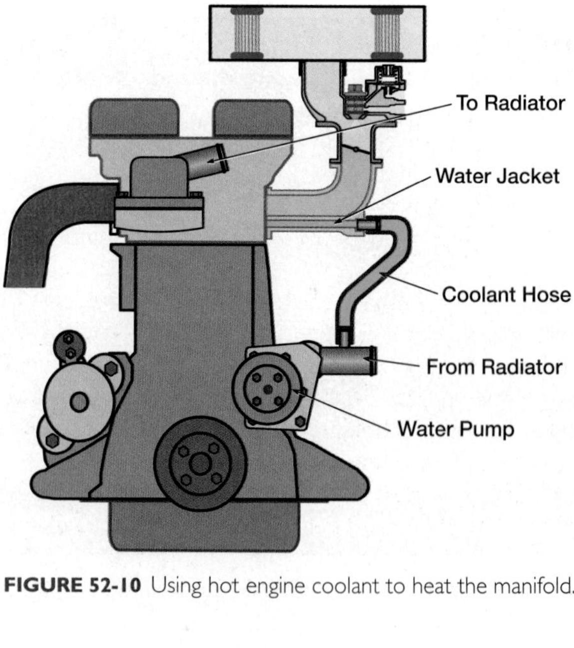

FIGURE 52-10 Using hot engine coolant to heat the manifold.

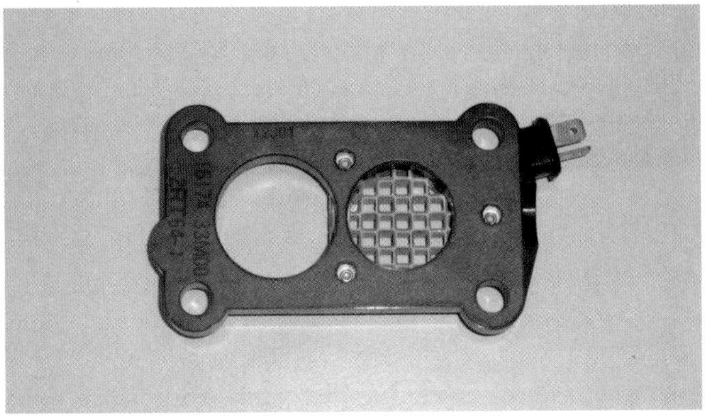

FIGURE 52-11 An electric heater.

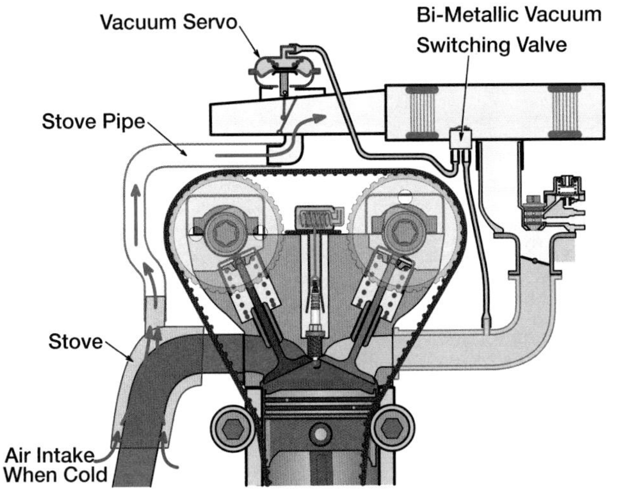

FIGURE 52-12 The heated air intake system, found on carbureted and throttle body–injection systems, is used to ensure that fuel does not collect on the cold intake manifold.

There is another method used to help vaporize the fuel in the intake manifold on carbureted and throttle body–injected engines. It is by warming the incoming air by passing it through a shroud around the exhaust manifold. This method was called an early fuel evaporation (EFE) system (**FIGURE 52-12**). An enormous amount of heat passes through the exhaust manifold and is used to heat the incoming air. The heated air then enters the air cleaner. It proceeds through the throttle body and intake manifold where it more easily mixes with the fuel.

Once an engine is hot, the incoming air can become too hot. So the temperature of the air entering the engine has to be controlled. On these systems, air cleaners use a blend door to control how much hot air enters the air cleaner. These systems are called heated air cleaners (HAC) and heated air systems (HAS). When the engine is started, only heated air is used. As engine temperature rises, the blend door opens and blends cold ambient air with the hot air. This ensures that the temperature of the air supplied to the engine stays fairly constant.

On most vehicles, the blend door is controlled by a vacuum diaphragm and a thermostatic control switch. The diaphragm is attached to the blend door and controls its movement. The thermostatic control switch controls vacuum to the diaphragm.

FIGURE 52-13 The port fuel injection manifold has a large open area, called the plenum chamber, and tubes that deliver air individually to each cylinder called intake runner tubes.

When a cold engine starts, vacuum from the intake manifold passes through the control switch and moves the diaphragm. It opens the air door and allows hot air to flow into the air cleaner. At the same time, it closes the ambient air intake.

As the air temperature in the air cleaner rises, the control valve slowly closes, reducing the flow of hot air and blending in cooler air. When carburetors and the throttle body–injection units were no longer installed in vehicles, heated intake air was no longer needed.

With a multipoint injection setup, the intake manifolds carry air only. So heating of the intake manifold is not needed. Also, the cross-sectional area of the intake runners can be larger. Because more air can flow, the engine will produce more power.

Fuel is injected directly into the intake ports of the cylinder head. The other possible fuel delivery method is the gasoline direct injection system. It directs fuel straight into the combustion chamber. This system uses an intake similar to the multipoint injection setup, flowing air only through the intake. The fuel-injected engine manifold has a **plenum chamber** that provides a reservoir of air. It also helps prevent interference with the flow of air between individual branches. The plenum chamber is a large portion of the intake manifold after the throttle plate but before the intake runner tubes (**FIGURE 52-13**).

▶ Intake Manifolds

LO 52-04 Describe differences in intake manifold designs.

Intake manifolds used to be made from cast iron, and later from light aluminum castings. In modern vehicles, the intake manifold is made from special heat-resistant polymers and plastics (**FIGURE 52-14**). Manifolds using this type of construction are up to 50% lighter. This means that they contribute to higher fuel efficiency. Plastic intake manifolds are commonly molded from a glass fiber–reinforced grade of crystalline polymer. This type of material is suited for replacing metal in underhood applications. Its strength, stiffness, and chemical resistance perform well under high-temperature operating conditions. The plastic can also be recycled, like cast iron and aluminum. The use of plastic has enabled more

FIGURE 52-14 Plastic intake manifold.

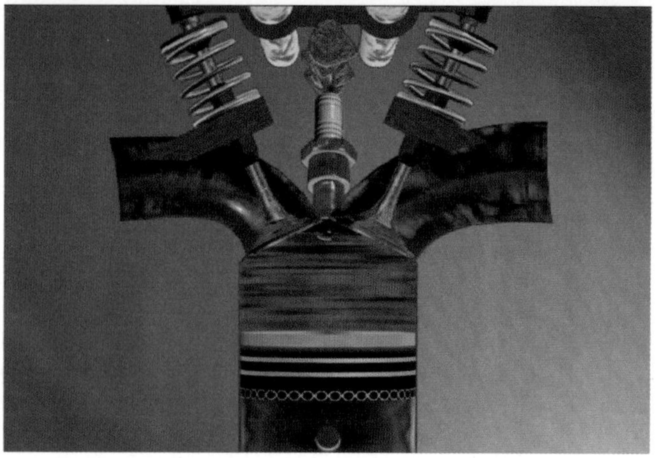

FIGURE 52-15 Cross-flow head.

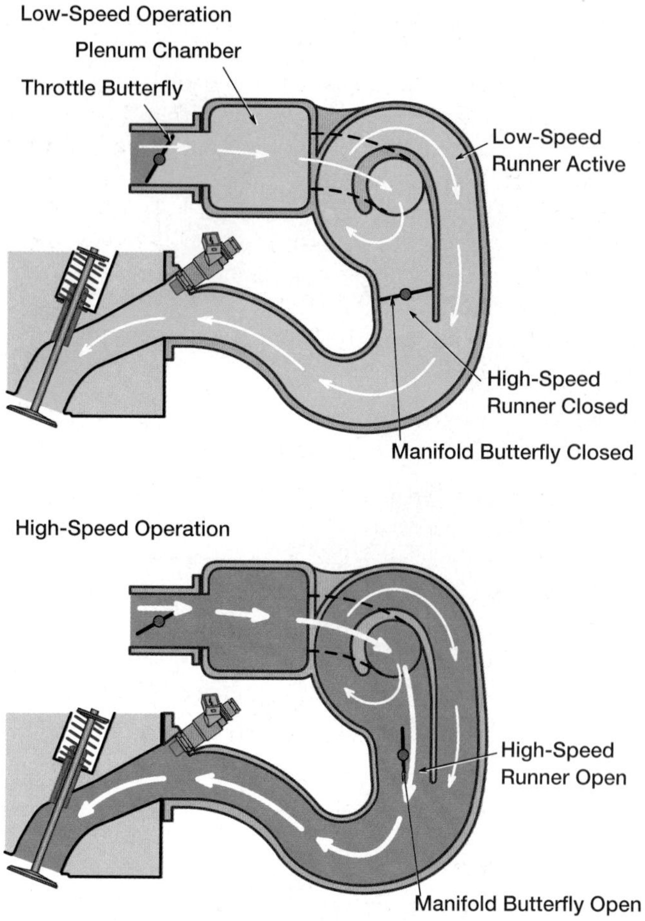

FIGURE 52-16 Typical variable intake manifold with alternate runners.

efficient airflow into the engine. This is due to smoother surfaces of the manifold. It has also allowed for easier shaping of the manifold. This then creates better airflow characteristics.

Cylinder heads that have intake and exhaust manifolds on opposite sides of the engine are known as cross-flow heads (**FIGURE 52-15**). This design tends to produce more power because the flow of air moves more easily across the head rather than coming in and back out on the same side. Most modern engines are set up with a cross-flow cylinder head. The cross-flow head contains individual branches or ports to carry air and fuel into the combustion chamber. In past designs, two cylinders could share an intake port. But this design did not allow for free flow of air. So it was replaced with individual ports to allow for more power.

▶ Variable Intake Systems

The intake manifold for fuel-injected engines normally has long branches of equal length. The long branches increase the pulsing effect of the airflow at lower engine rpm and help charge the cylinders. The more air drawn into the cylinder, the denser the air–fuel mixture is when the intake valve closes. The increased density of the air–fuel mixture determines how much pressure develops in the combustion chamber. This increases the amount of force on the piston to turn the crankshaft.

The design of the intake system largely influences the mass of air that can be drawn into a cylinder at a given engine speed. Thus, the intake system largely influences the engine's torque curve. In general, long intake manifold runners produce high torque at lower engine rpm. This is because of the increased amount of time between the air pulses due to the inertia of the long column of air. The increased time between the pulses matches the longer time between intake strokes at lower rpm operation. Shorter intake manifolds produce higher torque at higher engine rpm. This is because of the shorter amount of time between the pulses due to the inertia of a shorter column of air. This ram air effect allows for more total airflow at and near the tuned rpm.

Some manifolds are designed to respond to changes in engine load and speed by changing their effective length. They generally do this in two or three stages and are called variable inertia, or variable intake, manifold systems. The primary section is long and narrow for the low range of rpm. The secondary section is shorter and wider for the high range of rpm (**FIGURE 52-16**). This combination maintains a high-speed airflow in the system. The three-stage manifold extends the torque curve even more. This is because the torque curves overlap each other as advantageously as possible.

The different manifold stages are controlled by butterfly-type valves. They are moved by a stepper motor or vacuum actuator (**FIGURE 52-17**). The stepper motor or supply of vacuum is controlled by the engine control module. The system is called intake manifold runner control (IMRC). In this way, air is forced to run through the desired set of runners.

Some manufacturers are now using sealed pulse chambers for each intake runner. The chambers are opened and closed by butterfly valves. The pulse chambers are designed to time the return of intake pressure pulses back to the intake valves above a certain rpm (**FIGURE 52-18**). At this rpm, the butterfly valves open, and the pressure pulses assist in filling the cylinders with more air. Below this range, the butterfly valves are closed, and the pressure pulses are timed to happen at a lower rpm range. This helps to maintain a broader and more powerful torque range across the operating rpm of the engine.

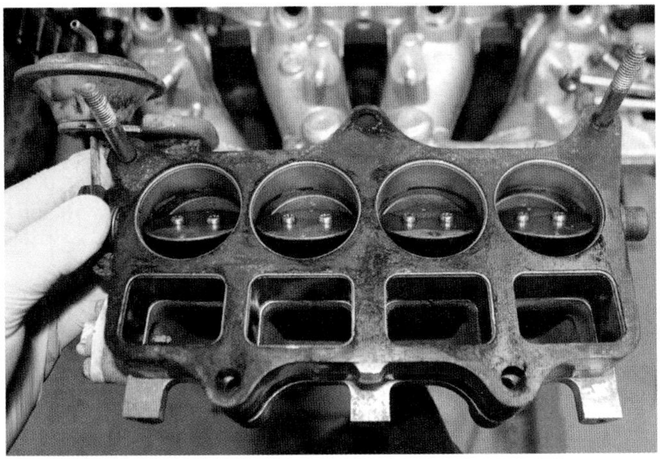

FIGURE 52-17 Different intake manifold runner lengths can be obtained by opening and closing alternate runners.

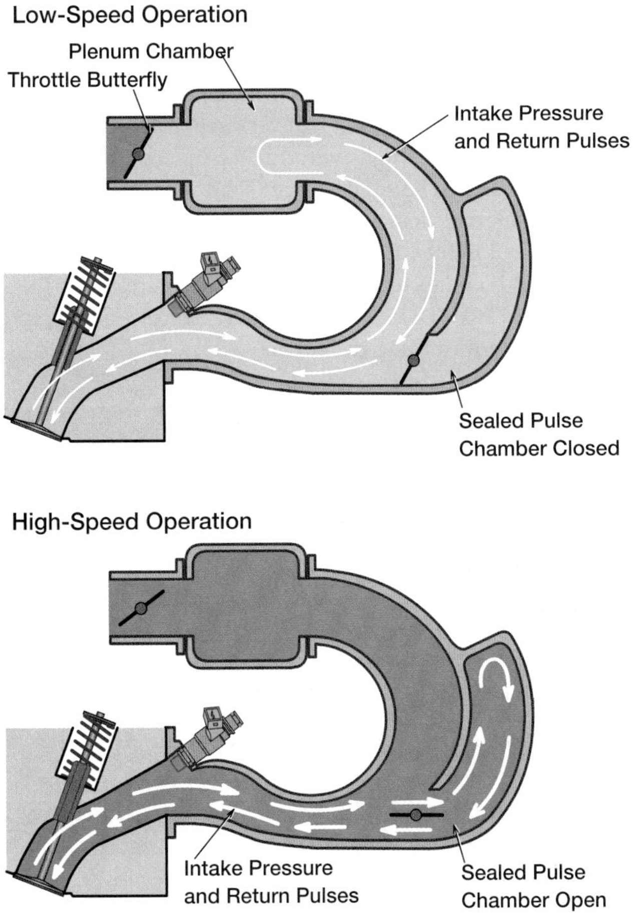

FIGURE 52-18 Typical operation of a pulse chamber intake manifold.

▶ Volumetric Efficiency

LO 52-05 Describe volumetric efficiency and forced induction.

Volumetric efficiency compares the volume of air entering a cylinder to the total piston displacement of the cylinder. It is usually expressed as a percentage (**FIGURE 52-19**). In a naturally aspirated engine, one without forced induction, volumetric efficiency can almost never be 100%. This is because of the resistance to airflow that occurs at the throttle body, manifold, and intake valve. However, some highly modified engines can exceed 100%. This is due to the valve overlap being matched to the intake and exhaust manifold sizes and lengths. These factors maximize the ram air effect at the tuned rpm.

To understand volumetric efficiency, you must have a good understanding of the physics of airflow. Air likes to move in a straight manner and is affected by the bends in the intake manifold. Sharp bends force the air to move to the opposite side of the tube and force the air to pile up, slowing the airflow. Also, air near the surface of the manifold slows as friction of the tube wall resists its motion. And yet, air in the center of the airstream tends to speed up as it moves over top of the slower air. All of these factors slow the airflow and reduce volumetric efficiency.

With forced induction, the incoming air is compressed by some sort of pump or compressor. Because of this greater pressure, a greater volume and mass of air is forced into the cylinder during the intake stroke. This increases volumetric efficiency to well above 100%. It therefore increases the engine's power output proportionately.

Volumetric efficiency is also affected by back pressure. Back pressure in an exhaust system refers to a buildup of pressure in the system that interferes with the outward flow of exhaust gases. This area of high pressure acts as a kind of wall to stop gas flow. It can be caused by a blockage in a muffler or a similar restriction.

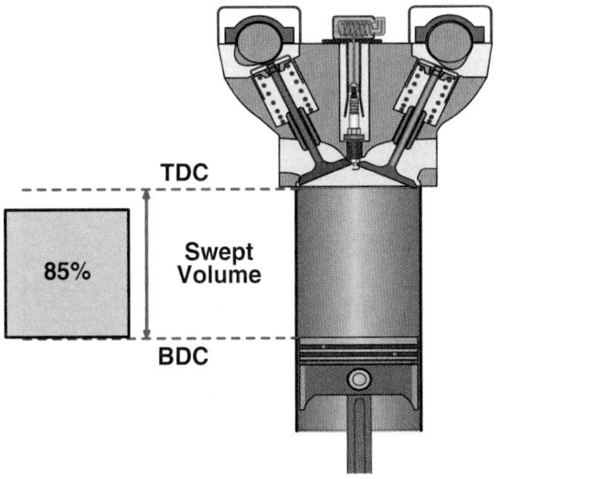

FIGURE 52-19 Volumetric efficiency of a normally aspirated engine.

One common cause of excessive backpressure is a catalytic converter that has melted down and created a restriction in the exhaust stream. Catalytic converters become hotter as the combustible mixture inside them is increased, such as from a misfire. The catalyst in the converter will melt if it gets too hot. It will then start to block off the small passageways.

Backpressure typically should be no more than 3 pounds per square inch (psi), or 20.7 kilopascals (kPa). If it is greater, it can create performance issues. This is because exhaust gases will back up into the combustion chamber minimizing the amount of fresh air (combustible mixture) flowing into the engine. Volumetric efficiency and engine power output are then reduced.

Forced Induction

One way to improve engine output is to increase the amount of the air–fuel mixture that is burned in the cylinder. This is due to increasing the volumetric efficiency. Volumetric efficiency can be improved the most by what is called **forced induction**. Forced induction increases air pressure in the intake manifold above atmospheric pressure. Thus, an engine using forced induction can have a volumetric efficiency well above 100% (**FIGURE 52-20**).

One way to achieve forced induction is by using a supercharger. A supercharger, turned by the crankshaft, compresses the air and forces it into the engine. The faster the engine turns, the more air is moved by the supercharger. Another way to achieve forced induction is to use a turbocharger. A turbocharger uses energy that is normally wasted through the exhaust system to spin a compressor that forces air into the engine.

Supercharger Systems

Power is produced when a mixture of air and fuel is burned inside an engine cylinder. If more air is forced into the cylinder, then more fuel can be burned and more power produced with each power stroke. A **supercharger** compresses the air in the intake system to above atmospheric pressure. This increases the density of the air entering the engine. Naturally aspirated engines operate with uncompressed air at atmospheric pressure, 14.7 psi (101 kPa). But a supercharger boosts that pressure another 6 psi (41.4 kPa) or higher.

In a supercharged system, there is a greater air mass flow rate. That is, a higher density and speed of airflow. Air pressure is increased by the compressor on the way into the engine. More fuel is added to the air, which creates more cylinder pressure and power. The increased cylinder pressure causes the exhaust gases to exit much more rapidly, making the exhaust sizing less important. Also, any residual gases tend to be pushed out of the cylinder by the pressurized airflow entering the combustion chamber.

Some of the extra power produced must be used to drive the supercharger. But the net result is more total power from the crankshaft (**FIGURE 52-21**). The supercharger may include a **bypass valve** system that allows the air to be returned back to the supercharger inlet. It is used to prevent excessive boost. The bypass valve can be mounted remotely or directly onto the intake port.

Several types of superchargers are manufactured for use on vehicles. The oldest version is the Roots-style supercharger. It uses lobes on two shafts to move air into the engine. This type of supercharger does not draw air between the two shafts. Rather, air is pushed around the sides of the housing and into the intake (**FIGURE 52-22**). The lobes on both shafts mesh together in the middle, so air cannot exit between them.

Another version of the supercharger is the twin-screw type. The twin-screw supercharger compresses air between the screws and forces it into the intake (**FIGURE 52-23**). This compressor design is very efficient, but very expensive.

Many aftermarket high-performance superchargers are the centrifugal design. The supercharger shaft is turned by a drive belt from the crankshaft. It spins a compressor wheel, similar to what is used in the turbocharger (**FIGURE 52-24**). The supercharger shaft is connected by a set of gears to increase the speed of the compressor. The compressor increases the air pressure, which forces more air into the engine.

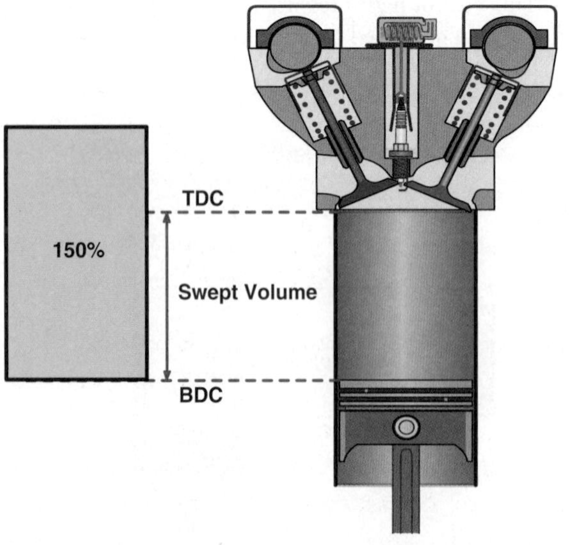

FIGURE 52-20 Volumetric efficiency of a forced induction engine.

FIGURE 52-21 The supercharger uses power from the crankshaft to force additional air and fuel into the engine, which increases the engine's power output.

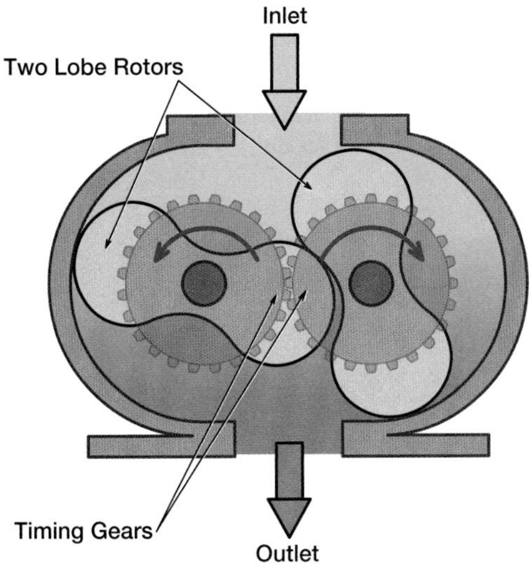

FIGURE 52-22 A Roots-style supercharger moves air around the sides of the lobes.

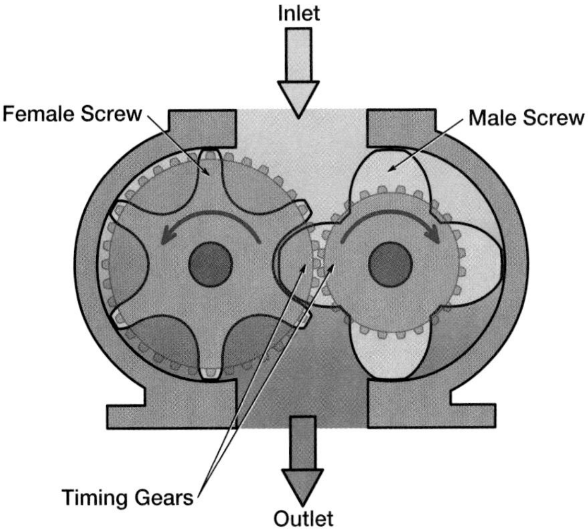

FIGURE 52-23 Typical twin-screw supercharger.

Because compressing air heats the air up, many supercharged engines use a heat exchanger. It is also called an intercooler and is located between the supercharger and the intake manifold. The heat exchanger is like a radiator that the air flows through. It cools the air before it enters the engine. This helps prevent engine-damaging detonation or overheating. Intercoolers are covered in detail later in the chapter.

▶ Turbocharger Systems

LO 52-06 Describe turbocharger and intercooler operation.

A **turbocharger** is a type of forced induction system. It uses wasted kinetic energy from the exhaust gases to increase the volume of air entering the engine (**FIGURE 52-25**). At the same time, it creates back pressure in the exhaust system. This means that valves, cam timing, and exhaust system design must be closely matched to the turbocharger and engine.

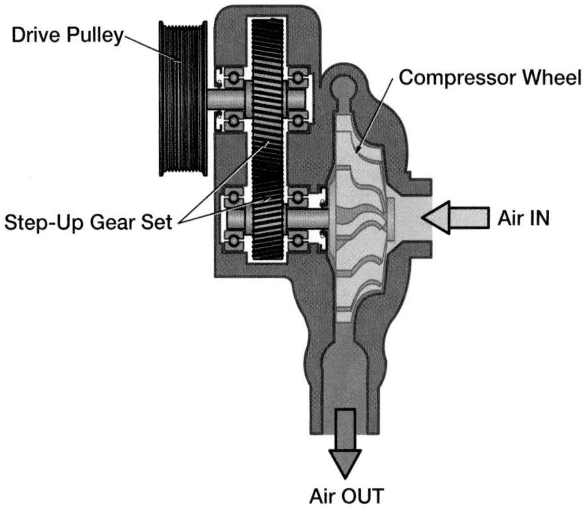

FIGURE 52-24 Typical centrifugal supercharger.

FIGURE 52-25 A typical turbocharger removed from the vehicle.

The turbocharger uses exhaust gases to turn a turbine fan. A shaft connects the turbine rigidly to a centrifugal compressor (**FIGURE 52-26**). The compressor compresses the air and forces it under pressure into the intake manifold. The turbine is turned by exhaust gases, so it runs at very high temperatures. It, along with the compressor, can rotate at well over 100,000 rpm. Because of this, the turbocharger needs a good supply of clean oil to lubricate the bearings. It also carries excess heat away from the turbocharger. Some engines also supply coolant to the turbocharger body to improve cooling.

Higher engine speeds mean increased cylinder pressure and exhaust gas volume. That makes the turbocharger spin faster and force more air into the cylinders. Pressures left unchecked can damage the engine if allowed to get too high. When pressure increases in an engine, so does cylinder temperature. As the temperature increases, the possibility of detonation also increases. This is because the gasoline will ignite before the piston reaches the proper position in the compression stroke.

To control this risk of detonation, a device called a **wastegate** is installed on the exhaust inlet of the turbocharger. When intake manifold pressure reaches a preset level, the

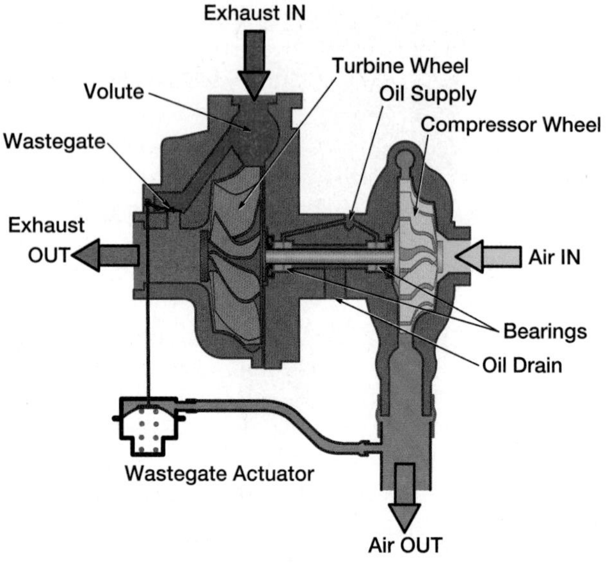

FIGURE 52-26 The turbocharger compresses air to feed to the intake and is powered by exhaust gases.

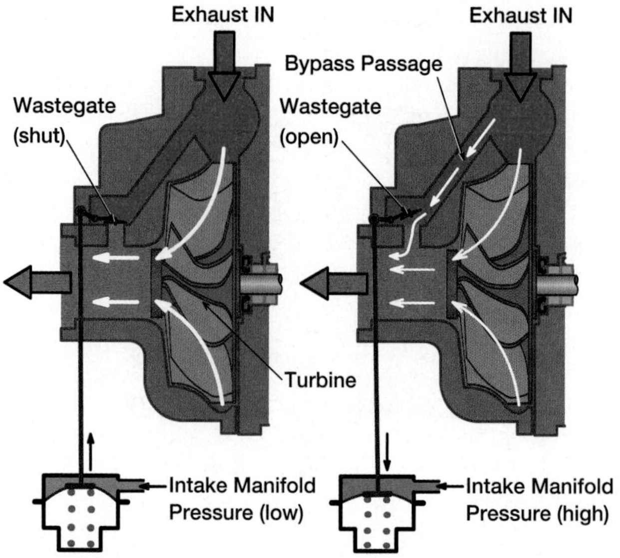

FIGURE 52-28 Typical mechanical wastegate operation.

FIGURE 52-27 The turbocharger wastegate actuator controls the amount of boost pressure going to the intake manifold.

FIGURE 52-29 The blow-off valve is a spring-loaded pressure relief valve that releases excessive pressure in the air intake tube going to the intake manifold.

wastegate opens. It allows the exhaust gases to bypass the turbine (**FIGURE 52-27**). This prevents intake pressure from building further. The wastegate can also be computer-controlled so that intake air pressure can be reduced in the case of detonation.

On a mechanically operated wastegate, the wastegate actuator uses a spring-loaded pressure diaphragm. The spring pressure in the diaphragm controls the pressure at which the wastegate opens. The manifold pressure pushes on the diaphragm. It overcomes the spring force at the specified point. This causes the control linkage to open the wastegate. Exhaust gases then bypass the turbine wheel. This reduces the boost pressure (**FIGURE 52-28**). As intake manifold pressure drops, the wastegate closes and allows exhaust to flow to the turbine wheel again.

The wastegate can be PCM-controlled. The manifold absolute pressure sensor reports the manifold pressure to the PCM. When needed, the PCM signals the wastegate actuator to open the wastegate and reduce turbocharger boost. The actuator can be vacuum-operated or electric motor–operated.

A pressure relief valve known as a **blow-off valve** may also be installed. It prevents excessive manifold pressure if the wastegate should fail. A blow-off valve works against spring pressure. As boost pressure rises above spring pressure, the pressure opens the valve. This allows excess pressure to "blow off" into the atmosphere.

The blow-off valve may also vent the pressure to the air cleaner box. This reduces the pressure release noise. The blow-off valve is located in the air intake tubing connected to the intake manifold. It is also used to ensure that the intake pressure does not rise excessively high when the throttle is abruptly shut (**FIGURE 52-29**).

Because the turbocharger uses the energy of the exhaust gases, there is a short delay between when a driver opens the throttle and when maximum power is available. This delay is called turbo lag. On larger engines, it can be quite noticeable if the turbocharger is not sized correctly.

Another new development in turbocharger technology is used on the BMW engine. The engine design uses reverse flow cylinder heads. The center of the V-type engine, which normally was for intake, is now for the exhaust passages (**FIGURE 52-30**). Exhaust leaves the head and moves into the twin turbochargers, one per engine bank. The intake manifold is now on the outside of the head. The different positioning of the turbochargers keeps exhaust temperatures high. This produces more turbine speed and increases the efficiency of the turbocharger. The catalytic converter also gets up to temperature faster, reducing emissions.

A turbocharger recycles heat energy that would otherwise be lost. So a turbocharged engine can increase an engine's efficiency and fuel economy. That is, as long as the vehicle is being driven conservatively. But, even though a turbocharger may seem to be offering more energy for nothing, it can introduce problems of its own. The extra heat and power it generates can put an extra load on the engine's cooling and lubrication systems. This is why most turbocharged vehicles have shorter service intervals for oil changes than non-turbocharged vehicles.

As the turbocharger compresses the air, it heats up. Hot air is less dense than cool air, so it tries to expand again. So some of the benefits of compressing it are lost. To stop this expansion and improve efficiency, some engines use an intercooler to cool the compressed air. It fits between the turbocharger and the engine. We look at this topic next.

Intercoolers

A turbocharger or supercharger is used to increase the volume of air entering the engine. It does so by compressing the air above atmospheric pressure. However, when air is compressed, it heats up. This causes the volume of the air to increase, lowering its density. Hot air under pressure in a cylinder contains fewer oxygen molecules than cooler air at the same pressure in the same volume.

The purpose of an intercooler is to reduce the intake air temperature. It can do so up to a few hundred degrees Fahrenheit before it enters the intake manifold. Decreasing air temperature increases the density of the pressurized air and improves engine efficiency. It also lowers the chance of engine-damaging detonation.

The intercooler is most efficient at removing heat when the turbocharger's boost pressure is near its maximum and the compressed air is the hottest. This is typically above 15 psi, or 100 kPa. Most intercoolers operate on an air-to-air principle. Compressed air is fed from the turbocharger through the intercooler and then into the intake manifold.

The intercooler works like a radiator. Inside it, the air passes through small tubes with thin fins attached. The heated compressed air flowing through the intercooler gives up its heat to the fins and the tubes (**FIGURE 52-31**). As the vehicle moves forward, the cool outside air flowing across the fins pulls heat away from the tubes and fins. This heat transfer occurs constantly during engine operation.

Some larger engine applications have a liquid-operated intercooler. In this system, the air is fed through small tubes in a heat exchanger. Engine coolant surrounds the tubes. The coolant absorbs the heat and transfers it to the engine cooling system.

▶ The Exhaust System

LO 52-07 Describe the exhaust system and exhaust pipe.

The exhaust system is made up of several components that work together to perform four main functions:

- remove exhaust gases from the engine,
- quiet the exhaust noise,
- ensure that poisonous exhaust gases do not enter the passenger compartment, and
- reduce harmful emissions in the exhaust stream (**FIGURE 52-32**).

Exhaust flow is described as follows. Burned gases exit the cylinder through the exhaust port and pass into the exhaust manifold. The first pipe is usually called the engine pipe or down pipe. The down pipe is connected to the outlet of the manifold.

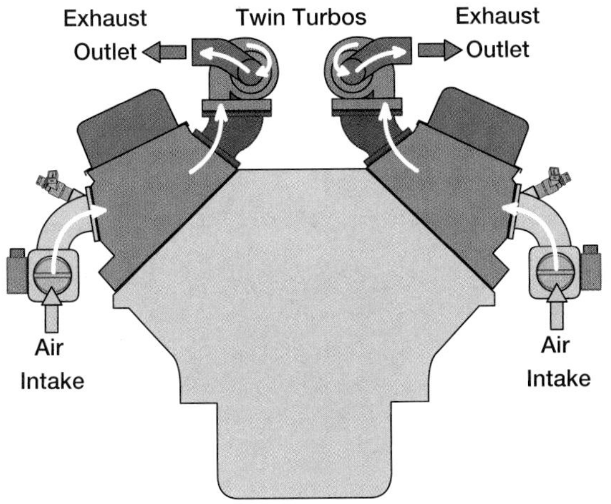

FIGURE 52-30 BMW's reverse flow cylinder heads with twin turbos in the center of the V.

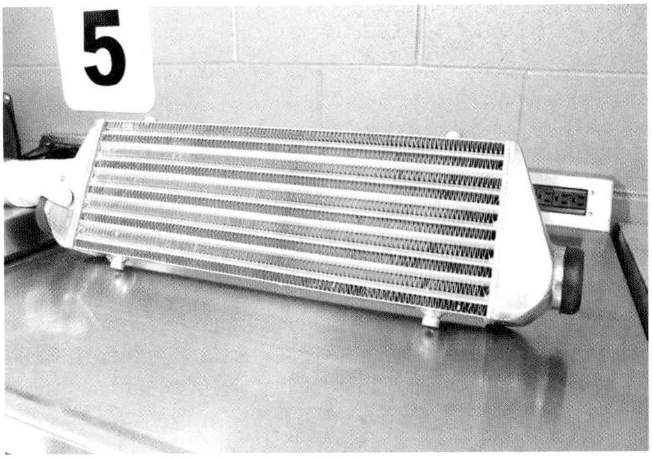

FIGURE 52-31 A typical air-to-air intercooler lowers the temperature of the compressed air before it enters the intake manifold.

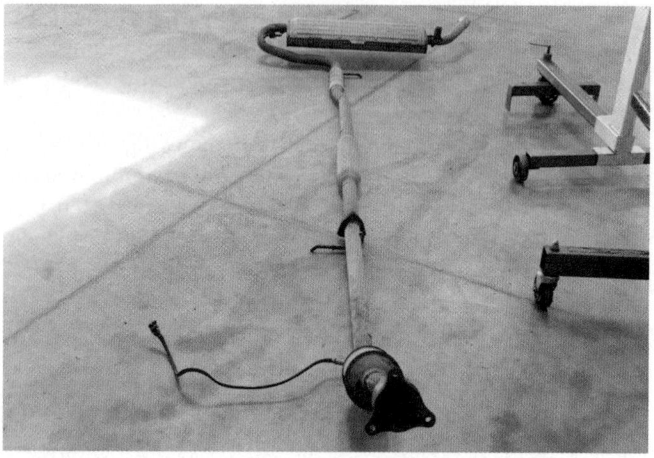

FIGURE 52-32 The exhaust system removes exhaust gases from the engine. It also prevents harmful exhaust gases from entering the passenger compartment. And it reduces noise, and reduces harmful emissions in the exhaust stream.

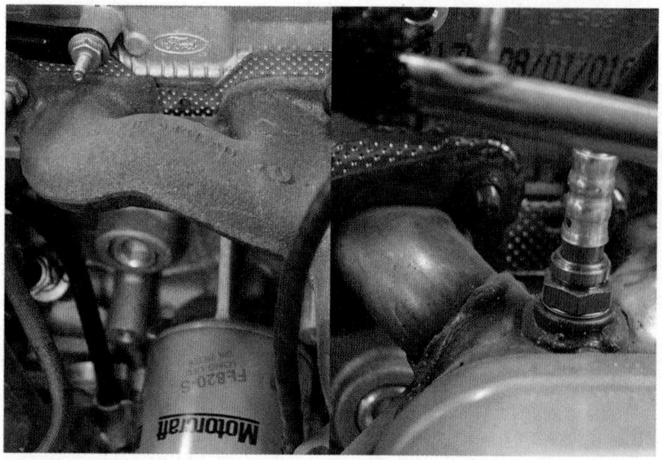

FIGURE 52-33 An exhaust manifold is typically made of cast iron, and a header is made of tubing of equal length.

It carries the exhaust gases to the catalytic converter. Catalytic converters were added to vehicles in the mid-1970s. They were designed to reduce exhaust emissions. The exhaust exits the converter and continues on through an intermediate pipe to the muffler, which reduces exhaust noise. Exhaust gases are then either passed through a resonator or simply discharged to the atmosphere through a tailpipe. Discharge is usually at the rear, to the side, or above the vehicle.

During engine operation, each time an exhaust valve opens, a pulse of hot exhaust gases is forced into the exhaust manifold. These hot gases produce a lot of noise, some of it at very high frequency. The muffler and resonator reduce the sound level of the exhaust. Even while quieting the exhaust noise, the exhaust system can be designed to enhance engine operation and efficiency. In fact, a well-designed system can improve drivability and performance while still reducing the noise.

▶ Exhaust System Components

Exhaust Manifold

The exhaust manifold is bolted to the engine's cylinder head. It also typically provides a mounting place for the oxygen sensor. It operates better when positioned as close to the cylinders as possible. This position helps it warm up faster during a cold start. The exhaust manifold is usually made from either cast iron or stainless steel. Because of the extreme temperatures generated at the exhaust manifold, heat shields can be installed. They protect other vehicle components from heat damage.

On many current vehicles, the exhaust manifold is often replaced with a header. The **header** is an exhaust manifold made of mandrel-formed tubes. They are of equal length and join at a common collector. **Mandrel forming** uses a special pipe-bending tool to ensure that piping bends in a smooth arc that is not collapsed. This avoids creating a partial restriction. The equal length of the pipes ensures that the exhaust pulses create a more equal exhaust flow out of each cylinder. Exhaust manifolds can be restricting, forcing the pistons to work harder to push the exhaust gases out. A header is freer flowing, allowing gases to leave the engine quickly (**FIGURE 52-33**).

The headers also provide a **scavenging** effect to help remove exhaust gases from the cylinders. The outgoing pulse from one cylinder is timed to arrive at the junction at exactly the right time to help draw out the pulse from another cylinder. This setup is called **tuned exhaust**, and the lengths of the header tubes determine the rpm range they are tuned to. Tuned exhaust is widely used on high-performance vehicles and race cars. It has also found its way into regular production vehicles as well. Performance gains can be realized by ensuring that exhaust gases flow freely. This is so that the engine can pull air in and push exhaust out efficiently.

Applied Science

AS-40: Decibels/Intensity: The technician can demonstrate an understanding of how sound intensity can be measured.

The degree of loudness is measured in units called decibels (dB). The average human can hear sounds between 0 and 120 dB. One-time exposure to noises above 120 dB, or sustained exposure above 85 dB, can damage the ears. Decibels can be measured with a sound pressure level meter. These meters are available in analog and digital styles.

As a baseline, the decibel level of some common sounds are as follows:

- Conversational speech at a distance of 3 feet: 60 dB
- A diesel truck engine at a distance of 30 feet: 90 dB
- A chain saw at a distance of 3 feet: 110 dB
- A jet aircraft at takeoff at a distance of 150 feet: 150 dB

Technicians should be concerned about the decibel rating, as well as the exposure time, of sounds in the workplace. For example, the CDC recommends that a person should be exposed to a sound pressure level of 100 dB for no more than 15 minutes. Ear protection should be worn to protect your hearing, based on the decibel rating and exposure time.

Engine Pipe

The engine pipe, or down pipe, is attached to the exhaust manifold and connects to the catalytic converter. The engine pipe is usually made of a nickel chromium material, which resists rust and corrosion to ensure it is long-lasting. Some exhaust down pipes may also use stainless steel to ensure long life. The engine pipe may be attached to the exhaust manifold by spring-loaded bolts. They allow the exhaust to move slightly as the engine moves in its mounts. Some engine pipes also have flexible connectors that allow movement between the engine pipe and the intermediate pipe (**FIGURE 52-34**). The flexible connector is used to close the gap. Its main functions are to allow engine movement—especially in front-wheel drive vehicles—and to reduce vibration without passing it along the exhaust (**FIGURE 52-35**).

Exhaust Brackets

The exhaust components are supported along the length of the vehicle by brackets suspended from the underbody. The supports are usually rubber mounted. This helps them to isolate the vibrations of the exhaust from the main body of the vehicle. Rubber is preferred because of its natural dampening effect (**FIGURE 52-36**).

FIGURE 52-34 The engine pipe connects the exhaust manifold to the catalytic converter and may contain a flexible connector.

FIGURE 52-35 The flexible connector is used to allow the pipe to flex as the engine moves.

▶ Catalytic Converter

LO 52-08 Describe catalytic converters and connecting pipes.

A catalytic converter converts unacceptable exhaust pollutants into less dangerous substances. Common pollutants are carbon monoxide, hydrocarbons, and oxides of nitrogen. Three-way converters convert oxides of nitrogen back into nitrogen and oxygen. They also convert the hydrocarbons and carbon monoxide into water and carbon dioxide. Older, two-way catalytic converters only converted hydrocarbons and carbon monoxide into water and carbon dioxide. They were not able to convert the oxides of nitrogen into nitrogen and oxygen.

A catalytic converter is fitted in line with the exhaust system. It is located close to the exhaust manifold. This is so that it can reach its operating temperature as soon as possible (**FIGURE 52-37**). Some manufacturers install a catalytic converter in the base of the exhaust manifold. They install another one downstream before the muffler.

Catalytic converters can become contaminated by lead or silicone. Leaded fuel must not be used in an engine with a catalytic converter. This is because lead coats the catalyst and

FIGURE 52-36 The exhaust is supported by rubber mounts that ensure that vibration is not felt by the driver.

FIGURE 52-37 The catalytic converter is located after the engine pipe and before the muffler. It changes harmful gases into nonharmful gases to be released to the atmosphere.

prevents it from interacting with the pollutants. Some types of silicone sealer also coat the catalyst. Once coated, the converter will most likely have to be replaced.

The catalytic converter operates by creating and then maintaining a chemical reaction while converting the exhaust pollutants. It usually creates heat while converting the harmful gases to less harmful ones, so it can get extremely hot. Because of this, it has a heat shield to prevent heat from radiating to bodywork and other parts. The catalytic converter is covered in greater detail in the Emission Control chapter.

Intermediate Pipe

The intermediate pipe connects the catalytic converter to the muffler (**FIGURE 52-38**). This pipe can be made to be clamped inside of the pipe of the catalytic converter. An exhaust clamp is typically used to hold and seal the joint in place. The pipe may also have flanges on the ends that bolt to a flange on the catalytic converter. It typically has a gasket between the bolted flanges to seal the exhaust gases in.

Tailpipe

The tailpipe takes the exhaust gases away from the vehicle. Its exit point must not allow any of the exhaust gases to enter the vehicle. The tailpipe is connected to the muffler or resonator and is held by flexible exhaust mounts. It is made of a noncorrosive and rust-resistant material to ensure long life. It also may include an exhaust tip for appearance purposes. And some exhaust tips include built-in resonators (**FIGURE 52-39**).

Exhaust Gaskets

Exhaust components are sometimes bolted together. If they are, they may need a gasket to seal them. Exhaust gaskets can often be found between:

- the engine cylinder head and the exhaust manifold,
- the engine pipe and the catalytic converter,
- and possibly the catalytic converter and the muffler.

The exhaust is extremely hot, so the gaskets must withstand the temperatures without burning (**FIGURE 52-40**). Exhaust gaskets can be multilayered high-temperature alloys that are formed with graphite or mica. They provide a good seal as exhaust components move with expansion and contraction. Donut gaskets are another type of exhaust gasket (**FIGURE 52-41**). They use spiral-wound steel and filler material, which is shaped into a round design. When installed in ball-shaped pipe, a seal is created while still allowing the pipes to be joined even if they are slightly out of line. Some of these gaskets also allow for slight movement of the engine.

FIGURE 52-38 A typical intermediate pipe between the catalytic converter and muffler.

FIGURE 52-39 Exhaust tips are typically used for appearance purposes, but some include resonator tips.

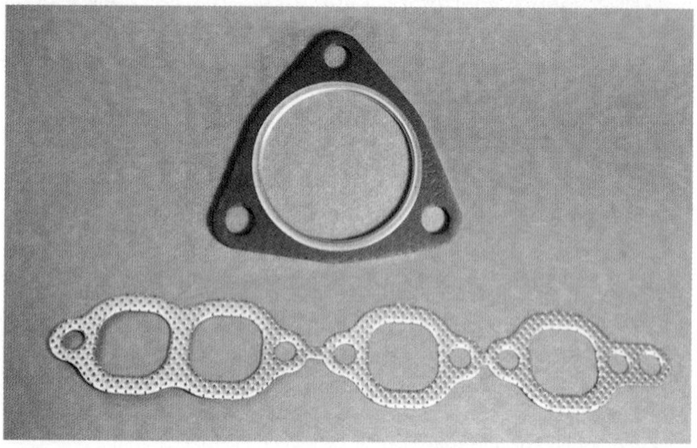

FIGURE 52-40 Exhaust gaskets come in many forms and are used in the extreme temperature of the exhaust system.

FIGURE 52-41 Donut gaskets are used between the exhaust manifold and down pipe on some vehicles.

▶ The Muffler System

LO 52-09 Describe mufflers and resonators.

The purpose of a vehicle's muffler is to manage the sound coming from the exhaust system. Typically, this means as much noise reduction as possible. But some customers desire a certain exhaust tone, so the muffler system is designed to provide that sound. Exhaust sounds originate from the combustion process within the engine. Exhaust noise becomes an issue as vehicle systems become generally quieter and as the number of vehicles on our roads increases.

To understand the operation of modern exhaust noise reduction systems, it is helpful to understand what sound is. We sense sound with our eardrum, located within the ear. The eardrum is made to move by variations in air pressure. Variations in air pressure can be created when a force is placed upon an object. An example is clapping your hands. As the two hands collide, they push the air surrounding them away. The moving air creates a wave of air pressure, or a sound wave. This sound wave moves your eardrum, which is interpreted as sound by your brain.

The engine makes noise, but why? It is because each combustion process is a rapid burning of air and fuel. It's like a controlled explosion. These explosions create a great deal of noise if they are not absorbed or canceled. ***Noise absorption*** refers to using sound-deadening materials to absorb sound waves. This happens in some mufflers by putting a sound-deadening material around a perforated pipe that the exhaust gases flow through (**FIGURE 52-42**). This is like placing noise-absorbing materials inside the walls of a house to make the rooms quieter.

Noise cancellation is a system that prevents the sound waves from leaving the exhaust system by canceling them out inside the muffler. These systems create gas pressures that are equal in force but opposite in direction to the noise source. These generated pressures are known as anti-noise. Any remaining sound is referred to as residual noise.

The muffler is designed to quiet the noises of combustion without restricting exhaust flow to the point of adversely affecting performance. The goal is to produce a vehicle that is smooth and quiet as well as powerful. A muffler may use a dissipative technique, which is sound absorption, or a noise-canceling technique, or both. Exhaust noise can be reduced by various means. This includes baffles and chambers, variable-flow exhaust, and electronic mufflers.

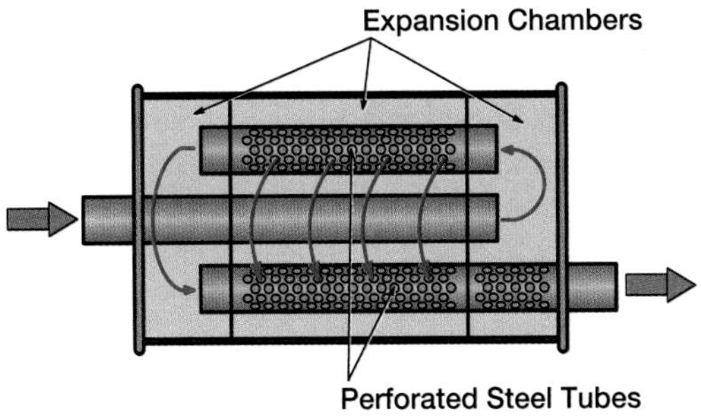

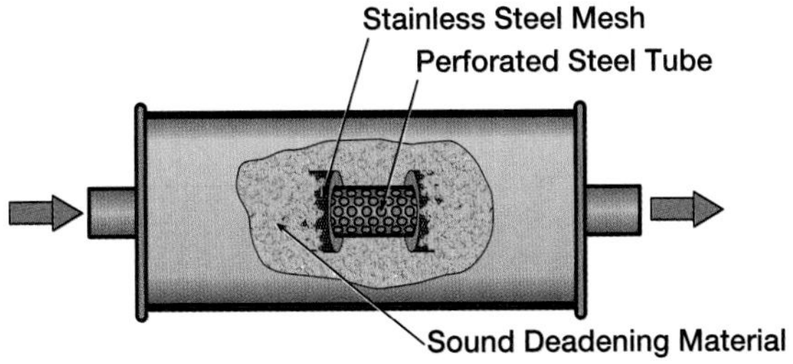

FIGURE 52-42 The muffler is one of two types of designs, either the absorptive type or the noise-canceling type.

Applied Science

AS-43: Noise/Acoustics: The technician can demonstrate an understanding of why the acoustics of the vehicle affect specific noises.

The acoustics of a vehicle is related to the behavior of sound waves in an enclosed space. Noise can enter the passenger compartment from a number of sources. Engine noise, tire noise, wind noise, and the noise of other vehicles are a few sources. Dampening material around the vehicle cabin is a very good method of noise reduction. Insulation is placed in trunk panels, rear wheel wells, doors, and the roof, as well as under the carpet. All of this insulation reduces unwanted environmental noise.

On some vehicles, to reduce weight and be more price-competitive, insulation and dampening materials have been reduced. In comparison, luxury vehicles may have more than 100 pounds of sound-deadening material. It is placed carefully in the most needed areas to ensure a quiet environment.

Technicians should understand that noises may not be the same in all vehicles due to these variations. The acoustics of a particular vehicle may differ widely as compared with another. On one type of vehicle, it may be typical for the driver to hear a squealing water pump as it failed. On another, the engine covers and multiple layers of dampening material between the engine and the passenger compartment may prevent the driver from hearing the same sound.

AS-44: Overtones/Harmonics: The technician can explain that the presence of overtones may indicate changes in vibration in systems.

A technician in a luxury car dealership has replaced a number of water pumps on a certain model of a popular vehicle. The published labor rate for this water pump replacement is 11.2 hours. This greatly exceeds the average water pump replacement labor rate by three or four times.

The technician has observed that the water pump on this particular vehicle tends to have a slightly noisy bearing before it fails. As a preventive measure, the technician uses a stethoscope to listen to the sound of the water pump bearing when vehicles of this model are in for basic service. The pump sounds are checked with the stethoscope at idle, at 1,000 rpm, and again at 1,500 rpm. Over the years, the technician has developed the listening skills with his stethoscope. The technician can evaluate the condition of the water pump bearing on this particular vehicle. When an abnormal condition exists during the stethoscope evaluation, the technician is aware of the situation. The presence of an overtone, which is a higher tone above the average tone, is a clue. This condition may indicate the first sign of a bearing with excessive vibration. The technician will record the findings on the repair order to alert the owner of this concern before the water pump fails.

▶ TECHNICIAN TIP

Pedestrians are faced with the opposite problem when dealing with hybrid and electric vehicles. The vehicles are too quiet. It is easy for pedestrians who are visually or hearing impaired, or who just aren't paying close attention, to walk out in front of one of these vehicles. While operating on electric power, they are very quiet, and get hit. Federal guidelines have just been released that all hybrid and electric vehicles will have to be equipped with a new alert system by September 1, 2019. The regulation also specifies that half of a manufacturer's new hybrid and electric vehicles will have to be compliant one year before that.

▶ Variable-Flow Exhaust

A moveable valve built within the exhaust system is used to change the path for exhaust to flow. It also changes the amount of exhaust backpressure. This system is used on many high-end performance vehicles to ensure they meet noise restrictions. The system operates in two stages. In the first stage, when the engine rpm and throttle position are low, the exhaust takes a longer route through two or more mufflers/resonators. It also provides a small amount of backpressure. This further lowers the exhaust noise as well as the hydrocarbon emissions during valve overlap at low rpm. When the throttle is opened and engine speed increases, the second stage is activated. A valve opens and allows the exhaust to bypass some of the silencers in the system. This makes the exhaust path more free-flowing (**FIGURE 52-43**). The valve can be operated by the following means:

- Exhaust gas pressure
- Vacuum diaphragm
- Electronic actuator

Adding a variable-flow exhaust to the baffle or chamber system reduces exhaust noise during normal driving. But it increases power and sound during performance driving. This is because the system can respond to changes in engine speed and load. Variable-flow exhaust was typically limited to high-end exotics. Examples include Ferraris, Aston Martins, or Lamborghinis. This can now be found on domestic vehicles such as Corvettes, Camaros, and Challengers.

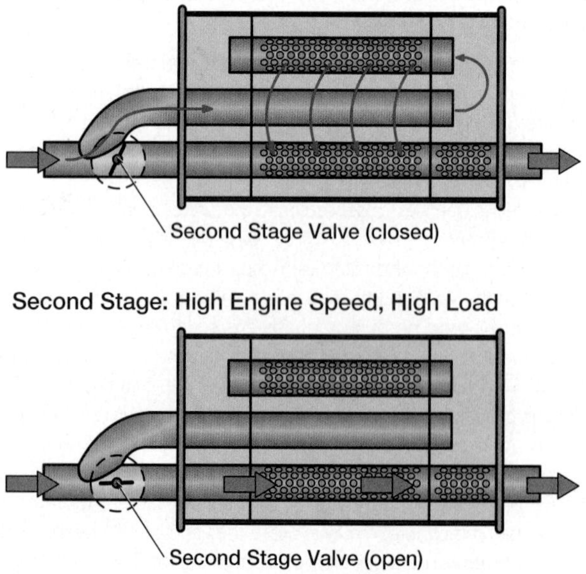

FIGURE 52-43 A typical variable-flow exhaust system.

Electronic Mufflers

Any restriction to exhaust flow in the exhaust system creates backpressure. Some backpressure can be good. Too much backpressure reduces an engine's volumetric efficiency. This in turn reduces engine efficiency. Electronic mufflers are designed to produce anti-noise which does not restrict exhaust flow. This computer-controlled system uses a microphone. It detects the sound waves produced within the exhaust system. A computer-driven loudspeaker generates equal but opposite sound waves. The sound waves cancel the exhaust noises. This use of opposing sound waves is sometimes called anti-noise.

The anti-noise is applied to the exhaust stream in an electronic muffler. The result is a virtually silent exhaust. With the added benefit or reducing unwanted backpressure across all engine operating conditions. This system increases fuel economy and reduces exhaust emissions. The electronic muffler is in research at the moment and has never been released on a production vehicle. But look for it to debut in the near future.

▶ TECHNICIAN TIP

Some vehicles use the audio system to cancel noise inside the passenger compartment in order to combat engine noise. This is done in two ways. The first is to use the entertainment system to actively cancel engine and road noise. The second way is to use a speed-sensitive entertainment system that adjusts the volume of the entertainment system according to vehicle speed. Low volume at low speed, and higher volume at higher speed. This helps overcome objectionable noise.

Resonator

Some manufacturers use a resonator in the exhaust system. It is located between the muffler and the exhaust outlet. Its function is to reduce any resonance levels that the muffler cannot adequately suppress (**FIGURE 52-44**). Some manufacturers may

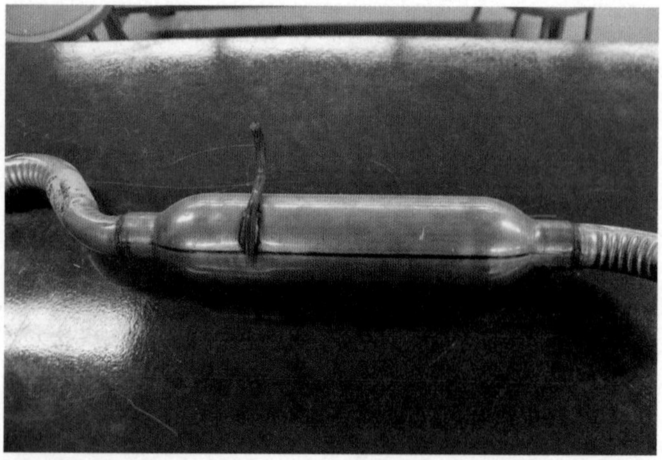

FIGURE 52-44 The resonator is an assistant to the muffler to ensure engine noises are canceled adequately.

Applied Science

AS-45: Pitch/Frequency: The technician can explain the relationship of pitch to frequency.

Heinrich Hertz, a German physicist, was the first to prove the existence of electromagnetic waves. He did so by engineering a radio wave transmitter. In honor of his work, the unit for measuring frequency, hertz, is named after him. Because it is named after a person, its abbreviation (Hz) begins with a capital letter.

Frequency is the number of sound waves produced in a given time. The frequency of a sound wave determines its pitch, described in terms of how "high" or "low" the sound is. Humans can perceive sound waves ranging from 20 Hz (lowest pitch) to 20,000 Hz (highest pitch).

The concept of sound is based on the principle that a sound wave begins with a vibrating object. The frequency of a sound wave, or Hz, can be expressed as 1 Hz = 1 vibration per second.

AS-46: Resonance: The technician can demonstrate an understanding of what happens when an object resonates.

In physics, resonance refers to the amplitude at which an object vibrates at a given frequency. There are a number of different types of resonance, including mechanical resonance. When a large group of soldiers are marching across a bridge, they are told to break step. This is to avoid damaging the bridge's structure. If the soldiers' footsteps were to fall simultaneously, extreme vibrations could result, which the bridge might not be able to withstand.

In automotive applications, resonance issues are often experienced in exhaust systems. A droning noise is produced from vibration. Vehicle manufacturers may add mass (weight) to the various exhaust components. This changes the resonance frequency.

use more than one resonator to quiet the combustion noises even further. Some resonators work on the Helmholtz principle. Namely, that air blown across a tube connected to a rounded container creates a vibration. Noise is produced, similar to air blown across a bottle. This principle is then used to cancel noise by bouncing waves off one another. This reduces the intensity of waves, which lowers the exhaust noise.

▶ Inspecting the Air Filter

LO 52-10 Inspect air filters, housings, and duct work.

The engine needs a free flow of clean air in order to operate correctly. Dust and grit in the air can be very abrasive and will severely shorten the life of the engine if not filtered out. If the filter element is not fitted correctly and does not seal properly, dirty air can bypass the filter and enter the engine.

The location of the air filter varies depending on the type of fuel system on the vehicle. Check the service information for the exact procedure. Some air filters are mounted to the top of the engine, usually found on older vehicles using a carburetor or throttle body fuel injection. The air cleaner on a multiport fuel-injected vehicle is typically located in a rectangular box. It is housed near the beginning of the air induction system. While inspecting the air filter, take a look at the air cleaner housing and ductwork for cracks or holes. These conditions would allow unfiltered air to enter the engine.

▶ TECHNICIAN TIP

The paper filter element actually becomes more efficient at filtering dirt particles the more it is used. This is because the passageways become smaller as dirt is caught in them. Smaller and smaller dirt particles are caught over time. However, if the filter becomes too clogged, it will restrict air, which reduces engine power output.

Once the air filter is removed, it is fairly easy to inspect. First inspect it for any damage to the sealing surfaces. If they are bent or damaged, they won't seal dirt out and require replacement. If the filter is in good shape, then inspect it for clogging. This is best done by holding the filter up to light and looking through it. If it is bright, then it is not clogged. If little to no light comes through, then it is clogged and needs to be replaced. To inspect and change the air filter, follow the steps in **SKILL DRILL 52-1**.

SKILL DRILL 52-1 Inspecting and Changing an Air Filter

1. On fuel-injected engines, unlatch or unscrew the filter housing fasteners to remove the air filter. It may be necessary to loosen the clamps and hoses on the induction tubing to remove the filter housing cover.

2. On carbureted or throttle body–injected engines, remove the top of the air filter by unscrewing the wing nut, and remove the air filter.

3. Inspect the air cleaner element by holding the filter element up to light and looking through it. If it is bright with no tears or cracks, it can be reused. If it is dark or damaged in any way, it needs to be replaced.

SKILL DRILL 52-1 Inspecting and Changing an Air Filter (Continued)

4. Clean the inside of the air filter housing, and inspect it and any ducts for cracks. If the air filter is being replaced, obtain a new air filter and compare it with the old one to ensure that they are exactly the same.

5. Place the new air filter inside the filter housing, making sure it is aligned properly on both sides.

6. Replace the cover of the air filter housing and tighten the latches, screws, or wing nut until completely closed. Reinstall any induction tubing or clamps.

▶ TECHNICIAN TIP

Just opening the filter housing up and looking at the top of the filter is not a proper way to inspect the filter. In most vehicles, the air flows UP through the filter, so it is the bottom side that is the dirty side. Looking only at the top side of the filter will give you inaccurate information about the filter. Remove it and hold it up to the light.

▶ Inspecting the Induction System

LO 52-11 Inspect the intake system for leaks.

The throttle body and the induction system control the airflow into the cylinders. If air leaks are present in the induction system downstream of the MAF sensor, then the sensor will not accurately measure the air coming into the engine. This additional air is sometimes referred to as "false air." As a result, the PCM will not deliver enough fuel, based on the MAF signal. This will cause the engine to run lean.

There are several ways to test for air leaks. They include using a substitute fuel to artificially enrich the mixture at the point of the leak. Another method uses a smoke machine to fill the intake system and show any external leaks. The third method uses an electronic stethoscope (or piece of heater hose) to listen for and pinpoint the source of the leak.

The use of a substitute fuel such as acetylene, propane, or carburetor cleaner is a quick and accurate method. But you risk catching the engine on fire. The technician carefully directs a small amount of fuel toward any suspected leaking components (**FIGURE 52-45**). If the fuel is passed over an intake leak, the vacuum will pull the gas into the intake manifold. This richens the mixture up, and will be noted in one of two ways. Either the engine will smooth out, or the engine rpm will increase.

If the engine has an idle control system, it may prevent the rpm from increasing. But it will likely stumble or even die as you pull the fuel away from the leak. You may need to disable the idle control system or hold the throttle open just enough that it doesn't control the idle. You may also be able to use a scan tool while applying and removing the acetylene or propane. Watch the oxygen sensor, injector pulse width, and short-term fuel trim reading for changes while doing that.

FIGURE 52-45 To find a leaking intake manifold gasket, a technician can use acetylene, propane, or carburetor cleaner to enrich the mixture.

Leaks can also happen in the ductwork, hoses, and vacuum diaphragms. Since they are connected to the intake manifold, they can create a false air condition. Although the acetylene and propane may still be able to find the leak, a smoke machine is invaluable. The smoke machine pushes smoke at low pressure into the intake manifold. The smoke machine can then be used to fill the entire intake system with smoke. If there are any leaks, the smoke will leak from inside the system to the outside. So anywhere the smoke is leaking from is an air leak. A high-intensity light can be used to make the smoke more visible.

SAFETY TIP

Be careful using carburetor cleaner, or any flammable liquid, on a hot engine. Flammable chemicals can start a fire on a hot engine.

SKILL DRILL 52-2 Inspecting the Throttle Body, Air Induction System, and Intake Manifold

1. Start the engine and let the idle stabilize. Using an acetylene- or propane-testing tool, place the hose near the suspected leak area.

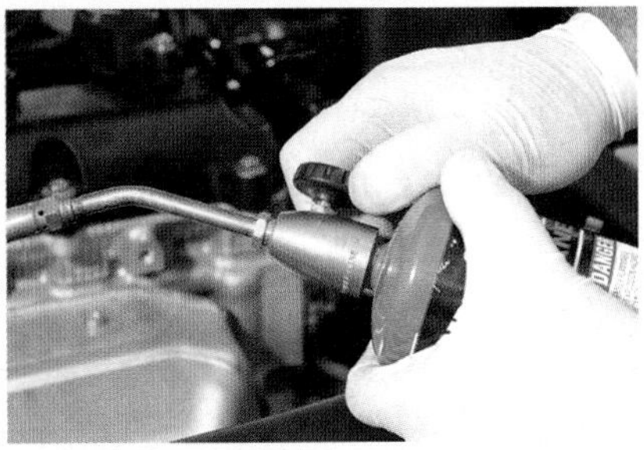

2. Open the fuel valve to slowly release the acetylene or propane.

3. On vehicles without an idle control system, observe the rpm and smoothness of the engine. If the engine speed increases or smoothes out, then a vacuum leak is present. Determine any necessary repairs.

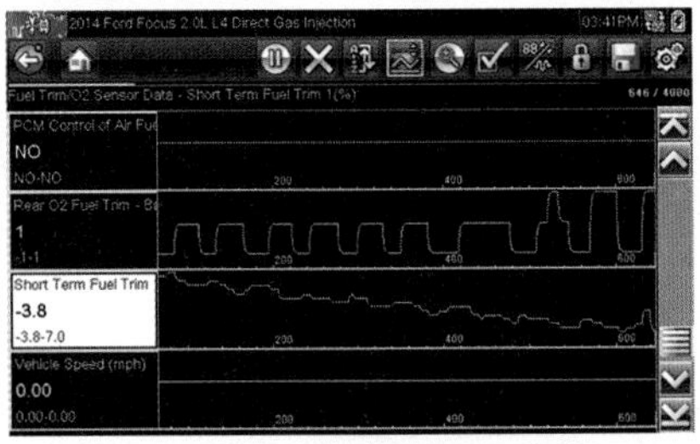

4. On vehicles with an idle control system, connect a scan tool and select data stream. Observe the oxygen, short-term fuel trims, and injector pulse width as you are moving the acetylene or propane around.

5. Shut off the engine and connect a smoke machine to the intake manifold. Start the smoke machine and inject smoke into the intake manifold.

6. Using a bright light, look around the engine compartment for traces of smoke. Any smoke coming from the air cleaner inlet is normal. Determine any necessary action(s).

To inspect the throttle body, air induction system, and intake manifold for leaks, follow the steps in **SKILL DRILL 52-2**.

▶ Inspecting the Exhaust System

LO 52-12 Inspect integrity of exhaust system components.

Inspection of the exhaust system is a necessary task as an automotive technician. Exhaust inspection is part of any mandatory state inspection process. This is because a leak will result in possible injury or death to the passengers due to carbon monoxide poisoning. Special care should be taken to ensure that the exhaust system is not leaking. Exhaust leaks leave a black carbon marking. Also, most municipalities have noise ordinances limiting the amount of noise an exhaust can emit.

A quick test for exhaust leaks is to use your shoe and a rag to block off the exhaust flow coming from the tailpipe. If the exhaust system is not leaking, pressure will build and either leak past the rag or cause the engine to die. If there are leaks in the exhaust, then hissing will be heard from the exhaust system through the leaks. It may be necessary to have an assistant block off the tailpipe with a rag while the technician listens for exhaust leaks. A small hose can help pinpoint small leaks while this test is being performed. Put one end of the hose in your ear. Then move the hose around the exhaust system listening for a hissing sound.

Another testing option is to use a smoke machine with an exhaust pipe adapter. This allows you to fill the exhaust system with smoke. With the engine off, any signs of smoke will pinpoint leaks in the exhaust system. Remember you may see smoke coming from the air cleaner assembly due to the valves in one or more cylinders being on overlap. This is considered normal and should not be confused as a leak from the exhaust.

The exhaust pipes can rust out over time. The water and corrosive chemicals in the exhaust stream corrode the pipe. A good way to test the integrity of the pipe is to use a large pair of adjustable pliers to squeeze the pipe. If it is springy or crushes easily, the pipe must be replaced. Make sure you check any low spots in the system along with bends in the pipe, as these are typically the weakest spots.

Also, check the exhaust system hangers, brackets, clamps, and heat shields. They wear out or break over time. Look for torn rubber on the hangers. On the brackets, clamps, and heat shields, look for damage, excessive rust, or completely missing components. Replace as needed.

To inspect the exhaust system for leaks, follow the steps in **SKILL DRILL 52-3**.

SKILL DRILL 52-3 Inspecting the Exhaust System for Leaks

1. Safely lift and secure the vehicle on a hoist. Inspect for leaks or rust holes in the exhaust system, including the exhaust manifolds. Inspect brackets, hangers, clamps, and heat shields.

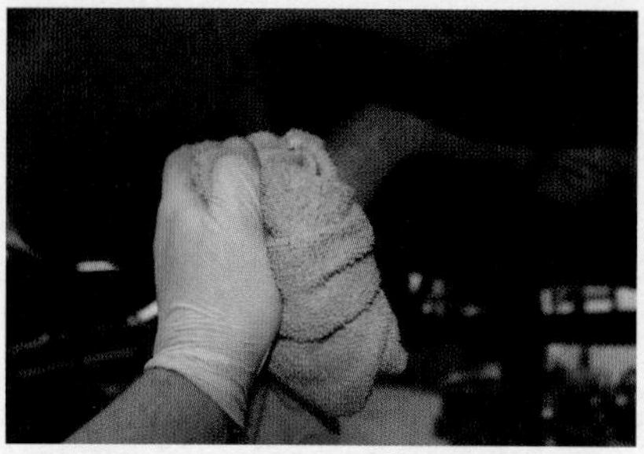

2. Have a helper hold a rag against the exhaust pipe(s) while the engine is idling to increase the pressure in the system. Listen and feel for any leaks.

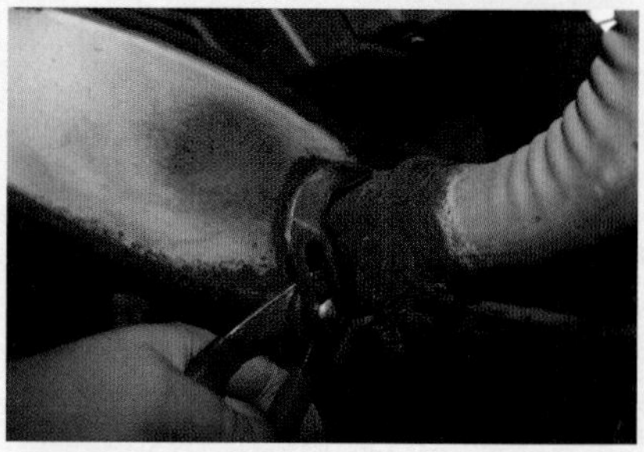

3. Use large adjustable pliers to test the integrity of the pipes by moderately squeezing the pipes. Determine necessary repairs.

▶ Wrap-Up

Ready for Review

- The intake system provides a sealed passageway to the combustion chambers and ensures that no contaminants leak into the system. Primary components of the intake system are: intake manifold, throttle body, and the air induction system.
- The air cleaner assembly filters the incoming air and muffles the noise of the intake pulses. Types of air filter systems are: pleated paper, oil impregnated cloth or felt, and oil bath air cleaner.
- Heated air intake systems provide greater vaporization of the fuel and ensure that fuel does not collect on the cold walls to create a lean mixture. Methods used to heat intake manifolds include: hot engine coolant, hot exhaust gases, and electric heaters.
- Variable intake systems are designed to respond to changes in engine load and speed by changing their effective length. The primary section is long and narrow for the low range of rpm and the secondary section is shorter and wider for the high range of rpm. Sealed pulse chambers are designed to time the return of intake pressure pulses back to the intake valves above a certain rpm.
- Volumetric efficiency compares the volume of air entering a cylinder to the total piston displacement of the cylinder and is expressed as a percentage. Forced induction increases air pressure in the intake manifold above atmospheric pressure and increases volumetric efficiency to well above 100%.
  - A supercharger or a turbocharger can be used to achieve forced induction.
- A turbocharger uses wasted kinetic energy from the exhaust gases to increase the volume of air entering the engine. Valves, cam timing, and exhaust system design must be closely matched to the turbocharger and engine to reduce backpressure. An intercooler reduces the intake air temperature to increase the density of the pressurized air and improve engine efficiency.
- An exhaust system functions to remove exhaust gases from the engine, quiet the exhaust noise, ensure that poisonous exhaust gases do not enter the passenger compartment, and reduce harmful emissions. The components of an exhaust system include: exhaust manifold, engine pipe, and exhaust brackets.
- A catalytic converter operates by creating and then maintaining a chemical reaction while converting the exhaust pollutants into less dangerous substances. The catalytic converter is located after the engine pipe and before the muffler.
- The muffler is designed to quiet the noises of combustion without restricting exhaust flow to the point of adversely affecting performance. A resonator is located between the muffler and the exhaust outlet and reduces any resonance levels that the muffler cannot adequately suppress.
- Check the service information for the exact location and procedure for an air filter service. The air cleaner on a multiport fuel-injected vehicle is typically located in a rectangular box housed near the beginning of the air induction system. Air filter inspection should include checks for damage to sealing surfaces.
- Different methods to test for air leaks in an intake system include the use of: a substitute fuel to artificially enrich the mixture at the point of the leak, a smoke machine, or an electronic stethoscope.
- Exhaust leaks can be tested by blocking off the exhaust flow coming from the tailpipe and listening for hissing

sounds or by using a smoke machine. The exhaust system hangers, brackets, clamps, and heat shields should be checked for wear.

- Exhaust leaks leave a black carbon marking.

Key Terms

blow-off valve Allows for excess pressure that is built up in the intake track to exit the intake after the pressure overcomes the spring pressure on this valve.

bypass valve A valve used in filter bases or near coolers that will open to provide flow when restriction become too great.

forced induction The pressurization of airflow going into the cylinder through the use of a turbocharger or supercharger.

header Version of an exhaust manifold, made of tubular bent steel, that increases exhaust velocity, thus increasing performance of the engine.

Helmholtz resonator A device that uses the principle of noise cancellation through the collision of sound waves. When necessary, the resonator is used in addition to the muffler to cancel additional sounds. This resonator may also be used on the induction system to muffle noise of airflow through the induction system. It is named after physicist Hermann von Helmholtz.

mandrel forming Using a pipe bender with mandrels to allow for very tight bends without creating kinks or reducing the size of the pipe.

plenum chamber A large portion of the intake manifold after the throttle plate and before the intake runner tubes. The plenum provides a reservoir of air and helps prevent interference with the flow of air between individual branches.

scavenging The process of removing burned gases from the cylinder through the use of moving airflow pulling or extracting the gases out.

supercharger Component that uses external power to turn the compressor wheel to compress the air as it enters the intake system. The supercharger is usually driven by a belt off the crankshaft, or directly from an electric motor.

tuned exhaust Exhaust system in which the exhaust pulse from one cylinder is timed to arrive at the right time to help draw out the pulse from another cylinder.

turbocharger Forced induction device driven by exhaust gas pressure and used to compress intake air so that more air can be pushed into the cylinders, thus creating more power.

wastegate A pressure regulator device that allows control of the pressure produced by the turbocharger. The wastegate opens to allow exhaust gases to bypass the turbine wheel of the turbocharger.

Review Questions

1. What type of engines have only air, and no fuel flowing through the intake manifold?
   a. Throttle body injection
   b. Electronically controlled carburetion
   c. Gasoline direct injection
   d. Carbureted
2. If an engine consumes 10 gallons of gas, approximately how many gallons of air will have passed through the air filter?
   a. 10 gallons
   b. 1,000 gallons
   c. 14,700 gallons
   d. 100,000 gallons
3. Why are carbureted or throttle body injected engine intake manifolds heated?
   a. To prevent them from getting too cold and cracking on start-up
   b. To provide better fuel vaporization and prevent fuel from collecting on walls
   c. To help cool the engine coolant while under heavy loads
   d. To lower emissions by directing exhaust gases into the intake so that it can be burned
4. How are most variable intake system valves controlled?
   a. Stepper motors or vacuum actuators
   b. Electric solenoids with a control module
   c. A cable attached to the throttle body
   d. Centrifugal force because of engine RPM
5. All the following affect an engine's volumetric efficiency EXCEPT:
   a. The throttle body
   b. The intake manifold
   c. The fuel pressure
   d. The intake valves
6. How does a wastegate control turbocharger boost pressure?
   a. A spring pops under pressure and releases boost to atmosphere
   b. A door closes and blocks the exhaust flow from leaving the engine
   c. The wastegate moves and exhaust gas bypasses the turbine wheel
   d. The wastegate moves and blocks flow to the intercooler
7. What is the purpose of the spring-loaded bolts on an exhaust downpipe?
   a. To prevent the bolts from loosening and falling out over time
   b. To quiet the exhaust gases as they leave the engine
   c. To make removing the downpipe easier for technicians
   d. To allow some movement between the downpipe and the manifold
8. Why should leaded fuel never be used in a vehicle equipped with a catalytic converter?
   a. The lead coats the catalyst and prevents it from converting pollutants
   b. The engine will burn too hot and damage the catalyst permanently
   c. The fuel will cause the catalyst to catch fire inside the exhaust system
   d. The lead will cause the catalyst to plug up and the engine will stall

9. Where is the air cleaner normally located on multiport fuel-injected vehicle?
   a. In a large round housing on top of the throttle body
   b. In a rectangular box behind the fender well
   c. In a rectangular box at the beginning of the induction system
   d. In a round container under the hood in an accessible location
10. What is the recommended method for testing induction systems for leaks?
   a. Spray it with soapy water and check for bubbles
   b. Inject UV dye and inspect it with a black light
   c. Install a smoke machine and look for smoke with a light
   d. Inject an alternative fuel and watch for a check engine light

ASE Technician A/Technician B Style Questions

1. A long-life air filtration system on a Ford Focus is being discussed. Technician A states that the system can capture hydrocarbons being released when the engine is turned off. Technician B states that the air cleaner and filter are an assembly that should last 150,000 miles. Who is correct?
   a. Technician A only
   b. Technician B only
   c. Both Technicians A and B
   d. Neither Technician A nor B
2. Intake manifolds are being discussed. Technician A states that an intake manifold on one side of a cylinder head and an exhaust manifold on the other side indicate the head is a cross-flow design. Technician B states that, generally, variable intake systems make maximum high RPM power when the longer intake runners are used. Who is correct?
   a. Technician A only
   b. Technician B only
   c. Both Technicians A and B
   d. Neither Technician A nor B
3. Volumetric efficiency is being discussed. Technician A states that most engines achieve 100% volumetric efficiency. Technician B states that superchargers and turbochargers can cause an engine to exceed 100%. Who is correct?
   a. Technician A only
   b. Technician B only
   c. Both Technicians A and B
   d. Neither Technician A nor B
4. A turbocharger system is being discussed. Technician A states that the turbocharger bearings require a reliable flow of clean, cool oil since they can spin at over 100,000 RPMs. Technician B states that the turbocharger blow-off valve is PCM controlled. Who is correct?
   a. Technician A only
   b. Technician B only
   c. Both Technicians A and B
   d. Neither Technician A nor B
5. An exhaust system is being discussed. Technician A states that exhaust manifolds remain cool so they don't crack. Technician B states that headers are only installed on racecars and do not come on brand new cars. Who is correct?
   a. Technician A only
   b. Technician B only
   c. Both Technicians A and B
   d. Neither Technician A nor B
6. A catalytic converter is being discussed. Technician A states that some silicone sealants can damage the catalytic converter. Technician B states that older catalytic converters were three-way type, but newer ones are two-way type. Who is correct?
   a. Technician A only
   b. Technician B only
   c. Both Technicians A and B
   d. Neither Technician A nor B
7. A muffler system is being discussed. Technician A states that some high-end performance cars have variable flow mufflers that can adjust noise level and backpressure. Technician B states that one type of muffler contains an anti-noise speaker. Who is correct?
   a. Technician A only
   b. Technician B only
   c. Both Technicians A and B
   d. Neither Technician A nor B
8. Air filter maintenance is being discussed. Technician A states that unfiltered air can contain dust and grit which will severely shorten engine life. Technician B states that an air filter needs to be changed when you can see a lot of light through it. Who is correct?
   a. Technician A only
   b. Technician B only
   c. Both Technicians A and B
   d. Neither Technician A nor B
9. An induction system is being discussed. Technician A states that air leaks will cause the MAP sensor to read higher vacuum (lower pressure) than normal. Technician B states that air leaks can be found by spraying a substitute fuel but may cause a fire so other methods are recommended. Who is correct?
   a. Technician A only
   b. Technician B only
   c. Both Technicians A and B
   d. Neither Technician A nor B
10. Inspecting an exhaust system is being discussed. Technician A states that a smoke machine can be used to find leaks. Technician B states that many exhaust leaks will leave black carbon marks around the leak area. Who is correct?
   a. Technician A only
   b. Technician B only
   c. Both Technicians A and B
   d. Neither Technician A nor B

CHAPTER 53
Emission Control

Learning Objectives

- **LO 53-01** Describe pollutants and the composition of air.
- **LO 53-02** Describe hydrocarbon and carbon monoxide emissions.
- **LO 53-03** Describe NO_x, SO_2, and particulate emissions.
- **LO 53-04** Describe how engine design reduces emissions.
- **LO 53-05** Describe how the air–fuel ratio can be controlled to affect emissions.
- **LO 53-06** Describe precombustion–postcombustion exhaust treatment and the evolution of emission controls.
- **LO 53-07** Explain the types and operation of catalytic converters.
- **LO 53-08** Explain the types and operation of the positive crankcase ventilation system.
- **LO 53-09** Explain the operation of exhaust gas recirculation system.
- **LO 53-10** Explain the operation of evaporative emission system.
- **LO 53-11** Explain the operation of heated air intake systems.
- **LO 53-12** Inspect and test the PCV system.

ASE Education Foundation Tasks

See Appendix A to view the 2017 ASE Education Foundation Automobile Accreditation Task List Correlation Guide.

▶ Introduction

LO 53-01 Describe pollutants and the composition of air.

Emissions are the release of substances into the atmosphere. They can occur naturally or be human-made. Natural sources are from forest fires, volcanoes, and the decomposition of natural materials. Common human-made sources are vehicles and industrial sites (**FIGURE 53-1**). In recent automotive applications, most emissions are the by-product of combustion. Because of this, they are primarily emitted from the exhaust system.

Not all emissions from combustion are considered hazardous. Those emissions from the automobile that are hazardous can be managed. This is done in a few ways. The first is by careful design of the engine. The second way is by accurately controlling the air–fuel ratio (AFR). The third way is by converting the harmful gases to nonharmful or less harmful gases. This is done through the use of add-on emission control devices. Today's vehicle technology has reduced the harmful emissions a vehicle produces to almost zero. This chapter details the methods and control devices designed to control those harmful emission gases.

FIGURE 53-1 Emissions occur naturally or are human-made, including stationary and mobile sources.

▶ Composition of Air

The air we breathe is composed mainly of two gases: nitrogen and oxygen. Nitrogen by far makes up the largest percentage of air (78%). It is an inert gas, which means it does not react easily with other compounds. Oxygen (21%) is the second most abundant element in air. It is a highly reactive gas, which means it combines readily with almost all other elements. In many cases, oxygen reacts through the process of combustion, which is also called oxidation. These two gases account for approximately 99% of the content of air. That leaves 1% for all of the other elements found in air, including carbon dioxide and argon.

The level of oxygen in the atmosphere is critical to humans. The gases in our atmosphere provide a balanced environment for plant, animal, and human life. Humans breathe in oxygen and exhale carbon dioxide. Trees and plants take in carbon dioxide and give back oxygen. This process is part of the oxygen cycle (**FIGURE 53-2**). There is also a nitrogen cycle, in which nitrogen from our atmosphere is converted to usable food for plants by bacteria in the ground. There is also a carbon cycle. Carbon fuels are burned which release carbon dioxide in the air, which is then converted to carbon and oxygen by plants. The carbon is used by the plants to grow, while oxygen is given back to the air.

Although nitrogen and oxygen make up 99% of the air, the other 1% contains a variety of gases (**TABLE 53-1**). Some of these gases can be useful in different ways. As a technician, while welding, you may have to use carbon dioxide, argon, or

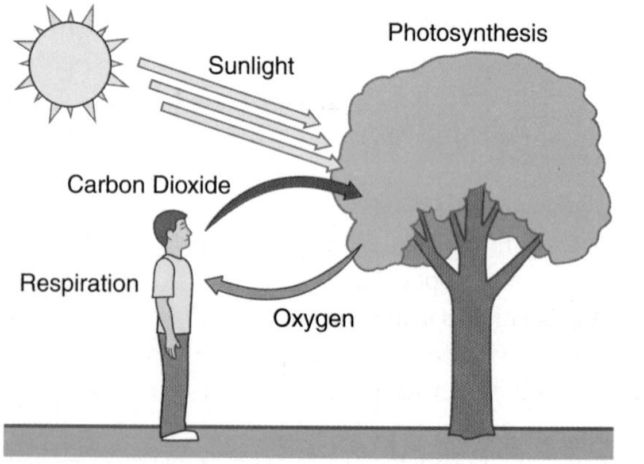

FIGURE 53-2 Humans breathe in oxygen and exhale carbon dioxide; trees and plants take in carbon dioxide and give back oxygen.

You Are the Automotive Technician

A customer brings a car into your shop with the malfunction indicator lamp (MIL) illuminated. You explain to the customer that they did the right thing by bringing it in. Some issues can cause further damage if left unchecked. The customer authorizes you to perform the diagnosis. You first plug in the scan tool and pull the active diagnostic trouble codes (DTCs). There are several codes: P0133-O2 Sensor Circuit Slow Response B1S1; P0401-EGR Flow Insufficient Detected; P0420-Catalyst Efficiency below Threshold-Bank 1. You know that a slow-responding oxygen sensor can affect the catalytic converter. So you start by diagnosing that DTC and then diagnose the other two DTCs.

1. How can a slow-responding O_2 sensor affect the catalytic converter?
2. Under what conditions should the EGR system operate?
3. Describe the operation of the reducing catalyst and the oxidizing catalyst.
4. How does an "EGR flow insufficient" condition affect emissions?

TABLE 53-1 Percentages of Gases Contained in the Atmosphere

Gas	Percentage
Nitrogen (N_2)	78.08
Oxygen (O_2)	20.95
Water (H_2O)	0 to 4, variable
Argon (Ar)	0.93
Carbon dioxide (CO_2)	0.0410, variable
Neon (Ne)	0.0018
Helium (He)	0.0005
Methane (CH_4)	0.00017, variable
Hydrogen (H_2)	0.00005
Oxides of nitrogen (NO_X)	0.00003, variable
Ozone (O_3)	0.000004

helium to keep oxygen away from the weld. These gases are used because they are inert. If oxygen is allowed to come in contact with the molten metal of the weld, oxidation will occur. The weld will be weak and could fail. Neon gas is drawn from the atmosphere. It produces light when exposed to electricity. It is used as an accent light in some show cars.

▶ Hydrocarbons

LO 53-02 Describe hydrocarbon and carbon monoxide emissions.

Gasoline, diesel, LPG, and natural gas are all hydrocarbon (HC) compounds. HC emissions are created in two ways. The first is when raw fuel evaporates into the atmosphere, such as when filling the gas tank. The second is when some or all of it does not burn in the combustion chamber and is exhausted out the tailpipe. In other words, HC emissions are unburned fuel that escapes to the atmosphere.

Gasoline must evaporate easily. This is so that it can mix with air to burn properly in the combustion chamber. But this property also means it can evaporate easily into the atmosphere. For example, when a vehicle is being refueled, HC vapors can escape from the filler neck into the atmosphere. Also, when the vehicle is left in the sun, the temperature of the fuel increases. This causes it to evaporate even more creating pressure in the tank. Fuel can then escape from the tank if not controlled.

> **▶ TECHNICIAN TIP**
>
> HC emissions react with oxides of nitrogen (NO_x) in the presence of sunlight to produce **photochemical smog**. This is the brown haze that hangs in the sky, typically seen over large cities. Smog is a major health issue to humans because it affects lung tissue. HCs are a major component of photochemical smog.

▶ Carbon Monoxide

Carbon monoxide (CO) is a colorless, odorless, tasteless, flammable, and highly toxic gas. It is also heavier than air, so it collects in low spaces such as lube pits. It is a product of the incomplete combustion of carbon and oxygen. The incomplete combustion can be caused by either too much fuel (carbon) or not enough air (oxygen). In these conditions, the carbon bonds with only one oxygen molecule, creating CO (**FIGURE 53-3**). Other causes of high CO emissions are improper ignition timing, low combustion temperature, or low cylinder compression. CO emissions have been reduced in modern vehicles as a result of:

- better engine design,
- better fuel management, and
- the use of catalytic converters in the exhaust system.

Most engine fuels, except pure hydrogen, use carbon-based fuel. In the real-world, internal combustion engines operating on a carbon-based fuel tend to produce some CO emissions. This is especially true when the combustion temperature is low, such as during a cold start; or when there is not enough oxygen present to completely burn the fuel; or when there is insufficient time in the combustion chamber for complete combustion.

▶ Toxicity

CO is an extremely poisonous gas. Because you cannot see it or smell it, it is very dangerous. Inhaling CO in a confined space can be lethal. Because it comes from the exhaust system, it is important not to run any engine inside a shop without venting the exhaust directly to the open air outside.

Victims of **carbon monoxide poisoning** can sometimes look healthy and pink-cheeked. This is because CO gives the blood a brighter red color than normal. Hemoglobin is the part of the blood that carries life-giving oxygen around the body. CO binds much more easily than oxygen with hemoglobin. If CO attaches to hemoglobin, it never leaves. As the hemoglobin becomes filled with CO, it can't carry oxygen. When the brain does not have oxygen, the victim becomes unconscious. This results in brain damage or death. Early signs of CO poisoning include headaches, nausea, weakness, and irritability.

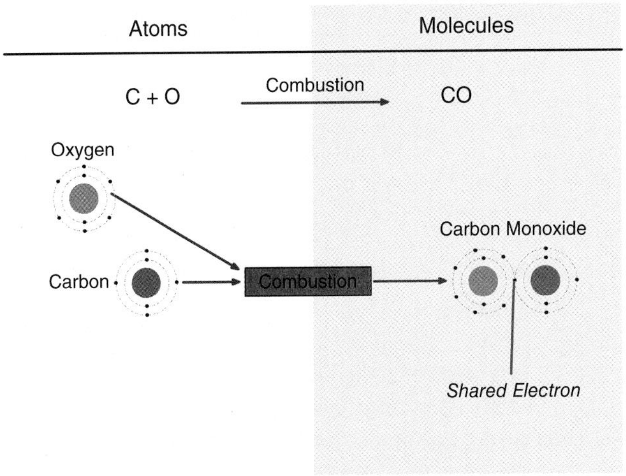

FIGURE 53-3 The incomplete combustion that creates CO can result from either too much fuel (carbon) or not enough air (oxygen), causing the carbon to be able to bond with only one oxygen molecule.

Applied Science

AS-24: Chemical Reactions: The technician can demonstrate an understanding of the chemical reactions that occur in the automotive engine that are related to the combustion of fuels and the operation of the catalytic converter.

An automotive teacher asks a student to use a hacksaw to cut apart a failed catalytic converter for evaluation purposes. When the converter is cut apart, it is evident that the converter had melted down. The students find six large pieces of material that resemble rocks. The ceramic catalyst was damaged because of a super-heated condition. It was the result of unburned fuel entering the converter. This teaching aid provides an introduction to the catalytic converter and how it works.

The stoichiometric ratio is the air–fuel ratio at which all of the oxygen in the air and all of the fuel are completely burned. This ratio is 14.7 parts of air to 1 part of fuel by weight. Maintaining close control of the air–fuel ratio allows an engine to run cleaner than ever. In many cases, precise control of the air–fuel ratio has made the job of hydrocarbon (HC)– and carbon monoxide (CO)–related emission control devices much easier. This is because they are dealing with much smaller volumes of the pollutants. Catalytic converters use a base structure of ceramic material coated with precious metals. They react with and break up or combine gases in the exhaust system. The catalytic converter cleans up the pollutants left over from combustion, which reduces tailpipe emissions.

In the case of the failed converter that was cut apart for evaluation, the engine had a misfiring cylinder as a result of a faulty spark plug. The vehicle owner ignored the check engine light that was illuminated on the dash for about one month, resulting in the failure of the converter. The unburned fuel from the misfiring cylinder was ignited in the converter. This caused overheating of the ceramic monolith.

Because CO is absorbed so easily by red blood cells, the following agencies set limits for the permissible exposure limit (PEL) for the workplace:

- OSHA PEL is 50 parts per million (ppm) for an eight-hour shift.
- The National Institute for Occupational Safety and Health (NIOSH) PEL is 35 ppm for an eight-hour shift, with a ceiling of 200 ppm for any length of time.
- The American Conference of Governmental Industrial Hygienists (ACGIH) PEL is 25 ppm for an eight-hour shift and a 40-hour workweek.

Exposure levels of 400 ppm may be fatal in as little as 30 minutes. As you can see, it doesn't take much CO in the air to be hazardous. To further help you understand the smallness of the PELs, consider that 1% CO equals 10,000 ppm of CO. Vehicles with an emission-related problem can run perfectly well and still easily have CO tailpipe readings between 3% and 5%. This equates to 30,000–50,000 ppm—more than 1000 times the allowable PELs given in the preceding list.

SAFETY TIP

Be sure when running a vehicle in a shop environment that you provide adequate ventilation. Use an exhaust extraction device such as a fan system and exhaust hoses. They are designed to draw exhaust gases from the vehicle's exhaust pipe and push them outside of the shop. Small amounts of carbon monoxide can kill if proper ventilation is not provided.

First Aid

First aid for victims of CO poisoning starts with recognizing the signs of exposure. Those include shortness of breath, nausea, headache, dizziness, or light-headedness. Awareness will allow you to make the decision to remove the other person, or yourself before severe poisoning occurs. If the victim of CO poisoning is unresponsive or demonstrates severe symptoms, call for qualified help. Then assess the environment to determine if you can safely remove the person from further exposure. Remember, CO is heavier than air, so it will collect in low spaces such as lube pits. If you are trained, you may need to perform CPR. If available, give first aid oxygen as well. Continue to treat the victim until medical help arrives.

SAFETY TIP

Well-meaning rescuers routinely die trying to save people who are unconscious in pits. They see a coworker or friend who is unresponsive in a pit and immediately enter it to help. What they don't know is that the air is bad and quickly are overcome themselves in the process. This can easily end up with multiple deaths.

▶ Oxides of Nitrogen

LO 53-03 **Describe NO_x, SO_2, and particulate emissions.**

Air that is drawn from the atmosphere into an engine contains almost 80% nitrogen. It is normally considered to be an inert gas, meaning it does not react easily. But under the high temperature and pressure of combustion, nitrogen may combine with oxygen. This produces oxides of nitrogen (NO_x) (**FIGURE 53-4**). It is

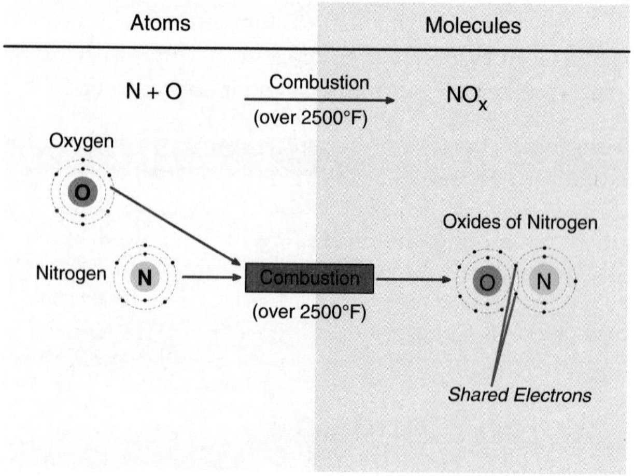

FIGURE 53-4 Under the high temperature and pressure of combustion, nitrogen may combine with oxygen to produce oxides of nitrogen.

typically produced only when combustion temperatures of around 2500°F (1400°C) and above occur. The higher the temperature, the greater the NO_x created. A lean mixture creates high levels of oxides of nitrogen. This is because lean mixtures burn faster than rich mixtures, producing excessive combustion temperature. The increase of available oxygen can oxidize the nitrogen at those high temperatures.

Oxides of nitrogen irritate the eyes, nose, and throat. In extreme cases, such as regular exposure to high concentrations, coughing and lung damage can occur. Oxides of nitrogen mix with HCs and sunlight to create photochemical smog. The brown-looking cloud that often hangs over large cities is photochemical smog.

▶ Sulfur Dioxide

Gasoline and diesel fuels contain sulfur as part of their chemical makeup. When sulfur burns in the combustion chamber, one sulfur molecule combines with two oxygen molecules. This creates sulfur dioxide. Sulfur dioxide mixes with water vapor formed during the combustion process and produces sulfuric acid (**FIGURE 53-5**). This corrosive compound is emitted into the atmosphere through the exhaust. Sulfuric acid is a major environmental pollutant. It comes back to earth in contaminated rainwater, called acid rain. This acid rain has been responsible for degrading vast areas of arable land. As a result, the reduction or removal of sulfur from motor fuels has become a major part of most countries' vehicle emission control programs.

High sulfur levels in fuel, when combined with water vapor, can also cause corrosive wear on valve guides and cylinder liners. This can lead to premature engine failure. Using proper lubricants and draining oil at the correct intervals help combat this effect. Many manufacturers have been using stainless steel in their exhaust pipes and mufflers. This helps to resist corrosion in these components.

Sulfur reduces catalyst efficiency in modern vehicles. Therefore, consider vehicles operating with higher sulfur gasoline have higher emissions than those operating on lower sulfur gasoline. At the same time, small amounts of sulfur in gasoline reduce the performance of oxygen sensors. This would then add to higher levels of all tailpipe emissions.

Regulations have reduced the permissible levels of sulfur in fuel. But, there are some side effects from using low-sulfur diesel fuel. The refining process used to reduce the sulfur level can reduce the natural lubricating properties of the diesel fuel. This is important for the lubrication and operation of diesel fuel system components. Examples include fuel pumps and fuel injectors. These components then must be manufactured with higher quality materials. Or special lubricants have to be used to provide the required level of protection.

To reduce the sulfur content, oil companies change the overall chemical composition of the fuel. Doing this can affect fuel pump seals, engine seals, and O-rings. Some of these tend to react to changes in fuel composition by swelling or shrinking. This problem can be avoided by manufacturers replacing the seals with ones that are less susceptible to swelling and shrinking. It also helps to perform regular maintenance on the fuel system.

Particulates

Particulates are small particles of solid matter that are suspended in the air. Particulates from modern engines are usually small particles of carbon (**FIGURE 53-6**). In spark-ignition engines, particulates are caused by incomplete combustion of rich air-fuel mixtures. In this case, the hydrogen burns away from the HC molecule. It leaves the carbon by itself.

In compression–ignition engines, particulates are caused by either too rich of a mixture or a poorly mixed air–fuel mixture. This produces very rich areas in the combustion chamber. This happens even though the overall mixture could be very lean. Newer compression–ignition engines use a particulate filter. It catches and holds the particulates in the exhaust stream. In some systems, the particulate filter can be regenerated periodically. This occurs by burning off the carbon particles. The powertrain control module (PCM) controls the regeneration process automatically. Also, a

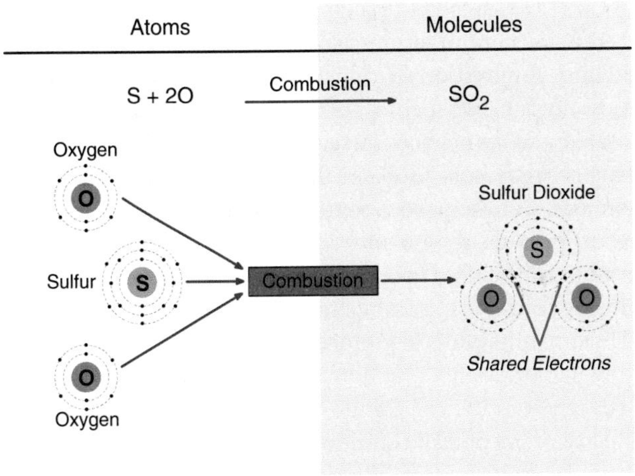

FIGURE 53-5 Sulfur dioxide mixes with water vapor formed during the combustion process and produces sulfuric acid.

FIGURE 53-6 Particulates from modern engines are usually small particles of carbon.

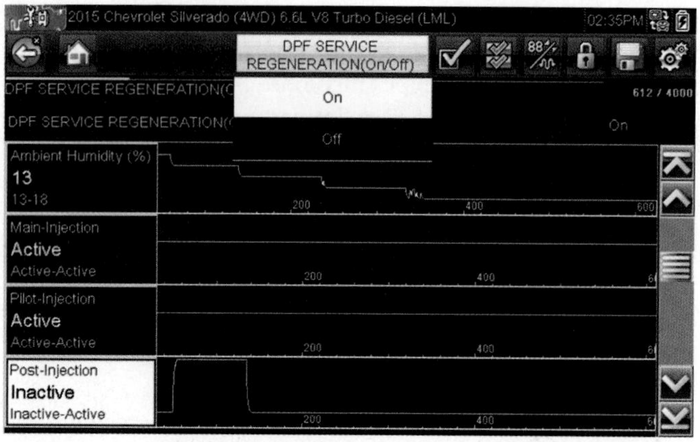

FIGURE 53-7 A technician using a scan tool to command regeneration of the particulate filter.

technician can command regeneration with a scan tool connected to the PCM (FIGURE 53-7).

Smog

Smog is produced in the atmosphere when unburned HCs and oxides of nitrogen react chemically with sunlight. Smog appears reddish brown on the horizon and can be seen over major cities (FIGURE 53-8). Smog is also known as ground-level ozone. Ozone blocks harmful ultraviolet light from the sun in the upper atmosphere. However, in the lower atmosphere, it creates irritation of the lungs and is believed to contribute to asthma.

FIGURE 53-8 Smog over a city.

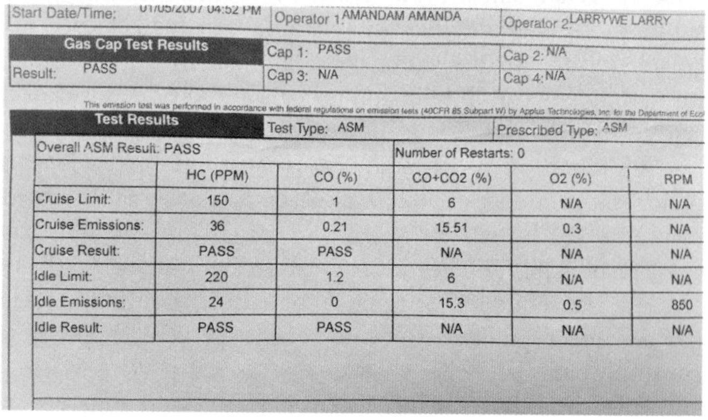

Start Date/Time: 01/05/2007 04:52 PM		Operator 1: AMANDAM AMANDA		Operator 2: LARRYWE LARRY	
Gas Cap Test Results		Cap 1: PASS		Cap 2: N/A	
Result: PASS		Cap 3: N/A		Cap 4: N/A	

This emission test was performed in accordance with federal regulations on emission tests (40CFR 85 Subpart W) by Applus Technologies, Inc. for the Department of Eco

Test Results — Test Type: ASM — Prescribed Type: ASM

Overall ASM Result: PASS — Number of Restarts: 0

	HC (PPM)	CO (%)	CO+CO2 (%)	O2 (%)	RPM
Cruise Limit:	150	1	6	N/A	N/A
Cruise Emissions:	36	0.21	15.51	0.3	N/A
Cruise Result:	PASS	PASS	N/A	N/A	N/A
Idle Limit:	220	1.2	6	N/A	N/A
Idle Emissions:	24	0	15.3	0.5	850
Idle Result:	PASS	PASS	N/A	N/A	N/A

FIGURE 53-9 One indication that an engine is in good working order is a clean emission test.

▶ Controlling Emissions

LO 53-04 Describe how engine design reduces emissions.

Manufacturers are usually free to determine how they will meet emission standards. This is as long as they do meet them. In designing systems and processes that effectively and cost-efficiently meet the standards, manufacturers use a couple of main strategies:

- The first is to design the engine in such a way to reduce emissions as much as possible. Examples are building turbulence-enhancing strategies and precisely controlling valve and spark timing.
- The second strategy is to closely control the air–fuel mixture under all driving conditions. This minimizes emission output.
- The third strategy is to warm up the engine as quickly as possible. And then operate it at higher temperatures than older engines.
- The last strategy is to integrate precombustion and post-combustion add-on devices. These devices are designed to further reduce emission output.

We look next at each of these strategies in more depth.

▶ TECHNICIAN TIP

Manufacturers continue to produce cleaner and more efficient vehicles. Most vehicles use similar emission controls. But, you should always refer to the service information. There, you will find a description and operation of the emission system. Read this before attempting to diagnose or repair it.

▶ Importance of Controlled Combustion

Most emissions are by-products of combustion. So, accurately controlling combustion within the cylinders can minimize emission output. For fuel to burn efficiently, it has to be fully evaporated and completely mixed with the right proportion of air. This is pretty easy to do in the controlled environment of a laboratory. But doing so repeatedly at a high rate of speed in very short periods of time under a huge variety of conditions is much more difficult. These are the conditions we are dealing with inside automotive combustion chambers.

Thus, maintaining low emission output is a challenge. It requires all of the engine's systems and components to be working properly for efficient combustion to occur. In fact, a good general test of the overall condition of the engine and its control systems is an emission test. If the vehicle passes the emission test with clean results, that is a good indication that all of the engine systems and the engine itself are operating well (FIGURE 53-9).

▶ Combustion Chamber Design

Combustion chamber design affects the combustion process, which also affects the level of emissions. In the combustion chamber where surface temperatures are low, the combustion flame can be **quenched** in those areas (**FIGURE 53-10**). The flame temperature drops so low that it goes out, or is quenched. Fuel left unburned in these zones is then sent out through the exhaust as HC and CO emissions.

If the spark plug is positioned near the center of the combustion chamber, the **flame front** travels evenly through the combustion chamber. This allows for more complete combustion. The flame front is the rapid burning of the air–fuel mixture that moves outward from the spark plug across the cylinder. Combustion chambers where the flame front is not blocked by protrusions or sharp corners help ensure complete combustion. Proper placement of the spark plug helps to minimize this effect.

The designs of intake ports and cylinder heads can aid in swirling the air–fuel mixture as it travels to the combustion chamber. Swirling of the air and fuel helps to fully mix the two. If fuel does not stay mixed in the air when being drawn into the combustion chamber, the mixture will be distributed unevenly. This creates drivability problems and excessive emissions. It also shows us how important proper movement of air is to combustion.

One factor that affects the ability of air to hold fuel is gas flow rate. Fast-moving air tends to hold fuel better than slow-moving air, which allows fuel to fall out of the airstream (**FIGURE 53-11**). Fast-moving air more easily causes swirling of the air and fuel. Gas flow rate can be increased by the use of smaller intake ports. This design works well at keeping good airflow at low engine revolutions per minute (rpm). But at high rpm, it can restrict the airflow too much and reduce engine power. Some manufacturers overcome this problem by using two or more intake valves in each cylinder. With the extra valves open at high rpm, the overall size of the intake port opening is then increased, and a good gas flow rate is maintained. The more ideal the gas flow rate of air entering the combustion chamber, the more efficient that engine. With better efficiency comes lower emissions.

Changing valve timing also alters the combustion process. Reducing the valve overlap reduces the **scavenging effect.** Exhaust gases moving out of the combustion chamber create a low-pressure area in the exhaust manifold. This helps to pull air and fuel into the cylinder. If the valve overlap is excessive, unburned air and fuel can be pulled out past the exhaust valve. Increased HC emissions will result (**FIGURE 53-12**).

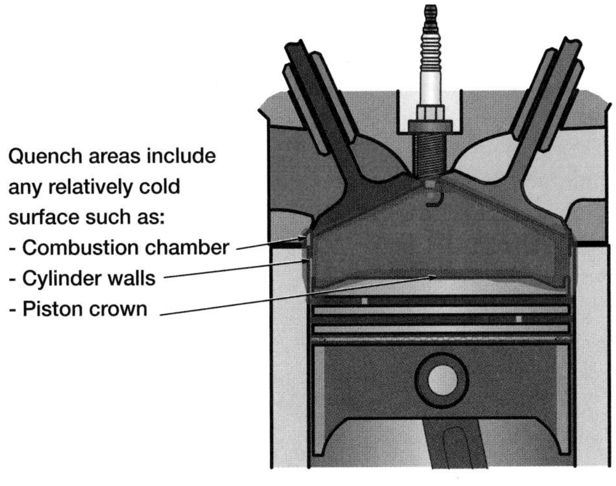
Quench areas include
any relatively cold
surface such as:
- Combustion chamber
- Cylinder walls
- Piston crown

FIGURE 53-10 Quench areas in a combustion chamber.

▶ Stoichiometric Ratio

LO 53-05 Describe how the air–fuel ratio can be controlled to affect emissions.

The stoichiometric ratio is also referred to as **lambda**. It is the ratio of air to fuel at which all of the oxygen in the air and all of the fuel are completely burned. As you can imagine, complete combustion reduces HC and CO emissions. The typical stoichiometric ratio for gasoline is 14.7:1 (14.7 parts air to 1 part gasoline by weight or mass). This ratio equates to a lambda

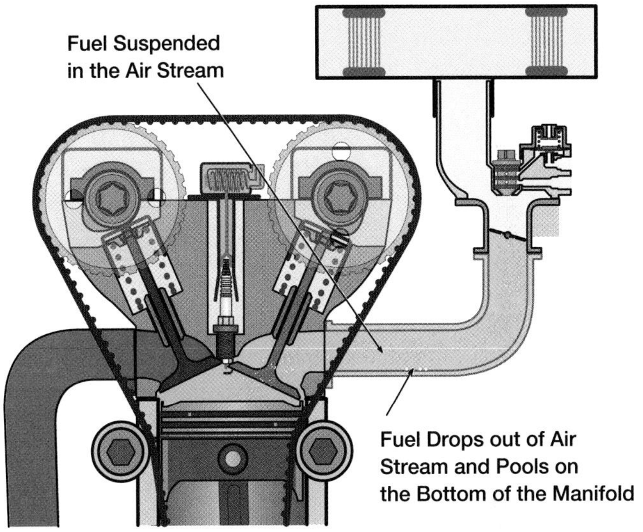
Fuel Suspended
in the Air Stream
Fuel Drops out of Air
Stream and Pools on
the Bottom of the Manifold

FIGURE 53-11 Gas flow rate affects the ability of air to hold fuel. Fast-moving air tends to hold fuel better than slow-moving air, which allows fuel to fall out of the airstream.

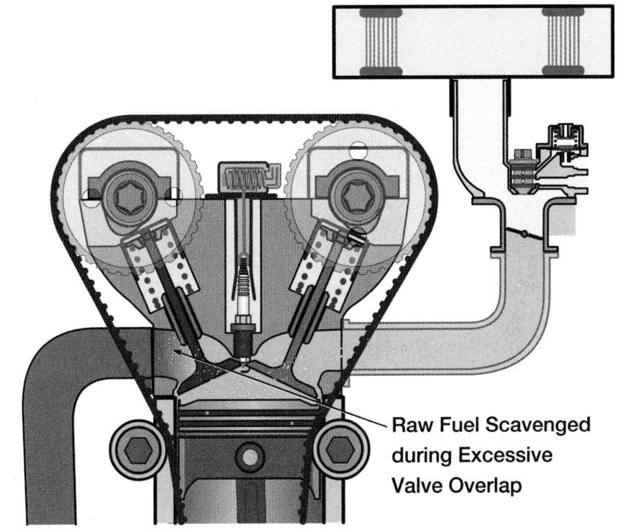
Raw Fuel Scavenged
during Excessive
Valve Overlap

FIGURE 53-12 Excessive valve overlap can cause raw fuel to be scavenged out of the cylinder, raising HC emissions.

TABLE 53-2 Stoichiometric Ratios for Various Fuel Types

Gasoline	14.7
No. 2 Diesel	14.7
Methanol	6.45
Ethanol	9.00
Propane	15.7
Compressed natural gas	17.2
Hydrogen	34.3

Source: Department of Energy website. Stoichiometric ratio is the ratio of air to fuel, expressed as 14.7:1.

reading of 1.0. Each fuel has its own stoichiometric ratio (**TABLE 53-2**). As the AFR moves away from the stoichiometric point, the HCs and CO emissions increase. In many cases, precise control of the AFR has made the job of HC and CO emission control devices much easier. This is because they are dealing with much smaller amounts of the pollutants.

A rich condition means more fuel is introduced than the engine can completely burn. This is anything lower (richer) than 14.7:1 or less than 1.0 lambda. When this condition occurs, HCs can only partially burn because of the lack of oxygen. This rich condition creates large amounts of CO. If excessively rich, some of the HCs will not burn at all. This leaves some HCs to pass through the engine unburned. Elevated amounts of CO in the exhaust are a good indicator that the engine is running rich.

A lean condition means more air is introduced than the engine can completely burn. This is anything above (leaner than) 14.7:1 or greater than 1.0 lambda. When this condition occurs, not all of the HCs burn. This is because the flame cannot reach some of the HC molecules due to the excessive amount of oxygen and nitrogen molecules. Thus, the HC molecules are pushed out of the exhaust pipe unburned. Elevated HC readings (with low CO readings) are a good indicator of a lean condition.

Carbon dioxide is a good indicator of whether the fuel is burning completely. For example, high readings of 12% to 15% of carbon dioxide indicate proper combustion efficiency. A smaller percentage of carbon dioxide in the exhaust indicates that combustion is not complete. This would be evidence of a rich or lean mixture, ignition problems, lowered compression, or an exhaust system leak.

▶ TECHNICIAN TIP

Carbon dioxide is a gas that manufacturers are trying to reduce. It is believed to contribute to global warming. Laws are underway that will require manufacturers to reduce its creation. Some of the ways that carbon dioxide can be reduced are listed here:

- Making vehicles smaller, lighter, and more aerodynamic
- Using smaller, more fuel-efficient engines
- Shutting down the engine when the vehicle is decelerating or stopped at a stop light
- Using regenerative braking systems to capture waste energy during braking

▶ Controlling Air–Fuel Ratios

Precise control of the AFR is one of the main strategies for minimizing a vehicle's emission output. If the AFR is close to the stoichiometric point, very little HC and CO are created. Electronic fuel injection and engine management systems can adjust the air–fuel mixture within certain limits. This needs to happen as parts wear. The proper control of the AFR can then be maintained over a broad range of conditions. These systems closely control the AFR entering each cylinder. They also ensure the ignition timing matches the operating conditions.

Sensors around the engine send information about many conditions. These include airflow, coolant temperature, throttle position, engine speed, and other inputs to the PCM. The PCM uses this information to set fuel and ignition settings, which change as sensor data change. The PCM has a programmed memory that works like a reference chart. The manufacturer's engineers program the PCM to constantly adjust the fuel and spark, depending on the values received from the sensors. For example, as airflow rises, the fuel increases proportionately.

Fuel delivery is controlled by the fuel injector. The typical fuel injector is simply an electric solenoid. When electrical current flows in the fuel injector winding, a magnetic field is produced (electromagnet). This causes the fuel injector to open and allow fuel to flow through it into the intake manifold. The PCM opens the fuel injector for a longer or shorter period of time. This creates more or less fuel delivery. The longer the fuel is delivered, the richer the mixture, and vice versa.

▶ TECHNICIAN TIP

The pulse width of the fuel injector is a piece of data that can be read on a scan tool or lab scope. It assists the technician in testing fuel delivery. The pulse width is the amount of time the fuel injector is turned on and fuel is sprayed into the intake manifold. It is measured in milliseconds. The longer the pulse width, the more fuel is delivered to the engine.

Changes in operating conditions can change mixture conditions. For instance, an engine is being driven at moderate speed and the throttle is suddenly closed. Any fuel condensed on the intake manifold walls or intake port of the cylinder head will be drawn into the cylinders. Also, a smaller amount of air moves into the combustion chamber with the throttle plates closed. So turbulence tends to be poor. This all leads to incomplete combustion and the release of partially burned and unburned gases into the exhaust.

To reduce this effect, the PCM detects that the throttle is being closed through the use of the throttle position sensor (TPS). It reduces the fuel spray by reducing the on-time of the injectors. If the engine is being used to slow the vehicle, the injectors can be completely shut off to save fuel and reduce pollution. This is called fuel shutoff mode.

Newer engines with electronic throttle control can move the throttle plate free of driver input. This feature allows the computer to control the throttle opening. It does this along with controlling the fuel injection pulse width. The proper mixture is thus maintained during all driving conditions.

▶ Precombustion/Postcombustion Treatment

LO 53-06 Describe precombustion–postcombustion exhaust treatment and the evolution of emission controls.

The pollution emitted from the internal combustion engine can be reduced. But it means taking action. This can happen before combustion and after combustion. Precombustion action looks to making sure that combustion happens in a controlled and complete way. This can lower postcombustion pollutant gases. There are two primary precombustion emission control systems:

- the heated air intake system, and
- the **exhaust gas recirculation (EGR)**system.

The heated air intake system was used on throttle body–injected (and carbureted) engines. It was designed to keep the air entering the engine at a constant temperature of approximately 100–110°F (38–43°C). This helped vaporize the fuel in the intake manifold. The better vaporized the fuel, the more likely it could be distributed evenly within the air. And thus, the cleaner the combustion process. This system is discussed further in the Emission Control System Evolution section.

Most port fuel-injection systems and gas direct-injection systems do not need a heated air intake system. This is because the fuel is either sprayed into the intake manifold right next to the intake valve, or into the combustion chamber itself. The injector sprays a fine mist of fuel, which is easier to vaporize. Plus it does not have to sit or travel along a cold intake manifold, where it can condense out of the air. Most new vehicles monitor the intake air temperature and use that to help determine the mass of air entering the engine. This helps the PCM to know exactly how much fuel to inject.

The second precombustion emission control system is the EGR system. This system recirculates some of the inert exhaust gases back into the combustion chamber. This happens when combustion temperatures are high enough to create oxides of nitrogen. The inert gases help lower peak combustion temperatures. It is important to lower them below the threshold when oxides of nitrogen are created. The next Emission Control Systems section discusses this in more detail.

Postcombustion emission control systems deal with the remaining pollutants that are left over from the combustion process. There are three primary postcombustion systems:

- the **positive crankcase ventilation (PCV)** system,
- the secondary air injection system, and
- the catalyst system.

The PCV system draws the blowby gases out of the crankcase and routes them to the intake manifold. From there, they are drawn into the cylinders and burnt in the combustion chamber. This effectively converts them into carbon dioxide and water.

The catalyst system reduces oxides of nitrogen in the exhaust stream. It converts them back into nitrogen and oxygen. It also oxidizes HCs and CO, converting them into carbon dioxide and water.

The secondary air injection system, if equipped, helps to complete the burning of hot HCs and CO in the exhaust system. It does so by adding oxygen to the exhaust stream. The oxygen in the air reacts with the HCs and CO. In doing so, it converts them into carbon dioxide and water.

All three of these systems reduce emissions to near zero if the engine and all of the systems are operating as designed. Each of them is covered in greater depth in the following section.

▶ Emission Control System Evolution

Emission control systems have been created by manufacturers. They minimize the amount of pollution a vehicle releases into the atmosphere. Before the days of emission control systems, vehicles and engines produced very large amounts of pollutants. This is evidenced by the large number of smog days in many big cities. Even though there are now many more vehicles and many more miles driven, the number of pollution days in a year has been reduced substantially. And in most areas, it continues to decline.

One of the earliest control devices was the PCV system. The system was first used around 1961. It later became a mandatory emissions device. PCV systems stop the release of HCs from the engine's crankcase to the atmosphere. Another control device is the evaporative control system, known as the **evaporative emission (EVAP) system**. This system was used beginning around 1972 and later became a mandatory emission system. It is used to stop the release of HCs from the fuel tank to the atmosphere.

Another early emission control device from the 1960s was the early fuel evaporation system. It was designed to quickly heat up the induction system and intake air on cold engine starts. There were several different ways of doing this, from heat riser valves to electric heater grids.

EGR systems began to appear on the automobile around 1972. It too became a mandatory part of emission controls. These systems reduce the levels of oxides of nitrogen created by controlling the temperature of the combustion process. The exhaust gases being relatively inert slowed down the speed at which the air–fuel burned. Because it burned slower, its temperature didn't peak as high. In the 2000s, variable valve timing (VVT) started to be used. It allowed some exhaust gas to remain in the cylinder after the exhaust stroke. The air–fuel mixture was then diluted with exhaust gases similar to an EGR system. So, VVT replaced the EGR valve on some vehicles.

The use of a system called secondary air injection began in the late 1960s. It was designed to limit HC release from the engine during cold engine operation. The system injects air into the exhaust stream near the exhaust valve. The oxygen mixes with the HCs and aid in the conversion process as the exhaust exits the combustion chambers. Later versions of this system sent air down to the catalytic converter once the engine warmed up.

Catalytic converters found their way onto the automobile around 1973. Catalytic converters reduce any pollutants that are left in the exhaust stream before they leave the tailpipe. As catalytic converters were introduced, leaded gas was phased out. Lead in the gas would contaminate (coat) the catalyst in the converter and render it inoperative.

▶ Catalytic Converters

LO 53-07 Explain the types and operation of catalytic converters.

Catalytic converters are a primary emission control device. They are installed near the front of the exhaust system (**FIGURE 53-13**). They perform a final cleanup of the tailpipe emissions. But they can only do so on a properly running engine equipped with functioning emission control systems. Catalytic converters are not designed to handle the emission output of a poorly running engine or one with faulty emission control systems. Catalytic converters create heat as they convert harmful gases into less harmful gases. Forcing converters to deal with more gases than they are designed to handle can cause them to overheat. In some cases, they may even melt down internally, which ruins them.

Catalytic converters use a base structure of ceramic material. It is coated with precious metals to react with and break up or combine gases in the exhaust system. The term "catalyst" refers to a material that causes a chemical reaction without itself being consumed or changed in the reaction. Specific precious metals have the ability to act as a catalyst. They can convert harmful exhaust gases to nonharmful or less harmful gases. Catalytic converters come in two types: oxidizing and reduction. Each type is used to react with particular gases.

The reduction catalyst contains a platinum and rhodium coating. It helps to reduce the oxides of nitrogen molecules into their base compounds. When a nitric oxide or nitrogen dioxide molecule comes into contact with the catalyst coating, the coating strips the nitrogen atom out of the molecule and retains it. This frees up the one or two oxygen atoms in the molecule. They combine in pairs to form molecules of oxygen. The nitrogen atoms bond with other nitrogen atoms that are retained in the catalyst and form molecules of nitrogen. Thus, two molecules of nitric oxide become one molecule of nitrogen and one molecule of oxygen. Or two molecules of nitrogen dioxide become one molecule of nitrogen and two molecules of oxygen (**FIGURE 53-14**).

The second type of catalyst is the oxidizing catalyst. It oxidizes any unburned HCs and CO as they travel over the platinum and palladium coating. This causes the CO and HCs to react with any remaining oxygen in the exhaust gas (**FIGURE 53-15**). Each CO molecule combines with an oxygen (O) molecule. They make one less harmful carbon dioxide (CO_2) molecule. And two hydrogen (H) atoms combine with one oxygen (O) molecule. They form one nonharmful water (H_2O) molecule.

Catalytic converters are referred to as either two-way or three-way catalytic converters. The **two-way catalytic converter** was the first type designed. It is an oxidizing catalyst. It converts two gases, HCs and CO, to carbon dioxide and water. The

FIGURE 53-13 Catalytic converters use a base structure of ceramic material coated with precious metals to react with and break up or combine gases in the exhaust system.

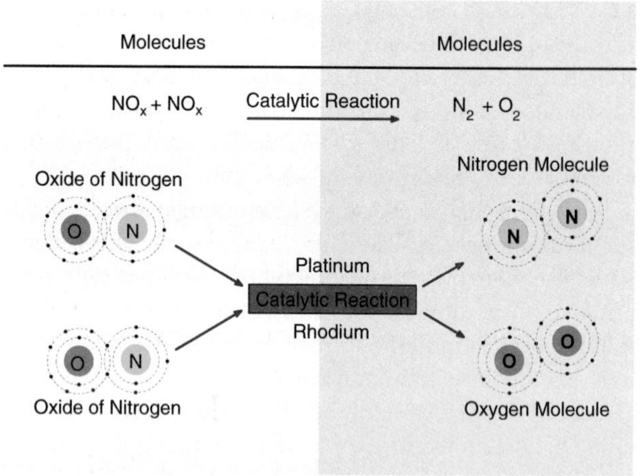

FIGURE 53-14 The reduction catalyst splits nitric oxide and nitrogen dioxide molecules into nitrogen atoms and oxygen atoms. The nitrogen atoms bond to form molecules of nitrogen (N_2), and the oxygen atoms bond to form oxygen (O_2).

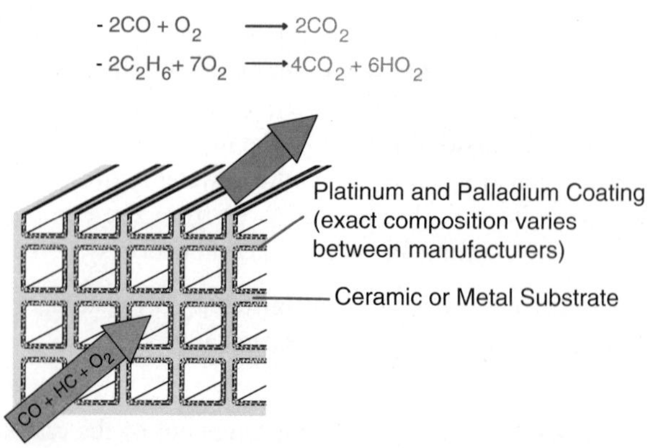

FIGURE 53-15 The oxidizing catalyst oxidizes any unburned HCs and carbon monoxide as they travel over the platinum and palladium coating.

early two-way catalytic converters used a bed of catalytic pellets (pellet-style converter). The exhaust passed through them.

There are more recent versions of the two-way (and three-way) catalytic converters. They are designed with a honeycomb catalytic element. This style provides long narrow passageways for the exhaust gases to flow through. It helps the gases to come in contact with the catalyst materials on the sides of the honeycomb walls. This contact helps to oxidize the HCs and CO once the catalyst reaches its operating temperature.

SAFETY TIP

Catalytic converters must be above approximately 500°F (260°C) to begin conversion. Typically, they operate at 900–1600°F (482–871°C). Catalytic converters can cause serious burns. So use gloves and heat protection sleeves when working around a hot catalytic converter.

▶ TECHNICIAN TIP

Catalytic converters may contain rhodium, palladium, platinum, or cerium. These materials are more valuable than gold and should be recycled. Catalytic converters have become a target for thieves. They cut them out of the exhaust system while the vehicle is parked in a parking lot or driveway.

Modern vehicles using petroleum-based fuels have three-way catalytic converters. **Three-way catalytic converters** contain both a reduction catalyst and an oxidizing catalyst. It is "three-way" because of the three regulated emissions it can convert. They are CO, HCs, and oxides of nitrogen. The reduction catalyst is in front. It reduces the oxides of nitrogen, nitric oxide, and nitrogen dioxide. It converts them back into harmless nitrogen and oxygen molecules (**FIGURE 53-16**). The oxidizing catalyst then uses the oxygen from the reduction catalyst. It oxidizes the HCs and CO into water and carbon dioxide.

Because of strict emission requirements, vehicles with three-way catalytic converters require a **feedback system** on the fuel system. The PCM monitors the oxygen content in the exhaust stream by using an exhaust gas oxygen (EGO) sensor. It is also known as a lambda sensor and is mounted in the exhaust manifold. This sensor tells the PCM how much oxygen is in the exhaust. The computer uses this information to control the pulse width of the fuel injectors.

The PCM can increase or decrease the amount of oxygen in the exhaust by adjusting the AFR. The ECU ensures that the AFR cycles just above and below the stoichiometric point in normal driving conditions. This ensures that there is always sufficient oxygen in the exhaust stream. That way, the oxidization catalyst can deal with unburned HCs and CO.

▶ TECHNICIAN TIP

Each style of catalytic converter is designed for a specific vehicle; one size does not fit all. A converter intended for the treatment of a four-cylinder engine will not be efficient on an eight-cylinder engine. Be sure to install the correct catalytic converter in the correct vehicle. Always check the manufacturer's specifications before making a repair. Be careful with aftermarket catalytic converters. They may not use as much catalyst agent (precious metals) as the emission system requires.

Catalyst Monitoring

Any vehicle produced after 1996 monitors the efficiency of the catalyst. It does so by comparing the readings of two oxygen sensors (**FIGURE 53-17**). One oxygen sensor is in front of the catalytic converter, as in the past. It indicates the amount of oxygen in the exhaust stream as it leaves the cylinders. The reading from this sensor normally toggles rich and lean as the PCM is controlling the AFR. The second oxygen sensor is located after the catalytic converter. It measures the oxygen in the exhaust stream after it leaves the catalytic converter.

The converter uses oxygen to oxidize the HCs and CO. Therefore, the oxygen content is normally lower after the converter. This results in a higher and steadier oxygen sensor voltage. The catalyst monitor tracks both oxygen sensor readings.

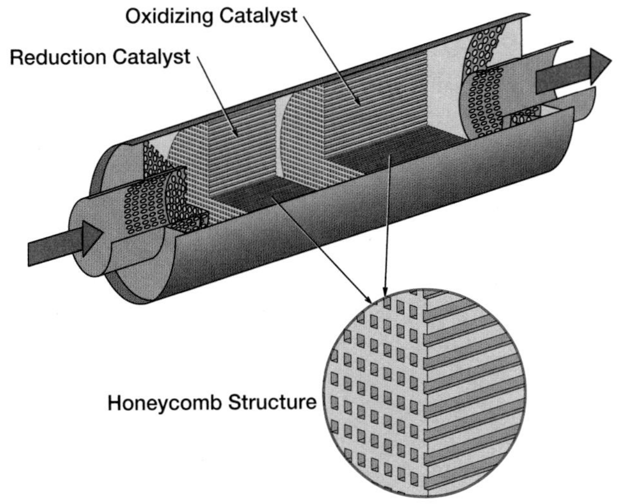

FIGURE 53-16 A typical three-way catalyst.

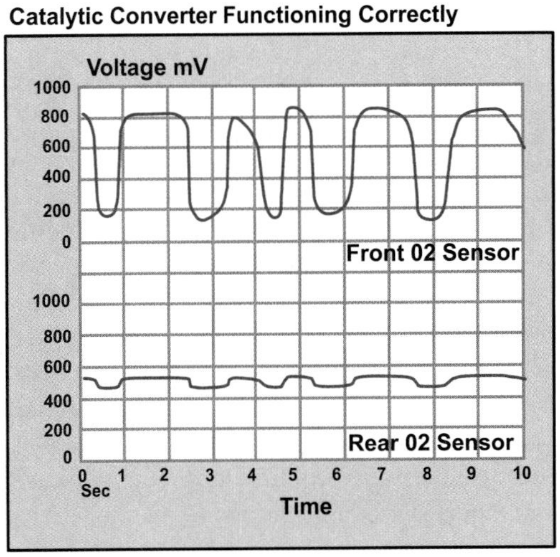

FIGURE 53-17 Any vehicle produced after 1996 monitors catalyst efficiency by using two oxygen sensors.

Applied Science

AS-89: Catalytic Converter: The technician can explain the principles by which a catalytic converter modifies emission gases at the atomic level to provide a lower level of HC, CO, and NOx in the final exhaust.

One of the greatest emission devices to be installed on vehicles is the catalytic converter. This component will last for the life of the vehicle if given proper care. Proper care for the converter includes maintaining the engine according to the manufacturer's service information. If the engine is operating normally, the converter won't have to work hard to accomplish its task.

Catalytic converters provide control of three important emissions. They are: hydrocarbons (HC), carbon monoxide (CO), and oxides of nitrogen (NO_x). Emissions pass through the converter's catalytic inlaid materials. These materials include rhodium, palladium, platinum, or cerium. All these special materials are precious metals. The reduction catalyst is in front. It converts the oxides of nitrogen, nitric oxide, and nitrogen dioxide. These convert back into harmless nitrogen and oxygen molecules. The oxidizing catalyst then uses the oxygen from the reduction catalyst. It oxidizes the hydrocarbons and carbon monoxide into water and carbon dioxide.

This ensures that the gases are being effectively converted by the catalytic converter. As the catalytic converter fails, the rear sensor tends to track the oscillations of the front oxygen sensor.

The PCM monitors these two oxygen sensor signals. Once the catalyst efficiency degrades to a predetermined point, the PCM turns on the MIL and sets a DTC. Access to this information allows the technician to verify the condition of the catalytic converter when it is suspect. Note that newer vehicles use an AFR sensor (mounted before the catalytic converter). It replaces the typical zirconia O_2 sensor, and has a wider range of detection. This allows the ECM to better control the air–fuel mixture.

▶ Crankcase Emission Control

LO 53-08 Explain the types and operation of the positive crankcase ventilation system.

While the engine is running, some gases from the combustion chamber leak past the piston rings and the cylinder walls. They leak down into the crankcase. This leakage is called **blowby**. To prevent pressure buildup during operation, the crankcase must be ventilated. Unburned fuel (HCs) and water from condensation also find their way into the crankcase. When the engine reaches its full operating temperature, the water and fuel in the crankcase evaporate. In older vehicles, crankcase vapors were vented directly to the atmosphere through a **breather tube** or road-draft tube. It was shaped so that air flowing past it while the vehicle was being driven helped draw the vapors from the crankcase. This resulted in blowby gases being vented directly to the atmosphere.

Modern vehicles are required to direct crankcase blowby gases and vapors back into the intake manifold. There, they can be returned to the combustion chamber to be burned. A common method of directing blowby gases back to the intake is through the **PCV** system (**FIGURE 53-18**). The PCV system regulates the flow of blowby gases between the crankcase and the intake manifold.

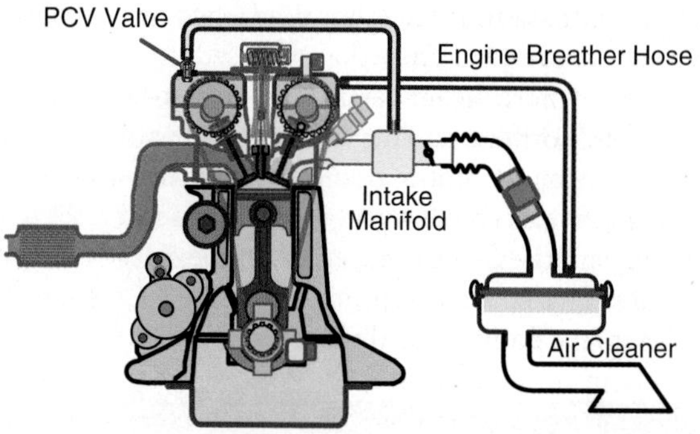

FIGURE 53-18 PCV systems draw blowby gases out of the crankcase to be burned in the combustion chambers.

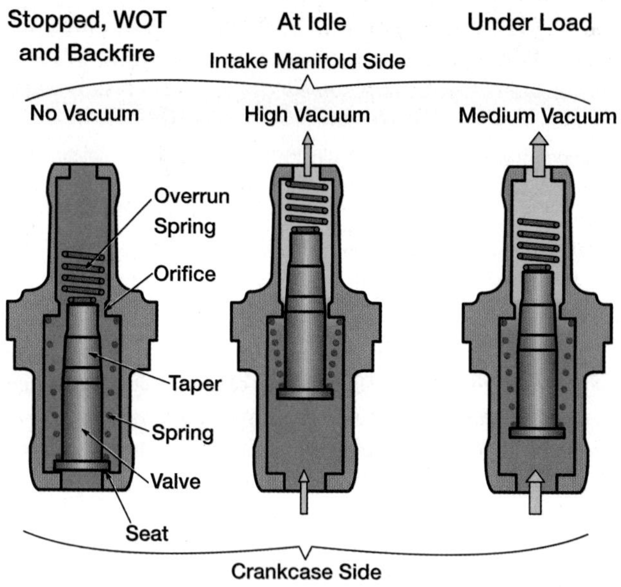

FIGURE 53-19 A variable-orifice PCV valve meters the amount of blowby gases, depending on engine load.

▶ Types of PCV Systems

There are three main types of PCV systems: the **variable orifice**, **fixed orifice**, and **separator** types. All three types do the same job. They ventilate blowby gases back to the intake manifold to be burned in the combustion chamber.

In a variable orifice–type system, a replaceable, spring-loaded **PCV valve** regulates gas flow (**FIGURE 53-19**). The position of the PCV valve is controlled by the pressure in the manifold. With the engine off, the spring holds the PCV valve in the closed position, and air cannot enter the inlet manifold. This allows the engine to start. At idle, relatively high intake manifold vacuum draws the PCV valve to the other end of the PCV valve's housing. There, it allows only a small, measured amount of blowby gases and air past the PCV valve. At wider throttle openings, the PCV valve plunger position allows maximum flow through the PCV valve's housing. This gives maximum crankcase ventilation. It tends to match the higher amount of blowby gases under that condition.

The PCV system is designed to remove more air than just blowby gases. So there should almost always be more ventilation capacity than the amount of blowby. The additional air that the system removes comes from a fresh air intake hose or tube. It is usually attached to the air cleaner assembly. The fresh air intake hose directs filtered air to one side or end of the crankcase. This intake point is usually as far as possible from the PCV valve. This ensures that as much of the blowby gases are pulled from the crankcase as possible, giving good ventilation.

The PCV system has another purpose. It prevents a buildup of pressure in the crankcase. Modern PCV systems are the closed (sealed) type. This means both ends of the PCV system are connected to the intake system. The PCV valve is connected directly to the intake manifold below the throttle plate. Whereas, the fresh air hose is connected to the air intake above the throttle plate. Having a closed system means any gases that cannot be handled through the vacuum side of the system are directed back through the fresh air connection to the air cleaner assembly. There, they are drawn into the intake airstream and burned in the combustion chamber (**FIGURE 53-20**).

The fixed orifice–type PCV system usually involves a screw-in fitting with a small hole drilled in it. The hole creates a predetermined vacuum leak. It draws a predetermined amount of crankcase vapors from the crankcase. Engineers have factored the vacuum leak into the engine management system to ensure the correct air–fuel mixture. The fixed orifice is typically mounted on or in the intake manifold (**FIGURE 53-21**).

The separator type of PCV system involves a valve that is hooked to the pressure side of the crankcase. An oil return line is connected to the bottom of the valve. And a suction line is connected on the other side. Oil that is mixed with blowby gases enters the separator. The heavy oil tends to fall out of the mixture to the bottom of the valve and return to the crankcase (**FIGURE 53-22**). The separator type is used on turbocharged applications. This is because turbocharged systems pressurize the intake system and would close off a standard PCV valve.

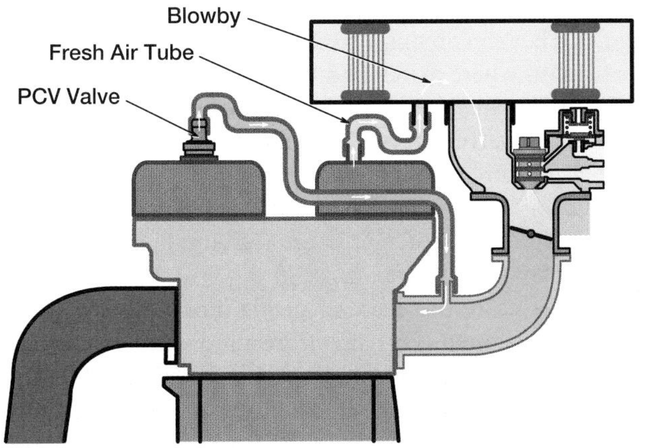

FIGURE 53-20 During heavy acceleration, excess blowby gases vent through the fresh air tube to the air cleaner assembly.

▶ Exhaust Gas Recirculation System

LO 53-09 Explain the operation of exhaust gas recirculation system.

The EGR system controls oxides of nitrogen. They are created in large amounts once combustion temperatures reach approximately 2500°F (1371°C). This can happen when the vehicle is under moderate to heavy loads and the AFR is at stoichiometric or leaner. EGR systems were designed to allow a measured amount of exhaust gases to enter the intake manifold during those times (**FIGURE 53-23**). As with all vehicle technology, EGR systems have evolved over time. Instead of using an EGR valve, some newer engines use VVT. They

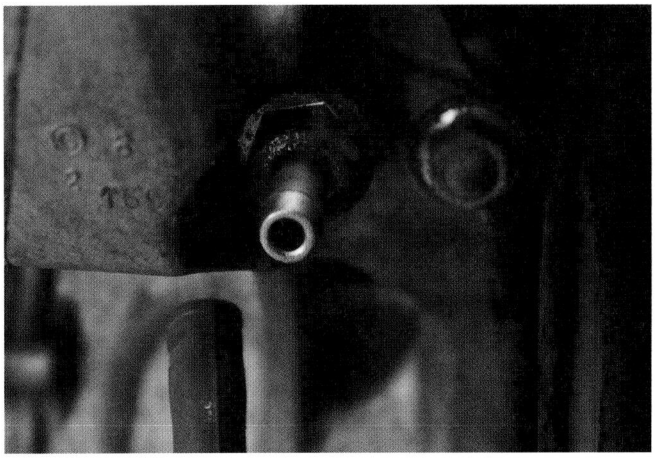

FIGURE 53-21 Typical fixed orifice passageway for that type of PCV system.

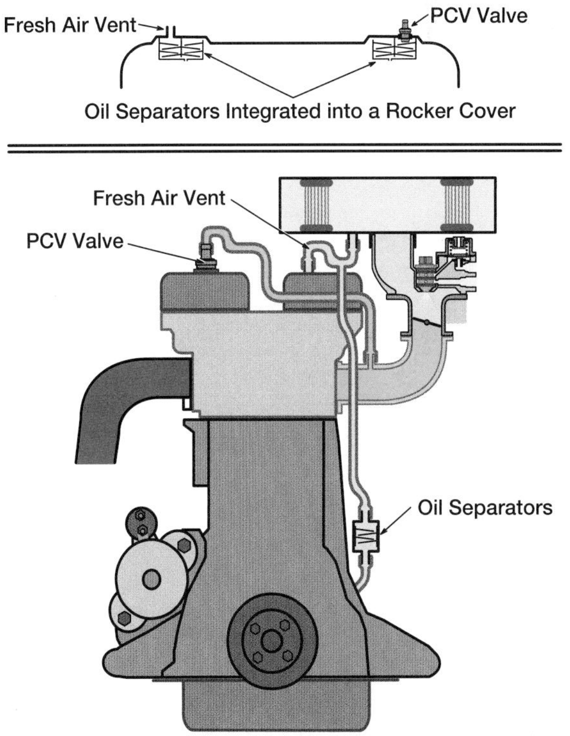

FIGURE 53-22 Separator type of PCV system.

FIGURE 53-23 The exhaust gas recirculation (EGR) system was designed by automotive engineers in the 1970s to control the emission of oxides of nitrogen. Nitrogen is oxidized in large amounts once combustion temperatures reach approximately 2500°F (1371°C).

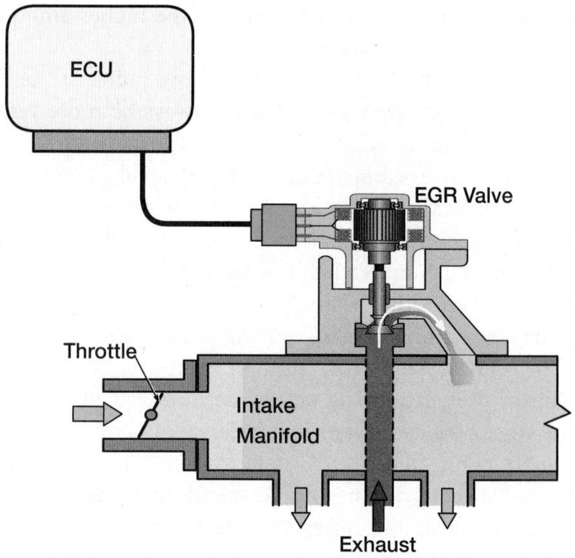

FIGURE 53-24 Typical computer-controlled EGR system.

draw some exhaust gases back into the cylinder through the exhaust valve, which is held open a little longer into the intake stroke than is typical.

TECHNICIAN TIP

In the combustion chamber, higher combustion temperatures tend to create more power. But high temperatures also create very high amounts of oxides of nitrogen. One way to overcome this is by understanding the difference between high peak combustion temperature and high average combustion temperature. High peak temperatures last for very little time. They produce a sharp spike in pressure. But it does not last long. Also, it tends to give a hammer blow to the piston, which is not very effective in pushing it down the cylinder. High average combustion temperature maintains high average pressure, which is good for a longer, smoother push of the piston. It produces good power while minimizing the creation of oxides of nitrogen.

Purpose and Operation

An EGR valve connects a passage from the exhaust port, or manifold, to the intake manifold (**FIGURE 53-24**). If engine operating conditions are likely to produce oxides of nitrogen, the EGR valve opens. This lets some burned exhaust gases pass from the exhaust into the intake system. During combustion, an amount of fresh air–fuel mixture is displaced by inert exhaust gases that do three things:

- First, they make it so there is not quite as much burnable mixture. This helps keep the temperature from rising as high as it otherwise would.
- Second, because they are inert and mix thoroughly with the rest of the air–fuel mixture, they cause the mixture to burn a bit more slowly. Therefore, it burns cooler, as the flame front must travel farther to burn around the inert molecules.
- Third, the inert gases absorb some of the heat of the burning gases.

These three situations lower the peak combustion temperature, thus reducing the formation of the oxides of nitrogen.

The EGR valve only opens under the conditions that create oxides of nitrogen. Oxides of nitrogen are not created in a cold engine. So, the EGR valve is kept from opening until after the engine achieves its normal operating temperature. The EGR valve also is not opened at idle. This is because there is only a small amount of air–fuel mixture being burned. So the combustion temperature is well below 2500°F (1371°C).

There is one condition when oxides of nitrogen are being created in large amounts and the EGR valve may not operate. This is when the vehicle is operated at wide-open throttle (WOT). In this condition, the EGR system is typically shut off. This way, full engine power is available for emergency needs such as avoiding an accident. But most passenger vehicles do not operate very often or for very long at WOT. So the overall amount of oxides of nitrogen created is minimal.

Some manufacturers have been able to accomplish the same reduction in oxides of nitrogen by using VVT on their engines. They can do this instead of the traditional EGR system. In situations where oxides of nitrogen can be created, the PCM delays the exhaust camshaft. This causes the exhaust valves to be held open during the first part of the intake stroke. The lower cylinder pressure pulls some of the exhaust gases back into the combustion chamber (**FIGURE 53-25**). These exhaust gases are used to dilute the air–fuel mixture with inert gases just like an external EGR system does.

In this case, the PCM controls how long the exhaust valve remains open. This affects the amount of exhaust gases that are retained in the cylinder. This is sometimes referred to as an internal EGR system. And the beauty is that it uses only the VVT components, which are used for other fuel economy and drivability reasons. So there are fewer components to install and go bad.

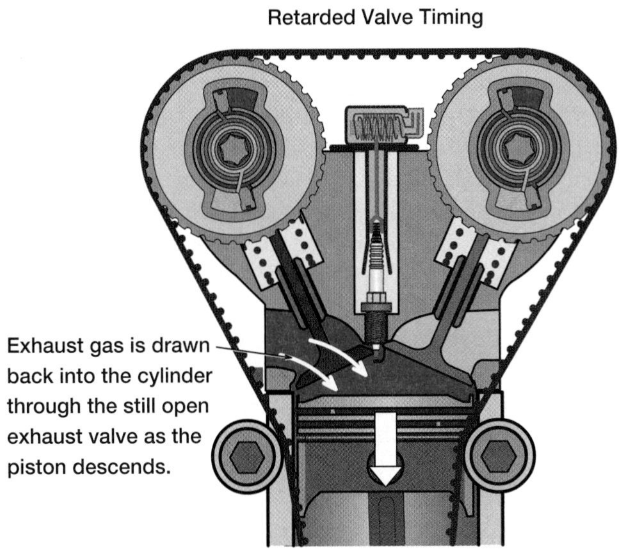

FIGURE 53-25 Variable valve timing can be used instead of an EGR valve to draw exhaust gases back into the combustion chamber.

▶ TECHNICIAN TIP

Many people believe that all emission control devices kill engine power and use extra gasoline. Although there may have been a small amount of truth to that when some of the systems were first introduced, that is almost never the case any longer. In the case of the EGR system, exhaust gases are typically recirculated only during part throttle operation. If any power is lost at that time, the throttle can be opened slightly to make up for it. And when the throttle is wide open, the EGR system typically does not operate, so there is no impact at that time. Finally, it could be argued that the extra inert gases from the EGR system help maintain a higher average cylinder pressure, which makes up for less air–fuel entering the cylinder.

▶ Evaporative Emission Control

LO 53-10 Explain the operation of evaporative emission system.

The EVAP system is designed to ensure that HCs are not released into the atmosphere. This occurs when fuel in the fuel tank begins to vaporize and build pressure. The EVAP system uses a charcoal canister and a series of hoses and valves. They capture the HC vapors that would normally be lost from the fuel tank. The system then delivers them to the intake manifold to be burned in the engine's combustion chambers (**FIGURE 53-26**).

The EVAP system stores the fuel vapors in the charcoal canister. The PCM determines when they can be burned in the engine without affecting the drivability of the vehicle. This is typically when the vehicle is driving down the road between certain speeds and within a certain engine operating temperature range. Once the PCM determines that the conditions are right for burning off the vapors, it will control the rate the vapors are released to the engine. The EVAP system also incorporates a monitoring system. It ensures that there are no leaks in the fuel storage system. It also verifies that the vapors are being delivered, as designed, to be burned.

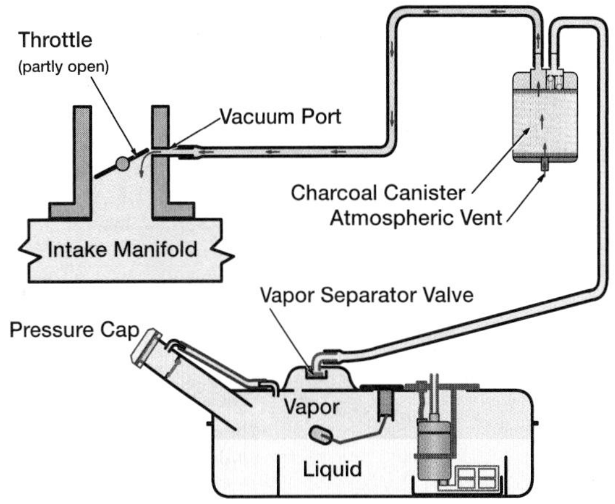

FIGURE 53-26 EVAP system.

Fuel Storage

Before 1971, vehicles vented the fuel tank through the filler cap into the atmosphere. As fuel in the tank vaporized, vaporous HCs would escape from the filler cap and the carburetor. This is a waste of perfectly good fuel as well as a pollutant of the air we breathe. Non-vented filler caps were designed to help prevent the release of these vapors. But the system has to accommodate changes in fuel tank level. Today, a **vacuum relief valve** somewhere in the system allows air to enter the fuel tank. It relieves the low pressure as the fuel is drawn out of the tank or when the fuel contracts as the temperature drops. The vacuum relief valve also stops the fuel tank from collapsing in these situations.

The fuel cap may also incorporate a **pressure relief valve**. If the fuel tank's internal pressure exceeds the set value of the relief valve, the pressure relief valve releases a small amount of pressure. This prevents the fuel tank from rupturing. Some modern caps have no valves at all. They are completely sealed to stop the entry of air and water into the fuel tank. They also stop the release of fuel vapor. Modern tanks also contain an **expansion volume**. It is an air chamber that allows for the expansion of liquid fuel on hot days (see Figure 53-26). It can be either directly built into the top of the fuel tank. Or it can be included as a separate chamber connected to the fuel tank by tubing.

A liquid–vapor separator may be connected to the fuel tank by a number of tubes. This separator allows a space for liquid fuel to separate from the vapors and return to the fuel tank (**FIGURE 53-27**). This separation ensures that the liquid fuel does not get pushed into the charcoal canister and ruin it. A **vapor line** is connected to the vapor space in the fuel tank or the liquid vapor separator. It carries fuel vapors from the fuel tank to a storage container, called a charcoal canister. This vapor line can incorporate a check valve. It is a valve with a steel or plastic ball in it. The check ball moves into the passageway to block fuel if the vehicle is tilted too far from the horizontal and rolls over. The check valve stops liquid fuel from leaking and creating a fire hazard. Thus, it is sometimes called a rollover valve.

Charcoal Storage

A charcoal canister is a canister filled with **activated charcoal** (**FIGURE 53-28**). "Activated" means the charcoal is highly

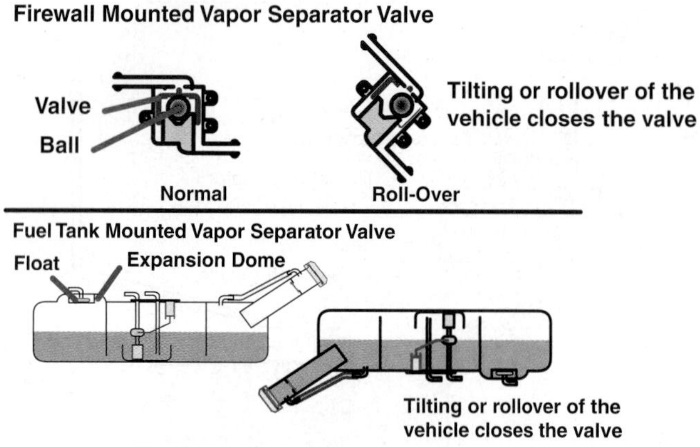

FIGURE 53-27 Liquid vapor separator.

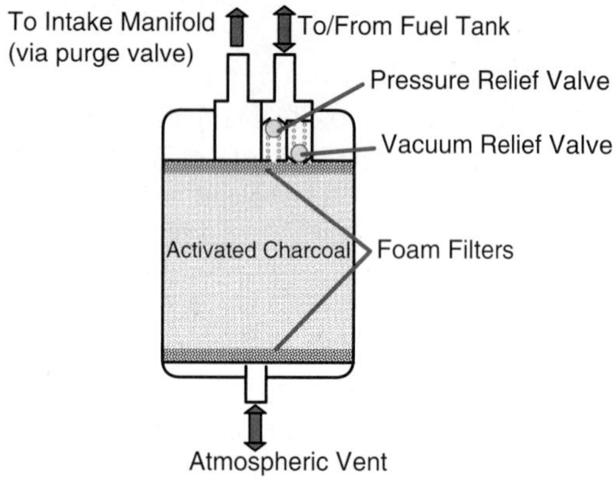

FIGURE 53-28 Charcoal canister.

porous. It has a very large surface area that can absorb and store large quantities of fuel vapor. The charcoal canister is connected to the fuel tank vapor line. The charcoal absorbs excess fuel vapor from the fuel tank, primarily when the engine is not running. In some earlier carbureted engine designs, the canister also has a vapor line connected to the **carburetor float bowl**. It can then absorb any fuel vapors from it.

The charcoal canister is designed to store the fuel vapors for a limited time before it must be purged. **Purging** is the process of drawing fresh air up through the activated charcoal (**FIGURE 53-29**). The purpose is to pull the fuel vapors from the charcoal and then burn them in the engine. Purging is designed to normally occur when the engine is above a minimum temperature and has run for a specified time. This system prevents the engine's AFR from being thrown rich or lean by the flow of purged gases while the engine is warming up. Purging is an important process, as it prepares the canister to absorb fuel vapors the next time the vehicle is not running. Because all air that enters the engine must be filtered, most charcoal canisters have an air filter installed on their air intake. This filter is replaceable on some charcoal canisters.

The rates of evaporation are not the same for all fuels. Evaporation is higher with gasoline than diesel because gasoline is more volatile. If a fuel is more volatile, it is able to evaporate quickly and easily. Volatility is important for cold weather operation but causes excess vapor in warmer weather. For this reason, gasoline fuels are rated for winter or summer use, as the winter fuel must be more volatile due to the cooler atmospheric air temperature. Thus, a problem arises when a vehicle has a winter blend of gasoline in the tank and an unexpectedly hot spring day happens. The charcoal canister can then become overwhelmed with more vapor than it can handle. It may then allow some of the vapor to escape past the canister to the atmosphere.

Manufacturers size charcoal canisters to a particular amount of fuel tank capacity and fuel volatility. In fact, some vehicles with large fuel tanks require the use of two or more charcoal canisters. This is needed to provide enough fuel vapor capacity for the vehicle.

▶ Heated Air Intake Systems

LO 53-11 Explain the operation of heated air intake systems.

Because liquid fuel will not burn, fuel must be in a vapor form for it to burn. So the challenge is to get liquid fuel to fully vaporize by the time it has to be ignited, but not much sooner, and certainly not later. Intake manifolds carry the air–fuel mixture. Fuel tends to condense on the inside surface of the cold manifold.

One method of minimizing this problem is to use a **heated air intake system**. This system collects hot air from around the exhaust manifold. Then, it mixes it with outside air entering the air cleaner assembly (**FIGURE 53-30**). The system maintains a specified air temperature of the air entering the engine. This simple system uses a temperature-sensitive valve inside the air cleaner. It operates a flap that blends the hot air with cool air so that the intake manifold receives air at about 100–110°F (38–43°C), regardless of the outside air temperature. Maintenance of this consistent temperature assists in the proper **vaporization** of the fuel, particularly when the engine is cold.

Vaporization is also assisted in some engines, such as in the Toyota Prius, by circulating stored hot liquid coolant

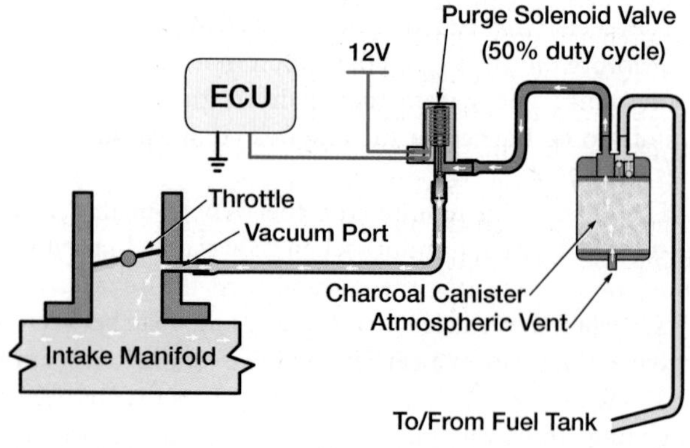

FIGURE 53-29 Purging the canister occurs when air is drawn through the charcoal to pick up fuel vapors on the way to the combustion chamber.

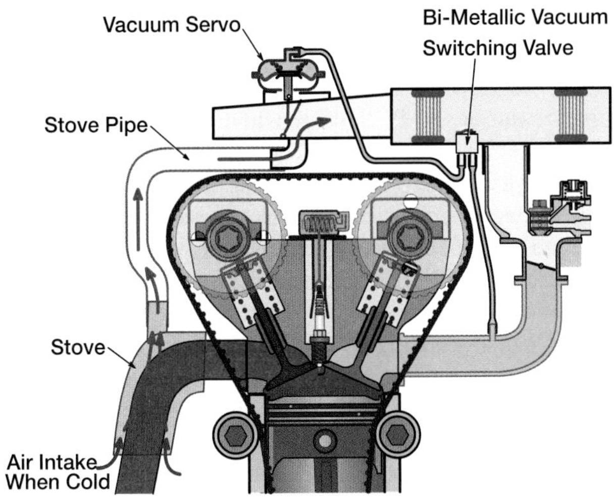

FIGURE 53-30 Heated air intake systems use the air around hot exhaust manifolds to blend with outside air, to supply warm air at a predetermined temperature.

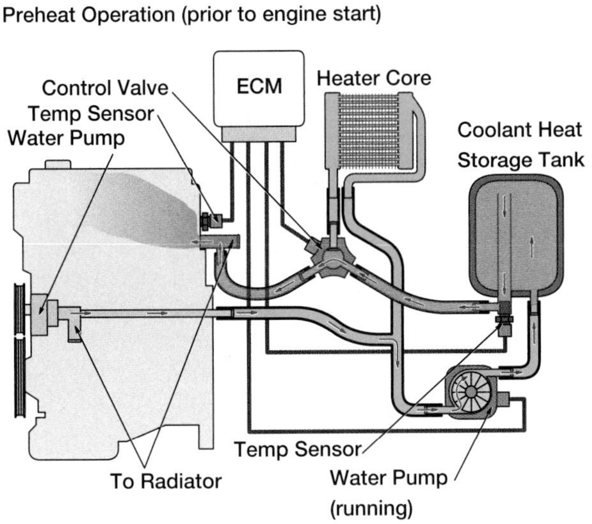

FIGURE 53-31 The Toyota coolant heat recovery system is used to preheat the engine during cold start.

through passages in the intake manifold and cylinder head. This preheats those parts right before the engine is started (**FIGURE 53-31**). Another past method of preheating fuel and air mixtures was to use a heated grid under the throttle body. It assisted in heating air in order to promote the vaporization of the fuel (**FIGURE 53-32**). Fuel will not mix easily in cold air temperatures, so the use of a heating grid assures a more combustible fuel mixture.

▶ Diagnosing PCV-Related Concerns

LO 53-12 Inspect and test the PCV system.

A PCV system is used to remove blowby gases from the crankcase in an environmentally friendly manner. If blowby gases

FIGURE 53-32 An electric grid under the throttle body was used on some vehicles to heat the incoming air.

are not removed from the crankcase, pressure can build up and cause oil seals and gaskets to leak oil. Replacement of seals or gaskets might temporarily fix the problem. But excessive pressure will push the seal or gasket out of place again. Proper testing of the PCV valve system is critical to get to the root cause of the concern and avoid repeat failures.

To diagnose PCV-related concerns, first locate the PCV valve. Refer to the service information for location of the PCV valve and hoses. If the PCV valve is located in the valve cover, remove the PCV valve from the cover. Ensure that a strong vacuum is coming through the valve with the engine running. This ensures the PCV valve and hose are clear. Reinstall the PCV valve into the valve cover, and remove the breather tube from the air cleaner assembly. With the engine still running, check that vacuum builds up in the breather tube (and crankcase). This indicates that the breather tube is clear and there are no excessive blowby gases. Inspect all hoses for softening, brittleness, cracks, kinks, and holes. Cracks in hoses can result in vacuum leaks, which can create a poor idle situation. Inspect the PCV valve and ensure it has the correct part number for the vehicle. If the incorrect part is found or the valve is sludged up or sticky, replace it with a new PCV valve.

▶ Inspecting and Servicing the PCV System

The PCV system draws crankcase gases out of the crankcase. It burns them in the engine to reduce crankcase emissions. It is an important system that has to be serviced on a regular maintenance interval. This is determined by the manufacturer and published in the service information.

PCV failures can occur and potentially create drivability problems for the customer. The most common problem is when the PCV system has a vacuum leak. The hose connecting the PCV valve to the intake manifold can crack and create a vacuum leak. Vacuum leaks tend to cause a lean air–fuel mixture and create a rough idle. This is most noticeable when the vehicle is stopped at a stop light. It also causes the long-term fuel trim (LTFT) to increase, the worse the leak.

PCV systems can also become clogged. Over time, especially if the vehicle's oil changes are neglected, the PCV valve or hose can plug up with sludge. If this happens, the PCV system cannot ventilate the crankcase. Blowby gases will contaminate the oil further, which creates more sludge throughout the engine. Another common failure from a clogged PCV valve or hose is oil being pushed up into the air filter housing. If oil is found in the air filter housing, the PCV system is either being overwhelmed by the blowby gases (the piston rings are worn out). Or the suction side of the PCV system is clogged/restricted. This causes pressure to build up in the crankcase. Oil mist is then forced up the fresh air hose to the air cleaner housing. In this case, check that there is strong vacuum present at the PCV valve when the engine is idling and that the valve is not restricted. Also, test the crankcase pressure, using a blowby gauge.

Repeat engine oil leaks may be the result of excessive pressure in the crankcase. If this pressure cannot be released, it will continue to build. Eventually it will either push oil past the seals or in some cases even push oil seals and gaskets out of position, causing an oil leak. In the event of multiple gasket or seal failures or repeat failures, be sure to check the PCV system. Make sure it is operating correctly by testing for the proper PCV system flow.

To test the PCV system, follow the steps in **SKILL DRILL 53-1**.

SKILL DRILL 53-1 Inspecting and Servicing the PCV System

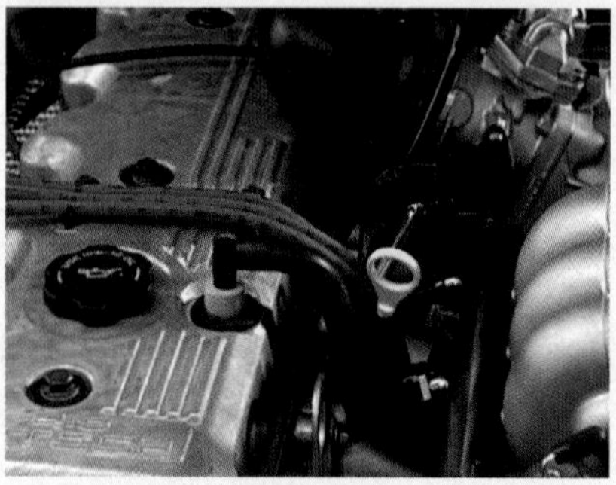

1. Locate the PCV valve. With the engine idling, remove the PCV valve, and check for the presence of a strong vacuum.

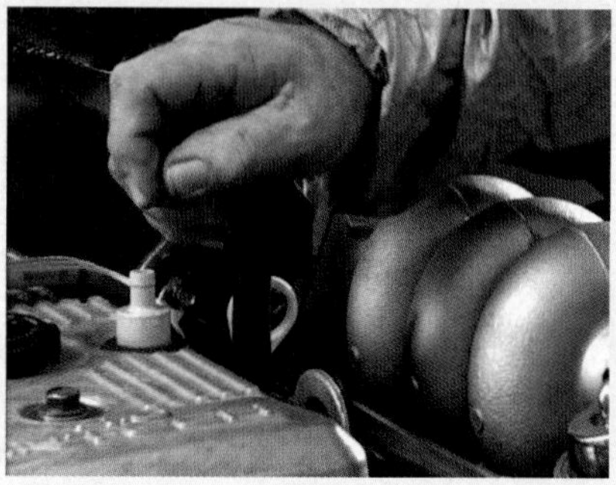

2. Remove the hose, and check that it is still pliable and not clogged with sludge deposits.

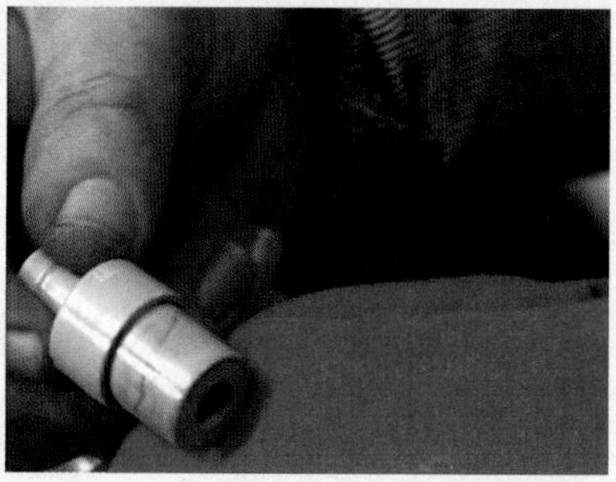

3. Remove the PCV valve, and inspect it for deposits. If there are any issues, replace it with a new one of the same type. If fixed orifice style, make sure the passageway is clean.

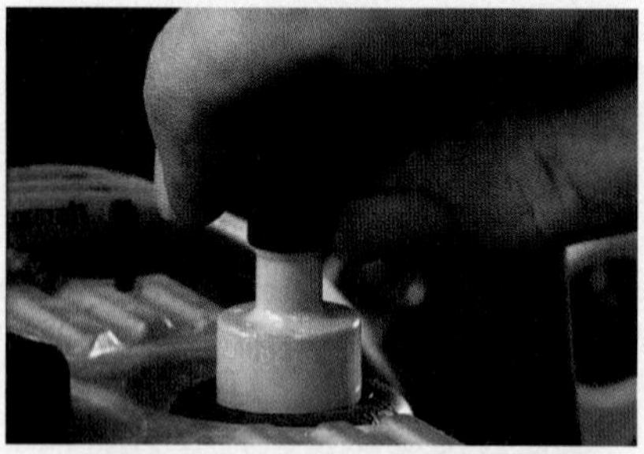

4. Reinstall the PCV valve into the valve cover.

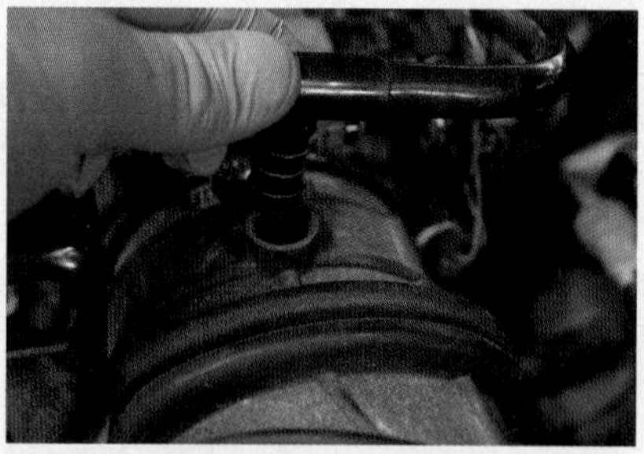

5. Remove the breather hose from the air cleaner assembly.

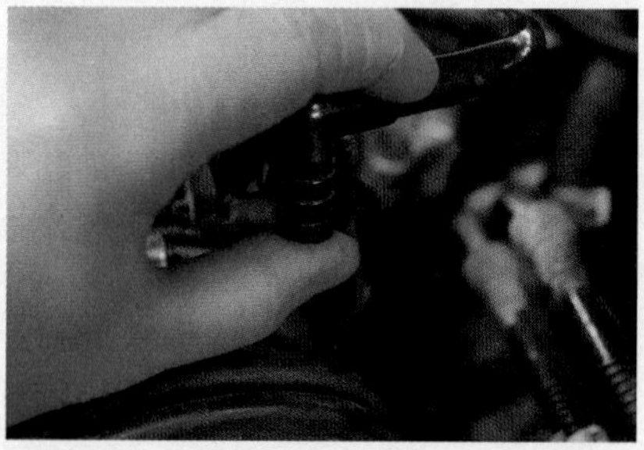

6. With the engine idling, block off the breather hose and feel for vacuum building up in the breather hose and crankcase. This indicates that the PCV system can handle the amount of blowby gases. There should also be a slight amount of measurable vacuum in the crankcase at idle, typically measurable at the dipstick tube as 0.5"–2" H_2O. This would indicate that the PCV system is operating and adequate for the blowby produced by the engine.

Wrap-Up

Ready for Review

- Atmospheric air is composed mainly of two gases: nitrogen and oxygen. Emissions can be due to forest fires, volcanoes, decomposition of natural materials, vehicles, or industrial sites.
- HC emissions are created when raw fuel evaporates into the atmosphere and when some or all of raw fuel does not burn in the combustion chamber and is exhausted out the tailpipe. Carbon monoxide is heavier than air and collects in low spaces such as lube pits. Incomplete combustion caused by either too much fuel or not enough air created carbon monoxide.
- Oxides of nitrogen are produced when combustion temperatures of around 2500°F (1400°C) and above occur. A lean mixture creates high levels of oxides of nitrogen. Sulfur dioxide mixes with water vapor formed during the combustion process and produces sulfuric acid, a corrosive compound. In spark-ignition engines, particulates are caused by incomplete combustion of rich air–fuel mixtures. In compression–ignition engines, particulates are caused by either too rich of a mixture or a poorly mixed air–fuel mixture.
- The following strategies can be used to reduce emissions: engine design to reduce emissions, air–fuel mixture control under all driving conditions, quick engine warm-up, integration of precombustion and postcombustion.
- Air-fuel ratio that is close to the stoichiometric point, produces very little HC and CO emissions. Electronic fuel injection and engine management systems can adjust the air–fuel mixture based on the information from various sensors.
- Two primary precombustion emission control systems are: the heated air intake system and the exhaust gas recirculation (EGR)system. Three primary postcombustion systems are: the positive crankcase ventilation (PCV) system, the secondary air injection system, and the catalyst system.
- Catalytic converters perform a final cleanup of the tailpipe emissions. Types of catalyst include: reduction catalyst and oxidizing catalyst. Two-way catalytic converter converts HCs and CO to carbon dioxide and water. Three-way catalytic converters contain both a reduction catalyst and an oxidizing catalyst.
- The PCV system regulates the flow of blowby gases between the crankcase and the intake manifold. Three main types of PCV systems: the variable orifice, fixed orifice, and separator types. All three types ventilate blowby gases back to the intake manifold to be burned in the combustion chamber.
- An EGR valve connects a passage from the exhaust port to the intake manifold to control oxides of nitrogen. The EGR valve opens when engine operating conditions are likely to produce oxides of nitrogen.
- The EVAP system uses a charcoal canister and a series of hoses and valves to capture the HC vapors that would normally be lost from the fuel tank. The EVAP system also incorporates a monitoring system that ensures that there are no leaks in the fuel storage system. It also verifies that the vapors are being delivered to be burned.
- Heated air intake system collects hot air from around the exhaust manifold and mixes it with outside air entering the air cleaner assembly. The system uses a temperature-sensitive valve inside the air cleaner.
- Inspection of PCV systems should include checks for clogs, oil leak around seals and gaskets, and correct part number. The hose connecting the PCV valve to the intake manifold should be checked for cracks in case of a vacuum leak.

Key Terms

activated charcoal Charcoal that will absorb large amounts of vapor; used in charcoal canisters to store and release fuel vapors.

blowby Pressure that leaks past the compression rings during compression and combustion.

blowby gas The result of combustion gases leaking past the compression rings and getting into the crankcase.

breather tube A tube used in the PCV system to allow fresh air into the engine crankcase.

carbon monoxide poisoning Exposure to higher than tolerable levels of carbon monoxide, resulting in headaches, fatigue, or loss of consciousness, eventually resulting in death.

carburetor float bowl The part of the carburetor that holds fuel to be burned in the engine; the bowl is at a constant level of fuel to ensure adequate fuel is present during driving.

EGR (exhaust gas recirculation) system A system that recirculates a portion of burned gases back into the combustion chamber to displace air and fuel and cool combustion temperatures.

evaporative emission (EVAP) system A system used to capture vapors or gases from an evaporating liquid.

expansion volume An engineered space that allows for growth of the volume of a liquid as it heats and expands.

feedback system A system that uses feedback to adjust what it is doing. Typically, a module gives a command and waits to see if the command was followed by use of a sensor.

fixed orifice tube A restriction device with a metered hole in it and a metal or plastic screen to keep out any debris.

flame front The front edge of the burning air-fuel mixture in the combustion chamber.

heated air intake system A system that uses hot air from around the exhaust manifold to warm the air going into the intake manifold.

lambda The ratio of air to fuel at which all of the oxygen in the air and all of the fuel are completely burned; see also *stoichiometric ratio.*

PCV valve See *positive crankcase ventilation valve.*

photochemical smog A brown haze that hangs in the sky, typically seen over large cities. Smog is a major health issue to humans because it affects lung tissue.

positive crankcase ventilation (PCV) system A system that draws blowby gases from the crankcase into the intake, to be burned.

pressure relief valve A valve that is designed to release pressure if it gets above a calibrated pressure; this limits pressure in one part of a circuit or a complete system.

purging The process of pulling stored fuel vapors from the charcoal canister and moving them into the engine to be burned.

quenched The state in a combustion chamber in which the flame cannot burn because of cold surfaces or poor distribution of the fuel mixture.

scavenging effect A condition caused by moving columns of air, which create a low-pressure area behind them, that results in a pulling force that is used to pull the remaining burned gases from the combustion chamber. Valve timing affects the amount of scavenging effect an engine has.

three-way catalytic converter A converter that changes hydrocarbons, carbon monoxide, and oxides of nitrogen into harmless elements.

two-way catalytic converter A converter that changes only hydrocarbons and carbon monoxide into harmless elements.

vacuum relief valve A mechanical valve used on the gas cap that prevents a low-pressure condition from occurring. It also ensures that the fuel tank will not collapse due to fuel usage or contraction of fuel on cooldown.

vapor line A rubber or plastic line that carries vapors from the fuel tank to the charcoal canister.

vaporization Change of state from a liquid to a gas.

variable orifice PCV system A suste, om wjocj a replaceable, spring-loaded PCV valve regulates gas flow. The position of the valve is controlled by the pressure in the manifold.

separator PCV system A PCV system with a device that uses gravity to allow oil to fall to the bottom and be returned to the crankcase; the device prevents liquid from traveling to the intake manifold.

Review Questions

1. What percentage of oxygen is in the atmosphere?
   a. 20.95%
   b. 78.08%
   c. 0.93%
   d. 0.0410%
2. Which exhaust gas is a result of incomplete combustion, is colorless, odorless, tasteless, flammable, and highly toxic?
   a. HC
   b. CO
   c. CO_2
   d. NO_x
3. If nitrogen reacts under the high temperature and pressure of internal combustion, it can become _______.
   a. CO_2
   b. HC
   c. NO_x
   d. PM
4. What is the best location for a spark plug when manufacturers are designing combustion chambers for the lowest possible emissions?
   a. The intake side
   b. The exhaust side
   c. The wedge corner
   d. The center
5. A stoichiometric air/fuel ratio of 14.7:1 is equal to ________.
   a. 1.0 Lambda
   b. 100% Lambda
   c. 128 Lambda
   d. 0 Lambda
6. Which emissions system is designed to draw out blowby gases from the crankcase and send them back through the intake manifold?
   a. EGR
   b. PCV
   c. VVT
   d. EVAP
7. Inside a three-way catalytic converter, what is incoming NO_x converted to as it passes through the reduction catalyst?
   a. Nitrogen and oxygen
   b. Hydrocarbons and carbon monoxide
   c. Nitrogen and hydrocarbons
   d. Carbon monoxide and oxygen
8. In a typical fixed-orifice PCV system, what pulls the vapors from the crankcase?
   a. The vacuum pump
   b. The oil separator extractor
   c. Intake manifold vacuum
   d. Exhaust backpressure
9. How does the Exhaust Gas Recirculation system lower NO_x emissions?
   a. The catalytic converter is cooled which increases its chemical efficiency
   b. The exhaust gas is inert which slows and reduces the peak combustion temperature
   c. The exhaust backpressure is relieved which is more efficient for emissions
   d. The engine burns the fuel twice which makes it cleaner
10. In an EVAP system, where are fuel vapors stored?
   a. In the fuel tank
   b. In the intake container
   c. In the purge tank
   d. In the charcoal canister

ASE Technician A/Technician B Style Questions

1. Emissions as a result of combustion is being discussed. Technician A states that all of the byproducts of combustion are harmful. Technician B states that of all the gases in the atmosphere, the most common is oxygen. Who is correct?
   a. Technician A only
   b. Technician B only
   c. Both Technicians A and B
   d. Neither Technician A nor B
2. Controlling emissions is being discussed. Technician A states that one way to control emissions is to design an engine in a way that emissions are reduced as much as possible. Technician B states that engines should warm up as quickly as possible to lower emissions. Who is correct?
   a. Technician A only
   b. Technician B only
   c. Both Technicians A and B
   d. Neither Technician A nor B
3. The stoichiometric air/fuel ratio is being discussed. Technician A states that the stoichiometric ratio requires 14.7 parts of air to 1 part of fuel by volume. Technician B states that a ratio of 15.0:1 by weight is a lean mixture. Who is correct?
   a. Technician A only
   b. Technician B only
   c. Both Technicians A and B
   d. Neither Technician A nor B
4. Air–fuel ratio is being discussed. Technician A states that a mixture with too much air, and not enough fuel, is known as being rich. Technician B states that engines use many sensors to detect conditions that require adjustment of the fuel injector spray time. Who is correct?
   a. Technician A only
   b. Technician B only
   c. Both Technicians A and B
   d. Neither Technician A nor B
5. The catalytic converter is being discussed. Technician A states that the catalytic converter is designed to clean up the emissions of a poorly running engine. Technician B states that a modern catalytic converter's operation is monitored by an oxygen sensor before the catalyst, and one after the catalyst. Who is correct?
   a. Technician A only
   b. Technician B only
   c. Both Technicians A and B
   d. Neither Technician A nor B
6. A positive crankcase ventilation system is being discussed. Technician A states that some PCV systems contain an oil separator with an oil drain pipe leading back to the crankcase. Technician B states that older vehicles used to simply vent harmful blowby gases straight to the atmosphere. Who is correct?
   a. Technician A only
   b. Technician B only
   c. Both Technicians A and B
   d. Neither Technician A nor B
7. An EGR system is being discussed. Technician A states that the EGR valve is normally open on start up, and closes as the engine gets warm. Technician B states that some manufacturers have used VVT to accomplish the same reduction in NO_x as an EGR system. Who is correct?
   a. Technician A only
   b. Technician B only
   c. Both Technicians A and B
   d. Neither Technician A nor B
8. An evaporative emissions system is being discussed. Technician A states that the EVAP system protects the environment against evaporating hydrocarbons. Technician B states that the fuel caps used in today's vehicles are vented which allows air to flow in and out of the gas tank freely. Who is correct?
   a. Technician A only
   b. Technician B only
   c. Both Technicians A and B
   d. Neither Technician A nor B
9. A heated intake system is being discussed. Technician A states that the system uses hot engine oil to warm the intake manifold. Technician B states that engine coolant can be used to heat the throttle body. Who is correct?
   a. Technician A only
   b. Technician B only
   c. Both Technicians A and B
   d. Neither Technician A nor B
10. A PCV system is being discussed. Technician A states that PCV hoses should be visually inspected for softening, brittleness, cracks, kinks, and holes. Technician B states that a symptom of a clogged PCV valve could be various leaking engine oil seals and gaskets. Who is correct?
    a. Technician A only
    b. Technician B only
    c. Both Technicians A and B
    d. Neither Technician A nor B

CHAPTER 54

Alternative Fuel Systems

Learning Objectives

- **LO 54-01** Identify the types of alternative fuels.
- **LO 54-02** Describe the problems alternative fuels address.
- **LO 54-03** Describe vehicle emissions and emission standards.
- **LO 54-04** Describe battery electric vehicles.
- **LO 54-05** Describe the types of hybrid and electric vehicles.
- **LO 54-06** Describe hybrid drive configurations, enhancements, and operation.
- **LO 54-07** Describe hybrid and electric vehicle service precautions, PPE, and tools.
- **LO 54-08** Disable the high-voltage system, service 12-volt battery.

ASE Education Foundation Tasks

See Appendix A to view the 2017 ASE Education Foundation Automobile Accreditation Task List Correlation Guide.

▶ Introduction

LO 54-01 Identify the types of alternative fuels.

The use of alternative fuels in the internal combustion engine (ICE) is nothing new. Back between 1890 and 1896, Henry Ford designed one of his first vehicles—the Quadricycle. It ran on ethanol derived from local farm crop waste (Ford's Model T also ran on ethanol) (**FIGURE 54-1**). Some of the earliest automobiles and trucks operated on electricity. Many used external combustion steam engines fueled with paraffin. Others used ICEs running on "town gas" produced from coal.

Petroleum replaced almost all other options for motor fuel used in light-duty road vehicles. There were many reasons (both political and practical) for this as we will see below. Petroleum-based fuels have dominated the transportation sector for the last century. This means that people are used to the long-standing use of gasoline. Some people wonder why we would ever switch back to alternative fuels for powering our road vehicles.

▶ Alternative Fuels

What are "alternative fuels," alternative fuel vehicles (AFVs), and alternative energy sources? An **alternative fuel** is essentially anything other than a petroleum-based motor fuel (gasoline or diesel fuel) that is used to propel a motorized vehicle. Alternative fuels may be liquid or gas (or electric). They contain a variety of latent heat energy for use in the ICE. An AFV, then, is a vehicle powered by something other than petroleum. One example is an electric vehicle. It is fueled by electricity generated in a variety of ways (more on this later). Here is a list of some of the recognized AFVs currently produced:

- Flexible-fuel vehicles (FFVs)
- Dedicated and bi-fuel compressed natural gas vehicles (CNGVs)
- Liquefied natural gas (LNG) vehicles
- Liquid petroleum gas (LPG, also called propane) vehicles
- Hydrogen-powered ICE vehicles
- **Battery-electric vehicles (BEVs)**
- Hybrid electric vehicles (HEVs)
- Hydrogen fuel cell (electric) vehicles (FCVs)

Some of the alternative fuels include:

- Ethanol (alcohol)
- Methanol (alcohol)
- Biodiesel
- Methane: CNG and LNG
- Liquefied petroleum gas (LPG, also called propane) from methane
- Synthetic fuels
- Electricity (not a fuel, per se, but an energy source)

What has prompted the renewed interest in alternative fuels? Reasons for making the switch from time-proven petroleum to "alternatives" are widely debated and include the following:

- Enhancing energy security by using domestically sourced fuels/energy
- Reducing the export of funds to overseas oil producers
- Reducing pollution from vehicle emissions
- Stimulating and bolstering the national and local economies (agri-business)

Boiled down, alternative fuels advance the following "3E" causes:

- Energy security
- Environmental concerns
- Economy

Let's look at each of these in more detail.

FIGURE 54-1 One of the first alternative fueled vehicles—Henry Ford's Quadricycle.

You Are the Automotive Technician

A Toyota Prius hybrid, which is a high-voltage vehicle, is brought into the shop for a regular 120,000-mile service. It requires replacement of the spark plugs and other maintenance items. You know that you need to follow the latest safety precautions when servicing a high-voltage vehicle. First, you do some research on the Toyota website and locate the service information for this make and model. Next, you create a service plan to prepare the workspace for safety. You then determine which tools are needed and which technicians are trained to perform the work.

1. Why is it important to read the Toyota service information for the hybrid vehicle prior to servicing?
2. Why is it necessary to inform or work with someone else while you are servicing a high-voltage vehicle?
3. What is the standard personal protection equipment (PPE) and equipment needed to work on high-voltage vehicles?
4. Why must the main high-voltage battery pack be disconnected before servicing a high-voltage vehicle?

▶ Energy Security

LO 54-02 Describe the problems alternative fuels address.

National security is a key issue in the United States, especially after the 9/11 terrorist attacks. For years, the United States has been importing oil from other countries. Oil importation for transportation has been as high as 70%. President George W. Bush said in his 2006 State of the Union Address that "Americans are addicted to oil" (**FIGURE 54-2**). Foreign petroleum stakeholders have large petroleum energy reserves. Some of whom are waging war against the United States. In an attempt to shift the playing field after the oil embargo of the 1970s, the United States Congress took action. In 1992, Congress passed the **Energy Policy Act (EPAct92).** The intent of which was to advance, through mandates, the use of vehicles capable of using alternatives to gasoline and diesel. EPAct92 considered the previously listed vehicles (and fuels) as legitimate AFVs. In 1996, EP Act was amended to include synthetically derived fuels such as coal-to-gas liquids (derived from the Fischer–Tropsch method), bio-based fuels, and more.

Unfortunately, EPAct92 failed to account for a very important fact. An alternative fuel **infrastructure** did not, for the most part, exist in the United States. So, federal, municipal, and private fleets were mandated to include AFVs as a certain percentage of their new vehicle purchases. But alternative fuels were not widely available for use in these flexible-fueled vehicles. Examples of these alternative fuels are ethanol and methanol. This means that they can use either alcohol or gasoline as their primary fuel. The Renewable Fuels Standard and other legislation helped to get more alternative fuel options. These options were available for many vehicle types. These include on-road and off-road vehicles (agricultural, railroads, etc.) and for marine and aviation applications. Today the U.S. military is a driving force in the move toward alternative fuels.

It is also important to consider this next point. Petroleum is not considered a renewable resource. That means that there is a finite amount of it. According to peak oil advocates, eventually it will be hard to extract it from proven easily accessible reserves. Then, the cost will become prohibitive. As sweet light crude oil becomes hard to find, crude oil will become more expensive. Developing countries like India and China are seeing an increased number of automobiles on their roads. So, the worldwide demand for oil has caused sharp rises in the price of crude oil. This has recently been tempered by the shale oil discoveries in the United States. Oil has become a valuable "black gold" commodity on the world market, making energy a strategic resource.

▶ Environmental Concerns

Consider the pollution created by the oil industry. They need to drill for oil, refine it, and transport the end products to the local filling station for sale. Once burned in the ICE, the resulting tailpipe emissions must be regulated. "Well-to-wheels" emissions are now being studied and regulated, to some degree or another, in most countries of the world.

Greenhouse gas emissions are a worldwide concern. Global climate change is widely discussed. Many countries regulate pollutants. Photochemical smog has created health issues for the young as well as for the elderly (**FIGURE 54-3**). Automobiles burning gasoline or petrodiesel fuel create pollutants. They are responsible for harm to humans and the environment. So transitioning to cleaner alternative energy and fuel choices can help reduce the effects of pollution. We explore the various pollutants linked with petroleum burned in automobile engines later on. We will then see why alternative choices make sense.

▶ Economic Concerns

To offset the unfavorable balance of U.S. dollars sent overseas for petroleum, "home-grown" bio-based fuels are being used. Congress has passed legislation requiring the use of alternatives to gasoline. And some states require that ethanol be blended at the pump with gasoline. In this way, it helps their farm economies. Nationwide, gasoline is widely blended with 10% ethanol (mostly corn-based ethanol). Yet, it still does not meet targets of nonpetroleum usage. Congressional targets for alternative fuel usage are difficult to reach. Producers have hit the "blend wall." This means that ethanol supplies now outweigh market demand.

The Environmental Protection Agency (EPA) has approved the use of 15% ethanol for 2001 and newer vehicles. This should

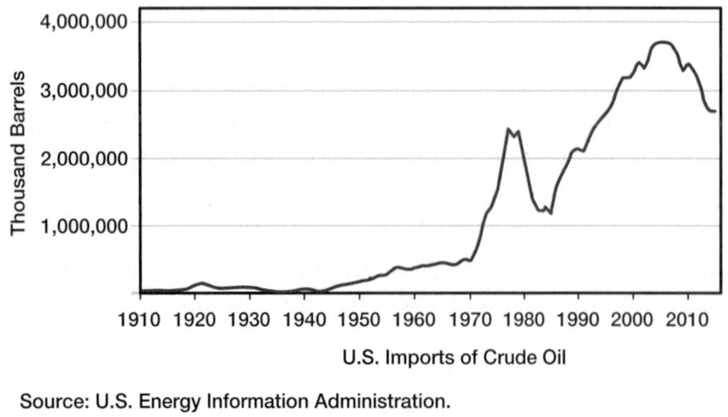

FIGURE 54-2 U.S. imports of crude oil have gone down as domestic oil production and alternative fuel production have gone up.

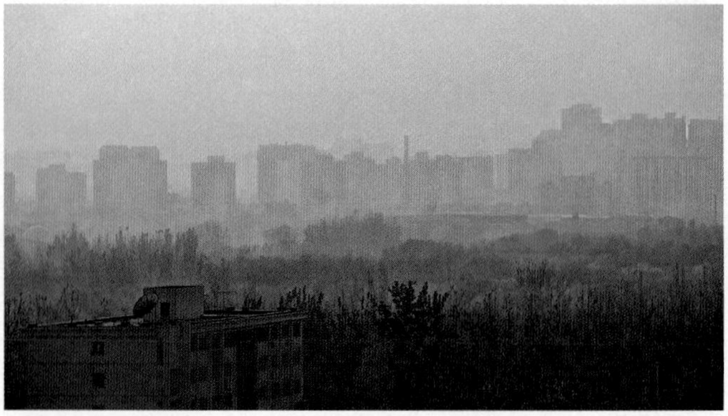

FIGURE 54-3 Smog is still a problem in parts of the world. Alternative fuels help reduce the problem.

help both ethanol suppliers and the environment. Corn prices per bushel have risen due to increased demand for fuel and for export overseas. The coproducts of ethanol production, such as dried distiller grains, are sold as feed for livestock.

Other alternatives likewise are helping local economies to prosper. Soybeans and other U.S.-grown feedstocks intended for use as biodiesel are finding their way to Midwestern U.S. markets. Some states mandate the use of 10% or more blends in biodiesel. This also encourages and assures local economies of a market for their products.

▶ Vehicle Emissions and Standards

LO 54-03 Describe vehicle emissions and emission standards.

As clean as they are by today's standards, left unchecked, ICEs and fuel systems emit pollutants both from without and from within the tailpipe. The pollutants regulated by EPA emission standards are:

- **Carbon monoxide (CO)**: Partially burned fuel. It displaces oxygen in the bloodstream and causes asphyxiation.
- **Hydrocarbons (HC)**: Volatile organic compounds (VOCs) of unburned fuel. They add to photochemical smog.
- **Carbon dioxide (CO_2)**: Although harmless in small amounts, too much carbon dioxide is an atmospheric concern. It is considered a greenhouse gas, although there are disagreements on how damaging it is.
- **Nitrogen oxides (NO_x)**: The various compounds of nitrogen contribute to the creation of smog in the presence of sunlight.
- **Particulate matter (PM)**: Mainly produced by diesel engines, PM is literally fine soot (measured in microns, or millionths of an inch). PM becomes trapped in the lungs and causes health issues. At one time, PM-10 (10 microns large) was thought to be harmless; today, even PM-2.5 is considered dangerous.

▶ Emission Standards

Various countries, states, and even localities set emission standards. They affect the amount of pollutants that vehicles may release into the environment. Different countries focus on different pollutants. It depends on what environmental issues they are dealing with. Around the world, most nations' standards have become stricter. Now, tailpipe exhaust is sometimes cleaner than what goes into the engine's intake system.

In the United States, the Environmental Protection Agency (EPA) controls federally mandated emission standards. These are mostly following the lead of California's clean air regulations. Some states, like California, have enacted stricter emission standards than those of the EPA. The California Air Resources Board (CARB) is a department within California's EPA. California is the only state that is permitted to have such a regulatory agency. This is because it is the only state that had one before the passage of the federal Clean Air Act. Other states are permitted to follow CARB standards or use the federal ones. But they cannot set their own.

We will not go into the actual cutoff points. There are cutoffs for each of the pollutants, vehicles, and locations. We will also not discuss when they became effective. The regulations are far too varied to list here. In general, however, regardless of where clean air regulations are regulated and enforced, pollution cutoff points have become progressively more stringent over time.

Low-Emission Vehicle Program Emission Standards

California used unique terms to describe vehicle compliance with their clean air regulations. Window stickers seen on vehicles showed their compliance with California's clean air limits. For example, from 1994 to 1999, California's regulations were stricter than the U.S. national "tier" regulations. The **Low-Emission Vehicle (LEV) standard** included six major emission categories. Each of them has several targets depending on vehicle weight and cargo capacity. Vehicles with a test weight up to 14,000 lb (6,350 kg) were covered by these regulations. The major emission categories were as follows:

- TLEV: Transitional Low-Emission Vehicle
- LEV: Low-Emission Vehicle
- ULEV: Ultra-Low-Emission Vehicle
- SULEV: Super-Ultra-Low-Emission Vehicle
- ZEV: Zero-Emission Vehicle

The last category is largely restricted to electric vehicles and hydrogen vehicles. However, such vehicles are usually partly reliant on grid-supplied power. This is derived from coal, oil, natural gas, hydro (water), or nuclear sources.

▶ Battery Electric Vehicles

LO 54-04 Describe battery electric vehicles.

BEVs are powered entirely by batteries that need recharging when they are low. In the past, lead-acid or nickel–metal hydride batteries were typically used. Now lithium-ion and other types of batteries are used for greater range. Most of the early electric vehicles have been reclaimed and scrapped by the manufacturers. Now, a whole new generation of electric vehicles have become available, such as the Tesla, Chevy Bolt, and Nissan Leaf (**FIGURE 54-4**). Watch for more BEVs to be released soon.

BEV Development

BEV principles were used more than a century ago. In fact, they were used before the advent of gasoline-powered internal combustion vehicles. With modern-day improvements in vehicle systems and battery technology, electric vehicles are returning to the marketplace. Because more than 75% of daily commuters drive no more than 40 miles (about 65 km) a day, electric vehicles are becoming viable. They can now deliver that range without having to consume any gasoline or diesel fuel at all. Therefore, they produce no emissions.

BEVs first enjoyed a brief resurgence during the late 1990s in the United States. Several manufacturers offered them for lease. These included GM's EV1, Toyota's RAV4

FIGURE 54-4 Nissan Leaf.

FIGURE 54-5 Early attempt by manufacturers to develop BEVs.

EV, Ford's Electric Ranger pickup, and Nissan's Altra station wagon/crossover vehicle (**FIGURE 54-5**). These vehicles were preproduction or limited production test vehicles. They were mainly for California commuters, police or postal fleets, and other limited-range applications.

The BEV experiment was limited. For example, it lacked a recharging infrastructure and a single-charge port configuration. Various conductive hookups were used along with the innovative inductive "paddle." The vehicles also suffered from disappointing range issues. Lead-acid batteries offered a limited range-to-weight ratio. But, the nickel–metal hydride batteries proved expensive and somewhat unreliable. So, the manufacturers recalled and crushed most BEVs of that period. But, a lot was learned from those electric vehicles. It propelled a new generation of HEVs and BEVs forward to the marketplace. There has been improved developments in battery technology and propulsion system management algorithms. Now, many manufacturers have entered the BEV market.

BEVs Today

BEVs have been on a resurgence lately. When coupled with sustainable or renewable energy sources, the BEV can truly serve as an emissions-free vehicle. Sustainable energy may be provided by unlimited sources of supply, including the following:

- Solar panels and reflectors (to drive steam turbines)
- Wind generators
- Tidal and wave-power generators
- Geothermal sources

These sources and others offer limitless energy sustainability without the use of fossil fuels like petroleum or coal. The BEV itself normally uses only one power source—the battery. The battery powers traction motors to propel the vehicle. The vehicle is returned to a charging station to maintain the energy (**FIGURE 54-6**). With a BEV, the average distance is limited to the storage capacity of the battery and the efficiency of its motor. Most BEVs, such as the Nissan Leaf, can travel about 60–100 miles (129–161 km) on a single charge, with some exceptions. Because of this, the Leaf has been one of the bestselling BEVs in history, with about 250,000 vehicles sold since it was introduced in 2010.

FIGURE 54-6 Ford Focus Electric being charged.

The Tesla Roadster was reportedly the first production automobile to use lithium-ion battery cells and the first production BEV (all-electric) to travel more than 200 miles (322 km) per charge. Tesla has followed that up with the Model S sedan and the Model X SUV (**FIGURE 54-7**). The Model 3 became available recently that sells for $35,000, competing with the Nissan Leaf and Chevy Bolt.

Recently, there have been promising battery and vehicle technologies. They will likely extend BEV range to more than 300 miles (483 km) in the not-too-distant future. In fact, Elon Musk, owner of Tesla says, "Future models may reach a 500-mile (800 km) range. This is partially because of a new patented battery system, pairing metal-air and lithium-ion batteries."

Another BEV, the Chevy Bolt, became available in 2016. It uses a "nickel-rich lithium-ion" battery pack that tolerates higher cell temperatures. The battery pack is rated at 60 kW of electrical storage and has a range of approximately 240 miles (380 km).

The Chevy Volt is known as an extended-range electric vehicle (EREV). It uses a lithium-ion battery that can be recharged from a domestic electrical power outlet (110 or 220 volts AC). This vehicle, however, is not a pure electric vehicle. It is a form of hybrid vehicle, as it also has a small gasoline/petrol-powered

FIGURE 54-7 A. Tesla Model S. **B.** Tesla Model X.

engine that drives a generator to provide electric power, should the vehicle travel beyond its battery-only range.

▶ Batteries

Battery technology has advanced in recent years. BEVs and HEVs are now viable for many more people. In the early days, there were lead-acid batteries that only provided a range of 20–30 miles. Now, lithium-ion batteries have been improved to the point that they provide over 200 miles of range. Although current battery technology is adequate for many people, it is still expensive, putting this technology out of the reach of the masses. But the progress being made in battery technology indicates that the price will continue to decrease, and the capacity increase. This will hasten the day when electric vehicles become a majority of vehicles on the road. Here is the progression of battery types used in production vehicles.

- Lead-acid battery—Variation of standard storage batteries used in regular passenger vehicles. It has thicker lead plates. These allow the battery to better withstand the heavy cycling. This cycling is common in electric vehicles. The batteries are called deep cycle batteries. Advantages include availability, cost, and simplicity of charging equipment required. Disadvantages include weight, size, and storage capacity.
- Nickel-cadmium (Ni-Cd)—Rechargeable battery. Has a higher energy density than lead-acid batteries. There are many advantages. The battery can withstand high discharge rates and deep discharges for long periods of time. It can withstand more discharge/charge cycles and is lighter and smaller than lead-acid batteries. There are disadvantages too. The batteries have a higher self-discharge rate. They also have a higher cost than lead-acid (equivalent cost to NiMH but less capacity). Also, the cadmium is highly toxic, and Ni-Cd requires a special charger because the batteries' internal resistance falls as the cell temperature rises.
- Nickel–metal hydride (NiMH)—Rechargeable battery like Ni-Cd, but with the cadmium replaced with a metal hydride structure. There are advantages. The batteries don't have highly toxic materials. It has two to three times the energy density of Ni-Cd. It is able to withstand high discharge rates. It is less expensive than lithium-ion batteries, and it is in mainstream use in the production of EVs and HEVs. Disadvantages are that it requires a special charger, has a moderately high self-discharge, and can be damaged if discharged too deeply (polarity reversal is possible).
- Lithium-ion (Li-ion)—Rechargeable battery with the highest energy density currently in use. Li-ion batteries are made from a variety of materials besides the lithium salt electrolyte. There are several advantages. They have a high energy density. They are minimally self-discharging. They have the ability to withstand large amount of discharge/charge cycles. It has relatively nontoxic materials. Disadvantages are that it has the highest cost of current battery options, is a possible thermal runaway and fire hazard, and requires a special charger.

Many high-voltage battery packs have to operate within a specified temperature range. This is especially true of Li-ion batteries. This requires special cooling for the battery module. Cooling can happen in a couple of ways: The first is with outside air being drawn into the battery pack by a computer-controlled fan (**FIGURE 54-8**). The other way is with a separate cooling system connected into the vehicle's HVAC system.

In many cases, the high-voltage controller also has to be cooled. This is typically cooled by a separate cooling system from the ICE's cooling system (**FIGURE 54-9**). However, they may share separate sections of a common radiator. The coolant is circulated by an electric water pump as needed, which is controlled by the HV controller. Typically, this coolant is the same type as the ICE coolant but has to be flushed separately.

▶ Types of Hybrids and Electric Vehicles

LO 54-05 Describe the types of hybrid and electric vehicles.

There are various kinds of electric and hybrid vehicles on the market today, and they are growing in popularity (**FIGURE 54-10**). They may be classified by body style, but how is the *degree* of

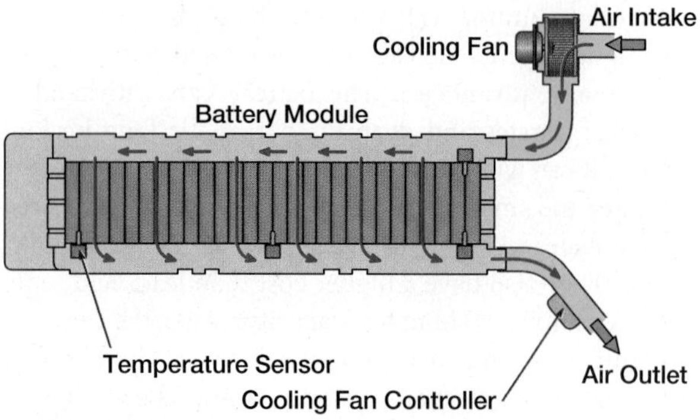

FIGURE 54-8 Typical battery pack cooling system.

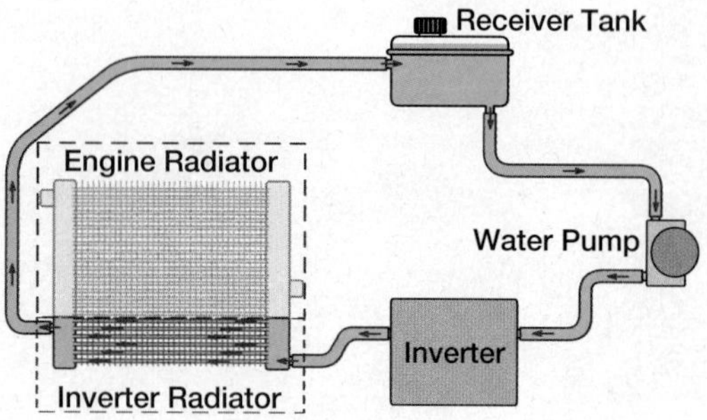

FIGURE 54-9 The high-voltage controller may have its own cooling system to prevent it from overheating.

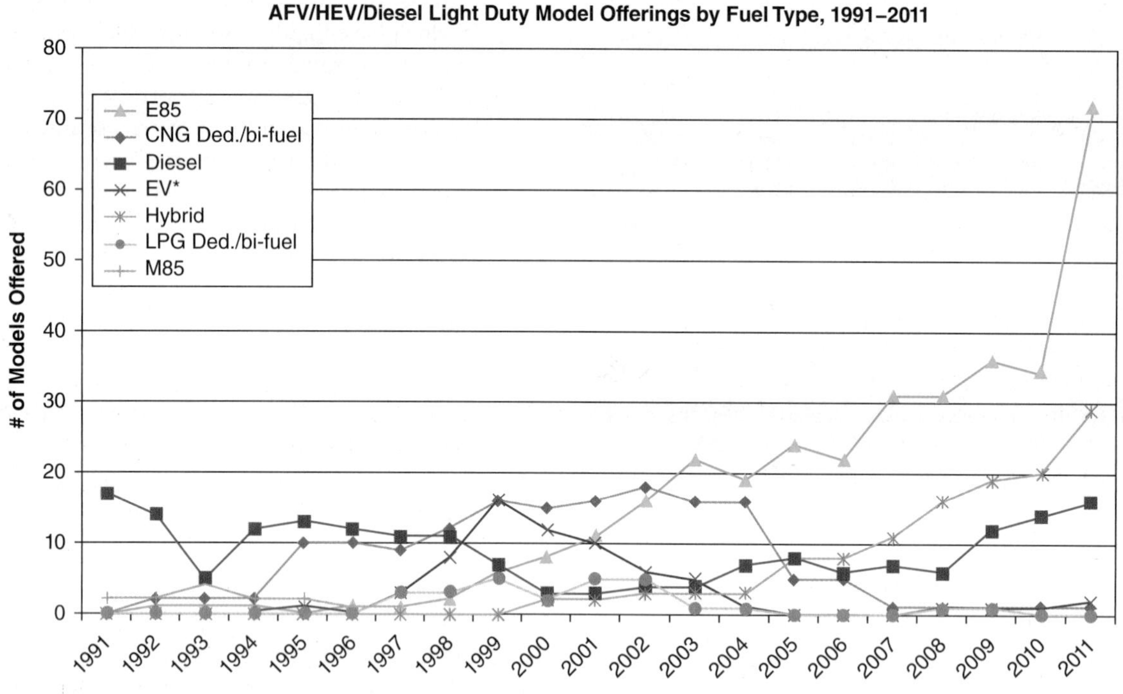

FIGURE 54-10 Notice that starting from 1999, there are close to 3 million hybrids. They rank the second highest among AFVs in the United States. These numbers will continue to climb in the coming years. Note also that as of 2011, there are more than 7 million E85 (85% ethanol) FFVs on U.S. roads. Most owners of these FFVs are unaware that they can handle alcohol fuel.

hybridization described? There seems to be some confusion. HEVs may be classified by the type of drive configuration used—such as planetary gear sets, multiple electric machines, and continuously variable transmissions. Other descriptions are thrown around with abandon. The following are some widely used descriptions you should be familiar with. But don't be surprised if other designations (often created for marketing appeal) are used for these hybrid vehicles.

▶ Power Sources

HEVs have two power sources, typically a gasoline ICE and high-voltage batteries working in both series and parallel. HEVs are not typically electric grid rechargeable. However, some have been converted to plug-in HEVs (see the next section) for more than 100 mpg (43 km/L). HEV types include micro, partial, and full hybrid vehicles.

- Micro-hybrids include GM's Silverado and Sierra with start/stop features.
- Partial hybrids include Honda's Civic and Insight, in which the engine (ICE) does not shut off while the vehicle is underway.
- Full hybrids include the Toyota Prius Highlander, and Camry models, the Ford Escape and Mercury Mariner hybrid, and the GM vehicles with their two-mode hybrid system. These and many others on the market are considered full hybrids. They comprise about perhaps 90% of the market. They can be driven short distances and at low speeds using only the (high-voltage) batteries.

FIGURE 54-11 More plug-in hybrid vehicles are coming to market.

FIGURE 54-12 The Chevy Volt is an extended-range electric vehicle.

▶ Plug-in Hybrid Electric Vehicles

Many manufacturers are entering the **plug-in hybrid electric vehicle (PHEV)** market. A PHEV is different in that only one power source, the battery, is used to propel the vehicle for a certain distance. This is limited by the storage capacity of the battery and the efficiency of the motor. A PHEV-20, for example, is designed to go 20 miles (32 km) before the batteries need charging and the ICE is needed. A PHEV-40 can go 40 miles (64 km), and so on (**FIGURE 54-11**). The ICE may also start up for added power before the batteries need a recharge.

Ford, Toyota, and other light-duty vehicle OEMs have PHEVs on test. PHEV automobiles achieve great fuel economy, with some getting more than 100 mpg (43 km/L). Medium-duty vehicles from Mercedes-Benz and others offer PHEVs for commercial use. International Truck and Engine sells a PHEV school bus. Combined with FFV capability, PHEVs offer a good miles per gallon of gasoline equivalent (MPG-e).

▶ Extended-Range Electric Vehicles

EREVs are plug-in electric vehicles that can drive a moderate distance on electric-only power. But they have a relatively small auxiliary ICE to drive a generator for charging the batteries and extending the vehicle's range. The Chevy Volt is a series EREV (**FIGURE 54-12**).

▶ Fuel Cell (Electric) Vehicles

FCVs are considered hybrid vehicles because they have two power sources. The sources typically are a hydrogen fuel cell "stack" and a storage device from which electric power is drawn. Regenerative braking electrical energy may be stored in the vehicle's high-voltage battery pack. Or as with Honda's early FCX and others, in ultra- or supercapacitors. They can more quickly be charged or provide acceleration power as needed.

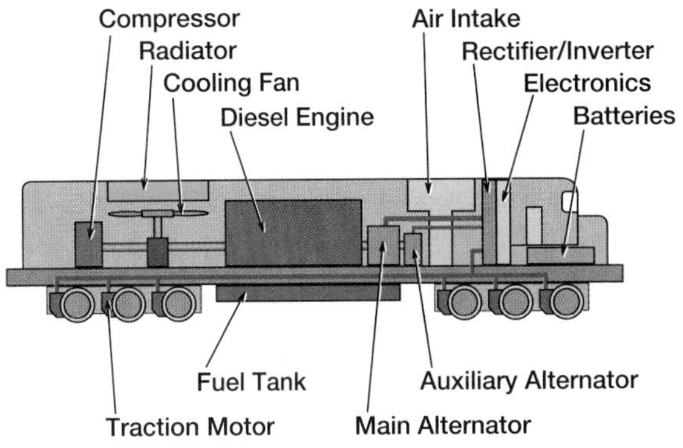

FIGURE 54-13 Hybrid drives have been used for about a hundred years in some applications.

▶ Hybrid Electric Vehicles

The hybrid drive concept is anything but new (**FIGURE 54-13**). Think about diesel–electric submarines or diesel–electric locomotives. Light- and medium-duty hybrids now come in a variety of configurations. Power from the ICE combines with auxiliary backup or primary power. They may run on gasoline, natural gas, diesel, or propane. The electric power comes from a variety of sources. They are electric batteries, ultra- or supercapacitors, or hydraulic-pneumatic accumulators. They may even be flywheels operating at super-high revolutions per minute (rpm). Other power options may include micro-turbines. They are in non-road special-purpose hybrids. At this time, U.S. light-duty hybrids are almost only gasoline–electric.

Hybrid electric vehicles (HEVs) use a combination of electric power and an ICE. They combine the different advantages of ICEs and electric motors. This provides a vehicle that operates efficiently. Usually, the engine drives a generator to provide the electricity. It is stored in batteries that then drive an electric motor. The engine may also drive the wheels directly. Hybrid vehicles are designed to drive like conventionally powered vehicles while producing significantly lower emissions and increased fuel economy.

Some conventional vehicles have only an ICE. In these, the engine must be large enough and powerful enough to produce momentary torque sufficient for rapid acceleration at low speeds or from a stationary position. Such an engine is larger than necessary for highway cruising, which requires far less power. The parasitic losses of the larger engine take their toll on fuel economy. A vehicle that has only an electric motor has its own complications. It must carry a large volume of heavy batteries to give it sufficient range to be useful, and these must be recharged or replaced when discharged.

Electric motors produce their maximum torque at stall speed. But, a gasoline engine is more efficient at higher rpm. Combining a small but efficient gasoline engine with an electric motor is both effective and efficient. This is because the engine is designed to operate at its optimum peak torque rpm. And the electric motor works best at slow speed. It can also double as a generator to recapture braking energy.

▶ Hybrid Drive Configurations

LO 54-06 Describe hybrid drive configurations, enhancements, and operation.

There are three basic kinds of hybrid drive configurations (**FIGURE 54-14**):

- **Series hybrid**—The gasoline engine is used only to drive a generator and charge a battery that produces power to drive an electric motor. The electric motor then drives the wheels of the vehicle. The gasoline engine is really only a battery charger inside an electric vehicle.
- **Parallel hybrid**—The engine always drives the wheels and keeps the batteries charged. The electric motor acts like a large starter motor, helping the engine to crank over when more power is needed, particularly when starting from rest or when accelerating rapidly. This allows the engine to be smaller and lighter and more fuel-efficient than if it were the sole source of power in the vehicle.
- **Series-parallel hybrid**—This design allows both the engine and the motor to drive the vehicle. This can be either one by itself, under the control of an onboard computer that determines the optimum combination of power delivered to the wheels at any time. The engine can shut down completely when the batteries are charged. When needed, the gasoline engine fires up again to assist in driving the wheels or to recharge the battery. The engine can be optimized for efficiency within a narrow rpm range. And the overall combination of both power sources takes up little more room than a conventional all-purpose gasoline engine. Yet, they can deliver lower emissions and much greater fuel economy. This design is used by the Toyota Prius and Ford Escape.

Almost all hybrids (except the Chevy Volt) transfer motive power to the wheels in either the parallel or the series-parallel drive configuration.

A variety of transmissions, clutches, and other features are used. This depends on the desired amount of performance (or fuel economy). It also depends on the selling price. Some hybrids come with manual transmissions. Others come with automatic transmissions (up to eight speeds) or planetary gear

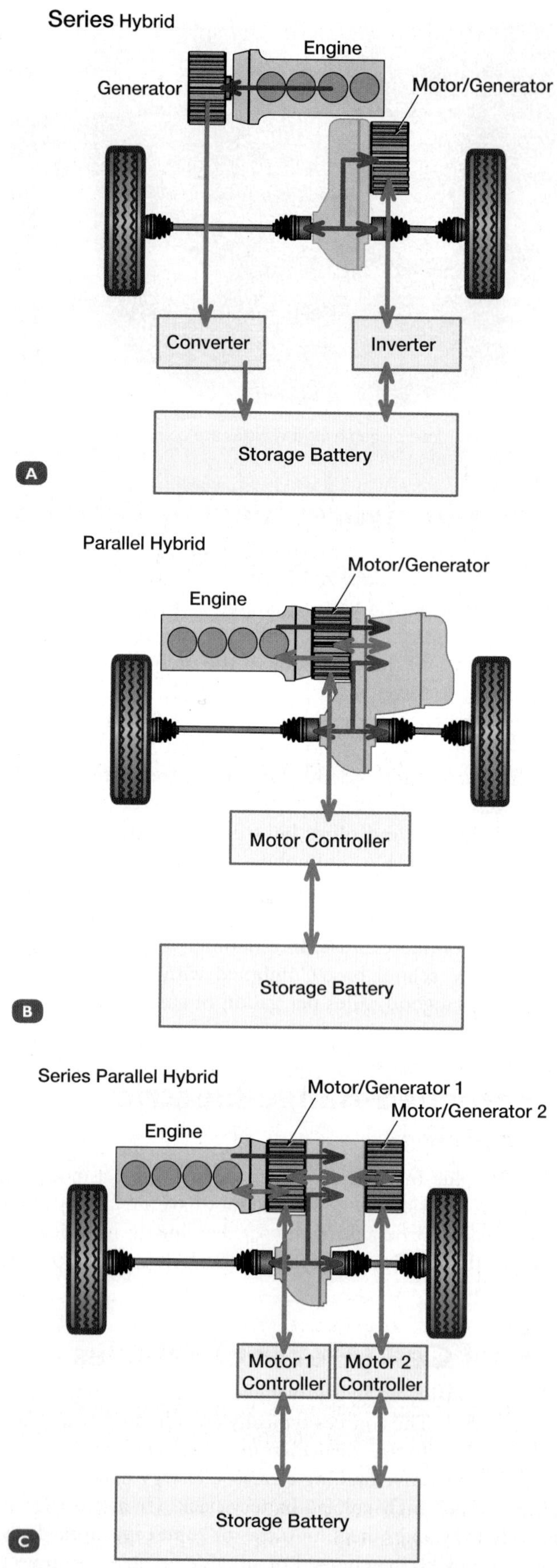

FIGURE 54-14 The three types of hybrid drive configurations.

set transmissions. Still, others come with continuously variable transmissions (some with sporty "paddle"-type shifters).

High-voltage controls vary also, along with the types of batteries and capacities, the type of battery cooling, and so on. There may be one, two, or more high-voltage **electric machines** (drive motors/generators) used. These may be three-phase AC permanent magnet or inductive motors. Usually, the ICE provides power when the vehicle is traveling at higher speeds. The electric motor is used when slowing to a halt and at lower speeds.

▶ Hybrid Vehicle Efficiency Enhancements

HEVs are able to offset ICE inefficiency. They use a smaller volumetric engine size and run the ICE at the ideal torque and speed. Efficiency is also achieved through regenerative braking. It can also be done by using displacement-on-demand cylinder deactivation, power-on-demand, and idle stop.

Regenerative braking occurs when the drive motor(s) act as generators to recharge the traction batteries during deceleration or when braking. This feature reportedly saves around 10–15% on fuel consumption. **Displacement-on-demand** deactivates cylinders when operating under cruise or coasting conditions to save fuel. The **power-on-demand** feature shuts off the ICE when it is not needed. This happens when idling or coasting. Drive may also be provided instead only by the electric motor, without burning fuel. This saves energy and reduces emissions.

Idle stop turns off the ICE when the vehicle is at a stop. Expect these features to be used more as newer vehicles hit the market. There are other fuel-saving features. They include:

- variable valve timing,
- Atkinson cycle (five-cycle) engines,
- Miller cycle (five-cycle with low rpm supercharger boost) engines,
- off-set pistons, and
- other interesting engine innovations.

Refer to the Motive Power Types chapter for more on these engines.

But engine technology is not the only thing that improves efficiency on hybrid vehicles. Many of them use:

- advanced aerodynamic technologies to reduce wind resistance. This can be due to shaping the body and exterior components to slip through the air easier.
- smoothing out the underside of the vehicle with panels that reduce drag.
- low-drag tires that are specially made for hybrid vehicles.
- electric power steering, electric air conditioning, and electric water pump to reduce engine load.

As you can see, making vehicles more efficient involves all aspects of the vehicle. And it involves understanding how each of the new technologies operates as well as how to diagnose and service them.

Hybrid vehicle types are classified by body style and by the degree of performance and luxury they offer. But technically speaking, they are also classified by how far they can be driven on batteries alone. For example, a "full hybrid" can cruise a limited distance without the engine running. A "partial hybrid" runs on ICE power at all times except when stopped. A large amount of U.S.-registered full HEVs use drive systems with one or more planetary gear sets licensed from, or similar to, Toyota's Hybrid Synergy Drive design. A number of other OEMs are using the GM-originated two-mode hybrid drive system. See the Hybrid and Continuously Variable Transmission chapter for more information on these drive systems.

▶ Hybrid Vehicle Operation

Let's look at the operation of a typical Toyota Prius. When the vehicle is moving from a stationary position, and when traveling at low-to-moderate speeds, the main electric motor/generator (MG2) drives the vehicle. At these speeds, the ICE is less efficient and is normally used only to charge the battery.

During normal driving, the ICE starts and drives the generator (MG1) and the power splitter. Power from the generator (MG1) is used to drive the electric motor (MG2). The motor control unit controls the **power splitter** so that the drive remains at its most efficient. When accelerating and when using power from the ICE, the control unit draws power from the battery and directs it to both electric motor/generators, providing more power to the wheels than the ICE could supply on its own.

During deceleration and braking, the ICE is turned off. Inertia from the wheels drives one or both electric motor/generators. The current produced is used to charge the high-voltage battery. Normally this energy would be wasted as heat, but in the hybrid, it is recovered and reused. The **retarding effect** (vehicle slowing) caused by this system provides a moderate amount of deceleration. So gentle application of the brake pedal does not apply the brakes. The braking effect is achieved by using the electric motor/generator. It converts the vehicle's inertial momentum into electrical energy to charge the high-voltage battery. This is called "regenerative braking."

When the vehicle is stationary and the brakes are applied, the ICE is stopped. Power is not applied to the electric motor/generator. No fuel is burned and no emissions are produced. Releasing the brakes applies a small amount of power to the electric motor. Moving the accelerator pedal further, more power is sent to the electric motor. If required, the ICE is started to increase the power further.

▶ Hybrid and Electric Vehicle Service Precautions

LO 54-07 Describe hybrid and electric vehicle service precautions, PPE, and tools.

Maintenance needs for HEVs, PHEVs, and all-electric vehicles (EVs) are like those of conventional vehicles. But, these vehicles have high-voltage electrical systems. The systems range from 100 to 600 volts. They also may start up at any time if not fully shut down. So, servicing them requires some particular precautions.

Don't attempt to perform service operations on any hybrid vehicle right away. First, read the manufacturer's service information for the vehicle you are about to service. The service information will instruct you in how to properly follow all safety procedures. The most important rule to remember is, never work on a high-voltage vehicle without first notifying someone who is trained in dealing with high-voltage electrocution. You need a responsible person in your work environment. This person needs to check on you as you work around high voltage.

The matter of servicing these vehicles safely is to be taken seriously. Always follow the manufacturer's safety and service procedures to keep out of trouble. There are many safety features built into hybrid and electric vehicles to prevent accidental shock. But once you decide to venture into high-voltage areas, be prepared to spend some money on safety equipment. This is no time to pinch pennies! Some of the standard equipment for servicing HEV systems is listed here:

- OEM or equivalent scan tools
- A high-quality (three-phase CAT III or CAT IV) digital multimeter (DMM) with appropriate high-voltage leads (**FIGURE 54-15**)
- High-voltage insulated gloves rated for 5000 volts and certified to 1000 volts (Class 0) and air-checked on a daily basis; also approved glove covers (**FIGURE 54-16**)
- High-voltage, insulated shoes
- Insulated tools meeting 1000 V/300 A specifications (**FIGURE 54-17**)
- Safety cones and safety tape (to mark off service areas) and other such items
- A retrieval hook in case someone becomes disabled from electrical shock (**FIGURE 54-18**)
- ABC-type fire extinguisher

If you decide to service electric and hybrid vehicles, get the proper training on the high-voltage areas. Also have a full understanding of how they operate. Stay alert and continually practice situational awareness.

Besides the high-voltage dangers of a hybrid or an electric vehicle, there are other safety precautions you need to be aware of. One of the first is that the vehicle should be completely powered down before working on it. This is because the ICE can start at any time the vehicle's PCM deems necessary.

Never assume that the vehicle is turned off because it is silent or still. The vehicle may appear to be turned off because the ICE has been shut down, and the vehicle is silent. But the

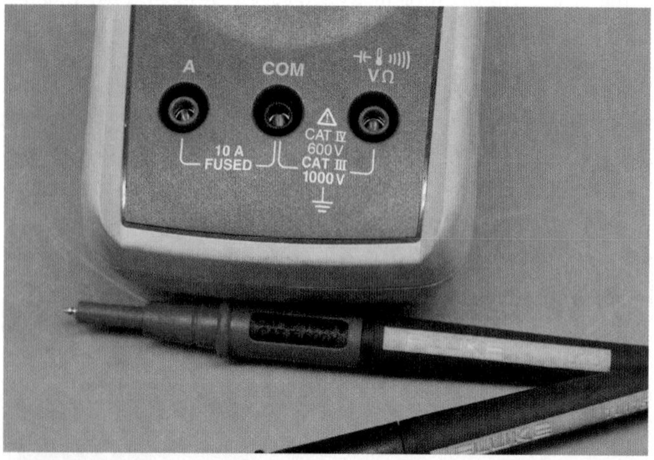

FIGURE 54-15 CAT III DMM.

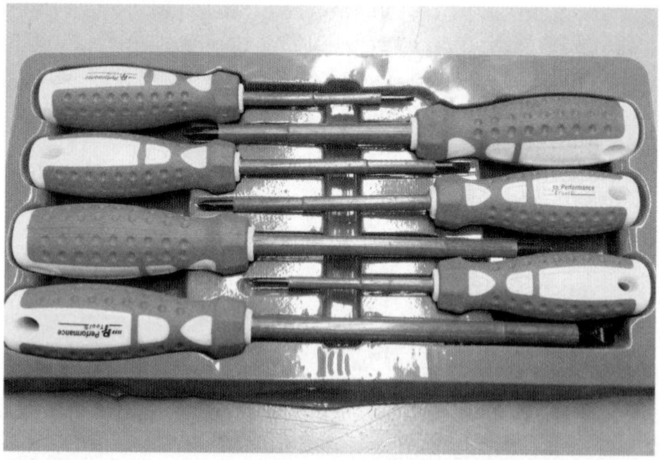
FIGURE 54-17 Insulated tools with high-voltage rating.

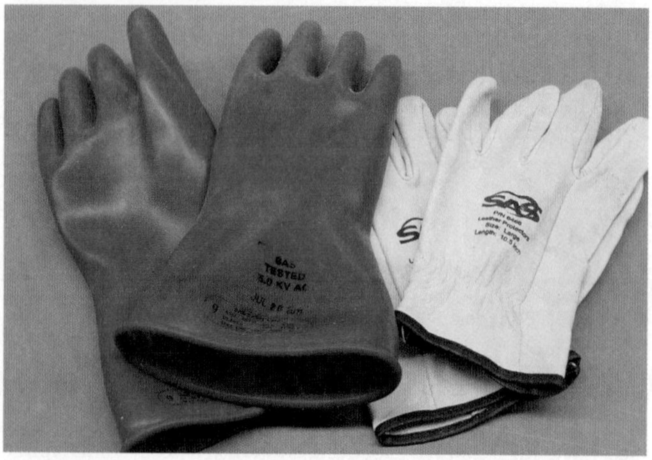
FIGURE 54-16 Class 0 HV insulated gloves and approved covers.

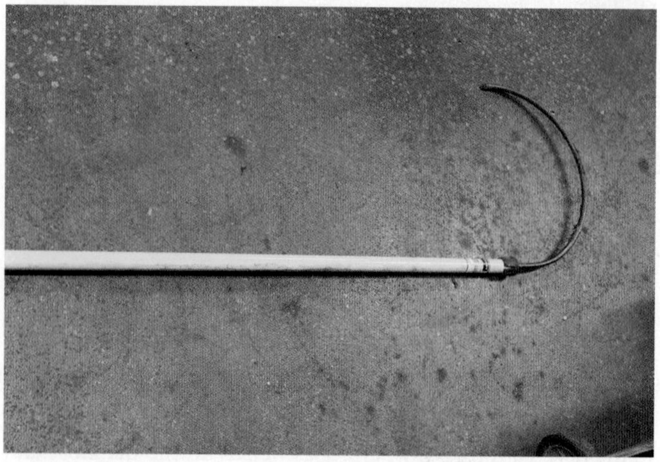
FIGURE 54-18 Retrieval hook for pulling a person away from high voltage.

system can still be in "ready" mode (most vehicle models show this on the vehicle dash display panel). In "ready' mode, it is capable of engaging the ICE or the drive motor at any time (**FIGURE 54-19**). If a technician is changing a belt or working around the ICE and the engine starts up, he/she could be injured or killed. Also, prior to service, the vehicle should be in Park, with the parking brake applied and the power button powered down. The keyless fob, if the vehicle is so equipped, should be more than 15" (4.6 m) from the vehicle. Also remove the hybrid high-voltage service plug.

SAFETY TIP

Always make sure there is someone in your work environment who is trained in the safety procedures about high-voltage electrocution. And have them nearby to check on you whenever you work around high-voltage systems.

▶ Identifying and Disabling the High-Voltage System

LO 54-08 Disable the high-voltage system, service 12-volt battery.

There are plenty of opportunities to service hybrid and electric vehicles without venturing into the high-voltage areas of the vehicle, such as changing oil. But just because you are not working on the main traction motor doesn't mean that you won't come into contact with other high-voltage circuits. In many hybrid and electric vehicles, high voltage is used to operate a variety of accessories. Some examples are power steering and air-conditioning. So you need to be aware that high-voltage wires can be located nearly anywhere in the vehicle. These high-voltage wires are recognized by orange (usually) convolute or wiring (**FIGURE 54-20**). Some vehicles have used other colors, such as blue, to designate high-voltage wiring. So always check the service information.

It should be noted that manufacturers of HEVs, PHEVs, and EVs design these vehicles with safety features. In most cases, they deactivate the high-voltage electrical system when they detect a collision, short circuit, or certain HV system faults. Typically, the deactivation occurs in the high-voltage battery itself. It can also occur in the battery junction box by one to three special relays. Toyota calls them system main relays (SMRs). They open the high-voltage circuit (**FIGURE 54-21**). The relays are controlled by the high-voltage power management control ECU. It switches them off if a fault is detected.

One method manufacturers use to detect shorts in the high-voltage wiring is by building electrical leakage detectors within the insulating material. If a leak is detected, the power management controller shuts down the high-voltage system. So never pierce a high-voltage wire when testing for voltage. You may inadvertently shut down the system and ruin the cable.

Only technicians who are trained to work on these high-voltage areas of a hybrid vehicle should do so. You first need to have the proper training, service information, and experienced help on hand. Also, the main high-voltage battery pack must be disconnected before service is performed on or around the high-voltage parts. Every manufacturer has

FIGURE 54-19 A vehicle in "ready" mode as indicated by the "ready" status.

FIGURE 54-20 High-voltage wiring.

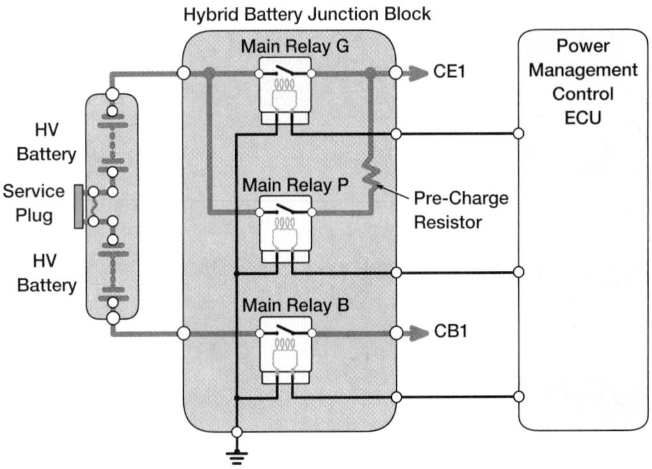

FIGURE 54-21 Typical high-voltage battery safety system on a Toyota.

a different way of doing this. Many systems use a service plug that is removed from or near the battery. This will open up the high-voltage circuit inside the battery (**FIGURE 54-22**). Other systems use a battery module switch that is turned to the Off position. This also disconnects the battery from the high-voltage system.

Consider what to do before you disable the high-voltage system. Make sure you are wearing the appropriate PPE. Also, make sure you are following the manufacturer's procedure. The following is an example of disconnecting the high-voltage battery. This example uses a third-generation Prius. Follow these general guidelines:

1. Locate the high-voltage battery disconnect (service plug).
2. Release the lock, release the lever, and remove the service plug (**FIGURE 54-23**).
3. Store it in a safe place, where it cannot be reinserted accidentally. Some manufacturers suggest that you keep it in your pocket so that it can't get reinstalled by anyone.
4. Wait for the specified time for the residual electricity to bleed off. It typically takes 5 to 15 minutes for this to occur.
5. Then, verify that the system is no longer live. To do that, locate the high-voltage access panel, and remove the bolts that secure the cover, using insulated tools (**FIGURE 54-24**).
6. Next, remove the cover and, using the correctly rated meter, measure that there is no voltage at the specified terminals (**FIGURE 54-25**).

To power down the high-voltage system on a Nissan Leaf, there are three possible methods: primary method, alternate method number 1, and alternate method number 2. We cover the primary method here, because the alternative methods are only used in the event the "Ready" indicator won't shut off:

- Verify that the "Ready" indicator is off.
- Keep the key at least 16 feet away from the vehicle.
- Open the hood and disconnect the 12-volt battery's negative cable.
- Insulate the negative battery cable with insulated tape.
- Wait at least 10 minutes for the high-voltage capacitor to discharge.

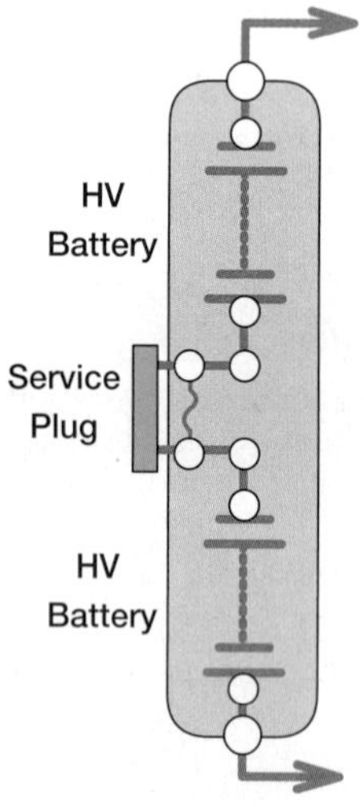

FIGURE 54-22 Removing the service plug opens the high-voltage circuit in the battery.

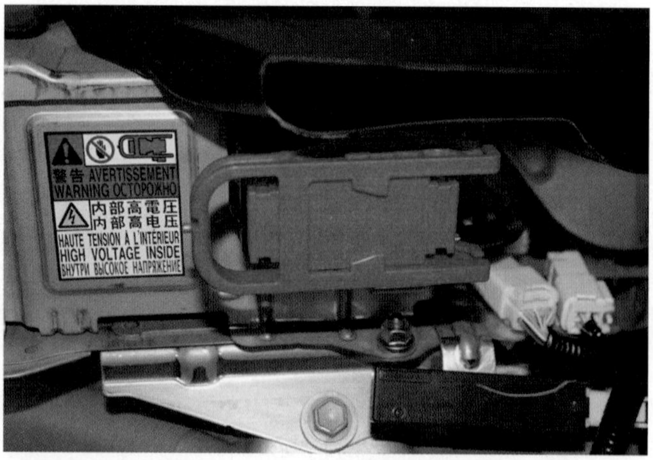

FIGURE 54-23 Locate the high-voltage battery disconnect (service plug), release the lock, release the lever, and remove the service plug.

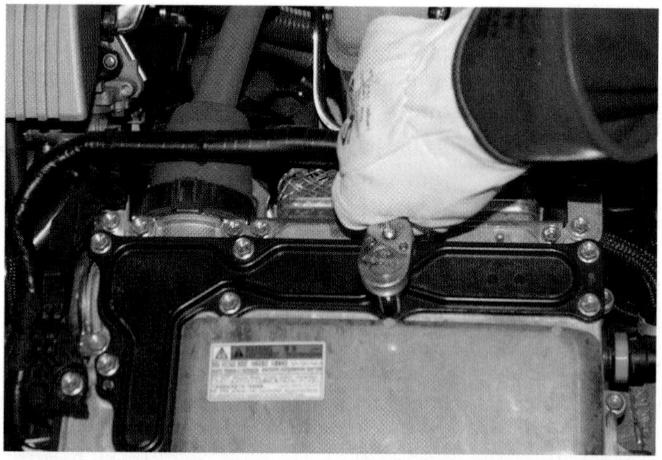

FIGURE 54-24 Locate the high-voltage access panel, and remove the bolts that secure the cover, using high-voltage insulated tools.

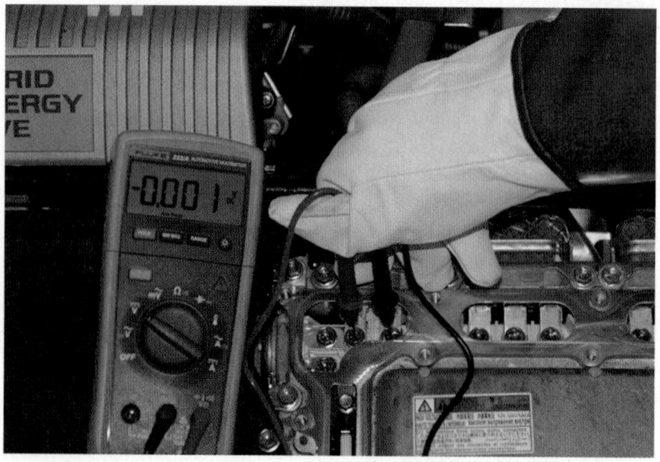

FIGURE 54-25 Remove the cover, and using the correctly rated meter, measure that there is no voltage at the specified terminals.

▶ Hybrid Auxiliary (12 V) Battery Service

On most hybrid vehicles, the 12-volt auxiliary battery is used to power up the vehicle's computers and run virtually all of the accessories. Because it powers up the computers, it is critical to starting the vehicle. This is even though it is generally not used to crank the engine over. That is typically done by one of the traction motors and a high-voltage battery. This means that the 12-V battery in a hybrid vehicle is generally of smaller capacity and size than in a standard vehicle. So it is more susceptible to parasitic current draws, as they don't have as much current capacity. It is also located separately from the traction batteries, typically either under the hood or in the trunk (**FIGURE 54-26**). In most cases, the 12-V battery is charged from an inverter built into the hybrid electronic controller, instead of a separate alternator (**FIGURE 54-27**).

In most hybrids, the 12-V battery is the absorbed glass mat (AGM) construction. This battery is sensitive to overcharging, which can damage it. So a special smart battery charger is required whenever charging is needed. Some manufacturers recommend removing the 12-V battery from the vehicle while it is charging, to prevent any damage to the vehicle's electrical system. But this can erase the computer and entertainment memories. It can also activate the antitheft system, so always follow the manufacturer's procedure when disconnecting any battery. Also, most manufacturers prohibit hybrid vehicles from being used to jump-start another vehicle. This is because of the reduced size of the auxiliary battery. There is also the potential to damage to the hybrid electrical system. The 12-V battery is best tested with a conductance tester, as described in the Battery System chapter (**FIGURE 54-28**).

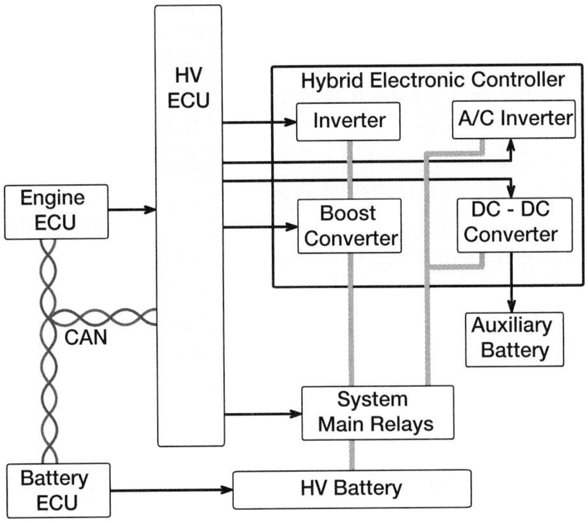

FIGURE 54-27 The 12-volt battery is typically charged from a DC–DC converter built into the hybrid electronic controller.

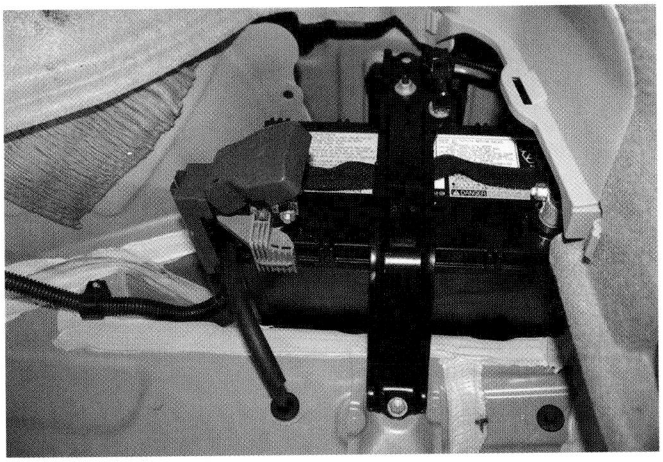

FIGURE 54-26 A 12 V Prius battery installed.

FIGURE 54-28 The 12-V auxiliary battery is best tested with a conductance tester.

▶ Wrap-Up

Ready for Review

- An alternative fuel may be liquid, gas, or electric, and the main reasons for their use are energy security, environmental concerns, and economy.
- The government actively promotes use of vehicles capable of using alternative fuels because oil is a strategic resource, pollution caused by both the oil industry and vehicles using gasoline or petrodiesel fuel, and to offset the unfavourable balance of U.S. dollars sent overseas for petroleum.
- The Environmental Protection Agency (EPA) sets emission standards and regulates the release of carbon monoxide, hydrocarbons, carbon dioxide, nitrogen oxides, and particulate matter. Emission targets vary with vehicle weight and cargo capacity.

- When combined with sustainable or renewable energy sources, the battery electric vehicle is fully emissions free. Among the different types of batteries, the lithium-ion battery is currently widely used. Battery packs operate within a specified temperature range so cooling is critical.
- Hybrid electric vehicles (HEVs) use a combination of electric power and an ICE to varying degrees. Electric motors produce their maximum torque at stall speed, but a gasoline engine is more efficient at higher rpm. So combining a small but efficient gasoline engine with an electric motor is both effective and efficient.
- There are three types of hybrid drive configurations, and vehicle efficiencies happen through displacement-on-demand cylinder deactivation, power-on-demand, and idle stop. During normal driving, the ICE starts and drives the generator (MG1) and the power splitter, and during deceleration and braking, the ICE is turned off.
- Before performing service on hybrid vehicles read the manufacturer's service information, notify someone trained in dealing with high-voltage electrocution, wear the proper safety equipment, have proper training and understanding of their operation, and ensure the vehicle is completely powered down before working on it.
- Every manufacturer has a different way of disconnecting the high-voltage battery pack so always refer to the service information before doing it. In most hybrids, the 12-V battery is the absorbed glass mat (AGM) construction so a special smart battery charger is required whenever charging is needed.

Key Terms

alternative fuel A non-petroleum-based motor fuel.

battery electric vehicle A vehicle powered by battery only.

displacement-on-demand A feature that allows cylinders to be taken off-line when not needed, such as during vehicle cruise.

electric machine Another name for a traction motor with regenerative capability.

Energy Policy Act Regulations enacted in 1992 by the US Congress to promote the use of alternative fuel vehicles and fuels.

hybrid electric vehicle (HEV) A vehicle that uses two power sources for propulsion, one of which is electricity.

idle stop A feature that turns off the internal combustion engine when the vehicle is at a standstill.

infrastructure The system for supplying fuel for transportation use; includes ships, pipelines, trucks, and so forth.

Low-Emission Vehicle (LEV) standard A program initiated in California to improve vehicle emission standards.

plug-in hybrid electric vehicle (PHEV) A hybrid electric vehicle in which only one power source, the battery, is used to propel the vehicle for a certain distance, limited by the storage capacity of the battery and the efficiency of the motor.

power splitter A device that receives power from an internal combustion engine and electric machine to power a hybrid electric vehicle.

power-on-demand A feature that shuts down the engine when not needed to save fuel.

regenerative braking A type of braking in which the kinetic energy of the vehicle's motion is captured rather than being lost to heat as it is in a conventional braking system. This is accomplished by using the drive motors as generators, which recharge the traction batteries.

retarding effect The result of retarding (slowing) the vehicle.

Review Questions

1. Alternative fuel vehicle is defined as a vehicle that runs on anything other than _____.
   a. Gasoline
   b. Battery
   c. Diesel
   d. Petroleum
2. Why is petroleum not considered a renewable resource?
   a. Because it pollutes too much
   b. Because it comes from oil
   c. Because it causes global warming
   d. Because it has a finite quantity
3. At gas stations nationwide, gasoline is mixed with what percentage of ethanol?
   a. 5%
   b. 10%
   c. 20%
   d. 75%
4. Which chemical is a result of unburned fuel?
   a. CO2
   b. NOx
   c. HC
   d. CO
5. What is the lowest vehicle emissions category currently?
   a. ULEV
   b. LEV
   c. TLEV
   d. ZEV
6. What makes a vehicle a hybrid?
   a. A vehicle with a high-voltage battery
   b. A vehicle that can drive with the engine off
   c. A vehicle that can be charged from an outlet
   d. A vehicle that has more than one power source
7. Which vehicle is a hybrid, without an ICE?
   a. FCV
   b. EREV
   c. PHEV
   d. HEV
8. Almost all hybrids transfer power to the wheels in either a parallel or series parallel drive configuration; the exception is the _________
   a. Toyota Prius
   b. Honda Insight
   c. Chevy Volt
   d. Tesla model S

9. What safety concern could happen to a hybrid vehicle that is in "ready mode"?
   a. The battery cooling fans can run
   b. The engine can start at any time
   c. The doors can lock with the keys inside
   d. The parking brake can release at any time
10. How can high-voltage cables be identified on most vehicles?
   a. They are orange
   b. They have stripes on them
   c. They are red
   d. They are green

ASE Technician A/Technician B Style Questions

1. Alternative fuel vehicles are being discussed. Technician A states that alternative fuels can include natural gas in liquid or compressed gas form. Technician B states that one of the earliest alternative fuel vehicles was a Tesla. Who is correct?
   a. Technician A only
   b. Technician B only
   c. Both Technicians A and B
   d. Neither Technician A nor B
2. The problems alternative fuels address are being discussed. Technician A states that some alternative fuels are intended to help the environment. Technician B states that some alternative fuels are used to reduce our country's dependence on foreign oil. Who is correct?
   a. Technician A only
   b. Technician B only
   c. Both Technicians A and B
   d. Neither Technician A nor B
3. Emissions are being discussed. Technician A states that particulate matter contains nitrogen and contributes to the formation of smog with the presence of sunlight. Technician B states that the tailpipe exhaust is sometimes cleaner than what goes into the intake system. Who is correct?
   a. Technician A only
   b. Technician B only
   c. Both Technicians A and B
   d. Neither Technician A nor B
4. Emissions levels are being discussed. Technician A states that emissions standards are set by the US, but some individual states have even higher standards. Technician B states that only electric vehicles fall into the zero-emission vehicle category. Who is correct?
   a. Technician A only
   b. Technician B only
   c. Both Technicians A and B
   d. Neither Technician A nor B
5. Battery electric vehicle technology is being discussed. Technician A states that BEV technology has only been studied since the late 1990s. Technician B states that lithium-ion batteries have high energy density but also are a thermal runaway and fire hazard. Who is correct?
   a. Technician A only
   b. Technician B only
   c. Both Technicians A and B
   d. Neither Technician A nor B
6. BEVs are being discussed. Technician A states that the Ford Ranger Electric was the first production vehicle to travel over 200 miles per charge. Technician B states that the Tesla Roadster was the first to use a lithium-ion rechargeable battery. Who is correct?
   a. Technician A only
   b. Technician B only
   c. Both Technicians A and B
   d. Neither Technician A nor B
7. Hybrid types are being discussed. Technician A states that partial hybrid engines do not shut off while the vehicle is moving. Technician B states that full hybrids can be driven at low speeds and short distances using only the HV battery. Who is correct?
   a. Technician A only
   b. Technician B only
   c. Both Technicians A and B
   d. Neither Technician A nor B
8. Hybrid drive types are being discussed. Technician A states that a series hybrid can use the engine, electric motor, or both to drive the wheels. Technician B states that a series-parallel hybrid example is the Toyota Prius. Who is correct?
   a. Technician A only
   b. Technician B only
   c. Both Technicians A and B
   d. Neither Technician A nor B
9. A hybrid high-voltage system is being discussed. Technician A states that when testing, HV cables insulation must never be pierced. Technician B states that piercing HV cable insulation, while dangerous, could also cause the controller to shut down. Who is correct?
   a. Technician A only
   b. Technician B only
   c. Both Technicians A and B
   d. Neither Technician A nor B
10. An auxiliary 12 V is being discussed. Technician A states that its main purpose is to turn the electric motors in the transaxle. Technician B states that it is typically charged by the DC–DC converter. Who is correct?
   a. Technician A only
   b. Technician B only
   c. Both Technicians A and B
   d. Neither Technician A nor B

APPENDIX A

2017 ASE EDUCATION FOUNDATION AUTOMOBILE ACCREDITATION TASK LIST CORRELATION GUIDE

Task List	MLR Priority	Chapter	LO
I. ENGINE REPAIR			
A. General			
1. Research vehicle service information, including fluid type, vehicle service history, service precautions, and technical service bulletins.	P-1	5	5-01
2. Verify operation of the instrument panel engine warning indicators.	P-1	10	10-01
3. Inspect engine assembly for fuel, oil, coolant, and other leaks; determine necessary action.	P-1	10; 15	10-04; 15-03
4. Install engine covers using gaskets, seals, and sealers as required.	P-1	17	17-05
5. Verify engine mechanical timing.	P-2	48	48-08
6. Perform common fastener and thread repair, to include: remove broken bolt, restore internal and external threads, and repair internal threads with thread insert.	P-1	8	8-04
7. Identify service precautions related to service of the internal combustion engine of a hybrid vehicle.	P-2	3; 54	3-05; 54-07
B. Cylinder Head and Valve Train			
1. Adjust valves (mechanical or hydraulic lifters).	P-3	6	6-07
2. Identify components of the cylinder head and valve train.	P-1	12	12-05
C. Lubrication and Cooling Systems			
1. Perform cooling system pressure and dye tests to identify leaks; check coolant condition and level; inspect and test radiator, pressure cap, coolant recovery tank, heater core, and galley plugs; determine necessary action.	P-1	17	17-08
2. Inspect, replace, and/or adjust drive belts, tensioners, and pulleys; check pulley and belt alignment.	P-1	17	17-03
3. Remove, inspect, and replace thermostat and gasket/seal.	P-1	17	17-05
4. Inspect and test coolant; drain and recover coolant; flush and refill cooling system; use proper fluid type per manufacturer specification; bleed air as required.	P-1	17	17-02
5. Perform engine oil and filter change; use proper fluid type per manufacturer specification; reset maintenance reminder as required.	P-1	15	15-02
6. Identify components of the lubrication and cooling systems.	P-1	14; 16	14-0-3; 16-03 to 08
II. AUTOMATIC TRANSMISSION AND TRANSAXLE			
A. General			
1. Research vehicle service information including fluid type, vehicle service history, service precautions, and technical service bulletins.	P-1	5	5-01
2. Check fluid level in a transmission or a transaxle equipped with a dip-stick.	P-1	19	19-01
3. Check fluid level in a transmission or a transaxle not equipped with a dip-stick.	P-1	19	19-01
4. Check transmission fluid condition; check for leaks.	P-2	19	10-01 & 02
5. Identify drive train components and configuration.	P-1	23	23-01 to 05
B. In-Vehicle Transmission/Transaxle			
1. Inspect, adjust, and/or replace external manual valve shift linkage, transmission range sensor/switch, and/or park/neutral position switch.	P-2	19	19-03
2. Inspect for leakage at external seals, gaskets, and bushings.	P-1	19	19-03

Task List	MLR Priority	Chapter	LO
3. Inspect, replace, and/or align power train mounts.	P-2	19	19-03
4. Drain and replace fluid and filter(s); use proper fluid type per manufacturer specification.	P-1	19	19-02
C. Off-Vehicle Transmission and Transaxle			
1. Describe the operational characteristics of a continuously variable transmission (CVT).	P-3	20	20-03
2. Describe the operational characteristics of a hybrid vehicle drive train.	P-3	20; 54	20-01 & 02; 54-06
III. MANUAL DRIVE TRAIN AND AXLES			
A. General			
1. Research vehicle service information including fluid type, vehicle service history, service precautions, and technical service bulletins.	P-1	5	5-01
2. Drain and refill manual transmission/transaxle and final drive unit; use proper fluid type per manufacturer specification.	P-1	21	21-07
3. Check fluid condition; check for leaks.	P-2	21	21-07
4. Identify manual drive train and axle components and configuration.	P-1	21	21-01 - 06
B. Clutch			
1. Check and adjust clutch master cylinder fluid level; use proper fluid type per manufacturer specification.	P-1	22	22-06
2. Check for hydraulic system leaks.	P-1	22	22-06
C. Transmission/Transaxle			
1. Describe the operational characteristics of an electronically controlled manual transmission/transaxle.	P-2	18	18-01
D. Drive Shaft, Half Shafts, Universal Joints and Constant-Velocity (CV) Joints (Front, Rear, All, and Four-Wheel Drive)			
1. Inspect, remove, and/or replace bearings, hubs, and seals.	P-2	23; 38	23-04; 38-06 & 07
2. Inspect, service, and/or replace shafts, yokes, boots, and universal/CV joints.	P-2	23	23-04
3. Inspect locking hubs.	P-3	38	38-06
4. Check for leaks at drive assembly and transfer case seals; check vents; check fluid level; use proper fluid type per manufacturer specification.	P-2	10; 23	10-04; 23-04
E. Differential Case Assembly			
1. Clean and inspect differential case; check for leaks; inspect housing vent.	P-1	10; 23	10-04; 23-04
2. Check and adjust differential case fluid level; use proper fluid type per manufacturer specification.	P-1	10; 23	10-04; 23-04
3. Drain and refill differential housing.	P-1	10; 23	10-04; 23-04
4. Inspect and replace drive axle wheel studs.	P-1	23; 35	23-04; 35-05
IV. SUSPENSION AND STEERING SYSTEMS			
A. General			
1. Research vehicle service information including fluid type, vehicle service history, service precautions, and technical service bulletins.	P-1	5	5-01
2. Disable and enable supplemental restraint system (SRS); verify indicator lamp operation.	P-1	27; 46	27-07; 46-09
3. Identify suspension and steering system components and configurations.	P-1	26; 28	26-01 to 09; 28-01 to 09
B. Related Suspension and Steering Service			
1. Inspect rack and pinion steering gear inner tie rod ends (sockets) and bellows boots.	P-1	27	27-04
2. Inspect power steering fluid level and condition.	P-1	10; 27	10-02; 27-03

(Continued)

Task List	MLR Priority	Chapter	LO
3. Flush, fill, and bleed power steering system; use proper fluid type per manufacturer specification.	P-2	27	27-03
4. Inspect for power steering fluid leakage.	P-1	10; 27	10-04; 27-02
5. Remove, inspect, replace, and/or adjust power steering pump drive belt.	P-1	17	17-03
6. Inspect and replace power steering hoses and fittings.	P-2	27	27-04
7. Inspect pitman arm, relay (centerlink/intermediate) rod, idler arm, mountings, and steering linkage damper.	P-1	27	27-05
8. Inspect tie rod ends (sockets), tie rod sleeves, and clamps.	P-1	27	27-05
9. Inspect upper and lower control arms, bushings, and shafts.	P-1	29	29-07
10. Inspect and replace rebound bumpers.	P-1		
11. Inspect track bar, strut rods/radius arms, and related mounts and bushings.	P-1	29	29-10
12. Inspect upper and lower ball joints (with or without wear indicators).	P-1	29	29-04; 29-07
13. Inspect suspension system coil springs and spring insulators (silencers).	P-1	29	20-06
14. Inspect suspension system torsion bars and mounts.	P-1	29	29-10
15. Inspect and/or replace front/rear stabilizer bar (sway bar) bushings, brackets, and links.	P-1	29	29-05
16. Inspect, remove, and/or replace strut cartridge or assembly; inspect mounts and bushings.	P-2	29	20-09
17. Inspect front strut bearing and mount.	P-1	29	29-09
18. Inspect rear suspension system lateral links/arms (track bars), control (trailing) arms.	P-1	29	29-10
19. Inspect rear suspension system leaf spring(s), spring insulators (silencers), shackles, brackets, bushings, center pins/bolts, and mounts.	P-1	29	29-10
20. Inspect, remove, and/or replace shock absorbers; inspect mounts and bushings.	P-1	29	29-05
21. Inspect electric power steering assist system.	P-2	27	27-06
22. Identify hybrid vehicle power steering system electrical circuits and safety precautions.	P-2	27	27-06
23. Describe the function of suspension and steering control systems and components (i.e., active suspension and stability control).	P-3	28	28-09
C. Wheel Alignment			
1. Perform prealignment inspection; measure vehicle ride height.	P-1	30	30-07
2. Describe alignment angles (camber, caster, and toe).	P-1	30	30-01
D. Wheels and Tires			
1. Inspect tire condition; identify tire wear patterns; check for correct tire size, application (load and speed ratings), and air pressure as listed on the tire information placard/label.	P-1	25	25-01 to 03
2. Rotate tires according to manufacturer's recommendations including vehicles equipped with tire pressure monitoring systems (TPMS).	P-1	25	25-03
3. Dismount, inspect, and remount tire on wheel; balance wheel and tire assembly.	P-1	25	25-05
4. Dismount, inspect, and remount tire on wheel equipped with tire pressure monitoring system sensor.	P-1	25	25-06
5. Inspect tire and wheel assembly for air loss; determine necessary action.	P-1	25	25-09
6. Repair tire following vehicle manufacturer approved procedure.	P-1	25	25-09
7. Identify indirect and direct tire pressure monitoring systems (TPMS); calibrate system; verify operation of instrument panel lamps.	P-1	24; 25	24-05; 25-08
8. Demonstrate knowledge of steps required to remove and replace sensors in a tire pressure monitoring system (TPMS) including relearn procedure.	P-1	25	25-07 & 08

Task List V. BRAKES A. General	MLR Priority	Chapter	LO
1. Research vehicle service information including fluid type, vehicle service history, service precautions, and technical service bulletins.	P-1	5	5-01
2. Describe procedure for performing a road test to check brake system operation, including an anti-lock brake system (ABS).	P-1	35	35-01
3. Install wheel and torque lug nuts.	P-1	35; 37	35-05; 37-06
4. Identify brake system components and configuration.	P-1	31;32; 34, 36	31-06; 32-02 to 11; 34-1 to 05; 36-1 to 07
B. Hydraulic System			
1. Describe proper brake pedal height, travel, and feel.	P-1	33	33-04
2. Check master cylinder for external leaks and proper operation.	P-1	33	33-04 & 05
3. Inspect brake lines, flexible hoses, and fittings for leaks, dents, kinks, rust, cracks, bulging, wear, and loose fittings/supports.	P-1	33	33-08
4. Select, handle, store, and fill brake fluids to proper level; use proper fluid type per manufacturer specification.	P-1	33	33-02
5. Identify components of hydraulic brake warning light system.	P-3	32; 33	32-09; 33-09
6. Bleed and/or flush brake system.	P-1	33	33-03
7. Test brake fluid for contamination.	P-1	33	33-02
C. Drum Brakes			
1. Remove, clean, and inspect brake drum; measure brake drum diameter; determine serviceability.	P-1	37	37-02
2. Refinish brake drum and measure final drum diameter; compare with specification.	P-1	37	37-03
3. Remove, clean, inspect, and/or replace brake shoes, springs, pins, clips, levers, adjusters/self-adjusters, other related brake hardware, and backing support plates; lubricate and reassemble.	P-1	37	37-04
4. Inspect wheel cylinders for leaks and proper operation; remove and replace as needed.	P-2	37	37-05
5. Pre-adjust brake shoes and parking brake; install brake drums or drum/hub assemblies and wheel bearings; make final checks and adjustments.	P-1	37	37-06
D. Disc Brakes			
1. Remove and clean caliper assembly; inspect for leaks and damage/wear; determine necessary action.	P-1	35	35-02
2. Inspect caliper mounting and slides/pins for proper operation, wear, and damage; determine necessary action.	P-1	35	35-02
3. Remove, inspect, and/or replace brake pads and retaining hardware; determine necessary action.	P-1	35	35-02
4. Lubricate and reinstall caliper, brake pads, and related hardware; seat brake pads and inspect for leaks.	P-1	35	35-02
5. Clean and inspect rotor and mounting surface, measure rotor thickness, thickness variation, and lateral runout; determine necessary action.	P-1	35	35-03
6. Remove and reinstall/replace rotor.	P-1	35	35-03
7. Refinish rotor on vehicle; measure final rotor thickness and compare with specification.	P-1	35	35-04
8. Refinish rotor off vehicle; measure final rotor thickness and compare with specification.	P-1	35	35-04
9. Retract and re-adjust caliper piston on an integral parking brake system.	P-2	35	35-02
10. Check brake pad wear indicator; determine necessary action.	P-1	35	35-02
11. Describe importance of operating vehicle to burnish/break-in replacement brake pads according to manufacturer's recommendation.	P-1	35	35-05

(Continued)

Task List	MLR Priority	Chapter	LO
E. Power-Assist Units			
1. Check brake pedal travel with, and without, engine running to verify proper power booster operation.	P-2	33	33-04
2. Identify components of the brake power assist system (vacuum and hydraulic); check vacuum supply (manifold or auxiliary pump) to vacuum-type power booster.	P-1	32; 33	32-10 & 11; 33-07
F. Related Systems (i.e. Wheel Bearings, Parking Brakes, Electrical)			
1. Remove, clean, inspect, repack, and install wheel bearings; replace seals; install hub and adjust bearings.	P-1	38	38-06
2. Check parking brake system components for wear, binding, and corrosion; clean, lubricate, adjust, and/or replace as needed.	P-2	38	38
3. Check parking brake operation and parking brake indicator light system operation; determine necessary action.	P-1	10; 33	10-01; 33-09
4. Check operation of brake stop light system.	P-1	10; 33	10-05; 33-09
5. Replace wheel bearing and race.	P-2	38	38-06
6. Inspect and replace wheel studs.	P-1	35	35-05
G. Electronic Brake, Traction Control, and Stability Control Systems			
1. Identify traction control/vehicle stability control system components.	P-3	39	39-06 & 07
2. Describe the operation of a regenerative braking system.	P-3	20	20-01
VI. ELECTRICAL/ELECTRONIC SYSTEMS			
A. General			
1. Research vehicle service information including vehicle service history, service precautions, and technical service bulletins.	P-1	5	5-01
2. Demonstrate knowledge of electrical/electronic series, parallel, and series-parallel circuits using principles of electricity (Ohm's Law).	P-1	40	40-04, 07, 09 & 10
3. Use wiring diagrams to trace electrical/electronic circuits.	P-1	41	41-08
4. Demonstrate proper use of a digital multimeter (DMM) when measuring source voltage, voltage drop (including grounds), current flow, and resistance.	P-1	42	42-01 to 13
5. Demonstrate knowledge of the causes and effects from shorts, grounds, opens, and resistance problems in electrical/electronic circuits.	P-1	42	42-14
6. Use a test light to check operation of electrical circuits.	P-2	42	42-15
7. Use fused jumper wires to check operation of electrical circuits.	P-2	42	42-15
8. Measure key-off battery drain (parasitic draw).	P-1	43	43-12
9. Inspect and test fusible links, circuit breakers, and fuses; determine necessary action.	P-1	42	42-16
10. Repair and/or replace connectors, terminal ends, and wiring of electrical/electronic systems (including solder repair).	P-1	41	41-09 & 10
11. Identify electrical/electronic system components and configuration.	P-1	41	41-01 to 07
B. Battery Service			
1. Perform battery state-of-charge test; determine necessary action.	P-1	43	43-09
2. Confirm proper battery capacity for vehicle application; perform battery capacity and load test; determine necessary action.	P-1	43	43-10
3. Maintain or restore electronic memory functions.	P-1	43	43-11
4. Inspect and clean battery; fill battery cells; check battery cables, connectors, clamps, and hold-downs.	P-1	43	43-07
5. Perform slow/fast battery charge according to manufacturer's recommendations.	P-1	43	43-08
6. Jump-start vehicle using jumper cables and a booster battery or an auxiliary power supply.	P-1	43	43-08

Task List	MLR Priority	Chapter	LO
7. Identify safety precautions for high voltage systems on electric, hybrid-electric, and diesel vehicles.	P-2	3; 54	3-05; 54-07
8. Identify electrical/electronic modules, security systems, radios, and other accessories that require reinitialization or code entry after reconnecting vehicle battery.	P-1	43	43-11
9. Identify hybrid vehicle auxiliary (12v) battery service, repair, and test procedures.	P-2	43; 54	43-10; 54-08
C. Starting System			
1. Perform starter current draw test; determine necessary action.	P-1	44	44-07
2. Perform starter circuit voltage drop tests; determine necessary action.	P-1	44	44-08 & 09
3. Inspect and test starter relays and solenoids; determine necessary action.	P-2	44	44-10
4. Remove and install starter in a vehicle.	P-1	44	44-11
5. Inspect and test switches, connectors, and wires of starter control circuits; determine necessary action.	P-2	44	44-09
6. Demonstrate knowledge of an automatic idle-stop/start-stop system.	P-3	44	44-12
D. Charging System			
1. Perform charging system output test; determine necessary action.	P-1	44	44-18
2. Inspect, adjust, and/or replace generator (alternator) drive belts; check pulleys and tensioners for wear; check pulley and belt alignment.	P-1	17	17-03
3. Remove, inspect, and/or replace generator (alternator).	P-2	44	44-20
4. Perform charging circuit voltage drop tests; determine necessary action.	P-2	44	44-19
E. Lighting, Instrument Cluster, Driver Information, and Body Electrical Systems			
1. Inspect interior and exterior lamps and sockets including headlights and auxiliary lights (fog lights/driving lights); replace as needed.	P-1	45	45-10
2. Aim headlights.	P-2	45	45-11
3. Identify system voltage and safety precautions associated with high-intensity discharge headlights.	P-2	45	45-09
4. Disable and enable supplemental restraint system (SRS); verify indicator lamp operation.	P-1	46	46-09
5. Remove and reinstall door panel.	P-1	46	46-05
6. Describe the operation of keyless entry/remote-start systems.	P-3	46	46-06
7. Verify operation of instrument panel gauges and warning/indicator lights; reset maintenance indicators.	P-1	10	10-01
8. Verify windshield wiper and washer operation; replace wiper blades.	P-1	46	46-08
VII. HEATING, VENTILATION, AND AIR CONDITIONING (HVAC)			
A. General			
1. Research vehicle service information, including refrigerant/oil type, vehicle service history, service precautions, and technical service bulletins.	P-1	5	5-01
2. Identify heating, ventilation and air conditioning (HVAC) components and configuration.	P-1	47	47-01 to 07
B. Refrigeration System Components			
1. Inspect and replace A/C compressor drive belts, pulleys, and tensioners; visually inspect A/C components for signs of leaks; determine necessary action.	P-1	17; 47	17-03; 47-08
2. Identify hybrid vehicle A/C system electrical circuits and the service/safety precautions.	P-2	3; 54	3-05; 54-07
3. Inspect A/C condenser for airflow restrictions; determine necessary action.	P-1	47	47-08
C. Heating, Ventilation, and Engine Cooling Systems			
1. Inspect engine cooling and heater systems hoses and pipes; determine necessary action.	P-1	17	17-04

(Continued)

Task List	MLR Priority	Chapter	LO
D. Operating Systems and Related Controls			
1. Inspect A/C-heater ducts, doors, hoses, cabin filters, and outlets; determine necessary action.	P-1	47	47-08
2. Identify the source of A/C system odors.	P-2	47	47-08
VIII. ENGINE PERFORMANCE			
A. General			
1. Research vehicle service information, including fluid type, vehicle service history, service precautions, and technical service bulletins.	P-1	5	5-01
2. Perform engine absolute manifold pressure tests (vacuum/boost); document results.	P-2	13	13-03
3. Perform cylinder power balance test; document results.	P-2	13	13-4
4. Perform cylinder cranking and running compression tests; document results.	P-2	13	13-05
5. Perform cylinder leakage test; document results.	P-2	13	13-06
6. Verify engine operating temperature.	P-1	17	17-08
7. Remove and replace spark plugs; inspect secondary ignition components for wear and damage.	P-1	48	48-11
B. Computerized Controls			
1. Retrieve and record diagnostic trouble codes (DTC), OBD monitor status, and freeze frame data; clear codes when applicable.	P-1	51	51-07
2. Describe the use of the OBD monitors for repair verification.	P-1	51	51-05 & 07
C. Fuel, Air Induction, and Exhaust Systems			
1. Replace fuel filter(s) where applicable.	P-2	49	49-16
2. Inspect, service, or replace air filters, filter housings, and intake duct work.	P-1	10; 52	10-03; 52-10
3. Inspect integrity of the exhaust manifold, exhaust pipes, muffler(s), catalytic converter(s), resonator(s), tail pipe(s), and heat shields; determine necessary action.	P-1	52	52-12
4. Inspect condition of exhaust system hangers, brackets, clamps, and heat shields; determine necessary action.	P-1	52	52-12
5. Check and refill diesel exhaust fluid (DEF).	P-2	10	10-02
D. Emissions Control Systems			
1. Inspect, test, and service positive crankcase ventilation (PCV) filter/breather, valve, tubes, orifices, and hoses; perform necessary action.	P-2	53	53-12

APPENDIX B

ASE EDUCATION FOUNDATION INTEGRATED APPLIED ACADEMIC SKILLS CORRELATION GUIDE

Applied Math	Chapter	Applied Science	Chapter	Applied Communication	Chapter
AM-1	10, 14, 15, 18, 24, 26, 30, 34, 40, 44	AS-1	14, 17, 20, 25, 30, 33, 34, 38, 40	AC-1	11
AM-2	24, 30, 40, 44	AS-2	3, 14, 17, 33, 40	AC-2	11
AM-3	30, 40, 42	AS-3	3, 14, 40	AC-3	11
AM-4	30, 40, 43	AS-4	3, 40	AC-4	11
AM-5	30, 43	AS-5	3, 40	AC-5	11
AM-6	42	AS-6	13, 40	AC-6	11
AM-7	43	AS-7	5, 40	AC-7	11
AM-8		AS-8	40	AC-8	11
AM-9	42	AS-9	40, 47	AC-9	11
AM-10	43	AS-10	50	AC-10	11
AM-11		AS-11	5	AC-11	11
AM-12	42	AS-12	31	AC-12	11
AM-13	44	AS-13		AC-13	11
AM-14		AS-14		AC-14	11
AM-15	50	AS-15		AC-15	11
AM-16	13	AS-16		AC-16	11
AM-17		AS-17		AC-17	11
AM-18	49	AS-18	51	AC-18	11
AM-19		AS-19		AC-19	11
AM-20		AS-20		AC-20	11
AM-21		AS-21		AC-21	11
AM-22		AS-22		AC-22	11
AM-23		AS-23	47	AC-23	5
AM-24		AS-24	53	AC-24	11
AM-25		AS-25		AC-25	11
AM-26		AS-26		AC-26	11
AM-27		AS-27	16	AC-27	11
AM-28		AS-28	47	AC-28	11
AM-29		AS-29	47	AC-29	11
AM-30		AS-30	16	AC-30	11
AM-31		AS-31	16	AC-31	11
AM-32		AS-32	16	AC-32	11
AM-33	47	AS-33	45	AC-33	11
AM-34		AS-34		AC-34	11
AM-35		AS-35		AC-35	11
AM-36		AS-36		AC-36	11
AM-37		AS-37		AC-37	

(Continued)

Applied Math	Chapter	Applied Science	Chapter	Applied Communication	Chapter
AM-38		AS-38	52	AC-38	
AM-39		AS-39	52	AC-39	
AM-40	47	AS-40	52	AC-40	
AM-41		AS-41		AC-41	
AM-42		AS-42		AC-42	
AM-43		AS-43	52	AC-43	
AM-44		AS-44	52	AC-44	
AM-45	42	AS-45	52	AC-45	
AM-46		AS-46	52	AC-46	
AM-47	42	AS-47		AC-47	
AM-48		AS-48		AC-48	
AM-49		AS-49		AC-49	
AM-50		AS-50	12	AC-50	
AM-51		AS-51	31	AC-51	
AM-52		AS-52		AC-52	
AM-53		AS-53		AC-53	
AM-54		AS-54	43	AC-54	
		AS-55	43		
		AS-56	43		
		AS-64	44		
		AS-65	48		
		AS-66	46		
		AS-67	46		
		AS-69	41		
		AS-72	44		
		AS-73	43		
		AS-74	43		
		AS-75	44		
		AS-76	48		
		AS-78	41		
		AS-83	50		
		AS-84	47		
		AS-89	53		
		AS-91	31		
		AS-99	12		

GLOSSARY

abrasive discs (cutting wheels) Abrasive wheels or flat discs fitted to bench, pedestal, and portable grinders.

ABS master cylinder Master cylinder used in anti-lock brake systems.

absorbed glass mat Batteries that have the electrolyte absorbed within a mat of fine glass fibers.

AC See alternating current.

acceleration An increase in a vehicle's speed.

accelerator pedal The foot-operated pedal used by the driver to increase and decrease the amount of power the engine develops.

accelerator pump circuit The carburetor circuit involved with heavy acceleration, directly connected to the accelerator pump. Also used to prime the engine during cold starting.

accelerator pump A small pump, usually located inside the carburetor, that sprays an extra amount of fuel into the carburetor air horn during acceleration.

accumulator A device placed between the evaporator and the compressor to collect liquid refrigerant and prevent it from entering the compressor.

Ackermann principle or angle The Ackermann principle angles the steering arms toward the center of the vehicle such that imaginary lines drawn from the center of the steering knuckle pivot points, through the center of the outer tie-rod ends, intersect at the center of the rear axle or at a point halfway between the two axles of a tandem drive.

activated charcoal Charcoal that will absorb large amounts of vapor; used in charcoal canisters to store and release fuel vapors.

active control A system of providing constant feedback from sensors in the vehicle to the control unit.

actuating The act of making something move or work.

actuators Electrical devices that are used to control fan speed, heater core coolant flow, air-conditioning clutch activation, compressor displacement, blend doors, and fresh-air/recirculation doors.

adaptive air suspension A suspension system that uses rubber bags or bladders filled with air to support the weight of the vehicle.

adaptive learning Software that can learn and change strategy based on different factors.

adequate ventilation Ventilation system or procedures that are designed to keep the air in the shop at a safe level.

adjustable bushing A brace or nylon part that pushes against the rack to adjust the mesh of the rack teeth to the pinion teeth.

adjusting nut The nut used to adjust the end play or preload of a wheel bearing.

advance mechanism A device used to trigger an earlier spark based on engine conditions.

aerate The tendency to create air bubbles in a fluid.

aftermarket A company other than the original manufacturer that produces equipment or provides services.

air driers A device fitted to compressed air lines to remove moisture.

air drill A compressed air–powered drill.

air gap The space or clearance between two components, such as the space between the tone wheel and the pickup coil in a wheel speed sensor.

air hammer A tool powered by compressed air with various hammer, cutting, punching, or chisel attachments; also called an air chisel.

air impact wrench An impact tool powered by compressed air, designed to undo tight fasteners.

air nozzle A compressed air device that emits a fine stream of compressed air for drying or cleaning parts.

air ratchet A ratchet tool for use with sockets, powered by compressed air.

air spring A part that provides the springing action or auxiliary spring. It is typically used in air suspension systems or heavy truck applications.

air spring Consists of a flexible rubber bladder, which seals the outside of the upper and lower halves of the shock absorber. When inflated, the bladder pushes the halves apart.

air supply system Equipment in a motor vehicle that delivers air to the engine.

Allen wrenches A tool that fits into a fastener with an internal hexagonal recess.

all-wheel drive (AWD) A drivetrain arrangement in which all of the wheels drive the vehicle.

alternative fuel A non-petroleum-based motor fuel.

ambient Relates to the immediate surroundings. For example, the temperature of the atmospheric air surrounding a vehicle.

American Wire Gauge (AWG) Standardized wire gauge used in North America. The higher the number, the smaller the wire is, and the lower the current-carrying capacity.

ammeter A device used to measure current flow.

amp An abbreviation for amperes, the unit for current measurement.

anchor pin A component of the backing plate that takes all of the braking force from the brake shoes.

angle grinder A portable grinder for grinding or cutting metal.

antifoaming agents Oil additives that keep oil from foaming.

antifriction bearing A bearing that uses rolling elements to reduce friction.

antilock brake system (ABS) A safety measure for the braking system that uses a computer to monitor the speed of each wheel and control the hydraulic pressure to each wheel to prevent wheel lockup.

applied force Pressure placed on something.

arc joint pliers Pliers with parallel slip jaws that can increase in size; also called Channellocks.

armature The rotating wire coils in motors and generators. It is also the moving part of a solenoid or relay, and the pole piece in a permanent magnet generator.

ASE Education Foundation The portion of ASE that evaluates and accredits entry-level automotive technology programs. It also brings together education, workplace experience, and mentorship.

aspect ratio The ratio of sidewall height to section width of a tire.

asymmetric tread pattern A tread pattern that differs on each side and therefore is usually directional.

atmospheric pressure (atm) The pressure of the air surrounding everything, caused by gravity and the weight of air. The higher the altitude from sea level, the lower the atmospheric pressure.

automatic brake self-adjuster A system on drum brakes that automatically adjusts the brakes to maintain a specified amount of running clearance between the shoes and drum.

automatic load-adjustable shock absorber Typically, an air shock absorber used in an automatic load-sensing system that adjusts ride height (ground clearance) automatically, such as when additional weight is added to the vehicle; also called self-leveling.

automatic oiler A device fitted to compressed air systems to oil air tools.

Automotive Service Excellence (ASE) An independent, nonprofit organization dedicated to the improvement of vehicle repair through the testing and certification of automotive professionals.

Automotive Youth Educational Systems (AYES) An independent, nonprofit organization that is a partnership among automotive manufacturers, their dealerships, and affiliated secondary automotive programs.

available voltage test Measurement of voltage at various points in a circuit, with the black meter lead on ground and the red lead probing the circuit.

available voltage The maximum amount of voltage that the induction coil secondary is capable of putting out.

aviation snips A scissorlike tool for cutting sheet metal.

axial load The load applied in line with a shaft. It can be controlled with thrust bearings.

axle A shaft connected to wheels that transmits the driving torque to the wheels.

BA (brake assist) An enhanced safety system built in to some ABS systems that anticipates a panic stop and applies maximum braking force to slow the vehicle as quickly as possible.

background color Background color used on a sign to indicate the type of warning

backing plate A metal plate to which the brake lining is fixed.

backlash The required clearance between two meshing gears.

baffles A part used to prevent sloshing of liquids. Used in oil pans on high performance vehicles.

ball joint A swivel connection mounted in the outer end of the front control arm. These swivels are typically constructed with a ball and socket to allow pivoting.

ballast A device that increases lighting voltage substantially and controls the current to the bulb.

ballast resistor Used to limit the amount of current flowing in the ignition primary circuit.

ball-peen (engineer's) hammer A hammer with one flat face and one rounded face.

band A metal band with friction material bonded to one side. The band is contracted around a drum to stop the drum from spinning.

band brake A type of brake that utilizes a steel band lined with friction material that wraps around a brake drum to slow the drum.

barometric pressure (BARO) sensor A sensor that measures atmospheric pressure.

barrier cream A cream that looks and feels like a moisturizing cream but has a specific formula to provide extra protection from chemicals and oils.

bars A unit of measure of pressure. Short for barometric. One bar equals 14.7 psi

batteries Device used to store electricity in chemical form.

battery charger A device that charges a battery, reversing the discharge process.

battery electric vehicle A vehicle powered by battery only.

battery terminal configuration The placement of positive and negative battery terminals.

bead seat The part of the wheel that the tire seals against.

beam axle Suspension system in which one set of wheels is connected laterally by a single beam or shaft.

bearing races Hardened metal surfaces that roller or ball bearings fit into when a bearing is properly assembled.

belt alternator starter (BAS) A type of hybrid drive system that uses a belt-driven alternator/starter that operates on 42 volts.

belt routing label A label that lists a diagram of the routing of the belt(s) for the engine accessories.

bench grinder (pedestal grinder) A grinder that is fixed to a bench or pedestal.

bench vice Tool used to hold parts while working on them.

bendable tangs Small tabs on the brake pad backing plate that are crimped on to the caliper, creating a secure fit and reducing noise.

bias-ply A tire constructed in a latticed, crisscrossing structure, with alternate plies crossing over one another and laid with the cord angles in opposite directions.

bidirectional control The ability to command different solenoids and actuators on and off to check their operation.

bidirectional scanners Scanners used to cause various components and systems to operate for test purposes.

bleeder screw A screw that allows air and brake fluid to be bled out of a hydraulic brake system when it is loosened, and seals the brake fluid in when it is tightened.

bleeding The process of removing air from a hydraulic braking system.

blind rivet A rivet that can be installed from its insertion side.

blink code A method of providing fault code data for a specific system, which involves counting the number of flashes from a warning lamp and observing longer pauses between the light blinks.

blowby Pressure that leaks past the compression rings during compression and combustion.

blowby gas The result of combustion gases leaking past the compression rings and getting into the crankcase.

blower motor An electric motor, usually the permanent magnet type, that moves air over the air-conditioning evaporator and heater core.

blow-off valve, intake Allows for excess pressure that is built up in the intake track to exit the intake after the pressure overcomes the spring pressure on this valve.

body control module (BCM) An onboard computer that controls many vehicle functions, including the vehicle interior and exterior lighting, horn, door locks, power seats, and windows.

boiling point The temperature at which a substance begins to change from a liquid to a gas.

boiling point test A test to determine the boiling point of brake fluid.

bolt A type of threaded fastener with a thread on one end and a hexagonal head on the other.

bonded linings Brake linings that are essentially glued to the brake pad backing plate; they are more common on light-duty vehicles.

boost valve A valve located in the HCU that is controlled by the EBCM; it allows brake fluid under high pressure to flow into the HCU hydraulic circuits, to apply the brakes when commanded.

bottoming tap A thread-cutting tap designed to cut threads to the bottom of a blind hole.

bouncy ride/excessive body movement Up and down movement that isn't being controlled by the shock absorbers/struts.

box-end wrench A wrench or spanner with a closed or ring end to grip bolts and nuts.

brake booster A vacuum- or hydraulically operated device that increases the driver's braking effort.

brake drum A short, wide, hollow cylinder that is capped on one end and bolted to a vehicle's wheel; it has an inner friction surface that the brake shoe is forced against.

brake fade The reduction in stopping power caused by a change in the brake system such as overheating, water, or overheated brake fluid.

brake hose A flexible section of the brake lines between the body and suspension that allows for steering and suspension movement.

brake lathe A tool used to refinish the friction surface of a drum or rotor surface by removing a small amount of metal and returning it to a concentric, nondirectional finish.

brake lines Made of seamless, double-walled steel and able to transmit over 1000 psi (6895 kPa) of hydraulic pressure through the hydraulic brake system.

brake lining thickness gauge A tool used to measure the thickness of the brake lining.

brake pad shims and guides Small pieces of metal that cushion the brake pad and absorb some of the vibration, helping to cut down on unwanted noise.

brake pedal The foot-operated lever which provides leverage to the brake system.

brake pedal emulator A brake pedal assembly used in electronically controlled braking systems to send the driver's braking intention to the computer; it mimics the feel of a standard brake pedal.

brake shoe The arched metal shoes and friction material in a drum brake system.

brake shoe adjustment gauge An adjustable tool used to pre-adjust the brake shoes to the diameter of the brake drum.

brake spoon A tool used to adjust the brake lining–to-drum clearance when the drum is installed on the vehicle.

brake spring pliers A tool used for removing and installing brake return springs.

brake switch The electrical switch that is activated by the brake pedal; it turns on the brake lights and signals the EBCM that the brakes are being applied.

brake wash station A piece of equipment designed to safely clean brake dust from drum and disc brake components.

brake-by-wire system A braking system that uses no mechanical connection between the brake pedal and each brake unit. The system uses electrically actuated motors or a separate hydraulic system to apply brake force.

brakes A system made up of hydraulic and mechanical components designed to slow or stop a vehicle.

braking torque The torque acting to twist the axle housing around its center during braking.

breaker plate The movable plate that the breaker points are mounted on that pivots as the vacuum advance pulls on it.

breather tube A tube used in the PCV system to allow fresh air into the engine crankcase.

British thermal unit (Btu) A measure of heat energy. It takes 1 Btu to raise the temperature of 1 pound of water 1°F.

brushless DC motor An electric motor that does not have any brushes and is sometimes called an electronically commutated motor. In this type of motor, an electronic control module replaces the brushes and commutator.

bump steer The undesired condition produced when hitting a bump, where the vehicle darts to one side as the steering linkage is pushed or pulled as a result of the travel of the suspension.

bushings A part that allows movement in a pivot point. It can also cushion suspension fulcrum points.

butt connector A crimp or solder joint that creates a permanent connection.

bypass filter An oil filter system that only filters some of the oil.

bypass valve, lubrication system A valve used in filter bases or near coolers that will open to provide flow when restriction become too great.

caliper A hydraulic device that uses pressure from the master cylinder to apply the brake pads against the rotor.

caliper dust boot seal driver set A set of drivers used to install metal-backed caliper dust boot seals.

caliper piston pliers A tool used to grip caliper pistons while removing them.

caliper piston retracting tool A tool used to retract caliper pistons on integrated parking brake systems.

cam lobes Raised areas or protrusions on an otherwise round shaft.

camber The side-to-side vertical tilt of the wheel. It is viewed from the front of the vehicle and measured in degrees. Negative camber is when the top of the tire is closer to the center of the vehicle than the bottom of the tire.

camshaft position (CMP) sensor A detection device that signals to the PCM the rotational position of the camshaft.

camshaft position sensor A sensor mounted near the camshaft and used to send camshaft and valve position information to the PCM.

CANbus circuit A two-wire communication network that transmits status and command signals between control modules in a vehicle.

carbon monoxide poisoning Exposure to higher than tolerable levels of carbon monoxide, resulting in headaches, fatigue, or loss of consciousness, eventually resulting in death.

carburetor float bowl The part of the carburetor that holds fuel to be burned in the engine; the bowl is at a constant level of fuel to ensure adequate fuel is present during driving.

carrier Part of the throw-out bearing assembly. The part that the bearing is pressed on to.

casing plies Network of cords that give the tire shape and strength; also known as casing cords.

caster The angle formed through the wheel pivot points when viewed from the side in comparison to a vertical line through the wheel.

catalyst heating mode An operating mode in GDI systems where fuel is injected near the top of the compression stroke; the fuel is ignited; and then an additional amount of fuel is injected during the power stroke to create heat later in the cycle so that the catalytic convert can be warmed up faster.

Cause A concise, but complete, description of the root cause of the customer's concern.

C-clamp A clamp shaped like the letter C; it comes in various sizes and can clamp various items.

center electrode The electrode located in the center of a spark plug. It is the hottest part of the spark plug.

center punch Less sharp than a prick punch, the center punch makes a bigger indentation that centers a drill bit at the point where a hole is required to be drilled.

centerline A real or imaginary line drawn through the center of something.

centrifugal advance mechanism An ignition timing device, located above or beneath the distributor base plate, that rotates with the distributor cam and is used to advance the spark. As engine speed rises, the flyweights on the advance mechanism are thrown outward by centrifugal force. Because the distributor cam is able to rotate on the distributor shaft, the weights act against their springs and move the distributor cam forward.

centrifugal force The apparent force by which a rotating mass tries to move outward, away from its axis of rotation.

centrifugal switch A switch in the sensor of a TPMS that allows the sensor to go to sleep when the vehicle stops, which extends the TPMS's battery life.

channel The number of wheel speed sensor circuits and hydraulic circuits the EBCM monitors and controls.

charcoal canister A device used to trap the fuel vapors. The fuel vapors adhere to the charcoal until the engine is started, and engine vacuum can be used to draw (purge) the vapors into the engine so that they can be burned along with the air-fuel mixture.

chassis The main support frame in a vehicle. It includes the running gear, such as suspension, the engine, and the drivetrain.

chassis and brake technicians Person who services the chassis and brake systems.

chlorofluorocarbon (CFC) A chlorine-based fluorocarbon compound used as refrigerant in older air-conditioning systems.

choke A device that provides a rich air-fuel mixture until the engine warms up by restricting the flow of air at the entrance to the carburetor, before the venturi.

circuit breaker A device that trips and opens a circuit, preventing excessive current flow in a circuit. It can be reset to allow for reuse.

circuit or schematic diagram A pictorial representation or road map of the wiring and electrical components.

CKP (crankshaft position) sensor A sensor used by the PCM to monitor engine speed. It can be one of three types of sensors—Hall effect, magnetic pickup, or optical.

Clean Air Act (CAA) A policy signed into law in 1990 that sets standards for air pollution to eliminate ozone-depleting elements.

cleaning gun A device with a nozzle controlled by a trigger fitted to the outlet of pressure cleaners.

clock spring A special rotary electrical connector located between the steering wheel and the steering column that maintains a constant electrical connection with the wiring system while the vehicle's steering wheel is being turned.

closed-loop system A refrigerant system where nothing else enters or exits the system.

clutch disc The center component of the clutch assembly, with friction material riveted on each side. Also called a clutch plate or friction disc.

clutch fork The part of the clutch linkage that operates the throw-out bearing.

clutch pedal The foot-operated pedal used by the driver to engage and disengage the clutch.

clutch system A mechanically operated assembly that connects and disconnects the engine from the transmission.

codes Another name for a diagnostic trouble code.

coefficient of friction (CoF) The amount of force required to move an object while in contact with another, divided by its weight.

coil spring Spring steel wire, heated and wound into a coil, that is used to support the weight of a vehicle.

coil spring pressure plate A type of pressure plate that uses coil springs to provide the clamping force.

cold chisel The most common type of chisel, used to cut cold metals. The cutting end is tempered and hardened so that it is harder than the metals that need to be cut.

cold cranking amps (CCA) The load in amps that a battery can deliver for 30 seconds while maintaining a voltage of 1.2 volts per cell (7.2 volts for a 12-volt battery) or higher at 0°F (−18°C).

combination wrench A type of wrench that has an open end on one end and a closed-end wrench on the other.

common bore When a single cylinder is used for two pistons. A tandem master cylinder would be an example of two pistons in one bore.

commutator A device made on armatures of electric generators and motors to control the direction of current flow in the armature windings.

companion flange A splined flange attached to a vehicle component, such as a drive axle pinion shaft, that bolts to a flange yoke on a driveshaft.

compensating port Connects the brake fluid reservoir to the master cylinder bore when the piston is fully retracted, allowing for expansion and contraction of the brake fluid.

compliance bushing A rubber bushing with a voided section molded in it that allows component movement under torque application. It is typically used in control arms on FWD vehicles to minimize torque steer issues.

compression ratio (CR) The volume of the cylinder with the piston at bottom dead center as compared to the volume of the cylinder at top dead center, given in a ratio such as 9:1 CR.

compression stroke The stroke of the piston during which air and fuel is being compressed into a small area prior to ignition.

compressor A belt- or electrically driven device designed to increase refrigerant pressure and cause refrigerant to travel through the air-conditioning system.

computer-controlled carburetor A carburetor that changes and sets air-fuel ratio based on commands from a PCM that uses memory and input from sensors.

concentric interrupter rings Two interrupter rings that have the same center.

Concern The first C in the three Cs, documenting the original concern that the customer came into the shop with. This documentation will go on the repair order, invoice, and service history.

condensation Change of state from a vapor to a liquid, such as the moisture that collects on a cool surface.

condenser A component of the HVAC system that transfers heat from the system to the atmosphere.

conduction The process of transferring heat through matter by the movement of heat energy through solids from one particle to another.

conductors A material that allows electricity to flow through it easily. It is made up of atoms with one to three valance ring electrons.

connector The plastic housing on the end of a wiring harness that holds the wire terminals in place. It can also refer to a type of wire terminal that connects wires together or to a common point such as a bolt.

conservation of energy A physical law that states that energy cannot be created or destroyed.

constant mesh A term used to describe two or more parts, such as gears, that are in constant contact with each other.

constant-velocity (CV) joints A flexible joint used to transmit torque in axles and driveshafts. This style maintains a constant velocity no matter the angle.

contact breaker point ignition system A type of ignition system that uses a mechanical means of turning the primary circuit on and off.

contact breaker points A mechanically operated electrical switch that is fixed to the distributor base plate and opened and closed by the distributor cam with the rotation of the engine. The contacts normally form a self-contained unit, fixed to the base plate by a retaining screw engaged in a slot in the fixed contact.

continuous monitoring A term that describes OBD II monitors that run continuously throughout the drive cycle.

control arm The primary load-bearing element of a vehicle's suspension system, commonly referred to as an A-arm or wishbone. These arms may be used as an upper and lower pivot point for the wheel assembly. They attach to the chassis with rubber bushings that allow up-and-down movement of the tire and wheel assembly.

control unit Any device that controls another object, such as a computer.

convection The process of transferring heat by the circulatory movement that occurs in a gas or fluid as areas of differing temperatures exchange places due to variations in density and the action of gravity.

conventional oil Type of oil that is processed from crude oil to the desired viscosity, after which additives are added to increase wear resistance.

conventional theory The theory that electrons flow from positive to negative.

convertible A vehicle that converts from having an enclosed top to having an open top by a roof that can be removed, retracted, or folded away.

coolant label A label that lists the type of coolant installed in the cooling system.

cored solder Solder that is in the form of a hollow wire. The center is filled with flux, which is used as a cleaning agent while the solder is being applied to the metal surfaces.

cornering force The force between the tread and the road surface as a vehicle turns.

Correction The third C in the three Cs, documenting the repair that solved the vehicle fault. This documentation goes on the repair order, invoice, and service history and must include the procedure used as well as a brief description of the correction.

corrosion inhibitors Oil additives that keep acid from forming in the oil.

cotter pin A single-use, soft, metal pin that can be bent into shape and is used to retain bearing adjusting nuts.

counter-electromotive force (CEMF) An electromagnetic force produced by the spinning magnetic field of the armature, which induces current in the opposite direction of battery current through the motor.

coupe A two-door vehicle that has seating for two people and may have a small rear seat.

crank angle position The position of the crankshaft, measured in degrees.

cranking amps (CA) A standard similar to CCA, but that measures the battery's function at a higher temperature (32°F [0°C]).

crankshaft position (CKP) sensor A sensor used by the PCM to monitor engine speed. It can be one of three types of sensors—Hall effect, magnetic pickup, or optical.

crankshaft A vehicle engine component that transfers the reciprocating movement of pistons into rotary motion.

crescent pump An oil pump that uses a crescent-shaped part to separate the oil pump gears from each other, allowing oil to be moved from one side of the pump to the other.

cross-arm A description for an arm that is set at right angles, or 90 degrees, to another component.

cross-cut chisel A type of chisel for metal work that cleans out or cuts key ways.

cross-flow radiator Type of radiator that has the coolant flow from side to side and is more conformable to the low hood designs of today's vehicles.

crude oil Oil that originates from the ground; it is then refined into a usable substance.

current clamp A device that clamps around a conductor to measure current flow. It is often used in conjunction with a digital multimeter (DMM).

current DTC Condition indicating that the test failed during the last ignition cycle (for two-trip codes, also on the previous cycle).

current flow The flow of electrons, typically within a circuit or component.

curved file A type of file that has a curved surface for filing holes.

CV joints A joint used to transmit torque through wider angles and without the change of velocity that occurs in U-joints.

CVT (continuously variable transmission) A type of transmission that lacks the fixed gears found in conventional transmissions; it adjusts gear ratios infinitely within the design of the transmission.

cylinder bore The diameter of the hole that the piston moves in; it is one of the main components that is included in the size of the engine.

cylindrical roller bearing assembly A type of wheel bearing with races and rollers that are cylindrical in shape and roll between inner and outer races, which are parallel to each other.

data link connector (DLC) The under-dash connector through which the scan tool communicates to the vehicle's computers. Connected to the DLC, the scan tool displays the readings from the various sensors and can retrieve trouble codes, freeze-frame data, and system monitor data.

DC See direct current.

dead axles An axle that only supports the wheel. It does not drive it.

dead blow hammer A type of hammer that has a cushioned head to reduce the amount of head bounce.

deceleration The process of decreasing a vehicle's speed.

deep dish wheel A wheel with negative offset, which gives the outside of the wheel a deep dish appearance. Deep dish refers to the side of the wheel that is farthest from the drop center.

deflecting force A force that moves an object in a different direction or into a different shape.

depth micrometer A micrometer that measures the depth of an item such as how far a piston is below the surface of the block.

detergents Reduce carbon deposits on parts such as piston rings and valves.

detonation The condition in which the remaining fuel charge fires or burns too rapidly after the initial combustion of the air-fuel mixture. It is audible through the combustion chamber walls as a knocking noise.

diagonal cutting pliers Cutting pliers for small wire or cable.

dial bore gauge A gauge that is used to measure the inside diameter of bores with a high degree of accuracy and speed.

dial indicator A device for precision measurements used to measure small variations, such as end play, movement in a bearing, or run-out.

diaphragm pressure plate A slightly conical, spring steel plate used to provide the clamping force for the clutch assembly.

dichlorodifluoromethane An inert, colorless gas that can be used as a refrigerant. It is stored in white containers.

die stock Tool used to hold a die during use.

die Used to cut external threads on a metal shank or bolt.

diesel exhaust fluid (DEF) A mixture of urea and water that is injected into the exhaust system of a late-model diesel-powered vehicle to reduce exhaust oxides of nitrogen emissions.

dieseling A condition in which the engine continues to run after the ignition key is turned off. Also referred to as run-on.

differential gears A gear arrangement that splits the available torque equally between two wheels while allowing them to turn at different speeds when required.

digital multimeter (DMM)-galvanic reaction test A test using a DMM to measure the voltage created in worn out brake fluid or engine coolant.

dipper A part used to splash oil inside the crankcase for lubrication purposes.

direct ignition system May refer to a waste spark ignition or a coil-on-plug ignition system, in which the coils are directly attached to the spark plugs.

direct TPMS A type of automated tire pressure monitoring system that measures tire pressure and possibly temperature via a sensor installed inside each wheel.

direct-acting telescopic shock absorber A shock absorber designed to reduce spring oscillations.

directional and asymmetric tread pattern A tread pattern that is both directional and asymmetric, which means the tire is designed to rotate in only one direction and has one side that must face outward to ensure that the tire performs as designed under operating conditions.

directional tread pattern A tread pattern designed to pump water out from under the tire; each tire must be placed in a particular spot on the vehicle.

disc brake pads Brake pads that consist of a friction material bonded or riveted to a steel backing plate; designed to wear out over time.

disc brake rotor micrometer A specially designed micrometer used to measure the thickness of a rotor.

disc brakes A type of brake system that forces stationary brake pads against the outside of a rotating brake rotor.

dispersants Oil additives that keep contaminants held in suspension in the oil, to be removed by the filter or when the oil is changed.

displacement-on-demand A feature that allows cylinders to be taken off-line when not needed, such as during vehicle cruise.

distributor The part of an ignition system that distributes the spark to the spark plugs in the correct sequence and at the correct time. It includes a distributor cap, rotor, shaft, and usually a switching device.

distributor base plate A round metal plate near the top of the distributor that is attached to a distributor housing; also called a breaker plate.

distributor cap The top portion of a distributor, used to make a connection between the spinning rotor and the high-tension leads.

distributorless ignition system An ignition system that does not include a distributor. It uses signals from the crankshaft position sensor and the camshaft position sensor sent to the PCM to determine when to send a signal to the ignition module.

doors and gates Used to separate areas and control access.

double Cardan joint A type of joint that uses two Cardan joints housed in a short carrier and that reduces the change in velocity of a single Cardan joint by using the second joint to cancel out the changes in velocity of the first joint.

double flare A seal that is made at the end of metal tubing or pipe.

double insulated Tools or appliances that are designed in such a way that no single failure can result in a dangerous voltage coming into contact with the outer casing of the device.

double-row ball bearing assembly A single ball bearing assembly using two rows of ball bearings riding in two channels in the races.

down-flow radiator A radiator in which the coolant flows from the top to the bottom.

drag link A steel or iron rod that transfers movement of the pitman arm to a relay lever.

drain line A wire included in a harness, with one end grounded to reduce interference or noise being induced into the harness.

drain plug Typically a threaded plug used to remove fluid from a system.

drawing-in method A method for replacing wheel studs that uses the lug nut to draw the wheel stud into the hub or flange.

drift punch A type of punch used to start pushing roll pins to prevent them from spreading.

drill press A device that incorporates a fixed drill with multiple speeds and an adjustable worktable. It can be free standing or fixed to a bench.

drill vice Tool used to hold parts while they are being drilled on a drill press.

drivability technician A technician who diagnoses and identifies mechanical and electrical faults that affect vehicle performance and emissions.

drive axle assembly The components that make up the drive axle, including the axles, final drive assembly, bearings, and axle housing.

drive axles A very strong part used to turn the wheels.

drive cycle A series of prescribed automobile operating conditions during which emissions testing is performed.

driveline angularity The angles at the universal joints.

driven center plate The friction disc that is held firmly against the flywheel by a pressure plate and that transfers power from the flywheel to the transmission input shaft.

driveshaft The shaft or tube fitted with universal couplings that is connected between the transmission and other drivetrain components, to transmit torque and rotation.

drivetrain A term used to identify the engine, transmission/transaxle, differential, axles, and wheels.

driving thrust The force transferred from the tire contact patch through the axle housing or suspension components that pushes the vehicle along the road.

drum brake micrometer A tool used for measuring the inside diameter of the brake drum.

drum brakes A type of brake system that forces brake shoes against the inside of a brake drum.

drum-style parking brake A mechanically operated drum brake that can be set while the vehicle is not moving, to serve as a parking brake.

dual-clutch transmission A transmission with two input shafts controlled by two separate clutches.

dual-shaft transmission An automatic transmission that more closely resembles a manual transmission, as it does not use planetary gearsets.

duck A heavy plain woven cotton fabric used in some shop clothing.

duo-servo drum brake system A system that uses servo action in both the forward and reverse direction.

dwell angle The amount of time that the primary circuit is energized, measured in degrees of distributor rotation.

dynamic imbalance A tire imbalance that causes the wheel assembly to turn inward and outward with each half revolution.

ear protection Protective gear worn when the sound levels exceed 85 decibels, when working around operating machinery for any period of time or when the equipment you are using produces loud noise.

edge code A two-digit code printed on the edge of a friction lining that describes its coefficient of friction.

EGR (exhaust gas recirculation) system A system that recirculates a portion of burned gases back into the combustion chamber to displace air and fuel and cool combustion temperatures.

EGR valve A valve that controls recirculation of a portion of burned gases back into the combustion chamber to displace air and fuel and cool combustion temperatures.

EH2 rim The specialized rim design that is used with some run-flat tires.

elasticity The amount of stretch or give a material has.

electric machine Another name for a traction motor with regenerative capability.

electric power steering system (EPS) A steering system that uses an electric motor and sensors to provide feedback to the vehicle's computer systems in order to decrease steering effort.

electrical capacity The ability of a circuit or component to carry electrical loads.

electrical power A measurement of the rate at which electricity is consumed or created.

electrical resistance A material's property that reduces voltage and amperage in an electrical current.

electrical technician A technician who diagnoses, replaces, maintains, identifies fault with, and repairs electrical wiring and computer-based equipment in vehicles.

electrically assisted steering (EAS) A power-assist system that uses an electric motor to replace the hydraulic pump to decrease steering effort.

electrically powered hydraulic steering (EPHS) A steering system that uses an electric motor to produce hydraulic assist for steering.

electrolysis The process of pulling metals apart by using electricity or by creating electricity through the use of chemicals and dissimilar metals.

electromagnet A conductor wound in a coil that produces a magnetic field when current flows through it.

electromagnetic induction The production of an electrical current in a conductor when it moves through a magnetic field or a magnetic field moves past it.

electromotive force An electrical pressure or voltage.

electron theory The theory that electrons, being negatively charged, repel other electrons and are attracted to positively charged objects; thus electrons flow from negative to positive.

electronic brake control (EBC) system A hydraulic brake system that has integrated electronic components for the purpose of closely controlling hydraulic pressure in the brake system.

electronic continuously variable transmission A type of hybrid transmission that often uses two electric motors in combination with an ICE. The two electric motors and the ICE transfer power through a planetary gearset, allowing an infinite number of gear ratios.

electronic fuel injection (EFI) An injection system in which fuel delivery is controlled electronically, allowing continuous adjustments to the air-fuel ratio.

electronic ignition system—distributor type An ignition system that uses a distributor but replaces the contact points with an electronic triggering device and control module.

electronic ignition system An ignition system that uses a nonmechanical (electronic) method of triggering the ignition coil's primary circuit.

electronic stability control (ESC) system A computer-controlled system added to ABS and TCS to assist the driver in maintaining vehicle stability while steering.

electronically controlled throttle A system that uses electronic, instead of mechanical, signals to control the throttle. Sometimes called drive-by-wire.

element The replaceable portion of a filter, such as an air filter element or oil filter element.

emission analyzer A service bay or lab device used for detecting/measuring vehicle tailpipe emissions.

emissions Tailpipe and volatile organic compound pollutants emitted by the automobile.

enabling criteria The operation conditions required before a monitor is allowed to run.

enameled copper wire Wire that uses a thin layer of enamel as insulating material. The thinness of the insulation allows the wire to be closely wound in a coil, creating a dense magnetic field when current flows through it.

end play Fore and aft movement between mating parts.

Energy Policy Act Regulations enacted in 1992 by the U.S. Congress to promote the use of alternative fuel vehicles and fuels.

energy The ability to do work.

engine configuration The way engine cylinders are arranged—for example, V, flat, or in-line.

engine control module (ECM) A computer that controls the ignition and fuel control and emissions control systems on an engine; also called the electronic control unit (ECU) or powertrain control module (PCM).

engine hoist A small crane used to lift engines.

engineering and work practice controls Systems and procedures required by OSHA and put in place by employers to protect their employees from hazards.

Environmental Protection Agency (EPA) A U.S. federal government agency that deals with issues related to environmental safety.

ethylene glycol A chemical used as antifreeze that provides the lower freezing point of coolant and raises the boiling point. It is a toxic antifreeze.

evaporative emission (EVAP) system A system used to capture vapors or gases from an evaporating liquid.

evaporator The cold heat exchanger of the air-conditioning system that absorbs heat from a cab or vehicle.

event cylinder The cylinder that is on compression and ready for the spark to ignite the air-fuel mixture on a waste spark ignition system.

exhaust gas recirculation (EGR) valve A valve that allows a controlled amount of exhaust gas into the intake manifold during a certain period of engine operation. Used to lower oxides of nitrogen exhaust emissions.

expansion volume An engineered space that allows for growth of the volume of a liquid as it heats and expands.

Extended Mobility Technology (EMT) Tires with thick sidewalls that allow the tire to be driven on even when it has no air pressure.

extension housing A component of the automatic transmission housing that covers the output shaft of the transmission. The extension housing also supports the end of the driveshaft and may hold components such as the vehicle speed sensor, speedometer drive assembly, and governor assembly.

external driveshaft A shaft used to transfer power from the transmission to the live axle.

extreme loading Large pressure placed on two bearing surfaces. Extreme loading will try to press oil from between bearing surfaces.

extreme-pressure additives coat parts with a protective layer so that the oil resists being forced out under heavy load.

eye ring terminal A type of crimp or solder terminal that has an enclosed eyelet to connect the terminal with a bolt or screw.

fast chargers A type of battery charger that charges batteries quickly.

fasteners Devices that securely hold items together, such as screws, cotter pins, rivets, and bolts.

fault codes An alphanumeric code system used to identify potential problems in a vehicle system.

feedback system A system that uses feedback to adjust what it is doing. Typically, a module gives a command and waits to see if the command was followed by use of a sensor.

feeler gauge Flat metal strips used to measure the width of gaps, such as the clearance between valves and rocker arms. Also called feeler blades.

fill plug Usually a threaded plug that can be removed to allow the level of a fluid to be checked and filled. This could also be a rubber snap fit plug.

filter sock The first line of defense in the fuel supply system. The sock typically consists of a fine mesh, which prevents most small particles from being drawn into the fuel pump and sent through the rest of the fuel system.

final drive assembly An assembly used to power the drive wheels and allow the wheels to rotate at different speeds as the vehicle turns.

final drive A component that provides a final gear reduction and allows for the difference in speed of each wheel when cornering.

finished rivet A rivet after the completion of the riveting process.

fixed caliper A type of brake caliper bolted firmly to the steering knuckle or axle housing, having at least one piston on each side of the rotor.

fixed orifice tube A restriction device with a metered hole in it and a metal or plastic screen to keep out any debris.

fixed orifice tube system A system with a fixed orifice tube that uses an accumulator between the evaporator and the compressor.

fixed resistor A resistor that has a fixed value.

fixed-type joint A joint that does not slide to allow for shaft lengthening or shortening; it simply allows for angle changes as the suspension moves.

flame front The front edge of the burning air-fuel mixture in the combustion chamber.

flare nut wrench A type of box-end wrench that has a slot in the box section to allow the wrench to slip through a tube or pipe. Also called a flare tubing wrench.

flasher can A mechanical device that switches the vehicle's turn signal and hazard flasher bulbs on and off.

flat seat lug nuts Lug nuts with a flat seat. The sides of the lug nut center the wheel to the axle flange.

flat-blade screwdriver A screwdriver that has a flat tip or blade.

flat-nose pliers Pliers that are flat and square at the end of the nose.

float chamber A chamber that holds a quantity of fuel at atmospheric pressure, ready for use.

floor jacks and safety stands Hydraulic jack used to lift the vehicle, and mechanical safety stands to support it once lifted.

fluid coupler A type of hydraulic coupling used on vintage vehicles to connect and transfer power from the engine to the transmission.

flux A material that is used during brazing and soldering operations to prevent oxidation and remove impurities from the metals.

flywheel A heavy, round metal disc attached to the end of the crankshaft to smooth out vibrations from the crankshaft assembly and provide one of the friction surfaces for a clutch disc used on a manual transmission.

forced induction The pressurization of airflow going into the cylinder through the use of a turbocharger or supercharger.

forces capability map data Data preprogrammed into the electronic control unit's memory by the manufacturer and used to determine how much power assistance is needed based on input from the vehicle's speed sensor and steering sensor.

forcing screw The center screw on a gear, bearing, or pulley puller. Also called a jacking screw.

four-post lift Vehicle lift with four lifting posts. One near each corner of the vehicle.

four-wheel drive (4WD) A drivetrain layout in which the engine can drive either two wheels or four wheels, depending on which mode is selected by the driver.

four-wheel steering. Vehicle that has the ability to steer all four wheels.

free electron An electron located on the outer ring, called the valence ring, that is only loosely held by the nucleus and that is free to move from one atom to another when an electrical potential (pressure) is applied.

free play The amount of movement between two mating parts.

freeze-frame data Refers to snapshots that are automatically stored in a vehicle's powertrain control module (PCM) when a fault occurs (only available on model year 1996 and newer).

frequency The number of events or cycles that occur in a period, usually 1 second.

friction The relative resistance to motion between any two bodies in contact with each other.

front bearing retainer sleeve The part that the throw-out bearing assembly rides on. It usually retains the input shaft bearing on a manual transmission.

front-wheel drive (FWD) A drivetrain layout in which the engine drives the front wheels.

fuel filter A device that removes impurities from the fuel before they reach the carburetor or injection system. Filters may be made of metal or plastic screen; paper, or gauze.

fuel filler neck The upper end of the fuel filler tube leading down to the fuel tank, which accepts the fuel hose nozzle at the gas station pump.

fuel gauge sending unit A device used for creating a signal to control a fuel gauge.

fuel metering system Equipment in a motor vehicle that delivers the proper amount of fuel to each cylinder.

fuel pressure regulator A system that controls the pressure of fuel entering the injectors.

fuel pump relay A relay to turn on or off the high-amperage circuit of the fuel pump.

fuel rail Tubing that connects several injectors to the main fuel line.

fuel shutoff mode A condition where the fuel injectors are shut off during certain driving conditions such as coasting.

fuel supply system Components in a motor vehicle that deliver fuel to the engine.

fuel system Components in a motor vehicle that deliver fuel to the engine.

fulcrum ring A steel ring that is used as a pivot point for the diaphragm spring in the pressure plate.

fulcrum A half-round bearing that the rocker moves on as a bearing surface.

full floating axle An axle where the axle only carries the twisting force for the wheels. The axle does NOT carry the weight of the vehicle.

full power mode An operating condition in a GDI fuel system where the fuel is injected on the intake stroke in a slightly rich condition and allowed to fully mix with the air in the cylinder. This creates maximum power and reduces detonation.

full-flow filter Type of filters that are designed to filter all of the oil before delivering it to the lubricated components.

fuses A circuit protection device with a conductive metal strip that melts when excessive current flows through it.

galleries Passageways drilled or cast into the engine block or head(s), which carry pressurized lubricating oil to various moving parts in the engine, such as the camshaft bearings.

garter spring A coiled spring that is fitted to the inside of the sealing lip of many seals, used to hold the lip in contact with the shaft.

gas welding goggles Protective gear designed for gas welding; they provide protection against foreign particles entering the eye and are tinted to reduce the glare of the welding flame.

gasket scraper A broad, sharp, flat blade to assist in removing gaskets and glue.

gasket A rubber, cork, or paper spacer that goes between two parts to seal the gap between the parts.

gasoline direct injection (GDI) A fuel injection system in which fuel is sprayed directly into the combustion chamber.

gasoline A volatile, flammable liquid mixture of hydrocarbons, obtained from crude oil and used as fuel for internal combustion engines.

gassing When gas escapes the battery; caused by overcharging or rapid charging of a battery.

gear pullers Tool used to remove press-fit gears from shafts.

gear ratio The relationship between two gears in mesh as a comparison to input versus output.

gear set Gears that are in mesh with each other.

gear synchronizer An assembly in the transmission that is used to bring two unequally spinning shafts or gears to the same speed when upshifting or downshifting.

gear A relatively round, rotating part with internal or external teeth that are designed to mesh with another gear for the purpose of transmitting torque.

geared oil pump An oil pump that has two gears running side by side together to move oil from one side of the pump gears to the other.

gelling The thickening of oil to a point that it will not flow through the engine; it becomes close to a solid in extreme cold temperatures.

grease A lubricating liquid thickened to make it suitable for use with many wheel bearings.

grease seal A component that is designed to keep grease from leaking out and contaminants from leaking in.

grinding wheels and discs Abrasive wheels or flat discs fitted to bench, pedestal, and portable grinders.

ground The return path for electrical current in a vehicle chassis, other metal of the vehicle, or dedicated wire.

guide pins Pins that allow the caliper to move in and out as the brakes operate and as the brake pads wear.

half-shaft An axle that has CV joints on each end and that fits between the transaxle and wheel. Typically, one is used on each side of a vehicle.

Hall effect An electrical effect where electrons tend to flow on one side of a special material when exposed to a magnetic field, causing a difference in voltage across the special material. When the magnetic field is removed, the electrons flow normally, and there is no difference of voltage across the special material. This effect can be used to determine the position or speed of an object.

Hall-effect sensor A sensor commonly used to measure the rotational speed of a shaft; they have the advantage of producing a digital signal square waveform and have strong signal strength at low shaft rotational speeds.

halogen lamp A type of incandescent lamp that is filled with a halogen gas such as bromine or iodine.

handrails Used to separate walkways from hazardous areas.

hard rubber mallet A special-purpose tool with a head made of hard rubber; often used for moving things into place where it is important not to damage the item being moved.

hard steering When it is difficult for the driver to steer the vehicle. There may be a problem in the power steering system, too much positive caster, or binding in the steering system.

hatchback A vehicle that has a shared passenger and cargo area; it typically is available in three- and five-door arrangements.

hazard Anything that could hurt you or someone else.

hazardous environment A place where hazards exist.

hazardous material Any material that poses an unreasonable risk of damage or injury to persons, property, or the environment if it is not properly controlled during handling, storage, manufacture, processing, packaging, use and disposal, or transportation.

header Version of an exhaust manifold, made of tubular bent steel, that increases exhaust velocity, thus increasing performance of the engine.

headgear Protective gear that includes items like hairnets, caps, or hard hats.

heat buildup A dangerous condition that occurs when the glove can no longer absorb or reflect heat, and heat is transferred to the inside of the glove.

heat dissipation The spreading of heat over a large area to increase heat transfer.

heat fade Brake fade caused by the buildup of heat in braking surfaces, which get so hot they cannot create any additional heat, leading to a loss of friction.

heat range The rating of a spark plug's operating temperature.

heat transfer The flow of heat from a hotter part to a cooler part; it can occur in solids, liquids, or gases.

heated air intake system A system that uses hot air from around the exhaust manifold to warm the air going into the intake manifold.

heavy line technician A technician who undertakes major engine, transmission, and differential overhaul and repair.

helical-cut gears Gears that are cut on a helix, or spiral.

helical-geared limited slip differential A type of differential that responds very quickly to changes in traction and that does not bind from friction in turns or lose its effectiveness, because there are no clutches.

Helmholtz resonator A device that uses the principle of noise cancellation through the collision of sound waves. When necessary, the resonator is used in addition to the muffler to cancel additional sounds. This resonator may also be used on the induction system to muffle noise of airflow through the induction system. It is named after physicist Hermann von Helmholtz.

high-intensity discharge (HID) A type of lighting that produces light with an electric arc rather than a glowing filament.

high-pressure accumulator A storage container designed to contain high-pressure liquids such as brake fluid.

high-tension leads The heavy insulated wires used to connect the distributor cap terminals to the spark plugs, and the ignition coil to the distributor cap; on waste spark systems, they connect the coils to the spark plugs.

history codes A fault code that has occurred but is not current. It is saved in the PCM's memory for 40 warm up cycles.

hold function A setting on a DMM to store the present reading.

hold-down spring tool A tool used for removing and installing hold-down springs.

hold-in winding The winding that is responsible for holding the solenoid in the "on" position; typically draws less current than the pull-in winding.

hole theory The theory that as electrons flow from negative to positive, holes flow from positive to negative.

hollow punch A punch with a center hollow for cutting circles in thin materials such as gaskets.

homogeneous mixture An air-fuel mixture evenly dispersed throughout the combustion chamber.

Hooke joint A joint with four trunnions and four bearing caps; also known as a Cardan joint or a universal joint.

horizontally opposed engine An engine with two banks of cylinders, 180 degrees apart, on opposite sides of the crankshaft. It is also called a flat engine or a boxer engine.

hot junction The heating point of a thermocouple.

hybrid drive system A drive system that uses two or more propulsion systems, such as electric motors and an ICE.

hybrid electric vehicle (HEV) A vehicle that uses two power sources for propulsion, one of which is electricity.

hydraulic control unit (HCU) An assembly that houses electrically operated solenoid valves used in electronic braking systems; also called a modulator.

hydraulic fade Brake fade caused by boiling brake fluid; causes a spongy brake pedal.

hydraulic jack A type of vehicle jack that uses oil under pressure to lift vehicles.

hydraulic press method A method for replacing wheel studs that uses a press to force the wheel stud into the flange until it bottoms out.

hydrocracking Refining crude oil with hydrogen, resulting in a base oil that has the higher performance characteristics of synthetic oils.

hydrogenating A process used during refining of crude oil. Hydrogen is added to crude oil to create a chemical reaction to take out impurities such as sulfur.

hygroscopic A property of a substance or liquid that causes it to attract and absorb moisture (water), as a sponge absorbs water. Brake fluid absorbs water out of the air; thus it is hygroscopic.

hypoid bevel gears A special design of spiral bevel gears where the centerline of the pinion is below the centerline of the ring gear.

hypoid gearing A type of spiral bevel gearset that mounts the pinion gear below the centerline of the ring gear.

ICE Internal combustion engine. A device that burns fuel inside itself to generate power.

ICM (ignition control module) A general control unit of some electronic ignition systems, usually with current and dwell angle control; driver and output stage; and, in some cases, electronic spark timing functions.

idle stop A feature that turns off the internal combustion engine when the vehicle is at a standstill.

idle The speed at which an engine runs without any throttle applied.

ignition advance The means of causing the spark to occur earlier within the compression stroke for better performance and fuel economy during changing engine conditions.

ignition coil A device used to amplify an input voltage into the much higher voltage needed to jump the electrodes of a spark plug.

ignition module An electronic component that electronically controls the ignition coil or coils.

ignition switch A switch operated by a key or start/stop button and used to turn a vehicle's electrical and ignition system on or off.

IMA (integrated motor assist) A Honda hybrid drive system that uses a moderate-sized electric motor installed between the engine and the transmission.

impact driver A tool that is struck with a hammer to provide an impact turning force to remove tight fasteners.

incandescent lamp The traditional bulb that uses a heated filament to produce light.

included angle The angle of camber added or subtracted to the SAI angle. This is the angle of the steering knuckle pivot points in relation to the camber angle of the wheel; also referred to as a diagnostic angle.

independent rear suspension (IRS) A type of suspension system where each rear wheel is capable of moving independently of the other.

independent suspension drive axle A type of suspension that allows each wheel on a drive axle to move independently of the other.

independent suspension A system for allowing the up-and-down movement of one tire without affecting the other tire on that axle.

indirect fuel injection Any fuel injection system where fuel is not sprayed directly into the combustion chamber.

indirect TPMS A type of automated tire pressure monitoring system that uses the antilock braking system of a vehicle to measure the difference in the rotational speed of the four wheels in order to determine tire pressure.

induced voltage The creation of voltage in a conductor by movement of a magnetic field that is near that conductor.

induction coil An electrical transformer that uses magnetic fields to produce high-voltage pulses from low-voltage direct current.

induction-type system A type of ignition system that uses a magnetic pulse generator to trigger the spark.

inductive current The current that has been created across a conductor by moving it through a magnetic field.

inductive-type sensor A sensor mounted on the crankcase housing that is used to sense the movement of the ring gear teeth on the flywheel, or a toothed disc on the crank pulley.

inertia The resistance to a change in motion.

infrastructure The system for supplying fuel for transportation use; includes ships, pipelines, trucks, and so forth.

injector A valve that is controlled by a solenoid or spring pressure to inject fuel into the engine.

inlet port Connects the brake fluid reservoir to the master cylinder bore around the piston and between the piston seals. Aids in recuperation.

in-line engine An engine in which the cylinders are arranged side by side in a single row.

inside micrometer A micrometer that measures inside dimensions.

inspection certificate Label applied to lifting equipment showing when it was last inspected and certified.

instrument panel warning lamps Lamps that illuminate to warn a driver of a fault in a system.

insulator A material that holds electrons tightly and prevents electron movement.

intake air temperature (IAT) sensor A sensor that measures the temperature of the incoming air through the air filtration system.

integral ABS system A brake system in which the master cylinder, power booster, and HCU are all combined in a common unit.

integrated circuit (IC) A complete circuit that is formed at a micro level. Processor chips include thousands of integrated circuits.

interference fit A condition in which two parts are held together by friction because the outside diameter of the inner component is slightly larger than the inside diameter of the outer component.

intermediate shaft A steel rod positioned at an angle from the steering column to the steering gear that functions in transferring movement from one to the other.

intermediate tap One of a series of taps designed to cut an internal thread. Also called a plug tap.

International Standards Organization (ISO) flare method A method for joining brake lines, also called a bubble flare. Created by flaring the line slightly out and then back in, leaving the line bubbled near the end.

interrupter ring A ferrous metal ring shaped like a very shallow cup with slits or windows cut into it at evenly spaced intervals. The ring has the same number of blades and windows as engine cylinders and is rotated by the engine moving the blades through an air gap. The purpose of an interrupter ring is to systematically block, and expose, the magnetic field in a Hall-effect sensor in order to turn the primary ignition circuit on and off.

inverted double flare A method for joining brake lines that forms a secure, leakproof connection.

ion An atom that has fewer electrons than protons (positive) or that has more electrons than protons (negative).

ISO flare Metric system flare commonly called a bubble flare.

ISO flare method A method for joining brake lines, also called a bubble flare. Created by flaring the line slightly out and then back in, leaving the line bubbled near the end.

isolation valve The valve in the HCU that either allows or blocks brake fluid that comes from the master cylinder and goes to the wheel brake unit.

jack stands Metal stands with a mechanical height adjustment to hold a vehicle once it has been jacked up.

keep alive memory (KAM) Memory that is retained by the ECM when the key is off.

keyed lock washer The washer that fits between the adjusting nut and the locknut; the face of the washer is drilled with a series of holes that mate to a short pin from the adjusting nut, locking it to the spindle.

keyed washer The washer that fits between the adjusting nut and the wheel bearing and that has the center hole keyed to fit a slot on the spindle or axle tube.

kinetic energy The energy of an object in motion; it doubles with weight and increases by the square of the speed.

kinetic friction The friction between two surfaces that are sliding against each other.

Kirchhoff's current law An electrical law stating that the sum of the current flowing into a junction is the same as the current flowing out of the junction.

knock sensor An engine sensor that detects pre-ignition, detonation, and knocking.

knocking A noise heard when the air-fuel mixture spontaneously ignites before the spark plug is fired at the optimum ignition moment.

knuckle The part that contains the wheel hub or spindle and attaches to the suspension components.

labor guide A guide that provides information to make estimates for repairs.

lambda The ratio of air to fuel at which all of the oxygen in the air and all of the fuel are completely burned; see also *stoichiometric ratio*.

latent heat of condensation The amount of heat removal necessary to change the state from a gas to a liquid without changing the actual gauge temperature.

latent heat of evaporation The amount of heat required to change the state from a liquid to a gas without raising the actual gauge temperature.

latent heat of freezing The amount of heat removal required to change the state from a liquid to a solid without changing the actual gauge temperature.

lateral runout The side-to-side movement of the rotor surfaces as the rotor turns; also called warpage.

leading shoes Brake shoes that are installed so that they are applied in the same direction as the forward rotation of the drum and thus are self-energizing.

leading/trailing shoe drum brake system Type of brake shoe arrangement where one shoe is positioned in a leading manner, and the other shoe in a trailing manner.

leaf spring A spring, made of one or more flat, tempered steel springs bracketed together, that is used in the suspension system to support the weight of the vehicle.

leather Shop PPE made from animal skins. Used to protect from dry heat and flame.

LED See *light-emitting diode*.

lemon law buyback A vehicle that has been bought back by the manufacturer due to inability to repair the same fault within a specified number of attempts.

lemon law A consumer protection law used in some states to identify a new vehicle that has undergone several unsuccessful attempts to repair the same fault.

lever A simple machine that can allow a large object to be moved with less force.

light line technician A technician who diagnoses and replaces the mechanical and electrical components of motor vehicles.

light-emitting diode (LED) Diodes that produces light when current flows across the P–N junction.

limited slip differential assembly A differential assembly that uses a clutch assembly or gear assembly to allow a limited amount of slip between the two axles. It is used to increase drive wheel traction in slippery conditions.

linear motion Movement in a straight line.

lithium soap A thickening agent for grease to give it the proper consistency.

live axle The axle that drives the machine by turning the power from the driveshaft 90 degrees to deliver it to the wheels and providing the final gear reduction in the drivetrain; also known as a drive axle.

load transfer Weight transfer from one set of wheels to the other set of wheels during braking, acceleration, or cornering.

lock cage The stamped sheet metal cap that fits over the bearing adjustment nut and is secured by a cotter pin going through it and the spindle/axle.

locking pliers A type of pliers where the jaws can be set and locked into position.

locknut The nut that holds the adjusting nut from turning; usually tightened much tighter than the adjusting nut.

Lockout/tag-out A safety tag system to ensure that faulty equipment or equipment in the middle of repair is not used.

longitudinal The orientation of the engine in which the front of the engine is facing the front of the vehicle. It is most commonly found in rear-wheel-drive vehicles.

lot attendant Person who is responsible for keeping vehicles in salable condition on a car lot.

low-drag caliper A caliper designed to maintain a larger brake pad–to-rotor clearance by retracting the pistons farther than normal.

Low-Emission Vehicle (LEV) standard A program initiated in California to improve vehicle emission standards.

lube technician A technician who carries out scheduled maintenance activities on a range of mechanical and related vehicle components.

lubricating oil Processed crude oil with additives to help it perform well in the engine.

lubrication system A system of parts that work together to deliver lubricating oil to the various moving parts of the engine.

lug A flange that is shaped to assist with aligning objects on other objects.

lug nuts Fasteners that secure the wheel onto the wheel studs.

lug wrench A tool designed to remove wheel lugs nuts and commonly shaped like a cross.

machinery guard Used to help prevent accidental contact with dangerous objects.

MacPherson strut A strut used on an independent suspension where the spring and shock are joined together; used on most front-wheel-drive vehicles.

magnetism The force that attracts or repels magnetic charges; the property of a material to respond to a magnetic field.

magneto-resistive Type of wheel speed sensor that uses an effect similar to a Hall effect sensor to create a digital signal.

magneto-rheological fluid A fluid that has the unique characteristic of changing viscosity when exposed to a magnetic field.

male and female terminal A crimp or solder terminal on which the male and female ends join to create a removable low-resistance connection.

mandrel forming Using a pipe bender with mandrels to allow for very tight bends without creating kinks or reducing the size of the pipe.

mandrel head The head of the pop rivet that connects to the shaft and causes the rivet body to flare.

mandrel The shaft of a pop rivet.

manual adjustable-rate shock absorber A shock absorber that allows manual adjustment of the dampening rate.

manual bleeding A bleeding method where one person manually operates the brake pedal while the other person opens and closes the bleeder screws on the wheel brake units to allow the air and old brake fluid to be pushed out.

manually adjustable air spring A rubber air bag placed inside coil springs to increase the spring's load carrying ability. It is filled manually through a valve similar to a tire valve stem.

master cylinder Converts the brake pedal force into hydraulic pressure, which is then transmitted via brake lines and hoses to one or more pistons at each wheel brake unit.

match mounting The process of matching up the tire's highest point with the rim's lowest point for the purpose of reducing the tire's radial runout.

measuring tape A thin measuring blade that rolls up and is contained in a spring-loaded dispenser.

mechanical advantage Occurs when we give up either speed or torque to increase either torque or speed through a machine.

mechanical disadvantage When the load distance on a lever is greater than the effort distance, which means the effort required to move the load is greater than the load itself.

mechanical jack A type of vehicle jack that uses mechanical leverage to lift a vehicle.

memory saver (memory minder) Battery backup device for vehicle computer systems.

meshed pinion A pinion when it is mated with the rack.

metering jet A calibrated orifice in a carburetor for fuel to flow through. Often it is replaceable for performance or economy desires.

metering valve A valve used on vehicles equipped with older rear drum/front disc brakes to delay application of the front disc brakes until the rear drum brakes are applied. Located in line with the front disc brakes.

micrometer An accurate measuring device for internal and external dimensions. Commonly abbreviated as *mic.*

min/max setting A setting on a DMM to display the maximum and minimum readings.

mineral oil Base stock processed from crude oil in a refinery, used as the base material of all conventional oil.

minivan A lighter-duty van used for carrying six to eight occupants or light cargo.

modules An electronic computer or circuit board that controls specific functions.

molybdenum thickening agent A compound used in some greases to give them the needed consistency.

monitor An OBD II test run to ensure that a specific component or system is working properly.

Morse taper A system for quickly changing and securing drill bits to drills.

multidisc clutch A type of holding or driving device used by automatic transmissions to stop or drive one component of a planetary gearset. It uses several thin friction discs and thin steel plates that are squeezed together when hydraulic pressure is applied to a piston in the clutch.

multimeter A test instrument used to measure volts, ohms, and amps. A digital multimeter may also be called a digital volt-ohmmeter (DVOM).

multiplexed The carrying of multiple signals on one wiring circuit. Digital signals are muptiplexed on a dedicated network; also referred to by the abbreviation MUX.

multipoint fuel injection (MPFI) An injection system in which fuel is injected into the intake ports just upstream of each cylinder's intake valve, rather than at a central point within an intake manifold. Also called multiport injection.

Mylar tape Polyester film that may be metalized and incorporated into a wiring harness to provide electrical shielding.

National Lubricating Grease Institute (NLGI) An organization that grades the thickness of automotive and industrial grease.

NC (normally closed) An electrical contact that is closed in the at-rest position.

needle-nose pliers Pliers with long tapered jaws for gripping small items and getting into tight spaces.

negative camber Condition where the top of the tire leans toward the centerline of the vehicle.

negative caster Forward tilt of the steering knuckle pivot points from the vertical line.

negative offset A condition in which the plane of the hub mounting surface is positioned toward the brake side or back of the wheel centerline.

negative scrub radius A condition in which the camber line is inside the steering axis centerline or when they intersect above the road surface.

negative temperature coefficient (NTC) A characteristic of materials whereby resistance decreases as temperature increases.

neutral steer A condition in which both the front and the rear tires of a vehicle are experiencing the same slip angle.

Newton's first law of motion A physical law that states that an object will stay at rest or uniform speed unless it is acted upon by an outside force.

nippers Another name for end-cutting pliers.

NO (normally open) An electrical contact that is open in the at-rest position.

noncontinuous monitor A monitor that runs only once per drive cycle.

nondirectional tread pattern A tread pattern that is nonspecific, allowing the tire to be placed on any wheel of a vehicle.

non-integral ABS systems A brake system in which the master cylinder, power booster, and HCU are all separate units.

number-one cylinder Typically the cylinder located farthest forward on the engine. It is the first cylinder in the firing order.

nut A fastener with a hexagonal head and internal threads for screwing on bolts.

OBD I The first generation of onboard diagnostic systems that originated for California vehicles.

OBD II The second generation of onboard diagnostic systems, which have been in effect for all U.S. vehicles since 1996.

Occupational Safety and Health Administration (OSHA) The agency that ensures safe and healthy working conditions by setting and enforcing standards and by providing training, outreach, education, and assistance.

octane rating A standard measure of the performance of a motor or aviation fuel. The higher the octane number, the more heat the fuel can withstand before self-igniting.

off-car brake lathe A tool used to machine (refinish) drums and rotors after they have been removed from the vehicle.

offset screwdriver A screwdriver with a 90-degree bend in the shaft for working in tight spaces.

offset vice Tool used to hold long components such as pipes and axles vertically.

Ohm's law A law that defines the relationship among current, resistance, and voltage.

ohm The unit for measuring electrical resistance.

oil consumption test A test designed to measure the amount of oil used in a certain amount of miles.

oil cooler A device that transfers heat away from oil by passing it near either engine coolant or outside air. Cooling the oil helps to keep it from overheating and breaking down.

oil galleries Main passageways for transporting lubrication in an engine.

oil monitoring system A system that monitors or calculates the condition of the oil and alerts the driver when it is time to be changed. These systems have to be reset for the customer after an oil change is performed.

oil pan The metal pan located at the bottom of the engine; it usually covers the crankshaft and rods, commonly where the oil sump is located on a wet sump oil system.

oil pressure regulator A spring operated valve which limits the maximum pressure in an engine lubrication system.

oil pressure relief valve A calibrated, spring-loaded valve that allows for pressure bleed-off if the oil pump creates too much pressure.

oil pump A positive-displacement pump that produces oil flow within the engine and lubricates the internal components.

oil pump strainer A screen located on the oil pump pickup that keeps debris from being picked up by the oil pump.

oil slinger A device used to fling oil up onto moving engine parts.

oil spurt holes Holes drilled into the connecting rod that spray oil up onto the cylinder walls and the piston wrist pins.

oil sump The lower part of the oil pan that collects and holds lubricating oil for the engine. The oil pickup screen sits in this low point.

oil-filter wrench A wrench used to grip and loosen an oil filter. Not to be used for tightening an oil filter.

on-car brake lathe A tool used to machine (refinish) rotors while they are still attached to the vehicle.

one-way clutch A type of holding device that allows free rotation in one direction but will lock up in the opposite direction; also called an over-running clutch.

open differential assembly A differential assembly that allows both axles to turn at their own speed when turning a corner, but is dependent on the traction of the tires to deliver torque to the ground. If one wheel has no traction, all of the engine's torque will be used at that wheel, causing it to simply spin.

open loop When the computer system feedback loop is not active.

open-end wrench A wrench with open jaws to allow side entry to a nut or bolt.

optical sensor A sensor that generates a voltage when excited by a beam of light.

oscillation The fluctuation of an object between two states. With regard to suspension springs, it refers to the uncontrolled compression and decompression of the spring following overshoot.

outlet port The port leaving a cylinder or pump.

outside micrometer A precision measuring instrument meant to measure the outside dimensions of components. It is usually accurate to 0.0001 inch (0.0025 mm).

overflow tank A tank used to catch any coolant that is released from the radiator cap (works like a catch can).

overshoot The amount a spring extends (springs back) past its original length following compression.

oversteer A condition when the front end steers more sharply than desired because the rear wheels lose adhesion during cornering. This causes the rear of the vehicle to swing out. The vehicle is said to be loose.

owner's manual An informational guide supplied by the manufacturer; it contains basic vehicle operating information.

oxidation inhibitor An oil additive that helps keep hot oil from combining with oxygen to produce sludge or tar, which clogs oil galleries and drain-back passages.

oxygen sensor (before and after catalytic convertor) An exhaust sensor used to measure the amount of oxygen in the exhaust gases fuel trim, not so much on spark timing is used to determine fuel mixture.

painted line Used to indicate a safe walking distance from a piece of equipment.

panhard rod A metal rod that is mounted between the body or frame of the vehicle and the axle. It controls suspension movement side to side; also known as a transverse torque rod.

paper-like fiber Used in disposable suits for protection against dust and splashes.

parallax error A visual error caused by viewing measurement markers at an incorrect angle.

parallel hybrid drivetrain A type of hybrid transmission in which power can flow from a gasoline engine, an electric motor, or any combination of the two.

parallelism Both surfaces of the rotor should be perfectly parallel to each other so that brake pulsations do not occur. Also called thickness variation.

parallelogram steering system A non-rack-and-pinion system that uses a series of parts, consisting of the pitman arm, idler arm, center link, and tie-rod assemblies, to relay movement from the steering gearbox to the wheel assembly.

parasitic draw The unwanted current draw that occurs once the vehicle has been turned off and the systems have gone to sleep.

parking brake A brake system used for holding the vehicle when it is stationary.

parking brake cable A mechanism used to transmit force from the parking brake actuating lever to the brake unit.

parking brake cable pliers A tool used to install parking brake cables.

parking brake cable removal tool A tool used to compress the spring steel fingers of the parking brake cable so that the cable can be removed from the backing plate.

parking brake mechanism A mechanism that operates the brake shoes or pads to hold the vehicle stationary when the parking brake is applied.

parts program A computer software program for identifying and ordering replacement vehicle parts.

parts specialist The person who serves customers at the parts counter.

Pascal's law The law of physics that states that pressure applied to a fluid in one part of a closed system will be transmitted equally to all other areas of the system.

PCV valve See *positive crankcase ventilation valve.*

peening A term used to describe the action of flattening a rivet through a hammering action.

pending codes Codes that have not been validated by failing a second test.

permanent magnet type An electric motor in which the magnetic field in the casing is produced by permanent magnets, and the armature has an electromagnetic field generated by passing electrical current through loops or winding, thereby producing the motor action.

personal protective equipment (PPE) Safety equipment designed to protect the technician, such as safety boots, gloves, clothing, protective eyewear, and hearing protection.

phenolic resin A very dense material used to create some brake pistons that is very resistant to corrosion and heat transfer.

Phillips head screwdriver A type of screwdriver that fits a head shaped like a cross in screws. Also called a Phillips screwdriver.

photochemical smog A brown haze that hangs in the sky, typically seen over large cities. Smog is a major health issue to humans because it affects lung tissue.

photovoltaic (PV) effect The conversion of sunlight into electricity.

pickle fork U-shaped wedge used for separating tie-rod ends that is operated by hammer or air hammer. This tool usually destroys the dust boot during the process, so it is not the best tool to use on tie-rods that will be reused.

pickup A vehicle that carries cargo; it has stronger chassis components and suspension than a sedan.

pickup assembly A component with a wire coil wrapped around a ferrous metal core; it is used to generate an electrical signal when a magnetic field passes through it.

pickup tube A tube connected to the oil pump that acts like a straw for the oil pump to pull oil from the sump of the oil pan.

Pictorial message Symbol used on a sign to convey a warning.

piezoelectric energy A type of electricity in which a material such as a quartz crystal produces voltage when mechanical pressure distorts it.

pilot bearing The bearing or bushing that supports the front of the transmission input shaft. It is mounted in the flywheel or the rear of the crankshaft.

pin punch A type of punch in various sizes with a straight or parallel shaft.

pipe wrench A wrench that grips pipes and can exert a lot of force to turn them. Because the handle pivots slightly, the more pressure put on the handle to turn the wrench, the more the grip tightens.

piston engine An internal combustion engine that uses cylindrical pistons, moving back and forth in a cylinder, to extract mechanical energy from chemical energy.

pitch Movement around the vehicle's y-axis. Or on a spring, the distance from the center of one coil to the center of the adjacent coil.

pitch The angle of something.

pitman arm puller A heavy-duty puller made specially for removing the pressed-on pitman arm from the sector shaft.

PKE (passive keyless entry) An automatic system that senses the proximity of a fob and locks or unlocks the vehicle.

planet carrier The device that holds the planet gears in place, keeping them equally spaced.

planet gears Gears that mesh with the sun gear and ring gear in planetary gear sets.

plenum chamber A large portion of the intake manifold after the throttle plate and before the intake runner tubes. The plenum provides a reservoir of air and helps prevent interference with the flow of air between individual branches.

pliers A hand tool with gripping jaws.

plug-in hybrid electric vehicle (PHEV) A hybrid electric vehicle in which only one power source, the battery, is used to propel the vehicle for a certain distance, limited by the storage capacity of the battery and the efficiency of the motor.

plunge-type joint The inner joint on the half-shaft that allows for changes in shaft length.

pneumatic jack A type of vehicle jack that uses compressed gas or air to lift a vehicle.

polarity The state of charge, positive or negative.

policy A guiding principle that sets the shop direction.

pollutant A potential threat to human health or the environment, resulting from excessive amounts of chemicals and waste.

polyalphaolefin (PAO) oil An artificially made base stock (synthetic) that is not refined from crude oil. Oil used in R-12 systems and those converted from R-12. Oil molecules are more consistent in size, and no impurities are found in this oil, as it is made in a lab.

polymeric positive temperature coefficient (PPTC) device A thermistor-like electronic device used to protect against circuit overloads. Also called resettable fuse.

pop rivet gun A hand tool for installing pop rivets.

poppet valve A valve that controls the flow of brake fluid at usually preset pressures.

positive camber Condition where the top of the tire leans away from the centerline of the vehicle.

positive caster Backward tilt of the steering knuckle pivot points from the vertical line.

positive crankcase ventilation (PCV) system A system that draws blowby gases from the crankcase into the intake, to be burned.

positive offset A condition in which the plane of the hub mounting surface is positioned toward the outside or front of the wheel centerline.

positive scrub radius A condition in which the camber line is outside of the SAI line or where they intersect below the road surface.

potentiometer A three-terminal resistive device with one terminal connected to the input of the resistor, one terminal connected to the output of the resistor, and the third terminal connected to a movable wiper arm that moves up and down the resistor, creating a varying voltage signal. Also called a pot, or variable resistor.

pour point depressants Oil additives that keep wax crystals from forming and causing the oil to gel during cold operation.

power The rate at which work is done; electrical power is measured in watts.

power assist unit Assembly used to assist the driver in turning the steering wheel.

power booster Device used to increase the brake pedal force on the master cylinder. Can be vacuum or hydraulic types.

power flow The path that power takes from the beginning of an assembly to the end.

power-on-demand A feature that shuts down some engine cylinders when not needed to save fuel.

power section A chamber in the rack, where pressurized fluid acts upon pistons that assist in steering.

power splitter A device that receives power from an internal combustion engine and electric machine to power a hybrid electric vehicle.

power steering pump A small hydraulic pump that provides assistance to the driver when turning the steering wheel.

power unit A belt- or gear-driven pump that produces hydraulic pressure for use in the steering box or rack.

power-splitting transmission (PST) A type of hybrid transmission that splits the power flow going to the wheels from one or more electric motors and an internal combustion engine.

powertrain control module (PCM) A computer that controls the ignition, fuel, and emissions control systems on an engine; also called the electronic control unit (ECU) or engine control module (ECM).

powertrain mount A rubber or metal bracket used to secure the engine and transmission into the vehicle. Some vehicles use hydraulic or electrohydraulic powertrain mounts.

preload Further pressure applied to bearing-supported parts after all the free play is taken up.

preloaded A part that is already compressed from pressure.

pressure bleeding A bleeding method that uses clean brake fluid under pressure from an auxiliary tool or piece of equipment to force the air and old brake fluid from the hydraulic braking system.

pressure differential valve A valve that monitors any pressure difference between the two separate hydraulic brake circuits; it usually contains a switch to turn on the brake warning light when there is a pressure difference.

pressure plate The friction surface of the clutch cover and the plate that squeezes the clutch disc against the flywheel.

pressure relief valve A valve that is designed to release pressure if it gets above a calibrated pressure; this limits pressure in one part of a circuit or a complete system.

pressure transducer An electrical device that creates an electrical signal based on a pressure input and displays it graphically on a lab scope.

pressure washer/cleaner A cleaning machine that boosts low-pressure tap water to a high-pressure output.

pressure, or force-feed, lubrication system A lubrication system that has a pump to pressurize the lubricating oil and push it through the engine to moving parts.

Pressure The force per unit area applied to the surface of an object.

prick punch A pinch with a sharp point for accurately marking a point on metal.

primary circuit The low-voltage circuit that turns the coil on and off.

primary cup A seal that holds pressure in the master cylinder when force is applied to the piston.

primary piston A brake piston in the master cylinder moved directly by the pushrod or the power booster; it generates hydraulic pressure to move the secondary piston.

primary winding The coil of wire in the low-voltage circuit that creates the magnetic field in a step-up transformer.

probing technique The way in which test probes are connected to a circuit.

procedure A list of the steps required to get the same result each time a task or activity is performed.

progressive rate of deflection The change in deflection rate that occurs as the weight of the vehicle changes. The greater the weight, the lower the rate of deflection due to increased resistance.

proportioning valves Valves used mostly on older vehicles equipped with rear drum brakes to reduce rear wheel hydraulic brake pressure under hard braking or light loads. Located in line with the rear brakes.

propylene glycol A chemical used as antifreeze. It is labeled as a nontoxic antifreeze.

pry bar A high-strength carbon steel rod with offsets for levering and prying. Also called a crowbar.

pry bar A high-strength carbon steel rod with offsets for levering and prying. Also called a crowbar.

PTO Power take-off. Used on some trucks to power accessories such as a dump bed.

pull Felt when the driver feels the steering wheel wanting to go to one side. It can mean the air pressure in one or more of the front tires is low. A simple tire pressure gauge is used to check the tire pressures.

puller Tool used to grab and pull press-fit parts apart.

pull-in winding A high-current winding found in starter solenoids that pulls the solenoid plunger into the activated position.

pulse width modulation (PWM) A digital on/off electrical signal. PWM is a very precise control method for an output device on a varying frequency. Usually, only fuel injectors are operated in this format, as the software programming and related circuits are very complex. Most other components are duty-cycled on a fixed frequency.

punches A generic term to describe high-strength carbon steel shafts with a blunt point for driving. Center and prick punches are exceptions and have a sharp point for marking or making an indentation.

purging The process of pulling stored fuel vapors from the charcoal canister and moving them into the engine to be burned.

push-on spade terminal A disconnectable type of crimp or solder terminal used to terminate electrical wires.

pushrod A tubular rod that stands between the tappet and the rocker arm in an overhead valve engine; the pushrod transfers cam motion to the rocker arm. In the braking system, a mechanism used to transmit force from the brake pedal to the master cylinder.

push-type clutch A typical clutch system, used in modern vehicles, where the clutch fork pushes the release bearing forward to release the friction facing from the pressure plate.

quadrant ratchet The device used in some cable-operated clutches to provide self-adjustment as the clutch disc wears. Some quadrant ratchets adjust if you lift up on the clutch pedal.

quench The quench in a combustion chamber in which the flame cannot burn because of cold surfaces or poor distribution of the fuel mixture.

quick take-up master cylinders Cylinders used on disc brake systems that are equipped with low-drag brake calipers to quickly move the brake pads into contact with the brake rotors.

quick take-up valve A valve used to release excess pressure from the larger piston in a quick take-up master cylinder once the brake pads have contacted the brake rotors.

rack-and-pinion steering system A steering system composed of a steering wheel, a main shaft, universal joints, and an intermediate shaft. When the steering wheel is turned, movement is transferred by the main shaft and intermediate shaft to the pinion.

radial load The load that is perpendicular to a shaft, usually controlled by bearings or bushings.

radial Type of tire where the casing plies run side-to-side from bead to bead.

radiation The movement of energy through space, such as the movement of energy from the sun to the earth.

radiator hoses Rubber hoses that connect the radiator to the engine. Because they are subject to pressure, they are reinforced with a layer of fabric, typically nylon. Some radiator hoses use coiled wire inside them to prevent hose collapse as the coolant cools.

radiator A device that transfers heat from a fluid within to a location outside.

ratchet A generic term to describe a handle for sockets that allows the user to select direction of rotation. It can turn sockets in restricted areas without the user having to remove the socket from the fastener.

ratcheting box-end wrench A wrench with an inner piece that is able to rotate within the outer housing, allowing it to be repositioned without being removed.

ratcheting screwdriver A screwdriver with a selectable ratchet mechanism built into the handle that allows the screwdriver tip to ratchet as it is being used.

ratcheting screwdriver A screwdriver with a selectable ratchet mechanism built into the handle that allows the screwdriver tip to ratchet as it is being used.

rattle gun A term used to describe an air impact wrench, based on the noise it makes.

reaction force A force that acts in the opposite direction to another force.

rear-wheel drive (RWD) A drivetrain layout in which the engine drives the rear wheels.

rebound clip A metal strap that is wrapped around the leaf spring to prevent excessive flexing of the main leaf during rebound.

recirculating ball steering box A worm gear steering box in which the worm rides on ball bearings.

rectification A process of converting alternating current (AC) into direct current (DC).

recuperation Process by which brake fluid moves from the reservoir past the edges of the seal into the chamber in front of the piston. This prevents air from being drawn into the hydraulic system caused by low pressure when the brake pedal is released quickly.

reed switch A type of speed sensor that uses a magnetic field to open and close a movable set of contacts. It is used with a rotating magnet to measure rpm of a shaft and send the signal to the PCM.

reed valve A small flexible metal plate that covers the inlet port of a two-stroke engine and opens and closes to let air and fuel into the crankcase.

refrigerant label A label that lists the type and total capacity of refrigerant that is installed in the A/C system.

refrigerant The name given to a chemical compound designed to meet the needs of the refrigeration system.

regenerative braking A type of braking in which the kinetic energy of the vehicle's motion is captured rather than being lost to heat as it is in a conventional braking system. This is accomplished by using the drive motors as generators, which recharge the traction batteries.

relay An electromechanical switching device whereby the magnetism from a coil winding acts on a lever that switches a set of contacts.

release mechanisms Components that operate the clutch. Usually included are the throw-out bearing and the clutch fork. Some manufacturers include the operating system.

reluctor A rotating, toothed wheel that changes the reluctance of a material to conduct magnetic lines of force.

remote keyless entry (RKE) A system that remotely unlocks and locks the vehicle without the use of a traditional key.

repair order A form used by shops to collect information regarding a vehicle coming in for repair; also referred to as a work order.

required voltage The amount of voltage needed to push current across the electrodes of a spark plug located in the combustion chamber.

reserve capacity (RC) Refers to the length of time, measured in minutes, that a new, fully charged 12-volt battery discharges under a specified load of 25 amps at 80°F (26.6°C) before battery cell voltage drops below 1.75 volts per cell (10.5 volts for a 12-volt battery).

residual pressure valve (residual check valve) In drum brake systems, a valve that maintains pressure in the wheel cylinders slightly above atmospheric pressure so that air does not enter the system through the seals in the wheel cylinders.

resistor A component designed to have a fixed resistance.

restriction A blockage that partially stops or slows the flow of a material such as refrigerant.

retarding effect The result of retarding (slowing) the vehicle.

return springs Springs that retract the brake shoes to their released position.

returnless fuel injection system A type of injection system in which no hot fuel is returned to the tank, thus keeping the fuel in the tank relatively cool and minimizing vaporization.

rheostat A variable resistor constructed of a fixed input terminal and a variable output terminal, which vary current flow by passing current through a long resistive tightly coiled wire.

ribbon cable A type of flat harness in which cables are insulated from one another but joined together side by side.

ride height The amount of ground clearance a vehicle has, measured from a point on the body or frame, depending on the manufacturer; also known as trim height.

rigid dead axle suspension A type of dead axle suspension system that is nonindependent and uses a beam or solid axle.

rigid spring hanger The rigid part typically welded to the body or frame of the vehicle to which the front of the leaf spring is attached.

rigid-axle coil-spring suspension A dead axle that uses a coil spring.

rim The outer circular lip of the metal on which the inside edge of the tire is mounted.

rim flanges The outside edges of the wheel that help keep the tire from popping off the wheel.

ring gear The outer gear in a planetary gear. Also the gear that meshes with the pinion gear in a final drive.

riveted linings Brake linings riveted to the brake pad backing plate with metal rivets and used on heavier-duty or high-performance vehicles.

road force imbalance Occurs when the wheel or tire is not concentric or when the tire's sidewall has uneven stiffness.

rods Straight (or precisely formed) pieces of steel used to control motion within the vehicle's suspension system.

roll bar Another type of pry bar, with one end used for prying and the other end for aligning larger holes, such as engine motor mounts.

roll Movement of a vehicle around its x-axis (the imaginary line down the center of the vehicle from front to back). It is commonly referred to as body roll or lean; when cornering, the body tries to move to the outside of the corner against the suspension.

roll-rate sensor A sensor that measures the amount of roll around the vehicle's horizontal axis that a vehicle is experiencing.

rosin A type of liquid or paste (flux) that, in solid form, is contained within solder and is used to prevent oxidization.

rotary engine An engine that uses a triangular rotor turning in a housing instead of conventional pistons.

rotary flow A type of fluid flow in a torque converter in which fluid flows around the centerline of the torque converter in a circle.

rotational force The force created by the rotating wheel when the brakes are applied; it causes the brake components to twist the brake support, and ultimately the vehicle, in the direction of wheel rotation.

rotational speed The speed at which an object rotates, measured in revolutions per minute (rpm).

rotor A high-voltage rotating switch that transfers voltage from the distributor cap's center terminal to the outer terminals.

rotor arm The portion of the rotor that extends toward, but not touching, the outer distributor cap terminals.

rotor housing Houses the rotors in a Wankel/rotary engine. This is the base for the engine, similar to the engine block.

rotor lobes Lobes or rounded edges on rotors that squeeze oil and create pressure.

rotor-type oil pump A pump with an inner rotor driving an outer one; as they turn, the volume between them increases. The larger volume created between the rotors lowers the pressure at the pump inlet, drawing fluid in and filling the spaces. As the lobes of the inner rotor move into the spaces in the outer rotor, oil is squeezed out through the outlet.

rubber-bonded bushing A bushing that has a steel outer housing and inner sleeve with rubber inside; also known as a metalastic bushing.

rubber, rubberized fabrics, neoprene, and plastics Shop PPE made from these materials. Used to protect from chemicals.

run-flat technology A tire design that allows the vehicle to keep moving under driver control following a puncture or rapid loss of pressure.

running clearance The amount of space between wheel bearing components while in operation.

Rzeppa joint A type of fixed constant-velocity joint that has an inner race, six steel ball bearings, a bearing cage, and an outer race.

SAE J1930 An SAE standard for across-the-board standardization of parts and systems nomenclature.

SAE J2012 An SAE standard for across-the-board identification of generic DTCs.

safe working load (SWL) The maximum safe lifting load for lifting equipment.

safety data sheet (SDS) A sheet that provides information about handling, use, and storage of a material that may be hazardous. Previously called material safety data sheets.

salvage title A record that a vehicle has been severely damaged or deemed a total loss by an insurance company; also called a branded title.

sand or bead blasters A cleaning system that uses high pressure and fine particles of glass bead or sand.

sander/polisher A power tool with a rotating disc or head to which polishing or sanding discs can be attached.

scavenge pump A pump used with a dry sump oiling system to pull oil from the dry sump pan and move it to an oil tank outside the engine.

scavenging The process of removing burned gases from the cylinder through the use of moving airflow pulling or extracting the gases out.

scavenging effect A condition caused by moving columns of air, which create a low-pressure area behind them, that results in a pulling force that is used to pull the remaining burned gases from the combustion chamber. Valve timing affects the amount of scavenging effect an engine has.

Schrader valve A one-way valve used in a valve stem.

scratcher A thin, spring steel wear indicator that is fixed to the backing plate of the brake pad; it emits a high-pitched squeal when the brakes are applied if the brake pads have become too thin.

screw extractor A tool for removing broken screws or bolts.

scrub brakes A brake system that uses leverage to force a friction block against one or more wheels.

scrub radius The distance between two imaginary lines—the camber line through the center of the tire, and the SAI line—or the point where they intersect above or below the surface of the road.

sealed bearings Wheel bearings that are assembled by the manufacturer, with the proper lubrication, and sealed for life; cannot normally be disassembled.

secondary circuit The part of an ignition system that operates on higher voltage and delivers the necessary high voltage to the spark plugs.

secondary cup A seal that prevents loss of fluid from the rear of each piston in the master cylinder.

secondary piston A piston that is moved by hydraulic pressure generated by the primary piston in the master cylinder.

secondary winding The coil of wire in which high voltage is induced in a step-up transformer.

sedan A vehicle configuration that has an enclosed body, with a maximum of four doors to allow access to the passenger compartment.

self-energizing The property of drum brakes that assists the driver in applying the brakes; when brake shoes come into contact with the moving drum, the friction tends to wedge the shoes against the drum, thus increasing the braking force.

self-leveling A vehicle with automatic load-adjustable shock absorbers.

self-sealing tire A tire constructed with a flexible and malleable lining inside the tire around the inner tubeless membrane. The lining can seal small tread-area punctures instantly and permanently.

semiconductor A device, usually made from silicon, that has been doped with boron and phosphorus to create two or more distinct layers. The layers can be joined to create components like diodes and transistors. A semiconductor has four valence electrons, and the doping creates a layer that has more positive charge carriers and a layer with more negative charge carriers. Current and voltage are used to manipulate the charge to provide an insulating or conducting function in a semiconductor.

semi-floating axle An axle with the bearing placed between the axle and the axle housing; therefore, it carries the load of the vehicle on its outer end.

sending unit The device that changes a temperature, fluid level, or pressure into a varying resistance. This varying resistance is in series with the related instrument cluster gauge, and the resulting voltage drop across the sending unit is what causes the needle on the gauge to indicate the related temperature, fluid level, or pressure.

separator PCV system A PCV system with a device that uses gravity to allow oil to fall to the bottom and be returned to the crankcase; the device prevents liquid from traveling to the intake manifold.

series hybrid drivetrain A type of hybrid transmission in which power flows from the engine through an electric motor. The electric motor supplements the power from the engine to the wheels.

series-parallel circuit A circuit that has both a series and a parallel circuit combined into one circuit.

series-parallel hybrid drivetrain A type of hybrid drivetrain that can function as both a series hybrid and parallel hybrid, meaning that the gasoline engine can turn a generator that can be used to power an electric motor. The gasoline engine can also drive the vehicle directly through the transmission, and the electric motor can work in parallel with the gasoline engine to drive the vehicle.

service brake A brake system that is operated while the vehicle is moving, in order to slow or stop the vehicle. Typically applied by foot.

service campaign and recall A corrective measure conducted by manufacturers when a safety issue is discovered with a particular vehicle.

service consultant/advisor A customer service worker who works with both customers and technicians; the first point of contact for customers seeking vehicle repairs.

service history A complete list of all the servicing and repairs that have been performed on a vehicle.

serviceable bearings Wheel bearings that can be disassembled, cleaned, inspected, packed, reinstalled, and adjusted.

servo action A drum brake design where one brake shoe, when activated, applies an increased activating force to the other brake shoe, in proportion to the initial activating force; further enhances the self-energizing feature of some drum brakes.

setback The distance one wheel is set back from the wheel on the opposite side of the axle.

shaft The long, narrow component that carries one or more gears or has gears machined into it.

shielded wiring harness A wiring harness that has shielding built into it to protect it from induced electrical interference.

shimmy/shake Vibration or wiggle in the steering or suspension system.

shock absorber A device on a vehicle designed to absorb bumps and jolts caused from driving on irregular surfaces; it also dampens body movement.

shop foreman The supervisor in a shop who oversees the work of technicians and staff, and communicates with customers and external suppliers.

shop or service manual Manufacturer or aftermarket information on the repair and service of vehicles.

short circuit A condition in which the current flows along an unintended route; also called a short.

short to ground Fault conditions in a circuit where the circuit is unintentionally contacting a grounded component or wire. This may result in a short, in the case of a power wire, or it could cause a circuit to stay live, as in the case of a switched ground circuit.

short to power A condition in which current flows from one circuit into another.

shroud A steel or plastic cover placed over the shock rod.

side force The pressure on the wheel that pushes it toward the outside or inside of the rim as the vehicle makes a turn.

side gears A gear that is splined to the axle shaft and meshes with the spider gears, and that allows the axles to rotate at their own speeds when cornering and turning.

signal word Specific words used on signs in a shop to indicate a danger, warning, or caution.

sine wave The shape of an AC waveform as it changes from positive to negative, graphed as a function of time.

single flare A sealing system made on the end of metal tubing.

single-piston master cylinder A master cylinder with a single piston that creates hydraulic pressure for all wheel units. If there is a leak in the system, there is a loss of pressure for all wheel units.

sintered Solid part used as a bearing surface. It is made of small particles that are bonded together with heat and pressure.

six-ply rating Tire casing made up of six casing plies. Or a tire casing having the strength of six casing plies.

SLA suspension See *short-/long-arm suspension.*

slave cylinder The component in a hydraulically operated clutch that converts hydraulic pressure to mechanical movement at the clutch fork.

sledgehammer A heavy hammer with two flat faces.

sliding or floating calipers A type of brake caliper that only has pistons on the inboard side of the rotor. The caliper is free to slide or float, thus pulling the outboard brake pad into the rotor when braking force is applied.

sliding spline driveshaft A two-piece driveshaft that is joined in the middle with splines. The driveshaft can slide on itself to increase or decrease in length.

sliding T-handle A handle, fitted at 90 degrees to the main body, that can be slid from side to side.

slip angle The angle between which the tire is pointing, and the vehicle is moving.

slip yoke Part of a two-piece driveshaft that is splined and allows for a change in length of the shaft as the suspension compresses and rebounds.

slow charger A battery charger that charges at low current.

smart charger A battery charger with microprocessor-controlled charging rates and times.

snap ring pliers A pair of pliers for installing and removing internal or external snap rings.

snap ring The spring steel, C-shaped ring that is fitted in a groove and holds gears, bearings, and shafts in place.

socket An enclosed metal tube, commonly with 6 or 12 points, used to remove and install bolts and nuts.

solder A metal with a low melting temperature that is used to fuse metal components.

soldering irons A heating tool to heat solder and wires to produce a low-resistance joint.

solder-type terminal A terminal that requires soldering to fasten the terminal to the cable or wire, instead of being crimped.

solenoid An electromagnet with a moving iron core that is used to cause mechanical motion.

solenoid valve A type of electromechanically operated valve that uses an electric current to control fluid flow.

solid axle A single piece of steel that provides a simple means of mounting the hub and wheel units; also called beam axle or straight axle.

solid rotor A type of brake rotor made of solid metal, not ventilated.

solid-state relay A relay that performs the function of a mechanical relay but uses only electronic components.

solvent tank A tank containing solvents to clean vehicle parts.

sound-insulated rooms A room designed to isolate noise from the rest of the shop.

spark plug A device that provides a gap for the high-voltage spark to occur in each cylinder.

spark plug reach The length of the spark plug from the seat to the end of the threads.

spark timing The point at which a spark occurs at the spark plug relative to the position of the piston.

specialty springs Springs used to return links and levers on the parking brake system or the self-adjuster mechanism.

speed brace A U-shaped socket wrench that allows high-speed operation; also called a speeder handle.

speedy sleeve An aftermarket repair kit that consists of a thin metal sleeve that fits tightly over the seal surface of the axle, providing a new, undamaged surface for the seal to ride against.

splash lubrication A lubrication system that relies on oil being splashed onto moving parts by rotating engine parts striking the oil. These systems are typically used in small engines.

spline Ridges or teeth on a shaft that mesh with grooves in a mating piece and transfer torque to it, maintaining the angular correspondence between them.

splined section A flat key made into a shaft to accommodate changes in shaft length due to movement in wheel camber with suspension action.

splined Typically, a shaft and gear that have parallel grooves machined in them so they mate with each other and lock together rotationally.

split ball gauge A gauge that is good for accurately measuring small holes where telescoping gauges cannot fit; also called a small hole gauge.

sport utility vehicle (SUV) A passenger vehicle built on a light-truck chassis; it is usually equipped with four-wheel drive and capable of hauling heavier loads than typical passenger vehicles.

spray wash cabinet A cleaning cabinet that sprays cleaning solution, under pressure, to clean vehicle parts.

spring eyes Rolled ends of some springs used to mount springs to the chassis.

spring shackle bushing A bushing that is positioned in the shackle to which the leaf spring mounts. Bushings allow the spring shackle to move as the leaf spring dimensions change over bumps.

spring-loaded rack guide yoke A spring-containing yoke that pushes on the back side of the rack to help reduce the play between the rack and the pinion while still allowing for relative movement.

springs Elastic component that supports the body of the vehicle. It allows the suspension system to flex with road irregularities.

springs and clips Various devices that hold the brake shoes in place or return them to their proper place.

spur gear A gear with teeth cut parallel to its axis of rotation.

square file A type of file with a square cross section.

square thread A thread type with square shoulders used to translate rotational to lateral movement.

square-cut O-ring An O-ring with a square cross section that is used to seal the pistons in disc brake calipers.

squib Used to ignite a device, like the propellant in an airbag module. While the device does not explode, it does create a very aggressive chemical reaction that creates the heat to expand and deploy the airbag from within its storage packaging.

standard torque converter A hydraulic coupling device consisting of an impeller, turbine, stator, and housing; located between the engine and the transmission.

starter ring gear Ring gear on a flywheel or flexplate that meshes with the starter drive gear.

static imbalance Assumes the imbalance is centered across the width of the tire. A tire with static imbalance tends to vibrate vertically, with the heavy area slapping the road surface with each turn of the wheel.

static toe A setting, designed to compensate for slight wear in steering components, that may cause the wheels to turn slightly outward or inward while the vehicle is in motion.

station wagon A vehicle configuration with four doors, with a roof line that continues into the rear cargo area and a rear door for access.

stationary winding An extended length of wire wrapped into a circle. These windings are fixed, as opposed to some types, which are meant to spin.

stator Portion of an electronic ignition system that is mounted to the base of the distributor. It has a circular permanent magnet with a number of projections or teeth corresponding to the number of engine cylinders, and a stationary coil of fine enameled copper wire wound on a plastic reel and positioned inside the magnet.

steel hammer A hammer with a head made of hardened steel.

steel rule An accurate measuring ruler made of steel, or stainless steel.

steering angle sensor A sensor that measures the amount of turning a driver desires. This information is used by the ESC system to know the driver's directional intent.

steering arm An arm that extends from the steering knuckle. The tie rods connect to these arms in order to steer the wheels.

steering axis inclination (SAI) The angle formed by an imaginary line running through the upper and lower steering pivots relative to vertical as viewed from the front.

steering box The assembly which converts the rotary motion of the steering wheel into the linear motion of the steering linkage.

steering column Transmits the driver's steering effort from the steering wheel down to the steering box, usually made to collapse during a crash.

steering damper A device used to prevent shocks from irregular roads from being transmitted through the steering linkage and back to the steering wheel.

steering knuckle The knuckle located either between the lower control arm and MacPherson strut or upper control arm. It has either a spindle formed or bolted onto it, to which the wheel hub is attached, or it provides the location of the sealed front wheel bearing on front drive axles.

steering linkage Steel rods that connect the steering box to the steering arms on the steering knuckle.

steering return concerns Steering that catches, binds, or doesn't return properly.

steering sensor A torque sensor that converts steering torque input and direction into voltage signals for the power steering control module (PSCM) to monitor. Also a steering angle or position sensor, it converts the rotation speed and direction into voltage signals for the PSCM to monitor.

steering system A term used to describe all of the components and parts involved in steering a vehicle.

steering wheel position sensor (SWPS) A sensor that signals to the EBCM both the position of the steering wheel and the speed at which it is being turned.

step-down transformer A component that converts high-voltage, low-current AC power from a wall outlet (or engine) to a lower-voltage, higher-current AC or DC output.

stepper motor A specialized DC motor that has a rotor that is operated by a series of coils that surround the rotor. The rotor is stepped, or moved incrementally, by pulsing the coils in sequence, causing the rotor to move in a specific direction and amount of rotation. The coils can be pulsed in either direction, so the rotor can move clockwise or counterclockwise. These motors are often used to move a component a very specific amount for precise control of the related output system, like the doors in heating and air conditioning duct work.

step-up transformer A transformer used to increase the voltage from a lower input voltage to a higher output, such as an ignition coil.

stoichiometric ratio The optimum ratio of air to fuel for combustion—14.7 parts air to 1 part fuel by weight.

stop A rubber part used to control the movement of control arms (suspension arms).

straight edge A measuring device, generally made of steel, to check how flat a surface is.

straight grinder A powered grinder with the wheel set at 90 degrees to the shaft.

strategy-based diagnosis A best practice diagnostic process that utilizes the same process every time.

stratified charge A layering of the air-fuel mixture in the combustion chamber. Used when the fuel injector places fuel near the spark plug, but not the surrounding space.

strut A rigid shock absorber assembly used on a MacPherson strut–type suspension.

stub axle An axle used for one wheel.

stub-axle carrier The body of the stub-axle knuckle.

stud A type of threaded fastener with a thread cut on each end rather than having a bolt head on one end.

sulfuric acid A type of acid that, when mixed with pure water, forms the basis of battery acid or electrolyte.

sun gear The center gear of a planetary gearset around which the other gears rotate.

supercharger Component that uses external power to turn the compressor wheel to compress the air as it enters the intake system. The supercharger is usually driven by a belt off the crankshaft, or directly from an electric motor.

supplemental restraint system (SRS) A passenger safety system, such as airbags and seat belt pretensioners.

supporting statement A statement that urges the speaker to elaborate on a particular topic.

surge protector An electrical protection device for preventing electrical surges.

surge tank A pressurized tank that is piped into the cooling system. Coolant constantly moves through it. It is used when the radiator is not the highest part of the cooling system. (Remember, air collects at the highest point in the cooling system.)

suspension action Movement of the chassis up and down.

suspension system A vehicle system designed to isolate the vehicle body from road bumps and vibrations.

sway bar A part used in vehicles as a stabilizer, or antiroll, bar. It is connected to the chassis in the center, and each end is connected to one side of the suspension system. It is typically installed on the front, and sometimes the rear, suspension.

swinging shackle A shackle connected to the rear of the multileaf spring that allows the leaf spring to move downward when a load is placed on the rear of the vehicle.

switch An electrical device with contacts that turns current flow on and off.

symmetric tread pattern A tread pattern with the same tread pattern on both sides of the tire; typically nondirectional.

synchromesh transmission A modern transmission that uses gear synchronizers to match the speeds of gears and shafts during upshifts and downshifts.

synthetic blend A blend of conventional engine oil and pure synthetic oil.

synthetic oil Synthetic oil that, in its pure form, uses artificially made base stocks and is not derived from crude oil. This oil lasts longer and performs better than normal oil. The base stock additives are similar to those in conventional oils.

tandem master cylinder A master cylinder that has two pistons that operate separate braking circuits, so if a leak develops in one circuit, the other circuit can still operate.

tap handle A tool designed to securely hold taps for cutting internal threads.

taper tap A tap with a taper; it is usually the first of three taps used when cutting internal threads.

tapered roller bearing A type of wheel bearing with races and rollers that are tapered in such a manner that all of the tapered angles meet at a common point, which allows them to roll freely and yet control thrust.

tapered seat lug nuts Lug nuts with a tapered seat. The tapered seat is used to center the wheel on the axle flange.

taps and dies Tools used to create internal and external threads.

technical service bulletins (TSBs) A bulletin from the vehicle manufacturer that includes commonly found faults in a particular system.

telescoping gauge A gauge that expands and locks to the internal diameter of bores; a caliper or outside micrometer is used to measure its size.

temperature grade a representation of a tire's ability to resist and dissipate heat. From highest to lowest: A, B, and C.

temporary barrier Used to provide specific protection for specific areas, such as a welding screen.

tensile strength The amount of force required before a material deforms or breaks when being pulled apart.

terminal Metal connectors that are attached to wire ends. They are used to create electrical connections that can be disconnected and reconnected.

test strip A chemically reactive strip that turns color when exposed to a reactive fluid.

tetrafluoroethane An inert, colorless gas that can be used as a refrigerant. It is stored in light blue containers.

text Information on a warning sign to provide additional information.

thermal expansion valve (TXV) system A system with a valve designed to sense evaporator outlet temperature and vary the inlet orifice size accordingly.

thermal runaway During thermal runaway, the high heat of the failing cell will propagate to neighboring cells, causing them to become thermally unstable as well. When lithium-ion batteries enter the thermal runaway, extreme overheating—and in some cases, fire—can be expected. Thermal runway is also referred to as venting the flame.

thermistor A variable resistor that changes its resistance based on temperature. Most thermistors have a negative temperature coefficient, meaning that their resistance decreases as temperature increases. Commonly used to measure coolant, oil, fuel, and air temperatures.

thermo-control switch A temperature-sensitive switch that is mounted into a coolant passage on the engine or into the radiator to control electric fan operation.

thermocouple A temperature-sensing component that consists of two dissimilar metals that produce voltage proportional to temperature.

thermostat Regulates coolant flow to the radiator. It opens at a predetermined temperature to allow coolant flow to the radiator for cooling. It also enables the engine to reach operating temperature more quickly for reduced emissions and wear.

thixotropy The ability of a semisolid grease to flow when agitated or stressed.

thread chaser A device similar to a die that cleans up rusty or damaged threads.

thread file A type of file that cleans clogged or distorted threads on bolts and studs.

thread pitch The coarseness or fineness of a thread as measured by either the threads per inch or the distance from the peak of one thread to the next. Metric fasteners are measured in millimeters.

thread pitch The coarseness or fineness of a thread as measured by either the threads per inch or the distance from the peak of one thread to the next. Metric fasteners are measured in millimeters.

thread repair A generic term to describe a number of processes that can be used to repair threads.

threaded fasteners Bolts, studs, and nuts designed to secure parts that are under various tension and sheer stresses.

three-quarter floating axle An axle on which there is only one wheel bearing that bears the weight of the vehicle, but the axle prevents the wheel from tipping inward or outward.

three-way catalytic converter A converter that changes hydrocarbons, carbon monoxide, and oxides of nitrogen into harmless elements.

throttle A device used to produce acceleration by controlling the amount of air entering the engine.

throttle body injection (TBI) A fuel injection system that uses one or more fuel injectors mounted above or in the throttle body itself; also called single-point injection.

throw-out bearing The part of the clutch release mechanism that imparts clutch pedal force to the rotating pressure plate levers.

thrust angle The angle formed between the perpendicular centerline of the rear axle in comparison to the centerline of the vehicle.

thrust line The imaginary line drawn perpendicular to the rear axle.

thrust washers Washers that provide a bearing surface between parts.

thrust-type angular-contact ball bearing A type of bearing that uses a deep groove in the bearing races where the ball bearings ride; this design is for thrust conditions.

tie-rod assembly The part that fits between the rack and the steering arms and transfers the movement of the rack.

tie-rod end puller Used to pull the tapered shaft on a tie-rod end from its mating steering component.

tie-rod sleeve adjusting tool Has a tab designed to grab the slot in the sleeve and is used to turn the sleeve when adjusting the toe setting.

time/mileage A reference to the method of determining when a particular service should be performed.

tin snips Cutting device for sheet metal; works in a similar fashion to scissors.

tire inflation pressure Level of air in the tire that provides it with load-carrying capacity and affects overall vehicle performance.

tire pressure gauge A tool used to check the air pressure in tires.

tire pressure monitoring system (TPMS) A federally mandated system to provide a means of reliable and continuous monitoring of vehicle tire pressure. It is designed to increase safety, decrease fuel consumption, and improve vehicle performance. A TPMS monitors the tires for low air pressure and alerts the driver when one or more tires are lower than (or in some cases, higher than) the designated thresholds. This alert can be an illuminated warning lamp or a chime.

title history A detailed account of a vehicle's past.

toe setting Setting of the toe-in or toe-out of the tires to the centerline of the vehicle.

toe-in When the front of the wheels, as seen from above, are closer together than the rear of the wheels.

toe-out on turns (TOOT) The difference in turning angle of the inside tire in comparison to the outside tire. This angle difference allows the tires to roll through the corner rather than the inside tire dragging. Also referred to as Ackermann angle.

toe-out When the rear of the wheels, as seen from above, are closer together than the front of the wheels.

tone wheel The part of the wheel speed sensor that has ribs and valleys used to create an electrical signal inside of the pickup assembly.

top hat parking brake A drum brake that is located inside a disc brake rotor in order to act as a parking brake.

toroidal CVT A type of CVT that uses moveable rollers in contact with input and output drive discs. The rollers transfer power from one drive disc to the other. Their position determines the effective gear ratio.

torque Twisting force applied to a shaft that may or may not result in motion.

torque angle A method of tightening bolts or nuts based on angles of rotation.

torque assist Use of an electric motor to supplement the engine's torque whenever additional torque is needed, allowing for a smaller ICE to be used.

torque converter A type of fluid coupling that is also capable of multiplying torque. It is turned by the crankshaft and transmits torque to the input shaft of an automatic transmission.

torque sensor A device used to measure the load on the steering wheel.

torque smoothing A process that uses an electric motor to smooth out engine power pulses when an ICE is operating at low rpm or when the vehicle is using fuel management techniques such as cylinder deactivation.

torque specifications Supplied by manufacturers, who describe the amount of twisting force required for a fastener or provide a specification showing the twisting force from an engine crankshaft.

torque steer A condition in which the vehicle pulls to one side during hard acceleration. It can be the result of unequal axle lengths as designed, which cannot be repaired if it is by design.

torque wrench A tool used to measure the rotational or twisting force applied to fasteners.

torque-to-yield (TTY) A method of tightening bolts close to their yield point or the point at which they will not return to their original length.

torque-to-yield (TTY) bolts Bolts that are tightened using the torque-to-yield method.

torsion bar Spring steel rod used in power steering systems to allow relative movement between the steering wheel and steering gear. This allows the hydraulic control valve to direct power steering fluid pressure as needed.

torsional load A force that is applied by clamping one end of an object to another object that is then twisted.

toxic dust Any dust that may contain fine particles that could be harmful to humans or the environment.

track rod On forward-control vehicles, it connects the relay lever to the idler arm.

traction control system (TCS) A computer-controlled system added to ABS to help prevent loss of traction while the vehicle is accelerating.

traction grade A standardized grading system that indicates how well a tire will maintain contact with the road surface when wet.

trailing arm suspension A type of suspension system that uses upper and lower control arms.

trailing shoes Brake shoes installed so that they are applied in the opposite direction to the forward rotation of the brake drum; not self-energizing and less efficient at developing braking force.

transaxle A type of transmission, typically used in FWD vehicles, in which the transmission also includes the differential and final drive gear assembly.

transfer case A gearbox arrangement that allows the torque from the transmission to be split between the front and rear driving axles of a vehicle.

transformer action The transfer of electrical energy from one coil to another through induction in a transformer.

transmission An assembly that houses a variety of gearsets that allow the vehicle to be driven at a wider range of speeds and terrain conditions than would be possible without a transmission.

transmission input shaft The shaft that brings engine torque into the transmission.

transmission mounted parking brake A drum brake that is mounted on the drive shaft, just after the transmission, to serve as a parking brake.

transmission specialist A technician who diagnoses, overhauls, and repairs transmissions.

transverse A term used to describe the side-to-side engine orientation when mounted in the engine compartment.

tread wear grade The number imprinted on the sidewall of a tire by the manufacturer, as required by the National Highway Traffic Safety Administration (NHTSA), that indicates the tread life of a tire's tread.

treated wool and cotton Fire-resistant material used in some shop clothing.

triangular file A type of file with three sides so it can get into internal corners.

trim height The height from the ground to a specified part of the vehicle; also known as ride height or curb height.

truck A large, heavy vehicle for carrying cargo.

tube flaring tool A tool that makes a sealing flare on the end of metal tubing.

tubing cutter A hand tool for cutting pipe or tubing squarely.

tulip/tripod joint A constant-velocity joint that has three equally spaced fingers shaped like a star. This configuration enables in-and-out movement of the shaft while also allowing flexing.

tuned exhaust Exhaust system in which the exhaust pulse from one cylinder is timed to arrive at the right time to help draw out the pulse from another cylinder.

turbocharger Forced induction device driven by exhaust gas pressure and used to compress intake air so that more air can be pushed into the cylinders, thus creating more power.

turn signal switch A switch that turns the left and right turn signal lights on and off.

turning radius A measure of how small a circle the outside front wheel (or the outside front corner of the vehicle body) can rotate around when the steering wheel is turned to the limit.

twin leading shoe drum brake system Brake shoe arrangement in which both brake shoes are self-energizing in the forward direction.

twisted pair Two conductors that are twisted together to reduce electrical interference.

two-mode hybrid A type of hybrid drive system in which there are two distinct modes of operation. In one mode, the electric motor can propel the vehicle and be used for regenerative braking; in the second mode, the electric motors can be used to assist the engine while the engine uses fuel management techniques such as cylinder deactivation.

two-post lifts Vehicle lift with two centrally located lifting posts.

two-stroke engine An engine that uses only two strokes to complete its running cycle.

two-way catalytic converter A converter that changes only hydrocarbons and carbon monoxide into harmless elements.

U-joint See *universal joint.*

UNC (Unified National Coarse) Used to describe thread pitch.

understeer A condition in which the vehicle's front wheels are turned more sharply than the vehicle's actual direction because the front tires lose adhesion during cornering. The vehicle is said to be "pushing" in the corners.

unibody design A vehicle design that does not use a rigid frame to support the body. The body panels are designed to provide the strength for the vehicle.

Unified National Coarse Thread (UNC) Used in the standard bolt system. A bolt with fewer threads per inch for a given diameter.

Unified National Fine Thread (UNF) Used in the standard bolt system. A bolt with more threads per inch for a given diameter.

Unified National Fine Used to describe thread pitch.

uniform pitch A spring whose pitch (the distance from the center of one coil to the center of the adjacent coil) is the same distance throughout.

Uniform Tire Quality Grading (UTQG) A standardized grading system, established by the National Highway Traffic Safety Administration (NHTSA), designed to provide tire buyers with a comparative measure of a tire's tread life, traction, and temperature characteristics.

unitized wheel bearing hub An assembly consisting of the hub, wheel bearing(s), and possibly the wheel flange, which is preassembled and ready to be installed on a vehicle.

universal joints A flexible cross-shaped joint used to transmit torque.

unsprung weight See *unsprung mass.*

U.S. Department of Transportation (DOT) A federal agency that regulates transportation safety in the United States, including vehicles' wheels and tires. The DOT requires a code—a series of letters and numbers—to be stamped into the sidewall of every tire made for public use in the United States. These codes contain information such as the date of manufacture and the plant where the tire was manufactured.

V blocks Tools used to set round objects in while measuring them.

V engine A term used to describe an engine configuration that has two banks of cylinders sitting side by side in a V arrangement and sharing a common crankshaft.

vacuum A pressure in an enclosed area that is lower than atmospheric pressure.

vacuum bleeding Bleeding process that uses a vacuum bleeder to pull the air and old brake fluid from the system.

vacuum relief valve A mechanical valve used on the gas cap that prevents a low-pressure condition from occurring. It also ensures that the fuel tank will not collapse due to fuel usage or contraction of fuel on cooldown.

vacuum tube fluorescent (VTF) A type of lighting used for instrumentation displays on vehicle instrument panel clusters. This type of lighting emits a very bright light with high contrast and can display in various colors; also called vacuum fluorescent display (VFD).

validating statement A statement that shows common interest in the topic being discussed.

valve core The one-way spring-loaded valve that screws into the valve stem; it allows air to be pumped into a tire and prevents it from flowing out.

valve overlap The portion of time that both valves are open at the same time.

valve stem cap A cap that fits tightly onto the valve stem to prevent debris from clogging it and acts as a secondary seal.

valve stem Part which allows air to be added or removed from a tire.

vapor line A rubber or plastic line that carries vapors from the fuel tank to the charcoal canister.

vapor lock A situation in which vapor forms in the fuel line, and the bubbles of vapor block the flow of fuel and stop the engine.

vaporization Change of state from a liquid to a gas.

variable orifice PCV system A system in which a replaceable, spring-loaded PCV valve regulates gas flow. The position of the valve is controlled by the pressure in the manifold.

variable reluctance sensor A sensor that uses the principle of magnetic induction to create its signal. It is used to measure rotational speed, including wheel speed, machine speed, engine speed, and camshaft and crankshaft position.

variable resistor A component that has a mechanism for varying resistance.

variable-diameter pulley (VDP) A type of CVT that uses two pulleys with moveable sheaves, allowing the effective diameter of the pulleys to change, resulting in variable gear ratios.

variable-diameter pulley A pulley that can change its diameter by moving closer or further apart.

vehicle emission control information (VECI) label A label used by technicians to identify engine and emission control information for the vehicle.

vehicle identification number (VIN) A unique serial number that is assigned to each vehicle produced.

vehicle inspection pit A trench permanently fitted into the floor of the shop to allow easy work access to the vehicle's underside.

vehicle jack A tool for lifting a vehicle.

vehicle lifts Equipment designed to lift the entire vehicle off the ground.

vehicle safety certification (VSC) label A label certifying that the vehicle meets the Federal Motor Vehicle Safety, Bumper, and Theft Prevention Standards in effect at the time of manufacture.

vehicle speed sensor (VSS) A sensor used by the PCM to measure vehicle speed. It is often located in the transmission extension housing. The output signal may be analog or digital.

vehicle wander Occurs when the vehicle is not driving in exactly the direction the driver is steering it. This tends to happen when the caster angle is off or there is looseness in the steering/suspension components.

ventilated rotor A type of brake rotor with passages between the rotor surfaces that are used to improve heat transfer to the atmosphere.

venturi A restriction (narrowed area) in the air horn.

viscosity index improver An oil additive that resists a change in viscosity over a range of temperatures.

viscosity The measurement of how easily a liquid flows; the most common organization that rates lubricating fluids is SAE.

viscous coupler Called a fan clutch, a hub that connects the water pump drive to the cooling fan, using a temperature-sensitive viscous fluid to cause the fan to turn faster as the temperature of the air pulled through the radiator increases.

viscous coupling A silicone clutch assembly used in AWD differentials to provide a slight amount of differential action for control of axle rotational speeds.

volatile organic compound (VOC) The hydrocarbons in petroleum products that contribute to combustion.

volt The unit used to measure potential difference or electrical pressure.

voltage The electrical pressure that causes current to flow in a circuit.

voltage drop The amount of potential difference between two points in a circuit.

voltage drop test Measurement of the difference in voltage between two points in a circuit: the black lead on the end point being tested and the red lead on the beginning point.

vortex flow State in which the fluid in the torque converter is traveling from the impeller, through the turbine, through the stator, and back to the impeller.

VR engine A term used to describe an engine configuration that uses a single bank of cylinders staggered at a shallow 15-degree V.

W engine A term used to describe an engine configuration consisting of two VR cylinder banks in a deep V arrangement.

wad punch A type of punch that is hollow for cutting circular shapes in soft materials such as gaskets.

warding file A type of thin, flat file with a tapered end.

warm-up cycle When the engine is started and run until the engine operating temperature reaches 160°F (71.1°C), has increased in temperature at least 40°F (22°C), and the engine is turned off again.

waste cylinder The cylinder in a waste spark ignition system that receives a spark near the top of its exhaust stroke.

waste spark ignition system An ignition system in which each ignition coil served two cylinders, with each end of the secondary winding attached by a high-tension lead to a spark plug. The spark is used to ignite the air-fuel mixture in one cylinder and has no effect on the other cylinder.

wastegate A pressure regulator device that allows control of the pressure produced by the turbocharger. The wastegate opens to allow exhaust gases to bypass the turbine wheel of the turbocharger.

water fade Brake fade caused by water-soaked brake linings.

water jacket Passages surrounding the cylinders and head on the engine where coolant can flow to pick up excess heat. They are sealed by replaceable core plugs.

watt's linkage Another name for a rigid-axle coil-spring suspension that uses two bars similar to a Panhard rod and a pivot point on the axle, to keep the axle from moving in turns.

watt The unit for measuring electrical power.

weight matching The process of matching the tire's lightest point with the rim's heaviest point (generally at the valve stem) for the purpose of reducing the tire's radial imbalance.

weight transfer Weight moving from one set of wheels to the other set of wheels during braking, acceleration, or cornering.

welding helmet Protective gear designed for arc welding; it provides protection against foreign particles entering the eye, and the lens is tinted to reduce the glare of the welding arc.

wheel alignment The practice of aligning the wheels of the vehicle to the centerline of the vehicle and to one another. It ensures that the vehicle will handle correctly and gives best tire wear.

wheel cylinder A hydraulic cylinder with one or two pistons, seals, dust boots, and a bleeder screw that pushes the brake shoes into contact with the brake drum to slow or stop the vehicle.

wheel cylinder piston clamp A tool that prevents the pistons from being pushed out of the wheel cylinders while the brake shoes are being replaced.

wheel flange The center portion of a wheel. Usually a formed disc which is welded to the rim.

wheel speed sensor A device that creates an analog or digital signal according to the speed of the wheel.

wheel studs Threaded fasteners that are pressed into the wheel hub flange and used to bolt the wheel onto the vehicle.

wheel width The distance between the bead seats on the wheel.

windage tray Component usually made out of sheet metal or plastic that bolts onto the bottom of the main bearing saddles; it prevents the churning of the oil by the rotation of the crankshaft.

wire A conductor usually made of multistranded copper with an external insulated coating; used to transmit electricity within circuits.

wiring diagram A schematic drawing and symbol representation of the wiring and components; also called an electrical schematic.

wiring harness connector A plug that contains multiple terminals with male and female ends.

wiring harness The network of wires, connectors, and terminals, pre-formed into bundles, that carry current within electrical circuits.

wishbone control arm Another term for an A-arm.

work The result of force creating movement; or the transformation of energy from one type to another.

worm A gear with a helical, threaded shaft that is attached to the steering column and meshes with a worm wheel that transfers motion from the steering wheel to the steering linkage.

worm gear steering A steering box consisting of a worm and worm gear.

worm shaft The protrusion of the worm gear that serves as the point of attachment to the steering column.

wrap leaf A spring containing spring eyes.

wrench A generic term to describe tools that tighten and loosen fasteners with hexagonal heads.

yaw Movement around the z-axis (vertical axis), felt when the vehicle deviates from its straight path, as when the rear wheels slide out during drifting.

yaw sensor A sensor that measures the amount a vehicle is turning around its vertical axis. This information is used by the ESC system to know how much a vehicle is turning.

yield point The point at which a bolt is stretched so much that it will not return to its original length when loosened; it is measured in pounds per square inch or kilopascals of bolt cross-section.

zero camber A tire with no tilt, or zero camber angle.

zero offset A condition in which the plane of the hub mounting surface is even with the centerline of the wheel.

zero scrub radius A condition in which the camber line through the center of the tire, intersects the SAI line at the road surface.

INDEX

A

abnormal noises, 1162
ABS. *See* antilock brake systems (ABS)
ABS electronic brake control module (EBCM), 940
ABS master cylinder, 784, 930
  operation of, 932–933
  portless ABS, 933f
  purpose of, 932–933
absorbed glass mat, 1038
AC generators, 1077
acceleration, 764
accelerator pedal position sensor, 1241
accelerator pump circuit, 1224
acid rain, 1303
Ackermann principle, 736
activated charcoal, 1313
active systems, 938
adaptive air suspension systems, 704
adaptive learning, 1255
adaptive suspension systems
  active and, 704–705
  adaptive air suspension, 705, 705f
adequate ventilation, 51
adjustable bushing, 636
adjustable proportioning valves
  diagonally split system, 792f
  electronic brake, 793
  front brake circuit failure, 792f
  load-sensitive, 792–793, 793f
adjustable shock absorbers
  air shock, 690f
  automatic load-adjustable suspension system, 692f
  coil-over shock, 690f
  electromagnetic, 691f
  electronic, 691–692
  load, 690
  manual, 690
  shock plunger rod, 691f
adjusting nut, 917
advance mechanisms, 1183
advanced batteries
  disadvantages of lithium-ion, 1039
  lithium-ion, 1038
  nickel-cadmium, 1038
  typical hybrid vehicle battery stack, 1039f
advanced brake systems
  brake assist, 760–761
  brake-by-wire, 760–761, 761f
  disc, 760f
aiming headlights, 1111–1112
air cleaner
  air filter systems, 1279f
  cyclone-type pre-cleaner, 1280f
  electronically fuel-injected systems, 1279
  filter service indicator, 1280f
  function of, 1279
  heavy-duty diesel engines, 1279, 1280
  long-life air filter, 1280, 1280f
air door operating mechanisms, 427f
air doors and actuators, 427
air drill, 180
air filter inspection, 1294–1295
air gap, 937
air hammer, 180
air impact wrench, 179
air measurement, 1246
air ratchet, 180
air spring, 690
air supply
  hot-wire MAF sensors, 1247
  intake runners branch off, 1246f
  MAF sensor, 1247
  mass of the air entering engine, 1247, 1247f
  mass sensor, 1246–1248
  measurement, 1246
  MPFI and GDI systems, 1246
  speed density method, 1246
  vane-type airflow meter, 1247f
  vortex-generating rod, 1247, 1247f
air supply system, 1202
air tools
  air automatic oiler, 179f
  air drills, 181–182
  air hammers, 180f, 182–183
  air impact wrenches, 181
  air ratchet, 180f
  air safety, 178
  air tools, 179–180
  chiller-type air drier, 178f
  driers and automatic oilers, 178–179
  hydraulic press, 183–184, 184f
  lubricating air tools, 179f
  OSHA-approved air nozzle, 180f
  quick disconnect fitting, 179f
  twist drill bit and drill chuck, 184f
  using air drills, 182–183
  using air hammers, 183
  using air impact wrenches, 181–182
  using air nozzles, 181f
  water trap air drier, 178f
airbags, 1136
air-conditioning components
  accumulator's job, 1155f
  automotive, 1154–1155
  evaporator transforms low-pressure liquid, 1154
  fixed orifice tube system, 1155, 1155f
  fluid flowing, 1154
  and operation, 1153
  refrigerant oils, 1156–1158, 1157f
  refrigerant types, 1156f
  service ports, 1157f
  TXV systems, 1155f
air-conditioning system odors, 1163
air-fuel ratio (AFR), 1302
Allen wrenches, 136
all-wheel drive (AWD) layout, 29, 550
alternating current, 1078
alternative fuel systems
  batteries, 1327
  battery electric vehicles, 1325–1327
  economic concerns, 1324–1325
  emission standards, 1325
  energy security, 1324
  environmental concerns, 1324
  extended-range electric vehicles, 1329
  fuel cell (electric) vehicles, 1329
  high-voltage system, 1333–1334
  hybrid and electric vehicle service precautions, 1331–1333
  hybrid auxiliary (12 v) battery service, 1335
  hybrid drive configurations, 1330–1331
  hybrid electric vehicles, 1329–1330
  hybrid vehicle efficiency enhancements, 1331
  hybrid vehicle operation, 1331
  hybrids and electric vehicles, 1327–1328
  overview of, 1323
  plug-in hybrid electric vehicles, 1329
  power source, 1328
  vehicle emissions and standards, 1325
alternative fuels
  defined, 1323
  3E causes, 1323
  liquid or gas, 1323
alternator component
  alternating north and south magnetic fields, 1079f
  brush assembly, 1079–1080
  brushless alternators, 1080
  overview of, 1079
  regular alternator, 1080f
  rotor, 1079
  slip rings and brushes aid, 1079–1080, 1080f
alternator cooling fan
  ADP, 1081–1082
  fan and pulley, 1081f
  freewheel, 1081
  overrunning alternator decoupler (OAD), 1081f
  and pulley, 1081

alternator decoupling pulleys (ADPs), 1081
alternator end frames and bearings, 1081, 1081f
alternator replacement, 1088–1089
alternator ripple, 1083, 1084
American Petroleum Institute (API), 375
American Society of Automotive Engineers (SAE), 376
American wire gauge (AWG), 983
ammeter, 956
analog signals
analog and digital waveforms, 1238f
brake on/off switch, 1239f
data transmission, 1239
and digital, 1238
duty cycle, 1239f
exhaust gas recirculation (EGR), 1239f
higher frequency, 1240
lower average current flow, 1239
on/off signal, 1238
speed and deceleration, 1240
anchor pin, 877
angle grinder, 186
antilock brake systems (ABS), 760
ABS master cylinder, 930
brake switch, 930
hydraulic control unit (HCU), 930, 930f
oscilloscope pattern, 931f
overview of, 930
power booster, 930
typical ABS system, 930f
wheel speed sensor and tone wheel, 931f
applied force, 682
arc joint pliers, 134
armature windings
and commutator, 1065
conductor loop, 1065
electromagnetic fields-with commutator and brushes, 1065, 1065f
single-loop motor, 1065f
switching point of commutator, 1066, 1066f
ASE. *See* Automotive Service Excellence (ASE)
ASE Education Foundation, 14
assembly parts, 646f
Association des Constructeurs Européensd' Automobiles (ACEA), 377
Atkinson and Miller cycle engines
crankshaft offset, 322f
scavenging, 322, 322f
atmospheric pressure, 1151
atomization, 1202
automatic brake self-adjuster, 872
automatic load-adjustable shock absorbers, 692
automatic oilers, 179
automatic transmission fundamentals
functions of, 461–462
holding/driving gears, 472–476
lock-up converters, 468–472
overview of, 461
types of, 462–468
automatic transmissions types
engine and the transmission, 465f
fluid flow, 467–468, 467f
Ford Focus dual-clutch transmission. Gears, 463f
Honda, main shaft and countershaft, 464f
impeller brazed, 466f
light-vehicle torque converter, 465f
planetary gear, 463
power-splitting transmission (PST), 463f
pulleys and metal belt of CVT transmission, 463f
rotary fluid flow, vortex flow, and spiral flow, 468f
stator removal, 466f
torque converter components, 465–466
torque converter operations, 466–467
torque converter principles, 464–465
torque converters, 464
torque multiplication, 467
transaxle, 462
transmission, 462
turbine mounted, 466f
typical torque converter, 465f
automatic transmission/transaxle
fluid and filters replacement, 482–485
in-vehicle transmission repair, 486–490
maintenance, 481–482
automobile history
Fully electric Tesla Model X vehicle, 7f
high-tech equipment, 6f
mass production of vehicles, 6f
technology, 6
vehicle manufacturing, 6
automotive heating
add-on luxury unit, 1148f
and cooling, 1147
EPA 611 license, 1148f
history, 1147
HVAC regulation, 1148
licensure, 1148
automotive industry certification
ongoing training, 15
special certification, 15
training class, 15f
automotive safety
hazardous materials safety, 61–71
high-voltage safety, 72–77
overview of, 41
safety overview, 42–46
shop safety inspections, 55–61
standard safety measures, 48
work environment, 47–48
automotive sector
careers in, 7
chassis and brake technician, 9–10, 10f
electrical technicians, 10f
electrical/drivability technician, 10, 10f, 11, 11f
heavy line technician, 9, 10f
light line technician, 7–9, 9f
lot attendant, 7, 9f
lube technician, 7
new car lots, 9f
performance of shop, service manager, 12f
service consultants, 11–12, 12f
service manager, 12
shop foreman, 11, 12f
transmission specialist, 11
vehicle scheduled maintenance, 9f
Automotive Service Excellence (ASE), 14
automotive technology
body designs, 19–22
drivetrain layouts, 28–33
engine classification, 33–37
overview of, 19
vehicle chassis, 22–23
vehicle system, 24–28
automotive technology careers
certification, 14–15
history, 5–6
sectors, 7–12
shop types, 12–14
Automotive Youth Educational Systems (AYES), 8f, 15
available voltage, 1003, 1008, 1174
AWG. *See* American wire gauge (AWG)
AWG system, 983f
AWG wire sizes, 984t
axial loads, 517
axial resistors, 981
axle flange measurement, 627, 630
axle inspection and repair
driveline and, 559
fluid leakage, 559
AYES. *See* Automotive Youth Educational Systems (AYES)

B

backing plate, 872
backlash, 556
backup lights, 1100, 1100f
bad compression, 347
baking principles
advanced, 760–761
factors affecting, 763
friction, 765–767
history of, 759–760
kinetic energy, 763–765
rotational force, 767–769
service and parking, 761–762
types, 770–772
ball joint play measuring, 716
ball joint press tool, 694
ballast resistor in ignition circuit, 1182, 1183f
ballistic pretensioner, 1137
band brake, 759
banked injection, 1217
BARO sensor, 1248f
barometric pressure (BARO) sensor, 1248
barrier cream, 87
BAS. *See* belt alternator starter (BAS)

basic hand tools
  anatomy of a socket, 129f
  box-end wrenches, 126f
  combination wrenches, 127f
  Crow's foot wrenches, 130f
  dial torque wrench reading torque, 133f
  digital torque wrench displaying torque, 134f
  flare nut wrench, 126f
  flexible extensions, 132f
  lug wrench, 132f
  oil filter wrenches, 128f
  open-end adjustable wrench, 127f
  pipe wrench, 128f
  ratcheting box-end wrench, 128f
  ratcheting open-end wrench, 128f
  six- and 12-point sockets, 130f
  sockets
    drivers, 131–132
    torque, 132–133
  standard and metric designations, 125–126
  standard wall socket, 130f
  torque wrench, 132f
  wrenches
    special, 128
    using correctly, 129
battery
  leave-acid battery, 1327
  lithium-ion (Li-ion)-rechargeable, 1327
  nickel-cadmium (Ni-Cd)-rechargeable, 1327
  nickel-metal hydride (NiMH)-rechargeable, 1327
  temperature range, 1327
battery and cables
  cleaning, 1046
  filling, 1046
  inspection, 1046
  replacing, 1046
battery cables
  displacement engine, 1041f
  layout types, 1040f
  lug terminal, 1040f
  posts, 1040f
  with sealed end, 1041f
  side-post, 1040f
  tapered-post, 1040f
  and terminals, 1040
battery charging, 171
  batteries connected in parallel, 173f
  battery disconnection, 172f
  high-voltage battery packs, 172f
  jumper cables, 175–176
  jump-starting vehicles, 174–177, 1047
  memory savor, 173f
  method, 172–173
  negative battery terminal, 1047
  smart charger, 171f
  testing, 1047
battery configurations
  allocation, 1039
  terminal, 1040
  underhood placement, 1039f
  under-seat placement, 1039f
battery discharging
  and charging cycle, 1042, 1042f
  discharging cycle, 1042, 1042f
battery electric vehicles
  Chevy Volt, 1326
  development, 1325–1326
  experiment, 1326
  Ford Focus Electric, 1326f
  Nissan Leaf, 1326f
  Tesla Model S, 1327f
  Tesla Model X, 1327f
  today, 1326–1327
battery life
  shorten/lengthen the life, 1042
  shorting out cell, 1043
  sulfated, 1043
battery life factors, lifetime, 1044
battery load testing, 1052–1053
battery maintenance
  baking soda and water, 1045
  battery state-of-charge (SOC), 1045
  clean battery post, 1045f
  DMM measure the voltage, 1044f
  electrolyte level, 1044f
  heavily oxidized battery post, 1045f
  purpose of, 1044
  surface of, 1044
  types of, 1044
battery ratings, 1041–1042
battery recycling, 1045–1046
battery removal
  identifying modules, 1053
  lose initialization, 1053
  memory saver and cable, 1053, 1054f
battery service precautions, 1045
battery state of charge
  acid concentration drops, 1050
  specific gravity, 1050, 1050t
  testing, 1050, 1051
battery systems
  cables and terminals, 1040–1041
  charge and specific gravity, 1050–1051
  charging, 1042, 1047–1048
  cleaning, 1046
  conductance testing for capacity, 1052
  discharging, 1042
  filling, 1046
  inspection, 1046
  jump-starting, 1048–1050
  lead-acid flooded cell, 1037
  life factors, 1042, 1044
  load testing, 1052–1053
  losing initialization, 1053–1054
  low-maintenance, 1037
  maintenance-free, 1037–1040, 1044–1046
  overview of, 1036–1037
  parasitic draw measurement, 1054–1056
  ratings, 1041–1042
  replacing, 1046
  testing capacity, 1052
battery terminal configuration, 1040
battery-electric vehicles (BEVs), 1323
BCM. *See* body control module (BCM); body control module (BCM)
bead seat, 580
beam axle, 639
bearing races, 863
before top dead center (BTDC), 1184
bell crank, 539
belt alternator starter (BAS), 497
belt routing label, 110
belt tension, 424
bench bleeding, 817
bench grinder, 185
bendable tangs, 840
BEVs. *See* battery-electric vehicles (BEVs)
bidirectional scanners, 1269
bleeder screw, 834
bleeding brake systems
  flushing, 813
  manual, 810, 811
  pressure, 810, 812
  vacuum, 810, 811
blink codes, 940
blowby gas, 327
blower motor, 1125
  and circuits, 1125, 1125f
blower motor and squirrel cage fan, 1160f
blower motor resistor and switch schematic, 1161f
blower motor resistor packs, 1160, 1160f
blow-off valve, 1287
body control module (BCM), 255, 930, 1069, 1118
body design, automotive technology
  convertible, 21
  coupe, 20, 20f
  four-door versions, 22f
  hardtop convertibles, 21, 21f
  hatchback, 20–21
  minivan, 22, 22f
  pickup, 21
  rear seats in hatchbacks, 21f
  sedan, 20, 20f
  sport utility vehicle, 22
  station wagon, 21, 21f
  steel ladder-frame chassis, 22f
  unibody design, 22f
body electrical system
  blower motor and circuits, 1125
  brush and brushless, 1124
  communication errors, 1122
  cooling fans and circuits, 1126
  door panel removing, 1128
  electric accessory motors, 1124
  electric lock, 1128–1129
  electronic modules, 1122
  flash reprogramming, 1122
  horn systems, 1130–1131
  keyless entry systems, 1128–1129
  multiplexing, 1118–1119
  networking, 1118–1119
  permanent magnet, 1124
  power door locks, 1127–1128

principles of operation, 1130
scan tool, 1122
software transfers, 1122
stepper type, 1124
supplemental restraint systems, 1135–1140
testing electric locks, 1128
testing electric motors, 1126–1127
types, 1129
vehicle communications networks, 1119–1122
wiper/washer system, 1131–1134
boiling point and pressure, 408
boiling point test, 808
bolts, studs, and nuts
bolt diameter and length, 206f
bolt failure, 208f
ductility, 209
fatigue strength, 208–209
grade 8.8 metric bolt, 208f
grade 5 standard bolt, 207f
grading of bolts, 207–208
metric thread pitches, 207f
proof load, 208
shear strength, 208
sizing bolts, 206
strength of bolts, 208
tensile strength, 208
thread pitch, 206–207
thread pitch gauge, 207, 207f
torsional strength, 209
toughness, 209
UNC and UNF standard bolt, 207f
variety of nuts, 209f
bonded linings, 838
boost valves, 942
bottoming tap, 145
bouncy ride, 713
box-end wrench, 126
brake assist (BA), 760
brake bleeder wrenches, 806f
brake drum, 872
brake fade, 839, 872
brake fluid handling, 807
brake fluid testing, 807f
boiling point, 808
digital multimeter (DMM)-galvanic reaction, 807–808
DMM measuring, 808f
DMM-galvanic reaction, 809
safety meter boils brake fluid, 808f
test strip, 808, 808f
testing brake fluid with, 810
time/mileage, 807
brake hoses, 789
flexible construction, 789f
materials
abrasion, 789
blockage of passageway, 789
bulges, 789
cracks, 789
kinks, 789
moveable brake unit, 789f
brake lights, 796
center high-mount stop light, 1100f
and CHMSL, 1099
brake line flaring, 822
brake line replacing
brake lines, brake hoses, and associated hardware inspection, 823
brake lines, hoses, fittings, and supports replacement, 823
fabricating brake lines, using the inverted double flare method, 823–824
fittings, 821
flaring, 822
hoses, 821
performing the ISO flare method, 824
supports, 821
vacuum-type power booster unit for leaks and inspecting the check valve, 822
brake lines
double-wrench method, 789f
flares
International Standards Organization (ISO) flare, 788
inverted double flare, 788
flexible brake shoe, 787f
inverted double-flared line, 788f
materials, 788
steel, 787f
brake lines and hoses, 821
brake lining thickness gauges, 851
brake pedal, 514
free play, 813, 814, 814f
height measurement, 814
inspection, 813
reading brake pedal height, 815f
reserve pedal, 815f
travel measuring, 815
brake pedal emulator, 760
brake pedals
adjustable pedal system, 786f
assembly, 785f
brake light switch, 786f
nonslip cover, 785f
return spring, 785f
brake repair legal standards
forgotten steps, 806f
improper brake repair, 805f
and technician liability, 805
vacuum bleeder tool, 805f
brake shoe adjustment gauge, 891
brake shoes, 872
bonded brake shoe lining, 881f
common installation error, 880f, 881f
drum, 881–882
and linings, 879, 879f
primary, 880
riveted and bonded friction materials, 881
riveted brake shoe lining, 881f
secondary, 880
temperature, coefficient of friction, 880f
worn down to metal, 882f
brake spoon, 891
brake spring pliers, 891
brake switch, 930, 938–940
brake system types
adjustment methods, 772f
drum-style parking brake, 771f
foot-operated parking brake mechanism, 772f
hand-operated parking brake, 772f
hydraulic, 770, 770f
integral parking brake, 771f
parking
apply mechanism, 771–772
brake adjustment, 772
cables, 771
parking brake cable, 771f
parking brake-transmission style, 771f
power booster, 770f
top hat design parking brake, 770f
brake warning lamp
brake fuse with a test light, 825f
CANbus circuit, 826f
diagnosing, 825
light circuit, 825f
light system in a non-CANbus system, 826–827
parking brake and indicator light system operation, 827
brake warning light
brake light switch operation, 797f
brake warning light circuit, 796f
braking system issue, 796f
CANbus brake warning light circuit, 796f
center high mount stop lamp (CHMSL), 797f
stop lights, 796–797
brake warning light system, 826
brake wash station, 851, 891
brake-by-wire system, 760
brakes
drum, 7580
emergency, 759f
history of, 759
old style, 759f
scrub, 759f
braking torque, 684
breaker plate assembly, 1189
breather tube, 1312
British thermal units (Btus), 1151
brush assembly, 1079
brushless alternators, 1080
brushless DC motor, 1124
brushless electric motor, 1124f
bucket lifter, 329
bug sprayers, 1201, 1202
bump steer, 639, 713
burned open fuse, 1030f
bus, 1118
bushings, 695
bypass circuit, 1182
bypass valve system, 1285

C

CAA. *See* Clean Air Act (CAA)
cabin air filter, 1163–1164, 1164f
caliper dust boot seal driver set, 853
caliper piston pliers, 851
caliper piston retracting tool, 851
calipers disassembling, 856
calipers inspection
brake pad and wear indicators, 854–855
brake pads checking, 855
mountings, 854
and removal, 853–854
slides and pins, 854
cam lobes, 1181
cam sensors
concentric interrupter rings, 1189
control variable valve timing, 1190f
and crank, 1189
interrupter rings, 1190f
monitors, 1189
cam-in-block engines, 329
camshaft position (CMP) sensor, 1189, 1243
cam-style adjuster, 885
CAN. *See* Controller Area Network (CAN)
carbon monoxide poisoning, 1303
carbureted fuel systems
air-fuel mixture, 1222
carburetors and mechanical fuel pumps, 1221
diaphragm-type pump, 1221
float bowl and float, 1221f
side-draft carburetors and updraft, 1222
types, 1222f
venturi shape, 1222
carburetor barrels
barrel, 1224f
increase power, 1224
large-capacity V8 engines, 1225
mechanically controlled secondaries, 1225f
progressive, 1225f
vacuum-operated secondaries, 1226f
carburetor body, 1223
carburetor circuits, 1223
carburetor float bowl, 1314
carburetor operation
air and fuel, 1222
air-fuel mixture, 1223
low-pressure area, 1222f
casing plies, 586
catalyst monitoring, 1311
catalytic converter
donut gaskets, 1291, 1291f
exhaust gaskets, 1291
exhaust system, 1290
extreme temperature of the exhaust system, 1291f
hydrocarbons and carbon monoxide, 1290
intermediate pipe, 1291, 1291f
tailpipe, 1291
catalytic converters
catalyst monitoring, 1311–1312
ceramic material coated, 1310, 1310f
nitric oxide and nitrogen dioxide, 1310f
oxygen sensor signals, 1312
PCM monitors, 1311
primary emission control device, 1310
three-way catalyst, 1311, 1311f
two oxygen sensors, 1311f
CCAs. *See* cold cranking amps (CCAs)
CEMF. *See* counter-electromotive force (CEMF)
center electrode, 1179
central locking systems, 1127
centrifugal advance units, 1184–1185
centrifugal force, 410
centrifugal switch, 591
charcoal canister, 1223, 1313
charging system circuit voltage drop, 1086–1087
charging system output testing, 1085–1086
charging systems
alternator installed on engine, 1077f
charging system, 1078
DC generators, 1077f
electronics and circuitry, 1077
external voltage regulator, 1078f
"generator" and "alternator," 1077
newer, 1078
rotating magnetic field in stationary wires, 1077f
voltage regulator circuit, 1078
chassis and brake technicians, 9
chassis ear, 710
checking lighting
common symbols, 1108t–1109t
and peripheral systems, 1107–1110
checking shock absorbers, 715
chlorofluorocarbons (CFCs), 1147
Choke, 1224
CI engines, 313
circuit breakers, 975
circuit protection devices
blade fuse color codes for common fuse sizes, 975t
circuit breaker, 976f
fuses, 975f
fusible links, 976f
maxi-fuses, 976f
typical underhood fuse box, 976f
circuit with a test light, 1029
circuits with fused jumper leads, 1029
Clean Air Act (CAA), 1148
cleaning gun, 193
cleaning tools
brake washers, 197–198
engine compartment, 193f
pressure washers and cleaners, 193–195
sand or bead blasters, 198–199
spray wash cabinets, 195–197, 195f
using brake washers, 198–199
using pressure washers, 194–195
using solvent tanks, 197–198
using spray wash cabinets, 196
clear flood mode, 1257
clock spring, 642, 664, 1130
closed-loop system, 1154, 1256
clutch components
the clamping force and reconnects the engine, 532f
dual-mass flywheels, 533f
flywheel, 532
operation of, 531
springs or diaphragm and frees, 532f
standard light vehicle clutch, 532f
stepped flywheels, 533f
types of flywheels, 533–534
clutch disc
components, 535f
multiplate clutch assembly, 535–536, 535f
waved springs, 535f
clutch maintenance
bleeding a hydraulic clutch system using the pressure, 543
bleeding/flushing hydraulic clutch system using the gravity method, 542
brake and clutch dust wash station, 540f
brake fluid for excess moisture, 540f
checking, adjusting, and bleeding a hydraulic clutch, 541
checking and adjusting a mechanical clutch, 540
clutch safety and hazards, 539
flushing a hydraulic clutch system using the manual method, 543
hydraulic clutch system bleeding, 542
preventive maintenance, 539–540
clutch operating mechanisms
bearings are located in transmission, 537f
cable, 537–538
cable-operated clutch with manual adjustment, 538f
cable-operated clutch with quadrant ratchet, 538f
clutch master cylinder, cutaway view of, 538f
hydraulic clutch control, 538f
hydraulic clutch mechanisms, 538–539
linkage-operated systems, 539
clutch pedal, 514
clutch principles
dry clutch, 530f
engages and disengages, 530f
input and output shaft speed, 531
transmission output shaft, 531
types, 530f
clutch system, 514, 517–518
clutch components, 532–534
clutch disc, 535–536
clutch maintenance, 539–543
clutch operating mechanisms, 537–539
operation components, 531–532
pilot bearing, 536–537
pressure plates, 534–535
principles, 530–532
throw-out bearing and clutch fork, 536
coefficient of friction, 473, 765
coil saturation, 1182

coil spring pressure plate, 534
coil-on-plug ignition system
 coil-near-plug, 1191f
 coils partially removed, 1191f
 COP and CNP systems, 1190, 1192
 four-cylinder engine, 1191f
 physical configurations, 1192
coil-over shock, 690
cold chisel, 142
cold cranking amps (CCAs), 1041
combination valve, 795
combination wrench, 126
combustion chamber design
 excessive valve overlap, 1307f
 gas flow rate, 1307, 1307f
 quench areas in, 1307, 1307f
common torque sequence for lug nuts, 867f
communication and employability skills
 accident report, 303, 303f
 active listening, 282
 appearance and environment, 290–291
 art of speaking, 285–286
 asking questions, 286
 business letter, 300f
 closed questions, 286
 communication in team, 287–288
 completing a repair order, 300–301
 comprehending reading method, 294f
 customer concern sheet, 293f
 customer satisfaction index report, 292f
 customer service and communication, 292f
 defective equipment report, 301–302, 301f
 description and operation section, 296f
 educational training sources, 298
 effective reading, 293–294
 effective writing, 299
 empathy, 283
 employability skills, 288
 employment requirements, 289–290
 giving and receiving instructions, 287
 good driving report, 289f
 identifying customers' needs, 292–293
 listening process, 282–283
 lockout/tagout, 302
 nonverbal contact, 284f
 nonverbal feedback, 283–284
 online repair order, 299f
 open questions, 286
 overview of, 282
 paper and online training materials, 298f
 personal space, 284f
 primary and secondary information, 297f
 problem solving, 297
 reading comprehension, 294–295
 reading method, 295f
 reading skills, 293f
 researching and using information sources, 295–297
 researching information, 295f
 service information, 296f
 shop safety inspection form, 301
 starting work at appointed time, 291f
 tagging out and securing defective tools, 301f
 technical assistance services, 298–299
 telephone skills, 287
 testing and diagnosis section, 296f
 time management, 291–292
 tools and equipment, 297f
 vehicle inspection form, 303–304, 304f
 vehicle service information, 297–298
 verbal feedback, 284
 writing business correspondence, 299–300
 yes-or-no questions, 286
commutator, 979
companion flange, 551
compensating port, 781
compliance bushing, 696
composite rotor, 842
composition of air
 oxygen and exhale carbon dioxide, 1302, 1302f
 oxygen level in atmosphere, 1302
 percentages of gases, 1303t
compression ratio, 321
compression stroke, 1171
compressor, 1153
computer-controlled carburetors, 1226
computer-controlled EGR system, 1314f
concentric slave cylinders, 536
condenser, 1153
condenser construction, 1182f
condenser for airflow restrictions, 1162
conductance testing for capacity, 1052
conduction, 403, 1150
conductors, 952, 953
cone-style synchronizer, 512f
connectors, 984, 1031
conservation of energy, 765
constant mesh, 636
constant-velocity joints
 lubrication and inspection, 569
 Rzeppa, 568f
 tulip tripod, 568f
constant-velocity (CV) joints, 515
contact breaker point ignition systems, 1172
 aluminum foil, 1181
 contact points, 1181
 mechanically operated, 1180
 point distributor, 1181f
 point operation, 1181f
 primary circuit voltage on an oscilloscope, 1181f
 primary ignition circuit, 1181
 secondary voltage, 1181
contact breaker points, 1172
contaminants fuel, 1230
Continuously Variable Transmission (CVT)
 changing sizes of the input and output pulleys, 502f
 electronic continuously variable transmission (ECVT), 501f
 low and reverse, 503
 low gear ratio, 504f
 steel CVT belt, 503f
 toroidal CVT, 504, 504f
 types of, 501–502
 variable-diameter pulley CVT, 502–503
 variable-diameter pulley or Reeves drive CVT, 501f
 VDP CVT, 502f
continuously variable transmissions (CVTs), 462
control arms
 ball joints, 693–695, 694f
 bushings, 695–696
 control arm bushing, 695f
 loaded *versus* follower ball joints, 695f
 primarily compression applications, 695f
 and rods, 692, 693
 rubber-bonded bushing, 696
 single-point control arm, 693f
 steering knuckle, 693, 693f
 straight stud ball joint, 694f
 strut rod bushings, 693f, 695f
 tapered stud ball joint, 694f
 wishbone control arm, 692f
control modules
 CKP sensor, 1254
 coil-on-plug ignition module, 1254, 1254f
 PCM timing signal, 1254
 PCM-controlled ignition control, 1254
 timing signal, 1253–1254
 waste spark ignition module circuit, 1254, 1254f
controlled combustion, 1304
controlled devices, 1252
Controller Area Network (CAN), 1119
controlling air-fuel ratios, 1308
convection, 1150
conventional oil, 365
coolant flow
 normal and reverse flow, 410
 in normal flow system, 411f
 in reverse flow system, 412f
 reverse-flow designs, 412
 surge tank, 412f
coolant label, 110
coolant system pressure tester, 453
coolant types
 corrosion inhibitors, 409
 HOAT coolant, 410f
 IAT coolant, 410f
 liquid-cooled engines, 409
 OAT coolant such as Dex-Cool, 410f
 POAT coolant, 410f
cooling fan
 bimetallic strip, 421f
 clutch fan, 419
 electric cooling, 420f
 engine-driven fan, 419
 flex fan, 419f
 hydraulically operated cooling fan, 421f
 one electric fan, 421f
 PCM-controlled, 421f
 viscous fan clutch, 420f

cooling fans and circuits, 1126
cooling system components
auxiliary coolers, 414
cooling fan, 413
drive belts, 413
heater core, 414
heater hoses, 413
radiator hoses, 413
recovery system, 413
surge tank, 413
temperature indicators, 413
thermostat, 413
water jackets, 414
water pump, 413
cooling system diagnosing
detect coolant leakage, 452f
engine operating temperature, 454
heater core leak, 452f
for leakage, 454
loose belt, 452f
pressure testing the cooling system, 455
temperature differences, 452f
tools, 453
verifying engine operating temperature, 455–456
visual inspection, 453
cooling system theory
air doors and actuators, 427
belt tension, 424
boiling point and pressure, 408
centrifugal force, 410
coolant types, 409–410
cooling fan, 419–421
core plugs, 426
drive belts, 423–424
electrolysis, 409
gasoline engine, 404f
heat transfer, 403
heat transfer in combustion engine, 404
heater control valve, 426
heater core, 426
hoses, 423
normal and reverse flow, 410–412
overview of, 403, 413–414
principles, 404–405
radiator hoses, 421–422
radiator pressure cap, 414–415, 416
radiator shrouding, 415
recovery system, 416
rotary engine, 412
simple cooling system, 404f
surge tank, 416
temperature indicators, 427–428
tensioners, 424–425
thermostat and housing, 416–418
thermostat controls, flow of coolant, 405f
vehicle coolant, 405–408
water jackets, 425
water pump, 418–419
COP boots, 1194
core plugs, 426, 426f
cored solder, 192
cornering force, 578, 684
cornering lights, 1099
cotter pin, 915
counter-electromotive force (CEMF), 1064
courtesy lights, 1107
crank angle position, 1188
crankcase emission control, 1312
cranking amps (CAs), 1041
cranking and running compression testing
cranking compression testing, 354
good compression, 353f
performing, 353–354
running compression testing performance, 354–355
crankshaft position sensor, 1176f, 1189
CKP, 1242
crankshaft speed and position, 1243
Hall-effect CKP sensor, 1243
polarity changes, 1243
crankshaft position (CKP) sensor, 1242
crescent pump, 370
cross-caster, 735
cross-flow heads, 1283
cross-flow radiator, 414
crude oil, 363
current and magnetic fields, 1010–1011
current clamp, 1055
current flow, 952, 954
current thread, 205f
customer pickup, 232
CV boot, 569
CV joint issues diagnosing
boot, 569f
inspecting half-shaft components, 570–571
CV joints, 521
CVTs. *See* continuously variable transmissions (CVTs)
cylinder block
connecting rod and piston, 325–327, 325f
and crankshaft, 324
crankshaft counterweights, 325f
crankshaft parts, 324f
crankshaft rod journals, 324f
flywheel, 324
forward and backward movement, 324f
piston coatings, 327f
piston crown shapes, 326f
powdered metal fracture-split rod, 326f
rod connected to a piston, 325–327
single part of engine, 324f
cylinder bore, 878
cylinder head
combustion chambers, 328f
poppet valves in, 328f
valves, 327–328, 328f
cylinder leakage testing
compression tester hose, 357f
cylinder leakage testing, 358–359
on good engine, 356f
leakage pass, 357f
oil fill hole, 356f
overview of, 355
performing, 358
cylinder power balance test
cylinder power balance testing, 352
disable cylinders, 352f
engine vacuum using a pressure transducer, 350
fuel injector harness, 351f
good engine, 350f
misfires, 351f
overview of, 349–350
performing, 351–352
scan tool can retrieve DTCs, 351f
test light and vacuum hose, 352f
cylindrical roller bearing assemblies, 907

D

damping, 683
data link-connector (DLC), 345, 1120, 1264f, 1265f
data stream, 1265
daytime running lights, 1106–1107
dead blow hammer, 142
deceleration, 764
deep dish wheels, 581
DEF. *See* diesel exhaust fluid (DEF)
deflecting force, 682
defroster, 1160
detonation, 1205
diagnosing engine lubrication system
common issues, 398–399
oil leakage, 398–399
oil pressure sending unit, 398
diagnosing steering systems
power steering fluid leakage, 663
power steering gear issues, 663
rack-and-pinion, 663–664
steering column issues, 664
diagnosing stop lights
non-CANbus stop light circuit, 827f
stop light system, 828
diagnosing suspension systems
body sway, 714
common issues
bouncy ride/excessive body movement, 713
bump steer, 713
hard steering, 713
noises, 713
pull, 712–713
shimmy/shake, 713
steering return concerns, 713
torque steer, 713
vehicle wander, 712
excessive body sway, 714f
suspension noises and vibrations noises, 713f
tools, 711f
universal strut nut kit, 712f
diagnostic trouble codes
decoding a diagnostic trouble code (DTC), 1266f
DTC, 1265
engine's coolant temperature, 1265

malfunction indicator lamp (MIL), 1265f
OBD II codes, 1266
scan tool displaying pending codes, 1266f
storage, 1265
warm-up cycle, 1266f
dial bore gauge, 159
dial indicators, 660, 851
diaphragm pressure plate, 534
dichlorodifluoromethane, 1155
diesel exhaust fluid (DEF), 262
dieseling, 1205
differential and final drive
adjusting final drive, 523–524
axle rotating different speed, 520f
changing manual transmission and final drive fluid, 524
checking and adjusting the differential/transfer case fluid level, 525
dead axles hold wheels, 521f
drive axle, 521
final drive assembly, 520f
fluid level checking, 522f
fluid level of a manual transmission, 523
fluid loss cause, 523
fluid loss in a transmission, cause of, 524
gear lube used in manual transmissions, 522f
independent rear axle assembly, 520f
independent suspension drive axle, 521f
lead-based grease, 522f
live axles drive wheels, 521f
lubrication, 522
manual transmission, proper fluid level for a, 523f
solid axle housing, 521f
solid rear axle assembly, 520f
transmission fluid changing, 525
transmission fluid checking, 522–523
differential assembly, 519, 555
differential gear set, 32, 515
digital meters, 998
digital multimeter (DMM), 807
digital multimeter (DMM)-galvanic reaction, 807–808
digital multimeter layout
and accessories, 1000
body, 1000f
digital multimeter purpose, 999
digital multimeters
CAT rating, 999t
diagnosing problems, 998, 998f
meter CAT ratings, 999t
meter fuses installed in a meter, 999t
test equipment, 998, 998f
direct current (DC), 954
direct ignition system, 1172
direct-acting telescopic shock absorbers, 688
directional tread patterns, 586
disc brake calipers
corroded caliper piston bore, 837f
fixed and floating, 834f
fixed caliper application on solid rotor, 835f
fixed calipers with multiple pistons, 835f
floating caliper and guide pins, 835f
low-drag caliper, 837f
O-rings, 836–838, 836f
phenolic piston, 837f
sliding or floating caliper application, 835f, 836f
square-cut O-ring brake application, 836f
disc brake diagnosing
concern, 851
overview of, 850–851
disc brake pads, 833
antinoise measuring, 840–841
bonded and riveted brake pads, 838, 838f
brake friction materials, 838–840, 838f
brake lining coefficient of friction, 838t
brake lining edge code (FF), 840f
brake lining grooves and contouring, 841f
brake pad locating lugs, 838, 838f
brake pad shims and guides, 840f
brake pad wear indicator system, 842f
ceramic materials, 839f
and friction materials, 838
low-metallic NAO materials, 839f
non-asbestos organic, 839f
scratcher brake wear indicator, 841f
semi-metallic materials, 839f
spring-loaded brake pad retainers, 840f
types, 841–842
wear indicators, 841
disc brake repair
causes of, 851t
dust boot seal/bushing driver set, 853f
and maintenance, 851
on-car brake lathe, 853f
tools, 852f
brake lining thickness, 851
brake wash station, 851
caliper piston pliers, 851
caliper piston retracting tool, 851
C-clamp, 851
dial indicator, 851
disc brake rotor micrometer, 851
disc brake rotor micrometer, 851
disc brake rotors
composite rotor, 842
directional ventilated rotor, 844f
disk thickness variation, 843f
lateral runout, 842–843, 843f
rotor thickness and heat capacity, 844f
solid and ventilated rotors, 843–844
typical dust shield, 843f
disc brake system
advantages, 834
disadvantages, 834
hub-style and hubless rotors, 833f
operation, 832f, 833
outboard caliper, 834f
pedal force into hydraulic pressure, 833f
disc brake system theory
calipers, 834–838
and friction materials, 838–842
operation, 833–834
overview of, 832
parking and disc, 845–846
primary components, 832
rotors, 842–845
dismounting, inspecting, and remounting a tire with a TPMS sensor, 617–619
distributor base plate, 1185
distributor cap, 1183
distributorless ignition systems, 1189
distributorless-type systems
coil-on-plug, 1189f
components of, 1189
types of, 1189
waste spark, 1189f
distributors
coil mounted on top, 1183
the high-voltage pulses, 1183
insulated rotor arm, 1183
main function of, 1183
spark plug wires, 1183f
distributor-type electronic ignition systems
under coils, 1186f
in distributor, 1186f
induction-type pickup assemblies, 1186–1187
modules, 1186
side of distributor, 1186f
stationary winding, 1186
strength of magnetic field, 1186
divided hydraulic systems
types, 786–787
vertical, diagonal, L-split, 787f
DLC. *See* data link connector (DLC); data link-connector (DLC)
DMM ranges, 1000t
DMM values, 1000t
door panel removing, 1128
DOT tire date manufacturing code, 590f
double Cardan joint, 564
double-acting wheel cylinder, 874, 875
double-row ball bearing, 909
down-flow radiator, 414
drain lines, 985
drain plug, 911
drift punch, 142
drill press, 185
drivability technicians, 11
drive axle assembly, 521
drive axle flange runout measuring, and shaft end play, 562–563
drive belts
coolant lines, 423f
multiple flexible coolant hoses, 423f
stretch-fit, 423f, 424
thermostat bypass hose, 423f
toothed, 423f, 424
V-type, 423, 423f
drive cycles, 1267
drive wheels, 30
driveline angularity, 552

driveline subassemblies
angularity, 552f
checking driveshaft joints, phasing, angles, balance, and runout, 553
and components, 550
driveline angle gauge, 552f
driveshaft joints and phasing angles, balance, and runout, 552
driveshaft yoke alignment, 552f
driveshafts-rear-wheel drive, 550–552
final drive pinion shaft, 551f
inspecting and servicing center support bearings, 552, 553–555
two-piece driveshaft, 551f
U-joint yokes, 551f
driven center plate, 535
driver's side airbag, 641
driveshafts, axles and final drives
all-wheel drive (AWD) layout, 550
axles and half-shafts, 560–561
constant-velocity joints, 568–569
CV joint issues diagnosing, 569–571
drive axle flange runout measuring, 562
driveline and axle inspection and repair, 559
driveline subassemblies and components, 550–555
final drives/differentials, 555–559
four-wheel drive layout, 549–550
front-wheel drive layout, 549
FWD/AWD axles/half-shafts, 561–562
inspecting and replacing lug nuts, 563
inspecting and replacing wheel studs, 563
joints and couplings, 563–568
rear-wheel drive layout, 548–549
shaft end play measuring, 562
driveshafts-rear-wheel drive, 550
drivetrain layouts
bold applied, torque, 33f
differential, axles, and wheels encompasses, 28f
engage and disengage the engine, 32f
engine, transmission encompasses, 28f
final drive assembly, 32f
four-wheel drive vehicle, 31f
front-engine, rear-transaxle arrangement, 30f
light vehicles, 30f
live and dead axles, 29–30
live axles location, 30f
longitudinal engine orientation, 29f
modified transaxle, 31f
rear-wheel drive arrangement, 31f
three engine-mounting positions, 28f
torque converter, 32–33, 32f
torque measurement, 33
transaxle, 29f
transmission and axle configurations, 29f
transmissions and final drives, 32
drivetrain system, 511
driving lights, 1098
driving thrust, 683
drum brake components
anchor pins, 877f
backing plate, 876–877, 876f
brake drums, 875–876
with cooling fin, 875f
hubless drum, 876f
hub-style drum, 876f
labyrinth seal, 877f
maximum diameter stamped, 876f
types, 876
without cooling fins, 875f
worn brake shoe contact pads, 877f
drum brake micrometer, 891
drum brake noises, 881
drum brake springs
brake springs and hardware, 882f
hold-down springs, 883, 883f
new return spring with tight coils, 882f
return springs, 882–883
shoe to rest position, 882f
specialty springs, 883, 883f
worn return spring with stretched coils, 882f
drum brake system overview
automatic brake self-adjuster, 872
backing plate, 872
brake drum, 872
brake shoes, 872
components of, 872f
drum removal, 872f
operation
self-energizing, 873–874
servo action, 873–874
rotating drum, 873f
self-energizing, 873f
springs and clips, 872
wheel cylinder, 872
drum brake system theory
brake shoes and linings, 879–882
components, 875–877
cylinders, 877–879
fundamentals, 872
overview of, 872–874
parking, 886
self-adjusters, 883–886
springs, 882–883
types, 874–875
drum brake system types
duo-servo, 875, 875f
forward direction, 874f
leading/trailing, 874–875, 874f
twin leading, 874
drum-style parking brake, 770
dry sump systems
drain plug, 369f
oil pan, 369f
wet sump, 368, 368f
windage tray, 369f
dual overhead cam (DOHC), 329
dual-clutch transmissions, 462
dual-diaphragm boosters, 799
dual-shaft transmissions, 464
ducting
cancel air noise as air flows, 1280f
MAF sensor, 1280
minimizes the noise, 1280f
duo-servo drum brake systems, 875
duty cycle, 1239, 1253
dwell angle, 1182
dynamic imbalance wheel, 610

E

ear protection, 93
early fuel evaporation (EFE), 1282
EAS. *See* electrically assisted steering (EAS)
EBCM. *See* electronic brake control module (EBCM)
economic concerns, 1324–1325
ECVT. *See* electronic continuously variable transmission (ECVT)
effects of electricity
chemical, 960
electromagnetic, 960–961
electromagnetic force, 960f
heating, 959–960
LED, light created by, 960f
light, 960
EFI. *See* electronic fuel injection (EFI)
EFI systems types
basic systems, 1216f
characteristics, 1215–1216
MPFI system, 1217f
multipoint fuel injection systems, 1216–1217
sequential fuel injection, 1217
simultaneous fuel injection, 1217
six-cylinder engine, 1217f
throttle body injection systems, 1216
EGR valve, 1248
elasticity, 682
electric accessory motors, 1124
electric cooling fan circuit, 1126
electric lock
door panel removing, 1129
and keyless entry systems, 1128–1129
wireless key fob with buttons, 1129f
electric motor actuators, 1255
electric motor diagram, 979f
electric power steering
column-assist EPS, 652f
direct-drive EPS, 652f
electronic power steering rack-and-pinion assembly, 652f
EPS light illuminated on the dash, 653f
front wheels, 654f
higher voltage electrically assisted, 653–654
high-voltage system, 654f
on hybrid vehicle, 653f
hydraulic steering pump and motor, 651f
operation, 653
primary types of, 652
rack-assist steering type, 652f
electric power steering (EPS) system, 652
electric power tools
abrasive wheel with a dressing tool, 186f
acid core solder, 190f
angle grinder, 188f
bench and angle grinders
using angle grinders, 188–189
using bench grinders, 186–187
butane soldering iron, 192f

desoldering tool, 192f
drills and drill bits, 184–185
hand-held cutoff wheel, 186f
heat shrink tubing, 192f
lead-free solder, 190f
rosin core solder, 190f
rosin flux, 190f
soldering tools, 189–193
soldering using a heat dam, 191f
tin-lead solder, 190f
electric power-assist steering
high-voltage circuits, 673f
hybrid vehicle power steering system electrical circuits, 673
inspection, 673
testing, 673
electric vehicle service precautions
Cat III DMM, 1332f
gloves and approved covers, 1332f
high voltage, 1332f
hybrid and, 1331–1332
servicing HEV systems, 1332
tools with high-voltage rating, 1332f
vehicle safety, 1332
electrical capacity, 1041
electrical circuit testing
battery voltage with the headlights on, 1021f
excessive voltage drop, dirty and loose ground connection, 1021f
ground side measurement indicates a fault, 1021f
Ohm's Law to Diagnosing Circuits, 1018–1022
resistance decreases, amperage must increase, 1019f
resistance increases, amperage must decrease, 1019f
resistance stays the same, amperage must decrease, 1019f
resistance stays the same, amperage must increase, 1019f
voltage at the dim headlight, 1021f
voltage is high or the resistance is low, 1020f
voltage low, resistance high, 1020f
electrical components
connectors are commonly marked, 974f
switches, 974–975
switches are wired into the circuit, 975f
typical automotive switches, 975f
electrical components and repair
circuit protection devices, 975–977
diagram fundamentals, 986–988
drain lines, 985
fixed resistors, 981
flasher can/control, 977
ignition coils and transformers, 980
motors, 979–980
Mylar tape, 985
potentiometers, 981–982
relays, 977–978
resistors, 980–981
rheostats, 981
shielding, 985
sizes, 982–984
soldering wires and terminals, 991–993
solderless terminal installing, 990–991
solenoids, 978–979
stripping wire insulation, 990
terminals and connectors, 984–985
thermistors, 982
twisted pair, 985
using wiring diagrams, 988–989
variable resistors, 981
wire maintenance and repair, 982, 990
wiring harnesses, 984
electrical concepts to solve problems, 968
electrical fundamentals
atom parts, 952f
basic electricity, 952–953
current flow (amps) in electrical circuit, 953f, 956f
current flow, resistance to, 954
electrical circuits, 954–955
excess of electrons, 954f
free electrons, movement of, 953–954
ions, 952f
piles of electrons, 955f
power and ground, 956
pressure in a water system, 955f
resistance (ohms) in electrical circuit, 956f
unlike charges attract, 952f
volts, amps, and ohms, 955–956
electrical power, 963
electrical resistance, 954
electrical schematics, 984
electrical systems principles
basic electrical, 968–969
electricity effects, 959–961
electricity sources, 957–959
fundamentals, 951–957
Ohm's Law, 961–968
electrical technicians, 10
electrically assisted steering (EAS), 652
electrically powered hydraulic steering (EPHS), 651
electricity sources
electrical charge, 958f
electrical current flow, 958f
electrochemical energy, 957–958
electromagnetic induction, 958–959, 959f
electrostatic energy, 957
photovoltaic effect, 958f
photovoltaic energy, 958
piezoelectric energy, 958, 959f
rubbing two insulators, 957f
thermoelectric energy, 957
electric-only propulsion, 496
electrochemical energy, 957
electrohydraulic brake boosters, 800
electrohydraulic power brake booster, 800f
electrolysis, 409, 436
electromagnetic induction, 958, 1078f, 1078
electromotive force, 954
electronic brake control
ABS electronic brake control module (EBCM), 940
ABS master cylinder, 932–933
antilock braking system operation, 930–932
brake switch, 938–940
electronic stability control (ESC), 943–944
evolution of, 928
HCU-control valve operation, 933–936
operations of, 928–930
traction control system (TCS), 940–942
wheel speed sensors, 936–938
electronic brake control module (EBCM), 1269
electronic brake control systems
driver avoid accidents, 929f
operation of, 928–929
oversteer condition, 929f
understeer condition, 929f
wheel is slipping, the TCS system, 929f
electronic brake control (EBC) systems, 928
electronic continuously variable transmission (ECVT), 502
electronic fuel injection (EFI), 1201
electronic ignition system-distributor type, 1172, 1185
electronic mufflers, 1293
electronic stability control (ESC)
directional control, 943
of ESC assist desired, 944f
ESC system, 943
operation, 944
overview of, 943
typical ESC schematic, 943f
vehicle control, 943f
electronic stability control (ESC) system, 929
electronically commutated motor, 1124
electronically controlled throttle, 1255
electrostatic energy, 957
eliminating air-conditioning system odors, 1164
emission analyzers, 1270
emission control, 1304
carbon monoxide, 1303
catalytic converters, 1310–1312
combustion chamber design, 1307
composition of air, 1302–1303
controlling air-fuel ratios, 1308–1309
crankcase emission control, 1312
design reduces, 1304
evaporative emission control, 1313–1314
evolution of, 1309–1310
exhaust gas recirculation system, 1313–1314
heated air intake systems, 1314–1315
hydrocarbons, 1303
importance, 1304
overview of, 1302
oxides of nitrogen, 1302–1303
PCV system inspection, 1315–16
PCV system servicing, 1315–1316
PCV system types, 1312–1313
PCV-related concerns, 1315
postcombustion treatment, 1309
purpose and operation, 1312–1313
stoichiometric ratio, 1307–1308
sulfur dioxide, 1303–1304
toxicity, 1303–1304

emission control system evolution, 1309–1310
emission standards, 1325
enameled copper wire, 1177
Energy Policy Act (EPAct92), 1324
energy security, 1324
energy transformation, 963
engine cam
  cam-in-block arrangement, 329f
  camshaft lobe control, 328, 330f
  camshaft specifications, 329–330
  lobes on camshaft, 329f
  overhead cam, 329f
  valve lift and duration, 330f
engine classification
  four- and six-cylinder configurations, 35f
  horizontally opposed, 34f, 35
  individual crankshaft journals, 36f
  in-line engines of equivalent capacity, 34f, 35, 35f
  piston, 33–35
  rotary engines, 36–37, 36f
  V, 35–36
  VR, 34f
  VR and W, 36
  VR cylinder arrangement, 36f
engine configuration, 33
engine displacement, 320
engine drive belts
  cabin air filter, 266–267
  heater hose, 265f
  hoses, 265
  inspecting and changing an air filter, 266–267
  inspecting the air filter, 266
  oil-soaked belt, 264f
  radiator hose, 265f
  serpentine belt, 264f
  serpentine belt wear tool, 265f
  V-type belt, 264f
engine load factor, 318
engine management sensors, 1240
engine management system
  accelerator pedal position, 1241
  air supply, 1246–1248
  barometric pressure sensor, 1248
  camshaft position sensor, 1243
  control modules, 1253–1255
  controlled devices, 1252
  crankshaft position sensor, 1242–1243
  digital and analog signals, 1238–1240
  electric motor actuators, 1255
  electronically controlled throttle, 1255
  feedback and looping, 1255–1256
  fuel pressure sensor, 1249
  fuel shutoff mode and clear flood mode, 1257–1258
  ignition pickup-style position sensor, 1243–1244
  knock sensor, 1249–1250
  manifold absolute pressure sensor, 1248
  oxygen sensor, 1244–1246
  potentiometers, 1240
  powertrain control module, 1250–1252
  relays, 1252–1253
  sensors, 1240
  short- and long-term fuel trim, 1257
  solenoids, 1253
  switches, 1250
  thermistors, 1241–1242
  throttle position sensor, 1240–1241
  vehicle speed sensor (VSS), 1244
engine mechanical testing
  compression tester, 344f
  cranking and running compression testing, 353–355
  cylinder leakage testing, 345f, 355–358
  cylinder power balance testing, 349–352
  intake manifold vacuum testing, 348–349
  overview of, 343–347
  pressure transducer and lab scope, 345f
  scan tool, 345f
  strategy-based diagnosing, 344f
  testing tools, 344–345
  vacuum gauge, 344f
  vacuum testing, 347–348
engine noise and vibrations
  cranking sound, 346
  diagnosing cranking sound, 345–347
engine oil checking
  cartridge filters, 395f
  cracked head, 390f
  disposable gloves, 392f
  draining engine oil, 394
  "full" and "add" marks, 390f
  instrument panel, 390f
  lubrication system maintenance, 389f
  oil analysis, 392
  oil and filter change, 392–394, 395
  oil drain plug, 393f
  oil filter oil ring and groove, 395f
  refilling engine oil, 397
  replacing spin-on filter, 395–396
  single-use drain plug gaskets, 393f
  spin-on and cartridge filters, 395
engine operation
  compressed gas increasing temperature, 314f
  cylinder leak, 313f
  pressure and temperature, 313–314
  pressure and volume, 314–315
  pressure temperature, 313f
  principles of, 313
  temperature and energy, 314
  temperature changes pressure, 314f
  volume affects pressure, 314f
engine spark timing, 1184
engine starting system
  current flow higher, 1062
  electric starter bolted to transmission, 1061f
  hand-cranked, 1061f
  high-amperage circuit, 1061
  two electrical circuits, 1061
engine vacuum testing
  cranking sound diagnosis, 347
  good idle vacuum, 347f
  reading, 347t
engineering controls, 81
environmental concerns, 1324
Environmental Protection Agency (EPA), 44, 1148
EPA. *See* Environmental Protection Agency (EPA)
EPHS. *See* electrically powered hydraulic steering (EPHS)
ethylene glycol, 407, 433
evaporative emission control
  charcoal canister, 1316f
  charcoal storage, 1316
  EVAP system, 1315
  fuel storage, 1315
  pressure relief valve, 1315
  purging the canister, 1316f
evaporative emission (EVAP) system, 1309
evaporator, 1154
evaporator housing water drain, 1162–1163
exhaust gas recirculation system, 1309, 1313–1314
exhaust gaskets, 1291
exhaust manifold, 334
exhaust system
  air-to-air intercooler, 1288f
  burned gas, 1288–1289
  components, 1288
  exhaust system removes exhaust gas, 1289f
exhaust system components
  engine pipe, 1290
  exhaust brackets, 1290
  flexible connector, 1290f
  manifold, 1289
  pipe to flex the engine, 1290f
  rubber mounts, 1290, 1290f
  scavenging effect, 1289
exhaust system inspection, 1296–1297
expansion volume, 1315
extended-range electric vehicle (EREV), 1326, 1329
extension housing, 475
exterior lights
  inspecting, 275
  shock absorbers, 276
  windshield wiper blades, 277–278
  wiper blades checking, 275–276
  wiper blades replacing, 275–276
exterior vehicle inspections
  performing, 274
  shock absorbers, 275
  vehicle's exterior, 275
external driveshafts, 703
extreme loading, 364
eye ring terminals, 984

F

factors affecting brakes
  braking performance, 763f
  different load, 763f
  load on wheel during stopping, 763
  road conditions, 763

road surface, 763
tires on vehicle, 763
vehicle height, 763
vehicle weight, 763
false air, 1248
Faraday's Law
conductor and magnetic field, 1173
induced by current, voltage, 1173
magnetic field collapsed, 1173–1174
secondary voltage, maximum value of, 1174
secondary windings of coil, 1173f
voltage in conductor, 1173
fastener standardization
grade number, 205f
standard bolts, 205–206
fault codes, 940
fault wire terminals, 990f
fault-tolerant CAN, 1120
feedback
air-fuel ratio, 1256, 1256f
closed loop, 1256f
and looping, 1255
MAF system, 1256
nominal battery voltage, 1255, 1255f
output voltage changes, 1257f
oxygen sensor voltage signal, 1256
feedback system, 1311
field coils, 1124
field diodes, 1084
filter sock, 1211
final drive assembly, 32, 514
final drives/differentials
assembly, 556f
clutch-style limited-slip differential, 558f
different speeds during cornering, 557f
differential gears, 555f
differential pinion gear assembly, 557f
FWD differentials, 558–559
FWD, transaxle case, 559f
hypoid gear arrangement, 556f
limited-slip, 558
rear-wheel final drive, 556
Torsen helical gear limited-slip, 558f
Torsen style, 558
fixed calipers, 834
fixed orifice tube system, 1154, 1312
fixed resistors, 981
fixed-type joint, 568
flare nut wrenches, 127, 806f
flare tubing wrench, 127
flaring tools
components of, 150f
riveting tools, 152–153
tubin cutter, 151f
using, 151–152
using riveting tools, 153
flash reprogramming on modules, 1123
flasher can/control, 977
flat seat lug nuts, 582
flat-blade screwdriver, 136
flat-head engines, 329
flat-nose pliers, 134
float bowl, float, and needle and seat, 1223f
float chamber, 1223
float circuit, 1223
floating calipers, 834
flooded cell battery, 1037
fluid and filter replacement, 484
fluid level and inspecting fluid loss, 483
flushing brake systems, 813
flywheel, 26
fog lights, 1099
forced induction, 1285
forces capability map data, 653
forcing screw, 149
Ford EEC-IV ignition pickup assembly, 1244f
four-stroke spark-ignition engines
compression ratio of an engine, 321f
conventional engine, 321f
engine displacement, 320f
engine measurement-size, 320–321
four-stroke operation, 319–320, 319f
Miller cycle and Atkinson cycle engines, 321f
piston movement, 318f
VR and W types, 319f
four-wheel drive layout, 29, 549–550
four-wheel steering systems, 654
rear actuator for, 655, 655f
turning radius, 655, 655f
yaw forces, 655f
free electrons, 953
freeze-frame data, 1267, 1270
frequency, 1239
friction brakes
brake fade, 766–767
coefficient of friction, 765f
heat fade, 767f
heat transfer, 766
heat transfer in disc brake, 766f
heat transfer in drum brake, 766f
hydraulic fade, 768f
water fade, 767f
friction facings, 531
front suspension
coil spring, 700f
double-wishbone modified, 699, 699f
front-wheel drive vehicles, 701f
modified, 699, 699f
SLA, 699–700
steering knuckles, 700f
strut, 698–699, 698f
twin I-beam suspension, 700, 700f
front-wheel drive layout
longitudinally mounted engine, powerflow in, 549f
transverse engine, powerflow in, 549f
fuel
absolute pressure *versus* gauge pressure, 1206f
air, 1205–1206
atmospheric pressure, 1207f
earth's atmosphere, 1206
gasoline engines, 1207
at idle, 1207f
partial throttle, 1207f
pressure and vacuum, 1206–1207
stoichiometric ratio, 1206
fuel burn controlling
air-fuel mixture, 1204
air-fuel ratio, 1204
anti-knock Index, 1204f
combustion, 1205
detonation/preignition, 1205
incomplete burning, 1204
rich and lean air-fuel ratios, 1204f
fuel cell (electric) vehicles, 1329
fuel delivery system
components, 1207
delivering gasoline, 1207
evaporative emissions system, 1208, 1208f
fuel filler neck, 1208, 1208f
fuel tank, 1208, 1208f
gas cap, 1209
new capless system, 1209, 1209f
tightened until they click, 1209f
fuel filler neck, 1208, 1210
fuel filter, maintenance and repair
engine idling, 1229
flexible hoses, 1227
inspecting and testing fuel pumps, 1229–1230
metal or plastic lines, 1226
replacing, 1226–1228
test drive, 1229
fuel gauge sending unit, 1208
fuel injectors, 979f
design, 1215
long service life, 1215
operations, 1214
O-rings and plastic cap, 1216f
PCM grounds injector, 1216f
port fuel injector, 1215f
sprays, 1215f
fuel injectors testing, 1231–1232
fuel lines, 1211
fuel metering system, 1203
fuel pressure regulation
circulation types, 1213
constant pressure drop across fuel injector, 1214f
delivered, 1213
diaphragm-operated valve, 1213f
electronically controlled fuel pressure, 1215f
manifold vacuum, 1214
mechanical returnless, 1214f
mounted, 1213
TBI systems, 1214
fuel pressure regulator, 1213
fuel pressure sensor, 1249
fuel pump
assemblies, 1210f
electrically operated and electronically, 1209
permanent-magnet electric motor, 1209f
pump chamber types, 1210

fuel pump relay, 1210, 1210f
fuel rail
 line connections, 1212, 1212f
 pressure test port, 1213f
 top of the fuel injectors, 1212, 1212f
fuel shutoff mode, 1257, 1308
fuel supply system, 1202
fuel system types
 EFI function, 1201–1202
 fuel delivery, 1201
 gasoline direct injection, 1202f
 mechanically operated, 1201
 port fuel injection, 1202f
 types, 1201
fuel tank sending unit
 changing, 1210
 and evaporative emission lines, 1211f
 filter sock, 1211
 float mechanism, 1211f
 sock installation, 1211f
 strainer, pump, filter, and sending unit, 1211f
 three jobs, 1210–1211
fuel trim, 1257
fulcrum, 769
full floating axle, 560
full power mode, 1218
full-flow filter systems, 372
fundamental wheel alignment
 camber, 733–735, 733f
 caster, 734–735, 734f
 checking the alignment, 733f
 positive caster, 734f
 secondary angles, 733
 toe is viewed from above the tire, 735f
 toe-in and toe-out, 735
fundamentals of manual transmissions
 gear ratios, 513–514
 mechanical advantage, 512–513
 power flow transmission, 514, 514f
 two gears with different ratio, 513f
fused jumper leads, 1028
FWD/AWD axles/half-shafts, 561–562

G

garter spring, 910
gas welding goggles, 93
gasket scraper, 139
gasoline direct injection (GDI), 1201
 direct and indirect, 1218f, 1219f
 full power mode, 1218, 1219f
 GDI engines, 1219
 GDI fuel systems, 1219
 GDI system layout, 1220f
 high-pressure GDI systems, 1220
 indirect, 1218f
 multi-pulse injection, 1219, 1219f
 operation, 1217
 stoichiometric mode, 1218, 1218f
 stratified charge mode, 1218
gasoline fuel, 1203
 from crude oil, 1203f
 higher octane ratings, 1204
 octane rating, 1203
 oxygenated fuel sticker, 1203f
 RON rating, 1204
 unleaded gasoline, 1204
 very volatile, 1203
gasoline fuel system principles
 air supply system, 1202–1203
 liquid fuel, 1202
 metering system, 1202–1203
 supply system, 1202
gasoline fuel systems
 accelerator pump circuit, 1224
 carbureted fuel systems, 1221–1222
 carburetor barrels, 1224–1226
 carburetor circuits, 1223
 carburetor operation, 1222–1223
 the choke, 1224
 computer-controlled carburetors, 1226
 controlling fuel burn, 1204–1205
 EFI system types, 1215–1217
 float circuit, 1223
 fuel, 1205–1207
 fuel delivery system components, 1207–1209
 fuel filter, 1211–1212
 fuel for contaminants and quality, 1230
 fuel injectors, 121–1215
 fuel lines, 1211
 fuel pressure regulation, 1213–1214
 fuel pump relay, 1209–1210
 fuel rail, 1212–1213
 fuel tank sending unit, 1210–1211
 fuel/air requirements for internal combustion, 1205
 gasoline direct injection systems, 1217–1220
 gasoline fuel, 1203–1204
 GDI drawbacks, 1220–1221
 idle and off-idle circuits, 1223
 inspection fuel pumps, 1229–1230
 main metering circuit, 1223–1224
 maintenance and repair, 1226–1228
 overview of, 1201–1202
 power circuit, 1224
 principles, 1202–1203
 testing fuel injectors, 1231–1232
 testing fuel pumps, 1229–1230
gassing, 1045
gauge pressure, 1206
GDI drawbacks
 carbon buildup reducing, 1221
 cylinder head removing, 1221
 induction cleaning, 1221
 intake removing, 1221
 poor-quality oil droplets, 1221
 positive quality, 1220
gear reduction, 513
gear set, 513, 517
gear synchronizer, 512
geared oil pump, 370
general safety guidelines
 flaring tools, 150–153
 hand tools, 125f, 134–150
 handling tools and equipment, 123
 lockout/tag-out, 124
 manufacturer's special tools, 125f
 safe handling and use of tools, 122
 safe procedures for handling tools and equipment, 123
 tool storage, 124–125
 tools, 121–122, 123
 typical technician's toolbox, 124f
 work safe and stay safe, 121
general safety guidelines safe handling, 121–122
good compression, 347
grading of bolts, 207
grease seals
 and axle seals, 910
 coated, 910f
 components of, 910f
grinding wheels and discs, 185
ground-level ozone, 1306
guide pins, 835

H

half-shafts, 561
 axle flange with lug studs, 561f
 axles and, 560
 ¾-floating axle, 560f
 full floating axle, 560f
 RWD solid axles
  flanges, 560–561
  seals, 560–561
 semi-floating axle, 560f
 single-lip and double-lip seals, 561f
 speedy sleeve, 561f
 typical half-shaft and CV joints, 561f
Hall-effect sensors
 interrupter ring, 1188
 operation, 1188f
 and operation, 1187
 optical sensor assembly, 1188, 1188f
 optical-type, 1188, 1188f
 pickup coil scope pattern, 1187f
 window is aligned with assembly, 1188
Hall-effect voltage, 1187
halogen lamps, 1094
hand and measuring tools
 basic, 125–133
 flaring tools, 150–153
 general safety guidelines, 121–125
 precision measuring tools, 154–165
 tools, 134–150
hand protection
 chemical gloves, 86
 heat-resistant sleeve, 87f
 leather gloves, 86–87
 light-duty gloves, 87f
hand tools
 Allen wrenches, 136–138, 136f
 bench vice, 148f
 C-clamp, 148f
 clamps, vices, and pullers, 147–149
 drill vice on a drill worktable, 148f
 flat-blade screwdrivers, 137f
 hammers and struck tools

automatic center punch, 144f
chisels, 142–143
file card, 141f
number and letter punches, 144f
punches, 143–144
thread file, 141f
wad punch, 144f
magnetic pickup tools and fingers
bolt cutters, 139f
cutting tools, 138
files, 140–141
gasket scrapers, 139–140
metal gasket scraper, 139f
pry bars, 138–139
rough, 140
offset screwdriver, 137f
Phillips screwdrivers, 137f
pliers, 134–136
arc joint, 134f
diagonal side, 135f
flat-nose snap, 135f
locking, 135f
needle-nose, 135f
snap ring, 135f
variety, 134f
proper use of taps and dies, 145–147
screw extractors, 147
straight-sided screw extractor, 147f
tap drill chart, 147
taps and dies, 145, 145f, 146f
using gear pullers, 149–150
hard rubber mallet, 142
hard steering, 713
hardtop convertibles, 21
hazard warning lights, 1102
hazardous materials safety
cleaning brake dust, 62–71
engine oil and fluids, 70–72
environmentally friendly way, 71f
safety data sheets, 61–62
Zamboni, 71f
HCU-control valve operation
ABS channels, 934f
dual, two-position valves, 935f
high-pressure accumulators, 936
hydraulic control operation, 934
low-pressure accumulator holds brake fluid, 935f
single three-position valve, 935f
types, 934–936
head protection
containing hair, 89f
ear protection, 93
eye and face protection, 89–90
full-face shield, 92
gas welding goggles, 93, 93f
hard hats prevent head wounds, 90f
headgear, 89
hearing protector, 94f
noise exposures, 93t
noise level of common sounds, 93t
protection against radiant energy, 92t
safety glasses, 90–91
safety goggles, 91–92
welding helmet, 92, 92f
headlight brightness, 1112–1113
headlight bulb
changing, 1111, 1112
checking, 1111, 1112
halogen bulb, 1111
headlight types
beam shaping, 1104
HID headlight assembly, 1106, 1106f
HID lights, 1104
high voltage, 1104
night vision, 1106, 1106f
projector bulb assembly, 1106, 1106f
replaceable halogen bulb, 1106f
semi-sealed beam, 1104
headlights
domestic-type turn signals, 1103f
foot-operated beam selector switch, 1105f
high- and low-beam filament, 1102–1104, 1104f
high-beam switch, 1105f
integrated hazard warning lights, 1104, 1104f
low-beam filament, 1104, 1104f
heat buildup, 86
heat dissipation, 413
heat engines, 311
heat range, 1180
heat transfer, 1150
heat-conducting resin coil, 980f
heated air cleaners (HAC), 1282
heated air intake systems, 1316, 1316f
heated air systems (HAS), 1282
heater control valve, 426
heater core, 426
heating and air conditioning systems
abnormal noises, 1162
automotive heating and cooling, 1147–1148
blower motors, 1160
cabin air filter, 1163–1164
condenser for airflow restrictions, 1162
defroster, 1160
evaporator housing water drain, 1162–1163
HVAC principles, 1148–1151
odors, 1163
operation, 1153–1155
overview of, 1147
performance testing, 1161
refrigerant principles, 1151–1153
refrigerant types, 1155–1158
resistor packs, 1160–1161
and ventilation system, 1158–1160
heavy line technicians, 9
helical springs, 684
helical-cut gears, 471
helical-geared limited-slip differentials, 558
Helmholtz resonator, 1280
HEVs. *See* hybrid electric vehicles (HEVs)
hex keys, 136
HID. *See* high-intensity discharge (HID)
HID safety precautions, 1107
hidden heat, 1151
high resistance
continuity between the output of the fuse and ground, 1027f
excessive voltage drop, 1026f
locating opens, 1025
open circuit fault starts, 1026f
short to power, 1028f
shorts and grounds, 1025
test light, 1027f
tested with an ohmmeter, 1027f
high-intensity discharge (HID), 1095
high-pressure accumulators, 936
high-tension leads, 1177–1178, 1178f
high-voltage safety
AC components, 76f
accidental SRS deployment, 73–74
air conditioning components, 76f
automatic controls safety, 72–73
automatic transmission, 76f
brake components, 75f
cooling system components, 75f
hot and cause severe burns, 75f
negative battery terminal, 74f
pressurized systems safely, 74–75
SRS injury reduction devices, 73f
SRS warning light, 74f
system-ready indicator, 72f
wiring, 72f
working near extreme temperatures, 75–77
high-voltage system
disabling, 1333
high-voltage battery safety system, 1333, 1333f
high-voltage system on Nissan Leaf, 1334, 1334f
high-voltage wiring, 1333, 1333f
identifying, 1333
no voltage the specified terminals, 1334, 1334f
third-generation Prius, 1334, 1334f
history codes, 1266
hold function, 1001
hold position, 799
HOLD reading, 1001f
hold-down spring tool, 891
hold-in winding, 1067
holding/driving gears
automatic transmission fluids, 473–474
case, extension housing, and pan, 475
compound planetary, 472f
dog clutches, 473f
flexplate and ring gear, 474, 474f
FWD transaxle, 475f
gaskets and seals, 475–476
one-way clutches, 473f
parking pawl assembly, 476
reusable gaskets, 475f
seals used automatic transmission, 476f
two multi-disc clutches, 472f
typical transmission band, 472f
welded-on ring gear, 474f

hollow punches, 142
homogeneous mixture, 1218, 1219
Hooke's joint, 564
horizontally opposed engines, 35
horn system testing, 1131
horn systems
  clock spring, 1131f
  horn, relay, switch, and clock spring, 1130–1131
  horn system, 1131
  inside components of a horn, 1130f
  typical horn circuit, 1131f
hub runout measurement, 627, 630
hubless rotor, 833
hubless-style drum, 876
hub-style drum, 876
hub-style rotor, 833
HVAC principles
  cold and hot, 1150–1151
  conduction, 1150f
  convection, 1150f
  heat energy, 1151
  liquids and gases, 1151
  metal coils, 1151
  principles of physics, 1149
  radiation, 1150f
  thermal and chemical transformations, 1149f
HVAC system, 1147, 1149
hybrid and continuously variable transmissions
  continuously variable transmission (CVT), 501–504
  hybrid drive systems, 494–497
hybrid auxiliary (12 v) battery, 1335, 1335f
hybrid drive configurations
  series-parallel, 1330
  types, 1330f
hybrid drive systems, 494
  driver information center, 495f
  driver information screen, 496f
  electric-only propulsion, 496
  idle stop, 494–495
  regenerative braking, 495–496
  torque assist, 496
  torque smoothing, 495
hybrid electric vehicle models
  BAS system installed on a Chevrolet Malibu, 497f
  belt alternator starter, 497
  Ford Motor Company Hybrids, 500
  Honda IMA system, 498f
  Honda integrated motor assist, 497–498
  Honda intelligent Multi-Mode Drive (i-MMD), 498–499
  Honda two-motor hybrid powertrain schematic, 498f
  M/G2 and the ICE, 499f
  planetary operation as engine, 499f
  ring gear, 499f
  Toyota and Lexus Hybrids, 499–500
  Toyota Prius, 497f
  two-mode hybrid, 500–501, 501f
hybrid electric vehicles (HEVs), 1329–1330
hybrid vehicle efficiency enhancements, 1331
hybrid vehicle operation, 1331
hybrids
  electric vehicles, 1327–1328
  types of, 1327
hydraulic brake booster, 800
hydraulic braking system control
  pressure-sensitive proportioning valve, 791, 791f
  proportioning valves, 790–791
hydraulic components
  brake fluid types and characteristics, 779–780
  DOT 3, DOT 4, DOT 5, DOT 5, 780f
  DOT ratings, 780t
hydraulic control unit (HCU), 930
hydraulic fade, 767
hydraulic jacks, 235
hydraulic modulator, 770
hydraulic power brake booster, 800f
hydraulic press method, 864
hydraulic system and power brake servicing
  bleeding brake systems, 810–813
  brake fluid handling, 807
  brake fluid testing, 807–809
  brake lines and hoses, 821
  brake pedal inspection, 813–815
  brake repair legal standards and technician liability, 805–806
  brake warning lamp, 825–827
  diagnosing stop lights, 827–828
  fittings, and supports, 821–824
  hydraulic system component diagnosing, 816
  hydraulic system maintenance, 806
  master cylinder inspection, 816–817
  master cylinder pushrod length, 817–818
  master cylinder service and bench bleeding, 817
  power brake systems, 818–821
hydraulic system component diagnosing, 816
hydraulic system maintenance tools, 806
hydraulic valve train
  cam drives, 333f
  cam in-block timing chain and gears, 333f
  freewheeling engine *vs.* interference engine, 333f
  mechanical and, 330
  roller lifters, 332f
  roller rockers and lifters, 331
  standard rocker, 331f
  timing chain, gears, and tensioner system, 333f
  valve adjustments, 332f
  valve clearance, 331–332
  valve train drives, 332–333
hydraulics and power brakes theory
  ABS master cylinders, 784
  adjustable proportioning valves, 791–793
  brake hoses, 789–790
  brake lines, 787–789
  brake pedals, 785–786
  brake warning light and stop light, 796–797
  braking system control, 790–791
  divided hydraulic systems, 786–787
  electrohydraulic brake boosters, 800–801
  hydraulic brake booster, 800
  hydraulic components, 779–780
  master cylinder, 780–784
  metering valve operation, 793–796
  power brakes, 797–799
  principles, 777–779
  quick take-up master cylinders, 784
  sealing washers and fittings, 790
hydraulics principles
  hydraulic pressure and force
    input force, 778
    output force, 778
    working pressure, 778
  mechanical force in hydraulic braking systems, 778f
  pistons and the input force, 779f
  pressure transmission, 778f
  schematic view, 777f
hydrocarbons, 1203, 1303
hydrocracking, 375
hydrometer, 453
hypoid bevel gears, 556
hypoid gears, 471

I

ICE. *See* internal combustion engine (ICE)
idle and off-idle circuits, 1223, 1223f
idle circuit, 1223
idle-stop/start-stop systems, 1076–1077
ignition advance, 1185
ignition coils, 1177
  cutway basic, 1177f
  secondary winding, 1177
  standard, 1177
ignition coils and transformers, 980
ignition control module (ICM), 1254
ignition modules, 1186
ignition pickup-style position sensor, 1243
ignition principles
  air-fuel mixture, 1171f
  breaker point, 1172f
  coil-on-plug, 1172f
  electronic switching device, 1172
  four-cylinder engine, 1172
  ignition system circuit, 1171f
  low voltage and high current, 1171
  system-distributor type, 1172f
  waste spark, 1172f
ignition switch, 1176
  accessory, 1176
  lock, 1176
  off, 1176
  on/run, 1176
  start/crank, 1176–1177
ignition system maintenance
  broken-off high-thread spark plugs, 1192f
  high-tension insulated wire puller, 1192f

plug gapping tool, 1192f
tools and safety practices, 1192
ignition systems, 1193
ballast resistor, 1182–1183
bypass circuit, 1182–1183
centrifugal advance units, 1184–1185
coil-on-plug, 1177, 1190–1192
common, 1176
contact breaker point, 1180–1182
COP boots, 1193
crank and cam sensors, 1189–1190
direct ignition system, 1172
distributorless-type systems, 1189
distributors, 1183–1184
distributor-type electronic ignition systems, 1186–1187
dwell angle, 1182
electronic switching device, 1172, 1185–1186
engine spark timing, 1184
Faraday's law, 1173–1174
hall-effect sensors, 1187–1188
heat range, 1180
high-tension leads, 1177–1178
maintenance, 1192
optical-type sensors, 1188–1189
overview of, 1171
primary and secondary circuits, 1172–1173
principles, 1171–1172
required *vs.* available, 1174
secondary ignition components removing, 1193
secondary ignition components replacing, 1193
service information, 1193
spark plug design, 1178–1179
spark plug reach, 1179–1180
spark plug size, 1179
spark plug wires, 1194–1195
spark timing, 1175–1176
switch, 1176–1177
types, 1180
vacuum advance units, 1185
waste spark ignition system, 1190
ignition systems types, 1180
IMA. *See* integrated motor assist (IMA)
impact driver, 137
incandescent lamps, 1094
included angle, 737
independent axle, 520
independent rear suspension (IRS), 833
independent suspension drive axle, 521
indirect fuel injection, 1217
induction and exhaust
air cleaner, 1278–1280
air filter, 1294
catalytic converter, 1290–1291
ducting, 1280
exhaust system components, 1288–1290
exhaust system inspection, 1296–1297
induction system, 1295–1296
intake air heating, 1281–1282
intake modules, 1282–1283
intake system, 1277–1278
muffler system, 1292
overview of, 1277
turbocharger systems, 1286–1288
variable intake systems, 1283–1284
variable-flow exhaust, 1293–1294
volumetric efficiency, 1284–1286
induction coil, 1172
induction system inspection, 1295–1296
induction-type systems, 1186
inductive current, 1181
inductive-type sensors, 1242
infrared temperature sensor, 453
injectors, 1213
in-line engines, 35
inline fuel filter, 1212f
inner tie-rod end tool, 660
inspecting and replacing wheel studs, 866–867
inspecting and testing relays and solenoids, 1075
inspecting the evaporator housing water drain, 1163
inspecting torsion bar suspension, 729
inspecting upper and lower ball joints, 722
inspection certificate, 235
inspection fuel injectors, 1231–1232
installing stabilizer components
inspecting, 716
removing and inspecting sway bar end links, 718
removing and replacing shock absorbers, 719
stabilizer bar bushings and mount brackets, 718
installing upper and lower ball joints, 722
instrument panel warning lamps, 253
insulators, 952, 953
intake air heating
electric heaters, 1281, 1282f
heated air intake system, 1282f
hot engine coolant, 1281f
methods of, 1281
multipoint injection setup, 1282
port fuel injection manifold, 1282f
thermostatic control switch, 1282
V-type engines, 1281
intake air temperature (IAT), 1242
intake manifold runner control (IMRC), 1284
intake manifold vacuum testing
pressure transducer, 349
testing engine vacuum, 349
vacuum gauge, testing engine vacuum using, 348–349
intake manifolds
cross-flow head, 1283f
designs, 1282–1283
plastic, 1283f
intake system
air filtration device, 1279f
automotive, 1277
fuel-injected and carbureted engines, 1277
intake manifold, 1277f
port fuel injection, 1278, 1278f
throttle body, 1278, 1278f
throttle plates, 1278f
in-tank fuel filter, 1212f
integral ABS systems, 932
integrated circuit, 1187
integrated motor assist (IMA), 497
interference fit, 906
intermediate pipe, 1291
intermediate shaft, 562f, 640
intermediate tap, 145
internal combustion engine (ICE), 311
International Lubricant Standardization and Approval Committee (ILSAC), 377
International Standards Organization (ISO), 788
interrupter ring, 1188
in-vehicle inspections
checking the brake pedal, 253
horn, 255
instrument panel warning lamps, 253, 254
interior lights, 255
parking brake, 253
retrieving and recording DTCs, 254
retrieving diagnostic trouble codes, 253
in-vehicle transmission repair
adjusting the shift linkage, 488
hydraulic-style powertrain mount, 488f
inspecting and repairing powertrain mounts, 487–488
inspecting and replacing gaskets and seals, 486
inspecting, replacing, and aligning powertrain mounts, 488–490
powertrain control module, 487
rubber style powertrain mounts, 488
transmission control module, 487
transmission range sensor, 488
inverted double flare, 788
iridium spark plug, 1193f
IRS. *See* independent rear suspension (IRS)
ISO flare method, 150, 824
isolation valve, 935

J

Japanese Automotive Standards Organization (JASO), 378
JASO diesel engine oil rating, 378f
joints
and couplings, 563
cross-and-roller U-joint, 564f
replacing U-joints, 564–568
typical double Cardan, 564
universal, 564
jounce stops, 682
jump-starting the vehicle
boost of electrical energy, 1049
damaging the electronics, 1048
spring-loaded clamps, 1049

K

KAM. *See* keep alive memory (KAM)
keep alive memory (KAM), 1053, 1265
keyed lock washer, 917
keyless starting, 1070
  driver to start the engine with key, 1070, 1070f
  remote start button, 1070, 1070f
kinetic energy, 763
  acceleration, 764
  deceleration, 764
  energy transformation, 765
  vehicle brakes, 764f
  weight doubles, 764f
kinetic friction, 578
Kirchhoff's Current Law, 968
knock sensor, 1249
  detection, 1249
  engine vibrations, 1250f
  function of, 1250
  mounted on engine, 1249f
  overheated engine, 1249
  technology, 1250
knocking, 1203

L

labor guide
  past program, 106
  typical exploded parts diagram, 106f
  using parts program, 106
labor guides, 105
lambda sensors, 1206, 1256, 1311
lamp types
  halogen bulb, 1094, 1094f
  HID headlamp assembly, 1095f
  HID headlight with lens cleaning system, 1095, 1095f
  high-intensity discharge, 1095–1096
  incandescent, 1094
  incandescent bulb with single filament, 1094f
  LED lights, 1094f
  light-emitting diode, 1094–1095
  sizes of, 1093
  vacuum fluorescent display, 1095f
  vacuum tube fluorescent, 1095
lamp/lightbulb configurations
  bayonet-style bulbs, 1096, 1096f
  built-in plastic base, 1097f
  dual-filament bulb, 1096, 1096f
  identifying numbers on them, 1096f
latent heat of condensation, 1151
latent heat of evaporation, 1151
latent heat of freezing, 1151
laws of condensation, 1151
laws of vaporization, 1151
lazy axle, 30
lead-acid flooded cell batteries, 1037
leading shoes, 874
leading/trailing shoe drum brake system, 874
leaf springs, 682
leaf springs inspection, 727, 729
LED. *See* light-emitting diode (LED)
lemon law buyback, 107
light line technicians, 7
light-duty gloves
  barrier cream, 87–88, 88f
  cleaning your hands, 88
  cloth gloves, 88f
  general-purpose cloth, 87
light-emitting diodes (LEDs), 188, 960, 1094, 1188
lighting circuit testing
  HID lamp, 1107
  HID safety precautions, 1107
  and service, 1107
  system wiring diagrams, 1107
lighting system types
  cornering, 1099, 1099f
  driving lights, 1098–1099
  fog, 1099
  license lights, 1098f
  marker lights, 1097, 1098f
  park, 1097
  park/tail/marker/license, 1097
  tail lamps, 1097, 1098f
  vehicle with factory fog lights, 1099f
lighting system wiring diagrams, 1107
lighting systems
  aiming headlights, 1111–1112
  backup lights, 1100
  brake lights, 1099–1100
  changing an exterior lightbulb, 1110–1111
  CHMSL, 1099–1100
  circuit testing and service, 1107
  courtesy lights, 1107
  daytime running, 1106–1107
  domestic and import systems, 1102
  hazard warning, 1102
  headlight brightness, 1112–1113
  headlight bulb, 1102–1104, 1111
  headlight types, 1104–1106
  lamp types, 1093–1096
  lamp/lightbulb configurations, 1096–1097
  overview, 1093
  peripheral systems, 1107–1110
  turn signal, 1100–1102
  types, 1097–1099
limited slip differential assembly, 520
LIN. *See* local interconnect network (LIN)
linear motion, 634
lining, 873
liquid-vapor separator, 1208
lithium soap, 912
load transfer, 791
local interconnect network (LIN), 1119
lock cage, 917
locking pliers, 135
lock-up converters
  absorb power pulses, 468f
  deep sump finned transmission pan, 469f
  external transmission cooler, 469f
  friction material, 468f
  front housing of torque, 468f
  gear ratio/torque multiplication, 470–471
  gear set styles, 471
  gear train, 470
  heat exchanger, 469
  hypoid gear arrangement, 471f
  planetary gear sets, 471–472
  simple planetary gear set, 470f, 471f
  spur, helical, and hypoid gears, 471f
  transmission fluid warmer, 469f
looping, 1256
lot attendant, 7
low power consumption, 1106
low-drag calipers, 837
Low-Emission Vehicle (LEV) standard, 1325
lower control arms and components removing, 721
low-pressure accumulators, 935
low-speed preignition, 1221
lubricating a steering system, 725–726
lubricating a suspension system, 725–726
lubricating oil, 363
lubricating steering, 722
lubrication
  axle bearings, 911f
  bearing grease, 911f
  bearing with grease by hand, 912f
  gear lube, 911f
  lithium soap grease, 912f
  molybdenum grease, 912f
  NLGI Rating System, 913t
  rear axle assembly, 911f
lubrication system components, 363, 368, 368f
lubrication system theory
  American Petroleum Institute (API), 375–376
  American Society of Automotive Engineers (SAE), 376–377
  Association des Constructeurs Européensd' Automobiles (ACEA), 377
  components, 368
  International Lubricant Standardization and Approval Committee (ILSAC), 377
  Japanese Automotive Standards Organization (JASO), 378
  lubricating oil, 364
  OEM-Specific Standards, 378
  oil coolers, 374
  oil filters, 372–374
  oil indicators, 378–380
  oil monitoring systems, 380–383
  oil pressure relief valve, 372
  oil pump, 370–372
  oil types, 364–367
  oil-certifying body, 374
  oil-certifying body and rating standards, 374–375
  pickup tube, 370
  spurt holes and oil galleries, 374
  wet sump and dry sump systems, 368–369
lug nuts replacement, 563

parallelogram steering linkage service
inspecting and replacing the center link, 671
inspecting and replacing the idler arm, 670, 671
inspecting and replacing the steering linkage damper, 672
inspecting and replacing tie-rod ends, 672
mountings, and steering linkage damper, 670
performing, 668
pitman arm, relay, 670
rod, idler arm, 670
tie-rod ends, tie-rod sleeves, and clamps, 670
parallelogram steering system, 635
parasitic draw, 1054
parking brake cable pliers, 851, 893
parking brake cable removal tool, 893
parking brake mechanism, 873
parking brake systems, 886
parking brakes on disc brakes
electric parking brake, 846
electrically integrated parking brake caliper, 846f
hand-operated, 845f
hat rotor and drum parking brake assembly, 845f
integrated mechanical calipers, 845
integrated parking brake operation, 845f
top hat design, 845–846
types, 845
parts programs, 106
parts specialists, 106
Pascal's law, 777
passive keyless entry (PKE), 1129
passive system, 654, 937
PCM. *See* powertrain control module (PCM)
PCV systems
air cleaner assembly, 1313f
failures, 1317
fixed orifice passageway, 1313f
fixed orifice-type, 1313
inspection, 1317
oil leakage, 1318
separator type of, 1313, 1313f
servicing, 1317
types, 1312
valve plunger position, 1312
PCV systems draw blowby gas, 1312f
PCV valve, 1312
PCV-related concerns, 1317
pedestrian airbag, 1137
pending codes, 1266
perform rack-and-pinion service
inspecting mounting bushings and brackets, 668
inspecting power steering hoses and fittings, 667
rack-and-pinion mounting bushings and brackets, 668
rack-and-pinion steering gear inner tie-rod ends and bellows boots, 668, 669
replacing power steering hoses and fittings, 667
performance testing, 1161
performing a wheel alignment
readings, 742, 742f
thrust angle alignment, reading, 742, 742f
types, 742
performing current measurements, 1009
circuit with single resistor, 1010f
two equal-resistance resistors in series, 1010f
two unequal-resistance resistors, 1010f
performing four-wheel alignment, 748–749
performing master cylinder service and bench bleeding, 818
performing ride height diagnosis, 714–715
permanent magnet type, 1124
personal protective equipment (PPE), 41
personal safety
hand protection, 86–87
head protection, 88–95
light-duty gloves, 87–88
overview of, 81–82
protective clothing, 82–86
phenolic resin, 837
PHEV. *See* plug-in hybrid electric vehicle (PHEV)
Phillips head screwdriver, 137
photovoltaic energy, 958
pickle fork, 660
pickup assembly, 937, 1185
pickup coil, 1187
pickup tube, 370
piezoelectric energy, 958
pilot bearing and input shaft, 536–537, 537f
piston displacement, 320
piston engine, 33
piston stroke, 320
pitman arm puller, 660
PKE. *See* passive keyless entry (PKE)
planet carrier, 470
plastic-retained U-joints, 564
platinum spark plug, 1193f
plenum chamber, 1282
plug-in hybrid electric vehicles, 1329
plunge-type joint, 568
pneumatic jacks, 235
pneumatic tires, 577
pop rivet guns, 152
ported, manifold, and venturi, 1185f
positive camber, 733
positive caster, 734
positive crankcase ventilation (PCV), 1309
positive ion, 952
positive scrub radius, 737
positive temperature coefficient (PTC), 975
postcombustion treatment, 1309
potential difference, 963
potentiometers, 981, 982f, 1240
power assist unit, 652
power balance testing, 349
power booster, 930
power brake systems
checking vacuum supply to vacuum-type power booster, 820
checking vacuum-type power booster unit for leaks, 821
testing, 819–820
testing pedal free travel and performance testing the vacuum booster, 820
vacuum supply, power booster, 820
power brakes
booster check valve, 798f
diaphragm and return spring, 798f
dual diaphragm brake booster, 799f
dual-diaphragm boosters, 799
hydraulic-assist, 797f
vacuum booster, 797
vacuum brake booster in apply position, 799f
vacuum brake booster in released position, 798f
vacuum-assist, 797f
power circuit, 1224
power door locks
DC motor type, 1127f
lock actuator-solenoid type, 1127, 1127f
permanent magnet plunger, 1127, 1127f
power door lock circuit, 1128f
remove door panels, 1127, 1127f
power sources, 1328–1329
power splitter, 1331
power steering
filter mounted, 648f
fluid and hoses, 648–649
fluid type, 648f
high-pressure line, 648f
high-pressure power steering hose, 649f
hydraulically assisted, 648
idle speed strategy, 650–651
O-ring, 649f
power-assisted rack-and-pinion system, 649f
power-assisted recirculating ball gearbox, 650f
slight flexing, 651f
steering process, 649–650
steering rotary valve and torsion bar, 650f
power steering fluid maintenance
adjusting power steering fluid, 664
checking and adjusting power steering fluid, 665
checking power steering fluid, 664
flushing, 665, 666
greased-for-life tie-rod end, 664f
inline power steering fluid filter, 665f
replacing pump filter, 665
reservoir style filter, 665f
zerk fitting used to lubricate the joint, 664f
power steering pump filter, 648, 665
power steering system, 665

power tools and equipment
air tools, 177–184
battery charging and jump-starting, 171–177
cleaning tools, 193–199
electric power tools, 184–193
power unit, 653
power-on-demand, 1331
power-splitting transmission (PST), 463
powertrain control module (PCM)
data information, 1250
data processor and drivers, 1251
electronic, 1252
input signal processing, 1251
memory and storage section, 1251
output signals to actuators, 1251, 1251f
section of, 1251f
powertrain mounts, 487
PPE. *See* personal protective equipment (PPE)
pre-alignment inspection, 747
precision measuring tools
bent feeler gauge set, 163f
dial bore gauge set, 159, 159f
dial indicators, 161–162, 161f
digital caliper, 161f
feeler gauge set, 163–165, 163f
markings on thimble, 156f
outside, inside, and depth micrometers, 155–157, 155f
split ball gauges, 158–159, 159f
stepped feeler gauge set, 164f
straight edges, 163
tapes, 154
telescoping gauges, 158, 158f
tipping a steel rule, 155f
using dial bore gauges, 160
using micrometers, 157, 158
Vernier calipers, 159–161
Vernier scale, 157f
pressure brake bleeder, 807f
pressure differential valve, 794
pressure plates
coil spring pressure plate, 534–535, 534f
diaphragm pressure plate, 534
diaphragm spring, 534f
pressure relief valve, 1315
pressure washers, 193
preventive maintenance, cooling system
accessory drive belt replacement, 441, 442
adjusting coolant freeze protection level, 437
bottomed-out serpentine belt, 440f
bottomed-out V-belt, 440f
checking and adjusting coolant, 436, 437
checking and replacing a coolant hose, 444
coolant level on a clear reservoir, 434f
coolant pH testing, 435
coolant stored for recycling, 434f
cooling fan wiring diagram, 448f
cracked hose, 444f
draining and refilling coolant, 438
electrolysis testing, 436
EPA guidelines, 433
fan clutch testing, 447f
flushing the coolant, 438–439
freeze protection, measuring, 434
glazed belt, 440f
hardened/brittle hose, 443f
hydrometer to test freeze point of the coolant, 434
inspecting and adjusting an accessory drive belt, 439–441
inspecting and replacing a coolant hose, 443–444
inspection and testing fans, fan clutch, fan shroud, and air dams, 447–449
inspection and testing the cooling fan, 445–447
inspection the thermostat bypass, 445
level checking, 433
performing electrolysis testing, 436
refractometer to test freeze point of the coolant, 435
removal and replacing a thermostat, 445
removing and replacing a thermostat, 446
rubber hose style, 446f
stretch fit belt replacement, 441
stretch fit drive belt, 442–443
swollen hose due to broken cords, 443f
temperature warning lamp illuminated, 446f
torn belt, 440f
water pump replacement, 449–451
prick punch, 142
primary and secondary circuit components, 1173t
primary and secondary circuits, 172
primary winding, 980
probes and probing techniques, 1003
probing techniques, 1003
production date code, 107
progressive carburetor, 1225f
progressive rate of deflection, 685
projector bulb assembly, 1106f
proper tire inflation
adjusting tire pressure, 603–604
green valve stem cap, 601f
nitrogen fill, advantages for, 59
tire cage, 600f
tire pressure gauges, 601–602, 602f
typical tire placard, 600f
using a tire pressure gauge, 602–603
proportioning valve/metering valve gauge sets, 790, 807f
propylene glycol, 407
protective clothing
care of clothing, 84
chemical exposure, 86f
duck material work clothes, 83f
footwear, 84
jewelry, 85
leather apron, 83f
neoprene, 83f
one-piece coverall, 84f
steel-toed boots, 85f
treated wool and cotton uniform, 83f
work clothing, 83–84
work shoes, 84f
protective covers
fender covers, 231
floor mats, 232
seat, 231
steering wheel covers, 232
pry bar, 660
PST. *See* power-splitting transmission (PST)
PTC. *See* positive temperature coefficient (PTC)
PTC thermistor resistance, 982f
pull-in winding, 1067
purging, 1316
push-on spade terminals, 984

Q

quadrant ratchet, 538
quick take-up master cylinders, 784
quick take-up valve, 784

R

rack bearing, 644
rack-and-pinion gearbox
center take-off rack, 644f
end take-off rack, 644f
knuckle and steering arm, 645f
rack housing, 645f
spring-loaded rack, 644f
variable-ratio rack, 644f
rack-and-pinion steering linkage
inner tie rod or socket, 637
inner tie-rod end or socket, 637f
outer tie-rod end, 637, 637f
pinion and rack, 637f
rack-and-pinion assembly, 637f
rubber bellows, 637
rack-and-pinion steering system, 636
radial loads, 517
radiation, 403, 1151
radiator
cooling tubes, 414
cross-flow, 414f
cross-flow and the down-flow designs, 414
dissipate heat, 414f
drain plugs, 415f
radiator hoses, 421
hose clamps, 422f
improper installation of hose clamp, 422f
installation of hose clamp, 422f
silicone coolant, 422f
radiator pressure cap, 416
radiator shrouding, 415, 415f
ratcheting box-end wrench, 127
ratcheting screwdriver, 137
rat-tail file, 141
reaction force, 680
rear suspension
control axle movement, 702f
Ford's Control Blade IRS, 703f

independent rear suspension, 704f
live independent rear suspension, 703f
lower control arms, 702f
MacPherson strut-type independent rear suspension, 703f
rear independent, 703
rear-wheel drive independent, 703–704
rigid dead axle, 702–703
rigid-axle coil-spring, 701–702, 701f
rigid-axle leaf-spring, 701, 701f
torsion beam axle, 703f
Watt's linkage, 702f
rear-wheel drive (RWD), 29
centerline of the transmission, 549f
-independent rear suspension, 548, 548f
ring and pinion gear set, 549f
typical RWD solid axle assembly, 548, 548f
reassembling caliper
inboard pad lines up, 858f
and pad assembly, 857
readjusting pistons, 858
retracting pistons, 858
rebound clips, 686
recirculating ball, 647
recovery system, 416
rectification, 1082
AC into DC, 1082
magnet rotator, 1082
north and south magnetic fields, 1082
positive and negative diode, 1083f
rectifier assembly, 1083
single phase in reverse direction, 1082f
single phase in the forward direction, 1082f
three phases-not rectified, 1082f
three phases-rectified, 1082f
rectifier assembly, 1083
recuperation, 781
reed switch, 1244
refinishing rotors off vehicle, 863, 865
refinishing rotors on vehicle, 963
reflash process, 1251
refrigerant label, 110
refrigerant principles
boiling point, 1152
boiling points of R-134a, 1152f
differential gas pressure, 1153
high-side components, 1153f
low-side components, 1153f
refrigerant falls, 1152
vaporization and condensation, 1151
vaporization and condensing points, 1152
water vapor condenses on cold surfaces, 1152f
regenerative braking, 495, 1331
reinstalling SLA suspension components, 723–724
relay inspection, 1074
relay test adapter, 1074f
relay testing, 1074
relays
fuel pump, 1252
PCM activation, 1253f
reset button on top, 1253f
spring-loaded trigger, 1253
release mechanism, 531
remote key locking, 1128
remote keyless entry (RKE), 1129
remote starting, 1070
removing and installing a starter, 1075–1076
removing and reinstalling sealed wheel bearings, 922
removing, inspecting, and installing upper and lower ball joints, 722
removing upper and lower ball joints, 722
repair order, 110
repair order information
electronic repair order, 111
repair order completion, 111–112
replacing fluid and filters
checking fluid level and inspecting fluid loss, 483
draining and replacing fluid and filter, 484
inspecting and cleaning the transmission cooler, 485
inspecting and flushing cooler lines, 485
required voltage *versus* available voltage, 1174
reserve capacity (RC), 1041
residual pressure valve, 781
resistance exercises, 1008
resistor ratings, 981
resistors, 980–981
resonator, 1293
restoring electronic memory functions, 1054
retarding effect, 1331
return springs, 882
returnless fuel injection systems, 1214
reverse lights, 1100
rheostats, 981
ribbon cable, 983, 983f
ride height measurement, 714, 741
rigid dead axle suspension, 702
rigid spring hanger, 686
rigid-axle coil-spring suspensions, 701
rim flange, 580
ring and pinion gears, 556
riveted linings, 838
RKE. *See* remote keyless entry (RKE)
road force imbalance, 610
rollover valve, 1316
roll-rate sensor, 943
rotary combustion spark-ignition engine
components of, 338–339
cutaway of a rotary engine, 337f
principles of, 337–338
rotary engine housing, 338f
rotary engine operation, 338f
rotary engine rotor and seals, 338f
stationary gear and internal rotor gear, 338f
Wankel rotary engine, 337f
rotary engine coolant passages, 413f
rotary engine cooling system, 413f
rotary engines, 36
rotary flow, 467
rotating components
circumference of circle, radiance, 316f
power in, 315
reciprocating motion, 316f
rotational force, 767
braking, 768f
first order lever, 769f
lever parts, 769f
levers and mechanical advantage, 769
second order lever, 769f
third order lever, 769f
weight transfer during braking, 768f
rotational speed, 513
rotor arm, 1183
rotor circuit control
alternator charging, 1084f
diode trio supplies power, 1084f
positive battery terminal, 1084
slip rings and rotor field winding, 1084
voltage regulator, 1085
rotor housing, 412
rotor lobes, 370
rotors removing
inner bearing and grease seal, 861
reinstalling, 860
removing and reinstalling rotor, 862
rotor-type oil pump, 370
rubber press-fit valve stem, 619f
rubber-bonded bushings, 696
run-flat technology, 592
running clearance, 908
Rzeppa joint, 569

S

safe edge files, 141
safe working load (SWL), 235
safety data sheets (SDS), 61
safety overview
accidents and injuries, any time, 43
accidents and injuries are avoidable, 43
accidents by mistakes, 43f
air quality, 45
correct chemical storage, 44f
don't underestimate the dangers, 42
exhaust hoses, 46f
OSHA and EPA, 43–44
OSHA inspection, 44f
poor lifting and handling techniques, 43f
recovering broken arm, 42f
running engines, 45
shop policies and procedures, 44–45
salvage title, 107
scan tools
bidirectional capability, 1269–1270
bidirectional scanners, 1270
cost and complexity, 1269
purpose, 1269
smartphone, 1269f
training and experience, 1270
scavenge pump, 369
scavenging effect, 322, 1289, 1307
schematic diagrams, 974
Schrader valve, 585, 1213

screw-in valve stem, 619f
screws
  anti-seize compound, 212f
  bolt torque chart, 218f
  Christmas tree type, 217f
  clips, 217
  cotter pin holding, 214f
  cotter pins, 216f
  external circlip, 215f
  internal snap ring holding, 214f
  machine screw heads, 212, 212f
  non-threaded, 213
  paddle-style snap ring, 214f
  pins, 216–217
  plastic retainers, 217f
  push clip holding, 214f
  screws variety, 212f
  self-tapping, 213, 213f
  sheet metal, 213, 213f
  snap ring pliers, 213–216, 215f
  thread-locking compound, 212f
  tie-rod nut, 216f
  torque and torque wrenches, 217–218
  torque charts, 218
  trim, 213, 213f
  U-joint retainer clip, 214f
scrub brakes, 759
scrub radius, 737
SDS. *See* safety data sheets (SDS)
sealed wheel bearings, 909
sealing washers and fittings, 790
seat belt pretensioners, 1136
240-second drive cycle, 1268f
secondary air injection, 1309
secondary alignment angles
  checking rear wheel thrust angle, 750
  checking toe-out on turns, 750
  front and/or rear cradle alignment, 751
  rear wheel thrust, 749–750
  SAI and included angle, 751
  toe-out on turns, 750f
secondary ignition components, 1193
secondary winding, 980
sector shaft, 646
self-adjusters
  duo-servo brakes, 884
  duo-servo-style brake, 884f
  ratchet-style adjuster, 885f
  star wheel assembly, 884f
  types of, 884
self-energizing, 873
self-induction, 1173
self-resetting circuit breakers, 977
self-sealing tires, 593
sellers thread, 204f
semiconductors, 953
semi-floating axle, 560
sending unit, 1210
sensible heat, 1151
series circuit measurements
  battery supply, fuse, switch, and lightbulb, 1012f
  lightbulbs of equal value connected, 1012f
  performing, 1011
  three unequal-resistance lightbulbs in series, 1013f
  two unequal lightbulbs connected, 1012f
  two unequal-resistance lightbulbs reversed, 1013f
series strings, 1094
series-parallel circuit measurements
  formed by bulbs, 1016f
  unwanted resistance, 1015
series-parallel circuits, 967
series-parallel hybrid drivetrain, 463
service brakes
  ABS brakes, 762f
  cylinder size, 762f
  outside brake disc, 762f
  parking brake, 761, 762f
  regenerative, 761
  vehicle display, 761f
  vehicle service brake, 761f
service campaigns and recalls, 104
service consultants, 11
service history, 112–113
service manuals, 99
serviceable bearings and sealed bearings, 907
serviceable wheel bearings maintenance
  adjustment, 916–917
  bearing packer, 916f, 917f
  cotter pin, 915f
  cotter pin removal tool, 916f
  cotter pin-style, 917f
  dust cap pliers, 916f
  installing the locking mechanism, 921
  installing wheel bearings, 920
  packing grease by hand, 918–919
  packing grease with a bearing packer, 919
  removing, cleaning, and inspecting wheel bearings, 918f
  repacking and adjusting wheel bearings, 917–920
  replacing wheel bearings and races, 921, 922
  seal puller, 916f
  tools, 916
  typical locknut-style, 917f
  wheel bearing locknut sockets, 916f
  wheel bearing race/seal installer set, 916f
servicing cooling system
  cooling system diagnosing, 451–456
  preventive maintenance, 433–451
servicing disc brakes
  calipers removal and inspection, 853–855
  diagnosing, 850–851
  disassembling calipers, 856–857
  final checking, 867–868
  installing wheels, 867–868
  maintain and repair disc brakes, 851–853
  measuring disc brake rotors, 858–860
  pad assembling, 857–858
  reassembling caliper, 857–858
  removing and reinstalling rotors, 860–862
  rotors off vehicle, refinishing, 863–864
  rotors on vehicle, refinishing, 863
  torquing lug nuts, 867–868
  wheel studs inspection, 864–867
  wheel studs replacement, 864–867
servicing drum brakes
  brake shoes and hardware, 896–898
  diagnosing, 890–891
  dissembling, cleaning, inspecting, and reassembling a non-servo brake, 899–900
  installing wheels, torquing lug nuts, and making final checks, 902
  maintaining
    brake shoe adjustment gauge, 891
    brake spoon, 891
    brake spring pliers, 891
    brake wash station, 891
    drum brake micrometer, 891
    drum brake tools, 892f
    hold-down spring tool, 891
    wheel cylinder piston clamp, 891
  pre-adjusting brakes and installing drums, 900–901
  refinishing, 894–896
  removing, cleaning, inspecting, 893–894
  removing, inspecting, and installing wheel cylinders, 898–901
  repairing, 891–902
servicing lubrication system
  diagnosing, 398–399
  engine oil checking, 388–397
  maintenance and repairing, 388
servicing steering systems
  diagnosing steering systems, 663–664
  disabling the SRS, 674–676
  enabling the SRS, 674–676
  inspecting and testing electric power-assist steering, 673
  maintenance and repair, 664–666
  overview of, 660–662
  parallelogram steering linkage service, 668–672
  perform rack-and-pinion service, 667–668
servicing suspension system, 727–728
  diagnosing, 710–714
  inspecting strut rods and bushings, 727
  inspecting torsion bar suspension, 728–729
  install SLA suspension components, 722
  installing stabilizer components, 716
  overview of, 710
  removing, inspecting, and installing upper and lower ball joints, 722
  removing stabilizer components, 716
  ride height measuring, 714
  shock absorbers, 714–715
  stabilizer components inspection, 716
  strut cartridge or assembly, 722–727
  suspension to measure ball joint play, 716
servicing wheels
  balance, 608–612
  dismounting, inspecting, and remounting a tire with a TPMS sensor, 617–619

dismounting tire, 613–614
inspecting and remounting tire, 616–617
mounting, 614–616
proper tire inflation, 600–604
tire diagnosing, 624–626
tire maintenance preliminaries, 598–600
tire rotation pattern, 606–608
tire wear patterns, 604–606
and tires, 598
TPMS sensors, 616–617, 620–624
valve stem replacement, 619–620
wheel, tire, axle flange, and hub runout measuring, 626–630
sheaves, 502
shielded wiring harnesses, 985
shielding, 985
shock absorbers, 714–715
cutaway view of a strut, 689f
driving on irregular surfaces, 688f
gas-pressurized, 689–690, 689f
strut-type, 688, 689f
valves, 688f
shop foreman, 11
shop manual
after-market, 100f
computer databases, 102f
factory, 100f
service information program, 100–103
using, 100
shop safety inspections
disposable eyewash packs, 56
draining fuel, 58
fire blankets, 61
fire classifications, 59
fire extinguisher, 58, 58f
fire extinguisher types, 59
fire prevention, 56–57
fire triangle, 59f
fixed eyewash station, 55–56
fuel retriever, 58f
fuel vapor, 57
operating fire extinguisher, 60f
spill response kit, 58f
spillage risks, 57–58
spill-proof gas can, 57f
shop types
dealership technicians, 13f
fleet shop, 14, 14f
independent shops, 13, 13f
lube franchise business, 13, 13f
services, 13, 13f
short- and long-term fuel trim
PCM watches, 1257
STFT and LTFT, 1257
short to ground, 1027
short to power, 1027
short-term fuel trim (STFT), 1257
side force, 578
simultaneous injection, 1217
sine wave, 967
single overhead cam (SOHC) engine, 329
single-acting wheel cylinder, 874
single-cell hydrometer float, 1038
single-phase AC generator, 1078
single-piston master cylinders, 781
single-point injection, 1215
six-ply rating, 585
sizing bolts, 206
SLA suspension coil springs, 719
SLA suspension components, 722
SLA suspension system coil springs, 720
slave cylinder, 538
sledgehammer, 142
sliding spline driveshaft, 549
sliding T-handle, 131
slip angle, 578
slip rings, 1079
slip yoke, 548
smart chargers, 171
snap ring pliers, 135, 516
software transfers, 1123
software updates, 1123
soldering wires
cleaning, 992f
and connectors, 993
electric gun, 992f
heat shrink tubing shrinks, 993f
rosin core, 992f
and terminals, 991
tinning, 992f
solder-type terminals, 985, 990
solenoid operation
magnetic field stopping, 1068f
move toward cap, 1068f
pull-in and hold-in windings, 1068
pull-in winding, 1067
soft iron plunger, 1067
starter contacts and starter drive linkage, 1067, 1067f
starter control circuit, 1067
two electrical windings, 1067, 1067f
solenoid solid-state devices, 1031
solenoid valves, 928, 933
solenoids, 978–979, 1074, 1253
solenoid-type air control valve, 1253f
solid axle, 521
solid rotors, 844
solid-state relay, 977
solvent tank, 196
spark duration, 1181
spark plug design
combustion chamber installation, 1178
installing plugs, 1179
metal gasket seal, 1178f
multiple side electrodes, 1179f
required voltage, 1179
tapered seat design, 1178f
spark plug reach, 1179–1180
spark plug servicing, 1193
spark plug size, 1179
spark plug wires, 1176, 1194, 1195
spark plugs replacing, 1194
spark timing, 1175
electronically triggered, 1175–1176
engine speed and load, 1175
higher engine speeds, 1175f
low engine revolutions per minute, 1175
spark-ignition engine
components of, 323
long block assembly, 323f
short block and long block, 323
short block assembly, 323f
vehicle power, 323f
specialty springs, 883
speed controls, 1160
spider gears, 556
spike-protected relays, 978f
spiral cable, 642, 1130
splash lubrication, 382
spray wash cabinets, 195
spring eyes, 686
spring insulators, 719
spring shackle bushings, 696
spring-loaded rack guide yoke, 644
springs and clips, 872
spur gears, 471
spurt holes, 374
square-cut O-ring, 836
squirrel cage, 1160
SRS. *See* supplemental restraint system (SRS)
avoid accidental deployment, 674f
disabling, 674
enabling, 674
steering wheel and center, 675, 676
time the coil, 675, 676
standard diameter, 876
standard safety measures
cordless LED light, 54f
electrical safety, 52
emergency showers, 54–55
eyewash stations, 54–55
fluorescent droplight, 54f
ground prong, 53f
identifying hazardous environments, 48–50
portable electrical equipment, 52–53
portable shop lights, 53
safety equipment
adequate ventilation, 51
doors and gates, 51
exhaust extraction and shop ventilation equipment, 51f
handrails, 50
machinery guards, 50
painted lines, 50
painted safety lines, 51f
sound-insulated rooms, 50–51, 51f
temporary barriers, 51–52
shop layout, 54
signs, 48
background color, 48
pictorial message, 48
signal word, 48
text, 48
standard torque converter, 462

starter control circuit, 1072–1073
clutch switch circuit, 1069f
neutral safety switch circuit, 1069f
operation, 1068
PCM-controlled starter circuit, 1069f
solenoid windings, 1068
starter draw testing, 1070
full cranking speed, 1070
starter test equipment, 1070–1071
starter drives
pinion gear, 1066
and ring gear, 1066
starter drive one-way clutch, 1067, 1067f
starter high-current circuit voltage drop, 1072
starter installing, 1074
starter magnet types
electromagnetic fields, 1064, 1064f
series wound motor, 1064, 1064f
series-parallel-wound motor, 1064, 1064f
starter motor construction
brushes transfer electricity to the rotating commutator, 1063, 1063f
cutway, 1063, 1063f
power flow, 1063, 1063f
starter motor engagement
flywheel ring gear, 1065f
pinion drive, 1064
slight armature rotation, 1065
starter switch, 1065
starter motor principles
electrical energy to mechanical energy, 1062
PCM signals, 1062
solenoid plunger, 1062
starter motor, solenoid, and starter drive, 1062f
starter motor solenoid, 979f
starter no-load test, 1070
starter removal, 1074
starter ring gear, 532
starter types
armature and starter drive, 1062f
direct-drive, 1062
gear reduction, 1062
gear reduction starter, 1062f
higher speed with lower current, 1062
start-stop systems, 1076
static imbalance wheel, 610
stationary winding, 1187
stator, 1080, 1081f
steel hammer, 142
steering angle sensor, 943
steering arm, 634
steering axis inclination (SAI)
changing wheel offset, 739f
included angle, 737
negative setting, 739f
positive camber angle, 738f
scrub radius, 737–739, 738f
self-centering effect, 738f
steering boxes, 634, 643
steering columns, 634
clock spring arrangement, 642f
collapsible, 640f
driver's side airbag, 641–642, 642f
electric motor and sensors, 641f
flexible joint, 641f
intermediate shaft, 640f
multifunction switch, 641f
steering wheel entertainment controls, 641f
tilt steering wheel mechanism, 642f
tilt/telescoping mechanism, 641
steering damper, 639
steering gear ratio, 644
steering geometry, 635
Ackermann principle, 636f
and suspension system, 635f
turn sharper than the outside tire, 635f
wheel scrub, 636f
steering knuckle assembly, 720
steering knuckles, 635
steering linkage, 634
steering offset, 737
steering sensor, 653
steering system overview
components, 634f
parallelogram, 635f
rack-and-pinion, 634f
typical steering knuckle, 635f
steering system service preliminaries
dial indicators, 660
floor jacks and safety stands, 660
inner tie-rod end tool, 660
pickle fork, 660
pitman arm puller, 660
pry bar, 660
steering system specialty tools, 661f
tie-rod end puller, 660
tie-rod sleeve adjusting tool, 660
steering system theory
boxes, 643
columns, 640–642
electric power steering, 651–654
four-wheel steering systems, 654–655
geometry, 635–636
overview of, 634–635
parallelogram steering linkage, 637–640
power, 647–651
rack-and-pinion gearbox, 643–645
rack-and-pinion steering linkage, 636–637
worm gearbox, 645–647
steering wheel position sensor, 929
step-down transformers, 980
stepper motors, 1125
stoichiometric ratio, 1206
fuel types, 1308t
good indicator, 1308
HC and CO emissions, 1307
stop lights, 1099
straight grinder, 186
strategy-based diagnosing
measuring tire pressure, 114f
service information, 114f
tire being plug patched, 115f
tire being testing, 114f
TPMS warning system, 114f
strategy-based diagnosis, 113
stratified charge, 1218
street cams, 330
stripping wire insulation, 990
strut assembly removal, 727
strut cartridge
front strut bearing and mount, 726
inspecting, 726
removing, inspecting, and installing a insulators, 727
removing, inspecting, and installing a strut coil spring, 727
removing, inspecting, and installing a upper strut bearing mount, 727
removing, inspecting, and installing the strut cartridge, 727
strut rod, 692
stub axle, 693
sulfur dioxide
carbon particles, 1303f
fuel levels, 1303
particulates, 1303
regeneration of particulate filter, 1304f
smog, 1304
water vapor form, 1303f
sun gear, 470
supercharger, 1285
supplemental restraint system (SRS), 1135
airbag, 1135f
airbags and pretensioners, 1136–1137
airbags parts, 1137f
during collision, 1135f
crash sensor bolted, 1138, 1138f
disabling, 1139, 1140
driver's side airbag, 1136f
electric motor-style seat belt pretensioner assembly, 1138f
enabling, 1139, 1140
injury reduction devices, 1135f
knee airbag, 1136f
large vent holes, 1137f
operation, 1135
passenger's side airbag, 1136f
pedestrian airbag, 1137f
purpose of, 1135
rip stitching rips, 1138f
safing sensor, 1139f
seat belt pretensioner, 1135f, 1138f
sensors, control module, and circuitry, 1138–1139
shorting bar, 1139f
side airbag, 1136f
surge protector, 175
surge tank, 412, 416
suspension action, 688
suspension system components
coil springs, 684–686
conical and barrel-shaped springs, 686f
constant rate *versus* a progressive rate spring, 685f
helical springs, 685f

- leaf springs, 686f
- long alloy-steel bar, 686f
- rubber stops, 687
- springs, 684
- sway bar, 687f
- sway bars, 687
- torsion bars, 686–687

suspension system principles
- common forces acting on, 683f
- controlling forces in, 683–684
- function, 681–683
- hitting a bump and pushing the body upward, 683f
- rubber stops, 682f
- speed bump, 681f
- spring at normal height, 683f
- spring overshoot on rebound, 683f
- spring types, 682f
- sprung and unsprung weight, 680–681
- unsprung weight, 681f
- yaw, pitch, and roll, 684

suspension systems, 722
suspension systems types
- "dead axle" and "live axle," 697f
- dead axle/live axle, 696–697
- independent suspension system, 698, 698f
- live rear axle, 697f
- MacPherson strut, 698f
- SLA suspension system, 698f
- solid axle, 697, 697f
- trailers, 697f

suspension theory system
- active and adaptive, 704–705
- adjustable shock absorbers, 690–692
- components, 684–687
- control arms and rods, 692–696
- front suspension, 698–700
- principles, 680–684
- rear suspension, 700–704
- shock absorbers and struts, 688–690
- types, 696–698

sway bar, 687
swinging shackle, 686
switch VSS, 1244f
switches, 1250
SWL. *See* safe working load (SWL)
synchromesh transmission, 512
synthetic blends, 364, 365
synthetic oil, 365
system charging and starting
- alternator component, 1079–1080
- alternator cooling fan and pulley, 1081–1082
- alternator end frames and bearings, 1081
- alternator replacement, 1088–1089
- armature windings and commutator, 1065–1066
- charging system circuit voltage drop, 1086–1087
- charging system output testing, 1085–1086
- charging systems, 1077–1078
- direct-drive, 1062
- electromagnetic induction in alternators, 1078
- engine, 1061–1062
- gear reduction, 1062
- idle-stop/start-stop systems, 1076–1077
- inspection relays and solenoids, 1074
- inspection starter control circuit, 1072–1073
- keyless starting/remote starting, 1070
- overview of, 1061
- rectification, 1082–1083
- rotor circuit control, 1084–1085
- solenoid operation, 1067–1068
- starter control circuit, 1068–1069
- starter draw testing, 1070–1071
- starter drives, 1066–1067
- starter high-current circuit voltage drop, 1072
- starter magnet types, 1063–1064
- starter motor construction, 1063
- starter motor engagement, 1064–1065
- starter motor principles, 1062
- starter, removing and installing, 1074–1076
- stator, 1080
- testing relays and solenoids, 1074
- testing the starter control circuit, 1072–1073
- undercharge and overcharging, diagnosing, 1087–1088
- vehicle immobilization systems, 1069–1070
- voltage regulation, 1083–1084
- voltage regulator, 1085

system readiness monitors
- continuous component, 1268
- fuel system, 1268
- lower priority faults, 1268
- misfire, 1268
- PCM runs, 1269

T

tailpipe, 1291
tandem master cylinders, 781
tap handle, 146
tapered roller bearings, 908
tapered seat lug nuts
TCS. *See* traction control system (TCS)
technical service bulletins (TSBs), 103
- manufacturer recall announcement, 104f
- service campaign information, 104
- service campaigns and recalls, 104
- using, 103–104

telescoping gauges, 158
temperature gauge wiring diagram, 428f
temperature grade, 589
temperature indicators, 427
temperature tire ratings, 589
temperature warning light diagram, 428f
temporary tire, 590
tensile strength, 585
tensioners
- manual, 425, 425f
- oil-actuated style, 425f
- spring-loaded automatic, 425, 425f
- timing belt, 425f

terminals, 984
test leads
- "-" in front of the reading, 1002f
- meter leads, slots for, 1002f
- positive battery post, 1002f

test light circuits, 1028
testing battery capacity preliminaries, 1052
testing circuit protection devices, 1028, 1030
testing electric locks, 1128
testing electric motors, 1126
testing starter circuit voltage drop, 1073
testing switches, 1031
tetrafluoroethane, 1152, 1156
the battery
- automotive storage, 1036f
- components, 1036f
- deep cycle, 1037f
- electrical to chemical energy, 1036
- starting, 1037f

thermal expansion valve (TXV), 1154
thermal runaway, 1039
thermistors, 982, 1241
- air-fuel mixture, 1242
- control unit, 1242
- coolant temperature sensor, 1241f
- engine coolant temperature (ECT) sensor, 1242f
- engine temperature sensors, 1242
- intake air temperature (IAT) sensor, 1242f
- operations, 1241
- PTC and NTC, 1241f

thermo-control switch, 420
thermodynamic heat engine
- compression engine, 313f
- external combustion engine, 312f
- SI engine, 312f
- steam engine, 311–312

thermoelectric energy, 957
thermostat and housing, 416
- coolant bleed valve, 418f
- forces the valve open, 417f
- with jiggle valve, 417f
- spring-loaded valve, 417f
- surge tank removes air and gases, 417f
- thermostats block, 417f

thixotropy, 522
thread chasers, 145
thread pitch,
thread pitch gauge, 157, 206, 207
thread repairing and fasteners
- fastener standardization, 205–206
- nuts, 209–211
- overview of, 204–205
- screws, 212–218
- threaded fasteners and torque, 218–219
- threadlocker and anti-seize, 212
- torque-to-yield and torque angle, 220–225

threaded fasteners and torque, 204, 204f
- bolt under sheer stresses, 219f
- torque sticks, 219
- torque wrench calibration, 219f
- using torque wrenches, 219

threadlocker and anti-seize, 212

3 C's
customer symptom questionnaire, 117f
fault correction, 117f
repair documentation, 117
repair order, 115, 115f
written up on repair order, 117f
three-phase AC signal, 1079f
three-quarter floating axle, 560
three-way catalytic converters, 1311
throttle body injection (TBI), 1201
throttle position sensor, 1240–1241
throttle valves, 1278
throw-out bearing and clutch fork, 536
thrust angle, 739
thrust line, 739
thrust washers, 471
thrust-type angular-contact ball bearings, 536
tie-rod assembly, 645
tie-rod end puller, 660
time/mileage, 807
tire construction
bias ply, 587, 585f
radial, 586
tread designs
asymmetric, 586
directional and asymmetric, 586
directional tread patterns, 586
nondirectional, 586
symmetric, 586
types of, 585
tire date of manufacture coding, 590
tire diagnosing
faults, 624
tire assembly in water and watch, 625f
tire repair, 626
wheel assembly for air loss, using the dunk tank method, 625–626
wheel assembly for air loss, using the spray bottle method, 626
tire dismounting
without TPMS, 613–614
tire inflation pressure, 600
tire maintenance preliminaries
repair tools and equipment, 599f
tire air pressure, 598f
tools, 598–599
and wheel issues
air loss, 599
an out-of-balance tire or wheel, 599
excessive lateral runout on the tire, 599
excessive loaded radial runout on the tire, 599
wheel trim imbalance, 600
tire markings
information, sidewall, 587f
maximum pressure, 588f
typical sidewall, 588f
tire measurement, 626
tire mountation, 614–615
tire patching, 627–628
tire pressure monitoring system (TPMS), 590
tire remountation inspection, 616
tire rotation pattern
five-tire forward-cross, 607f
five-tire rearward-cross, 607f
forward-cross pattern, 607f
front-to-rear, 607f
rearward-cross pattern, 607f
side-to-side, 607f
typical lug nut torque patterns, 608f
X pattern, 607f
tire runout measurement, 628–629
tire safety features
band-type TPMS sensor, 591f
collapsible space-saver spare and inflation canister, 593f
double-extended safety humps, 592f
indirect TPMS system, 592f
low tire pressure warning, 591f
pressure monitoring systems, 590–592
pressure sensors used in TPMS, 591f
run-flat tire, 592f
run-flat tires, 592–593
self-sealing tires, 593f
space-saver spare, 593
tire pressure monitoring system (TPMS), 591f
typical temporary tire, 593f
tire sizes
and designations, 588
low-profile tire, 589f
standard-profile tire, 589f
tire wear patterns
center wear, 605f
checking for, 604, 606
cupped wear have dips, 605f
edge wear, 605f
feathering wear, 604f
one-sided wear, 605f
wear indicator bars on, 605f
tires
caps, 585
cores, 585
tread, sidewalls, inner liner, and bead, 585f
valve stems, 585
toe-out, 735
toe-out on turns, 635, 735
Ackermann angle, 736f
adjusted on some vehicles, 737f
own true arc, 735f
turning radius, 736–737, 736f
tone wheel, 937
top hat parking brake, 770
torque assist, 496
torque converter, 32
torque sensor, 652
torque smoothing, 495
torque steer, 713
torque *versus* horsepower, 317
torque-to-yield
broken bolt, 223f
broken fasteners, 220–222
conducting thread repair, 225
left-hand lug nuts, 220f
New TTY bolt, 220f
oxyacetylene torch, 222f
penetrating oil, 222f
removing a bolt, 223f
repaired thread, 222f
thread repair, 222
thread-restoring tool, 223f
torque angle gauge with a torque wrench, 221
TTY bolt, 220f
types of thread repair, 222–225
torque-to-yield (TTY) bolts, 220
torquing lug nuts, 867
torsion beam axle, 702
torsional load, 686
toxicity
carbon monoxide poisoning, 1303
first aid, 1304
permissible exposure limit (PEL), 1304
Toyota coolant heat recovery system, 1315f
TPMS. *See* tire pressure monitoring system (TPMS)
TPMS sensor valve stem, 619f
TPMS sensors
band-style, 622f
grommets on screw-in valve stems, 622f
nickel-plated valve stem caps, 623f
non-replaceable battery, 621f
not chrome-plated caps, 623f
press-fit valve stems, 622f
relearn on TPMS, 623–624
replaceable battery, 621f
servicing, 620
tighten the valve core, 622f, 623f
valve stems, 622f
track rod, 638
traction control system (TCS), 929
deactivating, 943f
operation, 942
traction grade, 589
traction tire ratings, 589
trailing arm suspension, 687
trailing shoes, 874
transmission brake, 771
transmission input shaft, 531
transmission maintenance
additional fluid spills out, 482f
checking transmission fluid, 481–482
locating leakage, 482
transmission mounted parking brake, 771
transmission specialists, 11
transmission/transaxle
four-wheel drive (4WD), 519f
transfer case, 519
typical FWD transaxle, 518f
typical RWD transmission, 518f
transverse arms, 692
tread wear grade, 589
tread wear tire ratings, 589
trim height, 741
TSBs. *See* technical service bulletins (TSBs)
tube flaring tool, 150
tube-type tires, 577

tulip/tripod joint, 568
tuned exhaust, 1289
turbo lag, 1288
turbocharger systems
  blow-off valve, 1287, 1287f
  BMW's reverse flow cylinder, 1288f
  compresses air, 1287f
  higher engine speeds, 1286
  intercoolers, 1288
  removed from, 1286
  shaft connection, 1286
  wastegate actuator, 1287, 1287f
turn signal lights
  BCM, 1102f
  CAN/Bus backup light circuit, 1101f
  cancelling mechanism, 1100–1101
  domestic and import systems, 1102
  domestic style flashes the brake light, 1102, 1102f
  pulsing current, 1101
  turn signals, 1102, 1102f
  typical flasher can, 1102f
turning radius, 736
twin knock sensors, 1250
twin leading shoe drum brake system, 874
twisted pair, 985
two-mode hybrid, 500
two-stroke cycle engine
  crankcase compression, 335f
  principles, 335–336
  two stroke-crankcase intake, 335f
  two-stroke engine, 335f
two-stroke intake system
  four- and two-stroke engine differences, 336
  valve operation, 336f
two-stroke spark-ignition engines, 335
two-way catalytic converter, 1310
two-wheel drive (2WD) vehicles, 29
typical brush-type motor, 1124f
typical electronic relay, 978f
typical permanent magnet electric motor, 1124f

U

UNC. *See* Unified National Coarse Thread (UNC)
undercharge diagnosing, 1087
underhood fluid inspection
  automatic transmissions, 261f
  automatic transmission/transaxle fluid, 260–262
  brake and clutch fluid, 258–259
  checking fluid appearance, 256f
  engine coolant, 258
  engine oil, 256
  fluids, 255–256
  freeze protection level, 258
  manual transmission/transaxle fluid
    adding DEF to the reservoir, 262f
    diesel exhaust fluid, 262–263
    fill until transmission fluid, 262f
    windshield washer fluid, 263
  observe the markings, 261f
  oil dipsticks, 257f
  oil-life monitor, 257f
  parts damaged by lack of lubrication, 256f
  power steering fluid, 260, 260f
  transmission dipstick, 261f
  wipe off the dipstick, 261f
under-vehicle inspection
  checking differential fluid, 269
  checking for fluid leaks, 269
  checking manual transmission, 269
  differential fluid level, 270
  drive axles and driveshafts, 272
  engine, transmission, and exhaust, 273
  four-digit DOT date code, 269f
  locate and identify fluid leaks, 271
  manual transmission, 269f
  maximum tire inflation pressure, 268f
  other items, 273
  steering and suspension inspection, 272
  tire inspection, 268–269
  torn CV boot, 272f
  typical tire placard, 268f
  under-vehicle inspection, 273
  wear patterns, damage, and tread depth, 268f
  Wiggle one front wheel, 272f
  worn steering components, 272f
UNF. *See* Unified National Fine Thread (UNF)
unibody design, 22
Unified National Coarse Thread (UNC), 206
Unified National Fine Thread (UNF), 206
uniform pitch, 685
Uniform Tire Quality Grading (UTQG), 589
unitized wheel bearing hub, 909
universal joints, 515, 549
unleaded gasoline, 1204, 1206
unsprung mass, 680
unsprung weight, 680
U.S. Department of Transportation (DOT), 590
using service information, 1193
using wiring diagrams
  color coding, 988f
  diagnosing electrical circuits, 989
UTQG. *See* Uniform Tire Quality Grading (UTQG)
UTQG ratings, 589f

V

vacuum booster, 798
vacuum brake bleeder, 807f
vacuum fluorescent display (VFD), 1095
vacuum relief valve, 1313
vacuum tube fluorescent (VTF), 1095
vacuum-type power booster, 820
valence ring, 953
valve core, 585
valve margin, 328
valve overlap, 322
valve stem replacement, 619
valve-regulated lead-acid batteries, 1038
vapor lock, 1203
vaporization, 1314
variable intake systems
  air-fuel mixture, 1283
  butterfly-type valves, 1283f
  engine load and speed, 1283
  manifold with alternate runners, 1283f
  opening and closing alternate runners, 1284f
  pulse chambers, 1284
variable orifice, 1312
variable reluctance sensor, 931
variable reluctor sensors, 1186
variable resistors, 981, 1018
variable valve timing, 1313f
variable-flow exhaust
  electronic mufflers, 1293
  engine noises are cancel, 1293f
  hydrocarbon emissions, 1293
  resonator, 1293–1294
  typical, 1293f
variable-orifice PCV valve meters, 1312f
vehicle chassis
  closure designs, 23
  engine compartment release, 23f
  holding hoods open, 23f, 24f
vehicle communications networks
  configurations and terminal resistors, 1120
  controller area network (CAN), 1119–1120
  controller area network with flexible data-rate (CAN FD), 1120
  ethernet network systems, 1121–1122, 1121f
  fiber optic and electric communication systems, 1121f
  flexray network with a dual-channel system, 1120–1121, 1121f
  information act, 1119
  LIN network, 1120f
  local interconnect network (LIN), 1119
  master node, 1120f
  media-oriented systems transport networks, 1121
  multiplexed system, 1119
  network configurations, 1121f
  terminating resistors, 1120f
  vehicle speed, 1119
vehicle coolant
  crack blocks and heads, 406f
  ethylene glycol, 407f
  freezing point of antifreeze, 407f
  heat sink, 405
  Toyota coolant heat storage system, 406f
vehicle emissions and standards, 1325
vehicle for alignment tools
  brake pedal depressor, 747f
  common tools and equipment, 746f
  steering angle reset tool, 747f
  tie-rod assembly, 745, 745f
  tie-rod on a rack-and-pinion steering system, 745, 745f

vehicle immobilization systems, 1069–1070
vehicle information labels, 107
coolant label, 110f
other labels, 110
refrigerant label, 110f
Vehicle Emission Control Information (VECI), 109
Vehicle Safety Certification (VSC), 110
vehicle maintenance inspection
engine drive belts, 263–267
exterior vehicle inspections, 273–275
inspecting the exterior lights, 275–278
in-vehicle inspection, 252–255
overview, 251–252
underhood fluid inspection, 255–263
under-vehicle inspection, 267–273
vehicle protection and jack lift safety
corrosives and greases, 230
customer pickup, 232–233
engine hoists and stands, 240
engine stand, 240
fender, seat, carpet, and steering wheel covers, 230–231
fender, seat, carpet steering wheel covers absorbent materials, 231f
floor mats, 231
four-post lift, 244–245
four-post lifts to lift a vehicle, 246
jacks and jack stands, 235–237
lifting equipment, 234–235
moving and road testing vehicles, 233
overview of, 229
preventing vehicle damage, 229–230
ratings and inspections, 243
return vehicle to customer, 233
safety locks, 243
scissor jack, 237
single-post lift, 243
steering wheel covers, 231
tall jack stands, 237
two-post lift, 244
using engine hoists and stands, 240–241
using four-post lifts, 244–245
using lifting equipment, 238
using protective covers, 231–232
using two-post lifts, 244–245
using vehicle inspection pits, 247–248
using vehicle jacks and stands, 238–240
vehicle for customer, 232f
vehicle inspection pits, 244–245
vehicle lifts, 242–243
vehicle using a two-post lift, 245–246
vehicle walk-around, 229
vehicle safety certification (VSC), 110
vehicle service information
3 C's, 115–117
date code, and vehicle information labels, 107–109
information labels, 109–110
labor guide, 105–106
overview of, 98
owner's manual, 98–99
repair order information, 110–112
service history, 112–113
shop manual, 99–103
strategy-based diagnosis, 113–115
technical service bulletins, 103–105
VIN, 107
vehicle speed sensor (VSS), 940, 1244
vehicle systems
electrical system, 27f
engine management system, 26f
hybrid engine with internal combustion engine, 25f
internal combustion engine, 25f, 26f
operation and passenger safety, 24
overview of, 24–26
rear axle assembly, 27f
starter cranks the engine, 26f
steering system, 27f
suspension system absorbs road shock, 27f
torque converter, 26f
transmissions engine power, 27f
variety of systems, 25f
vehicle wander, 712
ventilation system
amount of heat delivered, 1158–1159
blend door, 1159f
boiling point of, 1158
driver and passenger safety, 1159
evaporator cooling, 1159
heater control valve, 1159f
heater core, 1158f
and heating, 1158
mode switch on an HVAC system, 1159f
temperature control on an HVAC system, 1159f
valve, 1159
VFD. *See* vacuum fluorescent display (VFD)
VIN
decoding, 108–109
locating, 108
North American, decoding, 109
viscosity, 366, 912
viscosity index improver, 367
viscous coupler, 31, 419
VOCs. *See* volatile organic compounds (VOCs)
volatile organic compounds (VOCs), 1262
voltage and voltage drop
available voltage test, 1022f
DMM, 1022
excessive voltage drops, 1023f
measuring, 1022
voltage drop on, circuit, 1023f
voltage drop testing, 964, 1008
full battery voltage, 1024f
horn circuit, 1024
negative side of the circuit, 1025f
performing, 1023
reversing the meter leads, 1025f
worn horn relay contacts, 1024f
voltage exercises
measuring, 1006–1008
ranges, 1006
two resistors of unequal resistance, 1007f
two unequal resistors connected in series, 1007f
unequal resistors connected in series, 1007f
using various ranges, 1006f
voltage splits up, 1007f
voltage regulation
alternator's output, 1083
A-type or B-type regulating circuits, 1083–1084
full-fielding an A-type regulator circuit, 1084f
output voltage, 1083
regulator switching, 1083
voltage regulator, 1085
volts, 955
volts measuring
amps, 1003
disconnected ends, 1005f
input and output terminals, 1005f
Ohms, 1003
series circuit, 1004
voltage drop tests, 1004f
volumetric efficiency
back pressure, 1284
engine's power output, 1285f
forced induction engine, 1285, 1285f
normally aspirated engine, 1284f
roots-style supercharger, 1286f
sharp bends force, 1284
supercharger systems, 1285–1286
supercharger types, 1285
twin-screw supercharger, 1286f
vortex flow, 467
VR engine, 36
VSC. *See* vehicle safety certification (VSC)
VSS. *See* vehicle speed sensor (VSS)
VSS magnetic pickup sensor, 1244, 1244f
VTF. *See* vacuum tube fluorescent (VTF)

W

W engine, 36
wad punches, 142
warm-up cycle, 1266
warpage, 842
waste cylinder, 1190
waste spark ignition system, 1172
effect on engine operation, 1190
ignition coil serves two cylinders, 1190
secondary circuit, 1190
waste cylinder, 1190
water fade, 766
water jackets
core plugs, 426f
cylinder head, 426f
water pump
head gaskets, 418f
and impeller, 418f
timing belts, 418f
and weep hole, 419f

Watt's linkage, 702
4WD vehicles, 29
weight matching, 614
weight transfer, 768
welding helmet, 92
wheel alignment, 733
  fundamentals, 733–735
  performing four-wheel, 748–749
  performing pre-alignment, 747
  preliminaries, 743–745
  ride height, 741
  secondary angles, 749–751
  steering axis inclination, 737–739
  thrust angle, centerlines, and setback, 739–741
  toe-out on turns, 735–737
  tools, 745–746
wheel alignment preliminaries
  adjustable ball joint sleeve, 745f
  adjustment methods, 743–745
  adjustment point for caster, 745f
  eccentric bolt adjustment, 744f
  rear alignment, 744f
  recalibrate a steering angle sensor, 743f
  shim-type adjustment, 744f
wheel balance
  dynamic balancing a tire, 611–612
  dynamic imbalance, 610, 610f
  off-car balancer, 610, 610f
  road force imbalance, 611f
  static imbalance, 610, 610f
  tire rotation, 609
wheel bearing arrangements
  full floating axle, 913f
  for rear drive axles, 913
  semi-floating axle, 913f
  three-quarter floating axle, 913f
wheel bearing diagnosis
  and failure analysis, 914–915
  failure chart, 915f
  transmission noise, 914
wheel bearing theory
  components of, 906f
  overview of, 906–907
  sealed, 907f
  serviceable bearings, 907
  technician servicing, 906f
wheel bearing types
  ball, 908–909
  cone, 908f
  control thrust in both directions, 908f
  cup, 908f
  cylindrical roller, 907, 907f
  double-row ball bearing assembly, 909f
  sealed, 909–910
  tapered roller, 908, 908f
  unitized wheel bearing hub assembly, 909f
wheel bearings, 906
  axle seals, 910
  diagnosing, 914–915
  grease seals, 910
  lubrication, 910–913
  maintenance and repairing, 915–922
  rear driveaxles, arrangements, 913–914
  sealed wheel bearings, 922–923
  theory, 906–907
  types, 907–910
wheel cylinder piston clamp, 891
wheel cylinders, 872
  on backing plate, 877f
  bleeder screw allows air and fluid, 879f
  brake linings, 877–878
  components, 878f
  cutaway, 878f
  sealing surface of, 878
  types, 879, 879f
wheel flange, 580, 584
wheel measurement, 626
wheel offset, 737
wheel physics
  center of gravity, 579
  font slip angle, 579f
  rear slip angle, 579f
  slip angle, 578f
  solid tires, 577f
  stretched tight by air pressure, 578f
  tire distortion, 578–579, 578f
  vehicle and road surface, 577f
wheel runout measurement, 629
wheel speed sensors
  Hall-effect operation, 939f
  Hall-effect wheel speed sensor assembly, 938f
  types of, 937–938
  variable reluctance wheel speed sensor assembly, 937f
  wheel speed sensor sine wave, 937f
wheel studs inspection, 864
wheel studs replacement, 563
wheel studs replacing, 864
wheel width, 580
wheels
  bolt pattern, 584
  center of gravity, 579f
  construction, 579–582
  deep dish wheel, 582f
  drop-center rim, 581f
  left-hand lug nut and stud, 583f
  lug nuts or wheel studs on the, 584f
  offset, 582
    negative, 582
    positive, 582
    zero, 582
  pitch circle diameter gauge, 584f
  rim of the wheel, 580f
  rim width and diameter, 580f
  safety ridge on a wheel, 581f
  separate washer and wheel, 583f
  studs and lug nuts, 582–584
  tapered seat lug nut and wheel, 583f
  types, 581t
  washer and, 583f
  wheel flange and rim, 580f
  zero, positive, and negative offset, 582f
wheels and tires theory
  physics, 577–579
  tire markings, 586–588
  tire safety features, 590–593
  tire size and designations, 588–589
  tires, 584–585
  traction and temperature, tire ratings for, 589–590
  tread wear, tire ratings for, 589–590
  types of tires, 585–586
  wheels, 579–584
Whitworth thread, 204f
windage tray, 369
wiper/washer system
  and delay circuits, 1131–1133
  dynamic braking, 1132
  headlights use a wiper and washer, 1133f
  rain sensor, 1133f
  testing, 1134
  testing the horn system, 1132
  wiper circuit diagram, 1132–1133, 1133f
  wiper motors, three brushes, 1133f
  wiper system testing, 1133–1134
wire maintenance and repair, 990
wiring diagram fundamentals, 986
  black wire with a white tracer, 987, 987f
  electrical device and component, 986, 986f
  heater blower motor circuit, 988, 988f
  identification codes, 987, 987f
  power at top, 988, 988f
  typical, 986, 986f
wiring harness connectors, 984f, 985
wiring looms, 984
wishbone control arm, 692
work, 315
work environment
  evacuation routes, 47, 47f
  relatively safe shop, 47f
  unsafe shop, 47f
work practice controls, 81
worm gear steering, 645
worm gearbox
  -and-roller gearbox, 646–647
  -and-roller steering, 646f
  -and-sector, 646, 646f
  assembly parts, 646f
  rack-and-pinion gear, 646f
  recirculating ball, 647
  road shock, 646f
  sector gear adjustment screw, 647f
worm wheel, 645
worm-and-roller gearbox, 646
worm-and-sector gearbox, 646
wrap leaf, 686

X

xenon headlamps, 1095

Z

zero camber, 734
zero scrub radius, 737